T0260134

Annals of Mathematics Studies
Number 210

Global Nonlinear Stability of Schwarzschild Spacetime under Polarized Perturbations

Sergiu Klainerman
Jérémie Szeftel

PRINCETON UNIVERSITY PRESS
PRINCETON AND OXFORD
2020

Published by Princeton University Press,
41 William Street, Princeton, New Jersey 08540
6 Oxford Street, Woodstock, Oxfordshire OX20 1TR

press.princeton.edu

Library of Congress Control Number: 2020948579

ISBN 9780691212432
ISBN (pbk.) 9780691212425
ISBN (e-book) 9780691218526

British Library Cataloging-in-Publication Data is available

Editorial: Susannah Shoemaker and Kristen Hop
Production Editorial: Nathan Carr
Production: Brigid Ackerman
Publicity: Matthew Taylor and Amy Stewart

The publisher would like to acknowledge the authors of this volume for providing the camera-ready copy from which this book was printed.

This book has been composed in LaTeX

Printed on acid-free paper. ∞

Printed in the United States of America

10 9 8 7 6 5 4 3 2 1

Contents

List of Figures

Acknowledgments

This work would be inconceivable without the remarkable advances made in the last sixty years on black holes. The works of Regge-Weeler, Israel, Carter, Teukolsky, Chandrasekhar, Wald, etc., made during the so-called *golden age* of black hole physics in the sixties and seventies, have greatly influenced our understanding of invariant quantities and the wave equations they satisfy. The advances made in the last fifteen years, which have led to the development of new mathematical methods to derive the decay of waves on black hole spacetimes, are even more immediately relevant to our work. In particular we would like to single out the direct influence of Dafermos-Holzegel-Rodnianski [26] in the gestation of our own ideas in this book. Finally, the work on the nonlinear stability of the Minkowski space in [20], a milestone in the mathematical GR, has significantly instructed our work here. We would like to thank E. Giorgi for her careful proofreading of various sections of the manuscript. Various discussions we had with S. Aksteiner were very useful. Finally we thank our wives Anca and Emilie for their incredible patience, understanding and support during our many years of work on this project.

The first author has been supported by the NSF grant DMS 1362872. He would like to thank the mathematics departments of Paris 6, Cergy-Pontoise and IHES for their hospitality during his many visits in the last six years. The second author is supported by the ERC grant ERC-2016 CoG 725589 EPGR.

Global Nonlinear Stability of Schwarzschild Spacetime under Polarized Perturbations

Chapter One

Introduction

1.1 BASIC NOTIONS IN GENERAL RELATIVITY

We provide a quick review of the basic concepts of general relativity relevant to this work. For a proper introduction to the subject we refer to the books by R. Wald [66] and S. Caroll [15].

1.1.1 Spacetime and causality

The main object of Einstein's general relativity is the spacetime. To define a spacetime, consider a four dimensional Lorentzian manifold $(\mathcal{M}, \mathbf{g})$, with $\mathbf{g}$ denoting a Lorentzian metric of signature $(-,+,+,+)$. Two Lorentzian manifolds $(\mathcal{M}, \mathbf{g})$, $(\mathcal{M}', \mathbf{g}')$ are equivalent if there exists a diffeomorphism $\Phi : \mathcal{M} \to \mathcal{M}'$ such that $\mathbf{g} = \Phi^{\#}(\mathbf{g}')$. A spacetime is simply a class of equivalence of such Lorentzian manifolds.

A Lorentzian metric divides vectors X in a tangent space $T_p(\mathcal{M})$ into *timelike*, *null* and *spacelike* according to whether $\mathbf{g}(X, X)$ is, respectively, negative, zero or positive. A curve $\gamma(t)$ is said to be timelike, respectively null, if its tangent vector $\dot\gamma(t)$ is timelike or null. It is called *causal* if it is either timelike or null.

Remark 1.1. *Observers in general relativity are identified to timelike curves, and freely moving observers correspond to timelike geodesics. Points of $\mathcal{M}$ are referred to as events and the proper time of an observer $\gamma(t)$ between the events $\gamma(t_1), \gamma(t_2)$ is the integral,*

$$\int_{t_1}^{t_2} \sqrt{-\mathbf{g}\big(\dot\gamma(t), \dot\gamma(t)\big)}\, dt.$$

Massless particles, on the other hand, follow null geodesics. The proper time of such a particle, i.e., the proper time of the corresponding null geodesic, is the affine parameter of the geodesic vectorfield associated to the curve.

Given a set $S \subset \mathcal{M}$, we denote by $\mathcal{I}^+(S)$ the set of all points in $\mathcal{M}$ which can be reached by future directed timelike curves[1] originating at S, called the *future set* of S. The set $\mathcal{J}^+(S)$, consisting of points which can be reached by future directed causal curves from S, is called the *causal future* of S. One defines in the same manner the past and causal pasts $\mathcal{I}^-(S)$ and $\mathcal{J}^-(S)$.

A hypersurface Σ is called *spacelike* or *null*, if the direction normal to it is timelike, respectively null. Typical spacelike hypersurfaces are given by the level surfaces of *time functions* t, i.e., non-degenerate functions on $\mathcal{M}$ ($dt \neq 0$) such that

[1] We assume the spacetime to be time oriented, i.e., there exists a globally defined non-degenerate timelike vectorfield T. In particular, a causal vectorfield X is future oriented if $\mathbf{g}(T, X) < 0$.

its gradient $-\mathbf{g}^{\mu\nu}\partial_\mu t\partial_\nu$ is timelike. Typical null hypersurfaces are given by level surfaces of *optical functions* u, i.e., non-degenerate functions $u : \mathcal{M} \to \mathcal{R}$ verifying

$$\mathbf{g}^{\mu\nu}\partial_\mu u\partial_\nu u = 0, \qquad du \neq 0. \tag{1.1.1}$$

In that case the gradient $L := -\mathbf{g}^{\mu\nu}\partial_\mu u\partial_\nu$ is both null and geodesic, i.e., $\mathbf{g}(L, L) = 0$ and $\mathbf{D}_L L = 0$.

A spacelike hypersurface Σ is said to be a *Cauchy hypersurface* in $\mathcal{M}$ if any in-extendible causal curve intersects Σ at precisely one point. Spacetimes which admit such hypersurfaces rule out causal pathologies such as the presence of closed timelike curves. A spacetime is called *globally hyperbolic* if it possesses such a hypersurface and, in addition, all sets of the form $\mathcal{J}^+(p) \cap \mathcal{J}^-(q)$ are compact.

1.1.2 The initial value formulation for Einstein equations

Let $(\mathcal{M}, \mathbf{g})$ a spacetime. *Einstein equations* are given by

$$\mathbf{R}_{\alpha\beta} - \frac{1}{2}\mathbf{g}_{\alpha\beta}\mathbf{R} = \mathbf{T}_{\alpha\beta} \tag{1.1.2}$$

with $\mathbf{R}_{\alpha\beta}$ the Ricci curvature of $\mathbf{g}$, $\mathbf{R}$ the scalar curvature of $\mathbf{g}$, and $\mathbf{T}_{\alpha\beta}$ the energy-momentum tensor of some matterfield defined on $(\mathcal{M}, \mathbf{g})$. An initial data set consists of a 3 dimensional manifold $\Sigma_{(0)}$, a complete Riemannian metric $g_{(0)}$, a symmetric 2-tensor $k_{(0)}$, and a well specified set of initial conditions corresponding to the matterfields under consideration. These have to verify a well known set of constraint equations. We restrict the discussion to *asymptotically flat* initial data sets, i.e., outside a sufficiently large compact set K, $\Sigma_{(0)} \setminus K$ is diffeomorphic to the complement of the unit ball in $\mathbb{R}^3$ and admits a system of coordinates in which $g_{(0)}$ is asymptotically euclidean, and $k_{(0)}$ vanishes asymptotically at appropriate order. A *Cauchy development* of an initial data set is a globally hyperbolic spacetime $(\mathcal{M}, \mathbf{g})$, verifying the Einstein equations (1.1.2) in the presence of a matterfield with energy-momentum $\mathbf{T}$ and an embedding $i : \Sigma \to \mathcal{M}$ such that $i_*(g_{(0)}), i_*(k_{(0)})$ are the first and second fundamental forms of $i(\Sigma_{(0)})$ in $\mathcal{M}$.

We restrict our attention to the Einstein vacuum equations (EVE), i.e., the case when the energy-momentum tensor vanishes identically and the equations take the purely geometric form,

$$\mathbf{R}_{\alpha\beta} = 0. \tag{1.1.3}$$

In that case, the constraint equations mentioned above take the form

$$\text{div}\, k_{(0)} - \nabla\, \text{tr} k_{(0)} = 0, \qquad R_{(0)} - |k_{(0)}|^2 + (\text{tr} k_{(0)})^2 = 0. \tag{1.1.4}$$

Here ∇ denotes the covariant derivative on $\Sigma_{(0)}$, div the usual divergence of a symmetric 2-tensor, defined with respect to ∇, and $R_{(0)}$ the scalar curvature of the metric $g_{(0)}$. Moreover $|k_{(0)}|$ and $\text{tr} k_{(0)}$ are the Riemannian norm and trace of $k_{(0)}$ with respect to $g_{(0)}$.

The most basic question concerning the initial value problem, solved in a satisfactory way for very large classes of evolution equations, is that of local existence and uniqueness of solutions. For the Einstein equations, this type of result was first established by Y. Choquet-Bruhat [13] with the help of wave coordinates.[2] According to this result any smooth initial data set admits a smooth, unique (up to an isometry) globally hyperbolic Cauchy development.[3] In the case of nonlinear systems of partial differential equations, the local existence and uniqueness result leads, through a straightforward extension argument, to a result concerning the maximal time interval of existence. The formulation of the same type of result for the Einstein equations is a little more subtle; something similar was achieved in [14], see also [60] for a modern version of the result.

Theorem 1.2 (Bruhat-Geroch). *For each smooth initial data set there exists a unique, smooth, maximal future globally hyperbolic development (MFGHD).*

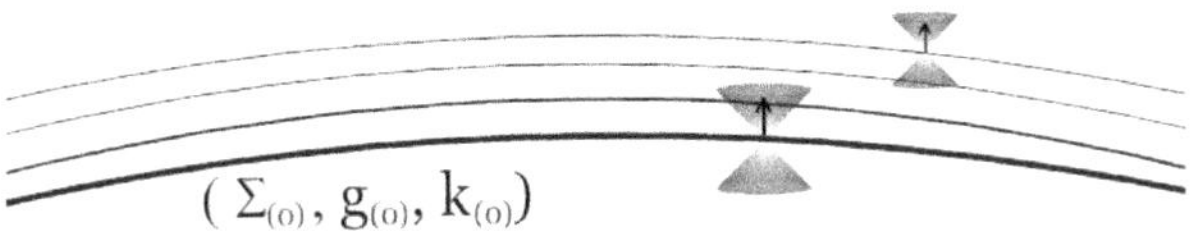

Figure 1.1: The initial value problem for Einstein vacuum equations

1.1.3 Special solutions

1.1.3.1 Minkowski space

The Minkowski space consists of the manifold $\mathbb{R}^{1+3}$ together with a Lorentzian metric $\mathbf{m}$ and a distinguished system of coordinates x^α, $\alpha = 0, 1, 2, 3$, called inertial, relative to which the metric has the diagonal form $\mathbf{m}_{\alpha\beta} = \mathrm{diag}(-1, 1, 1, 1)$. We write, splitting the spacetime coordinates x^α into the time component $x^0 = t$ and space components $x = x^1, x^2, x^3$,

$$\mathbf{m} = -dt^2 + (dx^1)^2 + (dx^2)^2 + (dx^3)^2.$$

In polar coordinates (t, r, θ, φ),

$$\mathbf{m} = -dt^2 + dr^2 + r^2 d\sigma_{\mathbb{S}^2}, \qquad d\sigma_{\mathbb{S}^2} := d\theta^2 + \sin^2\theta d\varphi^2.$$

The standard optical functions in $\mathbb{R}^{1+3}$ are given by $u = t - r$, $\underline{u} = t + r$, often called retarded and advanced time coordinates. One can compactify the Minkowski space by constructing a map $P : (u, \underline{u}, \omega) \to (U, \underline{U}, \omega)$, $\omega \in \mathbb{S}^2$, where

$$u = \tan U, \qquad \underline{u} = \tan \underline{U}, \qquad -\frac{\pi}{2} < U \le \underline{U} < \frac{\pi}{2}.$$

[2]These allow one to cast the Einstein vacuum equations in the form of a system of nonlinear wave equations for which classical local existence results can be applied.

[3]The precise result requires some minimal regularity for the initial data set. The optimal known result, the bounded L^2 curvature theorem, see [46], requires L^2 bounds for the curvature of the initial data set.

The map P establishes a conformal isometry[4] between the Minkowski space $\mathbb{R}^{1+3}$ and its image onto the Einstein cylinder $\mathbb{E}^{1+3} = \mathbb{R} \times \mathbb{S}^3$ with metric

$$\widetilde{\mathbf{m}} = -dU d\underline{U} + \frac{1}{4}\sin^2(\underline{U} - U) d\sigma_{\mathbb{S}^2}.$$

More precisely

$$P^{\#}(\widetilde{\mathbf{m}}) = \Omega^2 \mathbf{m}, \qquad \Omega = \cos U \cos \underline{U} = \frac{1}{(1+u^2)^{1/2}(1+\underline{u}^2)^{1/2}} \tag{1.1.5}$$

where $P^{\#}(\widetilde{\mathbf{m}})$ is the pullback by P of the metric $\widetilde{\mathbf{m}}$.

Definition 1.3. *The boundary of $P(\mathbb{R}^{1+3})$ in $\mathbb{E}^{1+3}$ is given by*

$$\partial P(\mathbb{R}^{1+3}) = \mathcal{I}^+ \cup \mathcal{I}^- \cup i^0 \cup i^+ \cup i^-.$$

The sets

$$\mathcal{I}^+ := \left\{\underline{U} = \frac{\pi}{2}, \, -\frac{\pi}{2} < U < \frac{\pi}{2}\right\}, \qquad \mathcal{I}^- := \left\{U = -\frac{\pi}{2}, \, -\frac{\pi}{2} < \underline{U} < \frac{\pi}{2}\right\},$$

are called the future and past null infinities of Minkowski space. The sets

$$i^0 := \left\{U = -\frac{\pi}{2}, \, \underline{U} = \frac{\pi}{2}\right\}, \qquad i^+ := \left\{U = \underline{U} = \frac{\pi}{2}\right\}, \qquad i^- := \left\{U = \underline{U} = -\frac{\pi}{2}\right\},$$

are called, respectively, spacelike, timelike future, and timelike past infinities.

Note that all timelike geodesics of Minkowski space begin at i^- and end at i^+, all spacelike geodesics begin and end at i^0 and all null geodesics start on $\mathcal{I}^-$ and end on $\mathcal{I}^+$. We also note that $\mathcal{I}^-, \mathcal{I}^+$ are complete null hypersurfaces, along which $d\Omega \neq 0$. One can also show that the boundary $\partial P(\mathbb{R}^{1+3})$ is of class C^2 at i^0 and real analytic, everywhere else.

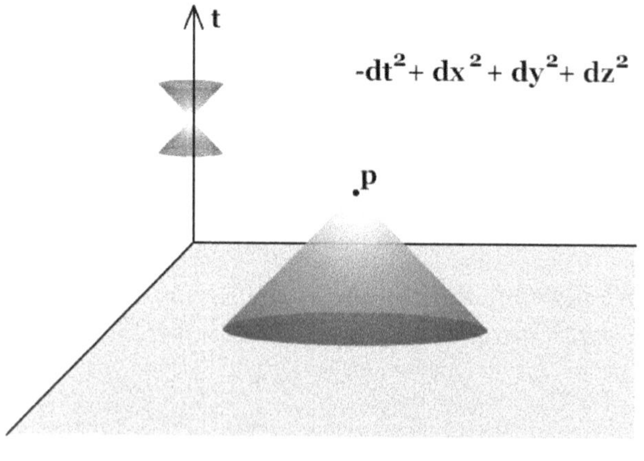

Figure 1.2: Minkowski in standard coordinates

[4]For a comprehensive discussion of conformal infinity, see section 11.1 in [66].

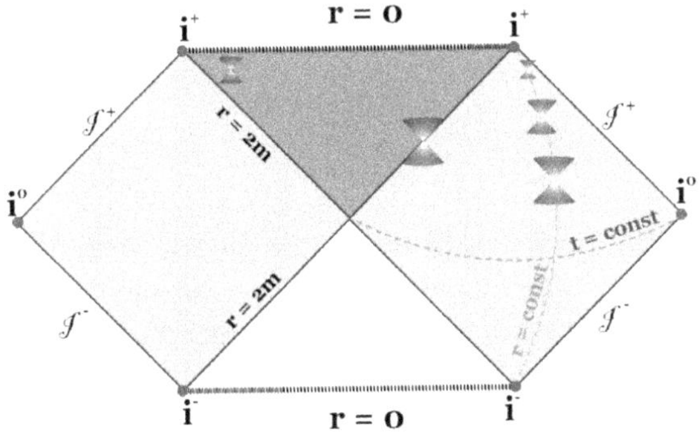

Figure 1.5: Complete Penrose diagram of Schwarzschild

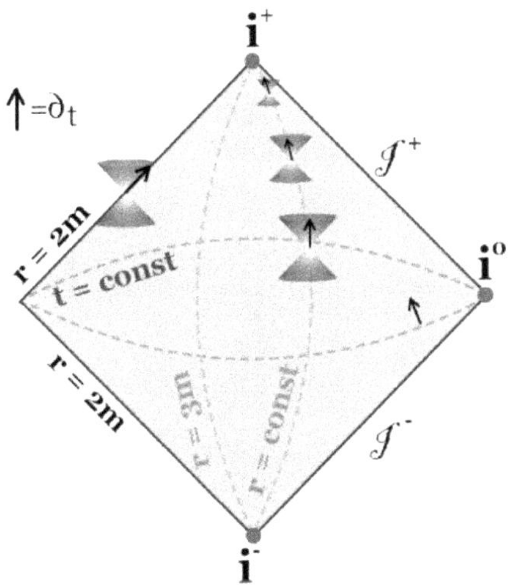

Figure 1.6: Exterior region of Schwarzschild

Similarly the white hole region is the complement of the future of past null infinity $\mathcal{J}^+(\mathcal{I}^-)$. The null hypersurface $\mathcal{H} = \{r = 2m\}$, called the *event horizon*, is the boundary of the black hole and of the white hole. In figure 1.6, representing one connected component of DOC, we note the presence of the timelike hypersurface $r = 3m$ on which null geodesics can be trapped.

1.1.3.3 Kerr space

The Schwarzschild family is included in a larger two parameter family of solutions $\mathcal{K}(a, m)$ discovered by Kerr. A given Kerr spacetime, with $0 \leq |a| \leq m$, has a well defined domain of outer communication $r > r_+ := m + (m^2 - a^2)^{1/2}$. In Boyer-Lindquist coordinates, well adapted to $r > r_+$, the Kerr metric has the form

$$\mathbf{g}_K = -\frac{(\Delta - a^2 \sin^2\theta)}{q^2} dt^2 - \frac{4amr}{q^2}\sin^2\theta dt d\varphi + \frac{q^2}{\Delta}dr^2 + q^2 d\theta^2 + \frac{\Sigma^2}{q^2}\sin^2\theta d\varphi^2$$

with $q^2 = r^2 + a^2\cos^2\theta$, $\Delta = r^2 + a^2 - 2mr$, $\Sigma^2 = (r^2+a^2)^2 - a^2(\sin\theta)^2\Delta$. Note that $\Delta(r_+) = 0$.

As in the Schwarzschild case, the exterior Kerr metric extends smoothly across the hypersurface $r = r_+$. The future and past sets of any point in the domain of outer communication intersect any timelike curve, passing through points of arbitrary large values of r, in finite time as measured relative to proper time along the curve. This fact is violated by points in the region $r \le r_+$, which consists of the union between a *black hole* region, extended towards the future, and a *white hole* region to the past. Thus physical signals (i.e., future timelike or null geodesics) which initiate at points in $r \le r_+$ cannot be registered by far away observers.[5] The domain of outer communication $\{r > r_+\}$ is real analytic. The boundary of the domain of outer communication $\{r = r_+\}$ is called the *event horizon*. In the non-degenerate case, $|a| < m$, the event horizon consists of two null hypersurfaces intersecting transversally on a compact 2-sphere. The Kerr solution can also be conformally compactified in the same manner as Minkowski and Schwarzschild. We can thus talk about the future and past null infinities $\mathcal{I}^+, \mathcal{I}^-$ as well as i^0, i^+, i^-. As before, $\mathcal{I}^+$ is a complete null hypersurface, smooth away from i^0.

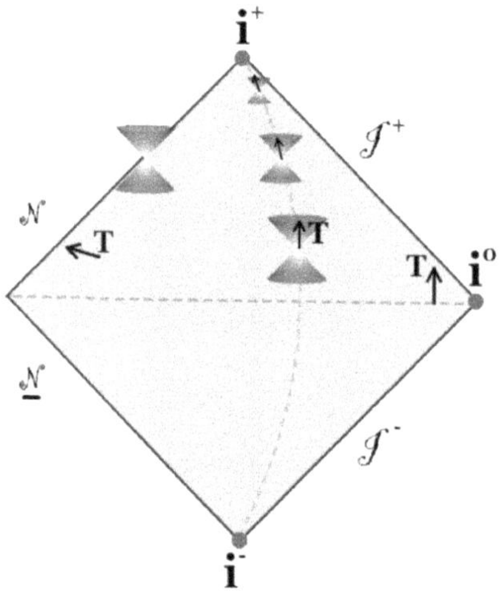

Figure 1.7: Exterior region of Kerr

The exterior Kerr metrics are *stationary*, which means, roughly, that the coefficients of the metric are independent of the time variable t. One can reformulate

[5]They must end in the singularity at $r = 0$, in Schwarzschild spacetime. Their behavior in Kerr is more complicated due to the presence of a Cauchy horizon at $r = r_-$ along which the spacetime remains smooth.

this by saying that the vectorfield $T = \partial_t$ is Killing[6] (everywhere in the domain of outer communication) and timelike at points with r large, i.e., the so-called *asymptotic region* (where the spacetime is close to flat). One can also easily check that T is tangent to the horizon $\mathcal{H} = \mathcal{N} \cup \underline{\mathcal{N}}$, which is itself a null hypersurface, i.e., the restriction of the metric to the tangent space to $\mathcal{H}$ is degenerate (see figure 1.7). In addition to being stationary, the coefficients of the Kerr metric are independent of the coordinate φ. Thus Kerr is stationary and *axially symmetric*. It has been conjectured that all asymptotically flat stationary solutions of the Einstein vacuum equations must be Kerr solutions. The conjecture has been verified only if additional assumptions are made, see [35] for a recent survey of known results.

The Schwarzschild metrics, corresponding to $a = 0$, are not just axially symmetric but spherically symmetric, which means that the metric is left invariant by the whole rotation group of the standard sphere $\mathbb{S}^2$. A well known theorem of Birkhoff shows that they are the only such solutions of the Einstein vacuum equations. Another peculiarity of a Schwarzschild metric, not true in the case of Kerr, is that the stationary Killing vectorfield $T = \partial_t$ is orthogonal to the hypersurface $t = 0$. A stationary spacetime which has this property is called *static*. This is also equivalent to the fact that the Schwarzschild metric is invariant with respect to the reflection $t \to -t$. Moreover, T is timelike for all $r > 2m$ and null along the Schwarzschild horizon $\mathcal{H} = \{r = 2m\}$. This is not the case for Kerr solutions in which case $T = \partial_t$ is only timelike for $r > m + (m^2 - a^2\cos^2\theta)^{1/2}$, null for $r = m + (m^2 - a^2\cos^2\theta)^{1/2}$ and spacelike in the region between r_+ and $r = m + (m^2 - a^2\cos^2\theta)^{1/2}$, called the *ergosphere*. Finally we remark that the Kerr family is not physically relevant for $|a| > m$, hence the restriction to $|a| \leq m$.

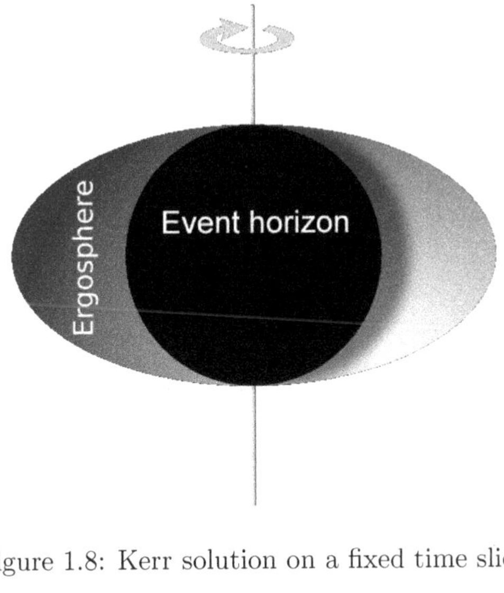

Figure 1.8: Kerr solution on a fixed time slice

To summarize:

1. The Kerr family $\mathcal{K}(a, m)$, $0 \leq |a| \leq m$, provides a two parameter family of asymp-

[6]A vectorfield X is said to be Killing if its associated 1-parameter flow consists of isometries of $\mathbf{g}$, i.e., the Lie derivative of the metric $\mathbf{g}$ with respect to X vanishes, $\mathcal{L}_X\mathbf{g} = 0$.

totically flat solutions of the Einstein vacuum equations exhibiting a smooth domain of outer communication and its complement, separated by the event horizon $\{r = r_+\}$. For $|a| < m$, the event horizon consists of two null hypersurfaces intersecting transversally on a compact 2-sphere.
2. All Kerr solutions are stationary, i.e., they admit a Killing vectorfield T which is timelike in the *asymptotic region*. The Schwarzschild spacetime (i.e., $a = 0$) is also static. Moreover the Kerr family is axially symmetric, i.e., it admits another Killing vectorfield Z which vanishes on the axis of symmetry. The Schwarzschild spacetime is spherically symmetric.
3. The stationary vectorfield T is tangent along the horizon and spacelike for all $0 < |a| \le m$. It remains spacelike in a small region of DOC called ergoregion. In the particular case $a = 0$, T is null along the horizon and timelike everywhere in DOC.
4. In all cases $0 \le |a| \le m$, DOC contains trapped null geodesics, i.e., null geodesics which are entirely contained in a region of DOC with a bounded value of r. In the case $a = 0$, all trapped null geodesics are either tangent to the timelike surface $\{r = 3m\}$ or asymptotic to it.
5. All physically acceptable Kerr solutions, i.e., $|a| \le m$, have complete future and past null infinities corresponding to $r = \infty$.

Here are some other important properties of the Kerr family.

- The Kerr solution has a remarkable algebraic feature, encoded in the so-called Petrov type D property, according to which it admits, at every point a pair of null vectors $(l, \underline{l})$, normalized by the condition $\mathbf{g}(l, \underline{l}) = -2$, called principal null vectors, such that all components of the Riemann curvature tensor vanish identically except for the two independent components
$$\mathbf{R}(l, \underline{l}, l, \underline{l}), \quad {}^\star\mathbf{R}(l, \underline{l}, l, \underline{l}),$$
with ${}^\star\mathbf{R}$ the Hodge dual of $\mathbf{R}$.
- In addition to the symmetries provided by the Killing vectorfields $\mathbf{T}$ and $\mathbf{Z}$, the Kerr solution possesses a nontrivial Killing tensor, i.e., a symmetric 2-covariant tensor $\mathbf{C}$ (the Carter tensor) verifying
$$\mathbf{D}_{(\alpha}\mathbf{C}_{\beta\gamma)} = 0.$$
- The Kerr family is distinguished among all stationary solutions of EVE by the vanishing of a four tensor called the Mars-Simon tensor, see [51].

1.1.4 Stability of Minkowski space

The Minkowski space $(\mathbb{R}^{1+3}, \mathbf{m})$ is the simplest solution of the Einstein vacuum equations. Note that it belongs to the Kerr family and corresponds to the particular case $a = m = 0$. Among all Kerr solutions, the Minkowski space is the only one free of pathologies such as singular boundaries, or the presence of Cauchy horizons. In particular, it is geodesically complete, i.e., any freely moving observer in $\mathcal{M}$ can be extended indefinitely, as measured relative to its proper time. Such a spacetime is said to have a regular MFGHD. Does this property persist under small perturbations?

The result stated below is a rough version of the global stability of Minkowski.

The complete result also provides very precise information about the decay of the curvature tensor along null and timelike directions as well as much other geometric information concerning the causal structure of the corresponding spacetime, see [20], as well as [42], [49] and [7]. Of particular interest are *peeling properties*, i.e., the precise decay rates of various components of the curvature tensor along future null geodesics.

Theorem 1.4 (Global stability of Minkowski). *The maximal future development of an asymptotically flat initial data set, sufficiently close to that of Minkowski space, in an appropriate topology, is geodesically complete and converges to the Minkowski space.*

Here are, very schematically, some of the main ideas in the proof of the stability of Minkowski space.

1. Perturbations radiate and decay *sufficiently fast* (just fast enough!) to insure convergence.
2. Interpret the Bianchi identities as a Maxwell like system. This is an effective, *invariant*, way to treat the hyperbolic character of the equations.
3. Rely on four important PDE advances of late last century:
   a) Vectorfield approach to get decay based on *approximate* Killing and conformal Killing symmetries of the equations, see [39], [40], [41], [19].
   b) Generalized energy estimates using both the Bianchi identities and the approximate Killing and conformal Killing vectorfields.
   c) The *null condition* identifies the deep mechanism for nonlinear stability, i.e., the specific structure of the nonlinear terms enables stability despite the slow decay rate of the perturbations, see [38], [40], [18].
   d) Involved bootstrap argument according to which one makes educated assumptions about the behavior of the spacetime and then proceeds to show that they are in fact satisfied. This amounts to a *conceptual linearization*, i.e., a method by which the equations become, essentially, linear[7] without actually linearizing them.

1.1.5 Cosmic censorship

Unlike the situation described in Theorem 1.4, we expect maximal developments of typical, nonsmall, initial data sets to be incomplete, with singular boundaries. As shown by D. Christodoulou [21], trapped surfaces can form in evolution starting with regular initial conditions.[8] Together with the well known singularity theorem of R. Penrose, these results show that there exists a large class of regular initial data whose MFGHD is incomplete.

The unavoidable presence of singularities, for sufficiently large initial data sets, as well as the analysis of explicit examples (such as Schwarzschild and Kerr) have led Penrose to formulate two fundamental conjectures, concerning the character of general solutions to the Einstein equations. Here we restrict our discussion only to the so-called weak cosmic censorship conjecture (WCC), which is the only one relevant to the problem of stability. To understand the statement of WCC, consider

[7]With quadratic and higher order terms satisfying the null condition on the right-hand side.
[8]That is, free of trapped surfaces. See also more recent results in [45], [44] and [2].

the different behavior of null rays in Schwarzschild and Minkowski spacetimes. In Minkowski space, light originating at any point $p = (t_0, x_0)$ propagates, towards future, along the null rays of the null cone $t - t_0 = |x - x_0|$. Any free observer in $\mathbb{R}^{1+3}$, following a straight timelike line, will necessarily meet this light cone in finite time, thus experiencing the event p. On the other hand, any point p in the trapped region $r < 2m$ of the Schwarzschild space is such that all null rays initiating at p remain trapped in the region $r < 2m$. In particular events causally connected to the singularity at $r = 0$ cannot influence events in the domain of outer communication $r > 2m$, which is thus entirely free of singularities. The same holds true in any Kerr solution with $0 \leq |a| \leq m$.

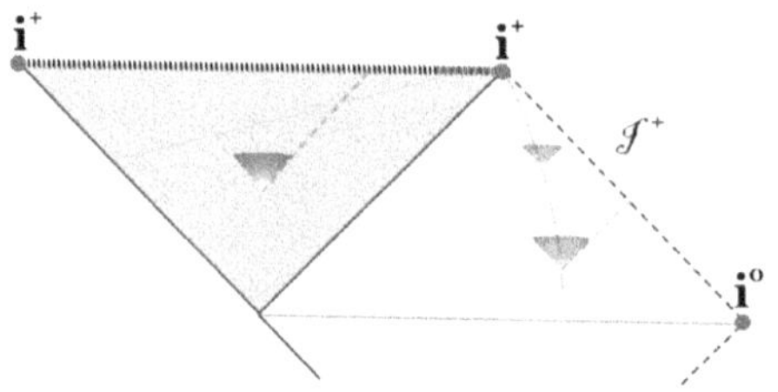

Figure 1.9: Behavior of null geodesics outside and inside the black hole

WCC is an optimistic extension of this fact to future developments of general, asymptotically flat initial data sets. The desired conclusion of the conjecture is that any such development, with the possible exception of a non-generic set of initial conditions, has the property that any *sufficiently distant observer* will not encounter singularities. To make this more precise, one needs to define what a sufficiently distant observer means. This is typically done by introducing the notion of future null infinity $\mathcal{I}^+$ which provides end points for the null geodesics that propagate to asymptotically large distances. As in the cases analyzed above, future null infinity is constructed by conformally embedding the physical spacetime $(\mathcal{M}, \mathbf{g})$ to a larger spacetime $(\widetilde{\mathcal{M}}, \widetilde{\mathbf{g}})$ such that $\widetilde{\mathbf{g}} = \Omega^2 \mathbf{g}$ in $\mathcal{M}$, with a null boundary $\mathcal{I}^+$ (where $\Omega = 0, d\Omega \neq 0$).

Definition 1.5. *The future null infinity $\mathcal{I}^+$ is said to be complete*[9] *if any future null geodesic along it can be indefinitely extended relative to an affine parameter.*[10]

Conjecture (Weak Cosmic Censorship conjecture). *Generic asymptotically flat initial data sets have maximal future developments possessing a complete future null infinity.*

Once the completeness of future null infinity has been established, one can then define the black hole region $\mathcal{B}$ to be the complement of the causal past of null

[9] A more precise definition of complete future null infinity, which avoids the technical and murky issue of the precise degree of smoothness of the conformal compactification, was proposed by Christodoulou in [17].

[10] This can be informally reformulated, for MFGHD spaces, by stating that there exists a sequence of relatively compact sets K_n exhausting the initial hypersurface $\Sigma_{(0)}$ such that the proper future time of observers starting in $K_{n+1} \setminus K_n$ tends to infinity as $n \to \infty$.

infinity

$$\mathcal{B} := \mathcal{M} \setminus \mathcal{J}^{-}(\mathcal{I}^{+}). \tag{1.1.7}$$

The boundary $\mathcal{H}^+$ of $\mathcal{B}$ is called the event horizon of the black hole.

1.2 STABILITY OF KERR CONJECTURE

The nonlinear stability of the Kerr family is one of the most pressing issues in mathematical GR today. Roughly, the problem is to show that all spacetime developments of initial data sets, sufficiently close to the initial data set of a Kerr spacetime, behave in the large like a (typically another) Kerr solution. This is not only a deep mathematical question but one with serious astrophysical implications. Indeed, if the Kerr family would be unstable under perturbations, black holes would be nothing more than mathematical artifacts. Here is a more precise formulation of the conjecture.

Conjecture (Stability of Kerr conjecture). *Vacuum initial data sets, sufficiently close to Kerr initial data, have a maximal development with complete future null infinity and with domain of outer communication*[11] *which approaches (globally) a nearby Kerr solution.*

There are four, related, major obstacles in passing from the stability of Minkowski to that of the Kerr family.

1. The first can be understood in the general framework of nonlinear hyperbolic or dispersive equations. Given a nonlinear equation $\mathcal{N}[\phi] = 0$ and a stationary solution ϕ_0 we have two notions of stability, *orbital stability*, according to which small perturbations of ϕ_0 lead to solutions ϕ which remain close, in some norm (typically L^2 based) for all time, and *asymptotical stability*, according to which the perturbed solutions converge, as $t \to \infty$, to a nearby stationary solution. Note that the second notion is far stronger, and much more precise, than the first and that orbital stability can only be established (without appealing to the stronger version) only for equations with very weak nonlinearities. For quasilinear equations, such as the Einstein field equations, a proof of stability requires, necessarily, a proof of asymptotic stability. This must then be based on a detailed understanding of the decay properties of the linearized[12] equations.
One is thus led to study the linearized equations $\mathcal{N}'[\phi_0]\psi = 0$, with $\mathcal{N}'[\phi_0]$ the Fréchet derivative of $\mathcal{N}$ at ϕ_0, which, in many important cases, are hyperbolic[13] systems with variable coefficients that typically present instabilities. In the exceptional situation, when nonlinear stability can ultimately be established, one can tie all the instability modes of the linearized system to two properties of the

[11]This presupposes the existence of an event horizon. Note that the existence of such an event horizon can only be established upon the completion of the proof of the conjecture.

[12]It is irrelevant whether a specific linearization procedure needs to be implemented; what is important here is to identify the linear mechanism for decay, such as the Maxwell system in the case of the stability of Minkowski space mentioned above.

[13]In the case of EVE the linearized equations are linear hyperbolic only after we mod out the linearized version of general coordinate transformations.

nonlinear equation:

a) The presence of a continuous[14] family of other stationary solutions of $\mathcal{N}[\phi] = 0$ near ϕ_0.
b) The presence of a continuous family of diffeomorphisms[15] of the background manifold which map, by pullback, solutions to solutions.

For a typical stationary solution ϕ_0, both properties exist and generate nontrivial solutions of the linearized equation $\mathcal{N}'[\phi_0]\psi = 0$. In the case of relatively simple scalar nonlinear equations, where the symmetry group of the equation is small, an effective strategy of dealing with this problem (known under the name of modulation theory) has been developed, see for example [53], [55]. In the case of the Einstein equations this problem is compounded by the large invariance group of the equations, i.e., all diffeomorphisms of the spacetime manifold. To deal with both problems and establish stability one has to:

- Track the parameters (a_f, m_f) of the final Kerr spacetime.
- Track the coordinate system (gauge condition) relative to which we have decay for all linearized quantities. Such a coordinate system cannot be imposed a priori, it has to emerge dynamically in the construction of the spacetime.

2. As described earlier, the fundamental insight in the stability of the Minkowski space was that we can treat the Bianchi identities as a Maxwell system in a slightly perturbed Minkowski space by using the vectorfield method. This cannot work for perturbations of Kerr due to the fact that some of the null components of the curvature tensor[16] are nontrivial in Kerr.
3. Even if we can establish a useful version of linearization (i.e., one which addresses the above mentioned problems), there are still major obstacles in understanding their decay properties. Indeed, when one considers the simplest, relevant, linear equation on a fixed Kerr background, i.e., the wave equation $\square_{\mathbf{g}}\psi = 0$ (often referred to as the *poor man's linearization* of EVE), one encounters serious difficulties even to prove the boundedness of solutions for the most reasonable, smooth, compactly supported data. Below is a very short description of these.
   - *The problem of trapped null geodesics.* This concerns the existence of null geodesics[17] neither crossing the event horizon nor escaping to null infinity, along which solutions can concentrate for arbitrary long times. This leads to degenerate energy estimates which require a very delicate analysis.
   - *The trapping properties of the horizon.* The horizon itself is ruled by null geodesics, which do not communicate with null infinity and can thus concentrate energy. This problem was solved by understanding the so-called redshift effect associated to the event horizon, which more than counteracts this type of trapping.
   - *The problem of superradiance.* This is essentially the failure of the stationary Killing field $\mathbf{T} = \partial_t$ to be everywhere timelike in the domain of outer communications and, thus, the failure of the associated conserved energy to be positive. Note that this problem is absent in Schwarzschild and, in general,

[14]This is responsible for the fact that a small perturbation of the fixed stationary solution ϕ_0 may not converge to ϕ_0 but to another nearby stationary solution. In the particular case of the stability of Kerr we have a two parameter family of solutions $\mathcal{K}(a, m)$.

[15]In the case of EVE, any diffeomorphism has that property.

[16]With respect to the so-called principal null directions.

[17]In the Schwarzschild case, these geodesics are located on the so-called photon sphere $r = 3m$.

for axially symmetric solutions.

- *Superposition problem.* This is the problem of combining the estimates in the near region, close to the horizon (including the ergoregion and trapping), with estimates in the asymptotic region, where the spacetime looks Minkowskian.

4. The full linearized system of EVE around Kerr, usually referred to as the linearized gravity system (LGS), whatever its formulation, presents far more difficulties beyond those mentioned above concerning the poor man's linear scalar wave equation on Kerr, see the discussion below.

Historically, two versions of LGS have been considered.

a) At the level of the metric itself, i.e., if $\mathbf{G}$ denotes the Einstein tensor, $\mathbf{G}_{\alpha\beta} = \mathbf{R}_{\alpha\beta} - \frac{1}{2}\mathbf{R}\mathbf{g}_{\alpha\beta}$,

$$\mathbf{G}'(\mathbf{g}_0)\,\delta\mathbf{g} = 0. \tag{1.2.1}$$

b) Via the Newman-Penrose (NP) formalism, based on null frames.

In what follows we review the main known results concerning solutions to the linearized equations on a Kerr background.

1.2.1 Formal mode analysis

The first important results concerning both items 3 and 4 above were obtained by physicists based on the classical method of separation of variables and formal mode analysis. In the particular case where $\mathbf{g}_0$ is the Schwarzschild metric, the linearized equations (1.2.1) can be formally decomposed into modes, by associating t-derivatives with multiplication by $i\omega$ and angular derivatives with multiplication by l, i.e., the eigenvalues of the spherical Laplacian. A similar decomposition, using oblate spheroidal harmonics, can be done in Kerr. The formal study of fixed modes from the point of view of *metric perturbations* as in (1.2.1) was initiated by Regge-Wheeler [58] who discovered the master Regge-Wheeler equation for odd-parity perturbations. This study was completed by Vishveshwara [65] and Zerilli [69]. A gauge-invariant formulation of *metric perturbations* was then given by Moncrief [56]. An alternative approach via the Newman-Penrose (NP) formalism was first undertaken by Bardeen-Press [6]. This latter type of analysis was later extended to the Kerr family by Teukolsky [64] who made the important discovery that the extreme curvature components, relative to a principal null frame, satisfy decoupled, separable, wave equations. These extreme curvature components also turn out to be *gauge invariant* in the sense that small perturbations of the frame lead to quadratic errors in their expression. The full extent of what could be done by mode analysis, in both approaches, can be found in Chandrasekhar's book [11]. Chandrasekhar also introduced (see [12]) a transformation theory relating the two approaches. More precisely, he exhibits a transformation which connects the Teukolsky equations to the Regge-Wheeler one. This transformation was further elucidated and extended by R. Wald [67] and recently by Aksteiner et al. [1]. The full mode stability, i.e., lack of exponentially growing modes, for the Teukolsky equation on Kerr is due to Whiting [68] (see also [61] for a stronger quantitive version).

1.2.2 Vectorfield method

Note that mode stability is far from establishing even boundedness of solutions to the linearized equations. To achieve that and, in addition, to derive realistic decay estimates one needs an entirely different approach based on a far-reaching extension of the classical vectorfield method[18] used in the proof of the nonlinear stability of Minkowski [20]. The new vectorfield method compensates for the lack of enough Killing and conformal Killing vectorfields on a Schwarzschild or Kerr background by introducing new vectorfields whose deformation tensors have coercive properties in different regions of spacetime, not necessarily causal. The new method has emerged in the last fifteen years in connection to the study of boundedness and decay for the scalar wave equation in the Kerr space $\mathcal{K}(a,m)$,

$$\square_{\mathbf{g}_{a,m}}\psi = 0. \tag{1.2.2}$$

The starting and most demanding part of the new method is the derivation of a global, simultaneous, *Energy-Morawetz* estimate which degenerates in the trapping region. This task is somewhat easier in Schwarzschild, or for axially symmetric solutions in Kerr, where the trapping region is restricted to a smooth hypersurface. The first such estimates, in Schwarzschild, were proved by Blue and Soffer in [8], [9], followed by a long sequence of further improvements in [10], [22], [54], etc. See also [36] and [62] for a vectorfield method treatment of the axially symmetric case in Kerr with applications to nonlinear equations. In the absence of axial symmetry the derivation of an Energy-Morawetz estimate in $Kerr(a,m)$, $|a/m| \ll 1$ requires a more refined analysis involving either Fourier decompositions, see [24], [63], or a systematic use of the second order Carter operator, see [3]. The derivation of such an estimate in the full sub-extremal case $|a| < m$ is even more subtle and was recently achieved by Dafermos, Rodnianski and Shlapentokh-Rothman [28] by combining mode decomposition with the vectorfield method.

Once an Energy-Morawetz estimate is established one can commute with the time translation vectorfield and the so-called *redshift* vectorfield,[19] first introduced in [22], to derive uniform bounds for solutions. The most efficient way to also get decay, and solve the *superposition problem*, is due to Dafermos and Rodnianski, see [23], based on the presence of a family of r^p*-weighted*, quasi-conformal vectorfields defined in the far r region of spacetime.[20]

[18]Method based on the symmetries of Minkowski space to derive uniform, robust decay for nonlinear wave equations, see [39], [40], [41], [19].

[19]Note that the redshift vectorfield is also used as a multiplier in the derivation of the Energy-Morawetz estimate.

[20]These replace the scaling and inverted time translation vectorfields used in [39] or their corresponding deformations used in [20]. A recent improvement of the method, relevant to our work here, allowing one to derive higher order decay can be found in [5]. See also [57] for further extensions of this method.

1.3 NONLINEAR STABILITY OF SCHWARZSCHILD UNDER POLARIZED PERTURBATIONS

1.3.1 Bare-bones version of our theorem

The goal of the book is to prove the nonlinear stability of the Schwarzschild spacetime under axially symmetric polarized perturbations, i.e., solutions of the Einstein vacuum equations (1.1.3) for asymptotically flat $1+3$ dimensional Lorentzian metrics which admit a hypersurface orthogonal spacelike Killing vectorfield $\mathbf{Z}$ with closed orbits. This class of perturbations allows us to restrict our analysis to the case when the final state of evolution is itself a Schwarzschild spacetime. This is not the case in general, as a typical perturbation of Schwarzschild may approach a member of the Kerr family with small angular momentum.

The simplest version of our main theorem can be stated as follows.

Theorem 1.6 (Main Theorem (first version)). *The future globally hyperbolic development of an* axially symmetric, polarized,[21] *asymptotically flat initial data set, sufficiently close (in a specified topology) to a Schwarzschild initial data set of mass* $m_0 > 0$, *has a complete future null infinity* $\mathcal{I}^+$ *and converges in its causal past* $\mathcal{J}^{-1}(\mathcal{I}^+)$ *to another nearby Schwarzschild solution of mass* m_∞ *close to* m_0.

Our theorem is an important step in the long-standing effort to prove the full nonlinear stability of Kerr spacetimes $\mathcal{K}(a,m)$, in the sub-extremal regime $|a| < m$. We give a succinct review below of some of the most important results which have been obtained so far in this direction.

1.3.2 Linear stability of the Schwarzschild spacetime

A first quantitative (i.e., which provides precise decay estimates) proof of the linear stability of Schwarzschild spacetime has recently been established[22] by Dafermos, Holzegel and Rodnianksi in [26], via the NP formalism (expressed in a double null foliation[23]). It is important to note that while the Teukolsky equation (in the NP formalism) is separable, and thus amenable to mode analysis, it is not Lagrangian and thus cannot be treated by direct energy type estimates. To overcome this difficulty [26] relies on a new physical space version of the Chandrasekhar transformation [12], which takes solutions of the Teukolsky equations to solutions of Regge-Wheeler, which is manifestly both Lagrangian and coercive. After quantitative decay has been established for this latter equation, based on the new vectorfield method, the physical space form of the transformation allows one to derive quantitative decay for solutions of the original Teukolsky equation. Once decay estimates for the Teukolsky equation have been established, the remaining work in [26] is to bound all other curvature and Ricci coefficients associated to the double null foliation. This last step requires carefully chosen gauge conditions along the

[21] See section 2.1.1 for a precise definition of axial symmetry and polarization. This property is preserved by the Einstein equations, i.e., if the data is axially symmetric, polarized, so is its development.

[22] A somewhat weaker version of linear stability of Schwarzschild was subsequently proved in [34] by using the original, direct, Regge-Wheeler, Zerilli approach combined with the vectorfield method and adapted gauge choices. See also [37] for an alternate proof of linear stability of Schwarzschild using wave coordinates.

[23] This is possible in Schwarzschild where the principal null directions are integrable.

event horizon of the fixed Schwarzschild background. This final gauge is itself then quantitatively bounded in terms of the initial data, thus giving a comprehensive statement of linear stability.

1.3.3 Main ideas in the proof of Theorem 1.6

In the passage from linear to nonlinear stability of Schwarzschild one has to overcome major new difficulties. Some are similar to those encountered in the stability of Minkowski [20], such as:

1. Need of an appropriate geometric setting which takes into account the decay and peeling properties of the curvature. In [20] this was achieved with the help of the foliation of the perturbed spacetime given by two optical functions ${}^{(int)}u$ and ${}^{(ext)}u$ and a maximal time function t. The exterior optical function ${}^{(ext)}u$, which was initialized at infinity, was essential to derive the decay and peeling properties along null directions while ${}^{(int)}u$, initialized on a timelike axis, was responsible for covering the interior, non-radiative, back scattering decay.
2. The peeling and decay estimates have to be derived by some version of the geometric vectorfield method which relates decay to generalized energy type estimates.
3. The peeling and decay estimates mentioned above should be sufficiently strong to be able to deal with the error terms generated by the vectorfield method. For this to happen, the error terms need to exhibit an appropriate null structure.

The new main difficulties are as follows:

1. One needs a procedure which allows one to take into account the change of mass and detect its final value. Note also that we need to restrict the nature of the perturbations to insure that the final state of a perturbation of Schwarzschild is still Schwarzschild.
2. While in the stability of Minkowski space all components of the curvature tensor were expected to approach zero, this is no longer true. Indeed, the middle curvature component (relative to an adapted null frame) ought to converge to its respective value in the final Schwarzschild spacetime. This statement is unfortunately hard to quantify since that value depends both on the final mass and on the corresponding Schwarzschild coordinates. Moreover, some of the other curvature components, which are expected to converge to zero, are also ill defined since a small change of the null frame can produce small linear distortion to the basic equation which these curvature components verify. Note that this difficulty was absent in the stability of the Minkowski space where small changes in the frame produce only quadratic errors.
3. The classical vectorfield method used in the nonlinear stability of Minkowski space was based on the construction, together with the spacetime, of an adequate family of approximate Killing and conformal Killing vectorfields which mimic the role played by the corresponding vectorfields in Minkowski space in establishing uniform decay estimates. The Schwarzschild space, however, has a much more limited set of Killing vectorfields and no useful conformal Killing ones. As mentioned above, this problem appears already in the analysis of the standard scalar linear wave equation in Schwarzschild.
4. As in the stability of the Minkowski space, one needs to make gauge conditions to insure that we are measuring decay relative to an appropriate center of mass

frame. Yet, as we saw above, it is no longer true that small perturbations of the null frame produce only quadratic errors for the curvature, as was the case in the stability of Minkowski space. In fact, the center of mass frame of the perturbed black hole continuously changes in response to incoming radiation. This, the so-called *recoil problem*, does not occur in linear theory.

Here is a very short summary of how we solve these new challenges in our work.

1. We resolve the first difficulty by restricting our analysis to axially symmetric, polarized perturbations and by tracking the mass using a quantity, called the quasi-local Hawking mass, for which we derive simple propagation equations which establish monotonicity of the mass up to errors which are quadratic with respect to the perturbations.
2. We resolve the second difficulty by making use of the fact that the extreme components of the curvature are, up to quadratic terms, invariant under null frame transformations. As in [26], we also make use of a transformation, similar to that of Chandrasekhar mentioned above, which maps the extreme components of the curvature to a new quantity $\mathfrak{q}$, defined up to quadratic errors, that verifies a Regge-Wheeler type equation. Once we manage to control $\mathfrak{q}$, i.e., to derive quantitative decay estimates for it, we can also control, in principle,[24] the two extreme curvature invariants α and $\underline{\alpha}$, the first by inverting the Chandrasekhar transformation and the second by using a variant of the Teukolsky-Starobinsky identities. One is then left with the arduous task of recovering[25] all other null components of the curvature tensor and all connection coefficients.
3. The third difficulty manifests itself in the most sensitive part of the entire argument, i.e., in the task of deriving quantitative decay estimates for $\mathfrak{q}$ by making use of the Regge-Wheeler type equation it verifies. To do this we rely on the new vectorfield method as outlined in section 1.2.2 above. The main new difficulties are:
   a) The vectorfield method introduces new error terms, not present in linear theory. To estimate these terms we need precise decay information, off the final Schwarzschild space, for all connection coefficients and curvature of the perturbation.
   b) The most difficult terms are those due to the quadratic errors made in the derivation of the Regge-Wheeler equation for $\mathfrak{q}$. As in the proof of the stability of the Minkowski space, the precise rates of decay for various curvature and connection coefficients, i.e., the peeling properties of the perturbation, and the precise structure of these error terms is of fundamental importance.
4. We solve the fourth and most important new difficulty by a procedure we call *General Covariant Modulation* (GCM). This procedure, which takes advantage of the full covariance of the Einstein equations, allows us to construct the perturbed spacetime by a continuity argument involving finite GCM admissible spacetimes $\mathcal{M}$ as represented in figure 1.10. The past boundaries $\underline{\mathcal{C}}_1 \cup \mathcal{C}_1$ are incoming and outgoing null hypersurfaces on which the initial perturbation is prescribed. The future boundaries consist of the union $\mathcal{A} \cup \underline{\mathcal{C}}_* \cup \mathcal{C}_* \cup \Sigma_*$ where $\mathcal{A}$ and Σ_* are spacelike, $\underline{\mathcal{C}}_*$ is incoming null, $\mathcal{C}_*$ outgoing null. The boundary $\mathcal{A}$ is chosen

[24] Provided that one can deal with the nonlinear terms.
[25] In the linear setting this was partially achieved in [25].

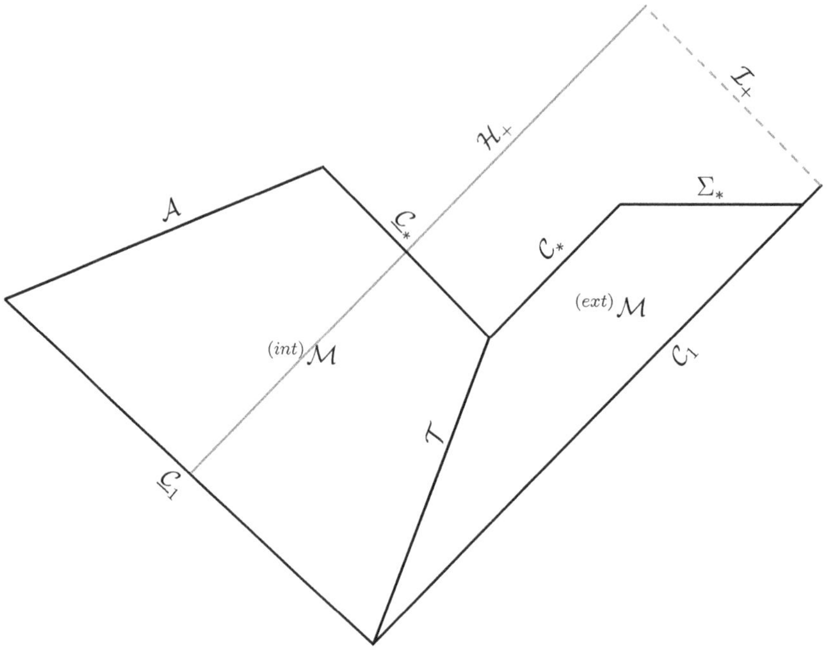

Figure 1.10: The GCM admissible spacetime $\mathcal{M}$

so that, in the limit when $\mathcal{M}$ converges to the final state, it is included in the perturbed black hole. The spacelike boundary Σ_* plays a fundamental role in our construction as seen below. The spacetime $\mathcal{M}$ also contains a timelike hypersurface $\mathcal{T}$ which divides $\mathcal{M}$ into an exterior region we call $^{(ext)}\mathcal{M}$ and an interior one $^{(int)}\mathcal{M}$. We say that $\mathcal{M}$ is a GCM admissible spacetime if it verifies the following properties.

a) The far region $^{(ext)}\mathcal{M}$ is foliated by a geodesic foliation induced by an outgoing optical function u initialized on Σ_*
b) The near region $^{(int)}\mathcal{M}$ is foliated by a geodesic foliation induced by an incoming optical function $\underline{u}$ initialized at $\mathcal{T}$ such that its level sets on $\mathcal{T}$ coincide with those of u.
c) The foliation induced on Σ_* is such that specific geometric quantities take Schwarzschildian values. We refer to these as GCM conditions. These conditions are dynamically reset in the continuation process on which our proof is based.
d) The area radius $r(u)$ of the spheres of constant u along Σ_* is far greater than the corresponding value of u. This condition allows us to simplify somewhat the null structure and Bianchi equations induced on Σ_* and corresponds to the expectation that the spacelike hypersurfaces Σ_* converges to the null infinity of the final state of the perturbation.

5. The GCM conditions together with the control derived on $\mathfrak{q}$, α and $\underline{\alpha}$ mentioned earlier allow us to control all null connection and curvature coefficients along on Σ_*, i.e., to derive appropriated decay estimates for them. These estimates can then be transported to $^{(ext)}\mathcal{M}$ using the full scope of the null structure and null Bianchi identities associated to the outgoing geodesic foliation.

6. The decay estimates in ${}^{(ext)}\mathcal{M}$ can then be used as initial conditions along the timelike hypersurface $\mathcal{T}$ for the incoming foliation of ${}^{(int)}\mathcal{M}$. These allow us to also derive appropriate decay estimates for all null connection and curvature coefficients of the foliation induced by $\underline{u}$.
7. The precise decay estimates derived in 5 are sufficiently strong to allow us to control all error terms generated in the process of estimating $\mathfrak{q}$, as mentioned in 3.

Note that in figure 1.10, one starts with initial conditions on the union of null hypersurfaces $\mathcal{C}_1 \cup \underline{\mathcal{C}}_1$ rather than an initial spacelike hypersurface $\Sigma_{(0)}$. One can justify this simplification based on the results of [42], [43], see Remark 3.12. The full red line $\mathcal{H}_+$ represents the future event horizon of the perturbed Schwarzschild. The line $\mathcal{T}$ represents the timelike hypersurface separating ${}^{(int)}\mathcal{M}$ from ${}^{(ext)}\mathcal{M}$. In deriving decay estimates the precise choice of $\mathcal{T}$ is irrelevant. A choice, however, needs to be made in order to avoid a derivative loss for our top energy estimates.[26]

The spacetime is constructed by a continuity argument, i.e., we assume that the spacetime terminating at $\mathcal{C}_* \cup \underline{\mathcal{C}}_*$ saturates a given bootstrap assumption (**BA**) and show, by a long sequence of a priori estimates which take advantage of the smallness of the initial perturbation, that (**BA**) can be improved and the spacetime extended past $\mathcal{C}_* \cup \underline{\mathcal{C}}_* \cup \Sigma_*$.

Our work here is the first to prove the nonlinear stability of Schwarzschild in a restricted class of nontrivial perturbations, i.e., perturbations for which new ideas, such as our GCM procedure are needed. To a large extent, the restriction to this class of perturbations is only needed to ensure that the final state of evolution is another Schwarzschild space. We are thus confident that our procedure may apply in a more general setting. We would like to single out two other recent important contributions to nonlinear stability of black holes. In the context of asymptotically flat Einstein vacuum equations the result of Dafermos-Holzegel-Rodnianski [25] constructs a class of Kerr black hole solutions starting from future infinity while Hintz-Vasy [31][27] prove the nonlinear stability of Kerr-de Sitter, for small angular momentum, in the context of the Einstein vacuum equations with a nontrivial positive cosmological constant. Though the two results are very different they share in common the fact that the perturbations they treat decay exponentially. This makes the analysis significantly easier than in our case when the decay is barely enough to control the nonlinear terms.

1.3.4 Beyond polarization

While we believe that the general strategy outlined in this work can be extended to the general case of perturbations of Kerr, there are a few major conceptual roadblocks, all connected to our symmetry assumption, which have to be overcome.[28] We describe the main effects played by polarization in our work as follows:

1. Polarization plays a crucial role in our nonlinear version of the Chandrasekhar transformation allowing us to pass from the Teukolsky equation for α to the

[26] See [20] for a similar situation.
[27] See also [32] for the stability of Kerr-Newman de Sitter.
[28] Note nevertheless that many steps in the proof of Theorem 1.6 do not depend at all on polarization.

Regge-Wheeler equation for $\mathfrak{q}$ mentioned above in section 1.3.3.
2. Polarization is also used in the derivation of the Morawetz type estimates for $\mathfrak{q}$ in Chapter 10, where we take advantage of the simpler nature of the trapped set for Schwarzschild metrics.
3. The GCM construction, which plays a fundamental role in this book, also makes essential use of polarization. There are fundamental conceptual difficulties to pass from the polarized case to the general case which will require new ideas.
4. A full stability result has to identify not only the final value of the mass m, but also the final value of the angular momentum. In our work, the final value of m is tracked with the help of the Hawking mass. Though there exists in the literature several interesting proposals[29] for a quasi-local definition of angular-momentum, in the spirit of the Hawking mass, none of them seem suitable to the dynamical approach developed here.

1.3.5 Note added in proof

We would like to report on additional progress made on the Kerr stability problem since we submitted our work, and dealing with the issues raised in section 1.3.4.[30]

- A notable development is the extension of the Chandrasekhar transformation to the case of linear Spin-2 equations in Kerr, as well as a methodology to derive, based on it, boundedness and decay estimates for the Teukolsky variable α. This was done independently by Ma [50] and Dafermos-Holzegel-Rodnianski [27]. In collaboration with E. Giorgi, we have extended, see [29], the derivation of the generalized Regge-Wheeler equation to realistic perturbations of Kerr, i.e., perturbations consistent with the decay and boundedness properties established in this work. Our derivation is based on a new geometric formalism which takes into account the lack of integrability of the principal null frames of Kerr.
- The most important advance, achieved in [47], [48], concerns the removal of any symmetry assumption in the construction of GCM spheres.
- The construction of intrinsic GCM spheres in [48] also leads to a canonical definition of the angular momentum on such spheres.

1.4 ORGANIZATION

The book is organized as follows. In Chapter 2 we introduce the main quantities, equations and basic tools needed later. It is our main reference kit providing all main null structure and null Bianchi equations, in general null frames, in the context of axially symmetric polarized spacetimes. Though we work with the reduced equations, i.e., the equations reduced by the symmetries, most of the work in the book does not really depend of the reduction. Besides insuring that the final state is a Schwarzschild space the reduction only plays a significant role in the GCM construction.

Chapter 3, the heart of the book, contains the precise version of our main theorem, its main conclusions as well as a full strategy of its proof, divided in nine

[29] See, for example, [59] or [16].

[30] Though not directly connected to our strategy, we would like also to refererence the work of Andersson-Bäckdahl-Blue-Ma [4] and Häfner-Hintz-Vasy [30] on the linear stability of Kerr.

supporting intermediate results, Theorems M0–M8. We also give a short description of the proof of each theorem.

In the other chapters of this book we give complete proofs of Theorems M0–M8 and a full description of our GCM procedure.

The reader versed in the formalism of null structure and Bianchi equations, as discussed in [20], is encouraged to glance fast over Chapter 2, to get familiarized with the notation, and then move directly to Chapter 3.

Chapter Two

Preliminaries

2.1 AXIALLY SYMMETRIC POLARIZED SPACETIMES

2.1.1 Axial symmetry

We consider vacuum, four dimensional, simply connected, axially symmetric spacetimes $(\mathcal{M}, \mathbf{g}, \mathbf{Z})$ with $\mathbf{g}$ Lorentzian and $\mathbf{Z}$ an axial Killing vectorfield on $\mathcal{M}$. We denote by $\mathfrak{A}$ the axis of symmetry, i.e., the points on $\mathcal{M}$ for which $X := \mathbf{g}(\mathbf{Z}, \mathbf{Z}) = 0$. In the case of interest for us we assume $dX \neq 0$ and that $\mathfrak{A}$ is a smooth manifold of codimension 2. The Ernst potential of the spacetime is given by

$$\sigma_\mu \quad := \quad \mathbf{D}_\mu(-\mathbf{Z}^\alpha \mathbf{Z}_\alpha) - i \in_{\mu\beta\gamma\delta} \mathbf{Z}^\beta \mathbf{D}^\gamma \mathbf{Z}^\delta.$$

The 1-form $\sigma_\mu dx^\mu$ is closed and thus there exists a function $\sigma : \mathcal{M} \to \mathbb{C}$, called the $\mathbf{Z}$-Ernst potential, such that $\sigma_\mu = \mathbf{D}_\mu \sigma$. Note also that $\mathbf{D}_\mu \mathbf{g}(\mathbf{Z}, \mathbf{Z}) = 2\mathbf{G}_{\mu\lambda}\mathbf{Z}^\lambda = -\Re(\sigma_\mu)$ where $G_{\mu\nu} = D_\mu Z_\nu$. Hence we can choose the potential σ such that $\Re\sigma = -X$. By a standard calculation one can show that

$$\Box\sigma = -X^{-1}\mathbf{D}_\mu\sigma\mathbf{D}^\mu\sigma.$$

Definition 2.1. *An axially symmetric Lorentzian manifold $(\mathcal{M}, \mathbf{g}, \mathbf{Z})$ is said to be polarized if the Ernst potential σ is real, i.e., $\sigma = -X$. In that case the metric $\mathbf{g}$ can be written in the form*

$$\mathbf{g} = X d\varphi^2 + g_{ab} dx^a dx^b \tag{2.1.1}$$

where X and g are independent of φ. We refer to the orbit space $\mathcal{M}/\mathbf{Z}$ as the reduced space and the metric $g = g_{ab}dx^a dx^b$ as the reduced metric. Note that the reduced space $(\mathcal{M}/\mathbf{Z}, g)$ is smooth away from the axis $\mathfrak{A}$. Moreover the scalar X verifies the wave equation

$$\Box_{\mathbf{g}} X = X^{-1}\mathbf{D}_\mu X \mathbf{D}^\mu X. \tag{2.1.2}$$

We denote by $\mathbf{R}$, resp. R, the curvature tensor of the spacetime metric $\mathbf{g}$, respectively g, and by $\Box_{\mathbf{g}}$, resp $\Box_g$, the d'Alembertian with respect to $\mathbf{g}$ and resp. the reduced metric g. We also denote by $\mathbf{\Gamma}$ the Christoffel symbols of $\mathbf{g}$ and by Γ the ones of g. Note that the only nonvanishing Christoffel symbols are:

$$\mathbf{\Gamma}^\varphi_{\varphi b} = \frac{1}{2}X^{-1}\partial_b X, \qquad \mathbf{\Gamma}^a_{\varphi\varphi} = -\frac{1}{2}g^{as}\partial_s X, \qquad \mathbf{\Gamma}^a_{bc} = \Gamma^a_{bc}. \tag{2.1.3}$$

One can easily prove the following.

Proposition 2.2. *The scalar curvature R of the reduced metric g of an axially* *symmetric polarized Einstein vacuum spacetime vanishes identically.*[1] *Moreover, setting* $\Phi := \frac{1}{2}\log X$ *we find*

$$R_{ab} = D_a D_b \Phi + D_a\Phi D_b\Phi, \qquad \square_g \Phi = -D^a\Phi D_a\Phi. \tag{2.1.4}$$

Also,

$$\begin{aligned}
\mathbf{R}_{a\varphi b}{}^{\varphi} &= -\frac{1}{2}X^{-1}D_aD_bX + \frac{1}{4}X^{-2}D_aXD_bX = -R_{ab},\\
\mathbf{R}_{acb}{}^{\varphi} &= 0,\\
\mathbf{R}_{abc}{}^{d} &= R_{abc}{}^{d},
\end{aligned} \tag{2.1.5}$$

and,

$$R_{abcd} = g_{ac}R_{bd} + g_{bd}R_{ac} - g_{ad}R_{bc} - g_{bc}R_{ad}. \tag{2.1.6}$$

Finally, when applied to **Z***-invariant functions,*

$$\square_{\mathbf{g}} = \square_g + g^{ab}\partial_a\Phi\partial_b. \tag{2.1.7}$$

Remark 2.3. *The wave equation in* (2.1.4) *is equivalent to*

$$\square_{\mathbf{g}}\Phi = 0. \tag{2.1.8}$$

Remark 2.4. *Schwarzschild spacetime is axially symmetric polarized with*

$$X = r^2(\sin\theta)^2, \qquad \Phi = \log(r) + \log(\sin\theta).$$

2.1.2 Z-frames

We consider orthonormal frames $e_0, e_1, e_\theta = e_2, e_\varphi = X^{-1/2}\mathbf{Z}$, with $X := \mathbf{g}(\mathbf{Z},\mathbf{Z})$, which are **Z**-equivariant, i.e., $[\mathbf{Z}, e_\alpha] = 0$. From now on, the index φ is referring to the frame rather than the coordinates.

Lemma 2.5. *Setting* $(\Lambda_\alpha)_{\beta\gamma} := \mathbf{g}(D_\alpha e_\gamma, e_\beta)$ *we have*

$$(\Lambda_\varphi)_{a\varphi} = -D_a\Phi, \quad (\Lambda_\varphi)_{ab} = (\Lambda_a)_{b\varphi} = 0, \qquad \forall a = 0,1,2, \tag{2.1.9}$$

and,

$$\begin{aligned}
\mathbf{D}_a e_b &= D_a e_b,\\
\mathbf{D}_a e_\varphi &= 0,\\
\mathbf{D}_\varphi e_a &= (\Lambda_\varphi)_{\varphi a} e_\varphi = (D_a\Phi) e_\varphi,\\
\mathbf{D}_\varphi e_\varphi &= (\Lambda_\varphi)_{a\varphi} e_a = -D^a\Phi e_a.
\end{aligned} \tag{2.1.10}$$

Proof. Straightforward verification. □

[1] *This is an easy consequence of the equation* (2.1.2).

Lemma 2.6. *We have*

$$\begin{aligned}
\mathbf{D}_s\mathbf{R}_{abcd} &= D_s R_{abcd},\\
\mathbf{D}_s\mathbf{R}_{\varphi bcd} &= 0,\\
\mathbf{D}_s\mathbf{R}_{\varphi b\varphi d} &= -D_s R_{bd},\\
\mathbf{D}_\varphi\mathbf{R}_{abcd} &= 0,\\
\mathbf{D}_\varphi\mathbf{R}_{\varphi bcd} &= D^s\Phi R_{sbcd} + D_c\Phi R_{bd} - D_d\Phi R_{bc},\\
\mathbf{D}_\varphi\mathbf{R}_{\varphi b\varphi d} &= 0.
\end{aligned}$$

Proof. Straightforward verification. □

Definition 2.7. *We say that a spacetime tensor* $\mathbf{U}$ *is* $\mathbf{Z}$*-invariant if* $\mathcal{L}_{\mathbf{Z}}U = 0$ *and* $\mathbf{Z}$*-invariant polarized if its contractions to an odd number of* $e_\varphi = X^{-1/2}\mathbf{Z}$ *vanish identically.*

Proposition 2.8. *All higher covariant derivatives of the Riemann curvature tensor* $\mathbf{R}$ *of an axially symmetric polarized spacetime* $(\mathbf{M}, \mathbf{g}, \mathbf{Z})$ *are* $\mathbf{Z}$*-invariant, polarized.*

Proof. The statement has been already verified above for both $\mathbf{R}$ and $\mathbf{DR}$. It suffices to show that, given an arbitrary $\mathbf{Z}$-invariant, polarized tensor $\mathbf{U}$, its covariant derivative $\mathbf{DU}$ is also $\mathbf{Z}$-invariant, polarized. The invariance is immediate. To show polarization we consider all frame components of $\mathbf{DU}$ with respect to our adapted equivariant frame e_1, e_2, e_3, e_φ. Assume first that the components of $\mathbf{DU}$ contain only one e_φ. These are

$$\mathbf{D}_\varphi\mathbf{U}_a, \qquad \mathbf{D}_a\mathbf{U}_{b\varphi c}$$

with various combinations of horizontal indices a, b, c. Now, in view of the polarization property of $\mathbf{U}$ and the relations $\mathbf{D}_a e_b = D_a e_b$, $\mathbf{D}_a e_\varphi = 0$ we easily deduce

$$\mathbf{D}_a\mathbf{U}_{b\varphi c} = e_a\mathbf{U}_{b\varphi c} - \mathbf{U}_{\mathbf{D}_a b\varphi c} - \mathbf{U}_{b\mathbf{D}_a\varphi c} - \mathbf{U}_{b\varphi c} = 0.$$

Similarly, since $e_\varphi(U_a) = X^{-1/2}\mathbf{Z}(U_a) = X^{-1/2}\mathcal{L}_{\mathbf{Z}}U_a = 0$ and $\mathbf{D}_\varphi e_a$ is proportional to e_φ,

$$\mathbf{D}_\varphi\mathbf{U}_a = e_\varphi(\mathbf{U}_a) - \mathbf{U}_{\mathbf{D}_\varphi}e_a = 0.$$

Similarly we can check that the contraction of $\mathbf{DU}$ with any odd number of e_φ must be zero. □

In what follows we shall refer to $\mathbf{Z}$-invariant, polarized tensors as simply $\mathbf{Z}$-polarized.

2.1.3 Axis of symmetry

We denote by $\mathfrak{A}$ the axis of symmetry of $\mathbf{Z}$, i.e., the set of zeroes of $X = \mathbf{g}(\mathbf{Z}, \mathbf{Z})$. Since we assume $dX \neq 0$, $\mathfrak{A}$ is a smooth timelike submanifold of dimension 2. In view of the definition of axial symmetry every trajectory of $\mathbf{Z}$ is closed and intersects $\mathfrak{A}$ at one point. The following regularity result at $\mathfrak{A}$ holds true.

Lemma 2.9. *At the axis of symmetry $\mathfrak{A}$ we have*

$$\frac{\mathbf{g}^{\mu\nu}\partial_\mu X \partial_\nu X}{4X} = e^{2\Phi}\mathbf{g}^{\mu\nu}\partial_\mu\Phi\partial_\nu\Phi \longrightarrow 1. \tag{2.1.11}$$

Proof. This is a classical result, see for example [52]. We provide a proof for the convenience of the reader. We introduce a coordinates system (x^0, x^1, x^2, x^3) centered at a point $q = (0,0,0,0)$ on the axis such that the Christoffel symbols of the metric vanish at q and $\partial_{x^0}|_q$ and $\partial_{x^1}|_q$ are tangent to the axis at q. In particular, in this coordinates system, the matrix $\partial_\alpha \mathbf{Z}^\mu(q)$ is given by

$$\partial_\alpha \mathbf{Z}^\mu(q) = \begin{pmatrix} 0 & 0 \\ 0 & A \end{pmatrix},$$

where A is an antisymmetric matrix. Note that we used the fact that $\mathbf{Z}$ vanishes on the axis, that q belongs to the axis, and that $\partial_\alpha \mathbf{Z}^\mu(q)$ is antisymmetric since $\mathbf{Z}$ is Killing. Now, if $x(\varphi)$ denotes an orbit of $\mathbf{Z}$ close to q, and $y = (x^2, x^3)$, we have in particular from the Taylor formula

$$\frac{dy}{d\varphi} = Ay + O(y^2).$$

Hence

$$\exp(-\varphi A)y(\varphi) = y(0) + O(\varphi y^2)$$

and since $y(2\pi) = y(0)$ in view of the 2π-periodicity of the orbits of $\mathbf{Z}$, we infer

$$\exp(-2\pi A)y(0) = y(0) + O(y^2).$$

As $y(0)$ can be taken arbitrarily small, we infer that $\exp(2\pi A)$ is the 2×2 identity matrix. Since A is antisymmetric and nonzero, its eigenvalues necessarily are i and $-i$, and hence $A^T A = I$. This yields

$$A^\alpha{}_\mu A_\gamma{}^\nu A_{\alpha\nu} = A^\alpha{}_\mu (A^T A)_{\gamma\alpha} = A_{\gamma\mu}$$

and hence

$$\partial^\alpha \mathbf{Z}_\mu(q)\partial_\gamma \mathbf{Z}^\nu(q)\partial_\alpha \mathbf{Z}_\nu(q) = \partial_\gamma \mathbf{Z}_\mu(q).$$

Finally, since $\mathbf{Z}$ vanishes on the axis, and since the coordinates system we use in this lemma has vanishing Christoffel symbols at q, we have as $|x|$ goes to 0

$$\begin{aligned}\frac{\mathbf{g}^{\mu\nu}\partial_\mu X \partial_\nu X}{4X} &= \frac{\mathbf{Z}^\mu \mathbf{D}^\alpha \mathbf{Z}_\mu \mathbf{Z}^\nu \mathbf{D}_\alpha \mathbf{Z}_\nu}{\mathbf{Z}^\mu \mathbf{Z}_\mu} \\ &= \frac{\partial_\beta \mathbf{Z}^\mu(q) x^\beta \partial^\alpha \mathbf{Z}_\mu(q) \partial_\gamma \mathbf{Z}^\nu(q) x^\gamma \partial_\alpha \mathbf{Z}_\nu(q)}{\partial_\beta \mathbf{Z}^\mu(q) x^\beta \partial_\gamma \mathbf{Z}_\mu(q) x^\gamma} + O(x).\end{aligned}$$

Together with the previous identity, we infer near any point q on the axis

$$\frac{\mathbf{g}^{\mu\nu}\partial_\mu X \partial_\nu X}{4X} \longrightarrow 1.$$

This concludes the proof of the lemma. □

We note that $\mathbf{Z}$-polarized, smooth vectorfields are automatically tangent to $\mathfrak{A}$. This is the content of the following.

Lemma 2.10. *Any, regular (i.e., smooth)* $\mathbf{Z}$*-polarized vectorfield* $\mathbf{U}$ *is tangent to the axis* $\mathfrak{A}$.

Proof. Let $\mathbf{U}$ a polarized $\mathbf{Z}$-invariant regular vectorfield. Since it is $\mathbf{Z}$-invariant, we have

$$0 = [\mathbf{Z}, \mathbf{U}] = \mathbf{Z}^\alpha \mathbf{D}_\alpha \mathbf{U} - \mathbf{U}^\alpha \mathbf{D}_\alpha \mathbf{Z}.$$

Since $\mathbf{Z} = 0$ on the axis and $\mathbf{U}$ is regular (hence bounded on the axis) we infer that $\mathbf{U}^\alpha \mathbf{D}_\alpha \mathbf{Z} = 0$ on $\mathfrak{A}$. In view of (2.1.10),

$$\mathbf{U}^\alpha \mathbf{D}_\alpha \mathbf{Z} = \mathbf{U}(e^\phi) e_\varphi,$$

and since e_φ is unitary, we infer that

$$\mathbf{U}(X^{1/2}) = \mathbf{U}(e^\phi) = 0 \text{ on } \mathfrak{A}$$

and hence $\mathbf{U}(X) = 0$ when $X = 0$. □

Corollary 2.11. *Let u be a smooth regular optical function, i.e.,* $\mathbf{g}^{\alpha\beta}\mathbf{D}_\alpha u \mathbf{D}_\beta u = 0$, *which is* $\mathbf{Z}$*-invariant, i.e.,* $\mathbf{Z}(u) = 0$. *Then its associated null geodesic generator* $L = -\mathbf{g}^{\alpha\beta}\partial_\alpha u \partial_\beta$ *is* $\mathbf{Z}$*-invariant, polarized, tangent to the axis of symmetry* $\mathfrak{A}$.

Proof. It is easy to check that L is $\mathbf{Z}$-invariant, polarized. It must therefore be tangent to $\mathfrak{A}$ in view of Lemma 2.10. □

2.1.4 Z-polarized S-surfaces

Throughout our work we shall deal with various $\mathbf{Z}$-polarized, S-foliations, i.e., foliations given by compact 2-surfaces S with induced metrics of the form

$$\not{g} = \gamma d\theta^2 + X d\varphi^2, \qquad \gamma = \gamma(\theta) > 0, \qquad \theta \in [0, \pi]. \tag{2.1.12}$$

Here γ and X are independent of φ, and e^Φ vanishes on the poles $\theta = 0$ and $\theta = \pi$, where $\Phi = \frac{1}{2} \log X$.

The regularity condition (2.1.11) takes the form

$$\lim_{\sin\theta \to 0} \Big(e_\theta(e^\Phi)\Big)^2 = 1 \tag{2.1.13}$$

where e_θ is the unit vector

$$e_\theta := \gamma^{-1/2} \partial_\theta.$$

We denote the induced covariant derivative $\nabla\!\!\!/$ and define the volume radius of S by the formula

$$|S| = 4\pi r^2$$

where $|S|$ is the volume of the surface using the volume form of the metric $\not g$. Note also that the area element on S is given by

$$\sqrt{\gamma}e^{\Phi}d\theta d\varphi.$$

In this section we record some basic general formulas concerning these surfaces. We consider adapted orthonormal frames

$$e_\theta, e_\varphi = X^{-1/2}\mathbf{Z} = X^{-1/2}\partial_\varphi.$$

Note that in view of (2.1.10) we have

$$\not\nabla_\varphi e_\varphi = -(e_\theta\Phi)e_\theta, \qquad \not\nabla_\varphi e_\theta = (e_\theta\Phi)e_\varphi, \qquad \not\nabla_\theta e_\theta = \not\nabla_\theta e_\varphi = 0. \tag{2.1.14}$$

In what follows, we consider **Z**-invariant polarized tensors tangent to S or simply polarized k-tensors on S.

In view of Lemma 2.10, a regular **Z**-polarized tensor on S must vanish on the axis of symmetry, i.e., at $\theta = 0$ and $\theta = \pi$. More precisely we have the following:

Lemma 2.12. *The following facts hold true for* **Z***-polarized tensors on* S.

1. *If U is a 1-form then, on the axis of symmetry*[2] *of* **Z** *(i.e., for $\theta = 0$ and $\theta = \pi$),*

$$U_\theta := U(e_\theta) = 0$$

2. *For a covariant 2-tensor, then, on the axis of symmetry*[3] *of* **Z** *(i.e., for $\theta = 0$ and $\theta = \pi$),*

$$U_{\theta\theta} = U_{\varphi\varphi} = 0.$$

Similar statements can be deduced for higher order tensors.

Proof. Immediate consequence of Lemma 2.10. □

Lemma 2.13. *The Gauss curvature K of the metric* (2.1.12) *can be expressed in terms of the polar function $\Phi := \frac{1}{2}\log X$ by the formula*

$$\not\Delta\Phi = -K. \tag{2.1.15}$$

Proof. Direct calculation using the form of the $\not g$ metric in (2.1.12). □

2.1.4.1 Basic operators on S

We recall (see [20] Chapter 2) the following operations which preserve the space of fully symmetric traceless tensors.

Definition 2.14. *We denote by $\mathcal{S}_k$ the set of k-covariant tensors which are fully*

[2] Note that the component U_φ must automatically vanish on S.

[3] Note that the components $U_{\theta\varphi}, U_{\varphi\theta}$ must automatically vanish on S.

symmetric and traceless, i.e., which verify[4]

$$f_{A_1\ldots A_k} = f_{(A_1\ldots A_k)}, \qquad \not{g}^{A_1A_2} f_{A_1A_2\ldots A_k} = 0.$$

Also, we define the Hodge dual of f by

$$^\star f_{A_1\ldots A_k} = \in_{A_1}{}^{C} f_{CA_2\ldots A_k}.$$

Remark 2.15. *Note*[5] *that if $f \in \mathcal{S}_k$, then there also holds $^\star f \in \mathcal{S}_k$.*

Definition 2.16. *We define the following operators on $\mathcal{S}_k$-tensors.*

1. *The operator $\not{\mathcal{D}}_k$ which takes $\mathcal{S}_k$ into $\mathcal{S}_{k-1}$ is the divergence operator,*

$$(\not{\mathcal{D}}_k f)_{A_2\ldots A_k} \;:=\; (\not{d}iv f)_{A_2\ldots A_k} = \not{g}^{AB}\not{\nabla}_B f_{AA_2\ldots A_k}.$$

2. *The operator $\not{\mathcal{C}}_k$ which takes $\mathcal{S}_k$ into $\mathcal{S}_{k-1}$ is the curl operator,*

$$(\not{\mathcal{C}}_k f)_{A_2\ldots A_k} \;:=\; \in^{AB} \not{\nabla}_A f_{BA_2\ldots A_k} = (\not{\mathcal{D}}_k(^\star f))_{A_2\ldots A_k}.$$

3. *The operator $\not{\mathcal{D}}^\star_k$ which takes $\mathcal{S}_{k-1}$ into $\mathcal{S}_k$ is the fully symmetrized, traceless, covariant derivative operator,*[6]

$$\begin{aligned}(\not{\mathcal{D}}^\star_k f)_{A_1\ldots A_k} \;:=\;& -\frac{1}{k}\sum_{i=1}^{k} \not{\nabla}_{A_i} f_{A_1\ldots \hat{A}_i\ldots A_k} \\ &+\frac{1}{k(k-1)}\sum_{1\le i<j\le k} \not{g}_{A_iA_j}(\not{d}iv f)_{A_1\ldots\hat{A}_i\ldots\hat{A}_j\ldots A_k}.\end{aligned}$$

4. *The operator $\not{\Delta}_k$ takes $\mathcal{S}_k$ into $\mathcal{S}_k$,*

$$(\not{\Delta}_k f)_{A_1\ldots A_k} := \not{g}^{BC}\not{\nabla}_B\not{\nabla}_C f_{A_1\ldots A_k}.$$

Lemma 2.17. *Given $f \in \mathcal{S}_k$, $k \ge 1$, we can express its covariant derivative*[7] *$\not{\nabla} f$ as a linear combination of the tensors $\not{\mathcal{D}}^\star_{k+1} f$, $(\not{g} \otimes \not{\mathcal{D}}_k)f$ and $(\in \otimes \not{\mathcal{C}}_k)f$.*

Proof. Consider for example the case $k = 2$. If f in $\mathcal{S}_2$, we have[8]

$$\begin{aligned}&3\not{\nabla}_B f_{A_1A_2}\\ =\;& (\not{\nabla}_B f_{A_1A_2} + \not{\nabla}_{A_1} f_{A_2B} + \not{\nabla}_{A_2} f_{BA_1}) + (\not{\nabla}_B f_{A_1A_2} - \not{\nabla}_{A_1} f_{BA_2})\\ &+(\not{\nabla}_B f_{A_1A_2} - \not{\nabla}_{A_2} f_{A_1B})\\ =\;& (\not{\nabla}_B f_{A_1A_2} + \not{\nabla}_{A_1} f_{A_2B} + \not{\nabla}_{A_2} f_{BA_1}) + \in_{BA_1} (\not{\mathcal{C}}_2 f)_{A_2} + \in_{BA_2} (\not{\mathcal{C}}_2 f)_{A_1}\\ =\;& -3(\not{\mathcal{D}}^\star_3 f)_{BA_1A_2} + \frac{1}{2}\Big(\not{g}_{A_1A_2}(\not{\mathcal{D}}_2 f)_B + \not{g}_{A_2B}(\not{\mathcal{D}}_2 f)_{A_1} + \not{g}_{A_1B}(\not{\mathcal{D}}_2 f)_{A_2}\Big)\\ &+ \in_{BA_1} (\not{\mathcal{C}}_2 f)_{A_2} + \in_{BA_2} (\not{\mathcal{C}}_2 f)_{A_1}\end{aligned}$$

---

[4] For an arbitrary k-tensor, $f_{(A_1\ldots A_k)} = \frac{1}{k!}\sum_{\sigma\in\mathfrak{S}_k} f_{A_{\sigma(1)}\ldots A_{\sigma(k)}}$.

[5] The result is easily verified for $k = 2$, see [20]. The general case follows by induction on k.

[6] In the particular case when $k = 1$, we get $(\not{\mathcal{D}}^\star_1 f)_A = -\not{\nabla}_A f$, and when $k = 2$, we get the familiar operator $\not{\mathcal{D}}^\star_2 f_{AB} = -\frac{1}{2}(\not{\nabla}_A f_B + \not{\nabla}_B f_A - \not{g}_{AB}\not{d}iv f)$.

[7] As a $k+1$ covariant tensor.

[8] Note that $\not{\nabla}_B f_{A_1A_2} - \not{\nabla}_{A_1} f_{BA_2} = \in_{BA_1} (\not{\mathcal{C}}_2 f)_{A_2}$.

which immediately leads the statement for $k = 2$. The general case can be verified in the same manner. □

Lemma 2.18. *The operators $\not{\mathcal{D}}$ and $\not{\mathcal{D}}^\star$ take polarized tensors to polarized tensors.*

Proof. We check the proof for tensors $f \in \mathcal{S}_2$. We start by showing that $\not{\mathcal{D}}_2 f$ is polarized

$$\begin{aligned} \mathbf{Z}^A(\not{\mathcal{D}}_2 f)_A &= \nabla^B(\mathbf{Z}^A f_{AB}) - \nabla^B \mathbf{Z}^A f_{AB} \\ &= \nabla^B(\mathbf{Z}^A f_{AB}) - \frac{1}{2}(\nabla_B \mathbf{Z}_A + \nabla_A \mathbf{Z}_B) f^{AB} = 0. \end{aligned}$$

To show that $\not{\mathcal{D}}_3^\star f$ is polarized, we calculate, relative to the frame e_θ, e_φ with the help of (2.1.14),

$$\begin{aligned} -3(\not{\mathcal{D}}_3^\star f)_{\varphi\theta\theta} &= \nabla_\varphi f_{\theta\theta} + \nabla_\theta f_{\theta\varphi} + \nabla_\theta f_{\theta\varphi} - \frac{1}{2} g_{\theta\theta}(\text{div } f)_\varphi = -\frac{1}{2}(\text{div } f)_\varphi \\ &= -\frac{1}{2}(\nabla_\theta f_{\theta\varphi} + \nabla_\varphi f_{\varphi\varphi}) = 0 \\ -3(\not{\mathcal{D}}_3^\star f)_{\varphi\varphi\varphi} &= 3\nabla_\varphi f_{\varphi\varphi} - \frac{3}{2}(\text{div } f)_\varphi = \frac{3}{2}(\nabla_\varphi f_{\varphi\varphi} - \nabla_\theta f_{\theta\varphi}) = 0. \end{aligned}$$

□

Remark 2.19. *The operations of taking dual and curl do not preserve polarization. On the other hand, if f is a polarized k-tensor, then $\nabla\mkern-13mu/\, f$ is a $k+1$ polarized tensor. Indeed, for example, for f in $\mathcal{S}_2$, we have, using* (2.1.14) *and the polarization property of f,*

$$\nabla\mkern-13mu/\,_\varphi f_{\theta\theta} = 0, \qquad \nabla\mkern-13mu/\,_\theta f_{\theta\varphi} = 0, \qquad \nabla\mkern-13mu/\,_\varphi f_{\varphi\varphi} = 0.$$

We can easily check that $\not{\mathcal{D}}_k^\star$ is the formal adjoint of $\not{\mathcal{D}}_k$, i.e.,

$$\int_S (\not{\mathcal{D}}_k f) g = \int_S f(\not{\mathcal{D}}_k^\star g).$$

It is also easy to check that the kernels of $\not{\mathcal{D}}_k$ are trivial for all $k \geq 1$ (see also Chapter 2 in [20]). The kernel of $\not{\mathcal{D}}_1^\star : \mathcal{S}_0 \to \mathcal{S}_1$ consists of constants on S while the kernel of $\not{\mathcal{D}}_2^\star$ consists of constant multiple of covectors f with $f_\theta = Ce^\Phi$. Moreover,

$$\begin{aligned} \not{\mathcal{D}}_1^\star \cdot \not{\mathcal{D}}_1 &= -\not\Delta_1 + K, \qquad & \not{\mathcal{D}}_1 \cdot \not{\mathcal{D}}_1^\star &= -\not\Delta_0, \\ \not{\mathcal{D}}_2^\star \cdot \not{\mathcal{D}}_2 &= -\frac{1}{2}\not\Delta_2 + K, \qquad & \not{\mathcal{D}}_2 \cdot \not{\mathcal{D}}_2^\star &= -\frac{1}{2}(\not\Delta_1 + K). \end{aligned} \tag{2.1.16}$$

Similar identities also hold for higher k. Using (2.1.16) one can also prove the following (see also Chapter 2 in [20]).

Proposition 2.20. *Let $(S, \not g)$ be a compact manifold with Gauss curvature K. We have:*

i.) *The following identity holds for vectorfields $f \in \mathcal{S}_1$,*

$$\int_S (|\nabla\mkern-13mu/\, f|^2 + K|f|^2) = \int_S |\not{\mathcal{D}}_1 f|^2.$$

ii.) *The following identity holds for symmetric, traceless tensors in $\mathcal{S}_2$,*

$$\int_S (|\nabla f|^2 + 2K|f|^2) = 2\int_S |\mathcal{D}_2 f|^2.$$

iii.) *The following identity holds for scalars $f \in \mathcal{S}_0$,*

$$\int_S |\nabla f|^2 = \int_S |\mathcal{D}_1^\star f|^2.$$

iv.) *The following identity holds for vectors $f \in \mathcal{S}_1$,*

$$\int_S (|\nabla f|^2 - K|f|^2) = 2\int_S |\mathcal{D}_2^\star f|^2.$$

Proof. All statements appear in [20]. □

Proposition 2.21. *We have for $f \in \mathcal{S}_0$,*

$$\int_S (|\nabla^2 f|^2 + K|\nabla f|^2) = \int_S |\triangle_0 f|^2.$$

Moreover, under mild assumptions on the curvature such as

$$K = \frac{1}{r^2} + O\left(\frac{\epsilon}{r^2}\right), \qquad re_\theta(K) = O\left(\frac{\epsilon}{r^2}\right),$$

for any $f \in \mathcal{S}_k$, $k \geq 1$,

$$\int_S (|\nabla^2 f|^2 + r^{-2}|\nabla f|^2) \lesssim \int_S |\triangle_k f|^2 + O(\epsilon)r^{-4}\int_S |f|^2.$$

Proof. Follows from the standard Bochner identity on S. □

2.1.4.2 Reduced picture

Lemma 2.22. *The following relations hold true between the spacetime picture and the reduced one.*

1. *Let ${}^{(1+3)}f \in \mathcal{S}_k$ such that ${}^{(1+3)}f_{\theta\ldots\theta} = f$. Then,*

$$(\mathcal{D}_k\, {}^{(1+3)}f)_{\theta\ldots\theta} = e_\theta(f) + ke_\theta(\Phi)f. \tag{2.1.17}$$

2. *If $f \in \mathcal{S}_0$ we have*

$$\not{d}_1^\star f = -e_\theta(f).$$

3. *If ${}^{(1+3)}f \in \mathcal{S}_{k-1}$, $k \geq 2$, such that ${}^{(1+3)}f_{\theta\ldots\theta} = f$ we have*

$$2(\mathcal{D}_k^\star\, {}^{(1+3)}f)_{\theta\ldots\theta} = -e_\theta(f) + (k-1)e_\theta(\Phi)f. \tag{2.1.18}$$

4. Let ${}^{(1+3)}f \in \mathcal{S}_k$ *such that* ${}^{(1+3)}f_{\theta\ldots\theta} = f$. *Then,*

$$\not{\triangle}_k\, {}^{(1+3)}f_{\theta_1\ldots\theta_k} \;=\; e_\theta(e_\theta f) + e_\theta(\Phi)e_\theta f - k^2\big(e_\theta(\Phi)\big)^2 f.$$

Proof. The proof follows easily from the definitions of $\not{\mathcal{D}}_k$, $\not{\mathcal{D}}_k^\star$, $\not{\triangle}_k$ and the formulae (2.1.14). We check below the formula (2.1.17).

$$\begin{aligned} -(\not{\mathcal{D}}_k^\star\, {}^{(1+3)}f)_{\theta\ldots\theta} &= e_\theta f - \frac{1}{2}(\not{\mathcal{D}}_{k-1} f)_{\theta\ldots\theta} = e_\theta(f) - \frac{1}{2}(e_\theta f + (k-1)e_\theta(\Phi)f) \\ &= \frac{1}{2}\,(e_\theta f - (k-1)e_\theta(\Phi)f) \end{aligned}$$

as desired. □

Definition 2.23. *We say that a scalar f is a reduced k-scalar on S if there is a* $\mathbf{Z}$*-invariant, polarized, k-covector* ${}^{(1+3)}f \in \mathcal{S}_k$ *such that*

$$f = {}^{(1+3)}f_{\theta\ldots\theta}.$$

We denote by $\mathfrak{s}_k$ the set of k-reduced scalars.

- *Given a k-reduced scalar f, reduced from ${}^{(1+3)}f$ we define*

$$|\nabla f|^2 = |\nabla\, {}^{(1+3)}f|^2, \qquad |\nabla^l f|^2 = |\nabla^l\, {}^{(1+3)}f|^2.$$

- *Given a k-reduced scalar f on S we define*

$$\not{d}_k f \;:= e_\theta(f) + k e_\theta(\Phi) f.$$

- *Given a $(k-1)$-reduced scalar $f \in S_{k-1}$ we define*

$$\not{d}_k^\star f \;:= -e_\theta(f) + (k-1)e_\theta(\Phi) f.$$

- *Given a k-reduced scalar $f \in \mathfrak{s}_k$ we define*

$$\not{\triangle}_k f \;:= e_\theta(e_\theta f) + e_\theta(\Phi)e_\theta f - k^2\big(e_\theta(\Phi)\big)^2 f.$$

In view of Lemma 2.22 we have

$$\not{d}_k f \;=\; (\not{\mathcal{D}}_k\, {}^{(1+3)}f)_{\theta\ldots\theta}$$

and

$$\not{d}_k^\star f = \begin{cases} (\not{\mathcal{D}}_k^\star\, {}^{(1+3)}f)_{\theta\ldots\theta}, & k=1, \\ 2(\not{\mathcal{D}}_k^\star\, {}^{(1+3)}f)_{\theta\ldots\theta}, & k \geq 2. \end{cases}$$

Clearly $\not{d}_k$ takes k-reduced scalars into $(k-1)$-reduced scalars, $\not{d}_k^\star$ takes $(k-1)$-reduced ones into k-reduced and $\not{\triangle}_k$ takes k-reduced scalars into k-reduced scalars.

Remark 2.24. *Note that, in view of Lemma 2.12, any reduced scalar in $\mathfrak{s}_k$, for $k \geq 1$, must vanish on the axis of symmetry of* $\mathbf{Z}$*, i.e., at the two poles.*

Remark 2.25. *The operator $\not{d}_k$ and $\not{d}_k^\star$ can only be applied to k-reduced, resp. $(k-1)$-reduced, scalars. Thus whenever we write a sequence of operators involving*

$\not{d}_k$, $\not{d}_k^\star$ *we understand from the context to which type of k-reduced scalars they are applied, see, for example, the proposition below. The same remark applies to* $\not{\Delta}_k$.

Remark 2.26. *Note that for given reduced scalar* $f \in \mathfrak{s}_k$ *and* $h \in \mathfrak{s}_1$ *we can write*

$$he_\theta(f) = \frac{1}{2} h \left(\not{d}_k f - \not{d}_{k+1}^\star f \right).$$

The term $h \not{d}_k f$ *is the reduced form of a tensor product of* ${}^{(1+3)}h$ *with* $\not{\mathcal{D}}_k {}^{(1+3)}f$ *while* $h \not{d}_{k+1}^\star f$ *is the reduced form of a contraction between* ${}^{(1+3)}h$ *and* $\not{\mathcal{D}}_{k+1}^\star {}^{(1+3)}f$. *This can be formalized precisely using Lemma 2.17. The remark will be useful in what follows, for example, in Lemma 2.69.*

Remark 2.27. *The duality between the operators* $\not{d}_k$ *and* $\not{d}_k^\star$ *follows in view of the duality of* $\not{\mathcal{D}}_k$ *and* $\not{\mathcal{D}}_k^\star$. *It can also be interpreted directly in terms of the area element* $\sqrt{\gamma} e^\Phi d\theta d\varphi$,

$$\begin{aligned} \int_S (\not{d}_k f g - f \not{d}_k^\star g) da_S &= \int_S e_\theta(fg) + e_\theta(\Phi) fg \\ &= \int_0^\pi \int_0^{2\pi} \left(e_\theta(fg) + e_\theta(\Phi) fg\right) \sqrt{\gamma} e^\Phi d\theta d\varphi \\ &= \int_0^\pi \int_0^{2\pi} \partial_\theta (e^\Phi fg) d\theta d\varphi = 0. \end{aligned}$$

Proposition 2.28. *The following identities hold true,*

$$\begin{aligned} \not{d}_k^\star \not{d}_k &= -\not{\Delta}_k + kK, \\ \not{d}_k \not{d}_k^\star &= -\not{\Delta}_{k-1} - (k-1)K. \end{aligned} \tag{2.1.19}$$

In particular for $k = 1, 2$

$$\not{d}_1^\star \not{d}_1 = -\not{\Delta}_1 + K, \quad \not{d}_1 \not{d}_1^\star = -\not{\Delta}_0, \quad \not{d}_2^\star \not{d}_2 = -\not{\Delta}_2 + 2K, \quad \not{d}_2 \not{d}_2^\star = -\not{\Delta}_1 - K.$$

Moreover, note the following commutation formulas

$$\begin{aligned} \not{d}_k \not{d}_k^\star - \not{d}_{k-1}^\star \not{d}_{k-1} &= -2(k-1)K, \\ -\not{d}_k \not{\Delta}_k + \not{\Delta}_{k-1} \not{d}_k &= -(2k-1)K \not{d}_k - k e_\theta(K), \\ -\not{d}_k^\star \not{\Delta}_{k-1} + \not{\Delta}_k \not{d}_k^\star &= (2k-1)K \not{d}_k^\star + (k-1) e_\theta(K). \end{aligned}$$

Proof. We have, for a k-reduced scalar f,

$$\begin{aligned} -\not{d}_k^\star \not{d}_k f &= (e_\theta - (k-1)e_\theta(\Phi))(e_\theta(f) + k e_\theta(\Phi) f) \\ &= e_\theta(e_\theta(f)) + k e_\theta(\Phi) e_\theta f + k(e_\theta e_\theta \Phi) f - (k-1) e_\theta(\Phi)(e_\theta(f) + k e_\theta(\Phi) f) \\ &= e_\theta(e_\theta(f)) + e_\theta(\Phi) e_\theta f + k(e_\theta e_\theta \Phi) f - k(k-1)\big(e_\theta(\Phi)\big)^2. \end{aligned}$$

In view of Lemma 2.13 we have, since Φ is a scalar,

$$-K = \not{\Delta}\Phi = e_\theta e_\theta(\Phi) + \big(e_\theta(\Phi)\big)^2.$$

Therefore,

$$\begin{aligned}
-\not{d}_k^{\star}\not{d}_k f &= e_\theta(e_\theta(f)) + e_\theta(\Phi)e_\theta f + k\left(-K - \big(e_\theta(\Phi)\big)^2\right) f - k(k-1)\big(e_\theta(\Phi)\big)^2 \\
&= e_\theta(e_\theta(f)) + e_\theta(\Phi)e_\theta f - kKf - k^2\big(e_\theta(\Phi)\big)^2 \\
&= \not{\Delta}_k f - kKf.
\end{aligned}$$

Similarly, for a $(k-1)$-reduced f,

$$\begin{aligned}
-\not{d}_k\not{d}_k^{\star} f &= (e_\theta + ke_\theta(\Phi))(e_\theta(f) - (k-1)e_\theta(\Phi)f) \\
&= e_\theta(e_\theta(f)) + ke_\theta(\Phi)e_\theta f - (k-1)(e_\theta e_\theta \Phi)f - (k-1)e_\theta(\Phi)e_\theta(f) \\
&- k(k-1)\big(e_\theta(\Phi)\big)^2 f \\
&= e_\theta(e_\theta(f)) + e_\theta(\Phi)e_\theta f - (k-1)\left(-K - \big(e_\theta(\Phi)\big)^2\right) f - k(k-1)\big(e_\theta(\Phi)\big)^2 f \\
&= e_\theta(e_\theta(f)) + e_\theta(\Phi)e_\theta f + (k-1)Kf - (k-1)^2\big(e_\theta(\Phi)\big)^2 f \\
&= \not{\Delta}_{k-1} f + (k-1)Kf.
\end{aligned}$$

Next, we check the commutation formulas. We have

$$\begin{aligned}
\not{d}_k\not{d}_k^{\star} - \not{d}_{k-1}^{\star}\not{d}_{k-1} &= -\not{\Delta}_{k-1} - (k-1)K - \Big(-\not{\Delta}_{k-1} + (k-1)K\Big) \\
&= -2(k-1)K
\end{aligned}$$

from which we infer

$$\begin{aligned}
\not{d}_k(-\not{\Delta}_k) &= \not{d}_k(\not{d}_k^{\star}\not{d}_k - kK) \\
&= \not{d}_k\not{d}_k^{\star}\not{d}_k - kK\not{d}_k - ke_\theta(K) \\
&= \Big(\not{d}_{k-1}^{\star}\not{d}_{k-1} - 2(k-1)K\Big)\not{d}_k - kK\not{d}_k - ke_\theta(K) \\
&= \Big(-\not{\Delta}_{k-1} - (k-1)K\Big)\not{d}_k - kK\not{d}_k - ke_\theta(K)
\end{aligned}$$

and hence

$$-\not{d}_k\not{\Delta}_k + \not{\Delta}_{k-1}\not{d}_k = -(2k-1)K\not{d}_k - ke_\theta(K).$$

Also, we have

$$\begin{aligned}
\not{d}_k^{\star}(-\not{\Delta}_{k-1}) &= \not{d}_k^{\star}(\not{d}_k\not{d}_k^{\star} + (k-1)K) \\
&= \not{d}_k^{\star}\not{d}_k\not{d}_k^{\star} + (k-1)K\not{d}_k^{\star} + (k-1)e_\theta(K) \\
&= \Big(-\not{\Delta}_k + kK\Big)\not{d}_k^{\star} + (k-1)K\not{d}_k^{\star} + (k-1)e_\theta(K)
\end{aligned}$$

and hence

$$-\not{d}_k^{\star}\not{\Delta}_{k-1} + \not{\Delta}_k\not{d}_k^{\star} = (2k-1)K\not{d}_k^{\star} + (k-1)e_\theta(K)$$

as desired. □

2.1.4.3 A remarkable identity

First, note the following observation which follows immediately from the form of $\not{d}_2^\star$.

Lemma 2.29. *The kernel of $\not{d}_2^\star$ is spanned by e^Φ.*

The above lemma, in connection with a Poincaré inequality for $\not{d}_2^\star$, see (2.1.34), will result in the need of a specific treatment for the projection of some of the quantities on the kernel of $\not{d}_2^\star$. This motivates the following definition.

Definition 2.30 (The $\ell = 1$ mode). *For a 1-reduced scalar f, the $\ell = 1$ mode denotes its projection on the kernel of $\not{d}_2^\star$, i.e.,*

$$\int_S fe^\Phi.$$

For a 0-reduced scalar f, the $\ell = 1$ mode denotes the projection of $e_\theta(f)$ on the kernel of $\not{d}_2^\star$, i.e.,

$$\int_S e_\theta(f)e^\Phi.$$

Remark 2.31. *The above definition is motivated by the fact that, in Schwarzschild, this corresponds to the projection on the $\ell = 1$ spherical harmonic.*[9]

We are now ready to state the following remarkable identity which will play a crucial role later in the book.

Lemma 2.32 (Vanishing of the $\ell = 1$ mode of the Gauss curvature). *The $\ell = 1$ mode of K vanishes identically, i.e.,*

$$\int_S e_\theta(K)e^\Phi = 0. \tag{2.1.20}$$

Proof. To prove (2.1.20) we write

$$-\int_S e_\theta(K)e^\Phi = \int_S \not{d}_1^\star(K)e^\Phi = \int_S K\not{d}_1(e^\Phi) = 2\int_S Ke_\theta(\Phi)e^\Phi.$$

Thus, in view of (2.1.15), using in addition $\not{\Delta}\Phi = e_\theta(e_\theta(\Phi)) + e_\theta(\Phi)^2$

$$\begin{aligned}\int_S e_\theta(K)e^\Phi &= 2\int_S \not{\Delta}\Phi e_\theta(\Phi)e^\Phi = 2\int_S \big(e_\theta(e_\theta(\Phi)) + e_\theta(\Phi)^2\big)e_\theta(\Phi)e^\Phi \\ &= \int_S \not{d}_2\big((e_\theta\Phi)^2\big)e^\Phi = \int_S (e_\theta\Phi)^2\not{d}_2^\star(e^\Phi) = 0\end{aligned}$$

as desired.[10] □

---

[9] In general, there are three spherical harmonics corresponding to $\ell = 1$, but only one is axially symmetric. This is why we have only one projection instead of three in our case.

[10] Note that the boundary term which appears from the last integration by parts has the form $(\partial_\theta\Phi)^2e^{2\Phi}(\pi) - (\partial_\theta\Phi)^2e^{2\Phi}(0)$ and hence vanishes in view of the regularity condition (2.1.13), see also the computation in Remark 2.27.

2.1.4.4 Poincaré inequalities on 2-spheres

Proposition 2.20 takes the following reduced form:

Proposition 2.33. *The following identities hold true for reduced k-scalars $f \in \mathfrak{s}_k$.*

i.) *If $f \in \mathfrak{s}_1$,*

$$\int_S \left(|\nabla f|^2 + Kf^2\right) = \int_S |{d\!\!/}_1 f|^2. \tag{2.1.21}$$

ii.) *If $f \in \mathfrak{s}_2$,*

$$\int_S \left(|\nabla f|^2 + 4Kf^2\right) = 2\int_S |{d\!\!/}_2 f|^2. \tag{2.1.22}$$

iii.) *If $f \in \mathfrak{s}_0$,*

$$\int_S |\nabla f|^2 = \int_S |{d\!\!/}_1^\star f|^2. \tag{2.1.23}$$

iv.) *If $f \in \mathfrak{s}_1$,*

$$\int_S \left(|\nabla f|^2 - Kf^2\right) = \int_S |{d\!\!/}_2^\star f|^2. \tag{2.1.24}$$

v.) *If $f \in \mathfrak{s}_0$,*

$$\int_S |\nabla^2 f|^2 + \int_S K|\nabla f|^2 = \int_S |\Delta_0 f|^2. \tag{2.1.25}$$

Under mild assumptions on the Gauss curvature K, such as

$$K = \frac{1}{r^2} + O\left(\frac{\epsilon}{r^2}\right), \qquad re_\theta(K) = O\left(\frac{\epsilon}{r^2}\right).$$

We also have for $f \in \mathfrak{s}_k$, $k \geq 1$,

$$\|\nabla^2 f\|^2_{L^2(S)} + r^{-2}\|\nabla f\|^2_{L^2(S)} \lesssim \|\Delta_k f\|^2_{L^2(S)} + \epsilon r^{-4}\|f\|^2_{L^2(S)}. \tag{2.1.26}$$

Proof. The proof of the above statements can be either derived from their spacetime version or checked directly. □

Lemma 2.34. *The following relations hold between **Z**-polarized S-tensors and reduced scalars.*[11]

- *If $f \in \mathfrak{s}_0$*

$$\begin{aligned} |\nabla f|^2 &= |e_\theta f|^2, \\ |\nabla^2 f|^2 &= |e_\theta(e_\theta f)|^2 + |e_\theta \Phi e_\theta f|^2. \end{aligned}$$

- *If $f \in \mathfrak{s}_1$,*

$$|\nabla f|^2 = |e_\theta f|^2 + |e_\theta(\Phi)|^2|f|^2.$$

[11]Note that the expressions on the left of the inequalities below should be interpreted as applying to the spacetime tensor from which f is reduced.

- *If* $f \in \mathfrak{s}_2$,

$$|\nabla\!\!\!/ f|^2 = 2\left(|e_\theta f|^2 + 4|e_\theta(\Phi)|^2|f|^2\right).$$

Proof. If $f \in \mathfrak{s}_0$,

$$|\nabla\!\!\!/^2 f|^2 \;=\; \nabla_A \nabla\!\!\!/_B f \nabla\!\!\!/^A \nabla\!\!\!/^B f = |\nabla\!\!\!/_\theta \nabla\!\!\!/_\theta f|^2 + |\nabla\!\!\!/_\varphi \nabla\!\!\!/_\varphi f|^2 = |e_\theta(e_\theta f)|^2 + |e_\theta \Phi e_\theta f|^2.$$

If $f \in \mathfrak{s}_1$ is reduced from a **Z**-invariant, polarized vector F,

$$\begin{aligned}|\nabla\!\!\!/ f|^2 &= \nabla_A F_B \nabla\!\!\!/^A F^B = |\nabla\!\!\!/_\theta F_\theta|^2 + |\nabla\!\!\!/_\varphi F_\varphi|^2\\ &= |e_\theta f|^2 + |e_\theta \Phi f|^2.\end{aligned}$$

If $f \in \mathfrak{s}_2$ is reduced from a symmetric, traceless **Z**-invariant, polarized tensor $F = {}^{(1+3)}f$ we have

$$\begin{aligned}|\nabla\!\!\!/ f|^2 &= |\nabla\!\!\!/_\theta F_{\theta\theta}|^2 + 2|\nabla\!\!\!/_\theta F_{\varphi\theta}|^2 + |\nabla\!\!\!/_\theta F_{\varphi\varphi}|^2 + |\nabla\!\!\!/_\varphi F_{\theta\theta}|^2 + 2|\nabla\!\!\!/_\varphi F_{\varphi\theta}|^2 + |\nabla\!\!\!/_\varphi F_{\varphi\varphi}|^2\\ &= |\nabla\!\!\!/_\theta F_{\theta\theta}|^2 + |\nabla\!\!\!/_\theta F_{\varphi\varphi}|^2 + 2|\nabla\!\!\!/_\varphi F_{\varphi\theta}|^2\end{aligned}$$

and

$$\begin{aligned}\nabla\!\!\!/_\theta F_{\theta\theta} &= e_\theta f = -\nabla\!\!\!/_\theta F_{\varphi\varphi},\\ \nabla\!\!\!/_\varphi F_{\varphi\theta} &= e_\theta \Phi F_{\theta\theta} - e_\theta \Phi F_{\varphi\varphi} = 2 e_\theta \Phi f.\end{aligned}$$

Thus,

$$|\nabla\!\!\!/ f|^2 = 2|e_\theta f|^2 + 8(e_\theta \Phi f)^2$$

as desired. □

Proposition 2.35 (Poincaré). *The following inequalities hold for k-reduced scalars.*

1. *If* $f \in \mathfrak{s}_0$,

$$\int_S |\nabla\!\!\!/^2 f|^2 \;\geq\; \int_S K|\not{d}_1^\star f|^2. \tag{2.1.27}$$

2. *If* $f \in \mathfrak{s}_1$,

$$\int_S |\nabla\!\!\!/ f|^2 \;\geq\; \int_S K f^2. \tag{2.1.28}$$

3. *If* $f \in \mathfrak{s}_2$,

$$\int_S |\nabla\!\!\!/ f|^2 \;\geq\; 4\int_S K f^2. \tag{2.1.29}$$

Proof. We first prove the result for $f \in \mathfrak{s}_2$. According to Lemma 2.34,

$$\begin{aligned}2^{-1}|\nabla\!\!\!/ f|^2 &= |e_\theta f|^2 + 4|e_\theta(\Phi)|^2|f|^2 = (e_\theta f - 2e_\theta(\Phi) f)^2 + 4f(e_\theta f)e_\theta(\Phi)\\ &= (e_\theta f - 2e_\theta(\Phi) f)^2 + 2e_\theta(f^2)e_\theta(\Phi).\end{aligned}$$

Hence,

$$2^{-1}\int_S |\nabla\!\!\!/ F|^2 da_S = \int_S (e_\theta f - 2e_\theta(\Phi)f)^2 da_S + 2\int_S e_\theta(f^2)e_\theta(\Phi)\sqrt{\gamma}e^\Phi d\theta d\varphi.$$

Now,

$$\begin{aligned}\int_S e_\theta(f^2)e_\theta(\Phi)\sqrt{\gamma}e^\Phi d\theta d\varphi &= \int_0^\pi\int_0^{2\pi} \partial_\theta(f^2)e_\theta(\Phi)e^\Phi d\theta d\varphi\\ &= -\int_0^\pi\int_0^{2\pi} f^2 e_\theta(e^\Phi e_\theta\Phi)\sqrt{\gamma}d\theta d\varphi\\ &= -\int_0^\pi\int_0^{2\pi} \left(e_\theta e_\theta\Phi + (e_\theta\Phi)^2\right) f^2 e^\Phi\sqrt{\gamma}d\theta d\varphi\\ &= \int_S Kf^2 da_S.\end{aligned}$$

Hence,

$$\int_S |\nabla\!\!\!/ f|^2 \geq 4\int_S Kf^2$$

as desired.

The result for $f \in \mathfrak{s}_1$ is proved in the same way.

If $f \in \mathfrak{s}_0$ we write, according to Lemma 2.34,

$$\begin{aligned}|\nabla\!\!\!/^2 f|^2 &= |e_\theta(e_\theta f)|^2 + |e_\theta\Phi e_\theta f|^2 = |e_\theta h|^2 + |e_\theta\Phi|^2|e_\theta f|^2\\ &= (e_\theta e_\theta f - e_\theta(\Phi)e_\theta f)^2 + e_\theta[(e_\theta f)^2]\, e_\theta(\Phi).\end{aligned}$$

Integrating by parts as before,

$$\int_S e_\theta[(e_\theta f)^2]\, e_\theta(\Phi)da_S = -\int_0^\pi\int_0^{2\pi}(e_\theta f)^2 e_\theta(e^\Phi e_\theta\Phi)\sqrt{\gamma}d\theta d\varphi = \int_S K(e_\theta f)^2 da_S.$$

Thus,

$$\int_S |\nabla\!\!\!/^2 f|^2 \geq \int_S K(e_\theta f)^2.$$

□

As a corollary we deduce the following:

Corollary 2.36. *The following hold true for reduced scalars:*

1. *If $f \in \mathfrak{s}_1$,*

$$\int_S |d\!\!\!/_1 f|^2 \geq \int_S 2Kf^2. \tag{2.1.30}$$

2. *If $f \in \mathfrak{s}_2$,*

$$\int_S |d\!\!\!/_2 f|^2 \geq 4\int_S Kf^2. \tag{2.1.31}$$

Proof. According to (2.1.21),

$$\int_S (|\nabla f|^2 + Kf^2) = \int_S |\not{d}_1 f|^2.$$

We deduce

$$\int_S |\not{d}_1 f|^2 \ \geq \ 2\int_S Kf^2.$$

According to (2.1.22)

$$\int_S (|\nabla f|^2 + 4Kf^2) = 2\int_S |\not{d}_2 f|^2.$$

Hence,

$$2\int_S |\not{d}_2 f|^2 \ \geq \ 8\int_S Kf^2$$

as desired. □

Corollary 2.37. *Under the following mild assumptions on the Gauss curvature*

$$K = \frac{1}{r^2} + O\left(\frac{\epsilon}{r^2}\right), \qquad re_\theta(K) = O\left(\frac{\epsilon}{r^2}\right),$$

the following holds.

1. *If $f \in \mathfrak{s}_0$ is orthogonal to the kernel of $\not{d}_1^\star$, i.e., $\int_S f = 0$, then, we have*

$$\int_S |\not{d}_1^\star f|^2 \ \geq \ 2\int_S (1+O(\epsilon))Kf^2. \tag{2.1.32}$$

2. *If $f \in \mathfrak{s}_1$ is orthogonal to the kernel of $\not{d}_2^\star$, i.e., $\int_S fe^\Phi = 0$, then, we have*

$$\int_S |\not{d}_1 f|^2 \geq 6\int_S (1+O(\epsilon))Kf^2 \ \text{and} \ \int_S |\not{d}_2^\star f|^2 \geq 4\int_S (1+O(\epsilon))Kf^2. \tag{2.1.33}$$

Proof. We start with the first assertion. If $f \in \mathfrak{s}_0$ satisfies $\int_S f = 0$ then, f is orthogonal to 1 which generates the kernel of $\not{d}_1^\star$, and hence, f is in the image of $\not{d}_1$, i.e., there exists $h \in \mathfrak{s}_1$ such that

$$f = \not{d}_1 h.$$

We deduce

$$\int_S (\not{d}_1^\star(f))^2 \ = \ \int_S (\not{d}_1^\star \not{d}_1 h)^2 = \int_S (\not{d}_1^\star \not{d}_1)^2 hh.$$

Now, the above Poincaré inequality for $\not{d}_1$ and the assumption on K implies a lower bound for the spectrum of the self-adjoint operator $\not{d}_1^\star \not{d}_1$ by $2K(1+O(\epsilon))$,

and hence

$$\begin{aligned}\int_S (\not{d}_1^\star(f))^2 &\geq 2\int_S K(1+O(\epsilon))(\not{d}_1^\star \not{d}_1)hh\\ &\geq 2\int_S (1+O(\epsilon))K(\not{d}_1 h)^2\\ &\geq 2\int_S (1+O(\epsilon))Kf^2\end{aligned}$$

which yields the first assertion.

Assume now that $f \in \mathfrak{s}_1$ satisfies $\int_S fe^{\Phi} = 0$, i.e., f is orthogonal to e^{Φ} which generates the kernel of $\not{d}_2^\star$, and hence, f is in the image of $\not{d}_2$, i.e., there exists $h \in \mathfrak{s}_2$ such that

$$f = \not{d}_2 h.$$

We deduce

$$\begin{aligned}\int_S (\not{d}_1 f)^2 &= \int_S (\not{d}_1 \not{d}_2 h)^2 \int_S \not{d}_1^\star \not{d}_1 \not{d}_2 h \not{d}_2 h = \int_S (\not{d}_2 \not{d}_2^\star + 2K)\not{d}_2 h \not{d}_2 h\\ &= \int_S (\not{d}_2^\star \not{d}_2)^2 hh + \int 2K(\not{d}_2 h)^2.\end{aligned}$$

Now, the above Poincaré inequality for $\not{d}_2$ and the assumption on K implies a lower bound for the spectrum of the self-adjoint operator $\not{d}_2^\star \not{d}_2$ by $4K(1+O(\epsilon))$, and hence

$$\begin{aligned}\int_S (\not{d}_1 f)^2 &\geq 4\int_S K(1+O(\epsilon))(\not{d}_2^\star \not{d}_2)hh + \int 2K(\not{d}_2 h)^2\\ &\geq 6\int_S (1+O(\epsilon))K(\not{d}_2 h)^2\\ &\geq 6\int_S (1+O(\epsilon))Kf^2.\end{aligned}$$

Together with the fact that

$$\int_S (\not{d}_2^\star f)^2 = \int_S \not{d}_2 \not{d}_2^\star ff = \int_S (\not{d}_1^\star \not{d}_1 - 2K)ff = \int_S (\not{d}_1 f)^2 - 2\int_S Kf^2,$$

this yields the second assertion and concludes the proof of the corollary. □

Lemma 2.38. *Assume that*

$$K = \frac{1}{r^2} + O\left(\frac{\epsilon}{r^2}\right), \qquad re_\theta(K) = O\left(\frac{\epsilon}{r^2}\right), \qquad \int_S e^{2\Phi} = r^4\left(\frac{8\pi}{3} + O(\epsilon)\right).$$

Then, if $f \in \mathfrak{s}_1$, we have the estimate

$$\int_S |f|^2 \lesssim r^2 \int_S |\not{d}_2^\star f|^2 + r^{-4}\left|\int_S e^{\Phi} f\right|^2. \tag{2.1.34}$$

More precisely,

$$f = \frac{\int_S f e^{\Phi}}{\int_S e^{2\Phi}} e^{\Phi} + f^{\perp} \tag{2.1.35}$$

with

$$\int_S |f^{\perp}|^2 \lesssim r^2 \int_S |\not{d}_2^\star f|^2.$$

Proof. According to Corollary 2.37, see 2.1.33, if $f \in \mathfrak{s}_1$ is orthogonal to the kernel of $\not{d}_2^\star$, i.e., $\int_S f e^{\Phi} = 0$, then, we have

$$\int_S |\not{d}_2^\star f|^2 \geq 4 \int_S (1 + O(\epsilon)) K f^2.$$

As a consequence $f^{\perp} = f - \left(\frac{\int_S f e^{\Phi}}{\int_S e^{2\Phi}} \right) e^{\Phi}$ verifies

$$r^{-2} \int_S |f^{\perp}|^2 \lesssim \int_S |\not{d}_2^\star (f^{\perp})|^2 = \int_S |\not{d}_2^\star f|^2$$

from which we derive

$$\int_S \left| f - \left(\frac{\int_S f e^{\Phi}}{\int_S e^{2\Phi}} \right) e^{\Phi} \right|^2 \lesssim r^2 \int_S |\not{d}_2^\star f|^2$$

or

$$\begin{aligned} \int_S |f|^2 &\lesssim r^2 \int_S |\not{d}_2^\star f|^2 + \left| \int_S e^{\Phi} f \right|^2 \frac{1}{\int_S e^{2\Phi}} \\ &\lesssim r^2 \int_S |\not{d}_2^\star f|^2 + r^{-4} \left| \int_S e^{\Phi} f \right|^2 \end{aligned}$$

as desired. □

2.1.4.5 Higher derivative operators and spaces

Definition 2.39. *Given f a k-reduced scalar and s a positive integer we define*

$$\not{\mathfrak{d}}^s f = \begin{cases} r^{2p} \not{\Delta}_k^p f, & \text{if } s = 2p, \\ r^{2p+1} \not{d}_k \not{\Delta}_k^p f, & \text{if } s = 2p + 1. \end{cases} \tag{2.1.36}$$

We also define the norms

$$\|f\|_{\mathfrak{h}_s(S)} := \sum_{i=0}^{s} \| \not{\mathfrak{d}}^i f \|_{L^2(S)}. \tag{2.1.37}$$

Lemma 2.40. *Assume the Gauss curvature K of S verifies the condition*

$$K = \frac{1}{r^2} + O(\epsilon), \qquad |r^i \nabla^i K| = O(\epsilon), \qquad 1 \leq i \leq [s/2] + 1.$$

Then, the following holds.

1. *If f is a k-scalar, reduced from ${}^{(1+3)}f$, we have*

$$\|f\|_{\mathfrak{h}_s(S)} \sim \sum_{j=0}^{s} r^j \|\nabla\!\!\!\!/^j f\|_{L^2(S)} \tag{2.1.38}$$

   where $\nabla\!\!\!\!/$ denotes the usual covariant derivative operator on S.
2. *Equivalently, the norm $r^{-s}\|f\|_{\mathfrak{h}_s(S)}$ of a reduced scalar $f \in \mathfrak{s}_s(S)$ can be defined as the sum of L^2 norms of any allowable sequence of Hodge operators $d\!\!\!/_a$, $d\!\!\!/_a^\star$ applied to f.*

Proof. For $s = 1, 2$ the proof of the first part follows immediately from Proposition 2.33. For higher s the proof follows, step by step, by a simple commutation argument between covariant derivatives and $\Delta\!\!\!\!/_k$ and applications of Proposition 2.33. The proof of the second part follows from our reduced elliptic estimates and definition of the reduced Hodge operators. □

As a consequence of the lemma we can derive the reduced form of the standard Sobolev and product Sobolev inequalities. Before stating the result we pause to define the product of two reduced scalars.

Definition 2.41. *Let $f \in \mathfrak{s}_a$ be reduced from an $\mathcal{S}_a$ tensor and $g \in \mathfrak{s}_b$ reduced from an $\mathcal{S}_b$ tensor. We define the product $f \cdot g$ to be the reduction of any product between the corresponding tensors on S, i.e., any contraction of the tensor product between them. Thus $f \cdot g \in \mathfrak{s}_{a+b-2c}$ where c denotes the number of indices affected by the contraction.*

Examples. Here are the most relevant examples for us.

- $f \in \mathfrak{s}_0$, $g \in \mathfrak{s}_k$ in which case $f \cdot g \in \mathfrak{s}_k$ and equals fg.
- $f \in \mathfrak{s}_1$, $g \in \mathfrak{s}_k$ in which case $f \cdot g \in \mathfrak{s}_{k-1}$ or $f \cdot g \in \mathfrak{s}_{k+1}$ and in both cases $f \cdot g = fg$ as simple product of the reduced scalars.
- $f \in \mathfrak{s}_2$, $g \in \mathfrak{s}_k$ in which case $f \cdot g \in \mathfrak{s}_{k-2}$ or $f \cdot g \in \mathfrak{s}_k$ or $f \cdot g \in \mathfrak{s}_{k+2}$. In the first case $f \cdot g = 2fg$. In the second case and third cases $f \cdot g = fg$ as simple product of the reduced scalars.

Lemma 2.42. *Let $f \in \mathfrak{s}_a(S)$, $g \in \mathfrak{s}_b(S)$, $a \geq b$, $a > 0$, and $f \cdot g \in \mathfrak{s}_{a+b-2c}$ where $0 \leq c \leq \frac{1}{2}(a-b)$ denotes the order of contraction. Then,*

$$\begin{aligned} d\!\!\!/_{a+b-2c}(fg) &= f\, d\!\!\!/_b g + g\left(\left(1 - \frac{c}{a}\right) d\!\!\!/_a f - \frac{c}{a}\, d\!\!\!/_{a+1}^\star f\right), \\ d\!\!\!/_{a+b-2c+1}^\star(fg) &= f\, d\!\!\!/_{b+1}^\star g + g\left(-\frac{c}{a}\, d\!\!\!/_a f - \left(-1 + \frac{c}{a}\right) d\!\!\!/_{a+1}^\star f\right). \end{aligned} \tag{2.1.39}$$

Proof. Assume $a \geq b$ and $c \leq \frac{a-b}{2}$. We write

$$d\!\!\!/_{a+b-2c}(fg) \quad = \quad f\, d\!\!\!/_b g + g\,(e_\theta f + (a-2c) e_\theta(\Phi) f).$$

We look for reals A, B with $A + B = 1$ such that

$$e_\theta f + (a-2c) e_\theta(\Phi) f \quad = \quad A\, d\!\!\!/_a f - B\, d\!\!\!/_{a+1}^\star f = e_\theta f + a(A-B) e_\theta \Phi f.$$

Therefore,

$$a(1-2B)=a-2c,$$

i.e., $B=\frac{c}{a}$, $A=1-\frac{c}{a}$ and we derive

$$\not{d}_{a+b-2c}(fg) \;=\; f\,\not{d}_b g+g\left(\left(1-\frac{c}{a}\right)\not{d}_a f-\frac{c}{a}\,\not{d}^\star_{a+1}f\right).$$

Also,

$$\not{d}^\star_{a+b-2c+1}(fg) \;=\; f\,\not{d}^\star_{b+1}g+g\left(-e_\theta(f)+(a-2c)e_\theta(\Phi)f\right).$$

As before we write, with $A+B=-1$

$$-e_\theta(f)+(a-2c)e_\theta(\Phi)f \;=\; A\,\not{d}_a f-B\,\not{d}^\star_{a+1}f=-e_\theta f+a(A-B)e_\theta\Phi f.$$

Hence,

$$a(-1-2B)=a-2c,$$

i.e., $B=-1+\frac{c}{a}$, $A=-\frac{c}{a}$. Hence,

$$\not{d}^\star_{a+b-2c+1}(fg) \;=\; f\,\not{d}^\star_{b+1}g+g\left(-\frac{c}{a}\,\not{d}_a f-\left(-1+\frac{c}{a}\right)\not{d}^\star_{a+1}f\right)$$

as desired. □

Proposition 2.43. *The following results hold true for k-reduced scalars on S:*

1. *If $f\in\mathfrak{s}_k$ we have*

$$\|f\|_{L^\infty(S)} \;\lesssim\; r^{-1}\|f\|_{\mathfrak{h}_2(S)}.$$

2. *Given two reduced scalars f,g we have*

$$\|f\cdot g\|_{\mathfrak{h}_s(S)} \;\lesssim\; r^{-1}\left(\|f\|_{\mathfrak{h}_{[s/2]+2}(S)}\|g\|_{\mathfrak{h}_s(S)}+\|g\|_{\mathfrak{h}_{[s/2]+2}(S)}\|f\|_{\mathfrak{h}_s(S)}\right)$$

where $[s/2]$ denotes the largest integer smaller than $s/2$.

Proof. Both statements are classical for $\mathcal{S}_k(S)$ tensors with respect to the norm on the right-hand side of (2.1.38). A direct proof can also be derived using Lemma 2.42 and the equivalence definition of the $\mathfrak{h}_s(S)$ norms. □

2.1.4.6 S-averages

Definition 2.44. *Given any $f\in\mathfrak{s}_0$ we denote its average by*

$$\overline{f}: \;=\; \frac{1}{|S|}\int_S f, \qquad \check{f}:=f-\overline{f}.$$

The following follows immediately from the definition.

Lemma 2.45. *For any two reduced scalars f and g in $\mathfrak{s}_0$ we have*

$$\overline{fg} = \overline{f}\,\overline{g} + \overline{\check{f}\check{g}},$$

and

$$fg - \overline{fg} = \check{f}\,\overline{g} + \overline{f}\check{g} + \left(\check{f}\check{g} - \overline{\check{f}\check{g}}\right).$$

Remark 2.46. *In view of the notations above, we may rewrite the Poincaré inequality for $\not{d}_1^\star$ as follows. Under mild assumptions on the Gauss curvature, i.e., $K = r^{-2} + O(\epsilon r^{-2})$ and $re_\theta(K) = O(\epsilon r^{-2})$, we have for any $f \in \mathfrak{s}_0$*

$$\int_S |\not{d}_1^\star f|^2 \geq 2\int_S (1 + O(\epsilon))K(\check{f})^2.$$

2.1.5 Invariant S-foliations

In this section we record the main equations associated to general, $\mathbf{Z}$-invariant Einstein vacuum spacetimes $(\mathcal{M}, \mathbf{g})$. We start by recalling the spacetime framework of [20] and then we show how the null structure and Bianchi identities simplify in the reduced picture. Throughout this section we consider given an invariant S-foliation[12] and a fixed adapted null pair e_3, e_4, i.e., future directed $\mathbf{Z}$-invariant, polarized, null vectors orthogonal to the leaves S of the foliation normalized by $\mathbf{g}(e_3, e_4) = -2$.

Definition 2.47. *We denote by $\mathcal{S}_k(\mathcal{M})$ the set of k-covariant polarized tensors on $\mathcal{M}$ tangent to the S-foliation and which restrict to $\mathcal{S}_k(S)$ on any S-surface of the foliation and by $\mathfrak{s}_k(\mathcal{M})$ their corresponding reductions.*

2.1.5.1 Spacetime null decompositions

Following [20] we define the spacetime Ricci coefficients,

$$\begin{aligned}
{}^{(1+3)}\chi_{AB} &:= \mathbf{g}(\mathbf{D}_A e_4, e_B), & {}^{(1+3)}\xi_A &:= \frac{1}{2}\mathbf{g}(\mathbf{D}_4 e_4, e_A),\\
{}^{(1+3)}\eta_A &:= \frac{1}{2}\mathbf{g}(\mathbf{D}_3 e_4, e_A), & {}^{(1+3)}\zeta_A &:= \frac{1}{2}\mathbf{g}(\mathbf{D}_A e_4, e_3),\\
{}^{(1+3)}\omega &:= \frac{1}{4}\mathbf{g}(\mathbf{D}_4 e_4, e_3),
\end{aligned} \tag{2.1.40}$$

and interchanging e_3, e_4,

$$\begin{aligned}
{}^{(1+3)}\underline{\chi}_{AB} &:= \mathbf{g}(\mathbf{D}_A e_3, e_B), & {}^{(1+3)}\underline{\xi}_A &:= \frac{1}{2}\mathbf{g}(\mathbf{D}_3 e_3, e_A),\\
{}^{(1+3)}\underline{\eta}_A &:= \frac{1}{2}\mathbf{g}(\mathbf{D}_4 e_3, e_A), & {}^{(1+3)}\zeta_A &:= -\frac{1}{2}\mathbf{g}(\mathbf{D}_A e_3, e_4),\\
{}^{(1+3)}\underline{\omega} &:= \frac{1}{4}\mathbf{g}(D_3 e_3, e_4).
\end{aligned} \tag{2.1.41}$$

[12] From now on, an invariant S-foliation is automatically assumed to be a $\mathbf{Z}$-invariant polarized foliation.

We also define the spacetime null curvature components,

$$
\begin{aligned}
{}^{(1+3)}\alpha_{AB} &:= \mathbf{R}_{A4B4}, & {}^{(1+3)}\beta_A &:= \frac{1}{2}\mathbf{R}_{A434}, & {}^{(1+3)}\rho &:= \frac{1}{4}\mathbf{R}_{3434},\\
{}^{(1+3)}\underline{\alpha}_{AB} &:= \mathbf{R}_{A3B3}, & {}^{(1+3)}\underline{\beta}_A &:= \frac{1}{2}\mathbf{R}_{A334}, & {}^{(1+3)\star}\rho &:= \frac{1}{4}{}^{\star}\mathbf{R}_{3434}.
\end{aligned}
\tag{2.1.42}
$$

2.1.5.2 Reduced null decompositions

We define the spacetime Ricci coefficients as follows:

Definition 2.48 (Ricci coefficients)**.** *Let (e_3, e_4, e_θ) be a reduced null frame. The following scalars*

$$
\begin{aligned}
\chi &= g(D_\theta e_4, e_\theta), & \underline{\chi} &= g(D_\theta e_3, e_\theta),\\
\eta &= \frac{1}{2}g(D_3 e_4, e_\theta), & \underline{\eta} &= \frac{1}{2}g(D_4 e_3, e_\theta),\\
\xi &= \frac{1}{2}g(D_4 e_4, e_\theta), & \underline{\xi} &= \frac{1}{2}g(D_3 e_3, e_\theta),\\
\omega &= \frac{1}{4}g(D_4 e_4, e_3), & \underline{\omega} &= \frac{1}{4}g(D_3 e_3, e_4),\\
\zeta &= \frac{1}{2}g(D_\theta e_4, e_3),
\end{aligned}
\tag{2.1.43}
$$

are called the Ricci coefficients associated to our canonical null pair.

Lemma 2.49. *The following lemma follows easily from the definitions,*

$$
\begin{aligned}
D_4 e_4 &= -2\omega e_4 + 2\xi e_\theta, & D_3 e_3 &= -2\underline{\omega} e_3 + 2\underline{\xi} e_\theta,\\
D_4 e_3 &= 2\omega e_3 + 2\underline{\eta} e_\theta, & D_3 e_4 &= 2\underline{\omega} e_4 + 2\eta e_\theta,\\
D_4 e_\theta &= \underline{\eta} e_4 + \xi e_3, & D_3 e_\theta &= \underline{\xi} e_4 + \eta e_3,\\
D_\theta e_4 &= -\zeta e_4 + \chi e_\theta, & D_\theta e_3 &= \zeta e_3 + \underline{\chi} e_\theta,\\
D_\theta e_\theta &= \frac{1}{2}\underline{\chi} e_4 + \frac{1}{2}\chi e_3.
\end{aligned}
\tag{2.1.44}
$$

Definition 2.50. *The null components of the Ricci curvature tensor*[13] *of the metric g are denoted by*

$$
R_{33} = \underline{\alpha}, \quad R_{44} = \alpha, \quad R_{3\theta} = \underline{\beta}, \quad R_{4\theta} = \beta, \quad R_{\theta\theta} = R_{34} = \rho, \quad R_{34} = \rho.
$$

2.1.5.3 Comparison to the spacetime frame

Let e_3, e_4, e_θ be a null frame for the reduced metric g and $e_3, e_4, e_\theta, e_\varphi = X^{-1/2}\partial_\varphi$ the augmented adapted $3+1$ frame for $\mathbf{g}$. Recall that we have denoted

$$
{}^{(1+3)}\chi,\ {}^{(1+3)}\xi,\ {}^{(1+3)}\eta,\ {}^{(1+3)}\underline{\eta},\ {}^{(1+3)}\zeta,\ {}^{(1+3)}\omega,\ {}^{(1+3)}\underline{\chi},\ {}^{(1+3)}\underline{\xi},\ {}^{(1+3)}\underline{\omega},
$$

[13] Recall that the scalar curvature of the reduced metric g vanishes, $R = 0$, and hence $R_{34} = R_{\theta\theta}$.

the standard (as defined in [20]) spacetime Ricci coefficients and by

$$^{(1+3)}\alpha,\ ^{(1+3)}\beta,\ ^{(1+3)}\rho,\ ^{(1+3)\star}\rho,\ ^{(1+3)}\underline{\beta},\ ^{(1+3)}\underline{\alpha},$$

the null decomposition of the curvature tensor $\mathbf{R}$.

Proposition 2.51. *The following relations between the spacetime and reduced Ricci and curvature null components hold true.*

- *We have*

$$\begin{aligned}
^{(1+3)}\chi_{\theta\theta} &= \chi, & ^{(1+3)}\chi_{\theta\varphi} &= 0, & ^{(1+3)}\chi_{\varphi\varphi} &= e_4(\Phi),\\
^{(1+3)}\underline{\chi}_{\theta\theta} &= \underline{\chi}, & ^{(1+3)}\underline{\chi}_{\theta\varphi} &= 0, & ^{(1+3)}\underline{\chi}_{\varphi\varphi} &= e_3(\Phi),\\
^{(1+3)}\alpha_{\theta\theta} &= \alpha, & ^{(1+3)}\alpha_{\theta\varphi} &= 0, & ^{(1+3)}\alpha_{\varphi\varphi} &= -\alpha,\\
^{(1+3)}\underline{\alpha}_{\theta\theta} &= \underline{\alpha}, & ^{(1+3)}\underline{\alpha}_{\theta\varphi} &= 0, & ^{(1+3)}\underline{\alpha}_{\varphi\varphi} &= -\underline{\alpha}.
\end{aligned}$$

- *All e_φ components of $^{(1+3)}\eta$, $^{(1+3)}\underline{\eta}$, $^{(1+3)}\zeta$, $^{(1+3)}\xi$, $^{(1+3)}\underline{\xi}$, $^{(1+3)}\beta$, $^{(1+3)}\underline{\beta}$ vanish and*

$$^{(1+3)}\eta_\theta = \eta,\ ^{(1+3)}\underline{\eta}_\theta = \underline{\eta},\ ^{(1+3)}\zeta_\theta = \zeta,\ ^{(1+3)}\xi_\theta = \xi,\ ^{(1+3)}\underline{\xi}_\theta = \underline{\xi},$$

as well as[14]

$$^{(1+3)}\beta_\theta = \beta,\ ^{(1+3)}\underline{\beta}_\theta = -\underline{\beta}.$$

Also,

$$^{(1+3)}\omega = \omega, \qquad ^{(1+3)}\underline{\omega} = \underline{\omega}, \qquad ^{(1+3)}\rho = \rho, \qquad ^{(1+3)\star}\rho = 0.$$

- *We have*

$$^{(1+3)}tr\chi = \chi + e_4(\Phi), \qquad ^{(1+3)}tr\underline{\chi} = \underline{\chi} + e_3(\Phi).$$

Recalling, see definition in [20],

$$\begin{aligned}
^{(1+3)}\widehat{\chi}_{AB} &= {}^{(1+3)}\chi_{AB} - \frac{1}{2}(\,^{(1+3)}tr\chi)\not{g}_{AB},\\
^{(1+3)}\widehat{\underline{\chi}}_{AB} &= {}^{(1+3)}\underline{\chi}_{AB} - \frac{1}{2}(\,^{(1+3)}tr\underline{\chi})\not{g}_{AB},
\end{aligned}$$

we have

$$^{(1+3)}\widehat{\chi}_{\theta\theta} = \frac{1}{2}\left(\chi - e_4(\Phi)\right), \qquad ^{(1+3)}\widehat{\underline{\chi}}_{\theta\theta} = \frac{1}{2}\left(\underline{\chi} - e_3(\Phi)\right).$$

Proof. We check only the less obvious relations such as those involving the null

[14]Note the change of sign for the $\underline{\beta}$ component.

components of curvature. Using (2.1.5) and (2.1.6) we deduce

$$
\begin{aligned}
{}^{(1+3)}\alpha_{\theta\theta} &= \mathbf{R}_{\theta 4\theta 4} = R_{\theta 4\theta 4} = g_{\theta\theta}R_{44} = \alpha,\\
{}^{(1+3)}\alpha_{\theta\varphi} &= \mathbf{R}_{\theta 4\varphi 4} = 0,\\
{}^{(1+3)}\alpha_{\varphi\varphi} &= \mathbf{R}_{\varphi 4\varphi 4} = -R_{44} = -\alpha,\\
2\,{}^{(1+3)}\beta_{\theta} &= \mathbf{R}_{\theta 434} = R_{\theta 434} = -g_{34}R_{\theta 4} = 2\beta,\\
2\,{}^{(1+3)}\beta_{\varphi} &= \mathbf{R}_{\varphi 343} = 0,\\
4\,{}^{(1+3)}\rho &= \mathbf{R}_{3434} = R_{3434} = -2g_{34}R_{34} = 4\rho,\\
4\,{}^{(1+3)\star}\rho &= {}^{\star}\mathbf{R}_{3434} = 0,\\
2\,{}^{(1+3)}\underline{\beta}_{\theta} &= \mathbf{R}_{\theta 334} = R_{\theta 334} = g_{34}R_{\theta 3} = -2\underline{\beta},\\
2\,{}^{(1+3)}\underline{\beta}_{\varphi} &= \mathbf{R}_{\varphi 334} = 0.
\end{aligned}
$$

□

Definition 2.52. *We introduce the notation,*

$$
\begin{aligned}
\vartheta &:= \chi - e_4(\Phi), \qquad \kappa := {}^{(1+3)}tr\chi = \chi + e_4(\Phi),\\
\underline{\vartheta} &:= \underline{\chi} - e_3(\Phi), \qquad \underline{\kappa} := {}^{(1+3)}tr\underline{\chi} = \underline{\chi} + e_3(\Phi).
\end{aligned}
$$

Thus,

$$
{}^{(1+3)}\widehat{\chi}_{\theta\theta} = {}^{(1+3)}\widehat{\chi}_{\varphi\varphi} = \frac{1}{2}\vartheta, \qquad {}^{(1+3)}\underline{\widehat{\chi}}_{\theta\theta} = {}^{(1+3)}\underline{\widehat{\chi}}_{\varphi\varphi} = \frac{1}{2}\underline{\vartheta}.
$$

In particular, $\chi = \frac{1}{2}(\vartheta + \kappa)$ *and* $\underline{\chi} = \frac{1}{2}(\underline{\vartheta} + \underline{\kappa})$.

Remark 2.53. *In view of Proposition 2.51 we have:*

1. *The quantities* $\kappa, \underline{\kappa}, \omega, \underline{\omega}, \rho$ *are reduced scalars in* $\mathfrak{s}_0$.
2. *The quantities* $\eta, \underline{\eta}, \zeta, \xi, \underline{\xi}, \beta, \underline{\beta}$ *are reduced scalars in* $\mathfrak{s}_1$.
3. *The quantities* $\vartheta, \underline{\vartheta}, \alpha, \underline{\alpha}$ *are reduced scalars in* $\mathfrak{s}_2$.

2.1.5.4 Commutation identities

We record first the commutation relations between the elements of the frame,

$$
\begin{aligned}
[e_\theta, e_3] &= \frac{1}{2}(\underline{\kappa} + \underline{\vartheta})e_\theta + (\zeta - \eta)e_3 - \underline{\xi}e_4,\\
[e_\theta, e_4] &= \frac{1}{2}(\kappa + \vartheta)e_\theta - (\zeta + \underline{\eta})e_4 - \xi e_3,\\
[e_3, e_4] &= 2\underline{\omega}e_4 - 2\omega e_3 + 2(\eta - \underline{\eta})e_\theta.
\end{aligned}
$$

Lemma 2.54. *The following commutation formulae hold true for reduced scalars.*

1. If $f \in \mathfrak{s}_k$,

$$\begin{aligned}
[\not{d}_k, e_3]f &= \frac{1}{2}\underline{\kappa}\not{d}_k f + \underline{Com}_k(f),\\
\underline{Com}_k(f) &= -\frac{1}{2}\underline{\vartheta}\not{d}_{k+1}^\star f + (\zeta-\eta)e_3 f - k\eta e_3\Phi f\\
&\quad - \underline{\xi}(e_4 f + k e_4(\Phi) f) - k\underline{\beta} f,\\
[\not{d}_k, e_4]f &= \frac{1}{2}\kappa\not{d}_k f + Com_k(f),\\
Com_k(f) &= -\frac{1}{2}\vartheta\not{d}_{k+1}^\star f - (\zeta+\underline{\eta})e_4 f - k\underline{\eta} e_4\Phi f\\
&\quad - \xi(e_3 f + k e_3(\Phi) f) - k\beta f.
\end{aligned} \tag{2.1.45}$$

2. If $f \in \mathfrak{s}_{k-1}$,

$$\begin{aligned}
[\not{d}_k^\star, e_3]f &= \frac{1}{2}\underline{\kappa}\not{d}_k^\star f + \underline{Com}_k^*(f),\\
\underline{Com}_k^*(f) &= -\frac{1}{2}\underline{\vartheta}\not{d}_{k-1} f - (\zeta-\eta)e_3 f - (k-1)\eta e_3\Phi f\\
&\quad + \underline{\xi}(e_4 f - (k-1)e_4(\Phi) f) - (k-1)\underline{\beta} f,\\
[\not{d}_k^\star, e_4]f &= \frac{1}{2}\kappa\not{d}_k^\star f + Com_k^*(f),\\
Com_k^*(f) &= -\frac{1}{2}\vartheta\not{d}_{k-1} f + (\zeta+\underline{\eta})e_4 f - (k-1)\underline{\eta} e_4\Phi f\\
&\quad + \xi(e_3 f - (k-1)e_3(\Phi) f) - (k-1)\beta f.
\end{aligned} \tag{2.1.46}$$

Proof. We write

$$[e_\theta + k e_\theta(\Phi), e_3]f \quad = \quad [e_\theta, e_3]f - k(e_3 e_\theta\Phi)f.$$

Recall that (see (2.1.4)), $D_a D_b\Phi = R_{ab} - D_a\Phi D_b\Phi$. Hence,

$$e_3 e_\theta\Phi - \underline{\xi} e_4\Phi - \eta e_3\Phi = D_3 D_\theta\Phi \quad = \quad R_{3\theta} - D_3\Phi D_\theta\Phi = \underline{\beta} - e_3\Phi e_\theta\Phi.$$

Thus,

$$e_3 e_\theta\Phi \quad = \quad \underline{\beta} - e_3\Phi e_\theta\Phi + \eta e_3(\Phi) + \underline{\xi} e_4\Phi.$$

We deduce, since $e_3\Phi = \frac{1}{2}(\underline{\kappa} - \underline{\vartheta})$,

$$\begin{aligned}
[e_\theta + k e_\theta(\Phi), e_3]f &= [e_\theta, e_3]f - k\left(\underline{\beta} - e_3\Phi e_\theta\Phi + \eta e_3(\Phi) + \underline{\xi} e_4\Phi\right) f\\
&= \frac{1}{2}(\underline{\kappa} + \underline{\vartheta})e_\theta f + (\zeta-\eta)e_3 f - \underline{\xi} e_4 f\\
&\quad -k\left(\underline{\beta} - e_3\Phi e_\theta\Phi + \eta e_3(\Phi) + \underline{\xi} e_4\Phi\right) f\\
&= \frac{1}{2}(\underline{\kappa} + \underline{\vartheta})e_\theta f + k\frac{1}{2}(\underline{\kappa} - \underline{\vartheta})e_\theta\Phi f\\
&\quad + (\zeta-\eta)e_3 f - k\eta e_3\Phi f - \underline{\xi}(e_4 f + k e_4(\Phi) f) - k\underline{\beta} f\\
&= \frac{1}{2}\underline{\kappa}\not{d}_k f + \frac{1}{2}\underline{\vartheta}(e_\theta f - k e_\theta\Phi f)\\
&\quad + (\zeta-\eta)e_3 f - k\eta e_3\Phi f - \underline{\xi}(e_4 f + k e_4(\Phi) f) - k\underline{\beta} f,
\end{aligned}$$

i.e., recalling the definition of $\not{d}^\star_{k+1}$,

$$\begin{aligned}
[e_\theta + k e_\theta(\Phi), e_3] f &= \frac{1}{2}\underline{\kappa}\, \not{d}_k f + \underline{Com}_k(f),\\
\underline{Com}_k(f) &= -\frac{1}{2}\underline{\vartheta}\, \not{d}^\star_{k+1} f + (\zeta - \eta) e_3 f - k\eta e_3 \Phi f\\
&\quad -\underline{\xi}(e_4 f + k e_4(\Phi) f) - k\underline{\beta} f.
\end{aligned}$$

The other commutation formulae are proved in the same manner. □

2.1.6 Schwarzschild spacetime

In standard coordinates the Schwarzschild metric has the form

$$ds^2 = -\Upsilon dt^2 + \Upsilon^{-1} dr^2 + r^2 d\theta^2 + X^2 d\varphi^2, \tag{2.1.47}$$

where

$$\Upsilon := 1 - \frac{2m}{r}, \qquad X = r^2 \sin^2\theta.$$

We denote by $\mathbf{T}$ the stationary Killing vectorfield $\mathbf{T} = \partial_t$ and by $\mathbf{Z} = \partial_\varphi$ the axial symmetric one. Recall the regular, $\mathbf{Z}$-invariant optical functions in the exterior region $r \geq 2m$ of Schwarzschild

$$u = t - r_*, \qquad \underline{u} = t + r_*, \qquad \frac{dr_*}{dr} = \Upsilon^{-1}, \tag{2.1.48}$$

with $r_* = r + 2m\log(\frac{r}{2m} - 1)$. The corresponding null geodesic generators are

$$\underline{L} := -g^{ab}\partial_a \underline{u}\,\partial_b = \Upsilon^{-1}\partial_t - \partial_r, \qquad L := -g^{ab}\partial_a u\,\partial_b = \Upsilon^{-1}\partial_t + \partial_r. \tag{2.1.49}$$

Clearly,

$$g(L, L) = g(\underline{L}, \underline{L}) = 0, \qquad g(L, \underline{L}) = -2\Upsilon^{-1}, \qquad D_L L = D_{\underline{L}} \underline{L} = 0.$$

Definition 2.55. *We can use the null geodesic generators $L, \underline{L}$ to define the following canonical null pairs. In all cases all curvature components vanish identically except*

$$^{(1+3)}\rho = -\frac{2m}{r^3}. \tag{2.1.50}$$

1. *The null frame (e_3, e_4) for which e_3 is geodesic (which is regular towards the future for all $r > 0$) is given by*

$$e_3 = \underline{L} = \Upsilon^{-1}\partial_t - \partial_r, \qquad e_4 = \Upsilon L = \partial_t + \Upsilon\partial_r, \qquad \Upsilon = 1 - \frac{2m}{r}. \tag{2.1.51}$$

   All Ricci coefficients vanish except

$$\chi = \frac{\Upsilon}{r}, \qquad \underline{\chi} = -\frac{1}{r}, \qquad \omega = -\frac{m}{r^2}, \qquad \underline{\omega} = 0.$$

2. The null frame (e_3, e_4) for which e_4 is geodesic is given by

$$e_4 = L = \Upsilon^{-1}\partial_t + \partial_r, \qquad e_3 = \Upsilon \underline{L} = \partial_t - \Upsilon\partial_r.$$

All Ricci coefficients vanish except

$$\chi = \frac{1}{r}, \qquad \underline{\chi} = -\frac{\Upsilon}{r}, \qquad \omega = 0, \qquad \underline{\omega} = \frac{m}{r^2}.$$

Note that the null pair (2.1.51) is regular along the future event horizon as can be easily seen by studying the behavior[15] of future directed ingoing null geodesics near $r = 2m$.

2.2 MAIN EQUATIONS

In this section we translate the null structure and null Bianchi identities associated to an S-foliation in the reduced picture. We start with general, $\mathbf{Z}$-invariant, S-foliation. We then consider the special case of geodesic foliations.

2.2.1 Main equations for general S-foliations

We consider a fixed $\mathbf{Z}$-invariant S-foliation with a fixed $\mathbf{Z}$-invariant null frame e_3, e_4.

2.2.1.1 Null structure equations

We simply translate the well known spacetime null structure equations (see[16] proposition 7.4.1 in [20]) in the reduced picture. Thus the spacetime equation[17]

$$\nabla\!\!\!/_3 \underline{\widehat{\chi}} + \mathrm{tr}\underline{\chi}\, \underline{\widehat{\chi}} = \nabla\!\!\!/ \widehat{\otimes} \underline{\xi} - 2\underline{\omega}\underline{\widehat{\chi}} + (\eta + \underline{\eta} - 2\zeta)\widehat{\otimes}\underline{\xi} - \underline{\alpha}$$

becomes[18]

$$e_3(\underline{\vartheta}) + \underline{\kappa}\,\underline{\vartheta} = 2(e_\theta(\underline{\xi}) - e_\theta(\Phi)\underline{\xi}) - 2\underline{\omega}\,\underline{\vartheta} + 2(\eta + \underline{\eta} - 2\zeta)\,\underline{\xi} - 2\underline{\alpha}. \tag{2.2.1}$$

The spacetime equation

$$e_3(\mathrm{tr}\underline{\chi}) + \frac{1}{2}\mathrm{tr}\underline{\chi}^2 = 2\mathrm{d}\!\!\!/\mathrm{iv}\,\underline{\xi} - 2\underline{\omega}\mathrm{tr}\underline{\chi} + 2\underline{\xi}\cdot(\eta + \underline{\eta} - 2\zeta) - \underline{\widehat{\chi}}\cdot\underline{\widehat{\chi}}$$

becomes

$$e_3(\underline{\kappa}) + \frac{1}{2}\underline{\kappa}^2 + 2\underline{\omega}\,\underline{\kappa} = 2(e_\theta\underline{\xi} + e_\theta(\Phi)\underline{\xi}) + 2(\eta + \underline{\eta} - 2\zeta)\underline{\xi} - \frac{1}{2}\underline{\vartheta}\,\underline{\vartheta}. \tag{2.2.2}$$

[15]That is, the null geodesics in the direction of $\underline{L}$ reach the horizon in finite proper time. Note that, on the other hand, the past null geodesics in the direction of L still meet the horizon in infinite proper time.

[16]Note however that the notations in [20] are different, see section 7.3 for the definitions.

[17]For convenience we drop the $^{(1+3)}$ labels in what follows.

[18]Recall that $^{(1+3)}\underline{\widehat{\chi}}_{\theta\theta} = \frac{1}{2}\underline{\vartheta}$.

The spacetime equation

$$\slashed{\nabla}_4\widehat{\underline{\chi}} + \frac{1}{2}\mathrm{tr}\chi\,\widehat{\underline{\chi}} = \slashed{\nabla}\widehat{\otimes}\underline{\eta} + 2\omega\widehat{\underline{\chi}} - \frac{1}{2}\mathrm{tr}\underline{\chi}\widehat{\chi} + \xi\widehat{\otimes}\underline{\xi} + \underline{\eta}\widehat{\otimes}\underline{\eta}$$

becomes

$$e_4\underline{\vartheta} + \frac{1}{2}\kappa\,\underline{\vartheta} - 2\omega\underline{\vartheta} = 2(e_\theta\underline{\eta} - e_\theta(\Phi)\underline{\eta}) - \frac{1}{2}\mathrm{tr}\underline{\chi}\,\vartheta + 2(\xi\,\underline{\xi} + \underline{\eta}^2).$$

The spacetime equation

$$\slashed{\nabla}_4\mathrm{tr}\underline{\chi} + \frac{1}{2}\mathrm{tr}\chi\,\mathrm{tr}\underline{\chi} = 2\slashed{\mathrm{div}}\,\underline{\eta} + 2\rho + 2\omega\,\mathrm{tr}\underline{\chi} - \widehat{\chi}\cdot\widehat{\underline{\chi}} + 2(\xi\cdot\underline{\xi} + \underline{\eta}\cdot\underline{\eta})$$

becomes

$$e_4(\underline{\kappa}) + \frac{1}{2}\kappa\,\underline{\kappa} - 2\omega\underline{\kappa} = 2(e_\theta\underline{\eta} + e_\theta(\Phi)\underline{\eta}) + 2\rho - \frac{1}{2}\vartheta\,\underline{\vartheta} + 2(\xi\,\underline{\xi} + \underline{\eta}\,\underline{\eta}).$$

The spacetime equation

$$\slashed{\nabla}_3\zeta = -\underline{\beta} - 2\slashed{\nabla}\underline{\omega} - \widehat{\underline{\chi}}\cdot(\zeta+\eta) - \frac{1}{2}\mathrm{tr}\underline{\chi}(\zeta+\eta) + 2\underline{\omega}(\zeta-\eta) + (\widehat{\chi} + \frac{1}{2}\mathrm{tr}\chi)\underline{\xi} + 2\omega\underline{\xi}$$

becomes (note that ${}^{(1+3)}\underline{\beta} = -\underline{\beta}$!)

$$e_3\zeta + \frac{1}{2}\underline{\kappa}(\zeta+\eta) - 2\underline{\omega}(\zeta-\eta) = \underline{\beta} - 2e_\theta(\underline{\omega}) + 2\omega\underline{\xi} + \frac{1}{2}\kappa\,\underline{\xi} - \frac{1}{2}\underline{\vartheta}(\zeta+\eta) + \frac{1}{2}\vartheta\,\underline{\xi}.$$

The spacetime equation

$$\slashed{\nabla}_4\underline{\xi} - \slashed{\nabla}_3\underline{\eta} = -\underline{\beta} + 4\omega\underline{\xi} + \widehat{\underline{\chi}}\cdot(\underline{\eta}-\eta) + \frac{1}{2}\mathrm{tr}\underline{\chi}(\underline{\eta}-\eta)$$

becomes[19]

$$e_4(\underline{\xi}) - e_3(\underline{\eta}) = \underline{\beta} + 4\omega\underline{\xi} + \frac{1}{2}\underline{\kappa}(\underline{\eta}-\eta) + \frac{1}{2}\underline{\vartheta}(\underline{\eta}-\eta).$$

The spacetime equation

$$\slashed{\nabla}_4\underline{\omega} + \slashed{\nabla}_3\omega = \rho + 4\omega\underline{\omega} + \xi\cdot\underline{\xi} + \zeta\cdot(\eta-\underline{\eta}) - \eta\cdot\underline{\eta}$$

becomes

$$e_4\underline{\omega} + e_3\omega = \rho + 4\omega\underline{\omega} + \xi\,\underline{\xi} + \zeta(\eta-\underline{\eta}) - \eta\,\underline{\eta}.$$

The spacetime Codazzi equation

$${}^{(1+3)}\slashed{\mathrm{div}}\,{}^{(1+3)}\widehat{\underline{\chi}} = {}^{(1+3)}\underline{\beta} + \frac{1}{2}\left({}^{(1+3)}\slashed{\nabla}\,{}^{(1+3)}\mathrm{tr}\underline{\chi} - {}^{(1+3)}\mathrm{tr}\underline{\chi}\,{}^{(1+3)}\zeta\right) + {}^{(1+3)}\widehat{\underline{\chi}}\cdot{}^{(1+3)}\zeta$$

---

[19]Note that ${}^{(1+3)}\underline{\beta} = -\underline{\beta}$ and $\slashed{d}_1^{\star} f = -e_\theta(f)$.

becomes[20]

$$\frac{1}{2}(e_\theta(\underline{\vartheta}) + 2e_\theta(\Phi)\underline{\vartheta}) = -\underline{\beta} + \frac{1}{2}(e_\theta(\underline{\kappa}) - \underline{\kappa}\zeta) + \frac{1}{2}\underline{\vartheta}\zeta.$$

The Gauss equation

$$K = -\frac{1}{4}\,{}^{(1+3)}\mathrm{tr}\chi\,{}^{(1+3)}\mathrm{tr}\underline{\chi} + \frac{1}{2}\,{}^{(1+3)}\widehat{\chi}\,{}^{(1+3)}\widehat{\underline{\chi}} - {}^{(1+3)}\rho$$

becomes

$$K = -\frac{1}{4}\kappa\underline{\kappa} + \frac{1}{4}\vartheta\underline{\vartheta} - \rho.$$

We summarize the results in the following proposition.

Proposition 2.56.

$$\begin{aligned}
e_3(\underline{\vartheta}) + \underline{\kappa}\,\underline{\vartheta} + 2\underline{\omega}\,\underline{\vartheta} &= -2\underline{\alpha} - 2\,\not{d}_2^\star\,\underline{\xi} + 2(\eta + \underline{\eta} - 2\zeta)\,\underline{\xi},\\
e_3(\underline{\kappa}) + \frac{1}{2}\underline{\kappa}^2 + 2\underline{\omega}\,\underline{\kappa} &= 2\,\not{d}_1\underline{\xi} + 2(\eta + \underline{\eta} - 2\zeta)\underline{\xi} - \frac{1}{2}\underline{\vartheta}^2,\\
e_4\underline{\vartheta} + \frac{1}{2}\kappa\,\underline{\vartheta} - 2\omega\underline{\vartheta} &= -2\,\not{d}_2^\star\,\underline{\eta} - \frac{1}{2}\underline{\kappa}\,\vartheta + 2(\xi\,\underline{\xi} + \underline{\eta}^2),\\
e_4(\underline{\kappa}) + \frac{1}{2}\kappa\,\underline{\kappa} - 2\omega\underline{\kappa} &= 2\,\not{d}_1\underline{\eta} + 2\rho - \frac{1}{2}\vartheta\,\underline{\vartheta} + 2(\xi\,\underline{\xi} + \eta\,\underline{\eta}),\\
e_3\zeta + \frac{1}{2}\underline{\kappa}(\zeta + \eta) - 2\underline{\omega}(\zeta - \eta) &= \underline{\beta} + 2\,\not{d}_1^\star\,\underline{\omega} + 2\omega\underline{\xi} + \frac{1}{2}\kappa\,\underline{\xi}\\
&\quad - \frac{1}{2}\underline{\vartheta}(\zeta + \eta) + \frac{1}{2}\vartheta\,\underline{\xi},\\
e_4(\underline{\xi}) - 4\omega\underline{\xi} - e_3(\underline{\eta}) &= \underline{\beta} + \frac{1}{2}\underline{\kappa}(\underline{\eta} - \eta) + \frac{1}{2}\underline{\vartheta}(\underline{\eta} - \eta),\\
e_4\underline{\omega} + e_3\omega &= \rho + 4\omega\underline{\omega} + \xi\,\underline{\xi} + \zeta(\eta - \underline{\eta}) - \eta\,\underline{\eta}.
\end{aligned} \tag{2.2.3}$$

[20]Note that ${}^{(1+3)}\underline{\beta}_\theta = -\underline{\beta}$, ${}^{(1+3)}\widehat{\underline{\chi}} = \frac{1}{2}\underline{\vartheta}$.

In view of the symmetry $e_3 - e_4$, we also derive

$$\begin{aligned}
e_4(\vartheta) + \kappa\,\vartheta + 2\omega\vartheta &= -2\alpha - 2\,{\not{d}}_2^\star\,\xi + 2(\underline{\eta} + \eta + 2\zeta)\xi,\\
e_4(\kappa) + \frac{1}{2}\kappa^2 + 2\omega\,\kappa &= 2\,{\not{d}}_1\xi + 2(\underline{\eta} + \eta + 2\zeta)\xi - \frac{1}{2}\vartheta^2,\\
e_3\vartheta + \frac{1}{2}\underline{\kappa}\,\vartheta - 2\underline{\omega}\vartheta &= -2\,{\not{d}}_2^\star\,\eta - \frac{1}{2}\kappa\,\underline{\vartheta} + 2(\underline{\xi}\,\xi + \eta^2),\\
e_3(\kappa) + \frac{1}{2}\underline{\kappa}\,\kappa - 2\underline{\omega}\kappa &= 2\,{\not{d}}_1\eta + 2\rho - \frac{1}{2}\underline{\vartheta}\,\vartheta + 2(\underline{\xi}\,\xi + \eta\,\eta),\\
-e_4\zeta + \frac{1}{2}\kappa(-\zeta + \underline{\eta}) + 2\omega(\zeta + \underline{\eta}) &= \beta + 2\,{\not{d}}_1^\star\omega + 2\underline{\omega}\xi + \frac{1}{2}\underline{\kappa}\,\xi\\
&\quad - \frac{1}{2}\vartheta(-\zeta + \underline{\eta}) + \frac{1}{2}\underline{\vartheta}\,\xi,\\
e_3(\xi) - e_4(\eta) &= \beta + 4\underline{\omega}\xi + \frac{1}{2}\kappa(\eta - \underline{\eta}) + \frac{1}{2}\vartheta(\eta - \underline{\eta}),\\
e_4\underline{\omega} + e_3\omega &= \rho + 4\omega\underline{\omega} + \xi\,\underline{\xi} + \zeta(\eta - \underline{\eta}) - \eta\,\underline{\eta}.
\end{aligned} \tag{2.2.4}$$

We also have the Codazzi equations

$$\begin{aligned}
{\not{d}}_2\underline{\vartheta} &= -2\underline{\beta} - {\not{d}}_1^\star\,\underline{\kappa} - \zeta\underline{\kappa} + \underline{\vartheta}\,\zeta,\\
{\not{d}}_2\vartheta &= -2\beta - {\not{d}}_1^\star\,\kappa + \zeta\kappa - \vartheta\,\zeta,
\end{aligned}$$

and the Gauss equation

$$K = -\rho - \frac{1}{4}\kappa\,\underline{\kappa} + \frac{1}{4}\vartheta\,\underline{\vartheta}.$$

2.2.2 Null Bianchi identities

We now translate the spacetime null Bianchi identities of [20] (see proposition 7.3.2) in the reduced picture. The spacetime equation (note that ${\not{\mathcal{D}}}_2^\star\beta := -\frac{1}{2}\,{}^{(1+3)}\nabla\!\!\!/ \otimes \beta)$

$$\nabla\!\!\!/_3\alpha + \frac{1}{2}\mathrm{tr}\underline{\chi}\,\alpha \quad = \quad -2\,{\not{\mathcal{D}}}_2^\star\,\beta + 4\underline{\omega}\alpha - 3(\widehat{\chi}\rho + {}^\star\widehat{\chi}\,{}^\star\rho) + (\zeta + 4\eta)\otimes\beta$$

becomes (note that ${}^\star\rho = 0$)

$$e_3\alpha + \frac{1}{2}\underline{\kappa}\alpha \quad = \quad (e_\theta(\beta) - (e_\theta\Phi)\beta) + 4\underline{\omega}\alpha - \frac{3}{2}\vartheta\rho + (\zeta + 4\eta)\beta. \tag{2.2.5}$$

The spacetime equation

$$\nabla\!\!\!/_4\beta + 2\mathrm{tr}\chi\beta \quad = \quad \mathrm{d\!\!/iv}\,\alpha - 2\omega\beta + (2\zeta + \underline{\eta})\cdot\alpha + 3(\xi\rho + {}^\star\xi\,{}^\star\rho)$$

becomes

$$e_4\beta + 2\kappa\beta \quad = \quad (e_\theta\alpha + 2(e_\theta\Phi)\alpha) - 2\omega\beta + (2\zeta + \underline{\eta})\alpha + 3\xi\rho. \tag{2.2.6}$$

The spacetime equation

$$\nabla\!\!\!/_3\beta + \mathrm{tr}\underline{\chi}\beta \quad = \quad {\not{\mathcal{D}}}_1^\star(-\rho, {}^\star\rho) + 2\widehat{\chi}\cdot\underline{\beta} + 2\underline{\omega}\,\beta + \underline{\xi}\cdot\alpha + 3(\eta\rho + {}^\star\eta\,{}^\star\rho)$$

becomes (recall ${}^{(1+3)}\underline{\beta}_\theta = -\underline{\beta}$)

$$e_3\beta + \underline{\kappa}\beta = e_\theta(\rho) + 2\underline{\omega}\beta + 3\eta\rho - \vartheta\underline{\beta} + \underline{\xi}\alpha. \tag{2.2.7}$$

The spacetime equation

$$e_4\rho + \frac{3}{2}\mathrm{tr}\chi\rho = \mathrm{d\!\!/iv}\,\beta - \frac{1}{2}\underline{\widehat{\chi}}\cdot\alpha + \zeta\cdot\beta + 2(\underline{\eta}\cdot\beta - \xi\cdot\underline{\beta})$$

becomes

$$e_4\rho + \frac{3}{2}\kappa\rho = (e_\theta(\beta) + (e_\theta\Phi)\beta) - \frac{1}{2}\underline{\vartheta}\,\alpha + \zeta\,\beta + 2(\underline{\eta}\,\beta + \xi\,\underline{\beta}). \tag{2.2.8}$$

Indeed note that

$${}^{(1+3)}\underline{\widehat{\chi}}\cdot{}^{(1+3)}\alpha = 2\,{}^{(1+3)}\underline{\widehat{\chi}}_{\theta\theta}\,{}^{(1+3)}\alpha_{\theta\theta} = \underline{\vartheta}\alpha.$$

All other equations in the proposition below are derived using the $e_3 - e_4$ symmetry. We summarize the results in the following proposition.

Proposition 2.57.

$$\begin{aligned}
e_3\alpha + \frac{1}{2}\underline{\kappa}\alpha &= -\,d\!\!/_2^\star\,\beta + 4\underline{\omega}\alpha - \frac{3}{2}\vartheta\rho + (\zeta + 4\eta)\beta,\\
e_4\beta + 2\kappa\beta &= d\!\!/_2\alpha - 2\omega\beta + (2\zeta + \underline{\eta})\alpha + 3\xi\rho,\\
e_3\beta + \underline{\kappa}\beta &= -\,d\!\!/_1^\star\rho + 2\underline{\omega}\beta + 3\eta\rho - \vartheta\underline{\beta} + \underline{\xi}\alpha,\\
e_4\rho + \frac{3}{2}\kappa\rho &= d\!\!/_1\beta - \frac{1}{2}\underline{\vartheta}\,\alpha + \zeta\,\beta + 2(\underline{\eta}\,\beta + \xi\,\underline{\beta}),\\
\\
e_3\rho + \frac{3}{2}\underline{\kappa}\rho &= d\!\!/_1\underline{\beta} - \frac{1}{2}\vartheta\,\underline{\alpha} - \zeta\,\underline{\beta} + 2(\eta\,\underline{\beta} + \underline{\xi}\,\beta),\\
e_4\underline{\beta} + \kappa\underline{\beta} &= -\,d\!\!/_1^\star\rho + 2\omega\underline{\beta} + 3\underline{\eta}\rho - \underline{\vartheta}\beta + \xi\underline{\alpha},\\
e_3\underline{\beta} + 2\underline{\kappa}\,\underline{\beta} &= d\!\!/_2\underline{\alpha} - 2\underline{\omega}\,\underline{\beta} + (-2\zeta + \eta)\underline{\alpha} + 3\underline{\xi}\rho,\\
e_4\underline{\alpha} + \frac{1}{2}\kappa\,\underline{\alpha} &= -\,d\!\!/_2^\star\,\underline{\beta} + 4\omega\underline{\alpha} - \frac{3}{2}\underline{\vartheta}\rho + (-\zeta + 4\underline{\eta})\underline{\beta}.
\end{aligned} \tag{2.2.9}$$

2.2.2.1 Mass aspect functions

We define the mass aspect functions

$$\begin{aligned}
\mu &:= -\,d\!\!/_1\zeta - \rho + \frac{1}{4}\vartheta\underline{\vartheta},\\
\underline{\mu} &:= d\!\!/_1\zeta - \rho + \frac{1}{4}\vartheta\underline{\vartheta}.
\end{aligned} \tag{2.2.10}$$

One can derive useful propagation equations, in the e_4 direction for μ and in the e_3 direction for $\underline{\mu}$, by using the null structure and null Bianchi equations, see [20] and [42]. In the next section we will do this in the context of null geodesic foliations.

2.2.3 Hawking mass

Definition 2.58. *The Hawking mass $m = m(S)$ of S is defined by the formula*

$$\frac{2m}{r} = 1 + \frac{1}{16\pi}\int_S \kappa\underline{\kappa}. \tag{2.2.11}$$

Proposition 2.59. *The following identities hold true.*

1. *The average of ρ is given by the formulas*

$$\overline{\rho} = -\frac{2m}{r^3} + \frac{1}{16\pi r^2}\int_S \vartheta\underline{\vartheta}. \tag{2.2.12}$$

2. *The average of the mass aspect function is*

$$\overline{\mu} = \overline{\underline{\mu}} = \frac{2m}{r^3}. \tag{2.2.13}$$

3. *The average of κ and $\underline{\kappa}$ are related by*

$$\overline{\kappa}\,\overline{\underline{\kappa}} = -\frac{4\Upsilon}{r^2} - \overline{\check{\kappa}\underline{\check{\kappa}}} \tag{2.2.14}$$

where $\Upsilon = 1 - \frac{2m}{r}$.

Proof. We have from the Gauss equation

$$K = -\frac{1}{4}\kappa\underline{\kappa} + \frac{1}{4}\vartheta\underline{\vartheta} - \rho.$$

Integrating on S and using the Gauss Bonnet formula, we infer

$$4\pi = -\frac{1}{4}\int_S \kappa\underline{\kappa} + \frac{1}{4}\int_S \vartheta\underline{\vartheta} - \int_S \rho.$$

Together with the definition of the Hawking mass, we infer

$$\begin{aligned}\int_S \rho &= -4\pi\left(1 + \frac{1}{16\pi}\int_S \kappa\underline{\kappa}\right) + \frac{1}{4}\int_S \vartheta\underline{\vartheta}\\ &= -\frac{8\pi m}{r} + \frac{1}{4}\int_S \vartheta\underline{\vartheta}\end{aligned}$$

and hence

$$\overline{\rho} = -\frac{2m}{r^3} + \frac{1}{16\pi r^2}\int_S \vartheta\underline{\vartheta}$$

which proves our first identity. The second identity follows easily from the definition of $\mu, \underline{\mu}$ and the first formula. Thus, for example,

$$\overline{\mu} = \frac{1}{|S|}\int_S \mu = \frac{1}{|S|}\int_S \left(-\,\not{d}_1\zeta - \rho + \frac{1}{4}\vartheta\underline{\vartheta}\right) = -\overline{\rho} + \frac{1}{4|S|}\int_S \vartheta\underline{\vartheta} = \frac{2m}{r^3}.$$

To prove the last identity we remark that, in view of the definition of the Hawking mass,

$$-\Upsilon = \frac{2m}{r} - 1 = \frac{1}{16\pi}\int_S \kappa\underline{\kappa} = \frac{1}{16\pi}\left(|S|\overline{\kappa}\,\overline{\underline{\kappa}} + \int_S \check{\kappa}\check{\underline{\kappa}}\right)$$

and hence

$$\begin{aligned}\overline{\kappa}\,\overline{\underline{\kappa}} &= -\frac{16\pi\Upsilon}{|S|} - \frac{1}{|S|}\int_S \check{\kappa}\check{\underline{\kappa}}\\ &= -\frac{4\Upsilon}{r^2} - \overline{\check{\kappa}\check{\underline{\kappa}}}.\end{aligned}$$

This concludes the proof of the proposition. □

2.2.4 Outgoing geodesic foliations

We restrict our attention to geodesic foliations, i.e., geodesic foliations by **Z**-invariant optical functions.

2.2.4.1 Basic definitions

Assume given an outgoing optical function u, i.e., **Z**-invariant solution of the equation

$$\mathbf{g}^{\alpha\beta}\partial_\alpha u \partial_\beta u = g^{ab}\partial_a u \partial_b u = 0$$

and $L = -g^{ab}\partial_b u \partial_a$ its null geodesic generator. We choose e_4 such that

$$e_4 = \varsigma L, \qquad L(\varsigma) = 0. \tag{2.2.15}$$

Remark 2.60. *In our definition of a GCM admissible spacetime, see section 3.1, we initialize ς on the spacelike hypersurface Σ_*.*

We then choose s such that

$$e_4(s) = 1. \tag{2.2.16}$$

The functions u, s generate what is called an outgoing geodesic foliation. Let $S_{u,s}$ be the 2-surfaces of intersection between the level surfaces of u and s. We choose e_3 the unique **Z**-invariant null vectorfield orthogonal to $S_{u,s}$ and such that $g(e_3, e_4) = -2$. We then let e_θ be unit tangent to $S_{u,s}$, **Z**-invariant and orthogonal to **Z**. We also introduce

$$\underline{\Omega} := e_3(s). \tag{2.2.17}$$

Lemma 2.61. *We have*

$$\omega = \xi = 0, \qquad \underline{\eta} = -\zeta, \tag{2.2.18}$$

$$\begin{aligned} \varsigma &= \frac{2}{e_3(u)}, \\ e_4(\varsigma) &= 0, \\ e_\theta(\log \varsigma) &= \eta - \zeta, \\ e_\theta(\underline{\Omega}) &= -\underline{\xi} - (\eta - \zeta)\underline{\Omega}, \\ e_4(\underline{\Omega}) &= -2\underline{\omega}. \end{aligned} \tag{2.2.19}$$

Proof. Recall that L is geodesic, $e_4 = \varsigma L$ and $L(\varsigma) = 0$. This immediately implies that e_4 is geodesic, and hence we have

$$\omega = \xi = 0.$$

Applying the vectorfield

$$[e_4, e_\theta] = (\underline{\eta} + \zeta)e_4 + \xi e_3 - \chi e_\theta$$

to s, and since $e_4(s) = 1$ and $e_\theta(s) = 0$, we derive

$$\underline{\eta} + \zeta = 0.$$

Next, note that

$$e_3(u) = g(e_3, -L) = -\varsigma^{-1} g(e_3, e_4) = \frac{2}{\varsigma}$$

and hence

$$\varsigma = \frac{2}{e_3(u)}.$$

Applying the vectorfield

$$[e_3, e_\theta] = \underline{\xi} e_4 + (\eta - \zeta)e_3 - \underline{\chi} e_\theta$$

to u and making use of the relation $e_4(u) = e_\theta(u) = 0$ we deduce

$$(\eta - \zeta)e_3(u) = e_3(e_\theta u) - e_\theta e_3(u) = -e_\theta e_3(u)$$

which together with the identity $\varsigma = 2/e_3(u)$ yields

$$\eta - \zeta = -e_\theta \log(e_3 u) = -e_\theta \log\left(\frac{2}{\varsigma}\right) = e_\theta(\log \varsigma)$$

and hence

$$e_\theta(\log \varsigma) = \eta - \zeta.$$

Applying the vectorfield

$$[e_3, e_\theta] = \underline{\xi} e_4 + (\eta - \zeta)e_3 - \underline{\chi} e_\theta$$

to s we deduce, since $e_4(s)=1$, $e_\theta(s)=0$ and $e_3(s)=\underline{\Omega}$,

$$e_\theta(\underline{\Omega}) = -\underline{\xi}-(\eta-\zeta)\underline{\Omega}.$$

Finally, applying

$$[e_4,e_3] = -2\underline{\omega}e_4 - 2(\eta-\underline{\eta})e_\theta + 2\omega e_3$$

to s, and using $e_4(s)=1$ and $e_\theta(s)=0$, we infer $e_4(e_3(s))=-2\underline{\omega}$, i.e., $e_4(\underline{\Omega})=-2\underline{\omega}$ as desired. □

Remark 2.62. *In the particular case when ς is constant we have $\eta=\zeta=-\underline{\eta}$. In Schwarzschild, relative to the standard outgoing geodesic frame, we have*

$$\varsigma=1, \qquad \underline{\Omega}=-\Upsilon=-\left(1-\frac{2m}{r}\right).$$

2.2.4.2 Basic equations

Proposition 2.63. *Relative to an outgoing geodesic foliation we have:*

1. *The reduced null structure equations take the form*

$$\begin{aligned}
e_4(\vartheta)+\kappa\,\vartheta &= -2\alpha,\\
e_4(\kappa)+\frac{1}{2}\kappa^2 &= -\frac{1}{2}\vartheta^2,\\
e_4\zeta+\kappa\zeta &= -\beta-\vartheta\zeta,\\
e_4(\eta-\zeta)+\frac{1}{2}\kappa(\eta-\zeta) &= -\frac{1}{2}\vartheta(\eta-\zeta),\\
e_4\underline{\vartheta}+\frac{1}{2}\kappa\,\underline{\vartheta} &= 2\,{\not{d}}_2^\star\,\zeta-\frac{1}{2}\underline{\kappa}\,\vartheta+2\zeta^2,\\
e_4(\underline{\kappa})+\frac{1}{2}\kappa\,\underline{\kappa} &= -2\,{\not{d}}_1\zeta+2\rho-\frac{1}{2}\vartheta\,\underline{\vartheta}+2\zeta^2,\\
e_4\underline{\omega} &= \rho+\zeta(2\eta+\zeta),\\
e_4(\underline{\xi}) &= -e_3(\zeta)+\underline{\beta}-\frac{1}{2}\underline{\kappa}(\zeta+\eta)-\frac{1}{2}\underline{\vartheta}(\zeta+\eta),
\end{aligned}$$

and

$$\begin{aligned}
e_3(\underline{\vartheta})+\underline{\kappa}\,\underline{\vartheta}+2\underline{\omega}\,\underline{\vartheta} &= -2\underline{\alpha}-2\,{\not{d}}_2^\star\,\underline{\xi}+2(\eta-3\zeta)\,\underline{\xi},\\
e_3(\underline{\kappa})+\frac{1}{2}\underline{\kappa}^2+2\underline{\omega}\,\underline{\kappa} &= 2\,{\not{d}}_1\underline{\xi}+2(\eta-3\zeta)\underline{\xi}-\frac{1}{2}\underline{\vartheta}^2,\\
e_3\zeta+\frac{1}{2}\underline{\kappa}(\zeta+\eta)-2\underline{\omega}(\zeta-\eta) &= \underline{\beta}+2\,{\not{d}}_1^\star\,\underline{\omega}+\frac{1}{2}\kappa\,\underline{\xi}-\frac{1}{2}\underline{\vartheta}(\zeta+\eta)+\frac{1}{2}\vartheta\,\underline{\xi},\\
e_3\vartheta+\frac{1}{2}\underline{\kappa}\,\vartheta-2\underline{\omega}\vartheta &= -2\,{\not{d}}_2^\star\,\eta-\frac{1}{2}\kappa\,\underline{\vartheta}+2\eta^2,\\
e_3(\kappa)+\frac{1}{2}\underline{\kappa}\,\kappa-2\underline{\omega}\kappa &= 2\,{\not{d}}_1\eta+2\rho-\frac{1}{2}\underline{\vartheta}\,\vartheta+2\eta^2,
\end{aligned}$$

and

$$
\begin{aligned}
\not{d}_2\underline{\vartheta} &= -2\underline{\beta} - \not{d}_1^\star\,\underline{\kappa} - \zeta\underline{\kappa} + \underline{\vartheta}\,\zeta,\\
\not{d}_2\vartheta &= -2\beta - \not{d}_1^\star\,\kappa + \zeta\kappa - \vartheta\,\zeta,\\
K &= -\rho - \frac{1}{4}\kappa\,\underline{\kappa} + \frac{1}{4}\vartheta\,\underline{\vartheta}.
\end{aligned}
$$

2. *The null Bianchi identities are given in this case by*

$$
\begin{aligned}
e_3\alpha + \frac{1}{2}\underline{\kappa}\alpha &= -\not{d}_2^\star\,\beta + 4\underline{\omega}\alpha - \frac{3}{2}\vartheta\rho + (\zeta + 4\eta)\beta,\\
e_4\beta + 2\kappa\beta &= \not{d}_2\alpha + \zeta\alpha,\\
e_3\beta + \underline{\kappa}\beta &= -\not{d}_1^\star\rho + 2\underline{\omega}\beta + 3\eta\rho - \vartheta\underline{\beta} + \xi\alpha,\\
e_4\rho + \frac{3}{2}\kappa\rho &= \not{d}_1\beta - \frac{1}{2}\underline{\vartheta}\,\alpha - \zeta\beta,
\end{aligned}
$$

$$
\begin{aligned}
e_3\rho + \frac{3}{2}\underline{\kappa}\rho &= \not{d}_1\underline{\beta} - \frac{1}{2}\vartheta\,\underline{\alpha} - \zeta\,\underline{\beta} + 2(\eta\,\underline{\beta} + \underline{\xi}\,\beta),\\
e_4\underline{\beta} + \kappa\underline{\beta} &= -\not{d}_1^\star\rho - 3\zeta\rho - \underline{\vartheta}\beta,\\
e_3\underline{\beta} + 2\underline{\kappa}\,\underline{\beta} &= \not{d}_2\underline{\alpha} - 2\underline{\omega}\,\underline{\beta} + (-2\zeta + \eta)\underline{\alpha} + 3\underline{\xi}\rho,\\
e_4\underline{\alpha} + \frac{1}{2}\kappa\,\underline{\alpha} &= -\not{d}_2^\star\,\underline{\beta} - \frac{3}{2}\underline{\vartheta}\rho - 5\zeta\underline{\beta}.
\end{aligned}
$$

3. *The mass aspect function* $\mu = -\not{d}_1\zeta - \rho + \frac{1}{4}\vartheta\underline{\vartheta}$, *defined in* (2.2.10), *verifies the transport equation*

$$
\begin{aligned}
e_4(\mu) + \frac{3}{2}\kappa\mu &= Err[e_4\mu],\\
Err[e_4\mu] : &= \frac{1}{2}\kappa\zeta^2 + e_\theta(\kappa)\zeta + \not{d}_1(\vartheta\zeta) - \frac{1}{8}\underline{\kappa}\vartheta^2.
\end{aligned}
$$

Proof. Concerning the null structure equations we only need to derive the equation for $\eta - \zeta$. According to Proposition 2.56 we have

$$
e_3(\xi) - e_4(\eta) = \beta + 4\underline{\omega}\xi + \frac{1}{2}\kappa(\eta - \underline{\eta}) + \frac{1}{2}\vartheta(\eta - \underline{\eta})
$$

which becomes

$$
e_4\eta = -\beta - \frac{1}{2}\kappa(\eta - \underline{\eta}) - \frac{1}{2}\vartheta(\eta - \underline{\eta})
$$

and

$$
-e_4\zeta + \frac{1}{2}\kappa(-\zeta + \underline{\eta}) + 2\omega(\zeta + \underline{\eta}) = \beta + 2\not{d}_1^\star\omega + 2\underline{\omega}\xi + \frac{1}{2}\underline{\kappa}\,\xi - \frac{1}{2}\vartheta(-\zeta + \underline{\eta}) - \frac{1}{2}\underline{\vartheta}\,\xi
$$

which becomes

$$
e_4\zeta = -\kappa\zeta - \beta - \vartheta\zeta.
$$

Hence,

$$
\begin{aligned}
e_4(\zeta-\eta) &= -\kappa\zeta - \vartheta\zeta + \frac{1}{2}\kappa(\eta-\underline{\eta}) + \frac{1}{2}\vartheta(\eta-\underline{\eta})\\
&= \kappa\left(-\zeta+\frac{1}{2}(\eta-\underline{\eta})\right) + \vartheta\left(-\zeta+\frac{1}{2}(\eta-\underline{\eta})\right).
\end{aligned}
$$

Since $\zeta = -\underline{\eta}$ we deduce $-\zeta + \frac{1}{2}(\eta - \underline{\eta}) = \frac{1}{2}(-\zeta + \eta)$ and thus,

$$
e_4(\zeta-\eta) = -\kappa(\zeta-\eta) - \vartheta(\zeta-\eta)
$$

as desired.

The Bianchi equations follow immediately from the general equations derived in the previous section. It only remains to check the equation verified by the mass aspect function μ. We have

$$
\begin{aligned}
e_4(\mu) &= -[e_4, \not{d}_1]\zeta - \not{d}_1 e_4(\zeta) - e_4(\rho) + \frac{1}{4}e_4(\vartheta\underline{\vartheta})\\
&= \frac{1}{2}\kappa\not{d}_1\zeta - \frac{1}{2}\vartheta\not{d}_2^\star\zeta + e_4(\Phi)\zeta^2 - \beta\zeta - \not{d}_1(-\kappa\zeta - \beta - \vartheta\zeta)\\
&\quad -\left(-\frac{3}{2}\kappa\rho + \not{d}_1\beta - \frac{1}{2}\underline{\vartheta}\alpha - \zeta\beta\right)\\
&\quad +\frac{1}{4}\vartheta\left(-\frac{1}{2}\kappa\underline{\vartheta} + 2\not{d}_2^\star\zeta - \frac{1}{2}\underline{\kappa}\vartheta + 2\zeta^2\right) + \frac{1}{4}\underline{\vartheta}(-\kappa\vartheta - 2\alpha)\\
&= \frac{3}{2}\kappa\left(\not{d}_1\zeta + \rho - \frac{1}{4}\vartheta\underline{\vartheta}\right) - \frac{1}{2}\vartheta\not{d}_2^\star\zeta + \frac{1}{2}(\kappa-\vartheta)\zeta^2 + e_\theta(\kappa)\zeta + \not{d}_1(\vartheta\zeta)\\
&\quad +\frac{1}{4}\vartheta\left(2\not{d}_2^\star\zeta - \frac{1}{2}\underline{\kappa}\vartheta + 2\zeta^2\right)
\end{aligned}
$$

and hence

$$
e_4(\mu) + \frac{3}{2}\kappa\mu = \frac{1}{2}\kappa\zeta^2 + e_\theta(\kappa)\zeta + \not{d}_1(\vartheta\zeta) - \frac{1}{8}\underline{\kappa}\vartheta^2
$$

as desired. This concludes the proof of the proposition. □

2.2.4.3 Transport equations for S-averages

Proposition 2.64. *For any scalar function f, we have*

$$
\begin{aligned}
e_4\left(\int_S f\right) &= \int_S (e_4(f) + \kappa f),\\
e_3\left(\int_S f\right) &= \int_S (e_3(f) + \underline{\kappa} f) + Err\left[e_3\left(\int_S f\right)\right],
\end{aligned}
\tag{2.2.20}
$$

where the error term is given by the formula

$$\begin{aligned} Err\left[e_3\left(\int_S f\right)\right]: &= -\varsigma^{-1}\check{\varsigma}\int_S (e_3(f)+\underline{\kappa}f)+\varsigma^{-1}\int_S \check{\varsigma}(e_3(f)+\underline{\kappa}f)\\ &+ (\check{\underline{\Omega}}+\varsigma^{-1}\overline{\underline{\Omega}}\check{\varsigma})\int_S (e_4 f+\kappa f)-\varsigma^{-1}\overline{\underline{\Omega}}\int_S \check{\varsigma}(e_4 f+\kappa f)\\ &- \varsigma^{-1}\int_S \check{\underline{\Omega}}\,\varsigma(e_4 f+\kappa f). \end{aligned}$$

In particular, we have

$$e_4(r)=\frac{r}{2}\overline{\kappa}, \qquad e_3(r)=\frac{r}{2}\left(\overline{\underline{\kappa}}+\underline{A}\right) \tag{2.2.21}$$

where

$$\underline{A}: \;=\; -\varsigma^{-1}\overline{\underline{\kappa}}\check{\varsigma}+\overline{\kappa}\left(\check{\underline{\Omega}}+\varsigma^{-1}\overline{\underline{\Omega}}\check{\varsigma}\right)+\varsigma^{-1}\overline{\check{\varsigma}\underline{\check{\kappa}}}-\varsigma^{-1}\overline{\underline{\Omega}}\,\overline{\check{\varsigma}\check{\kappa}}-\varsigma^{-1}\overline{\check{\underline{\Omega}}\varsigma\kappa}. \tag{2.2.22}$$

Proof. See section A.1. □

Corollary 2.65. *For a reduced scalar f, we have*

$$e_4\left(\int_S fe^{\Phi}\right) = \int_S \left(e_4(f)+\left(\frac{3}{2}\kappa-\frac{1}{2}\vartheta\right)f\right)e^{\Phi}$$

and

$$e_3\left(\int_S fe^{\Phi}\right) = \int_S \left(e_3(f)+\left(\frac{3}{2}\underline{\kappa}-\frac{1}{2}\underline{\vartheta}\right)f\right)e^{\Phi}+Err\left[e_3\left(\int_S fe^{\Phi}\right)\right].$$

Proof. In view of Proposition 2.64, we have

$$\begin{aligned} e_4\left(\int_S fe^{\Phi}\right) &= \int_S \left(e_4(fe^{\Phi})+\kappa fe^{\Phi}\right)\\ &= \int_S \left(e_4(f)+(\kappa+e_4\Phi)f\right)e^{\Phi}\\ &= \int_S \left(e_4(f)+\left(\frac{3}{2}\kappa-\frac{1}{2}\vartheta\right)f\right)e^{\Phi} \end{aligned}$$

as desired.

Also, using again Proposition 2.64, we have

$$\begin{aligned} e_3\left(\int_S fe^{\Phi}\right) &= \int_S \left(e_3(fe^{\Phi})+\underline{\kappa}fe^{\Phi}\right)+\mathrm{Err}\left[e_3\left(\int_S fe^{\Phi}\right)\right]\\ &= \int_S \left(e_3(f)+(\underline{\kappa}+e_3\Phi)f\right)e^{\Phi}+\mathrm{Err}\left[e_3\left(\int_S fe^{\Phi}\right)\right]\\ &= \int_S \left(e_3(f)+\left(\frac{3}{2}\underline{\kappa}-\frac{1}{2}\underline{\vartheta}\right)f\right)e^{\Phi}+\mathrm{Err}\left[e_3\left(\int_S fe^{\Phi}\right)\right] \end{aligned}$$

as desired. □

Corollary 2.66. *Given a scalar function f, we have*

$$
\begin{aligned}
e_4(\overline{f}) &= \overline{e_4(f)} + \overline{\check{\kappa}\,\check{f}},\\
e_4(\check{f}) &= e_4(f) - \overline{e_4(f)} - \overline{\check{\kappa}\,\check{f}},
\end{aligned}
\tag{2.2.23}
$$

and

$$
\begin{aligned}
e_3\left(\overline{f}\right) &= \overline{e_3(f)} + Err[e_3\overline{f}],\\
e_3(\check{f}) &= e_3(f) - \overline{e_3(f)} - Err[e_3(\overline{f})],
\end{aligned}
\tag{2.2.24}
$$

where

$$
\begin{aligned}
Err[e_3(\overline{f})] = {} & -\varsigma^{-1}\check{\varsigma}\left(\overline{e_3 f + \underline{\kappa} f} - \overline{\underline{\kappa}}\overline{f}\right) + \varsigma^{-1}\left(\overline{\check{\varsigma}(e_3 f + \underline{\kappa} f)} - \overline{\check{\varsigma}\check{\underline{\kappa}}}\,\overline{f}\right)\\
& + \left(\check{\underline{\Omega}} + \varsigma^{-1}\overline{\underline{\Omega}}\check{\varsigma}\right)\left(\overline{e_4 f + \kappa f} - \overline{\kappa}\,\overline{f}\right)\\
& - \varsigma^{-1}\overline{\underline{\Omega}}\left(\overline{\check{\varsigma}(e_4 f + \kappa f)} - \overline{\check{\varsigma}\check{\kappa}}\,\overline{f}\right) - \varsigma^{-1}\left(\overline{\check{\underline{\Omega}}\varsigma(e_4 f + \kappa f)} - \overline{\check{\underline{\Omega}}\varsigma\,\kappa}\overline{f}\right) + \overline{\check{\underline{\kappa}}\check{f}}.
\end{aligned}
\tag{2.2.25}
$$

Proof. We have, recalling Lemma 2.45 and $|S| = 4\pi r^2$,

$$
\begin{aligned}
e_4(\overline{f}) &= e_4\left(\frac{\int_S f}{|S|}\right) = \frac{1}{|S|}\int_S (e_4(f) + \kappa f) - \frac{e_4(|S|)}{|S|}\overline{f} = \overline{e_4(f) + \kappa f} - 2\frac{e_4 r}{r}\overline{f}\\
&= \overline{e_4(f)} + \overline{\kappa\, f} - \overline{\kappa}\overline{f} = \overline{e_4(f)} + \overline{\check{\kappa}\,\check{f}}.
\end{aligned}
$$

This also yields

$$
e_4(\check{f}) = e_4(f) - e_4(\overline{f}) = e_4(f) - \overline{e_4(f)} - \overline{\check{\kappa}\,\check{f}}
$$

as desired.

Similarly,

$$
\begin{aligned}
e_3(\overline{f}) &= e_3\left(\frac{\int_S f}{|S|}\right) = \frac{1}{|S|}e_3\left(\int_{\mathbf{S}} f\right) - \frac{2e_3(r)}{r}\overline{f}\\
&= \frac{1}{|S|}\int_S (e_3 f + \underline{\kappa} f) + \frac{1}{|S|}\mathrm{Err}\left[e_3\left(\int_S f\right)\right] - (\overline{\underline{\kappa}} + \underline{A})\overline{f}\\
&= \overline{e_3(f)} + \overline{\underline{\kappa} f} - \overline{\underline{\kappa}}\overline{f} + \frac{1}{|S|}\mathrm{Err}\left[e_3\left(\int_S f\right)\right] - \underline{A}\overline{f}\\
&= \overline{e_3(f)} + \overline{\check{\underline{\kappa}}\check{f}} + \frac{1}{|S|}\mathrm{Err}\left[e_3\left(\int_S f\right)\right] - \underline{A}\overline{f}.
\end{aligned}
$$

We deduce

$$
e_3(\overline{f}) = \overline{e_3(f)} + \mathrm{Err}[e_3(\overline{f})]
$$

where, recalling the definitions of the error terms $\mathrm{Err}\left[e_3\left(\int_S f\right)\right]$ and $\underline{A}$,

$$\begin{aligned}
\mathrm{Err}[e_3(\overline{f})] &= \overline{\check{\underline{\kappa}}\check{f}} + \frac{1}{|S|}\mathrm{Err}\left[e_3\left(\int_S f\right)\right] - \underline{A}\overline{f}\\
&= \overline{\check{\underline{\kappa}}\check{f}} - \varsigma^{-1}\check{\varsigma}\,\overline{e_3 f + \underline{\kappa} f} + \varsigma^{-1}\overline{\check{\varsigma}(e_3 f + \underline{\kappa} f)} + (\check{\underline{\Omega}} + \varsigma^{-1}\overline{\underline{\Omega}}\check{\varsigma})\,\overline{e_4 f + \kappa f}\\
&- \varsigma^{-1}\overline{\underline{\Omega}}\,\overline{\check{\varsigma}(e_4 f + \kappa f)} - \varsigma^{-1}\,\overline{\check{\underline{\Omega}}\varsigma(e_4 f + \kappa f)}\\
&- \overline{f}\left(-\varsigma^{-1}\overline{\underline{\kappa}}\check{\varsigma} + \varsigma^{-1}\overline{\check{\varsigma}\check{\underline{\kappa}}} + \overline{\kappa}\left(\check{\underline{\Omega}} + \varsigma^{-1}\overline{\underline{\Omega}}\check{\varsigma}\right) - \varsigma^{-1}\overline{\underline{\Omega}}\,\overline{\check{\varsigma}\check{\kappa}} - \varsigma^{-1}\overline{\check{\underline{\Omega}}\varsigma\kappa}\right),
\end{aligned}$$

i.e.,

$$\begin{aligned}
\mathrm{Err}[e_3(\overline{f})] &= \overline{\check{\underline{\kappa}}\check{f}} - \varsigma^{-1}\check{\varsigma}\left(\overline{e_3 f + \underline{\kappa} f} - \overline{\underline{\kappa}}\overline{f}\right) + \varsigma^{-1}\left(\overline{\check{\varsigma}(e_3 f + \underline{\kappa} f)} - \overline{\check{\varsigma}\check{\underline{\kappa}}}\,\overline{f}\right)\\
&+ \left(\check{\underline{\Omega}} + \varsigma^{-1}\overline{\underline{\Omega}}\check{\varsigma}\right)\left(\overline{e_4 f + \kappa f} - \overline{\kappa}\,\overline{f}\right) - \varsigma^{-1}\overline{\underline{\Omega}}\left(\overline{\check{\varsigma}(e_4 f + \kappa f)} - \overline{\check{\varsigma}\check{\kappa}}\,\overline{f}\right)\\
&- \varsigma^{-1}\left(\overline{\check{\underline{\Omega}}\varsigma(e_4 f + \kappa f)} - \overline{\check{\underline{\Omega}}\varsigma\kappa}\,\overline{f}\right)
\end{aligned}$$

as stated. Finally

$$e_3(\check{f}) = e_3 f - e_3(\overline{f}) = e_3 f - \overline{e_3(f)} - \mathrm{Err}[e_3\overline{f}]$$

which ends the proof of the corollary. □

The following is also an immediate application of Proposition 2.64.

Corollary 2.67. *If f verifies the scalar equation*

$$e_4(f) + \frac{p}{2}\overline{\kappa} f = F,$$

then

$$e_4(r^p f) = r^p F.$$

2.2.4.4 Commutation identities revisited

We revisit the general commutation identities of Lemma 2.54 in an outgoing geodesic foliation.

Lemma 2.68. *The following commutation formulae holds true:*

1. *If $f \in \mathfrak{s}_k$,*

$$\begin{aligned}
[r\,\not{d}_k, e_4]f &= r\left[Com_k(f) + \frac{1}{2}\check{\kappa}\,\not{d}_k f\right],\\
[r\,\not{d}_k, e_3]f &= r\left[\underline{Com}_k(f) + \frac{1}{2}(-\underline{A} + \check{\underline{\kappa}})\,\not{d}_k f\right].
\end{aligned} \tag{2.2.26}$$

2. *If $f \in \mathfrak{s}_{k-1}$,*

$$
\begin{aligned}
[r \,{\mathscr{d}}_k^\star, e_4] f &= r\left[Com_k^*(f) + \frac{1}{2}\check{\kappa}\,{\mathscr{d}}_k^\star f\right],\\
[r \,{\mathscr{d}}_k^\star, e_3] f &= r\left[\underline{Com}_k^*(f) + \frac{1}{2}(-\underline{A} + \check{\underline{\kappa}})\,{\mathscr{d}}_k^\star f\right].
\end{aligned}
\tag{2.2.27}
$$

Also, we have

$$
\begin{aligned}
\underline{Com}_k(f) &= -\frac{1}{2}\underline{\vartheta}\,{\mathscr{d}}_{k+1}^\star f + (\zeta - \eta) e_3 f - k\eta e_3 \Phi f - \underline{\xi}(e_4 f + k e_4(\Phi) f) - k\underline{\beta} f,\\
Com_k(f) &= -\frac{1}{2}\vartheta\,{\mathscr{d}}_{k+1}^\star f + k\zeta e_4 \Phi f - k\beta f,\\
\underline{Com}_k^*(f) &= -\frac{1}{2}\underline{\vartheta}\,{\mathscr{d}}_{k-1} f - (\zeta - \eta) e_3 f - (k-1)\eta e_3 \Phi f + \underline{\xi}(e_4 f - (k-1) e_4(\Phi) f)\\
&- (k-1)\underline{\beta} f,\\
Com_k^*(f) &= -\frac{1}{2}\vartheta\,{\mathscr{d}}_{k-1} f + (k-1)\zeta e_4 \Phi f - (k-1)\beta f.
\end{aligned}
$$

Proof. We make use of the commutation Lemma 2.54 and the definition of $\underline{A}$, see Proposition 2.64, to write, for $f \in \mathfrak{s}_k$,

$$
\begin{aligned}
[r\,{\mathscr{d}}_k, e_4] f &= r[{\mathscr{d}}_k, e_4] f - e_4(r)\,{\mathscr{d}}_k f\\
&= \frac{1}{2} r\kappa\,{\mathscr{d}}_k f + r\mathrm{Com}_k(f) - \frac{r}{2}\overline{\kappa}\,{\mathscr{d}}_k f\\
&= r\left[\mathrm{Com}_k(f) + \frac{1}{2}\check{\kappa}\,{\mathscr{d}}_k f\right]\\
[r\,{\mathscr{d}}_k, e_3] f &= r[{\mathscr{d}}_k, e_4] f - e_3(r)\,{\mathscr{d}}_k f\\
&= \frac{1}{2} r\underline{\kappa}\,{\mathscr{d}}_k f + r\underline{Com}_k(f) - \frac{r}{2}(\underline{A} + \overline{\underline{\kappa}})\,{\mathscr{d}}_k f\\
&= r\left[\underline{Com}_k(f) + \frac{1}{2}(-\underline{A} + \check{\underline{\kappa}})\,{\mathscr{d}}_k f\right].
\end{aligned}
$$

The remaining formulae are proved in the same manner. Also, the form of $\underline{Com}_k(f)$, $\mathrm{Com}_k(f)$, $\underline{Com}_k^*(f)$ and $\mathrm{Com}_k^*(f)$ follows from Lemma 2.54 and the fact that we have $\xi = \underline{\eta} + \zeta = 0$ in an outgoing geodesic foliation. □

We also record here for future use the following lemma.

Lemma 2.69. *Let* $\mathbf{T} = \frac{1}{2}(e_3 + \Upsilon e_4)$, *with* $\Upsilon = 1 - \frac{2m}{r}$. *We have,*

$$
\begin{aligned}
[\mathbf{T}, e_4] &= \left(\left(\underline{\omega} - \frac{m}{r^2}\right) - \frac{m}{2r}\left(\overline{\kappa} - \frac{2}{r}\right) + \frac{e_4(m)}{r}\right) e_4 + (\eta + \zeta) e_\theta,\\
[\mathbf{T}, e_3] &= \left(-\Upsilon\left(\underline{\omega} - \frac{m}{r^2}\right) - \frac{m}{2r}\left(\overline{\underline{\kappa}} + \frac{2\Upsilon}{r}\right) - \frac{m}{2r}\underline{A} + \frac{e_3(m)}{r}\right) e_4 - (\eta + \zeta)\Upsilon e_\theta.
\end{aligned}
\tag{2.2.28}
$$

Proof. Recall that $[e_3, e_4] = 2\underline{\omega} e_4 + 2(\eta+\zeta)e_\theta$. Thus,

$$\begin{aligned}
[\mathbf{T}, e_4] &= \frac{1}{2}[e_3 + \Upsilon e_4, e_4] = \frac{1}{2}\left(2\underline{\omega} e_4 + 2(\eta+\zeta)e_\theta - e_4\left(1-\frac{2m}{r}\right)e_4\right)\\
&= \left(\underline{\omega} - \frac{m}{r^2}e_4(r) + \frac{e_4(m)}{r}\right)e_4 + (\eta+\zeta)e_\theta\\
&= \left(\left(\underline{\omega} - \frac{m}{r^2}\right) - \frac{m}{r^2}(e_4(r)-1) + \frac{e_4(m)}{r}\right)e_4 + (\eta+\zeta)e_\theta\\
&= \left(\left(\underline{\omega} - \frac{m}{r^2}\right) - \frac{m}{r^2}\left(\frac{r}{2}\overline{\kappa}-1\right) + \frac{e_4(m)}{r}\right)e_4 + (\eta+\zeta)e_\theta\\
&= \left(\left(\underline{\omega} - \frac{m}{r^2}\right) - \frac{m}{2r}\left(\overline{\kappa}-\frac{2}{r}\right) + \frac{e_4(m)}{r}\right)e_4 + (\eta+\zeta)e_\theta
\end{aligned}$$

and,

$$\begin{aligned}
[\mathbf{T}, e_3] &= \frac{1}{2}[e_3 + \Upsilon e_4, e_3] = \frac{1}{2}\left(\Upsilon\left(-2\underline{\omega} e_4 - 2(\eta+\zeta)e_\theta\right) - e_3\left(1-\frac{2m}{r}\right)e_4\right)\\
&= \left(-\Upsilon\underline{\omega} - \frac{m}{r^2}e_3(r) + \frac{e_3(m)}{r}\right)e_4 - \Upsilon(\eta+\zeta)e_\theta\\
&= \left(-\Upsilon\left(\underline{\omega} - \frac{m}{r^2}\right) - \Upsilon\frac{m}{r^2} - \frac{m}{r^2}\frac{r}{2}\left(\overline{\underline{\kappa}} + \underline{A}\right) + \frac{e_3(m)}{r}\right)e_4 - \Upsilon(\eta+\zeta)e_\theta\\
&= \left(-\Upsilon\left(\underline{\omega} - \frac{m}{r^2}\right) - \frac{m}{2r}\left(\overline{\underline{\kappa}} + \frac{2\Upsilon}{r}\right) - \frac{m}{2r}\underline{A} + \frac{e_3(m)}{r}\right)e_4 - \Upsilon(\eta+\zeta)e_\theta
\end{aligned}$$

which concludes the proof of the lemma. □

Remark 2.70. *When applying the formulas of Lemma 2.69 to a k-reduced scalar $f \in \mathfrak{s}_k$, the term $(\eta+\zeta)e_\theta(f)$ should correspond to a reduced scalar. In fact, recalling Remark 2.26, we can write*

$$\zeta e_\theta(f) = \frac{1}{2}\zeta\left(\not{d}_k f - \not{d}^\star_{k+1} f\right)$$

which can indeed be shown to be a k-reduced scalar in $\mathfrak{s}_k$.

2.2.4.5 Derivatives of the Hawking mass

Proposition 2.71 (Derivatives of the Hawking mass)**.** *We have the following identities for the Hawking mass,*

$$e_4(m) = \frac{r}{32\pi}\int_S Err_1, \qquad (2.2.29)$$

and

$$\begin{aligned}
e_3(m) &= \left(1-\varsigma^{-1}\check{\varsigma}\right)\frac{r}{32\pi}\int_S \underline{Err}_1 + \left(\check{\underline{\Omega}}+\varsigma^{-1}\overline{\underline{\Omega}}\check{\varsigma}\right)\frac{r}{32\pi}\int_S Err_1\\
&+\varsigma^{-1}\frac{r}{32\pi}\int_S \check{\varsigma}\left(2\overline{\rho}\check{\underline{\kappa}}+2\check{\rho}\overline{\underline{\kappa}}+2\underline{\kappa}\,{\not d}_1\eta+2\kappa\,{\not d}_1\underline{\xi}+\underline{Err}_2\right)\\
&-\varsigma^{-1}\frac{r}{32\pi}\int_S (\overline{\underline{\Omega}}\check{\varsigma}+\check{\underline{\Omega}}\varsigma)\left(2\overline{\rho}\check{\kappa}+2\check{\rho}\overline{\kappa}-2\kappa\,{\not d}_1\zeta+Err_2\right)\\
&-\frac{m}{r}\varsigma^{-1}\left[-\overline{\check{\varsigma}\check{\underline{\kappa}}}+\overline{\underline{\Omega}}\,\overline{\check{\varsigma}\check{\kappa}}+\overline{\check{\underline{\Omega}}\varsigma\kappa}\right], \qquad (2.2.30)
\end{aligned}$$

where

$$\begin{aligned}
Err_1 &:= 2\check{\kappa}\check{\rho}+2e_\theta(\kappa)\zeta-\frac{1}{2}\underline{\kappa}\vartheta^2-\frac{1}{2}\check{\kappa}\vartheta\underline{\vartheta}+2\kappa\zeta^2,\\
\underline{Err}_1 &:= 2\check{\rho}\check{\underline{\kappa}}-2e_\theta(\underline{\kappa})\eta-2e_\theta(\kappa)\underline{\xi}-\frac{1}{2}\check{\underline{\kappa}}\vartheta\underline{\vartheta}+2\underline{\kappa}\eta^2+2\kappa(\eta-3\zeta)\underline{\xi}-\frac{1}{2}\kappa\underline{\vartheta}^2,\\
Err_2 &:= 2\check{\rho}\check{\kappa}-\frac{1}{2}\underline{\kappa}\vartheta^2-\frac{1}{2}\kappa\vartheta\underline{\vartheta}+2\kappa\zeta^2,\\
\underline{Err}_2 &:= 2\check{\rho}\check{\underline{\kappa}}+\underline{\kappa}\left(2\eta^2-\frac{1}{2}\vartheta\underline{\vartheta}\right)+2\kappa(\eta-3\zeta)\underline{\xi}-\frac{1}{2}\kappa\underline{\vartheta}^2.
\end{aligned}$$

Proof. The proof relies on the definition of the Hawking mass m given by the formula $\frac{2m}{r}=1+\frac{1}{16\pi}\int_S\kappa\underline{\kappa}$, Proposition 2.64, and the null structure equations for $e_4(\kappa)$, $e_4(\underline{\kappa})$, $e_3(\kappa)$ and $e_3(\underline{\kappa})$ provided by Proposition 2.63. We refer to section A.2 for the details. □

2.2.4.6 *Transport equations for main averaged quantities*

Lemma 2.72. *The following equations hold true:*

$$\begin{aligned}
e_4\left(\overline{\kappa}-\frac{2}{r}\right)+\frac{1}{2}\overline{\kappa}\left(\overline{\kappa}-\frac{2}{r}\right) &= -\frac{1}{4}\overline{\vartheta^2}+\frac{1}{2}\overline{\check{\kappa}^2},\\
e_4\left(\overline{\underline{\omega}}-\frac{m}{r^2}\right) &= \overline{\rho}+\frac{2m}{r^3}+\frac{m}{r^2}\left(\overline{\kappa}-\frac{2}{r}\right)-\frac{e_4(m)}{r^2}+3\overline{\zeta(2\eta+\zeta)}+\overline{\check{\kappa}\check{\underline{\omega}}}, \qquad (2.2.31)
\end{aligned}$$

and

$$\begin{aligned}
&e_3\left(\overline{\kappa}-\frac{2}{r}\right)+\frac{1}{2}\overline{\underline{\kappa}}\left(\overline{\kappa}-\frac{2}{r}\right) \qquad (2.2.32)\\
&= 2\overline{\underline{\omega}}\left(\overline{\kappa}-\frac{2}{r}\right)+\frac{4}{r}\left(\overline{\underline{\omega}}-\frac{m}{r^2}\right)+2\left(\overline{\rho}+\frac{2m}{r^3}\right)-\varsigma^{-1}\left(-\frac{1}{2}\overline{\underline{\kappa}}\,\overline{\kappa}+2\overline{\underline{\omega}}\,\overline{\kappa}+2\overline{\rho}\right)\check{\varsigma}\\
&-\frac{1}{2}\overline{\kappa}^2\left(\check{\underline{\Omega}}+\varsigma^{-1}\overline{\underline{\Omega}}\check{\varsigma}\right)-\frac{1}{r}\varsigma^{-1}\overline{\underline{\kappa}}\check{\varsigma}+\frac{1}{r}\overline{\kappa}\left(\check{\underline{\Omega}}+\varsigma^{-1}\overline{\underline{\Omega}}\check{\varsigma}\right)+Err\left[e_3\left(\overline{\kappa}-\frac{2}{r}\right)\right],
\end{aligned}$$

where

$$
\begin{aligned}
Err\left[e_3\left(\overline{\kappa}-\frac{2}{r}\right)\right] \quad :=\quad & 2\overline{\eta^2}+2\overline{\check{\underline{\omega}}\,\check{\kappa}}-\frac{1}{2}\overline{\check{\kappa}\,\check{\underline{\kappa}}}-\frac{1}{2}\overline{\vartheta\underline{\vartheta}}+\frac{1}{r}\varsigma^{-1}\overline{\check{\varsigma}\check{\underline{\kappa}}}-\frac{1}{r}\varsigma^{-1}\overline{\underline{\Omega}}\,\overline{\check{\varsigma}\check{\kappa}}\\
&-\frac{1}{r}\varsigma^{-1}\overline{\check{\underline{\Omega}}\varsigma\kappa}-\varsigma^{-1}\check{\varsigma}\,\overline{\left(\frac{1}{2}\check{\kappa}\check{\underline{\kappa}}+2\check{\underline{\omega}}\check{\kappa}-\frac{1}{2}\vartheta\underline{\vartheta}+2\eta^2\right)}\\
&+\varsigma^{-1}\left[\overline{\check{\varsigma}\left(\frac{1}{2}\kappa\underline{\kappa}+2\underline{\omega}\kappa+2\check{\rho}+2\,\not{d}_1\eta-\frac{1}{2}\vartheta\underline{\vartheta}+2\eta^2\right)}\right.\\
&\left.-\overline{\check{\varsigma}\check{\underline{\kappa}}}\,\overline{\kappa}\right]+\left(\check{\underline{\Omega}}+\varsigma^{-1}\overline{\underline{\Omega}}\check{\varsigma}\right)\overline{\left(\frac{1}{2}\check{\kappa}^2-\frac{1}{4}\vartheta^2\right)}\\
&-\varsigma^{-1}\overline{\underline{\Omega}}\left(\overline{\check{\varsigma}\left(\frac{1}{2}\kappa^2-\frac{1}{4}\vartheta^2\right)}-\overline{\check{\varsigma}\check{\kappa}}\,\overline{\kappa}\right)\\
&-\varsigma^{-1}\left(\overline{\check{\underline{\Omega}}\varsigma\left(\frac{1}{2}\kappa^2-\frac{1}{4}\vartheta^2\right)}-\overline{\check{\underline{\Omega}}\varsigma\,\kappa}\,\overline{\kappa}\right)+\overline{\check{\underline{\kappa}}\check{\kappa}}.
\end{aligned}
\tag{2.2.33}
$$

Proof. The proof relies on Corollary 2.66 and the null structure equations for $e_4(\kappa)$ and $e_3(\kappa)$ provided by Proposition 2.63. We refer to section A.3 for the details. □

2.2.4.7 *Transport equations for main checked quantities*

Proposition 2.73 (Transport equations for checked quantities). *We have the following transport equations in the e_4 direction:*

$$
\begin{aligned}
e_4\check{\kappa}+\overline{\kappa}\,\check{\kappa} &= Err[e_4\check{\kappa}],\\
Err[e_4\check{\kappa}] &:= -\frac{1}{2}\check{\kappa}^2-\frac{1}{2}\overline{\check{\kappa}^2}-\frac{1}{2}\left(\vartheta^2-\overline{\vartheta^2}\right),\\
e_4\check{\underline{\kappa}}+\frac{1}{2}\overline{\kappa}\check{\underline{\kappa}}+\frac{1}{2}\check{\kappa}\overline{\underline{\kappa}} &= -2\,\not{d}_1\zeta+2\check{\rho}+Err[e_4\check{\underline{\kappa}}],\\
Err[e_4\check{\underline{\kappa}}] &:= -\frac{1}{2}\check{\kappa}\check{\underline{\kappa}}-\frac{1}{2}\overline{\check{\kappa}\check{\underline{\kappa}}}+\left(-\frac{1}{2}\vartheta\underline{\vartheta}+2\zeta^2\right)-\overline{\left(-\frac{1}{2}\vartheta\underline{\vartheta}+2\zeta^2\right)},\\
e_4\check{\underline{\omega}} &= \check{\rho}+Err[e_4\check{\underline{\omega}}],\\
Err[e_4\check{\underline{\omega}}] &:= -\overline{\check{\kappa}\check{\underline{\omega}}}+\left(\zeta(2\eta+\zeta)-\overline{\zeta(2\eta+\zeta)}\right),
\end{aligned}
\tag{2.2.34}
$$

$$
\begin{aligned}
e_4\check{\rho}+\frac{3}{2}\overline{\kappa}\check{\rho}+\frac{3}{2}\overline{\rho}\check{\kappa} &= \not{d}_1\beta+Err[e_4\check{\rho}],\\
Err[e_4\check{\rho}] &:= -\frac{3}{2}\check{\kappa}\check{\rho}+\frac{1}{2}\overline{\check{\kappa}\check{\rho}}-\left(\frac{1}{2}\underline{\vartheta}\alpha+\zeta\beta\right)+\overline{\left(\frac{1}{2}\underline{\vartheta}\alpha+\zeta\beta\right)},\\
e_4\check{\mu}+\frac{3}{2}\overline{\kappa}\check{\mu}+\frac{3}{2}\overline{\mu}\check{\kappa} &= Err[e_4\check{\mu}],\\
Err[e_4\check{\mu}] &:= -\frac{3}{2}\check{\kappa}\check{\mu}+\frac{1}{2}\overline{\check{\kappa}\check{\mu}}+Err[e_4\mu]-\overline{Err[e_4\mu]},\\
e_4(\check{\underline{\Omega}}) &= -2\check{\underline{\omega}}+\overline{\check{\kappa}\check{\underline{\Omega}}}.
\end{aligned}
\tag{2.2.35}
$$

Also in the e_3 direction,

$$
\begin{aligned}
e_3(\check{\kappa}) &= 2\,\not{d}_1\eta + 2\check{\rho} - \frac{1}{2}\left(\underline{\kappa}\check{\kappa} + \kappa\check{\underline{\kappa}}\right) + 2\left(\underline{\omega}\check{\kappa} + \kappa\check{\underline{\omega}}\right)\\
&\quad + \varsigma^{-1}\left(-\frac{1}{2}\overline{\underline{\kappa}}\,\overline{\kappa} + 2\overline{\underline{\omega}}\,\overline{\kappa} + 2\overline{\rho}\right)\check{\varsigma} + \frac{1}{2}\overline{\kappa}^2\left(\check{\underline{\Omega}} + \varsigma^{-1}\overline{\underline{\Omega}}\check{\varsigma}\right) + Err[e_3\check{\kappa}],\\
e_3(\check{\underline{\kappa}}) + \underline{\kappa}\,\check{\underline{\kappa}} &= 2\,\not{d}_1\underline{\xi} - 2\left(\check{\underline{\omega}}\,\underline{\kappa} + \underline{\omega}\,\check{\underline{\kappa}}\right) + \varsigma^{-1}\check{\varsigma}\left(-\frac{1}{2}\overline{\underline{\kappa}}^2 - 2\overline{\underline{\omega}}\,\overline{\underline{\kappa}}\right)\\
&\quad - \left(\check{\underline{\Omega}} + \varsigma^{-1}\overline{\underline{\Omega}}\check{\varsigma}\right)\left(-\frac{1}{2}\overline{\kappa}\,\overline{\underline{\kappa}} + 2\overline{\rho}\right) + Err[e_3(\check{\underline{\kappa}})],\\
e_3\check{\rho} + \frac{3}{2}\overline{\underline{\kappa}}\check{\rho} &= -\frac{3}{2}\overline{\rho}\check{\underline{\kappa}} + \not{d}_1\underline{\beta} - \frac{3}{2}\overline{\underline{\kappa}}\,\overline{\rho}\varsigma^{-1}\check{\varsigma} + \frac{3}{2}\overline{\kappa}\,\overline{\rho}\left(\check{\underline{\Omega}} + \varsigma^{-1}\overline{\underline{\Omega}}\check{\varsigma}\right) + Err[e_3\check{\rho}],
\end{aligned}
\tag{2.2.36}
$$

with error terms given by

$$
\begin{aligned}
Err[e_3\check{\kappa}] &:= 2\left(\eta^2 - \overline{\eta^2}\right) - \frac{1}{2}\overline{\check{\kappa}\check{\underline{\kappa}}} + 2\overline{\check{\underline{\omega}}\check{\kappa}} - \frac{1}{2}\left(\vartheta\underline{\vartheta} - \overline{\vartheta\underline{\vartheta}}\right)\\
&\quad + \varsigma^{-1}\check{\varsigma}\left(\overline{\frac{1}{2}\check{\kappa}\check{\underline{\kappa}} + 2\check{\underline{\omega}}\check{\kappa} - \frac{1}{2}\vartheta\underline{\vartheta} + 2\eta^2}\right)\\
&\quad - \varsigma^{-1}\left(\overline{\check{\varsigma}\left(\frac{1}{2}\kappa\underline{\kappa} + 2\underline{\omega}\kappa + 2\check{\rho} + 2\,\not{d}_1\eta - \frac{1}{2}\vartheta\underline{\vartheta} + 2\eta^2\right)} - \overline{\check{\varsigma}}\,\overline{\check{\underline{\kappa}}}\,\overline{\kappa}\right)\\
&\quad - \left(\check{\underline{\Omega}} + \varsigma^{-1}\overline{\underline{\Omega}}\check{\varsigma}\right)\left(\overline{\frac{1}{2}\check{\kappa}^2 - \frac{1}{4}\vartheta^2}\right) + \varsigma^{-1}\overline{\underline{\Omega}}\left(\overline{\check{\varsigma}\left(\frac{1}{2}\kappa^2 - \frac{1}{4}\vartheta^2\right)} - \overline{\check{\varsigma}}\,\overline{\check{\kappa}}\,\overline{\kappa}\right)\\
&\quad + \varsigma^{-1}\left(\overline{\check{\underline{\Omega}}\varsigma\left(\frac{1}{2}\kappa^2 - \frac{1}{4}\vartheta^2\right)} - \overline{\check{\underline{\Omega}}\varsigma}\,\kappa\,\overline{\kappa}\right) - \overline{\check{\kappa}\check{\kappa}},
\end{aligned}
\tag{2.2.37}
$$

$$
\begin{aligned}
Err[e_3(\check{\underline{\kappa}})] &:= -\frac{1}{2}\overline{\check{\underline{\kappa}}^2} - 2\overline{\check{\underline{\omega}}\,\check{\underline{\kappa}}} + 2(\eta - 3\zeta)\underline{\xi} - \overline{2(\eta - 3\zeta)\underline{\xi}} - \frac{1}{2}\left(\underline{\vartheta}^2 - \overline{\underline{\vartheta}^2}\right)\\
&\quad - \varsigma^{-1}\left(\overline{\check{\varsigma}\left(\frac{1}{2}\underline{\kappa}^2 - 2\underline{\omega}\,\underline{\kappa} + 2\,\not{d}_1\underline{\xi} + 2(\eta - 3\zeta)\underline{\xi} - \frac{1}{2}\underline{\vartheta}^2\right)} - \overline{\check{\varsigma}}\,\overline{\check{\underline{\kappa}}}\,\overline{\underline{\kappa}}\right)\\
&\quad + \varsigma^{-1}\overline{\underline{\Omega}}\left(\overline{\check{\varsigma}\left(\frac{1}{2}\kappa\,\underline{\kappa} - 2\,\not{d}_1\zeta + 2\rho - \frac{1}{2}\vartheta\,\underline{\vartheta} + 2\zeta^2\right)} - \overline{\check{\varsigma}}\,\overline{\check{\kappa}}\,\overline{\underline{\kappa}}\right)\\
&\quad + \varsigma^{-1}\left(\overline{\check{\underline{\Omega}}\varsigma\left(\frac{1}{2}\kappa\,\underline{\kappa} - 2\,\not{d}_1\zeta + 2\rho - \frac{1}{2}\vartheta\,\underline{\vartheta} + 2\zeta^2\right)} - \overline{\check{\underline{\Omega}}\varsigma}\,\kappa\,\overline{\underline{\kappa}}\right) - \overline{\check{\underline{\kappa}}^2},
\end{aligned}
\tag{2.2.38}
$$

and

$$
\begin{aligned}
Err[e_3\check{\rho}] :=& -\left(\frac{1}{2}\vartheta\underline{\alpha}+\zeta\underline{\beta}-2\eta\underline{\beta}-2\underline{\xi}\beta\right)+\overline{\left(\frac{1}{2}\vartheta\underline{\alpha}+\zeta\underline{\beta}-2\eta\underline{\beta}-2\underline{\xi}\beta\right)}-\frac{3}{2}\underline{\check{\kappa}}\check{\rho}\\
&+\varsigma^{-1}\check{\varsigma}\left(\overline{-\frac{1}{2}\underline{\check{\kappa}}\check{\rho}-\frac{1}{2}\vartheta\,\underline{\alpha}-\zeta\,\underline{\beta}+2(\eta\,\underline{\beta}+\underline{\xi}\,\beta)}\right)\\
&-\varsigma^{-1}\left(\overline{\check{\varsigma}\left(-\frac{1}{2}\underline{\kappa}\rho+\not{d}_1\underline{\beta}-\frac{1}{2}\vartheta\,\underline{\alpha}-\zeta\,\underline{\beta}+2(\eta\,\underline{\beta}+\underline{\xi}\,\beta)\right)}-\overline{\check{\varsigma}\underline{\check{\kappa}}}\,\overline{\rho}\right)\\
&-\left(\underline{\check{\Omega}}+\varsigma^{-1}\overline{\underline{\Omega}}\check{\varsigma}\right)\left(\overline{-\frac{1}{2}\check{\kappa}\check{\rho}-\frac{1}{2}\underline{\vartheta}\,\alpha-\zeta\beta}\right)\\
&+\varsigma^{-1}\overline{\underline{\Omega}}\left(\overline{\check{\varsigma}\left(-\frac{1}{2}\kappa\rho+\not{d}_1\beta-\frac{1}{2}\underline{\vartheta}\,\alpha-\zeta\beta\right)}-\overline{\check{\varsigma}\check{\kappa}}\,\overline{\rho}\right)\\
&+\varsigma^{-1}\left(\overline{\underline{\check{\Omega}}\varsigma\left(-\frac{1}{2}\kappa\rho+\not{d}_1\beta-\frac{1}{2}\underline{\vartheta}\,\alpha-\zeta\beta\right)}-\overline{\underline{\check{\Omega}}\varsigma\,\kappa}\right)-\overline{\underline{\check{\kappa}}\check{\rho}}.
\end{aligned}
\tag{2.2.39}
$$

Proof. The proof relies on Corollary 2.66 and the null structure equations of Proposition 2.63. We refer to section A.4 for the details. □

2.2.5 Additional equations

We derive below additional equations for $\underline{\omega}, \eta, \underline{\xi}$.

Proposition 2.74. *The following identities hold true for a general forward geodesic foliation.*

- *The scalar* $\underline{\omega}$ *verifies*

$$
\begin{aligned}
2\,\not{d}_1^\star\underline{\omega} &= -\frac{1}{2}\kappa\underline{\xi}+\left(\frac{1}{2}\underline{\kappa}+2\underline{\omega}+\frac{1}{2}\underline{\vartheta}\right)\eta+e_3(\zeta)-\underline{\beta}\\
&\quad+\frac{1}{2}\underline{\kappa}\zeta-2\underline{\omega}\zeta+\frac{1}{2}\underline{\vartheta}\zeta-\frac{1}{2}\vartheta\underline{\xi}.
\end{aligned}
$$

- *The reduced* 1*-form* η *verifies*

$$
\begin{aligned}
2\,\not{d}_2\,\not{d}_2^\star\eta &= \kappa\left(-e_3(\zeta)+\underline{\beta}\right)-e_3(e_\theta(\kappa))-\kappa\left(\frac{1}{2}\underline{\kappa}\zeta-2\underline{\omega}\zeta\right)+6\rho\eta-\underline{\kappa}e_\theta\kappa\\
&\quad-\frac{1}{2}\kappa e_\theta(\underline{\kappa})+2\underline{\omega}e_\theta(\kappa)+2e_\theta(\rho)+Err[\,\not{d}_2\,\not{d}_2^\star\eta],\\
Err[\,\not{d}_2\,\not{d}_2^\star\eta] &= \left(2\,\not{d}_1\eta-\frac{1}{2}\kappa\underline{\vartheta}+2\eta^2\right)\eta+2e_\theta(\eta^2)-\kappa\left(\frac{1}{2}\underline{\vartheta}\zeta-\frac{1}{2}\vartheta\underline{\xi}\right)-\frac{1}{2}\underline{\vartheta}e_\theta(\kappa)\\
&\quad-\left(2\,\not{d}_1\eta-\frac{1}{2}\vartheta\underline{\vartheta}+2\eta^2\right)\zeta-\frac{1}{2}e_\theta(\underline{\vartheta}\,\vartheta)-\frac{1}{2}\vartheta^2\underline{\xi}-\frac{3}{2}\vartheta\underline{\vartheta}\eta.
\end{aligned}
$$

- *The reduced* 1*-form* $\underline{\xi}$ *verifies*

$$
\begin{aligned}
2\,\not{d}_2\,\not{d}_2^\star\underline{\xi} &= -e_3(e_\theta(\underline{\kappa})) + \underline{\kappa}\left(e_3(\zeta) - \underline{\beta}\right) + \underline{\kappa}^2\zeta - \frac{3}{2}\underline{\kappa}e_\theta\underline{\kappa} + 6\rho\underline{\xi} - 2\underline{\omega}e_\theta(\underline{\kappa})\\
&+ Err[\not{d}_2\,\not{d}_2^\star\underline{\xi}],\\
Err[\not{d}_2\,\not{d}_2^\star\underline{\xi}] &= \left(2\,\not{d}_1\underline{\xi} + \frac{1}{2}\underline{\kappa}\,\underline{\vartheta} + 2\eta\underline{\xi} - \frac{1}{2}\underline{\vartheta}^2\right)\eta + 2e_\theta(\eta\underline{\xi}) - \frac{1}{2}e_\theta(\underline{\vartheta}^2)\\
&+ \underline{\kappa}\left(\frac{1}{2}\underline{\vartheta}\zeta - \frac{1}{2}\vartheta\underline{\xi}\right) - \frac{1}{2}\underline{\vartheta}e_\theta\underline{\kappa} - \frac{1}{2}\vartheta\underline{\vartheta}\underline{\xi}\\
&- \zeta\left(2\,\not{d}_1\underline{\xi} + 2(\eta - 3\zeta)\underline{\xi} - \frac{1}{2}\underline{\vartheta}^2\right) + \underline{\xi}\Big(-\vartheta\underline{\vartheta} - 2\,\not{d}_1\zeta + 2\zeta^2\Big)\\
&- 6\eta\zeta\underline{\xi} - 6e_\theta(\zeta\underline{\xi}).
\end{aligned}
$$

Proof. The proof relies on the null structure equations of Proposition 2.63, in particular the ones for $e_3(\zeta)$, $e_3(\kappa)$ and $e_3(\underline{\kappa})$. We refer to section A.5 for the details. □

2.2.6 Ingoing geodesic foliation

All the equations of section 2.2.4 for outgoing geodesic foliations have a counterpart for ingoing geodesic foliations. The corresponding equations can be easily deduced from the ones in section 2.2.4 by performing the following substitutions:

$$
\begin{aligned}
&u \to \underline{u}, \quad s \to s, \quad \mathcal{C}_u \to \mathcal{C}_{\underline{u}}, \quad S_{u,s} \to S_{\underline{u},s}, \quad r \to r, \quad m \to m,\\
&e_4 \to e_3, \quad e_3 \to e_4, \quad e_\theta \to e_\theta, \quad e_4(s) = 1 \to e_3(s) = -1,\\
&\alpha \to \underline{\alpha}, \quad \beta \to, \underline{\beta}, \quad \rho \to \rho, \quad \mu \to \underline{\mu}, \quad \underline{\beta} \to \beta, \quad \underline{\alpha} \to \alpha,\\
&\xi \to \underline{\xi}, \quad \omega \to \underline{\omega}, \quad \kappa \to \underline{\kappa}, \quad \vartheta \to \underline{\vartheta}, \quad \eta \to \underline{\eta}, \quad \underline{\eta} \to \eta, \quad \zeta \to -\zeta, \quad \underline{\kappa} \to \kappa,\\
&\underline{\vartheta} \to \vartheta, \quad \underline{\omega} \to \omega, \quad \underline{\xi} \to \xi, \quad \underline{\Omega} = e_3(s) \to \Omega = e_4(s), \quad \varsigma = \frac{2}{e_3(u)} \to \underline{\varsigma} = \frac{2}{e_4(\underline{u})},\\
&\overline{\kappa} - \frac{2}{r} \to \underline{\overline{\kappa}} + \frac{2}{r}, \quad \underline{\overline{\kappa}} + \frac{2\Upsilon}{r} \to \overline{\kappa} - \frac{2\Upsilon}{r}, \quad \underline{\overline{\omega}} - \frac{m}{r^2} \to \overline{\omega} + \frac{m}{r^2},\\
&\overline{\rho} + \frac{2m}{r^3} \to \overline{\rho} + \frac{2m}{r^3}, \quad \overline{\mu} - \frac{2m}{r^3} \to \underline{\overline{\mu}} - \frac{2m}{r^3}, \quad \underline{\overline{\Omega}} + \Upsilon \to \overline{\Omega} - \Upsilon, \quad \overline{\varsigma} - 1 \to \underline{\overline{\varsigma}} - 1,\\
&\underline{A} = \frac{2}{r}e_3(r) - \underline{\overline{\kappa}} \to A = \frac{2}{r}e_4(r) - \overline{\kappa}.
\end{aligned}
$$

2.2.7 Adapted coordinates systems

2.2.7.1 (u, s, θ, φ) *coordinates*

Proposition 2.75. *Consider, in addition to the functions* u, s, φ*, an additional* **Z***-invariant function* θ*. Then, relative to the coordinates system* (u, s, θ, φ)*, the following hold true:*

1. *The spacetime metric takes the form*

$$
g = -2\varsigma du ds + \varsigma^2\underline{\Omega}du^2 + \gamma\left(d\theta - \frac{1}{2}\varsigma(\underline{b} - \underline{\Omega}b)du - bds\right)^2 + e^{2\Phi}(d\varphi)^2 \qquad (2.2.40)
$$

where

$$\underline{\Omega} = e_3(s), \qquad b = e_4(\theta), \qquad \underline{b} = e_3(\theta), \qquad \gamma^{-1} = e_\theta(\theta)^2. \tag{2.2.41}$$

2. *In these coordinates the reduced frame takes the form*

$$\begin{aligned} \partial_s &= e_4 - b\sqrt{\gamma}e_\theta, \qquad \partial_u = \varsigma\left(\frac{1}{2}e_3 - \frac{1}{2}\underline{\Omega}e_4 - \frac{1}{2}\sqrt{\gamma}(\underline{b} - b\underline{\Omega})e_\theta\right), \\ \partial_\theta &= \sqrt{\gamma}e_\theta. \end{aligned} \tag{2.2.42}$$

3. *In the particular case when $b = e_4(\theta) = 0$ we have*

$$e_4(\gamma) = 2\chi\gamma, \qquad e_4(\underline{b}) = -2(\zeta + \eta)\gamma^{-1/2}. \tag{2.2.43}$$

Proof. First, from the fact that (e_3, e_4, e_θ) forms a null frame, we easily verify that (2.2.42) holds. Then, (2.2.40) immediately follows from (2.2.42) and the fact that (e_3, e_4, e_θ) forms a null frame.

To prove the last statement, when $b = e_4(\theta) = 0$, we start with

$$[e_4, e_3] = 2\omega e_3 - 2\underline{\omega}e_4 + 2(\underline{\eta} - \eta)e_\theta = -2(\zeta + \eta)e_\theta - 2\underline{\omega}e_4.$$

Applying this to θ we derive

$$[e_4, e_3](\theta) = (-2(\zeta + \eta)e_\theta - 2\underline{\omega}e_4)(\theta) = -2(\zeta + \eta)e_\theta(\theta) = -2(\zeta + \eta)\gamma^{-1/2}.$$

We deduce

$$e_4(\underline{b}) = e_4(e_3(\theta)) = -2(\zeta + \eta)\gamma^{-1/2}.$$

To prove the equation for γ we make use of

$$[e_4, e_\theta] = (\underline{\eta} + \zeta)e_4 + \xi e_3 - \chi e_\theta = -\chi e_\theta$$

so that

$$e_4 e_\theta(\theta) = [e_4, e_\theta](\theta) \quad = \quad -\chi e_\theta(\theta) = -\chi\gamma^{-1/2}.$$

Thus

$$e_4(\gamma^{-1/2}) = -\chi\gamma^{-1/2}$$

from which

$$e_4(\gamma) = 2\chi\gamma.$$

This concludes the proof of the lemma. □

Remark 2.76. *In Schwarzschild, relative to the above coordinate system, we have*

$$\varsigma = 1, \qquad \underline{\Omega} = -\Upsilon, \qquad b = \underline{b} = 0, \qquad \gamma = r^2, \qquad e^\Phi = r\sin\theta,$$

so that we obtain outgoing Eddington-Finkelstein coordinates.

Remark 2.77. *The (u, s, θ, φ) coordinates system, with the choice $b = 0$ (i.e., θ is* *transported by $e_4(\theta) = 0$), will be used in section 3.7 and Chapter 9 in connection with our GCM procedure.*

2.2.7.2 (u, r, θ, φ) coordinates

Proposition 2.78. *Consider, in addition to the functions u, r, φ, an additional* **Z**-*invariant function θ. Relative to the coordinates (u, r, θ, φ) the following hold true:*

1. *The spacetime metric takes the form*

$$g = -\frac{4\varsigma}{r\overline{\kappa}}dudr + \frac{\varsigma^2(\underline{\overline{\kappa}} + \underline{A})}{\underline{\overline{\kappa}}}du^2 + \gamma\left(d\theta - \frac{1}{2}\varsigma\underline{b}du - \frac{b}{2}\Theta\right)^2 \tag{2.2.44}$$

where

$$b = e_4(\theta), \qquad \underline{b} = e_3(\theta), \qquad \gamma = \frac{1}{(e_\theta(\theta))^2} \tag{2.2.45}$$

and

$$\Theta := \frac{4}{r\overline{\kappa}}dr - \varsigma\left(\frac{\underline{\overline{\kappa}} + \underline{A}}{\overline{\kappa}}\right)du.$$

2. *The reduced coordinates derivatives take the form*

$$\begin{aligned}
\partial_r &= \frac{2}{r\overline{\kappa}}e_4 - \frac{2\sqrt{\gamma}}{r\overline{\kappa}}be_\theta, \\
\partial_\theta &= \sqrt{\gamma}e_\theta, \\
\partial_u &= \varsigma\left[\frac{1}{2}e_3 - \frac{1}{2}\frac{\underline{\overline{\kappa}} + \underline{A}}{\overline{\kappa}}e_4 - \frac{1}{2}\sqrt{\gamma}\left(\underline{b} - \left(\frac{\underline{\overline{\kappa}} + \underline{A}}{\overline{\kappa}}\right)b\right)e_\theta\right].
\end{aligned} \tag{2.2.46}$$

3. *To control e^Φ, we will rely on the following transport equation*

$$e_4\left(\frac{e^\Phi}{r\sin\theta} - 1\right) = \frac{e^\Phi}{2r\sin\theta}\left(\check{\kappa} - \vartheta\right). \tag{2.2.47}$$

Proof. First, from the fact that (e_3, e_4, e_θ) forms a null frame, we easily verify that (2.2.46) holds. Then, (2.2.44) immediately follows from (2.2.46) and the fact that (e_3, e_4, e_θ) forms a null frame.

It remains to prove (2.2.47). It follows from

$$\begin{aligned}
e_4\left(\frac{e^\Phi}{r\sin\theta} - 1\right) &= \frac{e^\Phi}{r\sin\theta}\left(e_4(\Phi) - \frac{e_4(r)}{r}\right) = \frac{e^\Phi}{r\sin\theta}\left(\frac{1}{2}(\kappa - \vartheta) - \frac{\overline{\kappa}}{2}\right) \\
&= \frac{e^\Phi}{2r\sin\theta}\left(\check{\kappa} - \vartheta\right)
\end{aligned}$$

which concludes the proof of the lemma. □

Remark 2.79. *In Schwarzschild, relative to the above coordinate system, we have*

$$\overline{\kappa} = \frac{2}{r}, \quad \underline{\overline{\kappa}} = -\frac{2\Upsilon}{r}, \quad \varsigma = 1, \quad \underline{A} = 0, \quad b = \underline{b} = 0, \quad \gamma = r^2, \quad e^\Phi = r\sin\theta,$$

so that we obtain outgoing Eddington-Finkelstein coordinates.

Remark 2.80. *The (u, r, θ, φ) coordinates system, with the choice* (2.2.52) *for θ introduced below, will be used in Proposition 3.17 to prove the convergence to the outgoing Eddington-Finkelstein coordinates of Schwarzschild.*

2.2.7.3 $(\underline{u}, r, \theta, \varphi)$ coordinates

We easily deduce an analog statement relative to $(\underline{u}, r, \theta, \varphi)$ coordinates.

Proposition 2.81. *Consider, in addition to the functions $\underline{u}, r, \varphi$, an additional $\mathbf{Z}$-invariant function θ. Relative to the coordinates $(\underline{u}, r, \theta, \varphi)$ the following hold true:*

1. *The spacetime metric takes the form*

$$g = -\frac{4\underline{\varsigma}}{r\underline{\overline{\kappa}}}d\underline{u}dr + \frac{\underline{\varsigma}^2(\overline{\kappa}+A)}{\underline{\overline{\kappa}}}d\underline{u}^2 + \gamma\left(d\theta - \frac{1}{2}\underline{\varsigma}\underline{b}d\underline{u} - \frac{b}{2}\underline{\Theta}\right)^2 \tag{2.2.48}$$

where

$$b = e_4(\theta), \qquad \underline{b} = e_3(\theta), \qquad \gamma = \frac{1}{(e_\theta(\theta))^2} \tag{2.2.49}$$

and

$$\underline{\Theta} := \frac{4}{r\underline{\overline{\kappa}}}dr - \underline{\varsigma}\left(\frac{\overline{\kappa}+A}{\underline{\overline{\kappa}}}\right)d\underline{u}.$$

2. *The reduced coordinates derivatives take the form*

$$\begin{aligned} \partial_r &= \frac{2}{r\underline{\overline{\kappa}}}e_3 - \frac{2\sqrt{\gamma}}{r\underline{\overline{\kappa}}}\underline{b}e_\theta, \\ \partial_\theta &= \sqrt{\gamma}e_\theta, \\ \partial_{\underline{u}} &= \underline{\varsigma}\left[\frac{1}{2}e_4 - \frac{1}{2}\frac{\overline{\kappa}+A}{\underline{\overline{\kappa}}}e_3 - \frac{1}{2}\sqrt{\gamma}\left(b - \left(\frac{\overline{\kappa}+A}{\underline{\overline{\kappa}}}\right)\underline{b}\right)e_\theta\right]. \end{aligned} \tag{2.2.50}$$

3. *To control e^Φ, we will rely on the following transport equation*

$$e_3\left(\frac{e^\Phi}{r\sin\theta} - 1\right) = \frac{e^\Phi}{2r\sin\theta}\left(\underline{\check{\kappa}} - \underline{\vartheta}\right). \tag{2.2.51}$$

Remark 2.82. *In Schwarzschild, relative to the above coordinate system, we have*

$$\overline{\kappa} = \frac{2}{r}, \quad \underline{\overline{\kappa}} = -\frac{2\Upsilon}{r}, \quad \underline{\varsigma} = 1, \quad A = 0, \quad b = \underline{b} = 0, \quad \gamma = r^2, \quad e^\Phi = r\sin\theta,$$

so that we obtain ingoing Eddington-Finkelstein coordinates.

Remark 2.83. *The $(\underline{u}, r, \theta, \varphi)$ coordinates system, with the choice* (2.2.52) *for θ introduced below, will be used in Proposition 3.18 to prove the convergence to the ingoing Eddington-Finkelstein coordinates of Schwarzschild.*

2.2.7.4 Initialization of θ

We now introduce the coordinate function θ that will be used for the (u, r, θ, φ) coordinates system and for the $(\underline{u}, r, \theta, \varphi)$ coordinates system, see Remarks 2.80 and 2.83.

Lemma 2.84. *Let $\theta \in [0, \pi]$ be the* **Z***-invariant scalar on $\mathcal{M}$ defined by*

$$\theta := \cot^{-1}(re_\theta(\Phi)). \tag{2.2.52}$$

Then,

$$\frac{e^\Phi}{r\sin\theta} = \sqrt{1+\mathfrak{a}} \tag{2.2.53}$$

where

$$\mathfrak{a} := \frac{e^{2\Phi}}{r^2} + (e_\theta(e^\Phi))^2 - 1. \tag{2.2.54}$$

Moreover, we have in an outgoing geodesic foliation

$$\begin{aligned}
re_\theta(\theta) &= 1 + \frac{r^2(K - \frac{1}{r^2})}{1 + (re_\theta(\Phi))^2},\\
e_3(\theta) &= -\frac{r\underline{\beta} + \frac{r}{2}\left(-\underline{\check{\kappa}} + \underline{A} + \underline{\vartheta}\right)e_\theta(\Phi) + r\underline{\xi}e_4(\Phi) + r\eta e_3(\Phi)}{1 + (re_\theta(\Phi))^2},\\
e_4(\theta) &= -\frac{r\beta + \frac{r}{2}\left(-\check{\kappa} + \vartheta\right)e_\theta(\Phi) - r\zeta e_3(\Phi)}{1 + (re_\theta(\Phi))^2},
\end{aligned}$$

and analog identities hold for an ingoing geodesic foliation.

Proof. In view of the definition of θ, we have $\theta \in [0, \pi]$, $\sin\theta \geq 0$ and

$$\begin{aligned}
\sin\theta &= \frac{1}{\sqrt{1+\cot\theta^2}} = \frac{1}{\sqrt{1+(re_\theta(\Phi))^2}} = \frac{e^\Phi}{\sqrt{e^{2\Phi} + (re_\theta(e^\Phi))^2}}\\
&= \frac{e^\Phi}{r\sqrt{\frac{e^{2\Phi}}{r^2} + (e_\theta(e^\Phi))^2}} = \frac{e^\Phi}{r\sqrt{1+\mathfrak{a}}}.
\end{aligned}$$

Hence

$$\frac{e^\Phi}{r\sin\theta} = \sqrt{\frac{e^{2\Phi}}{r^2} + (e_\theta(e^\Phi))^2} = \sqrt{1+\mathfrak{a}}.$$

Also, we compute

$$re_\theta(\theta) = -\frac{r^2 e_\theta e_\theta(\Phi)}{1 + (re_\theta(\Phi))^2}.$$

Next, recall that we have

$$e_\theta e_\theta(\Phi) = -K - (e_\theta(\Phi))^2.$$

We infer

$$re_\theta(\theta) = \frac{r^2(K+(e_\theta(\Phi))^2)}{1+(re_\theta(\Phi))^2} = 1+\frac{r^2(K-\frac{1}{r^2})}{1+(re_\theta(\Phi))^2}$$

as desired.

Also, we have in an outgoing geodesic foliation

$$\begin{aligned} e_4(\theta) &= -\frac{re_4e_\theta(\Phi)+e_4(r)e_\theta(\Phi)}{1+(re_\theta(\Phi))^2} \\ &= -\frac{r(D_4D_\theta\Phi+D_{D_4e_\theta}\Phi)+e_4(r)e_\theta(\Phi)}{1+(re_\theta(\Phi))^2} \\ &= -\frac{r\beta+r\left(\frac{e_4(r)}{r}-e_4(\Phi)\right)e_\theta(\Phi)-r\zeta e_4(\Phi)}{1+(re_\theta(\Phi))^2} \\ &= -\frac{r\beta+\frac{r}{2}\left(-\check{\kappa}+\vartheta\right)e_\theta(\Phi)-r\zeta e_4(\Phi)}{1+(re_\theta(\Phi))^2}. \end{aligned}$$

Finally, we compute in an outgoing geodesic foliation

$$\begin{aligned} e_3(\theta) &= -\frac{re_3e_\theta(\Phi)+e_3(r)e_\theta(\Phi)}{1+(re_\theta(\Phi))^2} \\ &= -\frac{r(D_3D_\theta\Phi+D_{D_3e_\theta}\Phi)+e_3(r)e_\theta(\Phi)}{1+(re_\theta(\Phi))^2} \\ &= -\frac{r\underline{\beta}+r\left(\frac{e_3(r)}{r}-e_3(\Phi)\right)e_\theta(\Phi)+r\underline{\xi}e_4(\Phi)+r\eta e_3(\Phi)}{1+(re_\theta(\Phi))^2} \\ &= -\frac{r\underline{\beta}+\frac{r}{2}\left(-\underline{\check{\kappa}}+\underline{A}+\underline{\vartheta}\right)e_\theta(\Phi)+r\underline{\xi}e_4(\Phi)+r\eta e_3(\Phi)}{1+(re_\theta(\Phi))^2}. \end{aligned}$$

This concludes the proof of the lemma. □

In view of (2.2.53), we will need to control the quantity $\mathfrak{a}$ defined in (2.2.54). To this end, we will need the following lemma.

Lemma 2.85. *The quantity $\mathfrak{a}$ defined in (2.2.54) vanishes on the axis of symmetry and verifies the following identities in an outgoing geodesic foliation:*

$$\begin{aligned} e_4(\mathfrak{a}) &= \frac{(\check{\kappa}-\vartheta)e^{2\Phi}}{r^2}+2e_\theta(e^\Phi)\Big(\beta-e_4(\Phi)\zeta\Big)e^\Phi, \\ e_\theta(\mathfrak{a}) &= 2e_\theta(\Phi)e^{2\Phi}\left(\left(\rho+\frac{2m}{r^3}\right)+\frac{1}{4}\left(\kappa\underline{\kappa}+\frac{4\Upsilon}{r^2}\right)-\frac{1}{4}\vartheta\underline{\vartheta}\right), \\ e_3(\mathfrak{a}) &= \frac{\Big(\underline{\check{\kappa}}-\underline{A}-\underline{\vartheta}\Big)e^{2\Phi}}{r^2}+2e_\theta(e^\Phi)\Big(\underline{\beta}+e_3(\Phi)\eta+\underline{\xi}e_4(\Phi)\Big)e^\Phi, \end{aligned}$$

and analog identities hold in an ingoing geodesic foliation.

Proof. The vanishing on the axis follow easily from the fact that both $e^{2\Phi}$ and $(e_\theta(e^\Phi))^2-1$ vanish on the axis (see (2.1.13)). To prove the second part of the lemma we recall that, with respect to the reduced metric (see equation (2.1.4)),

$$R_{ab} = D_a D_b \Phi + D_a \Phi D_b \Phi,$$

and (see Definition 2.50)

$$R_{3\theta} = \underline{\beta}, \quad R_{4\theta} = \beta, \quad R_{\theta\theta} = R_{34} = \rho, \quad R_{34} = \rho.$$

Starting with the definition $\mathfrak{a} = \frac{e^{2\Phi}}{r^2} + (e_\theta(e^\Phi))^2 - 1$, we compute in an outgoing geodesic foliation

$$\begin{aligned} e_4(\mathfrak{a}) &= \frac{2e_4(\Phi)e^{2\Phi}}{r^2} - \frac{2e_4(r)e^{2\Phi}}{r^3} + 2e_\theta(e^\Phi)e_4(e_\theta(e^\Phi)) \\ &= \frac{(\kappa - \vartheta)e^{2\Phi}}{r^2} - \frac{\overline{\kappa}e^{2\Phi}}{r^2} + 2e_\theta(e^\Phi)\Big(e_4(e_\theta(\Phi)) + e_\theta(\Phi)e_4(\Phi)\Big)e^\Phi \\ &= \frac{(\check{\kappa} - \vartheta)e^{2\Phi}}{r^2} + 2e_\theta(e^\Phi)\Big(\beta - e_4(\Phi)\zeta\Big)e^\Phi. \end{aligned}$$

Also

$$\begin{aligned} e_\theta(\mathfrak{a}) &= \frac{2e_\theta(\Phi)e^{2\Phi}}{r^2} + 2e_\theta(e^\Phi)e_\theta(e_\theta(e^\Phi)) \\ &= \frac{2e_\theta(\Phi)e^{2\Phi}}{r^2} + 2e_\theta(e^\Phi)\Big(e_\theta(e_\theta(\Phi)) + e_\theta(\Phi)^2\Big)e^\Phi \\ &= \frac{2e_\theta(\Phi)e^{2\Phi}}{r^2} + 2e_\theta(e^\Phi)\Big(\rho + D_{D_\theta e_\theta}\Phi\Big)e^\Phi \\ &= \frac{2e_\theta(\Phi)e^{2\Phi}}{r^2} + 2e_\theta(e^\Phi)\Big(\rho + \frac{1}{2}\chi e_3\Phi + \frac{1}{2}\underline{\chi}e_4\Phi\Big)e^\Phi \\ &= \frac{2e_\theta(\Phi)e^{2\Phi}}{r^2} + 2e_\theta(e^\Phi)\Big(\rho + \frac{1}{4}\kappa\underline{\kappa} - \frac{1}{4}\vartheta\underline{\vartheta}\Big)e^\Phi \\ &= 2e_\theta(\Phi)e^{2\Phi}\left(\left(\rho + \frac{2m}{r^3}\right) + \frac{1}{4}\left(\kappa\underline{\kappa} + \frac{4\Upsilon}{r^2}\right) - \frac{1}{4}\vartheta\underline{\vartheta}\right). \end{aligned}$$

Finally, we have in an outgoing geodesic foliation

$$\begin{aligned} e_3(\mathfrak{a}) &= \frac{2e_3(\Phi)e^{2\Phi}}{r^2} - \frac{2e_3(r)e^{2\Phi}}{r^3} + 2e_\theta(e^\Phi)e_3(e_\theta(e^\Phi)) \\ &= \frac{(\underline{\kappa} - \underline{\vartheta})e^{2\Phi}}{r^2} - \frac{\Big(\overline{\underline{\kappa}} + \underline{A}\Big)e^{2\Phi}}{r^2} + 2e_\theta(e^\Phi)\Big(e_3(e_\theta(\Phi)) + e_\theta(\Phi)e_3(\Phi)\Big)e^\Phi \\ &= \frac{\Big(\check{\underline{\kappa}} - \underline{A} - \underline{\vartheta}\Big)e^{2\Phi}}{r^2} + 2e_\theta(e^\Phi)\Big(\underline{\beta} + e_3(\Phi)\eta + \underline{\xi}e_4(\Phi)\Big)e^\Phi. \end{aligned}$$

This concludes the proof of the lemma. □

Remark 2.86. *The function θ defined by (2.2.52) defines*

- *together with the functions (u, r, φ), a regular coordinates system with the axis of symmetry corresponding to $\theta = 0, \pi$,*
- *together with the functions $(\underline{u}, r, \varphi)$, a regular coordinates system with the axis of symmetry corresponding to $\theta = 0, \pi$.*

2.3 PERTURBATIONS OF SCHWARZSCHILD AND INVARIANT QUANTITIES

Recall that in Schwarzschild all Ricci coefficients $\xi, \underline{\xi}, \vartheta, \underline{\vartheta}, \eta, \underline{\eta}, \zeta$ and curvature components $\alpha, \underline{\alpha}, \beta, \underline{\beta}$ vanish identically. In addition the check quantities $\check{\kappa}, \check{\underline{\kappa}}, \check{\omega}, \check{\underline{\omega}}$ and $\check{\rho}$ also vanish. Thus, roughly, we expect that in perturbations of Schwarzschild these quantities stay small, i.e., of order $O(\epsilon)$ for a sufficiently small ϵ. More precisely we say that a smooth, vacuum, $\mathbf{Z}$-invariant, polarized spacetime is an $O(\epsilon)$-perturbation of Schwarzschild, or simply $O(\epsilon)$-Schwarzschild, if the following are true relative to a $\mathbf{Z}$-invariant null frame e_3, e_4, e_θ:

$$\xi,\ \underline{\xi},\ \vartheta,\ \underline{\vartheta},\ \eta,\ \underline{\eta},\ \zeta,\ \check{\kappa},\ \check{\underline{\kappa}},\ \check{\omega},\ \check{\underline{\omega}} \qquad \alpha,\ \underline{\alpha},\ \beta,\ \underline{\beta}, \check{\rho} = O(\epsilon) \tag{2.3.1}$$

Moreover,

$$e_3(r) - \frac{r}{2}\underline{\overline{\kappa}} = O(\epsilon), \qquad e_4(r) - \frac{r}{2}\overline{\kappa} = O(\epsilon), \tag{2.3.2}$$

where r is the area radius of the 2-spheres generated by e_θ, e_φ, see (2.1.12).

In reality, of course, we expect that small perturbations of Schwarzschild remain not only close to the original Schwarzschild but also converge to a nearby Schwarzschild solution but for the discussion below this will suffice.

2.3.1 Null frame transformations

Our definition of $O(\epsilon)$-Schwarzschild perturbations does not specify a particular frame. In what follows we investigate how the main Ricci and curvature quantities change relative to frame transformations, i.e., linear transformations which take null frames into null frames.

Lemma 2.87. *A general null transformation can be written in the form*

$$\begin{aligned}
e_4' &= \lambda\left(e_4 + f e_\theta + \frac{1}{4}f^2 e_3\right),\\
e_\theta' &= \left(1+\frac{1}{2}f\underline{f}\right)e_\theta + \frac{1}{2}\underline{f}e_4 + \frac{1}{2}f\left(1+\frac{1}{4}f\underline{f}\right)e_3,\\
e_3' &= \lambda^{-1}\left(\left(1+\frac{1}{2}f\underline{f}+\frac{1}{16}f^2\underline{f}^2\right)e_3 + \underline{f}\left(1+\frac{1}{4}f\underline{f}\right)e_\theta + \frac{1}{4}\underline{f}^2 e_4\right).
\end{aligned} \tag{2.3.3}$$

Proof. It is straightforward to check that the transformation (2.3.3) takes null frames into null frames. One can also check that it can be written in the form type(3) $\circ$ type(1) $\circ$ type(2) where the type 1 transformations fix e_3, i.e., ($\lambda = 1, \underline{f} = 0$), type 2 transformations fix e_4, i.e., ($\lambda = 1, f = 0$) and type 3 transformations keep the directions of e_3, e_4, i.e., ($f = \underline{f} = 0$). □

Remark 2.88. *Note that $f, \underline{f}$ are reduced from spacetime 1 forms while λ is reduced from a scalar.*

Remark 2.89. *A transformation consistent with $O(\epsilon)$-Schwarzschild spacetimes must have $f, \underline{f} = O(\epsilon)$ and $a := \log\lambda = O(\epsilon)$.*

Proposition 2.90 (Transformation formulas). *Under a general transformation of type* (2.3.3), *the Ricci coefficients and curvature components transform as follows:*

$$\begin{aligned}
\xi' &= \lambda^2\left(\xi + \frac{1}{2}\lambda^{-1}e_4'(f) + \omega f + \frac{1}{4}f\kappa\right) + \lambda^2 Err(\xi,\xi'),\\
Err(\xi,\xi') &= \frac{1}{4}f\vartheta + l.o.t.,\\
\underline{\xi}' &= \lambda^{-2}\left(\underline{\xi} + \frac{1}{2}\lambda e_3'(\underline{f}) + \underline{\omega}\,\underline{f} + \frac{1}{4}\underline{f}\,\underline{\kappa}\right) + \lambda^{-2} Err(\underline{\xi},\underline{\xi}'),\\
Err(\underline{\xi},\underline{\xi}') &= -\frac{1}{8}\lambda\underline{f}^2 e_3'(f) + \frac{1}{4}\underline{f}\,\underline{\vartheta} + l.o.t.,
\end{aligned} \tag{2.3.4}$$

$$\begin{aligned}
\zeta' &= \zeta - e_\theta'(\log(\lambda)) + \frac{1}{4}(-f\underline{\kappa} + \underline{f}\kappa) + \underline{f}\omega - f\underline{\omega} + Err(\zeta,\zeta'),\\
Err(\zeta,\zeta') &= \frac{1}{2}\underline{f}e_\theta'(f) + \frac{1}{4}(-f\underline{\vartheta} + \underline{f}\vartheta) + l.o.t.,\\
\eta' &= \eta + \frac{1}{2}\lambda e_3'(f) + \frac{1}{4}\kappa\underline{f} - f\underline{\omega} + Err(\eta,\eta'),\\
Err(\eta,\eta') &= \frac{1}{4}\underline{f}\vartheta + l.o.t.,\\
\underline{\eta}' &= \underline{\eta} + \frac{1}{2}\lambda^{-1}e_4'(\underline{f}) + \frac{1}{4}\underline{\kappa}f - \underline{f}\omega + Err(\underline{\eta},\underline{\eta}'),\\
Err(\underline{\eta},\underline{\eta}') &= -\frac{1}{8}\underline{f}^2\lambda^{-1}e_4'(f) + \frac{1}{4}f\underline{\vartheta} + l.o.t.,
\end{aligned} \tag{2.3.5}$$

$$\begin{aligned}
\kappa' &= \lambda\left(\kappa + \slashed{d}_1{}'(f)\right) + \lambda Err(\kappa,\kappa'),\\
Err(\kappa,\kappa') &= f(\zeta+\eta) + \underline{f}\xi - \frac{1}{4}f^2\underline{\kappa} + f\underline{f}\omega - f^2\underline{\omega} + l.o.t.,\\
\underline{\kappa}' &= \lambda^{-1}\left(\underline{\kappa} + \slashed{d}_1{}'(\underline{f})\right) + \lambda^{-1}Err(\underline{\kappa},\underline{\kappa}'),\\
Err(\underline{\kappa},\underline{\kappa}') &= -\frac{1}{4}\underline{f}^2 e_\theta'(f) + \underline{f}(-\zeta+\underline{\eta}) + f\underline{\xi} - \frac{1}{4}\underline{f}^2\kappa + f\underline{f}\underline{\omega} - \underline{f}^2\omega + l.o.t.,
\end{aligned} \tag{2.3.6}$$

$$\begin{aligned}
\vartheta' &= \lambda\left(\vartheta - \slashed{d}_2^\star{}'(f)\right) + \lambda Err(\vartheta,\vartheta'),\\
Err(\vartheta,\vartheta') &= f(\zeta+\eta) + \underline{f}\xi + \frac{1}{4}f\underline{f}\kappa + f\underline{f}\omega - f^2\underline{\omega} + l.o.t.\\
\underline{\vartheta}' &= \lambda^{-1}\left(\underline{\vartheta} - \slashed{d}_2^\star{}'(\underline{f})\right) + \lambda^{-1}Err(\underline{\vartheta},\underline{\vartheta}'),\\
Err(\underline{\vartheta},\underline{\vartheta}') &= -\frac{1}{4}\underline{f}^2 e_\theta'(f) + \underline{f}(-\zeta+\underline{\eta}) + f\underline{\xi} + \frac{1}{4}f\underline{f}\underline{\kappa} + f\underline{f}\underline{\omega} - \underline{f}^2\omega + l.o.t.,
\end{aligned} \tag{2.3.7}$$

$$
\begin{aligned}
\omega' &= \lambda\left(\omega - \frac{1}{2}\lambda^{-1}e_4'(\log(\lambda))\right) + \lambda Err(\omega,\omega'),\\
Err(\omega,\omega') &= \frac{1}{4}\underline{f}e_4'(f) + \frac{1}{2}\omega f\underline{f} - \frac{1}{2}f\underline{\eta} + \frac{1}{2}\underline{f}\xi + \frac{1}{2}f\zeta - \frac{1}{8}\underline{\kappa}f^2 + \frac{1}{8}f\underline{f}\kappa\\
&\quad - \frac{1}{4}\underline{\omega}f^2 + l.o.t.,\\
\underline{\omega}' &= \lambda^{-1}\left(\underline{\omega} + \frac{1}{2}\lambda e_3'(\log(\lambda))\right) + \lambda^{-1}Err(\underline{\omega},\underline{\omega}'),\\
Err(\underline{\omega},\underline{\omega}') &= -\frac{1}{4}\underline{f}e_3'(f) + \underline{\omega}f\underline{f} - \frac{1}{2}\underline{f}\eta + \frac{1}{2}f\underline{\xi} - \frac{1}{2}\underline{f}\zeta - \frac{1}{8}\kappa\underline{f}^2 + \frac{1}{8}f\underline{f}\underline{\kappa}\\
&\quad - \frac{1}{4}\omega\underline{f}^2 + l.o.t.
\end{aligned}
\tag{2.3.8}
$$

The lower order terms we denote by l.o.t. are linear with respect to the Ricci coefficients $\Gamma = \{\xi, \underline{\xi}, \vartheta, \kappa, \eta, \underline{\eta}, \zeta, \underline{\kappa}, \underline{\vartheta}\}$ *and quadratic or higher order in* $f, \underline{f}$, *and do not contain derivatives of these latter.*

Also,

$$
\begin{aligned}
\alpha' &= \lambda^2\alpha + \lambda^2 Err(\alpha,\alpha'),\\
Err(\alpha,\alpha') &= 2f\beta + \frac{3}{2}f^2\rho + l.o.t.,\\
\beta' &= \lambda\left(\beta + \frac{3}{2}\rho f\right) + \lambda Err(\beta,\beta'),\\
Err(\beta,\beta') &= \frac{1}{2}\underline{f}\alpha + l.o.t.,\\
\rho' &= \rho + Err(\rho,\rho'),\\
Err(\rho,\rho') &= \frac{3}{2}\rho f\underline{f} + \underline{f}\beta + f\underline{\beta} + l.o.t.,\\
\underline{\beta}' &= \lambda^{-1}\left(\underline{\beta} + \frac{3}{2}\rho\underline{f}\right) + \lambda^{-1}Err(\underline{\beta},\underline{\beta}'),\\
Err(\underline{\beta},\underline{\beta}') &= \frac{1}{2}f\underline{\alpha} + l.o.t.,\\
\underline{\alpha}' &= \lambda^{-2}\underline{\alpha} + \lambda^{-2}Err(\underline{\alpha},\underline{\alpha}'),\\
Err(\underline{\alpha},\underline{\alpha}') &= 2\underline{f}\,\underline{\beta} + \frac{3}{2}\underline{f}^2\rho + l.o.t.
\end{aligned}
\tag{2.3.9}
$$

The lower order terms we denote by l.o.t. are linear with respect to the curvature quantities $\alpha, \beta, \rho, \underline{\beta}, \underline{\alpha}$ *and quadratic or higher order in* $f, \underline{f}$, *and do not contain derivatives of these latter.*

Proof. See section A.6. □

Lemma 2.91. *In the particular case when* $\lambda = 1, \underline{f} = 0$, *we have*

$$
\begin{aligned}
e_4' &= e_4 + fe_\theta + \frac{1}{4}f^2e_3,\\
e_\theta' &= e_\theta + \frac{1}{2}fe_3,\\
e_3' &= e_3,
\end{aligned}
$$

and

$$
\begin{aligned}
\xi' &= \xi + \frac{1}{2}e_4'f + \frac{1}{4}\kappa f + f\omega + \frac{1}{4}f\vartheta + \frac{1}{4}f^2\eta - \frac{1}{4}f^2\underline{\eta} + \frac{1}{2}f^2\zeta - \frac{1}{16}f^3\underline{\kappa}\\
&\quad -\frac{1}{4}f^3\underline{\omega} - \frac{1}{16}f^3\underline{\vartheta} - \frac{1}{16}f^4\underline{\xi},\\
\omega' &= \omega + \frac{1}{2}f\zeta - \frac{1}{2}\underline{\eta}f - \frac{1}{4}f^2\underline{\omega} - \frac{1}{8}f^2\underline{\kappa} - \frac{1}{8}f^2\underline{\vartheta} - \frac{1}{8}f^3\underline{\xi},\\
\zeta' &= \zeta - \left(\frac{1}{4}\underline{\kappa} + \underline{\omega}\right)f - f\left(\frac{1}{4}\underline{\vartheta} + \frac{1}{2}f\underline{\xi}\right),\\
\eta' &= \eta + \frac{1}{2}e_3'(f) - f\underline{\omega} - \frac{1}{4}f^2\underline{\xi}.
\end{aligned}
$$

Proof. The proof follows from Proposition 2.90 by setting $\lambda = 1, \underline{f} = 0$. Since we need precise formulas for the error terms, we provide a proof in section A.9. □

Lemma 2.92 (Transport equations for $(f, \underline{f}, \lambda)$)**.** *Assume that we have in the new null frame (e_3', e_4', e_θ') of type* (2.3.3)

$$\xi' = 0, \quad \omega' = 0, \quad \zeta' + \underline{\eta}' = 0.$$

Then, $(\underline{f}, f, \log(\lambda))$ satisfy the following transport equations

$$
\begin{aligned}
\lambda^{-1}e_4'(f) + \left(\frac{\kappa}{2} + 2\omega\right)f &= -2\xi + E_1(f,\Gamma),\\
\lambda^{-1}e_4'(\log(\lambda)) &= 2\omega + E_2(f,\Gamma),\\
\lambda^{-1}e_4'(\underline{f}) + \frac{\kappa}{2}\underline{f} &= -2(\zeta + \underline{\eta}) + 2e_\theta'(\log(\lambda)) + 2f\underline{\omega} + E_3(f,\underline{f},\Gamma),
\end{aligned}
$$

where E_1, E_2 and E_3 are given by

$$
\begin{aligned}
E_1(f,\Gamma) &= -\frac{1}{2}\vartheta f + l.o.t.,\\
E_2(f,\Gamma) &= f\zeta - \frac{1}{2}f^2\underline{\omega} - \underline{\eta}f - \frac{1}{4}f^2\underline{\kappa} + l.o.t.,\\
E_3(f,\underline{f},\Gamma) &= -\underline{f}e_\theta'(f) - \frac{1}{2}\underline{f}\vartheta + l.o.t.
\end{aligned}
$$

Here, l.o.t. denote terms which are cubic or higher order in $f, \underline{f}$ (or in f only in the case of E_1 and E_2) and $\check{\Gamma}$ and do not contain derivatives of these quantities, where Γ and $\check{\Gamma}$ denotes the Ricci coefficients and renormalized Ricci coefficients w.r.t. the original null frame (e_3, e_4, e_θ).

Proof. See section A.7. □

To avoid a potential log loss for the third equation in Lemma 2.92, i.e., the transport equation for $\underline{f}$, we state the following renormalized version of the lemma.

Corollary 2.93. *Assume given a null frame (e_3, e_4, e_θ) associated to an outgoing geodesic foliation as in section 2.2.4, and let r denote the corresponding area radius. Assume that we have in the new null frame (e_3', e_4', e_θ') of type* (2.3.3)

$$\xi' = 0, \quad \omega' = 0, \quad \zeta' + \underline{\eta}' = 0.$$

Then, $(\underline{f}, f, \log(\lambda))$ satisfy the following transport equations

$$
\begin{aligned}
\lambda^{-1}e_4'(rf) &= E_1'(f,\Gamma),\\
\lambda^{-1}e_4'(\log(\lambda)) &= E_2'(f,\Gamma),\\
\lambda^{-1}e_4'\Big(r\underline{f} - 2r^2e_\theta'(\log(\lambda)) + rf\underline{\Omega}\Big) &= E_3'(f,\underline{f},\lambda,\Gamma),
\end{aligned}
$$

where

$$
\begin{aligned}
E_1'(f,\Gamma) &= -\frac{r}{2}\check{\kappa}f - \frac{r}{2}\vartheta f + l.o.t.,\\
E_2'(f,\Gamma) &= f\zeta - \frac{1}{2}f^2\underline{\omega} - \underline{\eta}f - \frac{1}{4}f^2\underline{\kappa} + l.o.t.,\\
E_3'(f,\underline{f},\lambda,\Gamma) &= -\frac{r}{2}\check{\kappa}\underline{f} + r^2\left(\check{\kappa} - \left(\overline{\kappa} - \frac{2}{r}\right)\right)e_\theta'(\log(\lambda))\\
&\quad +r^2\Big(\not{d}_1'(f) + \lambda^{-1}\vartheta'\Big)e_\theta'(\log(\lambda)) - \frac{r}{2}\check{\kappa}\underline{\Omega}f + rE_3(f,\underline{f},\Gamma)\\
&\quad -2r^2e_\theta'(E_2(f,\Gamma)) + r\underline{\Omega}E_1(f,\Gamma),
\end{aligned}
$$

and where E_1, E_2 and E_3 are given in Lemma 2.92.

Proof. See section A.8. □

2.3.2 Schematic notation Γ_g and Γ_b

Many of the identities which we present below contain a huge number of $O(\epsilon^2)$ terms. In what follows we introduce schematic notation meant to keep track of the most important error terms. Note that the decomposition below between the terms Γ_g and Γ_b is consistent with our main bootstrap assumptions **BA-E** on energy and **BA-D** on decay, see section 3.4.1.

Definition 2.94. *We divide the small connection coefficient terms (relative to an arbitrary null frame) into*[21]

$$
\Gamma_g^{(0)} = \left\{r\xi,\ \vartheta,\ \zeta,\ \underline{\eta},\ \frac{2}{r}e_4(r) - \kappa,\ \frac{1}{r}e_\theta(r)\right\}, \qquad \Gamma_b^{(0)} = \left\{\eta,\ \underline{\vartheta},\ \underline{\xi},\ \frac{2}{r}e_3(r) - \underline{\kappa}\right\}.
$$

For higher derivatives we introduce

$$
\Gamma_g^{(1)} = \left\{\mathfrak{d}\Gamma_g^{(0)},\ r^2e_\theta(\omega),\ re_\theta(\kappa),\ re_\theta(\underline{\kappa})\right\}, \qquad \Gamma_b^{(1)} = \left\{\mathfrak{d}\Gamma_b^{(0)},\ re_\theta(\underline{\omega})\right\},
$$

and for $s \geq 2$,

$$
\Gamma_g^{(s)} = \mathfrak{d}^{\leq s}\Gamma_g, \qquad \Gamma_b^{(s)} = \mathfrak{d}^{\leq s}\Gamma_b,
$$

where we have introduced the notations

$$
\mathfrak{d} = \{e_3, re_4, \not{\mathfrak{d}}\},
$$

[21] In the frames we are using, we have in fact $\xi = 0$ for $r \geq 4m_0$ so that it behaves in fact better than the other components of $\Gamma_g^{(0)}$.

with angular derivatives $\not{\mathfrak{d}}$ *of reduced scalars in* $\mathfrak{s}_k$ *defined by* (2.1.36).

Remark 2.95. *According to the main bootstrap assumptions* ***BA-E***, ***BA-D*** *(see section 3.4.1), the terms* Γ_b *behave worse in powers of* r *than the terms in* Γ_g. *Thus, in the calculations below, we replace the terms of the form* $\Gamma_g^{(s)} + \Gamma_b^{(s)}$ *by* $\Gamma_b^{(s)}$. *Given the form of the bootstrap assumptions, we may also replace* $r^{-1}\Gamma_b^{(s)}$ *by* $\Gamma_g^{(s)}$. *We will denote l.o.t. the cubic and higher error terms in* $\check{\Gamma}, \check{R}$. *We also include in l.o.t. terms which decay faster in powers of* r *than the main quadratic terms.*

2.3.3 The invariant quantity $\mathfrak{q}$

Note from the transformation formulas of Proposition 2.90 that the only quantities which remain invariant up to quadratic or higher order error terms are α, $\underline{\alpha}$ and ρ. Among these only $\alpha, \underline{\alpha}$ vanish in Schwarzschild. We call such quantities $O(\epsilon^2)$ invariant. In what follows we show that, in addition to these two invariants, there exist other important invariants.

Lemma 2.96. *The expression*

$$e_3(e_3(\alpha)) + (2\underline{\kappa} - 6\underline{\omega})e_3(\alpha) + \left(-4e_3(\underline{\omega}) + 8\underline{\omega}^2 - 8\underline{\omega}\,\underline{\kappa} + \frac{1}{2}\underline{\kappa}^2\right)\alpha$$

is an $O(\epsilon^2)$ *invariant. It is also a conformal invariant, i.e., invariant under transformations* (2.3.3) *with* $f = \underline{f} = 0$.

Proof. Clearly the quantity vanishes in Schwarzschild and is an $O(\epsilon^2)$ invariant. For a conformal transformation, the result follows by a straightforward application of the transformation properties of Proposition 2.90 in the particular case where $f = \underline{f} = 0$. □

Remark 2.97. *Alternatively one can also define the corresponding quantity obtained by interchanging* e_3, e_4, *i.e.,*

$$e_4(e_4(\underline{\alpha})) + (2\kappa - 6\omega)e_4(\underline{\alpha}) + \left(-4e_4(\omega) + 8\omega^2 - 8\omega\kappa + \frac{1}{2}\kappa^2\right)\underline{\alpha}.$$

Note that it differs by $O(\epsilon^2)$ *from the previous one.*

Definition 2.98. *Given a general null frame* (e_4, e_3, e_θ), *and given a scalar function* r *satisfying the assumptions for section 2.3.2, i.e.,*

$$\frac{2}{r}e_4(r) - \kappa \in \Gamma_g, \quad \frac{1}{r}e_\theta(r) \in \Gamma_g, \quad \frac{2}{r}e_3(r) - \underline{\kappa} \in \Gamma_b,$$

we define our main quantity $\mathfrak{q}$ *as*

$$\begin{aligned}\mathfrak{q} :=& r^4\Bigg[e_3(e_3(\alpha)) + (2\underline{\kappa} - 6\underline{\omega})e_3(\alpha)\\ &+ \left(-4e_3(\underline{\omega}) + 8\underline{\omega}^2 - 8\underline{\omega}\,\underline{\kappa} + \frac{1}{2}\underline{\kappa}^2\right)\alpha\Bigg].\end{aligned} \tag{2.3.10}$$

2.3.4 Several identities for $\mathfrak{q}$

In this section, we state three identities involving the quantity $\mathfrak{q}$ defined by (2.3.10). All calculations are made in a general frame.

Proposition 2.99. *We have*

$$\mathfrak{q} \;=\; r^4\left(\,\not{d}_2^\star\,\not{d}_1^\star\rho+\frac{3}{4}\underline{\kappa}\rho\vartheta+\frac{3}{4}\kappa\rho\underline{\vartheta}\right)+Err[\mathfrak{q}] \tag{2.3.11}$$

with error term written schematically in the form

$$Err[\mathfrak{q}] \;=\; r^4 e_3\eta\cdot\beta+r^2\mathfrak{d}^{\leq 1}(\Gamma_b\cdot\Gamma_g). \tag{2.3.12}$$

Proof. See section A.10 □

The following consequence of Proposition 2.99 will prove to be very useful in the sequel.

Proposition 2.100. *We have*

$$\begin{aligned} e_3(r\mathfrak{q}) \;=\;& r^5\left\{\not{d}_2^\star\,\not{d}_1^\star\,\not{d}_1\underline{\beta}-\frac{3}{2}\rho\,\not{d}_2^\star\,\not{d}_1^\star\underline{\kappa}-\frac{3}{2}\underline{\kappa}\rho\,\not{d}_2^\star\zeta-\frac{3}{2}\kappa\rho\underline{\alpha}+\frac{3}{4}(2\rho^2-\kappa\underline{\kappa}\rho)\underline{\vartheta}\right\}\\ &+Err[e_3(r\mathfrak{q})], \end{aligned} \tag{2.3.13}$$

where the error term $Err[e_3(r\mathfrak{q})]$ is given schematically by

$$Err[e_3(r\mathfrak{q})] \;=\; r\Gamma_b\mathfrak{q}+r^5\mathfrak{d}^{\leq 1}(e_3\eta\cdot\beta)+r^3\mathfrak{d}^{\leq 2}(\Gamma_b\cdot\Gamma_g). \tag{2.3.14}$$

Proof. See section A.11. □

We deduce from Proposition 2.100 the following nonlinear version of the Teukolsky-Starobinsky identity.

Proposition 2.101. *The following identity holds true in ${}^{(int)}\mathcal{M}$:*

$$e_3(r^2e_3(r\mathfrak{q}))+2\underline{\omega}r^2e_3(r\mathfrak{q})=r^7\left\{\not{d}_2^\star\,\not{d}_1^\star\,\not{d}_1\,\not{d}_2\underline{\alpha}+\frac{3}{2}\rho\Big(\underline{\kappa}e_4-\kappa e_3\Big)\underline{\alpha}\right\}+Err[TS], \tag{2.3.15}$$

where the error term $Err[TS]$ is given schematically by

$$\begin{aligned} Err[TS] \;=\;& r^4\big(\not{\mathfrak{d}}\Gamma_b+r\Gamma_b\cdot\Gamma_b\big)\cdot\underline{\alpha}+r^2\Big(\Gamma_b e_3(r\mathfrak{q})+(\mathfrak{d}^{\leq 1}\Gamma_b)r\mathfrak{q}\Big)\\ +\;& r^7\mathfrak{d}^{\leq 2}(e_3\eta\cdot\beta)+r^5\mathfrak{d}^{\leq 3}(\Gamma_b\cdot\Gamma_g). \end{aligned}$$

Proof. See section A.12. □

2.4 INVARIANT WAVE EQUATIONS

In this section, we write wave equations for the invariant quantities α, $\underline{\alpha}$ and $\mathfrak{q}$.

2.4.1 Preliminaries

Lemma 2.102. *With respect to a general S-foliation we have, for a reduced scalar* $\psi \in \mathfrak{s}_0$,

$$\Box_{\mathbf{g}}\psi = -\frac{1}{2}\left(e_3 e_4 + e_4 e_3\right)\psi + \triangle\!\!\!\!/\,\psi + \left(\underline{\omega} - \frac{1}{2}\underline{\kappa}\right)e_4\psi + \left(\omega - \frac{1}{2}\kappa\right)e_3\psi \qquad (2.4.1)$$
$$+ (\eta + \underline{\eta})e_\theta\psi.$$

Also,

$$\Box_{\mathbf{g}}\psi = -e_3 e_4\psi + \triangle\!\!\!\!/\,\psi + \left(2\underline{\omega} - \frac{1}{2}\underline{\kappa}\right)e_4\psi - \frac{1}{2}\kappa e_3\psi + 2\eta e_\theta\psi,$$

$$\Box_{\mathbf{g}}\psi = -e_4 e_3\psi + \triangle\!\!\!\!/\,\psi + \left(2\omega - \frac{1}{2}\kappa\right)e_3\psi - \frac{1}{2}\underline{\kappa} e_4\psi + 2\underline{\eta} e_\theta\psi.$$

Proof. We calculate, in spacetime,

$$\Box_{\mathbf{g}}\psi = \mathbf{g}^{34}\mathbf{D}_3\mathbf{D}_4\psi + \mathbf{g}^{43}\mathbf{D}_4\mathbf{D}_3\psi + \delta^{AB}\mathbf{D}_A\mathbf{D}_B\psi$$
$$= -\frac{1}{2}(\mathbf{D}_3\mathbf{D}_4 + \mathbf{D}_4\mathbf{D}_3)\psi + \mathbf{g}^{AB}\mathbf{D}_A\mathbf{D}_B\psi.$$

Now,

$$\delta^{AB}\mathbf{D}_A\mathbf{D}_B\psi = \triangle\!\!\!\!/\,\psi - \frac{1}{2}{}^{(1+3)}\mathrm{tr}\chi e_3\psi - \frac{1}{2}{}^{(1+3)}\mathrm{tr}\underline{\chi} e_4\psi,$$
$$\mathbf{D}_3\mathbf{D}_4\psi = e_3 e_4\psi - 2\underline{\omega} e_4\psi - 2\eta e_\theta\psi,$$
$$\mathbf{D}_4\mathbf{D}_3\psi = e_4 e_3\psi - 2\omega e_3\psi - 2\underline{\eta} e_\theta\psi.$$

Hence,

$$\Box_{\mathbf{g}}\psi = -\frac{1}{2}(e_3 e_4 + e_4 e_3)\psi + \triangle\!\!\!\!/\,\psi - \frac{1}{2}{}^{(1+3)}\mathrm{tr}\chi e_3\psi - \frac{1}{2}{}^{(1+3)}\mathrm{tr}\underline{\chi} e_4\psi$$
$$+ \underline{\omega} e_4\psi + \eta e_\theta\psi + \omega e_3\psi + \underline{\eta} e_\theta\psi$$
$$= -\frac{1}{2}(e_3 e_4 + e_4 e_3)\psi + \triangle\!\!\!\!/\,\psi + \left(\underline{\omega} - \frac{1}{2}{}^{(1+3)}\mathrm{tr}\underline{\chi}\right)e_4\psi$$
$$+ \left(\omega - \frac{1}{2}{}^{(1+3)}\mathrm{tr}\chi\right)e_3\psi + (\eta + \underline{\eta})e_\theta\psi.$$

Since

$$\frac{1}{2}e_4 e_3\psi = \frac{1}{2}e_3 e_4\psi + \omega e_3\psi - \underline{\omega} e_4\psi + (\underline{\eta} - \eta)e_\theta\psi$$

we also have

$$\Box_{\mathbf{g}}\psi = -e_3 e_4\psi + \triangle\!\!\!\!/\,\psi + \left(2\underline{\omega} - \frac{1}{2}{}^{(1+3)}\mathrm{tr}\underline{\chi}\right)e_4\psi - \frac{1}{2}{}^{(1+3)}\mathrm{tr}\chi e_3\psi + 2\eta e_\theta\psi.$$

Since $\kappa = {}^{(1+3)}\mathrm{tr}\chi$, $\underline{\kappa} = {}^{(1+3)}\mathrm{tr}\underline{\chi}$, this concludes the proof of the lemma. $\square$

Definition 2.103. *Given a reduced k-scalar $\psi \in \mathfrak{s}_k$ we define*

$$\begin{aligned}\square_k\psi = &-\frac{1}{2}\left(e_3e_4+e_4e_3\right)\psi + {\not\Delta}_k\psi + (\underline{\omega}-\frac{1}{2}tr\underline{\chi})e_4\psi + (\omega-\frac{1}{2}tr\chi)e_3\psi\\ &+(\eta+\underline{\eta})e_\theta\psi.\end{aligned} \tag{2.4.2}$$

Equivalently, we have

$$\begin{aligned}\square_k\psi &= -e_3e_4\psi + {\not\Delta}_k\psi + \left(2\underline{\omega}-\frac{1}{2}\underline{\kappa}\right)e_4\psi - \frac{1}{2}\kappa e_3\psi + 2\eta e_\theta\psi,\\ \square_k\psi &= -e_4e_3\psi + {\not\Delta}_k\psi + \left(2\omega-\frac{1}{2}\kappa\right)e_3\psi - \frac{1}{2}\underline{\kappa}e_4\psi + 2\underline{\eta}e_\theta\psi.\end{aligned}$$

Remark 2.104. *Not that the terms $\eta e_\theta\psi, \underline{\eta}e_\theta\psi$ have to be interpreted as in Remark 2.26, i.e.,*

$$\eta e_\theta\psi = \frac{1}{2}\eta\left({\not d}_k\psi - {\not d}^\star_{k+1}\psi\right).$$

The term $\eta\,{\not d}_k\psi$ is the reduced form of a tensor product of ${}^{(1+3)}\eta$ with ${\not D}_k\,{}^{(1+3)}\psi$ while $\eta\,{\not d}^\star_{k+1}\psi$ is the reduced form of a contraction between the 1-form ${}^{(1+3)}\eta$ and $k+1$ tensor ${\not D}^\star_{k+1}\,{}^{(1+3)}\psi$.

Remark 2.105. *Recall that (see Definition 2.23),*

$${\not\Delta}_k f := e_\theta(e_\theta f) + e_\theta(\Phi)e_\theta f - k^2\left(e_\theta(\Phi)\right)^2 f.$$

Thus, for a $\psi \in \mathfrak{s}_k$, we have,

$${\not\Delta}_k\psi = {\not\Delta}\psi - k^2\left(e_\theta(\Phi)\right)^2\psi.$$

2.4.1.1 Spacetime interpretation of Definition 2.103

The linearized equation verified by our main quantity $\mathfrak{q}$, which will be derived in the next section, has the form

$$\square_2\psi = V\psi \tag{2.4.3}$$

with V a scalar potential. In what follows we give simple spacetime interpretation of the equation (see Appendix D for more details).

Given a mixed spacetime tensor in $\mathbf{T}^k\mathbf{M}\otimes\mathbf{T}^l_S\mathbf{M}$ of the form $U_{\mu_1\ldots\mu_k,A_1\ldots A_L}$ where e_μ is an orthonormal frame on $\mathcal{M}$ with $(e_A)_{A=1,2}$ tangent to S. We define

$$\begin{aligned}\dot{\mathbf{D}}_\mu U_{\nu_1\ldots\nu_k,A_1\ldots A_L} &= e_\mu U_{\nu_1\ldots\nu_k,A_1\ldots A_l} - U_{\mathbf{D}_\mu\nu_1\ldots\nu_k,A_1\ldots A_l} - \ldots - U_{\nu_1\ldots\mathbf{D}_\mu\nu_k,A_1\ldots A_l}\\ &- U_{\nu_1\ldots\nu_k,\dot{\mathbf{D}}_\mu A_1\ldots A_l} - U_{\nu_1\ldots\nu_k,A_1\ldots\dot{\mathbf{D}}_\mu A_l}\end{aligned}$$

with $\dot{\mathbf{D}}_\mu A$ denoting the projection of $\mathbf{D}_{e_\mu}e_A$ on S. One can easily check the commutator formulae

$$\begin{aligned}(\dot{\mathbf{D}}_\mu\dot{\mathbf{D}}_\nu - \dot{\mathbf{D}}_\nu\dot{\mathbf{D}}_\mu)\Psi_A &= \mathbf{R}_A{}^B{}_{\mu\nu}\Psi_B,\\ (\dot{\mathbf{D}}_\mu\dot{\mathbf{D}}_\nu - \dot{\mathbf{D}}_\nu\dot{\mathbf{D}}_\mu)\Psi_{\lambda A} &= \mathbf{R}_\lambda{}^\sigma{}_{\mu\nu}\Psi_{\sigma A} + \mathbf{R}_A{}^B{}_{\mu\nu}\Psi_{\lambda B}.\end{aligned}$$

Define

$$\dot{\square}_{\mathbf{g}}\Psi := \mathbf{g}^{\mu\nu}\dot{\mathbf{D}}_\mu\dot{\mathbf{D}}_\nu\Psi.$$

Consider the following Lagrangian for $\Psi = \Psi_{AB} \in \mathcal{S}_2$:

$$\mathcal{L}[\Psi] \quad = \quad \not{g}^{A_1B_1}\not{g}^{A_2B_2}\Big(\mathbf{g}^{\mu\nu}\dot{\mathbf{D}}_\mu\Psi_{A_1A_2}\dot{\mathbf{D}}_\mu\Psi_{B_1B_2} + V\Psi_{A_1A_2}\Psi_{B_1B_2}\Big).$$

Proposition 2.106. *The Euler-Lagrange equations for the Lagrangian $\mathcal{L}[\Psi]$ above are given by*

$$\dot{\square}\Psi = V\Psi \tag{2.4.4}$$

and its reduced form $\psi = \Psi_{\theta\theta}$ is precisely (2.4.3).

Proof. Straightforward verification. □

2.4.2 Wave equations for α, $\underline{\alpha}$, and $\mathfrak{q}$

We start with the wave equations for α and $\underline{\alpha}$, which are derived in a general null frame.

Proposition 2.107. *The following identities hold true.*

1. *The invariant quantity $\alpha \in \mathfrak{s}_2$ verifies the Teukolsky wave equation,*

$$\begin{aligned}\square_2\alpha &= -4\underline{\omega}e_4(\alpha) + (4\omega + 2\kappa)e_3(\alpha) + V\alpha + Err[\square_{\mathbf{g}}\alpha],\\ V &= -4\rho - 4e_4(\underline{\omega}) - 8\omega\underline{\omega} + 2\omega\,\underline{\kappa} - 10\kappa\,\underline{\omega} + \frac{1}{2}\kappa\,\underline{\kappa},\end{aligned} \tag{2.4.5}$$

where $Err[\square_{\mathbf{g}}\alpha]$ is given schematically by

$$Err(\square_{\mathbf{g}}\alpha) \quad = \quad \Gamma_g e_3(\alpha) + r^{-1}\mathfrak{d}^{\leq 1}\Big((\eta, \Gamma_g)(\alpha, \beta)\Big) + \xi(e_3(\beta), r^{-1}\mathfrak{d}\check{\rho}) + l.o.t.$$

where l.o.t. denote terms which are quadratic and enjoy better decay properties or are higher order and decay at least as good.

2. *The invariant quantity $\underline{\alpha} \in \mathfrak{s}_2$ verifies the Teukolsky wave equation,*

$$\begin{aligned}\square_2\underline{\alpha} &= -4\omega e_3(\underline{\alpha}) + (4\underline{\omega} + 2\underline{\kappa})e_4(\underline{\alpha}) + \underline{V}\,\underline{\alpha} + Err[\square_{\mathbf{g}}\underline{\alpha}],\\ \underline{V} &= -4\rho - 4e_3(\omega) - 8\omega\underline{\omega} + 2\underline{\omega}\kappa - 10\underline{\kappa}\,\omega + \frac{1}{2}\kappa\,\underline{\kappa},\end{aligned} \tag{2.4.6}$$

where

$$Err(\square_{\mathbf{g}}\underline{\alpha}) \quad = \quad r^{-1}\mathfrak{d}(\Gamma_b\underline{\alpha}) + \mathfrak{d}(\Gamma_b\underline{\beta}) + l.o.t.$$

Proof. See section A.13. □

We may now state the wave equation satisfied by $\mathfrak{q}$.

Theorem 2.108. *The invariant scalar quantity $\mathfrak{q}$ defined in* (2.3.10) *verifies the* *equation*

$$\square_2 \mathfrak{q} + \kappa \underline{\kappa}\, \mathfrak{q} = Err[\square_2 \mathfrak{q}] \tag{2.4.7}$$

where $Err[\square_2 \mathfrak{q}]$ *is* $O(\epsilon^2)$.

If $\mathfrak{q}$ *is defined relative to a null frame satisfying, in addition to the assumptions of section 2.3.2, that* $\eta \in \Gamma_g$ *and* $\xi = 0$ *for* $r \geq 4m_0$, *the error term is then given schematically by*

$$\begin{aligned} Err[\square_2 \mathfrak{q}] &= r^2 \mathfrak{d}^{\leq 2}(\Gamma_g \cdot (\alpha, \beta)) + e_3\Big(r^3 \mathfrak{d}^{\leq 2}(\Gamma_g \cdot (\alpha, \beta))\Big) \\ &\quad + \mathfrak{d}^{\leq 1}(\Gamma_g \cdot \mathfrak{q}) + l.o.t. \end{aligned} \tag{2.4.8}$$

Proof. See section A.14. □

Remark 2.109. *Note that the main frame used in this book is an outgoing geodesic null frame in* $r \geq 4m_0$ *so that* $\xi = 0$, *but unfortunately, as it turns out,* $\eta \in \Gamma_b$. *This would not allow us to control the error term appearing in* (2.4.7). *To overcome this problem, we are forced to define* $\mathfrak{q}$ *relative to a different frame where* $\xi = 0$ *still holds for* $r \geq 4m_0$ *and for which we have in addition* $\eta \in \Gamma_g$, *see Proposition 3.26 for the existence of such a frame. See also the discussion at the beginning of section 3.4.6.*

The remark above leads us to the following.

Remark 2.110. *The quantity* $\mathfrak{q}$ *we will be working with for the rest of the book is defined, according to equation* (2.3.10), *relative to the global frame of Proposition 3.26 for which* $\eta \in \Gamma_g$. *It is only in such a frame that* $\mathfrak{q}$ *verifies the correct decay estimates.*

Chapter Three

Main Theorem

3.1 GENERAL COVARIANT MODULATED ADMISSIBLE SPACETIMES

Note that all definitions below are consistent with the framework of **Z**-invariant polarized spacetimes.

3.1.1 Initial data layer

Recall that $m_0 > 0$ is given as the mass of the Schwarzschild solution to which the initial data is ϵ_0 close. Let $\delta_{\mathcal{H}} > 0$ be a sufficiently small constant which will be specified later.

Definition 3.1 (Initial data layer). *We consider a spacetime region $(\mathcal{L}_0, g)$, sketched below in figure 3.1, where*

- *The metric g is a reduced metric from a Lorentzian spacetime metric $\mathbf{g}$ close to Schwarzschild in a suitable topology.*[1]
- $\mathcal{L}_0 = {}^{(ext)}\mathcal{L}_0 \cup {}^{(int)}\mathcal{L}_0$.
- *The intersection ${}^{(ext)}\mathcal{L}_0 \cap {}^{(int)}\mathcal{L}_0$ is nontrivial.*

Furthermore, our initial data layer $(\mathcal{L}_0, g)$ satisfies the following:

1. ***Boundaries.*** *The future and past boundaries of $\mathcal{L}_0$ are given by*

$$\begin{aligned} \partial^+\mathcal{L}_0 &= \mathcal{A}_0 \cup \underline{\mathcal{C}}_{(2,\mathcal{L}_0)} \cup \mathcal{C}_{(2,\mathcal{L}_0)}, \\ \partial^-\mathcal{L}_0 &= \mathcal{C}_{(0,\mathcal{L}_0)} \cup \underline{\mathcal{C}}_{(0,\mathcal{L}_0)}, \end{aligned}$$

   where

   a) The past outgoing null boundary of the far region ${}^{(ext)}\mathcal{L}_0$ is denoted by $\mathcal{C}_{(0,\mathcal{L}_0)}$.
   b) The past incoming null boundary of the near region ${}^{(int)}\mathcal{L}_0$ is denoted by $\underline{\mathcal{C}}_{(0,\mathcal{L}_0)}$.
   c) ${}^{(ext)}\mathcal{L}_0$ is unbounded in the future outgoing null directions.
   d) The future outgoing null boundary of the far region ${}^{(ext)}\mathcal{L}_0$ is denoted by $\mathcal{C}_{(2,\mathcal{L}_0)}$.
   e) The future incoming null boundary of the near region ${}^{(int)}\mathcal{L}_0$ is denoted by $\underline{\mathcal{C}}_{(2,\mathcal{L}_0)}$.
   f) The future spacelike boundary of the near region ${}^{(int)}\mathcal{L}_0$ is denoted by $\mathcal{A}_0$.

2. ***Foliations of $\mathcal{L}_0$ and adapted null frames.*** *The spacetime $\mathcal{L}_0 = {}^{(ext)}\mathcal{L}_0 \cup$*

[1] This topology will be specified in our initial data layer assumptions, see (3.3.5) as well as section 3.2.4.

${}^{(int)}\mathcal{L}_0$ *is foliated as follows:*

a) The far region ${}^{(ext)}\mathcal{L}_0$ *is foliated by two functions* $(u_{\mathcal{L}_0}, {}^{(ext)}s_{\mathcal{L}_0})$ *such that*

- $u_{\mathcal{L}_0}$ *is an outgoing optical function on* ${}^{(ext)}\mathcal{L}_0$ *whose leaves are denoted by* $\mathcal{C}_{(u_{\mathcal{L}_0}, \mathcal{L}_0)}$.
- ${}^{(ext)}s_{\mathcal{L}_0}$ *is an affine parameter along the level hypersurfaces of* $u_{\mathcal{L}_0}$, *i.e.,*

$$ {}^{(ext)}L_0({}^{(ext)}s_{\mathcal{L}_0}) = 1 \text{ where } {}^{(ext)}L_0 := -g^{ab}\partial_b(u_{\mathcal{L}_0})\partial_a. $$

- *We denote by* $({}^{(ext)}(e_0)_3, {}^{(ext)}(e_0)_4, {}^{(ext)}(e_0)_\theta)$ *the null frame adapted to the outgoing geodesic foliation* $(u_{\mathcal{L}_0}, {}^{(ext)}s_{\mathcal{L}_0})$ *on* ${}^{(ext)}\mathcal{L}_0$.
- *Let* ${}^{(ext)}r_{\mathcal{L}_0}$ *denote the area radius of the 2-spheres* $S(u_{\mathcal{L}_0}, {}^{(ext)}s_{\mathcal{L}_0})$ *of this foliation.*
- *The outgoing future null boundary* $\mathcal{C}_{(2,\mathcal{L}_0)}$ *corresponds precisely to* $u_{\mathcal{L}_0} = 2$ *and the outgoing past null boundary* $\mathcal{C}_{(0,\mathcal{L}_0)}$ *corresponds to* $u_{\mathcal{L}_0} = 0$.
- *The foliation by* $u_{\mathcal{L}_0}$ *of* ${}^{(ext)}\mathcal{L}_0$ *terminates at the timelike boundary*

$$ \left\{ {}^{(ext)}r_{\mathcal{L}_0} = 2m_0\left(1 + \frac{\delta_{\mathcal{H}}}{4}\right)\right\}. $$

b) The near region ${}^{(int)}\mathcal{L}_0$ *is foliated by two functions* $(\underline{u}_{\mathcal{L}_0}, {}^{(int)}s_{\mathcal{L}_0})$ *such that*

- $\underline{u}_{\mathcal{L}_0}$ *is an ingoing optical function on* ${}^{(int)}\mathcal{L}_0$ *whose leaves are denoted by* $\underline{\mathcal{C}}_{(\underline{u}_{\mathcal{L}_0}, \mathcal{L}_0)}$.
- ${}^{(int)}s_{\mathcal{L}_0}$ *is an affine parameter along the level hypersurfaces of* $\underline{u}_{\mathcal{L}_0}$, *i.e.,*

$$ {}^{(int)}\underline{L}_0({}^{(int)}s_{\mathcal{L}_0}) = -1 \text{ where } {}^{(ext)}\underline{L}_0 := -g^{ab}\partial_b(\underline{u}_{\mathcal{L}_0})\partial_a. $$

- *We denote by* $({}^{(int)}(e_0)_3, {}^{(int)}(e_0)_4, {}^{(int)}(e_0)_\theta)$ *the null frame adapted to the outgoing geodesic foliation* $(u_{\mathcal{L}_0}, {}^{(int)}s_{\mathcal{L}_0})$ *on* ${}^{(int)}\mathcal{L}_0$.
- *Let* ${}^{(int)}r_{\mathcal{L}_0}$ *denote the area radius of the 2-spheres* $S(\underline{u}_{\mathcal{L}_0}, {}^{(int)}s_{\mathcal{L}_0})$ *of this foliation.*
- *The* $(\underline{u}_{\mathcal{L}_0}, {}^{(int)}s)$ *foliation is initialized on* ${}^{(ext)}r_{\mathcal{L}_0} = 2m_0(1+\frac{\delta_{\mathcal{H}}}{2})$ *as it will be made precise below.*
- *The foliation by* $\underline{u}_{\mathcal{L}_0}$ *of* ${}^{(int)}\mathcal{L}_0$ *terminates at the spacelike boundary*

$$ \mathcal{A}_0 = \left\{ {}^{(int)}r_{\mathcal{L}_0} = 2m_0(1 - 2\delta_{\mathcal{H}})\right\} $$

where m_0 *and* $\delta_{\mathcal{H}}$ *have been defined above.*
- *The ingoing future null boundary* $\underline{\mathcal{C}}_{(2,\mathcal{L}_0)}$ *corresponds precisely to* $\underline{u}_{\mathcal{L}_0} = 2$ *and the ingoing past null boundary* $\underline{\mathcal{C}}_{(0,\mathcal{L}_0)}$ *corresponds to* $\underline{u}_{\mathcal{L}_0} = 0$.
- *The foliation by* $\underline{u}_{\mathcal{L}_0}$ *of* ${}^{(int)}\mathcal{L}_0$ *terminates at the timelike boundary*

$$ \left\{ {}^{(int)}r_{\mathcal{L}_0} = 2m_0\,(1 + 2\delta_{\mathcal{H}})\right\}. $$

3. ***Initializations of the*** $(\underline{u}_{\mathcal{L}_0}, {}^{(int)}s_{\mathcal{L}_0})$ ***foliation.*** *The* $(\underline{u}_{\mathcal{L}_0}, {}^{(int)}s_{\mathcal{L}_0})$ *foliation is initialized on* ${}^{(ext)}r_{\mathcal{L}_0} = 2m_0(1 + \frac{\delta_{\mathcal{H}}}{2})$ *by setting*

$$ \underline{u}_{\mathcal{L}_0} = u_{\mathcal{L}_0}, \qquad {}^{(int)}s_{\mathcal{L}_0} = {}^{(ext)}s_{\mathcal{L}_0} $$

and, with $\lambda_0 = {}^{(ext)}\lambda_0 = 1 - \frac{2m_0}{{}^{(ext)}r_{\mathcal{L}_0}}$,

$$ {}^{(int)}(e_0)_4 = \lambda_0 \, {}^{(ext)}(e_0)_4, \quad {}^{(int)}(e_0)_3 = \lambda_0^{-1} \, {}^{(ext)}(e_0)_3, \quad {}^{(int)}(e_0)_\theta = {}^{(ext)}(e_0)_\theta. $$

4. ***Coordinates system on*** ${}^{(ext)}\mathcal{L}_0({}^{(ext)}r_{\mathcal{L}_0} \geq 4m_0)$. *In* ${}^{(ext)}\mathcal{L}_0({}^{(ext)}r_{\mathcal{L}_0} \geq 4m_0)$, *there exists adapted coordinates* $(u_{\mathcal{L}_0}, {}^{(ext)}s_{\mathcal{L}_0}, \theta_{\mathcal{L}_0}, \varphi)$ *with* $b = 0$, *see Proposition 2.75, such that the spacetime metric* $\mathbf{g}$ *takes the form*

$$ \mathbf{g} = -2du_{\mathcal{L}_0} d\left({}^{(ext)}s_{\mathcal{L}_0}\right) + \underline{\Omega}_{\mathcal{L}_0}(du_{\mathcal{L}_0})^2 + \gamma_{\mathcal{L}_0}\left(d\theta_{\mathcal{L}_0} - \frac{1}{2}\underline{b}_{\mathcal{L}_0} du_{\mathcal{L}_0}\right)^2 + e^{2\Phi}d\varphi^2. \tag{3.1.1} $$

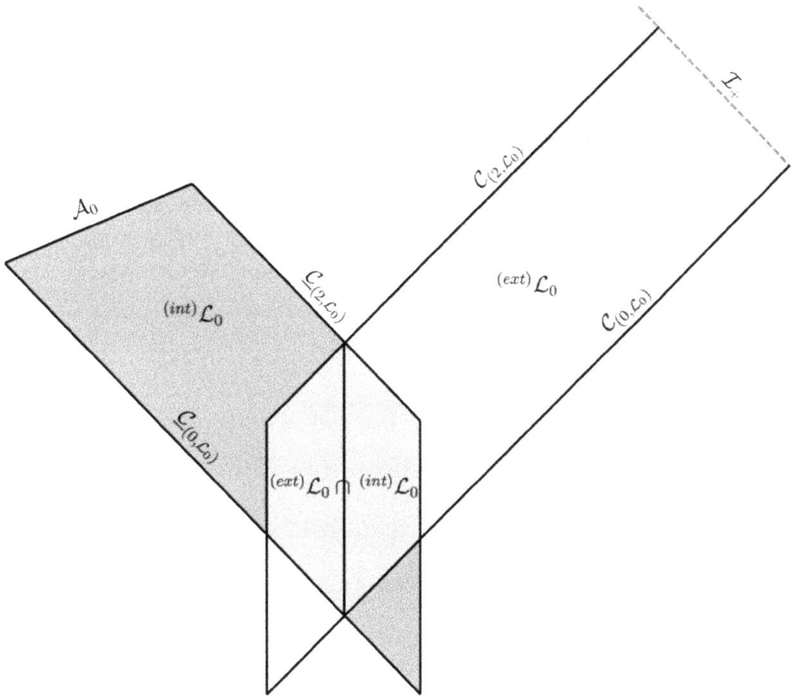

Figure 3.1: The initial data layer $\mathcal{L}_0$

3.1.2 Main definition

Recall that $m_0 > 0$ is given as the mass of the Schwarzschild solution to which the initial data is ϵ_0 close, and that $\delta_{\mathcal{H}} > 0$ is a sufficiently small constant which will be specified later.

Definition 3.2 (GCM admissible spacetime). *We consider a spacetime* $(\mathcal{M}, g)$, *sketched in figure 3.2, where*

- *The metric* g *is a reduced metric from a Lorentzian spacetime metric* $\mathbf{g}$ *close to*

Schwarzschild in a suitable topology.[2]
- $\mathcal{M} = {}^{(ext)}\mathcal{M} \cup {}^{(int)}\mathcal{M}$.
- $\mathcal{T} = {}^{(ext)}\mathcal{M} \cap {}^{(int)}\mathcal{M}$ *is a timelike hypersurface.*

$(\mathcal{M}, g)$ *is called a general covariant modulated admissible (or shortly GCM admissible) spacetime if it is defined as follows.*

1. ***Boundaries.*** *The future and past boundaries of* $\mathcal{M}$ *are given by*

$$\begin{aligned} \partial^+\mathcal{M} &= \mathcal{A} \cup \underline{\mathcal{C}}_* \cup \mathcal{C}_* \cup \Sigma_*, \\ \partial^-\mathcal{M} &= \mathcal{C}_1 \cup \underline{\mathcal{C}}_1, \end{aligned}$$

where

a) *The past boundary* $\mathcal{C}_1 \cup \underline{\mathcal{C}}_1$ *is included in the initial data layer* $\mathcal{L}_0$*, defined in section 3.1.1, in which the metric on* $\mathcal{M}$ *is specified to be a small perturbation of the Schwarzschild data.*
b) *The future spacelike boundary of the far region* ${}^{(ext)}\mathcal{M}$ *is denoted by* Σ_*.
c) *The future outgoing null boundary of the far region* ${}^{(ext)}\mathcal{M}$ *is denoted by* $\mathcal{C}_*$.
d) *The future incoming null boundary of the near region* ${}^{(int)}\mathcal{M}$ *is denoted by* $\underline{\mathcal{C}}_*$.
e) *The future spacelike boundary of the near region* ${}^{(int)}\mathcal{M}$ *is denoted by* $\mathcal{A}$.
f) *The timelike boundary* $\mathcal{T}$*, separating* ${}^{(ext)}\mathcal{M}$ *from* ${}^{(int)}\mathcal{M}$*, starts at* $\underline{\mathcal{C}}_1 \cap \mathcal{C}_1$ *and terminates at* $\underline{\mathcal{C}}_* \cap \mathcal{C}_*$.

2. ***Foliations of*** $\mathcal{M}$ ***and adapted null frames.*** *The spacetime* $\mathcal{M} = {}^{(ext)}\mathcal{M} \cup {}^{(int)}\mathcal{M}$ *is foliated as follows:*

a) *The far region* ${}^{(ext)}\mathcal{M}$ *is foliated by two functions* $(u, {}^{(ext)}s)$ *such that*
- u *is an outgoing optical function on* ${}^{(ext)}\mathcal{M}$*, initialized on* Σ_**, whose leaves are denoted by* $\mathcal{C}(u)$.
- ${}^{(ext)}s$ *is an affine parameter along the level hypersurfaces of* u*, i.e.,*

$$L({}^{(ext)}s) = 1 \text{ where } L := -g^{ab}\partial_b u \partial_a.$$

- *The* $(u, {}^{(ext)}s)$ *foliation is initialized on* Σ_* *as it will be made precise below.*
- *We denote by* $({}^{(ext)}e_3, {}^{(ext)}e_4, {}^{(ext)}e_\theta)$ *the null frame adapted to the outgoing geodesic foliation* $(u, {}^{(ext)}s)$ *on* ${}^{(ext)}\mathcal{M}$ *where* ${}^{(ext)}e_4 = L$.
- *Let* ${}^{(ext)}r$ *and* ${}^{(ext)}m$ *respectively the area radius and the Hawking mass of the 2-spheres* $S(u, {}^{(ext)}s)$ *of this foliation.*
- *The outgoing future null boundary* $\mathcal{C}_*$ *corresponds precisely to* $u = u_*$ *and the outgoing past null boundary* $\mathcal{C}_1$ *corresponds to* $u = 1$.
- *The foliation by* u *of* ${}^{(ext)}\mathcal{M}$ *terminates at the timelike boundary*

$$\mathcal{T} = \left\{ {}^{(ext)}r = r_{\mathcal{T}} \right\}$$

[2]This topology will be specified in our bootstrap assumptions, see (3.3.6) as well as section 3.2.

where $r_{\mathcal{T}}$ satisfies[3]

$$2m_0\left(1+\frac{\delta_{\mathcal{H}}}{2}\right) \leq r_{\mathcal{T}} \leq 2m_0\left(1+\frac{3\delta_{\mathcal{H}}}{2}\right).$$

b) The near region ${}^{(int)}\mathcal{M}$ is foliated by two functions $(\underline{u}, {}^{(int)}s)$ such that

- *$\underline{u}$ is an ingoing optical function on ${}^{(int)}\mathcal{M}$, initialized on $\mathcal{T}$, whose leaves are denoted by $\underline{\mathcal{C}}(\underline{u})$.*
- *${}^{(int)}s$ is an affine parameter along the level hypersurfaces of $\underline{u}$, i.e.,*

$$\underline{L}({}^{(int)}s) = -1 \text{ where } \underline{L} := -g^{ab}\partial_b\underline{u}\partial_a.$$

- *The $(\underline{u}, {}^{(int)}s)$ foliation is initialized on $\mathcal{T}$ as it will be made precise below.*
- *We denote by $({}^{(int)}e_3, {}^{(int)}e_4, {}^{(int)}e_\theta)$ the null frame adapted to the outgoing geodesic foliation $(u, {}^{(int)}s)$ on ${}^{(int)}\mathcal{M}$ where ${}^{(int)}e_3 = \underline{L}$.*
- *Let ${}^{(int)}r$ and ${}^{(int)}m$ respectively the area radius and the Hawking mass of the 2-spheres $S(\underline{u}, {}^{(int)}s)$ of this foliation.*
- *The foliation by $\underline{u}$ of ${}^{(int)}\mathcal{M}$ terminates at the spacelike boundary*

$$\mathcal{A} = \left\{ {}^{(int)}r = 2m_0(1-\delta_{\mathcal{H}})\right\}$$

  where m_0 and $\delta_{\mathcal{H}}$ have been defined above.
- *The ingoing future null boundary $\underline{\mathcal{C}}_*$ corresponds precisely to $\underline{u} = u_*$ and the ingoing past null boundary $\underline{\mathcal{C}}_1$ corresponds to $\underline{u} = 1$.*

3. ***GCM foliation of Σ_*.*** *The $(u, {}^{(ext)}s)$ foliation of ${}^{(ext)}\mathcal{M}$ restricted to the spacelike hypersurface Σ_* has the following properties:*

a) There exists a constant c_{Σ_*} such that

$$\Sigma_* := \{u + {}^{(ext)}r = c_{\Sigma_*}\}.$$

b) We have[4]

$$r \gg u_* \text{ on } \Sigma_*. \tag{3.1.2}$$

c) ${}^{(ext)}s$ satisfies[5]

$${}^{(ext)}s = {}^{(ext)}r \quad \text{on } \Sigma_*.$$

d) We say that Σ_* is a general covariant modulated hypersurface[6] *(or shortly GCM hypersurface) if relative to the above defined null frame of ${}^{(ext)}\mathcal{M}$, the*

[3] A specific choice of $r_{\mathcal{T}}$ will be made in section 3.8.9, see (3.8.8), in the context of a Lebesgue point argument needed to recover the top order derivatives.

[4] See (3.3.4) for the precise condition.

[5] Recall that ${}^{(ext)}s$ satisfies on ${}^{(ext)}\mathcal{M}$ the transport equation $L({}^{(ext)}s) = 1$ and thus needs to be initialized on a hypersurface transversal to L, chosen here to be Σ_*.

[6] More generally, a GCM hypersurface is one with the property that we can specify, using the full covariance of the Einstein equations, a number of vanishing conditions (equal to the number of degrees of freedom of the diffeomorphism group) for well-chosen components of $\check{\Gamma}$.

following conditions hold[7] *along* Σ_*:

$$\kappa = \frac{2}{r}, \quad \not{d}_2^\star \not{d}_1^\star \underline{\kappa} = 0, \quad \not{d}_2^\star \not{d}_1^\star \mu = 0,$$
$$\int_S \eta e^\Phi = 0, \quad \int_S \underline{\xi} e^\Phi = 0, \quad a\big|_{SP} = -1 - \frac{2m}{r}, \tag{3.1.3}$$

where a *is the unique scalar function such that* $\nu = e_3 + a e_4$ *is tangent to* Σ_*, *and* SP *denotes the south poles of the spheres on* Σ_*. *Moreover we also assume*

$$\int_{S_*} \beta e^\Phi = 0, \quad \int_{S_*} e_\theta(\underline{\kappa}) e^\Phi = 0, \quad \text{with } S_* := \Sigma_* \cap \mathcal{C}_*. \tag{3.1.4}$$

Note that the role of the GCM foliation of Σ_* *is to initialize the* $(u, {}^{(ext)}s)$ *foliation of* ${}^{(ext)}\mathcal{M}$.

e) *In view of the definition of* ν *and* ς, *we have* $\nu(u) = e_3(u) + a e_4(u) = 2/\varsigma$. ν *being tangent to* Σ_*, u *is thus transported along* Σ_*, *and hence defined up to a constant. To calibrate* u *on* Σ_*, *we fix the value* $u = 1$ *as follows:*

$$S_1 = \Sigma_* \cap \{u = 1\} \text{ is such that } S_1 \cap \mathcal{C}_{(1,\mathcal{L}_0)} \cap SP \neq \emptyset, \tag{3.1.5}$$

i.e., S_1 *is the unique sphere of* Σ_* *such that its south pole intersects the south pole of one of the spheres of the outgoing null cone* $\mathcal{C}_{(1,\mathcal{L}_0)}$ *of the initial data layer.*

4. ***Initialization the*** $(\underline{u}, {}^{(int)}s)$ ***foliation on*** $\mathcal{T}$. *The* $(\underline{u}, {}^{(int)}s)$ *foliation is initialized on* $\mathcal{T}$ *such that*

$$\underline{u} = u, \qquad {}^{(int)}s = {}^{(ext)}s.$$

In particular, the 2-spheres $S(\underline{u}, {}^{(int)}s)$ *coincide on* $\mathcal{T}$ *with* $S(u, {}^{(ext)}s)$ *and* ${}^{(int)}r = {}^{(ext)}r$. *Moreover, the null frame* $({}^{(int)}e_3, {}^{(int)}e_4, {}^{(int)}e_\theta)$ *is defined on* $\mathcal{T}$ *by the following renormalization,*

$${}^{(int)}e_4 = \lambda\, {}^{(ext)}e_4, \qquad {}^{(int)}e_3 = \lambda^{-1}\, {}^{(ext)}e_3, \qquad {}^{(int)}e_\theta = {}^{(ext)}e_\theta \quad \text{on } \mathcal{T}$$

where

$$\lambda = {}^{(ext)}\lambda = 1 - \frac{2\, {}^{(ext)}m}{{}^{(ext)}r}.$$

Remark 3.3. *In Schwarzschild,* $u = t - r_*$, $\underline{u} = t + r_*$, *with* $\frac{dr_*}{dr} = \Upsilon^{-1}$, *and*

$$\begin{aligned} {}^{(ext)}e_4 &= \Upsilon^{-1}\partial_t + \partial_r, & {}^{(ext)}e_3 &= \partial_t - \Upsilon\partial_r, \\ {}^{(int)}e_4 &= \partial_t + \Upsilon\partial_r, & {}^{(int)}e_3 &= \Upsilon^{-1}\partial_t - \partial_r. \end{aligned}$$

[7] The existence of such hypersurfaces is an essential part of our construction.

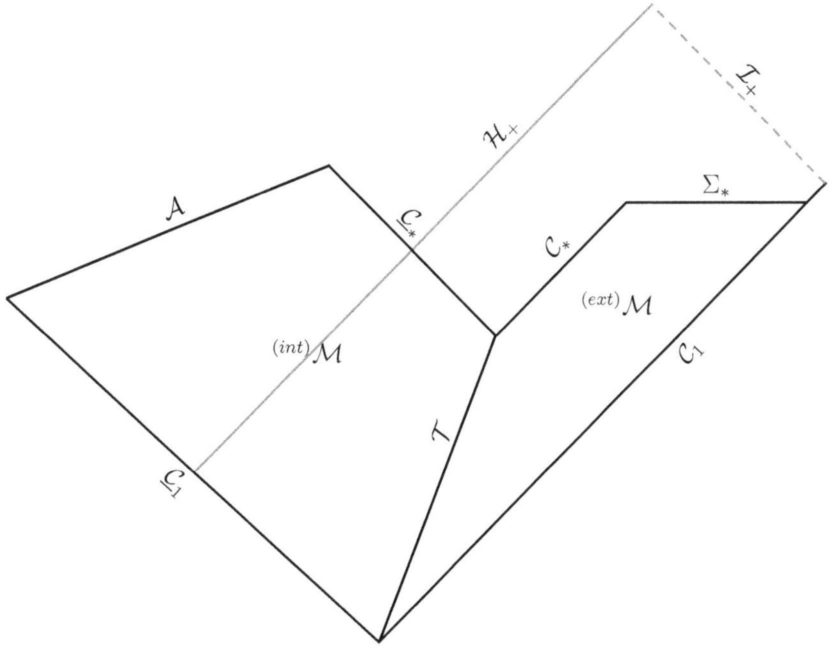

Figure 3.2: The GCM admissible spacetime $\mathcal{M}$

3.1.3 Renormalized curvature components and Ricci coefficients

For convenience, we introduce in this section a notation for renormalized curvature components and Ricci coefficients.

Definition 3.4 (Renormalized curvature and Ricci coefficients in ${}^{(ext)}\mathcal{M}$). *We introduce the following notations in* ${}^{(ext)}\mathcal{M}$

$$
{}^{(ext)}\check{R} = \Big\{\alpha,\, \beta,\, \check{\rho},\, \check{\mu},\, \underline{\beta},\, \underline{\alpha}\Big\}, \qquad {}^{(ext)}\check{\Gamma} = \Big\{\check{\kappa},\, \vartheta,\, \zeta,\, \eta,\, \check{\underline{\kappa}},\, \underline{\vartheta},\, \check{\underline{\omega}},\, \underline{\xi}\Big\},
$$

where, recall,

$$
\check{\rho} = \rho - \overline{\rho}, \quad \check{\mu} = \mu - \overline{\mu}, \quad \check{\kappa} = \kappa - \overline{\kappa}, \quad \check{\underline{\kappa}} = \underline{\kappa} - \overline{\underline{\kappa}}, \quad \check{\underline{\omega}} = \underline{\omega} - \overline{\underline{\omega}},
$$

and

$$
\xi = \omega = 0, \quad \underline{\eta} = -\zeta.
$$

Note that all the above quantities are defined with respect to the outgoing geodesic foliation of ${}^{(ext)}\mathcal{M}$ *(see section 2.2.4), and that the averages are taken with respect to the corresponding 2-spheres.*

Definition 3.5 (Renormalized curvature and Ricci coefficients in ${}^{(int)}\mathcal{M}$). *We introduce the following notations in* ${}^{(int)}\mathcal{M}$

$$
\begin{aligned}
{}^{(int)}\check{R} &= \Big\{\alpha,\, \beta,\, \check{\rho},\, \check{\underline{\mu}},\, \underline{\beta},\, \underline{\alpha}\Big\}, \\
{}^{(int)}\check{\Gamma} &= \Big\{\xi,\, \check{\omega},\, \check{\kappa},\, \vartheta,\, \zeta,\, \underline{\eta},\, \check{\underline{\kappa}},\, \underline{\vartheta}\Big\},
\end{aligned}
$$

where we have defined

$$\check{\rho}=\rho-\overline{\rho},\quad \check{\underline{\mu}}=\underline{\mu}-\overline{\underline{\mu}},\quad \check{\kappa}=\kappa-\overline{\kappa},\quad \check{\underline{\kappa}}=\underline{\kappa}-\overline{\underline{\kappa}},\quad \check{\omega}=\omega-\overline{\omega},$$

and we recall that

$$\underline{\xi}=\underline{\omega}=0,\quad \eta=\zeta,\quad \overline{\underline{\mu}}-\frac{2m}{r^3}=0.$$

Note that all the above quantities are defined with respect to the ingoing geodesic foliation of ${}^{(int)}\mathcal{M}$ *(see section 2.2.6), and that the averages are taken with respect to the corresponding 2-spheres.*

Remark 3.6. *In Schwarzschild, we have*

$${}^{(ext)}\check{R}=0,\quad {}^{(int)}\check{R}=0,\quad {}^{(ext)}\check{\Gamma}=0,\quad {}^{(int)}\check{\Gamma}=0.$$

3.2 MAIN NORMS

3.2.1 Main norms in ${}^{(ext)}\mathcal{M}$

All quantities appearing in this section are defined relative to the ${}^{(ext)}\mathcal{M}$ frame adapted to the $(u, {}^{(ext)}s)$ foliation. In particular, recall that with respect to this frame, we have

$$\xi=\omega=0,\qquad \underline{\eta}=-\zeta.$$

Recall the definition (2.1.36) of higher order angular derivatives $\not{\mathfrak{d}}^s$ of reduced scalars in $\mathfrak{s}_k$. We introduce the notations

$$\mathfrak{d}=\{e_3, re_4, \not{\mathfrak{d}}\}.$$

Definition 3.7. *We introduce the vectorfield* $\mathbf{T}$ *defined on* ${}^{(ext)}\mathcal{M}$ *as*

$$\mathbf{T}:=\frac{1}{2}\left(\left(1-\frac{2m}{r}\right)e_4+e_3\right). \tag{3.2.1}$$

We also introduce the vectorfield $\mathbf{N}$ *defined on* ${}^{(ext)}\mathcal{M}$ *by*

$$\mathbf{N}:=\frac{1}{2}\left(\left(1-\frac{2m}{r}\right)e_4-e_3\right). \tag{3.2.2}$$

Remark 3.8. *In Schwarzschild, we have*

$$\mathbf{T}=\partial_t,\quad \mathbf{N}=\left(1-\frac{2m_0}{r}\right)\partial_r$$

in the standard (t,r,θ,φ) *coordinates.*

We are ready to introduce our norms in ${}^{(ext)}\mathcal{M}$.

3.2.1.1 L^2 curvature norms in ${}^{(ext)}\mathcal{M}$

Let $\delta_B > 0$ a small constant to be specified later. We introduce the weighted curvature norms,

$$\begin{aligned}
\left({}^{(ext)}\mathfrak{R}_0^{\geq 4m_0}[\check{R}]\right)^2 &:= \sup_{1\leq u\leq u_*}\int_{\mathcal{C}_u(r\geq 4m_0)}\left(r^{4+\delta_B}\alpha^2+r^4\beta^2\right)\\
&+\int_{\Sigma_*}\left(r^{4+\delta_B}(\alpha^2+\beta^2)+r^4(\check{\rho})^2+r^2\underline{\beta}^2+\underline{\alpha}^2\right)\\
&+\int_{{}^{(ext)}\mathcal{M}(r\geq 4m_0)}\Big(r^{3+\delta_B}(\alpha^2+\beta^2)+r^{3-\delta_B}(\check{\rho})^2+r^{1-\delta_B}\underline{\beta}^2\\
&\qquad +r^{-1-\delta_B}\underline{\alpha}^2\Big),
\end{aligned}$$

$$\left({}^{(ext)}\mathfrak{R}_0^{\leq 4m_0}[\check{R}]\right)^2 := \int_{{}^{(ext)}\mathcal{M}(r\leq 4m_0)}\left(1-\frac{3m}{r}\right)^2|\check{R}|^2,$$

and

$${}^{(ext)}\mathfrak{R}_0[\check{R}] := {}^{(ext)}\mathfrak{R}_0^{\geq 4m_0}[\check{R}] + {}^{(ext)}\mathfrak{R}_0^{\leq 4m_0}[\check{R}].$$

For any nonzero integer k, we introduce the following higher derivatives norms

$$\begin{aligned}
\left({}^{(ext)}\mathfrak{R}_k[\check{R}]\right)^2 &:= \left({}^{(ext)}\mathfrak{R}_0[\mathfrak{d}^{\leq k}\check{R}]\right)^2\\
&+\int_{{}^{(ext)}\mathcal{M}(r\leq 4m_0)}\left(|\mathfrak{d}^{\leq k-1}\mathbf{N}\check{R}|^2+|\mathfrak{d}^{\leq k-1}\check{R}|^2\right).
\end{aligned}$$

Remark 3.9. *Note that the derivative in the* $\mathbf{N}$ *direction, unlike all other first derivatives of* $\check{R}$*, appears in the spacetime integral* $\int_{{}^{(ext)}\mathcal{M}(r\leq 4m_0)}$ *with top number of derivatives. This reflects the fact the* $\mathbf{N}$*-derivatives do not degenerate at* $r = 3m$ *in the Morawetz estimate.*

3.2.1.2 L^2 Ricci coefficients norms in ${}^{(ext)}\mathcal{M}$

For any $k \geq 2$, we introduce the following norms

$$\begin{aligned}
\left({}^{(ext)}\mathfrak{G}_k^{\geq 4m_0}[\check{\Gamma}]\right)^2 &:= \int_{\Sigma_*}\Big[r^2\left((\mathfrak{d}^{\leq k}\vartheta)^2+(\mathfrak{d}^{\leq k}\check{\kappa})^2+(\mathfrak{d}^{\leq k}\zeta)^2+(\mathfrak{d}^{\leq k}\underline{\check{\kappa}})^2\right)\\
&+ (\mathfrak{d}^{\leq k}\underline{\vartheta})^2+(\mathfrak{d}^{\leq k}\eta)^2+(\mathfrak{d}^{\leq k}\underline{\check{\omega}})^2+(\mathfrak{d}^{\leq k}\underline{\xi})^2\Big]\\
&+ \sup_{\lambda\geq 4m_0}\Bigg(\int_{\{r=\lambda\}}\Big[\lambda^2\left((\mathfrak{d}^{\leq k}\vartheta)^2+(\mathfrak{d}^{\leq k}\check{\kappa})^2+(\mathfrak{d}^{\leq k}\zeta)^2\right)\\
&+ \lambda^{2-\delta_B}(\mathfrak{d}^{\leq k}\underline{\check{\kappa}})^2+(\mathfrak{d}^{\leq k}\underline{\vartheta})^2+(\mathfrak{d}^{\leq k}\eta)^2+(\mathfrak{d}^{\leq k}\underline{\check{\omega}})^2\\
&+ \lambda^{-\delta_B}(\mathfrak{d}^{\leq k}\underline{\xi})^2\Big]\Bigg),
\end{aligned}$$

$$\left({}^{(ext)}\mathfrak{G}_k^{\leq 4m_0}\left[\check{\Gamma}\right]\right)^2 := \int_{{}^{(ext)}\mathcal{M}(\leq 4m_0)} \left|\mathfrak{d}^{\leq k}\left(\check{\Gamma}\right)\right|^2,$$

and

$${}^{(ext)}\mathfrak{G}_k\left[\check{\Gamma}\right] := {}^{(ext)}\mathfrak{G}_k^{\leq 4m_0}\left[\check{\Gamma}\right] + {}^{(ext)}\mathfrak{G}_k^{\geq 4m_0}\left[\check{\Gamma}\right].$$

3.2.1.3 Decay norms in ${}^{(ext)}\mathcal{M}$

Let $\delta_{dec} > 0$ a small constant to be specified later. We define

$$\begin{aligned}
{}^{(ext)}\mathfrak{D}_0[\alpha] &:= \sup_{{}^{(ext)}\mathcal{M}} \left(r^2(2r+u)^{1+\delta_{dec}} + r^3(2r+u)^{\frac{1}{2}+\delta_{dec}}\right)|\alpha|,\\
{}^{(ext)}\mathfrak{D}_0[\beta] &:= \sup_{{}^{(ext)}\mathcal{M}} \left(r^2(2r+u)^{1+\delta_{dec}} + r^3(2r+u)^{\frac{1}{2}+\delta_{dec}}\right)|\beta|,\\
{}^{(ext)}\mathfrak{D}_0[\check{\rho}] &:= \sup_{{}^{(ext)}\mathcal{M}} \left(r^2u^{1+\delta_{dec}} + r^3u^{\frac{1}{2}+\delta_{dec}}\right)|\check{\rho}|,\\
{}^{(ext)}\mathfrak{D}_0[\check{\mu}] &:= \sup_{{}^{(ext)}\mathcal{M}} r^3u^{1+\delta_{dec}}|\check{\mu}|,\\
{}^{(ext)}\mathfrak{D}_0[\underline{\beta}] &:= \sup_{{}^{(ext)}\mathcal{M}} r^2u^{1+\delta_{dec}}|\underline{\beta}|,\\
{}^{(ext)}\mathfrak{D}_0[\underline{\alpha}] &:= \sup_{{}^{(ext)}\mathcal{M}} ru^{1+\delta_{dec}}|\underline{\alpha}|,
\end{aligned}$$

and

$$\begin{aligned}
{}^{(ext)}\mathfrak{D}_0[\check{R}] &:= {}^{(ext)}\mathfrak{D}_0[\alpha] + {}^{(ext)}\mathfrak{D}_0[\beta] + {}^{(ext)}\mathfrak{D}_0[\check{\rho}] + {}^{(ext)}\mathfrak{D}_0[\check{\mu}]\\
&\quad + {}^{(ext)}\mathfrak{D}_0[\underline{\beta}] + {}^{(ext)}\mathfrak{D}_0[\underline{\alpha}].
\end{aligned}$$

Also, we introduce the following higher derivatives norms

$$\begin{aligned}
{}^{(ext)}\mathfrak{D}_1[\check{R}] &:= {}^{(ext)}\mathfrak{D}_0[\check{R}] + {}^{(ext)}\mathfrak{D}_0[\mathfrak{d}\check{R}]\\
&+ \sup_{{}^{(ext)}\mathcal{M}} \left(r^3(2r+u)^{1+\delta_{dec}} + r^4(2r+u)^{\frac{1}{2}+\delta_{dec}}\right)|e_3(\alpha)|\\
&+ \sup_{{}^{(ext)}\mathcal{M}} \left(r^3u^{1+\delta_{dec}} + r^4u^{\frac{1}{2}+\delta_{dec}}\right)|e_3(\beta)| + \sup_{{}^{(ext)}\mathcal{M}} r^3u^{1+\delta_{dec}}|e_3(\check{\rho})|,
\end{aligned}$$

and for any integer $k \geq 2$

$${}^{(ext)}\mathfrak{D}_k[\check{R}] := {}^{(ext)}\mathfrak{D}_1[\mathfrak{d}^{\leq k-1}\check{R}].$$

Also, we define

$$
\begin{aligned}
{}^{(ext)}\mathfrak{D}_0[\check{\kappa}] &:= \sup_{{}^{(ext)}\mathcal{M}} r^2 u^{1+\delta_{dec}}|\check{\kappa}|, \\
{}^{(ext)}\mathfrak{D}_0[\vartheta] &:= \sup_{{}^{(ext)}\mathcal{M}} \left(ru^{1+\delta_{dec}} + r^2 u^{\frac{1}{2}+\delta_{dec}}\right)|\vartheta|, \\
{}^{(ext)}\mathfrak{D}_0[\zeta] &:= \sup_{{}^{(ext)}\mathcal{M}} \left(ru^{1+\delta_{dec}} + r^2 u^{\frac{1}{2}+\delta_{dec}}\right)|\zeta|, \\
{}^{(ext)}\mathfrak{D}_0[\underline{\check{\kappa}}] &:= \sup_{{}^{(ext)}\mathcal{M}} \left(ru^{1+\delta_{dec}} + r^2 u^{\frac{1}{2}+\delta_{dec}}\right)|\underline{\check{\kappa}}|, \\
{}^{(ext)}\mathfrak{D}_0[\underline{\vartheta}] &:= \sup_{{}^{(ext)}\mathcal{M}} ru^{1+\delta_{dec}}|\underline{\vartheta}|, \\
{}^{(ext)}\mathfrak{D}_0[\eta] &:= \sup_{{}^{(ext)}\mathcal{M}} ru^{1+\delta_{dec}}|\eta| + \left(\int_{\Sigma_*} u^{2+2\delta_{dec}}\eta^2\right)^{\frac{1}{2}}, \\
{}^{(ext)}\mathfrak{D}_0[\underline{\check{\omega}}] &:= \sup_{{}^{(ext)}\mathcal{M}} ru^{1+\delta_{dec}}|\underline{\check{\omega}}|, \\
{}^{(ext)}\mathfrak{D}_0[\underline{\xi}] &:= \sup_{{}^{(ext)}\mathcal{M}} ru^{1+\delta_{dec}}|\underline{\xi}|,
\end{aligned}
$$

and

$$
\begin{aligned}
{}^{(ext)}\mathfrak{D}_0[\check{\Gamma}] := {}& {}^{(ext)}\mathfrak{D}_0[\check{\kappa}] + {}^{(ext)}\mathfrak{D}_0[\vartheta] + {}^{(ext)}\mathfrak{D}_0[\zeta] + {}^{(ext)}\mathfrak{D}_0[\underline{\check{\kappa}}] + {}^{(ext)}\mathfrak{D}_0[\underline{\vartheta}] \\
& + {}^{(ext)}\mathfrak{D}_0[\eta] + {}^{(ext)}\mathfrak{D}_0[\underline{\check{\omega}}] + {}^{(ext)}\mathfrak{D}_0[\underline{\xi}].
\end{aligned}
$$

Also, we introduce the following higher derivatives norms

$$
{}^{(ext)}\mathfrak{D}_1[\check{\Gamma}] := {}^{(ext)}\mathfrak{D}_0[\mathfrak{d}\check{\Gamma}] + \sup_{{}^{(ext)}\mathcal{M}} r^2 u^{1+\delta_{dec}}|e_3(\vartheta, \zeta, \underline{\check{\kappa}})|
$$

and for any integer $k \geq 2$

$$
{}^{(ext)}\mathfrak{D}_k[\check{\Gamma}] := {}^{(ext)}\mathfrak{D}_1[\mathfrak{d}^{\leq k-1}\check{\Gamma}].
$$

Remark 3.10. *The integral bootstrap assumption on Σ_* for η will only be needed in the proof of Proposition 3.20 and recovered in Proposition 7.22. In fact, other components satisfy an analog integral estimate on Σ_*: this is the case of $\underline{\vartheta}$, $\underline{\xi}$ and $r\underline{\beta}$, see Proposition 7.22. But η is the only component for which we need to make this type of bootstrap assumption.*

3.2.2 Main norms in ${}^{(int)}\mathcal{M}$

All quantities appearing in this section are defined relative to the ${}^{(int)}\mathcal{M}$ frame adapted to the $(\underline{u}, {}^{(int)}s)$ foliation.

3.2.2.1 L^2 based norms in ${}^{(int)}\mathcal{M}$

We introduce the curvature norms

$$
\left({}^{(int)}\mathfrak{R}_0[\check{R}]\right)^2 := \int_{{}^{(int)}\mathcal{M}} |\check{R}|^2.
$$

For any nonzero integer k, we introduce the following higher derivatives norms

$$ {}^{(int)}\mathfrak{R}_k[\check{R}] := {}^{(int)}\mathfrak{R}_0[\mathfrak{d}^{\leq k}\check{R}]. $$

For any $k \geq 0$, we introduce the following norms

$$ \left({}^{(int)}\mathfrak{G}_k[\check{\Gamma}]\right)^2 := \int_{{}^{(int)}\mathcal{M}} |\mathfrak{d}^{\leq k}\check{\Gamma}|^2. $$

3.2.2.2 Decay norms in ${}^{(int)}\mathcal{M}$

We define

$$ {}^{(int)}\mathfrak{D}_0[\check{R}] := \sup_{{}^{(int)}\mathcal{M}} \underline{u}^{1+\delta_{dec}}|\check{R}|, \quad {}^{(int)}\mathfrak{D}_0[\check{\Gamma}] := \sup_{{}^{(int)}\mathcal{M}} \underline{u}^{1+\delta_{dec}}|\check{\Gamma}|. $$

Also, we introduce the following higher derivatives norms for any integer $k \geq 1$

$$ {}^{(int)}\mathfrak{D}_k[\check{R}] := {}^{(int)}\mathfrak{D}_0[\mathfrak{d}^{\leq k}\check{R}], \quad {}^{(int)}\mathfrak{D}_k[\check{\Gamma}] := {}^{(int)}\mathfrak{D}_0[\mathfrak{d}^{\leq k}\check{\Gamma}]. $$

3.2.3 Combined norms

We define the following norms $\mathcal{M}$ by combining our above norms on ${}^{(ext)}\mathcal{M}$ and ${}^{(int)}\mathcal{M}$

$$ \begin{aligned} \mathfrak{N}_k^{(En)} &:= {}^{(ext)}\mathfrak{R}_k[\check{R}] + {}^{(ext)}\mathfrak{G}_k[\check{\Gamma}] + {}^{(int)}\mathfrak{R}_k[\check{R}] + {}^{(int)}\mathfrak{G}_k[\check{\Gamma}], \\ \mathfrak{N}_k^{(Dec)} &:= {}^{(ext)}\mathfrak{D}_k[\check{R}] + {}^{(ext)}\mathfrak{D}_k[\check{\Gamma}] + {}^{(int)}\mathfrak{D}_k[\check{R}] + {}^{(int)}\mathfrak{D}_k[\check{\Gamma}]. \end{aligned} $$

3.2.4 Initial layer norm

Recall the notations of section 3.1.1 concerning the initial data layer $\mathcal{L}_0$. Recall that the constant $m_0 > 0$ is the mass of the initial Schwarzschild spacetime relative to which our initial perturbation is measured. We define the initial layer norm to be[8]

$$ \mathfrak{I}_k := {}^{(ext)}\mathfrak{I}_k + {}^{(int)}\mathfrak{I}_k + \mathfrak{I}'_k $$

[8]Recall that the initial data layer foliations satisfy $\underline{\eta} + \zeta = 0$, as well as $\xi = \omega = 0$ on ${}^{(ext)}\mathcal{L}_0$, and $\eta = \zeta$ as well as $\underline{\xi} = \underline{\omega} = 0$ on ${}^{(int)}\mathcal{L}_0$.

where

$$
\begin{aligned}
{}^{(ext)}\mathfrak{I}_0 \;:=\;& \sup_{{}^{(ext)}\mathcal{L}_0}\left[r^{\frac{7}{2}+\delta_B}\left(|\alpha|+|\beta|\right)+r^3\left|\rho+\frac{2m_0}{r^3}\right|+r^2|\underline{\beta}|+r|\underline{\alpha}|\right]\\
&+\sup_{{}^{(ext)}\mathcal{L}_0} r^2\left(|\vartheta|+\left|\kappa-\frac{2}{r}\right|+|\zeta|+\left|\underline{\kappa}+\frac{2\left(1-\frac{2m_0}{r}\right)}{r}\right|\right)\\
&+\sup_{{}^{(ext)}\mathcal{L}_0} r\left(|\underline{\vartheta}|+\left|\underline{\omega}-\frac{m_0}{r^2}\right|+|\underline{\xi}|\right)\\
&+\sup_{{}^{(ext)}\mathcal{L}_0\left({}^{(ext)}r_0\geq 4m_0\right)}\left[r\left|\frac{\gamma}{r^2}-1\right|+r|\underline{b}|+|\underline{\Omega}+\Upsilon|+|\varsigma-1|\right.\\
&\qquad\left.+r\left|\frac{e^\Phi}{r\sin\theta}-1\right|\right],
\end{aligned}
$$

$$
\begin{aligned}
{}^{(int)}\mathfrak{I}_0 \;:=\;& \sup_{{}^{(int)}\mathcal{L}_0}\left(|\underline{\alpha}|+|\underline{\beta}|+\left|\rho+\frac{2m_0}{r^3}\right|+|\beta|+|\alpha|\right)\\
&+\sup_{{}^{(int)}\mathcal{L}_0}\left[|\vartheta|+\left|\kappa-\frac{2\left(1-\frac{2m_0}{r}\right)}{r}\right|+|\zeta|+\left|\underline{\kappa}+\frac{2}{r}\right|+|\underline{\vartheta}|\right.\\
&\qquad\left.+\left|\omega+\frac{m_0}{r^2}\right|+|\xi|\right],
\end{aligned}
$$

$$
\mathfrak{I}'_0 \;:=\; \sup_{{}^{(int)}\mathcal{L}_0\cap{}^{(ext)}\mathcal{L}_0}\left(|f|+|\underline{f}|+|\log(\lambda_0^{-1}\lambda)|\right),\quad \lambda_0={}^{(ext)}\lambda_0=1-\frac{2m_0}{{}^{(ext)}r_{\mathcal{L}_0}},
$$

with $\mathfrak{I}_k$ the corresponding higher derivatives norms obtained by replacing each component by $\mathfrak{d}^{\leq k}$ of it. In the definition of $\mathfrak{I}'_0$ above, $(f,\underline{f},\lambda)$ denote the transition functions of Lemma 2.87 from the frame of the outgoing part ${}^{(ext)}\mathcal{L}_0$ of the initial data layer to the frame of the ingoing part ${}^{(int)}\mathcal{L}_0$ of the initial data layer in the region ${}^{(int)}\mathcal{L}_0\cap{}^{(ext)}\mathcal{L}_0$.

Remark 3.11. *Note that in the definition of ${}^{(ext)}\mathfrak{I}_k$ we allow a higher power of r in front of α, β and their derivatives than what is consistent with the results of [20] and [42]. The additional r^{δ_B} power, for δ_B small, is consistent instead with the result of [43].*

3.3 MAIN THEOREM

3.3.1 Smallness constants

Before stating our main theorem, we first introduce the following constants that will be involved in its statement.

- The constant $m_0>0$ is the mass of the initial Schwarzschild spacetime relative to which our initial perturbation is measured.
- The integer k_{large} corresponds to the maximum number of derivatives of the

solution.

- The size of the initial data layer norm is measured by $\epsilon_0 > 0$.
- The size of the bootstrap assumption norms are measured by $\epsilon > 0$.
- $\delta_{\mathcal{H}} > 0$ measures the width of the region $|r - 2m_0| \leq 2m_0\delta_{\mathcal{H}}$ where the redshift estimate holds and which includes in particular the region ${}^{(int)}\mathcal{M}$.
- δ_{dec} is tied to decay estimates in u, $\underline{u}$ for $\check{\Gamma}$ and $\check{R}$.
- δ_B is involved in the r-power of the r^p-weighted estimates for curvature.

In what follows, m_0 is a fixed constant, $\delta_{\mathcal{H}}$, δ_B, and δ_{dec} are fixed, sufficiently small, universal constants, and k_{large} is a fixed, sufficiently large, universal constant, chosen such that

$$0 < \delta_{\mathcal{H}},\ \delta_{dec},\ \delta_B \ll \min\{m_0, 1\}, \qquad \delta_B > 2\delta_{dec}, \qquad k_{large} \gg \frac{1}{\delta_{dec}}. \tag{3.3.1}$$

Then, ϵ and ϵ_0 are chosen such that

$$\epsilon_0, \epsilon \ll \min\left\{\delta_{\mathcal{H}}, \delta_{dec}, \delta_B, \frac{1}{k_{large}}, m_0, 1\right\} \tag{3.3.2}$$

and

$$\epsilon = \epsilon_0^{\frac{2}{3}}. \tag{3.3.3}$$

Using the definition of ϵ_0, we may now specify the behavior (3.1.2) of r on Σ_*

$$\inf_{\Sigma_*} r = \epsilon_0^{-\frac{2}{3}} u_*^{1+\delta_{dec}}. \tag{3.3.4}$$

From now on, in the rest of the book, $\lesssim$ means bounded by a constant depending only on geometric universal constants (such as Sobolev embeddings, elliptic estimates,...) as well as the constants

$$m_0,\ \delta_{\mathcal{H}},\ \delta_{dec},\ \delta_B,\ k_{large}$$

but not on ϵ and ϵ_0.

3.3.2 Statement of the main theorem

We are now ready to give the following precise version of our main theorem.

Main Theorem (Main theorem, version 2)**.** *There exists a sufficiently large integer k_{large} and a sufficiently small constant $\epsilon_0 > 0$ such that given an initial layer defined as in section 3.1.1 and satisfying the bound*

$$\mathfrak{I}_{k_{large}+5} \leq \epsilon_0^{\frac{5}{3}}, \tag{3.3.5}$$

there exists a globally hyperbolic development with a complete future null infinity $\mathcal{I}_+$ and a future horizon $\mathcal{H}_+$ together with foliations and adapted null frames verifying the admissibility conditions of section 3.1.2 such that the following bound is satisfied

$$\mathfrak{N}^{(En)}_{k_{large}} + \mathfrak{N}^{(Dec)}_{k_{small}} \leq C\epsilon_0 \tag{3.3.6}$$

where C is a large enough universal constant and where k_{small} is given by

$$k_{small} = \left\lfloor \frac{1}{2} k_{large} \right\rfloor + 1. \tag{3.3.7}$$

In particular,

- *On ${}^{(ext)}\mathcal{M}$, we have*

$$\begin{aligned}
|\alpha|, |\beta| &\lesssim \min\left\{ \frac{\epsilon_0}{r^3(u+2r)^{\frac{1}{2}+\delta_{dec}}}, \frac{\epsilon_0}{r^2(u+2r)^{1+\delta_{dec}}} \right\}, \\
|\check{\rho}| &\lesssim \min\left\{ \frac{\epsilon_0}{r^3 u^{\frac{1}{2}+\delta_{dec}}}, \frac{\epsilon_0}{r^2 u^{1+\delta_{dec}}} \right\}, \\
|\underline{\beta}| &\lesssim \frac{\epsilon_0}{r^2 u^{1+\delta_{dec}}}, \\
|\underline{\alpha}| &\lesssim \frac{\epsilon_0}{r u^{1+\delta_{dec}}},
\end{aligned}$$

  and

$$\begin{aligned}
|\check{\kappa}| &\lesssim \frac{\epsilon_0}{r^2 u^{1+\delta_{dec}}}, \\
|\vartheta|, |\zeta|, |\check{\underline{\kappa}}| &\lesssim \min\left\{ \frac{\epsilon_0}{r^2 u^{\frac{1}{2}+\delta_{dec}}} \frac{\epsilon_0}{r u^{1+\delta_{dec}}} \right\}, \\
|\underline{\eta}|, |\underline{\vartheta}|, |\check{\underline{\omega}}|, |\underline{\xi}| &\lesssim \frac{\epsilon_0}{r u^{1+\delta_{dec}}}.
\end{aligned}$$

- *On ${}^{(int)}\mathcal{M}$ we have, with $\check{\Gamma} = \{\check{\underline{\kappa}}, \underline{\vartheta}, \zeta, \underline{\eta}, \check{\kappa}, \vartheta, \check{\omega}, \xi\}$, $\check{R} = \{\alpha, \beta, \check{\rho}, \underline{\beta}, \underline{\alpha}\}$,*

$$|\check{\Gamma}, \check{R}| \lesssim \frac{\epsilon_0}{\underline{u}^{1+\delta_{dec}}}.$$

- *The Bondi mass converges as $u \to +\infty$ along $\mathcal{I}_+$ to the final Bondi mass which we denote by m_∞. The final Bondi mass verifies the estimate*

$$\left| \frac{m_\infty}{m_0} - 1 \right| \lesssim \epsilon_0.$$

  In particular $m_\infty > 0$.
- *The Hawking mass m satisfies*

$$\frac{|m - m_\infty|}{m_0} \lesssim \begin{cases} \dfrac{\epsilon_0}{u^{1+\delta_{dec}}} & \text{on } {}^{(ext)}\mathcal{M}, \\ \dfrac{\epsilon_0}{\underline{u}^{1+\delta_{dec}}} & \text{on } {}^{(int)}\mathcal{M}. \end{cases}$$

- *The location of the future horizon $\mathcal{H}_+$ satisfies*

$$r = 2m_\infty + O\left(\frac{\sqrt{\epsilon_0}}{\underline{u}^{1+\frac{\delta_{dec}}{2}}} \right) \quad \text{on } \mathcal{H}_+.$$

- *On* ${}^{(ext)}\mathcal{M}$, *we have*

$$\begin{aligned}
\left|\rho+\frac{2m_\infty}{r^3}\right| &\lesssim \min\left\{\frac{\epsilon_0}{r^3u^{\frac{1}{2}+\delta_{dec}}}, \frac{\epsilon_0}{r^2u^{1+\delta_{dec}}}\right\},\\
\left|\kappa-\frac{2}{r}\right| &\lesssim \frac{\epsilon_0}{r^2u^{1+\delta_{dec}}},\\
\left|\underline{\kappa}+\frac{2\left(1-\frac{2m_\infty}{r}\right)}{r}\right| &\lesssim \min\left\{\frac{\epsilon_0}{r^2u^{\frac{1}{2}+\delta_{dec}}} \frac{\epsilon_0}{ru^{1+\delta_{dec}}}\right\},\\
\left|\underline{\omega}-\frac{m_\infty}{r^2}\right| &\lesssim \frac{\epsilon_0}{ru^{1+\delta_{dec}}}.
\end{aligned}$$

- *On* ${}^{(int)}\mathcal{M}$, *we have*

$$\left|\rho+\frac{2m_\infty}{r^3}\right|, \left|\underline{\kappa}+\frac{2}{r}\right|, \left|\kappa-\frac{2\left(1-\frac{2m_\infty}{r}\right)}{r}\right|, \left|\omega+\frac{m_\infty}{r^2}\right| \lesssim \frac{\epsilon_0}{\underline{u}^{1+\delta_{dec}}}.$$

- *On* ${}^{(ext)}\mathcal{M}$, *the spacetime metric* $\mathbf{g}$ *is given in the* (u,r,θ,φ) *coordinates system by*

$$\mathbf{g} = \mathbf{g}_{m_\infty,{}^{(ext)}\mathcal{M}} + O\left(\frac{\epsilon_0}{u^{1+\delta_{dec}}}\right)\Big((dr,du,rd\theta)^2, r^2(\sin\theta)^2(d\varphi)^2\Big)$$

where $\mathbf{g}_{m_\infty,{}^{(ext)}\mathcal{M}}$ *denotes the Schwarzschild metric of mass* $m_\infty>0$ *in outgoing Eddington-Finkelstein coordinates, i.e.,*

$$\mathbf{g}_{m_\infty,{}^{(ext)}\mathcal{M}} := -2dudr-\left(1-\frac{2m_\infty}{r}\right)(du)^2+r^2\Big((d\theta)^2+(\sin\theta)^2(d\varphi)^2\Big).$$

- *On* ${}^{(int)}\mathcal{M}$, *the spacetime metric* $\mathbf{g}$ *is given in the* $(\underline{u},r,\theta,\varphi)$ *coordinates system by*

$$\mathbf{g} = \mathbf{g}_{m_\infty,{}^{(int)}\mathcal{M}} + O\left(\frac{\epsilon_0}{\underline{u}^{1+\delta_{dec}}}\right)\Big((dr,d\underline{u},rd\theta)^2, r^2(\sin\theta)^2(d\varphi)^2\Big)$$

where $\mathbf{g}_{m_\infty,{}^{(ext)}\mathcal{M}}$ *denotes the Schwarzschild metric of mass* $m_\infty>0$ *in ingoing Eddington-Finkelstein coordinates, i.e.,*

$$\mathbf{g}_{m_\infty,{}^{(int)}\mathcal{M}} := 2d\underline{u}dr-\left(1-\frac{2m_\infty}{r}\right)(d\underline{u})^2+r^2\Big((d\theta)^2+(\sin\theta)^2(d\varphi)^2\Big).$$

Note that analog statements of the above estimates also hold for $\mathfrak{d}^k$ *derivatives with* $k\leq k_{small}$.

Remark 3.12. *In this book, we choose to specify the closeness to Schwarzschild of our initial data in the context of the Characteristic Cauchy problem. Note that the conclusions of our main theorem can be immediately extended to the case where the data are specified to be close to Schwarzschild on a spacelike hypersurface* Σ. *Indeed, one can reduce this latter case to our situation by invoking*

- *The results in [42], [43] which allow us to control the causal region between* Σ

and the outgoing part of the initial data layer.[9]

- *A standard local existence result which controls the finite causal region between Σ and the ingoing part of the initial data layer.*

Remark 3.13. *In the context of the previous remark, we note that the constant $m_0 > 0$ appearing in the initial data layer norm of the assumption* (3.3.5) *of our main theorem does not necessarily coincide with the ADM mass of the corresponding initial data set on the spacelike hypersurface Σ. With respect to this ADM mass, we would recover the well known inequality stating that the final Bondi mass is smaller than the ADM mass.*

Remark 3.14. *For most of the proof, it is sufficient to assume the following weaker analog of* (3.3.5) *for the initial data layer*

$$\mathfrak{I}_{k_{large}+5} \leq \epsilon_0.$$

The only place where we need the stronger assumption (3.3.5) *on the initial data layer is in section 8.1, see Remark 8.1.*

3.4 BOOTSTRAP ASSUMPTIONS AND FIRST CONSEQUENCES

3.4.1 Main bootstrap assumptions

We assume that the combined norms $\mathfrak{N}_k^{(En)}$ and $\mathfrak{N}_k^{(Dec)}$ defined in section 3.2 verify the following bounds:

BA-E *(Bootstrap Assumptions on energies and weighted energies)*

$$\mathfrak{N}_{k_{large}}^{(En)} \leq \epsilon, \tag{3.4.1}$$

BA-D *(Bootstrap Assumptions on decay)*

$$\mathfrak{N}_{k_{small}}^{(Dec)} \leq \epsilon. \tag{3.4.2}$$

In the remainder of section 3.4.1, we state several simple consequences of the bootstrap assumptions which will be proved in Chapter 4.

3.4.2 Control of the initial data

While the smallness constant involved in the bootstrap assumptions is $\epsilon > 0$, we need the smallness constant involved in the control of the initial data to be $\epsilon_0 > 0$. This is achieved in the theorem below.

Theorem M0. *Assume that the initial data layer $\mathcal{L}_0$, as defined in section 3.1.1, satisfies*

$$\mathfrak{I}_{k_{large}+5} \leq \epsilon_0.$$

[9]Note that the results of [43] are consistent with our initial data layer assumptions.

Then under the bootstrap assumptions ***BA-D*** *on decay, the following holds true on the initial data hypersurface* $\mathcal{C}_1 \cup \underline{\mathcal{C}}_1$*:*

$$\max_{0\leq k\leq k_{large}} \left\{ \sup_{\mathcal{C}_1} \left[r^{\frac{7}{2}+\delta_B} \left(|\mathfrak{d}^k\, {}^{(ext)}\alpha| + |\mathfrak{d}^k\, {}^{(ext)}\beta| \right) + r^{\frac{9}{2}+\delta_B} |\mathfrak{d}^{k-1} e_3({}^{(ext)}\alpha)| \right] \right.$$
$$\left. + \sup_{\mathcal{C}_1} \left[r^3 \left| \mathfrak{d}^k \left({}^{(ext)}\rho + \frac{2m_0}{r^3} \right) \right| + r^2 |\mathfrak{d}^k\, {}^{(ext)}\underline{\beta}| + r |\mathfrak{d}^k\, {}^{(ext)}\underline{\alpha}| \right] \right\} \lesssim \epsilon_0,$$

$$\max_{0\leq k\leq k_{large}} \sup_{\underline{\mathcal{C}}_1} \left[|\mathfrak{d}^k\, {}^{(int)}\alpha| + |\mathfrak{d}^k\, {}^{(int)}\beta| + \left| \mathfrak{d}^k \left({}^{(int)}\rho + \frac{2m_0}{r^3} \right) \right| \right.$$
$$\left. + |\mathfrak{d}^k\, {}^{(int)}\underline{\beta}| + |\mathfrak{d}^k\, {}^{(int)}\underline{\alpha}| \right] \lesssim \epsilon_0,$$

and

$$\sup_{\mathcal{C}_1\cup\underline{\mathcal{C}}_1} \left| \frac{m}{m_0} - 1 \right| \lesssim \epsilon_0.$$

3.4.3 Control of averages and of the Hawking mass

The following two lemma are simple consequences of the bootstrap assumptions and will be proved in section 4.2.

Lemma 3.15 (Control of averages). *Assume given a GCM admissible spacetime* $\mathcal{M}$ *as defined in section 3.1.2 verifying the bootstrap assumption for some sufficiently small* $\epsilon > 0$*. Then, we have*

$$\sup_{{}^{(ext)}\mathcal{M}} u^{1+\delta_{dec}} \left(r^3 \left| \mathfrak{d}^{\leq k_{small}} \left(\overline{\kappa} - \frac{2}{r} \right) \right| + r^3 \left| \mathfrak{d}^{\leq k_{small}} \left(\overline{\rho} + \frac{2m}{r^3} \right) \right| \right) \lesssim \epsilon_0,$$
$$\sup_{{}^{(ext)}\mathcal{M}} u^{1+\delta_{dec}} \left(r^2 \left| \mathfrak{d}^{\leq k_{small}} \left(\overline{\underline{\kappa}} + \frac{2\Upsilon}{r} \right) \right| + r^2 \left| \mathfrak{d}^{\leq k_{small}} \left(\overline{\underline{\omega}} - \frac{m}{r^2} \right) \right| \right) \lesssim \epsilon_0,$$

$$\sup_{{}^{(ext)}\mathcal{M}} u^{\frac{1}{2}+\delta_{dec}} \left(r^3 \left| \mathfrak{d}^{\leq k_{large}} \left(\overline{\kappa} - \frac{2}{r} \right) \right| + r^3 \left| \mathfrak{d}^{\leq k_{large}} \left(\overline{\rho} + \frac{2m}{r^3} \right) \right| \right) \lesssim \epsilon_0,$$
$$\sup_{{}^{(ext)}\mathcal{M}} u^{\frac{1}{2}+\delta_{dec}} \left(r^2 \left| \mathfrak{d}^{\leq k_{large}} \left(\overline{\underline{\kappa}} + \frac{2\Upsilon}{r} \right) \right| + r^2 \left| \mathfrak{d}^{\leq k_{large}} \left(\overline{\underline{\omega}} - \frac{m}{r^2} \right) \right| \right) \lesssim \epsilon_0,$$

$$\sup_{{}^{(int)}\mathcal{M}} u^{1+\delta_{dec}} \left(\left| \mathfrak{d}^{\leq k_{small}} \left(\overline{\kappa} - \frac{2\Upsilon}{r} \right) \right| + \left| \mathfrak{d}^{\leq k_{small}} \left(\overline{\rho} + \frac{2m}{r^3} \right) \right| \right) \lesssim \epsilon_0,$$
$$\sup_{{}^{(int)}\mathcal{M}} u^{1+\delta_{dec}} \left(\left| \mathfrak{d}^{\leq k_{small}} \left(\overline{\underline{\kappa}} + \frac{2}{r} \right) \right| + \left| \mathfrak{d}^{\leq k_{small}} \left(\overline{\omega} + \frac{m}{r^2} \right) \right| \right) \lesssim \epsilon_0,$$

$$\sup_{^{(int)}\mathcal{M}} u^{\frac{1}{2}+\delta_{dec}}\left(\left|\mathfrak{d}^{\leq k_{large}}\left(\overline{\kappa}-\frac{2\Upsilon}{r}\right)\right|+\left|\mathfrak{d}^{\leq k_{large}}\left(\overline{\rho}+\frac{2m}{r^3}\right)\right|\right) \lesssim \epsilon_0,$$

$$\sup_{^{(int)}\mathcal{M}} u^{\frac{1}{2}+\delta_{dec}}\left(\left|\mathfrak{d}^{\leq k_{large}}\left(\underline{\overline{\kappa}}+\frac{2}{r}\right)\right|+\left|\mathfrak{d}^{\leq k_{large}}\left(\overline{\omega}+\frac{m}{r^2}\right)\right|\right) \lesssim \epsilon_0.$$

Also, we have

$$\sup_{^{(ext)}\mathcal{M}}\left(u^{1+\delta_{dec}}r\left|\mathfrak{d}^{\leq k_{small}}\left(\underline{\overline{\Omega}}+\Upsilon\right)\right|+u^{\frac{1}{2}+\delta_{dec}}r\left|\mathfrak{d}^{\leq k_{large}}\left(\underline{\overline{\Omega}}+\Upsilon\right)\right|\right)\lesssim\epsilon_0,$$

$$\sup_{^{(int)}\mathcal{M}}\left(u^{1+\delta_{dec}}\left|\mathfrak{d}^{\leq k_{small}}\left(\overline{\Omega}-\Upsilon\right)\right|+u^{\frac{1}{2}+\delta_{dec}}\left|\mathfrak{d}^{\leq k_{large}}\left(\overline{\Omega}-\Upsilon\right)\right|\right)\lesssim\epsilon_0.$$

Finally, recall that $\overline{\mu}$ *and* $\underline{\overline{\mu}}$ *are given by the following formula*

$$\overline{\mu}=\frac{2m}{r^3}\quad on\ ^{(ext)}\mathcal{M},\qquad \underline{\overline{\mu}}=\frac{2m}{r^3}\quad on\ ^{(int)}\mathcal{M}.$$

Lemma 3.16 (Control of the Hawking mass)**.** *Assume given a GCM admissible spacetime* $\mathcal{M}$ *as defined in section 3.1.2 verifying the bootstrap assumption for some sufficiently small* $\epsilon>0$. *Then, we have*

$$\max_{0\leq k\leq k_{large}}\sup_{^{(ext)}\mathcal{M}} u^{1+\delta_{dec}}\Big(|\mathfrak{d}^k e_3(m)|+r|\mathfrak{d}^k e_4(m)|\Big) \lesssim \epsilon_0,$$

$$\max_{0\leq k\leq k_{large}}\sup_{^{(int)}\mathcal{M}} \underline{u}^{1+\delta_{dec}}\Big(|\mathfrak{d}^k e_3(m)|+|\mathfrak{d}^k e_4(m)|\Big) \lesssim \epsilon_0.$$

The e_4 *derivatives behave better in powers of* r,

$$\max_{0\leq k\leq k_{small}}\sup_{^{(ext)}\mathcal{M}} r^2u^{1+\delta_{dec}}|\mathfrak{d}^k e_4(m)| \lesssim \epsilon_0,$$

$$\max_{0\leq k\leq k_{large}}\sup_{^{(ext)}\mathcal{M}} r^2u^{\frac{1}{2}+\delta_{dec}}|\mathfrak{d}^k e_4(m)| \lesssim \epsilon_0.$$

Moreover,

$$\sup_{\mathcal{M}}\left|\frac{m}{m_0}-1\right| \lesssim \epsilon_0.$$

3.4.4 Control of coordinates system

The following two propositions on the existence of a suitable coordinates system both in $^{(ext)}\mathcal{M}$ and in $^{(int)}\mathcal{M}$ are also consequences of the bootstrap assumptions and will be proved in section 4.3.

Proposition 3.17 (Control of a coordinates system on $^{(ext)}\mathcal{M}$)**.** *Let* $\theta\in[0,\pi]$ *be the* **Z***-invariant scalar on* $\mathcal{M}$ *defined by* (2.2.52), *i.e.,*

$$\theta=\cot^{-1}\left(re_\theta(\Phi)\right). \tag{3.4.3}$$

Consider the (u,r,θ,φ) *coordinates system introduced in Proposition 2.78. Then, relative to these* (u,r,θ,φ) *coordinates,*

1. *The spacetime metric takes the form*

$$g = -\frac{4\varsigma}{r\overline{\kappa}}dudr + \frac{\varsigma^2(\underline{\overline{\kappa}} + \underline{A})}{\underline{\overline{\kappa}}}du^2 + \gamma\left(d\theta - \frac{1}{2}\varsigma\underline{b}du - \frac{b}{2}\Theta\right)^2 \tag{3.4.4}$$

where

$$b = e_4(\theta), \qquad \underline{b} = e_3(\theta), \qquad \gamma = \frac{1}{(e_\theta(\theta))^2} \tag{3.4.5}$$

and

$$\Theta = \frac{4}{r\overline{\kappa}}dr - \varsigma\left(\frac{\underline{\overline{\kappa}} + \underline{A}}{\overline{\kappa}}\right)du.$$

2. *The reduced coordinates derivatives take the form*

$$\begin{aligned}
\partial_r &= \frac{2}{r\overline{\kappa}}e_4 - \frac{2\sqrt{\gamma}}{r\overline{\kappa}}be_\theta, \\
\partial_\theta &= \sqrt{\gamma}e_\theta, \\
\partial_u &= \varsigma\left[\frac{1}{2}e_3 - \frac{1}{2}\frac{\underline{\overline{\kappa}} + \underline{A}}{\overline{\kappa}}e_4 - \frac{1}{2}\sqrt{\gamma}\left(\underline{b} - \left(\frac{\underline{\overline{\kappa}} + \underline{A}}{\overline{\kappa}}\right)b\right)e_\theta\right].
\end{aligned} \tag{3.4.6}$$

3. *The following estimates hold true:*

$$\begin{aligned}
\max_{0\le k\le k_{small}} \sup_{{}^{(ext)}\mathcal{M}} \left(ru^{\frac{1}{2}+\delta_{dec}} + u^{1+\delta_{dec}}\right)\left(\left|\mathfrak{d}^k\left(\frac{\gamma}{r^2} - 1\right)\right| + r\left|\mathfrak{d}^k b\right|\right) &\lesssim \epsilon, \\
\max_{0\le k\le k_{small}} \sup_{{}^{(ext)}\mathcal{M}} u^{1+\delta_{dec}}\left(\left|\mathfrak{d}^k\check{\underline{\Omega}}\right| + \left|\mathfrak{d}^k(\varsigma - 1)\right| + r\left|\mathfrak{d}^k\underline{b}\right|\right) &\lesssim \epsilon.
\end{aligned}$$

Also, e^Φ *satisfies*

$$\max_{0\le k\le k_{small}} \sup_{{}^{(ext)}\mathcal{M}} \left(ru^{\frac{1}{2}+\delta_{dec}} + u^{1+\delta_{dec}}\right)\left|\mathfrak{d}^k\left(\frac{e^\Phi}{r\sin\theta} - 1\right)\right| \lesssim \epsilon.$$

Proposition 3.18 (Control of a coordinates system on ${}^{(int)}\mathcal{M}$). *Let* $\theta \in [0, \pi]$ *be the* **Z**-*invariant scalar on* $\mathcal{M}$ *defined by* (3.4.3). *Consider the* $(\underline{u}, r, \theta, \varphi)$ *coordinates system introduced in Proposition 2.81. Then, relative to these* $(\underline{u}, r, \theta, \varphi)$ *coordinates,*

1. *The spacetime metric takes the form*

$$g = -\frac{4\underline{\varsigma}}{r\underline{\overline{\kappa}}}d\underline{u}dr + \frac{\underline{\varsigma}^2(\overline{\kappa} + A)}{\overline{\kappa}}d\underline{u}^2 + \gamma\left(d\theta - \frac{1}{2}\underline{\varsigma}bd\underline{u} - \frac{\underline{b}}{2}\underline{\Theta}\right)^2 \tag{3.4.7}$$

where

$$b = e_4(\theta), \qquad \underline{b} = e_3(\theta), \qquad \gamma = \frac{1}{(e_\theta(\theta))^2} \tag{3.4.8}$$

and

$$\underline{\Theta} := \frac{4}{r\underline{\overline{\kappa}}}dr - \underline{\varsigma}\left(\frac{\overline{\kappa} + A}{\underline{\overline{\kappa}}}\right)d\underline{u}.$$

2. *The reduced coordinates derivatives take the form*

$$\begin{aligned} \partial_r &= \frac{2}{r\underline{\kappa}}e_3 - \frac{2\sqrt{\gamma}}{r\underline{\kappa}}\underline{b}e_\theta, \\ \partial_\theta &= \sqrt{\gamma}e_\theta, \\ \partial_{\underline{u}} &= \varsigma\left[\frac{1}{2}e_4 - \frac{1}{2}\frac{\overline{\kappa}+A}{\underline{\kappa}}e_3 - \frac{1}{2}\sqrt{\gamma}\left(b - \left(\frac{\overline{\kappa}+A}{\underline{\kappa}}\right)\underline{b}\right)e_\theta\right]. \end{aligned} \tag{3.4.9}$$

3. *The following estimates hold true:*

$$\max_{0\le k\le k_{small}} \sup_{^{(int)}\mathcal{M}} \underline{u}^{1+\delta_{dec}}\Bigg[\left|\mathfrak{d}^k\check{\Omega}\right| + \left|\mathfrak{d}^k(\varsigma-1)\right| + \left|\mathfrak{d}^k\left(\frac{\gamma}{r^2}-1\right)\right| + \left|\mathfrak{d}^k b\right| + \left|\mathfrak{d}^k\underline{b}\right|\Bigg] \lesssim \epsilon.$$

Also, e^Φ *satisfies*

$$\max_{0\le k\le k_{small}} \sup_{^{(int)}\mathcal{M}} \underline{u}^{1+\delta_{dec}}\left|\mathfrak{d}^k\left(\frac{e^\Phi}{r\sin\theta}-1\right)\right| \lesssim \epsilon.$$

3.4.5 Pointwise bounds for higher order derivatives

We will need later to interpolate between the estimates provided by the bootstrap assumptions on decay and the bootstrap assumptions on energy. To this end, we will need the following consequence of the bootstrap assumptions on weighted energies.

Proposition 3.19. *The Ricci coefficients and curvature components satisfy the following pointwise estimates on* $\mathcal{M}$*:*

$$\begin{aligned} \max_{k\le k_{large}-5}\sup_{\mathcal{M}} \Big\{ & r^{\frac{7}{2}+\frac{\delta_B}{2}}\left(|\mathfrak{d}^k\alpha| + |\mathfrak{d}^k\beta|\right) + r^3\left(|\mathfrak{d}^k\mu| + |\mathfrak{d}^k\check{\rho}|\right) \\ & + r^2\left(|\mathfrak{d}^k\check{\kappa}| + |\mathfrak{d}^k\zeta| + |\mathfrak{d}^k\vartheta| + |\mathfrak{d}^k\check{\underline{\kappa}}| + |\mathfrak{d}^k\underline{\beta}|\right) \\ & + r\left(|\mathfrak{d}^k\eta| + |\mathfrak{d}^k\underline{\vartheta}| + |\mathfrak{d}^k\check{\underline{\omega}}| + |\mathfrak{d}^k\underline{\xi}| + |\mathfrak{d}^k\underline{\alpha}|\right)\Big\} \lesssim \epsilon. \end{aligned}$$

3.4.6 Construction of a second frame in $^{(ext)}\mathcal{M}$

Recall that the quantity $\mathfrak{q}$ satisfies the following wave equation, see (2.4.7),

$$\square_2\mathfrak{q} + \kappa\underline{\kappa}\mathfrak{q} = \mathrm{Err}[\square_2\mathfrak{q}]$$

where the nonlinear term $\mathrm{Err}[\square_2\mathfrak{q}]$ has the schematic structure exhibited in (2.4.8). Also, recall that according to our bootstrap assumption on decay and Proposition 3.19, η satisfies on $^{(ext)}\mathcal{M}$

$$|\mathfrak{d}^{\le k_{small}}\eta| \le \frac{\epsilon}{ru^{1+\delta_{dec}}}, \qquad |\mathfrak{d}^{\le k_{large}-5}\eta| \lesssim \frac{\epsilon}{r}.$$

As discussed in Remark 2.109, this decay in r^{-1} is too weak to derive suitable decay for $\mathfrak{q}$. We thus need to provide another frame for $^{(ext)}\mathcal{M}$. This is the aim of the following proposition.

Proposition 3.20. *Let an integer k_{loss} and a small constant $\delta_0 > 0$ satisfying*[10]

$$16 \leq k_{loss} \leq \frac{\delta_{dec}}{3}(k_{large} - k_{small}), \qquad \delta_0 := \frac{k_{loss}}{k_{large} - k_{small}}. \tag{3.4.10}$$

Let (e_4, e_3, e_θ) the outgoing geodesic null frame of ${}^{(ext)}\mathcal{M}$. There exists another frame (e_4', e_3', e_θ') of ${}^{(ext)}\mathcal{M}$ provided by

$$\begin{aligned}
e_4' &= e_4 + f e_\theta + \frac{1}{4} f^2 e_3, \\
e_\theta' &= e_\theta + \frac{1}{2} f e_3, \\
e_3' &= e_3,
\end{aligned}$$

such that the Ricci coefficients and curvature components with respect to that frame satisfy

$$\xi' = 0,$$

$$\begin{aligned}
\max_{0 \leq k \leq k_{small} + k_{loss}} \sup_{{}^{(ext)}\mathcal{M}} \quad & \Bigg\{ \left(r^2 u^{\frac{1}{2} + \delta_{dec} - 2\delta_0} + r u^{1 + \delta_{dec} - 2\delta_0} \right) |\mathfrak{d}^k \Gamma_g'| \\
& + r u^{1 + \delta_{dec} - 2\delta_0} |\mathfrak{d}^k \Gamma_b'| \\
& + r^2 u^{1 + \delta_{dec} - 2\delta_0} \left| \mathfrak{d}^{k-1} e_3' \left(\kappa' - \frac{2}{r}, \underline{\kappa}' + \frac{2\Upsilon}{r}, \vartheta', \zeta', \underline{\eta}', \eta' \right) \right| \\
& + \left(r^{\frac{7}{2} + \frac{\delta_B}{2}} + r^3 u^{\frac{1}{2} + \delta_{dec} - 2\delta_0} + r^2 u^{1 + \delta_{dec} - 2\delta_0} \right) |\mathfrak{d}^k (\alpha', \beta')| \\
& + \left(r^{\frac{9}{2} + \frac{\delta_B}{2}} + r^3 u^{1 + \delta_{dec}} + r^4 u^{\frac{1}{2} + \delta_{dec} - 2\delta_0} \right) |\mathfrak{d}^{k-1} e_3'(\alpha')| \\
& + \left(r^3 u^{1 + \delta_{dec}} + r^4 u^{\frac{1}{2} + \delta_{dec} - 2\delta_0} \right) |\mathfrak{d}^{k-1} e_3'(\beta')| \\
& + \left(r^3 u^{\frac{1}{2} + \delta_{dec} - 2\delta_0} + r^2 r u^{1 + \delta_{dec} - 2\delta_0} \right) |\mathfrak{d}^k \check{\rho}'| \\
& + u^{1 + \delta_{dec} - 2\delta_0} \left(r^2 |\mathfrak{d}^k \underline{\beta}'| + r |\mathfrak{d}^k \underline{\alpha}'| \right) \Bigg\} \lesssim \epsilon,
\end{aligned}$$

---

[10]Recall from (3.3.1) and (3.3.7) that we have

$$0 < \delta_{dec} \ll 1, \qquad \delta_{dec} k_{large} \gg 1, \qquad k_{small} = \left\lfloor \frac{1}{2} k_{large} \right\rfloor + 1.$$

In particular, we have $\delta_{dec}(k_{large} - k_{small}) \gg 1$ and hence there exists an integer k_{loss} satisfying the required constraints.

where we have used the notation[11]

$$\Gamma'_g = \left\{ r\omega',\, \kappa' - \frac{2}{r},\, \vartheta',\, \zeta',\, \eta',\, \underline{\eta}',\, \underline{\kappa}' + \frac{2\Upsilon}{r},\, r^{-1}(e'_4(r) - 1), r^{-1}e'_\theta(r),\, e'_4(m) \right\},$$
$$\Gamma'_b = \left\{ \underline{\vartheta}', \underline{\omega}' - \frac{m}{r^2}, \underline{\xi}',\, r^{-1}(e'_3(r) + \Upsilon),\, r^{-1}e'_3(m) \right\}.$$

Furthermore, f satisfies on ${}^{(ext)}\mathcal{M}$

$$\begin{aligned} |\mathfrak{d}^k f| &\lesssim \frac{\epsilon}{ru^{\frac{1}{2}+\delta_{dec}-2\delta_0} + u^{1+\delta_{dec}-2\delta_0}}, \quad \textit{for } k \leq k_{small} + k_{loss} + 2, \\ |\mathfrak{d}^{k-1} e'_3 f| &\lesssim \frac{\epsilon}{ru^{1+\delta_{dec}-2\delta_0}} \quad \textit{for } k \leq k_{small} + k_{loss} + 2. \end{aligned} \tag{3.4.11}$$

Remark 3.21. *The crucial point of Proposition 3.20 is that in the new null frame (e'_4, e'_3, e'_θ) of ${}^{(ext)}\mathcal{M}$, η' belongs to Γ'_g and thus displays a better decay in r^{-1} than η corresponding to the outgoing geodesic frame (e_4, e_3, e_θ) of ${}^{(ext)}\mathcal{M}$.*

3.5 GLOBAL NULL FRAMES

In this section, we construct 2 smooth global frames on $\mathcal{M}$ by matching the frame of ${}^{(int)}\mathcal{M}$ on the one hand with a renormalization of the frame on ${}^{(ext)}\mathcal{M}$, and on the other hand, with a renormalization of the second frame of ${}^{(ext)}\mathcal{M}$ given by Proposition 3.4.6.

3.5.1 Extension of frames

To construct the first global null frame, we need to extend the null frame of ${}^{(int)}\mathcal{M}$, i.e., $({}^{(int)}e_4, {}^{(int)}e_3, {}^{(int)}e_\theta)$, slightly into ${}^{(ext)}\mathcal{M}$, and the null frame of ${}^{(ext)}\mathcal{M}$, i.e., $({}^{(ext)}e_4, {}^{(ext)}e_3, {}^{(ext)}e_\theta)$, slightly into ${}^{(int)}\mathcal{M}$. We keep the same labels for the extended frame, i.e., $({}^{(int)}e_4, {}^{(int)}e_3, {}^{(int)}e_\theta)$ represents the extended frame of ${}^{(int)}\mathcal{M}$ in ${}^{(ext)}\mathcal{M}$ and vice versa. This convention also applies to the Ricci coefficients, curvature components, area radius and Hawking mass of the extended frames.

Note that these extensions require, in addition to the initialization of the frames on $\mathcal{T}$, to initialize

1. $({}^{(ext)}e_4, {}^{(ext)}e_3, {}^{(ext)}e_\theta)$ on $\underline{\mathcal{C}}_*$ by
$$({}^{(ext)}e_4, {}^{(ext)}e_3, {}^{(ext)}e_\theta) = (({}^{(int)}\Upsilon)^{-1}{}^{(int)}e_4, {}^{(int)}\Upsilon{}^{(int)}e_3, {}^{(int)}e_\theta).$$
2. $({}^{(int)}e_4, {}^{(int)}e_3, {}^{(int)}e_\theta)$ on $\mathcal{C}_*$ by
$$({}^{(int)}e_4, {}^{(int)}e_3, {}^{(int)}e_\theta) = ({}^{(ext)}\Upsilon{}^{(ext)}e_4, ({}^{(ext)}\Upsilon)^{-1}{}^{(ext)}e_3, {}^{(ext)}e_\theta).$$

[11]Here, r and m denote respectively the area radius and the Hawking mass of the outgoing geodesic foliation of ${}^{(ext)}\mathcal{M}$, i.e., $r = {}^{(ext)}r$ and $m = {}^{(ext)}m$. In particular, while $e_\theta(r) = e_\theta(m) = 0$, we have in general $e'_\theta(r) \neq 0$ and $e'_\theta(m) \neq 0$.

3.5.2 Construction of the first global frame

We start with the definition of the region where the frame of ${}^{(int)}\mathcal{M}$ and a conformal renormalization of the frame of ${}^{(ext)}\mathcal{M}$ will be matched.

Definition 3.22. *We define the matching region as the spacetime region*

$$\begin{aligned} Match \;\; := \;\; & \left({}^{(ext)}\mathcal{M}\cap\left\{{}^{(int)}r\leq 2m_0\left(1+\frac{3}{2}\delta_{\mathcal{H}}\right)\right\}\right) \\ & \cup\left({}^{(int)}\mathcal{M}\cap\left\{{}^{(int)}r\geq 2m_0\left(1+\frac{1}{2}\delta_{\mathcal{H}}\right)\right\}\right), \end{aligned}$$

where, as explained in the previous section, ${}^{(int)}r$ denotes the area radius of the ingoing geodesic foliation of ${}^{(int)}\mathcal{M}$ and its extension to ${}^{(ext)}\mathcal{M}$.

Here is our main proposition concerning our first global frame.

Proposition 3.23. *There exists a global null frame defined on ${}^{(int)}\mathcal{M}\cup{}^{(ext)}\mathcal{M}$ and denoted by $({}^{(glo)}e_4, {}^{(glo)}e_3, {}^{(glo)}e_\theta)$ such that*

a) In ${}^{(ext)}\mathcal{M}\setminus Match$, we have

$$({}^{(glo)}e_4, {}^{(glo)}e_3, {}^{(glo)}e_\theta) = \left({}^{(ext)}\Upsilon\,{}^{(ext)}e_4, {}^{(ext)}\Upsilon^{-1}{}^{(ext)}e_3, {}^{(ext)}e_\theta\right).$$

b) In ${}^{(int)}\mathcal{M}\setminus Match$, we have

$$({}^{(glo)}e_4, {}^{(glo)}e_3, {}^{(glo)}e_\theta) = \left({}^{(int)}e_4, {}^{(int)}e_3, {}^{(int)}e_\theta\right).$$

c) In the matching region, we have

$$\begin{aligned} \max_{0\leq k\leq k_{small}-2}\sup_{Match\cap{}^{(int)}\mathcal{M}} \underline{u}^{1+\delta_{dec}}\left|\mathfrak{d}^k({}^{(glo)}\check{\Gamma}, {}^{(glo)}\check{R})\right| &\lesssim \epsilon, \\ \max_{0\leq k\leq k_{small}-2}\sup_{Match\cap{}^{(ext)}\mathcal{M}} u^{1+\delta_{dec}}\left|\mathfrak{d}^k({}^{(glo)}\check{\Gamma}, {}^{(glo)}\check{R})\right| &\lesssim \epsilon, \\ \max_{0\leq k\leq k_{large}-1}\left(\int_{Match}\left|\mathfrak{d}^k({}^{(glo)}\check{\Gamma}, {}^{(glo)}\check{R})\right|^2\right)^{\frac{1}{2}} &\lesssim \epsilon, \end{aligned}$$

where ${}^{(glo)}\check{R}$ and ${}^{(glo)}\check{\Gamma}$ are given by

$$\begin{aligned} {}^{(glo)}\check{R} &= \left\{\alpha, \beta, \rho+\frac{2m}{r^3}, \underline{\beta}, \underline{\alpha}\right\}, \\ {}^{(glo)}\check{\Gamma} &= \left\{\xi, \omega+\frac{m}{r^2}, \kappa-\frac{2\Upsilon}{r}, \vartheta, \zeta, \eta, \underline{\eta}, \underline{\kappa}+\frac{2}{r}, \underline{\vartheta}, \underline{\omega}, \underline{\xi}\right\}. \end{aligned}$$

d) Furthermore, we may also choose the global frame such that, in addition, one of the following two possibilities hold:

i. We have on all ${}^{(ext)}\mathcal{M}$

$$({}^{(glo)}e_4, {}^{(glo)}e_3, {}^{(glo)}e_\theta) = \left({}^{(ext)}\Upsilon\,{}^{(ext)}e_4, {}^{(ext)}\Upsilon^{-1}{}^{(ext)}e_3, {}^{(ext)}e_\theta\right).$$

ii. We have on all ${}^{(int)}\mathcal{M}$

$$({}^{(glo)}e_4, {}^{(glo)}e_3, {}^{(glo)}e_\theta) = \left({}^{(int)}e_4, {}^{(int)}e_3, {}^{(int)}e_\theta\right).$$

Remark 3.24. *The global frame on* $\mathcal{M}$ *of Proposition 3.23 will be used to construct the second global frame in the next section, see Proposition 3.26. It will also be used to recover higher order derivatives in Theorem M8 (stated in section 3.6.2), see section 8.3.2.*

3.5.3 Construction of the second global frame

We start with the definition of the region where the first global frame of $\mathcal{M}$ (i.e., the one of Proposition 3.23) and a conformal renormalization of the second frame of ${}^{(ext)}\mathcal{M}$ (i.e., the one of Proposition 3.20) will be matched.

Definition 3.25. *We define the matching region as the spacetime region*

$$Match' := {}^{(ext)}\mathcal{M} \cap \left\{\frac{7m_0}{2} \le {}^{(ext)}r \le 4m_0\right\},$$

where ${}^{(ext)}r$ *denotes the area radius of the outgoing geodesic foliation of* ${}^{(ext)}\mathcal{M}$.

Here is our main proposition concerning our second global frame.

Proposition 3.26. *Let an integer* k_{loss} *and a small constant* $\delta_0 > 0$ *satisfying (3.4.10). There exists a global null frame* $({}^{(glo')}e_4, {}^{(glo')}e_3, {}^{(glo')}e_\theta)$ *defined on* ${}^{(int)}\mathcal{M} \cup {}^{(ext)}\mathcal{M}$ *such that*

a) In ${}^{(ext)}\mathcal{M} \cap \{{}^{(ext)}r \ge 4m_0\}$*, we have*

$$({}^{(glo')}e_4, {}^{(glo')}e_3, {}^{(glo')}e_\theta) = \left({}^{(ext)}\Upsilon\, {}^{(ext)}e'_4, {}^{(ext)}\Upsilon^{-1}\, {}^{(ext)}e'_3, {}^{(ext)}e'_\theta\right),$$

where $({}^{(ext)}e'_4, {}^{(ext)}e'_3, {}^{(ext)}e'_\theta)$ *denotes the second frame of* ${}^{(ext)}\mathcal{M}$*, i.e., the frame of Proposition 3.20.*

b) In ${}^{(int)}\mathcal{M} \cup ({}^{(ext)}\mathcal{M} \cap \{{}^{(ext)}r \le \frac{7m_0}{2}\})$*, we have*

$$({}^{(glo')}e_4, {}^{(glo')}e_3, {}^{(glo')}e_\theta) = ({}^{(glo)}e_4, {}^{(glo)}e_3, {}^{(glo)}e_\theta),$$

where $({}^{(glo)}e_4, {}^{(glo)}e_3, {}^{(glo)}e_\theta)$ *denotes the first global frame of* $\mathcal{M}$*, i.e., the frame of Proposition 3.23.*

c) In the matching region, we have

$$\max_{0\le k\le k_{small}+k_{loss}} \sup_{Match'} u^{1+\delta_{dec}-2\delta_0} \left|\mathfrak{d}^k({}^{(glo')}\check{\Gamma}, {}^{(glo')}\check{R})\right| \lesssim \epsilon,$$

where ${}^{(glo')}\check{R}$ *and* ${}^{(glo')}\check{\Gamma}$ *are given by*

$$\begin{aligned} {}^{(glo')}\check{R} &= \left\{\alpha, \beta, \rho + \frac{2m}{r^3}, \underline{\beta}, \underline{\alpha}\right\}, \\ {}^{(glo')}\check{\Gamma} &= \left\{\xi, \omega + \frac{m}{r^2}, \kappa - \frac{2\Upsilon}{r}, \vartheta, \zeta, \eta, \underline{\eta}, \underline{\kappa} + \frac{2}{r}, \underline{\vartheta}, \underline{\omega}, \underline{\xi}\right\}, \end{aligned}$$

with the Ricci coefficients and curvature components being the one associated to the frame $({}^{(glo')}e_4, {}^{(glo')}e_3, {}^{(glo')}e_\theta)$.

d) *Furthermore, we may also choose the global frame such that, in addition, one of the following two possibilities hold:*

   i. *We have on* ${}^{(ext)}\mathcal{M} \cap \{{}^{(ext)}r \geq \frac{15m_0}{4}\}$

$$({}^{(glo')}e_4, {}^{(glo')}e_3, {}^{(glo')}e_\theta) = \left({}^{(ext)}\Upsilon\, {}^{(ext)}e_4', {}^{(ext)}\Upsilon^{-1}\, {}^{(ext)}e_3', {}^{(ext)}e_\theta'\right).$$

   ii. *We have on* ${}^{(int)}\mathcal{M} \cup ({}^{(ext)}\mathcal{M} \cap \{{}^{(ext)}r \leq \frac{15m_0}{4}\})$

$$({}^{(glo')}e_4, {}^{(glo')}e_3, {}^{(glo')}e_\theta) = ({}^{(glo)}e_4, {}^{(glo)}e_3, {}^{(glo)}e_\theta).$$

Remark 3.27. *The global frame on* $\mathcal{M}$ *of Proposition 3.26 will be needed to derive decay estimates for the quantity* $\mathfrak{q}$ *in Theorem M1 (stated in section 3.6.1).*

3.6 PROOF OF THE MAIN THEOREM

3.6.1 Main intermediate results

We are ready to state our main intermediary results.

Theorem M1. *Assume given a GCM admissible spacetime* $\mathcal{M}$ *as defined in section 3.1.2 verifying the bootstrap assumptions*[12] ***BA-E*** *and* ***BA-D*** *for some sufficiently small* $\epsilon > 0$. *Then, if* $\epsilon_0 > 0$ *is sufficiently small, there exists* $\delta_{extra} > \delta_{dec}$ *such that we have the following estimates in* $\mathcal{M}$:

$$\max_{0\leq k\leq k_{small}+20} \sup_{{}^{(ext)}\mathcal{M}} \left\{\left(ru^{\frac{1}{2}+\delta_{extra}} + u^{1+\delta_{extra}}\right)|\mathfrak{d}^k\mathfrak{q}| + ru^{1+\delta_{extra}}|\mathfrak{d}^k e_3\mathfrak{q}|\right\} + \max_{0\leq k\leq k_{small}+20} \sup_{{}^{(int)}\mathcal{M}} \underline{u}^{1+\delta_{extra}}|\mathfrak{d}^k\mathfrak{q}| \lesssim \epsilon_0.$$

Moreover, $\mathfrak{q}$ *also satisfies the following estimate:*

$$\max_{0\leq k\leq k_{small}+21} \underline{u}^{2+2\delta_{extra}} \int_{{}^{(int)}\mathcal{M}(\geq \underline{u})} |\mathfrak{d}^k\mathfrak{q}|^2 + \max_{0\leq k\leq k_{small}+20} u^{2+2\delta_{extra}} \int_{\Sigma_*(\geq u)} |\mathfrak{d}^k e_3\mathfrak{q}|^2 \lesssim \epsilon_0^2.$$

Theorem M2. *Under the same assumptions as above we have the following decay estimates for* ${}^{(ext)}\alpha$:

$$\max_{0\leq k\leq k_{small}+20} \sup_{{}^{(ext)}\mathcal{M}} \left(\frac{r^2(2r+u)^{1+\delta_{extra}}}{\log(1+u)} + r^3(2r+u)^{\frac{1}{2}+\delta_{extra}}\right) \times \left(|\mathfrak{d}^k\, {}^{(ext)}\alpha| + r|\mathfrak{d}^k e_3\, {}^{(ext)}\alpha|\right) \lesssim \epsilon_0.$$

[12] Recall in particular that the conclusions of Theorem M0 hold under the bootstrap assumptions **BA-E** and **BA-D**.

Theorem M3. *Under the same assumptions as above we have the following decay estimates for $\underline{\alpha}$:*

$$ {}^{(int)}\mathfrak{D}_{k_{small}+16}[\underline{\alpha}] \lesssim \epsilon_0, \qquad \max_{0\leq k\leq k_{small}+18}\int_{\Sigma_*} u^{2+2\delta_{extra}}|\mathfrak{d}^k\underline{\alpha}|^2 \lesssim \epsilon_0^2. $$

Theorem M4. *Under the same assumptions as above we also have the following decay estimates in ${}^{(ext)}\mathcal{M}$:*

$$ {}^{(ext)}\mathfrak{D}_{k_{small}+8}[\check{R}] + {}^{(ext)}\mathfrak{D}_{k_{small}+8}[\check{\Gamma}] \lesssim \epsilon_0. $$

Theorem M5. *Under the same assumptions as above we also have the following decay estimates for $\check{R}$ and $\check{\Gamma}$ in ${}^{(int)}\mathcal{M}$:*

$$ {}^{(int)}\mathfrak{D}_{k_{small}+5}[\check{R}] + {}^{(int)}\mathfrak{D}_{k_{small}+5}[\check{\Gamma}] \lesssim \epsilon_0. $$

Note that, as an immediate consequence of Theorem M2 to Theorem M5, we have obtained, under the same assumptions as above, the following improvement of our bootstrap assumptions on decay:

$$ \mathfrak{N}^{(Dec)}_{k_{small}+5} \lesssim \epsilon_0. \tag{3.6.1} $$

3.6.2 End of the proof of the main theorem

Definition 3.28 (Definition of $\aleph(u_*)$). *Let $\epsilon_0 > 0$ and $\epsilon > 0$ be given small constants satisfying the constraint* (3.3.3). *Let $\aleph(u_*)$ be the set of all GCM admissible spacetimes $\mathcal{M}$ defined in section 3.1.2 such that*

- *u_* is the value of u on the last outgoing slice $\mathcal{C}_*$,*
- *u_* satisfies* (3.3.4),
- *the bootstrap assumptions* (3.4.1) *and* (3.4.2) *hold true, i.e., relative to the combined norms defined in section 3.2.3, we have*

$$ \mathfrak{N}^{(En)}_{k_{large}} \leq \epsilon, \quad \mathfrak{N}^{(Dec)}_{k_{small}} \leq \epsilon. $$

Definition 3.29. *Let $\mathcal{U}$ be the set of all values of $u_* \geq 0$ such that the spacetime $\aleph(u_*)$ exists.*

The following theorem shows that $\mathcal{U}$ is not empty.

Theorem M6. *There exists $\delta_0 > 0$ small enough such that for sufficiently small constants $\epsilon_0 > 0$ and $\epsilon > 0$ satisfying the constraints* (3.3.3) *and* (3.3.4), *we have $[1, 1+\delta_0] \subset \mathcal{U}$.*

In view of Theorem M6, we may define U_* as the supremum over all value of u_* that belongs to $\mathcal{U}$.

$$ U_* := \sup_{u_*\in\mathcal{U}} u_*. $$

Assume by contradiction that

$$ U_* < +\infty. $$

Then, by the continuity of the flow, $U_* \in \mathcal{U}$. Furthermore, according to the consequence (3.6.1) of Theorem M2 to Theorem M5, the bootstrap assumptions on decay (3.4.2) on any spacetime of $\aleph(U_*)$ are improved by

$$\mathfrak{N}^{(Dec)}_{k_{small}+5} \lesssim \epsilon_0.$$

To reach a contradiction, we still need an extension procedure for spacetimes in $\aleph(u_*)$ to larger values of u, as well as to improve our bootstrap assumptions on weighted energies (3.4.1). This is done in two steps.

Theorem M7. *Any GCM admissible spacetime in $\aleph(u_*)$ for some $0 < u_* < +\infty$ such that*

$$\mathfrak{N}^{(Dec)}_{k_{small}+5} \lesssim \epsilon_0,$$

has a GCM admissible extension (satisfying (3.3.4)), i.e., $u'_* > u_*$, initialized by Theorem M0, which verifies

$$\mathfrak{N}^{(Dec)}_{k_{small}} \lesssim \epsilon_0.$$

Remark 3.30. *Recall that the definition of a GCM admissible spacetime in section 3.1.2 is such that $\mathcal{T} = \{r = r_{\mathcal{T}}\}$ for some $r_{\mathcal{T}}$ satisfying*

$$2m_0\left(1 + \frac{\delta_{\mathcal{H}}}{2}\right) \leq r_{\mathcal{T}} \leq 2m_0\left(1 + \frac{3\delta_{\mathcal{H}}}{2}\right). \tag{3.6.2}$$

All results obtained so far, in particular Theorems M0–M7, hold for any choice of $r_{\mathcal{T}}$ satisfying (3.6.2), see Remark 8.2 for a more precise statement. It is at this stage, in Theorem M8 below, that we need to make a specific choice of $r_{\mathcal{T}}$ in the context of a Lebesgue point argument required for the control of top order derivatives. This choice will be made in (8.3.2).

Theorem M8. *There exists a choice of $r_{\mathcal{T}}$ satisfying (3.6.2) such that the GCM admissible spacetime exhibited in Theorem M7 satisfies in addition*

$$\mathfrak{N}^{(En)}_{k_{large}} \lesssim \epsilon_0$$

and therefore belongs to $\aleph(u'_*)$. In particular u'_* belongs to $\mathcal{U}$.

In view of Theorem M8, we have reached a contradiction, and hence

$$U_* = +\infty$$

so that the spacetime may be continued forever. This concludes the proof of the main theorem.

3.6.3 Conclusions

3.6.3.1 The Penrose diagram of $\mathcal{M}$

Complete future null infinity. We first deduce from our estimate that our spacetime $\mathcal{M}$ has a complete future null infinity $\mathcal{I}_+$. The portion of null infinity of

$\mathcal{M}$ corresponds to the limit $r \to +\infty$ along the leaves $\mathcal{C}_u$ of the outgoing geodesic foliation of ${}^{(ext)}\mathcal{M}$. As $\mathcal{C}_u$ exists for all $u \geq 0$ with suitable estimates, it suffices to prove that u is an affine parameter of $\mathcal{I}_+$. To this end, recall from our main theorem that the estimates $\mathfrak{N}^{(Dec)}_{k_{small}} \lesssim \epsilon_0$ hold which implies in particular[13]

$$\sup_{{}^{(ext)}\mathcal{M}} ru^{1+\delta_{dec}}\left(|\underline{\xi}| + \left|\underline{\omega} - \frac{m}{r^2}\right| + r^{-1}|\varsigma - 1|\right) \lesssim \epsilon_0. \tag{3.6.3}$$

As $|m - m_0| \lesssim \epsilon_0 m_0$, see Lemma 3.16, m is bounded. We infer that

$$\lim_{\mathcal{C}_u, r\to+\infty} \underline{\xi}, \underline{\omega} = 0 \text{ for all } 1 \leq u < \infty.$$

In view of the identity

$$D_3 e_3 = -2\underline{\omega} e_3 + 2\underline{\xi} e_\theta,$$

we infer that e_3 is a null geodesic generator of $\mathcal{I}_+$. Since we have $e_3(u) = \frac{2}{\varsigma}$ with $|\varsigma - 1| \lesssim \epsilon_0$ in view of (3.6.3), u is an affine parameter of $\mathcal{I}_+$ so that $\mathcal{I}_+$ is indeed complete.

Existence of a future event horizon. Next, note that the estimates $\mathfrak{N}^{(Dec)}_{k_{small}} \lesssim \epsilon_0$ also imply

$$\sup_{{}^{(int)}\mathcal{M}} \underline{u}^{1+\delta_{dec}}\left(\left|\underline{\kappa} + \frac{2}{r}\right| + \left|\kappa - \frac{2\left(1 - \frac{2m}{r}\right)}{r}\right|\right) \lesssim \epsilon_0.$$

In particular, considering the spacetime region $r \leq 2m_0(1 - \delta_{\mathcal{H}}/2)$ of ${}^{(int)}\mathcal{M}$, and in view of the estimate $|m - m_0| \lesssim \epsilon_0 m_0$, we infer, for all $r \leq 2m_0(1 - \delta_{\mathcal{H}}/2)$, that

$$\begin{aligned}\kappa &\leq 2\frac{r - 2m}{r^2} + O(\epsilon_0) \leq \frac{2}{r^2}(r - 2m_0 + 2m_0 - 2m) + O(\epsilon_0)\\ &\leq \frac{2m_0}{r^2}(-\delta_{\mathcal{H}} + \epsilon_0) + O(\epsilon_0).\end{aligned}$$

Thus, since $0 < \epsilon_0 \ll \delta_{\mathcal{H}} \ll 1$, we deduce,

$$\begin{aligned}\sup_{{}^{(int)}\mathcal{M}\left(r \leq 2m_0\left(1 - \frac{\delta_{\mathcal{H}}}{2}\right)\right)} \kappa &\leq -\frac{\delta_{\mathcal{H}}}{2m_0\left(1 - \frac{\delta_{\mathcal{H}}}{2}\right)^2} + O(\epsilon_0)\\ &\leq -\frac{\delta_{\mathcal{H}}}{4m_0}.\end{aligned}$$

Thus, all 2-spheres $S(\underline{u}, s)$ of the ingoing geodesic foliation of ${}^{(int)}\mathcal{M}$ which are located in the spacetime region $r \leq 2m_0(1 - \delta_{\mathcal{H}}/2)$ of ${}^{(int)}\mathcal{M}$ are trapped. This implies that the past of $\mathcal{I}_+$ in $\mathcal{M}$ does not contain this region, and hence $\mathcal{M}$ contains the event horizon $\mathcal{H}_+$ of a black hole in its interior. Moreover, since the timelike hypersurface $\mathcal{T}$ is foliated by the outgoing null cones $\mathcal{C}_u$ of ${}^{(ext)}\mathcal{M}$, it is in the past of $\mathcal{I}_+$. Hence, since $\mathcal{T}$ is one of the boundaries of ${}^{(int)}\mathcal{M}$, $\mathcal{H}_+$ is actually located in the interior of the region ${}^{(int)}\mathcal{M}$.

[13] Using also Proposition 3.17 for the control of ς.

Asymptotic stationarity of $\mathcal{M}$. Recall that we have introduced a vectorfield $\mathbf{T}$ in ${}^{(ext)}\mathcal{M}$ as well as one in ${}^{(int)}\mathcal{M}$ by

$$\mathbf{T} = e_3 + \Upsilon e_4 \text{ in } {}^{(ext)}\mathcal{M}, \quad \mathbf{T} = e_4 + \Upsilon e_3 \text{ in } {}^{(int)}\mathcal{M}.$$

We can easily express all components of ${}^{(T)}\pi$ in terms of $\check{\Gamma}$, $e_3(m), e_4 m$. Thus, making use of the estimate $\mathfrak{N}^{(Dec)}_{k_{small}} \lesssim \epsilon_0$ of our main theorem, we deduce

$$|{}^{(\mathbf{T})}\pi| \lesssim \frac{\epsilon_0}{ru^{1+\delta_{dec}}} \text{ in } {}^{(ext)}\mathcal{M} \text{ and } |{}^{(\mathbf{T})}\pi| \lesssim \frac{\epsilon_0}{\underline{u}^{1+\delta_{dec}}} \text{ in } {}^{(int)}\mathcal{M}.$$

In particular, $\mathbf{T}$ is an asymptotically Killing vectorfield and hence our spacetime $\mathcal{M}$ is asymptotically stationary.

The above conclusions regarding $\mathcal{I}_+$ and $\mathcal{H}_+$ allow us to draw the Penrose diagram of $\mathcal{M}$, see figure 3.3.

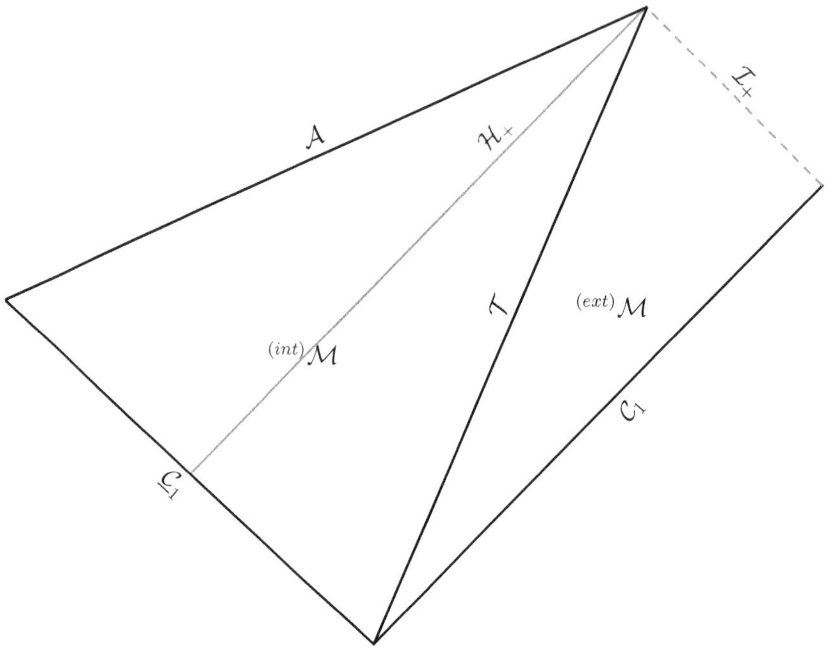

Figure 3.3: The Penrose diagram of the spacetime $\mathcal{M}$

3.6.3.2 Limits at null infinity and Bondi mass

Recall the following formula for the derivative of the Hawking mass in ${}^{(ext)}\mathcal{M}$, see Proposition 2.71,

$$e_4(m) = \frac{r}{32\pi}\int_S \left(-\frac{1}{2}\underline{\kappa}\vartheta^2 - \frac{1}{2}\check{\kappa}\vartheta\underline{\vartheta} + 2\check{\kappa}\check{\rho} + 2e_\theta(\kappa)\zeta + 2\kappa\zeta^2\right).$$

As a simple corollary of the decay estimates of our main theorem, i.e., $\mathfrak{N}^{(Dec)}_{k_{small}} \lesssim \epsilon_0$, we deduce

$$|e_4(m)| \lesssim \frac{\epsilon_0^2}{r^2 u^{1+2\delta_{dec}}}. \tag{3.6.4}$$

Since r^{-2} is integrable, we infer the existence of a limit to m as $r \to +\infty$ along $\mathcal{C}_u$

$$M_B(u) = \lim_{r\to+\infty} m(u,r) \text{ for all } 1 \le u < +\infty$$

where $M_B(u)$ is the so-called Bondi mass.

Next, we recall the following formula in ${}^{(ext)}\mathcal{M}$, see Proposition 2.63,

$$e_4(\underline{\vartheta}) + \frac{1}{2}\kappa\underline{\vartheta} = 2\,{\not{d}}_2^\star\zeta - \frac{1}{2}\underline{\kappa}\vartheta + 2\zeta^2.$$

In view of $\mathfrak{N}^{(Dec)}_{k_{small}} \lesssim \epsilon_0$, we deduce

$$|e_4(r\underline{\vartheta})| \lesssim \frac{\epsilon_0}{r^2 u^{\frac{1}{2}+\delta_{dec}}}.$$

Since r^{-2} is integrable, we infer the existence of a limit to $r\underline{\vartheta}$ as $r \to +\infty$ along $\mathcal{C}_u$

$$\underline{\Theta}(u,\cdot) = \lim_{r\to+\infty} r\underline{\vartheta}(r,u,\cdot) \text{ for all } 1 \le u < +\infty.$$

On the other hand, in view of $\mathfrak{N}^{(Dec)}_{k_{small}} \lesssim \epsilon_0$ again,

$$r|\underline{\vartheta}| \lesssim \frac{\epsilon_0}{u^{1+\delta_{dec}}}, \quad \text{on} \quad {}^{(ext)}\mathcal{M}.$$

We infer that

$$|\underline{\Theta}(u,\cdot)| \lesssim \frac{\epsilon_0}{u^{1+\delta_{dec}}} \text{ for all } 1 \le u < +\infty.$$

3.6.3.3 The spheres at null infinity are round

The Gauss curvature is given by the formula

$$K = -\rho - \frac{1}{4}\kappa\underline{\kappa} + \frac{1}{4}\vartheta\underline{\vartheta}.$$

Thus, in view of our estimates in ${}^{(ext)}\mathcal{M}$,

$$\left|K - \frac{1}{r^2}\right| \lesssim \frac{\epsilon_0}{r^3 u^{\frac{1}{2}+\delta_{dec}}}$$

so that

$$\lim_{r\to+\infty} r^2 K = 1.$$

In particular the spheres at null infinity are round.

3.6.3.4 A Bondi mass formula

Using the formula for $e_3(m)$ in ${}^{(ext)}\mathcal{M}$, see Proposition 2.71, together with the estimates $\mathfrak{N}^{(Dec)}_{k_{small}} \lesssim \epsilon_0$, we deduce

$$\left| e_3(m) + \frac{r}{64\pi}\int_S \kappa\underline{\vartheta}^2 \right| \lesssim \frac{\epsilon_0^2}{ru^{\frac{3}{2}+2\delta_{dec}}}$$

and hence

$$\left| e_3(m) + \frac{1}{8|S|}\int_S (r\underline{\vartheta})^2 \right| \lesssim \frac{\epsilon_0^2}{ru^{\frac{3}{2}+2\delta_{dec}}}.$$

Letting $r \to +\infty$ along $\mathcal{C}_u$, and using that the spheres at null infinity are round, we infer in view of the definition of M_B and $\underline{\Theta}$

$$e_3(M_B)(u) = -\frac{1}{8}\int_{\mathbb{S}^2} \underline{\Theta}^2(u,\cdot) \text{ for all } 1 \le u < +\infty.$$

Since $e_3(u) = \frac{2}{\varsigma}$ and e_3 is orthogonal to the spheres foliating $\mathcal{I}_+$, we infer $e_3 = \frac{2}{\varsigma}\partial_u$. Thus, we obtain the following Bondi mass type formula

$$\partial_u M_B(u) = -\frac{\varsigma}{16}\int_{\mathbb{S}^2} \underline{\Theta}^2(u,\cdot) \text{ for all } 1 \le u < +\infty,$$

with ς satisfying (3.6.3).

3.6.3.5 Final Bondi mass

In view of the estimate

$$|\underline{\Theta}(u,\cdot)| \lesssim \frac{\epsilon_0}{u^{1+\delta_{dec}}} \text{ for all } 1 \le u < +\infty,$$

and the control for ς in (3.6.3), we infer that

$$|\partial_u M_B(u)| \lesssim \frac{\epsilon_0^2}{u^{2+2\delta_{dec}}} \text{ for all } 1 \le u < +\infty.$$

In particular, since $u^{-2-2\delta_{dec}}$ is integrable, the limit along $\mathcal{I}_+$ exists

$$M_B(+\infty) = \lim_{u\to+\infty} M_B(u)$$

and is the so-called final Bondi mass. We denote it as m_∞, i.e., $m_\infty = M_B(+\infty)$.

Control of $m - m_\infty$. We have as a consequence of the above estimate for $\partial_u M_B$ and the definition of m_∞

$$|M_B(u) - m_\infty| \lesssim \frac{\epsilon_0^2}{u^{1+2\delta_{dec}}} \text{ for all } 1 \le u < +\infty.$$

Also, recall from (3.6.4) that we have obtained in ${}^{(ext)}\mathcal{M}$

$$|e_4(m)| \lesssim \frac{\epsilon_0^2}{r^2u^{1+2\delta_{dec}}}$$

which yields, together with the definition of $M_B(u)$, by integration in r at fixed u

$$|m(r,u) - M_B(u)| \lesssim \frac{\epsilon_0^2}{ru^{1+2\delta_{dec}}} \text{ in } {}^{(ext)}\mathcal{M}.$$

We infer

$$\sup_{{}^{(ext)}\mathcal{M}} u^{1+2\delta_{dec}}|m - m_\infty| \lesssim \epsilon_0^2. \tag{3.6.5}$$

Also, recall the following formula for the derivative of the Hawking mass in ${}^{(int)}\mathcal{M}$, see Proposition 2.71 in the context of an outgoing geodesic foliation,

$$e_3(m) = \frac{r}{32\pi}\int_S\left(-\frac{1}{2}\kappa\underline{\vartheta}^2 - \frac{1}{2}\underline{\check{\kappa}}\vartheta\underline{\vartheta} + 2\underline{\check{\kappa}}\check{\rho} - 2e_\theta(\underline{\kappa})\zeta + 2\underline{\kappa}\zeta^2\right).$$

Together with the estimates $\mathfrak{N}^{(Dec)}_{k_{small}} \lesssim \epsilon_0$, we deduce

$$|e_3(m)| \lesssim \frac{\epsilon_0^2}{\underline{u}^{2+2\delta_{dec}}} \text{ on } {}^{(int)}\mathcal{M}$$

and hence by integration in r at fixed $\underline{u}$, for $r \in [2m_0(1-\delta_{\mathcal{H}}), r_{\mathcal{T}}]$,

$$\left|m(r,\underline{u}) - m\Big(r_{\mathcal{T}},\underline{u}\Big)\right| \lesssim \frac{\epsilon_0^2}{\underline{u}^{2+2\delta_{dec}}}m_0\delta_{\mathcal{H}} \qquad \text{on } {}^{(int)}\mathcal{M}.$$

According to (3.6.5), since $\{r = r_{\mathcal{T}}\} = \mathcal{T} = {}^{(ext)}\mathcal{M} \cap {}^{(int)}\mathcal{M} \subset {}^{(ext)}\mathcal{M}$, and since $\underline{u} = u$ in $\mathcal{T}$ by the initialization of $\underline{u}$,

$$\underline{u}^{1+2\delta_{dec}}\left|m\Big(r_{\mathcal{T}},\underline{u}\Big) - m_\infty\right| \lesssim \epsilon_0^2.$$

We deduce

$$\sup_{{}^{(int)}\mathcal{M}} \underline{u}^{1+2\delta_{dec}}|m - m_\infty| \lesssim \epsilon_0^2. \tag{3.6.6}$$

Combining (3.6.5) and (3.6.6) with the estimate

$$\sup_{\mathcal{M}}|m - m_0| \lesssim \epsilon_0 m_0,$$

in the statement of our main theorem (see also Lemma 3.16), we infer that

$$|m_\infty - m_0| \lesssim \epsilon_0 m_0.$$

In particular we deduce that $m_\infty > 0$ since ϵ_0 can be made arbitrarily small.

3.6.3.6 Coordinates systems on ${}^{(ext)}\mathcal{M}$ and ${}^{(int)}\mathcal{M}$

In view of Proposition 3.17, and together with the control of the averages $\overline{\kappa}$, $\overline{\underline{\kappa}}$ provided by Lemma 3.15, the control of $\check{\kappa}$ provided by the estimates $\mathfrak{N}^{(Dec)}_{k_{small}} \lesssim \epsilon_0$, and the control of $m - m_\infty$ obtained in (3.6.5), we infer for the spacetime metric $\mathbf{g}$ on ${}^{(ext)}\mathcal{M}$ in the (u, r, θ, φ) coordinates system

$$\mathbf{g} \quad = \quad \mathbf{g}_{m_\infty, {}^{(ext)}\mathcal{M}} + O\left(\frac{\epsilon_0}{u^{1+\delta_{dec}}}\right)\Big((dr, du, rd\theta)^2, r^2(\sin\theta)^2(d\varphi)^2\Big)$$

where $\mathbf{g}_{m_\infty, {}^{(ext)}\mathcal{M}}$ denotes the Schwarzschild metric of mass $m_\infty > 0$ in outgoing Eddington-Finkelstein coordinates, i.e.,

$$\mathbf{g}_{m_\infty, {}^{(ext)}\mathcal{M}} \quad = \quad -2dudr - \left(1 - \frac{2m_\infty}{r}\right)(du)^2 + r^2\Big((d\theta)^2 + (\sin\theta)^2(d\varphi)^2\Big).$$

Also, in view of Proposition 3.18, and together with the control of the averages $\overline{\kappa}$, $\overline{\underline{\kappa}}$ provided by Lemma 3.15, the control of $\check{\underline{\kappa}}$ provided by the estimates $\mathfrak{N}^{(Dec)}_{k_{small}} \lesssim \epsilon_0$, and the control of $m - m_\infty$ obtained in (3.6.6), we infer for the spacetime metric $\mathbf{g}$ on ${}^{(int)}\mathcal{M}$ in the $(\underline{u}, r, \theta, \varphi)$ coordinates system

$$\mathbf{g} \quad = \quad \mathbf{g}_{m_\infty, {}^{(int)}\mathcal{M}} + O\left(\frac{\epsilon_0}{\underline{u}^{1+\delta_{dec}}}\right)\Big((dr, d\underline{u}, rd\theta)^2, r^2(\sin\theta)^2(d\varphi)^2\Big)$$

where $\mathbf{g}_{m_\infty, {}^{(ext)}\mathcal{M}}$ denotes the Schwarzschild metric of mass $m_\infty > 0$ in ingoing Eddington-Finkelstein coordinates, i.e.,

$$\mathbf{g}_{m_\infty, {}^{(int)}\mathcal{M}} \quad = \quad 2d\underline{u}dr - \left(1 - \frac{2m_\infty}{r}\right)(d\underline{u})^2 + r^2\Big((d\theta)^2 + (\sin\theta)^2(d\varphi)^2\Big).$$

Asymptotic of the future event horizon. We show below that $\mathcal{H}_+$ is located in the following region of ${}^{(int)}\mathcal{M}$

$$2m\left(1 - \frac{\sqrt{\epsilon_0}}{\underline{u}^{1+\delta_{dec}}}\right) \leq r \leq 2m\left(1 + \frac{\sqrt{\epsilon_0}}{\underline{u}^{1+\frac{\delta_{dec}}{2}}}\right) \text{ on } \mathcal{H}_+ \text{ for any } 1 \leq \underline{u} < +\infty. \tag{3.6.7}$$

Note first that the lower bound follows from the fact that

$$\begin{aligned}\sup_{{}^{(int)}\mathcal{M}\left(r \leq 2m\left(1 - \frac{\sqrt{\epsilon_0}}{\underline{u}^{1+\delta_{dec}}}\right)\right)} \kappa \quad &\leq \quad -\frac{\frac{\sqrt{\epsilon_0}}{\underline{u}^{1+\delta_{dec}}}}{m\left(1 - \frac{\sqrt{\epsilon_0}}{\underline{u}^{1+\delta_{dec}}}\right)^2} + O\left(\frac{\epsilon_0}{\underline{u}^{1+\delta_{dec}}}\right)\\ &\leq \quad -\frac{\sqrt{\epsilon_0}}{2m_0\underline{u}^{1+\delta_{dec}}} < 0.\end{aligned}$$

Concerning the upper bound, we need to show that any 2-sphere

$$S(\underline{u}_1) := S\left(\underline{u}_1, r = 2m\left(1 + \frac{\sqrt{\epsilon_0}}{\underline{u}_1^{1+\frac{\delta_{dec}}{2}}}\right)\right), \quad 1 \leq \underline{u}_1 < +\infty \tag{3.6.8}$$

is in the past of $\mathcal{I}_+$. Since ${}^{(ext)}\mathcal{M}$ is in the past of $\mathcal{I}_+$, it suffices to show that the forward outgoing null cone emanating from any 2-sphere (3.6.8) reaches ${}^{(ext)}\mathcal{M}$ in finite time.

Assume, by contradiction, that there exists an outgoing null geodesic, denoted by γ, perpendicular to $S(\underline{u}_1)$, that does not reach ${}^{(ext)}\mathcal{M}$ in finite time. Let e_4' be the geodesic generator of γ. In view of Lemma 2.87 on general null frame transformation, and denoting by (e_4, e_3, e_θ) the null frame[14] of ${}^{(int)}\mathcal{M}$, we look for e_4' under the form

$$e_4' = \lambda\left(e_4 + fe_\theta + \frac{1}{4}f^2 e_3\right),$$

and the fact that e_4' is geodesic implies the following transport equations along γ for f and λ in view of Lemma 2.92 (applied[15] with $\underline{f} = 0$):

$$\begin{aligned} \lambda^{-1}e_4'(f) + \left(\frac{\kappa}{2} + 2\omega\right) f &= -2\xi + E_1(f, \Gamma), \\ \lambda^{-1}e_4'(\log(\lambda)) &= 2\omega + E_2(f, \Gamma), \end{aligned}$$

where E_1 and E_2 are given schematically by

$$\begin{aligned} E_1(f, \Gamma) &= -\frac{1}{2}\vartheta f + \text{l.o.t.}, \\ E_2(f, \Gamma) &= f\zeta - \frac{1}{2}f^2\underline{\omega} - \underline{\eta} f - \frac{1}{4}f^2\underline{\kappa} + \text{l.o.t.} \end{aligned}$$

Here, l.o.t. denote terms which are cubic or higher order in f and Γ denotes the Ricci coefficients w.r.t. the original null frame (e_3, e_4, e_θ) of ${}^{(int)}\mathcal{M}$.

We then proceed as follows.

1. First, we initialize f and λ as follows on the $\gamma \cap S(\underline{u}_1)$:

$$f = 0, \quad \lambda = 1 \text{ on } \gamma \cap S(\underline{u}_1).$$

2. Then, we initiate a continuity argument by assuming for some

$$\underline{u}_1 < \underline{u}_2 < \underline{u}_1 + \left(\frac{\underline{u}_1}{\epsilon_0}\right)^{\frac{\delta_{dec}}{2}}$$

that we have

$$|f| \le \frac{\sqrt{\epsilon_0}}{\underline{u}_1^{\frac{1}{2}+\delta_{dec}}}, \quad \Upsilon \ge \frac{\sqrt{\epsilon_0}}{2\underline{u}_1^{1+\frac{\delta_{dec}}{2}}}, \quad 0 < \lambda < +\infty \text{ on } \gamma(\underline{u}_1, \underline{u}_2) \cap {}^{(int)}\mathcal{M} \quad (3.6.9)$$

where $\gamma(\underline{u}_1, \underline{u}_2)$ denotes the portion of γ in $\underline{u}_1 \le \underline{u} \le \underline{u}_2$.

3. We have

$$\lambda^{-1}e_4'(\underline{u}) = e_4(\underline{u}) + \frac{1}{4}f^2 e_3(\underline{u}) = \frac{2}{\underline{\varsigma}}.$$

Relying on our control of the ingoing geodesic foliation of ${}^{(int)}\mathcal{M}$, the above

[14] Recall that we assume by contradiction that γ does not reach ${}^{(ext)}\mathcal{M}$ and hence stays in ${}^{(int)}\mathcal{M}$.

[15] That is, we keep the direction of e_3 fixed.

assumption for f and the transport equation for f, we obtain on $\gamma(\underline{u}_1, \underline{u}_2) \cap {}^{(int)}\mathcal{M}$

$$\begin{aligned}\sup_{\gamma(\underline{u}_1,\underline{u}_2)\cap {}^{(int)}\mathcal{M}} |f| &\lesssim \frac{\epsilon_0}{\underline{u}_1^{1+\delta_{dec}}}(\underline{u}_2 - \underline{u}_1)\\ &\lesssim \frac{\epsilon_0^{1-\frac{\delta_{dec}}{2}}}{\underline{u}_1^{1+\frac{\delta_{dec}}{2}}}\end{aligned}$$

which improves our assumption in (3.6.9) on f.

4. We have in view of the control of f

$$\lambda^{-1}e_4'(r) = e_4(r) + \frac{1}{4}f^2 e_3(r) = \Upsilon + O\left(\frac{\epsilon_0}{\underline{u}_1^{1+\delta_{dec}}}\right).$$

This yields

$$\begin{aligned}\lambda^{-1}e_4'(\log(\Upsilon)) &= \frac{\frac{2m}{r^2}e_4(r) - \frac{2}{r}\lambda^{-1}e_4'(m)}{\Upsilon}\\ &= \frac{\frac{2m}{r^2}\Upsilon + O\left(\frac{\epsilon_0}{\underline{u}_1^{1+\delta_{dec}}}\right)}{\Upsilon}.\end{aligned}$$

Thanks to our assumption on the lower bound of Υ, we infer

$$\lambda^{-1}e_4'(\log(\Upsilon)) = \frac{2m}{r^2}(1 + O(\sqrt{\epsilon_0}))$$

and since we are in ${}^{(int)}\mathcal{M}$

$$\lambda^{-1}e_4'(\log(\Upsilon)) \geq \frac{1}{3m_0}.$$

Integrating from $\underline{u} = \underline{u}_1$, we deduce

$$\Upsilon \geq \frac{\sqrt{\epsilon_0}}{(1+\sqrt{\epsilon_0})\underline{u}_1^{1+\frac{\delta_{dec}}{2}}}\exp\left(\frac{\underline{u} - \underline{u}_1}{3m_0}\right)$$

which is an improvement of our assumption in (3.6.9) on Υ.

5. In view of the control of f and of the ingoing geodesic foliation of ${}^{(int)}\mathcal{M}$, we rewrite the transport equation for λ as

$$\begin{aligned}\lambda^{-1}e_4'(\log(\lambda)) &= 2\omega + E_2(f, \Gamma)\\ &= -\frac{2m}{r^2} + O\left(\frac{\epsilon_0}{\underline{u}_1^{1+\delta_{dec}}}\right).\end{aligned}$$

On the other hand, since we have obtained above

$$\lambda^{-1}e_4'(\log(\Upsilon)) = \frac{2m}{r^2}(1 + O(\sqrt{\epsilon_0}))$$

we immediately infer

$$\lambda^{-1}e_4'(\log(\lambda)\Upsilon^2) > 0, \quad \lambda^{-1}e_4'(\log(\lambda)\sqrt{\Upsilon}) < 0.$$

Integrating from $\underline{u} = \underline{u}_1$, this yields

$$\left(\frac{\sqrt{\epsilon_0}}{(1+\sqrt{\epsilon_0})\underline{u}^{1+\frac{\delta_{dec}}{2}}}\right)^2 \Upsilon^{-2} \leq \lambda \leq \left(\frac{\sqrt{\epsilon_0}}{(1+\sqrt{\epsilon_0})\underline{u}^{1+\frac{\delta_{dec}}{2}}}\right)^{\frac{1}{2}} \Upsilon^{-\frac{1}{2}}.$$

Since Υ has an explicit lower bound in view of our previous estimate, as well as an explicit upper bound since we are in ${}^{(int)}\mathcal{M}$, this yields an improvement of our assumptions in (3.6.9) for λ.

6. Since we have improved all our bootstrap assumptions (3.6.9), we infer by a continuity argument the following bound

$$\Upsilon \geq \frac{\sqrt{\epsilon_0}}{(1+\sqrt{\epsilon_0})\underline{u}_1^{1+\frac{\delta_{dec}}{2}}} \exp\left(\frac{\underline{u}-\underline{u}_1}{3m_0}\right) \text{ on } \gamma \left(\underline{u}_1, \underline{u}_1 + \left(\frac{\underline{u}_1}{\epsilon_0}\right)^{\frac{\delta_{dec}}{2}}\right) \cap {}^{(int)}\mathcal{M}.$$

Now, in this $\underline{u}$ interval, we may choose

$$\underline{u}_3 := \underline{u}_1 + 3m_0\left(1+\frac{\delta_{dec}}{2}\right)\log\left(\frac{\underline{u}_1}{\epsilon_0}\right)$$

for which we have $\Upsilon \geq 1$. This is a contradiction since $\Upsilon = O(\delta_{\mathcal{H}})$ in ${}^{(int)}\mathcal{M}$. Thus, we deduce that γ reaches ${}^{(ext)}\mathcal{M}$ before $\underline{u} = \underline{u}_3$, a contradiction to our assumption on γ. This concludes the proof of (3.6.7).

3.7 THE GENERAL COVARIANT MODULATION PROCEDURE

The role of this section is to give a short description of the results concerning our General Covariant Modulation (GCM) procedure, which is at the heart of our proof. We will apply it in ${}^{(ext)}\mathcal{M}$ under our main bootstrap assumptions **BA-E** and **BA-D**. The results stated in this section will be proved in Chapter 9.

3.7.1 Spacetime assumptions for the GCM procedure

To state our results, which are local in nature, it is convenient to consider axially symmetric polarized spacetime regions $\mathcal{R}$ foliated by two functions (u, s) such that

- On $\mathcal{R}$, (u, s) defines an outgoing geodesic foliation as in section 2.2.4.
- We denote by (e_3, e_4, e_θ) the null frame adapted to the outgoing geodesic foliation (u, s) on $\mathcal{R}$.
- We denote by $\overset{\circ}{S}$ a fixed sphere of $\mathcal{R}$

$$\overset{\circ}{S} \quad := \quad S(\overset{\circ}{u}, \overset{\circ}{s}) \tag{3.7.1}$$

and by $\overset{\circ}{r}$ the area radius of $\overset{\circ}{S}$, where $S(u, s)$ denote the 2-spheres of the outgoing geodesic foliation (u, s) on $\mathcal{R}$.

- In adapted coordinates (u, s, θ, φ) with $b = 0$, see Proposition 2.75, the spacetime metric $\mathbf{g}$ in $\mathcal{R}$ takes the form, with $\underline{\Omega} = e_3(s)$, $\underline{b} = e_3(\theta)$,

$$\mathbf{g} = -2\varsigma duds + \varsigma^2\underline{\Omega}du^2 + \gamma\left(d\theta - \frac{1}{2}\varsigma\underline{b}du\right)^2 + e^{2\Phi}d\varphi^2, \tag{3.7.2}$$

where θ is chosen such that $b = e_4(\theta) = 0$.

- The spacetime metric induced on $S(u, s)$ is given by

$$\not{g} = \gamma d\theta^2 + e^{2\Phi}d\varphi^2. \tag{3.7.3}$$

- The relation between the null frame and coordinate system is given by

$$e_4 = \partial_s, \qquad e_3 = \frac{2}{\varsigma}\partial_u + \underline{\Omega}\partial_s + \underline{b}\partial_\theta, \qquad e_\theta = \gamma^{-1/2}\partial_\theta. \tag{3.7.4}$$

- We denote the induced metric on $\overset{\circ}{S}$ by

$$\overset{\circ}{\not{g}} = \overset{\circ}{\gamma} d\theta^2 + e^{2\Phi}d\varphi^2.$$

Definition 3.31. *Let $0 < \overset{\circ}{\delta} \leq \overset{\circ}{\epsilon}$ two sufficiently small constants. Let $(\overset{\circ}{u}, \overset{\circ}{s})$ real numbers so that*

$$1 \leq \overset{\circ}{u} < +\infty, \quad 4m_0 \leq \overset{\circ}{s} < +\infty. \tag{3.7.5}$$

We define $\mathcal{R} = \mathcal{R}(\overset{\circ}{\delta}, \overset{\circ}{\epsilon})$ to be the region

$$\mathcal{R} := \left\{|u - \overset{\circ}{u}| \leq \delta_{\mathcal{R}}, \quad |s - \overset{\circ}{s}| \leq \delta_{\mathcal{R}}\right\}, \qquad \delta_{\mathcal{R}} := \overset{\circ}{\delta}(\overset{\circ}{\epsilon})^{-\frac{1}{2}}, \tag{3.7.6}$$

such that assumptions ***A1–A3*** *below, with constant $\overset{\circ}{\epsilon}$ on the background foliation of $\mathcal{R}$, are verified. The smaller constant $\overset{\circ}{\delta}$ controls the size of the GCM quantities as it will be made precise below.*

Consider the renormalized Ricci and curvature components associated to the (u, s) geodesic foliation of $\mathcal{R}$:

$$\check{\Gamma} := \left\{\check{\kappa},\, \vartheta,\, \zeta,\, \eta,\, \overline{\kappa} - \frac{2}{r},\, \overline{\underline{\kappa}} + \frac{2\Upsilon}{r},\, \check{\underline{\kappa}},\, \underline{\vartheta},\, \underline{\xi},\, \check{\underline{\omega}},\, \overline{\underline{\omega}} - \frac{m}{r^2},\, \check{\underline{\Omega}},\, (\overline{\underline{\Omega}} + \Upsilon),\, (\overline{\varsigma} + 1)\right\},$$

$$\check{R} := \left\{\alpha,\, \beta,\, \check{\rho},\, \overline{\rho} + \frac{2m}{r^3},\, \underline{\beta},\, \underline{\alpha}\right\}.$$

Since our foliation is outgoing geodesic we also have

$$\xi = \omega = 0, \quad \underline{\eta} + \zeta = 0. \tag{3.7.7}$$

We decompose $\check{\Gamma} = \Gamma_g \cup \Gamma_b$ where

$$\Gamma_g = \left\{\check{\kappa},\, \vartheta,\, \zeta,\, \check{\underline{\kappa}},\, \overline{\kappa} - \frac{2}{r},\, \overline{\underline{\kappa}} + \frac{2\Upsilon}{r}\right\},$$
$$\Gamma_b = \left\{\eta,\, \underline{\vartheta},\, \underline{\xi},\, \check{\underline{\omega}},\, \overline{\underline{\omega}} - \frac{m}{r^2},\, r^{-1}\check{\underline{\Omega}},\, r^{-1}\check{\varsigma}, r^{-1}(\overline{\underline{\Omega}} + \Upsilon),\, r^{-1}(\overline{\varsigma} - 1)\right\}. \tag{3.7.8}$$

Given an integer s_{max}, we assume the following:[16]

A1. For $k \le s_{max}$, we have on $\mathcal{R}$

$$\|\Gamma_g\|_{k,\infty} \lesssim \mathring{\epsilon} r^{-2},$$
$$\|\Gamma_b\|_{k,\infty} \lesssim \mathring{\epsilon} r^{-1}, \tag{3.7.9}$$

and

$$\|\alpha, \beta, \check{\rho}, \check{\mu}\|_{k,\infty} \lesssim \mathring{\epsilon} r^{-3},$$
$$\|e_3(\alpha, \beta)\|_{k-1,\infty} \lesssim \mathring{\epsilon} r^{-4},$$
$$\|\underline{\beta}\|_{k,\infty} \lesssim \mathring{\epsilon} r^{-2},$$
$$\|\underline{\alpha}\|_{k,\infty} \lesssim \mathring{\epsilon} r^{-1}. \tag{3.7.10}$$

A2. We have, with m_0 denoting the mass of the unperturbed spacetime,

$$\sup_{\mathcal{R}} \left|\frac{m}{m_0} - 1\right| \lesssim \mathring{\epsilon}. \tag{3.7.11}$$

A3. The metric coefficients are assumed to satisfy the following assumptions in $\mathcal{R}$, for all $k \le s_{max}$

$$r\left\|\left(\frac{\gamma}{r^2} - 1,\, \underline{b},\, \frac{e^\Phi}{r\sin\theta} - 1\right)\right\|_{\infty,k} + \|\underline{\Omega} + \Upsilon\|_{\infty,k} + \|\varsigma - 1\|_{\infty,k} \lesssim \mathring{\epsilon}. \tag{3.7.12}$$

We will assume, in addition, that there exists scalar functions $\underline{C} = \underline{C}(u,s)$, $M = M(u,s)$ such that the following small GCM conditions hold true on $\mathcal{R}$,

$$\left|\overline{\kappa} - \frac{2}{r}\right| + \left|\mathfrak{d}^k\check{\kappa}\right| + r\left|\mathfrak{d}^{k-1}(\not{d}_1^\star\underline{\kappa} - \underline{C}e^\Phi)\right|$$
$$+ r^2\left|\mathfrak{d}^{k-1}(\not{d}_1^\star\mu - Me^\Phi)\right| \lesssim \mathring{\delta} r^{-2} \text{ for all } k \le s_{max}, \tag{3.7.13}$$

$$r^{-2}\left|\int_S \eta e^\Phi\right| \lesssim \mathring{\delta}, \qquad r^{-2}\left|\int_S \underline{\xi} e^\Phi\right| \lesssim \mathring{\delta}. \tag{3.7.14}$$

[16]In applications, $s_{max} = k_{small} + 4$ in Theorem M7, and $s_{max} = k_{large} + 5$ in Theorem M0 and Theorem M6.

Also,

$$\left| \left(\frac{2}{\varsigma} + \underline{\Omega} \right) \Big|_{SP} - 1 - \frac{2m}{r} \right| \lesssim \overset{\circ}{\delta}. \tag{3.7.15}$$

Additionally we may assume on $\mathcal{R}$

$$r \left| \int_S \beta e^{\Phi} \right| \lesssim \overset{\circ}{\delta}, \qquad r \left| \int_S e_\theta(\kappa) e^{\Phi} \right| \lesssim \overset{\circ}{\delta}, \qquad r \left| \int_S e_\theta(\underline{\kappa}) e^{\Phi} \right| \lesssim \overset{\circ}{\delta}. \tag{3.7.16}$$

3.7.2 Deformations of surfaces

Definition 3.32. *We say that* $\mathbf{S}$ *is an* $O(\overset{\circ}{\epsilon})$ $\mathbf{Z}$*-polarized deformation of* $\overset{\circ}{S}$ *if there exists a map* $\Psi : \overset{\circ}{S} \to \mathbf{S}$ *of the form*

$$\Psi(\overset{\circ}{u}, \overset{\circ}{s}, \theta, \varphi) = \left(\overset{\circ}{u} + U(\theta), \overset{\circ}{s} + S(\theta), \theta, \varphi \right) \tag{3.7.17}$$

where U, S *are smooth functions defined on the interval* $[0, \pi]$ *of amplitude at most* $\overset{\circ}{\epsilon}$. *We denote by* ψ *the reduced map defined on the interval* $[0, \pi]$,

$$\psi(\theta) = (\overset{\circ}{u} + U(\theta), \overset{\circ}{s} + S(\theta), \theta). \tag{3.7.18}$$

We restrict ourselves to deformations which fix the south pole, i.e.,

$$U(0) = S(0) = 0. \tag{3.7.19}$$

3.7.3 Adapted frame transformations

We consider general null transformations introduced in Lemma 2.87,

$$\begin{aligned}
e_4' &= \lambda \left(e_4 + f e_\theta + \frac{1}{4} f^2 e_3 \right), \\
e_\theta' &= \left(1 + \frac{1}{2} f \underline{f} \right) e_\theta + \frac{1}{2} \underline{f} e_4 + \frac{1}{2} f \left(1 + \frac{1}{4} f \underline{f} \right) e_3, \\
e_3' &= \lambda^{-1} \left(\left(1 + \frac{1}{2} f \underline{f} + \frac{1}{16} f^2 \underline{f}^2 \right) e_3 + \underline{f} \left(1 + \frac{1}{4} f \underline{f} \right) e_\theta + \frac{1}{4} \underline{f}^2 e_4 \right).
\end{aligned} \tag{3.7.20}$$

Definition 3.33. *Given a deformation* $\Psi : \overset{\circ}{S} \to \mathbf{S}$ *we say that a new frame* (e_3', e_4', e_θ')*, obtained from the standard frame* (e_3, e_4, e_θ) *via the transformation* (3.7.20)*, is* $\mathbf{S}$*-adapted if we have*

$$e_\theta' = e_\theta^{\mathbf{S}} = \frac{1}{(\gamma^{\mathbf{S}})^{1/2}} \psi_\#(\partial_\theta) \tag{3.7.21}$$

where $\psi_\#(\partial_\theta)$ *is the push-forward defined by the deformation map* ψ.

The condition translates into the following relations between the functions U, S

defining the deformation and the transition functions $(f,\underline{f})$.

$$
\begin{aligned}
\varsigma^\# \partial_\theta U &= \left((\gamma^{\mathbf{S}})^\#\right)^{1/2} f^\# \left(1+\frac{1}{4}(f\underline{f})^\#\right),\\
\partial_\theta S - \frac{\varsigma^\#}{2}\underline{\Omega}^\# \partial_\theta U &= \frac{1}{2}\left((\gamma^{\mathbf{S}})^\#\right)^{1/2} \underline{f}^\#,\\
(\gamma^{\mathbf{S}})^\# &= \gamma^\# + (\varsigma^\#)^2\left(\underline{\Omega}+\frac{1}{4}\underline{b}^2\gamma\right)^\# (\partial_\theta U)^2\\
&\quad - 2\varsigma^\#\partial_\theta U \partial_\theta S - (\gamma\varsigma\underline{b})^\#\partial_\theta U,\\
U(0) &= S(0) = 0.
\end{aligned}
\tag{3.7.22}
$$

3.7.4 GCM results

Theorem 3.34 (GCMS-I). *Consider the region $\mathcal{R}$ as above, verifying the assumptions **A1–A3** and the small GCM conditions*[17] (3.7.13). *Let $\overset{\circ}{S}$ denote the sphere $\overset{\circ}{S} = S(\overset{\circ}{u},\overset{\circ}{s})$. For any fixed $\Lambda, \underline{\Lambda} \in \mathbb{R}$ verifying,*

$$
|\Lambda|,\ |\underline{\Lambda}| \ \lesssim\ \overset{\circ}{\delta}(\overset{\circ}{r})^2, \tag{3.7.23}
$$

1. *There exists a unique GCM sphere $\mathbf{S} = \mathbf{S}^{(\Lambda,\underline{\Lambda})}$, which is a deformation*[18] *of $\overset{\circ}{S}$, and an adapted null frame $e_3^{\mathbf{S}}, e_\theta^{\mathbf{S}}, e_4^{\mathbf{S}}$, such that the following GCM conditions are verified*[19]

$$
\not{d}_2^{\mathbf{S},\star}\not{d}_1^{\mathbf{S},\star}\underline{\kappa}^{\mathbf{S}} = \not{d}_2^{\mathbf{S},\star}\not{d}_1^{\mathbf{S},\star}\mu^{\mathbf{S}} = 0, \qquad \kappa^{\mathbf{S}} = \frac{2}{r^{\mathbf{S}}}. \tag{3.7.24}
$$

In addition

$$
\int_{\mathbf{S}} f e^\Phi = \Lambda, \qquad \int_{\mathbf{S}} \underline{f} e^\Phi = \underline{\Lambda}, \tag{3.7.25}
$$

where $(f,\underline{f})$ belong to the triplet $(f,\underline{f},\lambda = e^a)$ which denote the change of frame coefficients from the frame of $\overset{\circ}{S}$ to the one of $\mathbf{S}$.

2. *The transition functions $(f,\underline{f},\log\lambda)$ verify*

$$
\left\|(f,\underline{f},\log\lambda)\right\|_{\mathfrak{h}_k(\mathbf{S})} \lesssim \overset{\circ}{\delta}, \qquad k \le s_{max}+1. \tag{3.7.26}
$$

3. *The area radius $r^{\mathbf{S}}$ and Hawking mass $m^{\mathbf{S}}$ of $\mathbf{S}$ verify*

$$
\left|r^{\mathbf{S}} - \overset{\circ}{r}\right| \lesssim \overset{\circ}{\delta}, \qquad \left|m^{\mathbf{S}} - \overset{\circ}{m}\right| \lesssim \overset{\circ}{\delta}. \tag{3.7.27}
$$

The precise version of Theorem 3.34 and its proof are given in section 9.4.

The next result requires stronger assumptions for Γ_b than those made in **A1**.

---

[17] Here, the other assumptions (3.7.14) and (3.7.15) are not needed.

[18] In the sense of Definition 3.32.

[19] $\Gamma^{\mathbf{S}}$, $R^{\mathbf{S}}$ denote the Ricci and curvature components with respect to the adapted frame on $\mathbf{S}$.

A1-Strong. For $k \leq s_{max}$,

$$\Big\|\Gamma_g\Big\|_{k,\infty} \lesssim \overset{\circ}{\epsilon} r^{-2}, \qquad \Big\|\Gamma_b\Big\|_{k,\infty} \lesssim \overset{\circ}{\epsilon} r^{-1}, \qquad \Big\|\Gamma_b\Big\|_{k,\infty} \lesssim (\overset{\circ}{\epsilon})^{\frac{1}{3}} r^{-2}. \tag{3.7.28}$$

Theorem 3.35 (GCMS-II). *In addition to the assumptions of Theorem 3.34 we also assume that **A1-Strong** and* (3.7.16) *hold true. Then,*

1. *There exists a unique GCM sphere* $\mathbf{S}$*, which is a deformation of* $\overset{\circ}{S}$*, such that in addition to* (3.7.24) *the following GCM conditions also hold true on* $\mathbf{S}$*.*

$$\int_{\mathbf{S}} \beta^{\mathbf{S}} e^{\Phi} = 0, \qquad \int_{\mathbf{S}} e_\theta^{\mathbf{S}}(\underline{\kappa}^{\mathbf{S}}) e^{\Phi} = 0. \tag{3.7.29}$$

2. *The transition functions* $(f, \underline{f}, \log\lambda)$ *verify the estimates* (3.7.26).
3. *The area radius* $r^{\mathbf{S}}$ *and Hawking mass* $m^{\mathbf{S}}$ *of* $\mathbf{S}$ *verify* (3.7.27).

The precise version of Theorem 3.35 and its proof are given in 9.7.

Theorem 3.36 (GCMH). *Consider the region* $\mathcal{R}$ *as above, verifying the assumptions **A1–A3** and the small GCM conditions* (3.7.13)–(3.7.15). *Let the sphere* $\mathbf{S}_0 = \mathbf{S}_0[\overset{\circ}{u}, \overset{\circ}{s}, \Lambda_0, \underline{\Lambda}_0]$ *be the deformation of* $\overset{\circ}{S}$ *constructed in Theorem GCMS-I above.*

There exists a smooth spacelike hypersurface $\Sigma_0 \subset \mathcal{R}$ *passing through* $\mathbf{S}_0$*, a scalar function* $u^{\mathbf{S}}$ *defined on* Σ_0*, whose level surfaces are topological spheres denoted by* $\mathbf{S}$*, and a smooth collection of constants* $\Lambda^{\mathbf{S}}, \underline{\Lambda}^{\mathbf{S}}$ *verifying*

$$\Lambda^{\mathbf{S}_0} = \Lambda_0, \qquad \underline{\Lambda}^{\mathbf{S}_0} = \underline{\Lambda}_0,$$

such that the following conditions are verified:

1. *The surfaces* $\mathbf{S}$ *of constant* $u^{\mathbf{S}}$ *verify all the properties stated in Theorem GCMS-I for the prescribed constants* $\Lambda^{\mathbf{S}}, \underline{\Lambda}^{\mathbf{S}}$*. In particular they come endowed with null frames* $(e_4^{\mathbf{S}}, e_\theta^{\mathbf{S}}, e_3^{\mathbf{S}})$ *such that*
   i. *For each* $\mathbf{S}$ *the GCM conditions* (3.7.24), (3.7.25) *hold with* $\Lambda = \Lambda^{\mathbf{S}}, \underline{\Lambda} = \underline{\Lambda}^{\mathbf{S}}$.
   ii. *The transversality conditions hold true on each* $\mathbf{S}$.

$$\xi^{\mathbf{S}} = 0, \qquad \omega^{\mathbf{S}} = 0, \qquad \underline{\eta}^{\mathbf{S}} + \zeta^{\mathbf{S}} = 0. \tag{3.7.30}$$

2. *We have, for some constant* c_{Σ_0},

$$u^{\mathbf{S}} + r^{\mathbf{S}} = c_{\Sigma_0}, \qquad along \quad \Sigma_0. \tag{3.7.31}$$

3. *Let* $\nu^{\mathbf{S}}$ *be the unique vectorfield tangent to the hypersurface* Σ_0*, normal to* $\mathbf{S}$*, and normalized by* $g(\nu^{\mathbf{S}}, e_4^{\mathbf{S}}) = -2$*. There exists a unique scalar function* $a^{\mathbf{S}}$ *on* Σ_0 *such that* $\nu^{\mathbf{S}}$ *is given by*

$$\nu^{\mathbf{S}} = e_3^{\mathbf{S}} + a^{\mathbf{S}} e_4^{\mathbf{S}}.$$

The following normalization condition holds true at the south pole SP *of every sphere* $\mathbf{S}$, *i.e., at* $\theta = 0$,

$$a^{\mathbf{S}}\Big|_{SP} = -1 - \frac{2m^{\mathbf{S}}}{r^{\mathbf{S}}}. \tag{3.7.32}$$

4. *Under the additional transversality condition*[20] *on* Σ_0

$$e_4^{\mathbf{S}}(u^{\mathbf{S}}) = 0, \qquad e_4(r^{\mathbf{S}}) = \frac{r^{\mathbf{S}}}{2}\overline{\kappa^{\mathbf{S}}} = 1, \tag{3.7.33}$$

the Ricci coefficients $\eta^{\mathbf{S}}, \underline{\xi}^{\mathbf{S}}$ *are well defined and verify*

$$\int_{\mathbf{S}} \eta^{\mathbf{S}} e^{\Phi} = \int_{\mathbf{S}} \underline{\xi}^{\mathbf{S}} e^{\Phi} = 0. \tag{3.7.34}$$

5. *The transition functions* $(f, \underline{f}, \log\lambda)$ *verify the estimates* (3.7.26).
6. *The area radius* $r^{\mathbf{S}}$ *and Hawking mass* $m^{\mathbf{S}}$ *of* $\mathbf{S}$ *verify* (3.7.27).

The precise version of Theorem 3.36 and its proof are given in section 9.8.

3.7.5 Main ideas

Both theorems GCMS-II and GCMH are based on Theorem GSMS-I. They are heavily based on the transformation formulas for the Ricci and curvature coefficients recorded in Proposition 2.90.

3.7.5.1 Sketch of the proof of Theorems GSMS-I and GSMS-II

A given deformation $\Psi : \overset{\circ}{S} \to \mathbf{S}$ is fixed by the parameters U, S and transition functions $F = (f, \underline{f}, \lambda)$ connected by the system (3.7.22). Making use of the transformation formulas one can show that the GCM conditions (3.7.24)-(3.7.25) hold true if and only if the transition functions F verify a coercive nonlinear elliptic Hodge system of the form $\mathcal{D}_\Psi F = B(\Psi)$, where the operator $\mathcal{D}_\Psi$ depends on the deformation Ψ and the right-hand side B depends on both Ψ and the background foliation (see Proposition 9.33 for the precise form of the system). To find a desired GSMS deformation we have to solve a coupled system between the transport type equations in (3.7.22) and the elliptic coercive system $\mathcal{D}_\Psi F = 0$ of Proposition 9.33.

The actual proof is thus based on an iteration procedure for a sequence of deformation spheres $\mathbf{S}(n)$ of $\overset{\circ}{S}$ given by the maps $\Psi^{(n)} = (U^{(n)}, S^{(n)}) : \overset{\circ}{S} \to \mathbf{S}(n)$ and the corresponding transition functions $(f^{(n)}, \underline{f}^{(n)}, \lambda^{(n)})$. The iteration procedure for the quintets $Q^{(n)} = (U^{(n)}, S^{(n)}, f^{(n)}, \underline{f}^{(n)}, \lambda^{(n)})$, starting with the trivial quintet $Q^{(0)}$ corresponding to the zero deformation, is described in section 9.4.3. The main steps in the proof are as follows.

1. Given the triplet $(f^{(n)}, \underline{f}^{(n)}, \lambda^{(n)})$ the pair $(U^{(n)}, S^{(n)})$ defines the deformation sphere $\mathbf{S}(n)$ and the corresponding pullback map $\#_n : \overset{\circ}{S} \to \mathbf{S}(n)$ according to

[20]Here the average of $\kappa^{\mathbf{S}}$ is taken on $\mathbf{S}$. In view of the GCM conditions (3.7.24) we deduce $e_4^{\mathbf{S}}(r^{\mathbf{S}}) = 1$.

the equation (3.7.22).

2. Given the pair $\Psi^{(n)} = (U^{(n)}, S^{(n)})$ and the deformation sphere $\mathbf{S}(n)$ we define the triplet $(f^{(n+1)}, \underline{f}^{(n+1)}, \lambda^{(n+1)})$ by solving the corresponding elliptic system

$$\mathcal{D}_{\Psi(n)} F^{(n+1)} = B(\Psi^{(n)}).$$

This step is based on the crucial a priori estimates of section 9.4.1.

3. Given the new pair $(f^{(n+1)}, \underline{f}^{(n+1)})$ we make use of the equations (3.7.22) to find a unique new map $(U^{(n+1)}, S^{(n+1)})$ and thus the new deformation sphere $\mathbf{S}(n+1)$.
4. The convergence of the iterates $Q^{(n)}$, described in section 9.4.5 in the boundedness Proposition 9.42 and the contraction Proposition 9.43. The latter requires us to carefully compare the iterates $Q^{(n)}$, $Q^{(n+1)}$ by pulling them back to $\overset{\circ}{S}$. One has to be particularly careful with the behavior of the iterates on the axis of symmetry.

Theorem GSMS-II, which is an easy consequence of Theorem GSMS-I is proved in section 9.7 and the transformation formulas which relate $\int_{\mathbf{S}} \beta^{\mathbf{S}} e^{\Phi}$ to $\Lambda = \int_{\mathbf{S}} f e^{\Phi}$ and $\int_{\mathbf{S}} e_\theta^{\mathbf{S}}(\underline{\kappa}^{\mathbf{S}}) e^{\Phi}$ to $\underline{\Lambda} = \int_{\mathbf{S}} \underline{f} e^{\Phi}$. One can show that there exist choices of $\Lambda, \underline{\Lambda}$ such that $\int_{\mathbf{S}} \beta^{\mathbf{S}} e^{\Phi} = \int_{\mathbf{S}} e_\theta^{\mathbf{S}}(\underline{\kappa}^{\mathbf{S}}) e^{\Phi} = 0$.

3.7.5.2 Sketch of the proof of Theorem GCMH

The proof of Theorem GCMH makes use of Theorem GCMS-I to construct Σ_0 as a union of GCMS spheres.

Step 1. Theorem GCMS-I allows us to construct, for every value of the parameters (u, s) in $\mathcal{R}$ (i.e., such that the background spheres $S(u, s) \subset \mathcal{R}$) and every real numbers $(\Lambda, \underline{\Lambda})$, a unique GCM sphere $\mathbf{S}[u, s, \Lambda, \underline{\Lambda}]$, as a $\mathbf{Z}$-polarized deformation of $S(u, s)$. In particular (3.7.24) and (3.7.25) are verified and $\mathbf{S}_0 = \mathbf{S}_0[\overset{\circ}{u}, \overset{\circ}{s}, \Lambda_0, \underline{\Lambda}_0]$.

Step 2. We look for functions $\Psi(s), \Lambda(s), \underline{\Lambda}(s)$ such that

1. We have

$$\Psi(\overset{\circ}{s}) = \overset{\circ}{u}, \qquad \Lambda(\overset{\circ}{s}) = \Lambda_0, \qquad \underline{\Lambda}(\overset{\circ}{s}) = \underline{\Lambda}_0.$$

2. The resulting hypersurface $\Sigma_0 = \cup_s \mathbf{S}[\Psi(s), s, \Lambda(s), \underline{\Lambda}(s)]$ verifies

$$u^{\mathbf{S}} + r^{\mathbf{S}} = c_{\Sigma_0}, \quad \text{along} \quad \Sigma_0.$$

3. The additional GCM conditions (3.7.32) and (3.7.34) of Theorem GCMH are verified.

These conditions lead to a first order differential system for $\Psi(s), \Lambda(s), \underline{\Lambda}(s)$, with prescribed initial conditions at $\overset{\circ}{s}$ which allows us to determine the desired surface. The proof is given in detail in section 9.8.

3.8 OVERVIEW OF THE PROOF OF THEOREMS M0–M8

In this section, we provide a brief overview of the proof of Theorems M0–M8. In addition to the null frame adapted to the outgoing foliation of ${}^{(ext)}\mathcal{M}$ and to the null frame adapted to the ingoing foliation of ${}^{(int)}\mathcal{M}$, we have also introduced two global frames on $\mathcal{M} = {}^{(int)}\mathcal{M} \cup {}^{(ext)}\mathcal{M}$ as well as associated scalars r and m in section 3.5. Unless otherwise specified, when we discuss a particular spacetime region, i.e., ${}^{(ext)}\mathcal{M}$, ${}^{(int)}\mathcal{M}$ or $\mathcal{M}$, it should be understood that the frame as well as r and m are the ones corresponding to that region.

3.8.1 Discussion of Theorem M0

Step 1. Recall our GCM conditions on $S_* = \Sigma_* \cap \mathcal{C}_*$

$$\int_{S_*} e_\theta(\underline{\kappa})e^\Phi = 0, \qquad \int_{S_*} \beta e^\Phi = 0.$$

Recall that $\nu = e_3 + a_* e_4$ is the unique tangent vectorfield to Σ_* which is orthogonal to e_θ and normalized by $\mathbf{g}(\nu, e_4) = -2$. Using the null structure equation for $e_3(\underline{\kappa})$ and $e_3(\beta)$, as well as $e_4(\underline{\kappa})$ and $e_4(\beta)$, we obtain transport equations along Σ_* in the ν direction for

$$\int_S e_\theta(\underline{\kappa})e^\Phi \text{ and } \int_S \beta e^\Phi = 0.$$

Integrating these transport equations in ν, we propagate the control on S_* to Σ_*. In particular, we obtain the following estimates on $S_1 = \Sigma_* \cap \mathcal{C}_1$,

$$\left|\int_{S_1} e_\theta(\underline{\kappa})e^\Phi\right| + r\left|\int_{S_1} \beta e^\Phi\right| \lesssim \epsilon^2 + \frac{\epsilon}{r} \lesssim \epsilon_0, \tag{3.8.1}$$

where we used in the last inequality the dominance condition of r on Σ_*, see (3.3.4).

Step 2. We consider the transition functions $(f, \underline{f}, \lambda)$ from the frame of the initial data layer to the frame of ${}^{(ext)}\mathcal{M}$. Since

- S_1 is a sphere of ${}^{(ext)}\mathcal{M}$ in the initial data layer,
- S_1 is a sphere of the GCM hypersurface Σ_*,
- the estimate (3.8.1) holds on S_1,

we can invoke a corollary of the GCM procedure of section 3.7.4 to obtain a first improved bound for $(f, \underline{f}, \lambda)$ on S_1 with $O(\epsilon_0)$ smallness constant. After further improvements, leading in particular to a r^{-1} gain for f compared to $\underline{f}$ and λ, this ultimately leads to

$$\sup_{S_1}\left(r|\mathfrak{d}^{\leq k_{large}+4} f| + |\mathfrak{d}^{\leq k_{large}+4}(\underline{f}, \log\lambda)| + |m - m_0|\right) \lesssim \epsilon_0. \tag{3.8.2}$$

Step 3. Relying on the transport equations[21] in e_4 for $(f, \underline{f}, \lambda)$, see Corollary

[21]The control of $\underline{f}$ on $\mathcal{C}_1$ requires in fact a more subtle treatment, see Step 10 and Step 11 of the proof of Theorem M0.

2.93, and Proposition 2.71 for m, we propagate (3.8.2) to $\mathcal{C}_1$, and then, proceeding similarly in the e_3 direction to propagate the estimates to $\underline{\mathcal{C}}_1$, we finally obtain

$$\sup_{\mathcal{C}_1 \cup \underline{\mathcal{C}}_1} \left(r|\mathfrak{d}^{\leq k_{large}+1} f| + |\mathfrak{d}^{\leq k_{large}+1}(\underline{f}, \log \lambda)| + |m - m_0| \right) \lesssim \epsilon_0. \tag{3.8.3}$$

Together with the control of the initial data layer foliation and the transformation formulas of Proposition 2.90, we then obtain the desired estimates on $\mathcal{C}_1 \cup \underline{\mathcal{C}}_1$ for the curvature components.

Remark 3.37. *The fact that $\underline{f}$ and λ display a r loss compared to f in* (3.8.3) *does not affect the desired estimates for the curvature components on $\mathcal{C}_1 \cup \underline{\mathcal{C}}_1$, see Remark 4.4. See also Remark 4.5 for a heuristic explanation of this a priori anomalous behavior.*

3.8.2 Discussion of Theorem M1

Here are the main steps in the proof of Theorem M1.

Step 1. Consider the global frame on $\mathcal{M}$ constructed in Proposition 3.26 and the definition of $\mathfrak{q}$ on $\mathcal{M}$ with respect to that frame, see section 2.3.3 for the definition of $\mathfrak{q}$ with respect to any null frame. According to Theorem 2.108 we have

$$\square_2 \mathfrak{q} + V\mathfrak{q} = N, \quad V = \kappa\underline{\kappa} \tag{3.8.4}$$

where the nonlinear term $N = \mathrm{Err}[\square_{\mathbf{g}}\mathfrak{q}]$ is a long expression of terms quadratic, or higher order, in $\check{\Gamma}, \check{R}$ involving various powers of r. Making use of the symbolic notation introduced in Definition 2.94 we have, see (2.4.8),

$$\mathrm{Err}[\square_2 \mathfrak{q}] \;=\; r^2 \mathfrak{d}^{\leq 2}(\Gamma_g \cdot (\alpha, \beta)) + e_3\Big(r^3 \mathfrak{d}^{\leq 2}(\Gamma_g \cdot (\alpha, \beta))\Big) + \mathfrak{d}^{\leq 1}(\Gamma_g \cdot \mathfrak{q}) + \text{l.o.t.}$$

where the terms denoted by l.o.t. are higher order in $(\check{\Gamma}, \check{R})$.

Remark 3.38. *Recall from Remark 2.109 that the above good structure of the error term $\mathrm{Err}[\square_2 \mathfrak{q}]$ only holds in a frame for which $\xi = 0$ for $r \geq 4m_0$ and $\eta \in \Gamma_g$. This is why, in Theorem M1, $\mathfrak{q}$ is defined relative to the global frame of Proposition 3.26, see also Remark 2.110.*

Step 2. We follow the Dafermos-Rodnianski version of the vectorfield method to derive desired decay estimates. We recall that, in the context of a wave equation of the form $\square_{(Sch)}\psi = 0$ on Schwarzschild spacetime, their strategy consists in the following:

- Start by deriving Morawetz energy type estimates for ψ with nondegenerate flux energies and the usual degeneracy of bulk integrals at $r = 3m$.
- Derive r^p-weighted estimates for $0 < p < 2$ and use them, in conjunction with the Morawetz estimates, to derive decay estimates.
- The decay estimates obtained by using the standard r^p-weighted approach are too weak to be useful in our nonlinear approach. We improve them by making use of a recent variation of the Dafermos-Rodnianski approach due to Angelopoulos, Aretakis and Gajic [5] which is based on first commuting the wave equation

$\square_{(Sch)}\psi = 0$ with $r^2(e_4 + r^{-1})$ and then repeating the process described for the resulting new equation. This procedure allows us to derive the improved decay estimates consistent with our decay norms.
- Derive estimates for higher derivatives by commuting with $\mathbf{T}$, $r\not{\mathfrak{d}}$, the redshift vectorfield, and re_4.

Step 3. The estimates mentioned in Step 2 have to be adapted to the case of our equation (3.8.4). There are four main differences to take into account:

- The application of the vectorfield method to our context produces various nontrivial commutator terms which have to be absorbed. This is taken care of by our bootstrap assumption for $\check{\Gamma}, \check{R}$, as well as, in some cases, by integration by parts.
- The presence of the potential V is mostly advantageous but various modifications have to be nevertheless made, especially near the trapping region.[22]
- The presence of the nonlinear term N is the most important complication. The precise null structure of N is essential and various integrations by parts are needed.
- The quadratic terms involving η in N can only be treated provided the definition of $\mathfrak{q}$ is done with respect to the global frame on $\mathcal{M}$ constructed in Proposition 3.26, for with η behaves better in powers of r^{-1}.

3.8.3 Discussion of Theorem M2

Recall from section 2.3.3 that $\mathfrak{q}$ is defined with respect to a general null frame as follows:

$$\mathfrak{q} = r^4\left(e_3(e_3(\alpha)) + (2\underline{\kappa} - 6\underline{\omega})e_3(\alpha) + \left(-4e_3(\underline{\omega}) + 8\underline{\omega}^2 - 8\underline{\omega}\,\underline{\kappa} + \frac{1}{2}\underline{\kappa}^2\right)\alpha\right)$$

which yields the following transport equation for α

$$e_3(e_3(\alpha)) + (2\underline{\kappa} - 6\underline{\omega})e_3(\alpha) + \left(-4e_3(\underline{\omega}) + 8\underline{\omega}^2 - 8\underline{\omega}\,\underline{\kappa} + \frac{1}{2}\underline{\kappa}^2\right)\alpha \;=\; \frac{\mathfrak{q}}{r^4}.$$

Recall also that $\mathfrak{q}$, controlled in Theorem M1, is defined w.r.t. the global frame of Proposition 3.26 whose normalization is such that, in particular, $\underline{\omega}$ is a small quantity. Also, since we have

$$e_3(r) = \frac{r}{2}\underline{\kappa} + \text{l.o.t.}$$

we infer

$$e_3(e_3(r^2\alpha)) \;=\; \frac{\mathfrak{q}}{r^2} + \text{l.o.t.}$$

Integrating twice this transport equation from $\mathcal{C}_1$ where we control the initial data — and in particular α — in view of Theorem M0, and using the decay for $\mathfrak{q}$ provided

[22]At the linear level, on a Schwarzschild spacetime, this step was also treated (minus the improved decay) in the paper [26].

by Theorem M1, we deduce[23]

$$\sup_{^{(ext)}\mathcal{M}} \left(\frac{r^2(2r+u)^{1+\delta_{extra}}}{\log(1+u)} + r^3(2r+u)^{\frac{1}{2}+\delta_{extra}}\right) |\mathfrak{d}^{\leq k_{small}+20}\alpha| \lesssim \epsilon_0,$$
$$\sup_{^{(ext)}\mathcal{M}} \left(\frac{r^3(2r+u)^{1+\delta_{extra}}}{\log(1+u)} + r^4(2r+u)^{\frac{1}{2}+\delta_{extra}}\right) |\mathfrak{d}^{\leq k_{small}+19}e_3(\alpha)| \lesssim \epsilon_0.$$

Now that we control α in the global frame of Proposition 3.26, we need to go back to the frame of $^{(ext)}\mathcal{M}$. By invoking the relationships between our various frames of $^{(ext)}\mathcal{M}$, see Proposition 3.26 and Proposition 3.20, and the transformation formula for α, we infer

$$^{(ext)}\mathfrak{D}_{k_{small}+20}\left[{}^{(ext)}\alpha\right] \lesssim \epsilon_0$$

and hence the conclusion of Theorem M2.

3.8.4 Discussion of Theorem M3

Here are the main steps in the proof of Theorem M3.

Step 1. To derive decay estimates for $\underline{\alpha}$ in $\mathcal{M}$, we first recall the following Teukolsky-Starobinsky identity, see (2.3.15),

$$e_3(r^2e_3(r\mathfrak{q})) + 2\underline{\omega}r^2e_3(r\mathfrak{q}) = r^7\left\{\not{d}_2^\star\not{d}_1^\star\not{d}_1\not{d}_2\underline{\alpha} + \frac{3}{2}\underline{\kappa}\rho e_4\underline{\alpha} - \frac{3}{2}\kappa\rho e_3(\underline{\alpha})\right\} + \text{l.o.t.}$$

where l.o.t. denotes terms which are quadratic of higher, and where all quantities are defined w.r.t. the global frame of Proposition 3.26. Then, introducing the vectorfield

$$\widetilde{T} = e_4 - \frac{1}{\overline{\underline{\kappa}}}\left(\overline{\kappa} + \overline{\underline{\kappa}\Omega} - \overline{\check{\underline{\kappa}}\check{\Omega}}\right)e_3,$$

we rewrite the identity as

$$6m\widetilde{T}\underline{\alpha} + r^4\not{d}_2^\star\not{d}_1^\star\not{d}_1\not{d}_2\underline{\alpha} = \frac{1}{r^3}\Big(e_3(r^2e_3(r\mathfrak{q})) + 2\underline{\omega}r^2e_3(r\mathfrak{q})\Big) + \text{l.o.t.} \quad (3.8.5)$$

As it turns out, see Remark 6.11, this is a forward parabolic equation on each hypersurface of constant r in $^{(int)}\mathcal{M}$.

Step 2. Thanks to

- the control in $^{(int)}\mathcal{M}$ of the RHS of (3.8.5) which follows from the decay estimates of Theorem M1 for $\mathfrak{q}$, as well as the bootstrap assumptions for the quadratic and higher order terms,
- the control of $\underline{\alpha}$ on $\underline{\mathcal{C}}_1$ — i.e., of the initial data of (3.8.5) — provided by Theorem M0,
- parabolic estimates for the forward parabolic equation (3.8.5),

[23]Recall that δ_{extra} has been introduced in Theorem M1 and satisfies $\delta_{extra} > \delta_{dec}$.

we obtain the desired decay estimates for $\underline{\alpha}$ in ${}^{(int)}\mathcal{M}$.

Step 3. It remains to control $\underline{\alpha}$ on Σ_*. Recall that ν denotes the unique tangent vectorfield to Σ_* which can be written as $\nu = e_3 + ae_4$. The Teukolsky-Starobinsky identity of Step 1 can then be written as

$$6m\nu\underline{\alpha} + r^4 \not{d}_2^\star \not{d}_1^\star \not{d}_1 \not{d}_2\underline{\alpha} \quad = \quad \frac{1}{r^3}\Big(e_3(r^2 e_3(r\mathfrak{q})) + 2\underline{\omega}r^2 e_3(r\mathfrak{q})\Big) + \text{l.o.t.} \quad (3.8.6)$$

where l.o.t. denotes terms which are quadratic of higher, as well as terms which are linear but display additional decay in r. This is a forward parabolic equation along Σ_*. To obtain the desired decay for $\underline{\alpha}$ along Σ_*, one then proceeds as in Step 2, using in addition, for the linear term with extra decay in r, the behavior (3.3.4) of r on Σ_*.

3.8.5 Discussion of Theorem M4

Here are the main steps in the proof of Theorem M4.

Step 1. We derive decay estimates for the spacelike GCM hypersurface Σ_*. More precisely, thanks to

- the GCM conditions on Σ_*

$$\kappa = \frac{2}{r}, \quad \not{d}_2^\star \not{d}_1^\star \underline{\kappa} = 0, \quad \not{d}_2^\star \not{d}_1^\star \mu = 0, \quad \int_S \eta e^\Phi = 0, \quad \int_S \underline{\xi} e^\Phi = 0,$$

- the control of $\mathfrak{q}$ in ${}^{(ext)}\mathcal{M}$, established in Theorem M1, and hence in particular on Σ_*,
- the control of α of the outgoing geodesic foliation in ${}^{(ext)}\mathcal{M}$, established in Theorem M2, and hence in particular on Σ_*,
- the control of $\underline{\alpha}$ on Σ_*, established in Theorem M3,
- the dominance condition (3.3.4) of r on Σ_*

$$r\big|_{\Sigma_*} \geq \epsilon_0^{-\frac{2}{3}} u^{1+\delta_{dec}},$$

- the identity (2.3.11) relating $\mathfrak{q}$ to derivatives of ρ, i.e.,

$$\mathfrak{q} \quad = \quad r^4\left(\not{d}_2^\star \not{d}_1^\star \rho + \frac{3}{4}\rho\underline{\kappa}\vartheta + \frac{3}{4}\rho\kappa\underline{\vartheta} + \cdots\right),$$

- elliptic estimates for Hodge operators on the 2-spheres foliating Σ_*,

we infer the control with improved decay of all Ricci and curvature components on the spacelike hypersurface Σ_*.

Step 2. We derive decay estimates for the outgoing geodesic foliation of ${}^{(ext)}\mathcal{M}$. More precisely:

- First, we propagate the estimates involving only $u^{-\frac{1}{2}-\delta_{dec}}$ decay in u from Σ_* to ${}^{(ext)}\mathcal{M}$.
- We then focus on the harder to recover estimates, i.e., the ones involving $u^{-1-\delta_{dec}}$

decay in u. We proceed as follows.

- We first propagate the main GCM quantities $\check{\kappa}$, $\check{\mu}$, and a renormalized quantity involving $\check{\underline{\kappa}}$ (see the quantity Ξ in Lemma 7.36) from Σ_* to ${}^{(ext)}\mathcal{M}$.
- We then recover the estimates involving $u^{-1-\delta_{dec}}$ decay in u on $\mathcal{T}$. To this end, we use that we control the main GCM quantities, $\underline{\alpha}$ from Theorem M3 (since $\mathcal{T}$ belongs both to ${}^{(ext)}\mathcal{M}$ and ${}^{(int)}\mathcal{M}$), $\mathfrak{q}$ and α from Theorem M1–M2, and the estimates are then derived somewhat in the spirit of the ones on Σ_*, in particular by relying on elliptic estimates for Hodge operators on the 2-spheres foliating $\mathcal{T}$.
- To recover the remaining estimates in ${}^{(ext)}\mathcal{M}$ involving $u^{-1-\delta_{dec}}$ decay in u, we integrate the transport equations in e_4 forward from $\mathcal{T}$, which concludes the proof of Theorem M4.

3.8.6 Discussion of Theorem M5

Here are the main steps in the proof of Theorem M5.

Step 1. We first derive decay estimates for the ingoing geodesic foliation of ${}^{(int)}\mathcal{M}$ on the timelike hypersurface $\mathcal{T}$. More precisely, thanks to

- the fact that the null frame of ${}^{(int)}\mathcal{M}$ is defined on $\mathcal{T}$ as a simple conformal renormalization of the null frame of ${}^{(ext)}\mathcal{M}$ in view of its initialization, see section 3.1.2,
- the control of the outgoing geodesic foliation of ${}^{(ext)}\mathcal{M}$ on $\mathcal{T}$ obtained in Theorem M4,

this allows us to transfer the decay estimates for $({}^{(ext)}\check{R}, {}^{(ext)}\check{\Gamma})$ to $({}^{(int)}\check{R}, {}^{(int)}\check{\Gamma})$ on $\mathcal{T}$.

Step 2. We derive on ${}^{(int)}\mathcal{M}$ decay estimates for the ingoing geodesic foliation of ${}^{(int)}\mathcal{M}$. More precisely, thanks to

- the improved decay estimates for $\underline{\alpha}$ in ${}^{(int)}\mathcal{M}$ derived in Theorem M3,
- the improved decay estimates for $\check{\Gamma}$ and $\check{R}$ on $\mathcal{T}$ derived in Step 1,
- the null structure equations and Bianchi identities,

we infer $O(\epsilon_0 u^{-1-\delta_{dec}})$ decay estimates for $\check{\Gamma}$ and $\check{R}$ corresponding to the ingoing geodesic foliation of ${}^{(int)}\mathcal{M}$ which concludes the proof of Theorem M5.

3.8.7 Discussion of Theorem M6

Step 1. Using

a) The control of the initial data layer,
b) Theorem GCMS-II of section 3.7.4,
c) Theorem GCMH of section 3.7.4,

we produce a smooth hypersurface Σ_* in the initial data layer starting from a GCM sphere S_*, and satisfying all the required properties for the future spacelike boundary of a GCM admissible spacetime, according to item 3 of Definition 3.2.

Step 2. We then consider the outgoing geodesic foliation initialized on Σ_* which foliates the region we denote ${}^{(ext)}\mathcal{M}$, to the past of Σ_*, and included in the outgoing part ${}^{(ext)}\mathcal{L}_0$ of the initial data layer. In order to control it, we consider the transition functions $(f, \underline{f}, \lambda)$ from the background frame of the initial data layer to the frame of ${}^{(ext)}\mathcal{M}$. These functions satisfy transport equations in e_4 with the right-hand side depending on $(f, \underline{f}, \lambda)$ and the Ricci coefficients of the background foliation. Integrating the transport equations from Σ_*, where $(f, \underline{f}, \lambda)$ are under control as a by-product of the use of Theorem GCMH in Step 1, we obtain the control of $(f, \underline{f}, \lambda)$ in ${}^{(ext)}\mathcal{M}$. Using the transformation formulas of Proposition 2.90, and using the control of the initial data layer, we then infer the desired control (i.e., with ϵ_0 smallness constant and suitable r-weights) for the Ricci coefficients and curvature components of the foliation of ${}^{(ext)}\mathcal{M}$.

Step 3. ${}^{(ext)}\mathcal{M}$ terminates on a timelike hypersurface $\mathcal{T}$ of constant area radius.[24] We then consider the ingoing geodesic foliation initialized on $\mathcal{T}$ according to item 4 of Definition 3.2, which foliates the region we denote ${}^{(int)}\mathcal{M}$, included in the ingoing part ${}^{(int)}\mathcal{L}_0$ of the initial data layer. Proceeding as in Step 2, relying on transport equations in e_3 instead of e_4, we then derive the desired control (i.e., with ϵ_0 smallness constant) for the Ricci coefficients and curvature components of the foliation of ${}^{(int)}\mathcal{M}$, thus concluding the proof of Theorem M6.

3.8.8 Discussion of Theorem M7

From the assumptions of Theorem M7 we are given a GCM admissible spacetime $\mathcal{M} = \mathcal{M}(u_*) \in \aleph(u_*)$ verifying the following improved bounds, for a universal constant $C > 0$,

$$\mathfrak{N}^{(Dec)}_{k_{small}+5}(\mathcal{M}) \leq C\epsilon_0$$

provided by Theorems M1–M5. We then proceed as follows.

Step 1. We extend $\mathcal{M}$ by a local existence argument, to a strictly larger spacetime $\mathcal{M}^{(extend)}$, with a naturally extended foliation and the following slightly increased bounds

$$\mathfrak{N}^{(Dec)}_{k_{small}+5}(\mathcal{M}^{(extend)}) \leq 2C\epsilon_0$$

but which may not verify our admissibility criteria.

Step 2. Using

a) The control of the extended spacetime $\mathcal{M}^{(extend)}$,
b) Theorem GCMS-II of section 3.7.4,
c) Theorem GCMH of section 3.7.4,

we produce a small piece of smooth GCM hypersurface $\widetilde{\Sigma}_*$ in $\mathcal{M}^{(extend)} \setminus \mathcal{M}$ starting from a GCM sphere $\widetilde{S}_*$.

[24]With respect to the foliation of ${}^{(ext)}\mathcal{M}$.

Step 3. By a continuity argument based on a priori estimates, we extend $\widetilde{\Sigma}_*$ all the way to the initial data layer, while ensuring that it remains in $\mathcal{M}^{(extend)} \setminus \mathcal{M}$ and satisfying all the required properties for the future spacelike boundary of a GCM admissible spacetime, according to item 3 of Definition 3.2.

Step 4. We then consider the outgoing geodesic foliation initialized on $\widetilde{\Sigma}_*$ which foliates the region we denote ${}^{(ext)}\widetilde{\mathcal{M}}$, included in the outgoing part of $\mathcal{M}^{(extend)}$. In order to control it, we consider the transition functions $(f, \underline{f}, \lambda)$ from the background frame of the initial data layer to the frame of ${}^{(ext)}\widetilde{\mathcal{M}}$. These functions satisfy transport equations in e_4 with the right-hand side depending on $(f, \underline{f}, \lambda)$ and the Ricci coefficients of the background foliation. Integrating the transport equations from $\widetilde{\Sigma}_*$, where $(f, \underline{f}, \lambda)$ are under control as a by-product of the use of Theorem GCMH in Step 2, we obtain the control of $(f, \underline{f}, \lambda)$ in ${}^{(ext)}\widetilde{\mathcal{M}}$. Using the transformation formulas of Proposition 2.90, and using the control of the initial data layer, we then derive the desired control (i.e., with ϵ_0 smallness constant and suitable u and r weights) for the Ricci coefficients and curvature components of the foliation of ${}^{(ext)}\widetilde{\mathcal{M}}$.

Step 5. ${}^{(ext)}\widetilde{\mathcal{M}}$ terminates on a timelike hypersurface $\widetilde{\mathcal{T}}$ of constant area radius.[25] We then consider the ingoing geodesic foliation initialized on $\widetilde{\mathcal{T}}$ according to item 4 of Definition 3.2, which foliates the region we denote ${}^{(int)}\widetilde{\mathcal{M}}$, included in the ingoing part of $\mathcal{M}^{(extend)}$. Proceeding as in Step 4, relying on transport equations in e_3 instead of e_4, we then derive the desired control (i.e., with ϵ_0 smallness constant and suitable $\underline{u}$-weights) for the Ricci coefficients and curvature components of the foliation of ${}^{(int)}\widetilde{\mathcal{M}}$, thus concluding the proof of Theorem M7.

3.8.9 Discussion of Theorem M8

So far, we have only improved our bootstrap assumptions on decay estimates. We now improve our bootstrap assumptions on energies and weighted energies for $\check{R}$ and $\check{\Gamma}$ relying on an iterative procedure recovering derivatives one by one.[26]

Step 0. Let $I_{m_0,\delta_{\mathcal{H}}}$ the interval of $\mathbb{R}$ defined by

$$I_{m_0,\delta_{\mathcal{H}}} := \left[2m_0\left(1+\frac{\delta_{\mathcal{H}}}{2}\right), 2m_0\left(1+\frac{3\delta_{\mathcal{H}}}{2}\right)\right]. \tag{3.8.7}$$

Recall that $\mathcal{T} = \{r = r_{\mathcal{T}}\}$, where $r_{\mathcal{T}} \in I_{m_0,\delta_{\mathcal{H}}}$, and note, see also Remark 3.30, that the results of Theorems M0–M7 hold for any $r_{\mathcal{T}} \in I_{m_0,\delta_{\mathcal{H}}}$.

It is at this stage that we need to make a specific choice of $r_{\mathcal{T}}$ in the context of a Lebesgue point argument. More precisely, we choose $r_{\mathcal{T}}$ such that we have

$$\int_{\{r=r_{\mathcal{T}}\}} |\mathfrak{d}^{\leq k_{large}}\check{R}|^2 = \inf_{r_0\in I_{m_0,\delta_{\mathcal{H}}}} \int_{\{r=r_0\}} |\mathfrak{d}^{\leq k_{large}}\check{R}|^2. \tag{3.8.8}$$

---

[25] With respect to the foliation of ${}^{(ext)}\widetilde{\mathcal{M}}$.

[26] See also [33] for a related strategy to recover higher order derivatives from the control of lower order ones.

In view of this definition, and since $\mathcal{T} = \{r = r_{\mathcal{T}}\}$, we infer that

$$\int_{\mathcal{T}} |\mathfrak{d}^{\leq k_{large}} \check{R}|^2 \lesssim \int_{{}^{(ext)}\mathcal{M}\left(r \in I_{m_0, \delta_{\mathcal{H}}}\right)} |\mathfrak{d}^{\leq k_{large}} \check{R}|^2. \tag{3.8.9}$$

Remark 3.39. *From now on, we may thus assume that the spacetime $\mathcal{M}$ satisfies the conclusions of Theorem M0 and Theorem M7, as well as* (3.8.9)*, and our goal is to prove Theorem M8, i.e., to prove that $\mathfrak{N}^{(En)}_{k_{large}} \lesssim \epsilon_0$ holds.*

Step 1. The $O(\epsilon_0)$ decay estimates derived in Theorem M7 imply in particular the following (non-sharp) consequence

$$\mathfrak{N}^{(En)}_{k_{small}} \lesssim \epsilon_0,$$

where we recall[27]

$$\mathfrak{N}^{(En)}_k = {}^{(ext)}\mathfrak{R}_k[\check{R}] + {}^{(ext)}\mathfrak{G}_k[\check{\Gamma}] + {}^{(int)}\mathfrak{R}_k[\check{R}] + {}^{(int)}\mathfrak{G}_k[\check{\Gamma}].$$

This allows us to initialize our iteration scheme in the next step.

Step 2. Next, for J such that $k_{small} \leq J \leq k_{large} - 1$, consider the iteration assumption

$$\mathfrak{N}^{(En)}_J \lesssim \epsilon_{\mathcal{B}}[J], \tag{3.8.10}$$

where

$$\epsilon_{\mathcal{B}}[J] := \sum_{j=k_{small}-2}^{J} (\epsilon_0)^{\ell(j)} \mathcal{B}^{1-\ell(j)} + \epsilon_0^{\ell(J)}\mathcal{B}, \quad \ell(j) := 2^{k_{small}-2-j}, \tag{3.8.11}$$

$$\mathcal{B} := \left(\int_{{}^{(ext)}\mathcal{M}\left(r \in I_{m_0, \delta_{\mathcal{H}}}\right)} |\mathfrak{d}^{\leq k_{large}} \check{R}|^2 \right)^{\frac{1}{2}}. \tag{3.8.12}$$

In view of Step 1, (3.8.10) holds for $J = k_{small}$. From now on, we assume that (3.8.10) holds for J such that $k_{small} \leq J \leq k_{large} - 2$, and our goal is to show that this also holds for $J+1$ derivatives.

Step 3. Using the Teukolsky wave equations for α and $\underline{\alpha}$, as well as a wave equation for $\check{\rho}$, see Proposition 8.14, we derive Morawetz type estimates for $J+1$ derivatives of these quantities in terms of $O(\epsilon_{\mathcal{B}}[J] + \epsilon_0 \mathfrak{N}^{(En)}_{J+1})$.

Step 4. Relying on Bianchi identities, we also derive Morawetz type estimates for $J+1$ derivatives for β and $\underline{\beta}$. As a consequence, we obtain Morawetz type estimates for $J+1$ derivatives of all curvature components in terms of $O(\epsilon_{\mathcal{B}}[J] + \epsilon_0 \mathfrak{N}^{(En)}_{J+1})$.

[27]See sections 3.2.1 and 3.2.2 for the definition of our norms measuring energies for curvature components and Ricci coefficients.

Step 5. As a consequence of Step 4, we immediately obtain, for any $r_0 \geq 4m_0$,

$$^{(int)}\mathfrak{R}_{J+1}[\check{R}] + {}^{(ext)}\mathfrak{R}_{J+1}[\check{R}] \leq {}^{(ext)}\mathfrak{R}^{\geq r_0}_{J+1}[\check{R}] + O(r_0^{10}(\epsilon_{\mathcal{B}}[J] + \epsilon_0 \mathfrak{N}^{(En)}_{J+1})).$$

Step 6. Relying on the Bianchi identities, we derive r^p-weighted estimates for $J+1$ derivatives of curvature on $r \geq r_0$ with $r_0 \geq 4m_0$. We obtain

$$^{(ext)}\mathfrak{R}^{\geq r_0}_{J+1}[\check{R}] \lesssim \frac{1}{r_0^{\delta_B}} {}^{(ext)}\mathfrak{G}^{\geq r_0}_{k}[\check{\Gamma}] + r_0^{10}(\epsilon_{\mathcal{B}}[J] + \epsilon_0 \mathfrak{N}^{(En)}_{J+1}).$$

Step 7. Next, we estimate the Ricci coefficients of $^{(ext)}\mathcal{M}$. To control them, we rely on the null structure equations in $^{(ext)}\mathcal{M}$. Using the null structure equations in $^{(ext)}\mathcal{M}$ and the GCM conditions on Σ_*, we derive the following weighted estimates for $J+1$ derivatives of the Ricci coefficients:

$$^{(ext)}\mathfrak{G}_{J+1}[\check{\Gamma}] \lesssim {}^{(ext)}\mathfrak{R}_{J+1}[\check{R}] + \epsilon_{\mathcal{B}}[J] + \epsilon_0 \mathfrak{N}^{(En)}_{J+1}.$$

Together with the estimates of Step 5 and Step 6, we infer for a large enough choice of r_0

$$^{(ext)}\mathfrak{G}_{J+1}[\check{\Gamma}] + {}^{(int)}\mathfrak{R}_{J+1}[\check{R}] + {}^{(ext)}\mathfrak{R}_{J+1}[\check{R}] \lesssim \epsilon_{\mathcal{B}}[J] + \epsilon_0 \mathfrak{N}^{(En)}_{J+1}.$$

Step 8. Next, we estimate the Ricci coefficients of $^{(int)}\mathcal{M}$. Using the information on $\mathcal{T}$ induced by Step 7 and the null structure equations in $^{(int)}\mathcal{M}$, we derive

$$^{(int)}\mathfrak{G}_{J+1}[\check{\Gamma}] \lesssim {}^{(int)}\mathfrak{R}_{J+1}[\check{R}] + \epsilon_{\mathcal{B}}[J] + \epsilon_0 \mathfrak{N}^{(En)}_{J+1} + \left(\int_{\mathcal{T}} |\mathfrak{d}^{J+1}({}^{(ext)}\check{R})|^2\right)^{\frac{1}{2}}.$$

We need to deal with the last term. Relying on a trace theorem in the spacetime region $^{(ext)}\mathcal{M}(r \in I_{m_0,\delta_{\mathcal{H}}})$, and the fact that $J+2 \leq k_{large}$, we obtain

$$\left(\int_{\mathcal{T}} |\mathfrak{d}^{J+1}({}^{(ext)}\check{R})|^2\right)^{\frac{1}{2}} \lesssim \left(\int_{{}^{(ext)}\mathcal{M}\left(r\in I_{m_0,\delta_{\mathcal{H}}}\right)} |\mathfrak{d}^{k_{large}}\check{R}|^2\right)^{\frac{1}{4}} \left({}^{(ext)}\mathfrak{R}_{J+1}[\check{R}]\right)^{\frac{1}{2}} + {}^{(ext)}\mathfrak{R}_{J+1}[\check{R}].$$

Step 9. The last estimate of Step 7 and the two estimates of Step 8 yield, for $\epsilon_0 > 0$ small enough,

$$\mathfrak{N}^{(En)}_{J+1} \lesssim \epsilon_{\mathcal{B}}[J] + \left(\int_{{}^{(ext)}\mathcal{M}\left(r\in I_{m_0,\delta_{\mathcal{H}}}\right)} |\mathfrak{d}^{k_{large}}\check{R}|^2\right)^{\frac{1}{4}} \left(\epsilon_{\mathcal{B}}[J] + \epsilon_0 \mathfrak{N}^{(En)}_{J+1}\right)^{\frac{1}{2}}.$$

In view of the definition (3.8.11) of $\epsilon_{\mathcal{B}}[J]$, we infer that

$$\mathfrak{N}^{(En)}_{J+1} \lesssim \epsilon_{\mathcal{B}}[J+1]$$

which is the iteration assumption (3.8.10) for $J+1$ derivatives. We deduce that (3.8.10) holds for all $J \leq k_{large} - 1$, and hence

$$\mathfrak{N}^{(En)}_{k_{large}-1} \lesssim \epsilon_{\mathcal{B}}[k_{large} - 1].$$

Step 10. Relying on the conclusion of Step 9, and arguing as in Step 3 to Step 7, we obtain the conclusion of Step 7 for $J = k_{large} - 1$, i.e.,

$$ {}^{(ext)}\mathfrak{G}_{k_{large}}[\check{\Gamma}] + {}^{(int)}\mathfrak{R}_{k_{large}}[\check{R}] + {}^{(ext)}\mathfrak{R}_{k_{large}}[\check{R}] \lesssim \epsilon_{\mathcal{B}}[k_{large} - 1] + \epsilon_0 \mathfrak{N}^{(En)}_{k_{large}}. $$

We then infer that

$$ \epsilon_{\mathcal{B}}[k_{large} - 1] \lesssim \epsilon_0 + \epsilon_0 \mathfrak{N}^{(En)}_{k_{large}} $$

which yields, together with the last estimate of Step 9,

$$ {}^{(ext)}\mathfrak{G}_{k_{large}}[\check{\Gamma}] + {}^{(int)}\mathfrak{R}_{k_{large}}[\check{R}] + {}^{(ext)}\mathfrak{R}_{k_{large}}[\check{R}] \lesssim \epsilon_0 + \epsilon_0 \mathfrak{N}^{(En)}_{k_{large}}. $$

Step 11. It remains to recover ${}^{(int)}\mathfrak{G}_{k_{large}}[\check{\Gamma}]$. Arguing as for the first estimate of Step 8 with $J = k_{large} - 1$, we have

$$ {}^{(int)}\mathfrak{G}_{k_{large}}[\check{\Gamma}] \lesssim {}^{(int)}\mathfrak{R}_{k_{large}}[\check{R}] + \epsilon_{\mathcal{B}}[k_{large} - 1] + \epsilon_0 \mathfrak{N}^{(En)}_{k_{large}} + \left(\int_{\mathcal{T}} |\mathfrak{d}^{k_{large}}({}^{(ext)}\check{R})|^2 \right)^{\frac{1}{2}}. $$

Thanks to the outcome of Step 10, we deduce that

$$ {}^{(int)}\mathfrak{G}_{k_{large}}[\check{\Gamma}] \lesssim \epsilon_0 + \epsilon_0 \mathfrak{N}^{(En)}_{k_{large}} + \left(\int_{\mathcal{T}} |\mathfrak{d}^{k_{large}}({}^{(ext)}\check{R})|^2 \right)^{\frac{1}{2}} $$

and hence, for $\epsilon_0 > 0$ small enough, using again the last estimate of Step 10,

$$ \mathfrak{N}^{(En)}_{k_{large}} \lesssim \epsilon_0 + \left(\int_{\mathcal{T}} |\mathfrak{d}^{k_{large}}({}^{(ext)}\check{R})|^2 \right)^{\frac{1}{2}}. $$

It remains to estimate the last term of the RHS of the previous inequality. It is at this stage that we use the choice of $r_{\mathcal{T}}$, or rather its consequence (3.8.9), which implies

$$ \left(\int_{\mathcal{T}} |\mathfrak{d}^{k_{large}}({}^{(ext)}\check{R})|^2 \right)^{\frac{1}{2}} \lesssim \epsilon_0 + \epsilon_0 \mathfrak{N}^{(En)}_{k_{large}} $$

so that we finally obtain, for $\epsilon_0 > 0$ small enough,

$$ \mathfrak{N}^{(En)}_{k_{large}} \lesssim \epsilon_0 $$

which concludes the proof of Theorem M8.

3.9 STRUCTURE OF THE REST OF THE BOOK

The rest of this book is devoted to the proof of Theorems M0–M8, as well as our GCM procedure. More precisely,

1. Theorem M0, together with other first consequences of the bootstrap assump-

tions, is proved in Chapter 4.

2. Theorem M1 is proved in Chapter 5.
3. Theorems M2 and M3 are proved in Chapter 6.
4. Theorems M4 and M5 are proved in Chapter 7.
5. Theorems M6, M7 and M8 are proved in Chapter 8.
6. Our GCM procedure is described in details in Chapter 9.
7. Chapter 10 contains estimates for Regge-Wheeler type wave equations used in Theorem M1.
8. Many of the long calculations are to be found in the appendices.

Chapter Four

Consequences of the Bootstrap Assumptions

4.1 PROOF OF THEOREM M0

According to the statement of Theorem M0 we consider given the initial layer $\mathcal{L}_0 = {}^{(ext)}\mathcal{L}_0 \cup {}^{(int)}\mathcal{L}_0$ as defined in Definition 3.1. We also assume that the initial layer norm verifies

$$\sup_{k \leq k_{large}+5} \mathfrak{J}_k \lesssim \epsilon_0 \tag{4.1.1}$$

where $\mathfrak{J}_k = {}^{(ext)}\mathfrak{J}_k + {}^{(int)}\mathfrak{J}_k + \mathfrak{J}'_k$ and

$$\begin{aligned}
{}^{(ext)}\mathfrak{J}_0 &= \sup_{{}^{(ext)}\mathcal{L}_0} \left[r^{\frac{7}{2}+\delta_B} \left(|\alpha| + |\beta|\right) + r^3 \left|\rho + \frac{2m_0}{r^3}\right| + r^2|\underline{\beta}| + r|\underline{\alpha}| \right] \\
&+ \sup_{{}^{(ext)}\mathcal{L}_0} r^2 \left(|\vartheta| + \left|\kappa - \frac{2}{r}\right| + |\zeta| + \left|\underline{\kappa} + \frac{2\left(1 - \frac{2m_0}{r}\right)}{r}\right| \right) \\
&+ \sup_{{}^{(ext)}\mathcal{L}_0} r \left(|\underline{\vartheta}| + \left|\underline{\omega} - \frac{m_0}{r^2}\right| + |\underline{\xi}| \right) \\
&+ \sup_{{}^{(ext)}\mathcal{L}_0\left({}^{(ext)}r_0 \geq 4m_0\right)} \Bigg[r\left|\frac{\gamma}{r^2} - 1\right| + r|\underline{b}| + |\underline{\Omega} + \Upsilon| + |\varsigma - 1| \\
&\qquad + r\left|\frac{e^{\Phi}}{r\sin\theta} - 1\right| \Bigg],
\end{aligned}$$

$$\begin{aligned}
{}^{(int)}\mathfrak{J}_0 &= \sup_{{}^{(int)}\mathcal{L}_0} \left(|\underline{\alpha}| + |\underline{\beta}| + \left|\rho + \frac{2m_0}{r^3}\right| + |\beta| + |\alpha| \right) \\
&+ \sup_{{}^{(int)}\mathcal{L}_0} \Bigg[|\vartheta| + \left|\kappa - \frac{2\left(1 - \frac{2m_0}{r}\right)}{r}\right| + |\zeta| + \left|\underline{\kappa} + \frac{2}{r}\right| + |\underline{\vartheta}| \\
&\qquad + \left|\omega + \frac{m_0}{r^2}\right| + |\xi| \Bigg],
\end{aligned}$$

$$\mathfrak{J}'_0 = \sup_{{}^{(int)}\mathcal{L}_0 \cap {}^{(ext)}\mathcal{L}_0} \left(|f| + |\underline{f}| + |\log(\lambda_0^{-1}\lambda)|\right), \quad \lambda_0 = {}^{(ext)}\lambda_0 = 1 - \frac{2m_0}{{}^{(ext)}r_{\mathcal{L}_0}},$$

with $\mathfrak{J}_k$ the corresponding higher derivatives norms obtained by replacing each component by $\mathfrak{d}^{\leq k}$ of it. In the definition of $\mathfrak{J}'_0$ above, $(f, \underline{f}, \lambda)$ denote the transition functions of Lemma 2.87 from the frame of the outgoing part ${}^{(ext)}\mathcal{L}_0$ of the initial data layer to the frame of the ingoing part ${}^{(int)}\mathcal{L}_0$ of the initial data layer in the region ${}^{(int)}\mathcal{L}_0 \cap {}^{(ext)}\mathcal{L}_0$.

We divide the proof of Theorem M0 in the following steps.

Step 1. We have the following lemma.

Lemma 4.1. *We have on* ${}^{(ext)}\mathcal{M}$

$$
\begin{aligned}
e_4\left(\int_S e_\theta(\underline{\kappa})e^\Phi\right) &= \int_S\Big(-\underline{\kappa}e_\theta(\kappa)+4K\zeta-\vartheta e_\theta(\underline{\kappa})+2e_\theta(\zeta^2)\Big)e^\Phi,\\
e_4\left(\int_S \beta e^\Phi\right) &= \int_S\left(-\frac{1}{2}\kappa\beta+\zeta\alpha-\frac{1}{2}\vartheta\beta\right)e^\Phi,\\
e_3\left(\int_S \beta e^\Phi\right) &= -\frac{1}{4}\int_S e_\theta(\kappa\underline{\kappa})e^\Phi+3\overline{\rho}\int_S\eta e^\Phi+\frac{1}{4}\int_S e_\theta(\vartheta\underline{\vartheta})e^\Phi\\
&\quad+\int_S\left(\frac{1}{2}\underline{\kappa}\beta+2\underline{\omega}\beta+3\eta\check{\rho}-\vartheta\underline{\beta}+\underline{\xi}\alpha-\frac{1}{2}\underline{\vartheta}\beta\right)e^\Phi\\
&\quad+Err\left[e_3\left(\int_S\beta e^\Phi\right)\right],
\end{aligned}
$$

and

$$
\begin{aligned}
e_3\left(\int_S e_\theta(\underline{\kappa})e^\Phi\right) &= \underline{\kappa}e_3\left(\int_S\zeta e^\Phi\right)-\overline{\underline{\kappa}}\int_S\underline{\beta}e^\Phi+\int_S\Bigg\{-\check{\underline{\kappa}}\underline{\beta}-\frac{1}{2}\underline{\kappa}^2\zeta+6\rho\underline{\xi}\\
&\quad-2\underline{\omega}e_\theta(\underline{\kappa})-\frac{1}{2}\underline{\vartheta}(e_\theta(\underline{\kappa})-\underline{\kappa}\zeta)+Err[\not{d}_2\not{d}_2^\star\underline{\xi}]\Bigg\}e^\Phi\\
&\quad+Err\left[e_3\left(\int_S e_\theta(\underline{\kappa})e^\Phi\right)\right]+\int_S\check{\underline{\kappa}}\left(e_3(\zeta)+\left(\frac{3}{2}\underline{\kappa}-\frac{1}{2}\underline{\vartheta}\right)\zeta\right)e^\Phi\\
&\quad-\check{\underline{\kappa}}\int_S\left(e_3(\zeta)+\left(\frac{3}{2}\underline{\kappa}-\frac{1}{2}\underline{\vartheta}\right)\zeta\right)e^\Phi-\underline{\kappa}Err\left[e_3\left(\int_S\zeta e^\Phi\right)\right].
\end{aligned}
$$

Proof. We have in ${}^{(ext)}\mathcal{M}$, see Proposition 2.63,

$$e_4(\underline{\kappa})+\frac{1}{2}\kappa\,\underline{\kappa} \;=-2\not{d}_1\zeta+2\rho-\tfrac{1}{2}\vartheta\,\underline{\vartheta}+2\zeta^2.$$

Together with the following commutation relation

$$[e_\theta,e_4] \;=\; \frac{1}{2}(\kappa+\vartheta)e_\theta,$$

we infer

$$e_4(e_\theta(\underline{\kappa}))+\kappa e_\theta(\underline{\kappa})+\frac{1}{2}\vartheta e_\theta(\underline{\kappa})+\frac{1}{2}\underline{\kappa}e_\theta(\kappa)=2\not{d}_1^\star\not{d}_1\zeta+2e_\theta(\rho)-\frac{1}{2}e_\theta(\vartheta\,\underline{\vartheta})+2e_\theta(\zeta^2).$$

Also, we have in view of Proposition 2.74 the following identity

$$e_3(e_\theta(\underline{\kappa}))-\underline{\kappa}e_3(\zeta)=-2\not{d}_2\not{d}_2^\star\underline{\xi}-\underline{\kappa}\underline{\beta}+\underline{\kappa}^2\zeta-\frac{3}{2}\underline{\kappa}e_\theta\underline{\kappa}+6\rho\underline{\xi}-2\underline{\omega}e_\theta(\underline{\kappa})+\mathrm{Err}[\not{d}_2\not{d}_2^\star\underline{\xi}].$$

Next, in view of Corollary 2.65, we have in ${}^{(ext)}\mathcal{M}$

$$
\begin{aligned}
e_4\left(\int_S e_\theta(\underline{\kappa})e^\Phi\right) &= \int_S \left(e_4(e_\theta(\underline{\kappa})) + \left(\frac{3}{2}\kappa - \frac{1}{2}\vartheta\right)e_\theta(\underline{\kappa})\right)e^\Phi\\
e_4\left(\int_S \beta e^\Phi\right) &= \int_S \left(e_4(\beta) + \left(\frac{3}{2}\kappa - \frac{1}{2}\vartheta\right)\beta\right)e^\Phi,\\
e_3\left(\int_S \beta e^\Phi\right) &= \int_S \left(e_3(\beta) + \left(\frac{3}{2}\underline{\kappa} - \frac{1}{2}\underline{\vartheta}\right)\beta\right)e^\Phi + \mathrm{Err}\left[e_3\left(\int_S \beta e^\Phi\right)\right],
\end{aligned}
$$

and

$$
\begin{aligned}
& e_3\left(\int_S e_\theta(\underline{\kappa})e^\Phi\right) - \underline{\kappa}e_3\left(\int_S \zeta e^\Phi\right)\\
=& \int_S \left(e_3(e_\theta(\underline{\kappa})) + \left(\frac{3}{2}\underline{\kappa} - \frac{1}{2}\underline{\vartheta}\right)e_\theta(\underline{\kappa})\right)e^\Phi + \mathrm{Err}\left[e_3\left(\int_S e_\theta(\underline{\kappa})e^\Phi\right)\right]\\
& -\underline{\kappa}\int_S\left(e_3(\zeta) + \left(\frac{3}{2}\underline{\kappa} - \frac{1}{2}\underline{\vartheta}\right)\zeta\right)e^\Phi - \underline{\kappa}\mathrm{Err}\left[e_3\left(\int_S \zeta e^\Phi\right)\right]\\
=& \int_S \left(e_3(e_\theta(\underline{\kappa})) - \underline{\kappa}e_3(\zeta) + \left(\frac{3}{2}\underline{\kappa} - \frac{1}{2}\underline{\vartheta}\right)(e_\theta(\underline{\kappa}) - \underline{\kappa}\zeta)\right)e^\Phi\\
& + \mathrm{Err}\left[e_3\left(\int_S e_\theta(\underline{\kappa})e^\Phi\right)\right] + \int_S \check{\underline{\kappa}}\left(e_3(\zeta) + \left(\frac{3}{2}\underline{\kappa} - \frac{1}{2}\underline{\vartheta}\right)\zeta\right)e^\Phi\\
& - \check{\underline{\kappa}}\int_S\left(e_3(\zeta) + \left(\frac{3}{2}\underline{\kappa} - \frac{1}{2}\underline{\vartheta}\right)\zeta\right)e^\Phi - \underline{\kappa}\mathrm{Err}\left[e_3\left(\int_S \zeta e^\Phi\right)\right].
\end{aligned}
$$

Together with the above identities for $e_4(e_\theta(\underline{\kappa}))$ and $e_3(e_\theta(\underline{\kappa}))$, as well as the Bianchi identities of Proposition 2.63 for $e_4(\beta)$ and $e_3(\beta)$, we infer

$$
\begin{aligned}
e_4\left(\int_S e_\theta(\underline{\kappa})e^\Phi\right) &= \int_S \Bigg(\frac{1}{2}\kappa e_\theta(\underline{\kappa}) - \frac{1}{2}\underline{\kappa}e_\theta(\kappa) + 2 \not{d}_1^\star \not{d}_1\zeta + 2e_\theta(\rho) - \frac{1}{2}e_\theta(\vartheta\,\underline{\vartheta})\\
&\quad -\vartheta e_\theta(\underline{\kappa}) + 2e_\theta(\zeta^2)\Bigg)e^\Phi,\\
e_4\left(\int_S \beta e^\Phi\right) &= \int_S \left(-\frac{1}{2}\kappa\beta + \not{d}_2\alpha + \zeta\alpha - \frac{1}{2}\vartheta\beta\right)e^\Phi,\\
e_3\left(\int_S \beta e^\Phi\right) &= \int_S \left(\frac{1}{2}\underline{\kappa}\beta + e_\theta(\rho) + 2\underline{\omega}\beta + 3\eta\rho - \vartheta\underline{\beta} + \underline{\xi}\alpha - \frac{1}{2}\underline{\vartheta}\beta\right)e^\Phi\\
&\quad +\mathrm{Err}\left[e_3\left(\int_S \beta e^\Phi\right)\right],
\end{aligned}
$$

and

$$
\begin{aligned}
& e_3\left(\int_S e_\theta(\underline{\kappa})e^\Phi\right) - \underline{\kappa} e_3\left(\int_S \zeta e^\Phi\right) \\
= & \int_S \Bigg(- 2\, d\!\!/_2\, d\!\!/_2^\star\underline{\xi} - \underline{\kappa}\underline{\beta} - \frac{1}{2}\underline{\kappa}^2\zeta + 6\rho\underline{\xi} - 2\underline{\omega} e_\theta(\underline{\kappa}) - \frac{1}{2}\underline{\vartheta}(e_\theta(\underline{\kappa}) - \underline{\kappa}\zeta) \\
& + \mathrm{Err}[d\!\!/_2\, d\!\!/_2^\star\underline{\xi}]\Bigg)e^\Phi + \mathrm{Err}\left[e_3\left(\int_S e_\theta(\underline{\kappa})e^\Phi\right)\right] \\
& + \int_S \check{\underline{\kappa}}\left(e_3(\zeta) + \left(\frac{3}{2}\underline{\kappa} - \frac{1}{2}\underline{\vartheta}\right)\zeta\right)e^\Phi - \check{\underline{\kappa}}\int_S\left(e_3(\zeta) + \left(\frac{3}{2}\underline{\kappa} - \frac{1}{2}\underline{\vartheta}\right)\zeta\right)e^\Phi \\
& - \underline{\kappa}\mathrm{Err}\left[e_3\left(\int_S \zeta e^\Phi\right)\right].
\end{aligned}
$$

Using in particular the fact that $d\!\!/_2^\star(e^\Phi) = 0$, that $d\!\!/_2^\star$ is the adjoint of $d\!\!/_2$, and the identity $d\!\!/_1^\star\, d\!\!/_1 = d\!\!/_2\, d\!\!/_2^\star + 2K$, we deduce

$$
\begin{aligned}
e_4\left(\int_S e_\theta(\underline{\kappa})e^\Phi\right) &= \int_S \Bigg(\frac{1}{2}\kappa e_\theta(\underline{\kappa}) - \frac{1}{2}\underline{\kappa} e_\theta(\kappa) + 4K\zeta + 2e_\theta(\rho) - \frac{1}{2}e_\theta(\vartheta\,\underline{\vartheta}) \\
& \quad - \vartheta e_\theta(\underline{\kappa}) + 2e_\theta(\zeta^2)\Bigg)e^\Phi, \\
e_4\left(\int_S \beta e^\Phi\right) &= \int_S \left(-\frac{1}{2}\kappa\beta + \zeta\alpha - \frac{1}{2}\vartheta\beta\right)e^\Phi, \\
e_3\left(\int_S \beta e^\Phi\right) &= \int_S e_\theta(\rho)e^\Phi + 3\overline{\rho}\int_S \eta e^\Phi + \int_S \Bigg(\frac{1}{2}\underline{\kappa}\beta + 2\underline{\omega}\beta + 3\eta\check{\rho} \\
& \quad - \vartheta\underline{\beta} + \underline{\xi}\alpha - \frac{1}{2}\underline{\vartheta}\beta\Bigg)e^\Phi + \mathrm{Err}\left[e_3\left(\int_S \beta e^\Phi\right)\right],
\end{aligned}
$$

and

$$
\begin{aligned}
e_3\left(\int_S e_\theta(\underline{\kappa})e^\Phi\right) &= \underline{\kappa} e_3\left(\int_S \zeta e^\Phi\right) - \overline{\underline{\kappa}}\int_S \underline{\beta} e^\Phi + \int_S \Bigg(- \check{\underline{\kappa}}\underline{\beta} - \frac{1}{2}\underline{\kappa}^2\zeta + 6\rho\underline{\xi} \\
& \quad - 2\underline{\omega} e_\theta(\underline{\kappa}) - \frac{1}{2}\underline{\vartheta}(e_\theta(\underline{\kappa}) - \underline{\kappa}\zeta) + \mathrm{Err}[d\!\!/_2\, d\!\!/_2^\star\underline{\xi}]\Bigg)e^\Phi \\
& \quad + \mathrm{Err}\left[e_3\left(\int_S e_\theta(\underline{\kappa})e^\Phi\right)\right] + \int_S \check{\underline{\kappa}}\left(e_3(\zeta) + \left(\frac{3}{2}\underline{\kappa} - \frac{1}{2}\underline{\vartheta}\right)\zeta\right)e^\Phi \\
& \quad - \check{\underline{\kappa}}\int_S\left(e_3(\zeta) + \left(\frac{3}{2}\underline{\kappa} - \frac{1}{2}\underline{\vartheta}\right)\zeta\right)e^\Phi - \underline{\kappa}\mathrm{Err}\left[e_3\left(\int_S \zeta e^\Phi\right)\right].
\end{aligned}
$$

Finally, from the identity (2.1.20) for $e_\theta(K)$ and the formula for K, we have

$$
\int_S e_\theta(\rho)e^\Phi = -\frac{1}{4}\int_S e_\theta(\kappa\underline{\kappa})e^\Phi + \frac{1}{4}\int_S e_\theta(\vartheta\underline{\vartheta})e^\Phi.
$$

We deduce

$$e_4\Big(\int_S e_\theta(\underline{\kappa})e^\Phi\Big) = \int_S\Big(-\underline{\kappa}e_\theta(\kappa)+4K\zeta-\vartheta e_\theta(\underline{\kappa})+2e_\theta(\zeta^2)\Big)e^\Phi,$$

and

$$\begin{aligned}
e_3\Big(\int_S \beta e^\Phi\Big) &= -\frac{1}{4}\int_S e_\theta(\kappa\underline{\kappa})e^\Phi+3\overline{\rho}\int_S\eta e^\Phi+\frac{1}{4}\int_S e_\theta(\vartheta\underline{\vartheta})e^\Phi\\
&+\int_S\left(\frac{1}{2}\underline{\kappa}\beta+2\underline{\omega}\beta+3\eta\check{\rho}-\vartheta\underline{\beta}+\underline{\xi}\alpha-\frac{1}{2}\underline{\vartheta}\beta\right)e^\Phi\\
&+\mathrm{Err}\left[e_3\Big(\int_S\beta e^\Phi\Big)\right]
\end{aligned}$$

which concludes the proof of Lemma 4.1. □

Step 2. Using the transport equations of Lemma 4.1 and the bootstrap assumptions on decay for $k=0,1$ derivatives in ${}^{(ext)}\mathcal{M}$, we infer in that region, and in particular on Σ_*

$$\begin{aligned}
\left|e_4\Big(\int_S e_\theta(\underline{\kappa})e^\Phi\Big)\right| &\lesssim \frac{1}{r}\left|\int_S e_\theta(\kappa)e^\Phi\right|+\frac{1}{r^2}\left|\int_S\zeta e^\Phi\right|+\frac{\epsilon^2}{r^2u^{1+\delta_{dec}}},\\
\left|e_4\Big(\int_S \beta e^\Phi\Big)\right| &\lesssim \frac{1}{r}\left|\int_S\beta e^\Phi\right|+\frac{\epsilon^2}{r^2u^{1+\delta_{dec}}},\\
\left|e_3\Big(\int_S \beta e^\Phi\Big)\right| &\lesssim \left|\int_S e_\theta(\kappa\underline{\kappa})e^\Phi\right|+r^{-3}\left|\int_S\eta e^\Phi\right|+\frac{1}{r}\left|\int_S\beta e^\Phi\right|+\frac{\epsilon^2}{ru^{1+\delta_{dec}}},\\
\left|e_3\Big(\int_S e_\theta(\underline{\kappa})e^\Phi\Big)\right| &\lesssim \frac{1}{r}\left|e_3\Big(\int_S\zeta e^\Phi\Big)\right|+\frac{1}{r}\left|\int_S\underline{\beta}e^\Phi\right|\\
&+\frac{1}{r^3}\left|\int_S\underline{\xi}e^\Phi\right|+\frac{1}{r^2}\left|\int_S\zeta e^\Phi\right|+\frac{1}{r^2}\left|\int_S e_\theta(\underline{\kappa})e^\Phi\right|+\frac{\epsilon^2}{u^{2+2\delta_{dec}}}.
\end{aligned}$$

Recall the following GCM conditions

$$\kappa=\frac{2}{r},\qquad \int_S\eta e^\Phi=0,\qquad \int_S\underline{\xi}e^\Phi=0\quad\text{on }\Sigma_*.$$

We deduce on Σ_*

$$\begin{aligned}
\left|e_4\Big(\int_S e_\theta(\underline{\kappa})e^\Phi\Big)\right| &\lesssim \frac{1}{r^2}\left|\int_S\zeta e^\Phi\right|+\frac{\epsilon^2}{r^2u^{1+\delta_{dec}}},\\
\left|e_4\Big(\int_S \beta e^\Phi\Big)\right| &\lesssim \frac{1}{r}\left|\int_S\beta e^\Phi\right|+\frac{\epsilon^2}{r^2u^{1+\delta_{dec}}},\\
\left|e_3\Big(\int_S \beta e^\Phi\Big)\right| &\lesssim \frac{1}{r}\left|\int_S e_\theta(\underline{\kappa})e^\Phi\right|+\frac{1}{r}\left|\int_S\beta e^\Phi\right|+\frac{\epsilon^2}{ru^{1+\delta_{dec}}},\\
\left|e_3\Big(\int_S e_\theta(\underline{\kappa})e^\Phi\Big)\right| &\lesssim \frac{1}{r}\left|e_3\Big(\int_S\zeta e^\Phi\Big)\right|+\frac{1}{r}\left|\int_S\underline{\beta}e^\Phi\right|\\
&+\frac{1}{r^2}\left|\int_S\zeta e^\Phi\right|+\frac{1}{r^2}\left|\int_S e_\theta(\underline{\kappa})e^\Phi\right|+\frac{\epsilon^2}{u^{2+2\delta_{dec}}}.
\end{aligned}$$

Also, projecting both Codazzi on e^Φ, using $\slashed{d}_2^\star(e^\Phi) = 0$ and the fact that $\slashed{d}_2^\star$ is the adjoint of $\slashed{d}_2$, and using also the GCM condition for κ on Σ_*, we have on Σ_*

$$\begin{aligned}\int_S \underline{\beta} e^\Phi &= -\frac{1}{2}\int_S \slashed{d}_1^\star \underline{\kappa} e^\Phi - \frac{1}{2}\int_S \zeta\underline{\kappa} e^\Phi + \frac{1}{2}\int_S \underline{\vartheta}\zeta e^\Phi,\\ \int_S \zeta e^\Phi &= r\int_S \beta e^\Phi + \frac{r}{2}\int_S \vartheta\zeta e^\Phi.\end{aligned}$$

Together with the bootstrap assumptions on decay for $k = 0, 1$ derivatives in ${}^{(ext)}\mathcal{M}$, we infer on Σ_*

$$\begin{aligned}\left|e_3\left(\int_S \zeta e^\Phi\right)\right| &\lesssim r\left|e_3\left(\int_S \beta e^\Phi\right)\right| + \left|\int_S \beta e^\Phi\right| + \frac{\epsilon^2}{u^{1+\delta_{dec}}},\\ \left|\int_S \zeta e^\Phi\right| &\lesssim r\left|\int_S \beta e^\Phi\right| + \frac{\epsilon^2}{u^{1+\delta_{dec}}},\\ \left|\int_S \underline{\beta} e^\Phi\right| &\lesssim \left|\int_S e_\theta(\underline{\kappa}) e^\Phi\right| + \frac{1}{r}\left|\int_S \zeta e^\Phi\right| + \frac{\epsilon^2}{u^{\frac{3}{2}+\delta_{dec}}}\\ &\lesssim \left|\int_S e_\theta(\underline{\kappa}) e^\Phi\right| + \left|\int_S \beta e^\Phi\right| + \frac{\epsilon^2}{u^{\frac{3}{2}+\delta_{dec}}} + \frac{\epsilon^2}{r u^{1+\delta_{dec}}}.\end{aligned}$$

We have thus on Σ_*

$$\begin{aligned}\left|e_4\left(\int_S e_\theta(\underline{\kappa}) e^\Phi\right)\right| &\lesssim \frac{1}{r}\left|\int_S \beta e^\Phi\right| + \frac{\epsilon^2}{r^2 u^{1+\delta_{dec}}},\\ \left|e_4\left(\int_S \beta e^\Phi\right)\right| &\lesssim \frac{1}{r}\left|\int_S \beta e^\Phi\right| + \frac{\epsilon^2}{r^2 u^{1+\delta_{dec}}},\\ \left|e_3\left(\int_S \beta e^\Phi\right)\right| &\lesssim \frac{1}{r}\left|\int_S e_\theta(\underline{\kappa}) e^\Phi\right| + \frac{1}{r}\left|\int_S \beta e^\Phi\right| + \frac{\epsilon^2}{r u^{1+\delta_{dec}}},\\ \left|e_3\left(\int_S e_\theta(\underline{\kappa}) e^\Phi\right)\right| &\lesssim \left|e_3\left(\int_S \beta e^\Phi\right)\right| + \frac{1}{r}\left|\int_S \beta e^\Phi\right| + \frac{1}{r}\left|\int_S e_\theta(\underline{\kappa}) e^\Phi\right|\\ &\quad + \frac{\epsilon^2}{r u^{1+\delta_{dec}}} + \frac{\epsilon^2}{u^{2+2\delta_{dec}}}.\end{aligned} \tag{4.1.2}$$

In view of the behavior (3.3.4) of r on Σ_*, and plugging the third equation in the fourth, we infer on Σ_*

$$\begin{aligned}\left|e_4\left(\int_S e_\theta(\underline{\kappa}) e^\Phi\right)\right| &\lesssim \frac{\epsilon_0^{\frac{2}{3}}}{u_*^{1+\delta_{dec}}}\left|\int_S \beta e^\Phi\right| + \frac{\epsilon^2}{r^2 u^{1+\delta_{dec}}},\\ \left|e_4\left(\int_S \beta e^\Phi\right)\right| &\lesssim \frac{\epsilon_0^{\frac{2}{3}}}{u_*^{1+\delta_{dec}}}\left|\int_S \beta e^\Phi\right| + \frac{\epsilon^2}{r^2 u^{1+\delta_{dec}}},\\ \left|e_3\left(\int_S \beta e^\Phi\right)\right| &\lesssim \frac{1}{r}\left|\int_S e_\theta(\underline{\kappa}) e^\Phi\right| + \frac{\epsilon_0^{\frac{2}{3}}}{u_*^{1+\delta_{dec}}}\left|\int_S \beta e^\Phi\right| + \frac{\epsilon^2}{r u^{1+\delta_{dec}}},\\ \left|e_3\left(\int_S e_\theta(\underline{\kappa}) e^\Phi\right)\right| &\lesssim \frac{\epsilon_0^{\frac{2}{3}}}{u_*^{1+\delta_{dec}}}\left|\int_S \beta e^\Phi\right| + \frac{\epsilon_0^{\frac{2}{3}}}{u_*^{1+\delta_{dec}}}\left|\int_S e_\theta(\underline{\kappa}) e^\Phi\right| + \frac{\epsilon^2}{u^{2+2\delta_{dec}}}.\end{aligned}$$

Step 3. Let ν_* the unique tangent vector to Σ_* which can be written as

$$\nu_* = e_3 + a e_4$$

where a is a scalar function on Σ_*. Recall that there exists a constant c_* such that $\Sigma_* = \{u + r = c_*\}$. We infer $\nu_*(u+r) = 0$ and hence

$$0 = e_3(u+r) + a e_4(u+r) = \frac{2}{\varsigma} + \frac{r}{2}(\overline{\underline{\kappa}} + \underline{A}) + a\frac{r}{2}\overline{\kappa}$$

which yields

$$a = -\frac{\frac{2}{\varsigma} + \frac{r}{2}(\overline{\underline{\kappa}} + \underline{A})}{\frac{r}{2}\overline{\kappa}}.$$

In view of the GCM condition on κ and the definition of the Hawking mass m, we have on Σ_*

$$\overline{\kappa} = \frac{2}{r}, \qquad \overline{\underline{\kappa}} = -\frac{2\Upsilon}{r}$$

and hence, we have on Σ_*

$$a = -\frac{2}{\varsigma} + \Upsilon - \frac{r}{2}\underline{A}.$$

The bootstrap assumptions on decay for $k = 0$ derivatives in ${}^{(ext)}\mathcal{M}$, the definition (2.2.22) of $\underline{A}$, and the estimates for ς and $\underline{\Omega}$ yield the rough estimate[1]

$$|a| \lesssim 1.$$

Together with the fact that $\nu_* = e_3 + a e_4$ and the estimates of Step 2, we infer

$$\left|\nu_*\left(\int_S e_\theta(\underline{\kappa})e^\Phi\right)\right| \lesssim \frac{\epsilon_0^{\frac{2}{3}}}{u_*^{1+\delta_{dec}}}\left|\int_S \beta e^\Phi\right| + \frac{\epsilon_0^{\frac{2}{3}}}{u_*^{1+\delta_{dec}}}\left|\int_S e_\theta(\underline{\kappa})e^\Phi\right| + \frac{\epsilon_0}{u^{2+2\delta_{dec}}},$$

$$\left|\nu_*\left(\int_S \beta e^\Phi\right)\right| \lesssim \frac{\epsilon_0^{\frac{2}{3}}}{u_*^{1+\delta_{dec}}}\left|\int_S \beta e^\Phi\right| + \frac{1}{r}\left|\int_S e_\theta(\underline{\kappa})e^\Phi\right| + \frac{\epsilon_0}{r u^{1+\delta_{dec}}}.$$

Step 4. We assume on Σ_* the following bootstrap assumptions recovered at the end of this step

$$u^{1+\delta_{dec}}\left|\int_S e_\theta(\underline{\kappa})e^\Phi\right| + r u^{\delta_{dec}}\left|\int_S \beta e^\Phi\right| \leq \epsilon. \tag{4.1.3}$$

[1]The estimates for $\underline{\Omega}$ and ς are proved later in Proposition 3.17. Since the proof does not rely on Theorem M0, we may use it here.

This implies, using also the behavior (3.3.4) of r on Σ_*, and the fact that $\epsilon = \epsilon_0^{\frac{2}{3}}$,

$$\begin{aligned}\left|\nu_*\left(\int_S e_\theta(\underline{\kappa})e^\Phi\right)\right| &\lesssim \frac{\epsilon_0}{u^{2+2\delta_{dec}}},\\ \left|\nu_*\left(\int_S \beta e^\Phi\right)\right| &\lesssim \frac{1}{r}\left|\int_S e_\theta(\underline{\kappa})e^\Phi\right| + \frac{\epsilon_0}{ru^{1+\delta_{dec}}}.\end{aligned} \tag{4.1.4}$$

Now, recall that we have the following GCM on the last sphere $S_* = \Sigma_* \cap \mathcal{C}_*$ of Σ_*

$$\int_{S_*} e_\theta(\underline{\kappa})e^\Phi = \int_{S_*} \beta e^\Phi = 0.$$

Integrating backward from S_* the estimate for $e_\theta(\underline{\kappa})$ in (4.1.4) yields on Σ_*

$$\left|\int_S e_\theta(\underline{\kappa})e^\Phi\right| \lesssim \frac{\epsilon_0}{u^{1+\delta_{dec}}}.$$

Plugging in the estimate for β in (4.1.4), we infer on Σ_*

$$\left|\nu_*\left(\int_S \beta e^\Phi\right)\right| \lesssim \frac{\epsilon_0}{ru^{1+\delta_{dec}}}.$$

Integrating backward from S_* yields on Σ_*

$$\left|\int_S \beta e^\Phi\right| \lesssim \frac{\epsilon_0}{ru^{\delta_{dec}}}.$$

We have therefore obtained

$$u^{1+\delta_{dec}}\left|\int_S e_\theta(\underline{\kappa})e^\Phi\right| + ru^{\delta_{dec}}\left|\int_S \beta e^\Phi\right| \lesssim \epsilon_0$$

which is an improvement of the bootstrap assumptions (4.1.3). In particular, at the first sphere $S_1 = \Sigma_* \cap \mathcal{C}_1$ of Σ_*, we have obtained

$$\left|\int_{S_1} e_\theta(\underline{\kappa})e^\Phi\right| + r\left|\int_{S_1} \beta e^\Phi\right| \lesssim \epsilon_0. \tag{4.1.5}$$

Remark 4.2. *Note that the only bootstrap assumption used in the proof of Theorem M0 is the bootstrap assumption* ***BA-D*** *on decay for $k = 0, 1$ derivatives. Indeed, to obtain* (4.1.5)*, we have only used, in Steps 1–4, the bootstrap assumption* ***BA-D*** *on decay for $k = 0, 1$ derivatives, while, from now on, we will only rely on* (4.1.5) *and the assumptions* (4.1.1) *on the initial data layer. This observation will allow us to use the conclusions of Theorem M0, not only for the bootstrap spacetime $\mathcal{M}$ in Theorem M1–M5, but also for the extended spacetime in the proof of Theorem M8, where the only assumption is the one on decay (which is established for the extended spacetime in Theorem M7).*

Step 5. On the sphere $S_1 = \Sigma_* \cap \mathcal{C}_1$ of Σ_*, we have in view of the GCM conditions of Σ_* and (4.1.5)

$$\kappa = \frac{2}{r}, \quad \not{d}_2^\star \not{d}_1^\star \underline{\kappa} = 0, \quad \not{d}_2^\star \not{d}_1^\star \mu = 0, \quad \left|\int_{S_1} e_\theta(\underline{\kappa}) e^\Phi\right| + r\left|\int_{S_1} \beta e^\Phi\right| \lesssim \epsilon_0. \tag{4.1.6}$$

We consider the transition functions $(f, \underline{f}, \lambda)$ from the frame of the outgoing part ${}^{(ext)}\mathcal{L}_0$ of the initial data layer to the frame of ${}^{(ext)}\mathcal{M}$. We assume the following bootstrap assumptions along $\mathcal{C}_1$

$$\sup_{S\subset\mathcal{C}_1} \left(\|f\|_{\mathfrak{h}_4(S)} + r^{-1}\|(\underline{f}, \log(\lambda))\|_{\mathfrak{h}_4(S)}\right) \leq \epsilon. \tag{4.1.7}$$

In particular, the estimate (4.1.7) allows us to apply Lemma 9.15 with $\delta_1 = \epsilon$ which yields

$$\begin{aligned}\sup_{S_1} \Big(({}^{(ext)}r)^{-1}|{}^{(ext)}r - {}^{(ext)}r_{\mathcal{L}_0}| + |u - u_{\mathcal{L}_0}| \\ + ({}^{(ext)}r)^{-1}|{}^{(ext)}s - {}^{(ext)}s_{\mathcal{L}_0}|\Big) \quad \lesssim \quad \epsilon.\end{aligned} \tag{4.1.8}$$

In particular, since $u = 1$ on S_1, and ${}^{(ext)}r = {}^{(ext)}s$ on Σ_* verifies the dominant condition in r, we infer

$$\sup_{S_1} |u_{\mathcal{L}_0} - 1| \lesssim \epsilon, \qquad \inf_{S_1} {}^{(ext)}s_{\mathcal{L}_0} \geq \frac{1}{2}\epsilon_0^{-\frac{2}{3}}.$$

Since ${}^{(ext)}\mathcal{L}_0$ contains the region $\{4m_0 \leq {}^{(ext)}s_{\mathcal{L}_0} < +\infty\} \cup \{0 \leq u_{\mathcal{L}_0} \leq 2\}$, we infer that the sphere S_1 is included in ${}^{(ext)}\mathcal{L}_0$.

We will not only improve the bootstrap assumption (4.1.7), but also gain derivatives iteratively. To this end, for $4 \leq j \leq k_{large} + 5$, we consider the following iteration assumption

$$\|f\|_{\mathfrak{h}_j(S_1)} + r^{-1}\|(\underline{f}, \log(\lambda))\|_{\mathfrak{h}_j(S_1)} \quad \leq \quad \epsilon. \tag{4.1.9}$$

Note that (4.1.9) holds true for $j = 4$ in view of (4.1.7), and our goal is to show that (4.1.9) holds with j replaced by $j + 1$.

Since

- S_1 is a sphere of ${}^{(ext)}\mathcal{M}$ in ${}^{(ext)}\mathcal{L}_0$,
- S_1 is a sphere of the GCM hypersurface Σ_*,
- the estimate (4.1.6) holds on S_1,
- the estimate (4.1.9) holds on S_1,

we can invoke Corollary 9.51 with the choice $\overset{\circ}{\epsilon} = \overset{\circ}{\delta} = \epsilon_0$, $\delta_1 = \epsilon$, $s_{max} = j$, and with the background foliation being the one of the outgoing part ${}^{(ext)}\mathcal{L}_0$ of the initial data layer. We obtain

$$r^{-1}\|(f, \underline{f}, \lambda - \overline{\lambda}^{S_1})\|_{\mathfrak{h}_{j+1}(S_1)} \quad \lesssim \quad \epsilon_0 \tag{4.1.10}$$

and

$$|\overline{\lambda}^{S_1} - 1| \quad \lesssim \quad \epsilon_0 + r^{-1}\sup_{S_1}\left|{}^{(ext)}r - {}^{(ext)}r_{\mathcal{L}_0}\right|. \tag{4.1.11}$$

Remark 4.3. *In order to prove the iteration assumption (4.1.9) with j replaced by $j+1$, we need in particular to improve the estimate for f in (4.1.10) by r^{-1}. Obtaining this improvement is the focus of Steps 6 to 8.*

Remark 4.4. *The anomalous behavior for $\underline{f}$ and λ in (4.1.7), i.e., the fact that they display a r loss compared to f, does not affect the desired estimates for the curvature components, see (4.1.22). This is due to the fact that, in the change of frame formulas for the curvature components, λ and $\underline{f}$ are multiplied by terms that decay faster in r.*

Remark 4.5. *In view of (4.1.8), while $|u-u_{\mathcal{L}_0}| \lesssim \epsilon$ on S_1, we have $|s-s_{\mathcal{L}_0}| \lesssim r\epsilon$ on S_1. This, as well as the anomalous behavior of $\underline{f}$ mentioned above, shows that the sphere S_1 is a large deformation, along the outgoing direction, of spheres of the initial data layer ${}^{(ext)}\mathcal{L}_0$. This reflects the fact that S_1 (and Σ_*) captures the center of mass frame of the limiting Schwarzschild solution, while the initial data layer foliation captures the center of mass frame of the initial Schwarzschild solution. The behavior of $s-s_{\mathcal{L}_0}$, as well as the one of $\underline{f}$, is consistent with the presence of a Lorentz boost between these two center of mass frames.*

From now on, we denote the frame, Ricci coefficients and curvature components associated to the frame of ${}^{(ext)}\mathcal{M}$ with a prime, while the frame, Ricci coefficients and curvature components associated to the frame of ${}^{(ext)}\mathcal{L}_0$ are unprimed. From the following transformation formula of Proposition 2.90,

$$\beta' = \lambda\left(\beta + \frac{3}{2}\rho f + \frac{1}{2}\underline{f}\alpha + \text{l.o.t.}\right),$$

together with the estimate (4.1.10) for f to estimate the linear term ρf, the estimate (4.1.9) for $(f, \underline{f}, \lambda)$ to estimate the other terms, and the estimates (4.1.1) for the outgoing part ${}^{(ext)}\mathcal{L}_0$ of the initial data layer,[2] we have, since $j \le k_{large}+5$,

$$\max_{k\le j-1} r^2 \|\not{\mathfrak{d}}'^k \beta'\|_{L^2(S_1)} \lesssim \epsilon_0. \tag{4.1.12}$$

Also, we have

$$\rho' = \rho + \frac{3}{2}\rho f\underline{f} + \underline{f}\beta + f\underline{\beta} + \text{l.o.t.}$$

Differentiating with respect to e'_θ, and using the decomposition of e'_θ, we infer

$$\begin{aligned} e'_\theta(\rho') &= \left(\left(1+\frac{1}{2}f\underline{f}\right)e_\theta + \frac{1}{2}\underline{f}e_4 + \frac{1}{2}f\left(1+\frac{1}{4}f\underline{f}\right)e_3\right)\rho \\ &\quad + e'_\theta\left(\frac{3}{2}\rho f\underline{f} + \underline{f}\beta + f\underline{\beta} + \text{l.o.t.}\right) \\ &= e_\theta(\rho) + \frac{1}{2}\underline{f}e_4(\rho) + \frac{1}{2}fe_3(\rho) + e'_\theta\left(\frac{3}{2}\rho f\underline{f} + \underline{f}\beta + f\underline{\beta}\right) + \text{l.o.t.} \end{aligned}$$

[2] We use, here and in the remainder of the proof, property 6 of Lemma 9.15 to control the $\mathfrak{h}_j(S_1)$ norm of the Ricci coefficients and curvature components of the initial data foliation of ${}^{(ext)}\mathcal{L}_0$ in terms of their sup norm.

Together with the estimate (4.1.10) for f and $\underline{f}$ to estimate the linear terms $\underline{f}e_4(\rho)$ and $fe_3(\rho)$, the estimate (4.1.9) for $(f, \underline{f}, \lambda)$ to estimate the other terms, and the estimates (4.1.1) for the curvature components and the Ricci coefficients of the outgoing part ${}^{(ext)}\mathcal{L}_0$ of the initial data layer, we have, using also the behavior (3.3.4) of r on Σ_* and the fact that $S_1 \subset \Sigma_*$, as well as an elliptic estimate and the fact that $j \leq k_{large} + 5$,

$$\max_{k \leq j-1} r^2 \| \mathfrak{d}'^k \check{\rho}' \|_{L^2(S_1)} \lesssim \epsilon_0. \tag{4.1.13}$$

Step 6. Recall the definition of the mass aspect function μ':

$$\mu' = -\not{d}_1{}'\zeta' - \rho' + \frac{1}{4}\vartheta'\underline{\vartheta}'.$$

Together with the GCM conditions $\not{d}_2^\star{}'\not{d}_1^\star{}'\mu' = 0$ on Σ_*, and the fact that $S_1 \subset \Sigma_*$, we infer

$$\not{d}_2^\star{}'\not{d}_1^\star{}'\not{d}_1{}'\zeta' = -\not{d}_2^\star{}'\not{d}_1^\star{}'\rho' + \frac{1}{4}\not{d}_2^\star{}'\not{d}_1^\star{}'(\vartheta'\underline{\vartheta}').$$

In view of the identity $\not{d}_1^\star{}'\not{d}_1{}' = \not{d}_2{}'\not{d}_2^\star{}' + 2K'$, we infer

$$(\not{d}_2^\star{}'\not{d}_2{}' + 2K')\not{d}_2^\star{}'\zeta' = -\not{d}_2^\star{}'\not{d}_1^\star{}'\rho' + \frac{1}{4}\not{d}_2^\star{}'\not{d}_1^\star{}'(\vartheta'\underline{\vartheta}') + 2e_\theta'(K')\zeta'.$$

Using the estimate for ρ' of Step 5 and an elliptic estimate,

$$\max_{k \leq j-2} r^2 \| \mathfrak{d}'^k \not{d}_2^\star{}'\zeta' \|_{L^2(S_1)} \lesssim \epsilon_0. \tag{4.1.14}$$

Note that the quadratic terms involving $\vartheta'\underline{\vartheta}'$ and $e_\theta'(K')\zeta'$ are estimated using the transformation formulas,[3] the estimates (4.1.9) for $(f, \underline{f}, \lambda)$, and the estimates (4.1.1) for the curvature components and the Ricci coefficients of the outgoing part ${}^{(ext)}\mathcal{L}_0$ of the initial data layer.

Step 7. Recall Codazzi for ϑ'

$$\not{d}_2{}'\vartheta' = -2\beta' - \not{d}_1^\star\kappa' + \zeta'\kappa' - \vartheta'\zeta'.$$

We differentiate w.r.t. $\not{d}_2^\star{}'$ and use the GCM condition $\kappa' = 2/r'$ which holds on Σ_* and $S_1 \subset \Sigma_*$ to deduce

$$\not{d}_2^\star{}'\not{d}_2{}'\vartheta' = -2\not{d}_2^\star{}'\beta' + \kappa'\not{d}_2^\star{}'\zeta' - \not{d}_2^\star{}'(\vartheta'\zeta').$$

Together with the estimate of Step 5 for β', the estimate of Step 6 for $\not{d}_2^\star{}'\zeta'$, dealing with the quadratic terms as above, and using an elliptic estimate, we infer

$$\max_{k \leq j} r \| \mathfrak{d}'^k \vartheta' \|_{L^2(S_1)} \lesssim \epsilon_0.$$

[3]In fact, in view of the identity $K' = -\rho' - \frac{1}{4}\kappa'\underline{\kappa}' + \frac{1}{4}\vartheta'\underline{\vartheta}'$, the GCM conditions for κ' and $\underline{\kappa}'$, and the control of ρ' in Step 5, we only need the transformation formulas for ϑ' and $\underline{\vartheta}'$. These formulas involve at most one angular derivative of f and $\underline{f}$, and no transversal derivative.

Next, recall the transformation formula

$$\vartheta' = \lambda\left(\vartheta - {}^\star\!\not{d}_2'(f) + f(\zeta+\eta) + \underline{f}\xi + \frac{1}{4}f\underline{f}\kappa + f\underline{f}\omega - f^2\underline{\omega} + \text{l.o.t.}\right).$$

Together with the above estimate for ϑ', the estimate (4.1.9) for $(f, \underline{f}, \lambda)$, and the estimates (4.1.1) for the Ricci coefficients of the outgoing part ${}^{(ext)}\mathcal{L}_0$ of the initial data layer, we infer

$$\max_{k\leq j} \|r\, {}^\star\!\not{d}_2'(\not{\mathfrak{d}}'^k f)\|_{L^2(S_1)} \lesssim \epsilon_0 + \epsilon^2 \lesssim \epsilon_0.$$

Together with a Poincaré inequality, we infer

$$\max_{k\leq j+1} \|\not{\mathfrak{d}}'^k f\|_{L^2(S_1)} \lesssim \epsilon_0 + r^{-2}\left|\int_{S_1} fe^\Phi\right|. \tag{4.1.15}$$

Step 8. In view of the last estimate of Step 7, we need to control the $\ell = 1$ mode of f. Recall from Lemma 4.1

$$e_4'\left(r'\int_S \beta' e^\Phi\right) = \int_S\left(-\frac{1}{2}\check{\kappa}'\beta' + \zeta'\alpha' - \frac{1}{2}\vartheta'\beta'\right)e^\Phi.$$

Transporting along $\mathcal{C}_1$ from S_1, using the control of the $\ell = 1$ mode of β' in (4.1.6) on S_1, and using the bootstrap assumptions on ${}^{(ext)}\mathcal{M}$, we infer

$$\sup_{S\subset\mathcal{C}_1} r\left|\int_S \beta' e^\Phi\right| \lesssim \epsilon_0 + \epsilon^2 \lesssim \epsilon_0.$$

In particular, consider the sphere $S_{4m_0} = \mathcal{C}_1 \cap \{r' = 4m_0\}$. Then

$$\left|\int_{S_{4m_0}} \beta' e^\Phi\right| \lesssim \epsilon_0.$$

Together with the transformation formula

$$\beta' = \lambda\left(\beta + \frac{3}{2}\rho f + \frac{1}{2}\underline{f}\alpha + \text{l.o.t.}\right),$$

which we rewrite, multiply by e^Φ, and integrate on S_{4m_0},

$$\begin{aligned}\frac{3m'}{r'^3}\int_{S_{4m_0}} fe^\Phi &= -\int_{S_{4m_0}} \beta' e^\Phi + \int_{S_{4m_0}}\left(\frac{3m'}{r'^3} - \frac{3m}{r^3}\right)fe^\Phi \\ &+\frac{3}{2}\int_{S_{4m_0}}(\lambda-1)\rho fe^\Phi + \frac{3}{2}\int_{S_{4m_0}}\left(\rho + \frac{2m}{r^3}\right)fe^\Phi \\ &+\int_{S_{4m_0}}\lambda\left(\beta + \frac{1}{2}\underline{f}\alpha + \text{l.o.t.}\right)e^\Phi,\end{aligned}$$

the bootstrap assumptions (4.1.7) for $(f, \underline{f}, \lambda)$, and the control of the initial data layer, we infer

$$\begin{aligned}\left|\int_{S_{4m_0}} fe^{\Phi}\right| &\lesssim \epsilon_0+\epsilon^2+\epsilon\sup_{S_{4m_0}}\big(|r'-r|+|m'-m|\big)\\ &\lesssim \epsilon_0+\epsilon\sup_{S_{4m_0}}\big(|r'-r|+|m'-m|\big).\end{aligned}$$

Now, the bootstrap assumptions (4.1.7) for $(f, \underline{f}, \lambda)$, together with the estimate for $r-r'$ in Lemma 9.15 with $\delta_1=\epsilon$, and the one for $m'-m$ in Corollary 9.19 with $\overset{\circ}{\epsilon}=\epsilon$, yields

$$\sup_{S_{4m_0}}\big(|r'-r|+|m'-m|\big) \lesssim \epsilon$$

and hence

$$\left|\int_{S_{4m_0}} fe^{\Phi}\right| \lesssim \epsilon_0+\epsilon^2\lesssim\epsilon_0.$$

Next, recall from Corollary 2.93 that f satisfies the following transport equations along $\mathcal{C}_1$

$$\lambda^{-1}e_4'(r'f) = E_1'(f,\Gamma).$$

We deduce from Corollary 2.65 that

$$\begin{aligned}e_4\left(r'^{-2}\int_S fe^{\Phi}\right) &= r'^{-2}\int_S\left(e_4'(f)+\left(\frac{1}{2}\kappa'+\check{\kappa}'-\frac{1}{2}\vartheta'\right)f\right)e^{\Phi}\\ &= r'^{-2}\int_S\left(r'^{-1}e_4'(r'f)+\left(\frac{3}{2}\check{\kappa}'-\frac{1}{2}\vartheta'\right)f\right)e^{\Phi}\\ &= r'^{-2}\int_S\left(r'^{-1}\lambda' E_1'(f,\Gamma)+\left(\frac{3}{2}\check{\kappa}'-\frac{1}{2}\vartheta'\right)f\right)e^{\Phi}.\end{aligned}$$

In view of the form of E_1' in Corollary 2.93, the bootstrap assumption (4.1.7) for f, and the estimates (4.1.1) for the Ricci coefficients of the outgoing part ${}^{(ext)}\mathcal{L}_0$ of the initial data layer, we have

$$r^2|E_1'(f,\Gamma)| \lesssim \epsilon_0+\epsilon^2\lesssim\epsilon_0 \text{ on } \mathcal{C}_1.$$

We deduce

$$\left|e_4\left(r'^{-2}\int_S fe^{\Phi}\right)\right| \lesssim \frac{\epsilon_0}{r^2}+\sup_{S\subset\mathcal{C}_1}\left[r^{-1}\Big(\|\check{\kappa}'\|_{L^2(S)}+\|\vartheta'\|_{L^2(S)}\Big)\|f\|_{L^2(S)}\right].$$

Using the bootstrap assumption (4.1.7) for f, and the bootstrap assumption on decay on ${}^{(ext)}\mathcal{M}$ for $\check{\kappa}'$ and ϑ', we infer

$$\left|e_4\left(r'^{-2}\int_S fe^{\Phi}\right)\right| \lesssim \frac{\epsilon_0+\epsilon^2}{r^2}\lesssim\frac{\epsilon_0}{r^2}.$$

Integrating forward from $r=4m_0$, and using the above estimate for the $\ell=1$ mode of f on S_{4m_0}, we obtain

$$\sup_{S\subset\mathcal{C}_1} r^{-2}\left|\int_S fe^{\Phi}\right| \lesssim \epsilon_0.$$

Together with the estimate for $\not{\mathfrak{d}}'^k f$ of Step 7, and since $S_1\subset\mathcal{C}_1$, we deduce

$$\max_{k\leq j+1} \|\not{\mathfrak{d}}'^k f\|_{L^2(S_1)} \lesssim \epsilon_0.$$

Together with (4.1.10), we infer

$$\|f\|_{\mathfrak{h}_{j+1}(S_1)} + r^{-1}\|(\underline{f}, \lambda - \overline{\lambda}^{S_1})\|_{\mathfrak{h}_{j+1}(S_1)} \lesssim \epsilon_0.$$

In particular, the above estimate for $(f, \underline{f})$ allows us to use Lemma 9.15 with $\delta_1 = \epsilon_0$ which yields

$$\sup_{S_1}\left|\frac{r'}{r} - 1\right| \lesssim \epsilon_0.$$

Together with (4.1.11), we infer

$$\|f\|_{\mathfrak{h}_{j+1}(S_1)} + r^{-1}\|(\underline{f}, \log\lambda)\|_{\mathfrak{h}_{j+1}(S_1)} \lesssim \epsilon_0.$$

This implies the iteration assumption (4.1.9) for $j+1$, for all $4\leq j\leq k_{large}+5$. Thus, we have obtained

$$\|f\|_{\mathfrak{h}_{k_{large}+6}(S_1)} + r^{-1}\|(\underline{f}, \log\lambda)\|_{\mathfrak{h}_{k_{large}+6}(S_1)} \lesssim \epsilon_0.$$

In view of the above estimate for $(f, \underline{f}, \lambda)$, and since $S_1\subset\Sigma_*$, we may apply Corollary 9.53 with $\overset{\circ}{\delta} = \epsilon_0$ and $s_{max} = k_{large}+5$ which yields

$$\begin{aligned}
\|\mathfrak{d}^{\leq k_{large}+6} f\|_{L^2(S_1)} + r^{-1}\|\mathfrak{d}^{\leq k_{large}+6}(\underline{f}, \log\lambda)\|_{L^2(S_1)}&\\
+\|\mathfrak{d}^{\leq k_{large}+5} e_3'(\underline{f}, \log\lambda)\|_{L^2(S_1)} &\lesssim \epsilon_0.
\end{aligned}$$

The above control of $(f, \underline{f})$, together with Lemma 9.15 for $\delta_1 = \epsilon_0$, and Corollary 9.19 with $\delta_1 = \overset{\circ}{\epsilon} = \epsilon_0$, implies

$$\sup_{S_1}\left(\left|\frac{m'}{m_0} - 1\right| + \left|\frac{r'}{r} - 1\right|\right) \lesssim \epsilon_0.$$

We have thus obtained on S_1

$$\begin{aligned}
\|\mathfrak{d}^{\leq k_{large}+6} f\|_{L^2(S_1)} + r^{-1}\|\mathfrak{d}^{\leq k_{large}+6}(\underline{f}, \log(\lambda))\|_{L^2(S_1)}&\\
+\|\mathfrak{d}^{\leq k_{large}+5} e_3'(\underline{f}, \log(\lambda))\|_{L^2(S_1)} + \sup_{S_1}\left(\left|\frac{m'}{m_0} - 1\right| + \left|\frac{r'}{r} - 1\right|\right) &\lesssim \epsilon_0.
\end{aligned} \tag{4.1.16}$$

Finally, we will also need the following estimates on S_1

$$\begin{aligned} r\left\|\mathfrak{d}^{\leq k_{large}+5}\left(\overline{\kappa}'-\frac{2}{r'},\check{\kappa}',\vartheta'\right)\right\|_{L^2(S_1)} & \\ +r^{-1}\left\|\mathfrak{d}^{\leq k_{large}+5}\left(\frac{r'}{r}-1\right)\right\|_{L^2(S_1)} & \lesssim \epsilon_0. \end{aligned} \tag{4.1.17}$$

The estimates for κ' in (4.1.17) follow from the GCM condition on κ', as well as Raychaudhuri for transversal derivatives. The estimate for ϑ' in (4.1.17) follows from the transformation formula, the control (4.1.16) of $(f,\underline{f},\lambda)$, and the control of the initial data layer. We obtain similarly the control of $\underline{\xi}'$, $\underline{\omega}'$ and η' on S_1, which in turn yields the control of $\underline{\Omega}'$ and ς' on S_1 in view of Lemma 2.61, and finally the control of $\frac{r'}{r}$ in (4.1.17) relying on (4.1.16) and (2.2.21).

Step 9. Recall from Corollary 2.93 that $(f,\log(\lambda))$ satisfy the following transport equations along $\mathcal{C}_1$

$$\begin{aligned} \lambda^{-1}e_4'(rf) &= E_1'(f,\Gamma),\\ \lambda^{-1}e_4'(\log(\lambda)) &= E_2'(f,\Gamma), \end{aligned}$$

where, in view of the form of E_1', E_2' in Corollary 2.93 and the estimates (4.1.1) for the Ricci coefficients of the outgoing part ${}^{(ext)}\mathcal{L}_0$ of the initial data layer, we have

$$|\mathfrak{d}^kE_1'(f,\Gamma)|+|\mathfrak{d}^kE_2'(f,\Gamma)| \lesssim \frac{\epsilon_0}{r^2}+|\mathfrak{d}^{\leq k}f|^2 \text{ for } k\leq k_{large}+5 \text{ on } \mathcal{C}_1.$$

Next, recall from Lemma 2.69 the following commutator identity

$$[\mathbf{T},e_4] = \left(\left(\underline{\omega}-\frac{m}{r^2}\right)-\frac{m}{2r}\left(\overline{\kappa}-\frac{2}{r}\right)+\frac{e_4(m)}{r}\right)e_4+(\eta+\zeta)e_\theta$$

while from Lemma 2.68, we have schematically

$$[\not{\mathfrak{d}},e_4] = (\check{\kappa},\vartheta)\not{\mathfrak{d}}+(\zeta,r\beta).$$

Together with the fact that

$$\lambda^{-1}e_4' = e_4+fe_\theta+\frac{f^2}{4}e_3,$$

the commutator above identities for $[\mathbf{T},e_4]$ and $[\not{\mathfrak{d}},e_4]$, as well as the estimates (4.1.1) for the Ricci coefficients and curvature components of the outgoing part ${}^{(ext)}\mathcal{L}_0$ of the initial data layer, we infer, for $k\leq k_{large}+5$,

$$\begin{aligned} &|\mathfrak{d}^k[\mathbf{T},\lambda^{-1}e_4']h|+|\mathfrak{d}^k[\not{\mathfrak{d}},\lambda^{-1}e_4']h|\\ \lesssim\ & \frac{\epsilon_0}{r^2}|\mathfrak{d}^{\leq k+1}h|+\frac{1}{r}|\mathfrak{d}^{\leq k}(f\mathfrak{d}h)|+\frac{1}{r}|\mathfrak{d}^{\leq k}(h\mathfrak{d}f)|+|\mathfrak{d}^{\leq k}(f^2\mathfrak{d}h)|+|\mathfrak{d}^{\leq k}(hf\mathfrak{d}f)|. \end{aligned}$$

By commuting first the transport equations in $\lambda^{-1}e_4'$ with $(\mathbf{T},\not{\mathfrak{d}})^k$, and by using these transport equations to recover the e_4 derivatives, we deduce

$$\begin{aligned}\lambda^{-1}e_4'(r\mathfrak{d}^k f) &= E'_{1,k}(f,\Gamma),\\ \lambda^{-1}e_4'(\mathfrak{d}^k\log(\lambda)) &= E'_{2,k}(f,\Gamma),\end{aligned}$$

where we have

$$|E'_{1,k}(f,\Gamma)|+|E'_{2,k}(f,\Gamma)| \lesssim \frac{\epsilon_0}{r^2}+|\mathfrak{d}^{\leq k}f|^2 \text{ for } k\leq k_{large}+5 \text{ on } \mathcal{C}_1.$$

This allows us to propagate the estimates for (f,λ) in (4.1.10) on S_1 to any sphere on $\mathcal{C}_1$, and hence

$$\sup_{S\subset\mathcal{C}_1}\Big(\|\mathfrak{d}^{\leq k_{large}+5}f\|_{L^2(S)}+r^{-1}\|\mathfrak{d}^{\leq k_{large}+5}\log\lambda\|_{L^2(S)}\Big) \lesssim \epsilon_0. \quad (4.1.18)$$

Step 10. Our next goal is to control $\underline{f}$ along $\mathcal{C}_1$. We cannot proceed along the same lines as the control of (f,λ) in Step 9. Indeed, we cannot rely on the last transport equation along $\lambda^{-1}e_4'$ of Corollary 2.93, as it is not consistent with the control of $\underline{f}$ on S_1 derived in Step 8. Instead, we first control α', κ' and ϑ'.

Recall the following transformation formula

$$\alpha' = \lambda^2\Big(\alpha+2f\beta+\frac{3}{2}f^2\rho+\text{l.o.t.}\Big)$$

which does not depend on $\underline{f}$. Together with the control of (f,λ) of Step 9 and the control of the initial data layer, we infer

$$\sup_{S\subset\mathcal{C}_1} r^{\frac{5}{2}+\delta_B}\|\mathfrak{d}^{\leq k_{large}+5}\alpha'\|_{L^2(S)} \lesssim \epsilon_0.$$

Next, recall

$$e_4'\left(\overline{\kappa'}-\frac{2}{r'}\right)+\frac{1}{2}\overline{\kappa'}\left(\overline{\kappa'}-\frac{2}{r'}\right) = -\frac{1}{4}\overline{\vartheta'^2}+\frac{1}{2}\overline{\check{\kappa}'^2},$$

$$e_4'(\check{\kappa}')+\overline{\kappa'}\,\check{\kappa}' = -\frac{1}{2}(\check{\kappa}')^2-\frac{1}{2}\overline{(\check{\kappa}')^2}-\frac{1}{2}(\vartheta'^2-\overline{\vartheta'^2}),$$

and

$$e_4'(\vartheta')+\kappa'\vartheta' = -2\alpha'.$$

Proceeding as in Step 9, we commute first these transport equations with $(\mathbf{T},\not{\mathfrak{d}})^k$, and use the transport equations to recover the e_4 derivatives. By integrating the resulting transport equations from S_1 where $\overline{\kappa'}$, $\check{\kappa}'$ and ϑ' are under control in view of (4.1.17), and using the above control of α', we infer

$$\sup_{S\subset\mathcal{C}_1} r\left\|\mathfrak{d}^{\leq k_{large}+5}\left(\overline{\kappa'}-\frac{2}{r'},\,\check{\kappa}',\,\vartheta'\right)\right\|_{L^2(S)} \lesssim \epsilon_0. \quad (4.1.19)$$

Also, we have, using in particular (2.2.21),

$$\begin{aligned}
\lambda^{-1}e_4'\left(\log\left(\frac{r'}{r}\right)\right) &= \frac{\lambda^{-1}e_4'(r')}{r} - \frac{e_4(r)}{r} - \frac{f^2}{4}\frac{e_3(r)}{r}\\
&= \frac{1}{2}\Big(\lambda^{-1}\kappa' - \kappa\Big) - \frac{\lambda^{-1}}{2}\check{\kappa}' + \frac{1}{2}\check{\kappa} - \frac{rf^2}{8}(\overline{\underline{\kappa}} + \underline{A})\\
&= \frac{1}{2}\Big(\not{d}_1'(f) + \mathrm{Err}(\kappa,\kappa')\Big) - \frac{\lambda^{-1}}{2}\check{\kappa}' + \frac{1}{2}\check{\kappa} - \frac{rf^2}{8}(\overline{\underline{\kappa}} + \underline{A})
\end{aligned}$$

where we have also used the change of frame formula for κ'. Proceeding as in Step 9, we commute first these transport equations with $(\mathbf{T}, \not{\partial})^k$, and use the transport equations to recover the e_4 derivatives. By integrating the resulting transport equations from S_1 where $\frac{r'}{r}$ is under control in view of (4.1.17), and using the estimate[4] of Step 9 for f and λ, the estimate of Step 10 for $\check{\kappa}'$, and the estimate for the initial data foliation layer on ${}^{(ext)}\mathcal{L}_0$, we infer

$$\sup_{S\subset\mathcal{C}_1} r^{-1}\left\|\mathfrak{d}^{\leq k_{large}+5}\log\left(\frac{r'}{r}\right)\right\|_{L^2(S)} \lesssim \epsilon_0. \tag{4.1.20}$$

Step 11. Recall Codazzi for ϑ'

$$\not{d}_2{}'\vartheta' = -2\beta' - \not{d}_1^{\star}\kappa' + \zeta'\kappa' - \vartheta'\zeta'.$$

This yields

$$\zeta' = \frac{r'}{2}\left(2\beta' + \not{d}_2{}'\vartheta' + \not{d}_1^{\star}\kappa' + \vartheta'\zeta' - \zeta'\left(\kappa' - \frac{2}{r'}\right)\right).$$

Together with the control of κ', ϑ' and r' of Step 10, we infer

$$\begin{aligned}
\sup_{S\subset\mathcal{C}_1}\|\mathfrak{d}^{\leq k_{large}+4}\zeta'\|_{L^2(S)} &\lesssim \sup_{S\subset\mathcal{C}_1} r\|\mathfrak{d}^{\leq k_{large}+4}\beta'\|_{L^2(S)} + \epsilon_0\\
&\quad +\epsilon_0 \sup_{S\subset\mathcal{C}_1}\|\mathfrak{d}^{\leq k_{large}+4}\zeta'\|_{L^2(S)}
\end{aligned}$$

and hence, for ϵ_0 small enough,

$$\sup_{S\subset\mathcal{C}_1}\|\mathfrak{d}^{\leq k_{large}+4}\zeta'\|_{L^2(S)} \lesssim \sup_{S\subset\mathcal{C}_1} r\|\mathfrak{d}^{\leq k_{large}+4}\beta'\|_{L^2(S)} + \epsilon_0.$$

Recall the transformation formulas

$$\begin{aligned}
\beta' &= \lambda\left(\beta + \frac{3}{2}\rho f + \frac{1}{2}\underline{f}\alpha + \mathrm{l.o.t.}\right),\\
\zeta' &= \zeta - e_\theta'(\log(\lambda)) + \frac{1}{4}(-f\underline{\kappa} + \underline{f}\kappa) + \underline{f}\omega - f\underline{\omega} + \frac{1}{2}\underline{f}e_\theta'(f)\\
&\quad + \frac{1}{4}(-f\underline{\vartheta} + \underline{f}\vartheta) + \mathrm{l.o.t.}
\end{aligned}$$

Together with the control of f and λ from Step 9, the control of the initial data

[4] Note that the RHS of the transport equation does not depend on $\underline{f}$.

foliation layer on ${}^{(ext)}\mathcal{L}_0$, and the above control of ζ', we infer

$$\begin{aligned}\sup_{S\subset\mathcal{C}_1} r^{-1}\|\mathfrak{d}^{\leq k_{large}+4}\underline{f}\|_{L^2(S)} &\lesssim \sup_{S\subset\mathcal{C}_1}\|\mathfrak{d}^{\leq k_{large}+4}\zeta'\|_{L^2(S)}+\epsilon_0\\ &\lesssim \sup_{S\subset\mathcal{C}_1} r\|\mathfrak{d}^{\leq k_{large}+4}\beta'\|_{L^2(S)}+\epsilon_0\\ &\lesssim \epsilon_0+\epsilon_0\sup_{S\subset\mathcal{C}_1} r^{-1}\|\mathfrak{d}^{\leq k_{large}+4}\underline{f}\|_{L^2(S)}\end{aligned}$$

and hence, for ϵ_0 small enough,

$$\sup_{S\subset\mathcal{C}_1} r^{-1}\|\mathfrak{d}^{\leq k_{large}+4}\underline{f}\|_{L^2(S)} \lesssim \epsilon_0.$$

Together with the control of f and λ from Step 9, we have in particular

$$\sup_{S\subset\mathcal{C}_1}\Big(\|\mathfrak{d}^{\leq k_{large}+4}f\|_{L^2(S)}+r^{-1}\|\mathfrak{d}^{\leq k_{large}+4}(\log(\lambda),\underline{f})\|_{L^2(S)}\Big) \lesssim \epsilon_0.$$

Note that this concludes the improvement of the bootstrap assumptions (4.1.7) on $(f,\underline{f},\lambda)$. Also, using Sobolev, we infer

$$\sup_{\mathcal{C}_1}\Big(r|\mathfrak{d}^{\leq k_{large}+2}f|+|\mathfrak{d}^{\leq k_{large}+2}(\underline{f},\log(\lambda))|\Big) \lesssim \epsilon_0. \tag{4.1.21}$$

Step 12. In view of (4.1.21), the change of frame formulas of Proposition 2.90, and the estimates (4.1.1) for the curvature components of the outgoing part ${}^{(ext)}\mathcal{L}_0$ of the initial data layer, we obtain

$$\begin{aligned}\max_{0\leq k\leq k_{large}}\Bigg\{&\sup_{\mathcal{C}_1}\Big[r^{\frac{7}{2}+\delta_B}\Big(|\mathfrak{d}^k\,{}^{(ext)}\alpha|+|\mathfrak{d}^k\,{}^{(ext)}\beta|\Big)+r^{\frac{9}{2}+\delta_B}|\mathfrak{d}^{k-1}e_3({}^{(ext)}\alpha)|\Big] \qquad (4.1.22)\\ &+\sup_{\mathcal{C}_1}\Big[r^3\,\Big|\mathfrak{d}^k\Big({}^{(ext)}\rho+\frac{2m_0}{r^3}\Big)\Big|+r^2|\mathfrak{d}^k\,{}^{(ext)}\underline{\beta}|+r|\mathfrak{d}^k\,{}^{(ext)}\underline{\alpha}|\Big]\Bigg\} \lesssim \epsilon_0.\end{aligned}$$

Also, according to Proposition 2.71, we have in ${}^{(ext)}\mathcal{M}$

$$\begin{aligned}{}^{(ext)}e_4({}^{(ext)}m) &= \frac{{}^{(ext)}r}{32\pi}\int_S\Big(2\,{}^{(ext)}\check{\kappa}\,{}^{(ext)}\check{\rho}+2\,{}^{(ext)}e_\theta({}^{(ext)}\kappa)\,{}^{(ext)}\zeta\\ &\quad -\frac{1}{2}\,{}^{(ext)}\underline{\kappa}({}^{(ext)}\vartheta)^2-\frac{1}{2}\,{}^{(ext)}\check{\kappa}\,{}^{(ext)}\vartheta\,{}^{(ext)}\underline{\vartheta}+2\,{}^{(ext)}\kappa({}^{(ext)}\zeta)^2\Big),\end{aligned}$$

which together with the bootstrap assumptions on ${}^{(ext)}\mathcal{M}$ yields

$$\sup_{\mathcal{C}_1} r^2\Big|{}^{(ext)}e_4({}^{(ext)}m)\Big| \lesssim \epsilon_0^2.$$

This allows us to propagate the estimates for ${}^{(ext)}m$ in (4.1.16) on S_1 to any sphere on $\mathcal{C}_1$, and hence

$$\sup_{\mathcal{C}_1}\left|\frac{{}^{(ext)}m}{m_0}-1\right| \lesssim \epsilon_0. \tag{4.1.23}$$

Also, in view of the control $\frac{r'}{r}$ of Step 10, we have

$$\sup_{\mathcal{C}_1}\left|\frac{{}^{(ext)}r}{{}^{(ext)}r_{\mathcal{L}_0}}-1\right|\lesssim\epsilon_0.$$

Step 13. Recall that

- $({}^{(ext)}e_4, {}^{(ext)}e_3, {}^{(ext)}e_\theta)$ denotes the null frame of ${}^{(ext)}\mathcal{M}$,
- $({}^{(int)}e_4, {}^{(int)}e_3, {}^{(ext)}e_\theta)$ denotes the null frame of ${}^{(int)}\mathcal{M}$,
- $({}^{(ext)}(e_0)_3, {}^{(ext)}(e_0)_4, {}^{(ext)}(e_0)_\theta)$ denotes the null frame of ${}^{(ext)}\mathcal{L}_0$,
- $({}^{(int)}(e_0)_3, {}^{(int)}(e_0)_4, {}^{(int)}(e_0)_\theta)$ denotes the null frame of ${}^{(int)}\mathcal{L}_0$.

Also, recall that the timelike hypersurface $\mathcal{T}$ is given by

$$\mathcal{T}=\{{}^{(ext)}r=r_{\mathcal{T}}\}\text{ where }2m_0\left(1+\frac{\delta_{\mathcal{H}}}{2}\right)\leq r_{\mathcal{T}}\leq 2m_0\left(1+\frac{3\delta_{\mathcal{H}}}{2}\right)$$

to that $\mathcal{T}\subset{}^{(int)}\mathcal{L}_0\cap{}^{(ext)}\mathcal{L}_0$, and recall that the frame of ${}^{(int)}\mathcal{M}$ is initialed on $\mathcal{T}$ as follows

$${}^{(int)}e_4=\lambda\,{}^{(ext)}e_4,\qquad {}^{(int)}e_3=\lambda^{-1}\,{}^{(ext)}e_3,\qquad {}^{(int)}e_\theta={}^{(ext)}e_\theta\quad\text{on }\mathcal{T}$$

where

$$\lambda={}^{(ext)}\lambda=1-\frac{2\,{}^{(ext)}m}{{}^{(ext)}r}.$$

Denoting

- by $(f,\underline{f},\lambda)$ the transition functions from the frame of the outgoing part ${}^{(ext)}\mathcal{L}_0$ of the initial data layer to the frame of ${}^{(ext)}\mathcal{M}$ as in Steps 5 to 12,
- by $(f',\underline{f}',\lambda')$ the transition functions from the frame of the ingoing part ${}^{(int)}\mathcal{L}_0$ of the initial data layer to the frame of ${}^{(int)}\mathcal{M}$,
- by $(\tilde{f},\tilde{\underline{f}},\tilde{\lambda})$ the transition functions on ${}^{(int)}\mathcal{L}_0\cap{}^{(ext)}\mathcal{L}_0$ from the frame outgoing part ${}^{(ext)}\mathcal{L}_0$ of the initial data layer to the frame of the ingoing part ${}^{(int)}\mathcal{L}_0$ of the initial data layer,

we obtain, using also that $\mathcal{C}_1\cap\underline{\mathcal{C}}_1\subset\mathcal{T}$,

$$\begin{aligned}\sup_{\mathcal{C}_1\cap\underline{\mathcal{C}}_1}\left(|\mathfrak{d}^{\leq k_{large}+2}(f',\underline{f}',\log(\lambda'))|\right)\lesssim&\sup_{\mathcal{C}_1\cap\underline{\mathcal{C}}_1}\left(|\mathfrak{d}^{\leq k_{large}+2}(f,\underline{f},\log(\lambda))|\right)\\&+\sup_{\mathcal{C}_1\cap\underline{\mathcal{C}}_1}\left(|\mathfrak{d}^{\leq k_{large}+2}(\tilde{f},\tilde{\underline{f}},\log(\Upsilon_0^{-1}\tilde{\lambda}))|\right)\\&+\sup_{\mathcal{C}_1\cap\underline{\mathcal{C}}_1}\left(|\mathfrak{d}^{\leq k_{large}+2}\log(\Upsilon_0^{-1}\Upsilon)|\right)\end{aligned}$$

where we have denoted

$$\Upsilon_0=1-\frac{2m_0}{{}^{(ext)}r_{\mathcal{L}_0}},\qquad \Upsilon=1-\frac{2\,{}^{(ext)}m}{{}^{(ext)}r}.$$

Together with the control of $(\tilde{f},\tilde{\underline{f}},\log(\Upsilon_0^{-1}\tilde{\lambda}))$ provided on ${}^{(int)}\mathcal{L}_0\cap{}^{(ext)}\mathcal{L}_0$ by the estimates (4.1.1), the estimates (4.1.21) for $(f, \underline{f}, \lambda)$, and the estimates ${}^{(ext)}m - m_0$ and ${}^{(ext)}r - {}^{(ext)}r_{\mathcal{L}_0}$ obtained in Step 12, we infer

$$\sup_{\mathcal{C}_1 \cap \underline{\mathcal{C}}_1} \left(|\mathfrak{d}^{\leq k_{large}+2}(f', \underline{f}', \log(\lambda'))| \right) \lesssim \epsilon_0.$$

Step 14. We propagate the estimate for $(f', \underline{f}', \log(\lambda'))$ on $\mathcal{C}_1 \cap \underline{\mathcal{C}}_1$ provided by Step 8 to $\underline{\mathcal{C}}_1$ using the analog of Corollary 2.93 in the ingoing direction e_3. We obtain the following estimate

$$\sup_{\underline{\mathcal{C}}_1} \left(|\mathfrak{d}^{\leq k_{large}+2}(\underline{f}, \log \lambda)| \right) + \sup_{\mathcal{C}_1} |\mathfrak{d}^{\leq k_{large}+1} f| \lesssim \epsilon_0.$$

Together with the change of frame formulas of Proposition 2.90, and the estimates (4.1.1) for the curvature components of the ingoing part ${}^{(int)}\mathcal{L}_0$ of the initial data layer, we obtain

$$\max_{0 \leq k \leq k_{large}} \sup_{\underline{\mathcal{C}}_1} \Bigg[|\mathfrak{d}^k \, {}^{(int)}\alpha| + |\mathfrak{d}^k \, {}^{(int)}\beta| + \left| \mathfrak{d}^k \left({}^{(int)}\rho + \frac{2m_0}{r^3} \right) \right| + |\mathfrak{d}^k \, {}^{(int)}\underline{\beta}| + |\mathfrak{d}^k \, {}^{(int)}\underline{\alpha}| \Bigg] \lesssim \epsilon_0. \tag{4.1.24}$$

Also, since we have as a consequence of the initialization on $\mathcal{T}$ of the ingoing geodesic foliation of ${}^{(int)}\mathcal{M}$

$${}^{(int)}m = {}^{(ext)}m \quad \text{on} \quad \mathcal{C}_1 \cap \underline{\mathcal{C}}_1$$

we infer from the control of ${}^{(ext)}m$ provided by Step 12

$$|{}^{(int)}m - m_0| \lesssim \epsilon_0 \text{ on } \mathcal{C}_1 \cap \underline{\mathcal{C}}_1.$$

We then propagate, similarly to Step 12, this bound to $\underline{\mathcal{C}}_1$ and obtain

$$\sup_{\underline{\mathcal{C}}_1} \left| {}^{(int)}m - m_0 \right| \lesssim \epsilon_0.$$

Together with (4.1.22), (4.1.23) and (4.1.24), this concludes the proof of Theorem M0.

4.2 CONTROL OF AVERAGES AND OF THE HAWKING MASS

In this section, we prove Lemma 3.15 and Lemma 3.16.

4.2.1 Proof of Lemma 3.15

Step 1. We start with the control of $\overline{\rho}$ on $\mathcal{M}$. Recall the identity (2.2.12)

$$\overline{\rho} + \frac{2m}{r^3} = \frac{1}{4}\overline{\vartheta\underline{\vartheta}}.$$

Thus, in view of the bootstrap assumptions **BA-D**, **BA-E**, we have,

$$\begin{aligned}\left|\overline{\rho}+\frac{2m}{r^3}\right| &\lesssim \epsilon^2\min\{r^{-3}u^{-\frac{3}{2}-\delta_{dec}}, r^{-2}u^{-2-2\delta_{dec}}\} \qquad \text{in } {}^{(ext)}\mathcal{M},\\ \left|\overline{\rho}+\frac{2m}{r^3}\right| &\lesssim \epsilon^2\underline{u}^{-2-2\delta_{dec}} \qquad \text{in } {}^{(int)}\mathcal{M}.\end{aligned}$$

Differentiating the equation with respect to e_3, e_4 we derive

$$\begin{aligned}e_4\left(\overline{\rho}+\frac{2m}{r^3}\right) &= \frac{1}{4}\overline{e_4(\vartheta)\underline{\vartheta}+\vartheta e_4(\underline{\vartheta})}+\text{l.o.t.},\\ e_3\left(\overline{\rho}+\frac{2m}{r^3}\right) &= \frac{1}{4}\overline{e_3(\vartheta)\underline{\vartheta}+\vartheta e_3(\underline{\vartheta})}+\text{l.o.t.},\\ e_\theta\left(\overline{\rho}+\frac{2m}{r^3}\right) &= 0.\end{aligned}$$

Taking higher derivatives in e_3, e_4 and making use of the bootstrap assumptions **BA-D**, **BA-E**, we derive, in ${}^{(ext)}\mathcal{M}$,

$$\begin{aligned}\left|\mathfrak{d}^{\le k_{small}}\left(\overline{\rho}+\frac{2m}{r^3}\right)\right| &\lesssim \epsilon^2\min\{r^{-3}u^{-\frac{3}{2}-\delta_{dec}}, r^{-2}u^{-2-2\delta_{dec}}\},\\ \left|\mathfrak{d}^{\le k_{large}}\left(\overline{\rho}+\frac{2m}{r^3}\right)\right| &\lesssim r^{-3}u^{-1/2-\delta_{dec}},\end{aligned}$$

and in ${}^{(int)}\mathcal{M}$,

$$\begin{aligned}\left|\mathfrak{d}^{\le k_{small}}\left(\overline{\rho}+\frac{2m}{r^3}\right)\right| &\lesssim \epsilon^2\underline{u}^{-2-2\delta_{dec}},\\ \left|\mathfrak{d}^{\le k_{large}}\left(\overline{\rho}+\frac{2m}{r^3}\right)\right| &\lesssim \epsilon^2\underline{u}^{-1-\delta_{dec}}.\end{aligned}$$

In particular,

$$\begin{aligned}&\sup_{{}^{(ext)}\mathcal{M}} u^{\frac{3}{2}+\delta_{dec}}r^3\left|\mathfrak{d}^{\le k_{small}}\left(\overline{\rho}+\frac{2m}{r^3}\right)\right|\\ &+\sup_{{}^{(ext)}\mathcal{M}} u^{\frac{1}{2}+\delta_{dec}}r^3\left|\mathfrak{d}^{\le k_{large}}\left(\overline{\rho}+\frac{2m}{r^3}\right)\right| \lesssim \epsilon_0,\\ &\sup_{{}^{(int)}\mathcal{M}} \underline{u}^{\frac{3}{2}+\delta_{dec}}\left|\mathfrak{d}^{\le k_{small}}\left(\overline{\rho}+\frac{2m}{r^3}\right)\right|\\ &+\sup_{{}^{(int)}\mathcal{M}} \underline{u}^{\frac{1}{2}+\delta_{dec}}\left|\mathfrak{d}^{\le k_{large}}\left(\overline{\rho}+\frac{2m}{r^3}\right)\right| \lesssim \epsilon_0.\end{aligned}$$

Step 2. Next, we proceed with the control of $\overline{\kappa}$ in ${}^{(ext)}\mathcal{M}$. Recalling Lemma 2.72, we start with

$$e_4\left(\overline{\kappa}-\frac{2}{r}\right)+\frac{1}{2}\overline{\kappa}\left(\overline{\kappa}-\frac{2}{r}\right) = -\frac{1}{4}\overline{\vartheta^2}+\frac{1}{2}\overline{\check{\kappa}^2}. \tag{4.2.1}$$

In view of Corollary 2.67 we deduce, from the first equation,

$$e_4\left(r\left(\overline{\kappa}-\frac{2}{r}\right)\right) \quad = -r\left(\tfrac{1}{4}\overline{\vartheta^2}+\tfrac{1}{2}\overline{\check{\kappa}^2}\right). \tag{4.2.2}$$

Making use of the GCM condition

$$\kappa \quad = \quad \frac{2}{r} \text{ on } \Sigma_*,$$

which yields

$$\overline{\kappa} \quad = \quad \frac{2}{r} \text{ on } \Sigma_*,$$

we deduce, integrating (4.2.2) with respect to r along C_u from Σ_*,

$$\sup_{(ext)\mathcal{M}} u^{1+\delta_{dec}} r^3 \left|\overline{\kappa}-\frac{2}{r}\right| \quad \lesssim \quad \epsilon^2 \lesssim \epsilon_0.$$

Also, making use of the bootstrap assumptions **BA-D**, **BA-E** we easily deduce

$$\begin{aligned}
\sup_{(ext)\mathcal{M}} u^{1+\delta_{dec}} r^3 \left|\mathfrak{d}_{\nearrow}^{\leq k_{small}+1}\left(\overline{\kappa}-\frac{2}{r}\right)\right| \quad &\lesssim \quad \epsilon^2 \lesssim \epsilon_0,\\
\sup_{(ext)\mathcal{M}} u^{\frac{1}{2}+\delta_{dec}} r^3 \left|\mathfrak{d}_{\nearrow}^{\leq k_{large}+1}\left(\overline{\kappa}-\frac{2}{r}\right)\right| \quad &\lesssim \quad \epsilon^2 \lesssim \epsilon_0.
\end{aligned}$$

We next commute (4.2.2) with e_3 and derive

$$\begin{aligned}
e_4 e_3\left(r\left(\overline{\kappa}-\frac{2}{r}\right)\right) \quad &= \quad e_3\left(r\left(\frac{1}{4}\overline{\vartheta^2}+\frac{1}{2}\overline{\check{\kappa}^2}\right)\right) - [e_3, e_4]\left(r\left(\overline{\kappa}-\frac{2}{r}\right)\right)\\
&= \quad e_3\left(r\left(\frac{1}{4}\overline{\vartheta^2}+\frac{1}{2}\overline{\check{\kappa}^2}\right)\right) - 2\underline{\omega}\left(r\left(\overline{\kappa}-\frac{2}{r}\right)\right)\\
&\qquad -2\zeta\left(r\left(\overline{\kappa}-\frac{2}{r}\right)\right).
\end{aligned}$$

It is thus easy to see that we can prove estimates of the type

$$\begin{aligned}
\sup_{(ext)\mathcal{M}} u^{1+\delta_{dec}} r^3 \left|\mathfrak{d}^{\leq k_{small}+1}\left(\overline{\kappa}-\frac{2}{r}\right)\right| \quad &\lesssim \quad \epsilon^2 \lesssim \epsilon_0,\\
\sup_{(ext)\mathcal{M}} u^{\frac{1}{2}+\delta_{dec}} r^3 \left|\mathfrak{d}^{\leq k_{large}+1}\left(\overline{\kappa}-\frac{2}{r}\right)\right| \quad &\lesssim \quad \epsilon^2 \lesssim \epsilon_0,
\end{aligned}$$

provided that we can check that

$$\begin{aligned}
\sup_{(ext)\mathcal{M}} u^{1+\delta_{dec}} r^3 \left|e_3^{\leq k_{small}+1}\left(\overline{\kappa}-\frac{2}{r}\right)\right| \quad &\lesssim \quad \epsilon^2 \lesssim \epsilon_0,\\
\sup_{(ext)\mathcal{M}} u^{\frac{1}{2}+\delta_{dec}} r^3 \left|e_3^{\leq k_{large}+1}\left(\overline{\kappa}-\frac{2}{r}\right)\right| \quad &\lesssim \quad \epsilon^2 \lesssim \epsilon_0.
\end{aligned}$$

The difficulty in this case is to make sure that we can control terms of the type

$$e_3^{k+1}\left(r\left(\frac{1}{4}e_3^{k+1}(\overline{\vartheta^2})+\frac{1}{2}e_3^{k+1}(\overline{\check{\kappa}^2})\right)\right)$$

using only at most k derivatives of $\check{\Gamma}, \check{R}$. To see this we note that

$$\begin{aligned}
e_3(\overline{\vartheta^2}) &= \overline{e_3\vartheta^2} - (\check{\underline{\Omega}}\,\overline{\kappa} - \overline{\check{\underline{\Omega}}\,\check{\kappa}})\overline{\vartheta^2} + \overline{\check{\underline{\kappa}}\vartheta^2},\\
e_3(\overline{\check{\kappa}^2}) &= \overline{e_3\check{\kappa}^2} - (\check{\underline{\Omega}}\,\overline{\kappa} - \overline{\check{\underline{\Omega}}\,\check{\kappa}})\overline{\check{\kappa}^2} + \overline{\check{\underline{\kappa}}\check{\kappa}^2},
\end{aligned}$$

and

$$\begin{aligned}
e_3(\vartheta) + \frac{1}{2}\underline{\kappa}\vartheta - 2\underline{\omega}\vartheta &= -2\,{\not{d}}_2^\star\zeta - \frac{1}{2}\kappa\underline{\vartheta} + 2\zeta^2,\\
e_3\check{\kappa} + \frac{1}{2}\overline{\underline{\kappa}}\,\check{\kappa} &= -2\check{\mu} - \frac{1}{2}\overline{\kappa}\check{\underline{\kappa}} + 2(\check{\underline{\omega}}\overline{\kappa} + \overline{\underline{\omega}}\check{\kappa}) + \check{\underline{\Omega}}\overline{\kappa}\,\overline{\underline{\kappa}} + \mathrm{Err}[e_3\check{\kappa}], \qquad (4.2.3)\\
\mathrm{Err}[e_3\check{\kappa}] :&= 2(\zeta^2 - \overline{\zeta^2}) + 2(\check{\underline{\omega}}\check{\kappa} - \overline{\check{\underline{\omega}}\check{\kappa}}) - \frac{1}{2}\check{\underline{\kappa}}\,\check{\kappa} - \frac{1}{2}\overline{\check{\underline{\kappa}}\,\check{\kappa}} - \overline{\check{\underline{\Omega}}\check{\kappa}}\,\overline{\underline{\kappa}}.
\end{aligned}$$

We thus derive

$$\begin{aligned}
&\sup_{{}^{(ext)}\mathcal{M}} u^{1+\delta_{dec}} r^3 \left|\mathfrak{d}^{\leq k_{small}+1}\left(\overline{\kappa} - \frac{2}{r}\right)\right|\\
&+ \sup_{{}^{(ext)}\mathcal{M}} u^{\frac{1}{2}+\delta_{dec}} r^3 \left|\mathfrak{d}^{\leq k_{large}+1}\left(\overline{\kappa} - \frac{2}{r}\right)\right| \lesssim \epsilon_0.
\end{aligned}$$

Step 3. We next estimate $\overline{\underline{\kappa}}$ in ${}^{(ext)}\mathcal{M}$ making use of the identity (2.2.14) derived in connection to the Hawking mass

$$\overline{\underline{\kappa}} + \frac{2\Upsilon}{r} = \frac{2\Upsilon}{r\overline{\kappa}}\left(\overline{\kappa} - \frac{2}{r}\right) - \frac{1}{\overline{\kappa}}\overline{\check{\kappa}\check{\underline{\kappa}}}.$$

Thus, in view of the estimates for $\overline{\kappa}$ derived in Step 2 we easily infer that

$$\sup_{{}^{(ext)}\mathcal{M}} u^{\frac{3}{2}+\delta_{dec}} r^2 \left|\mathfrak{d}^{\leq k_{small}}\left(\overline{\underline{\kappa}} + \frac{2\Upsilon}{r}\right)\right| + \sup_{{}^{(ext)}\mathcal{M}} u^{\frac{1}{2}+\delta_{dec}} r^2 \left|\mathfrak{d}^{\leq k_{large}}\left(\overline{\underline{\kappa}} + \frac{2\Upsilon}{r}\right)\right| \lesssim \epsilon_0$$

as desired.

Step 4. We estimate $\overline{\underline{\omega}}$ in ${}^{(ext)}\mathcal{M}$ based on the following identity in Lemma 2.72

$$\begin{aligned}
e_3\left(\overline{\kappa} - \frac{2}{r}\right) + \frac{1}{2}\overline{\underline{\kappa}}\left(\overline{\kappa} - \frac{2}{r}\right) &= 2\overline{\underline{\omega}}\left(\overline{\kappa} - \frac{2}{r}\right) + \frac{4}{r}\left(\overline{\underline{\omega}} - \frac{m}{r^2}\right) + 2\left(\overline{\rho} + \frac{2m}{r^3}\right)\\
&\quad -\frac{1}{2}\overline{\kappa}\left(\overline{\kappa} - \frac{2}{r}\right)\check{\underline{\Omega}} + 2\overline{\check{\underline{\omega}}\check{\kappa}} - \frac{1}{2}\overline{\vartheta\underline{\vartheta}} + 2\overline{\zeta^2}\\
&\quad +\frac{1}{2}\check{\underline{\Omega}}\Big(-\overline{\vartheta^2} + \overline{\check{\kappa}^2}\Big) - \overline{\check{\underline{\Omega}}(e_4(\check{\kappa}) + \kappa\check{\kappa})} + \frac{1}{2}\overline{\check{\underline{\kappa}}\check{\kappa}}\\
&\quad -\frac{1}{r}\overline{\check{\underline{\Omega}}\check{\kappa}},
\end{aligned}$$

which we rewrite as

$$\begin{aligned}\overline{\underline{\omega}} - \frac{m}{r^2} &= \frac{r}{4}\Bigg\{ e_3\left(\overline{\kappa} - \frac{2}{r}\right) + \frac{1}{2}\overline{\underline{\kappa}}\left(\overline{\kappa} - \frac{2}{r}\right) - 2\overline{\underline{\omega}}\left(\overline{\kappa} - \frac{2}{r}\right) - 2\left(\overline{\rho} + \frac{2m}{r^3}\right)\\ &+\frac{1}{2}\overline{\kappa}\left(\overline{\kappa} - \frac{2}{r}\right)\check{\underline{\Omega}} - 2\overline{\check{\underline{\omega}}\check{\kappa}} + \frac{1}{2}\overline{\vartheta\underline{\vartheta}} - 2\overline{\zeta^2} - \frac{1}{2}\check{\underline{\Omega}}\Big(-\overline{\vartheta^2} + \overline{\check{\kappa}^2}\Big)\\ &+\overline{\check{\underline{\Omega}}(e_4(\check{\kappa}) + \kappa\check{\kappa})} - \frac{1}{2}\overline{\check{\underline{\kappa}}\check{\kappa}} + \frac{1}{r}\overline{\check{\underline{\Omega}}\check{\kappa}}\Bigg\}. \end{aligned} \tag{4.2.4}$$

Using the estimates of $\overline{\rho}$ in Step 1, the estimates for $\overline{\kappa}$ in Step 2, as well as our bootstrap assumptions on decay and energy, we easily derive

$$\sup_{{}^{(ext)}\mathcal{M}} u^{1+\delta_{dec}} r^2 \left|\mathfrak{d}^{\leq k_{small}}\left(\overline{\underline{\omega}} - \frac{m}{r^2}\right)\right| + \sup_{{}^{(ext)}\mathcal{M}} u^{\frac{1}{2}+\delta_{dec}} r^2 \left|\mathfrak{d}^{\leq k_{large}}\left(\overline{\underline{\omega}} - \frac{m}{r^2}\right)\right| \lesssim \epsilon_0.$$

Remark 4.6. *It is to estimate k_{large} derivatives of $\overline{\underline{\omega}} - mr^{-2}$ that we had to control $k_{large}+1$ derivatives of $\overline{\kappa} - 2/r$ in Step 2.*

Step 5. We estimate $\overline{\underline{\Omega}}$ in ${}^{(ext)}\mathcal{M}$. First we need the control of $\overline{\underline{\Omega}}$ on Σ_*. To this end, we recall that s is initialized on Σ_* by $s = r$ so that

$$\nu(s-r) \;=\; 0 \text{ on } \Sigma_*, \qquad \nu = e_3 + ae_4,$$

where the scalar function a is such that the vectorfield ν is tangent to Σ_*. On the other hand, we have $e_4(s) = 1$ and

$$e_4(r) = \frac{r}{2}\overline{\kappa} = 1 \text{ on } \Sigma_*$$

where we used the GCM condition $\kappa = 2/r$ on Σ_*. We infer $e_3(s) = e_3(r)$ on Σ_* and hence

$$\underline{\Omega} \;=\; e_3(r) \text{ on } \Sigma_*.$$

This yields

$$\overline{\underline{\Omega}} \;=\; \overline{e_3(r)} = \frac{r\overline{\underline{\kappa}}}{2} + \frac{r}{2}\overline{\underline{A}},$$

and hence, in view of the estimate for $\overline{\underline{\kappa}}$ of Step 3, the fact that $\overline{\underline{A}}$ contains only quadratic terms in view of the formula for $\underline{A}$, and in view of the bootstrap assumptions on decay and energy, we infer

$$\sup_{\Sigma_*} u^{1+\delta_{dec}} r \left|\mathfrak{d}^{\leq k_{small}}\left(\overline{\underline{\Omega}} - \frac{m}{r^2}\right)\right| + \sup_{\Sigma_*} u^{\frac{1}{2}+\delta_{dec}} r \left|\mathfrak{d}^{\leq k_{large}}\left(\overline{\underline{\Omega}} - \frac{m}{r^2}\right)\right| \lesssim \epsilon_0.$$

Then, we use $e_4(\underline{\Omega}) = -2\underline{\omega}$ and Corollary 2.66 to obtain

$$e_4(\overline{\underline{\Omega}}) \;=\; -2\overline{\underline{\omega}} + \overline{\check{\kappa}\,\check{\underline{\Omega}}}$$

and hence

$$e_4(\overline{\underline{\Omega}} + \Upsilon) \quad = \quad -2\left(\overline{\underline{\omega}} - \frac{m}{r^2}\right) + \frac{m}{r}\left(\overline{\kappa} - \frac{2}{r}\right) + \overline{\check{\kappa}\check{\underline{\Omega}}} - \frac{2e_4(m)}{r}.$$

Commuting with $\mathfrak{d}$, integrating from Σ_* where we have controlled $\overline{\underline{\Omega}}$ above, and using the estimates of Step 2 for $\overline{\kappa}$, Step 4 for $\overline{\underline{\omega}}$, the bootstrap assumptions, and the estimates for $e_4(m)$ of Lemma 3.16 (which do not depend on the control of $\overline{\underline{\Omega}}$), we infer

$$\sup_{^{(ext)}\mathcal{M}} u^{1+\delta_{dec}} r \left|\mathfrak{d}^{\leq k_{small}}\left(\overline{\underline{\Omega}} - \frac{m}{r^2}\right)\right| + \sup_{^{(ext)}\mathcal{M}} u^{\frac{1}{2}+\delta_{dec}} r \left|\mathfrak{d}^{\leq k_{large}}\left(\overline{\underline{\Omega}} - \frac{m}{r^2}\right)\right| \quad \lesssim \quad \epsilon_0.$$

Step 6. Next, we control ${}^{(int)}\overline{\underline{\kappa}}$ on the cylinder $\mathcal{T}$. From the initialization of the frame of ${}^{(int)}\mathcal{M}$ on $\mathcal{T}$, we have

$${}^{(int)}r = {}^{(ext)}r, \quad {}^{(int)}\overline{\kappa} = \Upsilon\, {}^{(ext)}\overline{\kappa}, \quad {}^{(int)}\overline{\underline{\kappa}} = \Upsilon^{-1}\, {}^{(ext)}\overline{\underline{\kappa}} \text{ on } \mathcal{T}.$$

Also, making use of the identity (2.2.14) derived in connection to the Hawking mass, we have

$${}^{(ext)}\overline{\underline{\kappa}} + \frac{2\Upsilon}{r} \quad = \quad \frac{2\Upsilon}{r\, {}^{(ext)}\overline{\kappa}}\left({}^{(ext)}\overline{\kappa} - \frac{2}{r}\right) - \frac{1}{{}^{(ext)}\overline{\kappa}}\overline{{}^{(ext)}\check{\kappa}\, {}^{(ext)}\check{\underline{\kappa}}}.$$

We deduce

$$\begin{aligned}{}^{(int)}\overline{\underline{\kappa}} + \frac{2}{r} \quad &= \quad \Upsilon^{-1}\left({}^{(ext)}\overline{\underline{\kappa}} + \frac{2\Upsilon}{r}\right) = \frac{2}{r\, {}^{(ext)}\overline{\kappa}}\left({}^{(ext)}\overline{\kappa} - \frac{2}{r}\right)\\ &\quad - \frac{\Upsilon^{-1}}{{}^{(ext)}\overline{\kappa}}\overline{{}^{(ext)}\check{\kappa}\, {}^{(ext)}\check{\underline{\kappa}}} \text{ on } \mathcal{T}.\end{aligned}$$

To derive higher tangential derivatives along $\mathcal{T}$ we remark that the vectorfield

$$T_{\mathcal{T}} = e_4 - \frac{e_4(r)}{e_3(r)}e_3 = e_4 - \frac{\overline{\kappa} + A}{\overline{\underline{\kappa}}}e_3,$$

together with e_θ, spans the tangent space to $\mathcal{T}$. The transversal derivatives, on the other hand, can be determined with help of the equation

$$e_3\left(\overline{\underline{\kappa}} + \frac{2}{r}\right) + \frac{1}{2}\overline{\underline{\kappa}}\left(\overline{\underline{\kappa}} + \frac{2}{r}\right) \quad = \quad -\frac{1}{4}\overline{\underline{\vartheta}^2} + \frac{1}{2}\overline{\check{\underline{\kappa}}^2}$$

adapted to the ${}^{(int)}\mathcal{M}$ foliation. Making use of the estimates for ${}^{(ext)}\overline{\kappa}$ in ${}^{(ext)}\mathcal{M}$ derived in Step 2 and the bootstrap assumptions, we infer that

$$\begin{aligned}&\sup_{\mathcal{T}} u^{\frac{3}{2}+\delta_{dec}}\left|\mathfrak{d}^{\leq k_{small}+1}\left({}^{(int)}\overline{\underline{\kappa}} + \frac{2}{r}\right)\right| + \sup_{\mathcal{T}} u^{\frac{1}{2}+\delta_{dec}}\left|\mathfrak{d}^{\leq k_{large}+1}\left({}^{(int)}\overline{\underline{\kappa}} + \frac{2}{r}\right)\right|\\ &\lesssim \quad \epsilon_0 + \sup_{\mathcal{T}} u^{\frac{1}{2}+\delta_{dec}}\left|\mathfrak{d}^{k_{large}+1}\left(\overline{{}^{(ext)}\check{\kappa}\, {}^{(ext)}\check{\underline{\kappa}}}\right)\right|.\end{aligned}$$

Now, in view of the transport equations involving ${}^{(ext)}e_4({}^{(ext)}\check{\kappa})$, ${}^{(ext)}e_3({}^{(ext)}\check{\kappa})$,

${}^{(ext)}e_4({}^{(ext)}\check{\underline{\kappa}})$ and ${}^{(ext)}e_3({}^{(ext)}\check{\kappa})$, as well as the bootstrap assumptions, we have

$$\begin{aligned}
&\sup_{\mathcal{T}} u^{\frac{1}{2}+\delta_{dec}} \left|\mathfrak{d}^{k_{large}+1}\left(\overline{{}^{(ext)}\check{\kappa}\,{}^{(ext)}\check{\underline{\kappa}}}\right)\right| \\
\lesssim\ & \epsilon_0 + \sup_{\mathcal{T}} u^{\frac{1}{2}+\delta_{dec}} \left|\overline{{}^{(ext)}\check{\underline{\kappa}}\mathfrak{d}^{k_{large}} \not{d}_1({}^{(ext)}\zeta)}\right| \\
&+ \sup_{\mathcal{T}} u^{\frac{1}{2}+\delta_{dec}} \left|\overline{{}^{(ext)}\check{\kappa}\mathfrak{d}^{k_{large}} \not{d}_1({}^{(ext)}\zeta)}\right| + \sup_{\mathcal{T}} u^{\frac{1}{2}+\delta_{dec}} \left|\overline{{}^{(ext)}\check{\kappa}\mathfrak{d}^{k_{large}} \not{d}_1({}^{(ext)}\underline{\xi})}\right| \\
\lesssim\ & \epsilon_0 + \sup_{\mathcal{T}} u^{\frac{1}{2}+\delta_{dec}} \left|\overline{\not{d}_1^\star\,{}^{(ext)}\check{\underline{\kappa}}\mathfrak{d}^{k_{large}}\,{}^{(ext)}\zeta}\right| + \sup_{\mathcal{T}} u^{\frac{1}{2}+\delta_{dec}} \left|\overline{\not{d}_1^\star\,{}^{(ext)}\check{\kappa}\mathfrak{d}^{k_{large}}\,{}^{(ext)}\zeta}\right| \\
&+ \sup_{\mathcal{T}} u^{\frac{1}{2}+\delta_{dec}} \left|\overline{\not{d}_1^\star\,{}^{(ext)}\check{\kappa}\mathfrak{d}^{k_{large}}\,{}^{(ext)}\underline{\xi}}\right| \\
\lesssim\ & \epsilon_0
\end{aligned}$$

where we have integrated $\not{d}_1$ by parts and used that $\not{d}_1^\star$ is its adjoint. We infer

$$\sup_{\mathcal{T}} \underline{u}^{\frac{3}{2}+\delta_{dec}} \left|\mathfrak{d}^{\leq k_{small}+1}\left({}^{(int)}\overline{\underline{\kappa}} + \frac{2}{r}\right)\right| + \sup_{\mathcal{T}} \underline{u}^{\frac{1}{2}+\delta_{dec}} \left|\mathfrak{d}^{\leq k_{large}+1}\left({}^{(int)}\overline{\underline{\kappa}} + \frac{2}{r}\right)\right| \lesssim \epsilon_0.$$

Step 7. From now on, we only work with the frame of ${}^{(int)}\mathcal{M}$. Starting with the equation

$$e_3\left(\overline{\underline{\kappa}} + \frac{2}{r}\right) + \frac{1}{2}\overline{\underline{\kappa}}\left(\overline{\underline{\kappa}} + \frac{2}{r}\right) = -\frac{1}{4}\overline{\underline{\vartheta}^2} + \frac{1}{2}\overline{\check{\underline{\kappa}}^2}$$

and using the estimates of Step 5, we can then proceed precisely as in Step 2 (using the ${}^{(int)}\mathcal{M}$ counterpart of the equations (4.2.3)) to derive

$$\sup_{{}^{(int)}\mathcal{M}} \underline{u}^{1+\delta_{dec}} \left|\mathfrak{d}^{\leq k_{small}+1}\left(\overline{\underline{\kappa}} + \frac{2}{r}\right)\right| + \sup_{{}^{(int)}\mathcal{M}} \underline{u}^{\frac{1}{2}+\delta_{dec}} \left|\mathfrak{d}^{\leq k_{large}+1}\left(\overline{\underline{\kappa}} + \frac{2}{r}\right)\right| \lesssim \epsilon_0.$$

Step 8. Finally, we estimate the remaining averages in ${}^{(int)}\mathcal{M}$, i.e., $\overline{\kappa}$ and $\overline{\underline{\omega}}$. To estimate $\overline{\kappa}$ we make use once more of the identity

$$\overline{\kappa} - \frac{2\Upsilon}{r} = -\frac{2\Upsilon}{r\overline{\underline{\kappa}}}\left(\overline{\underline{\kappa}} + \frac{2}{r}\right) - \frac{1}{\overline{\underline{\kappa}}}\overline{\check{\kappa}\check{\underline{\kappa}}}.$$

Making use of the estimates of $\overline{\underline{\kappa}}$ in Step 5 as well as the bootstrap assumptions for $\check{\kappa}$ and $\check{\underline{\kappa}}$ we easily derive

$$\sup_{{}^{(int)}\mathcal{M}} \underline{u}^{1+\delta_{dec}} \left|\mathfrak{d}^{\leq k_{small}}\left(\overline{\kappa} - \frac{2\Upsilon}{r}\right)\right| + \sup_{{}^{(int)}\mathcal{M}} \underline{u}^{\frac{1}{2}+\delta_{dec}} \left|\mathfrak{d}^{\leq k_{large}}\left(\overline{\kappa} - \frac{2\Upsilon}{r}\right)\right| \lesssim \epsilon_0.$$

Step 9. To estimate $\overline{\omega}$ we proceed as in Step 4 by making use of the identity

$$\begin{aligned}\overline{\omega}+\frac{m}{r^2} &= \frac{r}{4}\Bigg\{e_4\left(\underline{\overline{\kappa}}+\frac{2}{r}\right)+\frac{1}{2}\overline{\kappa}\left(\underline{\overline{\kappa}}+\frac{2}{r}\right)-2\overline{\omega}\left(\underline{\overline{\kappa}}+\frac{2}{r}\right)-2\left(\overline{\rho}+\frac{2m}{r^3}\right)\\ &+\frac{1}{2}\underline{\overline{\kappa}}\left(\underline{\overline{\kappa}}+\frac{2}{r}\right)\check{\Omega}-2\overline{\check{\omega}\underline{\check{\kappa}}}+\frac{1}{2}\overline{\vartheta\underline{\vartheta}}-2\overline{\zeta^2}-\frac{1}{2}\check{\Omega}\Big(-\overline{\underline{\vartheta}^2}+\overline{\underline{\check{\kappa}}^2}\Big)\\ &+\overline{\check{\Omega}(e_4(\underline{\check{\kappa}})+\kappa\underline{\check{\kappa}})}-\frac{1}{2}\overline{\check{\kappa}\underline{\check{\kappa}}}+\frac{1}{r}\overline{\check{\Omega}\underline{\check{\kappa}}}\Bigg\}.\end{aligned}$$

Thus, in view of the estimates of $\overline{\rho}$ in Step 1, the estimates for $\underline{\overline{\kappa}}$ in Step 5, the estimates of $\overline{\kappa}$ above,[5] as well as the bootstrap assumptions **BA-D** and **BA-E**, we deduce

$$\sup_{^{(int)}\mathcal{M}} \underline{u}^{1+\delta_{dec}}\left|\mathfrak{d}^{\leq k_{small}}\left(\overline{\omega}+\frac{m}{r^2}\right)\right|+\sup_{^{(int)}\mathcal{M}} \underline{u}^{\frac{1}{2}+\delta_{dec}}\left|\mathfrak{d}^{\leq k_{large}}\left(\overline{\omega}+\frac{m}{r^2}\right)\right| \lesssim \epsilon_0.$$

Step 10. It remains to estimate $\overline{\Omega}$ in $^{(int)}\mathcal{M}$. First we need the control of $\overline{\Omega}$ on $\mathcal{T}$. To this end, we recall that s is initialized on $\mathcal{T}$ by $s=r$ so that

$$T_{\mathcal{T}}(s-r) = 0 \text{ on } \mathcal{T}, \qquad T_{\mathcal{T}}=e_4-\frac{\overline{\kappa}+A}{\underline{\overline{\kappa}}}e_3,$$

where the vectorfield has been introduced above and is tangent to $\mathcal{T}$. On the other hand, we have $e_3(s)=-1$ and $e_3(r)=r\underline{\overline{\kappa}}/2$, and hence

$$\Omega = e_4(r)+\frac{\overline{\kappa}+A}{\underline{\overline{\kappa}}}(-1-e_3(r))=\frac{r}{2}(\overline{\kappa}+A)\left(1+\frac{\underline{\overline{\kappa}}-\frac{2}{r}}{\underline{\overline{\kappa}}}\right) \text{ on } \mathcal{T}.$$

This yields

$$\overline{\Omega} = \frac{r}{2}(\overline{\kappa}+\overline{A})\left(1+\frac{\underline{\overline{\kappa}}-\frac{2}{r}}{\underline{\overline{\kappa}}}\right) \text{ on } \mathcal{T},$$

and hence, in view of the estimate for $\underline{\overline{\kappa}}$ of Step 7, the estimate for $\overline{\kappa}$ of Step 8, the fact that $\overline{A}$ contains only quadratic terms in view of the formula for A, and in view of the bootstrap assumptions on decay and energy, we infer

$$\sup_{\mathcal{T}}\underline{u}^{1+\delta_{dec}}\left|\mathfrak{d}^{\leq k_{small}}\left(\overline{\Omega}-\Upsilon\right)\right|+\sup_{\mathcal{T}}\underline{u}^{\frac{1}{2}+\delta_{dec}}\left|\mathfrak{d}^{\leq k_{large}}\left(\overline{\Omega}-\Upsilon\right)\right| \lesssim \epsilon_0.$$

Then, we use the analog of the transport equation used to estimate $\underline{\overline{\Omega}}$ in $^{(ext)}\mathcal{M}$, i.e.,

$$e_3(\overline{\Omega}-\Upsilon) = 2\left(\overline{\omega}+\frac{m}{r^2}\right)-\frac{m}{r}\left(\underline{\overline{\kappa}}+\frac{2}{r}\right)+\overline{\underline{\check{\kappa}}\check{\Omega}}+\frac{2e_3(m)}{r}.$$

Commuting with $\mathfrak{d}$, integrating from $\mathcal{T}$ where we have controlled $\overline{\Omega}$ above, and using the estimates of Step 2 for $\overline{\kappa}$, Step 4 for $\underline{\overline{\omega}}$, the bootstrap assumptions, and

[5] It is to estimate k_{large} derivatives of $\overline{\omega}+m/r^2$ that we made sure to control $k_{large}+1$ derivatives of $\underline{\overline{\kappa}}+2/r$.

the estimates for $e_4(m)$ of Lemma 3.16 (which do not depend on the control of $\overline{\underline{\Omega}}$), we infer

$$\sup_{{}^{(int)}\mathcal{M}} \underline{u}^{1+\delta_{dec}} \left|\mathfrak{d}^{\leq k_{small}} \left(\overline{\underline{\Omega}} - \Upsilon\right)\right| + \sup_{{}^{(int)}\mathcal{M}} \underline{u}^{\frac{1}{2}+\delta_{dec}} \left|\mathfrak{d}^{\leq k_{large}} \left(\overline{\underline{\Omega}} - \Upsilon\right)\right| \lesssim \epsilon_0.$$

This concludes the proof of Lemma 3.15.

4.2.2 Proof of Lemma 3.16

Step 1. We start with the control of $e_3(m)$ and $e_4(m)$ in ${}^{(ext)}\mathcal{M}$. According to Proposition 2.71 we have in ${}^{(ext)}\mathcal{M}$

$$e_4(m) = \frac{r}{32\pi}\int_S \mathrm{Err}_1, \tag{4.2.5}$$

and

$$\begin{aligned} e_3(m) = {} & \left(1-\varsigma^{-1}\check{\varsigma}\right)\frac{r}{32\pi}\int_S \underline{\mathrm{Err}}_1 + \left(\check{\underline{\Omega}} + \varsigma^{-1}\overline{\underline{\Omega}}\check{\varsigma}\right)\frac{r}{32\pi}\int_S \mathrm{Err}_1 \\ & + \varsigma^{-1}\frac{r}{32\pi}\int_S \check{\varsigma}\left(2\overline{\rho}\underline{\check{\kappa}} + 2\check{\rho}\overline{\underline{\kappa}} + 2\underline{\kappa}\,\not{d}_1\eta + 2\kappa\,\not{d}_1\underline{\xi} + \underline{\mathrm{Err}}_2\right) \\ & - \varsigma^{-1}\frac{r}{32\pi}\int_S \left(\overline{\underline{\Omega}}\check{\varsigma} + \check{\underline{\Omega}}\varsigma\right)\left(2\overline{\rho}\check{\kappa} + 2\check{\rho}\overline{\kappa} - 2\kappa\,\not{d}_1\zeta + \mathrm{Err}_2\right) \\ & - \frac{m}{r}\varsigma^{-1}\left[-\overline{\check{\varsigma}\underline{\check{\kappa}}} + \overline{\underline{\Omega}}\,\overline{\check{\varsigma}\check{\kappa}} + \overline{\check{\underline{\Omega}}\varsigma\kappa}\right], \end{aligned} \tag{4.2.6}$$

where

$$\begin{aligned} \mathrm{Err}_1 &:= 2\check{\kappa}\check{\rho} + 2e_\theta(\kappa)\zeta - \frac{1}{2}\underline{\kappa}\vartheta^2 - \frac{1}{2}\check{\kappa}\vartheta\underline{\vartheta} + 2\kappa\zeta^2, \\ \underline{\mathrm{Err}}_1 &:= 2\check{\rho}\underline{\check{\kappa}} - 2e_\theta(\underline{\kappa})\eta - 2e_\theta(\kappa)\underline{\xi} - \frac{1}{2}\underline{\check{\kappa}}\vartheta\underline{\vartheta} + 2\underline{\kappa}\eta^2 + 2\kappa(\eta - 3\zeta)\underline{\xi} - \frac{1}{2}\kappa\underline{\vartheta}^2, \\ \mathrm{Err}_2 &:= 2\check{\rho}\check{\kappa} - \frac{1}{2}\underline{\kappa}\vartheta^2 - \frac{1}{2}\kappa\vartheta\underline{\vartheta} + 2\kappa\zeta^2, \\ \underline{\mathrm{Err}}_2 &:= 2\check{\rho}\underline{\check{\kappa}} + \underline{\kappa}\left(2\eta^2 - \frac{1}{2}\vartheta\underline{\vartheta}\right) + 2\kappa(\eta - 3\zeta)\underline{\xi} - \frac{1}{2}\kappa\underline{\vartheta}^2. \end{aligned}$$

Thus, according to the bootstrap assumption **BA-D** on decay, we deduce

$$\begin{aligned} |e_4(m)| &\lesssim \epsilon^2 r^{-2} u^{-1-\delta_{dec}}, \\ |e_3(m)| &\lesssim \epsilon^2 u^{-2-2\delta_{dec}}. \end{aligned}$$

Moreover, differentiating the equations with respect to e_3, e_4 and making use of both bootstrap assumptions **BA-D** and **BA-E** on decay and energy, and integrating by part once the e_θ derivative for the terms involving $e_\theta(\kappa)$ and $e_\theta(\underline{\kappa})$ when they contain top order derivatives, we infer that

$$\begin{aligned} \max_{0\leq k\leq k_{small}} \sup_{{}^{(ext)}\mathcal{M}} r^2 u^{1+\delta_{dec}} |\mathfrak{d}^k e_4(m)| &\lesssim \epsilon_0, \\ \max_{0\leq k\leq k_{large}} \sup_{{}^{(ext)}\mathcal{M}} \left(r^2 u^{\frac{1}{2}+\delta_{dec}} + r u^{1+\delta_{dec}}\right) |\mathfrak{d}^k e_4(m)| &\lesssim \epsilon_0, \end{aligned}$$

as well as

$$\max_{0\leq k\leq k_{small}} \sup_{^{(ext)}\mathcal{M}} u^{2+2\delta_{dec}}|\mathfrak{d}^k e_3(m)| \lesssim \epsilon_0,$$
$$\max_{0\leq k\leq k_{large}} \sup_{^{(ext)}\mathcal{M}} u^{1+\delta_{dec}}|\mathfrak{d}^k e_3(m)| \lesssim \epsilon_0,$$

consistent with the statement of the lemma.

Step 2. We derive the estimates on $^{(int)}\mathcal{M}$. According to the analogue of Proposition 2.71 in the situation of the incoming geodesic foliations of $^{(int)}\mathcal{M}$, and proceeding as in Step 1, we easily derive

$$\max_{0\leq k\leq k_{large}} \sup_{^{(int)}\mathcal{M}} \underline{u}^{1+\delta_{dec}}\Big(|\mathfrak{d}^k e_3(m)| + |\mathfrak{d}^k e_4(m)|\Big) \lesssim \epsilon^2 \lesssim \epsilon_0. \tag{4.2.7}$$

Step 3. We estimate $m - m_0$ in $^{(ext)}\mathcal{M}$. First, recall from Theorem M0 that we have

$$\sup_{\mathcal{C}_1\cup\underline{\mathcal{C}}_1} |m - m_0| \lesssim \epsilon_0 m_0. \tag{4.2.8}$$

We start with the control in $^{(ext)}\mathcal{M}$. Note that $^{(ext)}\mathcal{M}$ is covered by integral curves of e_3 starting from $\mathcal{C}_1$. Thus, integrating the $e_3 m$ equation and making use of the estimate $\sup_{\mathcal{C}_1}|m - m_0| \lesssim \epsilon_0 m_0$ as well as the fact that $e_3(u) = 2$, we easily deduce that

$$\sup_{^{(ext)}\mathcal{M}} |m - m_0| \lesssim \epsilon_0 m_0 + \epsilon^2 \lesssim \epsilon_0 m_0.$$

Step 4. We estimate $|m - m_0|$ on $\mathcal{T}$. In view of our initialization of the ingoing geodesic foliation of $^{(int)}\mathcal{M}$ on $\mathcal{T}$,

$$^{(int)}\kappa\, ^{(int)}\underline{\kappa} = {}^{(ext)}\kappa\, ^{(ext)}\underline{\kappa} \text{ on } \mathcal{T}.$$

Since the spheres of both foliations agree on $\mathcal{T}$, we infer from the definition of the Hawking mass

$$^{(int)}m = {}^{(ext)}m \text{ on } \mathcal{T}.$$

Using the estimate for $^{(ext)}m$ we infer that

$$\sup_{\mathcal{T}} |{}^{(int)}m - m_0| \lesssim \epsilon_0 m_0.$$

Step 5. We estimate $|m - m_0|$ on $^{(int)}\mathcal{M}$. Note first that in $^{(int)}\mathcal{M}$

$$e_3(r) + 1 = \frac{r}{2}\overline{\underline{\kappa}} + 1 = \frac{r}{2}\left(\overline{\underline{\kappa}} + \frac{2}{r}\right).$$

Thus, in view of the estimate for $\overline{\underline{\kappa}} + \frac{2}{r}$ derived in Lemma 3.15

$$\sup_{^{(int)}\mathcal{M}} |e_3(r) + 1| \lesssim \epsilon^2.$$

Thus integrating the estimate (4.2.7) in $r \in [2m_0(1-\delta_{\mathcal{H}}), r_{\mathcal{T}}]$, where we recall that $r_{\mathcal{T}} \leq 2m_0(1+2\delta_{\mathcal{H}})$, we derive

$$\sup_{{}^{(int)}\mathcal{M}} |m - m_0| \lesssim \epsilon_0 m_0.$$

Since $\mathcal{M} = {}^{(ext)}\mathcal{M} \cup {}^{(int)}\mathcal{M}$ we infer that

$$\sup_{\mathcal{M}} |m - m_0| \lesssim \epsilon_0 m_0.$$

This concludes the proof of Lemma 3.16.

4.3 CONTROL OF COORDINATES SYSTEMS

The goal of this section is to prove Propositions 3.17 and 3.18. In both cases, the first two claims, on the form of the spacetime metric in the corresponding coordinates system as well as on the expression of the coordinates vectorfield with respect to the null frame (e_4, e_3, e_θ), are already proved in Propositions 2.78 and 2.81. So we only focus on the third claim, i.e., on estimating $\check{\underline{\Omega}}$, $\check{\Omega}$, ς, $\underline{\varsigma}$, γ, b, $\underline{b}$ and e^Φ. The proof of Propositions 3.17 and 3.18 thus reduces to the proof of the following lemma.

Lemma 4.7. *Let $\theta \in [0, \pi]$ be the $\mathbf{Z}$-invariant scalar on $\mathcal{M}$ defined by (2.2.52), i.e.,*

$$\theta = \cot^{-1}\left(r e_\theta(\Phi)\right). \tag{4.3.1}$$

Let

$$b = e_4(\theta), \qquad \underline{b} = e_3(\theta), \qquad \gamma = \frac{1}{(e_\theta(\theta))^2}. \tag{4.3.2}$$

Then, we have

$$\max_{0\leq k\leq k_{small}} \sup_{{}^{(ext)}\mathcal{M}} \left(r u^{\frac{1}{2}+\delta_{dec}} + u^{1+\delta_{dec}}\right)\left(\left|\mathfrak{d}^k\left(\frac{\gamma}{r^2}-1\right)\right| + r\left|\mathfrak{d}^k b\right|\right) \lesssim \epsilon,$$

$$\max_{0\leq k\leq k_{small}} \sup_{{}^{(ext)}\mathcal{M}} u^{1+\delta_{dec}}\left(\left|\mathfrak{d}^k\check{\underline{\Omega}}\right| + \left|\mathfrak{d}^k(\varsigma-1)\right| + r\left|\mathfrak{d}^k\underline{b}\right|\right) \lesssim \epsilon,$$

$$\max_{0\leq k\leq k_{small}} \sup_{{}^{(int)}\mathcal{M}} \underline{u}^{1+\delta_{dec}}\Bigg(\left|\mathfrak{d}^k\check{\Omega}\right| + \left|\mathfrak{d}^k(\underline{\varsigma}-1)\right| + \left|\mathfrak{d}^k\left(\frac{\gamma}{r^2}-1\right)\right| + \left|\mathfrak{d}^k b\right| + \left|\mathfrak{d}^k\underline{b}\right|\Bigg) \lesssim \epsilon.$$

Also, e^Φ satisfies

$$\max_{0\leq k\leq k_{small}} \sup_{{}^{(ext)}\mathcal{M}} \left(r u^{\frac{1}{2}+\delta_{dec}} + u^{1+\delta_{dec}}\right)\left|\mathfrak{d}^k\left(\frac{e^\Phi}{r\sin\theta}-1\right)\right| \lesssim \epsilon,$$

$$\max_{0\leq k\leq k_{small}} \sup_{{}^{(int)}\mathcal{M}} \underline{u}^{1+\delta_{dec}}\left|\mathfrak{d}^k\left(\frac{e^\Phi}{r\sin\theta}-1\right)\right| \lesssim \epsilon.$$

Proof. We prove the estimates in ${}^{(ext)}\mathcal{M}$. The proof in ${}^{(int)}\mathcal{M}$ is similar and left to the reader.

Step 1. We start with the estimate for $\underline{\check{\Omega}}$. Recall that

$$\not{d}_1^\star \underline{\check{\Omega}} = \underline{\xi}$$

so that the bootstrap assumptions for $\underline{\xi}$ imply on any 2-sphere of the foliation of ${}^{(ext)}\mathcal{M}$ and for any $k \leq k_{small}$

$$r^{\frac{1}{2}}\|\mathfrak{d}^k r\not{d}_1^\star \underline{\check{\Omega}}\|_{L^4(S)} + \|\mathfrak{d}^k r\not{d}_1^\star \underline{\check{\Omega}}\|_{L^2(S)} \lesssim r^2 \sup_S |\mathfrak{d}^k \underline{\xi}| \lesssim \epsilon r u^{-1-\delta_{dec}}.$$

In view of the commutation formulas of Lemma 2.68 and of Proposition 2.28, together with the bootstrap assumptions, we infer any $k \leq k_{small}$, schematically,

$$[\mathfrak{d}^k, r\not{d}_1^\star] = O(\epsilon)\mathfrak{d}^{\leq k} + O(1)\mathfrak{d}^{\leq k-1},$$

and hence,

$$\begin{aligned} & r^{\frac{1}{2}}\|r\not{d}_1^\star\mathfrak{d}^k\underline{\check{\Omega}}\|_{L^4(S)} + \|r\not{d}_1^\star\mathfrak{d}^k\underline{\check{\Omega}}\|_{L^2(S)} \\ \lesssim\ & \epsilon r u^{-1-\delta_{dec}} + \epsilon\|\mathfrak{d}^{\leq k}\underline{\check{\Omega}}\|_{L^2(S)} + \epsilon r^{\frac{1}{2}}\|\mathfrak{d}^{\leq k}\underline{\check{\Omega}}\|_{L^4(S)} \\ & + \|\mathfrak{d}^{\leq k-1}\underline{\check{\Omega}}\|_{L^2(S)} + r^{\frac{1}{2}}\|\mathfrak{d}^{\leq k-1}\underline{\check{\Omega}}\|_{L^4(S)} \\ \lesssim\ & \epsilon r u^{-1-\delta_{dec}} + \epsilon\|r\not{d}_1^\star\mathfrak{d}^{\leq k}\underline{\check{\Omega}}\|_{L^2(S)} + \epsilon\|\mathfrak{d}^{\leq k}\underline{\check{\Omega}}\|_{L^2(S)} \\ & + \|\mathfrak{d}^{\leq k-1}\underline{\check{\Omega}}\|_{L^2(S)} + \|\mathfrak{d}^{\leq k}\underline{\check{\Omega}}\|^{\frac{1}{2}}_{L^2(S)}\|\mathfrak{d}^{\leq k-1}\underline{\check{\Omega}}\|^{\frac{1}{2}}_{L^2(S)}, \end{aligned}$$

where we used Gagliardo-Nirenberg on S. Together with the Poincaré inequality of Corollary 2.37 for $\not{d}_1^\star$, we deduce

$$r^{\frac{1}{2}}\|r\not{d}_1^\star\mathfrak{d}^k\underline{\check{\Omega}}\|_{L^4(S)} + \|r\not{d}_1^\star\mathfrak{d}^k\underline{\check{\Omega}}\|_{L^2(S)} + \|\mathfrak{d}^k\underline{\check{\Omega}}\|_{L^2(S)} \lesssim \epsilon r u^{-1-\delta_{dec}} + \|\mathfrak{d}^{\leq k-1}\underline{\check{\Omega}}\|_{L^2(S)}.$$

By iteration, and using again Gagliardo-Nirenberg on S, we infer on any 2-sphere of the foliation of ${}^{(ext)}\mathcal{M}$ and for any $k \leq k_{small}$

$$\|r\not{d}_1^\star\mathfrak{d}^k\underline{\check{\Omega}}\|_{L^4(S)} + \|\mathfrak{d}^k\underline{\check{\Omega}}\|_{L^4(S)} \lesssim \epsilon r^{\frac{1}{2}} u^{-1-\delta_{dec}},$$

and thus, by Sobolev embedding

$$\max_{0\leq k\leq k_{small}} \sup_{{}^{(ext)}\mathcal{M}} u^{1+\delta_{dec}}|\mathfrak{d}^k\underline{\check{\Omega}}| \lesssim \epsilon$$

which is the desired estimate for $\underline{\check{\Omega}}$.

Step 2. Next, we estimate ς. First, recall that we have

$$e_\theta(\log\varsigma) = \eta - \zeta.$$

Since the bootstrap assumptions for $\eta - \zeta$ are at least as good as for $\underline{\xi}$, we obtain, arguing as in Step 1, the following analog of the above estimate for $\check{\underline{\Omega}}$

$$\max_{0\leq k\leq k_{small}} \sup_{(ext)\mathcal{M}} u^{1+\delta_{dec}}|\mathfrak{d}^k\check{\varsigma}| \lesssim \epsilon.$$

Now that we control $\check{\varsigma}$, we turn to the estimate for ς. First, recall from the GCM on Σ_* that we have $u+r=c_{\Sigma_*}$ and $a\big|_{SP}=-1-\frac{2m}{r}$, where $\nu=e_3+ae_4$ and ν is tangent to Σ_*, with c_{Σ_*} a constant, and SP denoting the south pole of the spheres of Σ_*. We deduce on the south poles of Σ_*

$$0 = \nu(u+r)=e_3(u)+e_3(r)+ae_4(r)=\frac{2}{\varsigma}+e_3(r)-\left(1+\frac{2m}{r}\right)e_4(r)$$

and hence

$$\frac{2}{\varsigma}-2 = -\frac{r}{2}\left(\left(\underline{\overline{\kappa}}+\frac{2\Upsilon}{r}\right)+\underline{A}-\left(1+\frac{2m}{r}\right)\left(\overline{\kappa}-\frac{2}{r}\right)\right) \quad \text{on } SP\cap\Sigma_*.$$

Together with the fact that $\overline{\varsigma}=\varsigma-\check{\varsigma}$, the above control of $\check{\varsigma}$, the control of $\overline{\kappa}$ and $\underline{\overline{\kappa}}$ provided by Lemma 3.15, the formula for $\underline{A}$, the control for $\check{\underline{\Omega}}$ in Step 1, the bootstrap assumptions on decay, and the fact that $\overline{\varsigma}$ is constant on the sphere, we infer

$$\max_{0\leq k\leq k_{small}} \sup_{\Sigma_*} u^{1+\delta_{dec}}|\mathfrak{d}^k(\overline{\varsigma}-1)| \lesssim \epsilon.$$

Using $\varsigma=\overline{\varsigma}+\check{\varsigma}$ and the above estimates for $\overline{\varsigma}$ and $\check{\varsigma}$, we obtain

$$\max_{0\leq k\leq k_{small}} \sup_{\Sigma_*} u^{1+\delta_{dec}}|\mathfrak{d}^k(\varsigma-1)| \lesssim \epsilon.$$

Finally, recall

$$e_4(\varsigma) = 0.$$

Commuting with $\mathfrak{d}$, using the bootstrap assumptions on decay and the above control for $\varsigma-1$ on Σ_*, we infer

$$\max_{0\leq k\leq k_{small}} \sup_{(ext)\mathcal{M}} u^{1+\delta_{dec}}|\mathfrak{d}^k(\varsigma-1)| \lesssim \epsilon.$$

Remark 4.8. *In ${}^{(int)}\mathcal{M}$, we analogously transport $\underline{\varsigma}$ from the timelike hypersurface $\mathcal{T}$. To estimate $\underline{\varsigma}$ on $\mathcal{T}$, one uses the following identity (in the frame of ${}^{(int)}\mathcal{M}$)*

$$\frac{2}{\underline{\varsigma}}-1 = -\frac{\overline{\kappa}+A}{\Upsilon\underline{\overline{\kappa}}}\left(\frac{2}{\varsigma}-1\right)-\frac{A}{\Upsilon\underline{\overline{\kappa}}}-\frac{\left(\overline{\kappa}-\frac{2\Upsilon}{r}\right)+\Upsilon\left(\underline{\overline{\kappa}}+\frac{2}{r}\right)}{\Upsilon\underline{\overline{\kappa}}} \quad \textit{on } \mathcal{T}.$$

This identity follows from the definition of ς and $\underline{\varsigma}$, the identity for $e_3(r)$ and $e_4(r)$ in ${}^{(int)}\mathcal{M}$, the fact that $\underline{u}=u$ on $\mathcal{T}$, and that $\mathcal{T}=\{r=r_{\mathcal{T}}\}$ so that the vectorfield

$$T_{\mathcal{T}}=e_4-\frac{e_4(r)}{e_3(r)}e_3=e_4-\frac{\overline{\kappa}+A}{\underline{\overline{\kappa}}}e_3$$

is tangent to $\mathcal{T}$.

Step 3. We make the auxiliary bootstrap assumption which will be recovered at the end of Step 5

$$\left|e^{\Phi}\right| \leq 2r, \qquad \left|e_\theta(e^{\Phi})\right| \leq 2. \tag{4.3.3}$$

We start with the estimate for e^{Φ}. Recall from (2.2.53) that the following identity holds

$$\frac{e^{\Phi}}{r\sin\theta} = \sqrt{1+\mathfrak{a}} \tag{4.3.4}$$

where $\mathfrak{a}$ has been introduced in (2.2.54) by

$$\mathfrak{a} = \frac{e^{2\Phi}}{r^2} + (e_\theta(e^{\Phi}))^2 - 1.$$

In order to estimate e^{Φ}, it thus suffices to estimate $\mathfrak{a}$.

Step 4. Now, recall from Lemma 2.85 that $\mathfrak{a}$ verifies the following identities on ${}^{(ext)}\mathcal{M}$,

$$\begin{aligned}
e_4(\mathfrak{a}) &= \frac{(\check{\kappa}-\vartheta)e^{2\Phi}}{r^2} + 2e_\theta(e^{\Phi})\Big(\beta - e_4(\Phi)\zeta\Big)e^{\Phi},\\
e_\theta(\mathfrak{a}) &= 2e_\theta(\Phi)e^{2\Phi}\left(\left(\rho + \frac{2m}{r^3}\right) + \frac{1}{4}\left(\kappa\underline{\kappa} + \frac{4\Upsilon}{r^2}\right) - \frac{1}{4}\vartheta\underline{\vartheta}\right),\\
e_3(\mathfrak{a}) &= \frac{\Big(\underline{\check{\kappa}} - \underline{A} - \underline{\vartheta}\Big)e^{2\Phi}}{r^2} + 2e_\theta(e^{\Phi})\Big(\underline{\beta} + e_3(\Phi)\zeta + \underline{\xi}e_4(\Phi)\Big)e^{\Phi}.
\end{aligned}$$

Together with our bootstrap assumptions on decay for in ${}^{(ext)}\mathcal{M}$ for $\check{\kappa}$, ϑ, $\underline{\check{\kappa}}$, $\underline{\vartheta}$, β, $\underline{\beta}$, ρ, ζ, $\underline{\xi}$ and $\underline{\check{\Omega}}$ and the bootstrap assumption (4.3.3), we infer

$$\max_{1\leq k\leq k_{small}} \sup_{{}^{(ext)}\mathcal{M}} \left(ru^{\frac{1}{2}+\delta_{dec}} + u^{1+\delta_{dec}}\right)\left|\mathfrak{d}^k\mathfrak{a}\right| \lesssim \epsilon.$$

In particular, we deduce

$$\sup_{{}^{(ext)}\mathcal{M}} \left(ru^{\frac{1}{2}+\delta_{dec}} + u^{1+\delta_{dec}}\right)|\check{\mathfrak{a}}| \lesssim \epsilon.$$

Step 5. To estimate $\overline{\mathfrak{a}}$ we make use of equation (2.1.13) according to which

$$\Big(e_\theta(e^{\Phi})\Big)^2 = 1 \qquad \text{on the axis of symmetry.}$$

Since $e^{2\Phi}$ also vanishes there we infer that $\mathfrak{a} = 0$ on the axis. Therefore, on the axis, $\check{\mathfrak{a}} = -\overline{\mathfrak{a}}$, i.e.,

$$\overline{\mathfrak{a}} = -\check{\mathfrak{a}}|_{\text{axis}}$$

and therefore,

$$|\overline{\mathfrak{a}}| \lesssim |\check{\mathfrak{a}}| \lesssim \frac{\epsilon}{ru^{\frac{1}{2}+\delta_{dec}} + u^{1+\delta_{dec}}}.$$

We conclude that

$$\max_{0\leq k\leq k_{small}} \sup_{{}^{(ext)}\mathcal{M}} \left(ru^{\frac{1}{2}+\delta_{dec}} + u^{1+\delta_{dec}}\right) \left|\mathfrak{d}^k \mathfrak{a}\right| \lesssim \epsilon. \tag{4.3.5}$$

In view of (4.3.4) and (4.3.5), we immediately infer

$$\max_{0\leq k\leq k_{small}} \sup_{{}^{(ext)}\mathcal{M}} \left(ru^{\frac{1}{2}+\delta_{dec}} + u^{1+\delta_{dec}}\right) \left|\mathfrak{d}^k \left(\frac{e^\Phi}{r\sin\theta} - 1\right)\right| \lesssim \epsilon.$$

Together with (4.3.5) and the definition of $\mathfrak{a}$, this implies

$$\begin{aligned} \left|e^\Phi\right| &= (1+O(\epsilon))r\sin\theta \leq \frac{3r}{2}, \\ \left|e_\theta(e^\Phi)\right| &= \sqrt{1-\frac{e^{2\Phi}}{r^2}+\mathfrak{a}} \leq |\cos\theta| + O(\epsilon) \leq \frac{3}{2}, \end{aligned} \tag{4.3.6}$$

which is an improvement of the bootstrap assumption (4.3.3) which hence holds everywhere on ${}^{(ext)}\mathcal{M}$.

Step 6. We now prove the estimates for b, $\underline{b}$ and γ. Recall from Lemma 2.84 that θ defined by (4.3.1) satisfies

$$\begin{aligned} re_\theta(\theta) &= 1 + \frac{r^2(K-\frac{1}{r^2})}{1+(re_\theta(\Phi))^2}, \\ e_3(\theta) &= -\frac{r\underline{\beta} + \frac{r}{2}\left(-\underline{\check{\kappa}} + \underline{A} + \underline{\vartheta}\right)e_\theta(\Phi) + r\underline{\xi}e_4(\Phi) + r\zeta e_3(\Phi)}{1+(re_\theta(\Phi))^2}, \\ e_4(\theta) &= -\frac{r\beta + \frac{r}{2}\left(-\check{\kappa} + \vartheta\right)e_\theta(\Phi) - r\zeta e_3(\Phi)}{1+(re_\theta(\Phi))^2}. \end{aligned}$$

In view of the definition of b, $\underline{b}$ and γ, we infer

$$\begin{aligned} \frac{r}{\sqrt{\gamma}} &= 1 + \frac{r^2(K-\frac{1}{r^2})}{1+(re_\theta(\Phi))^2}, \\ \underline{b} &= -\frac{r\underline{\beta} + \frac{r}{2}\left(-\underline{\check{\kappa}} + \underline{A} + \underline{\vartheta}\right)e_\theta(\Phi) + r\underline{\xi}e_4(\Phi) + r\zeta e_3(\Phi)}{1+(re_\theta(\Phi))^2}, \\ b &= -\frac{r\beta + \frac{r}{2}\left(-\check{\kappa} + \vartheta\right)e_\theta(\Phi) - r\zeta e_3(\Phi)}{1+(re_\theta(\Phi))^2}. \end{aligned}$$

Also, we have in view of the definition of $\mathfrak{a}$

$$1+(re_\theta(\Phi))^2 = 1 + \frac{(e_\theta(e^\Phi))^2}{\frac{e^{2\Phi}}{r^2}} = \frac{r^2}{e^{2\Phi}}(1+\mathfrak{a})$$

and hence

$$
\begin{aligned}
\frac{r}{\sqrt{\gamma}} &= 1+\frac{e^{2\Phi}}{r^2}\left(\frac{r^2(K-\frac{1}{r^2})}{1+\mathfrak{a}}\right),\\
\underline{b} &= -\frac{e^{2\Phi}}{r^2}\left(\frac{r\underline{\beta}+\frac{r}{2}\left(-\underline{\check{\kappa}}+\underline{A}+\underline{\vartheta}\right)e_\theta(\Phi)+r\underline{\xi}e_4(\Phi)+r\zeta e_3(\Phi)}{1+\mathfrak{a}}\right),\\
b &= -\frac{e^{2\Phi}}{r^2}\left(\frac{r\beta+\frac{r}{2}\left(-\check{\kappa}+\vartheta\right)e_\theta(\Phi)-r\zeta e_3(\Phi)}{1+\mathfrak{a}}\right).
\end{aligned}
$$

The bootstrap assumptions on decay in ${}^{(ext)}\mathcal{M}$ for $\check{\kappa}$, ϑ, $\underline{\check{\kappa}}$, $\underline{\vartheta}$, β, $\underline{\beta}$, ζ, $\underline{\xi}$ and $\underline{\check{\Omega}}$, the estimate (4.3.5) for $\mathfrak{a}$, the estimate (4.3.6), and the identity

$$
\begin{aligned}
K-\frac{1}{r^2} &= -\frac{1}{4}\kappa\underline{\kappa}+\frac{1}{4}\vartheta\underline{\vartheta}-\rho-\frac{1}{r^2}\\
&= -\frac{1}{4}\left(\kappa\underline{\kappa}+\frac{4\Upsilon}{r^2}\right)-\left(\rho+\frac{2m}{r^3}\right)+\frac{1}{4}\vartheta\underline{\vartheta}
\end{aligned}
$$

imply

$$
\max_{0\leq k\leq k_{small}}\sup_{{}^{(ext)}\mathcal{M}}\left(ru^{\frac{1}{2}+\delta_{dec}}+u^{1+\delta_{dec}}\right)\left(\left|\mathfrak{d}^k\left(\frac{r}{\sqrt{\gamma}}-1\right)\right|+r\left|\mathfrak{d}^k b\right|\right) \lesssim \epsilon,
$$

and

$$
\max_{0\leq k\leq k_{small}}\sup_{{}^{(ext)}\mathcal{M}} ru^{1+\delta_{dec}}\left|\mathfrak{d}^k\underline{b}\right| \lesssim \epsilon.
$$

In particular, we also have

$$
\max_{0\leq k\leq k_{small}}\sup_{{}^{(ext)}\mathcal{M}}\left(ru^{\frac{1}{2}+\delta_{dec}}+u^{1+\delta_{dec}}\right)\left|\mathfrak{d}^k\left(\frac{\gamma}{r^2}-1\right)\right| \lesssim \epsilon.
$$

These are the desired estimates for b, $\underline{b}$ and γ in ${}^{(ext)}\mathcal{M}$. This concludes the proof of the lemma. □

In this section, we also prove two useful lemmas concerning estimates on 2-spheres of ${}^{(ext)}\mathcal{M}$ and ${}^{(int)}\mathcal{M}$.

Lemma 4.9. *Let $\theta\in[0,\pi]$ be the $\mathbf{Z}$-invariant scalar on $\mathcal{M}$ defined by (2.2.52). Then, we have on $\mathcal{M}$*

$$
re_\theta(\Phi) = \frac{\varpi}{\sin\theta}
$$

where ϖ is a reduced 1-scalar satisfying

$$
\sup_{\mathcal{M}}|\varpi| \leq 2.
$$

Also, we have

$$
\frac{1}{\sin\theta} \leq 2|re_\theta(\Phi)|+2 \text{ on } \mathcal{M}.
$$

Proof. The proof is similar on ${}^{(ext)}\mathcal{M}$ and ${}^{(int)}\mathcal{M}$ so we focus on ${}^{(ext)}\mathcal{M}$. Recall from (4.3.6) that

$$\left|e_\theta(e^\Phi)\right| \leq \frac{3}{2}.$$

Furthermore, in view of Proposition 3.17, we have in particular

$$\sup_{{}^{(ext)}\mathcal{M}}\left|\frac{e^\Phi}{r\sin\theta}-1\right| \lesssim \epsilon.$$

Since we have

$$\varpi = r\sin\theta e_\theta(\Phi),$$

we deduce

$$|\varpi| = \frac{r\sin\theta}{e^\Phi}|e_\theta(e^\Phi)| \leq \frac{3}{2}(1+O(\epsilon)) \leq 2,$$

which is the desired estimate for ϖ.

We now consider the upper bound for $(\sin\theta)^{-1}$. Recall the definition (2.2.54) of $\mathfrak{a}$

$$\mathfrak{a} = \frac{e^{2\Phi}}{r^2} + (e_\theta(e^\Phi))^2 - 1.$$

We infer

$$\begin{aligned} r^2 e_\theta(\Phi)^2 &= \frac{r^2(e_\theta(e^\Phi))^2}{e^{2\Phi}} \\ &= \frac{1+\mathfrak{a}}{\frac{e^{2\Phi}}{r^2}} - 1 \\ &= \frac{1+\mathfrak{a}-(\sin\theta)^2\left(1+\left(\frac{e^{2\Phi}}{r^2(\sin\theta)^2}-1\right)\right)}{(\sin\theta)^2\left(1+\left(\frac{e^{2\Phi}}{r^2(\sin\theta)^2}-1\right)\right)} \\ &= \frac{(\cos\theta)^2+\mathfrak{a}-(\sin\theta)^2\left(\frac{e^{2\Phi}}{r^2(\sin\theta)^2}-1\right)}{(\sin\theta)^2\left(1+\left(\frac{e^{2\Phi}}{r^2(\sin\theta)^2}-1\right)\right)} \end{aligned}$$

and hence

$$\sin\theta|re_\theta(\Phi)| = \frac{\sqrt{(\cos\theta)^2+\mathfrak{a}-(\sin\theta)^2\left(\frac{e^{2\Phi}}{r^2(\sin\theta)^2}-1\right)}}{\sqrt{1+\left(\frac{e^{2\Phi}}{r^2(\sin\theta)^2}-1\right)}}.$$

Now, in view of (4.3.5), $\mathfrak{a}$ satisfies in particular

$$\sup_{{}^{(ext)}\mathcal{M}}|\mathfrak{a}| \lesssim \epsilon.$$

Together with

$$\sup_{{}^{(ext)}\mathcal{M}}\left|\frac{e^\Phi}{r\sin\theta}-1\right| \lesssim \epsilon,$$

we infer

$$\sin\theta|re_\theta(\Phi)| = \frac{\sqrt{(\cos\theta)^2+O(\epsilon)}}{\sqrt{1+O(\epsilon)}}.$$

Thus, we deduce

$$\sin\theta|re_\theta(\Phi)| \geq \frac{\sqrt{2}}{2}(1+O(\epsilon)) \geq \frac{1}{2} \text{ for } 0\leq\theta\leq\frac{\pi}{4} \text{ and } \frac{3\pi}{4}\leq\theta\leq\pi.$$

On the other hand, we have

$$\sin\theta \geq \frac{\sqrt{2}}{2} \text{ on } \frac{\pi}{4}\leq\theta\leq\frac{3\pi}{4}$$

and hence

$$\frac{1}{\sin\theta} \leq 2|re_\theta(\Phi)|+2 \text{ on } 0\leq\theta\leq\pi$$

which is the desired estimate. This concludes the proof of the lemma. □

Lemma 4.10. *Let $\theta\in[0,\pi]$ be the $\mathbf{Z}$-invariant scalar on $\mathcal{M}$ defined by (2.2.52). Then, for any reduced 1-scalar h, we have on any 2-sphere S on ${}^{(ext)}\mathcal{M}$ and of ${}^{(int)}\mathcal{M}$*

$$\sup_S\frac{|h|}{e^\Phi}\lesssim r^{-1}\sup_S(|h|+|\not{\mathfrak{d}}h|) \quad \textit{and} \quad \left\|\frac{h}{e^\Phi}\right\|_{L^2(S)}\lesssim r^{-1}\|h\|_{\mathfrak{h}_1(S)}.$$

Proof. The proof is similar on ${}^{(ext)}\mathcal{M}$ and ${}^{(int)}\mathcal{M}$ so we focus on ${}^{(ext)}\mathcal{M}$. Recall that the 2-surface S is parametrized by the coordinate $\theta\in[0,\pi]$, and that the axis corresponds to the two poles $\theta=0$ and $\theta=\pi$. In view of

$$\sup_{{}^{(ext)}\mathcal{M}}\left|\frac{e^\Phi}{r\sin\theta}-1\right| \lesssim \epsilon,$$

we have

$$\sup_{S\cap\{\frac{\pi}{4}\leq\theta\leq\frac{3\pi}{4}\}}\frac{|h|}{e^\Phi}\lesssim r^{-1}\sup_S|h| \quad \text{and} \quad \left\|\frac{h}{e^\Phi}\right\|_{L^2(S\cap\{\frac{\pi}{4}\leq\theta\leq\frac{3\pi}{4}\})}\lesssim r^{-1}\|h\|_{L^2(S)}$$

which is the desired estimate for $\pi/4\leq\theta\leq3\pi/4$.

It remains to consider the portions $0\leq\theta\leq\pi/4$ and $3\pi/4\leq\theta\leq\pi$ of S. These regions can be treated analogously, so we focus on $0\leq\theta\leq\pi/4$. Recall from Remark 2.24 that any reduced scalar in $\mathfrak{s}_k$, for $k\geq1$, must vanish on the axis of symmetry of $\mathbf{Z}$, i.e., at the two poles. In particular, h must vanish at $\theta=0$. We deduce

$$\frac{h}{e^\Phi} = \frac{he^\Phi}{e^{2\Phi}} = \frac{\int_0^\theta \partial_\theta(e^\Phi h)}{e^{2\Phi}} = \frac{\int_0^\theta \sqrt{\gamma^{\mathbf{S}}} e_\theta(e^\Phi h)}{e^{2\Phi}} = \frac{\int_0^\theta \sqrt{\gamma} e^\Phi \not{d}_1 h}{e^{2\Phi}}.$$

Since we have $|\gamma| \lesssim r$, we infer

$$\frac{|h|}{e^\Phi} \lesssim \frac{\int_0^\theta e^\Phi |\not{\partial} h|}{e^{2\Phi}}$$

and since

$$\sup_{^{(ext)}\mathcal{M}} \left| \frac{e^\Phi}{r\sin\theta} - 1 \right| \lesssim \epsilon,$$

we deduce

$$\frac{|h|}{e^\Phi} \lesssim r^{-1} \frac{\int_0^\theta \sin(\theta')|\not{\partial} h| d\theta'}{(\sin\theta)^2}.$$

This yields

$$\sup_{S\cap\{0\le\theta\le\frac{\pi}{4}\}} \frac{|h|}{e^\Phi} \lesssim r^{-1} \sup_S |\not{\partial} h|$$

which is the desired sup norm estimate for $0 \le \theta \le \pi/4$.

It remains to control the L^2 norm on $0 \le \theta \le \pi/4$. We have, in view of the above,

$$\begin{aligned}
\left\| \frac{h}{e^\Phi} \right\|^2_{L^2(S\cap\{0\le\theta\le\frac{\pi}{4}\})} &\lesssim r^{-2} \int_0^{\frac{\pi}{4}} \frac{\left(\int_0^\theta \sin(\theta')|\not{\partial} h| d\theta'\right)^2}{(\sin\theta)^4} e^\Phi d\theta \\
&\lesssim r^{-1} \int_0^{\frac{\pi}{4}} \left(\int_0^\theta (\sin(\theta'))^2 |\not{\partial} h|^2 d\theta' \right) \frac{d\theta}{(\sin\theta)^2} \\
&\lesssim r^{-1} \int_0^{\frac{\pi}{4}} (\sin\theta)^2 |\not{\partial} h|^2 \left(\int_\theta^{\frac{\pi}{4}} \frac{d\theta'}{(\sin(\theta'))^2} \right) d\theta \\
&\lesssim r^{-1} \int_0^{\frac{\pi}{4}} |\not{\partial} h|^2 \sin\theta d\theta \\
&\lesssim r^{-2} \int_0^{\frac{\pi}{4}} |\not{\partial} h|^2 e^\Phi d\theta \\
&\lesssim r^{-2} \|\not{\partial} h\|^2_{L^2(S)}
\end{aligned}$$

and hence

$$\left\| \frac{h}{e^\Phi} \right\|_{L^2(S\cap\{0\le\theta\le\frac{\pi}{4}\})} \lesssim r^{-1} \|\not{\partial} h\|_{L^2(S)}$$

which is the desired $L^2(S)$ estimate for $0 \le \theta \le \pi/4$. This concludes the proof of the lemma. □

4.4 POINTWISE BOUNDS FOR HIGHER ORDER DERIVATIVES

The goal of this section is to prove Proposition 3.19. We deal first with the region $r \le 4m_0$ as follows:

1. The curvature components and Ricci coefficients satisfy in view of the bootstrap assumptions on energy

$$\max_{k\le k_{large}} \int_{^{(int)}\mathcal{M}} \left(|\check{R}|^2+|\check{\Gamma}|^2\right) + \max_{k\le k_{large}-1} \int_{^{(ext)}\mathcal{M}(r\le 4m_0)} \left(|\check{R}|^2+|\check{\Gamma}|^2\right) \le \epsilon^2.$$

2. We first take the trace on the ingoing null cones foliating $^{(int)}\mathcal{M}$ and the outgoing null cones foliating $^{(ext)}\mathcal{M}(r\le 4m_0)$ which loses one derivative. We thus obtain

$$\begin{aligned}&\max_{k\le k_{large}-1}\sup_{1\le \underline{u}\le u_*}\int_{\underline{\mathcal{C}}_{\underline{u}}}\left(|\check{R}|^2+|\check{\Gamma}|^2\right)\\ &+\max_{k\le k_{large}-2}\sup_{1\le u\le u_*}\int_{\mathcal{C}_u(r\le 4m_0)}\left(|\check{R}|^2+|\check{\Gamma}|^2\right) \lesssim \epsilon^2.\end{aligned}$$

3. We then take the trace on the 2-spheres S foliating the null cones in $^{(int)}\mathcal{M}$ and $^{(ext)}\mathcal{M}(r\le 4m_0)$ to infer

$$\begin{aligned}&\max_{k\le k_{large}-2}\sup_{^{(int)}\mathcal{M}}\left(\|\check{R}\|_{L^2(S)}+\|\check{\Gamma}\|_{L^2(S)}\right)\\ &+\max_{k\le k_{large}-3}\sup_{^{(ext)}\mathcal{M}(r\le 4m_0)}\left(|\check{R}|^2+|\check{\Gamma}|^2\right) \lesssim \epsilon.\end{aligned}$$

4. Finally, using the Sobolev embedding on the 2-sphere S, which loses 2 derivatives, we deduce

$$\max_{k\le k_{large}-4}\sup_{^{(int)}\mathcal{M}}\left(|\check{R}|+|\check{\Gamma}|\right) + \max_{k\le k_{large}-5}\sup_{^{(ext)}\mathcal{M}(r\le 4m_0)}\left(|\check{R}|+|\check{\Gamma}|\right) \lesssim \epsilon,$$

which is the desired estimate in the region $^{(int)}\mathcal{M}\cup {}^{(ext)}\mathcal{M}(r\le 4m_0)$.

It remains to consider the region $^{(ext)}\mathcal{M}(r\ge 4m_0)$. We proceed as follows.

Step 1. The Ricci coefficients satisfy, in view of the bootstrap assumptions on energy,

$$\begin{aligned}&\max_{k\le k_{large}}\int_{\Sigma_*}\Big[r^2\Big((\mathfrak{d}^{\le k}\vartheta)^2+(\mathfrak{d}^{\le k}\check{\kappa})^2+(\mathfrak{d}^{\le k}\zeta)^2+(\mathfrak{d}^{\le k}\underline{\check{\kappa}})^2\Big)+(\mathfrak{d}^{\le k}\underline{\vartheta})^2\\ &+\ (\mathfrak{d}^{\le k}\eta)^2+(\mathfrak{d}^{\le k}\underline{\check{\omega}})^2+(\mathfrak{d}^{\le k}\underline{\xi})^2\Big]\\ &+\ \sup_{\lambda\ge 4m_0}\Bigg(\int_{\{r=\lambda\}}\Big[\lambda^2\Big((\mathfrak{d}^{\le k}\vartheta)^2+(\mathfrak{d}^{\le k}\check{\kappa})^2+(\mathfrak{d}^{\le k}\zeta)^2\Big)\\ &+\ \lambda^{2-\delta_B}(\mathfrak{d}^{\le k}\underline{\check{\kappa}})^2+(\mathfrak{d}^{\le k}\underline{\vartheta})^2+(\mathfrak{d}^{\le k}\eta)^2+(\mathfrak{d}^{\le k}\underline{\check{\omega}})^2+\lambda^{-\delta_B}(\mathfrak{d}^{\le k}\underline{\xi})^2\Big]\Bigg) \le \epsilon^2.\end{aligned}$$

We take the trace on the 2-spheres S foliating the timelike cylinders $\{r = r_0\}$, for $r_0 \geq 4m_0$, which loses a derivative, and infer in particular

$$\max_{k\leq k_{large}-1} \sup_{^{(ext)}\mathcal{M}(r\geq 4m_0)} \Big\{r\big(\|\mathfrak{d}^k\check{\kappa}\|_{L^2(S)} + \|\mathfrak{d}^k\zeta\|_{L^2(S)} + \|\mathfrak{d}^k\vartheta\|_{L^2(S)}\big) + r^{1-\frac{\delta_B}{2}}\|\mathfrak{d}^k\underline{\check{\kappa}}\|_{L^2(S)} + \|\mathfrak{d}^k\eta\|_{L^2(S)} + \|\mathfrak{d}^k\underline{\vartheta}\|_{L^2(S)} + \|\mathfrak{d}^k\underline{\check{\omega}}\|_{L^2(S)} + r^{-\frac{\delta_B}{2}}\|\mathfrak{d}^k\underline{\xi}\|_{L^2(S)}\Big\} \lesssim \epsilon.$$

Also, we take the trace on the 2-spheres S foliating the spacelike hypersurface Σ_*, which loses a derivative, and infer in particular

$$\max_{k\leq k_{large}-1} \sup_{\Sigma_*} r\|\mathfrak{d}^k\underline{\check{\kappa}}\|_{L^2(S)} \lesssim \epsilon.$$

Step 2. On can easily prove the following trace theorem

$$\max_{k\leq k_{large}-1}\left(\sup_{r\geq 4m_0} r^{5+\delta_B}\int_S(\mathfrak{d}^k\alpha)^2\right) \lesssim \sup_{1\leq u\leq u_*}\int_{\mathcal{C}_u} r^{4+\delta_B}(\mathfrak{d}^{\leq k_{large}}\alpha)^2,$$

which together with the bootstrap assumptions on energy for α in $^{(ext)}\mathcal{M}(r \geq 4m_0)$ implies

$$\max_{k\leq k_{large}-1}\left(\sup_{r\geq 4m_0} r^{5+\delta_B}\int_S(\mathfrak{d}^k\alpha)^2\right) \lesssim \epsilon^2.$$

Step 3. Using the trace theorem

$$\max_{k\leq k_{large}-1}\left(\sup_{r\geq 4m_0} r^{5}\int_S(\mathfrak{d}^k\beta)^2\right) \lesssim \sup_{1\leq u\leq u_*}\int_{\mathcal{C}_u} r^{4}(\mathfrak{d}^{\leq k_{large}}\beta)^2,$$

we infer, together with the bootstrap assumptions on energy for β in the region $^{(ext)}\mathcal{M}(r \geq 4m_0)$,

$$\max_{k\leq k_{large}-1}\left(\sup_{r\geq 4m_0} r^{5}\int_S(\mathfrak{d}^k\beta)^2\right) \lesssim \epsilon^2. \tag{4.4.1}$$

The power of r of the above estimate is not strong enough. To upgrade the estimate, recall that we have the Bianchi identity

$$e_4(\beta) + 2\kappa\beta = \not{d}_2\alpha + \zeta\alpha.$$

This yields

$$\begin{aligned} e_4\left(r^{5+\delta_B}\int_S\beta^2\right) &= \int_S r^{5+\delta_B}\left(2\beta e_4(\beta) + \kappa\beta^2 + b\frac{e_4(r)}{r}\beta^2\right) \\ &= \int_S r^{5+\delta_B}\left(-\frac{1-\delta_B}{2}\kappa\beta^2 + 2\beta(r^{-1}\not{\mathfrak{d}}\alpha + \zeta\alpha) - \frac{5+\delta_B}{2}\check{\kappa}\beta^2\right) \end{aligned}$$

and hence

$$\begin{aligned}
& e_4\left(r^{5+\delta_B}\int_S \beta^2\right)+\frac{1-\delta_B}{2}\int_S r^{5+\delta_B}\kappa\beta^2 \\
= & \int_S r^{4+\delta_B}\left(2\beta(\not{\mathfrak{d}}\alpha+r\zeta\alpha)-\frac{5+\delta_B}{2}\check{\kappa}\beta^2\right) \\
\lesssim & \left(\int_S r^{4+\delta_B}(\mathfrak{d}^{\leq 1}\alpha)^2\right)^{\frac{1}{2}}\left(\int_S r^{4+\delta_B}\beta^2\right)^{\frac{1}{2}}+\epsilon\int_S r^{4+\delta_B}\beta^2
\end{aligned}$$

where we used the pointwise estimates of Step 1 for $\check{\kappa}$ and ζ. We infer

$$e_4\left(r^{5+\delta_B}\int_S \beta^2\right)+\int_S r^{4+\delta_B}\beta^2 \lesssim \int_S r^{4+\delta_B}(\mathfrak{d}^{\leq 1}\alpha)^2.$$

Integrating, from $r\geq 6m_0$, we deduce

$$\begin{aligned}
& \sup_{r\geq 6m_0} r^{5+\delta_B}\int_S \beta^2+\sup_{1\leq u\leq u_*}\int_{\mathcal{C}_u(r\geq 6m_0)} r^{4+\delta_B}\beta^2 \\
\lesssim & \sup_{1\leq u\leq u_*}\int_{\mathcal{C}_u} r^{4+\delta_B}(\mathfrak{d}^{\leq 1}\alpha)^2+\int_{S_{r=6m_0}}\beta^2 \\
\lesssim & \epsilon^2,
\end{aligned}$$

where we used the bootstrap assumptions on energy for α in ${}^{(ext)}\mathcal{M}(r\geq 4m_0)$ and the non-sharp estimate (4.4.1) for β. Using again (4.4.1), we obtain

$$\sup_{r\geq 4m_0} r^{5+\delta_B}\int_S \beta^2+\sup_{1\leq u\leq u_*}\int_{\mathcal{C}_u(r\geq 4m_0)} r^{4+\delta_B}\beta^2 \lesssim \epsilon^2.$$

To discuss higher order derivatives, recall from Lemma 2.68 the following commutator, written in schematic form,

$$[\not{\mathfrak{d}}, e_4] = (\check{\kappa},\vartheta)\not{\mathfrak{d}}+(\zeta, r\beta).$$

Also, recall from Lemma 2.69 the following commutator,

$$[\mathbf{T}, e_4] = \left(\left(\underline{\omega}-\frac{m}{r^2}\right)-\frac{m}{2r}\left(\overline{\kappa}-\frac{2}{r}\right)+\frac{e_4(m)}{r}\right)e_4+(\eta+\zeta)e_\theta.$$

In view of the estimates of Step 1 for $k_{large}-1$ derivatives of $\check{\kappa}$, ϑ, ζ, η, $\check{\underline{\omega}}$, the pointwise estimates for β in (4.4.1), the control of $\overline{\kappa}$ in Lemma 3.15, and the control of $e_4(m)$ in Lemma 3.16, we infer, schematically,

$$\left\|\mathfrak{d}^k\Big([\not{\mathfrak{d}}, e_4]\beta,\ [\mathbf{T}, e_4]\beta\Big)\right\|_{L^2(S)} \lesssim O(\epsilon r^{-2})\|\mathfrak{d}^{\leq k+1}\beta\|_{L^2(S)} \text{ for } k\leq k_{large}-2.$$

Thus, commuting the Bianchi identity for $e_4(\beta)$ with $\mathbf{T}$ and $\not{\mathfrak{d}}$ together with the above commutator estimate, using the Bianchi identity to recover the e_4 derivatives, we obtain for higher order derivatives

$$
\begin{aligned}
& \max_{k\leq k_{large}-1}\left(\sup_{r\geq 4m_0} r^{5+\delta_B}\int_S(\mathfrak{d}^k\beta)^2+\sup_{1\leq u\leq u_*}\int_{\mathcal{C}_u(r\geq 4m_0)} r^{4+\delta_B}(\mathfrak{d}^k\beta)^2\right)\\
\lesssim\ & \sup_{1\leq u\leq u_*}\int_{\mathcal{C}_u(r\geq 4m_0)} r^{4+\delta_B}(\mathfrak{d}^{\leq k_{large}}\alpha)^2\\
\lesssim\ & \epsilon^2.
\end{aligned}
$$

Step 4. Recall from Proposition 2.73 that we have

$$
\begin{aligned}
e_4\check{\rho}+\frac{3}{2}\overline{\kappa}\check{\rho}+\frac{3}{2}\overline{\rho}\check{\kappa} &= \not{d}_1\beta+\mathrm{Err}[e_4\check{\rho}],\\
\mathrm{Err}[e_4\check{\rho}] &= -\frac{3}{2}\check{\kappa}\check{\rho}+\frac{1}{2}\overline{\check{\kappa}\check{\rho}}-\left(\frac{1}{2}\underline{\vartheta}\alpha+\zeta\beta\right)+\overline{\left(\frac{1}{2}\underline{\vartheta}\alpha+\zeta\beta\right)}.
\end{aligned}
$$

This yields

$$
\begin{aligned}
e_4\left(r^4\int_S(\check{\rho})^2\right) &= \int_S r^4\left(2\check{\rho}e_4(\check{\rho})+\kappa\check{\rho}^2+4\frac{e_4(r)}{r}\check{\rho}^2\right)\\
&= \int_S r^4\Big(-3\overline{\rho}\check{\kappa}\check{\rho}+2\check{\rho}(r^{-1}\not{\mathfrak{d}}\beta+\mathrm{Err}[e_4\check{\rho}])+\check{\kappa}\check{\rho}^2\Big)
\end{aligned}
$$

and hence

$$
e_4\left[\left(r^4\int_S(\check{\rho})^2\right)^{\frac{1}{2}}\right] \lesssim \left[\int_S r^4\Big((\overline{\rho}\check{\kappa})^2+(r^{-1}\not{\mathfrak{d}}\beta)^2+(\mathrm{Err}[e_4\check{\rho}])^2+\check{\kappa}^2\check{\rho}^2\Big)\right]^{\frac{1}{2}}.
$$

Using the estimates of Steps 1, 2 and 3 for $\check{\kappa}$, ζ, $\underline{\vartheta}$, α and β, and the control of $\overline{\rho}$ in Lemma 3.15, we infer

$$
e_4\left[\left(r^4\int_S(\check{\rho})^2\right)^{\frac{1}{2}}\right] \lesssim \frac{\epsilon}{r^{\frac{3}{2}+\frac{\delta_B}{2}}}+\frac{\epsilon}{r^2}\left(r^4\int_S(\check{\rho})^2\right)^{\frac{1}{2}}.
$$

Integrating from $r=4m_0$, we control $\|\check{\rho}\|_{L^2(S)}$ from the control in $r\leq 4m_0$, we infer

$$
\sup_{r\geq 4m_0} r^4\int_S\check{\rho}^2 \lesssim \epsilon^2.
$$

Next, commuting the equation for $e_4(\check{\rho})$ with $\mathbf{T}$ and $\not{\mathfrak{d}}$ together with the commutator estimate of Step 3, using the equation for $e_4(\check{\rho})$ to recover the e_4 derivatives, we obtain similarly for higher order derivatives

$$
\max_{k\leq k_{large}-2}\sup_{r\geq 4m_0} r^4\int_S(\mathfrak{d}^k\check{\rho})^2 \lesssim \epsilon^2.
$$

Step 5. Recall from Proposition 2.73 that we have the following transport equations in the e_4 direction,

$$\begin{aligned} e_4\underline{\check{\kappa}} + \frac{1}{2}\overline{\kappa}\underline{\check{\kappa}} + \frac{1}{2}\check{\kappa}\overline{\underline{\kappa}} &= -2\,\not{d}_1\zeta + 2\check{\rho} + \mathrm{Err}[e_4\underline{\check{\kappa}}], \\ \mathrm{Err}[e_4\underline{\check{\kappa}}] &= -\frac{1}{2}\check{\kappa}\underline{\check{\kappa}} - \frac{1}{2}\overline{\check{\kappa}\underline{\check{\kappa}}} + \left(-\frac{1}{2}\vartheta\underline{\vartheta} + 2\zeta^2\right) - \overline{\left(-\frac{1}{2}\vartheta\underline{\vartheta} + 2\zeta^2\right)}. \end{aligned}$$

This yields

$$\begin{aligned} e_4\left(r\int_S(\underline{\check{\kappa}})^2\right) &= \int_S r\left(2\underline{\check{\kappa}}e_4(\underline{\check{\kappa}}) + \kappa\underline{\check{\kappa}}^2 + \frac{e_4(r)}{r}\underline{\check{\kappa}}^2\right) \\ &= \int_S r\left(2\underline{\check{\kappa}}\left(-\frac{1}{2}\check{\kappa}\overline{\underline{\kappa}} - 2\,\not{d}_1\zeta + 2\check{\rho} + \mathrm{Err}[e_4\underline{\check{\kappa}}]\right) + \check{\kappa}\underline{\check{\kappa}}^2\right) \end{aligned}$$

and hence, using the estimates of Steps 1 and 4 for $\check{\kappa}$, ζ, $\underline{\vartheta}$ and $\check{\rho}$, and the control of $\overline{\kappa}$ and $\overline{\underline{\kappa}}$ in Lemma 3.15, we infer

$$e_4\left(r\int_S(\underline{\check{\kappa}})^2\right) \lesssim \frac{\epsilon}{r^2}\int_S r\underline{\check{\kappa}}^2 + \epsilon r^{-\frac{3}{2}}\left(\int_S r\underline{\check{\kappa}}^2\right)^{\frac{1}{2}}$$

and hence

$$e_4\left(\left(r\int_S(\underline{\check{\kappa}})^2\right)^{\frac{1}{2}}\right) \lesssim \frac{\epsilon}{r^2}\left(r\int_S(\underline{\check{\kappa}})^2\right)^{\frac{1}{2}} + \epsilon r^{-\frac{3}{2}}.$$

Integrating backward from Σ_*, where $\underline{\check{\kappa}}$ is under control in view of Step 1, we infer

$$\sup_{r\geq 4m_0} r^2\int_S \underline{\check{\kappa}}^2 \lesssim \epsilon^2.$$

Next, commuting the equation for $e_4(\underline{\check{\kappa}})$ with $\mathbf{T}$ and $\not{\partial}$ together with the commutator estimate of Step 3, using the equation for $e_4(\underline{\check{\kappa}})$ to recover the e_4 derivatives, we obtain similarly for higher order derivatives

$$\max_{k\leq k_{large}-2}\sup_{r\geq 4m_0} r^4\int_S(\mathfrak{d}^k\underline{\check{\kappa}})^2 \lesssim \epsilon^2.$$

Step 6. In view of Codazzi for $\underline{\vartheta}$, and the estimates of Step 1 on ζ and $\underline{\vartheta}$, and of Step 3 on $\underline{\check{\kappa}}$ in ${}^{(ext)}\mathcal{M}(r\geq 4m_0)$, we infer

$$\max_{k\leq k_{large}-2}\sup_{{}^{(ext)}\mathcal{M}(r\geq 4m_0)} r\|\mathfrak{d}^k\underline{\beta}\|_{L^2(S)} \lesssim \epsilon.$$

Step 7. In view of the null structure equation for $e_3(\underline{\kappa})$, and the estimates of Step 1 on $\underline{\check{\omega}}$, ζ, η and $\underline{\vartheta}$, and of Step 3 on $\underline{\check{\kappa}}$ in ${}^{(ext)}\mathcal{M}(r\geq 4m_0)$, we infer

$$\max_{k\leq k_{large}-3}\sup_{{}^{(ext)}\mathcal{M}(r\geq 4m_0)} \|\mathfrak{d}^k\underline{\xi}\|_{L^2(S)} \lesssim \epsilon.$$

Step 8. In view of the Bianchi identity for $e_3(\underline{\beta})$, and the estimates of Step 1 on $\underline{\check{\omega}}$, ζ, and η, the estimates of Step 2 on $\check{\rho}$, of Step 3 on $\underline{\check{\kappa}}$ and of Step 5 on $\underline{\xi}$ in

${}^{(ext)}\mathcal{M}(r\geq 4m_0)$, we infer

$$\max_{k\leq k_{large}-3}\sup_{{}^{(ext)}\mathcal{M}(r\geq 4m_0)}\|\mathfrak{d}^k\underline{\alpha}\|_{L^2(S)} \lesssim \epsilon.$$

Step 9. Gathering the estimates for Step 1 to Step 8, we have obtained

$$\begin{aligned}
&\max_{k\leq k_{large}-1}\sup_{{}^{(ext)}\mathcal{M}(r\geq 4m_0)}\Big\{r^{\frac{5}{2}+\frac{\delta_B}{2}}\big(\|\mathfrak{d}^k\alpha\|_{L^2(S)}+\|\mathfrak{d}^k\beta\|_{L^2(S)}\big)\\
&+\ r\big(\|\mathfrak{d}^k\check{\kappa}\|_{L^2(S)}+\|\mathfrak{d}^k\zeta\|_{L^2(S)}+\|\mathfrak{d}^k\vartheta\|_{L^2(S)}\big)+\|\mathfrak{d}^k\eta\|_{L^2(S)}\\
&+\|\mathfrak{d}^k\underline{\vartheta}\|_{L^2(S)}+\|\mathfrak{d}^k\underline{\check{\omega}}\|_{L^2(S)}\Big\}\\
&+\ \max_{k\leq k_{large}-2}\sup_{{}^{(ext)}\mathcal{M}(r\geq 4m_0)}\Big\{r^2\big(\|\mathfrak{d}^k\mu\|_{L^2(S)}+\|\mathfrak{d}^k\check{\rho}\|_{L^2(S)}\big)\\
&+r\big(\|\mathfrak{d}^k\underline{\check{\kappa}}\|_{L^2(S)}+\|\mathfrak{d}^k\underline{\beta}\|_{L^2(S)}\big)\Big\}\\
&+\ \max_{k\leq k_{large}-3}\sup_{{}^{(ext)}\mathcal{M}(r\geq 4m_0)}\Big\{\|\mathfrak{d}\underline{\xi}\|_{L^2(S)}+\|\mathfrak{d}^k\underline{\alpha}\|_{L^2(S)}\Big\}\lesssim\epsilon.
\end{aligned}$$

Using the Sobolev embedding on the 2-sphere S which loses 2 derivatives, and in view of the previous estimate on ${}^{(ext)}\mathcal{M}(r\leq 4m_0)$, we infer

$$\begin{aligned}
\max_{k\leq k_{large}-5}\sup_{\mathcal{M}}\quad&\Big\{r^{\frac{7}{2}+\frac{\delta_B}{2}}\big(|\mathfrak{d}^k\alpha|+|\mathfrak{d}^k\beta|\big)+r^3\big(|\mathfrak{d}^k\check{\mu}|+|\mathfrak{d}^k\check{\rho}|\big)\\
&+r^2\big(|\mathfrak{d}^k\check{\kappa}|+|\mathfrak{d}^k\zeta|+|\mathfrak{d}^k\vartheta|+|\mathfrak{d}^k\underline{\check{\kappa}}|+|\mathfrak{d}^k\underline{\beta}|\big)\\
&+r\big(|\mathfrak{d}^k\eta|+|\mathfrak{d}^k\underline{\vartheta}|+|\mathfrak{d}^k\underline{\check{\omega}}|+|\mathfrak{d}\underline{\xi}|+|\mathfrak{d}^k\underline{\alpha}|\big)\Big\}\lesssim\epsilon
\end{aligned}$$

which is the desired estimate on ${}^{(ext)}\mathcal{M}(r\geq 4m_0)$. This concludes the proof of Proposition 3.19.

4.5 PROOF OF PROPOSITION 3.20

Let (e_4, e_3, e_θ) the outgoing geodesic null frame of ${}^{(ext)}\mathcal{M}$. We will exhibit another frame (e_4', e_3', e_θ') of ${}^{(ext)}\mathcal{M}$ provided by

$$\begin{aligned}
e_4' &= e_4+fe_\theta+\frac{1}{4}f^2e_3,\\
e_\theta' &= e_\theta+\frac{1}{2}fe_3,\\
e_3' &= e_3,
\end{aligned}\tag{4.5.1}$$

where f is such that

$$f=0 \text{ on } \Sigma_*\cap\mathcal{C}_*,\quad \eta'=0 \text{ on } \Sigma_*,\quad \xi'=0 \text{ on } {}^{(ext)}\mathcal{M}.\tag{4.5.2}$$

The desired estimates for the Ricci coefficients and curvature components with respect to the new frame (e_4', e_3', e_θ') of ${}^{(ext)}\mathcal{M}$ will be obtained using

- the change of frame formulas of Proposition 2.90, applied to the change of frame from (e_4, e_3, e_θ) to (e_4', e_3', e_θ'),

- the estimates for f on ${}^{(ext)}\mathcal{M}$,
- the estimates for the Ricci coefficients and curvature components with respect to the outgoing geodesic frame (e_4, e_3, e_θ) of ${}^{(ext)}\mathcal{M}$ provided by the bootstrap assumptions on decay and Proposition 3.19.

Step 1. We start by deriving an equation for f on ${}^{(ext)}\mathcal{M}$. In view of the condition $\xi' = 0$ on ${}^{(ext)}\mathcal{M}$, see (4.5.2), in view of $\xi = \omega = 0$ and $\underline{\eta} = -\zeta$ satisfied by the outgoing geodesic foliation of ${}^{(ext)}\mathcal{M}$, and in view of Lemma 2.91, we have

$$\begin{aligned} e_4'(f) + \frac{1}{2}\kappa f &= -\frac{1}{2}f\vartheta - \frac{1}{2}f^2\eta - \frac{3}{2}f^2\zeta \\ &\quad + \frac{1}{8}f^3\underline{\kappa} + \frac{1}{2}f^3\underline{\omega} + \frac{1}{8}f^3\underline{\vartheta} + \frac{1}{8}f^4\underline{\xi} \quad \text{on } {}^{(ext)}\mathcal{M}. \end{aligned} \tag{4.5.3}$$

We also derive an equation for f on Σ_*. In view of the condition $\eta' = 0$ on Σ_*, see (4.5.2), and in view of Lemma 2.91, we have

$$e_3'(f) = -2\eta + 2f\underline{\omega} + \frac{1}{2}f^2\underline{\xi} \quad \text{on } \Sigma_*. \tag{4.5.4}$$

Now, since $u + r$ is constant on Σ_*, the following vectorfield

$$\nu_{\Sigma_*}' := e_3' + a'e_4', \qquad a' := -\frac{e_3'(u+r)}{e_4'(u+r)},$$

is tangent to Σ_*. We compute in view of the above

$$\begin{aligned} \nu_{\Sigma_*}'(f) &= e_3'(f) + a'e_4'(f) \\ &= -2\eta + 2f\underline{\omega} + \frac{1}{2}f^2\underline{\xi} + a'\Bigg\{ -\frac{1}{2}\kappa f - \frac{1}{2}f\vartheta - \frac{1}{2}f^2\eta - \frac{3}{2}f^2\zeta \\ &\quad + \frac{1}{8}f^3\underline{\kappa} + \frac{1}{2}f^3\underline{\omega} + \frac{1}{8}f^3\underline{\vartheta} + \frac{1}{8}f^4\underline{\xi}\Bigg\}. \end{aligned}$$

Using (4.5.1), we have

$$\begin{aligned} a' &= -\frac{e_3'(u+r)}{e_4'(u+r)} \\ &= -\frac{e_3(u+r)}{\left(e_4 + fe_\theta + \frac{1}{4}f^2 e_3\right)(u+r)} \\ &= -\frac{\frac{2}{\varsigma} + \frac{r}{2}(\overline{\underline{\kappa}} + \underline{A})}{\frac{r}{2}\overline{\kappa} + \frac{1}{4}f^2\left(\frac{2}{\varsigma} + \frac{r}{2}(\overline{\underline{\kappa}} + \underline{A})\right)} \end{aligned}$$

and hence

$$\begin{aligned}\nu'_{\Sigma_*}(f) &= -2\eta + 2f\underline{\omega} + \frac{1}{2}f^2\underline{\xi}\\ &\quad -\frac{\frac{2}{\varsigma}+\frac{r}{2}(\overline{\kappa}+\underline{A})}{\frac{r}{2}\overline{\kappa}+\frac{1}{4}f^2\left(\frac{2}{\varsigma}+\frac{r}{2}(\overline{\kappa}+\underline{A})\right)}\Bigg\{-\frac{1}{2}\kappa f - \frac{1}{2}f\vartheta\\ &\quad -\frac{1}{2}f^2\eta - \frac{3}{2}f^2\zeta + \frac{1}{8}f^3\underline{\kappa} + \frac{1}{2}f^3\underline{\omega} + \frac{1}{8}f^3\underline{\vartheta} + \frac{1}{8}f^4\underline{\xi}\Bigg\} \quad \text{on } {}^{(ext)}\mathcal{M}.\end{aligned} \tag{4.5.5}$$

Step 2. Next, we estimate f on Σ_*. Introducing an integer k_{loss} and a small constant $\delta_0 > 0$ satisfying

$$16 \leq k_{loss} \leq \frac{\delta_{dec}}{3}(k_{large} - k_{small}), \qquad \delta_0 = \frac{k_{loss}}{k_{large} - k_{small}},$$

we assume the following local bootstrap assumption

$$|\mathfrak{d}^{\leq k_{small}+k_{loss}+2}f| \leq \frac{\sqrt{\epsilon}}{ru^{\frac{1}{2}+\delta_{dec}-2\delta_0}} \quad \text{on } u_1 \leq u \leq u_* \tag{4.5.6}$$

where

$$1 \leq u_1 < u_*.$$

Since $f = 0$ on $\Sigma_* \cap \mathcal{C}_*$ in view of (4.5.2), (4.5.6) holds for u_1 close enough to u_*, and our goal is to prove that we may in fact choose $u_1 = 1$ and replace $\sqrt{\epsilon}$ with ϵ in (4.5.6).

In view of the estimates for the Ricci coefficients and curvature components with respect to the outgoing geodesic frame (e_4, e_3, e_θ) of ${}^{(ext)}\mathcal{M}$ provided by Proposition 3.19, (4.5.5) yields

$$\nu'_{\Sigma_*}(f) = -2\eta + h, \qquad |\mathfrak{d}^k h| \lesssim r^{-1}(|\mathfrak{d}^{\leq k}f| + |\mathfrak{d}^{\leq k}f|^4) \text{ for } k \leq k_{large} - 5.$$

Using commutator identities, using also (4.5.3) and (4.5.4), and in view of (4.5.6), we infer

$$|\nu'_{\Sigma_*}(\not{\mathfrak{d}}^k f)| \lesssim |\not{\mathfrak{d}}^{\leq k}\eta| + \frac{\sqrt{\epsilon}}{r^2u^{\frac{1}{2}+\delta_{dec}-2\delta_0}} \text{ for } k \leq k_{small} + k_{loss} + 2, \; u_1 \leq u \leq u_*.$$

Since $f = 0$ on $\Sigma_* \cap \mathcal{C}_*$ in view of (4.5.2), and since ν'_{Σ_*} is tangent to Σ_*, we deduce on Σ_*, integrating along the integral curve of ν'_{Σ_*}

$$\begin{aligned}|\not{\mathfrak{d}}^k f| &\lesssim \int_u^{u_*} |\not{\mathfrak{d}}^{\leq k}\eta| + \frac{\sqrt{\epsilon}}{u^{\frac{1}{2}+\delta_{dec}-2\delta_0}}\int_u^{u_*}\frac{1}{\nu'_{\Sigma_*}(u')r^2}\\ &\quad \text{for } k \leq k_{small} + k_{loss} + 2, \; u_1 \leq u \leq u_*.\end{aligned}$$

Since

$$\begin{aligned}\nu'_{\Sigma_*}(u) &= e'_3(u) + a'e'_4(u)\\ &= e_3(u) - \frac{\frac{2}{\varsigma} + \frac{r}{2}(\overline{\kappa} + \underline{A})}{\frac{r}{2}\overline{\kappa} + \frac{1}{4}f^2\left(\frac{2}{\varsigma} + \frac{r}{2}(\overline{\kappa} + \underline{A})\right)}\left(e_4 + fe_\theta + \frac{1}{4}f^2 e_3\right)u\\ &= \frac{2}{\varsigma} - \frac{f^2}{2\varsigma}\frac{\frac{2}{\varsigma} + \frac{r}{2}(\overline{\kappa} + \underline{A})}{\frac{r}{2}\overline{\kappa} + \frac{1}{4}f^2\left(\frac{2}{\varsigma} + \frac{r}{2}(\overline{\kappa} + \underline{A})\right)}\end{aligned}$$

we have

$$\nu'_{\Sigma_*}(u) = 2 + O(\epsilon)$$

and hence

$$|\not{\mathfrak{d}}^k f| \lesssim \int_u^{u_*} |\not{\mathfrak{d}}^{\leq k}\eta| + \frac{\sqrt{\epsilon}}{u^{\frac{1}{2}+\delta_{dec}-2\delta_0}}\int_u^{u_*}\frac{1}{r^2}$$
$$\text{for } k \leq k_{small} + k_{loss} + 2,\ u_1 \leq u \leq u_*.$$

Together with the behavior (3.3.4) of r on Σ_*, we infer

$$|\not{\mathfrak{d}}^k f| \lesssim \int_u^{u_*} |\not{\mathfrak{d}}^{\leq k}\eta| + \frac{\epsilon}{ru^{\frac{1}{2}+\delta_{dec}-2\delta_0}} \text{ for } k \leq k_{small} + k_{loss} + 2,\ u_1 \leq u \leq u_*.$$

Next, we estimate η. We have by interpolation, since $k_{loss} \leq k_{large} - k_{small}$,

$$\|\not{\mathfrak{d}}^{\leq k_{small}+k_{loss}+4}\eta\|_{L^2(S)} \lesssim \|\not{\mathfrak{d}}^{\leq k_{small}}\eta\|_{L^2(S)}^{1-\frac{k_{loss}+4}{k_{large}-k_{small}}}\|\not{\mathfrak{d}}^{\leq k_{large}}\eta\|_{L^2(S)}^{\frac{k_{loss}+4}{k_{large}-k_{small}}},$$

and hence, using $\delta_0 > 0$, we have

$$\begin{aligned}&\int_u^{u_*}\|\not{\mathfrak{d}}^{\leq k_{small}+k_{loss}+4}\eta\|_{L^2(S)}\\ &\lesssim \left(\int_{\Sigma_*(\geq u)} u'^{1+\delta_0}|\not{\mathfrak{d}}^{\leq k_{small}+k_{loss}+4}\eta|^2\right)^{\frac{1}{2}}\\ &\lesssim \frac{1}{u^{\frac{1}{2}+\delta_{dec}-2\delta_0}}\left(\int_{\Sigma_*} u'^{2+2\delta_{dec}}|\not{\mathfrak{d}}^{\leq k_{small}}\eta|^2\right)^{\frac{1}{2}-\frac{k_{loss}+4}{2(k_{large}-k_{small})}}\\ &\quad\times\left(\int_{\Sigma_*}|\not{\mathfrak{d}}^{\leq k_{large}}\eta|^2\right)^{\frac{k_{loss}+4}{2(k_{large}-k_{small})}}\end{aligned}$$

where we have used the fact that

$$\frac{k_{loss}+4}{k_{large}-k_{small}}(1+\delta_{dec}) + \frac{\delta_0}{2} = \left(\left(1+\frac{4}{k_{loss}}\right)(1+\delta_{dec}) + \frac{1}{2}\right)\delta_0 \leq 2\delta_0$$

and

$$\frac{1}{2} + \delta_{dec} - 2\delta_0 = \frac{1}{2} + \delta_{dec} - \frac{4k_{loss}}{k_{large}-k_{small}} \geq \delta_{dec} > 0$$

since $16 \leq k_{loss} \leq \frac{1}{8}(k_{large} - k_{small})$ and $\delta_{dec} > 0$ is small. Now, recall from the bootstrap assumptions on decay and energy for η along Σ_* that we have

$$\int_{\Sigma_*} u^{2+2\delta_{dec}} |\mathfrak{d}^{\leq k_{small}} \eta|^2 + \int_{\Sigma_*} |\mathfrak{d}^{\leq k_{large}} \eta|^2 \leq \epsilon^2.$$

We deduce

$$\int_u^{u_*} \| \not{\mathfrak{d}}^{\leq k_{small}+k_{loss}+4} \eta \|_{L^2(S)} \lesssim \frac{\epsilon}{u^{\frac{1}{2}+\delta_{dec}-2\delta_0}}.$$

Together with the Sobolev embedding on the 2-spheres S foliating Σ_*, as well as the behavior (3.3.4) of r on Σ_*, we infer

$$\int_u^{u_*} | \not{\mathfrak{d}}^{\leq k_{small}+k_{loss}+2} \eta | \lesssim \frac{\epsilon}{r u^{\frac{1}{2}+\delta_{dec}-2\delta_0}}.$$

Plugging in the above estimate for f, we infer

$$|\not{\mathfrak{d}}^k f| \lesssim \frac{\epsilon}{r u^{\frac{1}{2}+\delta_{dec}-2\delta_0}} \text{ for } k \leq k_{small} + k_{loss} + 2, \ u_1 \leq u \leq u_*.$$

Together with (4.5.3) and (4.5.4), we recover e_4 and e_3 derivatives to deduce

$$|\mathfrak{d}^k f| \lesssim \frac{\epsilon}{r u^{\frac{1}{2}+\delta_{dec}-2\delta_0}} \text{ for } k \leq k_{small} + k_{loss} + 2, \ u_1 \leq u \leq u_*.$$

This is an improvement of the bootstrap assumption (4.5.6). Thus, we may choose $u_1 = 1$, and f satisfies the following estimate

$$|\mathfrak{d}^k f| \lesssim \frac{\epsilon}{r u^{\frac{1}{2}+\delta_{dec}-2\delta_0}} \text{ for } k \leq k_{small} + k_{loss} + 2 \text{ on } \Sigma_*.$$

Together with (4.5.4), as well as the behavior (3.3.4) of r on Σ_*, we infer

$$\begin{aligned} |\mathfrak{d}^{k-1} e_3' f| &\lesssim |\mathfrak{d}^{k-1} \eta| + \frac{\epsilon}{r^2} \\ &\lesssim \frac{\epsilon}{r u^{1+\delta_{dec}-2\delta_0}} \text{ for } k \leq k_{small} + k_{loss} + 2 \text{ on } \Sigma_*. \end{aligned}$$

Collecting the two above estimates, we obtain

$$\begin{aligned} |\mathfrak{d}^k f| &\lesssim \frac{\epsilon}{r u^{\frac{1}{2}+\delta_{dec}-2\delta_0}} && \text{for } k \leq k_{small} + k_{loss} + 2 \text{ on } \Sigma_*, \\ |\mathfrak{d}^{k-1} e_3' f| &\lesssim \frac{\epsilon}{r u^{1+\delta_{dec}-2\delta_0}} && \text{for } k \leq k_{small} + k_{loss} + 2 \text{ on } \Sigma_*. \end{aligned} \tag{4.5.7}$$

Step 3. Next, we estimate f on ${}^{(ext)}\mathcal{M}$. We assume the following local bootstrap assumption

$$|\mathfrak{d}^{\leq k_{small}+k_{loss}+2} f| \leq \frac{\sqrt{\epsilon}}{r u^{\frac{1}{2}+\delta_{dec}-2\delta_0} + u^{1+\delta_{dec}-2\delta_0}} \quad \text{on } r \geq r_1 \tag{4.5.8}$$

where $r_1 \geq 4m_0$. In view of the control of f on Σ_* provided by (4.5.7), (4.5.8) holds for r_1 sufficiently large, and our goal is to prove that we may in fact choose $r_1 = 4m_0$ and replace $\sqrt{\epsilon}$ with ϵ in (4.5.8).

Recall (4.5.3)

$$\begin{aligned} e_4'(f)+\frac{1}{2}\kappa f &= -\frac{1}{2}f\vartheta-\frac{1}{2}f^2\eta-\frac{3}{2}f^2\zeta \\ &\quad +\frac{1}{8}f^3\underline{\kappa}+\frac{1}{2}f^3\underline{\omega}+\frac{1}{8}f^3\underline{\vartheta}+\frac{1}{8}f^4\underline{\xi} \text{ on } {}^{(ext)}\mathcal{M}. \end{aligned}$$

In view of the estimates for the Ricci coefficients and curvature components with respect to the outgoing geodesic frame (e_4, e_3, e_θ) of ${}^{(ext)}\mathcal{M}$ provided by Proposition 3.19,

$$\begin{aligned} &\left|\mathfrak{d}^k\left(-\frac{1}{2}f\vartheta-\frac{1}{2}f^2\eta-\frac{3}{2}f^2\zeta+\frac{1}{8}f^3\underline{\kappa}+\frac{1}{2}f^3\underline{\omega}+\frac{1}{8}f^3\underline{\vartheta}+\frac{1}{8}f^4\underline{\xi}\right)\right| \\ &\lesssim \epsilon r^{-2}u^{-\frac{1}{2}}|\mathfrak{d}^{\leq k}f|+r^{-1}(|\mathfrak{d}^{\leq k}f|^2+|\mathfrak{d}^{\leq k}f|^4) \text{ for } k\leq k_{large}-5. \end{aligned}$$

Using commutator identities, using also (4.5.3), and in view of (4.5.8), we infer[6]

$$e_4'\Big((\not{\mathfrak{d}},T)^kf\Big)+\frac{1}{2}\kappa(\not{\mathfrak{d}},T)^kf \leq \frac{\epsilon}{r^3u^{1+\delta_{dec}-2\delta_0}} \text{ for } k\leq k_{small}+k_{loss}+2,\ r\geq r_1.$$

Integrating backwards from Σ_* where we have (4.5.7), we deduce[7]

$$|(\not{\mathfrak{d}},T)^kf| \leq \frac{\epsilon}{ru^{\frac{1}{2}+\delta_{dec}-2\delta_0}+u^{1+\delta_{dec}-2\delta_0}} \text{ for } k\leq k_{small}+k_{loss}+2,\ r\geq r_1.$$

Together with (4.5.3), we recover the e_4 derivatives and obtain

$$|\mathfrak{d}^kf| \leq \frac{\epsilon}{ru^{\frac{1}{2}+\delta_{dec}-2\delta_0}+u^{1+\delta_{dec}-2\delta_0}} \text{ for } k\leq k_{small}+k_{loss}+2,\ r\geq r_1.$$

This is an improvement of the bootstrap assumption (4.5.8). Thus, we may choose $r_1=4m_0$, and we have

$$|\mathfrak{d}^kf| \lesssim \frac{\epsilon}{ru^{\frac{1}{2}+\delta_{dec}-2\delta_0}+u^{1+\delta_{dec}-2\delta_0}} \text{ for } k\leq k_{small}+k_{loss}+2 \text{ on } {}^{(ext)}\mathcal{M}.$$

Also, commuting once (4.5.3) with e_3', using the commutator identity $[e_3', e_4'] = 2\underline{\omega}'e_4'-2\omega'e_3'+(\eta'-\underline{\eta}')e_\theta'$, and proceeding as above to integrate backward from Σ_* where $e_3'f$ is under control from (4.5.7), we also obtain

$$|\mathfrak{d}^{k-1}e_3'f| \lesssim \frac{\epsilon}{ru^{1+\delta_{dec}-2\delta_0}} \text{ for } k\leq k_{small}+k_{loss}+2 \text{ on } {}^{(ext)}\mathcal{M}.$$

[6] Note that

$$\delta_{dec}-2\delta_0=\delta_{dec}-\frac{2k_{loss}}{k_{large}-k_{small}}\geq\frac{\delta_{dec}}{3}>0$$

where we have used the definition of δ_0 and the upper bound on k_{loss}.

[7] Note that (4.5.7) yields

$$|\mathfrak{d}^kf| \lesssim \frac{\epsilon}{u^{1+\delta_{dec}-2\delta_0}} \text{ for } k\leq k_{small}+k_{loss}+2 \text{ on } \Sigma_*$$

in view of the behavior (3.3.4) of r on Σ_*.

Collecting the two above estimates, we obtain

$$
\begin{aligned}
|\mathfrak{d}^k f| &\lesssim \frac{\epsilon}{ru^{\frac{1}{2}+\delta_{dec}-2\delta_0}+u^{1+\delta_{dec}-2\delta_0}} \quad \text{for } k \le k_{small}+k_{loss}+2 \text{ on } {}^{(ext)}\mathcal{M},\\
|\mathfrak{d}^{k-1} e_3' f| &\lesssim \frac{\epsilon}{ru^{1+\delta_{dec}-2\delta_0}} \quad \text{for } k \le k_{small}+k_{loss}+2 \text{ on } {}^{(ext)}\mathcal{M},
\end{aligned}
\tag{4.5.9}
$$

which is the desired estimate for f.

Step 4. In view of Proposition 2.90 applied to our particular case, i.e., a triplet $(f, , \underline{f}, \lambda)$ with $\underline{f}=0$ and $\lambda = 1$, and the fact that the frame (e_4, e_3, e_θ) is outgoing geodesic, we have

$$
\begin{aligned}
\underline{\xi}' &= \underline{\xi},\\
\zeta' &= \zeta - \frac{1}{4}f\underline{\kappa} - f\underline{\omega} - \frac{1}{4}f\underline{\vartheta} + \text{l.o.t.},\\
\eta' &= \eta + \frac{1}{2}e_3'(f) - f\underline{\omega} + \text{l.o.t.},\\
\underline{\eta}' &= -\zeta + \frac{1}{4}\underline{\kappa}f + \frac{1}{4}f\underline{\vartheta} + \text{l.o.t.},
\end{aligned}
$$

$$
\begin{aligned}
\kappa' &= \kappa + \not{d}_1{}'(f) + f(\zeta+\eta) - \frac{1}{4}f^2\underline{\kappa} - f^2\underline{\omega} + \text{l.o.t.},\\
\underline{\kappa}' &= \underline{\kappa} + f\underline{\xi} + \text{l.o.t.},\\
\vartheta' &= \vartheta - \not{d}_2^{\star}{}'(f) + f(\zeta+\eta) - f^2\underline{\omega} + \text{l.o.t.},\\
\underline{\vartheta}' &= \underline{\vartheta} + f\underline{\xi} + \text{l.o.t.},
\end{aligned}
$$

$$
\begin{aligned}
\omega' &= f\zeta - \frac{1}{8}\underline{\kappa}f^2 - \frac{1}{4}\underline{\omega}f^2 + \text{l.o.t.},\\
\underline{\omega}' &= \underline{\omega} + \frac{1}{2}f\underline{\xi},
\end{aligned}
$$

and

$$
\begin{aligned}
\alpha' &= \alpha + 2f\beta + \frac{3}{2}f^2\rho + \text{l.o.t.},\\
\beta' &= \beta + \frac{3}{2}\rho f + \text{l.o.t.},\\
\rho' &= \rho + f\underline{\beta} + \text{l.o.t.},\\
\underline{\beta}' &= \underline{\beta} + \frac{1}{2}f\underline{\alpha},\\
\underline{\alpha}' &= \underline{\alpha},
\end{aligned}
\tag{4.5.10}
$$

where the lower order terms denoted by l.o.t. are linear with respect to ξ, $\underline{\xi}$, ϑ, κ, η, $\underline{\eta}$, ζ, $\underline{\kappa}$, $\underline{\vartheta}$ and α, β, ρ, $\underline{\beta}$, $\underline{\alpha}$, and quadratic or higher order in f, and do not contain derivatives of the latter. Together with the estimates (4.5.9) for f on ${}^{(ext)}\mathcal{M}$, and the estimates for the Ricci coefficients and curvature components with respect to the outgoing geodesic frame (e_4, e_3, e_θ) of ${}^{(ext)}\mathcal{M}$ provided by the bootstrap assumptions on decay and Proposition 3.19, we immediately infer

$$
\begin{aligned}
\max_{0\leq k\leq k_{small}+k_{loss}+1}\sup_{{}^{(ext)}\mathcal{M}}\Bigg\{&\Big(r^2u^{\frac{1}{2}+\delta_{dec}-2\delta_0}+ru^{1+\delta_{dec}-2\delta_0}\Big)|\mathfrak{d}^k(\Gamma'_g\setminus\{\eta'\})|\\
&+ru^{1+\delta_{dec}-2\delta_0}|\mathfrak{d}^k\Gamma'_b|\\
&+r^2u^{1+\delta_{dec}-2\delta_0}\left|\mathfrak{d}^{k-1}e'_3\left(\kappa'-\frac{2}{r},\underline{\kappa}'+\frac{2\Upsilon}{r},\vartheta',\zeta',\underline{\eta}'\right)\right|\\
&+\Big(r^3(u+2r)^{\frac{1}{2}+\delta_{dec}-2\delta_0}+r^2(u+2r)^{1+\delta_{dec}-2\delta_0}\Big)|\mathfrak{d}^k(\alpha',\beta')|\\
&+\Big(r^3(2r+u)^{1+\delta_{dec}}+r^4(2r+u)^{\frac{1}{2}+\delta_{dec}-2\delta_0}\Big)|\mathfrak{d}^{k-1}e'_3(\alpha')|\\
&+\Big(r^3u^{1+\delta_{dec}}+r^4u^{\frac{1}{2}+\delta_{dec}-2\delta_0}\Big)|\mathfrak{d}^{k-1}e'_3(\beta')|\\
&+\Big(r^3u^{\frac{1}{2}+\delta_{dec}-2\delta_0}+r^2ru^{1+\delta_{dec}-2\delta_0}\Big)|\mathfrak{d}^k\check{\rho}'|\\
&+u^{1+\delta_{dec}-2\delta_0}\Big(r^2|\mathfrak{d}^k\underline{\beta}'|+r|\mathfrak{d}^k\underline{\alpha}'|\Big)\Bigg\}\lesssim\epsilon
\end{aligned}
\tag{4.5.11}
$$

where we have introduced the notation

$$
\Gamma'_g\setminus\{\eta'\} \;=\; \left\{r\omega',\,\kappa'-\frac{2}{r},\,\vartheta',\,\zeta',\,\underline{\eta}',\,\underline{\kappa}'+\frac{2\Upsilon}{r},\,r^{-1}(e'_4(r)-1),r^{-1}e'_\theta(r),\,e'_4(m)\right\}.
$$

Note also, in view of the above transformation formula for ω', i.e.,

$$
\omega' \;=\; f\zeta-\frac{1}{8}\underline{\kappa}f^2-\frac{1}{4}\underline{\omega}f^2+\text{l.o.t.},
$$

that we have in fact a gain of r^{-1} for ω' compared to (4.5.11), i.e.,

$$
\max_{0\leq k\leq k_{small}+k_{loss}+1}\sup_{{}^{(ext)}\mathcal{M}}\Big(r^3u^{\frac{1}{2}+\delta_{dec}-2\delta_0}+r^2u^{1+\delta_{dec}-2\delta_0}\Big)|\mathfrak{d}^k\omega'|\lesssim\epsilon. \tag{4.5.12}
$$

We now focus on estimating η'. Proceeding as for the other Ricci coefficients would yield for η' the same behavior as η and hence a loss of r^{-1} compared to the desired estimate. Instead, we rely on the following null structure equation which follows from Proposition 2.56 and the fact that $\xi'=0$

$$
e'_4(\eta'-\zeta')+\frac{1}{2}\kappa'(\eta'-\zeta') \;=\; 2\,{\not{d}}_1^{\star}\omega'-\frac{1}{2}\vartheta'(\eta'-\zeta').
$$

Next,

- we commute with $\not{\mathfrak{d}}'$ and T', and we rely on the corresponding commutator identities,
- we use the above equation for $e'_4(\eta')$ to recover the e'_4 derivatives,
- we rely on the estimates (4.5.11), as well as the estimate (4.5.12) for ω',

which allows us to derive

$$\left| e_4'(\mathfrak{d}^k(\eta' - \zeta')) + \frac{1}{2}\kappa'\mathfrak{d}^k(\eta' - \zeta') \right| \lesssim \frac{\epsilon}{r^4 u^{\frac{1}{2}+\delta_{dec}-2\delta_0} + r^3 u^{1+\delta_{dec}-2\delta_0}} + \frac{\epsilon}{r^2}|\mathfrak{d}^{\leq k}(\eta' - \zeta')|, \qquad k \leq k_{small} + k_{loss}.$$

Integrating backwards from Σ_* where $\eta' = 0$ in view of (4.5.2), and using the control ζ' provided by (4.5.11), we infer

$$\begin{aligned} & \max_{0 \leq k \leq k_{small}+k_{loss}} \sup_{^{(ext)}\mathcal{M}} \left(r^2 u^{\frac{1}{2}+\delta_{dec}-2\delta_0} + r u^{1+\delta_{dec}-2\delta_0} \right) |\mathfrak{d}^k \eta'| \\ \lesssim \; & \epsilon + \max_{0 \leq k \leq k_{small}+k_{loss}} \sup_{^{(ext)}\mathcal{M}} \left(r^2 u^{\frac{1}{2}+\delta_{dec}-2\delta_0} + r u^{1+\delta_{dec}-2\delta_0} \right) |\mathfrak{d}^k \zeta'| \\ \lesssim \; & \epsilon. \end{aligned}$$

Also, commuting first the equation for $e_4'(\eta' - \zeta')$ with e_3', using the commutator identity $[e_3', e_4'] = 2\underline{\omega}' e_4' - 2\omega' e_3' + (\eta' - \underline{\eta}')e_\theta'$, and proceeding as above to integrate backward from Σ_*, we also obtain

$$\begin{aligned} & \max_{0 \leq k \leq k_{small}+k_{loss}} \sup_{^{(ext)}\mathcal{M}} r^2 u^{1+\delta_{dec}-2\delta_0} |\mathfrak{d}^{k-1} e_3' \eta'| \\ \lesssim \; & \epsilon + \max_{0 \leq k \leq k_{small}+k_{loss}} \sup_{^{(ext)}\mathcal{M}} r^2 u^{1+\delta_{dec}-2\delta_0} |\mathfrak{d}^{k-1} e_3' \zeta'| \\ \lesssim \; & \epsilon. \end{aligned}$$

Thus, together with (4.5.11), we infer

$$\begin{aligned} \max_{0 \leq k \leq k_{small}+k_{loss}} \sup_{^{(ext)}\mathcal{M}} \Bigg\{ & \left(r^2 u^{\frac{1}{2}+\delta_{dec}-2\delta_0} + r u^{1+\delta_{dec}-2\delta_0} \right) |\mathfrak{d}^k \Gamma_g'| \\ & + r u^{1+\delta_{dec}-2\delta_0} |\mathfrak{d}^k \Gamma_b'| \\ & + r^2 u^{1+\delta_{dec}-2\delta_0} \left| \mathfrak{d}^{k-1} e_3' \left(\kappa' - \frac{2}{r}, \underline{\kappa}' + \frac{2\Upsilon}{r}, \vartheta', \zeta', \underline{\eta}', \eta' \right) \right| \\ & + \left(r^3 (u + 2r)^{\frac{1}{2}+\delta_{dec}-2\delta_0} + r^2 (u + 2r)^{1+\delta_{dec}-2\delta_0} \right) |\mathfrak{d}^k (\alpha', \beta')| \\ & + \left(r^3 (2r + u)^{1+\delta_{dec}} + r^4 (2r + u)^{\frac{1}{2}+\delta_{dec}-2\delta_0} \right) |\mathfrak{d}^{k-1} e_3'(\alpha')| \\ & + \left(r^3 u^{1+\delta_{dec}} + r^4 u^{\frac{1}{2}+\delta_{dec}-2\delta_0} \right) |\mathfrak{d}^{k-1} e_3'(\beta')| \\ & + \left(r^3 u^{\frac{1}{2}+\delta_{dec}-2\delta_0} + r^2 r u^{1+\delta_{dec}-2\delta_0} \right) |\mathfrak{d}^k \check{\rho}'| \\ & + u^{1+\delta_{dec}-2\delta_0} \left(r^2 |\mathfrak{d}^k \underline{\beta}'| + r |\mathfrak{d}^k \underline{\alpha}'| \right) \Bigg\} \lesssim \epsilon. \end{aligned}$$

Together with the fact that $\xi' = 0$ in view of (4.5.2), this concludes the proof of Proposition 3.20.

4.6 EXISTENCE AND CONTROL OF THE GLOBAL FRAMES

4.6.1 Proof of Proposition 3.23

To match the frame of ${}^{(int)}\mathcal{M}$ and a conformal renormalization of the frame of ${}^{(ext)}\mathcal{M}$, we will need to introduce a cut-off function.

Definition 4.11. *Let $\psi : \mathbb{R} \to \mathbb{R}$ a smooth cut-off function such that $0 \leq \psi \leq 1$, $\psi = 0$ on $(-\infty, 0]$ and $\psi = 1$ on $[1, +\infty)$. We define $\psi_{m_0,\delta_{\mathcal{H}}}$ as follows:*

$$\psi_{m_0,\delta_{\mathcal{H}}}(r) = \begin{cases} 1 & \text{if } \ r \geq 2m_0\left(1 + \frac{3}{2}\delta_{\mathcal{H}}\right), \\ 0 & \text{if } \ r \leq 2m_0\left(1 + \frac{1}{2}\delta_{\mathcal{H}}\right), \end{cases}$$

and

$$\psi_{m_0,\delta_{\mathcal{H}}}(r) = \psi\left(\frac{r - 2m_0\left(1 + \frac{1}{2}\delta_{\mathcal{H}}\right)}{2m_0\delta_{\mathcal{H}}}\right) \quad \text{on } 2m_0\left(1 + \frac{1}{2}\delta_{\mathcal{H}}\right) \leq r \leq 2m_0\left(1 + \frac{3}{2}\delta_{\mathcal{H}}\right).$$

We are now ready to define the global frame of the statement of Proposition 3.23.

Definition 4.12 (Definition of the global frame)**.** *We introduce a global null frame defined on ${}^{(ext)}\mathcal{M} \cup {}^{(int)}\mathcal{M}$ and denoted by $({}^{(glo)}e_4, {}^{(glo)}e_3, {}^{(glo)}e_\theta)$. The global frame is defined as follows:*

1. *In ${}^{(ext)}\mathcal{M} \setminus Match$, we have*

$$({}^{(glo)}e_4, {}^{(glo)}e_3, {}^{(glo)}e_\theta) = \left({}^{(ext)}\Upsilon\, {}^{(ext)}e_4, {}^{(ext)}\Upsilon^{-1}\, {}^{(ext)}e_3, {}^{(ext)}e_\theta\right).$$

2. *In ${}^{(int)}\mathcal{M} \setminus Match$, we have*

$$({}^{(glo)}e_4, {}^{(glo)}e_3, {}^{(glo)}e_\theta) = \left({}^{(int)}e_4, {}^{(int)}e_3, {}^{(int)}e_\theta\right).$$

3. *It remains to define the global frame on the matching region. We denote by $(f, \underline{f}, \lambda)$ the reduced scalars such that we have in the matching region*

$$\begin{aligned}
{}^{(ext)}e_4 &= \lambda\left({}^{(int)}e_4 + f\,{}^{(int)}e_\theta + \frac{1}{4}f^2\,{}^{(int)}e_3\right), \\
{}^{(ext)}e_\theta &= \left(1 + \frac{1}{2}f\underline{f}\right){}^{(int)}e_\theta + \frac{\underline{f}}{2}\,{}^{(int)}e_4 + \frac{f}{2}\left(1 + \frac{f\underline{f}}{4}\right){}^{(int)}e_3, \\
{}^{(ext)}e_3 &= \lambda^{-1}\Bigg(\left(1 + \frac{1}{2}f\underline{f} + \frac{1}{16}f^2\underline{f}^2\right){}^{(int)}e_3 + \underline{f}\left(1 + \frac{f\underline{f}}{4}\right){}^{(int)}e_\theta \\
&\quad + \frac{\underline{f}^2}{4}\,{}^{(int)}e_4\Bigg),
\end{aligned}$$

where we recall that the frame of ${}^{(ext)}\mathcal{M}$ has been extended to ${}^{(int)}\mathcal{M}$, see section

3.5.1. Then, in the matching region, the global frame is given by

$$
\begin{aligned}
{}^{(glo)}e_4 &= \lambda'\left({}^{(int)}e_4 + f'\,{}^{(int)}e_\theta + \frac{1}{4}f'^2\,{}^{(int)}e_3\right),\\
{}^{(glo)}e_\theta &= \left(1+\frac{1}{2}f'\underline{f}'\right){}^{(int)}e_\theta + \frac{\underline{f}'}{2}{}^{(int)}e_4 + \frac{f'}{2}\left(1+\frac{f'\underline{f}'}{4}\right){}^{(int)}e_3,\\
{}^{(glo)}e_3 &= \lambda'^{-1}\Bigg(\left(1+\frac{1}{2}f'\underline{f}' + \frac{1}{16}f'^2\underline{f}'^2\right){}^{(int)}e_3 + \underline{f}'\left(1+\frac{f'\underline{f}'}{4}\right){}^{(int)}e_\theta\\
&\quad + \frac{\underline{f}'^2}{4}{}^{(int)}e_4\Bigg),
\end{aligned}
$$

where

$$
\begin{aligned}
& f' = \psi_{m_0,\delta_{\mathcal{H}}}({}^{(int)}r)f, \quad \underline{f}' = \psi_{m_0,\delta_{\mathcal{H}}}({}^{(int)}r)\underline{f},\\
& \lambda' = 1 - \psi_{m_0,\delta_{\mathcal{H}}}({}^{(int)}r) + \psi_{m_0,\delta_{\mathcal{H}}}({}^{(int)}r)\,{}^{(ext)}\Upsilon\lambda.
\end{aligned} \tag{4.6.1}
$$

Remark 4.13. *Recall that the smooth cut-off function ψ in Definition 3.22, allowing to define $\psi_{m_0,\delta_{\mathcal{H}}}$, is such that we have in particular $\psi = 0$ on $(-\infty, 0]$ and $\psi = 1$ on $[1, +\infty)$. The following two special cases correspond to the properties (d) i. and (d) ii. of Proposition 3.23.*

- *If the cut-off ψ in Definition 3.22 is such that $\psi = 1$ on $[1/2, +\infty)$, then*

$$({}^{(glo)}e_4, {}^{(glo)}e_3, {}^{(glo)}e_\theta) = \left({}^{(ext)}\Upsilon\,{}^{(ext)}e_4, {}^{(ext)}\Upsilon^{-1}\,{}^{(ext)}e_3, {}^{(ext)}e_\theta\right) \quad \text{on } {}^{(ext)}\mathcal{M}.$$

- *If the cut-off ψ in Definition 3.22 is such that $\psi = 0$ on $(-\infty, 1/2]$, then*

$$({}^{(glo)}e_4, {}^{(glo)}e_3, {}^{(glo)}e_\theta) = \left({}^{(int)}e_4, {}^{(int)}e_3, {}^{(int)}e_\theta\right) \quad \text{on } {}^{(int)}\mathcal{M}.$$

Definition 4.14 (Global area radius and Hawking mass)**.** *We define an area radius and a Hawking mass on ${}^{(ext)}\mathcal{M} \cup {}^{(int)}\mathcal{M}$ as follows:*

- *On ${}^{(ext)}\mathcal{M} \setminus Match$, we have*

$${}^{(glo)}r = {}^{(ext)}r, \quad {}^{(glo)}m = {}^{(ext)}m.$$

- *On ${}^{(int)}\mathcal{M} \setminus Match$, we have*

$${}^{(glo)}r = {}^{(int)}r, \quad {}^{(glo)}m = {}^{(int)}m.$$

- *On the matching region, we have*

$$
\begin{aligned}
{}^{(glo)}r &= (1-\psi_{m_0,\delta_{\mathcal{H}}}({}^{(int)}r))\,{}^{(int)}r + \psi_{m_0,\delta_{\mathcal{H}}}({}^{(int)}r)\,{}^{(ext)}r,\\
{}^{(glo)}m &= (1-\psi_{m_0,\delta_{\mathcal{H}}}({}^{(int)}r))\,{}^{(int)}m + \psi_{m_0,\delta_{\mathcal{H}}}({}^{(int)}r)\,{}^{(ext)}m.
\end{aligned}
$$

The following two lemmas provide the main properties of the global frame.

Lemma 4.15. *We have in ${}^{(ext)}\mathcal{M} \setminus Match$ the following relations between the* *quantities in the respective frames:*

$$
\begin{aligned}
&{}^{(glo)}\alpha = \Upsilon^{2\,(ext)}\alpha, \quad {}^{(glo)}\beta = \Upsilon^{(ext)}\beta, \quad {}^{(glo)}\rho + \frac{2m}{r^3} = {}^{(ext)}\rho + \frac{2m}{r^3},\\
&{}^{(glo)}\underline{\beta} = \Upsilon^{-1\,(ext)}\underline{\beta}, \quad {}^{(glo)}\underline{\alpha} = \Upsilon^{-2\,(ext)}\underline{\alpha}, \quad {}^{(glo)}\xi = 0, \quad {}^{(glo)}\underline{\xi} = \Upsilon^{-2\,(ext)}\underline{\xi},\\
&{}^{(glo)}\zeta = -{}^{(glo)}\underline{\eta} = {}^{(ext)}\zeta, \quad {}^{(glo)}\eta = {}^{(ext)}\eta,\\
&{}^{(glo)}\omega + \frac{m}{r^2} = -\frac{m}{2r}\left(\overline{{}^{(ext)}\kappa} - \frac{2}{r}\right) + \frac{e_4(m)}{r},\\
&{}^{(glo)}\underline{\omega} = \Upsilon^{-1}\Bigg({}^{(ext)}\underline{\omega} - \frac{m}{r^2} + \frac{m}{2\Upsilon r}\left(\overline{{}^{(ext)}\underline{\kappa}} - \frac{2\Upsilon}{r}\right)\\
&\qquad + \frac{m}{2\Upsilon r}\left({}^{(ext)}\check{\underline{\Omega}}\,\overline{{}^{(ext)}\kappa} - \overline{{}^{(ext)}\check{\underline{\Omega}}\,{}^{(ext)}\check{\kappa}}\right) - \frac{e_3(m)}{\Upsilon r}\Bigg),\\
&{}^{(glo)}\kappa - \frac{2\Upsilon}{r} = \Upsilon\left({}^{(ext)}\kappa - \frac{2}{r}\right), \quad {}^{(glo)}\underline{\kappa} + \frac{2}{r} = \Upsilon^{-1}\left({}^{(ext)}\underline{\kappa} + \frac{2\Upsilon}{r}\right),\\
&{}^{(glo)}\check{\kappa} = \Upsilon^{(ext)}\check{\kappa}, \quad {}^{(glo)}\check{\underline{\kappa}} = \Upsilon^{-1\,(ext)}\check{\underline{\kappa}}, \quad {}^{(glo)}\vartheta = \Upsilon^{(ext)}\vartheta, \quad {}^{(glo)}\underline{\vartheta} = \Upsilon^{-1\,(ext)}\underline{\vartheta}.
\end{aligned}
$$

Proof. The proof follows immediately from the change of frame formula with the choice ($f = 0, \underline{f} = 0, \lambda = \Upsilon$), the fact that $e_\theta(\Upsilon) = 0$, and the fact that the frame of ${}^{(ext)}\mathcal{M}$ is outgoing geodesic and thus satisfies in particular $\xi = \omega = 0$ and $\underline{\eta} = -\zeta$. □

Lemma 4.16 (Control of the global frame in the matching region). *In the matching region, the following estimates hold for the global frame:*[8]

$$
\begin{aligned}
&\max_{0\leq k\leq k_{small}-2}\Bigg[\sup_{Match\cap\,{}^{(int)}\mathcal{M}} \underline{u}^{1+\delta_{dec}}\left|\mathfrak{d}^k\left({}^{(glo)}\check{\Gamma}, {}^{(glo)}\check{R}\right)\right|\\
&\qquad + \sup_{Match\cap\,{}^{(ext)}\mathcal{M}} u^{1+\delta_{dec}}\left|\mathfrak{d}^k\left({}^{(glo)}\check{\Gamma}, {}^{(glo)}\check{R}\right)\right|\Bigg]\\
&\qquad + \max_{0\leq k\leq k_{large}-1}\left(\int_{Match}\left|\mathfrak{d}^k\left({}^{(glo)}\check{\Gamma}, {}^{(glo)}\check{R}\right)\right|^2\right)^{\frac{1}{2}} \lesssim \epsilon
\end{aligned}
$$

and

$$
\left(\int_{Match}\left|\mathfrak{d}^{k_{large}}\left({}^{(glo)}\check{\Gamma}, {}^{(glo)}\check{R}\right)\right|^2\right)^{\frac{1}{2}} \lesssim \epsilon + \left(\int_{\mathcal{T}}\left|\mathfrak{d}^{k_{large}}\left({}^{(ext)}\check{R}\right)\right|^2\right)^{\frac{1}{2}}.
$$

Remark 4.17. *The quantities associated to the global frame can be estimated as follows:*

- *In ${}^{(int)}\mathcal{M}\setminus Match$, the global frame coincides with the frame of ${}^{(int)}\mathcal{M}$, and hence, the quantities associated to the global frame satisfy the same estimates as the bootstrap assumptions for the frame of ${}^{(int)}\mathcal{M}$.*

[8]We only need the first estimate for the proof of Proposition 3.23, but the second estimate will be needed in the proof of Theorem M8.

- *In* $^{(ext)}\mathcal{M} \setminus$ *Match, estimates for the quantities associated to the global frame follow from the identities of Lemma 4.15 together with the bootstrap assumptions for the frame of* $^{(ext)}\mathcal{M}$.
- *In Match, the estimates for the quantities associated to the global frame are provided by Lemma 4.16.*

The proof of Proposition 3.23 easily follows from Definition 4.12, Remark 4.13, and Lemma 4.16. Thus, from now on, we focus on the proof of Lemma 4.16 which is carried out in the next section.

4.6.2 Proof of Lemma 4.16

In this section, we prove Lemma 4.16. To ease the exposition, the quantities associated to the frame of $^{(int)}\mathcal{M}$ are unprimed, the quantities associated to the frame of $^{(ext)}\mathcal{M}$ are primed, and the quantities associated to the global frame are double-primed.

Step 1. Let (e_3, e_θ, e_4) denote the null frame of $^{(int)}\mathcal{M}$ (and its extension) and (e_3', e_θ', e_4') the null frame of $^{(ext)}\mathcal{M}$ (and its extension). We denote by $(f, \underline{f}, \lambda)$ the reduced scalars such that

$$
\begin{aligned}
e_4' &= \lambda\left(e_4 + f e_\theta + \frac{1}{4}f^2 e_3\right),\\
e_\theta' &= \left(1+\frac{1}{2}f\underline{f}\right)e_\theta + \frac{\underline{f}}{2}e_4 + \frac{f}{2}\left(1+\frac{f\underline{f}}{4}\right)e_3,\\
e_3' &= \lambda^{-1}\left(\left(1+\frac{1}{2}f\underline{f}+\frac{1}{16}f^2\underline{f}^2\right)e_3 + \left(\underline{f}+\frac{1}{4}\underline{f}^2 f\right)e_\theta + \frac{\underline{f}^2}{4}e_4\right).
\end{aligned}
$$

Together with the initialization of the frame of $^{(ext)}\mathcal{M}$ and $^{(int)}\mathcal{M}$ on $\mathcal{T}$ in section 3.1.2 (where the spheres coincide), we have in particular

$$
f = \underline{f} = 0, \quad \lambda = \Upsilon^{-1} \text{ on } \mathcal{T}. \tag{4.6.2}
$$

Also, recall from section 3.5.1 that in order for (e_3', e_θ', e_4') to be defined everywhere on $^{(int)}\mathcal{M} \cap \text{Match}$, we need — in addition to the above initialization of $(f, \underline{f}, \lambda)$ on $\mathcal{T}$ — to initialize it also on $\underline{\mathcal{C}}_* \cap \text{Match}$ by

$$
f = \underline{f} = 0, \quad \lambda = \Upsilon^{-1} \text{ on } \underline{\mathcal{C}}_* \cap \text{Match}. \tag{4.6.3}
$$

Step 2. Next, we control the change of frame $(f, \underline{f}, \lambda)$ from (e_3, e_θ, e_4) to (e_3', e_θ', e_4') in the region $^{(int)}\mathcal{M} \cap \text{Match}$. To this end, we rely on the transport equation of Lemma 2.92 together with the fact that $\omega' = \xi' = \zeta' + \underline{\eta}' = 0$. Then, $(\underline{f}, f, \log(\lambda))$ satisfy the following transport equations:

$$
\begin{aligned}
\lambda^{-1}e_4'(f) + \left(\frac{\kappa}{2}+2\omega\right)f &= -2\xi + E_1(f,\Gamma),\\
\lambda^{-1}e_4'(\log(\lambda)) &= 2\omega + E_2(f,\Gamma),\\
\lambda^{-1}e_4'(\underline{f}) + \frac{\kappa}{2}\underline{f} &= -2(\zeta+\underline{\eta}) + 2e_\theta'(\log(\lambda)) + 2f\underline{\omega} + E_3(f,\underline{f},\Gamma),
\end{aligned}
$$

where E_1, E_2 and E_3 are given by

$$\begin{aligned}
E_1(f,\Gamma) &= -\frac{1}{2}\vartheta f + \text{l.o.t.},\\
E_2(f,\Gamma) &= f\zeta - \frac{1}{2}f^2\underline{\omega} - \underline{\eta}f - \frac{1}{4}f^2\underline{\kappa} + \text{l.o.t.},\\
E_3(f,\underline{f},\Gamma) &= -\underline{f}e'_\theta(f) - \frac{1}{2}\underline{f}\vartheta + \text{l.o.t.}
\end{aligned}$$

Here, l.o.t. denotes terms which are cubic or higher order in $f, \underline{f}$ (or in f only in the case of E_1 and E_2) and $\check{\Gamma}$ and do not contain derivatives of these quantities, where Γ and $\check{\Gamma}$ denote the Ricci coefficients and renormalized Ricci coefficients w.r.t. the original null frame (e_3, e_4, e_θ). We rewrite the transport equation for $\log(\lambda)$ as

$$\begin{aligned}
&\lambda^{-1}e'_4\left(\log\left(\Upsilon\lambda\right)\right)\\
&= \lambda^{-1}e'_4(\log(\lambda)) + \lambda^{-1}e'_4(\log(\Upsilon))\\
&= 2\omega + E_2(f,\Gamma) + \frac{1}{\Upsilon}\left(e_4 + fe_\theta + \frac{1}{4}f^2e_3\right)\Upsilon\\
&= 2\left(\omega + \frac{m}{r^2}\right) + E_2(f,\Gamma) + \frac{2}{\Upsilon}\frac{m(e_4(r)-\Upsilon)}{r^2} - \frac{2}{\Upsilon}\frac{e_4(m)}{r}\\
&\quad -\frac{1}{\Upsilon}\left(fe_\theta + \frac{1}{4}f^2e_3\right)\Upsilon.
\end{aligned}$$

In view of the above transport equations for f, $\underline{f}$ and λ, the initialization (4.6.2) and (4.6.3) for $(f, \underline{f}, \lambda)$ on $\mathcal{T} \cup (\underline{\mathcal{C}}_* \cap \text{Match})$, and the control of Γ induced by the bootstrap assumptions on ${}^{(int)}\mathcal{M}$, we easily deduce

$$\begin{aligned}
&\max_{0\le k\le k_{small}}\sup_{{}^{(int)}\mathcal{M}\cap\text{Match}} \underline{u}^{1+\delta_{dec}}\left|\mathfrak{d}^k(f,\log(\Upsilon\lambda))\right|\\
&\quad + \max_{0\le k\le k_{small}-1}\sup_{{}^{(int)}\mathcal{M}\cap\text{Match}} \underline{u}^{1+\delta_{dec}}\left|\mathfrak{d}^k\underline{f}\right| \lesssim \epsilon,\\
&\max_{0\le k\le k_{large}}\left(\int_{{}^{(int)}\mathcal{M}\cap\text{Match}}\left|\mathfrak{d}^k(f,\log(\Upsilon\lambda))\right|^2\right)^{\frac{1}{2}}\\
&\quad + \max_{0\le k\le k_{large}-1}\left(\int_{{}^{(int)}\mathcal{M}\cap\text{Match}}\left|\mathfrak{d}^k\underline{f}\right|^2\right)^{\frac{1}{2}} \lesssim \epsilon.
\end{aligned}$$

Step 3. We need to improve the number of derivatives in the top order estimate for $(f, \underline{f}, \log(\lambda))$. To this end, note first in view of the transformation formulas of Proposition 2.90 and the control of $(f, \underline{f}, \log(\lambda))$ provided by Step 2, we have in particular

$$\max_{0\le k\le k_{large}-1}\left(\int_{{}^{(int)}\mathcal{M}\cap\text{Match}}\left|\mathfrak{d}^k\check{R}'\right|^2\right)^{\frac{1}{2}} \lesssim \epsilon.$$

Relying on this estimate, the control of the Ricci coefficients associated to the outgoing null frame (e'_4, e'_3, e'_θ) on $\mathcal{T} \cup ({}^{(int)}\mathcal{M} \cap \text{Match})$, and the null structure equations, we infer

$$\max_{0\le k\le k_{large}-1}\left(\int_{{}^{(int)}\mathcal{M}\cap\mathrm{Match}}\left|\mathfrak{d}^k\check{\Gamma}'\right|^2\right)^{\frac{1}{2}} \lesssim \epsilon.$$

We refer to section 8.9 for a completely analogous proof where the Ricci coefficients are recovered in ${}^{(int)}\mathcal{M}$ based on the control of the curvature components.

In view of the transformation formulas of Proposition 2.90, which can be written schematically as

$$\partial\Big(f,\underline{f},\log(\lambda)\Big) = F(f,\underline{f},\lambda,\check{\Gamma}),$$

the control of $(f,\underline{f},\log(\lambda))$ provided by Step 2, and the above control of Γ', we infer

$$\max_{0\le k\le k_{large}}\left(\int_{{}^{(int)}\mathcal{M}\cap\mathrm{Match}}\left|\mathfrak{d}^k(f,\underline{f},\log(\Upsilon\lambda))\right|^2\right)^{\frac{1}{2}} \lesssim \epsilon.$$

Step 4. We still need to control one more derivative of $(f,\underline{f},\log(\lambda))$. Repeating the process of Step 3, we use again the transformation formulas of Proposition 2.90 and then the final estimate of Step 3 for $(f,\underline{f},\log(\lambda))$ yields the following control for the curvature components:

$$\max_{0\le k\le k_{large}}\left(\int_{{}^{(int)}\mathcal{M}\cap\mathrm{Match}}\left|\mathfrak{d}^k\check{R}'\right|^2\right)^{\frac{1}{2}} \lesssim \epsilon.$$

Arguing as in Step 3, we infer[9]

$$\max_{0\le k\le k_{large}}\left(\int_{{}^{(int)}\mathcal{M}\cap\mathrm{Match}}\left|\mathfrak{d}^k\check{\Gamma}'\right|^2\right)^{\frac{1}{2}} \lesssim \epsilon+\left(\int_{\mathcal{T}}\left|\mathfrak{d}^{k_{large}}\left({}^{(ext)}\check{R}\right)\right|^2\right)^{\frac{1}{2}}.$$

Using again the transformation formulas of Proposition 2.90, this yields the following control for $(f,\underline{f},\log(\lambda))$

$$\max_{0\le k\le k_{large}+1}\left(\int_{{}^{(int)}\mathcal{M}\cap\mathrm{Match}}\left|\mathfrak{d}^k(f,\underline{f},\log(\Upsilon\lambda))\right|^2\right)^{\frac{1}{2}} \lesssim \epsilon.$$

We have finally obtained for $(f,\underline{f},\lambda)$ in ${}^{(int)}\mathcal{M}\cap\mathrm{Match}$

$$\max_{0\le k\le k_{small}-1}\ \sup_{{}^{(int)}\mathcal{M}\cap\mathrm{Match}} \underline{u}^{1+\delta_{dec}}\left|\mathfrak{d}^k(f,\underline{f},\log(\Upsilon\lambda))\right| \lesssim \epsilon,$$

$$\max_{0\le k\le k_{large}}\left(\int_{{}^{(int)}\mathcal{M}\cap\mathrm{Match}}\left|\mathfrak{d}^k(f,\underline{f},\log(\Upsilon\lambda))\right|^2\right)^{\frac{1}{2}} \lesssim \epsilon,$$

---

[9]In Step 3, there is no term corresponding to the one integrated on $\mathcal{T}$. This is due to the fact that for $k\le k_{large}-1$, we have thanks to the bootstrap assumptions on energy and a trace estimate

$$\max_{0\le k\le k_{large}-1}\left(\int_{\mathcal{T}}\left|\mathfrak{d}^k\left({}^{(ext)}\check{R}\right)\right|^2\right)^{\frac{1}{2}}\lesssim\epsilon.$$

and

$$\left(\int_{^{(int)}\mathcal{M}\cap\mathrm{Match}} \left|\mathfrak{d}^{k_{large}+1}(f,\underline{f},\log(\Upsilon\lambda))\right|^2\right)^{\frac{1}{2}} \lesssim \epsilon+\left(\int_{\mathcal{T}}\left|\mathfrak{d}^{k_{large}}({}^{(ext)}\check{R})\right|^2\right)^{\frac{1}{2}}.$$

Step 5. In addition to the estimate of $(f,\underline{f},\lambda)$ in ${}^{(int)}\mathcal{M}\cap$ Match of Step 4, we need to estimate $(f,\underline{f},\lambda)$ in ${}^{(ext)}\mathcal{M}\cap$ Match. To this end, we first control in ${}^{(ext)}\mathcal{M}\cap$ Match the reduced scalar $(f',\underline{f}',\lambda')$ satisfying

$$\begin{aligned}
e_3 &= \lambda'\left(e_3'+\underline{f}'e_\theta'+\frac{1}{4}\underline{f}'^2e_4'\right),\\
e_\theta &= \left(1+\frac{1}{2}f'\underline{f}'\right)e_\theta'+\frac{1}{2}f'e_3'+\frac{1}{2}\left(\underline{f}'+\frac{1}{4}f\underline{f}'^2\right)e_4',\\
e_4 &= \lambda'^{-1}\left(\left(1+\frac{1}{2}f'\underline{f}'+\frac{1}{16}f'^2\underline{f}'^2\right)e_4'+\left(f'+\frac{1}{4}f'^2\underline{f}'\right)e_\theta'+\frac{1}{4}f'^2e_3'\right).
\end{aligned}$$

Together with the initialization of the frame of ${}^{(ext)}\mathcal{M}$ and ${}^{(int)}\mathcal{M}$ on $\mathcal{T}$ in section 3.1.2 (where the spheres coincide), we have in particular

$$f'=\underline{f}'=0,\quad \lambda'=\Upsilon^{-1}\text{ on }\mathcal{T}.$$

Also, recall from section 3.5.1 that in order for (e_3,e_θ,e_4) to be defined everywhere on ${}^{(ext)}\mathcal{M}\cap$ Match, we need — in addition to the above initialization of $(f,\underline{f},\lambda)$ on $\mathcal{T}$ — to initialize it also on $\mathcal{C}_*\cap$ Match by

$$f'=\underline{f}'=0,\quad \lambda'=\Upsilon^{-1}\text{ on }\mathcal{C}_*\cap\mathrm{Match}. \tag{4.6.4}$$

Arguing similarly to Steps 1–4, we estimate $(f',\underline{f}',\lambda')$ and $(\check{\Gamma},\check{R})$ in ${}^{(ext)}\mathcal{M}\cap$Match. We obtain

$$\begin{aligned}
\max_{0\leq k\leq k_{small}-2}\sup_{^{(ext)}\mathcal{M}\cap\mathrm{Match}} u^{1+\delta_{dec}}\left|\mathfrak{d}^k(\check{\Gamma},\check{R})\right| &\\
+\max_{0\leq k\leq k_{large}-1}\left(\int_{^{(ext)}\mathcal{M}\cap\mathrm{Match}}\left|\mathfrak{d}^k(\check{\Gamma},\check{R})\right|^2\right)^{\frac{1}{2}} &\lesssim \epsilon,\\
\left(\int_{^{(ext)}\mathcal{M}\cap\mathrm{Match}}\left|\mathfrak{d}^{k_{large}}\check{R}\right|^2\right)^{\frac{1}{2}} &\lesssim \epsilon,\\
\max_{0\leq k\leq k_{small}-1}\sup_{^{(ext)}\mathcal{M}\cap\mathrm{Match}} u^{1+\delta_{dec}}\left|\mathfrak{d}^k(f',\underline{f}',\log(\Upsilon'\lambda'))\right| &\lesssim \epsilon,\\
\max_{0\leq k\leq k_{large}}\left(\int_{^{(ext)}\mathcal{M}\cap\mathrm{Match}}\left|\mathfrak{d}^k(f',\underline{f}',\log(\Upsilon'\lambda'))\right|^2\right)^{\frac{1}{2}} &\lesssim \epsilon,
\end{aligned}$$

and

$$\begin{aligned}
\left(\int_{^{(ext)}\mathcal{M}\cap\mathrm{Match}}\left|\mathfrak{d}^{k_{large}}\check{\Gamma}\right|^2\right)^{\frac{1}{2}} &\lesssim \epsilon+\left(\int_{\mathcal{T}}\left|\mathfrak{d}^{k_{large}}({}^{(ext)}\check{R})\right|^2\right)^{\frac{1}{2}},\\
\left(\int_{^{(ext)}\mathcal{M}\cap\mathrm{Match}}\left|\mathfrak{d}^{k_{large}+1}(f',\underline{f}',\log(\Upsilon'\lambda'))\right|^2\right)^{\frac{1}{2}} &\lesssim \epsilon+\left(\int_{\mathcal{T}}\left|\mathfrak{d}^{k_{large}}({}^{(ext)}\check{R})\right|^2\right)^{\frac{1}{2}}.
\end{aligned}$$

Step 6. As mentioned above, in addition to the estimate of $(f, \underline{f}, \lambda)$ of Step 4 in ${}^{(int)}\mathcal{M} \cap \text{Match}$, we need to estimate $(f, \underline{f}, \lambda)$ in ${}^{(ext)}\mathcal{M} \cap \text{Match}$. To this end, we derive simple algebraic relations between $(f, \underline{f}, \lambda)$ and $(f', \underline{f}', \lambda')$ of Step 5. On the one hand, we have from the definition of $(f, \underline{f}, \lambda)$

$$g(e_4', e_3) = -2\lambda, \quad g(e_4', e_\theta) = \lambda f, \quad g(e_\theta', e_4) = -f\left(1 + \frac{f\underline{f}}{4}\right), \quad g(e_\theta', e_3) = -\underline{f},$$
$$g(e_3', e_4) = -2\lambda^{-1}\left(1 + \frac{f\underline{f}}{2} + \frac{1}{16}f^2\underline{f}^2\right), \quad g(e_3', e_\theta) = \lambda^{-1}\underline{f}\left(1 + \frac{f\underline{f}}{4}\right).$$

On the other hand, we have from the definition of $(f', \underline{f}', \lambda')$

$$g(e_3, e_4') = -2\lambda', \quad g(e_3, e_\theta') = \lambda'\underline{f}', \quad g(e_\theta, e_4') = -f',$$
$$g(e_\theta, e_3') = -\underline{f}'\left(1 + \frac{f'\underline{f}'}{4}\right), \quad g(e_4, e_3') = -2\lambda'^{-1}\left(1 + \frac{f'\underline{f}'}{2} + \frac{1}{16}f'^2\underline{f}'^2\right),$$
$$g(e_4, e_\theta') = \lambda'^{-1}f'\left(1 + \frac{f'\underline{f}'}{4}\right).$$

We immediately infer

$$\lambda' = \lambda, \quad f' = -\lambda f, \quad \underline{f}' = -\lambda^{-1}\underline{f}.$$

In view of the estimates of Step 5, we infer

$$\max_{0 \le k \le k_{small}-1} \sup_{{}^{(ext)}\mathcal{M} \cap \text{Match}} u^{1+\delta_{dec}} \left|\mathfrak{d}^k(f, \underline{f}, \log(\Upsilon\lambda))\right| \lesssim \epsilon,$$
$$\max_{0 \le k \le k_{large}} \left(\int_{{}^{(ext)}\mathcal{M} \cap \text{Match}} \left|\mathfrak{d}^k(f, \underline{f}, \log(\Upsilon\lambda))\right|^2\right)^{\frac{1}{2}} \lesssim \epsilon,$$

and

$$\left(\int_{{}^{(ext)}\mathcal{M} \cap \text{Match}} \left|\mathfrak{d}^{k_{large}+1}(f, \underline{f}, \log(\Upsilon\lambda))\right|^2\right)^{\frac{1}{2}} \lesssim \epsilon.$$

Together with Step 4, this yields

$$\max_{0 \le k \le k_{small}-1} \sup_{{}^{(int)}\mathcal{M} \cap \text{Match}} \underline{u}^{1+\delta_{dec}} \left|\mathfrak{d}^k(f, \underline{f}, \log(\Upsilon\lambda))\right| \lesssim \epsilon,$$
$$\max_{0 \le k \le k_{small}-1} \sup_{{}^{(ext)}\mathcal{M} \cap \text{Match}} u^{1+\delta_{dec}} \left|\mathfrak{d}^k(f, \underline{f}, \log(\Upsilon\lambda))\right| \lesssim \epsilon,$$
$$\max_{0 \le k \le k_{large}} \left(\int_{\text{Match}} \left|\mathfrak{d}^k(f, \underline{f}, \log(\Upsilon\lambda))\right|^2\right)^{\frac{1}{2}} \lesssim \epsilon,$$

and

$$\left(\int_{\text{Match}} \left|\mathfrak{d}^{k_{large}+1}(f, \underline{f}, \log(\Upsilon\lambda))\right|^2\right)^{\frac{1}{2}} \lesssim \epsilon + \left(\int_{\mathcal{T}} \left|\mathfrak{d}^{k_{large}}({}^{(ext)}\check{R})\right|^2\right)^{\frac{1}{2}}.$$

Step 7. Next, we estimate $r' - r$ and $m' - m$. Note first that in view of the initialization of the foliations of ${}^{(ext)}\mathcal{M}$ and ${}^{(int)}\mathcal{M}$ on $\mathcal{T}$, as well as the initializations (4.6.3) on $\underline{\mathcal{C}}_* \cap \text{Match}$ and (4.6.4) on $\mathcal{C}_* \cap \text{Match}$, we have

$$r' = r, \quad m' = m \text{ on } \mathcal{T} \cup \text{Match}. \tag{4.6.5}$$

We start with the region ${}^{(int)}\mathcal{M} \cap \text{Match}$. We have

$$\begin{aligned} e_4'(r') &= \frac{r'}{2}\overline{\kappa}' = 1 + \frac{r'}{2}\left(\overline{\kappa}' - \frac{2}{r'}\right), \\ e_3'(r') &= \frac{r'}{2}(\underline{\overline{\kappa}}' + \underline{A}') = -\Upsilon' + \frac{r'}{2}\left(\underline{\overline{\kappa}}' + \frac{2\Upsilon'}{r'}\right) + \frac{r'}{2}\underline{A}', \end{aligned}$$

which together with the identities for $e_4'(m')$ and $e_3'(m')$ in the outgoing foliation of ${}^{(ext)}\mathcal{M}$ and the control of the foliation of ${}^{(ext)}\mathcal{M}$ in ${}^{(int)}\mathcal{M} \cap \text{Match}$ established in Step 4 yields, using also $e_\theta'(r') = e_\theta'(m') = 0$,

$$\begin{aligned} \max_{0\leq k\leq k_{small}-2} \sup_{{}^{(int)}\mathcal{M}\cap\text{Match}} \underline{u}^{1+\delta_{dec}} \Big| \mathfrak{d}^k \big(e_4'(r') - 1, e_3'(r') + \Upsilon', e_\theta'(r'), & \\ e_4'(m'), e_3'(m'), e_\theta'(m')\big)\Big| &\lesssim \epsilon, \\ \max_{0\leq k\leq k_{large}-1} \left(\int_{{}^{(int)}\mathcal{M}\cap\text{Match}} \Big| \mathfrak{d}^k \big(e_4'(r') - 1, e_3'(r') + \Upsilon', e_\theta'(r'), e_4'(m'), \right. & \\ \left. e_3'(m'), e_\theta'(m')\big)\Big|^2\right)^{\frac{1}{2}} &\lesssim \epsilon. \end{aligned}$$

On the other hand, we have in view of the decomposition of e_4', e_3' and e_θ' of Step 1

$$\begin{aligned} e_4'(r) &= \lambda\left(e_4 + f e_\theta + \frac{1}{4}f^2 e_3\right) r \\ &= \lambda\left(\frac{r}{2}(\overline{\kappa} + A) + \frac{1}{4}f^2 e_3(r)\right) \\ &= 1 + \Big(\lambda\Upsilon - 1\Big) + \lambda\left(\frac{r}{2}\left(\overline{\kappa} - \frac{2\Upsilon}{r}\right) + \frac{r}{2}A + \frac{1}{4}f^2 e_3(r)\right), \end{aligned}$$

$$\begin{aligned} e_4'(m) &= \lambda\left(e_4 + f e_\theta + \frac{1}{4}f^2 e_3\right) m \\ &= \lambda\left(e_4(m) + \frac{1}{4}f^2 e_3(m)\right), \end{aligned}$$

$$
\begin{aligned}
e_3'(r) &= \lambda^{-1}\left(\left(1+\frac{1}{2}f\underline{f}+\frac{1}{16}f^2\underline{f}^2\right)e_3+\left(\underline{f}+\frac{1}{4}\underline{f}^2 f\right)e_\theta+\frac{\underline{f}^2}{4}e_4\right)r\\
&= \lambda^{-1}\left(e_3(r)+\left(\frac{1}{2}f\underline{f}+\frac{1}{16}f^2\underline{f}^2\right)e_3(r)+\frac{\underline{f}^2}{4}e_4(r)\right)\\
&= -\Upsilon+\lambda^{-1}(\lambda\Upsilon-1)+\lambda^{-1}\left(\frac{r}{2}\left(\overline{\underline{\kappa}}+\frac{2}{r}\right)+\left(\frac{1}{2}f\underline{f}+\frac{1}{16}f^2\underline{f}^2\right)e_3(r)\right.\\
&\quad\left.+\frac{\underline{f}^2}{4}e_4(r)\right),
\end{aligned}
$$

$$
\begin{aligned}
e_3'(m) &= \lambda^{-1}\left(\left(1+\frac{1}{2}f\underline{f}+\frac{1}{16}f^2\underline{f}^2\right)e_3+\left(\underline{f}+\frac{1}{4}\underline{f}^2 f\right)e_\theta+\frac{\underline{f}^2}{4}e_4\right)m\\
&= \lambda^{-1}\left(\left(1+\frac{1}{2}f\underline{f}+\frac{1}{16}f^2\underline{f}^2\right)e_3(m)+\frac{\underline{f}^2}{4}e_4(m)\right),
\end{aligned}
$$

$$
\begin{aligned}
e_\theta'(r) &= \left(\left(1+\frac{1}{2}f\underline{f}\right)e_\theta+\frac{f}{2}e_4+\frac{\underline{f}}{2}\left(1+\frac{f\underline{f}}{4}\right)e_3\right)r\\
&= \frac{f}{2}e_4(r)+\frac{\underline{f}}{2}\left(1+\frac{f\underline{f}}{4}\right)e_3(r),
\end{aligned}
$$

and

$$
\begin{aligned}
e_\theta'(r) &= \left(\left(1+\frac{1}{2}f\underline{f}\right)e_\theta+\frac{f}{2}e_4+\frac{\underline{f}}{2}\left(1+\frac{f\underline{f}}{4}\right)e_3\right)m\\
&= \frac{f}{2}e_4(m)+\frac{\underline{f}}{2}\left(1+\frac{f\underline{f}}{4}\right)e_3(m).
\end{aligned}
$$

Together with the identities for $e_4(m)$ and $e_3(m)$ in the ingoing foliation of ${}^{(int)}\mathcal{M}$, the final estimates of Step 6 for f and λ, and the bootstrap assumptions for the foliation of ${}^{(int)}\mathcal{M}$, we infer

$$
\begin{aligned}
\max_{0\le k\le k_{small}-2}\ \sup_{{}^{(int)}\mathcal{M}\cap\mathrm{Match}} \underline{u}^{1+\delta_{dec}}\Big|\mathfrak{d}^k(e_4'(r)-1, e_3'(r)+\Upsilon, e_\theta'(r), e_4'(m),&\\
e_3'(m), e_\theta'(m))\Big| &\lesssim \epsilon,\\
\max_{0\le k\le k_{large}-1}\left(\int_{{}^{(int)}\mathcal{M}\cap\mathrm{Match}}\Big|\mathfrak{d}^k(e_4'(r)-1, e_3'(r)+\Upsilon, e_\theta'(r), e_4'(m),\right.&\\
\left. e_3'(m), e_\theta'(m))\Big|^2\right)^{\frac{1}{2}} &\lesssim \epsilon.
\end{aligned}
$$

We deduce

$$\max_{0\leq k\leq k_{small}-2}\ \sup_{{}^{(int)}\mathcal{M}\cap\mathrm{Match}} \underline{u}^{1+\delta_{dec}}\left|\mathfrak{d}^k(e_4'(r'-r), e_\theta'(r-r'), \mathfrak{d}(m'-m))\right| \lesssim \epsilon,$$

$$\max_{0\leq k\leq k_{large}-1}\left(\int_{{}^{(int)}\mathcal{M}\cap\mathrm{Match}}\left|\mathfrak{d}^k(e_4'(r'-r), e_\theta'(r-r'), \mathfrak{d}(m'-m))\right|^2\right)^{\frac{1}{2}} \lesssim \epsilon.$$

In particular, we have

$$\sup_{{}^{(int)}\mathcal{M}\cap\mathrm{Match}} \underline{u}^{1+\delta_{dec}}\left|(e_4'(r'-r), e_4'(m'-m))\right| \lesssim \epsilon,$$

and together with the initialization (4.6.5), we integrate the transport equation from $\mathcal{T}\cup({}^{(int)}\mathcal{M}\cap\mathrm{Match})$ and obtain

$$\sup_{{}^{(int)}\mathcal{M}\cap\mathrm{Match}} \underline{u}^{1+\delta_{dec}}\left|(r'-r, m'-m)\right| \lesssim \epsilon.$$

Together with the above estimates, and recovering the $e_3'(r'-r)$ using

$$e_3'(r'-r) = \Big(e_3'(r')+\Upsilon'\Big)-\Big(e_3'(r)+\Upsilon\Big)+2\left(\frac{m'}{r'}-\frac{m}{r}\right),$$

we infer

$$\max_{0\leq k\leq k_{small}-1}\ \sup_{{}^{(int)}\mathcal{M}\cap\mathrm{Match}} \underline{u}^{1+\delta_{dec}}\left|\mathfrak{d}^k(r'-r, m'-m)\right| \lesssim \epsilon,$$

$$\max_{0\leq k\leq k_{large}}\left(\int_{{}^{(int)}\mathcal{M}\cap\mathrm{Match}}\left|\mathfrak{d}^k(r'-r, m'-m)\right|^2\right)^{\frac{1}{2}} \lesssim \epsilon.$$

Finally, arguing similarly in the region ${}^{(ext)}\mathcal{M}\cap\mathrm{Match}$, we infer

$$\max_{0\leq k\leq k_{small}-1}\ \sup_{{}^{(ext)}\mathcal{M}\cap\mathrm{Match}} u^{1+\delta_{dec}}\left|\mathfrak{d}^k(r'-r, m'-m)\right| \lesssim \epsilon,$$

$$\max_{0\leq k\leq k_{large}}\left(\int_{{}^{(ext)}\mathcal{M}\cap\mathrm{Match}}\left|\mathfrak{d}^k(r'-r, m'-m)\right|^2\right)^{\frac{1}{2}} \lesssim \epsilon,$$

and hence

$$\max_{0\leq k\leq k_{small}-1}\ \sup_{{}^{(int)}\mathcal{M}\cap\mathrm{Match}} \underline{u}^{1+\delta_{dec}}\left|\mathfrak{d}^k(r'-r, m'-m)\right| \lesssim \epsilon,$$

$$\max_{0\leq k\leq k_{small}-1}\ \sup_{{}^{(ext)}\mathcal{M}\cap\mathrm{Match}} u^{1+\delta_{dec}}\left|\mathfrak{d}^k(r'-r, m'-m)\right| \lesssim \epsilon,$$

$$\max_{0\leq k\leq k_{large}}\left(\int_{\mathrm{Match}}\left|\mathfrak{d}^k(r'-r, m'-m)\right|^2\right)^{\frac{1}{2}} \lesssim \epsilon.$$

Step 8. Recall from Definition 4.12 that we have defined the global null frame $(e_4'', e_3'', e_\theta'')$ as

- In ${}^{(int)}\mathcal{M}\setminus\mathrm{Match}$, $(e_4'', e_3'', e_\theta'') = (e_4, e_3, e_\theta)$.
- In ${}^{(ext)}\mathcal{M}\setminus\mathrm{Match}$, $(e_4'', e_3'', e_\theta'') = (\Upsilon e_4', \Upsilon^{-1}e_3', e_\theta')$.
- In Match, $(e_4'', e_3'', e_\theta'')$ is given by the change of frame formula starting from

(e_4, e_3, e_θ) and with change of frame coefficients $(f'', \underline{f}'', \lambda'')$ given by

$$f'' = \psi f, \quad \underline{f}'' = \psi \underline{f}, \quad \lambda'' = 1 - \psi + \psi \Upsilon' \lambda,$$

see (4.6.1).

Also, recall that we have defined r'' and m'' as

$$r'' = (1-\psi)r + \psi r', \quad m'' = (1-\psi)m + \psi m'.$$

Step 9. In view of the transformation formulas of Proposition 2.90, we have schematically

$$(\check{\Gamma}'', \check{R}'') = (\check{\Gamma}, \check{R}) + \mathfrak{d}(f'', \underline{f}'', \lambda'' - 1) + f'' + \underline{f}'' + (\lambda'' - 1) + (r'' - r) + (m'' - m).$$

In view of the definition of $(f'', \underline{f}'', \lambda'')$ and (r'', m'') in Step 8, we infer

$$(\check{\Gamma}'', \check{R}'') = (\check{\Gamma}, \check{R}) + \mathfrak{d}(f, \underline{f}, \Upsilon\lambda - 1) + f + \underline{f} + (\Upsilon\lambda - 1) + (r' - r) + (m' - m).$$

Together with the bootstrap assumptions in ${}^{(int)}\mathcal{M}$ for $(\Gamma, \check{R})$, the estimates for $(\Gamma, \check{R})$ in ${}^{(ext)}\mathcal{M}$ provided by Step 5, the estimates for $(f, \underline{f}, \lambda)$ provided by Step 6 in Match, and the estimates for $r' - r$ and $m' - m$ provided by Step 7, we deduce

$$\begin{aligned}
\max_{0 \leq k \leq k_{small}-2} \sup_{{}^{(int)}\mathcal{M} \cap \text{Match}} \underline{u}^{1+\delta_{dec}} \left| \mathfrak{d}^k (\check{\Gamma}'', \check{R}'') \right| &\lesssim \epsilon, \\
\max_{0 \leq k \leq k_{small}-2} \sup_{{}^{(ext)}\mathcal{M} \cap \text{Match}} u^{1+\delta_{dec}} \left| \mathfrak{d}^k (\check{\Gamma}'', \check{R}'') \right| &\lesssim \epsilon, \\
\max_{0 \leq k \leq k_{large}-1} \left(\int_{\text{Match}} \left| \mathfrak{d}^k (\check{\Gamma}'', \check{R}'') \right|^2 \right)^{\frac{1}{2}} &\lesssim \epsilon, \\
\left(\int_{\text{Match}} \left| \mathfrak{d}^{k_{large}} (\check{\Gamma}'', \check{R}'') \right|^2 \right)^{\frac{1}{2}} &\lesssim \epsilon + \left(\int_{\mathcal{T}} \left| \mathfrak{d}^{k_{large}} ({}^{(ext)}\check{R}) \right|^2 \right)^{\frac{1}{2}}.
\end{aligned}$$

Since the double-primed quantities correspond to the quantities associated to the global frame, this concludes the proof of Lemma 4.16.

4.6.3 Proof of Proposition 3.26

To match the first global frame of $\mathcal{M}$ of Proposition 3.26 with a conformal renormalization of the second frame of ${}^{(ext)}\mathcal{M}$ of Proposition 3.20, we will need to introduce a cut-off function.

Definition 4.18. *Let $\psi : \mathbb{R} \to \mathbb{R}$ a smooth cut-off function such that $0 \leq \psi \leq 1$, $\psi = 0$ on $(-\infty, 0]$ and $\psi = 1$ on $[1, +\infty)$. We define ψ_{m_0} as follows:*

$$\psi_{m_0}(r) = \begin{cases} 1 & \text{if} \quad r \geq 4m_0, \\ 0 & \text{if} \quad r \leq \frac{7m_0}{2}, \end{cases}$$

and

$$\psi_{m_0}(r) = \psi\left(\frac{2\left(r-\frac{7m_0}{2}\right)}{m_0}\right) \quad on \ \frac{7m_0}{2} \leq {}^{(ext)}r \leq 4m_0.$$

We are now ready to define the second global frame, i.e., the global frame of the statement of Proposition 3.26.

Definition 4.19 (Definition of the second global frame). *We introduce a global null frame defined on ${}^{(ext)}\mathcal{M} \cup {}^{(int)}\mathcal{M}$ and denoted by $({}^{(glo')}e_4, {}^{(glo')}e_3, {}^{(glo')}e_\theta)$. The second global frame is defined as follows:*

1. *In ${}^{(ext)}\mathcal{M} \cap \{{}^{(ext)}r \geq 4m_0\}$, we have*

$$({}^{(glo')}e_4, {}^{(glo')}e_3, {}^{(glo')}e_\theta) = \left({}^{(ext)}\Upsilon e_4', {}^{(ext)}\Upsilon^{-1}e_3', e_\theta'\right),$$

where (e_4', e_3', e_θ') denotes the second frame of ${}^{(ext)}\mathcal{M}$, i.e., the one constructed in Proposition 3.20.

2. *In ${}^{(int)}\mathcal{M} \cup ({}^{(ext)}\mathcal{M} \cap \{{}^{(ext)}r \leq \frac{7m_0}{2}\})$, we have*

$$({}^{(glo')}e_4, {}^{(glo')}e_3, {}^{(glo')}e_\theta) = \left({}^{(glo)}e_4, {}^{(glo)}e_3, {}^{(glo)}e_\theta\right),$$

where $\left({}^{(glo)}e_4, {}^{(glo)}e_3, {}^{(glo)}e_\theta\right)$ denotes the first global frame of $\mathcal{M}$ of Proposition 3.26.

3. *It remains to define the global frame on the matching region Match'. We denote by f the reduced scalar introduced in Proposition 3.20 such that we have in ${}^{(ext)}\mathcal{M}$*

$$\begin{aligned} e_4' &= {}^{(ext)}e_4 + f\,{}^{(ext)}e_\theta + \frac{1}{4}f^2\,{}^{(ext)}e_3, \\ e_\theta' &= {}^{(ext)}e_\theta + \frac{f}{2}\,{}^{(ext)}e_3, \\ e_3' &= {}^{(ext)}e_3. \end{aligned}$$

Then, in the matching region Match', the second global frame of $\mathcal{M}$ is given by

$$\begin{aligned} {}^{(glo')}e_4 &= \Upsilon'\left(\Upsilon'^{-1}\,{}^{(glo)}e_4 + f'\,{}^{(glo)}e_\theta + \frac{1}{4}f'^2\Upsilon'\,{}^{(glo)}e_3\right), \\ {}^{(glo')}e_\theta &= {}^{(glo)}e_\theta + \frac{f'}{2}\Upsilon'\,{}^{(glo)}e_3, \\ {}^{(glo')}e_3 &= {}^{(glo)}e_3, \end{aligned}$$

where

$$f' = \psi_{m_0}({}^{(ext)}r)f, \qquad \Upsilon' = 1 - \psi_{m_0}({}^{(ext)}r) + \psi_{m_0}({}^{(ext)}r)\,{}^{(ext)}\Upsilon. \tag{4.6.6}$$

Remark 4.20. *Recall that the smooth cut-off function ψ in Definition 3.25, allowing to define $\psi_{m_0,\delta_\mathcal{H}}$, is such that we have in particular $\psi = 0$ on $(-\infty, 0]$ and $\psi = 1$ on $[1, +\infty)$. The following two special cases correspond to the properties (d) i. and (d) ii. of Proposition 3.26.*

- *If the cut-off ψ in Definition 3.25 is such that $\psi = 1$ on $[1/2, +\infty)$, then*

$$
\begin{aligned}
({}^{(glo')}e_4, {}^{(glo')}e_3, {}^{(glo')}e_\theta) &= \left({}^{(ext)}\Upsilon e_4', {}^{(ext)}\Upsilon^{-1}e_3', e_\theta'\right) \\
&\quad \text{on } {}^{(ext)}\mathcal{M}\left({}^{(ext)}r \geq \frac{15m_0}{4}\right).
\end{aligned}
$$

- *If the cut-off ψ in Definition 3.25 is such that $\psi = 0$ on $(-\infty, 1/2]$, then*

$$
\begin{aligned}
({}^{(glo')}e_4, {}^{(glo')}e_3, {}^{(glo')}e_\theta) &= \left({}^{(glo)}e_4, {}^{(glo)}e_3, {}^{(glo)}e_\theta\right) \\
&\quad \text{on } {}^{(int)}\mathcal{M} \cup {}^{(ext)}\mathcal{M}\left({}^{(ext)}r \leq \frac{15m_0}{4}\right).
\end{aligned}
$$

Remark 4.21. *When dealing with the second global frame $({}^{(glo')}e_4, {}^{(glo')}e_3, {}^{(glo')}e_\theta)$, the area radius and Hawking mass that we use are the ones corresponding to the first global frame, i.e., ${}^{(glo)}r$ and ${}^{(glo)}m$.*

The following two lemmas provide the main properties of the second global frame of $\mathcal{M}$.

Lemma 4.22. *We have in ${}^{(ext)}\mathcal{M}(r \geq 4m_0)$ the following relations between the quantities in the second global frame of $\mathcal{M}$, i.e., $({}^{(glo')}e_4, {}^{(glo')}e_3, {}^{(glo')}e_\theta)$, and the second frame of ${}^{(ext)}\mathcal{M}$, i.e., (e_4', e_3', e_θ'),*

$$
\begin{aligned}
&{}^{(glo')}\alpha = \Upsilon^2\alpha', \quad {}^{(glo')}\beta = \Upsilon\beta', \quad {}^{(glo')}\rho + \frac{2m}{r^3} = \rho' + \frac{2m}{r^3}, \quad {}^{(glo')}\underline{\beta} = \Upsilon^{-1}\underline{\beta}', \\
&{}^{(glo')}\underline{\alpha} = \Upsilon^{-2}\underline{\alpha}', \quad {}^{(glo')}\xi = 0, \quad {}^{(glo')}\underline{\xi} = \Upsilon^{-2}\underline{\xi}', \quad {}^{(glo')}\zeta = -{}^{(glo')}\underline{\eta} = \zeta', \\
&{}^{(glo')}\eta = \eta', \quad {}^{(glo')}\omega + \frac{m}{r^2} = \Upsilon\omega' + \frac{m}{r^2}(1 - e_4'(r)) + \frac{e_4'(m)}{r}, \\
&{}^{(glo')}\underline{\omega} = \Upsilon^{-1}\left(\underline{\omega}' - \frac{m}{r^2} + \frac{m}{r^2}\left(1 - \frac{e_3'(r)}{\Upsilon}\right) - \frac{e_3'(m)}{\Upsilon r}\right), \\
&{}^{(glo')}\kappa - \frac{2\Upsilon}{r} = \Upsilon\left(\kappa' - \frac{2}{r}\right), \quad {}^{(glo')}\underline{\kappa} + \frac{2}{r} = \Upsilon^{-1}\left(\underline{\kappa}' + \frac{2\Upsilon}{r}\right), \\
&{}^{(glo')}\vartheta = \Upsilon\vartheta', \quad {}^{(glo')}\underline{\vartheta} = \Upsilon^{-1}\underline{\vartheta}'.
\end{aligned}
$$

Proof. The proof follows immediately from the change of frame formula with the choice $(f = 0, \underline{f} = 0, \lambda = \Upsilon)$, the fact that $e_\theta(\Upsilon) = 0$, and the fact that the frame (e_4', e_3', e_θ') is such that $\xi' = 0$ and $\underline{\eta}' = -\zeta'$. □

Lemma 4.23 (Control of the second global frame in the matching region)**.** *In the matching region, the following estimate holds for the second global frame*

$$
\max_{0 \leq k \leq k_{small}+k_{loss}} \sup_{Match'} u^{1+\delta_{dec}-2\delta_0}\left|\mathfrak{d}^k\left({}^{(glo')}\check{\Gamma}, {}^{(glo')}\check{R}\right)\right| \lesssim \epsilon.
$$

Remark 4.24. *The quantities associated to the second global frame can be estimated as follows:*

- *In ${}^{(int)}\mathcal{M} \cup {}^{(ext)}\mathcal{M}({}^{(ext)}r \leq \frac{7m_0}{2})$, the second global frame coincides with the first global frame, and hence, the quantities associated to the second global frame* *satisfy the same estimates as the corresponding quantities for the first global frame.*

- *In* $^{(ext)}\mathcal{M}(^{(ext)}r \geq 4m_0)$*, estimates for the quantities associated to the second global frame follow from the identities of Lemma 4.22 together with the estimates of Proposition 3.20 for the second frame of* $^{(ext)}\mathcal{M}$*.*
- *In Match′, the estimates for the quantities associated to the global frame are provided by Lemma 4.23.*

The proof of Proposition 3.26 easily follows from Definition 4.19, Remark 4.20, and Lemma 4.23. Thus, from now on, we focus on the proof of Lemma 4.23 which is carried out below.

Proof of Lemma 4.23. Recall from Definition 4.19 that we have in the matching region Match′

$$\begin{aligned}
^{(glo')}e_4 &= \Upsilon'\left(\Upsilon'^{-1}\,{}^{(glo)}e_4 + f'\,{}^{(glo)}e_\theta + \frac{1}{4}f'^2\Upsilon'\,{}^{(glo)}e_3\right),\\
^{(glo')}e_\theta &= {}^{(glo)}e_\theta + \frac{f'}{2}\Upsilon'\,{}^{(glo)}e_3,\\
^{(glo')}e_3 &= {}^{(glo)}e_3,
\end{aligned}$$

where

$$f' = \psi_{m_0}(^{(ext)}r)f, \qquad \Upsilon' = 1 - \psi_{m_0}(^{(ext)}r) + \psi_{m_0}(^{(ext)}r)\,{}^{(ext)}\Upsilon.$$

Now, since $^{(ext)}r \geq \frac{7m_0}{2}$ on Match′, we also have in that region

$$(^{(glo)}e_4, {}^{(glo)}e_3, {}^{(glo)}e_\theta) = (^{(ext)}\Upsilon\,{}^{(ext)}e_4, (^{(ext)}\Upsilon)^{-1}\,{}^{(ext)}e_3, {}^{(ext)}e_\theta).$$

We deduce on Match′

$$\begin{aligned}
^{(glo')}e_4 &= {}^{(ext)}\Upsilon\left({}^{(ext)}e_4 + f''\,{}^{(ext)}e_\theta + \frac{1}{4}f''^2\,{}^{(ext)}e_3\right),\\
^{(glo')}e_\theta &= {}^{(ext)}e_\theta + \frac{f''}{2}\,{}^{(ext)}e_3,\\
^{(glo')}e_3 &= (^{(ext)}\Upsilon)^{-1}\,{}^{(ext)}e_3,
\end{aligned}$$

where

$$\begin{aligned}
f'' &= \Upsilon'(^{(ext)}\Upsilon)^{-1}f'\\
&= \left(1 - \psi_{m_0}(^{(ext)}r) + \psi_{m_0}(^{(ext)}r)\,{}^{(ext)}\Upsilon\right)(^{(ext)}\Upsilon)^{-1}\psi_{m_0}(^{(ext)}r)f.
\end{aligned}$$

In view of the transformation formulas of Proposition 2.90, we deduce, schematically,

$$\left(^{(glo')}\check{\Gamma}, {}^{(glo')}\check{R}\right) = \left(^{(ext)}\check{\Gamma}, {}^{(ext)}\check{R}\right) + \mathfrak{d}f + f.$$

Together with the bootstrap assumptions on decay and Proposition 3.19 for $^{(ext)}\check{\Gamma}$

and ${}^{(ext)}\check{R}$, and the estimate (3.4.11) for f, we infer

$$\max_{0\leq k\leq k_{small}+k_{loss}}\sup_{\text{Match}'} u^{1+\delta_{dec}-2\delta_0}\left|\mathfrak{d}^k\left({}^{(glo')}\check{\Gamma}, {}^{(glo')}\check{R}\right)\right| \lesssim \epsilon$$

which concludes the proof of Lemma 4.23. □

Chapter Five

Decay Estimates for $\mathfrak{q}$ (Theorem M1)

The goal of this chapter is to prove Theorem M1, i.e., to derive decay estimates for the quantity $\mathfrak{q}$ for $k \leq k_{small} + 20$ derivatives. To this end, we will make use of the wave equation satisfied by $\mathfrak{q}$ (see (2.4.7))

$$\square_2 \mathfrak{q} + \kappa\underline{\kappa}\,\mathfrak{q} = N, \tag{5.0.1}$$

where N contains only quadratic or higher order terms. Now, in order to have a suitable right-hand side N, recall from the discussion in Remarks 2.109 and 2.110 that $\mathfrak{q}$ is defined relative to the global null frame of Proposition 3.26 for which $\xi = 0$ for $r \geq 4m_0$ and $\eta \in \Gamma_g$. For such a global fame, N is given schematically by, see (2.4.8),

$$N = r^2 \mathfrak{d}^{\leq 2}(\Gamma_g \cdot (\alpha, \beta)) + e_3\Big(r^3 \mathfrak{d}^{\leq 2}(\Gamma_g \cdot (\alpha, \beta))\Big) + \mathfrak{d}^{\leq 1}(\Gamma_g \cdot \mathfrak{q}) + \text{l.o.t.} \tag{5.0.2}$$

5.1 PRELIMINARIES

5.1.0.1 Smallness constants

Recall from the beginning of section 3.3.2 the constant m_0 and the main small constants $\delta_{\mathcal{H}}$, δ_B, δ_{dec}, ϵ and ϵ_0 such that

- The constant $m_0 > 0$ is the mass of the initial Schwarzschild spacetime relative to which our initial perturbation is measured.
- The integer k_{large} which corresponds to the maximum number of derivatives of the solution.
- The size of the initial data layer norm is measured by $\epsilon_0 > 0$.
- The size of the bootstrap assumption norms are measured by $\epsilon > 0$.
- $\delta_{\mathcal{H}} > 0$ measures the width of the region $|r - 2m_0| \leq 2m_0\delta_{\mathcal{H}}$ where the redshift estimate holds and which includes in particular the region ${}^{(int)}\mathcal{M}$.
- δ_{dec} is tied to decay estimates in u, $\underline{u}$ for $\check{\Gamma}$ and $\check{R}$.
- δ_B is involved in the r-power of the r^p-weighted estimates for curvature.

Recall also that these constants satisfy in view of (3.3.1), (3.3.2), and (3.3.3)

$$0 < \delta_{\mathcal{H}},\ \delta_{dec},\ \delta_B \ll \min\{m_0, 1\}, \qquad \delta_B > 2\delta_{dec}, \qquad k_{large} \gg \frac{1}{\delta_{dec}},$$

$$\epsilon_0, \epsilon \ll \min\{\delta_{\mathcal{H}}, \delta_{dec}, \delta_B, m_0, 1\},$$

and

$$\epsilon = \epsilon_0^{\frac{2}{3}}.$$

We will need the following additional small constants in this chapter:

- $\delta_{extra} > 0$, tied to the decay of $\mathfrak{q}$, and is chosen such that $\delta_{extra} > \delta_{dec}$,
- $\delta > 0$ for various degeneracies,
- $\delta_0 > 0$ which comes from interpolating between $k \le k_{small}$ derivatives of $(\check{\Gamma}, \check{R})$ and $k \le k_{large}$ derivatives of $(\check{\Gamma}, \check{R})$, see Lemma 5.1,
- $q_0 > 0$ which will allow us to recover the fact that the decay for $\mathfrak{q}$ in Theorem M1 has an extra gain $u^{-(\delta_{extra}-\delta_{dec})}$ compared to the expected behavior inferred from the bootstrap assumptions.

We will choose δ_{extra} such that

$$\delta_{dec} < \delta_{extra} < 2\delta_{dec}, \quad \delta_B \ge 2\delta_{extra},$$

δ and δ_0 such that

$$0 < \epsilon, \epsilon_0 \ll \delta, \delta_0 \ll \delta_{dec}, \delta_{extra}, \delta_{\mathcal{H}}, m_0, 1, \tag{5.1.1}$$

and q_0 such that[1]

$$2\delta_{dec} < q_0 < 4\delta_{dec} - 4\delta_0 - 4\delta. \tag{5.1.2}$$

5.1.1 The foliation of $\mathcal{M}$ by τ

Recall that the spacetime $\mathcal{M}$ is decomposed as $\mathcal{M} = {}^{(int)}\mathcal{M} \cup {}^{(ext)}\mathcal{M}$ and that u is an outgoing optical function on ${}^{(ext)}\mathcal{M}$ while $\underline{u}$ is an ingoing optical function. In this chapter, we rely on the global frame $(e_3, e_4, e_\theta, e_\varphi)$ defined in section 3.5, and r and m denote the corresponding scalar functions associated to it. Also, we define the trapping region $\mathcal{M}_{trap}$ as

$$\mathcal{M}_{trap} \quad := \quad \left\{ \frac{5m_0}{2} \le r \le \frac{7m_0}{2} \right\}. \tag{5.1.3}$$

Also, let ${}^{(\not{trap})}\mathcal{M} = \mathcal{M} \setminus {}^{(trap)}\mathcal{M}$ the complement of ${}^{(trap)}\mathcal{M}$ in $\mathcal{M}$.

We foliate our spacetime domain $\mathcal{M}$ by $\mathbf{Z}$-invariant hypersurfaces $\Sigma(\tau)$ which are:

- Incoming null in ${}^{(int)}\mathcal{M}$, with e_3 as null incoming generator. We denote this portion ${}^{(int)}\Sigma(\tau)$.
- Strictly spacelike in ${}^{(trap)}\mathcal{M}$. We denote this portion by ${}^{(trap)}\Sigma$.
- Outgoing null in $\mathcal{M}_{>4m_0}$. We denote this portion by $\Sigma_{>4m_0}(\tau)$.

[1]This will allow us to choose in the proof of Theorem M1, see (5.2.10),

$$\delta_{extra} = \frac{q_0 - \delta}{2}$$

which satisfies the desired estimate $\delta_{extra} > \delta_{dec}$ for $\delta > 0$ small enough.

- The parameter τ of $\Sigma(\tau)$ can be chosen, smoothly, such that

$$\tau := \begin{cases} u & \text{in } \mathcal{M}_{>4m_0}, \\ u+r & \text{in } \mathcal{M}_{trap}, \\ \underline{u} & \text{in } {}^{(int)}\mathcal{M}. \end{cases} \tag{5.1.4}$$

- In particular, the unit normal in the region $\mathcal{M}_{trap}$, i.e., the normal to ${}^{(trap)}\Sigma$, satisfies[2]

$$-2 \leq g(N_\Sigma, e_4) \leq -1, \quad -2 \leq g(N_\Sigma, e_3) \leq -1 \text{ on } \mathcal{M}_{trap}. \tag{5.1.5}$$

5.1.2 Assumptions for Ricci coefficients and curvature

Recall from Remark 2.110 that $\mathfrak{q}$ is defined, according to equation (2.3.10) in Lemma 2.96, relative to the global frame of Proposition 3.26 for which $\eta \in \Gamma_g$ with the notation

$$\begin{aligned} \Gamma_g &= \Gamma_g^{(0)} = \left\{ \xi,\, \vartheta,\, \omega + \frac{m}{r^2},\, \kappa - \frac{2\Upsilon}{r},\, \eta,\, \underline{\eta},\, \zeta,\, A \right\}, \\ \Gamma_b &= \Gamma_b^{(0)} = \left\{ \underline{\vartheta},\, \underline{\kappa} + \frac{2}{r},\, \underline{A},\, \underline{\omega},\, \underline{\xi} \right\}, \end{aligned}$$

where we recall that

$$\Upsilon = 1 - \frac{2m}{r}, \quad A = \frac{2}{r} e_4(r) - \kappa, \quad \underline{A} = \frac{2}{r} e_3(r) - \underline{\kappa}.$$

Note also that ξ vanishes in ${}^{(ext)}\mathcal{M}$ away from the matching region of Proposition 3.26, and in particular for $r \geq 4m_0$.

For higher derivatives we write

$$\begin{aligned} \Gamma_g^{(1)} &= \left\{ \mathfrak{d}\xi,\, \mathfrak{d}\vartheta,\, re_\theta\omega,\, re_\theta(\kappa),\, \mathfrak{d}\eta,\, \mathfrak{d}\underline{\eta}, \mathfrak{d}\zeta,\, \mathfrak{d}A \right\}, \\ \Gamma_b^{(1)} &= \left\{ \mathfrak{d}\underline{\vartheta}, re_\theta(\underline{\kappa}), \mathfrak{d}\underline{\xi}, \mathfrak{d}\underline{A},\, re_\theta\underline{\omega}, \mathfrak{d}\underline{\xi} \right\}, \end{aligned}$$

and for $s \geq 2$,

$$\Gamma_g^{(s)} = \mathfrak{d}^{s-1}\Gamma_g^{(1)}, \qquad \Gamma_b^{(s)} = \mathfrak{d}^{s-1}\Gamma_b^{(1)}.$$

Moreover we denote

$$\Gamma_g^{\leq s} = \left\{ \Gamma_g^{(0)}, \Gamma_g^{(1)}, \dots, \Gamma_g^{(s)} \right\}, \qquad \Gamma_b^{\leq s} = \left\{ \Gamma_b^{(0)}, \Gamma_b^{(1)}, \dots, \Gamma_b^{(s)} \right\}.$$

---

[2] N_Σ is given in view of its definition by

$$\begin{aligned} N_\Sigma &= \frac{1}{\sqrt{2}\sqrt{e_4(r)(e_3(u)+e_3(r))}} \Big(e_4(r)e_3 + (e_3(u)+e_3(r))e_4 \Big) \\ &= \frac{1}{\sqrt{2}\sqrt{2-\Upsilon+O(\epsilon)}} \Big((1+O(\epsilon))e_3 + (2-\Upsilon+O(\epsilon))e_4 \Big) \end{aligned}$$

where we used the bootstrap assumptions.

With these notations, we may now state the estimates satisfied by the Ricci coefficients and curvature components.

Lemma 5.1. *Consider the global frame of Proposition 3.26 and the above definition*[3] *of* Γ_g *and* Γ_b. *Let an integer* k_{loss} *and a small constant* $\delta_0 > 0$ *satisfying*[4]

$$16 \leq k_{loss} \leq \frac{\delta_{dec}}{3}(k_{large} - k_{small}), \qquad \delta_0 = \frac{k_{loss}}{k_{large} - k_{small}}. \tag{5.1.6}$$

Then, the Ricci coefficients and curvature components with respect to the global frame of Proposition 3.26 satisfy

$$\xi = 0 \text{ on } r \geq 4m_0,$$

$$\begin{aligned}\max_{0\leq k\leq k_{small}+k_{loss}} \sup_{\mathcal{M}} \Bigg\{ &\left(r^2\tau^{\frac{1}{2}+\delta_{dec}-2\delta_0} + r\tau^{1+\delta_{dec}-2\delta_0}\right)|\mathfrak{d}^k\Gamma_g| + r\tau^{1+\delta_{dec}-2\delta_0}|\mathfrak{d}^k\Gamma_b| \\ &+\left(r^{\frac{7}{2}+\frac{\delta_B}{2}} + r^3\tau^{\frac{1}{2}+\delta_{dec}-2\delta_0} + r^2\tau^{1+\delta_{dec}-2\delta_0}\right)\left(|\mathfrak{d}^k\alpha| + |\mathfrak{d}^k\beta|\right) \\ &+\left(r^3\tau^{\frac{1}{2}+\delta_{dec}-2\delta_0} + r^2\tau^{1+\delta_{dec}-2\delta_0}\right)|\mathfrak{d}^k\check{\rho}| \\ &+\tau^{1+\delta_{dec}-2\delta_0}\left(r^2|\mathfrak{d}^k\underline{\beta}| + r|\mathfrak{d}^k\underline{\alpha}|\right)\Bigg\} \lesssim \epsilon,\end{aligned}$$

$$\begin{aligned}\max_{0\leq k\leq k_{small}+k_{loss}} \sup_{\mathcal{M}} \Bigg\{ &r^2\tau^{1+\delta_{dec}-2\delta_0}|\mathfrak{d}^{k-1}e_3(\Gamma_g)| \\ &+r^3(\tau+2r)^{1+\delta_{dec}-2\delta_0}\left(|\mathfrak{d}^{k-1}e_3(\alpha)| + |\mathfrak{d}^k e_3(\beta)|\right)\Bigg\} \lesssim \epsilon.\end{aligned}$$

Proof. In $r \geq 4m_0$, the global frame of Proposition 3.26 coincides with a conformal renormalization of the second frame of ${}^{(ext)}\mathcal{M}$, see Proposition 3.20. The estimates there follow immediately from the ones of Proposition 3.20. In the matching region $7/2m_0 \leq r \leq 4m_0$, the estimates are stated in Proposition 3.26. Finally, for ${}^{(ext)}\mathcal{M}(r \leq 7/2m_0)$ and ${}^{(int)}\mathcal{M}$, the estimates follow directly from interpolation between the bootstrap assumptions on decay for $k \leq k_{small}$ and the pointwise estimates of Proposition 3.19 for $k \leq k_{large} - 5$. □

5.1.3 Structure of nonlinear terms

The following lemma will be important in what follows.

[3]Recall in particular that the global frame of Proposition 3.26 is such that $\eta \in \Gamma_g$.

[4]Recall that we have

$$0 < \delta_{dec} \ll 1, \qquad \delta_{dec}k_{large} \gg 1, \qquad k_{small} = \left\lfloor \frac{1}{2}k_{large}\right\rfloor + 1.$$

In particular, we have $\delta_{dec}(k_{large} - k_{small}) \gg 1$ and hence there exists an integer k_{loss} satisfying the required constraints. We will in fact choose $k_{loss} = 33$, see (5.2.3).

Lemma 5.2. *For the solution $\mathfrak{q}$ to the wave equation* (5.0.1)*, the structure of the error term N can be written schematically as follows:*

$$N = N_g + e_3(rN_g) + N_m[\mathfrak{q}] \tag{5.1.7}$$

where

$$\begin{aligned} N_g &= r^2 \mathfrak{d}^{\leq 2}(\Gamma_g \cdot (\alpha, \beta)), \\ N_m[\mathfrak{q}] &= \mathfrak{d}^{\leq 1}(\Gamma_g \cdot \mathfrak{q}). \end{aligned} \tag{5.1.8}$$

Moreover, for every $k \leq k_{large} - 3$ we have, schematically,

$$\mathfrak{d}^k N = \mathfrak{d}^{\leq k} N_g + e_3(\mathfrak{d}^k(rN_g)) + \mathfrak{d}^k N_m[\mathfrak{q}]. \tag{5.1.9}$$

Remark 5.3. *In fact,* (5.1.7) *and* (5.1.9) *also contain lower order terms which are strictly better in powers of r and contain at most the same number of derivatives. For convenience, we drop them in the rest of the proof of Theorem M1.*

Proof. For $k = 0$, this is an immediate consequence of (5.0.2). For the higher derivatives we write

$$\mathfrak{d}^k(e_3(rN_g)) = e_3(\mathfrak{d}^k(rN_g)) + [\mathfrak{d}^k, e_3](rN_g).$$

In view of the formula for $[e_3, \not{\mathfrak{d}}]$ of Lemma 2.68, and the commutator formula for $[e_3, e_4]$, we have, schematically,

$$[e_3, e_3] = 0, \quad [\not{\mathfrak{d}}, e_3] = \Gamma_b \mathfrak{d} + \Gamma_b, \quad [re_4, e_3] = \left(\frac{1}{r} + \Gamma_b\right)\mathfrak{d}.$$

In view of our assumptions

$$\left|\mathfrak{d}^i(\Gamma_b)\right| \leq r^{-1}\epsilon, \qquad i \leq k_{large} - 4,$$

Γ_b is at least as good as r^{-1}, and hence, we deduce, schematically,

$$[\mathfrak{d}, e_3] = \frac{1}{r}\mathfrak{d} + \frac{1}{r}.$$

On the other hand, we have, schematically,

$$[\mathfrak{d}, r] = r$$

and hence, for $k \leq k_{large} - 3$,

$$\begin{aligned} [\mathfrak{d}^k, e_3](rN_g) &= \sum_{i+j \leq k-1} \mathfrak{d}^i\left(\frac{1}{r}\mathfrak{d} + \frac{1}{r}\right)\mathfrak{d}^j(rN_g) \\ &= \mathfrak{d}^{\leq k} N_g \end{aligned}$$

as desired. ∎

5.1.4 Main quantities

We restrict our attention to the region $\mathcal{M}(\tau_1,\tau_2)=\mathcal{M}\cap\{\tau_1\le\tau\le\tau_2\}$. For a given $\psi\in\mathfrak{s}_2(\mathcal{M})$ we introduce the following quantities, for $0\le\tau_1<\tau_2\le\tau_*$.

5.1.4.1 Morawetz bulk quantities

Consider the vectorfields,

$$T:=\frac{1}{2}\left(e_4+\Upsilon e_3\right),\qquad R:=\frac{1}{2}\left(e_4-\Upsilon e_3\right). \tag{5.1.10}$$

Let θ a smooth bump function equal 1 on $|\Upsilon|\le\delta_{\mathcal{H}}^{\frac{1}{10}}$ vanishing for $|\Upsilon|\ge 2\delta_{\mathcal{H}}^{\frac{1}{10}}$ and define the modified vectorfields,

$$\begin{aligned}\breve{R}&:=\theta\frac{1}{2}(e_4-e_3)+(1-\theta)\Upsilon^{-1}R=\frac{1}{2}\left[\breve{\theta}e_4-e_3\right],\\ \breve{T}&:=\theta\frac{1}{2}(e_4+e_3)+(1-\theta)\Upsilon^{-1}T=\frac{1}{2}\left[\breve{\theta}e_4+e_3\right],\end{aligned} \tag{5.1.11}$$

where $\breve{\theta}=\theta+\Upsilon^{-1}(1-\theta)$. Note that

$$\breve{\theta}=\begin{cases}1 & \text{for}\quad |\Upsilon|\le\delta_{\mathcal{H}}^{\frac{1}{10}},\\ \Upsilon^{-1} & \text{for}\quad |\Upsilon|\ge 2\delta_{\mathcal{H}}^{\frac{1}{10}}.\end{cases} \tag{5.1.12}$$

Remark 5.4. *Note that*

$$\begin{aligned}&\breve{R}+\breve{T}=e_4, &&-\breve{R}+\breve{T}=e_3, &&\text{in } {}^{(int)}\mathcal{M},\\ &\breve{R}+\breve{T}=\Upsilon^{-1}e_4, &&-\breve{R}+\breve{T}=e_3, &&\text{in } \mathcal{M}_{>4m_0}.\end{aligned}$$

We define the quantities

$$\begin{aligned}\mathrm{Mor}[\psi](\tau_1,\tau_2):&=\int_{\mathcal{M}(\tau_1,\tau_2)}\frac{1}{r^3}|\breve{R}\psi|^2+\frac{1}{r^4}|\psi|^2\\ &+\left(1-\frac{3m}{r}\right)^2\frac{1}{r}\left(|\nabla\!\!\!/\psi|^2+\frac{1}{r^2}|\breve{T}\psi|^2\right),\\ \mathrm{Morr}[\psi](\tau_1,\tau_2):&=\mathrm{Mor}[\psi](\tau_1,\tau_2)+\int_{\mathcal{M}_{>4m_0}(\tau_1,\tau_2)}r^{-1-\delta}|e_3(\psi)|^2,\end{aligned} \tag{5.1.13}$$

with $m=m(\tau,r)=m(u,r)$ the Hawking mass in $\mathcal{M}$. The constant $\delta>0$ is a sufficiently small quantity. An equivalent definition for $\mathrm{Morr}[\psi](\tau_1,\tau_2)$ is given below:

$$
\begin{aligned}
\mathrm{Morr}[\psi](\tau_1,\tau_2) = \int_{{}^{(trap)}\mathcal{M}(\tau_1,\tau_2)} & \Bigg[|R\psi|^2 + r^{-2}|\psi|^2 \\
& + \left(1-\frac{3m}{r}\right)^2 \left(|\nabla\!\!\!/\psi|^2 + \frac{1}{r^2}|T\psi|^2\right)\Bigg] \\
+ \int_{{}^{(tr\not{a}p)}\mathcal{M}(\tau_1,\tau_2)} & \Big[r^{-3}\left(|e_4\psi|^2 + r^{-1}|\psi|^2\right) + r^{-1}|\nabla\!\!\!/\psi|^2 \\
& + r^{-1-\delta}|e_3\psi|^2 \Big]
\end{aligned} \tag{5.1.14}
$$

where ${}^{(tr\not{a}p)}\mathcal{M}$ denotes the complement of ${}^{(trap)}\mathcal{M}$.

5.1.4.2 Weighted bulk quantities

Define, for $0 < p < 2$,

$$
\dot{B}_{p\,;\,R}[\psi](\tau_1,\tau_2) : = \int_{\mathcal{M}_{\geq R}(\tau_1,\tau_2)} r^{p-1}\left(p|\check{e}_4(\psi)|^2 + (2-p)|\nabla\!\!\!/\psi|^2 + r^{-2}|\psi|^2\right), \tag{5.1.15}
$$

$$
B_p[\psi](\tau_1,\tau_2) : = \mathrm{Morr}[\psi](\tau_1,\tau_2) + \dot{B}_{p\,;\,4m_0}[\psi](\tau_1,\tau_2).
$$

The bulk quantity $B_p[\psi](\tau_1,\tau_2)$ is equivalent to[5]

$$
B_p[\psi](\tau_1,\tau_2) \simeq \int_{\tau_1}^{\tau_2} M_{p-1}[\psi](\tau)d\tau
$$

where

$$
\begin{aligned}
M_{p-1}[\psi](\tau) &= \int_{\Sigma_{\leq 4m_0}(\tau)} |\breve{R}\psi|^2 + r^{-2}|\psi|^2 + \left(1-\frac{3m}{r}\right)^2\left(|\nabla\!\!\!/\psi|^2 + \frac{m^2}{r^2}|\breve{T}\psi|^2\right) \\
&+ \int_{\Sigma_{\geq 4m_0}(\tau)} r^{p-1}\left(p|e_4(\psi)|^2 + (2-p)|\nabla\!\!\!/\psi|^2 + r^{-2}|\psi|^2\right) \\
&+ \int_{\Sigma_{\geq 4m_0}(\tau)} r^{-1-\delta}|e_3\psi|^2.
\end{aligned}
$$

Remark 5.5. *Note that, for $\delta \leq p \leq 2-\delta$,*

$$
B_p[\psi](\tau_1,\tau_2) : \;=\; Morr[\psi](\tau_1,\tau_2) + \dot{B}_{p\,;\,4m_0}[\psi](\tau_1,\tau_2)
$$

[5]This equivalence follows from the coarea formula and the fact that the lapse of the τ-foliation is controlled uniformly from above and below.

is equivalent to

$$\begin{aligned} B_p[\psi](\tau_1,\tau_2) &\simeq Morr[\psi](\tau_1,\tau_2) \\ &+\int_{\mathcal{M}_{\geq 4m_0}(\tau_1,\tau_2)} r^{p-1}\left(|\check{e}_4(\psi)|^2+|\nabla\psi|^2+r^{-2}|e_3\psi|^2+r^{-2}|\psi|^2\right). \end{aligned}$$

Indeed,

$$\int_{\mathcal{M}_{\geq 4m_0}(\tau_1,\tau_2)} r^{p-3}|e_3\psi|^2 \lesssim \int_{\mathcal{M}_{\geq 4m_0}(\tau_1,\tau_2)} r^{-1-\delta}|e_3\psi|^2.$$

Therefore, since $r^2\left(|\check{e}_4(\psi)|^2+|\nabla\psi|^2\right)\lesssim|\mathfrak{d}\psi|^2$, *we have*

$$\begin{aligned} B_p[\psi](\tau_1,\tau_2) &\simeq Morr[\psi](\tau_1,\tau_2) \\ &+\int_{\mathcal{M}_{\geq 4m_0}(\tau_1,\tau_2)} r^{p-3}\left(|\mathfrak{d}\psi|^2+|\psi|^2\right). \end{aligned} \tag{5.1.16}$$

5.1.4.3 Basic energy-flux quantity

The basic energy-flux quantity on a hypersurface $\Sigma(\tau)$ is defined by

$$\begin{aligned} E[\psi](\tau) &= \int_{\Sigma(\tau)}\bigg(\frac{1}{2}(N_\Sigma,e_3)^2\,|e_4\psi|^2+\frac{1}{2}(N_\Sigma,e_4)^2\,|e_3\psi|^2 \\ &\qquad +|\nabla\psi|^2+r^{-2}|\psi|^2\bigg). \end{aligned} \tag{5.1.17}$$

Here N_Σ denotes a choice for the normal to Σ so that in particular we have

$$N_\Sigma=\begin{cases} N_\Sigma=e_3 & \text{on } {}^{(int)}\Sigma, \\ N_\Sigma=e_4 & \text{on } {}^{(ext)}\Sigma, \end{cases} \tag{5.1.18}$$

and, in view of (5.1.5),

$$(N_\Sigma,e_3)\leq -1 \text{ and } (N_\Sigma,e_4)\leq -1 \qquad \text{on } {}^{(trap)}\Sigma. \tag{5.1.19}$$

5.1.4.4 Weighted energy-flux type quantities

We have

$$\dot{E}_{p\,;\,R}[\psi](\tau) := \begin{cases} \displaystyle\int_{\Sigma_{\geq R}(\tau)} r^p\Big(|\check{e}_4\psi|^2+r^{-2}|\psi|^2\Big) & \text{for } p\leq 1-\delta, \\ \displaystyle\int_{\Sigma_{\geq R}(\tau)} r^p\Big(|\check{e}_4\psi|^2+r^{-p-1-\delta}|\psi|^2\Big) & \text{for } p>1-\delta, \end{cases} \tag{5.1.20}$$

and

$$E_p[\psi](\tau) := E[\psi](\tau)+\dot{E}_{p\,;\,4m_0}[\psi](\tau). \tag{5.1.21}$$

Here $\check{e}_4$ denotes the first order operator

$$\check{e}_4\psi = r^{-1}\Upsilon^{-1}e_4(r\psi). \tag{5.1.22}$$

Remark 5.6. *To control the weighted quantities* (5.1.21), *it will be convenient to introduce in* ${}^{(ext)}\mathcal{M}(r \geq 4m_0)$ *the following renormalized frame*

$$e_4' = \Upsilon^{-1}e_4, \quad e_3' = \Upsilon e_3, \quad e_\theta' = e_\theta.$$

In particular, this yields

$$\check{e}_4\psi = r^{-1}e_4'(r\psi).$$

Note also that we have the following alternate form

$$\check{e}_4\psi = e_4'\psi + r^{-1}\psi + \frac{e_4'(r)-1}{r}\psi$$

where $e_4'(r) - 1 = \Upsilon^{-1}e_4(r) - 1 = O(\epsilon r^{-1})$ *in view of our assumption on* Γ_g.

5.1.4.5 Flux quantities

The boundary of $\mathcal{M}(\tau_1,\tau_2)$ is given by

$$\partial\mathcal{M}(\tau_1,\tau_2) \quad = \quad \Sigma(\tau_1)\cup\Sigma(\tau_2)\cup\mathcal{A}(\tau_1,\tau_2)\cup\Sigma_*(\tau_1,\tau_2).$$

Our basic flux quantity along the spacelike hypersurfaces $\mathcal{A}$ and Σ_* is given by

$$\begin{aligned} F[\Psi](\tau_1,\tau_2) \quad := \quad & \int_{\mathcal{A}(\tau_1,\tau_2)}\Big(\delta_{\mathcal{H}}^{-1}|e_4\Psi|^2 + \delta_{\mathcal{H}}|e_3\Psi|^2 + |\nabla\!\!\!/\Psi|^2 + r^{-2}|\Psi|^2\Big)\\ & + \int_{\Sigma_*(\tau_1,\tau_2)}\Big(|e_4\Psi|^2 + |e_3\Psi|^2 + |\nabla\!\!\!/\Psi|^2 + r^{-2}|\Psi|^2\Big), \end{aligned} \tag{5.1.23}$$

with $\mathcal{A}(\tau_1,\tau_2) = \mathcal{A}\cap\mathcal{M}(\tau_1,\tau_2)$ and $\Sigma_*(\tau_1,\tau_2) = \Sigma_*\cap\mathcal{M}(\tau_1,\tau_2)$.

5.1.4.6 Weighted flux quantities

$$\begin{aligned} \dot{F}_p[\psi](\tau_1,\tau_2) &:= \int_{\Sigma_*(\tau_1,\tau_2)} r^p\Big(|e_4\psi|^2 + |\nabla\!\!\!/\psi|^2 + r^{-2}|\psi|^2\Big),\\ F_p[\psi](\tau_1,\tau_2) &:= F[\psi](\tau_1,\tau_2) + \dot{F}_p[\psi](\tau_1,\tau_2). \end{aligned} \tag{5.1.24}$$

5.1.4.7 Weighted quantities for the inhomogeneous term N

Recall the decomposition (5.1.7) for the inhomogeneous term N

$$N \quad = \quad N_g + e_3(rN_g) + N_m[\mathfrak{q}].$$

We define, for $p \geq \delta$,

$$
\begin{aligned}
I_p[N_g](\tau_1, \tau_2) &= \left(\int_{\tau_1}^{\tau_2} d\tau \|N_g\|_{L^2({}^{(trap)}\Sigma(\tau))} \right)^2 + \int_{{}^{(tr\not{p})}\mathcal{M}(\tau_1,\tau_2)} r^{1+p}|N_g|^2 \\
&+ \int_{{}^{(tr\not{p})}\mathcal{M}(\tau_1,\tau_2)} r^{2+p}|N_g||e_3(N_g)| + \sup_{\tau\in[\tau_1,\tau_2]} \int_{\Sigma(\tau)} r^{p+2}\big|N_g\big|^2 \\
&+ \int_{{}^{(tr\not{p})}\mathcal{M}(\tau_1,\tau_2)} r^{3+\delta}|e_3(N_g)|^2. \qquad (5.1.25)
\end{aligned}
$$

Remark 5.7. *While $N_m[\mathfrak{q}]$ is present in the decomposition of the inhomogeneous term N, (5.1.25) only contains a norm for N_g. In fact, $N_m[\mathfrak{q}]$ will always be absorbed by the left-hand side wherever it appears.*

5.1.4.8 Higher order derivative quantities

We define the higher order derivative quantities $E^s[\psi], \mathrm{Mor}^s[\psi], \mathrm{Morr}^s[\psi], E^s_p[\psi]$, $B^s_b[\psi]$, $M^s_p[\psi]$, $F^s[\psi]$, $F^s_p[\psi]$, $I^s_p[N_g]$ by the obvious procedure,

$$
Q^s[\psi] = \sum_{k\leq s} Q[\mathfrak{d}^k\psi].
$$

Remark 5.8. *Note that in view of Remark 5.5 we can also write, equivalently, for $p < 2-\delta$,*

$$
B^s_p[\psi](\tau_1, \tau_2) = Morr^s[\psi](\tau_1, \tau_2) + \int_{\mathcal{M}_{>4m_0}(\tau_1,\tau_2)} r^{p-3}|\mathfrak{d}^{\leq 1+s}\psi|^2. \qquad (5.1.26)
$$

5.1.4.9 Decay norms

We introduce

$$
\begin{aligned}
\mathcal{E}^s_{p,d}[\psi] :&= \sup_{0\leq\tau\leq\tau_*} (1+\tau)^d E^s_p[\psi](\tau), \\
\mathcal{B}^s_{p,d}[\psi] :&= \sup_{0\leq\tau\leq\tau_*} (1+\tau)^d B^s_p[\psi](\tau, \tau_*), \\
&\simeq \sup_{0\leq\tau\leq\tau_*} (1+\tau)^d \int_{\tau}^{\tau_*} M^s_{p-1}[\psi](\tau')d\tau', \qquad (5.1.27) \\
\mathcal{F}^s_{p,d}[\psi] :&= \sup_{0\leq\tau\leq\tau_*} (1+\tau)^d F^s_p[\psi](\tau, \tau_*), \\
\mathcal{I}^s_{p,d}[N_g] :&= \sup_{0\leq\tau\leq\tau_*} (1+\tau)^d I^s_p[N_g](\tau, \tau_*).
\end{aligned}
$$

5.2 PROOF OF THEOREM M1

Recall that we have to prove for $k \leq k_{small} + 20$

$$\begin{aligned} |\mathfrak{d}^k \mathfrak{q}| &\lesssim \epsilon_0 r^{-1}(1+\tau)^{-\frac{1}{2}-\delta_{extra}}, \\ |\mathfrak{d}^k \mathfrak{q}| &\lesssim \epsilon_0 r^{-\frac{1}{2}}(1+\tau)^{-1-\delta_{extra}}, \\ |\mathfrak{d}^k e_3(\mathfrak{q})| &\lesssim \epsilon_0 r^{-1}(1+\tau)^{-1-\delta_{extra}}, \end{aligned}$$

and

$$\int_{{}^{(int)}\mathcal{M}(\tau,\tau_*)} |\mathfrak{d}^k e_3 \mathfrak{q}|^2 + \int_{\Sigma_*(\tau,\tau_*)} |\mathfrak{d}^k e_3 \mathfrak{q}|^2 \lesssim \epsilon_0^2 (1+\tau)^{-2-2\delta_{extra}},$$

for some constant δ_{extra} such that $\delta_{dec} < \delta_{extra} < 2\delta_{dec}$.

5.2.1 Flux decay estimates for $\mathfrak{q}$

The following result establishes decay of flux estimates for $\mathfrak{q}$.

Theorem 5.9. *Let $0 < q_0 < 1$ be a fixed number and $s \leq k_{small} + 25$. Then, for all $\delta > 0$, with a constant C depending only on s, δ and q_0 such that for all $\delta \leq p \leq 2 - \delta$, we have*

$$\begin{aligned} & \mathcal{E}^s_{p,2+q_0-p}[\mathfrak{q}] + \mathcal{B}^s_{p,2+q_0-p}[\mathfrak{q}] + \mathcal{F}^s_{p,2+q_0-p}[\mathfrak{q}] \\ \lesssim \;& \mathcal{E}^{s+2}_{q_0}[\check{\mathfrak{q}}](0) + \mathcal{E}^{s+4}_{2-\delta}[\mathfrak{q}](0) + \mathcal{I}^{s+5}_{q_0+2,0}[N_g] + \mathcal{I}^{s+5}_{\delta,2+q_0-\delta}[N_g], \end{aligned} \tag{5.2.1}$$

where we recall that the decay norms $\mathcal{I}^s_{p,d}[N_g]$ are defined by

$$\mathcal{I}^s_{p,d}[N_g] = \sup_{0 \leq \tau \leq \tau_*} (1+\tau)^d I^s_p[N_g](\tau, \tau_*).$$

Theorem 5.9 will be proved in section 5.4.3.

To prove Theorem M1 we have to eliminate the norms $\mathcal{I}^s_{p,d}[N_g]$ on the right-hand side of Theorem 5.9.

Proposition 5.10. *Let $s \leq k_{small} + 30$ and assume*

$$q_0 < 4\delta_{dec} - 4\delta_0 \tag{5.2.2}$$

where

$$\delta_0 = \frac{33}{k_{large} - k_{small}} = \frac{33}{k_{large} - \lfloor \frac{k_{large}}{2} \rfloor - 1} \tag{5.2.3}$$

is the small constant appearing in Lemma 5.1. Then, the following estimates hold true,

$$\mathcal{I}^s_{q_0+2,0}[N_g] + \mathcal{I}^s_{\delta,2+q_0-\delta}[N_g] \lesssim \epsilon^4.$$

The proof of Proposition 5.10 is postponed to section 5.2.3. Together with Theorem 5.9, Proposition 5.10 immediately yields the proof of the following corollary.

Corollary 5.11. *In addition to the assumptions of Theorem 5.9 we assume*

$$2\delta_{dec} < q_0 < 4\delta_{dec} - 4\delta_0 \tag{5.2.4}$$

where $\delta_0 > 0$ *is given by* (5.2.3). *Then for a sufficiently small bootstrap constant* $\epsilon > 0$, *for all* $s \le k_{small} + 25$ *and for all* $\delta \le p \le 2 - \delta$, *we have*

$$\mathcal{E}^s_{p,2+q_0-p}[\mathfrak{q}] + \mathcal{B}^s_{p,2+q_0-p}[\mathfrak{q}] + \mathcal{F}^s_{p,2+q_0-p}[\mathfrak{q}] \lesssim \mathcal{E}^{s+2}_{q_0}[\check{\mathfrak{q}}](0) + \mathcal{E}^{s+4}_{2-\delta}[\mathfrak{q}](0) + \epsilon^4.$$

5.2.2 Proof of Theorem M1

Since $\epsilon = \epsilon_0^{2/3}$, and in view of the control on $\mathfrak{q}$ at $\tau = 0$ provided by Theorem M0, we immediately deduce from Corollary 5.11, for all $0 < q \le q_0$, $\delta \le p \le 2 - \delta$, and $s \le k_{small} + 25$,

$$\mathcal{E}^s_{p,2+q_0-p}[\mathfrak{q}] + \mathcal{B}^s_{p,2+q_0-p}[\mathfrak{q}] + \mathcal{F}^s_{p,2+q_0-p}[\mathfrak{q}] \lesssim \epsilon_0^2. \tag{5.2.5}$$

We will also need the following two propositions concerning L^2 estimates on spheres.

Proposition 5.12. *On any* $S = S(\tau, r) \subset \Sigma(\tau)$, *for* $s \le k_{small} + 25$,

$$(1+\tau)^{1+q_0} \int_{S_r} |\mathfrak{q}^{(s)}|^2 \lesssim \left(\mathcal{E}^s_{1+\delta,1+q_0-\delta}[\mathfrak{q}]\right)^{\frac{1}{2}} \left(\mathcal{E}^s_{1-\delta,1+q_0+\delta}[\mathfrak{q}]\right)^{\frac{1}{2}} \tag{5.2.6}$$

and

$$r^{-1}(1+\tau)^{2+q_0-\delta} \int_{S_r} |\mathfrak{q}^{(s)}|^2 \lesssim \mathcal{E}^s_{\delta,2+q_0-\delta}[\mathfrak{q}]. \tag{5.2.7}$$

Proposition 5.13. *We have for* $s \le k_{small} + 25$

$$(1+\tau)^{2+q_0-\delta} \int_{\Sigma_*(\tau,\tau_*)} |e_3 \mathfrak{d}^{\le s} \mathfrak{q}|^2 \lesssim \mathcal{F}^s_{\delta,2+q_0-\delta}[\mathfrak{q}]. \tag{5.2.8}$$

Also, on any $S = S(\tau, r) \subset \Sigma(\tau)$, *for* $s \le k_{small} + 23$, *we have*

$$(1+\tau)^{2+q_0-\delta} \int_{S_r} |e_3 \mathfrak{d}^{\le s} \mathfrak{q}|^2 \lesssim \epsilon_0^2 + \mathcal{F}^{s+1}_{\delta,2+q_0-\delta}[\mathfrak{q}] + \mathcal{E}^{s+2}_{\delta,2+q_0-\delta}[\mathfrak{q}]. \tag{5.2.9}$$

The proof of Proposition 5.12 is postponed to section 5.4.4, and the proof of Proposition 5.13 is postponed to section 5.4.5.

We now conclude the proof of Theorem M1. Indeed, in view of (5.2.5), Proposition 5.12 and Proposition 5.13, we infer for $s \leq k_{small} + 25$

$$
\begin{aligned}
(1+\tau)^{2+q_0-\delta} \int_{{}^{(int)}\mathcal{M}(\tau,\tau_*)} |\mathfrak{d}^{\leq s+1}\mathfrak{q}|^2 &\lesssim \epsilon_0^2,\\
(1+\tau)^{1+q_0} \int_{S_r} |\mathfrak{q}^{(s)}|^2 &\lesssim \epsilon_0^2,\\
r^{-1}(1+\tau)^{2+q_0-\delta} \int_{S_r} |\mathfrak{q}^{(s)}|^2 &\lesssim \epsilon_0^2,\\
(1+\tau)^{2+q_0-\delta} \int_{\Sigma_*(\tau,\tau_*)} |e_3\mathfrak{d}^{\leq s}\mathfrak{q}|^2 &\lesssim \epsilon_0^2,
\end{aligned}
$$

and for $s \leq k_{small} + 23$

$$
(1+\tau)^{2+q_0-\delta} \int_S |\mathfrak{d}^s e_3\mathfrak{q}|^2 \lesssim \epsilon_0^2.
$$

In view of the standard Sobolev inequality on the 2-surfaces S, i.e.,

$$
\|\psi\|_{L^\infty(S)} \lesssim r^{-1}\|(r\nabla)^{\leq 2}\psi\|_{L^2(S)},
$$

we immediately infer for $s \leq k_{small} + 23$

$$
\begin{aligned}
|\mathfrak{q}^{(s)}| &\lesssim \epsilon_0 r^{-1}(1+\tau)^{-\frac{1}{2}-\frac{q_0}{2}},\\
|\mathfrak{q}^{(s)}| &\lesssim \epsilon_0 r^{-\frac{1}{2}}(1+\tau)^{-1-\frac{q_0-\delta}{2}},
\end{aligned}
$$

and for $s \leq k_{small} + 21$

$$
|\mathfrak{d}^s e_3(\mathfrak{q})| \lesssim \epsilon_0 r^{-1}(1+\tau)^{-1-\frac{q_0-\delta}{2}}.
$$

Recall that $q_0 > 2\delta_{dec}$ and that $\delta > 0$ can be chosen arbitrarily small so that we have $q_0 - \delta > 2\delta_{dec}$. In particular, we may choose

$$
\delta_{extra} := \frac{q_0-\delta}{2}, \quad \delta_{extra} > \delta_{dec}, \tag{5.2.10}
$$

which together with the above estimates for $\mathfrak{q}$ implies for $s \leq k_{small} + 25$

$$
\begin{aligned}
(1+\tau)^{2+q_0-\delta} \int_{{}^{(int)}\mathcal{M}(\tau,\tau_*)} |\mathfrak{d}^{\leq s+1}\mathfrak{q}|^2 &\lesssim \epsilon_0^2,\\
(1+\tau)^{2+2\delta_{extra}} \int_{\Sigma_*(\tau,\tau_*)} |e_3\mathfrak{d}^{\leq s}\mathfrak{q}|^2 &\lesssim \epsilon_0^2,
\end{aligned}
$$

for $s \leq k_{small} + 23$

$$
\begin{aligned}
|\mathfrak{q}^{(s)}| &\lesssim \epsilon_0 r^{-1}(1+\tau)^{-\frac{1}{2}-\delta_{extra}},\\
|\mathfrak{q}^{(s)}| &\lesssim \epsilon_0 r^{-\frac{1}{2}}(1+\tau)^{-1-\delta_{extra}},
\end{aligned}
$$

and for $s \leq k_{small} + 21$

$$
|\mathfrak{d}^s e_3(\mathfrak{q})| \lesssim \epsilon_0 r^{-1}(1+\tau)^{-1-\delta_{extra}}
$$

as desired. This concludes the proof of Theorem M1.

5.2.3 Proof of Proposition 5.10

Recall that

$$\begin{aligned} I_p[N_g](\tau_1,\tau_2) &= \left(\int_{\tau_1}^{\tau_2} d\tau \|N_g\|_{L^2({}^{(trap)}\Sigma(\tau))}\right)^2 + \int_{{}^{(tr\!\!/p)}\mathcal{M}(\tau_1,\tau_2)} r^{1+p}|N_g|^2 \\ &+ \int_{{}^{(tr\!\!/p)}\mathcal{M}(\tau_1,\tau_2)} r^{2+p}|N_g||e_3(N_g)| + \sup_{\tau\in[\tau_1,\tau_2]}\int_{\Sigma(\tau)} r^{p+2}|N_g|^2 \\ &+ \int_{{}^{(tr\!\!/p)}\mathcal{M}(\tau_1,\tau_2)} r^{3+\delta}|e_3(N_g)|^2 \end{aligned}$$

and

$$\mathcal{I}^s_{p,d}[N_g] = \sup_{0\le\tau\le\tau_*}(1+\tau)^d I^s_p[N_g](\tau,\tau_*).$$

Since we have

$$r^\delta(1+\tau)^{2+q_0-\delta} \lesssim r^{2+q_0}+(1+\tau)^{2+q_0},$$

and

$$\begin{aligned} \int_{{}^{(tr\!\!/p)}\mathcal{M}(\tau,\tau_*)} r^2|\mathfrak{d}^{\le s}e_3(N_g)||\mathfrak{d}^{\le s}N_g| &\lesssim \int_{{}^{(tr\!\!/p)}\mathcal{M}(\tau,\tau_*)} r|\mathfrak{d}^{\le s}N_g|^2 \\ &+ \int_{{}^{(tr\!\!/p)}\mathcal{M}(\tau,\tau_*)} r^3|\mathfrak{d}^{\le s}e_3(N_g)|^2, \end{aligned}$$

we infer

$$\begin{aligned} &\mathcal{I}^s_{q_0+2,0}[N_g]+\mathcal{I}^s_{\delta,2+q_0-\delta}[N_g] \qquad (5.2.11)\\ \lesssim& \sup_{0\le\tau\le\tau_*}\Bigg[\int_{{}^{(tr\!\!/p)}\mathcal{M}(\tau,\tau_*)} r^{4+q_0}|\mathfrak{d}^{\le s+1}N_g|^2 + \sup_{\tau'\in[\tau,\tau_*]}\int_{\Sigma(\tau')} r^{4+q_0}|\mathfrak{d}^{\le s}N_g|^2 \\ &+(1+\tau)^{2+q_0}\Bigg(\int_{{}^{(tr\!\!/p)}\mathcal{M}(\tau,\tau_*)} r|\mathfrak{d}^{\le s}N_g|^2 + \int_{{}^{(tr\!\!/p)}\mathcal{M}(\tau,\tau_*)} r^{3+\delta}|\mathfrak{d}^{\le s}e_3(N_g)|^2 \\ &+\sup_{\tau'\in[\tau,\tau_*]}\int_{\Sigma(\tau')} r^2|\mathfrak{d}^{\le s}N_g|^2\Bigg) \\ &+(1+\tau)^{2+q_0}\left(\int_\tau^{\tau_*} d\tau'\|\mathfrak{d}^{\le s}N_g\|_{L^2({}^{(trap)}\Sigma(\tau'))}\right)^2\Bigg]. \end{aligned}$$

In order to prove Proposition 5.10, it suffices to estimate the right-hand side of (5.2.11). To this end, we will estimate separately the terms with highest power of r, i.e., the first two terms, and the terms with highest power of τ, i.e., the four last terms.

5.2.3.1 Terms with highest power of r in (5.2.11)

We estimate the first two terms of (5.2.11). Recall from Lemma 5.2 that

$$N_g = r^2\mathfrak{d}^{\leq 2}(\Gamma_g \cdot (\alpha, \beta)).$$

Recall from Lemma 5.1 we have

$$\max_{0\leq k\leq k_{large}-3} \sup_{{}^{(ext)}\mathcal{M}(r\geq 4m_0)} r^{\frac{7}{2}+\frac{\delta_B}{2}}\left(|\mathfrak{d}^k\alpha| + |\mathfrak{d}^k\beta|\right) \lesssim \epsilon.$$

We infer for $s \leq k_{large} - 6$

$$|\mathfrak{d}^{\leq s+1}N_g| \lesssim \epsilon r^{-\frac{7}{2}-\frac{\delta_B}{2}}|r^2\mathfrak{d}^{\leq s+3}\Gamma_g|$$

and hence, for $s \leq k_{large} - 6$, we deduce

$$\begin{aligned}
&\int_{{}^{(tr\not{p})}\mathcal{M}(\tau,\tau_*)} r^{4+q_0}|\mathfrak{d}^{\leq s+1}N_g|^2 + \sup_{\tau'\in[\tau,\tau_*]}\int_{\Sigma(\tau')} r^{4+q_0}\left|\mathfrak{d}^{\leq s}N_g\right|^2\\
\lesssim\ & \epsilon^2\int_{{}^{(tr\not{p})}\mathcal{M}(\tau,\tau_*)} r^{-3-\delta_B+q_0}(r^2\mathfrak{d}^{\leq s+3}\Gamma_g)^2\\
&+\epsilon^2\sup_{\tau'\in[\tau,\tau_*]}\int_{\Sigma(\tau')} r^{-3-\delta_B+q_0}(r^2\mathfrak{d}^{\leq s+2}\Gamma_g)^2.
\end{aligned}$$

Since we also have for $s \leq k_{large} - 6$

$$\begin{aligned}
&\sup_{r_0\geq 4m_0}\int_{\{r=r_0\}}(r^2\mathfrak{d}^{\leq s+3}\Gamma_g)^2 \lesssim \epsilon^2, \quad \int_{\mathcal{M}_{r\leq 4m_0}}(\mathfrak{d}^{\leq s+3}\Gamma_g)^2 \lesssim \epsilon^2,\\
&\sup_{\mathcal{M}}|r^2\mathfrak{d}^{\leq s+2}\Gamma_g| \lesssim \epsilon,
\end{aligned}$$

we deduce

$$\begin{aligned}
&\int_{{}^{(tr\not{p})}\mathcal{M}(\tau,\tau_*)} r^{4+q_0}|\mathfrak{d}^{\leq s+1}N_g|^2 + \sup_{\tau'\in[\tau,\tau_*]}\int_{\Sigma(\tau')} r^{4+q_0}\left|\mathfrak{d}^{\leq s}N_g\right|^2\\
\lesssim\ & \epsilon^4\left(1+\int_{r\geq 4m_0}\frac{dr}{r^{1+\delta_B-q_0}}\right).
\end{aligned}$$

Since $q_0 < 4\delta_{dec}$ and $\delta_B \geq 4\delta_{dec}$, we have $q_0 < \delta_B$ and hence, we obtain for $s \leq k_{large} - 6$

$$\int_{{}^{(tr\not{p})}\mathcal{M}(\tau,\tau_*)} r^{4+q_0}|\mathfrak{d}^{\leq s+1}N_g|^2 + \sup_{\tau'\in[\tau,\tau_*]}\int_{\Sigma(\tau')} r^{4+q_0}\left|\mathfrak{d}^{\leq s}N_g\right|^2 \lesssim \epsilon^4.$$

This is the desired control of the terms with highest power of r in (5.2.11).

5.2.3.2 Terms with highest power of τ in (5.2.11)

We estimate the four last terms of (5.2.11). In view of Lemma 5.1 with $k_{loss} = 33$, so that

$$\delta_0 = \frac{33}{k_{large} - k_{small} - 2} = \frac{33}{k_{large} - \lfloor \frac{k_{large}}{2} \rfloor - 3},$$

we have

$$\begin{aligned}
\left|\mathfrak{d}^{\leq k_{small}+33}\Gamma_g\right| &\lesssim \epsilon r^{-2}\tau^{-1/2-\delta_{dec}+2\delta_0},\\
\left|\mathfrak{d}^{\leq k_{small}+33}\Gamma_g\right| &\lesssim \epsilon r^{-1}\tau^{-1-\delta_{dec}+2\delta_0},\\
\left|\mathfrak{d}^{\leq S+32}e_3\Gamma_g\right| &\lesssim \epsilon r^{-2}[\tau^{-1-\delta_{dec}}]^{1-\delta_0} \lesssim \epsilon r^{-2}\tau^{-1-\delta_{dec}+2\delta_0},\\
\left|\mathfrak{d}^{\leq k_{small}+33}(\alpha,\beta)\right| &\lesssim \epsilon r^{-3}(\tau+r)^{-1/2-\delta_{dec}+2\delta_0},\\
\left|\mathfrak{d}^{\leq k_{small}+33}(\alpha,\beta)\right| &\lesssim \epsilon r^{-2}(\tau+r)^{-1-\delta_{dec}+2\delta_0},\\
\left|\mathfrak{d}^{\leq S+32}e_3(\alpha,\beta)\right| &\lesssim \epsilon r^{-3-\frac{1}{2}\delta_0}[\tau^{-1-\delta_{dec}}]^{1-\delta_0} \lesssim \epsilon r^{-3}\tau^{-1-\delta_{dec}+2\delta_0}.
\end{aligned}$$

In particular, together with the bootstrap assumption for $k \leq k_{small}$, the pointwise bound

$$|\mathfrak{d}^{\leq k_{large}-5}\alpha| + |\mathfrak{d}^{\leq k_{large}-5}\beta| \lesssim \epsilon r^{-\frac{7}{2}-\frac{\delta_B}{2}}$$

and since $N_g = r^2\mathfrak{d}^{\leq 2}(\Gamma_g \cdot (\alpha,\beta))$, we infer for $s \leq k_{small} + 30$

$$\begin{aligned}
|\mathfrak{d}^s N_g| &\lesssim \epsilon^2 r^{-3}\tau^{-1-2\delta_{dec}+2\delta_0}\\
|\mathfrak{d}^s N_g| &\lesssim \epsilon^2 r^{-1}\tau^{-2-2\delta_{dec}+2\delta_0},\\
|\mathfrak{d}^s e_3(N_g)| &\lesssim \epsilon^2 r^{-3}\tau^{-\frac{3}{2}-2\delta_{dec}+2\delta_0},\\
|\mathfrak{d}^s e_3(N_g)| &\lesssim \epsilon^2 r^{-\frac{7}{2}-\frac{\delta_B}{2}}\tau^{-1-\delta_{dec}+2\delta_0}.
\end{aligned} \tag{5.2.12}$$

Using these 4 bounds and interpolation, we infer for $\delta > 0$

$$
\begin{aligned}
& (1+\tau)^{2+q_0}\left(\int_{^{(trap)}\mathcal{M}(\tau,\tau_*)} r|\mathfrak{d}^{\leq s}N_g|^2 + \int_{^{(trap)}\mathcal{M}(\tau,\tau_*)} r^{3+\delta}|\mathfrak{d}^{\leq s}e_3(N_g)|^2\right.\\
& \left. + \sup_{\tau'\in[\tau,\tau_*]}\int_{\Sigma(\tau')} r^2\left|\mathfrak{d}^{\leq s}N_g\right|^2\right)\\
& +(1+\tau)^{2+q_0}\left(\int_\tau^{\tau_*} d\tau'\|\mathfrak{d}^{\leq s}N_g\|_{L^2(^{(trap)}\Sigma(\tau'))}\right)^2\\
\lesssim\ & \epsilon^4(1+\tau)^{2+q_0}\int_{^{(trap)}\mathcal{M}(\tau,\tau_*)} r(r^{-3}\tau'^{-1-2\delta_{dec}+2\delta_0})^{1+\delta}(r^{-1}\tau'^{-2-2\delta_{dec}+2\delta_0})^{1-\delta}\\
& +\epsilon^4(1+\tau)^{2+q_0}\int_{^{(trap)}\mathcal{M}(\tau,\tau_*)} r^{3+\delta}(r^{-3}\tau'^{-\frac{3}{2}-2\delta_{dec}+2\delta_0})^{2-2\delta}\\
& \times(r^{-\frac{7}{2}-\frac{\delta_B}{2}}\tau'^{-1-\delta_{dec}+2\delta_0})^{2\delta}\\
& +\epsilon^4(1+\tau)^{2+q_0}\sup_{\tau'\in[\tau,\tau_*]}\int_{\Sigma(\tau')} r^2(r^{-3}\tau'^{-1-2\delta_{dec}+2\delta_0})^2\\
& +\epsilon^4(1+\tau)^{2+q_0}\left(\int_\tau^{\tau_*}\tau'^{-2-2\delta_{dec}+2\delta_0}d\tau'\right)^2\\
\lesssim\ & \epsilon^4(1+\tau)^{2+q_0}\int_{^{(trap)}\mathcal{M}(\tau,\tau_*)} r^{-3-\delta\delta_B}\tau'^{-3-4\delta_{dec}+\delta+4\delta_0+2\delta\delta_{dec}}\\
& +\epsilon^4(1+\tau)^{2+q_0}\sup_{\tau'\in[\tau,\tau_*]}\int_{\Sigma(\tau')} r^{-4}\tau'^{-2-4\delta_{dec}+4\delta_0}\\
& +\epsilon^4(1+\tau)^{2+q_0}\left(\int_\tau^{\tau_*}\tau'^{-2-2\delta_{dec}+2\delta_0}d\tau'\right)^2
\end{aligned}
$$

and since $\delta > 0$, we obtain

$$
\begin{aligned}
& (1+\tau)^{2+q_0}\left(\int_{^{(trap)}\mathcal{M}(\tau,\tau_*)} r|\mathfrak{d}^{\leq s}N_g|^2 + \int_{^{(trap)}\mathcal{M}(\tau,\tau_*)} r^{3+\delta}|\mathfrak{d}^{\leq s}e_3(N_g)|^2\right.\\
& \left. + \sup_{\tau'\in[\tau,\tau_*]}\int_{\Sigma(\tau')} r^2\left|\mathfrak{d}^{\leq s}N_g\right|^2\right) + (1+\tau)^{2+q_0}\left(\int_\tau^{\tau_*} d\tau'\|\mathfrak{d}^{\leq s}N_g\|_{L^2(^{(trap)}\Sigma(\tau'))}\right)^2\\
\lesssim\ & \epsilon^4(1+\tau)^{q_0-4\delta_{dec}+\delta+4\delta_0+2\delta\delta_{dec}}.
\end{aligned}
$$

As we have $q_0 < 4\delta_{dec} - 4\delta_0$, there exists $\delta > 0$ small enough such that

$$
q_0 - 4\delta_{dec} + \delta + 4\delta_0 + 2\delta\delta_{dec} \leq 0,
$$

and hence

$$(1+\tau)^{2+q_0}\left(\int_{{}^{(tr\not{p})}\mathcal{M}(\tau,\tau_*)} r|\mathfrak{d}^{\leq s}N_g|^2 + \int_{{}^{(tr\not{p})}\mathcal{M}(\tau,\tau_*)} r^{3+\delta}|\mathfrak{d}^{\leq s}e_3(N_g)|^2 + \sup_{\tau'\in[\tau,\tau_*]}\int_{\Sigma(\tau')} r^2\left|\mathfrak{d}^{\leq s}N_g\right|^2\right)$$
$$+(1+\tau)^{2+q_0}\left(\int_\tau^{\tau_*} d\tau'\|\mathfrak{d}^{\leq s}N_g\|_{L^2({}^{(trap)}\Sigma(\tau'))}\right)^2 \lesssim \epsilon^4.$$

This is the desired control of the terms with highest power of τ in (5.2.11). Together with (5.2.11) and the above control of the terms with highest power of r, we infer

$$\begin{aligned}
&\mathcal{I}^s_{q_0+2,0}[N_g] + \mathcal{I}^s_{\delta,2+q_0-\delta}[N_g]\\
\lesssim\ & \sup_{0\leq\tau\leq\tau_*}\Bigg[\int_{{}^{(tr\not{p})}\mathcal{M}(\tau,\tau_*)} r^{4+q_0}|\mathfrak{d}^{\leq s+1}N_g|^2 + \sup_{\tau'\in[\tau,\tau_*]}\int_{\Sigma(\tau')} r^{4+q_0}\left|\mathfrak{d}^{\leq s}N_g\right|^2\\
&+(1+\tau)^{2+q_0}\left(\int_{{}^{(tr\not{p})}\mathcal{M}(\tau,\tau_*)} r|\mathfrak{d}^{\leq s}N_g|^2 + \int_{{}^{(tr\not{p})}\mathcal{M}(\tau,\tau_*)} r^{3+\delta}|\mathfrak{d}^{\leq s}e_3(N_g)|^2\right.\\
&\left.+\sup_{\tau'\in[\tau,\tau_*]}\int_{\Sigma(\tau')} r^2\left|\mathfrak{d}^{\leq s}N_g\right|^2\right)\\
&+(1+\tau)^{2+q_0}\left(\int_\tau^{\tau_*} d\tau'\|\mathfrak{d}^{\leq s}N_g\|_{L^2({}^{(trap)}\Sigma(\tau'))}\right)^2\Bigg]\\
\lesssim\ & \epsilon^4
\end{aligned}$$

which is the desired estimate. This concludes the proof of Proposition 5.10.

5.3 IMPROVED WEIGHTED ESTIMATES

The goal of this section is to prove the two following theorems on improved weighted estimates.

Theorem 5.14. *Assume $\mathfrak{q}$ verifies the following wave equation, see* (5.0.1),

$$\square_2\mathfrak{q} + \kappa\underline{\kappa}\,\mathfrak{q} = N$$

with N given, in view of Lemma 5.2, by

$$N = N_g + e_3(rN_g) + N_m[\mathfrak{q}].$$

Then, for any $\delta\leq p\leq 2-\delta$, $0\leq s\leq k_{small}+30$,

$$\begin{aligned}
&\sup_{\tau\in[\tau_1,\tau_2]} E^s_p[\mathfrak{q}](\tau) + B^s_p[\mathfrak{q}](\tau_1,\tau_2) + F^s_p[\mathfrak{q}](\tau_1,\tau_2)\\
\lesssim\ & E^s_p[\mathfrak{q}](\tau_1) + I^{s+1}_p[N_g](\tau_1,\tau_2). \qquad (5.3.1)
\end{aligned}$$

The next result deals with weighted estimates for the quantity

$$\check{\mathfrak{q}} = f_2 \check{e}_4 \mathfrak{q}, \tag{5.3.2}$$

where f_2 is a fixed smooth function of r defined as follows,

$$f_2(r) = \begin{cases} r^2 & \text{for} \quad r \geq 6m_0, \\ 0 & \text{for} \quad r \leq 4m_0. \end{cases} \tag{5.3.3}$$

Theorem 5.15. *Assume* $\mathfrak{q}$ *verifies equation, see* (5.0.1),

$$\square_2 \mathfrak{q} + \kappa \underline{\kappa}\, \mathfrak{q} = N$$

with

$$N = N_g + e_3(rN_g) + N_m[\mathfrak{q}]$$

as in Lemma 5.2. Then, for any $-1+\delta < q \leq 1-\delta$, $0 \leq s \leq k_{small} + 29$,

$$\begin{aligned} &\sup_{\tau \in [\tau_1, \tau_2]} E^s_q[\check{\mathfrak{q}}](\tau) + B^s_q[\check{\mathfrak{q}}](\tau_1, \tau_2) \\ &\lesssim E^s_q[\check{\mathfrak{q}}](\tau_1) + E^{s+1}_{q+1}[\mathfrak{q}](\tau_1) + I^{s+2}_{q+2}[N_g](\tau_1, \tau_2). \end{aligned} \tag{5.3.4}$$

Remark 5.16. *Note that in* (5.3.1) *and* (5.3.4)*, the term* $N_m[\mathfrak{q}]$ *does not appear in the right-hand side since it turns out that it can be absorbed by the left-hand side.*

The proof of Theorem 5.14 is postponed to section 5.3.2, and the proof of Theorem 5.15 is postponed to section 5.3.3. These proofs will rely on weighted energy flux estimates introduced in the next section.

5.3.1 Basic and higher weighted estimates for wave equations

Assume given a spacetime $\mathcal{M}$ verifying the bootstrap assumptions with small constant $\epsilon > 0$. The proof of Theorem 5.14 and Theorem 5.15 will rely on estimates stated below for solutions $\psi \in \mathfrak{s}_2(\mathcal{M})$ of the equation

$$\square_2 \psi = V\psi + N, \qquad V = -\kappa\underline{\kappa}. \tag{5.3.5}$$

5.3.1.1 *Basic weighted estimates*

Theorem 5.17. *Recall the definitions in* (5.1.21), (5.1.15). *The following holds for any* $0 \leq s \leq k_{small} + 30$. *For all* $\delta \leq p \leq 2-\delta$, *we have*

$$\begin{aligned} &\sup_{\tau \in [\tau_1, \tau_2]} E^s_p[\psi](\tau) + B^s_p[\psi](\tau_1, \tau_2) + F^s_p[\psi](\tau_1, \tau_2) \\ &\lesssim E^s_p[\psi](\tau_1) + J^s_p[\psi, N](\tau_1, \tau_2), \end{aligned} \tag{5.3.6}$$

where, for $p \geq \delta$, we have introduced the notation

$$J_{p,R}[\psi, N](\tau_1, \tau_2) := \left| \int_{\mathcal{M}_{\geq R}(\tau_1,\tau_2)} r^p \check{e}_4 \psi N \right|,$$

$$J_p[\psi, N](\tau_1, \tau_2) := \left(\int_{\tau_1}^{\tau_2} d\tau \|N\|_{L^2({}^{(trap)}\Sigma(\tau))} \right)^2 + \int_{{}^{(tr\!\!\not{p})}\mathcal{M}(\tau_1,\tau_2)} r^{1+\delta}|N|^2 + J_{p,4m_0}[\psi, N](\tau_1, \tau_2), \tag{5.3.7}$$

and

$$J_p^s[\psi, N](\tau_1, \tau_2) := \sum_{k \leq s} J_p[\mathfrak{d}^k \psi, \mathfrak{d}^k N](\tau_1, \tau_2).$$

The proof of Theorem 5.17 is postponed to section 10.4.5.

5.3.1.2 Higher weighted estimates

The next result deals with weighted estimates for the quantity

$$\check{\psi} = f_2 \check{e}_4 \psi, \tag{5.3.8}$$

where f_2 is a fixed smooth function of r defined as follows,

$$f_2(r) = \begin{cases} r^2 & \text{for } r \geq 6m_0, \\ 0 & \text{for } r \leq 4m_0. \end{cases} \tag{5.3.9}$$

Theorem 5.18. *The following holds for any $-1+\delta < q \leq 1-\delta$, $0 \leq s \leq k_{small}+29$,*

$$\sup_{\tau \in [\tau_1, \tau_2]} E_q^s[\check{\psi}](\tau) + B_q^s[\check{\psi}](\tau_1, \tau_2) \lesssim E_q^s[\check{\psi}](\tau_1) + \check{J}_q^s[\check{\psi}, N](\tau_1, \tau_2) + E^{s+1}_{\max(q,\delta)}[\psi](\tau_1) + J^{s+1}_{\max(q,\delta)}[\psi, N], \tag{5.3.10}$$

where we have introduced the notation

$$\check{J}_q[\check{\psi}, N](\tau_1, \tau_2) := J_{q,4m_0}\left[\check{\psi}, r^2 \left(e_4 N + \frac{3}{r} N\right)\right](\tau_1, \tau_2) = \int_{\mathcal{M}_{\geq 4m_0}(\tau_1,\tau_2)} r^{q+2}(\check{e}_4 \check{\psi}) \cdot \left(e_4 N + \frac{3}{r} N\right),$$

and

$$\check{J}_q^s[\check{\psi}, N](\tau_1, \tau_2) := \sum_{k \leq s} \check{J}_q[\mathfrak{d}^k \check{\psi}, \mathfrak{d}^k N](\tau_1, \tau_2).$$

The proof of Theorem 5.18 is postponed to section 10.4.6.

We now proceed to the proof of Theorem 5.14 and Theorem 5.15 in the next two sections. The proofs will follow from the structure of the nonlinear term N of $\mathfrak{q}$ provided by Lemma 5.2 and the use of Theorem 5.17 and Theorem 5.18.

5.3.2 Proof of Theorem 5.14

Applying Theorem 5.17 to the equation for $\mathfrak{q}$, with N given by Lemma 5.2, we derive corresponding estimates with the norm $J_p^s[\mathfrak{q}, N](\tau_1, \tau_2)$ on the right-hand side, i.e., for $0 \le s \le k_{small} + 30$, and for $\delta \le p \le 2 - \delta$,

$$\begin{aligned} & \sup_{\tau\in[\tau_1,\tau_2]} E_p^s[\mathfrak{q}](\tau) + B_p^s[\mathfrak{q}](\tau_1, \tau_2) + F_p^s[\mathfrak{q}](\tau_1, \tau_2) \\ \lesssim \; & E_p^s[\mathfrak{q}](\tau_1) + J_p^s[\mathfrak{q}, N](\tau_1, \tau_2). \end{aligned} \tag{5.3.11}$$

To prove Theorem 5.14, it suffices, in view of (5.3.11), to estimate $J_p^s[\mathfrak{q}, N](\tau_1, \tau_2)$. Recall that, see (5.3.7) and (5.1.25),

$$\begin{aligned} I_p[N](\tau_1, \tau_2) &= \left(\int_{\tau_1}^{\tau_2} d\tau \|N\|_{L^2({}^{(trap)}\Sigma(\tau))}\right)^2 + \int_{{}^{(tr\not{p})}\mathcal{M}(\tau_1,\tau_2)} r^{1+p}|N|^2 \\ &\quad + \int_{{}^{(tr\not{p})}\mathcal{M}(\tau_1,\tau_2)} r^{2+p}|N_g||e_3(N_g)| + \sup_{\tau\in[\tau_1,\tau_2]} \int_{\Sigma(\tau)} r^{p+2}|N|^2 \\ &\quad + \int_{{}^{(tr\not{p})}\mathcal{M}(\tau_1,\tau_2)} r^{3+\delta}|e_3(N_g)|^2 \end{aligned}$$

and

$$\begin{aligned} J_{p,R}[\mathfrak{q}, N] &= \left|\int_{\mathcal{M}_{\ge R}(\tau_1,\tau_2)} r^p \check{e}_4(\mathfrak{q}) N\right|, \\ J_p[\mathfrak{q}, N](\tau_1, \tau_2) &= \left(\int_{\tau_1}^{\tau_2} d\tau \|N\|_{L^2({}^{(trap)}\Sigma(\tau))}\right)^2 + \int_{{}^{(tr\not{p})}\mathcal{M}(\tau_1,\tau_2)} r^{1+\delta}|N|^2 \\ &\quad + J_{p,4m_0}^s[\mathfrak{q}, N](\tau_1, \tau_2), \\ J_p^s[\mathfrak{q}, N](\tau_1, \tau_2) &= \sum_{k\le s} J_p[\mathfrak{d}^k \mathfrak{q}, \mathfrak{d}^k N]. \end{aligned}$$

Recall also from (5.1.9)

$$\mathfrak{d}^k N = \mathfrak{d}^{\le k} N_g + e_3(\mathfrak{d}^k(r N_g)) + \mathfrak{d}^k N_m[\mathfrak{q}] \tag{5.3.12}$$

and consider separately the three terms.

Case of $N_m[\mathfrak{q}]$. Recall that $N_m[\mathfrak{q}] = \mathfrak{d}^{\le 1}(\Gamma_g \cdot \mathfrak{q})$. We have, schematically,

$$\mathfrak{d}^k N_m[\mathfrak{q}] = \mathfrak{d}^{1+k}(\Gamma_g \cdot \mathfrak{q}) = \sum_{i+j=k+1} \mathfrak{d}^{\le i}\Gamma_g \mathfrak{d}^{\le j}\mathfrak{q}.$$

We make use of the following consequence of the bootstrap assumptions, valid for $k \le k_{large} - 5$,

$$\left|\mathfrak{d}^{\le k}\Gamma_g\right| \le \epsilon r^{-2}$$

to deduce

$$\left|\mathfrak{d}^k N_m[\mathfrak{q}]\right| \lesssim \epsilon r^{-2}\left|\mathfrak{d}^{\le k+1}\mathfrak{q}\right|. \tag{5.3.13}$$

We deduce

$$
\begin{aligned}
J^s_{p,4m_0}[\mathfrak{q}, N_m[\mathfrak{q}]](\tau_1,\tau_2) &\lesssim \sum_{k\le s}\int_{\mathcal{M}_{\ge 4m_0}(\tau_1,\tau_2)} r^p\big|\check{e}_4\mathfrak{q}^{(k)}\big|\,\big|\mathfrak{d}^k N_m[\mathfrak{q}]\big|\\
&\lesssim \epsilon\sum_{k\le s}\int_{\mathcal{M}_{\ge 4m_0}(\tau_1,\tau_2)} r^{p-3}\big|\mathfrak{d}^{1+k}\mathfrak{q}\big|^2.
\end{aligned}
$$

Thus, recalling Remark 5.8, we infer

$$
J^s_{p,4m_0}[\mathfrak{q}, N_m[\mathfrak{q}]](\tau_1,\tau_2) \lesssim \epsilon B^s_p[\mathfrak{q}](\tau_1,\tau_2). \tag{5.3.14}
$$

Next, we estimate in view of (5.3.13)

$$
\int_{{}^{(trap)}\!\!\!\!\!/\,\mathcal{M}(\tau_1,\tau_2)} r^{1+\delta}|\mathfrak{d}^k N_m[\mathfrak{q}]|^2 \lesssim \epsilon\int_{{}^{(trap)}\!\!\!\!\!/\,\mathcal{M}(\tau_1,\tau_2)} r^{\delta-3}|\mathfrak{d}^{\le k+1}\mathfrak{q}|^2
$$

which yields, using again Remark 5.8,

$$
\int_{{}^{(trap)}\!\!\!\!\!/\,\mathcal{M}(\tau_1,\tau_2)} r^{1+\delta}|\mathfrak{d}^k N_m[\mathfrak{q}]|^2 \lesssim \epsilon B^s_\delta[\mathfrak{q}](\tau_1,\tau_2). \tag{5.3.15}
$$

We next estimate the integral

$$
\int_{\tau_1}^{\tau_2} d\tau\|\mathfrak{d}^k N_m[\mathfrak{q}]\|_{L^2({}^{(trap)}\Sigma(\tau))}.
$$

In view of the definition of $N_m[\mathfrak{q}] = \mathfrak{d}^{\le 1}(\Gamma_g\cdot\mathfrak{q})$,

$$
\begin{aligned}
\mathfrak{d}^k N_m[\mathfrak{q}] &= \mathfrak{d}^{\le k+1}(\Gamma_g\cdot\mathfrak{q}) = \sum_{i+j=k+1}\mathfrak{d}^{\le i}\Gamma_g\,\mathfrak{d}^{\le j}\mathfrak{q}\\
&= \mathfrak{d}^{j\le(k+1)/2}\Gamma_g\,\mathfrak{d}^{\le k+1}\mathfrak{q} + \mathfrak{d}^{j\le(k+1)/2}\mathfrak{q}\,\mathfrak{d}^{\le k+1}\Gamma_g = J_1 + J_2.
\end{aligned}
$$

Now, since $\frac{k+1}{2}\le k_{small}$ we have

$$
\left|\mathfrak{d}^{j\le(k+1)/2}\Gamma_g\right| \lesssim \epsilon(1+\tau)^{-1-\delta_{dec}}.
$$

Hence,

$$
\begin{aligned}
\|J_1\|^2_{L^2({}^{(trap)}\Sigma(\tau))} &= \int_{{}^{(trap)}\Sigma(\tau)}\left|\mathfrak{d}^{j\le(k+1)/2}\Gamma_g\right|^2\left|\mathfrak{d}^{\le k+1}\mathfrak{q}\right|^2\\
&\lesssim \epsilon^2(1+\tau)^{-2-2\delta_{dec}}E^s[\mathfrak{q}](\tau),
\end{aligned}
$$

i.e.,

$$
\|J_1\|_{L^2({}^{(trap)}\Sigma(\tau))} \lesssim \epsilon(1+\tau)^{-1-\delta_{dec}}\left(E^s[\mathfrak{q}](\tau)\right)^{1/2}.
$$

For J_2 we write

$$\begin{aligned}\|J_2\|^2_{L^2({}^{(trap)}\Sigma(\tau))} &= \int_{{}^{(trap)}\Sigma(\tau)} \left|\mathfrak{d}^{j\leq(k+1)/2}\mathfrak{q}\right|^2 \left|\mathfrak{d}^{\leq k+1}\Gamma_g\right|^2\\ &\lesssim \left(\sup_{{}^{(trap)}\Sigma(\tau)}\left|\mathfrak{d}^{j\leq(k+1)/2}\mathfrak{q}\right|\right)^2 \int_{{}^{(trap)}\Sigma(\tau)}\left|\mathfrak{d}^{\leq k+1}\Gamma_g\right|^2\\ &\lesssim \int_{{}^{(trap)}\Sigma(\tau)}\left|\mathfrak{d}^{\leq(k+1)/2+2}\mathfrak{q}\right|^2 \int_{{}^{(trap)}\Sigma(\tau)}\left|\mathfrak{d}^{\leq k+1}\Gamma_g\right|^2\end{aligned}$$

or, since $(k+1)/2+2\leq s$,

$$\|J_2\|_{L^2({}^{(trap)}\Sigma(\tau))} \lesssim \left[\int_{{}^{(trap)}\Sigma(\tau)}\left|\mathfrak{d}^{\leq s}\mathfrak{q}\right|^2\right]^{1/2}\left[\int_{{}^{(trap)}\Sigma(\tau)}\left|\mathfrak{d}^{\leq k+1}\Gamma_g\right|^2\right]^{1/2}.$$

In view of the above estimates for J_1 and J_2, we deduce, for all $k\leq s\leq k_{large}-5$,

$$\begin{aligned}&\int_{\tau_1}^{\tau_2} d\tau\|\mathfrak{d}^k N_m[\mathfrak{q}]\|_{L^2({}^{(trap)}\Sigma(\tau))}\\ \lesssim\;& \epsilon \sup_{\tau_1\leq\tau\leq\tau_2}\left(E^s[\mathfrak{q}](\tau)\right)^{1/2}\\ &+\int_{\tau_1}^{\tau_2}d\tau\left[\int_{{}^{(trap)}\Sigma(\tau)}\left|\mathfrak{d}^{\leq s}\mathfrak{q}\right|^2\right]^{1/2}\left[\int_{{}^{(trap)}\Sigma(\tau)}\left|\mathfrak{d}^{\leq s}\Gamma_g\right|^2\right]^{1/2}\\ \lesssim\;& \epsilon \sup_{\tau_1\leq\tau\leq\tau_2}\left(E^s[\mathfrak{q}](\tau)\right)^{1/2}+\left(\int_{{}^{(tr\not{a}p)}\mathcal{M}(\tau_1,\tau_2)}\left|\mathfrak{d}^{\leq s}\mathfrak{q}\right|^2\right)^{\frac{1}{2}}\left(\int_{\mathcal{M}_{r\leq 4m_0}}\left|\mathfrak{d}^{\leq s}\Gamma_g\right|^2\right)^{1/2}.\end{aligned}$$

Making use of the following consequence of the bootstrap assumptions

$$\left(\int_{\mathcal{M}_{r\leq 4m_0}}\left|\mathfrak{d}^{\leq s}\Gamma_g\right|^2\right)^{1/2}\lesssim\epsilon,$$

as well as the fact that

$$\int_{{}^{(tr\not{a}p)}\mathcal{M}(\tau_1,\tau_2)}\left|\mathfrak{d}^{\leq s}\mathfrak{q}\right|^2 \lesssim \mathrm{Morr}^s[\mathfrak{q}](\tau_1,\tau_2),$$

we deduce

$$\begin{aligned}&\left(\int_{\tau_1}^{\tau_2}d\tau\|\mathfrak{d}^k N_m[\mathfrak{q}]\|_{L^2({}^{(trap)}\Sigma(\tau))}\right)^2\\ &\lesssim \epsilon^2\sup_{\tau_1\leq\tau\leq\tau_2}E^s[\mathfrak{q}](\tau)+\epsilon^2\mathrm{Morr}^s[\mathfrak{q}](\tau_1,\tau_2)\end{aligned}\qquad(5.3.16)$$

which together with (5.3.15) and (5.3.14) yields for any $p\geq\delta$

$$J_p^s[\mathfrak{q},N_m[\mathfrak{q}]](\tau_1,\tau_2) \lesssim \epsilon^2\sup_{\tau_1\leq\tau\leq\tau_2}E^s[\mathfrak{q}](\tau)+\epsilon B_p^s[\mathfrak{q}](\tau_1,\tau_2).\qquad(5.3.17)$$

Case of N_g. We write, as before,

$$\begin{aligned} J^s_{p,4m_0}[\mathfrak{q}, N_g](\tau_1,\tau_2) &\lesssim \sum_{k\leq s}\int_{\mathcal{M}_{\geq 4m_0}(\tau_1,\tau_2)} r^p\big|\check{e}_4\mathfrak{q}^{(k)}\mathfrak{d}^k N_g\big| \\ &\lesssim \sum_{k\leq s}\left(\int_{\mathcal{M}_{\geq 4m_0}(\tau_1,\tau_2)} r^{p-1}\big|\check{e}_4\mathfrak{q}^{(k)}\big|^2\right)^{1/2} \\ &\quad\times\left(\int_{\mathcal{M}_{\geq 4m_0}(\tau_1,\tau_2)} r^{p+1}\big|\mathfrak{d}^k N_g\big|^2\right)^{1/2}. \end{aligned}$$

Therefore,

$$\begin{aligned} J^s_{p,4m_0}[\mathfrak{q}, N_g](\tau_1,\tau_2) &\lesssim \left(B^s_p[\mathfrak{q}](\tau_1,\tau_2)\right)^{1/2}\left(I^s_p[N_g](\tau_1,\tau_2)\right)^{1/2} \\ &\lesssim \delta_1 B^s_p[\mathfrak{q}](\tau_1,\tau_2) + \delta_1^{-1} I^s_p[N_g](\tau_1,\tau_2) \end{aligned}$$

where $\delta_1 > 0$ is chosen sufficiently small so that we can later absorb the term $\delta_1 B^s_p[\mathfrak{q}](\tau_1,\tau_2)$ by the left-hand side of our main estimate.

Also, we have in view of the definition of $I^s_p[N](\tau_1,\tau_2)$ and the fact that $p\geq\delta$

$$\left(\int_{\tau_1}^{\tau_2} d\tau\|\mathfrak{d}^{\leq s}N_g\|_{L^2({}^{(trap)}\Sigma(\tau))}\right)^2 + \int_{{}^{(tr\!\!/p)}\mathcal{M}(\tau_1,\tau_2)} r^{1+\delta}|\mathfrak{d}^{\leq s}N_g|^2 \lesssim I^s_p[N_g](\tau_1,\tau_2).$$

Therefore,

$$\begin{aligned} J^s_p[\mathfrak{q}, N_g](\tau_1,\tau_2) &= \left(\int_{\tau_1}^{\tau_2} d\tau\|\mathfrak{d}^{\leq s}N_g\|_{L^2({}^{(trap)}\Sigma(\tau))}\right)^2 \\ &\quad + \int_{{}^{(tr\!\!/p)}\mathcal{M}(\tau_1,\tau_2)} r^{1+\delta}|\mathfrak{d}^{\leq s}N_g|^2 + J^s_{p,4m_0}[\mathfrak{q}, N_g](\tau_1,\tau_2) \\ &\lesssim I^s_\delta[N_g](\tau_1,\tau_2) + \delta_1^{-1}I^s_p[N_g](\tau_1,\tau_2) + \delta_1 B^s_p[\mathfrak{q}](\tau_1,\tau_2), \end{aligned}$$

i.e.,

$$J^s_p[\mathfrak{q}, N_g](\tau_1,\tau_2) \lesssim \delta_1^{-1}I^s_p[N_g](\tau_1,\tau_2) + \delta_1 B^s_p[\mathfrak{q}](\tau_1,\tau_2). \tag{5.3.18}$$

Case of $e_3(rN_g)$. First, note that we have

$$\begin{aligned} &\left(\int_{\tau_1}^{\tau_2} d\tau\|\mathfrak{d}^{\leq s}e_3(rN_g)\|_{L^2({}^{(trap)}\Sigma(\tau))}\right)^2 + \int_{{}^{(tr\!\!/p)}\mathcal{M}(\tau_1,\tau_2)} r^{1+\delta}|\mathfrak{d}^{\leq s}e_3(rN_g)|^2 \\ \lesssim\ &\left(\int_{\tau_1}^{\tau_2} d\tau\|\mathfrak{d}^{\leq s+1}N_g\|_{L^2({}^{(trap)}\Sigma(\tau))}\right)^2 + \int_{{}^{(tr\!\!/p)}\mathcal{M}(\tau_1,\tau_2)} r^{1+\delta}|\mathfrak{d}^{\leq s}N_g|^2 \\ &+ \int_{{}^{(tr\!\!/p)}\mathcal{M}(\tau_1,\tau_2)} r^{3+\delta}|\mathfrak{d}^{\leq s}e_3(N_g)|^2 \end{aligned}$$

where we used the fact that $|\mathfrak{d}^{\leq s}e_3(r)|\lesssim 1$ and $|\mathfrak{d}^{\leq s}r|\lesssim r$. Hence, we infer in view of the definition of $I^s_p[N](\tau_1,\tau_2)$ and the fact that $p\geq\delta$

$$\left(\int_{\tau_1}^{\tau_2}d\tau\|\mathfrak{d}^{\leq s}e_3(rN_g)\|_{L^2({}^{(trap)}\Sigma(\tau))}\right)^2+\int_{{}^{(tr\not{p})}\mathcal{M}(\tau_1,\tau_2)}r^{1+\delta}|\mathfrak{d}^{\leq s}e_3(rN_g)|^2$$
$$\lesssim I^{s+1}_p[N_g](\tau_1,\tau_2). \tag{5.3.19}$$

We then estimate

$$J_{p,4m_0}[\mathfrak{q}^{(k)},e_3(\mathfrak{d}^k(rN_g))](\tau_1,\tau_2),\qquad k\leq s.$$

To this end, we introduce a smooth cut-off function ϕ_0 vanishing for $r\leq 4m_0$ and equal to 1 for $r\geq 8m_0$. Then, we have

$$\begin{aligned}J_{p,4m_0}[\mathfrak{q}^{(k)},\mathfrak{d}^k(rN_g)](\tau_1,\tau_2) &= \left|\int_{\mathcal{M}(\tau_1,\tau_2)}r^p\check{e}_4\mathfrak{q}^{(k)}\,e_3\mathfrak{d}^k(rN_g)\right|\\ &\lesssim J_{p,4m_0}[\mathfrak{q}^{(k)},\phi_0\mathfrak{d}^k(rN_g)](\tau_1,\tau_2)\\ &\quad+J_{p,4m_0}[\mathfrak{q}^{(k)},(1-\phi_0)rN_g](\tau_1,\tau_2).\end{aligned}\tag{5.3.20}$$

In view of the fact that $1-\phi_0$ is supported in $r\leq 8m_0$, we easily obtain

$$\begin{aligned}&J_{p,4m_0}[\mathfrak{q}^{(k)},(1-\phi_0)rN_g](\tau_1,\tau_2)\\ &\lesssim\left(\sup_{\tau_1\leq\tau\leq\tau_2}E^s[\mathfrak{q}](\tau)+B^s_p[\mathfrak{q}](\tau_1,\tau_2)\right)^{1/2}\left(I^{s+1}_p[N_g](\tau_1,\tau_2)\right)^{\frac{1}{2}}\end{aligned}$$

and hence

$$\begin{aligned}&J_{p,4m_0}[\mathfrak{q}^{(k)},(1-\phi_0)rN_g](\tau_1,\tau_2)\\ &\lesssim\delta_1\Big(\sup_{\tau_1\leq\tau\leq\tau_2}E^s[\mathfrak{q}](\tau)+B^s_p[\mathfrak{q}](\tau_1,\tau_2)\Big)+\delta_1^{-1}I^{s+1}_p[N_g](\tau_1,\tau_2)\end{aligned}\tag{5.3.21}$$

where $\delta_1>0$ is chosen sufficiently small so that we can later absorb the terms $\delta_1\sup_{\tau_1\leq\tau\leq\tau_2}E^s[\mathfrak{q}](\tau)$ and $\delta_1B^s_p[\mathfrak{q}](\tau_1,\tau_2)$ by the left-hand side of our main estimate.

It remains to estimate the terms

$$J_{p,4m_0}[\mathfrak{q}^{(k)},\phi_0e_3(\mathfrak{d}^k(rN_g))](\tau_1,\tau_2),\qquad k\leq s$$

which is supported for $r\geq 4m_0$. Note that $e_3(rN_g)$ behaves like rN_g and therefore the same sequence of estimates as for N_g would lead to a loss of r^{-1}. For this reason we need to integrate by parts in e_3.

Proposition 5.19. *The following estimate holds true, for all $k\leq s\leq k_{large}-5$,*

$$\begin{aligned}&\sum_{k\leq s}J_{p,4m_0}[\mathfrak{q}^{(k)},\phi_0e_3(\mathfrak{d}^k(rN_g))](\tau_1,\tau_2)\\ &\lesssim\delta_1B^s_p[\mathfrak{q}](\tau_1,\tau_2)+\delta_1^{-1}I^{s+1}_p[N_g](\tau_1,\tau_2)\end{aligned}\tag{5.3.22}$$

for a sufficiently small $\delta_1>0$.

We postponed the proof of Proposition 5.19 to the end of the section. We are now in position to conclude the proof of Theorem 5.14.

Proof of Theorem 5.14. (5.3.21) and (5.3.22) yield

$$\sum_{k\leq s} J_{p,4m_0}[\mathfrak{q}^{(k)}, e_3(\mathfrak{d}^k(rN_g))](\tau_1,\tau_2) \lesssim \delta_1 B^s_p[\mathfrak{q}](\tau_1,\tau_2) + \delta_1^{-1} I^{s+1}_p[N_g](\tau_1,\tau_2).$$

Together with (5.3.17), (5.3.18) and (5.3.19), we infer

$$J^s_p[\mathfrak{q}, N](\tau_1,\tau_2) \lesssim (\delta_1+\epsilon) B^s_p[\mathfrak{q}](\tau_1,\tau_2) + \delta_1^{-1} I^{s+1}_p[N_g](\tau_1,\tau_2) + \epsilon^2 \sup_{\tau_1\leq\tau\leq\tau_2} E^s[\mathfrak{q}](\tau).$$

In view of (5.3.11), this concludes the proof of Theorem 5.14. □

The proof of Proposition 5.19 will rely in particular on the following identity.

Lemma 5.20. *The following hold true for any $\psi \in \mathfrak{s}_2$:*

- *We have, schematically,*

$$e_3e_4(r\psi) = -r\square_2\psi + r\Delta\!\!\!/_2\psi + r^{-1}\mathfrak{d}\psi. \tag{5.3.23}$$

- *The following identity holds true, schematically,*

$$e_3e_4(r\mathfrak{d}^k\psi) = -\mathfrak{d}^{\leq k}(r\square_2\psi) + r\Delta\!\!\!/_2(\mathfrak{d}^{\leq k}\psi) + r^{-1}\mathfrak{d}^{\leq k+1}\psi. \tag{5.3.24}$$

Proof. We start with the following identity for $\psi \in \mathfrak{s}_2$, see Definition 2.103,

$$\square_2\psi = -e_3e_4\psi + \Delta\!\!\!/_2\psi + \left(2\underline{\omega} - \frac{1}{2}\underline{\kappa}\right)e_4\psi - \frac{1}{2}\kappa e_3\psi + 2\eta e_\theta\psi$$

from which we deduce

$$r\square_2\psi = -re_3e_4\psi + r\left(\Delta\!\!\!/_2\psi + \left(2\underline{\omega} - \frac{1}{2}\underline{\kappa}\right)e_4\psi - \frac{1}{2}\kappa e_3\psi + 2\eta e_\theta\psi\right).$$

On the other hand,

$$\begin{aligned} re_3e_4\psi &= e_3(re_4\psi) - (e_3r)e_4\psi = e_3(e_4(r\psi) - e_4(r)\psi) - (e_3r)e_4\psi \\ &= e_3e_4(r\psi) - e_4(r)e_3\psi - (e_3r)e_4\psi - (e_3e_4r)\psi. \end{aligned}$$

Hence,

$$\begin{aligned} r\square_2\psi &= -e_3e_4(r\psi) + e_4(r)e_3\psi + (e_3r)e_4\psi + (e_3e_4r)\psi + r\Delta\!\!\!/_2\psi \\ &+ r\left(2\underline{\omega} - \frac{1}{2}\underline{\kappa}\right)e_4\psi - \frac{1}{2}r\kappa e_3\psi + 2r\eta e_\theta\psi \\ &= -e_3e_4(r\psi) + r\Delta\!\!\!/\psi + \left(e_4r - \frac{1}{2}r\kappa\right)e_3\psi + \left(e_3r - \frac{1}{2}r\underline{\kappa} + 2r\underline{\omega}\right)e_4\psi \\ &+2r\eta e_\theta\psi \\ &= -e_3e_4(r\psi) + r\Delta\!\!\!/\psi + \frac{r}{2}Ae_3\psi + \frac{r}{2}\left(\underline{A} + 4\underline{\omega}\right)e_4\psi + 2r\eta e_\theta\psi, \end{aligned}$$

i.e.,

$$e_3e_4(r\psi) \;=\; -r\square_2\psi + r\not{\Delta}_2\psi + \frac{r}{2}Ae_3\psi + \frac{r}{2}\left(\underline{A} + 4\underline{\omega}\right)e_4\psi + 2r\eta e_\theta\psi$$

or, schematically, in view of the definition of $\mathfrak{d}\psi$ and of $|\underline{\omega}| + r|\Gamma_g| + |\Gamma_b| \lesssim r^{-1}$,

$$\begin{aligned} e_3e_4(r\psi) &= -r\square_2\psi + r\not{\Delta}_2\psi + \left(r\Gamma_g + \Gamma_b + r^{-1}\right)e_3\psi \\ &= -r\square_2\psi + r\not{\Delta}_2\psi + r^{-1}\mathfrak{d}\psi \end{aligned}$$

which is (5.3.23).

To derive the identity for higher derivatives we write, schematically,

$$\mathfrak{d}^k e_3e_4(r\psi) \;=\; -\mathfrak{d}^k(r\square_2\psi) + \mathfrak{d}^k(r\not{\Delta}_2\psi) + \mathfrak{d}^k(r\Gamma_g\mathfrak{d}\psi).$$

We write

$$\begin{aligned} \mathfrak{d}^k e_3e_4(r\psi) &= e_3e_4(r\mathfrak{d}^k\psi) + [\mathfrak{d}^k, e_3e_4r]\psi = e_3e_4(r\mathfrak{d}^k\psi) + [\mathfrak{d}^k, e_3]\mathfrak{d}\psi + e_3[\mathfrak{d}^k, e_4r]\psi, \\ \mathfrak{d}^k(r\not{\Delta}_2\psi) &= r\not{\Delta}_2\mathfrak{d}^k\psi + [\mathfrak{d}^k, r\not{\Delta}]\psi = r\not{\Delta}_2\mathfrak{d}^k\psi + [\mathfrak{d}^k, r^{-1}]\mathfrak{d}^2\psi + r^{-1}[\mathfrak{d}^k, r^2\not{\Delta}]\psi. \end{aligned}$$

In view of the identities for $[e_3, \not{\mathfrak{d}}]$ and $[e_4, \not{\mathfrak{d}}]$ of Lemma 2.68, the identities of Proposition 2.28 for commutation formulas involving $\not{d}_k$ and $\not{d}_k^\star$ derivatives, and the commutator formula for $[e_3, e_4]$, we have schematically

$$\begin{aligned} &[e_3, e_3] = 0, \quad [\not{\mathfrak{d}}, r^2\not{\Delta}] = \not{\mathfrak{d}} + 1, \quad [e_3, e_4r] = (r^{-1} + \Gamma_g)\mathfrak{d}, \\ &[e_3, \not{\mathfrak{d}}] = \Gamma_b\mathfrak{d} + \Gamma_b, \quad [e_4r, \not{\mathfrak{d}}] = (r^2\xi + r\Gamma_g)\mathfrak{d} + r\Gamma_g. \end{aligned}$$

In view of the estimates for Γ_g, Γ_b, and the fact that $\xi = 0$ for $r \geq 4m_0$, we infer

$$[\mathfrak{d}^k, e_3] = r^{-1}\mathfrak{d}^{\leq k}, \quad [\mathfrak{d}^k, r^2\not{\Delta}] = \mathfrak{d}^{\leq k+1}, \quad [\mathfrak{d}^k, r^{-1}] = r^{-1}\mathfrak{d}^{\leq k-1}$$

and hence

$$\begin{aligned} \mathfrak{d}^k e_3e_4(r\psi) &= e_3e_4(r\mathfrak{d}^k\psi) + e_3[\mathfrak{d}^k, e_4r]\psi + r^{-1}\mathfrak{d}^{\leq k+1}\psi, \\ \mathfrak{d}^k(r\not{\Delta}_2\psi) &= r\not{\Delta}_2\mathfrak{d}^k\psi + r^{-1}\mathfrak{d}^{\leq k+1}\psi. \end{aligned}$$

Also, we have

$$[re_4, e_4r] = [re_4, e_4]r + e_4[re_4, r] = -e_4(r)e_4r - e_4re_4(r) = -2e_4r + r^{-1}\mathfrak{d}$$

and we infer by induction, schematically,

$$[(re_4)^j, e_4r] = e_4r(re_4)^{\leq j-1} + r^{-1}\mathfrak{d}^{\leq j}$$

so that, together with

$$[\mathfrak{d}_{\searrow}^{k-j}, e_4r] = r^{-1}\mathfrak{d}^{\leq k-j},$$

we infer

$$[\mathfrak{d}^k, e_4r] \;=\; [(re_4)^j\mathfrak{d}_{\searrow}^{k-j}, e_4r] = e_4r(re_4)^{\leq j-1}\mathfrak{d}_{\searrow}^{k-j} + r^{-1}\mathfrak{d}^{\leq k}.$$

We deduce

$$\begin{aligned} e_3e_4(r(re_4)^j\mathfrak{d}_{\searrow}^{k-j}\psi) &= -(re_4)^j\mathfrak{d}_{\searrow}^{k-j}(r\square_2\psi)+r\not{A}_2(\mathfrak{d}^k\psi)+r^{-1}\mathfrak{d}^{\leq k+1}\psi \\ &\quad +e_4r(re_4)^{\leq j-1}\mathfrak{d}_{\searrow}^{k-j}\psi. \end{aligned}$$

We infer by induction on j

$$e_3e_4(r(re_4)^j\mathfrak{d}_{\searrow}^{k-j}\psi) = -(re_4)^{\leq j}\mathfrak{d}_{\searrow}^{k-j}(r\square_2\psi)+r\not{A}_2(\mathfrak{d}^{\leq k}\psi)+r^{-1}\mathfrak{d}^{\leq k+1}\psi$$

and hence

$$e_3e_4(r\mathfrak{d}^k\psi) = -\mathfrak{d}^{\leq k}(r\square_2\psi)+r\not{A}_2(\mathfrak{d}^{\leq k}\psi)+r^{-1}\mathfrak{d}^{\leq k+1}\psi$$

which is (5.3.24). This concludes the proof of Lemma 5.20. □

We now are in position to prove Proposition 5.19.

Proof of Proposition 5.19. We integrate by parts

$$\begin{aligned} J_{p,4m_0}[\mathfrak{q}^{(k)},\phi_0\,\mathfrak{d}^k(rN_g)](\tau_1,\tau_2) &\lesssim \left|\int_{\mathcal{M}(\tau_1,\tau_2)} e_3\left(\phi_0(r)r^p\check{e}_4\mathfrak{q}^{(k)}\right)\mathfrak{d}^k(rN_g)\right| \\ &\quad +|B_p^k(\tau_1)|+|B_p^k(\tau_2)| \\ &\quad +\left|\int_{\mathcal{M}(\tau_1,\tau_2)} \mathbf{Div}(e_3)\phi_0(r)r^p\check{e}_4\mathfrak{q}^{(k)}\mathfrak{d}^k(rN_g)\right| \end{aligned} \tag{5.3.25}$$

where $\mathbf{Div}(e_3)$ denotes the spacetime divergence of e_3, and where the boundary terms are bounded by

$$\begin{aligned} |B_p^k(\tau_1)| &\lesssim \int_{\Sigma(\tau_1)} r^p|\check{e}_4\mathfrak{q}^{(k)}|\,|\mathfrak{d}^k(rN_g)|, \\ |B_p^k(\tau_2)| &\lesssim \int_{\Sigma(\tau_2)} r^p|\check{e}_4\mathfrak{q}^{(k)}|\,|\mathfrak{d}^k(rN_g)|. \end{aligned}$$

We estimate

$$\begin{aligned} |B_p^k(\tau)| &\lesssim \int_{\Sigma(\tau)} r^p|\check{e}_4\mathfrak{q}^{(k)}|\,|\mathfrak{d}^k(rN_g)| \\ &\lesssim \left(\int_{\Sigma(\tau)} r^p|\check{e}_4\mathfrak{q}^{(k)}|^2\right)^{1/2}\left(\int_{\Sigma(\tau)} r^p|\mathfrak{d}^k(rN_g)|^2\right)^{1/2} \\ &\lesssim \left(E_p^k[\mathfrak{q}](\tau)\right)^{1/2}\left(\int_{\Sigma(\tau)} r^{p+2}|\mathfrak{d}^kN_g|^2\right)^{1/2}. \end{aligned}$$

We deduce, with $\delta_1>0$ a sufficiently small constant, for any $\tau\in[\tau_1,\tau_2]$,

$$\begin{aligned} \left|B_p^k(\tau_1)\right| &\lesssim \delta_1\sup_{\tau_1\leq\tau\leq\tau_2}E_p^k[\mathfrak{q}](\tau)+\delta_1^{-1}\sup_{\tau_1\leq\tau\leq\tau_2}\int_{\Sigma(\tau)} r^{p+2}|N_g^{\leq k}|^2, \\ \left|B_p^k(\tau_2)\right| &\lesssim \delta_1\sup_{\tau_1\leq\tau\leq\tau_2}E_p^k[\mathfrak{q}](\tau)+\delta_1^{-1}\sup_{\tau_1\leq\tau\leq\tau_2}\int_{\Sigma(\tau)} r^{p+2}|N_g^{\leq k}|^2. \end{aligned} \tag{5.3.26}$$

Next, notice that $\mathbf{Div}(e_3)=\underline{\kappa}-2\underline{\omega}$ so that

$$|\mathbf{Div}(e_3)| \lesssim r^{-1}.$$

Together with the fact that $e_3(\Phi_0(r))$ is supported in $4m_0\leq r\leq 8m_0$, the fact that $|e_3(r)|\lesssim 1$ and

$$r\check{e}_4\mathfrak{q}^{(k)} = e_4(r\mathfrak{q}^{(k)})+O(r^{-1})e_4(\mathfrak{q}^{(k)}),$$

we infer

$$\begin{aligned}
&\left|\int_{\mathcal{M}(\tau_1,\tau_2)} e_3\left(\phi_0(r)r^p\check{e}_4\mathfrak{q}^{(k)}\right)\mathfrak{d}^k(rN_g)\right|\\
&+\left|\int_{\mathcal{M}(\tau_1,\tau_2)}\mathbf{Div}(e_3)\phi_0(r)r^p\check{e}_4\mathfrak{q}^{(k)}\mathfrak{d}^k(rN_g)\right|\\
\lesssim\ &\left|\int_{\mathcal{M}(\tau_1,\tau_2)}\phi_0(r)r^{p-1}e_3e_4(r\mathfrak{q}^{(k)})\mathfrak{d}^k(rN_g)\right|\\
&+\int_{\mathcal{M}_{\geq 4m_0}(\tau_1,\tau_2)} r^{p-1}|\check{e}_4(\mathfrak{q}^{(k)})||\mathfrak{d}^k(rN_g)|\\
&+\int_{\mathcal{M}_{4m_0\leq r\leq 8m_0}(\tau_1,\tau_2)}|\check{e}_4(\mathfrak{q}^{(k)})||\mathfrak{d}^k(rN_g)|\\
\lesssim\ &\left|\int_{\mathcal{M}(\tau_1,\tau_2)}\phi_0(r)r^{p-1}e_3e_4(r\mathfrak{q}^{(k)})\mathfrak{d}^k(rN_g)\right|\\
&+\left(\int_{\mathcal{M}_{\geq 4m_0}(\tau_1,\tau_2)} r^{p-1}|\check{e}_4(\mathfrak{q}^{(k)})|^2\right)^{\frac{1}{2}}\left(\int_{\mathcal{M}_{\geq 4m_0}(\tau_1,\tau_2)} r^{p+1}|\mathfrak{d}^{\leq k}N_g|^2\right)^{\frac{1}{2}}
\end{aligned}$$

and hence

$$\begin{aligned}
&\left|\int_{\mathcal{M}(\tau_1,\tau_2)} e_3\left(\phi_0(r)r^p\check{e}_4\mathfrak{q}^{(k)}\right)\mathfrak{d}^k(rN_g)\right|\\
&+\left|\int_{\mathcal{M}(\tau_1,\tau_2)}\mathbf{Div}(e_3)\phi_0(r)r^p\check{e}_4\mathfrak{q}^{(k)}\mathfrak{d}^k(rN_g)\right|\\
\lesssim\ &\left|\int_{\mathcal{M}(\tau_1,\tau_2)}\phi_0(r)r^{p-1}e_3e_4(r\mathfrak{q}^{(k)})\mathfrak{d}^k(rN_g)\right|+\left(B^s_p[\mathfrak{q}](\tau_1,\tau_2)\right)^{1/2}\left(I^s_p[N_g](\tau_1,\tau_2)\right)^{\frac{1}{2}}
\end{aligned}$$

which yields

$$\begin{aligned}
&\left|\int_{\mathcal{M}(\tau_1,\tau_2)} e_3\left(\phi_0(r)r^p\check{e}_4\mathfrak{q}^{(k)}\right)\mathfrak{d}^k(rN_g)\right|\\
&+\left|\int_{\mathcal{M}(\tau_1,\tau_2)}\mathbf{Div}(e_3)\phi_0(r)r^p\check{e}_4\mathfrak{q}^{(k)}\mathfrak{d}^k(rN_g)\right|\\
\lesssim\ &\left|L^k\right|+\delta_1 B^s_p[\mathfrak{q}](\tau_1,\tau_2)+\delta_1^{-1}I^s_p[N_g](\tau_1,\tau_2) \qquad (5.3.27)
\end{aligned}$$

where $\delta_1 > 0$ is chosen sufficiently small so that we can later absorb the term $\delta_1 B_p^s[\mathfrak{q}](\tau_1,\tau_2)$ by the left-hand side of our main estimate, and where we have introduced the notation

$$L^k := \int_{\mathcal{M}(\tau_1,\tau_2)} \phi_0(r) r^{p-1} e_3 e_4\big(r\mathfrak{q}^{(k)}\big)\mathfrak{d}^k(rN_g). \tag{5.3.28}$$

It remains to estimate the term L^k. Making use of Lemma 5.20, we deduce

$$\begin{aligned} L^k &= \int_{\mathcal{M}(\tau_1,\tau_2)} \phi_0(r) r^{p-1} e_3 e_4(r\mathfrak{q}^{(k)})\mathfrak{d}^k(rN_g) \\ &= -\int_{\mathcal{M}(\tau_1,\tau_2)} \phi_0(r) r^{p-1}\mathfrak{d}^{\leq k}(r\square_2\mathfrak{q})\mathfrak{d}^k(rN_g) \\ &+ \int_{\mathcal{M}(\tau_1,\tau_2)} \phi_0(r) r^{p} \not{\Delta}_2(\mathfrak{d}^{\leq k}\mathfrak{q})\mathfrak{d}^k(rN_g) \\ &+ \int_{\mathcal{M}(\tau_1,\tau_2)} \phi_0(r) r^{p-2}\mathfrak{d}^{\leq k+1}\mathfrak{q}\,\mathfrak{d}^k(rN_g) \\ &= L_1^k + L_2^k + L_3^k. \end{aligned}$$

We first estimate L_3^k as follows:

$$\begin{aligned} |L_3^k| &\lesssim \int_{\mathcal{M}_{\geq 4m_0}(\tau_1,\tau_2)} r^{p-2}|\mathfrak{d}^{\leq k+1}\mathfrak{q}|\,|\mathfrak{d}^k(rN_g)| \\ &\lesssim \left(\int_{\mathcal{M}_{\geq 4m_0}(\tau_1,\tau_2)} r^{p-3}\big|\mathfrak{d}^{\leq k+1}\mathfrak{q}\big|^2\right)^{1/2}\left(\int_{\mathcal{M}_{\geq 4m_0}(\tau_1,\tau_2)} r^{p+1}|\mathfrak{d}^{\leq k}N_g|^2\right)^{1/2}. \end{aligned}$$

In view of Remark 5.8 we thus deduce

$$\begin{aligned} |L_3^k| &\lesssim \big(B_p^k[\mathfrak{q}]\big)^{1/2}\left(\int_{\mathcal{M}_{\geq 4m_0}(\tau_1,\tau_2)} r^{p+1}|\mathfrak{d}^{\leq k}N_g|^2\right)^{1/2} \\ &\lesssim \big(B_p^k[\mathfrak{q}](\tau_1,\tau_2)\big)^{1/2}\big(I_p^k[N_g](\tau_1,\tau_2)\big)^{1/2} \end{aligned}$$

and hence

$$|L_3^k| \lesssim \delta_1 B_p^s[\mathfrak{q}](\tau_1,\tau_2) + \delta_1^{-1} I_p^s[N_g](\tau_1,\tau_2) \tag{5.3.29}$$

where $\delta_1 > 0$ is chosen sufficiently small so that we can later absorb the term $\delta_1 B_p^s[\mathfrak{q}](\tau_1,\tau_2)$ by the left-hand side of our main estimate.

We now estimate the term

$$L_2^k = \int_{\mathcal{M}(\tau_1,\tau_2)} \phi_0(r) r^{p} \not{\Delta}_2(\mathfrak{d}^k\mathfrak{q})\mathfrak{d}^k(rN_g)$$

by first performing another integration by parts in the angular directions

$$
\begin{aligned}
|L_2^k| &\lesssim \int_{\mathcal{M}_{\geq 4m_0}(\tau_1,\tau_2)} r^{p-2}|\mathfrak{d}^{k+1}\mathfrak{q}||\mathfrak{d}^{k+1}(rN_g)| \\
&\lesssim \left(\int_{\mathcal{M}_{\geq 4m_0}(\tau_1,\tau_2)} r^{p-3}|\mathfrak{d}^{k+1}\mathfrak{q}|^2\right)^{1/2}\left(\int_{\mathcal{M}_{\geq 4m_0}(\tau_1,\tau_2)} r^{p+1}|\mathfrak{d}^{\leq k+1}N_g|^2\right)^{1/2} \\
&\lesssim \left(B_p^k[\mathfrak{q}](\tau_1,\tau_2)\right)^{1/2}\left(I_p^{k+1}[N_g](\tau_1,\tau_2)\right)^{1/2}.
\end{aligned}
$$

Hence,

$$
|L_2^k| \lesssim \delta_1 B_p^s[\mathfrak{q}](\tau_1,\tau_2) + \delta_1^{-1} I_p^{s+1}[N_g](\tau_1,\tau_2) \tag{5.3.30}
$$

where $\delta_1 > 0$ is chosen sufficiently small so that we can later absorb the term $\delta_1 B_p^s[\mathfrak{q}](\tau_1,\tau_2)$ by the left-hand side of our main estimate.

It remains to estimate the term

$$
L_1^k = -\int_{\mathcal{M}(\tau_1,\tau_2)} \phi_0(r) r^{p-1}\mathfrak{d}^{\leq k}(r\square_2\mathfrak{q})\mathfrak{d}^k(rN_g).
$$

Making use of the equation verified by $\mathfrak{q}$, i.e., $\square_2\mathfrak{q} = -\kappa\underline{\kappa}\mathfrak{q} + N$, we deduce

$$
\mathfrak{d}^k(r\square_2\mathfrak{q}) = -\mathfrak{d}^k(r\kappa\underline{\kappa}\mathfrak{q}) + \mathfrak{d}^k(rN).
$$

Recall (5.1.9)

$$
\mathfrak{d}^k N = \mathfrak{d}^{\leq k}N_g + e_3(\mathfrak{d}^k(rN_g)) + \mathfrak{d}^k N_m[\mathfrak{q}].
$$

We infer

$$
\begin{aligned}
\mathfrak{d}^{\leq k}(rN) &= r\mathfrak{d}^{\leq k}N + \mathfrak{d}^{\leq k-1}N \\
&= r\mathfrak{d}^{\leq k}N_g + re_3(\mathfrak{d}^{\leq k}(rN_g)) + r\mathfrak{d}^{\leq k}N_m[\mathfrak{q}]
\end{aligned}
$$

and hence

$$
\begin{aligned}
|\mathfrak{d}^k(r\square_2\mathfrak{q})| &\lesssim r^{-1}|\mathfrak{d}^{\leq k}\mathfrak{q}| + r|\mathfrak{d}^{\leq k}N_g| + r^2|\mathfrak{d}^{\leq k}e_3(N_g)| + r|\mathfrak{d}^k N_m[\mathfrak{q}]| \\
&\lesssim r^{-1}|\mathfrak{d}^{\leq k+1}\mathfrak{q}| + r|\mathfrak{d}^{\leq k}N_g| + r^2|\mathfrak{d}^{\leq k}e_3(N_g)|.
\end{aligned} \tag{5.3.31}
$$

Note that we have used in the last inequality the form of $N_m[\mathfrak{q}] = \mathfrak{d}^{\leq 1}(\Gamma_g\mathfrak{q})$ and the fact that $|\Gamma_g| \leq \epsilon r^{-2}$. We deduce, using (5.3.31),

$$
\begin{aligned}
|L_1^k| &\lesssim \int_{\mathcal{M}_{\geq 4m_0}(\tau_1,\tau_2)} r^{p-1}|\mathfrak{d}^{\leq k+1}\mathfrak{q}||\mathfrak{d}^{\leq k}N_g| + \int_{\mathcal{M}_{\geq 4m_0}(\tau_1,\tau_2)} r^{p+1}|\mathfrak{d}^{\leq k}N_g|^2 \\
&\quad + \int_{\mathcal{M}_{\geq 4m_0}(\tau_1,\tau_2)} r^{p+2}|\mathfrak{d}^{\leq k}e_3(N_g)||\mathfrak{d}^{\leq k}N_g| \\
&\lesssim \left(B_p^k[\mathfrak{q}](\tau_1,\tau_2)\right)^{1/2}\left(I_p^k[N_g](\tau_1,\tau_2)\right)^{1/2} + I_p^k[N_g](\tau_1,\tau_2).
\end{aligned}
$$

We deduce

$$
|L_1^k| \lesssim \delta_1 B_p^s[\mathfrak{q}](\tau_1,\tau_2) + \delta_1^{-1} I_p^s[N_g](\tau_1,\tau_2) \tag{5.3.32}
$$

where $\delta_1 > 0$ is chosen sufficiently small so that we can later absorb the term $\delta_1 B_p^s[\mathfrak{q}](\tau_1,\tau_2)$ by the left-hand side of our main estimate.

Together with (5.3.29) and (5.3.30) we deduce

$$|L^k| \lesssim \delta_1 B_p^k[\mathfrak{q}](\tau_1,\tau_2) + \delta_1^{-1} I_p^k[N_g](\tau_1,\tau_2). \tag{5.3.33}$$

Together with (5.3.25), (5.3.26) and (5.3.28), we infer

$$\sum_{k\le s} J_{p,4m_0}[\mathfrak{q}^{(k)}, \phi_0\, \mathfrak{d}^k(rN_g)](\tau_1,\tau_2) \lesssim \delta_1 B_p^s[\mathfrak{q}](\tau_1,\tau_2) + \delta_1^{-1} I_p^{s+1}[N_g](\tau_1,\tau_2)$$

which concludes the proof of Proposition 5.19. □

5.3.3 Proof of Theorem 5.15

We apply Theorem 5.18 to the case when $\psi = \mathfrak{q}$. Hence,

$$\begin{aligned} E_q^s[\check{\mathfrak{q}}](\tau_2) + B_q^s[\check{\mathfrak{q}}](\tau_1,\tau_2) &\lesssim E_q^s[\check{\mathfrak{q}}](\tau_1) + \check{J}_q^s[\check{\mathfrak{q}}, N](\tau_1,\tau_2) \\ &\quad + E^{s+1}_{\max(q,\delta)}[\mathfrak{q}](\tau_1) + J^{s+1}_{\max(q,\delta)}[\mathfrak{q}, N](\tau_1,\tau_2). \end{aligned} \tag{5.3.34}$$

Also, recall that

$$\check{\mathfrak{q}} = f_2 \check{e}_4 \mathfrak{q},$$

where f_2 is a fixed smooth function of r defined as follows,

$$f_2(r) = \begin{cases} r^2 & \text{for } \quad r \ge 6m_0, \\ 0 & \text{for } \quad r \le 4m_0. \end{cases} \tag{5.3.35}$$

In particular, $\check{\mathfrak{q}}$ is supported in $r \ge 4m_0$, and hence, in view of Remark 5.5,

$$B_q[\check{\mathfrak{q}}](\tau_1,\tau_2) \simeq \int_{\mathcal{M}_{\ge 4m_0}(\tau_1,\tau_2)} r^{q-3} |\mathfrak{d}\check{\mathfrak{q}}|^2, \tag{5.3.36}$$

where we have used the fact that $-1+\delta \le q \le 1-\delta$.

First, notice that the proof of Theorem 5.14 yields

$$\begin{aligned} J^{s+1}_{\max(q,\delta)}[\mathfrak{q}, N](\tau_1,\tau_2) &\lesssim \sup_{\tau_1\le\tau\le\tau_2} E^{s+1}[\mathfrak{q}](\tau) + B^{s+1}_{\max(q,\delta)}[\mathfrak{q}](\tau_1,\tau_2) \\ &\quad + I^{s+2}_{\max(q,\delta)}[N_g](\tau_1,\tau_2). \end{aligned}$$

Hence, using Theorem 5.14, together with the fact that $\max(q,\delta) \le 1-\delta$, we infer

$$J^{s+1}_{\max(q,\delta)}[\mathfrak{q}, N](\tau_1,\tau_2) \lesssim E^{s+1}_{\max(q,\delta)}[\mathfrak{q}](\tau_1) + I^{s+2}_{\max(q,\delta)}[N_g](\tau_1,\tau_2).$$

Since $q \ge -1+\delta$, we have $\max(q,\delta) \le \delta \le q+1$ and thus

$$J^{s+1}_{\max(q,\delta)}[\mathfrak{q}, N](\tau_1,\tau_2) \lesssim E^{s+1}_{q+1}[\mathfrak{q}](\tau_1) + I^{s+2}_{q+1}[N_g](\tau_1,\tau_2). \tag{5.3.37}$$

It only remains to estimate the term

$$\check{J}_q^s[\check{\mathfrak{q}}, N](\tau_1, \tau_2) = \sum_{k \le s} \check{J}_q[\mathfrak{d}^k \check{\mathfrak{q}}, \mathfrak{d}^k N](\tau_1, \tau_2)$$

with

$$\begin{aligned} \check{J}_q[\check{\mathfrak{q}}, N](\tau_1, \tau_2) &= J_{q,4m_0}\left[\check{\mathfrak{q}}, r^2\left(e_4 N + \frac{3}{r}N\right)\right](\tau_1, \tau_2) \\ &= \int_{\mathcal{M}_{\ge 4m_0}(\tau_1,\tau_2)} r^{q+2}(\check{e}_4 \check{\mathfrak{q}}) \cdot \left(e_4 N + \frac{3}{r}N\right). \end{aligned}$$

We rewrite in the equivalent form

$$\check{J}_q[\mathfrak{d}^k \check{\mathfrak{q}}, \mathfrak{d}^k N](\tau_1, \tau_2) = \left| \int_{\mathcal{M}_{\ge 4m_0}(\tau_1,\tau_2)} r^q \left(r \check{e}_4 \mathfrak{d}^k \check{\mathfrak{q}}\right) \left(\mathfrak{d}^{k+1} N\right) \right|. \tag{5.3.38}$$

Using the identity (5.1.9), we have

$$\mathfrak{d}^{k+1} N = \mathfrak{d}^{\le k+1} N_g + e_3(\mathfrak{d}^{\le k+1} r N_g) + \mathfrak{d}^{k+1} N_m[\mathfrak{q}].$$

The integral due to $\mathfrak{d}^{\le k+1} N_g$ is treated as follows:

$$\begin{aligned} \check{J}_q[\mathfrak{d}^k \check{\mathfrak{q}}, \mathfrak{d}^k N_g](\tau_1, \tau_2) &\lesssim \int_{\mathcal{M}_{\ge 4m_0}(\tau_1,\tau_2)} r^q \left| r \check{e}_4 \mathfrak{d}^k \check{\mathfrak{q}} \right| \left| \mathfrak{d}^{\le k+1} N_g \right| \\ &\lesssim \left(\int_{\mathcal{M}_{\ge 4m_0}(\tau_1,\tau_2)} r^{q-3} \left| r \check{e}_4 \mathfrak{d}^k \check{\mathfrak{q}} \right|^2 \right)^{1/2} \\ &\quad \times \left(\int_{\mathcal{M}_{\ge 4m_0}(\tau_1,\tau_2)} r^{q+3} \left| \mathfrak{d}^{\le k+1} N_g \right|^2 \right)^{1/2}. \end{aligned}$$

Therefore,

$$\begin{aligned} \check{J}_q^s[\check{\mathfrak{q}}, N_g](\tau_1, \tau_2) &\lesssim \left(B_q^s[\check{\mathfrak{q}}](\tau_1, \tau_2)\right)^{1/2} \left(I_{q+2}^{s+1}[N_g](\tau_1, \tau_2)\right)^{1/2} \\ &\lesssim \delta_1 B_q^s[\check{\mathfrak{q}}](\tau_1, \tau_2) + \delta_1^{-1} I_{q+2}^{s+1}[N_g](\tau_1, \tau_2) \end{aligned} \tag{5.3.39}$$

where $\delta_1 > 0$ is chosen sufficiently small so that we can later absorb the term $\delta_1 B_q^s[\mathfrak{q}](\tau_1, \tau_2)$ by the left-hand side of our main estimate.

The integral due to $\mathfrak{d}^{k+1}N_m[\mathfrak{q}]$ is treated as follows:

$$\begin{aligned}
&\check{J}_q[\mathfrak{d}^k\check{\mathfrak{q}}, \mathfrak{d}^k N_m[\mathfrak{q}]](\tau_1,\tau_2)\\
\lesssim\; & \int_{\mathcal{M}_{\geq 4m_0}(\tau_1,\tau_2)} r^q\left|r\check{e}_4\mathfrak{d}^k\check{\mathfrak{q}}\right|\left|\mathfrak{d}^{k+1}N_m[\mathfrak{q}]\right|\\
\lesssim\; & \int_{\mathcal{M}_{\geq 4m_0}(\tau_1,\tau_2)} r^{q+1}\left|\check{e}_4\mathfrak{d}^k\check{\mathfrak{q}}\right|\left|\mathfrak{d}^{\leq k+2}\mathfrak{q}\right|\left|\mathfrak{d}^{\leq k+2}\Gamma_g\right|\\
\lesssim\; & \epsilon\int_{\mathcal{M}_{\geq 4m_0}(\tau_1,\tau_2)} r^{q-1}\tau^{-\frac{1}{2}-\delta_{dec}+2\delta_0}\left|\check{e}_4\mathfrak{d}^k\check{\mathfrak{q}}\right|\left|\mathfrak{d}^{\leq k+2}\mathfrak{q}\right|\\
\lesssim\; & \epsilon\left(\int_{\mathcal{M}_{\geq 4m_0}(\tau_1,\tau_2)} r^q\tau^{-1-2\delta_{dec}+4\delta_0}\left|\check{e}_4\mathfrak{d}^k\check{\mathfrak{q}}\right|^2\right)^{\frac{1}{2}}\left(\int_{\mathcal{M}_{\geq 4m_0}\tau_1,\tau_2)} r^{q-2}\left|\mathfrak{d}^{\leq k+2}\mathfrak{q}\right|^2\right)^{\frac{1}{2}}\\
\lesssim\; & \epsilon\left(\sup_{\tau_1\leq\tau\leq\tau_2} E^s_q[\check{\mathfrak{q}}](\tau)\right)^{\frac{1}{2}}\left(B^{s+1}_{q+1}[\mathfrak{q}](\tau_1,\tau_2)\right)^{\frac{1}{2}}
\end{aligned}$$

where we have used $|\Gamma_g| \lesssim \epsilon r^{-2}\tau^{-1/2-\delta_{dec}+2\delta_0}$ and $2\delta_0 < \delta_{dec}$. Since $\delta \leq q+1 \leq 2-\delta$ and $s \leq k_{small}+29$, we have in view of Theorem 5.14

$$B^{s+1}_{q+1}[\mathfrak{q}](\tau_1,\tau_2) \;\lesssim\; E^{s+1}_{q+1}[\mathfrak{q}](\tau_1) + I^{s+2}_{q+1}[N_g](\tau_1,\tau_2).$$

We infer

$$\begin{aligned}
&\sum_{k\leq s}\check{J}_q[\mathfrak{d}^k\check{\mathfrak{q}}, \mathfrak{d}^k N_m[\mathfrak{q}]](\tau_1,\tau_2)\\
\lesssim\; & \epsilon^2\sup_{\tau_1\leq\tau\leq\tau_2} E^s_q[\check{\mathfrak{q}}](\tau) + E^{s+1}_{q+1}[\mathfrak{q}](\tau_1) + I^{s+2}_{q+1}[N_g](\tau_1,\tau_2). \qquad (5.3.40)
\end{aligned}$$

It remains to estimate the integral due to $e_3(\mathfrak{d}^{\leq k+1}rN_g)$. We proceed as in Proposition 5.19 by integration by parts, and obtain in particular the following analog of (5.3.27):

$$\check{J}_q[\mathfrak{d}^k\check{\mathfrak{q}}, \mathfrak{d}^k e_3(rN_g)](\tau_1,\tau_2) \lesssim \left|P^k\right| + \delta_1 B^s_q[\check{\mathfrak{q}}](\tau_1,\tau_2) + \delta_1^{-1}I^{s+1}_{q+2}[N_g](\tau_1,\tau_2) \qquad (5.3.41)$$

where $\delta_1 > 0$ is chosen sufficiently small so that we can later absorb the term $\delta_1 B^s_q[\check{\mathfrak{q}}](\tau_1,\tau_2)$ by the left-hand side of our main estimate, and where we have introduced the notation P^k for the analog of L^k in (5.3.28), i.e.,[6]

$$P^k := \int_{\mathcal{M}(\tau_1,\tau_2)} r^q e_3 e_4(r\mathfrak{d}^k\check{\mathfrak{q}})\mathfrak{d}^{\leq k+1}(rN_g). \qquad (5.3.42)$$

As in Lemma 5.20,

$$e_3e_4(r\mathfrak{d}^k\check{\mathfrak{q}}) \;=\; -\mathfrak{d}^{\leq k}(r\square_2\check{\mathfrak{q}}) + r\Delta\!\!\!\!/_2(\mathfrak{d}^{\leq k}\check{\mathfrak{q}}) + r^{-1}\mathfrak{d}^{\leq k+1}\check{\mathfrak{q}}. \qquad (5.3.43)$$

[6] Recall that $\check{\mathfrak{q}}$ is localized in $r \geq 4m_0$ so that we don't need in (5.3.42) the cut-off function $\phi_0(r)$ introduced in Proposition 5.19.

We infer

$$\begin{aligned}
P^k &= \int_{\mathcal{M}(\tau_1,\tau_2)} r^q e_3 e_4 \big(r\mathfrak{d}^k\check{\mathfrak{q}}\big)\mathfrak{d}^{\leq k+1}(rN_g) \\
&= -\int_{\mathcal{M}(\tau_1,\tau_2)} r^q\mathfrak{d}^{\leq k}(r\square_2\check{\mathfrak{q}})\mathfrak{d}^{\leq k+1}(rN_g) \\
&+ \int_{\mathcal{M}(\tau_1,\tau_2)} r^{q+1}\not{\Delta}_2(\mathfrak{d}^{\leq k}\check{\mathfrak{q}})\mathfrak{d}^{\leq k+1}(rN_g) \\
&+ \int_{\mathcal{M}(\tau_1,\tau_2)} r^{q-1}\mathfrak{d}^{\leq k+1}\check{\mathfrak{q}}\ \mathfrak{d}^{\leq k+1}(rN_g) \\
&= P_1^k + P_2^k + P_3^k.
\end{aligned}$$

The last two terms on the right can be treated exactly as the corresponding terms in the treatment of L^k. This yields to the following analog of (5.3.29) and (5.3.30)

$$\begin{aligned}
\big|P_3^k\big| &\lesssim \delta_1 B_q^s[\check{\mathfrak{q}}](\tau_1,\tau_2) + \delta_1^{-1} I_{q+2}^{s+1}[N_g](\tau_1,\tau_2), \\
\big|P_2^k\big| &\lesssim \delta_1 B_q^s[\check{\mathfrak{q}}](\tau_1,\tau_2) + \delta_1^{-1} I_{q+2}^{s+2}[N_g](\tau_1,\tau_2),
\end{aligned} \tag{5.3.44}$$

where $\delta_1 > 0$ is chosen sufficiently small so that we can later absorb the term $\delta_1 B_p^s[\check{\mathfrak{q}}](\tau_1,\tau_2)$ by the left-hand side of our main estimate.

It thus only remains to consider the term analogous to L_1^k, i.e.,

$$P_1^k = \int_{\mathcal{M}(\tau_1,\tau_2)} r^q\mathfrak{d}^k(r\square_2\check{\mathfrak{q}})\mathfrak{d}^{\leq k+1}(rN_g).$$

Now, in view of Proposition 10.47, $\mathfrak{q}$ verifies, schematically,

$$\square_2\check{\mathfrak{q}} \quad = \quad r^{-2}\mathfrak{d}^{\leq 1}\check{\mathfrak{q}} + r^{-2}\mathfrak{d}^{\leq 2}\mathfrak{q} + r\mathfrak{d}^{\leq 1}N$$

so that

$$\begin{aligned}
\mathfrak{d}^k(r\square_2\check{\mathfrak{q}}) &= r^{-1}\mathfrak{d}^{\leq k+1}\check{\mathfrak{q}} + r^{-1}\mathfrak{d}^{\leq k+2}\mathfrak{q} + r^2\mathfrak{d}^{\leq k+1}N \\
&= r^{-1}\mathfrak{d}^{\leq k+1}\check{\mathfrak{q}} + r^{-1}\mathfrak{d}^{\leq k+2}\mathfrak{q} + r^2\mathfrak{d}^{\leq k+1}N_g + r^2\mathfrak{d}^{\leq k+1}N_m[\mathfrak{q}] \\
&\quad + r^2\mathfrak{d}^{\leq k+1}e_3(rN_g).
\end{aligned}$$

We infer the following decomposition of P_1^k:

$$\begin{aligned}
P_1^k &= \int_{\mathcal{M}(\tau_1,\tau_2)} r^{q-1}\Big(\mathfrak{d}^{\leq k+1}\check{\mathfrak{q}} + \mathfrak{d}^{\leq k+2}\mathfrak{q}\Big)\mathfrak{d}^{\leq k+1}(rN_g) \\
&+ \int_{\mathcal{M}(\tau_1,\tau_2)} r^{q+2}\mathfrak{d}^{\leq k+1}N_m[\mathfrak{q}]\mathfrak{d}^{\leq k+1}(rN_g) \\
&+ \int_{\mathcal{M}(\tau_1,\tau_2)} r^{q+2}\Big(\mathfrak{d}^{\leq k+1}N_g + \mathfrak{d}^{\leq k+1}e_3(rN_g)\Big)\mathfrak{d}^{\leq k+1}(rN_g) \\
&= P_{11}^k + P_{12}^k + P_{13}^k.
\end{aligned}$$

P^k_{11} is estimated as $\check{J}^s_q[\check{\mathfrak{q}}, N_g](\tau_1,\tau_2)$, see (5.3.39), and hence

$$\begin{aligned}|P^k_{11}| &\lesssim \left(B^s_q[\check{\mathfrak{q}}](\tau_1,\tau_2)\right)^{1/2}\left(I^{s+1}_{q+2}[N_g](\tau_1,\tau_2)\right)^{1/2}\\ &\lesssim \delta_1 B^s_q[\check{\mathfrak{q}}](\tau_1,\tau_2) + B^s_{\max(q,\delta)}[\mathfrak{q}](\tau_1,\tau_2) + \delta_1^{-1} I^{s+1}_{q+2}[N_g](\tau_1,\tau_2)\end{aligned}$$

which in view of Theorem 5.14 yields

$$|P^k_{11}| \lesssim \delta_1 B^s_q[\check{\mathfrak{q}}](\tau_1,\tau_2) + E^{s+1}_{\max(q,\delta)}[\mathfrak{q}](\tau_1) + \delta_1^{-1} I^{s+1}_{q+2}[N_g](\tau_1,\tau_2). \quad (5.3.45)$$

Next, P^k_{12} is estimated as follows:

$$\begin{aligned}|P^k_{12}| &\lesssim \int_{\mathcal{M}_{\geq 4m_0}(\tau_1,\tau_2)} r^{q+3}|\mathfrak{d}^{\leq k+1}N_m[\mathfrak{q}]||\mathfrak{d}^{\leq k+1}N_g|\\ &\lesssim \int_{\mathcal{M}_{\geq 4m_0}(\tau_1,\tau_2)} r^{q+3}|\mathfrak{d}^{\leq k+2}\Gamma_g||\mathfrak{d}^{\leq k+2}\mathfrak{q}||\mathfrak{d}^{\leq k+1}N_g|\\ &\lesssim \epsilon\int_{\mathcal{M}_{\geq 4m_0}(\tau_1,\tau_2)} r^{q+1}\tau^{-\frac{1}{2}-\delta_{dec}+2\delta_0}|\mathfrak{d}^{\leq k+2}\mathfrak{q}||\mathfrak{d}^{\leq k+1}N_g|\\ &\lesssim \epsilon\left(\int_{\mathcal{M}_{\geq 4m_0}(\tau_1,\tau_2)} r^{q-2}|\mathfrak{d}^{\leq k+2}\mathfrak{q}|^2\right)^{\frac{1}{2}}\\ &\quad\times\left(\int_{\mathcal{M}_{\geq 4m_0}(\tau_1,\tau_2)} r^{q+4}\tau^{-1-2\delta_{dec}+4\delta_0}|\mathfrak{d}^{\leq k+1}N_g|^2\right)^{\frac{1}{2}}\\ &\lesssim \epsilon\Big(B^{s+1}_{q+1}[\mathfrak{q}](\tau_1,\tau_2)\Big)^{\frac{1}{2}}\left(\sup_{\tau\in[\tau_1,\tau_2]}\int_{\Sigma(\tau)} r^{q+4}|\mathfrak{d}^{\leq k+1}N_g|^2\right)^{\frac{1}{2}}\end{aligned}$$

where we have used $|\Gamma_g| \lesssim \epsilon r^{-2}\tau^{-1/2-\delta_{dec}+2\delta_0}$ and $2\delta_0 < \delta_{dec}$. We infer

$$|P^k_{12}| \lesssim B^{s+1}_{q+1}[\mathfrak{q}](\tau_1,\tau_2) + I^{s+1}_{q+2}[N_g](\tau_1,\tau_2). \quad (5.3.46)$$

Finally, P^k_{13} is estimated as follows:

$$\begin{aligned}|P^k_{13}| &\lesssim \int_{\mathcal{M}_{\geq 4m_0}(\tau_1,\tau_2)} r^{q+3}\Big(|\mathfrak{d}^{\leq k+1}N_g| + |\mathfrak{d}^{\leq k+1}e_3(rN_g)|\Big)|\mathfrak{d}^{\leq k+1}N_g|\\ &\lesssim \int_{\mathcal{M}_{\geq 4m_0}(\tau_1,\tau_2)} r^{q+3}|\mathfrak{d}^{\leq k+1}N_g|^2\\ &\quad+\int_{\mathcal{M}_{\geq 4m_0}(\tau_1,\tau_2)} r^{q+4}|\mathfrak{d}^{\leq k+1}e_3(N_g)||\mathfrak{d}^{\leq k+1}N_g|\\ &\lesssim I^{s+1}_{q+2}[N_g](\tau_1,\tau_2).\end{aligned}$$

Together with (5.3.45) and (5.3.46), we infer

$$\begin{aligned}|P^k_1| &\leq |P^k_{11}| + |P^k_{12}| + |P^k_{13}|\\ &\lesssim \delta_1 B^s_q[\check{\mathfrak{q}}](\tau_1,\tau_2) + E^{s+1}_{\max(q,\delta)}[\mathfrak{q}](\tau_1) + \delta_1^{-1} I^{s+1}_{q+2}[N_g](\tau_1,\tau_2) + B^{s+1}_{q+1}[\mathfrak{q}](\tau_1,\tau_2).\end{aligned}$$

Together with (5.3.44), we deduce

$$\begin{aligned} |P^k| &\leq |P_1^k| + |P_2^k| + |P_3^k| \\ &\lesssim \delta_1 B_q^s[\check{\mathfrak{q}}](\tau_1, \tau_2) + E^{s+1}_{\max(q,\delta)}[\mathfrak{q}](\tau_1) + \delta_1^{-1} I^{s+2}_{q+2}[N_g](\tau_1, \tau_2) + B^{s+1}_{q+1}[\mathfrak{q}](\tau_1, \tau_2). \end{aligned}$$

Together with (5.3.34), (5.3.37), (5.3.39), (5.3.40) and (5.3.41), this concludes the proof of Theorem 5.15.

5.4 DECAY ESTIMATES

In this section we prove the decay estimates. In particular:

- In section 5.4.1, we prove first flux decay estimates for $\mathfrak{q}$.
- In section 5.4.2, we prove flux decay estimates for $\check{\mathfrak{q}}$.
- In section 5.4.3, we prove Theorem 5.9.
- In section 5.4.4, we prove Proposition 5.12 on pointwise decay estimates for $\mathfrak{q}$.
- In section 5.4.5, we prove Proposition 5.13 on flux estimates on Σ_* and on improved pointwise estimates for $e_3(\mathfrak{q})$.

The decay estimates rely on the norms (5.1.27) which we recall below.

$$\begin{aligned} \mathcal{E}^s_{p,d}[\psi] &= \sup_{0\leq\tau\leq\tau_*} (1+\tau)^d E^s_p[\psi](\tau), \\ \mathcal{B}^s_{p,d}[\psi] &= \sup_{0\leq\tau\leq\tau_*} (1+\tau)^d \int_\tau^{\tau_*} M^s_{p-1}[\psi](\tau) d\tau, \\ \mathcal{F}^s_{p,d}[\psi] &= \sup_{0\leq\tau\leq\tau_*} (1+\tau)^d F^s_p[\psi](\tau), \\ \mathcal{I}^s_{p,d}[N_g] &= \sup_{0\leq\tau\leq\tau_*} (1+\tau)^d I^s_p[N_g](\tau, \tau_*). \end{aligned}$$

5.4.1 First flux decay estimates

The goal of this section is to prove the following flux decay estimates for $\mathfrak{q}$.

Theorem 5.21. *Assume $\mathfrak{q}$ verifies all the estimates of Theorem 5.14. Then the following estimates hold true for all $s \leq k_{small} + 30$ and for all $\delta \leq p \leq 2 - \delta$:*

$$\begin{aligned} &\mathcal{E}^{s-[2-\delta-p]}_{p,2-\delta-p}[\mathfrak{q}] + \mathcal{B}^{s-[2-\delta-p]}_{p,2-\delta-p}[\mathfrak{q}] + \mathcal{F}^{s-[2-\delta-p]}_{p,2-\delta-p}[\mathfrak{q}] \\ \lesssim\ & \mathcal{E}^s_{2-\delta}[\mathfrak{q}](0) + \mathcal{I}^{s+1}_{2-\delta,0}[N_g] + \mathcal{I}^{s+1}_{\delta,2-2\delta}[N_g]. \end{aligned} \tag{5.4.1}$$

Here, $[x]$ denotes the least integer greater or equal to x.

Proof. We make use of Theorem 5.14 according to which we have, for $\delta \leq p \leq 2-\delta$, and $0 \leq k \leq k_{small} + 30$,

$$E^s_p[\mathfrak{q}](\tau_2) + B^s_p[\mathfrak{q}](\tau_1, \tau_2) + F^s_p[\mathfrak{q}](\tau_1, \tau_2) \lesssim E^s_p[\mathfrak{q}](\tau_1) + I^{s+1}_p[N_g](\tau_1, \tau_2)$$

which we write in the form

$$E^s_p(\tau_2) + \int_{\tau_1}^{\tau_2} M^s_{p-1}(\tau) d\tau \lesssim E^s_p(\tau_1) + I^{s+1}_p[N_g](\tau_1, \tau_2), \quad \delta \leq p \leq 2-\delta. \tag{5.4.2}$$

In particular,

$$E^s_{2-\delta}(\tau) + \int_{\tau/2}^{\tau} M^s_{1-\delta}(\lambda)d\lambda \lesssim E^s_{2-\delta}(\tau/2) + \mathcal{I}^{s+1}_{2-\delta,\,0}[N_g].$$

By the mean value theorem there exists $\tau_0 \in [\tau/2, \tau]$ such that

$$M^s_{1-\delta}(\tau_0) \lesssim \frac{1}{\tau}\left(E^s_{2-\delta}(\tau/2) + \mathcal{I}^{s+1}_{2-\delta,\,0}[N_g]\right).$$

Since[7]

$$E^{s-1}_{1-\delta}(\tau) \lesssim M^s_{1-\delta}(\tau),$$

we deduce

$$E^{s-1}_{1-\delta}(\tau_0) \lesssim \frac{1}{\tau}\left(E^s_{2-\delta}(\tau/2) + \mathcal{I}^{s+1}_{2-\delta,\,0}[N_g]\right).$$

Moreover, applying (5.4.2) again for $p = 1-\delta$, we deduce

$$\begin{aligned} E^{s-1}_{1-\delta}(\tau) + \int_{\tau_0}^{\tau} M^{s-1}_{-\delta}(\lambda)d\lambda &\lesssim E^{s-1}_{1-\delta}(\tau_0) + (1+\tau)^{-1}\mathcal{I}^s_{1-\delta,1}[N_g] \\ &\lesssim (1+\tau)^{-1}\left(E^s_{2-\delta}(\tau/2) + \mathcal{I}^{s+1}_{2-\delta,\,0}[N_g] + \mathcal{I}^s_{1-\delta,1}[N_g]\right). \end{aligned}$$

In particular,

$$E^{s-1}_{1-\delta}(\tau) \lesssim (1+\tau)^{-1}\left(E^s_{2-\delta}(\tau/2) + \mathcal{I}^{s+1}_{2-\delta,\,0}[N_g] + \mathcal{I}^s_{1-\delta,1}[N_g]\right). \quad (5.4.3)$$

Interpolating with

$$E^s_{2-\delta}(\tau) \lesssim E^s_{2-\delta}(\tau/2) + \mathcal{I}^{s+1}_{2-\delta,\,0}[N_g]$$

by using

$$E^s_p \lesssim (E^s_{p_1})^{\frac{p_2-p}{p_2-p_1}}(E^s_{p_2})^{\frac{p-p_1}{p_2-p_1}}, \quad p_1 \le p \le p_2,$$

we deduce

$$\begin{aligned} E^{s-1}_1(\tau) &\lesssim (E^{s-1}_{1-\delta}(\tau))^{1-\delta}(E^{s-1}_{2-\delta}(\tau))^{\delta} \\ &\lesssim (1+\tau)^{-1+\delta}\left(E^s_{2-\delta}(\tau/2) + \mathcal{I}^{s+1}_{2-\delta,\,0}[N_g] + \mathcal{I}^s_{1-\delta,1}[N_g]\right). \end{aligned}$$

The same inequality holds for τ replaced by $\tau/2$, i.e.,

$$E^{s-1}_1(\tau/2) \lesssim (1+\tau)^{-1+\delta}\left(E^s_{2-\delta}(\tau/4) + \mathcal{I}^{s+1}_{2-\delta,\,0}[N_g] + \mathcal{I}^s_{1-\delta,1}[N_g]\right). \quad (5.4.4)$$

We now repeat the procedure starting this time with the inequality (5.4.2) for

[7] Note that the loss of derivative is due to the degeneracy of the bulk integral in the trapping region.

$p = 1$,

$$E_1^{s-1}(\tau) + \int_{\tau/2}^{\tau} M_0^{s-1}(\lambda)d\lambda \lesssim E_1^{s-1}(\tau/2) + I_1^s[N_g](\tau/2, \tau)$$
$$\lesssim E_1^{s-1}(\tau/2) + (1+\tau)^{-1+\delta}\mathcal{I}_{1,1-\delta}^s[N_g].$$

Thus, in view of (5.4.4),

$$\int_{\tau/2}^{\tau} M_0^{s-1}(\lambda)d\lambda$$
$$\lesssim (1+\tau)^{-1+\delta}\Big(E_{2-\delta}^s(\tau/4) + \mathcal{I}_{2-\delta,\,0}^{s+1}[N_g] + \mathcal{I}_{1-\delta,1}^s[N_g] + \mathcal{I}_{1,1-\delta}^s[N_g]\Big)$$

or, since

$$E_{2-\delta}^s(\tau/4) \lesssim \mathcal{E}_{2-\delta}^s(0) + \mathcal{I}_{2-\delta,0}^{s+1}[N_g],$$

we infer that

$$\int_{\tau/2}^{\tau} M_0^{s-1}(\lambda)d\lambda \lesssim B(1+\tau)^{-1+\delta}$$

where

$$B : = \mathcal{E}_{2-\delta}^s(0) + \mathcal{I}_{2-\delta,\,0}^{s+1}[N_g] + \mathcal{I}_{1-\delta,1}^s[N_g] + \mathcal{I}_{1,1-\delta}^s[N_g]. \tag{5.4.5}$$

Repeating the mean value argument, we can find $\tau_1 \in [\tau/2, \tau]$ such that

$$M_0^{s-1}(\tau_1) \lesssim \frac{1}{\tau}\int_{\tau/2}^{\tau} M_0^{s-1}(\lambda)d\lambda \lesssim B(1+\tau)^{-2+\delta}.$$

We now make use of the fact that the energy norm E^{s-1} is comparable with M_0^{s-1} everywhere except in the trapping region where we lose a derivative. Thus

$$E^{s-2}(\tau_1) \lesssim M_0^{s-1}(\tau_1)$$

and therefore,

$$E^{s-2}(\tau_1) \lesssim B(1+\tau)^{-2+\delta}. \tag{5.4.6}$$

We would like now to compare $E^{s-2}(\tau)$ with $E^{s-2}(\tau_1)$ using the usual version of the energy inequality and thus derive a similar estimate for the former. Unfortunately,[8] we don't have a closed energy inequality for E and we therefore have instead to rely on E_δ for which we have the inequality

$$E_\delta^{s-2}(\tau) \lesssim E_\delta^{s-2}(\tau_1) + I_\delta^{s-1}[N_g](\tau_1, \tau). \tag{5.4.7}$$

[8]The loss of δ is due to the fact that we are on a perturbation of Schwarzschild rather than on Schwarzschild.

We also have in view of (5.4.3)

$$E^{s-2}_{1-\delta}(\tau_1) \lesssim (1+\tau)^{-1}\left(E^s_{2-\delta}(0) + \mathcal{I}^{s+1}_{2-\delta,0}[N_g] + \mathcal{I}^s_{1-\delta,1}[N_g]\right).$$

Interpolating this last inequality with (5.4.6) we deduce, for $\delta > 0$ sufficiently small,

$$\begin{aligned} E^{s-2}_{\delta}(\tau_1) &\lesssim \left(E^{s-2}(\tau_1)\right)^{\frac{1-2\delta}{1-\delta}} \left(E^{s-2}_{1-\delta}(\tau_1)\right)^{\frac{\delta}{1-\delta}} \\ &\lesssim (1+\tau)^{-2+2\delta}(B + \mathcal{E}^{s-2}_{1-\delta}(0) + \mathcal{I}^{s-1}_{1-\delta,0}[N_g]) \\ &\lesssim (1+\tau)^{-2+2\delta}B. \end{aligned}$$

Thus, in view of (5.4.7),

$$E^{s-2}_{\delta}(\tau) \lesssim E^{s-2}_{\delta}(\tau_1) + I^{s-1}_{\delta}[N_g](\tau_1,\tau) \lesssim (1+\tau)^{-2+2\delta}(B + \mathcal{I}^{s-1}_{\delta,2-2\delta}[N_g]),$$

i.e.,

$$\begin{aligned} & E^{s-2}_{\delta}(\tau) \\ \lesssim\ & (1+\tau)^{-2+2\delta}(\mathcal{E}^s_{2-\delta}(0) + \mathcal{I}^{s+1}_{2-\delta,0}[N_g] + \mathcal{I}^s_{1-\delta,1}[N_g] + \mathcal{I}^s_{1,1-\delta}[N_g] + \mathcal{I}^{s-1}_{\delta,2-2\delta}[N_g]). \end{aligned}$$

We infer

$$\mathcal{E}^{s-2}_{\delta,2-2\delta} \lesssim \left(\mathcal{E}^s_{2-\delta}(0) + \mathcal{I}^{s+1}_{2-\delta,0}[N_g] + \mathcal{I}^s_{1-\delta,1}[N_g] + \mathcal{I}^s_{1,1-\delta}[N_g] + \mathcal{I}^{s-1}_{\delta,2-2\delta}[N_g]\right)$$

which can be written in the shorter form (by interpolation of the middle terms)

$$\mathcal{E}^{s-2}_{\delta,2-2\delta} \lesssim \mathcal{E}^s_{2-\delta}(0) + \mathcal{I}^{s+1}_{2-\delta,0}[N_g] + \mathcal{I}^{s+1}_{\delta,2-2\delta}[N_g]. \tag{5.4.8}$$

Also, (5.4.3) yields

$$\begin{aligned} \mathcal{E}^{s-1}_{1-\delta,1} &\lesssim \mathcal{E}^s_{2-\delta}(0) + \mathcal{I}^{s+1}_{2-\delta,0}[N_g] + \mathcal{I}^s_{1-\delta,1}[N_g] \\ &\lesssim \mathcal{E}^s_{2-\delta}(0) + \mathcal{I}^{s+1}_{2-\delta,0}[N_g] + \mathcal{I}^{s+1}_{\delta,2-2\delta}[N_g], \end{aligned} \tag{5.4.9}$$

while from Theorem 5.14, we have

$$\mathcal{E}^s_{2-\delta,0} \lesssim \mathcal{E}^s_{2-\delta}(0) + \mathcal{I}^{s+1}_{2-\delta,0}[N_g]. \tag{5.4.10}$$

Interpolating (5.4.8) and (5.4.9), as well as (5.4.9) and (5.4.10), we infer for all $s \le k_{small} + 30$ and for all $\delta \le p \le 2-\delta$

$$\mathcal{E}^{s-[2-\delta-p]}_{p,2-\delta-p}[\mathfrak{q}] \lesssim \mathcal{E}^s_{2-\delta}[\mathfrak{q}](0) + \mathcal{I}^{s+1}_{2-\delta,0}[N_g] + \mathcal{I}^{s+1}_{\delta,2-2\delta}[N_g]. \tag{5.4.11}$$

Finally, making use of Theorem 5.14 between τ and τ_*, we have in particular

$$\begin{aligned} & B^{s-[2-\delta-p]}_p[\mathfrak{q}](\tau,\tau_*) + F^{s-[2-\delta-p]}_p[\mathfrak{q}](\tau,\tau_*) \\ \lesssim\ & E^{s-[2-\delta-p]}_p[\mathfrak{q}](\tau) + I^{s+1-[2-\delta-p]}_p[N_g](\tau,\tau_*) \\ \lesssim\ & (1+\tau)^{-(2-\delta-p)}\left(\mathcal{E}^{s-[2-\delta-p]}_{p,2-\delta-p}[\mathfrak{q}] + \mathcal{I}^{s+1}_{p,2-\delta-p}[N_g]\right) \end{aligned}$$

and hence, we infer for all $s \leq k_{small} + 30$ and for all $\delta \leq p \leq 2-\delta$

$$\mathcal{B}^{s-[2-\delta-p]}_{p,2-\delta-p}[\mathfrak{q}] + \mathcal{F}^{s-[2-\delta-p]}_{p,2-\delta-p}[\mathfrak{q}] \lesssim \mathcal{E}^{s-[2-\delta-p]}_{p,2-\delta-p}[\mathfrak{q}] + \mathcal{I}^{s+1}_{2-\delta,0}[N_g] + \mathcal{I}^{s+1}_{\delta,2-2\delta}[N_g].$$

Together with (5.4.11), this concludes the proof of Theorem 5.21. □

5.4.2 Flux decay estimates for $\check{\mathfrak{q}}$

The goal of this section is to prove the following flux decay estimates for $\check{\mathfrak{q}}$.

Theorem 5.22. *The following estimates hold for all $q_0 - 1 \leq q \leq q_0$, where q_0 is a fixed number $\delta < q_0 \leq 1-\delta$, and $s \leq k_{small} + 28$:*

$$\mathcal{E}^s_{q,q_0-q}[\check{\mathfrak{q}}] + \mathcal{B}^s_{q,q_0-q}[\check{\mathfrak{q}}] \lesssim \mathcal{E}^s_{q_0}[\check{\mathfrak{q}}](0) + \mathcal{E}^{s+2}_{2-\delta}[\mathfrak{q}](0) + \mathcal{I}^{s+3}_{q_0+2,0}[N_g] + \mathcal{I}^{s+3}_{\delta,2+q_0-\delta}[N_g].$$

Proof. Since $\delta < q_0 \leq 1-\delta$, according to Theorem 5.15, $\check{\mathfrak{q}} = f_2 \check{e}_4 \mathfrak{q}$ verifies, for any $q_0 - 1 \leq q \leq q_0$ and any $s \leq k_{small} + 29$,

$$E^s_q[\check{\mathfrak{q}}](\tau_2) + B^s_q[\check{\mathfrak{q}}](\tau_1,\tau_2) \lesssim E^s_q[\check{\mathfrak{q}}](\tau_1) + E^{s+1}_{q+1}[\mathfrak{q}](\tau_1) + I^{s+2}_{q+2}[N_g](\tau_1,\tau_2).$$

According to the definition of our decay norms above we have

$$I^{s+2}_{q+2}[N_g](\tau_1,\tau_2) \lesssim (1+\tau_1)^{q-q_0} \mathcal{I}^{s+2}_{q+2,q_0-q}[N_g]. \tag{5.4.12}$$

Also, according to the definition 5.1.27 for the decay norms for $\mathfrak{q}$ we also have

$$E^{s+1}_{q+1}[\mathfrak{q}](\tau_1) \lesssim (1+\tau_1)^{q-q_0} \mathcal{E}^{s+2}_{q+1,q_0-q}[\mathfrak{q}].$$

We deduce,[9] for all $q_0 - 1 \leq q \leq q_0$,

$$E^s_q[\check{\mathfrak{q}}](\tau_2) + \int_{\tau_1}^{\tau_2} M^s_q[\check{\mathfrak{q}}](\tau) \lesssim E^s_q[\check{\mathfrak{q}}](\tau_1) + (1+\tau_1)^{q-q_0} \tilde{\mathcal{E}}^s_{q,q_0-q} \tag{5.4.13}$$

where

$$\tilde{\mathcal{E}}^s_{q,q_0-q} := \mathcal{E}^{s+1}_{q+1,q_0-q}[\mathfrak{q}] + \mathcal{I}^{s+2}_{q+2,q_0-q}[N_g]. \tag{5.4.14}$$

In particular,

$$E^s_{q_0}[\check{\mathfrak{q}}](\tau_2) + \int_{\tau_1}^{\tau_2} M^s_{q_0-1}[\check{\mathfrak{q}}](\tau) d\tau \lesssim E^s_{q_0}[\check{\mathfrak{q}}](\tau_1) + \tilde{\mathcal{E}}^s_{q_0,0}. \tag{5.4.15}$$

By the mean value theorem we deduce that there exists $\tau_0 \in [\tau_1, \tau_2]$ such that

$$M^s_{q_0-1}[\check{\mathfrak{q}}](\tau_0) \lesssim \frac{1}{\tau_2-\tau_1}\left(E^s_{q_0}[\check{\mathfrak{q}}](\tau_1) + \tilde{\mathcal{E}}^s_{q_0,0}\right) \lesssim \frac{1}{\tau_2-\tau_1}\left(\mathcal{E}^s_{q_0,0}[\check{\mathfrak{q}}] + \tilde{\mathcal{E}}^s_{q_0,0}\right).$$

[9]Note that it is important in what follows that the r^q weighted estimates hold also for negative values of q.

Thus also,

$$E^s_{q_0-1}[\check{\mathfrak{q}}](\tau_0) \lesssim \frac{1}{\tau_2-\tau_1}\left(\mathcal{E}^s_{q_0,0}[\check{\mathfrak{q}}] + \tilde{\mathcal{E}}^s_{q_0,0}\right). \qquad (5.4.16)$$

We now make use of (5.4.13) to compare the quantities $E_q[\check{\mathfrak{q}}]$ for negative weights ($q = q_0 - 1$) at different values of τ.

$$E^s_{q_0-1}[\check{\mathfrak{q}}](\tau_2) \lesssim E^s_{q_0-1}[\check{\mathfrak{q}}](\tau_0) + (1+\tau_0)^{-1}\tilde{\mathcal{E}}^s_{q_0-1,1}.$$

Combining this with (5.4.16) we deduce

$$E^s_{q_0-1}[\check{\mathfrak{q}}](\tau_2) \lesssim \frac{1}{\tau_2-\tau_1}\left(\mathcal{E}^s_{q_0,0}[\check{\mathfrak{q}}] + \tilde{\mathcal{E}}^s_{q_0,0}\right) + (1+\tau_0)^{-1}\tilde{\mathcal{E}}^s_{q_0-1,1}.$$

Applying this inequality for $\tau_2 = \tau \leq \tau_*$, $\tau_1 = \frac{1}{2}\tau$, $\tau_0 \in [\tau_1, \tau_2]$, we deduce

$$E^s_{q_0-1}[\check{\mathfrak{q}}](\tau) \lesssim (1+\tau)^{-1}\left(\mathcal{E}^s_{q_0,0}[\check{\mathfrak{q}}] + \tilde{\mathcal{E}}^s_{q_0,0} + \tilde{\mathcal{E}}^s_{q_0-1,1}\right). \qquad (5.4.17)$$

We now interpolate this last inequality with the following immediate consequence of (5.4.15):

$$E^s_{q_0}[\check{\mathfrak{q}}](\tau) \lesssim \mathcal{E}^s_{q_0,0}[\check{\mathfrak{q}}] + \tilde{\mathcal{E}}^s_{q_0,0}$$

to deduce, for all $q_0 - 1 \leq q \leq q_0$,

$$E^s_q[\check{\mathfrak{q}}](\tau) \lesssim (1+\tau)^{q-q_0}\left(\mathcal{E}^s_{q_0,0}[\check{\mathfrak{q}}] + \tilde{\mathcal{E}}^s_{q_0,0} + \tilde{\mathcal{E}}^s_{q_0-1,1}\right),$$

i.e.,

$$\mathcal{E}^s_{q,q_0-q}[\check{\mathfrak{q}}] \lesssim \mathcal{E}^s_{q_0,0}[\check{\mathfrak{q}}] + \tilde{\mathcal{E}}^s_{q_0,0} + \tilde{\mathcal{E}}^s_{q_0-1,1}.$$

In view of the definition of $\tilde{\mathcal{E}}^s_{q,q_0-q}$, this yields, for all $q_0 - 1 \leq q \leq q_0$,

$$\mathcal{E}^s_{q,q_0-q}[\check{\mathfrak{q}}] \lesssim \mathcal{E}^s_{q_0,0}[\check{\mathfrak{q}}] + \mathcal{E}^{s+1}_{q_0+1,0}[\mathfrak{q}] + \mathcal{E}^{s+1}_{q_0,1}[\mathfrak{q}] + \mathcal{I}^{s+2}_{q_0+2,0}[N_g] + \mathcal{I}^{s+2}_{q_0+1,1}[N_g].$$

On the other hand, we have, in view of Theorem 5.15,

$$\mathcal{E}^s_{q_0,0}[\check{\mathfrak{q}}] \lesssim \mathcal{E}^s_{q_0}[\check{\mathfrak{q}}](0) + \mathcal{E}^{s+1}_{q_0+1,0}[\mathfrak{q}] + \mathcal{I}^{s+2}_{q_0+2,0}[N_g]$$

and hence

$$\mathcal{E}^s_{q,q_0-q}[\check{\mathfrak{q}}] \lesssim \mathcal{E}^s_{q_0}[\check{\mathfrak{q}}](0) + \mathcal{E}^{s+1}_{q_0+1,0}[\mathfrak{q}] + \mathcal{E}^{s+1}_{q_0,1}[\mathfrak{q}] + \mathcal{I}^{s+2}_{q_0+2,0}[N_g] + \mathcal{I}^{s+2}_{q_0+1,1}[N_g].$$

Now, since $\delta < q_0 \leq 1 - \delta$, we have $\delta < q_0 < q_0 + 1 \leq 2 - \delta$ and thus, we may apply Theorem 5.21 to obtain for all $q_0 - 1 \leq q \leq q_0$

$$\mathcal{E}^{s+1}_{q+1,q_0-q}[\mathfrak{q}] \lesssim \mathcal{E}^{s+2}_{2-\delta}[\mathfrak{q}](0) + \mathcal{I}^{s+3}_{2-\delta,0}[N_g] + \mathcal{I}^{s+3}_{\delta,2-2\delta}[N_g]. \qquad (5.4.18)$$

We thus infer

$$\begin{aligned}
&\mathcal{E}^s_{q,q_0-q}[\check{\mathfrak{q}}]\\
\lesssim\;& \mathcal{E}^s_{q_0}[\check{\mathfrak{q}}](0)+\mathcal{E}^{s+2}_{2-\delta}[\mathfrak{q}](0)+\mathcal{I}^{s+3}_{q_0+2,0}[N_g]+\mathcal{I}^{s+3}_{q_0+1,1}[N_g]+\mathcal{I}^{s+3}_{2-\delta,0}[N_g]+\mathcal{I}^{s+3}_{\delta,2-2\delta}[N_g]
\end{aligned}$$

and hence, for all $q_0-1\leq q\leq q_0$,

$$\mathcal{E}^s_{q,q_0-q}[\check{\mathfrak{q}}] \;\lesssim\; \mathcal{E}^s_{q_0}[\check{\mathfrak{q}}](0)+\mathcal{E}^{s+2}_{2-\delta}[\mathfrak{q}](0)+\mathcal{I}^{s+3}_{q_0+2,0}[N_g]+\mathcal{I}^{s+3}_{\delta,2-2\delta}[N_g]. \quad (5.4.19)$$

Finally, making use of Theorem 5.15 between τ and τ_*, we have in particular

$$\begin{aligned}
&B^s_q[\check{\mathfrak{q}}](\tau,\tau_*)\\
\lesssim\;& E^s_q[\check{\mathfrak{q}}](\tau)+E^{s+1}_{q+1}[\mathfrak{q}](\tau)+I^{s+2}_{q+2}[N_g](\tau,\tau_*)\\
\lesssim\;& (1+\tau)^{-(q_0-q)}\Big(\mathcal{E}^s_{q,q_0-q}[\check{\mathfrak{q}}]+\mathcal{E}^{s+1}_{q+1,q_0-q}[\mathfrak{q}]+\mathcal{I}^{s+2}_{q+2,q_0-q}[N_g]\Big)\\
\lesssim\;& (1+\tau)^{-(q_0-q)}\Big(\mathcal{E}^s_{q,q_0-q}[\check{\mathfrak{q}}]+\mathcal{E}^{s+2}_{2-\delta}[\mathfrak{q}](0)+\mathcal{I}^{s+3}_{q_0+2,0}[N_g]+\mathcal{I}^{s+3}_{\delta,2-2\delta}[N_g]\Big)
\end{aligned}$$

where we used (5.4.18) in the last inequality. Hence, we infer for all $s\leq k_{small}+28$ and for all $q_0-1\leq q\leq q_0$

$$\mathcal{B}^s_{q,q_0-q}[\check{\mathfrak{q}}] \;\lesssim\; \mathcal{E}^s_{q,q_0-q}[\check{\mathfrak{q}}]+\mathcal{E}^{s+2}_{2-\delta}[\mathfrak{q}](0)+\mathcal{I}^{s+3}_{q_0+2,0}[N_g]+\mathcal{I}^{s+3}_{\delta,2-2\delta}[N_g].$$

Together with (5.4.19), this concludes the proof of Theorem 5.22. □

5.4.3 Proof of Theorem 5.9

In this section, we prove Theorem 5.9 by making use of Theorem 5.21 and Theorem 5.22. We start with the main estimate of Theorem 5.22 with $q=-\delta$ which we write in the form

$$E^s_{-\delta}[\check{\mathfrak{q}}] \;\lesssim (1+\tau)^{-q_0-\delta}C^s_{q_0}$$

where

$$C^s_{q_0} \;:=\; \mathcal{E}^s_{q_0}[\check{\mathfrak{q}}](0)+\mathcal{E}^{s+2}_{2-\delta}[\mathfrak{q}](0)+\mathcal{I}^{s+3}_{q_0,0}[N_g]+\mathcal{I}^{s+3}_{\delta,q_0+2-\delta}[N_g].$$

In view of the definition (5.1.21) of $E^s_{-\delta}[\check{\mathfrak{q}}]$ and since $\check{\mathfrak{q}}=f_2\check{e}_4\mathfrak{q}$,

$$\int_{\Sigma_{\geq 4m_0}(\tau)} r^{-\delta}\left(|\check{e}_4\check{\mathfrak{q}}|^2+r^{-2}|\check{\mathfrak{q}}|^2\right) \;\lesssim (1+\tau)^{-q_0-\delta}C^s_{q_0}.$$

Hence,

$$\dot{E}^s_{2-\delta,4m_0}[\mathfrak{q}]=\int_{\Sigma_{\geq 4m_0}(\tau)} r^{2-\delta}|\check{e}_4\mathfrak{q}|^2 \;\lesssim (1+\tau)^{-q_0-\delta}C^s_{q_0}. \quad (5.4.20)$$

In view of the decay estimates (5.4.1) for $\mathfrak{q}$ established in Theorem 5.21 we have

$$\begin{aligned}
E^s(\tau) \;&\lesssim\; (1+\tau)^{-2+2\delta}B^{2+s}_{2-\delta},\\
B^{2+s}_{2-\delta}: \;&=\; \mathcal{E}^{s+2}_{2-\delta}[\mathfrak{q}](0)+\mathcal{I}^{s+3}_{2-\delta,0}[N_g]+\mathcal{I}^{s+3}_{\delta,2-2\delta}[N_g].
\end{aligned}$$

Thus, the quantity

$$E^s_{2-\delta} = E^s_{2-\delta}[\mathfrak{q}](\tau) = \dot{E}^s_{2-\delta,4m_0}[\mathfrak{q}] + E^s[\mathfrak{q}]$$

verifies

$$E^s_{2-\delta} \lesssim (1+\tau)^{-q_0-\delta}\left(C^s_{q_0} + B^{2+s}_{2-\delta}\right). \tag{5.4.21}$$

On the other hand, $E^s_{2-\delta}$ verifies (5.4.2) for $p = 2-\delta$, i.e.,

$$E^s_{2-\delta}(\tau_2) + \int_{\tau_1}^{\tau_2} M^s_{1-\delta}(\tau)d\tau \lesssim E^s_{2-\delta}(\tau_1) + I^{s+1}_{2-\delta}[N_g](\tau_1,\tau_2).$$

Since

$$I^{s+1}_{2-\delta}[N_g](\tau_1,\tau_2) \lesssim (1+\tau_1)^{-q_0-\delta}\mathcal{I}^{s+1}_{2-\delta,q_0+\delta}[N_g],$$

we infer

$$\begin{aligned}
&E^s_{2-\delta}(\tau) + \int_{\tau/2}^{\tau} M^s_{1-\delta}(\tau')d\tau' \\
\lesssim\ & E^s_{2-\delta}(\tau/2) + I^{s+1}_{2-\delta}[N_g](\tau/2,\tau) \\
\lesssim\ & (1+\tau)^{-q_0-\delta}\left(C^s_{q_0} + B^{2+s}_{2-\delta} + \mathcal{I}^{s+1}_{2-\delta,q_0+\delta}[N_g]\right).
\end{aligned} \tag{5.4.22}$$

Following the same arguments as in the proof of Theorem 5.21 we deduce, for a $\tau_0 \in [\tau/2,\tau]$,

$$E^{s-1}_{1-\delta}(\tau_0) \lesssim (1+\tau)^{-q_0-1-\delta}\left(C^s_{q_0} + B^{2+s}_{2-\delta} + \mathcal{I}^{s+1}_{2-\delta,q_0+\delta}[N_g]\right)$$

and since

$$E^s_{1-\delta}(\tau) \lesssim E^s_{1-\delta}(\tau_0) + I^{s+1}_{1-\delta}(\tau_0,\tau)[N_g],$$

we infer that

$$\begin{aligned}
&E^s_{1-\delta}(\tau) \\
\lesssim\ & (1+\tau)^{-q_0-1-\delta}\left(C^{s+1}_{q_0} + B^{3+s}_{2-\delta} + \mathcal{I}^{s+2}_{2-\delta,q_0+\delta}[N_g] + \mathcal{I}^{s+1}_{1-\delta,1+q_0+\delta}[N_g]\right).
\end{aligned} \tag{5.4.23}$$

Interpolating with (5.4.21), i.e.,

$$E^s_{2-\delta} \lesssim (1+\tau)^{-q_0-\delta}\left(C^s_{q_0} + B^{2+s}_{2-\delta}\right)$$

we deduce

$$\begin{aligned}
E^s_1 &\lesssim (E^s_{1-\delta})^{1-\delta}(E^s_{2-\delta})^{\delta} \\
&\lesssim (1+\tau)^{-q_0-1}\left(C^{s+1}_{q_0} + B^{3+s}_{2-\delta} + \mathcal{I}^{s+2}_{2-\delta,q_0+\delta}[N_g] + \mathcal{I}^{s+1}_{1-\delta,1+q_0+\delta}[N_g]\right).
\end{aligned}$$

Hence,

$$E^s_1 \lesssim (1+\tau)^{-q_0-1}\left(C^{s+1}_{q_0} + B^{3+s}_{2-\delta} + \mathcal{I}^{s+2}_{2-\delta,q_0+\delta} + \mathcal{I}^{s+1}_{1-\delta,1+q_0+\delta}\right). \tag{5.4.24}$$

As in the proof of Theorem 5.21 we repeat the procedure starting with the inequality (5.4.2) for $p = 1$,

$$\begin{aligned}
& E_1^s(\tau) + \int_{\tau/2}^{\tau} M_0^s(\lambda) d\lambda \\
\lesssim \; & E_1^s(\tau/2) + I_1^{s+1}[N_g](\tau/2, \tau) \\
\lesssim \; & (1+\tau)^{-q_0-1} \left(C_{q_0}^{s+1} + B_{2-\delta}^{3+s} + \mathcal{I}_{2-\delta, q_0+\delta}^{s+2}[N_g] + \mathcal{I}_{1-\delta,1+q_0+\delta}^{s+1}[N_g] \right) \\
& +(1+\tau)^{-1-q_0} \mathcal{I}_{1,1+q_0}^{s+1}[N_g] \\
\lesssim \; & (1+\tau)^{-q_0-1} \left(C_{q_0}^{s+1} + B_{2-\delta}^{3+s} + \mathcal{I}_{2-\delta, q_0+\delta}^{s+s}[N_g] + \mathcal{I}_{1-\delta,1+q_0+\delta}^{s+1}[N_g] + \mathcal{I}_{1,1+q_0}^{s+1}[N_g] \right)
\end{aligned}$$

from which we infer that, for a $\tau_0 \in [\tau/2, \tau]$,

$$\begin{aligned}
E^s(\tau_0) \lesssim \; & (1+\tau)^{-q_0-2} \Big(C_{q_0}^{s+2} + B_{2-\delta}^{s+4} + \mathcal{I}_{2-\delta, q_0+\delta}^{s+3}[N_g] + \mathcal{I}_{1-\delta,1+q_0+\delta}^{s+2}[N_g] \\
& + \mathcal{I}_{1,1+q_0}^{s+2}[N_g] \Big).
\end{aligned} \tag{5.4.25}$$

Interpolating (5.4.23) and (5.4.25) we deduce, for $\delta > 0$ sufficiently small,

$$\begin{aligned}
E_\delta^s(\tau_0) \lesssim \; & \left(E^s(\tau_0)\right)^{\frac{1-2\delta}{1-\delta}} \left(E_{1-\delta}^s(\tau_0)\right)^{\frac{\delta}{1-\delta}} \\
\lesssim \; & (1+\tau)^{-2-q_0+\delta} \Big(C_{q_0}^{s+2} + B_{2-\delta}^{s+4} + \mathcal{I}_{2-\delta, q_0+\delta}^{s+3}[N_g] \\
& + \mathcal{I}_{1-\delta,1+q_0+\delta}^{s+2}[N_g] + \mathcal{I}_{1,1+q_0}^{s+2}[N_g] \Big).
\end{aligned}$$

Thus, since we have, as in (5.4.7),

$$E_\delta^s(\tau) \lesssim E_\delta^s(\tau_0) + I_\delta^{s+1}[N_g](\tau_0, \tau),$$

we deduce

$$\begin{aligned}
E_\delta^s(\tau) \lesssim \; & (1+\tau)^{-2-q_0+\delta} \Big(C_{q_0}^{s+2} + B_{2-\delta}^{s+4} + \mathcal{I}_{2-\delta, q_0+\delta}^{s+3}[N_g] + \mathcal{I}_{1-\delta,1+q_0+\delta}^{s+2}[N_g] \\
& + \mathcal{I}_{1,1+q_0}^{s+2}[N_g] \Big) + (1+\tau)^{-2-q_0+\delta} \mathcal{I}_{\delta, 2+q_0-\delta}^{s+1}[N_g],
\end{aligned}$$

i.e.,

$$\begin{aligned}
E_\delta^s(\tau) \lesssim \; & (1+\tau)^{-2-q_0+\delta} \Big(C_{q_0}^{s+2} + B_{2-\delta}^{s+4} + \mathcal{I}_{2-\delta, q_0+\delta}^{s+3}[N_g] + \mathcal{I}_{1-\delta,1+q_0+\delta}^{s+2}[N_g] \\
& + \mathcal{I}_{1,1+q_0}^{s+2}[N_g] + \mathcal{I}_{\delta, 2+q_0-\delta}^{s+1}[N_g] \Big).
\end{aligned}$$

By interpolating the middle terms we write

$$E_\delta^s(\tau) \lesssim (1+\tau)^{-2-q_0+\delta} \left(C_{q_0}^{s+2} + B_{2-\delta}^{s+4} + \mathcal{I}_{2-\delta, q_0+\delta}^{s+3}[N_g] + \mathcal{I}_{\delta, 2+q_0-\delta}^{s+3}[N_g] \right).$$

We now recall

$$\begin{aligned}
C_{q_0}^s & := \mathcal{E}_{q_0}^s[\check{\mathfrak{q}}](0) + \mathcal{E}_{2-\delta}^{s+2}[\mathfrak{q}](0) + \mathcal{I}_{q_0+2,0}^{s+3}[N_g] + \mathcal{I}_{\delta, q_0+2-\delta}^{s+3}[N_g] \\
B_{2-\delta}^{2+s} : & = \mathcal{E}_{2-\delta}^{s+2}[\mathfrak{q}](0) + \mathcal{I}_{2-\delta,0}^{s+3}[N_g] + \mathcal{I}_{\delta, 2-2\delta}^{s+3}[N_g].
\end{aligned}$$

Hence,

$$\begin{aligned} & C_{q_0}^{s+2} + B_{2-\delta}^{s+4} + \mathcal{I}_{2-\delta,q_0+\delta}^{s+3}[N_g] + \mathcal{I}_{\delta,2+q_0-\delta}^{s+3}[N_g] \\ = \ & \mathcal{E}_{q_0}^{s+2}[\check{\mathfrak{q}}](0) + \mathcal{E}_{2-\delta}^{s+4}[\mathfrak{q}](0) + \mathcal{I}_{q_0+2,0}^{s+5}[N_g] + \mathcal{I}_{\delta,q_0+2-\delta}^{s+5}[N_g] \\ + \ & \mathcal{E}_{2-\delta}^{s+4}[\mathfrak{q}](0) + \mathcal{I}_{2-\delta,0}^{s+5}[N_g] + \mathcal{I}_{\delta,2-2\delta}^{s+5}[N_g] + \mathcal{I}_{2-\delta,q_0+\delta}^{s+3}[N_g] + \mathcal{I}_{\delta,2+q_0-\delta}^{s+3}[N_g]. \end{aligned}$$

We deduce

$$\mathcal{E}_{\delta,2+q_0-\delta}^{s}[\mathfrak{q}] \lesssim \mathcal{E}_{q_0}^{s+2}[\check{\mathfrak{q}}](0) + \mathcal{E}_{2-\delta}^{s+4}[\mathfrak{q}](0) + \mathcal{I}_{q_0+2,0}^{s+5}[N_g] + \mathcal{I}_{\delta,2+q_0-\delta}^{s+5}[N_g]. \quad (5.4.26)$$

We can also simplify the right-hand side of (5.4.24),

$$\begin{aligned} & C_{q_0}^{s+1} + B_{2-\delta}^{3+s} + \mathcal{I}_{2-\delta,q_0+\delta}^{s+2}[N_g] + \mathcal{I}_{1-\delta,1+q_0+\delta}^{s+1}[N_g] \\ & \lesssim \mathcal{E}_{q_0}^{s+2}[\check{\mathfrak{q}}](0) + \mathcal{E}_{2-\delta}^{s+4}[\mathfrak{q}](0) + \mathcal{I}_{q_0+2,0}^{s+5}[N_g] + \mathcal{I}_{\delta,2+q_0-\delta}^{s+5}[N_g]. \end{aligned}$$

Thus, (5.4.23) becomes

$$\mathcal{E}_{1-\delta,1+q_0+\delta}^{s} \lesssim \mathcal{E}_{q_0}^{s+2}[\check{\mathfrak{q}}](0) + \mathcal{E}_{2-\delta}^{s+4}[\mathfrak{q}](0) + \mathcal{I}_{q_0+2,0}^{s+5}[N_g] + \mathcal{I}_{\delta,2+q_0-\delta}^{s+5}[N_g]. \quad (5.4.27)$$

Similarly, (5.4.21) yields

$$\mathcal{E}_{2-\delta,q_0-\delta}^{s} \lesssim \mathcal{E}_{q_0}^{s+2}[\check{\mathfrak{q}}](0) + \mathcal{E}_{2-\delta}^{s+4}[\mathfrak{q}](0) + \mathcal{I}_{q_0+2,0}^{s+5}[N_g] + \mathcal{I}_{\delta,2+q_0-\delta}^{s+5}[N_g]. \quad (5.4.28)$$

Interpolating (5.4.26) and (5.4.27), as well as (5.4.27) and (5.4.28), we infer for all $s \leq k_{small} + 25$ and for all $\delta \leq p \leq 2-\delta$

$$\mathcal{E}_{p,2+q_0-p}^{s}[\mathfrak{q}] \lesssim \mathcal{E}_{q_0}^{s+2}[\check{\mathfrak{q}}](0) + \mathcal{E}_{2-\delta}^{s+4}[\mathfrak{q}](0) + \mathcal{I}_{q_0+2,0}^{s+5}[N_g] + \mathcal{I}_{\delta,2+q_0-\delta}^{s+5}[N_g]. \quad (5.4.29)$$

Finally, making use of Theorem 5.14 between τ and τ_*, we have in particular

$$\begin{aligned} B_p^s[\mathfrak{q}](\tau,\tau_*) + F_p^s[\mathfrak{q}](\tau,\tau_*) & \lesssim E_p^s[\mathfrak{q}](\tau) + I_p^{s+1}[N_g](\tau,\tau_*) \\ & \lesssim (1+\tau)^{-(2+q_0-p)}\Big(\mathcal{E}_{p,2+q_0-p}^{s}[\mathfrak{q}] + I_{p,2+q_0-p}^{s+1}[N_g]\Big) \end{aligned}$$

and hence, we infer for all $s \leq k_{small} + 25$ and for all $\delta \leq p \leq 2-\delta$

$$\mathcal{B}_{p,2+q_0-p}^{s}[\mathfrak{q}] + \mathcal{F}_{p,2+q_0-p}^{s}[\mathfrak{q}] \lesssim \mathcal{E}_{p,2+q_0-p}^{s}[\mathfrak{q}] + \mathcal{I}_{q_0+2,0}^{s+5}[N_g] + \mathcal{I}_{\delta,2+q_0-\delta}^{s+5}[N_g].$$

Together with (5.4.29), this concludes the proof of Theorem 5.9.

5.4.4 Proof of Proposition 5.12

Let χ be a smooth cut-off function vanishing for $r \le 4m_0$ and equal to 1 for $r \ge 6m_0$. To prove estimate (5.2.6) we consider the identity,

$$\begin{aligned} & e_4\left(\int_{S_r}\chi(\mathfrak{q}^{(s)})^2\right) \\ = & \int_{S_r}\left(e_4(\chi(\mathfrak{q}^{(s)})^2)+\kappa\chi(\mathfrak{q}^{(s)})^2\right) \\ = & \int_{S_r}\left(\chi(2\mathfrak{q}^{(s)}e_4\mathfrak{q}^{(s)}+2r^{-1}(\mathfrak{q}^{(s)})^2)+\chi'(\mathfrak{q}^{(s)})^2+\chi(\kappa-2r^{-1})|\mathfrak{q}^{(s)}|^2\right) \\ = & \int_{S_r}\left(2\chi\mathfrak{q}^{(s)}\check{e}_4\mathfrak{q}^{(s)}+\chi'(\mathfrak{q}^{(s)})^2+O(r^{-2})|\mathfrak{q}^{(s)}|^2\right). \end{aligned}$$

Integrating between $4m_0$ and r for a fixed $r \ge 6m_0$, we deduce, in view of the definitions of $E[\mathfrak{q}^{(s)}](\tau)$ and of $E_p[\mathfrak{q}^{(s)}](\tau)$,

$$\begin{aligned} \int_{S_r}|\mathfrak{q}^{(s)}|^2 & \lesssim \int_{\Sigma(\tau)\ge 4m_0}|\mathfrak{q}^{(s)}||\check{e}_4\mathfrak{q}^{(s)}|+E[\mathfrak{q}^{(s)}](\tau) \\ & \lesssim \left(\int_{\Sigma(\tau)\ge 4m_0} r^{1+\delta}|\check{e}_4\mathfrak{q}^{(s)}|^2\right)^{1/2}\left(\int_{\Sigma(\tau)\ge 4m_0} r^{-1-\delta}|\mathfrak{q}^{(s)}|^2\right)^{1/2} \\ & \quad +E[\mathfrak{q}^{(s)}](\tau) \\ & \lesssim \left(E_{1+\delta}[\mathfrak{q}^{(s)}](\tau)\right)^{1/2}\left(E_{1-\delta}[\mathfrak{q}^{(s)}](\tau)\right)^{1/2}. \end{aligned}$$

Clearly, this estimate also holds for $r \le 6m_0$. Together with the definition (5.1.27) of $\mathcal{E}^s_{p,d}[\mathfrak{q}^{(s)}]$, we immediately infer

$$(1+\tau)^{1+q_0}\int_{S_r}|\mathfrak{q}^{(s)}|^2 \lesssim \left(\mathcal{E}^s_{1+\delta,1+q_0-\delta}[\mathfrak{q}]\right)^{\frac12}\left(\mathcal{E}^s_{1-\delta,1+q_0+\delta}[\mathfrak{q}]\right)^{\frac12}$$

which is the desired estimate (5.2.6).

To prove (5.2.7) we start instead with the identity,

$$\begin{aligned} & e_4\left(r^{-1}\int_{S_r}\chi(\mathfrak{q}^{(s)})^2\right) \\ = & \int_{S_r} r^{-1}\left(e_4(\chi(\mathfrak{q}^{(s)})^2)+\kappa\chi(\mathfrak{q}^{(s)})^2\right)-\frac{e_4(r)}{r^2}\int_{S_r}\chi(\mathfrak{q}^{(s)})^2 \\ = & \int_{S_r} r^{-1}\left(\chi(2\mathfrak{q}^{(s)}e_4\mathfrak{q}^{(s)}+r^{-1}(\mathfrak{q}^{(s)})^2)+\chi'(\mathfrak{q}^{(s)})^2+\chi(\kappa-2r^{-1})|\mathfrak{q}^{(s)}|^2\right) \\ & -\frac{e_4(r)-1}{r^2}\int_{S_r}\chi(\mathfrak{q}^{(s)})^2 \\ = & \int_{S_r}\left(2r^{-1}\chi e_4(\mathfrak{q}^{(s)})\mathfrak{q}^{(s)}+r^{-1}\chi'(\mathfrak{q}^{(s)})^2+O(r^{-2})|\mathfrak{q}^{(s)}|^2\right). \end{aligned}$$

Integrating between $4m_0$ and r for a fixed $r \ge 6m_0$, we deduce, in view of the definitions of $E[\mathfrak{q}^{(s)}](\tau)$ and of $E_p[\mathfrak{q}^{(s)}](\tau)$,

$$\begin{aligned} r^{-1}\int_{S_r}|\psi|^2 &\lesssim \int_{\Sigma(\tau)_{\geq 4m_0}} r^{-1}|\mathfrak{q}^{(s)}||e_4(\mathfrak{q}^{(s)})| + E[\mathfrak{q}^{(s)}](\tau) \\ &\lesssim 2\left(\int_{\Sigma(\tau)_{\geq 4m_0}}|e_4(\mathfrak{q}^{(s)})|^2\right)^{1/2}\left(\int_{\Sigma(\tau)_{\geq 4m_0}} r^{-2}|\mathfrak{q}^{(s)}|^2\right)^{1/2} \\ &\quad + E[\mathfrak{q}^{(s)}](\tau) \\ &\lesssim E[\mathfrak{q}^{(s)}](\tau) \\ &\lesssim E_\delta[\mathfrak{q}^{(s)}](\tau). \end{aligned}$$

Clearly, this estimate also holds for $r \leq 6m_0$. Together with the definition (5.1.27) of $\mathcal{E}^s_{p,d}[\mathfrak{q}^{(s)}]$, we immediately infer

$$r^{-1}(1+\tau)^{2+q_0-\delta}\int_{S_r}|\mathfrak{q}^{(s)}|^2 \lesssim \mathcal{E}^s_{\delta,2+q_0-\delta}[\mathfrak{q}]$$

which is the desired estimate (5.2.7). This concludes the proof of Proposition 5.12.

5.4.5 Proof of Proposition 5.13

Recall the following definitions:

$$\begin{aligned} F[\psi](\tau_1,\tau_2) &= \int_{\mathcal{A}(\tau_1,\tau_2)}\left(\delta_{\mathcal{H}}^{-1}|e_4\Psi|^2 + \delta_{\mathcal{H}}|e_3\Psi|^2 + |\nabla\Psi|^2 + r^{-2}|\Psi|^2\right) \\ &\quad + \int_{\Sigma_*(\tau_1,\tau_2)}\left(|e_4\Psi|^2 + |e_3\Psi|^2 + |\nabla\Psi|^2 + r^{-2}|\Psi|^2\right), \\ \dot{F}_p[\psi](\tau_1,\tau_2) &= \int_{\Sigma_*(\tau_1,\tau_2)} r^p\left(|e_4\psi|^2 + |\nabla\psi|^2 + r^{-2}|\psi|^2\right), \\ F_p[\psi](\tau_1,\tau_2) &= F[\psi](\tau_1,\tau_2) + \dot{F}_p[\psi](\tau_1,\tau_2), \\ F^s[\psi](\tau_1,\tau_2) &= \sum_{k\leq s} F[\mathfrak{d}^k\psi](\tau_1,\tau_2), \\ F^s_p[\psi](\tau_1,\tau_2) &= \sum_{k\leq s} F_p[\mathfrak{d}^k\psi](\tau_1,\tau_2), \\ \mathcal{F}^s_{p,d}[\psi] &= \sup_{0\leq\tau\leq\tau_*}(1+\tau)^d F^s_p[\psi](\tau,\tau_*). \end{aligned}$$

We deduce

$$F^s[\mathfrak{q}](\tau,\tau_*) \leq F^s_\delta[\mathfrak{q}](\tau,\tau_*) \leq (1+\tau)^{-2-q_0+\delta}\mathcal{F}^s_{\delta,2+q_0-\delta}[\mathfrak{q}]$$

and hence in particular

$$(1+\tau)^{2+q_0-\delta}\int_{\Sigma_*(\tau,\tau_*)}\left(|e_3\mathfrak{d}^{\leq s}\mathfrak{q}|^2 + r^{-2}|\mathfrak{d}^{\leq s}\mathfrak{q}|^2\right) \lesssim \mathcal{F}^s_{\delta,2+q_0-\delta}[\mathfrak{q}] \quad (5.4.30)$$

which yields the desired estimate (5.2.8).

Next, we focus on the proof of (5.2.9). We start with the following trace estimate

$$\sup_{\Sigma_*(\tau,\tau_*)} \|e_3\mathfrak{d}^{\leq s}\mathfrak{q}\|_{L^2(S)} \lesssim \|\nu e_3\mathfrak{d}^{\leq s}\mathfrak{q}\|_{L^2(\Sigma_*(\tau,\tau_*))} + \|e_3\mathfrak{d}^{\leq s}\mathfrak{q}\|_{L^2(\Sigma_*(\tau,\tau_*))}$$

where we recall that ν is tangent to Σ_*, orthogonal to e_θ and given by

$$\nu = e_3 + ae_4, \quad -2 \leq a \leq -\frac{1}{2}.$$

We infer

$$\begin{aligned}
\sup_{\Sigma_*(\tau,\tau_*)} \|e_3\mathfrak{d}^{\leq s}\mathfrak{q}\|_{L^2(S)} &\lesssim \|e_3e_3\mathfrak{d}^{\leq s}\mathfrak{q}\|_{L^2(\Sigma_*(\tau,\tau_*))} + \|e_4e_3\mathfrak{d}^{\leq s}\mathfrak{q}\|_{L^2(\Sigma_*(\tau,\tau_*))} \\
&\quad +\|e_3\mathfrak{d}^{\leq s}\mathfrak{q}\|_{L^2(\Sigma_*(\tau,\tau_*))} \\
&\lesssim \|e_3\mathfrak{d}^{\leq s+1}\mathfrak{q}\|_{L^2(\Sigma_*(\tau,\tau_*))} + \|r^{-1}\mathfrak{d}^{\leq s+1}\mathfrak{q}\|_{L^2(\Sigma_*(\tau,\tau_*))} \\
&\quad +\|[e_4, e_3]\mathfrak{d}^{\leq s}\mathfrak{q}\|_{L^2(\Sigma_*(\tau,\tau_*))} \\
&\lesssim \|e_3\mathfrak{d}^{\leq s+1}\mathfrak{q}\|_{L^2(\Sigma_*(\tau,\tau_*))} + \|r^{-1}\mathfrak{d}^{\leq s+1}\mathfrak{q}\|_{L^2(\Sigma_*(\tau,\tau_*))}.
\end{aligned}$$

In view of (5.4.30), we deduce

$$\sup_{\Sigma_*}\left\{(1+\tau)^{2+q_0-\delta}\|e_3\mathfrak{d}^{\leq s}\mathfrak{q}\|^2_{L^2(S)}\right\} \lesssim \mathcal{F}^{s+1}_{\delta,2+q_0-\delta}[\mathfrak{q}]. \tag{5.4.31}$$

Next, we extend (5.4.31) to $r \geq 4m_0$. In view of (5.3.24), we have schematically

$$\begin{aligned}
e_3e_4(r\mathfrak{d}^k\mathfrak{q}) &= -\mathfrak{d}^{\leq k}(r\square_2\mathfrak{q}) + r\not{\Delta}_2(\mathfrak{d}^{\leq k}\mathfrak{q}) + r^{-1}\mathfrak{d}^{\leq k+1}\mathfrak{q} \\
&= -\mathfrak{d}^{\leq k}(r\square_2\mathfrak{q}) + r^{-1}\mathfrak{d}^{\leq k+2}\mathfrak{q}.
\end{aligned}$$

Also, we have

$$e_4(re_3(\mathfrak{d}^k\mathfrak{q})) = e_3e_4(r\mathfrak{d}^k\mathfrak{q}) + [e_4, e_3](r\mathfrak{d}^k\mathfrak{q}) - e_4(e_3(r)\mathfrak{d}^k\mathfrak{q})$$

and hence, we infer schematically

$$e_4(re_3(\mathfrak{d}^k\mathfrak{q})) = -\mathfrak{d}^{\leq k}(r\square_2\mathfrak{q}) + r^{-1}\mathfrak{d}^{\leq k+2}\mathfrak{q}.$$

Now, recall (5.3.31)

$$|\mathfrak{d}^k(r\square_2\mathfrak{q})| \lesssim r^{-1}|\mathfrak{d}^{\leq k+1}\mathfrak{q}| + r|\mathfrak{d}^{\leq k}N_g| + r^2|\mathfrak{d}^{\leq k}e_3(N_g)|.$$

We deduce

$$|e_4(re_3(\mathfrak{d}^k\mathfrak{q}))| \lesssim r^{-1}|\mathfrak{d}^{\leq k+2}\mathfrak{q}| + r|\mathfrak{d}^{\leq k}N_g| + r^2|\mathfrak{d}^{\leq k}e_3(N_g)|.$$

Now, we have

$$
\begin{aligned}
& e_4\left(r^{-2}\int_S (re_3(\mathfrak{d}^{\leq s}\mathfrak{q}))^2\right)\\
= \;& r^{-2}\int_S\Big(2e_4(re_3(\mathfrak{d}^{\leq s}\mathfrak{q}))re_3(\mathfrak{d}^{\leq s}\mathfrak{q})+\kappa(re_3(\mathfrak{d}^{\leq s}\mathfrak{q}))^2\Big)\\
& -\frac{2e_4(r)}{r}r^{-2}\int_S (re_3(\mathfrak{d}^{\leq s}\mathfrak{q}))^2\\
= \;& r^{-2}\int_S\Big(2e_4(re_3(\mathfrak{d}^{\leq s}\mathfrak{q}))re_3(\mathfrak{d}^{\leq s}\mathfrak{q})+(\kappa-2r^{-1})(re_3(\mathfrak{d}^{\leq s}\mathfrak{q}))^2\Big)\\
& -2\frac{e_4(r)-1}{r}r^{-2}\int_S (re_3(\mathfrak{d}^{\leq s}\mathfrak{q}))^2
\end{aligned}
$$

and hence

$$
\begin{aligned}
& \left|e_4\left(r^{-2}\int_S (re_3(\mathfrak{d}^{\leq s}\mathfrak{q}))^2\right)\right|\\
\lesssim \;& r^{-2}\int_S\bigg\{\Big(r^{-1}|\mathfrak{d}^{\leq s+2}\mathfrak{q}|+r|\mathfrak{d}^{\leq s}N_g|+r^2|\mathfrak{d}^{\leq s}e_3(N_g)|\Big)|re_3(\mathfrak{d}^{\leq s}\mathfrak{q})|\\
& +r^{-2}(re_3(\mathfrak{d}^{\leq s}\mathfrak{q}))^2\bigg\}\\
\lesssim \;& r^{-2}\int_S\Big(r^{-\frac{1}{2}}|\mathfrak{d}^{\leq s+2}\mathfrak{q}|^2+r^2|\mathfrak{d}^{\leq s}N_g|^2+r^4|\mathfrak{d}^{\leq s}e_3(N_g)|^2\Big)\\
& +r^{-\frac{7}{2}}\int_S (re_3(\mathfrak{d}^{\leq s}\mathfrak{q}))^2.
\end{aligned}
$$

Together with (5.2.7), this yields

$$
\begin{aligned}
& \left|e_4\left(r^{-2}\int_S (re_3(\mathfrak{d}^{\leq s}\mathfrak{q}))^2\right)\right|\\
\lesssim \;& r^{-2}\int_S\Big(r^{\frac{7}{2}}|\mathfrak{d}^{\leq s}N_g|^2+r^{\frac{11}{2}}|\mathfrak{d}^{\leq s}e_3(N_g)|^2\Big)\\
& +r^{-4}\int_S (re_3(\mathfrak{d}^{\leq s}\mathfrak{q}))^2+r^{-\frac{3}{2}}(1+\tau)^{-2-q_0+\delta}\mathcal{E}^{s+2}_{\delta,2+q_0-\delta}[\mathfrak{q}].
\end{aligned}
$$

Now, recall from (5.2.12) that we have for $s\leq k_{small}+30$

$$
\begin{aligned}
|\mathfrak{d}^s N_g| &\lesssim \epsilon^2 r^{-3}\tau^{-1-2\delta_{dec}+2\delta_0}\\
|\mathfrak{d}^s N_g| &\lesssim \epsilon^2 r^{-1}\tau^{-2-2\delta_{dec}+2\delta_0},\\
|\mathfrak{d}^s e_3(N_g)| &\lesssim \epsilon^2 r^{-3}\tau^{-\frac{3}{2}-2\delta_{dec}+2\delta_0},\\
|\mathfrak{d}^s e_3(N_g)| &\lesssim \epsilon^2 r^{-\frac{7}{2}-\frac{\delta_B}{2}}\tau^{-1-\delta_{dec}+2\delta_0}.
\end{aligned}
$$

By interpolation, we infer

$$\begin{aligned}
r^{-2}\int_S\left(r^{\frac{7}{2}}|\mathfrak{d}^{\leq s}N_g|^2+r^{\frac{11}{2}}|\mathfrak{d}^{\leq s}e_3(N_g)|^2\right) &\lesssim \epsilon^4 r^{-\frac{3}{2}}\tau^{-\frac{5}{2}-4\delta_{dec}+4\delta_0}\\
&\quad +\epsilon^4 r^{-1-\frac{\delta_B}{2}}\tau^{-\frac{5}{2}-3\delta_{dec}+4\delta_0}\\
&\lesssim \epsilon_0^2 r^{-1-\frac{\delta_B}{2}}\tau^{-\frac{5}{2}-3\delta_{dec}+4\delta_0}
\end{aligned}$$

and hence

$$\begin{aligned}
\left|e_4\left(r^{-2}\int_S(re_3(\mathfrak{d}^{\leq s}\mathfrak{q}))^2\right)\right| &\lesssim r^{-4}\int_S(re_3(\mathfrak{d}^{\leq s}\mathfrak{q}))^2+\epsilon_0^2 r^{-1-\frac{\delta_B}{2}}\tau^{-\frac{5}{2}-3\delta_{dec}+4\delta_0}\\
&\quad +r^{-\frac{3}{2}}(1+\tau)^{-2-q_0+\delta}\mathcal{E}^{s+2}_{\delta,2+q_0-\delta}[\mathfrak{q}].
\end{aligned}$$

We integrate from Σ_*. By Gronwall, and in view of (5.4.31), we deduce for $r\geq 4m_0$

$$(1+\tau)^{2+q_0-\delta}\int_{S_r}(e_3\mathfrak{d}^{\leq s}\mathfrak{q})^2 \lesssim \epsilon_0^2+\mathcal{F}^{s+1}_{\delta,2+q_0-\delta}[\mathfrak{q}]+\mathcal{E}^{s+2}_{\delta,2+q_0-\delta}[\mathfrak{q}].$$

On the other hand, we have by a trace estimate for $r\leq 4m_0$

$$(1+\tau)^{2+q_0-\delta}\int_{S_r}(e_3\mathfrak{d}^{\leq s}\mathfrak{q})^2 \lesssim \mathcal{E}^{s+2}_{0,2+q_0-\delta}[\mathfrak{q}].$$

We finally deduce on $\mathcal{M}$

$$(1+\tau)^{2+q_0-\delta}\int_{S_r}(e_3\mathfrak{d}^{\leq s}\mathfrak{q})^2 \lesssim \epsilon_0^2+\mathcal{F}^{s+1}_{\delta,2+q_0-\delta}[\mathfrak{q}]+\mathcal{E}^{s+2}_{\delta,2+q_0-\delta}[\mathfrak{q}]$$

which is the desired estimate (5.2.9). This concludes the proof of Proposition 5.13.

Chapter Six

Decay Estimates for α and $\underline{\alpha}$ (Theorems M2, M3)

In this chapter, we rely on the decay of $\mathfrak{q}$ to prove the decay estimates for α and $\underline{\alpha}$. More precisely, we rely on the results of Theorem M1 to prove Theorem M2 and M3.

6.1 PROOF OF THEOREM M2

6.1.1 A renormalized frame on ${}^{(ext)}\mathcal{M}$

In Theorem M1, decay estimates are derived for $\mathfrak{q}$ defined with respect to the global frame constructed in Proposition 3.26. We have the following control for the Ricci coefficients in that frame.

Lemma 6.1. *Consider the global null frame (e_3, e_4, e_θ) constructed in Proposition 3.26. Then, the Ricci coefficients satisfy the following estimates:*

$$\max_{0\leq k\leq k_{small}+20}\sup_{{}^{(ext)}\mathcal{M}} u^{\frac{1}{2}}\Bigg(r^2\left|\mathfrak{d}^k\left(\omega+\frac{m}{r^2}, \kappa-\frac{2\Upsilon}{r}, \vartheta, \zeta, \eta, \underline{\eta}\right)\right| + r\left|\mathfrak{d}^k\left(\underline{\xi}, \underline{\omega}, \underline{\kappa}+\frac{2}{r}, \underline{\vartheta}\right)\right| + \left|\mathfrak{d}^k\left(e_4(r)-\Upsilon, e_3(r)+1\right)\right|\Bigg) \lesssim \epsilon.$$

Proof. This follows immediately from the stronger estimates of Lemma 5.1 with the choice $k_{loss} = 20$. □

6.1.2 A transport equation for α

To recover α from $\mathfrak{q}$, we derive below a transport equation for α where $\mathfrak{q}$ is on the RHS. We are careful to avoid terms of the type $e_3(\underline{\omega})$ as they are anomalous w.r.t. decay in r. Indeed, they only decay linearly in r^{-1} while all comparable terms decay like r^{-2} in r.

Lemma 6.2. *We have*

$$\begin{aligned}
&\underline{\kappa}^2 e_3\left\{e_3\left(\frac{\alpha}{\underline{\kappa}^2}\right)-\left(8\underline{\omega}-\frac{2}{\underline{\kappa}}\left(2\,\not{d}_1\underline{\xi}-\frac{1}{2}\underline{\vartheta}^2\right)\right)\frac{\alpha}{\underline{\kappa}^2}\right\}\\
=\ &\frac{\mathfrak{q}}{r^4}+\left\{10\underline{\omega}+\frac{4}{\underline{\kappa}}\left(-\not{d}_1\underline{\xi}-2(\eta-3\zeta)\underline{\xi}+\frac{1}{4}\underline{\vartheta}^2\right)\right\}e_3\alpha\\
&+\Bigg\{-2\,\not{d}_1\underline{\xi}+\left(6\underline{\kappa}-24\underline{\omega}+\frac{8}{\underline{\kappa}}\left(2\,\not{d}_1\underline{\xi}+2(\eta-3\zeta)\underline{\xi}-\frac{1}{2}\underline{\vartheta}^2\right)\right)\underline{\omega}+\frac{1}{2}\underline{\vartheta}^2\\
&-\frac{4}{\underline{\kappa}}e_3((\eta-3\zeta)\underline{\xi})+\left(16+\frac{48}{\underline{\kappa}}\underline{\omega}-\frac{24}{\underline{\kappa}^2}\left(2\,\not{d}_1\underline{\xi}+2(\eta-3\zeta)\underline{\xi}-\frac{1}{2}\underline{\vartheta}^2\right)\right)\zeta\underline{\xi}\Bigg\}\alpha.
\end{aligned}$$

Proof. We compute

$$\begin{aligned}
e_3e_3\left(\frac{\alpha}{\underline{\kappa}^2}\right) &= e_3\left(\frac{e_3\alpha}{\underline{\kappa}^2}-\frac{2e_3(\underline{\kappa})\alpha}{\underline{\kappa}^3}\right)\\
&= \frac{1}{\underline{\kappa}^2}\left(e_3e_3\alpha-4\frac{e_3(\underline{\kappa})}{\underline{\kappa}}e_3\alpha-2\underline{\kappa}^2e_3\left(\frac{e_3(\underline{\kappa})}{\underline{\kappa}^3}\right)\alpha\right).
\end{aligned}$$

Now, recall the following null structure equation

$$e_3(\underline{\kappa})+\frac{1}{2}\underline{\kappa}^2+2\underline{\omega}\,\underline{\kappa}=2\,\not{d}_1\underline{\xi}+2(\eta-3\zeta)\underline{\xi}-\frac{1}{2}\underline{\vartheta}^2.$$

We infer

$$\frac{e_3(\underline{\kappa})}{\underline{\kappa}}=-\frac{1}{2}\underline{\kappa}-2\underline{\omega}+\frac{1}{\underline{\kappa}}\left(2\,\not{d}_1\underline{\xi}+2(\eta-3\zeta)\underline{\xi}-\frac{1}{2}\underline{\vartheta}^2\right)$$

and

$$\begin{aligned}
e_3\left(\frac{e_3(\underline{\kappa})}{\underline{\kappa}^3}\right) &= e_3\left(-\frac{1}{2\underline{\kappa}}-2\frac{\underline{\omega}}{\underline{\kappa}^2}+\frac{1}{\underline{\kappa}^3}\left(2\,\not{d}_1\underline{\xi}+2(\eta-3\zeta)\underline{\xi}-\frac{1}{2}\underline{\vartheta}^2\right)\right)\\
&= \frac{e_3(\underline{\kappa})}{2\underline{\kappa}^2}+e_3\left(-2\frac{\underline{\omega}}{\underline{\kappa}^2}+\frac{1}{\underline{\kappa}^3}\left(2\,\not{d}_1\underline{\xi}+2(\eta-3\zeta)\underline{\xi}-\frac{1}{2}\underline{\vartheta}^2\right)\right)\\
&= -\frac{1}{4}+\frac{1}{2\underline{\kappa}}\left(-2\underline{\omega}+\frac{1}{\underline{\kappa}}\left(2\,\not{d}_1\underline{\xi}+2(\eta-3\zeta)\underline{\xi}-\frac{1}{2}\underline{\vartheta}^2\right)\right)\\
&\quad +e_3\left(-2\frac{\underline{\omega}}{\underline{\kappa}^2}+\frac{1}{\underline{\kappa}^3}\left(2\,\not{d}_1\underline{\xi}+2(\eta-3\zeta)\underline{\xi}-\frac{1}{2}\underline{\vartheta}^2\right)\right)
\end{aligned}$$

and hence

$$\begin{aligned}
\underline{\kappa}^2 e_3 e_3\left(\frac{\alpha}{\underline{\kappa}^2}\right) &= e_3e_3\alpha - 4\frac{e_3(\underline{\kappa})}{\underline{\kappa}}e_3\alpha - 2\underline{\kappa}^2 e_3\left(\frac{e_3(\underline{\kappa})}{\underline{\kappa}^3}\right)\alpha\\
&= e_3e_3\alpha + 2\underline{\kappa}e_3\alpha + \frac{1}{2}\underline{\kappa}^2\alpha\\
&\quad +\left(8\underline{\omega} - \frac{4}{\underline{\kappa}}\left(2\,\not{d}_1\underline{\xi} + 2(\eta-3\zeta)\underline{\xi} - \frac{1}{2}\underline{\vartheta}^2\right)\right)e_3\alpha\\
&\quad +\Bigg\{-\underline{\kappa}\left(-2\underline{\omega} + \frac{1}{\underline{\kappa}}\left(2\,\not{d}_1\underline{\xi} + 2(\eta-3\zeta)\underline{\xi} - \frac{1}{2}\underline{\vartheta}^2\right)\right)\\
&\quad -2\underline{\kappa}^2 e_3\left(-2\frac{\underline{\omega}}{\underline{\kappa}^2} + \frac{1}{\underline{\kappa}^3}\left(2\,\not{d}_1\underline{\xi} + 2(\eta-3\zeta)\underline{\xi} - \frac{1}{2}\underline{\vartheta}^2\right)\right)\Bigg\}\alpha.
\end{aligned}$$

Next, recall from section 2.3.3 that $\mathfrak{q}$ is defined with respect to a general null frame as follows:

$$\mathfrak{q} = r^4\left(e_3(e_3(\alpha)) + (2\underline{\kappa} - 6\underline{\omega})e_3(\alpha) + \left(-4e_3(\underline{\omega}) + 8\underline{\omega}^2 - 8\underline{\omega}\,\underline{\kappa} + \frac{1}{2}\underline{\kappa}^2\right)\alpha\right).$$

We infer

$$\begin{aligned}
\underline{\kappa}^2 e_3 e_3\left(\frac{\alpha}{\underline{\kappa}^2}\right) &= \frac{\mathfrak{q}}{r^4} + \left(14\underline{\omega} - \frac{4}{\underline{\kappa}}\left(2\,\not{d}_1\underline{\xi} + 2(\eta-3\zeta)\underline{\xi} - \frac{1}{2}\underline{\vartheta}^2\right)\right)e_3\alpha\\
&\quad +\Bigg\{4e_3(\underline{\omega}) - 8\underline{\omega}^2 + 10\underline{\omega}\,\underline{\kappa} - \left(2\,\not{d}_1\underline{\xi} + 2(\eta-3\zeta)\underline{\xi} - \frac{1}{2}\underline{\vartheta}^2\right)\\
&\quad -2\underline{\kappa}^2 e_3\left(-2\frac{\underline{\omega}}{\underline{\kappa}^2} + \frac{1}{\underline{\kappa}^3}\left(2\,\not{d}_1\underline{\xi} + 2(\eta-3\zeta)\underline{\xi} - \frac{1}{2}\underline{\vartheta}^2\right)\right)\Bigg\}\alpha.
\end{aligned}$$

We rewrite the following terms

$$\begin{aligned}
&\left\{4e_3(\underline{\omega}) - 2\underline{\kappa}^2 e_3\left(-2\frac{\underline{\omega}}{\underline{\kappa}^2} + \frac{1}{\underline{\kappa}^3}\left(2\,\not{d}_1\underline{\xi} - \frac{1}{2}\underline{\vartheta}^2\right)\right)\right\}\alpha\\
&= \underline{\kappa}^2 e_3\left\{\left(8\frac{\underline{\omega}}{\underline{\kappa}^2} - \frac{2}{\underline{\kappa}^3}\left(2\,\not{d}_1\underline{\xi} - \frac{1}{2}\underline{\vartheta}^2\right)\right)\alpha\right\}\\
&\quad -4\underline{\kappa}^2\underline{\omega}e_3\left(\frac{\alpha}{\underline{\kappa}^2}\right) + \frac{2}{\underline{\kappa}}\left(2\,\not{d}_1\underline{\xi} - \frac{1}{2}\underline{\vartheta}^2\right)e_3(\alpha)\\
&= \underline{\kappa}^2 e_3\left\{\left(8\frac{\underline{\omega}}{\underline{\kappa}^2} - \frac{2}{\underline{\kappa}^3}\left(2\,\not{d}_1\underline{\xi} - \frac{1}{2}\underline{\vartheta}^2\right)\right)\alpha\right\}\\
&\quad +\left(-4\underline{\omega} + \frac{2}{\underline{\kappa}}\left(2\,\not{d}_1\underline{\xi} - \frac{1}{2}\underline{\vartheta}^2\right)\right)e_3\alpha - 4\underline{\kappa}^2\underline{\omega}e_3\left(\frac{1}{\underline{\kappa}^2}\right)\alpha
\end{aligned}$$

so that we obtain

$$\begin{aligned}
&\underline{\kappa}^2 e_3\left\{e_3\left(\frac{\alpha}{\underline{\kappa}^2}\right)-\left(8\frac{\underline{\omega}}{\underline{\kappa}^2}-\frac{2}{\underline{\kappa}^3}\left(2\,{\not d}_1\underline{\xi}-\frac{1}{2}\underline{\vartheta}^2\right)\right)\alpha\right\}\\
=\ &\frac{\mathfrak{q}}{r^4}+\left\{10\underline{\omega}-\frac{4}{\underline{\kappa}}\left({\not d}_1\underline{\xi}+2(\eta-3\zeta)\underline{\xi}-\frac{1}{4}\underline{\vartheta}^2\right)\right\}e_3\alpha+\bigg\{-8\underline{\omega}^2+10\underline{\omega}\,\underline{\kappa}\\
&-\left(2\,{\not d}_1\underline{\xi}+2(\eta-3\zeta)\underline{\xi}-\frac{1}{2}\underline{\vartheta}^2\right)-4\underline{\kappa}^2e_3\left(\frac{1}{\underline{\kappa}^3}(\eta-3\zeta)\underline{\xi}\right)-4\underline{\kappa}^2\underline{\omega}e_3\left(\frac{1}{\underline{\kappa}^2}\right)\bigg\}\alpha
\end{aligned}$$

which we rewrite as

$$\begin{aligned}
&\underline{\kappa}^2 e_3\left\{e_3\left(\frac{\alpha}{\underline{\kappa}^2}\right)-\left(8\underline{\omega}-\frac{2}{\underline{\kappa}}\left(2\,{\not d}_1\underline{\xi}-\frac{1}{2}\underline{\vartheta}^2\right)\right)\frac{\alpha}{\underline{\kappa}^2}\right\}\\
=\ &\frac{\mathfrak{q}}{r^4}+\left\{10\underline{\omega}-\frac{4}{\underline{\kappa}}\left({\not d}_1\underline{\xi}+2(\eta-3\zeta)\underline{\xi}-\frac{1}{4}\underline{\vartheta}^2\right)\right\}e_3\alpha\\
&+\bigg\{-8\underline{\omega}^2+10\underline{\omega}\,\underline{\kappa}-\left(2\,{\not d}_1\underline{\xi}+2(\eta-3\zeta)\underline{\xi}-\frac{1}{2}\underline{\vartheta}^2\right)-\frac{4}{\underline{\kappa}}e_3((\eta-3\zeta)\underline{\xi})\\
&+12\frac{e_3(\underline{\kappa})}{\underline{\kappa}^2}(\eta-3\zeta)\underline{\xi}+8\frac{e_3(\underline{\kappa})}{\underline{\kappa}}\underline{\omega}\bigg\}\alpha.
\end{aligned}$$

Now, recall from above that we have

$$\frac{e_3(\underline{\kappa})}{\underline{\kappa}}=-\frac{1}{2}\underline{\kappa}-2\underline{\omega}+\frac{1}{\underline{\kappa}}\left(2\,{\not d}_1\underline{\xi}+2(\eta-3\zeta)\underline{\xi}-\frac{1}{2}\underline{\vartheta}^2\right).$$

We finally deduce

$$\begin{aligned}
&\underline{\kappa}^2 e_3\left\{e_3\left(\frac{\alpha}{\underline{\kappa}^2}\right)-\left(8\underline{\omega}-\frac{2}{\underline{\kappa}}\left(2\,{\not d}_1\underline{\xi}-\frac{1}{2}\underline{\vartheta}^2\right)\right)\frac{\alpha}{\underline{\kappa}^2}\right\}\\
=\ &\frac{\mathfrak{q}}{r^4}+\left\{10\underline{\omega}+\frac{4}{\underline{\kappa}}\left(-{\not d}_1\underline{\xi}-2(\eta-3\zeta)\underline{\xi}+\frac{1}{4}\underline{\vartheta}^2\right)\right\}e_3\alpha\\
&+\bigg\{-2\,{\not d}_1\underline{\xi}+\left(6\underline{\kappa}-24\underline{\omega}+\frac{8}{\underline{\kappa}}\left(2\,{\not d}_1\underline{\xi}+2(\eta-3\zeta)\underline{\xi}-\frac{1}{2}\underline{\vartheta}^2\right)\right)\underline{\omega}+\frac{1}{2}\underline{\vartheta}^2\\
&-\frac{4}{\underline{\kappa}}e_3((\eta-3\zeta)\underline{\xi})+\left(16+\frac{48}{\underline{\kappa}}\underline{\omega}-\frac{24}{\underline{\kappa}^2}\left(2\,{\not d}_1\underline{\xi}+2(\eta-3\zeta)\underline{\xi}-\frac{1}{2}\underline{\vartheta}^2\right)\right)\zeta\underline{\xi}\bigg\}\alpha.
\end{aligned}$$

This concludes the proof of the lemma. □

6.1.3 Estimates for transport equations in e_3

The following lemma will be useful to integrate the transport equations in e_3.

Lemma 6.3. *Let $p\in{}^{(ext)}\mathcal{M}$. Let $\gamma[p]$ the unique integral curve of e_3 starting from* *a point on $\mathcal{C}_1$ terminating at p. Then, we have for $l \geq 1$*

$$\int_{\gamma[p]} \frac{1}{r'^{2+l} u'^{\frac{1}{2}+\delta_{extra}} + r'^{1+l} u'^{1+\delta_{extra}}} \lesssim \frac{1}{r^{1+l}(2r+u)^{\frac{1}{2}+\delta_{extra}} + r^l (2r+u)^{1+\delta_{extra}}}$$

and

$$\int_{\gamma[p]} \frac{1}{r'^{2} u'^{\frac{1}{2}+\delta_{extra}} + r' u'^{1+\delta_{extra}}} \lesssim \frac{1}{r(2r+u)^{\frac{1}{2}+\delta_{extra}} + \frac{1}{\log(1+u)}(2r+u)^{1+\delta_{extra}}}$$

where (u, r) correspond to p and (r', u') to a point on $\gamma[p]$, and where the integration along $\gamma[p]$ relies on a parametrization of $\gamma[p]$ normalized with respect to e_3.

Proof. Note first from the construction of ${}^{(ext)}\mathcal{M}$ that $\gamma[p]$ exists for any $p \in {}^{(ext)}\mathcal{M}$ (i.e., any point p can be joined to $\mathcal{C}_1$ by an integral curve of e_3), and $\gamma[p]$ is included in ${}^{(ext)}\mathcal{M}$.

Next, recall that the integration along $\gamma[p]$ relies on a parametrization of $\gamma[p]$ normalized with respect to e_3. To parametrize the integration by u or r, we will thus have to derive an upper bound for the corresponding Jacobian of the change of variable, i.e., for

$$\frac{1}{|e_3(u)|}, \quad \frac{1}{|e_3(r)|}.$$

To this end, note that we have on ${}^{(ext)}\mathcal{M}$

$$e_3(u) = \frac{2}{\varsigma\Upsilon} \geq \frac{2+O(\epsilon)}{\Upsilon} \geq \frac{1}{\Upsilon} \geq 1$$

since $\Upsilon \leq 1$ by definition. Also, we have on ${}^{(ext)}\mathcal{M}$ in view of Lemma 6.1

$$\begin{aligned} |e_3(r)| &\geq 1 - |e_3(r) + 1| \\ &= 1 + O(\epsilon) \\ &\geq \frac{1}{2}. \end{aligned}$$

Hence, we have obtained on ${}^{(ext)}\mathcal{M}$

$$\frac{1}{|e_3(u)|} \leq \frac{1}{2}, \qquad \frac{1}{|e_3(r)|} \leq 1. \tag{6.1.1}$$

Next, since $e_3(u) > 0$ and $e_3(r) < 0$ in ${}^{(ext)}\mathcal{M}$, we have $r' \geq r$ and $1 \leq u' \leq u$. We start with the proof of the first inequality. We consider two cases:

- If $r \geq u$, we have

$$\begin{aligned} &\int_{\gamma[p]} \frac{1}{r'^{2+l} u'^{\frac{1}{2}+\delta_{extra}} + r'^{1+l} u'^{1+\delta_{extra}}} \\ \leq\ & \frac{1}{r^{2+l}} \int_0^u \frac{1}{|e_3(u')|} \frac{du'}{u'^{\frac{1}{2}+\delta_{extra}}} \lesssim \frac{u^{\frac{1}{2}-\delta_{extra}}}{r^{2+l}} \\ \lesssim\ & \frac{1}{r^{1+l}(2r+u)^{\frac{1}{2}+\delta_{extra}} + r^l(2r+u)^{1+\delta_{extra}}}, \end{aligned}$$

where we used (6.1.1).

- If $r \leq u$, we separate the integral in $r' \geq u$, which coincides with $1 \leq u' \leq u$, and $r \leq r' \leq u$ and compute

$$\begin{aligned}
&\int_{\gamma[p]} \frac{1}{r'^{2+l}u'^{\frac{1}{2}+\delta_{extra}} + r'^{1+l}u'^{1+\delta_{extra}}}\\
= & \int_0^u \frac{1}{r'^{2+l}u'^{\frac{1}{2}+\delta_{extra}} + r'^{1+l}u'^{1+\delta_{extra}}}\frac{du'}{|e_3(u')|}\\
&+\int_r^u \frac{1}{r'^{2+l}u'^{\frac{1}{2}+\delta_{extra}} + r'^{1+l}u'^{1+\delta_{extra}}}\frac{dr'}{|e_3(r')|}\\
\lesssim & \frac{1}{u^{2+l}}\int_0^u \frac{1}{|e_3(u')|}\frac{du'}{u'^{\frac{1}{2}+\delta_{extra}}}\\
&+\min\left(\frac{1}{u^{\frac{1}{2}+\delta_{extra}}}\int_r^u \frac{1}{|e_3(r')|}\frac{dr'}{r'^{2+l}}, \frac{1}{u^{1+\delta_{extra}}}\int_r^u \frac{1}{|e_3(r')|}\frac{dr'}{r'^{1+l}}\right)\\
\lesssim & \frac{1}{u^{\frac{5}{2}+\delta_{extra}}} + \min\left(\frac{1}{u^{\frac{1}{2}+\delta_{extra}}}\frac{1}{r^{1+l}}, \frac{1}{u^{1+\delta_{extra}}}\frac{1}{r^l}\right)\\
\lesssim & \frac{1}{r^{1+l}(2r+u)^{\frac{1}{2}+\delta_{extra}} + r^l(2r+u)^{1+\delta_{extra}}},
\end{aligned}$$

where we used (6.1.1).

This proves the first inequality.

The second inequality is obtained similarly as follows:

- If $r \geq u$, we have

$$\begin{aligned}
\int_{\gamma[p]} \frac{1}{r'^2u'^{\frac{1}{2}+\delta_{extra}} + r'u'^{1+\delta_{extra}}} &\leq \frac{1}{r^2}\int_0^u \frac{1}{|e_3(u')|}\frac{du'}{u'^{\frac{1}{2}+\delta_{extra}}}\\
&\lesssim \frac{u^{\frac{1}{2}-\delta_{extra}}}{r^2}\\
&\lesssim \frac{1}{r(2r+u)^{\frac{1}{2}+\delta_{extra}} + (2r+u)^{1+\delta_{extra}}},
\end{aligned}$$

where we used (6.1.1).

- If $r \leq u$, we separate the integral in $r' \geq u$, which coincides with $1 \leq u' \leq u$, and $r \leq r' \leq u$ and compute

$$\begin{aligned}
&\int_{\gamma[p]} \frac{1}{r'^2 u'^{\frac{1}{2}+\delta_{extra}} + r' u'^{1+\delta_{extra}}} \\
= &\int_0^u \frac{1}{r'^2 u'^{\frac{1}{2}+\delta_{extra}} + r' u'^{1+\delta_{extra}}} \frac{du'}{|e_3(u')|} \\
&+ \int_r^u \frac{1}{r'^2 u'^{\frac{1}{2}+\delta_{extra}} + r' u'^{1+\delta_{extra}}} \frac{dr'}{|e_3(r')|} \\
\lesssim &\frac{1}{u^3} \int_0^u \frac{1}{|e_3(u')|} \frac{du'}{u'^{\frac{1}{2}+\delta_{extra}}} \\
&+ \min\left(\frac{1}{u^{\frac{1}{2}+\delta_{extra}}} \int_r^u \frac{1}{|e_3(r')|} \frac{dr'}{r'^2}, \frac{1}{u^{1+\delta_{extra}}} \int_r^u \frac{1}{|e_3(r')|} \frac{dr'}{r'} \right) \\
\lesssim &\frac{1}{u^{\frac{5}{2}+\delta_{extra}}} + \min\left(\frac{1}{u^{\frac{1}{2}+\delta_{extra}}} \frac{1}{r}, \frac{1}{u^{1+\delta_{extra}}} \int_{\frac{r}{u}}^1 \frac{dr''}{r''} \right) \\
\lesssim &\frac{1}{r(2r+u)^{\frac{1}{2}+\delta_{extra}} + \frac{(2r+u)^{1+\delta_{extra}}}{\log(1+u)}},
\end{aligned}$$

where we used (6.1.1).

This concludes the proof of the lemma. $\square$

Corollary 6.4. *Let ψ a solution of the following transport equation*

$$e_3(\psi) = h \text{ on } {}^{(ext)}\mathcal{M}.$$

Let also $0 < u_1 \leq u_*$. Then:

- *If h and ψ satisfy for $l \geq 1$*

$$\begin{aligned}
|h| &\lesssim \frac{\epsilon_0}{r^{2+l} u^{\frac{1}{2}+\delta_{extra}} + r^{1+l} u^{1+\delta_{extra}}} \text{ on } {}^{(ext)}\mathcal{M}(u \leq u_1), \\
|\psi| &\lesssim \frac{\epsilon_0}{r^{l+\frac{3}{2}+\delta_{extra}}} \text{ on } \mathcal{C}_1,
\end{aligned}$$

  we have

$$\sup_{{}^{(ext)}\mathcal{M}(u\leq u_1)} \left(r^{1+l}(2r+u)^{\frac{1}{2}+\delta_{extra}} + r^l (2r+u)^{1+\delta_{extra}} \right) |\psi| \lesssim \epsilon_0.$$

- *If h and ψ satisfy*

$$\begin{aligned}
|h| &\lesssim \frac{\epsilon_0}{r^2 u^{\frac{1}{2}+\delta_{extra}} + r u^{1+\delta_{extra}}} \text{ on } {}^{(ext)}\mathcal{M}(u \leq u_1), \\
|\psi| &\lesssim \frac{\epsilon_0}{r^{\frac{3}{2}+\delta_{extra}}} \text{ on } \mathcal{C}_1,
\end{aligned}$$

  we have

$$\sup_{{}^{(ext)}\mathcal{M}(u\leq u_1)} \left(r(2r+u)^{\frac{1}{2}+\delta_{extra}} + \frac{(2r+u)^{1+\delta_{extra}}}{\log(1+u)} \right) |\psi| \lesssim \epsilon_0.$$

Proof. This follows immediately from Lemma 6.3. □

6.1.4 Decay estimates for α

We start with an estimate for α on $\mathcal{C}_1$.

Lemma 6.5. *We have*

$$\max_{0\leq k\leq k_{small}+22}\sup_{\mathcal{C}_1} r^{\frac{7}{2}+\delta_{extra}}|\mathfrak{d}^k\alpha| + \max_{0\leq k\leq k_{small}+21}\sup_{\mathcal{C}_1} r^{\frac{9}{2}+\delta_{extra}}|\mathfrak{d}^k e_3\alpha| \lesssim \epsilon_0.$$

Proof. Recall that on $\mathcal{C}_1$, we have obtained in Theorem M0

$$\max_{0\leq k\leq k_{large}}\Bigg\{\sup_{\mathcal{C}_1}\left[r^{\frac{7}{2}+\delta_B}\left(|\mathfrak{d}^{k}\,{}^{(ext)}\alpha| + |\mathfrak{d}^{k}\,{}^{(ext)}\beta|\right) + r^{\frac{9}{2}+\delta_B}|\mathfrak{d}^{k-1}e_3({}^{(ext)}\alpha)|\right]$$
$$+\sup_{\mathcal{C}_1}\left[r^3\left|\mathfrak{d}^k\left({}^{(ext)}\rho+\frac{2m_0}{r^3}\right)\right| + r^2|\mathfrak{d}^{k}\,{}^{(ext)}\underline{\beta}| + r|\mathfrak{d}^{k}\,{}^{(ext)}\underline{\alpha}|\right]\Bigg\} \lesssim \epsilon_0.$$

Since we have chosen $\delta_B\geq\delta_{extra}$, we deduce

$$\max_{0\leq k\leq k_{large}}\sup_{\mathcal{C}_1}\left[r^{\frac{7}{2}+\delta_{extra}}|\mathfrak{d}^{k}\,{}^{(ext)}\alpha| + r^{\frac{9}{2}+\delta_{extra}}|\mathfrak{d}^{k-1}e_3({}^{(ext)}\alpha)|\right] \lesssim \epsilon_0.$$

Next, recall that $\mathfrak{q}$ is defined with respect to the global frame constructed in Proposition 3.26. In view of Proposition 3.26 and Proposition 3.20, and the change of frame formula for α in Proposition 2.90, we have

$$\alpha = ({}^{(ext)}\Upsilon)^2\left({}^{(ext)}\alpha + 2f\,{}^{(ext)}\beta + \frac{3}{2}f^2\,{}^{(ext)}\rho + \text{l.o.t.}\right) \tag{6.1.2}$$

where f satisfies,[1] see (3.4.11),

$$\begin{aligned} |\mathfrak{d}^k f| &\lesssim \frac{\epsilon}{ru^{\frac{1}{2}}+u}, \quad \text{for } k\leq k_{small}+22 \text{ on } {}^{(ext)}\mathcal{M},\\ |\mathfrak{d}^{k-1}e_3 f| &\lesssim \frac{\epsilon}{ru} \quad \text{for } k\leq k_{small}+22 \text{ on } {}^{(ext)}\mathcal{M}. \end{aligned} \tag{6.1.3}$$

We easily infer

$$\max_{0\leq k\leq k_{small}+22}\sup_{\mathcal{C}_1} r^{\frac{7}{2}+\delta_{extra}}|\mathfrak{d}^k\alpha| + \max_{0\leq k\leq k_{small}+21}\sup_{\mathcal{C}_1} r^{\frac{9}{2}+\delta_{extra}}|\mathfrak{d}^k e_3\alpha| \lesssim \epsilon_0.$$

This concludes the proof of the lemma. □

Next, let $0<u_1\leq u_*$. We introduce the following bootstrap assumption for α

[1] Here we use (3.4.11) with $k_{loss}=20$. Note also that the estimates we claim here for f are slightly weaker that those in (3.4.11).

on ${}^{(ext)}\mathcal{M}(u \leq u_1)$:

$$\max_{0\leq k\leq k_{small}+20} \sup_{{}^{(ext)}\mathcal{M}(u\leq u_1)} \left(\frac{r^2(2r+u)^{1+\delta_{extra}}}{\log(1+u)} + r^3(2r+u)^{\frac{1}{2}+\delta_{extra}}\right) \times\left(|\mathfrak{d}^k\alpha| + r|\mathfrak{d}^k e_3\alpha|\right) \leq \epsilon. \tag{6.1.4}$$

The goal of this section will be the following proposition, i.e., the improvement of these bootstrap assumptions.

Proposition 6.6. *We have*

$$\max_{0\leq k\leq k_{small}+20} \sup_{{}^{(ext)}\mathcal{M}(u\leq u_1)} \left(\frac{r^2(2r+u)^{1+\delta_{extra}}}{\log(1+u)} + r^3(2r+u)^{\frac{1}{2}+\delta_{extra}}\right) \times\left(|\mathfrak{d}^k\alpha| + r|\mathfrak{d}^k e_3\alpha|\right) \lesssim \epsilon_0.$$

Proposition 6.6 will be proved at the end of this section.

Based on the bootstrap assumptions (6.1.4), we estimate the RHS of the transport equation for α.

Lemma 6.7. *We have*

$$e_3\left\{e_3\left(\frac{\alpha}{\underline{\kappa}^2}\right) - F_1\right\} = F_2$$

where F_1 and F_2 satisfy

$$\max_{0\leq k\leq k_{small}+20} \sup_{{}^{(ext)}\mathcal{M}(u\leq u_1)} \left(r(2r+u)^{1+\delta_{extra}} + r^2(2r+u)^{\frac{1}{2}+\delta_{extra}}\right)|\mathfrak{d}^k F_1| + \max_{0\leq k\leq k_{small}+20} \sup_{{}^{(ext)}\mathcal{M}(u\leq u_1)} \left(r^2u^{1+\delta_{extra}} + r^3u^{\frac{1}{2}+\delta_{extra}}\right)|\mathfrak{d}^k F_2| \lesssim \epsilon_0.$$

Proof. Recall that we have

$$\begin{aligned}
&\underline{\kappa}^2 e_3\left\{e_3\left(\frac{\alpha}{\underline{\kappa}^2}\right) - \left(8\underline{\omega} - \frac{2}{\underline{\kappa}}\left(2\,\not{d}_1\underline{\xi} - \frac{1}{2}\underline{\vartheta}^2\right)\right)\frac{\alpha}{\underline{\kappa}^2}\right\}\\
&= \frac{\mathfrak{q}}{r^4} + \left\{10\underline{\omega} + \frac{4}{\underline{\kappa}}\left(-\not{d}_1\underline{\xi} - 2(\eta-3\zeta)\underline{\xi} + \frac{1}{4}\underline{\vartheta}^2\right)\right\}e_3\alpha\\
&\quad+\Bigg\{-2\,\not{d}_1\underline{\xi} + \left(6\underline{\kappa} - 24\underline{\omega} + \frac{8}{\underline{\kappa}}\left(2\,\not{d}_1\underline{\xi} + 2(\eta-3\zeta)\underline{\xi} - \frac{1}{2}\underline{\vartheta}^2\right)\right)\underline{\omega} + \frac{1}{2}\underline{\vartheta}^2\\
&\quad-\frac{4}{\underline{\kappa}}e_3((\eta-3\zeta)\underline{\xi}) + \left(16 + \frac{48}{\underline{\kappa}}\underline{\omega} - \frac{24}{\underline{\kappa}^2}\left(2\,\not{d}_1\underline{\xi} + 2(\eta-3\zeta)\underline{\xi} - \frac{1}{2}\underline{\vartheta}^2\right)\right)\zeta\underline{\xi}\Bigg\}\alpha
\end{aligned}$$

which we rewrite as

$$e_3\left\{e_3\left(\frac{\alpha}{\underline{\kappa}^2}\right) - F_1\right\} = F_2$$

where F_1 and F_2 are defined by

$$F_1 \;:=\; \left(8\underline{\omega} - \frac{2}{\underline{\kappa}}\left(2\,\not{d}_1\underline{\xi} - \frac{1}{2}\underline{\vartheta}^2\right)\right)\frac{\alpha}{\underline{\kappa}^2}$$

and

$$\begin{aligned}
F_2 \;:=\;& \frac{\mathfrak{q}}{r^4\underline{\kappa}^2} + \frac{1}{\underline{\kappa}^2}\left\{10\underline{\omega} + \frac{4}{\underline{\kappa}}\left(-\not{d}_1\underline{\xi} - 2(\eta-3\zeta)\underline{\xi} + \frac{1}{4}\underline{\vartheta}^2\right)\right\}e_3\alpha \\
&+\frac{1}{\underline{\kappa}^2}\Bigg\{-2\,\not{d}_1\underline{\xi} + \left(6\underline{\kappa} - 24\underline{\omega} + \frac{8}{\underline{\kappa}}\left(2\,\not{d}_1\underline{\xi} + 2(\eta-3\zeta)\underline{\xi} - \frac{1}{2}\underline{\vartheta}^2\right)\right)\underline{\omega} + \frac{1}{2}\underline{\vartheta}^2 \\
&-\frac{4}{\underline{\kappa}}e_3((\eta-3\zeta)\underline{\xi}) + \left(16 + \frac{48}{\underline{\kappa}}\underline{\omega} - \frac{24}{\underline{\kappa}^2}\left(2\,\not{d}_1\underline{\xi} + 2(\eta-3\zeta)\underline{\xi} - \frac{1}{2}\underline{\vartheta}^2\right)\right)\zeta\underline{\xi}\Bigg\}\alpha.
\end{aligned}$$

In view of the bootstrap assumptions (6.1.4) for α, the estimates of Lemma 6.1 for the Ricci coefficients, and using Theorem M2 to estimate $\mathfrak{q}$, we easily infer

$$\begin{aligned}
&\max_{0\leq k\leq k_{small}+20}\;\sup_{{}^{(ext)}\mathcal{M}(u\leq u_1)}\left(r(2r+u)^{1+\delta_{extra}} + r^2(2r+u)^{\frac{1}{2}+\delta_{extra}}\right)|\mathfrak{d}^k F_1| \\
\lesssim\;& \epsilon\max_{0\leq k\leq k_{small}+20}\;\sup_{{}^{(ext)}\mathcal{M}(u\leq u_1)}\left(\frac{r^2(2r+u)^{1+\delta_{extra}}}{\log(1+u)} + r^3(2r+u)^{\frac{1}{2}+\delta_{extra}}\right) \\
&\times\left(|\mathfrak{d}^k\alpha| + r|\mathfrak{d}^k e_3\alpha|\right) \\
\lesssim\;& \epsilon^2 \lesssim \epsilon_0
\end{aligned}$$

and

$$\begin{aligned}
&\max_{0\leq k\leq k_{small}+20}\;\sup_{{}^{(ext)}\mathcal{M}(u\leq u_1)}\left(r^2u^{1+\delta_{extra}} + r^3u^{\frac{1}{2}+\delta_{extra}}\right)|\mathfrak{d}^k F_2| \\
\lesssim\;& \max_{0\leq k\leq k_{small}+20}\;\sup_{{}^{(ext)}\mathcal{M}(u\leq u_1)}\left(u^{1+\delta_{extra}} + ru^{\frac{1}{2}+\delta_{extra}}\right)|\mathfrak{d}^k\mathfrak{q}| \\
&+\epsilon\max_{0\leq k\leq k_{small}+20}\;\sup_{{}^{(ext)}\mathcal{M}(u\leq u_1)}\left(\frac{r^2(2r+u)^{1+\delta_{extra}}}{\log(1+u)} + r^3(2r+u)^{\frac{1}{2}+\delta_{extra}}\right) \\
&\times\left(|\mathfrak{d}^k\alpha| + r|\mathfrak{d}^k e_3\alpha|\right) \\
\lesssim\;& \epsilon_0 + \epsilon^2 \lesssim \epsilon_0.
\end{aligned}$$

This concludes the proof of the lemma. □

Lemma 6.8. *For $0 \leq k+j \leq k_{small}+20$, we have*

$$e_3\left\{e_3\,\not{\mathfrak{d}}^k e_4^j\left(\frac{\alpha}{\underline{\kappa}^2}\right) - F_{1,\not{\mathfrak{d}}^k,e_4^j}\right\} \;=\; F_{2,\not{\mathfrak{d}}^k,e_4^j}$$

where

$$\max_{0\leq l\leq k_{small}+20-k}\sup_{^{(ext)}\mathcal{M}(u\leq u_1)}\Big(r(2r+u)^{1+\delta_{extra}}+r^2(2r+u)^{\frac{1}{2}+\delta_{extra}}\Big)|\mathfrak{d}^l F_{1,\not{\mathfrak{d}}^k}|$$
$$+\max_{0\leq l\leq k_{small}+20-k}\sup_{^{(ext)}\mathcal{M}(u\leq u_1)}\Big(r^2u^{1+\delta_{extra}}+r^3u^{\frac{1}{2}+\delta_{extra}}\Big)|\mathfrak{d}^l F_{2,\not{\mathfrak{d}}^k}|$$
$$\lesssim \epsilon_0,$$

and for $j\geq 1$

$$\max_{0\leq l\leq k_{small}+20-k-j}\sup_{^{(ext)}\mathcal{M}(u\leq u_1)}\Big(r^{1+j}(2r+u)^{1+\delta_{extra}}+r^{2+j}(2r+u)^{\frac{1}{2}+\delta_{extra}}\Big)$$
$$\times|\mathfrak{d}^l F_{1,\not{\mathfrak{d}}^k,e_4^j}|$$
$$+\max_{0\leq l\leq k_{small}+20-k-j}\sup_{^{(ext)}\mathcal{M}(u\leq u_1)}\Big(r^{2+j}u^{1+\delta_{extra}}+r^{3+j}u^{\frac{1}{2}+\delta_{extra}}\Big)|\mathfrak{d}^l F_{2,\not{\mathfrak{d}}^k,e_4^j}|$$
$$\lesssim \epsilon_0+\max_{0\leq j+k\leq k_{small}+20}\sup_{^{(ext)}\mathcal{M}(u\leq u_1)}\Big(\frac{r^2(2r+u)^{1+\delta_{extra}}}{\log(1+u)}+r^3(2r+u)^{\frac{1}{2}+\delta_{extra}}\Big)r$$
$$\times\Big(|\not{\mathfrak{d}}^k(re_4)^{j-1}e_3\alpha|+|\not{\mathfrak{d}}^k(re_4)^{j-2}e_3^2\alpha|\Big).$$

Proof. Recall from Lemma 6.7 that we have

$$e_3\left\{e_3\left(\frac{\alpha}{\underline{\kappa}^2}\right)-F_1\right\} \;=\; F_2$$

where F_1 and F_2 satisfy

$$\max_{0\leq k\leq k_{small}+20}\sup_{^{(ext)}\mathcal{M}}\Big(r(2r+u)^{1+\delta_{extra}}+r^2(2r+u)^{\frac{1}{2}+\delta_{extra}}\Big)|\mathfrak{d}^k F_1|$$
$$+\max_{0\leq k\leq k_{small}+20}\sup_{^{(ext)}\mathcal{M}}\Big(r^2u^{1+\delta_{extra}}+r^3u^{\frac{1}{2}+\delta_{extra}}\Big)|\mathfrak{d}^k F_2| \;\lesssim\; \epsilon_0.$$

Differentiating with $\not{\mathfrak{d}}^k$, this yields

$$e_3\left\{e_3\not{\mathfrak{d}}^k\left(\frac{\alpha}{\underline{\kappa}^2}\right)+[\not{\mathfrak{d}}^k,e_3]\left(\frac{\alpha}{\underline{\kappa}^2}\right)-\not{\mathfrak{d}}^kF_1\right\} \;=\; \not{\mathfrak{d}}^kF_2-[\not{\mathfrak{d}}^k,e_3]\left\{e_3\left(\frac{\alpha}{\underline{\kappa}^2}\right)-F_1\right\}$$

and hence

$$e_3\left\{e_3\not{\mathfrak{d}}^k\left(\frac{\alpha}{\underline{\kappa}^2}\right)-F_{1,\not{\mathfrak{d}}^k}\right\} \;=\; F_{2,\not{\mathfrak{d}}^k}$$

where

$$F_{1,\not{\mathfrak{d}}^k}:=\not{\mathfrak{d}}^kF_1-[\not{\mathfrak{d}}^k,e_3]\left(\frac{\alpha}{\underline{\kappa}^2}\right),\quad F_{2,\not{\mathfrak{d}}^k}:=\not{\mathfrak{d}}^kF_2-[\not{\mathfrak{d}}^k,e_3]\left\{e_3\left(\frac{\alpha}{\underline{\kappa}^2}\right)-F_1\right\}.$$

In view of Lemma 2.68, we have schematically

$$[\not{\mathfrak{d}},e_4] \;=\; \Gamma_g\mathfrak{d}+\Gamma_g+r\beta,$$
$$[\not{\mathfrak{d}},e_3] \;=\; \Gamma_b\mathfrak{d}+\Gamma_b+r\underline{\beta}.$$

Together with the estimates of Lemma 6.1 for the Ricci coefficients and curvature components as well as the bootstrap assumptions (6.1.4) for α on ${}^{(ext)}\mathcal{M}$, we infer

$$\max_{0\leq j\leq k_{small}+20-k}\sup_{{}^{(ext)}\mathcal{M}}\Big(r(2r+u)^{1+\delta_{extra}}+r^2(2r+u)^{\frac{1}{2}+\delta_{extra}}\Big)|\mathfrak{d}^j F_{1,\not{\mathfrak{d}}^k}|$$
$$+\max_{0\leq j\leq k_{small}+20-k}\sup_{{}^{(ext)}\mathcal{M}}\Big(r^2u^{1+\delta_{extra}}+r^3u^{\frac{1}{2}+\delta_{extra}}\Big)|\mathfrak{d}^j F_{2,\not{\mathfrak{d}}^k}| \ \lesssim\ \epsilon_0.$$

Next, we consider the case $j\geq 1$. We have the commutator

$$[e_4, e_3] \ = \ 2\omega e_3 - 2\underline{\omega} e_4 - 4\zeta e_\theta.$$

In view of the estimates of Lemma 6.1 for the Ricci coefficients, and in view of the bootstrap assumptions (6.1.4) for α, we infer after commutation by e_4^j for $0\leq k+j\leq k_{small}+20$

$$e_3\left\{e_3\not{\mathfrak{d}}^k e_4^j\left(\frac{\alpha}{\underline{\kappa}^2}\right)-F_{1,\not{\mathfrak{d}}^k,e_4^j}\right\} \ = \ F_{2,\not{\mathfrak{d}}^k,e_4^j}$$

where

$$\max_{0\leq l\leq k_{small}+20-k-j}\sup_{{}^{(ext)}\mathcal{M}(u\leq u_1)}\Big(r^{1+j}(2r+u)^{1+\delta_{extra}}+r^{2+j}(2r+u)^{\frac{1}{2}+\delta_{extra}}\Big)$$
$$\times|\mathfrak{d}^l F_{1,\not{\mathfrak{d}}^k,e_4^j}|$$
$$+\max_{0\leq l\leq k_{small}+20-k-j}\sup_{{}^{(ext)}\mathcal{M}(u\leq u_1)}\Big(r^{2+j}u^{1+\delta_{extra}}+r^{3+j}u^{\frac{1}{2}+\delta_{extra}}\Big)|\mathfrak{d}^l F_{2,\not{\mathfrak{d}}^k,e_4^j}|$$
$$\lesssim\ \epsilon_0+\max_{0\leq j+k\leq k_{small}+20}\sup_{{}^{(ext)}\mathcal{M}(u\leq u_1)}\left(\frac{r^2(2r+u)^{1+\delta_{extra}}}{\log(1+u)}+r^3(2r+u)^{\frac{1}{2}+\delta_{extra}}\right)r$$
$$\times\Big(|\not{\mathfrak{d}}^k(re_4)^{j-1}e_3\alpha|+|\not{\mathfrak{d}}^k(re_4)^{j-2}e_3^2\alpha|\Big)\lesssim\epsilon_0.$$

This concludes the proof of the lemma. □

We are now ready to prove Proposition 6.6.

Step 1. For $0\leq k\leq k_{small}+20$, recall from the above lemma with $j=0$ that we have

$$e_3\left\{e_3\not{\mathfrak{d}}^k\left(\frac{\alpha}{\underline{\kappa}^2}\right)-F_{1,\not{\mathfrak{d}}^k}\right\} \ = \ F_{2,\not{\mathfrak{d}}^k}$$

where

$$\max_{0\leq j\leq k_{small}+20-k}\sup_{{}^{(ext)}\mathcal{M}(u\leq u_1)}\Big(r^2u^{1+\delta_{extra}}+r^3u^{\frac{1}{2}+\delta_{extra}}\Big)|\mathfrak{d}^j F_{2,\not{\mathfrak{d}}^k}| \ \lesssim\ \epsilon_0.$$

Also, we have in view of Lemma 6.5

$$\max_{0\leq k\leq k_{large}-4}\sup_{\mathcal{C}_1} r^{\frac{5}{2}+\delta_{extra}}\left|e_3\not{\mathfrak{d}}^k\left(\frac{\alpha}{\underline{\kappa}^2}\right)-F_{1,\not{\mathfrak{d}}^k}\right| \ \lesssim\ \epsilon_0.$$

In view of Corollary 6.4, we immediately infer for any $0 \le k \le k_{small} + 20$

$$\max_{0\le k\le k_{small}+20} \sup_{{}^{(ext)}\mathcal{M}(u\le u_1)} \Big(r^2(2r+u)^{\frac{1}{2}+\delta_{extra}} + r(2r+u)^{1+\delta_{extra}}\Big) \times \left| e_3 \not{\mathfrak{d}}^k\left(\frac{\alpha}{\underline{\kappa}^2}\right) - F_{1,\not{\mathfrak{d}}^k}\right| \lesssim \epsilon_0.$$

Since we have from the above lemma that

$$\max_{0\le j\le k_{small}+20-k} \sup_{{}^{(ext)}\mathcal{M}(u\le u_1)} \Big(r(2r+u)^{1+\delta_{extra}} + r^2(2r+u)^{\frac{1}{2}+\delta_{extra}}\Big)|\mathfrak{d}^j F_{1,\not{\mathfrak{d}}^k}| \lesssim \epsilon_0,$$

we deduce that we have for any $0 \le k \le k_{small} + 20$

$$\max_{0\le k\le k_{small}+20} \sup_{{}^{(ext)}\mathcal{M}(u\le u_1)} \Big(r^2(2r+u)^{\frac{1}{2}+\delta_{extra}} + r(2r+u)^{1+\delta_{extra}}\Big) \times \left| e_3 \not{\mathfrak{d}}^k\left(\frac{\alpha}{\underline{\kappa}^2}\right)\right| \lesssim \epsilon_0. \tag{6.1.5}$$

Step 2. Next, note that we have in view of Lemma 6.5

$$\max_{0\le k\le k_{large}-3} \sup_{\mathcal{C}_1} r^{\frac{3}{2}+\delta_{extra}} \left|\not{\mathfrak{d}}^k\left(\frac{\alpha}{\underline{\kappa}^2}\right)\right| \lesssim \epsilon_0.$$

Together with the transport equation (6.1.5), and in view of Corollary 6.4, we infer

$$\max_{0\le k\le k_{small}+20} \sup_{{}^{(ext)}\mathcal{M}(u\le u_1)} \left(r(2r+u)^{\frac{1}{2}+\delta_{extra}} + \frac{(2r+u)^{1+\delta_{extra}}}{\log(1+u)}\right)\left|\not{\mathfrak{d}}^k\left(\frac{\alpha}{\underline{\kappa}^2}\right)\right| \lesssim \epsilon_0.$$

In view of the control of $\underline{\kappa}$ provided by Lemma 6.1, we easily deduce

$$\max_{0\le k\le k_{small}+20} \sup_{{}^{(ext)}\mathcal{M}(u\le u_1)} \left(r^3(2r+u)^{\frac{1}{2}+\delta_{extra}} + \frac{r^2(2r+u)^{1+\delta_{extra}}}{\log(1+u)}\right)|\not{\mathfrak{d}}^k\alpha| \lesssim \epsilon_0.$$

Together with (6.1.5), we infer

$$\max_{0\le k\le k_{small}+20} \sup_{{}^{(ext)}\mathcal{M}(u\le u_1)} \left(r^3(2r+u)^{\frac{1}{2}+\delta_{extra}} + \frac{r^2(2r+u)^{1+\delta_{extra}}}{\log(1+u)}\right) \times\Big(|\not{\mathfrak{d}}^k\alpha| + r|\not{\mathfrak{d}}^k e_3\alpha|\Big) \lesssim \epsilon_0.$$

Step 3. Next, recall from section 2.3.3 that $\mathfrak{q}$ is defined with respect to a general null frame as follows:

$$\mathfrak{q} = r^4\left(e_3(e_3(\alpha)) + (2\underline{\kappa} - 6\underline{\omega})e_3(\alpha) + \left(-4e_3(\underline{\omega}) + 8\underline{\omega}^2 - 8\underline{\omega}\,\underline{\kappa} + \frac{1}{2}\underline{\kappa}^2\right)\alpha\right).$$

We infer

$$e_3(e_3(\alpha)) = \frac{\mathfrak{q}}{r^4} - (2\underline{\kappa} - 6\underline{\omega})e_3(\alpha) - \left(-4e_3(\underline{\omega}) + 8\underline{\omega}^2 - 8\underline{\omega}\,\underline{\kappa} + \frac{1}{2}\underline{\kappa}^2\right)\alpha.$$

Together with the above estimate for α and $e_3\alpha$, we infer by iteration

$$\max_{0\leq k\leq k_{small}+20}\sup_{^{(ext)}\mathcal{M}(u\leq u_1)}\left(\frac{r^2(2r+u)^{1+\delta_{extra}}}{\log(1+u)}+r^3(2r+u)^{\frac{1}{2}+\delta_{extra}}\right)\times\Big(|(\not{\mathfrak{d}},e_3)^k\alpha|+r|(\not{\mathfrak{d}},e_3)^k e_3\alpha|\Big)\lesssim\epsilon_0.$$

Step 4. Arguing as for Step 1, but with $j\geq 1$, we infer the following analog of (6.1.5):

$$\begin{aligned}&\max_{0\leq j+k\leq k_{small}+20}\sup_{^{(ext)}\mathcal{M}(u\leq u_1)}\left(r^2(2r+u)^{\frac{1}{2}+\delta_{extra}}+r(2r+u)^{1+\delta_{extra}}\right)\\&\times\left|e_3\not{\mathfrak{d}}^k(re_4)^j\left(\frac{\alpha}{\underline{\kappa}^2}\right)\right|\\\lesssim\;&\epsilon_0+\max_{0\leq j+k\leq k_{small}+20}\sup_{^{(ext)}\mathcal{M}(u\leq u_1)}\left(\frac{r^2(2r+u)^{1+\delta_{extra}}}{\log(1+u)}+r^3(2r+u)^{\frac{1}{2}+\delta_{extra}}\right)r\\&\times\Big(|\not{\mathfrak{d}}^k(re_4)^{j-1}e_3\alpha|+|\not{\mathfrak{d}}^k(re_4)^{j-2}e_3^2\alpha|\Big).\end{aligned}$$

Step 5. Arguing as for Step 2, but with $j\geq 1$, we infer the following analog of the last estimate of Step 2:

$$\begin{aligned}&\max_{0\leq j\leq k_{small}+20}\sup_{^{(ext)}\mathcal{M}(u\leq u_1)}\left(r^3(2r+u)^{\frac{1}{2}+\delta_{extra}}+\frac{r^2(2r+u)^{1+\delta_{extra}}}{\log(1+u)}\right)\\&\times\Big(|\not{\mathfrak{d}}^k(re_4)^j\alpha|+r|\not{\mathfrak{d}}^k(re_4)^je_3\alpha|\Big)\\\lesssim\;&\epsilon_0+\max_{0\leq j+k\leq k_{small}+20}\sup_{^{(ext)}\mathcal{M}(u\leq u_1)}\left(\frac{r^2(2r+u)^{1+\delta_{extra}}}{\log(1+u)}+r^3(2r+u)^{\frac{1}{2}+\delta_{extra}}\right)r\\&\times\Big(|\not{\mathfrak{d}}^k(re_4)^{j-1}e_3\alpha|+|\not{\mathfrak{d}}^k(re_4)^{j-2}e_3^2\alpha|\Big).\end{aligned}$$

Step 6. Arguing as for Step 3, but with $j\geq 1$, we infer the following analog of the last estimate of Step 3:

$$\begin{aligned}&\max_{0\leq j+k\leq k_{small}+20}\sup_{^{(ext)}\mathcal{M}(u\leq u_1)}\left(\frac{r^2(2r+u)^{1+\delta_{extra}}}{\log(1+u)}+r^3(2r+u)^{\frac{1}{2}+\delta_{extra}}\right)\\&\times\Big(|(\not{\mathfrak{d}},e_3)^k(re_4)^j\alpha|+r|(\not{\mathfrak{d}},e_3)^k(re_4)^je_3\alpha|\Big)\\\lesssim\;&\epsilon_0+\max_{0\leq j+k\leq k_{small}+20}\sup_{^{(ext)}\mathcal{M}(u\leq u_1)}\left(\frac{r^2(2r+u)^{1+\delta_{extra}}}{\log(1+u)}+r^3(2r+u)^{\frac{1}{2}+\delta_{extra}}\right)r\\&\times\Big(|(\not{\mathfrak{d}},e_3)^k(re_4)^{j-1}e_3\alpha|+r|(\not{\mathfrak{d}},e_3)^k(re_4)^{j-2}e_3^2\alpha|\Big).\end{aligned}$$

Step 7. Arguing by iteration on j, noticing that the last estimate of Step 3 corresponds to the desired estimate for $j=0$, and in view of the estimate derived in Step 6, we finally obtain

$$\max_{0\leq j+k\leq k_{small}+20}\sup_{{}^{(ext)}\mathcal{M}(u\leq u_1)}\left(\frac{r^2(2r+u)^{1+\delta_{extra}}}{\log(1+u)}+r^3(2r+u)^{\frac{1}{2}+\delta_{extra}}\right)$$
$$\times\Big(|(\not{\partial}, e_3)^k(re_4)^j\alpha|+r|(\not{\partial}, e_3)^k(re_4)^je_3\alpha|\Big) \lesssim \epsilon_0$$

and hence

$$\max_{0\leq k\leq k_{small}+20}\sup_{{}^{(ext)}\mathcal{M}(u\leq u_1)}\left(\frac{r^2(2r+u)^{1+\delta_{extra}}}{\log(1+u)}+r^3(2r+u)^{\frac{1}{2}+\delta_{extra}}\right)$$
$$\times\Big(|\mathfrak{d}^k\alpha|+r|\mathfrak{d}^ke_3\alpha|\Big) \lesssim \epsilon_0.$$

This concludes the proof of Proposition 6.6.

6.1.5 End of the proof of Theorem M2

First, note in view of the estimates for α on $\mathcal{C}_1$ provided by Lemma 6.5 that the bootstrap assumptions (6.1.4) for α hold by continuity for some sufficiently small $u_1>0$. Then, we may in view of Proposition 6.6 choose $u_1=u_*$. We deduce therefore

$$\max_{0\leq k\leq k_{small}+20}\sup_{{}^{(ext)}\mathcal{M}}\left(\frac{r^2(2r+u)^{1+\delta_{extra}}}{\log(1+u)}+r^3(2r+u)^{\frac{1}{2}+\delta_{extra}}\right)$$
$$\times\Big(|\mathfrak{d}^k\alpha|+r|\mathfrak{d}^ke_3\alpha|\Big) \lesssim \epsilon_0.$$

Next, recall from (6.1.2) and (6.1.3) that we have

$$\alpha = ({}^{(ext)}\Upsilon)^2\left({}^{(ext)}\alpha+2f\,{}^{(ext)}\beta+\frac{3}{2}f^2\,{}^{(ext)}\rho+\text{l.o.t.}\right)$$

where f satisfies

$$|\mathfrak{d}^kf| \lesssim \frac{\epsilon}{ru^{\frac{1}{2}}+u}, \quad \text{for } k\leq k_{small}+22 \text{ on } {}^{(ext)}\mathcal{M},$$
$$|\mathfrak{d}^{k-1}e_3f| \lesssim \frac{\epsilon}{ru} \quad \text{for } k\leq k_{small}+22 \text{ on } {}^{(ext)}\mathcal{M}.$$

Together with bootstrap assumptions for ${}^{(ext)}\beta$ and ${}^{(ext)}\rho$, we easily infer

$$\max_{0\leq k\leq k_{small}+20}\sup_{{}^{(ext)}\mathcal{M}}\left(\frac{r^2(2r+u)^{1+\delta_{extra}}}{\log(1+u)}+r^3(2r+u)^{\frac{1}{2}+\delta_{extra}}\right)$$
$$\times\Big(|\mathfrak{d}^k\,{}^{(ext)}\alpha|+r|\mathfrak{d}^ke_3\,{}^{(ext)}\alpha|\Big) \lesssim \epsilon_0.$$

This concludes the proof of Theorem M2.

6.2 PROOF OF THEOREM M3

Theorem M3 contains decay estimates for $\underline{\alpha}$ in ${}^{(int)}\mathcal{M}$ and on Σ_*. We first proceed with the estimate on ${}^{(int)}\mathcal{M}$ before moving to ${}^{(ext)}\mathcal{M}$.

6.2.1 Estimate for $\underline{\alpha}$ in ${}^{(int)}\mathcal{M}$

Recall that $\mathfrak{q}$, controlled in Theorem M1, is defined with respect to the global frame of Proposition 3.26. Recall also that we may choose the global null frame to coincide with the ingoing geodesic null frame of ${}^{(int)}\mathcal{M}$ in ${}^{(int)}\mathcal{M}$ (see property (b) in Proposition 3.26 together with property (d) ii. in Proposition 3.23). Thus, in this section, as we only work on ${}^{(int)}\mathcal{M}$, the null frame (e_4, e_3, e_θ) denotes both the frame of ${}^{(int)}\mathcal{M}$ and the global frame with respect to which $\mathfrak{q}$ is defined. We start with the following definition.

Definition 6.9. *In ${}^{(int)}\mathcal{M}$, we define with respect to the ingoing geodesic frame of ${}^{(int)}\mathcal{M}$*

$$\widetilde{T} := e_4 - \frac{1}{\underline{\overline{\kappa}}}\Big(\overline{\kappa} + A\Big)e_3. \tag{6.2.1}$$

The estimate for $\underline{\alpha}$ in ${}^{(int)}\mathcal{M}$ relies on the following proposition.

Proposition 6.10. *Let $0 \le k \le k_{small} + 17$. Then, $\underline{\alpha}$ satisfies in ${}^{(int)}\mathcal{M}$*

$$6m\widetilde{T}(\mathfrak{d}^k\underline{\alpha}) + r^4\, \not{d}_2^\star\, \not{d}_1^\star\, \not{d}_1\, \not{d}_2(\mathfrak{d}^k\underline{\alpha}) \quad = \quad F_k$$

where F_k satisfies

$$\max_{0\le k\le k_{small}+17}\int_{{}^{(int)}\mathcal{M}} \underline{u}^{2+2\delta_{dec}}|\mathfrak{d}^{\le 1}F_k|^2 \quad \lesssim \quad \epsilon_0^2.$$

Remark 6.11. *In view of the definition of $\widetilde{T}$, we have*

$$\widetilde{T}(r) = e_4(r) - \frac{1}{\underline{\overline{\kappa}}}\Big(\overline{\kappa} + A\Big)e_3(r) = 0$$

so that $\widetilde{T}$ is tangent to the hypersurfaces of constant r. In particular, $(\widetilde{T}, e_\theta)$ spans the tangent space of hypersurfaces of constant r. Therefore, in view of Proposition 6.10, $\underline{\alpha}$ and its derivatives satisfy on each hypersurface of constant r in ${}^{(int)}\mathcal{M}$, i.e., on $\{r = r_0\}$ for $2m_0(1-\delta_{\mathcal{H}}) \le r \le r_{\mathcal{T}}$, a forward parabolic equation. Furthermore, since we have $\widetilde{T}(\underline{u}) = 2/\underline{\varsigma} = 2 + O(\epsilon)$, $\underline{u}$ plays the role of time in this forward parabolic equation.

We also derive estimates for the control of the parabolic equation appearing in the statement of Proposition 6.10.

Lemma 6.12. *Let f and h reduced 2-scalars such that*

$$\Big(6m\widetilde{T} + r^4\, \not{d}_2^\star\, \not{d}_1^\star\, \not{d}_1\, \not{d}_2\Big)f \quad = \quad h.$$

Then, for any real number $n \ge 0$ and any r_0 such that $2m_0(1-\delta_{\mathcal{H}}) \le r_0 \le r_{\mathcal{T}}$, we *have*

$$\sup_{1\le \underline{u}\le u_*}\int_{S(r=r_0,\underline{u})}(1+\underline{u}^n)f^2 \lesssim_n \int_{S(r=r_0,1)} f^2+\epsilon^2\int_1^{u_*}\int_{S(r=r_0,\underline{u})}(1+\underline{u}^{n-2})(\mathfrak{d}f)^2 + \int_{^{(int)}\mathcal{M}}(1+\underline{u}^n)(\mathfrak{d}^{\le 1}h)^2.$$

We are now in position to control $\underline{\alpha}$ in $^{(int)}\mathcal{M}$. Recall from Proposition 6.10 that $\underline{\alpha}$ satisfies in $^{(int)}\mathcal{M}$ for $0\le k\le k_{small}+17$

$$6m\widetilde{T}(\mathfrak{d}^k\underline{\alpha})+r^4\,{\not{d}}_2^\star\,{\not{d}}_1^\star\,{\not{d}}_1\,{\not{d}}_2(\mathfrak{d}^k\underline{\alpha}) = F_k.$$

Applying Lemma 6.12 with $n=2+2\delta_{dec}$, $f=\mathfrak{d}^k\underline{\alpha}$ and $h=F_k$, we infer for any r_0 such that $2m_0(1-\delta_{\mathcal{H}})\le r_0\le r_{\mathcal{T}}$

$$\begin{aligned}&\sup_{1\le \underline{u}\le u_*}\int_{S(r=r_0,\underline{u})}(1+\underline{u}^{2+2\delta_{dec}})(\mathfrak{d}^k\underline{\alpha})^2\\ \lesssim\;& \int_{S(r=r_0,1)}(\mathfrak{d}^k\underline{\alpha})^2+\epsilon^2\int_1^{u_*}\int_{S(r=r_0,\underline{u})}(1+\underline{u}^{2\delta_{dec}})(\mathfrak{d}^{k+1}\underline{\alpha})^2\\ &+\int_{^{(int)}\mathcal{M}}(1+\underline{u}^{2+2\delta_{dec}})(\mathfrak{d}^{\le 1}F_k)^2.\end{aligned}$$

Together with the bounds for $\underline{\alpha}$ on $\underline{\mathcal{C}}_1$ provided by Theorem M0, the bootstrap assumptions on decay and energy for $\underline{\alpha}$ in $^{(int)}\mathcal{M}$, and the bound for F_k provided by Proposition 6.10, we infer for $0\le k\le k_{small}+17$ in $^{(int)}\mathcal{M}$

$$\sup_{1\le \underline{u}\le u_*}\int_{S(r=r_0,\underline{u})}(1+\underline{u}^{2+2\delta_{dec}})(\mathfrak{d}^k\underline{\alpha})^2 \lesssim \epsilon_0^2.$$

In particular, we have obtained

$$\max_{0\le k\le k_{small}+17}\sup_{^{(int)}\mathcal{M}}\underline{u}^{1+\delta_{dec}}\|\mathfrak{d}^k\underline{\alpha}\|_{L^2(S)} \lesssim \epsilon_0.$$

Using the Sobolev embedding on 2-surface and the fact that r is bounded on $^{(int)}\mathcal{M}$, we infer

$$\max_{0\le k\le k_{small}+15}\sup_{^{(int)}\mathcal{M}}\underline{u}^{1+\delta_{dec}}|\mathfrak{d}^k\underline{\alpha}| \lesssim \epsilon_0$$

and hence

$$^{(int)}\mathfrak{D}_{k_{small}+15}[\underline{\alpha}] \lesssim \epsilon_0 \tag{6.2.2}$$

which is the desired estimate for $\underline{\alpha}$ in $^{(int)}\mathcal{M}$.

The proof of Proposition 6.10 will be given in section 6.2.3, and the proof of Lemma 6.12 in section 6.2.4. But first we show, in the next section, how to conclude the proof of Theorem M3 by controlling $\underline{\alpha}$ on Σ_*.

6.2.2 Estimate for $\underline{\alpha}$ on Σ_*

Recall that $\mathfrak{q}$, controlled in Theorem M1, is defined with respect to the global frame of Proposition 3.26. We will first control $\underline{\alpha}$ in this frame, before coming back to ${}^{(ext)}\mathcal{M}$ at the end of the argument. We start with the following definition.

Definition 6.13. *In Σ_*, we define, with respect to the global frame of Proposition 3.26,*

$$\widetilde{\nu} := e_3 + a e_4, \tag{6.2.3}$$

where the scalar function a is uniquely defined so that $\widetilde{\nu}$ is tangent to Σ_*.

The estimate for $\underline{\alpha}$ on Σ_* relies on the following proposition.

Proposition 6.14. *Let $0 \leq k \leq k_{small} + 18$. Then, $\underline{\alpha}$ satisfies on Σ_**

$$6m\widetilde{\nu}(\mathfrak{d}^k \underline{\alpha}) + r^4 \, \not{d}_2^\star \, \not{d}_1^\star \, \not{d}_1 \, \not{d}_2(\mathfrak{d}^k \underline{\alpha}) \quad = \quad F_k$$

where F_k satisfies

$$\max_{0\leq k\leq k_{small}+18} \int_{\Sigma_*} u^{2+2\delta_{dec}} |F_k|^2 \quad \lesssim \quad \epsilon_0^2.$$

Remark 6.15. *Since $\widetilde{\nu}$ is tangent to Σ_*, and since $(\widetilde{\nu}, e_\theta)$ spans the tangent space of Σ_*, in view of Proposition 6.14, $\underline{\alpha}$ and its derivatives satisfy on Σ_* a forward parabolic equation. Furthermore, since we have $\widetilde{\nu}(u) = 2 + O(\epsilon)$, u plays the role of time in this forward parabolic equation.*

We also derive estimates for the control of the parabolic equation appearing in the statement of Proposition 6.10.

Lemma 6.16. *Let f and h reduced 2-scalars such that*

$$\Big(6m\widetilde{\nu} + r^4 \, \not{d}_2^\star \, \not{d}_1^\star \, \not{d}_1 \, \not{d}_2\Big) f \quad = \quad h.$$

Then, for any real number $n \geq 0$, we have

$$\int_{\Sigma_*} (1+u^n) f^2 \quad \lesssim_n \quad \int_{\Sigma_* \cap \mathcal{C}_1} f^2 + \epsilon^2 \int_{\Sigma_*} (1+u^{n-2})(\mathfrak{d} f)^2 + \int_{\Sigma_*} (1+u^n) h^2.$$

Using the lemma we are in position to control $\underline{\alpha}$ on Σ_*. According to Proposition 6.14 $\underline{\alpha}$ satisfies in Σ_*, for $0 \leq k \leq k_{small} + 18$,

$$6m\widetilde{\nu}(\mathfrak{d}^k \underline{\alpha}) + r^4 \, \not{d}_2^\star \, \not{d}_1^\star \, \not{d}_1 \, \not{d}_2(\mathfrak{d}^k \underline{\alpha}) \quad = \quad F_k.$$

Applying Lemma 6.16 with $n = 2 + 2\delta_{dec}$, $f = \mathfrak{d}^k \underline{\alpha}$ and $h = F_k$, we infer

$$\begin{aligned} \int_{\Sigma_*} (1+u^{2+2\delta_{dec}})(\mathfrak{d}^k \underline{\alpha})^2 \quad \lesssim \quad & \int_{\Sigma_* \cap \mathcal{C}_1} (\mathfrak{d}^k \underline{\alpha})^2 + \epsilon^2 \int_{\Sigma_*} (1+u^{2\delta_{dec}})(\mathfrak{d}^{k+1} \underline{\alpha})^2 \\ & + \int_{\Sigma_*} (1+u^{2+2\delta_{dec}})(F_k)^2. \end{aligned}$$

Together with the bounds for $\underline{\alpha}$ on $\underline{\mathcal{C}}_1$ provided by Theorem M0, the bootstrap assumptions on decay and energy for $\underline{\alpha}$ in ${}^{(ext)}\mathcal{M}$, and the bound for F_k provided by Proposition 6.14, we infer

$$\int_{\Sigma_*}(1+u^{2+2\delta_{dec}})(\mathfrak{d}^k\underline{\alpha})^2 \lesssim \epsilon_0^2.$$

In particular, we have obtained

$$\max_{0\leq k\leq k_{small}+18}\int_{\Sigma_*}(1+u^{2+2\delta_{dec}})(\mathfrak{d}^k\underline{\alpha})^2 \lesssim \epsilon_0.$$

Now, recall that $\underline{\alpha}$ in the above estimate is defined with respect to the global frame of Proposition 3.26. In view of Proposition 3.26 and Proposition 3.20, and the change of frame formula for $\underline{\alpha}$ in Proposition 2.90, we have

$$\underline{\alpha} = \left({}^{(ext)}\Upsilon\right)^{-2}\,{}^{(ext)}\underline{\alpha}.$$

Hence, we immediately infer

$$\max_{0\leq k\leq k_{small}+18}\int_{\Sigma_*}(1+u^{2+2\delta_{dec}})(\mathfrak{d}^k\,{}^{(ext)}\underline{\alpha})^2 \lesssim \epsilon_0$$

which is the desired estimate in Σ_*. Together with (6.2.2), this concludes the proof of Theorem M3.

The proof of Proposition 6.14 will be given in section 6.2.5, and the proof of Lemma 6.16 will be given in section 6.2.6.

6.2.3 Proof of Proposition 6.10

In this section we derive as corollary of the Teukolsky-Starobinsky identity, see Proposition 2.101, a parabolic equation for $\underline{\alpha}$.

Corollary 6.17. *The quantity $\underline{\alpha}$ satisfies in ${}^{(int)}\mathcal{M}$ the following equation:*

$$\begin{aligned}
&6m\widetilde{T}\underline{\alpha}+r^4\,{\not{d}}_2^\star\,{\not{d}}_1^\star\,{\not{d}}_1\,{\not{d}}_2\underline{\alpha}\\
=\;&\frac{1}{r^3}\Big(e_3(r^2e_3(r\underline{\mathfrak{q}}))+2\underline{\omega}r^2e_3(r\underline{\mathfrak{q}})\Big)-r^{-3}Err[TS]\\
&-\left\{\frac{3}{2}r^4\left(\rho+\frac{2m}{r^3}\right)\underline{\kappa}-3mr\left(\underline{\kappa}+\frac{2}{r}\right)\right\}e_4\underline{\alpha}\\
&-\left\{-\frac{3}{2}r^4\left(\rho+\frac{2m}{r^3}\right)\kappa+\frac{3mr}{\overline{\underline{\kappa}}}\left(\overline{\underline{\kappa}}+\frac{2}{r}\right)\overline{\kappa}+3mr\check{\kappa}+\frac{6m}{\overline{\underline{\kappa}}}A\right\}e_3\underline{\alpha}
\end{aligned}$$

where the vectorfield $\widetilde{T}$ is defined by (6.2.1).

Proof. According to Proposition 2.3.15, we have

$$e_3(r^2e_3(r\underline{\mathfrak{q}}))+2\underline{\omega}r^2e_3(r\underline{\mathfrak{q}}) = r^7\left\{{\not{d}}_2^\star\,{\not{d}}_1^\star\,{\not{d}}_1\,{\not{d}}_2\underline{\alpha}+\frac{3}{2}\rho\Big(\underline{\kappa}e_4-\kappa e_3\Big)\underline{\alpha}\right\}+\mathrm{Err}[TS].$$

This yields

$$\frac{3}{2}r^4\rho\Big(\underline{\kappa}e_4-\kappa e_3\Big)\underline{\alpha}+r^4\,\not{d}_2^\star\,\not{d}_1^\star\,\not{d}_1\,\not{d}_2\underline{\alpha} = \frac{1}{r^3}\Big(e_3(r^2e_3(r\mathfrak{q}))+2\underline{\omega}r^2e_3(r\mathfrak{q})\Big) -r^{-3}\mathrm{Err}[TS].$$

Now, we have in view of the definition of $\widetilde{T}$

$$\begin{aligned}&\frac{3}{2}r^4\rho\Big(\underline{\kappa}e_4-\kappa e_3\Big)-6m\widetilde{T}\\ =\;& \left\{\frac{3}{2}r^4\left(\rho+\frac{2m}{r^3}\right)\underline{\kappa}-3mr\left(\underline{\kappa}+\frac{2}{r}\right)\right\}e_4\\ &+\left\{-\frac{3}{2}r^4\left(\rho+\frac{2m}{r^3}\right)\kappa+\frac{3mr}{\overline{\underline{\kappa}}}\left(\overline{\underline{\kappa}}+\frac{2}{r}\right)\overline{\kappa}+3mr\check{\kappa}+\frac{6m}{\overline{\underline{\kappa}}}A\right\}e_3.\end{aligned}$$

We infer

$$\begin{aligned}&6m\widetilde{T}\underline{\alpha}+r^4\,\not{d}_2^\star\,\not{d}_1^\star\,\not{d}_1\,\not{d}_2\underline{\alpha}\\ =\;&\frac{1}{r^3}\Big(e_3(r^2e_3(r\mathfrak{q}))+2\underline{\omega}r^2e_3(r\mathfrak{q})\Big)-r^{-3}\mathrm{Err}[TS]\\ &-\left\{\frac{3}{2}r^4\left(\rho+\frac{2m}{r^3}\right)\underline{\kappa}-3mr\left(\underline{\kappa}+\frac{2}{r}\right)\right\}e_4\underline{\alpha}\\ &-\left\{-\frac{3}{2}r^4\left(\rho+\frac{2m}{r^3}\right)\kappa+\frac{3mr}{\overline{\underline{\kappa}}}\left(\overline{\underline{\kappa}}+\frac{2}{r}\right)\overline{\kappa}+3mr\check{\kappa}+\frac{6m}{\overline{\underline{\kappa}}}A\right\}e_3\underline{\alpha}.\end{aligned}$$

This concludes the proof of the corollary. □

Corollary 6.18. *$\underline{\alpha}$ satisfies in ${}^{(int)}\mathcal{M}$*

$$6m\widetilde{T}\underline{\alpha}+r^4\,\not{d}_2^\star\,\not{d}_1^\star\,\not{d}_1\,\not{d}_2\underline{\alpha} = F$$

where F satisfies

$$\max_{0\leq k\leq k_{small}+18}\int_{{}^{(int)}\mathcal{M}}\underline{u}^{2+2\delta_{dec}}|\mathfrak{d}^kF|^2 \lesssim \epsilon_0^2.$$

Proof. In view of Corollary 6.17, $\underline{\alpha}$ satisfies

$$6m\widetilde{T}\underline{\alpha}+r^4\,\not{d}_2^\star\,\not{d}_1^\star\,\not{d}_1\,\not{d}_2\underline{\alpha} = F$$

with

$$\begin{aligned}F &:= \frac{1}{r^3}\Big(e_3(r^2e_3(r\mathfrak{q}))+2\underline{\omega}r^2e_3(r\mathfrak{q})\Big)+F_1,\\ F_1 &:= -r^{-3}\mathrm{Err}[TS]-\left\{\frac{3}{2}r^4\left(\rho+\frac{2m}{r^3}\right)\underline{\kappa}-3mr\left(\underline{\kappa}+\frac{2}{r}\right)\right\}e_4\underline{\alpha}\\ &\quad-\left\{-\frac{3}{2}r^4\left(\rho+\frac{2m}{r^3}\right)\kappa+\frac{3mr}{\overline{\underline{\kappa}}}\left(\overline{\underline{\kappa}}+\frac{2}{r}\right)\overline{\kappa}+3mr\check{\kappa}+\frac{6m}{\overline{\underline{\kappa}}}A\right\}e_3\underline{\alpha}.\end{aligned}$$

Using the bootstrap assumptions in ${}^{(int)}\mathcal{M}$ for decay and energies, and in view of the fact that F_1 contains only quadratic or higher order terms, we easily derive

$$\max_{0\leq k\leq k_{small}+18}\sup_{^{(int)}\mathcal{M}} \underline{u}^{\frac{3}{2}+\frac{3}{2}\delta_{dec}}|\mathfrak{d}^k F_1| \lesssim \epsilon^2\lesssim \epsilon_0.$$

In view of the definition of F, this yields

$$\max_{0\leq k\leq k_{small}+18}\int_{^{(int)}\mathcal{M}} \underline{u}^{2+2\delta_{dec}}|\mathfrak{d}^k F|^2 \lesssim \epsilon_0^2+\max_{0\leq k\leq k_{small}+20}\int_{^{(int)}\mathcal{M}} \underline{u}^{2+2\delta_{dec}}|\mathfrak{d}^k \mathfrak{q}|^2.$$

Together with Theorem M1, and the fact that $\delta_{extra}>\delta_{dec}$, we infer

$$\max_{0\leq k\leq k_{small}+18}\int_{^{(int)}\mathcal{M}} \underline{u}^{2+2\delta_{dec}}|\mathfrak{d}^k F|^2 \lesssim \epsilon_0^2.$$

This concludes the proof of the corollary. □

We are now ready to prove Proposition 6.10. In view of Corollary 6.18, $\underline{\alpha}$ satisfies

$$6m\widetilde{T}\underline{\alpha}+r^4\,\not{d}_2^\star\,\not{d}_1^\star\,\not{d}_1\,\not{d}_2\underline{\alpha} = F.$$

Commuting with $\mathfrak{d}^k$, we infer

$$6m\widetilde{T}(\mathfrak{d}^k\underline{\alpha})+r^4\,\not{d}_2^\star\,\not{d}_1^\star\,\not{d}_1\,\not{d}_2(\mathfrak{d}^k\underline{\alpha}) = F_k$$

where F_k is defined by

$$\begin{aligned}F_k := & -6m[\mathfrak{d}^k,\widetilde{T}]\underline{\alpha}-6\sum_{j=1}^k\mathfrak{d}^j(m)\mathfrak{d}^{k-j}\widetilde{T}\underline{\alpha}-[\mathfrak{d}^k,r\,\not{d}_2^\star]r\,\not{d}_1^\star r\,\not{d}_1 r\,\not{d}_2\underline{\alpha}\\ & -r\,\not{d}_2^\star[\mathfrak{d}^k,r\,\not{d}_1^\star]r\,\not{d}_1 r\,\not{d}_2\underline{\alpha}-r\,\not{d}_2^\star r\,\not{d}_1^\star[\mathfrak{d}^k,r\,\not{d}_1]r\,\not{d}_2\underline{\alpha}-r\,\not{d}_2^\star r\,\not{d}_1^\star r\,\not{d}_1[\mathfrak{d}^k,r\,\not{d}_2]\underline{\alpha}\\ & +\mathfrak{d}^kF.\end{aligned}$$

Note that we have schematically

$$[\mathfrak{d},\,\not{\mathfrak{d}}]=\check{\Gamma}\mathfrak{d},\quad [\widetilde{T},\,\not{\mathfrak{d}}]=\left(\mathfrak{d}\check{\Gamma}+\check{\Gamma}\right)\mathfrak{d},$$

as well as

$$
\begin{aligned}
[\widetilde{T}, re_4] &= e_4(r)e_4 - \frac{1}{\underline{\kappa}}\Big(\overline{\kappa}+A\Big)[e_3, re_4] + re_4\left(\frac{1}{\underline{\kappa}}\Big(\overline{\kappa}+A\Big)\right)e_3\\
&= \frac{r}{2}\Big(\overline{\kappa}+A\Big)e_4 - \frac{1}{\underline{\kappa}}\Big(\overline{\kappa}+A\Big)\left(\frac{r}{2}\underline{\kappa}e_4 + r\Big(-2\omega e_3 + 4\zeta e_\theta\Big)\right)\\
&\quad + re_4\left(\frac{\overline{\kappa}}{\underline{\kappa}}\right)e_3 + re_4\left(\frac{1}{\underline{\kappa}}A\right)e_3\\
&= \left\{\frac{2r}{\underline{\kappa}}\Big(\overline{\kappa}+A\Big)\omega + re_4\left(\frac{\overline{\kappa}}{\underline{\kappa}}\right) + re_4\left(\frac{1}{\underline{\kappa}}A\right)\right\}e_3\\
&\quad -\frac{4r}{\underline{\kappa}}\Big(\overline{\kappa}+A\Big)\zeta e_\theta\\
&= \Bigg\{-\frac{2m}{r}\left(\frac{\overline{\kappa}}{\underline{\kappa}}+\Upsilon\right) + \frac{2r}{\underline{\kappa}}\overline{\kappa}\left(\omega+\frac{m}{r}\right) - \frac{2m(e_4(r)-\Upsilon)}{r} + 2e_4(m)\\
&\quad + re_4\left(\frac{\overline{\kappa}}{\underline{\kappa}}+\Upsilon\right) + \frac{2r}{\underline{\kappa}}A\omega + re_4\left(\frac{1}{\underline{\kappa}}A\right)\Bigg\}e_3 - \frac{4r}{\underline{\kappa}}\Big(\overline{\kappa}+A\Big)\zeta e_\theta\\
&= \Big(\mathfrak{d}\check{\Gamma}+\check{\Gamma}\Big)\mathfrak{d},
\end{aligned}
$$

and

$$
\begin{aligned}
[\widetilde{T}, e_3] &= [e_4, e_3] + e_3\left(\frac{1}{\underline{\kappa}}\Big(\overline{\kappa}+A\Big)\right)e_3\\
&= \left\{2\omega + e_3\left(\frac{\overline{\kappa}}{\underline{\kappa}}\right) + e_3\left(\frac{1}{\underline{\kappa}}A\right)\right\}e_3 - 4\zeta e_\theta\\
&= \Bigg\{2\left(\omega+\frac{m}{r^2}\right) - \frac{2m(e_3(r)+1)}{r^2} + \frac{2e_3(m)}{r} + e_3\left(\frac{\overline{\kappa}}{\underline{\kappa}}+\Upsilon\right)\\
&\quad + e_3\left(\frac{1}{\underline{\kappa}}A\right)\Bigg\}e_3 - 4\zeta e_\theta\\
&= \Big(\mathfrak{d}\check{\Gamma}+\check{\Gamma}\Big)\mathfrak{d}.
\end{aligned}
$$

Together with the bootstrap assumptions in ${}^{(int)}\mathcal{M}$ for decay and energies, and in view of the fact that F_1 contains only quadratic or higher order terms, we easily derive

$$
\begin{aligned}
\max_{0\leq k\leq k_{small}+18}\sup_{{}^{(int)}\mathcal{M}} u^{\frac{3}{2}+\frac{3}{2}\delta_{dec}} \Bigg| & -6m[\mathfrak{d}^k, \widetilde{T}]\underline{\alpha} - 6\sum_{j=1}^{k}\mathfrak{d}^j(m)\mathfrak{d}^{k-j}\widetilde{T}\underline{\alpha}\\
& -[\mathfrak{d}^k, r\,{\not{d}}_2^\star]r\,{\not{d}}_1^\star r\,{\not{d}}_1 r\,{\not{d}}_2\underline{\alpha} - r\,{\not{d}}_2^\star[\mathfrak{d}^k, r\,{\not{d}}_1^\star]r\,{\not{d}}_1 r\,{\not{d}}_2\underline{\alpha}\\
& -r\,{\not{d}}_2^\star r\,{\not{d}}_1^\star[\mathfrak{d}^k, r\,{\not{d}}_1]r\,{\not{d}}_2\underline{\alpha}\\
& -r\,{\not{d}}_2^\star r\,{\not{d}}_1^\star r\,{\not{d}}_1[\mathfrak{d}^k, r\,{\not{d}}_2]\underline{\alpha}\Bigg| \lesssim \epsilon^2.
\end{aligned}
$$

In view of the definition of F_k, this yields

$$\max_{0\le k\le k_{small}+18}\int_{^{(int)}\mathcal{M}}\underline{u}^{2+2\delta_{dec}}|F_k|^2 \quad\lesssim\quad \epsilon^4+\max_{0\le k\le k_{small}+18}\int_{^{(int)}\mathcal{M}}\underline{u}^{2+2\delta_{dec}}|\mathfrak{d}^kF|^2.$$

Together with the estimate for F of Corollary 6.18, we infer

$$\max_{0\le k\le k_{small}+18}\int_{^{(int)}\mathcal{M}}\underline{u}^{2+2\delta_{dec}}|F_k|^2 \quad\lesssim\quad \epsilon^4+\epsilon_0^2\lesssim\epsilon_0^2.$$

This concludes the proof of Proposition 6.10.

6.2.4 Proof of Lemma 6.12

In this section we prove Lemma 6.12, i.e., we derive estimates for the control of the parabolic equation appearing in the statement of Proposition 6.10. To this end, we first start with a Poincaré inequality.

Lemma 6.19. *We have*

$$\int_S f\,{\not{d}}_2^\star\,{\not{d}}_1^\star\,{\not{d}}_1\,{\not{d}}_2 f \quad\ge\quad 24\int_S(1+O(\epsilon))K^2f^2.$$

Proof. We have

$$\begin{aligned}
{\not{d}}_2^\star\,{\not{d}}_1^\star\,{\not{d}}_1\,{\not{d}}_2 &= {\not{d}}_2^\star(-{\not{\Delta}}_1+K)\,{\not{d}}_2\\
&= -{\not{d}}_2^\star{\not{\Delta}}_1\,{\not{d}}_2+K\,{\not{d}}_2^\star\,{\not{d}}_2+{\not{d}}_1^\star(K)\,{\not{d}}_2\\
&= -{\not{\Delta}}_2\,{\not{d}}_2^\star\,{\not{d}}_2+\Big({\not{\Delta}}_2\,{\not{d}}_2^\star-{\not{d}}_2^\star{\not{\Delta}}_1\Big)\,{\not{d}}_2+K\,{\not{d}}_2^\star\,{\not{d}}_2+{\not{d}}_1^\star(K)\,{\not{d}}_2\\
&= ({\not{d}}_2^\star\,{\not{d}}_2-2K)\,{\not{d}}_2^\star\,{\not{d}}_2+\Big(3K\,{\not{d}}_2^\star-{\not{d}}_1^\star(K)\Big)\,{\not{d}}_2+K\,{\not{d}}_2^\star\,{\not{d}}_2+{\not{d}}_1^\star(K)\,{\not{d}}_2\\
&= ({\not{d}}_2^\star\,{\not{d}}_2)^2+2K\,{\not{d}}_2^\star\,{\not{d}}_2.
\end{aligned}$$

Recall also the Poincaré inequality for ${\not{d}}_2$ which holds for any reduced 2-scalar f

$$\int_S|{\not{d}}_2f|^2\ge 4\int_S Kf^2.$$

Then, we easily infer

$$\begin{aligned}
\int_S f\,{\not{d}}_2^\star\,{\not{d}}_1^\star\,{\not{d}}_1\,{\not{d}}_2 f &= \int_S f({\not{d}}_2^\star\,{\not{d}}_2)^2f+\int_S 2Kf\,{\not{d}}_2^\star\,{\not{d}}_2 f\\
&\ge 4^2\int_S(1+O(\epsilon))K^2f^2+8\int_S(1+O(\epsilon))K^2f^2\\
&\ge 24\int_S(1+O(\epsilon))K^2f^2
\end{aligned}$$

where we also used the estimates for the Gauss curvature

$$K=\frac{1}{r^2}+O\left(\frac{\epsilon}{r^2}\right),\quad re_\theta(K)=O\left(\frac{\epsilon}{r^2}\right),$$

which follow from the bootstrap assumptions. □

The following identity will be useful.

Lemma 6.20. *We have for any reduced scalar f*

$$\begin{aligned}\widetilde{T}\left(\int_S f^2\right) &= 2\int_S f\widetilde{T}f + \int_S\left(\frac{2}{\underline{\overline{\kappa}}}Afe_3(f) + \check{\kappa}f^2\right) - \frac{\overline{\kappa}}{\underline{\overline{\kappa}}}\left(\int_S \check{\underline{\kappa}}f^2\right)\\ &\quad -\frac{1}{\underline{\overline{\kappa}}}A\left(\int_S(2fe_3(f) + \underline{\kappa}f^2)\right) + Err\left[e_4\left(\int_S f^2\right)\right].\end{aligned}$$

Proof. Recall from the definition of $\widetilde{T}$ that

$$\widetilde{T} = e_4 - \frac{1}{\underline{\overline{\kappa}}}\left(\overline{\kappa} + A\right)e_3.$$

We infer, in view of the analog of Proposition 2.64 for an ingoing geodesic foliation,

$$\begin{aligned}&\widetilde{T}\left(\int_S f^2\right)\\ &= e_4\left(\int_S f^2\right) - \frac{1}{\underline{\overline{\kappa}}}\left(\overline{\kappa} + A\right)e_3\left(\int_S f^2\right)\\ &= \int_S(2fe_4(f) + \kappa f^2) + \mathrm{Err}\left[e_4\left(\int_S f^2\right)\right] - \frac{1}{\underline{\overline{\kappa}}}\left(\overline{\kappa} + A\right)\left(\int_S(2fe_3(f) + \underline{\kappa}f^2)\right)\\ &= \int_S\left(2f\widetilde{T}f + \frac{2}{\underline{\overline{\kappa}}}\left(\overline{\kappa} + A\right)fe_3(f) + \kappa f^2\right) + \mathrm{Err}\left[e_4\left(\int_S f^2\right)\right]\\ &\quad -\frac{1}{\underline{\overline{\kappa}}}\left(\overline{\kappa} + A\right)\left(\int_S(2fe_3(f) + \underline{\kappa}f^2)\right)\\ &= 2\int_S f\widetilde{T}f + \int_S\left(\frac{2}{\underline{\overline{\kappa}}}Afe_3(f) + \check{\kappa}f^2\right) - \frac{\overline{\kappa}}{\underline{\overline{\kappa}}}\left(\int_S \check{\underline{\kappa}}f^2\right)\\ &\quad -\frac{1}{\underline{\overline{\kappa}}}A\left(\int_S(2fe_3(f) + \underline{\kappa}f^2)\right) + \mathrm{Err}\left[e_4\left(\int_S f^2\right)\right].\end{aligned}$$

This concludes the proof of the lemma. □

We are now ready to prove Lemma 6.12. Recall from Lemma 6.20 that we have

$$\begin{aligned}\widetilde{T}\left(\int_S f^2\right) &= 2\int_S f\widetilde{T}f + \int_S\left(\frac{2}{\underline{\overline{\kappa}}}Afe_3(f) + \check{\kappa}f^2\right) - \frac{\overline{\kappa}}{\underline{\overline{\kappa}}}\left(\int_S \check{\underline{\kappa}}f^2\right)\\ &\quad -\frac{1}{\underline{\overline{\kappa}}}A\left(\int_S(2fe_3(f) + \underline{\kappa}f^2)\right) + \mathrm{Err}\left[e_4\left(\int_S f^2\right)\right].\end{aligned}$$

In view of the equation satisfied by f, we infer

$$\begin{aligned}\widetilde{T}\left(\int_S f^2\right) &= -\frac{1}{3m}\left(\int_S r^4 f \,{}^{*}\!\!\not{d}_2\, {}^{*}\!\!\not{d}_1 \not{d}_1 \not{d}_2 f\right) + \frac{1}{3m}\left(\int_S hf\right)\\ &\quad + \int_S\left(\frac{2}{\underline{\overline{\kappa}}}Afe_3(f) + \check{\kappa}f^2\right) - \frac{\overline{\kappa}}{\underline{\overline{\kappa}}}\left(\int_S \check{\underline{\kappa}}f^2\right)\\ &\quad -\frac{1}{\underline{\overline{\kappa}}}A\left(\int_S(2fe_3(f) + \underline{\kappa}f^2)\right) + \mathrm{Err}\left[e_4\left(\int_S f^2\right)\right].\end{aligned}$$

Now, from the definition of $\widetilde{T}$, we have $\widetilde{T}(\underline{u}) = 2/\underline{\varsigma}$. We deduce

$$\begin{aligned}
&\widetilde{T}\left(\underline{u}^n \int_S f^2\right) + \frac{\underline{u}^n}{3m}\left(\int_S r^4 f \, {d \mkern-8mu /}_2^\star {d \mkern-8mu /}_1^\star {d \mkern-8mu /}_1 {d \mkern-8mu /}_2 f\right) \\
= \; & \frac{\underline{u}^n}{3m}\left(\int_S hf\right) + \underline{u}^n \int_S \left(\frac{2}{\overline{\underline{\kappa}}} A f e_3(f) + \check{\kappa} f^2\right) - \underline{u}^n \frac{\overline{\kappa}}{\overline{\underline{\kappa}}}\left(\int_S \check{\underline{\kappa}} f^2\right) \\
& - \frac{\underline{u}^n}{\overline{\underline{\kappa}}} A\left(\int_S (2 f e_3(f) + \underline{\kappa} f^2)\right) + \underline{u}^n \mathrm{Err}\left[e_4\left(\int_S f^2\right)\right] + \frac{2}{\underline{\varsigma}} n \underline{u}^{n-1} \int_S f^2.
\end{aligned}$$

This yields in view of the bootstrap assumptions

$$\begin{aligned}
&\widetilde{T}\left(\underline{u}^n \int_S f^2\right) + \frac{\underline{u}^n}{3m}\left(\int_S r^4 f \, {d \mkern-8mu /}_2^\star {d \mkern-8mu /}_1^\star {d \mkern-8mu /}_1 {d \mkern-8mu /}_2 f\right) \\
\lesssim \; & \frac{\underline{u}^n}{3m}\|h\|_{L^2(S)}\|f\|_{L^2(S)} + \epsilon \underline{u}^{n-1} \int_S |f||\mathfrak{d}^{\leq 1} f| + n \underline{u}^{n-1} \int_S f^2.
\end{aligned}$$

Next, we rely on the Poincaré inequality of Lemma 6.19 to deduce

$$\widetilde{T}\left(\underline{u}^n \int_S f^2\right) + \underline{u}^n \int_S f^2 \;\lesssim\; \underline{u}^n \int_S h^2 + \epsilon^2 \underline{u}^{n-2} \int_S (\mathfrak{d} f)^2 + n \underline{u}^{n-1} \int_S f^2.$$

Integrating in $\underline{u}$ between 1 and u_*, and recalling that $\widetilde{T}(\underline{u}) = 2/\underline{\varsigma}$, we infer for any r_0 such that $2m_0(1 - \delta_{\mathcal{H}}) \leq r_0 \leq r_{\mathcal{T}}$

$$\begin{aligned}
&\int_{S(r=r_0,\underline{u})} \underline{u}^n f^2 + \int_1^{\underline{u}}\left(\int_{S(r=r_0,\underline{u}')} \underline{u}'^n f^2\right) d\underline{u}' \\
\lesssim \; & \int_{S(r=r_0,1)} f^2 + \int_1^{\underline{u}}\left(\int_{S(r=r_0,\underline{u}')} \underline{u}'^n h^2\right) d\underline{u}' \\
& + \epsilon^2 \int_1^{\underline{u}}\left(\int_{S(r=r_0,\underline{u}')} \underline{u}'^{n-2} (\mathfrak{d} f)^2\right) d\underline{u}' + n \int_0^{\underline{u}}\left(\int_{S(r=r_0,\underline{u}')} \underline{u}'^{n-1} f^2\right) d\underline{u}'.
\end{aligned}$$

In particular, we have for $n = 0$

$$\begin{aligned}
&\sup_{1 \leq \underline{u} \leq u_*} \int_{S(r=r_0,\underline{u})} f^2 + \int_1^{u_*}\left(\int_{S(r=r_0,\underline{u})} f^2\right) d\underline{u} \\
\lesssim \; & \int_{S(r=r_0,1)} f^2 + \int_1^{u_*}\left(\int_{S(r=r_0,\underline{u})} h^2\right) d\underline{u} + \epsilon^2 \int_1^{u_*}\left(\int_{S(r=r_0,\underline{u})} \underline{u}^{-2} (\mathfrak{d} f)^2\right) d\underline{u}.
\end{aligned}$$

Then, starting from the case $n = 0$ and arguing by iteration on the largest integer below n, one immediately deduces for any real $n \geq 0$

$$
\begin{aligned}
&\sup_{1\leq \underline{u}\leq u_*}\int_{S(r=r_0,\underline{u})}(1+\underline{u}^n)f^2+\int_1^{u_*}\left(\int_{S(r=r_0,\underline{u})}(1+\underline{u}^n)f^2\right)d\underline{u}\\
\lesssim\ & \int_{S(r=r_0,1)}f^2+\int_1^{u_*}\left(\int_{S(r=r_0,\underline{u})}(1+\underline{u}^n)h^2\right)d\underline{u}\\
&+\epsilon^2\int_1^{u_*}\left(\int_{S(r=r_0,\underline{u})}(1+\underline{u}^{n-2})(\mathfrak{d}f)^2\right)d\underline{u}.
\end{aligned}
$$

Now, a simple trace estimate yields

$$
\int_{S(r=r_0,\underline{u})}(1+\underline{u}^n)h^2 \lesssim \int_{\underline{\mathcal{C}}_{\underline{u}}}(1+\underline{u}^n)\Big(|h|^2+|e_3(h)|^2\Big)
$$

so that

$$
\begin{aligned}
\int_1^{u_*}\left(\int_{S(r=r_0,\underline{u})}(1+\underline{u}^n)h^2\right)d\underline{u} &\lesssim \int_1^{u_*}\int_{\underline{\mathcal{C}}_{\underline{u}}}(1+\underline{u}^n)\Big(|h|^2+|e_3(h)|^2\Big)d\underline{u}\\
&\lesssim \int_{{}^{(int)}\mathcal{M}}(1+\underline{u}^n)(\mathfrak{d}^{\leq 1}h)^2.
\end{aligned}
$$

We deduce

$$
\begin{aligned}
&\sup_{1\leq \underline{u}\leq u_*}\int_{S(r=r_0,\underline{u})}(1+\underline{u}^n)f^2+\int_1^{u_*}\left(\int_{S(r=r_0,\underline{u})}(1+\underline{u}^n)f^2\right)d\underline{u}\\
\lesssim\ & \int_{S(r=r_0,1)}f^2+\int_{{}^{(int)}\mathcal{M}}(1+\underline{u}^n)(\mathfrak{d}^{\leq 1}h)^2\\
&+\epsilon^2\int_1^{u_*}\left(\int_{S(r=r_0,\underline{u})}(1+\underline{u}^{n-2})(\mathfrak{d}f)^2\right)d\underline{u}
\end{aligned}
$$

which concludes the proof of Lemma 6.12.

6.2.5 Proof of Proposition 6.14

In this section, we infer from the Teukolsky-Starobinsky identity, see Proposition 2.101, a parabolic equation for $\underline{\alpha}$.

Corollary 6.21. *$\underline{\alpha}$ satisfies on Σ_* the following equation:*

$$
\begin{aligned}
&6m\widetilde{\nu}\underline{\alpha}+r^4\,{\not{d}}_2^\star\,{\not{d}}_1^\star\,{\not{d}}_1\,{\not{d}}_2\underline{\alpha}\\
=\ & \frac{1}{r^3}\Big(e_3(r^2e_3(r\underline{\mathfrak{q}}))+2\underline{\omega}r^2e_3(r\underline{\mathfrak{q}})\Big)-r^{-3}Err[TS]-\left\{\frac{3}{2}r^4\rho\underline{\kappa}-6am\right\}e_4\underline{\alpha}\\
&-\left\{-\frac{3}{2}r^4\left(\rho+\frac{2m}{r^3}\right)\kappa+3mr\left(\kappa-\frac{2\Upsilon}{r}\right)-\frac{12m}{r}\right\}e_3\underline{\alpha}
\end{aligned}
$$

where the vectorfield $\widetilde{\nu}$ is defined by (6.2.3).

Proof. Recall from (2.3.15) that we have

$$e_3(r^2 e_3(r\mathfrak{q})) + 2\underline{\omega} r^2 e_3(r\mathfrak{q}) = r^7 \left\{ \not{d}_2^\star \not{d}_1^\star \not{d}_1 \not{d}_2 \underline{\alpha} + \frac{3}{2}\rho\Big(\underline{\kappa} e_4 - \kappa e_3\Big)\underline{\alpha}\right\} + \mathrm{Err}[TS].$$

This yields

$$\begin{aligned}\frac{3}{2}r^4\rho\Big(\underline{\kappa} e_4 - \kappa e_3\Big)\underline{\alpha} + r^4 \not{d}_2^\star \not{d}_1^\star \not{d}_1 \not{d}_2 \underline{\alpha} &= \frac{1}{r^3}\Big(e_3(r^2 e_3(r\mathfrak{q})) + 2\underline{\omega} r^2 e_3(r\mathfrak{q})\Big)\\ &\quad -r^{-3}\mathrm{Err}[TS].\end{aligned}$$

Now, we have in view of the definition of $\widetilde{\nu}$

$$\begin{aligned}&\frac{3}{2}r^4\rho\Big(\underline{\kappa} e_4 - \kappa e_3\Big) - 6m\widetilde{\nu}\\ &= \left\{\frac{3}{2}r^4\rho\underline{\kappa} - 6am\right\} e_4 + \left\{-\frac{3}{2}r^4\left(\rho + \frac{2m}{r^3}\right)\kappa + 3mr\left(\kappa - \frac{2\Upsilon}{r}\right) - \frac{12m}{r}\right\} e_3.\end{aligned}$$

We infer

$$\begin{aligned}&6m\widetilde{\nu}\underline{\alpha} + r^4 \not{d}_2^\star \not{d}_1^\star \not{d}_1 \not{d}_2 \underline{\alpha}\\ &= \frac{1}{r^3}\Big(e_3(r^2 e_3(r\mathfrak{q})) + 2\underline{\omega} r^2 e_3(r\mathfrak{q})\Big) - r^{-3}\mathrm{Err}[TS] - \left\{\frac{3}{2}r^4\rho\underline{\kappa} - 6am\right\} e_4\underline{\alpha}\\ &\quad - \left\{-\frac{3}{2}r^4\left(\rho + \frac{2m}{r^3}\right)\kappa + 3mr\left(\kappa - \frac{2\Upsilon}{r}\right) - \frac{12m}{r}\right\} e_3\underline{\alpha}.\end{aligned}$$

This concludes the proof of the corollary. □

Corollary 6.22. *$\underline{\alpha}$ satisfies on Σ_**

$$6m\widetilde{\nu}\underline{\alpha} + r^4 \not{d}_2^\star \not{d}_1^\star \not{d}_1 \not{d}_2 \underline{\alpha} = F$$

where F satisfies

$$\max_{0\le k\le k_{small}+18} \int_{\Sigma_*} \underline{u}^{2+2\delta_{dec}} |\mathfrak{d}^k F|^2 \lesssim \epsilon_0^2.$$

Proof. In view of Corollary 6.17, $\underline{\alpha}$ satisfies

$$6m\widetilde{\nu}\underline{\alpha} + r^4 \not{d}_2^\star \not{d}_1^\star \not{d}_1 \not{d}_2 \underline{\alpha} = F$$

with

$$\begin{aligned}F &:= e_3(e_3(\mathfrak{q})) + F_1,\\ F_1 &:= \frac{1}{r^3}\Big(e_3(r^2 e_3(r)\mathfrak{q}) + e_3(r^3)e_3(\mathfrak{q}) + 2\underline{\omega} r^2 e_3(r\mathfrak{q})\Big) - r^{-3}\mathrm{Err}[TS]\\ &\quad - \left\{\frac{3}{2}r^4\rho\underline{\kappa} - 6am\right\} e_4\underline{\alpha}\\ &\quad - \left\{-\frac{3}{2}r^4\left(\rho + \frac{2m}{r^3}\right)\kappa + 3mr\left(\kappa - \frac{2\Upsilon}{r}\right) - \frac{12m}{r}\right\} e_3\underline{\alpha}.\end{aligned}$$

Recall also that Err$[TS]$ is given schematically by, see Proposition 2.101,

$$\begin{aligned}\mathrm{Err}[TS] &= r^4\big(\not{\mathfrak{d}}\Gamma_b + r\Gamma_b\cdot\Gamma_b\big)\cdot\underline{\alpha} + r^2\big(\Gamma_b e_3(r\mathfrak{q}) + (\mathfrak{d}^{\leq 1}\Gamma_b)r\mathfrak{q}\big)\\ &+ r^7\mathfrak{d}^{\leq 2}\big(e_3\eta\cdot\beta\big) + r^5\mathfrak{d}^{\leq 3}\big(\Gamma_b\cdot\Gamma_g\big).\end{aligned}$$

We infer that F_1 is given schematically by

$$\begin{aligned}F_1 &= r\big(\not{\mathfrak{d}}\Gamma_b + r\Gamma_b\cdot\Gamma_b\big)\cdot\underline{\alpha} + r^{-1}\big(\Gamma_b e_3(r\mathfrak{q}) + (\mathfrak{d}^{\leq 1}\Gamma_b)r\mathfrak{q}\big)\\ &+ r^4\mathfrak{d}^{\leq 2}\big(e_3\eta\cdot\beta\big) + r^2\mathfrak{d}^{\leq 3}\big(\Gamma_b\cdot\Gamma_g\big) + r^{-1}\Gamma_b\\ &= r^{-1}\Gamma_b + r^2\mathfrak{d}^{\leq 3}\big(\Gamma_b\cdot\Gamma_g\big) + r^4\mathfrak{d}^{\leq 3}\big(\Gamma_g\cdot\beta\big)\end{aligned}$$

where we have used:

- The fact we are working here with the global frame of Proposition 3.26 which has the property that $\eta\in\Gamma_g$.
- The fact that Γ_b behave better than $r\Gamma_g$.
- The fact that $\mathfrak{q}\in r\Gamma_g$.
- The fact that $\underline{\alpha}$ and $e_3(\mathfrak{q})$ behaves at least as good as Γ_b.
- The fact that $\rho+\frac{2m}{r^3}$ behaves as good as $r^{-1}\Gamma_g$.
- The fact that $e_3(r)+1$ belongs to $r\Gamma_b$.

Now, recall from Lemma 5.1 that the global frame of Proposition 3.26 satisfies in particular[2]

$$\max_{0\leq k\leq k_{small}+22}\sup_{\mathcal{M}}\Big\{r^{\frac{7}{2}+\delta_{dec}-2\delta_0}|\mathfrak{d}^k\beta| + r^2u^{\frac{1}{2}+\delta_{dec}-2\delta_0}|\mathfrak{d}^k\Gamma_g| + ru^{1+\delta_{dec}-2\delta_0}|\mathfrak{d}^k\Gamma_b|\Big\} \lesssim \epsilon. \quad (6.2.4)$$

Together with the schematic form of F_1 and the behavior (3.3.4) of r on Σ_*, and the fact that δ_0 can be chosen to satisfy[3] $8\delta_0\leq\delta_{dec}$, we infer

$$\max_{0\leq k\leq k_{small}+18}\sup_{\Sigma_*} ru^{\frac{3}{2}+\frac{3}{2}\delta_{dec}}|\mathfrak{d}^k F_1| \lesssim \epsilon u_*^{\frac{1}{2}+\delta_{dec}}\sup_{\Sigma_*}(r^{-1}) + \epsilon^2 \lesssim \epsilon_0.$$

In view of the definition of F, this yields

$$\max_{0\leq k\leq k_{small}+18}\int_{\Sigma_*} u^{2+2\delta_{dec}}|\mathfrak{d}^k F|^2 \lesssim \epsilon_0^2 + \max_{0\leq k\leq k_{small}+19}\int_{\Sigma_*} u^{2+2\delta_{dec}}|\mathfrak{d}^k e_3(\mathfrak{q})|^2.$$

Together with Theorem M1, and the fact that $\delta_{extra}>\delta_{dec}$, we infer

$$\max_{0\leq k\leq k_{small}+18}\int_{\Sigma_*} u^{2+2\delta_{dec}}|\mathfrak{d}^k F|^2 \lesssim \epsilon_0^2.$$

[2] Here we use (3.4.11) with $k_{loss}=22$.

[3] Recall from Lemma 5.1 that we have

$$\delta_0 = \frac{k_{loss}}{k_{large}-k_{small}}.$$

Since we have here $k_{loss}=22$, and since we have $2k_{small}\leq k_{large}+1$ and $k_{large}\delta_{dec}\gg 1$, we deduce $\delta_0\ll\delta_{dec}$ and we have indeed $8\delta_0\leq\delta_{dec}$.

This concludes the proof of the corollary. □

We are now ready to prove Proposition 6.14. In view of Corollary 6.22, $\underline{\alpha}$ satisfies

$$6m\widetilde{\nu}\underline{\alpha} + r^4 \,{\not{d}}_2^\star {\not{d}}_1^\star {\not{d}}_1 {\not{d}}_2\underline{\alpha} = F.$$

Commuting with $\mathfrak{d}^k$, we infer

$$6m\widetilde{\nu}(\mathfrak{d}^k\underline{\alpha}) + r^4 \,{\not{d}}_2^\star {\not{d}}_1^\star {\not{d}}_1 {\not{d}}_2(\mathfrak{d}^k\underline{\alpha}) = F_k$$

where F_k is defined by

$$\begin{aligned}
F_k := & -6m[\mathfrak{d}^k, \widetilde{\nu}]\underline{\alpha} - 6\sum_{j=1}^{k} \mathfrak{d}^j(m)\mathfrak{d}^{k-j}\widetilde{\nu}\underline{\alpha} - [\mathfrak{d}^k, r\,{\not{d}}_2^\star]r\,{\not{d}}_1^\star r\,{\not{d}}_1 r\,{\not{d}}_2\underline{\alpha}\\
& - r\,{\not{d}}_2^\star[\mathfrak{d}^k, r\,{\not{d}}_1^\star]r\,{\not{d}}_1 r\,{\not{d}}_2\underline{\alpha} - r\,{\not{d}}_2^\star r\,{\not{d}}_1^\star[\mathfrak{d}^k, r\,{\not{d}}_1]r\,{\not{d}}_2\underline{\alpha} - r\,{\not{d}}_2^\star r\,{\not{d}}_1^\star r\,{\not{d}}_1[\mathfrak{d}^k, r\,{\not{d}}_2]\underline{\alpha}\\
& + \mathfrak{d}^k F.
\end{aligned}$$

Note that we have schematically

$$\begin{aligned}
& [\mathfrak{d}, \not{\mathfrak{d}}] = r\Gamma_b\mathfrak{d}, \qquad [\widetilde{\nu}, \not{\mathfrak{d}}] = \Big(O(r^{-1}) + r\Gamma_b\Big)\mathfrak{d},\\
& [\widetilde{\nu}, re_4] = O(r^{-1})\mathfrak{d}, \qquad [\widetilde{\nu}, e_3] = O(r^{-1})\mathfrak{d}.
\end{aligned}$$

Together with the fact that $\underline{\alpha}$ behaves at least as good as Γ_b, we infer, schematically,

$$F_k = \mathfrak{d}^k F + r^{-1}\mathfrak{d}^{\leq k+4}\Gamma_b + r\mathfrak{d}^{\leq k+4}(\Gamma_b^2).$$

In view of (6.2.4) and the behavior (3.3.4) of r on Σ_*, we have

$$\begin{aligned}
\max_{0\leq k\leq k_{small}+18} \sup_{\Sigma_*} ru^{\frac{3}{2}+\frac{3}{2}\delta_{dec}}|r^{-1}\mathfrak{d}^{\leq k+4}\Gamma_b + r\mathfrak{d}^{\leq k+4}(\Gamma_b^2)| & \lesssim \epsilon u_*^{\frac{1}{2}+\delta_{dec}} \sup_{\Sigma_*} r^{-1} + \epsilon^2\\
& \lesssim \epsilon_0.
\end{aligned}$$

This yields

$$\max_{0\leq k\leq k_{small}+18} \int_{\Sigma_*} u^{2+2\delta_{dec}}|F_k|^2 \lesssim \epsilon_0^2 + \max_{0\leq k\leq k_{small}+18} \sup_{\Sigma_*} u^{2+2\delta_{dec}}|\mathfrak{d}^k F|^2.$$

Together with the estimate for F of Corollary 6.22, we infer

$$\max_{0\leq k\leq k_{small}+18} \int_{\Sigma_*} u^{2+2\delta_{dec}}|F_k|^2 \lesssim \epsilon^4 + \epsilon_0^2 \lesssim \epsilon_0^2.$$

This concludes the proof of Proposition 6.14.

6.2.6 Proof of Lemma 6.16

In this section we prove Lemma 6.16, i.e., we derive estimates for the control of the parabolic equation appearing in the statement of Proposition 6.14. The following identity will be useful.

Lemma 6.23. *We have for any reduced scalar f*

$$\begin{aligned}\widetilde{\nu}\left(\int_S f^2\right) &= 2\int_S f\widetilde{\nu}(f)+\int_S(-2afe_4(f)+\underline{\kappa}f^2)+a\int_S(2fe_4(f)+\kappa f^2)\\&\quad+Err\left[e_3\left(\int_S f^2\right)\right].\end{aligned}$$

Proof. Recall from the definition of $\widetilde{\nu}$ that

$$\widetilde{\nu} = e_3 + ae_4.$$

We infer, in view of Proposition 2.64,

$$\begin{aligned}&\widetilde{\nu}\left(\int_S f^2\right)\\&= e_3\left(\int_S f^2\right)+ae_4\left(\int_S f^2\right)\\&= \int_S(2fe_3(f)+\underline{\kappa}f^2)+\mathrm{Err}\left[e_3\left(\int_S f^2\right)\right]+a\int_S(2fe_4(f)+\kappa f^2)\\&= 2\int_S f\widetilde{\nu}(f)+\int_S(-2afe_4(f)+\underline{\kappa}f^2)+a\int_S(2fe_4(f)+\kappa f^2)\\&\quad+\mathrm{Err}\left[e_3\left(\int_S f^2\right)\right].\end{aligned}$$

This concludes the proof of the lemma. □

We are now ready to prove Lemma 6.16. Recall from Lemma 6.23 that we have

$$\begin{aligned}\widetilde{\nu}\left(\int_S f^2\right) &= 2\int_S f\widetilde{\nu}(f)+\int_S(-2afe_4(f)+\underline{\kappa}f^2)+a\int_S(2fe_4(f)+\kappa f^2)\\&\quad+\mathrm{Err}\left[e_3\left(\int_S f^2\right)\right].\end{aligned}$$

In view of the equation satisfied by f, we infer

$$\begin{aligned}\widetilde{\nu}\left(\int_S f^2\right) &= -\frac{1}{3m}\left(\int_S r^4 f\, \not{d}_2^\star\, \not{d}_1^\star\, \not{d}_1\, \not{d}_2 f\right)+\frac{1}{3m}\left(\int_S hf\right)\\&\quad+\int_S(-2afe_4(f)+\underline{\kappa}f^2)+a\int_S(2fe_4(f)+\kappa f^2)\\&\quad+\mathrm{Err}\left[e_3\left(\int_S f^2\right)\right].\end{aligned}$$

Now, from the definition of $\widetilde{\nu}$, we have $\widetilde{\nu}(u) = 2/\varsigma$. We deduce

$$\begin{aligned}
&\widetilde{\nu}\left(u^n \int_S f^2\right) + \frac{u^n}{3m}\left(\int_S r^4 f\, {\not d}_2^\star {\not d}_1^\star {\not d}_1 {\not d}_2 f\right)\\
= \; & \frac{u^n}{3m}\left(\int_S hf\right) + u^n \int_S (-2afe_4(f) + \underline{\kappa} f^2) + au^n \int_S (2fe_4(f) + \kappa f^2)\\
&+u^n \mathrm{Err}\left[e_3\left(\int_S f^2\right)\right] + \frac{2}{\varsigma} n u^{n-1} \int_S f^2.
\end{aligned}$$

This yields in view of the bootstrap assumptions

$$\begin{aligned}
&\widetilde{\nu}\left(u^n \int_S f^2\right) + \frac{u^n}{3m}\left(\int_S r^4 f\, {\not d}_2^\star {\not d}_1^\star {\not d}_1 {\not d}_2 f\right)\\
\lesssim \; & \frac{u^n}{3m}\|h\|_{L^2(S)}\|f\|_{L^2(S)} + \left(\frac{1}{r} + \epsilon u^{-1}\right) u^n \int_S |f||\mathfrak{d}^{\leq 1} f| + n u^{n-1}\int_S f^2\\
\lesssim \; & \frac{u^n}{3m}\|h\|_{L^2(S)}\|f\|_{L^2(S)} + \epsilon u^{n-1}\int_S |f||\mathfrak{d}^{\leq 1} f| + n u^{n-1}\int_S f^2
\end{aligned}$$

where we have used in the last inequality the behavior (3.3.4) of r on Σ_*. Next, we rely on the Poincaré inequality of Lemma 6.19 to deduce

$$\widetilde{\nu}\left(u^n \int_S f^2\right) + u^n \int_S f^2 \;\lesssim\; u^n \int_S h^2 + \epsilon^2 u^{n-2} \int_S (\mathfrak{d} f)^2 + n u^{n-1} \int_S f^2.$$

Integrating in u between 1 and u_*, and recalling that $\widetilde{\nu}(u) = 2/\varsigma$, we infer

$$\int_{\Sigma_*} u^n f^2 \;\lesssim\; \int_{\Sigma_* \cap \mathcal{C}_1} f^2 + \int_{\Sigma_*} u^n h^2 + \epsilon^2 \int_{\Sigma_*} u^{n-2}(\mathfrak{d} f)^2 + n \int_{\Sigma_*} u^{n-1} f^2.$$

In particular, we have for $n = 0$

$$\int_{\Sigma_*} f^2 \;\lesssim\; \int_{\Sigma_* \cap \mathcal{C}_1} f^2 + \int_{\Sigma_*} h^2 + \epsilon^2 \int_{\Sigma_*} u^{-2}(\mathfrak{d} f)^2.$$

Then, starting from the case $n = 0$ and arguing by iteration on the largest integer below n, one immediately deduces for any real $n \geq 0$

$$\int_{\Sigma_*} (1 + u^n) f^2 \;\lesssim\; \int_{\Sigma_* \cap \mathcal{C}_1} f^2 + \int_{\Sigma_*} (1 + u^n) h^2 + \epsilon^2 \int_{\Sigma_*} (1 + u^{n-2})(\mathfrak{d} f)^2$$

which concludes the proof of Lemma 6.16.

Chapter Seven

Decay Estimates (Theorems M4, M5)

In this chapter, we rely on the decay of $\mathfrak{q}$, α and $\underline{\alpha}$ to prove the decay estimates for all the other quantities. More precisely, we rely on the results of Theorems M1, M2 and M3 to prove Theorems M4 and M5.

7.1 PRELIMINARIES TO THE PROOF OF THEOREM M4

In what follows we give a detailed proof of Theorem M4, which, we recall, provides the main decay estimates in ${}^{(ext)}\mathcal{M}$. The proof makes use of the bootstrap assumptions **BA-D**, **BA-E**, and the results of Theorems M1, M2, M3 and Lemmas 3.15, 3.16. In this section, we start with some preliminaries.

7.1.1 Geometric structure of Σ_*

The proof of Theorem M4 depends in a fundamental way on the geometric properties of the GCM hypersuface Σ_*, the spacelike future boundary of ${}^{(ext)}\mathcal{M}$ introduced in section 3.1.2. For the convenience of the reader, we recall below its main features.

1. The affine parameter s is initialized on Σ_* such that $s = r$.
2. There exists a constant c_* such that

$$\Sigma_* := \{u + r = c_*\}.$$

3. Let $\nu_* = e_3 + a_* e_4$ be the unique vectorfield tangent to the hypersurface Σ_*, perpendicular to the foliation $S(u)$ induced on Σ_* and normalized by the condition $g(\nu_*, e_4) = -2$. The following normalization condition holds true at the south pole SP of every sphere S,

$$a_*\Big|_{SP} = -1 - \frac{2m}{r}. \tag{7.1.1}$$

4. We have

$$r \geq \epsilon_0^{-\frac{2}{3}} u_*^{1+\delta_{dec}} \qquad \text{on } \Sigma_*. \tag{7.1.2}$$

5. The following GCM conditions hold on Σ_*:

$$\kappa = \frac{2}{r}, \qquad {\not{d}}_2^\star {\not{d}}_1^\star \underline{\kappa} = 0, \qquad {\not{d}}_2^\star {\not{d}}_1^\star \mu = 0, \tag{7.1.3}$$

$$\int_S \eta e^{\Phi} = \int_S \underline{\xi} e^{\Phi} = 0. \tag{7.1.4}$$

Moreover on $S_*=\Sigma_*\cap\mathcal{C}_*$,

$$\int_{S_*}\beta e^{\Phi}=0, \qquad \int_{S_*}e_\theta(\underline{\kappa})e^{\Phi}=0. \tag{7.1.5}$$

6. According to the definition of the Hawking mass, i.e., $1-\frac{2m}{r}=-\frac{r^2}{4}\overline{\kappa\underline{\kappa}}$, and the GCM assumption for κ we also have

$$\overline{\underline{\kappa}} \;=\; -\frac{r}{2}\left(1-\frac{2m}{r}\right). \tag{7.1.6}$$

Thus on Σ_*,

$$e_3(r)=\frac{r}{2}\left(\overline{\underline{\kappa}}+\underline{A}\right)=-\Upsilon+\frac{r}{2}\underline{A}, \qquad e_4(r)=1. \tag{7.1.7}$$

7. In view of the definition of ν_* and that of ς we easily deduce[1] the following relation between a_* and ς on Σ_*:

$$a_* \;=\; -\frac{2}{\varsigma}+\Upsilon-\frac{r}{2}\underline{A}. \tag{7.1.8}$$

8. Since on Σ_* we have $r=s$ we deduce

$$\underline{\Omega}=e_3(r)=-\Upsilon+\frac{r}{2}\underline{A} \quad \text{on } \Sigma_*. \tag{7.1.9}$$

7.1.2 Main assumptions

We reformulate below the main bootstrap assumptions[2] in the form needed in the proof of Theorem M4.

Definition 7.1. *We make use of the following norms on $S=S(u,r)\subset {}^{(ext)}\mathcal{M}$,*

$$\begin{aligned}
\|f\|_\infty(u,r) &:= \|f\|_{L^\infty\left(S(u,r)\right)}, & \|f\|_2(u,r) &:= \|f\|_{L^2\left(S(u,r)\right)},\\
\|f\|_{\infty,k}(u,r) &:= \sum_{i=0}^k \|\mathfrak{d}^i f\|_\infty(u,r), & \|f\|_{2,k}(u,r) &:= \sum_{i=0}^k \|\mathfrak{d}^i f\|_2(u,r).
\end{aligned} \tag{7.1.10}$$

To simplify the exposition it also helps to introduce the following schematic notation for the connection coefficients (recall $\omega,\xi=0$ and $\zeta=-\underline{\eta}$),

$$\begin{aligned}
\Gamma_g &= \left\{\check{\kappa},\vartheta,\underline{\eta},\zeta,\check{\underline{\kappa}}\right\}\cup\left\{\overline{\kappa}-\frac{2}{r},\overline{\underline{\kappa}}+\frac{2\Upsilon}{r}\right\},\\
\Gamma_b &= \left\{\underline{\vartheta},\eta,\check{\underline{\omega}},\underline{\xi},\underline{A},r^{-1}\check{\varsigma},r^{-1}\check{\underline{\Omega}},\right\}\cup\left\{\overline{\underline{\omega}}-\frac{m}{r^2},r^{-1}(\overline{\varsigma}-1),r^{-1}(\overline{\underline{\Omega}}+\Upsilon)\right\}.
\end{aligned} \tag{7.1.11}$$

Remark 7.2. *It is important to note that η belongs to Γ_b rather than Γ_g as it may have been expected. Note also that $\underline{A}\in\Gamma_b$ in view of Proposition 2.64 and the fact*

[1]Indeed, since ν_* is tangent to Σ_* along which $u=-r+c_*$, using also (7.1.7), $\frac{2}{\varsigma}=e_3(u)=\nu_*(u)=-\nu_*(r)=-e_3(r)-a_*e_4(r)=-a_*+\Upsilon-\frac{r}{2}\underline{A}$.

[2]Based on bootstrap assumptions **BA-D**, **BA-E**, Theorems M1, M2, M3, and Lemmas 3.15, 3.16.

that $(\check{\varsigma}, \check{\underline{\Omega}}) \in r\Gamma_b$. *We also note that the averaged quantities* $\left\{\overline{\kappa} - \frac{2}{r}, \overline{\underline{\kappa}} + \frac{2\Upsilon}{r}\right\}$ *and* $\left\{\overline{\underline{\omega}} - \frac{m}{r^2}, r^{-1}(\overline{\varsigma} - 1), r^{-1}(\overline{\underline{\Omega}} + \Upsilon)\right\}$ *are actually better behaved in view of Lemmas 3.15, 3.16.*

Ref 1. According to our bootstrap assumptions **BA-D**, and the pointwise estimates of Proposition 3.19, which themselves follow from **BA-E**, as well as the control of averages in Lemma 3.15 and the control of the Hawking mass in Lemma 3.16, we have on ${}^{(ext)}\mathcal{M}$,

1. For $0 \le k \le k_{small}$,

$$\begin{aligned} \|\Gamma_g\|_{\infty,k} &\lesssim \epsilon \min\left\{r^{-2}u^{-\frac{1}{2}-\delta_{dec}},\ r^{-1}u^{-1-\delta_{dec}}\right\}, \\ \|e_3\Gamma_g\|_{\infty,k-1} &\lesssim \epsilon r^{-2}u^{-1-\delta_{dec}}, \\ \|\Gamma_b\|_{\infty,k} &\lesssim \epsilon r^{-1}u^{-1-\delta_{dec}}. \end{aligned} \tag{7.1.12}$$

2. For $k \le k_{large} - 5$,

$$\|\Gamma_g\|_{\infty,k} \ \lesssim \ \epsilon r^{-2}, \qquad \|\Gamma_b\|_{\infty,k} \lesssim \epsilon r^{-1}. \tag{7.1.13}$$

Ref 2. The quantity[3] $\mathfrak{q}$ satisfies on ${}^{(ext)}\mathcal{M}$, for all $0 \le k \le k_{small} + 20$,

$$\begin{aligned} \|\mathfrak{q}\|_{\infty,k} &\lesssim \epsilon_0 \min\left\{u^{-1-\delta_{extra}}, r^{-1}u^{-\frac{1}{2}-\delta_{extra}}\right\}, \\ \|e_3\mathfrak{q}\|_{\infty,k-1} &\lesssim \epsilon_0 r^{-1}u^{-1-\delta_{extra}}. \end{aligned} \tag{7.1.14}$$

In addition, on the last slice Σ_*, for all $k \le k_{small} + 20$,

$$\int_{\Sigma_*(\tau,\tau_*)} |e_3\mathfrak{d}^k\mathfrak{q}|^2 + |e_4\mathfrak{d}^k\mathfrak{q}|^2 + r^{-2}|\mathfrak{q}|^2 \ \lesssim \epsilon_0^2(1+\tau)^{-2-2\delta_{dec}}. \tag{7.1.15}$$

According to Theorem M2 we have on ${}^{(ext)}\mathcal{M}$, for all $0 \le k \le k_{small} + 20$,

$$\begin{aligned} \|\alpha\|_{\infty,k} &\lesssim \epsilon_0 \min\left\{r^{-3}(u+2r)^{-\frac{1}{2}-\delta_{extra}}, \log(1+u)r^{-2}(u+2r)^{-1-\delta_{extra}}\right\}, \\ \|e_3\alpha\|_{\infty,k-1} &\lesssim \epsilon_0 \min\left\{r^{-4}(u+2r)^{-\frac{1}{2}-\delta_{extra}},\right. \\ &\qquad\qquad \left.\log(1+u)r^{-3}(u+2r)^{-1-\delta_{extra}}\right\}. \end{aligned} \tag{7.1.16}$$

According to Theorem M3, the component $\underline{\alpha}$ verifies the following estimate[4] holds on $\mathcal{T}$, for $0 \le k \le k_{small} + 16$,

$$\sup_{\mathcal{T}} u^{1+\delta_{dec}}|\mathfrak{d}^k\underline{\alpha}| \ \lesssim \ \epsilon_0, \tag{7.1.17}$$

[3]Recall (see Remark 2.110) that the quantity $\mathfrak{q}$ we are working with is defined relative to the global frame of Proposition 3.26.

[4]In fact, the corresponding estimate in Theorem M3 holds on ${}^{(int)}\mathcal{M}$, and hence in particular on $\mathcal{T}$ since $\mathcal{T} \subset {}^{(int)}\mathcal{M}$.

and on the last slice Σ_* for all $k \leq k_{small}+18$

$$\int_{\Sigma_*(\tau,\tau_*)} |\mathfrak{d}^k\underline{\alpha}|^2 \lesssim \epsilon_0^2(1+\tau)^{-2-2\delta_{dec}}. \tag{7.1.18}$$

Ref 3. In view of the bootstrap assumptions **BA-D** and the pointwise estimates of Proposition 3.19 for the curvature components, which themselves follow from **BA-E**, we have in ${}^{(ext)}\mathcal{M}$,

i. For all $0 \leq k \leq k_{small}$,

$$\begin{aligned}
\|\beta\|_{\infty,k} &\lesssim \epsilon\min\left\{r^{-3}(u+2r)^{-\frac{1}{2}-\delta_{dec}},\, r^{-2}(u+2r)^{-1-\delta_{dec}}\right\},\\
\|e_3\beta\|_{\infty,k-1} &\lesssim \epsilon\min\left\{r^{-4}(u+2r)^{-\frac{1}{2}-\delta_{dec}},\, r^{-3}(u+2r)^{-1-\delta_{dec}}\right\},\\
\left\|\left(\check{\rho},\overline{\rho}+\frac{2m}{r^3}\right)\right\|_{\infty,k} &\lesssim \epsilon\min\left\{r^{-3}u^{-\frac{1}{2}-\delta_{dec}},\, r^{-2}u^{-1-\delta_{dec}}\right\},\\
\left\|e_3\left(\check{\rho},\overline{\rho}+\frac{2m}{r^3}\right)\right\|_{\infty,k-1} &\lesssim \epsilon r^{-3}u^{-1-\delta_{dec}},\\
\left\|\check{\mu},\overline{\mu}-\frac{2m}{r^3}\right\|_{\infty,k} &\lesssim \epsilon r^{-3}u^{-1-\delta_{dec}},\\
\|\underline{\beta}\|_{\infty,k} &\lesssim \epsilon r^{-2}u^{-1-\delta_{dec}}.
\end{aligned} \tag{7.1.19}$$

Since $K=-\rho-\frac{1}{4}\kappa\underline{\kappa}+\frac{1}{4}\vartheta\underline{\vartheta}=\frac{1}{r^2}-(\rho-\overline{\rho})-\frac{1}{4}(\kappa\underline{\kappa}-\overline{\kappa\underline{\kappa}})+\text{l.o.t.}$ we also deduce for all $0 \leq k \leq k_{small}$,

$$\left\|K-\frac{1}{r^2}\right\|_{\infty,k} \lesssim \epsilon\min\left\{r^{-3}u^{-\frac{1}{2}-\delta_{dec}},\, r^{-2}u^{-1-\delta_{dec}}\right\}.$$

ii. For all $k \leq k_{large}-5$,

$$\begin{aligned}
r^{\frac{7}{2}+\frac{\delta_B}{2}}\Big(\|\alpha\|_{\infty,k}+\|\beta\|_{\infty,k}\Big) &\lesssim \epsilon,\\
r^3\|\check{\rho}\|_{\infty,k}+r^2\|\underline{\beta}\|_{\infty,k}+r\|\underline{\alpha}\|_{\infty,k} &\lesssim \epsilon.
\end{aligned} \tag{7.1.20}$$

Remark 7.3. *In view of the control of averages Lemma 3.15 we have in fact better estimates for the scalars,*

$$\overline{\kappa}-\frac{2}{r},\quad \overline{\underline{\kappa}}+\frac{2\Upsilon}{r},\quad \overline{\underline{\omega}}-\frac{m}{r^2},\quad \overline{\rho}+\frac{2m}{r^3}.$$

*In particular they can be estimated by ϵ replaced by ϵ_0 in **Ref 1**.*

Remark 7.4. *Note that $r(\check{\rho},\overline{\rho}+\frac{2m}{r^3}), r(K-\frac{1}{r^2})$ behave as Γ_g. For convenience we shall just simply add them to Γ_g. Similarly $(r\underline{\beta},\underline{\alpha})$ behave as Γ_b. Thus, our extended Γ_g,Γ_b are*

$$\begin{aligned}
\Gamma_g &= \left\{\check{\kappa},\vartheta,\underline{\eta},\zeta,\check{\underline{\kappa}},r\check{\rho}\right\}\cup\left\{\overline{\kappa}-\frac{2}{r},\overline{\underline{\kappa}}+\frac{2\Upsilon}{r},r\left(\overline{\rho}+\frac{2m}{r^3}\right)\right\},\\
\Gamma_b &= \left\{\underline{\vartheta},\eta,\check{\underline{\omega}},\underline{\xi},\underline{A},r^{-1}\check{\varsigma},r^{-1}\check{\underline{\Omega}},r\underline{\beta},\underline{\alpha}\right\}\cup\left\{\overline{\underline{\omega}}-\frac{m}{r^2},r^{-1}(\overline{\varsigma}-1),r^{-1}(\overline{\underline{\Omega}}+\Upsilon)\right\}.
\end{aligned}$$

Note also that we can write $e_3(\Gamma_g) = r^{-1}\mathfrak{d}\Gamma_b$.

7.1.3 Basic lemmas

7.1.3.1 Commutation identities

Lemma 7.5. *We have, schematically,*

$$\begin{aligned} [\not{\mathfrak{d}}, e_4]\psi &= \Gamma_g \mathfrak{d}^{\nearrow}\psi + l.o.t., \\ [\not{\mathfrak{d}}, e_3]\psi &= r\Gamma_b e_3\psi + \Gamma_b^{\leq 1}\mathfrak{d}^{\nearrow}\psi + l.o.t. \end{aligned} \tag{7.1.21}$$

Proof. Follows from Lemma 2.68 and the symbolic notation introduced in (7.1.11), see also Remark 7.4. □

7.1.3.2 Interpolation and product estimates

We estimate quadratic error terms with the help of the following lemma.

Lemma 7.6. *Let* $k_{loss} = 25$. *Then, the following interpolation estimates hold true for all* $0 \leq k \leq k_{small} + k_{loss}$*:*

$$\begin{aligned} \|\Gamma_g\|_{\infty,k} + r\left\|\left(\check{\rho}, \beta, \alpha\right)\right\|_{\infty,k} &\lesssim \epsilon r^{-2}u^{-\frac{1}{2}-\frac{\delta_{dec}}{2}}, \\ \|(\Gamma_b, \underline{\alpha})\|_{\infty,k} + r\|\underline{\beta}\|_{\infty,k} &\lesssim \epsilon r^{-1}u^{-1-\frac{\delta_{dec}}{2}}. \end{aligned} \tag{7.1.22}$$

Also, the following product estimates hold true for all $0 \leq k \leq k_{small} + k_{loss}$*:*

$$\begin{aligned} \|\Gamma_g \cdot \Gamma_g\|_{\infty,k} + r\left\|\left(\check{\rho}, \beta, \alpha\right)\cdot\Gamma_g\right\|_{\infty,k} &\lesssim \epsilon_0 r^{-4}u^{-1-\delta_{dec}}, \\ \left\|\Gamma_g \cdot \left(\Gamma_b, \underline{\alpha}\right)\right\|_{\infty,k} + r\|\Gamma_g\cdot\underline{\beta}\|_{\infty,k} + r\left\|\left(\check{\rho}, \beta, \alpha\right)\cdot\Gamma_b\right\|_{\infty,k} &\lesssim \epsilon_0 r^{-3}u^{-\frac{3}{2}-\delta_{dec}} \\ \left\|\left(\beta, \alpha\right)\cdot\Gamma_b\right\|_{\infty,k} &\lesssim \epsilon_0 r^{-\frac{9}{2}}u^{-1-\delta_{dec}}, \\ \left\|\left(\Gamma_b, \underline{\alpha}\right)\cdot\Gamma_b\right\|_{\infty,k} + r\|\underline{\beta}\cdot\Gamma_b\|_{\infty,k} &\lesssim \epsilon_0 r^{-2}u^{-2-\delta_{dec}}. \end{aligned} \tag{7.1.23}$$

Proof. All estimates are easy to prove in the range $0 \leq k \leq k_{small}$. We shall thus assume that $k_{small} \leq k \leq k_{small} + k_{loss}$. Since $k_{loss} < k_{small}$ we have $k/2 < k_{small}$ for all k in that range.

For simplicity of notation we write $L := k_{large} - 5$, $S := k_{small}$. By standard interpolation inequalities, for all $S \leq k \leq L$,

$$\begin{aligned} \|\Gamma_g\|_{\infty,k} &\lesssim \|\Gamma_g\|_{\infty,L}^{\frac{k-S}{L-S}}\|\Gamma_g\|_{\infty,S}^{\frac{L-k}{L-S}} \lesssim \epsilon r^{-2}\left[u^{-\frac{1}{2}-\delta_{dec}}\right]^{\frac{L-k}{L-S}} \\ &\lesssim \epsilon r^{-2}u^{-\frac{1}{2}-\delta_{dec}}\left[u^{\frac{1}{2}+\delta_{dec}}\right]^{\frac{k-S}{L-S}}, \\ \|\Gamma_b\|_{\infty,k} &\lesssim \|\Gamma_b\|_{\infty,L}^{\frac{k-S}{L-S}}\|\Gamma_b\|_{\infty,S}^{\frac{L-k}{L-S}} \lesssim \epsilon r^{-1}\left[u^{-1-\delta_{dec}}\right]^{\frac{L-k}{L-S}} \\ &\lesssim \epsilon r^{-1}u^{-1-\delta_{dec}}\left[u^{1+\delta_{dec}}\right]^{\frac{k-S}{L-S}}. \end{aligned}$$

Now, we may assume that k_{loss} satisfies[5]

$$k_{loss} \leq \frac{\delta_{dec}}{3}(k_{large} - k_{small}).$$

Thus, for $k_{small} \leq k \leq k_{small} + k_{loss}$, we have

$$\left[u^{\frac{1}{2}+\delta_{dec}}\right]^{\frac{k-S}{L-S}} + \left[u^{1+\delta_{dec}}\right]^{\frac{k-S}{L-S}} \lesssim \left[u^{1+\delta_{dec}}\right]^{\frac{k_{loss}}{k_{large}-5-k_{small}}} \lesssim \left[u^{1+\delta_{dec}}\right]^{\frac{\delta_{dec}}{3}} \lesssim u^{\frac{\delta_{dec}}{2}}$$

and hence

$$\begin{aligned} \|\Gamma_g\|_{\infty,k} &\lesssim \epsilon r^{-2}u^{-\frac{1}{2}-\frac{\delta_{dec}}{2}}, \\ \|\Gamma_b\|_{\infty,k} &\lesssim \epsilon r^{-1}u^{-1-\frac{\delta_{dec}}{2}}. \end{aligned}$$

Since $r\check{\rho}$ satisfies the same estimates as Γ_g and $r\beta$ and $r\alpha$ satisfy even better estimates, and that $\underline{\alpha}$ and $r\underline{\beta}$ satisfy the same estimate as Γ_b, we infer

$$\|\Gamma_g\|_{\infty,k} + r\left\|\left(\check{\rho}, \beta, \alpha\right)\right\|_{\infty,k} \lesssim \epsilon r^{-2}u^{-\frac{1}{2}-\frac{\delta_{dec}}{2}},$$
$$\|(\Gamma_b, \underline{\alpha})\|_{\infty,k} + r\|\underline{\beta}\|_{\infty,k} \lesssim \epsilon r^{-1}u^{-1-\frac{\delta_{dec}}{2}},$$

which is the desired interpolation bound.

The first, second and last product estimates follow immediately from the above interpolation bound. Finally, the third product estimate follows from the above interpolation estimate for Γ_b together with the following interpolation estimate for $k_{small} \leq k \leq k_{small} + k_{loss}$:

$$\begin{aligned} \|(\beta, \alpha)\|_{\infty,k} &\lesssim \|(\beta, \alpha)\|_{\infty,L}^{\frac{k-S}{L-S}} \|(\beta, \alpha)\|_{\infty,S}^{\frac{L-k}{L-S}} \\ &\lesssim \epsilon\left[r^{-\frac{7}{2}-\frac{\delta_B}{2}}\right]^{\frac{k-S}{L-S}}\left[\min\left(r^{-\frac{7}{2}-\frac{\delta_B}{2}}, r^{-3}u^{-\frac{1}{2}-\delta_{dec}}\right)\right]^{\frac{L-k}{L-S}} \\ &\lesssim \epsilon\left[r^{-\frac{7}{2}-\frac{\delta_B}{2}}\right]^{1-\delta_{dec}}\left[r^{-3}u^{-\frac{1}{2}-\delta_{dec}}\right]^{\delta_{dec}} \\ &\lesssim \epsilon r^{-\frac{7}{2}}u^{-\frac{\delta_{dec}}{2}} \end{aligned}$$

where we have used in the last inequality the fact that $\delta_B > 2\delta_{dec}$. □

7.1.3.3 Elliptic estimates

We shall often make use of the results of Proposition 2.33 and Lemma 2.38 which we rewrite as follows with respect to the L^2 based $\mathfrak{h}_k(S)$ spaces introduced in Definition 2.39.

Lemma 7.7. *Under the assumptions* ***Ref 1–Ref 3****, the following elliptic estimates*

---

[5]Recall that we have

$$0 < \delta_{dec} \ll 1, \qquad \delta_{dec}\, k_{large} \gg 1, \qquad k_{small} = \left\lfloor \frac{1}{2}k_{large}\right\rfloor + 1.$$

In particular, we have $\delta_{dec}(k_{large} - k_{small}) \gg 1$ and hence we may indeed assume that $k_{loss} = 25$ satisfies the required constraints.

hold true for the Hodge operators $\not{d}_1$, $\not{d}_2$, $\not{d}_1^\star$, $\not{d}_2^\star$, *for all* $k \le k_{small} + 20$.

1. *If* $f \in \mathfrak{s}_1(S)$,

$$\|\not{\partial} f\|_{\mathfrak{h}_k(S)} + \|f\|_{\mathfrak{h}_k(S)} \lesssim r\|\not{d}_1 f\|_{\mathfrak{h}_k(S)}.$$

2. *If* $f \in \mathfrak{s}_2(S)$,

$$\|\not{\partial} f\|_{\mathfrak{h}_k(S)} + \|f\|_{\mathfrak{h}_k(S)} \lesssim r\|\not{d}_2 f\|_{\mathfrak{h}_k(S)}.$$

3. *If* $f \in \mathfrak{s}_0(S)$,

$$\|\not{\partial} f\|_{\mathfrak{h}_k(S)} \lesssim r\|\not{d}_1^\star f\|_{\mathfrak{h}_k(S)}.$$

4. *If* $f \in \mathfrak{s}_1(S)$,

$$\|f\|_{\mathfrak{h}_{k+1}(S)} \lesssim r\|\not{d}_2^\star f\|_{\mathfrak{h}_k(S)} + r^{-2}\left|\int_S e^\Phi f\right|.$$

5. *If* $f \in \mathfrak{s}_1(S)$,

$$\left\|f - \frac{\int_S f e^\Phi}{\int_S e^{2\Phi}} e^\Phi\right\|_{\mathfrak{h}_{k+1}(S)} \lesssim r\|\not{d}_2^\star f\|_{\mathfrak{h}_k(S)}.$$

7.1.4 Main equations

The proof of Theorem M4 relies heavily on the null structure and null Bianchi identities derived in section 2.2.4, see Proposition 2.63. We also rely on Proposition 2.73 for equations verified by the check quantities. We rewrite them below in a schematic form.

Proposition 7.8 (Transport equations for checked quantities). *We have the following transport equations in the* e_4 *direction,*

$$\begin{aligned}
e_4\check{\kappa} + \overline{\kappa}\,\check{\kappa} &= \Gamma_g \cdot \Gamma_g,\\
e_4\underline{\check{\kappa}} + \frac{1}{2}\overline{\kappa}\underline{\check{\kappa}} + \frac{1}{2}\check{\kappa}\overline{\underline{\kappa}} &= -2\not{d}_1\zeta + 2\check{\rho} + \Gamma_g \cdot \Gamma_b,\\
e_4\underline{\check{\omega}} &= \check{\rho} + \Gamma_g \cdot \Gamma_b,\\
e_4\check{\rho} + \frac{3}{2}\overline{\kappa}\check{\rho} + \frac{3}{2}\overline{\rho}\check{\kappa} &= \not{d}_1\beta + \Gamma_b \cdot \alpha + \Gamma_g \cdot \beta + \check{\kappa} \cdot \check{\rho},\\
e_4\check{\mu} + \frac{3}{2}\overline{\kappa}\check{\mu} + \frac{3}{2}\overline{\mu}\check{\kappa} &= r^{-1}\Gamma_g \cdot \not{\partial}^{\le 1}\Gamma_g.
\end{aligned} \tag{7.1.24}$$

Also, we have in the e_3 *direction,*

$$\begin{aligned}
e_3\underline{\check{\kappa}} &= r^{-1}\not{\partial}^{\le 1}\Gamma_b + \Gamma_b \cdot \not{\partial}^{\le 1}\Gamma_b,\\
e_3\check{\rho} &= r^{-2}\not{\partial}^{\le 1}\Gamma_b + r^{-1}\Gamma_b \cdot \not{\partial}^{\le 1}\Gamma_b.
\end{aligned} \tag{7.1.25}$$

Proof. The statements follow from the precise formulas of Proposition 2.73 and the symbolic notation in (7.1.11). We also use the convention made in Remark 7.4 according to which we write $r\check{\rho}, r\check{\mu} \in \Gamma_g$, $(r\underline{\beta}, \underline{\alpha}) \in \Gamma_b$ and $e_3(\Gamma_g) = r^{-1}(\mathfrak{d}\Gamma_b)$. □

7.1.5 Equations involving $\mathfrak{q}$

Recall that our main quantity $\mathfrak{q}$ has been introduced in Definition 2.98 with respect to the global frame of Proposition 3.26 (see Remark 2.110). The passage from the geodesic frame (e_3, e_θ, e_4) of ${}^{(ext)}\mathcal{M}$ to the global frame (e_3', e_θ', e_4') is given by

$$e_4' = \Upsilon\left(e_4 + fe_\theta + \frac{1}{4}f^2 e_3\right), \qquad e_\theta' = e_\theta + \frac{1}{2}fe_3, \qquad e_3' = \Upsilon^{-1}e_3, \quad (7.1.26)$$

with a reduced scalar f which was constructed in Proposition 3.20. We recall below the main relevant statements of Proposition 3.20 in connection to the construction of the global frame.

Proposition 7.9. *Under assumptions **Ref 1–2** on ${}^{(ext)}\mathcal{M}$ there exists a frame transformation of the form (7.1.26) verifying the following properties:*[6]

1. *Everywhere in ${}^{(ext)}\mathcal{M}$ we have $\xi' = 0$.*
2. *The transition function f verifies, relative to the background frame (e_3, e_θ, e_4), the estimates*[7]

$$\begin{aligned} |\mathfrak{d}^k f| &\lesssim \frac{\epsilon}{ru^{\frac{1}{2}+\delta_{dec}-2\delta_0} + u^{1+\delta_{dec}-2\delta_0}}, \quad \textit{for } k \leq k_{small} + 20 \textit{ on } {}^{(ext)}\mathcal{M}, \\ |\mathfrak{d}^{k-1} e_3' f| &\lesssim \frac{\epsilon}{ru^{1+\delta_{dec}-2\delta_0}} \quad \textit{for } k \leq k_{small} + 20 \textit{ on } {}^{(ext)}\mathcal{M}. \end{aligned} \quad (7.1.27)$$

3. *The primed Ricci coefficients and curvature components verify*[8]

$$\begin{aligned} \max_{0\leq k\leq k_{small}+k_{loss}} \sup_{{}^{(ext)}\mathcal{M}} \Bigg\{ & \left(r^2 u^{\frac{1}{2}+\delta_{dec}-2\delta_0} + ru^{1+\delta_{dec}-2\delta_0}\right)|\mathfrak{d}^k \Gamma_g'| \\ & + ru^{1+\delta_{dec}-2\delta_0}|\mathfrak{d}^k \Gamma_b'| \\ & + r^2 u^{1+\delta_{dec}-2\delta_0} \left|\mathfrak{d}^{k-1} e_3'\left(\kappa' - \frac{2\Upsilon}{r}, \underline{\kappa}' + \frac{2}{r}, \vartheta', \zeta', \underline{\eta}', \eta'\right)\right| \\ & + \left(r^3(u+2r)^{\frac{1}{2}+\delta_{dec}-2\delta_0} + r^2(u+2r)^{1+\delta_{dec}-2\delta_0}\right)|\mathfrak{d}^k(\alpha', \beta')| \\ & + \left(r^3(2r+u)^{1+\delta_{dec}} + r^4(2r+u)^{\frac{1}{2}+\delta_{dec}-2\delta_0}\right)|\mathfrak{d}^{k-1}e_3'(\alpha')| \\ & + \left(r^3 u^{1+\delta_{dec}} + r^4 u^{\frac{1}{2}+\delta_{dec}-2\delta_0}\right)|\mathfrak{d}^{k-1}e_3'(\beta')| \\ & + \left(r^3 u^{\frac{1}{2}+\delta_{dec}-2\delta_0} + r^2 r u^{1+\delta_{dec}-2\delta_0}\right)|\mathfrak{d}^k \check{\rho}'| \\ & + u^{1+\delta_{dec}-2\delta_0}\left(r^2|\mathfrak{d}^k \underline{\beta}'| + r|\mathfrak{d}^k \underline{\alpha}'|\right)\Bigg\} \lesssim \epsilon. \end{aligned}$$

We have the following analog of Proposition 2.99.

---

[6] We denote by primes the Ricci and curvature components w.r.t. the primed frame.

[7] In fact, the estimates hold for $k_{small} + k_{loss}$, see Proposition 3.20, and we choose here $k_{loss} = 20$.

[8] Note that u and r here are the outgoing optical function and area radius of the foliation of ${}^{(ext)}\mathcal{M}$.

Proposition 7.10. *We have, relative to the background frame of* ${}^{(ext)}\mathcal{M}$,

$$r^4\left(\not{d}_2^\star \not{d}_1^\star \rho + \frac{3}{4}\underline{\kappa}\rho\vartheta + \frac{3}{4}\kappa\rho\underline{\vartheta}\right) = \mathfrak{q} + Err \tag{7.1.28}$$

with error term expressed schematically in the form

$$Err = r^2 \not{\mathfrak{d}}^{\leq 2}(\Gamma_b \cdot \Gamma_g). \tag{7.1.29}$$

Proof. We make use of Proposition 2.99. Recall (see Remark 2.110) that the quantity $\mathfrak{q}$ we are working with is defined relative to the global frame of Proposition 3.26. We thus write[9]

$$\begin{aligned}\mathfrak{q} &= r^4\left((\not{d}_2^\star)'(\not{d}_1^\star)'\rho' + \frac{3}{4}\underline{\kappa}'\rho'\vartheta' + \frac{3}{4}\kappa'\rho'\underline{\vartheta}'\right) + \mathrm{Err}',\\ \mathrm{Err}' &= r^4 e_3'\eta'\cdot\beta' + r^2\mathfrak{d}^{\leq 1}(\Gamma_b\cdot\Gamma_g),\end{aligned}$$

where the primes refer to the global frame in which $\mathfrak{q}$ was defined. Since in that frame $e_3'\eta' \in r^{-1}\mathfrak{d}\Gamma_b$ and $\beta' \in r^{-1}\Gamma_g$ we can simplify and write

$$\mathrm{Err}' = r^2\mathfrak{d}^{\leq 1}(\Gamma_b\cdot\Gamma_g).$$

We also have in view of Proposition 2.90

$$\begin{aligned}\rho' &= \rho + f\underline{\beta} + O(f^2\underline{\alpha}),\\ \underline{\beta}' &= \underline{\beta} + \frac{1}{2}f\underline{\alpha},\\ \underline{\alpha}' &= \underline{\alpha},\\ \underline{\kappa}' &= \underline{\kappa} + f\underline{\xi},\\ \kappa' &= \kappa + \not{d}_1'(f) + f(\zeta+\eta) + O(r^{-1}f^2),\\ \vartheta' &= \vartheta - \not{d}_2^\star{}'(f) + f(\zeta+\eta) + O(r^{-1}f^2),\\ \underline{\vartheta}' &= \underline{\vartheta} + f\underline{\xi}.\end{aligned}$$

Note that

$$(\not{d}_1^\star)'\rho = -e_\theta'(\rho) = -e_\theta\rho - \frac{1}{2}fe_3\rho = \not{d}_1^\star\rho - \frac{1}{2}fe_3\rho.$$

We deduce

$$\begin{aligned}(\not{d}_2^\star)'(\not{d}_1^\star)'\rho' &= (\not{d}_2^\star)'(\not{d}_1^\star)'\rho + (\not{d}_2^\star)'(\not{d}_1^\star)'(\Gamma_b\cdot\Gamma_g) + \text{l.o.t.}\\ &= (\not{d}_2^\star)'\left(\not{d}_1^\star - \frac{1}{2}fe_3\right)\rho + r^{-2}\not{\mathfrak{d}}^{\leq 2}(\Gamma_b\cdot\Gamma_g)\\ &= \not{d}_2^\star\left(\not{d}_1^\star - \frac{1}{2}fe_3\right)\rho + r^{-2}\not{\mathfrak{d}}^{\leq 2}(\Gamma_b\cdot\Gamma_g)\\ &= \not{d}_2^\star\not{d}_1^\star\rho - \frac{1}{2}\not{d}_2^\star fe_3\rho + r^{-2}\not{\mathfrak{d}}^{\leq 2}(\Gamma_b\cdot\Gamma_g).\end{aligned}$$

[9]The values of r and r' differ only by lower order terms which do not affect the result.

Similarly,

$$\begin{aligned}\underline{\kappa}'\rho'\vartheta' &= \rho\underline{\kappa}(\vartheta - \not{d}_2^\star f) + r^{-3}\not{\mathfrak{d}}^{\leq 1}(\Gamma_g\cdot\Gamma_g),\\ \kappa'\rho'\underline{\vartheta}' &= \kappa\rho\underline{\vartheta} + r^{-3}\not{\mathfrak{d}}^{\leq 1}(\Gamma_b\cdot\Gamma_g).\end{aligned}$$

We deduce

$$\begin{aligned}(\not{d}_2^\star)'(\not{d}_1^\star)'\rho' + \frac{3}{4}\underline{\kappa}'\rho'\vartheta' + \frac{3}{4}\kappa'\rho'\underline{\vartheta}' &= \not{d}_2^\star\not{d}_1^\star\rho + \frac{3}{4}\underline{\kappa}\rho\vartheta + \frac{3}{4}\kappa\rho\underline{\vartheta}\\ &- \frac{1}{2}\not{d}_2^\star f\left(e_3\rho + \frac{3}{2}\underline{\kappa}\rho\right) + r^{-2}\not{\mathfrak{d}}^{\leq 2}(\Gamma_b\cdot\Gamma_g).\end{aligned}$$

Note that

$$\not{d}_2^\star f\left(e_3\rho + \frac{3}{2}\underline{\kappa}\rho\right) = \not{d}_2^\star f\left(\not{d}_1\underline{\beta} - \frac{1}{2}\vartheta\underline{\alpha} + \text{l.o.t.}\right) = r^{-2}\not{\mathfrak{d}}^{\leq 2}(\Gamma_g\cdot\Gamma_b).$$

Hence

$$(\not{d}_2^\star)'(\not{d}_1^\star)'\rho' + \frac{3}{4}\underline{\kappa}'\rho'\vartheta' + \frac{3}{4}\kappa'\rho'\underline{\vartheta}' = \not{d}_2^\star\not{d}_1^\star\rho + \frac{3}{4}\underline{\kappa}\rho\vartheta + \frac{3}{4}\kappa\rho\underline{\vartheta} + r^{-2}\not{\mathfrak{d}}^{\leq 2}(\Gamma_b\cdot\Gamma_g).$$

This concludes the proof of Proposition 7.10. □

We shall also need the following analogue of Proposition 2.100.

Proposition 7.11. *The following identity holds true in* $^{(ext)}\mathcal{M}$*, with respect to its background frame*

$$\begin{aligned}e_3(r\mathfrak{q}) = r^5\Bigg\{&\not{d}_2^\star\not{d}_1^\star\not{d}_1\underline{\beta} - \frac{3}{2}\rho\not{d}_2^\star\not{d}_1^\star\underline{\kappa} - \frac{3}{2}\underline{\kappa}\rho\not{d}_2^\star\zeta - \frac{3}{2}\kappa\rho\underline{\alpha} + \frac{3}{4}(2\rho^2 - \kappa\underline{\kappa}\rho)\underline{\vartheta}\Bigg\} \qquad (7.1.30)\\ &+ Err[e_3(r\mathfrak{q})],\end{aligned}$$

where

$$Err[e_3(r\mathfrak{q})] = r^3\mathfrak{d}^{\leq 3}(\Gamma_b\cdot\Gamma_g). \qquad (7.1.31)$$

Proof. We start with the result of Proposition 2.100 which we write in the form

$$\begin{aligned}(r')^{-5}e_3'(r'\mathfrak{q}) &= (\not{d}_2^\star\not{d}_1^\star\not{d}_1)'\underline{\beta}' - \frac{3}{2}\kappa'\rho'\underline{\alpha}' - \frac{3}{2}\rho'(\not{d}_2^\star\not{d}_1^\star)'\underline{\kappa}' - \frac{3}{2}\underline{\kappa}'\rho('\not{d}_2^\star)'\zeta'\\ &+ \frac{3}{4}(2(\rho')^2 - \kappa'\underline{\kappa}'\rho')\underline{\vartheta}' + (r')^{-5}\mathrm{Err}[e_3'(r'\mathfrak{q})]\\ \mathrm{Err}'[e_3'(r'\mathfrak{q})] &= r'\Gamma_b\mathfrak{q} + r^5\mathfrak{d}'^{\leq 1}(e_3'\eta'\cdot\beta') + r'^3\mathfrak{d}^{\leq 2}(\Gamma_b\cdot\Gamma_g).\end{aligned}$$

Since $e_3'\eta' \in r^{-1}\Gamma_b$ and $\mathfrak{q} \in \Gamma_b$, we deduce

$$\mathrm{Err}'[e_3'(r'\mathfrak{q})] = r^3\mathfrak{d}^{\leq 2}(\Gamma_b\cdot\Gamma_g).$$

Now, in view of Proposition 2.90,

$$(\not{d}_2^\star\not{d}_1^\star\not{d}_1)'\underline{\beta}' = (\not{d}_2^\star\not{d}_1^\star\not{d}_1)'\left(\underline{\beta} + \frac{1}{2}f\underline{\alpha}\right) = (\not{d}_2^\star\not{d}_1^\star\not{d}_1)'\underline{\beta} + r^{-2}\not{\mathfrak{d}}^3(\Gamma_b\cdot\Gamma_g).$$

Proceeding in the same manner with all other terms we find

$$
\begin{aligned}
&(\,\not{d}_2^\star \not{d}_1^\star \not{d}_1)'\underline{\beta}' - \frac{3}{2}\kappa'\rho'\underline{\alpha}' - \frac{3}{2}\rho'(\,\not{d}_2^\star \not{d}_1^\star)'\underline{\kappa}' - \frac{3}{2}\underline{\kappa}'\rho('\,\not{d}_2^\star)'\zeta' + \frac{3}{4}(2(\rho')^2 - \kappa'\underline{\kappa}'\rho')\underline{\vartheta}' \\
&= \not{d}_2^\star \not{d}_1^\star \not{d}_1\underline{\beta} - \frac{3}{2}\kappa\rho\underline{\alpha} - \frac{3}{2}\rho\,\not{d}_2^\star \not{d}_1^\star\underline{\kappa} - \frac{3}{2}\underline{\kappa}\rho\,\not{d}_2^\star\zeta + \frac{3}{4}(2\rho^2 - \kappa\underline{\kappa}\rho)\underline{\vartheta} \\
&+r^{-2}\,\not{\mathfrak{d}}^{\leq 3}(\Gamma_b\cdot\Gamma_g)
\end{aligned}
$$

from which the result easily follows. □

7.1.6 Additional equations

The following proposition is an immediate corollary of Proposition 2.74.

Proposition 7.12. *We have, schematically,*

$$
\begin{aligned}
2\,\not{d}_1^\star\underline{\omega} &= \left(\frac{1}{2}\underline{\kappa} + 2\underline{\omega}\right)\eta + e_3(\zeta) - \underline{\beta} - \frac{1}{2}\kappa\underline{\xi} + r^{-1}\Gamma_g + \Gamma_b\cdot\Gamma_b, \\
2\,\not{d}_2\,\not{d}_2^\star\eta &= \kappa\left(-e_3(\zeta) + \underline{\beta}\right) - e_3(e_\theta(\kappa)) + r^{-2}\,\not{\mathfrak{d}}^{\leq 1}\Gamma_g + r^{-1}\,\not{\mathfrak{d}}^{\leq 1}(\Gamma_b\cdot\Gamma_b), \\
2\,\not{d}_2\,\not{d}_2^\star\underline{\xi} &= \underline{\kappa}\left(e_3(\zeta) - \underline{\beta}\right) - e_3(e_\theta(\underline{\kappa})) + r^{-2}\,\not{\mathfrak{d}}^{\leq 1}\Gamma_g + r^{-1}\,\not{\mathfrak{d}}^{\leq 1}(\Gamma_b\cdot\Gamma_b).
\end{aligned}
$$

Remark 7.13. *Note that in fact* $\Gamma_g = \{\check{\kappa}, \vartheta, \zeta, \check{\underline{\kappa}}, r\check{\rho}\}$ *and* $\Gamma_b = \{\underline{\vartheta}, \eta, \underline{\xi}, \check{\underline{\omega}}, r\underline{\beta}, \underline{\alpha}\}$ *in the derivation of this proposition. It is important to note also that the terms denoted schematically by* $\not{\mathfrak{d}}(\Gamma_b\cdot\Gamma_b)$ *do not contain derivatives of* $\check{\underline{\omega}}$.

The following corollary of Proposition 7.12 will be very useful later on.

Proposition 7.14. *The following identities hold true on* Σ_*.

$$
\begin{aligned}
2\,\not{d}_2^\star \not{d}_1^\star \not{d}_1 \not{d}_2 \not{d}_2^\star\eta = \kappa\Big(e_3(\,\not{d}_2^\star \not{d}_1^\star\mu) + 2\,\not{d}_2^\star \not{d}_1^\star \not{d}_1\underline{\beta}\Big) - \not{d}_2^\star \not{d}_1^\star \not{d}_1 e_3(e_\theta(\kappa)) \\
+\, r^{-5}\,\not{\mathfrak{d}}^{\leq 4}\Gamma_g + r^{-4}\,\not{\mathfrak{d}}^{\leq 4}(\Gamma_b\cdot\Gamma_b) + l.o.t.
\end{aligned}
\tag{7.1.32}
$$

$$
\begin{aligned}
2\,\not{d}_2^\star \not{d}_1^\star \not{d}_1 \not{d}_2 \not{d}_2^\star\underline{\xi} = e_3\Big((\,\not{d}_2^\star \not{d}_2 + 2K)\,\not{d}_2^\star \not{d}_1^\star\underline{\kappa}\Big) - \underline{\kappa}\Big(e_3(\,\not{d}_2^\star \not{d}_1^\star\mu) + 2\,\not{d}_2^\star \not{d}_1^\star \not{d}_1\underline{\beta}\Big) \\
+\, r^{-5}\,\not{\mathfrak{d}}^{\leq 4}\Gamma_g + r^{-4}\,\not{\mathfrak{d}}^{\leq 4}(\Gamma_b\cdot\Gamma_b) + l.o.t.
\end{aligned}
\tag{7.1.33}
$$

Remark 7.15. *Here, as in Remark 7.13, we have* $\Gamma_g = \{\check{\kappa}, \vartheta, \zeta, \check{\underline{\kappa}}, r\check{\rho}\}$ *and* $\Gamma_b = \{\underline{\vartheta}, \eta, \underline{\xi}, \check{\underline{\omega}}, r\underline{\beta}, \underline{\alpha}\}$. *The quadratic terms denoted l.o.t. are lower order both in terms of decay in* r, u *as well in terms of number of derivatives. They also contain only angular derivatives* $\not{\mathfrak{d}}$ *and not* e_3 *nor* e_4.

Proof. We make use of Proposition 7.12 . We shall also make use of the conventions mentioned in Remark 7.4, i.e., $\check{\rho}, \check{\mu} \in r^{-1}\Gamma_g$, $\underline{\beta} \in r^{-1}\Gamma_b$, $\underline{\alpha} \in \Gamma_b$.

We start with

$$
2\,\not{d}_2\,\not{d}_2^\star\eta = \kappa\left(-e_3(\zeta) + \underline{\beta}\right) - e_3(e_\theta(\kappa)) + r^{-2}\,\not{\mathfrak{d}}^{\leq 1}\Gamma_g + r^{-1}\,\not{\mathfrak{d}}(\Gamma_b\cdot\Gamma_b).
$$

We apply $\not{d}_1^\star \not{d}_1$ to derive

$$\begin{aligned} 2 \not{d}_1^\star \not{d}_1 \not{d}_2 \not{d}_2^\star \eta &= \kappa \left(- \not{d}_1^\star \not{d}_1 e_3(\zeta) + \not{d}_1^\star \not{d}_1 \underline{\beta} \right) - \not{d}_1^\star \not{d}_1 e_3(e_\theta(\kappa)) + r^{-4} \not{\mathfrak{d}}^{\leq 3} \Gamma_g \\ & \quad + r^{-3} \not{\mathfrak{d}}^3 (\Gamma_b \cdot \Gamma_b) \\ &= \kappa \left(-e_3(\not{d}_1^\star \not{d}_1(\zeta) + \not{d}_1^\star \not{d}_1 \underline{\beta} \right) - \not{d}_1^\star \not{d}_1 e_3(e_\theta(\kappa)) \\ &- \kappa [\not{d}_1^\star \not{d}_1, e_3] \zeta + r^{-4} \not{\mathfrak{d}}^{\leq 3} \Gamma_g + r^{-3} \not{\mathfrak{d}}^3 (\Gamma_b \cdot \Gamma_b). \end{aligned}$$

Making use of the commutation formula, see Lemma 7.5, and the null structure equations for $e_3 \zeta, e_4 \zeta$,

$$[\not{d}_1, e_3] \zeta = - \eta e_3 \zeta + r^{-2} \not{\mathfrak{d}} \zeta + \Gamma_b e_4 \zeta + \text{l.o.t.} = r^{-1} \Gamma_b \cdot \Gamma_b + r^{-2} \not{\mathfrak{d}} \Gamma_g + \text{l.o.t.},$$

we deduce, schematically,

$$\begin{aligned} [\not{d}_1^\star \not{d}_1, e_3] \zeta &= \not{d}_1^\star [\not{d}_1, e_3] \zeta + [\not{d}_1^\star, e_3] \not{d}_1 \zeta \\ &= r^{-1} \not{\mathfrak{d}} \left(r^{-1} \Gamma_b \cdot \Gamma_b + r^{-2} \not{\mathfrak{d}} \zeta + \text{l.o.t.} \right) + \Gamma_b e_3 \not{d}_1 \zeta + r^{-2} \not{d}_1 \zeta + \text{l.o.t.} \\ &= r^{-2} \not{\mathfrak{d}} (\Gamma_b \cdot \Gamma_b) + r^{-3} \not{\mathfrak{d}}^2 \zeta + \Gamma_b \left(\not{d}_1 e_3 \zeta + \Gamma_b e_3 \zeta + r^{-2} \not{\mathfrak{d}} \zeta \right) + \text{l.o.t.} \\ &= r^{-2} \not{\mathfrak{d}} (\Gamma_b \cdot \Gamma_b) + r^{-2} \Gamma_b \not{\mathfrak{d}} (\mathfrak{d} \Gamma_b) + r^{-1} \Gamma_b \cdot \Gamma_b \cdot \Gamma_b + r^{-4} \not{\mathfrak{d}}^2 \Gamma_g \\ &= r^{-2} \not{\mathfrak{d}} (\Gamma_b \mathfrak{d}^{\leq 1} \Gamma_b) + r^{-4} \not{\mathfrak{d}}^2 \Gamma_g + \text{l.o.t.} \end{aligned}$$

Hence,

$$\begin{aligned} 2 \not{d}_1^\star \not{d}_1 \not{d}_2 \not{d}_2^\star \eta = \kappa \Big(- e_3(\not{d}_1^\star \not{d}_1 \zeta) + \not{d}_1^\star \not{d}_1 \underline{\beta} \Big) - \not{d}_1^\star \not{d}_1 e_3(e_\theta(\kappa)) \\ + r^{-4} \not{\mathfrak{d}}^{\leq 3} \Gamma_g + r^{-3} \not{\mathfrak{d}}^2 (\Gamma_b \cdot \not{\mathfrak{d}} \Gamma_b). \end{aligned} \tag{7.1.34}$$

Since $\mu = - \not{d}_1 \zeta - \rho + \frac{1}{4} \vartheta \underline{\vartheta}$, we deduce

$$\begin{aligned} \not{d}_1^\star \mu &= - \not{d}_1^\star \not{d}_1 \zeta - \not{d}_1^\star \rho + \frac{1}{4} \not{d}_1^\star (\vartheta \underline{\vartheta}), \\ e_3 \not{d}_1^\star \mu &= -e_3(\not{d}_1^\star \not{d}_1 \zeta) - e_3 \not{d}_1^\star \rho + \frac{1}{4} e_3 \not{d}_1^\star (\vartheta \underline{\vartheta}) \\ &= -e_3(\not{d}_1^\star \not{d}_1 \zeta) - \not{d}_1^\star e_3 \rho - [\not{d}_1^\star, e_3] \rho + \frac{1}{4} \not{d}_1^\star e_3 (\vartheta \underline{\vartheta}) + \frac{1}{4} [e_3, \not{d}_1^\star] (\vartheta \cdot \underline{\vartheta}). \end{aligned}$$

Making use of the equations for $e_3 \rho = \not{d}_1 \underline{\beta} - \frac{3}{2} \underline{\kappa} \rho + \Gamma_g \cdot \Gamma_b$ and also the equations for[10] $e_4 \rho, e_3 \vartheta, e_3 \underline{\vartheta}, e_4 \vartheta, e_4 \underline{\vartheta}$ (and writing $\not{d}_1 \beta = r^{-1} \not{\mathfrak{d}} \beta = r^{-2} \not{\mathfrak{d}} \Gamma_b$)

$$\begin{aligned} [e_3, \not{d}_1^\star] \rho &= \Gamma_b e_3 \rho + \Gamma_b e_4 \zeta + r^{-2} \not{\mathfrak{d}} \rho = r^{-2} \Gamma_b \not{\mathfrak{d}} \Gamma_b + r^{-3} \not{\mathfrak{d}} \Gamma_g + \text{l.o.t.}, \\ [e_3, \not{d}_1^\star] (\vartheta \cdot \underline{\vartheta}) &= \Gamma_b e_3 (\vartheta \cdot \underline{\vartheta}) + \Gamma_b e_4 (\vartheta \cdot \underline{\vartheta}) + r^{-2} \not{\mathfrak{d}} (\vartheta \cdot \underline{\vartheta}) = r^{-2} \not{\mathfrak{d}} \big(\Gamma_b \cdot \Gamma_g \big) + \text{l.o.t.} \end{aligned}$$

[10] This is to avoid the presence of e_3, e_4 derivatives in the error terms.

We deduce, ignoring the lower order terms,

$$\begin{aligned} e_3 \not{d}_1^\star \mu &= -e_3(\not{d}_1^\star \not{d}_1 \zeta) - \not{d}_1^\star \left(\not{d}_1 \underline{\beta} - \frac{3}{2}\underline{\kappa}\rho + \Gamma_g \cdot \Gamma_b \right) + r^{-2}\Gamma_b \not{\mathfrak{d}}\Gamma_b + r^{-2} \not{\mathfrak{d}}(\Gamma_b\Gamma_g) \\ &\quad + r^{-3} \not{\mathfrak{d}}\Gamma_g \\ &= -e_3(\not{d}_1^\star \not{d}_1 \zeta) - \not{d}_1^\star \not{d}_1 \underline{\beta} + \frac{3}{2}\underline{\kappa} \not{d}_1^\star \rho + r^{-3} \not{\mathfrak{d}}\Gamma_g + r^{-2} \not{\mathfrak{d}}^{\leq 1}(\Gamma_b \cdot \Gamma_b). \end{aligned}$$

Hence,

$$e_3(\not{d}_1^\star \not{d}_1 \zeta) = -e_3(\not{d}_1^\star \mu) - \not{d}_1^\star \not{d}_1 \underline{\beta} + r^{-3} \not{\mathfrak{d}}\Gamma_g + r^{-2} \not{\mathfrak{d}}^{\leq 2}(\Gamma_b \cdot \Gamma_b) + \text{l.o.t.} \tag{7.1.35}$$

and thus, back to (7.1.34),

$$\begin{aligned} 2 \not{d}_1^\star \not{d}_1 \not{d}_2 \not{d}_2^\star \eta = \kappa\Big(e_3(\not{d}_1^\star \mu) + 2 \not{d}_1^\star \not{d}_1 \underline{\beta} \Big) - \not{d}_1^\star \not{d}_1 e_3\big(e_\theta(\kappa)\big) \\ + r^{-4} \not{\mathfrak{d}}^{\leq 3}\Gamma_g + r^{-3} \not{\mathfrak{d}}^{\leq 3}(\Gamma_b \cdot \Gamma_b) + \text{l.o.t.} \end{aligned} \tag{7.1.36}$$

Applying $\not{d}_2^\star$ and commuting once more with e_3, i.e.,

$$\begin{aligned} 2 \not{d}_2^\star \not{d}_1^\star \not{d}_1 \not{d}_2 \not{d}_2^\star \eta = \kappa\Big(e_3(\not{d}_2^\star \not{d}_1^\star \mu) + 2 \not{d}_2^\star \not{d}_1^\star \not{d}_1 \underline{\beta} \Big) - \not{d}_2^\star \not{d}_1^\star \not{d}_1 e_3\big(e_\theta(\kappa)\big) \\ + \kappa[\not{d}_2^\star, e_3] \not{d}_1^\star \mu + r^{-1} \not{\mathfrak{d}}\Gamma_g \cdot \Big(e_3(\not{d}_1^\star \mu) + 2 \not{d}_1^\star \not{d}_1 \underline{\beta} \Big) \\ + r^{-5} \not{\mathfrak{d}}^{\leq 4}\Gamma_g + r^{-4} \not{\mathfrak{d}}^{\leq 4}(\Gamma_b \cdot \Gamma_b). \end{aligned} \tag{7.1.37}$$

Note that, in view of (7.1.36), we can write

$$\begin{aligned} e_3(\not{d}_1^\star \mu) = 2\kappa^{-1} \not{d}_2^\star \not{d}_1^\star \not{d}_1 e_3\big(e_\theta(\kappa)\big) - 2 \not{d}_1^\star \not{d}_1 \underline{\beta} + 2\kappa^{-1} \not{d}_1^\star \not{d}_1 \not{d}_2 \not{d}_2^\star \eta \\ = r^{-3} \not{\mathfrak{d}}^{\leq 4}\Gamma_b + \text{l.o.t.} \end{aligned} \tag{7.1.38}$$

Hence,

$$r^{-1} \not{\mathfrak{d}}\Gamma_g \cdot \Big(e_3(\not{d}_1^\star \mu) + 2 \not{d}_1^\star \not{d}_1 \underline{\beta} \Big) = r^{-4} \not{\mathfrak{d}}\Gamma_g \cdot \not{\mathfrak{d}}^{\leq 4}\Gamma_b.$$

Similarly,

$$\begin{aligned} [\not{d}_2^\star, e_3] \not{d}_1^\star \mu &= \Gamma_b \cdot e_3 \not{d}_1^\star \mu + \Gamma_b e_4 \not{d}_1^\star \mu + r^{-3} \not{\mathfrak{d}}^2 \mu + \text{l.o.t.} \\ &= r^{-3}\Gamma_b \cdot \not{\mathfrak{d}}^{\leq 4}\Gamma_b + \Gamma_b\big(\not{d}_1^\star e_4 \mu + [e_4, \not{d}_1^\star]\mu \big) + r^{-4} \not{\mathfrak{d}}^2\Gamma_g + \text{l.o.t.} \end{aligned}$$

Thus, making use of the equation for $e_4\mu$ and combining with the estimate above,

$$\kappa[\not{d}_2^\star, e_3] \not{d}_1^\star \mu + r^{-1} \not{\mathfrak{d}}\Gamma_g \cdot \Big(e_3(\not{d}_1^\star \mu) + 2 \not{d}_1^\star \not{d}_1 \underline{\beta} \Big) = r^{-4}\Gamma_b \cdot \not{\mathfrak{d}}^{\leq 4}\Gamma_b + r^{-5} \not{\mathfrak{d}}^{\leq 2}\Gamma_g.$$

Back to (7.1.37) we deduce

$$\begin{aligned} 2 \not{d}_2^\star \not{d}_1^\star \not{d}_1 \not{d}_2 \not{d}_2^\star \eta = \kappa\Big(e_3(\not{d}_2^\star \not{d}_1^\star \mu) + 2 \not{d}_2^\star \not{d}_1^\star \not{d}_1 \underline{\beta} \Big) - \not{d}_2^\star \not{d}_1^\star \not{d}_1 e_3\big(e_\theta(\kappa)\big) \\ + r^{-5} \not{\mathfrak{d}}^{\leq 4}\Gamma_g + r^{-4} \not{\mathfrak{d}}^{\leq 4}(\Gamma_b \cdot \Gamma_b) \end{aligned}$$

as desired.

To prove the second part we start with the formula for $\not{d}_2 \not{d}_2^\star \underline{\xi}$ in Corollary 7.12

$$2\not{d}_2\not{d}_2^\star\underline{\xi} = \underline{\kappa}\left(e_3(\zeta)-\underline{\beta}\right) - e_3(e_\theta(\underline{\kappa})) + r^{-2}\not{\mathfrak{d}}^{\leq 1}\Gamma_g + r^{-1}\not{\mathfrak{d}}(\Gamma_b\cdot\Gamma_b).$$

Applying $\not{d}_1^\star \not{d}_1$ and proceeding exactly as before in the derivation of (7.1.34) we derive

$$\begin{aligned}2\not{d}_1^\star\not{d}_1\not{d}_2\not{d}_2^\star\underline{\xi} &= -e_3(\not{d}_1^\star\not{d}_1 e_\theta(\underline{\kappa})) + \underline{\kappa}\left(e_3(\not{d}_1^\star\not{d}_1\zeta) - \not{d}_1^\star\not{d}_1\underline{\beta}\right)\\ &+ r^{-4}\not{\mathfrak{d}}^{\leq 3}\Gamma_g + r^{-3}\not{\mathfrak{d}}^2(\Gamma_b\cdot\mathfrak{d}\Gamma_b).\end{aligned} \tag{7.1.39}$$

Making use of (7.1.35) we deduce, as in (7.1.36),

$$\begin{aligned}2\not{d}_1^\star\not{d}_1\not{d}_2\not{d}_2^\star\underline{\xi} &= -e_3(\not{d}_1^\star\not{d}_1 e_\theta(\underline{\kappa})) + \underline{\kappa}\Big(-e_3(\not{d}_1^\star\mu) - 2\not{d}_1^\star\not{d}_1\underline{\beta}\Big)\\ &+ r^{-4}\not{\mathfrak{d}}^{3}\Gamma_g + r^{-3}\not{\mathfrak{d}}^{\leq 2}(\Gamma_b\cdot\mathfrak{d}\Gamma_b) + \text{l.o.t.}\end{aligned} \tag{7.1.40}$$

Applying $\not{d}_2^\star$ and proceeding as in the derivation of (7.1.37), by making use of (7.1.39) and (7.1.38) we obtain

$$\begin{aligned}2\not{d}_2^\star\not{d}_1^\star\not{d}_1\not{d}_2\not{d}_2^\star\underline{\xi} &= -e_3(\not{d}_2^\star\not{d}_1^\star\not{d}_1 e_\theta(\underline{\kappa})) - \underline{\kappa}\Big(e_3(\not{d}_2^\star\not{d}_1^\star\mu) + 2\not{d}_2^\star\not{d}_1^\star\not{d}_1\underline{\beta}\Big)\\ &+ r^{-5}\not{\mathfrak{d}}^{\leq 4}\Gamma_g + r^{-4}\not{\mathfrak{d}}^{\leq 4}(\Gamma_b\cdot\Gamma_b) + \text{l.o.t.}\end{aligned}$$

The identity $\not{d}_1^\star\not{d}_1 = \not{d}_2\not{d}_2^\star + 2K$ yields, together with the bootstrap assumptions,

$$\begin{aligned}2\not{d}_2^\star\not{d}_1^\star\not{d}_1\not{d}_2\not{d}_2^\star\underline{\xi} &= -e_3((\not{d}_2^\star\not{d}_2 + 2K)\not{d}_2^\star e_\theta(\underline{\kappa})) - \underline{\kappa}\Big(e_3(\not{d}_2^\star\not{d}_1^\star\mu) + 2\not{d}_2^\star\not{d}_1^\star\not{d}_1\underline{\beta}\Big)\\ &+ r^{-5}\not{\mathfrak{d}}^{\leq 4}\Gamma_g + r^{-4}\not{\mathfrak{d}}^{\leq 4}(\Gamma_b\cdot\Gamma_b) + \text{l.o.t.}\\ &= e_3((\not{d}_2^\star\not{d}_2 + 2K)\not{d}_2^\star\not{d}_1^\star(\underline{\kappa})) - \underline{\kappa}\Big(e_3(\not{d}_2^\star\not{d}_1^\star\mu) + 2\not{d}_2^\star\not{d}_1^\star\not{d}_1\underline{\beta}\Big)\\ &+ r^{-5}\not{\mathfrak{d}}^{\leq 4}\Gamma_g + r^{-4}\not{\mathfrak{d}}^{\leq 4}(\Gamma_b\cdot\Gamma_b) + \text{l.o.t.}\end{aligned}$$

as desired. □

7.2 STRUCTURE OF THE PROOF OF THEOREM M4

We rephrase the statement of Theorem M4 as follows.

Theorem 7.16. *Let $\mathcal{M} = {}^{(int)}\mathcal{M} \cup {}^{(ext)}\mathcal{M}$ be a GCM admissible spacetime.*[11] *Under the basic bootstrap assumptions and the results of Theorems M1–M4 (all encoded in **Ref 1–Ref 4**) the following estimates*[12] *hold true, for all $k \leq k_{small}+8$, everywhere on ${}^{(ext)}\mathcal{M}$,*

$$\begin{aligned}\|\Gamma_g\|_{\infty,k} &\lesssim \epsilon_0 \min\left\{r^{-2}u^{-\frac{1}{2}-\delta_{dec}}, r^{-1}u^{-1-\delta_{dec}}\right\},\\ \|e_3\Gamma_g\|_{\infty,k-1} &\lesssim \epsilon_0 r^{-2}u^{-1-\delta_{dec}},\\ \|\Gamma_b\|_{\infty,k} &\lesssim \epsilon_0 r^{-1}u^{-1-\delta_{dec}},\end{aligned} \tag{7.2.1}$$

[11] In particular the conditions (7.1.1)–(7.1.5) hold on the spacelike boundary Σ_*.

[12] See Remark 7.4 for the definition of Γ_g, Γ_b used here.

and,

$$\begin{aligned}
\|\beta\|_{\infty,k} &\lesssim \epsilon_0 \min\left\{r^{-2}(u+2r)^{-1-\delta_{dec}}, r^{-3}(u+2r)^{-\frac{1}{2}-\delta_{dec}}\right\},\\
\|e_3\beta\|_{\infty,k-1} &\lesssim \epsilon_0 r^{-3}(u+2r)^{-1-\delta_{dec}},\\
\|\check{\rho}\|_{\infty,k} &\lesssim \epsilon_0 \min\left\{r^{-2}u^{-1-\delta_{dec}}, r^{-3}u^{-\frac{1}{2}-\delta_{dec}}\right\},\\
\|e_3\check{\rho}\|_{\infty,k} &\lesssim \epsilon_0 r^{-3}u^{-1-\delta_{dec}},\\
\|\check{\mu}\|_{\infty,k} &\lesssim \epsilon_0 r^{-3}u^{-1-\delta_{dec}},\\
\|\underline{\beta}\|_{\infty,k} &\lesssim \epsilon_0 r^{-2}u^{-1-\delta_{dec}}.
\end{aligned} \tag{7.2.2}$$

Moreover, everywhere in $^{(ext)}\mathcal{M}$,

$$\|\underline{\alpha}\|_{\infty,k} \lesssim \epsilon_0 r^{-1}u^{-1-\delta_{dec}}. \tag{7.2.3}$$

Here is a short sketch of the proof of the theorem.

1. *Estimates on* Σ_*. To start with, we only have good[13] estimates for $\mathfrak{q}$, α and $\underline{\alpha}$, according to **Ref 2**. To proceed we make use in an essential way of all the GCM conditions (7.1.3)–(7.1.5) on the spacelike boundary Σ_* to estimate all the Ricci and curvature coefficients along Σ_*. We also take full advantage of the dominance condition $r \geq \epsilon\epsilon_0^{-1}u_*^{1+\delta}$ on Σ_*. The main result is stated in Proposition 7.28. The proof is divided in the following intermediary steps.
   a) In Proposition 7.22, we derive flux type estimates along Σ_* for the quantities $\underline{\beta}$ and Γ_b. These estimates take advantage in an essential way of the improved flux estimates for $\mathfrak{q}$ and $\underline{\alpha}$, see (7.1.15) and (7.1.18). This step also makes use of Proposition 7.11 and the identities of Proposition 7.21 for η, $\underline{\xi}$. Moreover, as a byproduct of the flux estimates, we obtain the desired estimates on Σ_* for $\underline{\beta}$ and Γ_b.
   b) We next estimate the $\ell = 1$ modes of the Ricci and curvature coefficients in Proposition 7.26. Besides the information provided by the estimates for $\mathfrak{q}$, α, $\underline{\alpha}$ and the GCM conditions, an important ingredient in the proof is the vanishing of the $\ell = 1$ mode of $e_\theta(K)$, i.e., $\int e_\theta(K)e^\Phi = 0$. The flux estimates derived in Proposition 7.22 play an essential role in deriving the desired estimate for the $\ell = 1$ mode of β.
   c) We make use of the previous steps to complete the proof for the remaining desired estimates on Σ_* in Proposition 7.28. This step also uses, in addition to the GCM conditions, Proposition 7.10 relating $\mathfrak{q}$ to $\not{d}_2^\star \not{d}_1^\star \rho$, the Codazzi equations and elliptic estimates on 2-surfaces.
2. *First estimates in* $^{(ext)}\mathcal{M}$. We make use of the propagation equations in e_4 and the estimates on Σ_* to derive some of the desired estimates of Theorem 7.16, more precisely the better estimates in powers of r for the Γ_g quantities. Note that these estimates decay only like $u^{-1/2-\delta_{dec}}$ in powers of u.
   a) We first prove the desired estimates for $\check{\kappa}, \check{\mu}$ by simply integrating the corresponding e_4 equations. Note that these estimates are also well behaved in terms of powers of u. This is done in section 7.4.3.

[13] That is, estimates in terms of ϵ_0.

b) We derive spacetime estimates for all the $\ell = 1$ modes in Lemma 7.34. This is done by propagating them from the last slice in the e_4 direction, combined with Codazzi equations and the vanishing of the $\ell = 1$ mode of $e_\theta(K)$.

c) We provide all the optimal estimates in terms of powers[14] of r for the quantities $\vartheta, \zeta, \eta, \check{\underline{\kappa}}, \beta, \check{\rho}$. This is achieved in Proposition 7.33 with the help of the estimates on the last slice, the propagation equation for these quantities and the estimates for the $\ell = 1$ modes derived in the previous step.

3. *Optimal u-decay estimates in* $^{(ext)}\mathcal{M}$. We derive all the remaining estimates of Theorem 7.16 for all but the quantities $\underline{\xi}, \check{\underline{\omega}}, \check{\underline{\Omega}}, \check{\varsigma}$. The main remaining difficulty is to get the top decay in powers of u for $\vartheta, \zeta, \eta, \check{\underline{\kappa}}, \beta, \check{\rho}, \underline{\beta}$. The result is stated in Proposition 7.35. We proceed as follows.

   a) One would like to start with ϑ by using the equation $e_4\vartheta + \kappa\vartheta = -2\alpha$. This unfortunately cannot work by integration[15] starting from the last slice Σ_*. Similar problems occur for ζ, β, $\check{\rho}$. On the other hand the quantities $\check{\underline{\kappa}}$ and $\underline{\vartheta}$ could in principle be propagated using their corresponding e_4 equations from Σ_*, but unfortunately they are strongly coupled with the other quantities for which we don't yet have information. For example, we have,

$$e_4\check{\underline{\kappa}} + \frac{1}{2}\overline{\kappa}\check{\underline{\kappa}} + \frac{1}{2}\check{\kappa}\overline{\underline{\kappa}} = -2\,\mathrm{d}\!\!\!/_1\zeta + 2\check{\rho} + \Gamma_g \cdot \Gamma_b,$$

   and therefore we cannot derive the estimate for $\check{\underline{\kappa}}$, by integration, before estimating $\mathrm{d}\!\!\!/_1\zeta$ and $\check{\rho}$. To circumvent this difficulty we proceed by an indirect method as follows.

   b) We can derive optimal decay information on various mixed quantities. For example making use of the equation

$$e_3\alpha + \left(\frac{1}{2}\underline{\kappa} - 4\underline{\omega}\right)\alpha = -\mathrm{d}\!\!\!/_2^\star\beta - \frac{3}{2}\vartheta\rho + 5\zeta\beta,$$

   we infer the desired decay in u for the quantity $\mathrm{d}\!\!\!/_2^\star\beta - \frac{3}{2}\vartheta\rho$. Other such information can be derived from the Codazzi equations for $\vartheta, \underline{\vartheta}$, the Bianchi identity for $\underline{\beta}$ and the identity (7.1.28) of Lemma 7.10.

   c) We combine the control we have for $\alpha, \check{\kappa}, \check{\mu}$ with the control for the mixed quantities mentioned above with a propagation equation for an intermediary quantity,

$$\Xi := r^2\left(e_\theta(\underline{\kappa}) + 4r\,\mathrm{d}\!\!\!/_1^\star\,\mathrm{d}\!\!\!/_1\zeta - 2r^2\,\mathrm{d}\!\!\!/_1^\star\,\mathrm{d}\!\!\!/_1\beta\right).$$

   We show in the crucial Lemma 7.36 that Ξ is also a good mixed quantity, i.e., it has optimal decay in u. It is important to note that this estimate does not depend linearly on $\underline{\alpha}$ for which we only have information on the last slice and $\mathcal{T}$.

   d) We can combine the control of Ξ with all other available information mentioned above, to derive good estimates, simultaneously, for $\mathrm{d}\!\!\!/_2^\star\,\mathrm{d}\!\!\!/_1^\star\underline{\kappa}$, $\mathrm{d}\!\!\!/_2^\star\zeta$ and

[14] These estimates also provide weak decay in u, i.e., $u^{-\frac{1}{2}-\delta_{dec}}$ decay.

[15] It would work however if instead we would integrate from the interior, but we don't possess information about optimal u-decay in the interior, for example, on the timelike boundary $\mathcal{T}$ of $^{(ext)}\mathcal{M}$.

$\not{d}_2^\star \beta$. This is achieved in a sequence of crucial lemma in section 7.5.2. Unfortunately this step is heavily dependent on the estimate of **Ref 2** for $\underline{\alpha}$ and therefore the estimates we derive are only useful on $\mathcal{T}$.

e) We also show that we have good estimates for $\not{d}_2^\star\Big(\zeta, \not{d}_1^\star \underline{\check{\kappa}}, \beta, \underline{\beta}, \not{d}_1^\star \check{\rho}\Big)$. To estimate $\underline{\check{\kappa}}, \zeta, \beta, \underline{\beta}, \check{\rho}$ from $\not{d}_2^\star\Big(\zeta, \not{d}_1^\star \underline{\check{\kappa}}, \beta, \underline{\beta}, \not{d}_1^\star \check{\rho}\Big)$ we rely on the elliptic Hodge Lemma 7.7 and the control we have for the $\ell = 1$ modes from Lemma 7.34 derived earlier. We obtain estimates for $\eta, \vartheta, \underline{\vartheta}$ as well. This establishes all the estimates of Proposition 7.35 on $\mathcal{T}$.

f) The estimates mentioned above on $\mathcal{T}$ can now be propagated by integrating forward the e_4 null structure and null Bianchi equations. This ends the proof of Proposition 7.35 in ${}^{(ext)}\mathcal{M}$.

4. In Proposition 7.43 we derive improved decay estimates for $e_3(\beta, \vartheta, \zeta, \underline{\check{\kappa}}, \check{\rho})$ and estimates for $\underline{\xi}, \underline{\check{\omega}}, \underline{\check{\Omega}}, \check{\varsigma}$ in terms of $u^{-1-\delta_{dec}}$ decay. The estimates for $\underline{\check{\omega}}$ and $\underline{\xi}$ are propagated from the last slice using their e_4 propagation equations. The estimate for $\underline{\check{\omega}}$ can be easily derived by integrating $e_4(\underline{\check{\omega}}) = \check{\rho} + \Gamma_g \cdot \Gamma_b$ from the last slice Σ_*. The estimate for $\underline{\xi}$ follows by integrating $e_4(\underline{\xi}) = -e_3(\zeta) + \underline{\beta} - \underline{\kappa}\zeta + \Gamma_b \cdot \Gamma_b$ and making use of the previously derived estimates for $e_3\zeta, \beta, \zeta$. The estimates for $\underline{\check{\Omega}}, \check{\varsigma}$ follow easily from the equations (2.2.19).

7.3 DECAY ESTIMATES ON THE LAST SLICE Σ_*

7.3.1 Preliminaries

We shall make use of the following norms on Σ_*.

$$\|\psi\|^*_{\infty,k}(u,r) := \sum_{j\le k} \|\mathfrak{d}_*^j \psi\|_{L^\infty(S(u,r))}, \qquad \mathfrak{d}_*^j = \sum_{j_1+j_2\le j} \not{\mathfrak{d}}^{j_1} (\nu_*)^{j_2}. \tag{7.3.1}$$

Recall that $\nu_* = \nu\Big|_{\Sigma_*} = e_3 + a_* e_4$ is the tangent vector to Σ_* and (see (7.1.8) (7.1.9)), along Σ_*,

$$a_* \;=\; -\frac{2}{\varsigma} + \Upsilon - \frac{r}{2}\underline{A} = -\frac{2}{\varsigma} - \underline{\Omega}. \tag{7.3.2}$$

Since $\varsigma - 1$ and $\underline{\Omega} + \Upsilon$ belong to $r\Gamma_b$ in view of (7.1.11), we deduce

$$a_* + 1 + \frac{2m}{r} \in r\Gamma_b. \tag{7.3.3}$$

As immediate consequence of the commutation Corollary 7.5 we derive the following.

Lemma 7.17. *We have, schematically,*

$$[\not{\mathfrak{d}}, \nu_*]\psi \;=\; r\Gamma_b\,(\nu_*\psi) + \mathfrak{d}^{\le 1}\Gamma_b \cdot \mathfrak{d}\psi. \tag{7.3.4}$$

Proof. Indeed, see Lemma 7.5,

$$
\begin{aligned}
[\not{\mathfrak{d}}, e_4]\psi &= \Gamma_g \mathfrak{d}_{\nearrow} \psi, \\
[\not{\mathfrak{d}}, e_3]\psi &= r\Gamma_b e_3\psi + \Gamma_b \mathfrak{d}_{\nearrow} \psi + \text{l.o.t.}
\end{aligned} \tag{7.3.5}
$$

Hence, since $\not{\mathfrak{d}} a_* \in r\not{\mathfrak{d}}\Gamma_b$,

$$
\begin{aligned}
[\not{\mathfrak{d}}, \nu_*]\psi &= [\not{\mathfrak{d}}, e_3 + a_* e_4]\psi = r\Gamma_b e_3\psi + \Gamma_b \mathfrak{d}_{\nearrow}\psi + a_*\Gamma_g \mathfrak{d}_{\nearrow}\psi + \not{\mathfrak{d}} a_* e_4\psi \\
&= r\Gamma_b\left(\nu_*\psi - a_* e_4\psi\right) + a_*\Gamma_g\mathfrak{d}_{\nearrow}\psi + \not{\mathfrak{d}} a_* e_4\psi \\
&= r\Gamma_b\,\nu_*\psi - a_*\left(\Gamma_b\mathfrak{d}_{\nearrow}\psi + \Gamma_g\mathfrak{d}_{\nearrow}\psi\right) + \not{\mathfrak{d}}\Gamma_b\cdot\mathfrak{d}\psi \\
&= r\Gamma_b\,\nu_*\psi + \mathfrak{d}^{\leq 1}\Gamma_b\cdot\mathfrak{d}\psi
\end{aligned}
$$

as desired. □

To estimate derivatives of the $\ell = 1$ modes on Σ_* we make use of the following.

Lemma 7.18. *For every scalar function h we have the formula*

$$
\nu_*\left(\int_S h\right) = (\varsigma)^{-1}\int_S \varsigma\left(\nu_*(h) + (\underline{\kappa} + a_*\kappa)h\right). \tag{7.3.6}
$$

In particular

$$
\nu_*(r) = \frac{r}{2}(\varsigma)^{-1}\overline{\varsigma(\underline{\kappa} + a_*\kappa)}. \tag{7.3.7}
$$

Proof. We consider the coordinates u, θ along Σ_* with $\nu_*(\theta) = 0$. In these coordinates we have

$$
\nu_* = \frac{2}{\varsigma}\partial_u.
$$

The lemma follows easily by expressing the volume element of the surfaces $S \subset \Sigma_*$ with respect to the coordinates u, θ (see also the proof of Proposition 2.64). □

Lemma 7.19. *Given $\psi \in \mathfrak{s}_1$, we have the formula*

$$
\nu_*\left(\int_S \psi e^{\Phi}\right) = \int_S (\nu_*\psi)e^{\Phi} + \frac{3}{2}\left(\overline{\underline{\kappa}} - 2\overline{\kappa} - \overline{\underline{\Omega}}\,\overline{\kappa}\right)\int_S \psi e^{\Phi} + Err[\psi, \nu_*] \tag{7.3.8}
$$

with error term

$$
Err[\psi, \nu_*] = r^4\Gamma_b\nu_*(\psi) + r^3\Gamma_b\psi.
$$

Proof. We have

$$
\begin{aligned}
\nu_*\left(\int_S \psi e^{\Phi}\right) &= \varsigma^{-1}\int_S \varsigma\left(\nu_*(\psi e^{\Phi}) + (\underline{\kappa} + a_*\kappa)\psi e^{\Phi}\right) \\
&= \varsigma^{-1}\int_{\mathbf{S}} \varsigma\left(\nu_*\psi e^{\Phi} + e^{-\Phi}\nu_*(e^{\Phi}) + \underline{\kappa} + a_*\kappa\right)\psi e^{\Phi}.
\end{aligned}
$$

Recalling that $e_4(\Phi) = \frac{1}{2}(\kappa - \vartheta)$, $e_3(\Phi) = \frac{1}{2}(\underline{\kappa} - \underline{\vartheta})$ we deduce

$$e^{-\Phi}\nu_*(e^\Phi) + \underline{\kappa} + a_*\kappa = \frac{3}{2}(\underline{\kappa} + a_*\kappa) - \frac{1}{2}(\underline{\vartheta} - a_*\vartheta).$$

Hence, writing also $\varsigma a_* = -2 - \varsigma\underline{\Omega}$, $\varsigma = \overline{\varsigma} + \check{\varsigma}$, $\kappa = \overline{\kappa} + \check{\kappa}$, $\underline{\kappa} = \overline{\underline{\kappa}} + \check{\underline{\kappa}}$, and $\underline{\Omega} = \overline{\underline{\Omega}} + \check{\underline{\Omega}}$,

$$\begin{aligned}
\nu_*\left(\int_S \psi e^\Phi\right) &= \varsigma^{-1}\int_S \varsigma\left(\nu_*\psi + \frac{3}{2}(\underline{\kappa} + a_*\kappa)\psi\right)e^\Phi - \frac{1}{2}\varsigma^{-1}\int_S \varsigma(\underline{\vartheta} - a_*\vartheta)\psi e^\Phi \\
&= \varsigma^{-1}\overline{\varsigma}\int_S \left(\nu_*\psi + \frac{3}{2}(\underline{\kappa} - \underline{\Omega}\kappa)\psi\right)e^\Phi \\
&+ \varsigma^{-1}\int_S \check{\varsigma}\left(\nu_*\psi + \frac{3}{2}(\underline{\kappa} - \underline{\Omega}\kappa)\right)e^\Phi - 3\varsigma^{-1}\int_S \kappa\psi e^\Phi \\
&- \frac{1}{2}\varsigma^{-1}\int_S \varsigma(\underline{\vartheta} - a_*\vartheta)\psi e^\Phi \\
&= \int_S \left(\nu_*\psi + \frac{3}{2}(\underline{\kappa} - \underline{\Omega}\kappa)\psi\right)e^\Phi - 3\varsigma^{-1}\int_S \kappa\psi e^\Phi \\
&+ (\varsigma^{-1}\overline{\varsigma} - 1)\int_S \left(\nu_*\psi + \frac{3}{2}(\underline{\kappa} - \underline{\Omega}\kappa)\psi\right)e^\Phi \\
&+ \varsigma^{-1}\int_S \check{\varsigma}\left(\nu_*\psi + \frac{3}{2}(\underline{\kappa} - \underline{\Omega}\kappa)\psi\right)e^\Phi - \frac{1}{2}\varsigma^{-1}\int_S \varsigma(\underline{\vartheta} - a_*\vartheta)\psi e^\Phi \\
&= \int_S (\nu_*\psi)e^\Phi + \frac{3}{2}(\overline{\underline{\kappa}} - 2\overline{\kappa} - \overline{\underline{\Omega}}\,\overline{\kappa})\int_S \psi e^\Phi + \mathrm{Err}[\psi, \nu_*]
\end{aligned}$$

where,

$$\mathrm{Err}[\psi, \nu_*] = r^4\Gamma_b\nu_*(\psi) + r^3\Gamma_b\psi + r^3\Gamma_b\psi + r^3\Gamma_g\psi$$

and the conclusion follows from the fact that Γ_g behaves at least as good as Γ_b. □

Corollary 7.20. *Given $\psi \in \mathfrak{s}_1$ and $k \geq 1$, the following estimate holds true:*

$$\left|\int_S (\nu_*^k\psi)e^\Phi\right| \lesssim \sum_{j=0}^k \left|\nu_*^j\left(\int_S \psi e^\Phi\right)\right| + \left|\mathfrak{d}^{\leq k-1}\left(r^4\Gamma_b\nu_*(\psi) + r^3\Gamma_b\psi\right)\right|. \quad (7.3.9)$$

Proof. We prove (7.3.9) by iteration. First, (7.3.9) holds true for $k = 1$ in view of Lemma 7.19. Also, assuming (7.3.9) for $k \geq 1$, we apply it with ψ replaced by $\nu_*\psi$ which implies

$$\left|\int_S (\nu_*^{k+1}\psi)e^\Phi\right| \lesssim \sum_{j=0}^k \left|\nu_*^j\left(\int_S \nu_*\psi e^\Phi\right)\right| + \left|\mathfrak{d}^{\leq k-1}\left(r^4\Gamma_b\nu_*^2(\psi) + r^3\Gamma_b\psi\right)\right|.$$

Applying Lemma 7.19 with ψ replaced by $\nu_*\psi$ to all terms in the sum of the left-hand side, we infer (7.3.9) with k replaced by $k+1$ which shows that (7.3.9) holds indeed for all k by iteration. □

7.3.2 Differential identities involving GCM conditions on Σ_*

Recall our GCM conditions on Σ_*:

$$\kappa = \frac{2}{r}, \quad {\not{d}}_2^\star {\not{d}}_1^\star \mu = 0, \quad {\not{d}}_2^\star {\not{d}}_1^\star \underline{\kappa} = 0, \quad \int_S \eta e^\Phi = 0, \quad \int_S \underline{\xi} e^\Phi = 0. \tag{7.3.10}$$

Also, on S_*, the last cut of Σ_*,

$$\int_{S_*} \beta e^\Phi = 0, \quad \int_{S_*} e_\theta(\underline{\kappa}) e^\Phi = 0. \tag{7.3.11}$$

The goal of this section is to derive identities involving the GCM conditions which will be used later, see Lemma 7.25.

Proposition 7.21. *The following identities hold true on Σ_*.*

$$\begin{aligned} 2\,{\not{d}}_2^\star {\not{d}}_1^\star {\not{d}}_1 {\not{d}}_2 {\not{d}}_2^\star \eta &= \kappa\Big(\nu_*({\not{d}}_2^\star {\not{d}}_1^\star \mu) + 2\,{\not{d}}_2^\star {\not{d}}_1^\star {\not{d}}_1 \underline{\beta}\Big) - {\not{d}}_2^\star {\not{d}}_1^\star {\not{d}}_1 \nu_*(e_\theta(\kappa)) \\ &\quad + r^{-5}\,\mathfrak{d}^{\leq 4}\Gamma_g + r^{-4}\,\mathfrak{d}^{\leq 4}(\Gamma_b \cdot \Gamma_b) + l.o.t., \end{aligned} \tag{7.3.12}$$

$$\begin{aligned} 2\,{\not{d}}_2^\star {\not{d}}_1^\star {\not{d}}_1 {\not{d}}_2 {\not{d}}_2^\star \underline{\xi} &= \nu_*\Big(({\not{d}}_2^\star {\not{d}}_2 + 2K)\,{\not{d}}_2^\star {\not{d}}_1^\star \underline{\kappa}\Big) - \underline{\kappa}\Big(\nu_*({\not{d}}_2^\star {\not{d}}_1^\star \mu) + 2\,{\not{d}}_2^\star {\not{d}}_1^\star {\not{d}}_1 \underline{\beta}\Big) \\ &\quad + r^{-5}\,\mathfrak{d}^{\leq 4}\Gamma_g + r^{-4}\,\mathfrak{d}^{\leq 4}(\Gamma_b \cdot \Gamma_b) + l.o.t. \end{aligned} \tag{7.3.13}$$

The quadratic terms denoted l.o.t. are lower order both in terms of decay in r, u as well as in terms of number of derivatives.

In particular, if the GCM conditions (7.3.10) *are verified, we deduce*

$$\begin{aligned} {\not{d}}_2^\star {\not{d}}_1^\star {\not{d}}_1 {\not{d}}_2 {\not{d}}_2^\star \eta &= \kappa\,{\not{d}}_2^\star {\not{d}}_1^\star {\not{d}}_1 \underline{\beta} + r^{-5}\,\mathfrak{d}^{\leq 4}\Gamma_g + r^{-4}\,\mathfrak{d}^{\leq 4}(\Gamma_b \cdot \Gamma_b) + l.o.t., \\ {\not{d}}_2^\star {\not{d}}_1^\star {\not{d}}_1 {\not{d}}_2 {\not{d}}_2^\star \underline{\xi} &= -\underline{\kappa}\,{\not{d}}_2^\star {\not{d}}_1^\star {\not{d}}_1 \underline{\beta} + r^{-5}\,\mathfrak{d}^{\leq 4}\Gamma_g + r^{-4}\,\mathfrak{d}^{\leq 4}(\Gamma_b \cdot \Gamma_b) + l.o.t. \end{aligned} \tag{7.3.14}$$

Proof. The proof is a straightforward application of Proposition 7.14. Indeed according to (7.1.32) we have

$$\begin{aligned} 2\,{\not{d}}_2^\star {\not{d}}_1^\star {\not{d}}_1 {\not{d}}_2 {\not{d}}_2^\star \eta &= \kappa\Big(e_3({\not{d}}_2^\star {\not{d}}_1^\star \mu) + 2\,{\not{d}}_2^\star {\not{d}}_1^\star {\not{d}}_1 \underline{\beta}\Big) - {\not{d}}_2^\star {\not{d}}_1^\star {\not{d}}_1 e_3(e_\theta(\kappa)) \\ &\quad + r^{-5}\,\mathfrak{d}^{\leq 4}\Gamma_g + r^{-4}\,\mathfrak{d}^{\leq 4}(\Gamma_b \cdot \Gamma_b) + \text{l.o.t.} \end{aligned}$$

On the other hand since $\nu_* = e_3 + a_* e_4$ with $a_* = \frac{2}{\varsigma_*} - \Upsilon + \frac{r}{2}\underline{A}$, see (7.1.8),

$$\begin{aligned} e_3({\not{d}}_2^\star {\not{d}}_1^\star \mu) &= \nu_*\left({\not{d}}_2^\star {\not{d}}_1^\star \mu\right) - a_* e_4\left({\not{d}}_2^\star {\not{d}}_1^\star \mu\right) \\ &= \nu_*\left({\not{d}}_2^\star {\not{d}}_1^\star \mu\right) - a_*\left({\not{d}}_2^\star {\not{d}}_1^\star e_4 \mu + [e_4, {\not{d}}_2^\star {\not{d}}_1^\star]\check{\mu}\right). \end{aligned}$$

Also, in the same fashion,[16]

$$\begin{aligned}
\not{d}_2^\star \not{d}_1^\star \not{d}_1 e_3(e_\theta(\kappa)) &= \not{d}_2^\star \not{d}_1^\star \not{d}_1 \left[\nu_*(e_\theta(\kappa)) - a_* e_4 e_\theta \kappa\right] \\
&= \not{d}_2^\star \not{d}_1^\star \not{d}_1 \left[(\nu_*(e_\theta(\kappa)))\right] - a_* \not{d}_2^\star \not{d}_1^\star \not{d}_1 (e_4 e_\theta \kappa) \\
&\quad + r^{-3} \sum_{i+j=2} \not{\mathfrak{d}}^i a_* \not{\mathfrak{d}}^j (e_4 e_\theta \kappa) \\
&= \not{d}_2^\star \not{d}_1^\star \not{d}_1 \left[\nu_*(e_\theta(\kappa)) - a_* e_\theta e_4 \kappa - [e_4, e_\theta]\kappa\right] \\
&= r^{-2} \sum_{i+j=2} \not{\mathfrak{d}}^i \Gamma_b \not{\mathfrak{d}}^j \left(e_\theta(e_4\kappa) + [e_\theta, e_4]\kappa\right).
\end{aligned}$$

Thus, after using the transport equations for $e_4\mu, e_4\kappa$ and the commutator lemma applied to $[e_4, e_\theta]$ we easily deduce

$$\begin{aligned}
2 \not{d}_2^\star \not{d}_1^\star \not{d}_1 \not{d}_2 \not{d}_2^\star \eta &= \kappa \Big(\nu_*(\not{d}_2^\star \not{d}_1^\star \mu) + 2 \not{d}_2^\star \not{d}_1^\star \not{d}_1 \underline{\beta}\Big) - \not{d}_2^\star \not{d}_1^\star \not{d}_1 \nu_*(e_\theta(\kappa)) \\
&\quad + r^{-5} \not{\mathfrak{d}}^{\leq 4} \Gamma_g + r^{-4} \not{\mathfrak{d}}^{\leq 4} (\Gamma_b \cdot \Gamma_b) + \text{l.o.t.}
\end{aligned}$$

which confirms the first identity of the proposition.

The second part of the proposition can be derived in the same manner starting with the identity (7.1.33)

$$\begin{aligned}
2 \not{d}_2^\star \not{d}_1^\star \not{d}_1 \not{d}_2 \not{d}_2^\star \underline{\xi} &= e_3 \Big((\not{d}_2^\star \not{d}_2 + 2K) \not{d}_2^\star \not{d}_1^\star \underline{\kappa} \Big) - \underline{\kappa} \Big(e_3(\not{d}_2^\star \not{d}_1^\star \mu) + 2 \not{d}_2^\star \not{d}_1^\star \not{d}_1 \underline{\beta} \Big) \\
&\quad + r^{-5} \not{\mathfrak{d}}^{\leq 4} \Gamma_g + r^{-4} \not{\mathfrak{d}}^{\leq 4} (\Gamma_b \cdot \Gamma_b) + \text{l.o.t.}
\end{aligned}$$

This concludes the proof of the proposition. □

7.3.3 Control of the flux of some quantities on Σ_*

The goal of this section is to establish the following.

Proposition 7.22. *The following estimate holds true for all $k \leq k_{small} + 18$:*

$$\int_{\Sigma_*(u,u_*)} \left(r^2 \left| \not{\mathfrak{d}}^{\leq 3} \mathfrak{d}_*^k \underline{\beta} \right|^2 + \left| \not{\mathfrak{d}}^{\leq 4} \mathfrak{d}_*^k \Gamma_b \right|^2 \right) \lesssim \epsilon_0^2 u^{-2-2\delta_{dec}}. \tag{7.3.15}$$

We also have for $k \leq k_{small} + 17$

$$r \| \not{\mathfrak{d}}^{\leq 1} \underline{\beta} \|^*_{\infty,k} + \| \not{\mathfrak{d}}^{\leq 2} \Gamma_b \|^*_{\infty,k} \lesssim \epsilon_0 r^{-1} u^{-1-\delta_{dec}}. \tag{7.3.16}$$

Remark 7.23. *The flux estimates* (7.3.15) *will be used in the proof of Proposition 7.26 on the control of the $\ell = 1$ mode of various quantities. They also improve the bootstrap assumption on the flux estimate for η on Σ_* which is part of the decay norm ${}^{(ext)}\mathfrak{D}_k[\eta]$.*

Proof. Note that (7.3.16) follows immediately from (7.3.15) using the trace theorem and Sobolev. We thus concentrate our attention on deriving (7.3.15).

Step 1. We first prove the corresponding estimates for $\underline{\beta}$ away from its $\ell = 1$ mode.

[16]Note that in view of (7.3.3), we have $\not{\mathfrak{d}} a_* \in r\Gamma_b$.

More precisely we prove:

Lemma 7.24. *The following estimate holds true for all $k \le k_{small} + 18$:*

$$\int_{\Sigma_*(u,u_*)} r^4 \left| {\not{d}}_2^\star ({\not{\mathfrak{d}}}^{\le 2} \mathfrak{d}_*^k \underline{\beta}) \right|^2 \lesssim \epsilon_0^2 u^{-2-2\delta_{dec}}. \tag{7.3.17}$$

Proof. We make use of Proposition 7.11 according to which

$$\begin{aligned} e_3(r\mathfrak{q}) &= r^5 \left\{ {\not{d}}_2^\star {\not{d}}_1^\star {\not{d}}_1 \underline{\beta} - \frac{3}{2}\kappa\rho\underline{\alpha} - \frac{3}{2}\rho {\not{d}}_2^\star {\not{d}}_1^\star \underline{\kappa} - \frac{3}{2}\underline{\kappa}\rho {\not{d}}_2^\star \zeta + \frac{3}{4}(2\rho^2 - \kappa\underline{\kappa}\rho)\underline{\vartheta} \right\} \\ &+ \mathrm{Err}[e_3(r\mathfrak{q})] \end{aligned}$$

where

$$\mathrm{Err}[e_3(r\mathfrak{q})] = r^3 \mathfrak{d}^{\le 3}(\Gamma_b \cdot \Gamma_g).$$

In view of Lemma 7.6, we have for all $k \le k_{small} + 18$,

$$\|\mathrm{Err}[e_3(r\mathfrak{q})]\|_{\infty,k}(u,r) \lesssim \epsilon_0 u^{-\frac{3}{2}-\delta_{dec}}. \tag{7.3.18}$$

We can also check, making use of the estimates (7.1.13), and Lemma 7.6 for $\underline{\vartheta}$,

$$\|\rho {\not{d}}_2^\star {\not{d}}_1^\star \underline{\kappa},\ \underline{\kappa}\rho {\not{d}}_2^\star \zeta,\ \kappa\underline{\kappa}\rho\underline{\vartheta},\ \rho^2\underline{\vartheta}\|_{\infty,k} \lesssim \epsilon\left(r^{-7} + r^{-6}u^{-1-\frac{\delta_{dec}}{2}}\right).$$

In view of our assumption for r on Σ_* we have $r \ge \frac{\epsilon}{\epsilon_0} u^{1+\delta_{dec}}$, we thus deduce for all $k \le k_{small} + 18$

$$\begin{aligned} \left\| e_3(r\mathfrak{q}) - r^5 \left({\not{d}}_2^\star {\not{d}}_1^\star {\not{d}}_1 \underline{\beta} - \frac{3}{2}\kappa\rho\underline{\alpha} \right) \right\|_{\infty,k} &\lesssim \epsilon\left(r^{-2} + r^{-1}u^{-1-\frac{\delta_{dec}}{2}}\right) + \epsilon_0 u^{-\frac{3}{2}-\delta_{dec}} \\ &\lesssim \epsilon_0 u^{-\frac{3}{2}-\delta_{dec}}. \end{aligned}$$

We infer that

$$\begin{aligned} r^{-1}\|r^4 \mathfrak{d}_*^k {\not{d}}_2^\star {\not{d}}_1^\star {\not{d}}_1 \underline{\beta}\|_{L^2(S)} &\lesssim r^{-1}\|r^{-1}\mathfrak{d}_*^k e_3(r\mathfrak{q})\|_{L^2(S)} + r^{-1}\|\mathfrak{d}_*^k \underline{\alpha}\|_{L^2(S)} \\ &+ \epsilon_0 r^{-1} u^{-\frac{3}{2}-\delta_{dec}} \end{aligned}$$

where $\mathfrak{d}_*^k = \nu_*^{k_1} {\not{\mathfrak{d}}}^{k_2}$ denote the tangential derivatives to Σ_*. Thus integrating on the last slice Σ_* and making use of the assumptions (7.1.15) and (7.1.18), i.e.,

$$\int_{\Sigma_*(u,u_*)} |e_3 \mathfrak{d}^k \mathfrak{q}|^2 + r^{-2}|\mathfrak{q}|^2 + |\mathfrak{d}^k \underline{\alpha}|^2 \lesssim \epsilon_0^2 (1+u)^{-2-2\delta_{dec}}, \qquad k \le k_{small} + 18,$$

we deduce

$$\int_{\Sigma_*(u,u_*)} r^8 \left| \mathfrak{d}_*^k ({\not{d}}_2^\star {\not{d}}_1^\star {\not{d}}_1 \underline{\beta}) \right|^2 \lesssim \epsilon_0^2 u^{-2-2\delta_{dec}}.$$

Taking into account the commutator Lemma 7.17, as well as the product Lemma

7.6, we deduce, for $k \leq k_{small} + 18$,

$$\int_{\Sigma_*(u,u_*)} r^8 \big| \,\mathfrak{d}\!\!/_2^\star \,\mathfrak{d}\!\!/_1^\star \,\mathfrak{d}\!\!/_1 (\mathfrak{d}_*^k \underline{\beta})\big|^2 \lesssim \epsilon_0^2 u^{-2-2\delta_{dec}}. \tag{7.3.19}$$

Since

$$\mathfrak{d}\!\!/_1^\star \,\mathfrak{d}\!\!/_1 = \mathfrak{d}\!\!/_2 \,\mathfrak{d}\!\!/_2^\star + 2K,$$

we infer that

$$\int_{\Sigma_*(u,u_*)} r^8 \left|\left(\mathfrak{d}\!\!/_2^\star \,\mathfrak{d}\!\!/_2 + 2K\right) \mathfrak{d}\!\!/_2^\star (\mathfrak{d}_*^k \underline{\beta})\right|^2 \lesssim \epsilon_0^2 u^{-2-2\delta_{dec}}.$$

In view of the coercivity of $\mathfrak{d}\!\!/_2^\star \,\mathfrak{d}\!\!/_2 + 2K$ we deduce

$$\int_{\Sigma_*(u,u_*)} r^4 \left| \mathfrak{d}\!\!/_2^\star (\mathfrak{d}\!\!/^{\leq 2} \mathfrak{d}_*^k \underline{\beta})\right|^2 \lesssim \epsilon_0^2 u^{-2-2\delta_{dec}}, \qquad k \leq k_{small} + 18.$$

This conclusdes the proof of Lemma 7.24. □

Step 2. We make use of Lemma 7.24 to prove the desired estimate for $\underline{\vartheta}$, i.e.,

$$\int_{\Sigma_*(u,u_*)} | \mathfrak{d}\!\!/^{\leq 4} \mathfrak{d}_*^k \underline{\vartheta}|^2 \lesssim \epsilon_0^2 u^{-2-2\delta_{dec}}, \qquad k \leq k_{small} + 18. \tag{7.3.20}$$

Proof of (7.3.20). One starts with the Codazzi equation

$$\mathfrak{d}\!\!/_2 \underline{\vartheta} = -2\underline{\beta} - \mathfrak{d}\!\!/_1^\star (\underline{\kappa}) - \zeta \underline{\kappa} + \Gamma_g \cdot \Gamma_b.$$

Differentiating w.r.t. $\mathfrak{d}\!\!/_2^\star$ and then taking tangential derivatives $\mathfrak{d}\!\!/^{\leq 2} \mathfrak{d}_*^k$ we derive

$$\mathfrak{d}\!\!/^{\leq 2} \mathfrak{d}_*^k \,\mathfrak{d}\!\!/_2^\star \,\mathfrak{d}\!\!/_2 \underline{\vartheta} = -2 \,\mathfrak{d}\!\!/^{\leq 2} \mathfrak{d}_*^k \,\mathfrak{d}\!\!/_2^\star \underline{\beta} - \mathfrak{d}\!\!/^{\leq 2} \mathfrak{d}_*^k \,\mathfrak{d}\!\!/_2^\star \,\mathfrak{d}\!\!/_1^\star (\underline{\kappa}) - \mathfrak{d}\!\!/^{\leq 2} \mathfrak{d}_*^k \left[r^{-2} \mathfrak{d}\!\!/ \Gamma_g + r^{-1} \mathfrak{d}\!\!/ (\Gamma_g \cdot \Gamma_b)\right].$$

Making use of the GCM condition $\mathfrak{d}\!\!/_2^\star \,\mathfrak{d}\!\!/_1^\star \underline{\kappa} = 0$ along Σ_* and the interpolation estimates of Lemma 7.6, for all $k \leq k_{small} + 18$,

$$\begin{aligned} \mathfrak{d}\!\!/^{\leq 2} \mathfrak{d}_*^k \,\mathfrak{d}\!\!/_2^\star \,\mathfrak{d}\!\!/_2 \underline{\vartheta} &= -2 \,\mathfrak{d}\!\!/^{\leq 2} \mathfrak{d}_*^k \,\mathfrak{d}\!\!/_2^\star \underline{\beta} + r^{-2} \mathfrak{d}\!\!/^{\leq 2} \mathfrak{d}^{k+1} \Gamma_g + r^{-1} \mathfrak{d}\!\!/^{\leq 2} \mathfrak{d}^{k+1} (\Gamma_g \cdot \Gamma_b) \\ &= -2 \,\mathfrak{d}\!\!/^{\leq 2} \mathfrak{d}_*^k \,\mathfrak{d}\!\!/_2^\star \underline{\beta} + O\left(\epsilon r^{-4} u^{-\frac{1}{2} - \frac{\delta_{dec}}{2}}\right) \end{aligned}$$

or, since $r \geq \frac{\epsilon}{\epsilon_0} u^{1+\delta_{dec}}$,

$$\mathfrak{d}\!\!/^{\leq 2} \mathfrak{d}_*^k \,\mathfrak{d}\!\!/_2^\star \,\mathfrak{d}\!\!/_2 \underline{\vartheta} = -2 \,\mathfrak{d}\!\!/^{\leq 2} \mathfrak{d}_*^k \,\mathfrak{d}\!\!/_2^\star \underline{\beta} + O\left(\epsilon_0 r^{-3} u^{-\frac{3}{2} - \delta_{dec}}\right).$$

Moreover,

$$\mathfrak{d}\!\!/_2^\star \,\mathfrak{d}\!\!/_2 \,\mathfrak{d}\!\!/^{\leq 2} \mathfrak{d}_*^k \underline{\vartheta} = -2 \,\mathfrak{d}\!\!/^{\leq 2} \mathfrak{d}_*^k \,\mathfrak{d}\!\!/_2^\star \underline{\beta} + O\left(\epsilon_0 r^{-3} u^{-\frac{3}{2} - \delta_{dec}}\right) + [\mathfrak{d}\!\!/^{\leq 2} \mathfrak{d}_*^k, \,\mathfrak{d}\!\!/_2^\star \,\mathfrak{d}\!\!/_2] \underline{\vartheta}.$$

Using the commutator estimates of Lemma 7.17 and the interpolation estimates of

Lemma 7.6, we derive

$$\not{d}_2^\star \not{d}_2 \mathfrak{d}_*^k \underline{\vartheta} = -2\mathfrak{d}_*^k \not{d}_2^\star \underline{\beta} + O\left(\epsilon_0 r^{-3} u^{-\frac{3}{2}-\delta_{dec}}\right).$$

Integrating and using the previously derived estimate for $\underline{\beta}$ we deduce

$$\int_{\Sigma_*(u,u_*)} r^4 \left| \not{d}_2^\star \not{d}_2 \not{\mathfrak{d}}^{\leq 2} \mathfrak{d}_*^k \underline{\vartheta} \right|^2 \lesssim \epsilon_0^2 u^{-2-2\delta_{dec}}, \qquad k \leq k_{small} + 18.$$

In view of the coercivity of $\not{d}_2^\star \not{d}_2$ we infer that

$$\int_{\Sigma_*(u,u_*)} \left| \not{\mathfrak{d}}^{\leq 4} \mathfrak{d}_*^k \underline{\vartheta} \right|^2 \lesssim \epsilon_0^2 u^{-2-2\delta_{dec}}, \qquad k \leq k_{small} + 18$$

as desired. □

Step 3. We next derive a non-sharp, preliminary estimate for the $\ell = 1$ mode of $\underline{\beta}$ with the help of the Codazzi equation for $\underline{\vartheta}$,

$$2\underline{\beta} = -\not{d}_2 \underline{\vartheta} + e_\theta(\underline{\kappa}) - \underline{\overline{\kappa}}\zeta + \Gamma_g \cdot \Gamma_b = -\not{d}_2 \underline{\vartheta} + r^{-1} \not{\mathfrak{d}}^{\leq 1} \Gamma_g + \Gamma_g \cdot \Gamma_b.$$

Projecting on the $\ell = 1$ mode, this yields

$$2\int_S \underline{\beta} e^\Phi = r^2 \not{\mathfrak{d}}^{\leq 1} \Gamma_g + r^3 \Gamma_g \cdot \Gamma_b.$$

Differentiating, and using Lemma 7.6, we deduce

$$\left| \nu_*^k \left(\int_S \underline{\beta} e^\Phi \right) \right| \lesssim \epsilon u^{-\frac{1}{2}-\frac{\delta_{dec}}{2}}, \qquad k \leq k_{small} + 18. \tag{7.3.21}$$

Together with Corollary 7.20, we infer

$$\left| \int_S (\nu_*^k \underline{\beta}) e^\Phi \right| \lesssim \epsilon u^{-\frac{1}{2}-\frac{\delta_{dec}}{2}} + r^4 |\mathfrak{d}^k (\underline{\beta} \cdot \Gamma_b)|.$$

Together with product estimates of Lemma 7.6, and since $r \geq (\epsilon\epsilon_0^{-1}) u^{1+\delta_{dec}}$ on Σ_*, we deduce, for $k \leq k_{small} + 18$,

$$\left| \int_S (\nu_*^k \underline{\beta}) e^\Phi \right| \lesssim \epsilon_0 r u^{-\frac{3}{2}-\delta_{dec}}. \tag{7.3.22}$$

We combine the result of Lemma 7.24 with (7.3.22) to deduce

$$\int_{\Sigma_*(u,u_*)} r^2 \left| \not{\mathfrak{d}}^{\leq 3} \mathfrak{d}_*^k \underline{\beta} \right|^2 \lesssim \epsilon_0^2 u^{-2-2\delta_{dec}}, \qquad k \leq k_{small} + 18. \tag{7.3.23}$$

Indeed, according to the last elliptic estimate of Lemma 7.7 and (7.3.22), we have

$$
\begin{aligned}
\int_S r^2 |\not{\mathfrak{d}}^{\leq 3}\mathfrak{d}_*^k \underline{\beta}|^2 &\lesssim r^4 \int_S |\not{d}_2^\star(\not{\mathfrak{d}}^{\leq 2}\mathfrak{d}_*^k\underline{\beta})|^2 + r^{-2}\left|\int_S (\nu_*^k \underline{\beta})e^\Phi\right|^2\\
&\lesssim r^4 \int_S |\not{d}_2^\star(\not{\mathfrak{d}}^{\leq 2}\mathfrak{d}_*^k\underline{\beta})|^2 + \epsilon_0^2 u^{-3-2\delta_{dec}}.
\end{aligned}
$$

Thus, integrating and making use of estimate (7.3.17), we infer

$$
\int_{\Sigma(u,u_*)} r^2 |\not{\mathfrak{d}}^{\leq 3}\mathfrak{d}_*^k \underline{\beta}|^2 \lesssim \int_{\Sigma(u,u_*)} r^4 |\not{d}_2^\star(\not{\mathfrak{d}}^{\leq 2}\mathfrak{d}_*^k\underline{\beta})|^2 + \epsilon_0^2 u^{-2-2\delta_{dec}} \lesssim \epsilon_0^2 u^{-2-2\delta_{dec}}
$$

which concludes the proof of (7.3.23).

Step 4. Next, we establish the estimates for η and $\underline{\xi}$. We first estimate $\not{d}_2^\star(\eta, \underline{\xi})$.

Lemma 7.25. *We have for* $k \leq k_{small} + 18$

$$
\int_{\Sigma_*(u,u_*)} r^2\Big(|\not{d}_2^\star(\not{\mathfrak{d}}^{\leq 5}\mathfrak{d}_*^k\eta)|^2 + |\not{d}_2^\star(\not{\mathfrak{d}}^{\leq 5}\mathfrak{d}_*^k\underline{\xi})|^2\Big) \lesssim \epsilon_0^2 u^{-2-2\delta_{dec}}. \tag{7.3.24}
$$

Proof. We prove Lemma 7.25 based on the identities of Proposition 7.21. To derive the desired flux estimate for η we make use of the first part of Proposition 7.21 according to which we have

$$
\begin{aligned}
2\not{d}_2^\star \not{d}_1^\star \not{d}_1 \not{d}_2 \not{d}_2^\star \eta = \kappa\Big(\nu_*(\not{d}_2^\star \not{d}_1^\star \mu) + 2\not{d}_2^\star \not{d}_1^\star \not{d}_1 \underline{\beta}\Big) - \not{d}_2^\star \not{d}_1^\star \not{d}_1 \nu_*(e_\theta(\kappa))\\
+ r^{-5}\not{\mathfrak{d}}^{\leq 4}\Gamma_g + r^{-4}\not{\mathfrak{d}}^{\leq 4}(\Gamma_b\cdot\Gamma_b) + \text{l.o.t.}
\end{aligned}
$$

Since, $\not{d}_1^\star \not{d}_1 = \not{d}_2 \not{d}_2^\star + 2K$, we deduce

$$
\begin{aligned}
\not{d}_2^\star(\not{d}_2 \not{d}_2^\star + 2K)\not{d}_2 \not{d}_2^\star \eta &= \frac{1}{2}\Big[\kappa\nu_*(\not{d}_2^\star \not{d}_1^\star \mu) - \not{d}_2^\star \not{d}_1^\star \not{d}_1 \nu_*(e_\theta(\kappa))\Big]\\
&\quad + \kappa\not{d}_2^\star(\not{d}_2 \not{d}_2^\star + 2K)\underline{\beta} + r^{-5}\not{\mathfrak{d}}^{\leq 4}\Gamma_g + r^{-4}\not{\mathfrak{d}}^{\leq 4}(\Gamma_b\cdot\Gamma_b)\\
&\quad + \text{l.o.t.},
\end{aligned}
$$

or

$$
\begin{aligned}
(\not{d}_2^\star \not{d}_2 + 2K)\left[\not{d}_2^\star \not{d}_2 \not{d}_2^\star \eta - \kappa\not{d}_2^\star\underline{\beta}\right] &= \frac{1}{2}\Big[\kappa\nu_*(\not{d}_2^\star \not{d}_1^\star \mu) - \not{d}_2^\star \not{d}_1^\star \not{d}_1 \nu_*(e_\theta(\kappa))\Big]\\
&\quad + r^{-5}\not{\mathfrak{d}}^{\leq 4}\Gamma_g + r^{-4}\not{\mathfrak{d}}^{\leq 4}(\Gamma_b\cdot\Gamma_b) + \text{l.o.t.}
\end{aligned}
$$

Taking higher tangential derivatives and using our GCM assumptions on Σ_*

$$
\mathfrak{d}_*^k(\not{d}_2 \not{d}_2^\star + 2K)\Big[\not{d}_2^\star \not{d}_2 \not{d}_2^\star \eta - \kappa\not{d}_2^\star\underline{\beta}\Big] = \mathfrak{d}_*^k\Big[r^{-5}\not{\mathfrak{d}}^{\leq 4}\Gamma_g + r^{-4}\not{\mathfrak{d}}^{\leq 3}(\Gamma_b\cdot\mathfrak{d}\Gamma_b)\Big] + \text{l.o.t.}
$$

Making use of the commutation Lemma 7.17 we can rewrite

$$
\begin{aligned}
& r^2(\not{d}_2 \not{d}_2^\star + 2K)\Big[\not{d}_2^\star \not{d}_2 \not{d}_2^\star(\mathfrak{d}_*^k\eta) - \kappa\not{d}_2^\star(\mathfrak{d}_*^k\underline{\beta})\Big]\\
&= \sum_{j\leq k}\not{\mathfrak{d}}^{\leq 2}\Big[r^{-3}\not{\mathfrak{d}}^{\leq 2}\mathfrak{d}_*^j\Gamma_g + r^{-2}\not{\mathfrak{d}}^{\leq 1}\mathfrak{d}_*^j(\Gamma_b\cdot\mathfrak{d}\Gamma_b)\Big].
\end{aligned}
$$

Using the ellipticity of the operators $(\not{d}_2 \not{d}_2^\star + 2K)$ and $\not{d}_2^\star \not{d}_2$, Lemma 7.6 and the dominance condition $r \geq (\epsilon\epsilon_0^{-1})\, u^{1+\delta_{dec}}$ on Σ_*, we derive, for $k \leq k_{small}+18$,

$$\| \not{d}_2^\star(\not{\mathfrak{d}}^{\leq 4}\mathfrak{d}_*^k\eta)\|_{L^2(S)} \lesssim \quad r\| \not{d}_2^\star(\not{\mathfrak{d}}^{\leq 2}\mathfrak{d}_*^k\underline{\beta})\|_{L^2(S)} + \epsilon_0 r^{-1}u^{-\frac{3}{2}-\delta_{dec}}. \qquad (7.3.25)$$

Finally, squaring, integrating on Σ_* and taking into account the flux estimate for $\underline{\beta}$ in (7.3.23) we deduce, for $k \leq k_{small}+18$,

$$\int_{\Sigma_*(u,u_*)} r^2 \big| \not{d}_2^\star(\not{\mathfrak{d}}^{\leq 4}\mathfrak{d}_*^k\eta)\big|^2 \lesssim \int_{\Sigma_*(u,u_*)} r^4 \big| \not{d}_2^\star(\not{\mathfrak{d}}^{\leq 2}\mathfrak{d}_*^k\underline{\beta})\big|^2 + \epsilon_0 u^{-2-2\delta_{dec}} \lesssim \epsilon_0 u^{-2-2\delta_{dec}}$$

as stated. This completes the proof of Lemma 7.25 for η. The proof for $\underline{\xi}$ is very similar and left to the reader. □

Step 5. In this step, we derive the desired estimates for η and $\underline{\xi}$, i.e., we show

$$\int_{\Sigma_*(u,u_*)} \Big(| \not{\mathfrak{d}}^{\leq 5}\mathfrak{d}_*^k\eta|^2 + | \not{\mathfrak{d}}^{\leq 5}\mathfrak{d}_*^k\underline{\xi}|^2\Big) \lesssim \epsilon_0^2 u^{-2-2\delta_{dec}}, \qquad k \leq k_{small}+18. \quad (7.3.26)$$

To this end, we prove the following estimates for the $\ell=1$ mode of $\underline{\xi}$ and η:

$$\left|\int_S (\nu_*^k\eta)e^\Phi\right| + \left|\int_S (\nu_*^k\underline{\xi})e^\Phi\right| \quad\lesssim\quad \epsilon_0 r^2\, u^{-2-\delta_{dec}}, \quad k \leq k_{small}+18. \quad (7.3.27)$$

Then, (7.3.26) follows from (7.3.27) and Lemma 7.25 using a Poincaré inequality.

To prove (7.3.27), we apply Corollary 7.20 to η and $\underline{\xi}$. This yields for $k \geq 1$

$$\begin{aligned}\left|\int_S (\nu_*^k\underline{\xi})e^\Phi\right| + \left|\int_S (\nu_*^k\eta)e^\Phi\right| \quad&\lesssim\quad \sum_{j=0}^k \left(\left|\nu_*^j\left(\int_S \underline{\xi}e^\Phi\right)\right| + \left|\nu_*^j\left(\int_S \eta e^\Phi\right)\right|\right)\\ &\quad + r^4\left|\mathfrak{d}^k(\Gamma_b\cdot\Gamma_b)\right|.\end{aligned}$$

In view of the GCM condition for the $\ell=1$ mode of η and $\underline{\xi}$, we infer

$$\left|\int_S (\nu_*^k\underline{\xi})e^\Phi\right| + \left|\int_S (\nu_*^k\eta)e^\Phi\right| \quad\lesssim\quad r^4\left|\mathfrak{d}^k(\Gamma_b\cdot\Gamma_b)\right|.$$

For $k \leq k_{small}+18$, we infer, using the product Lemma 7.6,

$$\left|\int_S (\nu_*^k\underline{\xi})e^\Phi\right| + \left|\int_S (\nu_*^k\eta)e^\Phi\right| \quad\lesssim\quad \epsilon_0^2 u^{-2-2\delta_{dec}}$$

which concludes the proof of (7.3.27), and hence of (7.3.26).

Step 6. Next, we derive the flux estimates for $\underline{\check{\omega}}$, $\check{\varsigma}$, $\underline{\check{\Omega}}$ and $\underline{A}$. Adding κ times the first equation to the second equation of Proposition 7.12, we obtain

$$\begin{aligned}2\kappa \not{d}_1^\star\underline{\omega} \quad&=\quad -e_3(e_\theta(\kappa)) - 2\not{d}_2 \not{d}_2^\star\eta + \kappa\left(\frac{1}{2}\underline{\kappa} + 2\underline{\omega}\right)\eta - \frac{1}{2}\kappa^2\underline{\xi}\\ &\quad + r^{-2}\not{\mathfrak{d}}^{\leq 1}\Gamma_g + r^{-1}\not{\mathfrak{d}}^{\leq 1}(\Gamma_b\cdot\Gamma_b).\end{aligned}$$

In view of the GCM condition for κ, the fact that $\nu_* = e_3 + a_* e_4$, and Raychaudhuri, we have

$$-e_3(e_\theta(\kappa)) = a_* e_4(e_\theta(\kappa)) = r^{-1}\not{\mathfrak{d}}(\Gamma_g \cdot \Gamma_g)$$

and hence

$$r \not{d}_1^\star \underline{\omega} = -\frac{1}{2}r^2 \not{d}_2 \not{d}_2^\star \eta + \frac{r^2}{4}\kappa\left(\frac{1}{2}\underline{\kappa} + 2\underline{\omega}\right)\eta - \frac{r^2}{8}\kappa^2 \underline{\xi} + \not{\mathfrak{d}}^{\leq 1}\Gamma_g + r\not{\mathfrak{d}}^{\leq 1}(\Gamma_b \cdot \Gamma_b).$$

The flux estimate for $\check{\underline{\omega}}$ follows easily from the above identity, the flux estimates for η and $\underline{\xi}$ derived in Step 4 and Step 5, the interpolation estimate of Lemma 7.6 for ζ, as well as the dominance property of r on Σ_*.

The flux estimates for $\check{\varsigma}$ and $\check{\underline{\Omega}}$ follow easily from the equations

$$\begin{aligned} \varsigma^{-1} e_\theta(\check{\varsigma}) &= \eta - \zeta, \\ e_\theta(\check{\underline{\Omega}}) &= -\underline{\xi} - (\eta - \zeta)\underline{\Omega}, \end{aligned}$$

the flux estimates for η and $\underline{\xi}$ derived in Step 4 and Step 5, the interpolation estimate of Lemma 7.6 for ζ, as well as the dominance property of r on Σ_*.

To estimate $\underline{A}$, note first that the flux estimate for $\underline{A} - \overline{\underline{A}}$ follows immediately from formula (7.1.9) and the above flux estimate for $\check{\underline{\Omega}}$. It then remains to control $\overline{\underline{A}}$. In view of (2.2.22), we have

$$\varsigma \underline{A} = -\overline{\underline{\kappa}}\check{\varsigma} + \overline{\kappa}\left(\varsigma\check{\underline{\Omega}} + \overline{\underline{\Omega}}\check{\varsigma}\right) + \overline{\check{\varsigma}\check{\underline{\kappa}}} - \overline{\underline{\Omega}}\,\overline{\check{\varsigma}\check{\kappa}} - \overline{\check{\underline{\Omega}}\varsigma\kappa}$$

and hence, taking the average, we infer

$$\overline{\underline{A}} = -(\overline{\varsigma} - 1)\overline{\underline{A}} - \overline{\check{\varsigma}\check{\underline{A}}} + \overline{\check{\varsigma}\check{\underline{\kappa}}} - \overline{\underline{\Omega}}\,\overline{\check{\varsigma}\check{\kappa}} - \overline{\check{\underline{\Omega}}\varsigma\check{\kappa}}.$$

The flux estimate for $\overline{\underline{A}}$ follows then from the product estimates of Lemma 7.6.

Step 7. It remains to derive the flux estimate for $\overline{\underline{\omega}}$, $\overline{\underline{\Omega}}$ and $\overline{\varsigma}$. Recall (4.2.4):

$$\begin{aligned} \overline{\underline{\omega}} - \frac{m}{r^2} = \frac{r}{4}\Bigg\{ & e_3\left(\overline{\kappa} - \frac{2}{r}\right) + \frac{1}{2}\overline{\underline{\kappa}}\left(\overline{\kappa} - \frac{2}{r}\right) - 2\overline{\underline{\omega}}\left(\overline{\kappa} - \frac{2}{r}\right) - 2\left(\overline{\rho} + \frac{2m}{r^3}\right) \\ & + \frac{1}{2}\overline{\kappa}\left(\overline{\kappa} - \frac{2}{r}\right)\check{\underline{\Omega}} - 2\overline{\check{\underline{\omega}}\check{\kappa}} + \frac{1}{2}\overline{\vartheta\underline{\vartheta}} - 2\overline{\zeta^2} - \frac{1}{2}\check{\underline{\Omega}}\Big(- \overline{\vartheta^2} + \overline{\check{\kappa}^2}\Big) \\ & + \overline{\check{\underline{\Omega}}(e_4(\check{\kappa}) + \kappa\check{\kappa})} - \frac{1}{2}\overline{\check{\underline{\kappa}}\check{\kappa}} + \frac{1}{r}\overline{\check{\underline{\Omega}}\check{\kappa}}\Bigg\}. \end{aligned}$$

Using the GCM condition for κ, the fact that $\nu_* = e_3 + a_* e_4$, Lemma 2.72 for $e_4(\overline{\kappa} - 2/r)$, and the identity (2.2.12) for $\overline{\rho}$, we infer

$$\overline{\underline{\omega}} - \frac{m}{r^2} = r\Gamma_b \cdot \Gamma_g$$

which together with Lemma 7.6 yields the flux estimate for $\overline{\underline{\omega}}$.

Next, taking the average of (7.1.9), we have

$$\overline{\underline{\Omega}} + \Upsilon = \frac{r}{2}\overline{\underline{A}}.$$

The flux estimate for $\overline{\underline{\Omega}}$ follows from the above identity and the flux estimate for $\underline{A}$ derived in Step 6.

Finally, we derive the flux estimate for $\overline{\varsigma}$. Recall equation (7.1.8)

$$a_* = -\frac{2}{\varsigma} + \Upsilon - \frac{r}{2}\underline{A}$$

and the GCM condition for a_*, see (7.1.1),

$$a_*\Big|_{SP} = -1 - \frac{2m}{r}.$$

We deduce

$$\frac{2}{\varsigma\Big|_{SP}} = 2 - \frac{r}{2}\underline{A}\Big|_{SP}.$$

Since

$$\overline{\varsigma} = \varsigma\Big|_{SP} + \check{\varsigma}\Big|_{SP},$$

we infer

$$\overline{\varsigma} - 1 = \check{\varsigma}\Big|_{SP} + \frac{1}{1 - \frac{r}{4}\underline{A}\Big|_{SP}} \frac{r}{4}\underline{A}\Big|_{SP}$$

and the flux estimate for $\overline{\varsigma}$ follows from the above identity and the flux estimates for $\check{\varsigma}$ and $\underline{A}$ of Step 6. This concludes the proof of Proposition 7.22. □

7.3.4 Estimates for some $\ell = 1$ modes on Σ_*

In this section, we control the $\ell = 1$ modes of $e_\theta(\underline{\kappa})$, $e_\theta(\rho)$, $e_\theta(\mu)$ and of β. We summarize the results in the following proposition.

Proposition 7.26. *The following estimates hold true:*

$$\begin{aligned}\left|\int_S e_\theta(\rho)e^\Phi\right| + \left|\int_S e_\theta(\mu)e^\Phi\right| + \max_{k\leq k_{small}+20}\left|\nu_*^k\left(\int_S \beta e^\Phi\right)\right| &\lesssim \epsilon_0 r^{-1}u^{-1-\delta_{dec}},\\ \left|\int_S e_\theta(\underline{\kappa})e^\Phi\right| &\lesssim \epsilon_0 u^{-2-\delta_{dec}}.\end{aligned} \tag{7.3.28}$$

Remark 7.27. *We note that the estimates for the $\ell = 1$ modes of $e_\theta(\underline{\kappa})$ and β are sharp.*[17] *During the proof we shall also need to derive sharp estimates for the $\ell = 1$ modes of ζ and $\underline{\beta}$, see* (7.3.30) *and* (7.3.32).

[17]Consistent in fact with strong peeling.

Proof. We will rely on the following auxiliary bootstrap assumptions:

$$\left|\int_S \beta e^\Phi\right| \lesssim \epsilon r^{-1}u^{-1-\delta_{dec}}, \qquad \left|\int_S e_\theta(\underline{\kappa})e^\Phi\right| \lesssim \epsilon u^{-2-\delta_{dec}}. \tag{7.3.29}$$

Step 1. We start with proving an intermediary estimate for the $\ell = 1$ mode of ζ. In view of the Codazzi equations and the GCM condition on κ,

$$\not{d}_2\vartheta = -2\beta + (e_\theta(\kappa) + \zeta\kappa) + \Gamma_g \cdot \Gamma_g = -2\beta + \frac{2}{r}\zeta + \Gamma_g \cdot \Gamma_g$$

and hence

$$\int_S \zeta e^\Phi = -r\int_S \beta e^\Phi + r^4\Gamma_g \cdot \Gamma_g.$$

Thus, using the product estimates of Lemma 7.6,

$$\left|\int_S \zeta e^\Phi\right| \lesssim r\left|\int_S \beta e^\Phi\right| + \epsilon_0 u^{-1-\delta_{dec}}.$$

In particular, in view of (7.3.29), we infer

$$\left|\int_S \zeta e^\Phi\right| \lesssim \epsilon u^{-1-\delta_{dec}}. \tag{7.3.30}$$

Step 2. Next, we establish an intermediary estimate for the $\ell = 1$ mode of $\underline{\beta}$. We start with the Codazzi equation for $\underline{\vartheta}$,

$$2\underline{\beta} = -\not{d}_2\underline{\vartheta} + e_\theta(\underline{\kappa}) + \frac{2\Upsilon}{r}\zeta + \Gamma_g \cdot \Gamma_b$$

and project on the $\ell = 1$ mode, i.e.,

$$2\int_S \underline{\beta}e^\Phi = \frac{2\Upsilon}{r}\int_S \zeta e^\Phi + \int_S e_\theta(\underline{\kappa})e^\Phi + \int_S \Gamma_g \cdot \Gamma_b e^\Phi. \tag{7.3.31}$$

We make use of the estimate (7.3.30) for ζ, the auxiliary estimate (7.3.29) for $e_\theta(\underline{\kappa})$, the dominance condition $r \geq \epsilon\epsilon_0^{-1}u^{1+\delta_{dec}}$ on Σ_*, and bootstrap assumptions for Γ_g to deduce

$$\begin{aligned} r^{-1}\left|\int_S \underline{\beta}e^\Phi\right| &\lesssim r^{-2}\left|\int_S \zeta e^\Phi\right| + r^{-1}\left|\int_S e_\theta(\underline{\kappa})e^\Phi\right| + \int_S |\Gamma_g \cdot \Gamma_b| \\ &\lesssim \epsilon r^{-2}u^{-1-\delta_{dec}} + \epsilon r^{-1}u^{-2-\delta_{dec}} + \epsilon r^{-2}u^{-\frac{1}{2}-\delta_{dec}}\int_S |\Gamma_b| \\ &\lesssim \epsilon_0 u^{-3-2\delta_{dec}} + \epsilon_0 u^{-\frac{3}{2}-2\delta_{dec}}\|\Gamma_b\|_{L^2(S)}. \end{aligned}$$

Thus, we infer

$$\int_u^{u_*} r^{-1}\left|\int_S \underline{\beta}e^\Phi\right| du \lesssim \epsilon_0 u^{-2-2\delta_{dec}} + \epsilon_0\left(\int_u^{u_*} u^{-3-4\delta_{dec}}\right)^{1/2}\left(\int_u^{u_*} \|\Gamma_b\|^2_{L^2(S)}\right)^{1/2}$$

which together with the flux estimate of Proposition 7.22 implies

$$\int_u^{u_*} r^{-1}\left|\int_S \underline{\beta} e^{\Phi}\right| du \lesssim \epsilon_0 u^{-2-2\delta_{dec}}. \tag{7.3.32}$$

Step 3. Next, we provide an intermediary estimate for the $\ell = 1$ mode of ρ. We start by differentiating the Gauss equation $K = -\rho - \frac{1}{4}\kappa\underline{\kappa} + \frac{1}{4}\vartheta\underline{\vartheta}$. Using the GCM condition for κ we derive

$$e_\theta(\rho) = -e_\theta(K) - \frac{1}{2r}e_\theta(\underline{\kappa}) + \frac{1}{4}e_\theta(\vartheta\underline{\vartheta}).$$

We make use of the vanishing of the $\ell = 1$ mode of $e_\theta(K)$ (see Lemma 2.32) to derive

$$\int_S e_\theta(\rho)e^{\Phi} = -\frac{1}{2r}\int_S e_\theta(\underline{\kappa})e^{\Phi} + \frac{1}{4}\int_S e_\theta(\vartheta\underline{\vartheta})e^{\Phi}. \tag{7.3.33}$$

Using the auxiliary estimate (7.3.29) for the $\ell = 1$ mode of $e_\theta(\underline{\kappa})$

$$\left|\int_S e_\theta(\rho)e^{\Phi}\right| \lesssim \epsilon r^{-1}u^{-2-\delta_{dec}} + \int_S |\not{\mathfrak{d}}^{\leq 1}(\Gamma_g\cdot\Gamma_b)|.$$

Making use of $r \geq \epsilon\epsilon_0^{-1}u^{1+\delta_{dec}}$ and the bootstrap assumptions on Γ_g, we deduce

$$\left|\int_S e_\theta(\rho)e^{\Phi}\right| \lesssim \epsilon_0 u^{-3-2\delta_{dec}} + \epsilon_0 u^{-\frac{3}{2}-2\delta_{dec}}\|\not{\mathfrak{d}}^{\leq 1}\Gamma_b\|_{L^2(S)}.$$

Integrating in u we derive

$$\int_u^{u_*}\left|\int_S e_\theta(\rho)e^{\Phi}\right| du \lesssim \epsilon_0 u^{-2-\delta_{dec}} + \epsilon_0\left(\int_u^{u_*} u^{-3-4\delta_{dec}}\right)^{\frac{1}{2}}\left(\int_u^{u_*}\|\not{\mathfrak{d}}^{\leq 1}\Gamma_b\|^2_{L^2(S)}\right)^{\frac{1}{2}}.$$

Using the flux estimate in Proposition 7.22, we infer

$$\int_u^{u_*}\left|\int_S e_\theta(\rho)e^{\Phi}\right| du \lesssim \epsilon_0 u^{-2-\delta_{dec}}. \tag{7.3.34}$$

Step 4. Next, we control the $\ell = 1$ mode of $e_\theta(\underline{\kappa})$ on Σ_*. To this end, we need the precise identity[18] of Proposition 2.74:

$$
\begin{aligned}
e_3(e_\theta(\underline{\kappa})) &= -2\,\not{d}_2\,\not{d}_2^\star\underline{\xi} + \underline{\kappa}\left(e_3(\zeta) - \underline{\beta}\right) + \underline{\kappa}^2\zeta - \frac{3}{2}\underline{\kappa}e_\theta\underline{\kappa} + 6\rho\underline{\xi} - 2\underline{\omega}e_\theta(\underline{\kappa})\\
&\quad + \mathrm{Err}[\not{d}_2\,\not{d}_2^\star\underline{\xi}],\\
\mathrm{Err}[\not{d}_2\,\not{d}_2^\star\underline{\xi}] &= \left(2\,\not{d}_1\underline{\xi} + \frac{1}{2}\underline{\kappa}\,\underline{\vartheta} + 2\eta\underline{\xi} - \frac{1}{2}\underline{\vartheta}^2\right)\eta + 2e_\theta(\eta\underline{\xi}) - \frac{1}{2}e_\theta(\underline{\vartheta}^2)\\
&\quad + \underline{\kappa}\left(\frac{1}{2}\underline{\vartheta}\zeta - \frac{1}{2}\vartheta\underline{\xi}\right) - \frac{1}{2}\underline{\vartheta}e_\theta\underline{\kappa} - \frac{1}{2}\vartheta\underline{\vartheta}\underline{\xi}\\
&\quad - \zeta\left(2\,\not{d}_1\underline{\xi} + 2(\eta - 3\zeta)\underline{\xi} - \frac{1}{2}\underline{\vartheta}^2\right) + \underline{\xi}\Big(-\vartheta\underline{\vartheta} - 2\,\not{d}_1\zeta + 2\zeta^2\Big)\\
&\quad - 6\eta\zeta\underline{\xi} - 6e_\theta(\zeta\underline{\xi}).
\end{aligned}
$$

The error term can be written schematically as

$$
\mathrm{Err}[\not{d}_2\,\not{d}_2^\star\underline{\xi}] = r^{-1}\,\not{\mathfrak{d}}^{\le 1}(\Gamma_b\cdot\Gamma_b) + r^{-1}\,\not{\mathfrak{d}}^{\le 1}(\Gamma_g\cdot\Gamma_b).
$$

Note also that we can write

$$
\begin{aligned}
&\underline{\kappa}\left(\frac{1}{2}\underline{\kappa}\zeta - 2\underline{\omega}\zeta\right) - \frac{3}{2}\underline{\kappa}e_\theta\underline{\kappa} + \zeta\left(\frac{1}{2}\underline{\kappa}^2 + 2\underline{\omega}\,\underline{\kappa}\right) + 6\rho\underline{\xi} - 2\underline{\omega}e_\theta(\underline{\kappa})\\
=\ &\frac{4\Upsilon^2}{r^2}\zeta + \frac{1}{r}\left(3 - \frac{8m}{r}\right)e_\theta(\underline{\kappa}) - \frac{12m}{r^3}\underline{\xi} + r^{-1}\,\not{\mathfrak{d}}^{\le 1}(\Gamma_g\cdot\Gamma_b).
\end{aligned}
$$

Also, using the transport equation for $e_4(\underline{\kappa})$ and the GCM condition for κ, we have

$$
\begin{aligned}
e_4(e_\theta(\underline{\kappa})) &= e_\theta e_4\underline{\kappa} + [e_4, e_\theta]\underline{\kappa}\\
&= e_\theta\left[-\frac{1}{2}\kappa\,\underline{\kappa} - 2\,\not{d}_1\zeta + 2\rho + \Gamma_g\cdot\Gamma_b\right] + \frac{1}{2}\kappa e_\theta\underline{\kappa} + \text{l.o.t.}\\
&= -\frac{1}{r}e_\theta\underline{\kappa} + 2\,\not{d}_1^\star\,\not{d}_1\zeta + 2e_\theta(\rho) + r^{-1}\,\not{\mathfrak{d}}(\Gamma_g\cdot\Gamma_b)\\
&= -\frac{1}{r}e_\theta\underline{\kappa} + 2(\not{d}_2\,\not{d}_2^\star + 2K)\zeta + 2e_\theta(\rho) + r^{-1}\,\not{\mathfrak{d}}(\Gamma_g\cdot\Gamma_b).
\end{aligned}
$$

We can also write, since $\nu_* = e_3 + a_* e_4$,

$$
\begin{aligned}
e_3(\zeta) &= \nu_*(\zeta) - a_* e_4(\zeta) = \nu_*(\zeta) - a_*\left(-\kappa\zeta - \beta + \Gamma_g\cdot\Gamma_g\right)\\
&= \nu_*(\zeta) - \left(1 + \frac{2m}{r}\right)\left(\frac{2}{r}\zeta + \beta\right) + \Gamma_b\cdot\Gamma_g.
\end{aligned}
$$

[18] Note that the schematic form of Proposition 7.12 is not suitable here.

Combining, we deduce

$$\nu_*(e_\theta(\underline{\kappa})) = -2\,\not{d}_2\,\not{d}_2^\star\underline{\xi} + \underline{\kappa}\nu_*(\zeta) + \frac{2\Upsilon}{r}\underline{\beta} + E_1 + E_2,$$
$$E_1 = \frac{4}{r^2}\left(1-\frac{6m}{r}\right)\zeta + \frac{2}{r}\left(2-\frac{3m}{r}\right)e_\theta(\underline{\kappa}) - \frac{12m}{r^3}\underline{\xi}$$
$$-2\left(1+\frac{2m}{r}\right)\not{d}_2\,\not{d}_2^\star\zeta + \frac{2\Upsilon}{r}\left(1+\frac{2m}{r}\right)\beta - 2\left(1+\frac{2m}{r}\right)e_\theta(\rho), \tag{7.3.35}$$
$$E_2 = r^{-1}\not{\mathfrak{d}}^{\le 1}(\Gamma_b\cdot\Gamma_b) + r^{-1}\not{\mathfrak{d}}^{\le 1}(\Gamma_g\cdot\Gamma_b).$$

We introduce the following notation

$$f := e_\theta(\underline{\kappa}) - \underline{\kappa}\zeta. \tag{7.3.36}$$

Using the fact that $\nu_* = e_3 + a_* e_4$, and the transport equation for $e_3(\underline{\kappa})$ and $e_4(\underline{\kappa})$, we have

$$\begin{aligned}
\nu_*(f) &= \nu_*(e_\theta(\underline{\kappa})) - \underline{\kappa}\nu_*(\zeta) - \nu_*(\underline{\kappa})\zeta\\
&= \nu_*(e_\theta(\underline{\kappa})) - \underline{\kappa}\nu_*(\zeta) - \big(e_3(\underline{\kappa}) + a_* e_4(\underline{\kappa})\big)\zeta\\
&= \nu_*(e_\theta(\underline{\kappa})) - \underline{\kappa}\nu_*(\zeta) + \frac{4}{r^2}\left(1-\frac{4m}{r}\right)\zeta + r^{-1}\not{\mathfrak{d}}^{\le 1}(\Gamma_g\cdot\Gamma_b).
\end{aligned}$$

Together with (7.3.35), we deduce, with a similar E_2,

$$\nu_*(f) = -2\,\not{d}_2\,\not{d}_2^\star\underline{\xi} + \frac{2\Upsilon}{r}\underline{\beta} + \frac{4}{r^2}\left(1-\frac{4m}{r}\right)\zeta + E_1 + E_2. \tag{7.3.37}$$

Projecting on the $\ell = 1$ mode and integrating $\not{d}_2\,\not{d}_2^\star\underline{\xi}$ by parts, we derive

$$\int_S \nu_*(f)e^\Phi = \frac{2\Upsilon}{r}\int_S \underline{\beta}e^\Phi + \frac{4}{r^2}\left(1-\frac{4m}{r}\right)\int_S \zeta e^\Phi + \int_S (E_1+E_2)e^\Phi.$$

In view of **Ref 1**, we have schematically

$$\int_S E_2 e^\Phi = O\Big(\|\not{\mathfrak{d}}^{\le 1}\Gamma_b\|^2_{L^2(S)} + \epsilon r^{-1}u^{-\frac{1}{2}-\delta_{dec}}\|\not{\mathfrak{d}}^{\le 1}\Gamma_b\|_{L^2(S)}\Big).$$

Also, using the GCM condition for the $\ell = 1$ mode of $\underline{\xi}$ and integrating $\not{d}_2^\star\,\not{d}_2\zeta$ by parts, we infer

$$\begin{aligned}
&\int_S\left(\frac{4}{r^2}\left(1-\frac{4m}{r}\right)\zeta + E_1\right)e^\Phi\\
&= \frac{8}{r^2}\left(1-\frac{5m}{r}\right)\int_S \zeta e^\Phi + \frac{2}{r}\left(2-\frac{3m}{r}\right)\int_S e_\theta(\underline{\kappa})e^\Phi\\
&\quad + \frac{2\Upsilon}{r}\left(1+\frac{2m}{r}\right)\int_S \beta e^\Phi - 2\left(1+\frac{2m}{r}\right)\int_S e_\theta(\rho)e^\Phi.
\end{aligned}$$

Hence, in view of the intermediate assumption 7.3.29 for the $\ell = 1$ modes of β and

$e_\theta(\underline{\kappa})$ as well as estimate (7.3.30) for the $\ell = 1$ mode of ζ,

$$\left|\int_S\left(\frac{4}{r^2}\left(1-\frac{4m}{r}\right)\zeta+E_1\right)e^\Phi\right| \lesssim \epsilon r^{-2}u^{-1-\delta_{dec}}+\epsilon r^{-1}u^{-2-2\delta_{dec}} + \left|\int_S e_\theta(\rho)e^\Phi\right|.$$

We deduce

$$\left|\int_S(\nu_* f)e^\Phi\right| \lesssim r^{-1}\left|\int_S\underline{\beta}e^\Phi\right|+\left|\int_S e_\theta(\rho)e^\Phi\right|+\|\not{\mathfrak{d}}^{\leq 1}\Gamma_b\|^2_{L^2(S)} +\epsilon r^{-1}u^{-\frac{1}{2}-\delta_{dec}}\|\not{\mathfrak{d}}^{\leq 1}\Gamma_b\|_{L^2(S)}+\epsilon r^{-2}u^{-1-\delta_{dec}}+\epsilon r^{-1}u^{-2-2\delta_{dec}},$$

or, making use of the assumption $r \geq \epsilon\epsilon_0^{-1}u^{1+\delta_{dec}}$,

$$\left|\int_S(\nu_* f)e^\Phi\right| \lesssim r^{-1}\left|\int_S\underline{\beta}e^\Phi\right|+\left|\int_S e_\theta(\rho)e^\Phi\right|+\|\not{\mathfrak{d}}^{\leq 1}\Gamma_b\|^2_{L^2(S)} +\epsilon_0 u^{-3-2\delta_{dec}}. \tag{7.3.38}$$

On the other hand, according to Lemma 7.19,

$$\nu_*\left(\int_S fe^\Phi\right)=\int_S(\nu_* f)e^\Phi+\frac{3}{2}\left(\overline{\underline{\kappa}}-2\overline{\kappa}-\overline{\underline{\Omega}\kappa}\right)\int_S fe^\Phi+\mathrm{Err}[f,\nu_*] \tag{7.3.39}$$

with error term

$$\mathrm{Err}[f,\nu_*] = r^4\Gamma_b\nu_*(f)+r^3\Gamma_b f. \tag{7.3.40}$$

Note that

$$\begin{aligned} r^{-1}\int_S fe^\Phi &= r^{-1}\int_S e_\theta(\underline{\kappa})e^\Phi - r^{-1}\int_S\underline{\kappa}\zeta e^\Phi \\ &= r^{-1}\int_S e_\theta(\underline{\kappa})e^\Phi+2r^{-2}\Upsilon\int_S\zeta e^\Phi-r^{-1}\int_S\check{\underline{\kappa}}\zeta e^\Phi. \end{aligned}$$

Thus in view of our auxiliary assumption (7.3.29) for $e_\theta(\underline{\kappa})$, estimate (7.3.30) of Step 1, and the dominance condition for r, we deduce

$$\left|r^{-1}\int_S fe^\Phi\right| \lesssim \epsilon r^{-1}u^{-2-2\delta_{dec}}+\epsilon r^{-2}u^{-1-\delta_{dec}}\lesssim\epsilon_0 u^{-3-3\delta_{dec}}. \tag{7.3.41}$$

Also, we have in view of the definition of f and (7.3.37)

$$f=r^{-1}\not{\mathfrak{d}}^{\leq 1}\Gamma_g, \qquad \nu_*(f)=r^{-2}\not{\mathfrak{d}}^{\leq 2}\Gamma_b.$$

Together with (7.3.40), we infer

$$\mathrm{Err}[f,\nu_*] = r^4\Gamma_b r^{-2}\not{\mathfrak{d}}^{\leq 2}\Gamma_b+r^2\Gamma_b\cdot\not{\mathfrak{d}}^{\leq 1}\Gamma_g.$$

We deduce, together with (7.3.38), (7.3.39) and (7.3.41),

$$\left|\nu_*\left(\int_S fe^\Phi\right)\right| \lesssim r^{-1}\left|\int_S \underline{\beta}e^\Phi\right| + \left|\int_S e_\theta(\rho)e^\Phi\right| + \|\not{\mathfrak{d}}^{\le 4}\Gamma_b\|^2_{L^2(S)} + \epsilon_0 u^{-3-2\delta_{dec}}$$

where we have also used the control of Γ_g and Sobolev.

Integrating in u, and making use of Proposition 7.22 on the flux estimates for Γ_b, as well as the estimate (7.3.32) for the $\ell = 1$ mode of $\underline{\beta}$ and the estimate (7.3.34) for the $\ell = 1$ mode of $e_\theta(\rho)$, we infer

$$\int_u^{u_*}\left|\nu_*\left(\int_S fe^\Phi\right)\right| du' \lesssim \epsilon_0 u^{-2-\delta_{dec}}.$$

We deduce

$$\begin{aligned}\left|\int_{S(u)} fe^\Phi\right| &\lesssim \left|\int_{S_*} fe^\Phi\right| + \int_u^{u_*}\left|\nu_*\left(\int_S fe^\Phi\right)\right| du' \\ &\lesssim \left|\int_{S_*} fe^\Phi\right| + \epsilon_0 u^{-2-\delta_{dec}}.\end{aligned}$$

Together with the definition of f and the GCM condition for the $\ell = 1$ mode of $e_\theta(\underline{\kappa})$ on S_*, this yields

$$\begin{aligned}\left|\int_{S(u)} e_\theta(\underline{\kappa})e^\Phi\right| &\lesssim \left|\int_{S(u)} \underline{\kappa}\zeta e^\Phi\right| + \left|\int_{S_*} \underline{\kappa}\zeta e^\Phi\right| + \epsilon_0 u^{-2-\delta_{dec}} \\ &\lesssim r^{-1}\left|\int_{S(u)} \zeta e^\Phi\right| + r^{-1}\left|\int_{S_*} \zeta e^\Phi\right| + r^3|\Gamma_g\cdot\Gamma_g| + \epsilon_0 u^{-2-\delta_{dec}}.\end{aligned}$$

Together with the estimate (7.3.30) for the $\ell = 1$ mode of ζ, the control of Γ_g, and the dominance condition of r on Σ_*, we obtain

$$\left|\int_S e_\theta(\underline{\kappa})e^\Phi\right| \lesssim \epsilon r^{-1}u^{-1-\delta_{dec}} + \epsilon_0 u^{-2-\delta_{dec}} \lesssim \epsilon_0 u^{-2-\delta_{dec}}$$

which improves the estimate for the $\ell = 1$ mode of $e_\theta(\underline{\kappa})$ in (7.3.29) and establishes the desired estimate for the $\ell = 1$ mode of $e_\theta(\underline{\kappa})$.

Step 5. We establish the desired estimate for the $\ell = 1$ mode of $e_\theta(\rho)$. In view of (7.3.33), we have

$$\left|\int_S e_\theta(\rho)e^\Phi\right| \lesssim r^{-1}\left|\int_S e_\theta(\underline{\kappa})e^\Phi\right| + \int_S |\not{\mathfrak{d}}^{\le 1}(\Gamma_g\cdot\Gamma_b)|. \tag{7.3.42}$$

Using the improved estimate of Step 4 for the $\ell = 1$ mode of $e_\theta(\underline{\kappa})$ and the control of Γ_b and Γ_g, we infer

$$\left|\int_S e_\theta(\rho)e^\Phi\right| \lesssim \epsilon_0 r^{-1}u^{-2-\delta_{dec}} + \epsilon^2 r^{-1}u^{-\frac{3}{2}-\delta_{dec}} \lesssim \epsilon_0 r^{-1}u^{-1-\delta_{dec}}$$

which is the desired estimate for the $\ell = 1$ mode of $e_\theta(\rho)$.

Step 6. To estimate the $\ell = 1$ mode of μ we differentiate the relation

$$\mu = -\text{div}\,\zeta - \rho + \Gamma_g \cdot \Gamma_b$$

by e_θ and obtain

$$\begin{aligned} e_\theta(\mu) &= \not{d}_1^\star \not{d}_1 \zeta - e_\theta(\rho) + r^{-1} \not{\mathfrak{d}}(\Gamma_g \cdot \Gamma_b) \\ &= (\not{d}_2 \not{d}_2^\star + 2K)\zeta - e_\theta(\rho) + r^{-1} \not{\mathfrak{d}}(\Gamma_g \cdot \Gamma_b) \\ &= \not{d}_2 \not{d}_2^\star \zeta + \frac{2}{r^2}\zeta - e_\theta(\rho) + r^{-1} \not{\mathfrak{d}}(\Gamma_g \cdot \Gamma_b) + \text{l.o.t.} \end{aligned}$$

Hence,

$$\int_S e_\theta(\mu) e^\Phi = 2r^{-2} \int_S \zeta e^\Phi - \int_S e_\theta(\rho) e^\Phi + r^2 \not{\mathfrak{d}}(\Gamma_g \cdot \Gamma_b). \tag{7.3.43}$$

Using the estimate (7.3.30) for the $\ell = 1$ mode of ζ and the estimate of Step 5 for the $\ell = 1$ mode of $e_\theta(\rho)$, we deduce, using also the dominant condition for r on Σ_*,

$$\left| \int_S e_\theta(\mu) e^\Phi \right| \lesssim \epsilon r^{-2} u^{-1-\delta_{dec}} + \epsilon_0 r^{-1} u^{-1-\delta_{dec}} \lesssim \epsilon_0 r^{-1} u^{-1-\delta_{dec}}$$

which is the desired estimate for the $\ell = 1$ mode of $e_\theta(\mu)$.

Step 7. It remains to estimate the $\ell = 1$ mode of β. We start with the $e_3 \beta$ equation

$$e_3 \beta + \underline{\kappa} \beta = -\not{d}_1^\star \rho + 2\underline{\omega}\beta + 3\eta\rho - \vartheta \underline{\beta} + \underline{\xi}\alpha.$$

Also, taking into account the e_4 equation for β and recalling that a_* satisfies the identity $a_* = -(1 + \frac{2m}{r}) + r\Gamma_b$,

$$\begin{aligned} \nu_*(\beta) &= e_3(\beta) + a_* e_4 \beta \\ &= -\underline{\kappa}\beta - \not{d}_1^\star \rho + 2\underline{\omega}\beta + 3\eta\rho - \vartheta\underline{\beta} + \underline{\xi}\alpha + a^* \left(-2\kappa\beta + \not{d}_2 \alpha + \zeta\alpha\right) \\ &= -\not{d}_1^\star \rho - \frac{6m}{r^3}\eta + \frac{6}{r}\left(1 - \frac{m}{r}\right)\beta - \left(1 + \frac{2m}{r}\right) \not{d}_2 \alpha + r^{-1}\Gamma_b \cdot \not{\mathfrak{d}}^{\leq 1}\Gamma_g. \end{aligned}$$

Projecting on the $\ell = 1$ mode, and using the GCM condition for the $\ell = 1$ mode of η,

$$\int_S \nu_*(\beta) e^\Phi = \int_S e_\theta(\rho) e^\Phi + \frac{6}{r}\left(1 - \frac{m}{r}\right) \int_S \beta e^\Phi + \int_S \Gamma_b \cdot \not{\mathfrak{d}}^{\leq 1}\Gamma_g. \tag{7.3.44}$$

On the other hand, making use of Lemma 7.19, the auxiliary assumption (7.3.29) for the $\ell = 1$ mode of β and $r \geq \epsilon \epsilon_0^{-1} u^{1+\delta_{dec}}$,

$$\begin{aligned} \nu_* \left(\int_S \beta e^\Phi \right) &= \int_S (\nu_* \beta) e^\Phi + \frac{3}{2}(\underline{\kappa} - 2\overline{\kappa} - \overline{\underline{\Omega}\,\kappa}) \int_S \beta e^\Phi + r^4 \Gamma_b \nu_*(\beta) + r^3 \Gamma_b \beta \\ &= \int_S (\nu_* \beta) e^\Phi - \frac{6}{r} \int_S \beta e^\Phi + r^2 \Gamma_b \cdot \mathfrak{d}^{\leq 1} \Gamma_g \end{aligned}$$

where we have used the fact that $\nu_*\beta = r^{-2}\Gamma_g$. Together with (7.3.44), we deduce

$$\left|\nu_*\left(\int_S \beta e^\Phi\right)\right| \lesssim \left|\int_S e_\theta(\rho)e^\Phi\right| + \frac{1}{r}\left|\int_S \beta e^\Phi\right| + r^2\Gamma_b\cdot\mathfrak{d}^{\leq 1}\Gamma_g.$$

Using (7.3.29) and (7.3.42), we infer

$$\begin{aligned}\left|\nu_*\left(\int_S \beta e^\Phi\right)\right| &\lesssim \frac{1}{r}\left|\int_S e_\theta(\underline{\kappa})e^\Phi\right| + \epsilon r^{-2}u^{-1-\delta_{dec}} + r^2|\Gamma_b\cdot\mathfrak{d}^{\leq 1}\Gamma_g| \\ &+ \int_S |\not{\mathfrak{d}}^{\leq 1}(\Gamma_g\cdot\Gamma_b)|.\end{aligned}$$

Using the control of the $\ell = 1$ mode for $e_\theta(\underline{\kappa})$ derived in Step 4, our control for Γ_g and Sobolev, as well as the dominance condition in r on Σ_*, we infer

$$\left|\nu_*\left(\int_S \beta e^\Phi\right)\right| \lesssim \epsilon_0 r^{-1}u^{-2-\delta_{dec}} + \epsilon r^{-1}u^{-\frac{1}{2}-\delta_{dec}}\|\not{\mathfrak{d}}^{\leq 2}\Gamma_b\|_{L^2(S)}. \tag{7.3.45}$$

Integrating (7.3.45) in u, and making use of Proposition 7.22 on the flux estimates for Γ_b, we infer

$$\begin{aligned}\int_u^{u_*}\left|\nu_*\left(\int_S \beta e^\Phi\right)\right| &\lesssim \epsilon_0 r^{-1}u^{-1-\delta_{dec}} \\ &+\epsilon r^{-1}\left(\int_u^{u_*} u'^{-1-2\delta_{dec}}\right)^{\frac{1}{2}}\left(\int_{\Sigma_*(u,u_*)}|\Gamma_b|^2\right)^{\frac{1}{2}} \\ &\lesssim \epsilon_0 r^{-1}u^{-1-\delta_{dec}}.\end{aligned}$$

In view of the GCM condition for the $\ell = 1$ mode of β on S_*, we infer on Σ_*

$$\left|\int_S \beta e^\Phi\right| \lesssim \epsilon_0 r^{-1}u^{-1-\delta_{dec}}$$

which is the desired estimate for $k = 0$. Also, coming back to (7.3.45), and using our control for Γ_b, we have

$$\left|\nu_*\left(\int_S \beta e^\Phi\right)\right| \lesssim \epsilon_0 r^{-1}u^{-1-\delta_{dec}}$$

which is our desired estimate for $k = 1$.

It remains to consider the case $2 \leq k \leq k_{small} + 20$. In view of Corollary 7.20, we easily derive the following estimate:

$$\begin{aligned}\left|\nu_*^k\left(\int_S \beta e^\Phi\right)\right| &\lesssim \left|\nu_*^{\leq k-2}\left(\int_S \nu_*^2\beta e^\Phi\right)\right| + \left|\nu_*\left(\int_S \beta e^\Phi\right)\right| + \left|\int_S \beta e^\Phi\right| \\ &+\left|\mathfrak{d}^{\leq k-1}(r^4\Gamma_b\nu_*(\beta) + r^3\Gamma_b\beta)\right|.\end{aligned}$$

Since $\nu_*\beta = r^{-2}\Gamma_g$, and using our product estimates, as well as the above improved estimates for the $\ell = 1$ mode of β for $k = 0$ and $k = 1$, we deduce the following, for

$2 \leq k \leq k_{small} + 20$,

$$\left|\nu_*^k\left(\int_S \beta e^\Phi\right)\right| \lesssim \left|\nu_*^{\leq k-2}\left(\int_S \nu_*^2\beta e^\Phi\right)\right| + \epsilon_0 r^{-1}u^{-1-\delta_{dec}}. \qquad (7.3.46)$$

In view of (7.3.46), we need to estimate $\nu_*^2\beta$. Recall from above that

$$\nu_*(\beta) = -\not{d}_1^\star\rho - \frac{6m}{r^3}\eta + \frac{6}{r}\left(1-\frac{m}{r}\right)\beta - \left(1+\frac{2m}{r}\right)\not{d}_2\alpha + r^{-1}\Gamma_b\cdot\not{\mathfrak{d}}^{\leq 1}\Gamma_g.$$

Differentiating again, and relying on the commutation formula of Lemma 7.17, we infer

$$\begin{aligned}\nu_*^2(\beta) &= -\not{d}_1^\star(\nu_*\check{\rho}) - \frac{6m}{r^3}\nu_*\eta + \frac{6}{r}\left(1-\frac{m}{r}\right)\nu_*\beta - \left(1+\frac{2m}{r}\right)\not{d}_2(\nu_*\alpha)\\ &\quad + r^{-1}\mathfrak{d}^{\leq 1}(\Gamma_b\cdot\not{\mathfrak{d}}^{\leq 1}\Gamma_g).\end{aligned}$$

Also, using the Bianchi identity for $e_3(\check{\rho})$ and $e_4(\check{\rho})$, as well as $\nu_* = e_3 + a_*e_4$, we have

$$\nu_*\check{\rho} = \not{d}_1\underline{\beta} + r^{-2}\mathfrak{d}^{\leq 1}\Gamma_g$$

and hence

$$\begin{aligned}\nu_*^2(\beta) &= -\not{d}_1^\star\not{d}_1\underline{\beta} - \frac{6m}{r^3}\nu_*\eta + \frac{6}{r}\left(1-\frac{m}{r}\right)\nu_*\beta - \left(1+\frac{2m}{r}\right)\not{d}_2(\nu_*\alpha)\\ &\quad + r^{-3}\mathfrak{d}^{\leq 2}\Gamma_g + r^{-1}\mathfrak{d}^{\leq 1}(\Gamma_b\cdot\not{\mathfrak{d}}^{\leq 1}\Gamma_g).\end{aligned}$$

This yields

$$\begin{aligned}\int_S \nu_*^2(\beta)e^\Phi &= -\int_S \not{d}_1^\star\not{d}_1\underline{\beta}e^\Phi - \frac{6m}{r^3}\int_S \nu_*\eta e^\Phi + \frac{6}{r}\left(1-\frac{m}{r}\right)\int_S \nu_*\beta e^\Phi\\ &\quad + \mathfrak{d}^{\leq 2}\Gamma_g + r^2\mathfrak{d}^{\leq 1}(\Gamma_b\cdot\not{\mathfrak{d}}^{\leq 1}\Gamma_g).\end{aligned}$$

Since $\not{d}_1^\star\not{d}_1 = \not{d}_2\not{d}_2^\star + 2K$, we infer

$$\begin{aligned}\int_S \nu_*^2(\beta)e^\Phi &= -\frac{2}{r^2}\int_S \underline{\beta}e^\Phi - \frac{6m}{r^3}\int_S \nu_*\eta e^\Phi + \frac{6}{r}\left(1-\frac{m}{r}\right)\int_S \nu_*\beta e^\Phi\\ &\quad + \mathfrak{d}^{\leq 2}\Gamma_g + r^2\mathfrak{d}^{\leq 1}(\Gamma_b\cdot\not{\mathfrak{d}}^{\leq 1}\Gamma_g).\end{aligned}$$

Also, applying Lemma 7.19 to the last two integrals of the RHS, and using again that $\nu_*\beta = r^{-2}\Gamma_g$, we obtain

$$\begin{aligned}\int_S \nu_*^2(\beta)e^\Phi &= -\frac{2}{r^2}\int_S \underline{\beta}e^\Phi - \frac{6m}{r^3}\nu_*\left(\int_S \eta e^\Phi\right) + \frac{6}{r}\left(1-\frac{m}{r}\right)\nu_*\left(\int_S \beta e^\Phi\right)\\ &\quad + \mathfrak{d}^{\leq 2}\Gamma_g + r^2\mathfrak{d}^{\leq 1}(\Gamma_b\cdot\not{\mathfrak{d}}^{\leq 1}\Gamma_g).\end{aligned}$$

Together with (7.3.46), the control of Γ_b and Γ_g, and the dominance of r on Σ_*, and arguing again by iteration for the β term, we deduce

$$\left|\nu_*^k\left(\int_S \beta e^\Phi\right)\right| \lesssim \frac{1}{r^2}\left|\nu_*^{k-2}\left(\int_S \underline{\beta} e^\Phi\right)\right| + \frac{1}{r^3}\left|\nu_*^{k-1}\left(\int_S \eta e^\Phi\right)\right| + \epsilon_0 r^{-1}u^{-1-\delta_{dec}}.$$

Together with the GCM condition for the $\ell = 1$ mode for η and the estimate (7.3.21), we infer for $2 \leq k \leq k_{small} + 20$

$$\left|\nu_*^k\left(\int_S \beta e^\Phi\right)\right| \lesssim \epsilon_0 r^{-1}u^{-1-\delta_{dec}}$$

as desired. This completes the proof of Proposition 7.26. □

7.3.5 Decay of Ricci and curvature components on Σ_*

Recall that:

- we have already derived improved pointwise estimates for α and $\underline{\alpha}$, respectively in Theorem M2 and Theorem M3,
- we have already derived improved pointwise estimates for $\underline{\beta}$ and Γ_b on Σ_*, see (7.3.16) in Proposition 7.22,
- $\check{\kappa} = 0$ on Σ_* in view of the GCM condition for κ.

In the following proposition, we derive improved pointwise estimates on Σ_* for the remaining quantities, i.e., $\check{\underline{\kappa}}$, $\check{\rho}$, ζ, $\check{\mu}$, ϑ and β.

Proposition 7.28. *The following estimates hold along Σ_* for all $k \leq k_{small} + 18$:*

$$\begin{aligned}\|\check{\underline{\kappa}}, r\check{\mu}\|^*_{\infty,k} &\lesssim \epsilon_0 r^{-2}u^{-1-\delta_{dec}},\\ \|\vartheta, \zeta, r\check{\rho}\|^*_{\infty,k} &\lesssim \epsilon_0 r^{-2}u^{-\frac{1}{2}-\delta_{dec}},\\ \|\beta\|^*_{\infty,k} &\lesssim \epsilon_0 r^{-3}(2r+u)^{-\frac{1}{2}-\delta_{dec}}.\end{aligned} \tag{7.3.47}$$

Also, for all $k \leq k_{small} + 17$

$$\begin{aligned}\left\|\nu_*(\vartheta, \zeta, r\check{\rho})\right\|^*_{\infty,k} &\lesssim \epsilon_0 r^{-2}u^{-1-\delta_{dec}},\\ \|\nu_*\beta\|^*_{\infty,k} &\lesssim \epsilon_0 r^{-4}u^{-\frac{1}{2}-\delta_{dec}}.\end{aligned} \tag{7.3.48}$$

Proof. We proceed in several steps.

Step 1. In this step we control $\check{\underline{\kappa}}$. First, note from Proposition 2.73 that we have

$$e_3(\check{\underline{\kappa}}) = r^{-1}\not{\mathfrak{d}}^{\leq 1}\Gamma_b + r^{-1}\Gamma_g + \Gamma_b\cdot\Gamma_b, \qquad e_4(\check{\underline{\kappa}}) = r^{-1}\not{\mathfrak{d}}^{\leq 1}\Gamma_g,$$

and hence

$$\nu_*(\check{\underline{\kappa}}) = r^{-1}\not{\mathfrak{d}}^{\leq 1}\Gamma_b + r^{-1}\not{\mathfrak{d}}^{\leq 1}\Gamma_g + \Gamma_b\cdot\Gamma_b.$$

Together with the improved control of Γ_b in (7.3.16), the control of Γ_g, and the dominance in r condition, we infer, for all $k \leq k_{small} + 17$,

$$\|\nu_*\check{\underline{\kappa}}\|^*_{\infty,k} \lesssim \epsilon_0 r^{-2}u^{-1-\delta_{dec}} + \epsilon r^{-3} \lesssim \epsilon_0 r^{-2}u^{-1-\delta_{dec}}.$$

It then remains to control $\not{\mathfrak{d}}^k\check{\underline{\kappa}}$. Since we have $\not{d}_2^\star \not{d}_1^\star \underline{\kappa} = 0$ in view of our GCM condition, we infer, using a Poincaré inequality,

$$r^{-1}\|\not{\mathfrak{d}}^k\check{\underline{\kappa}}\|_{L^2(S)} \lesssim r^{-2}\left|\int_S e_\theta(\underline{\kappa})e^\Phi\right| \lesssim \epsilon_0 r^{-2}u^{-1-\delta_{dec}}$$

where we have used Proposition 7.26 to estimate the $\ell = 1$ mode of $\underline{\kappa}$. Together with the above estimate for $\nu_*\check{\underline{\kappa}}$, we infer, for all $k \leq k_{small} + 18$, the desired estimate

$$\|\check{\underline{\kappa}}\|^*_{\infty,k} \lesssim \epsilon_0 r^{-2}u^{-1-\delta_{dec}}.$$

Step 2. Next, we estimate $\check{\rho}$. First, note from Proposition 2.73 that we have

$$e_3(\check{\rho}) = r^{-1}\not{\mathfrak{d}}^{\leq 1}\underline{\beta} + r^{-2}\Gamma_g + \Gamma_b\cdot\Gamma_g, \qquad e_4(\check{\rho}) = r^{-2}\not{\mathfrak{d}}^{\leq 1}\Gamma_g,$$

and hence

$$\nu_*(\check{\rho}) = r^{-1}\not{\mathfrak{d}}^{\leq 1}\underline{\beta} + r^{-2}\not{\mathfrak{d}}^{\leq 1}\Gamma_g + \Gamma_b\cdot\Gamma_g.$$

Together with the improved control of $\underline{\beta}$ in (7.3.16), the control of Γ_g, and the dominance in r condition, we infer, for all $k \leq k_{small} + 17$,

$$\|\nu_*\check{\rho}\|^*_{\infty,k} \lesssim \epsilon_0 r^{-3}u^{-1-\delta_{dec}} + \epsilon r^{-4} \lesssim \epsilon_0 r^{-3}u^{-1-\delta_{dec}}.$$

It then remains to control $\not{\mathfrak{d}}^k\check{\rho}$. In view of Proposition 7.10, we have, relative to the background frame of ${}^{(ext)}\mathcal{M}$,

$$r^4\left(\not{d}_2^\star \not{d}_1^\star\rho + \frac{3}{4}\underline{\kappa}\rho\vartheta + \frac{3}{4}\kappa\rho\underline{\vartheta}\right) = \mathfrak{q} + r^2\not{\mathfrak{d}}^{\leq 2}(\Gamma_b\cdot\Gamma_g) + \text{l.o.t.}$$

Using the improved control for $\mathfrak{q}$ in **Ref 2**, the improved control for $\underline{\vartheta}$ in (7.3.16), and the product Lemma 7.6, we have, for all $k \leq k_{small} + 17$,

$$\|\mathfrak{q}\|^*_{\infty,k} + \|\underline{\vartheta}\|^*_{\infty,k} + \|r^2\not{\mathfrak{d}}^{\leq 2}(\Gamma_b\cdot\Gamma_g)\|^*_{\infty,k} \lesssim \epsilon_0 r^{-1}u^{-\frac{1}{2}-\delta_{dec}}.$$

Also, using our condition on r along Σ_*

$$\|\vartheta\|^*_{\infty,k} \lesssim \epsilon r^{-2} \lesssim \epsilon_0 r^{-1}u^{-1-\delta_{dec}}.$$

We deduce, for all $k \leq k_{small} + 17$,

$$\|\not{d}_2^\star \not{d}_1^\star\check{\rho}\|^*_{\infty,k} \lesssim \epsilon_0 r^{-5}u^{-\frac{1}{2}-\delta_{dec}}.$$

We infer, using a Poincaré inequality, for all $k \leq k_{small} + 19$,

$$r^{-1}\|\not{\mathfrak{d}}^k\check{\rho}\|_{L^2(S)} \lesssim r^{-2}\left|\int_S e_\theta(\rho)e^\Phi\right| + \epsilon_0 r^{-3}u^{-\frac{1}{2}-\delta_{dec}} \lesssim \epsilon_0 r^{-3}u^{-\frac{1}{2}-\delta_{dec}}$$

where we have used Proposition 7.26 to estimate the $\ell = 1$ mode of ρ. Together with the above estimate for $\nu_*\check{\underline{\kappa}}$, we infer, for all $k \leq k_{small} + 18$, the desired estimate

$$\|\check{\rho}\|^*_{\infty,k} \lesssim \epsilon_0 r^{-3}u^{-\frac{1}{2}-\delta_{dec}}.$$

Step 3. Next, we control $\not{\mathfrak{d}}^k \check{\mu}$ and $\not{\mathfrak{d}}^k \zeta$. Since we have $\not{d}_2^\star \not{d}_1^\star \mu = 0$ in view of our GCM condition, we infer, using a Poincaré inequality,

$$r^{-1}\|\not{\mathfrak{d}}^k \check{\mu}\|_{L^2(S)} \lesssim r^{-2}\left|\int_S e_\theta(\mu) e^\Phi\right| \lesssim \epsilon_0 r^{-3} u^{-1-\delta_{dec}}$$

where we have used Proposition 7.26 to estimate the $\ell = 1$ mode of μ.

Also, from the definition of $\mu = -\not{d}_1 \zeta - \rho + \frac{1}{4}\vartheta\underline{\vartheta}$, we have

$$\not{d}_1 \zeta = -\check{\mu} - \check{\rho} + \Gamma_g \cdot \Gamma_b$$

and hence, using also the product Lemma 7.6, we infer

$$r^{-1}\|\not{\mathfrak{d}}^k \zeta\|_{L^2(S)} \lesssim \|\not{\mathfrak{d}}^k \check{\mu}\|_{L^2(S)} + \|\not{\mathfrak{d}}^k \check{\rho}\|_{L^2(S)} + \epsilon_0 r^{-2} u^{-1-\delta_{dec}}.$$

Together with the above improved estimates for $\check{\mu}$, and the improved estimates for $\check{\rho}$ of Step 2, we deduce

$$r^{-1}\|\not{\mathfrak{d}}^k \zeta\|_{L^2(S)} \lesssim \epsilon_0 r^{-2} u^{-\frac{1}{2}-\delta_{dec}}, \qquad r^{-1}\|\not{\mathfrak{d}}^k \check{\mu}\|_{L^2(S)} \lesssim \epsilon_0 r^{-3} u^{-1-\delta_{dec}}.$$

Step 4. We conclude in this step the control of ζ and $\check{\mu}$. From the null structure equations, we have

$$e_3(\zeta) = r^{-1}\not{\mathfrak{d}}^{\le 1}\Gamma_b + r^{-1}\Gamma_g + \Gamma_b \cdot \Gamma_b, \qquad e_4(\zeta) = r^{-1}\not{\mathfrak{d}}^{\le 1}\Gamma_g,$$

and hence

$$\nu_*(\zeta) = r^{-1}\not{\mathfrak{d}}^{\le 1}\Gamma_b + r^{-1}\not{\mathfrak{d}}^{\le 1}\Gamma_g + \Gamma_b \cdot \Gamma_b.$$

Together with the improved control of Γ_b in (7.3.16), the control of Γ_g, and the dominance in r condition, we infer, for all $k \le k_{small} + 17$,

$$\|\not{\mathfrak{d}}^{\le 1}\nu_* \zeta\|^*_{\infty,k} \lesssim \epsilon_0 r^{-2} u^{-1-\delta_{dec}} + \epsilon r^{-3} \lesssim \epsilon_0 r^{-2} u^{-1-\delta_{dec}}.$$

Using

$$\check{\mu} = -\not{d}_1 \zeta - \check{\rho} + \Gamma_g \cdot \Gamma_b,$$

we also have for all $k \le k_{small} + 17$, in view of the above improved estimate for $\nu_* \zeta$, the commutation formula of Lemma 7.17, and the improved estimate of $\check{\rho}$ of Step 2,

$$\|\nu_* \check{\rho}\|^*_{\infty,k} \lesssim \epsilon_0 r^{-3} u^{-1-\delta_{dec}}.$$

Together with the improved estimate for $\not{\mathfrak{d}}^k \zeta$ and $\not{\mathfrak{d}}^k \check{\rho}$ of Step 3, we infer, for all $k \le k_{small} + 18$, the desired estimates

$$\|\zeta\|^*_{\infty,k} \lesssim \epsilon_0 r^{-2} u^{-\frac{1}{2}-\delta_{dec}}, \qquad \|\check{\mu}\|^*_{\infty,k} \lesssim \epsilon_0 r^{-3} u^{-1-\delta_{dec}}.$$

Step 5. Next, we estimate ϑ. From the null structure equations, we have

$$e_3(\vartheta) = r^{-1}\not{\mathfrak{d}}^{\le 1}\Gamma_b + r^{-1}\Gamma_g + \Gamma_b \cdot \Gamma_b, \qquad e_4(\vartheta) = r^{-1}\not{\mathfrak{d}}^{\le 1}\Gamma_g,$$

and hence

$$\nu_*(\vartheta) \quad = \quad r^{-1}\mathfrak{d}^{\leq 1}\Gamma_b + r^{-1}\mathfrak{d}^{\leq 1}\Gamma_g + \Gamma_b\cdot\Gamma_b.$$

Together with the improved control of Γ_b in (7.3.16), the control of Γ_g, and the dominance in r condition, we infer, for all $k\leq k_{small}+17$,

$$\|\nu_*\vartheta\|^*_{\infty,k} \quad \lesssim \quad \epsilon_0 r^{-2}u^{-1-\delta_{dec}} + \epsilon r^{-3} \lesssim \epsilon_0 r^{-2}u^{-1-\delta_{dec}}.$$

It remains to control $\mathfrak{d}^k\vartheta$. We use Codazzi and the GCM equation for κ, which yields

$$\begin{aligned}\not{d}_2\vartheta \quad &= \quad -2\beta + (e_\theta(\kappa)+\zeta\kappa) + \Gamma_g\cdot\Gamma_g,\\ &= \quad -2\beta + \frac{2}{r}\zeta + \Gamma_g\cdot\Gamma_g.\end{aligned}$$

Making use of the bootstrap assumption for β and the estimate of Step 3 for ζ,

$$\begin{aligned}r^{-1}\big\|\mathfrak{d}^k\not{d}_2\vartheta\big\|_{L^2(S)} \quad &\lesssim \quad \|\beta\|_{\infty,k} + r^{-1}\|\zeta\|^*_{\infty,k} + \epsilon_0 r^{-3}u^{-1-\delta_{dec}}\\ &\lesssim \quad \epsilon r^{-\frac{7}{2}-\delta_{dec}} + \epsilon_0 r^{-3}u^{-1/2-\delta_{dec}}\end{aligned}$$

from which we derive, using also the condition $r\geq\epsilon\epsilon_0^{-1}u^{1+\delta_{dec}}$ and a Poincaré inequality,

$$r^{-1}\big\|\mathfrak{d}^k\vartheta\big\|_{L^2(S)} \quad \lesssim \quad \epsilon_0 r^{-2}u^{-1/2-\delta_{dec}}.$$

Together with the above estimate for $\nu_*\vartheta$, we infer, for all $k\leq k_{small}+18$,

$$\|\vartheta\|^*_{\infty,k} \quad \lesssim \quad \epsilon_0 r^{-2}u^{-1/2-\delta_{dec}}.$$

Step 6. Finally, we estimate β. From Bianchi, we have

$$e_3\beta = r^{-1}\mathfrak{d}\check{\rho} + r^{-1}\beta + r^{-3}\Gamma_b + r^{-1}\Gamma_b\cdot\Gamma_g, \quad e_4(\beta) = r^{-1}\mathfrak{d}\alpha + r^{-1}\beta + r^{-1}\Gamma_g\cdot\Gamma_g$$

and hence

$$\nu_*(\beta) \quad = \quad r^{-1}\mathfrak{d}\check{\rho} + r^{-1}\mathfrak{d}\alpha + r^{-1}\beta + r^{-3}\Gamma_b + r^{-1}\Gamma_b\cdot\Gamma_g.$$

Together with the improved control of Γ_b in (7.3.16), the bootstrap assumptions for α and β, the improved estimate for $\check{\rho}$ of Step 2, and the dominance in r condition, we infer, for all $k\leq k_{small}+17$,

$$\|\nu_*\beta\|^*_{\infty,k} \quad \lesssim \quad \epsilon_0 r^{-4}u^{-\frac{1}{2}-\delta_{dec}} + \epsilon r^{-\frac{9}{2}} \lesssim \epsilon_0 r^{-4}u^{-\frac{1}{2}-\delta_{dec}}.$$

It remains to control $\mathfrak{d}^k\beta$. We have from Bianchi

$$\not{d}_2^\star\beta \quad = \quad e_3\alpha + r^{-1}\alpha + r^{-3}\vartheta + \Gamma_b\cdot(\alpha,\beta) + r^{-1}\Gamma_g\cdot\Gamma_g.$$

Hence, using in particular the improved estimate for ϑ of Step 6, and the improved estimate for α and $e_3\alpha$ of **Ref 2**, we infer

$$\|\mathfrak{d}^k\not{d}_2^\star\beta\|_{L^2(S)} \quad \lesssim \quad \epsilon_0 r^{-4}(2r+u)^{-\frac{1}{2}-\delta_{dec}}.$$

This yields, using a Poincaré inequality,

$$r^{-1}\| \not{\mathfrak{d}}^k\beta\|_{L^2(S)} \lesssim r^{-3}\left|\int_S \beta e^{\Phi}\right| + \epsilon_0 r^{-3}(2r+u)^{-\frac{1}{2}-\delta_{dec}} \lesssim \epsilon_0 r^{-3}(2r+u)^{-\frac{1}{2}-\delta_{dec}}$$

where we have used Proposition 7.26 to estimate the $\ell=1$ mode of β. Together with the above estimate for $\nu_*\beta$, we infer, for all $k\le k_{small}+18$, the desired estimate

$$\|\beta\|^*_{\infty,k} \lesssim \epsilon_0 r^{-4}u^{-\frac{1}{2}-\delta_{dec}}.$$

This concludes the proof of Proposition 7.28. □

7.4 CONTROL IN ${}^{(ext)}\mathcal{M}$, PART I

7.4.1 Preliminaries

7.4.1.1 Commutation lemmas

Here and below we write schematically

$$\not{\mathfrak{d}} = r\not{d}, \quad \mathfrak{d}_{\nearrow} = \{re_4, \not{\mathfrak{d}}\}, \quad \mathfrak{d} = (e_3, re_4, r\not{d}), \quad \mathbf{T} = \frac{1}{2}(e_3+\Upsilon e_4).$$

Lemma 7.29. *We have, schematically,*

$$[\mathbf{T}, e_4] = r^{-1}\Gamma_b\mathfrak{d}_{\nearrow}, \qquad [\not{\mathfrak{d}}, e_4] = \check{\Gamma}_g\mathfrak{d}_{\nearrow} + \Gamma_g. \tag{7.4.1}$$

Also,

$$\mathbf{T}(r) = \frac{1}{2r}\underline{A} = r^{-1}\Gamma_b.$$

Proof. The identity for $[\not{\mathfrak{d}}, e_4]$ has already been discussed in Corollary 7.5. According to Lemma 2.69, we have

$$[\mathbf{T}, e_4] = \left(\left(\underline{\omega} - \frac{m}{r^2}\right) - \frac{m}{2r}\left(\overline{\kappa} - \frac{2}{r}\right) + \frac{e_4(m)}{r}\right)e_4 + (\eta+\zeta)e_\theta.$$

In view of **Ref 4** and bootstrap assumptions **Ref 2**, the factors of e_4 and e_θ, on the right-hand side, behave at worst like Γ_b. Thus schematically $[\mathbf{T}, e_4] = r^{-1}\Gamma_b\mathfrak{d}_{\nearrow}$. □

7.4.1.2 Transport lemmas

The following lemma will be used repeatedly in what follows.

Lemma 7.30. *If f verifies the transport equation*

$$e_4(f) + \frac{p}{2}\overline{\kappa}f = F,$$

we have for fixed u *and any* $r_0 \le r \le r_*$,

$$
\begin{aligned}
r^p \|f\|_\infty(u,r) &\lesssim r_0^p \|f\|_\infty(u, r_{\mathcal{H}}) + \int_{r_0}^{r} \lambda^p \|F\|_\infty(u,\lambda) d\lambda, \\
r^p \|f\|_\infty(u,r) &\lesssim r_*^p \|f\|_\infty(u, r_*) + \int_{r}^{r_*} \lambda^p \|F\|_\infty(u,\lambda) d\lambda,
\end{aligned}
\tag{7.4.2}
$$

where r *is the area radius at fixed* u.

Proof. According to Corollary 2.67 we have $e_4(r^p f) = r^p F$. The desired estimates follow easily by integration with respect to the affine parameter s, where we recall that $e_4(s) = 1$. □

Proposition 7.31. *The following inequalities hold true for all* $k \le k_{large} - 5$, $r_0 \le r \le r_*$*:*

$$
\begin{aligned}
r^p \|f\|_{\infty,k}(u,r) &\lesssim r_*^p \|f\|_{\infty,k}(u, r_*) + \int_{r}^{r_*} \lambda^p \|F\|_{\infty,k}(u,\lambda) d\lambda, \\
r^p \|f\|_{\infty,k}(u,r) &\lesssim r_0^p \|f\|_{\infty,k}(u, r_0) + \int_{r_0}^{r} \lambda^p \|F\|_{\infty,k}(u,\lambda) d\lambda.
\end{aligned}
\tag{7.4.3}
$$

Proof. Commuting the equation $e_4(r^p f) = r^p F$ with $\not{\mathfrak{d}}$, applying the commutation Lemma 7.29 and our bootstrap assumptions on Γ_g, we derive

$$
\begin{aligned}
e_4(r^p|\not{\mathfrak{d}} f|) &\lesssim r^p|\not{\mathfrak{d}} F| + r^p|\not{\mathfrak{d}} F| + \epsilon r^{-2} r^p\big(|\not{\mathfrak{d}} f| + r e_4(r^p f)\big) \\
&\lesssim r^p(|\not{\mathfrak{d}} F| + |F|) + \epsilon r^{-2} r^p(|\not{\mathfrak{d}} f| + |f|).
\end{aligned}
$$

Similarly, commuting with $\mathbf{T}$,

$$
e_4\big(\mathbf{T}(r^p f)\big) = \mathbf{T}(r^p F) - [\mathbf{T}, e_4](r^p f) = r^p F - p r^{p-1}\mathbf{T}(r) F - r^{-1}\Gamma_b \mathfrak{d}_{\nearrow} r^p f.
$$

Hence,

$$
e_4\big(r^p \mathbf{T} f\big) = r^p \mathbf{T}(F) - p r^{p-1}\mathbf{T}(r) F - r^{-1}\Gamma_b \mathfrak{d}_{\nearrow} r^p f - p e_4\big(r^{p-1}\mathbf{T}(r) f\big),
$$

i.e.,

$$
e_4\big(r^p|\mathbf{T} f|\big) \lesssim r^p\big(|\mathbf{T}F| + |F|\big) + O(\epsilon r^{-2})\big(r^p|F| + |\not{\mathfrak{d}} f| + |f|\big).
$$

Similarly, commuting the equation with $r e_4$ we derive

$$
e_4\big(r^p|r f|\big) \lesssim r^p\big(|r e_4 F| + |F|\big) + O(\epsilon r^{-2})\big(r^p|F| + |\not{\mathfrak{d}} f| + |f|\big).
$$

Integrating the inequalities,

$$
\begin{aligned}
e_4(r^p|\not{\mathfrak{d}} f|) &\lesssim r^p(|\not{\mathfrak{d}} F| + |F|) + \epsilon r^{-2} r^p(|\not{\mathfrak{d}} f| + |f|) \\
e_4\big(r^p|\mathbf{T} f|\big) &\lesssim r^p\big(|\mathbf{T}F| + |F|\big) + \epsilon r^{-2} r^p\big(|\not{\mathfrak{d}} f| + |f|\big) \\
e_4\big(r^p|r f|\big) &\lesssim r^p\big(|r e_4 F| + |F|\big) + \epsilon r^{-2}\big(r^p|F| + |\not{\mathfrak{d}} f| + |f|\big)
\end{aligned}
$$

and applying Gronwall, we derive the desired estimates in (7.4.3) for $k = 1$.

Repeating the procedure for $\widetilde{\mathfrak{d}}^k$, any combination of derivatives of the form

$\widetilde{\mathfrak{d}}^k = \mathbf{T}^{k_1} \not{\mathfrak{d}}^{k_2}$ with $k_1 + k_2 = k$, estimating the corresponding commutators using our assumptions **Ref 1**, we deduce for all $0 \leq k \leq k_{large} - 5$,

$$e_4(r^p|\widetilde{\mathfrak{d}}^{\leq k} f|) \quad \lesssim \quad r^p|\widetilde{\mathfrak{d}}^{\leq k} F| + \epsilon r^{-2} r^p|\widetilde{\mathfrak{d}}^{\leq k} f|$$

and the desired estimates follow by integration. □

7.4.1.3 Transport equations for $\ell = 1$ modes

To estimate $\ell = 1$ modes we make use of the following.

Lemma 7.32. *The following equation holds for reduced scalars $\psi \in \mathfrak{s}_1({}^{(ext)}\mathcal{M})$:*

$$e_4\left(\int_S \psi e^{\Phi}\right) = \int_S e_4(\psi) e^{\Phi} + \frac{3}{2}\overline{\kappa} \int_S \psi e^{\Phi} + \int_S \frac{1}{2}\left(3\check{\kappa} - \vartheta\right) \psi e^{\Phi}. \tag{7.4.4}$$

Proof. This is an immediate consequence of Proposition (2.64). Indeed according to it and $e_4\Phi = \frac{1}{2}(\kappa - \vartheta)$,

$$\begin{aligned}
e_4\left(\int_S \psi e^{\Phi}\right) &= \int_S (e_4(\psi e^{\Phi}) + \kappa\psi e^{\Phi}) = \int_S \left(e_4(\psi) + \frac{1}{2}(3\kappa - \vartheta)\psi\right) e^{\Phi} \\
&= \int_S \left(e_4(\psi) + \frac{3}{2}\overline{\kappa}\psi\right) e^{\Phi} + \int_S \frac{1}{2}\left(3\check{\kappa} - \vartheta\right) \psi e^{\Phi} \\
&= \int_S e_4(\psi) e^{\Phi} + \frac{3}{2}\overline{\kappa} \int_S \psi e^{\Phi} + \int_S \frac{1}{2}\left(3\check{\kappa} - \vartheta\right) \psi e^{\Phi}
\end{aligned}$$

as desired. □

7.4.2 Proposition 7.33

In what follows we prove the stronger estimates in terms of powers of r for the quantities $\check{\kappa}, \check{\mu}, \vartheta, \zeta, \underline{\check{\kappa}}, \beta, \check{\rho}$. More precisely we establish the following.

Proposition 7.33. *The following estimates hold in ${}^{(ext)}\mathcal{M}$ for all $k \leq k_{small} + 20$:*

$$\begin{aligned}
\|\check{\kappa}\|_{\infty,k} &\lesssim \epsilon_0 r^{-2} u^{-1-\delta_{dec}}, \\
\|\check{\mu}\|_{\infty,k-2} &\lesssim \epsilon_0 r^{-3} u^{-1-\delta_{dec}}.
\end{aligned} \tag{7.4.5}$$

Also, for all $k \leq k_{small} + 18$,

$$\begin{aligned}
\left\|\vartheta, \zeta, \underline{\check{\kappa}}, r\check{\rho}\right\|_{\infty,k} &\lesssim \epsilon_0 r^{-2} u^{-1/2-\delta_{dec}}, \\
\left\|\beta\right\|_{\infty,k} &\lesssim \epsilon_0 r^{-3}(2r+u)^{-1/2-\delta_{dec}}, \\
\left\|e_3\beta\right\|_{\infty,k-1} &\lesssim \epsilon_0 r^{-4} u^{-1/2-\delta_{dec}}, \\
\left\|e_\theta K\right\|_{\infty,k-1} &\lesssim \epsilon_0 r^{-4} u^{-1/2-\delta_{dec}}.
\end{aligned} \tag{7.4.6}$$

7.4.3 Estimates for $\check{\kappa}$, $\check{\mu}$ in ${}^{(ext)}\mathcal{M}$

Step 1. We prove the following estimates for $\check{\kappa}$ in ${}^{(ext)}\mathcal{M}$.

$$\|\check{\kappa}\|_{\infty,k} \lesssim \epsilon_0 r^{-5/2} u^{-1-\delta_{dec}}, \qquad k \le k_{small}+20. \tag{7.4.7}$$

We make use of the equation

$$e_4(\check{\kappa}) + \overline{\kappa}\,\check{\kappa} = F := -\frac{1}{2}\check{\kappa}^2 - \frac{1}{2}\overline{\check{\kappa}^2} - \frac{1}{2}\left(\vartheta^2 - \overline{\vartheta^2}\right).$$

In view of our assumptions **Ref 1–2** and Lemma 7.6

$$\|F\|_{\infty,k}(u,\lambda) \lesssim \epsilon_0 \lambda^{-7/2} u^{-1-\delta_{dec}}.$$

Applying Proposition 7.31, we deduce

$$\begin{aligned} r^2\|\check{\kappa}\|_{\infty,k}(u,r) &\lesssim r_*^2\|\check{\kappa}\|_{\infty,k}(u,r_*) + \epsilon_0 u^{-1-\delta_{dec}} \int_r^{r_*} \lambda^2\lambda^{-7/2} d\lambda \\ &\lesssim r_*^2\|\check{\kappa}\|_{\infty}(u,r_*) + \epsilon_0 r^{-1/2} u^{-1-\delta_{dec}}. \end{aligned}$$

In view of the control on the last slice we infer that, everywhere in ${}^{(ext)}\mathcal{M}$,

$$\|\check{\kappa}\|_{\infty,k}(u,r) \lesssim \epsilon_0 r^{-5/2} u^{-1-\delta_{dec}}.$$

Step 2. We prove the estimate

$$\|\check{\mu}\|_{\infty,k} \lesssim \epsilon_0 r^{-3} u^{-1-\delta_{dec}}, \qquad k \le k_{small}+18. \tag{7.4.8}$$

Recall that we have

$$\begin{aligned} e_4(\check{\mu}) + \frac{3}{2}\overline{\kappa}\check{\mu} &= -\frac{3}{2}\overline{\mu}\check{\kappa} + F, \\ F: &= -\frac{3}{2}\check{\mu}\check{\kappa} + \frac{1}{2}\overline{\check{\mu}\check{\kappa}} + \mathrm{Err}[e_4\check{\mu}] - \overline{\mathrm{Err}[e_4\check{\mu}]}, \\ \mathrm{Err}[e_4\check{\mu}] &= -\frac{1}{8}\underline{\kappa}\vartheta^2 - \vartheta\, \not{d}_2^\star\zeta - \vartheta\zeta^2 + \left(2e_\theta(\kappa) - 2\beta + \frac{3}{2}\kappa\zeta\right)\zeta. \end{aligned}$$

In view of Lemma 7.6 we check

$$\|F\|_{\infty,k}(u,\lambda) \lesssim \epsilon_0 \lambda^{-9/2} u^{-1-\delta_{dec}}.$$

Applying Proposition 7.31 and the estimates on the last slice for $\check{\mu}$ we deduce

$$\begin{aligned} r^3\|\widetilde{\mathfrak{d}}^k\check{\mu}\|_{\infty,k}(u,\lambda) &\lesssim r_*^3\|\widetilde{\mathfrak{d}}^k\check{\mu}\|_{\infty,k}(u,r_*) + \epsilon_0 u^{-1-\delta_{dec}} \int_r^{r_*} \lambda^3\lambda^{-9/2} \\ &\lesssim r_*^3\|\widetilde{\mathfrak{d}}^k\check{\mu}\|_{\infty,k}(u,r_*) + \epsilon_0 u^{-1-\delta_{dec}} r^{-1/2} \\ &\lesssim \epsilon_0 u^{-1-\delta_{dec}} \end{aligned}$$

from which the desired estimate (7.4.8) follows.

7.4.4 Estimates for the $\ell = 1$ modes in ${}^{(ext)}\mathcal{M}$

We extend the validity of Lemma 7.26 to the entire region ${}^{(ext)}\mathcal{M}$.

Lemma 7.34. *The following estimates hold true on ${}^{(ext)}\mathcal{M}$ for all $k \le k_{small}+19$:*

$$\begin{aligned}\left\|\int_S \beta e^{\Phi}\right\|_{\infty,k}(u,r) &\lesssim \epsilon_0 r^{-1}u^{-1-\delta_{dec}},\\ \left\|\int_S \zeta e^{\Phi}\right\|_{\infty,k}(u,r) &\lesssim \epsilon_0 r u^{-1-\delta_{dec}},\\ \left\|\int_S e_\theta(\rho) e^{\Phi}\right\|_{\infty,k}(u,r) &\lesssim \epsilon_0 r^{-1}u^{-1-\delta_{dec}},\\ \left\|\int_S e_\theta(\underline{\kappa}) e^{\Phi}\right\|_{\infty,k}(u,r) &\lesssim \epsilon_0 u^{-1-\delta_{dec}},\\ \left\|\int_S \underline{\beta} e^{\Phi}\right\|_{\infty,k}(u,r) &\lesssim \epsilon_0 u^{-1-\delta_{dec}}.\end{aligned} \tag{7.4.9}$$

Proof. We first note that the estimate for the $\ell = 1$ mode of $\check{\mu}$ is an immediate consequence of the estimate (7.4.8). To prove the remaining estimates we proceed in steps as follows.

Step 1. Observe that the estimates of Lemma 7.26 remain valid when we replace the norms $\| \|^*_{\infty,k}$ by $\| \|_{\infty,k}$. To show this it suffices to prove estimates for re_4 of all $\ell = 1$ modes. This can easily be achieved with the help of Lemma 7.32 and our e_4 transport equations for $\zeta, \check{\rho}, \check{\mu}, \check{\underline{\kappa}}, \underline{\beta}$.

Step 2. We establish the estimate

$$\left\|\int_S \beta e^{\Phi}\right\|_{\infty,k} \lesssim \epsilon_0 r^{-1}u^{-1-\delta_{dec}}, \qquad k \le k_{small}+20. \tag{7.4.10}$$

In view of (7.4.4) and the Bianchi identity for $e_4(\beta)$

$$\begin{aligned} e_4\left(\int_S \beta e^{\Phi}\right) &= \int_S e_4\beta e^{\Phi} + \frac{3}{2}\overline{\kappa}\int_S \beta e^{\Phi} + \int_S \frac{1}{2}(3\check{\kappa}-\vartheta)\beta e^{\Phi}\\ &= \int_S(-2\kappa\beta + \not{d}_2^{\star}\alpha + \zeta\alpha)e^{\Phi} + \frac{3}{2}\overline{\kappa}\int_S \beta e^{\Phi} + \int_S \frac{1}{2}(3\check{\kappa}-\vartheta)\beta e^{\Phi}\\ &= -\frac{\overline{\kappa}}{2}\int_S \beta e^{\Phi} + \int_S\left(\zeta\alpha + \frac{1}{2}(-\check{\kappa}+\vartheta)\beta\right)e^{\Phi},\end{aligned}$$

and hence

$$e_4\left(\int_S \beta e^{\Phi}\right) + \frac{\overline{\kappa}}{2}\int_S \beta e^{\Phi} = \int_S\left(\zeta\alpha + \frac{1}{2}(-\check{\kappa}+\vartheta)\beta\right)e^{\Phi}. \tag{7.4.11}$$

Recall that

$$|(\alpha,\beta)| \lesssim \epsilon r^{-3}(2r+u)^{-1/2-\delta_{dec}}.$$

We deduce

$$\begin{aligned}\left|e_4\left(r\int_S\beta e^\Phi\right)\right| &\lesssim r\epsilon_0 r^{-2}u^{-1/2-\delta_{dec}}r^{-3}(2r+u)^{-1/2-\delta_{dec}}\int_S|e^\Phi|\\ &\lesssim \epsilon_0 r^{-1}u^{-1/2-\delta_{dec}}(2r+u)^{-1/2-\delta_{dec}}\\ &\lesssim \epsilon_0 r^{-1-\delta}u^{-1-\delta_{dec}},\end{aligned}$$

i.e., in view of the estimate on Σ_*, everywhere on ${}^{(ext)}\mathcal{M}$,

$$\left|\int_S\beta e^\Phi\right| \lesssim \epsilon r^{-1}u^{-1-\delta_{dec}}. \tag{7.4.12}$$

Commuting with $\mathbf{T}$, $\not{\mathfrak{d}}$ and re_4 we also easily deduce

$$\left\|\widetilde{\mathfrak{d}}^{k_2}(re_4)^{k_1}\int_S\beta e^\Phi\right\|_\infty \lesssim \epsilon r^{-1}u^{-1-\delta_{dec}}, \qquad \forall\, k_1+k_2\le k_{small}+20$$

from which (7.4.10) follows.

Step 3. We prove the estimate

$$\left\|\int_S\zeta e^\Phi\right\|_{\infty,k} \lesssim \epsilon_0 r^{1/2}u^{-1-\delta_{dec}}, \qquad k\le k_{small}+19 \tag{7.4.13}$$

which is better than the desired estimate in Lemma 7.34. This follows, as for the corresponding estimate on Σ_*, by projecting the Codazzi equation for ϑ on the $\ell=1$ mode

$$\int_S\zeta e^\Phi = \frac{r}{2\Upsilon}\left(2\int_S\beta e^\Phi-\int_S e_\theta(\kappa)e^\Phi-\int_S\vartheta\zeta e^\Phi-\int_S\left(\kappa-\frac{2\Upsilon}{r}\right)\zeta e^\Phi\right).$$

Note that in view of the estimates for $\check{\kappa}$ in (7.4.7) already established[19] we have

$$\left\|\int_S e_\theta(\kappa)e^\Phi\right\|_{\infty,k} \lesssim \epsilon_0 r^{-1/2}u^{-1-\delta_{dec}}, \qquad k\le k_{small}+19.$$

Also, making use of (7.4.10),

$$\left\|\int_S\beta e^\Phi\right\|_{\infty,k} \lesssim \epsilon_0 r^{-1}u^{-1-\delta_{dec}}, \qquad k\le k_{small}+19.$$

Thus, since Υ is bounded away from zero in ${}^{(ext)}\mathcal{M}$, we easily deduce

$$\left\|\int_S\zeta e^\Phi\right\|_{\infty,k} \lesssim \epsilon_0 r^{1/2}u^{-1-\delta_{dec}}, \qquad k\le k_{small}+19.$$

[19]Note that the estimate for $\check{\kappa}$ is stronger in powers of r than the corresponding bootstrap assumption.

Step 4. We prove the estimate

$$\left\| \int_S e_\theta(\rho) e^\Phi \right\|_{\infty,k} \lesssim \epsilon_0 r^{-1} u^{-1-\delta_{dec}}, \qquad k \le k_{small} + 19. \tag{7.4.14}$$

We proceed as in Step 4 of the proof of Lemma 7.26. In view of the definition of μ and the identity ${\not{d}}_1^\star {\not{d}}_1 = {\not{d}}_2 {\not{d}}_2^\star + 2K$, we write

$$\begin{aligned}
\int_S e_\theta(\rho) e^\Phi &= -\int_S e_\theta(\mu) e^\Phi + \int_S {\not{d}}_1^\star {\not{d}}_1 \zeta e^\Phi + \frac{1}{4}\int_S e_\theta(\vartheta \underline{\vartheta}) e^\Phi \\
&= -\int_S e_\theta(\mu) e^\Phi + \int_S ({\not{d}}_2 {\not{d}}_2^\star + 2K)\zeta e^\Phi + \frac{1}{4}\int_S e_\theta(\vartheta \underline{\vartheta}) e^\Phi \\
&= -\int_S e_\theta(\mu) e^\Phi + \frac{2}{r^2}\int_S \zeta e^\Phi + 2\int_S \left(K - \frac{1}{r^2}\right)\zeta e^\Phi + \frac{1}{4}\int_S e_\theta(\vartheta \underline{\vartheta}) e^\Phi.
\end{aligned}$$

Together with the above estimate for the $\ell = 1$ mode of ζ, the estimate (7.4.8) for $\check{\mu}$ and the bootstrap assumptions, we infer that

$$\left\| \int_S e_\theta(\rho) e^\Phi \right\|_{\infty,k} \lesssim \epsilon_0 r^{-1} u^{-1-\delta_{dec}}.$$

Step 5. We prove the estimate

$$\left\| \int_S e_\theta(\underline{\kappa}) e^\Phi \right\|_{\infty,k} \lesssim \epsilon_0 u^{-1-\delta_{dec}}, \qquad k \le k_{small} + 19. \tag{7.4.15}$$

As in the corresponding estimate on the last slice we make use of the remarkable identity for the $\ell = 1$ mode of $e_\theta(K)$, i.e.,

$$\int_S e_\theta(\rho) e^\Phi + \frac{1}{4}\int_S e_\theta(\kappa\underline{\kappa}) e^\Phi - \frac{1}{4}\int_S e_\theta(\vartheta\underline{\vartheta}) e^\Phi = 0.$$

We infer

$$\begin{aligned}
\int_S e_\theta(\underline{\kappa}) e^\Phi = & -2r\int_S e_\theta(\rho) e^\Phi - \frac{r}{2}\int_S \underline{\kappa} e_\theta(\kappa) e^\Phi + \frac{r}{2}\int_S e_\theta(\vartheta\underline{\vartheta}) e^\Phi \\
& -\frac{r}{2}\int_S \left(\kappa - \frac{2}{r}\right) e_\theta(\underline{\kappa}) e^\Phi.
\end{aligned}$$

The estimate (7.4.15) follows easily from the above estimate for the $\ell = 1$ mode of $e_\theta(\rho)$, the estimate for $\check{\kappa}$ in (7.4.7) and the bootstrap assumptions.

Step 6. We prove the estimate

$$\left\| \int_S \underline{\beta} e^\Phi \right\|_{\infty,k} \lesssim \epsilon_0 u^{-1-\delta_{dec}}, \qquad k \le k_{small} + 19. \tag{7.4.16}$$

Projecting the Codazzi for $\underline{\vartheta}$ on the $\ell = 1$ mode, we have

$$-2\int_S \underline{\beta} e^\Phi + \int_S e_\theta(\underline{\kappa}) e^\Phi - \int_S \underline{\kappa}\zeta e^\Phi + \int_S \underline{\vartheta}\zeta e^\Phi = 0$$

and hence

$$\int_S \underline{\beta} e^\Phi = \frac{1}{2}\int_S e_\theta(\underline{\kappa})e^\Phi + \frac{\Upsilon}{r}\int_S \zeta e^\Phi - \frac{1}{2}\int_S \left(\underline{\kappa} + \frac{2\Upsilon}{r}\right)\zeta e^\Phi + \frac{1}{2}\int_S \underline{\vartheta}\zeta e^\Phi.$$

The desired estimate follows easily in view of the above estimates for the $\ell = 1$ mode of $e_\theta(\underline{\kappa})$, the $\ell = 1$ mode of ζ and the bootstrap assumptions. □

7.4.5 Completion of the proof of Proposition 7.33

We prove the second part of Proposition 7.33, i.e., we prove for all $k \le k_{small} + 18$

$$\begin{aligned} \big\|\vartheta, \zeta, \check{\underline{\kappa}}, r\check{\rho}\big\|_{\infty,k} &\lesssim \epsilon_0 r^{-2} u^{-1/2-\delta_{dec}}, \\ \big\|\beta\big\|_{\infty,k} &\lesssim \epsilon_0 r^{-3}(2r+u)^{-1/2-\delta_{dec}}, \\ \big\|e_3\beta\big\|_{\infty,k-1} &\lesssim \epsilon_0 r^{-4} u^{-1/2-\delta_{dec}}, \\ \big\|e_\theta K\big\|_{\infty,k-1} &\lesssim \epsilon_0 r^{-4} u^{-1/2-\delta_{dec}}. \end{aligned} \tag{7.4.17}$$

We also prove the stronger estimate for β

$$\|\beta\|_{\infty,k} \lesssim \epsilon_0 \log(1+u) r^{-3}(2r+u)^{-1/2-\delta_{extra}}. \tag{7.4.18}$$

Proof. We proceed in steps as follows.

Step 1. We derive the estimate

$$\big\|\vartheta\big\|_{\infty,k} \lesssim \epsilon_0 r^{-2} u^{-1/2-\delta_{dec}}, \qquad \forall k \le k_{small} + 19, \tag{7.4.19}$$

with the help of the equation $e_4\vartheta + \overline{\kappa}\vartheta = F := -2\alpha - \check{\kappa}\vartheta$ and the corresponding estimate on the last slice.

Note that

$$\big\|\alpha\big\|_{\infty,k} \lesssim \epsilon_0 r^{-3-\delta}(2r+u)^{-1/2-\delta_{dec}}$$

where $\delta > 0$ is a small constant, $\delta < \delta_{extra} - \delta_{dec}$. Thus, using also the product estimates of Lemma 7.6, we easily check that

$$\|F\|_{\infty,k} \lesssim \epsilon_0 r^{-3-\delta} u^{-1/2-\delta_{dec}} + \epsilon_0 r^{-7/2} u^{-1-\delta_{dec}}, \qquad k \le k_{small} + 20.$$

Making use of Proposition 7.31 we deduce, for all $k \le k_{small} + 19$,

$$r^2\|\mathfrak{d}^k\vartheta\|_{\infty,k}(u,r) \lesssim r_*^2\|\mathfrak{d}^k\vartheta\|_{\infty,k}(u,r_*) + \epsilon_0 u^{-1/2-\delta_{dec}}\int_r^{r_*}\lambda^{-1-\delta}d\lambda.$$

Thus, in view of the results on the last slice Σ_*, we deduce

$$\|\mathfrak{d}^k\vartheta\|_\infty(u,r) \lesssim r^{-2}u^{-1/2-\delta_{dec}}.$$

Step 2. We derive the estimate

$$\big\|\beta\big\|_{\infty,k} \lesssim \epsilon_0 r^{-3}(2r+u)^{-1/2-\delta_{dec}}, \qquad \forall k \le k_{small} + 19. \tag{7.4.20}$$

We proceed exactly as in the estimates for β on the last slice Σ_* by making use of the Bianchi identity $e_3\alpha+\left(\frac{1}{2}\underline{\kappa}-4\underline{\omega}\right)\alpha=-\,\not{d}_2^\star\beta-\frac{3}{2}\vartheta\rho+5\zeta\beta$, from which we deduce

$$\|\,\not{d}_2^\star\beta\|_{\infty,k-1} \lesssim \|e_3\alpha\|_{\infty,k-1}+r^{-1}\|\alpha\|_{\infty,k-1}+r^{-3}\|\vartheta\|_{\infty,k-1}+\epsilon_0 r^{-5}u^{-1-\delta_{dec}}.$$

Thus, in view of the above estimate for ϑ and **Ref 2** for α,

$$\|\,\not{d}_2^\star\beta\|_{\infty,k-1} \lesssim \epsilon_0 r^{-4}(2r+u)^{-1/2-\delta_{dec}}+\epsilon_0 r^{-5}u^{-1/2-\delta_{dec}}.$$

On the other hand we have, according to (7.4.10),

$$\left\|\int_S \beta e^\Phi\right\|_{\infty,k} \lesssim \epsilon_0 r^{-1}u^{-1-\delta_{dec}}, \qquad k\le k_{small}+20.$$

Estimate (7.4.20) follows then easily, according to part 4 of the elliptic Hodge Lemma 7.7.

We can prove a stronger estimate for β. Indeed we have, in view of **Ref 2.**

$$\begin{aligned}|\alpha| &\lesssim \log(1+u)r^{-3}(2r+u)^{-1/2-\delta_{extra}},\\ |e_3\alpha| &\lesssim r^{-4}(2r+u)^{-1/2-\delta_{extra}}.\end{aligned}$$

Hence, using the equation $e_3\alpha+\left(\frac{1}{2}\underline{\kappa}-4\underline{\omega}\right)\alpha=-\,\not{d}_2^\star\beta-\frac{3}{2}\vartheta\rho+5\zeta\beta$,

$$\|\,\not{d}_2^\star\beta\|_{\infty,k} \lesssim \epsilon_0\log(1+u)r^{-4}(2r+u)^{-1/2-\delta_{extra}}+\epsilon_0 r^{-5}u^{-1/2-\delta_{dec}}.$$

According to Lemma 7.7

$$\|\beta\|_{\mathfrak{h}_{k+1}(S)} \lesssim r\|\,\not{d}_2^\star\beta\|_{\mathfrak{h}_k(S)}+r^{-2}\left|\int_S e^\Phi\beta\right|$$

and thus, in view of the estimate (7.4.10) for the $\ell=1$ mode of β,

$$\begin{aligned}\|\beta\|_{\mathfrak{h}_{k+1}(S)} &\lesssim \epsilon_0\log(1+u)r^{-2}(2r+u)^{-1/2-\delta_{extra}}+\epsilon_0 r^{-3}u^{-1-\delta_{dec}}\\ &\lesssim \epsilon_0\log(1+u)r^{-2}(2r+u)^{-1/2-\delta_{extra}}.\end{aligned}$$

The estimates for the T and e_4 derivatives are derived in the same manner and hence

$$\|\beta\|_{\infty,k}\lesssim\epsilon_0\log(1+u)r^{-3}(2r+u)^{-1/2-\delta_{extra}}, \qquad \forall\, k\le k_{small}+19. \quad (7.4.21)$$

This improvement is needed in the next step.

Step 3. We derive the estimate

$$\|\zeta\|_{\infty,k} \lesssim \epsilon_0 r^{-2}u^{-1/2-\delta_{dec}}, \qquad \forall\, k\le k_{small}+19. \qquad (7.4.22)$$

For this we make use of the transport equation for ζ,

$$e_4\zeta+\overline{\kappa}\zeta=F:=-\beta+\Gamma_g\cdot\Gamma_g,$$

and the improved estimate for β in the previous step. Thus, making use of the product Lemma 7.6,

$$
\begin{aligned}
\|F\|_{\infty,k} &\lesssim \|\beta\|_{\infty,k} + \epsilon_0 r^{-7/2} u^{-1-\delta_{dec}} \\
&\lesssim \epsilon_0 \log(1+u) r^{-3}(2r+u)^{-1/2-\delta_{extra}} + \epsilon_0 r^{-7/2} u^{-1-\delta_{dec}} \\
&\lesssim \epsilon_0 r^{-3-\delta} u^{-1/2-\delta_{dec}} + \epsilon_0 r^{-7/2} u^{-1-\delta_{dec}}.
\end{aligned}
$$

Making use of Proposition 7.31 we deduce

$$
r^2 \|\mathfrak{d}^k \zeta\|_{\infty,k}(u,r) \lesssim r_*^2 \|\mathfrak{d}^k \zeta\|_{\infty,k}(u,r_*) + \epsilon_0 u^{-1/2-\delta_{dec}} \int_r^{r_*} \lambda^{-1-\delta} d\lambda.
$$

Thus, in view of the estimates on the last slice,

$$
r^2 \|\mathfrak{d}^k \zeta\|_{\infty}(u,r) \lesssim \epsilon_0 u^{-1/2-\delta_{dec}}, \qquad k \le k_{small} + 19
$$

as desired.

Step 4. We derive the estimate

$$
\|\check{\rho}\|_{\infty,k} \lesssim \epsilon_0 r^{-3} u^{-1/2-\delta_{dec}}, \qquad \forall\, k \le k_{small} + 18. \tag{7.4.23}
$$

We make use of the definition of μ from which we infer that

$$
\check{\mu} = -\not{d}_1 \zeta - \check{\rho} + \Gamma_g \cdot \Gamma_b.
$$

Hence, in view of the product lemma and the estimates already derived, for all $k \le k_{small} + 18$,

$$
\begin{aligned}
\|\check{\rho}\|_{\infty,k} &\lesssim r^{-1} \|\zeta\|_{\infty,k+1} + \|\check{\mu}\|_{\infty,k} + \epsilon_0 r^{-3} u^{-1-\delta_{dec}} \\
&\lesssim \epsilon_0 r^{-3} u^{-1/2-\delta_{dec}}
\end{aligned}
$$

as desired.

Step 5. We derive the estimate

$$
\|\check{\underline{\kappa}}\|_{\infty,k} \lesssim \epsilon_0 r^{-2} u^{-1/2-\delta_{dec}}, \qquad \forall\, k \le k_{small} + 18. \tag{7.4.24}
$$

We make use of the equation

$$
e_4 \check{\underline{\kappa}} + \frac{1}{2} \overline{\kappa} \check{\underline{\kappa}} = F := -2 \not{d}_1 \zeta - \frac{1}{2} \check{\kappa} \underline{\overline{\kappa}} + 2\check{\rho} + \Gamma_g \cdot \Gamma_b.
$$

In view of the previously derived estimates,

$$
\|F\|_{\infty,k} \lesssim \epsilon_0 r^{-3} u^{-1/2-\delta_{dec}}, \qquad k \le k_{small} + 18.
$$

Making use of Proposition 7.31 we deduce, for all $k \le k_{small} + 18$,

$$
\begin{aligned}
r \|\mathfrak{d}^k \check{\underline{\kappa}}\|_{\infty,k}(u,r) &\lesssim r_* \|\mathfrak{d}^k \check{\underline{\kappa}}\|_{\infty,k}(u,r_*) + \epsilon_0 u^{-1/2-\delta_{dec}} \int_r^{r_*} \lambda^{-2} d\lambda \\
&\lesssim r_* \|\mathfrak{d}^k \check{\underline{\kappa}}\|_{\infty,k}(u,r_*) + \epsilon_0 r^{-1} u^{-1/2-\delta_{dec}}.
\end{aligned}
$$

Thus, in view of the estimates on the last slice,

$$r\|\mathfrak{d}^k\underline{\check{\kappa}}\|_\infty(u,r) \lesssim \epsilon_0(r_*)^{-1}u^{-1/2-\delta_{dec}} + \epsilon_0 r^{-1}u^{-1/2-\delta_{dec}}$$

from which the desired estimate easily follows. □

Step 6. We derive the estimate

$$\|e_3\beta\|_{\infty,k} \lesssim \epsilon_0 r^{-4}u^{-1/2-\delta_{dec}}, \qquad \forall k \le k_{small}+17, \tag{7.4.25}$$

making use of the equation $e_3\beta + (\underline{\kappa} - 2\underline{\omega})\beta = -\,{}^{*}\!\!\not{d}_1\rho + 3\zeta\rho + \Gamma_g\underline{\beta} + \Gamma_b\alpha$ and the estimates derived above for β, ${}^{*}\!\!\not{d}_1\rho$, ζ. Hence,

$$\begin{aligned}\|e_3\beta\|_{\infty,k} &\lesssim r^{-1}\|\beta\|_{\infty,k} + \|{}^{*}\!\!\not{d}_1\rho\|_{\infty,k} + r^{-3}\|\zeta\|_{\infty,k} + \epsilon_0 r^{-4}u^{-1-\delta_{dec}}\\ &\lesssim \epsilon_0 r^{-4}u^{-1/2-\delta_{dec}}.\end{aligned}$$

Step 7. As a corollary of the above estimates (see also **Ref 4**) we also derive, in ${}^{(ext)}\mathcal{M}$,

$$\begin{aligned}\|K - \overline{K}\|_{\infty,k-1} &\lesssim \epsilon_0 r^{-3}u^{-1/2-\delta_{dec}}, \qquad k \le k_{small}+18,\\ \left\|K - \frac{1}{r^2}\right\|_{\infty,k-1} &\lesssim \epsilon_0 r^{-3}u^{-1/2-\delta_{dec}}, \qquad k \le k_{small}+18.\end{aligned} \tag{7.4.26}$$

In view of the definition of K we have,

$$e_\theta(K) = -e_\theta\left(\check{\rho} - \frac{1}{4}\kappa\underline{\check{\kappa}} - \frac{1}{4}\underline{\kappa}\check{\kappa} + \frac{1}{4}\vartheta\underline{\vartheta}\right).$$

Thus, in view of the above estimates, for all $k \le k_{small}+17$,

$$\|e_\theta K\|_{\infty,k} \lesssim \epsilon_0 r^{-4}u^{-1/2-\delta_{dec}}$$

from which the desired estimate easily follows.

7.5 CONTROL IN ${}^{(ext)}\mathcal{M}$, PART II

We derive the crucial decay estimates which imply, in particular, decay of order $u^{-1-\delta_{dec}}$ for all quantities in Γ and $\check{R}$ (except $\underline{\xi}, \underline{\check{\omega}}, \underline{\check{\Omega}}$ which will be treated separately) in the interior. More precisely we prove the following:

Proposition 7.35. *The following estimates hold in ${}^{(ext)}\mathcal{M}$, for all $k \le k_{small}+8$,*

$$\|\vartheta,\, \zeta,\, \eta,\, \underline{\check{\kappa}},\, \underline{\vartheta},\, r\beta,\, r\check{\rho},\, r\underline{\beta}, \underline{\alpha}\|_{\infty,k} \lesssim \epsilon_0 r^{-1}u^{-1-\delta_{dec}}. \tag{7.5.1}$$

To prove the proposition we make use of the fact that we already have good decay estimates in terms of powers of u for $\check{\kappa}, \check{\mu}$. We also derive below decay estimates for various renormalized quantities.

7.5.1 Estimate for η

We start with the following simple estimate for η in terms of ζ:

$$\|\eta\|_{\infty,k} \lesssim \|\zeta\|_{\infty,k} + \epsilon_0 r^{-1} u^{-1-\delta_{dec}}, \qquad k \le k_{small} + 17. \tag{7.5.2}$$

This can be derived by propagation from the last slice with the help of the equation

$$e_4(\eta - \zeta) + \frac{1}{2}\kappa(\eta - \zeta) = -\frac{1}{2}\vartheta(\eta - \zeta) = \Gamma_g \cdot \Gamma_b.$$

Note that

$$\|\Gamma_g \cdot \Gamma_b\|_{\infty,k} \lesssim \epsilon_0 r^{-3} u^{-1-\delta_{dec}}.$$

Thus making use of Proposition 7.31 we deduce

$$\begin{aligned} r\|\eta - \zeta\|_{\infty,k}(u,r) &\lesssim r_*\|\eta - \zeta\|_{\infty,k}(u,r_*) + \int_r^{r_*} \lambda \|\Gamma_g \cdot \Gamma_b\|_{\infty,k}(u,\lambda) \\ &\lesssim r_*\|\eta - \zeta\|_{\infty,k}(u,r_*) + \epsilon_0 u^{-1-\delta_{dec}} \end{aligned}$$

with r_* the value of r on $C(u) \cap \Sigma_*$. On the last slice we have derived the estimates, recorded in Proposition 7.22 and Proposition 7.33,

$$\begin{aligned} \|\eta\|^*_{\infty,k} &\lesssim \epsilon_0 r^{-1} u^{-1-\delta_{dec}}, \\ \|\zeta\|^*_{\infty,k} &\lesssim \epsilon_0 r^{-2} u^{-1/2-\delta_{dec}}. \end{aligned}$$

In view of the dominance condition on r on Σ_* we deduce

$$\|\eta - \zeta\|^*_{\infty,k}(u,r) \lesssim \epsilon_0 r^{-1} u^{-1-\delta_{dec}}$$

and therefore, also,

$$r_*\|\eta - \zeta\|_{\infty,k}(u,r_*) \lesssim \epsilon_0 u^{-1-\delta_{dec}}.$$

Therefore,

$$r\|\eta\|_{\infty,k}(u,r) \lesssim r\|\zeta\|_{\infty,k}(u,r) + \epsilon_0 u^{-1-\delta_{dec}}$$

as desired.

7.5.2 Crucial lemmas

We start with the following lemma.

Lemma 7.36. *The $\mathfrak{s}_1(\mathcal{M})$ reduced tensor*

$$\Xi := r^2 \left(e_\theta(\underline{\kappa}) + 4r \, \not{d}_1^\star \not{d}_1 \zeta - 2r^2 \not{d}_1^\star \not{d}_1 \beta\right) \tag{7.5.3}$$

verifies in ${}^{(ext)}\mathcal{M}$ the estimate

$$\|\Xi\|_{\infty,k} \lesssim \epsilon_0 u^{-1-\delta_{dec}}, \qquad \forall k \le k_{small} + 13. \tag{7.5.4}$$

Proof. To calculate $e_4\Xi$ we make use of the equations

$$\begin{aligned} e_4(\underline{\kappa}) + \frac{1}{2}\kappa\underline{\kappa} &= -2\,\not{d}_1\zeta + 2\rho - \frac{1}{2}\vartheta\underline{\vartheta} + 2\zeta^2, \\ e_4\zeta + \kappa\zeta &= -\beta - \vartheta\zeta, \\ e_4\beta + 2\kappa\beta &= \not{d}_2\alpha + \zeta\alpha. \end{aligned}$$

Since we already have an estimate for $\check{\mu}$ we re-express $\rho = -\mu - \not{d}_1\zeta + \frac{1}{4}\vartheta\underline{\vartheta}$ and derive

$$e_4(\underline{\kappa}) + \frac{1}{2}\kappa\underline{\kappa} = -2\mu - 4\,\not{d}_1\zeta + 2\zeta^2.$$

Commuting with $\not{d}_1^\star$ and making use of $[\not{d}_1^\star, e_4] = \frac{1}{2}(\kappa + \vartheta)\,\not{d}_1^\star$ we derive

$$e_4(\not{d}_1^\star\underline{\kappa}) + \kappa\,\not{d}_1^\star\underline{\kappa} + \frac{1}{2}\underline{\kappa}\,\not{d}_1^\star\kappa = -\not{d}_1^\star\check{\mu} - 4\,\not{d}_1^\star\,\not{d}_1\zeta + 2\,\not{d}_1^\star(\zeta^2) + \vartheta\,\not{d}_1^\star\underline{\kappa}. \quad (7.5.5)$$

Hence, since $e_4(r) = \frac{r}{2}\overline{\kappa}$,

$$\begin{aligned} e_4(r^2\,\not{d}_1^\star\underline{\kappa}) &= r^2(\kappa - \overline{\kappa})\,\not{d}_1^\star\underline{\kappa} - \frac{1}{2}r^2\underline{\kappa}\,\not{d}_1^\star\check{\kappa} - 4r^2\,\not{d}_1^\star\,\not{d}_1\zeta - r^2\,\not{d}_1^\star\check{\mu} \\ &\quad + r^2(\not{d}_1^\star(\zeta^2) + \vartheta\,\not{d}_1^\star\underline{\kappa}) \\ &= -\frac{1}{2}r^2\underline{\kappa}\,\not{d}_1^\star\check{\kappa} - 4r^2\,\not{d}_1^\star\,\not{d}_1\zeta + \mathrm{Err}_1 \end{aligned}$$

where

$$\mathrm{Err}_1 := -\frac{1}{2}r^2\underline{\kappa}\,\not{d}_1^\star\check{\kappa} - r^2\,\not{d}_1^\star\check{\mu} + r^2\left(\check{\kappa}\,\not{d}_1^\star\underline{\kappa} + \not{d}_1^\star(\zeta^2) + \vartheta\,\not{d}_1^\star\underline{\kappa}\right).$$

In view of the estimate already established for $\check{\kappa}$, $\check{\mu}$ and the product Lemma 7.6 we check

$$\|\mathrm{Err}_1\|_{\infty,k} \lesssim \epsilon_0 r^{-2}u^{-1-\delta_{dec}}, \qquad k \le k_{small} + 17.$$

To simplify notation we introduce the following.

Definition 7.37. *We say that a quantity $\psi \in \mathfrak{s}_k(\mathcal{M})$ is $r^{-p}Good_a$ provided that it verifies the estimate, everywhere in ${}^{(ext)}\mathcal{M}$,*

$$\|\psi\|_{\infty,k} \lesssim \epsilon_0 r^{-p}u^{-1-\delta_{dec}}, \qquad \forall k \le k_{small} + a. \quad (7.5.6)$$

Using this notation we write

$$e_4(r^2\,\not{d}_1^\star\underline{\kappa}) = -4r^2\,\not{d}_1^\star\,\not{d}_1\zeta + r^{-2}Good_{17}. \quad (7.5.7)$$

Using the same notation, the transport equation for ζ can be written in the form

$$e_4\zeta + \overline{\kappa}\zeta = -\beta - \vartheta\zeta - \check{\kappa}\zeta = -\beta + r^{-7/2}Good_{20}.$$

Commuting with $(r \not{d}_1^\star)(r \not{d}_1)$ (making use of Lemma 7.29) we derive

$$e_4(r^2 \not{d}_1^\star \not{d}_1 \zeta) + \overline{\kappa}(r^2 \not{d}_1^\star \not{d}_1 \zeta) = -r^2 \not{d}_1^\star \not{d}_1 \beta + r^{-7/2} Good_{18}.$$

Since $e_4(r) = \frac{1}{2} r\overline{\kappa}$ we deduce

$$e_4(r^3 \not{d}_1^\star \not{d}_1 \zeta) = -\frac{1}{2}\overline{\kappa} r^3 \not{d}_1^\star \not{d}_1 \zeta - r^3 \not{d}_1^\star \not{d}_1 \beta + r^{-5/2} Good_{18}. \tag{7.5.8}$$

Similarly the transport equation for β takes the form

$$e_4\beta + 2\overline{\kappa}\beta = \not{d}_2\alpha + \zeta\alpha - 2\check{\kappa}\beta = \not{d}_2\alpha + r^{-9/2} Good_{20}$$

and

$$e_4(r^2 \not{d}_1^\star \not{d}_1 \beta) + 2\overline{\kappa} r^2 \not{d}_1^\star \not{d}_1 \beta = r^2 \not{d}_1^\star \not{d}_1 \not{d}_2 \alpha + r^{-9/2} Good_{18}.$$

As before, since $e_4(r) = \frac{1}{2} r\overline{\kappa}$, we deduce

$$e_4(r^4 \not{d}_1^\star \not{d}_1 \beta) = -\overline{\kappa} r^4 \not{d}_1^\star \not{d}_1 \beta + r^4 \not{d}_1^\star \not{d}_1 \not{d}_2 \alpha + r^{-5/2} Good_{18}. \tag{7.5.9}$$

Combining (7.5.7)–(7.5.9) we deduce

$$\begin{aligned}
e_4\Xi &= e_4\Big[r^2\left(-\not{d}_1^\star \underline{\kappa} + 4r \not{d}_1^\star \not{d}_1 \zeta - 2r^2 \not{d}_1^\star \not{d}_1 \beta\right)\Big] \\
&= 4r^2 \not{d}_1^\star \not{d}_1 \zeta + 4\left(-\frac{1}{2}\overline{\kappa} r^3 \not{d}_1^\star \not{d}_1 \zeta - r^3 \not{d}_1^\star \not{d}_1 \beta\right) \\
&- 2\left(-\overline{\kappa} r^4 \not{d}_1^\star \not{d}_1 \beta + r^4 \not{d}_1^\star \not{d}_1 \not{d}_2 \alpha\right) + r^{-2} Good_{17} \\
&= -2\left(\overline{\kappa} - \frac{2}{r}\right) r^3 \not{d}_1^\star \not{d}_1 \zeta + 2r^4\left(\overline{\kappa} - \frac{2}{r}\right) \not{d}_1^\star \not{d}_1 \beta - 2r^4 \not{d}_1^\star \not{d}_1 \not{d}_2 \alpha + r^{-2} Good_{17}.
\end{aligned}$$

Making use of **Ref 4** estimates for $\overline{\kappa} - \frac{2}{r}$ and the estimates for α in **Ref 2**, i.e.,

$$r^4 |\not{d}_1^\star \not{d}_1 \not{d}_2 \alpha| \lesssim \epsilon_0 r^{-1} (2r + u)^{-1-\delta_{extra}} \lesssim \epsilon_0 r^{-1-\delta} u^{-1-\delta_{extra}+\delta}, \qquad 0 < \delta < \delta_{extra},$$

i.e.,

$$r^4 \not{d}_1^\star \not{d}_1 \not{d}_2 \alpha = r^{-1-\delta} Good_{13},$$

we thus deduce

$$e_4\Xi = r^{-1-\delta} Good_{13}.$$

We deduce

$$\|\Xi\|_{\infty,k}(u,r) \lesssim \|\Xi\|_{\infty,k}(u,r_*) + \epsilon_0 u^{-1-\delta_{dec}} \int_r^{r_*} \lambda^{-1-\delta} d\lambda, \qquad \forall k \le k_{small} + 13.$$

In view of the estimates on the last slice it is easy to check that

$$\|\Xi\|_{\infty,k}(u,r_*) \lesssim \epsilon_0 u^{-1-\delta_{dec}}, \qquad \forall k \le k_{small} + 13.$$

Indeed, on the last slice,

$$
\begin{aligned}
\| \slashed{d}_1^\star \underline{\kappa} \|_{\infty,k} &\lesssim \epsilon_0 r^{-3} u^{-1/2-\delta_{dec}}, \\
\| \slashed{d}_1^\star \slashed{d}_1 \zeta \|_{\infty,k} &\lesssim \epsilon_0 r^{-4} u^{-1/2-\delta_{dec}}, \\
\| \slashed{d}_1^\star \slashed{d}_1 \beta \|_{\infty,k} &\lesssim \epsilon_0 r^{-5} u^{-1/2-\delta_{dec}}.
\end{aligned}
$$

Hence, since $r \gg u$ on Σ_*,

$$
\|\Xi\|_{\infty,k}(u, r_*) \lesssim \epsilon_0 r^{-1} u^{-1/2-\delta_{dec}} \lesssim \epsilon_0 u^{-1-\delta_{dec}}.
$$

Thus everywhere on ${}^{(ext)}\mathcal{M}$,

$$
\|\Xi\|_{\infty,k} \lesssim \epsilon_0 u^{-1-\delta_{dec}}, \qquad \forall k \le k_{small} + 13
$$

as desired. □

In the following lemma, we make use of the control we have already established for $\mathfrak{q}, \alpha, \underline{\alpha}, \check{\kappa}, \check{\mu}$ in ${}^{(ext)}\mathcal{M}$ to derive two nontrivial relations between angular derivatives of $\zeta, \underline{\check{\kappa}}$ and β.

Remark 7.38. *According to Theorem M3 we only have good estimates for $\underline{\alpha}$ along $\mathcal{T}$ and on the last slice Σ_*. To keep track of this fact we denote by $r^{-p}Good_a(\underline{\alpha})$ those $r^{-p}Good_a$ terms which depend linearly on $\underline{\alpha}$ and their derivatives.*

Lemma 7.39. *Let A, B be the operators $A := \slashed{d}_2^\star \slashed{d}_2 - 3\overline{\rho}$, $B = \slashed{d}_2^\star \slashed{d}_2 + 2K$. The following identities hold true:*

$$
\begin{aligned}
AB\, \slashed{d}_2^\star \zeta - \frac{3}{4}\overline{\kappa}\,\overline{\rho}\, \slashed{d}_2^\star e_\theta(\underline{\kappa}) &\in r^{-6} Good_{14}(\underline{\alpha}) \\
A^2 B\, \slashed{d}_2^\star \beta + \frac{9}{8}\left(\overline{\kappa}\,\overline{\rho}\right)^2 \slashed{d}_2^\star e_\theta(\underline{\kappa}) &\in r^{-9} Good_9(\underline{\alpha})
\end{aligned}
\tag{7.5.10}
$$

Proof. In view of the improved control for α in Theorem M2, $\underline{\alpha}$ in Theorem M3, and $\mathfrak{q}$ in Theorem M1, the bootstrap assumptions and product lemma, and the control we have already derived for $\check{\kappa}$ and $\check{\mu}$ in ${}^{(ext)}\mathcal{M}$, we obtain

$$
\begin{aligned}
\slashed{d}_2 \vartheta + 2\beta - \overline{\kappa}\zeta &\in r^{-3} Good_{20}, && \text{Codazzi and control of } \check{\kappa}, \\
\slashed{d}_2 \underline{\vartheta} + 2\underline{\beta} - e_\theta(\underline{\kappa}) + \overline{\underline{\kappa}}\zeta &\in r^{-3} Good_{20}, && \text{Codazzi}, \\
\slashed{d}_2^\star \beta + \frac{3}{2}\overline{\rho}\vartheta &\in r^{-3} Good_{15}, && \text{Bianchi and control of } \alpha, \\
\slashed{d}_2^\star \underline{\beta} + \frac{3}{2}\overline{\rho}\underline{\vartheta} &\in r^{-2} Good_{15}(\underline{\alpha}), && \text{Bianchi and control of } \underline{\alpha}, \\
\slashed{d}_2^\star \slashed{d}_1^\star \rho + \frac{3}{4}\overline{\underline{\kappa}}\,\overline{\rho}\vartheta + \frac{3}{4}\overline{\kappa}\ \overline{\rho}\underline{\vartheta} &\in r^{-4} Good_{18}, && \text{(7.1.28) and control of } \mathfrak{q},
\end{aligned}
\tag{7.5.11}
$$

where we used Codazzi for the two first inequalities, Bianchi for the third and fourth inequalities, the definition of μ for the fifth one, and the identity relating $\mathfrak{q}$ and $\slashed{d}_2^\star \slashed{d}_1^\star \rho$ for the last one.

Combining the first statement with the third and the second with the fourth we infer that

$$\begin{aligned}(\not{d}_2^\star \not{d}_2 - 3\overline{\rho})\vartheta - \overline{\kappa}\,\not{d}_2^\star\zeta &\in r^{-3}Good_{14},\\ (\not{d}_2^\star \not{d}_2 - 3\overline{\rho})\underline{\vartheta} - \not{d}_2^\star e_\theta(\underline{\kappa}) + \overline{\underline{\kappa}}\,\not{d}_2^\star\zeta &\in r^{-2}Good_{14}(\underline{\alpha}),\end{aligned}$$

or, setting

$$A := \not{d}_2^\star \not{d}_2 - 3\overline{\rho},$$

$$\begin{aligned}A\vartheta - \overline{\kappa}\,\not{d}_2^\star\zeta &\in r^{-3}Good_{14},\\ A\underline{\vartheta} - \not{d}_2^\star e_\theta(\underline{\kappa}) + \overline{\underline{\kappa}}\,\not{d}_2^\star\zeta &\in r^{-2}Good_{14}(\underline{\alpha}).\end{aligned} \tag{7.5.12}$$

From the fifth equations we deduce

$$A\left(\not{d}_2^\star \not{d}_1^\star\rho + \frac{3}{4}\overline{\underline{\kappa}}\,\overline{\rho}\vartheta + \frac{3}{4}\overline{\kappa}\,\overline{\rho}\underline{\vartheta}\right) \in r^{-6}Good_{18},$$

i.e.,

$$A\,\not{d}_2^\star \not{d}_1^\star\rho + \frac{3}{4}\overline{\underline{\kappa}}\,\overline{\rho}\,A\vartheta + \frac{3}{4}\overline{\kappa}\,\overline{\rho}\,A\underline{\vartheta} \in r^{-6}Good_{18}.$$

Making use of (7.5.12) we deduce

$$A\,\not{d}_2^\star \not{d}_1^\star\rho + \frac{3}{4}\overline{\underline{\kappa}}\,\overline{\rho}\,(\overline{\kappa}\,\not{d}_2^\star\zeta) + \frac{3}{4}\overline{\kappa}\,\overline{\rho}\big(\not{d}_2^\star e_\theta(\underline{\kappa}) - \overline{\underline{\kappa}}\,\not{d}_2^\star\zeta\big) \in r^{-6}Good_{14}(\underline{\alpha}).$$

Hence, simplifying,

$$A\,\not{d}_2^\star \not{d}_1^\star\rho + \frac{3}{4}\overline{\kappa}\,\overline{\rho}\,\not{d}_2^\star e_\theta(\underline{\kappa}) \in r^{-6}Good_{14}(\underline{\alpha}). \tag{7.5.13}$$

Next, in view of the identity $\not{d}_1^\star \not{d}_1 = \not{d}_2 \not{d}_2^\star + 2K$,

$$\begin{aligned}(\not{d}_2^\star \not{d}_2 + 2K)\,\not{d}_2^\star\zeta &= \not{d}_2^\star \not{d}_2 \not{d}_2^\star\zeta + 2K\,\not{d}_2^\star\zeta\\ &= \not{d}_2^\star\,(\not{d}_2 \not{d}_2^\star\zeta + 2K\zeta) - 2\,\not{d}_2^\star K\zeta\\ &= \not{d}_2^\star \not{d}_1^\star \not{d}_1\zeta + r^{-9/2}Good_{19}.\end{aligned}$$

Recalling the definition of $\mu = -\not{d}_1\zeta - \rho + \frac{1}{4}\vartheta\underline{\vartheta}$ and the product lemma we write

$$\not{d}_1^\star \not{d}_1\zeta \;=\; -\not{d}_1^\star\mu - \not{d}_1^\star\rho + \frac{1}{4}\not{d}_1^\star(\vartheta\underline{\vartheta}) = -\not{d}_1^\star\mu - \not{d}_1^\star\rho + r^{-4}Good_{19}.$$

In view of the estimates for $\check{\mu}$ we have already established we deduce

$$\not{d}_1^\star \not{d}_1\zeta = -\not{d}_1^\star\rho + r^{-4}Good_{17}.$$

Thus,

$$(\not{d}_2^\star \not{d}_2 + 2K)\,\not{d}_2^\star\zeta \;=\; -\not{d}_2^\star \not{d}_1^\star\rho + r^{-5}Good_{16}. \tag{7.5.14}$$

Therefore, making use of (7.5.13),

$$\begin{aligned} A(\,\not{d}_2^\star\,\not{d}_2 + 2K)\,\not{d}_2^\star\zeta &= -A\,\not{d}_2^\star\,\not{d}_1^\star\rho + r^{-6}Good_{14}\\ &= \frac{3}{4}\overline{\kappa}\,\overline{\rho}\,\not{d}_2^\star e_\theta(\underline{\kappa}) + r^{-6}Good_{14}(\underline{\alpha}), \end{aligned}$$

i.e.,

$$A(\,\not{d}_2^\star\,\not{d}_2 + 2K)\,\not{d}_2^\star\zeta - \frac{3}{4}\overline{\kappa}\,\overline{\rho}\,\not{d}_2^\star e_\theta(\underline{\kappa}) = r^{-6}Good_{14}(\underline{\alpha}) \tag{7.5.15}$$

as desired.

To prove the second statement of the lemma we write, using (7.5.12),

$$(\,\not{d}_2^\star\,\not{d}_2 + 2K)A\vartheta = \overline{\kappa}\,(\,\not{d}_2^\star\,\not{d}_2 + 2K)\,\not{d}_2^\star\zeta + r^{-5}Good_{12}(\underline{\alpha}).$$

Hence applying A and making use of (7.5.15),

$$\begin{aligned} A(\,\not{d}_2^\star\,\not{d}_2 + 2K)A\vartheta &= \overline{\kappa}A(\,\not{d}_2^\star\,\not{d}_2 + 2K)\,\not{d}_2^\star\zeta + r^{-7}Good_{10}(\underline{\alpha})\\ &= \frac{3}{4}\overline{\kappa}^2\,\overline{\rho}\,\not{d}_2^\star e_\theta(\underline{\kappa}) + r^{-7}Good_{10}(\underline{\alpha}). \end{aligned}$$

Finally, making use of the relation $\not{d}_2^\star\beta + \frac{3}{2}\overline{\rho}\,\vartheta \in r^{-3}Good_{15}$, we have

$$\begin{aligned} A^2(\,\not{d}_2^\star\,\not{d}_2 + 2K)\,\not{d}_2^\star\beta &= A(\,\not{d}_2^\star\,\not{d}_2 + 2K)A\,\not{d}_2^\star\beta + r^{-9}Good_9(\underline{\alpha})\\ &= -\frac{3}{2}\overline{\rho}A(\,\not{d}_2^\star\,\not{d}_2 + 2K)A\vartheta + r^{-9}Good_9(\underline{\alpha})\\ &= -\frac{3}{2}\overline{\rho}\left(\frac{3}{4}\overline{\kappa}^2\,\overline{\rho}\,\not{d}_2^\star e_\theta(\underline{\kappa}) + r^{-8}Good_{10}(\underline{\alpha})\right) + r^{-9}Good_9\\ &= -\frac{9}{8}\overline{\kappa}^2\overline{\rho}^2\,\not{d}_2^\star e_\theta(\underline{\kappa}) + r^{-9}Good_9(\underline{\alpha}) \end{aligned}$$

as desired. This concludes the proof of the lemma. □

Corollary 7.40. *The $\mathfrak{s}_1(\mathcal{M})$ tensor $e_\theta(\underline{\kappa}) = -\,\not{d}_1^\star\underline{\check{\kappa}}$ verifies the following fifth order elliptic equation in ${}^{(ext)}\mathcal{M}$:*

$$A^2\,\not{d}_2^\star\,(e_\theta\underline{\kappa}) - \frac{12m}{r^3}A\,\not{d}_2^\star e_\theta(\underline{\kappa}) + \frac{36m^2}{r^6}\,\not{d}_2^\star e_\theta(\underline{\kappa}) \in r^{-7}Good_8(\underline{\alpha}). \tag{7.5.16}$$

Proof. According to Lemma 7.39

$$\begin{aligned} AB\,\not{d}_2^\star\zeta - \frac{3}{4}\overline{\kappa}\,\overline{\rho}\,\not{d}_2^\star e_\theta(\underline{\kappa}) &\in r^{-6}Good_{14}(\underline{\alpha}),\\ A^2B\,\not{d}_2^\star\beta + \frac{9}{8}\left(\overline{\kappa}\,\overline{\rho}\right)^2\,\not{d}_2^\star e_\theta(\underline{\kappa}) &\in r^{-9}Good_9(\underline{\alpha}), \end{aligned}$$

we have

$$\begin{aligned} A^2B\,\not{d}_2^\star\zeta &= \frac{3}{4}\overline{\kappa}\,\overline{\rho}\,A\,\not{d}_2^\star e_\theta(\underline{\kappa}) + r^{-8}Good_{12}(\underline{\alpha}),\\ A^2B\,\not{d}_2^\star\beta &= -\frac{9}{8}\left(\overline{\kappa}\,\overline{\rho}\right)^2\,\not{d}_2^\star e_\theta(\underline{\kappa}) + r^{-9}Good_9(\underline{\alpha}). \end{aligned} \tag{7.5.17}$$

In view of Lemma 7.36 we have, on ${}^{(ext)}\mathcal{M}$,

$$e_\theta(\underline{\kappa}) + 4r\,\not{d}_1^\star\,\not{d}_1\zeta - 2r^2\,\not{d}_1^\star\,\not{d}_1\beta \in r^{-2}Good_{13}. \tag{7.5.18}$$

Thus,

$$A^2\,\not{d}_2^\star\Big(e_\theta(\underline{\kappa}) + 4r\,\not{d}_1^\star\,\not{d}_1\zeta - 2r^2\,\not{d}_1^\star\,\not{d}_1\beta\Big) \in r^{-7}Good_8.$$

Making use of

$$\not{d}_2^\star\,\not{d}_1^\star\,\not{d}_1 \;=\; \not{d}_2^\star\Big(\,\not{d}_2\,\not{d}_2^\star + 2K\Big) = (\,\not{d}_2^\star\,\not{d}_2 + 2K)\,\not{d}_2^\star - e_\theta(K),$$

we deduce

$$\begin{aligned} A^2\,\not{d}_2^\star\,(e_\theta\underline{\kappa}) &= -4rA^2\,\not{d}_2^\star\,\not{d}_1^\star\,\not{d}_1\zeta + 2r^2A^2\,\not{d}_2^\star\,\not{d}_1^\star\,\not{d}_1\beta + r^{-7}Good_8\\ &= -4rA^2(\,\not{d}_2^\star\,\not{d}_2 + 2K)\,\not{d}_2^\star\zeta + 2r^2A^2(\,\not{d}_2^\star\,\not{d}_2 + 2K)\,\not{d}_2^\star\beta + r^{-7}Good_8\\ &= -4rA^2B\,\not{d}_2^\star\zeta + 2r^2A^2B\,\not{d}_2^\star\beta + r^{-7}Good_8. \end{aligned}$$

Thus, in view of the lemma,

$$\begin{aligned} A^2\,\not{d}_2^\star\,(e_\theta\underline{\kappa}) &= -3r\left(\overline{\kappa}\,\overline{\rho}A\,\not{d}_2^\star e_\theta(\underline{\kappa}) + r^{-8}Good_{12}\right)\\ &- \frac{9}{4}r^2\left(\left(\overline{\kappa}\,\overline{\rho}\right)^2\not{d}_2^\star e_\theta(\underline{\kappa}) + r^{-9}Good_9(\underline{\alpha})\right) + r^{-7}Good_8. \end{aligned}$$

We deduce

$$A^2\,\not{d}_2^\star e_\theta\underline{\kappa} + 3r\left(\overline{\kappa}\,\overline{\rho}\right)A\,\not{d}_2^\star e_\theta(\underline{\kappa}) + \frac{9}{4}r^2\left(\overline{\kappa}\,\overline{\rho}\right)^2\not{d}_2^\star e_\theta(\underline{\kappa}) \in r^{-7}Good_8(\underline{\alpha}).$$

Finally,

$$A^2\,\not{d}_2^\star e_\theta\underline{\kappa} - \frac{12m}{r^3}A\,\not{d}_2^\star e_\theta(\underline{\kappa}) + \frac{36m^2}{r^6}\,\not{d}_2^\star e_\theta(\underline{\kappa}) \in r^{-7}Good_8(\underline{\alpha})$$

as desired. □

Lemma 7.41. *We have the following Poincaré inequality on* ${}^{(ext)}\mathcal{M}$ *for* $f \in \mathfrak{s}_2(\mathcal{M})$ *with* $A = (\,\not{d}_2^\star\,\not{d}_2 - 3\overline{\rho})$*:*

$$\int_S f\left(A^2 - \frac{12m}{r^3}A + \frac{36m^2}{r^6}\right)f \;\geq\; \frac{1}{4r^2}\int_S(\,\not{d}_2 f)^2 + \frac{9}{r^4}\int_S f^2.$$

Proof. Recall that we have the following Poincaré inequality for $\not{d}_2$:

$$\int_S(\,\not{d}_2 f)^2 \geq 4\int_S Kf^2.$$

Since $\left|K - r^{-2}\right| \lesssim \epsilon r^{-2}$,

$$\begin{aligned}\int_S fAf &= \int_S f(\not{d}_2^\star \not{d}_2 - 3\overline{\rho})f \geq \int_S (4K - 3\overline{\rho})f^2 \\ &= \left(\frac{4}{r^2} + \frac{6m}{r^3} + O(r^{-2}\epsilon)\right)\int_S f^2.\end{aligned}$$

Since A is positive self-adjoint,

$$\begin{aligned}\int_S fA^2f &= \int_S (A^{1/2}f)A(A^{1/2}f) = \left(\frac{4}{r^2} + \frac{6m}{r^3} + O(r^{-2}\epsilon)\right)\int_S |A^{1/2}f|^2 \\ &= \left(\frac{4}{r^2} + \frac{6m}{r^3} + O(r^{-2}\epsilon)\right)\int_S fAf.\end{aligned}$$

This yields

$$\begin{aligned}\int_S f\left(A^2f - \frac{12m}{r^3}Af\right) &= \left(\frac{4}{r^2} + \frac{6m}{r^3} - \frac{12m}{r^3} + O(r^{-2}\epsilon)\right)\int_S fAf \\ &= \left(\frac{4}{r^2} - \frac{6m}{r^3} + O(r^{-2}\epsilon)\right)\int_S fAf,\end{aligned}$$

and therefore,

$$\begin{aligned}\int_S f\left(A^2 - \frac{12m}{r^3}A + \frac{36m^2}{r^6}\right) &\geq \left(\frac{4}{r^2} - \frac{6m}{r^3} + O(r^{-2}\epsilon)\right)\int_S fAf + \frac{36m^2}{r^6}\int_S f^2 \\ &= \left(\frac{4}{r^2}\left(1 - \frac{3m}{2r}\right) + O(r^{-2}\epsilon)\right)\int_S fAf \\ &\quad + \frac{36m^2}{r^6}\int_S f^2.\end{aligned}$$

Note that for $r > 2m$ we have

$$1 - \frac{3m}{2r} > \frac{1}{4}.$$

We deduce, for sufficiently small ϵ, everywhere in ${}^{(ext)}\mathcal{M}$,

$$\int_S f\left(A^2 - \frac{12m}{r^3}A + \frac{36m^2}{r^6}\right) > \frac{1}{r^2}\left(\int_S fAf + \frac{36m^2}{r^4}\int_S f^2\right).$$

Now, since $\left|\overline{\rho} + \frac{2m}{r^3}\right| \lesssim \epsilon_0 r^{-3}$,

$$\int_S fAf = \int_S f(\not{d}_2^\star \not{d}_2 - 3\overline{\rho})f = \int_S \left(|\not{d}_2 f|^2 + \left(\frac{6m}{r^3} + O(r^{-3}\epsilon_0)\right)|f|^2\right).$$

Hence,

$$\int_S fAf + \frac{36m^2}{r^4}\int_S f^2 > \int_S \left(|\not{d}_2 f|^2 + (\frac{6m}{r^3} + \frac{36m^2}{r^6})|f|^2\right) > \int_S |\not{d}_2 f|^2$$

or, since $\int_S (\not{d}_2 f)^2 \geq 4\int_S \frac{1}{r^2}\int_S |f|^2 + O(\epsilon_0 r^{-3})\int_S |f|^2$. We deduce

$$\int_S f\left(A^2 - \frac{12m}{r^3}A + \frac{36m^2}{r^6}\right) \geq \frac{1}{4r^2}\int_S (\not{d}_2 f)^2 + \frac{9}{r^4}\int_S f^2$$

as desired. This concludes the proof of the lemma. □

Applying the lemma to $f = \not{d}_2^\star e_\theta \underline{\kappa}$ in (7.5.16), i.e.,

$$A^2 \not{d}_2^\star (e_\theta \underline{\kappa}) - \frac{12m}{r^3} A \not{d}_2^\star e_\theta(\underline{\kappa}) + \frac{36m^2}{r^6} \not{d}_2^\star e_\theta(\underline{\kappa}) \in r^{-7} Good_8(\underline{\alpha})$$

or, in any region where

$$\|\underline{\alpha}\|_{2,k} \lesssim \epsilon_0 r^{-1} u^{-1-\delta_{dec}}, \qquad k \leq k_{small} + 16,$$

we have

$$\left\|A^2 \not{d}_2^\star (e_\theta \underline{\kappa}) - \frac{12m}{r^3} A \not{d}_2^\star e_\theta(\underline{\kappa}) + \frac{36m^2}{r^6} \not{d}_2^\star e_\theta(\underline{\kappa})\right\|_{2,k}$$
$$\lesssim \epsilon_0 r^{-6} u^{-1-\delta_{dec}}, \qquad k \leq k_{small} + 8.$$

We deduce, by L^2-elliptic estimates,

$$\|\not{d}_2^\star e_\theta \check{\underline{\kappa}}\|_{2,k} \lesssim \epsilon_0 r^{-2} u^{-1-\delta_{dec}}, \qquad k \leq k_{small} + 12. \tag{7.5.19}$$

Since we control the $\ell = 1$ mode of $e_\theta \check{\underline{\kappa}}$ we infer that

$$\|e_\theta \check{\underline{\kappa}}\|_{2,k} \lesssim \epsilon_0 r^{-1} u^{-1-\delta_{dec}}, \qquad k \leq k_{small} + 13,$$

i.e.,

$$\|\check{\underline{\kappa}}\|_{2,k} \lesssim \epsilon_0 u^{-1-\delta_{dec}}, \qquad k \leq k_{small} + 14.$$

Therefore, using the Sobolev embedding,

$$\|\check{\underline{\kappa}}\|_{\infty,k} \lesssim \epsilon_0 r^{-1} u^{-1-\delta_{dec}} \qquad k \leq k_{small} + 12.$$

This proves the following:

Proposition 7.42. *In any region of ${}^{(ext)}\mathcal{M}$ where*

$$\|\underline{\alpha}\|_{2,k} \lesssim \epsilon_0 r^{-1} u^{-1-\delta_{dec}}, \qquad k \leq k_{small} + 16,$$

we also have

$$\|\check{\underline{\kappa}}\|_{\infty,k} \lesssim \epsilon_0 r^{-1} u^{-1-\delta_{dec}}, \qquad k \leq k_{small} + 12. \tag{7.5.20}$$

7.5.3 Proof of Proposition 7.35, Part I

We first prove Proposition 7.35 in the region where the estimate

$$\|\underline{\alpha}\|_{2,k} \lesssim \epsilon_0 r^{-1} u^{-1-\delta_{dec}}, \qquad k \leq k_{small} + 16, \tag{7.5.21}$$

holds true.

Step 1. We prove the estimates

$$\|\zeta\|_{\infty,k} \lesssim \epsilon_0 r^{-1} u^{-1-\delta_{dec}}, \qquad k \le k_{small}+15,$$
$$\|\beta\|_{\infty,k} \lesssim \epsilon_0 r^{-2} u^{-1-\delta_{dec}}, \qquad k \le k_{small}+12. \tag{7.5.22}$$

According to (7.5.17)

$$A^2 B \,\not{d}_2^\star \zeta = \frac{3}{4}\overline{\kappa}\,\overline{\rho}\, A \,\not{d}_2^\star e_\theta(\underline{\kappa}) + r^{-8} Good_{12}(\underline{\alpha}),$$
$$A^2 B \,\not{d}_2^\star \beta = -\frac{9}{8}\left(\overline{\kappa}\,\overline{\rho}\right)^2 \not{d}_2^\star e_\theta(\underline{\kappa}) + r^{-9} Good_9(\underline{\alpha}).$$

In view of (7.5.19) we deduce, in L^2 norms,

$$\|A^2 B \,\not{d}_2^\star \zeta\|_{2,k} \lesssim r^{-4}\|A \,\not{d}_2^\star e_\theta \check{\kappa}\|_{2,k} + \epsilon_0 r^{-7} u^{-1-\delta_{dec}}, \qquad k \le k_{small}+12,$$
$$\|A^2 B \,\not{d}_2^\star \beta\|_{2,k} \lesssim r^{-8}\| \not{d}_2^\star e_\theta \check{\kappa}\|_{2,k} + \epsilon_0 r^{-8} u^{-1-\delta_{dec}}, \qquad k \le k_{small}+9.$$

Thus, in view of the estimates for $\check{\kappa}$ derived above,

$$\|A^2 B \,\not{d}_2^\star \zeta\|_{2,k} \lesssim \epsilon_0 r^{-7} u^{-1-\delta_{dec}}, \qquad k \le k_{small}+12,$$
$$\|A^2 B \,\not{d}_2^\star \beta\|_{2,k} \lesssim \epsilon_0 r^{-8} u^{-1-\delta_{dec}}, \qquad k \le k_{small}+9.$$

Thus, by elliptic estimates,

$$\| \not{d}_2^\star \zeta\|_{2,k} \lesssim \epsilon_0 r^{-1} u^{-1-\delta_{dec}}, \qquad k \le k_{small}+16,$$
$$\| \not{d}_2^\star \beta\|_{2,k} \lesssim \epsilon_0 r^{-2} u^{-1-\delta_{dec}}, \qquad k \le k_{small}+13.$$

In view of the estimates for the $\ell = 1$ modes of ζ, β we deduce

$$\|\zeta\|_{2,k} \lesssim \epsilon_0 u^{-1-\delta_{dec}}, \qquad k \le k_{small}+17,$$
$$\|\beta\|_{2,k} \lesssim \epsilon_0 r^{-1} u^{-1-\delta_{dec}}, \qquad k \le k_{small}+14.$$

Passing to L^∞ norms we derive

$$\|\zeta\|_{\infty,k} \lesssim \epsilon_0 r^{-1} u^{-1-\delta_{dec}}, \qquad k \le k_{small}+15,$$
$$\|\beta\|_{\infty,k} \lesssim \epsilon_0 r^{-2} u^{-1-\delta_{dec}}, \qquad k \le k_{small}+13. \tag{7.5.23}$$

Step 2. We prove the estimate

$$\|\eta\|_{\infty,k} \lesssim \epsilon_0 r^{-1} u^{-1-\delta_{dec}}, \qquad k \le k_{small}+15.$$

This follows immediately from the estimate from ζ and the previously derived estimate (7.5.2). Indeed,

$$\|\eta\|_{\infty,k} \lesssim \|\zeta\|_{\infty,k} + \epsilon_0 r^{-1} u^{-1-\delta_{dec}} \lesssim \epsilon_0 r^{-1} u^{-1-\delta_{dec}}.$$

Step 3. We derive the estimate

$$\|\vartheta\|_{\infty,k} \lesssim r^{-1} u^{-1-\delta_{dec}}, \qquad k \le k_{small}+11. \tag{7.5.24}$$

This follows easily in view of the equation (see (7.5.11))

$$\not{d}_2\vartheta + 2\beta - \overline{\kappa}\zeta \quad \in r^{-3}Good_{20}$$

from which, in view of Step 1,

$$\| \not{d}_2\vartheta\|_{2,k} \quad \lesssim \quad \epsilon_0 r^{-1}u^{-1-\delta_{dec}}, \qquad k \le k_{small} + 12.$$

The desired estimate follows by elliptic estimates and Sobolev.

Step 4. We derive the intermediate estimate for $\underline{\vartheta}$,

$$\|\underline{\vartheta}\|_{\infty,k} \quad \lesssim \quad \epsilon_0 u^{-1-\delta_{dec}}, \qquad k \le k_{small} + 12. \tag{7.5.25}$$

To show this we combine the equations (see (7.5.11))

$$\begin{aligned} \not{d}_2\underline{\vartheta} + 2\underline{\beta} - e_\theta(\underline{\kappa}) + \overline{\underline{\kappa}}\zeta &\in r^{-3}Good_{20}, \\ \not{d}_2^\star\underline{\beta} + \frac{3}{2}\overline{\rho}\underline{\vartheta} &\in r^{-2}Good_{15}, \end{aligned}$$

to deduce

$$\not{d}_2^\star\, \not{d}_2\underline{\vartheta} - 3\overline{\rho}\underline{\vartheta} \quad = \quad \not{d}_2^\star e_\theta\underline{\kappa} + \overline{\underline{\kappa}}\, \not{d}_2^\star\zeta + r^{-2}Good_{15},$$

and hence,

$$\|A\underline{\vartheta}\|_{2,k} \quad \lesssim \quad \epsilon_0 r^{-1}u^{-1-\delta_{dec}}, \qquad k \le k_{small} + 12.$$

Thus,

$$\|\underline{\vartheta}\|_{2,k} \quad \lesssim \quad \epsilon_0 r u^{-1-\delta_{dec}}, \qquad k \le k_{small} + 14$$

and hence,

$$\|\underline{\vartheta}\|_{\infty,k} \quad \lesssim \quad \epsilon_0 u^{-1-\delta_{dec}}, \qquad k \le k_{small} + 12$$

as desired.

Step 5. We derive the estimate

$$\|\check{\rho}\|_{\infty,k} \quad \lesssim \epsilon_0 r^{-2}u^{-1-\delta_{dec}}, \qquad k \le k_{small} + 14. \tag{7.5.26}$$

From

$$\not{d}_2^\star\, \not{d}_1^\star\rho + \frac{3}{4}\overline{\underline{\kappa}}\,\overline{\rho}\vartheta + \frac{3}{4}\overline{\kappa}\,\overline{\rho}\underline{\vartheta} \quad \in r^{-4}Good_{20},$$

we deduce

$$\begin{aligned} \| \not{d}_2^\star\, \not{d}_1^\star\rho\|_{2,k} &\lesssim r^{-4}\left(\|\theta\|_{2,k} + \|\underline{\theta}\|_{2,k}\right) + \epsilon_0 r^{-3}u^{-1-\delta_{dec}}, \qquad k \le k_{small} + 14 \\ &\lesssim \epsilon_0 r^{-3}u^{-1-\delta_{dec}}, \qquad k \le k_{small} + 14. \end{aligned}$$

Since we control the $\ell = 1$ mode of $\not{d}_1^\star \rho$ (see Lemma 7.34) we infer that

$$\|\check{\rho}\|_{2,k} \lesssim \epsilon_0 r^{-1} u^{-1-\delta_{dec}}, \qquad k \le k_{small} + 16,$$

i.e.,

$$\|\check{\rho}\|_{\infty,k} \lesssim \epsilon_0 r^{-2} u^{-1-\delta_{dec}}, \qquad k \le k_{small} + 14$$

as desired.

Step 6. We derive the estimate

$$\|\underline{\beta}\|_{\infty,k} \lesssim \epsilon_0 r^{-2} u^{-1-\delta_{dec}}, \qquad \forall k \le k_{small} + 9 \tag{7.5.27}$$

with the help of the identity

$$\begin{aligned} e_3(r\mathfrak{q}) &= r^5 \left\{ \not{d}_2^\star \not{d}_1^\star \not{d}_1 \underline{\beta} - \frac{3}{2}\kappa\rho\underline{\alpha} - \frac{3}{2}\rho \not{d}_2^\star \not{d}_1^\star \underline{\kappa} - \frac{3}{2}\underline{\kappa}\rho \not{d}_2^\star \zeta + \frac{3}{4}(2\rho^2 - \kappa\underline{\kappa}\rho)\underline{\vartheta} \right\} \\ &\quad + \mathrm{Err}[e_3(r\mathfrak{q})], \\ \mathrm{Err}[e_3(r\mathfrak{q})] &= r^4(e_3\Gamma_b) \cdot \not{\mathfrak{d}}^{\le 1}\beta + r\Gamma_b \cdot \mathfrak{q} + r^2 \not{\mathfrak{d}}^3(\Gamma_g \cdot \Gamma_b), \end{aligned}$$

of Proposition 7.11. In view of (7.3.18) we have

$$\|\mathrm{Err}[e_3(r\mathfrak{q})]\|_{\infty,k}(u,r) \lesssim \epsilon_0 u^{-1-\delta_{dec}}, \qquad k \le k_{small} + 16.$$

We can now make use of the estimates for $\check{\underline{\kappa}}, , \zeta, \vartheta, \underline{\vartheta}$ already derived and the **Ref 2** estimate for $e_3(\mathfrak{q})$ and $\underline{\alpha}$ to deduce, for all $k \le k_{small} + 10$,

$$\begin{aligned} \|\rho \not{d}_2^\star \not{d}_1^\star \underline{\kappa}\|_{\infty,k} &\lesssim \epsilon_0 r^{-6} u^{-1-\delta_{dec}}, \\ \|\underline{\kappa}\rho \not{d}_2^\star \zeta\|_{\infty,k} &\lesssim \epsilon_0 r^{-6} u^{-1-\delta_{dec}}, \\ \|\rho^2 \underline{\vartheta}\|_{\infty,k} &\lesssim \epsilon_0 r^{-6} u^{-1-\delta_{dec}}, \\ \|\kappa\underline{\kappa}\vartheta\|_{\infty,k} &\lesssim \epsilon_0 r^{-5} u^{-1-\delta_{dec}}, \\ \|\kappa\rho\underline{\alpha}\|_{\infty,k} &\lesssim \epsilon_0 r^{-5} u^{-1-\delta_{dec}}, \\ \|e_3(r\mathfrak{q})\|_{\infty,k} &\lesssim \epsilon_0 u^{-1-\delta_{dec}}. \end{aligned}$$

Therefore,

$$\|\not{d}_2^\star \not{d}_1^\star \not{d}_1 \underline{\beta}\|_{\infty,k} \lesssim \epsilon_0 r^{-5} u^{-1-\delta_{dec}}, \qquad k \le k_{small} + 10,$$

i.e.,

$$\|\not{d}_2^\star \not{d}_1^\star \not{d}_1 \underline{\beta}\|_{2,k} \lesssim \epsilon_0 r^{-4} u^{-1-\delta_{dec}}, \qquad k \le k_{small} + 10.$$

Making use of the identity

$$\not{d}_1^\star \not{d}_1 = \not{d}_2 \not{d}_2^\star + 2K,$$

we deduce

$$\left\|(\not{d}_2^\star \not{d}_2 + K) \not{d}_2^\star \underline{\beta}\right\|_{2,k} \lesssim \epsilon_0 r^{-4} u^{-1-\delta_{dec}}.$$

Since $\not{d}_2^\star \not{d}_2 + K$ is coercive we deduce

$$\| \not{d}_2^\star \underline{\beta}\|_{2,k} \lesssim \epsilon_0 r^{-2} u^{-1-\delta_{dec}}, \qquad \forall k \le k_{small} + 10.$$

Since we control the $\ell = 1$ mode of $\underline{\beta}$ (see Lemma 7.34) according to Lemma 7.26,

$$\|\underline{\beta}\|_{2,k} \lesssim \epsilon_0 r^{-1} u^{-1-\delta_{dec}}, \qquad \forall k \le k_{small} + 11.$$

Hence,

$$\|\underline{\beta}\|_{\infty,k} \lesssim \epsilon_0 r^{-2} u^{-1-\delta_{dec}}, \qquad \forall k \le k_{small} + 9. \tag{7.5.28}$$

Step 7. Using the above estimate for $\underline{\beta}$ we can improve the estimate for $\underline{\vartheta}$ derived in Step 4. We show, in the region where the estimate (7.5.21) for $\underline{\alpha}$ holds,

$$\|\underline{\vartheta}\|_{\infty,k} \lesssim \epsilon_0 r^{-1} u^{-1-\delta_{dec}}, \qquad k \le k_{small} + 9. \tag{7.5.29}$$

Indeed in view of the Codazzi equation

$$\not{d}_2 \underline{\vartheta} + 2\underline{\beta} - e_\theta(\underline{\kappa}) + \overline{\underline{\kappa}}\zeta \quad \in r^{-3} Good_{20},$$

we infer that, for all $k \le k_{small} + 11$,

$$\begin{aligned} \| \not{d}_2 \underline{\vartheta}\|_{2,k} &\lesssim \|\underline{\beta}\|_{2,k} + r^{-1}\|\check{\underline{\kappa}}\|_{2,k+1} + r^{-1}\|\zeta\|_{2,k} + \epsilon_0 r^{-2} u^{-1-\delta_{dec}} \\ &\lesssim \epsilon_0 r^{-2} u^{-1-\delta_{dec}}. \end{aligned}$$

Thus, for all $k \le k_{small} + 12$,

$$\|\underline{\vartheta}\|_{2,k} \lesssim \epsilon_0 r^{-1} u^{-1-\delta_{dec}}$$

and hence,

$$\|\underline{\vartheta}\|_{\infty,k} \lesssim \epsilon_0 r^{-2} u^{-1-\delta_{dec}}, \qquad k \le k_{small} + 10. \tag{7.5.30}$$

This ends the proof of Proposition 7.35 in the region for which the desired estimate (7.5.21) for $\underline{\alpha}$ holds true.

Since (7.5.21) for $\underline{\alpha}$ holds true on $\mathcal{T}$ in view of[20] Theorem M3, this ends the proof of Proposition 7.35 on $\mathcal{T}$.

7.5.4 Proof of Proposition 7.35, Part II

We extend the validity of Proposition 7.35 to all of ${}^{(ext)}\mathcal{M}$ propagating the estimates derived in the first part on $\mathcal{T}$. We also recall that we have good decay estimates for $\check{\kappa}$ and $\check{\mu}$ everywhere on ${}^{(ext)}\mathcal{M}$.

Step 1. We first derive estimates for ϑ in $\mathcal{M}_{ext}$ making use of the transport

[20] Recall that r is bounded on $\mathcal{T}$ and that $\mathcal{T} \subset {}^{(int)}\mathcal{M}$ so that (7.5.21) holds true for ${}^{(int)}\underline{\alpha}$ on $\mathcal{T}$ in view of Theorem M3. Then, since we have ${}^{(ext)}\underline{\alpha} = ({}^{(ext)}\Upsilon)^2\, {}^{(int)}\underline{\alpha}$ on $\mathcal{T}$, (7.5.21) holds indeed true for ${}^{(ext)}\underline{\alpha}$ on $\mathcal{T}$.

equation

$$e_4(\vartheta) + \overline{\kappa}\vartheta \quad = \quad -2\alpha - (\kappa - \overline{\kappa})\vartheta = -2\alpha + \Gamma_g \cdot \Gamma_g.$$

Making use of Proposition 7.31 we derive, for all $r \geq r_0 = r_{\mathcal{T}}$,

$$r^2\|\vartheta\|_{\infty,k}(u,r) \quad \lesssim \quad r_0^2\|\vartheta\|_{\infty,k}(u,r_0) + \int_{r_0}^r \lambda^2\|\alpha\|_{\infty,k}(u,\lambda)d\lambda + \epsilon_0 u^{-1-\delta_{dec}}.$$

We now make use of the estimate

$$\|\alpha\|_{\infty,k}(u,r) \lesssim \epsilon_0 r^{-2}u^{-1-\delta_{dec}}, \qquad k \leq k_{small} + 20$$

and

$$\|\vartheta\|_{\infty,k}(u,r_0) \quad \lesssim \quad \epsilon_0 u^{-1-\delta_{dec}}$$

derived above in (7.5.24), to derive

$$r^2\|\vartheta\|_{\infty,k}(u,r) \quad \lesssim \quad \epsilon_0 u^{-1-\delta_{dec}} + \epsilon_0 r u^{-1-\delta_{dec}}.$$

Therefore, everywhere on ${}^{(ext)}\mathcal{M}$,

$$\|\vartheta\|_{\infty,k}(u,r) \quad \lesssim \quad \epsilon_0 r^{-1}u^{-1-\delta_{dec}}. \tag{7.5.31}$$

Step 2. Next, we estimate β from the equation

$$e_4\beta + 2\overline{\kappa}\beta \quad = \quad \not{d}_2\alpha - (\kappa - \overline{\kappa})\beta + \Gamma_g \cdot \alpha = \not{d}_2\alpha + \Gamma_g \cdot (\alpha, \beta)$$

to deduce in the same manner

$$r^4\|\beta\|_{\infty,k}(u,r) \quad \lesssim \quad r_0^2\|\beta\|_{\infty,k}(u,r_0) + \int_{r_0}^r \lambda^4\|\not{d}_2\alpha\|_{\infty,k}(u,\lambda)d\lambda + \epsilon_0 r u^{-1-\delta_{dec}}.$$

Thus, in view of the estimates for α in (7.5.23) and the estimates for α in **Ref 2**, i.e., for $0 \leq k \leq k_{small} + 20$,

$$\|\alpha\|_{\infty,k} \lesssim \epsilon_0 \min\{r^{-2}\log(1+u)(u+2r)^{-1-\delta_{extra}}, r^{-3}(u+2r)^{-\frac{1}{2}-\delta_{extra}}\}.$$

Thus we have with $I(u,r) := \int_{r_0}^r \lambda^4\|\not{d}_2\alpha\|_{\infty,k}(u,\lambda)d\lambda$

$$I(u,r) \lesssim \epsilon_0 \min\left\{\log(1+u)\int_{r_0}^r \lambda(u+2\lambda)^{-1-\delta_{extra}}d\lambda, \int_{r_0}^r (u+2\lambda)^{-1/2-\delta_{extra}}d\lambda\right\}.$$

If $r \leq 2u$ we have

$$\int_{r_0}^r \lambda(u+2\lambda)^{-1-\delta_{extra}}d\lambda \lesssim r^2u^{-1-\delta_{extra}} \lesssim r^2(u+2r)^{-1-\delta_{extra}}$$

and

$$r^{-4}I(u,r) \quad \lesssim \quad \epsilon_0 r^{-2}\log(1+u)(u+2r)^{-1-\delta_{extra}}.$$

If $r \geq 2u$ we have

$$\int_{r_0}^{r} (u+2\lambda)^{-1/2-\delta_{extra}} \lesssim (u+2r)^{1/2+\delta_{extra}}$$

and

$$r^{-4}I(u,r) \lesssim r^{-4}(u+2r)^{1/2+\delta_{extra}} \lesssim r^{-2}(u+2r)^{-1-\delta_{extra}}.$$

We deduce

$$\|\beta\|_{\infty,k} \lesssim r^{-4}\|\beta\|_{\infty,k}(u,r_0) + \epsilon_0 r^{-2}\log(1+u)(u+2r)^{-1-\delta_{extra}}.$$

Thus, in view of (7.5.23),

$$\|\beta\|_{\infty,k} \lesssim \epsilon_0 r^{-2}\log(1+u)(u+2r)^{-1-\delta_{extra}}. \tag{7.5.32}$$

Step 3. We now estimate ζ using the equation

$$e_4(\zeta) + \kappa\zeta = -\beta + \Gamma_g \cdot \Gamma_g.$$

This can be done exactly as in Step 1 making use of the estimates already derived for β and the estimate (7.5.23) for ζ along $\mathcal{T}$. We thus derive

$$\|\zeta\|_{\infty,k} \lesssim \epsilon_0 r^{-1}u^{-1-\delta_{dec}}, \qquad k \leq k_{small+15}.$$

Step 4. We estimate $\check{\rho}$ using equation

$$\check{\rho} = -\,\not{d}_1\zeta - \check{\mu} + \Gamma_g \cdot \Gamma_b,$$

the previous estimate for ζ and $\check{\mu}$ in ${}^{(ext)}\mathcal{M}$. We deduce

$$\|\check{\rho}\|_{\infty,k} \lesssim \epsilon_0 r^{-2}u^{-1-\delta_{dec}}, \qquad k \leq k_{small} + 14. \tag{7.5.33}$$

Step 5. We estimate $\check{\underline{\kappa}}$ using the equation

$$e_4\check{\underline{\kappa}} + \frac{1}{2}\overline{\kappa}\check{\underline{\kappa}} + \frac{1}{2}\check{\kappa}\overline{\underline{\kappa}} = -2\,\not{d}_1\zeta + 2\check{\rho} + \Gamma_g \cdot \Gamma_b.$$

Making use of the estimates in ${}^{(ext)}\mathcal{M}$ for $\check{\kappa}$, ζ and $\check{\rho}$ as well as the estimates for $\check{\underline{\kappa}}$ on $\mathcal{T}$ in Proposition 7.42 we derive, everywhere on ${}^{(ext)}\mathcal{M}$,

$$\|\check{\underline{\kappa}}\|_{\infty,k} \lesssim \epsilon_0 r^{-1}u^{-1-\delta_{dec}}, \qquad k \leq k_{small} + 12. \tag{7.5.34}$$

Alternatively, we could make use of the estimate for the auxiliary quantity $\Xi = r^2\left(e_\theta(\underline{\kappa}) + 4r\,\not{d}_1^\star\,\not{d}_1\zeta - 2r^2\,\not{d}_1^\star\,\not{d}_1\beta\right)$ in Lemma 7.36, which holds everywhere on ${}^{(ext)}\mathcal{M}$, and the above estimates for ζ, β.

Step 6. We estimate $\underline{\beta}$ everywhere on ${}^{(ext)}\mathcal{M}$ with the help of the equation

$$e_4\underline{\beta} + \overline{\kappa}\underline{\beta} = -\,\not{d}_1\rho - 3\zeta\rho - \underline{\vartheta}\beta - (\kappa - \overline{\kappa})\underline{\beta}$$

together with the estimate (7.5.28) for $\underline{\beta}$ on $\mathcal{T}$ and the above derived estimates for

$\check{\rho}, \zeta$ in ${}^{(ext)}\mathcal{M}$ to infer that

$$\|\underline{\beta}\|_{\infty,k} \lesssim \epsilon_0 r^{-2} u^{-1-\delta_{dec}}, \qquad \forall k \leq k_{small}+9. \tag{7.5.35}$$

Step 7. We estimate $\underline{\vartheta}$ everywhere on ${}^{(ext)}\mathcal{M}$ by making use of the Codazzi equation for $\underline{\vartheta}$ in (7.5.11),

$$\not{d}_2\underline{\vartheta} + 2\underline{\beta} - e_\theta(\underline{\kappa}) + \overline{\underline{\kappa}}\zeta \quad \in r^{-3} Good_{20}.$$

Using the estimates already derived above, we infer that, for all $k \leq k_{small}+11$,

$$\begin{aligned} \|\not{d}_2\underline{\vartheta}\|_{2,k} &\lesssim \|\underline{\beta}\|_{2,k} + r^{-1}\|\underline{\check{\kappa}}\|_{2,k+1} + r^{-1}\|\zeta\|_{2,k} + \epsilon_0 r^{-2} u^{-1-\delta_{dec}} \\ &\lesssim \epsilon_0 r^{-2} u^{-1-\delta_{dec}}. \end{aligned}$$

Hence, everywhere in ${}^{(ext)}\mathcal{M}$,

$$\|\underline{\vartheta}\|_{2,k} \lesssim \epsilon_0 r^{-1} u^{-1-\delta_{dec}}, \qquad \text{for all } k \leq k_{small}+12,$$

and therefore,

$$\|\underline{\vartheta}\|_{\infty,k} \lesssim \epsilon_0 r^{-2} u^{-1-\delta_{dec}}, \qquad \text{for all } k \leq k_{small}+10.$$

Step 8. We estimate $\underline{\alpha}$ everywhere on ${}^{(ext)}\mathcal{M}$ by making use of the equation

$$e_4\underline{\alpha} + \frac{1}{2}\overline{\kappa}\underline{\alpha} = -\not{d}_2^\star\underline{\beta} - \frac{3}{2}\underline{\vartheta}\rho - 5\zeta\underline{\beta} - \frac{1}{2}(\kappa - \overline{\kappa})\underline{\alpha}$$

as well as the estimate (7.5.21) for $\underline{\alpha}$ on $\mathcal{T}$ and the above estimates in all ${}^{(ext)}\mathcal{M}$ for $\underline{\beta}$ and $\underline{\vartheta}$. Proceeding as before we derive

$$\|\underline{\alpha}\|_{\infty,k} \lesssim \epsilon_0 r^{-1} u^{-1-\delta_{dec}}, \qquad \text{for all } k \leq k_{small}+8. \tag{7.5.36}$$

This concludes the proof of Proposition 7.35.

7.6 CONCLUSION OF THE PROOF OF THEOREM M4

So far we have established the following estimates, for all $k \leq k_{small}+8$,

$$\begin{aligned} \|\check{\kappa}, r\check{\mu}\|_{\infty,k} &\lesssim \epsilon_0 r^{-2} u^{-1-\delta_{dec}}, \\ \|\vartheta,\ \zeta,\ \underline{\check{\kappa}}, r\check{\rho}\|_{\infty,k} &\lesssim \epsilon_0 r^{-2} u^{-1/2-\delta_{dec}}, \\ \|\vartheta,\ \zeta,\ \eta,\ \underline{\check{\kappa}},\ \underline{\vartheta},\ r\beta,\ r\check{\rho},\ r\underline{\beta},\ \underline{\alpha}\|_{\infty,k} &\lesssim \epsilon_0 r^{-1} u^{-1-\delta_{dec}}, \\ \|\beta, re_3\beta\|_{\infty,k} &\lesssim \epsilon_0 r^{-3}(2r+u)^{-1/2-\delta_{dec}}. \end{aligned} \tag{7.6.1}$$

It only remains to derive improved decay estimates for $e_3(\beta, \vartheta, \zeta, \underline{\check{\kappa}}, \check{\rho})$ and the estimates for $\underline{\xi}, \underline{\check{\omega}}, \check{\varsigma}, \underline{\check{\Omega}}$ as well as $\overline{\varsigma}+1$ and $\overline{\underline{\Omega}}+\Upsilon$ in terms of $u^{-1-\delta_{dec}}$ decay. More precisely it remains to prove the following.

Proposition 7.43. *The following estimates hold true on* ${}^{(ext)}\mathcal{M}$ *for all integers* k

such that $k \le k_{small} + 7$*:*

$$\begin{aligned}\big\|e_3(\vartheta,\,\zeta\,,\check{\underline{\kappa}}),\, re_3\beta,\, re_3\check{\rho}\big\|_{\infty,k} &\lesssim \epsilon_0 r^{-2}u^{-1-\delta_{dec}},\\ \big\|\underline{\xi},\, \check{\underline{\omega}}\big\|_{\infty,k} &\lesssim \epsilon_0 r^{-1}u^{-1-\delta_{dec}},\\ \big\|\check{\varsigma}, \check{\underline{\Omega}}, \overline{\varsigma}+1,\, \overline{\underline{\Omega}}+\Upsilon\big\|_{\infty,k} &\lesssim \epsilon_0 u^{-1-\delta_{dec}}.\end{aligned}$$

Proof. We proceed in steps as follows.

Step 1. We make use of the equation $e_3\vartheta = -\frac{1}{2}\underline{\kappa}\,\vartheta + 2\underline{\omega}\vartheta - 2\,{\not{d}}_2^\star\eta - \frac{1}{2}\kappa\,\underline{\vartheta} + 2\eta^2$ and the previously derived estimates to derive

$$\big\|e_3\vartheta\big\|_{\infty,k} \lesssim \epsilon_0 r^{-2}u^{-1-\delta_{dec}}, \qquad k \le k_{small}+9. \tag{7.6.2}$$

Step 2. We make use of the equation $e_3\beta + (\underline{\kappa} - 2\underline{\omega})\beta = -\,{\not{d}}_1^\star\rho + 3\eta\rho + \Gamma_g\underline{\beta} + \Gamma_b\alpha$ and the previously derived estimates for $\beta, \check{\rho}, \underline{\beta}$ to derive

$$\big\|e_3\beta\big\|_{\infty,k} \lesssim \epsilon_0 r^{-3}u^{-1-\delta_{dec}}, \qquad k \le k_{small}+9.$$

Step 3. To estimate $e_3\zeta$ in the next step we actually need a stronger estimate for $e_3\beta$ than the one derived above. At the same time we derive an improved estimate for β. We show in fact, for some $0 < \delta$,

$$\begin{aligned}\big\|\beta\big\|_{\infty,k} &\lesssim \epsilon_0 r^{-2-\delta}u^{-1-\delta_{dec}}, \qquad k \le k_{small}+10,\\ \big\|e_3\beta\big\|_{\infty,k-1} &\lesssim \epsilon_0 r^{-3-\delta}u^{-1-\delta_{dec}}, \qquad k \le k_{small}+10.\end{aligned} \tag{7.6.3}$$

This makes use of the equation

$$e_4\beta + 2\kappa\beta = {\not{d}}_2\alpha + \Gamma_g\cdot\alpha = F := {\not{d}}_2\alpha + \Gamma_g\cdot\alpha - 2\check{\kappa}\beta$$

and the estimates for α in **Ref 2.** Thus, for some $0 < \delta < \delta_{extra} - \delta_{dec}$,

$$\begin{aligned}\|F\|_{\infty,k} &\lesssim \log(1+u)r^{-3}(2r+u)^{-1-\delta_{extra}} + \epsilon_0 r^{-4}u^{-1-\delta_{dec}}\\ &\lesssim \epsilon_0 u^{-1-\delta_{dec}}r^{-3-\delta}.\end{aligned}$$

Integrating from $\mathcal{T}$, where $r = r_{\mathcal{T}} = r_0 \lesssim 1$, we deduce with the help of Proposition 7.31

$$\begin{aligned}r^4\|\beta\|_{\infty,k}(u,r) &\lesssim r_0{}^4\|\beta\|_{\infty,k}(u,r_0) + \int_{r_0}^{r}\lambda^4\|F\|_{\infty,k}(u,\lambda)d\lambda\\ &\lesssim \|\beta\|_{\infty,k}(u,r_0) + \epsilon_0\int_{r_0}^{r}\lambda^{1-\delta}d\lambda.\end{aligned}$$

Based on the previously derived estimate for β we have $\|\beta\|_{\infty,k}(u,r_{\mathcal{H}}) \lesssim \epsilon_0 u^{-1-\delta_{dec}}$. Hence,

$$\|\beta\|_{\infty,k}(u,r) \lesssim \epsilon_0 r^{-4}u^{-1-\delta_{dec}} + \epsilon_0 r^{-4}r^{2-\delta}u^{-1-\delta_{dec}} \lesssim \epsilon_0 r^{-2-\delta}u^{-1-\delta_{dec}}$$

as desired.

To prove the second estimate in (7.6.3) we commute the transport equation for

β with $\mathbf{T}$ and make use of the corresponding estimate for $\mathbf{T}\alpha$ (which follows from **Ref 2**)

$$\|\mathbf{T}\alpha\|_{\infty,k} \lesssim \epsilon_0 \log(1+u) r^{-4}(2r+u)^{-1-\delta_{extra}} \lesssim \epsilon_0 u^{-1-\delta_{dec}} r^{-4-\delta}$$

as well as the fact that we control $\mathbf{T}\beta$ on $\mathcal{T}$, i.e., $\|\mathbf{T}\beta\|_{\infty,k-1}(u,r_0) \lesssim \epsilon_0 u^{-1-\delta_{dec}}$.

Step 4. We make use of the equation $e_4\zeta + \overline{\kappa}\zeta = -\beta + \Gamma_g \cdot \Gamma_g$ to derive

$$\|e_3\zeta\|_{k,\infty} \lesssim \epsilon_0 r^{-2} u^{-1-\delta_{dec}}, \qquad k \le k_{small} + 9. \tag{7.6.4}$$

Indeed commuting the equation with $\mathbf{T}$ we derive

$$e_4\mathbf{T}\zeta + \overline{\kappa}\mathbf{T}\zeta = F := -\mathbf{T}\beta + [\mathbf{T}, e_4]\zeta + \zeta\mathbf{T}\overline{\kappa} + \mathbf{T}(\Gamma_g \cdot \Gamma_g).$$

It is easy to check, in view of the commutation Lemma 7.29,

$$\|F\|_{\infty,k-1} \lesssim \|\mathbf{T}\beta\|_{\infty,k-1} + \epsilon_0 r^{-4} u^{-1-\delta_{dec}}.$$

Thus, in view of the estimate for $e_3\zeta$ derived in Step 3 and the estimate for $e_4\zeta$ we infer that

$$\|F\|_{\infty,k-1} \lesssim \epsilon_0 r^{-3-\delta} u^{-1-\delta_{dec}}.$$

Integrating from $\mathcal{T}$, and relying also on the previously derived estimate for ζ, i.e., $\|\zeta\|_{k,\infty} \lesssim \epsilon_0 r^{-1} u^{-1-\delta_{dec}}$, we infer

$$\begin{aligned} r^2\|\mathbf{T}\zeta\|_{\infty,k-1} &\lesssim r_0{}^2\|\mathbf{T}\zeta\|_{\infty,k-1}(u,r_0) + \epsilon_0 u^{-1-\delta_{dec}} \int_{r_0}^{r} \lambda^{-1-\delta} d\lambda \\ &\lesssim \|\mathbf{T}\zeta\|_{\infty,k-1}(u,r_0) + \epsilon_0 r^{-\delta} u^{-1-\delta_{dec}} \lesssim \epsilon_0 u^{-1-\delta_{dec}}. \end{aligned}$$

Hence

$$\|\mathbf{T}\zeta\|_{\infty,k-1} \lesssim \epsilon_0 r^{-2} u^{-1-\delta_{dec}}$$

from which the desired estimate easily follows.

Step 5. We make use of the equation $e_4(\check{\underline{\omega}}) = \check{\rho} + \Gamma_g \cdot \Gamma_b$ and the previously derived estimates for $\check{\rho}$ as well as the estimates of $\check{\underline{\omega}}$ on the last slice (see Proposition 7.28) to derive the estimate

$$\|\check{\underline{\omega}}\|_{\infty,k} \lesssim \epsilon_0 r^{-1} u^{-1-\delta_{dec}}, \qquad k \le k_{small} + 9. \tag{7.6.5}$$

Indeed,

$$\|e_4\check{\underline{\omega}}\|_{\infty,k} \lesssim \|\check{\rho}\|_{\infty,k} + \epsilon_0 r^{-3} u^{-1-\delta_{dec}} \lesssim \epsilon_0 r^{-2} u^{-1-\delta_{dec}}.$$

Thus, applying Proposition 7.31, integrating from Σ_* and using the previously derived estimate for $\check{\underline{\omega}}$ on Σ_*,

$$\begin{aligned}\|\check{\underline{\omega}}\|_{\infty,k}(u,r) &\lesssim \|\check{\underline{\omega}}\|_{\infty,k}(u,r_*) + \epsilon_0 u^{-1-\delta_{dec}} \int_r^{r_*} \lambda^{-2} d\lambda \\ &\lesssim \epsilon_0 r^{-1} u^{-1-\delta_{dec}}\end{aligned}$$

as desired.

Step 6. We derive the estimate

$$\|\underline{\xi}\|_{\infty,k} \lesssim \epsilon_0 r^{-1} u^{-1-\delta_{dec}}, \qquad k \leq k_{small}+9 \tag{7.6.6}$$

by making use of the transport equation $e_4(\underline{\xi}) = F := -e_3(\zeta)+\underline{\beta}-\frac{1}{2}\underline{\kappa}(\zeta+\eta)+\Gamma_b\cdot\Gamma_b$. In view of the previously derived estimates for $e_3\zeta, \underline{\beta}, \zeta, \eta$ we derive

$$\|F\|_{\infty,k} \lesssim \epsilon_0 r^{-2} u^{-1-\delta_{dec}}.$$

Integrating from Σ_* and making use of the estimate for $\underline{\xi}$ on Σ_* (see Proposition 7.28) we derive

$$\|\underline{\xi}\|_{\infty,k}(u,r) \lesssim \|\underline{\xi}\|_{\infty,k}(u,r_*) + \epsilon_0 r^{-1} u^{-1-\delta_{dec}} \lesssim \epsilon_0 r^{-1} u^{-1-\delta_{dec}}.$$

Step 7. We derive the estimate

$$\|\check{\underline{\Omega}}\|_{\infty,k} \lesssim \epsilon_0 u^{-1-\delta_{dec}}, \qquad k \leq k_{small}+8. \tag{7.6.7}$$

This follows immediately from the equation $e_\theta(\underline{\Omega}) = -\underline{\xi} - (\eta-\zeta)\underline{\Omega}$, see (2.2.19), and the previous estimate for $\underline{\xi}$. Note that $\overline{\underline{\Omega}}$ has been estimated in Lemma 3.15.

Step 8. We derive the estimate

$$\|\varsigma - 1\|_{\infty,k} \lesssim \epsilon_0 u^{-1-\delta_{dec}}, \qquad k \leq k_{small}+8. \tag{7.6.8}$$

The estimate follows from the propagation equation $e_4(\varsigma)=0$ and the estimate for $\varsigma - 1$ on the last slice Σ_*.

Step 9. We derive the estimate

$$\|e_3\check{\rho}\|_{\infty,k} \lesssim \epsilon_0 r^{-3} u^{-1-\delta_{dec}}, \qquad k \leq k_{small}+8 \tag{7.6.9}$$

with the help of the equation (see Proposition 7.8)

$$e_3\check{\rho} = r^{-2}\mathfrak{d}^{\leq 1}\Gamma_b + r^{-1}\Gamma_b\cdot\Gamma_b$$

and the previously derived estimates for $\underline{\beta}, \check{\underline{\kappa}}, \check{\rho}, \check{\underline{\Omega}}, \check{\varsigma}$.

Step 10. We derive the estimate

$$\|e_3\check{\underline{\kappa}}\|_{\infty,k} \lesssim \epsilon_0 r^{-2} u^{-1-\delta_{dec}}, \qquad k \leq k_{small}+8 \tag{7.6.10}$$

using the equation (see Proposition 7.8)

$$e_3\check{\underline{\kappa}} \quad = r^{-1}\not{\mathfrak{d}}^{\leq 1}\Gamma_b + \Gamma_b \cdot \Gamma_b$$

and the previously derived estimates for $\check{\underline{\kappa}}, \underline{\xi}, \check{\underline{\omega}}, \check{\underline{\Omega}}, \check{\underline{\varsigma}}$. This ends the proof of Proposition 7.43 and Theorem M4. □

7.7 PROOF OF THEOREM M5

Recall from Theorem M3 that we have obtained the following estimate for ${}^{(int)}\underline{\alpha}$ in ${}^{(int)}\mathcal{M}$:

$$\max_{0\leq k\leq k_{small}+16} \sup_{{}^{(int)}\mathcal{M}} \underline{u}^{1+\delta_{dec}}|\mathfrak{d}^k\underline{\alpha}| \quad \lesssim \quad \epsilon_0. \tag{7.7.1}$$

Step 1. We consider the control of the other curvature components, as well as the Ricci components on $\mathcal{T}$. Recall that the $(\underline{u}, {}^{(int)}s)$ foliation is initialized on $\mathcal{T}$ as follows:

- $\underline{u}$ and ${}^{(int)}s$ are defined on $\mathcal{T}$ by

$$\underline{u} = u \text{ and } {}^{(int)}s = {}^{(ext)}s \text{ on } \mathcal{T}.$$

  In particular, on the hypersurface $\mathcal{T}$, the 2-spheres $\mathbf{S}(u, {}^{(int)}s)$ coincide with the 2-sphere $\mathbf{S}(u, {}^{(ext)}s)$.
- In view of the above initialization, and the fact that $\mathcal{T} = \{r = r_{\mathcal{T}}\}$, we infer that

$${}^{(int)}r = {}^{(ext)}r = r_{\mathcal{T}}, \qquad {}^{(int)}m = {}^{(ext)}m.$$

- The null frame $({}^{(int)}e_3, {}^{(int)}e_4, {}^{(int)}e_\theta)$ is defined on $\mathcal{T}$ by

$${}^{(int)}e_4 = {}^{(ext)}\lambda\, {}^{(ext)}e_4, \quad {}^{(int)}e_3 = ({}^{(ext)}\lambda)^{-1}\, {}^{(ext)}e_3, \quad {}^{(int)}e_\theta = {}^{(ext)}e_\theta \text{ on } \mathcal{T}$$

  where

$${}^{(ext)}\lambda = 1 - \frac{2\,{}^{(ext)}m}{{}^{(ext)}r}.$$

In particular, we deduce the following identities for the curvature components and Ricci coefficients on $\mathcal{T}$.

Lemma 7.44. *We have on* $\mathcal{T}$

$$\begin{aligned} {}^{(int)}\underline{\varsigma} &= -\frac{\overline{{}^{(ext)}\underline{\kappa} + {}^{(ext)}\underline{A}}}{\overline{{}^{(ext)}\underline{\kappa}}}\lambda^{-1}\, {}^{(ext)}\varsigma, \\ {}^{(int)}\Omega &= \lambda - \lambda^2\frac{\overline{{}^{(ext)}\kappa}}{\overline{{}^{(ext)}\underline{\kappa} + {}^{(ext)}\underline{A}}} - \lambda\frac{\overline{{}^{(ext)}\kappa}}{\overline{{}^{(ext)}\underline{\kappa} + {}^{(ext)}\underline{A}}}\, {}^{(ext)}\underline{\Omega}, \end{aligned}$$

where

$$\lambda = {}^{(ext)}\lambda = 1 - \frac{2\,{}^{(ext)}m}{{}^{(ext)}r}.$$

Moreover, we have on $\mathcal{T}$

$$ {}^{(int)}\alpha = \lambda^2\, {}^{(ext)}\alpha, \quad {}^{(int)}\beta = \lambda\, {}^{(ext)}\beta, \quad {}^{(int)}\rho = {}^{(ext)}\rho, \quad {}^{(int)}\underline{\beta} = \lambda^{-1}\, {}^{(ext)}\underline{\beta}, $$
$$ {}^{(int)}\underline{\alpha} = \lambda^{-2}\, {}^{(ext)}\underline{\alpha}, $$

$$ {}^{(int)}\underline{\xi} = 0, \quad {}^{(int)}\underline{\omega} = 0, \quad {}^{(int)}\zeta = {}^{(ext)}\zeta, \quad {}^{(int)}\underline{\eta} = -\, {}^{(ext)}\zeta, $$
$$ {}^{(int)}\kappa = \lambda\, {}^{(ext)}\kappa, \quad {}^{(int)}\vartheta = \lambda\, {}^{(ext)}\vartheta, \quad {}^{(int)}\underline{\kappa} = \lambda^{-1}\, {}^{(ext)}\underline{\kappa}, \quad {}^{(int)}\underline{\vartheta} = \lambda^{-1}\, {}^{(ext)}\underline{\vartheta}, $$

and

$$
\begin{aligned}
{}^{(int)}\xi &= \frac{\lambda^2\, \overline{{}^{(ext)}\kappa}}{\overline{{}^{(ext)}\underline{\kappa}} + {}^{(ext)}\underline{A}}\left({}^{(ext)}\zeta - {}^{(ext)}\eta\right),\\
{}^{(int)}\omega &= \frac{\lambda\, \overline{{}^{(ext)}\kappa}}{\overline{{}^{(ext)}\underline{\kappa}} + {}^{(ext)}\underline{A}}\, {}^{(ext)}\underline{\omega},\\
{}^{(int)}\underline{\eta} &= {}^{(ext)}\zeta - \frac{\overline{{}^{(ext)}\kappa}}{\overline{{}^{(ext)}\underline{\kappa}} + {}^{(ext)}\underline{A}}\, {}^{(ext)}\underline{\xi}.
\end{aligned}
$$

Proof. The following vectorfield is tangent to $\mathcal{T}$:

$$ \nu_{\mathcal{T}} := {}^{(ext)}e_3 - \frac{\overline{{}^{(ext)}\underline{\kappa}} + {}^{(ext)}\underline{A}}{\overline{{}^{(ext)}\kappa}}\, {}^{(ext)}e_4, $$

which can also be written as

$$ \nu_{\mathcal{T}} = \lambda\, {}^{(int)}e_3 - \frac{\overline{{}^{(ext)}\underline{\kappa}} + {}^{(ext)}\underline{A}}{\overline{{}^{(ext)}\kappa}}\lambda^{-1}\, {}^{(int)}e_4. $$

Since $\nu_{\mathcal{T}}$ is tangent to $\mathcal{T}$, and in view of the definition of $\underline{u}$ and ${}^{(int)}s$, we immediately infer

$$ \nu_{\mathcal{T}}(\underline{u}) = \nu_{\mathcal{T}}(u) \text{ and } \nu_{\mathcal{T}}({}^{(int)}s) = \nu_{\mathcal{T}}({}^{(ext)}s) \text{ on } \mathcal{T} $$

and hence, using the identities

$$ {}^{(ext)}e_4(u) = {}^{(int)}e_3(\underline{u}) = 0, \quad {}^{(ext)}e_4({}^{(ext)}s) = 1, \quad {}^{(int)}e_3({}^{(int)}s) = -1, $$

we deduce on $\mathcal{T}$

$$
\begin{aligned}
-\frac{\overline{{}^{(ext)}\underline{\kappa}} + {}^{(ext)}\underline{A}}{\overline{{}^{(ext)}\kappa}}\lambda^{-1}\, {}^{(int)}e_4(\underline{u}) &= {}^{(ext)}e_3(u),\\
-\lambda - \frac{\overline{{}^{(ext)}\underline{\kappa}} + {}^{(ext)}\underline{A}}{\overline{{}^{(ext)}\kappa}}\lambda^{-1}\, {}^{(int)}e_4({}^{(int)}s) &= {}^{(ext)}e_3({}^{(ext)}s) - \frac{\overline{{}^{(ext)}\underline{\kappa}} + {}^{(ext)}\underline{A}}{\overline{{}^{(ext)}\kappa}}.
\end{aligned}
$$

In view of the definition of ${}^{(ext)}\varsigma$, ${}^{(int)}\underline{\varsigma}$, ${}^{(ext)}\underline{\Omega}$ and ${}^{(int)}\Omega$, this yields

$$
\begin{aligned}
{}^{(int)}\underline{\varsigma} &= -\frac{\overline{{}^{(ext)}\underline{\kappa}+{}^{(ext)}\underline{A}}}{\overline{{}^{(ext)}\kappa}}\lambda^{-1}\,{}^{(ext)}\varsigma,\\
{}^{(int)}\Omega &= \lambda-\lambda^2\frac{\overline{{}^{(ext)}\kappa}}{\overline{{}^{(ext)}\underline{\kappa}+{}^{(ext)}\underline{A}}}-\lambda\frac{\overline{{}^{(ext)}\kappa}}{\overline{{}^{(ext)}\underline{\kappa}+{}^{(ext)}\underline{A}}}\,{}^{(ext)}\underline{\Omega}.
\end{aligned}
$$

Next, we consider the Ricci coefficients of ${}^{(int)}\mathcal{M}$ on $\mathcal{T}$. From

$$
{}^{(int)}e_4=\lambda\,{}^{(ext)}e_4,\qquad {}^{(int)}e_3=\lambda^{-1}\,{}^{(ext)}e_3,\qquad {}^{(int)}e_\theta={}^{(ext)}e_\theta\ \text{ on }\mathcal{T},
$$

the fact that λ is constant on $\mathcal{T}$, and the fact that ${}^{(ext)}e_\theta$ is tangent to $\mathcal{T}$, we infer on $\mathcal{T}$

$$
\begin{aligned}
&{}^{(int)}\alpha=\lambda^2\,{}^{(ext)}\alpha,\ \ {}^{(int)}\beta=\lambda\,{}^{(ext)}\beta,\ \ {}^{(int)}\rho={}^{(ext)}\rho,\ \ {}^{(int)}\underline{\beta}=\lambda^{-1}\,{}^{(ext)}\underline{\beta},\\
&{}^{(int)}\underline{\alpha}=\lambda^{-2}\,{}^{(ext)}\underline{\alpha},
\end{aligned}
$$

and

$$
\begin{aligned}
&{}^{(int)}\zeta={}^{(ext)}\zeta,\ \ {}^{(int)}\kappa=\lambda\,{}^{(ext)}\kappa,\ \ {}^{(int)}\vartheta=\lambda\,{}^{(ext)}\vartheta,\ \ {}^{(int)}\underline{\kappa}=\lambda^{-1}\,{}^{(ext)}\underline{\kappa},\\
&{}^{(int)}\underline{\vartheta}=\lambda^{-1}\,{}^{(ext)}\underline{\vartheta}.
\end{aligned}
$$

Also, since the foliation of ${}^{(int)}\mathcal{M}$ is ingoing geodesic, we have

$$
{}^{(int)}\underline{\xi}=0,\quad {}^{(int)}\underline{\omega}=0,\quad {}^{(int)}\underline{\eta}=-\,{}^{(int)}\zeta.
$$

It remains to find identities for ${}^{(int)}\xi$, ${}^{(int)}\omega$ and ${}^{(int)}\eta$. Since λ is constant on $\mathcal{T}$ and $\nu_{\mathcal{T}}$ tangent to $\mathcal{T}$, we have on $\mathcal{T}$

$$
D_{\nu_{\mathcal{T}}}\,{}^{(int)}e_4=\lambda D_{\nu_{\mathcal{T}}}\,{}^{(ext)}e_4,\quad D_{\nu_{\mathcal{T}}}\,{}^{(int)}e_3=\lambda^{-1}D_{\nu_{\mathcal{T}}}\,{}^{(ext)}e_3
$$

and hence

$$
\begin{aligned}
g(D_{\nu_{\mathcal{T}}}\,{}^{(int)}e_4,\,{}^{(int)}e_\theta) &= \lambda g(D_{\nu_{\mathcal{T}}}\,{}^{(ext)}e_4,\,{}^{(ext)}e_\theta),\\
g(D_{\nu_{\mathcal{T}}}\,{}^{(int)}e_4,\,{}^{(int)}e_3) &= g(D_{\nu_{\mathcal{T}}}\,{}^{(ext)}e_4,\,{}^{(ext)}e_3),\\
g(D_{\nu_{\mathcal{T}}}\,{}^{(int)}e_3,\,{}^{(int)}e_\theta) &= \lambda^{-1}g(D_{\nu_{\mathcal{T}}}\,{}^{(ext)}e_3,\,{}^{(ext)}e_\theta).
\end{aligned}
$$

We deduce

$$
\begin{aligned}
2\lambda\,{}^{(int)}\eta-2\frac{\overline{{}^{(ext)}\underline{\kappa}+{}^{(ext)}\underline{A}}}{\overline{{}^{(ext)}\kappa}}\lambda^{-1}\,{}^{(int)}\xi &= \lambda\left(2\,{}^{(ext)}\eta-2\frac{\overline{{}^{(ext)}\underline{\kappa}+{}^{(ext)}\underline{A}}}{\overline{{}^{(ext)}\kappa}}\,{}^{(ext)}\xi\right),\\
-4\lambda\,{}^{(int)}\underline{\omega}-4\frac{\overline{{}^{(ext)}\underline{\kappa}+{}^{(ext)}\underline{A}}}{\overline{{}^{(ext)}\kappa}}\lambda^{-1}\,{}^{(int)}\omega &= -4\,{}^{(ext)}\underline{\omega}-4\frac{\overline{{}^{(ext)}\underline{\kappa}+{}^{(ext)}\underline{A}}}{\overline{{}^{(ext)}\kappa}}\,{}^{(ext)}\omega,\\
2\lambda\,{}^{(int)}\underline{\xi}-2\frac{\overline{{}^{(ext)}\underline{\kappa}+{}^{(ext)}\underline{A}}}{\overline{{}^{(ext)}\kappa}}\lambda^{-1}\,{}^{(int)}\underline{\eta} &= \lambda^{-1}\Bigg(2\,{}^{(ext)}\underline{\xi}\\
&\qquad -2\frac{\overline{{}^{(ext)}\underline{\kappa}+{}^{(ext)}\underline{A}}}{\overline{{}^{(ext)}\kappa}}\,{}^{(ext)}\underline{\eta}\Bigg),
\end{aligned}
$$

and thus

$$
\begin{aligned}
{}^{(int)}\xi &= \frac{\lambda^2\overline{{}^{(ext)}\kappa}}{\overline{{}^{(ext)}\underline{\kappa}}+{}^{(ext)}\underline{A}}({}^{(ext)}\zeta - {}^{(ext)}\eta),\\
{}^{(int)}\omega &= \frac{\lambda\overline{{}^{(ext)}\kappa}}{\overline{{}^{(ext)}\underline{\kappa}}+{}^{(ext)}\underline{A}}{}^{(ext)}\underline{\omega},\\
{}^{(int)}\underline{\eta} &= {}^{(ext)}\zeta - \frac{\overline{{}^{(ext)}\kappa}}{\overline{{}^{(ext)}\underline{\kappa}}+{}^{(ext)}\underline{A}}{}^{(ext)}\underline{\xi}.
\end{aligned}
$$

This concludes the proof of the lemma. □

Remark 7.45. *Since the 2-spheres* $\mathbf{S}(u, {}^{(int)}s)$ *coincide on* $\mathcal{T}$ *with the 2-spheres* $\mathbf{S}(u, {}^{(ext)}s)$*, the above lemma immediately yields*

$$
\begin{aligned}
&{}^{(int)}\check{\rho} = {}^{(ext)}\check{\rho}, \quad {}^{(int)}\check{\kappa} = \lambda\,{}^{(ext)}\check{\kappa}, \quad {}^{(int)}\check{\underline{\kappa}} = \lambda^{-1}\,{}^{(ext)}\check{\underline{\kappa}},\\
&{}^{(int)}\check{\underline{\mu}} = -\,{}^{(ext)}\check{\mu} - 2\,{}^{(ext)}\check{\rho} + \frac{1}{2}\,{}^{(ext)}\vartheta\,{}^{(ext)}\underline{\vartheta} - \frac{1}{2}\overline{{}^{(ext)}\vartheta\,{}^{(ext)}\underline{\vartheta}},\\
&{}^{(int)}\check{\omega} = \lambda\overline{{}^{(ext)}\kappa}\left(\frac{{}^{(ext)}\underline{\omega}}{\overline{{}^{(ext)}\underline{\kappa}}+{}^{(ext)}\underline{A}} - \overline{\frac{{}^{(ext)}\underline{\omega}}{\overline{{}^{(ext)}\underline{\kappa}}+{}^{(ext)}\underline{A}}}\right),
\end{aligned}
$$

and

$$
\begin{aligned}
{}^{(int)}\check{\underline{\varsigma}} &= -\frac{1}{\lambda\overline{{}^{(ext)}\kappa}}\left(\left(\overline{{}^{(ext)}\underline{\kappa}}+{}^{(ext)}\underline{A}\right){}^{(ext)}\varsigma - \overline{\left(\overline{{}^{(ext)}\underline{\kappa}}+{}^{(ext)}\underline{A}\right){}^{(ext)}\varsigma}\right),\\
{}^{(int)}\check{\underline{\Omega}} &= -\lambda^2\overline{{}^{(ext)}\kappa}\left(\frac{1}{\overline{{}^{(ext)}\underline{\kappa}}+{}^{(ext)}\underline{A}} - \overline{\frac{1}{\overline{{}^{(ext)}\underline{\kappa}}+{}^{(ext)}\underline{A}}}\right)\\
&\quad -\lambda\overline{{}^{(ext)}\kappa}\left(\frac{{}^{(ext)}\underline{\Omega}}{\overline{{}^{(ext)}\underline{\kappa}}+{}^{(ext)}\underline{A}} - \overline{\frac{{}^{(ext)}\underline{\Omega}}{\overline{{}^{(ext)}\underline{\kappa}}+{}^{(ext)}\underline{A}}}\right).
\end{aligned}
$$

Together with the estimates on $\mathcal{T}$ for the outgoing geodesic foliation of ${}^{(ext)}\mathcal{M}$ derived in Theorem M4, we infer the control of tangential derivatives to $\mathcal{T}$, i.e., $(e_\theta, T_{\mathcal{T}})$ derivatives. Recovering the transversal derivative thanks to the transport equations in the direction e_3, we infer for the ingoing geodesic foliation of ${}^{(int)}\mathcal{M}$ on $\mathcal{T}$

$$
\begin{aligned}
\max_{0\le k\le k_{small}+8}\sup_{\mathcal{T}} u^{1+\delta_{dec}}\Big\|\mathfrak{d}^k\Big(&{}^{(int)}\alpha, {}^{(int)}\beta, {}^{(int)}\check{\rho}, {}^{(int)}\underline{\beta}, {}^{(int)}\check{\underline{\mu}}, {}^{(int)}\check{\kappa},\\
&{}^{(int)}\vartheta, {}^{(int)}\zeta, {}^{(int)}\underline{\eta}, {}^{(int)}\check{\underline{\kappa}}, {}^{(int)}\underline{\vartheta},\\
&{}^{(int)}\xi, {}^{(int)}\check{\omega}, {}^{(int)}\check{\underline{\varsigma}}, {}^{(int)}\check{\underline{\Omega}}\Big)\Big\|_{L^2(S)} \lesssim \epsilon_0.
\end{aligned}
$$

Step 2. Relying on the estimates of the ingoing geodesic foliation of ${}^{(int)}\mathcal{M}$ on $\mathcal{T}$ derived in Step 1, we propagate these estimates to ${}^{(int)}\mathcal{M}$ thanks to transport equations in the e_3 direction given by the null structure equations and Bianchi identities. Recalling that $\underline{\alpha}$ has already been estimated in Theorem M3, see (7.7.1), quantities are recovered in the following order:

1. We recover $\underline{\check{\kappa}}$, with a control of $k_{small}+8$ derivatives, from
$$e_3\underline{\check{\kappa}} + \overline{\underline{\kappa}}\,\underline{\check{\kappa}} = \mathrm{Err}[e_3\underline{\check{\kappa}}].$$

2. We recover $\underline{\vartheta}$, with a control of $k_{small}+8$ derivatives, from
$$e_3(\underline{\vartheta}) + \underline{\kappa}\,\underline{\vartheta} = -2\underline{\alpha}.$$

3. We recover $\underline{\beta}$, with a control of $k_{small}+8$ derivatives, from
$$e_3\underline{\beta} + 2\underline{\kappa}\,\underline{\beta} = \not{d}_2\underline{\alpha} - \zeta\underline{\alpha}.$$

4. We recover ζ, with a control of $k_{small}+8$ derivatives, from
$$e_3(\zeta) + \underline{\kappa}\zeta = \underline{\beta} - \underline{\vartheta}\zeta.$$

5. We recover $\underline{\eta}$, with a control of $k_{small}+8$ derivatives, from
$$e_3(\underline{\eta}+\zeta) + \frac{1}{2}\underline{\kappa}(\underline{\eta}+\zeta) = -\frac{1}{2}\underline{\vartheta}(\underline{\eta}+\zeta).$$

6. We recover $\underline{\check{\mu}}$, with a control of $k_{small}+8$ derivatives, from
$$e_3\underline{\check{\mu}} + \frac{3}{2}\overline{\underline{\kappa}}\,\underline{\check{\mu}} + \frac{3}{2}\overline{\underline{\mu}}\,\underline{\check{\kappa}} = \mathrm{Err}[e_3\underline{\check{\mu}}].$$

7. We recover $\check{\rho}$, with a control of $k_{small}+7$ derivatives, from
$$e_3\check{\rho} + \frac{3}{2}\overline{\underline{\kappa}}\check{\rho} + \frac{3}{2}\overline{\rho}\,\underline{\check{\kappa}} = \not{d}_1\underline{\beta} + \mathrm{Err}[e_3\check{\rho}].$$

8. We recover $\check{\kappa}$, with a control of $k_{small}+7$ derivatives, from
$$e_3\check{\kappa} + \frac{1}{2}\overline{\underline{\kappa}}\check{\kappa} + \frac{1}{2}\underline{\check{\kappa}}\overline{\kappa} = 2\not{d}_1\zeta + 2\check{\rho} + \mathrm{Err}[e_3\check{\kappa}].$$

9. We recover ϑ, with a control of $k_{small}+7$ derivatives, from
$$e_3\vartheta + \frac{1}{2}\underline{\kappa}\,\vartheta = -2\not{d}_2^\star\zeta - \frac{1}{2}\kappa\,\underline{\vartheta} + 2\zeta^2.$$

10. We recover β, with a control of $k_{small}+6$ derivatives, from
$$e_3\beta + \underline{\kappa}\beta = e_\theta(\rho) + 3\zeta\rho - \vartheta\underline{\beta}.$$

11. We recover α, with a control of $k_{small}+5$ derivatives, from
$$e_3\alpha + \frac{1}{2}\underline{\kappa}\alpha = -\not{d}_2^\star\beta - \frac{3}{2}\vartheta\rho + 5\zeta\beta.$$

12. We recover $\check{\omega}$, with a control of $k_{small}+7$ derivatives, from
$$e_3\check{\omega} = \check{\rho} + \mathrm{Err}[e_3\check{\omega}].$$

13. We recover $\check{\Omega}$, with a control of $k_{small} + 7$ derivatives, from

$$e_3(\check{\Omega}) = -2\check{\omega} + \overline{\check{\underline{\kappa}}\check{\Omega}}.$$

14. We recover ξ, with a control of $k_{small} + 6$ derivatives, from

$$e_3(\xi) = e_4(\zeta) + \beta + \frac{1}{2}\kappa(\zeta - \underline{\eta}) + \frac{1}{2}\vartheta(\zeta - \underline{\eta}).$$

15. We recover $\underline{\varsigma}$, with a control of $k_{small} + 8$ derivatives, from

$$e_3(\underline{\varsigma} - 1) = 0.$$

As the estimates are significantly simpler to derive[21] and in the same spirit as the corresponding ones in Theorem M4, we leave the details to the reader. This concludes the proof of Theorem M5.

[21]Note that r is bounded on ${}^{(int)}\mathcal{M}$ and that all quantities behave the same in ${}^{(int)}\mathcal{M}$.

Chapter Eight

Initialization and Extension (Theorems M6, M7, M8)

In this chapter, we prove M6 concerning initialization, Theorem M7 concerning extension, and Theorem M8 concerning the improvement of higher order weighted energies.

8.1 PROOF OF THEOREM M6

Step 1. Let r_0 such that

$$r_0 \quad := \quad d_0\epsilon_0^{-\frac{2}{3}}, \tag{8.1.1}$$

where the constant d_0 satisfies

$$\frac{1}{2} \leq d_0 \leq 2$$

and will be suitably chosen in Step 3. Also, let $\delta_0 > 0$ sufficiently small. Consider the unique sphere $\overset{\circ}{S}$ of the initial data layer on $\mathcal{C}_{(1+\delta_0, \mathcal{L}_0)}$ with area radius r_0. Then, denoting $S(u_{\mathcal{L}_0}, {}^{(ext)}s_{\mathcal{L}_0})$ the spheres of the outgoing portion of the initial data layer, we have

$$\overset{\circ}{S} = S(\overset{\circ}{u}, \overset{\circ}{s}), \qquad \overset{\circ}{u} = 1 + \delta_0, \qquad |\overset{\circ}{s} - r_0| \lesssim \epsilon_0.$$

Relying on the control of the initial data layer given by (3.3.5), i.e.,

$$\mathfrak{I}_{k_{large}+5} \leq \epsilon_0^{\frac{5}{3}},$$

we then invoke Theorem GCMS-II of section 3.7.4 with the choices

$$\overset{\circ}{\delta} = \overset{\circ}{\epsilon} = \epsilon_0, \quad s_{max} = k_{large} + 5,$$

to produce a unique GCM sphere S_*, which is a deformation of $\overset{\circ}{S}$, satisfying

$$\kappa^{S_*} = \frac{2}{r^{S_*}}, \qquad {\not{d}_2^\star}^{S_*} {\not{d}_1^\star}^{S_*} \underline{\kappa}^{S_*} = {\not{d}_2^\star}^{S_*} {\not{d}_1^\star}^{S_*} \mu^{S_*} = 0 \quad \text{ on } S_*,$$
$$\int_{S_*} \beta^{S_*} e^{\Phi} = 0, \qquad \int_{S_*} e_\theta^{S_*}(\underline{\kappa}^{S_*}) e^{\Phi} = 0.$$

Remark 8.1. *In order to apply Theorem GCMS-II to the above setting, one needs to check that the initial data foliation layer satisfies the assumptions of the theorem,* *and in particular*

$$|\mathfrak{d}^{\leq s_{max}}\Gamma_b| \lesssim (\overset{\circ}{\epsilon})^{\frac{1}{3}} r_0^{-2},$$
$$r_0\left|\int_S \beta e^{\Phi}\right| \lesssim \overset{\circ}{\delta}, \qquad r_0\left|\int_S e_\theta(\kappa) e^{\Phi}\right| \lesssim \overset{\circ}{\delta}, \qquad r_0\left|\int_S e_\theta(\underline{\kappa}) e^{\Phi}\right| \lesssim \overset{\circ}{\delta}.$$

Now, in view of the above choice for s_{max}, $\overset{\circ}{\delta}$, $\overset{\circ}{\epsilon}$ and r_0, this follows from

$$r|\mathfrak{d}^{\leq k_{large}+5}\Gamma_b| \lesssim \epsilon_0, \qquad r^3|\beta| + r^2|\not{\mathfrak{d}}\check{\kappa}| + r^2|\not{\mathfrak{d}}\check{\underline{\kappa}}| \lesssim \epsilon_0^{\frac{5}{3}}$$

and hence from

$$\mathfrak{I}_{k_{large}+5} \leq \epsilon_0^{\frac{5}{3}}$$

which is (3.3.5).

Step 2. Starting from S_* constructed in Step 1, and relying on the control of the initial data layer, we then invoke Theorem GCMH of section 3.7.4 to produce a smooth spacelike hypersurface Σ_* included in the initial data layer, passing through the sphere S_*, and a scalar function u defined on Σ_* such that

- The following GCM conditions hold:

$$\kappa = \frac{2}{r}, \quad \not{d}_2^\star \not{d}_1^\star \underline{\kappa} = \not{d}_2^\star \not{d}_1^\star \mu = 0, \quad \int_S \eta e^{\Phi} = \int_S \underline{\xi} e^{\Phi} = 0 \text{ on } \Sigma_*.$$

- We have, for some constant c_{Σ_*},

$$u + r = c_{\Sigma_*}, \quad \text{along} \quad \Sigma_*.$$

- The following normalization condition holds true at the south pole SP of every sphere S,

$$a\Big|_{SP} = -1 - \frac{2m}{r}$$

  where a is such that we have

$$\nu = e_3 + a e_4,$$

  with ν the unique vectorfield tangent to the hypersurface Σ_*, normal to S, and normalized by $g(\nu, e_4) = -2$.

Furthermore, we have[1]

$$\max_{k\leq k_{large}+4} \sup_{\Sigma_*} r\Big(|\mathfrak{d}^k f| + |\mathfrak{d}^k \underline{f}| + |\mathfrak{d}^k \log(\lambda)|\Big) \lesssim \epsilon_0, \tag{8.1.2}$$

[1]We have in fact

$$\max_{k\leq k_{large}+6} \sup_{\Sigma_*} \Big(\|\mathfrak{d}^k f\|_{L^2(S)} + \|\mathfrak{d}^k \underline{f}\|_{L^2(S)} + \|\mathfrak{d}^k \log(\lambda)\|_{L^2(S)}\Big) \lesssim \epsilon_0,$$

and then use the Sobolev embedding on the 2-spheres S foliating Σ_* to deduce (8.1.2).

and

$$\sup_{\Sigma_*}\Big(|m-m_0|+|r-r_0|\Big) \lesssim \epsilon_0, \tag{8.1.3}$$

where $(f, \underline{f}, \lambda)$ are the transition functions from the frame of the initial data layer to the frame of Σ_*.

Step 3. Provided $\delta_0 > 0$ has been chosen sufficiently small, the spacelike hypersurface Σ_* of Step 2 intersects the curve of the south poles of the spheres foliating the outgoing cone $\mathcal{C}_{(1,\mathcal{L}_0)}$ of the initial data layer. We then call S_1 the unique sphere of Σ_* such that its south pole coincides with the south pole of a sphere of $\mathcal{C}_{(1,\mathcal{L}_0)}$, and we calibrate u such that $u = 1$ on S_1. We then can compare $\mathring{u} = 1 + \delta_0$ to $u(S_*)$ and obtain

$$|u(S_*) - 1 - \delta_0| \lesssim \epsilon_0\delta_0,$$

so that

$$1 \le u \le u(S_*) \quad \text{ on } \Sigma_* \text{ where } 1 < u(S_*) < 1 + 2\delta_0.$$

Together with the estimate (8.1.3), and in view of the choice (8.1.1) for r_0, we have

$$\begin{aligned} \inf_{\Sigma_*} r &= r(S_*) = r_0 + O(\epsilon_0) = d_0\epsilon_0^{-\frac{2}{3}} + O(\epsilon_0) \\ &= \epsilon_0^{-\frac{2}{3}}(u(S_*))^{1+\delta_{dec}}\left(d_0 + O(\delta_0) + O\left(\epsilon_0^{\frac{5}{3}}\right)\right). \end{aligned}$$

Thus, we may choose the constant d_0 in the range $\frac{1}{2} \le d_0 \le 2$ such that

$$\inf_{\Sigma_*} r = \epsilon_0^{-\frac{2}{3}}(u(S_*))^{1+\delta_{dec}}$$

so that the dominant condition (3.3.4) for r is satisfied.

Step 4. In view of Step 1 to Step 3, Σ_* satisfies all the required properties for the future spacelike boundary of a GCM admissible spacetime, see item 3 of Definition 3.2. We now control the outgoing geodesic foliation initialized on Σ_* and covering the region we denote by ${}^{(ext)}\mathcal{M}$, which is included in the initial data layer. Let $(f, \underline{f}, \lambda)$ the transition functions from the frame of the outgoing part of the initial data layer to the frame of ${}^{(ext)}\mathcal{M}$. Since both frames are outgoing geodesic, we may apply Corollary 2.93 which yields for $(\underline{f}, f, \log(\lambda))$ the following transport equations:

$$\begin{aligned} \lambda^{-1}e_4'(rf) &= E_1'(f, \Gamma), \\ \lambda^{-1}e_4'(\log(\lambda)) &= E_2'(f, \Gamma), \\ \lambda^{-1}e_4'\Big(r\underline{f} - 2r^2e_\theta'(\log(\lambda)) + rf\underline{\Omega}\Big) &= E_3'(f, \underline{f}, \lambda, \Gamma), \end{aligned}$$

where

$$\begin{aligned}
E_1'(f,\Gamma) &= -\frac{r}{2}\check{\kappa}f - \frac{r}{2}\vartheta f + \text{l.o.t.},\\
E_2'(f,\Gamma) &= f\zeta - \frac{1}{2}f^2\underline{\omega} - \underline{\eta}f - \frac{1}{4}f^2\underline{\kappa} + \text{l.o.t.},\\
E_3'(f,\underline{f},\lambda,\Gamma) &= -\frac{r}{2}\check{\kappa}\underline{f} + r^2\left(\check{\kappa} - \left(\overline{\kappa} - \frac{2}{r}\right)\right)e_\theta'(\log(\lambda))\\
&\quad + r^2\left(\not{d}_1'(f) + \lambda^{-1}\vartheta'\right)e_\theta'(\log(\lambda)) - \frac{r}{2}\check{\kappa}\underline{\Omega}f + rE_3(f,\underline{f},\Gamma)\\
&\quad - 2r^2e_\theta'(E_2(f,\Gamma)) + r\underline{\Omega}E_1(f,\Gamma),
\end{aligned}$$

and where E_1, E_2 and E_3 are given in Lemma 2.92. Integrating these transport equations from Σ_*, using the control (8.1.2) of $(f,\underline{f},\lambda)$ on Σ_*, and together with the assumption (3.3.5) for the Ricci coefficients of the foliation of the initial data layer, we obtain

$$\sup_{{}^{(ext)}\mathcal{M}(r\geq 2m_0(1+\delta_{\mathcal{H}}))} r\Big(|\mathfrak{d}^{\leq k_{large}+4}(f,\log(\lambda))| + |\mathfrak{d}^{\leq k_{large}+3}\underline{f}|\Big) \lesssim \epsilon_0. \tag{8.1.4}$$

Then, let $\mathcal{T} = \{r = 2m_0(1+\delta_{\mathcal{H}})\}$, i.e., we choose $r_{\mathcal{T}} = 2m_0(1+\delta_{\mathcal{H}})$. We initialize the ingoing geodesic foliation of ${}^{(int)}\mathcal{M}$ on $\mathcal{T}$ using the outgoing geodesic foliation of ${}^{(ext)}\mathcal{M}$ as in item 4 of Definition 3.2. Using the control of $(f,\underline{f},\lambda)$ induced on $\mathcal{T}$ by (8.1.4), and using the analog of Corollary 2.93 in the e_3 direction for ingoing foliations, we obtain similarly

$$\sup_{{}^{(int)}\mathcal{M}} \Big(|\mathfrak{d}^{\leq k_{large}+3}(\underline{f},\log(\lambda))| + |\mathfrak{d}^{\leq k_{large}+2}f|\Big) \lesssim \epsilon_0. \tag{8.1.5}$$

Then, in view of (8.1.4), (8.1.5), and the assumption (3.3.5) for the Ricci coefficients and curvature components of the foliation of the initial data layer, and using the transformation formulas of Proposition 2.90, we deduce

$$\begin{aligned}
\max_{k\leq k_{large}} &\left\{ \sup_{{}^{(ext)}\mathcal{M}} \Big(r^{\frac{7}{2}+\delta_B}(|\mathfrak{d}^k\alpha| + |\mathfrak{d}^k\beta|) + r^3|\mathfrak{d}^k\check{\rho}| + r^2|\mathfrak{d}^k\underline{\beta}| + r|\mathfrak{d}^k\underline{\alpha}|\Big)\right.\\
&+ \sup_{{}^{(ext)}\mathcal{M}} r^2(|\mathfrak{d}^k\check{\kappa}| + |\mathfrak{d}^k\vartheta| + |\mathfrak{d}^k\zeta| + |\mathfrak{d}^k\underline{\check{\kappa}}|)\\
&\left.+ \sup_{{}^{(ext)}\mathcal{M}} r(|\mathfrak{d}^k\eta| + |\mathfrak{d}^k\underline{\vartheta}| + |\mathfrak{d}^k\underline{\check{\omega}}| + |\mathfrak{d}^k\underline{\xi}|)\right\} \lesssim \epsilon_0,
\end{aligned}$$

and

$$\max_{k\leq k_{large}} \sup_{{}^{(int)}\mathcal{M}} \Big(|\mathfrak{d}^k\check{R}| + |\mathfrak{d}^k\check{\Gamma}|\Big) \lesssim \epsilon_0.$$

In particular, we infer that

$$\mathfrak{N}^{(En)}_{k_{large}} + \mathfrak{N}^{(Dec)}_{k_{small}} \lesssim \epsilon_0$$

which concludes the proof of Theorem M6.

8.2 PROOF OF THEOREM M7

In view of the assumptions of Theorem M7, we are given a GCM admissible spacetime $\mathcal{M} = \mathcal{M}(u_*) \in \aleph(u_*)$ verifying the following improved bounds, for a universal constant $C > 0$,

$$\mathfrak{N}^{(Dec)}_{k_{small}+5}(\mathcal{M}) \leq C\epsilon_0 \tag{8.2.1}$$

provided by Theorems M1–M5. We then proceed as follows.

Step 1. We extend $\mathcal{M}$ by a local existence argument, to a strictly larger spacetime $\mathcal{M}^{(extend)}$, with a naturally extended foliation and the following slightly increased bounds

$$\mathfrak{N}^{(Dec)}_{k_{small}+5}(\mathcal{M}^{(extend)}) \leq 2C\epsilon_0,$$

but which may not verify our admissibility criteria.

Step 2. We then invoke Theorem GCMH of section 3.7.4 to extend the hypersurface Σ_* in $\mathcal{M}^{(extend)} \setminus \mathcal{M}$ as a smooth spacelike hypersurface $\Sigma_*^{(extend)}$, together with a scalar function $u^{(extend)}$, satisfying the same GCM conditions as Σ_*.

Step 3. We consider the outgoing geodesic foliation $(u^{(extend)}, s^{(extend)})$ initialized on $\Sigma_*^{(extend)}$ to the future of $\Sigma_*^{(extend)}$ in $\mathcal{M}^{(extend)}$. Note in particular that we have from the definition of Σ_* and $\Sigma_*^{(extend)}$

$$u^{(extend)} + s^{(extend)} = c_{\Sigma_*}.$$

We define the following spacetime region to the future of $\Sigma_*^{(extend)}$:

$$\widetilde{\mathcal{R}} \; := \; \Big\{ u_* \leq u^{(extend)} \leq u_* + \delta_{ext}, \quad c_{\Sigma_*} \leq u^{(extend)} + s^{(extend)} \leq c_{\Sigma_*} + \Delta_{ext} \Big\},$$

where

$$\Delta_{ext} := \frac{d_0 r_*}{u_*}\delta_{ext}, \qquad r_* := r(S_*), \qquad S_* := \Sigma_* \cap C_*,$$

with $\delta_{ext} > 0$ chosen sufficiently small so that $\widetilde{\mathcal{R}} \subset \mathcal{M}^{(extend)}$, and with d_0 a constant satisfying

$$\frac{1}{2} \leq d_0 \leq 1$$

which will be suitably chosen in Step 11 below. From now on, for convenience, we drop the index $(extend)$ and simply denote $u^{(extend)}$ and $s^{(extend)}$ by u and s.

Step 4. Since we have on $\Sigma_*^{(extend)}$ the GCM conditions $\not{d}_2^\star \not{d}_1^\star \underline{\kappa} = \not{d}_2^\star \not{d}_1^\star \mu = 0$, and since e^Φ generates the kernel of $\not{d}_2^\star$, we infer

$$\not{d}_1^\star \underline{\kappa} = -\frac{\int_S e_\theta(\underline{\kappa}) e^\Phi}{\int_S e^{2\Phi}} e^\Phi, \qquad \not{d}_1^\star \mu = -\frac{\int_S e_\theta(\mu) e^\Phi}{\int_S e^{2\Phi}} e^\Phi, \qquad \text{on } \Sigma_*^{(extend)}.$$

Thus, introducing the following two scalar functions

$$\underline{C}(u) := -\frac{\int_S e_\theta(\underline{\kappa})e^\Phi}{\int_S e^{2\Phi}}, \qquad M(u) := -\frac{\int_S e_\theta(\mu)e^\Phi}{\int_S e^{2\Phi}}, \qquad \text{on } \Sigma_*^{(extend)},$$

we rewrite the GCM conditions on $\Sigma_*^{(extend)}$ as follows:

$$\kappa = \frac{2}{r}, \qquad \not{d}_1^\star \underline{\kappa} = \underline{C}(u)e^\Phi, \qquad \not{d}_1^\star \mu = M(u)e^\Phi, \qquad \int_S \eta e^\Phi = \int_S \underline{\xi} e^\Phi = 0.$$

Propagating these GCM quantities in the e_4 direction from $\Sigma_*^{(extend)}$, and propagating the scalar functions $\underline{C}$ and M by $e_4(r^4\underline{C}) = 0$ and $e_4(r^5 M) = 0$ so that we have[2] $\underline{C} = \underline{C}(u,s)$ and $M = M(u,s)$ in $\widetilde{\mathcal{R}}$, we obtain for all $k \le k_{small} + 4$

$$\sup_{\widetilde{\mathcal{R}}} \left(r^2 \left|\mathfrak{d}^k \left(\kappa - \frac{2}{r}\right)\right| + r^3 \left|\mathfrak{d}^{k-1}\left(\not{d}_1^\star \underline{\kappa} - \underline{C}e^\Phi\right)\right| + r^4 \left|\mathfrak{d}^{k-1}\left(\not{d}_1^\star \mu - Me^\Phi\right)\right| \right)$$
$$\lesssim \frac{\epsilon_0}{r}\Delta_{ext}$$

and

$$\sup_{\widetilde{\mathcal{R}}} r^{-2}\left(\left|\int_S \underline{\xi} e^\Phi\right| + \left|\int_S \eta e^\Phi\right| \right) \lesssim \frac{\epsilon_0}{r}\Delta_{ext}.$$

Next, in view of (4.1.2) and the fact that $\nu = e_3 + ae_4$, we have on $\Sigma_*^{(extend)}$

$$\left|\nu\left(\int_S \beta e^\Phi\right)\right| \lesssim \frac{\epsilon_0}{ru^{1+\delta_{dec}}}, \qquad \left|\nu\left(\int_S e_\theta(\underline{\kappa})e^\Phi\right)\right| \lesssim \frac{\epsilon_0}{ru^{1+\delta_{dec}}} + \frac{\epsilon_0^2}{u^{2+2\delta_{dec}}}.$$

In particular, since $r(S_*) = \epsilon_0^{-\frac{2}{3}}(u(S_*))^{1+\delta_{dec}}$ in view of (3.3.4) and $u(S_*) = u_*$, we infer $r \sim \epsilon_0^{-\frac{2}{3}} u^{1+\delta_{dec}}$ on $\Sigma_*^{(extend)}(u_* \le u \le u_* + \delta_{ext})$ and hence

$$\left|\nu\left(\int_S \beta e^\Phi\right)\right| + \left|\nu\left(\int_S e_\theta(\underline{\kappa})e^\Phi\right)\right| \lesssim \frac{\epsilon_0}{ru^{1+\delta_{dec}}} \quad \text{on} \quad \Sigma_*^{(extend)}(u_* \le u \le u_* + \delta_{ext}).$$

We integrate from S_* where we have

$$\int_{S_*} \beta e^\Phi = \int_{S_*} e_\theta(\underline{\kappa})e^\Phi = 0$$

and obtain

$$\sup_{\Sigma_*^{(extend)}(u_* \le u \le u_* + \delta_{ext})} \left(r\left|\int_S \beta e^\Phi\right| + r\left|\int_S e_\theta(\underline{\kappa})e^\Phi\right| \right) \lesssim \frac{\epsilon_0}{u_*}\delta_{ext}.$$

We now integrate in the e_4 direction from $\Sigma_*^{(extend)}(u_* \le u \le u_* + \delta_{ext})$ where we

[2] More precisely, we have $\underline{C} = r^{-4}\widetilde{\underline{C}}$ and $M = r^{-5}\widetilde{M}$, with $\widetilde{\underline{C}}$ and $\widetilde{M}$ given by the restriction of $r^4\underline{C}$ and $r^5 M$ to $\Sigma_*^{(extend)}$ so that $\widetilde{\underline{C}} = \widetilde{\underline{C}}(u)$ and $\widetilde{M} = \widetilde{M}(u)$. Note also that $r = r(u,s)$.

have the above estimate as well as $e_\theta(\kappa)=0$. We obtain

$$\sup_{\widetilde{\mathcal{R}}\cap\{u\geq u_*\}}\left(r\left|\int_S \beta e^\Phi\right|+r\left|\int_S e_\theta(\kappa)e^\Phi\right|+r\left|\int_S e_\theta(\underline{\kappa})e^\Phi\right|\right) \lesssim \frac{\epsilon_0}{u_*}\delta_{ext}+\frac{\epsilon_0}{r}\Delta_{ext}$$
$$\lesssim \frac{\epsilon_0}{r}\Delta_{ext}.$$

Also, recall that $\nu=e_3+a_*e_4$ denotes the unique tangent vectorfield to Σ_* which is orthogonal to e_θ and normalized by $\mathbf{g}(\nu,e_4)=-2$. Then, one has, since $u+r$ is constant on Σ_* and $s=r$ on Σ_*,

$$0=\nu(u+s)=e_3(u)+ae_4(u)+e_3(s)+ae_4(s)=\frac{2}{\varsigma}+\underline{\Omega}+a$$

and hence

$$a = -\frac{2}{\varsigma}-\underline{\Omega} \text{ on } \Sigma_*.$$

Together with the GCM condition on a, we infer

$$\left(\frac{2}{\varsigma}+\underline{\Omega}\right)\Big|_{SP} = 1+\frac{2m}{r} \text{ on } \Sigma_*.$$

As above, propagating forward in e_4, we infer

$$\sup_{\widetilde{\mathcal{R}}}\left|\left(\frac{2}{\varsigma}+\underline{\Omega}\right)\Big|_{SP}-\left(1+\frac{2m}{r}\right)\right| \lesssim \frac{\epsilon_0}{r}\Delta_{ext}.$$

Finally, arguing as we did above on $\Sigma_*^{(extend)}(u_*\leq u\leq u_*+\delta_{ext})$, we have $r\sim\epsilon_0^{-\frac{2}{3}}u^{1+\delta_{dec}}$ on $\widetilde{\mathcal{R}}\cap\{u\geq u_*\}$ and hence

$$\sup_{\widetilde{\mathcal{R}}\cap\{u\geq u_*\}} r^2|\Gamma_b| \lesssim \sup_{\widetilde{\mathcal{R}}\cap\{u\geq u_*\}}\left(\frac{r\epsilon_0}{u^{1+\delta_{dec}}}\right)\lesssim\epsilon_0^{\frac{1}{3}}.$$

Step 5. We fix the following sphere of the $(u^{(extend)},s^{(extend)})$ foliation in the region $\widetilde{\mathcal{R}}\cap\{u\geq u_*\}$:

$$\overset{\circ}{S}:=S(\overset{\circ}{u},\overset{\circ}{s}), \qquad \overset{\circ}{u}:=u_*+\frac{\delta_{ext}}{2}, \qquad \overset{\circ}{s}:=r_*+\frac{3d_0r_*}{4u_*}\delta_{ext}. \tag{8.2.2}$$

Define

$$\overset{\circ}{\delta}:=\frac{\epsilon_0}{r}\Delta_{ext}=\frac{d_0\epsilon_0\delta_{ext}}{u_*}, \qquad \overset{\circ}{\epsilon}:=\epsilon_0,$$

and the small spacetime neighborhood of $\overset{\circ}{S}$

$$\mathcal{R}(\overset{\circ}{\epsilon},\overset{\circ}{\delta}):=\left\{|u-\overset{\circ}{u}|\leq\delta_{\mathcal{R}},\quad |s-\overset{\circ}{s}|\leq\delta_{\mathcal{R}}\right\}, \qquad \delta_{\mathcal{R}}=\overset{\circ}{\delta}(\overset{\circ}{\epsilon})^{-\frac{1}{2}}.$$

Note that $\mathcal{R}(\overset{\circ}{\epsilon},\overset{\circ}{\delta})\subset\widetilde{\mathcal{R}}$. In view of the estimates in Step 4, we are in position to apply Theorem GCMS-II of section 3.7.4, with $s_{max} = k_{small} + 4$, which yields the existence of a unique sphere $\widetilde{S}_*$, which is a deformation of $\overset{\circ}{S}$, is included in $\mathcal{R}(\overset{\circ}{\epsilon}, \overset{\circ}{\delta})$, and is such that the following GCM conditions hold on it:

$$\widetilde{\not{d}_2^\star} \widetilde{\not{d}_1^\star} \widetilde{\underline{\kappa}} = \widetilde{\not{d}_2^\star} \widetilde{\not{d}_1^\star} \widetilde{\mu} = 0, \qquad \widetilde{\kappa} = \frac{2}{\widetilde{r}}, \qquad \int_{\widetilde{S}_*} \widetilde{\beta} e^\Phi = \int_{\widetilde{S}_*} \widetilde{e}_\theta(\widetilde{\underline{\kappa}}) e^\Phi = 0,$$

where the tilde refer to the quantities and tangential operators on $\widetilde{S}_*$.

Step 6. Starting from $\widetilde{S}_*$ constructed in Step 5, and in view of the estimates in Step 4, we may apply Theorem GCMH of section 3.7.4, with $s_{max} = k_{small} + 4$, which yields the existence of a smooth small piece of spacelike hypersurface $\widetilde{\Sigma}_*$ starting from $\widetilde{S}_*$ towards the initial data layer, together with a scalar function $\widetilde{u}$ defined on $\widetilde{\Sigma}_*$, whose level surfaces are topological spheres denoted by $\widetilde{S}$, so that

- The following GCM conditions are verified on $\widetilde{\Sigma}_*$:

$$\widetilde{\not{d}_2^\star} \widetilde{\not{d}_1^\star} \widetilde{\underline{\kappa}} = \widetilde{\not{d}_2^\star} \widetilde{\not{d}_1^\star} \widetilde{\mu} = 0, \qquad \widetilde{\kappa} = \frac{2}{\widetilde{r}}, \qquad \int_{\widetilde{S}} \widetilde{\eta} e^\Phi = \int_{\widetilde{S}} \widetilde{\underline{\xi}} e^\Phi = 0,$$

  where the tilde refer to the quantities and tangential operators on $\widetilde{\Sigma}_*$.
- We have, for some constant $c_{\widetilde{\Sigma}_*}$,

$$\widetilde{u} + \widetilde{r} = c_{\widetilde{\Sigma}_*}, \quad \text{along} \quad \widetilde{\Sigma}_*.$$

- The following normalization condition holds true at the south pole SP of every sphere $\widetilde{S}$,

$$\widetilde{a}\Big|_{SP} = -1 - \frac{2\widetilde{m}}{\widetilde{r}}$$

  where $\widetilde{a}$ is such that we have

$$\widetilde{\nu} = \widetilde{e}_3 + \widetilde{a}\widetilde{e}_4,$$

  with $\widetilde{\nu}$ the unique vectorfield tangent to the hypersurface $\widetilde{\Sigma}_*$, normal to $\widetilde{S}$, and normalized by $g(\widetilde{\nu}, \widetilde{e}_4) = -2$.
- The transition functions $(f, \underline{f}, \lambda)$ from the frame of $\mathcal{M}^{(extend)}$ to the frame of $\widetilde{\Sigma}_*$

$$\|(f, \underline{f}, \log(\lambda))\|_{\mathfrak{h}_{k_{small}+5}} \lesssim \overset{\circ}{\delta}.$$

Step 7. The spacelike GCM hypersurface $\widetilde{\Sigma}_*$ has been constructed in Step 6 in a small neighborhood of $\widetilde{S}_*$. We now focus on proving that it in fact extends all the way to the initial data layer. To this end, we denote by u_1 with

$$1 \leq u_1 < \overset{\circ}{u},$$

the minimal value of u such that

- We have

$$\widetilde{\Sigma}_* \cap \mathcal{C}_u \neq \emptyset \text{ for any } u_1 \leq u \leq \overset{\circ}{u}. \tag{8.2.3}$$

- There exists a large constant $D \geq 1$ such that we have for any sphere $\widetilde{S}$ of $\widetilde{\Sigma}_*(u \geq u_1)$

$$\|(f, \underline{f}, \log(\lambda))\|_{\mathfrak{h}_{k_{small}+5}(\widetilde{S})} \leq Du_*\overset{\circ}{\delta}. \tag{8.2.4}$$

- For the same large constant $D \geq 1$ as above, we have along $\widetilde{\Sigma}_*(u \geq u_1)$

$$|\psi(s)| \leq Du_*\overset{\circ}{\delta}, \tag{8.2.5}$$

where the function $\psi(s)$ is such that the curve

$$\left(u = -s + c_{\widetilde{\Sigma}_*} + \psi(s),\ s,\ \theta = 0\right) \text{ with } \psi(\overset{\circ}{s}) = 0, \tag{8.2.6}$$

coincides with the south poles of the sphere $\widetilde{S}$ of $\widetilde{\Sigma}_*$ and the constant $c_{\widetilde{\Sigma}_*}$ is fixed by the condition $\psi(\overset{\circ}{s}) = 0$.

The fact that $\psi(\overset{\circ}{s}) = 0$ together with the bounds of Step 6 implies that (8.2.3), (8.2.4), (8.2.5) hold for $u_1 < \overset{\circ}{u}$ with u_1 close enough to $\overset{\circ}{u}$. By a continuity argument based on reapplying Theorem GCMH, it suffices to show that we may improve the bounds (8.2.4), (8.2.5) independently of the value of u_1.

Step 8. We now focus on improving the bounds (8.2.4), (8.2.5). We first prove that $\widetilde{\Sigma}_*(u \geq u_1)$ is included in $\widetilde{\mathcal{R}}$. Indeed, (8.2.4), (8.2.5) imply

$$\begin{aligned}
\sup_{\widetilde{\Sigma}_*(u\geq u_1)} |u + s - c_{\widetilde{\Sigma}_*}| &\lesssim \sup_{\widetilde{\Sigma}_*(u\geq u_1)} \left(|\psi| + r|f| + r|\underline{f}|\right) \\
&\lesssim Du_*\overset{\circ}{\delta} \\
&\lesssim \frac{Du_*}{r}\epsilon_0\Delta_{ext} \\
&\lesssim \epsilon_0^{\frac{2}{3}} D\epsilon_0\Delta_{ext} \\
&\lesssim \epsilon_0\Delta_{ext}.
\end{aligned}$$

On the other hand, by construction, $\psi(\overset{\circ}{s}) = 0$ and the south pole of $\overset{\circ}{S}$ and $\widetilde{S}_*$ coincide, so that we have

$$\begin{aligned}
c_{\widetilde{\Sigma}_*} &= \overset{\circ}{u} + \overset{\circ}{s} = u_* + r_* + \frac{\delta_{ext}}{2} + \frac{3d_0 r_*}{4u_*}\delta_{ext} \\
&= c_{\Sigma_*} + \frac{3}{4}\left(1 + \frac{2u_*}{3d_0 r_*}\right)\Delta_{ext}
\end{aligned}$$

and hence

$$\begin{aligned}\sup_{\widetilde{\Sigma}_*(u\geq u_1)}\left|u+s-c_{\Sigma_*}-\frac{3}{4}\Delta_{ext}\right| &\lesssim \left(\frac{u_*}{2d_0r_*}+\epsilon_0\right)\Delta_{ext}\\ &\lesssim \epsilon_0^{\frac{2}{3}}\Delta_{ext}.\end{aligned}$$

In view of the definition of $\widetilde{\mathcal{R}}$, we infer

$$\widetilde{\Sigma}_*(u\geq u_1)\subset\widetilde{\mathcal{R}} \tag{8.2.7}$$

as claimed.

Step 9. Since $\widetilde{\Sigma}_*(u\geq u_1)\subset\widetilde{\mathcal{R}}$, the bounds of Step 4 apply, and hence we have

$$\sup_{\widetilde{\mathcal{R}}}\left|\left(\frac{2}{\varsigma}+\underline{\Omega}\right)\Big|_{SP}-\left(1+\frac{2m}{r}\right)\right| \lesssim \frac{\epsilon_0}{r}\Delta_{ext}\lesssim\mathring{\delta},$$

and for all $k\leq k_{small}+4$

$$\sup_{\widetilde{\mathcal{R}}}\left(r^2\left|\mathfrak{d}^k\left(\kappa-\frac{2}{r}\right)\right|+r^2|\mathfrak{d}^{k-2}(r^2\,\not{d}_2^\star\not{d}_1^\star\underline{\kappa})|+r^3|\mathfrak{d}^{k-2}(r^2\,\not{d}_2^\star\not{d}_1^\star\mu)|\right)\lesssim\frac{\epsilon_0}{r}\Delta_{ext}\lesssim\mathring{\delta},$$

as well as

$$\sup_{\widetilde{\mathcal{R}}\cap\{u\geq u_*\}}\left(r\left|\int_S\beta e^{\Phi}\right|+r\left|\int_S e_\theta(\kappa)e^{\Phi}\right|+r\left|\int_S e_\theta(\underline{\kappa})e^{\Phi}\right|\right) \lesssim \frac{\epsilon_0}{r}\Delta_{ext}\lesssim\mathring{\delta}.$$

Together with the a priori estimates of Chapter 9 on the GCM construction, this yields

$$\begin{aligned}|\psi'(s)| &\lesssim \left|1+\frac{2\widetilde{m}}{\widetilde{r}}+\left(\underline{\Omega}+\frac{2}{\varsigma}\right)\Big|_{SP}\right|+|\lambda-1|\\ &\lesssim \left|\frac{\widetilde{m}}{\widetilde{r}}-\frac{m}{r}\right|+|\lambda-1|+\frac{\epsilon_0}{r}\Delta_{ext}.\end{aligned}$$

In view of (8.2.4), we have

$$|\widetilde{r}-r|+|\widetilde{m}-m| \lesssim \sup_{\widetilde{S}}r(|f|+|\underline{f}|)\lesssim Du_*\mathring{\delta} \tag{8.2.8}$$

and we infer

$$\begin{aligned}|\psi'(s)| &\lesssim \frac{Du_*}{r}\mathring{\delta}+\mathring{\delta}\\ &\lesssim \left(1+\epsilon_0^{\frac{2}{3}}D\right)\mathring{\delta}\\ &\lesssim \mathring{\delta}.\end{aligned}$$

Integrating from $\overset{\circ}{s}$ where $\psi(\overset{\circ}{s}) = 0$, we infer

$$\begin{aligned} |\psi(s)| &\lesssim |s - \overset{\circ}{s}|\overset{\circ}{\delta} \\ &\lesssim u_*\overset{\circ}{\delta} \end{aligned}$$

which improves (8.2.5) for $D \geq 1$ large enough.

Similarly, we obtain

$$\|(f, \underline{f}, \log(\lambda))\|_{\mathfrak{h}_{k_{small}+5}(\widetilde{S})} \lesssim r^{-2}\left(\left|\int_S fe^{\Phi}\right| + \left|\int_S \underline{f}e^{\Phi}\right|\right) + \overset{\circ}{\delta}$$

and

$$\left|\tilde{\nu}\left(\int_S fe^{\Phi}\right)\right| + \left|\tilde{\nu}\left(\int_S \underline{f}e^{\Phi}\right)\right| \lesssim r^2\overset{\circ}{\delta} + \frac{1}{r}\left(\left|\int_S fe^{\Phi}\right| + \left|\int_S \underline{f}e^{\Phi}\right|\right).$$

In view of (8.2.4), we infer

$$\left|\tilde{\nu}\left(\int_S fe^{\Phi}\right)\right| + \left|\tilde{\nu}\left(\int_S \underline{f}e^{\Phi}\right)\right| \lesssim r^2\overset{\circ}{\delta} + rDu_*\overset{\circ}{\delta}$$

and integrating from $\widetilde{S}_*$, we infer

$$\begin{aligned} r^{-2}\left(\left|\int_S fe^{\Phi}\right| + \left|\int_S \underline{f}e^{\Phi}\right|\right) &\lesssim u_*\overset{\circ}{\delta} + \frac{D(u_*)^2}{r}\overset{\circ}{\delta} \\ &\lesssim \left(1 + \epsilon_0^{\frac{2}{3}}D\right)u_*\overset{\circ}{\delta} \\ &\lesssim u_*\overset{\circ}{\delta}. \end{aligned}$$

This yields

$$\|(f, \underline{f}, \log(\lambda))\|_{\mathfrak{h}_{k_{small}+5}(\widetilde{S})} \lesssim u_*\overset{\circ}{\delta}$$

which improves (8.2.4) for $D \geq 1$ large enough. We thus conclude that $u_1 = 1$, $\widetilde{\Sigma}_*$ extends all the way to the initial data layer, $\widetilde{\Sigma}_* \subset \widetilde{\mathcal{R}}$, and we have the bounds

$$\|(f, \underline{f}, \log(\lambda))\|_{\mathfrak{h}_{k_{small}+5}(\widetilde{S})} \lesssim u_*\overset{\circ}{\delta}, \qquad |\psi(s)| \lesssim u_*\overset{\circ}{\delta}.$$

In view of the definition of $\overset{\circ}{\delta}$, we infer in particular for any sphere $\widetilde{S}$ of $\widetilde{\Sigma}_*$

$$\|(f, \underline{f}, \log(\lambda))\|_{\mathfrak{h}_{k_{small}+5}(\widetilde{S})} \lesssim \epsilon_0\delta_{ext}, \qquad |\psi(s)| \lesssim \epsilon_0\delta_{ext}. \tag{8.2.9}$$

Step 10. As $\widetilde{\Sigma}_*$ extends all the way to the initial data layer, this allows us to calibrate $\tilde{u}$ along $\widetilde{\Sigma}_*$ by fixing the value $\widetilde{u} = 1$ as in (3.1.5):

$$\widetilde{S}_1 = \widetilde{\Sigma}_* \cap \{\widetilde{u} = 1\} \text{ is such that } \widetilde{S}_1 \cap \mathcal{C}_{(1,\mathcal{L}_0)} \cap SP \neq \emptyset, \tag{8.2.10}$$

i.e., $\widetilde{S}_1$ is the unique sphere of $\widetilde{\Sigma}_*$ such that its south pole intersects the south pole of one of the spheres of the outgoing null cone $\mathcal{C}_{(1,\mathcal{L}_0)}$ of the initial data layer.

Now that $\widetilde{u}$ is calibrated, we define

$$\tilde{u}_* := \tilde{u}(\widetilde{S}_*). \tag{8.2.11}$$

For the proof of Theorem M7, we need in particular to prove that $\tilde{u}_* > u_*$. First, note that, since $\tilde{u} + \tilde{r}$ is constant along $\widetilde{\Sigma}_*$, we have

$$\widetilde{\Sigma}_* = \left\{\widetilde{u} + \widetilde{r} = 1 + \widetilde{r}(\widetilde{S}_1)\right\}. \tag{8.2.12}$$

Since $\widetilde{S}_* \subset \widetilde{\Sigma}_*$, and in view of (8.2.12), (8.2.2), (8.2.6), we infer

$$\begin{aligned}\left|\widetilde{u}(\widetilde{S}_*) - \left(u_* + \frac{\delta_{ext}}{2}\right)\right| &= \left|\widetilde{u}(\widetilde{S}_*) - u(\mathring{S})\right| \\ &= \left|1 + \widetilde{r}(\widetilde{S}_1) - \widetilde{r}(\widetilde{S}_*) - \left(-s(\mathring{S}) + c_{\widetilde{\Sigma}_*}\right)\right|.\end{aligned}$$

Next, note from

$$s = r \text{ on } \Sigma_*, \qquad e_4(r - s) = \frac{r}{2}\left(\overline{\kappa} - \frac{2}{r}\right)$$

that we have

$$\sup_{\widetilde{\mathcal{R}}} |r - s| \lesssim \frac{\epsilon_0}{r}\Delta_{ext} \lesssim \epsilon_0 \delta_{ext}. \tag{8.2.13}$$

Together with (8.2.8), this yields

$$\left|\widetilde{u}(\widetilde{S}_*) - \left(u_* + \frac{\delta_{ext}}{2}\right)\right| \lesssim \left|1 + \widetilde{r}(\widetilde{S}_1) - c_{\widetilde{\Sigma}_*}\right| + \epsilon_0 \delta_{ext}.$$

Since $c_{\widetilde{\Sigma}_*}$ in (8.2.6) is a constant, we have in particular

$$c_{\widetilde{\Sigma}_*} = u(\widetilde{S}_1) + r(\widetilde{S}_1) - \psi(s(\widetilde{S}_1))$$

and thus

$$\begin{aligned}\left|\widetilde{u}(\widetilde{S}_*) - \left(u_* + \frac{\delta_{ext}}{2}\right)\right| &\lesssim \left|1 + \widetilde{r}(\widetilde{S}_1) - u(\widetilde{S}_1) - r(\widetilde{S}_1) + \psi(s(\widetilde{S}_1))\right| + \epsilon_0 \delta_{ext} \\ &\lesssim \left|1 - u(\widetilde{S}_1)\right| + \left|\widetilde{r}(\widetilde{S}_1) - r(\widetilde{S}_1)\right| + \left|\psi(s(\widetilde{S}_1))\right| + \epsilon_0 \delta_{ext}.\end{aligned}$$

In view of (8.2.9) and (8.2.8), we infer

$$\left|\widetilde{u}(\widetilde{S}_*) - \left(u_* + \frac{\delta_{ext}}{2}\right)\right| \lesssim \left|1 - u(\widetilde{S}_1)\right| + \epsilon_0 \delta_{ext}.$$

Also, since (recall in particular (3.1.5))

$$u = 1 \text{ on } S_1 \cap SP, \qquad e_4^{\mathcal{L}_0}(u) = O\left(\frac{\epsilon_0}{r^2}\right),$$

and since the south pole of S_1 coincides with the one of the corresponding sphere of $\mathcal{C}_{\mathcal{L}_0,1}$, we infer

$$\sup_{\widetilde{\mathcal{R}}\cap\mathcal{C}_{\mathcal{L}_0,1}\cap SP} |u-1| \lesssim \Delta_{ext}\frac{\epsilon_0}{r^2} \lesssim \epsilon_0\delta_{ext}.$$

This yields

$$\left|\widetilde{u}(\widetilde{S}_*) - \left(u_* + \frac{\delta_{ext}}{2}\right)\right| \lesssim \epsilon_0\delta_{ext}. \tag{8.2.14}$$

In particular, we deduce, for ϵ_0 small enough,

$$\widetilde{u}(\widetilde{S}_*) > u_* \tag{8.2.15}$$

as desired.

Step 11. We would like to check that the dominant condition (3.3.4) for r holds on $\widetilde{\Sigma}_*$, i.e., we need to prove that there exists a choice of constant d_0 satisfying $\frac{1}{2} \le d_0 \le 1$ such that

$$\widetilde{r}(\widetilde{S}_*) = \epsilon_0^{-\frac{2}{3}}(\widetilde{u}(\widetilde{S}_*))^{1+\delta_{dec}}.$$

To this end, note that we have in view of (8.2.8), (8.2.13) and (8.2.14)

$$\begin{aligned}
&\widetilde{r}(\widetilde{S}_*) - \epsilon_0^{-\frac{2}{3}}(\widetilde{u}(\widetilde{S}_*))^{1+\delta_{dec}} \\
=\;& s(\overset{\circ}{S}) + O(\epsilon_0\delta_{ext}) - \epsilon_0^{-\frac{2}{3}}\left(u_* + \frac{\delta_{ext}}{2} + O(\epsilon_0\delta_{ext})\right)^{1+\delta_{dec}} \\
=\;& s(\overset{\circ}{S}) - \epsilon_0^{-\frac{2}{3}}(u_*)^{1+\delta_{dec}} - \frac{1+\delta_{dec}}{2}\epsilon_0^{-\frac{2}{3}}(u_*)^{\delta_{dec}}\delta_{ext} \\
&+\epsilon_0^{-\frac{2}{3}}(u_*)^{\delta_{dec}}\delta_{ext}O\left(\frac{\delta_{ext}}{u_*} + \epsilon_0\right) + O(\epsilon_0\delta_{ext}).
\end{aligned}$$

Together with (8.2.2), we infer

$$\begin{aligned}
&\widetilde{r}(\widetilde{S}_*) - \epsilon_0^{-\frac{2}{3}}(\widetilde{u}(\widetilde{S}_*))^{1+\delta_{dec}} \\
=\;& r_* + \frac{3d_0r_*}{4u_*}\delta_{ext} - \epsilon_0^{-\frac{2}{3}}(u_*)^{1+\delta_{dec}} - \frac{1+\delta_{dec}}{2}\epsilon_0^{-\frac{2}{3}}(u_*)^{\delta_{dec}}\delta_{ext} \\
&+\epsilon_0^{-\frac{2}{3}}(u_*)^{\delta_{dec}}\delta_{ext}O\left(\frac{\delta_{ext}}{u_*} + \epsilon_0\right) + O(\epsilon_0\delta_{ext}) \\
=\;& r_* - \epsilon_0^{-\frac{2}{3}}(u_*)^{1+\delta_{dec}} + \left(\frac{3d_0r_*}{4} - \frac{1+\delta_{dec}}{2}\epsilon_0^{-\frac{2}{3}}(u_*)^{1+\delta_{dec}}\right)\frac{\delta_{ext}}{u_*} \\
&+\epsilon_0^{-\frac{2}{3}}(u_*)^{\delta_{dec}}\delta_{ext}O\left(\frac{\delta_{ext}}{u_*} + \epsilon_0\right) + O(\epsilon_0\delta_{ext}).
\end{aligned}$$

Since we have by the condition (3.3.4) of r on Σ_*

$$r_* = \epsilon_0^{-\frac{2}{3}}u_*^{1+\delta_{dec}},$$

we deduce

$$\begin{aligned}
&\widetilde{r}(\widetilde{S}_*) - \epsilon_0^{-\frac{2}{3}}(\widetilde{u}(\widetilde{S}_*))^{1+\delta_{dec}} \\
= & \left(\frac{3d_0}{4} - \frac{1+\delta_{dec}}{2}\right)\frac{r_*\delta_{ext}}{u_*} + \epsilon_0^{-\frac{2}{3}}(u_*)^{\delta_{dec}}\delta_{ext}O\left(\frac{\delta_{ext}}{u_*}+\epsilon_0\right) + O(\epsilon_0\delta_{ext}) \\
= & \frac{3r_*\delta_{ext}}{4u_*}\left(d_0 - \frac{2+2\delta_{dec}}{3} + O\left(\epsilon_0 + \frac{\delta_{ext}}{u_*}\right)\right).
\end{aligned}$$

Thus, we may choose the contant d_0 such that $\frac{1}{2} \leq d_0 \leq 1$ and

$$\widetilde{r}(\widetilde{S}_*) = \epsilon_0^{-\frac{2}{3}}(\widetilde{u}(\widetilde{S}_*))^{1+\delta_{dec}}$$

as desired.

Step 12. We summarize the properties of $\widetilde{\Sigma}_*$ obtained so far:

- $\widetilde{\Sigma}_*$ is a spacelike hypersurface included in the spacetime region $\widetilde{\mathcal{R}}$.
- The scalar function $\widetilde{u}$ is defined on $\widetilde{\Sigma}_*$ and its level sets are topological 2-spheres denoted by $\widetilde{S}$.
- The following GCM conditions hold on $\widetilde{\Sigma}_*$:

$$\widetilde{\not{d}}_2^\star \widetilde{\not{d}}_1^\star \underline{\widetilde{\kappa}} = \widetilde{\not{d}}_2^\star \widetilde{\not{d}}_1^\star \widetilde{\mu} = 0, \qquad \widetilde{\kappa} = \frac{2}{\widetilde{r}}, \quad \int_{\widetilde{S}} \widetilde{\eta} e^\Phi = \int_{\widetilde{S}} \underline{\widetilde{\xi}} e^\Phi = 0.$$

- In addition, the following GCM conditions hold on the sphere $\widetilde{S}_*$ of $\widetilde{\Sigma}_*$:

$$\int_{\widetilde{S}_*} \widetilde{\beta} e^\Phi = \int_{\widetilde{S}_*} \widetilde{e}_\theta(\underline{\widetilde{\kappa}}) e^\Phi = 0.$$

- We have, for some constant $c_{\widetilde{\Sigma}_*}$,

$$\widetilde{u} + \widetilde{r} = c_{\widetilde{\Sigma}_*}, \quad \text{along} \quad \widetilde{\Sigma}_*.$$

- The following normalization condition holds true at the south pole SP of every sphere $\widetilde{S}$:

$$\widetilde{a}\Big|_{SP} = -1 - \frac{2\widetilde{m}}{\widetilde{r}}$$

  where $\widetilde{a}$ is such that we have

$$\widetilde{\nu} = \widetilde{e}_3 + \widetilde{a}\widetilde{e}_4,$$

  with $\widetilde{\nu}$ the unique vectorfield tangent to the hypersurface $\widetilde{\Sigma}_*$, normal to $\widetilde{S}$, and normalized by $g(\widetilde{\nu}, \widetilde{e}_4) = -2$.
- The dominant condition (3.3.4) for r holds on $\widetilde{\Sigma}_*$, i.e., we have

$$\widetilde{r}(\widetilde{S}_*) = \epsilon_0^{-\frac{2}{3}}(\widetilde{u}(\widetilde{S}_*))^{1+\delta_{dec}}.$$

- $\tilde{u}$ is calibrated along $\widetilde{\Sigma}_*$ by fixing the value $\widetilde{u}=1$:

$$\widetilde{S}_1 = \widetilde{\Sigma}_* \cap \{\widetilde{u}=1\} \text{ is such that } \widetilde{S}_1 \cap \mathcal{C}_{(1,\mathcal{L}_0)} \cap SP \neq \emptyset, \tag{8.2.16}$$

i.e., $\widetilde{S}_1$ is the unique sphere of $\widetilde{\Sigma}_*$ such that its south pole intersects the south pole of one of the sphere of the outgoing null cone $\mathcal{C}_{(1,\mathcal{L}_0)}$ of the initial data layer.

Thus $\widetilde{\Sigma}_*$ satisfies all the required properties for the future spacelike boundary of a GCM admissible spacetime, see item 3 of Definition 3.2. Furthermore, we have on $\widetilde{\Sigma}_*$

$$\widetilde{u}(\widetilde{S}_*) > u_*, \tag{8.2.17}$$

and $(f, \underline{f}, \lambda)$ satisfy in view of (8.2.9) and Corollary 9.53

$$\sup_{\widetilde{\Sigma}_*} \|\mathfrak{d}^{\leq k_{small}+5}(f, \underline{f}, \log(\lambda))\|_{L^2(\widetilde{S})} \lesssim \epsilon_0 \delta_{ext}.$$

Together with the Sobolev embedding on the spheres $\widetilde{S}$, we find

$$\sup_{\widetilde{\Sigma}_*} \widetilde{r}\, |\mathfrak{d}^{\leq k_{small}+3}(f, \underline{f}, \log(\lambda))| \lesssim \epsilon_0 \delta_{ext}.$$

Possibly reducing the size of $\delta_{ext}>0$, we deduce

$$\sup_{\widetilde{\Sigma}_*} \widetilde{r}\, \widetilde{u}^{\frac{1}{2}+\delta_{dec}} |\mathfrak{d}^{\leq k_{small}+3}(f, \underline{f}, \log(\lambda))| \lesssim \epsilon_0. \tag{8.2.18}$$

Step 13. We now control the outgoing geodesic foliation initialized on $\widetilde{\Sigma}_*$. We denote by ${}^{(ext)}\widetilde{\mathcal{M}}$ the region covered by this outgoing geodesic foliation. Let (e_4, e_3, e_θ) of ${}^{(ext)}\mathcal{M}$ be extended to the spacetime $\mathcal{M}^{(extend)}$, and satisfy, as discussed in Step 1 to Step 3,

$$\mathfrak{N}^{(Dec)}_{k_{small}+5}(\mathcal{M}^{(extend)}) \lesssim \epsilon_0. \tag{8.2.19}$$

Let $(f, \underline{f}, \lambda)$ the transition functions from the null frame (e_4, e_3, e_θ) to the null frame $(\widetilde{e}_4, \widetilde{e}_3, \widetilde{e}_\theta)$ of ${}^{(ext)}\widetilde{\mathcal{M}}$. Since both frames are outgoing geodesic, we may apply Corollary 2.93 which yields for $(\underline{f}, f, \log(\lambda))$ the following transport equations:

$$\begin{aligned} \lambda^{-1}e_4'(rf) &= E_1'(f,\Gamma), \\ \lambda^{-1}e_4'(\log(\lambda)) &= E_2'(f,\Gamma), \\ \lambda^{-1}e_4'\Big(r\underline{f} - 2r^2 e_\theta'(\log(\lambda)) + rf\underline{\Omega}\Big) &= E_3'(f,\underline{f},\lambda,\Gamma), \end{aligned}$$

where

$$\begin{aligned}
E_1'(f,\Gamma) &= -\frac{r}{2}\check{\kappa}f - \frac{r}{2}\vartheta f + \text{l.o.t.},\\
E_2'(f,\Gamma) &= f\zeta - \frac{1}{2}f^2\underline{\omega} - \underline{\eta}f - \frac{1}{4}f^2\underline{\kappa} + \text{l.o.t.},\\
E_3'(f,\underline{f},\lambda,\Gamma) &= -\frac{r}{2}\check{\kappa}\underline{f} + r^2\left(\check{\kappa} - \left(\overline{\kappa} - \frac{2}{r}\right)\right)e_\theta'(\log(\lambda))\\
&\quad + r^2\left(\not{d}_1'(f) + \lambda^{-1}\vartheta'\right)e_\theta'(\log(\lambda)) - \frac{r}{2}\check{\kappa}\underline{\Omega}f + rE_3(f,\underline{f},\Gamma)\\
&\quad -2r^2e_\theta'(E_2(f,\Gamma)) + r\underline{\Omega}E_1(f,\Gamma),
\end{aligned}$$

and where E_1, E_2 and E_3 are given in Lemma 2.92. Integrating these transport equations from $\widetilde{\Sigma}_*$, using the control (8.2.18) of $(f,\underline{f},\lambda)$ on $\widetilde{\Sigma}_*$, and together with the control (8.2.19) for the Ricci coefficients of the foliation of $\mathcal{M}^{(extend)}$, we obtain

$$\sup_{^{(ext)}\widetilde{\mathcal{M}}\left(\widetilde{r}\geq 2m_0(1+\frac{\delta_{\mathcal{H}}}{2})\right)} \left(\widetilde{r}\,\widetilde{u}^{\frac{1}{2}+\delta_{dec}} + \widetilde{u}^{1+\delta_{dec}}\right) \times\left(|\mathfrak{d}^{\leq k_{small}+3}(f,\log(\lambda))| + |\mathfrak{d}^{\leq k_{small}+2}\underline{f}|\right) \lesssim \epsilon_0. \tag{8.2.20}$$

Then, for any $r_{\mathcal{T}}$ in the interval

$$2m_0\left(1+\frac{\delta_{\mathcal{H}}}{2}\right) \leq r_{\mathcal{T}} \leq 2m_0\left(1+\frac{3\delta_{\mathcal{H}}}{2}\right), \tag{8.2.21}$$

we initialize the ingoing geodesic foliation of $^{(int)}\widetilde{\mathcal{M}}[r_{\mathcal{T}}]$ on $\widetilde{r} = r_{\mathcal{T}}$ using the outgoing geodesic foliation of $^{(ext)}\widetilde{\mathcal{M}}$ as in item 4 of Definition 3.2. Using the control of $(f,\underline{f},\lambda)$ induced on $\widetilde{r} = r_{\mathcal{T}}$ by (8.2.20), and using the analog of Corollary 2.93 in the e_3 direction for ingoing foliations, we obtain similarly, for any $r_{\mathcal{T}}$ in the interval (8.2.21),

$$\sup_{^{(int)}\widetilde{\mathcal{M}}[r_{\mathcal{T}}]} \widetilde{\underline{u}}^{1+\delta_{dec}}\left(|\mathfrak{d}^{\leq k_{small}+2}(\underline{f},\log(\lambda))| + |\mathfrak{d}^{\leq k_{small}+1}f|\right) \lesssim \epsilon_0. \tag{8.2.22}$$

Let now, for any $r_{\mathcal{T}}$ in the interval (8.2.21),

$$\mathcal{M}[r_{\mathcal{T}}] := {}^{(ext)}\widetilde{\mathcal{M}}(\widetilde{r} \geq r_{\mathcal{T}}) \cup {}^{(int)}\widetilde{\mathcal{M}}[r_{\mathcal{T}}].$$

Then, in view of (8.2.20), (8.2.22), and (8.2.19), and using the transformation formulas of Proposition 2.90, we deduce

$$\mathfrak{N}^{(Dec)}_{k_{small}}(\mathcal{M}[r_{\mathcal{T}}]) \lesssim \epsilon_0$$

which concludes the proof of Theorem M7.

8.3 PROOF OF THEOREM M8

So far, we have only improved our bootstrap assumptions on decay estimates. We now improve our bootstrap assumptions on energies and weighted energies for $\check{R}$

and $\check{\Gamma}$ relying on an iterative procedure which recovers derivatives one by one.[3]

Let $I_{m_0,\delta_{\mathcal{H}}}$ the interval of $\mathbb{R}$ defined by

$$I_{m_0,\delta_{\mathcal{H}}} := \left[2m_0\left(1+\frac{\delta_{\mathcal{H}}}{2}\right), 2m_0\left(1+\frac{3\delta_{\mathcal{H}}}{2}\right)\right]. \tag{8.3.1}$$

Remark 8.2. *Recall that the results of Theorems M0–M7 hold for any $r_{\mathcal{T}} \in I_{m_0,\delta_{\mathcal{H}}}$, see Remark 3.30. More precisely,*

- *they hold on ${}^{(ext)}\mathcal{M}(r \geq 2m_0(1+\frac{\delta_{\mathcal{H}}}{2}))$, and hence on ${}^{(ext)}\mathcal{M}(r \geq r_{\mathcal{T}})$ for any $r_{\mathcal{T}} \in I_{m_0,\delta_{\mathcal{H}}}$,*
- *they hold on ${}^{(int)}\mathcal{M}[r_{\mathcal{T}}]$ for any $r_{\mathcal{T}} \in I_{m_0,\delta_{\mathcal{H}}}$, where ${}^{(int)}\mathcal{M}[r_{\mathcal{T}}]$ is initialized on $\mathcal{T} = \{r = r_{\mathcal{T}}\}$ using ${}^{(ext)}\mathcal{M}(r \geq r_{\mathcal{T}})$ as in section 3.1.2.*

It is at this stage that we need to make a specific choice of $r_{\mathcal{T}}$ in the context of a Lebesgue point argument. More precisely, we choose $r_{\mathcal{T}}$ such that we have

$$\int_{\{r=r_{\mathcal{T}}\}} |\mathfrak{d}^{\leq k_{large}}\check{R}|^2 = \inf_{r_0 \in I_{m_0,\delta_{\mathcal{H}}}} \int_{\{r=r_0\}} |\mathfrak{d}^{\leq k_{large}}\check{R}|^2. \tag{8.3.2}$$

Remark 8.3. *In case the above infimum is achieved for several values of r, we choose $r_{\mathcal{T}}$ to be the largest of such values, so that $r_{\mathcal{T}}$ is uniquely defined. Note also that the infimum could a priori be infinite, and will only be shown to be finite — and more precisely $O(\epsilon_0)$ — at the end of the proof of Theorem M8, see section 8.3.4. This could be made rigorous in the context of a continuity argument.*

In view of the definition of $r_{\mathcal{T}}$, and since $\mathcal{T} = \{r = r_{\mathcal{T}}\}$, we have

$$\int_{\mathcal{T}} |\mathfrak{d}^{\leq k_{large}}\check{R}|^2 \leq \frac{1}{2m_0\delta_{\mathcal{H}}}\int_{I_{m_0,\delta_{\mathcal{H}}}} \left(\int_{\{r=r_0\}} |\mathfrak{d}^{\leq k_{large}}\check{R}|^2\right) dr_0$$

and hence[4]

$$\int_{\mathcal{T}} |\mathfrak{d}^{\leq k_{large}}\check{R}|^2 \lesssim \int_{{}^{(ext)}\mathcal{M}\left(r\in I_{m_0,\delta_{\mathcal{H}}}\right)} |\mathfrak{d}^{\leq k_{large}}\check{R}|^2. \tag{8.3.3}$$

From now on, we may thus assume that the spacetime $\mathcal{M}$ satisfies

[3]See also [33] for a related strategy to recover higher order derivatives from the control of lower order ones.

[4]We use the coarea formula, $d\mathcal{M} = \frac{1}{\sqrt{\mathbf{g}(\mathbf{D}r,\mathbf{D}r)}} d\{r = r_0\}dr_0$, and the fact that, for $r \in I_{m_0,\delta_{\mathcal{H}}}$, $\mathbf{g}(\mathbf{D}r,\mathbf{D}r) = -e_3(r)e_4(r) = \Upsilon + O(\epsilon) \geq \frac{\delta_{\mathcal{H}}}{2} + O(\epsilon + \delta_{\mathcal{H}}^2) \geq \frac{\delta_{\mathcal{H}}}{4}$. Note that $\lesssim$ here depends on $\delta_{\mathcal{H}}^{-1}$, see the convention for $\lesssim$ made at the end of section 3.3.1.

- the conclusions of Theorem M0, i.e.,

$$\begin{aligned}\max_{0\leq k\leq k_{large}} \Bigg\{ \sup_{\mathcal{C}_1} \Bigg[& r^{\frac{7}{2}+\delta_B}\left(|\mathfrak{d}^{k}\,{}^{(ext)}\alpha| + |\mathfrak{d}^{k}\,{}^{(ext)}\beta|\right) \\ & + r^{\frac{9}{2}+\delta_B}|\mathfrak{d}^{k-1}e_3({}^{(ext)}\alpha)|\Bigg] \\ + \sup_{\mathcal{C}_1}\Bigg[& r^3\left|\mathfrak{d}^{k}\left({}^{(ext)}\rho + \frac{2m_0}{r^3}\right)\right| + r^2|\mathfrak{d}^{k}\,{}^{(ext)}\underline{\beta}| \\ & + r|\mathfrak{d}^{k}\,{}^{(ext)}\underline{\alpha}|\Bigg]\Bigg\} \lesssim \epsilon_0 \end{aligned} \tag{8.3.4}$$

and

$$\begin{aligned}\max_{0\leq k\leq k_{large}} \sup_{\underline{\mathcal{C}}_1} \Bigg[& |\mathfrak{d}^{k}\,{}^{(int)}\alpha| + |\mathfrak{d}^{k}\,{}^{(int)}\beta| + \left|\mathfrak{d}^{k}\left({}^{(int)}\rho + \frac{2m_0}{r^3}\right)\right| \\ & + |\mathfrak{d}^{k}\,{}^{(int)}\underline{\beta}| + |\mathfrak{d}^{k}\,{}^{(int)}\underline{\alpha}|\Bigg] \lesssim \epsilon_0, \end{aligned} \tag{8.3.5}$$

- the conclusions of Theorem M7, i.e.,

$$\mathfrak{N}^{(Dec)}_{k_{small}} \lesssim \epsilon_0, \tag{8.3.6}$$

see section 3.2.3 for the definition of the combined norm on decay $\mathfrak{N}^{(Dec)}_{k}$,
- the estimate

$$\int_{\mathcal{T}} |\mathfrak{d}^{\leq k_{large}}\check{R}|^2 \lesssim \int_{{}^{(ext)}\mathcal{M}\left(r\in I_{m_0,\delta_{\mathcal{H}}}\right)} |\mathfrak{d}^{\leq k_{large}}\check{R}|^2. \tag{8.3.7}$$

The goal of this section is to prove Theorem M8, i.e., to prove that the following bound holds on $\mathcal{M}$ for the weighted energies:

$$\mathfrak{N}^{(En)}_{k_{large}} \lesssim \epsilon_0,$$

see section 3.2.3 for the definition of the combined norm on weighted energies $\mathfrak{N}^{(En)}_{k}$.

8.3.1 Main norms

We recall below our norms for measuring weighted energies for curvature components and Ricci coefficients, see sections 3.2.1 and 3.2.2. Let $r_0 \geq 4m_0$. Then, we have for ${}^{(ext)}\mathcal{M}$

$$
\begin{aligned}
\left({}^{(ext)}\mathfrak{R}_0^{\geq r_0}[\check{R}]\right)^2 &= \sup_{0\leq u\leq u_*}\int_{\mathcal{C}_u(r\geq r_0)}\left(r^{4+\delta_B}\alpha^2+r^4\beta^2\right)\\
&+\int_{\Sigma_*}\left(r^{4+\delta_B}(\alpha^2+\beta^2)+r^4(\check{\rho})^2+r^2\underline{\beta}^2+\underline{\alpha}^2\right)\\
&+\int_{{}^{(ext)}\mathcal{M}(r\geq r_0)}\Big(r^{3+\delta_B}(\alpha^2+\beta^2)+r^{3-\delta_B}(\check{\rho})^2+r^{1-\delta_B}\underline{\beta}^2\\
&\qquad +r^{-1-\delta_B}\underline{\alpha}^2\Big),
\end{aligned}
$$

$$
\left({}^{(ext)}\mathfrak{R}_0^{\leq r_0}[\check{R}]\right)^2 = \int_{{}^{(ext)}\mathcal{M}(r\leq r_0)}\left(1-\frac{3m}{r}\right)^2|\check{R}|^2,
$$

$$
{}^{(ext)}\mathfrak{R}_0[\check{R}] = {}^{(ext)}\mathfrak{R}_0^{\geq 4m_0}[\check{R}]+{}^{(ext)}\mathfrak{R}_0^{\leq 4m_0}[\check{R}],
$$

$$
\begin{aligned}
\left({}^{(ext)}\mathfrak{R}_k[\check{R}]\right)^2 = \left({}^{(ext)}\mathfrak{R}_0[\mathfrak{d}^{\leq k}\check{R}]\right)^2+\int_{{}^{(ext)}\mathcal{M}(r\leq 4m_0)}\Big(|\mathfrak{d}^{\leq k-1}\mathbf{N}\check{R}|^2\\
+|\mathfrak{d}^{\leq k-1}\check{R}|^2\Big), \qquad \text{for } k\geq 1,
\end{aligned}
$$

and

$$
\begin{aligned}
\left({}^{(ext)}\mathfrak{G}_k^{\geq r_0}[\check{\Gamma}]\right)^2 &= \int_{\Sigma_*}\Big[r^2\Big((\mathfrak{d}^{\leq k}\vartheta)^2+(\mathfrak{d}^{\leq k}\check{\kappa})^2+(\mathfrak{d}^{\leq k}\zeta)^2+(\mathfrak{d}^{\leq k}\underline{\check{\kappa}})^2\Big)\\
&+(\mathfrak{d}^{\leq k}\underline{\vartheta})^2+(\mathfrak{d}^{\leq k}\eta)^2+(\mathfrak{d}^{\leq k}\underline{\check{\omega}})^2+(\mathfrak{d}^{\leq k}\underline{\xi})^2\Big]\\
&+\sup_{\lambda\geq 4m_0}\Bigg(\int_{\{r=\lambda\}}\Big[\lambda^2\Big((\mathfrak{d}^{\leq k}\vartheta)^2+(\mathfrak{d}^{\leq k}\check{\kappa})^2+(\mathfrak{d}^{\leq k}\zeta)^2\Big)\\
&+\lambda^{2-\delta_B}(\mathfrak{d}^{\leq k}\underline{\check{\kappa}})^2+(\mathfrak{d}^{\leq k}\underline{\vartheta})^2+(\mathfrak{d}^{\leq k}\eta)^2+(\mathfrak{d}^{\leq k}\underline{\check{\omega}})^2\\
&+\lambda^{-\delta_B}(\mathfrak{d}^{\leq k}\underline{\xi})^2\Big]\Bigg),\\
\left({}^{(ext)}\mathfrak{G}_k^{\leq r_0}[\check{\Gamma}]\right)^2 &= \int_{{}^{(ext)}\mathcal{M}(\leq 4m_0)}\left|\mathfrak{d}^{\leq k}\left(\check{\Gamma}\right)\right|^2,\\
{}^{(ext)}\mathfrak{G}_k[\check{\Gamma}] &= {}^{(ext)}\mathfrak{G}_k^{\leq 4m_0}[\check{\Gamma}]+{}^{(ext)}\mathfrak{G}_k^{\geq 4m_0}[\check{\Gamma}].
\end{aligned}
$$

Also, we have for ${}^{(int)}\mathcal{M}$

$$
\left({}^{(int)}\mathfrak{R}_k[\check{R}]\right)^2 = \int_{{}^{(int)}\mathcal{M}}|\mathfrak{d}^{\leq k}\check{R}|^2,
$$

and

$$\left({}^{(int)}\mathfrak{G}_k[\check{\Gamma}] \right)^2 = \int_{{}^{(int)}\mathcal{M}} |\mathfrak{d}^{\leq k}\check{\Gamma}|^2.$$

Finally, we recall the following Morawetz type norms, see section 5.1.4. For $\delta > 0$, we have

$$\begin{aligned} & B_\delta[\psi](\tau_1, \tau_2) \\ = & \int_{{}^{(trap)}\mathcal{M}(\tau_1,\tau_2)} |R\psi|^2 + r^{-2}|\psi|^2 + \left(1 - \frac{3m}{r}\right)^2 \left(|\nabla\psi|^2 + \frac{1}{r^2}|T\psi|^2\right) \\ + & \int_{{}^{(tr\not{a}p)}\mathcal{M}(\tau_1,\tau_2)} r^{\delta-3} \left(|\mathfrak{d}\psi|^2 + |\psi|^2\right) \end{aligned}$$

where the scalar function τ and the spacetime region ${}^{(trap)}\mathcal{M}$ have been introduced in section 5.1.1, and where ${}^{(tr\not{a}p)}\mathcal{M}$ denotes the complement of ${}^{(trap)}\mathcal{M}$. Also, we have

$$\begin{aligned} E_\delta[\psi](\tau) &= \int_{\Sigma(\tau)} \left(\frac{1}{2}(N_\Sigma, e_3)^2 |e_4\psi|^2 + \frac{1}{2}(N_\Sigma, e_4)^2 |e_3\psi|^2 + |\nabla\psi|^2 + r^{-2}|\psi|^2 \right) \\ &\quad + \int_{\Sigma_{\geq 4m_0}(\tau)} r^\delta \left(|e_4\psi|^2 + r^{-2}|\psi|^2 \right). \end{aligned}$$

Here $\Sigma(\tau)$ denotes the level set of τ, see section 5.1.1, N_Σ denotes a choice for the normal to Σ, and recall that we have

$$N_\Sigma = \begin{cases} N_\Sigma = e_3 & \text{on} \quad {}^{(int)}\Sigma, \\ N_\Sigma = e_4 & \text{on} \quad {}^{(ext)}\Sigma, \end{cases}$$

with ${}^{(int)}\Sigma$ and ${}^{(ext)}\Sigma$ defined in section 5.1.1, and

$$(N_\Sigma, e_3) \leq -1 \text{ and } (N_\Sigma, e_4) \leq -1 \qquad \text{on} \quad {}^{(trap)}\Sigma.$$

Moreover, we have

$$\begin{aligned} F_\delta[\psi](\tau_1, \tau_2) &= \int_{\mathcal{A}(\tau_1,\tau_2)} \left(\delta_{\mathcal{H}}^{-1}|e_4\Psi|^2 + \delta_{\mathcal{H}}|e_3\Psi|^2 + |\nabla\Psi|^2 + r^{-2}|\Psi|^2 \right) \\ &\quad + \int_{\Sigma_*(\tau_1,\tau_2)} \left(|e_3\Psi|^2 + r^\delta \left(|e_4\psi|^2 + |\nabla\psi|^2 + r^{-2}|\psi|^2 \right) \right) \end{aligned}$$

with $\mathcal{A}(\tau_1, \tau_2) = \mathcal{A} \cap \mathcal{M}(\tau_1, \tau_2)$ and $\Sigma_*(\tau_1, \tau_2) = \Sigma_* \cap \mathcal{M}(\tau_1, \tau_2)$.

8.3.2 Control of the global frame

Some quantities will be controlled based on the wave equation they satisfy, and will thus need to be defined w.r.t. a global frame, i.e., a smooth frame on $\mathcal{M}$. To this end, we will rely on the global frame of section 3.5.2. We recall below the main properties of that global frame.

From Definition 3.22, the region where the frame of ${}^{(int)}\mathcal{M}$ and a conformal renormalization of the frame of ${}^{(ext)}\mathcal{M}$ are matched is given by

$$\begin{aligned} \text{Match} \; &:= \; \left({}^{(ext)}\mathcal{M} \cap \left\{ {}^{(int)}r \leq 2m_0\left(1+\frac{3}{2}\delta_{\mathcal{H}}\right)\right\}\right) \\ &\cup \left({}^{(int)}\mathcal{M} \cap \left\{ {}^{(int)}r \geq 2m_0\left(1+\frac{1}{2}\delta_{\mathcal{H}}\right)\right\}\right), \end{aligned}$$

where ${}^{(int)}r$ denotes the area radius of the ingoing geodesic foliation of ${}^{(int)}\mathcal{M}$ and its extension to ${}^{(ext)}\mathcal{M}$.

The following proposition concerning the global frame is an immediate consequence of Proposition 3.23 and the decay estimates (8.3.6).

Proposition 8.4. *Assume (8.3.6). Then, there exists a global null frame defined on ${}^{(int)}\mathcal{M} \cup {}^{(ext)}\mathcal{M}$ and denoted by $({}^{(glo)}e_4, {}^{(glo)}e_3, {}^{(glo)}e_\theta)$ such that*

a) In ${}^{(ext)}\mathcal{M} \setminus$ Match, we have

$$({}^{(glo)}e_4, {}^{(glo)}e_3, {}^{(glo)}e_\theta) = \left({}^{(ext)}\Upsilon\, {}^{(ext)}e_4, {}^{(ext)}\Upsilon^{-1}\, {}^{(ext)}e_3, {}^{(ext)}e_\theta\right).$$

b) In ${}^{(int)}\mathcal{M} \setminus$ Match, we have

$$({}^{(glo)}e_4, {}^{(glo)}e_3, {}^{(glo)}e_\theta) = \left({}^{(int)}e_4, {}^{(int)}e_3, {}^{(int)}e_\theta\right).$$

c) In the matching region, we have

$$\begin{aligned} \max_{0\leq k\leq k_{small}-2} \sup_{Match \cap {}^{(int)}\mathcal{M}} \underline{u}^{1+\delta_{dec}} \left|\mathfrak{d}^k({}^{(glo)}\check{\Gamma}, {}^{(glo)}\check{R})\right| &\lesssim \epsilon_0, \\ \max_{0\leq k\leq k_{small}-2} \sup_{Match \cap {}^{(ext)}\mathcal{M}} u^{1+\delta_{dec}} \left|\mathfrak{d}^k({}^{(glo)}\check{\Gamma}, {}^{(glo)}\check{R})\right| &\lesssim \epsilon_0, \end{aligned}$$

where ${}^{(glo)}\check{R}$ and ${}^{(glo)}\check{\Gamma}$ are given by

$$\begin{aligned} {}^{(glo)}\check{R} &= \left\{\alpha, \beta, \rho + \frac{2m}{r^3}, \underline{\beta}, \underline{\alpha}\right\}, \\ {}^{(glo)}\check{\Gamma} &= \left\{\xi, \omega + \frac{m}{r^2}, \kappa - \frac{2\Upsilon}{r}, \vartheta, \zeta, \eta, \underline{\eta}, \underline{\kappa} + \frac{2}{r}, \underline{\vartheta}, \underline{\omega}, \underline{\xi}\right\}. \end{aligned}$$

d) Furthermore, we may also choose the global frame such that, in addition, one of the following two possibilities hold:

i. We have on all ${}^{(ext)}\mathcal{M}$

$$({}^{(glo)}e_4, {}^{(glo)}e_3, {}^{(glo)}e_\theta) = \left({}^{(ext)}\Upsilon\, {}^{(ext)}e_4, {}^{(ext)}\Upsilon^{-1}\, {}^{(ext)}e_3, {}^{(ext)}e_\theta\right).$$

ii. We have on all ${}^{(int)}\mathcal{M}$

$$({}^{(glo)}e_4, {}^{(glo)}e_3, {}^{(glo)}e_\theta) = \left({}^{(int)}e_4, {}^{(int)}e_3, {}^{(int)}e_\theta\right).$$

8.3.3 Iterative procedure

Recall our norms for measuring energies for curvature components and Ricci coefficients which are given respectively by ${}^{(int)}\mathfrak{R}_k[\check{R}]$, ${}^{(ext)}\mathfrak{R}_k[\check{R}]$ and ${}^{(int)}\mathfrak{G}_k[\check{\Gamma}]$, ${}^{(ext)}\mathfrak{G}_k[\check{\Gamma}]$, see sections 3.2.1 and 3.2.2. Recall also our combined weighted energy norm

$$\mathfrak{N}_k^{(En)} = {}^{(ext)}\mathfrak{R}_k[\check{R}] + {}^{(ext)}\mathfrak{G}_k[\check{\Gamma}] + {}^{(int)}\mathfrak{R}_k[\check{R}] + {}^{(int)}\mathfrak{G}_k[\check{\Gamma}].$$

We also introduce the following norm controlling on the matching region the Ricci coefficients and curvature components of the global frame of Proposition 8.4:

$$\mathcal{N}_k^{(match)} := \left(\int_{\text{Match}} \left|\mathfrak{d}^{\leq k}({}^{(glo)}\check{\Gamma}, {}^{(glo)}\check{R})\right|^2\right)^{\frac{1}{2}}. \tag{8.3.8}$$

To initiate the iterative procedure, we rely on the following lemma.

Lemma 8.5. *We have*

$$\mathfrak{N}_{k_{small}}^{(En)} + \mathcal{N}_{k_{small}-2}^{(match)} \lesssim \epsilon_0. \tag{8.3.9}$$

Proof. The estimate (8.3.6) and Proposition 8.4 imply in particular

$$\begin{aligned}&{}^{(ext)}\widehat{\mathfrak{R}_{k_{small}}^{\geq 4m_0}}[\check{R}] + {}^{(ext)}\mathfrak{R}_{k_{small}}^{\leq 4m_0}[\check{R}] + {}^{(ext)}\mathfrak{G}_{k_{small}}[\check{\Gamma}]\\ &+ {}^{(int)}\mathfrak{R}_{k_{small}}[\check{R}] + {}^{(int)}\mathfrak{G}_{k_{small}}[\check{\Gamma}] + \mathcal{N}_{k_{small}-2}^{(match)} \lesssim \epsilon_0\end{aligned} \tag{8.3.10}$$

where the first term of the right-hand side is defined by

$$\begin{aligned}\left({}^{(ext)}\widehat{\mathfrak{R}_k^{\geq 4m_0}}[\check{R}]\right)^2 &:= \sup_{0\leq u\leq u_*}\int_{\mathcal{C}_u(r\geq 4m_0)} r^4|\mathfrak{d}^{\leq k}\beta|^2\\ &+ \int_{\Sigma_*}\left(r^4|\mathfrak{d}^{\leq k}\check{\rho}|^2 + r^2|\mathfrak{d}^{\leq k}\underline{\beta}|^2 + |\mathfrak{d}^{\leq k}\underline{\alpha}|^2\right)\\ &+ \int_{{}^{(ext)}\mathcal{M}(r\geq 4m_0)}\Big(r^{3-\delta_B}|\mathfrak{d}^{\leq k}\check{\rho}|^2 + r^{1-\delta_B}|\mathfrak{d}^{\leq k}\underline{\beta}|^2\\ &\quad + r^{-1-\delta_B}|\mathfrak{d}^{\leq k}\underline{\alpha}|^2\Big).\end{aligned}$$

In view of the definition of the combined weighted energy norm $\mathfrak{N}_k^{(En)}$, we infer

$$\begin{aligned}\mathfrak{N}_{k_{small}}^{(En)} + \mathcal{N}_{k_{small}-2}^{(match)} &\lesssim \epsilon_0 + \Bigg[\sup_{1\leq u\leq u_*}\int_{\mathcal{C}_u(r\geq 4m_0)} r^{4+\delta_B}|\mathfrak{d}^{\leq k_{small}}\alpha|^2\\ &+ \int_{\Sigma_*} r^{4+\delta_B}(|\mathfrak{d}^{\leq k_{small}}\alpha|^2 + |\mathfrak{d}^{\leq k_{small}}\beta|^2)\\ &+ \int_{{}^{(ext)}\mathcal{M}(r\geq 4m_0)} r^{3+\delta_B}(|\mathfrak{d}^{\leq k_{small}}\alpha|^2 + |\mathfrak{d}^{\leq k_{small}}\beta|^2)\Bigg]^{\frac{1}{2}}.\end{aligned} \tag{8.3.11}$$

Note that the terms on the RHS of the above estimate can not be estimated directly by (8.3.6) since $\delta_{dec} < \delta_B$.

Next we claim the estimate

$$\sup_{1\leq u\leq u_*}\int_{\mathcal{C}_u(r\geq 4m_0)} r^{4+\delta_B}|\mathfrak{d}^{\leq k_{small}}\alpha|^2+\int_{\Sigma_*} r^{4+\delta_B}(|\mathfrak{d}^{\leq k_{small}}\alpha|^2+|\mathfrak{d}^{\leq k_{small}}\beta|^2)$$
$$+\int_{{}^{(ext)}\mathcal{M}(r\geq 4m_0)} r^{3+\delta_B}(|\mathfrak{d}^{\leq k_{small}}\alpha|^2+|\mathfrak{d}^{\leq k_{small}}\beta|^2)$$
$$\lesssim \left({}^{(ext)}\mathfrak{R}^{\leq 4m_0}_{k_{small}}[\check{R}]\right)^2+\left({}^{(ext)}\mathfrak{G}_{k_{small}}[\check{\Gamma}]\right)^2+\epsilon_0^2+\epsilon_0^2(\mathfrak{N}^{(En)}_{k_{small}})^2. \tag{8.3.12}$$

The proof of (8.3.12) relies on r^p-weighted estimates for the Bianchi pair (α,β) and is postponed to section 8.7.3. Then (8.3.9) follows immediately from (8.3.10), (8.3.11) and (8.3.12) for $\epsilon_0>0$ small enough. □

Next, for J such that $k_{small}-2\leq J\leq k_{large}-1$, consider the iteration assumption

$$\mathfrak{N}^{(En)}_J+\mathcal{N}^{(match)}_J \lesssim \epsilon_{\mathcal{B}}[J], \tag{8.3.13}$$

where

$$\epsilon_{\mathcal{B}}[J] := \sum_{j=k_{small}-2}^{J}(\epsilon_0)^{\ell(j)}\,\mathcal{B}^{1-\ell(j)}+\epsilon_0^{\ell(J)}\mathcal{B}, \quad \ell(j):=2^{k_{small}-2-j},$$
$$\mathcal{B}:=\left(\int_{{}^{(ext)}\mathcal{M}\left(r\in I_{m_0,\delta_{\mathcal{H}}}\right)}|\mathfrak{d}^{\leq k_{large}}\check{R}|^2\right)^{\frac{1}{2}}. \tag{8.3.14}$$

Lemma 8.6. *The following estimate holds true for $\epsilon_{\mathcal{B}}[J]$ as defined above:*

$$\epsilon_{\mathcal{B}}[J]+\mathcal{B}^{\frac{1}{2}}(\epsilon_{\mathcal{B}}[J])^{\frac{1}{2}}+\epsilon_0\mathcal{B}\lesssim\epsilon_{\mathcal{B}}[J+1]. \tag{8.3.15}$$

Proof. We clearly have

$$\epsilon_{\mathcal{B}}[J]+\epsilon_0\mathcal{B}\lesssim\epsilon_{\mathcal{B}}[J+1]. \tag{8.3.16}$$

Also, we have, using $\ell(j)=2\ell(j+1)$,

$$\begin{aligned}
\mathcal{B}\epsilon_{\mathcal{B}}[J] &\lesssim \sum_{j=k_{small}-2}^{J}(\epsilon_0)^{\ell(j)}\,\mathcal{B}^{2-\ell(j)}+\epsilon_0^{\ell(J)}\mathcal{B}^2\\
&\lesssim \sum_{j=k_{small}-1}^{J+1}(\epsilon_0)^{2\ell(j)}\,\mathcal{B}^{2-2\ell(j)}+\epsilon_0^{2\ell(J+1)}\mathcal{B}^2\\
&\lesssim \left(\sum_{j=k_{small}-2}^{J+1}(\epsilon_0)^{\ell(j)}\,\mathcal{B}^{1-\ell(j)}+\epsilon_0^{\ell(J+1)}\mathcal{B}\right)^2\\
&= (\epsilon_{\mathcal{B}}[J+1])^2
\end{aligned}$$

which concludes the proof of the lemma. □

In view of (8.3.9), (8.3.13) holds for $J=k_{small}-2$. The propositions below will allow us to prove Theorem M8 in the next section.

Proposition 8.7. *Let J such that $k_{small} - 2 \leq J \leq k_{large} - 1$. Consider the global frame constructed in Proposition 8.4. In that frame, let*

$$\tilde{\rho} \quad := \quad r^2\rho + 2mr^{-1}. \tag{8.3.17}$$

Then, under the iteration assumption (8.3.13), *we have*

$$\sup_{\tau\in[1,\tau_*]} E^J_\delta[\tilde{\rho}](\tau) + B^J_\delta[\tilde{\rho}](1,\tau_*) + F^J_\delta[\tilde{\rho}](1,\tau_*) \lesssim (\epsilon_{\mathcal{B}}[J])^2 + \epsilon_0^2\Big(\mathfrak{N}^{(En)}_{J+1} + \mathcal{N}^{(match)}_{J+1}\Big)^2.$$

Proposition 8.8. *Let J such that $k_{small} - 2 \leq J \leq k_{large} - 1$. Consider the global frame constructed in Proposition 8.4. In that frame, under the iteration assumption* (8.3.13), *we have*

$$\begin{aligned}&\sup_{\tau\in[1,\tau_*]} E^J_\delta[\alpha + \Upsilon^2\underline{\alpha}](\tau) + B^J_\delta[\alpha + \Upsilon^2\underline{\alpha}](1,\tau_*) + F^J_\delta[\alpha + \Upsilon^2\underline{\alpha}](1,\tau_*)\\ &\lesssim \quad (\epsilon_{\mathcal{B}}[J])^2 + \epsilon_0^2\Big(\mathfrak{N}^{(En)}_{J+1} + \mathcal{N}^{(match)}_{J+1}\Big)^2.\end{aligned}$$

Proposition 8.9. *Let J such that $k_{small} - 2 \leq J \leq k_{large} - 1$. Consider the global frame constructed in Proposition 8.4. In that frame, under the iteration assumption* (8.3.13), *we have*

$$B^J_{-2}\Big[\check{\rho},\, \alpha,\, \underline{\alpha},\, \beta,\, \underline{\beta}\Big](1,\tau_*) \quad \lesssim \quad (\epsilon_{\mathcal{B}}[J])^2 + \epsilon_0^2\Big(\mathfrak{N}^{(En)}_{J+1} + \mathcal{N}^{(match)}_{J+1}\Big)^2.$$

Proposition 8.10. *Let J such that $k_{small} - 2 \leq J \leq k_{large} - 1$. Under the iteration assumption* (8.3.13), *we have for $r_0 \geq 4m_0$*

$$\begin{aligned}{}^{(int)}\mathfrak{R}_{J+1}[\check{R}] + {}^{(ext)}\mathfrak{R}_{J+1}[\check{R}] \quad \leq \quad & {}^{(ext)}\mathfrak{R}^{\geq r_0}_{J+1}[\check{R}]\\ & + O\left(r_0^{10}\left(\epsilon_{\mathcal{B}}[J] + \epsilon_0\Big(\mathfrak{N}^{(En)}_{J+1} + \mathcal{N}^{(match)}_{J+1}\Big)\right)\right)\end{aligned}$$

and

$${}^{(ext)}\mathfrak{R}^{\geq r_0}_{J+1}[\check{R}] \quad \lesssim \quad r_0^{-\delta_B}\, {}^{(ext)}\mathfrak{G}^{\geq r_0}_{J+1}[\check{\Gamma}] + r_0^{10}\left(\epsilon_{\mathcal{B}}[J] + \epsilon_0\left(\mathfrak{N}^{(En)}_{J+1} + \mathcal{N}^{(match)}_{J+1}\right)\right).$$

Proposition 8.11. *Let J such that $k_{small} - 2 \leq J \leq k_{large} - 1$. Under the iteration assumption* (8.3.13), *we have*

$${}^{(ext)}\mathfrak{G}_{J+1}[\check{\Gamma}] + {}^{(int)}\mathfrak{R}_{J+1}[\check{R}] + {}^{(ext)}\mathfrak{R}_{J+1}[\check{R}] \quad \lesssim \quad \epsilon_{\mathcal{B}}[J] + \epsilon_0\Big(\mathfrak{N}^{(En)}_{J+1} + \mathcal{N}^{(match)}_{J+1}\Big).$$

Proposition 8.12. *Let J such that $k_{small} - 2 \leq J \leq k_{large} - 1$. Under the iteration assumption* (8.3.13), *we have*

$${}^{(int)}\mathfrak{G}_{J+1}[\check{\Gamma}] \quad \lesssim \quad \epsilon_{\mathcal{B}}[J] + \epsilon_0\Big(\mathfrak{N}^{(En)}_{J+1} + \mathcal{N}^{(match)}_{J+1}\Big) + \left(\int_{\mathcal{T}}|\mathfrak{d}^{J+1}({}^{(ext)}\check{R})|^2\right)^{\frac{1}{2}}.$$

Proposition 8.13. *Let J such that $k_{small} - 2 \leq J \leq k_{large} - 1$. Under the iteration* *assumption* (8.3.13), *we have*

$$\mathcal{N}^{(match)}_{J+1} \lesssim \mathfrak{N}^{(En)}_{J+1} + \left(\int_{\mathcal{T}} |\mathfrak{d}^{J+1}({}^{(ext)}\check{R})|^2\right)^{\frac{1}{2}}.$$

The proofs of Propositions 8.7, 8.8, 8.9, 8.10, 8.11, 8.12 and 8.13 are postponed respectively to sections 8.4, 8.5, 8.6, 8.7, 8.8, 8.9 and 8.10.

8.3.4 End of the proof of Theorem M8

To prove Theorem M8, we rely on Propositions 8.11, 8.12 and 8.13. Note that among these propositions, only the last two involve the dangerous boundary term $\left(\int_{\mathcal{T}} |\mathfrak{d}^{J+1}({}^{(ext)}\check{R})|^2\right)^{\frac{1}{2}}$. We proceed as follows.

Step 1. As mentioned earlier, the estimate (8.3.9) trivially implies the iteration assumption (8.3.13) with $J = k_{small} - 2$. We assume that the iteration assumption (8.3.13) holds for any fixed J such that $k_{small} - 2 \leq J \leq k_{large} - 2$. In view of Proposition 8.12, we have

$$ {}^{(int)}\mathfrak{G}_{J+1}[\check{\Gamma}] \lesssim \epsilon_{\mathcal{B}}[J] + \epsilon_0\left(\mathfrak{N}^{(En)}_{J+1} + \mathcal{N}^{(match)}_{J+1}\right) + \left(\int_{\mathcal{T}} |\mathfrak{d}^{J+1}({}^{(ext)}\check{R})|^2\right)^{\frac{1}{2}}. \tag{8.3.18}$$

We need to deal with the last term in the RHS of (8.3.18). Relying on a trace theorem in the spacetime region ${}^{(ext)}\mathcal{M}(r \in I_{m_0,\delta_{\mathcal{H}}})$, as well as the fact that $J+2 \leq k_{large}$, we obtain

$$\left(\int_{\mathcal{T}} |\mathfrak{d}^{J+1}({}^{(ext)}\check{R})|^2\right)^{\frac{1}{2}} \lesssim \left(\int_{{}^{(ext)}\mathcal{M}\left(r\in I_{m_0,\delta_{\mathcal{H}}}\right)} |\mathfrak{d}^{k_{large}}\check{R}|^2\right)^{\frac{1}{4}} \left({}^{(ext)}\mathfrak{R}_{J+1}[\check{R}]\right)^{\frac{1}{2}} + {}^{(ext)}\mathfrak{R}_{J+1}[\check{R}]. \tag{8.3.19}$$

Proposition 8.11, (8.3.18) and (8.3.19) yield, for $\epsilon_0 > 0$ small enough so that we can absorb some of the terms to the left,

$$\begin{aligned}\mathfrak{N}^{(En)}_{J+1} \lesssim\ & \epsilon_{\mathcal{B}}[J] + \left(\int_{{}^{(ext)}\mathcal{M}\left(r\in I_{m_0,\delta_{\mathcal{H}}}\right)} |\mathfrak{d}^{k_{large}}\check{R}|^2\right)^{\frac{1}{4}} \\ & \times\left(\epsilon_{\mathcal{B}}[J] + \epsilon_0\left(\mathfrak{N}^{(En)}_{J+1} + \mathcal{N}^{(match)}_{J+1}\right)\right)^{\frac{1}{2}} + \epsilon_0\mathcal{N}^{(match)}_{J+1},\end{aligned}$$

and using also Proposition 8.13,

$$\begin{aligned}
\mathcal{N}_{J+1}^{(match)} &\lesssim \mathfrak{N}_{J+1}^{(En)} + \left(\int_{\mathcal{T}} |\mathfrak{d}^{J+1}({}^{(ext)}\check{R})|^2\right)^{\frac{1}{2}} \\
&\lesssim \epsilon_{\mathcal{B}}[J] + \left(\int_{{}^{(ext)}\mathcal{M}\left(r\in I_{m_0,\delta_{\mathcal{H}}}\right)} |\mathfrak{d}^{k_{large}}\check{R}|^2\right)^{\frac{1}{4}} \\
&\quad \times\left(\epsilon_{\mathcal{B}}[J] + \epsilon_0\left(\mathfrak{N}_{J+1}^{(En)} + \mathcal{N}_{J+1}^{(match)}\right)\right)^{\frac{1}{2}} + \epsilon_0\mathcal{N}_{J+1}^{(match)}.
\end{aligned}$$

For $\epsilon_0 > 0$ small enough, we infer, by absorbing the appropriate terms to the left,

$$\begin{aligned}
&\mathfrak{N}_{J+1}^{(En)} + \mathcal{N}_{J+1}^{(match)} \\
\lesssim\;& \epsilon_{\mathcal{B}}[J] + \left(\int_{{}^{(ext)}\mathcal{M}\left(r\in I_{m_0,\delta_{\mathcal{H}}}\right)} |\mathfrak{d}^{k_{large}}\check{R}|^2\right)^{\frac{1}{4}} \\
&\times\left(\epsilon_{\mathcal{B}}[J] + \epsilon_0\left(\mathfrak{N}_{J+1}^{(En)} + \mathcal{N}_{J+1}^{(match)}\right)\right)^{\frac{1}{2}} \\
\lesssim\;& \epsilon_{\mathcal{B}}[J] + \left(\int_{{}^{(ext)}\mathcal{M}\left(r\in I_{m_0,\delta_{\mathcal{H}}}\right)} |\mathfrak{d}^{k_{large}}\check{R}|^2\right)^{\frac{1}{4}} \left(\epsilon_{\mathcal{B}}[J]\right)^{\frac{1}{2}} \\
&+ \left(\int_{{}^{(ext)}\mathcal{M}\left(r\in I_{m_0,\delta_{\mathcal{H}}}\right)} |\mathfrak{d}^{k_{large}}\check{R}|^2\right)^{\frac{1}{4}} \left(\epsilon_0\left(\mathfrak{N}_{J+1}^{(En)} + \mathcal{N}_{J+1}^{(match)}\right)\right)^{\frac{1}{2}}
\end{aligned}$$

and hence

$$\begin{aligned}
\mathfrak{N}_{J+1}^{(En)} + \mathcal{N}_{J+1}^{(match)} &\lesssim \epsilon_{\mathcal{B}}[J] + \left(\int_{{}^{(ext)}\mathcal{M}\left(r\in I_{m_0,\delta_{\mathcal{H}}}\right)} |\mathfrak{d}^{k_{large}}\check{R}|^2\right)^{\frac{1}{4}} \left(\epsilon_{\mathcal{B}}[J]\right)^{\frac{1}{2}} \\
&\quad +\epsilon_0 \left(\int_{{}^{(ext)}\mathcal{M}\left(r\in I_{m_0,\delta_{\mathcal{H}}}\right)} |\mathfrak{d}^{k_{large}}\check{R}|^2\right)^{\frac{1}{2}}.
\end{aligned}$$

In view of Lemma 8.6, we deduce

$$\mathfrak{N}_{J+1}^{(En)} + \mathcal{N}_{J+1}^{(match)} \lesssim \epsilon_{\mathcal{B}}[J+1]$$

which is (8.3.13) for $J+1$ derivatives. We deduce that the estimate (8.3.13) holds for all $J \leq k_{large} - 1$, and hence

$$\mathfrak{N}_{k_{large}-1}^{(En)} + \mathcal{N}_{k_{large}-1}^{(match)} \lesssim \epsilon_{\mathcal{B}}[k_{large} - 1]. \tag{8.3.20}$$

Step 2. Next, Proposition 8.11 implies in view of (8.3.20)

$$ {}^{(ext)}\mathfrak{G}_{k_{large}}[\check{\Gamma}] + {}^{(int)}\mathfrak{R}_{k_{large}}[\check{R}] + {}^{(ext)}\mathfrak{R}_{k_{large}}[\check{R}] \lesssim \epsilon_{\mathcal{B}}[k_{large}-1] + \epsilon_0\Big(\mathfrak{N}^{(En)}_{k_{large}} + \mathcal{N}^{(match)}_{k_{large}}\Big). \tag{8.3.21} $$

In particular, we have

$$ \begin{aligned} \left(\int_{{}^{(ext)}\mathcal{M}\left(r\in I_{m_0,\delta_{\mathcal{H}}}\right)} |\mathfrak{d}^{\leq k_{large}}\check{R}|^2\right)^{\frac{1}{2}} &\leq {}^{(ext)}\mathfrak{R}_{k_{large}}[\check{R}] \\ &\lesssim \epsilon_{\mathcal{B}}[k_{large}-1] + \epsilon_0\Big(\mathfrak{N}^{(En)}_{k_{large}} + \mathcal{N}^{(match)}_{k_{large}}\Big). \end{aligned} $$

In view of the definition of $\epsilon_{\mathcal{B}}[k_{large}-1]$, we infer for $\epsilon_0 > 0$ small enough

$$ \left(\int_{{}^{(ext)}\mathcal{M}\left(r\in I_{m_0,\delta_{\mathcal{H}}}\right)} |\mathfrak{d}^{\leq k_{large}}\check{R}|^2\right)^{\frac{1}{2}} \lesssim \epsilon_0 + \epsilon_0\Big(\mathfrak{N}^{(En)}_{k_{large}} + \mathcal{N}^{(match)}_{k_{large}}\Big) $$

and hence

$$ \epsilon_{\mathcal{B}}[k_{large}-1] \lesssim \epsilon_0 + \epsilon_0\Big(\mathfrak{N}^{(En)}_{k_{large}} + \mathcal{N}^{(match)}_{k_{large}}\Big) $$

which yields, together with (8.3.21),

$$ \begin{aligned} & {}^{(ext)}\mathfrak{G}_{k_{large}}[\check{\Gamma}] + {}^{(int)}\mathfrak{R}_{k_{large}}[\check{R}] + {}^{(ext)}\mathfrak{R}_{k_{large}}[\check{R}] \\ \lesssim\ & \epsilon_0 + \epsilon_0\Big(\mathfrak{N}^{(En)}_{k_{large}} + \mathcal{N}^{(match)}_{k_{large}}\Big). \end{aligned} \tag{8.3.22} $$

Step 3. Next, Proposition 8.12 implies in view of (8.3.22)

$$ {}^{(int)}\mathfrak{G}_{k_{large}}[\check{\Gamma}] \lesssim \epsilon_0 + \epsilon_0\Big(\mathfrak{N}^{(En)}_{k_{large}} + \mathcal{N}^{(match)}_{k_{large}}\Big) + \left(\int_{\mathcal{T}} |\mathfrak{d}^{k_{large}}({}^{(ext)}\check{R})|^2\right)^{\frac{1}{2}} $$

and hence, for $\epsilon_0 > 0$ small enough, using again (8.3.22),

$$ \mathfrak{N}^{(En)}_{k_{large}} \lesssim \epsilon_0 + \epsilon_0\mathcal{N}^{(match)}_{k_{large}} + \left(\int_{\mathcal{T}} |\mathfrak{d}^{k_{large}}({}^{(ext)}\check{R})|^2\right)^{\frac{1}{2}}. $$

Together with Proposition 8.13, we infer for $\epsilon_0 > 0$ small enough

$$ \mathfrak{N}^{(En)}_{k_{large}} + \mathcal{N}^{(match)}_{k_{large}} \lesssim \epsilon_0 + \left(\int_{\mathcal{T}} |\mathfrak{d}^{J+1}({}^{(ext)}\check{R})|^2\right)^{\frac{1}{2}}. $$

Step 4. It remains to estimate the last term of the RHS of the previous inequality.

Now, in view of (8.3.7) and (8.3.22), we have

$$\begin{aligned}\left(\int_{\mathcal{T}}|\mathfrak{d}^{k_{large}}({}^{(ext)}\check{R})|^2\right)^{\frac{1}{2}} &\lesssim \left(\int_{{}^{(ext)}\mathcal{M}\left(r\in I_{m_0,\delta_{\mathcal{H}}}\right)}|\mathfrak{d}^{\leq k_{large}}\check{R}|^2\right)^{\frac{1}{2}}\\ &\lesssim {}^{(ext)}\mathfrak{R}_{k_{large}}[\check{R}]\\ &\lesssim \epsilon_0+\epsilon_0\mathfrak{N}^{(En)}_{k_{large}}\end{aligned}$$

so that we finally obtain, for $\epsilon_0>0$ small enough,

$$\mathfrak{N}^{(En)}_{k_{large}} \lesssim \epsilon_0.$$

This concludes the proof of Theorem M8.

8.4 PROOF OF PROPOSITION 8.7

8.4.1 A wave equation for $\tilde{\rho}$

Proposition 8.14. *The following wave equations hold true.*

1. *The curvature component ρ verifies the identity*

$$\square_{\mathbf{g}}\rho = \underline{\kappa}e_4\rho+\kappa e_3\rho+\frac{3}{2}\Big(\underline{\kappa}\kappa+2\rho\Big)\rho+Err[\square_{\mathbf{g}}\rho],$$

where

$$\begin{aligned}Err[\square_{\mathbf{g}}\rho] &:= \frac{3}{2}\rho\left(-\frac{1}{2}\underline{\vartheta}\,\vartheta+2(\underline{\xi}\,\xi+\eta\,\eta)\right)\\ &+\left(\frac{3}{2}\underline{\kappa}-2\underline{\omega}\right)\left(\frac{1}{2}\underline{\vartheta}\,\alpha-\zeta\,\beta-2(\underline{\eta}\,\beta+\xi\,\underline{\beta})\right)\\ &-\frac{1}{2}\underline{\vartheta}\,{\not{d}}_2^\star\beta+(\zeta-\eta)e_3\beta-\eta e_3(\Phi)\beta-\underline{\xi}(e_4\beta+e_4(\Phi)\beta)-\underline{\beta}\beta\\ &-e_3\left(-\frac{1}{2}\underline{\vartheta}\,\alpha+\zeta\,\beta+2(\underline{\eta}\,\beta+\xi\,\underline{\beta})\right)\\ &-{\not{d}}_1^\star(\underline{\kappa})\beta+2\,{\not{d}}_1^\star(\underline{\omega})\beta+3\eta\,{\not{d}}_1^\star(\rho)-{\not{d}}_1\Big(-\vartheta\underline{\beta}+\underline{\xi}\alpha\Big)-2\eta e_\theta\rho.\end{aligned}$$

2. *The small curvature quantity*

$$\tilde{\rho}:=r^2\left(\rho+\frac{2m}{r^3}\right)$$

verifies the wave equation

$$\begin{aligned}\square_{\mathbf{g}}(\tilde{\rho})+\frac{8m}{r^3}\tilde{\rho} &= -6m\frac{\square_{\mathbf{g}}(r)-(\frac{2}{r}-\frac{2m}{r^2})}{r^2}-\frac{3m}{r}\left(\kappa\underline{\kappa}+\frac{4\Upsilon}{r^2}\right)\\ &-\frac{3m}{r}\left(A\underline{\kappa}+\underline{A}\kappa\right)+Err[\square_g\tilde{\rho}],\end{aligned}$$

where

$$
\begin{aligned}
Err[\square_g \tilde{\rho}] \quad := \quad & -\frac{6m}{r} A\underline{A} + \frac{3}{r^2}\tilde{\rho}^2 + \frac{3}{2}\left(\frac{4}{3}A\frac{e_3(r)}{r} + \frac{4}{3}\underline{A}\frac{e_4(r)}{r}\right)\tilde{\rho} \\
& + \left(\frac{3}{2}\left(\kappa\underline{\kappa} - \frac{8m}{r^3} + \frac{2}{3r^2}\square_{\mathbf{g}}(r^2)\right) + \frac{8m}{r^3}\right)\tilde{\rho} \\
& - Ae_3(\tilde{\rho}) - \underline{A}e_4(\tilde{\rho}) + \frac{2}{r}Ae_3(m) + \frac{2}{r}\underline{A}e_4(m) \\
& + 4D^a(m)D_a\left(\frac{1}{r}\right) + \frac{2}{r}\square_{\mathbf{g}}(m) + 4r\,\not{d}_1^\star(r)\,\not{d}_1^\star(\rho) + r^2 Err[\square_{\mathbf{g}}\rho],
\end{aligned}
$$

and where we recall that

$$
A = \frac{2}{r}e_4(r) - \kappa, \qquad \underline{A} = \frac{2}{r}e_3(r) - \underline{\kappa}.
$$

Proof. See section B.1. □

8.4.2 Control of $\square_{\mathbf{g}}(r)$

Lemma 8.15. *Let r the function on $\mathcal{M}$ associated to the global frame constructed in Proposition 8.4, see Definition 4.14. Let J such that $k_{small} - 2 \leq J \leq k_{large} - 1$. Under the iteration assumption (8.3.13), we have*

$$
\begin{aligned}
& \int_{{}^{(int)}\mathcal{M}\cup{}^{(ext)}\mathcal{M}(r\leq 4m_0)} \left(\mathfrak{d}^J\left(\square_{\mathbf{g}}(r) - \left(\frac{2}{r} - \frac{2m}{r^2}\right)\right)\right)^2 \\
& + \sup_{r_0\geq 4m_0} \int_{\{r=r_0\}} \left(\mathfrak{d}^J\left(\square_{\mathbf{g}}(r) - \left(\frac{2}{r} - \frac{2m}{r^2}\right)\right)\right)^2 \\
\lesssim \quad & (\epsilon_{\mathcal{B}}[J])^2 + \epsilon_0^2\left(\mathfrak{N}^{(En)}_{J+1} + \mathcal{N}^{(match)}_{J+1}\right)^2
\end{aligned}
$$

and

$$
\begin{aligned}
& \int_{{}^{(trap)}\mathcal{M}} \left(\mathfrak{d}^J e_4\left(\square_{\mathbf{g}}({}^{(ext)}r) - \left(\frac{2}{{}^{(ext)}r} - \frac{2\,{}^{(ext)}m}{({}^{(ext)}r)^2}\right)\right)\right)^2 \\
\lesssim \quad & (\epsilon_{\mathcal{B}}[J])^2 + \epsilon_0^2\left(\mathfrak{N}^{(En)}_{J+1} + \mathcal{N}^{(match)}_{J+1}\right)^2.
\end{aligned}
$$

Proof. Recall that, according to Definition 4.14, r is defined on ${}^{(ext)}\mathcal{M} \cup {}^{(int)}\mathcal{M}$ as follows:

- on ${}^{(ext)}\mathcal{M} \setminus \text{Match}$, we have

$$
{}^{(glo)}r = {}^{(ext)}r,
$$

- on ${}^{(int)}\mathcal{M} \setminus \text{Match}$, we have

$$
{}^{(glo)}r = {}^{(int)}r,
$$

- on the matching region, we have

$$
{}^{(glo)}r = (1-\psi_{m_0,\delta_{\mathcal{H}}}({}^{(int)}r))\,{}^{(int)}r + \psi_{m_0,\delta_{\mathcal{H}}}({}^{(int)}r)\,{}^{(ext)}r,
$$

where the matching region of Proposition 8.4 is given by

$$
\begin{aligned}
\text{Match} &:= \left({}^{(ext)}\mathcal{M}\cap\left\{{}^{(int)}r\leq 2m_0\left(1+\frac{3}{2}\delta_{\mathcal{H}}\right)\right\}\right)\\
&\quad\cup\left({}^{(int)}\mathcal{M}\cap\left\{{}^{(int)}r\geq 2m_0\left(1+\frac{1}{2}\delta_{\mathcal{H}}\right)\right\}\right),
\end{aligned}
$$

and where $\psi_{m_0,\delta_{\mathcal{H}}}$ is given by

$$
\psi_{m_0,\delta_{\mathcal{H}}}(r) = \psi\left(\frac{r-2m_0\left(1+\frac{1}{2}\delta_{\mathcal{H}}\right)}{2m_0\delta_{\mathcal{H}}}\right) \text{ on } 2m_0\left(1+\frac{1}{2}\delta_{\mathcal{H}}\right)\leq r\leq 2m_0\left(1+\frac{3}{2}\delta_{\mathcal{H}}\right)
$$

with $\psi:\mathbb{R}\to\mathbb{R}$ a smooth cut-off function such that $0\leq\psi\leq 1$, $\psi=0$ on $(-\infty,0]$ and $\psi=1$ on $[1,+\infty)$.

We have on ${}^{(ext)}\mathcal{M}$

$$
\begin{aligned}
\square_{\mathbf{g}}({}^{(ext)}r) &= -e_3e_4({}^{(ext)}r) + \Delta\!\!\!\!/\,({}^{(ext)}r) + \left(2\underline{\omega}-\frac{1}{2}\underline{\kappa}\right)e_4({}^{(ext)}r)\\
&\quad -\frac{1}{2}\kappa e_3({}^{(ext)}r) + 2\eta e_\theta({}^{(ext)}r).
\end{aligned}
$$

Here, (e_4, e_3, e_θ) denotes the frame of ${}^{(ext)}\mathcal{M}$ and the Ricci coefficients are computed w.r.t. this frame, so we have

$$
e_4({}^{(ext)}r) = \frac{{}^{(ext)}r}{2}\overline{\kappa}, \qquad e_3({}^{(ext)}r) = \frac{{}^{(ext)}r}{2}(\overline{\underline{\kappa}}+\underline{A}), \qquad e_\theta({}^{(ext)}r) = 0
$$

and hence

$$
\begin{aligned}
\square_{\mathbf{g}}({}^{(ext)}r) &= -e_3\left(\frac{{}^{(ext)}r}{2}\overline{\kappa}\right) + \left(2\underline{\omega}-\frac{1}{2}\underline{\kappa}\right)\frac{{}^{(ext)}r}{2}\overline{\kappa} - \frac{1}{2}\kappa\frac{{}^{(ext)}r}{2}(\overline{\underline{\kappa}}+\underline{A})\\
&= -\frac{{}^{(ext)}r}{2}e_3(\overline{\kappa}) - \frac{e_3({}^{(ext)}r)}{2}\overline{\kappa} + \left(2\underline{\omega}-\frac{1}{2}\underline{\kappa}\right)\frac{{}^{(ext)}r}{2}\overline{\kappa} - \frac{{}^{(ext)}r}{4}\kappa\overline{\underline{\kappa}}\\
&\quad -\frac{{}^{(ext)}r}{4}\kappa\underline{A}\\
&= -\frac{{}^{(ext)}r}{2}e_3(\overline{\kappa}) - \frac{1}{2}\overline{\kappa}\frac{{}^{(ext)}r}{2}(\overline{\underline{\kappa}}+\underline{A}) + \left(2\underline{\omega}-\frac{1}{2}\underline{\kappa}\right)\frac{{}^{(ext)}r}{2}\overline{\kappa}\\
&\quad -\frac{{}^{(ext)}r}{4}\kappa\overline{\underline{\kappa}} - \frac{{}^{(ext)}r}{4}\kappa\underline{A}.
\end{aligned}
$$

Now, we have

$$\begin{aligned} e_3(\overline{\kappa}) &= \overline{e_3(\kappa)} + \mathrm{Err}[e_3\overline{\kappa}] \\ &= \overline{-\frac{1}{2}\kappa\underline{\kappa} + 2\underline{\omega}\kappa + 2\rho + 2\,{\not d}_1\eta - \frac{1}{2}\vartheta\underline{\vartheta} + 2\eta^2} + \mathrm{Err}[e_3\overline{\kappa}] \\ &= \overline{-\frac{1}{2}\kappa\underline{\kappa} + 2\underline{\omega}\kappa + 2\rho - \frac{1}{2}\vartheta\underline{\vartheta} + 2\eta^2} + \mathrm{Err}[e_3\overline{\kappa}] \end{aligned}$$

and hence

$$\begin{aligned} \square_{\mathbf{g}}({}^{(ext)}r) &= -\frac{{}^{(ext)}r}{2}\left(\overline{-\frac{1}{2}\kappa\underline{\kappa} + 2\underline{\omega}\kappa + 2\rho - \frac{1}{2}\vartheta\underline{\vartheta} + 2\eta^2} + \mathrm{Err}[e_3\overline{\kappa}]\right) \\ & -\frac{1}{2}\overline{\kappa}\frac{{}^{(ext)}r}{2}(\overline{\underline{\kappa}} + \underline{A}) + \left(2\underline{\omega} - \frac{1}{2}\underline{\kappa}\right)\frac{{}^{(ext)}r}{2}\overline{\kappa} - \frac{{}^{(ext)}r}{4}\overline{\kappa\underline{\kappa}} - \frac{{}^{(ext)}r}{4}\kappa\underline{A}. \end{aligned}$$

Together with (8.3.4) and the iteration assumption (8.3.13), we easily infer[5]

$$\begin{aligned} & \int_{{}^{(ext)}\mathcal{M}(r\le 4m_0)}\left(\mathfrak{d}^J\left(\square_{\mathbf{g}}({}^{(ext)}r) - \left(\frac{2}{{}^{(ext)}r} - \frac{2\,{}^{(ext)}m}{({}^{(ext)}r)^2}\right)\right)\right)^2 \\ & + \sup_{r_0\ge 4m_0}\int_{\{r=r_0\}}\left(\mathfrak{d}^J\left(\square_{\mathbf{g}}({}^{(ext)}r) - \left(\frac{2}{{}^{(ext)}r} - \frac{2\,{}^{(ext)}m}{({}^{(ext)}r)^2}\right)\right)\right)^2 \\ \lesssim\; & (\epsilon_{\mathcal{B}}[J])^2 + \epsilon_0^2\left(\mathfrak{N}^{(En)}_{J+1} + \mathcal{N}^{(match)}_{J+1}\right)^2. \end{aligned} \tag{8.4.1}$$

Also, using again (8.3.4) and the iteration assumption (8.3.13), we have

$$\begin{aligned} & \int_{{}^{(trap)}\mathcal{M}}\left(\mathfrak{d}^J e_4\left(\square_{\mathbf{g}}({}^{(ext)}r) - \left(\frac{2}{{}^{(ext)}r} - \frac{2\,{}^{(ext)}m}{({}^{(ext)}r)^2}\right)\right)\right)^2 \\ \lesssim\; & (\epsilon_{\mathcal{B}}[J])^2 + \epsilon_0^2\left(\mathfrak{N}^{(En)}_{J+1} + \mathcal{N}^{(match)}_{J+1}\right)^2, \end{aligned} \tag{8.4.2}$$

where we have used the null structure equations for $e_4(\kappa)$, $e_4(\underline{\kappa})$, $e_4(\underline{\omega})$, $e_4(\vartheta)$, $e_4(\underline{\vartheta})$, $e_4(\eta)$, the equations for $e_4(\underline{\Omega})$,$e_4(\varsigma)$, $e_4(r)$, and the Bianchi identity for $e_4(\rho)$.

Remark 8.16. *Note that we have used in the last estimate the following observations to avoid a potential loss of one derivative:*

$$\begin{aligned} e_4(\underline{\kappa}) &= -2\,{\not d}_1\zeta + \cdots = 2\left(\rho + \mu - \frac{1}{4}\vartheta\underline{\vartheta}\right) + \cdots, \\ \overline{e_4(\rho)} &= \overline{{\not d}_1\beta} + \cdots = \cdots, \\ e_4(Err[e_3\overline{\kappa}]) &= 2e_4(\varsigma^{-1}\overline{\check{\varsigma}\,{\not d}_1\eta}) + \cdots = 2\varsigma^{-1}\overline{\check{\varsigma}\,{\not d}_1 e_4\eta} + \cdots = -2\varsigma^{-1}\overline{e_\theta(\varsigma)e_4\eta} + \cdots. \end{aligned}$$

Note also that there is no term involving $\mathfrak{d}^J\rho$ (without average) as such a term appears only in the null structure equations for $e_4(\underline{\kappa})$, as well as $e_4(\underline{\omega})$, and vanishes

[5]Recall in particular that $\overline{\rho}$ is under control in view of Lemma 3.15.

due to the cancellation

$$\begin{aligned} e_4\left(2\underline{\omega}-\frac{1}{2}\underline{\kappa}\right) &= 2e_4(\underline{\omega})-\frac{1}{2}e_4(\underline{\kappa}) \\ &= 2\rho+\cdots-\frac{1}{2}(-2\,\mathcal{d}_1\zeta+2\rho)+\cdots \\ &= 2\mu+\cdots. \end{aligned}$$

This is important as such a term would otherwise violate (8.4.2) at $r=3m$.

Remark 8.17. *Recall that the global null frame constructed in Proposition 8.4*

- *coincides with the null frame of ${}^{(int)}\mathcal{M}$ in the region ${}^{(int)}\mathcal{M}\setminus Match$,*
- *coincides with a conformal renormalization of the null frame of ${}^{(ext)}\mathcal{M}$ in the region ${}^{(ext)}\mathcal{M}\setminus Match$.*

Thus, $J+1$ derivatives of its Ricci coefficients and curvature components are controlled

- *by $\mathcal{N}^{(match)}_{J+1}$ in Match,*
- *by $\mathfrak{N}^{(En)}_{J+1}$ in $\mathcal{M}\setminus Match$,*

and hence by $\mathfrak{N}^{(En)}_{J+1}+\mathcal{N}^{(match)}_{J+1}$ on $\mathcal{M}$. This explains the occurrence of the term $\mathfrak{N}^{(En)}_{J+1}+\mathcal{N}^{(match)}_{J+1}$ on the right-hand side of numerous estimates, see for example (8.4.1), (8.4.2).

Arguing similarly for ${}^{(int)}r$, we obtain the following analog of (8.4.1):

$$\begin{aligned} &\int_{{}^{(int)}\mathcal{M}}\left(\mathfrak{d}^J\left(\square_{\mathbf{g}}({}^{(int)}r)-\left(\frac{2}{{}^{(int)}r}-\frac{2\,{}^{(int)}m}{({}^{(int)}r)^2}\right)\right)\right)^2 \\ &\lesssim (\epsilon_{\mathcal{B}}[J])^2+\epsilon_0^2\left(\mathfrak{N}^{(En)}_{J+1}+\mathcal{N}^{(match)}_{J+1}\right)^2. \end{aligned} \tag{8.4.3}$$

Then, since

- on ${}^{(ext)}\mathcal{M}\setminus$ Match, we have

$$\square_{\mathbf{g}}(r)=\square_{\mathbf{g}}({}^{(ext)}r), \qquad m={}^{(ext)}m,$$

- on ${}^{(int)}\mathcal{M}\setminus$ Match, we have

$$\square_{\mathbf{g}}(r)=\square_{\mathbf{g}}({}^{(int)}r), \qquad m={}^{(int)}m,$$

we immediately infer from (8.4.1), (8.4.2) and (8.4.3)

$$\begin{aligned} &\int_{\left({}^{(int)}\mathcal{M}\cup{}^{(ext)}\mathcal{M}(r\le 4m_0)\right)\setminus\text{Match}}\left(\mathfrak{d}^J\left(\square_{\mathbf{g}}(r)-\left(\frac{2}{r}-\frac{2m}{r^2}\right)\right)\right)^2 \\ &+\sup_{r_0\ge 4m_0}\int_{\{r=r_0\}}\left(\mathfrak{d}^J\left(\square_{\mathbf{g}}(r)-\left(\frac{2}{r}-\frac{2m}{r^2}\right)\right)\right)^2 \\ &\lesssim (\epsilon_{\mathcal{B}}[J])^2+\epsilon_0^2\left(\mathfrak{N}^{(En)}_{J+1}+\mathcal{N}^{(match)}_{J+1}\right)^2 \end{aligned}$$

and

$$\int_{^{(trap)}\mathcal{M}} \left(\mathfrak{d}^J e_4\left(\square_{\mathbf{g}}(r) - \left(\frac{2}{r} - \frac{2m}{r^2}\right)\right)\right)^2 \lesssim (\epsilon_{\mathcal{B}}[J])^2 + \epsilon_0^2\left(\mathfrak{N}^{(En)}_{J+1} + \mathcal{N}^{(match)}_{J+1}\right)^2$$

which are the desired estimates outside of the matching region. Note that we have used the fact that ${}^{(trap)}\mathcal{M} \cap \text{Match} = \emptyset$.

It remains to derive the desired estimates in the matching region. To this end, we need to estimate ${}^{(ext)}r - {}^{(int)}r$ and ${}^{(int)}m - {}^{(ext)}m$ in the matching region. Step 7 or the proof of Lemma 4.16 in section 4.6.2 yields[6]

$$\int_{^{(int)}\mathcal{M}} \left(\mathfrak{d}^{J+1}\left({}^{(ext)}r - {}^{(int)}r,\ {}^{(ext)}m - {}^{(int)}m\right)\right)^2 \lesssim (\mathfrak{N}^{(En)}_J)^2 + (\mathcal{N}^{(match)}_J)^2.$$

We infer, in view of the iteration assumption (8.3.13),

$$\int_{^{(int)}\mathcal{M}} \left(\mathfrak{d}^{J+1}\left({}^{(ext)}r - {}^{(int)}r,\ {}^{(ext)}m - {}^{(int)}m\right)\right)^2 \lesssim (\epsilon_{\mathcal{B}}[J])^2. \tag{8.4.4}$$

Then, since we have on the matching region

$$\begin{aligned}
r &= (1 - \psi_{m_0,\delta_{\mathcal{H}}}({}^{(int)}r))\,{}^{(int)}r + \psi_{m_0,\delta_{\mathcal{H}}}({}^{(int)}r)\,{}^{(ext)}r,\\
m &= (1 - \psi_{m_0,\delta_{\mathcal{H}}}({}^{(int)}r))\,{}^{(int)}m + \psi_{m_0,\delta_{\mathcal{H}}}({}^{(int)}r)\,{}^{(ext)}m,\\
\square_{\mathbf{g}}(r) &= (1 - \psi_{m_0,\delta_{\mathcal{H}}}({}^{(int)}r))\square_{\mathbf{g}}({}^{(int)}r) + \psi_{m_0,\delta_{\mathcal{H}}}({}^{(int)}r)\square_{\mathbf{g}}({}^{(ext)}r)\\
&\quad +2\psi'_{m_0,\delta_{\mathcal{H}}}({}^{(int)}r)\mathbf{D}^\alpha({}^{(int)}r)\mathbf{D}_\alpha({}^{(ext)}r - {}^{(int)}r)\\
&\quad +({}^{(ext)}r - {}^{(int)}r)\square_{\mathbf{g}}(\psi_{m_0,\delta_{\mathcal{H}}}),
\end{aligned}$$

we deduce there

$$\begin{aligned}
&\square_{\mathbf{g}}(r) - \left(\frac{2}{r} - \frac{2m}{r^2}\right)\\
&= (1 - \psi_{m_0,\delta_{\mathcal{H}}}({}^{(int)}r))\left(\square_{\mathbf{g}}({}^{(int)}r) - \left(\frac{2}{{}^{(int)}r} - \frac{2\,{}^{(int)}m}{({}^{(int)}r)^2}\right)\right)\\
&\quad +\psi_{m_0,\delta_{\mathcal{H}}}({}^{(int)}r)\left(\square_{\mathbf{g}}({}^{(ext)}r) - \left(\frac{2}{{}^{(ext)}r} - \frac{2\,{}^{(ext)}m}{({}^{(ext)}r)^2}\right)\right)\\
&\quad +(1 - \psi_{m_0,\delta_{\mathcal{H}}}({}^{(int)}r))\left(\frac{2}{{}^{(int)}r} - \frac{2}{r} - \frac{2\,{}^{(int)}m}{({}^{(int)}r)^2} + \frac{2m}{r^2}\right)\\
&\quad +\psi_{m_0,\delta_{\mathcal{H}}}({}^{(int)}r)\left(\frac{2}{{}^{(ext)}r} - \frac{2}{r} - \frac{2\,{}^{(ext)}m}{({}^{(ext)}r)^2} + \frac{2m}{r^2}\right)\\
&\quad +2\psi'_{m_0,\delta_{\mathcal{H}}}({}^{(int)}r)\mathbf{D}^\alpha({}^{(int)}r)\mathbf{D}_\alpha({}^{(ext)}r - {}^{(int)}r)\\
&\quad +({}^{(ext)}r - {}^{(int)}r)\square_{\mathbf{g}}(\psi_{m_0,\delta_{\mathcal{H}}})
\end{aligned}$$

[6]The proof of Lemma 4.16 in section 4.6.2 is done in the particular case $J = k_{large} - 1$ but extends immediately to the case $k_{small} - 2 \leq J \leq k_{large} - 1$.

and thus, in view of (8.4.1), (8.4.3) and (8.4.4), we have on the matching region

$$\int_{\text{Match}} \left(\mathfrak{d}^J \left(\square_{\mathbf{g}}(r) - \left(\frac{2}{r} - \frac{2m}{r^2} \right) \right) \right)^2 \lesssim (\epsilon_{\mathcal{B}}[J])^2 + \epsilon_0^2 \left(\mathfrak{N}^{(En)}_{J+1} + \mathcal{N}^{(match)}_{J+1} \right)^2$$

as desired. This concludes the proof of the lemma. □

Corollary 8.18. *Let N_0 the RHS of the wave equation for $\tilde{\rho}$ provided by Proposition 8.14, i.e.,*

$$N_0 = -6m \frac{\square_{\mathbf{g}}(r) - \left(\frac{2}{r} - \frac{2m}{r^2}\right)}{r^2} - \frac{3m}{r}\left(\kappa\underline{\kappa} + \frac{4\Upsilon}{r^2}\right) - \frac{3m}{r}\left(A\underline{\kappa} + \underline{A}\kappa\right) + Err[\square_g \tilde{\rho}].$$

Then, $N_0 - Err[\square_g \tilde{\rho}]$ satisfies

$$\begin{aligned} &\int_{^{(int)}\mathcal{M} \cup \, ^{(ext)}\mathcal{M}(r \le 4m_0)} \left(\mathfrak{d}^J \left(N_0 - Err[\square_g \tilde{\rho}] \right) \right)^2 \\ &+ \sup_{r_0 \ge 4m_0} \int_{\{r=r_0\}} \left(\mathfrak{d}^J \left(N_0 - Err[\square_g \tilde{\rho}] \right) \right)^2 \\ \lesssim \;& (\epsilon_{\mathcal{B}}[J])^2 + \epsilon_0^2 \left(\mathfrak{N}^{(En)}_{J+1} + \mathcal{N}^{(match)}_{J+1} \right)^2 \end{aligned}$$

and

$$\mathfrak{d}^J e_4 \left(N_0 - Err[\square_g \tilde{\rho}] \right) = -\frac{12m\kappa}{r} \mathfrak{d}^J \rho + a_J \text{ on } ^{(trap)}\mathcal{M}$$

where a^J satisfies

$$\int_{^{(trap)}\mathcal{M}} \left(\mathfrak{d}^J e_4 a^J \right)^2 \lesssim (\epsilon_{\mathcal{B}}[J])^2 + \epsilon_0^2 \left(\mathfrak{N}^{(En)}_{J+1} + \mathcal{N}^{(match)}_{J+1} \right)^2.$$

Proof. The first estimate is an immediate consequence of Lemma 8.15, (8.3.4) and the iteration assumption (8.3.13).

Concerning the second estimate, note that the term $\mathfrak{d}^J \rho$ is due to the null structure equations for $e_4(\underline{\kappa})$, i.e.,

$$\begin{aligned} e_4(\underline{\kappa}) &= -2 \, \not{d}_1 \zeta + 2\rho + \cdots \\ &= 4\rho + \cdots . \end{aligned}$$

Then, the estimate for a_J follows from Lemma 8.15, (8.3.4) and the iteration assumption (8.3.13). □

8.4.3 End of the proof of Proposition 8.7

In view of Proposition 8.14, $\tilde{\rho}$ satisfies

$$(\square_0 + V_0)\tilde{\rho} = N_0, \qquad V_0 = \frac{8m}{r^3},$$

where

$$N_0 \ := \ -6m\frac{\square_{\mathbf{g}}(r) - (\frac{2}{r} - \frac{2m}{r^2})}{r^2} - \frac{3m}{r}\left(\kappa\underline{\kappa} + \frac{4\Upsilon}{r^2}\right) - \frac{3m}{r}\left(A\underline{\kappa} + \underline{A}\kappa\right) + \mathrm{Err}[\square_g\tilde{\rho}].$$

We may thus apply the estimate (10.5.2) of Theorem 10.67 with $\phi = \tilde{\rho}$ and $s = J$ to obtain for any $k_{small} \leq J \leq k_{large} - 1$

$$\begin{aligned}
&\sup_{\tau\in[1,\tau_*]} E^J_\delta[\tilde{\rho}](\tau) + B^J_\delta[\tilde{\rho}](1,\tau_*) + F^J_\delta[\tilde{\rho}](1,\tau_*)\\
&\lesssim \ E^J_\delta[\tilde{\rho}](1) + \sup_{\tau\in[1,\tau_*]} E^{J-1}_\delta[\tilde{\rho}](\tau) + B^{J-1}_\delta[\tilde{\rho}](1,\tau_*) + F^{J-1}_\delta[\tilde{\rho}](1,\tau_*)\\
&\quad + D_J[\Gamma]\left(\sup_{\mathcal{M}} r u_{trap}^{\frac{1}{2}+\delta_{dec}}|\mathfrak{d}^{\leq k_{small}}\tilde{\rho}|\right)^2 + \int_{\Sigma(\tau_*)}\frac{(\mathfrak{d}^{\leq J}\tilde{\rho})^2}{r^3}\\
&\quad + \int_{\mathcal{M}} r^{1+\delta}|\mathfrak{d}^{\leq J}N_0|^2 + \left|\int_{{}^{(trap)}\mathcal{M}} T(\mathfrak{d}^J\tilde{\rho})\mathfrak{d}^J N_0\right|,
\end{aligned}$$

where $D_J[\Gamma]$ is defined by

$$\begin{aligned}
D_J[\Gamma] \ := \ & \int_{{}^{(int)}\mathcal{M}\cup{}^{(ext)}\mathcal{M}(r\leq 4m_0)} (\mathfrak{d}^{\leq J}\check{\Gamma})^2\\
& + \sup_{r_0\geq 4m_0}\left(r_0\int_{\{r=r_0\}}|\mathfrak{d}^{\leq J}\Gamma_g|^2 + r_0^{-1}\int_{\{r=r_0\}}|\mathfrak{d}^{\leq J}\Gamma_b|^2\right).
\end{aligned}$$

Next we use the iteration assumption (8.3.13) which yields in particular

$$D_J[\Gamma] \ \lesssim \ (\epsilon_{\mathcal{B}}[J])^2.$$

Also, we have

$$\tilde{\rho} \ = \ r^2\left(\overline{\rho} - \frac{2m}{r^3}\right) + r^2\check{\rho}$$

and hence, using again the iteration assumption (8.3.13), as well as the control on averages provided by Lemma 3.15, we infer

$$\sup_{\tau\in[1,\tau_*]} E^{J-1}_\delta[\tilde{\rho}](\tau) + B^{J-1}_\delta[\tilde{\rho}](1,\tau_*) + F^{J-1}_\delta[\tilde{\rho}](1,\tau_*) + \int_{\Sigma(\tau_*)}\frac{(\mathfrak{d}^{\leq J}\tilde{\rho})^2}{r^3} \lesssim (\epsilon_{\mathcal{B}}[J])^2.$$

Together with the control of $\mathfrak{d}^{\leq k_{small}}\tilde{\rho}$ provided by the decay estimate (8.3.6), we infer from the above estimates

$$\begin{aligned}
&\sup_{\tau\in[1,\tau_*]} E^J_\delta[\tilde{\rho}](\tau) + B^J_\delta[\tilde{\rho}](1,\tau_*) + F^J_\delta[\tilde{\rho}](1,\tau_*)\\
&\lesssim \ E^J_\delta[\tilde{\rho}](1) + (\epsilon_{\mathcal{B}}[J])^2 + \int_{\mathcal{M}} r^{1+\delta}|\mathfrak{d}^{\leq J}N_0|^2 + \left|\int_{{}^{(trap)}\mathcal{M}} T(\mathfrak{d}^J\tilde{\rho})\mathfrak{d}^J N_0\right|.
\end{aligned}$$

Next, using the form of N_0, as well as Corollary 8.18, we derive

$$\int_{\mathcal{M}} r^{1+\delta}|\mathfrak{d}^{\leq J}N_0|^2 \ \lesssim \ (\epsilon_{\mathcal{B}}[J])^2 + \epsilon_0^2\left(\mathfrak{N}^{(En)}_{J+1} + \mathcal{N}^{(match)}_{J+1}\right)^2.$$

Also, decomposing T as a combination of R and e_4, integrating e_4 by parts, using again the form of N_0, as well as Corollary 8.18, we have

$$\begin{aligned}
&\left|\int_{^{(trap)}\mathcal{M}} T(\mathfrak{d}^J\tilde{\rho})\mathfrak{d}^J N_0\right| \\
\lesssim\ & \left|\int_{^{(trap)}\mathcal{M}} R(\mathfrak{d}^J\tilde{\rho})\mathfrak{d}^J(N_0 - \mathrm{Err}[\square_g\tilde{\rho}])\right| + \left|\int_{^{(trap)}\mathcal{M}} e_4(N_0 - \mathrm{Err}[\square_g\tilde{\rho}])\mathfrak{d}^J N_0\right| \\
&+ \int_{^{(trap)}\mathcal{M}} |T(\mathfrak{d}^J\tilde{\rho})||\mathfrak{d}^J\mathrm{Err}[\square_g\tilde{\rho}]| \\
\lesssim\ & \left|\int_{^{(trap)}\mathcal{M}} \mathfrak{d}^J\tilde{\rho}e_4(\mathfrak{d}^J(N_0 - \mathrm{Err}[\square_g\tilde{\rho}]))\right| \\
&+ \left(\epsilon_{\mathcal{B}}[J] + \epsilon_0\left(\mathfrak{N}^{(En)}_{J+1} + \mathcal{N}^{(match)}_{J+1}\right)\right)\left(\sup_{\tau\in[1,\tau_*]} E^J_\delta[\tilde{\rho}](\tau) + B^J_\delta[\tilde{\rho}](1,\tau_*)\right)^{\frac{1}{2}} \\
\lesssim\ & \int_{^{(trap)}\mathcal{M}} (\mathfrak{d}^J\tilde{\rho})^2 \\
&+ \left(\epsilon_{\mathcal{B}}[J] + \epsilon_0\left(\mathfrak{N}^{(En)}_{J+1} + \mathcal{N}^{(match)}_{J+1}\right)\right)\left(\sup_{\tau\in[1,\tau_*]} E^J_\delta[\tilde{\rho}](\tau) + B^J_\delta[\tilde{\rho}](1,\tau_*)\right)^{\frac{1}{2}}.
\end{aligned}$$

In view of the above, we infer

$$\begin{aligned}
&\sup_{\tau\in[1,\tau_*]} E^J_\delta[\tilde{\rho}](\tau) + B^J_\delta[\tilde{\rho}](1,\tau_*) + F^J_\delta[\tilde{\rho}](1,\tau_*) \\
\lesssim\ & \int_{^{(trap)}\mathcal{M}} (\mathfrak{d}^J\tilde{\rho})^2 + (\epsilon_{\mathcal{B}}[J])^2 + \epsilon_0^2\left(\mathfrak{N}^{(En)}_{J+1} + \mathcal{N}^{(match)}_{J+1}\right)^2.
\end{aligned}$$

Next, note that we have

$$\begin{aligned}
R(r-3m) &= \frac{1}{2}(e_4(r) - \Upsilon e_3(r)) - \frac{3}{2}(e_4(m) - \Upsilon e_3(m)) \\
&= \Upsilon + O(\epsilon_0) \geq \frac{1}{6} \text{ on } {}^{(trap)}\mathcal{M},
\end{aligned}$$

and hence, using also integration by parts,

$$\begin{aligned}
\int_{^{(trap)}\mathcal{M}} (\mathfrak{d}^J\tilde{\rho})^2 &\lesssim \int_{^{(trap)}\mathcal{M}} R(r-3m)(\mathfrak{d}^J\tilde{\rho})^2 \\
&\lesssim \int_{^{(trap)}\mathcal{M}} \left|1 - \frac{3m}{r}\right| |\mathfrak{d}^J\tilde{\rho}||R\mathfrak{d}^J\rho| \\
&\lesssim \epsilon_{\mathcal{B}}[J]\left(\sup_{\tau\in[1,\tau_*]} E^J_\delta[\tilde{\rho}](\tau) + B^J_\delta[\tilde{\rho}](1,\tau_*)\right)^{\frac{1}{2}}.
\end{aligned}$$

We deduce

$$\sup_{\tau\in[1,\tau_*]} E^J_\delta[\tilde{\rho}](\tau) + B^J_\delta[\tilde{\rho}](1,\tau_*) + F^J_\delta[\tilde{\rho}](1,\tau_*) \lesssim (\epsilon_{\mathcal{B}}[J])^2 + \epsilon_0^2\left(\mathfrak{N}^{(En)}_{J+1} + \mathcal{N}^{(match)}_{J+1}\right)^2$$

as desired. This concludes the proof of Proposition 8.7.

8.5 PROOF OF PROPOSITION 8.8

8.5.1 A wave equation for $\alpha+\Upsilon^2\underline{\alpha}$

Lemma 8.19. *We have*

$$
\begin{aligned}
\square_2(\alpha+\Upsilon^2\underline{\alpha}) &= \frac{4}{r}\left(1-\frac{3m}{r}\right)\Big(e_3(\alpha)-\Upsilon e_4(\underline{\alpha})\Big)+\left(-\frac{2}{r^2}+\frac{16m}{r^3}\right)\alpha\\
&\quad -\frac{2\Upsilon}{r^2}\left(1-\frac{2m}{r}-\frac{8m^2}{r^2}\right)\underline{\alpha}+Err\Big[\square_2(\alpha+\Upsilon^2\underline{\alpha})\Big]
\end{aligned}
$$

where

$$
\begin{aligned}
& Err\Big[\square_2(\alpha+\Upsilon^2\underline{\alpha})\Big]\\
=\; & \left(\Upsilon^2\underline{V}+\frac{4m}{r^2}\Upsilon\square_g(r)-\frac{8m}{r^3}\Upsilon\mathbf{D}^\alpha(r)\mathbf{D}_\alpha(r)+\frac{8m^2}{r^4}\mathbf{D}^\alpha(r)\mathbf{D}_\alpha(r)\right)\underline{\alpha}\\
& +\left(4\left(\omega+\frac{m}{r^2}\right)+2\left(\kappa-\frac{2\Upsilon}{r}\right)\right)e_3(\alpha)-4\underline{\omega}e_4(\alpha)\\
& -4\Upsilon\left(\Upsilon\left(\omega+\frac{m}{r^2}\right)+\frac{m}{r^2}(e_4(r)-1)-\frac{e_4(m)}{r}\right)e_3(\underline{\alpha})\\
& +\left(4\Upsilon^2\underline{\omega}+2\Upsilon^2\left(\underline{\kappa}+\frac{2}{r}\right)-4m\Upsilon\frac{(e_3(r)+1)}{r^2}+4\Upsilon\frac{e_3(m)}{r}\right)e_4(\underline{\alpha})\\
& +\Bigg\{\left(-4\rho-\frac{8m}{r^3}\right)+2\left(\omega\,\underline{\kappa}-\frac{2m}{r^3}\right)+\frac{1}{2}\left(\kappa\,\underline{\kappa}+\frac{4\Upsilon}{r^2}\right)-4e_4(\underline{\omega})-8\omega\underline{\omega}\\
& -10\kappa\,\underline{\omega}\Bigg\}\alpha+\Bigg\{\frac{8\Upsilon}{r^2}\mathbf{D}^\alpha(m)\mathbf{D}_\alpha(r)-\frac{4\Upsilon}{r}\square_g(m)-\frac{8m}{r^3}\mathbf{D}^\alpha(r)\mathbf{D}_\alpha(m)\\
& +\frac{8m}{r}\mathbf{D}^\alpha(m)\mathbf{D}_\alpha\left(\frac{m}{r}\right)\Bigg\}\underline{\alpha}+\Upsilon^2\Bigg\{\left(-4\rho-\frac{8m}{r^3}\right)-4\left(e_3(\omega)-\frac{2m}{r^3}\right)\\
& -10\left(\underline{\kappa}\,\omega-\frac{2m}{r^3}\right)+\frac{1}{2}\left(\kappa\,\underline{\kappa}+\frac{4\Upsilon}{r^2}\right)-8\omega\underline{\omega}+2\kappa\underline{\omega}\Bigg\}\underline{\alpha}\\
& +\frac{4m}{r^2}\Upsilon\left(\square_g(r)-\left(\frac{2}{r}-\frac{2m}{r^2}\right)\right)\underline{\alpha}\\
& -\frac{8m}{r^3}\left(1-\frac{3m}{r}\right)\Big(-e_4(r)e_3(r)-\Upsilon+(e_\theta(r))^2\Big)\underline{\alpha}+4\Upsilon e_\theta(\Upsilon)e_\theta(\underline{\alpha})\\
& +Err[\square_{\mathbf{g}}\alpha]+\Upsilon^2 Err[\square_{\mathbf{g}}\underline{\alpha}].
\end{aligned}
$$

Proof. Recall from Proposition 2.107 that the curvature components α and $\underline{\alpha}$ verify the following Teukolsky equations:

$$
\begin{aligned}
\square_2\alpha &= -4\underline{\omega}e_4(\alpha)+(4\omega+2\kappa)e_3(\alpha)+V\alpha+\mathrm{Err}[\square_{\mathbf{g}}\alpha],\\
V &= -4\rho-4e_4(\underline{\omega})-8\omega\underline{\omega}+2\omega\,\underline{\kappa}-10\kappa\,\underline{\omega}+\frac{1}{2}\kappa\,\underline{\kappa},
\end{aligned}
$$

where

$$
\begin{aligned}
&\mathrm{Err}(\square_{\mathbf{g}}\alpha)\\
= \;& \frac{1}{2}\vartheta e_3(\alpha)+\frac{3}{4}\vartheta^2\rho+e_\theta(\Phi)\vartheta\beta-\frac{1}{2}\kappa(\zeta+4\eta)\beta-(\zeta+\underline{\eta})e_4(\beta)-\xi e_3(\beta)\\
&+e_\theta(\Phi)(2\zeta+\underline{\eta})\alpha+\beta^2+e_4(\Phi)\underline{\eta}\beta+e_3(\Phi)\xi\beta-(\zeta+4\eta)e_4(\beta)\\
&-(e_4(\zeta)+4e_4(\eta))\beta-2(\kappa+\omega)(\zeta+4\eta)\beta+2e_\theta(\kappa+\omega)\beta-e_\theta((2\zeta+\underline{\eta})\alpha)\\
&-3\xi e_\theta(\rho)+2\underline{\eta}e_\theta(\alpha)+\frac{3}{2}\vartheta\,\not{d}_1\beta+3\rho(\underline{\eta}+\eta+2\zeta)\xi+\not{d}_1\underline{\eta}\alpha+\frac{1}{4}\underline{\kappa}\vartheta\alpha-2\underline{\omega}\vartheta\alpha\\
&-\frac{1}{2}\vartheta\underline{\vartheta}\alpha+\xi\underline{\xi}\alpha+\underline{\eta}^2\alpha+\frac{3}{2}\vartheta\zeta\beta+3\vartheta(\underline{\eta}\beta+\xi\underline{\beta})-\frac{1}{2}\vartheta(\zeta+4\eta)\beta,
\end{aligned}
$$

and

$$
\begin{aligned}
\square_2\underline{\alpha} &= -4\omega e_3(\underline{\alpha})+(4\underline{\omega}+2\underline{\kappa})e_4(\underline{\alpha})+\underline{V}\,\underline{\alpha}+\mathrm{Err}[\square_{\mathbf{g}}\underline{\alpha}],\\
\underline{V} &= -4\rho-4e_3(\omega)-8\omega\underline{\omega}+2\underline{\omega}\kappa-10\underline{\kappa}\,\omega+\frac{1}{2}\kappa\,\underline{\kappa},
\end{aligned}
$$

where

$$
\begin{aligned}
&\mathrm{Err}(\square_{\mathbf{g}}\underline{\alpha})\\
= \;& \frac{1}{2}\underline{\vartheta}e_4(\underline{\alpha})+\frac{3}{4}\underline{\vartheta}^2\rho+e_\theta(\Phi)\underline{\vartheta}\,\underline{\beta}-\frac{1}{2}\kappa(-\zeta+4\underline{\eta})\underline{\beta}-(-\zeta+\eta)e_3(\underline{\beta})-\underline{\xi}e_4(\underline{\beta})\\
&+e_\theta(\Phi)(-2\zeta+\eta)\underline{\alpha}+\underline{\beta}^2+e_3(\Phi)\eta\underline{\beta}+e_4(\Phi)\underline{\xi}\,\underline{\beta}-(-\zeta+4\underline{\eta})e_3(\underline{\beta})\\
&-(-e_3(\zeta)+4e_3(\underline{\eta}))\underline{\beta}-2(\underline{\kappa}+\underline{\omega})(-\zeta+4\underline{\eta})\underline{\beta}+2e_\theta(\underline{\kappa}+\underline{\omega})\underline{\beta}-e_\theta((-2\zeta+\eta)\underline{\alpha})\\
&-3\underline{\xi}e_\theta(\rho)+2\eta e_\theta(\underline{\alpha})+\frac{3}{2}\underline{\vartheta}\,\not{d}_1\underline{\beta}+3\rho(\eta+\underline{\eta}-2\zeta)\underline{\xi}+\not{d}_1\eta\underline{\alpha}+\frac{1}{4}\kappa\underline{\vartheta}\,\underline{\alpha}-2\omega\underline{\vartheta}\,\underline{\alpha}\\
&-\frac{1}{2}\underline{\vartheta}\vartheta\underline{\alpha}+\xi\underline{\xi}\underline{\alpha}+\eta^2\underline{\alpha}-\frac{3}{2}\underline{\vartheta}\zeta\underline{\beta}+3\underline{\vartheta}(\eta\underline{\beta}+\underline{\xi}\beta)-\frac{1}{2}\underline{\vartheta}(-\zeta+4\underline{\eta})\underline{\beta}.
\end{aligned}
$$

We infer from the above wave equations

$$
\begin{aligned}
&\square_2(\alpha+\Upsilon^2\underline{\alpha})\\
= \;& \square_2(\alpha)+\Upsilon^2\square_2(\underline{\alpha})+2\mathbf{D}^\mu(\Upsilon^2)\mathbf{D}_\mu(\underline{\alpha})+\square_0(\Upsilon^2)\underline{\alpha}\\
= \;& -4\underline{\omega}e_4(\alpha)+(4\omega+2\kappa)e_3(\alpha)\\
&+\Upsilon^2\Big(-4\omega e_3(\underline{\alpha})+(4\underline{\omega}+2\underline{\kappa})e_4(\underline{\alpha})\Big)-2\Upsilon e_3(\Upsilon)e_4(\underline{\alpha})-2\Upsilon e_4(\Upsilon)e_3(\underline{\alpha})\\
&+V\alpha+\Big(\Upsilon^2\underline{V}+\square_0(\Upsilon^2)\Big)\underline{\alpha}+4\Upsilon e_\theta(\Upsilon)e_\theta(\underline{\alpha})+\mathrm{Err}[\square_{\mathbf{g}}\alpha]+\Upsilon^2\mathrm{Err}[\square_{\mathbf{g}}\underline{\alpha}]
\end{aligned}
$$

and hence

$$\begin{aligned}
&\square_2(\alpha+\Upsilon^2\underline{\alpha})\\
= \;& \frac{4}{r}\left(1-\frac{3m}{r}\right)\Big(e_3(\alpha)-\Upsilon e_4(\underline{\alpha})\Big)\\
&+V\alpha+\left(\Upsilon^2\underline{V}+\frac{4m}{r^2}\Upsilon\square_g(r)-\frac{8m}{r^3}\Upsilon\mathbf{D}^\alpha(r)\mathbf{D}_\alpha(r)+\frac{8m^2}{r^4}\mathbf{D}^\alpha(r)\mathbf{D}_\alpha(r)\right)\underline{\alpha}\\
&+\left(4\left(\omega+\frac{m}{r^2}\right)+2\left(\kappa-\frac{2\Upsilon}{r}\right)\right)e_3(\alpha)-4\underline{\omega}e_4(\alpha)\\
&-4\Upsilon\left(\Upsilon\left(\omega+\frac{m}{r^2}\right)+\frac{m}{r^2}(e_4(r)-1)-\frac{e_4(m)}{r}\right)e_3(\underline{\alpha})\\
&+\left(4\Upsilon^2\underline{\omega}+2\Upsilon^2\left(\underline{\kappa}+\frac{2}{r}\right)-4m\Upsilon\frac{(e_3(r)+1)}{r^2}+4\Upsilon\frac{e_3(m)}{r}\right)e_4(\underline{\alpha})\\
&+\left(\frac{8\Upsilon}{r^2}\mathbf{D}^\alpha(m)\mathbf{D}_\alpha(r)-\frac{4\Upsilon}{r}\square_g(m)-\frac{8m}{r^3}\mathbf{D}^\alpha(r)\mathbf{D}_\alpha(m)\right.\\
&\left.+\frac{8m}{r}\mathbf{D}^\alpha(m)\mathbf{D}_\alpha\left(\frac{m}{r}\right)\right)\underline{\alpha}+4\Upsilon e_\theta(\Upsilon)e_\theta(\underline{\alpha})+\mathrm{Err}[\square_{\mathbf{g}}\alpha]+\Upsilon^2\mathrm{Err}[\square_{\mathbf{g}}\underline{\alpha}].
\end{aligned}$$

Next, we have in view of the formula for V

$$\begin{aligned}
V &= -4\rho-4e_4(\underline{\omega})-8\omega\underline{\omega}+2\omega\,\underline{\kappa}-10\kappa\,\underline{\omega}+\frac{1}{2}\kappa\,\underline{\kappa}\\
&= -\frac{2}{r^2}+\frac{16m}{r^3}+\left(-4\rho-\frac{8m}{r^3}\right)+2\left(\omega\,\underline{\kappa}-\frac{2m}{r^3}\right)+\frac{1}{2}\left(\kappa\,\underline{\kappa}+\frac{4\Upsilon}{r^2}\right)\\
&\quad -4e_4(\underline{\omega})-8\omega\underline{\omega}-10\kappa\,\underline{\omega}.
\end{aligned}$$

Also, we have in view of the formula for $\underline{V}$

$$\begin{aligned}
\underline{V} &= -4\rho-4e_3(\omega)-8\omega\underline{\omega}+2\underline{\omega}\kappa-10\underline{\kappa}\,\omega+\frac{1}{2}\kappa\,\underline{\kappa}\\
&= -\frac{2}{r^2}+\left(-4\rho-\frac{8m}{r^3}\right)-4\left(e_3(\omega)+\frac{2m}{r^3}\right)-10\left(\underline{\kappa}\,\omega-\frac{2m}{r^3}\right)\\
&\quad +\frac{1}{2}\left(\kappa\,\underline{\kappa}+\frac{4\Upsilon}{r^2}\right)-8\omega\underline{\omega}+2\kappa\underline{\omega}.
\end{aligned}$$

Moreover, we have

$$
\begin{aligned}
& \frac{4m}{r^2}\Upsilon\square_g(r) - \frac{8m}{r^3}\Upsilon\mathbf{D}^\alpha(r)\mathbf{D}_\alpha(r) + \frac{8m^2}{r^4}\mathbf{D}^\alpha(r)\mathbf{D}_\alpha(r) \\
= & \frac{4m\Upsilon}{r^2}\left(\frac{2}{r} - \frac{2m}{r^2}\right) + \frac{4m}{r^2}\Upsilon\left(\square_g(r) - \left(\frac{2}{r} - \frac{2m}{r^2}\right)\right) \\
& -\frac{8m}{r^3}\left(1 - \frac{3m}{r}\right)(-e_4(r)e_3(r) + (e_\theta(r))^2) \\
= & \frac{16m^2\Upsilon}{r^4} + \frac{4m}{r^2}\Upsilon\left(\square_g(r) - \left(\frac{2}{r} - \frac{2m}{r^2}\right)\right) \\
& -\frac{8m}{r^3}\left(1 - \frac{3m}{r}\right)\Big(- e_4(r)e_3(r) - \Upsilon + (e_\theta(r))^2\Big)
\end{aligned}
$$

and hence

$$
\begin{aligned}
& \Upsilon^2\underline{V} + \frac{4m}{r^2}\Upsilon\square_g(r) - \frac{8m}{r^3}\Upsilon\mathbf{D}^\alpha(r)\mathbf{D}_\alpha(r) + \frac{8m^2}{r^4}\mathbf{D}^\alpha(r)\mathbf{D}_\alpha(r) \\
= & -\frac{2\Upsilon}{r^2}\left(1 - \frac{2m}{r} - \frac{8m^2}{r^2}\right) \\
& +\Upsilon^2\left\{\left(-4\rho - \frac{8m}{r^3}\right) - 4\left(e_3(\omega) + \frac{2m}{r^3}\right) - 10\left(\underline{\kappa}\,\omega - \frac{2m}{r^3}\right)\right. \\
& \left. +\frac{1}{2}\left(\kappa\,\underline{\kappa} + \frac{4\Upsilon}{r^2}\right) - 8\omega\underline{\omega} + 2\kappa\underline{\omega}\right\} + \frac{4m}{r^2}\Upsilon\left(\square_g(r) - \left(\frac{2}{r} - \frac{2m}{r^2}\right)\right) \\
& -\frac{8m}{r^3}\left(1 - \frac{3m}{r}\right)\Big(- e_4(r)e_3(r) - \Upsilon + (e_\theta(r))^2\Big).
\end{aligned}
$$

We deduce

$$
\begin{aligned}
\square_2(\alpha + \Upsilon^2\underline{\alpha}) = & \frac{4}{r}\left(1 - \frac{3m}{r}\right)\Big(e_3(\alpha) - \Upsilon e_4(\underline{\alpha})\Big) + \left(-\frac{2}{r^2} + \frac{16m}{r^3}\right)\alpha \\
& -\frac{2\Upsilon}{r^2}\left(1 - \frac{2m}{r} - \frac{8m^2}{r^2}\right)\underline{\alpha} + \mathrm{Err}\Big[\square_2(\alpha + \Upsilon^2\underline{\alpha})\Big]
\end{aligned}
$$

where

$$
\begin{aligned}
&\mathrm{Err}\Big[\square_2(\alpha+\Upsilon^2\underline{\alpha})\Big]\\
=\ & \left(\Upsilon^2\underline{V}+\frac{4m}{r^2}\Upsilon\square_g(r)-\frac{8m}{r^3}\Upsilon\mathbf{D}^\alpha(r)\mathbf{D}_\alpha(r)+\frac{8m^2}{r^4}\mathbf{D}^\alpha(r)\mathbf{D}_\alpha(r)\right)\underline{\alpha}\\
&+\left(4\left(\omega+\frac{m}{r^2}\right)+2\left(\kappa-\frac{2\Upsilon}{r}\right)\right)e_3(\alpha)-4\underline{\omega}e_4(\alpha)\\
&-4\Upsilon\left(\Upsilon\left(\omega+\frac{m}{r^2}\right)+\frac{m}{r^2}(e_4(r)-1)-\frac{e_4(m)}{r}\right)e_3(\underline{\alpha})\\
&+\left(4\Upsilon^2\underline{\omega}+2\Upsilon^2\left(\underline{\kappa}+\frac{2}{r}\right)-4m\Upsilon\frac{(e_3(r)+1)}{r^2}+4\Upsilon\frac{e_3(m)}{r}\right)e_4(\underline{\alpha})\\
&+\Bigg\{\left(-4\rho-\frac{8m}{r^3}\right)+2\left(\omega\,\underline{\kappa}-\frac{2m}{r^3}\right)+\frac{1}{2}\left(\kappa\,\underline{\kappa}+\frac{4\Upsilon}{r^2}\right)-4e_4(\underline{\omega})-8\omega\underline{\omega}\\
&-10\kappa\,\underline{\omega}\Bigg\}\alpha+\Bigg\{\frac{8\Upsilon}{r^2}\mathbf{D}^\alpha(m)\mathbf{D}_\alpha(r)-\frac{4\Upsilon}{r}\square_g(m)-\frac{8m}{r^3}\mathbf{D}^\alpha(r)\mathbf{D}_\alpha(m)\\
&+\frac{8m}{r}\mathbf{D}^\alpha(m)\mathbf{D}_\alpha\left(\frac{m}{r}\right)\Bigg\}\underline{\alpha}+\Upsilon^2\Bigg\{\left(-4\rho-\frac{8m}{r^3}\right)-4\left(e_3(\omega)-\frac{2m}{r^3}\right)\\
&-10\left(\underline{\kappa}\,\omega-\frac{2m}{r^3}\right)+\frac{1}{2}\left(\kappa\,\underline{\kappa}+\frac{4\Upsilon}{r^2}\right)-8\omega\underline{\omega}+2\kappa\underline{\omega}\Bigg\}\underline{\alpha}\\
&+\frac{4m}{r^2}\Upsilon\left(\square_g(r)-\left(\frac{2}{r}-\frac{2m}{r^2}\right)\right)\underline{\alpha}\\
&-\frac{8m}{r^3}\left(1-\frac{3m}{r}\right)\Big(-e_4(r)e_3(r)-\Upsilon+(e_\theta(r))^2\Big)\underline{\alpha}+4\Upsilon e_\theta(\Upsilon)e_\theta(\underline{\alpha})\\
&+\mathrm{Err}[\square_{\mathbf{g}}\alpha]+\Upsilon^2\mathrm{Err}[\square_{\mathbf{g}}\underline{\alpha}]
\end{aligned}
$$

as desired. This concludes the proof of the lemma. □

Lemma 8.20. *We have*

$$
\begin{aligned}
e_3(\alpha) &= -\frac{1}{2}\underline{\kappa}\alpha-{\not d}_2^\star{\not d}_1^{-1}\Bigg\{e_4\left(\frac{\tilde{\rho}}{r^2}\right)+\frac{3}{2r^2}\kappa\tilde{\rho}-\frac{3m}{r^3}\left(\kappa-\frac{2\Upsilon}{r}\right)+\frac{6m(e_4(r)-\Upsilon)}{r^4}\\
&\quad-\frac{2e_4(m)}{r^3}+\frac{1}{2}\underline{\vartheta}\alpha-\zeta\beta-2(\underline{\eta}\beta+\xi\underline{\beta})\Bigg\}+4\underline{\omega}\alpha-\frac{3}{2}\vartheta\rho+(\zeta+4\eta)\beta
\end{aligned}
$$

and

$$
\begin{aligned}
e_4(\underline{\alpha}) &= -\frac{1}{2}\kappa\underline{\alpha}-{\not d}_2^\star{\not d}_1^{-1}\Bigg\{e_3\left(\frac{\tilde{\rho}}{r^2}\right)+\frac{3}{2r^2}\underline{\kappa}\tilde{\rho}-\frac{3m}{r^3}\left(\underline{\kappa}+\frac{2}{r}\right)+\frac{6m(e_3(r)+1)}{r^4}\\
&\quad-\frac{2e_3(m)}{r^3}+\frac{1}{2}\vartheta\underline{\alpha}+\zeta\underline{\beta}-2(\eta\underline{\beta}+\underline{\xi}\beta)\Bigg\}+4\omega\underline{\alpha}-\frac{3}{2}\underline{\vartheta}\rho+(-\zeta+4\underline{\eta})\underline{\beta}.
\end{aligned}
$$

Proof. Recall that we have

$$\begin{aligned}\square_2(\alpha+\Upsilon^2\underline{\alpha}) &= \frac{4}{r}\left(1-\frac{3m}{r}\right)\Big(e_3(\alpha)-\Upsilon e_4(\underline{\alpha})\Big)+\left(-\frac{2}{r^2}+\frac{16m}{r^3}\right)\alpha\\ &-\frac{2\Upsilon}{r^2}\left(1-\frac{2m}{r}-\frac{8m^2}{r^2}\right)\underline{\alpha}+\mathrm{Err}\Big[\square_2(\alpha+\Upsilon^2\underline{\alpha})\Big].\end{aligned}$$

We first express $e_3(\alpha)-\Upsilon e_4(\underline{\alpha})$ in terms of $\tilde{\rho}$, where we recall that $\tilde{\rho}=r^2\rho+\frac{2m}{r}$. Using Bianchi, we have

$$\begin{aligned}e_3(\alpha) &= -\frac{1}{2}\underline{\kappa}\alpha-\not{d}_2^\star\beta+4\underline{\omega}\alpha-\frac{3}{2}\vartheta\rho+(\zeta+4\eta)\beta,\\ \not{d}_1\beta &= e_4(\rho)+\frac{3}{2}\kappa\rho+\frac{1}{2}\underline{\vartheta}\alpha-\zeta\beta-2(\underline{\eta}\beta+\xi\underline{\beta})\\ &= e_4\left(\frac{\tilde{\rho}}{r^2}-\frac{2m}{r^3}\right)+\frac{3}{2}\kappa\rho+\frac{1}{2}\underline{\vartheta}\alpha-\zeta\beta-2(\underline{\eta}\beta+\xi\underline{\beta})\\ &= e_4\left(\frac{\tilde{\rho}}{r^2}\right)+\frac{3}{2r^2}\kappa\tilde{\rho}-\frac{3m}{r^3}\left(\kappa-\frac{2\Upsilon}{r}\right)+\frac{6m(e_4(r)-\Upsilon)}{r^4}-\frac{2e_4(m)}{r^3}\\ &+\frac{1}{2}\underline{\vartheta}\alpha-\zeta\beta-2(\underline{\eta}\beta+\xi\underline{\beta})\end{aligned}$$

and hence

$$\begin{aligned}e_3(\alpha) &= -\frac{1}{2}\underline{\kappa}\alpha-\not{d}_2^\star\not{d}_1^{-1}\left\{e_4\left(\frac{\tilde{\rho}}{r^2}\right)+\frac{3}{2r^2}\kappa\tilde{\rho}-\frac{3m}{r^3}\left(\kappa-\frac{2\Upsilon}{r}\right)+\frac{6m(e_4(r)-\Upsilon)}{r^4}\right.\\ &\left.-\frac{2e_4(m)}{r^3}+\frac{1}{2}\underline{\vartheta}\alpha-\zeta\beta-2(\underline{\eta}\beta+\xi\underline{\beta})\right\}+4\underline{\omega}\alpha-\frac{3}{2}\vartheta\rho+(\zeta+4\eta)\beta.\end{aligned}$$

Similarly, we have

$$\begin{aligned}e_4(\underline{\alpha}) &= -\frac{1}{2}\kappa\underline{\alpha}-\not{d}_2^\star\underline{\beta}+4\omega\underline{\alpha}-\frac{3}{2}\underline{\vartheta}\rho+(-\zeta+4\underline{\eta})\underline{\beta},\\ \not{d}_1\underline{\beta} &= e_3(\rho)+\frac{3}{2}\underline{\kappa}\rho+\frac{1}{2}\vartheta\underline{\alpha}+\zeta\underline{\beta}-2(\eta\underline{\beta}+\underline{\xi}\beta)\\ &= e_3\left(\frac{\tilde{\rho}}{r^2}-\frac{2m}{r^3}\right)+\frac{3}{2}\underline{\kappa}\rho+\frac{1}{2}\vartheta\underline{\alpha}+\zeta\underline{\beta}-2(\eta\underline{\beta}+\underline{\xi}\beta)\\ &= e_3\left(\frac{\tilde{\rho}}{r^2}\right)+\frac{3}{2r^2}\underline{\kappa}\tilde{\rho}-\frac{3m}{r^3}\left(\underline{\kappa}+\frac{2}{r}\right)+\frac{6m(e_3(r)+1)}{r^4}-\frac{2e_3(m)}{r^3}\\ &+\frac{1}{2}\vartheta\underline{\alpha}+\zeta\underline{\beta}-2(\eta\underline{\beta}+\underline{\xi}\beta)\end{aligned}$$

and hence

$$\begin{aligned}e_4(\underline{\alpha}) &= -\frac{1}{2}\kappa\underline{\alpha}-\not{d}_2^\star\not{d}_1^{-1}\left\{e_3\left(\frac{\tilde{\rho}}{r^2}\right)+\frac{3}{2r^2}\underline{\kappa}\tilde{\rho}-\frac{3m}{r^3}\left(\underline{\kappa}+\frac{2}{r}\right)+\frac{6m(e_3(r)+1)}{r^4}\right.\\ &\left.-\frac{2e_3(m)}{r^3}+\frac{1}{2}\vartheta\underline{\alpha}+\zeta\underline{\beta}-2(\eta\underline{\beta}+\underline{\xi}\beta)\right\}+4\omega\underline{\alpha}-\frac{3}{2}\underline{\vartheta}\rho+(-\zeta+4\underline{\eta})\underline{\beta}.\end{aligned}$$

This concludes the proof of the lemma. □

Corollary 8.21. *We have*

$$\begin{aligned}
&\square_2(\alpha+\Upsilon^2\underline{\alpha})-\frac{2}{r^2}\left(1+\frac{2m}{r}\right)(\alpha+\Upsilon^2\underline{\alpha})\\
=\;& -\frac{8}{r}\left(1-\frac{3m}{r}\right)\not{d}_2^\star\not{d}_1^{-1}R\left(\frac{\tilde{\rho}}{r^2}\right)-\frac{6}{r}\left(1-\frac{3m}{r}\right)(\vartheta-\Upsilon\underline{\vartheta})\rho\\
&-\frac{4}{r}\left(1-\frac{3m}{r}\right)\not{d}_2^\star\not{d}_1^{-1}\left\{\frac{3}{2r^2}\kappa\tilde{\rho}-\frac{3m}{r^3}\left(\kappa-\frac{2\Upsilon}{r}\right)+\frac{6m(e_4(r)-\Upsilon)}{r^4}\right\}\\
&+\frac{4\Upsilon}{r}\left(1-\frac{3m}{r}\right)\not{d}_2^\star\not{d}_1^{-1}\left\{\frac{3}{2r^2}\underline{\kappa}\tilde{\rho}-\frac{3m}{r^3}\left(\underline{\kappa}+\frac{2}{r}\right)+\frac{6m(e_3(r)+1)}{r^4}\right\}+Err_1,
\end{aligned}$$

where

$$\begin{aligned}
Err_1 \;:=\;& \frac{4}{r}\left(1-\frac{3m}{r}\right)\Bigg[4\underline{\omega}\alpha+(\zeta+4\eta)\beta-\Upsilon(-\zeta+4\underline{\eta})\underline{\beta}+[\Upsilon,\not{d}_2^\star\not{d}_1^{-1}]e_3\left(\frac{\tilde{\rho}}{r^2}\right)\\
&-\not{d}_2^\star\not{d}_1^{-1}\left\{-\frac{2e_4(m)}{r^3}+\frac{1}{2}\underline{\vartheta}\alpha-\zeta\beta-2(\underline{\eta}\beta+\xi\underline{\beta})\right\}\\
&+\Upsilon\not{d}_2^\star\not{d}_1^{-1}\left\{-\frac{2e_3(m)}{r^3}+\frac{1}{2}\vartheta\underline{\alpha}+\zeta\underline{\beta}-2(\eta\underline{\beta}+\underline{\xi}\beta)\right\}\Bigg]\\
&+\frac{4\Upsilon}{r}\left(1-\frac{3m}{r}\right)\left(\frac{1}{2}\left(\kappa-\frac{2\Upsilon}{r}\right)-4\left(\omega+\frac{m}{r^2}\right)\right)\underline{\alpha}\\
&-\frac{2}{r}\left(1-\frac{3m}{r}\right)\left(\underline{\kappa}+\frac{2}{r}\right)\alpha+Err\Big[\square_2(\alpha+\Upsilon^2\underline{\alpha})\Big].
\end{aligned}$$

Proof. Recall from Lemma 8.19 that we have

$$\begin{aligned}
\square_2(\alpha+\Upsilon^2\underline{\alpha}) \;=\;& \frac{4}{r}\left(1-\frac{3m}{r}\right)\Big(e_3(\alpha)-\Upsilon e_4(\underline{\alpha})\Big)+\left(-\frac{2}{r^2}+\frac{16m}{r^3}\right)\alpha\\
&-\frac{2\Upsilon}{r^2}\left(1-\frac{2m}{r}-\frac{8m^2}{r^2}\right)\underline{\alpha}+\mathrm{Err}\Big[\square_2(\alpha+\Upsilon^2\underline{\alpha})\Big].
\end{aligned}$$

In view of Lemma 8.20, we have

$$
\begin{aligned}
e_3(\alpha)-\Upsilon e_4(\underline{\alpha}) \quad = \quad & -2\,{\not{d}}_2^\star {\not{d}}_1^{-1} R\left(\frac{\tilde{\rho}}{r^2}\right)-\frac{1}{2}\underline{\kappa}\alpha+\Upsilon\left(\frac{1}{2}\kappa-4\omega\right)\underline{\alpha}-\frac{3}{2}(\vartheta-\Upsilon\underline{\vartheta})\rho\\
&-{\not{d}}_2^\star {\not{d}}_1^{-1}\left\{\frac{3}{2r^2}\kappa\tilde{\rho}-\frac{3m}{r^3}\left(\kappa-\frac{2\Upsilon}{r}\right)+\frac{6m(e_4(r)-\Upsilon)}{r^4}\right\}\\
&+\Upsilon\,{\not{d}}_2^\star {\not{d}}_1^{-1}\left\{\frac{3}{2r^2}\underline{\kappa}\tilde{\rho}-\frac{3m}{r^3}\left(\underline{\kappa}+\frac{2}{r}\right)+\frac{6m(e_3(r)+1)}{r^4}\right\}\\
&+4\underline{\omega}\alpha+(\zeta+4\eta)\beta-\Upsilon(-\zeta+4\underline{\eta})\underline{\beta}+[\Upsilon,\,{\not{d}}_2^\star {\not{d}}_1^{-1}]e_3\left(\frac{\tilde{\rho}}{r^2}\right)\\
&-{\not{d}}_2^\star {\not{d}}_1^{-1}\left\{-\frac{2e_4(m)}{r^3}+\frac{1}{2}\underline{\vartheta}\alpha-\zeta\beta-2(\underline{\eta}\beta+\xi\underline{\beta})\right\}\\
&+\Upsilon\,{\not{d}}_2^\star {\not{d}}_1^{-1}\left\{-\frac{2e_3(m)}{r^3}+\frac{1}{2}\vartheta\underline{\alpha}+\zeta\underline{\beta}-2(\eta\underline{\beta}+\underline{\xi}\beta)\right\}.
\end{aligned}
$$

We infer

$$
\begin{aligned}
&\square_2(\alpha+\Upsilon^2\underline{\alpha})\\
= \;& -\frac{8}{r}\left(1-\frac{3m}{r}\right){\not{d}}_2^\star {\not{d}}_1^{-1}R\left(\frac{\tilde{\rho}}{r^2}\right)-\frac{2}{r}\left(1-\frac{3m}{r}\right)\underline{\kappa}\alpha+\left(-\frac{2}{r^2}+\frac{16m}{r^3}\right)\alpha\\
&+\frac{4\Upsilon}{r}\left(1-\frac{3m}{r}\right)\left(\frac{1}{2}\kappa-4\omega\right)\underline{\alpha}-\frac{2\Upsilon}{r^2}\left(1-\frac{2m}{r}-\frac{8m^2}{r^2}\right)\underline{\alpha}\\
&-\frac{6}{r}\left(1-\frac{3m}{r}\right)(\vartheta-\Upsilon\underline{\vartheta})\rho\\
&-\frac{4}{r}\left(1-\frac{3m}{r}\right){\not{d}}_2^\star {\not{d}}_1^{-1}\left\{\frac{3}{2r^2}\kappa\tilde{\rho}-\frac{3m}{r^3}\left(\kappa-\frac{2\Upsilon}{r}\right)+\frac{6m(e_4(r)-\Upsilon)}{r^4}\right\}\\
&+\frac{4\Upsilon}{r}\left(1-\frac{3m}{r}\right){\not{d}}_2^\star {\not{d}}_1^{-1}\left\{\frac{3}{2r^2}\underline{\kappa}\tilde{\rho}-\frac{3m}{r^3}\left(\underline{\kappa}+\frac{2}{r}\right)+\frac{6m(e_3(r)+1)}{r^4}\right\}\\
&+\frac{4}{r}\left(1-\frac{3m}{r}\right)\Bigg[4\underline{\omega}\alpha+(\zeta+4\eta)\beta-\Upsilon(-\zeta+4\underline{\eta})\underline{\beta}+[\Upsilon,\,{\not{d}}_2^\star {\not{d}}_1^{-1}]e_3\left(\frac{\tilde{\rho}}{r^2}\right)\\
&-{\not{d}}_2^\star {\not{d}}_1^{-1}\left\{-\frac{2e_4(m)}{r^3}+\frac{1}{2}\underline{\vartheta}\alpha-\zeta\beta-2(\underline{\eta}\beta+\xi\underline{\beta})\right\}\\
&+\Upsilon\,{\not{d}}_2^\star {\not{d}}_1^{-1}\left\{-\frac{2e_3(m)}{r^3}+\frac{1}{2}\vartheta\underline{\alpha}+\zeta\underline{\beta}-2(\eta\underline{\beta}+\underline{\xi}\beta)\right\}\Bigg]+\mathrm{Err}\Big[\square_2(\alpha+\Upsilon^2\underline{\alpha})\Big].
\end{aligned}
$$

Since we have

$$
\begin{aligned}
&-\frac{2}{r}\left(1-\frac{3m}{r}\right)\underline{\kappa}\alpha+\frac{4\Upsilon}{r}\left(1-\frac{3m}{r}\right)\left(\frac{1}{2}\kappa-4\omega\right)\underline{\alpha}\\
= \;& \frac{4}{r^2}\left(1-\frac{3m}{r}\right)\alpha+\frac{4\Upsilon}{r^2}\left(1-\frac{3m}{r}\right)\left(1+\frac{2m}{r}\right)\underline{\alpha}-\frac{2}{r}\left(1-\frac{3m}{r}\right)\left(\underline{\kappa}+\frac{2}{r}\right)\alpha\\
&+\frac{4\Upsilon}{r}\left(1-\frac{3m}{r}\right)\left(\frac{1}{2}\left(\kappa-\frac{2\Upsilon}{r}\right)-4\left(\omega+\frac{m}{r^2}\right)\right)\underline{\alpha},
\end{aligned}
$$

this yields

$$
\begin{aligned}
&\square_2(\alpha+\Upsilon^2\underline{\alpha})\\
=\;& -\frac{8}{r}\left(1-\frac{3m}{r}\right)\not{d}_2^\star\not{d}_1^{-1}R\left(\frac{\tilde{\rho}}{r^2}\right)+\frac{2}{r^2}\left(1+\frac{2m}{r}\right)\alpha+\frac{2\Upsilon}{r^2}\left(1-\frac{4m^2}{r^2}\right)\underline{\alpha}\\
&-\frac{6}{r}\left(1-\frac{3m}{r}\right)(\vartheta-\Upsilon\underline{\vartheta})\rho\\
&-\frac{4}{r}\left(1-\frac{3m}{r}\right)\not{d}_2^\star\not{d}_1^{-1}\left\{\frac{3}{2r^2}\kappa\tilde{\rho}-\frac{3m}{r^3}\left(\kappa-\frac{2\Upsilon}{r}\right)+\frac{6m(e_4(r)-\Upsilon)}{r^4}\right\}\\
&+\frac{4\Upsilon}{r}\left(1-\frac{3m}{r}\right)\not{d}_2^\star\not{d}_1^{-1}\left\{\frac{3}{2r^2}\underline{\kappa}\tilde{\rho}-\frac{3m}{r^3}\left(\underline{\kappa}+\frac{2}{r}\right)+\frac{6m(e_3(r)+1)}{r^4}\right\}+\mathrm{Err}_1,
\end{aligned}
$$

where

$$
\begin{aligned}
\mathrm{Err}_1 \;=\;& \frac{4}{r}\left(1-\frac{3m}{r}\right)\Bigg[4\underline{\omega}\alpha+(\zeta+4\eta)\beta-\Upsilon(-\zeta+4\underline{\eta})\underline{\beta}+[\Upsilon,\not{d}_2^\star\not{d}_1^{-1}]e_3\left(\frac{\tilde{\rho}}{r^2}\right)\\
&-\not{d}_2^\star\not{d}_1^{-1}\left\{-\frac{2e_4(m)}{r^3}+\frac{1}{2}\underline{\vartheta}\alpha-\zeta\beta-2(\underline{\eta}\beta+\xi\underline{\beta})\right\}\\
&+\Upsilon\not{d}_2^\star\not{d}_1^{-1}\left\{-\frac{2e_3(m)}{r^3}+\frac{1}{2}\vartheta\underline{\alpha}+\zeta\underline{\beta}-2(\eta\underline{\beta}+\underline{\xi}\beta)\right\}\Bigg]\\
&+\frac{4\Upsilon}{r}\left(1-\frac{3m}{r}\right)\left(\frac{1}{2}\left(\kappa-\frac{2\Upsilon}{r}\right)-4\left(\omega+\frac{m}{r^2}\right)\right)\underline{\alpha}\\
&-\frac{2}{r}\left(1-\frac{3m}{r}\right)\left(\underline{\kappa}+\frac{2}{r}\right)\alpha+\mathrm{Err}\Big[\square_2(\alpha+\Upsilon^2\underline{\alpha})\Big].
\end{aligned}
$$

Now, since we have

$$
\frac{2}{r^2}\left(1+\frac{2m}{r}\right)\alpha+\frac{2\Upsilon}{r^2}\left(1-\frac{4m^2}{r^2}\right)\underline{\alpha} \;=\; \frac{2}{r^2}\left(1+\frac{2m}{r}\right)(\alpha+\Upsilon^2\underline{\alpha}),
$$

we infer

$$
\begin{aligned}
&\square_2(\alpha+\Upsilon^2\underline{\alpha})-\frac{2}{r^2}\left(1+\frac{2m}{r}\right)(\alpha+\Upsilon^2\underline{\alpha})\\
=\;& -\frac{8}{r}\left(1-\frac{3m}{r}\right)\not{d}_2^\star\not{d}_1^{-1}R\left(\frac{\tilde{\rho}}{r^2}\right)-\frac{6}{r}\left(1-\frac{3m}{r}\right)(\vartheta-\Upsilon\underline{\vartheta})\rho\\
&-\frac{4}{r}\left(1-\frac{3m}{r}\right)\not{d}_2^\star\not{d}_1^{-1}\left\{\frac{3}{2r^2}\kappa\tilde{\rho}-\frac{3m}{r^3}\left(\kappa-\frac{2\Upsilon}{r}\right)+\frac{6m(e_4(r)-\Upsilon)}{r^4}\right\}\\
&+\frac{4\Upsilon}{r}\left(1-\frac{3m}{r}\right)\not{d}_2^\star\not{d}_1^{-1}\left\{\frac{3}{2r^2}\underline{\kappa}\tilde{\rho}-\frac{3m}{r^3}\left(\underline{\kappa}+\frac{2}{r}\right)+\frac{6m(e_3(r)+1)}{r^4}\right\}+\mathrm{Err}_1,
\end{aligned}
$$

as desired. This concludes the proof of the corollary. □

8.5.2 End of the proof of Proposition 8.8

In view of Corollary 8.21, $\alpha+\Upsilon^2\underline{\alpha}$ satisfies

$$(\square_2+V_2)(\alpha+\Upsilon^2\underline{\alpha}) \quad = \quad N_2, \qquad V_2=-\frac{2}{r^2}\left(1+\frac{2m}{r}\right),$$

where

$$\begin{aligned}N_2 \quad := \quad & -\frac{8}{r}\left(1-\frac{3m}{r}\right)\mathscr{d}_2^\star\mathscr{d}_1^{-1}R\left(\frac{\tilde{\rho}}{r^2}\right)-\frac{6}{r}\left(1-\frac{3m}{r}\right)(\vartheta-\Upsilon\underline{\vartheta})\rho\\ & -\frac{4}{r}\left(1-\frac{3m}{r}\right)\mathscr{d}_2^\star\mathscr{d}_1^{-1}\left\{\frac{3}{2r^2}\kappa\tilde{\rho}-\frac{3m}{r^3}\left(\kappa-\frac{2\Upsilon}{r}\right)+\frac{6m(e_4(r)-\Upsilon)}{r^4}\right\}\\ & +\frac{4\Upsilon}{r}\left(1-\frac{3m}{r}\right)\mathscr{d}_2^\star\mathscr{d}_1^{-1}\left\{\frac{3}{2r^2}\underline{\kappa}\tilde{\rho}-\frac{3m}{r^3}\left(\underline{\kappa}+\frac{2}{r}\right)+\frac{6m(e_3(r)+1)}{r^4}\right\}\\ & +\mathrm{Err}_1.\end{aligned}$$

We may thus apply the estimate (10.5.1) of Theorem 10.67 with $\psi=\alpha+\Upsilon^2\underline{\alpha}$ and $s=J$ to obtain for any $k_{small}\leq J\leq k_{large}-1$

$$\begin{aligned}&\sup_{\tau\in[1,\tau_*]} E_\delta^J[\alpha+\Upsilon^2\underline{\alpha}](\tau)+B_\delta^J[\alpha+\Upsilon^2\underline{\alpha}](1,\tau_*)+F_\delta^J[\alpha+\Upsilon^2\underline{\alpha}](1,\tau_*)\\ \lesssim\ & E_\delta^J[\alpha+\Upsilon^2\underline{\alpha}](1)+\sup_{\tau\in[1,\tau_*]}E_\delta^{J-1}[\alpha+\Upsilon^2\underline{\alpha}](\tau)+B_\delta^{J-1}[\alpha+\Upsilon^2\underline{\alpha}](1,\tau_*)\\ & +F_\delta^{J-1}[\alpha+\Upsilon^2\underline{\alpha}](1,\tau_*)+D_J[\Gamma]\left(\sup_{\mathcal{M}} r u_{trap}^{\frac{1}{2}+\delta_{dec}}|\mathfrak{d}^{\leq k_{small}}(\alpha+\Upsilon^2\underline{\alpha})|\right)^2\\ & +\int_{\mathcal{M}} r^{1+\delta}|\mathfrak{d}^{\leq J}N_2|^2+\left|\int_{^{(trap)}\mathcal{M}} T(\mathfrak{d}^J(\alpha+\Upsilon^2\underline{\alpha}))\mathfrak{d}^J N_2\right|,\end{aligned}$$

where $D_J[\Gamma]$ is defined by

$$\begin{aligned}D_J[\Gamma] \quad := \quad & \int_{^{(int)}\mathcal{M}\cup{}^{(ext)}\mathcal{M}(r\leq 4m_0)}(\mathfrak{d}^{\leq J}\check{\Gamma})^2\\ & +\sup_{r_0\geq 4m_0}\left(r_0\int_{\{r=r_0\}}|\mathfrak{d}^{\leq J}\Gamma_g|^2+r_0^{-1}\int_{\{r=r_0\}}|\mathfrak{d}^{\leq J}\Gamma_b|^2\right).\end{aligned}$$

Next we use the iteration assumption (8.3.13) which yields in particular

$$D_J[\Gamma] \quad \lesssim \quad (\epsilon_{\mathcal{B}}[J])^2$$

and

$$\begin{aligned}&\sup_{\tau\in[1,\tau_*]} E_\delta^{J-1}[\alpha+\Upsilon^2\underline{\alpha}](\tau)+B_\delta^{J-1}[\alpha+\Upsilon^2\underline{\alpha}](1,\tau_*)\\ &\qquad\qquad +F_\delta^{J-1}[\alpha+\Upsilon^2\underline{\alpha}](1,\tau_*) \quad \lesssim \quad (\epsilon_{\mathcal{B}}[J])^2.\end{aligned}$$

Together with the control of $\mathfrak{d}^{\leq k_{small}}(\alpha+\Upsilon^2\underline{\alpha})$ provided by the decay estimate

(8.3.6), we infer from the above estimates

$$\begin{aligned}
&\sup_{\tau\in[1,\tau_*]} E^J_\delta[\alpha+\Upsilon^2\underline{\alpha}](\tau)+B^J_\delta[\alpha+\Upsilon^2\underline{\alpha}](1,\tau_*)+F^J_\delta[\alpha+\Upsilon^2\underline{\alpha}](1,\tau_*)\\
\lesssim\;& E^J_\delta[\alpha+\Upsilon^2\underline{\alpha}](1)+(\epsilon_{\mathcal{B}}[J])^2+\int_{\mathcal{M}} r^{1+\delta}|\mathfrak{d}^{\leq J}N_2|^2\\
&+\left|\int_{{}^{(trap)}\mathcal{M}} T(\mathfrak{d}^J(\alpha+\Upsilon^2\underline{\alpha}))\mathfrak{d}^J N_2\right|.
\end{aligned}$$

Next, using the form of N_2, as well as the control of $\tilde{\rho}$ provided by Proposition 8.7, we derive

$$\int_{\mathcal{M}} r^{1+\delta}|\mathfrak{d}^{\leq J}N_2|^2 \;\lesssim\; (\epsilon_{\mathcal{B}}[J])^2+\epsilon_0^2\Big(\mathfrak{N}^{(En)}_{J+1}+\mathcal{N}^{(match)}_{J+1}\Big)^2$$

and

$$\begin{aligned}
&\left|\int_{{}^{(trap)}\mathcal{M}} T(\mathfrak{d}^J(\alpha+\Upsilon^2\underline{\alpha}))\mathfrak{d}^J N_2\right|\\
\lesssim\;& \int_{{}^{(trap)}\mathcal{M}}\left|1-\frac{3m}{r}\right||T(\mathfrak{d}^J(\alpha+\Upsilon^2\underline{\alpha}))|\Big(|R(\mathfrak{d}^J\tilde{\rho})|+|\mathfrak{d}^J\tilde{\rho}|+|\mathfrak{d}^J\check{\Gamma}|\Big)\\
&+\int_{{}^{(trap)}\mathcal{M}}|T(\mathfrak{d}^J(\alpha+\Upsilon^2\underline{\alpha}))||\mathrm{Err}_1|\\
\lesssim\;& \Big(\epsilon_{\mathcal{B}}[J]+\epsilon_0\Big(\mathfrak{N}^{(En)}_{J+1}+\mathcal{N}^{(match)}_{J+1}\Big)\Big)\left(\sup_{\tau\in[1,\tau_*]} E^J_\delta[\tilde{\rho}](\tau)+B^J_\delta[\tilde{\rho}](1,\tau_*)\right)^{\frac{1}{2}}.
\end{aligned}$$

In view of the above, we infer

$$\begin{aligned}
&\sup_{\tau\in[1,\tau_*]} E^J_\delta[\alpha+\Upsilon^2\underline{\alpha}](\tau)+B^J_\delta[\alpha+\Upsilon^2\underline{\alpha}](1,\tau_*)+F^J_\delta[\alpha+\Upsilon^2\underline{\alpha}](1,\tau_*)\\
\lesssim\;& (\epsilon_{\mathcal{B}}[J])^2+\epsilon_0^2\Big(\mathfrak{N}^{(En)}_{J+1}+\mathcal{N}^{(match)}_{J+1}\Big)^2
\end{aligned}$$

as desired. This concludes the proof of Proposition 8.8.

8.6 PROOF OF PROPOSITION 8.9

8.6.1 Control of α and $\Upsilon^2\underline{\alpha}$

We initiate the proof of Proposition 8.9 by deriving a suitable control for α and $\Upsilon^2\underline{\alpha}$. Recall from Lemma 8.20 that we have

$$\begin{aligned}
e_4(\underline{\alpha}) \;=\;& -\frac{1}{2}\kappa\underline{\alpha}-{\not{d}}_2^\star{\not{d}}_1^{-1}\Bigg\{e_3\left(\frac{\tilde{\rho}}{r^2}\right)+\frac{3}{2r^2}\underline{\kappa}\tilde{\rho}-\frac{3m}{r^3}\left(\underline{\kappa}+\frac{2}{r}\right)+\frac{6m(e_3(r)+1)}{r^4}\\
&-\frac{2e_3(m)}{r^3}+\frac{1}{2}\vartheta\underline{\alpha}+\zeta\underline{\beta}-2(\eta\underline{\beta}+\underline{\xi}\beta)\Bigg\}+4\omega\underline{\alpha}-\frac{3}{2}\underline{\vartheta}\rho+(-\zeta+4\underline{\eta})\underline{\beta}.
\end{aligned}$$

We infer

$$
\begin{aligned}
& e_4(\alpha - \Upsilon^2\underline{\alpha})\\
= \;& e_4(\alpha + \Upsilon^2\underline{\alpha}) - 2e_4(\Upsilon^2\underline{\alpha})\\
= \;& e_4(\alpha + \Upsilon^2\underline{\alpha}) - 2\Upsilon^2 e_4(\underline{\alpha}) - 2e_4(\Upsilon^2)\underline{\alpha}\\
= \;& e_4(\alpha + \Upsilon^2\underline{\alpha}) + 2\Upsilon^2 \, {}^{*}\!\!\not{d}_2 \not{d}_1^{-1}\Bigg\{ e_3\left(\frac{\tilde{\rho}}{r^2}\right) + \frac{3}{2r^2}\underline{\kappa}\tilde{\rho} - \frac{3m}{r^3}\left(\underline{\kappa} + \frac{2}{r}\right)\\
& + \frac{6m(e_3(r)+1)}{r^4} - \frac{2e_3(m)}{r^3} + \frac{1}{2}\vartheta\underline{\alpha} + \zeta\underline{\beta} - 2(\eta\underline{\beta} + \xi\beta)\Bigg\} + \Upsilon^2\kappa\underline{\alpha} - 8\Upsilon^2\omega\underline{\alpha}\\
& + 3\Upsilon^2\underline{\vartheta}\rho - 2\Upsilon^2(-\zeta + 4\underline{\eta})\underline{\beta} - \frac{8m\Upsilon e_4(r)}{r^2}\underline{\alpha} + \frac{8\Upsilon e_4(m)}{r}\underline{\alpha}.
\end{aligned}
$$

Also, recall from Lemma 8.20 that we have

$$
\begin{aligned}
e_3(\alpha) \;=\;& -\frac{1}{2}\underline{\kappa}\alpha - {}^{*}\!\!\not{d}_2 \not{d}_1^{-1}\Bigg\{ e_4\left(\frac{\tilde{\rho}}{r^2}\right) + \frac{3}{2r^2}\kappa\tilde{\rho} - \frac{3m}{r^3}\left(\kappa - \frac{2\Upsilon}{r}\right) + \frac{6m(e_4(r)-\Upsilon)}{r^4}\\
& - \frac{2e_4(m)}{r^3} + \frac{1}{2}\underline{\vartheta}\alpha - \zeta\beta - 2(\underline{\eta}\beta + \xi\underline{\beta})\Bigg\} + 4\underline{\omega}\alpha - \frac{3}{2}\vartheta\rho + (\zeta + 4\eta)\beta.
\end{aligned}
$$

We infer

$$
\begin{aligned}
& e_3(\alpha - \Upsilon^2\underline{\alpha})\\
= \;& -e_3(\alpha + \Upsilon^2\underline{\alpha}) + 2e_3(\alpha)\\
= \;& -e_3(\alpha + \Upsilon^2\underline{\alpha}) - 2\, {}^{*}\!\!\not{d}_2 \not{d}_1^{-1}\Bigg\{ e_4\left(\frac{\tilde{\rho}}{r^2}\right) + \frac{3}{2r^2}\kappa\tilde{\rho} - \frac{3m}{r^3}\left(\kappa - \frac{2\Upsilon}{r}\right)\\
& + \frac{6m(e_4(r)-\Upsilon)}{r^4} - \frac{2e_4(m)}{r^3} + \frac{1}{2}\underline{\vartheta}\alpha - \zeta\beta - 2(\underline{\eta}\beta + \xi\underline{\beta})\Bigg\} - \underline{\kappa}\alpha + 8\underline{\omega}\alpha\\
& - 3\vartheta\rho + 2(\zeta + 4\eta)\beta.
\end{aligned}
$$

In view of the above identities for $e_4(\alpha - \Upsilon^2\underline{\alpha})$ and $e_3(\alpha - \Upsilon^2\underline{\alpha})$, and using the control for $\tilde{\rho}$ provided by Proposition 8.7 as well as the control for $\alpha + \Upsilon^2\underline{\alpha}$ provided by Proposition 8.8, and the iteration assumption (8.3.13), we obtain

$$
\begin{aligned}
& B_\delta^{J-1}[e_3(\alpha - \Upsilon^2\underline{\alpha})](1,\tau_*) + B_\delta^{J-1}[re_4(\alpha - \Upsilon^2\underline{\alpha})](1,\tau_*)\\
\lesssim \;& (\epsilon_{\mathcal{B}}[J])^2 + \epsilon_0^2\left(\mathfrak{N}_{J+1}^{(En)} + \mathcal{N}_{J+1}^{(match)}\right)^2.
\end{aligned}
$$

Also, using the Bianchi identity for $\not{d}_2\alpha$ and $\not{d}_1\beta$, we have

$$\begin{aligned}
\not{d}_1\not{d}_2\alpha &= \not{d}_1\Big(e_4\beta+2(\kappa+\omega)\beta-(2\zeta+\eta)\alpha-3\xi\rho\Big)\\
&= e_4(\not{d}_1\beta)+[\not{d}_1,e_4]\beta+\not{d}_1\Big(2(\kappa+\omega)\beta-(2\zeta+\eta)\alpha-3\xi\rho\Big)\\
&= e_4\left(e_4\rho+\frac{3}{2}\rho+\frac{1}{2}\underline{\vartheta}\alpha-\zeta\beta-2(\eta\beta+\xi\underline{\beta})\right)+[\not{d}_1,e_4]\beta\\
&\quad+\not{d}_1\Big(2(\kappa+\omega)\beta-(2\zeta+\eta)\alpha-3\xi\rho\Big)\\
&= e_4\left[e_4\left(\frac{\tilde{\rho}}{r^2}\right)+\frac{3}{2r^2}\kappa\tilde{\rho}-\frac{3m}{r^3}\left(\kappa-\frac{2\Upsilon}{r}\right)+\frac{6m(e_4(r)-\Upsilon)}{r^4}\right.\\
&\quad\left.-\frac{2e_4(m)}{r^3}+\frac{1}{2}\underline{\vartheta}\alpha-\zeta\beta-2(\underline{\eta}\beta+\xi\underline{\beta})\right]+[\not{d}_1,e_4]\beta\\
&\quad+\not{d}_1\Big(2(\kappa+\omega)\beta-(2\zeta+\eta)\alpha-3\xi\rho\Big).
\end{aligned}$$

Using the control for $\tilde{\rho}$ provided by Proposition 8.7 as well as the iteration assumption (8.3.13), we obtain

$$B_\delta^{J-2}[r^2\not{d}_1\not{d}_2\alpha](1,\tau_*) \lesssim (\epsilon_{\mathcal{B}}[J])^2+\epsilon_0^2\left(\mathfrak{N}_{J+1}^{(En)}+\mathcal{N}_{J+1}^{(match)}\right)^2.$$

Using the control for $\alpha+\Upsilon^2\underline{\alpha}$ provided by Proposition 8.8, we infer

$$\begin{aligned}
B_\delta^{J-2}[r^2\not{d}_1\not{d}_2(\alpha-\Upsilon^2\underline{\alpha})](1,\tau_*) &\lesssim B_\delta^{J-2}[r^2\not{d}_1\not{d}_2\alpha](1,\tau_*)\\
&\quad+B_\delta^{J-2}[r^2\not{d}_1\not{d}_2(\alpha+\Upsilon^2\underline{\alpha})](1,\tau_*)\\
&\lesssim (\epsilon_{\mathcal{B}}[J])^2+\epsilon_0^2\left(\mathfrak{N}_{J+1}^{(En)}+\mathcal{N}_{J+1}^{(match)}\right)^2.
\end{aligned}$$

Using a Poincaré inequality for $\not{d}_1$ and for $\not{d}_2$, we deduce

$$B_\delta^{J-2}[\not{\partial}^2(\alpha-\Upsilon^2\underline{\alpha})](1,\tau_*) \lesssim (\epsilon_{\mathcal{B}}[J])^2+\epsilon_0^2\left(\mathfrak{N}_{J+1}^{(En)}+\mathcal{N}_{J+1}^{(match)}\right)^2.$$

Together with the above estimate for $e_3(\alpha-\Upsilon^2\underline{\alpha})$ and $re_4(\alpha-\Upsilon^2\underline{\alpha})$, we deduce

$$B_\delta^{J}[\alpha-\Upsilon^2\underline{\alpha}](1,\tau_*) \lesssim (\epsilon_{\mathcal{B}}[J])^2+\epsilon_0^2\left(\mathfrak{N}_{J+1}^{(En)}+\mathcal{N}_{J+1}^{(match)}\right)^2.$$

Together with the control for $\alpha+\Upsilon^2\underline{\alpha}$ provided by Proposition 8.8, we finally obtain

$$B_\delta^{J}[\alpha](1,\tau_*)+B_\delta^{J}[\Upsilon^2\underline{\alpha}](1,\tau_*) \lesssim (\epsilon_{\mathcal{B}}[J])^2+\epsilon_0^2\left(\mathfrak{N}_{J+1}^{(En)}+\mathcal{N}_{J+1}^{(match)}\right)^2. \tag{8.6.1}$$

8.6.2 Control of $\underline{\alpha}$

(8.6.1) provides in particular the control of $\Upsilon^2\underline{\alpha}$. In this section, we infer a suitable control for $\underline{\alpha}$ using the wave equation satisfied by $\underline{\alpha}$ and the redshift vectorfield.

Let $Y_{(0)}$ the vectorfield given by

$$Y_{(0)} := \left(1+\frac{5}{4m}(r-2m)+\Upsilon\right)e_3+\left(1+\frac{5}{4m}(r-2m)\right)e_4,$$

where $Y_{(0)}$ has been introduced in Proposition 10.29 in connection with the redshift vectorfield.

Lemma 8.22. *We have*

$$\square_2\underline{\alpha} \;=\; \frac{4m}{r^2\left(1+\frac{5}{4m}(r-2m)+\Upsilon\right)}Y_{(0)}\underline{\alpha}+\widetilde{N}_2$$

where $\widetilde{N}_2$ is given by

$$\begin{aligned}\widetilde{N}_2 \;:=\; & -\frac{4}{r}\left(1+\frac{m\left(1+\frac{5}{4m}(r-2m)\right)}{r\left(1+\frac{5}{4m}(r-2m)+\Upsilon\right)}\right)\Bigg[-\frac{1}{2}\kappa\underline{\alpha}\\ & -{\not{d}}_2^{\star}{\not{d}}_1^{-1}\left\{e_3\left(\frac{\tilde{\rho}}{r^2}\right)+\frac{3}{2r^2}\underline{\kappa}\tilde{\rho}-\frac{3m}{r^3}\left(\underline{\kappa}+\frac{2}{r}\right)+\frac{6m(e_3(r)+1)}{r^4}-\frac{2e_3(m)}{r^3}\right.\\ & \left.+\frac{1}{2}\vartheta\underline{\alpha}+\zeta\underline{\beta}-2(\eta\underline{\beta}+\xi\beta)\right\}+4\omega\underline{\alpha}-\frac{3}{2}\underline{\vartheta}\rho+(-\zeta+4\underline{\eta})\underline{\beta}\Bigg]\\ & +\underline{V}\,\underline{\alpha}-4\left(\omega+\frac{m}{r^2}\right)e_3(\underline{\alpha})+\left(4\underline{\omega}+2\left(\underline{\kappa}+\frac{2}{r}\right)\right)e_4(\underline{\alpha})+Err[\square_{\mathbf{g}}\underline{\alpha}].\end{aligned}$$

Proof. Recall from Proposition 2.107 that $\underline{\alpha}$ verifies the following Teukolsky equation:

$$\begin{aligned}\square_2\underline{\alpha} &= -4\omega e_3(\underline{\alpha})+(4\underline{\omega}+2\underline{\kappa})e_4(\underline{\alpha})+\underline{V}\,\underline{\alpha}+\mathrm{Err}[\square_{\mathbf{g}}\underline{\alpha}],\\ \underline{V} &= -4\rho-4e_3(\omega)-8\omega\underline{\omega}+2\underline{\omega}\kappa-10\underline{\kappa}\,\omega+\frac{1}{2}\kappa\,\underline{\kappa},\end{aligned}$$

where

$$\begin{aligned}&\mathrm{Err}(\square_{\mathbf{g}}\underline{\alpha})\\ =\;& \frac{1}{2}\vartheta e_4(\underline{\alpha})+\frac{3}{4}\underline{\vartheta}^2\rho+e_\theta(\Phi)\underline{\vartheta}\,\underline{\beta}-\frac{1}{2}\kappa(-\zeta+4\underline{\eta})\underline{\beta}-(-\zeta+\eta)e_3(\underline{\beta})-\underline{\xi}e_4(\underline{\beta})\\ &+e_\theta(\Phi)(-2\zeta+\eta)\underline{\alpha}+\underline{\beta}^2+e_3(\Phi)\eta\underline{\beta}+e_4(\Phi)\underline{\xi}\,\underline{\beta}-(-\zeta+4\underline{\eta})e_3(\underline{\beta})\\ &-(-e_3(\zeta)+4e_3(\underline{\eta}))\underline{\beta}-2(\underline{\kappa}+\underline{\omega})(-\zeta+4\underline{\eta})\underline{\beta}+2e_\theta(\underline{\kappa}+\underline{\omega})\underline{\beta}-e_\theta((-2\zeta+\eta)\underline{\alpha})\\ &-3\underline{\xi}e_\theta(\rho)+2\eta e_\theta(\underline{\alpha})+\frac{3}{2}\underline{\vartheta}\,{\not{d}}_1\underline{\beta}+3\rho(\eta+\underline{\eta}-2\zeta)\underline{\xi}+{\not{d}}_1\eta\underline{\alpha}+\frac{1}{4}\kappa\underline{\vartheta}\,\underline{\alpha}-2\omega\underline{\vartheta}\,\underline{\alpha}\\ &-\frac{1}{2}\underline{\vartheta}\vartheta\underline{\alpha}+\xi\underline{\xi}\underline{\alpha}+\eta^2\underline{\alpha}-\frac{3}{2}\underline{\vartheta}\zeta\underline{\beta}+3\underline{\vartheta}(\eta\underline{\beta}+\underline{\xi}\beta)-\frac{1}{2}\underline{\vartheta}(-\zeta+4\underline{\eta})\underline{\beta}.\end{aligned}$$

We deduce

$$\begin{aligned}\square_2\underline{\alpha} \;=\;& \frac{4m}{r^2}e_3(\underline{\alpha})-\frac{4}{r}e_4(\underline{\alpha})+\underline{V}\,\underline{\alpha}-4\left(\omega+\frac{m}{r^2}\right)e_3(\underline{\alpha})\\ &+\left(4\underline{\omega}+2\left(\underline{\kappa}+\frac{2}{r}\right)\right)e_4(\underline{\alpha})+\mathrm{Err}[\square_{\mathbf{g}}\underline{\alpha}].\end{aligned}$$

In view of the definition of $Y_{(0)}$, we infer

$$\begin{aligned}
&\square_2\underline{\alpha}\\
&= \frac{4m}{r^2\left(1+\frac{5}{4m}(r-2m)+\Upsilon\right)}Y_{(0)}\underline{\alpha} - \frac{4}{r}\left(1+\frac{m\left(1+\frac{5}{4m}(r-2m)\right)}{r\left(1+\frac{5}{4m}(r-2m)+\Upsilon\right)}\right)e_4(\underline{\alpha})\\
&\quad +\underline{V}\,\underline{\alpha} - 4\left(\omega+\frac{m}{r^2}\right)e_3(\underline{\alpha}) + \left(4\underline{\omega}+2\left(\underline{\kappa}+\frac{2}{r}\right)\right)e_4(\underline{\alpha}) + \mathrm{Err}[\square_{\mathbf{g}}\underline{\alpha}].
\end{aligned}$$

Next, recall from Lemma 8.20 that we have

$$\begin{aligned}
e_4(\underline{\alpha}) &= -\frac{1}{2}\kappa\underline{\alpha} - \not{d}_2^\star \not{d}_1^{-1}\left\{e_3\left(\frac{\tilde{\rho}}{r^2}\right) + \frac{3}{2r^2}\underline{\kappa}\tilde{\rho} - \frac{3m}{r^3}\left(\underline{\kappa}+\frac{2}{r}\right) + \frac{6m(e_3(r)+1)}{r^4}\right.\\
&\quad \left. -\frac{2e_3(m)}{r^3} + \frac{1}{2}\vartheta\underline{\alpha} + \zeta\underline{\beta} - 2(\eta\underline{\beta}+\underline{\xi}\beta)\right\} + 4\omega\underline{\alpha} - \frac{3}{2}\underline{\vartheta}\rho + (-\zeta+4\underline{\eta})\underline{\beta}.
\end{aligned}$$

We infer

$$\square_2\underline{\alpha} = \frac{4m}{r^2\left(1+\frac{5}{4m}(r-2m)+\Upsilon\right)}Y_{(0)}\underline{\alpha} + \widetilde{N}_2$$

where $\widetilde{N}_2$ is given by

$$\begin{aligned}
\widetilde{N}_2 &= -\frac{4}{r}\left(1+\frac{m\left(1+\frac{5}{4m}(r-2m)\right)}{r\left(1+\frac{5}{4m}(r-2m)+\Upsilon\right)}\right)\left[-\frac{1}{2}\kappa\underline{\alpha}\right.\\
&\quad -\not{d}_2^\star \not{d}_1^{-1}\left\{e_3\left(\frac{\tilde{\rho}}{r^2}\right) + \frac{3}{2r^2}\underline{\kappa}\tilde{\rho} - \frac{3m}{r^3}\left(\underline{\kappa}+\frac{2}{r}\right) + \frac{6m(e_3(r)+1)}{r^4} - \frac{2e_3(m)}{r^3}\right.\\
&\quad \left.\left.+\frac{1}{2}\vartheta\underline{\alpha} + \zeta\underline{\beta} - 2(\eta\underline{\beta}+\underline{\xi}\beta)\right\} + 4\omega\underline{\alpha} - \frac{3}{2}\underline{\vartheta}\rho + (-\zeta+4\underline{\eta})\underline{\beta}\right]\\
&\quad +\underline{V}\,\underline{\alpha} - 4\left(\omega+\frac{m}{r^2}\right)e_3(\underline{\alpha}) + \left(4\underline{\omega}+2\left(\underline{\kappa}+\frac{2}{r}\right)\right)e_4(\underline{\alpha}) + \mathrm{Err}[\square_{\mathbf{g}}\underline{\alpha}].
\end{aligned}$$

This concludes the proof of the lemma. □

Lemma 8.23. *$\widetilde{N}_2$, in the RHS of the wave equation for $\underline{\alpha}$ introduced in Lemma 8.22, satisfies*

$$\int_{^{(int)}\mathcal{M}} |\mathfrak{d}^J \widetilde{N}_2|^2 \lesssim (\epsilon_{\mathcal{B}}[J])^2 + \epsilon_0^2\left(\mathfrak{N}^{(En)}_{J+1} + \mathcal{N}^{(match)}_{J+1}\right)^2.$$

Proof. The proof of the lemma follows immediately from the form of $\widetilde{N}_2$, see Lemma 8.22, as well as the control for $\tilde{\rho}$ provided by Proposition 8.7, (8.3.6), and the iteration assumption (8.3.13). □

In view of Lemma 8.22, we may apply Proposition 10.69 with

$$\psi = \underline{\alpha}, \qquad f_2(r,m) = \frac{4m}{r^2\left(1+\frac{5}{4m}(r-2m)+\Upsilon\right)}.$$

We infer

$$\int_{^{(int)}\mathcal{M}(1,\tau_*)} (\mathfrak{d}^{J+1}\underline{\alpha})^2 \lesssim E^J_\delta[\underline{\alpha}](\tau=1) + \int_{^{(ext)}\mathcal{M}_{r\leq\frac{5}{2}m_0}(1,\tau_*)} (\mathfrak{d}^{J+1}\underline{\alpha})^2$$
$$+D_J[\Gamma]\left(\sup_{^{(int)}\mathcal{M}(1,\tau_*)\cup{}^{(ext)}\mathcal{M}_{r\leq\frac{5}{2}m_0}} r|\mathfrak{d}^{\leq k_{small}}\underline{\alpha}|\right)^2$$
$$+\int_{^{(int)}\mathcal{M}(1,\tau_*)\cup{}^{(ext)}\mathcal{M}_{r\leq\frac{5}{2}m_0}} \Big((\mathfrak{d}^{\leq s}\underline{\alpha})^2 + (\mathfrak{d}^{\leq J+1}\widetilde{N}_2)^2\Big).$$

Next we use the iteration assumption (8.3.13) which yields in particular

$$D_J[\Gamma] \lesssim (\epsilon_{\mathcal{B}}[J])^2$$

together with the control of $\mathfrak{d}^{\leq k_{small}}\underline{\alpha}$ provided by the decay estimate (8.3.6), as well as the iteration assumption and the control for $\widetilde{N}_2$ provided by Lemma 8.23 to deduce

$$\int_{^{(int)}\mathcal{M}(1,\tau_*)} (\mathfrak{d}^{J+1}\underline{\alpha})^2 \lesssim (\epsilon_{\mathcal{B}}[J])^2 + \epsilon_0^2\Big(\mathfrak{N}^{(En)}_{J+1} + \mathcal{N}^{(match)}_{J+1}\Big)^2$$
$$+\int_{^{(ext)}\mathcal{M}_{r\leq\frac{5}{2}m_0}(1,\tau_*)} (\mathfrak{d}^{J+1}\underline{\alpha})^2.$$

Note that $\Upsilon^2 \gtrsim \delta_{\mathcal{H}}^2 > 0$ on ${}^{(ext)}\mathcal{M}$ and hence

$$\int_{^{(ext)}\mathcal{M}_{r\leq\frac{5}{2}m_0}(1,\tau_*)} (\mathfrak{d}^{J+1}\underline{\alpha})^2 \lesssim \int_{^{(ext)}\mathcal{M}_{r\leq\frac{5}{2}m_0}(1,\tau_*)} (\mathfrak{d}^{J+1}(\Upsilon^2\underline{\alpha}))^2$$

which together with the control of $\Upsilon^2\underline{\alpha}$ provided by (8.6.1) yields

$$\int_{^{(ext)}\mathcal{M}_{r\leq\frac{5}{2}m_0}(1,\tau_*)} (\mathfrak{d}^{J+1}\underline{\alpha})^2 \lesssim (\epsilon_{\mathcal{B}}[J])^2 + \epsilon_0^2\Big(\mathfrak{N}^{(En)}_{J+1} + \mathcal{N}^{(match)}_{J+1}\Big)^2$$

and hence

$$\int_{^{(int)}\mathcal{M}(1,\tau_*)} (\mathfrak{d}^{J+1}\underline{\alpha})^2 \lesssim (\epsilon_{\mathcal{B}}[J])^2 + \epsilon_0^2\Big(\mathfrak{N}^{(En)}_{J+1} + \mathcal{N}^{(match)}_{J+1}\Big)^2.$$

Since

$$B^J_\delta[\underline{\alpha}](1,\tau_*) \lesssim \int_{^{(int)}\mathcal{M}(1,\tau_*)} (\mathfrak{d}^{\leq J+1}\underline{\alpha})^2 + B^J_\delta[\Upsilon^2\underline{\alpha}](1,\tau_*),$$

using again (8.6.1), we finally obtain

$$B^J_\delta[\underline{\alpha}](1,\tau_*) \lesssim (\epsilon_{\mathcal{B}}[J])^2 + \epsilon_0^2\Big(\mathfrak{N}^{(En)}_{J+1} + \mathcal{N}^{(match)}_{J+1}\Big)^2. \tag{8.6.2}$$

8.6.3 End of the proof of Proposition 8.9

We have

$$\check{\rho} = \frac{\tilde{\rho}}{r^2} - \left(\overline{\rho} - \frac{2m}{r^3}\right).$$

Together with the control for $\tilde{\rho}$ provided by Proposition 8.7, as well as the control on averages provided by Lemma 3.15, we infer

$$B_\delta^J[\check{\rho}](1,\tau_*) \lesssim (\epsilon_{\mathcal{B}}[J])^2 + \epsilon_0^2\left(\mathfrak{N}_{J+1}^{(En)} + \mathcal{N}_{J+1}^{(match)}\right)^2.$$

Together with the control for α provided by (8.6.1) and the control for $\underline{\alpha}$ provided by (8.6.2), we infer

$$B_\delta^J[\alpha,\check{\rho},\underline{\alpha}](1,\tau_*) \lesssim (\epsilon_{\mathcal{B}}[J])^2 + \epsilon_0^2\left(\mathfrak{N}_{J+1}^{(En)} + \mathcal{N}_{J+1}^{(match)}\right)^2.$$

Together with the Bianchi identities for $e_4(\beta)$, $e_3(\beta)$, $\not{d}_1\beta$, $e_4(\underline{\beta})$, $e_3(\underline{\beta})$, $\not{d}_1\underline{\beta}$, as well as the iteration assumption (8.3.13), we infer

$$B_{\delta-2}^J[\beta,\underline{\beta}](1,\tau_*) \lesssim (\epsilon_{\mathcal{B}}[J])^2 + \epsilon_0^2\left(\mathfrak{N}_{J+1}^{(En)} + \mathcal{N}_{J+1}^{(match)}\right)^2$$

and hence

$$B_{-2}^J[\alpha,\beta,\check{\rho},\underline{\beta},\underline{\alpha}](1,\tau_*) \lesssim (\epsilon_{\mathcal{B}}[J])^2 + \epsilon_0^2\left(\mathfrak{N}_{J+1}^{(En)} + \mathcal{N}_{J+1}^{(match)}\right)^2$$

as desired. This concludes the proof of Proposition 8.9.

8.7 PROOF OF PROPOSITION 8.10

First, note that, by definition of the norms B_{-2}^J, ${}^{(int)}\mathfrak{R}_{J+1}[\check{R}]$ and ${}^{(ext)}\mathfrak{R}_{J+1}[\check{R}]$, we have for any $r_0 \geq 4m_0$

$${}^{(int)}\mathfrak{R}_{J+1}[\check{R}] + {}^{(ext)}\mathfrak{R}_{J+1}^{\leq r_0}[\check{R}] \lesssim r^{10} B_{-2}^J[\alpha,\beta,\check{\rho},\underline{\beta},\underline{\alpha}](1,\tau_*).$$

Together with Proposition 8.9, this implies

$${}^{(int)}\mathfrak{R}_{J+1}[\check{R}] + {}^{(ext)}\mathfrak{R}_{J+1}^{\leq r_0}[\check{R}] \lesssim r_0^{10}\left((\epsilon_{\mathcal{B}}[J])^2 + \epsilon_0^2\left(\mathfrak{N}_{J+1}^{(En)} + \mathcal{N}_{J+1}^{(match)}\right)^2\right).$$

Since we have

$${}^{(int)}\mathfrak{R}_{J+1}[\check{R}] + {}^{(ext)}\mathfrak{R}_{J+1}[\check{R}] = {}^{(int)}\mathfrak{R}_{J+1}[\check{R}] + {}^{(ext)}\mathfrak{R}_{J+1}^{\leq r_0}[\check{R}] + {}^{(ext)}\mathfrak{R}_{J+1}^{\geq r_0}[\check{R}],$$

we deduce for any $r_0 \geq 4m_0$

$$\begin{aligned} & {}^{(int)}\mathfrak{R}_{J+1}[\check{R}] + {}^{(ext)}\mathfrak{R}_{J+1}[\check{R}] \\ \leq\; & {}^{(ext)}\mathfrak{R}_{J+1}^{\geq r_0}[\check{R}] + O\left(r_0^{10}\left(\epsilon_{\mathcal{B}}[J] + \epsilon_0\left(\mathfrak{N}_{J+1}^{(En)} + \mathcal{N}_{J+1}^{(match)}\right)\right)\right). \end{aligned}$$

Thus, to prove Proposition 8.10, it suffices to establish the following inequality:

$$ {}^{(ext)}\mathfrak{R}^{\geq r_0}_{J+1}[\check{R}] \lesssim r_0^{-\delta_B} \, {}^{(ext)}\mathfrak{G}^{\geq r_0}_k[\check{\Gamma}] + r_0^{10}\left(\epsilon_{\mathcal{B}}[J] + \epsilon_0\left(\mathfrak{N}^{(En)}_{J+1} + \mathcal{N}^{(match)}_{J+1}\right)\right). $$

This will follow from r^p-weighted estimates for the curvature components.

8.7.1 r-weighted divergence identities for Bianchi pairs

Lemma 8.24. *Let $k \geq 1$, let $a_{(1)}$ and $a_{(2)}$ real numbers. We consider the following equations.*

- *If $\psi_{(1)}, h_{(1)} \in \mathfrak{s}_k$, $\psi_{(2)}, h_{(2)} \in \mathfrak{s}_{k-1}$, let $(\psi_{(1)}, \psi_{(2)})$ such that*

$$ \begin{cases} e_3(\psi_{(1)}) + a_{(1)}\underline{\kappa}\psi_{(1)} &= -\, \not{d}_k^\star \psi_{(2)} + h_{(1)}, \\ e_4(\psi_{(2)}) + a_{(2)}\kappa\psi_{(2)} &= \not{d}_k \psi_{(1)} + h_{(2)}. \end{cases} \tag{8.7.1} $$

- *If $\psi_{(1)}, h_{(1)} \in \mathfrak{s}_{k-1}$, $\psi_{(2)}, h_{(2)} \in \mathfrak{s}_k$, let $(\psi_{(1)}, \psi_{(2)})$ such that*

$$ \begin{cases} e_3(\psi_{(1)}) + a_{(1)}\underline{\kappa}\psi_{(1)} &= \not{d}_k \psi_{(2)} + h_{(1)}, \\ e_4(\psi_{(2)}) + a_{(2)}\kappa\psi_{(2)} &= -\, \not{d}_k^\star \psi_{(1)} + h_{(2)}. \end{cases} \tag{8.7.2} $$

Then, the pair $(\psi_{(1)}, \psi_{(2)})$ satisfies for any real number b

$$ \begin{aligned} &\mathbf{Div}\Big(r^b\psi_{(1)}^2 e_3\Big) + \mathbf{Div}\Big(r^b\psi_{(2)}^2 e_4\Big) - \frac{1}{2}r^b\underline{\kappa}\Big(-4a_{(1)} + b + 2\Big)\psi_{(1)}^2 \\ &+\frac{1}{2}r^b\kappa\Big(4a_{(2)} - b - 2\Big)\psi_{(2)}^2 \\ =\;& 2r^b \not{d}_1(\psi_{(1)}\psi_{(2)}) - 2r^b\underline{\omega}\psi_{(1)}^2 - 2r^b\omega\psi_{(2)}^2 + 2r^b\psi_{(1)}h_{(1)} + 2r^b\psi_{(2)}h_{(2)} \\ &+br^{b-1}\left(e_3(r) - \frac{r}{2}\underline{\kappa}\right)\psi_{(1)}^2 + br^{b-1}\left(e_4(r) - \frac{r}{2}\kappa\right)\psi_{(2)}^2. \end{aligned} \tag{8.7.3} $$

Remark 8.25. *Note that the Bianchi identities can be written as systems of equations of the type* (8.7.1), (8.7.2). *In particular*

- *the Bianchi pair (α, β) satisfies* (8.7.1) *with $k = 2$, $a_{(1)} = \frac{1}{2}$, $a_{(2)} = 2$,*
- *the Bianchi pair (β, ρ) satisfies* (8.7.1) *with $k = 1$, $a_{(1)} = 1$, $a_{(2)} = \frac{3}{2}$,*
- *the Bianchi pair $(\rho, \underline{\beta})$ satisfies* (8.7.2) *with $k = 1$, $a_{(1)} = \frac{3}{2}$, $a_{(2)} = 1$,*
- *the Bianchi pair $(\underline{\beta}, \underline{\alpha})$ satisfies* (8.7.2) *with $k = 2$, $a_{(1)} = 2$, $a_{(2)} = \frac{1}{2}$.*

Proof of Lemma 8.24. The proof being identical for (8.7.1) and (8.7.2), it suffices to prove it in the case where $(\psi_{(1)}, \psi_{(2)})$ satisfies (8.7.1).

We compute

$$ \begin{aligned} \mathbf{D}_\gamma e_4^\gamma &= -\frac{1}{2}\mathbf{g}(\mathbf{D}_4 e_4, e_3) - \frac{1}{2}\mathbf{g}(\mathbf{D}_3 e_4, e_4) + \mathbf{g}(D_\theta e_4, e_\theta) + \mathbf{g}(D_\varphi e_4, e_\varphi) \\ &= \kappa - 2\omega \end{aligned} $$

and

$$\begin{aligned}\mathbf{D}_\gamma e_3^\gamma &= -\frac{1}{2}\mathbf{g}(\mathbf{D}_4 e_3, e_3) - \frac{1}{2}\mathbf{g}(\mathbf{D}_3 e_3, e_4) + \mathbf{g}(D_\theta e_3, e_\theta) + \mathbf{g}(D_\varphi e_3, e_\varphi)\\ &= \underline{\kappa} - 2\underline{\omega}.\end{aligned}$$

We infer in view of (8.7.1)

$$\begin{aligned}&\mathbf{D}_\gamma\left(r^b\psi_{(1)}^2 e_3^\gamma\right)\\ =\;& 2r^b\psi_{(1)}e_3(\psi_{(1)}) + br^{b-1}e_3(r)\psi_{(1)}^2 + r^b\psi_{(1)}^2\mathbf{D}_\gamma e_3^\gamma\\ =\;& 2r^b\psi_{(1)}\Big(-a_{(1)}\underline{\kappa}\psi_{(1)} - \not{d}_k^\star\psi_{(2)} + h_{(1)}\Big) + br^{b-1}e_3(r)\psi_{(1)}^2 + r^b\psi_{(1)}^2(\underline{\kappa} - 2\underline{\omega})\\ =\;& -2r^b\psi_{(1)}\,\not{d}_k^\star\psi_{(2)} + r^b\Big(-2a_{(1)} + \frac{b}{2} + 1\Big)\underline{\kappa}\psi_{(1)}^2 + br^{b-1}\left(e_3(r) - \frac{r}{2}\underline{\kappa}\right)\psi_{(1)}^2\\ & -2\underline{\omega}r^b\psi_{(1)}^2 + 2r^b\psi_{(1)}h_{(1)}\end{aligned}$$

and

$$\begin{aligned}&\mathbf{D}_\gamma\left(r^b\psi_{(2)}^2 e_4^\gamma\right)\\ =\;& 2r^b\psi_{(2)}e_4(\psi_{(2)}) + br^{b-1}e_4(r)\psi_{(2)}^2 + r^b\psi_{(2)}^2\mathbf{D}_\gamma e_4^\gamma\\ =\;& 2r^b\psi_{(2)}\Big(-a_{(2)}\kappa\psi_{(2)} + \not{d}_k\psi_{(1)} + h_{(2)}\Big) + br^{b-1}e_4(r)\psi_{(2)}^2 + r^b\psi_{(2)}^2(\kappa - 2\omega)\\ =\;& 2r^b\psi_{(2)}\,\not{d}_k\psi_{(1)} + r^b\Big(-2a_{(2)} + \frac{b}{2} + 1\Big)\kappa\psi_{(2)}^2 + br^{b-1}\left(e_4(r) - \frac{r}{2}\kappa\right)\psi_{(2)}^2\\ & -2r^b\omega\psi_{(2)}^2 + 2r^b\psi_{(2)}h_{(2)}.\end{aligned}$$

We sum the two identities

$$\begin{aligned}&\mathbf{D}_\gamma\left(r^b\psi_{(1)}^2 e_3^\gamma\right) + \mathbf{D}_\gamma\left(r^b\psi_{(2)}^2 e_4^\gamma\right)\\ =\;& -2r^b\psi_{(1)}\,\not{d}_k^\star\psi_{(2)} + 2r^b\psi_{(2)}\,\not{d}_k\psi_{(1)} + r^b\Big(-2a_{(1)} + \frac{b}{2} + 1\Big)\underline{\kappa}\psi_{(1)}^2\\ & + r^b\Big(-2a_{(2)} + \frac{b}{2} + 1\Big)\kappa\psi_{(2)}^2 + br^{b-1}\left(e_3(r) - \frac{r}{2}\underline{\kappa}\right)\psi_{(1)}^2\\ & + br^{b-1}\left(e_4(r) - \frac{r}{2}\kappa\right)\psi_{(2)}^2 - 2r^b\underline{\omega}\psi_{(1)}^2 - 2r^b\omega\psi_{(2)}^2\\ & + 2r^b\psi_{(2)}h_{(2)} + 2r^b\psi_{(1)}h_{(1)}\end{aligned}$$

and hence

$$\begin{aligned}&\mathbf{D}_\gamma\left(r^b\psi_{(1)}^2 e_3^\gamma\right) + \mathbf{D}_\gamma\left(r^b\psi_{(2)}^2 e_4^\gamma\right) - r^b\underline{\kappa}\Big(-2a_{(1)} + \frac{b}{2} + 1\Big)\psi_{(1)}^2\\ & + r^b\kappa\Big(2a_{(2)} - \frac{b}{2} - 1\Big)\psi_{(2)}^2\\ =\;& 2r^b\,\not{d}_1(\psi_{(1)}\psi_{(2)}) + br^{b-1}\left(e_3(r) - \frac{r}{2}\underline{\kappa}\right)\psi_{(1)}^2 + br^{b-1}\left(e_4(r) - \frac{r}{2}\kappa\right)\psi_{(2)}^2\\ & -2r^b\underline{\omega}\psi_{(1)}^2 - 2r^b\omega\psi_{(2)}^2 + 2r^b\psi_{(1)}h_{(1)} + 2r^b\psi_{(2)}h_{(2)}.\end{aligned}$$

This concludes the proof of Lemma 8.24. □

To obtain r^p-weighted estimates for higher order derivatives of the curvature components, we will need several lemmas.

Lemma 8.26. *Let $k \geq 1$ and $s \geq 1$ two integers. Let $\psi_{(1)} \in \mathfrak{s}_k$ and $\psi_{(2)} \in \mathfrak{s}_{k-1}$. Then, we have*

$$-\not{\mathfrak{d}}^s\psi_{(1)}\,\not{\mathfrak{d}}^s\,\not{d}_k^\star\psi_{(2)} + \not{\mathfrak{d}}^s\psi_{(2)}\,\not{\mathfrak{d}}^s\,\not{d}_k\psi_{(1)} \quad = \quad \not{d}_1\Big(\not{\mathfrak{d}}^s\psi_{(1)}\,\not{\mathfrak{d}}^s\psi_{(2)}\Big) + E[\not{\mathfrak{d}}, s, k, \psi_{(1)}, \psi_{(2)}]$$

where

$$\begin{aligned}|E[\not{\mathfrak{d}}, s, k, \psi_{(1)}, \psi_{(2)}]| \quad &\lesssim \quad r|\not{\mathfrak{d}}^s\psi_{(1)}|\sum_{j=0}^{s-1}|\not{\mathfrak{d}}^{s-1-j}(\psi_{(2)})||\not{\mathfrak{d}}^j(K)| \\ &\qquad +r|\not{\mathfrak{d}}^s\psi_{(2)}|\sum_{j=0}^{s-1}|\not{\mathfrak{d}}^{s-1-j}(\psi_{(1)})||\not{\mathfrak{d}}^j(K)|.\end{aligned}$$

Proof. Recall our definition $\not{\mathfrak{d}}^s$ for higher angular derivatives. Given f a k-reduced scalar and s a positive integer we define

$$\not{\mathfrak{d}}^s f = \begin{cases} r^{2p}\not{\Delta}_k^p f, & \text{if} \quad s = 2p, \\ r^{2p+1}\,\not{d}_k\not{\Delta}_k^p f, & \text{if} \quad s = 2p+1. \end{cases}$$

We start with the case $s = 2p$, i.e., s is even. Since $\psi_{(1)} \in \mathfrak{s}_k$ and $\psi_{(2)} \in \mathfrak{s}_{k-1}$, we have

$$\begin{aligned}&-\not{\mathfrak{d}}^s\psi_{(1)}\,\not{\mathfrak{d}}^s\,\not{d}_k^\star\psi_{(2)} + \not{\mathfrak{d}}^s\psi_{(2)}\,\not{\mathfrak{d}}^s\,\not{d}_k\psi_{(1)} \\ = \quad & r^{4p}\Big(- \not{\Delta}_k^p\psi_{(1)}\not{\Delta}_k^p\,\not{d}_k^\star\psi_{(2)} + \not{\Delta}_{k-1}^p\psi_{(2)}\not{\Delta}_{k-1}^p\,\not{d}_k\psi_{(1)}\Big).\end{aligned}$$

Next, recall the commutation formulas

$$\begin{aligned}-\not{d}_k\not{\Delta}_k + \not{\Delta}_{k-1}\not{d}_k \quad &= \quad -(2k-1)K\,\not{d}_k - ke_\theta(K), \\ -\not{d}_k^\star\not{\Delta}_{k-1} + \not{\Delta}_k\,\not{d}_k^\star \quad &= \quad (2k-1)K\,\not{d}_k^\star + (k-1)e_\theta(K).\end{aligned}$$

We infer

$$\begin{aligned}\not{\Delta}_{k-1}^p\not{d}_k \quad &= \quad \not{d}_k\not{\Delta}_k^p + \sum_{j=1}^p \not{\Delta}_{k-1}^{p-j}\Big(\not{\Delta}_{k-1}\not{d}_k - \not{d}_k\not{\Delta}_k\Big)\not{\Delta}_k^{j-1} \\ &= \quad \not{d}_k\not{\Delta}_k^p + \sum_{j=1}^p \not{\Delta}_{k-1}^{p-j}\Big(- (2k-1)K\,\not{d}_k - ke_\theta(K)\Big)\not{\Delta}_k^{j-1}\end{aligned}$$

and

$$\begin{aligned}\not{\Delta}_k^p\,\not{d}_k^\star \quad &= \quad \not{d}_k^\star\not{\Delta}_{k-1}^p + \sum_{j=1}^p \not{\Delta}_k^{p-j}\Big(\not{\Delta}_k\,\not{d}_k^\star - \not{d}_k^\star\not{\Delta}_{k-1}\Big)\not{\Delta}_{k-1}^{j-1} \\ &= \quad \not{d}_k^\star\not{\Delta}_{k-1}^p + \sum_{j=1}^p \not{\Delta}_k^{p-j}\Big((2k-1)K\,\not{d}_k^\star + (k-1)e_\theta(K)\Big)\not{\Delta}_{k-1}^{j-1}.\end{aligned}$$

This yields

$$
\begin{aligned}
& -\not{\mathfrak{d}}^s\psi_{(1)}\not{\mathfrak{d}}^s\not{d}_k^\star\psi_{(2)}+\not{\mathfrak{d}}^s\psi_{(2)}\not{\mathfrak{d}}^s\not{d}_k\psi_{(1)}\\
= \; & r^{4p}\Bigg\{-\not{\Delta}_k^p\psi_{(1)}\not{d}_k^\star\not{\Delta}_{k-1}^p\psi_{(2)}\\
& -\sum_{j=1}^p\not{\Delta}_k^p\psi_{(1)}\not{\Delta}_k^{p-j}\Big((2k-1)K\not{d}_k^\star+(k-1)e_\theta(K)\Big)\not{\Delta}_{k-1}^{j-1}\psi_{(2)}\\
& +\not{\Delta}_{k-1}^p\psi_{(2)}\not{d}_k\not{\Delta}_k^p\psi_{(1)}\\
& +\sum_{j=1}^p\not{\Delta}_{k-1}^p\psi_{(2)}\not{\Delta}_{k-1}^{p-j}\Big(-(2k-1)K\not{d}_k-ke_\theta(K)\Big)\not{\Delta}_k^{j-1}\psi_{(1)}\Bigg\}\\
= \; & \not{d}_1\Big(\not{\mathfrak{d}}^s\psi_{(1)}\not{\mathfrak{d}}^s\psi_{(2)}\Big)\\
& -\sum_{j=1}^p\not{\mathfrak{d}}^s\psi_{(1)}\not{\mathfrak{d}}^{s-2j}\Big((2k-1)r^2K\not{d}_k^\star+(k-1)r^2e_\theta(K)\Big)\not{\mathfrak{d}}^{2j-2}\psi_{(2)}\\
& +\sum_{j=1}^p\not{\mathfrak{d}}^s\psi_{(2)}\not{\mathfrak{d}}^{s-2j}\Big(-(2k-1)r^2K\not{d}_k-kr^2e_\theta(K)\Big)\not{\mathfrak{d}}^{2j-2}\psi_{(1)}.
\end{aligned}
$$

Hence, we infer

$$
-\not{\mathfrak{d}}^s\psi_{(1)}\not{\mathfrak{d}}^s\not{d}_k^\star\psi_{(2)}+\not{\mathfrak{d}}^s\psi_{(2)}\not{\mathfrak{d}}^s\not{d}_k\psi_{(1)} \;=\; \not{d}_1\Big(\not{\mathfrak{d}}^s\psi_{(1)}\not{\mathfrak{d}}^s\psi_{(2)}\Big)+E[\not{\mathfrak{d}},s,k,\psi_{(1)},\psi_{(2)}]
$$

where

$$
\begin{aligned}
|E[\not{\mathfrak{d}},s,k,\psi_{(1)},\psi_{(2)}]| \;\lesssim\; & r^2|\not{\mathfrak{d}}^s\psi_{(1)}|\sum_{j=0}^{s-1}|\not{\mathfrak{d}}^{s-1-j}(\psi_{(2)})||\not{\mathfrak{d}}^j(K)|\\
& +r^2|\not{\mathfrak{d}}^s\psi_{(2)}|\sum_{j=0}^{s-1}|\not{\mathfrak{d}}^{s-1-j}(\psi_{(1)})||\not{\mathfrak{d}}^j(K)|.
\end{aligned}
$$

Next, we deal with the case $s=2p+1$, i.e., s odd. Since $\psi_{(1)}\in\mathfrak{s}_k$ and $\psi_{(2)}\in\mathfrak{s}_{k-1}$, we have

$$
\begin{aligned}
& -\not{\mathfrak{d}}^s\psi_{(1)}\not{\mathfrak{d}}^s\not{d}_k^\star\psi_{(2)}+\not{\mathfrak{d}}^s\psi_{(2)}\not{\mathfrak{d}}^s\not{d}_k\psi_{(1)}\\
= \; & r^{4p+2}\Big(-\not{d}_k\not{\Delta}_k^p\psi_{(1)}\not{d}_k\not{\Delta}_k^p\not{d}_k^\star\psi_{(2)}+\not{d}_{k-1}\not{\Delta}_{k-1}^p\psi_{(2)}\not{d}_{k-1}\not{\Delta}_{k-1}^p\not{d}_k\psi_{(1)}\Big).
\end{aligned}
$$

In view of the case $s = 2p$ above, we infer

$$
\begin{aligned}
& -\not{\partial}^s\psi_{(1)}\,\not{\partial}^s\,\not{d}_k^\star\psi_{(2)} + \not{\partial}^s\psi_{(2)}\,\not{\partial}^s\,\not{d}_k\psi_{(1)}\\
= \quad & r^{4p+2}\Bigg\{ -\not{d}_k\not{\Delta}_k^p\psi_{(1)}\,\not{d}_k\,\not{d}_k^\star\not{\Delta}_{k-1}^p\psi_{(2)}\\
& -\sum_{j=1}^p \not{d}_k\not{\Delta}_k^p\psi_{(1)}\,\not{d}_k\not{\Delta}_k^{p-j}\Big((2k-1)K\,\not{d}_k^\star + (k-1)e_\theta(K)\Big)\not{\Delta}_{k-1}^{j-1}\psi_{(2)}\\
& +\not{d}_{k-1}\not{\Delta}_{k-1}^p\psi_{(2)}\,\not{d}_{k-1}\,\not{d}_k\not{\Delta}_k^p\psi_{(1)}\\
& +\sum_{j=1}^p \not{d}_{k-1}\not{\Delta}_{k-1}^p\psi_{(2)}\,\not{d}_{k-1}\not{\Delta}_{k-1}^{p-j}\Big(-(2k-1)K\,\not{d}_k - ke_\theta(K)\Big)\not{\Delta}_k^{j-1}\psi_{(1)}\Bigg\}.
\end{aligned}
$$

Next, recall the commutation formula

$$
\not{d}_k\,\not{d}_k^\star - \not{d}_{k-1}^\star\,\not{d}_{k-1} \quad = \quad -2(k-1)K.
$$

We infer

$$
\begin{aligned}
& -\not{\partial}^s\psi_{(1)}\,\not{\partial}^s\,\not{d}_k^\star\psi_{(2)} + \not{\partial}^s\psi_{(2)}\,\not{\partial}^s\,\not{d}_k\psi_{(1)}\\
= \quad & r^{4p+2}\Bigg\{ -\not{d}_k\not{\Delta}_k^p\psi_{(1)}\Big(\not{d}_{k-1}^\star\,\not{d}_{k-1} - 2(k-1)K\Big)\not{\Delta}_{k-1}^p\psi_{(2)}\\
& -\sum_{j=1}^p \not{d}_k\not{\Delta}_k^p\psi_{(1)}\,\not{d}_k\not{\Delta}_k^{p-j}\Big((2k-1)K\,\not{d}_k^\star + (k-1)e_\theta(K)\Big)\not{\Delta}_{k-1}^{j-1}\psi_{(2)}\\
& +\not{d}_{k-1}\not{\Delta}_{k-1}^p\psi_{(2)}\,\not{d}_{k-1}\,\not{d}_k\not{\Delta}_k^p\psi_{(1)}\\
& +\sum_{j=1}^p \not{d}_{k-1}\not{\Delta}_{k-1}^p\psi_{(2)}\,\not{d}_{k-1}\not{\Delta}_{k-1}^{p-j}\Big(-(2k-1)K\,\not{d}_k - ke_\theta(K)\Big)\not{\Delta}_k^{j-1}\psi_{(1)}\Bigg\}\\
= \quad & \not{d}_1\Big(\not{\partial}^s\psi_{(1)}\,\not{\partial}^s\psi_{(2)}\Big)\\
& +2(k-1)r^2K\,\not{\partial}^s\psi_{(1)}\,\not{\partial}^{s-1}\psi_{(2)}\\
& -\sum_{j=1}^p \not{\partial}^s\psi_{(1)}\,\not{\partial}^{s-2j}\Big((2k-1)r^2K\,\not{d}_k^\star + (k-1)r^2e_\theta(K)\Big)\not{\partial}^{2j-2}\psi_{(2)}\\
& +\sum_{j=1}^p \not{\partial}^s\psi_{(2)}\,\not{\partial}^{s-2j}\Big(-(2k-1)r^2K\,\not{d}_k - kr^2e_\theta(K)\Big)\not{\partial}^{2j-2}\psi_{(1)}.
\end{aligned}
$$

Hence, we obtain

$$
-\not{\partial}^s\psi_{(1)}\,\not{\partial}^s\,\not{d}_k^\star\psi_{(2)} + \not{\partial}^s\psi_{(2)}\,\not{\partial}^s\,\not{d}_k\psi_{(1)} \quad = \quad \not{d}_1\Big(\not{\partial}^s\psi_{(1)}\,\not{\partial}^s\psi_{(2)}\Big) + E[\not{\partial}, s, k, \psi_{(1)}, \psi_{(2)}]
$$

where

$$|E[\not{\mathfrak{d}}, s, k, \psi_{(1)}, \psi_{(2)}]| \lesssim r^2|\not{\mathfrak{d}}^s\psi_{(1)}|\sum_{j=0}^{s-1}|\not{\mathfrak{d}}^{s-1-j}(\psi_{(2)})||\not{\mathfrak{d}}^j(K)| \\ +r^2|\not{\mathfrak{d}}^s\psi_{(2)}|\sum_{j=0}^{s-1}|\not{\mathfrak{d}}^{s-1-j}(\psi_{(1)})||\not{\mathfrak{d}}^j(K)|.$$

This concludes the proof of the lemma. □

Corollary 8.27. *Let $k \geq 1$, let $a_{(1)}$ and $a_{(2)}$ real numbers and let $0 \leq s \leq k_{large}$. Consider the outgoing geodesic foliation of ${}^{(ext)}\mathcal{M}$. We consider the following equations.*

- *If $\psi_{(1)} \in \mathfrak{s}_k$, $\psi_{(2)} \in \mathfrak{s}_{k-1}$, let $(\psi_{(1)}, \psi_{(2,s)})$ such that*

$$\begin{cases} e_3(\not{\mathfrak{d}}^s\psi_{(1)}) + a_{(1)}\underline{\kappa}\not{\mathfrak{d}}^s\psi_{(1)} &= -\not{\mathfrak{d}}^s \not{d}_k^\star\psi_{(2)} + h_{(1,s)}, \\ e_4(\not{\mathfrak{d}}^s\psi_{(2)}) + a_{(2)}\kappa\not{\mathfrak{d}}^s\psi_{(2)} &= \not{\mathfrak{d}}^s \not{d}_k\psi_{(1)} + h_{(2,s)}. \end{cases}$$

- *If $\psi_{(1)} \in \mathfrak{s}_{k-1}$, $\psi_{(2)}, h_{(2)} \in \mathfrak{s}_k$, let $(\psi_{(1)}, \psi_{(2)})$ such that*

$$\begin{cases} e_3(\not{\mathfrak{d}}^s\psi_{(1)}) + a_{(1)}\underline{\kappa}\not{\mathfrak{d}}^s\psi_{(1)} &= \not{\mathfrak{d}}^s \not{d}_k\psi_{(2)} + h_{(1,s)}, \\ e_4(\not{\mathfrak{d}}^s\psi_{(2)}) + a_{(2)}\kappa\not{\mathfrak{d}}^s\psi_{(2)} &= -\not{\mathfrak{d}}^s \not{d}_k^\star\psi_{(1)} + h_{(2,s)}. \end{cases}$$

Then, the pair $(\psi_{(1)}, \psi_{(2)})$ satisfies for any real number b

$$\begin{aligned} &\mathbf{Div}\Big(r^b(\not{\mathfrak{d}}^s\psi_{(1)})^2e_3\Big) + \mathbf{Div}\Big(r^b(\not{\mathfrak{d}}^s\psi_{(2)})^2e_4\Big) \\ &-\frac{1}{2}r^b\underline{\kappa}\Big(-4a_{(1)} + b + 2\Big)(\not{\mathfrak{d}}^s\psi_{(1)})^2 + \frac{1}{2}r^b\kappa\Big(4a_{(2)} - b - 2\Big)(\not{\mathfrak{d}}^s\psi_{(2)})^2 \\ = \;& 2r^b\not{d}_1\Big(\not{\mathfrak{d}}^s\psi_{(1)}\not{\mathfrak{d}}^s\psi_{(2)}\Big) + 2r^bE[\not{\mathfrak{d}}, s, k, \psi_{(1)}, \psi_{(2)}] - 2r^b\underline{\omega}(\not{\mathfrak{d}}^s\psi_{(1)})^2 \\ &+2r^b\not{\mathfrak{d}}^s\psi_{(1)}h_{(1,s)} + 2r^b\not{\mathfrak{d}}^s\psi_{(2)}h_{(2,s)} + br^{b-1}\left(e_3(r) - \frac{r}{2}\underline{\kappa}\right)(\not{\mathfrak{d}}^s\psi_{(1)})^2 \\ &+br^{b-1}\left(e_4(r) - \frac{r}{2}\kappa\right)(\not{\mathfrak{d}}^s\psi_{(2)})^2 \end{aligned}$$

where $E[\not{\mathfrak{d}}, s, k, \psi_{(1)}, \psi_{(2)}]$ has been introduced in Lemma 8.26.

Proof. The proof follows immediately from combining Lemma 8.24 and Lemma 8.26. □

Lemma 8.28. *Let j, k, l three integers. Consider a Bianchi $(\psi_{(1)}, \psi_{(2)})$ satisfying*

(8.7.1) *or* (8.7.2). *Then, the pair* $(\psi_{(1)}, \psi_{(2)})$ *satisfies for any real number* b

$$\begin{aligned}
&\mathbf{Div}\Big(r^b(\not\mathfrak{d}^j(re_4)^k\mathbf{T}^l\psi_{(1)})^2e_3\Big) + \mathbf{Div}\Big(r^b(\not\mathfrak{d}^j(re_4)^k\mathbf{T}^l\psi_{(2)})^2e_4\Big)\\
&-\frac{1}{2}r^b\underline{\kappa}\Big(-4a_{(1)}+2k+b+2\Big)(\not\mathfrak{d}^j(re_4)^k\mathbf{T}^l\psi_{(1)})^2\\
&+\frac{1}{2}r^b\kappa\Big(4a_{(2)}-2k-b-2\Big)(\not\mathfrak{d}^j(re_4)^k\mathbf{T}^l\psi_{(2)})^2\\
= \;& 2r^b\,\not{d}_1\Big(\not\mathfrak{d}^j(re_4)^k\mathbf{T}^l\psi_{(1)}\,\not\mathfrak{d}^j(re_4)^k\mathbf{T}^l\psi_{(2)}\Big)\\
&+2r^bE[\not\mathfrak{d}, j, k, (re_4)^k\mathbf{T}^l\psi_{(1)}, (re_4)^k\mathbf{T}^l\psi_{(2)}] - 2r^b\underline{\omega}(\not\mathfrak{d}^j(re_4)^k\mathbf{T}^l\psi_{(1)})^2\\
&+2r^b\not\mathfrak{d}^j(re_4)^k\mathbf{T}^l\psi_{(1)}h_{(1),j,k,l} + 2r^b\not\mathfrak{d}^j(re_4)^k\mathbf{T}^l\psi_{(2)}h_{(2),j,k,l}\\
&+br^{b-1}\Big(e_3(r)-\frac{r}{2}\underline{\kappa}\Big)(\not\mathfrak{d}^j(re_4)^k\mathbf{T}^l\psi_{(1)})^2\\
&+br^{b-1}\Big(e_4(r)-\frac{r}{2}\kappa\Big)(\not\mathfrak{d}^j(re_4)^k\mathbf{T}^l\psi_{(2)})^2
\end{aligned}$$

where $E[\not\mathfrak{d}, s, k, (re_4)^k\mathbf{T}^l\psi_{(1)}, (re_4)^k\mathbf{T}^l\psi_{(2)}]$ *has been introduced in Lemma 8.26, and where* $h_{(1),j,k,l}$ *and* $h_{(2),j,k,l}$ *are given, schematically, by*

$$\begin{aligned}
h_{(1),j,k,l} \;=\;& \not\mathfrak{d}^{\leq j+k+l}(h_{(1)}) + kr^{-1}\not\mathfrak{d}^{j+1}(re_4)^{k-1}\mathbf{T}^l\psi_{(2)}\\
&+r\mathfrak{d}^{j+k+l}\Big(\Gamma_g(\psi_{(1)},\psi_{(2)})\Big) + O(r^{-1})\mathfrak{d}^{\leq j+k+l-1}(\psi_{(1)},\psi_{(2)})
\end{aligned}$$

and

$$\begin{aligned}
h_{(2),j,k,l} \;=\;& \not\mathfrak{d}^{\leq j+k+l}(h_{(2)}) + kr^{-1}\not\mathfrak{d}^{j+1}(re_4)^{k-1}\mathbf{T}^l\psi_{(1)} + r\mathfrak{d}^{j+k+l}\Big(\Gamma_g(\psi_{(1)},\psi_{(2)})\Big)\\
&+O(r^{-1})\mathfrak{d}^{\leq j+k+l-1}(\psi_{(1)},\psi_{(2)}).
\end{aligned}$$

Proof. We have the following simple schematic consequences of the commutator identities:

$$\begin{aligned}
&[T, e_4],\ [T, e_3] = r^{-1}\Gamma_b\mathfrak{d}, && [T, \not{d}_k] = -\eta e_3 + \Gamma_g\mathfrak{d},\\
&[\not\mathfrak{d}, e_4] = \Gamma_g\mathfrak{d} + \Gamma_g, && [\not\mathfrak{d}, e_3] = -r\eta e_3 + r\Gamma_g\mathfrak{d},\\
&[re_4, e_4] = -\frac{r}{2}\kappa e_4 + \Gamma_g\mathfrak{d}, && [re_4, e_3] = -\frac{r}{2}\underline{\kappa}e_4 + \Gamma_b\mathfrak{d},\\
&[re_4, \not\mathfrak{d}_k] = r^{-1}\not\mathfrak{d} + \Gamma_g\mathfrak{d} + \Gamma_g.
\end{aligned}$$

Then, differentiating with $\not\mathfrak{d}^j(re_4)^k\mathbf{T}^l$ the equations

$$\begin{cases}
e_3(\psi_{(1)}) + a_{(1)}\underline{\kappa}\psi_{(1)} &= -\not{d}_k^\star\psi_{(2)} + h_{(1)},\\
e_4(\psi_{(2)}) + a_{(2)}\kappa\psi_{(2)} &= \not{d}_k\psi_{(1)} + h_{(2)},
\end{cases}$$

and using the above commutator identities we infer

$$\begin{cases}
e_3(\not\mathfrak{d}^j(re_4)^k\mathbf{T}^l\psi_{(1)}) + \left(a_{(1)} - \frac{k}{2}\right)\underline{\kappa}\not\mathfrak{d}^j(re_4)^k\mathbf{T}^l\psi_{(1)} &= -\not\mathfrak{d}^j\not{d}_k^\star((re_4)^k\mathbf{T}^l\psi_{(2)})\\
&\quad +h_{(1),j,k,l},\\
e_4(\not\mathfrak{d}^j(re_4)^k\mathbf{T}^l\psi_{(2)}) + \left(a_{(2)} - \frac{k}{2}\right)\kappa\not\mathfrak{d}^j(re_4)^k\mathbf{T}^l\psi_{(2)} &= \not\mathfrak{d}^j\not{d}_k((re_4)^k\mathbf{T}^l\psi_{(1)})\\
&\quad +h_{(2),j,k,l},
\end{cases}$$

where

$$\begin{aligned} h_{(1),j,k,l} &= \not{\mathfrak{d}}^j (re_4)^k \mathbf{T}^l (h_{(1)}) + kr^{-1} \not{\mathfrak{d}}^{j+1} (re_4)^{k-1} \mathbf{T}^l \psi_{(2)} + jr\eta \mathfrak{d}^{j+k+l-1} e_3 \psi_{(1)} \\ &\quad + r\mathfrak{d}^{j+k+l}\Big(\Gamma_g(\psi_{(1)}, \psi_{(2)})\Big) + O(r^{-1})\mathfrak{d}^{\leq j+k+l-1}(\psi_{(1)}, \psi_{(2)}) \end{aligned}$$

and

$$\begin{aligned} h_{(2),j,k,l} &= \not{\mathfrak{d}}^j (re_4)^k \mathbf{T}^l (h_{(2)}) + kr^{-1} \not{\mathfrak{d}}^{j+1} (re_4)^{k-1} \mathbf{T}^l \psi_{(1)} \\ &\quad + r\mathfrak{d}^{j+k+l}\Big(\Gamma_g(\psi_{(1)}, \psi_{(2)})\Big) + O(r^{-1})\mathfrak{d}^{\leq j+k+l-1}(\psi_{(1)}, \psi_{(2)}). \end{aligned}$$

Also, using the equation

$$e_3(\psi_{(1)}) = -a_{(1)}\underline{\kappa}\psi_{(1)} - \not{d}_k^\star \psi_{(2)} + h_{(1)},$$

we obtain

$$\begin{aligned} jr\eta\mathfrak{d}^{j+k+l-1} e_3 \psi_{(1)} &= r\mathfrak{d}^{j+k+l}\Big(\Gamma_g(\psi_{(1)}, \psi_{(2)})\Big) + O(r^{-1})\mathfrak{d}^{\leq j+k+l-1}(\psi_{(1)}, \psi_{(2)}) \\ &\quad + r\eta\mathfrak{d}^{k+j+l-1}(h_{(1)}) \end{aligned}$$

and hence,

$$\begin{aligned} h_{(1),j,k,l} &= \not{\mathfrak{d}}^{\leq j+k+l}(h_{(1)}) + kr^{-1} \not{\mathfrak{d}}^{j+1} (re_4)^{k-1} \mathbf{T}^l \psi_{(2)} \\ &\quad + r\mathfrak{d}^{j+k+l}\Big(\Gamma_g(\psi_{(1)}, \psi_{(2)})\Big) + O(r^{-1})\mathfrak{d}^{\leq j+k+l-1}(\psi_{(1)}, \psi_{(2)}). \end{aligned}$$

We have thus obtained the desired form for $h_{(1),j,k,l}$ and $h_{(2),j,k,l}$.

The divergence identity now follows from the equations

$$\begin{cases} e_3(\not{\mathfrak{d}}^j (re_4)^k \mathbf{T}^l \psi_{(1)}) + \left(a_{(1)} - \frac{k}{2}\right) \underline{\kappa} \not{\mathfrak{d}}^j (re_4)^k \mathbf{T}^l \psi_{(1)} = -\not{\mathfrak{d}}^j \not{d}_k^\star((re_4)^k \mathbf{T}^l \psi_{(2)}) \\ \qquad + h_{(1),j,k,l}, \\ e_4(\not{\mathfrak{d}}^j (re_4)^k \mathbf{T}^l \psi_{(2)}) + \left(a_{(2)} - \frac{k}{2}\right) \kappa \not{\mathfrak{d}}^j (re_4)^k \mathbf{T}^l \psi_{(2)} = \not{\mathfrak{d}}^j \not{d}_k((re_4)^k \mathbf{T}^l \psi_{(1)}) \\ \qquad + h_{(2),j,k,l}, \end{cases}$$

together with Corollary 8.27. This concludes the proof of the lemma. □

Corollary 8.29. *Let $r_0 \geq 4m_0$ and $1 \leq u_0 \leq u_*$. We introduce the spacetime region*

$$\mathcal{R}_{u_0} = {}^{(ext)}\mathcal{M} \cap \{r \geq 4m_0\} \cap \{1 \leq u \leq u_0\}.$$

Let j, k, l three integers. Assume that the frame of ${}^{(ext)}\mathcal{M}$ satisfies

$$\sup_{{}^{(ext)}\mathcal{M}} \left(\left| e_3(r) - \frac{r}{2}\underline{\kappa} \right| + r\left(|\underline{\omega}| + \left| e_4(r) - \frac{r}{2}\kappa \right| \right) \right) \lesssim \epsilon_0.$$

Consider a pair $(\psi_{(1)}, \psi_{(2)})$ satisfying (8.7.1) or (8.7.2). Then, $(\psi_{(1)}, \psi_{(2)})$ satisfies, for any real number b,

a) If

$$-4a_{(1)} + 2k + b + 2 > 0 \text{ and } 4a_{(2)} - 2k - b - 2 > 0,$$

then, we have

$$\begin{aligned}
&\int_{\mathcal{C}_{u_0}(r\geq r_0)} r^b(\not{\mathfrak{d}}^j(re_4)^k\mathbf{T}^l\psi_{(1)})^2\\
&+\int_{\Sigma_*(\leq u_0)} r^b\Big((\not{\mathfrak{d}}^j(re_4)^k\mathbf{T}^l\psi_{(1)})^2+(\not{\mathfrak{d}}^j(re_4)^k\mathbf{T}^l\psi_{(2)})^2\Big)\\
&+\int_{\mathcal{R}_{u_0}(r\geq r_0)} r^{b-1}\Big((\not{\mathfrak{d}}^j(re_4)^k\mathbf{T}^l\psi_{(1)})^2+(\not{\mathfrak{d}}^j(re_4)^k\mathbf{T}^l\psi_{(2)})^2\Big)\\
\lesssim &\int_{{}^{(ext)}\mathcal{M}(\frac{r_0}{2}\leq r\leq r_0)} r^{b-1}\Big((\not{\mathfrak{d}}^j(re_4)^k\mathbf{T}^l\psi_{(1)})^2+(\not{\mathfrak{d}}^j(re_4)^k\mathbf{T}^l\psi_{(2)})^2\Big)\\
&+\int_{\mathcal{R}_{u_0}(r\geq r_0)} r^{b+1}\Big((h_{(1),j,k,l})^2+(h_{(2),j,k,l})^2\Big)\\
&+\int_{\mathcal{R}_{u_0}(r\geq r_0)} r^bE[\not{\mathfrak{d}},j,k,(re_4)^k\mathbf{T}^l\psi_{(1)},(re_4)^k\mathbf{T}^l\psi_{(2)}].
\end{aligned}$$

b) If

$$-4a_{(1)}+2k+b+2\leq 0 \textit{ and } 4a_{(2)}-2k-b-2>0,$$

then, we have

$$\begin{aligned}
&\int_{\mathcal{C}_{u_0}(r\geq r_0)} r^b(\not{\mathfrak{d}}^j(re_4)^k\mathbf{T}^l\psi_{(1)})^2\\
&+\int_{\Sigma_*(\leq u_0)} r^b\Big((\not{\mathfrak{d}}^j(re_4)^k\mathbf{T}^l\psi_{(1)})^2+(\not{\mathfrak{d}}^j(re_4)^k\mathbf{T}^l\psi_{(2)})^2\Big)\\
&+\int_{\mathcal{R}_{u_0}(r\geq r_0)} r^{b-1}(\not{\mathfrak{d}}^j(re_4)^k\mathbf{T}^l\psi_{(2)})^2\\
\lesssim &\int_{{}^{(ext)}\mathcal{M}(\frac{r_0}{2}\leq r\leq r_0)} r^{b-1}\Big((\not{\mathfrak{d}}^j(re_4)^k\mathbf{T}^l\psi_{(1)})^2+(\not{\mathfrak{d}}^j(re_4)^k\mathbf{T}^l\psi_{(2)})^2\Big)\\
&+\int_{\mathcal{R}_{u_0}(r\geq r_0)} r^{b+1}\Big((h_{(1),j,k,l})^2+(h_{(2),j,k,l})^2\Big)\\
&+\int_{\mathcal{R}_{u_0}(r\geq r_0)} r^{b-1}(\not{\mathfrak{d}}^j(re_4)^k\mathbf{T}^l\psi_{(1)})^2\\
&+\int_{\mathcal{R}_{u_0}(r\geq r_0)} r^bE[\not{\mathfrak{d}},j,k,(re_4)^k\mathbf{T}^l\psi_{(1)},(re_4)^k\mathbf{T}^l\psi_{(2)}].
\end{aligned}$$

c) If

$$4a_{(2)}-2k-b-2=0,$$

then, we have

$$\begin{aligned}
&\int_{\mathcal{C}_{u_0}(r\geq r_0)} r^b(\not{\mathfrak{d}}^j(re_4)^k\mathbf{T}^l\psi_{(1)})^2 \\
&+\int_{\Sigma_*(\leq u_0)} r^b\Big((\not{\mathfrak{d}}^j(re_4)^k\mathbf{T}^l\psi_{(1)})^2+(\not{\mathfrak{d}}^j(re_4)^k\mathbf{T}^l\psi_{(2)})^2\Big) \\
\lesssim &\int_{{}^{(ext)}\mathcal{M}(\frac{r_0}{2}\leq r\leq r_0)} r^{b-1}\Big((\not{\mathfrak{d}}^j(re_4)^k\mathbf{T}^l\psi_{(1)})^2+(\not{\mathfrak{d}}^j(re_4)^k\mathbf{T}^l\psi_{(2)})^2\Big) \\
&+\int_{\mathcal{R}_{u_0}(r\geq r_0)} r^{b+1-\delta_B}(h_{(1),j,k,l})^2+\int_{\mathcal{R}_{u_0}(r\geq r_0)} r^{b+1+\delta_B}(h_{(2),j,k,l})^2 \\
&+\int_{\mathcal{R}_{u_0}(r\geq r_0)} r^{b-1+\delta_B}(\not{\mathfrak{d}}^j(re_4)^k\mathbf{T}^l\psi_{(1)})^2 \\
&+\int_{\mathcal{R}_{u_0}(r\geq r_0)} r^{b-1-\delta_B}(\not{\mathfrak{d}}^j(re_4)^k\mathbf{T}^l\psi_{(2)})^2 \\
&+\int_{\mathcal{R}_{u_0}(r\geq r_0)} r^b E[\not{\mathfrak{d}},j,k,(re_4)^k\mathbf{T}^l\psi_{(1)},(re_4)^k\mathbf{T}^l\psi_{(2)}].
\end{aligned}$$

d) If

$$-4a_{(1)}+2k+b+2>0,$$

then, we have

$$\begin{aligned}
&\int_{\mathcal{C}_{u_0}(r\geq r_0)} r^b(\not{\mathfrak{d}}^j(re_4)^k\mathbf{T}^l\psi_{(1)})^2 \\
&+\int_{\Sigma_*(\leq u_0)} r^b\Big((\not{\mathfrak{d}}^j(re_4)^k\mathbf{T}^l\psi_{(1)})^2+(\not{\mathfrak{d}}^j(re_4)^k\mathbf{T}^l\psi_{(2)})^2\Big) \\
&+\int_{\mathcal{R}_{u_0}(r\geq r_0)} r^{b-1}(\not{\mathfrak{d}}^j(re_4)^k\mathbf{T}^l\psi_{(1)})^2 \\
\lesssim &\int_{{}^{(ext)}\mathcal{M}(\frac{r_0}{2}\leq r\leq r_0)} r^{b-1}\Big((\not{\mathfrak{d}}^j(re_4)^k\mathbf{T}^l\psi_{(1)})^2+(\not{\mathfrak{d}}^j(re_4)^k\mathbf{T}^l\psi_{(2)})^2\Big) \\
&+\int_{\mathcal{R}_{u_0}(r\geq r_0)} r^{b+1}(h_{(1),j,k,l})^2+\int_{\mathcal{R}_{u_0}(r\geq r_0)} r^{b+1}(h_{(2),j,k,l})^2 \\
&+\int_{\mathcal{R}_{u_0}(r\geq r_0)} r^{b-1}(\not{\mathfrak{d}}^j(re_4)^k\mathbf{T}^l\psi_{(2)})^2 \\
&+\int_{\mathcal{R}_{u_0}(r\geq r_0)} r^b E[\not{\mathfrak{d}},j,k,(re_4)^k\mathbf{T}^l\psi_{(1)},(re_4)^k\mathbf{T}^l\psi_{(2)}].
\end{aligned}$$

Proof. We multiply the pair $(\psi_{(1)},\psi_{(2)})$ by a smooth cut-off function in r supported in $r\geq\frac{r_0}{2}$ and identically one for $r\geq r_0$. We obtain again a solution to (8.7.1) or (8.7.2) up to error terms that are supported in the region $\frac{r_0}{2}\leq r\leq r_0$. We then integrate the divergence identities of Lemma 8.28 on the region $\mathcal{R}_{u_0}$ and the corollary follows. □

8.7.2 End of the proof of Proposition 8.10

Let $r_0 \geq 4m_0$. Recall that, to prove Proposition 8.10, it suffices to establish the following inequality:

$$
{}^{(ext)}\mathfrak{R}^{\geq r_0}_{J+1}[\check{R}] \quad \lesssim \quad r_0^{-\delta_B}\, {}^{(ext)}\mathfrak{G}^{\geq r_0}_{k}[\check{\Gamma}] + r_0^{10}\left(\epsilon_{\mathcal{B}}[J] + \epsilon_0\left(\mathfrak{N}^{(En)}_{J+1} + \mathcal{N}^{(match)}_{J+1}\right)\right).
$$

To this end, we will rely on the r^p-weighted estimates derived in Corollary 8.29 applied to the Bianchi pairs, where we recall Remark 8.25.

Remark 8.30. *For the Bianchi pair (β, ρ), we replace the Bianchi identities for $e_4(\rho)$ by its analog for $e_4(\check{\rho})$, i.e.,*

$$
e_4\check{\rho} + \frac{3}{2}\overline{\kappa}\check{\rho} \quad = \quad \not{d}_1\beta - \frac{3}{2}\overline{\rho}\check{\kappa} + Err[e_4\check{\rho}],
$$

while for the Bianchi pair $(\rho, \underline{\beta})$, we replace the Bianchi identities for $e_3(\rho)$ by its analog for $e_3(\check{\rho})$, i.e.,

$$
e_3\check{\rho} + \frac{3}{2}\underline{\overline{\kappa}}\check{\rho} \quad = \quad \not{d}_1\underline{\beta} - \frac{3}{2}\overline{\rho}\underline{\check{\kappa}} - \frac{3}{2}\underline{\overline{\kappa}}\,\overline{\rho}\varsigma^{-1}\check{\varsigma} + \frac{3}{2}\overline{\kappa}\,\overline{\rho}\left(\underline{\check{\Omega}} + \varsigma^{-1}\underline{\overline{\Omega}}\check{\varsigma}\right) + Err[e_3\check{\rho}],
$$

see Proposition 2.73 for the derivation of these equations.

Let j, k, l three integers such that

$$
j + k + l = J + 1.
$$

To derive r^p-weighted curvature estimates for $\not{\mathfrak{d}}^j(re_4)^k\mathbf{T}^l$ derivatives in the region $r \geq r_0$, we proceed as follows.

Step 1. We start with the case $k = 0$, i.e., we derive r^p-weighted curvature estimates for $\not{\mathfrak{d}}^j\mathbf{T}^l$ derivatives with $j + l = J + 1$. First, we apply Corollary 8.29

- to the Bianchi pair (α, β) with the choice $b = 4 + \delta_B$,
- to the Bianchi pair (β, ρ) with the choice $b = 4 - \delta_B$,
- to the Bianchi pair $(\rho, \underline{\beta})$ with the choice $b = 2 - \delta_B$,
- to the Bianchi pair $(\underline{\beta}, \underline{\alpha})$ with the choice $b = -\delta_B$.

The above choices are such that we are in case (a) of Corollary 8.29 for the Bianchi pairs (α, β) and (β, ρ), and in case (b) of Corollary 8.29 for the last two

Bianchi pairs. In particular, we obtain

$$\begin{aligned}
&\sum_{j+l=J+1}\Bigg\{\sup_{1\leq u\leq u_*}\int_{\mathcal{C}_u(r\geq r_0)}\Big(r^{4+\delta_B}(\not{\mathfrak{d}}^j\mathbf{T}^l\alpha)^2+r^{4-\delta_B}(\not{\mathfrak{d}}^j\mathbf{T}^l\beta)^2\\
&+r^{2-\delta_B}(\not{\mathfrak{d}}^j\mathbf{T}^l\check{\rho})^2+r^{-\delta_B}(\not{\mathfrak{d}}^j\mathbf{T}^l\underline{\beta})^2\Big)+\int_{\Sigma_*}\Big(r^{4+\delta_B}\Big((\not{\mathfrak{d}}^j\mathbf{T}^l\alpha)^2+(\not{\mathfrak{d}}^j\mathbf{T}^l\beta)^2\Big)\\
&+r^{4-\delta_B}(\not{\mathfrak{d}}^j\mathbf{T}^l\check{\rho})^2+r^{2-\delta_B}(\not{\mathfrak{d}}^j\mathbf{T}^l\underline{\beta})^2+r^{-\delta_B}(\not{\mathfrak{d}}^j\mathbf{T}^l\underline{\alpha})^2\Big)\\
&+\int_{{}^{(ext)}\mathcal{M}(r\geq r_0)}\Big(r^{3+\delta_B}\Big((\not{\mathfrak{d}}^j\mathbf{T}^l\alpha)^2+(\not{\mathfrak{d}}^j\mathbf{T}^l\beta)^2\Big)+r^{3-\delta_B}(\not{\mathfrak{d}}^j\mathbf{T}^l\check{\rho})^2\\
&+r^{1-\delta_B}(\not{\mathfrak{d}}^j\mathbf{T}^l\underline{\beta})^2+r^{-1-\delta_B}(\not{\mathfrak{d}}^j\mathbf{T}^l\underline{\alpha})^2\Big)\Bigg\}\\
\lesssim\;& r_0^{8+\delta_B}\int_{{}^{(ext)}\mathcal{M}(\frac{r_0}{2}\leq r\leq r_0)}\frac{(\mathfrak{d}^{J+1}\check{R})^2}{r^5}+\int_{{}^{(ext)}\mathcal{M}(r\geq r_0)}\Bigg\{r^{-1+\delta_B}(\not{\mathfrak{d}}^j\mathbf{T}^l\vartheta)^2\\
&+r^{-1-\delta_B}\Big((\not{\mathfrak{d}}^j\mathbf{T}^l\eta)^2+(\not{\mathfrak{d}}^j\mathbf{T}^l\check{\kappa})^2\Big)+r^{-3-\delta_B}\Big((\not{\mathfrak{d}}^j\mathbf{T}^l\underline{\check{\kappa}})^2+(\not{\mathfrak{d}}^j\mathbf{T}^l\zeta)^2\Big)\\
&+r^{-5-\delta_B}\Big((\not{\mathfrak{d}}^j\mathbf{T}^l\underline{\xi})^2+(\not{\mathfrak{d}}^j\mathbf{T}^l\underline{\vartheta})^2+(\not{\mathfrak{d}}^j\mathbf{T}^l\check{\varsigma})^2+(\not{\mathfrak{d}}^j\mathbf{T}^l\underline{\check{\Omega}})^2\Big)\Bigg\}\\
&+(\epsilon_{\mathcal{B}}[J])^2+\epsilon_0^2(\mathfrak{N}^{(En)}_{J+1})^2.
\end{aligned}$$

Using Proposition 8.9 to bound the first term on the right-hand side, and using also the definition of the norm ${}^{(ext)}\mathfrak{G}_k^{\geq r_0}[\check{\Gamma}]$, we infer that

$$\begin{aligned}
&\sum_{j+l=J+1}\Bigg\{\sup_{1\leq u\leq u_*}\int_{\mathcal{C}_u(r\geq r_0)}\Big(r^{4+\delta_B}(\not{\mathfrak{d}}^j\mathbf{T}^l\alpha)^2+r^{4-\delta_B}(\not{\mathfrak{d}}^j\mathbf{T}^l\beta)^2\\
&+r^{2-\delta_B}(\not{\mathfrak{d}}^j\mathbf{T}^l\check{\rho})^2+r^{-\delta_B}(\not{\mathfrak{d}}^j\mathbf{T}^l\underline{\beta})^2\Big)+\int_{\Sigma_*}\Big(r^{4+\delta_B}\Big((\not{\mathfrak{d}}^j\mathbf{T}^l\alpha)^2+(\not{\mathfrak{d}}^j\mathbf{T}^l\beta)^2\Big)\\
&+r^{4-\delta_B}(\not{\mathfrak{d}}^j\mathbf{T}^l\check{\rho})^2+r^{2-\delta_B}(\not{\mathfrak{d}}^j\mathbf{T}^l\underline{\beta})^2+r^{-\delta_B}(\not{\mathfrak{d}}^j\mathbf{T}^l\underline{\alpha})^2\Big)\\
&+\int_{{}^{(ext)}\mathcal{M}(r\geq r_0)}\Big(r^{3+\delta_B}\Big((\not{\mathfrak{d}}^j\mathbf{T}^l\alpha)^2+(\not{\mathfrak{d}}^j\mathbf{T}^l\beta)^2\Big)\\
&+r^{3-\delta_B}(\not{\mathfrak{d}}^j\mathbf{T}^l\check{\rho})^2+r^{1-\delta_B}(\not{\mathfrak{d}}^j\mathbf{T}^l\underline{\beta})^2+r^{-1-\delta_B}(\not{\mathfrak{d}}^j\mathbf{T}^l\underline{\alpha})^2\Big)\Bigg\}\\
\lesssim\;&\left(\int_{r_0}^{+\infty}\frac{dr}{r^{1+\delta_B}}\right)({}^{(ext)}\mathfrak{G}^{\geq r_0}_{J+1}[\check{\Gamma}])^2+r_0^{10}\Big((\epsilon_{\mathcal{B}}[J])^2+\epsilon_0^2\Big(\mathfrak{N}^{(En)}_{J+1}+\mathcal{N}^{(match)}_{J+1}\Big)^2\Big)
\end{aligned}$$

and hence

$$\begin{aligned}
&\sum_{j+l=J+1}\Bigg\{\sup_{1\le u\le u_*}\int_{\mathcal{C}_u(r\ge r_0)}\Big(r^{4+\delta_B}(\not{\mathfrak{d}}^j\mathbf{T}^l\alpha)^2+r^{4-\delta_B}(\not{\mathfrak{d}}^j\mathbf{T}^l\beta)^2\\
&+r^{2-\delta_B}(\not{\mathfrak{d}}^j\mathbf{T}^l\check{\rho})^2+r^{-\delta_B}(\not{\mathfrak{d}}^j\mathbf{T}^l\underline{\beta})^2\Big)+\int_{\Sigma_*}\Big(r^{4+\delta_B}\Big((\not{\mathfrak{d}}^j\mathbf{T}^l\alpha)^2+(\not{\mathfrak{d}}^j\mathbf{T}^l\beta)^2\Big)\\
&+r^{4-\delta_B}(\not{\mathfrak{d}}^j\mathbf{T}^l\check{\rho})^2+r^{2-\delta_B}(\not{\mathfrak{d}}^j\mathbf{T}^l\underline{\beta})^2+r^{-\delta_B}(\not{\mathfrak{d}}^j\mathbf{T}^l\underline{\alpha})^2\Big)\\
&+\int_{{}^{(ext)}\mathcal{M}(r\ge r_0)}\Big(r^{3+\delta_B}\Big((\not{\mathfrak{d}}^j\mathbf{T}^l\alpha)^2+(\not{\mathfrak{d}}^j\mathbf{T}^l\beta)^2\Big)\\
&+r^{3-\delta_B}(\not{\mathfrak{d}}^j\mathbf{T}^l\check{\rho})^2+r^{1-\delta_B}(\not{\mathfrak{d}}^j\mathbf{T}^l\underline{\beta})^2+r^{-1-\delta_B}(\not{\mathfrak{d}}^j\mathbf{T}^l\underline{\alpha})^2\Big)\Bigg\}\\
\lesssim\;& r_0^{-\delta_B}({}^{(ext)}\mathfrak{G}^{\ge r_0}_{J+1}[\check{\Gamma}])^2+r_0^{10}\Big((\epsilon_{\mathcal{B}}[J])^2+\epsilon_0^2\Big(\mathfrak{N}^{(En)}_{J+1}+\mathcal{N}^{(match)}_{J+1}\Big)^2\Big). \qquad (8.7.4)
\end{aligned}$$

Step 2. We derive additional r^p-weighted curvature estimates for $\not{\mathfrak{d}}^j\mathbf{T}^l$ derivatives with $j+l=J+1$. To this end, we apply Corollary 8.29

- to the Bianchi pair (β,ρ) with the choice $b=4$,
- to the Bianchi pair $(\rho,\underline{\beta})$ with the choice $b=2$,
- to the Bianchi pair $(\underline{\beta},\underline{\alpha})$ with the choice $b=0$.

All the above choices are such that we are in case (c) of Corollary 8.29. In particular, we obtain

$$\begin{aligned}
&\sum_{j+l=J+1}\Bigg\{\sup_{1\le u\le u_*}\int_{\mathcal{C}_u(r\ge r_0)}\Big(r^4(\not{\mathfrak{d}}^j\mathbf{T}^l\beta)^2+r^2(\not{\mathfrak{d}}^j\mathbf{T}^l\check{\rho})^2+(\not{\mathfrak{d}}^j\mathbf{T}^l\underline{\beta})^2\Big)\\
&+\int_{\Sigma_*}\Big(r^4\Big((\not{\mathfrak{d}}^j\mathbf{T}^l\beta)^2+(\not{\mathfrak{d}}^j\mathbf{T}^l\check{\rho})^2\Big)+r^2(\not{\mathfrak{d}}^j\mathbf{T}^l\underline{\beta})^2+(\not{\mathfrak{d}}^j\mathbf{T}^l\underline{\alpha})^2\Big)\Bigg\}\\
\lesssim\;& r_0^8\int_{{}^{(ext)}\mathcal{M}(\frac{r_0}{2}\le r\le r_0)}\frac{(\mathfrak{d}^{J+1}\check{R})^2}{r^5}+\sum_{j+l=J+1}\Bigg\{\int_{{}^{(ext)}\mathcal{M}(r\ge r_0)}\Big(r^{3+\delta_B}(\not{\mathfrak{d}}^j\mathbf{T}^l\beta)^2\\
&+r^{3-\delta_B}(\not{\mathfrak{d}}^j\mathbf{T}^l\check{\rho})^2+r^{1-\delta_B}(\not{\mathfrak{d}}^j\mathbf{T}^l\underline{\beta})^2+r^{-1-\delta_B}(\not{\mathfrak{d}}^j\mathbf{T}^l\underline{\alpha})^2\Big)\Bigg\}\\
&+\int_{{}^{(ext)}\mathcal{M}(r\ge r_0)}\Bigg\{r^{-1-\delta_B}(\not{\mathfrak{d}}^j\mathbf{T}^l\eta)^2+r^{-1+\delta_B}(\not{\mathfrak{d}}^j\mathbf{T}^l\check{\kappa})^2\\
&+r^{-3+\delta_B}\Big((\not{\mathfrak{d}}^j\mathbf{T}^l\underline{\check{\kappa}})^2+(\not{\mathfrak{d}}^j\mathbf{T}^l\zeta)^2\Big)\\
&+r^{-5+\delta_B}\Big((\not{\mathfrak{d}}^j\mathbf{T}^l\underline{\xi})^2+(\not{\mathfrak{d}}^j\mathbf{T}^l\underline{\vartheta})^2+(\not{\mathfrak{d}}^j\mathbf{T}^l\check{\varsigma})^2+(\not{\mathfrak{d}}^j\mathbf{T}^l\underline{\check{\Omega}})^2\Big)\Bigg\}\\
&+(\epsilon_{\mathcal{B}}[J])^2+\epsilon_0^2(\mathfrak{N}^{(En)}_{J+1})^2.
\end{aligned}$$

Using Proposition 8.9 to bound the first term on the right-hand side, and using also the definition of the norm ${}^{(ext)}\mathfrak{G}_k^{\geq r_0}[\check{\Gamma}]$, we infer that

$$\begin{aligned}
&\sum_{j+l=J+1}\left\{\sup_{1\leq u\leq u_*}\int_{\mathcal{C}_u(r\geq r_0)}\Big(r^4(\not{\partial}^j\mathbf{T}^l\beta)^2+r^2(\not{\partial}^j\mathbf{T}^l\check{\rho})^2+(\not{\partial}^j\mathbf{T}^l\underline{\beta})^2\Big)\right.\\
&\left.+\int_{\Sigma_*}\Big(r^4\Big((\not{\partial}^j\mathbf{T}^l\beta)^2+(\not{\partial}^j\mathbf{T}^l\check{\rho})^2\Big)+r^2(\not{\partial}^j\mathbf{T}^l\underline{\beta})^2+(\not{\partial}^j\mathbf{T}^l\underline{\alpha})^2\Big)\right\}\\
\lesssim\;&\left(\int_{r_0}^{+\infty}\frac{dr}{r^{1+\delta_B}}\right)({}^{(ext)}\mathfrak{G}_{J+1}^{\geq r_0}[\check{\Gamma}])^2+r_0^{10}\Big((\epsilon_{\mathcal{B}}[J])^2+\epsilon_0^2\Big(\mathfrak{N}_{J+1}^{(En)}+\mathcal{N}_{J+1}^{(match)}\Big)^2\Big)\\
&+\sum_{j+l=J+1}\left\{\int_{{}^{(ext)}\mathcal{M}(r\geq r_0)}\Big(r^{3+\delta_B}(\not{\partial}^j\mathbf{T}^l\beta)^2+r^{3-\delta_B}(\not{\partial}^j\mathbf{T}^l\check{\rho})^2\right.\\
&\left.+r^{1-\delta_B}(\not{\partial}^j\mathbf{T}^l\underline{\beta})^2+r^{-1-\delta_B}(\not{\partial}^j\mathbf{T}^l\underline{\alpha})^2\Big)\right\}
\end{aligned}$$

and hence

$$\begin{aligned}
&\sum_{j+l=J+1}\left\{\sup_{1\leq u\leq u_*}\int_{\mathcal{C}_u(r\geq r_0)}\Big(r^4(\not{\partial}^j\mathbf{T}^l\beta)^2+r^2(\not{\partial}^j\mathbf{T}^l\check{\rho})^2+(\not{\partial}^j\mathbf{T}^l\underline{\beta})^2\Big)\right.\\
&\left.+\int_{\Sigma_*}\Big(r^4\Big((\not{\partial}^j\mathbf{T}^l\beta)^2+(\not{\partial}^j\mathbf{T}^l\check{\rho})^2\Big)+r^2(\not{\partial}^j\mathbf{T}^l\underline{\beta})^2+(\not{\partial}^j\mathbf{T}^l\underline{\alpha})^2\Big)\right\}\\
\lesssim\;&r_0^{-\delta_B}({}^{(ext)}\mathfrak{G}_{J+1}^{\geq r_0}[\check{\Gamma}])^2+r_0^{10}\Big((\epsilon_{\mathcal{B}}[J])^2+\epsilon_0^2\Big(\mathfrak{N}_{J+1}^{(En)}+\mathcal{N}_{J+1}^{(match)}\Big)^2\Big)\\
&+\sum_{j+l=J+1}\left\{\int_{{}^{(ext)}\mathcal{M}(r\geq r_0)}\Big(r^{3+\delta_B}(\not{\partial}^j\mathbf{T}^l\beta)^2+r^{3-\delta_B}(\not{\partial}^j\mathbf{T}^l\check{\rho})^2\right.\\
&\left.+r^{1-\delta_B}(\not{\partial}^j\mathbf{T}^l\underline{\beta})^2+r^{-1-\delta_B}(\not{\partial}^j\mathbf{T}^l\underline{\alpha})^2\Big)\right\}.
\end{aligned}$$

Together with (8.7.4), we deduce

$$\begin{aligned}
&\sum_{j+l=J+1}\left\{\sup_{1\leq u\leq u_*}\int_{\mathcal{C}_u(r\geq r_0)}\Big(r^{4+\delta_B}(\not{\partial}^j\mathbf{T}^l\alpha)^2+r^4(\not{\partial}^j\mathbf{T}^l\beta)^2+r^2(\not{\partial}^j\mathbf{T}^l\check{\rho})^2\right.\\
&+(\not{\partial}^j\mathbf{T}^l\underline{\beta})^2\Big)+\int_{\Sigma_*}\Big(r^{4+\delta_B}\Big((\not{\partial}^j\mathbf{T}^l\alpha)^2+(\not{\partial}^j\mathbf{T}^l\beta)^2\Big)+r^4(\not{\partial}^j\mathbf{T}^l\check{\rho})^2\\
&+r^2(\not{\partial}^j\mathbf{T}^l\underline{\beta})^2+(\not{\partial}^j\mathbf{T}^l\underline{\alpha})^2\Big)+\int_{{}^{(ext)}\mathcal{M}(r\geq r_0)}\Big(r^{3+\delta_B}\Big((\not{\partial}^j\mathbf{T}^l\alpha)^2+(\not{\partial}^j\mathbf{T}^l\beta)^2\Big)\\
&\left.+r^{3-\delta_B}(\not{\partial}^j\mathbf{T}^l\check{\rho})^2+r^{1-\delta_B}(\not{\partial}^j\mathbf{T}^l\underline{\beta})^2+r^{-1-\delta_B}(\not{\partial}^j\mathbf{T}^l\underline{\alpha})^2\Big)\right\}\\
\lesssim\;&r_0^{-\delta_B}({}^{(ext)}\mathfrak{G}_{J+1}^{\geq r_0}[\check{\Gamma}])^2+r_0^{10}\Big((\epsilon_{\mathcal{B}}[J])^2+\epsilon_0^2\Big(\mathfrak{N}_{J+1}^{(En)}+\mathcal{N}_{J+1}^{(match)}\Big)^2\Big).
\end{aligned}\tag{8.7.5}$$

Step 3. We now argue by iteration on k. For $0\leq k\leq J$, we consider the following iteration assumption:

$$\begin{aligned}
&\sum_{j+l=J+1-k}\Bigg\{\sup_{1\leq u\leq u_*}\int_{\mathcal{C}_u(r\geq r_0)}\Big(r^{4+\delta_B}(\not{\mathfrak{d}}^j(re_4)^k\mathbf{T}^l\alpha)^2+r^4(\not{\mathfrak{d}}^j(re_4)^k\mathbf{T}^l\beta)^2\\
&+r^2(\not{\mathfrak{d}}^j(re_4)^k\mathbf{T}^l\check{\rho})^2+(\not{\mathfrak{d}}^j(re_4)^k\mathbf{T}^l\underline{\beta})^2\Big)\\
&+\int_{\Sigma_*}\Big(r^{4+\delta_B}\Big((\not{\mathfrak{d}}^j(re_4)^k\mathbf{T}^l\alpha)^2+(\not{\mathfrak{d}}^j(re_4)^k\mathbf{T}^l\beta)^2\Big)+r^4(\not{\mathfrak{d}}^j(re_4)^k\mathbf{T}^l\check{\rho})^2\\
&+r^2(\not{\mathfrak{d}}^j(re_4)^k\mathbf{T}^l\underline{\beta})^2+(\not{\mathfrak{d}}^j(re_4)^k\mathbf{T}^l\underline{\alpha})^2\Big)\\
&+\int_{{}^{(ext)}\mathcal{M}(r\geq r_0)}\Big(r^{3+\delta_B}\Big((\not{\mathfrak{d}}^j(re_4)^k\mathbf{T}^l\alpha)^2+(\not{\mathfrak{d}}^j(re_4)^k\mathbf{T}^l\beta)^2\Big)\\
&+r^{3-\delta_B}(\not{\mathfrak{d}}^j(re_4)^k\mathbf{T}^l\check{\rho})^2+r^{1-\delta_B}(\not{\mathfrak{d}}^j(re_4)^k\mathbf{T}^l\underline{\beta})^2+r^{-1-\delta_B}(\not{\mathfrak{d}}^j(re_4)^k\mathbf{T}^l\underline{\alpha})^2\Big)\Bigg\}\\
\lesssim\;& r_0^{-\delta_B}({}^{(ext)}\mathfrak{G}^{\geq r_0}_{J+1}[\check{\Gamma}])^2+r_0^{10}\Big((\epsilon_{\mathcal{B}}[J])^2+\epsilon_0^2\Big(\mathfrak{N}^{(En)}_{J+1}+\mathcal{N}^{(match)}_{J+1}\Big)^2\Big). \qquad (8.7.6)
\end{aligned}$$

(8.7.6) holds true for $k=0$ in view of (8.7.5). We now assume that (8.7.6) holds true for k such that $0\leq k\leq J$, and our goal is to prove that it also holds for $k+1$.

First, note that the Bianchi identities for $e_4(\beta)$, $e_4(\check{\rho})$, $e_4(\underline{\beta})$ and $e_4(\underline{\alpha})$, together with (8.7.6), yield

$$\begin{aligned}
&\sum_{j+l=J+1-(k+1)}\Bigg\{\sup_{1\leq u\leq u_*}\int_{\mathcal{C}_u(r\geq r_0)}\Big(r^4(\not{\mathfrak{d}}^j(re_4)^{k+1}\mathbf{T}^l\beta)^2\\
&+r^2(\not{\mathfrak{d}}^j(re_4)^{k+1}\mathbf{T}^l\check{\rho})^2+(\not{\mathfrak{d}}^j(re_4)^{k+1}\mathbf{T}^l\underline{\beta})^2\Big)+\int_{\Sigma_*}\Big(r^{4+\delta_B}(\not{\mathfrak{d}}^j(re_4)^{k+1}\mathbf{T}^l\beta)^2\\
&+r^4(\not{\mathfrak{d}}^j(re_4)^{k+1}\mathbf{T}^l\check{\rho})^2+r^2(\not{\mathfrak{d}}^j(re_4)^{k+1}\mathbf{T}^l\underline{\beta})^2+(\not{\mathfrak{d}}^j(re_4)^{k+1}\mathbf{T}^l\underline{\alpha})^2\Big)\\
&+\int_{{}^{(ext)}\mathcal{M}(r\geq r_0)}\Big(r^{3+\delta_B}(\not{\mathfrak{d}}^j(re_4)^{k+1}\mathbf{T}^l\beta)^2+r^{3-\delta_B}(\not{\mathfrak{d}}^j(re_4)^{k+1}\mathbf{T}^l\check{\rho})^2\\
&+r^{1-\delta_B}(\not{\mathfrak{d}}^j(re_4)^{k+1}\mathbf{T}^l\underline{\beta})^2+r^{-1-\delta_B}(\not{\mathfrak{d}}^j(re_4)^{k+1}\mathbf{T}^l\underline{\alpha})^2\Big)\Bigg\}\\
\lesssim\;& r_0^{-\delta_B}({}^{(ext)}\mathfrak{G}^{\geq r_0}_{J+1}[\check{\Gamma}])^2+r_0^{10}\Big((\epsilon_{\mathcal{B}}[J])^2+\epsilon_0^2\Big(\mathfrak{N}^{(En)}_{J+1}+\mathcal{N}^{(match)}_{J+1}\Big)^2\Big). \qquad (8.7.7)
\end{aligned}$$

We still need to estimate $\not{\mathfrak{d}}^j(re_4)^{k+1}\mathbf{T}^l\alpha$. To this end, we apply Corollary 8.29 to the Bianchi pair (α,β) with the choice $b=4+\delta_B$. Since $k+1\geq 1$, we are in case (d) of Corollary 8.29. In particular, we obtain, arguing similarly as above,

$$\sum_{j+l=J+1-(k+1)}\Bigg\{\sup_{1\leq u\leq u_*}\int_{\mathcal{C}_u(r\geq r_0)} r^{4+\delta_B}(\not{\mathfrak{d}}^j(re_4)^{k+1}\mathbf{T}^l\alpha)^2$$
$$+\int_{\Sigma_*} r^{4+\delta_B}(\not{\mathfrak{d}}^j(re_4)^{k+1}\mathbf{T}^l\alpha)^2+\int_{{}^{(ext)}\mathcal{M}(r\geq r_0)} r^{3+\delta_B}(\not{\mathfrak{d}}^j(re_4)^{k+1}\mathbf{T}^l\alpha)^2\Bigg\}$$
$$\lesssim \sum_{j+l=J+1-(k+1)}\Bigg\{\int_{{}^{(ext)}\mathcal{M}(r\geq r_0)} r^{3+\delta_B}(\not{\mathfrak{d}}^j(re_4)^{k+1}\mathbf{T}^l\beta)^2\Bigg\}$$
$$+r_0^{-\delta_B}({}^{(ext)}\mathfrak{G}^{\geq r_0}_{J+1}[\check{\Gamma}])^2+r_0^{10}\Big((\epsilon_{\mathcal{B}}[J])^2+\epsilon_0^2\Big(\mathfrak{N}^{(En)}_{J+1}+\mathcal{N}^{(match)}_{J+1}\Big)^2\Big).$$

Together with (8.7.7), this implies (8.7.6) for $k+1$. Hence, by iteration, (8.7.6) holds for any $0\leq k\leq J+1$. This implies

$$\sum_{k\leq J+1}\Bigg\{\sup_{1\leq u\leq u_*}\int_{\mathcal{C}_u(r\geq r_0)}\Big(r^{4+\delta_B}(\mathfrak{d}^k\alpha)^2+r^4(\mathfrak{d}^k\beta)^2+r^2(\mathfrak{d}^k\check{\rho})^2+(\mathfrak{d}^k\underline{\beta})^2\Big)$$
$$+\int_{\Sigma_*}\Big(r^{4+\delta_B}\Big((\mathfrak{d}^k\alpha)^2+(\mathfrak{d}^k\beta)^2\Big)+r^4(\mathfrak{d}^k\check{\rho})^2+r^2(\mathfrak{d}^k\underline{\beta})^2+(\mathfrak{d}^k\underline{\alpha})^2\Big)$$
$$+\int_{{}^{(ext)}\mathcal{M}(r\geq r_0)}\Big(r^{3+\delta_B}\Big((\mathfrak{d}^k\alpha)^2+(\mathfrak{d}^k\beta)^2\Big)+r^{3-\delta_B}(\mathfrak{d}^k\check{\rho})^2+r^{1-\delta_B}(\mathfrak{d}^k\underline{\beta})^2$$
$$+r^{-1-\delta_B}(\mathfrak{d}^k\underline{\alpha})^2\Big)\Bigg\}$$
$$\lesssim r_0^{-\delta_B}({}^{(ext)}\mathfrak{G}^{\geq r_0}_{J+1}[\check{\Gamma}])^2+r_0^{10}\Big((\epsilon_{\mathcal{B}}[J])^2+\epsilon_0^2\Big(\mathfrak{N}^{(En)}_{J+1}+\mathcal{N}^{(match)}_{J+1}\Big)^2\Big).$$

Hence, we have obtained

$${}^{(ext)}\mathfrak{R}^{\geq r_0}_{J+1}[\check{R}] \lesssim r_0^{-\delta_B}\,{}^{(ext)}\mathfrak{G}^{\geq r_0}_{J+1}[\check{\Gamma}]+r_0^{10}\Big(\epsilon_{\mathcal{B}}[J]+\epsilon_0\Big(\mathfrak{N}^{(En)}_{J+1}+\mathcal{N}^{(match)}_{J+1}\Big)\Big)$$

which concludes the proof of Proposition 8.10.

8.7.3 Proof of (8.3.12)

To prove (8.3.12), we argue as in the proof of Proposition 8.10. Let j,k,l three integers such that

$$j+k+l\leq k_{small}.$$

To derive r^p-weighted curvature estimates for $\not{\mathfrak{d}}^j(re_4)^k\mathbf{T}^l$ derivatives of (α,β) in the region $r\geq 4m_0$, we proceed as follows.

Step 1. We start with the case $k=0$, i.e., we derive r^p-weighted curvature estimates for $\not{\mathfrak{d}}^j\mathbf{T}^l$ derivatives of (α,β) with $j+l\leq k_{small}$. First, we apply Corollary 8.29 to the Bianchi pair (α,β) with the choice $b=4+\delta_B$. This choice is such that we are in case (a) of Corollary 8.29. In particular, we obtain

$$\sum_{j+l\leq k_{small}}\Bigg\{\sup_{1\leq u\leq u_*}\int_{\mathcal{C}_u(r\geq 4m_0)} r^{4+\delta_B}(\not{\mathfrak{d}}^j\mathbf{T}^l\alpha)^2+\int_{\Sigma_*} r^{4+\delta_B}\Big((\not{\mathfrak{d}}^j\mathbf{T}^l\alpha)^2$$
$$+(\not{\mathfrak{d}}^j\mathbf{T}^l\beta)^2\Big)+\int_{{}^{(ext)}\mathcal{M}(r\geq 4m_0)} r^{3+\delta_B}\Big((\not{\mathfrak{d}}^j\mathbf{T}^l\alpha)^2+(\not{\mathfrak{d}}^j\mathbf{T}^l\beta)^2\Big)\Bigg\}$$
$$\lesssim \int_{{}^{(ext)}\mathcal{M}(\frac{7m_0}{2}\leq r\leq 4m_0)}\frac{(\mathfrak{d}^{J+1}\check{R})^2}{r^5}+\int_{{}^{(ext)}\mathcal{M}(r\geq 4m_0)}\Big\{r^{-1+\delta_B}(\not{\mathfrak{d}}^j\mathbf{T}^l\vartheta)^2\Big\}+\epsilon_0^2$$
$$+\epsilon_0^2(\mathfrak{N}^{(En)}_{k_{small}})^2.$$

We infer that

$$\sum_{j+l\leq k_{small}}\Bigg\{\sup_{1\leq u\leq u_*}\int_{\mathcal{C}_u(r\geq 4m_0)} r^{4+\delta_B}(\not{\mathfrak{d}}^j\mathbf{T}^l\alpha)^2+\int_{\Sigma_*} r^{4+\delta_B}\Big((\not{\mathfrak{d}}^j\mathbf{T}^l\alpha)^2$$
$$+(\not{\mathfrak{d}}^j\mathbf{T}^l\beta)^2\Big)+\int_{{}^{(ext)}\mathcal{M}(r\geq 4m_0)} r^{3+\delta_B}\Big((\not{\mathfrak{d}}^j\mathbf{T}^l\alpha)^2+(\not{\mathfrak{d}}^j\mathbf{T}^l\beta)^2\Big)\Bigg\}$$
$$\lesssim \Big({}^{(ext)}\mathfrak{R}^{\leq 4m_0}_{k_{small}}[\check{R}]\Big)^2+\Big({}^{(ext)}\mathfrak{G}_{k_{small}}[\check{\Gamma}]\Big)^2+\epsilon_0^2+\epsilon_0^2(\mathfrak{N}^{(En)}_{k_{small}})^2. \tag{8.7.8}$$

Step 2. We now argue by iteration on k. For $0\leq k\leq k_{small}-1$, we consider the following iteration assumption:

$$\sum_{j+l\leq k_{small}-k}\Bigg\{\sup_{1\leq u\leq u_*}\int_{\mathcal{C}_u(r\geq 4m_0)} r^{4+\delta_B}(\not{\mathfrak{d}}^j(re_4)^k\mathbf{T}^l\alpha)^2$$
$$+\int_{\Sigma_*} r^{4+\delta_B}\Big((\not{\mathfrak{d}}^j(re_4)^k\mathbf{T}^l\alpha)^2+(\not{\mathfrak{d}}^j(re_4)^k\mathbf{T}^l\beta)^2\Big)$$
$$+\int_{{}^{(ext)}\mathcal{M}(r\geq 4m_0)} r^{3+\delta_B}\Big((\not{\mathfrak{d}}^j(re_4)^k\mathbf{T}^l\alpha)^2+(\not{\mathfrak{d}}^j(re_4)^k\mathbf{T}^l\beta)^2\Big)\Bigg\}$$
$$\lesssim \Big({}^{(ext)}\mathfrak{R}^{\leq 4m_0}_{k_{small}}[\check{R}]\Big)^2+\Big({}^{(ext)}\mathfrak{G}_{k_{small}}[\check{\Gamma}]\Big)^2+\epsilon_0^2+\epsilon_0^2(\mathfrak{N}^{(En)}_{k_{small}})^2. \tag{8.7.9}$$

(8.7.9) holds true for $k=0$ in view of (8.7.8). We now assume that (8.7.9) holds true for k such that $0\leq k\leq k_{small}-1$, and our goal is to prove that it also holds for $k+1$.

First, note that the Bianchi identity for $e_4(\beta)$, together with (8.7.9), yields

$$\sum_{j+l\leq k_{small}-(k+1)}\Bigg\{\int_{\Sigma_*} r^{4+\delta_B}(\not{\mathfrak{d}}^j(re_4)^{k+1}\mathbf{T}^l\beta)^2$$
$$+\int_{{}^{(ext)}\mathcal{M}(r\geq 4m_0)} r^{3+\delta_B}(\not{\mathfrak{d}}^j(re_4)^{k+1}\mathbf{T}^l\beta)^2\Bigg\}$$
$$\lesssim \Big({}^{(ext)}\mathfrak{R}^{\leq 4m_0}_{k_{small}}[\check{R}]\Big)^2+\Big({}^{(ext)}\mathfrak{G}_{k_{small}}[\check{\Gamma}]\Big)^2+\epsilon_0^2+\epsilon_0^2(\mathfrak{N}^{(En)}_{k_{small}})^2. \tag{8.7.10}$$

We still need to estimate $\not{\mathfrak{d}}^j (re_4)^{k+1}\mathbf{T}^l\alpha$. To this end, we apply Corollary 8.29 to the Bianchi pair (α, β) with the choice $b = 4 + \delta_B$. Since $k+1 \geq 1$, we are in case (c) of Corollary 8.29. In particular, we obtain, arguing similarly as above,

$$\sum_{j+l\leq k_{small}-(k+1)} \left\{ \sup_{1\leq u\leq u_*} \int_{\mathcal{C}_u(r\geq 4m_0)} r^{4+\delta_B} (\not{\mathfrak{d}}^j (re_4)^{k+1}\mathbf{T}^l\alpha)^2 \right.$$
$$\left. + \int_{\Sigma_*} r^{4+\delta_B} (\not{\mathfrak{d}}^j (re_4)^{k+1}\mathbf{T}^l\alpha)^2 + \int_{{}^{(ext)}\mathcal{M}(r\geq 4m_0)} r^{3+\delta_B} (\not{\mathfrak{d}}^j (re_4)^{k+1}\mathbf{T}^l\alpha)^2 \right\}$$
$$\lesssim \sum_{j+l\leq k_{small}-(k+1)} \left\{ \int_{{}^{(ext)}\mathcal{M}(r\geq 4m_0)} r^{3+\delta_B} (\not{\mathfrak{d}}^j (re_4)^{k+1}\mathbf{T}^l\beta)^2 \right\}$$
$$+ \left({}^{(ext)}\mathfrak{R}^{\leq 4m_0}_{k_{small}}[\check{R}] \right)^2 + \left({}^{(ext)}\mathfrak{G}_{k_{small}}[\check{\Gamma}] \right)^2 + \epsilon_0^2 + \epsilon_0^2 (\mathfrak{N}^{(En)}_{k_{small}})^2.$$

Together with (8.7.10), this implies (8.7.9) for $k+1$. Hence, by iteration, (8.7.9) holds for any $0 \leq k \leq k_{small}$. Now, (8.7.9) for any $0 \leq k \leq k_{small}$ is equivalent to (8.3.12) which is the desired estimate.

8.8 PROOF OF PROPOSITION 8.11

To prove Proposition 8.11, we rely on the following three propositions.

Proposition 8.31. *Let J such that $k_{small} - 2 \leq J \leq k_{large} - 1$. Then, we have*

$$ {}^{(\Sigma_*)}\mathfrak{G}_{J+1}[\check{\Gamma}] + {}^{(\Sigma_*)}\mathfrak{G}'_{J+1}[\check{\Gamma}] \lesssim {}^{(\Sigma_*)}\mathfrak{R}_{J+1}[\check{R}] + {}^{(\Sigma_*)}\mathfrak{G}_J[\check{\Gamma}],$$

where we have introduced the notations

$$ {}^{(\Sigma_*)}\mathfrak{G}_k[\check{\Gamma}] := \int_{\Sigma_*} \Big[r^2 \Big((\mathfrak{d}^{\leq k}\vartheta)^2 + (\mathfrak{d}^{\leq k}\check{\kappa})^2 + (\mathfrak{d}^{\leq k}\zeta)^2 + (\mathfrak{d}^{\leq k}\underline{\check{\kappa}})^2 \Big) + (\mathfrak{d}^{\leq k}\underline{\vartheta})^2$$
$$ + (\mathfrak{d}^{\leq k}\eta)^2 + (\mathfrak{d}^{\leq k}\underline{\check{\omega}})^2 + (\mathfrak{d}^{\leq k}\underline{\xi})^2 \Big],$$

$$ {}^{(\Sigma_*)}\mathfrak{G}'_k[\check{\Gamma}] := \int_{\Sigma_*} r^2 \Big[(\mathfrak{d}^{k+1}\not{\mathfrak{d}}\check{\kappa})^2 + (\mathfrak{d}^{k+1}\check{\kappa})^2 + (\mathfrak{d}^{\leq k+1}\check{\mu})^2 + (\mathfrak{d}^{k+1}\underline{\check{\kappa}})^2 + (\mathfrak{d}^{k+1}\zeta)^2 \Big],$$

and

$$ {}^{(\Sigma_*)}\mathfrak{R}_k[\check{R}] := \int_{\Sigma_*} \Big(r^{4+\delta_B} \big((\mathfrak{d}^{\leq k}\alpha)^2 + (\mathfrak{d}^{\leq k}\beta)^2 \big) + r^4 (\mathfrak{d}^{\leq k}\check{\rho})^2 + r^2 (\mathfrak{d}^{\leq k}\underline{\beta})^2$$
$$ + (\mathfrak{d}^{\leq k}\underline{\alpha})^2 \Big).$$

Proposition 8.32. *Let J such that $k_{small} - 2 \leq J \leq k_{large} - 1$. Then, we have*

$$ {}^{(ext)}\mathfrak{G}^{\geq 4m_0}_{J+1}[\check{\Gamma}] + {}^{(ext)}\mathfrak{G}^{\geq 4m_0\,\prime}_{J+1}[\check{\Gamma}] \lesssim {}^{(\Sigma_*)}\mathfrak{G}_{J+1}[\check{\Gamma}] + {}^{(\Sigma_*)}\mathfrak{G}'_{J+1}[\check{\Gamma}]$$
$$ + {}^{(ext)}\mathfrak{R}_{J+1}[\check{R}] + {}^{(ext)}\mathfrak{G}_J[\check{\Gamma}],$$

where we have introduced the notation

$$
\begin{aligned}
{}^{(ext)}\mathfrak{G}_k^{\geq 4m_0\,'}[\check{\Gamma}] \;:=\; \sup_{\lambda\geq 4m_0}\Bigg(\int_{\{r=\lambda\}}\Bigg[&\lambda^6\Big(\mathfrak{d}^k\Big(\not{d}_1\not{d}_1^\star\kappa-\vartheta\Big(\not{d}_4\not{d}_3^\star\not{d}_2^{-1}+\not{d}_2^\star\Big)\not{d}_1^{-1}\check{\rho}\Big)\Big)^2\\
&+\lambda^2(\mathfrak{d}^{k+1}\check{\kappa})^2+\lambda^6\Big(\mathfrak{d}^k\Big(e_\theta(\mu)+\vartheta\not{d}_2\not{d}_2^\star(\not{d}_1^\star\not{d}_1)^{-1}\underline{\beta}+2\zeta\check{\rho}\Big)\Big)^2\\
&+\lambda^4(\mathfrak{d}^{\leq k}\check{\mu})^2+\lambda^2\Big(\mathfrak{d}^k\Big(e_\theta(\underline{\kappa})-4\underline{\beta}\Big)\Big)^2+\lambda^2\Big(\mathfrak{d}^k\Big(e_3(\zeta)+\underline{\beta}\Big)\Big)^2\Bigg]\Bigg).
\end{aligned}
$$

Proposition 8.33. *Let J such that $k_{small}-2\leq J\leq k_{large}-1$. Then, we have*

$$
\begin{aligned}
{}^{(ext)}\mathfrak{G}_{J+1}^{\leq 4m_0}[\check{\Gamma}]+{}^{(ext)}\mathfrak{G}_{J+1}^{\leq 4m_0\,'}[\check{\Gamma}] \;\lesssim\;& {}^{(ext)}\mathfrak{G}_{J+1}^{\geq 4m_0}[\check{\Gamma}]+{}^{(ext)}\mathfrak{G}_{J+1}^{\geq 4m_0\,'}[\check{\Gamma}]\\
&+{}^{(ext)}\mathfrak{R}_{J+1}[\check{R}]+{}^{(ext)}\mathfrak{G}_J[\check{\Gamma}],
\end{aligned}
$$

where we have introduced the notation

$$
\begin{aligned}
&{}^{(ext)}\mathfrak{G}_k^{\leq 4m_0\,'}[\check{\Gamma}]\\
:=\;& \sup_{r_{\mathcal{T}}\leq\lambda\leq 4m_0}\Bigg(\int_{\{r=\lambda\}}\Bigg[\lambda^6\Big(\mathfrak{d}^k\Big(\not{d}_1\not{d}_1^\star\kappa-\vartheta\Big(\not{d}_4\not{d}_3^\star\not{d}_2^{-1}+\not{d}_2^\star\Big)\not{d}_1^{-1}\check{\rho}\Big)\Big)^2\\
&+\lambda^2(\mathfrak{d}^{k+1}\check{\kappa})^2+\lambda^6\Big(\mathfrak{d}^k\Big(e_\theta(\mu)+\vartheta\not{d}_2\not{d}_2^\star(\not{d}_1^\star\not{d}_1)^{-1}\underline{\beta}+2\zeta\check{\rho}\Big)\Big)^2\\
&+\lambda^4(\mathfrak{d}^{\leq k}\check{\mu})^2+\lambda^2\Big(\mathfrak{d}^k\Big(e_\theta(\underline{\kappa})-4\underline{\beta}\Big)\Big)^2+\lambda^2\Big(\mathfrak{d}^{k-1}\mathbf{N}\Big(e_3(\zeta)+\underline{\beta}\Big)\Big)^2\Bigg]\Bigg).
\end{aligned}
$$

The proof of Proposition 8.31 is postponed to section 8.8.1, the proof of Proposition 8.32 is postponed to section 8.8.4, and the proof of Proposition 8.33 is postponed to section 8.8.5. The proof of the two latter propositions will rely in particular on basic weighted estimates for transport equations along e_4 in ${}^{(ext)}\mathcal{M}$ derived in section 8.8.2, as well as several renormalized identities derived in section 8.8.3.

We now conclude the proof of Proposition 8.11. In view of Propositions 8.31, 8.32 and 8.33, we have, for J such that $k_{small}-2\leq J\leq k_{large}-1$,

$$
{}^{(ext)}\mathfrak{G}_{J+1}[\check{\Gamma}] \;\lesssim\; {}^{(ext)}\mathfrak{R}_{J+1}[\check{R}]+{}^{(ext)}\mathfrak{G}_J[\check{\Gamma}],
$$

where we have used the fact that

$$
{}^{(\Sigma_*)}\mathfrak{R}_{J+1}[\check{R}]\leq{}^{(ext)}\mathfrak{R}_{J+1}[\check{R}],\qquad {}^{(\Sigma_*)}\mathfrak{G}_J[\check{\Gamma}]\leq{}^{(ext)}\mathfrak{G}_J[\check{\Gamma}].
$$

In view of the iteration assumption (8.3.13), we infer

$$
{}^{(ext)}\mathfrak{G}_{J+1}[\check{\Gamma}] \;\lesssim\; {}^{(ext)}\mathfrak{R}_{J+1}[\check{R}]+\epsilon_{\mathcal{B}}[J].
$$

Since the estimates in Proposition 8.32 are integrated from Σ_*, we obtain similarly, for any $r_0\geq 4m_0$,

$$
{}^{(ext)}\mathfrak{G}_{J+1}^{\geq r_0}[\check{\Gamma}] \;\lesssim\; {}^{(ext)}\mathfrak{R}_{J+1}^{\geq r_0}[\check{R}]+\epsilon_{\mathcal{B}}[J].
$$

On the other hand, we have in view of Proposition 8.10, for any $r_0 \geq 4m_0$,

$$ {}^{(ext)}\mathfrak{R}^{\geq r_0}_{J+1}[\check{R}] \lesssim r_0^{-\delta_B} \, {}^{(ext)}\mathfrak{G}^{\geq r_0}_{J+1}[\check{\Gamma}] + r_0^{10}\left(\epsilon_{\mathcal{B}}[J] + \epsilon_0\left(\mathfrak{N}^{(En)}_{J+1} + \mathcal{N}^{(match)}_{J+1}\right)\right) $$

and

$$ \begin{aligned} {}^{(int)}\mathfrak{R}_{J+1}[\check{R}] + {}^{(ext)}\mathfrak{R}_{J+1}[\check{R}] \leq & \; {}^{(ext)}\mathfrak{R}^{\geq r_0}_{J+1}[\check{R}] \\ & + O\left(r_0^{10}\left(\epsilon_{\mathcal{B}}[J] + \epsilon_0\left(\mathfrak{N}^{(En)}_{J+1} + \mathcal{N}^{(match)}_{J+1}\right)\right)\right). \end{aligned} $$

Choosing $r_0 \geq 4m_0$ large enough, we infer from the above estimates

$$ {}^{(ext)}\mathfrak{G}_{J+1}[\check{\Gamma}] + {}^{(int)}\mathfrak{R}_{J+1}[\check{R}] + {}^{(ext)}\mathfrak{R}_{J+1}[\check{R}] \lesssim \epsilon_{\mathcal{B}}[J] + \epsilon_0\left(\mathfrak{N}^{(En)}_{J+1} + \mathcal{N}^{(match)}_{J+1}\right). $$

This concludes the proof of Proposition 8.11.

8.8.1 Proof of Proposition 8.31

Step 1. We control κ on Σ_*. Recall the GCM conditions $\kappa = 2/r$ on Σ_*. Since ν_{Σ_*} and e_θ are tangent, we infer

$$ (\not{\mathfrak{d}}, \nu_{\Sigma_*})^k\left(\kappa - \frac{2}{r}\right) = 0. $$

Together with Raychaudhuri, we infer

$$ \begin{aligned} & \max_{k\leq J+2}\int_{\Sigma_*}\left(r^2\left(\mathfrak{d}^k\left(\kappa - \frac{2}{r}\right)\right)^2 + r^4\left(\mathfrak{d}^k e_\theta(\kappa)\right)^2\right) \\ \lesssim & \left({}^{(\Sigma_*)}\mathfrak{R}_{J+1}[\check{R}] + {}^{(\Sigma_*)}\mathfrak{G}_J[\check{\Gamma}] + \epsilon_0 \, {}^{(\Sigma_*)}\mathfrak{G}_{J+1}[\check{\Gamma}]\right)^2, \end{aligned} $$

where we have used the fact that e_3 is in the span of e_4 and ν_{Σ_*}. Note that we have used Codazzi for ϑ to control the term $\mathfrak{d}^{J+1}e_4(e_\theta(\kappa))$.

Step 2. We control the $\ell = 1$ modes on Σ_*. In view of the GCM conditions for κ, and projecting the Codazzi for ϑ on the $\ell = 1$ mode, we infer on Σ_*

$$ \int_S \zeta e^\Phi = r\int_S \beta e^\Phi + \frac{r}{2}\int_S \vartheta\zeta e^\Phi. $$

Since the vectorfield ν is tangent to Σ_*, we infer

$$ \begin{aligned} \nu^{J+2}\left(\int_S \zeta e^\Phi\right) &= r\int_S \nu^{J+2}\beta e^\Phi + \frac{r}{2}\int_S \nu^{J+2}(\vartheta\zeta)e^\Phi + \text{l.o.t.} \\ &= r\int_S \nu^{J+2}\beta e^\Phi + \frac{r}{2}\int_S \zeta\nu^{J+2}(\vartheta)e^\Phi + \frac{r}{2}\int_S \vartheta\nu^{J+2}(\zeta)e^\Phi + \text{l.o.t.} \end{aligned} $$

where l.o.t. denote, here and below, terms that

- either are linear and contain at most $J + 1$ derivatives of curvature components and J derivatives of Ricci coefficients,

- or are quadratic and contain at most $J+1$ derivatives of Ricci coefficients and curvature components.

Using Bianchi identities and the null structure equations, we deduce

$$
\begin{aligned}
& \nu^{J+2}\left(\int_S \zeta e^\Phi\right) \\
= \;& r\int_S \nu^{J+1}(\not{d}_2\alpha, \not{d}_1^\star\rho - 3\rho\eta)e^\Phi + \frac{r}{2}\int_S \zeta\nu^{J+1}\not{d}_2^\star\eta e^\Phi + \frac{r}{2}\int_S \vartheta\nu^{J+1}\not{d}_1^\star\underline{\omega}e^\Phi + \text{l.o.t.} \\
= \;& r\int_S (\not{d}_2\nu^{J+1}\alpha, \nu^J\not{d}_1^\star\not{d}_1(\beta,\underline{\beta}))e^\Phi - 3\overline{\rho}\int_S \nu^{J+1}\eta e^\Phi + \frac{r}{2}\int_S \zeta\not{d}_2^\star\nu^{J+1}\eta e^\Phi \\
& + \frac{r}{2}\int_S \vartheta\not{d}_1^\star\nu^{J+1}\underline{\omega}e^\Phi + \text{l.o.t.} \\
= \;& r\int_S (\not{d}_2\nu^{J+1}\alpha, \not{d}_1^\star\not{d}_1\nu^J(\beta,\underline{\beta}))e^\Phi + \frac{r}{2}\int_S \zeta\not{d}_2^\star\nu^{J+1}\eta e^\Phi + \frac{r}{2}\int_S \vartheta\not{d}_1^\star\nu^{J+1}\underline{\omega}e^\Phi \\
& + \text{l.o.t.},
\end{aligned}
$$

where we have used, in the last equality, a cancellation due to the fact that ν is tangent to Σ_* and $\int_S \eta e^\Phi = 0$ on Σ_*. Using the identity $\not{d}_1^\star\not{d}_1 = \not{d}_2\not{d}_2^\star + 2K$, integration by parts for all terms, and the fact that $\not{d}_2^\star(e^\Phi) = 0$ so that the top order linear term vanishes, we infer

$$
\nu^{J+2}\left(\int_S \zeta e^\Phi\right) = \text{l.o.t.}
$$

with the above convention for the lower order terms. Also, relying on the null equation for $e_4(\zeta)$, i.e.,

$$
e_4(\zeta) = -\kappa\zeta - \beta - \vartheta\zeta,
$$

we obtain, with more ease since this estimate is at one lower level of derivatives,

$$
(re_4, \nu)^{J+2}\left(\int_S \zeta e^\Phi\right) = \text{l.o.t.}
$$

We infer

$$
\max_{k\leq J+2}\int_1^{u_*} r^{-2}\left(\mathfrak{d}^k\left(\int_S \zeta e^\Phi\right)\right)^2 \lesssim \left({}^{(\Sigma_*)}\mathfrak{R}_{J+1}[\check{R}] + {}^{(\Sigma_*)}\mathfrak{G}_J[\check{\Gamma}] + \epsilon_0\,{}^{(\Sigma_*)}\mathfrak{G}_{J+1}[\check{\Gamma}]\right)^2.
$$

Next, we have, in view of the definition of μ and the identity $\not{d}_1^\star\not{d}_1 = \not{d}_2\not{d}_2^\star + 2K$,

$$
\begin{aligned}
\int_S e_\theta(\mu)e^\Phi &= \int_S \not{d}_1^\star\not{d}_1\zeta e^\Phi - \int_S e_\theta(\rho)e^\Phi + \frac{1}{4}\int_S e_\theta(\vartheta\underline{\vartheta})e^\Phi \\
&= 2\int_S K\zeta e^\Phi - \int_S e_\theta(\rho)e^\Phi + \frac{1}{4}\int_S e_\theta(\vartheta\underline{\vartheta})e^\Phi \\
&= \frac{2}{r^2}\int_S \zeta e^\Phi - \int_S e_\theta(\rho)e^\Phi + \int_S\left(K - \frac{2}{r^2}\right)\zeta e^\Phi + \frac{1}{4}\int_S e_\theta(\vartheta\underline{\vartheta})e^\Phi.
\end{aligned}
$$

To estimate the RHS, we use in particular the following.

- For the second term, in view of Bianchi, we have, schematically,

$$
\begin{aligned}
& (e_3, re_4)^{J+2} e_\theta(\rho) \\
= \; & (e_3, re_4)^{J+1} \, d\!\!/_1^\star \, d\!\!/_1 (r\beta, \underline{\beta}) - \frac{3}{2}(e_3, re_4)^{J+1} \, d\!\!/_1^\star (r\kappa\rho, \underline{\kappa}\rho) \\
& + (e_3, re_4)^{J+1} \, d\!\!/_1^\star \big(r\underline{\vartheta}\alpha, (r\zeta, \underline{\xi})\beta, (\zeta, \eta)\underline{\beta}, \vartheta\underline{\alpha} \big) + \text{l.o.t.} \\
= \; & (e_3, re_4)^{J+1} \, d\!\!/_2 \, d\!\!/_2^\star (r\beta, \underline{\beta}) + \frac{3}{2}\rho (e_3, re_4)^{J+1} (re_\theta(\kappa), e_\theta(\underline{\kappa})) \\
& + r\underline{\vartheta}(e_3, re_4)^{J+1} \, d\!\!/_1^\star \alpha + (r\zeta, \underline{\xi})(e_3, re_4)^{J+1} \, d\!\!/_1^\star \beta + (\zeta, \eta)(e_3, re_4)^{J+1} \, d\!\!/_1^\star \underline{\beta} \\
& + \vartheta (e_3, re_4)^{J+1} \, d\!\!/_1^\star \underline{\alpha} + r\alpha (e_3, re_4)^{J+1} \, d\!\!/_1^\star \underline{\vartheta} + \beta (e_3, re_4)^{J+1} \, d\!\!/_1^\star (r\zeta, \underline{\xi}) \\
& + \underline{\beta}(e_3, re_4)^{J+1} \, d\!\!/_1^\star (\zeta, \eta) + \underline{\alpha}(e_3, re_4)^{J+1} \, d\!\!/_1^\star \vartheta) + \text{l.o.t.} \\
= \; & d\!\!/_2 (e_3, re_4)^{J+1} \, d\!\!/_2^\star (r\beta, \underline{\beta}) + [(e_3, re_4)^{J+1}, \, d\!\!/_2] \, d\!\!/_2^\star (r\beta, \underline{\beta}) \\
& + \frac{3}{2}\overline{\rho}(e_3, re_4)^{J+1} (re_\theta(\kappa), e_\theta(\underline{\kappa})) - \frac{3}{2}\check{\rho} \, d\!\!/_1^\star (e_3, re_4)^{J+1} (r\check{\kappa}, \underline{\check{\kappa}}) \\
& + r\underline{\vartheta} \, d\!\!/_1^\star (e_3, re_4)^{J+1} \alpha + (r\zeta, \underline{\xi}) \, d\!\!/_1^\star (e_3, re_4)^{J+1} \beta + (\zeta, \eta) \, d\!\!/_1^\star (e_3, re_4)^{J+1} \underline{\beta} \\
& + \vartheta \, d\!\!/_1^\star (e_3, re_4)^{J+1} \underline{\alpha} + r\alpha \, d\!\!/_1^\star (e_3, re_4)^{J+1} \underline{\vartheta} + \beta \, d\!\!/_1^\star (e_3, re_4)^{J+1} (r\zeta, \underline{\xi}) \\
& + \underline{\beta} \, d\!\!/_1^\star (e_3, re_4)^{J+1} (\zeta, \eta) + \underline{\alpha} \, d\!\!/_1^\star (e_3, re_4)^{J+1} \vartheta) + \text{l.o.t.}
\end{aligned}
$$

- For the third term

$$
\begin{aligned}
& (e_3, re_4)^{J+2} \left(\left(K - \frac{2}{r^2} \right) \zeta \right) \\
= \; & \zeta (e_3, re_4)^{J+2} \left(K - \frac{2}{r^2} \right) + \left(K - \frac{2}{r^2} \right) (e_3, re_4)^{J+2} \zeta + \text{l.o.t.} \\
= \; & \zeta (e_3, re_4)^{J+1} \Big(d\!\!/_1 (r\beta, \underline{\beta}, \eta, r^{-1}\underline{\xi}) \Big) + \left(K - \frac{2}{r^2} \right) (e_3, re_4)^{J+1} e_\theta(\underline{\omega}) + \text{l.o.t.} \\
= \; & \zeta (e_3, re_4)^{J+1} \, d\!\!/_1 \left(r\beta, \underline{\beta}, \eta, r^{-1}\underline{\xi} \right) + \left(K - \frac{2}{r^2} \right) (e_3, re_4)^{J+1} e_\theta(\underline{\omega}) + \text{l.o.t.} \\
= \; & \zeta \, d\!\!/_1 (e_3, re_4)^{J+1} \left(r\beta, \underline{\beta}, \eta, r^{-1}\underline{\xi} \right) + \left(K - \frac{2}{r^2} \right) d\!\!/_1^\star (e_3, re_4)^{J+1} \underline{\check{\omega}} + \text{l.o.t.}
\end{aligned}
$$

- For the fourth term

$$
\begin{aligned}
(e_3, re_4)^{J+2} e_\theta(\vartheta\underline{\vartheta}) &= (e_3, re_4)^{J+1} e_\theta(\vartheta \, d\!\!/_2^\star (\underline{\xi}, r\zeta)) + (e_3, re_4)^{J+1} e_\theta(\underline{\vartheta} \, d\!\!/_2^\star \eta) + \text{l.o.t.} \\
&= \vartheta \, d\!\!/_1^\star \, d\!\!/_2^\star (e_3, re_4)^{J+1} (\underline{\xi}, r\zeta) + \underline{\vartheta} \, d\!\!/_1^\star \, d\!\!/_2^\star (e_3, re_4)^{J+1} \eta + \text{l.o.t.}
\end{aligned}
$$

We infer

$$
\begin{aligned}
&(e_3, re_4)^{J+2}\left(\int_S e_\theta(\mu)e^\Phi\right)\\
= \quad & \frac{2}{r^2}(e_3, re_4)^{J+2}\left(\int_S \zeta e^\Phi\right) + \int_S \not{d}_2 (e_3, re_4)^{J+1}\, \not{d}_2^\star(r\beta, \underline{\beta})e^\Phi\\
& + \int_S [(e_3, re_4)^{J+1}, \not{d}_2]\, \not{d}_2^\star(r\beta, \underline{\beta})e^\Phi + \frac{3}{2}\overline{\rho}\int_S (e_3, re_4)^{J+1}(re_\theta(\kappa), e_\theta(\underline{\kappa}))e^\Phi\\
& -\frac{3}{2}\int_S \check{\rho}\,\not{d}_1^\star(e_3, re_4)^{J+1}(r\check{\kappa}, \check{\underline{\kappa}})e^\Phi\\
& + \int_S \Big[r\underline{\vartheta}\,\not{d}_1^\star(e_3, re_4)^{J+1}\alpha + (r\zeta, \underline{\xi})\,\not{d}_1^\star(e_3, re_4)^{J+1}\beta + (\zeta, \eta)\,\not{d}_1^\star(e_3, re_4)^{J+1}\underline{\beta}\\
& + \vartheta\,\not{d}_1^\star(e_3, re_4)^{J+1}\underline{\alpha}\Big]e^\Phi + \int_S \Big[r\alpha\,\not{d}_1^\star(e_3, re_4)^{J+1}\underline{\vartheta} + \beta\,\not{d}_1^\star(e_3, re_4)^{J+1}(r\zeta, \underline{\xi})\\
& + \underline{\beta}\,\not{d}_1^\star(e_3, re_4)^{J+1}(\zeta, \eta) + \underline{\alpha}\,\not{d}_1^\star(e_3, re_4)^{J+1}\vartheta\Big]e^\Phi\\
& + \int_S \zeta\,\not{d}_1(e_3, re_4)^{J+1}\left(r\beta, \underline{\beta}, \zeta, r^{-1}\underline{\xi}\right)e^\Phi + \int_S \left(K - \frac{2}{r^2}\right)\not{d}_1^\star(e_3, re_4)^{J+1}\underline{\check{\omega}}e^\Phi\\
& + \int_S \vartheta\,\not{d}_1^\star\,\not{d}_2^\star(e_3, re_4)^{J+1}(\underline{\xi}, r\zeta)e^\Phi + \int_S \underline{\vartheta}\,\not{d}_1^\star\,\not{d}_2^\star(e_3, re_4)^{J+1}\zeta e^\Phi + \text{l.o.t.}
\end{aligned}
$$

and after integrations by parts and the fact that

$$
\not{d}_k(Fe^\Phi) = \not{d}_{k+1}(F)e^\Phi, \quad \not{d}_k^\star(Fe^\Phi) = \not{d}_{k-1}^\star(F)e^\Phi,
$$

we obtain

$$
\begin{aligned}
&(e_3, re_4)^{J+2}\left(\int_S e_\theta(\mu)e^\Phi\right)\\
= \quad & \frac{2}{r^2}(e_3, re_4)^{J+2}\left(\int_S \zeta e^\Phi\right) + \int_S \not{d}_1^\star\Big([(e_3, re_4)^{J+1}, \not{d}_2]\Big)(r\beta, \underline{\beta})e^\Phi\\
& + \frac{3}{2}\overline{\rho}\int_S (e_3, re_4)^{J+1}(re_\theta(\kappa), e_\theta(\underline{\kappa}))e^\Phi + \frac{3}{2}\int_S \not{d}_2\rho(e_3, re_4)^{J+1}(r\check{\kappa}, \check{\underline{\kappa}})e^\Phi\\
& + \int_S \Big[r\,\not{d}_2\underline{\vartheta}(e_3, re_4)^{J+1}\alpha + \not{d}_2(r\zeta, \underline{\xi})(e_3, re_4)^{J+1}\beta + \not{d}_2(\zeta, \eta)(e_3, re_4)^{J+1}\underline{\beta}\\
& + \not{d}_2\vartheta(e_3, re_4)^{J+1}\underline{\alpha}\Big]e^\Phi + \int_S \Big[r\,\not{d}_2\alpha(e_3, re_4)^{J+1}\underline{\vartheta} + \not{d}_2\beta(e_3, re_4)^{J+1}(r\zeta, \underline{\xi})\\
& + \not{d}_2\underline{\beta}(e_3, re_4)^{J+1}(\zeta, \eta) + \not{d}_2\underline{\alpha}(e_3, re_4)^{J+1}\vartheta\Big]e^\Phi\\
& + \int_S \not{d}_1\zeta(e_3, re_4)^{J+1}\left(r\beta, \underline{\beta}, \zeta, r^{-1}\underline{\xi}\right)e^\Phi + \int_S \not{d}_2\left(K - \frac{2}{r^2}\right)(e_3, re_4)^{J+1}\underline{\check{\omega}}e^\Phi\\
& + \int_S \not{d}_3\,\not{d}_2\vartheta(e_3, re_4)^{J+1}(\underline{\xi}, r\zeta)e^\Phi + \int_S \not{d}_3\,\not{d}_2\underline{\vartheta}(e_3, re_4)^{J+1}\zeta e^\Phi + \text{l.o.t.}
\end{aligned}
$$

Together with the above estimate for the $\ell = 1$ mode of ζ and the estimate of Step

1 for κ, we infer

$$\begin{aligned}
&\max_{k\leq J+2}\int_1^{u_*} r^2\left(\mathfrak{d}^k\left(\int_S e_\theta(\mu)e^\Phi\right)\right)^2\\
\lesssim\;& \left({}^{(\Sigma_*)}\mathfrak{R}_{J+1}[\check{R}]+{}^{(\Sigma_*)}\mathfrak{G}_J[\check{\Gamma}]+\epsilon_0\,{}^{(\Sigma_*)}\mathfrak{G}_{J+1}[\check{\Gamma}]\right)^2\\
&+\max_{k\leq J+1}\int_1^{u_*} r^{-4}\left(\mathfrak{d}^k\left(\int_S e_\theta(\underline{\kappa})e^\Phi\right)\right)^2.
\end{aligned}$$

In view of the dominant condition (3.3.4) for r on Σ_*, we infer

$$\begin{aligned}
&\max_{k\leq J+2}\int_1^{u_*} r^2\left(\mathfrak{d}^k\left(\int_S e_\theta(\mu)e^\Phi\right)\right)^2\\
\lesssim\;& \left({}^{(\Sigma_*)}\mathfrak{R}_{J+1}[\check{R}]+{}^{(\Sigma_*)}\mathfrak{G}_J[\check{\Gamma}]+\epsilon_0\,{}^{(\Sigma_*)}\mathfrak{G}_{J+1}[\check{\Gamma}]\right)^2\\
&+\epsilon_0\max_{k\leq J+1}\int_1^{u_*} \left(\mathfrak{d}^k\left(\int_S e_\theta(\underline{\kappa})e^\Phi\right)\right)^2.
\end{aligned}$$

Next, in view of the remarkable identity for the $\ell=1$ mode of $e_\theta(K)$, we have

$$-\int_S e_\theta(\rho)e^\Phi-\frac{1}{4}\int_S e_\theta(\kappa\underline{\kappa})e^\Phi+\frac{1}{4}\int_S e_\theta(\vartheta\underline{\vartheta})e^\Phi \;=\; 0$$

and hence

$$\begin{aligned}
\int_S e_\theta(\underline{\kappa})e^\Phi \;=\;& -2r\int_S e_\theta(\rho)e^\Phi-\frac{r}{2}\int_S\left(\kappa-\frac{2}{r}\right)e_\theta(\underline{\kappa})e^\Phi-\frac{r}{2}\int_S \underline{\kappa}e_\theta(\kappa)e^\Phi\\
&+\frac{r}{2}\int_S e_\theta(\vartheta\underline{\vartheta})e^\Phi.
\end{aligned}$$

Arguing as for the estimate of the $\ell=1$ mode of $e_\theta(\mu)$, and using the smallness of ϵ_0, we infer

$$\begin{aligned}
&\max_{k\leq J+2}\int_1^{u_*} r^2\left(\mathfrak{d}^k\left(\int_S e_\theta(\mu)e^\Phi\right)\right)^2+\max_{k\leq J+2}\int_1^{u_*}\left(\mathfrak{d}^k\left(\int_S e_\theta(\underline{\kappa})e^\Phi\right)\right)^2\\
\lesssim\;& \left({}^{(\Sigma_*)}\mathfrak{R}_{J+1}[\check{R}]+{}^{(\Sigma_*)}\mathfrak{G}_J[\check{\Gamma}]+\epsilon_0\,{}^{(\Sigma_*)}\mathfrak{G}_{J+1}[\check{\Gamma}]\right)^2.
\end{aligned}$$

We have thus obtained

$$\begin{aligned}
&\max_{k\leq J+2}\int_1^{u_*}\Bigg[r^{-2}\left(\mathfrak{d}^k\left(\int_S \zeta e^\Phi\right)\right)^2+r^2\left(\mathfrak{d}^k\left(\int_S e_\theta(\mu)e^\Phi\right)\right)^2\\
&\qquad\qquad+\left(\mathfrak{d}^k\left(\int_S e_\theta(\underline{\kappa})e^\Phi\right)\right)^2\Bigg]\\
\lesssim\;& \left({}^{(\Sigma_*)}\mathfrak{R}_{J+1}[\check{R}]+{}^{(\Sigma_*)}\mathfrak{G}_J[\check{\Gamma}]+\epsilon_0\,{}^{(\Sigma_*)}\mathfrak{G}_{J+1}[\check{\Gamma}]\right)^2.
\end{aligned}$$

Step 3. Recall the GCM conditions $\not{d}_2^\star \not{d}_1^\star \underline{\kappa} = \not{d}_2^\star \not{d}_1^\star \mu = 0$ on Σ_*. This yields on Σ_*

$$e_\theta(\mu) = \frac{\int_S e_\theta(\mu)e^\Phi}{\int_S e^{2\Phi}} e^\Phi, \quad e_\theta(\underline{\kappa}) = \frac{\int_S e_\theta(\underline{\kappa})e^\Phi}{\int_S e^{2\Phi}} e^\Phi.$$

Together with Step 2, we infer

$$\begin{aligned} &\max_{k\leq J+2} \int_{\Sigma_*} \left(r^4 \left((\not{\mathfrak{d}}, \nu_{\Sigma_*})^k \check{\mu} \right)^2 + r^2 \left((\not{\mathfrak{d}}, \nu_{\Sigma_*})^k \check{\underline{\kappa}} \right)^2 \right) \\ \lesssim & \left({}^{(\Sigma_*)}\mathfrak{R}_{J+1}[\check{R}] + {}^{(\Sigma_*)}\mathfrak{G}_J[\check{\Gamma}] + \epsilon_0 \, {}^{(\Sigma_*)}\mathfrak{G}_{J+1}[\check{\Gamma}] \right)^2. \end{aligned}$$

Then, in view of the null structure equations for $e_4(\check{\mu})$ and $e_4(\check{\underline{\kappa}})$,

$$\begin{aligned} e_4(\check{\mu}) &= -\frac{3}{2}\overline{\kappa}\check{\mu} - \frac{3}{2}\overline{\mu}\check{\kappa} + \mathrm{Err}[e_4\check{\mu}] \\ e_4(\check{\underline{\kappa}}) &= -\frac{1}{2}\overline{\kappa}\check{\underline{\kappa}} - \frac{1}{2}\overline{\underline{\kappa}}\check{\kappa} + 2\check{\mu} + 4\check{\rho} + \mathrm{Err}[e_4\check{\underline{\kappa}}], \end{aligned}$$

we infer, together with the control of $\check{\kappa}$ provided by Step 1,

$$\begin{aligned} &\max_{k\leq J+2} \int_{\Sigma_*} \left(r^4 \left(\mathfrak{d}^k \check{\mu} \right)^2 + r^2 \left(\mathfrak{d}^k \check{\underline{\kappa}} \right)^2 \right) \\ \lesssim & \left({}^{(\Sigma_*)}\mathfrak{R}_{J+1}[\check{R}] + {}^{(\Sigma_*)}\mathfrak{G}_J[\check{\Gamma}] + \epsilon_0 \, {}^{(\Sigma_*)}\mathfrak{G}_{J+1}[\check{\Gamma}] \right)^2. \end{aligned}$$

Step 4. Recall that we have

$$\not{d}_1 \zeta = -\check{\mu} - \check{\rho} + \frac{1}{4}\vartheta\underline{\vartheta}.$$

Differentiating, and using the Bianchi identities for $e_4(\check{\rho})$ and $e_3(\check{\rho})$, and the null structure equations for $e_4(\vartheta)$, $e_3(\vartheta)$, $e_4(\underline{\vartheta})$ and $e_3(\underline{\vartheta})$, we infer

$$\begin{aligned} \not{d}_1 \mathfrak{d}^k \zeta &= -\mathfrak{d}^k \check{\mu} - \mathfrak{d}^{k-1} \not{\mathfrak{d}} \left(\check{\rho}, \beta, r^{-1}\underline{\beta} \right) + \frac{1}{4}\mathfrak{d}^{k-1} \left(\not{\mathfrak{d}}(\vartheta\underline{\vartheta}), r^{-1}\underline{\vartheta}\not{\mathfrak{d}}\eta, \vartheta\not{\mathfrak{d}}\zeta, r^{-1}\vartheta\not{\mathfrak{d}}\underline{\xi} \right) \\ &\quad + \text{l.o.t.} \\ &= -\mathfrak{d}^k \check{\mu} - \not{\mathfrak{d}}\mathfrak{d}^{k-1} \left(\check{\rho}, \beta, r^{-1}\underline{\beta} \right) + \frac{1}{4}\not{\mathfrak{d}} \left(\underline{\vartheta}\mathfrak{d}^{k-1}(\vartheta, r^{-1}\eta) \right) \\ &\quad + \frac{1}{4}\not{\mathfrak{d}} \left(\vartheta\mathfrak{d}^{k-1}(\underline{\vartheta}, \zeta, r^{-1}\underline{\xi}) \right) + \text{l.o.t.} \end{aligned}$$

We infer, since $\not{d}_1$ is invertible in view of the corresponding Poincaré inequality,

$$\begin{aligned} \mathfrak{d}^k \zeta &= -r\not{\mathfrak{d}}^{-1}\mathfrak{d}^k \check{\mu} - \mathfrak{d}^{k-1} \left(\check{\rho}, \beta, r^{-1}\underline{\beta} \right) + \frac{1}{4} \left(\underline{\vartheta}\mathfrak{d}^{k-1}(\vartheta, r^{-1}\eta) \right) \\ &\quad + \frac{1}{4} \left(\vartheta\mathfrak{d}^{k-1}(\underline{\vartheta}, \zeta, r^{-1}\underline{\xi}) \right) + \text{l.o.t.} \end{aligned}$$

Together with the estimate for $\check{\mu}$ of Step 3, this yields

$$\max_{k\leq J+2} \int_{\Sigma_*} r^2 (\mathfrak{d}^k \zeta)^2 \lesssim \left({}^{(\Sigma_*)}\mathfrak{R}_{J+1}[\check{R}] + {}^{(\Sigma_*)}\mathfrak{G}_J[\check{\Gamma}] + \epsilon_0 \, {}^{(\Sigma_*)}\mathfrak{G}_{J+1}[\check{\Gamma}] \right)^2.$$

Step 5. Recall from the GCM condition that we have on Σ_*

$$\int_S \eta e^\Phi = 0.$$

Together with the transport equation

$$e_4(\eta - \zeta) = -\frac{1}{2}\kappa(\eta - \zeta) - \frac{1}{2}\vartheta(\eta - \zeta),$$

we infer in view of the estimates for ζ of Step 4,

$$\max_{k\leq J+1} \int_1^{u_*} r^{-4}\left(\mathfrak{d}^k\left(\int_S \eta e^\Phi\right)\right)^2 \lesssim \left({}^{(\Sigma_*)}\mathfrak{R}_{J+1}[\check{R}] + {}^{(\Sigma_*)}\mathfrak{G}_J[\check{\Gamma}] + \epsilon_0\, {}^{(\Sigma_*)}\mathfrak{G}_{J+1}[\check{\Gamma}]\right)^2.$$

Next, recall from Proposition 2.74 that η verifies

$$\begin{aligned} 2\,{\not{d}}_2\,{\not{d}}_2^\star\eta &= \kappa\left(-e_3(\zeta) + \underline{\beta}\right) - e_3(e_\theta(\kappa)) - \kappa\left(\frac{1}{2}\underline{\kappa}\zeta - 2\underline{\omega}\zeta\right) + 6\rho\eta - \underline{\kappa}e_\theta\kappa \\ &- \frac{1}{2}\kappa e_\theta(\underline{\kappa}) + 2\underline{\omega}e_\theta(\kappa) + 2e_\theta(\rho) + \mathrm{Err}[{\not{d}}_2\,{\not{d}}_2^\star\eta], \\ \mathrm{Err}[{\not{d}}_2\,{\not{d}}_2^\star\eta] &= \left(2\,{\not{d}}_1\eta - \frac{1}{2}\kappa\underline{\vartheta} + 2\eta^2\right)\eta + 2e_\theta(\eta^2) - \kappa\left(\frac{1}{2}\underline{\vartheta}\zeta - \frac{1}{2}\vartheta\underline{\xi}\right) - \frac{1}{2}\underline{\vartheta}e_\theta(\kappa) \\ &- \left(2\,{\not{d}}_1\eta - \frac{1}{2}\vartheta\underline{\vartheta} + 2\eta^2\right)\zeta - \frac{1}{2}e_\theta(\underline{\vartheta}\,\vartheta) - \frac{1}{2}\vartheta^2\underline{\xi} - \frac{3}{2}\vartheta\underline{\vartheta}\eta. \end{aligned}$$

Together with the estimates for κ of Step 1, the estimates for $\underline{\kappa}$ of Step 3, and the estimates for ζ of Step 4,

$$\begin{aligned} &\max_{k\leq J+1}\int_{\Sigma_*}\left(\mathfrak{d}^k\left(r^2\,{\not{d}}_2\,{\not{d}}_2^\star\eta - r^2e_\theta(\rho) - \frac{r^2}{2}\,{\not{d}}_2(\eta^2) - r^2e_\theta(\eta^2) + r^2\,{\not{d}}_1(\zeta\eta)\right.\right. \\ &\qquad\left.\left. + \frac{1}{4}r^2e_\theta(\underline{\vartheta}\,\vartheta)\right)\right)^2 \\ &\lesssim \max_{k\leq J+1}\int_{\Sigma_*} r^{-2}|\mathfrak{d}^k\eta|^2 + \left({}^{(\Sigma_*)}\mathfrak{R}_{J+1}[\check{R}] + {}^{(\Sigma_*)}\mathfrak{G}_J[\check{\Gamma}] + \epsilon_0\, {}^{(\Sigma_*)}\mathfrak{G}_{J+1}[\check{\Gamma}]\right)^2. \end{aligned}$$

In view of the dominant condition (3.3.4) for r on Σ_*, we infer

$$\begin{aligned} &\max_{k\leq J+1}\int_{\Sigma_*}\left(\mathfrak{d}^k\left(r^2\,{\not{d}}_2\,{\not{d}}_2^\star\eta - r^2e_\theta(\rho) - \frac{r^2}{2}\,{\not{d}}_2(\eta^2) - r^2e_\theta(\eta^2) + r^2\,{\not{d}}_1(\zeta\eta)\right.\right. \\ &\qquad\left.\left. + \frac{1}{4}r^2e_\theta(\underline{\vartheta}\,\vartheta)\right)\right)^2 \\ &\lesssim \epsilon_0^{\frac{2}{3}}\max_{k\leq J+1}\int_{\Sigma_*} |\mathfrak{d}^k\eta|^2 + \left({}^{(\Sigma_*)}\mathfrak{R}_{J+1}[\check{R}] + {}^{(\Sigma_*)}\mathfrak{G}_J[\check{\Gamma}] + \epsilon_0\, {}^{(\Sigma_*)}\mathfrak{G}_{J+1}[\check{\Gamma}]\right)^2. \end{aligned}$$

This yields

$$\begin{aligned}
&\max_{k\leq J+1}\int_{\Sigma_*}\Bigg(r^2 \not{d}_2 \not{d}_2^\star \mathfrak{d}^k\eta + r\not{d}_2^\star[\mathfrak{d}^k, r\not{d}_2]\eta + r\not{d}_2[\mathfrak{d}^k, r\not{d}_2^\star]\eta\\
&-r^2 e_\theta(\mathfrak{d}^k\rho) - \frac{r^2}{2}\not{d}_2\mathfrak{d}^k(\eta^2) - r^2 e_\theta\mathfrak{d}^k(\eta^2) + r^2\not{d}_1\mathfrak{d}^k(\zeta\eta) + \frac{1}{4}r^2 e_\theta\mathfrak{d}^k(\underline{\vartheta}\,\vartheta)\Bigg)^2\\
\lesssim\ & \epsilon_0^{\frac{2}{3}}\max_{k\leq J+1}\int_{\Sigma_*}|\mathfrak{d}^k\eta|^2 + \Big({}^{(\Sigma_*)}\mathfrak{R}_{J+1}[\check{R}] + {}^{(\Sigma_*)}\mathfrak{G}_J[\check{\Gamma}] + \epsilon_0\, {}^{(\Sigma_*)}\mathfrak{G}_{J+1}[\check{\Gamma}]\Big)^2.
\end{aligned}$$

We deduce, using a Poincaré inequality for $\not{d}_2$,

$$\begin{aligned}
\max_{k\leq J+1}\int_{\Sigma_*}\big(r\not{d}_2^\star\mathfrak{d}^k\eta\big)^2 \ \lesssim\ & \epsilon_0^{\frac{2}{3}}\max_{k\leq J+1}\int_{\Sigma_*}|\mathfrak{d}^k\eta|^2\\
&+\Big({}^{(\Sigma_*)}\mathfrak{R}_{J+1}[\check{R}] + {}^{(\Sigma_*)}\mathfrak{G}_J[\check{\Gamma}] + \epsilon_0\, {}^{(\Sigma_*)}\mathfrak{G}_{J+1}[\check{\Gamma}]\Big)^2.
\end{aligned}$$

Together with a Poincaré inequality for $r\not{d}_2^\star$ and the above control of the $\ell = 1$ mode of η, we infer

$$\begin{aligned}
\max_{k\leq J+1}\int_{\Sigma_*}\big(\mathfrak{d}^k\eta\big)^2 \ \lesssim\ & \epsilon_0^{\frac{2}{3}}\max_{k\leq J+1}\int_{\Sigma_*}|\mathfrak{d}^k\eta|^2\\
&+\Big({}^{(\Sigma_*)}\mathfrak{R}_{J+1}[\check{R}] + {}^{(\Sigma_*)}\mathfrak{G}_J[\check{\Gamma}] + \epsilon_0\, {}^{(\Sigma_*)}\mathfrak{G}_{J+1}[\check{\Gamma}]\Big)^2,
\end{aligned}$$

and hence, for ϵ_0 small enough,

$$\max_{k\leq J+1}\int_{\Sigma_*}\big(\mathfrak{d}^k\eta\big)^2 \ \lesssim\ \Big({}^{(\Sigma_*)}\mathfrak{R}_{J+1}[\check{R}] + {}^{(\Sigma_*)}\mathfrak{G}_J[\check{\Gamma}] + \epsilon_0\, {}^{(\Sigma_*)}\mathfrak{G}_{J+1}[\check{\Gamma}]\Big)^2.$$

Step 6. Recall from the GCM condition that we have on Σ_*

$$\int_S \underline{\xi} e^\Phi = 0.$$

Together with the transport equation

$$e_4(\underline{\xi}) \ = \ -e_3(\zeta) + \underline{\beta} - \underline{\kappa}\zeta - \zeta\underline{\vartheta},$$

we infer in view of the estimates for ζ of Step 4, the estimates for $\underline{\beta}$, and the bootstrap assumptions

$$\max_{k\leq J+1}\int_1^{u_*} r^{-4}\left(\mathfrak{d}^k\left(\int_S\underline{\xi}e^\Phi\right)\right)^2 \lesssim \Big({}^{(\Sigma_*)}\mathfrak{R}_{J+1}[\check{R}] + {}^{(\Sigma_*)}\mathfrak{G}_J[\check{\Gamma}] + \epsilon_0\, {}^{(\Sigma_*)}\mathfrak{G}_{J+1}[\check{\Gamma}]\Big)^2.$$

Next, from Proposition 2.74, we have

$$\begin{aligned}
2\,\mathscr{d}_2\,\mathscr{d}_2^\star\underline{\xi} &= -e_3(e_\theta(\underline{\kappa}))+\underline{\kappa}\left(e_3(\zeta)-\underline{\beta}\right)+\underline{\kappa}^2\zeta-\frac{3}{2}\underline{\kappa}e_\theta\underline{\kappa}+6\rho\underline{\xi}-2\underline{\omega}e_\theta(\underline{\kappa})\\
&\quad +\mathrm{Err}[\mathscr{d}_2\,\mathscr{d}_2^\star\underline{\xi}],\\
\mathrm{Err}[\mathscr{d}_2\,\mathscr{d}_2^\star\underline{\xi}] &= \left(2\,\mathscr{d}_1\underline{\xi}+\frac{1}{2}\underline{\kappa}\,\underline{\vartheta}+2\eta\underline{\xi}-\frac{1}{2}\underline{\vartheta}^2\right)\eta+2e_\theta(\eta\underline{\xi})-\frac{1}{2}e_\theta(\underline{\vartheta}^2)\\
&+\ \underline{\kappa}\left(\frac{1}{2}\underline{\vartheta}\zeta-\frac{1}{2}\vartheta\underline{\xi}\right)-\frac{1}{2}\underline{\vartheta}e_\theta\underline{\kappa}-\frac{1}{2}\vartheta\underline{\vartheta}\underline{\xi}\\
&\quad -\zeta\left(2\,\mathscr{d}_1\underline{\xi}+2(\eta-3\zeta)\underline{\xi}-\frac{1}{2}\underline{\vartheta}^2\right)\\
&+\ \underline{\xi}\Big(-\vartheta\underline{\vartheta}-2\,\mathscr{d}_1\zeta+2\zeta^2\Big)-6\eta\zeta\underline{\xi}-6e_\theta(\zeta\underline{\xi}).
\end{aligned}$$

Together with the estimates for κ of Step 1, the estimates for $\underline{\kappa}$ of Step 3, and the estimates for ζ of Step 4,

$$\begin{aligned}
&\max_{k\le J+1}\int_{\Sigma_*}\left(\mathfrak{d}^k\left(r^2\,\mathscr{d}_2\,\mathscr{d}_2^\star\underline{\xi}+\frac{1}{2}e_\theta(e_3(\check{\underline{\kappa}}))-\eta\,\mathscr{d}_1\underline{\xi}-e_\theta(\eta\underline{\xi})+\frac{1}{4}e_\theta(\underline{\vartheta}^2)+\mathscr{d}_2(\zeta\underline{\xi})\right.\right.\\
&\qquad\qquad\left.\left.+3e_\theta(\zeta\underline{\xi})\right)\right)^2\\
\lesssim\ &\max_{k\le J+1}\int_{\Sigma_*} r^{-2}|\mathfrak{d}^k\underline{\xi}|^2+\left({}^{(\Sigma_*)}\mathfrak{R}_{J+1}[\check{R}]+{}^{(\Sigma_*)}\mathfrak{G}_J[\check{\Gamma}]+\epsilon_0\,{}^{(\Sigma_*)}\mathfrak{G}_{J+1}[\check{\Gamma}]\right)^2.
\end{aligned}$$

In view of the dominant condition (3.3.4) for r on Σ_*, we infer

$$\begin{aligned}
&\max_{k\le J+1}\int_{\Sigma_*}\left(\mathfrak{d}^k\left(r^2\,\mathscr{d}_2\,\mathscr{d}_2^\star\underline{\xi}+\frac{1}{2}e_\theta(e_3(\check{\underline{\kappa}}))-\eta\,\mathscr{d}_1\underline{\xi}-e_\theta(\eta\underline{\xi})+\frac{1}{4}e_\theta(\underline{\vartheta}^2)+\mathscr{d}_2(\zeta\underline{\xi})\right.\right.\\
&\qquad\qquad\left.\left.+3e_\theta(\zeta\underline{\xi})\right)\right)^2\\
\lesssim\ &\epsilon_0^{\frac{2}{3}}\max_{k\le J+1}\int_{\Sigma_*}|\mathfrak{d}^k\underline{\xi}|^2+\left({}^{(\Sigma_*)}\mathfrak{R}_{J+1}[\check{R}]+{}^{(\Sigma_*)}\mathfrak{G}_J[\check{\Gamma}]+\epsilon_0\,{}^{(\Sigma_*)}\mathfrak{G}_{J+1}[\check{\Gamma}]\right)^2.
\end{aligned}$$

This yields

$$\begin{aligned}
&\max_{k\le J+1}\int_{\Sigma_*}\left(r^2\,\mathscr{d}_2\,\mathscr{d}_2^\star\mathfrak{d}^k\underline{\xi}+r\,\mathscr{d}_2^\star[\mathfrak{d}^k,r\,\mathscr{d}_2]\underline{\xi}+r\,\mathscr{d}_2[\mathfrak{d}^k,r\,\mathscr{d}_2^\star]\underline{\xi}+\frac{1}{2}e_\theta(\mathfrak{d}^k e_3(\check{\underline{\kappa}}))\right.\\
&\left.-\,\mathscr{d}_1(\eta\mathfrak{d}^k\underline{\xi})-e_\theta\mathfrak{d}^k(\eta\underline{\xi})+\frac{1}{4}e_\theta\mathfrak{d}^k(\underline{\vartheta}^2)+\mathscr{d}_2\mathfrak{d}^k(\zeta\underline{\xi})+3e_\theta\mathfrak{d}^k(\zeta\underline{\xi})\right)^2\\
\lesssim\ &\epsilon_0^{\frac{2}{3}}\max_{k\le J+1}\int_{\Sigma_*}|\mathfrak{d}^k\underline{\xi}|^2+\left({}^{(\Sigma_*)}\mathfrak{R}_{J+1}[\check{R}]+{}^{(\Sigma_*)}\mathfrak{G}_J[\check{\Gamma}]+\epsilon_0\,{}^{(\Sigma_*)}\mathfrak{G}_{J+1}[\check{\Gamma}]\right)^2.
\end{aligned}$$

We deduce, using a Poincaré inequality for $\not{d}_2$ and the estimates for $\underline{\kappa}$ of Step 3,

$$\max_{k\leq J+1}\int_{\Sigma_*}\left(r\not{d}_2^\star\mathfrak{d}^k\underline{\xi}\right)^2 \lesssim \epsilon_0^{\frac{2}{3}}\max_{k\leq J+1}\int_{\Sigma_*}|\mathfrak{d}^k\underline{\xi}|^2 +\left({}^{(\Sigma_*)}\mathfrak{R}_{J+1}[\check{R}]+{}^{(\Sigma_*)}\mathfrak{G}_J[\check{\Gamma}]+\epsilon_0\,{}^{(\Sigma_*)}\mathfrak{G}_{J+1}[\check{\Gamma}]\right)^2.$$

Together with a Poincaré inequality for $r\not{d}_2^\star$ and the above control of the $\ell=1$ mode of $\underline{\xi}$, we infer

$$\max_{k\leq J+1}\int_{\Sigma_*}\left(\mathfrak{d}^k\underline{\xi}\right)^2 \lesssim \epsilon_0^{\frac{2}{3}}\max_{k\leq J+1}\int_{\Sigma_*}|\mathfrak{d}^k\underline{\xi}|^2 +\left({}^{(\Sigma_*)}\mathfrak{R}_{J+1}[\check{R}]+{}^{(\Sigma_*)}\mathfrak{G}_J[\check{\Gamma}]+\epsilon_0\,{}^{(\Sigma_*)}\mathfrak{G}_{J+1}[\check{\Gamma}]\right)^2,$$

and hence, for ϵ_0 small enough,

$$\max_{k\leq J+1}\int_{\Sigma_*}\left(\mathfrak{d}^k\underline{\xi}\right)^2 \lesssim \left({}^{(\Sigma_*)}\mathfrak{R}_{J+1}[\check{R}]+{}^{(\Sigma_*)}\mathfrak{G}_J[\check{\Gamma}]+\epsilon_0\,{}^{(\Sigma_*)}\mathfrak{G}_{J+1}[\check{\Gamma}]\right)^2.$$

Step 7. Using the Codazzi for ϑ and $\underline{\vartheta}$, the transport equation for ϑ and $\underline{\vartheta}$ in the e_4 and e_3 direction, the control of $\check{\kappa}$ of Step 1, the control of $\check{\underline{\kappa}}$ of Step 3, the control of ζ of Step 4, the control of η of Step 5, the control of $\underline{\xi}$ of Step 6, and a Poincaré inequality for $\not{d}_2$, we infer

$$\max_{k\leq J+1}\int_{\Sigma_*}\left(r^2(\mathfrak{d}^k\vartheta)^2+(\mathfrak{d}^k\underline{\vartheta})^2\right)\lesssim\left({}^{(\Sigma_*)}\mathfrak{R}_{J+1}[\check{R}]+{}^{(\Sigma_*)}\mathfrak{G}_J[\check{\Gamma}]+\epsilon_0^{\frac{2}{3}}\,{}^{(\Sigma_*)}\mathfrak{G}_{J+1}[\check{\Gamma}]\right)^2.$$

Step 8. Recall from Proposition 2.74 that $\underline{\omega}$ verifies

$$\begin{aligned}2\not{d}_1^\star\underline{\omega} &= -\frac{1}{2}\kappa\underline{\xi}+\left(\frac{1}{2}\underline{\kappa}+2\underline{\omega}+\frac{1}{2}\underline{\vartheta}\right)\eta+e_3(\zeta)-\underline{\beta}\\ &\quad+\frac{1}{2}\underline{\kappa}\zeta-2\underline{\omega}\zeta+\frac{1}{2}\underline{\vartheta}\zeta-\frac{1}{2}\vartheta\underline{\xi}.\end{aligned}$$

Together with a Poincaré inequality for $\not{d}_1^\star$, the control of $\underline{\xi}$ from Step 6, the control of η from Step 5, and the control of ζ from Step 4, we infer

$$\max_{k\leq J+1}\int_{\Sigma_*}|\mathfrak{d}^k\check{\underline{\omega}}|^2 \lesssim \left({}^{(\Sigma_*)}\mathfrak{R}_{J+1}[\check{R}]+{}^{(\Sigma_*)}\mathfrak{G}_J[\check{\Gamma}]+\epsilon_0^{\frac{2}{3}}\,{}^{(\Sigma_*)}\mathfrak{G}_{J+1}[\check{\Gamma}]\right)^2.$$

Finally, gathering the estimates of Step 1 to Step 8, we infer

$${}^{(\Sigma_*)}\mathfrak{G}_{J+1}[\check{\Gamma}]+{}^{(\Sigma_*)}\mathfrak{G}'_{J+1}[\check{\Gamma}] \lesssim {}^{(\Sigma_*)}\mathfrak{R}_{J+1}[\check{R}]+{}^{(\Sigma_*)}\mathfrak{G}_J[\check{\Gamma}]+\epsilon_0^{\frac{2}{3}}\,{}^{(\Sigma_*)}\mathfrak{G}_{J+1}[\check{\Gamma}]$$

and hence, for ϵ_0 small enough,

$${}^{(\Sigma_*)}\mathfrak{G}_{J+1}[\check{\Gamma}]+{}^{(\Sigma_*)}\mathfrak{G}'_{J+1}[\check{\Gamma}] \lesssim {}^{(\Sigma_*)}\mathfrak{R}_{J+1}[\check{R}]+{}^{(\Sigma_*)}\mathfrak{G}_J[\check{\Gamma}]$$

as desired. This concludes the proof of Proposition 8.31.

8.8.2 Weighted estimates for transport equations along e_4 in ${}^{(ext)}\mathcal{M}$

Lemma 8.34. *Let the following transport equation in* ${}^{(ext)}\mathcal{M}$

$$e_4(f) + \frac{a}{2}\kappa f = h$$

where $a \in \mathbb{R}$ *is a given constant, and* f *and* h *are scalar functions. Also, let* $\delta_B > 0$. *Then,* f *satisfies*

$$\sup_{r_0 \geq 4m_0} \left(r_0^{2a-2} \int_{\{r=r_0\}} f^2 \right) \lesssim \int_{\Sigma_*} r^{2a-2} f^2 + \int_{{}^{(ext)}\mathcal{M}(\geq 4m_0)} r^{2a-1+\delta_B} h^2.$$

Proof. Multiply by f to obtain

$$\frac{1}{2} e_4(f^2) + \frac{a}{2}\kappa f^2 = hf.$$

Next, integrate over $S_{u,r}$ to obtain

$$\begin{aligned} \frac{1}{2} e_4\left(\int_{S_{u,r}} f^2 \right) &= \int_{S_{u,r}} \frac{1}{2}(e_4(f^2) + \kappa f^2) \\ &= -\int_{S_{u,r}} \frac{a-1}{2}\kappa f^2 + \int_{S_{u,r}} hf \\ &= -\frac{a-1}{2}\overline{\kappa} \int_{S_{u,r}} f^2 - \frac{a-1}{2} \int_{S_{u,r}} \check{\kappa} f^2 + \int_{S_{u,r}} hf \end{aligned}$$

and hence

$$\frac{1}{2} e_4\left(\int_{S_{u,r}} f^2 \right) + \frac{a-1}{2}\overline{\kappa} \int_{S_{u,r}} f^2 = -\frac{a-1}{2} \int_{S_{u,r}} \check{\kappa} f^2 + \int_{S_{u,r}} hf.$$

Also, we multiply by r^{2a-2} which yields

$$\frac{1}{2} e_4\left(r^{2a-2} \int_{S_{u,r}} f^2 \right) = -\frac{a-1}{2} r^{2a-2} \int_{S_{u,r}} \check{\kappa} f^2 + r^{2a-2} \int_{S_{u,r}} hf$$

where we used the fact that $2e_4(r) = r\overline{\kappa}$. This yields

$$-e_4\left(r^{2a-2} \int_{S_{u,r}} f^2 \right) \leq r^{2a-3-\delta_B} \int_{S_{u,r}} f^2 + \frac{1}{4} r^{2a-1+\delta_B} \int_{S_{u,r}} h^2$$

and hence

$$-e_4\left(e^{-\delta_B^{-1} r^{-\delta_B}} r^{2a-2} \int_{S_{u,r}} f^2 \right) \lesssim e^{-\delta_B^{-1} r^{-\delta_B}} r^{2a-1+\delta_B} \int_{S_{u,r}} h^2$$

where we used the fact that $2e_4(r) = r\overline{\kappa} = 2 + O(\epsilon_0)$. Integrating between $r = r_0$

and $r = r_*(u)$, where $r_*(u)$ is such that $S_{u,r_*(u)} \subset \Sigma_*$, we infer

$$r_0^{2a-2}\int_{S_{u,r_0}} f^2 \lesssim r_*(u)^{2a-2}\int_{S_{u,r_*(u)}} f^2 + \int_{r_0}^{r_*(u)} r^{2a-1+\delta_B}\int_{S_{u,r}} h^2. \tag{8.8.1}$$

Remark 8.35. *Note that we have the following consequences of the coarea formula*

$$d\Sigma_* = \varsigma\sqrt{\frac{2}{\varsigma} - \Upsilon + \frac{r}{2}\underline{A}}\; d\mu_{u,\Sigma_*}du, \qquad d\{r = r_0\} = \frac{\varsigma\sqrt{-\underline{\kappa} - \underline{A}}}{\sqrt{\overline{\kappa}}}\; d\mu_{u,r_0}du,$$

where we used in particular that $\Sigma_* = \{u + r = c_{\Sigma_*}\}$*. Also, we have in* ${}^{(ext)}\mathcal{M}$

$$d\mathcal{M} = \frac{4\varsigma^2}{r^2\overline{\kappa}^2}d\mu_{u,r}dudr.$$

We infer, in ${}^{(ext)}\mathcal{M}$*, using in particular the dominant condition of* r *on* Σ_**,*

$$d\Sigma_* = \left(1 + O\left(\epsilon_0^{\frac{2}{3}}\right)\right) d\mu_{u,\Sigma_*}du, \quad d\{r = r_0\} = \sqrt{1 - \frac{2m_0}{r_0}}(1 + O(\epsilon_0))\; d\mu_{u,r_0}du,$$

and

$$d\mathcal{M} = (1 + O(\epsilon_0))d\mu_{u,r}dudr.$$

Integrating (8.8.1) in $u \in [1, u_*]$, and relying on Remark 8.35 we deduce for $r_0 \geq 4m_0$

$$r_0^{2a-2}\int_{\{r=r_0\}} f^2 \lesssim \int_{\Sigma_*} r^{2a-2}f^2 + \int_{{}^{(ext)}\mathcal{M}(r\geq 4m_0)} r^{2a-\frac{1}{2}}h^2$$

as desired. This concludes the proof of the lemma. □

Corollary 8.36. *Let the following transport equation in* ${}^{(ext)}\mathcal{M}$

$$e_4(f) + \frac{a}{2}\kappa f = h$$

where $a \in \mathbb{R}$ *is a given constant, and* f *and* h *are scalar functions. Also, let* $\delta_B > 0$*. Then,* f *satisfies for* $5 \leq k \leq k_{large} + 1$

$$\begin{aligned}
&\sup_{r_0\geq 4m_0}\left(r_0^{2a-2}\int_{\{r=r_0\}}(\mathfrak{d}^k f)^2\right)\\
&\lesssim \int_{\Sigma_*} r^{2a-2}(\mathfrak{d}^{\leq k}f)^2 + \sup_{r_0\geq 4m_0}\left(r_0^{2a-2}\int_{\{r=r_0\}}(\mathfrak{d}^{\leq k-1}f)^2\right)\\
&\quad + \int_{{}^{(ext)}\mathcal{M}(\geq 4m_0)} r^{2a-1+\delta_B}(\mathfrak{d}^{\leq k}h)^2\\
&\quad + \left(\sup_{{}^{(ext)}\mathcal{M}(\geq 4m_0)}\left(r^a|\mathfrak{d}^{\leq k-5}f|\right)\right)^2\left({}^{(ext)}\mathfrak{G}_{k-1}^{\geq 4m_0}[\check{\Gamma}] + {}^{(ext)}\mathfrak{G}_k^{\geq 4m_0}[\check{\kappa}]\right)^2.
\end{aligned}$$

Proof. We first differentiate the equation for f with $(\not{\mathfrak{d}}, \mathbf{T})^l$ and obtain

$$\begin{aligned} e_4((\not{\mathfrak{d}}, \mathbf{T})^l f) + \frac{a}{2}\kappa(\not{\mathfrak{d}}, \mathbf{T})^l f &= h_l, \\ h_l &:= (\not{\mathfrak{d}}, \mathbf{T})^l h - [(\not{\mathfrak{d}}, \mathbf{T})^l, e_4] f - \frac{a}{2}[(\not{\mathfrak{d}}, \mathbf{T})^l, \kappa] f. \end{aligned}$$

In view of Lemma 8.34, we deduce

$$\sup_{r_0 \geq 4m_0} \left(r_0^{2a-2} \int_{\{r=r_0\}} ((\not{\mathfrak{d}}, \mathbf{T})^l f)^2 \right) \lesssim \int_{\Sigma_*} r^{2a-2} (\mathfrak{d}^l f)^2 + \int_{{}^{(ext)}\mathcal{M}(\geq 4m_0)} r^{2a-1+\delta_B} h_l^2.$$

Now, we have the following schematic commutation formulas

$$[\not{\mathfrak{d}}, e_4] = \Gamma_g \mathfrak{d} + \Gamma_g, \quad [\mathbf{T}, e_4] = r^{-1} \Gamma_b \mathfrak{d}.$$

Together with the definition of h_l and for $5 \leq l \leq k_{large} + 1$, we deduce

$$\begin{aligned} & \int_{{}^{(ext)}\mathcal{M}(\geq 4m_0)} r^{2a-1+\delta_B} h_l^2 \\ \lesssim & \int_{{}^{(ext)}\mathcal{M}(\geq 4m_0)} r^{2a-1+\delta_B} (\mathfrak{d}^l h)^2 + \epsilon_0^2 \int_{{}^{(ext)}\mathcal{M}} r^{2a-5+\delta_B} (\mathfrak{d}^{\leq l} f)^2 \\ & + \left(\sup_{{}^{(ext)}\mathcal{M}} (r^a |\mathfrak{d}^{\leq l-5} f|) \right)^2 \left({}^{(ext)}\mathfrak{G}_{l-1}[\check{\Gamma}] + {}^{(ext)}\mathfrak{G}_l[\check{\kappa}] \right)^2 \end{aligned}$$

and hence

$$\begin{aligned} & \sup_{r_0 \geq r_{\mathcal{T}}} \left(r_0^{2a-2} \int_{\{r=r_0\}} ((\not{\mathfrak{d}}, \mathbf{T})^l f)^2 \right) \\ \lesssim & \int_{\Sigma_*} r^{2a-2} (\mathfrak{d}^l f)^2 + \int_{{}^{(ext)}\mathcal{M}} r^{2a-1+\delta_B} (\mathfrak{d}^l h)^2 \\ & + \epsilon_0^2 \int_{{}^{(ext)}\mathcal{M}(\geq 4m_0)} r^{2a-5+\delta_B} (\mathfrak{d}^{\leq l} f)^2 \\ & + \left(\sup_{{}^{(ext)}\mathcal{M}(\geq 4m_0)} (r^a |\mathfrak{d}^{\leq l-5} f|) \right)^2 \left({}^{(ext)}\mathfrak{G}_{l-1}^{\geq 4m_0}[\check{\Gamma}] + {}^{(ext)}\mathfrak{G}_l^{\geq 4m_0}[\check{\kappa}] \right)^2 \end{aligned}$$

or

$$\begin{aligned} & \sup_{r_0 \geq 4m_0} \left(r_0^{2a-2} \int_{\{r=r_0\}} ((\not{\mathfrak{d}}, \mathbf{T})^l f)^2 \right) \\ \lesssim & \int_{\Sigma_*} r^{2a-2} (\mathfrak{d}^l f)^2 + \int_{{}^{(ext)}\mathcal{M}(\geq 4m_0)} r^{2a-1+\delta_B} (\mathfrak{d}^l h)^2 \\ & + \epsilon_0^2 \sup_{r_0 \geq 4m_0} \left(r_0^{2a-2} \int_{\{r=r_0\}} (\mathfrak{d}^{\leq l} f)^2 \right) \\ & + \left(\sup_{{}^{(ext)}\mathcal{M}(\geq 4m_0)} (r^a |\mathfrak{d}^{\leq l-5} f|) \right)^2 \left({}^{(ext)}\mathfrak{G}_{l-1}^{\geq 4m_0}[\check{\Gamma}] + {}^{(ext)}\mathfrak{G}_l^{\geq 4m_0}[\check{\kappa}] \right)^2. \end{aligned}$$

Together with the first equation which yields

$$re_4((\not{\mathfrak{d}}, \mathbf{T})^l f) + \frac{a}{2} r\kappa(\not{\mathfrak{d}}, \mathbf{T})^l f \quad = \quad rh_l,$$

and hence

$$(re_4)^j((\not{\mathfrak{d}}, \mathbf{T})^l f) + \frac{a}{2}(re_4)^{j-1}\Big(r\kappa(\not{\mathfrak{d}}, \mathbf{T})^l f\Big) \quad = \quad (re_4)^{j-1}(rh_l),$$

we infer, for $\epsilon_0 > 0$ small enough, and for $5 \leq k \leq k_{large} + 1$,

$$\begin{aligned}
&\sup_{r_0 \geq 4m_0} \left(r_0^{2a-2} \int_{\{r=r_0\}} (\mathfrak{d}^k f)^2 \right) \\
\lesssim \;& \int_{\Sigma_*} r^{2a-2}(\mathfrak{d}^{\leq k} f)^2 + \int_{{}^{(ext)}\mathcal{M}(\geq 4m_0)} r^{2a-1+\delta_B}(\mathfrak{d}^{\leq k} h)^2 \\
&+ \sup_{r_0 \geq 4m_0} \left(r_0^{2a-2} \int_{\{r=r_0\}} (\mathfrak{d}^{\leq k-1} h)^2 \right) + \sup_{r_0 \geq 4m_0} \left(r_0^{2a-2} \int_{\{r=r_0\}} (\mathfrak{d}^{\leq k-1} f)^2 \right) \\
&+ \left(\sup_{{}^{(ext)}\mathcal{M}(\geq 4m_0)} \left(r^a |\mathfrak{d}^{\leq k-5} f| \right) \right)^2 \left({}^{(ext)}\mathfrak{G}_{k-1}^{\geq 4m_0}[\check{\Gamma}] + {}^{(ext)}\mathfrak{G}_k^{\geq 4m_0}[\check{\kappa}] \right)^2.
\end{aligned}$$

Using a trace estimate, we infer

$$\begin{aligned}
&\sup_{r_0 \geq 4m_0} \left(r_0^{2a-2} \int_{\{r=r_0\}} (\mathfrak{d}^k f)^2 \right) \\
\lesssim \;& \int_{\Sigma_*} r^{2a-2}(\mathfrak{d}^{\leq k} f)^2 + \int_{{}^{(ext)}\mathcal{M}(\geq 4m_0)} r^{2a-1+\delta_B}(\mathfrak{d}^{\leq k} h)^2 \\
&+ \sup_{r_0 \geq 4m_0} \left(r_0^{2a-2} \int_{\{r=r_0\}} (\mathfrak{d}^{\leq k-1} f)^2 \right) \\
&+ \left(\sup_{{}^{(ext)}\mathcal{M}(\geq 4m_0)} \left(r^a |\mathfrak{d}^{\leq k-5} f| \right) \right)^2 \left({}^{(ext)}\mathfrak{G}_{k-1}^{\geq 4m_0}[\check{\Gamma}] + {}^{(ext)}\mathfrak{G}_k^{\geq 4m_0}[\check{\kappa}] \right)^2
\end{aligned}$$

as desired. This concludes the proof of the corollary. □

Lemma 8.37. *Let the following transport equation in* ${}^{(ext)}\mathcal{M}$

$$e_4(f) + \frac{a}{2}\kappa f = h$$

where $a \in \mathbb{R}$ *is a given constant, and* f *and* h *are scalar functions. Let* $b > 2a - 2$. *Then,* f *satisfies*

$$\begin{aligned}
&\sup_{r_0 \geq 4m_0} \left(r_0^b \int_{\{r=r_0\}} f^2 \right) + \int_{{}^{(ext)}\mathcal{M}(\geq 4m_0)} r^{b-1} f^2 \\
\lesssim \;& \int_{\Sigma_*} r^b f^2 + \int_{{}^{(ext)}\mathcal{M}(\geq 4m_0)} r^{b+1} h^2.
\end{aligned}$$

Proof. Recall from Lemma 8.34 the following identity:

$$\frac{1}{2}e_4\left(\int_{S_{u,r}} f^2\right) + \frac{a-1}{2}\overline{\kappa}\int_{S_{u,r}} f^2 = -\frac{a-1}{2}\int_{S_{u,r}} \check{\kappa} f^2 + \int_{S_{u,r}} hf.$$

We multiply by r^b which yields

$$\frac{1}{2}e_4\left(r^b\int_{S_{u,r}} f^2\right) + \frac{1}{2}\left(a-1-\frac{b}{2}\right)\overline{\kappa}\int_{S_{u,r}} r^b f^2 = -\frac{a-1}{2}r^b\int_{S_{u,r}} \check{\kappa} f^2 + r^b\int_{S_{u,r}} hf$$

where we used the fact that $2e_4(r) = r\overline{\kappa}$. We choose $b > 2a-2$ and integrate between $r = r_0$ and $r = r_*(u)$, where $r_*(u)$ is such that $S_{u,r_*(u)} \subset \Sigma_*$, which yields

$$\int_{S_{u,r_0}} r^b f^2 + \int_{r_0}^{r_*}\int_{S_{u,r}} r^{b-1}f^2 \lesssim \int_{S_{u,r_*}} r^b f^2 + \int_{r_0}^{r_*}\int_{S_{u,r}} r^{b+1}h^2.$$

Then, integrating in u in $u \in [1, u_*]$, and relying on Remark 8.35 we deduce for $r_0 \geq 4m_0$,

$$r_0^b\int_{\{r=r_0\}} f^2 + \int_{^{(ext)}\mathcal{M}\cap\{r\geq r_0\}} r^{b-1}f^2 \lesssim \int_{\Sigma_*} r^b f^2 + \int_{^{(ext)}\mathcal{M}(\geq 4m_0)} r^{b+1}h^2.$$

This concludes the proof of the lemma. □

Corollary 8.38. *Let the following transport equation in* $^{(ext)}\mathcal{M}$

$$e_4(f) + \frac{a}{2}\kappa f = h$$

where $a \in \mathbb{R}$ is a given constant, and f and h are scalar functions. Let $b > 2a-2$. Then, f satisfies for $5 \leq l \leq k_{large}+1$

$$\begin{aligned}
&\sup_{r_0\geq 4m_0}\left(r_0^b\int_{\{r=r_0\}}(\mathfrak{d}^k f)^2\right) + \int_{^{(ext)}\mathcal{M}(\geq 4m_0)} r^{b-1}(\mathfrak{d}^k f)^2\\
\lesssim\; & \int_{\Sigma_*} r^b(\mathfrak{d}^{\leq k}f)^2 + \sup_{r_0\geq 4m_0}\left(r_0^b\int_{\{r=r_0\}}(\mathfrak{d}^{\leq k-1}f)^2\right) + \int_{^{(ext)}\mathcal{M}(\geq 4m_0)} r^{b-1}(\mathfrak{d}^{\leq k}h)^2\\
&+\left(\sup_{^{(ext)}\mathcal{M}(\geq 4m_0)}\left(r^b|\mathfrak{d}^{\leq k-5}f|\right)\right)^2\left({}^{(ext)}\mathfrak{G}_{k-1}^{\geq 4m_0}[\check{\Gamma}] + {}^{(ext)}\mathfrak{G}_{k}^{\geq 4m_0}[\check{\kappa}]\right)^2.
\end{aligned}$$

Proof. The proof is based on Lemma 8.37. It is similar to the one of Corollary 8.36 and left to the reader. □

Lemma 8.39. *Let the following transport equation in* $^{(ext)}\mathcal{M}$

$$e_4(f) + \frac{a}{2}\kappa f = h$$

where $a \in \mathbb{R}$ is a given constant, and f and h are scalar functions. Then, f satisfies

$$\sup_{r_\mathcal{T}\leq r_0\leq 4m_0}\int_{\{r=r_0\}} f^2 \lesssim \int_{\{r=4m_0\}} f^2 + \int_{^{(ext)}\mathcal{M}(\leq 4m_0)} h^2.$$

Proof. Let $b > 2a - 2$. Recall from Lemma 8.37 the following identity:

$$\frac{1}{2}e_4\left(r^b\int_{S_{u,r}} f^2\right) + \frac{1}{2}\left(a-1-\frac{b}{2}\right)\overline{\kappa}\int_{S_{u,r}} r^b f^2 = -\frac{a-1}{2}r^b\int_{S_{u,r}}\check{\kappa}f^2 + r^b\int_{S_{u,r}} hf.$$

Choosing $b = 2a$, we obtain

$$\frac{1}{2}e_4\left(r^b\int_{S_{u,r}} f^2\right) - \frac{1}{2}\overline{\kappa}\int_{S_{u,r}} r^b f^2 = -\frac{a-1}{2}r^b\int_{S_{u,r}}\check{\kappa}f^2 + r^b\int_{S_{u,r}} hf.$$

Next, let $1 \leq u \leq u_*$ and $r_{\mathcal{T}} \leq r_0 \leq 4m_0$. We now integrate in $r_0 \leq r \leq 4m_0$ and along $\mathcal{C}_u$ in ${}^{(ext)}\mathcal{M}$. Since r is bounded on ${}^{(ext)}\mathcal{M}(r \leq 4m_0)$ from above and below, we obtain, for $\epsilon_0 > 0$ small enough,

$$\int_{S_{u,r_0}} f^2 \lesssim \int_{S_{u,4m_0}} f^2 + \int_{r_{\mathcal{T}}}^{4m_0}\int_{S_{\underline{u},r}} h^2.$$

We may now integrate in u to deduce

$$\int_1^{u_*}\int_{S_{u,r_0}} f^2 \lesssim \int_1^{u_*}\int_{S_{u,4m_0}} f^2 + \int_1^{u_*}\int_{r_{\mathcal{T}}}^{4m_0}\int_{S_{u,r}} h^2.$$

Relying on Remark 8.35 we deduce

$$\sup_{r_{\mathcal{T}}\leq r_0\leq 4m_0}\int_{\{r=r_0\}} f^2 \lesssim \int_{\{r=4m_0\}} f^2 + \int_{{}^{(ext)}\mathcal{M}(\leq 4m_0)} h^2$$

as desired. This concludes the proof of the lemma. □

Corollary 8.40. *Let the following transport equation in ${}^{(ext)}\mathcal{M}$*

$$e_4(f) + \frac{a}{2}\kappa f = h$$

where $a \in \mathbb{R}$ is a given constant, and f and h are scalar functions. Then, f satisfies for $5 \leq l \leq k_{large} + 1$

$$\begin{aligned}&\sup_{r_{\mathcal{T}}\leq r_0\leq 4m_0}\int_{\{r=r_0\}} (\mathfrak{d}^k f)^2\\ \lesssim\ & \int_{\{r=4m_0\}}(\mathfrak{d}^{\leq k} f)^2 + \sup_{r_{\mathcal{T}}\leq r_0\leq 4m_0}\int_{\{r=r_0\}}(\mathfrak{d}^{\leq k-1} f)^2 + \int_{{}^{(ext)}\mathcal{M}(\leq 4m_0)}(\mathfrak{d}^{\leq k} h)^2\\ &+\left(\sup_{{}^{(ext)}\mathcal{M}(r\leq 4m_0)}\left(|\mathfrak{d}^{\leq k-5} f|\right)\right)^2\left({}^{(ext)}\mathfrak{G}^{\leq 4m_0}_{k-1}[\check{\Gamma}] + {}^{(ext)}\mathfrak{G}^{\leq 4m_0}_{k}[\check{\kappa}]\right)^2.\end{aligned}$$

Proof. The proof is based on Lemma 8.39. It is similar to the one of Corollary 8.36 and left to the reader. □

8.8.3 Several identities

The goal of this section is to prove the identities below that will be used to avoid losing derivatives when controlling the weighted energies of the Ricci coefficients.

Lemma 8.41. *We have*

$$\begin{aligned}
& e_4\left(\not{d}_1\,\not{d}_1^\star\kappa-\vartheta\Big(\not{d}_4\,\not{d}_3^\star\,\not{d}_2^{-1}+\not{d}_2^\star\Big)\not{d}_1^{-1}\check{\rho}\right)\\
& +2\kappa\left(\not{d}_1\,\not{d}_1^\star\kappa-\vartheta\Big(\not{d}_4\,\not{d}_3^\star\,\not{d}_2^{-1}+\not{d}_2^\star\Big)\not{d}_1^{-1}\check{\rho}\right)\\
= & -\left(-\frac{1}{2}\vartheta\,\not{d}_2^\star+\zeta e_4(\Phi)-\beta\right)\not{d}_1^\star\kappa-\frac{1}{2}\vartheta\,\not{d}_1\,\not{d}_1^\star\kappa+\frac{1}{2}\,\not{d}_1^\star(\kappa+\vartheta)\,\not{d}_1^\star\kappa+(\,\not{d}_1^\star\kappa)^2\\
& -2\kappa\vartheta\Big(\not{d}_4\,\not{d}_3^\star\,\not{d}_2^{-1}+\not{d}_2^\star\Big)\not{d}_1^{-1}\check{\rho}+(\kappa\vartheta+2\alpha)\Big(\not{d}_4\,\not{d}_3^\star\,\not{d}_2^{-1}+\not{d}_2^\star\Big)\not{d}_1^{-1}\check{\rho}\\
& +\vartheta\left[\Big(\not{d}_4\,\not{d}_3^\star\,\not{d}_2^{-1}+\not{d}_2^\star\Big)\not{d}_1^{-1},e_4\right]\check{\rho}\\
& +\vartheta\Big(\not{d}_4\,\not{d}_3^\star\,\not{d}_2^{-1}+\not{d}_2^\star\Big)\not{d}_1^{-1}\left(\frac{3}{2}\overline{\kappa}\check{\rho}+\frac{3}{2}\overline{\rho}\check{\kappa}-Err[e_4\check{\rho}]\right)\\
& -\frac{1}{2}\vartheta\,\not{d}_4\,\not{d}_3^\star\,\not{d}_2^{-1}(-\,\not{d}_1^\star\kappa+\kappa\zeta-\vartheta\zeta)-\frac{1}{2}\vartheta\,\not{d}_2^\star(-\,\not{d}_1^\star\kappa+\kappa\zeta-\vartheta\zeta)\\
& +\frac{1}{2}\,\not{d}_3^\star\,\not{d}_2^{-1}(-2\beta-\,\not{d}_1^\star\kappa+\kappa\zeta-\vartheta\zeta)\,\not{d}_3^\star\vartheta+\frac{1}{2}(-2\beta-\,\not{d}_1^\star\kappa+\kappa\zeta-\vartheta\zeta)\,\not{d}_2\vartheta,
\end{aligned}$$

$$\begin{aligned}
& e_4\Big(e_\theta(\mu)+\vartheta\,\not{d}_2\,\not{d}_2^\star(\,\not{d}_1^\star\,\not{d}_1)^{-1}\underline{\beta}+2\zeta\check{\rho}\Big)+2\kappa\Big(e_\theta(\mu)+\vartheta\,\not{d}_2\,\not{d}_2^\star(\,\not{d}_1^\star\,\not{d}_1)^{-1}\underline{\beta}+2\zeta\check{\rho}\Big)\\
= & -\frac{3}{2}\mu e_\theta(\kappa)-\frac{1}{2}\vartheta e_\theta(\mu)-(\kappa\vartheta+2\alpha)\,\not{d}_2\,\not{d}_2^\star(\,\not{d}_1^\star\,\not{d}_1)^{-1}\underline{\beta}\\
& -\vartheta\Big[\not{d}_2\,\not{d}_2^\star(\,\not{d}_1^\star\,\not{d}_1)^{-1},e_4\Big]\underline{\beta}-\vartheta\,\not{d}_2\,\not{d}_2^\star(\,\not{d}_1^\star\,\not{d}_1)^{-1}\left(\kappa\underline{\beta}+3\rho\zeta+\underline{\vartheta}\beta\right)\\
& +\,\not{d}_3^\star\vartheta\,\not{d}_2^\star\,\not{d}_1^{-1}\check{\rho}-e_\theta\left(\vartheta\,\not{d}_2^\star\,\not{d}_1^{-1}\left(-\check{\mu}+\frac{1}{4}\vartheta\underline{\vartheta}\right)\right)-e_\theta\left(\vartheta\left(\frac{1}{8}\underline{\kappa}\vartheta+\zeta^2\right)\right)\\
& -2\zeta\,\not{d}_1\,\not{d}_1^\star\kappa-2e_\theta(\kappa)\,\not{d}_2^\star\zeta-2(\kappa\zeta+\beta+\vartheta\zeta)\check{\rho}-2\zeta\left(\frac{3}{2}\overline{\kappa}\check{\rho}+\frac{3}{2}\overline{\rho}\check{\kappa}-Err[e_4\check{\rho}]\right)\\
& +2\beta\,\not{d}_2^\star\zeta+\frac{3}{2}e_\theta\left(\kappa\zeta^2\right)+2\kappa\Big(\vartheta\,\not{d}_2\,\not{d}_2^\star(\,\not{d}_1^\star\,\not{d}_1)^{-1}\underline{\beta}+2\zeta\check{\rho}\Big),
\end{aligned}$$

and

$$\begin{aligned}
& e_4(e_\theta(\underline{\kappa})-4\underline{\beta})+\kappa(e_\theta(\underline{\kappa})-4\underline{\beta})\\
= & \ 2e_\theta(\mu)+12\rho\zeta-\frac{1}{2}\underline{\kappa}e_\theta(\kappa)+4\underline{\vartheta}\beta-\frac{1}{2}\vartheta e_\theta(\underline{\kappa})-e_\theta(\vartheta\underline{\vartheta})+2e_\theta(\zeta^2)\\
= & \ 2\Big(e_\theta(\mu)+\vartheta\,\not{d}_2\,\not{d}_2^\star(\,\not{d}_1^\star\,\not{d}_1)^{-1}\underline{\beta}+2\zeta\check{\rho}\Big)+12\rho\zeta-\frac{1}{2}\underline{\kappa}e_\theta(\kappa)-2\vartheta\,\not{d}_2\,\not{d}_2^\star(\,\not{d}_1^\star\,\not{d}_1)^{-1}\underline{\beta}\\
& +2\zeta\check{\rho}+4\underline{\vartheta}\beta-\frac{1}{2}\vartheta(e_\theta(\underline{\kappa})-4\underline{\beta})-2\vartheta\underline{\beta}-e_\theta(\vartheta\underline{\vartheta})+2e_\theta(\zeta^2).
\end{aligned}$$

Proof. Recall Raychaudhuri

$$e_4(\kappa)+\frac{1}{2}\kappa^2 \quad = \quad -\frac{1}{2}\vartheta^2.$$

We commute with $\not{d}_1\not{d}_1^\star$ which yields

$$\begin{aligned}
e_4(\not{d}_1\not{d}_1^\star\kappa)+2\kappa\not{d}_1\not{d}_1^\star\kappa &= -\left(-\frac{1}{2}\vartheta\not{d}_2^\star+\zeta e_4(\Phi)-\beta\right)\not{d}_1^\star\kappa-\frac{1}{2}\vartheta\not{d}_1\not{d}_1^\star\kappa\\
&\quad+\frac{1}{2}\not{d}_1^\star(\kappa+\vartheta)\not{d}_1^\star\kappa+(\not{d}_1^\star\kappa)^2-\frac{1}{2}\not{d}_1\not{d}_1^\star(\vartheta^2).
\end{aligned}$$

We have in view of Codazzi for ϑ

$$\begin{aligned}
&\frac{1}{2}\not{d}_1\not{d}_1^\star(\vartheta^2)\\
=\;& \not{d}_1(\vartheta\not{d}_1^\star\vartheta)\\
=\;& \frac{1}{2}\not{d}_1(\vartheta\not{d}_3^\star\vartheta-\vartheta\not{d}_2\vartheta)\\
=\;& \frac{1}{2}\not{d}_1\Big(\vartheta\not{d}_3^\star\not{d}_2^{-1}(-2\beta-\not{d}_1^\star\kappa+\kappa\zeta-\vartheta\zeta)-\vartheta(-2\beta-\not{d}_1^\star\kappa+\kappa\zeta-\vartheta\zeta)\Big)\\
=\;& \frac{1}{2}\vartheta\not{d}_4\not{d}_3^\star\not{d}_2^{-1}(-2\beta-\not{d}_1^\star\kappa+\kappa\zeta-\vartheta\zeta)+\frac{1}{2}\vartheta\not{d}_2^\star(-2\beta-\not{d}_1^\star\kappa+\kappa\zeta-\vartheta\zeta)\\
&-\frac{1}{2}\not{d}_3^\star\not{d}_2^{-1}(-2\beta-\not{d}_1^\star\kappa+\kappa\zeta-\vartheta\zeta)\not{d}_3^\star\vartheta-\frac{1}{2}(-2\beta-\not{d}_1^\star\kappa+\kappa\zeta-\vartheta\zeta)\not{d}_2\vartheta\\
=\;& -\vartheta\not{d}_4\not{d}_3^\star\not{d}_2^{-1}\beta-\vartheta\not{d}_2^\star\beta+\frac{1}{2}\vartheta\not{d}_4\not{d}_3^\star\not{d}_2^{-1}(-\not{d}_1^\star\kappa+\kappa\zeta-\vartheta\zeta)\\
&+\frac{1}{2}\vartheta\not{d}_2^\star(-\not{d}_1^\star\kappa+\kappa\zeta-\vartheta\zeta)-\frac{1}{2}\not{d}_3^\star\not{d}_2^{-1}(-2\beta-\not{d}_1^\star\kappa+\kappa\zeta-\vartheta\zeta)\not{d}_3^\star\vartheta\\
&-\frac{1}{2}(-2\beta-\not{d}_1^\star\kappa+\kappa\zeta-\vartheta\zeta)\not{d}_2\vartheta.
\end{aligned}$$

Together with Bianchi for $e_4(\check{\rho})$, we infer

$$\begin{aligned}
&\frac{1}{2}\not{d}_1\not{d}_1^\star(\vartheta^2)\\
=\;& -\vartheta\Big(\not{d}_4\not{d}_3^\star\not{d}_2^{-1}+\not{d}_2^\star\Big)\not{d}_1^{-1}\left(e_4(\check{\rho})+\frac{3}{2}\overline{\kappa}\check{\rho}+\frac{3}{2}\overline{\rho}\check{\kappa}-\mathrm{Err}[e_4\check{\rho}]\right)\\
&+\frac{1}{2}\vartheta\not{d}_4\not{d}_3^\star\not{d}_2^{-1}(-\not{d}_1^\star\kappa+\kappa\zeta-\vartheta\zeta)+\frac{1}{2}\vartheta\not{d}_2^\star(-\not{d}_1^\star\kappa+\kappa\zeta-\vartheta\zeta)\\
&-\frac{1}{2}\not{d}_3^\star\not{d}_2^{-1}(-2\beta-\not{d}_1^\star\kappa+\kappa\zeta-\vartheta\zeta)\not{d}_3^\star\vartheta-\frac{1}{2}(-2\beta-\not{d}_1^\star\kappa+\kappa\zeta-\vartheta\zeta)\not{d}_2\vartheta\\
=\;& -\vartheta e_4\left(\Big(\not{d}_4\not{d}_3^\star\not{d}_2^{-1}+\not{d}_2^\star\Big)\not{d}_1^{-1}\check{\rho}\right)-\vartheta\left[\Big(\not{d}_4\not{d}_3^\star\not{d}_2^{-1}+\not{d}_2^\star\Big)\not{d}_1^{-1},e_4\right]\check{\rho}\\
&-\vartheta\Big(\not{d}_4\not{d}_3^\star\not{d}_2^{-1}+\not{d}_2^\star\Big)\not{d}_1^{-1}\left(\frac{3}{2}\overline{\kappa}\check{\rho}+\frac{3}{2}\overline{\rho}\check{\kappa}-\mathrm{Err}[e_4\check{\rho}]\right)\\
&+\frac{1}{2}\vartheta\not{d}_4\not{d}_3^\star\not{d}_2^{-1}(-\not{d}_1^\star\kappa+\kappa\zeta-\vartheta\zeta)+\frac{1}{2}\vartheta\not{d}_2^\star(-\not{d}_1^\star\kappa+\kappa\zeta-\vartheta\zeta)\\
&-\frac{1}{2}\not{d}_3^\star\not{d}_2^{-1}(-2\beta-\not{d}_1^\star\kappa+\kappa\zeta-\vartheta\zeta)\not{d}_3^\star\vartheta-\frac{1}{2}(-2\beta-\not{d}_1^\star\kappa+\kappa\zeta-\vartheta\zeta)\not{d}_2\vartheta.
\end{aligned}$$

In view of the null structure equation for $e_4(\vartheta)$, we infer

$$
\begin{aligned}
&\frac{1}{2}\,{\not{d}}_1\,{\not{d}}_1^\star(\vartheta^2)\\
= \quad & -e_4\left(\vartheta\Big({\not{d}}_4\,{\not{d}}_3^\star\,{\not{d}}_2^{-1}+{\not{d}}_2^\star\Big)\,{\not{d}}_1^{-1}\check{\rho}\right)-(\kappa\vartheta+2\alpha)\Big({\not{d}}_4\,{\not{d}}_3^\star\,{\not{d}}_2^{-1}+{\not{d}}_2^\star\Big)\,{\not{d}}_1^{-1}\check{\rho}\\
&-\vartheta\left[\Big({\not{d}}_4\,{\not{d}}_3^\star\,{\not{d}}_2^{-1}+{\not{d}}_2^\star\Big)\,{\not{d}}_1^{-1},e_4\right]\check{\rho}\\
&-\vartheta\Big({\not{d}}_4\,{\not{d}}_3^\star\,{\not{d}}_2^{-1}+{\not{d}}_2^\star\Big)\,{\not{d}}_1^{-1}\left(\frac{3}{2}\overline{\kappa}\check{\rho}+\frac{3}{2}\overline{\rho}\check{\kappa}-\mathrm{Err}[e_4\check{\rho}]\right)\\
&+\frac{1}{2}\vartheta\,{\not{d}}_4\,{\not{d}}_3^\star\,{\not{d}}_2^{-1}(-{\not{d}}_1^\star\kappa+\kappa\zeta-\vartheta\zeta)+\frac{1}{2}\vartheta\,{\not{d}}_2^\star(-{\not{d}}_1^\star\kappa+\kappa\zeta-\vartheta\zeta)\\
&-\frac{1}{2}\,{\not{d}}_3^\star\,{\not{d}}_2^{-1}(-2\beta-{\not{d}}_1^\star\kappa+\kappa\zeta-\vartheta\zeta)\,{\not{d}}_3^\star\vartheta-\frac{1}{2}(-2\beta-{\not{d}}_1^\star\kappa+\kappa\zeta-\vartheta\zeta)\,{\not{d}}_2\vartheta.
\end{aligned}
$$

This yields

$$
\begin{aligned}
&e_4\left({\not{d}}_1\,{\not{d}}_1^\star\kappa-\vartheta\Big({\not{d}}_4\,{\not{d}}_3^\star\,{\not{d}}_2^{-1}+{\not{d}}_2^\star\Big)\,{\not{d}}_1^{-1}\check{\rho}\right)\\
&+2\kappa\left({\not{d}}_1\,{\not{d}}_1^\star\kappa-\vartheta\Big({\not{d}}_4\,{\not{d}}_3^\star\,{\not{d}}_2^{-1}+{\not{d}}_2^\star\Big)\,{\not{d}}_1^{-1}\check{\rho}\right)\\
= \quad & -\left(-\frac{1}{2}\vartheta\,{\not{d}}_2^\star+\zeta e_4(\Phi)-\beta\right){\not{d}}_1^\star\kappa-\frac{1}{2}\vartheta\,{\not{d}}_1\,{\not{d}}_1^\star\kappa+\frac{1}{2}\,{\not{d}}_1^\star(\kappa+\vartheta)\,{\not{d}}_1^\star\kappa+({\not{d}}_1^\star\kappa)^2\\
&-2\kappa\vartheta\Big({\not{d}}_4\,{\not{d}}_3^\star\,{\not{d}}_2^{-1}+{\not{d}}_2^\star\Big)\,{\not{d}}_1^{-1}\check{\rho}+(\kappa\vartheta+2\alpha)\Big({\not{d}}_4\,{\not{d}}_3^\star\,{\not{d}}_2^{-1}+{\not{d}}_2^\star\Big)\,{\not{d}}_1^{-1}\check{\rho}\\
&+\vartheta\left[\Big({\not{d}}_4\,{\not{d}}_3^\star\,{\not{d}}_2^{-1}+{\not{d}}_2^\star\Big)\,{\not{d}}_1^{-1},e_4\right]\check{\rho}\\
&+\vartheta\Big({\not{d}}_4\,{\not{d}}_3^\star\,{\not{d}}_2^{-1}+{\not{d}}_2^\star\Big)\,{\not{d}}_1^{-1}\left(\frac{3}{2}\overline{\kappa}\check{\rho}+\frac{3}{2}\overline{\rho}\check{\kappa}-\mathrm{Err}[e_4\check{\rho}]\right)\\
&-\frac{1}{2}\vartheta\,{\not{d}}_4\,{\not{d}}_3^\star\,{\not{d}}_2^{-1}(-{\not{d}}_1^\star\kappa+\kappa\zeta-\vartheta\zeta)-\frac{1}{2}\vartheta\,{\not{d}}_2^\star(-{\not{d}}_1^\star\kappa+\kappa\zeta-\vartheta\zeta)\\
&+\frac{1}{2}\,{\not{d}}_3^\star\,{\not{d}}_2^{-1}(-2\beta-{\not{d}}_1^\star\kappa+\kappa\zeta-\vartheta\zeta)\,{\not{d}}_3^\star\vartheta+\frac{1}{2}(-2\beta-{\not{d}}_1^\star\kappa+\kappa\zeta-\vartheta\zeta)\,{\not{d}}_2\vartheta.
\end{aligned}
$$

Next, recall that we have

$$
e_4\mu+\frac{3}{2}\kappa\mu \quad = \quad -\vartheta\,{\not{d}}_2^\star\zeta-\vartheta\left(\frac{1}{8}\underline{\kappa}\vartheta+\zeta^2\right)+\left(2e_\theta(\kappa)-2\beta+\frac{3}{2}\kappa\zeta\right)\zeta.
$$

We commute with e_θ which yields

$$
\begin{aligned}
& e_4(e_\theta(\mu)) + 2\kappa e_\theta(\mu)\\
= & -\frac{3}{2}\mu e_\theta(\kappa) - \frac{1}{2}\vartheta e_\theta(\mu) - e_\theta\left(\vartheta \, \not{d}_2^\star \not{d}_1^{-1}\left(-\check{\mu} - \check{\rho} + \frac{1}{4}\vartheta\underline{\vartheta}\right)\right)\\
& -e_\theta\left(\vartheta\left(\frac{1}{8}\underline{\kappa}\vartheta + \zeta^2\right)\right) + e_\theta\left(\left(2e_\theta(\kappa) - 2\beta + \frac{3}{2}\kappa\zeta\right)\zeta\right)\\
= & -\frac{3}{2}\mu e_\theta(\kappa) - \frac{1}{2}\vartheta e_\theta(\mu) + \vartheta \, \not{d}_2 \not{d}_2^\star(\not{d}_1^\star \not{d}_1)^{-1} \not{d}_1^\star \rho + \not{d}_3^\star \vartheta \, \not{d}_2^\star \not{d}_1^{-1}\check{\rho}\\
& -e_\theta\left(\vartheta \, \not{d}_2^\star \not{d}_1^{-1}\left(-\check{\mu} + \frac{1}{4}\vartheta\underline{\vartheta}\right)\right) - e_\theta\left(\vartheta\left(\frac{1}{8}\underline{\kappa}\vartheta + \zeta^2\right)\right)\\
& -2\zeta \, \not{d}_1 \not{d}_1^\star \kappa - 2e_\theta(\kappa) \not{d}_2^\star \zeta - 2\zeta \, \not{d}_1 \beta + 2\beta \, \not{d}_2^\star \zeta + \frac{3}{2}e_\theta\left(\kappa\zeta^2\right).
\end{aligned}
$$

Now, using the Bianchi identities for $e_4(\underline{\beta})$ and $e_4(\check{\rho})$, we have

$$
\begin{aligned}
& \vartheta \, \not{d}_2 \not{d}_2^\star(\not{d}_1^\star \not{d}_1)^{-1} \not{d}_1^\star \rho\\
= & -\vartheta \, \not{d}_2 \not{d}_2^\star(\not{d}_1^\star \not{d}_1)^{-1}\left(e_4\underline{\beta} + \kappa\underline{\beta} + 3\rho\zeta + \underline{\vartheta}\beta\right)\\
= & -e_4\Big(\vartheta \, \not{d}_2 \not{d}_2^\star(\not{d}_1^\star \not{d}_1)^{-1}\underline{\beta}\Big) + e_4(\vartheta) \not{d}_2 \not{d}_2^\star(\not{d}_1^\star \not{d}_1)^{-1}\underline{\beta}\\
& -\vartheta\Big[\not{d}_2 \not{d}_2^\star(\not{d}_1^\star \not{d}_1)^{-1}, e_4\Big]\underline{\beta} - \vartheta \, \not{d}_2 \not{d}_2^\star(\not{d}_1^\star \not{d}_1)^{-1}\left(\kappa\underline{\beta} + 3\rho\zeta + \underline{\vartheta}\beta\right)\\
= & -e_4\Big(\vartheta \, \not{d}_2 \not{d}_2^\star(\not{d}_1^\star \not{d}_1)^{-1}\underline{\beta}\Big) - (\kappa\vartheta + 2\alpha) \not{d}_2 \not{d}_2^\star(\not{d}_1^\star \not{d}_1)^{-1}\underline{\beta}\\
& -\vartheta\Big[\not{d}_2 \not{d}_2^\star(\not{d}_1^\star \not{d}_1)^{-1}, e_4\Big]\underline{\beta} - \vartheta \, \not{d}_2 \not{d}_2^\star(\not{d}_1^\star \not{d}_1)^{-1}\left(\kappa\underline{\beta} + 3\rho\zeta + \underline{\vartheta}\beta\right)
\end{aligned}
$$

and

$$
\begin{aligned}
\zeta \, \not{d}_1 \beta &= \zeta\left(e_4(\check{\rho}) + \frac{3}{2}\overline{\kappa}\check{\rho} + \frac{3}{2}\overline{\rho}\check{\kappa} - \mathrm{Err}[e_4\check{\rho}]\right)\\
&= e_4(\zeta\check{\rho}) - e_4(\zeta)\check{\rho} + \zeta\left(\frac{3}{2}\overline{\kappa}\check{\rho} + \frac{3}{2}\overline{\rho}\check{\kappa} - \mathrm{Err}[e_4\check{\rho}]\right)\\
&= e_4(\zeta\check{\rho}) + (\kappa\zeta + \beta + \vartheta\zeta)\check{\rho} + \zeta\left(\frac{3}{2}\overline{\kappa}\check{\rho} + \frac{3}{2}\overline{\rho}\check{\kappa} - \mathrm{Err}[e_4\check{\rho}]\right).
\end{aligned}
$$

We infer

$$
\begin{aligned}
& e_4\Big(e_\theta(\mu)) + \vartheta \, \not{d}_2 \not{d}_2^\star(\not{d}_1^\star \not{d}_1)^{-1}\underline{\beta} + 2\zeta\check{\rho}\Big) + 2\kappa e_\theta(\mu)\\
= & -\frac{3}{2}\mu e_\theta(\kappa) - \frac{1}{2}\vartheta e_\theta(\mu) - (\kappa\vartheta + 2\alpha) \not{d}_2 \not{d}_2^\star(\not{d}_1^\star \not{d}_1)^{-1}\underline{\beta}\\
& -\vartheta\Big[\not{d}_2 \not{d}_2^\star(\not{d}_1^\star \not{d}_1)^{-1}, e_4\Big]\underline{\beta} - \vartheta \, \not{d}_2 \not{d}_2^\star(\not{d}_1^\star \not{d}_1)^{-1}\left(\kappa\underline{\beta} + 3\rho\zeta + \underline{\vartheta}\beta\right)\\
& + \not{d}_3^\star \vartheta \, \not{d}_2^\star \not{d}_1^{-1}\check{\rho} - e_\theta\left(\vartheta \, \not{d}_2^\star \not{d}_1^{-1}\left(-\check{\mu} + \frac{1}{4}\vartheta\underline{\vartheta}\right)\right) - e_\theta\left(\vartheta\left(\frac{1}{8}\underline{\kappa}\vartheta + \zeta^2\right)\right)\\
& -2\zeta \, \not{d}_1 \not{d}_1^\star \kappa - 2e_\theta(\kappa) \not{d}_2^\star \zeta - 2(\kappa\zeta + \beta + \vartheta\zeta)\check{\rho} - 2\zeta\left(\frac{3}{2}\overline{\kappa}\check{\rho} + \frac{3}{2}\overline{\rho}\check{\kappa} - \mathrm{Err}[e_4\check{\rho}]\right)\\
& +2\beta \, \not{d}_2^\star \zeta + \frac{3}{2}e_\theta\left(\kappa\zeta^2\right)
\end{aligned}
$$

and hence

$$\begin{aligned}
& e_4\Big(e_\theta(\mu)+\vartheta\, \not{d}_2\, \not{d}_2^\star(\not{d}_1^\star\, \not{d}_1)^{-1}\underline{\beta}+2\zeta\check{\rho}\Big)+2\kappa\Big(e_\theta(\mu)+\vartheta\, \not{d}_2\, \not{d}_2^\star(\not{d}_1^\star\, \not{d}_1)^{-1}\underline{\beta}+2\zeta\check{\rho}\Big)\\
= & -\frac{3}{2}\mu e_\theta(\kappa)-\frac{1}{2}\vartheta e_\theta(\mu)-(\kappa\vartheta+2\alpha)\, \not{d}_2\, \not{d}_2^\star(\not{d}_1^\star\, \not{d}_1)^{-1}\underline{\beta}\\
& -\vartheta\Big[\not{d}_2\, \not{d}_2^\star(\not{d}_1^\star\, \not{d}_1)^{-1}, e_4\Big]\underline{\beta}-\vartheta\, \not{d}_2\, \not{d}_2^\star(\not{d}_1^\star\, \not{d}_1)^{-1}\left(\kappa\underline{\beta}+3\rho\zeta+\underline{\vartheta}\beta\right)\\
& +\not{d}_3^\star\vartheta\, \not{d}_2^\star\, \not{d}_1^{-1}\check{\rho}-e_\theta\left(\vartheta\, \not{d}_2^\star\, \not{d}_1^{-1}\left(-\check{\mu}+\frac{1}{4}\vartheta\underline{\vartheta}\right)\right)-e_\theta\left(\vartheta\left(\frac{1}{8}\underline{\kappa}\vartheta+\zeta^2\right)\right)\\
& -2\zeta\, \not{d}_1\, \not{d}_1^\star\kappa-2e_\theta(\kappa)\, \not{d}_2^\star\zeta-2(\kappa\zeta+\beta+\vartheta\zeta)\check{\rho}-2\zeta\left(\frac{3}{2}\overline{\kappa}\check{\rho}+\frac{3}{2}\overline{\rho}\check{\kappa}-\mathrm{Err}[e_4\check{\rho}]\right)\\
& +2\beta\, \not{d}_2^\star\zeta+\frac{3}{2}e_\theta\left(\kappa\zeta^2\right)+2\kappa\Big(\vartheta\, \not{d}_2\, \not{d}_2^\star(\not{d}_1^\star\, \not{d}_1)^{-1}\underline{\beta}+2\zeta\check{\rho}\Big).
\end{aligned}$$

Finally, recall that we have

$$\begin{aligned}
e_4(\underline{\kappa})+\frac{1}{2}\kappa\underline{\kappa} &= -2\, \not{d}_1\zeta+2\rho-\frac{1}{2}\vartheta\underline{\vartheta}+2\zeta^2\\
&= 2\mu+4\rho-\vartheta\underline{\vartheta}+2\zeta^2.
\end{aligned}$$

We commute with e_θ which yields

$$e_4(e_\theta(\underline{\kappa}))+\kappa e_\theta(\underline{\kappa}) = 2e_\theta(\mu)+4e_\theta(\rho)-\frac{1}{2}\underline{\kappa}e_\theta(\kappa)-\frac{1}{2}\vartheta e_\theta(\underline{\kappa})-e_\theta(\vartheta\underline{\vartheta})+2e_\theta(\zeta^2).$$

Together with Bianchi for $e_4(\underline{\beta})$, we infer

$$\begin{aligned}
& e_4(e_\theta(\underline{\kappa})-4\underline{\beta})+\kappa(e_\theta(\underline{\kappa})-4\underline{\beta})\\
= & 2e_\theta(\mu)+12\rho\zeta-\frac{1}{2}\underline{\kappa}e_\theta(\kappa)+4\underline{\vartheta}\beta-\frac{1}{2}\vartheta e_\theta(\underline{\kappa})-e_\theta(\vartheta\underline{\vartheta})+2e_\theta(\zeta^2)\\
= & 2\Big(e_\theta(\mu)+\vartheta\, \not{d}_2\, \not{d}_2^\star(\not{d}_1^\star\, \not{d}_1)^{-1}\underline{\beta}+2\zeta\check{\rho}\Big)+12\rho\zeta-\frac{1}{2}\underline{\kappa}e_\theta(\kappa)-2\vartheta\, \not{d}_2\, \not{d}_2^\star(\not{d}_1^\star\, \not{d}_1)^{-1}\underline{\beta}\\
& +2\zeta\check{\rho}+4\underline{\vartheta}\beta-\frac{1}{2}\vartheta(e_\theta(\underline{\kappa})-4\underline{\beta})-2\vartheta\underline{\beta}-e_\theta(\vartheta\underline{\vartheta})+2e_\theta(\zeta^2).
\end{aligned}$$

This concludes the proof of the lemma. $\square$

8.8.4 Proof of Proposition 8.32

We introduce the following notation which will constantly appear on the RHS of the equalities below:

$$\begin{aligned}
N^{\geq 4m_0}[J,\check{\Gamma},\check{R}] := & \, {}^{(\Sigma_*)}\mathfrak{G}_{J+1}[\check{\Gamma}]+{}^{(\Sigma_*)}\mathfrak{G}'_{J+1}[\check{\Gamma}]+{}^{(ext)}\mathfrak{R}_{J+1}[\check{R}]+{}^{(ext)}\mathfrak{G}_J[\check{\Gamma}]\\
& +\epsilon_0\Big({}^{(ext)}\mathfrak{G}^{\geq 4m_0}_{J+1}[\check{\Gamma}]+{}^{(ext)}\mathfrak{G}^{\geq 4m_0\,\prime}_{J+1}[\check{\Gamma}]\Big).
\end{aligned} \tag{8.8.2}$$

Step 1. Recall that

$$e_4(\vartheta)+\kappa\vartheta = -2\alpha.$$

In view of Corollary 8.36 with $a=2$, we have for any $r_0 \geq 4m_0$

$$\max_{k\leq J+1}\sup_{r_0\geq 4m_0} r_0^2\int_{\{r=r_0\}}(\mathfrak{d}^k\vartheta)^2 \lesssim \left(N^{\geq 4m_0}[J,\check{\Gamma},\check{R}]\right)^2.$$

Step 2. Next, recall that

$$e_4(\check{\kappa})+\overline{\kappa}\,\check{\kappa} = -\frac{1}{4}\vartheta^2+\frac{1}{4}\overline{\vartheta^2}-\overline{\check{\kappa}^2}.$$

In view of Corollary 8.36 with $a=2$, we have for any $r_0 \geq 4m_0$

$$\max_{k\leq J+2}\sup_{r_0\geq 4m_0} r_0^2\int_{\{r=r_0\}}(\mathfrak{d}^k\check{\kappa})^2 \lesssim \left(N^{\geq 4m_0}[J,\check{\Gamma},\check{R}]\right)^2$$

where we have used the null structure equations for $e_4(\vartheta)$, $e_3(\vartheta)$ and $\not{d}_2\vartheta$ to avoid a loss of one derivative for the RHS.

Step 3. Next, recall that

$$e_4(\zeta)+\kappa\zeta = -\beta-\vartheta\zeta.$$

In view of Corollary 8.36 with $a=2$, we have for any $r_0 \geq 4m_0$

$$\max_{k\leq J+1}\sup_{r_0\geq 4m_0} r_0^2\int_{\{r=r_0\}}(\mathfrak{d}^k\zeta)^2 \lesssim \left(N^{\geq 4m_0}[J,\check{\Gamma},\check{R}]\right)^2.$$

Step 4. Next, recall that

$$e_4\left(\check{\mu}\right)+\frac{3}{2}\overline{\kappa}\check{\mu} = -\frac{3}{2}\overline{\mu}\check{\kappa}+\mathrm{Err}[e_4\check{\mu}].$$

In view of Corollary 8.36 with $a=3$, commuting with $\not{\mathfrak{d}}$ and $\mathbf{T}$, we have for any $r_0 \geq 4m_0$

$$\max_{k\leq J+1}\sup_{r_0\geq 4m_0} r_0^4\int_{\{r=r_0\}}(\mathfrak{d}^k\check{\mu})^2 \lesssim \left(N^{\geq 4m_0}[J,\check{\Gamma},\check{R}]\right)^2$$

where we used the estimates for $\check{\kappa}$ on ${}^{(ext)}\mathcal{M}$ derived in Step 2.

Step 5. Next, recall that

$$e_4(\check{\underline{\kappa}})+\frac{1}{2}\overline{\kappa}\check{\underline{\kappa}} = -\frac{1}{2}\overline{\underline{\kappa}}\check{\kappa}+2\check{\rho}-2\not{d}_1\zeta+\mathrm{Err}[e_4\check{\underline{\kappa}}].$$

In view of Corollary 8.38 with $a=1$ and $b=2-\delta_B$ which satisfy the constraint $b>2a-2$, we have for any $r_0 \geq 4m_0$

$$\max_{k\leq J+1}\sup_{r_0\geq 4m_0}\left(r_0^{2-\delta_B}\int_{\{r=r_0\}}(\mathfrak{d}^k\check{\underline{\kappa}})^2\right)+\int_{{}^{(ext)}\mathcal{M}(r\geq 4m_0)} r^{1-\delta_B}(\mathfrak{d}^k\check{\underline{\kappa}})^2$$
$$\lesssim \left(N^{\geq 4m_0}[J,\check{\Gamma},\check{R}]\right)^2$$

where we used the estimates for $\check{\kappa}$ and $\check{\mu}$ on ${}^{(ext)}\mathcal{M}$ derived respectively in Step 2 and Step 4.

Step 6. Next, recall that

$$
\begin{aligned}
e_4(\underline{\vartheta}) + \frac{1}{2}\kappa\underline{\vartheta} &= 2\,{\not{d}}_2^\star\zeta - \frac{1}{2}\underline{\kappa}\vartheta + 2\zeta^2 \\
&= 2\,{\not{d}}_2^\star\,{\not{d}}_1^{-1}\left(-\mu - \rho + \frac{1}{4}\vartheta\underline{\vartheta}\right) - \frac{1}{2}\underline{\kappa}\vartheta + 2\zeta^2.
\end{aligned}
$$

In view of Corollary 8.36 with $a = 1$, we have for any $r_0 \geq 4m_0$

$$
\max_{k\leq J+1}\sup_{r_0\geq 4m_0}\int_{\{r=r_0\}}(\mathfrak{d}^k\underline{\vartheta})^2 \lesssim \left(N^{\geq 4m_0}[J,\check{\Gamma},\check{R}]\right)^2
$$

where we used the estimates for ϑ and $\check{\mu}$ on ${}^{(ext)}\mathcal{M}$ derived respectively in Step 1 and Step 4.

Step 7. Next, recall that

$$
e_4(\check{\underline{\omega}}) = \check{\rho} + 3\zeta^2 - 3\overline{\zeta^2} - \overline{\check{\kappa}\check{\underline{\omega}}}.
$$

In view of Corollary 8.38 with $a = 0$ and $b = 0$ which satisfy the required constraint $b > 2a - 2$, we have for any $r_0 \geq 4m_0$

$$
\max_{k\leq J+1}\sup_{r_0\geq 4m_0}\left(\int_{\{r=r_0\}}(\mathfrak{d}^k\check{\underline{\omega}})^2\right) + \int_{{}^{(ext)}\mathcal{M}(\geq 4m_0)} r^{-1}(\mathfrak{d}^k\check{\underline{\omega}})^2 \lesssim \left(N^{\geq 4m_0}[J,\check{\Gamma},\check{R}]\right)^2.
$$

Step 8. In order to estimate $\underline{\xi}$ in Step 9, we derive an estimate for $e_3(\zeta) + \underline{\beta}$. Recall that we have

$$
e_4(\zeta) + \kappa\zeta = -\beta - \vartheta\zeta.
$$

Commuting with e_3, we infer

$$
e_4(e_3(\zeta)) + [e_3, e_4]\zeta + \kappa e_3(\zeta) + e_3(\kappa)\zeta = -e_3(\beta) - \vartheta\zeta.
$$

In view of the null structure equation for $e_3(\kappa)$, the Bianchi identity for $e_3(\beta)$ and the commutator identity for $[e_3, e_4]$, we infer

$$
\begin{aligned}
& e_4(e_3(\zeta)) + \left(2\underline{\omega}e_4 + 4\zeta e_\theta\right)\zeta + \kappa e_3(\zeta) \\
& + \left(-\frac{1}{2}\kappa\underline{\kappa} + 2\underline{\omega}\kappa + 2\,{\not{d}}_1\zeta + 2\rho - \frac{1}{2}\vartheta\underline{\vartheta} + 2\zeta^2\right)\zeta \\
= \; & (\underline{\kappa} - 2\underline{\omega})\beta + {\not{d}}_1^\star\rho - 3\zeta\rho + \vartheta\underline{\beta} - \underline{\xi}\alpha - \vartheta\zeta.
\end{aligned}
$$

Together with the null structure equation for $e_4(\zeta)$, the Bianchi identity for $e_4(\underline{\beta})$

to get rid of the term $\not{d}_1^\star \rho$, and the definition of μ, we infer

$$
\begin{aligned}
& e_4(e_3(\zeta)) + 2\underline{\omega}(-\kappa\zeta - \beta - \vartheta\zeta) + 4\zeta \not{d}_1^\star \not{d}_1^{-1}\left(\check{\mu} + \check{\rho} - \frac{1}{4}\vartheta\underline{\vartheta} + \frac{1}{4}\overline{\vartheta\underline{\vartheta}}\right) \\
& + \kappa e_3(\zeta) + \left(-\frac{1}{2}\kappa\underline{\kappa} + 2\underline{\omega}\kappa - 2\mu + 2\zeta^2\right)\zeta \\
= \; & (\underline{\kappa} - 2\underline{\omega})\beta - e_4(\underline{\beta}) - \kappa\underline{\beta} - 3\zeta\rho - \underline{\vartheta}\beta - 3\zeta\rho + \vartheta\underline{\beta} - \underline{\xi}\alpha - \vartheta\zeta
\end{aligned}
$$

and hence

$$
\begin{aligned}
& e_4(e_3(\zeta) + \underline{\beta}) + \kappa(e_3(\zeta) + \underline{\beta}) \\
= \; & \underline{\kappa}\beta + \left(\frac{1}{2}\kappa\underline{\kappa} + 2\mu - 6\rho\right)\zeta \\
& - \underline{\vartheta}\beta + \vartheta\underline{\beta} - \underline{\xi}\alpha - \vartheta\zeta + 2\underline{\omega}\vartheta\zeta - 4\zeta \not{d}_1^\star \not{d}_1^{-1}\left(\check{\mu} + \check{\rho} - \frac{1}{4}\vartheta\underline{\vartheta} + \frac{1}{4}\overline{\vartheta\underline{\vartheta}}\right) - 2\zeta^3.
\end{aligned}
$$

In view of Corollary 8.36 with $a = 2$, we have for any $r_0 \geq 4m_0$

$$
\max_{k \leq J+1} \sup_{r_0 \geq 4m_0} r_0^2 \int_{\{r=r_0\}} (\mathfrak{d}^k(e_3(\zeta) + \underline{\beta}))^2 \lesssim \left(N^{\geq 4m_0}[J, \check{\Gamma}, \check{R}]\right)^2
$$

where we used the estimates for ζ derived in Step 3.

Step 9. Next, recall that we have

$$
\begin{aligned}
e_4(\underline{\xi}) &= -e_3(\zeta) + \underline{\beta} - \underline{\kappa}\zeta - \zeta\underline{\vartheta} \\
&= -(e_3(\zeta) + \underline{\beta}) + 2\underline{\beta} - \underline{\kappa}\zeta - \zeta\underline{\vartheta}.
\end{aligned}
$$

In view of Corollary 8.38 with $a = 0$ and $b = -\delta_B$ which satisfy the constraint $b > 2a - 2$, we have for any $r_0 \geq 4m_0$

$$
\begin{aligned}
& \max_{k \leq J+1} \sup_{r_0 \geq 4m_0} r_0^{-\delta_B}\left(\int_{\{r=r_0\}} (\mathfrak{d}^k\underline{\xi})^2\right) + \int_{{}^{(ext)}\mathcal{M}(r \geq 4m_0)} r^{-1-\delta_B}(\mathfrak{d}^k\underline{\xi})^2 \\
\lesssim \; & \left(N^{\geq 4m_0}[J, \check{\Gamma}, \check{R}]\right)^2
\end{aligned}
$$

where we used the estimates for ζ and $e_3(\zeta) + \underline{\beta}$ on ${}^{(ext)}\mathcal{M}$ derived respectively in Step 3 and Step 8.

Step 10. Recall that we have

$$\begin{aligned}
& e_4\left(\not{d}_1\not{d}_1^\star\kappa-\vartheta\left(\not{d}_4\not{d}_3^\star\not{d}_2^{-1}+\not{d}_2^\star\right)\not{d}_1^{-1}\check{\rho}\right)\\
& +2\kappa\left(\not{d}_1\not{d}_1^\star\kappa-\vartheta\left(\not{d}_4\not{d}_3^\star\not{d}_2^{-1}+\not{d}_2^\star\right)\not{d}_1^{-1}\check{\rho}\right)\\
= & -\left(-\frac{1}{2}\vartheta\not{d}_2^\star+\zeta e_4(\Phi)-\beta\right)\not{d}_1^\star\kappa-\frac{1}{2}\vartheta\not{d}_1\not{d}_1^\star\kappa+\frac{1}{2}\not{d}_1^\star(\kappa+\vartheta)\not{d}_1^\star\kappa+(\not{d}_1^\star\kappa)^2\\
& -2\kappa\vartheta\left(\not{d}_4\not{d}_3^\star\not{d}_2^{-1}+\not{d}_2^\star\right)\not{d}_1^{-1}\check{\rho}+(\kappa\vartheta+2\alpha)\left(\not{d}_4\not{d}_3^\star\not{d}_2^{-1}+\not{d}_2^\star\right)\not{d}_1^{-1}\check{\rho}\\
& +\vartheta\left[\left(\not{d}_4\not{d}_3^\star\not{d}_2^{-1}+\not{d}_2^\star\right)\not{d}_1^{-1},e_4\right]\check{\rho}\\
& +\vartheta\left(\not{d}_4\not{d}_3^\star\not{d}_2^{-1}+\not{d}_2^\star\right)\not{d}_1^{-1}\left(\frac{3}{2}\overline{\kappa}\check{\rho}+\frac{3}{2}\overline{\rho}\check{\kappa}-\mathrm{Err}[e_4\check{\rho}]\right)\\
& -\frac{1}{2}\vartheta\not{d}_4\not{d}_3^\star\not{d}_2^{-1}(-\not{d}_1^\star\kappa+\kappa\zeta-\vartheta\zeta)-\frac{1}{2}\vartheta\not{d}_2^\star(-\not{d}_1^\star\kappa+\kappa\zeta-\vartheta\zeta)\\
& +\frac{1}{2}\not{d}_3^\star\not{d}_2^{-1}(-2\beta-\not{d}_1^\star\kappa+\kappa\zeta-\vartheta\zeta)\not{d}_3^\star\vartheta+\frac{1}{2}(-2\beta-\not{d}_1^\star\kappa+\kappa\zeta-\vartheta\zeta)\not{d}_2\vartheta.
\end{aligned}$$

In view of Corollary 8.36 with $a=4$, we have

$$\begin{aligned}
& \max_{k\leq J+1}\sup_{r_0\geq 4m_0}\int_{\{r=r_0\}}r_0^6\left(\mathfrak{d}^k\left(\not{d}_1\not{d}_1^\star\kappa-\vartheta\left(\not{d}_4\not{d}_3^\star\not{d}_2^{-1}+\not{d}_2^\star\right)\not{d}_1^{-1}\check{\rho}\right)\right)^2\\
\lesssim & \left(N^{\geq 4m_0}[J,\check{\Gamma},\check{R}]\right)^2
\end{aligned}$$

where we have used

- the fact that $\not{\mathfrak{d}}\vartheta=\not{\mathfrak{d}}\not{d}_2^{-1}\not{d}_2\vartheta$ and Codazzi for ϑ to estimate the terms of the RHS with one angular derivative of ϑ,
- the estimates of Step 2 to estimate the terms of the RHS with one derivative of $\check{\kappa}$,
- the fact that $\not{\mathfrak{d}}\zeta=\not{\mathfrak{d}}\not{d}_1^{-1}\not{d}_1\zeta$ and the definition of μ to estimate terms of the RHS with one angular derivative of ζ,
- the identity

$$\begin{aligned}
\not{\mathfrak{d}}\not{d}_1^\star\kappa = & \not{\mathfrak{d}}\not{d}_1^{-1}\left(\vartheta\left(\not{d}_4\not{d}_3^\star\not{d}_2^{-1}+\not{d}_2^\star\right)\not{d}_1^{-1}\check{\rho}\right)\\
& +\not{\mathfrak{d}}\not{d}_1^{-1}\left(\not{d}_1\not{d}_1^\star\kappa-\vartheta\left(\not{d}_4\not{d}_3^\star\not{d}_2^{-1}+\not{d}_2^\star\right)\not{d}_1^{-1}\check{\rho}\right)
\end{aligned}$$

  to estimate the terms of the RHS with two angular derivatives of $\check{\kappa}$.

Step 11. Recall that we have

$$\begin{aligned}
& e_4\Big(e_\theta(\mu) + \vartheta \,\not{d}_2 \,\not{d}_2^\star(\not{d}_1^\star \not{d}_1)^{-1}\underline{\beta} + 2\zeta\check{\rho}\Big) + 2\kappa\Big(e_\theta(\mu) + \vartheta \,\not{d}_2 \,\not{d}_2^\star(\not{d}_1^\star \not{d}_1)^{-1}\underline{\beta} + 2\zeta\check{\rho}\Big) \\
= & -\frac{3}{2}\mu e_\theta(\kappa) - \frac{1}{2}\vartheta e_\theta(\mu) - (\kappa\vartheta + 2\alpha) \not{d}_2 \,\not{d}_2^\star(\not{d}_1^\star \not{d}_1)^{-1}\underline{\beta} \\
& -\vartheta\Big[\not{d}_2 \,\not{d}_2^\star(\not{d}_1^\star \not{d}_1)^{-1}, e_4\Big]\underline{\beta} - \vartheta \not{d}_2 \,\not{d}_2^\star(\not{d}_1^\star \not{d}_1)^{-1}\left(\kappa\underline{\beta} + 3\rho\zeta + \underline{\vartheta}\beta\right) \\
& + \not{d}_3^\star\vartheta \,\not{d}_2^\star \not{d}_1^{-1}\check{\rho} - e_\theta\left(\vartheta \,\not{d}_2^\star \not{d}_1^{-1}\left(-\check{\mu} + \frac{1}{4}\vartheta\underline{\vartheta}\right)\right) - e_\theta\left(\vartheta\left(\frac{1}{8}\underline{\kappa}\vartheta + \zeta^2\right)\right) \\
& -2\zeta \not{d}_1 \not{d}_1^\star\kappa - 2e_\theta(\kappa)\not{d}_2^\star\zeta - 2(\kappa\zeta + \beta + \vartheta\zeta)\check{\rho} - 2\zeta\left(\frac{3}{2}\overline{\kappa}\check{\rho} + \frac{3}{2}\overline{\rho}\check{\kappa} - \mathrm{Err}[e_4\check{\rho}]\right) \\
& +2\beta\not{d}_2^\star\zeta + \frac{3}{2}e_\theta\left(\kappa\zeta^2\right) + 2\kappa\Big(\vartheta \not{d}_2 \,\not{d}_2^\star(\not{d}_1^\star \not{d}_1)^{-1}\underline{\beta} + 2\zeta\check{\rho}\Big).
\end{aligned}$$

In view of Corollary 8.36 with $a = 4$, we have

$$\begin{aligned}
& \max_{k\le J+1}\sup_{r_0\ge 4m_0} r_0^6\int_{\{r=r_0\}}\left(\mathfrak{d}^k\left(e_\theta(\mu) + \vartheta \not{d}_2 \,\not{d}_2^\star(\not{d}_1^\star \not{d}_1)^{-1}\underline{\beta} + 2\zeta\check{\rho}\right)\right)^2 \\
\lesssim & \left(N^{\ge 4m_0}[J,\check{\Gamma},\check{R}]\right)^2 + \epsilon^2\int_{{}^{(ext)}\mathcal{M}(\ge 4m_0)}\left(\mathfrak{d}^{\le J+1}\left(e_\theta(\underline{\kappa}) - 4\underline{\beta}\right)\right)^2,
\end{aligned}$$

where we have used

- the fact that $\not{\mathfrak{d}}\vartheta = \not{\mathfrak{d}}\not{d}_2^{-1}\not{d}_2\vartheta$ and Codazzi for ϑ to estimate the terms of the RHS with one angular derivative of ϑ,
- the estimates of Step 2 to estimate the terms of the RHS with one derivative of $\check{\kappa}$,
- the fact that $\not{\mathfrak{d}}\zeta = \not{\mathfrak{d}}\not{d}_1^{-1}\not{d}_1\zeta$ and the definition of μ to estimate terms of the RHS with one angular derivative of ζ,
- the fact that $e_\theta(\underline{\kappa}) = (e_\theta(\underline{\kappa}) - 4\underline{\beta}) + 4\underline{\beta}$ to estimate the term with one angular derivative of $\underline{\kappa}$,
- the identity

$$\begin{aligned}
\not{\mathfrak{d}}\not{d}_1^\star\kappa = & \not{\mathfrak{d}}\not{d}_1^{-1}\left(\vartheta\left(\not{d}_4\not{d}_3^\star\not{d}_2^{-1} + \not{d}_2^\star\right)\not{d}_1^{-1}\rho\right) \\
& + \not{\mathfrak{d}}\not{d}_1^{-1}\left(\not{d}_1\not{d}_1^\star\kappa - \vartheta\left(\not{d}_4\not{d}_3^\star\not{d}_2^{-1} + \not{d}_2^\star\right)\not{d}_1^{-1}\check{\rho}\right)
\end{aligned}$$

  and the estimates of Step 10 to estimate the terms of the RHS with two angular derivatives of $\check{\kappa}$.

Step 12. Recall that we have

$$\begin{aligned}
& e_4(e_\theta(\underline{\kappa}) - 4\underline{\beta}) + \kappa(e_\theta(\underline{\kappa}) - 4\underline{\beta}) \\
= & 2e_\theta(\mu) + 12\rho\zeta - \frac{1}{2}\underline{\kappa}e_\theta(\kappa) + 4\underline{\vartheta}\beta - \frac{1}{2}\vartheta e_\theta(\underline{\kappa}) - e_\theta(\vartheta\underline{\vartheta}) + 2e_\theta(\zeta^2) \\
= & 2\Big(e_\theta(\mu) + \vartheta \not{d}_2 \,\not{d}_2^\star(\not{d}_1^\star \not{d}_1)^{-1}\underline{\beta} + 2\zeta\check{\rho}\Big) + 12\rho\zeta - \frac{1}{2}\underline{\kappa}e_\theta(\kappa) - 2\vartheta \not{d}_2 \,\not{d}_2^\star(\not{d}_1^\star \not{d}_1)^{-1}\underline{\beta} \\
& +2\zeta\check{\rho} + 4\underline{\vartheta}\beta - \frac{1}{2}\vartheta(e_\theta(\underline{\kappa}) - 4\underline{\beta}) - 2\vartheta\underline{\beta} - e_\theta(\vartheta\underline{\vartheta}) + 2e_\theta(\zeta^2).
\end{aligned}$$

In view of Corollary 8.36 with $a = 2$, we have

$$\begin{aligned}
&\max_{k\leq J+1}\sup_{r_0\geq 4m_0} r_0^2\int_{\{r=r_0\}}\left(\mathfrak{d}^k\left(e_\theta(\underline{\check{\kappa}}) - 4\underline{\beta}\right)\right)^2\\
\lesssim\; &\left(N^{\geq 4m_0}[J,\check{\Gamma},\check{R}]\right)^2\\
&+\int_{{}^{(ext)}\mathcal{M}(\geq 4m_0)} r^4\left(\mathfrak{d}^{\leq J+1}\left(e_\theta(\mu) + \vartheta\, \not{d}_2\, \not{d}_2^\star(\not{d}_1^\star\not{d}_1)^{-1}\underline{\beta} + 2\zeta\check{\rho}\right)\right)^2,
\end{aligned}$$

where we have used

- the fact that $\not{\mathfrak{d}}\vartheta = \not{\mathfrak{d}}\,\not{d}_2^{-1}\not{d}_2\vartheta$ and Codazzi for ϑ to estimate the terms of the RHS with one angular derivative of ϑ,
- the fact that $\not{\mathfrak{d}}\underline{\vartheta} = \not{\mathfrak{d}}\,\not{d}_2^{-1}\not{d}_2\underline{\vartheta}$ and Codazzi for $\underline{\vartheta}$ to estimate the terms of the RHS with one angular derivative of $\underline{\vartheta}$,
- the estimates of Step 2 to estimate the terms of the RHS with one derivative of $\check{\kappa}$,
- the fact that $\not{\mathfrak{d}}\zeta = \not{\mathfrak{d}}\,\not{d}_1^{-1}\not{d}_1\zeta$ and the definition of μ to estimate terms of the RHS with one angular derivative of ζ,
- the estimate for ζ of Step 3.

Together with the estimate of Step 11, we infer

$$\begin{aligned}
&\max_{k\leq J+1}\sup_{r_0\geq 4m_0} r_0^6\int_{\{r=r_0\}}\left(\mathfrak{d}^k\left(e_\theta(\mu) + \vartheta\,\not{d}_2\,\not{d}_2^\star(\not{d}_1^\star\not{d}_1)^{-1}\underline{\beta} + 2\zeta\check{\rho}\right)\right)^2\\
&+\max_{k\leq J+1}\sup_{r_0\geq 4m_0} r_0^2\int_{\{r=r_0\}}\left(\mathfrak{d}^k\left(e_\theta(\underline{\check{\kappa}}) - 4\underline{\beta}\right)\right)^2\\
\lesssim\; &\left(N^{\geq 4m_0}[J,\check{\Gamma},\check{R}]\right)^2.
\end{aligned}$$

Finally, we have obtained

$$\begin{aligned}
&\max_{k\leq J+1}\sup_{r_0\geq 4m_0}\Bigg(\int_{\{r=r_0\}}\Bigg(r_0^4(\mathfrak{d}^k\check{\mu})^2 + r_0^2(\mathfrak{d}^k\vartheta)^2 + r_0^2(\mathfrak{d}^k\zeta)^2\\
&+r_0^2(\mathfrak{d}^k(e_3(\zeta)+\underline{\beta}))^2 + r_0^{2-\delta_B}(\mathfrak{d}^k\underline{\check{\kappa}})^2 + (\mathfrak{d}^k\underline{\vartheta})^2 + (\mathfrak{d}^k\underline{\check{\omega}})^2 + r_0^{-\delta_B}(\mathfrak{d}^k\underline{\xi})^2\Bigg)\Bigg)\\
\lesssim\; &\left(N^{\geq 4m_0}[J,\check{\Gamma},\check{R}]\right)^2,
\end{aligned}$$

$$\max_{k\leq J+2}\sup_{r_0\geq 4m_0}\left(r_0^2\int_{\{r=r_0\}}\left(\mathfrak{d}^k\left(\kappa-\frac{2}{r}\right)\right)^2\right) \lesssim \left(N^{\geq 4m_0}[J,\check{\Gamma},\check{R}]\right)^2,$$

and

$$\max_{k\leq J+1}\sup_{r_{\mathcal{T}}\leq r_0\leq 4m_0}\left(\int_{\{r=r_0\}}\left\{r_0^6\left(\mathfrak{d}^k\left(\not{d}_1\not{d}_1^\star\kappa-\vartheta\left(\not{d}_4\not{d}_3^\star\not{d}_2^{-1}+\not{d}_2^\star\right)\not{d}_1^{-1}\check{\rho}\right)\right)^2\right.\right.$$
$$\left.\left.+r_0^6\left(\mathfrak{d}^k\left(e_\theta(\mu)+\vartheta\not{d}_2\not{d}_2^\star(\not{d}_1^\star\not{d}_1)^{-1}\underline{\beta}+2\zeta\check{\rho}\right)\right)^2+r_0^2\left(\mathfrak{d}^k\left(e_\theta(\underline{\kappa})-4\underline{\beta}\right)\right)^2\right\}\right)$$
$$\lesssim \left(N^{\geq 4m_0}[J,\check{\Gamma},\check{R}]\right)^2.$$

In view of the definition (8.8.2) of $N^{\geq 4m_0}[J,\check{\Gamma},\check{R}]$, and of the various norms, we infer

$$\begin{aligned}
&{}^{(ext)}\mathfrak{G}^{\geq 4m_0}_{J+1}[\check{\Gamma}]+{}^{(ext)}\mathfrak{G}^{\geq 4m_0\prime}_{J+1}[\check{\Gamma}]\\
\lesssim\ &{}^{(\Sigma_*)}\mathfrak{G}_{J+1}[\check{\Gamma}]+{}^{(\Sigma_*)}\mathfrak{G}'_{J+1}[\check{\Gamma}]+{}^{(ext)}\mathfrak{R}_{J+1}[\check{R}]+{}^{(ext)}\mathfrak{G}_J[\check{\Gamma}]\\
&+\epsilon_0\left({}^{(ext)}\mathfrak{G}^{\geq 4m_0}_{J+1}[\check{\Gamma}]+{}^{(ext)}\mathfrak{G}^{\geq 4m_0\prime}_{J+1}[\check{\Gamma}]\right)
\end{aligned}$$

and hence, for ϵ_0 small enough,

$$\begin{aligned}
&{}^{(ext)}\mathfrak{G}^{\geq 4m_0}_{J+1}[\check{\Gamma}]+{}^{(ext)}\mathfrak{G}^{\geq 4m_0\prime}_{J+1}[\check{\Gamma}]\\
\lesssim\ &{}^{(\Sigma_*)}\mathfrak{G}_{J+1}[\check{\Gamma}]+{}^{(\Sigma_*)}\mathfrak{G}'_{J+1}[\check{\Gamma}]+{}^{(ext)}\mathfrak{R}_{J+1}[\check{R}]+{}^{(ext)}\mathfrak{G}_J[\check{\Gamma}].
\end{aligned}$$

This concludes the proof of Proposition 8.32.

8.8.5 Proof of Proposition 8.33

In the proof below, we will repeatedly use the following estimate:

$$\begin{aligned}
&\max_{k\leq J+1}\int_{{}^{(ext)}\mathcal{M}(r\leq 4m_0)}(\mathfrak{d}^k f)^2\\
\lesssim\ &\max_{k\leq J}\int_{{}^{(ext)}\mathcal{M}(r\leq 4m_0)}\left((\mathfrak{d}^k f)^2+(\mathfrak{d}^k\mathbf{N}f)^2+(\mathfrak{d}^k e_4 f)^2+(\mathfrak{d}^k\not{\mathfrak{d}}f)^2\right) \qquad (8.8.3)
\end{aligned}$$

which follows from the fact that $\mathfrak{d}=(\not{\mathfrak{d}},re_4,e_3)$ and $e_3=\Upsilon e_4-2\mathbf{N}$, where we recall that

$$\mathbf{N} = \frac{1}{2}\left(\Upsilon e_4-e_3\right).$$

Also, we introduce the following notation which will constantly appear on the RHS of the equalities below:

$$\begin{aligned}
N^{\leq 4m_0}[J,\check{\Gamma},\check{R}] \ :=\ &{}^{(ext)}\mathfrak{G}^{\geq 4m_0}_{J+1}[\check{\Gamma}]+{}^{(ext)}\mathfrak{G}^{\geq 4m_0\prime}_{J+1}[\check{\Gamma}]+{}^{(ext)}\mathfrak{R}_{J+1}[\check{R}]+{}^{(ext)}\mathfrak{G}_J[\check{\Gamma}]\\
&+\epsilon_0\left({}^{(ext)}\mathfrak{G}^{\leq 4m_0}_{J+1}[\check{\Gamma}]+{}^{(ext)}\mathfrak{G}^{\leq 4m_0\prime}_{J+1}[\check{\Gamma}]\right). \qquad (8.8.4)
\end{aligned}$$

Step 1. Recall that

$$e_4(\check{\kappa})+\kappa\check{\kappa} = \mathrm{Err}[e_4\check{\kappa}].$$

In view of Corollary 8.40, we have

$$\max_{k\leq J+2}\sup_{r_{\mathcal{T}}\leq r_0\leq 4m_0}\int_{\{r=r_0\}}(\mathfrak{d}^k\check{\kappa})^2 \lesssim \Big(N^{\leq 4m_0}[J,\check{\Gamma},\check{R}]\Big)^2$$

where we have used the null structure equations for $e_4(\vartheta)$, $e_3(\vartheta)$ and $\not{d}_2\vartheta$ to avoid losing one derivative.

Step 2. Next, recall that

$$e_4(\check{\mu})+\frac{3}{2}\overline{\kappa}\check{\mu} = -\frac{3}{2}\overline{\mu}\check{\kappa}+\mathrm{Err}[e_4\check{\mu}].$$

In view of Corollary 8.40, we have

$$\max_{k\leq J+1}\sup_{r_{\mathcal{T}}\leq r_0\leq 4m_0}\int_{\{r=r_0\}}(\mathfrak{d}^k\check{\mu})^2 \lesssim \Big(N^{\leq 4m_0}[J,\check{\Gamma},\check{R}]\Big)^2$$

where we have used the estimates for $\check{\kappa}$ of Step 1.

Step 3. Next, recall that

$$e_4(\zeta)+\kappa\zeta = -\beta-\vartheta\zeta.$$

In view of Corollary 8.40, we have

$$\max_{k\leq J}\sup_{r_{\mathcal{T}}\leq r_0\leq 4m_0}\int_{\{r=r_0\}}\Big((\mathfrak{d}^k e_4\zeta)^2+(\mathfrak{d}^k\zeta)^2\Big) \lesssim \Big(N^{\leq 4m_0}[J,\check{\Gamma},\check{R}]\Big)^2.$$

Also, commuting first with $\mathbf{N}$, and proceeding analogously, we infer

$$\max_{k\leq J}\sup_{r_{\mathcal{T}}\leq r_0\leq 4m_0}\int_{\{r=r_0\}}(\mathfrak{d}^k\mathbf{N}\zeta)^2 \lesssim \Big(N^{\leq 4m_0}[J,\check{\Gamma},\check{R}]\Big)^2.$$

Furthermore, in view of the definition of μ and a Poincaré inequality for $\not{d}_1$, we have

$$\max_{k\leq J}\sup_{r_{\mathcal{T}}\leq r_0\leq 4m_0}\int_{\{r=r_0\}}(\mathfrak{d}^k\not{\mathfrak{d}}\zeta)^2 \lesssim \Big(N^{\leq 4m_0}[J,\check{\Gamma},\check{R}]\Big)^2$$

where we have used a trace estimate and the estimate for $\check{\mu}$ of Step 2. The above estimates, together with (8.8.3), imply

$$\max_{k\leq J+1}\sup_{r_{\mathcal{T}}\leq r_0\leq 4m_0}\int_{\{r=r_0\}}(\mathfrak{d}^k\zeta)^2 \lesssim \Big(N^{\leq 4m_0}[J,\check{\Gamma},\check{R}]\Big)^2.$$

Step 4. Recall that

$$e_4(\vartheta)+\kappa\vartheta = -2\alpha.$$

In view of Corollary 8.40, we have

$$\max_{k\leq J}\sup_{r_{\mathcal{T}}\leq r_0\leq 4m_0}\int_{\{r=r_0\}}\Big((\mathfrak{d}^k e_4\vartheta)^2+(\mathfrak{d}^k\vartheta)^2\Big) \lesssim \Big(N^{\leq 4m_0}[J,\check{\Gamma},\check{R}]\Big)^2.$$

Also, commuting first one time with $\mathbf{N}$, and proceeding analogously, we infer

$$\max_{k\leq J}\sup_{r_{\mathcal{T}}\leq r_0\leq 4m_0}\int_{\{r=r_0\}}(\mathfrak{d}^k\mathbf{N}\vartheta)^2 \lesssim \left(N^{\leq 4m_0}[J,\check{\Gamma},\check{R}]\right)^2.$$

Furthermore, in view of Codazzi for ϑ, and a Poincaré inequality for $\not{d}_2$, we have

$$\max_{k\leq J}\sup_{r_{\mathcal{T}}\leq r_0\leq 4m_0}\int_{\{r=r_0\}}(\mathfrak{d}^k\not{\mathfrak{d}}\vartheta)^2 \lesssim \left(N^{\leq 4m_0}[J,\check{\Gamma},\check{R}]\right)^2$$

where we have used a trace estimate, and the estimate for $\check{\kappa}$ and ζ respectively in Step 1 and Step 3. The above estimates, together with (8.8.3), imply

$$\max_{k\leq J+1}\sup_{r_{\mathcal{T}}\leq r_0\leq 4m_0}\int_{\{r=r_0\}}(\mathfrak{d}^k\vartheta)^2 \lesssim \left(N^{\leq 4m_0}[J,\check{\Gamma},\check{R}]\right)^2.$$

Step 5. Recall that we have

$$\begin{aligned}
e_4(\check{\underline{\kappa}})+\frac{1}{2}\overline{\kappa}\check{\underline{\kappa}} &= -\frac{1}{2}\overline{\underline{\kappa}}\check{\kappa}-2\not{d}_1\zeta+2\check{\rho}+\mathrm{Err}[e_4\check{\underline{\kappa}}]\\
&= -\frac{1}{2}\overline{\underline{\kappa}}\check{\kappa}+2\check{\mu}+4\check{\rho}-\frac{1}{2}\vartheta\underline{\vartheta}+\mathrm{Err}[e_4\check{\underline{\kappa}}].
\end{aligned}$$

In view of Corollary 8.40, we have

$$\max_{k\leq J}\sup_{r_{\mathcal{T}}\leq r_0\leq 4m_0}\int_{\{r=r_0\}}\left((\mathfrak{d}^k e_4\check{\underline{\kappa}})^2+(\mathfrak{d}^k\check{\underline{\kappa}})^2\right) \lesssim \left(N^{\leq 4m_0}[J,\check{\Gamma},\check{R}]\right)^2$$

where we have used the estimates for $\check{\kappa}$ and $\check{\mu}$ derived respectively in Step 1 and Step 2. Also, commuting first one time with $\mathbf{N}$, and proceeding analogously, we infer

$$\max_{k\leq J}\sup_{r_{\mathcal{T}}\leq r_0\leq 4m_0}\int_{\{r=r_0\}}(\mathfrak{d}^k\mathbf{N}\check{\underline{\kappa}})^2 \lesssim \left(N^{\leq 4m_0}[J,\check{\Gamma},\check{R}]\right)^2$$

where we have used the estimates for $\check{\kappa}$ and $\check{\mu}$ derived respectively in Step 1 and Step 2. Furthermore, commuting the equation for $e_4(\kappa)$ once with e_θ, we have

$$e_4(e_\theta(\underline{\kappa}))+\kappa e_\theta(\underline{\kappa}) = -\frac{1}{2}\underline{\kappa}e_\theta(\kappa)+2e_\theta(\mu)+4e_\theta(\rho)-e_\theta(\vartheta\underline{\vartheta})+2e_\theta(\zeta^2)-\frac{1}{2}\vartheta e_\theta(\underline{\kappa}).$$

Together with the Bianchi identity for $e_4(\underline{\beta})$, we infer

$$\begin{aligned}
e_4(e_\theta(\underline{\kappa})-4\underline{\beta})+\kappa(e_\theta(\underline{\kappa})-4\underline{\beta}) &= -\frac{1}{2}\underline{\kappa}e_\theta(\kappa)+2e_\theta(\mu)+12\rho\zeta\\
&\quad +4\underline{\vartheta}\beta-e_\theta(\vartheta\underline{\vartheta})+2e_\theta(\zeta^2)-\frac{1}{2}\vartheta e_\theta(\underline{\kappa}).
\end{aligned}$$

In view of Corollary 8.40, we have

$$\begin{aligned}
&\max_{k\leq J}\sup_{r_{\mathcal{T}}\leq r_0\leq 4m_0}\int_{\{r=r_0\}}\left((\mathfrak{d}^k e_4(e_\theta(\underline{\kappa})-4\underline{\beta}))^2+(\mathfrak{d}^k(e_\theta(\underline{\kappa})-4\underline{\beta}))^2\right)\\
&\lesssim \left(N^{\leq 4m_0}[J,\check{\Gamma},\check{R}]\right)^2
\end{aligned}$$

where we have used the estimates for $\check{\kappa}$, $\check{\mu}$ and ζ derived respectively in Step 1, Step 2 and Step 3.

The above estimates, together with (8.8.3), imply

$$\begin{aligned}
&\max_{k\leq J+1}\sup_{r_{\mathcal{T}}\leq r_0\leq 4m_0}\int_{\{r=r_0\}}(\mathfrak{d}^k\underline{\check{\kappa}})^2\\
\lesssim\;& \left(N^{\leq 4m_0}[J,\check{\Gamma},\check{R}]\right)^2+\max_{k\leq J}\sup_{r_{\mathcal{T}}\leq r_0\leq 4m_0}\int_{\{r=r_0\}}(\mathfrak{d}^k\underline{\beta})^2\\
\lesssim\;& \left(N^{\leq 4m_0}[J,\check{\Gamma},\check{R}]\right)^2
\end{aligned}$$

where we have used a trace estimate on $\{r=r_0\}$ for $r_{\mathcal{T}}\leq r_0\leq 4m_0$.

Step 6. Recall that we have

$$e_4(\underline{\check{\omega}}) \;=\; \check{\rho}+\mathrm{Err}[e_4\underline{\check{\omega}}].$$

In view of Corollary 8.40, we have

$$\max_{k\leq J}\sup_{r_{\mathcal{T}}\leq r_0\leq 4m_0}\int_{\{r=r_0\}}\left((\mathfrak{d}^k e_4\underline{\check{\omega}})^2+(\mathfrak{d}^k\underline{\check{\omega}})^2\right)\;\lesssim\;\left(N^{\leq 4m_0}[J,\check{\Gamma},\check{R}]\right)^2.$$

Also, commuting first one time with $\mathbf{N}$, and proceeding analogously, we infer

$$\max_{k\leq J}\sup_{r_{\mathcal{T}}\leq r_0\leq 4m_0}\int_{\{r=r_0\}}(\mathfrak{d}^k\mathbf{N}\underline{\check{\omega}})^2\;\lesssim\;\left(N^{\leq 4m_0}[J,\check{\Gamma},\check{R}]\right)^2.$$

Step 7. Recall that we have

$$\begin{aligned}
&e_4(e_3(\zeta)+\underline{\beta})+\kappa(e_3(\zeta)+\underline{\beta})\\
=\;&\underline{\kappa}\beta+\left(\frac{1}{2}\kappa\underline{\kappa}+2\mu-6\rho\right)\zeta\\
&-\underline{\vartheta}\beta+\vartheta\underline{\beta}-\underline{\xi}\alpha-\vartheta\zeta+2\underline{\omega}\vartheta\zeta-4\zeta\,{\not d}_1^\star{\not d}_1^{-1}\left(\check{\mu}+\check{\rho}-\frac{1}{4}\vartheta\underline{\vartheta}+\frac{1}{4}\overline{\vartheta\underline{\vartheta}}\right)-2\zeta^3.
\end{aligned}$$

Commuting first one time with $\mathbf{N}$, and in view of Corollary 8.40, we have

$$\max_{k\leq J}\sup_{r_{\mathcal{T}}\leq r_0\leq 4m_0}\int_{\{r=r_0\}}(\mathfrak{d}^k\mathbf{N}(e_3(\zeta)+\underline{\beta}))^2\;\lesssim\;\left(N^{\leq 4m_0}[J,\check{\Gamma},\check{R}]\right)^2$$

where we have used the estimate for ζ in Step 3.

Step 8. Recall that we have

$$e_4(\underline{\xi}) \;=\; -e_3(\zeta)+\underline{\beta}-\underline{\kappa}\zeta-\zeta\underline{\vartheta}.$$

In view of Corollary 8.40, we have

$$\max_{k\leq J}\sup_{r_{\mathcal{T}}\leq r_0\leq 4m_0}\int_{\{r=r_0\}}\left((\mathfrak{d}^k e_4\underline{\xi})^2+(\mathfrak{d}^k\underline{\xi})^2\right)\;\lesssim\;\left(N^{\leq 4m_0}[J,\check{\Gamma},\check{R}]\right)^2$$

where we have used the estimates for ζ derived in Step 3. Also, commuting first one time with $\mathbf{N}$, and proceeding analogously, we infer

$$\max_{k\leq J}\sup_{r_{\mathcal{T}}\leq r_0\leq 4m_0}\int_{\{r=r_0\}}(\mathfrak{d}^k\mathbf{N}\underline{\xi})^2 \lesssim \left(N^{\leq 4m_0}[J,\check{\Gamma},\check{R}]\right)^2$$

where we have used the estimates for $e_3(\zeta)+\underline{\beta}$ derived in Step 7.

Step 9. Recall

$$2\,\not{d}_1^\star\underline{\omega} = e_3\zeta+\underline{\kappa}\zeta-\underline{\beta}-\frac{1}{2}\kappa\underline{\xi}+\underline{\vartheta}\zeta-\frac{1}{2}\vartheta\underline{\xi}.$$

Using a Poincaré inequality for $\not{d}_1^\star$, we infer

$$\max_{k\leq J}\sup_{r_{\mathcal{T}}\leq r_0\leq 4m_0}\int_{\{r=r_0\}}(\mathfrak{d}^k\,\not{\mathfrak{d}}\check{\underline{\omega}})^2 \lesssim \left(N^{\leq 4m_0}[J,\check{\Gamma},\check{R}]\right)^2$$

where we have used a trace estimate and the estimate for ζ and $\underline{\xi}$ respectively in Step 3 and Step 8. The above estimates, together with the estimates for $\check{\underline{\omega}}$ of Step 6 and (8.8.3), imply

$$\max_{k\leq J+1}\sup_{r_{\mathcal{T}}\leq r_0\leq 4m_0}\int_{\{r=r_0\}}(\mathfrak{d}^k\check{\underline{\omega}})^2 \lesssim \left(N^{\leq 4m_0}[J,\check{\Gamma},\check{R}]\right)^2.$$

Step 10. Recall that we have

$$e_4(\check{\underline{\Omega}}) = -2\check{\underline{\omega}}+\overline{\check{\kappa}\check{\underline{\Omega}}}.$$

In view of Corollary 8.40, we have

$$\max_{k\leq J+1}\sup_{r_{\mathcal{T}}\leq r_0\leq 4m_0}\int_{\{r=r_0\}}(\mathfrak{d}^k\check{\underline{\Omega}})^2 \lesssim \left(N^{\leq 4m_0}[J,\check{\Gamma},\check{R}]\right)^2$$

where we have used the estimates for $\check{\underline{\omega}}$ derived in Step 9.

Step 11. Recall

$$2\,\not{d}_1\underline{\xi} = e_3(\check{\underline{\kappa}})+\overline{\underline{\kappa}}\,\check{\underline{\kappa}}+2\overline{\underline{\omega}}\,\check{\underline{\kappa}}+2\overline{\underline{\kappa}}\,\check{\underline{\omega}}-\left(\frac{1}{2}\overline{\kappa\underline{\kappa}}-2\overline{\rho}\right)\check{\underline{\Omega}}-\mathrm{Err}[e_3\check{\underline{\kappa}}].$$

Using a Poincaré inequality for $\not{d}_1$, we infer

$$\max_{k\leq J}\sup_{r_{\mathcal{T}}\leq r_0\leq 4m_0}\int_{\{r=r_0\}}(\mathfrak{d}^k\,\not{\mathfrak{d}}\underline{\xi})^2 \lesssim \left(N^{\leq 4m_0}[J,\check{\Gamma},\check{R}]\right)^2$$

where we have used the estimates for $\check{\underline{\kappa}}$, $\check{\underline{\omega}}$ and $\check{\underline{\Omega}}$ respectively in Step 5, Step 9 and Step 10. The above estimates, together with the estimates for $\underline{\xi}$ of Step 8 and (8.8.3), imply

$$\max_{k\leq J+1}\sup_{r_{\mathcal{T}}\leq r_0\leq 4m_0}\int_{\{r=r_0\}}(\mathfrak{d}^k\underline{\xi})^2 \lesssim \left(N^{\leq 4m_0}[J,\check{\Gamma},\check{R}]\right)^2.$$

Step 12. Recall that

$$\begin{aligned} e_4(\underline{\vartheta}) + \frac{1}{2}\kappa\underline{\vartheta} &= 2\,\mathrm{d}\!\!\!/_2^\star \underline{\zeta} - \frac{1}{2}\underline{\kappa}\vartheta + 2\zeta^2 \\ &= 2\,\mathrm{d}\!\!\!/_2^\star\, \mathrm{d}\!\!\!/_1^{-1}\left(-\check{\mu} - \check{\rho} + \frac{1}{4}\vartheta\underline{\vartheta} - \frac{1}{4}\overline{\vartheta\underline{\vartheta}}\right) - \frac{1}{2}\underline{\kappa}\vartheta + 2\zeta^2. \end{aligned}$$

In view of Corollary 8.40, we have

$$\max_{k\leq J}\sup_{r_{\mathcal{T}}\leq r_0\leq 4m_0}\int_{\{r=r_0\}}\left((\mathfrak{d}^k e_4\underline{\vartheta})^2 + (\mathfrak{d}^k\underline{\vartheta})^2\right) \lesssim \left(N^{\leq 4m_0}[J,\check{\Gamma},\check{R}]\right)^2$$

where we have used the estimate for $\check{\mu}$ and ϑ respectively in Step 2 and Step 4. Also, commuting first one time with $\mathbf{N}$, and proceeding analogously, we infer

$$\max_{k\leq J}\sup_{r_{\mathcal{T}}\leq r_0\leq 4m_0}\int_{\{r=r_0\}}(\mathfrak{d}^k\mathbf{N}\underline{\vartheta})^2 \lesssim \left(N^{\leq 4m_0}[J,\check{\Gamma},\check{R}]\right)^2$$

where we have used the estimate for $\check{\mu}$ and ϑ respectively in Step 2 and Step 4. Furthermore, in view of Codazzi for $\underline{\vartheta}$, and a Poincaré inequality for $\mathrm{d}\!\!\!/_2$, we have

$$\max_{k\leq J}\sup_{r_{\mathcal{T}}\leq r_0\leq 4m_0}\int_{\{r=r_0\}}(\mathfrak{d}^k\,\nabla\!\!\!\!/\,\underline{\vartheta})^2 \lesssim \left(N^{\leq 4m_0}[J,\check{\Gamma},\check{R}]\right)^2$$

where we have used a trace estimate and the estimate for $\underline{\check{\kappa}}$ and ζ respectively in Step 5 and Step 3. The above estimates, together with (8.8.3), imply

$$\max_{k\leq J+1}\sup_{r_{\mathcal{T}}\leq r_0\leq 4m_0}\int_{\{r=r_0\}}(\mathfrak{d}^k\underline{\vartheta})^2 \lesssim \left(N^{\leq 4m_0}[J,\check{\Gamma},\check{R}]\right)^2.$$

Step 13. Recall that we have

$$\begin{aligned} & e_4\left(\mathrm{d}\!\!\!/_1\,\mathrm{d}\!\!\!/_1^\star\kappa - \vartheta\left(\mathrm{d}\!\!\!/_4\,\mathrm{d}\!\!\!/_3^\star\,\mathrm{d}\!\!\!/_2^{-1} + \mathrm{d}\!\!\!/_2^\star\right)\mathrm{d}\!\!\!/_1^{-1}\check{\rho}\right) \\ & +2\kappa\left(\mathrm{d}\!\!\!/_1\,\mathrm{d}\!\!\!/_1^\star\kappa - \vartheta\left(\mathrm{d}\!\!\!/_4\,\mathrm{d}\!\!\!/_3^\star\,\mathrm{d}\!\!\!/_2^{-1} + \mathrm{d}\!\!\!/_2^\star\right)\mathrm{d}\!\!\!/_1^{-1}\check{\rho}\right) \\ = & -\left(-\frac{1}{2}\vartheta\,\mathrm{d}\!\!\!/_2^\star + \zeta e_4(\Phi) - \beta\right)\mathrm{d}\!\!\!/_1^\star\kappa - \frac{1}{2}\vartheta\,\mathrm{d}\!\!\!/_1\,\mathrm{d}\!\!\!/_1^\star\kappa + \frac{1}{2}\mathrm{d}\!\!\!/_1^\star(\kappa+\vartheta)\,\mathrm{d}\!\!\!/_1^\star\kappa + (\mathrm{d}\!\!\!/_1^\star\kappa)^2 \\ & -2\kappa\vartheta\left(\mathrm{d}\!\!\!/_4\,\mathrm{d}\!\!\!/_3^\star\,\mathrm{d}\!\!\!/_2^{-1} + \mathrm{d}\!\!\!/_2^\star\right)\mathrm{d}\!\!\!/_1^{-1}\check{\rho} + (\kappa\vartheta + 2\alpha)\left(\mathrm{d}\!\!\!/_4\,\mathrm{d}\!\!\!/_3^\star\,\mathrm{d}\!\!\!/_2^{-1} + \mathrm{d}\!\!\!/_2^\star\right)\mathrm{d}\!\!\!/_1^{-1}\check{\rho} \\ & +\vartheta\left[\left(\mathrm{d}\!\!\!/_4\,\mathrm{d}\!\!\!/_3^\star\,\mathrm{d}\!\!\!/_2^{-1} + \mathrm{d}\!\!\!/_2^\star\right)\mathrm{d}\!\!\!/_1^{-1}, e_4\right]\check{\rho} \\ & +\vartheta\left(\mathrm{d}\!\!\!/_4\,\mathrm{d}\!\!\!/_3^\star\,\mathrm{d}\!\!\!/_2^{-1} + \mathrm{d}\!\!\!/_2^\star\right)\mathrm{d}\!\!\!/_1^{-1}\left(\frac{3}{2}\overline{\kappa}\check{\rho} + \frac{3}{2}\overline{\rho}\check{\kappa} - \mathrm{Err}[e_4\check{\rho}]\right) \\ & -\frac{1}{2}\vartheta\,\mathrm{d}\!\!\!/_4\,\mathrm{d}\!\!\!/_3^\star\,\mathrm{d}\!\!\!/_2^{-1}(-\mathrm{d}\!\!\!/_1^\star\kappa + \kappa\zeta - \vartheta\zeta) - \frac{1}{2}\vartheta\,\mathrm{d}\!\!\!/_2^\star(-\mathrm{d}\!\!\!/_1^\star\kappa + \kappa\zeta - \vartheta\zeta) \\ & +\frac{1}{2}\mathrm{d}\!\!\!/_3^\star\,\mathrm{d}\!\!\!/_2^{-1}(-2\beta - \mathrm{d}\!\!\!/_1^\star\kappa + \kappa\zeta - \vartheta\zeta)\,\mathrm{d}\!\!\!/_3^\star\vartheta + \frac{1}{2}(-2\beta - \mathrm{d}\!\!\!/_1^\star\kappa + \kappa\zeta - \vartheta\zeta)\,\mathrm{d}\!\!\!/_2\vartheta. \end{aligned}$$

In view of Corollary 8.40, we have

$$\max_{k\leq J+1}\sup_{r_{\mathcal{T}}\leq r_0\leq 4m_0}\int_{\{r=r_0\}}\left(\mathfrak{d}^k\left(\not{d}_1\not{d}_1^\star\kappa-\vartheta\Big(\not{d}_4\not{d}_3^\star\not{d}_2^{-1}+\not{d}_2^\star\Big)\not{d}_1^{-1}\check{\rho}\right)\right)^2$$
$$\lesssim\left(N^{\leq 4m_0}[J,\check{\Gamma},\check{R}]\right)^2$$

where we have used

- the fact that $\not{\mathfrak{d}}\vartheta=\not{\mathfrak{d}}\not{d}_2^{-1}\not{d}_2\vartheta$ and Codazzi for ϑ to estimate the terms of the RHS with one angular derivative of ϑ,
- the estimates of Step 1 to estimate the terms of the RHS with one derivative of $\check{\kappa}$,
- the fact that $\not{\mathfrak{d}}\zeta=\not{\mathfrak{d}}\not{d}_1^{-1}\not{d}_1\zeta$ and the definition of μ to estimate terms of the RHS with one angular derivative of ζ,
- the identity

$$\begin{aligned}\not{\mathfrak{d}}\not{d}_1^\star\kappa &= \not{\mathfrak{d}}\not{d}_1^{-1}\left(\vartheta\Big(\not{d}_4\not{d}_3^\star\not{d}_2^{-1}+\not{d}_2^\star\Big)\not{d}_1^{-1}\check{\rho}\right)\\ &\quad+\not{\mathfrak{d}}\not{d}_1^{-1}\left(\not{d}_1\not{d}_1^\star\kappa-\vartheta\Big(\not{d}_4\not{d}_3^\star\not{d}_2^{-1}+\not{d}_2^\star\Big)\not{d}_1^{-1}\check{\rho}\right)\end{aligned}$$

  to estimate the terms of the RHS with two angular derivatives of $\check{\kappa}$.

Step 14. Recall that we have

$$\begin{aligned}&e_4\Big(e_\theta(\mu)+\vartheta\not{d}_2\not{d}_2^\star(\not{d}_1^\star\not{d}_1)^{-1}\underline{\beta}+2\zeta\check{\rho}\Big)+2\kappa\Big(e_\theta(\mu)+\vartheta\not{d}_2\not{d}_2^\star(\not{d}_1^\star\not{d}_1)^{-1}\underline{\beta}+2\zeta\check{\rho}\Big)\\ =\;&-\frac{3}{2}\mu e_\theta(\kappa)-\frac{1}{2}\vartheta e_\theta(\mu)-(\kappa\vartheta+2\alpha)\not{d}_2\not{d}_2^\star(\not{d}_1^\star\not{d}_1)^{-1}\underline{\beta}\\ &-\vartheta\Big[\not{d}_2\not{d}_2^\star(\not{d}_1^\star\not{d}_1)^{-1},e_4\Big]\underline{\beta}-\vartheta\not{d}_2\not{d}_2^\star(\not{d}_1^\star\not{d}_1)^{-1}\left(\kappa\underline{\beta}+3\rho\zeta+\underline{\vartheta}\beta\right)\\ &+\not{d}_3^\star\vartheta\not{d}_2^\star\not{d}_1^{-1}\check{\rho}-e_\theta\left(\vartheta\not{d}_2^\star\not{d}_1^{-1}\left(-\check{\mu}+\frac{1}{4}\vartheta\underline{\vartheta}\right)\right)-e_\theta\left(\vartheta\left(\frac{1}{8}\underline{\kappa}\vartheta+\zeta^2\right)\right)\\ &-2\zeta\not{d}_1\not{d}_1^\star\kappa-2e_\theta(\kappa)\not{d}_2^\star\zeta-2(\kappa\zeta+\beta+\vartheta\zeta)\check{\rho}-2\zeta\left(\frac{3}{2}\overline{\kappa}\check{\rho}+\frac{3}{2}\overline{\rho}\check{\kappa}-\mathrm{Err}[e_4\check{\rho}]\right)\\ &+2\beta\not{d}_2^\star\zeta+\frac{3}{2}e_\theta\left(\kappa\zeta^2\right)+2\kappa\Big(\vartheta\not{d}_2\not{d}_2^\star(\not{d}_1^\star\not{d}_1)^{-1}\underline{\beta}+2\zeta\check{\rho}\Big).\end{aligned}$$

In view of Corollary 8.40, we have

$$\max_{k\leq J+1}\sup_{r_{\mathcal{T}}\leq r_0\leq 4m_0}\int_{\{r=r_0\}}\left(\mathfrak{d}^k\left(e_\theta(\mu)+\vartheta\not{d}_2\not{d}_2^\star(\not{d}_1^\star\not{d}_1)^{-1}\underline{\beta}+2\zeta\check{\rho}\right)\right)^2$$
$$\lesssim\left(N^{\leq 4m_0}[J,\check{\Gamma},\check{R}]\right)^2+\epsilon^2\int_{{}^{(ext)}\mathcal{M}(\leq 4m_0)}\left(\mathfrak{d}^{\leq J+1}\left(e_\theta(\underline{\kappa})-4\underline{\beta}\right)\right)^2,$$

where we have used

- the fact that $\not{\mathfrak{d}}\vartheta=\not{\mathfrak{d}}\not{d}_2^{-1}\not{d}_2\vartheta$ and Codazzi for ϑ to estimate the terms of the RHS with one angular derivative of ϑ,
- the estimates of Step 1 to estimate the terms of the RHS with one derivative of $\check{\kappa}$,

- the fact that $\not{\partial}\zeta = \not{\partial}\,\not{d}_1^{-1}\not{d}_1\zeta$ and the definition of μ to estimate terms of the RHS with one angular derivative of ζ,
- the fact that $e_\theta(\underline{\kappa}) = (e_\theta(\underline{\kappa}) - 4\underline{\beta}) + 4\underline{\beta}$ to estimate the term with one angular derivative of $\underline{\kappa}$,
- the identity

$$\begin{aligned}\not{\partial}\,\not{d}_1^\star\kappa &= \not{\partial}\,\not{d}_1^{-1}\Big(\vartheta\Big(\not{d}_4\,\not{d}_3^\star\,\not{d}_2^{-1}+\not{d}_2^\star\Big)\not{d}_1^{-1}\check{\rho}\Big)\\ &\quad+\not{\partial}\,\not{d}_1^{-1}\Big(\not{d}_1\,\not{d}_1^\star\kappa-\vartheta\Big(\not{d}_4\,\not{d}_3^\star\,\not{d}_2^{-1}+\not{d}_2^\star\Big)\not{d}_1^{-1}\check{\rho}\Big)\end{aligned}$$

and the estimates of Step 13 to estimate the terms of the RHS with two angular derivatives of $\check{\kappa}$.

Step 15. Recall that we have

$$\begin{aligned}& e_4(e_\theta(\underline{\kappa})-4\underline{\beta})+\kappa(e_\theta(\underline{\kappa})-4\underline{\beta})\\ =\;& 2e_\theta(\mu)+12\rho\zeta-\frac{1}{2}\underline{\kappa}e_\theta(\kappa)+4\underline{\vartheta}\beta-\frac{1}{2}\vartheta e_\theta(\underline{\kappa})-e_\theta(\vartheta\underline{\vartheta})+2e_\theta(\zeta^2)\\ =\;& 2\Big(e_\theta(\mu)+\vartheta\,\not{d}_2\,\not{d}_2^\star(\not{d}_1^\star\,\not{d}_1)^{-1}\underline{\beta}+2\zeta\check{\rho}\Big)+12\rho\zeta-\frac{1}{2}\underline{\kappa}e_\theta(\kappa)-2\vartheta\,\not{d}_2\,\not{d}_2^\star(\not{d}_1^\star\,\not{d}_1)^{-1}\underline{\beta}\\ & +2\zeta\check{\rho}+4\underline{\vartheta}\beta-\frac{1}{2}\vartheta(e_\theta(\underline{\kappa})-4\underline{\beta})-2\vartheta\underline{\beta}-e_\theta(\vartheta\underline{\vartheta})+2e_\theta(\zeta^2).\end{aligned}$$

In view of Corollary 8.40, we have

$$\begin{aligned}& \max_{k\le J+1}\sup_{r_{\mathcal{T}}\le r_0\le 4m_0}\int_{\{r=r_0\}}\left(\mathfrak{d}^k\left(e_\theta(\underline{\kappa})-4\underline{\beta}\right)\right)^2\\ \lesssim\;& \left(N^{\le 4m_0}[J,\check{\Gamma},\check{R}]\right)^2\\ & +\int_{{}^{(ext)}\mathcal{M}(\le 4m_0)}\left(\mathfrak{d}^{\le J+1}\left(e_\theta(\mu)+\vartheta\,\not{d}_2\,\not{d}_2^\star(\not{d}_1^\star\,\not{d}_1)^{-1}\underline{\beta}+2\zeta\check{\rho}\right)\right)^2.\end{aligned}$$

where we have used

- the fact that $\not{\partial}\vartheta = \not{\partial}\,\not{d}_2^{-1}\not{d}_2\vartheta$ and Codazzi for ϑ to estimate the terms of the RHS with one angular derivative of ϑ,
- the fact that $\not{\partial}\underline{\vartheta} = \not{\partial}\,\not{d}_2^{-1}\not{d}_2\underline{\vartheta}$ and Codazzi for $\underline{\vartheta}$ to estimate the terms of the RHS with one angular derivative of $\underline{\vartheta}$,
- the estimates of Step 1 to estimate the terms of the RHS with one derivative of $\check{\kappa}$,
- the fact that $\not{\partial}\zeta = \not{\partial}\,\not{d}_1^{-1}\not{d}_1\zeta$ and the definition of μ to estimate terms of the RHS with one angular derivative of ζ,
- the estimate for ζ of Step 3.

Together with the estimate of Step 14, we infer

$$\begin{aligned}
&\max_{k\leq J+1}\sup_{r_{\mathcal{T}}\leq r_0\leq 4m_0}\int_{\{r=r_0\}}\left(\mathfrak{d}^k\left(e_\theta(\mu)+\vartheta\,\not{d}_2\,\not{d}_2^\star(\not{d}_1^\star\not{d}_1)^{-1}\underline{\beta}+2\zeta\check{\rho}\right)\right)^2\\
&+\max_{k\leq J+1}\sup_{r_{\mathcal{T}}\leq r_0\leq 4m_0}\int_{\{r=r_0\}}\left(\mathfrak{d}^k\left(e_\theta(\underline{\kappa})-4\underline{\beta}\right)\right)^2\\
\lesssim\;&\left(N^{\leq 4m_0}[J,\check{\Gamma},\check{R}]\right)^2.
\end{aligned}$$

In view of Step 1 to Step 15, of the definition (8.8.4) of $N^{\leq 4m_0}[J,\check{\Gamma},\check{R}]$, and of the various norms, we infer

$$\begin{aligned}
&{}^{(ext)}\mathfrak{G}^{\leq 4m_0}_{J+1}[\check{\Gamma}]+{}^{(ext)}\mathfrak{G}^{\leq 4m_0\,'}_{J+1}[\check{\Gamma}]\\
\lesssim\;&{}^{(ext)}\mathfrak{G}^{\geq 4m_0}_{J+1}[\check{\Gamma}]+{}^{(ext)}\mathfrak{G}^{\geq 4m_0\,'}_{J+1}[\check{\Gamma}]+{}^{(ext)}\mathfrak{R}_{J+1}[\check{R}]+{}^{(ext)}\mathfrak{G}_J[\check{\Gamma}]\\
&+\epsilon_0\left({}^{(ext)}\mathfrak{G}^{\leq 4m_0}_{J+1}[\check{\Gamma}]+{}^{(ext)}\mathfrak{G}^{\leq 4m_0\,'}_{J+1}[\check{\Gamma}]\right)
\end{aligned}$$

and hence, for ϵ_0 small enough,

$$\begin{aligned}
&{}^{(ext)}\mathfrak{G}^{\leq 4m_0}_{J+1}[\check{\Gamma}]+{}^{(ext)}\mathfrak{G}^{\leq 4m_0\,'}_{J+1}[\check{\Gamma}]\\
\lesssim\;&{}^{(ext)}\mathfrak{G}^{\geq 4m_0}_{J+1}[\check{\Gamma}]+{}^{(ext)}\mathfrak{G}^{\geq 4m_0\,'}_{J+1}[\check{\Gamma}]+{}^{(ext)}\mathfrak{R}_{J+1}[\check{R}]+{}^{(ext)}\mathfrak{G}_J[\check{\Gamma}].
\end{aligned}$$

This concludes the proof of Proposition 8.33.

8.9 PROOF OF PROPOSITION 8.12

To prove Proposition 8.12, we rely on the following proposition.

Proposition 8.42. *Let J such that $k_{small}-2\leq J\leq k_{large}-1$. Then, we have*

$$\begin{aligned}
{}^{(int)}\mathfrak{G}_{J+1}[\check{\Gamma}]+{}^{(int)}\mathfrak{G}'_{J+1}[\check{\Gamma}]\;\lesssim\;&{}^{(ext)}\mathfrak{G}_{J+1}[\check{\Gamma}]+{}^{(ext)}\mathfrak{G}'_{J+1}[\check{\Gamma}]+{}^{(int)}\mathfrak{R}_{J+1}[\check{R}]\\
&+\left(\int_{\mathcal{T}}|\mathfrak{d}^{J+1}({}^{(ext)}\check{R})|^2\right)^{\frac{1}{2}},
\end{aligned}$$

where the notation ${}^{(ext)}\mathfrak{G}'_{J+1}[\check{\Gamma}]$ has been introduced in Proposition 8.32, and where we have introduced the notation

$${}^{(int)}\mathfrak{G}'_k[\check{\Gamma}]\;:=\;\int_{{}^{(int)}\mathcal{M}}\left[\left(\mathfrak{d}^k e_\theta(\underline{\kappa})\right)^2+(\mathfrak{d}^{\leq k}\underline{\check{\mu}})^2+\left(\mathfrak{d}^k\left(e_4(\zeta)-\beta\right)\right)^2\right].$$

The proof of Proposition 8.42 is postponed to section 8.9.2. It will rely in particular on basic weighted estimates for transport equations along e_3 in ${}^{(int)}\mathcal{M}$ derived in section 8.9.1.

We now conclude the proof of Proposition 8.12. In view of Proposition 8.42, we have

$$
\begin{aligned}
{}^{(int)}\mathfrak{G}_{J+1}[\check{\Gamma}] \;\lesssim\; & {}^{(ext)}\mathfrak{G}_{J+1}[\check{\Gamma}] + {}^{(ext)}\mathfrak{G}'_{J+1}[\check{\Gamma}] + {}^{(int)}\mathfrak{R}_{J+1}[\check{R}] \\
& + \left(\int_{\mathcal{T}} |\mathfrak{d}^{J+1}({}^{(ext)}\check{R})|^2\right)^{\frac{1}{2}}.
\end{aligned}
$$

Also, we have in view of Proposition 8.31, Proposition 8.32 and the iteration assumption (8.3.13)

$$
{}^{(ext)}\mathfrak{G}_{J+1}[\check{\Gamma}] + {}^{(ext)}\mathfrak{G}'_{J+1}[\check{\Gamma}] \;\lesssim\; {}^{(ext)}\mathfrak{R}_{J+1}[\check{R}] + \epsilon_{\mathcal{B}}[J].
$$

We infer

$$
{}^{(int)}\mathfrak{G}_{J+1}[\check{\Gamma}] \;\lesssim\; {}^{(int)}\mathfrak{R}_{J+1}[\check{R}] + {}^{(ext)}\mathfrak{R}_{J+1}[\check{R}] + \epsilon_{\mathcal{B}}[J] + \left(\int_{\mathcal{T}} |\mathfrak{d}^{J+1}({}^{(ext)}\check{R})|^2\right)^{\frac{1}{2}}.
$$

Together with Proposition 8.11, we deduce

$$
{}^{(int)}\mathfrak{G}_{J+1}[\check{\Gamma}] \;\lesssim\; \epsilon_{\mathcal{B}}[J] + \epsilon_0\left(\mathfrak{N}^{(En)}_{J+1} + \mathcal{N}^{(match)}_{J+1}\right) + \left(\int_{\mathcal{T}} |\mathfrak{d}^{J+1}({}^{(ext)}\check{R})|^2\right)^{\frac{1}{2}}
$$

which concludes the proof of Proposition 8.12.

The rest of this section is dedicated to the proof of Proposition 8.42.

8.9.1 Weighted estimates for transport equations along e_3 in ${}^{(int)}\mathcal{M}$

Lemma 8.43. *Let the following transport equation in* ${}^{(int)}\mathcal{M}$

$$
e_3(f) + \frac{a}{2}\kappa f = h
$$

where $a \in \mathbb{R}$ *is a given constant, and* f *and* h *are scalar functions. Then,* f *satisfies*

$$
\int_{{}^{(int)}\mathcal{M}} f^2 \;\lesssim\; \int_{\mathcal{T}} f^2 + \int_{{}^{(int)}\mathcal{M}} h^2.
$$

Proof. Multiply by f to obtain

$$
\frac{1}{2}e_3(f^2) + \frac{a}{r}f^2 = hf.
$$

Next, integrate over $S_{\underline{u},r}$ to obtain

$$
\begin{aligned}
\frac{1}{2}e_3\left(\int_{S_{\underline{u},r}} f^2\right) &= \int_{S_{\underline{u},r}} \frac{1}{2}(e_3(f^2) + \underline{\kappa} f^2) \\
&= -\int_{S_{\underline{u},r}} \frac{a-1}{2}\underline{\kappa} f^2 + \int_{S_{\underline{u},r}} hf \\
&= -\frac{a-1}{2}\overline{\underline{\kappa}}\int_{S_{\underline{u},r}} f^2 - \frac{a-1}{2}\int_{S_{\underline{u},r}} \check{\underline{\kappa}} f^2 + \int_{S_{\underline{u},r}} hf
\end{aligned}
$$

and hence

$$\frac{1}{2}e_3\left(r^b\int_{S_{\underline{u},r}}f^2\right)+\frac{1}{2}\left(a+1+\frac{b}{2}\right)\overline{\underline{\kappa}}r^b\int_{S_{\underline{u},r}}f^2=-\frac{a-1}{2}r^b\int_{S_{u,r}}\check{\underline{\kappa}}f^2+r^b\int_{S_{\underline{u},r}}hf$$

where we used the fact that $2e_3(r)=r\overline{\underline{\kappa}}$. Also, choosing $b=-2a$, we obtain

$$\frac{1}{2}e_3\left(r^{-2a}\int_{S_{\underline{u},r}}f^2\right)+\frac{1}{2}\overline{\underline{\kappa}}r^{-2a}\int_{S_{\underline{u},r}}f^2=-\frac{a-1}{2}r^{-2a}\int_{S_{u,r}}\check{\underline{\kappa}}f^2+r^{-2a}\int_{S_{\underline{u},r}}hf.$$

Next, let $1\leq\underline{u}\leq u_*$. We now integrate in r and along $\mathcal{C}_{\underline{u}}$ in ${}^{(int)}\mathcal{M}$. Since r is bounded on ${}^{(int)}\mathcal{M}$ from above and below, we obtain, for $\epsilon_0>0$ small enough,

$$\int_{2m_0-2m_0\delta_0}^{r_\mathcal{T}}\int_{S_{\underline{u},r}}f^2 \lesssim \int_{S_{\underline{u},r_\mathcal{T}}}f^2+\int_{2m_0-2m_0\delta_0}^{r_\mathcal{T}}\int_{S_{\underline{u},r}}h^2.$$

We may now integrate in $\underline{u}$ to deduce

$$\int_1^{u_*}\int_{2m_0-2m_0\delta_0}^{r_\mathcal{T}}\int_{S_{\underline{u},r}}f^2\lesssim\int_1^{u_*}\int_{S_{\underline{u},r_\mathcal{T}}}f^2+\int_1^{u_*}\int_{2m_0-2m_0\delta_0}^{r_\mathcal{T}}\int_{S_{\underline{u},r}}h^2. \quad (8.9.1)$$

Remark 8.44. *Note that we have the following consequence of the coarea formula*

$$d\mathcal{T}=\frac{\varsigma\sqrt{\overline{\kappa}+A}}{\sqrt{-\overline{\underline{\kappa}}}}\,d\mu_{\underline{u}r_\mathcal{T}}d\underline{u},$$

where we used that $\mathcal{T}=\{r=r_\mathcal{T}\}$. *Also, we have in* ${}^{(int)}\mathcal{M}$

$$d\mathcal{M}=\frac{4\varsigma^2}{r^2\overline{\underline{\kappa}}^2}d\mu_{\underline{u},r}d\underline{u}dr.$$

We infer, in ${}^{(int)}\mathcal{M}$,

$$d\mathcal{T}=\sqrt{1-\frac{2m_0}{r_\mathcal{T}}(1+O(\epsilon_0))}\;d\mu_{\underline{u},r_0}d\underline{u},$$

and

$$d\mathcal{M}=(1+O(\epsilon_0))d\mu_{\underline{u},r}d\underline{u}dr.$$

Relying on Remark 8.44 we deduce from (8.9.1)

$$\int_{{}^{(int)}\mathcal{M}}f^2 \lesssim \int_\mathcal{T}f^2+\int_{{}^{(int)}\mathcal{M}}h^2$$

as desired. This concludes the proof of the lemma. □

Corollary 8.45. *Let the following transport equation in* ${}^{(int)}\mathcal{M}$

$$e_3(f)+\frac{a}{2}\underline{\kappa}f=h$$

where $a \in \mathbb{R}$ is a given constant, and f and h are scalar functions. Then, f satisfies for $5 \le l \le k_{large} + 1$

$$\int_{^{(int)}\mathcal{M}} (\mathfrak{d}^k f)^2 \lesssim \int_{\mathcal{T}} (\mathfrak{d}^{\le k} f)^2 + \int_{^{(int)}\mathcal{M}} (\mathfrak{d}^{\le k-1} f)^2 + \int_{^{(int)}\mathcal{M}} (\mathfrak{d}^{\le k} h)^2 + \left(\sup_{^{(int)}\mathcal{M}} |\mathfrak{d}^{\le k-5} f|\right)^2 \left({}^{(int)}\mathfrak{G}_{k-1}[\check{\Gamma}] + {}^{(int)}\mathfrak{G}_k[\check{\underline{\kappa}}]\right)^2.$$

Proof. The proof is based on Lemma 8.43. It is similar to the one of Corollary 8.36 and left to the reader. □

8.9.2 Proof of Proposition 8.42

We introduce the following notation which will constantly appear on the RHS of the equalities below:

$$\begin{aligned} N^{(int)}[J, \check{\Gamma}, \check{R}] &:= {}^{(ext)}\mathfrak{G}_{J+1}[\check{\Gamma}] + {}^{(ext)}\mathfrak{G}'_{J+1}[\check{\Gamma}] + {}^{(int)}\mathfrak{R}_{J+1}[\check{R}] \\ &+ \left(\int_{\mathcal{T}} |\mathfrak{d}^{J+1}({}^{(ext)}\check{R})|^2\right)^{\frac{1}{2}} \\ &+ \epsilon_0\left({}^{(int)}\mathfrak{G}_{J+1}[\check{\Gamma}] + {}^{(int)}\mathfrak{G}'_{J+1}[\check{\Gamma}]\right). \end{aligned} \tag{8.9.2}$$

Step 1. In view of Lemma 7.44 relating the Ricci coefficients and curvature components of ${}^{(int)}\mathcal{M}$ to the ones of ${}^{(ext)}\mathcal{M}$ on the timelike hypersurface $\mathcal{T}$, we have

$$\int_{\mathcal{T}} \left|\mathfrak{d}^{J+1}({}^{(int)}\check{\Gamma})\right|^2 \lesssim \int_{\mathcal{T}} |\mathfrak{d}^{J+1}({}^{(ext)}\check{\Gamma})|^2.$$

Also, using again Lemma 7.44, we have

$$\begin{aligned} &\int_{\mathcal{T}} \left|\mathfrak{d}^{J+1}\left({}^{(int)}e_\theta({}^{(int)}\underline{\kappa}), {}^{(int)}\check{\underline{\mu}}, {}^{(int)}e_4({}^{(int)}\zeta - {}^{(int)}\beta)\right)\right|^2 \\ \lesssim &\int_{\mathcal{T}} \left|\mathfrak{d}^{J+1}\left({}^{(ext)}e_\theta({}^{(int)}\underline{\kappa}) - 4{}^{(ext)}\underline{\beta}, {}^{(int)}\check{\mu}, {}^{(int)}e_3({}^{(int)}\zeta) + {}^{(int)}\underline{\beta}\right)\right|^2 \\ &+ \int_{\mathcal{T}} |\mathfrak{d}^{J+1}({}^{(ext)}\check{R})|^2. \end{aligned}$$

We deduce, using that $\mathcal{T} = \{r = r_{\mathcal{T}}\}$ and the definitions of the various norms on ${}^{(ext)}\mathcal{M}$,

$$\begin{aligned} &\int_{\mathcal{T}} \left|\mathfrak{d}^{J+1}({}^{(int)}\check{\Gamma})\right|^2 + \int_{\mathcal{T}} \left|\mathfrak{d}^{J+1}\left({}^{(int)}e_\theta({}^{(int)}\underline{\kappa}), {}^{(int)}\check{\underline{\mu}}, {}^{(int)}e_4({}^{(int)}\zeta - {}^{(int)}\beta)\right)\right|^2 \\ \lesssim\ &{}^{(ext)}\mathfrak{G}_{J+1}[\check{\Gamma}] + {}^{(ext)}\mathfrak{G}'_{J+1}[\check{\Gamma}] + \left(\int_{\mathcal{T}} |\mathfrak{d}^{J+1}({}^{(ext)}\check{R})|^2\right)^{\frac{1}{2}} \end{aligned}$$

and hence, in view of (8.9.2),

$$\int_{\mathcal{T}}\left|\mathfrak{d}^{J+1}({}^{(int)}\check{\Gamma})\right|^2+\int_{\mathcal{T}}\left|\mathfrak{d}^{J+1}\Big({}^{(int)}e_\theta({}^{(int)}\underline{\kappa}),\,{}^{(int)}\check{\underline{\mu}},\,{}^{(int)}e_4({}^{(int)}\zeta-{}^{(int)}\beta)\Big)\right|^2$$
$$\lesssim\ \left(N^{(int)}[J,\check{\Gamma},\check{R}]\right)^2.$$

From now on, we only consider the frame of ${}^{(int)}\mathcal{M}$. The previous estimate can be written as

$$\max_{k\le J+1}\left(\int_{\mathcal{T}}\Big((\mathfrak{d}^k\check{\underline{\mu}})^2+(\mathfrak{d}^k\zeta)^2+(\mathfrak{d}^k\check{\kappa})^2+(\mathfrak{d}^k\vartheta)^2+(\mathfrak{d}^k\check{\underline{\kappa}})^2+(\mathfrak{d}^k\underline{\vartheta})^2\right.$$
$$\left.+(\mathfrak{d}^k(e_4(\zeta)-\beta))^2+(\mathfrak{d}^k\xi)^2+(\mathfrak{d}^k\check{\omega})^2+(\mathfrak{d}^k\check{\Omega})^2\Big)\right)$$
$$\lesssim\ \left(N^{(int)}[J,\check{\Gamma},\check{R}]\right)^2$$

and

$$\max_{k\le J+1}\int_{\mathcal{T}}(\mathfrak{d}^k e_\theta(\underline{\kappa}))^2\ \lesssim\ \left(N^{(int)}[J,\check{\Gamma},\check{R}]\right)^2.$$

Step 2. We have obtained all the desired estimates on $\mathcal{T}$ for the foliation of ${}^{(int)}\mathcal{M}$ in Step 1. We now derive the desired estimates on ${}^{(int)}\mathcal{M}$. To this end, we rely on the transport equations in the e_3 directions which we estimate thanks to Corollary 8.45. The initial data on $\mathcal{T}$ is estimated thanks to Step 1. In particular, we proceed in the following order:

- From

$$e_3(\check{\underline{\kappa}})+\overline{\underline{\kappa}}\,\check{\underline{\kappa}}\ =\ \mathrm{Err}[e_3\check{\underline{\kappa}}]$$

  and the bootstrap assumptions, we infer

$$\max_{k\le J+1}\int_{{}^{(int)}\mathcal{M}}(\mathfrak{d}^k\check{\underline{\kappa}})^2\ \lesssim\ \left(N^{(int)}[J,\check{\Gamma},\check{R}]\right)^2.$$

- From

$$e_3(e_\theta(\underline{\kappa}))+\frac{3}{2}\underline{\kappa}e_\theta(\underline{\kappa})\ =\ -\frac{1}{2}\underline{\vartheta}e_\theta(\underline{\kappa})-\frac{1}{2}e_\theta(\underline{\vartheta}^2)$$

  and the bootstrap assumptions, we infer

$$\max_{k\le J+1}\int_{{}^{(int)}\mathcal{M}}(\mathfrak{d}^k e_\theta(\underline{\kappa}))^2\ \lesssim\ \left(N^{(int)}[J,\check{\Gamma},\check{R}]\right)^2.$$

- From

$$e_3(\check{\underline{\mu}})+\frac{3}{2}\overline{\underline{\kappa}}\,\check{\underline{\mu}}\ =\ -\frac{3}{2}\overline{\underline{\mu}}\,\check{\underline{\kappa}}+\mathrm{Err}[e_3\check{\underline{\mu}}],$$

  the above control of $\check{\underline{\kappa}}$ and $e_\theta(\underline{\kappa})$ (the control of $e_\theta(\underline{\kappa})$ is needed to estimate

$\mathrm{Err}[e_3\underline{\check{\mu}}]$), and the bootstrap assumptions, we infer

$$\max_{k\leq J+1}\int_{{}^{(int)}\mathcal{M}}(\mathfrak{d}^k\underline{\check{\mu}})^2 \lesssim \left(N^{(int)}[J,\check{\Gamma},\check{R}]\right)^2.$$

- From

$$e_3(\underline{\vartheta})+\underline{\kappa}\,\underline{\vartheta} = -2\underline{\alpha}$$

and the control of $\underline{\alpha}$, we infer

$$\max_{k\leq J+1}\int_{{}^{(int)}\mathcal{M}}(\mathfrak{d}^k\underline{\vartheta})^2 \lesssim \left(N^{(int)}[J,\check{\Gamma},\check{R}]\right)^2.$$

- From

$$e_3(\zeta)+\underline{\kappa}\zeta = \underline{\beta}-\underline{\vartheta}\zeta$$

the control of $\underline{\beta}$, and the bootstrap assumptions, we infer

$$\max_{k\leq J+1}\int_{{}^{(int)}\mathcal{M}}(\mathfrak{d}^k\zeta)^2 \lesssim \left(N^{(int)}[J,\check{\Gamma},\check{R}]\right)^2.$$

- From

$$\begin{aligned}
e_3(\check{\kappa})+\frac{1}{2}\overline{\underline{\kappa}}\check{\kappa} &= -\frac{1}{2}\overline{\kappa}\underline{\check{\kappa}}+2\,\not{d}_1\zeta+2\check{\rho}+\mathrm{Err}[e_3\check{\kappa}]\\
&= -\frac{1}{2}\overline{\kappa}\underline{\check{\kappa}}+2\underline{\check{\mu}}+4\check{\rho}-\frac{1}{2}\vartheta\underline{\vartheta}+\frac{1}{2}\overline{\vartheta\underline{\vartheta}}+\mathrm{Err}[e_3\check{\kappa}],
\end{aligned}$$

the control of $\check{\rho}$, the above control of $\underline{\check{\kappa}}$ and $\underline{\check{\mu}}$, and the bootstrap assumptions, we infer

$$\max_{k\leq J+1}\int_{{}^{(int)}\mathcal{M}}(\mathfrak{d}^k\check{\kappa})^2 \lesssim \left(N^{(int)}[J,\check{\Gamma},\check{R}]\right)^2.$$

- From

$$\begin{aligned}
e_3(\vartheta)+\frac{1}{2}\underline{\kappa}\vartheta &= 2\,\not{d}_2^\star\zeta-\frac{1}{2}\kappa\underline{\vartheta}+2\zeta^2\\
&= 2\,\not{d}_2^\star\,\not{d}_1^{-1}\left(\underline{\check{\mu}}+\check{\rho}-\frac{1}{4}\vartheta\underline{\vartheta}+\frac{1}{4}\overline{\vartheta\underline{\vartheta}}\right)-\frac{1}{2}\kappa\underline{\vartheta}+2\zeta^2,
\end{aligned}$$

the control of $\check{\rho}$, the above control of $\underline{\vartheta}$ and $\underline{\check{\mu}}$, and the bootstrap assumptions, we infer

$$\max_{k\leq J+1}\int_{{}^{(int)}\mathcal{M}}(\mathfrak{d}^k\vartheta)^2 \lesssim \left(N^{(int)}[J,\check{\Gamma},\check{R}]\right)^2.$$

- From

$$e_3(\check{\omega}) = \check{\rho}+\mathrm{Err}[e_3\check{\omega}],$$

the control of $\check{\rho}$, and the bootstrap assumptions, we infer

$$\max_{k\leq J+1}\int_{^{(int)}\mathcal{M}}(\mathfrak{d}^k\check{\omega})^2 \lesssim \left(N^{(int)}[J,\check{\Gamma},\check{R}]\right)^2.$$

- From

$$\begin{aligned}
& e_3(e_4(\zeta)-\beta)+\underline{\kappa}(e_4(\zeta)-\beta)\\
= & -\kappa\underline{\beta}+\left(\frac{1}{2}\kappa\underline{\kappa}+2\underline{\mu}-6\rho\right)\zeta\\
& +\vartheta\underline{\beta}-\underline{\vartheta}\beta+\xi\underline{\alpha}-\underline{\vartheta}\zeta+2\omega\underline{\vartheta}\zeta-4\zeta\,\not{d}_1^\star\,\not{d}_1^{-1}\left(\check{\underline{\mu}}+\check{\rho}-\frac{1}{4}\vartheta\underline{\vartheta}+\frac{1}{4}\overline{\vartheta\underline{\vartheta}}\right)-2\zeta^3,
\end{aligned}$$

the control of $\underline{\beta}$, the above control of ζ, and the bootstrap assumptions, we infer

$$\max_{k\leq J+1}\int_{^{(int)}\mathcal{M}}(\mathfrak{d}^k(e_4(\zeta)-\beta))^2 \lesssim \left(N^{(int)}[J,\check{\Gamma},\check{R}]\right)^2.$$

- From

$$e_3(\xi) = (e_4(\zeta)-\beta)+2\beta+\kappa\zeta+\vartheta\zeta,$$

the control of $\underline{\beta}$, the above control of $e_4(\zeta)-\beta$ and ζ, and the bootstrap assumptions, we infer

$$\max_{k\leq J+1}\int_{^{(int)}\mathcal{M}}(\mathfrak{d}^k\xi)^2 \lesssim \left(N^{(int)}[J,\check{\Gamma},\check{R}]\right)^2.$$

In view of the above estimates, of the definition (8.9.2) of $N^{\leq 4m_0}[J,\check{\Gamma},\check{R}]$, and of the various norms, we infer

$$\begin{aligned}
{}^{(int)}\mathfrak{G}_{J+1}[\check{\Gamma}]+{}^{(int)}\mathfrak{G}'_{J+1}[\check{\Gamma}] \lesssim\ & {}^{(ext)}\mathfrak{G}_{J+1}[\check{\Gamma}]+{}^{(ext)}\mathfrak{G}'_{J+1}[\check{\Gamma}]+{}^{(int)}\mathfrak{R}_{J+1}[\check{R}]\\
& +\left(\int_{\mathcal{T}}|\mathfrak{d}^{J+1}({}^{(ext)}\check{R})|^2\right)^{\frac{1}{2}}\\
& +\epsilon_0\left({}^{(int)}\mathfrak{G}_{J+1}[\check{\Gamma}]+{}^{(int)}\mathfrak{G}'_{J+1}[\check{\Gamma}]\right)
\end{aligned}$$

and hence, for ϵ_0 small enough,

$$\begin{aligned}
{}^{(int)}\mathfrak{G}_{J+1}[\check{\Gamma}]+{}^{(int)}\mathfrak{G}'_{J+1}[\check{\Gamma}] \lesssim\ & {}^{(ext)}\mathfrak{G}_{J+1}[\check{\Gamma}]+{}^{(ext)}\mathfrak{G}'_{J+1}[\check{\Gamma}]+{}^{(int)}\mathfrak{R}_{J+1}[\check{R}]\\
& +\left(\int_{\mathcal{T}}|\mathfrak{d}^{J+1}({}^{(ext)}\check{R})|^2\right)^{\frac{1}{2}}.
\end{aligned}$$

This concludes the proof of Proposition 8.42.

8.10 PROOF OF PROPOSITION 8.13

Lemma 4.16 corresponds to the particular case $J = k_{large}-1$ of Proposition 8.13. Its proof in section 4.6.2 extends immediately to the case $k_{small}-2\leq J\leq k_{large}-1$ which thus yields the proof of Proposition 8.13.

Chapter Nine

GCM Procedure

9.1 PRELIMINARIES

We consider an axially symmetric polarized spacetime region $\mathcal{R}$ foliated by two functions (u, s) such that:

- On $\mathcal{R}$, (u, s) defines an outgoing geodesic foliation as in section 2.2.4.
- We denote by (e_3, e_4, e_θ) the null frame adapted to the outgoing geodesic foliation (u, s) on $\mathcal{R}$.
- Let

$$\overset{\circ}{S} \quad := \quad S(\overset{\circ}{u}, \overset{\circ}{s}) \tag{9.1.1}$$

  and $\overset{\circ}{r}$ the area radius of $\overset{\circ}{S}$, where $S(u, s)$ denote the 2-spheres of the outgoing geodesic foliation (u, s) on $\mathcal{R}$.
- In adapted coordinates (u, s, θ, φ) with $b = 0$, see Proposition 2.75, the spacetime metric $\mathbf{g}$ in $\mathcal{R}$ takes the form, with $\underline{\Omega} = e_3(s)$, $\underline{b} = e_3(\theta)$,

$$\mathbf{g} = -2\varsigma du ds + \varsigma^2 \underline{\Omega} du^2 + \gamma \left(d\theta - \frac{1}{2}\varsigma \underline{b} du \right)^2 + e^{2\Phi} d\varphi^2, \tag{9.1.2}$$

  where θ is chosen such that $b = e_4(\theta) = 0$.
- The spacetime metric induced on $S(u, s)$ is given by

$$\not{g} = \gamma d\theta^2 + e^{2\Phi} d\varphi^2. \tag{9.1.3}$$

- The relation between the null frame and coordinate system is given by

$$e_4 = \partial_s, \qquad e_3 = \frac{2}{\varsigma}\partial_u + \underline{\Omega}\partial_s + \underline{b}\partial_\theta, \qquad e_\theta = \gamma^{-1/2}\partial_\theta. \tag{9.1.4}$$

- We denote the induced metric on $\overset{\circ}{S}$ by

$$\overset{\circ}{\not{g}} = \overset{\circ}{\gamma} d\theta^2 + e^{2\Phi} d\varphi^2.$$

Definition 9.1. *Let $0 < \overset{\circ}{\delta} \le \overset{\circ}{\epsilon}$ two sufficiently small constants. Let $(\overset{\circ}{u}, \overset{\circ}{s})$ real numbers so that*

$$1 \le \overset{\circ}{u} < +\infty, \quad 4m_0 \le \overset{\circ}{s} < +\infty. \tag{9.1.5}$$

We define $\mathcal{R} = \mathcal{R}(\overset{\circ}{\delta}, \overset{\circ}{\epsilon})$ to be the region

$$\mathcal{R} := \left\{ |u - \overset{\circ}{u}| \le \delta_{\mathcal{R}}, \quad |s - \overset{\circ}{s}| \le \delta_{\mathcal{R}} \right\}, \qquad \delta_{\mathcal{R}} := \overset{\circ}{\delta} (\overset{\circ}{\epsilon})^{-\frac{1}{2}}, \tag{9.1.6}$$

*such that assumptions **A1**–**A3** below with constant $\overset{\circ}{\epsilon}$ on the background foliation of $\mathcal{R}$ are verified. The smaller constant $\overset{\circ}{\delta}$ controls the size of the GCMS quantities as it will be made precise below.*

In this section we define the renormalized Ricci and curvature components,

$$\check{\Gamma} := \left\{ \check{\kappa},\, \vartheta,\, \zeta,\, \eta,\, \overline{\kappa} - \frac{2}{r},\, \overline{\underline{\kappa}} + \frac{2\Upsilon}{r},\, \check{\underline{\kappa}},\, \underline{\vartheta},\, \underline{\xi},\, \check{\underline{\omega}},\, \overline{\underline{\omega}} - \frac{m}{r^2},\, \check{\underline{\Omega}},\, (\overline{\underline{\Omega}} + \Upsilon),\, (\overline{\varsigma} + 1) \right\},$$

$$\check{R} := \left\{ \alpha,\, \beta,\, \check{\rho},\, \overline{\rho} + \frac{2m}{r^3},\, \underline{\beta},\, \underline{\alpha} \right\}.$$

Since our foliation is outgoing geodesic we also have

$$\xi = \omega = 0, \quad \underline{\eta} + \zeta = 0. \tag{9.1.7}$$

We decompose $\check{\Gamma} = \Gamma_g \cup \Gamma_b$ where

$$\begin{aligned} \Gamma_g &= \left\{ \check{\kappa},\, \vartheta,\, \zeta,\, \check{\underline{\kappa}},\, \overline{\kappa} - \frac{2}{r},\, \overline{\underline{\kappa}} + \frac{2\Upsilon}{r} \right\}, \\ \Gamma_b &= \left\{ \eta,\, \underline{\vartheta},\, \underline{\xi},\, \check{\underline{\omega}},\, \overline{\underline{\omega}} - \frac{m}{r^2},\, r^{-1}\check{\underline{\Omega}},\, r^{-1}\check{\varsigma}, r^{-1}(\overline{\underline{\Omega}} + \Upsilon),\, r^{-1}(\overline{\varsigma} - 1) \right\}. \end{aligned} \tag{9.1.8}$$

Given a p-reduced scalar $f \in \mathfrak{s}_p(\mathcal{M})$, with respect to the given geodesic foliation on $\mathcal{R}$, we consider the following norms on spheres $S = S(u,r) \subset \mathcal{R}$,

$$\begin{aligned} \|f\|_\infty(u,r) &:= \|f\|_{L^\infty(S(u,r))}, & \|f\|_2(u,r) &:= \|f\|_{L^2(S(u,r))}, \\ \|f\|_{\infty,k}(u,r) &= \sum_{i=0}^k \|\mathfrak{d}^i f\|_\infty(u,r), & \|f\|_{2,k}(u,r) &= \sum_{i=0}^k \|\mathfrak{d}^i f\|_2(u,r), \end{aligned} \tag{9.1.9}$$

where, we recall, that $\mathfrak{d}^i$ stands for any combination of length i of operators of the form $e_3, re_4, \not{\mathfrak{d}}$. Recall that

$$\not{\mathfrak{d}}^s f = \begin{cases} r^{2p}\not{\Delta}_k^p, & \text{if} \quad s = 2p, \\ r^{2p+1}\not{d}_k\not{\Delta}_k^p, & \text{if} \quad s = 2p+1. \end{cases} \tag{9.1.10}$$

On a given polarized surface $\mathbf{S} \subset \mathcal{R}$, not necessarily a leaf S of the given foliation, we define

$$\|f\|_{\mathfrak{h}_s^q(\mathbf{S})} : \; = \; \sum_{i=0}^s \| \left(\not{\mathfrak{d}}^{\mathbf{S}}\right)^i f\|_{L^q(\mathbf{S})} \tag{9.1.11}$$

where $\not{\mathfrak{d}}^{\mathbf{S}}$ is defined as above with respect to the intrinsic metric on $\mathbf{S}$. In the particular case when $q = 2$ we omit the upper index, i.e., $\mathfrak{h}_s(\mathbf{S}) = \mathfrak{h}_s^2(\mathbf{S})$.

9.1.1 Main assumptions

Given an integer s_{max}, we assume the following:[1]
A1. For all $k \leq s_{max}$, we have on $\mathcal{R}$

$$\begin{aligned} \|\Gamma_g\|_{k,\infty} &\lesssim \mathring{\epsilon} r^{-2}, \\ \|\Gamma_b\|_{k,\infty} &\lesssim \mathring{\epsilon} r^{-1}, \end{aligned} \tag{9.1.12}$$

and

$$\begin{aligned} \|\alpha, \beta, \check{\rho}, \check{\mu}\|_{k,\infty} &\lesssim \mathring{\epsilon} r^{-3}, \\ \|e_3(\alpha, \beta)\|_{k-1,\infty} &\lesssim \mathring{\epsilon} r^{-4}, \\ \|\underline{\beta}\|_{k,\infty} &\lesssim \mathring{\epsilon} r^{-2}, \\ \|\underline{\alpha}\|_{k,\infty} &\lesssim \mathring{\epsilon} r^{-1}. \end{aligned} \tag{9.1.13}$$

A2. We have, with m_0 denoting the mass of the unperturbed spacetime,

$$\sup_{\mathcal{R}} \left| \frac{m}{m_0} - 1 \right| \lesssim \mathring{\epsilon}. \tag{9.1.14}$$

A3. The metric coefficients are assumed to satisfy the following assumptions in $\mathcal{R}$, for all $k \leq s_{max}$,

$$r \left\| \left(\frac{\gamma}{r^2} - 1, \, \underline{b}, \, \frac{e^{\Phi}}{r \sin\theta} - 1 \right) \right\|_{\infty,k} + \|\underline{\Omega} + \Upsilon\|_{\infty,k} + \|\varsigma - 1\|_{\infty,k} \lesssim \mathring{\epsilon}. \tag{9.1.15}$$

Remark 9.2. *The above assumptions imply in particular the following:*

$$|e_4(r)|, \, |e_3(r)| \lesssim 1, \qquad e_4(s) = 1 + O(\mathring{\epsilon}), \qquad e_3(u) = 2 + O(\mathring{\epsilon}), \qquad e_4(u) = 0.$$

Hence, since $r = \mathring{r}$ at $(\mathring{u}, \mathring{s})$, we infer

$$|r - \mathring{r}| \lesssim |s - \mathring{s}| + |u - \mathring{u}|,$$

and thus, in view of the definition (9.1.6) *of $\mathcal{R}$,*

$$\sup_{\mathcal{R}} |r - \mathring{r}| \lesssim \mathring{\delta} (\mathring{\epsilon})^{-\frac{1}{2}}. \tag{9.1.16}$$

We will make use of the following lemma, see Lemmas 4.9 and 4.10.

Lemma 9.3. *Under the assumption **A3** for the metric coefficients we have*

$$r|e_\theta(\Phi)| \leq \frac{2}{sin\theta}, \qquad \frac{1}{\sin\theta} \leq 2\left(r|e_\theta \Phi| + 1\right). \tag{9.1.17}$$

[1]In applications, $s_{max} = k_{small} + 4$ in Theorem M7, and $s_{max} = k_{large} + 5$ in Theorem M0 and Theorem M6.

Moreover, for any reduced 1-scalar h, we have

$$\sup_S \frac{|h|}{e^\Phi} \lesssim r^{-1} \sup_S(|h| + |\not{\partial} h|), \qquad \|\tfrac{h}{e^\Phi}\|_{L^2(S)} \lesssim r^{-1} \|h\|_{\mathfrak{h}_1(S)}. \quad (9.1.18)$$

9.1.2 Elliptic Hodge lemma

We shall often make use of the results of Proposition 2.33 and Lemma 2.38 which we rewrite as follows.

Lemma 9.4. *Under the assumptions **A1, A3** the following elliptic estimates hold true for the Hodge operators $\not{d}_1$, $\not{d}_2$, $\not{d}_1^\star$, $\not{d}_2^\star$, for all $k \leq s_{max}$:*

1. *If $f \in \mathfrak{s}_1(S)$*

$$\|\not{\partial} f\|_{\mathfrak{h}_k(S)} + \|f\|_{\mathfrak{h}_k(S)} \lesssim r \|\not{d}_1 f\|_{\mathfrak{h}_k(S)}.$$

2. *If $f \in \mathfrak{s}_2(S)$*

$$\|\not{\partial} f\|_{\mathfrak{h}_k(S)} + \|f\|_{\mathfrak{h}_k(S)} \lesssim r \|\not{d}_2 f\|_{\mathfrak{h}_k(S)}.$$

3. *If $f \in \mathfrak{s}_0(S)$*

$$\|\not{\partial} f\|_{\mathfrak{h}_k(S)} \lesssim r \|\not{d}_1^\star f\|_{\mathfrak{h}_k(S)}.$$

4. *If $f \in \mathfrak{s}_1(S)$*

$$\|f\|_{\mathfrak{h}_{k+1}(S)} \lesssim r \|\not{d}_2^\star f\|_{\mathfrak{h}_k(S)} + r^{-2} \left| \int_S e^\Phi f \right|.$$

5. *If $f \in \mathfrak{s}_1(S)$*

$$\left\| f - \frac{\int_S f e^\Phi}{\int_S e^{2\Phi}} e^\Phi \right\|_{\mathfrak{h}_{k+1}(S)} \lesssim r \|\not{d}_2^\star f\|_{\mathfrak{h}_k(S)}.$$

We shall often make use of the following non-sharp product estimate on $\mathbf{S}$, see Proposition 2.43.

Lemma 9.5. *The following estimates hold true on a given polarized surface $\mathbf{S} \subset \mathcal{R}$, for any contraction between two reduced scalars ψ_1, ψ_2, $k \geq 2$,*

$$\|\psi_1 \cdot \psi_2\|_{\mathfrak{h}_k(\mathbf{S})} \lesssim r^{-1} \|\psi_1\|_{\mathfrak{h}_k(\mathbf{S})} \|\psi_2\|_{\mathfrak{h}_k(\mathbf{S})}.$$

9.2 DEFORMATIONS OF S SURFACES

9.2.1 Deformations

Recall that $\overset{\circ}{S} = S(\overset{\circ}{u}, \overset{\circ}{s})$ is a fixed sphere of the (u, s) outgoing geodesic foliation of a fixed spacetime region $\mathcal{R} = \mathcal{R}(\overset{\circ}{\epsilon}, \overset{\circ}{\delta})$.

Definition 9.6. *We say that* $\mathbf{S}$ *is an* $O(\overset{\circ}{\epsilon})$ $\mathbf{Z}$*-polarized deformation of* $\overset{\circ}{S}$ *if there* *exists a map* $\Psi : \overset{\circ}{S} \to \mathbf{S}$ *of the form*

$$\Psi(\overset{\circ}{u}, \overset{\circ}{s}, \theta, \varphi) = \left(\overset{\circ}{u} + U(\theta), \overset{\circ}{s} + S(\theta), \theta, \varphi\right) \tag{9.2.1}$$

where U, S *are functions defined on the interval* $[0, \pi]$ *of amplitude at most* $\overset{\circ}{\epsilon}$, *leading to a smooth surface* $\mathbf{S}$. *We denote by* ψ *the reduced map defined on the interval* $[0, \pi]$,

$$\psi(\theta) = (\overset{\circ}{u} + U(\theta), \overset{\circ}{s} + S(\theta), \theta). \tag{9.2.2}$$

We restrict ourselves to deformations which fix the south pole, i.e.,

$$U(0) = S(0) = 0. \tag{9.2.3}$$

9.2.2 Pullback map

We recall that given a scalar function f on $\mathbf{S}$ one defines its pullback on $\overset{\circ}{S}$ to be the function

$$f^\# := \Psi^\# f = f \circ \Psi.$$

On the other hand, given a vectorfield X on $\overset{\circ}{S}$ one defines its push-forward $\Psi_\# X$ to be the vectorfield on $\mathbf{S}$ defined by

$$\Psi_\# X(f) = X(\Psi^\# f) = X(f \circ \Psi).$$

Given a covariant tensor U on $\mathbf{S}$, one defines its pullback to $\overset{\circ}{S}$ to be the tensor

$$\Psi^\# U(X_1, \ldots, X_k) = U(\Psi_\# X_1, \ldots, \Psi_\# X_k).$$

Lemma 9.7. *Given a* $\mathbf{Z}$*-invariant deformation* $\Psi : \overset{\circ}{S} \to \mathbf{S}$, *we have:*

1. *Let* $\not{g}^{\mathbf{S}}$ *the induced metric on* $\mathbf{S}$ *and* $\not{g}^{\mathbf{S},\#} = \gamma^{\mathbf{S},\#} d\theta^2 + e^{2\Phi^\#} d\varphi^2$ *its pullback to* $\overset{\circ}{S}$. *The metric coefficients* $\gamma^{\mathbf{S}}$ *and* $\gamma^{\mathbf{S},\#}$ *are related by*

$$\gamma^{\mathbf{S},\#}(\theta) = \gamma^{\mathbf{S}}(\psi(\theta)) = \gamma^{\mathbf{S}}(\overset{\circ}{u} + U(\theta), \overset{\circ}{s} + S(\theta), \theta) \tag{9.2.4}$$

where $\gamma^{\mathbf{S}}$ *is defined implicitly by*

$$(\gamma^{\mathbf{S}})^\# = \gamma^\# + (\varsigma^\#)^2 \left(\underline{\Omega} + \frac{1}{4}\underline{b}^2\gamma\right)^\# (U')^2 - 2\varsigma^\# U'S' - (\gamma\varsigma\underline{b})^\# U', \tag{9.2.5}$$

that is,

$$\begin{aligned} \gamma^{\mathbf{S}}(\psi(\theta)) &= \gamma(\psi(\theta)) + \varsigma^2(\psi(\theta)) \left(\underline{\Omega}(\psi(\theta)) + \frac{1}{4}(\underline{b}(\psi(\theta)))^2\gamma(\psi(\theta))\right) (U'(\theta))^2 \\ &- 2\varsigma(\psi(\theta))U'(\theta)S'(\theta) - \gamma(\psi(\theta))\varsigma(\psi(\theta))\underline{b}(\psi(\theta))U'(\theta). \end{aligned}$$

2. *The* **Z**-*invariant vectorfield* $\partial_\theta^{\mathbf{S}} := \Psi_\#(\partial_\theta)$ *is tangent to* **S** *and*

$$\partial_\theta^{\mathbf{S}}|_{\Psi(p)} = \Big[\Big(\partial_\theta S - \frac{\varsigma}{2}\underline{\Omega}\partial_\theta U\Big)e_4 + \frac{\varsigma}{2}\partial_\theta U e_3 + \sqrt{\gamma}\Big(1 - \frac{\varsigma}{2}\underline{b}\partial_\theta U\Big)e_\theta\Big]\Big|_{\Psi(p)}. \quad (9.2.6)$$

3. *If* $f \in \mathfrak{s}_k(\mathbf{S})$ *and* $P^{\mathbf{S}}$ *is a geometric operator acting on* f, *then*

$$(P^{\mathbf{S}}[f])^\# = P^{\mathbf{S},\#}[f^\#] \quad (9.2.7)$$

where $P^{\mathbf{S},\#}$ *is the corresponding geometric operator on* $\overset{\circ}{S}$ *with respect to the metric* $\not{g}^{\mathbf{S},\#}$ *and* $f^\# = \psi^\# f$.

4. *The* L^2 *norm of* $f^\# = \psi^\# f$ *with respect to the metric* $\not{g}^{\mathbf{S},\#}$ *is the same as the* L^2 *norm of* f *with respect to the metric* $\not{g}^{\mathbf{S}}$, *i.e.,*

$$\int_{\overset{\circ}{S}} |f^\#|^2 da_{\not{g}^{\mathbf{S},\#}} = \int_{\mathbf{S}} |f|^2 da_{\not{g}^{\mathbf{S}}}.$$

5. *If* $f \in \mathfrak{h}_k(\mathbf{S})$ *and* $f^\#$ *is its pullback by* ψ *then*

$$\|f^\#\|_{\mathfrak{h}_k(\overset{\circ}{S}, \not{g}^{\mathbf{S},\#})} = \|f\|_{\mathfrak{h}_k(\mathbf{S})}.$$

Proof. If ∂_θ denotes the coordinate derivative $\partial_\theta = \frac{\partial}{\partial\theta}$ then, at every point $p \in \overset{\circ}{S}$,

$$\Psi_\#(\partial_\theta)|_{\Psi(p)} = \partial_\theta U \partial_u|_{\Psi(p)} + \partial_\theta S \partial_s|_{\Psi(p)} + \partial_\theta|_{\Psi(p)}, \qquad \Psi_\#(\partial_\varphi) = \partial_\varphi.$$

In view of (9.1.4) we have

$$\partial_s = e_4, \quad \partial_u = \frac{\varsigma}{2}\Big(e_3 - \underline{\Omega} e_4 - \underline{b}\gamma^{1/2} e_\theta\Big), \quad \partial_\theta = \sqrt{\gamma} e_\theta.$$

Hence, at a point $\Psi(p)$ on **S** we have

$$\Psi_\#(\partial_\theta) = \Big(\partial_\theta S - \frac{\varsigma}{2}\underline{\Omega}\partial_\theta U\Big)e_4 + \frac{\varsigma}{2}\partial_\theta U e_3 + \sqrt{\gamma}\Big(1 - \frac{\varsigma}{2}\underline{b}\partial_\theta U\Big)e_\theta.$$

We denote by $\not{g}^\# = \Psi^\#(\not{g}^{\mathbf{S}})$ the pullback to $\overset{\circ}{S}$ of the metric $\not{g}^{\mathbf{S}}$ on **S**, i.e., at any point $p \in \overset{\circ}{S}$,

$$\begin{aligned}
\not{g}^\#(\partial_\theta, \partial_\theta) &= \not{g}^{\mathbf{S}}(\Psi_\#\partial_\theta, \Psi_\#\partial_\theta) = \mathbf{g}(\partial_\theta U\partial_u + \partial_\theta S\partial_s + \partial_\theta, \partial_\theta U\partial_u + \partial_\theta S\partial_s + \partial_\theta) \\
&= (\partial_\theta U)^2 \mathbf{g}_{uu} + 2\partial_\theta U \partial_\theta S \mathbf{g}_{us} + 2\partial_\theta U \mathbf{g}_{u\theta} + \mathbf{g}_{\theta\theta}, \\
\not{g}^\#(\partial_\theta, \partial_\varphi) &= 0, \\
\not{g}^\#(\partial_\varphi, \partial_\varphi) &= e^{2\Phi^\#},
\end{aligned}$$

where

$$\mathbf{g}_{uu} = \varsigma^2\big(\underline{\Omega} + \frac{1}{4}\gamma\underline{b}^2\big), \quad \mathbf{g}_{us} = -\varsigma, \quad \mathbf{g}_{u\theta} = -\frac{\varsigma}{2}\gamma\underline{b}, \quad \mathbf{g}_{ss} = \mathbf{g}_{s\theta} = 0, \quad \mathbf{g}_{\theta\theta} = \gamma.$$

Hence the pullback metric $\Psi^\#(\not{g}^{\mathbf{S}})$ on $\overset{\circ}{S}$ is given by

$$\gamma^{\mathbf{S},\#} d\theta^2 + e^{2\Phi^\#} d\varphi^2$$

where

$$\gamma^{\mathbf{S},\#} = (\gamma^{\mathbf{S}})^{\#}, \tag{9.2.8}$$

with $\gamma^{\mathbf{S}}$ defined by

$$(\gamma^{\mathbf{S}})^{\#} = \gamma^{\#} + (\varsigma^{\#})^2 \left(\underline{\Omega} + \frac{1}{4}\underline{b}^2\gamma\right)^{\#} (U')^2 - 2\varsigma^{\#}U'S' - (\gamma\varsigma\underline{b})^{\#}U'. \tag{9.2.9}$$

Note that the vectorfield

$$e_\theta^{\mathbf{S}} := \frac{1}{(\gamma^{\mathbf{S}})^{1/2}}\Psi_{\#}(\partial_\theta)$$

is tangent, $\mathbf{Z}$-invariant and forms together with e_φ an orthonormal frame on $\mathbf{S}$. Note that we can also write

$$e_\theta^{\mathbf{S}} := \frac{(\overset{\circ}{\gamma})^{1/2}}{(\gamma^{\mathbf{S}})^{1/2}}\Psi_{\#}(e_\theta)$$

where $\overset{\circ}{\gamma}$ is the coefficient in front of $d\theta^2$ of the metric induced by $\mathbf{g}$ on $\overset{\circ}{S}$,

$$\overset{\circ}{\not{g}} = \overset{\circ}{\gamma}\, d\theta^2 + e^{2\Phi}d\varphi^2.$$

In general, any geometric calculation on $\mathbf{S}$ can be reduced to a geometric calculation on $\overset{\circ}{S}$ with respect to the metric $\not{g}^{\mathbf{S},\#}$. Moreover the L^2 norm on $\mathbf{S}$ with respect to the metric $\not{g}^{\mathbf{S}}$ is the same as the L^2 norm of $f^{\#} = \psi^{\#}f$ with respect to the norm $\not{g}^{\mathbf{S},\#}$. This concludes the proof of the lemma. □

9.2.3 Comparison of norms between deformations

Lemma 9.8. *Let* $\Psi : \overset{\circ}{S} \to \mathbf{S}$ *a* $\mathbf{Z}$*-invariant deformation in* $\mathcal{R}(\overset{\circ}{\epsilon}, \overset{\circ}{\delta})$ *with* U, S *verifying the bounds*

$$\sup_{0\le\theta\le\pi}\Big(|U'(\theta)| + |S'(\theta)|\Big) \lesssim \overset{\circ}{\delta}, \tag{9.2.10}$$

as well as the bound (9.1.15) for the coordinates system (u, s, θ, φ) *of* $\mathcal{R}$. *The following hold true:*

1. *We have*

$$\left|\gamma^{\mathbf{S},\#} - \overset{\circ}{\gamma}\right| \lesssim \overset{\circ}{\delta}\overset{\circ}{r}. \tag{9.2.11}$$

2. *For every* $f \in \mathfrak{s}_k(\mathbf{S})$ *we have*

$$\|f^{\#}\|_{L^2(\overset{\circ}{S}, \not{g}^{\mathbf{S},\#})} = \|f^{\#}\|_{L^2(\overset{\circ}{S}, \overset{\circ}{\not{g}})}\Big(1 + O(r^{-1}\overset{\circ}{\delta})\Big). \tag{9.2.12}$$

3. As a corollary of (9.2.12) *(choosing $f = 1$) we deduce*[2]

$$\frac{r^{\mathbf{S}}}{\overset{\circ}{r}} = 1 + O(\overset{\circ}{r}^{-1}\overset{\circ}{\delta}) \tag{9.2.13}$$

where $r^{\mathbf{S}}$ is the area radius of $\mathbf{S}$ and $\overset{\circ}{r}$ that of $\overset{\circ}{S}$.

Proof. Recall

$$\begin{aligned}\gamma^{\mathbf{S},\#}(\overset{\circ}{u},\overset{\circ}{s},\theta) &= \gamma(\psi(\theta)) + \varsigma^2(\psi(\theta))\left(\underline{\Omega}(\psi(\theta)) + \frac{1}{4}(\underline{b}(\psi(\theta)))^2\gamma(\psi(\theta))\right)(U'(\theta))^2 \\ &- 2\varsigma(\psi(\theta))U'(\theta)S'(\theta) - \gamma(\psi(\theta))\varsigma(\psi(\theta))\underline{b}(\psi(\theta))U'(\theta).\end{aligned}$$

In view of our assumptions on U' and S' as well as our estimates (9.1.15) for γ, $\underline{\Omega}$ and $\underline{b}$ and ς, we infer

$$|\gamma^{\mathbf{S},\#} - \gamma| \lesssim |\gamma^{\#} - \gamma| + \overset{\circ}{r}\overset{\circ}{\epsilon}^{1/2}\overset{\circ}{\delta}.$$

Also, we have

$$\begin{aligned}\gamma^{\#}(\overset{\circ}{u},\overset{\circ}{s},\theta) - \gamma(\overset{\circ}{u},\overset{\circ}{s},\theta) &= \gamma(\overset{\circ}{u} + U(\theta), \overset{\circ}{s} + S(\theta), \theta) - \gamma(\overset{\circ}{u},\overset{\circ}{s},\theta) \\ &= \int_0^1 \frac{d}{d\lambda}\left[\gamma(\overset{\circ}{u} + \lambda U(\theta), \overset{\circ}{s} + \lambda S(\theta), \theta)\right] d\lambda \\ &= U(\theta)\int_0^1 \partial_u\gamma(\overset{\circ}{u} + \lambda U(\theta), \overset{\circ}{s} + \lambda S(\theta), \theta)d\lambda \\ &\quad + S(\theta)\int_0^1 \partial_s\gamma(\overset{\circ}{u} + \lambda U(\theta), \overset{\circ}{s} + \lambda S(\theta), \theta)d\lambda.\end{aligned}$$

In view of our estimates (9.1.15) for γ, the assumption (9.2.10) on (U', S') and the fact that

$$\partial_s = e_4, \quad \partial_u = \frac{\varsigma}{2}\left(e_3 - \underline{\Omega}e_4 - \underline{b}\gamma^{1/2}e_\theta\right),$$

we infer[3]

$$|\gamma^{\#} - \gamma| \lesssim \overset{\circ}{r}\overset{\circ}{\delta}.$$

We have finally obtained

$$|\gamma^{\mathbf{S},\#} - \gamma| \lesssim |\gamma^{\#} - \gamma| + \overset{\circ}{r}\,\overset{\circ}{\delta} \lesssim \overset{\circ}{r}\,\overset{\circ}{\delta}.$$

---

[2]Recall also from (9.1.16) that $r - \overset{\circ}{r} = O(\overset{\circ}{\delta}(\overset{\circ}{\epsilon})^{-1/2})$.

[3]Note that we also use the assumption $U(0) = S(0) = 0$ to estimate (U, S) from (U', S').

To prove the second part of the lemma we write

$$\begin{aligned}\int_{\overset{\circ}{S}} |f^\#|^2 da_{g^{\mathbf{S},\#}} &= \int_{\overset{\circ}{S}} |f^\#|^2 \frac{\sqrt{\gamma^{\mathbf{S},\#}}}{\sqrt{\overset{\circ}{\gamma}}} da_{\overset{\circ}{g}}\\ &= \int_{\overset{\circ}{S}} |f^\#|^2 da_{\overset{\circ}{g}} + \int_{\overset{\circ}{S}} |f^\#|^2 \left(\frac{\sqrt{\gamma^{\mathbf{S},\#}}}{\sqrt{\overset{\circ}{\gamma}}} - 1\right) da_{\overset{\circ}{g}}\end{aligned}$$

which yields, in view of the first part,

$$\int_{\overset{\circ}{S}} |f^\#|^2 da_{g^{\mathbf{S},\#}} = \int_{\overset{\circ}{S}} |f^\#|^2 da_{\overset{\circ}{g}} \left(1 + O(\overset{\circ}{r}^{-1}\overset{\circ}{\delta})\right).$$

This conclude the proof of the lemma. □

Remark 9.9. *In view of* (9.2.13) *and* (9.1.16), $\overset{\circ}{r}, r^{\mathbf{S}}$ *and the value of* r *along* $\mathbf{S}$ *are all comparable.*

Corollary 9.10. *Under the assumptions of Lemma 9.8 the following estimate*[4] *holds true for an arbitrary scalar* $f \in \mathfrak{s}_0(\mathcal{R})$,

$$\left|\int_{\mathbf{S}} f - \int_{\overset{\circ}{S}} f\right| \lesssim \overset{\circ}{\delta}\overset{\circ}{r}\left(\sup_{\mathcal{R}} |\mathfrak{d}_{\nearrow}^{\leq 1} f| + \sup_{\mathcal{R}} r|e_3 f|\right).$$

Proof. We have

$$\int_{\mathbf{S}} f - \int_{\overset{\circ}{S}} f = \int_{\overset{\circ}{S}} f^\# \frac{\sqrt{\gamma^{\mathbf{S},\#}}}{\sqrt{\overset{\circ}{\gamma}}} - \int_{\overset{\circ}{S}} f = \int_{\overset{\circ}{S}} f^\# \left(\frac{\sqrt{\gamma^{\mathbf{S},\#}}}{\sqrt{\overset{\circ}{\gamma}}} - 1\right) + \int_{\overset{\circ}{S}} (f^\# - f).$$

Hence,

$$\left|\int_{\mathbf{S}} f - \int_{\overset{\circ}{S}} f\right| \lesssim \overset{\circ}{\delta}\overset{\circ}{r} \sup_{\mathbf{S}} |f| + \int_{\overset{\circ}{S}} |f^\# - f|.$$

Now, proceeding as in the proof of (9.2.11),

$$\begin{aligned}f(\overset{\circ}{u} + U(\theta), \overset{\circ}{s} + S(\theta)) - f(\overset{\circ}{u}, \overset{\circ}{s}) &\lesssim \int_0^1 \frac{d}{d\lambda}\left[f(\overset{\circ}{u} + \lambda U(\theta), \overset{\circ}{s} + \lambda S(\theta), \theta)\right] d\lambda\\ &= U(\theta)\int_0^1 \partial_u f(\overset{\circ}{u} + \lambda U(\theta), \overset{\circ}{s} + \lambda S(\theta), \theta) d\lambda\\ &\quad + S(\theta)\int_0^1 \partial_s f(\overset{\circ}{u} + \lambda U(\theta), \overset{\circ}{s} + \lambda S(\theta), \theta) d\lambda.\end{aligned}$$

[4]Recall that $\mathcal{R} := \{|u - \overset{\circ}{u}| \leq \delta_{\mathcal{R}}, \quad |s - \overset{\circ}{s}| \leq \delta_{\mathcal{R}}\}$, see (9.1.6).

Therefore,

$$\begin{aligned}\left|\int_{\mathbf{S}} f - \int_{\overset{\circ}{S}} f\right| &\lesssim \overset{\circ}{r}\overset{\circ}{\delta} \sup_{\mathbf{S}} |f| + \overset{\circ}{\delta}\overset{\circ}{r}\left(\sup_{\mathcal{R}} |\mathfrak{d}_{\nearrow} f| + \sup_{\mathcal{R}} r|e_3 f|\right)\\ &\lesssim \overset{\circ}{\delta}\overset{\circ}{r}\left(\sup_{\mathcal{R}} |\mathfrak{d}_{\nearrow}^{\leq 1} f| + \sup_{\mathcal{R}} r|e_3 f|\right)\end{aligned}$$

as stated. □

To compare higher order Sobolev spaces, we will need the following lemma.

Lemma 9.11. *Let $\overset{\circ}{S} \subset \mathcal{R} = \mathcal{R}(\overset{\circ}{\epsilon}, \overset{\circ}{\delta})$ as in Definition 9.1 verifying the assumptions* ***A1–A3****. Let $\Psi : \overset{\circ}{S} \to \mathbf{S}$ be* **Z***-invariant deformation. Assume the bound*

$$\|(U', S')\|_{L^\infty(\overset{\circ}{S})} + (\overset{\circ}{r})^{-1}\|(U', S')\|_{\mathfrak{h}_{s_{max}-1}(\overset{\circ}{S}, \overset{\circ}{\not g})} \lesssim \overset{\circ}{\delta}. \tag{9.2.14}$$

Then, we have for any reduced scalar h defined on $\mathcal{R}$

$$\|h\|_{\mathfrak{h}_s(\mathbf{S})} \lesssim r \sup_{\mathcal{R}} |\mathfrak{d}^{\leq s} h|, \qquad \textit{for } 0 \leq s \leq s_{max}.$$

Also, if $f \in \mathfrak{h}_s(\mathbf{S})$ and $f^\#$ is its pullback by ψ, we have

$$\|f\|_{\mathfrak{h}_s(\mathbf{S})} = \|f^\#\|_{\mathfrak{h}_s(\overset{\circ}{S}, \not g^{\mathbf{S},\#})} = \|f^\#\|_{\mathfrak{h}_s(\overset{\circ}{S}, \overset{\circ}{\not g})}(1 + O(r^{-1}\overset{\circ}{\delta})) \quad \textit{for } 0 \leq s \leq s_{max} - 1.$$

Proof. See section C.1. □

Corollary 9.12. *Under the same assumptions as Lemma 9.11, we have, for all $j, k \geq 0$ with $0 \leq j + k \leq s_{max}$,*

$$\begin{aligned}\|\mathfrak{d}^{\leq j}\Gamma_g\|_{\mathfrak{h}_k(\mathbf{S})} &\lesssim \overset{\circ}{\epsilon} r^{-1},\\ \|\mathfrak{d}^{\leq j}\Gamma_b\|_{\mathfrak{h}_k(\mathbf{S})} &\lesssim \overset{\circ}{\epsilon},\end{aligned} \tag{9.2.15}$$

$$\begin{aligned}\|\mathfrak{d}^{\leq j}(\alpha, \beta, \check{\rho}, \check{\mu})\|_{\mathfrak{h}_k(\mathbf{S})} &\lesssim \overset{\circ}{\epsilon} r^{-2},\\ \|\mathfrak{d}^{\leq j}\underline{\beta}\|_{\mathfrak{h}_k(\mathbf{S})} &\lesssim \overset{\circ}{\epsilon} r^{-1},\\ \|\mathfrak{d}^{\leq j}\underline{\alpha}\|_{\mathfrak{h}_k(\mathbf{S})} &\lesssim \overset{\circ}{\epsilon},\end{aligned} \tag{9.2.16}$$

$$\begin{aligned}\left\|\mathfrak{d}^{\leq j}\left(\frac{\gamma}{r^2} - 1, \underline{b}, \frac{e^\Phi}{r\sin\theta} - 1\right)\right\|_{\mathfrak{h}_k(\mathbf{S})} &\lesssim \overset{\circ}{\epsilon},\\ \left\|\mathfrak{d}^{\leq j}(\underline{\Omega} + \Upsilon)\right\|_{\mathfrak{h}_k(\mathbf{S})} + \left\|\mathfrak{d}^{\leq j}(\varsigma - 1)\right\|_{\mathfrak{h}_k(\mathbf{S})} &\lesssim \overset{\circ}{\epsilon} r.\end{aligned} \tag{9.2.17}$$

Proof. In view of Lemma 9.11 and assumptions **A1–A3** we have, for $j, k \geq 0$ with $0 \leq j + k \leq s_{max}$,

$$\left\|\mathfrak{d}^{\leq j}\Gamma_g\right\|_{\mathfrak{h}_k(\mathbf{S})} \lesssim r \sup_{\mathcal{R}} \left|\mathfrak{d}^{\leq k}\mathfrak{d}^{\leq j}\Gamma_g\right| \lesssim r \sup_{\mathcal{R}} \left|\mathfrak{d}^{\leq s_{max}}\Gamma_g\right| \lesssim r^{-1}\overset{\circ}{\epsilon}.$$

The other estimates are proved in the same manner. □

9.2.4 Adapted frame transformations

We consider general null transformations introduced in Lemma 2.87,

$$
\begin{aligned}
e_4' &= \lambda\left(e_4 + f e_\theta + \frac{1}{4} f^2 e_3\right),\\
e_\theta' &= \left(1 + \frac{1}{2} f\underline{f}\right) e_\theta + \frac{1}{2}\underline{f} e_4 + \frac{1}{2} f\left(1 + \frac{1}{4} f\underline{f}\right) e_3,\\
e_3' &= \lambda^{-1}\left(\left(1 + \frac{1}{2} f\underline{f} + \frac{1}{16} f^2\underline{f}^2\right) e_3 + \underline{f}\left(1 + \frac{1}{4} f\underline{f}\right) e_\theta + \frac{1}{4}\underline{f}^2 e_4\right).
\end{aligned}
\tag{9.2.18}
$$

Definition 9.13. *Given a deformation $\Psi : \overset{\circ}{S} \to \mathbf{S}$ we say that a new frame (e_3', e_4', e_θ'), obtained from the standard frame (e_3, e_4, e_θ) via the transformation (9.2.18), is $\mathbf{S}$-adapted if we have*

$$
e_\theta' = e_\theta^{\mathbf{S}} = \frac{1}{(\gamma^{\mathbf{S}})^{1/2}} \Psi_\#(\partial_\theta). \tag{9.2.19}
$$

Proposition 9.14. *Consider a deformation $\Psi : \overset{\circ}{S} \to \mathbf{S}$ in $\mathcal{R} = \mathcal{R}(\overset{\circ}{\epsilon}, \overset{\circ}{\delta})$ verifying the assumption **A3**. The following statements hold true.*

1. *A new frame e_3', e_θ', e_4' generated by $(f, \underline{f}, \lambda = e^a)$ according to (9.2.18) is adapted to $\mathbf{S} = \mathbf{S}(\overset{\circ}{u} + U, \overset{\circ}{s} + S)$ provided that, at all points $\theta \in [0, \pi]$,*

$$
\begin{aligned}
\sqrt{\gamma^\#}\left(1 - \frac{\varsigma^\#}{2}\underline{b}^\# U'\right) &= \left((\gamma^{\mathbf{S}})^\#\right)^{1/2}\left(1 + \frac{1}{2}(f\underline{f})^\#\right),\\
\varsigma^\# U' &= \left((\gamma^{\mathbf{S}})^\#\right)^{1/2} f^\#\left(1 + \frac{1}{4}(f\underline{f})^\#\right),\\
2\left(S' - \frac{\varsigma^\#}{2}\underline{\Omega}^\# U'\right) &= \left((\gamma^{\mathbf{S}})^\#\right)^{1/2} \underline{f}^\#,
\end{aligned}
\tag{9.2.20}
$$

   where

$$
(\gamma^{\mathbf{S}})^\# \;=\; \gamma^\# + (\varsigma^\#)^2\left(\underline{\Omega} + \frac{1}{4}\underline{b}^2\gamma\right)^\# (\partial_\theta U)^2 - 2\varsigma^\#\partial_\theta U \partial_\theta S - (\gamma\varsigma\underline{b})^\#\partial_\theta U
$$

   and $\#$ denotes the pullback by ψ of the corresponding reduced scalars, i.e., for example, $f^\#(\theta) = f(\overset{\circ}{u} + U(\theta), \overset{\circ}{s} + S(\theta), \theta)$.

2. *There exists a small enough constant[5] δ_1 such that for given $f, \underline{f}$ on $\mathcal{R}$ satisfying*

$$
\sup_{\mathcal{R}}\left(|f| + |\underline{f}|\right) \le r^{-1}\delta_1,
$$

[5] In later applications, we will have

$$
\sup_{\mathcal{R}}(|f| + |\underline{f}|) \lesssim r^{-1}\overset{\circ}{\delta}.
$$

we can uniquely solve the system (9.2.20) for U, S subject to the initial conditions,

$$U(0) = 0, \qquad S(0) = 0.$$

Thus, if $(\mathring{u}, \mathring{s}, 0)$ corresponds to the south pole of $\mathring{S}$ and $f, \underline{f}$ are given there exists a unique deformation $\mathbf{S} \subset \mathcal{R}$, given by $U, S : [0, \pi] \to \mathbb{R}$, adapted to frames generated by[6] *$(f, \underline{f})$ which passes through the same south pole. Moreover,*

$$\sup_{[0,\pi]} |(U', S')| \quad \lesssim \quad \mathring{r} \sup_{\mathbf{S}} \left(|f| + |\underline{f}|\right) \tag{9.2.21}$$

and, for $2 \le s \le s_{max} - 1$,

$$\|(U', S')\|_{L^\infty(\mathring{S})} + (\mathring{r})^{-1}\|(U', S')\|_{\mathfrak{h}_s(\mathring{S}, \mathring{\not g})} \quad \lesssim \quad \|f, \underline{f}\|_{\mathfrak{h}_s(\mathbf{S})} \tag{9.2.22}$$

with $\|f, \underline{f}\|_{\mathfrak{h}_s(\mathbf{S})} = \|f\|_{\mathfrak{h}_s(\mathbf{S})} + \|\underline{f}\|_{\mathfrak{h}_s(\mathbf{S})}$.

3. *As a consequence of (9.2.22) the deformation thus obtained verifies the conclusions of Lemmas 9.8–9.11 and Corollary 9.12. In particular,*

   a) We have

$$\left|\gamma^{\mathbf{S},\#} - \mathring{\gamma}\right| \quad \lesssim \quad \delta_1 \mathring{r}.$$

   b) We have

$$\left|\frac{r^{\mathbf{S}}}{\mathring{r}} - 1\right| \lesssim \mathring{r}^{-1} \delta_1.$$

Proof. In view of Lemma 9.7, The **Z**-invariant vectorfield $e^{\mathbf{S}}_\theta := \frac{1}{(\gamma^{\mathbf{S}})^{1/2}} \Psi_\#(\partial_\theta)$ can be expressed by the formula

$$e^{\mathbf{S}}_\theta \quad = \quad \frac{1}{(\gamma^{\mathbf{S}})^{1/2}}\Big[\left(\partial_\theta S - \frac{\varsigma}{2}\underline{\Omega}\partial_\theta U\right) e_4 + \frac{\varsigma}{2}\partial_\theta U e_3 + \sqrt{\gamma}\left(1 - \frac{\varsigma}{2}\underline{b}\partial_\theta U\right) e_\theta\Big]$$

where $\psi(p) = (\mathring{u} + U(\theta), \mathring{s} + S(\theta), \theta)$ and $U' = \partial_\theta U(\theta)$, $S' = \partial_\theta S(\theta)$. On the other hand, according to (9.2.18), at $\Psi(p) \in \mathbf{S}$,

$$e'_\theta = \left(1 + \frac{1}{2} f\underline{f}\right) e_\theta + \frac{1}{2} f \left(1 + \frac{1}{4} f\underline{f}\right) e_3 + \frac{1}{2}\underline{f} e_4.$$

We deduce, at every $\theta \in [0, \pi]$,

$$\begin{aligned}
\sqrt{\gamma^\#}\left(1 - \frac{\varsigma^\#}{2}\underline{b}^\# U'\right) \quad &= \quad ((\gamma^{\mathbf{S}})^\#)^{1/2}\left(1 + \frac{1}{2}(f\underline{f})^\#\right),\\
\varsigma^\# U' \quad &= \quad ((\gamma^{\mathbf{S}})^\#)^{1/2} f^\# \left(1 + \frac{1}{4}(f\underline{f})^\#\right),\\
2\left(S' - \frac{\varsigma^\#}{2}\underline{\Omega}^\# U'\right) \quad &= \quad ((\gamma^{\mathbf{S}})^\#)^{1/2} \underline{f}^\#,
\end{aligned}$$

[6] Note that a is not restricted in this result.

as desired.

To prove the second part of the lemma we first check for the compatibility of the three equations in (9.2.20). Note that, if we denote

$$A = 1 + \frac{1}{2}(f\underline{f})^\#, \quad B = f^\# \left(1 + \frac{1}{4}(f\underline{f})^\#\right), \quad C = \underline{f}^\#,$$

we have $A^2 - BC = 1$. Hence, squaring the first equation and subtracting the product of the other two we derive

$$\begin{aligned}
(\gamma^{\mathbf{S}})^\# &= \left(\sqrt{\gamma^\#}\left(1 - \frac{\varsigma^\#}{2}\underline{b}^\# U'\right)\right)^2 - 2U'\varsigma^\# \left(S' - \frac{\varsigma^\#}{2}\underline{\Omega}^\# U'\right) \\
&= \gamma^\# \left(1 - (\varsigma\underline{b})^\# U' + \frac{1}{4}\left((\varsigma\underline{b})^\# U'\right)^2\right) - 2\varsigma^\# U'S' + (\varsigma^\#)^2 \underline{\Omega}^\# (U')^2 \\
&= \gamma^\# + (\varsigma^\#)^2 \left(\underline{\Omega}^\# + \frac{1}{4}\left(\underline{b}^\#\right)^2 \gamma^\#\right)(U')^2 \\
&\quad -2\varsigma^\# U'S' - \gamma^\# \varsigma^\# \underline{b}^\# U' \qquad (9.2.23)
\end{aligned}$$

which coincides with the formula (9.2.5). It thus suffices to only consider the last two equations in (9.2.20) which we write in the form

$$\begin{aligned}
U' &= (\varsigma^\#)^{-1}((\gamma^{\mathbf{S}})^\#)^{1/2} f^\# \left(1 + \frac{1}{4}(f\underline{f})^\#\right), \\
S' &= \frac{1}{2}((\gamma^{\mathbf{S}})^\#)^{1/2} \underline{f}^\# + \frac{1}{2}\underline{\Omega}^\# ((\gamma^{\mathbf{S}})^\#)^{1/2} f^\# \left(\underline{f}^\# + \frac{1}{4}(f\underline{f})^\#\right),
\end{aligned} \qquad (9.2.24)$$

i.e.,

$$\begin{aligned}
U'(\theta) &= \left[\varsigma^{-1}(\gamma^{\mathbf{S}})^{1/2} f \left(1 + \frac{1}{4}(f\underline{f})\right)\right](\overset{\circ}{u} + U(\theta), \overset{\circ}{s} + S(\theta), \theta), \\
S'(\theta) &= \left[\frac{1}{2}(\gamma^{\mathbf{S}})^{1/2}\underline{f} + \frac{1}{2}\underline{\Omega}(\gamma^{\mathbf{S}})^{1/2} f \left(1 + \frac{1}{4}f\underline{f}\right)\right](\overset{\circ}{u} + U(\theta), \overset{\circ}{s} + S(\theta), \theta).
\end{aligned}$$

Thus under the assumption $\sup_{\mathcal{R}}(|f| + |\underline{f}|) \leq \overset{\circ}{r}^{-1} \delta_1$, with δ_1 sufficiently small, making use also of the expression (9.2.23) of γ^S, and the estimates (9.1.15) for $(\gamma, \underline{b}, \underline{\Omega})$, for $\overset{\circ}{\epsilon}$ sufficiently small, we can uniquely solve for U, S subject to the initial conditions

$$U(0) = 0, \qquad S(0) = 0.$$

Moreover the solution verifies

$$\sup_{[0,\pi]} |(U', S')| \lesssim \overset{\circ}{r} \sup_{\mathbf{S}} \left(|f| + |\underline{f}|\right)$$

according to Definition 9.6. Estimate (9.2.22) can be easily derived by taking higher derivatives and using Lemma 9.11 and **A1**–**A3**. This concludes the proof of the lemma. □

We now provide a lemma analogous to Proposition 9.14 in the particular case when $\underline{f}$ is only bounded in r, unlike the rest of the chapter where it decays like r^{-1}. This lemma is not needed for the construction of GCM spheres in this chapter. It is used in the proof of Theorem M0 in the region ${}^{(ext)}\mathcal{L}_0 \cap {}^{(ext)}\mathcal{M}$ of the initial data layer, see Step 8 in section 4.1.

Lemma 9.15. *There exists a small enough constant δ_1 such that for given $f, \underline{f}$ on $\mathcal{R}$ satisfying*

$$\|f\|_{\mathfrak{h}_{s_{max}-1}(\mathbf{S})} + (r^{\mathbf{S}})^{-1}\|\underline{f}\|_{\mathfrak{h}_{s_{max}-1}(\mathbf{S})} \leq \delta_1,$$

the following holds:

1. *We have*

$$\|U'\|_{L^\infty(\overset{\circ}{S})} + (\overset{\circ}{r})^{-1}\|S'\|_{L^\infty(\overset{\circ}{S})} + (\overset{\circ}{r})^{-1}\|U'\|_{\mathfrak{h}_{s_{max}-1}(\overset{\circ}{S},\overset{\circ}{\not{g}})} + (\overset{\circ}{r})^{-2}\|S'\|_{\mathfrak{h}_{s_{max}-1}(\overset{\circ}{S},\overset{\circ}{\not{g}})} \lesssim \delta_1.$$

In particular, we have

$$\sup_{\mathbf{S}} |u - \overset{\circ}{u}| \lesssim \delta_1, \qquad \sup_{\mathbf{S}} |s - \overset{\circ}{s}| \lesssim \overset{\circ}{r}\delta_1.$$

2. *We have*

$$\left|\gamma^{\mathbf{S},\#} - \overset{\circ}{\gamma}\right| \lesssim \delta_1(\overset{\circ}{r})^2.$$

3. *We have*

$$\left|\frac{r^{\mathbf{S}}}{\overset{\circ}{r}} - 1\right| + \sup_{\mathbf{S}}\left|\frac{r^{\mathbf{S}}}{r} - 1\right| \lesssim \delta_1.$$

4. *The following estimate holds true for an arbitrary scalar $h \in \mathfrak{s}_0(\mathcal{R})$,*

$$\left|h^\# - h\right| \lesssim \delta_1 \sup_{\mathcal{R}} |\mathfrak{d}^{\leq 1}h|.$$

5. *The following estimate holds true for an arbitrary scalar $h \in \mathfrak{s}_0(\mathcal{R})$,*

$$\left|\int_{\mathbf{S}} h - \int_{\overset{\circ}{S}} h\right| \lesssim \delta_1(\overset{\circ}{r})^2 \sup_{\mathcal{R}} |\mathfrak{d}^{\leq 1}h|.$$

6. *We have for any reduced scalar h defined on $\mathcal{R}$*

$$\|h\|_{\mathfrak{h}_s(\mathbf{S})} \lesssim r \sup_{\mathcal{R}} |\mathfrak{d}^{\leq s}h|, \qquad \text{for } 0 \leq s \leq s_{max}.$$

7. *If $h \in \mathfrak{h}_s(\mathbf{S})$ and $h^\#$ is its pullback by ψ, we have*

$$\|f\|_{\mathfrak{h}_s(\mathbf{S})} = \|f^\#\|_{\mathfrak{h}_s(\overset{\circ}{S},\not{g}^{\mathbf{S},\#})} = \|f^\#\|_{\mathfrak{h}_s(\overset{\circ}{S},\overset{\circ}{\not{g}})}(1 + O(\delta_1)) \quad \text{for } 0 \leq s \leq s_{max} - 1.$$

Proof. Recall from (9.2.20) that we have in particular

$$\varsigma^\# U' = \left((\gamma^{\mathbf{S}})^\#\right)^{1/2} f^\# \left(1 + \frac{1}{4}(f\underline{f})^\#\right),$$

$$2\left(S' - \frac{\varsigma^\#}{2}\underline{\Omega}^\# U'\right) = \left((\gamma^{\mathbf{S}})^\#\right)^{1/2} \underline{f}^\#.$$

In view of the assumptions on $(f, \underline{f})$, and the control of the background foliation of $\mathcal{R}$, we immediately obtain the first claim of the lemma concerning the control of (U, S). Note that the estimate for $u - \overset{\circ}{u}$ and $s - \overset{\circ}{s}$ follows then from

$$\begin{aligned}\sup_{\mathbf{S}} |u - \overset{\circ}{u}| &\lesssim \sup_{\overset{\circ}{S}} |U| \lesssim \sup_{\overset{\circ}{S}} |U'| \lesssim \delta_1,\\ \sup_{\mathbf{S}} |s - \overset{\circ}{s}| &\lesssim \sup_{\overset{\circ}{S}} |S| \lesssim \sup_{\overset{\circ}{S}} |S'| \lesssim \overset{\circ}{r}\delta_1.\end{aligned}$$

The first claim then yields the second and third claims by a straightforward adaptation of the proof of Proposition 9.14. Also, the fifth claim follows from the second and the fourth claims, by a simple adaptation of the proof of Corollary 9.10. The sixth and seventh claims follow from the other claims by a simple adaptation of Lemma 9.11.

Finally, we focus on the fourth claim. We have for an arbitrary scalar $h \in \mathfrak{s}_0(\mathcal{R})$,

$$\begin{aligned}h^\#(\overset{\circ}{u}, \overset{\circ}{s}, \theta) - h(\overset{\circ}{u}, \overset{\circ}{s}, \theta) &= h(\overset{\circ}{u} + U(\theta), \overset{\circ}{s} + S(\theta), \theta) - h(\overset{\circ}{u}, \overset{\circ}{s}, \theta)\\ &= \int_0^1 \frac{d}{d\lambda}\left[h(\overset{\circ}{u} + \lambda U(\theta), \overset{\circ}{s} + \lambda S(\theta), \theta)\right] d\lambda\\ &= U(\theta)\int_0^1 \partial_u h(\overset{\circ}{u} + \lambda U(\theta), \overset{\circ}{s} + \lambda S(\theta), \theta) d\lambda\\ &\quad + S(\theta)\int_0^1 \partial_s h(\overset{\circ}{u} + \lambda U(\theta), \overset{\circ}{s} + \lambda S(\theta), \theta) d\lambda.\end{aligned}$$

In view of our estimates (9.1.15) for γ, the assumption (9.2.10) on (U', V') and the fact that

$$\partial_s = e_4, \quad \partial_u = \frac{\varsigma}{2}\left(e_3 - \underline{\Omega} e_4 - \underline{b}\gamma^{1/2} e_\theta\right),$$

we infer[7] together with the first claim

$$\begin{aligned}|h^\# - h| &\lesssim \sup_{\overset{\circ}{S}} |U| \sup_{\mathcal{R}} |\mathfrak{d} h| + r^{-1} \sup_{\overset{\circ}{S}} |S| \sup_{\mathcal{R}} |r e_4(h)|\\ &\lesssim \delta_1 \sup_{\mathcal{R}} |\mathfrak{d} h|\end{aligned}$$

as desired. □

Lemma 9.15 yields the following corollaries.

[7]Note that we also use the assumption $U(0) = S(0) = 0$ to estimate (U, S) from (U', S').

Corollary 9.16. *Assume that there exists a small enough constant δ_1 such that we have*

$$\|U'\|_{L^\infty(\overset{\circ}{S})} + (\overset{\circ}{r})^{-1}\|S'\|_{L^\infty(\overset{\circ}{S})} + (\overset{\circ}{r})^{-1}\|U'\|_{\mathfrak{h}_{s_{max}-1}(\overset{\circ}{S},\overset{\circ}{\not g})} + (\overset{\circ}{r})^{-2}\|S'\|_{\mathfrak{h}_{s_{max}-1}(\overset{\circ}{S},\overset{\circ}{\not g})} \leq \delta_1.$$

Then, we have, for all $j, k \geq 0$ with $0 \leq j + k \leq s_{max}$,

$$\begin{aligned} \|\mathfrak{d}^{\leq j}\Gamma_g\|_{\mathfrak{h}_k(\mathbf{S})} &\lesssim \overset{\circ}{\epsilon} r^{-1}, \\ \|\mathfrak{d}^{\leq j}\Gamma_b\|_{\mathfrak{h}_k(\mathbf{S})} &\lesssim \overset{\circ}{\epsilon}, \end{aligned} \tag{9.2.25}$$

$$\begin{aligned} \|\mathfrak{d}^{\leq j}(\alpha, \beta, \check{\rho}, \check{\mu})\|_{\mathfrak{h}_k(\mathbf{S})} &\lesssim \overset{\circ}{\epsilon} r^{-2}, \\ \|\mathfrak{d}^{\leq j}\underline{\beta}\|_{\mathfrak{h}_k(\mathbf{S})} &\lesssim \overset{\circ}{\epsilon} r^{-1}, \\ \|\mathfrak{d}^{\leq j}\underline{\alpha}\|_{\mathfrak{h}_k(\mathbf{S})} &\lesssim \overset{\circ}{\epsilon}, \end{aligned} \tag{9.2.26}$$

$$\begin{aligned} \left\|\mathfrak{d}^{\leq j}\left(\frac{\gamma}{r^2} - 1,\, \underline{b},\, \frac{e^\Phi}{r\sin\theta} - 1\right)\right\|_{\mathfrak{h}_k(\mathbf{S})} &\lesssim \overset{\circ}{\epsilon}, \\ \left\|\mathfrak{d}^{\leq j}(\underline{\Omega} + \Upsilon)\right\|_{\mathfrak{h}_k(\mathbf{S})} + \left\|\mathfrak{d}^{\leq j}(\varsigma - 1)\right\|_{\mathfrak{h}_k(\mathbf{S})} &\lesssim \overset{\circ}{\epsilon} r. \end{aligned} \tag{9.2.27}$$

Proof. The proof is similar to the one of Corollary 9.12 and relies on property 6 of Lemma 9.15 and the control **A1**–**A3** of the background foliation. □

Corollary 9.17. *Let $3 \leq s \leq s_{max}$. There exists a small enough constant δ_1 such that given $f, \underline{f}$ on $\mathcal{R}$ satisfying*

$$\|f\|_{\mathfrak{h}_s(\mathbf{S})} + (r^{\mathbf{S}})^{-1}\|\underline{f}\|_{\mathfrak{h}_s(\mathbf{S})} \leq \delta_1,$$

then

$$\sup_{\mathbf{S}}\left|K^{\mathbf{S}} - \frac{1}{(r^{\mathbf{S}})^2}\right| \lesssim \frac{\delta_1 + \overset{\circ}{\epsilon}}{(r^{\mathbf{S}})^2}, \qquad \left\|K^{\mathbf{S}} - \frac{1}{(r^{\mathbf{S}})^2}\right\|_{\mathfrak{h}_{s-1}(\mathbf{S})} \lesssim \frac{\delta_1 + \overset{\circ}{\epsilon}}{r^{\mathbf{S}}},$$

and

$$\int_{\mathbf{S}} e^{2\Phi} = \frac{4\pi}{3}(r^{\mathbf{S}})^4(1 + O(\delta_1 + \overset{\circ}{\epsilon})).$$

Proof. Using

$$K^{\mathbf{S}} = -\rho^{\mathbf{S}} - \frac{1}{4}\kappa^{\mathbf{S}}\underline{\kappa}^{\mathbf{S}} + \frac{1}{4}\vartheta^{\mathbf{S}}\underline{\vartheta}^{\mathbf{S}}, \qquad K = -\rho - \frac{1}{4}\kappa\underline{\kappa} + \frac{1}{4}\vartheta\underline{\vartheta},$$

the change of frame formulas for $\rho^{\mathbf{S}}$, $\kappa^{\mathbf{S}}$, $\underline{\kappa}^{\mathbf{S}}$, $\vartheta^{\mathbf{S}}$ and $\underline{\vartheta}^{\mathbf{S}}$, and the assumptions

(9.4.23) for[8] $(f, \underline{f})$, we infer

$$\sup_{\mathbf{S}} \left| K^{\mathbf{S}} - K \right| \lesssim \frac{\delta_1}{(r^{\mathbf{S}})^2}, \qquad \left\| K^{\mathbf{S}} - K \right\|_{\mathfrak{h}_{s-1}(\mathbf{S})} \lesssim \frac{\delta_1}{r^{\mathbf{S}}}.$$

Together with the control **A1**–**A3** for the background foliation, we deduce

$$\sup_{\mathbf{S}} \left| K^{\mathbf{S}} - \frac{1}{r^2} \right| \lesssim \frac{\delta_1}{(r^{\mathbf{S}})^2} + \frac{\mathring{\epsilon}}{r^2}, \qquad \left\| K^{\mathbf{S}} - \frac{1}{r^2} \right\|_{\mathfrak{h}_{s-1}(\mathbf{S})} \lesssim \frac{\delta_1}{r^{\mathbf{S}}} + \frac{\mathring{\epsilon}}{r}.$$

Also, in view of the assumptions (9.4.23) for $(f, \underline{f})$, we may apply Lemma 9.15. Using property 3 of that lemma on the control of $r - r^{\mathbf{S}}$, we easily infer

$$\sup_{\mathbf{S}} \left| K^{\mathbf{S}} - \frac{1}{(r^{\mathbf{S}})^2} \right| \lesssim \frac{\delta_1 + \mathring{\epsilon}}{(r^{\mathbf{S}})^2}, \qquad \left\| K^{\mathbf{S}} - \frac{1}{(r^{\mathbf{S}})^2} \right\|_{\mathfrak{h}_{s-1}(\mathbf{S})} \lesssim \frac{\delta_1 + \mathring{\epsilon}}{r^{\mathbf{S}}}.$$

Also, using property 5 of that lemma, we have

$$\left| \int_{\mathbf{S}} e^{2\Phi} - \int_{\mathring{S}} e^{2\Phi} \right| \lesssim \delta_1 (r^{\mathbf{S}})^4$$

which together with the control **A3** for the background foliation implies

$$\int_{\mathbf{S}} e^{2\Phi} = \frac{4\pi}{3} (r^{\mathbf{S}})^4 (1 + O(\delta_1 + \mathring{\epsilon})).$$

This concludes the proof of the corollary. □

Corollary 9.18. *Let $2 \leq s \leq s_{max}$. There exists a small enough constant δ_1 such that given $f, \underline{f}$ on $\mathcal{R}$ satisfying*

$$\|f\|_{\mathfrak{h}_s(\mathbf{S})} + (r^{\mathbf{S}})^{-1} \|\underline{f}\|_{\mathfrak{h}_s(\mathbf{S})} \leq \delta_1,$$

then, for any scalar function $D = D(u, s)$ on $\mathcal{R}$ depending only on the coordinates (u, s) of the background foliation, we have

$$\left\| D - \overline{D}^{\mathbf{S}} \right\|_{\mathfrak{h}_s(\mathbf{S})} \lesssim r \Big(\|f\|_{\mathfrak{h}_s(\mathbf{S})} + r^{-1} \|\underline{f}\|_{\mathfrak{h}_s(\mathbf{S})} \Big) \sup_{\mathcal{R}} |\mathfrak{d}^{\leq s} D|.$$

Proof. We have, using a Poincaré inequality,

$$\left\| D - \overline{D}^{\mathbf{S}} \right\|_{\mathfrak{h}_s(\mathbf{S})} \lesssim \left\| r^{\mathbf{S}} e_\theta^{\mathbf{S}}(D) \right\|_{\mathfrak{h}_{s-1}(\mathbf{S})}, \quad s \geq 1.$$

Thus, we need to compute $e_\theta^{\mathbf{S}}(D)$. Decomposing $e_\theta^{\mathbf{S}}$ on the background frame, we

[8] Note that the change of frame formulas for $\rho^{\mathbf{S}}$, $\kappa^{\mathbf{S}} \underline{\kappa}^{\mathbf{S}}$ and $\vartheta^{\mathbf{S}} \underline{\vartheta}^{\mathbf{S}}$ do not involve λ, and involve at most one tangential derivative to $\mathbf{S}$ of $(f, \underline{f})$.

have

$$\begin{aligned} e_\theta^{\mathbf{S}}(D) &= \left(1+\frac{1}{2}f\underline{f}\right)e_\theta(D)+\frac{1}{2}\underline{f}e_4(D)+\frac{1}{2}f\left(1+\frac{1}{4}f\underline{f}\right)e_3(D)\\ &= \frac{1}{2}\underline{f}e_4(D)+\frac{1}{2}f\left(1+\frac{1}{4}f\underline{f}\right)e_3(D) \end{aligned}$$

where we have used in the last inequality $D = D(u, s)$ and $e_\theta(u) = e_\theta(s) = 0$. We infer, for $2 \le s \le s_{max}$,

$$\begin{aligned} \left\|D-\overline{D}^{\mathbf{S}}\right\|_{\mathfrak{h}_s(\mathbf{S})} &\lesssim r^{\mathbf{S}}\left\|\frac{1}{2}\underline{f}e_4(D)+f\left(1+\frac{1}{4}f\underline{f}\right)e_3(D)\right\|_{\mathfrak{h}_{s-1}(\mathbf{S})}\\ &\lesssim \|\underline{f}\|_{\mathfrak{h}_s(\mathbf{S})}\|e_4(D)\|_{\mathfrak{h}_{s-1}(S)}\\ &\quad+\|f\|_{\mathfrak{h}_s(\mathbf{S})}\left(1+r^{-2}\|f\|_{\mathfrak{h}_s(\mathbf{S})}\|\underline{f}\|_{\mathfrak{h}_s(\mathbf{S})}\right)\|e_3(D)\|_{\mathfrak{h}_{s-1}(S)}\\ &\lesssim r\left(\|f\|_{\mathfrak{h}_s(\mathbf{S})}+r^{-1}\|\underline{f}\|_{\mathfrak{h}_s(\mathbf{S})}\right)\sup_{\mathcal{R}}|\mathfrak{d}^{\le s}D|, \end{aligned}$$

where we have used in the last inequality the control on $(f, \underline{f})$, as well as property 6 of Lemma 9.15 with $h = e_4(D)$ and $h = e_3(D)$. □

Corollary 9.19. *Assume that $(f, \underline{f})$ given on $\mathcal{R}$ satisfy for a small enough constant δ_1*

$$\|f\|_{\mathfrak{h}_{s_{max}-1}(\mathbf{S})}+(r^{\mathbf{S}})^{-1}\|\underline{f}\|_{\mathfrak{h}_{s_{max}-1}(\mathbf{S})} \le \delta_1.$$

Then, we have

$$|m^{\mathbf{S}}-\overset{\circ}{m}| \lesssim \delta_1+(\overset{\circ}{\epsilon})^2.$$

Proof. According to the identity (2.2.12), we have

$$\begin{aligned} \int_{\mathbf{S}}\rho^{\mathbf{S}} &= -\frac{8\pi m^{\mathbf{S}}}{r^{\mathbf{S}}}+\frac{1}{4}\int_S \vartheta^{\mathbf{S}}\underline{\vartheta}^{\mathbf{S}},\\ \int_{\overset{\circ}{S}}\rho &= -\frac{8\pi\overset{\circ}{m}}{\overset{\circ}{r}}+\frac{1}{4}\int_S \vartheta\underline{\vartheta}. \end{aligned}$$

In view of the transformation formulas for $\vartheta^{\mathbf{S}}$ and $\underline{\vartheta}^{\mathbf{S}}$, and noticing that the product $\vartheta^{\mathbf{S}}\underline{\vartheta}^{\mathbf{S}}$ only involves $(f, \underline{f})$ but not λ, we infer from the assumptions **A1**–**A3** for the background foliation of $\mathcal{R}$, and the assumptions on $(f, \underline{f})$ that

$$|\vartheta^{\mathbf{S}}\underline{\vartheta}^{\mathbf{S}}|+|\vartheta\underline{\vartheta}| \lesssim \frac{(\overset{\circ}{\epsilon})^2}{r^3}.$$

We infer

$$\begin{aligned} \int_{\mathbf{S}}\rho^{\mathbf{S}} &= -\frac{8\pi m^{\mathbf{S}}}{r^{\mathbf{S}}}+O\left(\frac{(\overset{\circ}{\epsilon})^2}{r}\right),\\ \int_{\overset{\circ}{S}}\rho &= -\frac{8\pi\overset{\circ}{m}}{\overset{\circ}{r}}+O\left(\frac{(\overset{\circ}{\epsilon})^2}{r}\right), \end{aligned}$$

and hence

$$\left|m^{\mathbf{S}} - \overset{\circ}{m}\right| \lesssim \left|r^{\mathbf{S}} \int_{\mathbf{S}} \rho^{\mathbf{S}} - \overset{\circ}{r} \int_{\overset{\circ}{S}} \rho\right| + (\overset{\circ}{\epsilon})^2.$$

Next, we apply Lemma 9.15 and infer in particular

$$|r^{\mathbf{S}} - \overset{\circ}{r}| \lesssim \overset{\circ}{r}\delta_1, \qquad \left|\int_{\mathbf{S}} \rho - \int_{\overset{\circ}{S}} \rho\right| \lesssim \frac{\delta_1}{\overset{\circ}{r}}.$$

We deduce

$$\left|m^{\mathbf{S}} - \overset{\circ}{m}\right| \lesssim r\left|\int_{\mathbf{S}} (\rho^{\mathbf{S}} - \rho)\right| + \delta_1 + (\overset{\circ}{\epsilon})^2.$$

Together with the transformation formula for $\rho^{\mathbf{S}}$, which only involves $(f, \underline{f})$ but not λ, we infer from the assumptions **A1**–**A3** for the background foliation of $\mathcal{R}$, and the assumptions on $(f, \underline{f})$ that

$$\left|m^{\mathbf{S}} - \overset{\circ}{m}\right| \lesssim \delta_1 + (\overset{\circ}{\epsilon})^2$$

as desired. □

9.3 FRAME TRANSFORMATIONS

For the convenience of the reader we start by recalling the transformation formulas recorded in Proposition 2.90.

Proposition 9.20 (Transformation formulas-GCM)**.** *Under a general transformation of type* (9.2.18) *with* $\lambda = e^a$ *the Ricci coefficients and curvature components transform as follows:*

$$\begin{aligned}
\xi' &= \lambda^2\left(\xi + \frac{1}{2}\lambda^{-1}e_4'(f) + \omega f + \frac{1}{4}f\kappa\right) + \lambda^2 Err(\xi, \xi'),\\
Err(\xi, \xi') &= \frac{1}{4}f\vartheta + l.o.t.,\\
\underline{\xi}' &= \lambda^{-2}\left(\underline{\xi} + \frac{1}{2}\lambda e_3'(\underline{f}) + \underline{\omega}\,\underline{f} + \frac{1}{4}\underline{f}\,\underline{\kappa}\right) + \lambda^{-2}Err(\underline{\xi}, \underline{\xi}'),\\
Err(\underline{\xi}, \underline{\xi}') &= -\frac{1}{8}\lambda\underline{f}^2 e_3'(f) + \frac{1}{4}\underline{f}\,\underline{\vartheta} + l.o.t.,
\end{aligned} \tag{9.3.1}$$

$$
\begin{aligned}
\zeta' &= \zeta - e'_\theta(\log(\lambda)) + \frac{1}{4}(-f\underline{\kappa} + \underline{f}\kappa) + \underline{f}\omega - f\underline{\omega} + Err(\zeta, \zeta'),\\
Err(\zeta, \zeta') &= \frac{1}{2}\underline{f}e'_\theta(f) + \frac{1}{4}(-f\underline{\vartheta} + \underline{f}\vartheta) + l.o.t.,\\
\eta' &= \eta + \frac{1}{2}\lambda e'_3(f) + \frac{1}{4}\kappa\underline{f} - f\underline{\omega} + Err(\eta, \eta'),\\
Err(\eta, \eta') &= \frac{1}{4}\underline{f}\vartheta + l.o.t.,\\
\underline{\eta}' &= \underline{\eta} + \frac{1}{2}\lambda^{-1}e'_4(\underline{f}) + \frac{1}{4}\underline{\kappa}f - \underline{f}\omega + Err(\underline{\eta}, \underline{\eta}'),\\
Err(\underline{\eta}, \underline{\eta}') &= -\frac{1}{8}\underline{f}^2\lambda^{-1}e'_4(f) + \frac{1}{4}f\underline{\vartheta} + l.o.t.,
\end{aligned}
\tag{9.3.2}
$$

$$
\begin{aligned}
\kappa' &= \lambda\left(\kappa + \not{d}_1{}'(f)\right) + \lambda Err(\kappa, \kappa'),\\
Err(\kappa, \kappa') &= f(\zeta + \eta) + \underline{f}\xi - \frac{1}{4}f^2\underline{\kappa} + f\underline{f}\omega - f^2\underline{\omega} + l.o.t.,\\
\underline{\kappa}' &= \lambda^{-1}\left(\underline{\kappa} + \not{d}_1{}'(\underline{f})\right) + \lambda^{-1}Err(\underline{\kappa}, \underline{\kappa}'),\\
Err(\underline{\kappa}, \underline{\kappa}') &= -\frac{1}{4}\underline{f}^2 e'_\theta(f) + \underline{f}(-\zeta + \underline{\eta}) + f\underline{\xi} - \frac{1}{4}\underline{f}^2\kappa + f\underline{f}\underline{\omega} - \underline{f}^2\omega + l.o.t.,
\end{aligned}
\tag{9.3.3}
$$

$$
\begin{aligned}
\vartheta' &= \lambda\left(\vartheta - \not{d}_2^\star{}'(f)\right) + \lambda Err(\vartheta, \vartheta'),\\
Err(\vartheta, \vartheta') &= f(\zeta + \eta) + \underline{f}\xi + \frac{1}{4}f\underline{f}\kappa + f\underline{f}\omega - f^2\underline{\omega} + l.o.t.\\
\underline{\vartheta}' &= \lambda^{-1}\left(\underline{\vartheta} - \not{d}_2^\star{}'(\underline{f})\right) + \lambda^{-1}Err(\underline{\vartheta}, \underline{\vartheta}'),\\
Err(\underline{\vartheta}, \underline{\vartheta}') &= -\frac{1}{4}\underline{f}^2 e'_\theta(f) + \underline{f}(-\zeta + \underline{\eta}) + f\underline{\xi} + \frac{1}{4}f\underline{f}\underline{\kappa} + f\underline{f}\underline{\omega} - \underline{f}^2\omega + l.o.t.,
\end{aligned}
\tag{9.3.4}
$$

$$
\begin{aligned}
\omega' &= \lambda\left(\omega - \frac{1}{2}\lambda^{-1}e'_4(\log(\lambda))\right) + \lambda Err(\omega, \omega'),\\
Err(\omega, \omega') &= \frac{1}{4}\underline{f}e'_4(f) + \frac{1}{2}\omega f\underline{f} - \frac{1}{2}f\underline{\eta} + \frac{1}{2}\underline{f}\xi + \frac{1}{2}f\zeta - \frac{1}{8}\underline{\kappa}f^2 + \frac{1}{8}f\underline{f}\kappa - \frac{1}{4}\underline{\omega}f^2\\
&\quad + l.o.t.,\\
\underline{\omega}' &= \lambda^{-1}\left(\underline{\omega} + \frac{1}{2}\lambda e'_3(\log(\lambda))\right) + \lambda^{-1}Err(\underline{\omega}, \underline{\omega}'),\\
Err(\underline{\omega}, \underline{\omega}') &= -\frac{1}{4}\underline{f}e'_3(f) + \underline{\omega}f\underline{f} - \frac{1}{2}\underline{f}\eta + \frac{1}{2}f\underline{\xi} - \frac{1}{2}\underline{f}\zeta - \frac{1}{8}\kappa\underline{f}^2 + \frac{1}{8}f\underline{f}\underline{\kappa} - \frac{1}{4}\omega\underline{f}^2\\
&\quad + l.o.t.
\end{aligned}
\tag{9.3.5}
$$

The lower order terms we denote by l.o.t. are linear with respect to the Ricci coefficients $\{\xi, \underline{\xi}, \vartheta, \kappa, \eta, \underline{\eta}, \zeta, \underline{\kappa}, \underline{\vartheta}\}$ *of the background, and quadratic or higher order in* $f, \underline{f}$*, and do not contain derivatives of these latter.*

Also,

$$\begin{aligned}
\alpha' &= \lambda^2\alpha + \lambda^2 Err(\alpha,\alpha'),\\
Err(\alpha,\alpha') &= 2f\beta + \frac{3}{2}f^2\rho + l.o.t.,\\
\beta' &= \lambda\left(\beta + \frac{3}{2}\rho f\right) + \lambda Err(\beta,\beta'),\\
Err(\beta,\beta') &= \frac{1}{2}\underline{f}\alpha + l.o.t.,\\
\rho' &= \rho + Err(\rho,\rho'),\\
Err(\rho,\rho') &= \frac{3}{2}\rho f\underline{f} + \underline{f}\beta + f\underline{\beta} + l.o.t.,\\
\underline{\beta}' &= \lambda^{-1}\left(\underline{\beta} + \frac{3}{2}\rho\underline{f}\right) + \lambda^{-1}Err(\underline{\beta},\underline{\beta}'),\\
Err(\underline{\beta},\underline{\beta}') &= \frac{1}{2}f\underline{\alpha} + l.o.t.,\\
\underline{\alpha}' &= \lambda^{-2}\underline{\alpha} + \lambda^{-2}Err(\underline{\alpha},\underline{\alpha}'),\\
Err(\underline{\alpha},\underline{\alpha}') &= 2\underline{f}\,\underline{\beta} + \frac{3}{2}\underline{f}^2\rho + l.o.t.
\end{aligned} \tag{9.3.6}$$

The lower order terms we denote by l.o.t. are linear with respect to the curvature quantities $\alpha, \beta, \rho, \underline{\beta}, \underline{\alpha}$ and quadratic or higher order in $f, \underline{f}$, and do not contain derivatives of these latter.

In the following lemma we rewrite a subset of these transformations in a more useful form:

Lemma 9.21. *Under a general transformation of type (9.2.18) with $\lambda = e^a$ we have, in particular,*

$$\begin{aligned}
\zeta' &= \zeta - e_\theta'(a) - f\underline{\omega} + \underline{f}\omega - \frac{1}{2}f\underline{\chi} + \frac{1}{2}\underline{f}\chi + Err(\zeta,\zeta'),\\
Err(\zeta,\zeta') &= \frac{1}{2}\underline{f}\left(1 + \frac{1}{4}f\underline{f}\right)e_\theta'(f) - \frac{1}{16}\underline{f}^2 e_\theta'(f^2) + \frac{1}{4}(-f\underline{\vartheta} + \underline{f}\vartheta) + l.o.t.
\end{aligned} \tag{9.3.7}$$

$$\begin{aligned}
\kappa' &= e^a\left(\kappa + \not{d}_1' f\right) + e^a Err(\kappa,\kappa'),\\
Err(\kappa,\kappa') &= \frac{1}{2}f\underline{f}e_\theta'(f) - \frac{1}{4}\underline{f}e_\theta'(f^2) + f(\zeta+\eta) + \underline{f}\xi - \frac{1}{4}f^2\underline{\kappa}\\
&\quad + f\underline{f}\omega - f^2\underline{\omega} + l.o.t.
\end{aligned} \tag{9.3.8}$$

$$\begin{aligned}
\underline{\kappa}' &= e^{-a}\left(\underline{\kappa} + \not{d}_1'\underline{f}\right) + e^{-a}Err(\underline{\kappa},\underline{\kappa}'),\\
Err(\underline{\kappa},\underline{\kappa}') &= -\frac{1}{2}\underline{f}e_\theta'\left(f\underline{f} + \frac{1}{8}f^2\underline{f}^2\right) + \left(\frac{3}{4}f\underline{f} + \frac{1}{8}(f\underline{f})^2\right)e_\theta'(\underline{f})\\
&\quad + \frac{1}{4}\left(1 + \frac{1}{2}f\underline{f}\right)\underline{f}e_\theta'(f\underline{f}) - \frac{1}{4}f\left(1 + \frac{1}{4}f\underline{f}\right)e_\theta'(\underline{f}^2)\\
&\quad + \underline{f}(-\zeta + \underline{\eta}) + f\underline{\xi} - \frac{1}{4}\underline{f}^2\kappa + f\underline{f}\underline{\omega} - \underline{f}^2\omega + l.o.t.
\end{aligned} \tag{9.3.9}$$

Also,

$$
\begin{aligned}
\vartheta' &= \lambda\left(\vartheta - {\not{d}}_2^{\star\prime}(f)\right) + \lambda Err(\vartheta, \vartheta'),\\
Err(\vartheta, \vartheta') &= \frac{1}{2} f \underline{f} e'_\theta(f) - \frac{1}{4}\underline{f} e'_\theta(f^2) + f(\zeta+\eta) + \underline{f}\xi + \frac{1}{4} f\underline{f}\kappa\\
&\quad + f\underline{f}\omega - f^2\underline{\omega} + l.o.t.\\
\underline{\vartheta}' &= \lambda^{-1}\left(\underline{\vartheta} - {\not{d}}_2^{\star\prime}(\underline{f})\right) + \lambda^{-1} Err(\underline{\vartheta}, \underline{\vartheta}'),\\
Err(\underline{\vartheta}, \underline{\vartheta}') &= -\frac{1}{2}\underline{f} e'_\theta\left(f\underline{f} + \frac{1}{8} f^2\underline{f}^2\right) + \left(\frac{3}{4} f\underline{f} + \frac{1}{8}(f\underline{f})^2\right) e'_\theta(\underline{f})\\
&\quad + \frac{1}{4}\left(1 + \frac{1}{2} f\underline{f}\right)\underline{f} e'_\theta(f\underline{f}) - \frac{1}{4} f\left(1 + \frac{1}{4} f\underline{f}\right) e'_\theta(\underline{f}^2) + \underline{f}(-\zeta + \underline{\eta})\\
&\quad + f\underline{\xi} + \frac{1}{4} f\underline{f}\underline{\kappa} + f\underline{f}\underline{\omega} - \underline{f}^2\omega + l.o.t.
\end{aligned}
\tag{9.3.10}
$$

The lower order terms we denote by l.o.t. are cubic or higher order in the small quantities $\xi, \underline{\xi}, \vartheta, \eta, \underline{\eta}, \zeta, \underline{\vartheta}$ *as well as* $f, \underline{f}$*, and do not contain derivatives of these quantities.*

We also have

$$
\begin{aligned}
\beta' &= \lambda\left(\beta + \frac{3}{2}\rho f\right) + \lambda Err(\beta, \beta'),\\
Err(\beta, \beta') &= \frac{1}{2}\underline{f}\alpha + l.o.t.,\\
\rho' &= \rho + Err(\rho, \rho'),\\
Err(\rho, \rho') &= \frac{3}{2}\rho f\underline{f} + \underline{f}\beta + f\underline{\beta} + l.o.t.
\end{aligned}
\tag{9.3.11}
$$

The lower order terms above denoted by l.o.t. are cubic or higher order in the small quantities $\xi, \underline{\xi}, \vartheta, \eta, \underline{\eta}, \zeta, \underline{\vartheta}$ *as well as* $a, f, \underline{f}$*.*

Lemma 9.22. *The following transformation formula holds true:*

$$
\begin{aligned}
\mu' &= \mu + ({\not{d}}_1)'\left(-({\not{d}}_1^\star)' a + f\underline{\omega} - \underline{f}\omega + \frac{1}{4} f\underline{\kappa} - \frac{1}{4}\underline{f}\kappa\right) + Err(\mu, \mu'),\\
Err(\mu, \mu') &= -{\not{d}}_1' Err(\zeta, \zeta') - Err(\rho, \rho') + \frac{1}{4}\left(\vartheta'\underline{\vartheta}' - \vartheta\underline{\vartheta}\right).
\end{aligned}
$$

The error term $Err(\mu, \mu')$ *is quadratic or higher order with respect to* $(f, \underline{f}, a, \check{\Gamma}, \check{R})$ *and depends only on at most two angular derivatives* e'_θ *of* f *and one angular derivative* e'_θ *of* $a, \underline{f}$*.*

Proof. Recall that

$$
\mu = -{\not{d}}_1\zeta - \rho + \frac{1}{4}\vartheta\underline{\vartheta}.
$$

Therefore,

$$
\begin{aligned}
\mu' &= -\not{d}_1'\zeta' - \rho' + \frac{1}{4}\vartheta'\underline{\vartheta}' \\
&= -\not{d}_1'\Big(\zeta - e_\theta'(a) - f\underline{\omega} + \underline{f}\omega - \frac{1}{4}f\underline{\kappa} + \frac{1}{4}\underline{f}\kappa + \mathrm{Err}(\zeta,\zeta')\Big) - \rho - \mathrm{Err}(\rho,\rho') \\
&\quad + \frac{1}{4}\vartheta'\underline{\vartheta}' \\
&= -\not{d}_1'\zeta - \rho + \frac{1}{4}\vartheta\underline{\vartheta} - \not{d}_1'\Big((\not{d}_1^\star)'a - f\underline{\omega} + \underline{f}\omega - \frac{1}{4}f\underline{\kappa} + \frac{1}{4}\underline{f}\kappa\Big) \\
&\quad - \not{d}_1'\mathrm{Err}(\zeta,\zeta') - \mathrm{Err}(\rho,\rho') + \frac{1}{4}(\vartheta'\underline{\vartheta}' - \vartheta\underline{\vartheta}).
\end{aligned}
$$

Note that

$$
\begin{aligned}
-\not{d}_1'\zeta - \rho + \frac{1}{4}\vartheta\underline{\vartheta} &= -\not{d}_1\zeta - \rho + \frac{1}{4}\vartheta\underline{\vartheta} + fe_3\zeta + \underline{f}e_4\zeta + \text{l.o.t.} \\
&= \mu + fe_3\zeta + \underline{f}e_4\zeta + \text{l.o.t.}
\end{aligned}
$$

Hence,

$$
\mu' = \mu + \not{d}_1'\Big(-(\not{d}_1^\star)'a + f\underline{\omega} - \underline{f}\omega + \frac{1}{4}f\underline{\kappa} - \frac{1}{4}\underline{f}\kappa\Big) + \mathrm{Err}(\mu,\mu')
$$

where

$$
\mathrm{Err}(\mu,\mu') = -\not{d}_1'\mathrm{Err}(\zeta,\zeta') - \mathrm{Err}(\rho,\rho') + \frac{1}{4}(\vartheta'\underline{\vartheta}' - \vartheta\underline{\vartheta}) + fe_3\zeta + \underline{f}e_4\zeta + \text{l.o.t.}
$$

In view of the transformation formulas for $\vartheta, \underline{\vartheta}$ and the structure of the error terms $\mathrm{Err}(\zeta,\zeta')$, $\mathrm{Err}(\rho,\rho')$, $\mathrm{Err}(\vartheta,\vartheta')$, $\mathrm{Err}(\underline{\vartheta},\underline{\vartheta}')$ in Lemma 9.21 we easily deduce that the error term $\mathrm{Err}(\mu,\mu')$ depends only on at most two angular derivatives e_θ' of f and one angular derivative e_θ' of $a, \underline{f}$. □

We shall also make use of the following lemma.

Lemma 9.23. *We have the transformation equations,*

$$
\begin{aligned}
e_\theta'(\kappa') &= e_\theta\kappa + e_\theta'\not{d}_1'f + \kappa e_\theta' a - \frac{1}{4}\kappa(f\underline{\kappa} + \underline{f}\kappa) + \kappa(f\underline{\omega} - \omega\underline{f}) + f\rho \\
&\quad + Err(e_\theta'\kappa', e_\theta\kappa), \\
e_\theta'(\underline{\kappa}') &= e_\theta\underline{\kappa} + e_\theta'\not{d}_1'\underline{f} - \underline{\kappa}e_\theta' a - \frac{1}{4}\underline{\kappa}(f\underline{\kappa} + \underline{f}\kappa) + \underline{\kappa}(\underline{f}\omega - \underline{\omega}f) + \underline{f}\rho \\
&\quad + Err(e_\theta'\underline{\kappa}', e_\theta\underline{\kappa}), \\
e_\theta'(\mu') &= e_\theta\mu + e_\theta'(\not{d}_1)'\Big(-(\not{d}_1^\star)'a + f\underline{\omega} - \underline{f}\omega + \frac{1}{4}f\underline{\kappa} - \frac{1}{4}\underline{f}\kappa\Big) + \frac{3}{4}\rho(f\underline{\kappa} + \underline{f}\kappa) \\
&\quad + Err(e_\theta'\mu', e_\theta\mu),
\end{aligned}
\tag{9.3.12}
$$

where

$$\begin{aligned}
Err(e'_\theta\kappa', e_\theta\kappa) &= (e^a-1)\Big(e_\theta\kappa + e'_\theta\,{\not{d}}'_1 f + \frac{1}{2}\underline{f}e_4\kappa + \frac{1}{2}fe_3\kappa\Big)\\
&+ e^a\Bigg[e'_\theta\, Err(\kappa,\kappa') + e'_\theta(a)\Big({\not{d}}'_1 f + Err(\kappa,\kappa')\Big) + \frac{1}{2}f\underline{f}e_\theta\kappa\\
&+ \frac{1}{8}f^2\underline{f}e_3\kappa\Bigg] + \frac{1}{2}f\left(2\,{\not{d}}_1\eta - \frac{1}{2}\underline{\vartheta}\,\vartheta + 2(\xi\underline{\xi}+\eta^2)\right)\\
&+ \frac{1}{2}f\underline{f}e_\theta\kappa + \frac{1}{8}f^2\underline{f}e_3\kappa + \frac{1}{2}\underline{f}\left(2\,{\not{d}}_1\xi - \frac{1}{2}\vartheta^2 + 2(\eta+\underline{\eta}+2\zeta)\xi\right),
\end{aligned}$$

$$\begin{aligned}
Err(e'_\theta\underline{\kappa}', e_\theta\underline{\kappa}) &= (e^{-a}-1)\Big(e_\theta\underline{\kappa} + e'_\theta\,{\not{d}}'_1 \underline{f} + \frac{1}{2}fe_3\underline{\kappa} + \frac{1}{2}\underline{f}e_4\underline{\kappa}\Big)\\
&+ e^{-a}\Bigg[e'_\theta\, Err(\underline{\kappa},\underline{\kappa}') + e'_\theta(a)\Big({\not{d}}'_1 \underline{f} + Err(\underline{\kappa},\underline{\kappa}')\Big) + \frac{1}{2}f\underline{f}e_\theta\underline{\kappa}\\
&+ \frac{1}{8}f^2\underline{f}e_3\underline{\kappa}\Bigg] + \frac{1}{2}\underline{f}\left(2\,{\not{d}}_1\underline{\eta} - \frac{1}{2}\underline{\vartheta}\,\vartheta + 2(\xi\underline{\xi}+\underline{\eta}^2)\right)\\
&+ \frac{1}{2}f\underline{f}e_\theta\underline{\kappa} + \frac{1}{8}f^2\underline{f}e_3\underline{\kappa} + \frac{1}{2}f\left(2\,{\not{d}}_1\underline{\xi} - \frac{1}{2}\underline{\vartheta}^2 + 2(\eta+\underline{\eta}-2\zeta)\xi\right),
\end{aligned}$$

and

$$\begin{aligned}
Err(e'_\theta\mu', e_\theta\mu) &= e'_\theta Err(\mu,\mu') + \frac{1}{2}f\underline{f}e_\theta\mu + \frac{1}{8}f^2\underline{f}e_3\mu\\
&- \frac{1}{2}f\Big({\not{d}}_1\underline{\beta} - \frac{1}{2}\vartheta\,\underline{\alpha} - \zeta\,\underline{\beta} + 2(\eta\,\underline{\beta} + \underline{\xi}\,\beta)\Big)\\
&- \frac{1}{2}\underline{f}\Big({\not{d}}_1\beta - \frac{1}{2}\underline{\vartheta}\,\alpha + \zeta\,\beta + 2(\underline{\eta}\,\beta + \xi\,\underline{\beta})\Big)\\
&+ \frac{1}{2}\underline{f}e_4\left(-{\not{d}}_1\zeta + \frac{1}{4}\vartheta\underline{\vartheta}\right) + \frac{1}{2}fe_3\left(-{\not{d}}_1\zeta + \frac{1}{4}\vartheta\underline{\vartheta}\right).
\end{aligned}$$

Proof. Applying the vectorfield e'_θ to

$$\kappa' = e^a\left(\kappa + {\not{d}}'_1 f + \mathrm{Err}(\kappa,\kappa')\right)$$

we deduce

$$e'_\theta(\kappa') = e^a\Big(e'_\theta\kappa + e'_\theta\,{\not{d}}'_1 f + e'_\theta\mathrm{Err}(\kappa,\kappa')\Big) + e^a e'_\theta(a)\Big(\kappa + {\not{d}}'_1 f + \mathrm{Err}(\kappa,\kappa')\Big).$$

Hence,

$$e^{-a}e'_\theta(\kappa') = e'_\theta\kappa + e'_\theta\,{\not{d}}'_1 f + e'_\theta\mathrm{Err}(\kappa,\kappa') + e'_\theta(a)\Big(\kappa + {\not{d}}'_1 f + \mathrm{Err}(\kappa,\kappa')\Big)$$

and thus

$$e'_\theta(\kappa') = e_\theta\kappa + e'_\theta(a)\kappa + e'_\theta\,{\not{d}}'_1 f + \frac{1}{2}\underline{f}e_4\kappa + \frac{1}{2}fe_3\kappa + \mathrm{Err}_1[e_\theta(\kappa), e'_\theta(\kappa')]$$

with error term

$$\begin{aligned}\mathrm{Err}_1[e_\theta(\kappa), e'_\theta(\kappa')] &= (e^a-1)\Big(e_\theta\kappa + e'_\theta \,\mathit{d}'_1 f + \frac{1}{2}\underline{f}e_4\kappa + \frac{1}{2}fe_3\kappa\Big)\\ &+ e^a\Bigg[e'_\theta \mathrm{Err}(\kappa,\kappa') + e'_\theta(a)\Big(\mathit{d}'_1 f + \mathrm{Err}(\kappa,\kappa')\Big) + \frac{1}{2}f\underline{f}e_\theta\kappa\\ &+ \frac{1}{8}f^2\underline{f}e_3\kappa\Bigg].\end{aligned}$$

Now, making use of

$$\begin{aligned}e'_\theta\kappa &= \left(1+\frac{1}{2}f\underline{f}\right)e_\theta\kappa + \frac{1}{2}\underline{f}e_4\kappa + \frac{1}{2}f\left(1+\frac{1}{4}f\underline{f}\right)e_3\kappa\\ &= e_\theta\kappa + \frac{1}{2}\underline{f}e_4\kappa + \frac{1}{2}fe_3\kappa + \frac{1}{2}f\underline{f}e_\theta\kappa + \frac{1}{8}f^2\underline{f}e_3\kappa\end{aligned}$$

and the null structure equations,

$$\begin{aligned}e_3(\kappa) + \frac{1}{2}\underline{\kappa}\,\kappa - 2\underline{\omega}\kappa &= 2\,\mathit{d}_1\eta + 2\rho - \frac{1}{2}\underline{\vartheta}\,\vartheta + 2(\xi\underline{\xi} + \eta^2),\\ e_4\kappa + \frac{1}{2}\kappa^2 + 2\omega\kappa &= 2\,\mathit{d}_1\xi - \frac{1}{2}\vartheta^2 + 2(\eta + \underline{\eta} + 2\zeta)\xi,\end{aligned}$$

we deduce

$$\begin{aligned}e'_\theta\kappa &= e_\theta\kappa + \frac{1}{2}\underline{f}\left(-\frac{1}{2}\kappa^2 - 2\omega\kappa\right) + \frac{1}{2}f\left(-\frac{1}{2}\underline{\kappa}\,\kappa + 2\underline{\omega}\kappa + 2\rho\right)\\ &+ \frac{1}{2}f\underline{f}e_\theta\kappa + \frac{1}{8}f^2\underline{f}e_3\kappa + \frac{1}{2}\underline{f}\left(2\,\mathit{d}_1\xi - \frac{1}{2}\vartheta^2 + 2(\eta+\underline{\eta}+2\zeta)\xi\right)\\ &+ \frac{1}{2}f\left(2\,\mathit{d}_1\eta - \frac{1}{2}\underline{\vartheta}\,\vartheta + 2(\xi\underline{\xi}+\eta^2)\right).\end{aligned}$$

Hence,

$$\begin{aligned}e'_\theta(\kappa') &= e_\theta\kappa + e'_\theta(a)\kappa + e'_\theta\,\mathit{d}'_1 f + \kappa e'_\theta a - \frac{1}{4}\kappa(f\underline{\kappa} + \underline{f}\kappa) + \kappa(f\underline{\omega} - \omega\underline{f}) + f\rho\\ &+\mathrm{Err}(e'_\theta\kappa', e_\theta\kappa)\end{aligned}$$

where

$$\begin{aligned}\mathrm{Err}(e'_\theta\kappa', e_\theta\kappa) &= \mathrm{Err}_1(e'_\theta\kappa', e_\theta\kappa) + \frac{1}{2}f\left(2\,\mathit{d}_1\eta - \frac{1}{2}\underline{\vartheta}\,\vartheta + 2(\xi\underline{\xi}+\eta^2)\right)\\ &+ \frac{1}{2}f\underline{f}e_\theta\kappa + \frac{1}{8}f^2\underline{f}e_3\kappa + \frac{1}{2}\underline{f}\left(2\,\mathit{d}_1\xi - \frac{1}{2}\vartheta^2 + 2(\eta+\underline{\eta}+2\zeta)\xi\right)\end{aligned}$$

as desired. The formula for $e'_\theta(\underline{\kappa}')$ is easily derived by symmetry from the one on $e'_\theta(\kappa')$. Note however that a becomes $-a$ in the transformation.

Applying the operator $e'_\theta = \left(1+\frac{1}{2}f\underline{f}\right)e_\theta + \frac{1}{2}\underline{f}e_4 + \frac{1}{2}f\left(1+\frac{1}{4}f\underline{f}\right)e_3$ to the transformation formula for μ,

$$\mu' = \mu + (\not{d}_1)' \left(-(\not{d}_1^\star)' a + f\underline{\omega} - \underline{f}\omega + \frac{1}{4} f\underline{\kappa} - \frac{1}{4}\underline{f}\kappa \right) + \mathrm{Err}(\mu, \mu'),$$

we derive

$$\begin{aligned} e'_\theta(\mu') &= e'_\theta(\mu) + e'_\theta(\not{d}_1)' \left(-(\not{d}_1^\star)' a + f\underline{\omega} - \underline{f}\omega + \frac{1}{4} f\underline{\kappa} - \frac{1}{4}\underline{f}\kappa \right) + e'_\theta \mathrm{Err}(\mu, \mu') \\ &= e_\theta(\mu) + \frac{1}{2}\underline{f} e_4\mu + \frac{1}{2} f e_3 \mu + e'_\theta(\not{d}_1)' \left(-(\not{d}_1^\star)' a + f\underline{\omega} - \underline{f}\omega + \frac{1}{4} f\underline{\kappa} - \frac{1}{4}\underline{f}\kappa \right) \\ &+ e'_\theta \mathrm{Err}(\mu, \mu') + \frac{1}{2} f\underline{f} e_\theta \mu + \frac{1}{8} f^2 \underline{f} e_3 \mu. \end{aligned}$$

Recalling that $\mu = - \not{d}_1\zeta - \rho + \frac{1}{4}\vartheta\underline{\vartheta}$ we find

$$\begin{aligned} \frac{1}{2}\underline{f} e_4\mu + \frac{1}{2} f e_3\mu &= -\frac{1}{2}(f e_3 + \underline{f} e_4)\rho + \frac{1}{2}\underline{f} e_4 \left(- \not{d}_1\zeta + \frac{1}{4}\vartheta\underline{\vartheta} \right) \\ &+ \frac{1}{2} f e_3 \left(- \not{d}_1\zeta + \frac{1}{4}\vartheta\underline{\vartheta} \right). \end{aligned}$$

Recalling the Bianchi equations for $e_3\rho, e_4\rho$

$$\begin{aligned} e_4\rho + \frac{3}{2}\kappa\rho &= \not{d}_1\beta - \tfrac{1}{2}\underline{\vartheta}\,\alpha + \zeta\,\beta + 2(\underline{\eta}\,\beta + \xi\,\underline{\beta}), \\ e_3\rho + \frac{3}{2}\underline{\kappa}\rho &= \not{d}_1\underline{\beta} - \tfrac{1}{2}\vartheta\,\underline{\alpha} - \zeta\,\underline{\beta} + 2(\eta\,\underline{\beta} + \underline{\xi}\,\beta), \end{aligned}$$

we further deduce

$$\begin{aligned} \frac{1}{2}\underline{f} e_4\mu + \frac{1}{2} f e_3\mu &= \frac{3}{4}\rho(f\underline{\kappa} + \underline{f}\kappa) \\ &- \frac{1}{2} f \Big(\not{d}_1\underline{\beta} - \frac{1}{2}\vartheta\,\underline{\alpha} - \zeta\,\underline{\beta} + 2(\eta\,\underline{\beta} + \underline{\xi}\,\beta) \Big) \\ &- \frac{1}{2}\underline{f} \Big(\not{d}_1\beta - \frac{1}{2}\underline{\vartheta}\,\alpha + \zeta\,\beta + 2(\underline{\eta}\,\beta + \xi\,\underline{\beta}) \Big) \\ &+ \frac{1}{2}\underline{f} e_4 \left(- \not{d}_1\zeta + \frac{1}{4}\vartheta\underline{\vartheta} \right) + \frac{1}{2} f e_3 \left(- \not{d}_1\zeta + \frac{1}{4}\vartheta\underline{\vartheta} \right). \end{aligned}$$

Therefore,

$$\begin{aligned} e'_\theta(\mu') &= e_\theta(\mu) + \frac{3}{4}\rho(f\underline{\kappa} + \underline{f}\kappa) + e'_\theta(\not{d}_1)' \left(-(\not{d}_1^\star)' a + f\underline{\omega} - \underline{f}\omega + \frac{1}{4} f\underline{\kappa} - \frac{1}{4}\underline{f}\kappa \right) \\ &+ \mathrm{Err}(e_\theta\mu, e_\theta\mu) \end{aligned}$$

with

$$\begin{aligned}\mathrm{Err}(e_\theta'\mu', e_\theta\mu) &= e_\theta'\mathrm{Err}(\mu,\mu') + \frac{1}{2}f\underline{f}e_\theta\mu + \frac{1}{8}f^2\underline{f}e_3\mu\\ &- \frac{1}{2}f\Big(\not{d}_1\underline{\beta} - \frac{1}{2}\vartheta\,\underline{\alpha} - \zeta\,\underline{\beta} + 2(\eta\,\underline{\beta} + \underline{\xi}\,\beta)\Big)\\ &- \frac{1}{2}\underline{f}\Big(\not{d}_1\beta - \frac{1}{2}\underline{\vartheta}\,\alpha + \zeta\,\beta + 2(\underline{\eta}\,\beta + \xi\,\underline{\beta})\Big)\\ &+ \frac{1}{2}\underline{f}e_4\left(-\not{d}_1\zeta + \frac{1}{4}\vartheta\underline{\vartheta}\right) + \frac{1}{2}fe_3\left(-\not{d}_1\zeta + \frac{1}{4}\vartheta\underline{\vartheta}\right)\end{aligned}$$

as desired. □

Finally recalling the definition of the Hodge operators $\not{d}_1, \not{d}_1^\star, (\not{d}_1)', (\not{d}_1^\star)'$ and noticing that

$$\begin{aligned}(\not{d}_1^\star)'(\kappa') &= (\not{d}_1^\star)'(\check{\kappa}'), \qquad (\not{d}_1^\star)(\kappa) = (\not{d}_1^\star)(\check{\kappa}),\\ (\not{d}_1^\star)'(\underline{\kappa}') &= (\not{d}_1^\star)'(\check{\underline{\kappa}}'), \qquad (\not{d}_1^\star)(\underline{\kappa}) = (\not{d}_1^\star)(\check{\underline{\kappa}}),\\ (\not{d}_1^\star)'(\mu') &= (\not{d}_1^\star)'(\check{\mu}'), \qquad (\not{d}_1^\star)(\mu) = (\not{d}_1^\star)(\check{\mu}),\\ (\not{d}_1^\star)'(\underline{\mu}') &= (\not{d}_1^\star)'(\check{\underline{\mu}}'), \qquad (\not{d}_1^\star)(\underline{\mu}) = (\not{d}_1^\star)(\check{\underline{\mu}}),\end{aligned}$$

we recast the results of Lemma 9.23 in the following form.

Lemma 9.24. *We have the transformation equations,*

$$\begin{aligned}(\not{d}_1^\star)'(\check{\kappa}') &= \not{d}_1^\star(\check{\kappa}) + (\not{d}_1^\star)'(\not{d}_1)'f + \kappa(\not{d}_1^\star)'a - \rho f + \frac{1}{4}\kappa(f\underline{\kappa} + \underline{f}\kappa)\\ &\quad - \kappa(f\underline{\omega} - \underline{f}\omega) - Err_1,\\ (\not{d}_1^\star)'(\check{\underline{\kappa}}') &= \not{d}_1^\star(\check{\underline{\kappa}}) + (\not{d}_1^\star)'(\not{d}_1)'\underline{f} - \underline{\kappa}(\not{d}_1^\star)'a - \rho\underline{f} + \frac{1}{4}\underline{\kappa}(f\underline{\kappa} + \underline{f}\kappa)\\ &\quad - \underline{\kappa}(\underline{f}\omega - f\underline{\omega}) - Err_2,\\ (\not{d}_1^\star)'(\check{\mu}') &= \not{d}_1^\star(\check{\mu}) + (\not{d}_1^\star)'(\not{d}_1)'\left(-(\not{d}_1^\star)'a + f\underline{\omega} - \underline{f}\omega + \frac{1}{4}f\underline{\kappa} - \frac{1}{4}\underline{f}\kappa\right)\\ &\quad - \frac{3}{4}\rho\,(f\underline{\kappa} + \underline{f}\kappa) - Err_3,\end{aligned} \tag{9.3.13}$$

where

$$\begin{aligned}Err_1 &= Err(e_\theta'\kappa', e_\theta\kappa) = e_\theta'\,Err(\kappa,\kappa') + ae_\theta'\,\not{d}_1'f + e_\theta'(a)\,\not{d}_1'f\\ &\quad + a\left(e_\theta\kappa + \frac{1}{2}\left(\underline{f}e_4\kappa + fe_3\kappa\right)\right) + f\,\not{d}_1\eta + \underline{f}\,\not{d}_1\xi + l.o.t.,\\ Err_2 &= Err(e_\theta'\underline{\kappa}', e_\theta\underline{\kappa}) = e_\theta'\,Err(\underline{\kappa},\underline{\kappa}') - ae_\theta'\,\not{d}_1'\underline{f} - e_\theta'(a)\,\not{d}_1'\underline{f}\\ &\quad - a\left(e_\theta\underline{\kappa} + \frac{1}{2}\left(\underline{f}e_4\underline{\kappa} + fe_3\underline{\kappa}\right)\right) + \underline{f}\,\not{d}_1\underline{\eta} + f\,\not{d}_1\underline{\xi} + l.o.t.,\end{aligned} \tag{9.3.14}$$

$$\begin{aligned}Err_3 &= Err(e_\theta'\mu', e_\theta\mu) = e_\theta'\,Err(\mu,\mu') - \frac{1}{2}\left(\underline{f}\,\not{d}_1\beta + f\,\not{d}_1\underline{\beta}\right)\\ &\quad - \frac{1}{2}(fe_3 + \underline{f}e_4)\,\not{d}_1\zeta + l.o.t.,\end{aligned}$$

where the terms denoted by l.o.t. are cubic or higher order in $a, f, \underline{f}, \check{\Gamma}, \check{R}$ *and contain no derivatives of* $(a, f, \underline{f})$.

9.3.1 Main GCM equations

Given a deformation $\Psi : \overset{\circ}{S} \to \mathbf{S}$ and adapted frame (e_3', e_4', e_θ') with $e_\theta' = e_\theta^{\mathbf{S}}$ we derive an elliptic system for the transition parameters $(a, f, \underline{f})$. The system will later be used in the construction of GCM surfaces.

In what follows we denote by $\not{d}_1^{\mathbf{S}}, \not{d}_2^{\mathbf{S}}, \not{d}_1^{\mathbf{S},\star}, \not{d}_2^{\mathbf{S},\star}$ the basic Hodge operators on $\mathbf{S}$. Noting that the transformation formulae in (9.3.13)–(9.3.14) contain only the operators $(\not{d}_1)' = \not{d}_1^{\mathbf{S}}, (\not{d}_1^\star)' = \not{d}_1^{\mathbf{S},\star}$ applied to $a, f, \underline{f}$ we introduce the simplified notation,

$$\not{d}^{\mathbf{S}} := (\not{d}_1)', \qquad \not{d}^{\mathbf{S},\star} := (\not{d}_1^\star)', \qquad A^{\mathbf{S}} := \not{d}^{\mathbf{S},\star}\not{d}^{\mathbf{S}}, \qquad \not{d}^\star := \not{d}_1^\star. \tag{9.3.15}$$

With these notation (9.3.13) takes the following form:

$$\begin{aligned}
\not{d}^{\mathbf{S},\star}\check{\kappa}^{\mathbf{S}} &= \not{d}^\star\check{\kappa} + A^{\mathbf{S}}f + \kappa\not{d}^{\mathbf{S},\star}a - \rho f + \frac{1}{4}\kappa(f\underline{\kappa} + \underline{f}\kappa) - \kappa(f\underline{\omega} - \underline{f}\omega) - \mathrm{Err}_1,\\
\not{d}^{\mathbf{S},\star}\check{\underline{\kappa}}^{\mathbf{S}} &= \not{d}^\star\check{\underline{\kappa}} + A^{\mathbf{S}}\underline{f} - \underline{\kappa}\not{d}^{\mathbf{S},\star}a - \rho\underline{f} + \frac{1}{4}\underline{\kappa}(f\underline{\kappa} + \underline{f}\kappa) - \underline{\kappa}(\underline{f}\omega - f\underline{\omega}) - \mathrm{Err}_2,\\
\not{d}^{\mathbf{S},\star}\check{\mu}^{\mathbf{S}} &= \not{d}^\star\check{\mu} + A^{\mathbf{S}}\left(-\not{d}^{\mathbf{S},\star}a + f\underline{\omega} - \underline{f}\omega + \frac{1}{4}f\underline{\kappa} - \frac{1}{4}\underline{f}\kappa\right) - \frac{3}{4}\rho\left(f\underline{\kappa} + \underline{f}\kappa\right) - \mathrm{Err}_3,
\end{aligned}$$

or

$$\begin{aligned}
A^{\mathbf{S}}\left(-\not{d}^{\mathbf{S},\star}a + f\underline{\omega} - \underline{f}\omega + \frac{1}{4}f\underline{\kappa} - \frac{1}{4}\underline{f}\kappa\right) - \frac{3}{4}\rho(\kappa\underline{f} + \underline{\kappa}f) &= \not{d}^{\mathbf{S},\star}\check{\mu}^{\mathbf{S}} - \not{d}^\star\check{\mu}\\
&\quad + \mathrm{Err}_3,\\
A^{\mathbf{S}}f + \kappa\not{d}^{\mathbf{S},\star}a - \rho f + \frac{1}{4}\kappa(f\underline{\kappa} + \underline{f}\kappa) - \kappa(f\underline{\omega} - \underline{f}\omega) &= \not{d}^{\mathbf{S},\star}\check{\kappa}^{\mathbf{S}} - \not{d}^\star\check{\kappa}\\
&\quad + \mathrm{Err}_1,\\
A^{\mathbf{S}}\underline{f} - \underline{\kappa}\not{d}^{\mathbf{S},\star}a - \rho\underline{f} + \frac{1}{4}\underline{\kappa}(f\underline{\kappa} + \underline{f}\kappa) - \underline{\kappa}(\underline{f}\omega - f\underline{\omega}) &= \not{d}^{\mathbf{S},\star}\check{\underline{\kappa}}^{\mathbf{S}} - \not{d}^\star\check{\underline{\kappa}}\\
&\quad + \mathrm{Err}_2.
\end{aligned} \tag{9.3.16}$$

Since $A^{\mathbf{S}}$ is invertible[9] we can write, setting $z := \kappa\underline{f} + \underline{\kappa}f$,

$$\begin{aligned}
\not{d}^{\mathbf{S},\star}a &= f\underline{\omega} - \underline{f}\omega + \frac{1}{4}f\underline{\kappa} - \frac{1}{4}\underline{f}\kappa - \frac{3}{4}(A^{\mathbf{S}})^{-1}(\rho z)\\
&\quad + (A^{\mathbf{S}})^{-1}\big(-\not{d}^{\mathbf{S},\star}\check{\mu}^{\mathbf{S}} + \not{d}^\star\check{\mu} - \mathrm{Err}_3\big).
\end{aligned}$$

[9]We have $\int_{\mathbf{S}} fA^{\mathbf{S}}f = \int_{\mathbf{S}}(\not{d}^{\mathbf{S}}f)^2$ which in view of the identity (2.1.21) for $\not{d}_1^{\mathbf{S}}$ and the definition of $\not{d}^{\mathbf{S}}$ implies that $A^{\mathbf{S}}$ is invertible.

We can thus eliminate $\slashed{d}^{\mathbf{S},\star}a$ from the last two equations,

$$
\begin{aligned}
A^{\mathbf{S}}f+\left(\frac{1}{2}\kappa\underline{\kappa}-\rho\right)f-\frac{3}{4}\kappa(A^{\mathbf{S}})^{-1}(\rho z) &= \slashed{d}^{\mathbf{S},\star}\check{\kappa}^{\mathbf{S}}-\slashed{d}^{\star}\check{\kappa}\\
&\quad -\kappa(A^{\mathbf{S}})^{-1}\big(-\slashed{d}^{\mathbf{S},\star}\check{\mu}^{\mathbf{S}}+\slashed{d}^{\star}\check{\mu}\big)+\mathrm{Err}_4,\\
A^{\mathbf{S}}\underline{f}+\left(\frac{1}{2}\kappa\underline{\kappa}-\rho\right)\underline{f}+\frac{3}{4}\underline{\kappa}(A^{\mathbf{S}})^{-1}(\rho z) &= \slashed{d}^{\mathbf{S},\star}\check{\underline{\kappa}}^{\mathbf{S}}-\slashed{d}^{\star}\check{\underline{\kappa}}\\
&\quad +\underline{\kappa}(A^{\mathbf{S}})^{-1}\big(-\slashed{d}^{\mathbf{S},\star}\check{\mu}^{\mathbf{S}}+\slashed{d}^{\star}\check{\mu}\big)+\mathrm{Err}_5,
\end{aligned}
$$

where

$$
\mathrm{Err}_4 = \mathrm{Err}_1+\kappa(A^{\mathbf{S}})^{-1}\mathrm{Err}_3, \qquad \mathrm{Err}_5=\mathrm{Err}_2-\underline{\kappa}(A^{\mathbf{S}})^{-1}\mathrm{Err}_3.
$$

Therefore the system (9.3.16) is equivalent to the system

$$
\begin{aligned}
A^{\mathbf{S}}f+\left(\frac{1}{2}\kappa\underline{\kappa}-\rho\right)f-\frac{3}{4}\kappa(A^{\mathbf{S}})^{-1}(\rho z) &= \slashed{d}^{\mathbf{S},\star}\check{\kappa}^{\mathbf{S}}-\slashed{d}^{\star}\check{\kappa}\\
&\quad -\kappa(A^{\mathbf{S}})^{-1}\big(-\slashed{d}^{\mathbf{S},\star}\check{\mu}^{\mathbf{S}}+\slashed{d}^{\star}\check{\mu}\big)+\mathrm{Err}_4,\\
A^{\mathbf{S}}\underline{f}+\left(\frac{1}{2}\kappa\underline{\kappa}-\rho\right)\underline{f}+\frac{3}{4}\underline{\kappa}(A^{\mathbf{S}})^{-1}(\rho z) &= \slashed{d}^{\mathbf{S},\star}\check{\underline{\kappa}}^{\mathbf{S}}-\slashed{d}^{\star}\check{\underline{\kappa}}\\
&\quad +\underline{\kappa}(A^{\mathbf{S}})^{-1}\big(-\slashed{d}^{\mathbf{S},\star}\check{\mu}^{\mathbf{S}}+\slashed{d}^{\star}\check{\mu}\big)+\mathrm{Err}_5,\\
\slashed{d}^{\mathbf{S},\star}a+\frac{3}{4}(A^{\mathbf{S}})^{-1}(\rho z)-f\underline{\omega}+\underline{f}\omega&\\
-\frac{1}{4}f\underline{\kappa}+\frac{1}{4}\underline{f}\kappa &= (A^{\mathbf{S}})^{-1}\big(-\slashed{d}^{\mathbf{S},\star}\check{\mu}^{\mathbf{S}}+\slashed{d}^{\star}\check{\mu}\big)-(A^{\mathbf{S}})^{-1}\mathrm{Err}_3.
\end{aligned}
$$

We summarize the results of the above calculation in the following lemma.

Lemma 9.25. *The original system* (9.3.13) *in* $(a,f,\underline{f})$ *associated to a deformation sphere* $\mathbf{S}$ *is equivalent to the following:*

$$
\begin{aligned}
(A^{\mathbf{S}}+V)\,f &= \frac{3}{4}\kappa(A^{\mathbf{S}})^{-1}(\rho z)+\slashed{d}^{\mathbf{S},\star}\check{\kappa}^{\mathbf{S}}-\slashed{d}^{\star}\check{\kappa}\\
&\quad -\kappa(A^{\mathbf{S}})^{-1}\big(-\slashed{d}^{\mathbf{S},\star}\check{\mu}^{\mathbf{S}}+\slashed{d}^{\star}\check{\mu}\big)+Err_4,\\
(A^{\mathbf{S}}+V)\,\underline{f} &= -\frac{3}{4}\underline{\kappa}(A^{\mathbf{S}})^{-1}(\rho z)+\slashed{d}^{\mathbf{S},\star}\check{\underline{\kappa}}^{\mathbf{S}}-\slashed{d}^{\star}\check{\underline{\kappa}}\\
&\quad +\underline{\kappa}(A^{\mathbf{S}})^{-1}\big(-\slashed{d}^{\mathbf{S},\star}\check{\mu}^{\mathbf{S}}+\slashed{d}^{\star}\check{\mu}\big)+Err_5,\\
\slashed{d}^{\mathbf{S},\star}a &= -\frac{3}{4}(A^{\mathbf{S}})^{-1}(\rho z)+f\underline{\omega}-\underline{f}\omega+\frac{1}{4}f\underline{\kappa}-\frac{1}{4}\underline{f}\kappa\\
&\quad +(A^{\mathbf{S}})^{-1}\big(-\slashed{d}^{\mathbf{S},\star}\check{\mu}^{\mathbf{S}}+\slashed{d}^{\star}\check{\mu}\big)-(A^{\mathbf{S}})^{-1}Err_3,
\end{aligned}
\tag{9.3.17}
$$

where

$$
z:=\underline{\kappa}f+\kappa\underline{f}, \qquad V:=\frac{1}{2}\kappa\underline{\kappa}-\rho. \tag{9.3.18}
$$

The error terms are given by Err_1, Err_2, Err_3*, defined in Lemma 9.24, and*

$$
Err_4 = Err_1+\kappa(A^{\mathbf{S}})^{-1}Err_3, \qquad Err_5=Err_2-\underline{\kappa}(A^{\mathbf{S}})^{-1}Err_3. \tag{9.3.19}
$$

Remark 9.26. *We note the following remarks concerning the system* (9.3.17).

1. *The right-hand side of the equations is linear in the quantities*

$$\not{d}^{\mathbf{S},\star}\check{\kappa}^{\mathbf{S}}, \quad \not{d}^{\mathbf{S},\star}\check{\underline{\kappa}}^{\mathbf{S}}, \quad \not{d}^{\mathbf{S},\star}\check{\mu}^{\mathbf{S}}, \qquad \text{as well as} \qquad \not{d}\overline{\kappa}, \quad \not{d}\overline{\underline{\kappa}}, \quad \not{d}\overline{\mu}.$$

The first group is to be constrained by our GCM conditions in the next section while the second group depends on assumptions regarding the background foliation of $\mathcal{R}$.

2. *The error terms contain only* **S**-*angular derivatives of* $(a, f, \underline{f})$ *of order at most equal to the order of the corresponding operators on the left-hand sides, see Lemma 9.27 below. Thus the system is in a standard quasilinear elliptic system form.*
3. *In order to uniquely solve the equations for* f *and* $\underline{f}$, *we need to establish the coercivity of the operator* $A^{\mathbf{S}} + V$. *One can easily show that the potential* V *is positive for small values of* r, *i.e.,* r *near* $r_{\mathcal{H}} = 2m_0(1+\delta_{\mathcal{H}})$ *but negative for large* r. *In fact* $A^{\mathbf{S}} + V$ *has a nontrivial kernel for large* r *as one can easily see from the following calculation. Since*

$$A^{\mathbf{S}} = \not{d}_1^{\star \mathbf{S}}\, \not{d}_1{}^{\mathbf{S}} = \not{d}_2{}^{\mathbf{S}}\, \not{d}_2^{\star \mathbf{S}} + 2K, \qquad K = -\rho - \frac{1}{4}\kappa\underline{\kappa} + \frac{1}{4}\vartheta\underline{\vartheta}$$

we deduce

$$A^{\mathbf{S}} + V = A^{\mathbf{S}} + \frac{1}{2}\kappa\underline{\kappa} - \rho \quad = \quad \not{d}_2{}^{\mathbf{S}}\, \not{d}_2^{\star \mathbf{S}} - 3\rho + \frac{1}{2}\vartheta\underline{\vartheta}.$$

Thus for large enough r *the operator* $A^{\mathbf{S}} + V$ *behaves like* $\not{d}_2^{\mathbf{S}}\, \not{d}_2^{\mathbf{S},\star}$ *which has a nontrivial kernel.*

4. *To be able to correct for the lack of coercivity of the system, we need to prescribe the* $\ell = 1$ *modes of* $(f, \underline{f})$.
5. *The equations do not provide information on the average of* a. *For this we will need yet another equation derived in section 9.3.2.*

Lemma 9.27. *The error terms* $Err_1, \ldots, Err_5$ *can be written schematically as follows:*

$$\begin{aligned} r^2 Err_1 &= (\not{\mathfrak{d}}^{\mathbf{S}})^2\left((f,\underline{f},a)^2\right) + \not{\mathfrak{d}}^{\mathbf{S}}\left((f,\underline{f},a)(r\Gamma_g)\right) + l.o.t., \\ r\, Err_2 &= r^{-1}(\not{\mathfrak{d}}^{\mathbf{S}})^2\left((f,\underline{f},a)^2\right) + \not{\mathfrak{d}}^{\mathbf{S}}\left((f,\underline{f},a)(\check{\Gamma})\right) + l.o.t., \\ r^3 Err_3 &= (\not{\mathfrak{d}}^{\mathbf{S}})^3\left((f,\underline{f},a)^2\right) + (\not{\mathfrak{d}}^{\mathbf{S}})^2\left((f,\underline{f},a)(r\Gamma_g)\right) + l.o.t., \\ Err_4, Err_5 &= Err_1 + r^{-1}(A^{\mathbf{S}})^{-1}Err_3, \end{aligned} \tag{9.3.20}$$

where the lower order terms denoted l.o.t. are cubic with respect to $a, f, \underline{f}, \check{\Gamma}, \check{R}$ *and may involve fewer angular (along* **S***) derivatives of* $a, f, \underline{f}$.

Remark 9.28. *Note that* Err_2 *behaves worse in powers of* r *than* Err_1. *The reason is the presence of the terms* $f e_\theta \underline{\xi}, e_\theta(f\underline{\xi})$ *in the formula for* $e_\theta^{\mathbf{S}}(Err(\underline{\kappa}', \underline{\kappa}))$.

Proof. Note that in the spacetime region $\mathcal{R}$ of interest r and $r^{\mathbf{S}}$ are comparable.

Recall, see (9.3.14),

$$
\begin{aligned}
\mathrm{Err}_1 &= \mathrm{Err}(e'_\theta\kappa', e_\theta\kappa) = e'_\theta\,\mathrm{Err}(\kappa,\kappa') + ae'_\theta\,{\not{d}}'_1 f + e'_\theta(a)\,{\not{d}}'_1 f\\
&\quad + a\left(e_\theta\kappa + \frac{1}{2}\left(\underline{f}e_4\kappa + fe_3\kappa\right)\right) + f\,{\not{d}}_1\eta + \underline{f}\,{\not{d}}_1\xi + \text{l.o.t.},\\
\mathrm{Err}_2 &= \mathrm{Err}(e'_\theta\underline{\kappa}', e_\theta\underline{\kappa}) = e'_\theta\,\mathrm{Err}(\underline{\kappa},\underline{\kappa}') - ae'_\theta\,{\not{d}}'_1\underline{f} - e'_\theta(a)\,{\not{d}}'_1\underline{f}\\
&\quad - a\left(e_\theta\underline{\kappa} + \frac{1}{2}\left(\underline{f}e_4\underline{\kappa} + fe_3\underline{\kappa}\right)\right) + \underline{f}\,{\not{d}}_1\underline{\eta} + f\,{\not{d}}_1\underline{\xi} + \text{l.o.t.},\\
\mathrm{Err}_3 &= \mathrm{Err}(e'_\theta\mu', e_\theta\mu) = e'_\theta\mathrm{Err}(\mu,\mu') - \frac{1}{2}\left(\underline{f}\,{\not{d}}_1\beta + f\,{\not{d}}_1\underline{\beta}\right) - \frac{1}{2}(fe_3 + \underline{f}e_4)\,{\not{d}}_1\zeta\\
&\quad + \text{l.o.t.},
\end{aligned}
$$

and[10]

$$
\begin{aligned}
\mathrm{Err}(\kappa,\kappa') &= \frac{1}{2}f\underline{f}e'_\theta(f) - \frac{1}{4}\underline{f}e'_\theta(f^2) + f(\zeta+\eta) + \underline{f}\xi - \frac{1}{4}f^2\underline{\kappa} + f\underline{f}\omega - f^2\underline{\omega}\\
&\quad +\text{l.o.t.},\\
\mathrm{Err}(\underline{\kappa},\underline{\kappa}') &= -\frac{1}{2}\underline{f}e'_\theta\left(f\underline{f} + \frac{1}{8}f^2\underline{f}^2\right) + \left(\frac{3}{4}f\underline{f} + \frac{1}{8}(f\underline{f})^2\right)e'_\theta(\underline{f})\\
&\quad + \frac{1}{4}\left(1+\frac{1}{2}f\underline{f}\right)\underline{f}e'_\theta\left(f\underline{f}\right) - \frac{1}{4}f\left(1+\frac{1}{4}f\underline{f}\right)e'_\theta\left(\underline{f}^2\right)\\
&\quad + \underline{f}(-\zeta+\underline{\eta}) + f\underline{\xi} - \frac{1}{4}\underline{f}^2\kappa + f\underline{f}\underline{\omega} - \underline{f}^2\omega + \text{l.o.t.}
\end{aligned}
$$

Also,

$$
\begin{aligned}
\mathrm{Err}(\mu,\mu') &= -e'_\theta\mathrm{Err}(\zeta,\zeta') - \mathrm{Err}(\rho,\rho') + \frac{1}{4}(\vartheta'\underline{\vartheta}' - \vartheta\underline{\vartheta}),\\
\mathrm{Err}(\zeta,\zeta') &= \frac{1}{2}\underline{f}\left(1+\frac{1}{4}f\underline{f}\right)e'_\theta(f) - \frac{1}{16}\underline{f}^2e'_\theta(f^2) + \frac{1}{4}(-f\underline{\vartheta} + \underline{f}\vartheta) + \text{l.o.t.},\\
\mathrm{Err}(\rho,\rho') &= \frac{3}{2}\rho f\underline{f} + \underline{f}\beta + f\underline{\beta} + \text{l.o.t.}
\end{aligned}
$$

We write schematically[11]

$$
\begin{aligned}
\mathrm{Err}_1 &= (f,\underline{f},a)(r^{-2}\,\mathfrak{\not{d}}^{\mathbf{S}})^2(f,\underline{f},a) + (r^{-1}\,\mathfrak{\not{d}}^{\mathbf{S}}(f,\underline{f},a))^2 + r^{-1}\,\mathfrak{\not{d}}^{\mathbf{S}}\left((f,\underline{f},a)\Gamma_g\right)\\
&\quad + r^{-2}\,\mathfrak{\not{d}}^{\mathbf{S}}(f^2) + \frac{1}{2}a\left(\underline{f}e_4\kappa + fe_3\kappa\right) + \text{l.o.t.}
\end{aligned}
$$

Making use of

$$
\begin{aligned}
e_3(\kappa) + \frac{1}{2}\underline{\kappa}\,\kappa - 2\underline{\omega}\kappa &= 2\,{\not{d}}_1\eta + 2\rho - \frac{1}{2}\underline{\vartheta}\,\vartheta + 2(\xi\underline{\xi} + \eta^2),\\
e_4\kappa + \frac{1}{2}\kappa^2 + 2\omega\kappa &= 2\,{\not{d}}_1\xi - \frac{1}{2}\vartheta^2 + 2(\eta + \underline{\eta} + 2\zeta)\xi,
\end{aligned}
$$

[10]Recall also the outgoing geodesic conditions, i.e., $\xi = 0$, $\zeta + \underline{\eta} = 0$, $\zeta - \eta = 0$, $\omega = 0$.

[11]The last term $r^{-2}\,\mathfrak{\not{d}}^{\mathbf{S}}(f^2)$ on the right of the identity below is due to the term $e'_\theta(f^2\underline{\omega})$ in the expression of $e'_\theta\,\mathrm{Err}(\underline{\kappa},\underline{\kappa}')$.

and treating the curvature terms that appear as Γ_g, we easily derive

$$r^2\mathrm{Err}_1 = (\not{\mathfrak{d}}^{\mathbf{S}})^2\left((f,\underline{f},a)^2\right) + \not{\mathfrak{d}}^{\mathbf{S}}\left((f,\underline{f},a)(r\Gamma_g)\right).$$

We obtain a worse estimate for Err_2 because of the presence $e'_\theta(f\underline{\xi})$, since $\underline{\xi}\in\Gamma_b$. In fact,

$$r\mathrm{Err}_2 = r^{-1}(\not{\mathfrak{d}}^{\mathbf{S}})^2\left((f,\underline{f},a)^2\right) + \not{\mathfrak{d}}^{\mathbf{S}}\left((f,\underline{f},a)\check{\Gamma}\right).$$

For Err_3 we write similarly, treating the curvature terms that appear as Γ_g,

$$e_\theta(\mu,\mu') = r^{-3}(\not{\mathfrak{d}}^{\mathbf{S}})^3\left((f,\underline{f},a)^2\right) + r^{-3}(\not{\mathfrak{d}}^{\mathbf{S}})^2\left((f,\underline{f},a)\Gamma_g\right) + \text{l.o.t.}$$

Using the null structure equations for ζ we infer that

$$\begin{aligned}\mathrm{Err}_3 &= e_\theta(\mu,\mu') - \frac{1}{2}\left(\underline{f}\,\not{d}_1\beta + f\,\not{d}_1\underline{\beta}\right) - \frac{1}{2}(fe_3+\underline{f}e_4)\not{d}_1\zeta + \text{l.o.t.}\\ &= r^{-3}(\not{\mathfrak{d}}^{\mathbf{S}})^3\left((f,\underline{f},a)^2\right) + r^{-3}(\not{\mathfrak{d}}^{\mathbf{S}})^2\left((f,\underline{f},a)\Gamma_g\right) + \text{l.o.t.}\end{aligned}$$

as stated. □

Making use of the above lemma and the assumptions **A1**–**A3** we can derive the following.

Lemma 9.29. *Assume given a deformation $\Psi:\overset{\circ}{S}\to\mathbf{S}$ in $\mathcal{R}$ and adapted frame (e'_3,e'_4,e'_θ) with $e'_\theta=e^{\mathbf{S}}_\theta$ with transition parameters $a,f,\underline{f}$ defined on $\mathbf{S}$. Assume that there exists a small enough constant δ_1 such that the following holds true:*

$$(\overset{\circ}{r})^{-1}\|U'\|_{\mathfrak{h}_{s_{max}-1}(\overset{\circ}{S},\overset{\circ}{\not{g}})} + (\overset{\circ}{r})^{-2}\|S'\|_{\mathfrak{h}_{s_{max}-1}(\overset{\circ}{S},\overset{\circ}{\not{g}})} \lesssim \delta_1.$$

Then, for $5\le s\le s_{max}+1$,

$$\begin{aligned}\|Err_1,Err_2\|_{\mathfrak{h}_{s-2}(\mathbf{S})} &\lesssim r^{-2}\left\|(f,\underline{f},a)\right\|_{\mathfrak{h}_s(\mathbf{S})}\left(\overset{\circ}{\epsilon}+r^{-1}\left\|f,\underline{f},a\right\|_{\mathfrak{h}_{s-1}(\mathbf{S})}\right),\\ \|Err_3\|_{\mathfrak{h}_{s-3}(\mathbf{S})} &\lesssim r^{-3}\left\|(f,\underline{f},a)\right\|_{\mathfrak{h}_s(\mathbf{S})}\left(\overset{\circ}{\epsilon}+r^{-1}\left\|f,\underline{f},a\right\|_{\mathfrak{h}_{s-1}(\mathbf{S})}\right),\\ \|Err_4,Err_5\|_{\mathfrak{h}_{s-2}(\mathbf{S})} &\lesssim r^{-2}\left\|(f,\underline{f},a)\right\|_{\mathfrak{h}_s(\mathbf{S})}\left(\overset{\circ}{\epsilon}+r^{-1}\left\|f,\underline{f},a\right\|_{\mathfrak{h}_{s-1}(\mathbf{S})}\right).\end{aligned}\tag{9.3.21}$$

Proof. The proof follows easily from Lemma 9.27, Corollary 9.16, coercivity of $A^{\mathbf{S}}$ and obvious product estimates on $\mathbf{S}$. Consider for example the term

$$\mathrm{Err}_2 = r^{-2}(\not{\mathfrak{d}}^{\mathbf{S}})^2\left((f,\underline{f},a)^2\right) + r^{-1}\not{\mathfrak{d}}^{\mathbf{S}}\left((f,\underline{f},a)(\check{\Gamma})\right).$$

We write

$$(\not{\mathfrak{d}}^{\mathbf{S}})^k\mathrm{Err}_2 = r^{-2}(\not{\mathfrak{d}}^{\mathbf{S}})^{2+k}\left((f,\underline{f},a)^2\right) + r^{-1}(\not{\mathfrak{d}}^{\mathbf{S}})^{1+k}\left((f,\underline{f},a)(\check{\Gamma})\right) + \text{l.o.t.}$$

and

$$(\not{\mathfrak{d}}^{\mathbf{S}})^{2+k}\left((f,\underline{f},a)^2\right) = \sum_{i+j=k+2}\not{\mathfrak{d}}^i(f,\underline{f},a)\cdot\not{\mathfrak{d}}^j(f,\underline{f},a).$$

Thus, dividing the sum into terms with $i \geq [\frac{k+2}{2}]$ and $i < [\frac{k+2}{2}]$ and using Sobolev estimates for the terms involving fewer derivatives we derive, for $[\frac{k+2}{2}]+2 \leq k+1$,

$$\|(\not{\mathfrak{d}}^{\mathbf{S}})^{2+k}\left((f,\underline{f},a)^2\right)\|_{L^2(\mathbf{S})} \lesssim r^{-1}\|(a,f,\underline{f})\|_{\mathfrak{h}_{k+1}(\mathbf{S})}\|(a,f,\underline{f})\|_{\mathfrak{h}_{k+2}(\mathbf{S})}.$$

Similarly, making use of our assumptions for $\check{\Gamma}$,

$$\begin{aligned}(\not{\mathfrak{d}}^{\mathbf{S}})^{1+k}\left((f,\underline{f},a)(\check{\Gamma})\right) &\lesssim r^{-1}\|(a,f,\underline{f})\|_{\mathfrak{h}_{k+1}(\mathbf{S})}\|\check{\Gamma}\|_{\mathfrak{h}_{k+1}(\mathbf{S})}\\ &\lesssim \mathring{\epsilon}r^{-1}\|(a,f,\underline{f})\|_{\mathfrak{h}_{k+1}(\mathbf{S})}.\end{aligned}$$

Thus, for all $3 \leq k \leq s_{max}-1$

$$\|(\not{\mathfrak{d}}^{\mathbf{S}})^k \mathrm{Err}_2\|_{L^2(\mathbf{S})} \lesssim r^{-2}\,\|(f,\underline{f},a)\|_{\mathfrak{h}_{k+2}(\mathbf{S})}\left(\mathring{\epsilon}+r^{-1}\Big\|(f,\underline{f},a)\Big\|_{\mathfrak{h}_{k+1}(\mathbf{S})}\right),$$

i.e., for $5 \leq s \leq s_{max}+1$,

$$\|\mathrm{Err}_2\|_{\mathfrak{h}_{s-2}(\mathbf{S})} \lesssim r^{-2}\Big\|(f,\underline{f},a)\Big\|_{\mathfrak{h}_s(\mathbf{S})}\left(\mathring{\epsilon}+r^{-1}\Big\|f,\underline{f},a\Big\|_{\mathfrak{h}_{s-1}(\mathbf{S})}\right).$$

All other terms can be treated similarly. □

9.3.2 Equation for the average of a

In the proof of existence and uniqueness of GCMS, see Theorem 9.32, we will need, in addition to the equations derived so far, an equation for the average of a. To achieve this we make use of the transformation formula for κ of Lemma 9.21:

$$\begin{aligned}\kappa' &= e^a\left(\kappa+\not{d}_1' f\right)+e^a\mathrm{Err}(\kappa,\kappa'),\\ \mathrm{Err}(\kappa,\kappa') &= f(\zeta+\eta)+\underline{f}\xi+\frac{1}{2}\underline{f}e_4 f+\frac{1}{2}fe_3 f+\frac{1}{4}f\underline{f}\kappa+f\underline{f}\omega-\underline{\omega}f^2+\mathrm{l.o.t.}\end{aligned}$$

which we rewrite in the form

$$\kappa^{\mathbf{S}} = e^a\left(\frac{2}{r}+\check{\kappa}+\left(\overline{\kappa}-\frac{2}{r}\right)+\not{d}^{\mathbf{S}}f+\mathrm{Err}(\kappa,\kappa')\right).$$

We deduce

$$\begin{aligned}(e^a-1)\frac{2}{r} &= \kappa^{\mathbf{S}}-\frac{2}{r}-e^a\left(\check{\kappa}+\overline{\kappa}-\frac{2}{r}+\not{d}^{\mathbf{S}}f\right)-e^a\mathrm{Err}(\kappa,\kappa')\\ &= \kappa^{\mathbf{S}}-\frac{2}{r^{\mathbf{S}}}+\left(\frac{2}{r^{\mathbf{S}}}-\frac{2}{r}\right)-\left(\check{\kappa}+\overline{\kappa}-\frac{2}{r}+\not{d}^{\mathbf{S}}f\right)\\ &- e^a\mathrm{Err}(\kappa,\kappa')-(e^a-1)\left(\check{\kappa}+\overline{\kappa}-\frac{2}{r}+\not{d}^{\mathbf{S}}f\right)\end{aligned}$$

or

$$
\begin{aligned}
a\frac{2}{r} &= \kappa^{\mathbf{S}} - \frac{2}{r^{\mathbf{S}}} + \left(\frac{2}{r^{\mathbf{S}}} - \frac{2}{r}\right) - \left(\check{\kappa} + \overline{\kappa} - \frac{2}{r} + \not{d}^{\mathbf{S}} f\right)\\
&- e^a \mathrm{Err}(\kappa, \kappa') - (e^a - 1)\left(\check{\kappa} + \overline{\kappa} - \frac{2}{r} + \not{d}^{\mathbf{S}} f\right) - (e^a - 1 - a)\frac{2}{r}.
\end{aligned}
$$

We deduce

$$
\begin{aligned}
a &= \frac{r^{\mathbf{S}}}{2}\left(\kappa^{\mathbf{S}} - \frac{2}{r^{\mathbf{S}}}\right) + \left(1 - \frac{r^{\mathbf{S}}}{r}\right) - \frac{r^{\mathbf{S}}}{2}\left(\check{\kappa} + \overline{\kappa} - \frac{2}{r} + \not{d}^{\mathbf{S}} f\right) + \mathrm{Err}_6\\
\mathrm{Err}_6 &= -\frac{r^{\mathbf{S}}}{2}\left[e^a \mathrm{Err}(\kappa, \kappa') - (e^a - 1)\left(\check{\kappa} + \overline{\kappa} - \frac{2}{r} + \not{d}^{\mathbf{S}} f\right) - (e^a - 1 - a)\frac{2}{r}\right]\\
&- a\left(\frac{r^{\mathbf{S}}}{r} - 1\right).
\end{aligned}
$$

Taking the average on $\mathbf{S}$ we infer that

$$
\overline{a}^{\mathbf{S}} = \frac{r^{\mathbf{S}}}{2}\left(\overline{\kappa^{\mathbf{S}}}^{\mathbf{S}} - \frac{2}{r^{\mathbf{S}}}\right) + \overline{\left(1 - \frac{r^{\mathbf{S}}}{r}\right)}^{\mathbf{S}} - \frac{r^{\mathbf{S}}}{2}\overline{\left(\check{\kappa} + \overline{\kappa} - \frac{2}{r}\right)}^{\mathbf{S}} + \overline{\mathrm{Err}_6}^{\mathbf{S}} \tag{9.3.22}
$$

where $\overline{h}^{\mathbf{S}}$ denotes the average of h on $\mathbf{S}$.

9.3.3 Transversality conditions

Lemma 9.30. *Assume given a deformed sphere $\mathbf{S} \subset \mathcal{R}$ with adapted null frame $e_3^{\mathbf{S}}, e_4^{\mathbf{S}}, e_\theta^{\mathbf{S}}$ and transition functions $(a, f, \underline{f})$. We can extend $a, f, \underline{f}$, and thus the frame $e_3^{\mathbf{S}}, e_4^{\mathbf{S}}, e_\theta^{\mathbf{S}}$, in a small neighborhood of $\mathbf{S}$ such that the following hold true:*

$$
\xi^{\mathbf{S}} = 0, \qquad \omega^{\mathbf{S}} = 0, \qquad \underline{\eta}^{\mathbf{S}} + \zeta^{\mathbf{S}} = 0. \tag{9.3.23}
$$

Proof. According to Proposition 9.20 we have

$$
\begin{aligned}
\xi^{\mathbf{S}} &= e^{2a}\left(\xi + \frac{1}{2}e^{-a}e_4^{\mathbf{S}}(f) + \frac{1}{4}f\kappa + f\omega\right) + e^{2a}\mathrm{Err}(\xi, \xi^{\mathbf{S}}),\\
\zeta^{\mathbf{S}} &= \zeta - e_\theta^{\mathbf{S}}(a) - f\underline{\omega} + \underline{f}\omega - \frac{1}{4}f\underline{\kappa} + \frac{1}{4}\underline{f}\kappa + \mathrm{Err}(\zeta, \zeta^{\mathbf{S}}),\\
\underline{\eta}^{\mathbf{S}} &= \underline{\eta} + \frac{1}{2}e^{-a}e_4^{\mathbf{S}}\underline{f} - \underline{f}\omega + \frac{1}{4}\underline{f}\kappa + \mathrm{Err}(\underline{\eta}, \underline{\eta}^{\mathbf{S}}),\\
\omega^{\mathbf{S}} &= e^a\left(\omega - \frac{1}{2}e^{-a}e_4^{\mathbf{S}}a\right) + e^a\mathrm{Err}(\omega, \omega^{\mathbf{S}}).
\end{aligned}
$$

Clearly the conditions $\xi^{\mathbf{S}} = 0$, $\omega^{\mathbf{S}} = 0$ allow us to determine $e_4^{\mathbf{S}}f$ and $e_4^{\mathbf{S}}a$ on $\mathbf{S}$ while the condition $\underline{\eta}^{\mathbf{S}} + \zeta^{\mathbf{S}} = 0$ allows us to determine $e_4^{\mathbf{S}}\underline{f}$ on $\mathbf{S}$. □

Remark 9.31. *According to Proposition 9.20 we also have*

$$\underline{\xi}^{\mathbf{S}} = e^{-2a}\left(\underline{\xi} + \frac{1}{2}e^a e_3^{\mathbf{S}}(\underline{f}) + \frac{1}{4}\underline{f}\,\underline{\kappa} + \underline{f}\,\underline{\omega}\right) + e^{-2a}\,Err(\underline{\xi}, \underline{\xi}^{\mathbf{S}}),$$
$$\underline{\omega}^{\mathbf{S}} = e^{-a}\left(\underline{\omega} + \frac{1}{2}e^a e_3^{\mathbf{S}} a\right) + e^{-a}\,Err(\underline{\omega}, \underline{\omega}^{\mathbf{S}}),$$

so that we may impose, in addition, vanishing conditions on $\underline{\xi}^{\mathbf{S}}$ *and* $\underline{\omega}^{\mathbf{S}}$ *along* $\mathbf{S}$. *Indeed these are determined by* $e_3^{\mathbf{S}}\underline{f}$ *and* $e_3^{\mathbf{S}}a$.

9.4 EXISTENCE OF GCM SPHERES

We now impose the GCM conditions on the deformed sphere $\mathbf{S}$:

$$\begin{aligned} &\not{d}_2^{\mathbf{S},\star}\not{d}_1^{\mathbf{S},\star}\underline{\kappa}^{\mathbf{S}} = \not{d}_2^{\mathbf{S},\star}\not{d}_1^{\mathbf{S},\star}\mu^{\mathbf{S}} = 0, \qquad \kappa^{\mathbf{S}} = \frac{2}{r^{\mathbf{S}}}, \\ &\int_{\mathbf{S}} f e^{\Phi} = \Lambda, \qquad \int_{\mathbf{S}} \underline{f} e^{\Phi} = \underline{\Lambda}, \end{aligned} \tag{9.4.1}$$

where $(f, \underline{f})$ belong to the triplet $(f, \underline{f}, \lambda = e^a)$ which denotes the change of frame coefficients from the frame of $\overset{\circ}{S}$ to the one of $\mathbf{S}$.

We are ready to state the first main result of this chapter.

Theorem 9.32 (Existence of GCM spheres). *Let* $\overset{\circ}{S} = S(\overset{\circ}{u}, \overset{\circ}{s})$ *be a fixed sphere of the* (u, s) *outgoing geodesic foliation of a fixed spacetime region* $\mathcal{R}$. *Assume in addition to* ***A1–A3*** *that there exists scalar functions* $\underline{C} = \underline{C}(u, s)$, $M = M(u, s)$, *such that the following estimates hold true on* $\mathcal{R}$, *for all* $k \le s_{max}$, *with* $s_{max} \ge 6$,

$$\begin{aligned} \left|\mathfrak{d}^{k-1}\left(\not{d}_1^{\star}\underline{\kappa} - \underline{C}e^{\Phi}\right)\right| &\lesssim \overset{\circ}{\delta} r^{-3}, \\ \left|\mathfrak{d}^{k-1}\left(\not{d}_1^{\star}\mu - Me^{\Phi}\right)\right| &\lesssim \overset{\circ}{\delta} r^{-4}, \\ \left|\overline{\kappa} - \frac{2}{r}\right| + \left|\mathfrak{d}^k\check{\kappa}\right| &\lesssim \overset{\circ}{\delta} r^{-2}. \end{aligned} \tag{9.4.2}$$

For any fixed $\Lambda, \underline{\Lambda} \in \mathbb{R}$ *verifying*

$$|\Lambda|, |\underline{\Lambda}| \lesssim \overset{\circ}{\delta} r^2 \tag{9.4.3}$$

there exists a unique GCM sphere $\mathbf{S} = \mathbf{S}^{(\Lambda,\underline{\Lambda})}$, *which is a deformation of* $\overset{\circ}{S}$, *such that the GCM conditions* (9.4.1) *are verified. Moreover the following estimates hold true.*

1. *We have*

$$\left|\frac{r^{\mathbf{S}}}{\overset{\circ}{r}} - 1\right| \lesssim r^{-1}\overset{\circ}{\delta}. \tag{9.4.4}$$

In particular $r, \overset{\circ}{r}$ *and* $r^{\mathbf{S}}$ *are all comparable in* $\mathcal{R}$.

2. *The unique functions* $(\lambda, f, \underline{f})$ *on* $\mathbf{S}$*, which relate the original frame* e_3, e_4, e_θ *to the new frame on* $e_3^{\mathbf{S}}, e_4^{\mathbf{S}}, e_\theta^{\mathbf{S}}$ *according to* (9.2.18)*, verify the estimates*

$$\Big\|f, \underline{f}, \log\lambda\Big\|_{\mathfrak{h}_k(\mathbf{S})} \lesssim \overset{\circ}{\delta}, \qquad k \le s_{max}+1. \tag{9.4.5}$$

3. *The parameters* U, S *of the deformation, see Definition 9.6, verify the estimate*

$$\|(U', S')\|_{L^\infty(\overset{\circ}{S})} + \max_{0\le s\le s_{max}-1} r^{-1}\|(U', S')\|_{\mathfrak{h}_s(\overset{\circ}{S}, \overset{\circ}{\not{g}})} \lesssim \overset{\circ}{\delta}. \tag{9.4.6}$$

4. *The Hawking mass* $m^{\mathbf{S}}$ *verifies the estimate*

$$\left|m^{\mathbf{S}} - \overset{\circ}{m}\right| \lesssim \overset{\circ}{\delta}. \tag{9.4.7}$$

5. *The curvature components* $(\alpha^{\mathbf{S}}, \beta^{\mathbf{S}}, \rho^{\mathbf{S}}, \underline{\beta}^{\mathbf{S}}, \underline{\alpha}^{\mathbf{S}})$*, as well as* $\mu^{\mathbf{S}}$ *and the Ricci coefficients*[12] $(\kappa^{\mathbf{S}}, \vartheta^{\mathbf{S}}, \zeta^{\mathbf{S}}, \underline{\kappa}^{\mathbf{S}}, \underline{\vartheta}^{\mathbf{S}})$ *on* $\mathbf{S}$*, verify, for all* $k \le s_{max}$*,*

$$\begin{aligned}
\|\check{\kappa}^{\mathbf{S}}, \vartheta^{\mathbf{S}}, \zeta^{\mathbf{S}}, \check{\underline{\kappa}}^{\mathbf{S}}\|_{\mathfrak{h}_k(\mathbf{S})} &\lesssim \overset{\circ}{\epsilon} r^{-1},\\
\|\underline{\vartheta}^{\mathbf{S}}\|_{\mathfrak{h}_k(\mathbf{S})} &\lesssim \overset{\circ}{\epsilon},\\
\|\alpha^{\mathbf{S}}, \beta^{\mathbf{S}}, \check{\rho}^{\mathbf{S}}, \check{\mu}^{\mathbf{S}}\|_{\mathfrak{h}_k(\mathbf{S})} &\lesssim \overset{\circ}{\epsilon} r^{-2},\\
\|\underline{\beta}^{\mathbf{S}}\|_{\mathfrak{h}_k(\mathbf{S})} &\lesssim \overset{\circ}{\epsilon} r^{-1},\\
\|\underline{\alpha}^{\mathbf{S}}\|_{\mathfrak{h}_k(\mathbf{S})} &\lesssim \overset{\circ}{\epsilon}.
\end{aligned} \tag{9.4.8}$$

6. *The functions,* $(\lambda, f, \underline{f})$ *uniquely defined above, can be smoothly extended to a small neighborhood of* $\mathbf{S}$ *in such a way that the corresponding Ricci coefficients verify the following transversality conditions*

$$\xi^{\mathbf{S}} = 0, \qquad \omega^{\mathbf{S}} = 0, \qquad \underline{\eta}^{\mathbf{S}} + \zeta^{\mathbf{S}} = 0. \tag{9.4.9}$$

In that case, the following estimates hold[13] *for all* $k \le s_{max} - 1$

$$\|e_4^{\mathbf{S}}(f, \underline{f}, \log\lambda)\|_{\mathfrak{h}_k(\mathbf{S})} \lesssim r^{-1}\overset{\circ}{\delta} + r^{-3}\left(|\Lambda| + |\underline{\Lambda}|\right), \tag{9.4.10}$$

and

$$\begin{aligned}
\|e_4^{\mathbf{S}}(\check{\kappa}^{\mathbf{S}}, \vartheta^{\mathbf{S}}, \zeta^{\mathbf{S}}, \check{\underline{\kappa}}^{\mathbf{S}})\|_{\mathfrak{h}_k(\mathbf{S})} &\lesssim \overset{\circ}{\epsilon} r^{-2},\\
\|e_4^{\mathbf{S}}(\underline{\vartheta}^{\mathbf{S}})\|_{\mathfrak{h}_k(\mathbf{S})} &\lesssim \overset{\circ}{\epsilon} r^{-1},\\
\|e_4^{\mathbf{S}}\left(\alpha^{\mathbf{S}}, \beta^{\mathbf{S}}, \check{\rho}^{\mathbf{S}}, \check{\mu}^{\mathbf{S}}\right)\|_{\mathfrak{h}_k(\mathbf{S})} &\lesssim \overset{\circ}{\epsilon} r^{-3},\\
\|e_4^{\mathbf{S}}(\underline{\beta}^{\mathbf{S}})\|_{\mathfrak{h}_k(\mathbf{S})} &\lesssim \overset{\circ}{\epsilon} r^{-2},\\
\|e_4^{\mathbf{S}}(\underline{\alpha}^{\mathbf{S}})\|_{\mathfrak{h}_k(\mathbf{S})} &\lesssim \overset{\circ}{\epsilon} r^{-1}.
\end{aligned} \tag{9.4.11}$$

---

[12] All other Ricci coefficients involve the transversal derivatives $e_3^{\mathbf{S}}, e_4^{\mathbf{S}}$ of the frame.

[13] To be more precise one should replace r by $\overset{\circ}{r}$ in the estimates below. Of course r and $\overset{\circ}{r}$ are comparable in $\mathcal{R}$, in particular on $\mathbf{S}$.

To prove Theorem 9.32, it will be useful, using the fact that the kernel of $\not{d}_2^{\mathbf{S},\star}$ is spanned by e^{Φ}, to rewrite the GCM conditions (9.4.1) in the following form:

$$\begin{aligned}
&\not{d}_1^{\mathbf{S},\star}\underline{\kappa}^{\mathbf{S}} = \underline{C}^{\mathbf{S}}e^{\Phi}, \qquad \not{d}_1^{\mathbf{S},\star}\mu^{\mathbf{S}} = M^{\mathbf{S}}e^{\Phi}, \qquad \kappa^{\mathbf{S}} = \frac{2}{r^{\mathbf{S}}},\\
&\int_{\mathbf{S}} fe^{\Phi} = \Lambda, \qquad \int_{\mathbf{S}} \underline{f}e^{\Phi} = \underline{\Lambda},
\end{aligned} \tag{9.4.12}$$

where $\underline{C}^{\mathbf{S}}$ and $M^{\mathbf{S}}$ are constants.

Proposition 9.33. *Assume that there exists constants $\underline{C}^{\mathbf{S}}$, $M^{\mathbf{S}}$, such that the deformed sphere $\mathbf{S}$ verifies the GCM conditions (9.4.12). Then, the deformation parameters $(a, f, \underline{f})$ verify the system*

$$\begin{aligned}
(A^{\mathbf{S}} + V)\, f &= \frac{3}{4}\kappa(A^{\mathbf{S}})^{-1}(\rho z) - \not{d}^{\star}\check{\kappa} - \kappa(A^{\mathbf{S}})^{-1}\big(- M^{\mathbf{S}}e^{\Phi} + \not{d}^{\star}\check{\mu}\big)\\
&\quad + Err_4,\\
(A^{\mathbf{S}} + V)\, \underline{f} &= -\frac{3}{4}\underline{\kappa}(A^{\mathbf{S}})^{-1}(\rho z) + \underline{C}^{\mathbf{S}}e^{\Phi} - \not{d}^{\star}\check{\underline{\kappa}} + \underline{\kappa}(A^{\mathbf{S}})^{-1}\big(- M^{\mathbf{S}}e^{\Phi} + \not{d}^{\star}\check{\mu}\big)\\
&\quad + Err_5,\\
\not{d}^{\mathbf{S},\star}a &= -\frac{3}{4}(A^{\mathbf{S}})^{-1}(\rho z) + f\underline{\omega} - \underline{f}\omega + \frac{1}{4}f\underline{\kappa} - \frac{1}{4}\underline{f}\kappa\\
&\quad + (A^{\mathbf{S}})^{-1}\big(- M^{\mathbf{S}}e^{\Phi} + \not{d}^{\star}\check{\mu}\big) - (A^{\mathbf{S}})^{-1}Err_3,\\
\overline{a}^{\mathbf{S}} &= \overline{\left(1 - \frac{r^{\mathbf{S}}}{r}\right)}^{\mathbf{S}} - \frac{r^{\mathbf{S}}}{2}\overline{\left(\check{\kappa} + \overline{\kappa} - \frac{2}{r}\right)}^{\mathbf{S}} + \overline{Err_6}^{\mathbf{S}},
\end{aligned} \tag{9.4.13}$$

and

$$\int_{\mathbf{S}} e^{\Phi} f = \Lambda, \qquad \int_{\mathbf{S}} e^{\Phi}\underline{f} = \underline{\Lambda}, \tag{9.4.14}$$

where we recall that

$$z = \underline{\kappa} f + \kappa \underline{f}, \qquad V = \frac{1}{2}\kappa\underline{\kappa} - \rho,$$

$$Err_4 = Err_1 + \kappa(A^{\mathbf{S}})^{-1}Err_3, \qquad Err_5 = Err_2 - \underline{\kappa}(A^{\mathbf{S}})^{-1}Err_3,$$

$$\begin{aligned}
Err_6 &= -\frac{r^{\mathbf{S}}}{2}\left[e^a Err(\kappa, \kappa') - (e^a - 1)\left(\check{\kappa} + \overline{\kappa} - \frac{2}{r} + \not{d}^{\mathbf{S}}f\right) - (e^a - 1 - a)\frac{2}{r}\right]\\
&\quad - a\left(\frac{r^{\mathbf{S}}}{r} - 1\right)
\end{aligned}$$

with the error terms Err_1, Err_2, Err_3, defined in Lemma 9.24.

Conversely, if there exist constants $\underline{C}^{\mathbf{S}}$, $M^{\mathbf{S}}$, such that the deformation parameters $(a, f, \underline{f})$ verify the system (9.4.13), (9.4.14), then the deformed sphere $\mathbf{S}$ verifies the GCM conditions (9.4.12).

Proof. The first statement is an immediate consequence of Lemma 9.25 and (9.3.22).

We then focus on the second statement, i.e., we assume that the deformation parameters $(a, f, \underline{f})$ verify the system (9.4.13), (9.4.14) for some constants $\underline{C}^{\mathbf{S}}$, $M^{\mathbf{S}}$. Then, subtracting the first three equations of (9.4.13) from (9.3.17) and the last equation of (9.4.13) from (9.3.22), we obtain

$$\begin{aligned}
\not{d}^{\mathbf{S},\star}\check{\kappa}^{\mathbf{S}} - \kappa(A^{\mathbf{S}})^{-1}\big(- \not{d}^{\mathbf{S},\star}\check{\mu}^{\mathbf{S}} + M^{\mathbf{S}}e^{\Phi}\big) &= 0,\\
\not{d}^{\mathbf{S},\star}\underline{\check{\kappa}}^{\mathbf{S}} - \underline{C}^{\mathbf{S}}e^{\Phi} + \underline{\kappa}(A^{\mathbf{S}})^{-1}\big(- \not{d}^{\mathbf{S},\star}\check{\mu}^{\mathbf{S}} + M^{\mathbf{S}}e^{\Phi}\big) &= 0,\\
(A^{\mathbf{S}})^{-1}\big(- \not{d}^{\mathbf{S},\star}\check{\mu}^{\mathbf{S}} + M^{\mathbf{S}}e^{\Phi}\big) &= 0,\\
\frac{r^{\mathbf{S}}}{2}\left(\overline{\kappa^{\mathbf{S}}}^{\mathbf{S}} - \frac{2}{r^{\mathbf{S}}}\right) &= 0,
\end{aligned}$$

which, together with (9.4.14), immediately implies (9.4.12). □

Remark 9.34. *In view of Propositions 9.14 and 9.33, to find a GCM sphere amounts to solving the following coupled system:*

$$\begin{aligned}
\varsigma^{\#}U' &= \big((\gamma^{\mathbf{S}})^{\#}\big)^{1/2} f^{\#}\left(1 + \frac{1}{4}(f\underline{f})^{\#}\right),\\
S' - \frac{\varsigma^{\#}}{2}\underline{\Omega}^{\#}U' &= \frac{1}{2}\big((\gamma^{\mathbf{S}})^{\#}\big)^{1/2}\underline{f}^{\#},\\
(\gamma^{\mathbf{S}})^{\#} &= \gamma^{\#} + (\varsigma^{\#})^2\left(\underline{\Omega} + \frac{1}{4}\underline{b}^2\gamma\right)^{\#}(\partial_\theta U)^2 - 2\varsigma^{\#}\partial_\theta U\partial_\theta S\\
&\quad - (\gamma\varsigma\underline{b})^{\#}\partial_\theta U,\\
U(0) &= S(0) = 0,
\end{aligned} \tag{9.4.15}$$

where the inputs $(a, f, \underline{f})$ verify (9.4.13), (9.4.14)*. Recall that for a reduced scalar h defined on* $\mathbf{S}$ *we write*

$$h^{\#}(\mathring{u}, \mathring{s}, \theta) \quad = \quad h(\mathring{u} + U(\theta), \mathring{s} + S(\theta), \theta).$$

We will solve the coupled system of equations (9.4.13), (9.4.14), (9.4.15) *by an iteration argument which will be introduced below. Before doing this however it pays to observe that the system* (9.4.13) *can be interpreted as an elliptic system on a fixed surface* $\mathbf{S}$ *for $(a, f, \underline{f})$. In the next section we state a result which establishes the coercivity of the corresponding linearized system. The full proof of the theorem is detailed in section 9.4.3 to section 9.6.3.*

9.4.1 The linearized GCM system

We start with the following linearized version of the equations (9.4.13):

$$
\begin{aligned}
B^{\mathbf{S}} f &= -\frac{6m^{\mathbf{S}}}{(r^{\mathbf{S}})^5}(A^{\mathbf{S}})^{-1}(\underline{f} - \Upsilon^{\mathbf{S}} f) + \frac{2}{r^{\mathbf{S}}}\left(M^{\mathbf{S}} - \overline{M}^{\mathbf{S}}\right)(A^{\mathbf{S}})^{-1}e^{\Phi} + F_1,\\
B^{\mathbf{S}} \underline{f} &= -\frac{6m^{\mathbf{S}}\Upsilon^{\mathbf{S}}}{(r^{\mathbf{S}})^5}(A^{\mathbf{S}})^{-1}\left(\underline{f} - \Upsilon^{\mathbf{S}} f\right) + \left(\underline{C}^{\mathbf{S}} - \overline{\underline{C}}^{\mathbf{S}}\right)e^{\Phi}\\
&\quad + \frac{2\Upsilon^{\mathbf{S}}}{r^{\mathbf{S}}}\left(M^{\mathbf{S}} - \overline{M}^{\mathbf{S}}\right)(A^{\mathbf{S}})^{-1}e^{\Phi} + F_2,\\
\not{d}^{\mathbf{S},\star} a &= F_3,\\
\overline{a}^{\mathbf{S}} &= b_0,
\end{aligned}
\tag{9.4.16}
$$

where F_1, F_2, and F_3 are given reduced scalar on $\mathbf{S}$, b_0 is a given constant, and where we have introduced the notation

$$
B^{\mathbf{S}} \quad := \quad \not{d}_2^{\mathbf{S}} \not{d}_2^{\star\mathbf{S}} + \frac{6m^{\mathbf{S}}}{(r^{\mathbf{S}})^3}. \tag{9.4.17}
$$

Remark 9.35. *Recalling that we have* $A^{\mathbf{S}} + V = \not{d}_2^{\mathbf{S}} \not{d}_2^{\star\mathbf{S}} - 3\rho + \frac{1}{2}\vartheta\underline{\vartheta}$, *the GCM system* (9.4.13) *corresponds, in view of the definition of* $B^{\mathbf{S}}$, *to the linearized GCM system* (9.4.16) *with the following choices for* F_1, F_2, F_3, b_0*:*

$$
\begin{aligned}
F_1 \quad &= \quad -\frac{3m^{\mathbf{S}}}{(r^{\mathbf{S}})^4}(A^{\mathbf{S}})^{-1}\left(\left(\kappa - \frac{2}{r^{\mathbf{S}}}\right)\underline{f} + \left(\underline{\kappa} + \frac{2\Upsilon^{\mathbf{S}}}{r^{\mathbf{S}}}\right)f\right)\\
&\quad +\frac{3}{4}\left(\kappa - \frac{2}{r^{\mathbf{S}}}\right)(A^{\mathbf{S}})^{-1}(\rho z) - \frac{3}{2r^{\mathbf{S}}}(A^{\mathbf{S}})^{-1}\left(\left(\rho + \frac{2m^{\mathbf{S}}}{(r^{\mathbf{S}})^3}\right)z\right) - \not{d}^\star\check{\kappa}\\
&\quad -\left(\kappa - \frac{2}{r^{\mathbf{S}}}\right)(A^{\mathbf{S}})^{-1}(-\ M^{\mathbf{S}}e^{\Phi} + \not{d}^\star\check{\mu}) - \frac{2}{r^{\mathbf{S}}}(A^{\mathbf{S}})^{-1}\left(\not{d}^\star\check{\mu} - \overline{M}^{\mathbf{S}}e^{\Phi}\right)\\
&\quad +Err_4,
\end{aligned}
$$

$$
\begin{aligned}
F_2 \quad &= \quad -\frac{3m^{\mathbf{S}}\Upsilon^{\mathbf{S}}}{(r^{\mathbf{S}})^4}(A^{\mathbf{S}})^{-1}\left(\left(\kappa - \frac{2}{r^{\mathbf{S}}}\right)\underline{f} + \left(\underline{\kappa} + \frac{2\Upsilon^{\mathbf{S}}}{r^{\mathbf{S}}}\right)f\right)\\
&\quad -\frac{3}{4}\left(\underline{\kappa} + \frac{2\Upsilon^{\mathbf{S}}}{r^{\mathbf{S}}}\right)(A^{\mathbf{S}})^{-1}(\rho z) + \frac{3\Upsilon^{\mathbf{S}}}{2r^{\mathbf{S}}}(A^{\mathbf{S}})^{-1}\left(\left(\rho + \frac{2m^{\mathbf{S}}}{(r^{\mathbf{S}})^3}\right)z\right)\\
&\quad -\left(\not{d}^\star\check{\underline{\kappa}} - \underline{C}^{\mathbf{S}}e^{\Phi}\right) + \left(\underline{\kappa} + \frac{2\Upsilon^{\mathbf{S}}}{r^{\mathbf{S}}}\right)(A^{\mathbf{S}})^{-1}(-\ M^{\mathbf{S}}e^{\Phi} + \not{d}^\star\check{\mu})\\
&\quad -\frac{2\Upsilon^{\mathbf{S}}}{r^{\mathbf{S}}}(A^{\mathbf{S}})^{-1}\left(\not{d}^\star\check{\mu} - \overline{M}^{\mathbf{S}}e^{\Phi}\right) + Err_5,
\end{aligned}
$$

$$
\begin{aligned}
F_3 \quad &= \quad -\frac{3}{4}(A^{\mathbf{S}})^{-1}(\rho z) + f\underline{\omega} - \underline{f}\omega + \frac{1}{4}f\underline{\kappa} - \frac{1}{4}\underline{f}\kappa\\
&\quad +(A^{\mathbf{S}})^{-1}(-\ M^{\mathbf{S}}e^{\Phi} + \not{d}^\star\check{\mu}) - (A^{\mathbf{S}})^{-1}Err_3,\\
b_0 \quad &= \quad \overline{\left(1 - \frac{r^{\mathbf{S}}}{r}\right)}^{\mathbf{S}} - \frac{r^{\mathbf{S}}}{2}\overline{\left(\check{\kappa} + \overline{\kappa} - \frac{2}{r}\right)}^{\mathbf{S}} + \overline{Err_6}^{\mathbf{S}}.
\end{aligned}
$$

Remark 9.36. *To motivate the introduction of the system* (9.4.16), *let us note that the above particular choices for F_1 and F_2 in Remark 9.35 correspond to the terms in the first two equations of* (9.4.13) *which*[14]

- *either depend on $\check{\kappa}$, $\not{d}^\star\check{\underline{\kappa}} - \underline{C}e^\Phi$, and $\not{d}^\star\check{\mu} - Me^\Phi$,*
- *or are nonlinear.*

The following result plays a main role in the proof of Theorem 9.32.

Proposition 9.37. *Let a fixed spacetime region $\mathcal{R}$ verifying assumptions **A1–A3** and* (9.4.2). *Assume $\mathbf{S}$ is a given surface in $\mathcal{R}$ such that, for a small enough constant $\delta_1 > 0$ and for any $2 \le s \le s_{max}+1$,*

$$\sup_{\mathbf{S}}\left|K^{\mathbf{S}} - \frac{1}{(r^{\mathbf{S}})^2}\right| \le \frac{\delta_1}{(r^{\mathbf{S}})^2}, \quad \|K^{\mathbf{S}}\|_{\mathfrak{h}_{s-2}(\mathbf{S})} \lesssim \frac{1}{r^{\mathbf{S}}}, \quad \int_{\mathbf{S}} e^{2\Phi} = \frac{4\pi}{3}(r^{\mathbf{S}})^4(1+O(\delta_1)).$$

Then, for every $\Lambda, \underline{\Lambda}$,

- *Existence and uniqueness. There exists unique constants $(\underline{C}^{\mathbf{S}}, M^{\mathbf{S}})$ and a unique solution $(f, \underline{f}, \lambda)$ of the system* (9.4.16), (9.4.14) *verifying the estimates*

$$|\underline{C}^{\mathbf{S}} - \overline{\underline{C}}^{\mathbf{S}}| + r^{\mathbf{S}}|M^{\mathbf{S}} - \overline{M}^{\mathbf{S}}| \lesssim (r^{\mathbf{S}})^{-7}(|\Lambda| + |\underline{\Lambda}|) + (r^{\mathbf{S}})^{-2}\|F_1\|_{L^2(\mathbf{S})} + (r^{\mathbf{S}})^{-2}\|F_2\|_{L^2(\mathbf{S})}, \tag{9.4.18}$$

$$\|(f,\underline{f})\|_{\mathfrak{h}_s(\mathbf{S})} \lesssim (r^{\mathbf{S}})^{-2}(|\Lambda| + |\underline{\Lambda}|) + (r^{\mathbf{S}})^2\|F_1\|_{\mathfrak{h}_{s-2}(\mathbf{S})} + (r^{\mathbf{S}})^2\|F_2\|_{\mathfrak{h}_{s-2}(\mathbf{S})}, \tag{9.4.19}$$

$$\|\check{a}^{\mathbf{S}}\|_{\mathfrak{h}_s(\mathbf{S})} \lesssim r^{\mathbf{S}}\|F_3\|_{\mathfrak{h}_{s-1}(\mathbf{S})} \tag{9.4.20}$$

and

$$|\overline{a}^{\mathbf{S}}| \lesssim |b_0|. \tag{9.4.21}$$

- *A priori estimates. If $(f, \underline{f}, \lambda)$ verifies the system* (9.4.16), (9.4.14) *for some constant $(\underline{C}^{\mathbf{S}}, M^{\mathbf{S}})$, then $(M^{\mathbf{S}}, \underline{C}^{\mathbf{S}})$ satisfies* (9.4.18) *and $(f, \underline{f}, \lambda)$ satisfies* (9.4.19), (9.4.20), *and* (9.4.21).

As a corollary, we derive the following rigidity result for GCM spheres.

Corollary 9.38. *Let a fixed spacetime region $\mathcal{R}$ verifying assumptions **A1–A3** and* (9.4.2). *Assume that $\mathbf{S}$ is a deformed sphere in $\mathcal{R}$ which verifies the GCM conditions*

$$\kappa^{\mathbf{S}} = \frac{2}{r^{\mathbf{S}}}, \quad \not{d}_2^{\star\,\mathbf{S}}\not{d}_1^{\star\,\mathbf{S}}\underline{\kappa}^{\mathbf{S}} = \not{d}_2^{\star\,\mathbf{S}}\not{d}_1^{\star\,\mathbf{S}}\mu^{\mathbf{S}} = 0 \tag{9.4.22}$$

---

[14]Note that the terms $\not{d}^\star\check{\underline{\kappa}} - \overline{\underline{C}}^{\mathbf{S}}e^\Phi$ and $\not{d}^\star\check{\mu} - \overline{M}^{\mathbf{S}}e^\Phi$ can be decomposed as follows:

$$\not{d}^\star\check{\underline{\kappa}} - \overline{\underline{C}}^{\mathbf{S}}e^\Phi = \not{d}^\star\check{\underline{\kappa}} - \underline{C}e^\Phi + (\underline{C} - \overline{\underline{C}}^{\mathbf{S}})e^\Phi, \qquad \not{d}^\star\check{\mu} - \overline{M}^{\mathbf{S}}e^\Phi = \not{d}^\star\check{\mu} - Me^\Phi + (M - \overline{M}^{\mathbf{S}})e^\Phi,$$

where $\underline{C} - \overline{\underline{C}}^{\mathbf{S}}$ and $M - \overline{M}^{\mathbf{S}}$ are nonlinear in view of Corollary 9.18 applied to $D = \underline{C}$ and $D = M$.

and such that for a small enough constant $\delta_1 > 0$, the transition functions $(f, \underline{f}, \lambda)$ from the background frame of $\mathcal{R}$ to that of $\mathbf{S}$ verify, for some $4 \le s \le s_{max}$, the bound

$$\|f\|_{\mathfrak{h}_s(\mathbf{S})} + (r^{\mathbf{S}})^{-1}\|(\underline{f}, a)\|_{\mathfrak{h}_s(\mathbf{S})} \quad \le \quad \delta_1. \tag{9.4.23}$$

Then $(f, \underline{f}, \lambda)$ verify the estimates

$$\|(f, \underline{f}, \check{a}^{\mathbf{S}})\|_{\mathfrak{h}_{s+1}(\mathbf{S})} \quad \lesssim \quad \overset{\circ}{\delta} + r^{-2}\left(\left|\int_{\mathbf{S}} fe^{\Phi}\right| + \left|\int_{\mathbf{S}} \underline{f}e^{\Phi}\right|\right) + r\delta_1\left(\overset{\circ}{\epsilon} + \delta_1\right)$$

and

$$r|\overline{a}^{\mathbf{S}}| \quad \lesssim \quad \overset{\circ}{\delta} + r^{-2}\left(\left|\int_{\mathbf{S}} fe^{\Phi}\right| + \left|\int_{\mathbf{S}} \underline{f}e^{\Phi}\right|\right) + \sup_{\mathbf{S}} |r - r^{\mathbf{S}}|.$$

Remark 9.39. *As mentioned before Lemma 9.15, the anomalous behavior for $(\underline{f}, a)$ in the assumption (9.4.23) does not appear in the construction of GCM spheres in this chapter. It appears however in the proof of Theorem M0 in the region ${}^{(ext)}\mathcal{L}_0 \cap {}^{(ext)}\mathcal{M}$ of the initial data layer, see Step 8 in section 4.1.*

The proof of Proposition 9.37 will be given in section 9.5.1 while the proof of Corollary 9.38 will be given in section 9.5.2.

9.4.2 Comparison of the Hawking mass

We establish the estimate (9.4.7) concerning the Hawking mass $m^{\mathbf{S}}$. Recall that

$$\begin{aligned} m^{\mathbf{S}} &= \frac{r^{\mathbf{S}}}{2}\left(1 + \frac{1}{16\pi}\int_{\mathbf{S}} \kappa^{\mathbf{S}}\underline{\kappa}^{\mathbf{S}}\right), \\ \overset{\circ}{m} &= \frac{\overset{\circ}{r}}{2}\left(1 + \frac{1}{16\pi}\int_{\overset{\circ}{S}} \overset{\circ}{\kappa}\overset{\circ}{\underline{\kappa}}\right). \end{aligned}$$

We write

$$\begin{aligned} 2\left(\frac{m^{\mathbf{S}}}{r^{\mathbf{S}}} - \frac{\overset{\circ}{m}}{\overset{\circ}{r}}\right) &= \frac{1}{16\pi}\left[\int_{\mathbf{S}} (\kappa^{\mathbf{S}}\underline{\kappa}^{\mathbf{S}} - \kappa\underline{\kappa}) + \left(\int_{\mathbf{S}} \kappa\underline{\kappa} - \int_{\overset{\circ}{S}} \kappa\underline{\kappa}\right) - \int_{\overset{\circ}{S}} \left(\kappa\underline{\kappa} - \overset{\circ}{\kappa}\overset{\circ}{\underline{\kappa}}\right)\right] \\ &= I_1 + I_2 + I_3. \end{aligned}$$

In view of Proposition 9.14 we have $|r^{\mathbf{S}} - \overset{\circ}{r}| \lesssim \overset{\circ}{\delta}$ and $\left|\gamma^{\mathbf{S},\#} - \overset{\circ}{\gamma}\right| \lesssim \overset{\circ}{\delta}\overset{\circ}{r}$. Making use of Corollary 9.10 and the assumptions **A1**–**A3** for $\kappa, \underline{\kappa}$ we deduce

$$|I_2| = \left|\int_{\mathbf{S}} \kappa\underline{\kappa} - \int_{\overset{\circ}{S}} \kappa\underline{\kappa}\right| \quad \lesssim \quad \overset{\circ}{\delta} r^{-1}.$$

Similarly, taking into account the definition of $\mathcal{R} := \left\{|u - \overset{\circ}{u}| \le \overset{\circ}{\delta}, \quad |s - \overset{\circ}{s}| \le \overset{\circ}{\delta}\right\}$,

$$|I_3| \quad \lesssim \quad \overset{\circ}{\delta} r^{-1}.$$

Finally, making use also of the transformation formula from the original frame (e_4, e_3, e_θ) to the frame $(e_4^{\mathbf{S}}, e_3^{\mathbf{S}}, e_\theta^{\mathbf{S}})$ of $\mathbf{S}$

$$\kappa^{\mathbf{S}}\underline{\kappa}^{\mathbf{S}} = \big(\kappa + \not{d}^{\mathbf{S}} f + \mathrm{Err}(\kappa, \kappa^{\mathbf{S}})\big)\big(\underline{\kappa} + \not{d}^{\mathbf{S}}\underline{f} + \mathrm{Err}(\underline{\kappa}, \underline{\kappa}^{\mathbf{S}})\big)$$

and the estimates for $(f, \underline{f}, a = \log\lambda)$ we deduce

$$\left|\kappa^{\mathbf{S}}\underline{\kappa}^{\mathbf{S}} - \kappa\underline{\kappa}\right| \lesssim r^{-3}\mathring{\delta}.$$

Hence,

$$\left|I_1\right| \lesssim \mathring{\delta} r^{-1}.$$

We infer that

$$\left|\frac{m^{\mathbf{S}}}{r^{\mathbf{S}}} - \frac{\mathring{m}}{\mathring{r}}\right| \lesssim \mathring{\delta} r^{-1}$$

from which the desired estimate (9.4.7) easily follows.

9.4.3 Iteration procedure for Theorem 9.32

We solve the coupled system of equations (9.4.13), (9.4.14), (9.4.15) by an iteration argument as follows.

Starting with the septets

$$\begin{aligned} Q^{(0)} &:= (U^{(0)}, S^{(0)}, a^{(0)}, f^{(0)}, \underline{f}^{(0)}, \underline{C}^{(0)}, M^{(0)}) = (0, 0, 0, 0, 0, \underline{C}(\mathring{u}, \mathring{s}), M(\mathring{u}, \mathring{s})), \\ Q^{(1)} &:= (U^{(1)}, S^{(1)}, a^{(1)}, f^{(1)}, \underline{f}^{(1)}, \underline{C}^{(1)}, M^{(1)}) = (0, 0, 0, 0, 0, \underline{C}(\mathring{u}, \mathring{s}), M(\mathring{u}, \mathring{s})), \end{aligned}$$

corresponding to the undeformed sphere $\mathring{S}$, we define iteratively the quintet

$$Q^{(n+1)} = (U^{(n+1)}, S^{(n+1)}, a^{(n+1)}, f^{(n+1)}, \underline{f}^{(n+1)}, \underline{C}^{(n+1)}, M^{(n+1)})$$

from

$$\begin{aligned} Q^{(n-1)} &= (U^{(n-1)}, S^{(n-1)}, a^{(n-1)}, f^{(n-1)}, \underline{f}^{(n-1)}, \underline{C}^{(n-1)}, M^{(n-1)}), \\ Q^{(n)} &= (U^{(n)}, S^{(n)}, a^{(n)}, f^{(n)}, \underline{f}^{(n)}, \underline{C}^{(n)}, M^{(n)}), \end{aligned}$$

as follows.

1. The pair $(U^{(n)}, S^{(n)})$ defines the deformation sphere $\mathbf{S}(n)$ and the corresponding pullback map $\#_n$ given by the map $\Psi^{(n)} : \mathring{S} \to \mathbf{S}(n)$,

$$(\mathring{u}, \mathring{s}, \theta, \varphi) \longrightarrow (\mathring{u} + U^{(n)}(\theta), \mathring{s} + S^{(n)}(\theta), \theta, \varphi). \tag{9.4.24}$$

By induction we assume that the following estimates hold true:

$$
\begin{aligned}
r^4|\underline{C}^{(n)} - \overline{\underline{C}}^{\mathbf{S}(n-1)}| + r^5|M^{(n)} - \overline{M}^{\mathbf{S}(n-1)}| & \\
+ \left\|(a^{(n)}, f^{(n)}, \underline{f}^{(n)})\right\|_{\mathfrak{h}_{s_{max}-1}(\mathbf{S}(n-1))} & \lesssim \overset{\circ}{\delta},
\end{aligned}
\tag{9.4.25}
$$

and

$$
\begin{aligned}
& \|\partial_\theta \left(U^{(n-1)}, S^{(n-1)}\right)\|_{L^\infty(\overset{\circ}{S})} + r^{-1}\|\partial_\theta \left(U^{(n-1)}, S^{(n-1)}\right)\|_{\mathfrak{h}_{s_{max}-1}(\overset{\circ}{S}, \overset{\circ}{\not g})} \\
& + \|\partial_\theta \left(U^{(n)}, S^{(n)}\right)\|_{L^\infty(\overset{\circ}{S})} + r^{-1}\|\partial_\theta \left(U^{(n)}, S^{(n)}\right)\|_{\mathfrak{h}_{s_{max}-1}(\overset{\circ}{S}, \overset{\circ}{\not g})} \\
\lesssim & \ \overset{\circ}{\delta}.
\end{aligned}
\tag{9.4.26}
$$

2. We then define the quintet $(a^{(n+1)}, f^{(n+1)}, \underline{f}^{(n+1)}, \underline{C}^{(n+1)}, M^{(n+1)})$ by solving the system on $\mathbf{S}(n)$ consisting of the equations (9.4.27), (9.4.32) and (9.4.33) below.

$$
\begin{aligned}
B^{\mathbf{S}(n)} f^{(n+1)} &= -\frac{6m^{\mathbf{S}(n)}}{(r^{\mathbf{S}(n)})^5}(A^{\mathbf{S}(n)})^{-1}(\underline{f}^{(n+1)} - \Upsilon^{\mathbf{S}(n)} f^{(n+1)}) \\
&\quad + \frac{2}{r^{\mathbf{S}(n)}}(M^{(n+1)} - \overline{M}^{\mathbf{S}(n)})(A^{\mathbf{S}(n)})^{-1}e^{\Phi} + E^{(n+1)}, \\
B^{\mathbf{S}(n)} \underline{f}^{(n+1)} &= -\frac{6m^{\mathbf{S}(n)}\Upsilon^{\mathbf{S}(n)}}{(r^{\mathbf{S}(n)})^5}(A^{\mathbf{S}(n)})^{-1}\left(\underline{f}^{(n+1)} - \Upsilon^{\mathbf{S}(n)} f^{(n+1)}\right) \\
&\quad + (\underline{C}^{(n+1)} - \overline{\underline{C}}^{\mathbf{S}(n)})e^{\Phi} + \frac{2\Upsilon^{\mathbf{S}(n)}}{r^{\mathbf{S}}}(M^{(n+1)} - \overline{M}^{\mathbf{S}(n)})(A^{\mathbf{S}(n)})^{-1}e^{\Phi} \\
&\quad + \underline{E}^{(n+1)}, \\
\not{d}^{\mathbf{S}(n)} a^{(n+1)} &= \widetilde{E}^{(n+1)},
\end{aligned}
\tag{9.4.27}
$$

with

$$
\begin{aligned}
E^{(n+1)} :=& -\frac{3m^{\mathbf{S}(n)}}{(r^{\mathbf{S}(n)})^4}(A^{\mathbf{S}(n)})^{-1} \\
& \times \left(\left(\kappa - \frac{2}{r^{\mathbf{S}(n)}}\right)\underline{f}^{(n)}_{n-1} + \left(\underline{\kappa} + \frac{2\Upsilon^{\mathbf{S}(n)}}{r^{\mathbf{S}(n)}}\right)f^{(n)}_{n-1}\right) \\
& + \frac{3}{4}\left(\kappa - \frac{2}{r^{\mathbf{S}(n)}}\right)(A^{\mathbf{S}(n)})^{-1}(\rho z^{(n)}_{n-1}) \\
& - \frac{3}{2r^{\mathbf{S}(n)}}(A^{\mathbf{S}(n)})^{-1}\left(\left(\rho + \frac{2m^{\mathbf{S}(n)}}{(r^{\mathbf{S}(n)})^3}\right)z^{(n)}_{n-1}\right) - \not{d}^\star\check{\kappa} \\
& - \left(\kappa - \frac{2}{r^{\mathbf{S}(n)}}\right)(A^{\mathbf{S}(n)})^{-1}(-M^{(n)}e^{\Phi} + \not{d}^\star\check{\mu}) \\
& - \frac{2}{r^{\mathbf{S}(n)}}(A^{\mathbf{S}(n)})^{-1}(\not{d}^\star\check{\mu} - \overline{M}^{\mathbf{S}(n)}e^{\Phi}) + \mathrm{Err}_4^{(n+1)},
\end{aligned}
\tag{9.4.28}
$$

$$
\begin{aligned}
\underline{E}^{(n+1)} := & -\frac{3m^{\mathbf{S}(n)}\Upsilon^{\mathbf{S}(n)}}{(r^{\mathbf{S}(n)})^4}(A^{\mathbf{S}(n)})^{-1}\\
& \times\left(\left(\kappa-\frac{2}{r^{\mathbf{S}(n)}}\right)\underline{f}^{(n)}_{n-1}+\left(\underline{\kappa}+\frac{2\Upsilon^{\mathbf{S}(n)}}{r^{\mathbf{S}(n)}}\right)f^{(n)}_{n-1}\right)\\
& -\frac{3}{4}\left(\underline{\kappa}+\frac{2\Upsilon^{\mathbf{S}(n)}}{r^{\mathbf{S}(n)}}\right)(A^{\mathbf{S}(n)})^{-1}\big(\rho z^{(n)}_{n-1}\big)\\
& +\frac{3\Upsilon^{\mathbf{S}(n)}}{2r^{\mathbf{S}(n)}}(A^{\mathbf{S}(n)})^{-1}\left(\left(\rho+\frac{2m^{\mathbf{S}(n)}}{(r^{\mathbf{S}(n)})^3}\right)z^{(n)}_{n-1}\right)\\
& -\big({\not d}^\star\underline{\check{\kappa}}-\overline{\underline{C}}^{\mathbf{S}(n)}e^\Phi\big)+\left(\underline{\kappa}+\frac{2\Upsilon^{\mathbf{S}(n)}}{r^{\mathbf{S}(n)}}\right)(A^{\mathbf{S}(n)})^{-1}\big(-M^{(n)}e^\Phi+{\not d}^\star\check{\mu}\big)\\
& -\frac{2\Upsilon^{\mathbf{S}(n)}}{r^{\mathbf{S}(n)}}(A^{\mathbf{S}(n)})^{-1}\big({\not d}^\star\check{\mu}-\overline{M}^{\mathbf{S}(n)}e^\Phi\big)+\mathrm{Err}_5^{(n+1)},
\end{aligned}
\tag{9.4.29}
$$

$$
\begin{aligned}
\widetilde{E}^{(n+1)} := & -\frac{3}{4}(A^{\mathbf{S}(n)})^{-1}\big(\rho z^{(n+1)}\big)+f^{(n+1)}\underline{\omega}-\underline{f}^{(n+1)}\omega+\frac{1}{4}f^{(n+1)}\underline{\kappa}\\
& -\frac{1}{4}\underline{f}^{(n+1)}\kappa+(A^{\mathbf{S}(n)})^{-1}\big(-M^{(n+1)}e^\Phi+{\not d}^\star\check{\mu}\big)\\
& -(A^{\mathbf{S}(n)})^{-1}\big(\mathrm{Err}_3^{(n+1)}\big),
\end{aligned}
\tag{9.4.30}
$$

where

$$
f^{(n)}_{n-1}=f^{(n)}\circ\big(\Psi^{(n-1)}\circ(\Psi^{(n)})^{-1}\big),\qquad \underline{f}^{(n)}_{n-1}=\underline{f}^{(n)}\circ\big(\Psi^{(n-1)}\circ(\Psi^{(n)})^{-1}\big),
$$

$$
z^{(n+1)}:=\underline{\kappa}f^{(n+1)}+\kappa\underline{f}^{(n+1)},\qquad z^{(n)}_{n-1}:=\underline{\kappa}f^{(n)}_{n-1}+\kappa\underline{f}^{(n)}_{n-1},
$$

and the error terms,

$$
\mathrm{Err}_1^{(n+1)},\mathrm{Err}_2^{(n+1)},\mathrm{Err}_3^{(n+1)},\mathrm{Err}_4^{(n+1)},\mathrm{Err}_5^{(n+1)},
\tag{9.4.31}
$$

are obtained from the error terms Err_1, Err_2, Err_3, Err_4 and Err_5 by setting $(a,f,\underline{f})=(a^{(n)},f^{(n)},\underline{f}^{(n)})$ and their derivatives by the corresponding ones on $\mathbf{S}(n-1)$, and then composing by $\Psi^{(n-1)}\circ(\Psi^{(n)})^{-1}$ so that the error terms in (9.4.31) are well defined on $\mathbf{S}(n)$.
We also set

$$
\int_{\mathbf{S}(n)}e^\Phi f^{(n+1)}=\Lambda,\qquad \int_{\mathbf{S}(n)}e^\Phi\underline{f}^{(n+1)}=\underline{\Lambda},
\tag{9.4.32}
$$

and

$$
\overline{a^{(n+1)}}^{\mathbf{S}(n)}=\overline{\left(1-\frac{r^{\mathbf{S}(n)}}{r}\right)}^{\mathbf{S}(n)}-\frac{r^{\mathbf{S}(n)}}{2}\overline{\left(\check{\kappa}+\overline{\kappa}-\frac{2}{r}\right)}^{\mathbf{S}(n)}+\overline{\mathrm{Err}_6^{(n+1)}}^{\mathbf{S}(n)},
\tag{9.4.33}
$$

where $\mathrm{Err}_6^{(n+1)}$ is obtained from the error terms Err_6, as above in (9.4.31), by setting $(a,f,\underline{f})=(a^{(n)},f^{(n)},\underline{f}^{(n)})$ and their derivatives by the corresponding ones on the sphere $\mathbf{S}(n-1)$, and then composing by $\Psi^{(n-1)}\circ(\Psi^{(n)})^{-1}$ so that

$\mathrm{Err}_6^{(n+1)}$ is defined on $\mathbf{S}(n)$.

3. The system of equations (9.4.27), (9.4.32) and (9.4.33) admits a unique solution $(f^{(1+n)}, \underline{f}^{(1+n)}, a^{(n+1)}, \underline{C}^{(n+1)}, M^{(n+1)})$ according to Proposition 9.40 below.
4. We then use the new pair $(f^{(n+1)}, \underline{f}^{(n+1)})$ to solve the equations on $\overset{\circ}{S}$,

$$
\begin{aligned}
\varsigma^{\#_n}\partial_\theta U^{(n+1)} &= (\gamma^{(n)})^{1/2}(f^{(n+1)})^{\#_n} \\
&\quad \times \left(1+\frac{1}{4}\left(f^{(n+1)}\underline{f}^{(n+1)}\right)^{\#_n}\right), \\
\partial_\theta S^{(n+1)} - \frac{1}{2}\varsigma^{\#_n}\underline{\Omega}^{\#_n}\partial_\theta U^{(n+1)} &= \frac{1}{2}(\gamma^{(n)})^{1/2}(\underline{f}^{(n+1)})^{\#_n}, \\
\gamma^{(n)} &= \gamma^{\#_n} + (\varsigma^{\#_n})^2\left(\underline{\Omega}+\frac{1}{4}\underline{b}^2\gamma\right)^{\#_n}(\partial_\theta U^{(n)})^2 \\
&\quad - 2\varsigma^{\#_n}\partial_\theta U^{(n)}\partial_\theta S^{(n)} - \left(\gamma\varsigma\underline{b}\right)^{\#_n}\partial_\theta U^{(n)}, \\
U^{(n+1)}(0) &= S^{(n+1)}(0) = 0,
\end{aligned}
\tag{9.4.34}
$$

where, we repeat, the pullback $\#_n$ is defined with respect to the map

$$
\Psi^{(n)}(\overset{\circ}{u}, \overset{\circ}{s}, \theta) = (\overset{\circ}{u} + U^{(n)}(\theta), \overset{\circ}{s} + S^{(n)}(\theta), \theta),
$$

and

$$
\gamma^{(n)} := \gamma^{\mathbf{S}(n), \#_n}.
$$

The equation (9.4.34) admits a unique solution $(U^{(n+1)}, S^{(n+1)})$ according to Proposition 9.41 below. The new pair $(U^{(n+1)}, S^{(n+1)})$ defines the new polarized sphere $\mathbf{S}(n+1)$ and we can proceed with the next step of the iteration.

9.4.4 Existence and boundedness of the iterates

9.4.4.1 Existence and boundedness of $(f^{(n+1)}, \underline{f}^{(n+1)}, a^{(n+1)}, \underline{C}^{(n+1)}, M^{(n+1)})$

Proposition 9.40. *The system of equations* (9.4.27), (9.4.32) *and* (9.4.33) *admits a unique solution* $(f^{(1+n)}, \underline{f}^{(1+n)}, a^{(n+1)}, \underline{C}^{(n+1)}, M^{(n+1)})$ *verifying the estimates*

$$
\begin{aligned}
& r^4|\underline{C}^{(n+1)} - \overline{\underline{C}}^{\mathbf{S}(n)}| + r^5|M^{(n+1)} - \overline{M}^{\mathbf{S}(n)}| \\
& + \left\|\left(a^{(n+1)} - \overline{a^{(n+1)}}^{\mathbf{S}(n)}, f^{(n+1)}, \underline{f}^{(n+1)}\right)\right\|_{\mathfrak{h}_{s_{max}-1}(\mathbf{S}(n))} \lesssim \overset{\circ}{\delta}
\end{aligned}
$$

and

$$
r\left|\overline{a^{(n+1)}}^{\mathbf{S}(n)}\right| \lesssim \overset{\circ}{\delta} + \|\partial_\theta\left(U^{(n)}, S^{(n)}\right)\|_{L^\infty(\overset{\circ}{S})}
$$

uniformly for all $n \in \mathbb{N}$.

Proof. The system (9.4.27), (9.4.32) and (9.4.33) corresponds to the linearized GCM system (9.4.16), (9.4.14) with the following choice for F_1, F_2, F_3 and b_0:

$$F_1 = E^{(n+1)}, \qquad F_2 = \underline{E}^{(n+1)}, \qquad F_3 = \widetilde{E}^{(n+1)},$$
$$b_0 = \overline{\left(1 - \frac{r^{\mathbf{S}(n)}}{r}\right)}^{\mathbf{S}(n)} - \frac{r^{\mathbf{S}(n)}}{2}\overline{\left(\check{\kappa} + \overline{\kappa} - \frac{2}{r}\right)}^{\mathbf{S}(n)} + \overline{\mathrm{Err}_6^{(n+1)}}^{\mathbf{S}(n)}.$$

Also, the induction assumptions (9.4.26) for $(U^{(n)}, S^{(n)})$ together with Corollary 9.17 imply that the sphere $\mathbf{S}(n)$ satisfies in particular the assumptions of Proposition 9.37. We infer from that proposition the existence and uniqueness of the quintet solution $(f^{(1+n)}, \underline{f}^{(1+n)}, a^{(n+1)}, \underline{C}^{(n+1)}, M^{(n+1)})$ to (9.4.27), (9.4.32) and (9.4.33), as well as the following a priori estimate:

$$\begin{aligned}
& (r^{\mathbf{S}(n)})^4|\underline{C}^{(n+1)} - \overline{\underline{C}}^{\mathbf{S}(n)}| + (r^{\mathbf{S}(n)})^5|M^{(n+1)} - \overline{M}^{\mathbf{S}(n)}| \\
& + \left\|(f^{(n+1)}, \underline{f}^{(n+1)})\right\|_{\mathfrak{h}_{s_{max}-1}(\mathbf{S}(n))} \\
\lesssim\ & (r^{\mathbf{S}(n)})^{-2}\big(|\Lambda| + |\underline{\Lambda}|\big) + (r^{\mathbf{S}(n)})^2\|E^{(n+1)}\|_{\mathfrak{h}_{s_{max}-3}(\mathbf{S}(n))} \\
& +(r^{\mathbf{S}(n)})^2\|\underline{E}^{(n+1)}\|_{\mathfrak{h}_{s_{max}-3}(\mathbf{S}(n))},
\end{aligned} \tag{9.4.35}$$

$$\left\|a^{(n+1)} - \overline{a^{(n+1)}}^{\mathbf{S}(n)}\right\|_{\mathfrak{h}_{s_{max}-1}(\mathbf{S}(n))} \lesssim r^{\mathbf{S}(n)}\|\widetilde{E}^{(n+1)}\|_{\mathfrak{h}_{s_{max}-2}(\mathbf{S}(n))}, \tag{9.4.36}$$

and

$$\begin{aligned}
\left|\overline{a^{(n+1)}}^{\mathbf{S}(n)}\right| \lesssim\ & \left|\overline{\left(1 - \frac{r^{\mathbf{S}(n)}}{r}\right)}^{\mathbf{S}(n)}\right| + \left|\frac{r^{\mathbf{S}(n)}}{2}\overline{\left(\check{\kappa} + \overline{\kappa} - \frac{2}{r}\right)}^{\mathbf{S}(n)}\right| \\
& + \left|\overline{\mathrm{Err}_6^{(n+1)}}^{\mathbf{S}(n)}\right|.
\end{aligned} \tag{9.4.37}$$

We need to control the RHS of the inequalities (9.4.35), (9.4.36), (9.4.37). We start with the control of the error terms $\mathrm{Err}_j^{(n+1)}$, $j = 3, 4, 5, 6$. The induction assumptions (9.4.26) for $(U^{(n)}, S^{(n)})$ imply that the sphere $\mathbf{S}(n)$ satisfies in particular the assumptions of Lemma 9.29 with $\delta_1 = \overset{\circ}{\delta}$. We deduce from that lemma

$$\begin{aligned}
\|\mathrm{Err}_1^{(n+1)}, \mathrm{Err}_2^{(n+1)}\|_{\mathfrak{h}_{s_{max}-3}(\mathbf{S}(n))} \lesssim \overset{\circ}{\epsilon} r^{-2}\left\|(f^{(n)}, \underline{f}^{(n)}, a^{(n)})\right\|_{\mathfrak{h}_{s_{max}-1}(\mathbf{S}(n-1))} &\lesssim r^{-2}\overset{\circ}{\epsilon}\overset{\circ}{\delta}, \\
\left\|\mathrm{Err}_3^{(n+1)}\right\|_{\mathfrak{h}_{s_{max}-4}(\mathbf{S}(n))} \lesssim \overset{\circ}{\epsilon} r^{-3}\left\|(f^{(n)}, \underline{f}^{(n)}, a^{(n)})\right\|_{\mathfrak{h}_{s_{max}-1}(\mathbf{S}(n-1))} &\lesssim r^{-3}\overset{\circ}{\epsilon}\overset{\circ}{\delta}, \\
\|\mathrm{Err}_4^{(n+1)}, \mathrm{Err}_5^{(n+1)}\|_{\mathfrak{h}_{s_{max}-3}(\mathbf{S}(n))} \lesssim \overset{\circ}{\epsilon} r^{-2}\left\|(f^{(n)}, \underline{f}^{(n)}, a^{(n)})\right\|_{\mathfrak{h}_{s_{max}-1}(\mathbf{S}(n-1))} &\lesssim r^{-2}\overset{\circ}{\epsilon}\overset{\circ}{\delta},
\end{aligned}$$

where we have also used the induction assumptions (9.4.25) for $(f^{(n)}, \underline{f}^{(n)}, a^{(n)})$, as well as Lemma 9.11 which implies for a reduced scalar h on $\mathbf{S}(n-1)$

$$\|h \circ \big(\Psi^{(n-1)} \circ (\Psi^{(n)})^{-1}\big)\|_{\mathfrak{h}_s(\mathbf{S}(n))} = \|h\|_{\mathfrak{h}_s(\mathbf{S}(n-1))}(1 + O(r^{-1}\overset{\circ}{\delta})), \ 0 \leq s \leq s_{max} - 1.$$

Also, recall that $\mathrm{Err}_6^{(n+1)}$ is given by

$$\mathrm{Err}_6^{(n+1)} = \Bigg[-\frac{r^{\mathbf{S}(n)}}{2}\Bigg\{ e^{a^{(n)}}\mathrm{Err}(\kappa, \kappa') - (e^{a^{(n)}} - 1)\left(\check{\kappa} + \overline{\kappa} - \frac{2}{r} + \not{d}^{\mathbf{S}(n-1)} f^{(n)}\right) - \left(e^{a^{(n)}} - 1 - a^{(n)}\right)\frac{2}{r}\Bigg\} - a^{(n)}\left(\frac{r^{\mathbf{S}(n)}}{r} - 1\right)\Bigg] \circ \left(\Psi^{(n-1)} \circ (\Psi^{(n)})^{-1}\right)$$

which together with the control **A1**–**A3** of the background foliation, the induction assumptions (9.4.25) for $(f^{(n)}, \underline{f}^{(n)}, \lambda^{(n)})$, the control of $r - r^{\mathbf{S}(n)}$ following from the induction assumptions (9.4.26) for $(U^{(n)}, S^{(n)})$ and Lemma 9.8, and Sobolev, yields

$$\sup_{\mathbf{S}(n)} |\mathrm{Err}_6^{(n+1)}| \lesssim r^{-1}\overset{\circ}{\epsilon}\left\|(f^{(n)}, \underline{f}^{(n)}, a^{(n)})\right\|_{\mathfrak{h}_3(\mathbf{S}(n-1))} \lesssim r^{-1}\overset{\circ}{\epsilon}\overset{\circ}{\delta}.$$

In view of

- the definition (9.4.28), (9.4.29), (9.4.30) of $E^{(n+1)}$, $\underline{E}^{(n+1)}$ and $\widetilde{E}^{(n+1)}$,
- the control of the background foliation on $\mathbf{S}(n)$ provided by Corollary 9.12,
- the assumption (9.4.2) for $\check{\kappa}$, $\not{d}_1^\star\underline{\kappa} - \underline{C}e^\Phi$ and $\not{d}_1^\star\mu - Me^\Phi$,
- the control of $\underline{C} - \overline{\underline{C}}^{\mathbf{S}(n)}$ and $M - \overline{M}^{\mathbf{S}(n)}$ using Corollary 9.18, the control of the background foliation, as well as the induction assumptions (9.4.26) for $(U^{(n)}, S^{(n)})$,
- the control of $r - r^{\mathbf{S}(n)}$ following from the induction assumptions (9.4.26) for $(U^{(n)}, S^{(n)})$ and Lemma 9.8,
- the control of $m - m^{\mathbf{S}(n)}$ thanks to section 9.4.2 which uses the control of $\mathbf{S}(n)$ provided by the induction assumptions (9.4.26) for $(U^{(n)}, S^{(n)})$, as well as the induction assumptions (9.4.25) for $(f^{(n)}, \underline{f}^{(n)}, \lambda^{(n)})$,
- the above estimates for $\mathrm{Err}_j^{(n+1)}$, $j = 3, 4, 5, 6$,

we infer

$$\|E^{(n+1)}\|_{\mathfrak{h}_{s_{max}-3}(\mathbf{S}(n))} + \|\underline{E}^{(n+1)}\|_{\mathfrak{h}_{s_{max}-3}(\mathbf{S}(n))} \lesssim \max_{k\leq s_{max}-2}\sup_{\mathcal{R}}\left(\left|\mathfrak{d}^k\check{\kappa}\right| + r\left|\mathfrak{d}^{k-1}\left(\not{d}_1^\star\underline{\kappa} - \underline{C}e^\Phi\right)\right| + r^2\left|\mathfrak{d}^{k-1}\left(\not{d}_1^\star\mu - Me^\Phi\right)\right|\right) + r^{-2}\overset{\circ}{\epsilon}\overset{\circ}{\delta},$$

$$\|\widetilde{E}^{(n+1)}\|_{\mathfrak{h}_{s_{max}-2}(\mathbf{S}(n))} \lesssim r^3\max_{k\leq s_{max}-3}\sup_{\mathcal{R}}\left|\mathfrak{d}^{k-1}\left(\not{d}_1^\star\mu - Me^\Phi\right)\right| + r^4|M^{(n)} - \overline{M}^{\mathbf{S}(n-1)}| + r^{-1}\left\|(f^{(n)}, \underline{f}^{(n)})\right\|_{\mathfrak{h}_{s_{max}-1}(\mathbf{S}(n-1))} + r^{-1}\overset{\circ}{\epsilon}\overset{\circ}{\delta},$$

and

$$\left|\overline{\left(1-\frac{r^{\mathbf{S}(n)}}{r}\right)}^{\mathbf{S}(n)}\right|+\left|\frac{r^{\mathbf{S}(n)}}{2}\overline{\left(\check{\kappa}+\overline{\kappa}-\frac{2}{r}\right)}^{\mathbf{S}(n)}\right|+\left|\overline{\mathrm{Err}_6^{(n+1)}}^{\mathbf{S}(n)}\right|$$
$$\lesssim r\sup_{\mathcal{R}}\left(\left|\overline{\kappa}-\frac{2}{r}\right|+|\check{\kappa}|\right)+r^{-1}\sup_{\mathbf{S}(n)}|r-r^{\mathbf{S}(n)}|+r^{-1}\overset{\circ}{\epsilon}\overset{\circ}{\delta}.$$

Together with (9.4.35), (9.4.36) and (9.4.37), as well as the assumption (9.4.2) for $\check{\kappa}$, $\not{d}_1^\star\underline{\kappa}-\underline{C}e^\Phi$ and $\not{d}_1^\star\mu-Me^\Phi$, the induction assumptions (9.4.25) for $M^{(n)}$ and $(f^{(n)},\underline{f}^{(n)},\lambda^{(n)})$, and the control of $r-r^{\mathbf{S}(n)}$ following from the induction assumptions (9.4.26) for $(U^{(n)},S^{(n)})$ and Lemma 9.8, this implies, uniformly in n,

$$r^4|\underline{C}^{(n+1)}-\overline{\underline{C}}^{\mathbf{S}(n)}|+r^5|M^{(n+1)}-\overline{M}^{\mathbf{S}(n)}|$$
$$+\left\|\left(a^{(n+1)}-\overline{a^{(n+1)}}^{\mathbf{S}(n)},f^{(n+1)},\underline{f}^{(n+1)}\right)\right\|_{\mathfrak{h}_{s_{max}-1}(\mathbf{S}(n))}\lesssim\overset{\circ}{\delta}$$

and

$$r\left|\overline{a^{(n+1)}}^{\mathbf{S}(n)}\right|\lesssim\overset{\circ}{\delta}+\|\partial_\theta\left(U^{(n)},S^{(n)}\right)\|_{L^\infty(\overset{\circ}{S})}.$$

This concludes the proof of Proposition 9.40. □

9.4.4.2 Existence and boundedness of $(U^{(n+1)},S^{(n+1)})$

Proposition 9.41. *The equation* (9.4.34) *admits a unique solution* $U^{(1+n)},S^{(1+n)}$ *verifying the estimate*

$$\|\partial_\theta\left(U^{(n+1)},S^{(n+1)}\right)\|_{L^\infty(\overset{\circ}{S})}+r^{-1}\|\partial_\theta\left(U^{(n+1)},S^{(n+1)}\right)\|_{\mathfrak{h}_{s_{max}-1}(\overset{\circ}{S},\overset{\circ}{\not{g}})}\lesssim\overset{\circ}{\delta}$$

uniformly for all $n\in\mathbb{N}$.

Proof. The existence and uniqueness part of the proposition is an immediate consequence of the standard results for ODEs.

To prove the desired estimate, we use the equations for $(U^{(1+n)},S^{(1+n)})$ and infer, for $s=s_{max}-1$,

$$\|\partial_\theta U^{(n+1)}\|_{\mathfrak{h}_s(\overset{\circ}{S},\overset{\circ}{\not{g}})}$$
$$\lesssim\left\|(\gamma^{(n)})^{1/2}(\varsigma^{\#_n})^{-1}(f^{(1+n)})^{\#_n}\left(1+\frac{1}{4}\left(f^{(1+n)}\underline{f}^{(1+n)}\right)^{\#_n}\right)\right\|_{\mathfrak{h}_s(\overset{\circ}{S},\overset{\circ}{\not{g}})}.$$

Together with the non-sharp product estimate on $(\overset{\circ}{S},\overset{\circ}{\not{g}})$, see Lemma 9.5, we infer that, for $s = s_{max} - 1$,

$$\begin{aligned}
&\|\partial_\theta U^{(n+1)}\|_{\mathfrak{h}_s(\overset{\circ}{S},\overset{\circ}{\not g})} \\
\lesssim\ & r^{-1}\left\|(f^{(n+1)})^{\#_n},(\underline{f}^{(n+1)})^{\#_n}\right\|_{\mathfrak{h}_{s_{max}-2}(\overset{\circ}{S},\overset{\circ}{\not g})}\left\|\left(\varsigma^{\#_n}\right)^{-1}(\gamma^{(n)})^{1/2}\right\|_{\mathfrak{h}_s(\overset{\circ}{S},\overset{\circ}{\not g})} \\
&\times\left(1+\left\|(f^{(1+n)})^{\#_n},(\underline{f}^{(1+n)})^{\#_n}\right\|^2_{\mathfrak{h}_s(\overset{\circ}{S},\overset{\circ}{\not g})}\right).
\end{aligned}$$

In view of Lemma 9.11, Corollary 9.12, and the bound for $(f^{(n+1)},\underline{f}^{(n+1)})$ provided by Proposition 9.40, we deduce

$$\|\partial_\theta U^{(n+1)}\|_{\mathfrak{h}_s(\overset{\circ}{S},\overset{\circ}{\not g})} \lesssim \overset{\circ}{\delta} r^{-1}\left\|\left(\varsigma^{\#_n}\right)^{-1}(\gamma^{(n)})^{1/2}\right\|_{\mathfrak{h}_s(\overset{\circ}{S},\overset{\circ}{\not g})}.$$

We recall that

$$\gamma^{(n)} = \gamma^{\#_n}+\left(\varsigma^{\#_n}\right)^2\left(\underline{\Omega}+\tfrac{1}{4}\underline{b}^2\gamma\right)^{\#_n}(\partial_\theta U^{(n)})^2-2\varsigma^{\#_n}\partial_\theta U^{(n)}-\left(\gamma\varsigma\underline{b}\right)^{\#_n}\partial_\theta U^{(n)}.$$

In view of our assumptions on the Ricci coefficients and the non-sharp product estimates of Lemma 9.5

$$\left\|\left(\varsigma(\underline{\Omega}+\frac{1}{4}\underline{b}^2\gamma)\right)^{\#_n}\right\|_{\mathfrak{h}_{s_{max}-1}(\overset{\circ}{S},\overset{\circ}{\not g})}+\left\|\left(\gamma\underline{b}\right)^{\#_n}\right\|_{\mathfrak{h}_{s_{max}-1}(\overset{\circ}{S},\overset{\circ}{\not g})} \lesssim \overset{\circ}{\epsilon} r$$

we deduce

$$\left\|\left(\varsigma^{\#_n}\right)^{-1}\gamma^{(n)}\right\|_{\mathfrak{h}_{s_{max}-1}(\overset{\circ}{S},\overset{\circ}{\not g})} \lesssim \|\gamma^{\#_n}\|_{\mathfrak{h}_{s_{max}-1}(\overset{\circ}{S},\overset{\circ}{\not g})}+\overset{\circ}{\delta}r^2.$$

Together with Lemma 9.11 and Corollary 9.12, we deduce

$$\left\|\left(\varsigma^{\#_n}\right)^{-1}\left(\gamma^{(n)}\right)^{1/2}\right\|_{\mathfrak{h}_{s_{max}-1}(\overset{\circ}{S},\overset{\circ}{\not g})} \lesssim r^2$$

and therefore,

$$\|\partial_\theta U^{(n+1)}\|_{\mathfrak{h}_{s_{max}-1}(\overset{\circ}{S},\overset{\circ}{\not g})} \lesssim \overset{\circ}{\delta} r.$$

Proceeding in the same manner with equation

$$\partial_\theta S^{(1+n)}-\frac{1}{2}\varsigma^{\#_n}\underline{\Omega}^{\#_n}\partial_\theta U^{(1+n)} = \frac{1}{2}(\gamma^{(n)})^{1/2}(\underline{f}^{(1+n)})^{\#_n}$$

we infer that

$$r^{-1}\|\partial_\theta U^{(n+1)},\partial_\theta S^{(n+1)}\|_{\mathfrak{h}_{s_{max}-1}(\overset{\circ}{S},\overset{\circ}{\not g})} \lesssim \overset{\circ}{\delta}.$$

This, together with the Sobolev inequality, concludes the proof of Proposition 9.41. $\square$

9.4.5 Convergence of the iterates

To finish the proof of Theorem 9.32, it remains to prove convergence of the iterates.

Step 1. In order to prove the convergence of the iterative scheme, we introduce the following septets $P^{(n)}$:

$$P^{(0)} = (0,0,0,0,0,M(\overset{\circ}{u},\overset{\circ}{s}),\underline{C}(\overset{\circ}{u},\overset{\circ}{s})), \qquad P^{(1)} = (0,0,0,0,0,M(\overset{\circ}{u},\overset{\circ}{s}),\underline{C}(\overset{\circ}{u},\overset{\circ}{s})),$$
$$P^{(n)} = \Big(U^{(n)},S^{(n)},(a^{(n)})^{\#_{n-1}},(f^{(n)})^{\#_{n-1}},(\underline{f}^{(n)})^{\#_{n-1}},\underline{C}^{(n)},M^{(n)}\Big), \quad n\geq 2.$$

Since $(a^{(n)},f^{(n)},\underline{f}^{(n)})$ are defined on $\mathbf{S}(n-1)$, their respective pullback by $\Psi^{(n-1)}$ is defined on $\overset{\circ}{S}$ so that $P^{(n)}$ consists of a quintet of functions on $\overset{\circ}{S}$, together with two constants, for any n, and we may introduce the following norms to compare the elements of the sequence:

$$\begin{aligned}\|P^{(n)}\|_k := {} & r^{-1}\|\partial_\theta\left(U^{(n)},S^{(n)}\right)\|_{\mathfrak{h}_{k-1}(\overset{\circ}{S})} + r^4|\underline{C}^{(n)}-\overline{\underline{C}}^{\mathbf{S}(n-1)}| \\ & + r^5|M^{(n)}-\overline{M}^{\mathbf{S}(n-1)}| \\ & + \Big\|\Big((a^{(n)})^{\#_{n-1}},(f^{(n)})^{\#_{n-1}},(\underline{f}^{(n)})^{\#_{n-1}}\Big)\Big\|_{\mathfrak{h}_{k-1}(\overset{\circ}{S})}.\end{aligned} \tag{9.4.38}$$

Here are the steps needed to implement a convergence argument.

1. The quintets $P^{(n)}$ are bounded with respect to the norm (9.4.38) for the choice $k=s_{max}$.
2. The quintets $P^{(n)}$ are contractive with respect to the norm (9.4.38) for the choice $k=2$.

The precise statements are given in the following propositions.

Proposition 9.42. *We have, uniformly for all $n\in\mathbb{N}$,*

$$\|P^{(n)}\|_{s_{max}} \lesssim \overset{\circ}{\delta}.$$

Proof. The proof is an immediate consequence of Propositions 9.37, 9.41 and the estimate

$$\Big\|(\Psi^{(n-1)})^\#\left(f^{(n)},\underline{f}^{(n)},a^{(n)}\right)\Big\|_{\mathfrak{h}_{s_{max}-1}(\overset{\circ}{S})} \lesssim \Big\|\left(f^{(n)},\underline{f}^{(n)},a^{(n)}\right)\Big\|_{\mathfrak{h}_{s_{max}-1}(\mathbf{S}(n-1))}$$

which is a consequence of Lemma 9.11. □

Proposition 9.43. *We have, uniformly for all $n\in\mathbb{N}$, the contraction estimate*

$$\begin{aligned}&\|P^{(n+1)}-P^{(n)}\|_2 \\ \lesssim {} & \overset{\circ}{\epsilon}\Big(\|P^{(n)}-P^{(n-1)}\|_2+\|P^{(n-1)}-P^{(n-2)}\|_2+\|P^{(n-2)}-P^{(n-3)}\|_2\Big).\end{aligned}$$

The proof of Proposition 9.43 is postponed to section 9.6.

Step 2. In view of Proposition 9.43, we have

$$\|P^{(n+1)} - P^{(n)}\|_2 \lesssim (\overset{\circ}{\epsilon})^{\lfloor\frac{n-2}{3}\rfloor} \Big(\|P^{(3)} - P^{(2)}\|_2 + \|P^{(2)} - P^{(1)}\|_2 + \|P^{(1)} - P^{(0)}\|_2\Big), \quad n \geq 3,$$

which in view of Proposition 9.42 yields

$$\|P^{(n+1)} - P^{(n)}\|_2 \lesssim \overset{\circ}{\delta}\,(\overset{\circ}{\epsilon})^{\lfloor\frac{n}{3}\rfloor}, \quad n \geq 3.$$

Together with a simple interpolation argument on $\overset{\circ}{S}$ and Proposition 9.42, we infer

$$\|P^{(n+1)} - P^{(n)}\|_k \lesssim \overset{\circ}{\delta}\,\overset{\circ}{\epsilon}^{\left(\frac{s_{max}-k}{s_{max}-2}\right)\lfloor\frac{n}{3}\rfloor}, \quad 2 \leq k \leq s_{max}, \quad n \geq 3.$$

We infer the existence of a septet $P^{(\infty)}$ such that

$$\|P^{(\infty)}\|_{s_{max}} \lesssim \overset{\circ}{\delta} \tag{9.4.39}$$

and

$$\lim_{n\to+\infty} \|P^{(n)} - P^{(\infty)}\|_{s_{max}-1} = 0. \tag{9.4.40}$$

Also, we have

$$P^{(\infty)} = \Big(U^{(\infty)}, S^{(\infty)}, a_0^{(\infty)}, f_0^{(\infty)}, \underline{f}_0^{(\infty)}, \underline{C}^{(\infty)}, M^{(\infty)}\Big),$$

where the quintet of functions are defined on $\overset{\circ}{S}$ and $(\underline{C}^{(\infty)}, M^{(\infty)})$ are two constants. The functions $(U^{(\infty)}, S^{(\infty)})$ define a sphere $\mathbf{S}(\infty)$ and we introduce the map

$$\Psi^{(\infty)}(\overset{\circ}{u}, \overset{\circ}{s}, \theta, \varphi) = \Big(\overset{\circ}{u} + U^{(\infty)}(\theta), \overset{\circ}{s} + S^{(\infty)}(\theta), \theta, \varphi\Big)$$

so that $\Psi^{(\infty)}$ is a map from $\overset{\circ}{S}$ to $\mathbf{S}(\infty)$. Then, let

$$a^{(\infty)} = a_0^{(\infty)} \circ (\Psi^{(\infty)})^{-1}, \quad f^{(\infty)} = f_0^{(\infty)} \circ (\Psi^{(\infty)})^{-1}, \quad \underline{f}^{(\infty)} = \underline{f}_0^{(\infty)} \circ (\Psi^{(\infty)})^{-1}$$

so that $a^{(\infty)}, f^{(\infty)}, \underline{f}^{(\infty)}$ are defined on $\mathbf{S}(\infty)$ and

$$a_0^{(\infty)} = (a^{(\infty)})^{\#_\infty}, \quad f_0^{(\infty)} = (f^{(\infty)})^{\#_\infty}, \quad \underline{f}_0^{(\infty)} = (\underline{f}^{(\infty)})^{\#_\infty}.$$

From these definitions, the above control of $P^{(\infty)}$ and Lemma 9.11, we infer

$$r^{-1}\|(\partial_\theta U^{(\infty)}, \partial_\theta S^{(\infty)})\|_{\mathfrak{h}_{s_{max}-1}(\overset{\circ}{S})} + \|(a^{(\infty)}, f^{(\infty)}, \underline{f}^{(\infty)})\|_{\mathfrak{h}_{s_{max}-1}(\mathbf{S}(\infty))} \lesssim \overset{\circ}{\delta}.$$

In particular, applying Corollary 9.38 twice, first with $s = s_{max} - 1$, and then with $s = s_{max}$, we deduce

$$\|(a^{(\infty)}, f^{(\infty)}, \underline{f}^{(\infty)})\|_{\mathfrak{h}_{s_{max}+1}(\mathbf{S}(\infty))} \lesssim \overset{\circ}{\delta}.$$

Together with the above control for $(U^{(\infty)}, S^{(\infty)})$, we finally obtain

$$r^{-1}\|(U^{(\infty)\prime}, S^{(\infty)\prime})\|_{\mathfrak{h}_{s_{max}-1}(\overset{\circ}{S})} + \|(a^{(\infty)}, f^{(\infty)}, \underline{f}^{(\infty)})\|_{\mathfrak{h}_{s_{max}+1}(\mathbf{S}(\infty))} \lesssim \overset{\circ}{\delta}. \quad (9.4.41)$$

Step 3. We proceed to control the area radius $r^{\mathbf{S}(\infty)}$ and the Hawking mass $m^{\mathbf{S}(\infty)}$ of the sphere $\mathbf{S}(\infty)$. First, note from (9.4.41) and the Sobolev embedding on $\overset{\circ}{S}$ that we have

$$\|(U^{(\infty)}, S^{(\infty)})\|_{L^\infty(\overset{\circ}{S})} \lesssim \overset{\circ}{\delta}. \quad (9.4.42)$$

Together with Lemma 9.8, we infer that[15]

$$\left|\frac{r^{\mathbf{S}(\infty)}}{r} - 1\right| \lesssim \overset{\circ}{\delta}. \quad (9.4.43)$$

Next, we denote by $\Gamma^{\mathbf{S}(\infty)}$ the connection coefficients of $\mathbf{S}(\infty)$. We have in view of the transformation formula from the original frame (e_4, e_3, e_θ) to the frame $(e_4^{\mathbf{S}(\infty)}, e_3^{\mathbf{S}(\infty)}, e_\theta^{\mathbf{S}(\infty)})$ of $\mathbf{S}(\infty)$

$$\begin{aligned} \kappa^{\mathbf{S}(\infty)}\underline{\kappa}^{\mathbf{S}(\infty)} &= \left(\kappa + \not{d}^{\mathbf{S}(\infty)} f^{(\infty)} + \mathrm{Err}(\kappa, \kappa^{\mathbf{S}(\infty)})\right) \\ &\times \left(\underline{\kappa} + \not{d}^{\mathbf{S}(\infty)} \underline{f}^{(\infty)} + \mathrm{Err}(\underline{\kappa}, \underline{\kappa}^{\mathbf{S}(\infty)})\right). \end{aligned}$$

Together with the estimate (9.4.41) for $f^{(\infty)}$ and $\underline{f}^{(\infty)}$ and the assumptions **A1**–**A3** for $\check{\Gamma}$ corresponding to the original frame (e_4, e_3, e_θ), we infer

$$\left|\kappa^{\mathbf{S}(\infty)}\underline{\kappa}^{\mathbf{S}(\infty)} - \kappa\underline{\kappa}\right| \lesssim \overset{\circ}{\delta} r^{-3}.$$

Recall that (see (9.4.2))

$$\left|\kappa - \frac{2}{r}\right| \lesssim \overset{\circ}{\delta} r^{-2}, \qquad \left|\overline{\underline{\kappa}} + \frac{2\left(1 - \frac{2m}{r}\right)}{r}\right| \lesssim \overset{\circ}{\delta}.$$

Thus, since $\underline{\kappa} = \overline{\underline{\kappa}} + \check{\underline{\kappa}}$,

$$\kappa\underline{\kappa} = -\frac{4\left(1 - \frac{2m}{r}\right)}{r^2} + \frac{2}{r}\check{\underline{\kappa}} + O(\overset{\circ}{\delta})r^{-2}.$$

We deduce

$$\left|\kappa^{\mathbf{S}(\infty)}\underline{\kappa}^{\mathbf{S}(\infty)} + \frac{4\left(1 - \frac{2m}{r}\right)}{r^2} - \frac{2}{r}\check{\underline{\kappa}}\right| \lesssim \overset{\circ}{\delta} r^{-3}.$$

[15] Here, we also use the fact that, on $S^{(\infty)}$, we have

$$|r - \overset{\circ}{r}| \lesssim \|(U^{(\infty)}, S^{(\infty)})\|_{L^\infty(\overset{\circ}{S})} \lesssim \overset{\circ}{\delta}.$$

Thus, in view of (9.4.43),

$$\int_{\mathbf{S}(\infty)} \kappa^{\mathbf{S}(\infty)} \underline{\kappa}^{\mathbf{S}(\infty)} = -\int_{\mathbf{S}(\infty)} \frac{4\left(1-\frac{2m}{r}\right)}{r^2} + O(\overset{\circ}{\delta}) r^{-1}.$$

Making use of the definition of the Hawking mass, i.e.,

$$m^{\mathbf{S}(\infty)} = \frac{r^{\mathbf{S}(\infty)}}{2}\left(1+\frac{1}{16\pi}\int_{\mathbf{S}(\infty)} \kappa^{\mathbf{S}(\infty)} \underline{\kappa}^{\mathbf{S}(\infty)}\right),$$

we easily deduce[16]

$$\left|m^{\mathbf{S}(\infty)} - m\right| \lesssim \overset{\circ}{\delta}. \tag{9.4.44}$$

Step 4. We make use of Lemma 9.30 to extend $(a^{(\infty)}, f^{(\infty)}, \underline{f}^{(\infty)})$ as well as the frame $(e_3^{\mathbf{S}(\infty)}, e_4^{\mathbf{S}(\infty)}, e_\theta^{\mathbf{S}(\infty)})$ in a small neighborhood of $\mathbf{S}(\infty)$ such that we have

$$\xi^{\mathbf{S}(\infty)} = 0, \qquad \omega^{\mathbf{S}(\infty)} = 0, \qquad \underline{\eta}^{\mathbf{S}(\infty)} + \zeta^{\mathbf{S}(\infty)} = 0, \tag{9.4.45}$$

and then provide estimates for the corresponding Ricci coefficients and curvature components $\check{\Gamma}^{\mathbf{S}(\infty)}$, $\check{R}^{\mathbf{S}(\infty)}$. More precisely we make use of the assumption **A1**, the estimates in (9.4.41) for $(a^{(\infty)}, f^{(\infty)}, \underline{f}^{(\infty)})$, and the transformation formulae to derive the desired estimates (9.4.8) for s_{max} derivative of the Ricci coefficients and curvature components of $\mathbf{S}(\infty)$.

Step 5. Thanks to (9.4.40), we can pass to the limit in (9.4.27), (9.4.32), (9.4.33). In view of Remark 9.35, we deduce that equations (9.4.13), (9.4.14) hold true. Thus, we may apply Proposition 9.33 which implies that (9.4.12) holds true. In particular, the desired GCM conditions (9.4.1) hold true which concludes the proof of Theorem 9.32.

9.5 PROOF OF PROPOSITION 9.37 AND OF COROLLARY 9.38

9.5.1 Proof of Proposition 9.37

Step 1. We start with the proof of existence. Note first that the existence of $\overline{a}^{\mathbf{S}}$ and $\check{a}^{\mathbf{S}}$ is immediate in view of the last two equations of (9.4.16). We thus focus on the existence of $(f, \underline{f})$. In view of the first two equations of (9.4.16), we have

$$\begin{aligned} B^{\mathbf{S}} f &= -\frac{6m^{\mathbf{S}}}{(r^{\mathbf{S}})^5}(A^{\mathbf{S}})^{-1}(\underline{f} - \Upsilon^{\mathbf{S}} f) + \frac{2}{r^{\mathbf{S}}}(M^{\mathbf{S}} - \overline{M}^{\mathbf{S}})(A^{\mathbf{S}})^{-1} e^{\Phi} + F_1, \\ B^{\mathbf{S}} \underline{f} &= -\frac{6m^{\mathbf{S}}\Upsilon^{\mathbf{S}}}{(r^{\mathbf{S}})^5}(A^{\mathbf{S}})^{-1}(\underline{f} - \Upsilon^{\mathbf{S}} f) + (\underline{C}^{\mathbf{S}} - \overline{\underline{C}}^{\mathbf{S}}) e^{\Phi} \\ &\quad + \frac{2\Upsilon^{\mathbf{S}}}{r^{\mathbf{S}}}(M^{\mathbf{S}} - \overline{M}^{\mathbf{S}})(A^{\mathbf{S}})^{-1} e^{\Phi} + F_2. \end{aligned} \tag{9.5.1}$$

[16]See also section 9.4.2.

In particular, subtracting $\Upsilon^{\mathbf{S}}$ times the first equation to the second equation, we infer that the existence of $(f, \underline{f})$ is equivalent to the existence of

$$\begin{aligned} B^{\mathbf{S}}(\underline{f} - \Upsilon^{\mathbf{S}} f) &= F_2 - \Upsilon^{\mathbf{S}} F_1 + (\underline{C}^{\mathbf{S}} - \overline{\underline{C}}^{\mathbf{S}}) e^{\Phi}, \\ B^{\mathbf{S}} f &= -\frac{6m^{\mathbf{S}}}{(r^{\mathbf{S}})^5}(A^{\mathbf{S}})^{-1}(\underline{f} - \Upsilon^{\mathbf{S}} f) + \frac{2}{r^{\mathbf{S}}}(M^{\mathbf{S}} - \overline{M}^{\mathbf{S}})(A^{\mathbf{S}})^{-1} e^{\Phi} + F_1. \end{aligned} \tag{9.5.2}$$

Step 2. Next, we differentiate (9.5.2) w.r.t. $\not{d}_2^{\star\mathbf{S}}$ which yields the system

$$\begin{aligned} \left(\not{d}_2^{\star\mathbf{S}} \not{d}_2^{\mathbf{S}} + \frac{3m^{\mathbf{S}}}{(r^{\mathbf{S}})^3}\right) \not{d}_2^{\star\mathbf{S}}(\underline{f} - \Upsilon^{\mathbf{S}} f) &= \not{d}_2^{\star\mathbf{S}}\Big\{F_2 - \Upsilon^{\mathbf{S}} F_1\Big\}, \\ \left(\not{d}_2^{\star\mathbf{S}} \not{d}_2^{\mathbf{S}} + \frac{3m^{\mathbf{S}}}{(r^{\mathbf{S}})^3}\right) \not{d}_2^{\star\mathbf{S}} f &= \not{d}_2^{\star\mathbf{S}}\Bigg\{ -\frac{6m^{\mathbf{S}}}{(r^{\mathbf{S}})^5}(A^{\mathbf{S}})^{-1}(\underline{f} - \Upsilon^{\mathbf{S}} f) \\ &\quad + \frac{2}{r^{\mathbf{S}}}(M^{\mathbf{S}} - \overline{M}^{\mathbf{S}})(A^{\mathbf{S}})^{-1} e^{\Phi} + F_1\Bigg\}, \end{aligned} \tag{9.5.3}$$

where we have used the fact that $\not{d}_2^{\star\mathbf{S}} e^{\Phi} = 0$. Since the operator $\not{d}_2^{\star\mathbf{S}} \not{d}_2^{\mathbf{S}}$ is coercive and invertible, so is $\not{d}_2^{\star\mathbf{S}} \not{d}_2^{\mathbf{S}} + \frac{3m^{\mathbf{S}}}{(r^{\mathbf{S}})^3}$. Thus, using also the fact that e^{Φ} generates the kernel of $\not{d}_2^{\star\mathbf{S}}$ and that $\not{d}_2^{\star\mathbf{S}}$ is surjective, there exists $\underline{f} - \Upsilon^{\mathbf{S}} f$ solution to

$$\begin{aligned} \not{d}_2^{\star\mathbf{S}}(\underline{f} - \Upsilon^{\mathbf{S}} f) &= \left(\not{d}_2^{\star\mathbf{S}} \not{d}_2^{\mathbf{S}} + \frac{6m^{\mathbf{S}}}{(r^{\mathbf{S}})^3}\right)^{-1} \not{d}_2^{\star\mathbf{S}}\Big\{F_2 - \Upsilon^{\mathbf{S}} F_1\Big\}, \\ \int_{\mathbf{S}} (\underline{f} - \Upsilon^{\mathbf{S}} f) e^{\Phi} &= \underline{\Lambda} - \Upsilon^{\mathbf{S}} \Lambda. \end{aligned} \tag{9.5.4}$$

Step 3. Next, we have, using in particular the assumptions on $K^{\mathbf{S}}$,

$$\begin{aligned} A^{\mathbf{S}}(e^{\Phi}) &= \not{d}_1^{\star\mathbf{S}} \not{d}_1^{\mathbf{S}}(e^{\Phi}) = \left(\not{d}_2^{\mathbf{S}} \not{d}_2^{\star\mathbf{S}} + 2K^{\mathbf{S}}\right) e^{\Phi} = 2K^{\mathbf{S}} e^{\Phi} \\ &= \frac{2}{(r^{\mathbf{S}})^2} e^{\Phi} + \left(K^{\mathbf{S}} - \frac{2}{(r^{\mathbf{S}})^2}\right) e^{\Phi} \end{aligned}$$

and hence

$$(A^{\mathbf{S}})^{-1}(e^{\Phi}) = \frac{(r^{\mathbf{S}})^2}{2} e^{\Phi} - \frac{(r^{\mathbf{S}})^2}{2}(A^{\mathbf{S}})^{-1}\left[\left(K^{\mathbf{S}} - \frac{2}{(r^{\mathbf{S}})^2}\right) e^{\Phi}\right]. \tag{9.5.5}$$

In particular, we have, in view of the assumptions of the proposition,

$$\int_{\mathbf{S}} e^{2\Phi} = \frac{4\pi}{3}(r^{\mathbf{S}})^4(1 + O(\delta_1)), \quad \int_{\mathbf{S}} e^{\Phi}(A^{\mathbf{S}})^{-1}(e^{\Phi}) = \frac{2\pi}{3}(r^{\mathbf{S}})^6(1 + O(\delta_1)), \tag{9.5.6}$$

so that these quantities do not vanish. We may thus choose $\underline{C}^{\mathbf{S}}$ and $M^{\mathbf{S}}$ as follows:

$$
\begin{aligned}
\underline{C}^{\mathbf{S}} &= \overline{\underline{C}}^{\mathbf{S}} + \left\{\frac{6m^{\mathbf{S}}}{(r^{\mathbf{S}})^3}(\underline{\Lambda} - \Upsilon^{\mathbf{S}}\Lambda) + \int_{\mathbf{S}} \left[\Upsilon^{\mathbf{S}} F_1 - F_2\right] e^{\Phi}\right\} \left(\int_{\mathbf{S}} e^{2\Phi}\right)^{-1}, \\
M^{\mathbf{S}} &= \overline{M}^{\mathbf{S}} + \frac{r^{\mathbf{S}}}{2}\left\{\frac{6m^{\mathbf{S}}}{(r^{\mathbf{S}})^3}\Lambda + \int_{\mathbf{S}} \left[\frac{6m^{\mathbf{S}}}{(r^{\mathbf{S}})^5}(A^{\mathbf{S}})^{-1}(f - \Upsilon^{\mathbf{S}}\underline{f}) - F_1\right] e^{\Phi}\right\} \\
&\quad \times \left(\int_{\mathbf{S}} e^{\Phi}(A^{\mathbf{S}})^{-1}(e^{\Phi})\right)^{-1},
\end{aligned}
\tag{9.5.7}
$$

where $\underline{f} - \Upsilon^{\mathbf{S}} f$ appearing on the RHS of the above choice of $M^{\mathbf{S}}$ is the solution of (9.5.4).

Step 4. Next, with $\underline{f} - \Upsilon^{\mathbf{S}} f$ chosen as in (9.5.4) and $M^{\mathbf{S}}$ chosen as in (9.5.7), and arguing as in Step 2, there exists f solution to

$$
\begin{aligned}
{\not{d}}_2^{\star\mathbf{S}} f &= \left({\not{d}}_2^{\star\mathbf{S}} {\not{d}}_2^{\mathbf{S}} + \frac{6m^{\mathbf{S}}}{(r^{\mathbf{S}})^3}\right)^{-1} {\not{d}}_2^{\star\mathbf{S}} \Bigg\{ - \frac{6m^{\mathbf{S}}}{(r^{\mathbf{S}})^5}(A^{\mathbf{S}})^{-1}(\underline{f} - \Upsilon^{\mathbf{S}} f) \\
&\quad + \frac{2}{r^{\mathbf{S}}}(M^{\mathbf{S}} - \overline{M}^{\mathbf{S}})(A^{\mathbf{S}})^{-1} e^{\Phi} + F_1 \Bigg\},
\end{aligned}
\tag{9.5.8}
$$

$$
\int_{\mathbf{S}} f e^{\Phi} = \Lambda.
$$

Now, in view of

1. the fact that $(f, \underline{f})$ satisfies (9.5.3) in view of (9.5.4), (9.5.8),
2. the choice (9.5.7) for the constants $\underline{C}^{\mathbf{S}}$ and $M^{\mathbf{S}}$,
3. the fact that e^{Φ} generates the kernel of ${\not{d}}_2^{\star\mathbf{S}}$,

we infer that $(f, \underline{f})$ satisfies (9.5.2), and hence (9.5.1), which concludes the existence part of the proof.

Step 5. Next, we focus on the proof of the a priori estimates. Note first that the last two equations of (9.4.16) immediately yield the a priori estimates for $\overline{a}^{\mathbf{S}}$ and $\check{a}^{\mathbf{S}}$. We then focus on the a priori control of $(\underline{C}^{\mathbf{S}}, M^{\mathbf{S}})$ and $(f, \underline{f})$. We multiply the first two equations of (9.5.1) by e^{Φ} and integrate on $\mathbf{S}$. Using the fact that e^{Φ} generates the kernel of ${\not{d}}_2^{\star\mathbf{S}}$, and that ${\not{d}}_2^{\star\mathbf{S}}$ is the adjoint of ${\not{d}}_2^{\mathbf{S}}$, we deduce that the constants $\underline{C}^{\mathbf{S}}$ and $M^{\mathbf{S}}$ are given by (9.5.7). Together with (9.5.6), we infer the following control for the constants $\underline{C}^{\mathbf{S}}$ and $M^{\mathbf{S}}$:

$$
\begin{aligned}
|\underline{C}^{\mathbf{S}} - \overline{\underline{C}}^{\mathbf{S}}| &\lesssim (r^{\mathbf{S}})^{-7}(|\Lambda| + |\underline{\Lambda}|) + (r^{\mathbf{S}})^{-2}\|F_1\|_{L^2(\mathbf{S})} + (r^{\mathbf{S}})^{-2}\|F_2\|_{L^2(\mathbf{S})}, \\
|M^{\mathbf{S}} - \overline{M}^{\mathbf{S}}| &\lesssim (r^{\mathbf{S}})^{-8}(|\Lambda| + |\underline{\Lambda}|) + (r^{\mathbf{S}})^{-6}\|\underline{f} - \Upsilon^{\mathbf{S}} f\|_{L^2(\mathbf{S})} + (r^{\mathbf{S}})^{-3}\|F_1\|_{L^2(\mathbf{S})}.
\end{aligned}
\tag{9.5.9}
$$

Step 6. Next, we multiply the first equation of (9.5.2) by $(\underline{f} - \Upsilon^{\mathbf{S}} f)$, integrate on

$\mathbf{S}$, and integrate by parts the term $B^{\mathbf{S}}(\underline{f}-\Upsilon^{\mathbf{S}}f)$. We obtain

$$\begin{aligned}\|r^{\mathbf{S}}\,{\not{d}}_2^{\star\mathbf{S}}(\underline{f}-\Upsilon^{\mathbf{S}}f)\|^2_{L^2(\mathbf{S})} &\lesssim \Big((r^{\mathbf{S}})^2\|F_1\|_{L^2(\mathbf{S})}+(r^{\mathbf{S}})^2\|F_2\|_{L^2(\mathbf{S})}\Big)\|\underline{f}-\Upsilon^{\mathbf{S}}f\|_{L^2(\mathbf{S})}\\ &+(r^{\mathbf{S}})^2|\underline{C}^{\mathbf{S}}-\overline{\underline{C}}^{\mathbf{S}}|(|\Lambda|+|\underline{\Lambda}|).\end{aligned}$$

Together with a Poincaré inequality for ${\not{d}}_2^{\star\mathbf{S}}$ and the estimate for $\underline{C}^{\mathbf{S}}-\overline{\underline{C}}^{\mathbf{S}}$ in (9.5.9), we deduce

$$\|\underline{f}-\Upsilon^{\mathbf{S}}f\|_{\mathfrak{h}_1(\mathbf{S})}\lesssim (r^{\mathbf{S}})^{-2}(|\Lambda|+|\underline{\Lambda}|)+(r^{\mathbf{S}})^2\|F_1\|_{L^2(\mathbf{S})}+(r^{\mathbf{S}})^2\|F_2\|_{L^2(\mathbf{S})}. \tag{9.5.10}$$

In particular, together with (9.5.9), we infer

$$\begin{aligned}|\underline{C}^{\mathbf{S}}-\overline{\underline{C}}^{\mathbf{S}}|+r^{\mathbf{S}}|M^{\mathbf{S}}-\overline{M}^{\mathbf{S}}| &\lesssim (r^{\mathbf{S}})^{-7}(|\Lambda|+|\underline{\Lambda}|)+(r^{\mathbf{S}})^{-2}\|F_1\|_{L^2(\mathbf{S})}\\ &+(r^{\mathbf{S}})^{-2}\|F_2\|_{L^2(\mathbf{S})}\end{aligned} \tag{9.5.11}$$

which is the desired a priori estimate for $(\underline{C}^{\mathbf{S}}, M^{\mathbf{S}})$.

Step 7. Next, we multiply the second equation of (9.5.2) by f, integrate on $\mathbf{S}$, and integrate by parts the term $B^{\mathbf{S}}f$. We obtain

$$\begin{aligned}\|r^{\mathbf{S}}\,{\not{d}}_2^{\star\mathbf{S}}f\|^2_{L^2(\mathbf{S})} &\lesssim \Big((r^{\mathbf{S}})^{-1}\|\underline{f}-\Upsilon^{\mathbf{S}}f\|_{L^2(\mathbf{S})}+(r^{\mathbf{S}})^2\|F_1\|_{L^2(\mathbf{S})}\\ &+(r^{\mathbf{S}})^5|M^{\mathbf{S}}-\overline{M}^{\mathbf{S}}|\Big)\|f\|_{L^2(\mathbf{S})}\end{aligned}$$

which together with a Poincaré inequality for ${\not{d}}_2^{\star\mathbf{S}}$, (9.5.10), and (9.5.11) yields

$$\|f\|_{\mathfrak{h}_1(\mathbf{S})} \lesssim (r^{\mathbf{S}})^{-3}(|\Lambda|+|\underline{\Lambda}|)+(r^{\mathbf{S}})^2\|F_1\|_{L^2(\mathbf{S})}+(r^{\mathbf{S}})^2\|F_2\|_{L^2(\mathbf{S})}.$$

Together with (9.5.10), we obtain

$$\begin{aligned}\|f\|_{\mathfrak{h}_1(\mathbf{S})}+\|\underline{f}\|_{\mathfrak{h}_1(\mathbf{S})} &\lesssim (r^{\mathbf{S}})^{-2}(|\Lambda|+|\underline{\Lambda}|)+(r^{\mathbf{S}})^2\|F_1\|_{L^2(\mathbf{S})}\\ &+(r^{\mathbf{S}})^2\|F_2\|_{L^2(\mathbf{S})}.\end{aligned} \tag{9.5.12}$$

Step 8. Finally, using the identity ${\not{d}}_2^{\mathbf{S}}\,{\not{d}}_2^{\star\mathbf{S}}={\not{d}}_1^{\star\mathbf{S}}\,{\not{d}}_1^{\mathbf{S}}-2K^{\mathbf{S}}$, we rewrite (9.5.1) as follows:

$$\begin{aligned}{\not{d}}_1^{\star\mathbf{S}}\,{\not{d}}_1^{\mathbf{S}}f &= \left(2K^{\mathbf{S}}-\frac{6m^{\mathbf{S}}}{(r^{\mathbf{S}})^3}\right)f-\frac{6m^{\mathbf{S}}}{(r^{\mathbf{S}})^5}(A^{\mathbf{S}})^{-1}(\underline{f}-\Upsilon^{\mathbf{S}}f)\\ &+\frac{2}{r^{\mathbf{S}}}(M^{\mathbf{S}}-\overline{M}^{\mathbf{S}})(A^{\mathbf{S}})^{-1}e^{\Phi}+F_1,\\ {\not{d}}_1^{\star\mathbf{S}}\,{\not{d}}_1^{\mathbf{S}}\underline{f} &= \left(2K^{\mathbf{S}}-\frac{6m^{\mathbf{S}}}{(r^{\mathbf{S}})^3}\right)\underline{f}-\frac{6m^{\mathbf{S}}\Upsilon^{\mathbf{S}}}{(r^{\mathbf{S}})^5}(A^{\mathbf{S}})^{-1}\left(\underline{f}-\Upsilon^{\mathbf{S}}f\right)+(\underline{C}^{\mathbf{S}}-\overline{\underline{C}}^{\mathbf{S}})e^{\Phi}\\ &+\frac{2\Upsilon^{\mathbf{S}}}{r^{\mathbf{S}}}(M^{\mathbf{S}}-\overline{M}^{\mathbf{S}})(A^{\mathbf{S}})^{-1}e^{\Phi}+F_2.\end{aligned}$$

Together with (9.5.12), (9.5.11) and the assumptions for $K^{\mathbf{S}}$

$$\sup_{\mathbf{S}}|K^{\mathbf{S}}|\lesssim\frac{1}{(r^{\mathbf{S}})^2},\qquad \|K^{\mathbf{S}}\|_{\mathfrak{h}_{s-2}(\mathbf{S})}\lesssim\frac{1}{r^{\mathbf{S}}},$$

we deduce by iteration

$$\|(f,\underline{f})\|_{\mathfrak{h}_s(\mathbf{S})} \lesssim (r^{\mathbf{S}})^2\|F_1\|_{\mathfrak{h}_{s-2}(\mathbf{S})} + (r^{\mathbf{S}})^2\|F_2\|_{\mathfrak{h}_{s-2}(\mathbf{S})} + (r^{\mathbf{S}})^{-2}(|\Lambda| + |\underline{\Lambda}|)$$

which concludes the part on a priori estimates. The part on uniqueness follows from the linearity of the equations and the a priori estimates. This ends the proof of Proposition 9.37.

9.5.2 Proof of Corollary 9.38

Step 1. First, we introduce for convenience the notation

$$\Lambda := \int_{\mathbf{S}} f e^{\Phi}, \qquad \underline{\Lambda} := \int_{\mathbf{S}} \underline{f} e^{\Phi}.$$

Then, in view of the assumptions of Corollary 9.38, $(f, \underline{f}, \lambda)$ satisfies (9.4.1), and hence, there exist constants $(\underline{C}^{\mathbf{S}}, M^{\mathbf{S}})$ such that $(f, \underline{f}, \lambda)$ satisfies (9.4.12). In particular, from Proposition 9.33, $(f, \underline{f}, \lambda)$ satisfies (9.4.13), (9.4.14). In view of Remark 9.35, we deduce that $(f, \underline{f}, \lambda)$ satisfies the linearized GCM system (9.4.16) with the following choices for F_1, F_2, F_3, b_0,

$$\begin{aligned}
F_1 &= -\frac{3m^{\mathbf{S}}}{(r^{\mathbf{S}})^4}(A^{\mathbf{S}})^{-1}\left(\left(\kappa - \frac{2}{r^{\mathbf{S}}}\right)\underline{f} + \left(\underline{\kappa} + \frac{2\Upsilon^{\mathbf{S}}}{r^{\mathbf{S}}}\right) f\right)\\
&\quad + \frac{3}{4}\left(\kappa - \frac{2}{r^{\mathbf{S}}}\right)(A^{\mathbf{S}})^{-1}(\rho z) - \frac{3}{2r^{\mathbf{S}}}(A^{\mathbf{S}})^{-1}\left(\left(\rho + \frac{2m^{\mathbf{S}}}{(r^{\mathbf{S}})^3}\right) z\right) - {\not{d}}^\star\check{\kappa}\\
&\quad - \left(\kappa - \frac{2}{r^{\mathbf{S}}}\right)(A^{\mathbf{S}})^{-1}\left(- M^{\mathbf{S}} e^{\Phi} + {\not{d}}^\star\check{\mu}\right) - \frac{2}{r^{\mathbf{S}}}(A^{\mathbf{S}})^{-1}\left({\not{d}}^\star\check{\mu} - \overline{M}^{\mathbf{S}} e^{\Phi}\right) + \mathrm{Err}_4,
\end{aligned}$$

$$\begin{aligned}
F_2 &= -\frac{3m^{\mathbf{S}}\Upsilon^{\mathbf{S}}}{(r^{\mathbf{S}})^4}(A^{\mathbf{S}})^{-1}\left(\left(\kappa - \frac{2}{r^{\mathbf{S}}}\right)\underline{f} + \left(\underline{\kappa} + \frac{2\Upsilon^{\mathbf{S}}}{r^{\mathbf{S}}}\right) f\right)\\
&\quad - \frac{3}{4}\left(\underline{\kappa} + \frac{2\Upsilon^{\mathbf{S}}}{r^{\mathbf{S}}}\right)(A^{\mathbf{S}})^{-1}(\rho z) + \frac{3\Upsilon^{\mathbf{S}}}{2r^{\mathbf{S}}}(A^{\mathbf{S}})^{-1}\left(\left(\rho + \frac{2m^{\mathbf{S}}}{(r^{\mathbf{S}})^3}\right) z\right)\\
&\quad - \left({\not{d}}^\star\check{\underline{\kappa}} - \overline{\underline{C}}^{\mathbf{S}} e^{\Phi}\right) + \left(\underline{\kappa} + \frac{2\Upsilon^{\mathbf{S}}}{r^{\mathbf{S}}}\right)(A^{\mathbf{S}})^{-1}\left(- M^{\mathbf{S}} e^{\Phi} + {\not{d}}^\star\check{\mu}\right)\\
&\quad - \frac{2\Upsilon^{\mathbf{S}}}{r^{\mathbf{S}}}(A^{\mathbf{S}})^{-1}\left({\not{d}}^\star\check{\mu} - \overline{M}^{\mathbf{S}} e^{\Phi}\right) + \mathrm{Err}_5,
\end{aligned}$$

$$\begin{aligned}
F_3 &= -\frac{3}{4}(A^{\mathbf{S}})^{-1}(\rho z) + f\underline{\omega} - \underline{f}\omega + \frac{1}{4}f\underline{\kappa} - \frac{1}{4}\underline{f}\kappa\\
&\quad + (A^{\mathbf{S}})^{-1}\left(- M^{\mathbf{S}} e^{\Phi} + {\not{d}}^\star\check{\mu}\right) - (A^{\mathbf{S}})^{-1}\mathrm{Err}_3,\\
b_0 &= \overline{\left(1 - \frac{r^{\mathbf{S}}}{r}\right)}^{\mathbf{S}} - \frac{r^{\mathbf{S}}}{2}\overline{\left(\check{\kappa} + \overline{\kappa} - \frac{2}{r}\right)}^{\mathbf{S}} + \overline{\mathrm{Err}_6}^{\mathbf{S}}.
\end{aligned}$$

Step 2. In view of Corollary 9.17, we may apply Proposition 9.37. In particular, the following a priori estimates hold:

$$|\underline{C}^{\mathbf{S}} - \overline{\underline{C}}^{\mathbf{S}}| + r^{\mathbf{S}}|M^{\mathbf{S}} - \overline{M}^{\mathbf{S}}| \lesssim (r^{\mathbf{S}})^{-7}\big(|\Lambda| + |\underline{\Lambda}|\big) + (r^{\mathbf{S}})^{-2}\|F_1\|_{L^2(\mathbf{S})} + (r^{\mathbf{S}})^{-2}\|F_2\|_{L^2(\mathbf{S})}, \tag{9.5.13}$$

$$\|(f, \underline{f})\|_{\mathfrak{h}_{s+1}(\mathbf{S})} \lesssim (r^{\mathbf{S}})^{-2}\big(|\Lambda| + |\underline{\Lambda}|\big) + (r^{\mathbf{S}})^2\|F_1\|_{\mathfrak{h}_{s-1}(\mathbf{S})} + (r^{\mathbf{S}})^2\|F_2\|_{\mathfrak{h}_{s-1}(\mathbf{S})}, \tag{9.5.14}$$

$$\|\check{a}^{\mathbf{S}}\|_{\mathfrak{h}_{s+1}(\mathbf{S})} \lesssim r^{\mathbf{S}}\|F_3\|_{\mathfrak{h}_s(\mathbf{S})} \tag{9.5.15}$$

and

$$|\overline{a}^{\mathbf{S}}| \lesssim |b_0|, \tag{9.5.16}$$

where F_1, F_2, F_3 and b_0 are given in Step 1.

Step 3. In view of the a priori estimates of Step 2, we need to estimate F_1, F_2, F_3 and b_0. We start with the control of the error terms Err_j, $j = 3, 4, 5, 6$. In view of Lemma 9.29, we have, since $4 \leq s \leq s_{max}$,

$$\begin{aligned}
\|\mathrm{Err}_1, \mathrm{Err}_2\|_{\mathfrak{h}_{s-1}(\mathbf{S})} &\lesssim r^{-2}\|(f, \underline{f}, a)\|_{\mathfrak{h}_{s+1}(\mathbf{S})}\left(\overset{\circ}{\epsilon} + r^{-1}\|f, \underline{f}, a\|_{\mathfrak{h}_s(\mathbf{S})}\right),\\
\|\mathrm{Err}_3\|_{\mathfrak{h}_{s-2}(\mathbf{S})} &\lesssim r^{-3}\|(f, \underline{f}, a)\|_{\mathfrak{h}_{s+1}(\mathbf{S})}\left(\overset{\circ}{\epsilon} + r^{-1}\|f, \underline{f}, a\|_{\mathfrak{h}_s(\mathbf{S})}\right),\\
\|\mathrm{Err}_4, \mathrm{Err}_5\|_{\mathfrak{h}_{s-1}(\mathbf{S})} &\lesssim r^{-2}\|(f, \underline{f}, a)\|_{\mathfrak{h}_{s+1}(\mathbf{S})}\left(\overset{\circ}{\epsilon} + r^{-1}\|f, \underline{f}, a\|_{\mathfrak{h}_s(\mathbf{S})}\right).
\end{aligned}$$

In particular, in view of the assumptions (9.4.23) for $(f, \underline{f}, \lambda)$, we deduce

$$\begin{aligned}
\|\mathrm{Err}_1, \mathrm{Err}_2\|_{\mathfrak{h}_{s-1}(\mathbf{S})} &\lesssim r^{-2}\|(f, \underline{f}, a)\|_{\mathfrak{h}_{s+1}(\mathbf{S})}\left(\overset{\circ}{\epsilon} + \delta_1\right),\\
\|\mathrm{Err}_3\|_{\mathfrak{h}_{s-2}(\mathbf{S})} &\lesssim r^{-3}\|(f, \underline{f}, a)\|_{\mathfrak{h}_{s+1}(\mathbf{S})}\left(\overset{\circ}{\epsilon} + \delta_1\right),\\
\|\mathrm{Err}_4, \mathrm{Err}_5\|_{\mathfrak{h}_{s-1}(\mathbf{S})} &\lesssim r^{-2}\|(f, \underline{f}, a)\|_{\mathfrak{h}_{s+1}(\mathbf{S})}\left(\overset{\circ}{\epsilon} + \delta_1\right).
\end{aligned} \tag{9.5.17}$$

Also, recall that Err_6 is given by

$$\begin{aligned}
\mathrm{Err}_6 &= -\frac{r^{\mathbf{S}}}{2}\left[e^a \mathrm{Err}(\kappa, \kappa') - (e^a - 1)\left(\check{\kappa} + \overline{\kappa} - \frac{2}{r} + \,d\!\!/^{\mathbf{S}} f\right) - (e^a - 1 - a)\frac{2}{r}\right]\\
&- a\left(\frac{r^{\mathbf{S}}}{r} - 1\right)
\end{aligned}$$

which together with the control **A1**–**A3** of the background foliation, the assumptions (9.4.23) for $(f, \underline{f}, \lambda)$, the control of $r - r^{\mathbf{S}}$, and Sobolev, yields

$$\sup_{\mathbf{S}}|\mathrm{Err}_6| \lesssim r^{-1}(\overset{\circ}{\epsilon} + \delta_1)\|(f, \underline{f}, a)\|_{\mathfrak{h}_3(\mathbf{S})}.$$

Step 4. We now estimate F_1, F_2, F_3 and b_0. In view of

- the definition of F_1, F_2, F_3 and b_0 in Step 1,
- the control **A1**–**A3** for the background foliation,
- the assumption (9.4.2) for $\check{\kappa}$, $\not{d}_1^\star\underline{\kappa} - \underline{C}e^\Phi$ and $\not{d}_1^\star\mu - Me^\Phi$,
- the control of $\underline{C} - \overline{\underline{C}}^{\mathbf{S}}$ and $M - \overline{M}^{\mathbf{S}}$ using Corollary 9.18 and the control of the background foliation,
- the control of $r - r^{\mathbf{S}}$ in property 3 of Lemma 9.15,
- the control of $m - m^{\mathbf{S}}$ thanks to Corollary 9.19 and property 4 of Lemma 9.15,
- property 6 of Lemma 9.15,
- the estimates for Err_j, $j = 3, 4, 5, 6$ of Step 3,

we infer

$$\|F_1\|_{\mathfrak{h}_{s-1}(\mathbf{S})} + \|F_2\|_{\mathfrak{h}_{s-1}(\mathbf{S})} \lesssim r^{-2}\mathring{\delta} + \left[r^3|M^{\mathbf{S}} - \overline{M}^{\mathbf{S}}| + r^{-2}\|(f, \underline{f}, a)\|_{\mathfrak{h}_{s+1}(\mathbf{S})}\right] \times \left(\mathring{\epsilon} + \delta_1\right),$$

$$\|F_3\|_{\mathfrak{h}_s(\mathbf{S})} \lesssim r^{-1}\mathring{\delta} + r^4|M^{\mathbf{S}} - \overline{M}^{\mathbf{S}}| + r^{-1}\|(f, \underline{f})\|_{\mathfrak{h}_{s+1}(\mathbf{S})} + r^{-1}\left(\mathring{\epsilon} + \delta_1\right)\|a\|_{\mathfrak{h}_{s+1}(\mathbf{S})},$$

and

$$r|b_0| \lesssim \mathring{\delta} + \sup_{\mathbf{S}}|r - r^{\mathbf{S}}| + (\mathring{\epsilon} + \delta_1)\|(f, \underline{f}, a)\|_{\mathfrak{h}_3(\mathbf{S})}.$$

Step 5. In view of the estimates of Step 2 for $(\underline{C}^{\mathbf{S}}, M^{\mathbf{S}})$ and $(f, \underline{f}, \lambda)$, and the estimate for F_1, F_2, F_3 and b_0 in Step 4, we deduce

$$|\underline{C}^{\mathbf{S}} - \overline{\underline{C}}^{\mathbf{S}}| + r^{\mathbf{S}}|M^{\mathbf{S}} - \overline{M}^{\mathbf{S}}| \lesssim (r^{\mathbf{S}})^{-4}\mathring{\delta} + (r^{\mathbf{S}})^{-7}(|\Lambda| + |\underline{\Lambda}|) + \left(\mathring{\epsilon} + \delta_1\right)\left[r^{\mathbf{S}}|M^{\mathbf{S}} - \overline{M}^{\mathbf{S}}| + (r^{\mathbf{S}})^{-4}\|(f, \underline{f}, a)\|_{\mathfrak{h}_{s+1}(\mathbf{S})}\right],$$

$$\|(f, \underline{f})\|_{\mathfrak{h}_{s+1}(\mathbf{S})} \lesssim \mathring{\delta} + (r^{\mathbf{S}})^{-2}(|\Lambda| + |\underline{\Lambda}|) + \left(\mathring{\epsilon} + \delta_1\right)\left[(r^{\mathbf{S}})^5|M^{\mathbf{S}} - \overline{M}^{\mathbf{S}}| + \|(f, \underline{f}, a)\|_{\mathfrak{h}_{s+1}(\mathbf{S})}\right],$$

$$\|\check{a}^{\mathbf{S}}\|_{\mathfrak{h}_{s+1}(\mathbf{S})} \lesssim \mathring{\delta} + (r^{\mathbf{S}})^5|M^{\mathbf{S}} - \overline{M}^{\mathbf{S}}| + \|(f, \underline{f})\|_{\mathfrak{h}_{s+1}(\mathbf{S})} + \left(\mathring{\epsilon} + \delta_1\right)\|a\|_{\mathfrak{h}_{s+1}(\mathbf{S})}$$

and

$$r|\overline{a}^{\mathbf{S}}| \lesssim \mathring{\delta} + \sup_{\mathbf{S}}|r - r^{\mathbf{S}}| + (\mathring{\epsilon} + \delta_1)\|(f, \underline{f}, a)\|_{\mathfrak{h}_3(\mathbf{S})}.$$

The above estimates for $(\underline{C}^{\mathbf{S}}, M^{\mathbf{S}})$ and $(f, \underline{f})$ yields for δ_1 and $\mathring{\epsilon}$ small enough

$$\begin{aligned}&(r^{\mathbf{S}})^4|\underline{C}^{\mathbf{S}} - \overline{\underline{C}}^{\mathbf{S}}| + (r^{\mathbf{S}})^5|M^{\mathbf{S}} - \overline{M}^{\mathbf{S}}| + \|(f, \underline{f})\|_{\mathfrak{h}_{s+1}(\mathbf{S})}\\ \lesssim\ & \mathring{\delta} + (r^{\mathbf{S}})^{-2}(|\Lambda| + |\underline{\Lambda}|) + \left(\mathring{\epsilon} + \delta_1\right)\Big\|a\Big\|_{\mathfrak{h}_{s+1}(\mathbf{S})}.\end{aligned}$$

Plugging in the above estimate for $\check{a}^{\mathbf{S}}$, we infer for δ_1 and $\overset{\circ}{\epsilon}$ small enough

$$\begin{aligned}
&(r^{\mathbf{S}})^4|\underline{C}^{\mathbf{S}}-\overline{\underline{C}}^{\mathbf{S}}|+(r^{\mathbf{S}})^5|M^{\mathbf{S}}-\overline{M}^{\mathbf{S}}|+\|(f,\underline{f},\check{a}^{\mathbf{S}})\|_{\mathfrak{h}_{s+1}(\mathbf{S})}\\
\lesssim\ & \overset{\circ}{\delta}+(r^{\mathbf{S}})^{-2}\big(|\Lambda|+|\underline{\Lambda}|\big)+\left(\overset{\circ}{\epsilon}+\delta_1\right)r\left|\overline{a}^{\mathbf{S}}\right|.
\end{aligned}$$

Also, plugging in the above estimate for $\overline{a}^{\mathbf{S}}$, we infer for δ_1 and $\overset{\circ}{\epsilon}$ small enough

$$\begin{aligned}
&(r^{\mathbf{S}})^4|\underline{C}^{\mathbf{S}}-\overline{\underline{C}}^{\mathbf{S}}|+(r^{\mathbf{S}})^5|M^{\mathbf{S}}-\overline{M}^{\mathbf{S}}|+\|(f,\underline{f},\check{a}^{\mathbf{S}})\|_{\mathfrak{h}_{s+1}(\mathbf{S})}\\
\lesssim\ & \overset{\circ}{\delta}+(r^{\mathbf{S}})^{-2}\big(|\Lambda|+|\underline{\Lambda}|\big)+\left(\overset{\circ}{\epsilon}+\delta_1\right)\sup_{\mathbf{S}}|r-r^{\mathbf{S}}|
\end{aligned}$$

and

$$r|\overline{a}^{\mathbf{S}}| \ \lesssim\ \overset{\circ}{\delta}+(r^{\mathbf{S}})^{-2}\big(|\Lambda|+|\underline{\Lambda}|\big)+\sup_{\mathbf{S}}|r-r^{\mathbf{S}}|.$$

Using the third property of Lemma 9.15 in the first equation to bound $r-r^{\mathbf{S}}$, we finally obtain

$$\begin{aligned}
&(r^{\mathbf{S}})^4|\underline{C}^{\mathbf{S}}-\overline{\underline{C}}^{\mathbf{S}}|+(r^{\mathbf{S}})^5|M^{\mathbf{S}}-\overline{M}^{\mathbf{S}}|+\|(f,\underline{f},\check{a}^{\mathbf{S}})\|_{\mathfrak{h}_{s+1}(\mathbf{S})}\\
\lesssim\ & \overset{\circ}{\delta}+(r^{\mathbf{S}})^{-2}\big(|\Lambda|+|\underline{\Lambda}|\big)+r\delta_1\left(\overset{\circ}{\epsilon}+\delta_1\right)
\end{aligned}$$

and

$$r|\overline{a}^{\mathbf{S}}| \ \lesssim\ \overset{\circ}{\delta}+(r^{\mathbf{S}})^{-2}\big(|\Lambda|+|\underline{\Lambda}|\big)+\sup_{\mathbf{S}}|r-r^{\mathbf{S}}|$$

which are the desired estimates. This concludes the proof of Corollary 9.38.

9.6 PROOF OF PROPOSITION 9.43

9.6.1 Pullback of the main equations

According to Proposition 9.42 we may assume valid the uniform bounds for the quintets $P^{(n)}$. To establish a contraction estimate we need to compare the quantities,

$$\begin{aligned}
h^{(n)}:&= (\Psi^{(n-1)})^\# f^{(n)}, \quad \underline{h}^{(n)}:=(\Psi^{(n-1)})^\#\underline{f}^{(n)}, \quad w^{(n)}:=(\Psi^{(n-1)})^\# z^{(n)},\\
e^{(n)}:&= (\Psi^{(n-1)})^\# a^{(n)},
\end{aligned}$$

and

$$\begin{aligned}
h^{(n+1)}:&= (\Psi^{(n)})^\# f^{(n+1)}, \ \underline{h}^{(n+1)}:=(\Psi^{(n)})^\#\underline{f}^{(n+1)}, \ w^{(n+1)}:=(\Psi^{(n)})^\# z^{(n+1)},\\
e^{(n+1)}:&= (\Psi^{(n)})^\# a^{(n+1)}.
\end{aligned}$$

According to Lemma 9.7 we have

$$(\Psi^{(n)})^\# \left(\not{d}^{\mathbf{S}(n)} f^{(n+1)} \right) = \not{d}^{(n)} h^{(n+1)},$$
$$(\Psi^{(n)})^\# \left(A^{\mathbf{S}(n)} f^{(n+1)} \right) = A^{(n)} h^{(n+1)},$$
$$(\Psi^{(n)})^\# \left(B^{\mathbf{S}(n)} f^{(n+1)} \right) = B^{(n)} h^{(n+1)},$$

where $\not{d}^{(n)}$, $\not{d}^{\star(n)}$, $A^{(n)}, B^{(n)}$ are the corresponding Hodge operators on $\overset{\circ}{S}$ defined with respect to the metric $\not{g}^{(n)} := (\Psi^{(n)})^\#(\not{g}^{\mathbf{S}(n)})$ given by

$$\not{g}^{(n)} = \gamma^{(n)} d\theta^2 + e^{2\Phi^{\#_n}} d\varphi^2.$$

Consequently the system (9.4.27) takes the form

$$\begin{aligned}
B^{(n)} h^{(n+1)} &= -\frac{6m^{\mathbf{S}(n)}}{(r^{\mathbf{S}(n)})^5} (A^{(n)})^{-1} \big(\underline{h}^{(n+1)} - \Upsilon^{\mathbf{S}(n)} h^{(n+1)} \big) \\
&\quad + \frac{2}{r^{\mathbf{S}(n)}} (M^{(n+1)} - \overline{M}^{\mathbf{S}(n)}) (A^{(n)})^{-1} e^{\Phi^{\#_n}} + (\Psi^{(n)})^\# E^{(n+1)}, \\
B^{(n)} \underline{h}^{(n+1)} &= -\frac{6m^{\mathbf{S}(n)} \Upsilon^{\mathbf{S}(n)}}{(r^{\mathbf{S}(n)})^5} (A^{(n)})^{-1} \left(\underline{h}^{(n+1)} - \Upsilon^{\mathbf{S}(n)} h^{(n+1)} \right) \\
&\quad + (\underline{C}^{(n+1)} - \overline{\underline{C}}^{\mathbf{S}(n)}) e^{\Phi^{\#_n}} \\
&\quad + \frac{2\Upsilon^{\mathbf{S}(n)}}{r^{\mathbf{S}}} (M^{(n+1)} - \overline{M}^{\mathbf{S}(n)}) (A^{(n)})^{-1} e^{\Phi^{\#_n}} + (\Psi^{(n)})^\# \underline{E}^{(n+1)}, \\
\not{d}^{\star(n)} e^{(n+1)} &= (\Psi^{(n)})^\# \widetilde{E}^{(n+1)}.
\end{aligned} \tag{9.6.1}$$

Equation (9.4.32) takes the form

$$\int_{(\overset{\circ}{S}, \not{g}^{(n)})} e^{\Phi^{\#_n}} h^{(n+1)} = \Lambda, \qquad \int_{(\overset{\circ}{S}, \not{g}^{(n)})} e^{\Phi^{\#_n}} \underline{h}^{(n+1)} = \underline{\Lambda}. \tag{9.6.2}$$

Equation (9.4.33) takes the form

$$\begin{aligned}
\overline{e^{(n+1)}}^{\overset{\circ}{S}, \not{g}^{(n)}} &= \overline{\left(1 - \frac{r^{\mathbf{S}(n)}}{r}\right)^{(n)}}^{\overset{\circ}{S}, \not{g}^{(n)}} - \frac{r^{\mathbf{S}(n)}}{2} \overline{\left(\check{\kappa} + \overline{\kappa} - \frac{2}{r} \right)^{\#_n}}^{\overset{\circ}{S}, \not{g}^{(n)}} \\
&\quad + \overline{\mathrm{Err}_6^{(n+1)}}^{\overset{\circ}{S}, \not{g}^{(n)}}.
\end{aligned} \tag{9.6.3}$$

Finally the system (9.4.34) takes the form

$$\begin{aligned}
\varsigma^{\#_n}\partial_\theta U^{(1+n)} &= (\gamma^{(n)})^{1/2}h^{(1+n)}\left(1+\frac{1}{4}h^{(1+n)}\underline{h}^{(1+n)}\right),\\
\partial_\theta S^{(1+n)} - \frac{1}{2}\varsigma^{\#_n}\underline{\Omega}^{\#_n}\partial_\theta U^{(1+n)} &= \frac{1}{2}(\gamma^{(n)})^{1/2}\underline{h}^{(1+n)},\\
\gamma^{(n)} &= \gamma^{\#_n}\\
&\quad + (\varsigma^{\#_n})^2\left(\underline{\Omega}^{\#_n}+\frac{1}{4}(\underline{b}^{\#_n})^2\gamma^{\#_n}\right)(\partial_\theta U^{(n)})^2\\
&\quad - 2\varsigma^{\#_n}\partial_\theta U^{(n)}\partial_\theta S^{(n)} - \gamma^{\#_n}\varsigma^{\#_n}\underline{b}^{\#_n}\partial_\theta U^{(n)},\\
U^{(1+n)}(0) &= S^{(1+n)}(0) = 0.
\end{aligned} \tag{9.6.4}$$

We recall, see (9.4.38), the definition of the norm for the quintets $P^{(n)}$ in the particular case $k=2$

$$\begin{aligned}
\|P^{(n)}\|_2 := {} & r^{-1}\|\partial_\theta\left(U^{(n)}, S^{(n)}\right)\|_{\mathfrak{h}_1(\overset{\circ}{S})} + r^4|\underline{C}^{(n)} - \overline{\underline{C}}^{\mathbf{S}(n)}|\\
& + r^5|M^{(n)} - \overline{M}^{\mathbf{S}(n)}| + \left\|\left((a^{(n)})^{\#_{n-1}}, (f^{(n)})^{\#_{n-1}}, (\underline{f}^{(n)})^{\#_{n-1}}\right)\right\|_{\mathfrak{h}_1(\overset{\circ}{S})}.
\end{aligned}$$

To prove the estimate

$$\begin{aligned}
& \|P^{(n+1)} - P^{(n)}\|_2\\
\lesssim {} & \overset{\circ}{\epsilon}\Big(\|P^{(n)} - P^{(n-1)}\|_2 + \|P^{(n-1)} - P^{(n-2)}\|_2 + \|P^{(n-2)} - P^{(n-3)}\|_2\Big),
\end{aligned}$$

we set

$$\begin{aligned}
\delta w^{(n+1)} &= w^{(n+1)} - w^{(n)}, \quad \delta h^{(n+1)} = h^{(n+1)} - h^{(n)}, \quad \delta \underline{h}^{(n+1)} = \underline{h}^{(n+1)} - \underline{h}^{(n)},\\
\delta e^{(n+1)} &= e^{(n+1)} - e^{(n)}, \quad \delta U^{(n+1)} = U^{(n+1)} - U^{(n)}, \quad \delta S^{(n+1)} = S^{(n+1)} - S^{(n)},\\
\delta \underline{C}^{(n+1)} &= \underline{C}^{(n+1)} - \underline{C}^{(n)}, \quad \delta M^{(n+1)} = M^{(n+1)} - M^{(n)},
\end{aligned}$$

and

$$\begin{aligned}
\delta w^{(n)} &= w^{(n)} - w^{(n-1)}, \quad \delta h^{(n)} = h^{(n)} - h^{(n-1)}, \quad \delta \underline{h}^{(n)} = \underline{h}^{(n)} - \underline{h}^{(n-1)},\\
\delta e^{(n)} &= e^{(n)} - e^{(n-1)}, \quad \delta U^{(n)} = U^{(n)} - U^{(n-1)}, \quad \delta S^{(n)} = S^{(n)} - S^{(n-1)},\\
\delta \underline{C}^{(n)} &= \underline{C}^{(n)} - \underline{C}^{(n-1)}, \quad \delta M^{(n)} = M^{(n)} - M^{(n-1)}.
\end{aligned}$$

We will derive in section 9.6.3 the following estimates:

$$\begin{aligned}
& r^4|\delta\underline{C}^{(n+1)}| + r^5|\delta M^{(n+1)}|\\
& + \left\|\left(\delta h^{(n+1)}, \delta\underline{h}^{(n+1)}, \delta e^{(n+1)} - \overline{\delta e^{(n+1)}}^{\overset{\circ}{S},\not{g}^{(n)}}\right)\right\|_{\mathfrak{h}_1(\overset{\circ}{S})}\\
\lesssim {} & \overset{\circ}{\epsilon}\Big(\|P^{(n)} - P^{(n-1)}\|_2 + \|P^{(n-1)} - P^{(n-2)}\|_2\Big),
\end{aligned} \tag{9.6.5}$$

$$r\left|\overline{\delta e^{(n+1)}}^{\overset{\circ}{S},\not{g}^{(n)}}\right| \lesssim r^{-1}\|\partial_\theta(\delta U^{(n)},\delta S^{(n)})\|_{\mathfrak{h}_1(\overset{\circ}{S})} + \overset{\circ}{\epsilon}\Big(\|P^{(n)}-P^{(n-1)}\|_2+\|P^{(n-1)}-P^{(n-2)}\|_2\Big), \quad (9.6.6)$$

and

$$r^{-1}\|\partial_\theta(\delta U^{(n+1)},\delta S^{(n+1)})\|_{\mathfrak{h}_1(\overset{\circ}{S})} \lesssim \|\delta h^{(n+1)},\delta\underline{h}^{(n+1)}\|_{\mathfrak{h}_1(\overset{\circ}{S})} + \overset{\circ}{\epsilon}\|P^{(n)}-P^{(n-1)}\|_2. \quad (9.6.7)$$

Proposition 9.43 is then an immediate consequence of (9.6.5), (9.6.6), (9.6.7). Thus, from now on, we focus on the proof of (9.6.5), (9.6.6), (9.6.7). To this end, we will rely on the following lemmas.

9.6.2 Basic lemmas

Lemma 9.44. *Let F be a reduced scalar function defined in a neighborhood of $\overset{\circ}{S}$ in $\mathcal{R}$ and define its pullback $F^{(n)}=(\Psi^{(n)})^\# F$ to $\overset{\circ}{S}$, i.e.,*

$$\begin{aligned} F^{(n)}(\theta) &= F(\overset{\circ}{u}+U^{(n)}(\theta),\overset{\circ}{s}+S^{(n)}(\theta),\theta),\\ F^{(n-1)}(\theta) &= F(\overset{\circ}{u}+U^{(n-1)}(\theta),\overset{\circ}{s}+S^{(n-1)}(\theta),\theta). \end{aligned}$$

Then,[17] *for all $1\le p\le\infty$, with $\delta_n U=U^{(n+1)}-U^{(n)}$, $\delta_n S=S^{(n+1)}-S^{(n)}$*

$$\|\delta_n F\|_{L^p(\overset{\circ}{S})} \lesssim \left(\|\delta_n U\|_{L^p(\overset{\circ}{S})}+\|\delta_n S\|_{L^p(\overset{\circ}{S})}\right)\sup_{\mathcal{R}}\left(|\partial_s F|+|\partial_u F|\right). \quad (9.6.8)$$

Also,

$$\|\delta_n F\|_{\mathfrak{h}_1(\overset{\circ}{S})} \lesssim \left(\|\delta_n U\|_{\mathfrak{h}_1(\overset{\circ}{S})}+\|\delta_n S\|_{\mathfrak{h}_1(\overset{\circ}{S})}\right)\sup_{\mathcal{R}}\left(|\mathfrak{d}^{\le 1}\partial_s F|+|\mathfrak{d}^{\le 1}\partial_u F|\right) \quad (9.6.9)$$

where $\delta_n U=U^{(n+1)}-U^{(n)}$, $\delta_n S=S^{(n+1)}-S^{(n)}$.

Proof. We write

$$\begin{aligned} \delta_n F &:= F(u_0+U^{(n)}(\theta),s_0+S^{(n)}(\theta),\theta)-F(u_0+U^{(n-1)}(\theta),s_0+S^{(n-1)}(\theta),\theta)\\ &= \int_0^1\frac{d}{dt}F\Big(u_0+tU^{(n)}(\theta)+(1-t)U^{(n-1)}(\theta),\\ &\qquad\qquad s_0+tS^{(n)}(\theta)+(1-t)S^{(n-1)}(\theta),\theta\Big), \end{aligned}$$

[17]Recall $\partial_s=e_4$, $\partial_u=\frac{\varsigma}{2}\left(e_3-\underline{\Omega}e_4-\underline{b}\gamma^{1/2}e_\theta\right)$.

i.e., denoting $\delta_n U = U^{(n)} - U^{(n-1)}$, $\delta_n S = S^{(n)} - S^{(n-1)}$,

$$\begin{aligned} |\delta_n F| &\lesssim |\delta_n U| \int_0^1 \Big| \partial_u F\Big(u_0 + tU^{(n)}(\theta) + (1-t)U^{(n-1)}(\theta), \\ &\qquad s_0 + tS^{(n)}(\theta) + (1-t)S^{(n-1)}(\theta), \theta\Big)\Big| \\ &+ |\delta_n S| \int_0^1 \Big| \partial_s F\Big(u_0 + tU^{(n)}(\theta) + (1-t)U^{(n-1)}(\theta), \\ &\qquad s_0 + tS^{(n)}(\theta) + (1-t)S^{(n-1)}(\theta), \theta\Big)\Big|, \end{aligned}$$

i.e.,

$$|\delta_n F| \lesssim \left|U^{(n)}(\theta) - U^{(n-1)}(\theta)\right| \sup_{\overset{\circ}{S}+\delta\overset{\circ}{S}} |\partial_u F| + \left|S^{(n)}(\theta) - S^{(n-1)}(\theta)\right| \sup_{\overset{\circ}{S}+\epsilon\overset{\circ}{S}} |\partial_s F|$$

from which (9.6.8) easily follows.

Similarly,

$$\| \not{\partial}\delta_n F\|_{L^2(\overset{\circ}{S})} \lesssim \left(\|\delta_n U\|_{\mathfrak{h}_1(\overset{\circ}{S})} + \|\delta_n S\|_{\mathfrak{h}_1(\overset{\circ}{S})} \right) \sup_{\mathcal{R}} \left(\left|\mathfrak{d}^{\le 1}\partial_s F\right| + \left|\mathfrak{d}^{\le 1}\partial_u F\right| \right).$$

Hence,

$$\|\delta_n F\|_{\mathfrak{h}_1(\overset{\circ}{S})} \lesssim \left(\|\delta_n U\|_{\mathfrak{h}_1(\overset{\circ}{S})} + \|\delta_n S\|_{\mathfrak{h}_1(\overset{\circ}{S})} \right) \sup_{\mathcal{R}} \left(\left|\mathfrak{d}^{\le 1}\partial_s F\right| + \left|\mathfrak{d}^{\le 1}\partial_u F\right| \right)$$

as desired. ☐

Lemma 9.45. *Let* $\psi, h \in \mathfrak{s}_1(\overset{\circ}{S})$, *and* $\delta B^{(n)} = B^{(n)} - B^{(n-1)}$. *The following formula holds true.*

$$\left| \int_{(\overset{\circ}{S}, \not{g}^{(n)})} \psi \delta B^{(n)} h \right| \lesssim r^{-3} \|\partial_\theta \left(\Psi^{(n)} - \Psi^{(n-1)} \right) \|_{\mathfrak{h}_1(\overset{\circ}{S})} \|\psi\|_{\mathfrak{h}_1(\overset{\circ}{S})} \|h\|_{\mathfrak{h}_2(\overset{\circ}{S})}.$$

Proof. Recall that the metric $\not{g}^{(n)}$ is given by

$$\not{g}^{(n)} = \gamma^{(n)} d\theta^2 + e^{2\Phi^{\#_n}} d\varphi^2$$

so that the operator $B^{(n)} = \not{d}_2^{(n)} \not{d}_2^{\star\,(n)}$, applied to $\mathfrak{s}_1$ tensors h on $\overset{\circ}{S}$, is given by

$$B^{(n)} h = \frac{1}{\sqrt{\gamma^{(n)}}} (\partial_\theta + 2\partial_\theta(\Phi^{\#_n})) \left(\frac{1}{\sqrt{\gamma^{(n)}}} \left(-\partial_\theta h + \partial_\theta(\Phi^{\#_n}) h \right) \right).$$

This yields

$$
\begin{aligned}
& \delta B^{(n)} h \\
= \; & \left(\frac{1}{\sqrt{\gamma^{(n)}}}-\frac{1}{\sqrt{\gamma^{(n-1)}}}\right)\left(\partial_\theta+2 \partial_\theta\left(\Phi^{\#_n}\right)\right)\left(\frac{1}{\sqrt{\gamma^{(n)}}}\left(-\partial_\theta h+\partial_\theta\left(\Phi^{\#_n}\right) h\right)\right) \\
& +\frac{1}{\sqrt{\gamma^{(n-1)}}}\left(\partial_\theta+2 \partial_\theta\left(\Phi^{\#_n}\right)\right)\left(\left(\frac{1}{\sqrt{\gamma^{(n)}}}-\frac{1}{\sqrt{\gamma^{(n-1)}}}\right)\left(-\partial_\theta h+\partial_\theta\left(\Phi^{\#_n}\right) h\right)\right) \\
& +\frac{1}{\sqrt{\gamma^{(n-1)}}}\left(\partial_\theta+2 \partial_\theta\left(\Phi^{\#_n}\right)\right)\left(\frac{1}{\sqrt{\gamma^{(n-1)}}} \partial_\theta\left(\Phi^{\#_n}-\Phi^{\#_{n-1}}\right) h\right) \\
& +\frac{2}{\gamma^{(n-1)}} \partial_\theta\left(\Phi^{\#_n}-\Phi^{\#_{n-1}}\right) \psi \partial_\theta\left(\Phi^{\#_{n-1}}\right) h .
\end{aligned}
$$

Using the previous formula to integrate $\psi \delta B^{(n)} h$ on $\overset{\circ}{S}$ with the volume of $\not g^{(n)}$, and after integration by parts, we infer

$$
\begin{aligned}
& \int_{(\overset{\circ}{S}, \not g^{(n)})} \psi \delta B^{(n)} h \\
= \; & \int_{(\overset{\circ}{S}, \not g^{(n)})} \frac{1}{\sqrt{\gamma^{(n)}}}\left(-\partial_\theta+\partial_\theta\left(\Phi^{\#_n}\right)\right)\left(\left(1-\frac{\sqrt{\gamma^{(n)}}}{\sqrt{\gamma^{(n-1)}}}\right) \psi\right) \\
& \times \frac{1}{\sqrt{\gamma^{(n)}}}\left(-\partial_\theta h+\partial_\theta\left(\Phi^{\#_n}\right) h\right) \\
& +\int_{(\overset{\circ}{S}, \not g^{(n)})} \frac{1}{\sqrt{\gamma^{(n)}}}\left(-\partial_\theta+\partial_\theta\left(\Phi^{\#_n}\right)\right)\left(\frac{\sqrt{\gamma^{(n)}}}{\sqrt{\gamma^{(n-1)}}} \psi\right)\left(\frac{1}{\sqrt{\gamma^{(n)}}}-\frac{1}{\sqrt{\gamma^{(n-1)}}}\right) \\
& \times\left(\partial_\theta h+\partial_\theta\left(\Phi^{\#_n}\right) h\right) \\
& +\int_{(\overset{\circ}{S}, \not g^{(n)})} \frac{1}{\sqrt{\gamma^{(n)}}}\left(-\partial_\theta+\partial_\theta\left(\Phi^{\#_n}\right)\right)\left(\frac{\sqrt{\gamma^{(n)}}}{\sqrt{\gamma^{(n-1)}}} \psi\right) \\
& \times \frac{1}{\sqrt{\gamma^{(n-1)}}} \partial_\theta\left(\Phi^{\#_n}-\Phi^{\#_{n-1}}\right) h \\
& +\int_{(\overset{\circ}{S}, \not g^{(n)})} \frac{2}{\gamma^{(n-1)}} \partial_\theta\left(\Phi^{\#_n}-\Phi^{\#_{n-1}}\right) \psi \partial_\theta\left(\Phi^{\#_{n-1}}\right) h .
\end{aligned}
$$

We now make use of the bounds (9.1.15) for $(\underline{\Omega}, \underline{b}, \gamma)$ involved in the definition of $\gamma^{(n-1)}$ and $\gamma^{(n)}$, the uniform bound of $P^{(n)}$ provided by Proposition 9.42 and the Sobolev inequality to deduce

$$
\begin{aligned}
\left|\int_{(\overset{\circ}{S}, \not g^{(n)})} \psi \delta B^{(n)} h\right| \; \lesssim \; & r^{-5}\left\|\gamma^{(n)}-\gamma^{(n-1)}\right\|_{\mathfrak{h}_1(\overset{\circ}{S})}\|\psi\|_{\mathfrak{h}_1(\overset{\circ}{S})}\|h\|_{\mathfrak{h}_2(\overset{\circ}{S})} \\
& + r^{-2}\left\|\partial_\theta\left(\Phi^{\#_n}-\Phi^{\#_{n-1}}\right) h\right\|_{L^2(\overset{\circ}{S})}\|\psi\|_{\mathfrak{h}_1(\overset{\circ}{S})}, \quad (9.6.10)
\end{aligned}
$$

where we have also used Lemma 9.3 to estimate

$$
\left\|\partial_\theta\left(\Phi^{\#_{n-1}}\right) \psi\right\|_{L^2(\overset{\circ}{S}, \not g^{(n)})} \; \lesssim \; r\left\|\frac{\psi}{e^{\Phi}}\right\|_{L^2(\overset{\circ}{S}, \not g^{(n)})} \lesssim r\left\|\frac{\psi}{e^{\Phi}}\right\|_{L^2(\overset{\circ}{S})} \lesssim\|\psi\|_{\mathfrak{h}_1(\overset{\circ}{S})} .
$$

To estimate the term $\gamma^{(n)} - \gamma^{(n-1)}$ we recall that

$$\begin{aligned}\gamma^{(n)} &= \gamma^{\#_n} + (\varsigma^{\#_n})^2 \left(\underline{\Omega}^{\#_n} + \frac{1}{4}(\underline{b}^{\#_n})^2 \gamma^{\#_n}\right) (\partial_\theta U^{(n)})^2 \\ &\quad -2\varsigma^{\#_n} \partial_\theta U^{(n)} \partial_\theta S^{(n)} - \gamma^{\#_n} \varsigma^{\#_n} \underline{b}^{\#_n} \partial_\theta U^{(n)} \\ \gamma^{(n-1)} &= \gamma^{\#_{n-1}} + (\varsigma^{\#_{n-1}})^2 \left(\underline{\Omega}^{\#_{n-1}} + \frac{1}{4}(\underline{b}^{\#_{n-1}})^2 \gamma^{\#_{n-1}}\right) (\partial_\theta U^{(n-1)})^2 \\ &\quad -2\varsigma^{\#_{n-1}} \partial_\theta U^{(n-1)} \partial_\theta S^{(n-1)} - \gamma^{\#_{n-1}} \varsigma^{\#_{n-1}} \underline{b}^{\#_{n-1}} \partial_\theta U^{(n-1)}.\end{aligned}$$

The principal term $\gamma^{\#_n} - \gamma^{\#_{n-1}}$ can be estimated with the help of Lemma 9.44, the uniform bound of $P^{(n)}$ provided by Proposition 9.42, and the bounds provided[18] by **A3**. All other terms can be estimated in a similar fashion. We derive

$$\|\gamma^{(n)} - \gamma^{(n-1)}\|_{\mathfrak{h}_1(\overset{\circ}{S})} \lesssim r \|\partial_\theta \left(\Psi^{(n)} - \Psi^{(n-1)}\right)\|_{\mathfrak{h}_1(\overset{\circ}{S})} \tag{9.6.11}$$

where

$$\|\partial_\theta \left(\Psi^{(n)} - \Psi^{(n-1)}\right)\|_{\mathfrak{h}_1(\overset{\circ}{S})} := \|\partial_\theta (U^{(n)} - U^{(n-1)})\|_{\mathfrak{h}_1(\overset{\circ}{S})} + \|\partial_\theta (S^{(n)} - S^{(n-1)})\|_{\mathfrak{h}_1(\overset{\circ}{S})}.$$

We deduce

$$\begin{aligned}\left|\int_{(\overset{\circ}{S}, \not{g}^{(n)})} \psi \delta B^{(n)} h\right| &\lesssim r^{-4} \|\partial_\theta \left(\Psi^{(n)} - \Psi^{(n-1)}\right)\|_{\mathfrak{h}_1(\overset{\circ}{S})} \|\psi\|_{\mathfrak{h}_1(\overset{\circ}{S})} \|h\|_{\mathfrak{h}_2(\overset{\circ}{S})} \\ &\quad + r^{-2} \left\|\partial_\theta \left(\Phi^{\#_n} - \Phi^{\#_{n-1}}\right) h\right\|_{L^2(\overset{\circ}{S})} \|\psi\|_{\mathfrak{h}_1(\overset{\circ}{S})}.\end{aligned} \tag{9.6.12}$$

The proof of Lemma 9.45 is now an immediate consequence of the following lemma. □

Lemma 9.46. *The following estimate holds true for a reduced scalar $h \in \mathfrak{s}_1(\overset{\circ}{S})$:*

$$\left\|\partial_\theta \left(\Phi^{\#_n} - \Phi^{\#_{n-1}}\right) h\right\|_{L^2(\overset{\circ}{S})} \lesssim r^{-1} \|\partial_\theta \left(\Psi^{(n)} - \Psi^{(n-1)}\right)\|_{\mathfrak{h}_1(\overset{\circ}{S})} \|h\|_{\mathfrak{h}_2(\overset{\circ}{S})}. \tag{9.6.13}$$

[18]Note in particular that **A3** implies $\partial_u(\gamma) = \partial_u(r^2) + O(\overset{\circ}{\epsilon} r) = O(r)$ and $\partial_s(\gamma) = \partial_s(r^2) + O(\overset{\circ}{\epsilon} r) = O(r)$.

Proof. We write

$$
\begin{aligned}
&\partial_\theta\left(\Phi^{\#_n}-\Phi^{\#_{n-1}}\right)\\
&=\left\{\left(\partial_\theta S^{(n)}-\frac{1}{2}\underline{\Omega}\partial_\theta U^{(n)}\right)e_4\Phi+\frac{1}{2}\partial_\theta U^{(n)}e_3\Phi+\sqrt{\gamma}\left(1-\frac{1}{2}\underline{b}\partial_\theta U^{(n)}\right)e_\theta\Phi\right\}^{\#_n}\\
&-\Bigg\{\left(\partial_\theta S^{(n-1)}-\frac{1}{2}\underline{\Omega}\partial_\theta U^{(n-1)}\right)e_4\Phi+\frac{1}{2}\partial_\theta U^{(n-1)}e_3\Phi\\
&+\sqrt{\gamma}\left(1-\frac{1}{2}\underline{b}\partial_\theta U^{(n-1)}\right)e_\theta\Phi\Bigg\}^{\#_{n-1}}\\
&=\left(\partial_\theta S^{(n)}-\frac{1}{2}\underline{\Omega}^{\#_n}\partial_\theta U^{(n)}\right)(e_4\Phi)^{\#_n}+\frac{1}{2}\partial_\theta U^{(n)}(e_3\Phi)^{\#_n}\\
&+\sqrt{\gamma}^{\#_n}\left(1-\frac{1}{2}\underline{b}^{\#_n}\partial_\theta U^{(n)}\right)(e_\theta\Phi)^{\#_n}\\
&-\left(\partial_\theta S^{(n-1)}-\frac{1}{2}\underline{\Omega}^{\#_{n-1}}\partial_\theta U^{(n-1)}\right)(e_4\Phi)^{\#_{n-1}}-\frac{1}{2}\partial_\theta U^{(n-1)}(e_3\Phi)^{\#_{n-1}}\\
&-\sqrt{\gamma}^{\#_{n-1}}\left(1-\frac{1}{2}\underline{b}^{\#_{n-1}}\partial_\theta U^{(n-1)}\right)(e_\theta\Phi)^{\#_{n-1}},
\end{aligned}
$$

i.e., grouping the terms appropriately,

$$
\begin{aligned}
&\partial_\theta\left(\Phi^{\#_n}-\Phi^{\#_{n-1}}\right)=J_1+J_2+J_3,\\
J_1\quad&=\left(\partial_\theta S^{(n)}-\frac{1}{2}\underline{\Omega}^{\#_n}\partial_\theta U^{(n)}\right)(e_4\Phi)^{\#_n}\\
&-\left(\partial_\theta S^{(n-1)}-\frac{1}{2}\underline{\Omega}^{\#_n-1}\partial_\theta U^{(n-1)}\right)(e_4\Phi)^{\#_{n-1}},\\
J_2\quad&=\frac{1}{2}\partial_\theta U^{(n)}(e_3\Phi)^{\#_n}-\frac{1}{2}\partial_\theta U^{(n-1)}(e_3\Phi)^{\#_{n-1}},\\
J_3\quad&=\sqrt{\gamma}^{\#_n}\left(1-\frac{1}{2}\underline{b}^{\#_n}\partial_\theta U^{(n)}\right)(e_\theta\Phi)^{\#_n}\\
&-\sqrt{\gamma}^{\#_{n-1}}\left(1-\frac{1}{2}\underline{b}^{\#_{n-1}}\partial_\theta U^{(n-1)}\right)(e_\theta\Phi)^{\#_{n-1}},
\end{aligned}
$$

and

$$
\begin{aligned}
J_3 &= J_{31}+J_{32},\\
J_{31} &= (e_\theta\Phi)^{\#_{n-1}}\Bigg[\sqrt{\gamma}^{\#_n}\left(1-\frac{1}{2}\underline{b}^{\#_n}\partial_\theta U^{(n)}\right)\\
&-\sqrt{\gamma}^{\#_{n-1}}\left(1-\frac{1}{2}\underline{b}^{\#_{n-1}}\partial_\theta U^{(n-1)}\right)\Bigg],\\
J_{32} &= \sqrt{\gamma}^{\#_n}\left(1-\frac{1}{2}\underline{b}^{\#_n}\partial_\theta U^{(n)}\right)\left((e_\theta\Phi)^{\#_n}-(e_\theta\Phi)^{\#_{n-1}}\right).
\end{aligned}
$$

The contribution to the estimate of Lemma 9.46 given by J_1, J_2, J_{31} can be easily estimated by making use of the uniform bound of $P^{(n)}$ provided by Proposition

9.42, the bound (9.1.15) for $(\underline{\Omega}, \underline{b}, \gamma)$, and Lemma 9.11, as well as Lemma 9.44. We thus derive

$$\|(J_1, J_2, J_{31})h\|_{L^2(\overset{\circ}{S})} \lesssim r^{-1}\|\partial_\theta\Psi^{(n)} - \partial_\theta\Psi^{(n-1)}\|_{\mathfrak{h}_1(\overset{\circ}{S})}\|h\|_{\mathfrak{h}_2(\overset{\circ}{S})}.$$

It remains to estimate the term $\|J_{32}h\|_{L^2(\overset{\circ}{S})}$ which presents a difficulty at the axis of symmetry where $\sin\theta = 0$. Clearly, we have

$$\|J_{32}\, h\|_{L^2(\overset{\circ}{S})} \lesssim r\left\|\left((e_\theta\Phi)^{\#_n} - (e_\theta\Phi)^{\#_{n-1}}\right) h\right\|_{L^2(\overset{\circ}{S})}.$$

We are thus left to estimate the term $\left\|\left((e_\theta\Phi)^{\#_n} - (e_\theta\Phi)^{\#_{n-1}}\right) h\right\|_{L^2(\overset{\circ}{S})}$. Proceeding as in the proof of Lemma 9.44 we write, for $F = e_\theta\Phi$,

$$\begin{aligned}
|\delta_n F| \lesssim\ & |\delta_n U|\int_0^1 \Big|\partial_u F\Big(u_0 + tU^{(n)}(\theta) + (1-t)U^{(n-1)}(\theta),\\
&\qquad\qquad s_0 + tS^{(n)}(\theta) + (1-t)S^{(n-1)}(\theta), \theta\Big)\Big|\\
+\ & |\delta_n S|\int_0^1 \Big|\partial_s F\Big(u_0 + tU^{(n)}(\theta) + (1-t)U^{(n-1)}(\theta),\\
&\qquad\qquad s_0 + tS^{(n)}(\theta) + (1-t)S^{(n-1)}(\theta), \theta\Big)\Big|.
\end{aligned}$$

We need to pay special attention on the axis,[19] where $\sin\theta = 0$, to the integral term involving

$$\partial_u(e_\theta\Phi) = \frac{1}{2}\left(e_3 - \underline{\Omega}e_4 - \underline{b}\gamma^{1/2}e_\theta\right)e_\theta\Phi.$$

This leads us to consider the integral

$$\int_0^1 \left[\underline{b}e_\theta(e_\theta(\Phi))\right](\overset{\circ}{u} + tU^{(n)}(\theta) + (1-t)U^{(n-1)}(\theta), \overset{\circ}{s}, \theta)dt$$

and the L^2 norm of its product with h on $\overset{\circ}{S}$. We recall (see Lemma 2.13) that $\Delta\!\!\!\!/\,\Phi = -K$, and therefore, $|e_\theta(e_\theta\Phi)| \lesssim r^{-2} + |e_\theta\Phi|^2$. The contribution due to K does not present any difficulties on the axis therefore we are led to consider the integral

$$I(\theta) := \int_0^1 \left[\underline{b}(e_\theta(\Phi))^2\right](\overset{\circ}{u} + tU^{(n)}(\theta) + (1-t)U^{(n-1)}(\theta), \overset{\circ}{s}, \theta)dt$$

and the L^2 norm of its product with h on $\overset{\circ}{S}$. Making use of (9.1.17) and then the first estimate of (9.1.18) of Lemma 9.3 together with our assumption **A3** we derive

[19]Indeed the term $e_\theta(e_\theta\Phi)$ is quite singular on the axis.

the bound

$$\begin{aligned} r^2\big|I(\theta)h(\theta)\big| &\lesssim \frac{1}{\sin^2\theta}\left(\int_0^1 \Big|\underline{b}(\mathring{u}+tU^{(n)}(\theta)+(1-t)U^{(n-1)}(\theta),\mathring{s},\theta)\Big|\,dt\right)|h(\theta)| \\ &\lesssim \left|\frac{h(\theta)}{\sin\theta}\right|\sup_{\mathcal{R}}\left|\frac{\underline{b}}{\sin\theta}\right| \lesssim r^2\left|\frac{h(\theta)}{e^{\Phi}}\right|\sup_{\mathcal{R}}\left|\frac{\underline{b}}{e^{\Phi}}\right| \lesssim \mathring{\epsilon}\left|\frac{h(\theta)}{e^{\Phi}}\right|. \end{aligned}$$

Making use of the second estimate in (9.1.17) we then derive

$$\|I\cdot h\|_{L^2(\mathring{S})} \lesssim \mathring{\epsilon}r^{-2}\left\|\frac{h}{e^{\Phi}}\right\|_{L^2(S)} \lesssim r^{-3}\mathring{\epsilon}\|h\|_{\mathfrak{h}_1(S)}.$$

This shows that the behavior along the axis in (9.6.13) is not an issue. This ends the proof of Lemma 9.46. □

Lemma 9.47. *Let $\delta h^{(n+1)}$ and $\delta\underline{h}^{(n+1)}$ reduced scalars on $\mathring{S}$ satisfying*

$$\begin{aligned} B^{(n)}\delta h^{(n+1)} &= -\frac{6m^{\mathbf{S}(n)}}{(r^{\mathbf{S}(n)})^5}(A^{(n)})^{-1}(\delta\underline{h}^{(n+1)}-\Upsilon^{\mathbf{S}(n)}\delta h^{(n+1)}) \\ &\quad+\frac{2}{r^{\mathbf{S}(n)}}\delta M^{(n+1)}(A^{(n)})^{-1}e^{\Phi\#_n}-(\delta B^{(n)})h^{(n)}+H^{(n+1)}, \\ B^{(n)}\delta\underline{h}^{(n+1)} &= -\frac{6m^{\mathbf{S}(n)}\Upsilon^{\mathbf{S}(n)}}{(r^{\mathbf{S}(n)})^5}(A^{(n)})^{-1}\left(\delta\underline{h}^{(n+1)}-\Upsilon^{\mathbf{S}(n)}\delta h^{(n+1)}\right) \\ &\quad+\delta\underline{C}^{(n+1)}e^{\Phi\#_n}+\frac{2\Upsilon^{\mathbf{S}(n)}}{r^{\mathbf{S}}}\delta M^{(n+1)}(A^{(n)})^{-1}e^{\Phi\#_n} \\ &\quad-(\delta B^{(n)})\underline{h}^{(n)}+\underline{H}^{(n+1)}, \end{aligned} \tag{9.6.14}$$

as well as

$$\begin{aligned} \int_{\mathring{S},\not{g}^{(n)}}\delta h^{(n+1)}e^{\Phi\#_n} &= D^{(n+1)}, \\ \int_{\mathring{S},\not{g}^{(n)}}\delta\underline{h}^{(n+1)}e^{\Phi\#_n} &= \underline{D}^{(n+1)}. \end{aligned}$$

Also, assume the bounds

$$\|(h^{(n)},\underline{h}^{(n)})\|_{\mathfrak{h}_2(\mathring{S})} \lesssim \mathring{\delta}.$$

Then we have

$$\begin{aligned} &r^4|\delta\underline{C}^{(n+1)}|+r^5|\delta M^{(n+1)}|+\|(\delta h^{(n+1)},\delta\underline{h}^{(n+1)})\|_{\mathfrak{h}_1(\mathring{S})} \\ &\lesssim \mathring{\delta}r^{-1}\|\partial_\theta\left(\Psi^{(n)}-\Psi^{(n-1)}\right)\|_{\mathfrak{h}_1(\mathring{S})}+r^2\|H^{(n+1)}\|_{L^2(\mathring{S})}+r^2\|\underline{H}^{(n+1)}\|_{L^2(\mathring{S})} \\ &+r^{-2}\left(\left|D^{(n+1)}\right|+\left|\underline{D}^{(n+1)}\right|\right). \end{aligned} \tag{9.6.15}$$

Proof. We proceed exactly as for the a priori estimates in Step 5 to Step 7 of the proof of Proposition 9.37, see section 9.5.1, with the exception of the terms involving $\delta B^{(n)}$ for which we do not use Cauchy-Schwarz. We obtain the following analog of

(9.5.12), (9.5.9):

$$
\begin{aligned}
&\left(r^4|\delta\underline{C}^{(n+1)}|+r^5|\delta M^{(n+1)}|+\|(\delta h^{(n+1)},\delta\underline{h}^{(n+1)})\|_{\mathfrak{h}_1(\overset{\circ}{S})}\right)^2\\
\lesssim\ &\left\{r^2\|H^{(n+1)}\|_{L^2(\overset{\circ}{S})}+r^2\|\underline{H}^{(n+1)}\|_{L^2(\overset{\circ}{S})}+r^{-2}\left(\left|D^{(n+1)}\right|+\left|\underline{D}^{(n+1)}\right|\right)\right\}\\
&\times\|(\delta h^{(n+1)},\delta\underline{h}^{(n+1)})\|_{L^2(\overset{\circ}{S})}\\
&+r^2\left|\int_{(\overset{\circ}{S},\not{g}^{(n)})}(\delta\underline{h}^{(n+1)}-\Upsilon^{(\mathbf{S}(n)}\delta h^{(n+1)})\delta B^{(n)}(\underline{h}^{(n)}-\Upsilon^{\mathbf{S}(n)}h^{(n)})\right|\\
&+r^2\left|\int_{(\overset{\circ}{S},\not{g}^{(n)})}\delta h^{(n+1)}\delta B^{(n)}h^{(n)}\right|\\
&+\left(\left|\int_{(\overset{\circ}{S},\not{g}^{(n)})}e^{\Phi}(\delta B^{(n)}h^{(n)})\right|+\left|\int_{(\overset{\circ}{S},\not{g}^{(n)})}e^{\Phi}(\delta B^{(n)}\underline{h}^{(n)})\right|\right)^2.
\end{aligned}
$$

Next, we estimate the terms involving $\delta B^{(n)}$. Using Lemma 9.45 with the choices

- $\psi=\delta\underline{h}^{(n+1)}-\Upsilon^{(\mathbf{S}(n)}\delta h^{(n+1)}$ and $h=\underline{h}^{(n)}-\Upsilon^{\mathbf{S}(n)}h^{(n)}$,
- $\psi=\delta h^{(n+1)}$ and $h=h^{(n)}$,
- $\psi=e^{\Phi}$ and $h=h^{(n)}$,
- $\psi=e^{\Phi}$ and $h=\underline{h}^{(n)}$,

which yields, together with the assumption on the $\mathfrak{h}_2(\overset{\circ}{S})$ norm of $h^{(n)}$ and $\underline{h}^{(n)}$,

$$
\begin{aligned}
&\left|\int_{(\overset{\circ}{S},\not{g}^{(n)})}(\delta\underline{h}^{(n+1)}-\Upsilon^{(\mathbf{S}(n)}\delta h^{(n+1)})\delta B^{(n)}(\underline{h}^{(n)}-\Upsilon^{\mathbf{S}(n)}h^{(n)})\right|\\
&+\left|\int_{(\overset{\circ}{S},\not{g}^{(n)})}\delta h^{(n+1)}\delta B^{(n)}h^{(n)}\right|\\
\lesssim\ &r^{-3}\overset{\circ}{\delta}\|\partial_\theta\left(\Psi^{(n)}-\Psi^{(n-1)}\right)\|_{\mathfrak{h}_1(\overset{\circ}{S})}\|(\delta h^{(n+1)},\delta\underline{h}^{(n+1)})\|_{\mathfrak{h}_1(\overset{\circ}{S})}
\end{aligned}
$$

and

$$
\begin{aligned}
&\left(\left|\int_{(\overset{\circ}{S},\not{g}^{(n)})}e^{\Phi}(\delta B^{(n)}h^{(n)})\right|+\left|\int_{(\overset{\circ}{S},\not{g}^{(n)})}e^{\Phi}(\delta B^{(n)}\underline{h}^{(n)})\right|\right)^2\\
\lesssim\ &\left(r^{-1}\overset{\circ}{\delta}\|\partial_\theta\left(\Psi^{(n)}-\Psi^{(n-1)}\right)\|_{\mathfrak{h}_1(\overset{\circ}{S})}\right)^2.
\end{aligned}
$$

Plugging in the above estimate, we infer

$$\begin{aligned}
&\left(r^4|\delta\underline{C}^{(n+1)}|+r^5|\delta M^{(n+1)}|+\|(\delta h^{(n+1)},\delta\underline{h}^{(n+1)})\|_{\mathfrak{h}_1(\overset{\circ}{S})}\right)^2\\
\lesssim\ &\left\{r^2\|H^{(n+1)}\|_{L^2(\overset{\circ}{S})}+r^2\|\underline{H}^{(n+1)}\|_{L^2(\overset{\circ}{S})}+r^{-2}\left(\left|D^{(n+1)}\right|+\left|\underline{D}^{(n+1)}\right|\right)\right\}\\
&\times\|(\delta h^{(n+1)},\delta\underline{h}^{(n+1)})\|_{L^2(\overset{\circ}{S})}\\
&+r^{-1}\overset{\circ}{\delta}\|\partial_\theta\left(\Psi^{(n)}-\Psi^{(n-1)}\right)\|_{\mathfrak{h}_1(\overset{\circ}{S})}\|(\delta h^{(n+1)},\delta\underline{h}^{(n+1)})\|_{\mathfrak{h}_1(\overset{\circ}{S})}\\
&+\left(r^{-1}\overset{\circ}{\delta}\|\partial_\theta\left(\Psi^{(n)}-\Psi^{(n-1)}\right)\|_{\mathfrak{h}_1(\overset{\circ}{S})}\right)^2
\end{aligned}$$

and hence

$$\begin{aligned}
&r^4|\delta\underline{C}^{(n+1)}|+r^5|\delta M^{(n+1)}|+\|(\delta h^{(n+1)},\delta\underline{h}^{(n+1)})\|_{\mathfrak{h}_1(\overset{\circ}{S})}\\
&\lesssim\overset{\circ}{\delta}r^{-1}\|\partial_\theta\left(\Psi^{(n)}-\Psi^{(n-1)}\right)\|_{\mathfrak{h}_1(\overset{\circ}{S})}+r^2\|H^{(n+1)}\|_{L^2(\overset{\circ}{S})}+r^2\|\underline{H}^{(n+1)}\|_{L^2(\overset{\circ}{S})}\\
&+r^{-2}\left(\left|D^{(n+1)}\right|+\left|\underline{D}^{(n+1)}\right|\right)
\end{aligned}$$

as desired. □

9.6.3 Proof of the estimates (9.6.5), (9.6.6), (9.6.7)

We are now in position to prove (9.6.5), (9.6.6), (9.6.7).

Step 1. We start by estimating $\delta h^{(n+1)},\delta\underline{h}^{(n+1)}$. To this end, we need to apply Lemma 9.47 to the equations for $\delta h^{(n+1)}$, $\delta\underline{h}^{(n+1)}$, derived from the first two equations in (9.6.1) and (9.6.2), and estimate the corresponding terms $H^{(n+1)}$, $\underline{H}^{(n+1)}$, $D^{(n+1)}$ and $\underline{D}^{(n+1)}$ on the right-hand side. This is tedious but straightforward and one derives

$$\begin{aligned}
&r^2\|H^{(n+1)}\|_{L^2(\overset{\circ}{S})}+r^2\|\underline{H}^{(n+1)}\|_{L^2(\overset{\circ}{S})}+r^{-2}\left(\left|D^{(n+1)}\right|+\left|\underline{D}^{(n+1)}\right|\right)\\
\lesssim\ &\overset{\circ}{\epsilon}\left(\|P^{(n)}-P^{(n-1)}\|_2+\|P^{(n-1)}-P^{(n-2)}\|_2\right).
\end{aligned}$$

Remark 9.48. *Note that the presence of the inverse operators $(A^{(n)})^{-1}$ in the right-hand side of the equations for $\delta h^{(n+1)}$, $\delta\underline{h}^{(n+1)}$ do not create any difficulties when taking differences. Indeed, we can write*

$$(A^{(n)})^{-1}-(A^{(n-1)})^{-1}\ =\ (A^{(n)})^{-1}\left(A^{(n-1)}-A^{(n)}\right)(A^{(n-1)})^{-1}$$

and estimate the difference $\delta A^{(n)}=A^{(n)}-A^{(n-1)}$ similar to the estimate for $\delta B^{(n)}$ in the proof of Lemma 9.45.

We infer from Lemma 9.47 and the above estimates

$$
\begin{aligned}
& r^4|\delta\underline{C}^{(n+1)}| + r^5|\delta M^{(n+1)}| + \|(\delta h^{(n+1)}, \delta\underline{h}^{(n+1)})\|_{\mathfrak{h}_1(\overset{\circ}{S})} \\
\lesssim\ & \overset{\circ}{\epsilon}\Big(\|P^{(n)} - P^{(n-1)}\|_2 + \|P^{(n-1)} - P^{(n-2)}\|_2\Big). \qquad (9.6.16)
\end{aligned}
$$

Step 2. Next, we estimate $\not{d}^\star\delta e^{(n+1)}$. Recall (9.6.1)

$$
\not{d}^{\star(n)} e^{(n+1)} = (\Psi^{(n)})^\# \widetilde{E}^{(n+1)},
$$

where

$$
\begin{aligned}
\widetilde{E}^{(n+1)} &= -\frac{3}{4}(A^{\mathbf{S}(n)})^{-1}\big(\rho z^{(n+1)}\big) + f^{(n+1)}\underline{\omega} - \underline{f}^{(n+1)}\omega + \frac{1}{4}f^{(n+1)}\underline{\kappa} \\
& -\frac{1}{4}\underline{f}^{(n+1)}\kappa + (A^{\mathbf{S}(n)})^{-1}\big(- M^{(n+1)}e^\Phi + \not{d}^\star\check{\mu}\big) - (A^{\mathbf{S}(n)})^{-1}\big(\mathrm{Err}_3^{(n+1)}\big).
\end{aligned}
$$

This yields

$$
\not{d}^{\star(n)}\delta e^{(n+1)} = -\left(1 - \frac{\sqrt{\gamma^{(n)}}}{\sqrt{\gamma^{(n-1)}}}\right)\not{d}^{\star(n)} e^{(n)} + \widetilde{H}^{(n+1)}.
$$

The control of $\widetilde{H}^{(n+1)}$ is tedious but straightforward and one derives, using in particular Remark 9.48,

$$
\begin{aligned}
r\Big\|\widetilde{H}^{(n+1)}\Big\|_{L^2(\overset{\circ}{S})} \lesssim\ & r^5|\delta M^{(n+1)}| + \|(\delta h^{(n+1)}, \delta\underline{h}^{(n+1)})\|_{L^2(\overset{\circ}{S})} \\
& +\overset{\circ}{\epsilon}\Big(\|P^{(n)} - P^{(n-1)}\|_2 + \|P^{(n-1)} - P^{(n-2)}\|_2\Big).
\end{aligned}
$$

Also, in view of the boundedness of $e^{(n)}$ and $\gamma^{(n)}$, we have

$$
\begin{aligned}
r\left\|\left(1 - \frac{\sqrt{\gamma^{(n)}}}{\sqrt{\gamma^{(n-1)}}}\right)\not{d}^{\star(n)} e^{(n)}\right\|_{L^2(\overset{\circ}{S})} &\lesssim r^{-3}\overset{\circ}{\delta}\|\gamma^{(n)} - \gamma^{(n-1)}\|_{L^2(\overset{\circ}{S})} \\
&\lesssim r^{-2}\overset{\circ}{\delta}\|\partial_\theta\left(\Psi^{(n)} - \Psi^{(n-1)}\right)\|_{\mathfrak{h}_1(\overset{\circ}{S})}
\end{aligned}
$$

where we have used (9.6.11) in the last inequality. We deduce

$$
\begin{aligned}
r\Big\|\not{d}^{\star(n)}\delta e^{(n+1)}\Big\|_{L^2(\overset{\circ}{S})} \lesssim\ & r^5|\delta M^{(n+1)}| + \|(\delta h^{(n+1)}, \delta\underline{h}^{(n+1)})\|_{L^2(\overset{\circ}{S})} \\
& +\overset{\circ}{\epsilon}\Big(\|P^{(n)} - P^{(n-1)}\|_2 + \|P^{(n-1)} - P^{(n-2)}\|_2\Big)
\end{aligned}
$$

and hence, using a Poincaré inequality,

$$
\begin{aligned}
\left\|\delta e^{(n+1)} - \overline{\delta e^{(n+1)}}^{\overset{\circ}{S}, \not{g}^{(n)}}\right\|_{\mathfrak{h}_1(\overset{\circ}{S})} \lesssim\ & r^5|\delta M^{(n+1)}| + \|(\delta h^{(n+1)}, \delta\underline{h}^{(n+1)})\|_{L^2(\overset{\circ}{S})} \\
& + \overset{\circ}{\epsilon}\Big(\|P^{(n)} - P^{(n-1)}\|_2 + \|P^{(n-1)} - P^{(n-2)}\|_2\Big).
\end{aligned}
$$

Together with (9.6.16), we deduce

$$\begin{aligned}
& r^4|\delta\underline{C}^{(n+1)}| + r^5|\delta M^{(n+1)}| \\
& + \left\| \left(\delta h^{(n+1)}, \delta \underline{h}^{(n+1)}, \delta e^{(n+1)} - \overline{\delta e^{(n+1)}}^{\overset{\circ}{S}, \not{g}^{(n)}} \right) \right\|_{\mathfrak{h}_1(\overset{\circ}{S})} \\
\lesssim\ & \overset{\circ}{\epsilon}\Big(\|P^{(n)} - P^{(n-1)}\|_2 + \|P^{(n-1)} - P^{(n-2)}\|_2 \Big),
\end{aligned}$$

which is the desired estimate (9.6.5).

Step 3. Next, we estimate the average of $\delta e^{(n+1)}$. Recall from (9.6.3)

$$\overline{e^{(n+1)}}^{\overset{\circ}{S}, \not{g}^{(n)}} = \overline{\left(1 - \frac{r^{\mathbf{S}(n)}}{r}\right)^{(n)}}^{\overset{\circ}{S}, \not{g}^{(n)}} - \frac{r^{\mathbf{S}(n)}}{2} \overline{\left(\check{\kappa} + \overline{\kappa} - \frac{2}{r}\right)^{(n)}}^{\overset{\circ}{S}, \not{g}^{(n)}} + \overline{\mathrm{Err}_6^{(n+1)}}^{\overset{\circ}{S}, \not{g}^{(n)}}$$

and

$$\begin{aligned}
\overline{e^{(n)}}^{\overset{\circ}{S}, \not{g}^{(n)}} =\ & \overline{\left(1 - \frac{r^{\mathbf{S}(n-1)}}{r}\right)^{(n-1)}}^{\overset{\circ}{S}, \not{g}^{(n-1)}} - \frac{r^{\mathbf{S}(n-1)}}{2} \overline{\left(\check{\kappa} + \overline{\kappa} - \frac{2}{r}\right)^{(n)}}^{\overset{\circ}{S}, \not{g}^{(n-1)}} \\
& + \overline{\mathrm{Err}_6^{(n)}}^{\overset{\circ}{S}, \not{g}^{(n-1)}}.
\end{aligned}$$

Taking the difference, recalling that we have in the (θ, φ) coordinates system

$$dvol_{\not{g}^{(n)}} = \sqrt{\gamma^{(n)}} e^{\Phi^{\#n}} d\theta d\varphi, \quad 4\pi (r^{\mathbf{S}(n)})^2 = \int_0^{2\pi} \int_0^{\pi} \sqrt{\gamma^{(n)}} e^{\Phi^{\#n}} d\theta d\varphi$$

and

$$dvol_{\not{g}^{(n-1)}} = \sqrt{\gamma^{(n-1)}} e^{\Phi^{\#n-1}} d\theta d\varphi, \quad 4\pi (r^{\mathbf{S}(n-1)})^2 = \int_0^{2\pi} \int_0^{\pi} \sqrt{\gamma^{(n-1)}} e^{\Phi^{\#n-1}} d\theta d\varphi$$

and using the uniform bound of $P^{(n)}$ provided by Proposition 9.42 and the bounds **A1** for $\check{\Gamma}$, we infer

$$\begin{aligned}
r \left| \overline{\delta e^{(n+1)}}^{\overset{\circ}{S}} \right| \lesssim\ & r^{-2} \|\gamma^{(n)} - \gamma^{(n-1)}\|_{L^2(\overset{\circ}{S})} + r^{-1} \|e^{\Phi^{\#n}} - e^{\Phi^{\#n-1}}\|_{L^2(\overset{\circ}{S})} \\
& + \|\delta \mathrm{Err}_6^{(n+1)}\|_{L^2(\overset{\circ}{S})}.
\end{aligned}$$

Arguing as above, we deduce

$$\begin{aligned}
r \left| \overline{\delta e^{(n+1)}}^{\overset{\circ}{S}, \not{g}^{(n)}} \right| \lesssim\ & r^{-1} \|\partial_\theta (\delta U^{(n)}, \delta S^{(n)})\|_{\mathfrak{h}_1(\overset{\circ}{S})} \\
& + \overset{\circ}{\epsilon}\Big(\|P^{(n)} - P^{(n-1)}\|_2 + \|P^{(n-1)} - P^{(n-2)}\|_2 \Big)
\end{aligned}$$

which is the desired estimate (9.6.6).

Step 4. Finally, we focus on (9.6.7). Recall (9.6.4)

$$\begin{aligned}
\varsigma^{\#_n}\partial_\theta U^{(1+n)} &= (\gamma^{(n)})^{1/2} h^{(1+n)}\left(1+\frac{1}{4}h^{(1+n)}\underline{h}^{(1+n)}\right),\\
\partial_\theta S^{(1+n)} - \frac{1}{2}\varsigma^{\#_n}\underline{\Omega}^{\#_n}\partial_\theta U^{(1+n)} &= \frac{1}{2}(\gamma^{(n)})^{1/2}\underline{h}^{(1+n)},\\
\gamma^{(n)} &= \gamma^{\#_n} + \left(\varsigma^{\#_n}\right)^2\left(\underline{\Omega}^{\#_n} + \frac{1}{4}(\underline{b}^{\#_n})^2\gamma^{\#_n}\right)(\partial_\theta U^{(n)})^2\\
&\quad - 2\varsigma^{\#_n}\partial_\theta U^{(n)}\partial_\theta S^{(n)} - \gamma^{\#_n}\varsigma^{\#_n}\underline{b}^{\#_n}\partial_\theta U^{(n)},\\
U^{(1+n)}(0) &= S^{(1+n)}(0) = 0.
\end{aligned}$$

Taking the difference and arguing as above, we derive

$$\begin{aligned}
r^{-1}\|\partial_\theta\delta U^{(1+n)}\|_{\mathfrak{h}_1(\overset{\circ}{S})} &\lesssim \|\delta h^{(n+1)}, \delta\underline{h}^{(n+1)}\|_{\mathfrak{h}_1(\overset{\circ}{S})} + r^{-3}\overset{\circ}{\epsilon}\|\gamma^{(n)} - \gamma^{(n-1)}\|_{\mathfrak{h}_1(\overset{\circ}{S})}\\
&\quad + \overset{\circ}{\epsilon}\|P^{(n)} - P^{(n-1)}\|_2,\\
r^{-1}\|\partial_\theta\delta S^{(1+n)}\|_{\mathfrak{h}_1(\overset{\circ}{S})} &\lesssim \|\partial_\theta\delta U^{(1+n)}\|_{\mathfrak{h}_1(\overset{\circ}{S})} + \|\delta h^{(n+1)}, \delta\underline{h}^{(n+1)}\|_{\mathfrak{h}_1(\overset{\circ}{S})}\\
&\quad + r^{-3}\overset{\circ}{\epsilon}\|\gamma^{(n)} - \gamma^{(n-1)}\|_{\mathfrak{h}_1(\overset{\circ}{S})} + \overset{\circ}{\epsilon}\|P^{(n)} - P^{(n-1)}\|_2.
\end{aligned}$$

Estimating $\gamma^{(n)} - \gamma^{(n-1)}$ as above, we infer

$$r^{-1}\|\partial_\theta\left(\delta U^{(n+1)}, \delta S^{(n+1)}\right)\|_{\mathfrak{h}_1(\overset{\circ}{S})} \lesssim \|\delta h^{(n+1)}, \delta\underline{h}^{(n+1)}\|_{\mathfrak{h}_1(\overset{\circ}{S})} + \overset{\circ}{\epsilon}\|P^{(n)} - P^{(n-1)}\|_2$$

which is the desired estimate (9.6.7). This concludes the proof of Proposition 9.43.

9.7 A COROLLARY TO THEOREM 9.32

In what follows we prove an important corollary of Theorem 9.32 which makes use of the arbitrariness of $\Lambda, \underline{\Lambda}$ to ensure the vanishing of the $\ell = 1$ modes of β and $\check{\underline{\kappa}}$. The result requires stronger assumptions than those made in **A1**. Namely we assume that Γ_b has the same behavior as Γ_g, i.e.,

A1-Strong. For $k \le s_{max}$,

$$\Big\|\Gamma_g\Big\|_{k,\infty} \lesssim \overset{\circ}{\epsilon} r^{-2}, \qquad \Big\|\Gamma_b\Big\|_{k,\infty} \lesssim \overset{\circ}{\epsilon} r^{-1}, \qquad \Big\|\Gamma_b\Big\|_{k,\infty} \lesssim (\overset{\circ}{\epsilon})^{\frac{1}{3}} r^{-2}. \tag{9.7.1}$$

Theorem 9.49 (Existence of GCM spheres). *In addition to the assumptions of Theorem 9.32, we assume that* ***A1-Strong*** *holds, and that, for any background sphere S in $\mathcal{R}$,*

$$r\left|\int_S \beta e^\Phi\right| + r\left|\int_S e_\theta(\kappa)e^\Phi\right| + r\left|\int_S e_\theta(\underline{\kappa})e^\Phi\right| \lesssim \overset{\circ}{\delta}. \tag{9.7.2}$$

Then there exists a unique GCM sphere **S**, *which is a deformation of $\overset{\circ}{S}$, such that* *the following GCM conditions hold true:*

$$\begin{aligned} & \not{d}_2^{\mathbf{S},\star} \not{d}_1^{\mathbf{S},\star} \underline{\kappa}^{\mathbf{S}} = \not{d}_2^{\mathbf{S},\star} \not{d}_1^{\mathbf{S},\star} \mu^{\mathbf{S}} = 0, \qquad \kappa^{\mathbf{S}} = \frac{2}{r^{\mathbf{S}}}, \\ & \int_{\mathbf{S}} \beta^{\mathbf{S}} e^{\Phi} = 0, \qquad \int_{\mathbf{S}} e_\theta^{\mathbf{S}}(\kappa^{\mathbf{S}}) e^{\Phi} = 0. \end{aligned} \tag{9.7.3}$$

Moreover, all other estimates of Theorem 9.32 hold true.

Proof. The proof of the theorem follows easily in view of Theorem 9.32 and the following lemma.

Lemma 9.50. *Let* $\mathbf{S}$ *be a deformation of* $\overset{\circ}{S}$ *as in Theorem 9.32 with* $\Lambda = \int_{\mathbf{S}} f e^{\Phi}$ *and* $\underline{\Lambda} = \int_{\mathbf{S}} \underline{f} e^{\Phi}$. *The following identities hold true:*

$$\begin{aligned} \Lambda &= \frac{r^3}{3m}\left(\int_{\overset{\circ}{S}} \beta e^{\Phi} - \int_{\mathbf{S}} \beta^{\mathbf{S}} e^{\Phi}\right) + F_1(\Lambda, \underline{\Lambda}), \\ \underline{\Lambda} &= \frac{r^3}{6m}\left(\int_{\overset{\circ}{S}} e_\theta(\underline{\kappa}) e^{\Phi} - \int_{\mathbf{S}} (e_\theta^{\mathbf{S}} \underline{\kappa}^{\mathbf{S}}) e^{\Phi} - \Upsilon \int_{\overset{\circ}{S}} e_\theta(\kappa) e^{\Phi}\right) + \Upsilon\Lambda + F_2(\Lambda, \underline{\Lambda}), \end{aligned} \tag{9.7.4}$$

where F_1, F_2 *are continuous*[20] *in* $\Lambda, \underline{\Lambda}$, *verifying, provided* ***A1-Strong*** *holds, the estimates*

$$\begin{aligned} |F_1| + |F_2| &\lesssim (\overset{\circ}{\epsilon})^{\frac{1}{3}} \overset{\circ}{\delta} r^2, \\ |\partial_{\Lambda, \underline{\Lambda}} F_1| + |\partial_{\Lambda, \underline{\Lambda}} F_2| &\lesssim (\overset{\circ}{\epsilon})^{\frac{1}{3}} r^2. \end{aligned} \tag{9.7.5}$$

Proof. To prove (9.7.4), we start with the change of frame formula,

$$\begin{aligned} \beta^{\mathbf{S}} &= e^a\left(\beta + \frac{3}{2}\rho f\right) + e^a \mathrm{Err}(\beta, \beta^{\mathbf{S}}), \\ \mathrm{Err}(\beta, \beta^{\mathbf{S}}) &= \frac{1}{2}\underline{f}\alpha + \text{l.o.t.} \end{aligned}$$

We write[21]

$$\begin{aligned} \beta^{\mathbf{S}} &= \beta + \frac{3}{2}\rho f + (e^a - 1)\left(\beta + \frac{3}{2}\rho f\right) + e^a \mathrm{Err}(\beta, \beta^{\mathbf{S}}) \\ &= \beta + \frac{3}{2}\left(-\frac{2m}{r^3}\right) f + \frac{3}{2}\left(\rho + \frac{2m}{r^3}\right) f + (e^a - 1)\left(\beta + \frac{3}{2}\rho f\right) + e^a \mathrm{Err}(\beta, \beta^{\mathbf{S}}) \end{aligned}$$

and deduce

$$\beta^{\mathbf{S}} + \frac{3m^{\mathbf{S}}}{(r^{\mathbf{S}})^3} f = \beta + \mathrm{Err}'(\beta, \beta^{\mathbf{S}})$$

---

[20]In fact smooth.

[21]Here (r, m) represents the area radius and Hawking mass of $\overset{\circ}{S}$, while $(r^{\mathbf{S}}, m^{\mathbf{S}})$ represent the area radius and Hawking mass of $\mathbf{S}$. Since $|\frac{r^{\mathbf{S}}}{r} - 1| \lesssim \overset{\circ}{\delta}$ and $|m^{\mathbf{S}} - m| \lesssim \overset{\circ}{\delta}$, we can interchange freely $r^{\mathbf{S}}$ with r and $m^{\mathbf{S}}$ with m.

with error term $\mathrm{Err}'(\beta, \beta^{\mathbf{S}})$,

$$\mathrm{Err}'(\beta, \beta^{\mathbf{S}}) = \left(\frac{3m^{\mathbf{S}}}{(r^{\mathbf{S}})^3} - \frac{3m}{r^3}f\right) + \frac{3}{2}\left(\rho + \frac{2m}{r^3}\right)f + (e^a - 1)\left(\beta + \frac{3}{2}\rho f\right) + e^a \mathrm{Err}(\beta, \beta^{\mathbf{S}}).$$

Making use of the assumptions **A1**–**A3** , the estimates of Theorem 9.32 for $(f, \underline{f}, a)$ as well as the bounds for $\mathring{r} - r^{\mathbf{S}}$, $\mathring{m} - m^{\mathbf{S}}$, we deduce

$$\left|\mathrm{Err}'(\beta, \beta^{\mathbf{S}})\right| \lesssim r^{-1}\mathring{\delta}\mathring{\epsilon}.$$

Thus,

$$\begin{aligned}\frac{3m^{\mathbf{S}}}{(r^{\mathbf{S}})^3}\int_{\mathbf{S}} fe^{\Phi} &= \int_{\mathbf{S}} \beta e^{\Phi} - \int_{\mathbf{S}} \beta^{\mathbf{S}} e^{\Phi} + \int_{\mathbf{S}} \mathrm{Err}'(\beta, \beta^{\mathbf{S}})e^{\Phi}\\ &= \int_{\mathring{S}} \beta e^{\Phi} - \int_{\mathbf{S}} \beta^{\mathbf{S}} e^{\Phi} + \int_{\mathbf{S}} \mathrm{Err}'(\beta, \beta^{\mathbf{S}})e^{\Phi} + \left(\int_{\mathbf{S}} \beta e^{\Phi} - \int_{\mathring{S}} \beta e^{\Phi}\right)\end{aligned}$$

or

$$\begin{aligned}\frac{3m}{r^3}\int_{\mathbf{S}} fe^{\Phi} &= \int_{\mathring{S}} \beta e^{\Phi} - \int_{\mathbf{S}} \beta^{\mathbf{S}} e^{\Phi} + \int_{\mathbf{S}} \mathrm{Err}'(\beta, \beta^{\mathbf{S}})e^{\Phi} + \left(\int_{\mathbf{S}} \beta e^{\Phi} - \int_{\mathring{S}} \beta e^{\Phi}\right)\\ &+ \left(\frac{3m}{r^3} - \frac{3m^{\mathbf{S}}}{(r^{\mathbf{S}})^3}\right)\int_{\mathbf{S}} fe^{\Phi}.\end{aligned}$$

Clearly,

$$\left|\int_{\mathbf{S}} \mathrm{Err}'(\beta, \beta^{\mathbf{S}})e^{\Phi}\right| \lesssim r^{-1}\mathring{\delta}\mathring{\epsilon}.$$

Also, proceeding exactly as in Corollary 9.10 we deduce

$$\left|\int_{\mathbf{S}} \beta e^{\Phi} - \int_{\mathring{S}} \beta e^{\Phi}\right| \lesssim \mathring{\delta}\left(\sup_{\mathcal{R}} \mathring{r}\left|\mathfrak{d}_{\nearrow}^{\leq 1}(\beta e^{\Phi})\right| + \sup_{\mathcal{R}} \mathring{r}^2\left|e_3(\beta e^{\Phi})\right|\right).$$

Thus, in view of the assumptions **A1**–**A3**,

$$\left|\int_{\mathbf{S}} \beta e^{\Phi} - \int_{\mathring{S}} \beta e^{\Phi}\right| \lesssim \mathring{\delta}\mathring{\epsilon}r^{-1}. \tag{9.7.6}$$

We deduce

$$\Lambda = \frac{r^3}{3m}\left(\int_{\mathring{S}} \beta e^{\Phi} - \int_{\mathbf{S}} \beta^{\mathbf{S}} e^{\Phi}\right) + F_1(\Lambda, \underline{\Lambda})$$

where the error term $F_1(\Lambda, \underline{\Lambda})$ is a continuous function of $\Lambda, \underline{\Lambda}$ verifying the estimate

$$\left|F_1(\Lambda, \underline{\Lambda})\right| \lesssim \mathring{\epsilon}\mathring{\delta}r^2.$$

We also recall, see Lemma 9.23,

$$\begin{aligned} e^{\mathbf{S}}_\theta(\underline{\kappa}^{\mathbf{S}}) &= e_\theta\underline{\kappa} - \not{d}_1^{\mathbf{S},\star}\not{d}_1^{\mathbf{S}}\underline{f} - \underline{\kappa}e^{\mathbf{S}}_\theta a - \frac{1}{4}\underline{\kappa}(f\underline{\kappa}+\underline{f}\kappa) + \underline{\kappa}(\underline{f}\omega - \underline{\omega}f) + \underline{f}\rho \\ &\quad + \mathrm{Err}(e^{\mathbf{S}}_\theta\underline{\kappa}^{\mathbf{S}}, e_\theta\underline{\kappa}) \end{aligned}$$

where

$$\begin{aligned} \mathrm{Err}(e^{\mathbf{S}}_\theta\underline{\kappa}^{\mathbf{S}}, e_\theta\underline{\kappa}) &= (e^{-a}-1)\Big(e_\theta\underline{\kappa} - \not{d}^{\mathbf{S},\star}\not{d}_1^{\mathbf{S}}\underline{f} + \frac{1}{2}fe_3\underline{\kappa} + \frac{1}{2}\underline{f}e_4\underline{\kappa}\Big) \\ &+ e^{-a}\Bigg[e^{\mathbf{S}}_\theta\,\mathrm{Err}(\underline{\kappa},\underline{\kappa}^{\mathbf{S}}) + e^{\mathbf{S}}_\theta(a)\Big(\not{d}_1^{\mathbf{S}}\underline{f} + \mathrm{Err}(\underline{\kappa},\underline{\kappa}^{\mathbf{S}})\Big) + \frac{1}{2}f\underline{f}e_\theta\underline{\kappa} \\ &+ \frac{1}{8}f^2\underline{f}e_3\underline{\kappa}\Bigg] + \frac{1}{2}\underline{f}\left(2\not{d}_1\underline{\eta} - \frac{1}{2}\underline{\vartheta}\vartheta + 2(\xi\underline{\xi}+\underline{\eta}^2)\right) \\ &+ \frac{1}{2}f\underline{f}e_\theta\underline{\kappa} + \frac{1}{8}f^2\underline{f}e_3\underline{\kappa} + \frac{1}{2}f\left(2\not{d}_1\underline{\xi} - \frac{1}{2}\underline{\vartheta}^2 + 2(\eta+\underline{\eta}-2\zeta)\underline{\xi}\right). \end{aligned}$$

Making use of the identity $\not{d}_1^{\mathbf{S},\star}\not{d}_1^{\mathbf{S}} = \not{d}_2^{\mathbf{S}}\not{d}_2^{\mathbf{S},\star} + 2K^{\mathbf{S}}$ we deduce

$$\begin{aligned} e^{\mathbf{S}}_\theta(\underline{\kappa}^{\mathbf{S}}) + \left(\frac{1}{4}\kappa\underline{\kappa} - \rho + 2K\right)\underline{f} &= e_\theta\underline{\kappa} - \not{d}_2^{\mathbf{S}}\not{d}_2^{\mathbf{S},\star}\underline{f} - \underline{\kappa}e^{\mathbf{S}}_\theta a - \frac{1}{4}\underline{\kappa}^2 f - \underline{\kappa}\,\underline{\omega}f \\ &\quad -2(K^{\mathbf{S}}-K)\underline{f} + \mathrm{Err}(e^{\mathbf{S}}_\theta\underline{\kappa}^{\mathbf{S}}, e_\theta\underline{\kappa}). \end{aligned}$$

Using $\kappa = \frac{2}{r} + (\kappa - \frac{2}{r})$, $\underline{\kappa} = -\frac{2\Upsilon}{r} + (\underline{\kappa} + \frac{2\Upsilon}{r})$, $\rho = -\frac{2m}{r^3} + (\rho + \frac{2m}{r^3})$, $K = \frac{1}{r^2} + (K - \frac{1}{r^2})$, we have

$$\begin{aligned} \frac{1}{4}\kappa\underline{\kappa} - \rho + 2K &= \frac{1}{r^2} + \frac{4m}{r^3} + \frac{1}{2r}\left(\underline{\kappa} + \frac{2\Upsilon}{r}\right) - \frac{\Upsilon}{2r}\left(\kappa - \frac{2}{r}\right) + \left(\rho + \frac{2m}{r^3}\right) \\ &\quad +2\left(K - \frac{1}{r^2}\right) \\ &= \frac{1}{r^2} + \frac{4m}{r^3} + O(\overset{\circ}{\epsilon}r^{-3}). \end{aligned}$$

Also, using **A1-Strong**,

$$\begin{aligned} \underline{\kappa} &= -\frac{2\Upsilon}{r} + \left(\underline{\kappa} + \frac{2\Upsilon}{r}\right) = -\frac{2\Upsilon}{r} + O(r^{-2}\overset{\circ}{\epsilon}), \\ \underline{\kappa}\,\underline{\omega} &= -\frac{2m\Upsilon}{r^3} + O\left(r^{-3}(\overset{\circ}{\epsilon})^{\frac{1}{3}}\right), \end{aligned}$$

and, in view of **A1-Strong**, and since $a, f, \underline{f} = O(r^{-1}\overset{\circ}{\delta})$,

$$\mathrm{Err}(e^{\mathbf{S}}_\theta\underline{\kappa}^{\mathbf{S}}, e_\theta\underline{\kappa}) = O(r^{-4}\overset{\circ}{\delta}(\overset{\circ}{\epsilon})^{\frac{1}{3}}).$$

We deduce

$$e^{\mathbf{S}}_\theta(\underline{\kappa}^{\mathbf{S}}) + \left(\frac{1}{r^2} + \frac{4m}{r^3}\right)\underline{f} = e_\theta\underline{\kappa} - \not{d}_2\not{d}_2^\star\underline{f} + \frac{2\Upsilon}{r}e^{\mathbf{S}}_\theta a - \frac{\Upsilon\left(1-\frac{4m}{r}\right)}{r^2}f + \mathrm{Err}_1$$

with error term

$$\left|\text{Err}_1\right| \lesssim (\overset{\circ}{\epsilon})^{\frac{1}{3}}\overset{\circ}{\delta} r^{-4}.$$

Projecting on e^{Φ} and proceeding as before,

$$\begin{aligned}\left(\frac{1}{r^2}+\frac{4m}{r^3}\right)\int_{\mathbf{S}}\underline{f}e^{\Phi} &= \int_{\overset{\circ}{S}}(e_\theta\underline{\kappa})e^{\Phi}-\int_{\mathbf{S}}(e^{\mathbf{S}}_\theta\underline{\kappa}^{\mathbf{S}})e^{\Phi}-\frac{2\Upsilon}{r}\int_{\mathbf{S}}(e^{\mathbf{S}}_\theta a)e^{\Phi}\\ &-\frac{\Upsilon\left(1-\frac{4m}{r}\right)}{r^2}\int_{\mathbf{S}}fe^{\Phi}+I_1(\Lambda,\underline{\Lambda})\end{aligned} \tag{9.7.7}$$

where the error term I_1 is continuous in $\Lambda, \underline{\Lambda}$ and verifies the estimate

$$\left|I_1\right| \lesssim r^{-1}(\overset{\circ}{\epsilon})^{\frac{1}{3}}\overset{\circ}{\delta}.$$

We now calculate $\int_{\mathbf{S}}(e^{\mathbf{S}}_\theta a)e^{\Phi}$. Recall from Lemma 9.23

$$\begin{aligned}e^{\mathbf{S}}_\theta(\kappa^{\mathbf{S}}) &= e_\theta\kappa-\not{d}^{\star\mathbf{S}}_1\not{d}^{\mathbf{S}}_1 f+\kappa e^{\mathbf{S}}_\theta a-\frac{1}{4}\kappa(f\underline{\kappa}+\underline{f}\kappa)+\kappa(f\underline{\omega}-\omega\underline{f})+f\rho\\ &+\text{Err}(e^{\mathbf{S}}_\theta\kappa^{\mathbf{S}}, e_\theta\kappa),\end{aligned}$$

where

$$\begin{aligned}\text{Err}(e^{\mathbf{S}}_\theta\kappa^{\mathbf{S}}, e_\theta\kappa) &= (e^a-1)\Big(e_\theta\kappa+e^{\mathbf{S}}_\theta\not{d}^{\mathbf{S}}_1 f+\frac{1}{2}\underline{f}e_4\kappa+\frac{1}{2}fe_3\kappa\Big)\\ &+ e^a\left[e^{\mathbf{S}}_\theta\,\text{Err}(\kappa,\kappa^{\mathbf{S}})+e^{\mathbf{S}}_\theta(a)\Big(\not{d}^{\mathbf{S}}_1 f+\text{Err}(\kappa,\kappa^{\mathbf{S}})\Big)+\frac{1}{2}f\underline{f}e_\theta\kappa\right.\\ &+ \left.\frac{1}{8}f^2\underline{f}e_3\kappa\right]+\frac{1}{2}f\left(2\not{d}_1\eta-\frac{1}{2}\underline{\vartheta}\vartheta+2(\xi\underline{\xi}+\eta^2)\right)\\ &+ \frac{1}{2}f\underline{f}e_\theta\kappa+\frac{1}{8}f^2\underline{f}e_3\kappa+\frac{1}{2}\underline{f}\left(2\not{d}_1\xi-\frac{1}{2}\vartheta^2+2(\eta+\underline{\eta}+2\zeta)\xi\right).\end{aligned}$$

Using again the identity $\not{d}^{\mathbf{S},\star}_1\not{d}^{\mathbf{S}}_1 = \not{d}^{\mathbf{S}}_2\not{d}^{\mathbf{S},\star}_2+2K^{\mathbf{S}}$ and proceeding as above, we infer, using also the GCM condition for $\kappa^{\mathbf{S}}$ which yields $e^{\mathbf{S}}_\theta(\kappa^{\mathbf{S}})=0$,

$$\begin{aligned}0 &= e_\theta\kappa-\not{d}^{\mathbf{S}}_2\not{d}^{\star\mathbf{S}}_2 f+\frac{2}{r^{\mathbf{S}}}e^{\mathbf{S}}_\theta a-\frac{1}{4}\kappa^2\underline{f}+\left(\frac{1}{4}\kappa\underline{\kappa}+\kappa\underline{\omega}+3\rho\right)f\\ &+\left(\kappa-\frac{2}{r^{\mathbf{S}}}\right)e^{\mathbf{S}}_\theta a-2(K^{\mathbf{S}}-K)f-\frac{1}{2}\vartheta\underline{\vartheta}f+\text{Err}(e^{\mathbf{S}}_\theta\kappa^{\mathbf{S}}, e_\theta\kappa).\end{aligned}$$

Integrating over $\mathbf{S}$, we deduce

$$
\begin{aligned}
&\frac{2}{r^{\mathbf{S}}}\int_{\mathbf{S}} e_\theta^{\mathbf{S}}(a)e^\Phi \\
=&\ -\int_{\mathbf{S}} e_\theta(\kappa)e^\Phi + \frac{1}{4}\int_{\mathbf{S}}\kappa^2\underline{f}e^\Phi - \int_{\mathbf{S}}\left(\frac{1}{4}\kappa\underline{\kappa} + \kappa\underline{\omega} + 3\rho\right) fe^\Phi \\
&+\int_{\mathbf{S}}\left[\left(\kappa - \frac{2}{r^{\mathbf{S}}}\right)e_\theta^{\mathbf{S}}a - 2(K^{\mathbf{S}} - K)f - \frac{1}{2}\vartheta\underline{\vartheta}f + \mathrm{Err}(e_\theta^{\mathbf{S}}\kappa^{\mathbf{S}}, e_\theta\kappa)\right]e^\Phi \\
=&\ -\int_{\overset{\circ}{S}} e_\theta(\kappa)e^\Phi + \frac{1}{r^2}\underline{\Lambda} + \frac{1}{r^2}\left(1 + \frac{2m}{r}\right)\Lambda + I_2(\Lambda, \underline{\Lambda})
\end{aligned}
$$

where, using in particular **A1-Strong**,

$$
\left|I_2(\Lambda, \underline{\Lambda})\right| \lesssim r^{-1}(\overset{\circ}{\epsilon})^{\frac{1}{3}}\overset{\circ}{\delta}.
$$

Indeed, using once more Corollary 9.10, we note that

$$
\left|\int_{\mathbf{S}} e_\theta(\kappa)e^\Phi - \int_{\overset{\circ}{S}} e_\theta(\kappa)e^\Phi\right| \lesssim r^{-1}\overset{\circ}{\delta}\left(\overset{\circ}{\epsilon} + r^4\sup_{\mathcal{R}}|e_3(e_\theta(\kappa))|\right) \lesssim r^{-1}(\overset{\circ}{\epsilon})^{\frac{1}{3}}\overset{\circ}{\delta},
$$

where we used **A1-Strong**, the transport equation for $e_3(\kappa)$ and a commutator formula for $[e_3, e_\theta]$ to estimate $e_3(e_\theta(\kappa))$. All other error terms are easily estimated.

Back to (9.7.7) we deduce

$$
\begin{aligned}
\left(\frac{1}{r^2} + \frac{4m}{r^3}\right)\underline{\Lambda} =&\ \int_{\overset{\circ}{S}}(e_\theta\underline{\kappa})e^\Phi - \int_{\mathbf{S}}(e_\theta^{\mathbf{S}}\underline{\kappa}^{\mathbf{S}})e^\Phi + \frac{2\Upsilon}{r}\int_{\mathbf{S}}(e_\theta^{\mathbf{S}}a)e^\Phi - \frac{\Upsilon\left(1 - \frac{4m}{r}\right)}{r^2}\Lambda \\
&+I_1(\Lambda, \underline{\Lambda}) \\
=&\ \int_{\overset{\circ}{S}}(e_\theta\underline{\kappa})e^\Phi - \int_{\mathbf{S}}(e_\theta^{\mathbf{S}}\underline{\kappa}^{\mathbf{S}})e^\Phi - \frac{\Upsilon\left(1 - \frac{4m}{r}\right)}{r^2}\Lambda + I_1(\Lambda, \underline{\Lambda}) \\
&+ \Upsilon\left(-\int_{\overset{\circ}{S}} e_\theta(\kappa)e^\Phi + \frac{1}{r^2}\underline{\Lambda} + \frac{1}{r^2}\left(1 + \frac{2m}{r}\right)\Lambda + I_2(\Lambda, \underline{\Lambda})\right) \\
=&\ \int_{\overset{\circ}{S}}(e_\theta\underline{\kappa})e^\Phi - \int_{\mathbf{S}}(e_\theta^{\mathbf{S}}\underline{\kappa}^{\mathbf{S}})e^\Phi - \Upsilon\int_{\overset{\circ}{S}} e_\theta(\kappa)e^\Phi + \frac{\Upsilon}{r^2}\underline{\Lambda} + \frac{6m\Upsilon}{r^3}\Lambda \\
&+ I_1(\Lambda, \underline{\Lambda}) + \Upsilon I_2(\Lambda, \underline{\Lambda}).
\end{aligned}
$$

Hence,

$$
\begin{aligned}
\frac{6m}{r^3}\underline{\Lambda} =&\ \int_{\overset{\circ}{S}}(e_\theta\underline{\kappa})e^\Phi - \int_{\mathbf{S}}(e_\theta^{\mathbf{S}}\underline{\kappa}^{\mathbf{S}})e^\Phi - \Upsilon\int_{\overset{\circ}{S}} e_\theta(\kappa)e^\Phi + \frac{6m\Upsilon}{r^3}\Lambda \\
&+ I_1(\Lambda, \underline{\Lambda}) + \Upsilon I_2(\Lambda, \underline{\Lambda}).
\end{aligned}
$$

Thus,

$$
\underline{\Lambda} = \frac{r^3}{6m}\left(\int_{\overset{\circ}{S}}(e_\theta\underline{\kappa})e^\Phi - \int_{\mathbf{S}}(e_\theta^{\mathbf{S}}\underline{\kappa}^{\mathbf{S}})e^\Phi - \Upsilon\int_{\overset{\circ}{S}} e_\theta(\kappa)e^\Phi\right) + \Upsilon\Lambda + F_2(\Lambda, \underline{\Lambda})
$$

with error term $F_2(\Lambda, \underline{\Lambda})$ continuous in $\Lambda, \underline{\Lambda}$ and verifying the estimate

$$\left|F_2(\Lambda, \underline{\Lambda})\right| \lesssim (\overset{\circ}{\epsilon})^{\frac{1}{3}}\overset{\circ}{\delta} r^2.$$

To check the second part in (9.7.5) one needs to revisit the proof of Theorem 9.32 and check the dependence of $U, S, f, \underline{f}, \lambda$ on the parameters $\Lambda, \underline{\Lambda}$. It is tedious but standard to check the following estimates for the derivatives with respect to $\Lambda, \underline{\Lambda}$.

$$\left\|\partial_{\Lambda,\underline{\Lambda}}(f, \underline{f}, \log\lambda)\right\|_{\mathfrak{h}_k(\mathbf{S})} \lesssim 1, \qquad k \leq s_{max}, \tag{9.7.8}$$

$$\|\partial_{\Lambda,\underline{\Lambda}}(U', S')\|_{L^\infty(\overset{\circ}{S})} + \max_{0\leq s\leq s_{max}-1} r^{-1}\|\partial_{\Lambda,\underline{\Lambda}}(U', S')\|_{\mathfrak{h}_s(\overset{\circ}{S},\overset{\circ}{\not g})} \lesssim 1. \tag{9.7.9}$$

Using these estimates and taking into account the structure of the error terms F_1, F_2 we derive the second inequality in (9.7.5). This ends the proof of the lemma. □

Under the assumptions of the theorem, the system

$$\Lambda = \frac{r^3}{3m}\int_{\overset{\circ}{S}} \beta e^\Phi + F_1(\Lambda, \underline{\Lambda}),$$
$$\underline{\Lambda} = \frac{r^3}{6m}\left(\int_{\overset{\circ}{S}} e_\theta(\underline{\kappa})e^\Phi - \Upsilon\int_{\overset{\circ}{S}} e_\theta(\kappa)e^\Phi\right) + \Upsilon\Lambda + F_2(\Lambda, \underline{\Lambda}),$$

has a unique solution $\Lambda_0, \underline{\Lambda}_0$ verifying the estimate

$$|\Lambda_0| + |\underline{\Lambda}_0| \lesssim \overset{\circ}{\delta} r^2.$$

Taking $\Lambda = \Lambda_0, \underline{\Lambda} = \underline{\Lambda}_0$ in (9.7.4) we deduce

$$\int_{\mathbf{S}} \beta^{\mathbf{S}} e^\Phi = 0, \qquad \int_{\mathbf{S}} e_\theta^{\mathbf{S}}(\underline{\kappa}^{\mathbf{S}})e^\Phi = 0,$$

as stated. □

Corollary 9.51. *Let a fixed spacetime region $\mathcal{R}$ verifying assumptions* ***A1–A3*** *and (9.4.2), as well as, for any background sphere S in $\mathcal{R}$,*

$$\left|\int_S \beta e^\Phi\right| + \left|\int_S e_\theta(\underline{\kappa})e^\Phi\right| \lesssim \overset{\circ}{\delta}. \tag{9.7.10}$$

Assume that $\mathbf{S}$ is a sphere in $\mathcal{R}$ which verifies the GCM conditions

$$\kappa^{\mathbf{S}} = \frac{2}{r^{\mathbf{S}}}, \qquad \not{d}_2^{\star\,\mathbf{S}}\not{d}_1^{\star\,\mathbf{S}}\underline{\kappa}^{\mathbf{S}} = \not{d}_2^{\star\,\mathbf{S}}\not{d}_1^{\star\,\mathbf{S}}\mu^{\mathbf{S}} = 0 \tag{9.7.11}$$

and such that, for a small enough constant $\delta_1 > 0$, the transition functions $(f, \underline{f}, \lambda)$ from the background frame of $\mathcal{R}$ to that of $\mathbf{S}$ verifies, for some $4 \leq s \leq s_{max}$, the bound

$$\|f\|_{\mathfrak{h}_s(\mathbf{S})} + (r^{\mathbf{S}})^{-1}\|(\underline{f}, \log\lambda)\|_{\mathfrak{h}_s(\mathbf{S})} \leq \delta_1. \tag{9.7.12}$$

Assume in addition that we have

$$\left|\int_{\mathbf{S}}\beta^{\mathbf{S}}e^{\Phi}\right|+\left|\int_{\mathbf{S}}e_\theta^{\mathbf{S}}(\underline{\kappa}^{\mathbf{S}})e^{\Phi}\right| \lesssim \overset{\circ}{\delta}. \tag{9.7.13}$$

Then the transition functions $(f,\underline{f},\lambda)$ from the background frame of $\mathcal{R}$ to that of $\mathbf{S}$ *verify the estimates*

$$\|(f,\underline{f},\check{\lambda}^{\mathbf{S}})\|_{\mathfrak{h}_{s+1}(\mathbf{S})} \lesssim r\Big(\overset{\circ}{\delta}+\delta_1(\overset{\circ}{\epsilon}+\delta_1)\Big)$$

and

$$r|\overline{\lambda}^{\mathbf{S}}-1| \lesssim r\Big(\overset{\circ}{\delta}+\delta_1(\overset{\circ}{\epsilon}+\delta_1)\Big)+\sup_{\mathbf{S}}\left|r-r^{\mathbf{S}}\right|.$$

Proof. Applying Corollary 9.38, we have

$$\begin{aligned}\|(f,\underline{f},\check{\lambda}^{\mathbf{S}})\|_{\mathfrak{h}_{s_{max}+1}(\mathbf{S})} &\lesssim \overset{\circ}{\delta}+r^{-2}\left(|\Lambda|+|\underline{\Lambda}|\right)+r\delta_1(\overset{\circ}{\epsilon}+\delta_1),\\ r|\overline{\lambda}^{\mathbf{S}}-1| &\lesssim \overset{\circ}{\delta}+r^{-2}\left(|\Lambda|+|\underline{\Lambda}|\right)+\sup_{\mathbf{S}}\left|r-r^{\mathbf{S}}\right|.\end{aligned}$$

Thus, to conclude, it suffices to prove the estimate

$$|\Lambda|+|\underline{\Lambda}| \lesssim r^3\Big(\overset{\circ}{\delta}+\delta_1(\overset{\circ}{\epsilon}+\delta_1)\Big).$$

Now, revisiting the proof of Lemma 9.50 without assuming that **A1-Strong** holds, and using in particular (9.7.12), we obtain the following analog of (9.7.4):

$$\begin{aligned}\Lambda &= \frac{r^3}{3m}\left(\int_{\overset{\circ}{S}}\beta e^{\Phi}-\int_{\mathbf{S}}\beta^{\mathbf{S}}e^{\Phi}\right)+O\left(r^3\delta_1(\overset{\circ}{\epsilon}+\delta_1)\right),\\ \underline{\Lambda} &= \frac{r^3}{6m}\left(\int_{\overset{\circ}{S}}(e_\theta\underline{\kappa})e^{\Phi}-\int_{\mathbf{S}}(e_\theta^{\mathbf{S}}\underline{\kappa}^{\mathbf{S}})e^{\Phi}-\Upsilon\int_{\overset{\circ}{S}}e_\theta(\kappa)e^{\Phi}\right)+\Upsilon\Lambda+O\left(r^3\delta_1(\overset{\circ}{\epsilon}+\delta_1)\right).\end{aligned}$$

The desired estimate for $(\Lambda,\underline{\Lambda})$ follows then immediately from (9.4.2) for κ, (9.7.10) and (9.7.13). □

9.8 CONSTRUCTION OF GCM HYPERSURFACES

We are ready to state our main result concerning the construction of GCM hypersurfaces.

Theorem 9.52. *Let a fixed spacetime region $\mathcal{R}$ verifying assumptions **A1**–**A3** and* (9.4.2). *In addition we assume that*

$$\left|\int_{S(u,s)}\eta e^{\Phi}\right|\lesssim r^2\overset{\circ}{\delta}, \qquad \left|\int_{S(u,s)}\underline{\xi}e^{\Phi}\right|\lesssim r^2\overset{\circ}{\delta}, \tag{9.8.1}$$

and, everywhere on $\mathcal{R}$,

$$\left| \left(\frac{2}{\varsigma} + \Omega \right) \Big|_{SP} - 1 - \frac{2m}{r} \right| \lesssim \overset{\circ}{\delta} \tag{9.8.2}$$

where SP denotes the south pole, i.e., $\theta = 0$ relative to the adapted geodesic coordinates u, s, θ.

Let $\mathbf{S}_0 = \mathbf{S}_0[\overset{\circ}{u}, \overset{\circ}{s}, \Lambda_0, \underline{\Lambda}_0]$ be a fixed GCMS provided by Theorem 9.32. Then, there exists a unique, local,[22] *smooth, $\mathbf{Z}$-invariant spacelike hypersurface Σ_0 passing through $\mathbf{S}_0$, a scalar function $u^{\mathbf{S}}$ defined on Σ_0, whose level surfaces are topological spheres denoted by $\mathbf{S}$, and a smooth collection of constants $\Lambda^{\mathbf{S}}, \underline{\Lambda}^{\mathbf{S}}$ verifying*

$$\Lambda^{\mathbf{S}_0} = \Lambda_0, \qquad \underline{\Lambda}^{\mathbf{S}_0} = \underline{\Lambda}_0,$$

such that the following conditions are verified:

1. *The surfaces $\mathbf{S}$ of constant $u^{\mathbf{S}}$ verify all the properties stated in Theorem 9.32 for the prescribed constants $\Lambda^{\mathbf{S}}, \underline{\Lambda}^{\mathbf{S}}$. In particular they come endowed with null frames $(e_4^{\mathbf{S}}, e_\theta^{\mathbf{S}}, e_3^{\mathbf{S}})$ such that*
   a) *For each $\mathbf{S}$ the GCM conditions (9.4.1) with $\Lambda = \Lambda^{\mathbf{S}}, \underline{\Lambda} = \underline{\Lambda}^{\mathbf{S}}$, are verified.*
   b) *The transition functions $(f, \underline{f}, a = \log \lambda)$ verify the estimates (9.4.5).*
   c) *The transversality conditions (9.4.9) are verified.*
   d) *The corresponding Ricci and curvature coefficients verify the estimates (9.4.8) and (9.4.11).*
2. *Denoting $r^{\mathbf{S}}$ to be the area radius of the spheres $\mathbf{S}$ we have, for some constant c_*,*

$$u^{\mathbf{S}} + r^{\mathbf{S}} = c_*, \quad \textit{along} \quad \Sigma_0. \tag{9.8.3}$$

3. *Let $\nu^{\mathbf{S}}$ be the unique vectorfield tangent to the hypersurface Σ_0, normal to $\mathbf{S}$, and normalized by $g(\nu^{\mathbf{S}}, e_4^{\mathbf{S}}) = -2$. There exists a unique scalar function $a^{\mathbf{S}}$ on Σ_0 such that $\nu^{\mathbf{S}}$ is given by*

$$\nu^{\mathbf{S}} = e_3^{\mathbf{S}} + a^{\mathbf{S}} e_4^{\mathbf{S}}.$$

   The following normalization condition holds true at the south pole SP of every sphere $\mathbf{S}$, i.e., at $\theta = 0$,

$$a^{\mathbf{S}}\Big|_{SP} = -1 - \frac{2m^{\mathbf{S}}}{r^{\mathbf{S}}}. \tag{9.8.4}$$

4. *We extend $u^{\mathbf{S}}$ and $r^{\mathbf{S}}$ in a small neighborhood of Σ_0 such that the following transversality conditions are verified*[23] *on Σ_0,*

$$e_4^{\mathbf{S}}(u^{\mathbf{S}}) = 0, \qquad e_4^{\mathbf{S}}(r^{\mathbf{S}}) = \frac{r^{\mathbf{S}}}{2} \overline{\kappa^{\mathbf{S}}} = 1. \tag{9.8.5}$$

[22] That is, in a neighborhood of $\mathbf{S}_0$.

[23] Here the average of $\kappa^{\mathbf{S}}$ is taken on $\mathbf{S}$. In view of the GCM conditions (9.8.16) we deduce $e_4^{\mathbf{S}}(r^{\mathbf{S}}) = 1$.

5. *In view of* (9.8.5) *the Ricci coefficients* $\eta^{\mathbf{S}}, \underline{\xi}^{\mathbf{S}}$ *are well defined for every* $\mathbf{S} \subset \Sigma_0$ *and verify*

$$\int_{\mathbf{S}} \eta^{\mathbf{S}} e^{\Phi} = \int_{\mathbf{S}} \underline{\xi}^{\mathbf{S}} e^{\Phi} = 0. \tag{9.8.6}$$

6. *The following estimates hold true for all* $k \leq s_{max}$,

$$\|\eta^{\mathbf{S}}\|_{\mathfrak{h}_k(\mathbf{S})} \lesssim \overset{\circ}{\epsilon}, \tag{9.8.7}$$

$$\|\underline{\xi}^{\mathbf{S}}\|_{\mathfrak{h}_k(\mathbf{S})} \lesssim \overset{\circ}{\epsilon}, \tag{9.8.8}$$

$$\left\|a^{\mathbf{S}} + 1 + \frac{2m^{\mathbf{S}}}{r^{\mathbf{S}}}\right\|_{\mathfrak{h}_k(\mathbf{S})} \lesssim \overset{\circ}{\epsilon}. \tag{9.8.9}$$

The $e_3^{\mathbf{S}}$ *derivatives of* $\check{\kappa}^{\mathbf{S}}, \vartheta^{\mathbf{S}}, \zeta^{\mathbf{S}}, \underline{\check{\kappa}}^{\mathbf{S}}, \underline{\vartheta}^{\mathbf{S}}, \alpha^{\mathbf{S}}, \beta^{\mathbf{S}}, \check{\rho}^{\mathbf{S}}, \mu^{\mathbf{S}}, \underline{\beta}^{\mathbf{S}}$ *are well defined on* Σ_0 *and we have, for all* $k \leq s_{max} - 1$,

$$\begin{aligned}
\|e_3^{\mathbf{S}}(\check{\kappa}^{\mathbf{S}}, \vartheta^{\mathbf{S}}, \zeta^{\mathbf{S}}, \underline{\check{\kappa}}^{\mathbf{S}})\|_{\mathfrak{h}_k(\mathbf{S})} &\lesssim \overset{\circ}{\epsilon} r^{-1},\\
\|e_3^{\mathbf{S}}(\underline{\vartheta}^{\mathbf{S}})\|_{\mathfrak{h}_k(\mathbf{S})} &\lesssim \overset{\circ}{\epsilon},\\
\|e_3^{\mathbf{S}}\left(\alpha^{\mathbf{S}}, \beta^{\mathbf{S}}, \check{\rho}^{\mathbf{S}}, \mu^{\mathbf{S}}\right)\|_{\mathfrak{h}_k(\mathbf{S})} &\lesssim \overset{\circ}{\epsilon} r^{-2},\\
\|e_3^{\mathbf{S}}(\underline{\beta}^{\mathbf{S}})\|_{\mathfrak{h}_k(\mathbf{S})} &\lesssim \overset{\circ}{\epsilon} r^{-1},\\
\|e_3^{\mathbf{S}}(\underline{\alpha}^{\mathbf{S}})\|_{\mathfrak{h}_k(\mathbf{S})} &\lesssim \overset{\circ}{\epsilon}.
\end{aligned} \tag{9.8.10}$$

7. *The transition functions from the background foliation to that of* Σ_0 *verify*

$$\|\mathfrak{d}^{\leq s_{max}+1}(f, \underline{f}, \log\lambda)\|_{L^2(\mathbf{S})} \lesssim \overset{\circ}{\delta}. \tag{9.8.11}$$

Corollary 9.53. *Let a fixed spacetime region* $\mathcal{R}$ *verifying assumptions* ***A1–A3*** *and the small GCM conditions* (9.4.2). *Assume given a GCM hypersurface* $\Sigma_0 \subset \mathcal{R}$ *foliated by surfaces* $\mathbf{S}$ *such that*

$$\kappa^{\mathbf{S}} = \frac{2}{r^{\mathbf{S}}}, \quad {\not{d}}_2^{\star\,\mathbf{S}} {\not{d}}_1^{\star\,\mathbf{S}} \underline{\kappa}^{\mathbf{S}} = {\not{d}}_2^{\star\,\mathbf{S}} {\not{d}}_1^{\star\,\mathbf{S}} \mu^{\mathbf{S}} = 0, \quad \int_{\mathbf{S}} \eta^{\mathbf{S}} e^{\Phi} = 0, \quad \int_{\mathbf{S}} \underline{\xi}^{\mathbf{S}} e^{\Phi} = 0.$$

1. *If we assume in addition that for a specific sphere* $\mathbf{S}_0$ *on* Σ_0, *the transition functions* $f, \underline{f}$ *from the background foliation to* $\mathbf{S}_0$ *verify*

$$\|(f, \underline{f}, \log(\lambda))\|_{\mathfrak{h}_{s_{max}+1}(\mathbf{S}_0)} \lesssim \overset{\circ}{\delta}, \tag{9.8.12}$$

then, the following holds true:

$$\|\mathfrak{d}^{\leq s_{max}+1}(f, \underline{f}, \log(\lambda))\|_{L^2(\mathbf{S}_0)} \lesssim \overset{\circ}{\delta}. \tag{9.8.13}$$

2. *If we assume in addition that for a specific sphere* $\mathbf{S}_0$ *on* Σ_0, *the transition functions* $f, \underline{f}$ *from the background foliation to* $\mathbf{S}_0$ *verify*

$$\|f\|_{\mathfrak{h}_{s_{max}+1}(\mathbf{S}_0)} + (r^{\mathbf{S}_0})^{-1}\|(\underline{f}, \log\lambda)\|_{\mathfrak{h}_{s_{max}+1}(\mathbf{S}_0)} \lesssim \overset{\circ}{\delta}, \tag{9.8.14}$$

then, the following holds true:

$$\|\mathfrak{d}^{\leq s_{max}+1}f\|_{L^2(\mathbf{S}_0)}+r^{-1}\|\mathfrak{d}^{\leq s_{max}+1}(\underline{f},\log\lambda)\|_{L^2(\mathbf{S}_0)}$$
$$+\|\mathfrak{d}^{\leq s_{max}}e_3^{\mathbf{S}}(\underline{f},\log\lambda)\|_{L^2(\mathbf{S}_0)} \lesssim \overset{\circ}{\delta}. \quad (9.8.15)$$

We give below the proof of Theorem 9.52 and of Corollary 9.53.

9.8.1 Definition of Σ_0

As stated in the theorem, we assume given a region $\mathcal{R}=\{|u-\overset{\circ}{u}|\leq\delta_{\mathcal{R}},\ |s-\overset{\circ}{s}|\leq\delta_{\mathcal{R}}\}$ (see definition(9.1.6)) endowed with a background foliation such that the condition **A1**–**A3** hold true. We also assume given a deformation sphere

$$\mathbf{S}_0 \quad := \quad \mathbf{S}[\overset{\circ}{u},\overset{\circ}{s},\Lambda_0,\underline{\Lambda}_0]$$

of a given sphere $\overset{\circ}{S}=S(\overset{\circ}{u},\overset{\circ}{s})$ of the background foliation which verify the conclusions of Theorem 9.32. We then proceed to construct, in a small neighborhood of $\mathbf{S}_0$, a spacelike hypersurface Σ_0 initiating at $\mathbf{S}_0$ verifying all the desired properties mentioned above. In what follows we outline the main steps in the construction.

Step 1. According to Theorem 9.32, for every value of the parameters (u,s) in $\mathcal{R}$ (i.e., such that the background spheres $S(u,s)\subset\mathcal{R}$) and every real numbers $(\Lambda,\underline{\Lambda})$, there exists a unique GCM sphere $\mathbf{S}[u,s,\Lambda,\underline{\Lambda}]$, as a $\mathbf{Z}$-polarized deformation of $S(u,s)$. In particular the following are verified:

- $\mathbf{S}$ coincides with $S(u,s)$ at their south poles (i.e., for $\theta=0$ in the adapted coordinates).
- On $\mathbf{S}$, the following GCMS conditions hold:

$$\kappa^{\mathbf{S}}=\frac{2}{r^{\mathbf{S}}},\qquad \not{d}_2^{\mathbf{S},\star}\not{d}_1^{\mathbf{S},\star}\underline{\kappa}^{\mathbf{S}}=0,\qquad \not{d}_2^{\mathbf{S},\star}\not{d}_1^{\mathbf{S},\star}\mu^{\mathbf{S}}=0,\tag{9.8.16}$$

$$\int_{\mathbf{S}}fe^{\Phi}=\Lambda^{\mathbf{S}},\qquad \int_{\mathbf{S}}\underline{f}e^{\Phi}=\underline{\Lambda}^{\mathbf{S}},\tag{9.8.17}$$

where $(f,\underline{f},\lambda)$ are the transition parameters of the frame transformation from the background frame (e_3,e_θ,e_4) to the adapted frame $(e_3^{\mathbf{S}},e_\theta^{\mathbf{S}},e_4^{\mathbf{S}})$. The constants $\Lambda^{\mathbf{S}},\underline{\Lambda}^{\mathbf{S}}$ depend smoothly on the surfaces $\mathbf{S}$ and

$$\Lambda^{\mathbf{S}_0}=\Lambda_0,\qquad \underline{\Lambda}^{\mathbf{S}_0}=\underline{\Lambda}_0.$$

- There is a map $\Xi: S(u,s)\to\mathbf{S}$ given by

$$\Xi:(u,s,\theta)=\Big(u+U(\theta,u,\ s,\ \Lambda,\ \underline{\Lambda}),\ s+S(\theta,u,s,\Lambda,\underline{\Lambda}),\ \theta\Big)\tag{9.8.18}$$

  with U,S vanishing at $\theta=0$.
- The transversality conditions (9.4.9) hold, i.e., $\xi^{\mathbf{S}}=\omega^{\mathbf{S}}=\zeta^{\mathbf{S}}+\underline{\eta}^{\mathbf{S}}=0$. Note that these specify the $e_4^{\mathbf{S}}$ derivatives of $(f,\underline{f},\lambda)$ on $\mathbf{S}$.

- The Ricci coefficients[24] $\kappa^{\mathbf{S}}, \underline{\kappa}^{\mathbf{S}}, \vartheta^{\mathbf{S}}, \underline{\vartheta}^{\mathbf{S}}, \zeta^{\mathbf{S}}$ are well defined on each sphere $\mathbf{S}$ of Σ_0, and hence on Σ_0. The same holds true for all curvature coefficients $\alpha^{\mathbf{S}}, \beta^{\mathbf{S}}, \rho^{\mathbf{S}}$, $\underline{\beta}^{\mathbf{S}}, \underline{\alpha}^{\mathbf{S}}$. Taking into account our transversality condition we remark that the only ill defined Ricci coefficients are $\eta^{\mathbf{S}}, \underline{\xi}^{\mathbf{S}}, \underline{\omega}^{\mathbf{S}}$.
- Let $\nu^{\mathbf{S}}$ be the unique vectorfield tangent to the hypersurface Σ_0, normal to $\mathbf{S}$, and normalized by $g(\nu^{\mathbf{S}}, e_4^{\mathbf{S}}) = -2$. There exists a unique scalar function $a^{\mathbf{S}}$ on Σ_0 such that $\nu^{\mathbf{S}}$ is given by

$$\nu^{\mathbf{S}} = e_3^{\mathbf{S}} + a^{\mathbf{S}} e_4^{\mathbf{S}}.$$

We deduce that the quantities

$$\begin{aligned} g(D_{\nu^{\mathbf{S}}} e_4^{\mathbf{S}}, e_\theta^{\mathbf{S}}) &= 2\eta^{\mathbf{S}} + 2a^{\mathbf{S}}\xi^{\mathbf{S}} = 2\eta^{\mathbf{S}}, \\ g(D_{\nu^{\mathbf{S}}} e_3^{\mathbf{S}}, e_\theta^{\mathbf{S}}) &= 2\underline{\xi}^{\mathbf{S}} + 2a^{\mathbf{S}}\underline{\eta}^{\mathbf{S}} = 2(\underline{\xi}^{\mathbf{S}} - a^{\mathbf{S}}\zeta^{\mathbf{S}}), \\ g(D_{\nu^{\mathbf{S}}} e_3^{\mathbf{S}}, e_4^{\mathbf{S}}) &= 4\underline{\omega}^{\mathbf{S}} - 4a^{\mathbf{S}}\omega^{\mathbf{S}} = 4\underline{\omega}^{\mathbf{S}}, \end{aligned}$$

are well defined on Σ_0. Thus the scalar $a^{\mathbf{S}}$ allows us to specify the remaining Ricci coefficients, $\eta^{\mathbf{S}}, \underline{\xi}^{\mathbf{S}}, \underline{\omega}^{\mathbf{S}}$ along Σ_0, which we do below.

9.8.2 Extrinsic properties of Σ_0

We analyze the extrinsic properties of the hypersurfaces Σ_0 defined in Step 1.

Step 2. We define the scalar function $u^{\mathbf{S}}$ on Σ_0 as

$$u^{\mathbf{S}} := c_0 - r^{\mathbf{S}}, \tag{9.8.19}$$

where $r^{\mathbf{S}}$ is the area radius of $\mathbf{S}$ and the constant c_0 is such that $u^{\mathbf{S}}|_{\mathbf{S}_0} = \overset{\circ}{u}$, i.e., $c_0 = \overset{\circ}{u} + r^{\mathbf{S}}|_{\mathbf{S}_0}$.

Step 3. We extend $u^{\mathbf{S}}$ and $r^{\mathbf{S}}$ in a small neighborhood of Σ_0 such that the following transversality conditions are verified.

$$e_4^{\mathbf{S}}(u^{\mathbf{S}}) = 0, \qquad e_4^{\mathbf{S}}(r^{\mathbf{S}}) = \frac{r^{\mathbf{S}}}{2}\overline{\kappa^{\mathbf{S}}}, \tag{9.8.20}$$

where the average of $\kappa^{\mathbf{S}}$ is taken on $\mathbf{S}$. In view of the GCM conditions (9.8.16) we deduce $e_4^{\mathbf{S}}(r^{\mathbf{S}}) = 1$.

Step 4. Note that $e_3^{\mathbf{S}}(u^{\mathbf{S}}, r^{\mathbf{S}})$ remain undetermined. On the other hand, since

[24]Consequently the Hawking mass $m^{\mathbf{S}}$ is also well defined.

$e_\theta^{\mathbf{S}}(u^{\mathbf{S}}) = e_\theta^{\mathbf{S}}(r^{\mathbf{S}}) = 0$, we deduce in view of (9.8.20)

$$\begin{aligned} e_\theta^{\mathbf{S}}(e_3^{\mathbf{S}}(u^{\mathbf{S}})) &= [e_\theta^{\mathbf{S}}, e_3^{\mathbf{S}}]u^{\mathbf{S}} = \left[\frac{1}{2}(\underline{\kappa}^{\mathbf{S}} + \underline{\vartheta}^{\mathbf{S}})e_\theta^{\mathbf{S}} + (\zeta^{\mathbf{S}} - \eta^{\mathbf{S}})e_3^{\mathbf{S}} + \underline{\xi}^{\mathbf{S}} e_4^{\mathbf{S}}\right] u^{\mathbf{S}} \\ &= (\zeta^{\mathbf{S}} - \eta^{\mathbf{S}})e_3^{\mathbf{S}}(u^{\mathbf{S}}), \\ e_\theta^{\mathbf{S}}(e_3^{\mathbf{S}}(r^{\mathbf{S}})) &= [e_\theta^{\mathbf{S}}, e_3^{\mathbf{S}}]r^{\mathbf{S}} = \left[\frac{1}{2}(\underline{\kappa}^{\mathbf{S}} + \underline{\vartheta}^{\mathbf{S}})e_\theta^{\mathbf{S}} + (\zeta^{\mathbf{S}} - \eta^{\mathbf{S}})e_3^{\mathbf{S}} + \underline{\xi}^{\mathbf{S}} e_4^{\mathbf{S}}\right] r^{\mathbf{S}} \\ &= (\zeta^{\mathbf{S}} - \eta^{\mathbf{S}})e_3^{\mathbf{S}}(r^{\mathbf{S}}) + \underline{\xi}^{\mathbf{S}}. \end{aligned}$$

Thus introducing the scalars

$$\varsigma^{\mathbf{S}} := \frac{2}{e_3^{\mathbf{S}}(u^{\mathbf{S}})}, \tag{9.8.21}$$

and

$$\underline{A}^{\mathbf{S}} := \frac{2}{r^{\mathbf{S}}}(e_3^{\mathbf{S}}(r^{\mathbf{S}}) + \Upsilon^{\mathbf{S}}), \tag{9.8.22}$$

we deduce

$$e_\theta^{\mathbf{S}}(\log \varsigma^{\mathbf{S}}) = (\eta^{\mathbf{S}} - \zeta^{\mathbf{S}}), \tag{9.8.23}$$

$$e_\theta^{\mathbf{S}}(\underline{A}^{\mathbf{S}}) = -\frac{2\Upsilon^{\mathbf{S}}}{r^{\mathbf{S}}}(\zeta^{\mathbf{S}} - \eta^{\mathbf{S}}) - \frac{2}{r^{\mathbf{S}}}\underline{\xi}^{\mathbf{S}} + (\zeta^{\mathbf{S}} - \eta^{\mathbf{S}})\underline{A}^{\mathbf{S}}. \tag{9.8.24}$$

We infer that $e_\theta^{\mathbf{S}}(\log \varsigma^{\mathbf{S}})$ and $e_\theta^{\mathbf{S}}(\underline{A}^{\mathbf{S}})$ are determined in terms of $\eta, \underline{\xi}$.

Step 5. In view of the definition of $\nu^{\mathbf{S}}$ and $\varsigma^{\mathbf{S}}$ we make use of (9.8.20) to deduce

$$\nu^{\mathbf{S}}(u^{\mathbf{S}}) = e_3^{\mathbf{S}}(u^{\mathbf{S}}) + a^{\mathbf{S}} e_4^{\mathbf{S}}(u^{\mathbf{S}}) = \frac{2}{\varsigma^{\mathbf{S}}}.$$

On the other hand, since $u^{\mathbf{S}} := c_0 - r^{\mathbf{S}}$ along Σ_0,

$$\nu^{\mathbf{S}}(u^{\mathbf{S}}) = -\nu^{\mathbf{S}}(r^{\mathbf{S}}) = -e_3^{\mathbf{S}}(r^{\mathbf{S}}) - a^{\mathbf{S}} e_4^{\mathbf{S}}(r^{\mathbf{S}}) = \Upsilon^{\mathbf{S}} - \frac{r^{\mathbf{S}}}{2}\underline{A}^{\mathbf{S}} - a^{\mathbf{S}}$$

and therefore,

$$a^{\mathbf{S}} = -\frac{2}{\varsigma^{\mathbf{S}}} + \Upsilon^{\mathbf{S}} - \frac{r^{\mathbf{S}}}{2}\underline{A}^{\mathbf{S}} = -\frac{2}{\varsigma^{\mathbf{S}}} - \underline{\Omega}^{\mathbf{S}} \tag{9.8.25}$$

where

$$\underline{\Omega}^{\mathbf{S}} := e_3^{\mathbf{S}}(r^{\mathbf{S}}) = -\Upsilon^{\mathbf{S}} - \frac{r^{\mathbf{S}}}{2}\underline{A}^{\mathbf{S}}. \tag{9.8.26}$$

Step 6. The following lemma will be used, in particular,[25] to determine the $\overline{\underline{A}^{\mathbf{S}}}$.

[25]It will also be used below to derive equations for $\Lambda, \underline{\Lambda}$.

Lemma 9.54. *For every scalar function h we have the formula*

$$\nu^{\mathbf{S}}\left(\int_{\mathbf{S}} h\right) = (\varsigma^{\mathbf{S}})^{-1}\int_{\mathbf{S}} \varsigma^{\mathbf{S}}\left(\nu^{\mathbf{S}}(h) + (\underline{\kappa}^{\mathbf{S}} + a^{\mathbf{S}}\kappa^{\mathbf{S}})h\right). \tag{9.8.27}$$

In particular

$$\nu^{\mathbf{S}}(r^{\mathbf{S}}) = \frac{r^{\mathbf{S}}}{2}(\varsigma^{\mathbf{S}})^{-1}\overline{\varsigma^{\mathbf{S}}(\underline{\kappa}^{\mathbf{S}} + a^{\mathbf{S}}\kappa^{\mathbf{S}})}$$

where the average is with respect to $\mathbf{S}$.

Proof. We consider the coordinates $u^{\mathbf{S}}$, $\theta^{\mathbf{S}}$ along Σ_0 with $\nu^{\mathbf{S}}(\theta^{\mathbf{S}}) = 0$. In these coordinates we have

$$\nu^{\mathbf{S}} = \frac{2}{\varsigma^{\mathbf{S}}}\partial_{u^{\mathbf{S}}}.$$

The lemma follows easily by expressing the volume element of the surfaces $\mathbf{S} \subset \Sigma_0$ with respect to the coordinates $u^{\mathbf{S}}, \theta^{\mathbf{S}}$ (see also the proof of Proposition 2.64). □

Step 7. Note that the GCM condition $\kappa^{\mathbf{S}} = \frac{2}{r^{\mathbf{S}}}$ together with the definition of the Hawking mass implies that

$$\overline{\underline{\kappa}^{\mathbf{S}}} = -\frac{2\Upsilon^{\mathbf{S}}}{r^{\mathbf{S}}}, \qquad \Upsilon^{\mathbf{S}} = 1 - \frac{2m^{\mathbf{S}}}{r^{\mathbf{S}}},$$

where the average is taken with respect to $\mathbf{S}$. Thus in view of Lemma 9.54 we deduce

$$\begin{aligned}
e_3^{\mathbf{S}}(r^{\mathbf{S}}) + a^{\mathbf{S}} &= \nu^{\mathbf{S}}(r^{\mathbf{S}}) = \frac{r^{\mathbf{S}}}{2}(\varsigma^{\mathbf{S}})^{-1}\overline{\varsigma^{\mathbf{S}}(\underline{\kappa}^{\mathbf{S}} + a^{\mathbf{S}}\kappa^{\mathbf{S}})}\\
&= \frac{r^{\mathbf{S}}}{2}(\varsigma^{\mathbf{S}})^{-1}\left(\overline{\varsigma^{\mathbf{S}}}\,\overline{\underline{\kappa}^{\mathbf{S}}} + \overline{\check{\varsigma}^{\mathbf{S}}\check{\underline{\kappa}}^{\mathbf{S}}}\right) + (\varsigma^{\mathbf{S}})^{-1}\overline{\varsigma^{\mathbf{S}}a^{\mathbf{S}}}\\
&= -\Upsilon^{\mathbf{S}}(\varsigma^{\mathbf{S}})^{-1}\overline{\varsigma^{\mathbf{S}}} + \frac{r^{\mathbf{S}}}{2}(\varsigma^{\mathbf{S}})^{-1}\overline{\check{\varsigma}^{\mathbf{S}}\check{\underline{\kappa}}^{\mathbf{S}}} + (\varsigma^{\mathbf{S}})^{-1}\overline{\varsigma^{\mathbf{S}}a^{\mathbf{S}}}.
\end{aligned}$$

Since according to (9.8.22) $e_3^{\mathbf{S}}(r^{\mathbf{S}}) = -\Upsilon^{S} + \frac{r^{\mathbf{S}}}{2}\underline{A}^{\mathbf{S}}$, we deduce

$$\underline{A}^{\mathbf{S}} = \frac{2}{r^{\mathbf{S}}}\left(\Upsilon^{S} - a^{\mathbf{S}} - \Upsilon^{\mathbf{S}}(\varsigma^{\mathbf{S}})^{-1}\overline{\varsigma^{\mathbf{S}}} + \frac{r^{\mathbf{S}}}{2}(\varsigma^{\mathbf{S}})^{-1}\overline{\check{\varsigma}^{\mathbf{S}}\check{\underline{\kappa}}^{\mathbf{S}}} + (\varsigma^{\mathbf{S}})^{-1}\overline{\varsigma^{\mathbf{S}}a^{\mathbf{S}}}\right).$$

In particular, multiplying by $\varsigma^{\mathbf{S}}$ and taking the average, we infer

$$\overline{\varsigma^{\mathbf{S}}\underline{A}^{\mathbf{S}}} = \overline{\check{\varsigma}^{\mathbf{S}}\check{\underline{\kappa}}^{\mathbf{S}}},$$

and hence

$$\overline{\underline{A}^{\mathbf{S}}} = \frac{1}{\overline{\varsigma^{\mathbf{S}}}}\left(\overline{\check{\varsigma}^{\mathbf{S}}\check{\underline{\kappa}}^{\mathbf{S}}} - \overline{\check{\varsigma}^{\mathbf{S}}\check{\underline{A}}^{\mathbf{S}}}\right). \tag{9.8.28}$$

Step 8. We summarize the results in Steps 1–7 in the following.

Proposition 9.55. *Let Σ_0 be a smooth spacelike hypersurface foliated by framed*[26] *spheres $(\mathbf{S}, e_4^{\mathbf{S}}, e_\theta^{\mathbf{S}}, e_3^{\mathbf{S}})$ whose Ricci coefficients verify the GCM condition $\kappa^{\mathbf{S}} = \frac{2}{r^{\mathbf{S}}}$ and transversality condition (9.4.9). Define $u^{\mathbf{S}}$ as in (9.8.19) such that $u^{\mathbf{S}} + r^{\mathbf{S}}$ is constant on Σ_0 with $r^{\mathbf{S}}$ the area radius of the spheres $\mathbf{S}$. Extend $u^{\mathbf{S}}$ and $r^{\mathbf{S}}$ in a neighborhood of Σ_0 such that the transversality conditions (9.8.20) are verified. Then, defining the scalars $\varsigma^{\mathbf{S}}, \underline{A}^{\mathbf{S}}$ as in (9.8.21), (9.8.22) we establish the following relations between $\eta^{\mathbf{S}}, \underline{\xi}^{\mathbf{S}}$ and $\varsigma^{\mathbf{S}}, \underline{A}^{\mathbf{S}}$ and $a^{\mathbf{S}}$, where the latter scalar is defined in Step 1,*

$$\begin{aligned} e_\theta^{\mathbf{S}}(\log \varsigma^{\mathbf{S}}) &= (\eta^{\mathbf{S}} - \zeta^{\mathbf{S}}), \\ e_\theta^{\mathbf{S}}(\underline{A}^{\mathbf{S}}) &= -\frac{2\Upsilon^{\mathbf{S}}}{r^{\mathbf{S}}}(\zeta^{\mathbf{S}} - \eta^{\mathbf{S}}) - \frac{2}{r^{\mathbf{S}}}\underline{\xi}^{\mathbf{S}} + (\zeta^{\mathbf{S}} - \eta^{\mathbf{S}})\underline{A}^{\mathbf{S}}, \\ \overline{\underline{A}^{\mathbf{S}}} &= \frac{1}{\overline{\varsigma^{\mathbf{S}}}}\left(\overline{\check{\varsigma}^{\mathbf{S}}\check{\underline{\kappa}}^{\mathbf{S}}} - \overline{\check{\varsigma}^{\mathbf{S}}\check{\underline{A}}^{\mathbf{S}}}\right), \\ a^{\mathbf{S}} &= -\frac{2}{\varsigma^{\mathbf{S}}} + \Upsilon^{\mathbf{S}} - \frac{r^{\mathbf{S}}}{2}\underline{A}^{\mathbf{S}}. \end{aligned} \tag{9.8.29}$$

Remark 9.56. *Note that we lack equations for $\eta^{\mathbf{S}}, \underline{\xi}^{\mathbf{S}}$ and the average of $a^{\mathbf{S}}$. The latter can be fixed by fixing the value of $a^{\mathbf{S}}\big|_{SP}$ and observing that*

$$\overline{a}^{\mathbf{S}} = a^{\mathbf{S}}\big|_{SP} - \check{a}^{\mathbf{S}}\big|_{SP}. \tag{9.8.30}$$

In what follows we state a result which ties $\eta^{\mathbf{S}}, \underline{\xi}^{\mathbf{S}}$ to the other GCM conditions in (9.8.16)–(9.8.17).

Step 9. To state the proposition below we split the Ricci coefficients into the following groups.

$$\begin{aligned} \Gamma_g^{\mathbf{S}} &= \left\{\check{\kappa}^{\mathbf{S}}, \vartheta^{\mathbf{S}}, \zeta^{\mathbf{S}}, \check{\underline{\kappa}}^{\mathbf{S}}, r\check{\rho}^{\mathbf{S}}, \overline{\kappa^{\mathbf{S}}} - \frac{2}{r^{\mathbf{S}}}, \overline{\rho^{\mathbf{S}}} + \frac{2m^{\mathbf{S}}}{(r^{\mathbf{S}})^3}\right\}, \\ \Gamma_b^{\mathbf{S}} &= \left\{\eta^{\mathbf{S}}, \underline{\xi}^{\mathbf{S}}, \check{\underline{\omega}}^{\mathbf{S}}, \underline{\omega}^{\mathbf{S}} - \frac{m^{\mathbf{S}}}{(r^{\mathbf{S}})^2}, \underline{\beta}^{\mathbf{S}}, \underline{\alpha}^{\mathbf{S}}\right\}. \end{aligned}$$

Proposition 9.57. *The following statements hold true:*[27]

1. Under the same assumptions as in Proposition 9.55, the Ricci coefficients $\eta^{\mathbf{S}}$,

---

[26] That is, differentiable spheres $\mathbf{S}$ endowed with adapted null frames $(e_4^{\mathbf{S}}, e_\theta^{\mathbf{S}}, e_3^{\mathbf{S}})$.

[27] $r_{\mathbf{S}}$ here denotes $r^{\mathbf{S}}$ the area radius of $\mathbf{S}$.

$\underline{\xi}^{\mathbf{S}}$, $\underline{\omega}^{\mathbf{S}}$ *verify the following identities.*

$$
\begin{aligned}
2\,\not{d}_2^{\mathbf{S},\star}\,\not{d}_1^{\mathbf{S},\star}\,\not{d}_1^{\mathbf{S}}\,\not{d}_2^{\mathbf{S}}\,\not{d}_2^{\mathbf{S},\star}\eta^{\mathbf{S}} &= \kappa^{\mathbf{S}}\Big(C_1 + 2\,\not{d}_2^{\mathbf{S},\star}\,\not{d}_1^{\mathbf{S},\star}\,\not{d}_1^{\mathbf{S}}\underline{\beta}^{\mathbf{S}}\Big) - r_{\mathbf{S}}^{-3}\not{\mathfrak{d}}^3 C_2\\
&\quad + r_{\mathbf{S}}^{-5}(\not{\mathfrak{d}}^{\mathbf{S}})^{\leq 4}\Gamma_g^{\mathbf{S}} + r_{\mathbf{S}}^{-4}(\not{\mathfrak{d}}^{\mathbf{S}})^{\leq 4}(\Gamma_b^{\mathbf{S}}\cdot\Gamma_b^{\mathbf{S}}) + l.o.t.,\\
2\,\not{d}_2^{\mathbf{S},\star}\,\not{d}_1^{\mathbf{S},\star}\,\not{d}_1^{\mathbf{S}}\,\not{d}_2^{\mathbf{S}}\,\not{d}_2^{\mathbf{S},\star}\underline{\xi} &= C_3 - \underline{\kappa}^{\mathbf{S}}\Big(C_1 + 2\,\not{d}_2^{\mathbf{S},\star}\,\not{d}_1^{\mathbf{S},\star}\,\not{d}_1^{\mathbf{S}}\underline{\beta}^{\mathbf{S}}\Big) + r_{\mathbf{S}}^{-5}(\not{\mathfrak{d}}^{\mathbf{S}})^{\leq 4}\Gamma_g^{\mathbf{S}}\\
&\quad + r_{\mathbf{S}}^{-4}(\not{\mathfrak{d}}^{\mathbf{S}})^{\leq 4}(\Gamma_b^{\mathbf{S}}\cdot\Gamma_b^{\mathbf{S}}) + l.o.t.,\\
\not{d}_1^{\mathbf{S},\star}\underline{\omega}^{\mathbf{S}} &= \left(\frac{1}{4}\underline{\kappa}^{\mathbf{S}} + \underline{\omega}^{\mathbf{S}}\right)\eta^{\mathbf{S}} - (\kappa^{\mathbf{S}})^{-1}\,\not{d}_2^{\mathbf{S}}\,\not{d}_2^{\mathbf{S},\star}\eta^{\mathbf{S}}\\
&\quad + \frac{1}{4}\kappa^{\mathbf{S}}\underline{\xi}^{\mathbf{S}} - \frac{1}{2}(\kappa^{\mathbf{S}})^{-1}C_2 + r_{\mathbf{S}}^{-1}(\not{\mathfrak{d}}^{\mathbf{S}})^{\leq 1}\Gamma_g^{\mathbf{S}}\\
&\quad + \not{\mathfrak{d}}^{\mathbf{S}}(\Gamma_b^{\mathbf{S}}\cdot\Gamma_b^{\mathbf{S}}),
\end{aligned}
\tag{9.8.31}
$$

where

$$
\begin{aligned}
C_1 &= e_3^{\mathbf{S}}(\not{d}_2^{\mathbf{S},\star}\,\not{d}_1^{\mathbf{S},\star}\mu^{\mathbf{S}}),\\
C_2 &= e_3^{\mathbf{S}}(e_\theta^{\mathbf{S}}\kappa^{\mathbf{S}}),\\
C_3 &= e_3\Big((\not{d}_2^{\mathbf{S},\star}\,\not{d}_2^{\mathbf{S}} + 2K^{\mathbf{S}})\,\not{d}_2^{\mathbf{S},\star}\,\not{d}_1^{\mathbf{S},\star}\underline{\kappa}^{\mathbf{S}}\Big).
\end{aligned}
\tag{9.8.32}
$$

The quadratic terms denoted by l.o.t. are lower order both in terms of decay as well as in terms of number of derivatives. Also, they contain angular derivatives $\not{\mathfrak{d}}^{\mathbf{S}}$, *but neither* $e_3^{\mathbf{S}}$ *nor* $e_4^{\mathbf{S}}$ *derivatives. Moreover, we note that the error terms* $r_{\mathbf{S}}^{-5}(\not{\mathfrak{d}}^{\mathbf{S}})^{\leq 4}\Gamma_g^{\mathbf{S}}$ *and* $r_{\mathbf{S}}^{-4}(\not{\mathfrak{d}}^{\mathbf{S}})^{\leq 4}(\Gamma_b^{\mathbf{S}}\cdot\Gamma_b^{\mathbf{S}})$ *do in fact not contain more than three derivatives of* $\check{\underline{\omega}}^{\mathbf{S}}$.

2. *If in addition* (9.4.8) *of Theorem 9.32 hold true then, for* $k \leq s_{max} - 7$,

$$
\begin{aligned}
\|\not{d}_2^{\mathbf{S},\star}\eta^{\mathbf{S}}\|_{\mathfrak{h}_{4+k}(\mathbf{S})} &\lesssim r_{\mathbf{S}}^3\|C_1\|_{\mathfrak{h}_k(\mathbf{S})} + r\|C_2\|_{\mathfrak{h}_{3+k}(\mathbf{S})} + \mathring{\epsilon} r_{\mathbf{S}}^{-1}\\
&\quad + r_{\mathbf{S}}^{-1}\|\Gamma_g^{\mathbf{S}}\|_{\mathfrak{h}_{4+k}(\mathbf{S})} + \|\Gamma_b^{\mathbf{S}}\cdot\Gamma_b^{\mathbf{S}}\|_{\mathfrak{h}_{4+k}(\mathbf{S})} + l.o.t.,\\
\|\not{d}_2^{\mathbf{S},\star}\underline{\xi}^{\mathbf{S}}\|_{\mathfrak{h}_{4+k}(\mathbf{S})} &\lesssim r_{\mathbf{S}}^4\|C_3\|_{\mathfrak{h}_k(\mathbf{S})} + r_{\mathbf{S}}^3\|C_1\|_{\mathfrak{h}_{3+k}(\mathbf{S})} + \mathring{\epsilon} r_{\mathbf{S}}^{-1}\\
&\quad + r_{\mathbf{S}}^{-1}\|\Gamma_g^{\mathbf{S}}\|_{\mathfrak{h}_{4+k}(\mathbf{S})} + \|\Gamma_b^{\mathbf{S}}\cdot\Gamma_b^{\mathbf{S}}\|_{\mathfrak{h}_{4+k}(\mathbf{S})} + l.o.t.,\\
\|\not{d}_1^{\mathbf{S},\star}\underline{\omega}^{\mathbf{S}}\|_{\mathfrak{h}_{2+k}(\mathbf{S})} &\lesssim r_{\mathbf{S}}^{-1}\|\eta^{\mathbf{S}}\|_{\mathfrak{h}_{4+k}(\mathbf{S})} + r_{\mathbf{S}}^{-1}\|\underline{\xi}^{\mathbf{S}}\|_{\mathfrak{h}_{2+k}(\mathbf{S})} + r\|C_2\|_{\mathfrak{h}_{2+k}(S)}\\
&\quad + r_{\mathbf{S}}^{-1}\|\Gamma_g^{\mathbf{S}}\|_{\mathfrak{h}_{3+k}(\mathbf{S})} + \|\Gamma_b^{\mathbf{S}}\cdot\Gamma_b^{\mathbf{S}}\|_{\mathfrak{h}_{3+k}(\mathbf{S})} + l.o.t.
\end{aligned}
\tag{9.8.33}
$$

3. *If in addition the GCM conditions* (9.8.16) *hold true along* Σ_0 *and the estimates* (9.4.11) *are also verified then, for* $k \leq s_{max} - 7$,

$$
\begin{aligned}
\|C_1\|_{\mathfrak{h}_{k-2}(\mathbf{S})} &\lesssim \mathring{\epsilon} r^{-5}\left(|\overline{a^{\mathbf{S}}} + 1 + \frac{2m^{\mathbf{S}}}{r^{\mathbf{S}}}| + r^{-1}\|\check{a}^{\mathbf{S}}\|_{\mathfrak{h}_{k-2}(\mathbf{S})}\right),\\
\|C_2\|_{\mathfrak{h}_{k-1}(\mathbf{S})} &\lesssim \mathring{\epsilon} r^{-3}\left(|\overline{a^{\mathbf{S}}} + 1 + \frac{2m^{\mathbf{S}}}{r^{\mathbf{S}}}| + r^{-1}\|\check{a}^{\mathbf{S}}\|_{\mathfrak{h}_{k-1}(\mathbf{S})}\right),\\
\|C_3\|_{\mathfrak{h}_{k-4}(\mathbf{S})} &\lesssim \mathring{\epsilon} r^{-5}\left(|\overline{a^{\mathbf{S}}} + 1 + \frac{2m^{\mathbf{S}}}{r^{\mathbf{S}}}| + r^{-1}\|\check{a}^{\mathbf{S}}\|_{\mathfrak{h}_{k-4}(\mathbf{S})}\right),
\end{aligned}
\tag{9.8.34}
$$

where $a^{\mathbf{S}}$ *was defined in Step 1 and can be expressed in terms of* $\varsigma^{\mathbf{S}}$ *and* $\underline{A}^{\mathbf{S}}$ *by formula* (9.8.25).

Proof. The proof[28] of the first two identities in (9.8.31) was derived in Proposition 7.21 in connection to the proof[29] of Theorem M4, starting with the following:[30]

$$
\begin{aligned}
2\,\mathfrak{d}_1^{\mathbf{S},\star}\underline{\omega}^{\mathbf{S}} &= \left(\frac{1}{2}\underline{\kappa}^{\mathbf{S}} + 2\underline{\omega}^{\mathbf{S}}\right)\eta^{\mathbf{S}} + e_3^{\mathbf{S}}(\zeta^{\mathbf{S}}) - \underline{\beta}^{\mathbf{S}} + \frac{1}{2}\kappa\underline{\xi}^{\mathbf{S}} + r^{-1}\Gamma_g^{\mathbf{S}} + \Gamma_b^{\mathbf{S}}\cdot\Gamma_b^{\mathbf{S}},\\
2\,\mathfrak{d}_2^{\mathbf{S}}\,\mathfrak{d}_2^{\mathbf{S},\star}\eta^{\mathbf{S}} &= \kappa^{\mathbf{S}}\left(-e_3(\zeta^{\mathbf{S}}) + \underline{\beta}^{\mathbf{S}}\right) - e_3^{\mathbf{S}}(e_\theta^{\mathbf{S}}(\kappa^{\mathbf{S}})) + r_{\mathbf{S}}^{-2}(\mathfrak{d}^{\mathbf{S}})^{\leq 1}\Gamma_g^{\mathbf{S}}\\
&\quad + r_{\mathbf{S}}^{-1}\,\mathfrak{d}(\Gamma_b^{\mathbf{S}}\cdot\Gamma_b^{\mathbf{S}}),\\
2\,\mathfrak{d}_2^{\mathbf{S}}\,\mathfrak{d}_2^{\mathbf{S},\star}\underline{\xi}^{\mathbf{S}} &= \underline{\kappa}^{\mathbf{S}}\left(e_3(\zeta^{\mathbf{S}}) - \underline{\beta}^{\mathbf{S}}\right) - e_3^{\mathbf{S}}(e_\theta^{\mathbf{S}}(\underline{\kappa}^{\mathbf{S}})) + r_{\mathbf{S}}^{-2}(\mathfrak{d}^{\mathbf{S}})^{\leq 1}\Gamma_g^{\mathbf{S}}\\
&\quad + r_{\mathbf{S}}^{-1}\,\mathfrak{d}^{\mathbf{S}}(\Gamma_b^{\mathbf{S}}\cdot\Gamma_b^{\mathbf{S}}).
\end{aligned}
\tag{9.8.35}
$$

The last identity in (9.8.31) follows by combining the first two identities in (9.8.35).

To prove the estimates for $\eta^{\mathbf{S}}$ in the second part of the proposition we make use of the identity $\mathfrak{d}_1^{\mathbf{S},\star}\,\mathfrak{d}_1^{\mathbf{S}} = \mathfrak{d}_2^{\mathbf{S}}\,\mathfrak{d}_2^{\mathbf{S},\star} + 2K^{\mathbf{S}}$ to deduce

$$
\begin{aligned}
&\mathfrak{d}_2^{\mathbf{S},\star}(\mathfrak{d}_2^{\mathbf{S}}\,\mathfrak{d}_2^{\mathbf{S},\star} + 2K^{\mathbf{S}})\,\mathfrak{d}_2^{\mathbf{S}}\,\mathfrak{d}_2^{\mathbf{S},\star}\eta^{\mathbf{S}}\\
= \;& \frac{1}{2}\kappa^{\mathbf{S}}C_1 + \kappa^{\mathbf{S}}\,\mathfrak{d}_2^{\mathbf{S},\star}(\mathfrak{d}_2^{\mathbf{S}}\,\mathfrak{d}_2^{\mathbf{S},\star} + 2K^{\mathbf{S}})\underline{\beta}^{\mathbf{S}} - r_{\mathbf{S}}^{-3}(\mathfrak{d}^{\mathbf{S}})^3 C_2 + r_{\mathbf{S}}^{-5}(\mathfrak{d}^{\mathbf{S}})^{\leq 4}\Gamma_g^{\mathbf{S}}\\
&+ r_{\mathbf{S}}^{-4}(\mathfrak{d}^{\mathbf{S}})^{\leq 4}(\Gamma_b^{\mathbf{S}}\cdot\Gamma_b^{\mathbf{S}}) + \text{l.o.t.},
\end{aligned}
$$

i.e.,

$$
\begin{aligned}
&\mathfrak{d}_2^{\mathbf{S},\star}(\mathfrak{d}_2^{\mathbf{S}}\,\mathfrak{d}_2^{\mathbf{S},\star} + 2K^{\mathbf{S}})\Big(\mathfrak{d}_2^{\mathbf{S}}\,\mathfrak{d}_2^{\mathbf{S},\star}\eta^{\mathbf{S}} - \kappa^{\mathbf{S}}\underline{\beta}^{\mathbf{S}}\Big)\\
= \;& \frac{1}{2}\kappa^{\mathbf{S}}C_1 - \frac{1}{2}r_{\mathbf{S}}^{-3}(\mathfrak{d}^{\mathbf{S}})^3 C_2 + r_{\mathbf{S}}^{-5}(\mathfrak{d}^{\mathbf{S}})^{\leq 4}\Gamma_g^{\mathbf{S}} + r_{\mathbf{S}}^{-4}(\mathfrak{d}^{\mathbf{S}})^{\leq 4}(\Gamma_b^{\mathbf{S}}\cdot\Gamma_b^{\mathbf{S}}) + \text{l.o.t.}
\end{aligned}
$$

Similarly for $\underline{\xi}^{\mathbf{S}}$

$$
\begin{aligned}
&\mathfrak{d}_2^{\mathbf{S},\star}(\mathfrak{d}_2^{\mathbf{S}}\,\mathfrak{d}_2^{\mathbf{S},\star} + 2K^{\mathbf{S}})\Big(\mathfrak{d}_2^{\mathbf{S}}\,\mathfrak{d}_2^{\mathbf{S},\star}\underline{\xi}^{\mathbf{S}} + \underline{\kappa}^{\mathbf{S}}\underline{\beta}^{\mathbf{S}}\Big)\\
= \;& \frac{1}{2}C_3 - \frac{1}{2}\underline{\kappa}^{\mathbf{S}}C_1 + r_{\mathbf{S}}^{-5}(\mathfrak{d}^{\mathbf{S}})^{\leq 4}\Gamma_g^{\mathbf{S}} + r_{\mathbf{S}}^{-4}(\mathfrak{d}^{\mathbf{S}})^{\leq 4}(\Gamma_b^{\mathbf{S}}\cdot\Gamma_b^{\mathbf{S}}) + \text{l.o.t.}
\end{aligned}
$$

The desired estimates for $\eta^{\mathbf{S}}$ and $\underline{\xi}^{\mathbf{S}}$ follow then by making use of the coercivity of the operator $\mathfrak{d}_2^{\mathbf{S},\star}(\mathfrak{d}_2^{\mathbf{S}}\,\mathfrak{d}_2^{\mathbf{S},\star} + 2K^{\mathbf{S}})$ and the estimate for $\underline{\beta} = \underline{\beta}^{\mathbf{S}}$ in (9.4.11). The estimate for $\mathfrak{d}_1^{\mathbf{S},\star}\underline{\omega}^{\mathbf{S}}$ is straightforward from the last identity in (9.8.35).

To prove the last part of the proposition we make use of the GCM conditions

[28] The equations used in the derivation of these identities only require the transversality conditions (9.4.9).

[29] Strictly speaking Proposition 7.21 requires the e_3 Ricci and Bianchi identities of a geodesic foliation. It is easy to justify the application of these equations in our context by using the transversality conditions to generate a geodesic foliation in a neighborhood of Σ_0.

[30] These identities were recorded in Proposition 7.12. Note also that $\mathfrak{d}(\Gamma_b^{\mathbf{S}}\cdot\Gamma_b^{\mathbf{S}})$ does not contain derivatives of $\underline{\check{\omega}}$.

(9.8.16) on Σ_0 to deduce that

$$\nu^{\mathbf{S}}(\not{d}_2^{\mathbf{S},\star}\not{d}_1^{\mathbf{S},\star}\mu^{\mathbf{S}})=0, \quad \nu^{\mathbf{S}}(e_\theta^{\mathbf{S}}\kappa^{\mathbf{S}})=0, \quad \nu^{\mathbf{S}}\Big((\not{d}_2^{\mathbf{S},\star}\not{d}_2^{\mathbf{S}}+2K^{\mathbf{S}})\not{d}_2^{\mathbf{S},\star}\not{d}_1^{\mathbf{S},\star}\underline{\kappa}^{\mathbf{S}}\Big)=0.$$

Hence, the quantities C_1, C_2, C_3 in (9.8.32) can be expressed in the form

$$\begin{aligned} C_1 &= -a^{\mathbf{S}}e_4^{\mathbf{S}}(\not{d}_2^{\mathbf{S},\star}\not{d}_1^{\mathbf{S},\star}\mu^{\mathbf{S}}),\\ C_2 &= -a^{\mathbf{S}}e_4^{\mathbf{S}}(e_\theta^{\mathbf{S}}\kappa^{\mathbf{S}}),\\ C_3 &= a^{\mathbf{S}}e_4^{\mathbf{S}}\Big((\not{d}_2^{\mathbf{S},\star}\not{d}_2^{\mathbf{S}}+2K^{\mathbf{S}})\not{d}_2^{\mathbf{S},\star}\not{d}_1^{\mathbf{S},\star}\underline{\kappa}^{\mathbf{S}}\Big). \end{aligned}$$

Making use of our commutation formulas of Lemma 2.68 and the estimates (9.4.11) and (9.4.8) we easily deduce

$$\begin{aligned} \|e_4^{\mathbf{S}}(\not{d}_2^{\mathbf{S},\star}\not{d}_1^{\mathbf{S},\star}\mu^{\mathbf{S}})\|_{\mathfrak{h}_{k-2}(\mathbf{S})} &\lesssim \mathring{\epsilon}r^{-5},\\ \|e_4^{\mathbf{S}}(e_\theta^{\mathbf{S}}\kappa^{\mathbf{S}})\|_{\mathfrak{h}_{k-1}(\mathbf{S})} &\lesssim \mathring{\epsilon}r^{-3}. \end{aligned}$$

Similarly,

$$\Big\|e_4^{\mathbf{S}}\Big((\not{d}_2^{\mathbf{S},\star}\not{d}_2^{\mathbf{S}}+2K^{\mathbf{S}})\not{d}_2^{\mathbf{S},\star}\not{d}_1^{\mathbf{S},\star}\underline{\kappa}^{\mathbf{S}}\Big)\Big\|_{\mathfrak{h}_{k-4}(\mathbf{S})} \lesssim \mathring{\epsilon}r^{-5}.$$

Writing $a^{\mathbf{S}}=\overline{a^{\mathbf{S}}}+\check{a}^{\mathbf{S}}$ and making use of product estimates we deduce

$$\begin{aligned} \|C_1\|_{\mathfrak{h}_{k-2}(\mathbf{S})} &\lesssim \mathring{\epsilon}r^{-5}\left(\Big|\overline{a^{\mathbf{S}}}+1+\frac{2m^{\mathbf{S}}}{r^{\mathbf{S}}}\Big|+r^{-1}\|\check{a}^{\mathbf{S}}\|_{\mathfrak{h}_{k-2}(\mathbf{S})}\right),\\ \|C_2\|_{\mathfrak{h}_{k-1}(\mathbf{S})} &\lesssim \mathring{\epsilon}r^{-3}\left(\Big|\overline{a^{\mathbf{S}}}+1+\frac{2m^{\mathbf{S}}}{r^{\mathbf{S}}}\Big|+r^{-1}\|\check{a}^{\mathbf{S}}\|_{\mathfrak{h}_{k-1}(\mathbf{S})}\right),\\ \|C_3\|_{\mathfrak{h}_{k-4}(\mathbf{S})} &\lesssim \mathring{\epsilon}r^{-5}\left(\Big|\overline{a^{\mathbf{S}}}+1+\frac{2m^{\mathbf{S}}}{r^{\mathbf{S}}}\Big|+r^{-1}\|\check{a}^{\mathbf{S}}\|_{\mathfrak{h}_{k-4}(\mathbf{S})}\right), \end{aligned}$$

as stated. □

Step 10. Propositions 9.55 and 9.57 provide us with potential[31] estimates for $\not{d}_2^{\mathbf{S},\star}\eta^{\mathbf{S}}$, $\not{d}_2^{\mathbf{S},\star}\underline{\xi}^{\mathbf{S}}$, $\not{d}_1^{\mathbf{S},\star}\underline{\omega}^{\mathbf{S}}$, $\not{d}_1^{\mathbf{S},\star}\varsigma^{\mathbf{S}}$. To close we also need to control the $\ell=1$ modes of $\eta^{\mathbf{S}}, \underline{\xi}^{\mathbf{S}}$ the average of $\underline{\omega}^{\mathbf{S}}$ and the average[32] of $a^{\mathbf{S}}$. Note that the average of $\underline{\omega}^{\mathbf{S}}$ can in fact be derived from the equation

$$e_3^{\mathbf{S}}(\kappa^{\mathbf{S}})+\frac{1}{2}\underline{\kappa}^{\mathbf{S}}\kappa^{\mathbf{S}}-2\underline{\omega}^{\mathbf{S}}\kappa^{\mathbf{S}} \;=2\not{d}_1^{\mathbf{S}}\eta^{\mathbf{S}}+2\rho^{\mathbf{S}}-\tfrac{1}{2}\underline{\vartheta}^{\mathbf{S}}\vartheta^{\mathbf{S}}+2(\eta^{\mathbf{S}})^2$$

[31]We cannot close the estimates without also being able to estimate the $\ell=1$ modes of $\eta^{\mathbf{S}}, \underline{\xi}^{\mathbf{S}}, \overline{\underline{\omega}^{\mathbf{S}}}$ and the average $\overline{a}^{\mathbf{S}}$.

[32]The quantity $\check{a}^{\mathbf{S}}$ can be determined using Proposition 9.55.

in terms of $\overline{\underline{A}^{\mathbf{S}}}$ and $\eta^{\mathbf{S}}$. Indeed, making use of the GCM condition $\kappa^{\mathbf{S}}=\frac{2}{r^{\mathbf{S}}}$,

$$\begin{aligned}
\underline{\omega}^{\mathbf{S}} &= \frac{1}{2\kappa^{\mathbf{S}}}\left[e_3^{\mathbf{S}}(\kappa^{\mathbf{S}})+\frac{1}{2}\underline{\kappa}^{\mathbf{S}}\kappa^{\mathbf{S}}-2\,\mathscr{d}_1^{\mathbf{S}}\eta^{\mathbf{S}}-2\rho^{\mathbf{S}}+\frac{1}{2}\underline{\vartheta}^{\mathbf{S}}\vartheta^{\mathbf{S}}-2(\eta^{\mathbf{S}})^2\right]\\
&= -\frac{1}{2}e_3(r^{\mathbf{S}})-\frac{\Upsilon^{\mathbf{S}}}{2r^{\mathbf{S}}}+\frac{r^{\mathbf{S}}}{4}\left[-2\,\mathscr{d}_1^{\mathbf{S}}\eta^{\mathbf{S}}-2\rho^{\mathbf{S}}+\frac{1}{2}\underline{\vartheta}^{\mathbf{S}}\vartheta^{\mathbf{S}}-2(\eta^{\mathbf{S}})^2\right]\\
&= -\frac{1}{4}\underline{A}^{\mathbf{S}}+\frac{r^{\mathbf{S}}}{4}\left[-2\,\mathscr{d}_1^{\mathbf{S}}\eta^{\mathbf{S}}-2\rho^{\mathbf{S}}+\frac{1}{2}\underline{\vartheta}^{\mathbf{S}}\vartheta^{\mathbf{S}}-2(\eta^{\mathbf{S}})^2\right].
\end{aligned}$$

Thus, recalling the definition of $\mu^{\mathbf{S}}$,

$$\overline{\underline{\omega}^{\mathbf{S}}} = -\frac{1}{4}\overline{\underline{A}^{\mathbf{S}}}+\frac{r^{\mathbf{S}}}{2}\overline{\mu^{\mathbf{S}}-(\eta^{\mathbf{S}})^2}$$

or

$$\overline{\underline{\omega}^{\mathbf{S}}}-\frac{m^{\mathbf{S}}}{(r^{\mathbf{S}})^2} = -\frac{1}{4}\overline{\underline{A}^{\mathbf{S}}}+\frac{r^{\mathbf{S}}}{2}\left(\overline{\mu^{\mathbf{S}}-\frac{m^{\mathbf{S}}}{(r^{\mathbf{S}})^3}}-\overline{\eta^{\mathbf{S}}\cdot\eta^{\mathbf{S}}}\right). \tag{9.8.36}$$

Step 11. In view of the above we can determine $\eta^{\mathbf{S}},\underline{\xi}^{\mathbf{S}},\underline{\omega}^{\mathbf{S}},\varsigma^{\mathbf{S}},\underline{A}^{\mathbf{S}}$ provided that we control the $\ell=1$ modes of $\eta^{\mathbf{S}},\underline{\xi}^{\mathbf{S}}$ and the average of $\varsigma^{\mathbf{S}}$. For this reason we introduce,[33] along Σ_0,

$$B^{\mathbf{S}}=\int_{\mathbf{S}}\eta^{\mathbf{S}}e^{\Phi}, \qquad \underline{B}^{\mathbf{S}}=\int_{\mathbf{S}}\underline{\xi}^{\mathbf{S}}e^{\Phi}, \qquad D^{\mathbf{S}}=a^{\mathbf{S}}\Big|_{SP}+1+\frac{2m^{\mathbf{S}}}{r^{\mathbf{S}}}. \tag{9.8.37}$$

We are now ready to prove the following:

Proposition 9.58. *Let Σ_0 be a smooth spacelike hypersurface foliated by framed spheres $(\mathbf{S},e_4^{\mathbf{S}},e_\theta^{\mathbf{S}}.e_3^{\mathbf{S}})$ which verify the GCM conditions* (9.8.16), *transversality condition* (9.4.9) *and the estimates* (9.4.8)–(9.4.11) *of Theorem 9.32. Let $u^{\mathbf{S}}$ as in* (9.8.19) *such that $u^{\mathbf{S}}+r^{\mathbf{S}}$ is constant on Σ_0. Extend $u^{\mathbf{S}}$ and $r^{\mathbf{S}}$ in a neighborhood of Σ_0 such that the transversality conditions* (9.8.20) *are verified. As shown above these allow us to define $\eta^{\mathbf{S}},\underline{\xi}^{\mathbf{S}},\underline{\omega}^{\mathbf{S}},\varsigma^{\mathbf{S}},\underline{A}^{\mathbf{S}},a^{\mathbf{S}}$ and the constants $B^{\mathbf{S}},\underline{B}^{\mathbf{S}},D^{\mathbf{S}}$ as in* (9.8.37). *Finally we assume that*

$$r^{-2}\big(|B^{\mathbf{S}}|+|\underline{B}^{\mathbf{S}}|\big)+|D^{\mathbf{S}}|\le \overset{\circ}{\epsilon}^{1/2}. \tag{9.8.38}$$

Under these assumptions the following estimates hold true for all $k\le s_{max}-7$:

[33] Note that to prove our main theorem we have to construct our hypersurface Σ_0 such that in fact $B=\underline{B}=D=0$.

1. *The Ricci coefficients $\eta^{\mathbf{S}}, \underline{\xi}^{\mathbf{S}}, \underline{\omega}^{\mathbf{S}}$ verify*

$$\begin{aligned} \|\eta^{\mathbf{S}}\|_{\mathfrak{h}_{5+k}(\mathbf{S})} &\lesssim \mathring{\epsilon} + r_{\mathbf{S}}^{-2}|B^{\mathbf{S}}|, \\ \|\underline{\xi}^{\mathbf{S}}\|_{\mathfrak{h}_{5+k}(\mathbf{S})} &\lesssim \mathring{\epsilon} + r_{\mathbf{S}}^{-2}|\underline{B}^{\mathbf{S}}|, \\ \|\check{\underline{\omega}}^{\mathbf{S}}\|_{\mathfrak{h}_{3+k}(\mathbf{S})} &\lesssim \mathring{\epsilon} + r_{\mathbf{S}}^{-2}\left(|\underline{B}^{\mathbf{S}}| + |B^{\mathbf{S}}|\right), \\ \left|\overline{\underline{\omega}^{\mathbf{S}}} - \frac{m^{\mathbf{S}}}{(r^{\mathbf{S}})^2}\right| &\lesssim \mathring{\epsilon} + r_{\mathbf{S}}^{-2}\left(|\underline{B}^{\mathbf{S}}| + |B^{\mathbf{S}}|\right). \end{aligned} \tag{9.8.39}$$

2. *The scalar $a^{\mathbf{S}}$ verifies*

$$r_{\mathbf{S}}^{-1}\|\check{a}^{\mathbf{S}}\|_{\mathfrak{h}_{k+1}(\mathbf{S})} + \left|\overline{a}^{\mathbf{S}} + 1 + \frac{2m^{\mathbf{S}}}{r^{\mathbf{S}}}\right| \quad \lesssim \quad \mathring{\epsilon} + r_{\mathbf{S}}^{-2}|B^{\mathbf{S}}| + |D^{\mathbf{S}}|. \tag{9.8.40}$$

3. *We also have*

$$\begin{aligned} \|\underline{A}^{\mathbf{S}}\|_{\mathfrak{h}_{k+1}(\mathbf{S})} &\lesssim \mathring{\epsilon} + r_{\mathbf{S}}^{-2}\left(|\underline{B}^{\mathbf{S}}| + |B^{\mathbf{S}}|\right) + |D^{\mathbf{S}}|, \\ r_{\mathbf{S}}^{-1}\|\check{\varsigma}^{\mathbf{S}}\|_{\mathfrak{h}_{k+1}(\mathbf{S})} + |\overline{\varsigma}^{\mathbf{S}} - 1| &\lesssim \mathring{\epsilon} + r_{\mathbf{S}}^{-2}\left(|\underline{B}^{\mathbf{S}}| + |B^{\mathbf{S}}|\right) + |D^{\mathbf{S}}|. \end{aligned} \tag{9.8.41}$$

4. *We also have, for all $k \leq s_{max} - 4$,*

$$\begin{aligned} \|e_3^{\mathbf{S}}(\check{\kappa}^{\mathbf{S}}, \vartheta^{\mathbf{S}}, \zeta^{\mathbf{S}}, \check{\underline{\kappa}}^{\mathbf{S}})\|_{\mathfrak{h}_k(\mathbf{S})} &\lesssim \mathring{\epsilon} r_{\mathbf{S}}^{-1}, \\ \|e_3^{\mathbf{S}}(\underline{\vartheta}^{\mathbf{S}})\|_{\mathfrak{h}_k(\mathbf{S})} &\lesssim \mathring{\epsilon}, \\ \|e_3^{\mathbf{S}}\left(\alpha^{\mathbf{S}}, \beta^{\mathbf{S}}, \check{\rho}^{\mathbf{S}}, \mu^{\mathbf{S}}\right)\|_{\mathfrak{h}_k(\mathbf{S})} &\lesssim \mathring{\epsilon} r_{\mathbf{S}}^{-2}, \\ \|e_3^{\mathbf{S}}(\underline{\beta}^{\mathbf{S}})\|_{\mathfrak{h}_k(\mathbf{S})} &\lesssim \mathring{\epsilon} r_{\mathbf{S}}^{-1}, \\ \|e_3^{\mathbf{S}}(\underline{\alpha}^{\mathbf{S}})\|_{\mathfrak{h}_k(\mathbf{S})} &\lesssim \mathring{\epsilon}. \end{aligned}$$

Proof. To simplify the exposition below we make the auxiliary bootstrap assumptions,

$$\|\eta^{\mathbf{S}}\|_{\mathfrak{h}_{5+k}(\mathbf{S})} + \|\underline{\xi}^{\mathbf{S}}\|_{\mathfrak{h}_{5+k}(\mathbf{S})} \quad \lesssim \mathring{\epsilon}^{1/2}. \tag{9.8.42}$$

We start with the following lemma.

Lemma 9.59. *The following estimates hold true:*

$$r_{\mathbf{S}}^{-1}\|\check{a}^{\mathbf{S}}\|_{\mathfrak{h}_{k+1}(\mathbf{S})} + \left|\overline{a}^{\mathbf{S}} + 1 + \frac{2m^{\mathbf{S}}}{r^{\mathbf{S}}}\right| \lesssim |D^{\mathbf{S}}| + \|\eta^{\mathbf{S}}\|_{\mathfrak{h}_k(\mathbf{S})} + \|\underline{\xi}^{\mathbf{S}}\|_{\mathfrak{h}_k(\mathbf{S})} + \mathring{\epsilon}. \tag{9.8.43}$$

Proof. Since $a^{\mathbf{S}} = \overline{a}^{\mathbf{S}} + \check{a}^{\mathbf{S}}$ we deduce $a^{\mathbf{S}}\big|_{SP} = \overline{a}^{\mathbf{S}} + \check{a}^{\mathbf{S}}\big|_{SP}$. Hence,

$$\overline{a}^{\mathbf{S}} \quad = \quad D^{\mathbf{S}} - 1 - \frac{2m^{\mathbf{S}}}{r^{\mathbf{S}}} - \check{a}^{\mathbf{S}}\big|_{SP}. \tag{9.8.44}$$

We also have (see Proposition 9.55)

$$a^{\mathbf{S}} = -\frac{2}{\varsigma^{\mathbf{S}}} + \Upsilon^{\mathbf{S}} - \frac{r^{\mathbf{S}}}{2}\underline{A}^{\mathbf{S}}.$$

Hence,

$$a^{\mathbf{S}} = -\frac{2}{\overline{\varsigma^{\mathbf{S}}}+\check{\varsigma}^{\mathbf{S}}} + \Upsilon^{\mathbf{S}} - \frac{r^{\mathbf{S}}}{2}\underline{A}^{\mathbf{S}} = -\frac{2}{\overline{\varsigma^{\mathbf{S}}}}\left(1-\frac{\check{\varsigma}^{\mathbf{S}}}{\overline{\varsigma^{\mathbf{S}}}}+O\left(\frac{\check{\varsigma}^{\mathbf{S}}}{\overline{\varsigma^{\mathbf{S}}}}\right)^2\right) + \Upsilon^{\mathbf{S}} - \frac{r^{\mathbf{S}}}{2}\underline{A}^{\mathbf{S}}.$$

Taking the average on $\mathbf{S}$ we deduce

$$\overline{a^{\mathbf{S}}} = -\frac{2}{\overline{\varsigma^{\mathbf{S}}}} + \Upsilon^{\mathbf{S}} - \frac{r^{\mathbf{S}}}{2}\overline{\underline{A}^{\mathbf{S}}} + O\left(\frac{\check{\varsigma}^{\mathbf{S}}}{\overline{\varsigma^{\mathbf{S}}}}\right)^2. \tag{9.8.45}$$

Also, using (9.8.44),

$$\check{a}^{\mathbf{S}} = 2\check{\varsigma}^{\mathbf{S}} - \frac{r^{\mathbf{S}}}{2}\check{\underline{A}}^{\mathbf{S}} + \text{l.o.t.} \tag{9.8.46}$$

where l.o.t. denotes higher order terms in $\check{\varsigma}^{\mathbf{S}}$ and $\overline{\varsigma^{\mathbf{S}}}-1$. Indeed

$$\begin{aligned}
\check{a}^{\mathbf{S}} = a^{\mathbf{S}} - \overline{a^{\mathbf{S}}} &= -\frac{2}{\varsigma^{\mathbf{S}}} + \Upsilon^{\mathbf{S}} - \frac{r^{\mathbf{S}}}{2}\underline{A}^{\mathbf{S}} - \left(-\frac{2}{\overline{\varsigma^{\mathbf{S}}}} + \Upsilon^{\mathbf{S}} - \frac{r^{\mathbf{S}}}{2}\overline{\underline{A}^{\mathbf{S}}}\right)\\
&= -\frac{2}{\varsigma^{\mathbf{S}}} + \frac{2}{\overline{\varsigma^{\mathbf{S}}}} - \frac{r^{\mathbf{S}}}{2}\check{\underline{A}}^{\mathbf{S}} = \frac{2\check{\varsigma}^{\mathbf{S}}}{\varsigma^{\mathbf{S}}\overline{\varsigma^{\mathbf{S}}}} - \frac{r^{\mathbf{S}}}{2}\check{\underline{A}}^{\mathbf{S}} = 2\check{\varsigma}^{\mathbf{S}} - \frac{r^{\mathbf{S}}}{2}\check{\underline{A}}^{\mathbf{S}} + \text{l.o.t.}
\end{aligned}$$

Thus to estimate $\check{a}^{\mathbf{S}}$ and $\overline{a^{\mathbf{S}}}$ we first need to estimate $\underline{A}^{\mathbf{S}}$, $\check{\varsigma}^{\mathbf{S}}$ and $\overline{\varsigma^{\mathbf{S}}}$. Using the equations (see Proposition 9.55)

$$\begin{aligned}
e_\theta^{\mathbf{S}}(\underline{A}^{\mathbf{S}}) &= -\frac{2\Upsilon^{\mathbf{S}}}{r^{\mathbf{S}}}(\zeta^{\mathbf{S}}-\eta^{\mathbf{S}}) - \frac{2}{r^{\mathbf{S}}}\underline{\xi}^{\mathbf{S}} + (\zeta^{\mathbf{S}}-\eta^{\mathbf{S}})\underline{A}^{\mathbf{S}},\\
\overline{\underline{A}^{\mathbf{S}}} &= \frac{1}{\overline{\varsigma^{\mathbf{S}}}}\left(\overline{\check{\varsigma}^{\mathbf{S}}\underline{\check{\kappa}}^{\mathbf{S}}} - \overline{\check{\varsigma}^{\mathbf{S}}\check{\underline{A}}^{\mathbf{S}}}\right),
\end{aligned}$$

and the auxiliary assumption we derive

$$\|\underline{A}^{\mathbf{S}}\|_{\mathfrak{h}_{k+1}(\mathbf{S})} \lesssim \|\eta^{\mathbf{S}}\|_{\mathfrak{h}_k(\mathbf{S})} + \|\underline{\xi}^{\mathbf{S}}\|_{\mathfrak{h}_k(\mathbf{S})} + \mathring{\epsilon} r^{-1}\left(1 + \|\check{\varsigma}^{\mathbf{S}}\|_{\mathfrak{h}_k(\mathbf{S})} + \left|\overline{\varsigma^{\mathbf{S}}}-1\right|\right). \tag{9.8.47}$$

From the equation

$$e_\theta^{\mathbf{S}}(\log\varsigma^{\mathbf{S}}) = (\eta^{\mathbf{S}} - \zeta^{\mathbf{S}}),$$

we also derive

$$r_{\mathbf{S}}^{-1}\|\check{\varsigma}^{\mathbf{S}}\|_{\mathfrak{h}_{k+1}(\mathbf{S})} \lesssim \|\eta^{\mathbf{S}}\|_{\mathfrak{h}_k(\mathbf{S})} + \mathring{\epsilon} + \mathring{\epsilon}\left|\overline{\varsigma^{\mathbf{S}}}-1\right|. \tag{9.8.48}$$

To estimate $\overline{\varsigma^{\mathbf{S}}}-1$ we derive from (9.8.45) and (9.8.46)

$$\begin{aligned}
\frac{2}{\overline{\varsigma^{\mathbf{S}}}} &= -\overline{a^{\mathbf{S}}} + \Upsilon^{\mathbf{S}} - \frac{r^{\mathbf{S}}}{2}\overline{\underline{A}^{\mathbf{S}}} = -\left(D^{\mathbf{S}} - 1 - \frac{2m^{\mathbf{S}}}{r^{\mathbf{S}}} - \check{a}^{\mathbf{S}}\big|_{SP}\right) + \Upsilon^{\mathbf{S}} - \frac{r^{\mathbf{S}}}{2}\overline{\underline{A}^{\mathbf{S}}} + \text{l.o.t.}\\
&= -D^{\mathbf{S}} + 2 + \check{a}^{\mathbf{S}}\big|_{SP} - \frac{r^{\mathbf{S}}}{2}\overline{\underline{A}^{\mathbf{S}}} + \text{l.o.t.}\\
&= -D^{\mathbf{S}} + 2 + 2\check{\varsigma}^{\mathbf{S}}\big|_{SP} - \frac{r^{\mathbf{S}}}{2}\left(\overline{\underline{A}^{\mathbf{S}}} + \check{\underline{A}}^{\mathbf{S}}\big|_{SP}\right) + \text{l.o.t.}
\end{aligned}$$

and therefore,

$$\frac{2(1-\overline{\varsigma^{\mathbf{S}}})}{\overline{\varsigma^{\mathbf{S}}}} = -D^{\mathbf{S}} + 2\check{\varsigma}^{\mathbf{S}}\big|_{SP} - \frac{r^{\mathbf{S}}}{2}\left(\overline{\underline{A}}^{\mathbf{S}} + \check{\underline{A}}^{\mathbf{S}}\big|_{SP}\right) + \text{l.o.t.},$$

i.e.,

$$\overline{\varsigma^{\mathbf{S}}} - 1 = \frac{1}{2}D^{\mathbf{S}} - \check{\varsigma}^{\mathbf{S}}\big|_{SP} + \frac{r^{\mathbf{S}}}{4}\left(\overline{\underline{A}}^{\mathbf{S}} + \check{\underline{A}}^{\mathbf{S}}\big|_{SP}\right) + \text{l.o.t.}$$

where l.o.t. denote higher order terms in $\check{\varsigma}^{\mathbf{S}}$ and $\overline{\varsigma^{\mathbf{S}}} - 1$. Thus,

$$\left|\overline{\varsigma^{\mathbf{S}}} - 1\right| \lesssim |D^{\mathbf{S}}| + \|\check{\varsigma}^{\mathbf{S}}\|_{L^\infty(\mathbf{S})} + r^{\mathbf{S}}\|\underline{A}^{\mathbf{S}}\|_{L^\infty(\mathbf{S})}.$$

Hence, back to (9.8.48) we derive

$$r_{\mathbf{S}}^{-1}\|\check{\varsigma}^{\mathbf{S}}\|_{\mathfrak{h}_{k+1}(\mathbf{S})} + \left|\overline{\varsigma^{\mathbf{S}}} - 1\right| \lesssim |D^{\mathbf{S}}| + r^{\mathbf{S}}\|\underline{A}^{\mathbf{S}}\|_{L^\infty(\mathbf{S})} + r\|\eta^{\mathbf{S}}\|_{\mathfrak{h}_k(\mathbf{S})} + \overset{\circ}{\epsilon}.$$

Combining with (9.8.47) we deduce

$$\begin{aligned} \|\underline{A}^{\mathbf{S}}\|_{\mathfrak{h}_{k+1}(\mathbf{S})} &\lesssim \|\eta^{\mathbf{S}}\|_{\mathfrak{h}_k(\mathbf{S})} + \|\underline{\xi}^{\mathbf{S}}\|_{\mathfrak{h}_k(\mathbf{S})} + \overset{\circ}{\epsilon}, \\ r_{\mathbf{S}}^{-1}\|\check{\varsigma}^{\mathbf{S}}\|_{\mathfrak{h}_{k+1}(\mathbf{S})} + \left|\overline{\varsigma^{\mathbf{S}}} - 1\right| &\lesssim r\|\eta^{\mathbf{S}}\|_{\mathfrak{h}_k(\mathbf{S})} + \overset{\circ}{\epsilon}. \end{aligned} \tag{9.8.49}$$

In view of (9.8.46) we also deduce

$$\begin{aligned} r_{\mathbf{S}}^{-1}\|\check{a}^{\mathbf{S}}\|_{\mathfrak{h}_{k+1}(\mathbf{S})} &\lesssim r_{\mathbf{S}}^{-1}\|\check{\varsigma}^{\mathbf{S}}\|_{\mathfrak{h}_{k+1}(\mathbf{S})} + \|\underline{A}^{\mathbf{S}}\|_{\mathfrak{h}_{k+1}(\mathbf{S})} \\ &\lesssim \|\eta^{\mathbf{S}}\|_{\mathfrak{h}_k(\mathbf{S})} + \|\underline{\xi}^{\mathbf{S}}\|_{\mathfrak{h}_k(\mathbf{S})} + \overset{\circ}{\epsilon}. \end{aligned}$$

From (9.8.44) we further deduce

$$\left|\overline{a}^{\mathbf{S}} + 1 + \frac{2m^{\mathbf{S}}}{r^{\mathbf{S}}}\right| \lesssim |D^{\mathbf{S}}| + \|\check{a}^{\mathbf{S}}\|_{L^\infty(\mathbf{S})} \lesssim |D^{\mathbf{S}}| + \|\eta^{\mathbf{S}}\|_{\mathfrak{h}_k(\mathbf{S})} + \|\underline{\xi}^{\mathbf{S}}\|_{\mathfrak{h}_k(\mathbf{S})} + \overset{\circ}{\epsilon}.$$

Hence,

$$r_{\mathbf{S}}^{-1}\|\check{a}^{\mathbf{S}}\|_{\mathfrak{h}_{k+1}(\mathbf{S})} + \left|\overline{a}^{\mathbf{S}} + 1 + \frac{2m^{\mathbf{S}}}{r^{\mathbf{S}}}\right| \lesssim |D^{\mathbf{S}}| + \|\eta^{\mathbf{S}}\|_{\mathfrak{h}_k(\mathbf{S})} + \|\underline{\xi}^{\mathbf{S}}\|_{\mathfrak{h}_k(\mathbf{S})} + \overset{\circ}{\epsilon}$$

as stated. □

In view of the lemma above and the assumption $|D^{\mathbf{S}}| \lesssim \overset{\circ}{\epsilon}^{1/2}$ the estimates (9.8.34) become

$$\begin{aligned} \|C_1\|_{\mathfrak{h}_k(\mathbf{S})} &\lesssim \overset{\circ}{\epsilon} r_{\mathbf{S}}^{-4}\left(\|\eta^{\mathbf{S}}\|_{\mathfrak{h}_k(\mathbf{S})} + \|\underline{\xi}^{\mathbf{S}}\|_{\mathfrak{h}_k(\mathbf{S})} + \overset{\circ}{\epsilon}^{1/2}\right), \\ \|C_2\|_{\mathfrak{h}_{k+3}(\mathbf{S})} &\lesssim \overset{\circ}{\epsilon} r_{\mathbf{S}}^{-2}\left(\|\eta^{\mathbf{S}}\|_{\mathfrak{h}_{k+3}(\mathbf{S})} + \|\underline{\xi}^{\mathbf{S}}\|_{\mathfrak{h}_{k+3}(\mathbf{S})} + \overset{\circ}{\epsilon}^{1/2}\right), \\ \|C_3\|_{\mathfrak{h}_k(\mathbf{S})} &\lesssim \overset{\circ}{\epsilon} r_{\mathbf{S}}^{-4}\left(\|\eta^{\mathbf{S}}\|_{\mathfrak{h}_k(\mathbf{S})} + \|\underline{\xi}^{\mathbf{S}}\|_{\mathfrak{h}_k(\mathbf{S})} + \overset{\circ}{\epsilon}^{1/2}\right). \end{aligned} \tag{9.8.50}$$

To prove the desired estimate for $\eta^{\mathbf{S}}, \underline{\xi}^{\mathbf{S}}, \underline{\omega}^{\mathbf{S}}$ we make use of (9.8.33) and the following lemma.

Lemma 9.60. *The error term*

$$E_k = r_{\mathbf{S}}^{-1}\|\Gamma_g^{\mathbf{S}}\|_{\mathfrak{h}_{4+k}(\mathbf{S})} + \|\Gamma_b^{\mathbf{S}} \cdot \Gamma_b^{\mathbf{S}}\|_{\mathfrak{h}_{4+k}(\mathbf{S})}, \qquad k \le s_{max} - 7,$$

appearing in (9.8.33) *verifies the estimate*

$$E_k \lesssim r_{\mathbf{S}}^{-1}\overset{\circ}{\epsilon} + r_{\mathbf{S}}^{-1}\overset{\circ}{\epsilon}^{1/2}\Big(\|(\eta^{\mathbf{S}}, \underline{\xi}^{\mathbf{S}})\|_{\mathfrak{h}_{4+k}(\mathbf{S})} + \|\underline{\check{\omega}}^{\mathbf{S}}\|_{\mathfrak{h}_{k+3}(\mathbf{S})}\Big).$$

Proof. Since $\Gamma_g^{\mathbf{S}}$ contains only terms estimated by (9.4.8),

$$\|\Gamma_g^{\mathbf{S}}\|_{\mathfrak{h}_{4+k}(\mathbf{S})} \lesssim r_{\mathbf{S}}^{-1}\overset{\circ}{\epsilon},$$

$\Gamma_b^{\mathbf{S}}$ contains $\underline{\vartheta}^{\mathbf{S}}$, which is estimated by (9.4.8), as well as $\eta^{\mathbf{S}}, \underline{\xi}^{\mathbf{S}}, \underline{\check{\omega}}^{\mathbf{S}}, \overline{\omega}^{\mathbf{S}} - \frac{m^{\mathbf{S}}}{(r^{\mathbf{S}})^2}$. Thus, in view of the auxiliary estimates $\big\|\eta^{\mathbf{S}}, \underline{\xi}^{\mathbf{S}}\big\|_{\mathfrak{h}_{5+k}(\mathbf{S})} \lesssim \overset{\circ}{\epsilon}^{1/2}$ and the fact that the quadratic error terms contain one less derivative of $\underline{\check{\omega}}^{\mathbf{S}}$, we deduce

$$\|\Gamma_b^{\mathbf{S}} \cdot \Gamma_b^{\mathbf{S}}\|_{\mathfrak{h}_{4+k}(\mathbf{S})} \lesssim r_{\mathbf{S}}^{-1}\overset{\circ}{\epsilon}^{1/2}\left(\|\eta^{\mathbf{S}}, \underline{\xi}^{\mathbf{S}}\|_{\mathfrak{h}_{4+k}(\mathbf{S})} + \|\underline{\check{\omega}}^{\mathbf{S}}\|_{\mathfrak{h}_{k+3}(\mathbf{S})} + r^{\mathbf{S}}\left|\overline{\omega}^{\mathbf{S}} - \frac{m^{\mathbf{S}}}{(r^{\mathbf{S}})^2}\right|\right).$$

In view of equation (9.8.36), $\overline{\underline{\omega}^{\mathbf{S}}} - \frac{m^{\mathbf{S}}}{(r^{\mathbf{S}})^2} = -\frac{1}{4}\overline{\underline{A}^{\mathbf{S}}} + \frac{r^{\mathbf{S}}}{2}\Big(\overline{\mu^{\mathbf{S}} - \frac{m^{\mathbf{S}}}{(r^{\mathbf{S}})^3}} - \overline{\eta^{\mathbf{S}} \cdot \eta^{\mathbf{S}}}\Big)$,

$$\begin{aligned}
\left|\overline{\underline{\omega}^{\mathbf{S}}} - \frac{m^{\mathbf{S}}}{(r^{\mathbf{S}})^2}\right| &\lesssim |\overline{\underline{A}^{\mathbf{S}}}| + r^{\mathbf{S}}\left|\overline{\mu} - \frac{m^{\mathbf{S}}}{(r^{\mathbf{S}})^3}\right| + |\eta^{\mathbf{S}}|^2\\
&\lesssim r_{\mathbf{S}}^{-1}\overset{\circ}{\epsilon}|D^{\mathbf{S}}| + r^{-1}\overset{\circ}{\epsilon}^{1/2}\left(\|\eta^{\mathbf{S}}\|_{\mathfrak{h}_2(\mathbf{S})} + \|\underline{\xi}^{\mathbf{S}}\|_{\mathfrak{h}_2(\mathbf{S})}\right) + \overset{\circ}{\epsilon}r^{-2}\\
&\lesssim r_{\mathbf{S}}^{-1}\overset{\circ}{\epsilon}^{1/2}\left(\|\eta^{\mathbf{S}}\|_{\mathfrak{h}_2(\mathbf{S})} + \|\underline{\xi}^{\mathbf{S}}\|_{\mathfrak{h}_2(\mathbf{S})} + \overset{\circ}{\epsilon}\right).
\end{aligned}$$

Hence,

$$\|\Gamma_b^{\mathbf{S}} \cdot \Gamma_b^{\mathbf{S}}\|_{\mathfrak{h}_{4+k}(\mathbf{S})} \lesssim r_{\mathbf{S}}^{-1}\overset{\circ}{\epsilon}^{1/2}\left(\|\eta^{\mathbf{S}}, \underline{\xi}^{\mathbf{S}}\|_{\mathfrak{h}_{4+k}(\mathbf{S})} + \|\underline{\check{\omega}}^{\mathbf{S}}\|_{\mathfrak{h}_{k+3}(\mathbf{S})} + \overset{\circ}{\epsilon}\right)$$

and

$$\begin{aligned}
E_k &= r_{\mathbf{S}}^{-1}\|\Gamma_g^{\mathbf{S}}\|_{\mathfrak{h}_{4+k}(\mathbf{S})} + \|\Gamma_b^{\mathbf{S}} \cdot \Gamma_b^{\mathbf{S}}\|_{\mathfrak{h}_{4+k}(\mathbf{S})}\\
&\lesssim r_{\mathbf{S}}^{-1}\overset{\circ}{\epsilon} + r_{\mathbf{S}}^{-1}\overset{\circ}{\epsilon}^{1/2}\left(\|\eta^{\mathbf{S}}, \underline{\xi}^{\mathbf{S}}\|_{\mathfrak{h}_{4+k}(\mathbf{S})} + \|\underline{\check{\omega}}^{\mathbf{S}}\|_{\mathfrak{h}_{k+3}(\mathbf{S})}\right)
\end{aligned}$$

as stated. ◻

In view of the lemma and estimates (9.8.50) for C_1, C_2, C_3, the estimates (9.8.33) of Proposition 9.57 become

$$\begin{aligned}
\| \not{d}_2^{\mathbf{S},\star} \eta^{\mathbf{S}} \|_{\mathfrak{h}_{4+k}(\mathbf{S})} &\lesssim r_{\mathbf{S}}^{-1} \overset{\circ}{\epsilon} + r_{\mathbf{S}}^{-1} \overset{\circ}{\epsilon}^{1/2} \left(\| \eta^{\mathbf{S}}, \underline{\xi}^{\mathbf{S}} \|_{\mathfrak{h}_{4+k}(\mathbf{S})} + \| \underline{\check{\omega}}^{\mathbf{S}} \|_{\mathfrak{h}_{k+3}(\mathbf{S})} \right), \\
\| \not{d}_2^{\mathbf{S},\star} \underline{\xi}^{\mathbf{S}} \|_{\mathfrak{h}_{4+k}(\mathbf{S})} &\lesssim r_{\mathbf{S}}^{-1} \overset{\circ}{\epsilon} + r_{\mathbf{S}}^{-1} \overset{\circ}{\epsilon}^{1/2} \left(\| \eta^{\mathbf{S}}, \underline{\xi}^{\mathbf{S}} \|_{\mathfrak{h}_{4+k}(\mathbf{S})} + \| \underline{\check{\omega}}^{\mathbf{S}} \|_{\mathfrak{h}_{k+3}(\mathbf{S})} \right), \\
\| \not{d}_1^{\mathbf{S},\star} \underline{\omega}^{\mathbf{S}} \|_{\mathfrak{h}_{2+k}(\mathbf{S})} &\lesssim r_{\mathbf{S}}^{-1} \| \eta^{\mathbf{S}} \|_{\mathfrak{h}_{4+k}(\mathbf{S})} + r_{\mathbf{S}}^{-1} \| \underline{\xi}^{\mathbf{S}} \|_{\mathfrak{h}_{2+k}(\mathbf{S})} \\
&\quad + r_{\mathbf{S}}^{-1} \overset{\circ}{\epsilon} + r_{\mathbf{S}}^{-1} \overset{\circ}{\epsilon}^{1/2} \left(\| \eta^{\mathbf{S}}, \underline{\xi}^{\mathbf{S}} \|_{\mathfrak{h}_{3+k}(\mathbf{S})} + \| \underline{\check{\omega}}^{\mathbf{S}} \|_{\mathfrak{h}_{2+k}(\mathbf{S})} \right).
\end{aligned} \tag{9.8.51}$$

From the last equation we derive

$$\| \underline{\check{\omega}}^{\mathbf{S}} \|_{\mathfrak{h}_{3+k}(\mathbf{S})} \lesssim \| \eta^{\mathbf{S}} \|_{\mathfrak{h}_{4+k}(\mathbf{S})} + \| \underline{\xi}^{\mathbf{S}} \|_{\mathfrak{h}_{2+k}(\mathbf{S})} + \overset{\circ}{\epsilon}.$$

Thus the first two equations in (9.8.51) become

$$\begin{aligned}
r^{\mathbf{S}} \| \not{d}_2^{\mathbf{S},\star} \eta^{\mathbf{S}} \|_{\mathfrak{h}_{4+k}(\mathbf{S})} &\lesssim \overset{\circ}{\epsilon} + \overset{\circ}{\epsilon}^{1/2} \left(\| \eta^{\mathbf{S}} \|_{\mathfrak{h}_{4+k}(\mathbf{S})} + \| \underline{\xi}^{\mathbf{S}} \|_{\mathfrak{h}_{4+k}(\mathbf{S})} \right), \\
r^{\mathbf{S}} \| \not{d}_2^{\mathbf{S},\star} \underline{\xi}^{\mathbf{S}} \|_{\mathfrak{h}_{4+k}(\mathbf{S})} &\lesssim \overset{\circ}{\epsilon} + \overset{\circ}{\epsilon}^{1/2} \left(\| \eta^{\mathbf{S}} \|_{\mathfrak{h}_{4+k}(\mathbf{S})} + \| \underline{\xi}^{\mathbf{S}} \|_{\mathfrak{h}_{4+k}(\mathbf{S})} \right),
\end{aligned} \tag{9.8.52}$$

from which we deduce

$$\begin{aligned}
\| \eta^{\mathbf{S}} \|_{\mathfrak{h}_{5+k}(\mathbf{S})} &\lesssim \overset{\circ}{\epsilon} + r_{\mathbf{S}}^{-2} |B^{\mathbf{S}}|, \\
\| \underline{\xi}^{\mathbf{S}} \|_{\mathfrak{h}_{5+k}(\mathbf{S})} &\lesssim \overset{\circ}{\epsilon} + r_{\mathbf{S}}^{-2} |\underline{B}^{\mathbf{S}}|, \\
\| \underline{\check{\omega}}^{\mathbf{S}} \|_{\mathfrak{h}_{3+k}(\mathbf{S})} &\lesssim \overset{\circ}{\epsilon} + r_{\mathbf{S}}^{-2} \left(|\underline{B}^{\mathbf{S}}| + |B^{\mathbf{S}}| \right),
\end{aligned}$$

as stated. We can then go back to the preliminary estimates obtained above for $\varsigma^{\mathbf{S}}$, $\underline{A}^{\mathbf{S}}$ and $a^{\mathbf{S}}$ to derive the remaining statements (1–4) of Proposition 9.58. To prove the last part of the proposition we make use of the corresponding Ricci and Bianchi equations in the $e_3^{\mathbf{S}}$ direction. □

Corollary 9.61. *Under the same assumptions as in the proposition above we have the more precise estimates, with $d(\mathbf{S}) = \int_{\mathbf{S}} e^{2\Phi}$,*

$$\begin{aligned}
\left\| \eta^{\mathbf{S}} - \frac{1}{d(\mathbf{S})} B^{\mathbf{S}} e^{\Phi} \right\|_{\mathfrak{h}_{5+k}(\mathbf{S})} &\lesssim \overset{\circ}{\epsilon}, \\
\left\| \underline{\xi}^{\mathbf{S}} - \frac{1}{d(\mathbf{S})} \underline{B}^{\mathbf{S}} e^{\Phi} \right\|_{\mathfrak{h}_{5+k}(\mathbf{S})} &\lesssim \overset{\circ}{\epsilon}.
\end{aligned}$$

Note also that

$$d(\mathbf{S}) = (r^{\mathbf{S}})^4 \left(\frac{8\pi}{3} + O(\overset{\circ}{\epsilon}) \right).$$

Proof. In view of (9.8.52), (9.8.39) and auxiliary assumption (9.8.38), we deduce

$$\begin{aligned}\left\|\eta^{\mathbf{S}}-\left(\frac{\int_{\mathbf{S}}\eta^{\mathbf{S}}e^{\Phi}}{\int_{\mathbf{S}}e^{2\Phi}}\right)e^{\Phi}\right\|_{\mathfrak{h}_{5+k}(\mathbf{S})} &\lesssim r\|\not{d}_2^{\mathbf{S},\star}\eta^{\mathbf{S}}\|_{\mathfrak{h}_{4+k}(\mathbf{S})}\\ &\lesssim \mathring{\epsilon}+\mathring{\epsilon}^{1/2}\left(\|\eta^{\mathbf{S}}\|_{\mathfrak{h}_{4+k}(\mathbf{S})}+\|\underline{\xi}^{\mathbf{S}}\|_{\mathfrak{h}_{4+k}(\mathbf{S})}\right)\\ &\lesssim \mathring{\epsilon}+\mathring{\epsilon}^{1/2}\left(\mathring{\epsilon}+r^{-2}\left(|\underline{B}^{\mathbf{S}}|+|B^{\mathbf{S}}|\right)\right)\\ &\lesssim \mathring{\epsilon}.\end{aligned}$$

We deduce

$$\left\|\eta^{\mathbf{S}}-B^{\mathbf{S}}\frac{1}{\int_{\mathbf{S}}e^{2\Phi}}e^{\Phi}\right\|_{\mathfrak{h}_{5+k}(\mathbf{S})}\lesssim\mathring{\epsilon}.$$

Similarly,

$$\left\|\underline{\xi}^{\mathbf{S}}-\underline{B}^{\mathbf{S}}\frac{1}{\int_{\mathbf{S}}e^{2\Phi}}e^{\Phi}\right\|_{\mathfrak{h}_{5+k}(\mathbf{S})}\lesssim\mathring{\epsilon}$$

as desired. □

9.8.3 Construction of Σ_0

To construct the spacelike hypersurface of Theorem 9.52 we proceed as follows.

Step 12. Let $\Psi(s),\Lambda(s),\underline{\Lambda}(s)$ real valued functions that will be carefully chosen later. We look for the hypersurface Σ_0 in the form

$$\Sigma_0 = \bigcup_{s\geq\mathring{s}}\mathbf{S}[P(s)]=\bigcup_{s\geq\mathring{s}}\mathbf{S}[\Psi(s),s,\Lambda(s),\underline{\Lambda}(s)] \tag{9.8.53}$$

where $P(s)$ is a curve in the parameter space P given by

$$P(s)=(\Psi(s),s,\Lambda(s),\underline{\Lambda}(s)). \tag{9.8.54}$$

In order for Σ_0 to start at $\mathbf{S}_0=\mathbf{S}[\mathring{u},\mathring{s},\Lambda_0,\underline{\Lambda}_0]$ we impose the conditions

$$\Psi(\mathring{s})=\mathring{u},\quad \Lambda(\mathring{s})=\Lambda_0,\quad \underline{\Lambda}(\mathring{s})=\underline{\Lambda}_0. \tag{9.8.55}$$

Step 13. We expect Σ_0 to be a perturbation of the spacelike hypersurface $u+s=c_0$ for some constant c_0. We thus introduce the notation

$$\psi(s):=\Psi(s)+s-c_0,\ \text{so that}\ \Psi(s)=-s+c_0+\psi(s)$$

and expect $\psi(s)=O(\mathring{\delta})$.

Step 14. In view of (9.8.18) we can express the collection of spheres Σ_0 in the form

$$\Sigma_0 = \Big\{\Xi(s,\theta), \quad s \ge \overset{\circ}{s},\ \theta \in [0,\pi]\Big\} \tag{9.8.56}$$

where the map $\Xi(s,\theta) = \Xi(\Psi(s), s, \theta)$ is defined as

$$\Xi(s,\theta) := \Big(\Psi(s) + U(\theta, P(s)),\ s + S(\theta, P(s)),\ \theta\Big). \tag{9.8.57}$$

At the south pole, i.e., $\theta = 0$, where $U(0,P) = S(0,P) = 0$,

$$\Xi(s,0) \quad = \quad \Big(\Psi(s),\ s,\ 0\Big). \tag{9.8.58}$$

Clearly,

$$\begin{aligned}
\partial_s \Xi(s,\theta) &= \Big(\Psi'(s) + \partial_P U(\theta, P(s))P'(s),\ 1 + \partial_P S(\theta, P(s))P'(s),\ 0\Big),\\
\partial_\theta \Xi(s,\theta) &= \Big(\partial_\theta U(\theta, P(s)),\ \partial_\theta S(\theta, P(s)),\ 1\Big),
\end{aligned}$$

where

$$\begin{aligned}
\partial_P U(\cdot)P'(s) &= \Psi'(s)\partial_u U(\cdot) + \partial_s U(\cdot) + \Lambda'(s)\partial_\Lambda U(\cdot) + \underline{\Lambda}'(s)\partial_{\underline{\Lambda}} U(\cdot),\\
\partial_P S(\cdot)P'(s) &= \Psi'(s)\partial_u S(\cdot) + \partial_s S(\cdot) + \Lambda'(s)\partial_\Lambda S(\cdot) + \underline{\Lambda}'(s)\partial_{\underline{\Lambda}} S(\cdot).
\end{aligned}$$

Given f a function on Σ_0 we have

$$\begin{aligned}
\frac{d}{ds} f(\Xi(s,\theta)) &= \Big(\Psi'(s) + \partial_P U(\theta, P(s))P'(s)\Big)\partial_u f\\
&\quad + \Big(1 + \partial_P S(\theta, P(s))P'(s)\Big)\partial_s f\\
&= X_* f,\\
\frac{d}{d\theta} f(\Xi(s,\theta)) &= \partial_\theta U(\theta, P(s))\partial_s f + \partial_\theta S(\theta, P(s))\partial_s + \partial_\theta f\\
&= Y_* f,
\end{aligned}$$

where X_*, Y_* are the following tangent vectorfields along Σ_0:

$$\begin{aligned}
X_*(s,\theta) :&= \Big(\Psi'(s) + \partial_P U(\theta, P(s))P'(s)\Big)\partial_u + \Big(1 + \partial_P S(\theta, P(s))P'(s)\Big)\partial_s,\\
Y_*(s,\theta) :&= \partial_\theta U(\theta, P(s))\partial_s + \partial_\theta S(\theta, P(s))\partial_s + \partial_\theta,
\end{aligned} \tag{9.8.59}$$

or

$$\begin{aligned}
X_*(s,\theta) :&= \Big(\Psi'(s) + \breve{A}(s,\theta)\Big)\partial_u + \Big(1 + \breve{B}(s,\theta)P'(s)\Big)\partial_s,\\
Y_*(s,\theta) :&= \breve{C}(s,\theta)\partial_u + \breve{D}(s,\theta)\partial_s + \partial_\theta,
\end{aligned} \tag{9.8.60}$$

where

$$\begin{aligned}
\breve{A}(s,\theta) : &= \partial_P U(\theta, P(s))P'(s)\\
&= \partial_u U(\theta, P(s))\Psi'(s) + \partial_s U(\theta, P(s)) + \partial_\Lambda U(\theta, P(s))\Lambda'(s)\\
&\quad +\partial_{\underline{\Lambda}} U(\theta, P(s))\underline{\Lambda}'(s),\\
\breve{B}(s,\theta) : &= \partial_P S(\theta, P(s))P'(s)\\
&= \partial_u S(\theta, P(s))\Psi'(s) + \partial_s S(\theta, P(s)) + \partial_\Lambda U(\theta, P(s))\Lambda'(s)\\
&\quad +\partial_{\underline{\Lambda}} S(\theta, P(s))\underline{\Lambda}'(s),\\
\breve{C}(s,\theta) : &= \partial_\theta U(\theta, P(s)),\\
\breve{D}(s,\theta) : &= \partial_\theta S(\theta, P(s)).
\end{aligned}$$

Step 15. Define the vectorfield, along the south pole of each $\mathbf{S} \subset \Sigma_0$,

$$X_*\Big|_{SP} h = \frac{d}{ds} h\big(\Xi(s,0)\big). \tag{9.8.61}$$

Lemma 9.62. *At the south pole we have the relations (recall $\nu^{\mathbf{S}} = e_3^{\mathbf{S}} + a^{\mathbf{S}} e_4^{\mathbf{S}}$)*

$$X_*|_{SP} = \frac{1}{2\lambda}\varsigma\Psi'\,\nu^{\mathbf{S}}\Big|_{SP}, \tag{9.8.62}$$

$$a^{\mathbf{S}}|_{SP} = \frac{2\lambda^2}{\Psi'(s)\varsigma}(1 - \frac{1}{2}\Psi'(s)\varsigma\underline{\Omega})|_{SP}, \tag{9.8.63}$$

or, more precisely,

$$a^{\mathbf{S}}(\Psi(s), s, 0) = \frac{1}{\Psi'(s)}\frac{2\lambda^2}{\varsigma}\left(1 - \frac{1}{2}\Psi'(s)\varsigma\underline{\Omega}\right)(\Psi(s), s, 0).$$

Here $f, \underline{f}, \lambda$ are the transition functions and $\varsigma, \underline{\Omega}$ correspond to the background foliation.

Proof. Note that

$$\breve{A}(s,0) = \breve{B}(s,0) = \breve{C}(s,0) = \breve{D}(s,0) = 0.$$

Thus, at the south pole SP,

$$X_*(s,0) = \Psi'(s)\partial_u + \partial_s.$$

Recall that

$$\partial_s = e_4, \quad \partial_u = \frac{1}{2}\varsigma\left(e_3 - \underline{\Omega} e_4 - \underline{b}\gamma^{1/2} e_\theta\right), \quad \partial_\theta = \sqrt{\gamma} e_\theta,$$

or, since $\underline{b}$ vanishes at the south pole,

$$X_*(s,0) = \Psi'\frac{1}{2}\varsigma\,(e_3 - \underline{\Omega} e_4) + e_4 = (1 - \frac{1}{2}\Psi'(s)\varsigma\underline{\Omega})e_4 + \frac{1}{2}\Psi'(s)\varsigma e_3.$$

On the other hand, since the transition functions $f, \underline{f}$ vanish at the south pole,

$$e_4^{\mathbf{S}} = \lambda e_4, \qquad e_3^{\mathbf{S}} = \lambda^{-1} e_3.$$

Hence,

$$\begin{aligned}
X_*(s,0) &= \lambda\big(1 - \frac{1}{2}\Psi'(s)\varsigma\underline{\Omega}\big)e_4^{\mathbf{S}} + \frac{1}{2}\lambda^{-1}\Psi'(s)\varsigma e_3^{\mathbf{S}} \\
&= \frac{1}{2}\lambda^{-1}\Psi'(s)\varsigma\left(e_3^{\mathbf{S}} + \frac{2\lambda^2}{\Psi'(s)\varsigma}\big(1 - \frac{1}{2}\Psi'(s)\varsigma\underline{\Omega}\big)e_4^{\mathbf{S}}\right).
\end{aligned}$$

In view of the definition of $\nu^{\mathbf{S}}$ we deduce

$$X_*(s,0) = \frac{1}{2}\lambda^{-1}\Psi'(s)\varsigma\, \nu^{\mathbf{S}}\big|_{SP}$$

and[34]

$$a^{\mathbf{S}}(s,0) = \frac{2\lambda^2}{\Psi'(s)\varsigma}\left(1 - \frac{1}{2}\Psi'(s)\varsigma\underline{\Omega}\right)$$

as stated. □

Step 16. The transition functions $(f, \underline{f}, \lambda)$ are uniquely determined on $\mathbf{S}$ by the results of Theorem 9.32 in terms of $\Lambda, \underline{\Lambda}$. The same holds true for all curvature components and the Ricci coefficients $\kappa^{\mathbf{S}}, \vartheta^{\mathbf{S}}, \zeta^{\mathbf{S}}, \underline{\kappa}^{\mathbf{S}}, \underline{\vartheta}^{\mathbf{S}}$. One can easily see from the transformation formulas that the values of the $e_3^{\mathbf{S}}$ derivatives of $(f, \underline{f}, \lambda)$ are determined by the transversal Ricci coefficients $\eta^{\mathbf{S}}, \underline{\xi}^{\mathbf{S}}, \underline{\omega}^{\mathbf{S}}$. Indeed, schematically, from the transformation formulas for $\eta, \underline{\xi}, \underline{\omega}$ in Proposition 9.20,

$$\begin{aligned}
e_3^{\mathbf{S}} f &= 2(\eta^{\mathbf{S}} - \eta) - \frac{1}{2}\kappa\underline{f} + f\underline{\omega} + F\cdot\Gamma_b + \text{l.o.t.}, \\
e_3^{\mathbf{S}} \underline{f} &= 2(\underline{\xi}^{\mathbf{S}} - \underline{\xi}) - \frac{1}{2}\underline{f}(\underline{\kappa} + 4\underline{\omega}) + F\cdot\Gamma_b + \text{l.o.t.}, \qquad (9.8.64)\\
e_3^{\mathbf{S}}(\log\lambda) &= 2(\underline{\omega}^{\mathbf{S}} - \underline{\omega}) + \Gamma_b\cdot F + \text{l.o.t.},
\end{aligned}$$

where $F = (f, \underline{f}, \log\lambda)$ and l.o.t. denotes terms which are linear in Γ_g, Γ_b and linear and higher order in F. Recall also that the $e_4^{\mathbf{S}}$ derivatives of F are fixed by our transversality condition (9.4.9) More precisely we have

$$\begin{aligned}
e_4^{\mathbf{S}}(f) &= -\frac{1}{2}\kappa f + \text{l.o.t.}, \\
e_4^{\mathbf{S}}(\underline{f}) &= 2e_\theta^{\mathbf{S}}(\log\lambda) - \underline{f}\kappa + 2\left(\underline{\omega} + \frac{1}{4}\underline{\kappa}\right)f + \text{l.o.t.}, \\
e_4^{\mathbf{S}}(\log\lambda) &= \text{l.o.t.}
\end{aligned}$$

It follows that $\eta^{\mathbf{S}}, \underline{\xi}^{\mathbf{S}}, \underline{\omega}^{\mathbf{S}}$ can be determined by $\nu^{\mathbf{S}}(f, \underline{f}, \lambda)$ and the scalar $a^{\mathbf{S}}$. More

[34]Note that $a^{\mathbf{S}}(s,0) = a^{\mathbf{S}}(\Xi(s,0))$.

precisely,

$$
\begin{aligned}
\nu^{\mathbf{S}}(f) &= 2(\eta^{\mathbf{S}} - \eta) - \frac{1}{2}(\kappa \underline{f} + a^{\mathbf{S}} \underline{\kappa} f) + f \underline{\omega} + F \cdot \Gamma_b + \text{l.o.t.}, \\
\nu^{\mathbf{S}}(\underline{f}) &= 2(\underline{\xi}^{\mathbf{S}} - \underline{\xi}) - \frac{1}{2}(\underline{\kappa} + 4\underline{\omega})(\underline{f} - a^{\mathbf{S}} f) + a^{\mathbf{S}} \left(2 e_\theta^{\mathbf{S}}(\log \lambda) - \underline{f} \kappa\right) \\
&\quad + F \cdot \Gamma_b + \text{l.o.t.}
\end{aligned}
\tag{9.8.65}
$$

Step 17. We derive equations for $\Lambda(s) = \Lambda(\Psi(s), s, 0)$, $\underline{\Lambda}(s) = \underline{\Lambda}(\Psi(s), s, 0)$ as follows.

Lemma 9.63. *We have the following identities*

$$
\begin{aligned}
c(s) \frac{1}{\Psi'(s)} \Lambda'(s) &= \int_{\mathbf{S}} \nu^{\mathbf{S}}(f) e^{\Phi} - \frac{6}{r^{\mathbf{S}}} \Lambda(s) + E(s), \\
c(s) \frac{1}{\Psi'(s)} \underline{\Lambda}'(s) &= \int_{\mathbf{S}(s)} \nu^{\mathbf{S}}(\underline{f}) e^{\Phi} - \frac{6}{r^{\mathbf{S}}} \underline{\Lambda}(s) + \underline{E}(s),
\end{aligned}
\tag{9.8.66}
$$

where

$$
c(s) = \left(\frac{2\lambda}{\varsigma}\right)\Big|_{SP}(s) = \left(\frac{2\lambda}{\varsigma}\right)(\Psi(s), s, 0)
$$

and error terms

$$
\begin{aligned}
E(s) &= \frac{1}{2} \int_{\mathbf{S}(s)} \left(3\underline{\check{\kappa}}^{\mathbf{S}} - \underline{\vartheta}^{\mathbf{S}} - a^{\mathbf{S}} \vartheta^{\mathbf{S}} + \frac{3}{r^{\mathbf{S}}}\left(a^{\mathbf{S}} + \left(1 + \frac{2m^{\mathbf{S}}}{r^{\mathbf{S}}}\right)\right)\right) f e^{\Phi} \\
&+ (\varsigma^{\mathbf{S}})^{-1}\Big|_{SP} \int_{\mathbf{S}(s)} \left(\varsigma^{\mathbf{S}} - \varsigma^{\mathbf{S}}\Big|_{SP}\right) \left(\nu^{\mathbf{S}}(f) - \frac{6}{r^{\mathbf{S}}} \Lambda(s)\right) e^{\Phi} + l.o.t., \\
\underline{E}(s) &= \frac{1}{2} \int_{\mathbf{S}(s)} \left(3\underline{\check{\kappa}}^{\mathbf{S}} - \underline{\vartheta}^{\mathbf{S}} - a^{\mathbf{S}} \vartheta^{\mathbf{S}} + \frac{3}{r^{\mathbf{S}}}\left(a^{\mathbf{S}} + \left(1 + \frac{2m^{\mathbf{S}}}{r^{\mathbf{S}}}\right)\right)\right) \underline{f} e^{\Phi} \\
&+ (\varsigma^{\mathbf{S}})^{-1}\Big|_{SP} \int_{\mathbf{S}(s)} \left(\varsigma^{\mathbf{S}} - \varsigma^{\mathbf{S}}\Big|_{SP}\right) \left(\nu^{\mathbf{S}}(\underline{f}) - \frac{6}{r^{\mathbf{S}}} \underline{\Lambda}(s)\right) e^{\Phi} + l.o.t.
\end{aligned}
$$

Proof. According to Lemma 9.54 we have

$$
\nu^{\mathbf{S}}\left(\int_{\mathbf{S}} h\right) = (\varsigma^{\mathbf{S}})^{-1} \int_{\mathbf{S}} \varsigma^{\mathbf{S}} \left(\nu^{\mathbf{S}}(h) + (\underline{\kappa}^{\mathbf{S}} + a^{\mathbf{S}} \kappa^{\mathbf{S}}) h\right).
$$

Thus, applying the vectorfield $\nu^{\mathbf{S}}\big|_{SP} = \frac{2\lambda}{\varsigma \Psi'} X_*\big|_{SP}$ to the formulas (9.8.17),

$$
\begin{aligned}
&\frac{1}{\Psi'(s)} \left(\frac{2\lambda}{\varsigma}\right)\Big|_{SP} \frac{d}{ds} \Lambda(s) = \nu^{\mathbf{S}}\Big|_{SP}(\Lambda) = \nu^{\mathbf{S}}(\Lambda)\big|_{SP} = \nu^{\mathbf{S}}\Big(\int_{\mathbf{S}} f e^{\Phi}\Big)\Big|_{SP} \\
&= (\varsigma^{\mathbf{S}})^{-1}\Big|_{SP} \int_{\mathbf{S}(s)} \varsigma^{\mathbf{S}} \Big(\nu^{\mathbf{S}}(f e^{\Phi}) + (\underline{\kappa}^{\mathbf{S}} + a^{\mathbf{S}} \kappa^{\mathbf{S}}) f e^{\Phi}\Big).
\end{aligned}
$$

Introducing

$$
J(f) = e^{-\Phi} \nu^{\mathbf{S}}(f e^{\Phi}) + (\underline{\kappa}^{\mathbf{S}} + a^{\mathbf{S}} \kappa^{\mathbf{S}}) f
\tag{9.8.67}
$$

we deduce

$$\begin{aligned}
c(s)\frac{1}{\Psi'(s)} &= (\varsigma^{\bf S})^{-1}\Big|_{SP}\int_{{\bf S}(s)}\varsigma^{\bf S} J(f)e^{\Phi}\\
&= \int_{{\bf S}(s)} J(f)e^{\Phi} + (\varsigma^{\bf S})^{-1}\Big|_{SP}\int_{{\bf S}(s)}\left(\varsigma^{\bf S}-\varsigma^{\bf S}\Big|_{SP}\right)J(f).
\end{aligned}$$

On the other hand, since $e_3\Phi = \frac{1}{2}(\underline{\kappa}-\underline{\vartheta})$, $e_4\Phi = \frac{1}{2}(\kappa-\vartheta)$,

$$\begin{aligned}
J(f) &= \nu^{\bf S}(f) + \left(e_3^{\bf S}\Phi + a^{\bf S}e_4^{\bf S}\Phi + \underline{\kappa}^{\bf S} + a^{\bf S}\kappa^{\bf S}\right)f\\
&= \nu^{\bf S}(f) + \frac{1}{2}\left(3\underline{\kappa}^{\bf S} - \underline{\vartheta}^{\bf S} + a^{\bf S}(3\kappa^{\bf S} - \underline{\vartheta}^{\bf S})\right)f\\
&= \nu^{\bf S}(f) + \frac{3}{2}\left(\underline{\kappa}^{\bf S} + a^{\bf S}\kappa^{\bf S}\right) - \frac{1}{2}\left(\underline{\vartheta}^{\bf S} + a^{\bf S}\vartheta^{\bf S}\right).
\end{aligned}$$

Since $\kappa^{\bf S} = \frac{2}{r^{\bf S}}$ and $\underline{\kappa}^{\bf S} = \overline{\underline{\kappa}^{\bf S}} + \underline{\check{\kappa}}^{\bf S} = -\frac{2\Upsilon^{\bf S}}{r^{\bf S}} + \underline{\check{\kappa}}^{\bf S}$, we deduce

$$\begin{aligned}
J(f) &= \nu^{\bf S}(f) + \frac{3}{r^{\bf S}}\left(-\Upsilon^{\bf S} + a^{\bf S}\right)f + \frac{1}{2}\left(3\underline{\check{\kappa}}^{\bf S} - \underline{\vartheta}^{\bf S} - a^{\bf S}\vartheta^{\bf S}\right)f\\
&= \nu^{\bf S}(f) + \frac{3}{r^{\bf S}}\left(-\Upsilon^{\bf S} - (1+\frac{2m^{\bf S}}{r^{\bf S}})\right)f\\
&\quad +\frac{1}{2}\left(3\underline{\check{\kappa}}^{\bf S} - \underline{\vartheta}^{\bf S} - a^{\bf S}\vartheta^{\bf S} + \frac{3}{r^{\bf S}}\left(a^{\bf S} + \left(1+\frac{2m^{\bf S}}{r^{\bf S}}\right)\right)\right)f\\
&= \nu^{\bf S}(f) - \frac{6}{r^{\bf S}}\Lambda(s) + \frac{1}{2}\left(3\underline{\check{\kappa}}^{\bf S} - \underline{\vartheta}^{\bf S} - a^{\bf S}\vartheta^{\bf S} + \frac{3}{r^{\bf S}}\left(a^{\bf S} + \left(1+\frac{2m^{\bf S}}{r^{\bf S}}\right)\right)\right)f.
\end{aligned}$$

We deduce

$$c(s)\frac{1}{\Psi'(s)} = \int_{\bf S}\nu^{\bf S}(f)e^{\Phi} - \frac{6}{r^{\bf S}}\Lambda(s) + E(s)$$

where

$$\begin{aligned}
E(s) &= \frac{1}{2}\int_{{\bf S}(s)}\left(3\underline{\check{\kappa}}^{\bf S} - \underline{\vartheta}^{\bf S} - a^{\bf S}\vartheta^{\bf S} + \frac{3}{r^{\bf S}}\left(a^{\bf S} + \left(1+\frac{2m^{\bf S}}{r^{\bf S}}\right)\right)\right)fe^{\Phi}\\
&+ (\varsigma^{\bf S})^{-1}\Big|_{SP}\int_{{\bf S}(s)}\left(\varsigma^{\bf S}-\varsigma^{\bf S}\Big|_{SP}\right)J(f)\\
&= \frac{1}{2}\int_{{\bf S}(s)}\left(3\underline{\check{\kappa}}^{\bf S} - \underline{\vartheta}^{\bf S} - a^{\bf S}\vartheta^{\bf S} + \frac{3}{r^{\bf S}}\left(a^{\bf S} + \left(1+\frac{2m^{\bf S}}{r^{\bf S}}\right)\right)\right)fe^{\Phi}\\
&+ (\varsigma^{\bf S})^{-1}\Big|_{SP}\int_{{\bf S}(s)}\left(\varsigma^{\bf S}-\varsigma^{\bf S}\Big|_{SP}\right)\left(\nu^{\bf S}(f) - \frac{6}{r^{\bf S}}\Lambda(s)\right)e^{\Phi} + \text{l.o.t.}
\end{aligned}$$

The proof for $\underline{\Lambda}$ is exactly the same. □

Step 18. We make use of the estimates for $F = (f, \underline{f}, \log\lambda)$ and $e_4^{\bf S}(F)$ derived in Theorem 9.32 as well as the estimates for $a^{\bf S}, \varsigma^{\bf S}, \eta^{\bf S}, \underline{\xi}^{\bf S}, \underline{\omega}^{\bf S}$ derived in Proposition 9.58 to evaluate the right-hand sides of (9.8.66). Recall that in Proposition 9.58 we have made the auxiliary assumption (9.8.38), i.e.,

$$r_{\mathbf{S}}^{-2}\big(|B^{\mathbf{S}}|+|\underline{B}^{\mathbf{S}}|\big)+|D^{\mathbf{S}}|\le \mathring{\epsilon}^{1/2}.$$

Proposition 9.64. *The following equations hold true for the functions*[35]

$$\begin{array}{lll}\Lambda(s) &=& \Lambda(\Psi(s),s,0), \quad \underline{\Lambda}(s)=\underline{\Lambda}(\Psi(s),s,0),\\ B(s) &=& \Lambda(\Psi(s),s,0), \quad \underline{B}(s)=\Lambda(\Psi(s),s,0), \quad r(s)=r(\Psi(s),s,0),\end{array}$$

$$\begin{aligned}\frac{1}{-1+\psi'(s)}\Lambda'(s) &= B(s)-\frac{1}{2}r(s)^{-1}\underline{\Lambda}(s)-\frac{7}{2}r(s)^{-1}\Lambda(s)+O(r^{-1})\Lambda(s)\\ &\quad +N(B,\underline{B},D,\Lambda,\underline{\Lambda},\psi)(s),\\ \frac{1}{-1+\psi'(s)}\underline{\Lambda}'(s) &= \underline{B}(s)-\frac{7}{2}r(s)^{-1}\underline{\Lambda}(s)+\frac{1}{2}r(s)^{-1}\Lambda(s)\\ &\quad +O(r^{-1})\big(\Lambda(s)+\underline{\Lambda}(s)\big)+\underline{N}(B,\underline{B},D,\Lambda,\underline{\Lambda},\psi)(s).\end{aligned} \tag{9.8.68}$$

The expressions $N,\underline{N}$ verify the following properties.

- *They depend on $B,\underline{B},D,\Lambda,\underline{\Lambda},\psi,F=(f,\underline{f},\lambda-1)$, the background Ricci coefficients Γ_b,Γ_g and curvature $\check{R}=\{\alpha,\beta,\check{\rho},\underline{\beta},\underline{\alpha}\}$.*
- *$N,\underline{N}$ vanish at $(B,\underline{B},D,\Lambda,\underline{\Lambda},\psi)=(0,0,0,0,0,0)$. In fact,*

$$|N,\underline{N}| \lesssim r^2\mathring{\delta}.$$

- *The linear part in $B,\underline{B},D$ has $O(\mathring{\epsilon})$ coefficients, i.e., coefficients which depend on the quantities $\Gamma_b,\Gamma_g,\check{R},F$ and $\Lambda,\underline{\Lambda},\psi$.*
- *The linear part in $\Lambda,\underline{\Lambda},\psi$ has $O(\mathring{\epsilon})$ coefficients.*

Proof. To prove the desired result we make use of (9.8.65) to check the following:

$$\begin{aligned}\int_{\mathbf{S}(s)}\nu^{\mathbf{S}}(f)e^{\Phi} &= 2B(s)-r^{-1}\underline{\Lambda}(s)-r^{-1}\Lambda(s)+O(r^{-1})\Lambda(s)+O(r^2\mathring{\delta}),\\ \int_{\mathbf{S}(s)}\nu^{\mathbf{S}}(\underline{f})e^{\Phi} &= 2\underline{B}(s)-r^{-1}\underline{\Lambda}(s)+r^{-1}\Lambda(s)+O(r^{-1})\big(\Lambda(s)+\underline{\Lambda}(s)\big)+O(r^2\mathring{\delta}).\end{aligned} \tag{9.8.69}$$

Combining this with (9.8.66),

$$\begin{aligned}c(s)\frac{1}{\Psi'(s)}\Lambda'(s) &= \int_{\mathbf{S}}\nu^{\mathbf{S}}(f)e^{\Phi}-\frac{6}{r^{\mathbf{S}}}\Lambda(s)+E(s),\\ c(s)\frac{1}{\Psi'(s)}\underline{\Lambda}'(s) &= \int_{\mathbf{S}(s)}\nu^{\mathbf{S}}(\underline{f})e^{\Phi}-\frac{6}{r^{\mathbf{S}}}\underline{\Lambda}(s)+\underline{E}(s),\end{aligned}$$

and the following estimates for the error terms $E,\underline{E}$,

$$|E(s)|+|\underline{E}(s)| \lesssim r^2\mathring{\delta}, \tag{9.8.70}$$

[35]Note also that $r^{\mathbf{S}(s)}=r^{\mathbf{S}(s)}=r|_{SP(\mathbf{S}(s))}=r(s)$.

we deduce

$$\frac{1}{\Psi'(s)}\Lambda'(s) = \frac{1}{c(s)}\left(2B(s) - r^{-1}\underline{\Lambda}(s) - 7r^{-1}\Lambda(s) + O(r^{-1})\Lambda(s) + O(r^2\overset{\circ}{\delta})\right),$$

$$\frac{1}{\Psi'(s)}\underline{\Lambda}'(s) = \frac{1}{c(s)}\Bigg(2\underline{B}(s) - 7r^{-1}\underline{\Lambda}(s) + r^{-1}\Lambda(s) + O(r^{-1})\big(\Lambda(s) + \underline{\Lambda}(s)\big) + O(r^2\overset{\circ}{\delta})\Bigg).$$

According to our assumptions $\varsigma = 1 + O(\overset{\circ}{\epsilon})$. Also according to Theorem 9.32 $\lambda = 1 + O(r^{-1}\overset{\circ}{\epsilon})$. Thus,

$$c(s) = \left(\frac{2\lambda}{\varsigma}\right)\Big|_{SP}(s) = \frac{2(1 + O(r^{-1}\overset{\circ}{\epsilon}))}{1 + O(\overset{\circ}{\epsilon})} = 2 + O(\overset{\circ}{\epsilon}).$$

Hence,

$$\begin{aligned}\frac{1}{\Psi'(s)}\Lambda'(s) &= \frac{1}{2}\left(2B(s) - r^{-1}\underline{\Lambda}(s) - 7r^{-1}\Lambda(s) + O(r^{-1})\Lambda(s) + O(r^2\overset{\circ}{\delta})\right)\\ &= B(s) - \frac{1}{2}r^{-1}\underline{\Lambda}(s) - \frac{7}{2}r^{-1}\Lambda(s) + O(r^{-1})\Lambda(s) + O(r^2\overset{\circ}{\delta}).\end{aligned}$$

Setting $\Psi(s) = -s + \psi(s) + c_0$ and recalling the structure of the error terms we have denoted by $O(r^2\overset{\circ}{\delta})$

$$\begin{aligned}\frac{1}{-1+\psi'(s)}\Lambda'(s) &= B(s) - \frac{1}{2}r^{-1}\underline{\Lambda}(s) - \frac{7}{2}r^{-1}\Lambda(s) + O(r^{-1})\Lambda(s)\\ &+ N(B, \underline{B}, D, \Lambda, \underline{\Lambda}, \psi)(s)\end{aligned}$$

where N verifies the properties mentioned in the proposition. In the same manner we derive

$$\begin{aligned}\frac{1}{-1+\psi'(s)}\underline{\Lambda}'(s) &= \underline{B}(s) - \frac{7}{2}r^{-1}\underline{\Lambda}(s) + \frac{1}{2}r^{-1}\Lambda(s) + O(r^{-1})\big(\Lambda(s) + \underline{\Lambda}(s)\big)\\ &+ \underline{N}(B, \underline{B}, D, \Lambda, \underline{\Lambda}, \psi)(s)\end{aligned}$$

as stated in the proposition.

It remains to check (9.8.69) and (9.8.70). According to (9.8.65) and our assumptions on the Ricci coefficients $\kappa, \underline{\kappa}, \underline{\omega}$, we have along the sphere $\mathbf{S}$

$$\begin{aligned}\nu^{\mathbf{S}}(f) &= 2(\eta^{\mathbf{S}} - \eta) - \frac{1}{2}\left(\frac{2}{r}\underline{f} - a^{\mathbf{S}}\frac{2\Upsilon}{r}f\right) + f\frac{m}{r^2} + F \cdot \Gamma_b + \text{l.o.t.}\\ &= 2(\eta^{\mathbf{S}} - \eta) - r^{-1}\underline{f} + r^{-1}\left(\frac{m}{r} + a^{\mathbf{S}}\left(1 - \frac{2m}{r}\right)\right)f + F \cdot \Gamma.\end{aligned}$$

According to (9.8.43) and auxiliary assumption (9.8.38)

$$\left|a^{\mathbf{S}} + \left(1 + \frac{2m^{\mathbf{S}}}{r^{\mathbf{S}}}\right)\right| \lesssim \overset{\circ}{\epsilon} + r^{-2}\left(|\underline{B}^{\mathbf{S}}| + |B^{\mathbf{S}}|\right) + |D^{\mathbf{S}}| \lesssim \overset{\circ}{\epsilon}^{1/2}.$$

Thus,

$$\nu^{\mathbf{S}}(f) = 2(\eta^{\mathbf{S}}-\eta) - r^{-1}\underline{f} - r^{-1}\left(1-\frac{m}{r}-\frac{m^2}{r^2}\right)f + r^{-2}O\left(\overset{\circ}{\delta}\,\overset{\circ}{\epsilon}^{1/2}\right).$$

Since r and $r^{\mathbf{S}}$ are comparable along $\mathbf{S}$, i.e., $|r-r^{\mathbf{S}}|\leq\overset{\circ}{\delta}$, we deduce, recalling the definition of B,

$$\begin{aligned}\int_{\mathbf{S}(s)}\nu^{\mathbf{S}}(f)e^{\Phi} &= 2B(s) - 2\int_{\mathbf{S}(s)}\eta e^{\Phi} - r^{-1}\underline{\Lambda}(s) - r^{-1}\left(1-\frac{m}{r}-\frac{m^2}{r^2}\right)\Lambda(s)\\ &\quad + rO\left(\overset{\circ}{\delta}\,\overset{\circ}{\epsilon}^{1/2}\right).\end{aligned}$$

Making use of the assumption (9.8.1) for η as well as Corollary 9.10 we easily deduce

$$\left|\int_{\mathbf{S}(s)}\eta e^{\Phi}\right| \lesssim r^2\overset{\circ}{\delta}.$$

Hence,

$$\begin{aligned}\int_{\mathbf{S}(s)}\nu^{\mathbf{S}}(f)e^{\Phi} &= 2B(s) - r^{-1}\underline{\Lambda}(s) - r^{-1}(1+O(r^{-1}))\Lambda(s) + O(r^2\overset{\circ}{\delta})\\ &= 2B(s) - r^{-1}\underline{\Lambda}(s) - r^{-1}\Lambda(s) + O(r^{-1})\Lambda(s) + O(r^2\overset{\circ}{\delta}).\end{aligned}$$

Similarly, starting with

$$\nu^{\mathbf{S}}(\underline{f}) = 2(\underline{\xi}^{\mathbf{S}}-\underline{\xi}) - \tfrac{1}{2}(\underline{\kappa}+4\underline{\omega})(\underline{f}-a^{\mathbf{S}}f) + a^{\mathbf{S}}\left(2e_\theta^{\mathbf{S}}(\log\lambda) - \underline{f}\kappa\right) + F\cdot\Gamma_b + \text{l.o.t.}$$

we deduce

$$\begin{aligned}\int_{\mathbf{S}(s)}\nu^{\mathbf{S}}(\underline{f})e^{\Phi} &= 2\underline{B}(s) - 2\int_{\mathbf{S}(s)}\underline{\xi}e^{\Phi} + r^{-1}\left(1+\frac{8m}{r}\right)\underline{\Lambda}(s)\\ &+ r^{-1}\left(1-\frac{2m}{r}-\frac{8m^2}{r^2}\right)\Lambda(s)\\ &- 2\left(1+\frac{2m}{r}\right)\int_{\mathbf{S}(s)}e_\theta^{\mathbf{S}}(\log\lambda)e^{\Phi} + rO\left(\overset{\circ}{\delta}\,\overset{\circ}{\epsilon}^{1/2}\right).\end{aligned}$$

Making use of the assumption (9.8.1) for $\underline{\xi}$, as well as Corollary 9.10,

$$\left|\int_{\mathbf{S}(s)}\underline{\xi}e^{\Phi}\right| \lesssim r^2\overset{\circ}{\delta}. \tag{9.8.71}$$

Also, in view of the estimates of Theorem 9.32,

$$\left|\int_{\mathbf{S}(s)}e_\theta^{\mathbf{S}}(\log\lambda)e^{\Phi}\right| \lesssim r^2\overset{\circ}{\delta}.$$

We deduce

$$\int_{\mathbf{S}(s)} \nu^{\mathbf{S}}(\underline{f})e^{\Phi} = 2\underline{B}(s) - r^{-1}(1+O(r^{-1}))\underline{\Lambda}(s) + r^{-1}(1+O(r^{-1}))\Lambda(s) + O(r^2\overset{\circ}{\delta})$$

as stated. The estimates for $E, \underline{E}$ in (9.8.70) can also be easily checked. This ends the proof of Proposition 9.64. □

Step 19. We derive an equation for ψ. The main result is stated in the proposition below.

Proposition 9.65. *The function $\psi(s) = \Psi(s) + s - c_0$ defined in Step 13 verifies the following equation:*

$$\psi'(s) = -\tfrac{1}{2}D(s) + O(D(s)^2) + M(s) \tag{9.8.72}$$

where $M(s)$ is a function which depends only on Γ, R of the background foliation, ψ and $(f, \underline{f}, \lambda - 1)$, such that

$$|M(s)| \lesssim \overset{\circ}{\delta} r(s)^{-1}.$$

Proof. In view of (9.8.63) and the definition of $c(s) = \left(\frac{2\lambda}{\varsigma}\right)\Big|_{SP}(s)$ we have

$$\Psi'(s) = \frac{1}{a^{\mathbf{S}}\big|_{SP}(s)} \cdot \frac{2\lambda^2}{\varsigma}\left(1 - \frac{1}{2}\Psi'\varsigma\underline{\Omega}\right)\Big|_{SP}(s)$$

or

$$\Psi'(s) = \left[\frac{2\lambda^2}{\varsigma}\frac{1}{a^{\mathbf{S}} + \lambda^2\underline{\Omega}}\right]\Big|_{SP}. \tag{9.8.73}$$

Now, we have

$$\begin{aligned}\frac{2\lambda^2}{\varsigma}\frac{1}{a^{\mathbf{S}} + \lambda^2\underline{\Omega}} &= \frac{2}{\varsigma}\frac{1}{a^{\mathbf{S}} + \underline{\Omega}} + O(\lambda - 1)\\ &= -1 + \frac{a^{\mathbf{S}} + \frac{2}{\varsigma} + \underline{\Omega}}{a^{\mathbf{S}} + \underline{\Omega}} + O(\lambda - 1).\end{aligned}$$

Hence,

$$\begin{aligned}\psi'(s) &= \Psi'(s) + 1 = \left[\frac{a^{\mathbf{S}} + \frac{2}{\varsigma} + \underline{\Omega}}{a^{\mathbf{S}} + \underline{\Omega}}\right]\Big|_{SP} + O(\lambda - 1)\\ &= \left[\frac{a^{\mathbf{S}} + \frac{2}{\varsigma} + \underline{\Omega}}{a^{\mathbf{S}} + \underline{\Omega}}\right]\Big|_{SP} + O(r^{-1}\overset{\circ}{\delta}).\end{aligned}$$

We have, see (9.8.37),

$$a^{\mathbf{S}}\big|_{SP}(s) = D(s) - 1 - \frac{2m^{\mathbf{S}}}{r^{\mathbf{S}}}.$$

Hence,

$$\begin{aligned}(a^{\mathbf{S}}+\underline{\Omega})\big|_{SP}(s) &= D(s)-1-\frac{2m^{\mathbf{S}}}{r^{\mathbf{S}}}+\underline{\Omega}\big|_{SP}(s)\\ &= D(s)-1-\frac{2m^{\mathbf{S}}}{r^{\mathbf{S}}}-\left(1-\frac{2m}{r}\right)+O(\overset{\circ}{\epsilon})\\ &= D(s)-2+O(\overset{\circ}{\epsilon}).\end{aligned}$$

In view of the assumption (9.8.2)

$$\left|\left(\frac{2}{\varsigma}+\underline{\Omega}\right)\Big|_{SP}-1-\frac{2m}{r}\right| \lesssim r^{-1}\overset{\circ}{\delta}$$

we deduce

$$\begin{aligned}\left(a^{\mathbf{S}}+\frac{2}{\varsigma}+\underline{\Omega}\right)\Big|_{SP}(s) &= a^{\mathbf{S}}\big|_{SP}+1+\frac{2m}{r}+O(r^{-1}\overset{\circ}{\delta})\\ &= D(s)+\frac{2m}{r}-\frac{2m^{\mathbf{S}}}{r^{\mathbf{S}}}+O(r^{-1}\overset{\circ}{\delta})\\ &= D(s)+O(r^{-1}\overset{\circ}{\delta}).\end{aligned}$$

Hence,

$$\psi'(s) = \left[\frac{a^{\mathbf{S}}+\frac{2}{\varsigma}+\underline{\Omega}}{a^{\mathbf{S}}+\underline{\Omega}}\right]\Bigg|_{SP}+O(r^{-1}\overset{\circ}{\delta}) = -\frac{1}{2}D(s)+O(D(s)^2)+O(r^{-1}\overset{\circ}{\delta})$$

as stated. □

Step 20. We combine Propositions 9.64 and 9.65 to derive the closed system of equations in $\Lambda, \underline{\Lambda}, \psi$,

$$\begin{aligned}\frac{1}{-1+\psi'(s)}\Lambda'(s) &= B(s)-\frac{1}{2}r(s)^{-1}\underline{\Lambda}(s)-\frac{7}{2}r(s)^{-1}\Lambda(s)+O(r^{-1})\Lambda(s)\\ &\quad +N(B,\underline{B},D,\Lambda,\underline{\Lambda},\psi)(s),\\ \frac{1}{-1+\psi'(s)}\underline{\Lambda}'(s) &= \underline{B}(s)-\frac{7}{2}r(s)^{-1}\underline{\Lambda}(s)+\frac{1}{2}r(s)^{-1}\Lambda(s)\\ &\quad +O(r^{-1})\big(\Lambda(s)+\underline{\Lambda}(s)\big)+\underline{N}(B,\underline{B},D,\Lambda,\underline{\Lambda},\psi)(s),\\ \psi'(s) &= -\frac{1}{2}D(s)+O(D(s)^2)+M(s),\end{aligned} \tag{9.8.74}$$

with initial conditions

$$\psi(\overset{\circ}{s})=0, \qquad \Lambda(\overset{\circ}{s})=\Lambda_0, \qquad \underline{\Lambda}(\overset{\circ}{s})=\underline{\Lambda}_0. \tag{9.8.75}$$

Recall also that $r(s)$ is a smooth function of $\psi(s)$.

The system (9.8.74) is verified for all choices of $(\Lambda, \underline{\Lambda}, \Psi)$. We now make a suitable particular choice for $(\Lambda, \underline{\Lambda}, \Psi)$ as follows.

Consider in particular the system obtained from(9.8.74) by setting $B, \underline{B}, D$ to zero

$$
\begin{aligned}
\psi'(s) &= M(s),\\
\frac{1}{-1+\psi'(s)}\Lambda'(s) &= -\frac{1}{2}r(s)^{-1}\underline{\Lambda}(s) - \frac{7}{2}r(s)^{-1}\Lambda(s)\\
&\quad + O(r^{-1})\Lambda(s) + \widetilde{N}(\Lambda, \underline{\Lambda}, \psi)(s),\\
\frac{1}{-1+\psi'(s)}\underline{\Lambda}'(s) &= -\frac{7}{2}r(s)^{-1}\underline{\Lambda}(s) + \frac{1}{2}r(s)^{-1}\Lambda(s)\\
&\quad + O(r^{-1})\big(\Lambda(s) + \underline{\Lambda}(s)\big) + \underline{\widetilde{N}}(\Lambda, \underline{\Lambda}, \psi)(s),
\end{aligned}
\tag{9.8.76}
$$

where

$$
\begin{aligned}
\widetilde{N}(\Lambda, \underline{\Lambda}, \psi) &= N(0,0,0,\Lambda, \underline{\Lambda}, \psi),\\
\underline{\widetilde{N}}(\Lambda, \underline{\Lambda}, \psi) &= \underline{N}(0,0,0,\Lambda, \underline{\Lambda}, \psi).
\end{aligned}
$$

We initialize the system at $s = \overset{\circ}{s}$ as in (9.8.75), i.e.,

$$
\Lambda(\overset{\circ}{s}) = \psi(\overset{\circ}{s}) = 0, \qquad \Lambda_0, \qquad \underline{\Lambda}(\overset{\circ}{s}) = \underline{\Lambda}_0.
$$

The system admits a unique solution $\psi(s)$ defined in a small neighborhood $\overset{\circ}{I}$ of $\overset{\circ}{s}$. The function $\Psi(s) = -s + \psi(s) + c_0$ defines the desired hypersurface Σ_0.

Step 21. It remains to show that the function $B, \underline{B}, D$ vanish on the hypersurface Σ_0 defined above. Since the system (9.8.74) is verified for all functions $\Lambda, \underline{\Lambda}, \psi$ we deduce, along Σ_0,

$$
\begin{aligned}
D &= 0,\\
B &= N(B, \underline{B}, D, \Lambda, \underline{\Lambda}, \psi)(s) - N(0,0,0,\Lambda, \underline{\Lambda}, \psi)(s),\\
\underline{B} &= \underline{N}(B, \underline{B}, D, \Lambda, \underline{\Lambda}, \psi)(s) - \underline{N}(0,0,0,\Lambda, \underline{\Lambda}, \psi)(s).
\end{aligned}
$$

In view of the properties of $N, \underline{N}$ we deduce

$$
\begin{aligned}
\Big|N(B, \underline{B}, D, \Lambda, \underline{\Lambda}, \psi)(s) - N(0,0,0,\Lambda, \underline{\Lambda}, \psi)(s)\Big| &\lesssim \overset{\circ}{\epsilon} \sup_{\overset{\circ}{I}} \Big(|B(s)| + |\underline{B}(s)|\Big),\\
\Big|\underline{N}(B, \underline{B}, D, \Lambda, \underline{\Lambda}, \psi)(s) - \underline{N}(0,0,0,\Lambda, \underline{\Lambda}, \psi)(s)\Big| &\lesssim \overset{\circ}{\epsilon} \sup_{\overset{\circ}{I}} \Big(|B(s)| + |\underline{B}(s)|\Big).
\end{aligned}
$$

Hence,

$$
\sup_{\overset{\circ}{I}} |B(s)| + \sup_{\overset{\circ}{I}} |\underline{B}(s)| \lesssim \overset{\circ}{\delta}\big(\sup_{\overset{\circ}{I}} |B(s)| + \sup_{\overset{\circ}{I}} |\underline{B}(s)|\big).
$$

Hence $B, \underline{B}, D$ vanish identically on Σ_0.

Step 22. We have

$$
\left|\frac{dr}{ds} - 1\right| \lesssim \overset{\circ}{\epsilon}. \tag{9.8.77}
$$

Indeed, according to Step 15 and Lemma 9.54 we have

$$\begin{aligned}\frac{d}{ds}r(s) &= X_*\Big|_{SP} r^{\mathbf{S}} = \frac{1}{2\lambda}\varsigma\Psi'\nu^{\mathbf{S}}(r^{\mathbf{S}})\Big|_{SP} = (-1+\psi'(s))\frac{1}{2\lambda}\varsigma\nu^{\mathbf{S}}(r^{\mathbf{S}})\Big|_{SP}\\ &= (-1+\psi'(s))\left(\frac{1}{2\lambda}\varsigma\frac{r^{\mathbf{S}}}{2}(\varsigma^{\mathbf{S}})^{-1}\overline{\varsigma^{\mathbf{S}}(\underline{\kappa}^{\mathbf{S}}+a^{\mathbf{S}}\kappa^{\mathbf{S}})}\right)\Big|_{SP}.\end{aligned}$$

In view of Proposition 9.65, with $D=0$, $|\psi'|\lesssim r^{-1}\overset{\circ}{\delta}$. We deduce

$$\frac{d}{ds}r(s) = -\left(\frac{1}{2\lambda}\varsigma\frac{r^{\mathbf{S}}}{2}(\varsigma^{\mathbf{S}})^{-1}\overline{\varsigma^{\mathbf{S}}(\underline{\kappa}^{\mathbf{S}}+a^{\mathbf{S}}\kappa^{\mathbf{S}})}\right)\Big|_{SP} + O(\overset{\circ}{\delta}).$$

Step 23. Therefore the functions $B, \underline{B}, D$ vanish identically on the hypersurface Σ_0 defined by the function $\Psi(s) = -s+\psi(s)+c_0$ which accomplishes the main task of Theorem 9.52. More precisely we have produced a local hypersurface Σ_0, as defined in Step 12, foliated by the function $u^{\mathbf{S}}$, defined in Step 2 and extended in Step 3, such that the items 2–5 of the theorem are verified. The estimates in items 6–7 are an immediate consequence of Proposition 9.58. It only remains to prove the smoothness of the function $\Xi(s,\theta)$ in (9.8.56), Step 14 and the estimates for $F=(f,\underline{f},\log\lambda)$ in the last part of the theorem. To check the differentiability properties recall that

$$\begin{aligned}\partial_s\Xi(s,\theta) &= \Big(\Psi'(s)+\partial_P U(\theta,P(s))P'(s),\ 1+\partial_P S(\theta,P(s))P'(s),\ 0\Big),\\ \partial_\theta\Xi(s,\theta) &= \Big(\partial_\theta U(\theta,P(s)),\ \partial_\theta S(\theta,P(s)),\ 1\Big),\end{aligned}$$

where

$$\begin{aligned}\partial_P U(\cdot)P'(s) &= \Psi'(s)\partial_u U(\cdot)+\partial_s U(\cdot)+\Lambda'(s)\partial_\Lambda U(\cdot)+\underline{\Lambda}'(s)\partial_{\underline{\Lambda}}U(\cdot),\\ \partial_P S(\cdot)P'(s) &= \Psi'(s)\partial_u S(\cdot)+\partial_s S(\cdot)+\Lambda'(s)\partial_\Lambda S(\cdot)+\underline{\Lambda}'(s)\partial_{\underline{\Lambda}}S(\cdot).\end{aligned}$$

Thus to prove the smoothness of Ξ we need to appeal to the smoothness of U,S with respect to the parameters $\Lambda,\underline{\Lambda}$ and u,s. Though tedious, this can be easily done by appealing to the coupled system of equations (9.4.13), (9.4.14), (9.4.15), as in the proof of Theorem 9.32, and studying its dependence on these parameters.

Step 24. It only remains to derive the estimates (9.8.11) for the transition functions $F=(f,\underline{f},\log\lambda)$. To start with we have, in view of the construction of Σ_0 and the estimates for $F=(f,\underline{f},\log\lambda)$ of Theorem 9.32, for every $\mathbf{S}\subset\Sigma_0$

$$\|F\|_{\mathfrak{h}_{s_{\max}+1}(\mathbf{S})}\lesssim\overset{\circ}{\delta}. \tag{9.8.78}$$

To derive the remaining tangential derivatives of F along Σ_0 we commute the GCM system (9.4.13) of Proposition 9.33 with respect to $\nu=\nu^{\mathbf{S}}=e_3^{\mathbf{S}}+a^{\mathbf{S}}e_4^{\mathbf{S}}$ and then proceed, as in the proof of the a priori estimates of Theorem 9.37 to derive recursively the estimates, for $K = s_{max} + 1$,

$$\|\nu^l(F)\|_{\mathfrak{h}_{K-l}(\mathbf{S})} \lesssim \overset{\circ}{\delta} + r^{-2}\left(\left|\int_{\mathbf{S}} \nu^l(f)e^\Phi\right| + \left|\int_{\mathbf{S}} \nu^l(\underline{f})e^\Phi\right|\right) + \overset{\circ}{\delta} r^{-1}\left\|\nu^{\leq l-1} a^{\mathbf{S}}\right\|_{\mathfrak{h}_{K-l+1}(\mathbf{S})} + \|\nu^{\leq l-1} F\|_{\mathfrak{h}_{K-l+1}(\mathbf{S})}. \tag{9.8.79}$$

We already have estimates for the $\ell = 1$ modes of $F = (f, \underline{f})$. To estimate the $\ell = 1$ modes of $\nu^l(f, \underline{f})$, $l \geq 1$, we make use of the equations (9.8.65) and the vanishing of the $\ell = 1$ modes of $\eta^{\mathbf{S}}$, $\underline{\xi}^{\mathbf{S}}$ along Σ_0 to derive, recursively, for all $1 \leq l \leq K$,

$$\begin{aligned} r^{-2}\left(\left|\int_{\mathbf{S}} \nu^l(f)e^\Phi\right| + \left|\int_{\mathbf{S}} \nu^l(\underline{f})e^\Phi\right|\right) \lesssim\ & \overset{\circ}{\delta} + r^{-1}\overset{\circ}{\delta}\left\|\nu^{\leq l-1}\left(a, \underline{\Omega}^{\mathbf{S}}, \varsigma^{\mathbf{S}}\right)\right\|_{\mathfrak{h}_{K-l+1}(\mathbf{S})} \\ & + \overset{\circ}{\delta}\left\|\nu^{\leq l-1}\left(\underline{\xi}^{\mathbf{S}}, \eta^{\mathbf{S}}, \underline{\check{\omega}}^{\mathbf{S}}\right)\right\|_{\mathfrak{h}_{K-l+1}(\mathbf{S})} \\ & + r^{-1}\|\nu^{\leq l-1}(F)\|_{\mathfrak{h}_{K-l+1}(\mathbf{S})}. \end{aligned} \tag{9.8.80}$$

We can then proceed as in the proof of Proposition 9.58 to derive, recursively, the estimates

$$\begin{aligned} & r^{-1}\left\|\nu^{\leq l-1}\left(a^{\mathbf{S}}, \underline{\Omega}^{\mathbf{S}}, \varsigma^{\mathbf{S}}\right)\right\|_{\mathfrak{h}_{K-l+1}(\mathbf{S})} + \left\|\nu^{\leq l-1}\left(\underline{\xi}^{\mathbf{S}}, \eta^{\mathbf{S}}, \underline{\check{\omega}}^{\mathbf{S}}\right)\right\|_{\mathfrak{h}_{K-l+1}(\mathbf{S})} \\ \lesssim\ & 1 + r^{-1}\|F\|_{\mathfrak{h}_K(\mathbf{S})} + \sum_{j=1}^{l} \|\nu^j(F)\|_{\mathfrak{h}_{K-j}(\mathbf{S})}. \end{aligned} \tag{9.8.81}$$

Combining (9.8.79), (9.8.80), (9.8.81), we obtain

$$\|\nu^l(F)\|_{\mathfrak{h}_{K-l}(\mathbf{S})} \lesssim \overset{\circ}{\delta} + \sum_{j=0}^{l-1} \|\nu^j(F)\|_{\mathfrak{h}_{K-j}(\mathbf{S})},$$

which, together with (9.8.78), yields the desired estimate for all tangential derivatives

$$\sum_{j=0}^{K} \|\nu^j(F)\|_{\mathfrak{h}_{K-j}(\mathbf{S})} \lesssim \overset{\circ}{\delta}. \tag{9.8.82}$$

To complete the desired estimate for all derivatives we make use of the equations for $e_4^{\mathbf{S}}(F)$, due to the transversality conditions (9.4.9). The $e_3^{\mathbf{S}}$ derivatives can then be derived from $\nu^{\mathbf{S}} = e_3^{\mathbf{S}} + a^{\mathbf{S}} e_4^{\mathbf{S}}$ and the estimates for $a^{\mathbf{S}}$. This concludes the proof of Theorem 9.52.

Step 25. We now prove Corollary 9.53. Consider first the simpler case where

$$\|(f, \underline{f}, \log(\lambda))\|_{\mathfrak{h}_{s_{max}+1}(\mathbf{S}_0)} \lesssim \overset{\circ}{\delta},$$

so that the estimate (9.8.78) holds true for $\mathbf{S}_0$. We then proceed exactly as in Step 24 to derive the estimates (9.8.79), (9.8.80), (9.8.81) for our distinguished sphere $\mathbf{S}_0$. Note that $\mathbf{S}_0$ can be viewed as a deformation of the unique background sphere sharing the same south pole.

It remains to prove Corollary 9.53 in the more difficult case where

$$\|f\|_{\mathfrak{h}_{s_{max}+1}(\mathbf{S}_0)} + (r^{\mathbf{S}_0})^{-1}\|(\underline{f}, \log\lambda)\|_{\mathfrak{h}_{s_{max}+1}(\mathbf{S}_0)} \lesssim \overset{\circ}{\delta}.$$

In view of Lemma 9.15, with $\delta_1 = \overset{\circ}{\delta}$, we infer

$$\left|\frac{r^{\mathbf{S}_0}}{\overset{\circ}{r}} - 1\right| + \sup_{\mathbf{S}_0}\left|\frac{r^{\mathbf{S}_0}}{r} - 1\right| \lesssim \overset{\circ}{\delta}$$

so that r and $r^{\mathbf{S}_0}$ are comparable, and hence

$$\|f\|_{\mathfrak{h}_{s_{max}+1}(\mathbf{S}_0)} + r^{-1}\|(\underline{f}, \log(\lambda))\|_{\mathfrak{h}_{s_{max}+1}(\mathbf{S}_0)} \lesssim \overset{\circ}{\delta}. \tag{9.8.83}$$

Next, we introduce as in Step 24 the notation $K = s_{max} + 1$. We claim the following analog of (9.8.82):

$$\sum_{j=1}^{K}\|\nu^j(F)\|_{\mathfrak{h}_{K-j}(\mathbf{S}_0)} \lesssim \overset{\circ}{\delta}. \tag{9.8.84}$$

To complete the desired estimate for all derivatives we then make use, as in Step 24, of the equations for $e_4^{\mathbf{S}_0}(F)$, due to the transversality conditions (9.4.9), and recover the $e_3^{\mathbf{S}_0}$ derivatives from $\nu^{\mathbf{S}_0} = e_3^{\mathbf{S}_0} + a^{\mathbf{S}_0}e_4^{\mathbf{S}}$, which concludes the proof of Corollary 9.53.

It thus remains to prove (9.8.84). Note that $\mathbf{S}_0$ can be viewed as a deformation of the unique background sphere sharing the same south pole. We proceed exactly as in Step 24 to derive the estimates (9.8.80), (9.8.81) for our distinguished sphere $\mathbf{S}_0$, which yields, for all $1 \leq l \leq K$,

$$\begin{aligned} & r^{-2}\left(\left|\int_{\mathbf{S}}\nu^l(f)e^{\Phi}\right| + \left|\int_{\mathbf{S}}\nu^l(\underline{f})e^{\Phi}\right|\right) \\ \lesssim \;& \overset{\circ}{\delta} + r^{-1}\|F\|_{\mathfrak{h}_K(\mathbf{S})} + \sum_{j=1}^{l}\|\nu^j(F)\|_{\mathfrak{h}_{K-j}(\mathbf{S})} \end{aligned} \tag{9.8.85}$$

and

$$r^{-1}\left\|\nu^{\leq l-1}a^{\mathbf{S}}\right\|_{\mathfrak{h}_{K-l+1}(\mathbf{S})} \lesssim 1 + r^{-1}\|F\|_{\mathfrak{h}_K(\mathbf{S})} + \sum_{j=1}^{l}\|\nu^j(F)\|_{\mathfrak{h}_{K-j}(\mathbf{S})}. \tag{9.8.86}$$

We now claim the following sharpened version of (9.8.79):

$$\begin{aligned} \|\nu^l(F)\|_{\mathfrak{h}_{K-l}(\mathbf{S})} \lesssim\; & \overset{\circ}{\delta} + r^{-2}\left(\left|\int_{\mathbf{S}}\nu^l(f)e^{\Phi}\right| + \left|\int_{\mathbf{S}}\nu^l(\underline{f})e^{\Phi}\right|\right) \\ & + \overset{\circ}{\delta}r^{-1}\left\|\nu^{\leq l-1}a^{\mathbf{S}}\right\|_{\mathfrak{h}_{K-l+1}(\mathbf{S})} + \|f\|_{\mathfrak{h}_K(\mathbf{S})} \\ & + r^{-1}\|(\underline{f}, \log(\lambda))\|_{\mathfrak{h}_K(\mathbf{S})} + \sum_{j=1}^{l-1}\|\nu^j(F)\|_{\mathfrak{h}_{K-j}(\mathbf{S})}. \end{aligned} \tag{9.8.87}$$

Then, (9.8.85), (9.8.86) and (9.8.87) imply

$$\|\nu^l(F)\|_{\mathfrak{h}_{K-l}(\mathbf{S})} \lesssim \mathring{\delta} + \|f\|_{\mathfrak{h}_K(\mathbf{S})} + r^{-1}\|(\underline{f}, \log(\lambda))\|_{\mathfrak{h}_K(\mathbf{S})} + \sum_{j=1}^{l-1} \|\nu^j(F)\|_{\mathfrak{h}_{K-j}(\mathbf{S})}.$$

Together with (9.8.83), we deduce (9.8.84) by iteration.

Finally, it remains to prove (9.8.87). As for the proof of (9.8.79), we commute the GCM system (9.4.13) of Proposition 9.33 with respect to $\nu^{\mathbf{S}} = e_3^{\mathbf{S}} + a^{\mathbf{S}} e_4^{\mathbf{S}}$ and then proceed, as in the proof of the a priori estimates of Theorem 9.37 to derive (9.8.87) recursively. To obtain a stronger conclusion than (9.8.79), we need to analyze the differentiation w.r.t. $\nu^{\mathbf{S}}$ more carefully. First, note that the commutator $[\not{\mathfrak{d}}^{\mathbf{S}}, \nu^{\mathbf{S}}]F$ satisfies, in view of Lemma 2.68,

$$\|[\not{\mathfrak{d}}^{\mathbf{S}}, \nu^{\mathbf{S}}]F\|_{\mathfrak{h}_l(\mathbf{S})} \lesssim \mathring{\epsilon} r^{-1}\|F\|_{\mathfrak{h}_{l+1}(\mathbf{S})} + \mathring{\epsilon}\|\nu^{\mathbf{S}} F\|_{\mathfrak{h}_l(\mathbf{S})}$$

where the important observation is that the first term on the right-hand side gains a power of r^{-1} which is consistent with (9.8.87). It remains to analyze the differentiation of the error terms $\mathrm{Err}_1, \ldots, \mathrm{Err}_6$ of the GCM system w.r.t. $\nu^{\mathbf{S}}$. To this end, in what follows, we single out all the terms that lose one power of r^{-1} in view of the anomalous behavior of $(\underline{f}, \log(\lambda))$ compared to the one in Step 24, and denote by $\cdots$ all terms that behave as before. We have by direct check

$$\mathrm{Err}(\kappa, \kappa') = \cdots, \qquad \mathrm{Err}(\underline{\kappa}, \underline{\kappa}') = -\frac{1}{4}\kappa \underline{f}^2 + \cdots,$$
$$\mathrm{Err}(\zeta, \zeta') = \cdots, \qquad \mathrm{Err}(\rho, \rho') = \cdots, \qquad \mathrm{Err}(\mu, \mu') = \cdots.$$

Then, in view of the identities in Lemma 9.23

$$\begin{aligned}
\mathrm{Err}(e'_\theta \kappa', e_\theta \kappa) &= (e^a - 1)\frac{1}{2}\underline{f} e_4 \kappa + \cdots, \\
\mathrm{Err}(e'_\theta \underline{\kappa}', e_\theta \underline{\kappa}) &= (e^{-a} - 1)\left(e'_\theta \not{d}'_1 \underline{f} + \frac{1}{2}\underline{f} e_4 \underline{\kappa}\right) \\
&+ e^{-a}\left[e'_\theta\left(-\frac{1}{4}\kappa \underline{f}^2\right) + e'_\theta(a)\left(\not{d}'_1 \underline{f} + -\frac{1}{4}\kappa \underline{f}^2\right)\right] + \cdots, \\
\mathrm{Err}(e'_\theta \mu', e_\theta \mu) &= \cdots.
\end{aligned}$$

In view of (9.3.14), we deduce

$$\begin{aligned}
\mathrm{Err}_1 &= -\frac{1}{r^2}(e^a - 1)\underline{f} + \cdots, \\
\mathrm{Err}_2 &= \frac{1}{r^2}\left\{(e^{-a} - 1)\left((\not{\mathfrak{d}}^{\mathbf{S}})^2 \underline{f} + \underline{f}\right) + e^{-a}\left[\not{\mathfrak{d}}^{\mathbf{S}}(\underline{f}^2) + \not{\mathfrak{d}}^{\mathbf{S}}(a)\left(\not{\mathfrak{d}}^{\mathbf{S}} \underline{f} + \underline{f}^2\right)\right]\right\} + \cdots, \\
\mathrm{Err}_3 &= \cdots.
\end{aligned}$$

We infer, in view of the expression of Err_4 and Err_5 in section 9.3.1,

$$\begin{aligned}
\mathrm{Err}_4 &= -\frac{1}{r^2}(e^a - 1)\underline{f} + \cdots, \\
\mathrm{Err}_5 &= \frac{1}{r^2}\left\{(e^{-a} - 1)\left((\not{\mathfrak{d}}^{\mathbf{S}})^2 \underline{f} + \underline{f}\right) + e^{-a}\left[\not{\mathfrak{d}}^{\mathbf{S}}(\underline{f}^2) + \not{\mathfrak{d}}^{\mathbf{S}}(a)\left(\not{\mathfrak{d}}^{\mathbf{S}} \underline{f} + \underline{f}^2\right)\right]\right\} + \cdots.
\end{aligned}$$

Finally, in view of the expression of Err_6 in section 9.3.2, we have

$$\mathrm{Err}_6 = \frac{r^{\mathbf{S}}}{r}(e^a - 1 - a) - a\left(\frac{r^{\mathbf{S}}}{r} - 1\right) + \cdots.$$

Thus, to conclude the proof of (9.8.87), it suffices to show that all terms singled out in the above expression of $\mathrm{Err}_1, \ldots, \mathrm{Err}_6$ gain a power of r^{-1} when differentiated w.r.t. $\nu^{\mathbf{S}}$. Now, they are all quadratic expressions involving a, $\underline{f}$, r and $r^{\mathbf{S}}$. Since $\nu^{\mathbf{S}}(a)$ and $\nu^{\mathbf{S}}(\underline{f})$ gain r^{-1} compared to a and $\underline{f}$ in view of (9.8.84), the conclusion then follows from the straightforward estimate

$$\frac{|\nu^{\mathbf{S}}(r^{\mathbf{S}})|}{r^{\mathbf{S}}} + \frac{|\nu^{\mathbf{S}}(r)|}{r} \lesssim r^{-1}.$$

Chapter Ten

Regge-Wheeler Type Equations

The goal of this chapter is to prove Theorem 5.17 and Theorem 5.18 concerning the weighted estimates for the solution ψ to

$$\square_2\psi = V\psi + N, \qquad V = -\kappa\underline{\kappa}.$$

Recall that these theorems were used in Chapter 5 to prove Theorem M1.

The structure of the chapter is as follows.

- In section 10.1, we prove basic Morawetz estimates for ψ.
- In section 10.2, we prove r^p-weighted estimates in the spirit of Dafermos and Rodnianski [23] for ψ. In particular, we obtain as an immediate corollary the proof of Theorem 5.17 in the case $s = 0$ (i.e., without commutating the equation of ψ with derivatives).
- In section 10.3, we use a variation of the method of [5] to derive slightly stronger weighted estimates and prove Theorem 5.18 in the case $s = 0$ (i.e., without commutating the equation of $\check{\psi}$ with derivatives).
- In section 10.4, commuting the equation of ψ with derivatives, we complete the proof of Theorem 5.17 by controlling higher order derivatives of ψ, i.e., for $s \leq k_{small}+30$. Also, commuting the equation of $\check{\psi}$ with derivatives, we complete the proof of Theorem 5.18 by controlling higher order derivatives of $\check{\psi}$, i.e., for $s \leq k_{small} + 29$.

10.1 BASIC MORAWETZ ESTIMATES

Recall

- the definitions in section 5.1.1 of ${}^{(trap)}\mathcal{M}$, ${}^{(tr\not{p})}\mathcal{M}$, τ, $\Sigma(\tau)$ and ${}^{(trap)}\Sigma$,
- the main quantities involved in the energy and Morawetz estimates, e.g., $E[\psi](\tau)$, $\mathrm{Mor}[\psi](\tau_1,\tau_2)$, $\mathrm{Morr}[\psi](\tau_1,\tau_2)$, $F[\psi](\tau_1,\tau_2)$, $J_\delta[\psi,N](\tau_1,\tau_2)$ and $\dot{B}^s_{p;R}[\psi](\tau_1,\tau_2)$, introduced in section 5.1.4.

The following theorem claims basic Morawetz estimates for the solution ψ of the wave equation (5.3.5).

Theorem 10.1 (Morawetz). *Let ψ a reduced 2-scalar solution to*

$$\square_2\psi = V\psi + N, \qquad V = -\kappa\underline{\kappa}.$$

Let $\delta > 0$ be a fixed small constant verifying $0 < \epsilon \ll \delta$. The following estimates hold true in $\mathcal{M}(\tau_1,\tau_2)$, $0 \leq \tau_1 < \tau_2 \leq \tau_*$,

$$E[\psi](\tau_2) + Mor[\psi](\tau_1,\tau_2) + F[\psi](\tau_1,\tau_2) \lesssim E[\psi](\tau_1) + J_\delta[\psi,N](\tau_1,\tau_2) + O(\epsilon)\dot{B}_{\delta\,;\,4m_0}[\psi](\tau_1,\tau_2). \tag{10.1.1}$$

Also,

$$E[\psi](\tau_2) + Morr[\psi](\tau_1,\tau_2) + F[\psi](\tau_1,\tau_2) \lesssim E[\psi](\tau_1) + J_\delta[\psi,N](\tau_1,\tau_2) + \dot{B}_{\delta\,;\,4m_0}[\psi](\tau_1,\tau_2). \tag{10.1.2}$$

Remark 10.2. *Note that the bulk term $\dot{B}_{\delta\,;\,4m_0}[\psi](\tau_1,\tau_2)$ cannot yet be absorbed on the left-hand side of the inequality. To do that we will rely on the r^p-weighted estimates of Theorem 10.37.*

Remark 10.3. *In addition to ϵ and δ, the proof of Theorem 10.1 will involve several smallness constants: C^{-1}, $\widehat{\delta}$, δ_1, $\delta_{\mathcal{H}}$, $\epsilon_{\mathcal{H}}$, $\Lambda_{\mathcal{H}}^{-1}$ and Λ^{-1}. These smallness constants will be chosen such that*

$$0 < \epsilon \ll \widehat{\delta},\, \delta_{\mathcal{H}},\, \epsilon_{\mathcal{H}},\, \Lambda_{\mathcal{H}}^{-1},\, \Lambda^{-1} \ll \delta_1 \ll C^{-1}. \tag{10.1.3}$$

In addition, $\widehat{\delta}$, $\epsilon_{\mathcal{H}}$, $\Lambda_{\mathcal{H}}^{-1}$ and Λ^{-1} will in fact be chosen towards the end of the proof as explicit powers of $\delta_{\mathcal{H}}$, see (10.1.63), (10.1.65) *and Proposition 10.30.*

The goal of this section is to prove Theorem 10.1. This will be achieved in section 10.1.15.

10.1.1 Structure of the proof of Theorem 10.1

To prove Theorem 10.1, we proceed as follows:

- In section 10.1.2, we introduce a simplified set of assumptions of the Ricci coefficients which is sufficient in order to prove Theorem 10.1.
- In section 10.1.3, we discuss notations concerning functions depending on m and r.
- In section 10.1.4, we compute the deformation tensor of the vectorfields R, T, and $X = f(r,m)R$.
- In section 10.1.5, we introduce the basic integral identities for wave equations that will be used repeatedly in the proof of Theorem 10.1.
- In section 10.1.6, we derive the main Morawetz identity.
- In section 10.1.7, we derive a first estimate. This estimate is insufficient due to
  - a lack of positivity of the bulk in the region $3m \le r \le 4m$,
  - a log divergence of a suitable choice of vectorfield at $r = 2m$,
  - a degeneracy at $r = 2m$.
- In section 10.1.8, we add a correction and rely on a Poincaré inequality to obtain a positive estimate also on the region $3m \le r \le 4m$.
- In section 10.1.9, we perform a cut-off to remove above mentioned log divergence at $r = 2m$.
- In section 10.1.10, we introduce the redshift vectorfield to remove the above mentioned degeneracy at $r = 2m$.
- In section 10.1.11, we combine the previous estimates with the redshift vectorfield to obtain a bulk term suitable on the whole spacetime $\mathcal{M}$.
- In section 10.1.12, we prove the positivity of the boundary terms arising from adding a large multiple of the energy estimate to the Morawetz estimate.
- In section 10.1.13, combining the good properties of the bulk and of the bound-

ary terms established so far, we obtain a first Morawetz estimate providing in particular the control of the quantity $\mathrm{Mor}[\psi]$.
- In section 10.1.14, we analyze an error term appearing in the right-hand side of the above mentioned Morawetz estimate.
- Finally, in section 10.1.15, we add a correction to upgrade the control of $\mathrm{Mor}[\psi]$ to the control of the quantity $\mathrm{Morr}[\psi]$, hence concluding the proof of Theorem 10.1.

10.1.2 A simplified set of assumptions

To prove Theorem 10.1, it suffices to make a simplified set of assumptions. Define

$$u_{trap} = \begin{cases} 1+\tau & \text{for} \quad r \in [\frac{5m_0}{2}, \frac{7m_0}{2}], \\ 1 & \text{for} \quad r \notin [\frac{5m_0}{2}, \frac{7m_0}{2}]. \end{cases} \tag{10.1.4}$$

For $k = 0, 1$, we assume the following.

Mor1. The renormalized Ricci coefficients $\check{\Gamma}^{\leq k}$ verify, on $\mathcal{M} = {}^{(int)}\mathcal{M} \cup {}^{(ext)}\mathcal{M}$,

$$\begin{aligned} |\check{\Gamma}^{\leq k}| &\lesssim \epsilon r^{-1} u_{trap}^{-1-\delta_{dec}}, \\ \left|\mathfrak{d}^{\leq k}\left(\omega + \frac{m}{r^2}, \xi\right)\right| &\lesssim \epsilon r^{-2} u_{trap}^{-1-\delta_{dec}}. \end{aligned} \tag{10.1.5}$$

Mor2. The Gauss curvature K of S and ρ verify

$$\begin{aligned} \left|\mathfrak{d}^{\leq k}\left(\rho + \frac{2m}{r^3}\right)\right| &\lesssim \epsilon r^{-2}\, u_{trap}^{-1-\delta_{dec}}, \\ \left|\mathfrak{d}^{\leq k}\left(K - \frac{1}{r^2}\right)\right| &\lesssim \epsilon r^{-2}\, u_{trap}^{-1-\delta_{dec}}. \end{aligned} \tag{10.1.6}$$

Mor3. We also assume

$$\begin{aligned} |m - m_0| &\lesssim \epsilon m_0, \\ |\mathfrak{d}^{\leq k}(e_3 m, r^2 e_4 m)| &\lesssim \epsilon\, u_{trap}^{-1-\delta_{dec}}. \end{aligned} \tag{10.1.7}$$

Remark 10.4. *Note that in the case when the bootstrap constant $\epsilon = 0$, i.e., in Schwarzschild, the assumptions made above are consistent with the behavior relative to the regular frame (near horizon)*

$$e_3 = \Upsilon^{-1}\partial_t - \partial_r, e_4 = \partial_t + \Upsilon\partial_r.$$

10.1.3 Functions depending on m and r

In order to prove Theorem 10.1, we will adapt the derivation of the Morawetz estimate for the wave equation in Schwarzschild. In particular, we will need to consider various scalar functions, used to define suitable analogs of the vectorfields in Schwarzschild, which depend on m and r. Now, m is a scalar function unlike the Schwarzschild case where it is constant. To take this into account, we will rely on the following lemma.

Lemma 10.5. *Let $f = f(r,m)$ a C^1 function of r and m. Then, we have*

$$\begin{aligned} e_4(f(r,m)) &= \partial_r f(r,m) e_4(r) + O(\epsilon r^{-2} u_{trap}^{-1-\delta_{dec}} |\partial_m f|), \\ e_3(f(r,m)) &= \partial_r f(r,m) e_3(r) + O(\epsilon u_{trap}^{-1-\delta_{dec}} |\partial_m f|), \\ e_4(e_3(f(r,m))) &= \partial_r^2 f(r,m) e_4(r) e_3(r) + \partial_r f(r,m) e_4(e_3(r)) \\ &\quad + O(\epsilon r^{-2} u_{trap}^{-1-\delta_{dec}} (r|\partial_r \partial_m f| + |\partial_m^2 f|)), \\ e_3(e_4(f(r,m))) &= \partial_r^2 f(r,m) e_4(r) e_3(r) + \partial_r f(r,m) e_3(e_4(r)) \\ &\quad + O(\epsilon r^{-2} u_{trap}^{-1-\delta_{dec}} (r|\partial_r \partial_m f| + |\partial_m^2 f|)), \\ e_\theta(f(r,m)) &= 0. \end{aligned}$$

Proof. Straightforward verification using (10.1.7). □

Remark 10.6. *Note that in the sequel, $\partial_r f$ will not denote a spacetime coordinate vectorfield applied to f, but instead the partial derivative with respect to the variable r of the function $f(r,m)$.*

10.1.4 Deformation tensors of the vectorfields R, T, X

Recall the definition (5.1.10) of the regular vectorfields[1]

$$T = \frac{1}{2}(e_4 + \Upsilon e_3), \qquad R = \frac{1}{2}(e_4 - \Upsilon e_3).$$

Note that

$$-g(T,T) = g(R,R) = \Upsilon, \qquad g(T,R) = 0.$$

Note also that

$$R(r) = 1 - \frac{2m}{r} + O(\epsilon u_{trap}^{-1-\delta_{dec}}), \qquad T(r) = O(\epsilon u_{trap}^{-1-\delta_{dec}}).$$

Lemma 10.7. *The following hold true.*

1. *The components of the deformation tensor of $R = \frac{1}{2}(e_4 - \Upsilon e_3)$ are given by*

$$\begin{aligned} \left| {}^{(R)}\pi_{34} + \frac{4m}{r^2} \right| &\lesssim \epsilon r^{-1} u_{trap}^{-1-\delta_{dec}}, \\ \left| {}^{(R)}\pi(e_A, e_B) - \frac{2}{r}\Upsilon \delta_{AB} \right| &\lesssim \epsilon r^{-1} u_{trap}^{-1-\delta_{dec}}, \\ \left| {}^{(R)}\pi_{33} \right| &\lesssim \epsilon r^{-1} u_{trap}^{-1-\delta_{dec}}, \\ \left| {}^{(R)}\pi_{3\theta} \right| &\lesssim \epsilon r^{-1} u_{trap}^{-1-\delta_{dec}}, \\ \left| {}^{(R)}\pi_{4\theta} \right| &\lesssim \epsilon r^{-1} u_{trap}^{-1-\delta_{dec}}. \end{aligned}$$

[1] In Schwarzschild, in standard coordinates, we have $T = \partial_t$, $R = \Upsilon \partial_r$ which are regular near the horizon.

Moreover,

$$\left| {}^{(R)}\pi_{44} \right| \lesssim \epsilon r^{-2} u_{trap}^{-1-\delta_{dec}}.$$

2. *If* $V := -\kappa \underline{\kappa}$, *we have*

$$\begin{aligned} e_3(V) &= \frac{8}{r^3}\left(1 - \frac{3m}{r}\right) + O(\epsilon) r^{-3} u_{trap}^{-1-\delta_{dec}}, \\ e_4(V) &= -\frac{8\Upsilon}{r^3}\left(1 - \frac{3m}{r}\right) + O(\epsilon) r^{-3} u_{trap}^{-1-\delta_{dec}}, \end{aligned} \tag{10.1.8}$$

and

$$\begin{aligned} R(V) &= -\frac{8\Upsilon}{r^3}\left(1 - \frac{3m}{r}\right) + O(\epsilon) r^{-3} u_{trap}^{-1-\delta_{dec}}, \\ T(V) &= O(\epsilon) r^{-3} u_{trap}^{-1-\delta_{dec}}. \end{aligned}$$

3. *All components of the deformation tensor of* $T = \frac{1}{2}(e_4 + \Upsilon e_3)$ *can be bounded by* $O(\epsilon r^{-1} u_{trap}^{-1-\delta_{dec}})$. *Moreover,*

$$\left| {}^{(T)}\pi_{44} \right| \lesssim \epsilon r^{-2} u_{trap}^{-1-\delta_{dec}}.$$

Proof. We have

$$\begin{aligned} {}^{(R)}\pi_{44} &= \mathbf{g}(\mathbf{D}_4(e_4 - \Upsilon e_3), e_4) = 2e_4(\Upsilon) + 4\Upsilon\omega, \\ {}^{(R)}\pi_{34} &= \frac{1}{2}\mathbf{g}(\mathbf{D}_3(e_4 - \Upsilon e_3), e_4) + \frac{1}{2}\mathbf{g}(\mathbf{D}_4(e_4 - \Upsilon e_3), e_3) \\ &= e_3(\Upsilon) - 2\Upsilon\underline{\omega} + 2\omega, \\ {}^{(R)}\pi_{33} &= \mathbf{g}(\mathbf{D}_3(e_4 - \Upsilon e_3), e_3) = -4\underline{\omega}, \\ {}^{(R)}\pi_{AB} &= \frac{1}{2}\mathbf{g}(\mathbf{D}_A(e_4 - \Upsilon e_3), e_B) + \frac{1}{2}\mathbf{g}(\mathbf{D}_B(e_4 - \Upsilon e_3), e_A), \\ &= {}^{(1+3)}\chi_{AB} - \Upsilon\, {}^{(1+3)}\underline{\chi}_{AB} \\ &= \frac{1}{2}(\kappa - \Upsilon\underline{\kappa})\delta_{AB} + {}^{(1+3)}\widehat{\chi}_{AB} - \Upsilon\, {}^{(1+3)}\widehat{\underline{\chi}}_{AB}. \end{aligned}$$

Note that

$$\begin{aligned} e_3(\Upsilon) &= e_3\left(1 - \frac{2m}{r}\right) = \frac{2m}{r^2} e_3(r) - \frac{2e_3 m}{r} = \frac{m}{r}(\underline{\kappa} + \underline{A}) + O(\epsilon r^{-1} u_{trap}^{-1-\delta_{dec}}) \\ &= \frac{m}{r}\underline{\kappa} + O(\epsilon r^{-1} u_{trap}^{-1-\delta_{dec}}) = -\frac{2m}{r^2} + O(\epsilon r^{-1} u_{trap}^{-1-\delta_{dec}}), \\ e_4(\Upsilon) &= e_4\left(1 - \frac{2m}{r}\right) = \frac{2m}{r^2} e_4(r) - \frac{2e_4 m}{r} = \frac{m}{r}(\kappa + A) + O(\epsilon r^{-2} u_{trap}^{-1-\delta_{dec}}) \\ &= \frac{m}{r}\kappa + O(\epsilon r^{-2} u_{trap}^{-1-\delta_{dec}}) = \frac{2m}{r^2}\Upsilon + O(\epsilon r^{-2} u_{trap}^{-1-\delta_{dec}}). \end{aligned}$$

Thus

$$\begin{aligned}
{}^{(R)}\pi_{44} &= O(\epsilon r^{-2} u_{trap}^{-1-\delta_{dec}}),\\
{}^{(R)}\pi_{33} &= O(\epsilon r^{-1} u_{trap}^{-1-\delta_{dec}}),\\
{}^{(R)}\pi_{34} &= -4\frac{m}{r^2} + O(\epsilon r^{-1} u_{trap}^{-1-\delta_{dec}}),\\
{}^{(R)}\pi_{AB} &= \frac{2\Upsilon}{r}\delta_{AB} + O(\epsilon r^{-1} u_{trap}^{-1-\delta_{dec}}).
\end{aligned}$$

Also, in view of

$$\left|\xi, \underline{\xi}, \eta, \underline{\eta}, \zeta\right| \lesssim \epsilon r^{-1} u_{trap}^{-1-\delta_{dec}},$$

we deduce

$$\left|{}^{(R)}\pi_{3\theta}, {}^{(R)}\pi_{4\theta}\right| \lesssim \epsilon r^{-1} u_{trap}^{-1-\delta_{dec}}$$

as desired.

To prove the second part of the lemma we write

$$\begin{aligned}
e_3(V) &= -e_3(\kappa)\underline{\kappa} - \kappa e_3(\underline{\kappa})\\
&= -\left(-\frac{1}{2}\kappa\underline{\kappa} + 2\underline{\omega}\kappa + 2\rho + O(\epsilon r^{-2} u_{trap}^{-1-\delta_{dec}})\right)\underline{\kappa}\\
&\quad -\kappa\left(-\frac{1}{2}\underline{\kappa}^2 - 2\underline{\omega}\,\underline{\kappa} + O(\epsilon r^{-2} u_{trap}^{-1-\delta_{dec}})\right)\\
&= (\kappa\underline{\kappa} - 2\rho)\underline{\kappa} + O(\epsilon r^{-3} u_{trap}^{-1-\delta_{dec}}).
\end{aligned}$$

On the other hand,

$$\begin{aligned}
\kappa\underline{\kappa} - 2\rho &= -\left(\frac{2\Upsilon}{r} + O(\epsilon) r^{-1} u_{trap}^{-1-\delta_{dec}}\right)\left(\frac{2}{r} + O(\epsilon) r^{-1} u_{trap}^{-1-\delta_{dec}}\right) + \frac{4m}{r^3}\\
&\quad + O(\epsilon r^{-3} u_{trap}^{-1-\delta_{dec}})\\
&= -\frac{4}{r^2}\left(1 - \frac{3m}{r}\right) + O(\epsilon r^{-2} u_{trap}^{-1-\delta_{dec}}).
\end{aligned}$$

Hence,

$$\begin{aligned}
e_3(V) &= (\kappa\underline{\kappa} - 2\rho)\underline{\kappa} + O(\epsilon r^{-3} u_{trap}^{-1-\delta_{dec}})\\
&= \frac{8}{r^3}\left(1 - \frac{3m}{r}\right) + O(\epsilon r^{-3} u_{trap}^{-1-\delta_{dec}})
\end{aligned}$$

and similarly for $e_4(V)$. Thus,

$$\begin{aligned}
R(V) &= \frac{1}{2}(e_4 - \Upsilon e_3)V = -\frac{8\Upsilon}{r^3}\left(1 - \frac{3m}{r}\right) + O(\epsilon r^{-3} u_{trap}^{-1-\delta_{dec}}),\\
T(V) &= \frac{1}{2}(e_4 + \Upsilon e_3)V = O(\epsilon) r^{-3} u_{trap}^{-1-\delta_{dec}},
\end{aligned}$$

as desired.

To prove the last part of the lemma we write

$$\begin{aligned}
{}^{(T)}\pi_{44} &= \mathbf{g}\left(\mathbf{D}_4(e_4+\Upsilon e_3), e_4\right) = -2e_4(\Upsilon) - 4\Upsilon\omega,\\
{}^{(T)}\pi_{34} &= \frac{1}{2}\mathbf{g}(\mathbf{D}_3(e_4+\Upsilon e_3), e_4) + \frac{1}{2}\mathbf{g}(\mathbf{D}_4(e_4+\Upsilon e_3), e_3)\\
&= -e_3(\Upsilon) + 2\Upsilon\underline{\omega} + 2\omega,\\
{}^{(T)}\pi_{33} &= \mathbf{g}\left(\mathbf{D}_3(e_4+\Upsilon e_3), e_3\right) = -4\underline{\omega},\\
{}^{(T)}\pi_{AB} &= \frac{1}{2}\mathbf{g}\left(\mathbf{D}_A(e_4+\Upsilon e_3), e_B\right) + \frac{1}{2}\mathbf{g}\left(\mathbf{D}_B(e_4+\Upsilon e_3), e_A\right),\\
&= {}^{(1+3)}\chi_{AB} + \Upsilon\, {}^{(1+3)}\underline{\chi}_{AB}\\
&= \frac{1}{2}(\kappa + \Upsilon\underline{\kappa})\delta_{AB} + {}^{(1+3)}\widehat{\chi}_{AB} + \Upsilon\, {}^{(1+3)}\widehat{\underline{\chi}}_{AB},
\end{aligned}$$

and the proof continues as above in view of our assumptions. □

Consider now $X = f(r,m)R$ and ${}^{(X)}\pi$ its deformation tensor. We have the following lemma.

Lemma 10.8. *Let $X = f(r,m)R$ and ${}^{(X)}\pi$ its deformation tensor. We have*

$$ {}^{(X)}\pi = {}^{(X)}\dot{\pi} + \epsilon\, {}^{(X)}\ddot{\pi} $$

where[2]

- *The only nonvanishing components of ${}^{(X)}\dot{\pi}$ are*

$$\begin{aligned}
{}^{(X)}\dot{\pi}_{33} &= 2\partial_r f,\\
{}^{(X)}\dot{\pi}_{44} &= 2\partial_r f\Upsilon^2,\\
{}^{(X)}\dot{\pi}_{34} &= -\frac{4m}{r^2}f - 2\partial_r f\Upsilon,\\
{}^{(X)}\dot{\pi}_{AB} &= f\frac{2\Upsilon}{r}\delta_{AB}.
\end{aligned}$$

- *All components of ${}^{(X)}\ddot{\pi}$ verify*

$$ \left|{}^{(X)}\ddot{\pi}\right| \lesssim r^{-1}u_{trap}^{-1-\delta_{dec}}(|f| + r|\partial_m f| + r^2|\partial_r f|). $$

Moreover,

$$ \left|{}^{(X)}\ddot{\pi}_{44}\right| \lesssim r^{-2}u_{trap}^{-1-\delta_{dec}}(|f| + r|\partial_m f| + r^2|\partial_r f|). $$

Proof. Clearly,

$$ {}^{(X)}\pi_{\mu\nu} = f\,{}^{(R)}\pi_{\mu\nu} + e_\mu f R_\nu + e_\nu f R_\mu. $$

[2]Recall from Remark 10.6 that $\partial_r f$ does not denote a spacetime coordinate vectorfield applied to f, but instead the partial derivative with respect to the variable r of the function $f(r,m)$.

Therefore, since $\mathbf{g}(R, e_3) = -1$, $\mathbf{g}(R, e_4) = \Upsilon$ and

$$\Big|e_4(r) - \Upsilon, e_3(r) + 1\Big| \lesssim \epsilon\, u_{trap}^{-1-\delta_{dec}},$$

and using Lemma 10.5, we deduce

$$\begin{aligned}
{}^{(X)}\pi_{33} &= f\,{}^{(R)}\pi_{33} - 2e_3(f) = f\,{}^{(R)}\pi_{33} - 2\partial_r f e_3(r) - 2\partial_m f e_3(m)\\
&= 2\partial_r f + O\Big(\epsilon r^{-1} u_{trap}^{-1-\delta_{dec}}(|f| + r|\partial_m f| + r^2|\partial_r f|)\Big)\\
{}^{(X)}\pi_{44} &= f\,{}^{(R)}\pi_{44} + 2\Upsilon e_4(f) = f\,{}^{(R)}\pi_{44} + 2\Upsilon\partial_r f e_4(r) + 2\Upsilon\partial_m f e_4(m)\\
&= 2\partial_r f \Upsilon^2 + O\Big(\epsilon r^{-2} u_{trap}^{-1-\delta_{dec}}(|f| + r|\partial_m f| + r^2|\partial_r f|)\Big)\\
{}^{(X)}\pi_{34} &= f\,{}^{(R)}\pi_{34} + e_3(f)\Upsilon - e_4(f)\\
&= f\,{}^{(R)}\pi_{34} + (\partial_r f e_3(r) + \partial_m f e_3(m))\Upsilon - (\partial_r f e_4(r) + \partial_m f e_4(m))\\
&= -\frac{4m}{r^2} f - 2\partial_r f\Upsilon + O\Big(\epsilon r^{-1} u_{trap}^{-1-\delta_{dec}}(|f| + r|\partial_m f| + r^2|\partial_r f|)\Big).
\end{aligned}$$

This concludes the proof of the lemma. □

10.1.5 Basic integral identities

We recall, see section 2.4.1, that wave equations for $\psi \in \mathfrak{s}_2(\mathcal{M})$ of the form

$$\square_2\psi = V\psi + N[\psi], \qquad V = -\kappa\underline{\kappa}, \tag{10.1.9}$$

can be lifted to the spacetime version[3]

$$\dot{\square}\Psi = V\Psi + N[\Psi] \tag{10.1.10}$$

where $\Psi \in \mathcal{S}_2(\mathcal{M})$ and $N[\Psi] \in \mathcal{S}_2(\mathcal{M})$ are defined according to Proposition 2.106. In fact,

$$\begin{aligned}
\Psi_{\theta\theta} &= -\Psi_{\varphi\varphi} = \psi, & \Psi_{\theta\varphi} = 0.\\
N_{\theta\theta}[\Psi] &= N_{\varphi\varphi}[\Psi] = N(\psi), & N[\Psi]_{\theta\varphi} = 0.
\end{aligned}$$

All estimates for (10.1.10) derived in this section can be easily transferred to estimates for (10.1.9) and vice versa.

Consider wave equations of the form

$$\dot{\square}_{\mathbf{g}}\Psi = V\Psi + N \tag{10.1.11}$$

with $\Psi \in \mathcal{S}_2(\mathcal{M})$ and N a given symmetric traceless tensor, i.e., $\mathcal{N} \in \mathcal{S}_2(\mathcal{M})$.

Proposition 10.9. *Assume* $\Psi \in \mathcal{S}_2(\mathcal{M})$ *verifies* (10.1.10). *Then,*

[3]See section 2.4.1.1 and Appendix D for the precise definition of the covariant derivative $\dot{\mathbf{D}}$ and wave operator $\dot{\square}$ on $\mathcal{S}_2(\mathcal{M})$.

1. *The energy-momentum tensor $\mathcal{Q} = \mathcal{Q}[\Psi]$ given by*

$$\begin{aligned}\mathcal{Q}_{\mu\nu} : &= \dot{\mathbf{D}}_\mu \Psi \cdot \dot{\mathbf{D}}_\nu \Psi - \frac{1}{2}\mathbf{g}_{\mu\nu}\left(\dot{\mathbf{D}}_\lambda \Psi \cdot \dot{\mathbf{D}}^\lambda \Psi + V\Psi \cdot \Psi\right)\\ &= \dot{\mathbf{D}}_\mu \Psi \cdot \dot{\mathbf{D}}_\nu \Psi - \frac{1}{2}\mathbf{g}_{\mu\nu}\mathcal{L}(\Psi)\end{aligned}$$

verifies

$$\mathbf{D}^\nu \mathcal{Q}_{\mu\nu} = \dot{\mathbf{D}}_\mu \Psi \cdot \mathcal{N}[\Psi] + \dot{\mathbf{D}}^\nu \Psi^A \mathbf{R}_{AB\nu\mu}\Psi^B - \frac{1}{2}\mathbf{D}_\mu V \Psi \cdot \Psi.$$

2. *The null components of $\mathcal{Q}$ are given by*

$$\begin{aligned}\mathcal{Q}_{33} &= |e_3\Psi|^2,\\ \mathcal{Q}_{44} &= |e_4\Psi|^2,\\ \mathcal{Q}_{34} &= |\nabla\!\!\!/\,\Psi|^2 + V|\Psi|^2,\end{aligned}$$

and

$$\mathbf{g}^{\mu\nu}\mathcal{Q}_{\mu\nu} = -\mathcal{L}(\Psi) - V|\Psi|^2.$$

Also,

$$|\mathcal{L}(\Psi)| \lesssim |e_3\Psi|\,|e_4\Psi| + |\nabla\!\!\!/\,\Psi|^2 + V|\Psi|^2$$

and

$$\begin{aligned}|\mathcal{Q}_{AB}| &\leq |e_3\Psi||e_4\Psi| + |\nabla\!\!\!/\,\Psi|^2 + |V||\Psi|^2,\\ |\mathcal{Q}_{A3}| &\leq |e_3\Psi||\nabla\!\!\!/\,\Psi|,\\ |\mathcal{Q}_{A4}| &\leq |e_4\Psi||\nabla\!\!\!/\,\Psi|.\end{aligned}$$

3. *Introducing*

$$\widehat{\mathcal{Q}}_{34} := \mathcal{Q}_{34} - V|\Psi|^2 = |\nabla\!\!\!/\,\Psi|^2,$$

we have

$$-\widehat{\mathcal{Q}}_{34} + \mathcal{Q}_{\theta\theta} + \mathcal{Q}_{\varphi\varphi} = -\mathcal{L}(\Psi).$$

4. *Let $X = ae_3 + be_4$. Then, since $\mathbf{R}_{AB34} = 0$ in an axially symmetric polarized spacetime,*

$$\mathbf{D}^\mu(\mathcal{Q}_{\mu\nu}X^\nu) = \frac{1}{2}\mathcal{Q}\cdot{}^{(X)}\pi + X(\Psi)\cdot\mathcal{N}[\Psi] - \frac{1}{2}X(V)\Psi\cdot\Psi.$$

5. *Let $X = ae_3 + be_4$ as above, w a scalar function and M a 1-form. Define*

$$\mathcal{P}_\mu = \mathcal{P}_\mu[X, w, M] = \mathcal{Q}_{\mu\nu}X^\nu + \frac{1}{2}w\Psi\dot{\mathbf{D}}_\mu\Psi - \frac{1}{4}|\Psi|^2\partial_\mu w + \frac{1}{4}|\Psi|^2 M_\mu.$$

Then,

$$\mathbf{D}^\mu \mathcal{P}_\mu[X, w, M] = \frac{1}{2}\mathcal{Q}\cdot {}^{(X)}\pi - \frac{1}{2}X(V)\Psi\cdot\Psi + \frac{1}{2}w\mathcal{L}[\Psi] - \frac{1}{4}|\Psi|^2\square_{\mathbf{g}} w + \frac{1}{4}\dot{\mathbf{D}}^\mu(|\Psi|^2 M_\mu) + \left(X(\Psi) + \frac{1}{2}w\Psi\right)\cdot\mathcal{N}[\Psi]. \tag{10.1.12}$$

Proof. See sections D.1.4 and D.2 in the appendix. □

Notation. For convenience we introduce the notation

$$\mathcal{E}[X, w, M](\Psi) := \mathbf{D}^\mu \mathcal{P}_\mu[X, w, M] - \left(X(\Psi) + \frac{1}{2}w\Psi\right)\cdot\mathcal{N}[\Psi]. \tag{10.1.13}$$

Thus equation (10.1.12) becomes

$$\begin{aligned}\mathcal{E}[X, w, M](\Psi) &= \frac{1}{2}\mathcal{Q}\cdot {}^{(X)}\pi - \frac{1}{2}X(V)\Psi\cdot\Psi + \frac{1}{2}w\mathcal{L}[\Psi] \\ &\quad -\frac{1}{4}|\Psi|^2\square_{\mathbf{g}} w + \frac{1}{4}\dot{\mathbf{D}}^\mu(|\Psi|^2 M_\mu).\end{aligned} \tag{10.1.14}$$

When $M = 0$ we simply write $\mathcal{E}[X, w](\Psi)$.

10.1.6 Main Morawetz identity

Lemma 10.10. *Let $f(r, m)$ a function of r and m, and let X a vectorfield defined by $X = f(r, m)R$. Then, we have*[4]

$$\begin{aligned}\mathcal{Q}\cdot {}^{(X)}\dot{\pi} &= f\left(-\frac{2m}{r^2} + \frac{2\Upsilon}{r}\right)|\nabla\!\!\!/\Psi|^2 + 2\partial_r f|R\Psi|^2 - \left(\frac{2\Upsilon}{r}f + \Upsilon\partial_r f\right)\mathcal{L}\Psi \\ &\quad -\frac{2m}{r^2}fV|\Psi|^2\end{aligned}$$

where ${}^{(X)}\dot{\pi}$ has been defined in Lemma 10.8.

Proof. In view of Lemma 10.8, we have

$$\begin{aligned}\mathcal{Q}\cdot {}^{(X)}\dot{\pi} &= \frac{1}{2}\mathcal{Q}_{34}\dot{\pi}_{34} + \frac{1}{4}\mathcal{Q}_{44}\dot{\pi}_{33} + \frac{1}{4}\mathcal{Q}_{33}\dot{\pi}_{44} + \mathcal{Q}_{AB}\dot{\pi}_{AB} \\ &= -\frac{2m}{r^2}f\mathcal{Q}_{34} - \partial_r f\Upsilon\mathcal{Q}_{34} + \frac{1}{2}\mathcal{Q}_{44}\partial_r f + \frac{1}{2}\mathcal{Q}_{33}\Upsilon^2\partial_r f + \frac{2\Upsilon}{r}f\delta^{AB}\mathcal{Q}_{AB} \\ &= -\frac{2m}{r^2}f\mathcal{Q}_{34} + \frac{2\Upsilon}{r}f\delta^{AB}\mathcal{Q}_{AB} + \frac{1}{2}\partial_r f\left(\mathcal{Q}_{44} - 2\Upsilon\mathcal{Q}_{34} + \Upsilon^2\mathcal{Q}_{33}\right).\end{aligned}$$

Note that

$$\left(\mathcal{Q}_{44} - 2\Upsilon\mathcal{Q}_{34} + \Upsilon^2\mathcal{Q}_{33}\right) = 4\mathcal{Q}_{RR}$$

[4]Recall from Remark 10.6 that $\partial_r f$ does not denote a spacetime coordinate vectorfield applied to f, but instead the partial derivative with respect to the variable r of the function $f(r, m)$.

and, since $\mathbf{g}^{\mu\nu}\mathcal{Q}_{\mu\nu} = -\mathcal{L}(\Psi) - V|\Psi|^2$,

$$\delta^{AB}\mathcal{Q}_{AB} = \mathcal{Q}_{34} - \mathcal{L} - V|\Psi|^2 = \widehat{\mathcal{Q}}_{34} - \mathcal{L}.$$

Hence,

$$\begin{aligned}
\mathcal{Q}\cdot {}^{(X)}\dot{\pi} &= -\frac{2m}{r^2}f\mathcal{Q}_{34} + \frac{2\Upsilon}{r}f\left(\widehat{\mathcal{Q}}_{34} - \mathcal{L}\right) + 2\partial_r f\mathcal{Q}_{RR}\\
&= -\frac{2m}{r^2}f\left(\widehat{\mathcal{Q}}_{34} + V|\Psi|^2\right) + \frac{2\Upsilon}{r}f\left(\widehat{\mathcal{Q}}_{34} - \mathcal{L}\right) + 2\partial_r f\mathcal{Q}_{RR}\\
&= f\left(-\frac{2m}{r^2} + \frac{2\Upsilon}{r}\right)\widehat{\mathcal{Q}}_{34} + 2\partial_r f\mathcal{Q}_{RR} - \frac{2\Upsilon}{r}f\mathcal{L} - \frac{2m}{r^2}fV|\Psi|^2.
\end{aligned}$$

Finally,

$$\mathcal{Q}_{RR} = |R\Psi|^2 - \frac{1}{2}\mathbf{g}(R,R)\mathcal{L} = |R\Psi|^2 - \frac{1}{2}\Upsilon\mathcal{L}.$$

Hence,

$$\begin{aligned}
&\mathcal{Q}\cdot {}^{(X)}\dot{\pi}\\
&= 2f\left(-\frac{m}{r^2} + \frac{\Upsilon}{r}\right)\widehat{\mathcal{Q}}_{34} + 2\partial_r f\left(|R\Psi|^2 - \frac{1}{2}\Upsilon\mathcal{L}\right) - \frac{2\Upsilon}{r}f\mathcal{L} - \frac{2m}{r^2}fV|\Psi|^2\\
&= 2f\left(-\frac{m}{r^2} + \frac{\Upsilon}{r}\right)|\nabla\!\!\!\!/\,\Psi|^2 + 2\partial_r f|R\Psi|^2 - \left(\frac{2\Upsilon}{r}f + \partial_r f\Upsilon\right)\mathcal{L} - \frac{2m}{r^2}fV|\Psi|^2.
\end{aligned}$$

This concludes the proof of the lemma. □

We shall also make use of the following lemma.

Lemma 10.11. *If $f = f(r,m)$, then*

$$\begin{aligned}
\square_{\mathbf{g}}(f(r,m)) &= r^{-2}\partial_r(r^2\Upsilon\partial_r f) + O(\epsilon r^{-2}\, u_{trap}^{-1-\delta_{dec}})\Big[r^2|\partial_r^2 f(r,m)| + r|\partial_r f(r,m)|\\
&\quad + r|\partial_r\partial_m f(r,m)| + |\partial_m^2 f(r,m)|\Big].
\end{aligned}$$

Proof. Recall from Lemma 2.102 that, for a general scalar f,

$$\begin{aligned}
\square_{\mathbf{g}} f &= -\frac{1}{2}(e_3e_4 + e_4e_3)f + \Delta\!\!\!\!/\, f + \left({}^{(1+3)}\underline{\omega} - \frac{1}{2}{}^{(1+3)}\mathrm{tr}\underline{\chi}\right)e_4 f\\
&\quad + \left({}^{(1+3)}\omega - \frac{1}{2}{}^{(1+3)}\mathrm{tr}\chi\right)e_3 f.
\end{aligned}$$

Recall that

$${}^{(1+3)}\mathrm{tr}\chi = 2\chi - \vartheta, \quad {}^{(1+3)}\mathrm{tr}\underline{\chi} = 2\underline{\chi} - \underline{\vartheta}, \quad {}^{(1+3)}\omega = \omega, \quad {}^{(1+3)}\underline{\omega} = \underline{\omega}$$

and

$$\Delta\!\!\!\!/\, f = e_\theta e_\theta f + (e_\theta\Phi)^2 e_\theta f.$$

Using Lemma 10.5, we deduce, for a function $f = f(r,m)$,

$$
\begin{aligned}
\square_{\mathbf{g}} f &= -\frac{1}{2}(e_3e_4+e_4e_3)f + \left(\underline{\omega} - \frac{1}{2}\underline{\kappa}\right) e_4 f + \left(\omega - \frac{1}{2}\kappa\right) e_3 f \\
&= -\partial_r^2 f(r,m) e_3(r) e_4(r) - \frac{1}{2}\partial_r f(r,m)\left(e_3e_4(r) + e_4e_3(r)\right) \\
&- \frac{1}{2}\underline{\kappa}\partial_r f(r,m) e_4 r + \left(\omega - \frac{1}{2}\kappa\right)\partial_r f(r,m) e_3(r) + O(\epsilon r^{-2}\, u_{trap}^{-1-\delta_{dec}}) \\
&\quad \times\Big[r^2|\partial_r^2 f(r,m)| + r|\partial_r f(r,m)| + r|\partial_r\partial_m f(r,m)| + |\partial_m^2 f(r,m)|\Big] \\
&= -\partial_r^2 f(r,m)\left(-\Upsilon + O(\epsilon u_{trap}^{-1-\delta_{dec}})\right) + \partial_r f(r,m)\frac{m}{r^2} + \partial_r f(r,m)\frac{\Upsilon}{r} \\
&\quad + \frac{r-m}{r^2}\partial_r f(r,m) + O(\epsilon r^{-2}\, u_{trap}^{-1-\delta_{dec}}) \\
&\quad \times\Big[r^2|\partial_r^2 f(r,m)| + r|\partial_r f(r,m)| + r|\partial_r\partial_m f(r,m)| + |\partial_m^2 f(r,m)|\Big] \\
&= \Upsilon\partial_r^2 f(r,m) + \partial_r f(r,m)\left(\frac{2}{r} - \frac{2m}{r^2}\right) + O(\epsilon r^{-2}\, u_{trap}^{-1-\delta_{dec}}) \\
&\quad \times\Big[r^2|\partial_r^2 f(r,m)| + r|\partial_r f(r,m)| + r|\partial_r\partial_m f(r,m)| + |\partial_m^2 f(r,m)|\Big] \\
&= r^{-2}\partial_r(r^2\Upsilon\partial_r f) + O(\epsilon r^{-2}\, u_{trap}^{-1-\delta_{dec}}) \\
&\quad \times\Big[r^2|\partial_r^2 f(r,m)| + r|\partial_r f(r,m)| + r|\partial_r\partial_m f(r,m)| + |\partial_m^2 f(r,m)|\Big]
\end{aligned}
$$

as desired. $\square$

According to equation (10.1.14) we have

$$
\mathcal{E}[X,w](\Psi) = \frac{1}{2}\mathcal{Q}\cdot {}^{(X)}\pi - \frac{1}{2}X(V)|\Psi|^2 + \frac{1}{2}w\mathcal{L}(\Psi) - \frac{1}{4}|\Psi|^2\square_{\mathbf{g}} w.
$$

In the next proposition we choose X to be of the form $X = f(r,m)R$ and make a choice of w as a function of f.

Proposition 10.12. *Assume*

$$
X = f(r,m)R \text{ and } w(r,m) = r^{-2}\Upsilon\partial_r(r^2 f).
$$

Then,

$$
\mathcal{E}[X,w](\Psi) = \dot{\mathcal{E}}[X,w] + \mathcal{E}_\epsilon[X,w]
$$

where, with $\widehat{\mathcal{Q}}_{34} := \mathcal{Q}_{34} - V|\Psi|^2 = |\nabla\!\!\!/\Psi|^2$,

$$
\begin{aligned}
\dot{\mathcal{E}}[fR,w](\Psi) &= \frac{1}{r}\left(1 - \frac{3m}{r}\right) f\widehat{\mathcal{Q}}_{34} + \partial_r f|R(\Psi)|^2 - \frac{1}{4}r^{-2}\partial_r(r^2\Upsilon\partial_r w)|\Psi|^2 \\
&\quad + 4\Upsilon\frac{r-4m}{r^4} f|\Psi|^2, \\
\mathcal{E}_\epsilon[fR,w](\Psi) &= \epsilon\frac{1}{2}\mathcal{Q}\cdot {}^{(X)}\ddot{\pi} + O\Big(\epsilon r^{-3} u_{trap}^{-1-\delta_{dec}}\big(|f| + r^2|\partial_r w| + r^3|\partial_r^2 w| \\
&\quad + r^2|\partial_r\partial_m w| + r|\partial_m^2 w|\big)\Big)|\Psi|^2.
\end{aligned}
\tag{10.1.15}
$$

Proof. According to Lemma 10.8 and equation (10.1.14) we have

$$\mathcal{E}[X,w](\Psi) = \frac{1}{2}\mathcal{Q}\cdot({}^{(X)}\dot{\pi}+\epsilon\,{}^{(X)}\ddot{\pi})-\frac{1}{2}X(V)|\Psi|^2+\frac{1}{2}w\mathcal{L}(\Psi)-\frac{1}{4}|\Psi|^2\square_{\mathbf{g}}w.$$

Hence, in view of Lemmas 10.8 and 10.10,

$$\begin{aligned}\mathcal{E}[X,w](\Psi)-\epsilon\frac{1}{2}\mathcal{Q}\cdot{}^{(X)}\ddot{\pi} &= \frac{1}{2}\mathcal{Q}\cdot{}^{(X)}\dot{\pi}-\frac{1}{2}X(V)|\Psi|^2+\frac{1}{2}w\mathcal{L}(\Psi)-\frac{1}{4}|\Psi|^2\square_{\mathbf{g}}w\\ &= f\left(-\frac{m}{r^2}+\frac{\Upsilon}{r}\right)|\nabla\!\!\!/\Psi|^2+\partial_r f|R\Psi|^2\\ &- \left(\frac{\Upsilon}{r}f+\frac{1}{2}\Upsilon\partial_r f\right)\mathcal{L}(\Psi)-\frac{m}{r^2}fV|\Psi|^2\\ &- \frac{1}{2}X(V)|\Psi|^2+\frac{1}{2}w\mathcal{L}(\Psi)-\frac{1}{4}|\Psi|^2\square_{\mathbf{g}}w.\end{aligned}$$

Thus, assuming $w=r^{-2}\Upsilon\partial_r(r^2 f)=\frac{2\Upsilon}{r}f+\partial_r f\Upsilon$,

$$\begin{aligned}\mathcal{E}[X,w](\Psi)-\epsilon\frac{1}{2}\mathcal{Q}\cdot{}^{(X)}\ddot{\pi} &= r^{-1}f\left(1-\frac{3m}{r}\right)|\nabla\!\!\!/\Psi|^2+\partial_r f|R\Psi|^2\\ &- \left(\frac{m}{r^2}fV+\frac{1}{2}X(V)+\frac{1}{4}\square_{\mathbf{g}}w\right)|\Psi|^2.\end{aligned}$$

Note that, in view of Lemma 10.7,

$$X(V) = fR(V)=-8\Upsilon f\frac{r-3m}{r^4}+O(\epsilon r^{-3}u_{trap}^{-1-\delta_{dec}}|f|)$$

and

$$\begin{aligned}\frac{m}{r^2}fV+\frac{1}{2}X(V) &= f\left(\frac{4m}{r^4}\Upsilon-4\Upsilon\frac{r-3m}{r^4}\right)+O(\epsilon r^{-3}u_{trap}^{-1-\delta_{dec}}|f|)\\ &= -4f\Upsilon\frac{r-4m}{r^4}+O(\epsilon r^{-3}\,u_{trap}^{-1-\delta_{dec}}|f|).\end{aligned}$$

Note also that, in view of Lemma 10.11,

$$\begin{aligned}\square_{\mathbf{g}}(w) &= r^{-2}\partial_r(r^2\Upsilon\partial_r w)\\ &+O(\epsilon r^{-2}\,u_{trap}^{-1-\delta_{dec}})\Big[r^2|\partial_r^2 w|+r|\partial_r w|+r|\partial_r\partial_m w|+|\partial_m^2 w|\Big].\end{aligned}$$

Thus,

$$\begin{aligned}\frac{m}{r^2}fV+\frac{1}{2}X(V)+\frac{1}{4}\square_{\mathbf{g}}w &= -4\Upsilon\frac{r-4m}{r^4}f+\frac{1}{4}r^{-2}\partial_r(r^2\Upsilon\partial_r w)\\ &+O(\epsilon r^{-3}\,u_{trap}^{-1-\delta_{dec}})\Big[|f|+r^3|\partial_r^2 w|+r^2|\partial_r w|\\ &+r^2|\partial_r\partial_m w|+r|\partial_m^2 w|\Big]\end{aligned}$$

and hence

$$\begin{aligned}\mathcal{E}[X,w](\Psi)-\epsilon\frac{1}{2}\mathcal{Q}\cdot{}^{(X)}\ddot{\pi} &= r^{-1}f\left(1-\frac{3m}{r}\right)|\nabla\!\!\!\!/\,\Psi|^2+\partial_r f|R\Psi|^2\\ &- \frac{1}{4}|\Psi|^2 r^{-2}\partial_r(r^2\Upsilon\partial_r w)+4\Upsilon\frac{r-4m}{r^4}f|\Psi|^2\\ &+ O\Big(\epsilon r^{-3}\,u_{trap}^{-1-\delta_{dec}}\big(|f|+r^2|\partial_r w|+r^3|\partial_r^2 w|\\ &\qquad +r^2|\partial_r\partial_m w|+r|\partial_m^2 w|\big)\Big)|\Psi|^2\end{aligned}$$

as desired. □

10.1.7 A first estimate

We concentrate our attention on the principal term

$$\begin{aligned}\dot{\mathcal{E}}[fR,w](\Psi) &= \frac{1}{r}\left(1-\frac{3m}{r}\right)f\widehat{\mathcal{Q}}_{34}+\partial_r f|R(\Psi)|^2-\frac{1}{4}r^{-2}\partial_r(r^2\Upsilon\partial_r w)|\Psi|^2\\ &\quad +4\Upsilon\frac{r-4m}{r^4}f|\Psi|^2\end{aligned}$$

and choose $f=f(r,m)$ such that the right-hand side is positive definite.

Consider the quadratic forms,

$$\begin{aligned}\dot{\mathcal{E}}_0(\Psi) &:= A\widehat{\mathcal{Q}}_{34}+B|R\Psi|^2+r^{-2}W|\Psi|^2,\\ \dot{\mathcal{E}}(\Psi) &:= \dot{\mathcal{E}}_0(\Psi)+4\Upsilon\frac{r-4m}{r^4}f|\Psi|^2,\end{aligned}\tag{10.1.16}$$

with the coefficients

$$A:=r^{-1}f\left(1-\frac{3m}{r}\right),\qquad B:=\partial_r f,\qquad W:=-\frac{1}{4}\partial_r(r^2\Upsilon\partial_r w).\tag{10.1.17}$$

The goal is to show that there exist choices of f,w verifying the condition of Proposition 10.12, i.e., $w=r^{-2}\partial_r(r^2 f)$, which makes $\dot{\mathcal{E}}(\Psi)$ positive definite, for all smooth S-valued tensorfields Ψ defined in the region $r\geq 2m_0(1-\delta_{\mathcal{H}})$, which decay reasonably fast at infinity. We look first for choices of f,w such that the coefficient A,B,W are non-negative. Note in particular that f must be increasing as a function of r and $f=0$ on $r=3m$. Following J. Stogin [62] we choose w first to ensure that W is non-negative and then choose f, compatible with the equation

$$\partial_r(r^2 f)=\frac{r^2}{\Upsilon}w,\qquad f=0 \text{ on } r=3m.\tag{10.1.18}$$

To ensure that $A=r^{-2}f(r-3m)$ is positive we need a non-negative w which verifies (modulo error terms[5]) $W=-\frac{1}{4}\partial_r(r^2\Upsilon\partial_r w)\geq 0$. It is more difficult to choose w such that $B=\partial_r f$ is also non-negative.

Stogin defines w based on the following lemma.

[5]That is, terms which vanish in Schwarzschild.

Lemma 10.13. *The scalar function w defined by*

$$w(r,m)=\begin{cases}\frac{1}{4m}, & \text{if} \quad r\le 4m,\\ \frac{2\Upsilon}{r}, & \text{if} \quad r\ge 4m,\end{cases}$$

is C^1, non-negative and such that $W=-\frac{1}{4}\partial_r(r^2\Upsilon\partial_r w)$ verifies

$$W(r,m)=\begin{cases}0, & \text{if} \quad r<4m,\\ \frac{m}{r^2}\left(3-\frac{8m}{r}\right), & \text{if} \quad r>4m.\end{cases} \tag{10.1.19}$$

Proof. For $r\ge 4m$, we have

$$w(r,m)=\frac{2\Upsilon}{r}, \quad \partial_r w(r,m)=-\frac{2}{r^2}+\frac{8m}{r^3}, \quad \partial_r^2 w(r,m)=\frac{4}{r^3}-\frac{24m}{r^4}.$$

In particular, we have

$$w=\frac{1}{4m}, \qquad \partial_r w=0 \quad \text{at } r=4m$$

so that w is indeed C^1. Furthermore, we also have

$$\begin{aligned} r^{-2}\partial_r(r^2\Upsilon\partial_r w) &= \Upsilon\partial_r^2 w(r)+\partial_r w(r)\left(\frac{2}{r}-\frac{2m}{r^2}\right)\\ &= \Upsilon\left(\frac{4}{r^3}-\frac{24m}{r^4}\right)+\left(-\frac{2}{r^2}+\frac{8m}{r^3}\right)\left(\frac{2}{r}-\frac{2m}{r^2}\right)\\ &= -\frac{4m}{r^4}\left(3-\frac{8m}{r}\right)\end{aligned}$$

so that, for $r\ge 4m$,

$$W=-\frac{1}{4}\partial_r(r^2\Upsilon\partial_r w) \quad = \quad \frac{m}{r^2}\left(3-\frac{8m}{r}\right)$$

as desired. □

Once w is defined we can evaluate f as follows.

Lemma 10.14. *Let $w(r,m)$ defined as in Lemma 10.13. Then, the function $f(r,m)$ given by*

$$r^2 f(r,m):=\begin{cases}2m^2\log\left(\frac{r-2m}{m}\right)+(r-3m)\frac{r^2+6mr+30m^2}{12m}, & \text{for } r\le 4m,\\ C_* m^2+r^2-(4m)^2, & \text{for } r\ge 4m,\end{cases} \tag{10.1.20}$$

with the constant C_* given by[6]

$$C_* \quad := \quad 2\log(2)+\frac{35}{6}, \qquad C_*\sim 7.22,$$

[6] C_* is chosen so that f is continuous across $r=4m$.

is C^2 and satisfies (10.1.18), *i.e., we have*

$$\partial_r(r^2 f) = r^2\Upsilon^{-1}w, \qquad f = 0 \text{ on } r = 3m.$$

Proof. By direct check,[7] we have for $r \leq 4m$

$$\begin{aligned}
\partial_r(r^2 f)(r,m) &= \frac{2m^2}{(r-2m)} + \frac{r^2+6mr+30m^2}{12m} + (r-3m)\frac{2r+6m}{12m} \\
&= \frac{r^3}{4m(r-2m)} \\
&= \frac{r^2}{\Upsilon}\frac{1}{4m}
\end{aligned}$$

and for $r \geq 4m$

$$\partial_r(r^2 f)(r,m) = 2r,$$

as well as $f = 0$ on $r = 3m$ so that, in view of the definition of $w(r)$ in Lemma 10.13, we infer

$$\partial_r(r^2 f) = r^2\Upsilon^{-1}w, \qquad f = 0 \text{ on } r = 3m$$

as desired. Note also that w being C^1, f is thus indeed C^2. □

Next, we derive a lower bound on $\partial_r f$ for $r \leq 4m$.

Lemma 10.15. *We have for all r and m*

$$r^3\partial_r f \geq 16m^2.$$

Also, there exists a constant $C > 0$ such that for all r and m

$$\left(1-\frac{3m}{r}\right)f \geq C^{-1}\left(1-\frac{3m}{r}\right)^2.$$

Proof. We have

$$\partial_r(r^3\partial_r f) = \partial_r(r\partial_r(r^2 f)) - 2\partial_r(r^2 f).$$

Using the identity $\partial_r(r^2 f) = r^2\Upsilon^{-1}w$, we infer

$$\partial_r(r^3\partial_r f) = \partial_r(r^3\Upsilon^{-1}w) - 2r^2\Upsilon^{-1}w.$$

[7] Recall from Remark 10.6 that $\partial_r f$ does not denote a spacetime coordinate vectorfield applied to f, but instead the partial derivative with respect to the variable r of the function $f(r,m)$.

For $r \leq 4m$, we have $w = (4m)^{-1}$ and hence

$$\begin{aligned}\partial_r(r^3\partial_r f) &= \frac{1}{4m}\left(\partial_r(r^3\Upsilon^{-1}) - 2r^2\Upsilon^{-1}\right)\\ &= \frac{1}{4m}r^3\partial_r(\Upsilon^{-1})\\ &= -\frac{r}{2\Upsilon^2}.\end{aligned}$$

In particular, $r^3\partial_r f$ is decreasing in r on $r \leq 4m$ and hence

$$r^3\partial_r f \geq (4m)^3\partial_r f(r = 4m, m) \text{ on } r \leq 4m.$$

On the other hand, we have, in view of the definition of f in (10.1.20),

$$\begin{aligned}\partial_r(r^2 f)(r = 4m, m) &= (4m)^2\partial_r f(r = 4m, m) + 8mf(r = 4m, m)\\ &= (4m)^2\partial_r f(r = 4m, m) + \frac{m}{2}C_*\end{aligned}$$

and hence

$$(4m)^2\partial_r f(r = 4m, m) = \left(8 - \frac{C_*}{2}\right)m$$

so that

$$r^3\partial_r f \geq 2(16 - C_*)m^2 \text{ on } r \leq 4m.$$

Since $C_* \sim 7.22 < 8$, we deduce

$$r^3\partial_r f \geq 16m^2 \text{ on } r \leq 4m.$$

Also, for $r \geq 4m$, we have

$$f = 1 - \frac{(16 - C_*)m^2}{r^2}$$

so that

$$\partial_r f = \frac{2(16 - C_*)m^2}{r^3}.$$

Since $C_* \sim 7.22 < 8$, we deduce

$$r^3\partial_r f \geq 16m^2 \text{ on } r \geq 4m$$

which together with the case $r \leq 4m$ above yields for all r and m the desired estimate for $\partial_r f$

$$r^3\partial_r f \geq 16m^2.$$

In particular, $\partial_r f > 0$ and hence is strictly increasing. On the other hand, $f = 0$ on $r = 3$ and converges to 1 as $r \to +\infty$. We deduce the existence of a constant

$C > 0$ such that

$$\left(1 - \frac{3m}{r}\right) f \geq C^{-1} \left(1 - \frac{3m}{r}\right)^2$$

as desired. □

We summarize the results in the following.

Proposition 10.16. *There exist functions $f \in C^2, w \in C^1$ verifying the relation $w = r^{-2}\Upsilon\partial_r(r^2 f)$ and such that*

$$r^2 f = \begin{cases} 2m^2 \log\left(\frac{r-2m}{m}\right) + (r - 3m)\frac{r^2+6mr+30m^2}{12m}, & \text{for } r \leq 4m, \\ C_* m^2 + r^2 - (4m)^2, & \text{for } r \geq 4m, \end{cases} \tag{10.1.21}$$

where C_* is a constant satisfying $7 < C_* < 8$. In particular,

$$f = \begin{cases} \frac{2m^2}{r^2} \log\left(\frac{r-2m}{m}\right) + O(\frac{r-3m}{m}), & \text{for } \ r \leq 4m, \\ 1 + O(\frac{m^2}{r^2}), & \text{for } \ r \geq 4m, \end{cases} \tag{10.1.22}$$

and, for some $C > 0$ and all $r \geq 2m$,

$$\left(1 - \frac{3m}{r}\right) f \geq C^{-1} \left(1 - \frac{3m}{r}\right)^2, \qquad \partial_r f \geq \frac{16m^2}{r^3}. \tag{10.1.23}$$

Also, w is given by

$$w = \begin{cases} \frac{1}{4m}, & \text{for } \ r \leq 4m, \\ \frac{2}{r}\left(1 - \frac{2m}{r}\right), & \text{for } \ r \geq 4m. \end{cases} \tag{10.1.24}$$

Moreover $W = -\frac{1}{4}\partial_r(r^2\Upsilon\partial_r w)$ verifies

$$W = \begin{cases} 0, & \text{if } \ r < 4m, \\ \frac{m}{r^2}\left(3 - \frac{8m}{r}\right), & \text{if } \ r > 4m, \end{cases} \tag{10.1.25}$$

and

$$\begin{aligned} \dot{\mathcal{E}}_0[fR, w](\Psi) &= \partial_r f|R(\Psi)|^2 + r^{-2}W|\Psi|^2 + r^{-1}\left(1 - \frac{3m}{r}\right) f\widehat{\mathcal{Q}}_{34}, \\ \dot{\mathcal{E}}[fR, w](\Psi) &= \mathcal{E}_0[fR, w](\Psi) + 4\Upsilon\frac{r - 4m}{r^4} f|\Psi|^2. \end{aligned} \tag{10.1.26}$$

Recall also that

$$\widehat{\mathcal{Q}}_{34} \ = \ |\nabla\!\!\!/\Psi|^2.$$

Remark 10.17. *The estimates obtained so far have two major deficiencies.*

1. *The quadratic form $\dot{\mathcal{E}}_0[fR, w](\Psi) + 4\Upsilon\frac{r-4m}{r^4} f|\Psi|^2$ fails to be positive definite in the region $3m \leq r \leq 4m$ because of the potential term $\Upsilon\frac{r-4m}{r^4} f|\Psi|^2$.*
2. *The function f blows up logarithmically at $r = 2m$ in ${}^{(int)}\mathcal{M}$.*

In the next section we deal with the first issue. We handle the second problem in the following two sections.

10.1.8 Improved lower bound in ${}^{(ext)}\mathcal{M}$

Note that the term $4f\Upsilon\frac{r-4m}{r^4}$ is negative for $3m \leq r \leq 4m$ and positive everywhere else. An improvement can be obtained by using the following Poincaré inequality.

Lemma 10.18. *We have, for $\Psi \in \mathcal{S}_2(\mathcal{M})$,*

$$\int_S |\nabla\!\!\!\!/\,\Psi|^2 \;\geq\; 2r^{-2}\Big(1 - O(\epsilon)\Big)\int_S \Psi^2 da_S. \tag{10.1.27}$$

Proof. See Proposition 2.35. □

According to Proposition 10.16 we deduce

$$\begin{aligned}
&\int_S \dot{\mathcal{E}}[fR, w](\Psi) \geq \int_S \dot{\mathcal{E}}_1 - O(\epsilon r^{-3})\int_S \Psi^2 da_S,\\
&\dot{\mathcal{E}}_1 := \partial_r f|R(\Psi)|^2 + r^{-2}W|\Psi|^2 + 2r^{-3}\left(1 - \frac{3m}{r}\right)f|\Psi|^2 + 4\Upsilon\frac{r-4m}{r^4}f|\Psi|^2,
\end{aligned} \tag{10.1.28}$$

with W defined in (10.1.25). It is easy to see however that $\dot{\mathcal{E}}_1$ still fails to be positive for $3m < r < 4m$. To achieve positivity we also need to modify the original energy density $\mathcal{E}[fR, w](\Psi)$ by considering instead the modified energy density $\mathcal{E}[fR, w, M](\Psi)$ (see (10.1.12) and notation (10.1.13)) with $M = 2hR$ for a function $h = h(r, m)$ supported for $r \geq 3m$ and constant for $r \geq 4m$.

$$\begin{aligned}
\mathcal{E}[fR, w, M](\Psi) &= \mathcal{E}[fR, w](\Psi) + \frac{1}{4}\dot{\mathbf{D}}^\mu(|\Psi|^2 M_\mu)\\
&= \mathcal{E}[fR, w](\Psi) + \frac{1}{4}(\mathbf{D}^\mu M_\mu)|\Psi|^2 + \frac{1}{2}\Psi M(\Psi)\\
&= \mathcal{E}[fR, w](\Psi) + \frac{1}{2}\mathbf{D}^\mu(hR_\mu)|\Psi|^2 + h\Psi R(\Psi).
\end{aligned}$$

To take into account the additional terms in the modified $\mathcal{E}[fR, w, M](\Psi)$ we first derive the following.

Lemma 10.19. *Let $h(r, m)$ a C^1 function of r and m. We have*

$$\mathbf{D}^\mu(hR_\mu) = r^{-2}\partial_r(\Upsilon r^2 h) + O\Big(\epsilon r^{-1}u_{trap}^{-1-\delta_{dec}}\big(r|\partial_r h| + |h| + r|\partial_m h|\big)\Big). \tag{10.1.29}$$

Proof. In view of Lemma 10.7, which computes the components of ${}^{(R)}\pi$, as well as

Lemma 10.5 to compute $R(h)$, we calculate

$$\begin{aligned}
\mathbf{D}^\mu(hR_\mu) &= R(h)+h(\mathbf{D}^\mu R_\mu) = \frac{1}{2}(e_4(h)-\Upsilon e_3(h)) + h\frac{1}{2}\mathrm{tr}\,({}^{(R)}\pi)\\
&= \frac{1}{2}(e_4(r)-\Upsilon e_3(r))\partial_r h + O(\epsilon\, u_{trap}^{-1-\delta_{dec}}|\partial_m h|)\\
&\quad +\frac{1}{2}h\left(-{}^{(R)}\pi_{34}+{}^{(R)}\pi_{\theta\theta}+{}^{(R)}\pi_{\varphi\varphi}\right)\\
&= \Upsilon\partial_r h + \frac{1}{2}\left(\frac{4m}{r^2}+4\frac{\Upsilon}{r}\right)h + O\left(\epsilon\, u_{trap}^{-1-\delta_{dec}}(|\partial_r h|+r^{-1}|h|+|\partial_m h|)\right)\\
&= r^{-2}\partial_r(\Upsilon r^2 h) + O\left(\epsilon\, u_{trap}^{-1-\delta_{dec}}(|\partial_r h|+r^{-1}|h|+|\partial_m h|)\right)
\end{aligned}$$

as desired. □

In view of the lemma we write

$$\begin{aligned}
\mathcal{E}[fR,w,2hR](\Psi) &= \dot{\mathcal{E}}[fR,w,2hR](\Psi)+\mathcal{E}_\epsilon[fR,w,2hR](\Psi),\\
\dot{\mathcal{E}}[fR,w,2hR](\Psi) &:= \dot{\mathcal{E}}[fR,w](\Psi)+\frac{1}{2}r^{-2}\partial_r(\Upsilon r^2 h)|\Psi|^2 + h\Psi R(\Psi),\\
\mathcal{E}_\epsilon[fR,w,2hR](\Psi) &:= \mathcal{E}_\epsilon[fR,w](\Psi)\\
&\quad + O\left(r^{-1}\epsilon\, u_{trap}^{-1-\delta_{dec}}(r|\partial_r h|+|h|+r|\partial_m h|)\right)|\Psi|^2.
\end{aligned} \tag{10.1.30}$$

The main result of this section is stated below.

Proposition 10.20. *There exists a function $h=h(r,m)$ with bounded derivative h', supported in $r\geq 3m$ such that $h=O(r^{-2}), h'=O(r^{-3})$ for $r\geq 4m$ such that*

$$\begin{aligned}
\mathcal{E}[fR,w,2hR](\Psi) &= \dot{\mathcal{E}}[fR,w,2hR](\Psi)+\mathcal{E}_\epsilon[fR,w,2hR](\Psi),\\
\mathcal{E}_\epsilon[fR,w,2hR](\Psi) &= \epsilon\frac{1}{2}\mathcal{Q}\cdot{}^{(X)}\ddot{\pi} + O\Big(\epsilon r^{-3}u_{trap}^{-1-\delta_{dec}}(|f|+1)\Big)|\Psi|^2,
\end{aligned} \tag{10.1.31}$$

and, for sufficiently large universal constant $C>0$, in the region $r\geq\frac{5m}{2}$,

$$\int_S \dot{\mathcal{E}}[fR,w,2hR](\Psi) \geq C^{-1}\int_S\left(\frac{m^2}{r^3}|R(\Psi)|^2 + r^{-1}\left(1-\frac{3m}{r}\right)^2|\nabla\!\!\!/\Psi|^2 + \frac{m}{r^4}|\Psi|^2\right). \tag{10.1.32}$$

Proof. We first derive the weaker inequality

$$\int_S \dot{\mathcal{E}}[fR,w,2hR](\Psi)\geq C^{-1}\int_S\left(\frac{m^2}{r^3}|R(\Psi)|^2+\frac{m}{r^4}|\Psi|^2\right) \quad \text{on } r\geq\frac{5m}{2}$$

by making full use of the Poincaré inequality above, i.e.,

$$\int_S r^{-1}\left(1-\frac{3m}{r}\right)f(r,m)|\nabla\!\!\!/\Psi|^2 \geq \int_S (2-O(\epsilon))r^{-3}\left(1-\frac{3m}{r}\right)f(r,m)|\Psi|^2.$$

The result will easily follow by writing instead, with a sufficiently small $\mu > 0$,

$$\begin{aligned}
&\int_S r^{-1}\left(1-\frac{3m}{r}\right) f(r,m)|\nabla\Psi|^2 \\
= &\ \mu\int_S r^{-1}\left(1-\frac{3m}{r}\right) f(r,m)|\nabla\Psi|^2 + (1-\mu)\int_S r^{-1}\left(1-\frac{3m}{r}\right) f(r,m)|\nabla\Psi|^2 \\
\geq &\ \mu\int_S r^{-1}\left(1-\frac{3m}{r}\right) f(r,m)|\nabla\Psi|^2 + (1-\mu)\int_S 2r^{-3}\left(1-\frac{3m}{r}\right) f(r,m)|\Psi|^2
\end{aligned}$$

and then proceeding exactly as below.

We start with

$$\dot{\mathcal{E}}[fR,w,2hR](\Psi) \quad = \quad \dot{\mathcal{E}}[fR,w](\Psi) + \frac{1}{2}r^{-2}\partial_r(\Upsilon r^2 h)|\Psi|^2 + h\Psi R(\Psi).$$

Recalling the definition of $\dot{\mathcal{E}}_1$ in (10.1.28),

$$\dot{\mathcal{E}}_1 := \partial_r f|R(\Psi)|^2 + r^{-2}W|\Psi|^2 + 2r^{-3}\left(1-\frac{3m}{r}\right) f|\Psi|^2 + 4\Upsilon\frac{r-4m}{r^4}f|\Psi|^2,$$

and setting

$$\begin{aligned}
\dot{\mathcal{E}}_2 \quad := &\quad \dot{\mathcal{E}}_1 + \frac{1}{2}r^{-2}(\Upsilon r^2 h)'|\Psi|^2 + h\Psi R(\Psi) \qquad (10.1.33)\\
= &\quad \partial_r f|R(\Psi)|^2 + 2r^{-3}\left(1-\frac{3m}{r}\right) f|\Psi|^2 + 4\Upsilon\frac{r-4m}{r^4}f|\Psi|^2 + r^{-2}W|\Psi|^2 \\
+ &\quad \frac{1}{2}r^{-2}(\Upsilon r^2 h)'|\Psi|^2 + h\Psi R(\Psi),
\end{aligned}$$

we deduce, from (10.1.28),

$$\int_S \dot{\mathcal{E}}[fR,w,2hR](\Psi) \geq \int_S \dot{\mathcal{E}}_2 - O(\epsilon r^{-3})\int_S |\Psi|^2.$$

We now substitute

$$h \quad = \quad 4\Upsilon r^{-4}\widetilde{h}.$$

Hence,

$$\begin{aligned}
\frac{1}{2}r^{-2}\partial_r(\Upsilon r^2 h)|\Psi|^2 + h\Psi R(\Psi) \quad &= \quad \frac{1}{2}r^{-2}\partial_r(4\Upsilon^2 r^{-2}\widetilde{h})|\Psi|^2 + 4\Upsilon r^{-4}\widetilde{h}\Psi R(\Psi) \\
&= \quad \frac{1}{2}r^{-2}\partial_r(4\Upsilon^2 r^{-2})\widetilde{h}|\Psi|^2 + 2r^{-4}\Upsilon^2\partial_r\widetilde{h}|\Psi|^2 \\
&\quad\ +4\Upsilon r^{-4}\widetilde{h}\Psi R(\Psi)
\end{aligned}$$

or, since $\frac{1}{2}r^{-2}\partial_r(4\Upsilon^2 r^{-2}) = -4r^{-2}\Upsilon\frac{r-4m}{r^4}$,

$$\begin{aligned}
\frac{1}{2}r^{-2}\partial_r(\Upsilon r^2 h)|\Psi|^2 + h\Psi R(\Psi) \quad &= \quad -4r^{-2}\Upsilon\frac{r-4m}{r^4}\widetilde{h}|\Psi|^2 + 2r^{-4}\Upsilon^2\partial_r\widetilde{h}|\Psi|^2 \\
&\quad\ +4\Upsilon r^{-4}\widetilde{h}\Psi R(\Psi).
\end{aligned}$$

Thus we have

$$\begin{aligned}\dot{\mathcal{E}}_2 &= \partial_r f|R(\Psi)|^2 + 2r^{-3}\left(1-\frac{3m}{r}\right) f|\Psi|^2 + 4\Upsilon\frac{r-4m}{r^4}(f - r^{-2}\widetilde{h})|\Psi|^2 \\ &+ 2r^{-4}\Upsilon^2\partial_r\widetilde{h}|\Psi|^2 + 4\Upsilon r^{-4}\widetilde{h}\Psi R(\Psi) + r^{-2}W|\Psi|^2.\end{aligned}$$

We also express

$$4\Upsilon r^{-4}\widetilde{h}\Psi R(\Psi) = \frac{2\widetilde{h}}{r^3}(R(\Psi)+\Upsilon r^{-1}\Psi)^2 - \frac{2\widetilde{h}}{r^3}|R(\Psi)|^2 - \frac{2\widetilde{h}}{r^5}\Upsilon^2|\Psi|^2$$

and, therefore,

$$\begin{aligned}\dot{\mathcal{E}}_2 &= (\partial_r f - 2r^{-3}\widetilde{h})|R(\Psi)|^2 + \frac{2\widetilde{h}}{r^3}(R(\Psi)+\Upsilon r^{-1}\Psi)^2 + r^{-2}W|\Psi|^2 \\ &+ \left[2r^{-3}\left(1-\frac{3m}{r}\right) f + 4\Upsilon\frac{r-4m}{r^4}(f - r^{-2}\widetilde{h}) + 2r^{-4}\Upsilon\partial_r\widetilde{h} - 2r^{-5}\Upsilon^2\widetilde{h}\right]|\Psi|^2.\end{aligned}$$

We choose $\widetilde{h}(r,m)$ as the following continuous and piecewise C^1 function,

$$\widetilde{h} = \begin{cases} 0, & r \le \frac{5m}{2}, \\ \delta_{\widetilde{h}}\left(\frac{5m}{2} - r\right), & \frac{5m}{2} \le r \le \frac{11m}{4}, \\ \delta_{\widetilde{h}}(r-3m), & \frac{11m}{4} \le r \le 3m, \\ r^2 f, & 3m \le r \le 4m, \\ (4m)^2 f(4m,m), & r \ge 4m, \end{cases}$$

where the constant $\delta_{\widetilde{h}} > 0$ will be chosen small enough. We consider the following cases:

Case 1 ($\frac{5m}{2} \le r \le 3m$). In view of the definition of $\widetilde{h}$ and since $W = 0$, we deduce

$$\begin{aligned}\dot{\mathcal{E}}_2 &= \partial_r f|R(\Psi)|^2 + \Bigg[2r^{-3}\left(1-\frac{3m}{r}\right) f + 4\Upsilon\frac{r-4m}{r^4}(f - r^{-2}\widetilde{h}) \\ &+2\delta_{\widetilde{h}} r^{-4}\Upsilon^2\left(1_{\frac{11m}{4}\le r\le 3m} - 1_{\frac{5m}{2}\le r\le\frac{11m}{4}}\right)\Bigg]|\Psi|^2 + \delta_{\widetilde{h}} O(1)\Psi R(\Psi) 1_{\frac{5m}{2}\le r\le 3m}.\end{aligned}$$

In view of (10.1.23), we may assume, choosing for $\delta_{\widetilde{h}} > 0$ small enough, that

$$f - \widetilde{h} \le -\frac{1}{2}|f| \text{ on } r \le 3m. \tag{10.1.34}$$

We infer, using also that $f < 0$ on $r \le 3m$,

$$\begin{aligned}\dot{\mathcal{E}}_2 &\ge \partial_r f|R(\Psi)|^2 + \Bigg[2r^{-3}\left(1-\frac{3m}{r}\right) f + 2\Upsilon\frac{r-4m}{r^4}f \\ &+2\delta_{\widetilde{h}} r^{-4}\Upsilon^2\left(1_{\frac{11m}{4}\le r\le 3m} - 1_{\frac{5m}{2}\le r\le\frac{11m}{4}}\right)\Bigg]|\Psi|^2 + \delta_{\widetilde{h}} O(1)\Psi R(\Psi) 1_{\frac{5m}{2}\le r\le 3m}.\end{aligned}$$

Since we have

$$\partial_r f \gtrsim 1, \quad 2r^{-3}\left(1-\frac{3m}{r}\right)+4\Upsilon\frac{r-4m}{r^4} \lesssim -1, \quad f \lesssim -\left|1-\frac{3m}{r}\right| \qquad \text{on } r \le 3m,$$

where we have used in particular Lemma 10.15 and Proposition 10.16, we infer

$$\begin{aligned}\dot{\mathcal{E}}_2 &\gtrsim |R(\Psi)|^2+\left(\left|1-\frac{3m}{r}\right|+\delta_{\widetilde{h}}1_{\frac{11m}{4}\le r\le 3m}-O(1)\delta_{\widetilde{h}}1_{\frac{5m}{2}\le r\le\frac{11m}{4}}\right)|\Psi|^2\\ &\quad -\delta_{\widetilde{h}}O(1)\Psi R(\Psi)1_{\frac{5m}{2}\le r\le 3m}\\ &\ge \frac{1}{2}|R(\Psi)|^2+\Bigg(\left|1-\frac{3m}{r}\right|+\delta_{\widetilde{h}}\left(1-O(1)\delta_{\widetilde{h}}\right)1_{\frac{11m}{4}\le r\le 3m}\\ &\quad -O(1)\delta_{\widetilde{h}}1_{\frac{5m}{2}\le r\le\frac{11m}{4}}\Bigg)|\Psi|^2.\end{aligned}$$

Thus, for $\delta_{\widetilde{h}}>0$ small enough, there exists some large $C>0$ such that

$$\mathcal{E}_2 \ge C^{-1}\left[|R(\Psi)|^2+|\Psi|^2\right] \quad \text{on } \frac{5m}{2}\le r\le 3m. \tag{10.1.35}$$

Case 2 ($3m \le r \le 4m$). Since $\widetilde{h}=r^2f$ and $W=0$, using in particular $\widetilde{h}\ge 0$ on $3m\le r\le 4m$, we deduce

$$\begin{aligned}\dot{\mathcal{E}}_2 &\ge (\partial_r f-2r^{-3}(r^2f))|R(\Psi)|^2\\ &\quad +\left[2r^{-3}\left(1-\frac{3m}{r}\right)f+2r^{-4}\Upsilon\partial_r(r^2f)-2r^{-5}\Upsilon^2(r^2f)\right]|\Psi|^2\\ &= (\partial_r f-2r^{-1}f)|R(\Psi)|^2\\ &\quad +\left[2r^{-3}\left(1-\frac{3m}{r}\right)f+2r^{-4}\Upsilon^2(2rf+r^2\partial_r f)-2r^{-3}\Upsilon^2 f\right]|\Psi|^2\\ &= (\partial_r f-2r^{-1}f)|R(\Psi)|^2\\ &\quad +\left[2r^{-3}\left(1-\frac{3m}{r}\right)f+2r^{-2}\Upsilon^2\partial_r f+2r^{-3}\Upsilon^2 f\right]|\Psi|^2.\end{aligned}$$

Note that the second term is strictly positive. It remains to analyze the first term.

Lemma 10.21. *In the interval* $[3m,4m]$ *we have*

$$\partial_r f-2r^{-1}f>0.$$

Proof. Recall from Proposition 10.16 that $w=r^{-2}\Upsilon\partial_r(r^2f)=\frac{1}{4m}$ in the interval $[3m,4m]$. Using also $f=0$ on $r=3m$, we deduce

$$\partial_r(r^2f) = \frac{r^2}{\Upsilon}\frac{1}{4m}.$$

We compute

$$\begin{aligned}\partial_r\left(r^2 f - \frac{(r-3m)r^2}{4m\Upsilon}\right) &= -\frac{(r-3m)}{4m}\partial_r\left(\frac{r^2}{\Upsilon}\right)\\ &= -\frac{(r-3m)(r-4m)}{2m\Upsilon^2}\\ &\leq 0 \quad \text{on } 3m\leq r\leq 4m,\end{aligned}$$

so that the differentiated quantity decays in r on $[3m, 4m]$. Since it vanishes on $r = 3m$, we infer

$$f \leq \frac{(r-3m)}{4m\Upsilon} \text{ on } 3m \leq r \leq 4m.$$

Thus, we deduce, using again $\partial_r(r^2 f) = \frac{r^2}{\Upsilon}\frac{1}{4m}$,

$$\begin{aligned}\partial_r f - \frac{2}{r}f &= r^{-2}\Big(\partial_r(r^2 f) - 4rf\Big)\\ &= \frac{1}{4m\Upsilon} - \frac{4}{r}f\\ &\geq \frac{1}{4m\Upsilon} - \frac{(r-3m)}{rm\Upsilon}\\ &\geq \frac{1}{4m\Upsilon}\left(1-4\left(1-\frac{3m}{r}\right)\right)\\ &> 0 \text{ on } 3m\leq r< 4m.\end{aligned}$$

On the other hand, we have by direct check at $r = 4m$, using (10.1.21),

$$\left(\partial_r f - \frac{2}{r}f\right)_{r=4m} = \frac{1}{2m} - \frac{1}{m}f_{r=4m} = \frac{1}{2m}\left(1-\frac{C_*}{8}\right) > 0$$

since $C_* < 8$. Hence, we infer

$$\partial_r f - 2r^{-1}f > 0 \text{ on } 3m \leq r \leq 4m$$

as desired. □

We thus conclude, for some $C > 0$, in the interval $[3m, 4m]$

$$\dot{\mathcal{E}}_2 \geq C^{-1}\left[|R(\Psi)|^2 + |\Psi|^2\right]. \tag{10.1.36}$$

Case 3 ($r \geq 4m$). Since $\widetilde{h}$ is constant and positive on $r \geq 4m$, we deduce

$$\begin{aligned}\dot{\mathcal{E}}_2 &\geq (\partial_r f - 2r^{-3}\widetilde{h})|R(\Psi)|^2\\ &+ \left[2r^{-3}\left(1-\frac{3m}{r}\right)f + 4\Upsilon\frac{r-4m}{r^4}(f - r^{-2}\widetilde{h}) - 2r^{-5}\Upsilon^2\widetilde{h} + r^{-2}W\right]|\Psi|^2.\end{aligned}$$

We examine the first term. In view of the formula for f for $r \geq 4m$, see (10.1.21),

$$\partial_r f = \frac{2}{r^3}(16 - C_*)m^2, \qquad \widetilde{h} = (4m)^2 f(4m, m) = C_* m^2$$

and hence

$$\partial_r f - 2r^{-3}\widetilde{h} \;=\; \frac{2(16-2C_*)m^2}{r^3}$$

and hence, since $C_* < 8$, we have

$$\partial_r f - 2r^{-3}\widetilde{h} \;\gtrsim\; \frac{m^2}{r^3} \quad \text{for } r \geq 4m.$$

It remains to analyze the sign of

$$\begin{aligned}
& 2r^{-3}\left(1-\frac{3m}{r}\right)f + 4\Upsilon\frac{r-4m}{r^4}(f - r^{-2}\widetilde{h}) - 2r^{-5}\Upsilon^2\widetilde{h} \\
=& \left[2r^{-3}\left(1-\frac{3m}{r}\right) + 4\Upsilon\frac{r-4m}{r^4}\right](f - r^{-2}\widetilde{h}) \\
&+ \left[2r^{-3}\left(1-\frac{3m}{r}\right) - 2r^{-3}\Upsilon^2\right]r^{-2}\widetilde{h}.
\end{aligned}$$

The first term, which can be written in the form

$$\left[2r^{-3}\left(1-\frac{3m}{r}\right) + 4\Upsilon\frac{r-4m}{r^4}\right]r^{-2}\Big(r^2 f(r,m) - (4m)^2 f(4m,m)\Big),$$

is manifestly positive for $r \geq 4m$. To evaluate the sign of the second term we calculate

$$2r^{-3}\left(1-\frac{3m}{r}\right) - 2r^{-3}\Upsilon^2 = 2mr^{-5}(r-4m).$$

Thus, for $r \geq 4m$,

$$2r^{-3}\left(1-\frac{3m}{r}\right)f + 4\Upsilon\frac{r-4m}{r^4}(f - r^{-2}\widetilde{h}) - 2r^{-5}\Upsilon^2\widetilde{h} \;\geq\; 0.$$

Also, since $W = \frac{m}{r^2}\left(3 - \frac{8m}{r}\right)$, we have

$$r^{-2}W \;\gtrsim\; \frac{1}{r^4}.$$

Thus, in view of the above, we have, for some $C > 0$ and for $r \geq 4m$,

$$\dot{\mathcal{E}}_2 \;\geq C^{-1}\left[\tfrac{1}{r^3}|R(\Psi)|^2 + \tfrac{1}{r^4}|\Psi|^2\right]. \tag{10.1.37}$$

Gathering (10.1.35), (10.1.36) and (10.1.37), we infer for some $C > 0$,

$$\dot{\mathcal{E}}_2 \;\geq C^{-1}\left[\tfrac{1}{r^3}|R(\Psi)|^2 + \tfrac{1}{r^4}|\Psi|^2\right] \quad \text{on } r \geq \tfrac{5m}{2}.$$

Recalling

$$\int_S \dot{\mathcal{E}}[fR, w, 2hR](\Psi) \geq \int_S \dot{\mathcal{E}}_2 - O(\epsilon r^{-3})\int_S |\Psi|^2,$$

we infer

$$\int_S \dot{\mathcal{E}}[fR, w, 2hR](\Psi) \geq C^{-1}\int_S\left[\frac{1}{r^3}|R(\Psi)|^2+\frac{1}{r^4}|\Psi|^2\right] - O(\epsilon r^{-3})\int_S |\Psi|^2$$

and hence, for $\epsilon > 0$ small enough,

$$\int_S \dot{\mathcal{E}}[fR, w, 2hR](\Psi) \geq \frac{1}{2}C^{-1}\int_S\left[\frac{1}{r^3}|R(\Psi)|^2+\frac{1}{r^4}|\Psi|^2\right] \quad \text{on } r \geq \frac{5m}{2}$$

as desired.

It remains to analyze the error term,

$$\begin{aligned}
\mathcal{E}_\epsilon[fR, w, 2hR](\Psi) &= \mathcal{E}_\epsilon[fR, w](\Psi) \\
&\quad + O\left(r^{-1}\epsilon\, u_{trap}^{-1-\delta_{dec}}(r|\partial_r h| + |h| + r|\partial_m h|)\right)|\Psi|^2 \\
&= \epsilon\frac{1}{2}\mathcal{Q}^{(X)}\ddot{\pi} + O(\epsilon r^{-3}u_{trap}^{-1-\delta_{dec}}(|f| + r^2|\partial_r w| + r^3|\partial_r^2 w| \\
&\quad + r^2|\partial_r\partial_m w| + r|\partial_m^2 w|))|\Psi|^2 \\
&\quad + O\left(r^{-3}\epsilon\, u_{trap}^{-1-\delta_{dec}}(r^3|\partial_r h| + r^2|h| + r^3|\partial_m h|)\right)|\Psi|^2.
\end{aligned}$$

Recall that

$$w = \begin{cases} \frac{1}{4m}, & \text{for } \quad r \leq 4m, \\ \frac{2}{r}\left(1-\frac{2m}{r}\right), & \text{for } \quad r \geq 4m, \end{cases}$$

and $h = 4\Upsilon r^{-4}\widetilde{h}$, with

$$\widetilde{h} = \begin{cases} 0, & r \leq \frac{5m}{2}, \\ \delta_{\widetilde{h}}\left(\frac{5m}{2}-r\right), & \frac{5m}{2} \leq r \leq \frac{11m}{4}, \\ \delta_{\widetilde{h}}(r-3m), & \frac{11m}{4} \leq r \leq 3m, \\ r^2 f, & 3m \leq r \leq 4m, \\ (4m)^2 f(4m, m), & r \geq 4m. \end{cases}$$

We deduce

$$\mathcal{E}_\epsilon[fR, w, 2hR](\Psi) = \epsilon\mathcal{Q}^{(X)}\ddot{\pi} + O\Big(\epsilon r^{-3}u_{trap}^{-1-\delta_{dec}}(|f|+1)\Big)$$

which concludes the proof of Proposition 10.20. □

10.1.9 Cut-off correction in ${}^{(int)}\mathcal{M}$

So far we have found a triplet ($X = fR, w = r^{-2}\Upsilon\partial_r\left(r^2 f\right), M = 2hR$) with f defined in Proposition 10.16 and h in Proposition 10.20 allowing for the lower bound (10.1.32) on $\int_S \dot{\mathcal{E}}[fR, w, M](\Psi)$. The main problem which remains to be addressed is:

1. f blows up logarithmically near $r = 2m$.
2. The lower bound for $\int_S \dot{\mathcal{E}}[fR, w, 2hR](\Psi)$ does not control $e_3(\Psi)$ near $r = 2m$.

In this section, we deal with the first problem, while the second problem will be treated in section 10.1.10. To correct for the first problem, i.e., the fact that f blows up logarithmically near $r = 2m$, we have to modify our choice of f and w there. Introducing

$$u := r^2 f,$$

we have

$$f = r^{-2}u, \qquad w = r^{-2}\Upsilon\partial_r u. \tag{10.1.38}$$

Warning. The auxiliary function u introduced here, and used only in this section, has of course nothing to do with our previously defined optical function on ${}^{(ext)}\mathcal{M}$.

Definition 10.22. *For a given $\widehat{\delta} > 0$ we define the following functions of (r, m)*

$$u_{\widehat{\delta}} := -\frac{m^2}{\widehat{\delta}}F\left(-\frac{\widehat{\delta}}{m^2}u\right), \qquad f_{\widehat{\delta}} := r^{-2}u_{\widehat{\delta}},$$
$$w_{\widehat{\delta}} := r^{-2}\Upsilon\partial_r u_{\widehat{\delta}}, \qquad W_{\widehat{\delta}} := -\frac{1}{4}\partial_r\left(r^2\Upsilon\partial_r w_{\widehat{\delta}}\right),$$

where $F : \mathbb{R} \to \mathbb{R}$ is a fixed, increasing, smooth function such that

$$F(x) = \begin{cases} x & \text{for } x \leq 1, \\ 2 & \text{for } x \geq 3. \end{cases}$$

We now derive useful properties satisfied by $f_{\widehat{\delta}}$, $w_{\widehat{\delta}}$ and $W_{\widehat{\delta}}$.

Lemma 10.23. *Let $f_{\widehat{\delta}}$, $w_{\widehat{\delta}}$ and $W_{\widehat{\delta}}$ introduced in Definition 10.22. Then, we have $f_{\widehat{\delta}} \in C^2(r > 0)$, $w_{\widehat{\delta}} \in C^1(r > 0)$, and we have for $\widehat{\delta} > 0$ sufficiently small*

$$f_{\widehat{\delta}} = f \qquad w_{\widehat{\delta}} = w, \qquad W_{\widehat{\delta}} = W \quad \text{for } r \geq \frac{5m}{2}.$$

Also, we have for all $r > 0$

$$r^{-1}f_{\widehat{\delta}}\left(1 - \frac{3m}{r}\right) \geq C^{-1}r^{-1}\left(1 - \frac{3m}{r}\right)^2 \tag{10.1.39}$$

and

$$\partial_r(f_{\widehat{\delta}}) \geq \frac{16m^2}{r^3}. \tag{10.1.40}$$

Proof. Note first that

$$w_{\widehat{\delta}} = r^{-2}\Upsilon\partial_r u_{\widehat{\delta}} = r^{-2}\Upsilon F'\left(-\frac{\widehat{\delta}}{m^2}u\right)\partial_r u = wF'\left(-\frac{\widehat{\delta}}{m^2}u\right).$$

In view of the definition of $u_{\widehat{\delta}}$, $f_{\widehat{\delta}}$, $w_{\widehat{\delta}}$ and $W_{\widehat{\delta}}$, we have

$$u_{\widehat{\delta}} = u, \qquad f_{\widehat{\delta}} = f, \qquad w_{\widehat{\delta}} = w, \qquad W_{\widehat{\delta}} = W \qquad \text{for} \quad u \geq -\frac{m^2}{\widehat{\delta}},$$
$$u_{\widehat{\delta}} = -\frac{2m^2}{\widehat{\delta}}, \qquad f_{\widehat{\delta}} = -\frac{2m^2}{\widehat{\delta} r^2}, \qquad w_{\widehat{\delta}} = 0, \qquad W_{\widehat{\delta}} = 0 \qquad \text{for} \quad u \leq -\frac{3m^2}{\widehat{\delta}}.$$

Also, according to (10.1.22)

$$u = \begin{cases} 2m^2 \log \frac{r-2m}{m} + O(m(r-3m)), & \text{for} \quad r \leq 4m, \\ r^2 + O(m^2), & \text{for} \quad r \geq 4m, \end{cases}$$

and hence, for $\widehat{\delta} > 0$ sufficiently small

$$\left\{ r \geq 2m + e^{-\frac{1}{3\widehat{\delta}}} \right\} \cup \left\{ u \geq -\frac{m^2}{\widehat{\delta}} \right\}, \qquad \left\{ r \leq 2m + e^{-\frac{2}{\widehat{\delta}}} \right\} \subset \left\{ u \leq -\frac{3m^2}{\widehat{\delta}} \right\}.$$

This yields

$$f_{\widehat{\delta}} = f \qquad w_{\widehat{\delta}} = w, \qquad W_{\widehat{\delta}} = W \quad \text{for } r \geq \frac{5m}{2}.$$

Also, we have

$$f_{\widehat{\delta}} = \begin{cases} -\frac{2m^2}{\widehat{\delta} r^2}, & \text{for} \quad r \leq 2m + e^{-\frac{2}{\widehat{\delta}}}, \\ f, & \text{for} \quad r \geq 2m + e^{-\frac{1}{3\widehat{\delta}}}, \end{cases}$$

and

$$f_{\widehat{\delta}} \gtrsim \frac{1}{\widehat{\delta}} \quad \text{on} \quad 2m + e^{-\frac{2}{\widehat{\delta}}} \leq r \leq 2m + e^{-\frac{1}{3\widehat{\delta}}},$$

and thus, there exists $C > 0$ such that, for all $r > 0$,

$$r^{-1} f_{\widehat{\delta}} \left(1 - \frac{3m}{r}\right) \geq C^{-1} r^{-1} \left(1 - \frac{3m}{r}\right)^2$$

which is (10.1.39).

For $u \leq -\frac{3m^2}{\widehat{\delta}}$,

$$\partial_r(f_{\widehat{\delta}}) \quad = \quad \partial_r(r^{-2} u_{\widehat{\delta}}) = -2r^{-3} u_{\widehat{\delta}} + r^{-2} \partial_r(u_{\widehat{\delta}}) = \frac{4m^2}{\widehat{\delta}} r^{-3}.$$

For $-\frac{3m^2}{\widehat{\delta}} \leq u \leq -\frac{m^2}{\widehat{\delta}}$

$$\begin{aligned}\partial_r(f_{\widehat{\delta}}) &= \partial_r(r^{-2}u_{\widehat{\delta}}) = -2r^{-3}u_{\widehat{\delta}} + r^{-2}\partial_r(u_{\widehat{\delta}}) \\ &= -2r^{-3}u_{\widehat{\delta}} + r^{-2}F'\left(-\frac{\widehat{\delta}}{m^2}u\right)\partial_r u \\ &= -2r^{-3}u_{\widehat{\delta}} + r^{-2}F'\left(-\frac{\widehat{\delta}}{m^2}u\right)r^2\Upsilon^{-1}w,\end{aligned}$$

and since $w \geq 0$ and $F' \geq 0$, we deduce

$$\partial_r(f_{\widehat{\delta}}) \geq -2r^{-3}u_{\widehat{\delta}} \geq 2\widehat{\delta}^{-1}m^2r^{-3}.$$

For $u \geq -\frac{m^2}{\widehat{\delta}}$, using Lemma 10.15, we have

$$\partial_r(f_{\widehat{\delta}}) = \partial_r f \geq \frac{16m^2}{r^3}.$$

Hence, for all $r \geq 2m$, $\widehat{\delta} > 0$ sufficiently small,

$$\partial_r(f_{\widehat{\delta}}) \geq \frac{16m^2}{r^3}$$

which is (10.1.40). □

It remains to evaluate $W_{\widehat{\delta}}$. This is done in the following lemma.

Lemma 10.24. *Let*

$$\overline{W}_{\widehat{\delta}}(r,m) := 1_{r\leq\frac{5m}{2}}|W_{\widehat{\delta}}|. \tag{10.1.41}$$

Then, $\overline{W}_{\widehat{\delta}}$ *is supported, for* $\delta > 0$ *small enough, in the region*

$$2m + e^{-\frac{2}{\delta}} \leq r \leq \frac{9m}{4}.$$

Moreover its primitive,

$$\widetilde{W}_{\widehat{\delta}}(r,m) := \int_{2m}^{r} \overline{W}_{\widehat{\delta}}(r',m)dr', \tag{10.1.42}$$

verifies the pointwise estimate

$$\widetilde{W}_{\widehat{\delta}}(r,m) \lesssim \widehat{\delta}. \tag{10.1.43}$$

Proof. Recall that we have chosen $w = \frac{1}{4m}$ to be constant in the region $r \leq 4m$. Hence, in that region,

$$w_{\widehat{\delta}} = \frac{1}{4m}F'\left(-\frac{\widehat{\delta}}{m^2}u\right), \qquad \partial_r w_{\widehat{\delta}} = \frac{1}{4m}\partial_r\left(F'\left(-\frac{\widehat{\delta}}{m^2}u\right)\right).$$

Hence,

$$\begin{aligned} W_{\widehat{\delta}} &= -\frac{1}{4}r^{-2}\partial_r\left(\frac{1}{4m}r^2\Upsilon\partial_r\left(F'\left(-\frac{\widehat{\delta}}{m^2}u\right)\right)\right) \\ &= -\frac{1}{16m}r^{-2}\partial_r\left(r^2\Upsilon\partial_r\left(F'\left(-\frac{\widehat{\delta}}{m^2}u\right)\right)\right). \end{aligned}$$

Now, setting $\delta_0 = \frac{\widehat{\delta}}{m^2}$ for convenience below,

$$\begin{aligned} r^{-2}\partial_r\left(r^2\Upsilon\partial_r\left(F'\left(-\delta_0 u\right)\right)\right) &= -\delta_0 F''(-\delta_0 u)r^{-2}\partial_r\left(r^2\Upsilon\partial_r u\right) \\ &\quad +\delta_0^2 F'''(-\delta_0 u)\Upsilon(\partial_r u)^2. \end{aligned}$$

Note that, since $r^{-2}\Upsilon\partial_r u = w$ and $w = (4m)^{-1}$ is constant in r in the region of interest,

$$r^{-2}\partial_r\left(r^2\Upsilon\partial_r u\right) = r^{-2}\partial_r\left(r^4 r^{-2}\Upsilon\partial_r u\right) = r^{-2}\partial_r\left(\frac{r^4}{4m}\right) = \frac{r}{m}.$$

Hence,

$$\begin{aligned} r^{-2}\partial_r\left(r^2\Upsilon\partial_r\left(F'\left(-\delta_0 u\right)\right)\right) &= -\delta_0 F''(-\delta_0 u)r^{-2}\partial_r\left(r^2\Upsilon\partial_r u\right) \\ &\quad +\delta_0^2 F'''(-\delta_0 u)\Upsilon(\partial_r u)^2 \\ &= -\delta_0 F''(-\delta_0 u)\frac{r}{m} + \delta_0^2 F'''(-\delta_0 u)\Upsilon(\partial_r u)^2. \end{aligned}$$

Hence, for $r \leq 4m$, with $\delta_0 = \frac{\widehat{\delta}}{m^2}$,

$$|W_{\widehat{\delta}}| \lesssim \delta_0^2|\Upsilon||F'''(-\delta_0 u)|(\partial_r u)^2 + \delta_0|F''(-\delta_0 u)|$$

or, since $|\partial_r u| \lesssim \frac{1}{r-2m}$, in the region of interest,

$$|W_{\widehat{\delta}}| \lesssim \frac{\delta_0^2}{|r-2m|}|F'''(-\delta_0 u)| + \delta_0|F''(-\delta_0 u)|.$$

Since $F''(-\delta_0 u)$ and $F'''(-\delta_0 u)$ are supported in the region $1 \leq -\delta_0 u \leq 3$, i.e., in $-\frac{3}{\delta_0} \leq u \leq -\frac{1}{\delta_0}$, for $\widehat{\delta} > 0$ sufficiently small

$$e^{-\frac{2}{\delta}} \leq r - 2m \leq e^{-\frac{1}{3\delta}} \leq \frac{m}{4}.$$

Hence,

$$\overline{W}_{\widehat{\delta}} = 1_{r\leq\frac{5}{2}m}|W_{\widehat{\delta}}| \lesssim \widehat{\delta}\left(\frac{\delta}{r-2m}+1\right)\kappa_{\widehat{\delta}}(r-2m)$$

with $\kappa_{\widehat{\delta}}(x)$ the characteristic function of the interval $[e^{-\frac{2}{\delta}}, e^{-\frac{1}{3\delta}}]$. Note that the primitive of $\overline{W}_{\widehat{\delta}}$, i.e.,

$$\widetilde{W_{\widehat{\delta}}}(r,m) = \int_{2m}^{r}\overline{W}_{\widehat{\delta}}(r',m)dr',$$

is a positive, increasing function. Moreover,

$$\widetilde{W_{\widehat{\delta}}}(r) \lesssim \int_{2m}^{4m} \overline{W}_{\widehat{\delta}}(r)dr \lesssim \delta + \delta^2 \int_{e^{-\frac{2}{\delta}}}^{e^{-\frac{1}{3\delta}}} \frac{1}{x}dx \lesssim \delta$$

as desired. □

We now recall that, see (10.1.16),

$$\dot{\mathcal{E}}[fR,w](\Psi) = \dot{\mathcal{E}}_0[fR,w](\Psi) + 4\Upsilon\frac{r-4m}{r^4}f|\Psi|^2,$$
$$\dot{\mathcal{E}}_0[fR,w](\Psi) = \partial_r f|R(\Psi)|^2 + r^{-2}W|\Psi|^2 + r^{-1}\left(1-\frac{3m}{r}\right)f\widehat{\mathcal{Q}}_{34}.$$

Using the functions $f_{\widehat{\delta}}$, $w_{\widehat{\delta}}$ and $W_{\widehat{\delta}}$ introduced in Definition 10.22, we have

$$\dot{\mathcal{E}}_0[f_{\widehat{\delta}}R, w_{\widehat{\delta}}](\Psi) = \frac{1}{r}f_{\widehat{\delta}}\left(1-\frac{3m}{r}\right)\widehat{\mathcal{Q}}_{34} + \partial_r(f_{\widehat{\delta}})|R\Psi|^2 + W_{\widehat{\delta}}|\Psi|^2.$$

Note that in view of the estimates (10.1.39), (10.1.40), and Lemma 10.24, we immediately deduce the existence of a constant $C>0$ independent of $\widehat{\delta}$ such that

$$\begin{aligned}\dot{\mathcal{E}}[f_{\widehat{\delta}}R, w_{\widehat{\delta}}](\Psi) \geq\ & C^{-1}\left[|R\Psi|^2 + |\nabla\Psi|^2 + \Upsilon|\Psi|^2\right] \\ & -\overline{W}_{\widehat{\delta}}|\Psi|^2 \quad \text{on } r \leq \frac{5m}{2}\end{aligned} \tag{10.1.44}$$

where $\overline{W}_{\widehat{\delta}}$ is a non-negative potential supported in the region $2m + e^{-\frac{2}{\delta}} \leq r \leq \frac{9m}{4}$, whose primitive $\widetilde{W_{\widehat{\delta}}}(r) = \int_{2m}^{r}\overline{W}_{\widehat{\delta}}(r'm)dr'$ verifies $\widetilde{W_{\widehat{\delta}}} \lesssim \widehat{\delta}$. Combining this with estimates of the previous section we derive the following.

Proposition 10.25. *There exists a constant $C>0$, and for any small enough $\widehat{\delta}>0$, there exist functions $f_{\widehat{\delta}} \in C^2(r>0)$, $w_{\widehat{\delta}} \in C^1(r>0)$ and $h \in C^2(r>0)$ verifying, for all $r>0$,*

$$|f_{\widehat{\delta}}(r)| \lesssim \widehat{\delta}^{-1}, \qquad w_{\widehat{\delta}} \lesssim r^{-1}, \qquad h \lesssim r^{-4},$$

such that

$$\mathcal{E}[f_{\widehat{\delta}}R, w_{\widehat{\delta}}, 2hR](\Psi) = \dot{\mathcal{E}}[f_{\widehat{\delta}}R, w_{\widehat{\delta}}, 2hR] + \mathcal{E}_\epsilon[f_{\widehat{\delta}}R, w_{\widehat{\delta}}, 2hR](\Psi)$$

satisfies

$$\begin{aligned}\int_S \dot{\mathcal{E}}[f_{\widehat{\delta}}R, w_{\widehat{\delta}}, 2hR] \geq\ & C^{-1}\int_S\Bigg[\frac{m^2}{r^3}|R(\Psi)|^2 \\ & + r^{-1}\left(1-\frac{3m}{r}\right)^2\left(|\nabla\Psi|^2 + \frac{m^2}{r^2}|T\Psi|^2\right) + \frac{m}{r^4}|\Psi|^2\Bigg] \\ & - \int_S \overline{W}_{\widehat{\delta}}|\Psi|^2, \\ \mathcal{E}_\epsilon[f_{\widehat{\delta}}R, w_{\widehat{\delta}}, 2hR] =\ & \epsilon\frac{1}{2}\mathcal{Q}\cdot{}^{(f_{\widehat{\delta}}R)}\ddot{\pi} + O(r^{-3}u_{trap}^{-1-\delta_{dec}}(1+|f_{\widehat{\delta}}|))|\Psi|^2,\end{aligned}$$

where $\overline{W}_{\widehat{\delta}}$ *is non-negative, supported in the region* $2m + e^{-\frac{2}{\delta}} \le r \le \frac{9m}{4}$*, and such that its primitive* $\widetilde{W}_{\widehat{\delta}}(r) = \int_{2m}^{r} \overline{W}_{\widehat{\delta}}$ *verifies* $\widetilde{W}_{\widehat{\delta}} \lesssim \widehat{\delta}$.

Proof. We choose h to be the function of (r, m) introduced in Proposition 10.20, $f_{\widehat{\delta}}$ to be the function of (r, m) introduced in Definition 10.22, and $\overline{W}_{\widehat{\delta}}$, introduced in Lemma 10.24. Also, by an abuse of notation, we denote by $w_{\widehat{\delta},0}$ the function denoted by $w_{\widehat{\delta}}$ in Definition 10.22. Then, combining Proposition 10.20 in the region $r \ge \frac{5m}{2}$ with the estimate (10.1.44) in the region $r \le \frac{5m}{2}$, we immediately obtain

$$\begin{aligned}\int_S \dot{\mathcal{E}}[f_{\widehat{\delta}}R, w_{\widehat{\delta},0}, 2hR] \;\ge\;& C^{-1}\int_S \left[\frac{m^2}{r^3}|R(\Psi)|^2\right.\\ &\left.+r^{-1}\left(1-\frac{3m}{r}\right)^2|\nabla\!\!\!/\Psi|^2 + \frac{m}{r^4}|\Psi|^2\right]\\ &-\int_S \overline{W}_{\widehat{\delta}}|\Psi|^2. \end{aligned} \tag{10.1.45}$$

(10.1.45) corresponds to the desired estimate without the presence of the term $|T\Psi|^2$ on the right-hand side. To get the improved estimate of Proposition 10.25, we set

$$w_{\widehat{\delta}} \;:=\; w_{\widehat{\delta},0} - \delta_1 w_1, \tag{10.1.46}$$

for a small parameter $\delta_1 > 0$ to be chosen later, where $w_{\widehat{\delta},0}$ is our previous choice introduced in Definition 10.22, and where

$$w_1(r,m) := r^{-1}\frac{m^2}{r^2}\Upsilon\left(1-\frac{3m}{r}\right)^2. \tag{10.1.47}$$

We evaluate (modulo the same type of error terms as before which we include in $\mathcal{E}_\epsilon$)

$$\dot{\mathcal{E}}[f_{\widehat{\delta}}R, w_{\widehat{\delta}}, 2hR](\Psi) \;=\; \dot{\mathcal{E}}[X_{\widehat{\delta}}, w_{\widehat{\delta},0}, 2hR] - \frac{1}{2}\delta_1 w_1 \mathcal{L}(\Psi) + \frac{\delta_1}{4}|\Psi|^2 r^{-2}\partial_r(r^2\Upsilon\partial_r w_1).$$

Now, since

$$\begin{aligned}\mathcal{L}(\Psi) \;&=\; -e_3\Psi\cdot e_4\Psi + |\nabla\!\!\!/\Psi|^2 + V|\Psi|^2\\ &=\; \Upsilon^{-1}\left(-|T\Psi|^2 + |R\Psi|^2\right) + |\nabla\!\!\!/\Psi|^2 + V|\Psi|^2,\end{aligned}$$

we have

$$
\begin{aligned}
& -\frac{1}{2}\delta_1 w_1 \mathcal{L}(\Psi) + \frac{\delta_1}{4}|\Psi|^2 r^{-2}\partial_r(r^2\Upsilon\partial_r w_1) \\
= & -\frac{1}{2}\delta_1 r^{-1}\frac{m^2}{r^2}\Upsilon\left(1-\frac{3m}{r}\right)^2 \mathcal{L}(\Psi) \\
& +\frac{\delta_1}{4}|\Psi|^2 r^{-2}\partial_r\left(r^2\Upsilon\partial_r\left(r^{-1}\frac{m^2}{r^2}\Upsilon\left(1-\frac{3m}{r}\right)^2\right)\right) \\
= & \frac{1}{2}\delta_1 r^{-1}\left(1-\frac{3m}{r}\right)^2\frac{m^2}{r^2}|T\Psi|^2 \\
& +O(\delta_1)\left(\frac{m^2}{r^3}|R(\Psi)|^2 + r^{-1}\left(1-\frac{3m}{r}\right)^2|\nabla\Psi|^2 + \frac{m}{r^4}|\Psi|^2\right)
\end{aligned}
$$

and hence

$$
\begin{aligned}
& \dot{\mathcal{E}}[f_{\widehat{\delta}}R, w_{\widehat{\delta}}, 2hR](\Psi) \\
= & \dot{\mathcal{E}}[X_{\widehat{\delta}}, w_{\widehat{\delta},0}, 2hR] + \frac{1}{2}\delta_1 r^{-1}\left(1-\frac{3m}{r}\right)^2\frac{m^2}{r^2}|T\Psi|^2 \\
& +O(\delta_1)\left(\frac{m^2}{r^3}|R(\Psi)|^2 + r^{-1}\left(1-\frac{3m}{r}\right)^2|\nabla\Psi|^2 + \frac{m}{r^4}|\Psi|^2\right). \quad (10.1.48)
\end{aligned}
$$

The desired estimate now follows from (10.1.45) and (10.1.48) provided $\delta_1 > 0$ is chosen small enough compared to the constant $C > 0$ of (10.1.45) so that the last term $O(\delta_1)$ in the above identity can be absorbed. □

10.1.10 The redshift vectorfield

Note that the vectorfields T and R both become proportional to e_4 for $\Upsilon = 0$ which means that the estimate of Proposition 10.25 degenerates along $\Upsilon = 0$, i.e., it does not control $e_3(\Psi)$ there. In this section we make use of the Dafermos-Rodnianski redshift vectorfield to compensate for this degeneracy. The crucial ingredient here is the favorable sign of ω in a small neighborhood of $r = 2m$.

Lemma 10.26. *Let $\pi^{(3)}$, $\pi^{(4)}$ denote the deformation tensors of e_3, e_4. In the region $r \le 3m$ all components are $O(\epsilon)$ with the exception of*

$$
\begin{aligned}
\pi^{(3)}_{44} &= -8\omega = \frac{8m}{r^2} + O(\epsilon), & \pi^{(3)}_{\theta\theta} &= \underline{\kappa} + \underline{\vartheta} = -\frac{2}{r} + O(\epsilon), \\
\pi^{(3)}_{\varphi\varphi} &= \underline{\kappa} - \underline{\vartheta} = -\frac{2}{r} + O(\epsilon), \\
\pi^{(4)}_{34} &= 4\omega = -\frac{4m}{r^2} + O(\epsilon), & \pi^{(4)}_{\theta\theta} &= \kappa + \vartheta = \frac{2\Upsilon}{r} + O(\epsilon), \\
\pi^{(4)}_{\varphi\varphi} &= \kappa - \vartheta = \frac{2\Upsilon}{r} + O(\epsilon).
\end{aligned}
$$

Proof. Immediate verification in view of our assumptions. □

Lemma 10.27. *Given the vectorfield*

$$Y = a(r,m)e_3 + b(r,m)e_4, \tag{10.1.49}$$

and assuming

$$\sup_{r\le 3m}\Big(|a|+|\partial_r a|+|\partial_m a|+|b|+|\partial_r b|+|\partial_m b|\Big) \lesssim 1,$$

we have, for $r\le 3m$,

$$\begin{aligned}
\mathcal{Q}^{\alpha\beta}\,{}^{(Y)}\pi_{\alpha\beta} &= \left(\frac{2m}{r^2}a-\Upsilon\partial_r a\right)\mathcal{Q}_{33}+\partial_r b\mathcal{Q}_{44}+\left(\partial_r a-\frac{2m}{r^2}b-\Upsilon\partial_r b\right)\mathcal{Q}_{34}\\
&+ \frac{2}{r}(b\Upsilon-a)e_3\Psi\cdot e_4\Psi+8\frac{\Upsilon}{r^3}(a-\Upsilon b)|\Psi|^2\\
&+ O(\epsilon)\big(|\mathcal{Q}(\Psi)|+r^{-2}|\Psi|^2\big).
\end{aligned}$$

Moreover, with the notation (10.1.14),

$$\mathcal{E}[Y,0](\Psi)=\frac{1}{2}\mathcal{Q}^{\alpha\beta}\,{}^{(Y)}\pi_{\alpha\beta}+4\frac{r-3m}{r^4}(-a+b\Upsilon)|\Psi|^2+O(\epsilon)r^{-2}|\Psi|^2. \tag{10.1.50}$$

Proof. In view of

$$|e_4(r)-\Upsilon, e_3(r)+1| \lesssim \epsilon,$$

Lemma 10.5, and the assumptions on the derivatives of a and b w.r.t. (r,m), we have

$$\begin{aligned}
e_4(a) &= \Upsilon\partial_r a+O(\epsilon), \quad e_3(a)=-\partial_r a+O(\epsilon),\\
e_4(b) &= \Upsilon\partial_r b+O(\epsilon), \quad e_3(b)=-\partial_r b+O(\epsilon), \quad e_\theta(a)=e_\theta(b)=0.
\end{aligned}$$

We infer

$$\begin{aligned}
\mathcal{Q}^{\alpha\beta}\,{}^{(Y)}\pi_{\alpha\beta} &= a\mathcal{Q}^{\alpha\beta}\,{}^{(3)}\pi_{\alpha\beta}-(\mathcal{Q}_{33}e_4a+\mathcal{Q}_{43}e_3a)+b\mathcal{Q}^{\alpha\beta}\,{}^{(4)}\pi_{\alpha\beta}\\
&\quad -(\mathcal{Q}_{34}e_4b+\mathcal{Q}_{44}e_3b)+O(\epsilon)|\mathcal{Q}(\Psi)|\\
&= a\mathcal{Q}^{\alpha\beta}\,{}^{(3)}\pi_{\alpha\beta}+b\mathcal{Q}^{\alpha\beta}\,{}^{(4)}\pi_{\alpha\beta}-\mathcal{Q}_{33}\Upsilon\partial_r a-\mathcal{Q}_{34}\left(-\partial_r a+\Upsilon\partial_r b\right)\\
&\quad +\mathcal{Q}_{44}\partial_r b+O(\epsilon)|\mathcal{Q}(\Psi)|.
\end{aligned}$$

Note that

$$\mathcal{Q}_{\theta\theta}+\mathcal{Q}_{\varphi\varphi}=e_3\Psi\cdot e_4\Psi-V|\Psi|^2=e_3\Psi\cdot e_4\Psi-4\frac{\Upsilon}{r^2}|\Psi|^2+O(\epsilon)r^{-2}|\Psi|^2. \tag{10.1.51}$$

Hence,

$$\begin{aligned}
\mathcal{Q}^{\alpha\beta}\,\pi^{(3)}_{\alpha\beta} &= \mathcal{Q}^{44}\,\pi^{(3)}_{44} + \mathcal{Q}^{\theta\theta}\,\pi^{(3)}_{\theta\theta} + \mathcal{Q}^{\varphi\varphi}\,\pi^{(3)}_{\varphi\varphi} + O(\epsilon)|\mathcal{Q}(\Psi)| \\
&= \frac{1}{4}\mathcal{Q}_{33}\frac{8m}{r^2} - \frac{2}{r}\left(\mathcal{Q}_{\theta\theta} + \mathcal{Q}_{\varphi\varphi}\right) + O(\epsilon)|\mathcal{Q}(\Psi)| \\
&= \frac{2m}{r^2}\mathcal{Q}_{33} - \frac{2}{r}e_3\Psi\cdot e_4\Psi + 8\frac{\Upsilon}{r^3}|\Psi|^2 + O(\epsilon)\big(|\mathcal{Q}(\Psi)| + r^{-2}|\Psi|^2\big),
\end{aligned}$$

$$\begin{aligned}
\mathcal{Q}^{\alpha\beta}\,\pi^{(4)}_{\alpha\beta} &= 2\mathcal{Q}^{34}\,\pi^{(4)}_{34} + \mathcal{Q}^{\theta\theta}\,\pi^{(4)}_{\theta\theta} + \mathcal{Q}^{\varphi\varphi}\,\pi^{(4)}_{\varphi\varphi} + O(\epsilon)|\mathcal{Q}(\Psi)| \\
&= \frac{1}{2}\mathcal{Q}_{34}(-4\frac{m}{r^2}) + \frac{2\Upsilon}{r}\left(\mathcal{Q}_{\theta\theta} + \mathcal{Q}_{\varphi\varphi}\right) + O(\epsilon)|\mathcal{Q}(\Psi)| \\
&= -\frac{2m}{r^2}\mathcal{Q}_{34} + \frac{2\Upsilon}{r}e_3\Psi\cdot e_4\Psi - 8\frac{\Upsilon^2}{r^3}|\Psi|^2 + O(\epsilon)\big(|\mathcal{Q}(\Psi)| + r^{-2}|\Psi|^2\big).
\end{aligned}$$

Therefore,

$$\begin{aligned}
\mathcal{Q}^{\alpha\beta\,(Y)}\pi_{\alpha\beta} &= a\left[\frac{2m}{r^2}\mathcal{Q}_{33} - \frac{2}{r}e_3\Psi e_4\Psi + 8\frac{\Upsilon}{r^3}|\Psi|^2\right] \\
&\quad + b\left[-\frac{2m}{r^2}\mathcal{Q}_{34} + \frac{2\Upsilon}{r}e_3\Psi\cdot e_4\Psi - 8\frac{\Upsilon^2}{r^3}|\Psi|^2\right] - \mathcal{Q}_{33}\Upsilon\partial_r a \\
&\quad - \mathcal{Q}_{34}\left(-\partial_r a + \Upsilon\partial_r b\right) + \mathcal{Q}_{44}\partial_r b + O(\epsilon)\big(|\mathcal{Q}(\Psi)| + r^{-2}|\Psi|^2\big) \\
&= \left(\frac{2m}{r^2}a - \Upsilon\partial_r a\right)\mathcal{Q}_{33} + \mathcal{Q}_{44}\partial_r b + \frac{2}{r}(b\Upsilon - a)e_3\Psi e_4\Psi \\
&\quad + \left(\partial_r a - \frac{2m}{r^2}b - \Upsilon\partial_r b\right)\mathcal{Q}_{34} + 8\frac{\Upsilon}{r^3}(a - \Upsilon b)|\Psi|^2 \\
&\quad + O(\epsilon)\big(|\mathcal{Q}(\Psi)| + r^{-2}|\Psi|^2\big).
\end{aligned}$$

To prove the second part of the lemma we recall (see (10.1.14)),

$$\mathcal{E}[Y,0](\Psi) = \frac{1}{2}\mathcal{Q}^{\alpha\beta\,(Y)}\pi_{\alpha\beta} - \frac{1}{2}Y(V)|\Psi|^2$$

and, relying on Lemma 10.5, we have on $r \leq 3m$

$$Y(V) = (-a + b\Upsilon)\partial_r V + O(\epsilon) = (-a + b\Upsilon)\left(-8\frac{r-3m}{r^4}\right) + O(\epsilon)$$

which concludes the proof of the lemma. ☐

Corollary 10.28. *If we choose*

$$a(2m,m) = 1, \qquad b(2m,m) = 0, \qquad \partial_r a(2m,m) \geq \frac{1}{4m}, \qquad \partial_r b(2m,m) \geq \frac{5}{4m},$$

then, at $r = 2m$, we have

$$\mathcal{Q}^{\alpha\beta\,(Y)}\pi_{\alpha\beta} \geq \frac{1}{4m}\left(|e_3\Psi|^2 + |e_4\Psi|^2 + \mathcal{Q}_{34}\right) + O(\epsilon)\big(|\mathcal{Q}(\Psi)| + r^{-2}|\Psi|^2\big) \tag{10.1.52}$$

and

$$\begin{aligned}\mathcal{E}[Y,0](\Psi) &\geq \frac{1}{8m}\left(|e_3\Psi|^2+|e_4\Psi|^2+\mathcal{Q}_{34}+\frac{1}{m^2}|\Psi|^2\right)\\ &\quad +O(\epsilon)\big(|\mathcal{Q}(\Psi)|+r^{-2}|\Psi|^2\big). \qquad (10.1.53)\end{aligned}$$

Moreover the estimates remain true if we add to Y *a multiple of* $T=\frac{1}{2}(e_4+\Upsilon e_3)$.

Proof. Recall from Lemma 10.27 that we have, for $r\leq 3m$,

$$\begin{aligned}\mathcal{Q}^{\alpha\beta}\,{}^{(Y)}\pi_{\alpha\beta} &= \left(\frac{2m}{r^2}a-\Upsilon\partial_r a\right)\mathcal{Q}_{33}+\partial_r b\mathcal{Q}_{44}+\left(\partial_r a-\frac{2m}{r^2}b-\Upsilon\partial_r b\right)\mathcal{Q}_{34}\\ &+ \frac{2}{r}(b\Upsilon-a)e_3\Psi\cdot e_4\Psi+8\frac{\Upsilon}{r^3}(a-\Upsilon b)|\Psi|^2\\ &+ O(\epsilon)\big(|\mathcal{Q}(\Psi)|+r^{-2}|\Psi|^2\big).\end{aligned}$$

Hence, at $r=2m$, using $\Upsilon=0$, $a=1$, $b=0$, $\partial_r a\geq(4m)^{-1}$ and $\partial_r b\geq 5(4m)^{-1}$, we deduce

$$\begin{aligned}\mathcal{Q}^{\alpha\beta}\,{}^{(Y)}\pi_{\alpha\beta} &= \frac{1}{2m}\mathcal{Q}_{33}+\partial_r b\mathcal{Q}_{44}+\partial_r a\mathcal{Q}_{34}-\frac{1}{m}e_3\Psi\cdot e_4\Psi\\ &\quad +O(\epsilon)\big(|\mathcal{Q}(\Psi)|+r^{-2}|\Psi|^2\big)\\ &\geq \frac{1}{2m}|e_3(\Psi)|^2+\frac{5}{4m}|e_4(\Psi)|^2+\frac{1}{4m}\mathcal{Q}_{34}-\frac{1}{m}e_3\Psi\cdot e_4\Psi\\ &\quad +O(\epsilon)\big(|\mathcal{Q}(\Psi)|+r^{-2}|\Psi|^2\big)\end{aligned}$$

from which the desired lower bound in (10.1.52) follows.

Also, at $r=2m$, using (10.1.50), $\Upsilon=0$, $a=1$, and $b=0$, we have

$$\begin{aligned}\mathcal{E}[Y,0](\Psi) &= \frac{1}{2}\mathcal{Q}^{\alpha\beta}\,{}^{(Y)}\pi_{\alpha\beta}+4\frac{r-3m}{r^4}(-a+b\Upsilon)|\Psi|^2+O(\epsilon)r^{-2}|\Psi|^2\\ &= \frac{1}{2}\mathcal{Q}^{\alpha\beta}\,{}^{(Y)}\pi_{\alpha\beta}+\frac{1}{4m^3}|\Psi|^2\\ &\geq \frac{1}{8m}\left(|e_3\Psi|^2+|e_4\Psi|^2+\mathcal{Q}_{34}+\frac{1}{m^2}|\Psi|^2\right)+O(\epsilon)\big(|\mathcal{Q}(\Psi)|+r^{-2}|\Psi|^2\big)\end{aligned}$$

which yields (10.1.53). □

We are now ready to prove the following result.

Proposition 10.29. *Given a small parameter* $\delta_{\mathcal{H}}>0$ *there exists a smooth vectorfield* $Y_{\mathcal{H}}$ *supported in the region* $|\Upsilon|\leq 2\delta_{\mathcal{H}}^{\frac{1}{10}}$ *such that the following estimate holds:*

$$\begin{aligned}\mathcal{E}[Y_{\mathcal{H}},0](\Psi) &\geq \frac{1}{16m}1_{|\Upsilon|\leq\delta_{\mathcal{H}}^{\frac{1}{10}}}\left(|e_3\Psi|^2+|e_4\Psi|^2+\widehat{\mathcal{Q}}_{34}+m^{-2}|\Psi|^2\right)\\ &- \frac{1}{m}\delta_{\mathcal{H}}^{-\frac{1}{10}}1_{\delta_{\mathcal{H}}^{\frac{1}{10}}\leq\Upsilon\leq 2\delta_{\mathcal{H}}^{\frac{1}{10}}}\left(|e_3\Psi|^2+|e_4\Psi|^2+\widehat{\mathcal{Q}}_{34}+m^{-2}|\Psi|^2\right)\\ &+ O(\epsilon)1_{|\Upsilon|\leq 2\delta_{\mathcal{H}}^{\frac{1}{10}}}\big(|\mathcal{Q}(\Psi)|+m^{-2}|\Psi|^2\big).\end{aligned}$$

Moreover, for $|\Upsilon| \le \delta_{\mathcal{H}}^{\frac{1}{10}}$, *we have*

$$Y_{\mathcal{H}} = e_3 + e_4 + O(\delta_{\mathcal{H}}^{\frac{1}{10}})(e_3 + e_4).$$

Proof. We introduce the vectorfield

$$Y_{(0)} := ae_3 + be_4 + 2T, \quad a(r,m) := 1 + \frac{5}{4m}(r-2m), \quad b(r,m) := \frac{5}{4m}(r-2m),$$

with $T = \frac{1}{2}(e_4 + \Upsilon e_3)$. Also, we pick positive bump function $\kappa = \kappa(r)$, supported in the region in $[-2,2]$ and equal to 1 for $[-1,1]$ and define, for sufficiently small $\delta_{\mathcal{H}} > 0$,

$$Y_{\mathcal{H}} := \kappa_{\mathcal{H}} Y_{(0)}, \qquad \kappa_{\mathcal{H}} := \kappa\left(\frac{\Upsilon}{\delta_{\mathcal{H}}^{\frac{1}{10}}}\right). \tag{10.1.54}$$

We have

$$\begin{aligned}
\mathcal{E}[Y_{\mathcal{H}}, 0](\Psi) &= \mathcal{Q} \cdot {}^{(Y_{\mathcal{H}})}\pi - Y_{\mathcal{H}}(V)|\Psi|^2 \\
&= \kappa_{\mathcal{H}} \mathcal{Q} \cdot {}^{(Y_0)}\pi + \mathcal{Q}(Y_{(0)}, d\kappa_{\mathcal{H}}) + \kappa_{\mathcal{H}} Y_{(0)}(V)|\Psi|^2 \\
&= \kappa_{\mathcal{H}} \mathcal{E}[Y_{(0)}, 0](\Psi) \\
&\quad + O(\delta_{\mathcal{H}}^{-\frac{1}{10}}) 1_{\delta_{\mathcal{H}}^{\frac{1}{10}} \le \Upsilon \le 2\delta_{\mathcal{H}}^{\frac{1}{10}}} \left(|e_3\Psi|^2 + |e_4\Psi|^2 + \widehat{\mathcal{Q}}_{34} + m^{-2}|\Psi|^2\right).
\end{aligned}$$

Note from the definition of $Y_{(0)}$ and the choice of a and b that Corollary 10.28 applies to $Y_{(0)}$. In particular, we deduce from (10.1.53) for $\delta_{\mathcal{H}} > 0$ small enough,

$$\begin{aligned}
\mathcal{E}[Y_{\mathcal{H}}, 0](\Psi) &\ge \frac{1}{16m} 1_{|\Upsilon| \le \delta_{\mathcal{H}}^{\frac{1}{10}}} \left(|e_3\Psi|^2 + |e_4\Psi|^2 + \widehat{\mathcal{Q}}_{34} + m^{-2}|\Psi|^2\right) \\
&\quad - \frac{1}{m} \delta_{\mathcal{H}}^{-\frac{1}{10}} 1_{\delta_{\mathcal{H}}^{\frac{1}{10}} \le \Upsilon \le 2\delta_{\mathcal{H}}^{\frac{1}{10}}} \left(|e_3\Psi|^2 + |e_4\Psi|^2 + \widehat{\mathcal{Q}}_{34} + m^{-2}|\Psi|^2\right) \\
&\quad + O(\epsilon) 1_{|\Upsilon| \le 2\delta_{\mathcal{H}}^{\frac{1}{10}}} \left(|\mathcal{Q}(\Psi)| + m^{-2}|\Psi|^2\right)
\end{aligned}$$

as desired. □

10.1.11 Combined estimate

We consider the combined Morawetz triplet

$$(X, w, M) := (X_{\widehat{\delta}}, w_{\widehat{\delta}}, 2hR) + \epsilon_{\mathcal{H}}(Y_{\mathcal{H}}, 0, 0), \tag{10.1.55}$$

with $\epsilon_{\mathcal{H}} > 0$ sufficiently small to be determined later. Here $(X_{\widehat{\delta}} = f_{\widehat{\delta}} R, w_{\widehat{\delta}}, 2hR)$ is the triplet given by Proposition 10.25 and $Y_{\mathcal{H}}$ the vectorfield of Proposition 10.29.

Recall, see Proposition 10.25, that $\dot{\mathcal{E}}_{\widehat{\delta}} := \dot{\mathcal{E}}[f_{\widehat{\delta}}R, w_{\widehat{\delta}}, 2hR](\Psi)$ verifies

$$\begin{aligned}\int_S \dot{\mathcal{E}}_{\widehat{\delta}} &\geq C^{-1}\int_S \left[\frac{m^2}{r^3}|R(\Psi)|^2 + \left(1-\frac{3m}{r}\right)^2 r^{-1}\left(\widehat{\mathcal{Q}}_{34} + \frac{m^2}{r^2}|T\Psi|^2\right)\right.\\ &\left. +\Upsilon\frac{m}{r^4}|\Psi|^2\right] - \int_S \overline{W}_{\widehat{\delta}}|\Psi|^2.\end{aligned}$$

According to Proposition 10.29, we write, for $\mathcal{E}_{\mathcal{H}} = \mathcal{E}(Y_{\mathcal{H}}, 0, 0)(\Psi)$,

$$\begin{aligned}\mathcal{E}_{\mathcal{H}} &= \dot{\mathcal{E}}_{\mathcal{H}} + \mathcal{E}_{\mathcal{H},\epsilon},\\ \dot{\mathcal{E}}_{\mathcal{H}} &\geq \frac{1}{8m}1_{|\Upsilon|\leq\delta_{\mathcal{H}}^{\frac{1}{10}}}\left(|e_3\Psi|^2 + |e_4\Psi|^2 + \widehat{\mathcal{Q}}_{34} + m^{-2}|\Psi|^2\right)\\ &- \frac{1}{m}\delta_{\mathcal{H}}^{-\frac{1}{10}}1_{\delta_{\mathcal{H}}^{\frac{1}{10}}\leq\Upsilon\leq 2\delta_{\mathcal{H}}^{\frac{1}{10}}}\left(|e_3\Psi|^2 + |e_4\Psi|^2 + \widehat{\mathcal{Q}}_{34} + m^{-2}|\Psi|^2\right),\\ \mathcal{E}_{\mathcal{H},\epsilon} &= O(\epsilon)\left(|\mathcal{Q}(\Psi)| + m^{-2}|\Psi|^2\right)1_{|\Upsilon|\leq 2\delta_{\mathcal{H}}^{\frac{1}{10}}}.\end{aligned}$$

Note that, for $|\Upsilon| \geq \delta_{\mathcal{H}}^{\frac{1}{10}}$, we have

$$|R\Psi|^2 + |T\Psi|^2 = \frac{1}{2}(|e_4\Psi|^2 + \Upsilon^2|e_3\Psi|^2) \geq \frac{1}{2}\delta_{\mathcal{H}}^{\frac{1}{5}}(|e_4\Psi|^2 + |e_3\Psi|^2).$$

We now proceed to find a lower bound for the expression $\dot{\mathcal{E}}_{\widehat{\delta}} + \epsilon_{\mathcal{H}}\dot{\mathcal{E}}_{\mathcal{H}}$. For brevity the S integration is omitted below.

Region $\delta_{\mathcal{H}}^{\frac{1}{10}} \leq |\Upsilon| \leq 2\delta_{\mathcal{H}}^{\frac{1}{10}}$.

$$\begin{aligned}\dot{\mathcal{E}}_{\widehat{\delta}} + \epsilon_{\mathcal{H}}\dot{\mathcal{E}}_{\mathcal{H}} &\geq m^{-1}C^{-1}\left[\delta_{\mathcal{H}}^{\frac{1}{5}}(|e_4\Psi|^2 + |e_3\Psi|^2) + m^{-2}|\Psi|^2 + |\nabla\!\!\!\!/\,\Psi|^2\right] - \overline{W}_{\widehat{\delta}}|\Psi|^2\\ &- \epsilon_{\mathcal{H}}\frac{1}{m}\delta_{\mathcal{H}}^{-\frac{1}{10}}\left(|e_3\Psi|^2 + |e_4\Psi|^2 + |\nabla\!\!\!\!/\,\Psi|^2 + m^{-2}|\Psi|^2\right).\end{aligned}$$

Therefore, choosing $\epsilon_{\mathcal{H}} \leq (2C)^{-1}\delta_{\mathcal{H}}^{\frac{3}{10}}$, we deduce

$$\dot{\mathcal{E}}_{\widehat{\delta}} + \epsilon_{\mathcal{H}}\dot{\mathcal{E}}_{\mathcal{H}} \geq m^{-1}\delta_{\mathcal{H}}^{\frac{1}{5}}(2C)^{-1}\left(|e_4\Psi|^2 + |e_3\Psi|^2 + |\nabla\!\!\!\!/\,\Psi|^2 + m^{-2}|\Psi|^2\right) - \overline{W}_{\widehat{\delta}}|\Psi|^2.$$

Region $|\Upsilon| \leq \delta_{\mathcal{H}}^{\frac{1}{10}}$.

$$\epsilon_{\mathcal{H}}\dot{\mathcal{E}}_{\mathcal{H}} + \dot{\mathcal{E}}_{\widehat{\delta}} \geq \epsilon_{\mathcal{H}}\frac{1}{16m}\left(|e_3\Psi|^2 + |e_4\Psi|^2 + \widehat{\mathcal{Q}}_{34} + m^{-2}|\Psi|^2\right) - \overline{W}_{\widehat{\delta}}|\Psi|^2.$$

Region $\Upsilon \geq 2\delta_{\mathcal{H}}^{\frac{1}{10}}$. In this region $\dot{\mathcal{E}}_{\widehat{\delta}} + \epsilon_{\mathcal{H}}\dot{\mathcal{E}}_{\mathcal{H}} = \dot{\mathcal{E}}_{\widehat{\delta}}$. Hence (ignoring the S integration),

$$\begin{aligned}\dot{\mathcal{E}}_{\widehat{\delta}} + \epsilon_{\mathcal{H}}\dot{\mathcal{E}}_{\mathcal{H}} &\geq C^{-1}\left(\frac{m^2}{r^3}|R(\Psi)|^2 + \left(1-\frac{3m}{r}\right)^2 r^{-1}\left(\widehat{\mathcal{Q}}_{34} + \frac{m^2}{r^2}|T\Psi|^2\right) + \frac{m}{r^4}|\Psi|^2\right)\\ &- \overline{W}_{\widehat{\delta}}|\Psi|^2.\end{aligned}$$

To combine these three cases together we modify the vectorfields R, T near

$r = 2m$ according to (5.1.11), i.e.,

$$\breve{R} := \theta\frac{1}{2}(e_4 - e_3) + (1-\theta)\Upsilon^{-1}R = \frac{1}{2}\left[\breve{\theta}e_4 - e_3\right],$$
$$\breve{T} := \theta\frac{1}{2}(e_4 + e_3) + (1-\theta)\Upsilon^{-1}T = \frac{1}{2}\left[\breve{\theta}e_4 + e_3\right],$$

where θ a smooth bump function equal 1 on $|\Upsilon| \le \delta_{\mathcal{H}}^{\frac{1}{10}}$ vanishing for $|\Upsilon| \ge 2\delta_{\mathcal{H}}^{\frac{1}{10}}$, and where

$$\breve{\theta} = \theta + \Upsilon^{-1}(1-\theta) = \begin{cases} 1, & \text{for } |\Upsilon| \le \delta_{\mathcal{H}}^{\frac{1}{10}}, \\ \Upsilon^{-1}, & \text{for } |\Upsilon| \ge 2\delta_{\mathcal{H}}^{\frac{1}{10}}. \end{cases}$$

Note that

$$2(|\breve{R}\Psi|^2 + |\breve{T}\Psi|^2) = |e_3\Psi|^2 + \breve{\theta}^2|e_4\Psi|^2.$$

Thus in the region $|\Upsilon| \le \delta_{\mathcal{H}}^{\frac{1}{10}}$ we have $|e_3\Psi|^2 + |e_4\Psi|^2 = 2(|\breve{R}\Psi|^2 + |\breve{T}\Psi|^2)$ and therefore,

$$\begin{aligned}
\dot{\mathcal{E}}_{\widehat{\delta}} + \epsilon_{\mathcal{H}}\dot{\mathcal{E}}_{\mathcal{H}} &\ge \epsilon_{\mathcal{H}}\frac{1}{16m}\left(|e_3\Psi|^2 + |e_4\Psi|^2 + \widehat{\mathcal{Q}}_{34} + m^{-2}|\Psi|^2\right) - \overline{W}_{\widehat{\delta}}|\Psi|^2 \\
&= \epsilon_{\mathcal{H}}\frac{1}{16m}\left(|\breve{R}\Psi|^2 + |\breve{T}\Psi|^2 + \widehat{\mathcal{Q}}_{34} + m^{-2}|\Psi|^2\right) - \overline{W}_{\widehat{\delta}}|\Psi|^2.
\end{aligned}$$

In the region $\delta_{\mathcal{H}}^{\frac{1}{10}} \le |\Upsilon| \le 2\delta_{\mathcal{H}}^{\frac{1}{10}}$, we have $|\breve{R}\Psi|^2 + |\breve{T}\Psi|^2 \lesssim |e_3\Psi|^2 + \delta_{\mathcal{H}}^{-\frac{1}{5}}|e_4\Psi|^2$. Hence, for $\epsilon_{\mathcal{H}} \le (2C)^{-1}\delta_{\mathcal{H}}^{\frac{3}{10}}$, we deduce

$$\begin{aligned}
\dot{\mathcal{E}}_{\widehat{\delta}} + \epsilon_{\mathcal{H}}\dot{\mathcal{E}}_{\mathcal{H}} &\ge m^{-1}\delta_{\mathcal{H}}^{\frac{1}{5}}(2C)^{-1}\left(|e_4\Psi|^2 + |e_3\Psi|^2 + |\nabla\Psi|^2 + m^{-2}|\Psi|^2\right) - \overline{W}_{\widehat{\delta}}|\Psi|^2 \\
&\ge m^{-1}\delta_{\mathcal{H}}^{\frac{1}{5}}(2C)^{-1}\left(\delta_{\mathcal{H}}^{\frac{1}{5}}\left(|\breve{R}\Psi|^2 + |\breve{T}\Psi|^2\right) + |\nabla\Psi|^2 + m^{-2}|\Psi|^2\right) \\
&\quad -\overline{W}_{\widehat{\delta}}|\Psi|^2.
\end{aligned}$$

Finally, for $\Upsilon \ge 2\delta_{\mathcal{H}}^{\frac{1}{10}}$ we have $\breve{R} = \Upsilon^{-1}R$, $\breve{T} = \Upsilon^{-1}T$. Hence,

$$\begin{aligned}
\dot{\mathcal{E}}_{\widehat{\delta}} + \epsilon_{\mathcal{H}}\dot{\mathcal{E}}_{\mathcal{H}} &\ge C^{-1}\Bigg[\frac{m^2}{r^3}|R(\Psi)|^2 + \left(1 - \frac{3m}{r}\right)^2 r^{-1}\left(\widehat{\mathcal{Q}}_{34} + \frac{m^2}{r^2}|T\Psi|^2\right) \\
&\quad + \Upsilon\frac{m}{r^4}|\Psi|^2\Bigg] - \overline{W}_{\widehat{\delta}}|\Psi|^2 \\
&\ge C^{-1}\Bigg[\delta_{\mathcal{H}}^{\frac{1}{5}}\frac{m^2}{r^3}|\breve{R}(\Psi)|^2 + \left(1 - \frac{3m}{r}\right)^2 r^{-1}\left(\widehat{\mathcal{Q}}_{34} + \delta_{\mathcal{H}}^{\frac{2}{10}}\frac{m^2}{r^2}|\breve{T}\Psi|^2\right) \\
&\quad + \Upsilon\frac{m}{r^4}|\Psi|^2\Bigg] - \overline{W}_{\widehat{\delta}}|\Psi|^2.
\end{aligned}$$

We deduce the following.

Proposition 10.30. *Let $C > 0$ the constant of Proposition 10.25. Consider the* *combined Morawetz triplet*

$$(X, w, M) := (X_{\widehat{\delta}}, w_{\widehat{\delta}}, 2hR) + \epsilon_{\mathcal{H}}(Y_{\mathcal{H}}, 0, 0), \tag{10.1.56}$$

with $C^{-1}\delta_{\mathcal{H}}^{\frac{2}{5}} \le \epsilon_{\mathcal{H}} \le (2C)^{-1}\delta_{\mathcal{H}}^{\frac{3}{10}}$ *where, for given fixed* $\widehat{\delta} > 0$, $(X_{\widehat{\delta}}, w_{\widehat{\delta}}, 2hR)$ *is the triplet of Proposition 10.25 and* $Y_{\mathcal{H}}$ *the vectorfield of Proposition 10.29, supported in* $|\Upsilon| \le 2\delta_{\mathcal{H}}^{\frac{1}{10}}$ *with* $\delta_{\mathcal{H}} > 0$ *sufficiently small, independent of* $\widehat{\delta}$. *Let* $\dot{\mathcal{E}}_{\widehat{\delta}}, \dot{\mathcal{E}}_{\mathcal{H}}$ *be the principal parts of* $\mathcal{E}[f_{\widehat{\delta}}R, w_{\widehat{\delta}}, 2hR](\Psi)$ *and respectively* $\mathcal{E}_{\mathcal{H}}[Y_{\mathcal{H}}, 0, 0](\Psi)$ *and* $\mathcal{E}_{\widehat{\delta},\epsilon}$, $\mathcal{E}_{\mathcal{H},\epsilon}$ *the corresponding error terms, i.e.,*

$$\mathcal{E}[f_{\widehat{\delta}}R, w_{\widehat{\delta}}, 2hR](\Psi) = \dot{\mathcal{E}}_{\widehat{\delta}} + \mathcal{E}_{\widehat{\delta},\epsilon}, \qquad \mathcal{E}_{\mathcal{H}}[Y_{\mathcal{H}}, 0, 0](\Psi) = \dot{\mathcal{E}}_{\mathcal{H}} + \mathcal{E}_{\mathcal{H},\epsilon}.$$

Then, provided $\delta_{\mathcal{H}} > 0$ *is sufficiently small, we have:*

1. *In the region* $-2\delta_{\mathcal{H}}^{\frac{1}{10}} \le \Upsilon$, $r \le \frac{5m}{2}$, *we have with a constant* $\Lambda_{\mathcal{H}}^{-1} := C^{-1}\delta_{\mathcal{H}}^{\frac{2}{5}} > 0$

$$\begin{aligned}\int_S (\dot{\mathcal{E}}_{\widehat{\delta}} + \epsilon_{\mathcal{H}}\dot{\mathcal{E}}_{\mathcal{H}}) \;&\ge\; m^{-1}\Lambda_{\mathcal{H}}^{-1}\int_S \left(|\breve{R}(\Psi)|^2 + |\breve{T}\Psi|^2 + |\nabla\!\!\!\!/\,\Psi|^2 + m^{-2}|\Psi|^2\right)\\ &\quad - \int_S \overline{W}_{\widehat{\delta}}|\Psi|^2.\end{aligned}$$

2. *In the region* $r \ge \frac{5m}{2}$, *where* $\dot{\mathcal{E}}_{\widehat{\delta}} + \epsilon_{\mathcal{H}}\dot{\mathcal{E}}_{\mathcal{H}} = \dot{\mathcal{E}}_{\widehat{\delta}}$ *and* $\overline{W}_{\widehat{\delta}} = 0$, *we have the same estimate as in Proposition 10.25, i.e.,*

$$\begin{aligned}\int_S (\dot{\mathcal{E}}_{\widehat{\delta}} + \epsilon_{\mathcal{H}}\dot{\mathcal{E}}_{\mathcal{H}}) \;&\ge\; C^{-1}\int_S \Bigg[\frac{m^2}{r^3}|R(\Psi)|^2\\ &\quad + r^{-1}\left(1 - \frac{3m}{r}\right)^2\left(|\nabla\!\!\!\!/\,\Psi|^2 + \frac{m^2}{r^2}|T\Psi|^2\right) + \frac{m}{r^4}|\Psi|^2\Bigg].\end{aligned}$$

3. *The* ϵ*-error terms verify the upper bound estimate,*

$$\begin{aligned}\mathcal{E}_{\widehat{\delta},\epsilon} + \epsilon_{\mathcal{H}}\mathcal{E}_{\mathcal{H},\epsilon} \;&\lesssim\; C\epsilon\widehat{\delta}^{-1}u_{trap}^{-1-\delta_{dec}}\left[r^{-2}|e_3\Psi|^2 + r^{-1}(|e_4\Psi|^2 + |\nabla\!\!\!\!/\,\Psi|^2)\right]\\ &\quad + C\epsilon\widehat{\delta}^{-1}u_{trap}^{-1-\delta_{dec}}r^{-1}|e_3\Psi|\,(|e_4\Psi| + |\nabla\!\!\!\!/\,\Psi|)\\ &\quad + C\epsilon\widehat{\delta}^{-1}u_{trap}^{-1-\delta_{dec}}r^{-3}|\Psi|^2.\end{aligned}$$

Proof. It only remains to check the last part. In view of Proposition 10.20 we have

$$\mathcal{E}_{\widehat{\delta},\epsilon} = \mathcal{E}_\epsilon[f_{\widehat{\delta}}R, w_{\widehat{\delta}}, 2hR](\Psi) = \epsilon\frac{1}{2}\mathcal{Q}\cdot\,{}^{(X_{\widehat{\delta}})}\ddot{\pi} + O\big(\epsilon r^{-3}u_{trap}^{-1-\delta_{dec}}(|f_{\widehat{\delta}}| + 1)\big)|\Psi|^2$$

and $|f_{\widehat{\delta}}| \lesssim \widehat{\delta}^{-1}$. Hence,

$$\mathcal{E}_{\widehat{\delta},\epsilon} = \mathcal{E}_\epsilon[f_{\widehat{\delta}}R, w_{\widehat{\delta}}, 2hR](\Psi) = \epsilon\left(\frac{1}{2}\mathcal{Q}\cdot\,{}^{X_{\widehat{\delta}}}\ddot{\pi} + O(\widehat{\delta}^{-1}r^{-3}u_{trap}^{-1-\delta_{dec}})|\Psi|^2\right).$$

We write, with $\ddot{\pi} = {}^{(X_{\widehat{\delta}})}\ddot{\pi}$ for simplicity,

$$\mathcal{Q}\cdot\ddot{\pi} = \frac{1}{4}(\mathcal{Q}_{33}\ddot{\pi}_{44} + 2\mathcal{Q}_{34}\ddot{\pi}_{34} + \mathcal{Q}_{44}\dot{\pi}_{33}) - \frac{1}{2}(\mathcal{Q}_{A3}\ddot{\pi}_{A4} + \mathcal{Q}_{A4}\ddot{\pi}_{A3}) + \mathcal{Q}_{AB}\ddot{\pi}_{AB}.$$

Thus, recalling parts 1 and 2 of Proposition 10.9, and Lemma 10.8,

$$\begin{aligned}\mathcal{Q}\cdot\ddot{\pi} &\lesssim r^{-2}u_{trap}^{-1-\delta_{dec}}|e_3\Psi|^2 + r^{-1}u_{trap}^{-1-\delta_{dec}}\left(|e_4\Psi|^2 + |\nabla\!\!\!/\Psi|^2 + r^{-2}|\Psi|^2\right)\\ &+ r^{-1}u_{trap}^{-1-\delta_{dec}}|e_3\Psi|\left(|e_4\Psi| + |\nabla\!\!\!/\Psi|\right).\end{aligned}$$

Finally, since $r\sim 2m$ and $u_{trap}=1$ on $|\Upsilon|\le 2\delta_{\mathcal{H}}^{\frac{1}{10}}$, the error terms generated by the redshift vectorfield $Y_{\mathcal{H}}$,

$$\mathcal{E}_{\mathcal{H},\epsilon} = O(\epsilon)1_{|\Upsilon|\le 2\delta_{\mathcal{H}}^{\frac{1}{10}}}\left(|\mathcal{Q}(\Psi)| + m^{-2}|\Psi|^2\right),$$

can easily be absorbed on the right-hand side to derive the desired estimate. □

10.1.11.1 Elimination of $\overline{W}_{\widehat{\delta}}$

We now proceed to eliminate the potential $\overline{W}_{\widehat{\delta}}$ by a procedure analogous to that used in section 10.1.8. More precisely we set, in view of (10.1.14),

$$\mathcal{E}_{\widehat{\delta}} = \mathcal{E}[f_{\widehat{\delta}}R, w_{\widehat{\delta}}, 2hR](\Psi), \qquad \mathcal{E}'_{\widehat{\delta}} = \mathcal{E}[f_{\widehat{\delta}}R, w_{\widehat{\delta}}, 2(hR + h_2\breve{R})](\Psi),$$

and

$$\mathcal{E}'_{\widehat{\delta}} = \mathcal{E}_{\widehat{\delta}} + h_2\Psi\breve{R}\Psi + \frac{1}{2}\mathbf{D}^\mu(h_2\breve{R}_\mu)|\Psi|^2,$$

where h_2 is a smooth, compactly supported function supported[8] in the region $r\le\frac{9m}{4}$.

Thus, we have in view of Proposition 10.30, ignoring the integration on S,

$$\begin{aligned}\dot{\mathcal{E}}'_{\widehat{\delta}} + \epsilon_{\mathcal{H}}\dot{\mathcal{E}}_{\mathcal{H}} &= \dot{\mathcal{E}}_{\widehat{\delta}} + \epsilon_{\mathcal{H}}\dot{\mathcal{E}}_{\mathcal{H}} + h_2\Psi\breve{R}\Psi + \frac{1}{2}\mathbf{D}^\mu(h_2\breve{R}_\mu)|\Psi|^2 \qquad (10.1.57)\\ &\ge \mathcal{I}(\Psi) + m^{-1}\Lambda_{\mathcal{H}}^{-1}\left(\frac{1}{2}|\breve{R}(\Psi)|^2 + |\breve{T}\Psi|^2 + |\nabla\!\!\!/\Psi|^2 + m^{-2}|\Psi|^2\right)\end{aligned}$$

where

$$\mathcal{I}(\Psi) := \frac{1}{2}\Lambda_{\mathcal{H}}^{-1}m^{-1}|\breve{R}(\Psi)|^2 + \Psi h_2\breve{R}\Psi + \frac{1}{2}\mathbf{D}^\mu(h_2\breve{R}_\mu)|\Psi|^2 - \overline{W}_{\widehat{\delta}}|\Psi|^2$$

so that we have

$$\mathcal{I}(\Psi) \ge \frac{1}{2}\left[\mathbf{D}^\mu(h_2\breve{R}_\mu) - 2\overline{W}_{\widehat{\delta}} - m\Lambda_{\mathcal{H}}h_2^2\right]|\Psi|^2. \qquad (10.1.58)$$

We focus on the coefficient in front of $|\Psi|^2$ on the RHS of (10.1.58). Ignoring the error terms in ϵ (which can easily be incorporated in the upper bound for $\mathcal{E}_{\widehat{\delta},\epsilon} + \epsilon_{\mathcal{H}}\mathcal{E}_{\mathcal{H},\epsilon}$ of the previous proposition), we have

$$\mathbf{Div}\breve{R} = \frac{1}{2}\left(\mathbf{D}^\mu(\breve{\theta}(e_4)_\mu) - \mathbf{D}^\mu(e_3)_\mu\right) = \frac{1}{4}\left(\breve{\theta}\mathrm{tr}\pi^{(4)} - \mathrm{tr}\pi^{(3)}\right) + \frac{1}{2}e_4(\breve{\theta}) = O(\delta_{\mathcal{H}}^{-\frac{1}{10}})$$

[8]Recall that $\overline{W}_{\widehat{\delta}}$ is supported in the region $2m < r \le \frac{5m}{2}$.

and, using in particular Lemma 10.5,

$$\begin{aligned}
\mathbf{D}^\mu(h_2\breve{R}_\mu) &= \breve{R}h_2 + h_2\mathbf{Div}\breve{R} = \frac{1}{2}\partial_r h_2(\breve{\theta}e_4 r - e_3 r) + h_2\mathbf{Div}\breve{R}\\
&= \frac{1}{2}\partial_r h_2(\breve{\theta}\Upsilon + 1) + h_2 O(\delta_{\mathcal{H}}^{-\frac{1}{10}})\\
&\geq \frac{1}{2}\partial_r h_2 + h_2 O(\delta_{\mathcal{H}}^{-\frac{1}{10}}).
\end{aligned}$$

Together with (10.1.58), we infer

$$\mathcal{I}(\Psi) \geq \frac{1}{4}\Big[\partial_r h_2 - 4\overline{W}_{\widehat{\delta}} - 4m\Lambda_{\mathcal{H}}h_2^2 + h_2 O(\delta_{\mathcal{H}}^{-\frac{1}{10}})\Big]|\Psi|^2. \tag{10.1.59}$$

We now consider the choice of the function $h_2 = h_2(r,m)$. Recall (see Lemma 10.24) that $\overline{W}_{\widehat{\delta}}$ is supported in the region $2m+e^{-\frac{2}{\delta}} \leq r \leq \frac{9m}{4}$ and that its primitive $\widetilde{W}_{\widehat{\delta}}(r) := \int_{2m}^r \overline{W}_{\widehat{\delta}}$ verifies $\widetilde{W}_{\widehat{\delta}} \lesssim m^{-2}\widehat{\delta}$. We choose

$$h_2 =: \begin{cases} 4\widetilde{W}_{\widehat{\delta}}, & \text{for } r \leq \frac{9m}{4}\\ 0, & \text{for } r \geq \frac{5m}{2}\end{cases} \tag{10.1.60}$$

and since $\widetilde{W}_{\widehat{\delta}} \lesssim m^{-2}\widehat{\delta}$, we may extend h_2 in $\frac{9m}{4} \leq r \leq \frac{5m}{2}$ such that h_2 is C^1 and we have for all $r > 0$

$$|h_2| \lesssim m^{-2}\widehat{\delta}, \qquad |\partial_r h_2| \lesssim m^{-3}\widehat{\delta}. \tag{10.1.61}$$

In view of (10.1.59), this choice of h_2 yields

$$\mathcal{I}(\Psi) \geq -\frac{1}{4}O\Big(m^{-1}\widehat{\delta} + \Lambda_{\mathcal{H}}m^{-1}\widehat{\delta}^2 + \widehat{\delta}(\delta_{\mathcal{H}})^{-\frac{1}{10}}\Big)|\Psi|^2.$$

Hence, for $\widehat{\delta} \ll \delta_{\mathcal{H}}^{\frac{1}{10}}\Lambda_{\mathcal{H}}^{-1}$, i.e., $\widehat{\delta} \ll \delta_{\mathcal{H}}^{\frac{1}{2}}$ (recall that $\Lambda_{\mathcal{H}}^{-1} = C^{-1}\delta_{\mathcal{H}}^{\frac{2}{5}}$) and h_2 defined as above, we infer

$$\mathcal{I}(\Psi) \geq -\frac{1}{2}m^{-1}\Lambda_{\mathcal{H}}^{-1}m^{-2}|\Psi|^2$$

which together with (10.1.57) finally yields

$$\begin{aligned}
\int_S(\dot{\mathcal{E}}'_{\widehat{\delta}} + \epsilon_{\mathcal{H}}\dot{\mathcal{E}}_{\mathcal{H}}) &\geq \int_S \mathcal{I}(\Psi)\\
&\quad + m^{-1}\Lambda_{\mathcal{H}}^{-1}\int_S\left(\frac{1}{2}|\breve{R}(\Psi)|^2 + |\breve{T}\Psi|^2 + |\nabla\!\!\!/\Psi|^2 + m^{-2}|\Psi|^2\right)\\
&\geq m^{-1}\Lambda_{\mathcal{H}}^{-1}\int_S\left(\frac{1}{2}|\breve{R}(\Psi)|^2 + |\breve{T}\Psi|^2 + |\nabla\!\!\!/\Psi|^2 + \frac{1}{2}m^{-2}|\Psi|^2\right).
\end{aligned}$$

10.1.11.2 Summary of results so far

We summarize the result in the following.

Proposition 10.31. *Consider the combined Morawetz triplet*

$$(X, w, M) := (f_{\widehat{\delta}}R, w_{\widehat{\delta}}, 2hR) + \epsilon_{\mathcal{H}}(Y_{\mathcal{H}}, 0, 0) + (0, 0, 2h_2\breve{R}) \tag{10.1.62}$$

with $(f_{\widehat{\delta}}R, w_{\widehat{\delta}}, 2hR)$ *the triplet of Proposition 10.25,* $Y_{\mathcal{H}}$ *the redshift vectorfield of Proposition 10.29 (corresponding to the small parameter* $\delta_{\mathcal{H}}$*) and* h_2 *the* C^1 *function above satisfying* (10.1.60), (10.1.61)*. Let* $\dot{\mathcal{E}}[X, w, M]$ *the principal part of* $\mathcal{E}[X, w, M]$ *(independent of* ϵ*) and* $\mathcal{E}_\epsilon[X, w, M]$ *the error term in* ϵ *such that* $\mathcal{E} = \dot{\mathcal{E}} + \mathcal{E}_\epsilon$*.*

We choose the small strictly positive parameters $\epsilon_{\mathcal{H}}$*,* $\delta_{\mathcal{H}}, \delta$ *such that*[9]

$$\epsilon_{\mathcal{H}} = \delta_{\mathcal{H}}^{\frac{7}{20}}, \qquad \widehat{\delta} = \delta_{\mathcal{H}}^{\frac{3}{5}}. \tag{10.1.63}$$

Then, there holds[10]

$$\begin{aligned}\int_S \dot{\mathcal{E}}[X, w, M](\Psi) &\geq \delta_{\mathcal{H}}^{\frac{1}{2}} \int_S \Bigg[\frac{m^2}{r^3}|\breve{R}\Psi|^2 + r^{-1}\left(1 - \frac{3m}{r}\right)^2 \left(\widehat{\mathcal{Q}}_{34} + \frac{m^2}{r^2}|\breve{T}\Psi|^2\right) \\ &\quad + \frac{m}{r^4}|\Psi|^2\Bigg], \\ \mathcal{E}_\epsilon[X, w, M](\Psi) &\leq \delta_{\mathcal{H}}^{-1}\epsilon u_{trap}^{-1-\delta_{dec}}\left[r^{-2}|e_3\Psi|^2 + r^{-1}(|e_4\Psi|^2 + |\nabla\Psi|^2)\right] \\ &\quad + \delta_{\mathcal{H}}^{-1}\epsilon u_{trap}^{-1-\delta_{dec}} r^{-1}|e_3\Psi|\left(|e_4\Psi| + |\nabla\Psi|\right) \\ &\quad + \delta_{\mathcal{H}}^{-1}\epsilon u_{trap}^{-1-\delta_{dec}} r^{-3}|\Psi|^2.\end{aligned} \tag{10.1.64}$$

10.1.12 Lower bounds for $\mathcal{Q}$

In this section we prove lower bounds for $\mathcal{Q}(X + 2\Lambda T, e_3)$ and $\mathcal{Q}(X + 2\Lambda T, e_4)$ in the region $r_{\mathcal{H}} \leq r$, for $r_{\mathcal{H}}$ to be determined and Λ sufficiently large.

Proposition 10.32. *Under the assumptions of Proposition 10.31, and with the choice*

$$\Lambda := \frac{1}{4}\delta_{\mathcal{H}}^{-\frac{13}{20}}, \tag{10.1.65}$$

the following inequalities hold true for $r \geq 2m_0(1 - \delta_{\mathcal{H}})$*.*

1. *For the region such that* $r \geq 2m_0(1 - \delta_{\mathcal{H}})$ *and* $\Upsilon \leq \delta_{\mathcal{H}}^{\frac{1}{10}}$*, we have*

$$\begin{aligned}\mathcal{Q}(X + \Lambda T, e_3) &\geq \frac{1}{4}\epsilon_{\mathcal{H}}\mathcal{Q}_{33} + \frac{1}{2}\Lambda\mathcal{Q}_{34}, \\ \mathcal{Q}(X + \Lambda T, e_4) &\geq \frac{1}{4}\epsilon_{\mathcal{H}}\mathcal{Q}_{34} + \frac{1}{2}\Lambda\mathcal{Q}_{44}.\end{aligned}$$

---

[9]Note that (10.1.63) verifies all the restrictions we have encountered so far, i.e., $\delta_{\mathcal{H}}^{\frac{2}{5}} \ll \epsilon_{\mathcal{H}} \ll \delta_{\mathcal{H}}^{\frac{3}{10}}$ and $0 < \widehat{\delta} \ll \delta_{\mathcal{H}}^{\frac{1}{2}}$.

[10]Note that $\delta_{\mathcal{H}}^{\frac{1}{2}} \ll \Lambda_{\mathcal{H}}^{-1}$ (recall that $\Lambda_{\mathcal{H}}^{-1} = C^{-1}\delta_{\mathcal{H}}^{\frac{2}{5}}$) and $\delta_{\mathcal{H}}^{-1} \gg \widehat{\delta}^{-1}$ in view of (10.1.63).

2. For the region $\delta_{\mathcal{H}}^{\frac{1}{10}} \leq \Upsilon \leq \frac{1}{3}$*, we have*

$$\begin{aligned}\mathcal{Q}(X+\Lambda T, e_3) &\geq \delta_{\mathcal{H}}^{-\frac{1}{2}}\left(\mathcal{Q}_{33}+\mathcal{Q}_{34}\right),\\ \mathcal{Q}(X+\Lambda T, e_4) &\geq \delta_{\mathcal{H}}^{-\frac{1}{2}}\left(\mathcal{Q}_{44}+\mathcal{Q}_{34}\right).\end{aligned}$$

3. For the region $r \geq 3m$*, we have*

$$\begin{aligned}\mathcal{Q}(X+\Lambda T, e_3) &\geq \frac{1}{4}\Lambda\left(\mathcal{Q}_{33}+\mathcal{Q}_{34}\right),\\ \mathcal{Q}(X+\Lambda T, e_4) &\geq \frac{1}{4}\Lambda\left(\mathcal{Q}_{44}+\mathcal{Q}_{34}\right).\end{aligned}$$

4. The null components of $\mathcal{Q}$ *are given by (recall Proposition 10.9),*

$$\mathcal{Q}_{33} = |e_3\Psi|^2, \qquad \mathcal{Q}_{44} = |e_4\Psi|^2, \qquad \mathcal{Q}_{34} = |\nabla\!\!\!/\Psi|^2 + \frac{4\Upsilon}{r^2}(1+O(\epsilon))|\Psi|^2.$$

Proof. Since $X = f_{\widehat{\delta}}R + \epsilon_{\mathcal{H}}Y_{\mathcal{H}}$ and $T = \frac{1}{2}(e_4+\Upsilon e_3)$, $R = \frac{1}{2}(e_4 - \Upsilon e_3)$, we write

$$\begin{aligned}\mathcal{Q}(X+2\Lambda T, e_3) &= \mathcal{Q}(X, e_3) + \Lambda\mathcal{Q}(e_4+\Upsilon e_3, e_3) = \mathcal{Q}(X, e_3) + \Lambda\left(\mathcal{Q}_{34}+\Upsilon\mathcal{Q}_{33}\right)\\ &= \epsilon_{\mathcal{H}}\mathcal{Q}(Y_{\mathcal{H}}, e_3) + \Lambda\left(\mathcal{Q}_{34}+\Upsilon\mathcal{Q}_{33}\right) + \frac{1}{2}f_{\widehat{\delta}}\left(\mathcal{Q}_{34}-\Upsilon\mathcal{Q}_{33}\right).\end{aligned}$$

In the region $2m_0(1-\delta_{\mathcal{H}}) \leq r \leq 2m$ we have $Y_{\mathcal{H}} = e_3 + e_4 + O(\delta_{\mathcal{H}})(e_3+e_4)$, $\Upsilon \geq 0$ and $f_{\widehat{\delta}} < 0$. Hence, in that region,

$$\begin{aligned}\mathcal{Q}(X+2\Lambda T, e_3) &\geq \frac{1}{2}\epsilon_{\mathcal{H}}\left(\mathcal{Q}_{33}+\mathcal{Q}_{34}\right) + \left(\Lambda - \frac{1}{2}|f_{\widehat{\delta}}|\right)\mathcal{Q}_{34} - |\Upsilon|\left(\Lambda + \frac{1}{2}|f_{\widehat{\delta}}|\right)\mathcal{Q}_{33}\\ &\geq \left(\frac{1}{2}\epsilon_{\mathcal{H}} - |\Upsilon|\left(\Lambda+\frac{1}{2}|f_{\widehat{\delta}}|\right)\right)\mathcal{Q}_{33} + \left(\frac{1}{2}\epsilon_{\mathcal{H}} + \Lambda - \frac{1}{2}|f_{\widehat{\delta}}|\right)\mathcal{Q}_{34}.\end{aligned}$$

Thus, we need to choose Λ such that

$$\frac{1}{2}|f_{\widehat{\delta}}| \leq \Lambda \leq \frac{1}{4}\frac{\epsilon_{\mathcal{H}}}{\delta_{\mathcal{H}}} - \frac{1}{2}|f_{\widehat{\delta}}|.$$

Now, recall (10.1.63) as well as the fact that $|f_{\widehat{\delta}}|$ is of size $O((\widehat{\delta})^{-1})$. Thus it suffices to choose Λ such that

$$O(\delta_{\mathcal{H}}^{-\frac{3}{5}}) \leq \Lambda \leq \frac{1}{2}\delta_{\mathcal{H}}^{-\frac{13}{20}} - O(\delta_{\mathcal{H}}^{-\frac{3}{5}}),$$

i.e., it suffices to choose, for $\delta_{\mathcal{H}} > 0$ small enough,

$$\Lambda = \frac{1}{4}\delta_{\mathcal{H}}^{-\frac{13}{20}},$$

to deduce the inequality,

$$\mathcal{Q}(X+2\Lambda T, e_3) \geq \frac{1}{4}\epsilon_{\mathcal{H}}\mathcal{Q}_{33} + \frac{1}{2}\Lambda\mathcal{Q}_{34}.$$

Next, in the region $0 \le \Upsilon \le \delta_{\mathcal{H}}^{\frac{1}{10}}$, the sign of Υ is more favorable and we have

$$\begin{aligned}\mathcal{Q}(X+2\Lambda T,e_3) &\ge \frac{1}{2}\epsilon_{\mathcal{H}}(\mathcal{Q}_{33}+\mathcal{Q}_{34}) + \left(\Lambda - \frac{1}{2}|f_{\widehat{\delta}}|\right)\mathcal{Q}_{34} + |\Upsilon|\left(\Lambda + \frac{1}{2}|f_{\widehat{\delta}}|\right)\mathcal{Q}_{33}\\ &\ge \left(\frac{1}{2}\epsilon_{\mathcal{H}} + |\Upsilon|\left(\Lambda + \frac{1}{2}|f_{\widehat{\delta}}|\right)\right)\mathcal{Q}_{33} + \left(\frac{1}{2}\epsilon_{\mathcal{H}} + \Lambda - \frac{1}{2}|f_{\widehat{\delta}}|\right)\mathcal{Q}_{34}.\end{aligned}$$

In particular, we simply need $\Lambda \gg \widehat{\delta}^{-1}$, which is in particular satisfied by (10.1.65), to deduce the same inequality,

$$\mathcal{Q}(X+2\Lambda T,e_3) \ge \frac{1}{4}\epsilon_{\mathcal{H}}\mathcal{Q}_{33} + \frac{1}{2}\Lambda\mathcal{Q}_{34}.$$

In the region $\delta_{\mathcal{H}}^{\frac{1}{10}} \le \Upsilon \le \frac{1}{3}$, where $f_{\widehat{\delta}} \le 0$, and using the fact that $|f_{\widehat{\delta}}|$ is of size $O((\widehat{\delta})^{-1})$

$$\begin{aligned}\mathcal{Q}(X+2\Lambda T,e_3) &= \epsilon_{\mathcal{H}}\mathcal{Q}(Y_{\mathcal{H}},e_3) + \Lambda(\mathcal{Q}_{34}+\Upsilon\mathcal{Q}_{33}) + \frac{1}{2}f_{\widehat{\delta}}(\mathcal{Q}_{34}-\Upsilon\mathcal{Q}_{33})\\ &\ge \Lambda\left(\mathcal{Q}_{34} + \delta_{\mathcal{H}}^{\frac{1}{10}}\mathcal{Q}_{33}\right) - O(\widehat{\delta}^{-1})\mathcal{Q}_{34}.\end{aligned}$$

Hence, for the choice (10.1.65), and in view of (10.1.63), we infer

$$\mathcal{Q}(X+2\Lambda T,e_3) \ge \delta_{\mathcal{H}}^{-\frac{1}{2}}(\mathcal{Q}_{33}+\mathcal{Q}_{34}).$$

Finally, for $r \ge 3m$ where we have $0 \le f_{\widehat{\delta}} \lesssim 1$, $\frac{1}{3} \le \Upsilon \le 1$ and $Y_{\mathcal{H}} = 0$,

$$\begin{aligned}\mathcal{Q}(X+2\Lambda T,e_3) &= \Lambda(\mathcal{Q}_{34}+\Upsilon\mathcal{Q}_{33}) + \frac{1}{2}f_{\widehat{\delta}}(\mathcal{Q}_{34}-\Upsilon\mathcal{Q}_{33})\\ &\ge \Lambda\left(\mathcal{Q}_{34}+\frac{1}{3}\mathcal{Q}_{33}\right) - O(1)\mathcal{Q}_{33}\end{aligned}$$

and hence, (10.1.65) implies

$$\mathcal{Q}(X+2\Lambda T,e_3) \ge \frac{1}{4}\Lambda(\mathcal{Q}_{34}+\mathcal{Q}_{33})$$

as desired. The proof for $\mathcal{Q}(X+\Lambda T,e_4)$ is similar. □

10.1.13 First Morawetz estimate

We are now ready to state our first Morawetz estimate which is simply obtained by integrating the pointwise inequality in Proposition 10.30 on our spacetime domain $\mathcal{M} = {}^{(int)}\mathcal{M} \cup {}^{(ext)}\mathcal{M}$ described at the beginning of the section, with X replaced by $X+\Lambda T$ for $\Lambda > 0$ sufficiently large. In view of the choice of τ, note that we have

$$N_\Sigma = ae_3 + be_4, \qquad 0 \le a, b \le 1, \qquad a+b \ge 1, \tag{10.1.66}$$

with

$$b=0,\ a=1 \text{ on } {}^{(int)}\mathcal{M}, \quad a=0,\ b=1 \text{ on } \mathcal{M}_{r\geq 4m_0}, \quad a,\ b\geq \frac{1}{4} \text{ on } {}^{(trap)}\mathcal{M}.$$

We recall the following quantities for Ψ in regions $\mathcal{M}(\tau_1,\tau_2)\subset\mathcal{M}$ in the past of $\Sigma(\tau_2)$ and future of $\Sigma(\tau_1)$.

1. Morawetz bulk quantity

$$\begin{aligned}\mathrm{Mor}[\Psi](\tau_1,\tau_2) &= \int_{\mathcal{M}(\tau_1,\tau_2)}\Bigg[\frac{m^2}{r^3}|\breve{R}\psi|^2+\frac{m}{r^4}|\Psi|^2\\ &\quad+\left(1-\frac{3m}{r}\right)^2\frac{1}{r}\left(|\nabla\!\!\!/\psi|^2+\frac{m^2}{r^2}|\breve{T}\psi|^2\right)\Bigg].\end{aligned}$$

2. Basic energy quantity

$$E[\Psi](\tau)=\int_{\Sigma(\tau)}\left(\frac{1}{2}(N_\Sigma,e_3)^2\,|e_4\Psi|^2+\frac{1}{2}(N_\Sigma,e_4)^2\,|e_3\Psi|^2+|\nabla\!\!\!/\Psi|^2+r^{-2}|\Psi|^2\right).$$

3. Flux through $\mathcal{A}$ and Σ_*

$$\begin{aligned}F[\Psi](\tau_1,\tau_2) &= \int_{\mathcal{A}(\tau_1,\tau_2)}(\delta_{\mathcal{H}}^{-1}|e_4\Psi|^2+\delta_{\mathcal{H}}|e_3\Psi|^2+|\nabla\!\!\!/\Psi|^2+r^{-2}|\Psi|^2)\\ &\quad+\int_{\Sigma_*(\tau_1,\tau_2)}(|e_4\Psi|^2+|e_3\Psi|^2+|\nabla\!\!\!/\Psi|^2+r^{-2}|\Psi|^2),\end{aligned}$$

with $\mathcal{A}(\tau_1,\tau_2)=\mathcal{A}\cap\mathcal{M}(\tau_1,\tau_2)$ and $\Sigma_*(\tau_1,\tau_2)=\Sigma_*\cap\mathcal{M}(\tau_1,\tau_2)$.

The following theorem is our first Morawetz estimate.

Theorem 10.33. *Consider the equation* (10.1.10), *i.e.,* $\dot{\square}\Psi=V\Psi+\mathcal{N}$, *with the potential* $V=-\kappa\underline{\kappa}$ *and a domain* $\mathcal{M}(\tau_1,\tau_2)\subset\mathcal{M}$. *Then, we have*

$$\begin{aligned}E[\Psi](\tau_2)+Mor[\Psi](\tau_1,\tau_2)+F[\Psi](\tau_1,\tau_2) &\lesssim E[\Psi](\tau_1)+J[N,\Psi](\tau_1,\tau_2)\\ &\quad+Err_\epsilon(\tau_1,\tau_2)[\Psi],\\ J[N,\Psi](\tau_1,\tau_2) &:= \int_{\mathcal{M}(\tau_1,\tau_2)}\Big[|\breve{R}\Psi|+|\breve{T}\Psi|\\ &\quad+r^{-1}|\Psi|\Big]|N|,\\ Err_\epsilon[\Psi](\tau_1,\tau_2) &= \int_{\mathcal{M}(\tau_1,\tau_2)}\mathcal{E}_\epsilon[\Psi],\end{aligned}\tag{10.1.67}$$

where

$$\begin{aligned}\mathcal{E}_\epsilon[\Psi] &\lesssim \epsilon u_{trap}^{-1-\delta_{dec}}\Big[r^{-2}|e_3\Psi|^2+r^{-1}(|e_4\Psi|^2+|\nabla\!\!\!/\Psi|^2+r^{-2}|\Psi|^2\\ &\quad+|e_3\Psi|\,(|e_4\Psi|+|\nabla\!\!\!/\Psi|))\Big].\end{aligned}$$

Proof. Recall that, see (10.1.13),

$$\mathcal{E}[X,w,M](\Psi) \quad := \quad \mathbf{D}^\mu \mathcal{P}_\mu[X,w,M] - \left(X(\Psi) + \frac{1}{2}w\Psi\right) \cdot \mathcal{N}[\Psi]$$

where

$$\mathcal{P}_\mu \quad = \quad \mathcal{P}_\mu[X,w,M] = \mathcal{Q}_{\mu\nu}X^\nu + \frac{1}{2}w\Psi\dot{\mathbf{D}}_\mu\Psi - \frac{1}{4}|\Psi|^2\partial_\mu w + \frac{1}{4}|\Psi|^2 M_\mu$$

with triplet

$$(X,w,M) := (f_{\widehat{\delta}}R, w_{\widehat{\delta}}, 2hR) + \epsilon_{\mathcal{H}}(Y_{\mathcal{H}},0,0) + (0,0,2h_2\breve{R})$$

given in Proposition 10.30. Replacing X by $\check{X} = X + \Lambda T$ in the calculation above we deduce

$$\check{\mathcal{P}}_\mu \quad = \quad \mathcal{P}_\mu[\check{X},w,M] = \mathcal{Q}_{\mu\nu}\check{X}^\nu + \frac{1}{2}w\Psi\dot{\mathbf{D}}_\mu\Psi - \frac{1}{4}|\Psi|^2\partial_\mu w + \frac{1}{4}|\Psi|^2 M_\mu.$$

By the divergence theorem we have

$$\begin{aligned}&\int_{\mathcal{A}} \check{\mathcal{P}}\cdot N_{\mathcal{A}} + \int_{\Sigma_2} \check{\mathcal{P}}\cdot N_\Sigma + \int_{\mathcal{M}(\tau_1,\tau_2)} \mathcal{E} + \int_{\Sigma_*} \check{\mathcal{P}}\cdot N_{\Sigma_*} = \int_{\Sigma_1} \check{\mathcal{P}}\cdot N_\Sigma \\ &- \int_{\mathcal{M}(\tau_1,\tau_2)} (\check{X}(\Psi) + \frac{1}{2}w\Psi)N[\Psi]\end{aligned} \tag{10.1.68}$$

where $\mathcal{E} = \mathcal{E}[\check{X},w,M](\Psi)$. Now,

$$\mathcal{E}[\check{X},w,M](\Psi) \quad = \quad \mathcal{E}[X,w,M](\Psi) + \frac{1}{2}\Lambda\mathcal{Q}\cdot {}^{(T)}\pi - \frac{1}{2}T(V)|\Psi|^2.$$

According to Lemma 10.7, $T(V) = O(\epsilon)r^{-3}u_{trap}^{-1-\delta_{dec}}$, and all components of ${}^{(T)}\pi$ are $O(\epsilon r^{-1}u_{trap}^{-1-\delta_{dec}})$ except for ${}^{(T)}\pi_{44}$ which is $O(\epsilon r^{-2}u_{trap}^{-1-\delta_{dec}})$. We easily deduce

$$\Lambda|\mathcal{Q}\cdot {}^{(T)}\pi| + |T(V)||\Psi|^2 \quad \lesssim \quad \Lambda\mathcal{E}_\epsilon.$$

Thus in view of Proposition 10.31, we have[11]

$$\begin{aligned}\int_{\mathcal{M}(\tau_1,\tau_2)} \mathcal{E} \geq \delta_{\mathcal{H}}^{\frac{1}{2}} \int_{\mathcal{M}(\tau_1,\tau_2)} &\left[\frac{m^2}{r^3}|\breve{R}\Psi|^2 + r^{-1}\left(1-\frac{3m}{r}\right)^2\left(|\nabla\!\!\!/\Psi|^2 + \frac{m^2}{r^2}|\breve{T}\Psi|^2\right)\right. \\ &\left. + \frac{m}{r^4}|\Psi|^2\right] - O(\delta_{\mathcal{H}}^{-1})\int_{\mathcal{M}(\tau_1,\tau_2)} \mathcal{E}_\epsilon,\end{aligned}$$

i.e.,

$$\int_{\mathcal{M}(\tau_1,\tau_2)} \mathcal{E} \quad \geq \delta_{\mathcal{H}}^{\frac{1}{2}}\mathrm{Mor}[\Psi](\tau_1,\tau_2) - O(\delta_{\mathcal{H}}^{-1})\mathrm{Err}_\epsilon(\tau_1,\tau_2). \tag{10.1.69}$$

[11]Recall from (10.1.65) that we have $\Lambda = \frac{1}{4}\delta_{\mathcal{H}}^{-\frac{13}{20}} \ll \delta_{\mathcal{H}}^{-1}$.

We now analyze the boundary terms in (10.1.68).

10.1.13.1 Boundary term along $\mathcal{A}$

Along the spacelike hypersurface $\mathcal{A}$, i.e., $r = 2m_0(1-\delta_{\mathcal{H}})$, the unit normal $N_{\mathcal{A}}$ is given by

$$\begin{aligned} N_{\mathcal{A}} &= \frac{1}{2\sqrt{\frac{e_4(r)}{e_3(r)}}}\left(e_4 + \frac{e_4(r)}{e_3(r)}e_3\right) \\ &= \frac{1}{2\sqrt{\delta_{\mathcal{H}} + O(\epsilon)}}\Big(e_4 + (\delta_{\mathcal{H}} + O(\epsilon))e_3\Big), \end{aligned}$$

and we have $h, h_2 = 0$ as well as $w = -\delta_1 w_1$ where $\delta_1 > 0$ is a small constant and w_1 is given by (10.1.47)

$$w_1(r,m) = r^{-1}\frac{m^2}{r^2}\Upsilon\left(1 - \frac{3m}{r}\right)^2.$$

In particular, we have on $\mathcal{A}$ in view of the formula for w_1 and for $N_{\mathcal{A}}$

$$|w_1| \lesssim \delta_{\mathcal{H}}, \qquad |N_{\mathcal{A}}(w_1)| \lesssim \sqrt{\delta_{\mathcal{H}}}.$$

Hence,

$$\begin{aligned} \mathcal{P}\cdot N_{\mathcal{A}} &= \mathcal{Q}(X + \Lambda T, N_{\mathcal{A}}) - \frac{\delta_1}{2}w_1\Psi N_{\mathcal{A}}(\Psi) + \frac{\delta_1}{4}|\Psi|^2 N_{\mathcal{A}}(w_1) \\ &= \frac{2}{\sqrt{\delta_{\mathcal{H}} + O(\epsilon)}}\mathcal{Q}(X + \Lambda T, e_4) + 2\sqrt{\delta_{\mathcal{H}} + O(\epsilon)}\mathcal{Q}(X + \Lambda T, e_3) \\ &\quad -O(\sqrt{\delta_{\mathcal{H}}})\Psi e_4(\Psi) - O(\delta_{\mathcal{H}}^{\frac{3}{2}})\Psi e_3(\Psi) - O(\sqrt{\delta_{\mathcal{H}}})|\Psi|^2. \end{aligned}$$

Thus, in view of Proposition 10.32, we infer

$$\begin{aligned} \mathcal{P}\cdot N_{\mathcal{A}} &\geq \frac{2}{\sqrt{\delta_{\mathcal{H}} + O(\epsilon)}}\left(\frac{1}{4}\epsilon_{\mathcal{H}}\mathcal{Q}_{34} + \frac{1}{2}\Lambda\mathcal{Q}_{44}\right) \\ &\quad +2\sqrt{\delta_{\mathcal{H}} + O(\epsilon)}\left(\frac{1}{4}\epsilon_{\mathcal{H}}\mathcal{Q}_{33} + \frac{1}{2}\Lambda\mathcal{Q}_{34}\right) \\ &\quad -O(\sqrt{\delta_{\mathcal{H}}})\Psi e_4(\Psi) - O(\delta_{\mathcal{H}}^{\frac{3}{2}})\Psi e_3(\Psi) - O(\sqrt{\delta_{\mathcal{H}}})|\Psi|^2. \end{aligned}$$

Using in particular (10.1.63) and (10.1.65), we deduce

$$
\begin{aligned}
\mathcal{P}\cdot N_{\mathcal{A}} &\geq \frac{2}{\sqrt{\delta_{\mathcal{H}}+O(\epsilon)}}\left(\frac{1}{4}\delta_{\mathcal{H}}^{\frac{7}{20}}\left(|\nabla\Psi|^2+O(\delta_{\mathcal{H}})|\Psi|^2\right)+\frac{1}{8}\delta_{\mathcal{H}}^{-\frac{13}{20}}|e_4\Psi|^2\right)\\
&+2\sqrt{\delta_{\mathcal{H}}+O(\epsilon)}\left(\frac{1}{4}\delta_{\mathcal{H}}^{\frac{7}{20}}|e_3\Psi|^2+\frac{1}{8}\delta_{\mathcal{H}}^{-\frac{13}{20}}\left(|\nabla\Psi|^2+O(\delta_{\mathcal{H}})|\Psi|^2\right)\right)\\
&-O(\sqrt{\delta_{\mathcal{H}}})\Psi e_4(\Psi)-O(\delta_{\mathcal{H}}^{\frac{3}{2}})\Psi e_3(\Psi)-O(\sqrt{\delta_{\mathcal{H}}})|\Psi|^2\\
&\geq \frac{1}{2}\delta_{\mathcal{H}}^{-\frac{3}{20}}|\nabla\Psi|^2+\frac{1}{8}\delta_{\mathcal{H}}^{-\frac{23}{20}}|e_4\Psi|^2+\frac{1}{4}\delta_{\mathcal{H}}^{\frac{17}{20}}|e_3\Psi|^2\\
&-O(\sqrt{\delta_{\mathcal{H}}})\Psi e_4(\Psi)-O(\delta_{\mathcal{H}}^{\frac{3}{2}})\Psi e_3(\Psi)-O(\sqrt{\delta_{\mathcal{H}}})|\Psi|^2.
\end{aligned}
$$

Recalling the Poincaré inequality (10.1.27),

$$
\int_S|\nabla\Psi|^2 \;\geq\; 2r^{-2}(1-O(\epsilon))\int_S\Psi^2 da_S,
$$

we deduce, in this region,

$$
\int_{\mathcal{A}(\tau_1,\tau_2)}\mathcal{P}\cdot N_{\mathcal{A}}\geq\frac{1}{8}\int_{\mathcal{A}(\tau_1,\tau_2)}\left(\delta_{\mathcal{H}}^{-1}|e_4\Psi|^2+\delta_{\mathcal{H}}|e_3\Psi|^2+|\nabla\Psi|^2+r^{-2}|\Psi|^2\right)
$$

as desired in view of the definition of the flux along $\mathcal{A}$.

10.1.13.2 Boundary terms along $\Sigma(\tau_1),\Sigma(\tau_2)$

Along a hypersurface $\Sigma(\tau)$ with timelike unit future normal $N_{\Sigma(\tau)}=ae_3+be_4$, we have

$$
\mathcal{P}\cdot N_\Sigma=\mathcal{Q}(X+\Lambda T,N_\Sigma)+\frac{1}{2}w\Psi N_\Sigma(\Psi)-\frac{1}{4}N_\Sigma(w)|\Psi|^2+\frac{1}{2}N_\Sigma\cdot(hR+h_2\breve{R})|\Psi|^2
$$

and

$$
E[\Psi](\tau) \;=\; \int_{\Sigma(\tau)}\left(2b^2|e_4\Psi|^2+2a^2|e_3\Psi|^2+|\nabla\Psi|^2+r^{-2}|\Psi|^2\right).
$$

1. In the region $r\geq 2m_0(1-\delta_{\mathcal{H}})$, $\Upsilon\leq\delta_{\mathcal{H}}^{\frac{1}{10}}$ we have $h=0$, $h_2=O(\widehat{\delta})$ and $N_\Sigma=e_3$ (i.e., $a=1$, $b=0$). Also, we have $w=-\delta_1 w_1$, where $\delta_1>0$ is a small constant and w_1 is given by (10.1.47)

$$
w_1(r,m)=r^{-1}\frac{m^2}{r^2}\Upsilon\left(1-\frac{3m}{r}\right)^2.
$$

In particular, we have in the region of interest, in view of the formula for w_1 and for N_Σ

$$
|w_1|\lesssim\delta_{\mathcal{H}}^{\frac{1}{10}},\qquad |N_\Sigma(w_1)|=|e_3(w_1)|\lesssim 1.
$$

We infer

$$\begin{aligned}\mathcal{P}\cdot N_\Sigma &= \mathcal{Q}(X+\Lambda T, e_3)-\frac{\delta_1}{2}w_1\Psi e_3(\Psi)+\frac{\delta_1}{4}e_3(w_1)|\Psi|^2+\frac{1}{2}h_2 e_3\cdot \breve{R}|\Psi|^2\\ &= \mathcal{Q}(X+\Lambda T, e_3)-O(\delta_{\mathcal{H}}^{\frac{1}{10}})w_1\Psi e_3(\Psi)-O(1)|\Psi|^2\end{aligned}$$

where we used the fact that $\breve{R}=\frac{1}{2}(e_4-e_3)$ in the region of interest. Thus, according to Proposition 10.32,

$$\mathcal{P}\cdot N_\Sigma \geq \frac{1}{4}\epsilon_{\mathcal{H}}\mathcal{Q}_{33}+\frac{1}{2}\Lambda\mathcal{Q}_{34}-O(\delta_{\mathcal{H}}^{\frac{1}{10}})|\Psi||e_3(\Psi)|-O(1)|\Psi|^2.$$

Using in particular (10.1.63) and (10.1.65), we deduce

$$\mathcal{P}\cdot N_\Sigma \geq \frac{1}{4}\delta_{\mathcal{H}}^{\frac{7}{20}}|e_3\Psi|^2+\frac{1}{8}\delta_{\mathcal{H}}^{-\frac{13}{20}}(|\nabla\!\!\!/\Psi|^2+O(\epsilon)|\Psi|^2)-O(\delta_{\mathcal{H}}^{\frac{1}{10}})|\Psi||e_3\Psi|-O(1)|\Psi|^2.$$

Together with the Poincaré inequality (10.1.27), we deduce

$$\begin{aligned}&\int_{\Sigma_{r\geq 2m_0(1-\delta_{\mathcal{H}}),\ \Upsilon\leq\delta_{\mathcal{H}}^{\frac{1}{10}}}(\tau)}\mathcal{P}\cdot N_\Sigma\\ \geq\ & \frac{1}{8}\delta_{\mathcal{H}}^{\frac{7}{20}}\int_{\Sigma_{r\geq 2m_0(1-\delta_{\mathcal{H}}),\ \Upsilon\leq\delta_{\mathcal{H}}^{\frac{1}{10}}}(\tau)}\Big(|e_3\Psi|^2+|\nabla\!\!\!/\Psi|^2+r^{-2}|\Psi|^2\Big)\\ \geq\ & \frac{1}{8}\delta_{\mathcal{H}}^{\frac{7}{20}}E_{r\geq 2m_0(1-\delta_{\mathcal{H}}),\ \Upsilon\leq\delta_{\mathcal{H}}^{\frac{1}{10}}}[\Psi](\tau).\end{aligned}$$

2. In the region $\Upsilon\geq\delta_{\mathcal{H}}^{\frac{1}{10}}$, we have $w=O(r^{-1})$, $N_\Sigma(w)=O(r^{-2})$, $h=O(r^{-4})$ and $h_2=O(r^{-4})$. We infer

$$\begin{aligned}\mathcal{P}\cdot N_\Sigma &= a\mathcal{Q}(X+\Lambda T, e_3)+b\mathcal{Q}(X+\Lambda T, e_4)-O(r^{-1})|\Psi|(a|e_3\Psi|+b|e_4\Psi|)\\ &\quad -O(r^{-2})|\Psi|^2.\end{aligned}$$

Thus, according to Proposition 10.32,

$$\begin{aligned}\mathcal{P}\cdot N_\Sigma &\geq \delta_{\mathcal{H}}^{-\frac{1}{2}}(a\mathcal{Q}_{33}+b\mathcal{Q}_{44}+(a+b)\mathcal{Q}_{34})-O(1)(a^2|e_3\Psi|^2+b^2|e_4\Psi|^2)\\ &\quad -O(r^{-2})|\Psi|^2\\ &= \delta_{\mathcal{H}}^{-\frac{1}{2}}\left(a|e_3\Psi|^2+b|e_4\Psi|^3+(a+b)\left(|\nabla\!\!\!/\Psi|^2+\frac{4\Upsilon}{r^2}|\Psi|^2\right)\right)\\ &\quad -O(1)\Big(a^2|e_3\Psi|^2+b^2|e_4\Psi|^2+r^{-2}|\Psi|^2\Big)\\ &\geq \delta_{\mathcal{H}}^{-\frac{1}{2}}\left(a|e_3\Psi|^2+b|e_4\Psi|^3+(a+b)\left(|\nabla\!\!\!/\Psi|^2+\frac{4\delta_{\mathcal{H}}^{\frac{1}{10}}}{r^2}|\Psi|^2\right)\right)\\ &\quad -O(1)\Big(a^2|e_3\Psi|^2+b^2|e_4\Psi|^2+r^{-2}|\Psi|^2\Big).\end{aligned}$$

Hence, for $\delta_{\mathcal{H}}>0$ sufficiently small, and since $a^2\leq a$, $b^2\leq b$ and $a+b\geq 1$, we infer in this region

$$
\begin{aligned}
\mathcal{P}\cdot N_\Sigma &\geq \delta_{\mathcal{H}}^{-\frac{1}{5}}\int_{\Sigma_{\Upsilon\geq\delta_{\mathcal{H}}^{\frac{1}{10}}}(\tau)}\Big(2b^2\,|e_4\Psi|^2+2a^2\,|e_3\Psi|^2+|\nabla\!\!\!/\Psi|^2+r^{-2}|\Psi|^2\Big)\\
&= \delta_{\mathcal{H}}^{-\frac{1}{5}}E[\Psi]_{\Upsilon\geq\delta_{\mathcal{H}}^{\frac{1}{10}}}(\tau).
\end{aligned}
$$

In view of the above estimates in $r\geq 2m_0(1-\delta_{\mathcal{H}})$, $\Upsilon\leq\delta_{\mathcal{H}}^{\frac{1}{10}}$ and in $\Upsilon\geq\delta_{\mathcal{H}}^{\frac{1}{10}}$, we deduce, everywhere,

$$
\int_{\Sigma(\tau)}\mathcal{P}\cdot N_\Sigma\geq\frac{1}{8}\delta_{\mathcal{H}}^{\frac{7}{20}}E[\Psi](\tau). \tag{10.1.70}
$$

10.1.13.3 Boundary terms along Σ_*

On Σ_*, we have

$$
N_{\Sigma_*} = T+O\left(\epsilon+\frac{m}{r}\right)(e_3+e_4),
$$

$w=O(r^{-1})$, $N_{\Sigma_*}(w)=O(\epsilon r^{-2})$, $h=O(r^{-4})$ and $h_2=0$.

Proceeding as before, we have along Σ_*,

$$
\begin{aligned}
\mathcal{P}\cdot N_\Sigma &= \left(\frac{1}{2}+O\left(\epsilon+\frac{m}{r}\right)\right)\mathcal{Q}(X+\Lambda T,e_3)\\
&\quad+\left(\frac{1}{2}+O\left(\epsilon+\frac{m}{r}\right)\right)\mathcal{Q}(X+\Lambda T,e_4)\\
&\quad-O(r^{-1})|\Psi|(|e_3\Psi|+|e_4\Psi|)-O(r^{-2})|\Psi|^2\\
&\geq \frac{1}{4}\mathcal{Q}(X+\Lambda T,e_3)+\frac{1}{4}\mathcal{Q}(X+\Lambda T,e_4)-O\Big(|e_3\Psi|^2+|e_4\Psi|^2+r^{-2}|\Psi|^2\Big).
\end{aligned}
$$

Thus, according to Proposition 10.32, we have

$$
\begin{aligned}
\mathcal{P}\cdot N_\Sigma &\geq \frac{1}{16}\Lambda\left(\mathcal{Q}_{33}+\mathcal{Q}_{44}+2\mathcal{Q}_{34}\right)-O\Big(|e_3\Psi|^2+|e_4\Psi|^2+r^{-2}|\Psi|^2\Big)\\
&= \frac{1}{16}\Lambda\left(|e_3\Psi|^2+|e_4\Psi|^2+2\left(|\nabla\!\!\!/\Psi|^2+\frac{4\Upsilon}{r^2}|\Psi|^2\right)\right)\\
&\quad-O\Big(|e_3\Psi|^2+|e_4\Psi|^2+r^{-2}|\Psi|^2\Big)\\
&\geq \frac{1}{64}\delta_{\mathcal{H}}^{-\frac{13}{20}}\left(|e_3\Psi|^2+|e_4\Psi|^2+|\nabla\!\!\!/\Psi|^2+r^{-2}|\Psi|^2\right)\\
&\quad-O\Big(|e_3\Psi|^2+|e_4\Psi|^2+r^{-2}|\Psi|^2\Big)\\
&\geq \delta_{\mathcal{H}}^{-\frac{1}{2}}\Big(|e_3\Psi|^2+|e_4\Psi|^2+|\nabla\!\!\!/\Psi|^2+r^{-2}|\Psi|^2\Big)
\end{aligned}
$$

and hence

$$
\int_{\Sigma_*(\tau_1,\tau_2)}\mathcal{P}\cdot N_{\Sigma_*}\geq\delta_{\mathcal{H}}^{-\frac{1}{2}}\int_{\Sigma_*(\tau_1,\tau_2)}\Big(|e_3\Psi|^2+|e_4\Psi|^2+|\nabla\!\!\!/\Psi|^2+r^{-2}|\Psi|^2\Big). \tag{10.1.71}
$$

10.1.13.4 The inhomogeneous term $\int_{\mathcal{M}(\tau_1,\tau_2)}(\check{X}(\Psi)+\frac{1}{2}w\Psi)N[\Psi]$

Recall that $\check{X}=X+\Lambda T=f_{\widehat{\delta}}R+Y_{\mathcal{H}}+\Lambda T$. We easily check, recalling the properties of $f_{\widehat{\delta}}$, w and Λ and the definition of $J[N,\Psi]$,

$$\begin{aligned}
&\left|\int_{\mathcal{M}(\tau_1,\tau_2)}\left(\check{X}(\Psi)+\frac{1}{2}w\Psi\right)N[\Psi]\right|\\
\leq\;& \delta_{\mathcal{H}}^{-\frac{3}{4}}\int_{\mathcal{M}(\tau_1,\tau_2)}\left(|\check{R}\Psi|+|\check{T}\Psi|+r^{-2}|\Psi|^2\right)|N(\Psi)|\\
=\;& \delta_{\mathcal{H}}^{-\frac{3}{4}}J[N,\Psi](\tau_1,\tau_2).
\end{aligned}\tag{10.1.72}$$

Going back to (10.1.68) we deduce

$$E[\Psi](\tau_2)+\int_{\mathcal{M}(\tau_1,\tau_2)}\mathcal{E}\;+F[\Psi](\tau_1,\tau_2)\;\leq\;\delta_{\mathcal{H}}^{-\frac{7}{20}}\left(E[\Psi](\tau_1)+J[N,\Psi](\tau_1,\tau_2)\right).$$

In view of (10.1.69) we obtain

$$\begin{aligned}
E[\Psi](\tau_2)+\mathrm{Mor}[\Psi](\tau_1,\tau_2)+F[\Psi](\tau_1,\tau_2)\;\leq\;&\delta_{\mathcal{H}}^{-1}\left(E[\Psi](\tau_1)+J[N,\Psi](\tau_1,\tau_2)\right)\\
&+O\left(\delta_{\mathcal{H}}^{-\frac{3}{2}}\right)\mathrm{Err}_\epsilon(\tau_1,\tau_2).
\end{aligned}$$

This concludes the proof of Theorem 10.33. □

10.1.14 Analysis of the error term $\mathcal{E}_\epsilon$

Recall that $\mathrm{Err}_\epsilon(\tau_1,\tau_2)=\int_{\mathcal{M}(\tau_1,\tau_2)}\mathcal{E}_\epsilon$ where

$$\begin{aligned}
\mathcal{E}_\epsilon\;\lesssim\;&\epsilon u_{trap}^{-1-\delta_{dec}}\Big[r^{-2}|e_3\Psi|^2+r^{-1}\big(|e_4\Psi|^2+|\nabla\!\!\!/\Psi|^2+r^{-2}|\Psi|^2\\
&+|e_3\Psi|\left(|e_4\Psi|+|\nabla\!\!\!/\Psi|\right)\big)\Big].
\end{aligned}$$

- In the trapping region $\mathcal{M}_{trap}$, i.e., $\frac{5m}{2}\leq r\leq\frac{7m}{2}$, where $u_{trap}=1+\tau$ and $\Sigma(\tau)$ is strictly spacelike, we have

$$\begin{aligned}
\int_{\Sigma_{trap}(\tau)}\mathcal{E}_\epsilon\;\lesssim\;&\epsilon\tau_{trap}^{-1-\delta_{dec}}\int_{\Sigma_{trap}(\tau)}\left(|e_3\Psi|^2+|e_4\Psi|^2+|\nabla\!\!\!/\Psi|^2+|\Psi|^2\right)\\
\lesssim\;&\epsilon\tau_{trap}^{-1-\delta_{dec}}E[\Psi](\tau).
\end{aligned}$$

  Thus,

$$\begin{aligned}
\int_{\mathcal{M}_{trap}(\tau_1,\tau_2)}\mathcal{E}_\epsilon\;\lesssim\;&\epsilon\int_{\tau_1}^{\tau_2}\tau_{trap}^{-1-\delta_{dec}}E[\Psi](\tau)\\
\lesssim\;&\epsilon\left(\int_{\tau_1}^{\tau_2}(1+\tau)^{-1-\delta}\right)\sup_{\tau\in[\tau_1,\tau_2]}\mathcal{E}[\Psi](\tau)\\
\lesssim\;&\epsilon\sup_{\tau\in[\tau_1,\tau_2]}\mathcal{E}[\Psi](\tau)
\end{aligned}$$

  and therefore, for small $\epsilon>0$, the integral $\int_{\mathcal{M}_{trap}(\tau_1,\tau_2)}\mathcal{E}_\epsilon$ can be absorbed on the left-hand side of (10.1.67).

- In the non-trapping region $\mathcal{M}_{\not{trap}}$ we write, with a fixed $\delta > 0$,

$$\mathcal{E}_\epsilon \lesssim \epsilon r^{-1-\delta}|e_3\Psi|^2 + r^{-1+\delta}\left(|e_4\Psi|^2 + |\nabla\!\!\!/\Psi|^2 + r^{-2}|\Psi|^2\right).$$

Hence,

$$\begin{aligned}\int_{\mathcal{M}_{\not{trap}}(\tau_1,\tau_2)} \mathcal{E}_\epsilon &\lesssim \epsilon \int_{\mathcal{M}_{r\geq 4m_0}(\tau_1,\tau_2)} r^{-1-\delta}|e_3\Psi|^2 \\ &+ \epsilon \int_{\mathcal{M}_{r\geq 4m_0}(\tau_1,\tau_2)} r^{-1+\delta}\left(|e_4\Psi|^2 + |\nabla\!\!\!/\Psi|^2 + r^{-2}|\Psi|^2\right) \\ &+ \epsilon \int_{(\not{trap})\mathcal{M}_{r\leq 4m_0}(\tau_1,\tau_2)} \left(|e_3\Psi|^2 + |e_4\Psi|^2 + |\nabla\!\!\!/\Psi|^2 + |\Psi|^2\right).\end{aligned}$$

Note that for $\epsilon > 0$ sufficiently small, the last integral, on ${}^{(\not{trap})}\mathcal{M}_{r\leq 4m_0}$, can be absorbed by the left-hand side of (10.1.67).

As a consequence we deduce the following.

Corollary 10.34. *The statement of Theorem 10.33 remains true if we replace Err_ϵ in the statement of the theorem with*

$$\begin{aligned}Err_\epsilon &= \int_{\mathcal{M}_{r\geq 4m_0}(\tau_1,\tau_2)} \mathcal{E}_\epsilon, \\ \mathcal{E}_\epsilon &\lesssim \epsilon r^{-1-\delta}|e_3\Psi|^2 + \epsilon r^{-1+\delta}\left(|e_4\Psi|^2 + |\nabla\!\!\!/\Psi|^2 + |r^{-2}|\Psi|^2\right),\end{aligned}$$

for a fixed $\delta > 0$.

Remark 10.35. *Note that the error terms Err_ϵ cannot yet be absorbed to the left-hand side of* (10.1.67)*. In fact we need additional estimates. The Morawetz bulk quantity* (5.1.13)*,*

$$\begin{aligned}Mor[\Psi](\tau_1,\tau_2) &= \int_{\mathcal{M}(\tau_1,\tau_2)} \left[\frac{m^2}{r^3}|\breve{R}\Psi|^2 + \frac{m}{r^4}|\Psi|^2 \right.\\ &\left. + \left(1-\frac{3m}{r}\right)^2 \frac{1}{r}\left(|\nabla\!\!\!/\Psi|^2 + \frac{m^2}{r^2}|\breve{T}\Psi|^2\right)\right],\end{aligned}$$

is quite weak for r large with regard to the terms $|\breve{R}\Psi|^2$ and $|\breve{T}\Psi|^2$, while, using the Poincaré inequality, $Mor[\Psi]$ controls the term $\int_{\mathcal{M}_{r\geq 4m_0}(\tau_1,\tau_2)} r^{-1}\left(|\nabla\!\!\!/\Psi|^2 + r^{-2}|\Psi|^2\right)$. In the next section we show how we can estimate the term $\int_{\mathcal{M}_{\geq R_0}(\tau_1,\tau_2)} r^{-1-\delta}|e_3\Psi|^2$ by $\int_{\mathcal{M}_{\geq R_0}(\tau_1,\tau_2)} r^{-1-\delta}|e_4\Psi|^2$ and then, we provide estimates for the remaining terms. Note also that the weight $r^{-1-\delta}$ is optimal in estimating $e_3\Psi$ in the wave zone region.

10.1.15 Proof of Theorem 10.1

We are now ready to prove Theorem 10.1. Note that it suffices to improve the previous Morawetz estimate of Theorem 10.33 by replacing the quantity $\mathrm{Mor}[\Psi](\tau_1,\tau_2)$ with

$$\mathrm{Morr}[\Psi](\tau_1,\tau_2) := \mathrm{Mor}[\Psi](\tau_1,\tau_2) + \int_{\mathcal{M}_{far}(\tau_1,\tau_2)} r^{-1-\delta}|e_3(\Psi)|^2.$$

In view of the Morawetz estimate (10.1.67) and Corollary 10.34 we have

$$\begin{aligned} E[\Psi](\tau_2) + \mathrm{Mor}[\Psi](\tau_1,\tau_2) + F[\Psi](\tau_1,\tau_2) &\lesssim E[\Psi](\tau_2) + J[N,\Psi](\tau_1,\tau_2) \\ &\quad + \mathrm{Err}_\epsilon(\tau_1,\tau_2), \\ J[N,\Psi](\tau_1,\tau_2) : &= \int_{\mathcal{M}(\tau_1,\tau_2)} (|\breve{R}\Psi| + |\breve{T}\Psi| + r^{-1}|\Psi|)|N|, \end{aligned}$$

with error term

$$\mathrm{Err}_\epsilon \lesssim \epsilon \int_{\mathcal{M}_{\geq 4m_0}(\tau_1,\tau_2)} r^{-1-\delta}|e_3\Psi|^2 + r^{-1+\delta}\left(|e_4\Psi|^2 + |\nabla\!\!\!/\Psi|^2 + r^{-2}|\Psi|^2\right).$$

We divide $J[N] = J[N,\Psi]$ as follows:

$$J[N] = J[N]_{trap} + J[N]_{tr\not{a}p}$$

where

$$\begin{aligned} J[N]_{trap} : &= \int_{\mathcal{M}_{trap}} (|\breve{R}\Psi| + |\breve{T}\Psi| + r^{-1}|\Psi|)|N|, \\ J[N]_{tr\not{a}p} : &= \int_{{}^{(tr\not{a}p)}\mathcal{M}} (|\breve{R}\Psi| + |\breve{T}\Psi| + r^{-1}|\Psi|)|N|. \end{aligned}$$

For the trapping region, where the hypersurfaces $\Sigma(\tau)$ are strictly spacelike, we write

$$\begin{aligned} J[N]_{trap}(\tau_1,\tau_2) &= \int_{\tau_1}^{\tau_2} d\tau \int_{\Sigma_{trap}(\tau)} (|\breve{R}\Psi| + |\breve{T}\Psi| + r^{-1}|\Psi|)|N| \\ &\leq \int_{\tau_1}^{\tau_2} E[\Psi](\tau)^{1/2}\left(\int_{\Sigma_{trap}(\tau)} |N|^2\right)^{1/2} \\ &\leq \sup_{\tau\in[\tau_1,\tau_2]} E[\Psi](\tau)^{1/2}\int_{\tau_1}^{\tau_2}\left(\int_{\Sigma_{trap}(\tau)} |N|^2\right)^{1/2} \\ &\lesssim \lambda \sup_{\tau\in[\tau_1,\tau_2]} E[\Psi](\tau) + \lambda^{-1}\left(\int_{\tau_1}^{\tau_2} \|N\|_{L^2(\Sigma_{trap}(\tau))}\right)^2. \end{aligned}$$

Hence, for $\lambda > 0$ sufficiently small, we deduce

$$\begin{aligned}
&E[\Psi](\tau_2) + \mathrm{Mor}[\Psi](\tau_1,\tau_2) + F[\Psi](\tau_1,\tau_2)\\
\lesssim\;& E[\Psi](\tau_2) + \mathrm{Err}_\epsilon(\tau_1,\tau_2) + J_{tr\not{a}p}[N,\Psi](\tau_1,\tau_2) + \left(\int_{\tau_1}^{\tau_2} \|N\|_{L^2(\Sigma_{trap}(\tau))}\right)^2.
\end{aligned}$$

On the other hand we have

$$\begin{aligned}
J[N]_{tr\not{a}p}(\tau_1,\tau_2) &= \int_{\mathcal{M}_{tr\not{a}p}(\tau_1,\tau_2)} (|\breve{R}\Psi| + |\breve{T}\Psi| + r^{-1}|\Psi|)|N|\\
&\leq \lambda \int_{\mathcal{M}_{tr\not{a}p}} r^{-1-\delta}(|\breve{R}\Psi|^2 + |\breve{T}\Psi|^2 + r^{-2}|\Psi|^2)\\
&\quad +\lambda^{-1}\int_{\mathcal{M}_{tr\not{a}p}} r^{1+\delta}|N|^2.
\end{aligned}$$

The first integral on the right can be divided further into integrals for $r \leq 4m_0$ and $r \geq 4m_0$. The first integral can then be easily absorbed by the term $Mor[\Psi](\tau_1,\tau_2)$, if $\lambda > 0$ is sufficiently small. We are thus led to the estimate

$$\begin{aligned}
&E[\Psi](\tau_2) + \mathrm{Mor}[\Psi](\tau_1,\tau_2) + F[\Psi](\tau_1,\tau_2)\\
\lesssim\;& E[\Psi](\tau_2) + \mathrm{Err}_\epsilon(\tau_1,\tau_2) + \mathcal{I}_\delta[N](\tau_1,\tau_2)\\
&+\int_{\mathcal{M}_{r\geq 4m_0}} r^{-1-\delta}(|e_3\Psi|^2 + |e_4\Psi|^2 + r^{-2}|\Psi|^2)
\end{aligned}$$

where

$$\mathcal{I}_\delta[N](\tau_1,\tau_2) : = \int_{\mathcal{M}_{tr\not{a}p}(\tau_1,\tau_2)} r^{1+\delta}|N|^2 + \left(\int_{\tau_1}^{\tau_2} d\tau \|N\|_{L^2(\Sigma_{trap}(\tau))}\right)^2.$$

Recalling the definition of Err_ϵ in Corollary 10.34, we deduce

$$\begin{aligned}
&E[\Psi](\tau_2) + \mathrm{Mor}[\Psi](\tau_1,\tau_2) + F[\Psi](\tau_1,\tau_2)\\
\lesssim\;& E[\Psi](\tau_2) + \mathrm{Err}_\epsilon(\tau_1,\tau_2) + \mathcal{I}_\delta[N](\tau_1,\tau_2).
\end{aligned} \tag{10.1.73}$$

To eliminate the term in $e_3\Psi$ from the error term we appeal to the following proposition.

Proposition 10.36. *Assume $\square\Psi = V\Psi + N$ and consider the vectorfield $X = f_{-\delta}T$ with $f_{-\delta} := r^{-\delta}$ for $r \geq 4m_0$ and compactly supported in $r \geq \frac{7m_0}{2}$. With the notation of Proposition 10.9, let*

$$\begin{aligned}
\mathcal{P}_\mu[f_{-\delta}T,0,0] &= f_{-\delta}\mathcal{Q}_{\alpha\mu}T^\mu,\\
\mathcal{E}[f_{-\delta}T,0,0] &= \mathbf{D}^\mu\mathcal{P}_\mu[f_{-\delta}T,0,0] - f_{-\delta}T(\Psi)N.
\end{aligned}$$

Then:

1. *We have, for* $r \geq 4m_0$,

$$\begin{aligned}\mathcal{E}[f_{-\delta}T, 0, 0] &= \frac{\Upsilon^2}{4}\delta r^{-1-\delta}|e_3\Psi|^2 - \frac{1}{4}\delta r^{-1-\delta}|e_4\Psi|^2 \\ &+ O\left(\epsilon r^{-1-\delta}\left(|\mathbf{D}\Psi|^2 + r^{-2}|\Psi|^2\right)\right). \qquad (10.1.74)\end{aligned}$$

2. *We have*

$$\mathcal{P}[f_{-\delta}T, 0, 0]\cdot e_4 = f_{-\delta}\mathcal{Q}(T, e_4) \geq 0, \qquad \mathcal{P}[f_{-\delta}T, 0, 0]\cdot e_3 = f_{-\delta}\mathcal{Q}(T, e_3) \geq 0.$$

We postpone the proof of Proposition 10.36 and continue the proof of Theorem 10.1. By integration, the proposition provides a bound for[12]

$$\int_{\mathcal{M}_{\geq 4m_0}(\tau_1,\tau_2)} r^{-1-\delta}|e_3\Psi|^2$$

in terms of $E[\Psi](\tau_1)$, $\int_{\mathcal{M}_{\geq \frac{7m_0}{2}}(\tau_1,\tau_2)} r^{-1-\delta}|e_4\Psi|^2$ and $\int_{\mathcal{M}_{\geq \frac{7m_0}{2}}(\tau_1,\tau_2)} r^{-\delta}T(\Psi)N$, as well as the error terms. The second bulk integral involving the inhomogeneous term N can be estimated exactly like before. Thus combining the new estimate with that in Corollary 10.34 we derive the desired estimates, both (10.1.1) and (10.1.2), hence concluding the proof of Theorem 10.1.

10.1.15.1 Proof of Proposition 10.36

We consider vectorfields of the form $X = f(r)T$ with $T = \frac{1}{2}(\Upsilon e_3 + e_4)$. Recall, see Lemma 10.7, that all components of the deformation tensor ${}^{(T)}\pi$ can be bounded by $O(\epsilon r^{-1})$. Since $f = O(r^{-\delta})$, we deduce

$${}^{(X)}\pi_{\alpha\beta} = f\,{}^{(T)}\pi_{\alpha\beta} + \mathbf{D}_\alpha f T_\beta + \mathbf{D}_\beta f T_\alpha = \mathbf{D}_\alpha f T_\beta + \mathbf{D}_\beta f T_\alpha + O(\epsilon r^{-1-\delta}).$$

Also,

$$e_3(f) = f' e_3(r) = -f' + O(\epsilon r^{-1-\delta}), \qquad e_4(f) = f' e_4(r) = \Upsilon f' + O(\epsilon r^{-1-\delta}).$$

Thus, modulo error terms of the form $O(\epsilon)r^{-1-\delta}\left(|e_3\Psi|^2 + |e_4\Psi|^2 + |\nabla\!\!\!/\Psi|^2 + r^{-2}|\Psi|^2\right)$, we have

$$\begin{aligned}\mathcal{Q}\cdot{}^{(X)}\pi &= 2\mathcal{Q}^{\alpha\beta}T_\alpha\mathbf{D}_\beta f = 2\left(\mathcal{Q}^{3\beta}T_\beta e_3 f + \mathcal{Q}^{4\beta}T_\beta e_4 f\right) \\ &= -\mathcal{Q}(e_4, T)e_3 f - \mathcal{Q}(e_3, T)e_4 f \\ &= \frac{1}{2}\mathcal{Q}(e_4, e_4 + \Upsilon e_3)f' - \frac{1}{2}f'\Upsilon\mathcal{Q}(e_3, e_4 + \Upsilon e_3) \\ &= \frac{1}{2}f'\left(|e_4\Psi|^2 - \Upsilon^2|e_3\Psi|^2\right).\end{aligned}$$

We now apply Proposition 10.9, as well as (10.1.13), (10.1.14), with $X = f_{-\delta}(r)T$, $w = 0$, $M = 0$ so that

$$\mathcal{P}_\mu[f_{-\delta}T, 0, 0] = f_{-\delta}\mathcal{Q}_{\alpha\mu}T^\mu, \qquad \mathcal{E}[f_{-\delta}T, 0, 0] := \mathbf{D}^\mu\mathcal{P}_\mu[f_{-\delta}T, 0, 0] - f_{-\delta}T(\Psi)N$$

[12] Note that $\Upsilon^2 \geq \frac{1}{4}$ in $r \geq 4m_0$.

and

$$\begin{aligned}\mathcal{E}[f_{-\delta}T,0,0] &= \frac{1}{2}\mathcal{Q}\cdot{}^{(X)}\pi - \frac{1}{2}f_{-\delta}T(V)|\Psi|^2\\ &= \frac{1}{4}f'_{-\delta}(r)\left(|e_4\Psi|^2 - \Upsilon^2|e_3\Psi|^2\right) + O\left(\epsilon r^{-1-\delta}\left(|\mathbf{D}\Psi|^2 + r^{-2}|\Psi|^2\right)\right)\end{aligned}$$

with $|\mathbf{D}\Psi|^2 = |e_3\Psi|^2 + |e_4\Psi|^2 + |\nabla\Psi|^2 + r^{-2}|\Psi|^2$. Since $f_{-\delta}(r) = r^{-\delta}$ for $r \geq 4m_0$, we deduce, for $r \geq 4m_0$,

$$\begin{aligned}\mathcal{E}[f_{-\delta}T,0,0] &= \frac{\Upsilon^2}{4}\delta r^{-1-\delta}|e_3\Psi|^2 - \frac{1}{4}\delta r^{-1-\delta}|e_4\Psi|^2\\ &\quad +O\left(\epsilon r^{-1-\delta}\left(|\mathbf{D}\Psi|^2 + r^{-2}|\Psi|^2\right)\right).\end{aligned}$$

On the other hand,

$$\begin{aligned}\mathcal{P}[f_{-\delta}T,0,0]\cdot e_4 &= f_{-\delta}\mathcal{Q}(T,e_4) \geq 0,\\ \mathcal{P}[f_{-\delta}T,0,0]\cdot e_3 &= f_{-\delta}\mathcal{Q}(T,e_3) \geq 0,\end{aligned}$$

as desired. This concludes the proof of Proposition 10.36.

10.2 DAFERMOS-RODNIANSKI r^p-WEIGHTED ESTIMATES

For convenience, we work in this section with the renormalized frame (e'_3, e'_4) defined in (10.2.6) instead of the original frame (e_3, e_4). To simplify the exposition, we still denote it as (e_3, e_4). Recall that the two are frames are equivalent up to lower terms in m/r.

In this section we rely on the Morawetz estimates proved in the previous section to establish r^p-weighted estimates in the spirit of Dafermos-Rodnianski [23]. The following theorem claims r^p-weighted estimates for the solution ψ of the wave equation (5.3.5).

Theorem 10.37 (r^p-weighted estimates)**.** *Consider a fixed $\delta > 0$ and let $R \gg \frac{m_0}{\delta}$, $\epsilon \ll \delta$. The following estimates hold true for all $\delta \leq p \leq 2-\delta$,*

$$\begin{aligned}&\dot{E}_{p;R}[\psi](\tau_2) + \dot{B}_{p;R}[\psi](\tau_1,\tau_2) + \dot{F}_p[\psi](\tau_1,\tau_2)\\ \lesssim\ & E_p[\psi](\tau_1) + J_p[\psi,N](\tau_1,\tau_2).\end{aligned} \tag{10.2.1}$$

Remark 10.38. *Note that Theorem 10.1 on Morawetz estimates and Theorem 10.37 on r^p-weighted estimates immediately yield for all $\delta \leq p \leq 2-\delta$,*

$$\sup_{\tau\in[\tau_1,\tau_2]} E_p[\psi](\tau) + B_p[\psi](\tau_1,\tau_2) + F_p[\psi](\tau_1,\tau_2) \lesssim E_p[\psi](\tau_1) + J_p[\psi,N](\tau_1,\tau_2),$$

which corresponds to Theorem 5.17 in the case $s = 0$.

Theorem 10.37 will be proved in section 10.2.3. We will need in this section stronger assumptions in the region $r \geq 4m_0$, away from trapping, than those in (10.1.5)–(10.1.7) of the previous section. For convenience we express our conditions with respect to the weights[13]

$$w_{p,q}(u,r) = r^{-p}(1+\tau)^{-q-\delta_{dec}+2\delta_0}.$$

RP0. The assumptions **Mor1**–**Mor3** made in the previous section hold true.

RP1. The Ricci coefficients verify, for $r \geq 4m_0$,

$$\begin{aligned}
\left|\underline{\xi}, \vartheta, \underline{\vartheta}, \eta, \underline{\eta}, \zeta, \underline{\omega}\right| &\lesssim \epsilon w_{1,1},\\
\left|\underline{\kappa} + \frac{2}{r}\right|, \left|\underline{\chi} + \frac{1}{r}\right|, \left|e_3\Phi - \underline{\chi}\right| &\lesssim \epsilon w_{1,1},\\
\left|\kappa - \frac{2\Upsilon}{r}\right|, \left|\chi - \frac{\Upsilon}{r}\right|, \left|e_4\Phi - \chi\right| &\lesssim \epsilon \min\{w_{1,1}, w_{2,1/2}\},\\
\left|\omega + \frac{m}{r^2}\right|, |\xi| &\lesssim \epsilon \min\{w_{2,1}, w_{3,1/2}\}.
\end{aligned} \tag{10.2.2}$$

RP2. The derivatives of r verify, for $r \geq 4m_0$,

$$\begin{aligned}
\left|e_3(r) + 1\right| &\lesssim \epsilon w_{0,1},\\
\left|e_4(r) - \Upsilon\right| &\lesssim \epsilon w_{1,1},\\
\left|e_3e_4(r) + \frac{2m}{r^2},\ e_4e_3(r)\right| &\lesssim \epsilon w_{1,1}.
\end{aligned} \tag{10.2.3}$$

RP3. For $r \geq 4m_0$,

$$\begin{aligned}
\left|\rho + \frac{2m}{r^3}\right| &\lesssim \epsilon w_{3,1},\\
\left|K - \frac{1}{r^2}\right| &\lesssim \epsilon r^{-2},\\
\left|e_\theta(\Phi)\right| &\lesssim r^{-1}.
\end{aligned} \tag{10.2.4}$$

RP4. We also assume, for $r \geq 4m_0$,

$$\begin{aligned}
|m - m_0| &\lesssim \epsilon,\\
|e_3 m,\ r^2 e_4 m| &\lesssim \epsilon w_{0,1},\\
|e_3e_4(m),\ e_4e_3(m)| &\lesssim \epsilon w_{1,1}.
\end{aligned} \tag{10.2.5}$$

Since the estimates we are establishing are restricted to the far region $r > R$ it is convenient, in this section, to work with the renormalized frame

$$e_3' = \Upsilon e_3, \qquad e_4' = \Upsilon^{-1} e_4, \qquad e_\theta' = e_\theta. \tag{10.2.6}$$

Relative to the new frame (e_3', e_4', e_θ') we have

$$\xi' = \Upsilon^{-2}\xi, \quad \underline{\xi}' = \Upsilon^2\underline{\xi}, \quad \zeta' = \zeta, \quad \eta' = \eta, \quad \chi' = \Upsilon^{-1}\chi, \quad \underline{\chi}' = \Upsilon\underline{\chi}$$

[13]The assumptions are consistent with the global frame used in Theorem M1, see Lemma 5.1. In particular, $\delta_0 > 0$ is such that $\delta_{dec} - 2\delta_0 > 0$ which is the only needed property of $\delta_{dec} - 2\delta_0$ to derive the r^p-weighted estimates.

and

$$\begin{aligned}
\omega' &= \Upsilon^{-1}\left(\omega + \frac{1}{2}e_4(\log\Upsilon)\right) = \Upsilon^{-1}\left(\omega + \Upsilon^{-1}\frac{m}{r^2}e_4(r) - \Upsilon^{-1}\frac{e_4(m)}{r}\right)\\
&= \Upsilon^{-1}\left(\left(\omega + \frac{m}{r^2}\right) + \Upsilon^{-1}\frac{m}{r^2}(e_4(r) - \Upsilon) - \Upsilon^{-1}\frac{e_4(m)}{r}\right),\\
\underline{\omega}' &= \Upsilon\left(\underline{\omega} - \frac{1}{2}e_3(\log\Upsilon)\right) = \Upsilon\left(\underline{\omega} - \Upsilon^{-1}\frac{m}{r^2}e_3(r) + \Upsilon^{-1}\frac{e_3(m)}{r}\right)\\
&= \frac{m}{r^2} + \Upsilon\left(\underline{\omega} - \Upsilon^{-1}\frac{m}{r^2}(e_3(r)+1) + \Upsilon^{-1}\frac{e_3(m)}{r}\right).
\end{aligned}$$

Thus in the new frame we have, for $r \geq 4m_0$:

RP1'. The Ricci coefficients with respect to the null frame (e_3', e_4', e_θ') verify, for $r \geq 4m_0$:

$$\begin{aligned}
\left|\xi', \vartheta', \underline{\vartheta}, \eta', \underline{\eta}', \zeta'\right|, \left|\underline{\omega}' - \frac{m}{r^2}\right| &\lesssim \epsilon w_{1,1},\\
\left|\underline{\kappa}' + \frac{2\Upsilon}{r}\right|, \left|\underline{\chi}' + \frac{\Upsilon}{r}\right|, \left|e_3'\Phi - \underline{\chi}'\right| &\lesssim \epsilon w_{1,1},\\
\left|\kappa' - \frac{2}{r}\right|, \left|\chi' - \frac{1}{r}\right|, \left|e_4'\Phi - \chi'\right| &\lesssim \epsilon \min\{w_{1,1}, w_{2,1/2}\},\\
\left|\omega'\right|, \left|\xi'\right| &\lesssim \epsilon \min\{w_{2,1}, w_{3,1/2}\}.
\end{aligned} \tag{10.2.7}$$

RP2'. The derivatives of r verify

$$\begin{aligned}
\left|e_3'(r) + \Upsilon\right| &\lesssim \epsilon w_{0,1},\\
\left|e_4'(r) - 1\right| &\lesssim \epsilon w_{1,1},\\
\left|e_3'e_4'(r),\, e_4'e_3'(r) + \frac{2m}{r^2}\right| &\lesssim \epsilon w_{1,1}.
\end{aligned} \tag{10.2.8}$$

RP3'. The Gauss curvature K of S and ρ verify

$$\begin{aligned}
\left|\rho + \frac{2m}{r^3}\right| &\lesssim \epsilon r^{-3},\\
\left|K - \frac{1}{r^2}\right| &\lesssim \epsilon r^{-2}.
\end{aligned} \tag{10.2.9}$$

RP4'. We also assume

$$\begin{aligned}
|m - m_0| &\lesssim \epsilon,\\
|e_3'm,\, r^2 e_4'm| &\lesssim \epsilon w_{0,1},\\
|e_3'e_4'(m),\, e_4'e_3'(m)| &\lesssim \epsilon w_{1,1}.
\end{aligned} \tag{10.2.10}$$

Remark 10.39. *In the far region $r \geq 4m_0$ all norms we are using in our estimates are equivalent when expressed relative to the null frame (e_3, e_4, e_θ) or (e_3', e_4', e_θ').*

Convention. For the remainder of this section we shall do all calculations with respect to the renormalized frame (e_3', e_4', e_θ'). For convenience we shall drop the primes, throughout this section, since there is no danger of confusion. Note however that the main results, which include the interior region $r \leq R$, are always expressed with respect to the original frame.

10.2.1 Vectorfield $X = f(r)e_4$

Lemma 10.40. *Consider the vectorfield $X = f(r)e_4$.*

1. *We have the decomposition*

$$^{(X)}\pi = {}^{(X)}\Lambda \mathbf{g} + {}^{(X)}\widetilde{\pi}, \qquad {}^{(X)}\Lambda = \frac{2}{r}f$$

with symmetric tensor $^{(X)}\widetilde{\pi}$ which verifies

$$\begin{aligned} {}^{(X)}\widetilde{\pi}_{43} &= -2f' + \frac{4f}{r} + O(\epsilon)w_{1,1}\left(|f| + r|f'|\right) \\ {}^{(X)}\widetilde{\pi}_{33} &= 4f'\Upsilon - 4\Upsilon' + O(\epsilon)w_{1,1}(|f| + r|f'|) \\ {}^{(X)}\widetilde{\pi}_{4\theta} &= O(\epsilon)w_{2,1/2}|f| \\ {}^{(X)}\widetilde{\pi}_{AB} &= O(\epsilon)w_{2,1/2}|f| \\ {}^{(X)}\widetilde{\pi}_{3\theta} &= O(\epsilon)w_{1,1}|f|. \end{aligned} \tag{10.2.11}$$

2. *We have*

$$\square\, {}^{(X)}\Lambda \;=\; \frac{2}{r}f'' + O\left(\frac{m}{r^4} + \epsilon w_{3,1}\right)\left(|f| + r|f'| + r^2|f''|\right). \tag{10.2.12}$$

Proof. See Lemma D.14 in appendix. □

10.2.2 Energy densities for $X = f(r)e_4$

We start with the following proposition.

Proposition 10.41. *Assume Ψ verifies the equation $\dot{\square}_{\mathbf{g}}\Psi = V\Psi + N$ and let $X = fe_4$ and $w = {}^{(X)}\Lambda = \frac{2f}{r}$ and let $\mathcal{E} := \mathcal{E}[X, w] = \mathcal{E}[X = fe_4, w = \frac{2f}{r}]$.*

1. *We have*

$$\begin{aligned} \mathcal{E} \;=\; & \frac{1}{2}f'|e_4\Psi|^2 + \frac{1}{2}\left(-f' + \frac{2f}{r}\right)\left(|\nabla\!\!\!\!/\,\Psi|^2 + V|\Psi|^2\right) - \frac{1}{2r}f''|\Psi|^2 \\ & + Err\left(\epsilon, \frac{m}{r}, f\right)(\Psi) \end{aligned}$$

where

$$\begin{aligned} & Err\left(\epsilon, \frac{m}{r}, f\right)(\Psi) \\ = \;& O\left(\frac{m}{r^2}\right)(|f| + r|f'|)|e_4\Psi|^2 + O\left(\frac{m}{r^4} + \epsilon w_{3,1}\right)\left(|f| + r|f'| + r^2|f''|\right)|\Psi|^2 \\ + \;& O(\epsilon)w_{1,1}(|f| + r|f'|)\left(|e_4\Psi|^2 + |\nabla\!\!\!\!/\,\Psi|^2 + r^{-2}|\Psi|^2\right) \\ + \;& O(\epsilon)w_{2,1/2}|f|\Big(|e_3\Psi|(|e_4\Psi| + r^{-1}|\nabla\!\!\!\!/\,\Psi|) + |\nabla\!\!\!\!/\,\Psi|^2 + r^{-2}|\Psi|^2\Big). \end{aligned}$$

2. *The current*

$$\mathcal{P}_\mu \quad = \quad \mathcal{P}_\mu[X,w] = \mathcal{Q}_{\mu\nu}X^\nu + \frac{1}{2}w\Psi\cdot\mathbf{D}_\mu\Psi - \frac{1}{4}|\Psi|^2\partial_\mu w$$

verifies

$$\begin{aligned}
\mathcal{P}\cdot e_4 &= f\left|e_4\Psi + \frac{1}{r}\Psi\right|^2 - \frac{1}{2}r^{-2}e_4(rf|\Psi|^2) + O(\epsilon r^{-3})f\,|\Psi|^2,\\
\mathcal{P}\cdot e_3 &= f\mathcal{Q}_{34} + \frac{1}{2}r^{-2}e_3(rf\psi^2) + r^{-1}f'\psi^2 + O(mr^{-3}+\epsilon r^{-2})|rf'|\,|\Psi|^2.
\end{aligned}$$

3. *Let $\theta=\theta(r)$ supported for $r\geq R/2$ with $\theta=1$ for $r\geq R$ such that $f_p=\theta(r)r^p$. Let ${}^{(p)}\mathcal{P} := \mathcal{P}[f_pe_4, w_p]$. Then, for all $r\geq R$,*

$${}^{(p)}\mathcal{P}\cdot e_4 + \frac{p}{2}r^{-2}e_4(\theta r^{p+1}|\Psi|^2) \geq \frac{1}{8}r^{p-2}(p-1)^2|\Psi|^2.$$

Before proceeding with the proof of Proposition 10.41, we first establish the following lemma.

Lemma 10.42. *We have*

$$\begin{aligned}
\mathcal{Q}\cdot{}^{(X)}\widetilde{\pi} &= \left(f' + O\left(\frac{m}{r^2}\right)(|f|+r|f'|)\right)|e_4\Psi|^2 + \left(-f' + \frac{2f}{r}\right)(|\nabla\!\!\!/\Psi|^2 + V|\Psi|^2)\\
&+ O(\epsilon)w_{1,1}(|f|+r|f'|)\Big(|e_4\Psi|^2 + |\nabla\!\!\!/\Psi|^2 + r^{-2}|\Psi|^2\Big)\\
&+ O(\epsilon)w_{2,1/2}|f|\Big(|e_3\Psi|(|e_4\Psi| + r^{-1}|\nabla\!\!\!/\Psi|) + |\nabla\!\!\!/\Psi|^2 + r^{-2}|\Psi|^2\Big).
\end{aligned}$$

Proof. Recall from Proposition 10.9 that we have

$$\mathcal{Q}_{33} = |e_3\Psi|^2, \quad \mathcal{Q}_{44} = |e_4\Psi|^2, \quad \mathcal{Q}_{34} = |\nabla\!\!\!/\Psi|^2 + V|\Psi|^2,$$

and

$$|\mathcal{Q}_{AB}| \leq |e_3\Psi||e_4\Psi| + |\nabla\!\!\!/\Psi|^2 + |V||\Psi|^2, \quad |\mathcal{Q}_{A3}| \leq |e_3\Psi||\nabla\!\!\!/\Psi|, \quad |\mathcal{Q}_{A4}| \leq |e_4\Psi||\nabla\!\!\!/\Psi|.$$

Hence, in view of Lemma D.14 for ${}^{(X)}\widetilde{\pi}$, we have

$$\begin{aligned}
\mathcal{Q}\cdot{}^{(X)}\widetilde{\pi} &= \frac{1}{4}\mathcal{Q}_{44}\,{}^{(X)}\widetilde{\pi}_{33} + \frac{1}{2}\mathcal{Q}_{34}\,{}^{(X)}\widetilde{\pi}_{34} - \frac{1}{2}\mathcal{Q}_{4A}\,{}^{(X)}\widetilde{\pi}_{3A} - \frac{1}{2}\mathcal{Q}_{3A}\,{}^{(X)}\widetilde{\pi}_{4A}\\
&\quad +\mathcal{Q}_{AB}\,{}^{(X)}\widetilde{\pi}_{AB}\\
&= \Big(f'\Upsilon - \Upsilon'f + O(\epsilon)w_{1,1}(|f|+r|f'|)\Big)\mathcal{Q}_{44}\\
&+ \left(-f' + \frac{2f}{r} + O(\epsilon)w_{1,1}\,(|f|+r|f'|)\right)\mathcal{Q}_{34}\\
&+ O(\epsilon)w_{1,1}|f|\;\mathcal{Q}_{4A} + O(\epsilon)w_{2,1/2}|f|\,(\mathcal{Q}_{3A}+\mathcal{Q}_{AB})\\
&= \left(f' + O\left(\frac{m}{r^2}\right)(|f|+r|f'|)\right)\mathcal{Q}_{44} + \left(-f' + \frac{2f}{r}\right)\mathcal{Q}_{34}\\
&+ O(\epsilon)(|f|+r|f'|)w_{1,1}\,(\mathcal{Q}_{44}+\mathcal{Q}_{4A}) + O(\epsilon)w_{2,1/2}|f|\,(\mathcal{Q}_{AB}+\mathcal{Q}_{34})\\
&\quad +O(\epsilon)w_{3,1/2}|f|\mathcal{Q}_{3A}
\end{aligned}$$

from which we deduce

$$
\begin{aligned}
\mathcal{Q}\cdot {}^{(X)}\widetilde{\pi} &= \left(f' + O\left(\frac{m}{r^2}\right)(|f| + r|f'|)\right)|e_4\Psi|^2 + \left(-f' + \frac{2f}{r}\right)(|\nabla\!\!\!/\Psi|^2 + V|\Psi|^2)\\
&+ O(\epsilon)w_{1,1}(|f| + r|f'|)\Big(|e_4\Psi|^2 + |\nabla\!\!\!/\Psi|^2 + r^{-2}|\Psi|^2\Big)\\
&+ O(\epsilon)w_{2,1/2}|f|\Big(|e_3\Psi|(|e_4\Psi| + r^{-1}|\nabla\!\!\!/\Psi|) + |\nabla\!\!\!/\Psi|^2 + r^{-2}|\Psi|^2\Big)
\end{aligned}
$$

as desired. □

We are now ready to prove Proposition 10.41.

Proof of Proposition 10.41. If $\mathcal{Q} = \mathcal{Q}[\Psi]$ is the energy-momentum tensor of Ψ (recall $\dot{\square}\Psi = V\Psi + N$) and

$$
{}^{(X)}\pi = {}^{(X)}\Lambda\,\mathbf{g} + {}^{(X)}\widetilde{\pi}
$$

we deduce

$$
\mathcal{Q}\cdot {}^{(X)}\pi = {}^{(X)}\Lambda \mathrm{tr}\mathcal{Q} + \mathcal{Q}\cdot {}^{(X)}\widetilde{\pi} = {}^{(X)}\Lambda\left(-\mathcal{L}(\Psi) - V|\Psi|^2\right) + \mathcal{Q}\cdot {}^{(X)}\widetilde{\pi}.
$$

Hence, for $X = fe_4$ and $w = {}^{(X)}\Lambda = \frac{2f}{r}$,

$$
\frac{1}{2}\mathcal{Q}\cdot {}^{(X)}\pi + \frac{1}{2}w\mathcal{L}[\Psi] = -\frac{1}{2}{}^{(X)}\Lambda V|\Psi|^2 + \frac{1}{2}\mathcal{Q}\cdot {}^{(X)}\widetilde{\pi}.
$$

In view of (10.1.14), we infer

$$
\begin{aligned}
\mathcal{E} &:= \mathcal{E}[X, w = {}^{(X)}\Lambda, M = 0]\\
&= \frac{1}{2}\mathcal{Q}\cdot {}^{(X)}\widetilde{\pi} - \frac{1}{4}|\Psi|^2\square_{\mathbf{g}}\,{}^{(X)}\Lambda - \frac{1}{2}(X(V) + {}^{(X)}\Lambda V)|\Psi|^2.
\end{aligned}
$$

Recall that $V = -\kappa\underline{\kappa}$. Hence,

$$
\begin{aligned}
X(V) + {}^{(X)}\Lambda V &= fe_4(V) + \frac{2f}{r}V = -f\left(e_4(\kappa\underline{\kappa}) + \frac{2}{r}\kappa\underline{\kappa}\right)\\
&= f\left(\kappa^2\underline{\kappa} - 2\kappa\rho - \frac{2}{r}\kappa\underline{\kappa} + O(\epsilon)w_{3,1}\right)\\
&= O\left(\frac{m}{r^4} + \epsilon w_{3,1}\right)f.
\end{aligned}
$$

Hence, in view of the computation (10.2.12) of $\square\,{}^{(X)}\Lambda$,

$$
\begin{aligned}
&-\frac{1}{4}|\Psi|^2\square_{\mathbf{g}}\,{}^{(X)}\Lambda - \frac{1}{2}(X(V) + {}^{(X)}\Lambda V)|\Psi|^2\\
&= -\frac{1}{2r}f''|\Psi|^2 + O\left(\frac{m}{r^4} + \epsilon w_{3,1}\right)\left(|f| + r|f'| + r^2|f''|\right)|\Psi|^2.
\end{aligned}
$$

We deduce

$$
\mathcal{E} = \frac{1}{2}\mathcal{Q}\cdot {}^{(X)}\widetilde{\pi} - \frac{1}{2r}f''|\Psi|^2 + O\left(\frac{m}{r^4} + \epsilon w_{3,1}\right)\left(|f| + r|f'| + r^2|f''|\right)|\Psi|^2.
$$

Using Lemma 10.42, we deduce

$$\mathcal{E} = \frac{1}{2}f'|e_4\Psi|^2 + \frac{1}{2}\left(-f' + \frac{2f}{r}\right)(|\nabla\Psi|^2 + V|\Psi|^2) - \frac{1}{2r}f''|\Psi|^2 + \mathrm{Err}\left(\epsilon, \frac{m}{r}, f\right)(\Psi)$$

where

$$\begin{aligned}
\mathrm{Err}\left(\epsilon, \frac{m}{r}, f\right)(\Psi) &= O\left(\frac{m}{r^2}\right)(|f| + r|f'|)|e_4\Psi|^2 \\
&+ O\left(\frac{m}{r^4} + \epsilon w_{3,1}\right)\left(|f| + r|f'| + r^2|f''|\right)|\Psi|^2 \\
&+ O(\epsilon)w_{1,1}(|f| + r|f'|)\Big(|e_4\Psi|^2 + |\nabla\Psi|^2 + r^{-2}|\Psi|^2\Big) \\
&+ O(\epsilon)w_{2,1/2}|f|\Big(|e_3\Psi|(|e_4\Psi| + r^{-1}|\nabla\Psi|) + |\nabla\Psi|^2 + r^{-2}|\Psi|^2\Big)
\end{aligned}$$

which is the first part of Proposition 10.41.

To prove the second part of Proposition 10.41, we compute

$$\begin{aligned}
\mathcal{P}\cdot e_4 &= fQ_{44} + \frac{1}{r}f\Psi\cdot e_4\Psi - \frac{1}{2}e_4(r^{-1}f)|\Psi|^2 \\
&= f\left(|e_4\Psi|^2 + \frac{1}{r}\Psi\cdot e_4\Psi\right) - \frac{1}{2}e_4(r^{-1}f)|\Psi|^2 \\
&= f\left|e_4\Psi + \frac{1}{r}\Psi\right|^2 - \frac{1}{r}f\Psi\cdot e_4\Psi - r^{-2}f|\Psi|^2 - \frac{1}{2}e_4(r^{-1}f)|\Psi|^2 \\
&= f\left|e_4\Psi + \frac{1}{r}\Psi\right|^2 - \frac{1}{2}r^{-2}e_4(rf|\Psi|^2) + \frac{1}{2}r^{-2}e_4(rf)|\Psi|^2 - r^{-2}f|\Psi|^2 \\
&- \frac{1}{2}e_4(r^{-1}f)|\Psi|^2 \\
&= f\left|e_4\Psi + \frac{1}{r}\Psi\right|^2 - \frac{1}{2}r^{-2}e_4(rf|\Psi|^2) + r^{-2}(e_4(r) - 1)f|\Psi|^2.
\end{aligned}$$

Since

$$e_4(r) = \frac{r}{2}(\kappa + A),$$

we have[14]

$$e_4(r) - 1 = O(\epsilon r^{-1}).$$

Thus, as desired,

$$\mathcal{P}\cdot e_4 = f\left|e_4\Psi + \frac{1}{r}\Psi\right|^2 - \frac{1}{2}r^{-2}e_4(rf|\Psi|^2) + O(\epsilon r^{-3})f|\Psi|^2.$$

[14]Note that so far we have only used the weaker version $e_4(r) - 1 = O(\epsilon)$. This is the first time we need the stronger version of the estimate in this chapter.

Also,

$$\begin{aligned}
\mathcal{P}\cdot e_3 &= f\mathcal{Q}_{34}+r^{-1}f\Psi\cdot e_3\Psi-\frac{1}{2}e_3(r^{-1}f)|\Psi|^2\\
&= f\mathcal{Q}_{34}+\frac{1}{2}r^{-1}fe_3(|\Psi|^2)-\frac{1}{2}e_3(r^{-1}f)|\Psi|^2\\
&= f\mathcal{Q}_{34}+\frac{1}{2}\left[r^{-2}e_3(rf|\Psi|^2)-r^{-2}e_3(rf)|\Psi|^2\right]-\frac{1}{2}e_3(r^{-1}f)|\Psi|^2\\
&= f\mathcal{Q}_{34}+\frac{1}{2}r^{-2}e_3(rf|\Psi|^2)+r^{-1}f'\Upsilon|\Psi|^2-r^{-1}f'(e_3(r)+\Upsilon)|\Psi|^2\\
&= f\mathcal{Q}_{34}+\frac{1}{2}r^{-2}e_3(rf|\Psi|^2)+r^{-1}f'|\Psi|^2+O\left(mr^{-3}+\epsilon r^{-2}\right)(r|f'|)|\Psi|^2
\end{aligned}$$

as desired.

It remains to prove the last part of Proposition 10.41. We have, for $r\geq R$,

$$ {}^{(p)}\mathcal{P}\cdot e_4 = r^p|e_4\Psi|^2+r^{p-1}\Psi e_4\Psi-\frac{1}{2}e_4(r^{p-1})|\Psi|^2$$

and

$$\begin{aligned}
{}^{(p)}\mathcal{P}\cdot e_4+\frac{p}{2}r^{-2}e_4(\theta r^{p+1}|\Psi|^2) &= r^p|e_4\Psi|^2+r^{p-1}\Psi\cdot e_4\Psi-\frac{1}{2}e_4(r^{p-1})|\Psi|^2\\
&+ pr^{p-1}\Psi\cdot e_4\Psi+\frac{p(p+1)}{2}r^{p-2}e_4(r)|\Psi|^2\\
&= r^p|e_4\Psi|^2+(p+1)\,r^{p-1}\Psi\cdot e_4\Psi\\
&+ \frac{p^2+1}{2}e_4(r)r^{p-2}|\Psi|^2\\
&= r^p\left[\left|e_4\Psi+\frac{p+1}{2r}\Psi\right|^2+\frac{(p-1)^2}{4r^2}|\Psi|^2\right]\\
&+ \frac{p^2+1}{2}(e_4(r)-1)r^{p-2}|\Psi|^2\\
&\geq r^{p-2}\frac{(p-1)^2}{4}|\Psi|^2-O(\epsilon)\frac{p^2+1}{2}r^{p-3}|\Psi|^2.
\end{aligned}$$

This concludes the proof of Proposition 10.41. □

In applications we would like to apply Proposition 10.41 to $f=r^p$, $0<p<2$. We note however that the presence of the term $-\frac{1}{2}r^{-1}f''|\Psi|^2$ on the right-hand side of the $\mathcal{E}$ identity requires an additional correction if $p>1$. This additional correction is taken into account by the following proposition.

Proposition 10.43. *Assume Ψ verifies the equation $\square_{\mathbf{g}}\Psi=V\Psi+N$ and let $X=f(r)e_4$, $w={}^{(X)}\Lambda=\frac{2f}{r}$ and $M=2r^{-1}f'e_4$. Then:*

1. *We have, with $\check{e}_4=r^{-1}e_4(r\cdot)$,*

$$\mathcal{E}[X,w,M]=\frac{1}{2}f'|\check{e}_4(\Psi)|^2+\frac{1}{2}\left(\frac{2f}{r}-f'\right)\mathcal{Q}_{34}+Err\Big(\epsilon,\frac{m}{r};f\Big)[\Psi] \quad (10.2.13)$$

with error term

$$
\begin{aligned}
& Err\left(\epsilon, \frac{m}{r}, f\right)(\Psi) \\
= \;& O\left(\frac{m}{r^2}\right)(|f|+r|f'|)|e_4\Psi|^2 + O\left(\frac{m}{r^4}+\epsilon w_{3,1}\right)\left(|f|+r|f'|+r^2|f''|\right)|\Psi|^2 \\
+ \;& O(\epsilon)w_{1,1}(|f|+r|f'|)\left(|e_4\Psi|^2+|\nabla\!\!\!/\Psi|^2+r^{-2}|\Psi|^2\right) \\
+ \;& O(\epsilon)w_{2,1/2}|f|\Big(|e_3\Psi|(|e_4\Psi|+r^{-1}|\nabla\!\!\!/\Psi|)+|\nabla\!\!\!/\Psi|^2+r^{-2}|\Psi|^2\Big).
\end{aligned}
$$

2. *The current*

$$
\mathcal{P}_\mu \;=\; \mathcal{P}_\mu[X,w,M] = \mathcal{Q}_{\mu\nu}X^\nu + \frac{1}{2}w\Psi\mathbf{D}_\mu\Psi - \frac{1}{4}|\Psi|^2\partial_\mu w + \frac{1}{4}M_\mu|\Psi|^2
$$

verifies

$$
\begin{aligned}
\mathcal{P}\cdot e_4 &= f(\check{e}_4\Psi)^2 - \frac{1}{2}r^{-2}e_4(rf|\Psi|^2) + O(\epsilon r^{-1})f(|e_4\Psi|^2+r^{-2}|\Psi|^2), \\
\mathcal{P}\cdot e_3 &= f\mathcal{Q}_{34} + \frac{1}{2}r^{-2}e_3(rf|\Psi|^2) + O(mr^{-3}+\epsilon r^{-2})(|f|+r|f'|)|\Psi|^2.
\end{aligned}
$$

3. *Let $\theta=\theta(r)$ supported for $r\geq R/2$ with $\theta=1$ for $r\geq R$ such that $f_p=\theta(r)r^p$. Let ${}^{(p)}\mathcal{P} := \mathcal{P}[f_pe_4, w_p, M_p]$. Then, for all $r\geq R$,*

$$
{}^{(p)}\mathcal{P}\cdot e_4 + \frac{p}{2}r^{-2}e_4(\theta r^{p+1}|\Psi|^2) \geq \frac{1}{8}r^{p-2}(p-1)^2|\Psi|^2.
$$

Proof. We start with the first part of Proposition 10.43. To this end, we use

$$
{}^{(X)}\pi_{43} = -2e_4f + 4f\omega, \qquad {}^{(X)}\pi_{AB} = 2f\,{}^{(1+3)}\chi_{AB},
$$

so that

$$
\begin{aligned}
\mathrm{tr}\,{}^{(X)}\pi &= -\,{}^{(X)}\pi_{43} + \mathbf{g}^{AB}\,{}^{(X)}\pi_{AB} \\
&= 2e_4f - 4f\omega + 2f\kappa,
\end{aligned}
$$

and we compute

$$
\begin{aligned}
& \mathbf{D}^\mu M_\mu \\
=\;& \mathbf{D}^\mu(2r^{-1}f'e_4)_\mu = \mathbf{D}^\mu\left(\frac{2f'}{rf}X_\mu\right) = \frac{2f'}{rf}\mathbf{Div}X + X\left(\frac{2f'}{rf}\right) \\
=\;& \frac{f'}{rf}\mathrm{tr}\,{}^{(X)}\pi + X\left(\frac{2f'}{rf}\right) \\
=\;& \frac{f'}{rf}(2e_4f - 4f\omega + 2f\kappa) + 2fe_4\left(\frac{f'}{rf}\right) \\
=\;& \frac{f'}{rf}\left(2e_4(r)f' + \frac{4f}{r} + O\left(\frac{m}{r^3}+\epsilon w_{2,1}\right)\right) + 2f\left(\frac{f''rf - f'(f+rf')}{r^2f^2}\right)e_4(r) \\
=\;& \frac{4f'}{r^2} + \frac{2(f')^2}{rf} + 2\frac{f''}{r} - \frac{2f'}{r^2} - \frac{2(f')^2}{rf} + O\left(\frac{m}{r^4}+\epsilon w_{3,1}\right)(|f|+r|f'|+r^2|f''|) \\
=\;& \frac{2f'}{r^2} + \frac{2f''}{r} + O\left(\frac{m}{r^4}+\epsilon w_{3,1}\right)(|f|+r|f'|+r^2|f''|).
\end{aligned}
$$

We also have

$$\frac{1}{2}\Psi\cdot\mathbf{D}_\mu\Psi M^\mu \quad = \quad r^{-1}f'\Psi\cdot\mathbf{D}_4\Psi.$$

Since we have

$$\mathcal{E}[X,w,M] \quad = \quad \mathcal{E}[X,w]+\frac{1}{4}(\mathbf{D}^\mu M_\mu)|\Psi|^2+\frac{1}{2}\Psi\cdot\mathbf{D}_\mu\Psi M^\mu,$$

we infer

$$\begin{aligned}
&\mathcal{E}[X,w,M]\\
=\;&\mathcal{E}[X,w]+\left(\frac{f'}{2r^2}+\frac{f''}{2r}+O\left(\frac{m}{r^4}+\epsilon w_{3,1}\right)(|f|+r|f'|+r^2|f''|)\right)|\Psi|^2\\
&+r^{-1}f'\Psi\cdot\mathbf{D}_4\Psi.
\end{aligned}$$

Together with Proposition 10.41, this yields

$$\begin{aligned}
\mathcal{E}[X,w,M] \;=\;& \frac{1}{2}f'\left(|e_4\Psi|^2+2r^{-1}\Psi\cdot\mathbf{D}_4\Psi+r^{-2}|\Psi|^2\right)\\
&+\frac{1}{2}\left(-f'+\frac{2f}{r}\right)(|\nabla\!\!\!/\Psi|^2+V|\Psi|^2)\\
&+\mathrm{Err}\left(\epsilon,\frac{m}{r},f\right)(\Psi)+O\left(\frac{m}{r^4}+\epsilon w_{3,1}\right)(|f|+r|f'|+r^2|f''|)|\Psi|^2\\
=\;& \frac{1}{2}f'|e_4\Psi+r^{-1}\Psi|^2+\frac{1}{2}\left(-f'+\frac{2f}{r}\right)(|\nabla\!\!\!/\Psi|^2+V|\Psi|^2)\\
&+\mathrm{Err}\left(\epsilon,\frac{m}{r},f\right)(\Psi)+O\left(\frac{m}{r^4}+\epsilon w_{3,1}\right)(|f|+r|f'|+r^2|f''|)|\Psi|^2\\
=\;& \frac{1}{2}f'|\check{e}_4\Psi+r^{-1}(1-e_4(r))\Psi|^2+\frac{1}{2}\left(-f'+\frac{2f}{r}\right)(|\nabla\!\!\!/\Psi|^2+V|\Psi|^2)\\
&+\mathrm{Err}\left(\epsilon,\frac{m}{r},f\right)(\Psi)+O\left(\frac{m}{r^4}+\epsilon w_{3,1}\right)(|f|+r|f'|+r^2|f''|)|\Psi|^2
\end{aligned}$$

and hence

$$\mathcal{E}[X,w,M]=\frac{1}{2}f'|\check{e}_4\Psi|^2+\frac{1}{2}\left(-f'+\frac{2f}{r}\right)(|\nabla\!\!\!/\Psi|^2+V|\Psi|^2)+\mathrm{Err}\left(\epsilon,\frac{m}{r},f\right)(\Psi)$$

where

$$\begin{aligned}
\mathrm{Err}\left(\epsilon,\frac{m}{r},f\right)(\Psi) \;=\;& O\left(\frac{m}{r^2}\right)(|f|+r|f'|)|e_4\Psi|^2\\
+\;& O\left(\frac{m}{r^4}+\epsilon w_{3,1}\right)(|f|+r|f'|+r^2|f''|)|\Psi|^2\\
+\;& O(\epsilon)w_{1,1}(|f|+r|f'|)\left(|e_4\Psi|^2+|\nabla\!\!\!/\Psi|^2+r^{-2}|\Psi|^2\right)\\
+\;& O(\epsilon)w_{2,1/2}|f|\Big(|e_3\Psi|(|e_4\Psi|+r^{-1}|\nabla\!\!\!/\Psi|)+|\nabla\!\!\!/\Psi|^2+r^{-2}|\Psi|^2\Big).
\end{aligned}$$

This is the desired estimate (10.2.13).

Next, we consider the second part of Proposition 10.43.

$$\mathcal{P}_\mu[X,w,M] = \mathcal{P}_\mu[X,w] + \frac{1}{4}|\Psi|^2 M_\mu = \mathcal{P}_\mu[X,w] + \frac{1}{2}r^{-1}f'|\Psi|^2 e_4.$$

Hence, in view of the results in part 2 of Proposition 10.41,

$$\begin{aligned}
\mathcal{P}_4[X,w,M] &= \mathcal{P}_4[X,w] = f\Big|\check{e}_4\Psi + (1-e_4(r))\Psi\Big|^2 - \frac{1}{2}r^{-2}e_4(rf|\Psi|^2) \\
&\quad +O(\epsilon r^{-3})f|\Psi|^2 \\
&= f|\check{e}_4\Psi|^2 - \frac{1}{2}r^{-2}e_4(rf|\Psi|^2) + O(\epsilon r^{-1})f(|e_4\Psi|^2 + r^{-2}|\Psi|^2), \\
\mathcal{P}_3[X,w,M] &= \mathcal{P}_3[X,w] - r^{-1}f'|\Psi|^2 \\
&= f\mathcal{Q}_{34} + \frac{1}{2}r^{-2}e_3(rf|\Psi|^2) + r^{-1}f'|\Psi|^2 - r^{-1}f'|\Psi|^2 \\
&\quad + O(\epsilon r^{-2})(|f| + r|f'|)|\Psi|^2 \\
&= f\mathcal{Q}_{34} + \frac{1}{2}r^{-2}e_3(rf|\Psi|^2) + O(\epsilon r^{-2})(|f| + r|f'|)|\Psi|^2
\end{aligned}$$

as desired. The last part follows from the third part of Proposition 10.41. □

Lemma 10.44. *On Σ_*, we have*

$$\begin{aligned}
\mathcal{P}\cdot N_{\Sigma_*} &= \frac{1}{2}f\mathcal{Q}_{34} + \frac{1}{2}f(\check{e}_4\Psi)^2 + \frac{1}{2}div_{\,\Sigma_*}\Big(r^{-1}f|\Psi|^2\nu_{\Sigma_*}\Big) \\
&\quad +O(mr^{-1}+\epsilon)(|f| + r|f'|)(|e_4\Psi|^2 + |\nabla\!\!\!/\Psi|^2 + r^{-2}|\Psi|^2).
\end{aligned}$$

Proof. Recall that there exists a constant c_* such that $u+r = c_*$ on Σ_*. In particular, the unit normal N_{Σ_*} is collinear to

$$\begin{aligned}
-2\mathbf{g}^{\alpha\beta}\partial_\alpha(u+r)\partial_\beta &= e_4(u+r)e_3 + e_3(u+r)e_4 \\
&= e_4(r)e_3 + (e_3(u)+e_3(r))e_4
\end{aligned}$$

and since

$$\begin{aligned}
& g\Big(e_4(r)e_3 + (e_3(u)+e_3(r))e_4, e_4(r)e_3 + (e_3(u)+e_3(r))e_4\Big) \\
=& -4e_4(r)(e_3(u)+e_3(r)),
\end{aligned}$$

we infer

$$N_{\Sigma_*} = \frac{\sqrt{e_4(r)}}{2\sqrt{e_3(u)+e_3(r)}}e_3 + \frac{\sqrt{e_3(u)+e_3(r)}}{2\sqrt{e_4(r)}}e_4.$$

In particular, we have

$$\mathcal{P}\cdot N_{\Sigma_*} = \frac{\sqrt{e_4(r)}}{2\sqrt{e_3(u)+e_3(r)}}\mathcal{P}\cdot e_3 + \frac{\sqrt{e_3(u)+e_3(r)}}{2\sqrt{e_4(r)}}\mathcal{P}\cdot e_4.$$

Now, recall from Proposition 10.43 that we have

$$\begin{aligned}\mathcal{P}\cdot e_4 &= f(\check{e}_4\Psi)^2-\frac{1}{2}r^{-2}e_4(rf|\Psi|^2)+O(\epsilon r^{-1})f(|e_4\Psi|^2+r^{-2}|\Psi|^2),\\ \mathcal{P}\cdot e_3 &= f\mathcal{Q}_{34}+\frac{1}{2}r^{-2}e_3(rf|\Psi|^2)+O(mr^{-3}+\epsilon r^{-2})(|f|+r|f'|)|\Psi|^2.\end{aligned}$$

We deduce

$$\begin{aligned}\mathcal{P}\cdot N_{\Sigma_*} &= \frac{\sqrt{e_4(r)}}{2\sqrt{e_3(u)+e_3(r)}}\Bigg(f\mathcal{Q}_{34}+\frac{1}{2}r^{-2}e_3(rf|\Psi|^2)\\ &\quad +O(mr^{-3}+\epsilon r^{-2})(|f|+r|f'|)|\Psi|^2\Bigg)\\ &\quad +\frac{\sqrt{e_3(u)+e_3(r)}}{2\sqrt{e_4(r)}}\Bigg(f(\check{e}_4\Psi)^2-\frac{1}{2}r^{-2}e_4(rf|\Psi|^2)\\ &\quad +O(\epsilon r^{-1})f(|e_4\Psi|^2+r^{-2}|\Psi|^2)\Bigg)\\ &= \frac{\sqrt{e_4(r)}}{2\sqrt{e_3(u)+e_3(r)}}f\mathcal{Q}_{34}+\frac{\sqrt{e_3(u)+e_3(r)}}{2\sqrt{e_4(r)}}f(\check{e}_4\Psi)^2\\ &\quad +\frac{\sqrt{e_4(r)}}{2\sqrt{e_3(u)+e_3(r)}}\frac{1}{2}r^{-2}e_3(rf|\Psi|^2)-\frac{\sqrt{e_3(u)+e_3(r)}}{2\sqrt{e_4(r)}}\frac{1}{2}r^{-2}e_4(rf|\Psi|^2)\\ &\quad +O(mr^{-3}+\epsilon r^{-2})(|f|+r|f'|)|\Psi|^2+O(\epsilon r^{-1})f(|e_4\Psi|^2+r^{-2}|\Psi|^2)\\ &= \frac{1}{2\sqrt{2-\Upsilon}}(1+O(\epsilon))f\mathcal{Q}_{34}+\frac{\sqrt{2-\Upsilon}}{2}(1+O(\epsilon))f(\check{e}_4\Psi)^2\\ &\quad +\frac{1}{2}r^{-2}\nu_{\Sigma_*}(rf|\Psi|^2)+O(mr^{-3}+\epsilon r^{-2})(|f|+r|f'|)|\Psi|^2\\ &\quad +O(\epsilon r^{-1})f(|e_4\Psi|^2+r^{-2}|\Psi|^2)\\ &= \frac{1}{2}f\mathcal{Q}_{34}+\frac{1}{2}f(\check{e}_4\Psi)^2+\frac{1}{2}r^{-2}\nu_{\Sigma_*}(rf|\Psi|^2)\\ &\quad +O(mr^{-1}+\epsilon)(|f|+r|f'|)(|e_4\Psi|^2+|\nabla\!\!\!/\Psi|^2+r^{-2}|\Psi|^2)\end{aligned}$$

where we used

$$e_4(r)=1+O(\epsilon),\quad e_3(r)=-\Upsilon+O(\epsilon),\quad e_3(u)=2+O(\epsilon),$$

and where ν_{Σ_*} denotes the vectorfield

$$\nu_{\Sigma_*} = \frac{\sqrt{e_4(r)}}{2\sqrt{e_3(u)+e_3(r)}}e_3-\frac{\sqrt{e_3(u)+e_3(r)}}{2\sqrt{e_4(r)}}e_4.$$

Next, note from the formula that ν_{Σ_*} is unitary and orthogonal to N_{Σ_*} so that ν_{Σ_*} is a unit vectorfield, tangent to Σ_* and normal to e_θ. Furthermore, since $(\nu_{\Sigma_*}, e_\theta, e_\varphi)$ is an orthonormal frame of Σ_*, we have

$$\text{div}_{\Sigma_*}(\nu_{\Sigma_*}) = \mathbf{g}(\mathbf{D}_{\nu_{\Sigma_*}}\nu_{\Sigma_*},\nu_{\Sigma_*})+\mathbf{g}(\mathbf{D}_{e_\theta}\nu_{\Sigma_*},e_\theta)+\mathbf{g}(\mathbf{D}_{e_\varphi}\nu_{\Sigma_*},e_\varphi).$$

Since ν_{Σ_*} is a unit vector, the first term vanishes, and hence

$$\begin{aligned}
\text{div}_{\Sigma_*}(\nu_{\Sigma_*}) &= \mathbf{g}(\mathbf{D}_{e_\theta}\nu_{\Sigma_*}, e_\theta) + \mathbf{g}(\mathbf{D}_{e_\varphi}\nu_{\Sigma_*}, e_\varphi) \\
&= \frac{\sqrt{e_4(r)}}{2\sqrt{e_3(u)+e_3(r)}} g(D_\theta e_3, e_\theta) - \frac{\sqrt{e_3(u)+e_3(r)}}{2\sqrt{e_4(r)}} g(D_\theta e_4, e_\theta) \\
&\quad +\nu_{\Sigma_*}(\Phi) \\
&= \frac{\sqrt{e_4(r)}}{2\sqrt{e_3(u)+e_3(r)}}\underline{\kappa} - \frac{\sqrt{e_3(u)+e_3(r)}}{2\sqrt{e_4(r)}}\kappa.
\end{aligned}$$

In particular, we have

$$\begin{aligned}
\text{div}_{\Sigma_*}\Big(r^{-1}f|\Psi|^2\nu_{\Sigma_*}\Big) &= r^{-2}\nu_{\Sigma_*}(rf|\Psi|^2) + \nu_{\Sigma_*}(r^{-2})rf|\Psi|^2 \\
&\quad +\text{div}_{\Sigma_*}(\nu_{\Sigma_*})r^{-1}f|\Psi|^2 \\
&= r^{-2}\nu_{\Sigma_*}(rf|\Psi|^2) + \left(\text{div}_{\Sigma_*}(\nu_{\Sigma_*}) - \frac{2\nu_{\Sigma_*}(r)}{r}\right) r^{-1}f|\Psi|^2 \\
&= r^{-2}\nu_{\Sigma_*}(rf|\Psi|^2) + \left(\frac{\sqrt{e_4(r)}}{2\sqrt{e_3(u)+e_3(r)}}\left(\underline{\kappa} - \frac{2e_3(r)}{r}\right)\right. \\
&\quad \left. - \frac{\sqrt{e_3(u)+e_3(r)}}{2\sqrt{e_4(r)}}\left(\kappa - \frac{2e_4(r)}{r}\right)\right) r^{-1}f|\Psi|^2
\end{aligned}$$

and hence

$$r^{-2}\nu_{\Sigma_*}(rf|\Psi|^2) = \text{div}_{\Sigma_*}\Big(r^{-1}f|\Psi|^2\nu_{\Sigma_*}\Big) + O(\epsilon r^{-2})f|\Psi|^2.$$

We finally obtain

$$\begin{aligned}
\mathcal{P}\cdot N_{\Sigma_*} &= \frac{1}{2}fQ_{34} + \frac{1}{2}f(\check{e}_4\Psi)^2 + \frac{1}{2}r^{-2}\nu_{\Sigma_*}(rf|\Psi|^2) \\
&\quad +O(mr^{-1}+\epsilon)(|f|+r|f'|)(|e_4\Psi|^2 + |\nabla\!\!\!/\Psi|^2 + r^{-2}|\Psi|^2) \\
&= \frac{1}{2}fQ_{34} + \frac{1}{2}f(\check{e}_4\Psi)^2 + \frac{1}{2}\text{div}_{\Sigma_*}\Big(r^{-1}f|\Psi|^2\nu_{\Sigma_*}\Big) \\
&\quad +O(mr^{-1}+\epsilon)(|f|+r|f'|)(|e_4\Psi|^2 + |\nabla\!\!\!/\Psi|^2 + r^{-2}|\Psi|^2)
\end{aligned}$$

which concludes the proof of the lemma. □

10.2.3 Proof of Theorem 10.37

Consider the function $f_p = f_{p,R}$ defined by

$$f_p = \begin{cases} r^p, & \text{if } r \geq R, \\ 0, & \text{if } r \leq \frac{R}{2}, \end{cases} \tag{10.2.14}$$

where R is a fixed, sufficiently large constant which will be chosen in the proof. We also consider

$$X_p = f_p e_4, \quad w_p = \frac{2f_p}{r}, \quad M_p = \frac{2f_p'}{r}e_4.$$

The proof relies on Proposition 10.43.

Step 0. (Reduction to the region $r \geq R$) In view of the definition of $\mathcal{E}[X_p, w_p, M_p]$, see (10.1.13), and in view of the choice of X_p and w_p, we have

$$\mathbf{D}^\mu \mathcal{P}_\mu[X_p, w_p, M_p] \quad = \quad \mathcal{E}[X_p, w_p, M_p] + f_p(r)\check{e}_4\Psi \cdot N.$$

We integrate this identity on the domain $\mathcal{M}(\tau_1, \tau_2)$ to derive

$$\begin{aligned} & \int_{\Sigma(\tau_2)} \mathcal{P} \cdot N_\Sigma + \int_{\Sigma_*(\tau_1,\tau_2)} \mathcal{P} \cdot N_{\Sigma_*} + \int_{\mathcal{M}(\tau_1,\tau_2)} \mathcal{E} \\ = & \int_{\Sigma(\tau_1)} \mathcal{P} \cdot N_\Sigma - \int_{\mathcal{M}(\tau_1,\tau_2)} f_p \check{e}_4 \Psi \cdot N. \end{aligned}$$

Denoting the boundary terms,

$$\begin{aligned} K_{\geq R}(\tau_1, \tau_2) : &= \int_{\Sigma_{\geq R}(\tau_2)} \mathcal{P} \cdot e_4 + \int_{\Sigma_*(\tau_1,\tau_2)} \mathcal{P} \cdot N_{\Sigma_*} - \int_{\Sigma_{\geq R}(\tau_1)} \mathcal{P} \cdot e_4, \\ K_{\leq R}(\tau_1, \tau_2) : &= \int_{\Sigma_{\leq R}(\tau_1)} \mathcal{P} \cdot N_\Sigma - \int_{\Sigma_{\leq R}(\tau_2)} \mathcal{P} \cdot N_\Sigma, \end{aligned}$$

we write

$$K_{\geq R}(\tau_1, \tau_2) + \int_{\mathcal{M}_{\geq R}(\tau_1,\tau_2)} \mathcal{E} = K_{\leq R}(\tau_1, \tau_2) - \int_{\mathcal{M}_{\leq R}(\tau_1,\tau_2)} \mathcal{E} - \int_{\mathcal{M}(\tau_1,\tau_2)} f_p \check{e}_4 \Psi \cdot N.$$

We have the following lemma.

Lemma 10.45. *For $p \geq \delta$, we have*

$$\begin{aligned} K_{\geq R}(\tau_1, \tau_2) + \int_{\mathcal{M}_{\geq R}(\tau_1,\tau_2)} \mathcal{E} \quad \lesssim \quad & R^{p+2}\Big(E[\Psi](\tau_1) + J_p[N, \psi](\tau_1, \tau_2) \\ & + O(\epsilon)\dot{B}_{\delta\,;\,4m_0}[\psi](\tau_1, \tau_2)\Big). \end{aligned}$$

Proof of Lemma 10.45. The terms $\int_{\Sigma_{\leq R}(\tau)} \mathcal{P} \cdot N_\Sigma$ and $\int_{\mathcal{M}_{\leq R}(\tau_1,\tau_2)} \mathcal{E}$ on the right can be estimated as follows:

$$\begin{aligned} \left| \int_{\Sigma_{\leq R}(\tau_1)} \mathcal{P} \cdot N_\Sigma \right| \quad &\lesssim \quad R^p E[\Psi](\tau_1), \\ \left| \int_{\Sigma_{\leq R}(\tau_2)} \mathcal{P} \cdot N_\Sigma \right| \quad &\lesssim \quad R^p E[\Psi](\tau_2), \\ \left| \int_{\mathcal{M}_{\leq R}(\tau_1,\tau_2)} \mathcal{E} \right| \quad &\lesssim \quad R^{p+2} \mathrm{Mor}[\Psi](\tau_1, \tau_2). \end{aligned}$$

Hence,

$$K_{\leq R}(\tau_1, \tau_2) - \int_{\mathcal{M}_{\leq R}(\tau_1,\tau_2)} \mathcal{E} \quad \lesssim \quad R^{p+2} \left(E[\Psi](\tau_1) + E[\Psi](\tau_2) + \mathrm{Mor}[\Psi](\tau_1, \tau_2)\right).$$

In view of the improved Morawetz Theorem 10.1 we have, for fixed $\delta > 0$,

$$\begin{aligned} E[\Psi](\tau_2) + \mathrm{Morr}[\Psi](\tau_1,\tau_2) + F[\Psi](\tau_1,\tau_2) \;\lesssim\; & E[\Psi](\tau_1) + J_\delta[N,\psi](\tau_1,\tau_2) \\ & + O(\epsilon)\dot{B}_{\delta\,;\,4m_0}[\psi](\tau_1,\tau_2) \end{aligned}$$

which implies

$$\begin{aligned} & K_{\geq R}(\tau_1,\tau_2) + \int_{\mathcal{M}_{\geq R}(\tau_1,\tau_2)} \mathcal{E} \\ \lesssim\; & R^{p+2}\Big(E[\Psi](\tau_1) + J_\delta[N,\psi](\tau_1,\tau_2) + O(\epsilon)\dot{B}_{\delta\,;\,4m_0}[\psi](\tau_1,\tau_2)\Big) \\ & + \left|\int_{\mathcal{M}(\tau_1,\tau_2)} f_p \check{e}_4\Psi\cdot N\right|. \end{aligned}$$

Together with the definition (5.3.7) of J_p and the fact that $p \geq \delta$, we infer

$$\begin{aligned} & K_{\geq R}(\tau_1,\tau_2) + \int_{\mathcal{M}_{\geq R}(\tau_1,\tau_2)} \mathcal{E} \\ \lesssim\; & R^{p+2}\Big(E[\Psi](\tau_1) + J_p[N,\psi](\tau_1,\tau_2) + O(\epsilon)\dot{B}_{\delta\,;\,4m_0}[\psi](\tau_1,\tau_2)\Big) \end{aligned}$$

which concludes the proof of Lemma 10.45. □

The proof of Theorem 10.37 now proceeds according to the following steps.

Step 1. (Bulk terms for $r \geq R$) We prove the following lower bound for $\int_{\mathcal{M}_{\geq R}(\tau_1,\tau_2)} \mathcal{E}$.

Lemma 10.46. *Given a fixed $\delta > 0$ we have for all $\delta \leq p \leq 2-\delta$ and $R \gg \frac{m}{\delta}$, $\epsilon \ll \delta$,*

$$\begin{aligned} \int_{\mathcal{M}_{\geq R}(\tau_1,\tau_2)} \mathcal{E} \;\geq\; & \frac{1}{4}\int_{\mathcal{M}_{\geq R}(\tau_1,\tau_2)} r^{p-1}\Big(p|\check{e}_4(\Psi)|^2 + (2-p)(|\nabla\!\!\!/\Psi|^2 + r^{-2}|\Psi|^2)\Big) \\ & - O(\epsilon)\mathit{Morr}[\Psi](\tau_1,\tau_2). \end{aligned} \tag{10.2.15}$$

Proof of Lemma 10.46. We make use of Proposition 10.43 according to which

$$\begin{aligned} \mathcal{E}[X,w,M] &= \frac{1}{2}f_p'|\check{e}_4(\Psi)|^2 + \frac{1}{2}\left(\frac{2f_p}{r} - f_p'\right)\mathcal{Q}_{34} + \mathrm{Err}\left(\epsilon,\frac{m}{r};f_p\right)[\Psi] \\ &= r^{p-1}\left[\frac{p}{2}|\check{e}_4(\Psi)|^2 + \frac{1}{2}(2-p)(|\nabla\!\!\!/\Psi|^2 + V|\Psi|^2)\right] + \mathrm{Err}\left(\epsilon,\frac{m}{r};f_p\right)[\Psi] \\ &\geq r^{p-1}\left[\frac{p}{2}|\check{e}_4(\Psi)|^2 + \frac{1}{2}(2-p)(|\nabla\!\!\!/\Psi|^2 + r^{-2}|\Psi|^2)\right] \\ &\quad + \mathrm{Err}\left(\epsilon,\frac{m}{r};f_p\right)[\Psi] \end{aligned}$$

where

$$\begin{aligned}
\mathrm{Err}\left(\epsilon, \frac{m}{r}, f_p\right)(\Psi) &= r^p O\left(\frac{m}{r^2}\right)\left[|e_4\Psi|^2 + r^{-2}\Psi|^2\right] \\
&+ r^p O(\epsilon) w_{1,1}\left(|e_4\Psi|^2 + |\nabla\!\!\!/\Psi|^2 + r^{-2}|\Psi|^2\right) \\
&+ r^p O(\epsilon) w_{2,1/2}\Big(|e_3\Psi|(|e_4\Psi| + r^{-1}|\nabla\!\!\!/\Psi|) + |\nabla\!\!\!/\Psi|^2 + r^{-2}|\Psi|^2\Big) \\
&\lesssim \mathrm{Err}\left(\frac{m}{r}\right) + \mathrm{Err}(\epsilon), \\
\mathrm{Err}\left(\frac{m}{r}\right) &= O\left(\frac{m}{r}\right) r^{p-1}\left[|\check{e}_4\Psi|^2 + r^{-2}|\Psi|^2\right], \\
\mathrm{Err}(\epsilon) &= O(\epsilon) r^{p-1}\left(|\check{e}_4\Psi|^2 + |\nabla\!\!\!/\Psi|^2 + r^{-2}|\Psi|^2 + r^{-2}|e_3\Psi|^2\right).
\end{aligned}$$

Thus,

$$\begin{aligned}
\int_{\mathcal{M}_{\geq R}(\tau_1,\tau_2)} \mathcal{E} &\geq \int_{\mathcal{M}_{\geq R}(\tau_1,\tau_2)} r^{p-1}\Big(\frac{p}{2}|\check{e}_4(\Psi)|^2 + \frac{1}{2}(2-p)(|\nabla\!\!\!/\Psi|^2 + r^{-2}|\Psi|^2)\Big) \\
&- O\left(\frac{m}{R}\right)\int_{\mathcal{M}_{\geq R}(\tau_1,\tau_2)} r^{p-1}\left[|\check{e}_4\Psi|^2 + r^{-2}|\Psi|^2\right] \\
&- O(\epsilon)\int_{\mathcal{M}_{\geq R}(\tau_1,\tau_2)} r^{p-1}\left(|\check{e}_4\Psi|^2 + |\nabla\!\!\!/\Psi|^2 + r^{-2}|\Psi|^2\right) \\
&- O(\epsilon)\int_{\mathcal{M}_{\geq R}(\tau_1,\tau_2)} r^{p-3}|e_3\Psi|^2.
\end{aligned}$$

For $\delta \leq p \leq 2-\delta$, $R \gg \frac{m}{\delta}$ and $\epsilon \ll \delta$ we can absorb all error terms except the last, i.e.,

$$\begin{aligned}
\int_{\mathcal{M}_{\geq R}(\tau_1,\tau_2)} \mathcal{E} &\geq \frac{1}{4}\int_{\mathcal{M}_{\geq R}(\tau_1,\tau_2)} r^{p-1}\Big(p|\check{e}_4(\Psi)|^2 + (2-p)(|\nabla\!\!\!/\Psi|^2 + r^{-2}|\Psi|^2)\Big) \\
&- O(\epsilon)\int_{\mathcal{M}_{\geq R}(\tau_1,\tau_2)} r^{p-3}|e_3\Psi|^2.
\end{aligned}$$

Note also that for all $\delta \leq p \leq 2-\delta$ we have

$$\int_{\mathcal{M}_{\geq R}(\tau_1,\tau_2)} r^{p-3}|e_3\Psi|^2 \lesssim \mathrm{Morr}(\tau_1,\tau_2).$$

Hence, for all $\delta \leq p \leq 2-\delta$ and $R \gg \frac{m}{\delta}$, $\epsilon \ll \delta$,

$$\begin{aligned}
\int_{\mathcal{M}_{\geq R}(\tau_1,\tau_2)} \mathcal{E} &\geq \frac{1}{4}\int_{\mathcal{M}_{\geq R}(\tau_1,\tau_2)} r^{p-1}\Big(p|\check{e}_4(\Psi)|^2 + (2-p)(|\nabla\!\!\!/\Psi|^2 + r^{-2}|\Psi|^2)\Big) \\
&- O(\epsilon)\mathrm{Morr}[\Psi](\tau_1,\tau_2)
\end{aligned}$$

as desired. □

Combining (10.2.15) with Lemma 10.45, we deduce

$$\begin{aligned}
K_{\geq R}(\tau_1,\tau_2) + \dot{B}_{p,R}[\Psi](\tau_1,\tau_2) &\lesssim R^{p+2}\Big(E[\Psi](\tau_1) + J_p[N,\psi](\tau_1,\tau_2) \\
&+ O(\epsilon)\dot{B}_{\delta\,;\,4m_0}[\psi](\tau_1,\tau_2)\Big). \qquad (10.2.16)
\end{aligned}$$

Step 2. (Boundary terms for $r \geq R$.) Recall that, according to Proposition 10.43,

$$\mathcal{P}\cdot e_4 \quad = \quad f_p|\check{e}_4\Psi|^2 - \frac{1}{2}r^{-2}e_4(rf_p|\Psi|^2) + O(\epsilon r^{-1})f_p(|e_4\Psi|^2 + r^{-2}|\Psi|^2),$$

and according to Lemma 10.44

$$\begin{aligned}\mathcal{P}\cdot N_{\Sigma_*} \quad = \quad & \frac{1}{2}f\mathcal{Q}_{34} + \frac{1}{2}f(\check{e}_4\Psi)^2 + \frac{1}{2}\mathrm{div}\,_{\Sigma_*}\Big(r^{-1}f|\Psi|^2\nu_{\Sigma_*}\Big)\\ & + O(mr^{-1}+\epsilon)(|f| + r|f'|)(|e_4\Psi|^2 + |\nabla\!\!\!/\Psi|^2 + r^{-2}|\Psi|^2).\end{aligned}$$

Recalling the definition of

$$K_{\geq R} = \int_{\Sigma_{\geq R}(\tau_2)}\mathcal{P}\cdot e_4 + \int_{\Sigma_*(\tau_1,\tau_2)}\mathcal{P}\cdot N_{\Sigma_*} - \int_{\Sigma_{\geq R}(\tau_1)}\mathcal{P}\cdot e_4$$

we write

$$\begin{aligned}K_{\geq R} \quad = \quad & \int_{\Sigma_{\geq R}(\tau_2)} f_p|\check{e}_4\Psi|^2 + \frac{1}{2}\int_{\Sigma_*(\tau_1,\tau_2)} r^p\Big(\mathcal{Q}_{34} + (\check{e}_4\Psi)^2\Big) - \int_{\Sigma_{\geq R}(\tau_1)} f_p(\check{e}_4\Psi)^2\\ & - \frac{1}{2}\int_{\Sigma_{\geq R}(\tau_2)} r^{-2}e_4(rf_p|\Psi|^2) + \frac{1}{2}\int_{\Sigma_*(\tau_1,\tau_2)}\mathrm{div}\,_{\Sigma_*}\Big(r^{-1}f|\Psi|^2\nu_{\Sigma_*}\Big)\\ & + \frac{1}{2}\int_{\Sigma_{\geq R}(\tau_1)} r^{-2}e_4(rf_p|\Psi|^2)\\ & + \int_{\Sigma_*(\tau_1,\tau_2)} O(mr^{-1}+\epsilon)r^{p-2}(|e_4\Psi|^2 + |\nabla\!\!\!/\Psi|^2 + r^{-2}|\Psi|^2)\\ & + O(\epsilon)\Bigg(\int_{\Sigma_{\geq R}(\tau_2)} r^{p-1}(|e_4\Psi|^2 + r^{-2}|\Psi|^2)\\ & \qquad - \int_{\Sigma_{\geq R}(\tau_1)} r^{p-1}(|e_4\Psi|^2 + r^{-2}|\Psi|^2)\Bigg).\end{aligned}$$

Now, the following integrations by parts hold true:

$$\begin{aligned}& \int_{\Sigma_{\geq R}(\tau)} r^{-2}e_4(rf_p|\Psi|^2)\\ = \quad & \int_{r\geq R}\left(\int_{S_r} r^{-2}e_4(rf_p|\Psi|^2)\right)\frac{1}{e_4(r)}\\ = \quad & \int_{r\geq R}\frac{1}{e_4(r)}e_4\left(\int_{S_r} r^{-1}f_p|\Psi|^2\right) - \int_{\Sigma_{\geq R}(\tau)}\left(e_4\left(r^{-2}\right) + \kappa r^{-2}\right)rf_p|\Psi|^2\\ = \quad & \int_{S_*(\tau)} r^{p-1}|\Psi|^2 - \int_{S_R(\tau)} r^{p-1}|\Psi|^2 - \int_{\Sigma_{\geq R}(\tau)}\left(\kappa - \frac{2e_4(r)}{r}\right)r^{-1}f_p|\Psi|^2\\ = \quad & \int_{S_*(\tau)} r^{p-1}|\Psi|^2 - \int_{S_R(\tau)} r^{p-1}|\Psi|^2 + O(\epsilon)\int_{\Sigma_{\geq R}(\tau)} r^{p-3}|\Psi|^2\end{aligned}$$

and

$$\int_{\Sigma_*(\tau_1,\tau_2)}\mathrm{div}\,_{\Sigma_*}\Big(r^{-1}f|\Psi|^2\nu_{\Sigma_*}\Big) \quad = \quad \int_{S_*(\tau_2)} r^{p-1}|\Psi|^2 - \int_{S_*(\tau_1)} r^{p-1}|\Psi|^2$$

where $S_*(\tau)$ denotes the 2-sphere $\Sigma_* \cap \Sigma(\tau)$. Note that the boundary terms cancel, except the one on $r = R$, and hence

$$\begin{aligned}
K_{\geq R} &= \int_{\Sigma_{\geq R}(\tau_2)} f_p|\check{e}_4\Psi|^2 + \frac{1}{2}\int_{\Sigma_*(\tau_1,\tau_2)} r^p\Big(\mathcal{Q}_{34} + (\check{e}_4\Psi)^2\Big) - \int_{\Sigma_{\geq R}(\tau_1)} f_p|\check{e}_4\Psi|^2 \\
&+ \int_{\Sigma_*(\tau_1,\tau_2)} O(mr^{-1}+\epsilon)r^{p-2}(|e_4\Psi|^2 + |\nabla\!\!\!/\Psi|^2 + r^{-2}|\Psi|^2) \\
&+ O(\epsilon)\Bigg(\int_{\Sigma_{\geq R}(\tau_2)} r^{p-1}(|e_4\Psi|^2 + r^{-2}|\Psi|^2) \\
&\qquad - \int_{\Sigma_{\geq R}(\tau_1)} r^{p-1}(|e_4\Psi|^2 + r^{-2}|\Psi|^2)\Bigg) \\
&+ \frac{1}{2}\int_{S_R(\tau_2)} r^{p-1}|\Psi|^2 - \frac{1}{2}\int_{S_R(\tau_1)} r^{p-1}|\Psi|^2.
\end{aligned}$$

Using

$$\begin{aligned}
\mathcal{Q}_{34} + |\check{e}_4\Psi|^2 &= |\nabla\!\!\!/\Psi|^2 + \frac{4\Upsilon}{r^2}|\Psi|^2 + \left|e_4\Psi + \frac{1}{r}\Psi\right|^2 \\
&= |\nabla\!\!\!/\Psi|^2 + \frac{4\Upsilon}{r^2}|\Psi|^2 + (e_4\Psi)^2 + \frac{1}{r^2}|\Psi|^2 + \frac{2}{r}\Psi\cdot e_4(\Psi) \\
&\geq |\nabla\!\!\!/\Psi|^2 + \frac{4\Upsilon - 3}{r^2}|\Psi|^2 + \frac{2}{3}|e_4\Psi|^2
\end{aligned}$$

and the fact that $4\Upsilon \geq 3 + 2/3$ for $r \geq R$ and R large enough, we infer

$$\begin{aligned}
K_{\geq R} &\geq \frac{1}{2}\left(\int_{\Sigma_{\geq R}(\tau_2)} r^p|\check{e}_4\Psi|^2 - \int_{\Sigma_{\geq R}(\tau_1)} r^p|\check{e}_4\Psi|^2 + \dot{F}_p[\Psi](\tau_1,\tau_2)\right) \\
&+ O(\epsilon)\left(\int_{\Sigma_{\geq R}(\tau_2)} r^{p-3}|\Psi|^2 - \int_{\Sigma_{\geq R}(\tau_1)} r^{p-3}|\Psi|^2\right) \\
&+ \frac{1}{2}\int_{S_R(\tau_2)} r^{p-1}|\Psi|^2 - \frac{1}{2}\int_{S_R(\tau_1)} r^{p-1}|\Psi|^2. \qquad (10.2.17)
\end{aligned}$$

Next, recall that, according to Proposition 10.43, we have

$$\mathcal{P}\cdot e_4 \geq \frac{1}{8}r^{p-2}(p-1)^2|\Psi|^2 - \frac{p}{2}r^{-2}e_4(rf_p|\Psi|^2).$$

We infer

$$\int_{\Sigma_{\geq R}(\tau_2)} \mathcal{P}\cdot e_4 \geq \frac{1}{8}\int_{\Sigma_{\geq R}(\tau_2)} r^{p-2}(p-1)^2|\Psi|^2 - \frac{p}{2}\int_{\Sigma_{\geq R}(\tau_2)} r^{-2}e_4(rf_p|\Psi|^2).$$

Integrating by parts similarly as before, we infer

$$\begin{aligned}\int_{\Sigma_{\geq R}(\tau_2)} \mathcal{P}\cdot e_4 \;\geq\; & \frac{1}{8}\int_{\Sigma_{\geq R}(\tau_2)} r^{p-2}(p-1)^2|\Psi|^2 - \frac{p}{2}\int_{S_*(\tau)} r^{p-1}|\Psi|^2\\ & +\frac{p}{2}\int_{S_R(\tau)} r^{p-1}|\Psi|^2 + O(\epsilon)\int_{\Sigma_{\geq R}(\tau)} r^{p-3}|\Psi|^2.\end{aligned}$$

Arguing as for the proof of (10.2.17) except for the boundary term on $\Sigma_{\geq R}(\tau_2)$ for which we use the above estimate, we deduce

$$\begin{aligned}K_{\geq R} \;\geq\; & \frac{1}{8}\int_{\Sigma_{\geq R}(\tau_2)} r^{p-2}(p-1)^2|\Psi|^2 + \frac{1}{2}\left(-\int_{\Sigma_{\geq R}(\tau_1)} r^p|\check{e}_4\Psi|^2 + \dot{F}_p[\Psi](\tau_1,\tau_2)\right)\\ & + O(\epsilon)\Bigg(\int_{\Sigma_{\geq R}(\tau_2)} r^{p-1}(|e_4\Psi|^2 + r^{-2}|\Psi|^2)\\ & \qquad - \int_{\Sigma_{\geq R}(\tau_1)} r^{p-1}(|e_4\Psi|^2 + r^{-2}|\Psi|^2)\Bigg)\\ & + \frac{1-p}{2}\int_{S_*(\tau_2)} r^{p-1}|\Psi|^2 + \frac{p}{2}\int_{S_R(\tau_2)} r^{p-1}|\Psi|^2 - \frac{1}{2}\int_{S_R(\tau_1)} r^{p-1}|\Psi|^2.\end{aligned}$$

We first focus on the case $\delta\leq p\leq 1-\delta$, in which case the previous estimate yields

$$\begin{aligned}K_{\geq R} \;\geq\; & \frac{\delta^2}{8}\int_{\Sigma_{\geq R}(\tau_2)} r^{p-2}|\Psi|^2 + \frac{1}{2}\left(-\int_{\Sigma_{\geq R}(\tau_1)} r^p|\check{e}_4\Psi|^2 + \dot{F}_p[\Psi](\tau_1,\tau_2)\right)\\ & + O(\epsilon)\left(\int_{\Sigma_{\geq R}(\tau_2)} r^{p-1}(|e_4\Psi|^2 + r^{-2}|\Psi|^2) - \int_{\Sigma_{\geq R}(\tau_1)} r^{p-3}|\Psi|^2\right)\\ & + \frac{1}{2}\int_{S_R(\tau_2)} r^{p-1}|\Psi|^2 - \frac{1}{2}\int_{S_R(\tau_1)} r^{p-1}|\Psi|^2.\end{aligned}$$

Together with (10.2.17) and the fact that $\epsilon\ll\delta^2$ by assumption, we infer in view of the definition of $\dot{E}_{p,R}[\Psi]$ for $\delta\leq p\leq 1-\delta$,

$$\dot{E}_{p,R}[\Psi](\tau_2) + \dot{F}_p[\Psi](\tau_1,\tau_2) \;\lesssim\; K_{\geq R} + \dot{E}_{p,R}[\Psi](\tau_1) + \int_{S_R(\tau_2)} r^{p-1}|\Psi|^2.$$

Together with (10.2.16), we deduce for $\delta\leq p\leq 1-\delta$

$$\begin{aligned}& \dot{E}_{p,R}[\Psi](\tau_2) + \dot{F}_p[\Psi](\tau_1,\tau_2) + \dot{B}_{p,R}[\Psi](\tau_1,\tau_2)\\ \lesssim\; & \dot{E}_{p,R}[\Psi](\tau_1) + \int_{S_R(\tau_2)} r^{p-1}|\Psi|^2 + R^{p+2}\Big(E[\Psi](\tau_1) + J_p[N,\psi](\tau_1,\tau_2)\\ & + O(\epsilon)\dot{B}_{\delta\,;\,4m_0}[\psi](\tau_1,\tau_2)\Big).\end{aligned}$$

In view of the improved Morawetz Theorem 10.1, and thanks also to the term $\dot{B}_{p,R}[\Psi](\tau_1,\tau_2)$ on the left-hand side, we may absorb the term $O(\epsilon)\dot{B}_{\delta\,;\,4m_0}[\psi](\tau_1,\tau_2)$

and obtain

$$\begin{aligned}&\dot{E}_{p,R}[\Psi](\tau_2)+\dot{F}_p[\Psi](\tau_1,\tau_2)+\dot{B}_{p,R}[\Psi](\tau_1,\tau_2)\\ \lesssim\;& R^{p+2}\Big(E_p[\Psi](\tau_1)+J_p[N,\psi](\tau_1,\tau_2)\Big)\end{aligned}\tag{10.2.18}$$

which is the desired estimate in the case $\delta\le p\le 1-\delta$.

Finally, we focus on the remaining case, i.e., $1-\delta\le p\le 2-\delta$. Combining (10.2.17) and (10.2.16), arguing as in the proof of (10.2.18), and in view of the definition of $\dot{E}_{p,R}[\Psi]$ for $1-\delta\le p\le 2-\delta$, we obtain

$$\begin{aligned}&\dot{E}_{p,R}[\Psi](\tau_2)+\dot{F}_p[\Psi](\tau_1,\tau_2)+\dot{B}_{p,R}[\Psi](\tau_1,\tau_2)\\ \lesssim\;& R^{p+2}\left(E_p[\Psi](\tau_1)+J_p[N,\psi](\tau_1,\tau_2)\right)\\ &+O(\epsilon)\int_{\Sigma_{\ge R}(\tau_2)}r^{p-3}|\Psi|^2+\int_{\Sigma_{\ge R}(\tau_2)}r^{-1-\delta}|\Psi|^2\\ \lesssim\;& R^{p+2}\left(E_p[\Psi](\tau_1)+J_p[N,\psi](\tau_1,\tau_2)\right)+E_{1-\delta}[\Psi](\tau_2)\end{aligned}$$

where we also used the fact that $p\le 2-\delta$ so that $p-3\le -1-\delta$. Together with the fact that

$$\dot{E}_{p,R}[\Psi](\tau)\ge \dot{E}_{1-\delta,R}[\Psi](\tau)\text{ for }p\ge 1-\delta$$

and (10.2.18), we infer

$$\dot{E}_{p,R}[\Psi](\tau_2)+\dot{F}_p[\Psi](\tau_1,\tau_2)+\dot{B}_{p,R}[\Psi](\tau_1,\tau_2)\lesssim R^{p+2}\left(E_p[\Psi](\tau_1)+J_p[N,\psi](\tau_1,\tau_2)\right)$$

for all $\delta\le p\le 2-\delta$ as desired. This concludes the proof of Theorem 10.37.

10.3 HIGHER WEIGHTED ESTIMATES

We use a variation of the method of [5] to derive slightly stronger weighted estimates. This allows us to prove Theorem 5.18 for $s=0$ in section 10.4.6. The proof for higher order derivatives $s\le k_{small}+29$ will be provided in section 10.4.6.

As in the previous section we rely on the assumptions (10.2.7)–(10.2.10) to which we add:

RP5. The assumptions **RP0**–**RP4** hold true for one extra derivative with respect to $\mathfrak{d}$.

RP6. $e_4(m)$ satisfies the following improvement of **RP4**:

$$|\mathfrak{d}^{\le 2}e_4(m)|\;\lesssim\;\epsilon w_{2,1}.\tag{10.3.1}$$

10.3.1 Wave equation for $\check{\psi}$

Proposition 10.47. *Assume ψ verifies $\square_2\psi=-\kappa\underline{\kappa}\psi+N$. Then $\check{\psi}=f_2\check{e}_4\psi$ verifies:*

1. *In the region* $r \geq 6m_0$,

$$(\square_2 + \kappa\underline{\kappa})\check{\psi} = r^2\left(e_4(N) + \frac{3}{r}N\right) + \frac{2}{r}\left(1 - \frac{3m}{r}\right)e_4\check{\psi} + O(r^{-2})\mathfrak{d}^{\leq 1}\psi$$
$$+r\Gamma_b e_4\mathfrak{d}\psi + \mathfrak{d}^{\leq 1}(\Gamma_b)\mathfrak{d}^{\leq 1}\psi + r\mathfrak{d}^{\leq 1}(\Gamma_g)e_3\psi + \mathfrak{d}^{\leq 1}(\Gamma_g)\mathfrak{d}^2\psi.$$

2. *In the region* $4m_0 \leq r \leq 6m_0$,

$$(\square_2 + \kappa\underline{\kappa})\check{\psi} = f_2\left(e_4(N) + \frac{3}{r}N\right) + O(1)\mathfrak{d}^{\leq 2}\psi.$$

The proof of Proposition 10.47 is postponed to section D.4.

10.3.2 The r^p-weighted estimates for $\check{\psi}$

The goal of this section is to prove Theorem 5.18 in the case $s = 0$. The proof for higher order derivatives $s \leq k_{small} + 29$ will be provided in section 10.4.6.

Proof of Theorem 5.18 in the case $s = 0$. We write, in accordance with Proposition 10.47,

$$\square\check{\psi} - V\check{\psi} = \check{N} + f_2\left(e_4 + \frac{3}{r}\right)N$$

where

$$\check{N} = \begin{cases} \frac{2}{r}\left(1 - \frac{3m}{r}\right)e_4\check{\psi} + O(r^{-2})\mathfrak{d}^{\leq 1}\psi + r\Gamma_b e_4\mathfrak{d}\psi \\ +\mathfrak{d}^{\leq 1}(\Gamma_b)\mathfrak{d}^{\leq 1}\psi + r\mathfrak{d}^{\leq 1}(\Gamma_g)e_3\psi + \mathfrak{d}^{\leq 1}(\Gamma_g)\mathfrak{d}^2\psi, & r \geq 6m_0, \\ \\ O(1)\mathfrak{d}^{\leq 2}\psi, & 4m_0 \leq r \leq 6m_0. \end{cases} \tag{10.3.2}$$

We apply the first part of Proposition 10.43 to ψ replaced by $\check{\psi}$. This yields, using also (10.1.13),

$$\mathbf{Div}\mathcal{P}_q(\check{\psi}) = \mathcal{E}_q(\check{\psi}) + f_q\check{e}_4\check{\psi}\cdot\check{N} + f_q\check{e}_4\check{\psi}\cdot f_2\left(e_4 + \frac{3}{r}\right)N,$$

where, with $f = f_q$, $X_q = f_q e_4$, $w_q = \frac{2f_q}{r}$, $M_q = 2r^{-1}f_q' e_4$,

$$\begin{aligned}
\mathcal{E}_q(\check{\psi}) &= \mathcal{E}[X_q, w_q, M_q] = \frac{1}{2}f_q'|\check{e}_4(\check{\psi})|^2 + \frac{1}{2}\left(\frac{2f_q}{r} - f_q'\right)\mathcal{Q}_{34}(\check{\psi}) + \mathrm{Err}_q(\check{\psi}),\\
\mathrm{Err}_q(\check{\psi}) : &= \mathrm{Err}\left(\epsilon, \frac{m}{r}; f_q\right)[\check{\psi}]\\
&= O\left(\frac{m}{r^2}\right)r^q|e_4\check{\psi}|^2 + O\left(\frac{m}{r^4} + \epsilon w_{3,1}\right)r^q|\check{\psi}|^2\\
&+ O(\epsilon)w_{1,1}r^q\left(|e_4\check{\psi}|^2 + |\nabla\check{\psi}|^2 + r^{-2}|\check{\psi}|^2\right)\\
&+ O(\epsilon)w_{2,1/2}r^q\Big(\left(|e_4\check{\psi}| + r^{-1}|\nabla\check{\psi}|\right)|e_3\check{\psi}| + |\nabla\check{\psi}|^2 + r^{-2}|\check{\psi}|^2\Big),\\
\mathcal{P}_k(\check{\psi}) &= \mathcal{P}[X_q, w_q, M_q](\check{\psi}).
\end{aligned}$$

We then integrate on the domain $\mathcal{M}(\tau_1, \tau_2)$ to derive, exactly as in the proof of Theorem 10.37 (see section 10.2.3),

$$\begin{aligned}&\int_{\Sigma(\tau_2)} \mathcal{P}_q \cdot e_4 + \int_{\Sigma_*(\tau_1,\tau_2)} \mathcal{P}_q \cdot N_{\Sigma_*} + \int_{\mathcal{M}(\tau_1,\tau_2)} (\mathcal{E}_q + f_q \check{e}_4 \check{\psi} \check{N}) \\ &= \int_{\Sigma(\tau_1)} \mathcal{P}_q \cdot e_4 - \int_{\mathcal{M}(\tau_1,\tau_2)} f_q \check{e}_4 \check{\psi} \cdot f_2 \left(e_4 + \frac{3}{r} \right) N. \qquad (10.3.3)\end{aligned}$$

All terms can be treated exactly as in the proof of Theorem 10.37, except for the bulk term, i.e., we obtain the following analog of (10.2.1):

$$\begin{aligned}&\dot{E}_{q\,;\,R}[\check{\psi}](\tau_2) + \int_{\mathcal{M}(\tau_1,\tau_2)} (\mathcal{E}_q + r^q \check{e}_4(\check{\psi})\check{N}) + \dot{F}_q[\check{\psi}](\tau_1,\tau_2) \\ &\lesssim E_q[\check{\psi}](\tau_1) + J_q\left[\check{\psi}, f_2\left(e_4 + \frac{3}{r}\right)N\right](\tau_1,\tau_2).\end{aligned}$$

Since all terms for $r \leq R$ can be controlled by one derivative of ψ, we infer

$$\begin{aligned}&\dot{E}_q[\check{\psi}](\tau_2) + \mathrm{Morr}[\check{\psi}](\tau_1,\tau_2) + \int_{\mathcal{M}_{\geq R}(\tau_1,\tau_2)} (\mathcal{E}_q + r^q \check{e}_4(\check{\psi})\check{N}) + \dot{F}_q[\check{\psi}](\tau_1,\tau_2) \\ &\lesssim E_q[\check{\psi}](\tau_1) + \check{J}_q[\check{\psi}, N](\tau_1,\tau_2) + R^{q+3}(E^1[\psi](\tau_2) + \mathrm{Morr}^1[\psi](\tau_1,\tau_2)). \quad (10.3.4)\end{aligned}$$

Also, since $\delta \leq \max(q,\delta) \leq 1-\delta$, we have in view of Theorem 5.17 in the case $s = 1$[15]

$$\begin{aligned}&\sup_{\tau\in[\tau_1,\tau_2]} E^1_{\max(q,\delta)}[\psi](\tau) + B^1_{\max(q,\delta)}[\psi](\tau_1,\tau_2) \\ &\lesssim E^1_{\max(q,\delta)}[\psi](\tau_1) + J^1_{\max(q,\delta)}[\psi, N](\tau_1,\tau_2). \qquad (10.3.5)\end{aligned}$$

In view of (10.3.4) and (10.3.5), it thus only remains to estimate the integral

$$\int_{\mathcal{M}_{\geq R}(\tau_1,\tau_2)} (\mathcal{E}_q + r^q \check{e}_4(\check{\psi})\check{N}),$$

i.e., we need to derive the analog of (10.2.15) used in the proof of Theorem 10.37. This is achieved in Proposition 10.48 below, which together with (10.3.4) and (10.3.5) immediately yields the proof of Theorem 5.18 in the case $s = 0$. □

Proposition 10.48. *The following estimate holds true:*

$$\begin{aligned}\int_{\mathcal{M}_{\geq R}(\tau_1,\tau_2)} (\mathcal{E}_q + r^q \check{e}_4(\check{\psi})\check{N}) \geq \frac{1}{8} \int_{\mathcal{M}_{\geq R}(\tau_1,\tau_2)} & r^{q-1}\Big((2+q)|\check{e}_4\check{\psi}|^2 + (2-q)|\nabla\check{\psi}|^2 \\ & + 2r^{-2}|\check{\psi}|^2\Big) - O(\epsilon) \sup_{\tau_1\leq\tau\leq\tau_2} \dot{E}_{q,R}[\check{\psi}](\tau) \\ & - O(1)\left(E^1_{\max(q,\delta)}[\psi](\tau_1) + J^1_{\max(q,\delta)}[\psi, N]\right).\end{aligned}$$

[15]The proof of Theorem 5.17 for higher derivatives $s \geq 1$, even though proved later in section 10.4.5, is in fact independent of the proof of Theorem 5.18 and can thus be invoked here.

We now focus on the proof of Proposition 10.48. In view of the definition of $\check{N}$, we have for $r \geq R$

$$\begin{aligned}
\check{N} &= A_0 + A_1 + A_2, \\
A_0 &= \frac{2}{r} e_4 \check{\psi} = \frac{2}{r}(\check{e}_4 \check{\psi} - r^{-1} \check{\psi}), \\
A_1 &= -\frac{6m}{r^2} e_4 \check{\psi} + O(r^{-2}) \mathfrak{d}^{\leq 1}\psi, \\
A_2 &= \mathrm{Err}[\square_{\mathbf{g}} \check{\psi}], \\
\mathrm{Err}[\square_{\mathbf{g}} \check{\psi}] &= r^2 \Gamma_g e_4 \mathfrak{d}\psi + r \mathfrak{d}^{\leq 1}(\Gamma_g) \mathfrak{d}^{\leq 1}\psi + \mathfrak{d}^{\leq 1}(\Gamma_g)\mathfrak{d}^2\psi.
\end{aligned}$$

Also, recall that we have for $r \geq R$

$$\mathcal{E}_q(\check{\psi}) = \mathcal{E}[X_q, w_q, M_q] = \frac{q}{2} r^{q-1} |\check{e}_4(\check{\psi})|^2 + \frac{2-q}{2} r^{q-1} \mathcal{Q}_{34}(\check{\psi}) + \mathrm{Err}_q(\check{\psi}).$$

Consequently, we write

$$\begin{aligned}
(\mathcal{E}_q + r^q \check{e}_4(\check{\psi})\check{N}) &= I_0 + I_1 + I_2, \\
I_0 : &= \frac{1}{2} r^{q-1} \left(q|\check{e}_4 \check{\psi}|^2 + (2-q)|\nabla \check{\psi}|^2 + 4(2-q) r^{-2} \check{\psi}^2 \right) \\
&\quad + 2r^{q-1} \check{e}_4 \check{\psi}(\check{e}_4 \check{\psi} - r^{-1} \check{\psi}) \\
&= \frac{1}{2} r^{q-1} \Big((q+4)|\check{e}_4 \check{\psi}|^2 + (2-q)|\nabla \check{\psi}|^2 \\
&\quad + 4(2-q) r^{-2} \check{\psi}^2 - 4r^{-1} \check{e}_4 \check{\psi} \check{\psi} \Big), \qquad (10.3.6) \\
I_1 : &= r^{q-2} \check{e}_4(\check{\psi}) \left[-6m e_4 \check{\psi} + O(1) \mathfrak{d}^{\leq 1}\psi \right] + O\left(\frac{m}{r}\right) r^{q-3} (\check{\psi})^2, \\
I_2 : &= \mathrm{Err}_q(\check{\psi}) + r^q \check{e}_4(\check{\psi}) A_2.
\end{aligned}$$

We will rely on the following two lemmas.

Lemma 10.49. *The following lower bound estimate holds true for $q \leq 1 - \delta$ and $r \geq R$, where R is sufficiently large:*

$$\begin{aligned}
I_0 + I_1 &\geq \frac{1}{4} r^{q-1} \left((2+q)|\check{e}_4 \check{\psi}|^2 + (2-q)|\nabla \check{\psi}|^2 + 2r^{-2}|\check{\psi}|^2 \right) \\
&\quad - O(1) r^{q-3} (\mathfrak{d}^{\leq 1}\psi)^2. \qquad (10.3.7)
\end{aligned}$$

Lemma 10.50. *The following estimate holds true for the error term I_2:*

$$\begin{aligned}
\int_{\mathcal{M}_{\geq R}(\tau_1, \tau_2)} |I_2| &\lesssim \epsilon \sup_{\tau_1 \leq \tau \leq \tau_2} \dot{E}_{q,R}[\check{\psi}](\tau) + \left(\frac{m_0}{R} + \epsilon\right) \dot{B}_{q,R}[\check{\psi}](\tau_1, \tau_2) \\
&\quad + \epsilon \left(\sup_{\tau_1 \leq \tau \leq \tau_2} E_q^1[\psi](\tau) + B_q^1[\psi](\tau_1, \tau_2) + J_q[\psi, N](\tau_1, \tau_2) \right).
\end{aligned}$$

We postpone the proof of Lemma 10.49 and Lemma 10.50 to finish the proof of Proposition 10.48.

Proof of Proposition 10.48. In view of Lemma 10.49 and Lemma 10.50, we have

$$\begin{aligned}
&\int_{\mathcal{M}_{\geq R}(\tau_1,\tau_2)} (\mathcal{E}_q + r^q \check{e}_4(\check{\psi})\check{N}) \\
= &\int_{\mathcal{M}_{\geq R}(\tau_1,\tau_2)} (I_0 + I_1) + \int_{\mathcal{M}_{\geq R}(\tau_1,\tau_2)} I_2 \\
\geq &\frac{1}{4}\int_{\mathcal{M}_{\geq R}(\tau_1,\tau_2)} r^{q-1}\left((2+q)|\check{e}_4\check{\psi}|^2 + (2-q)|\nabla\check{\psi}|^2 + 2r^{-2}|\check{\psi}|^2\right) \\
- &O(1)\int_{\mathcal{M}_{\geq R}(\tau_1,\tau_2)} r^{q-3}(\mathfrak{d}^{\leq 1}\psi)^2 \\
- &O(\epsilon)\sup_{\tau_1\leq\tau\leq\tau_2} \dot{E}_{q,R}[\check{\psi}](\tau) + O\left(\frac{m_0}{R}+\epsilon\right)\dot{B}_{q,R}[\check{\psi}](\tau_1,\tau_2) \\
- &O(\epsilon)\left(\sup_{\tau_1\leq\tau\leq\tau_2} E^1_q[\psi](\tau) + B^1_q[\psi](\tau_1,\tau_2) + J_q[\psi,N](\tau_1,\tau_2)\right)
\end{aligned}$$

so that, since $1-\delta < q \leq 1-\delta$, and for R sufficiently large and small[16] ϵ,

$$\begin{aligned}
&\int_{\mathcal{M}_{\geq R}(\tau_1,\tau_2)} (\mathcal{E}_q + r^q \check{e}_4(\check{\psi})\check{N}) \\
\geq &\frac{1}{8}\int_{\mathcal{M}_{\geq R}(\tau_1,\tau_2)} r^{q-1}\left((2+q)|\check{e}_4\check{\psi}|^2 + (2-q)|\nabla\check{\psi}|^2 + 2r^{-2}|\check{\psi}|^2\right) \\
- &O(\epsilon)\sup_{\tau_1\leq\tau\leq\tau_2} \dot{E}_{q,R}[\check{\psi}](\tau) \\
- &O(\epsilon)\left(\sup_{\tau_1\leq\tau\leq\tau_2} E^1_q[\psi](\tau) + B^1_q[\psi](\tau_1,\tau_2) + J_q[\psi,N](\tau_1,\tau_2)\right).
\end{aligned}$$

In view of (10.3.5), we infer

$$\begin{aligned}
\int_{\mathcal{M}_{\geq R}(\tau_1,\tau_2)} (\mathcal{E}_q + r^q \check{e}_4(\check{\psi})\check{N}) \geq \frac{1}{8}\int_{\mathcal{M}_{\geq R}(\tau_1,\tau_2)} & r^{q-1}\Big((2+q)|\check{e}_4\check{\psi}|^2 \\
&+ (2-q)|\nabla\check{\psi}|^2 + 2r^{-2}|\check{\psi}|^2\Big) \\
&- O(\epsilon)\sup_{\tau_1\leq\tau\leq\tau_2} \dot{E}_{q,R}[\check{\psi}](\tau) \\
&- O(1)\left(E^1_{\max(q,\delta)}[\psi](\tau_1) + J^1_{\max(q,\delta)}[\psi,N]\right)
\end{aligned}$$

which concludes the proof. □

It finally remains to prove Lemma 10.49 and Lemma 10.50.

[16]Using also the second bound on Morr from Theorem 10.1 and the bound on $\dot{B}_{\delta,4m_0}$ from the r^p estimates of Theorem 10.37. See also Remark 10.38.

Proof of Lemma 10.49. Note that

$$\begin{aligned} & (q+4)|\check{e}_4 \check{\psi}|^2 - 4r^{-1}(\check{e}_4 \check{\psi})\check{\psi} + 4(2-q)r^{-2}|\check{\psi}|^2 \\ = & (q+2)|\check{e}_4 \check{\psi}|^2 + (6-4q)r^{-2}|\check{\psi}|^2 + 2\left(\check{e}_4 \check{\psi} - r^{-1}\check{\psi}\right)^2 \\ \geq & (q+2)|\check{e}_4 \check{\psi}|^2 + (6-4q)r^{-2}|\check{\psi}|^2 \\ \geq & (q+2)|\check{e}_4 \check{\psi}|^2 + 2r^{-2}|\check{\psi}|^2, \end{aligned}$$

where we used the fact that $q \leq 1-\delta$. Hence,

$$I_0 \geq \frac{1}{2} r^{q-1}\left((2+q)|\check{e}_4 \check{\psi}|^2 + (2-q)|\nabla \check{\psi}|^2 + 2r^{-2}|\check{\psi}|^2\right).$$

We also have

$$I_1 \leq O\left(\frac{m}{r}\right) r^{q-1}\left(|\check{e}_4 \check{\psi}|^2 + r^{-2}|\check{\psi}|^2\right) + O(1)\left(r^{q-1}(\check{e}_4 \check{\psi})^2\right)^{\frac{1}{2}}\left(r^{q-3}(\mathfrak{d}^{\leq 1}\psi)^2\right)^{\frac{1}{2}}.$$

Thus if m_0/R is sufficiently small, and since $q \leq 1-\delta$, we deduce, for $r \geq R$,

$$I_0 + I_1 \geq \frac{1}{4} r^{q-1}\left((2+q)|\check{e}_4 \check{\psi}|^2 + (2-q)|\nabla \check{\psi}|^2 + 2r^{-2}|\check{\psi}|^2\right) - O(1) r^{q-3}(\mathfrak{d}^{\leq 1}\psi)^2$$

as desired. □

Proof of Lemma 10.50. Recall that

$$\begin{aligned} I_2 &= \mathrm{Err}_q(\check{\psi}) + r^q \check{e}_4(\check{\psi}) A_2, \\ A_2 &= \mathrm{Err}[\square_{\mathbf{g}} \check{\psi}], \\ \mathrm{Err}[\square_{\mathbf{g}} \check{\psi}] &= r\Gamma_b e_4 \mathfrak{d}\psi + \mathfrak{d}^{\leq 1}(\Gamma_b)\mathfrak{d}^{\leq 1}\psi + r\mathfrak{d}^{\leq 1}(\Gamma_g) e_3 \psi + \mathfrak{d}^{\leq 1}(\Gamma_g)\mathfrak{d}^2\psi, \\ \mathrm{Err}_q(\check{\psi}) &= O\left(\frac{m}{r^2}\right) r^q |e_4 \check{\psi}|^2 + O\left(\frac{m}{r^4} + \epsilon w_{3,1}\right) r^q |\check{\psi}|^2 \\ &+ O(\epsilon) w_{1,1} r^q \left(|e_4 \check{\psi}|^2 + |\nabla \check{\psi}|^2 + r^{-2}|\check{\psi}|^2\right) \\ &+ O(\epsilon) w_{2,1/2} r^q \Big(\left(|e_4 \check{\psi}| + r^{-1}|\nabla \check{\psi}|\right)|e_3 \check{\psi}| + |\nabla \check{\psi}|^2 + r^{-2}|\check{\psi}|^2\Big). \end{aligned}$$

Hence,

$$\begin{aligned} |I_2| \lesssim\ & \epsilon r^q |\check{e}_4(\check{\psi})| \Big(\tau^{-1-\delta_{dec}}\left(|\check{e}_4 \mathfrak{d}^{\leq 1}\psi| + |\mathfrak{d}^{\leq 1}\psi|\right) \\ & + r^{-1}\tau^{-\frac{1}{2}-\delta_{dec}}\left(|e_3\psi| + r^{-1}|\mathfrak{d}^2\psi|\right)\Big) \\ & + \left(\frac{m}{r} + \epsilon\right) r^{q-1}\left(|\check{e}_4 \check{\psi}|^2 + |\nabla \check{\psi}|^2 + r^{-2}|\check{\psi}|^2\right) + \epsilon r^{q-2}\tau^{-\frac{1}{2}-\delta_{dec}}|e_4 \check{\psi}||e_3 \check{\psi}| \\ & + \epsilon r^{q-3}|\nabla \check{\psi}||e_3 \check{\psi}|. \end{aligned}$$

This yields, using $q \le 1-\delta$,

$$
\begin{aligned}
&\int_{\mathcal{M}_{\ge R}(\tau_1,\tau_2)} |I_2| \\
\lesssim\;& \epsilon\left(\sup_{\tau_1\le\tau\le\tau_2} \dot{E}_{q,R}[\check{\psi}](\tau)\right)^{\frac{1}{2}}\left(\sup_{\tau_1\le\tau\le\tau_2} E^1_q[\psi](\tau)+B^1_q[\psi](\tau_1,\tau_2)\right)^{\frac{1}{2}} \\
&+\left(\frac{m_0}{R}+\epsilon\right)\dot{B}_{q,R}[\check{\psi}](\tau_1,\tau_2) \\
&+\epsilon\left(\sup_{\tau_1\le\tau\le\tau_2} \dot{E}_{q,R}[\check{\psi}](\tau)+\dot{B}_q[\check{\psi}](\tau_1,\tau_2)\right)^{\frac{1}{2}}\left(\int_{\mathcal{M}_{\ge R}(\tau_1,\tau_2)} r^{q-4}|e_3\,\check{\psi}|^2\right)^{\frac{1}{2}} \\
\lesssim\;& \epsilon\sup_{\tau_1\le\tau\le\tau_2} \dot{E}_{q,R}[\check{\psi}](\tau)+\left(\frac{m_0}{R}+\epsilon\right)\dot{B}_{q,R}[\check{\psi}](\tau_1,\tau_2) \\
&+\epsilon\left(\sup_{\tau_1\le\tau\le\tau_2} E^1_q[\psi](\tau)+B^1_q[\psi](\tau_1,\tau_2)\right)+\epsilon\int_{\mathcal{M}_{\ge R}(\tau_1,\tau_2)} r^{q-4}|e_3\,\check{\psi}|^2.
\end{aligned}
$$

Next, we estimate the term involving $e_3\,\check{\psi}$. For this we need to appeal to the formula in Lemma 5.20 which we recall below:

$$
\square_2\psi = -e_3e_4\psi+\mathbb{A}_2\psi+\left(2\underline{\omega}-\frac{1}{2}\underline{\kappa}\right)e_4\psi-\frac{1}{2}\kappa e_3\psi+2\eta e_\theta\psi.
$$

We have for $r \ge 6m_0$

$$
\begin{aligned}
e_3\,\check{\psi} &= e_3(re_4(r\psi)) = re_3(re_4\psi)+e_3(r)e_4(r\psi) \\
&= r^2e_3e_4\psi+2re_3(r)e_4\psi+e_3(r)e_4(r)\psi \\
&= r^2\left(-\square_2\psi+\mathbb{A}_2\psi+\left(2\underline{\omega}-\frac{1}{2}\underline{\kappa}\right)e_4\psi-\frac{1}{2}\kappa e_3\psi+2\eta e_\theta\psi\right) \\
&\quad +2re_3(r)e_4\psi+e_3(r)e_4(r)\psi
\end{aligned}
$$

so that

$$
|e_3\,\check{\psi}| \lesssim r^2|N|+r|e_3\psi|+|\mathfrak{d}^{\le 2}\psi|
$$

and hence

$$
\int_{\mathcal{M}_{\ge R}(\tau_1,\tau_2)} r^{q-4}|e_3\,\check{\psi}|^2 \lesssim \int_{\mathcal{M}_{\ge R}(\tau_1,\tau_2)} r^{q-4}\Big(r^4|N|^2+r^2|e_3\psi|^2+|\mathfrak{d}^{\le 2}\psi|^2\Big).
$$

Since $q \le 1-\delta$, we infer

$$
\begin{aligned}
\int_{\mathcal{M}_{\ge R}(\tau_1,\tau_2)} r^{q-4}|e_3\,\check{\psi}|^2 &\lesssim \int_{{}^{(tr\not{p})}\mathcal{M}(\tau_1,\tau_2)} r^{1-\delta}|N|^2+B^1_q[\psi](\tau_1,\tau_2) \\
&\lesssim J_q[\psi,N](\tau_1,\tau_2)+B^1_q[\psi](\tau_1,\tau_2)
\end{aligned}
$$

and thus

$$\begin{aligned}\int_{\mathcal{M}_{\geq R}(\tau_1,\tau_2)}|I_2| \quad\lesssim\quad & \epsilon \sup_{\tau_1\leq\tau\leq\tau_2}\dot{E}_{q,R}[\check{\psi}](\tau)+\Big(\frac{m_0}{R}+\epsilon\Big)\dot{B}_{q,R}[\check{\psi}](\tau_1,\tau_2)\\ &+\epsilon\Big(\sup_{\tau_1\leq\tau\leq\tau_2}E^1_q[\psi](\tau)+B^1_q[\psi](\tau_1,\tau_2)\Big)\\ &+\epsilon\int_{\mathcal{M}_{\geq R}(\tau_1,\tau_2)}r^{q-4}|e_3\check{\psi}|^2\\ \lesssim\quad & \epsilon \sup_{\tau_1\leq\tau\leq\tau_2}\dot{E}_{q,R}[\check{\psi}](\tau)+\Big(\frac{m_0}{R}+\epsilon\Big)\dot{B}_{q,R}[\check{\psi}](\tau_1,\tau_2)\\ &+\epsilon\Big(\sup_{\tau_1\leq\tau\leq\tau_2}E^1_q[\psi](\tau)+B^1_q[\psi](\tau_1,\tau_2)+J_q[\psi,N](\tau_1,\tau_2)\Big)\end{aligned}$$

which concludes the proof of Lemma 10.50. □

10.4 HIGHER DERIVATIVE ESTIMATES

We have proved, respectively in section 10.2 and section 10.3.2, Theorem 5.17 on basic weighted estimates (see Remark 10.38) and Theorem 5.18 on higher weighted estimates only in the case $s = 0$. In this section, we conclude the proof of these theorems by recovering higher order derivatives[17] one by one.

10.4.1 Basic assumptions

Recall that any Ricci coefficient either belongs to Γ_g or Γ_b, where Γ_g and Γ_b are defined in section 5.1.2. We make use of the following non-sharp consequence of the estimates of Lemma 5.1. We assume, concerning the Ricci coefficients,

$$\begin{aligned}|\mathfrak{d}^k(\Gamma_g)| &\lesssim \frac{\epsilon}{r^2u_{trap}^{1+\delta_{dec}-2\delta_0}} && \text{for } k\leq k_{small}+30,\\ |\mathfrak{d}^k(\Gamma_b)| &\lesssim \frac{\epsilon}{ru_{trap}^{1+\delta_{dec}-2\delta_0}} && \text{for } k\leq k_{small}+30,\\ |\mathfrak{d}^k(\alpha,\beta,\check{\rho})| &\lesssim \frac{\epsilon}{r^3u_{trap}^{1+\delta_{dec}-2\delta_0}} && \text{for } k\leq k_{small}+30,\\ |\mathfrak{d}^k\underline{\alpha}|+r|\mathfrak{d}^k\underline{\beta}| &\lesssim \frac{\epsilon}{ru_{trap}^{1+\delta_{dec}-2\delta_0}} && \text{for } k\leq k_{small}+30,\end{aligned}$$

where we recall that δ_{dec} and δ_0 are such that we have in particular $0 < 2\delta_0 < \delta_{dec}$.

10.4.2 Strategy for recovering higher order derivatives

So far, we have proved Theorem 5.17 in the case $s = 0$[18] in section 10.2, and Theorem 5.18 on higher weighted estimates in the case $s = 0$ in section 10.3. We

[17] Respectively $s \leq k_{small}+30$ in the case of Theorem 5.17, and $s \leq k_{small}+29$ in the case of Theorem 5.18.

[18] Recall that Theorem 5.17 in the case $s = 0$ is obtained as a consequence of Theorem 10.1 on Morawetz and energy estimates, and Theorem 10.37 on r^p-weighted estimates, see Remark 10.38.

now conclude the proof of these theorems by recovering higher order derivatives one by one. Since going from $s=0$ to $s=1$ is analogous to going from s to $s+1$, we will in fact consider only the former. More precisely, we assume the following bounds proved respectively in section 10.2 and section 10.3:

$$\begin{aligned}&\sup_{\tau\in[\tau_1,\tau_2]} E_p[\psi](\tau)+B_p[\psi](\tau_1,\tau_2)+F_p[\psi](\tau_1,\tau_2)\\ &\lesssim\ E_p[\psi](\tau_1)+J_p[\psi,N](\tau_1,\tau_2),\end{aligned} \tag{10.4.1}$$

and

$$\begin{aligned}\sup_{\tau\in[\tau_1,\tau_2]} E_q[\check{\psi}](\tau)+B_q[\check{\psi}](\tau_1,\tau_2) \ \lesssim\ & E_q[\check{\psi}](\tau_1)+\check{J}_q[\check{\psi},N](\tau_1,\tau_2)\\ &+E^1_{\max(q,\delta)}[\psi](\tau_1)+J^1_{\max(q,\delta)}[\psi,N],\end{aligned} \tag{10.4.2}$$

and our goal is to prove the corresponding estimates for $s=1$. We will proceed as follows:

1. We first commute the wave equation for ψ and $\check{\psi}$ with T and derive (10.4.1) for $T\psi$ instead of ψ, and (10.4.2) for $T\check{\psi}$ instead of $\check{\psi}$.
2. We then commute the wave equation for ψ and $\check{\psi}$ with $r\,\mathfrak{d}\!\!/_2$ and derive (10.4.1) for $r\,\mathfrak{d}\!\!/_2\psi$ instead of ψ, and (10.4.2) for $r\,\mathfrak{d}\!\!/_2\check{\psi}$ instead of $\check{\psi}$.
3. We then use the wave equation satisfied by ψ to derive an estimate for $R^2\psi$ in $r\leq 6m_0$[19] with a degeneracy at $r=3m$.
4. We then commute the wave equation for ψ with R and remove the degeneracy at $r=3m$ for $R^2\psi$.
5. We then commute the wave equation for ψ with the redshift vectorfield $Y_{\mathcal{H}}$ and derive (10.4.1) for $Y_{\mathcal{H}}\psi$ instead of ψ.
6. We then commute the wave equation for ψ and $\check{\psi}$ with f_1e_4 and derive (10.4.1) for $re_4\psi$ instead of ψ, and (10.4.2) for $f_1e_4\check{\psi}$ instead of $\check{\psi}$, where $f_1=r$ for $r\geq 6m_0$ and $f_1=0$ for $r\leq 4m_0$.
7. We finally gather all estimates and conclude.

We will follow the above strategy in section 10.4.5 to prove Theorem 5.17, and in section 10.4.6 to prove Theorem 5.18. To this end, we first derive several commutator identities and estimates.

10.4.3 Commutation formulas with the wave equation

10.4.3.1 Commutation with T

Lemma 10.51. *We have, schematically, the following commutator formulae:*

$$[T,e_4]=\Gamma_g\mathfrak{d},\quad [T,e_3]=\Gamma_b\mathfrak{d},\quad [T,\mathfrak{d}\!\!/_k]=\Gamma_b\mathfrak{d}+\Gamma_b,\quad [T,\mathfrak{d}\!\!/_k^\star]=\Gamma_b\mathfrak{d}+\Gamma_b.$$

Proof. Recall that we have

$$[e_3,e_4]\ =\ 2\underline{\omega}e_4-2\omega e_3+2(\eta-\underline{\eta})e_\theta.$$

[19]Note that any finite region in r strictly containing the trapping region would suffice.

We infer

$$
\begin{aligned}
2[T, e_4] &= [e_4 + \Upsilon e_3, e_4] \\
&= \Upsilon[e_3, e_4] - e_4(\Upsilon)e_3 \\
&= \Upsilon\left(2\underline{\omega}e_4 - 2\omega e_3 + 2(\eta - \underline{\eta})e_\theta\right) - e_4(\Upsilon)e_3 \\
&= \Upsilon\left(2\underline{\omega}e_4 - 2\left(\omega + \frac{1}{2}\Upsilon^{-1}e_4(\Upsilon)\right)e_3 + 2(\eta - \underline{\eta})e_\theta\right) \\
&= (r^{-1}\Gamma_b + \Gamma_g)\mathfrak{d} \\
&= \Gamma_g\mathfrak{d},
\end{aligned}
$$

and

$$
\begin{aligned}
2[T, e_3] &= [e_4 + \Upsilon e_3, e_3] \\
&= [e_4, e_3] - e_3(\Upsilon)e_3 \\
&= -2\underline{\omega}e_4 + 2\omega e_3 - 2(\eta - \underline{\eta})e_\theta - e_3(\Upsilon)e_3 \\
&= -2\underline{\omega}e_4 + 2\left(\omega - \frac{1}{2}e_3(\Upsilon)\right)e_3 - 2(\eta - \underline{\eta})e_\theta \\
&= -2\underline{\omega}e_4 + 2\left(\omega + \frac{1}{2}\Upsilon^{-1}e_4(\Upsilon) - \frac{1}{2\Upsilon}T(\Upsilon)\right)e_3 - 2(\eta - \underline{\eta})e_\theta \\
&= (\Gamma_g + \Gamma_b)\mathfrak{d} \\
&= \Gamma_b\mathfrak{d}.
\end{aligned}
$$

Next, recall in view of Lemma 2.54, the following commutation formulae for reduced scalars.

1. If $f \in \mathfrak{s}_k$,

$$
\begin{aligned}
[\not{d}_k, e_3]f &= \frac{1}{2}\underline{\kappa}\,\not{d}_k f + \underline{Com}_k(f), \\
\underline{Com}_k(f) &= -\frac{1}{2}\underline{\vartheta}\,\not{d}_{k+1}^\star f + (\zeta - \eta)e_3 f - k\eta e_3\Phi f - \underline{\xi}(e_4 f + k e_4(\Phi)f) - k\underline{\beta}f, \\
[\not{d}_k, e_4] &= \frac{1}{2}\kappa\,\not{d}_k f + \mathrm{Com}_k(f), \\
\mathrm{Com}_k(f) &= -\frac{1}{2}\vartheta\,\not{d}_{k+1}^\star f - (\zeta + \underline{\eta})e_4 f - k\underline{\eta}e_4\Phi f - \xi(e_3 f + k e_3(\Phi)f) - k\beta f.
\end{aligned}
$$

2. If $f \in \mathfrak{s}_{k-1}$,

$$
\begin{aligned}
[\not{d}_k^\star, e_3]f &= \frac{1}{2}\underline{\kappa}\,\not{d}_k^\star f + \underline{Com}_k^*(f), \\
\underline{Com}_k^*(f) &= -\frac{1}{2}\underline{\vartheta}\,\not{d}_{k-1} f - (\zeta - \eta)e_3 f - (k-1)\eta e_3\Phi f + \underline{\xi}(e_4 f - (k-1)e_4(\Phi)f) \\
&\quad - (k-1)\underline{\beta}f, \\
[\not{d}_k^\star, e_4]f &= \frac{1}{2}\kappa\,\not{d}_k f + \mathrm{Com}_k^*(f), \\
\mathrm{Com}_k^*(f) &= -\frac{1}{2}\vartheta\,\not{d}_{k-1} f + (\zeta + \underline{\eta})e_4 f - (k-1)\underline{\eta}e_4\Phi f + \xi(e_3 f - (k-1)e_3(\Phi)f) \\
&\quad - (k-1)\beta f.
\end{aligned}
$$

We infer, schematically,

$$\begin{aligned} 2[T, \not{d}_k] &= [e_4 + \Upsilon e_3, \not{d}_k] \\ &= [e_4, \not{d}_k] + \Upsilon[e_3, \not{d}_k] - e_\theta(\Upsilon)e_3 \\ &= -\frac{1}{2}(\kappa + \Upsilon\underline{\kappa})\not{d}_k + \Gamma_b\mathfrak{d} + r^{-1}\Gamma_b + 2e_\theta\left(\frac{m}{r}\right)e_3 \\ &= \Gamma_b\mathfrak{d} + \Gamma_b. \end{aligned}$$

The estimate for $[T, \not{d}_k^\star]$ is similar and left to the reader. This concludes the proof of the lemma. □

Lemma 10.52. *We have*

$$T(\kappa) = \mathfrak{d}^{\leq 1}\Gamma_g, \qquad T\left(2\underline{\omega} - \frac{1}{2}\underline{\kappa}\right) = \mathfrak{d}^{\leq 1}\Gamma_b, \qquad T(K) = \mathfrak{d}^{\leq 1}\Gamma_g.$$

Proof. We have

$$\begin{aligned} 2T(\kappa) &= (e_4 + \Upsilon e_3)\kappa \\ &= -\frac{1}{2}\kappa^2 - 2\omega\kappa + 2\not{d}_1\xi + 2(\underline{\eta} + \eta + 2\zeta)\xi - \frac{1}{2}\vartheta^2 \\ &\quad + \Upsilon\left(-\frac{1}{2}\kappa\underline{\kappa} + 2\underline{\omega}\kappa + 2\not{d}_1\eta + 2\rho - \frac{1}{2}\vartheta\underline{\vartheta} + 2(\xi\underline{\xi} + \eta^2)\right) \\ &= -\frac{1}{2}\kappa^2 - 2\omega\kappa - \frac{1}{2}\underline{\kappa}\kappa\Upsilon + 2\underline{\omega}\kappa\Upsilon + 2\Upsilon\rho \\ &\quad + 2\not{d}_1\xi + 2\Upsilon\not{d}_1\eta + 2(\underline{\eta} + \eta + 2\zeta)\xi - \frac{1}{2}\vartheta^2 + \Upsilon\left(-\frac{1}{2}\vartheta\underline{\vartheta} + 2(\xi\underline{\xi} + \eta^2)\right) \\ &= r^{-1}\mathfrak{d}\Gamma_b + r^{-1}\Gamma_b \\ &= \mathfrak{d}^{\leq 1}\Gamma_g. \end{aligned}$$

Also, we have

$$T(r) = e_4(r) + \Upsilon e_3(r) = e_4(r) - \Upsilon + \Upsilon(e_3(r) - 1) \in r\Gamma_b.$$

We infer

$$\begin{aligned} T\left(2\underline{\omega} - \frac{1}{2}\underline{\kappa}\right) &= T\left(\frac{1}{r}\right) + \Gamma_b - \frac{1}{2}T\left(\underline{\kappa} + \frac{2}{r}\right) \\ &= -\frac{T(r)}{r^2} + \mathfrak{d}\Gamma_b \\ &= \mathfrak{d}^{\leq 1}\Gamma_b \end{aligned}$$

and

$$\begin{aligned} T(K) &= T\left(\frac{1}{r^2}\right) + T\left(K - \frac{1}{r^2}\right) \\ &= -\frac{2T(r)}{r^3} + r^{-1}\Gamma_b \\ &= r^{-1}\mathfrak{d}(\Gamma_b) + r^{-1}\Gamma_b \\ &= \mathfrak{d}^{\leq 1}\Gamma_g. \end{aligned}$$

This concludes the proof of the lemma. □

Corollary 10.53. *We have*

$$[T,\square_2]\psi = \mathfrak{d}^{\leq 1}(\Gamma_g)\mathfrak{d}^{\leq 2}\psi.$$

Proof. Recall that we have

$$\square_2\psi = -e_3e_4\psi+\not{\Delta}_2\psi+\left(2\underline{\omega}-\frac{1}{2}\underline{\kappa}\right)e_4\psi-\frac{1}{2}\kappa e_3\psi+2\eta e_\theta\psi.$$

We infer

$$\begin{aligned}
[T,\square_2]\psi &= -[T,e_3]e_4\psi-e_3[T,e_4]\psi+[T,\not{\Delta}_2]\psi+\left(2\underline{\omega}-\frac{1}{2}\underline{\kappa}\right)[T,e_4]\psi\\
&\quad +T\left(2\underline{\omega}-\frac{1}{2}\underline{\kappa}\right)e_4\psi-\frac{1}{2}\kappa[T,e_3]\psi-\frac{1}{2}T(\kappa)e_3\psi+2\eta[T,e_\theta]\psi\\
&\quad +2T(\eta)e_\theta\psi
\end{aligned}$$

and hence, using also $\not{\Delta}_2=-\not{d}_2^\star\not{d}_2+2K$,

$$\begin{aligned}
& [T,\square_2]\psi\\
= & -[T,e_3](r^{-1}\mathfrak{d}\psi)-\mathfrak{d}[T,e_4]\psi-r^{-1}\mathfrak{d}[T,\not{d}_2]\psi-[T,\not{d}_2^\star]r^{-1}\mathfrak{d}\psi+2T(K)\psi\\
& +\left(2\underline{\omega}-\frac{1}{2}\underline{\kappa}\right)[T,e_4]\psi+T\left(2\underline{\omega}-\frac{1}{2}\underline{\kappa}\right)r^{-1}\mathfrak{d}\psi-\frac{1}{2}\kappa[T,e_3]\psi-\frac{1}{2}T(\kappa)\mathfrak{d}\psi\\
& +2\eta[T,e_\theta]\psi+2T(\eta)r^{-1}\mathfrak{d}\psi.
\end{aligned}$$

In view of

$$[T,e_4]=\Gamma_g\mathfrak{d},\quad [T,e_3]=\Gamma_b\mathfrak{d},\quad [T,\not{d}_k]=\Gamma_b\mathfrak{d}+\Gamma_b,\quad [T,\not{d}_k^\star]=\Gamma_b\mathfrak{d}+\Gamma_b$$

and

$$T(\kappa)=\mathfrak{d}^{\leq 1}\Gamma_g,\qquad T\left(2\underline{\omega}-\frac{1}{2}\underline{\kappa}\right)=\mathfrak{d}^{\leq 1}\Gamma_b,\qquad T(K)=\mathfrak{d}^{\leq 1}\Gamma_g,$$

we deduce, schematically,

$$\begin{aligned}
[T,\square_2]\psi &= \mathfrak{d}^{\leq 1}(\Gamma_g)\mathfrak{d}^{\leq 2}\psi+r^{-1}\mathfrak{d}^{\leq 1}(\Gamma_b)\mathfrak{d}^{\leq 2}\psi\\
&= \mathfrak{d}^{\leq 1}(\Gamma_g)\mathfrak{d}^{\leq 2}\psi.
\end{aligned}$$

This concludes the proof of the corollary. □

10.4.3.2 Commutation with angular derivatives

Lemma 10.54. *We have, schematically,*

$$\begin{aligned}
&[r\not{d}_k,e_4]f,\ [r\not{d}_k^\star,e_4]f=\Gamma_g\mathfrak{d}^{\leq 1}f,\ [r^2\not{\Delta}_k,e_4]f=\mathfrak{d}^{\leq 1}(\Gamma_g)\mathfrak{d}^{\leq 2}f\\
&[r\not{d}_k,e_3]f=-r\eta e_3(f)+\Gamma_b\mathfrak{d}^{\leq 1}f,\ [r\not{d}_k^\star,e_3]f=r\eta e_3(f)+\Gamma_b\mathfrak{d}^{\leq 1}f.
\end{aligned}$$

Proof. Recall from Lemma 2.68 that the following commutation formulae holds true:

1. If $f \in \mathfrak{s}_k$,

$$[r \not{d}_k, e_4] = r \left[\text{Com}_k(f) - \frac{1}{2} A \not{d}_k f \right],$$
$$[r \not{d}_k, e_3] f = r \left[\underline{Com}_k(f) - \frac{1}{2} \underline{A} \not{d}_k f \right].$$

2. If $f \in \mathfrak{s}_{k-1}$,

$$[r \not{d}_k^\star, e_4] f = r \left[\text{Com}_k^*(f) - \frac{1}{2} A \not{d}_k^\star f \right],$$
$$[r \not{d}_k^\star, e_3] f = r \left[\underline{Com}_k^*(f) - \frac{1}{2} \underline{A} \not{d}_k^\star f \right],$$

where $A = 2/re_4(r) - \kappa$ and $\underline{A} = 2/re_3(r) - \underline{\kappa}$. Now, we have

$$\text{Com}_k(f) = r^{-1}\Gamma_g \mathfrak{d}^{\leq 1} f, \quad \text{Com}_k^*(f) = r^{-1}\Gamma_g \mathfrak{d}^{\leq 1} f,$$
$$\underline{Com}_k(f) = -\eta e_3(f) + r^{-1}\Gamma_b \mathfrak{d}^{\leq 1} f, \quad \underline{Com}_k^*(f) = \eta e_3(f) + r^{-1}\Gamma_b \mathfrak{d}^{\leq 1} f,$$

which together with the fact that $A \in \Gamma_g$ and $\underline{A} \in \Gamma_b$ implies, schematically,

$$[r \not{d}_k, e_4] f, \ [r \not{d}_k^\star, e_4] f = \Gamma_g \mathfrak{d}^{\leq 1} f,$$
$$[r \not{d}_k, e_3] f = -r\eta e_3(f) + \Gamma_b \mathfrak{d}^{\leq 1} f, \ [r \not{d}_k^\star, e_3] f = r\eta e_3(f) + \Gamma_b \mathfrak{d}^{\leq 1} f.$$

Since $\not{\Delta}_k = -\not{d}_k^\star \not{d}_k + kK$, we infer

$$\begin{aligned} [r^2 \not{\Delta}_k, e_4] &= [-r^2 \not{d}_k^\star \not{d}_k + kr^2 K, e_4] \\ &= -[r \not{d}_k^\star, e_4] r \not{d}_k - r \not{d}_k^\star [r \not{d}_k, e_4] - k e_4(r^2 K) \\ &= \mathfrak{d}^{\leq 1}(\Gamma_g) \mathfrak{d}^{\leq 2} f. \end{aligned}$$

This concludes the proof of the lemma. □

Corollary 10.55. *We have*

$$r \not{d}_2(\square_2 \psi) - (\square_1 - K)(r \not{d}_2 \psi) = -r\eta \square_2 \psi + \mathfrak{d}^{\leq 1}(\Gamma_g) \mathfrak{d}^{\leq 2} \psi$$

and

$$r \not{d}_2^\star(\square_1 \phi) - (\square_2 - 3K)(r \not{d}_2^\star \phi) = r\eta \square_1 \phi + \mathfrak{d}^{\leq 1}(\Gamma_g) \mathfrak{d}^{\leq 2} \phi.$$

Proof. Recall that we have

$$\square_2 \psi = -e_3 e_4 \psi + \not{\Delta}_2 \psi + \left(2\underline{\omega} - \frac{1}{2} \underline{\kappa} \right) e_4 \psi - \frac{1}{2} \kappa e_3 \psi + 2\eta e_\theta \psi$$

and

$$\square_1 \phi = -e_3 e_4 \phi + \not{\Delta}_2 \phi + \left(2\underline{\omega} - \frac{1}{2} \underline{\kappa} \right) e_4 \phi - \frac{1}{2} \kappa e_3 \phi + 2\eta e_\theta \phi.$$

We infer

$$\begin{aligned}
& r\,{d\!\!/}_2(\square_2\psi)-\square_1(r\,{d\!\!/}_2\psi)\\
= & -[r\,{d\!\!/}_2,e_3]e_4\psi-e_3[r\,{d\!\!/}_2,e_4]\psi+r({d\!\!/}_2{\triangle\!\!\!/}_2-{\triangle\!\!\!/}_1{d\!\!/}_2)\psi+[r,{\triangle\!\!\!/}_1]\,{d\!\!/}_2\psi\\
& +re_\theta\left(2\underline{\omega}-\frac{1}{2}\underline{\kappa}\right)e_4\psi+\left(2\underline{\omega}-\frac{1}{2}\underline{\kappa}\right)[r\,{d\!\!/}_2,e_4]\psi-\frac{1}{2}re_\theta(\kappa)e_3\psi\\
& -\frac{1}{2}\kappa[r\,{d\!\!/}_2,e_3]\psi+2r\,{d\!\!/}_2(\eta e_\theta\psi)-2\eta e_\theta(r\,{d\!\!/}_2\psi),
\end{aligned}$$

and

$$\begin{aligned}
& r\,{d\!\!/}_2^\star(\square_1\phi)-\square_2(r\,{d\!\!/}_2^\star\phi)\\
= & -[r\,{d\!\!/}_2^\star,e_3]e_4\phi-e_3[r\,{d\!\!/}_2^\star,e_4]\phi+r({d\!\!/}_2^\star{\triangle\!\!\!/}_1-{\triangle\!\!\!/}_2{d\!\!/}_2^\star)\phi+[r,{\triangle\!\!\!/}_2]\,{d\!\!/}_2^\star\phi\\
& -re_\theta\left(2\underline{\omega}-\frac{1}{2}\underline{\kappa}\right)e_4\phi+\left(2\underline{\omega}-\frac{1}{2}\underline{\kappa}\right)[r\,{d\!\!/}_2^\star,e_4]\phi+\frac{1}{2}re_\theta(\kappa)e_3\phi\\
& -\frac{1}{2}\kappa[r\,{d\!\!/}_2^\star,e_3]\phi+2r\,{d\!\!/}_2(\eta e_\theta\phi)-2\eta e_\theta(r\,{d\!\!/}_2\phi),
\end{aligned}$$

and hence, using also in particular the following identities from Proposition 2.28,

$$\begin{aligned}
{d\!\!/}_2{\triangle\!\!\!/}_2-{\triangle\!\!\!/}_1{d\!\!/}_2 &= -K\,{d\!\!/}_2+2e_\theta(K),\\
{d\!\!/}_2^\star{\triangle\!\!\!/}_1-{\triangle\!\!\!/}_2{d\!\!/}_2^\star &= -3K\,{d\!\!/}_2^\star-e_\theta(K),
\end{aligned}$$

we infer

$$\begin{aligned}
& r\,{d\!\!/}_2(\square_2\psi)-(\square_1-K)(r\,{d\!\!/}_2\psi)\\
= & -[r\,{d\!\!/}_2,e_3]e_4\psi-\mathfrak{d}[r\,{d\!\!/}_2,e_4]\psi+2re_\theta(K)\psi+[r,{\triangle\!\!\!/}_1](r^{-1}\mathfrak{d}\psi)\\
& +re_\theta\left(2\underline{\omega}-\frac{1}{2}\underline{\kappa}\right)r^{-1}\mathfrak{d}\psi+\left(2\underline{\omega}-\frac{1}{2}\underline{\kappa}\right)[r\,{d\!\!/}_2,e_4]\psi-\frac{1}{2}re_\theta(\kappa)\mathfrak{d}\psi\\
& -\frac{1}{2}\kappa[r\,{d\!\!/}_2,e_3]\psi+2\mathfrak{d}(r^{-1}\eta\mathfrak{d}\psi)-2r^{-1}\eta\mathfrak{d}(\mathfrak{d}\psi),
\end{aligned}$$

and

$$\begin{aligned}
& r\,{d\!\!/}_2^\star(\square_1\phi)-(\square_2-3K)(r\,{d\!\!/}_2^\star\phi)\\
= & -[r\,{d\!\!/}_2^\star,e_3]e_4\phi-\mathfrak{d}[r\,{d\!\!/}_2^\star,e_4]\phi-re_\theta(K)\phi+[r,{\triangle\!\!\!/}_2](r^{-1}\mathfrak{d}\phi)\\
& -re_\theta\left(2\underline{\omega}-\frac{1}{2}\underline{\kappa}\right)r^{-1}\mathfrak{d}\phi+\left(2\underline{\omega}-\frac{1}{2}\underline{\kappa}\right)[r\,{d\!\!/}_2^\star,e_4]\phi+\frac{1}{2}re_\theta(\kappa)\mathfrak{d}\phi\\
& -\frac{1}{2}\kappa[r\,{d\!\!/}_2^\star,e_3]\phi+2\mathfrak{d}(r^{-1}\eta\mathfrak{d}\phi)-2r^{-1}\eta\mathfrak{d}(\mathfrak{d}\phi).
\end{aligned}$$

This yields, schematically,

$$\begin{aligned}
& r\,{d\!\!/}_2(\square_2\psi)-(\square_1-K)(r\,{d\!\!/}_2\psi)\\
= & -[r\,{d\!\!/}_2,e_3]e_4\psi-\mathfrak{d}[r\,{d\!\!/}_2,e_4]\psi+r^{-1}[r\,{d\!\!/}_2,e_4]\psi-\frac{1}{2}\kappa[r\,{d\!\!/}_2,e_3]\psi+\mathfrak{d}^{\leq 1}(\Gamma_g)\mathfrak{d}^{\leq 2}\psi
\end{aligned}$$

and

$$
\begin{aligned}
& r\,{\not{d}}_2^\star(\square_1\phi) - (\square_2 - 3K)(r\,{\not{d}}_2^\star\phi) \\
= \; & -[r\,{\not{d}}_2^\star, e_3]e_4\phi - \mathfrak{d}[r\,{\not{d}}_2^\star, e_4]\phi + r^{-1}[r\,{\not{d}}_2^\star, e_4]\phi - \frac{1}{2}\kappa[r\,{\not{d}}_2^\star, e_3]\phi + \mathfrak{d}^{\leq 1}(\Gamma_g)\mathfrak{d}^{\leq 2}\phi
\end{aligned}
$$

where we used the fact that $r{-}1\mathfrak{d}^{\leq 1}\Gamma_b$ is at least as good as $\mathfrak{d}^{\leq 1}\Gamma_g$ and the fact that $r^{-1}e_\theta(r)$ is Γ_g.

Next, we rely on

$$
\begin{aligned}
& [r\,{\not{d}}_k, e_4]f,\ [r\,{\not{d}}_k^\star, e_4]f = \Gamma_g\mathfrak{d}^{\leq 1}f, \\
& [r\,{\not{d}}_k, e_3]f = -r\eta e_3(f) + \Gamma_b\mathfrak{d}^{\leq 1}f,\ [r\,{\not{d}}_k^\star, e_3]f = r\eta e_3(f) + \Gamma_b\mathfrak{d}^{\leq 1}f
\end{aligned}
$$

to infer

$$
r\,{\not{d}}_2(\square_2\psi) - (\square_1 - K)(r\,{\not{d}}_2\psi) \;=\; r\eta e_3e_4\psi + \frac{1}{2}r\kappa\eta e_3\psi + \mathfrak{d}^{\leq 1}(\Gamma_g)\mathfrak{d}^{\leq 2}\psi
$$

and

$$
r\,{\not{d}}_2^\star(\square_1\phi) - (\square_2 - 3K)(r\,{\not{d}}_2^\star\phi) \;=\; -r\eta e_3e_4\phi - \frac{1}{2}r\kappa\eta e_3\phi + \mathfrak{d}^{\leq 1}(\Gamma_g)\mathfrak{d}^{\leq 2}\phi.
$$

This yields

$$
\begin{aligned}
r\,{\not{d}}_2(\square_2\psi) - (\square_1 - K)(r\,{\not{d}}_2\psi) \;&=\; r\eta\left(-\square_2\psi + {\not{\Delta}}_2\psi + \left(2\underline{\omega} - \frac{1}{2}\underline{\kappa}\right)e_4\psi + 2\eta e_\theta\psi\right) \\
&\quad + \mathfrak{d}^{\leq 1}(\Gamma_g)\mathfrak{d}^{\leq 2}\psi \\
&=\; -r\eta\square_2\psi + \mathfrak{d}^{\leq 1}(\Gamma_g)\mathfrak{d}^{\leq 2}\psi
\end{aligned}
$$

and

$$
\begin{aligned}
& r\,{\not{d}}_2^\star(\square_1\phi) - (\square_2 - 3K)(r\,{\not{d}}_2^\star\phi) \\
= \; & -r\eta\left(-\square_1\phi + {\not{\Delta}}_2\phi + \left(2\underline{\omega} - \frac{1}{2}\underline{\kappa}\right)e_4\phi + 2\eta e_\theta\phi\right) + \mathfrak{d}^{\leq 1}(\Gamma_g)\mathfrak{d}^{\leq 2}\phi \\
= \; & r\eta\square_1\phi + \mathfrak{d}^{\leq 1}(\Gamma_g)\mathfrak{d}^{\leq 2}\phi,
\end{aligned}
$$

where we used the fact that $r{-}1\mathfrak{d}^{\leq 1}\Gamma_b$ is at least as good as $\mathfrak{d}^{\leq 1}\Gamma_g$. This concludes the proof of the corollary. □

10.4.3.3 Commutation with R in the region $r \leq r_0$

We derive in the following lemma commutator identities that are non-sharp as far as decay in r is concerned. This is sufficient for our needs since we will commute the wave equation with R only in the region $r \leq r_0$ for a fixed $r_0 \geq 4m_0$ large enough. We will use in particular the following estimate:

$$
\max_{k\leq k_{small}+30}|\mathfrak{d}(\Gamma_g)| \lesssim \frac{\epsilon}{r^2 u_{trap}^{1+\delta_{dec}-2\delta_0}}. \tag{10.4.3}
$$

Lemma 10.56. *We have*

$$[R, e_4] = O\left(\frac{\epsilon}{u_{trap}^{1+\delta_{dec}-2\delta_0}}\right)\mathfrak{d}, \qquad [R, e_3] = -\frac{2m}{r^2}e_3 + O\left(\frac{\epsilon}{u_{trap}^{1+\delta_{dec}-2\delta_0}}\right)\mathfrak{d},$$
$$[r\,\not{d}_k, R]f,\ [r\,\not{d}_k^\star, R]f = O\left(\frac{\epsilon}{u_{trap}^{1+\delta_{dec}-2\delta_0}}\right)\mathfrak{d}^{\leq 1}f,$$
$$[r^2\not{\Delta}_k, R]f = O\left(\frac{\epsilon}{u_{trap}^{1+\delta_{dec}-2\delta_0}}\right)\mathfrak{d}^{\leq 2}f.$$

Proof. Recall that R is defined by

$$R = \frac{1}{2}(e_4 - \Upsilon e_3)$$

and that we have

$$[e_3, e_4] \quad = \quad 2\underline{\omega}e_4 - 2\omega e_3 + 2(\eta - \underline{\eta})e_\theta.$$

We infer

$$\begin{aligned}[R, e_4] &= \frac{1}{2}[-\Upsilon e_3, e_4] = -\frac{\Upsilon}{2}[e_3, e_4] + \frac{1}{2}e_4(\Upsilon)e_3 \\ &= \left(\Upsilon\omega e_3 + \frac{m}{r^2}e_4(r)\right)e_3 + O\left(\frac{\epsilon}{u_{trap}^{1+\delta_{dec}-2\delta_0}}\right)\mathfrak{d},\end{aligned}$$

and

$$\begin{aligned}[R, e_3] &= \frac{1}{2}[e_4 - \Upsilon e_3, e_3] = \frac{1}{2}[e_4, e_3] + \frac{1}{2}e_3(\Upsilon)e_3 \\ &= \left(\omega e_3 + \frac{m}{r^2}e_3(r)\right)e_3 + O\left(\frac{\epsilon}{u_{trap}^{1+\delta_{dec}-2\delta_0}}\right)\mathfrak{d},\end{aligned}$$

and hence,

$$[R, e_4] = O\left(\frac{\epsilon}{u_{trap}^{1+\delta_{dec}-2\delta_0}}\right)\mathfrak{d}, \qquad [R, e_3] = -\frac{2m}{r^2}e_3 + O\left(\frac{\epsilon}{u_{trap}^{1+\delta_{dec}-2\delta_0}}\right)\mathfrak{d}.$$

Also, recall that we have

$$\begin{aligned}&[r\,\not{d}_k, e_4]f,\ [r\,\not{d}_k^\star, e_4]f = \Gamma_g\mathfrak{d}^{\leq 1}f,\ [r^2\not{\Delta}_k, e_4]f = \mathfrak{d}^{\leq 1}(\Gamma_g)\mathfrak{d}^{\leq 2}f,\\ &[r\,\not{d}_k, e_3]f = -r\eta e_3(f) + \Gamma_b\mathfrak{d}^{\leq 1}f,\ [r\,\not{d}_k^\star, e_3]f = r\eta e_3(f) + \Gamma_b\mathfrak{d}^{\leq 1}f.\end{aligned}$$

We infer

$$[r\,\phi\!\!\!d_k, e_4]f,\ [r\,\phi\!\!\!d_k^\star, e_4]f = O\left(\frac{\epsilon}{u_{trap}^{1+\delta_{dec}-2\delta_0}}\right)\mathfrak{d}^{\leq 1}f,$$

$$[r^2\triangle\!\!\!\!/_k, e_4]f = O\left(\frac{\epsilon}{u_{trap}^{1+\delta_{dec}-2\delta_0}}\right)\mathfrak{d}^{\leq 2}f,$$

$$[r\,\phi\!\!\!d_k, e_3]f = O\left(\frac{\epsilon}{u_{trap}^{1+\delta_{dec}-2\delta_0}}\right)\mathfrak{d}^{\leq 1}f,\ [r\,\phi\!\!\!d_k^\star, e_3]f = O\left(\frac{\epsilon}{u_{trap}^{1+\delta_{dec}-2\delta_0}}\right)\mathfrak{d}^{\leq 1}f.$$

Together with the definition for R, we deduce

$$[r\,\phi\!\!\!d_k, R]f,\ [r\,\phi\!\!\!d_k^\star, R]f = O\left(\frac{\epsilon}{u_{trap}^{1+\delta_{dec}-2\delta_0}}\right)\mathfrak{d}^{\leq 1}f,$$

$$[r^2\triangle\!\!\!\!/_k, R]f = O\left(\frac{\epsilon}{u_{trap}^{1+\delta_{dec}-2\delta_0}}\right)\mathfrak{d}^{\leq 2}f.$$

This concludes the proof of the lemma. □

Corollary 10.57. *We have in the region* $r \leq r_0$

$$\square_2(R\psi) = \left(1-\frac{3m}{r}\right)\mathfrak{d}^2\psi + O\left(\frac{\epsilon}{u_{trap}^{1+\delta_{dec}-2\delta_0}}\right)\mathfrak{d}^2\psi + O(1)\mathfrak{d}^{\leq 1}\psi + O(1)\mathfrak{d}^{\leq 1}N.$$

Proof. Recall that we have

$$\square_2\psi = -e_4(e_3(\psi)) + \triangle\!\!\!\!/_2\psi - \frac{1}{2}\underline{\kappa}e_4\psi + \left(-\frac{1}{2}\kappa + 2\omega\right)e_3\psi + 2\underline{\eta}e_\theta\psi.$$

Multiplying by r^2, we infer

$$r^2\square_2\psi = -r^2e_4(e_3(\psi)) + r^2\triangle\!\!\!\!/_2\psi - \frac{1}{2}r^2\underline{\kappa}e_4\psi + r^2\left(-\frac{1}{2}\kappa + 2\omega\right)e_3\psi + 2r\underline{\eta}re_\theta\psi$$

and hence

$$\begin{aligned} R(r^2\square_2\psi) = {} & r^2\square_2(R\psi) - [R, r^2e_4e_3]\psi + [R, r^2\triangle\!\!\!\!/_2]\psi - \frac{1}{2}R(r^2\underline{\kappa})e_4\psi \\ & -\frac{1}{2}r^2\underline{\kappa}[R, e_4]\psi + R\left(r^2\left(-\frac{1}{2}\kappa + 2\omega\right)\right)e_3\psi \\ & +r^2\left(-\frac{1}{2}\kappa + 2\omega\right)[R, e_3]\psi + 2R(r\underline{\eta})re_\theta\psi + 2r\underline{\eta}[R, re_\theta]\psi. \end{aligned}$$

Using the commutation identities of the previous lemma, we infer in the region $r \leq r_0$

$$R(r^2\square_2\psi) = r^2\square_2(R\psi) - [R, r^2e_4e_3]\psi + O\left(\frac{\epsilon}{u_{trap}^{1+\delta_{dec}-2\delta_0}}\right)\mathfrak{d}^2\psi + O(1)\mathfrak{d}\psi.$$

Also, since ψ satisfies $\square_2\psi = V\psi + N$, we infer in the region $r \leq r_0$

$$r^2\square_2(R\psi) = [R, r^2 e_4 e_3]\psi + O\left(\frac{\epsilon}{u_{trap}^{1+\delta_{dec}-2\delta_0}}\right)\mathfrak{d}^2\psi + O(1)\mathfrak{d}^{\leq 1}\psi + O(1)\mathfrak{d}^{\leq 1}N.$$

Next, recall that we have

$$[R, e_4] = O\left(\frac{\epsilon}{u_{trap}^{1+\delta_{dec}-2\delta_0}}\right)\mathfrak{d}, \qquad [R, e_3] = -\frac{2m}{r^2}e_3 + O(\epsilon)\mathfrak{d}.$$

We infer

$$\begin{aligned}
[R, r^2 e_4 e_3]\psi &= R(r^2)e_4 e_3 + r^2[R, e_4]e_3 + r^2 e_4[R, e_3] \\
&= \frac{1}{2}\Big(e_4(r^2) - \Upsilon e_3(r^2)\Big)e_4 e_3\psi + r^2 e_4\left(-\frac{2m}{r^2}e_3\psi\right) \\
&\quad +O\left(\frac{\epsilon}{u_{trap}^{1+\delta_{dec}-2\delta_0}}\right)\mathfrak{d}^2\psi \\
&= 2\Big(r - 3m\Big)e_4 e_3\psi + O\left(\frac{\epsilon}{u_{trap}^{1+\delta_{dec}-2\delta_0}}\right)\mathfrak{d}^2\psi + O(1)\mathfrak{d}\psi
\end{aligned}$$

and thus, in the region $r \leq r_0$,

$$\square_2(R\psi) = \left(1 - \frac{3m}{r}\right)\mathfrak{d}^2\psi + O\left(\frac{\epsilon}{u_{trap}^{1+\delta_{dec}-2\delta_0}}\right)\mathfrak{d}^2\psi + O(1)\mathfrak{d}^{\leq 1}\psi + O(1)\mathfrak{d}^{\leq 1}N$$

as desired. □

10.4.3.4 Commutation with the redshift vectorfield

Let a positive bump function $\kappa = \kappa(r)$, supported in the region in $[-2, 2]$ and equal to 1 for $[-1, 1]$. Recall that the redshift vectorfield is given by

$$Y_{\mathcal{H}} = \kappa_{\mathcal{H}} Y_{(0)}, \qquad \kappa_{\mathcal{H}} := \kappa\left(\frac{\Upsilon}{\delta_{\mathcal{H}}}\right)$$

where $Y_{(0)}$ is defined by

$$Y_{(0)} = a e_3 + b e_4 + 2T, \quad a = 1 + \frac{5}{4m}(r - 2m), \quad b = \frac{5}{4m}(r - 2m).$$

Lemma 10.58. *We have*

$$[\square_2, e_3]\psi = -2\omega e_3(e_3\psi) + \underline{\kappa} e_4(e_3\psi) + \underline{\kappa}\square_2\psi + \mathfrak{d}^{\leq 1}(\Gamma_g)\mathfrak{d}^2\psi + r^{-2}\mathfrak{d}^{\leq 1}\psi.$$

Proof. Recall that we have

$$\square_2\psi = -e_4(e_3(\psi)) + \Delta\!\!\!\!/_2\psi - \frac{1}{2}\underline{\kappa} e_4\psi + \left(-\frac{1}{2}\kappa + 2\omega\right)e_3\psi + 2\underline{\eta} e_\theta\psi.$$

Since we have

$$[e_4, e_3] = 2\omega e_3 + r^{-1}\Gamma_b \not{\mathfrak{d}}, \quad [\not{d}_k, e_3] = \frac{1}{2}\underline{\kappa}\not{d}_k + \Gamma_b\mathfrak{d} + r^{-1}\Gamma_b,$$
$$[\not{d}_k^\star, e_3] = \frac{1}{2}\underline{\kappa}\not{d}_k^\star + \Gamma_b\mathfrak{d} + r^{-1}\Gamma_b,$$

we infer

$$\begin{aligned}
[\square_2, e_3]\psi &= -[e_4, e_3](e_3(\psi)) + [\not{\Delta}_2, e_3] - \frac{1}{2}\underline{\kappa}[e_4, e_3](\psi)\\
&\quad +\frac{1}{2}e_3(\underline{\kappa})e_4(\psi) - e_3\left(-\frac{1}{2}\kappa + 2\omega\right)e_3(\psi) + 2\underline{\eta}[e_\theta, e_3]\psi - 2e_3(\underline{\eta})e_\theta(\psi)\\
&= -2\omega e_3(e_3\psi) + \underline{\kappa}\not{\Delta}_2\psi + \mathfrak{d}^{\leq 1}(\Gamma_g)\mathfrak{d}^2\psi + r^{-2}\mathfrak{d}^{\leq 1}\psi.
\end{aligned}$$

Using again

$$\square_2\psi = -e_4(e_3(\psi)) + \not{\Delta}_2\psi - \frac{1}{2}\underline{\kappa}e_4\psi + \left(-\frac{1}{2}\kappa + 2\omega\right)e_3\psi + 2\underline{\eta}e_\theta\psi,$$

we deduce

$$\begin{aligned}
[\square_2, e_3]\psi &= -2\omega e_3(e_3\psi) + \underline{\kappa}\Big(\square_2\psi + e_4(e_3\psi)\Big) + \mathfrak{d}^{\leq 1}(\Gamma_g)\mathfrak{d}^2\psi + r^{-2}\mathfrak{d}^{\leq 1}\psi\\
&= -2\omega e_3(e_3\psi) + \underline{\kappa}e_4(e_3\psi) + \underline{\kappa}\square_2\psi + \mathfrak{d}^{\leq 1}(\Gamma_g)\mathfrak{d}^2\psi + r^{-2}\mathfrak{d}^{\leq 1}\psi.
\end{aligned}$$

This concludes the proof of the lemma. □

Lemma 10.59. *There exists a scalar function d_0 satisfying the bound*

$$d_0 = \frac{1}{2m_0} + O(\delta_{\mathcal{H}}) \text{ on the support of } \kappa_{\mathcal{H}},$$

such that we have, schematically,

$$\begin{aligned}
[\square_2, Y_{\mathcal{H}}]\psi &= d_0 Y_{(0)}(Y_{\mathcal{H}}\psi) + 1_{\Upsilon\leq 2\delta_{\mathcal{H}}}\left(\square_2\psi + \mathfrak{d}T\psi + \mathfrak{d}^{\leq 1}(\Gamma_g)\mathfrak{d}^2\psi + \frac{1}{\delta_{\mathcal{H}}^2}\mathfrak{d}^{\leq 1}\psi\right)\\
&\quad +\frac{1}{\delta_{\mathcal{H}}}1_{\delta_{\mathcal{H}}\leq\Upsilon\leq 2\delta_{\mathcal{H}}}\mathfrak{d}^{\leq 2}\psi.
\end{aligned}$$

Proof. We have

$$\begin{aligned}
Y_{(0)} = ae_3 + be_4 + 2T &= ae_3 + b(2T - \Upsilon e_3) + 2T\\
&= (a - \Upsilon b)e_3 + 2(1+b)T.
\end{aligned}$$

Thus, in view of the commutator identities

$$\begin{aligned}
[T, \square_2]\psi &= \mathfrak{d}^{\leq 1}(\Gamma_g)\mathfrak{d}^{\leq 2}\psi,\\
[\square_2, e_3]\psi &= -2\omega e_3(e_3\psi) + \underline{\kappa}e_4(e_3\psi) + \underline{\kappa}\square_2\psi + \mathfrak{d}^{\leq 1}(\Gamma_g)\mathfrak{d}^2\psi + r^{-2}\mathfrak{d}^{\leq 1}\psi,
\end{aligned}$$

we deduce, schematically,

$$\begin{aligned}
[\square_2, Y_{(0)}]\psi &= [\square_2, (a-\Upsilon b)e_3]\psi + [\square_2, 2(1+b)T]\psi\\
&= (a-\Upsilon b)[\square_2, e_3]\psi + \mathbf{g}^{\alpha\beta}\mathbf{D}_\alpha(a)\mathbf{D}_\beta e_3\psi + 2(1+b)[\square_2, T]\psi\\
&\quad +2\mathbf{g}^{\alpha\beta}\mathbf{D}_\alpha(b)\mathbf{D}_\beta T\psi + \mathfrak{d}^{\leq 1}\psi\\
&= (a-\Upsilon b)\Big(-2\omega e_3(e_3\psi) + \underline{\kappa} e_4(e_3\psi) + \underline{\kappa}\square_2\psi\Big) - \frac{1}{2}e_3(a)e_4(e_3\psi)\\
&\quad -\frac{1}{2}e_4(a)e_3(e_3\psi) + \mathfrak{d}T\psi + \mathfrak{d}^{\leq 1}(\Gamma_g)\mathfrak{d}^2\psi + \mathfrak{d}^{\leq 1}\psi.
\end{aligned}$$

Since $e_4 = -\Upsilon e_3 + 2T$, we infer schematically

$$\begin{aligned}
[\square_2, Y_{(0)}]\psi &= \left((a-\Upsilon b)(-2\omega - \Upsilon\underline{\kappa}) + \frac{\Upsilon}{2}e_3(a) - \frac{1}{2}e_4(a)\right)e_3(e_3\psi)\\
&\quad +\square_2\psi + \mathfrak{d}T\psi + \mathfrak{d}^{\leq 1}(\Gamma_g)\mathfrak{d}^2\psi + \mathfrak{d}^{\leq 1}\psi.
\end{aligned}$$

We deduce

$$\begin{aligned}
[\square_2, Y_{\mathcal{H}}]\psi &= [\square_2, \kappa_{\mathcal{H}} Y_{(0)}]\psi\\
&= \kappa_{\mathcal{H}}[\square_2, Y_{(0)}]\psi + \kappa'_{\mathcal{H}}\mathfrak{d}^{\leq 2}\psi + \kappa''_{\mathcal{H}}\mathfrak{d}^{\leq 1}\psi\\
&= \kappa_{\mathcal{H}}\left((a-\Upsilon b)(-2\omega - \Upsilon\underline{\kappa}) + \frac{\Upsilon}{2}e_3(a) - \frac{1}{2}e_4(a)\right)e_3(e_3\psi)\\
&\quad +1_{\Upsilon\leq 2\delta_{\mathcal{H}}}\left(\square_2\psi + \mathfrak{d}T\psi + \mathfrak{d}^{\leq 1}(\Gamma_g)\mathfrak{d}^2\psi + \frac{1}{\delta_{\mathcal{H}}^2}\mathfrak{d}^{\leq 1}\psi\right)\\
&\quad +\frac{1}{\delta_{\mathcal{H}}}1_{\delta_{\mathcal{H}}\leq\Upsilon\leq 2\delta_{\mathcal{H}}}\mathfrak{d}^{\leq 2}\psi.
\end{aligned}$$

Now, we have

$$\begin{aligned}
\kappa_{\mathcal{H}} e_3(e_3\psi) &= \frac{1}{a-\Upsilon b}\kappa_{\mathcal{H}} Y_{(0)}(e_3\psi) + T\mathfrak{d}\psi\\
&= \frac{1}{(a-\Upsilon b)^2}\kappa_{\mathcal{H}} Y_{(0)}(Y_{(0)}\psi) + \mathfrak{d}T\psi + \mathfrak{d}^{\leq 1}\psi\\
&= \frac{1}{(a-\Upsilon b)^2}Y_{(0)}(Y_{\mathcal{H}}\psi) + \mathfrak{d}T\psi + \frac{1}{\delta_{\mathcal{H}}}\mathfrak{d}^{\leq 1}\psi
\end{aligned}$$

and hence

$$\begin{aligned}
[\square_2, Y_{\mathcal{H}}]\psi &= \frac{(a-\Upsilon b)(-2\omega - \Upsilon\underline{\kappa}) + \frac{\Upsilon}{2}e_3(a) - \frac{1}{2}e_4(a)}{(a-\Upsilon b)^2}Y_{(0)}(Y_{\mathcal{H}}\psi)\\
&\quad +1_{\Upsilon\leq 2\delta_{\mathcal{H}}}\left(\square_2\psi + \mathfrak{d}T\psi + \mathfrak{d}^{\leq 1}(\Gamma_g)\mathfrak{d}^2\psi + \frac{1}{\delta_{\mathcal{H}}^2}\mathfrak{d}^{\leq 1}\psi\right)\\
&\quad +\frac{1}{\delta_{\mathcal{H}}}1_{\delta_{\mathcal{H}}\leq\Upsilon\leq 2\delta_{\mathcal{H}}}\mathfrak{d}^{\leq 2}\psi.
\end{aligned}$$

Now, we have, in view of the definition of a and b,

$$\begin{aligned}\frac{(a-\Upsilon b)(-2\omega-\Upsilon\underline{\kappa})+\frac{\Upsilon}{2}e_3(a)-\frac{1}{2}e_4(a)}{(a-\Upsilon b)^2} &= \frac{(1+O(\Upsilon))(-2\omega+O(\Upsilon))+O(\Upsilon)}{(1+O(\Upsilon))^2}\\ &= \frac{1}{2m}+O(\epsilon)+O(\Upsilon)\\ &= \frac{1}{2m_0}+O(\epsilon)+O(\Upsilon)\end{aligned}$$

where we also used our assumptions on ω and m. Thus, we have on the support of $\kappa_{\mathcal{H}}$

$$\begin{aligned}\frac{(a-\Upsilon b)(-2\omega-\Upsilon\underline{\kappa})+\frac{\Upsilon}{2}e_3(a)-\frac{1}{2}e_4(a)}{(a-\Upsilon b)^2} &= \frac{1}{2m_0}+O(\epsilon+\delta_{\mathcal{H}})\\ &= \frac{1}{2m_0}+O(\delta_{\mathcal{H}})\end{aligned}$$

where we used the fact that $\epsilon \ll \delta_{\mathcal{H}}$ by assumption. Setting

$$d_0 \;:=\; \frac{(a-\Upsilon b)(-2\omega-\Upsilon\underline{\kappa})+\frac{\Upsilon}{2}e_3(a)-\frac{1}{2}e_4(a)}{(a-\Upsilon b)^2},$$

this concludes the proof of the lemma. □

10.4.3.5 Commutation with re_4

Lemma 10.60. *We have, schematically,*

$$\begin{aligned}[\square_2, re_4]\psi &= \frac{\Upsilon}{r}\left(1+\frac{2m}{r\Upsilon^2}\right)\check{e}_4(re_4\psi)+\square_2\psi+\Gamma_g\mathfrak{d}^2\psi+\frac{1}{\Upsilon}r^{-2}\mathfrak{d}T\psi\\ &\quad +r^{-2}\not{\mathfrak{d}}^2\psi+r^{-2}\mathfrak{d}\psi.\end{aligned}$$

Proof. Recall that we have

$$\square_2\psi \;=\; -e_3(e_4(\psi))+\not{\Delta}_2\psi+\left(2\underline{\omega}-\frac{1}{2}\underline{\kappa}\right)e_4\psi-\frac{1}{2}\kappa e_3\psi+2\eta e_\theta\psi.$$

Since we have

$$\begin{aligned}&[re_4, e_3] = 2r\omega e_3-\frac{r}{2}\underline{\kappa}e_4+\Gamma_b\mathfrak{d}, \quad [re_4, e_4] = -\frac{r}{2}\kappa e_4+\Gamma_g\mathfrak{d},\\ &[\not{d}_k, re_4] = \frac{1}{2}r\kappa\not{d}_k+\Gamma_g\mathfrak{d}+\Gamma_g, \quad [\not{d}_k^\star, re_4] = \frac{1}{2}r\kappa\not{d}_k^\star+\Gamma_g\mathfrak{d}+\Gamma_g,\end{aligned}$$

we infer, schematically,

$$
\begin{aligned}
[\square_2, re_4]\psi &= -[e_3, re_4]e_4\psi - e_3[e_4, re_4]\psi + [\not\Delta_2, re_4]\psi + \left(2\underline{\omega} - \frac{1}{2}\underline{\kappa}\right)[e_4, re_4]\psi \\
&\quad -re_4\left(2\underline{\omega} - \frac{1}{2}\underline{\kappa}\right)e_4\psi - \frac{1}{2}\kappa[e_3, re_4]\psi + \frac{1}{2}re_4(\kappa)e_3\psi \\
&\quad +2\eta[e_\theta, re_4]\psi - 2re_4(\eta)e_\theta\psi \\
&= \left(2r\omega e_3 - \frac{r}{2}\underline{\kappa}e_4\right)e_4\psi - \frac{1}{2}e_3\left(r\kappa e_4\psi\right) + r\kappa\not\Delta_2\psi + \frac{1}{2}re_4(\kappa)e_3\psi \\
&\quad +\Gamma_g\mathfrak{d}^2\psi + r^{-2}\mathfrak{d}\psi \\
&= \left(2r\omega e_3 - \frac{r}{2}\underline{\kappa}e_4\right)e_4\psi - \frac{1}{2}r\kappa e_3e_4\psi - \frac{1}{4}r\kappa^2 e_3\psi + \Gamma_g\mathfrak{d}^2\psi + r^{-2}\not\mathfrak{d}^2\psi \\
&\quad +r^{-2}\mathfrak{d}\psi.
\end{aligned}
$$

Using again

$$
\square_2\psi = -e_3(e_4(\psi)) + \not\Delta_2\psi + \left(2\underline{\omega} - \frac{1}{2}\underline{\kappa}\right)e_4\psi - \frac{1}{2}\kappa e_3\psi + 2\eta e_\theta\psi,
$$

we have

$$
-\frac{1}{2}r\kappa e_3e_4\psi - \frac{1}{4}r\kappa^2 e_3\psi = \frac{1}{2}r\kappa\square_2\psi + r^{-2}\not\mathfrak{d}^2\psi + r^{-2}\mathfrak{d}\psi
$$

and hence

$$
\begin{aligned}
[\square_2, re_4]\psi &= \left(2r\omega e_3 - \frac{r}{2}\underline{\kappa}e_4\right)e_4\psi + \square_2\psi + \Gamma_g\mathfrak{d}^2\psi + r^{-2}\not\mathfrak{d}^2\psi + r^{-2}\mathfrak{d}\psi \\
&= \left(2r\omega\frac{1}{\Upsilon}(2T - e_4) - \frac{r}{2}\underline{\kappa}e_4\right)e_4\psi + \square_2\psi + \Gamma_g\mathfrak{d}^2\psi + r^{-2}\not\mathfrak{d}^2\psi \\
&\quad +r^{-2}\mathfrak{d}\psi \\
&= \left(-2r\omega\frac{1}{\Upsilon} - \frac{r}{2}\underline{\kappa}\right)e_4(e_4\psi) + \square_2\psi + \Gamma_g\mathfrak{d}^2\psi + \frac{1}{\Upsilon}r^{-2}\mathfrak{d}T\psi + r^{-2}\not\mathfrak{d}^2\psi \\
&\quad +r^{-2}\mathfrak{d}\psi \\
&= \left(\Upsilon + \frac{2m}{r\Upsilon}\right)e_4(e_4\psi) + \square_2\psi + \Gamma_g\mathfrak{d}^2\psi + \frac{1}{\Upsilon}r^{-2}\mathfrak{d}T\psi + r^{-2}\not\mathfrak{d}^2\psi \\
&\quad +r^{-2}\mathfrak{d}\psi \\
&= \frac{\Upsilon}{r}\left(1 + \frac{2m}{r\Upsilon^2}\right)\breve{e}_4(re_4\psi) + \square_2\psi + \Gamma_g\mathfrak{d}^2\psi + \frac{1}{\Upsilon}r^{-2}\mathfrak{d}T\psi + r^{-2}\not\mathfrak{d}^2\psi \\
&\quad +r^{-2}\mathfrak{d}\psi.
\end{aligned}
$$

This concludes the proof of the lemma. $\square$

10.4.4 Some weighted estimates for wave equations

Recall from Corollary 10.55 that we have the following commutator identity:

$$
r\not{d}_2(\square_2\psi) - (\square_1 - K)(r\not{d}_2\psi) = -r\eta\square_2\psi + \mathfrak{d}^{\leq 1}(\Gamma_g)\mathfrak{d}^{\leq 2}\psi.
$$

In particular, to derive weighted estimates for $r\,\not{d}_2$, we need to derive weighted estimates for solutions ϕ to wave equations of the type

$$(\square_1 - V_1)\phi \quad = \quad N,$$

where ϕ is a reduced 1-scalar and the potential V_1 is given by $V_1 = V+K = -\kappa\underline{\kappa}+K$. This is done in the following theorem.

Theorem 10.61. *Let ϕ a reduced 1-scalar solution to*

$$(\square_1 - V_1)\phi \quad = \quad N, \qquad V_1 = -\kappa\underline{\kappa} + K.$$

Then, ϕ satisfies, for all $\delta \leq p \leq 2-\delta$,

$$\begin{aligned}
&\sup_{\tau\in[\tau_1,\tau_2]} E_p[\phi](\tau) + B_p[\phi](\tau_1,\tau_2) + F_p[\phi](\tau_1,\tau_2)\\
&\lesssim \; E_p[\phi](\tau_1) + J_p[\phi, N](\tau_1,\tau_2) + \int_{{}^{(trap)}\mathcal{M}(\tau_1,\tau_2)} \left|1-\frac{3m}{r}\right| |\phi|(|\phi|+|R\phi|)\\
&\quad + \int_{{}^{(tr\not{p})}\mathcal{M}(\tau_1,\tau_2)} r^{p-3}|\phi|(|\phi|+|\mathfrak{d}\phi|),
\end{aligned} \tag{10.4.4}$$

and $\check{\phi} = f_2\check{e}_4\phi$ satisfies for all $1-\delta < q \leq 1-\delta$,

$$\begin{aligned}
&\sup_{\tau\in[\tau_1,\tau_2]} E_q[\check{\phi}](\tau) + B_q[\check{\phi}](\tau_1,\tau_2)\\
&\lesssim \; E_q[\check{\phi}](\tau_1) + \check{J}_q[\check{\phi}, N](\tau_1,\tau_2) + E^1_{\max(q,\delta)}[\phi](\tau_1) + J^1_{\max(q,\delta)}[\phi, N]\\
&\quad + \int_{\mathcal{M}(\tau_1,\tau_2)} r^{q-3}|\check{\phi}|(|\check{\phi}|+|\mathfrak{d}\check{\phi}|).
\end{aligned} \tag{10.4.5}$$

Remark 10.62. *Although we will not need it, we expect that the last two terms in the right-hand side of* (10.4.4) *and the last term in the right-hand side of* (10.4.5) *could be removed.*

Proof. We start with the following observations.

- (10.4.5) is the analog of (10.4.1), i.e., of Theorem 5.18 in the case $s=0$, with V replaced by V_1, and with the reduced 2-scalar ψ replaced by the reduced 1-scalar ϕ. The proof is in fact significantly easier in view of the presence of the term

$$\int_{\mathcal{M}(\tau_1,\tau_2)} r^{q-3}|\check{\phi}|(|\check{\phi}|+|\mathfrak{d}\check{\phi}|)$$

  on the right-hand side of (10.4.5).
- (10.4.4) is the analog of (10.4.2), i.e., of Theorem 5.17 in the case $s=0$, with V replaced by V_1, and with the reduced 2-scalar ψ replaced by the reduced 1-scalar ϕ. The proof is in fact significantly easier in view of the presence of the terms

$$\int_{{}^{(trap)}\mathcal{M}(\tau_1,\tau_2)} \left|1-\frac{3m}{r}\right| |\phi|(|\phi|+|R\phi|) + \int_{{}^{(tr\not{p})}\mathcal{M}(\tau_1,\tau_2)} r^{p-3}|\phi|(|\phi|+|\mathfrak{d}\phi|)$$

  on the right-hand side of (10.4.4).

- The boundary terms can be treated as in the proof of (10.4.1) and (10.4.2) in view of the fact that V_1 is a positive potential.[20]
- The only place where there might be a potential difficulty concerns the proof of (10.4.5) in ${}^{(trap)}\mathcal{M}$ where the second-to-last term on the right-hand side is required to have a more precise structure.

In view of the above observations, and in particular of the last one, we focus on recovering the bulk term leading to (10.4.5) in ${}^{(trap)}\mathcal{M}$. To this end, we choose f and w as in Proposition 10.16. This yields[21]

$$\dot{\mathcal{E}}[fR,w](\Psi) \geq f'|R(\Psi)|^2 + r^{-1}\left(1-\tfrac{3m}{r}\right) f|\nabla\Psi|^2 + O\left(\tfrac{1-\frac{3m}{r}}{r^3}\right)|\Psi|^2.$$

We infer

$$\begin{aligned}\dot{\mathcal{E}}[fR,w,M=2hR](\Psi) \ \geq\ & f'|R(\Psi)|^2 + r^{-1}\left(1-\frac{3m}{r}\right) f|\nabla\Psi|^2 \\ & + O\left(\frac{1-\frac{3m}{r}}{r^3}\right)|\Psi|^2 + \frac{1}{2}r^{-2}(\Upsilon r^2 h)'|\Psi|^2 + h\Psi R(\Psi).\end{aligned}$$

We now choose a smooth h, compactly supported in $[5/2m_0, 7/2m_0]$, such that $h(3m)=0$ and $h'(3m)=1$.[22] We infer $r^{-2}(\Upsilon r^2 h)'(3m) = 1/3 > 0$ and hence

$$\begin{aligned}\dot{\mathcal{E}}[fR,w,M](\Psi) \ \geq\ & f'|R(\Psi)|^2 + r^{-1}\left(1-\frac{3m}{r}\right) f|\nabla\Psi|^2 + \frac{\Upsilon}{r^3}|\Psi|^2 \\ & + O\left(\frac{1-\frac{3m}{r}}{r^3}\right)|\Psi|(|\Psi|+|R(\Psi)|).\end{aligned}$$

In view of the choice of f in Proposition 10.16, we have

$$f' \gtrsim \frac{1}{r^3}, \qquad \left(1-\frac{3m}{r}\right) f \geq \left(1-\frac{3m}{r}\right)^2,$$

and hence, there exist two constants $c_0>0$ and $C_0>0$ such that

$$\begin{aligned}\dot{\mathcal{E}}[fR,w_0,M](\Psi) \ \geq\ & c_0\left(\frac{1}{r^3}|R(\Psi)|^2 + r^{-1}\left(1-\frac{3m}{r}\right)^2|\nabla\Psi|^2 + \frac{\Upsilon}{r^3}|\Psi|^2\right) \\ & - C_0\frac{\left|1-\frac{3m}{r}\right|}{r^3}|\Psi|(|\Psi|+|R(\Psi)|).\end{aligned}$$

The last term above is responsible for the second-to-last term on the right-hand

---

[20]We have

$$V_1 = -\kappa\underline{\kappa} + K = \frac{4\Upsilon + 1 + O(\epsilon)}{r^2}$$

in view of the assumptions so that V_1 is indeed a positive potential.

[21]Note that Proposition 10.16 does not use the particular form of the potential and the type of the reduced scalar ϕ and hence holds in our more general case.

[22]This differs from the choice of h in the proof of (10.4.1) in order to avoid using a Poincaré inequality (which depends on the type of the reduced scalar) and the particular form of the potential V_1.

side of (10.4.5). □

Next, we have the following consequence of (10.4.1) and Theorem 10.61.

Corollary 10.63. *Let ϕ be a reduced k-scalar for $k=1,2$ such that ϕ satisfies*[23]

$$(\square_k - W)\phi = O\left(\frac{\epsilon}{r^2 u_{trap}^{1+\delta_{dec}-2\delta_0}}\right)\mathfrak{d}\phi_1 + \phi_2$$

where ϕ_1 and ϕ_2 are given reduced scalars, and where $W=V$ in the case $k=2$ and $W=V_1$ in the case $k=1$. Then, ϕ satisfies, for all $\delta \le p \le 2-\delta$,

$$\begin{aligned}
&\sup_{\tau\in[\tau_1,\tau_2]} E_p[\phi](\tau) + B_p[\phi](\tau_1,\tau_2) + F_p[\phi](\tau_1,\tau_2)\\
\lesssim\ & E_p[\phi](\tau_1) + \epsilon^2\left(\sup_{[\tau_1,\tau_2]} E[\phi_1](\tau) + B_p[\phi_1](\tau_1,\tau_2)\right) + J_p[\phi,\phi_2](\tau_1,\tau_2)\\
&+\int_{{}^{(trap)}\mathcal{M}(\tau_1,\tau_2)}\left|1-\frac{3m}{r}\right||\phi|(|\phi|+|R\phi|) + \int_{{}^{(tr\!\!\not{}p)}\mathcal{M}(\tau_1,\tau_2)} r^{p-3}|\phi|(|\phi|+|\mathfrak{d}\phi|).
\end{aligned}$$

Proof. The wave equation for ϕ satisfies the assumptions of (10.4.1) and Theorem 10.61 with

$$N = O\left(\frac{\epsilon}{r^2 u_{trap}^{1+\delta_{dec}-2\delta_0}}\right)\mathfrak{d}\phi_1 + \phi_2.$$

We deduce

$$\begin{aligned}
&\sup_{\tau\in[\tau_1,\tau_2]} E_p[\phi](\tau) + B_p[\phi](\tau_1,\tau_2) + F_p[\phi](\tau_1,\tau_2)\\
\lesssim\ & E_p[\phi](\tau_1) + J_p\left[\phi, O\left(\frac{\epsilon}{r^2 u_{trap}^{1+\delta_{dec}-2\delta_0}}\right)\mathfrak{d}\phi_1 + \phi_2\right](\tau_1,\tau_2)\\
&+\int_{{}^{(trap)}\mathcal{M}(\tau_1,\tau_2)}\left|1-\frac{3m}{r}\right||\phi|(|\phi|+|R\phi|) + \int_{{}^{(tr\!\!\not{}p)}\mathcal{M}(\tau_1,\tau_2)} r^{p-3}|\phi|(|\phi|+|\mathfrak{d}\phi|).
\end{aligned}$$

Now, in view of the definition

$$\begin{aligned}
J_{p,R}[\psi,N](\tau_1,\tau_2) &= \left|\int_{\mathcal{M}_{\ge R}(\tau_1,\tau_2)} r^p \check{e}_4\psi N\right|,\\
J_p[\psi,N](\tau_1,\tau_2) &= \left(\int_{\tau_1}^{\tau_2} d\tau \|N\|_{L^2({}^{(trap)}\Sigma(\tau))}\right)^2 + \int_{{}^{(tr\!\!\not{}p)}\mathcal{M}(\tau_1,\tau_2)} r^{1+\delta}|N|^2\\
&\quad + J_{p,4m_0}[\psi,N](\tau_1,\tau_2),
\end{aligned}$$

[23]Recall that we have $\delta_0 \ll \delta_{dec}$ in view of (5.1.1), and hence $\delta_{dec} - 2\delta_0 > 0$.

we have

$$\begin{aligned} & J_p\left[\phi, O\left(\frac{\epsilon}{r^2 u_{trap}^{1+\delta_{dec}-2\delta_0}}\right)\mathfrak{d}\phi_1+\phi_2\right](\tau_1,\tau_2)\\ \lesssim\ & J_p\left[\phi, O\left(\frac{\epsilon}{r^2 u_{trap}^{1+\delta_{dec}-2\delta_0}}\right)\mathfrak{d}\phi_1\right](\tau_1,\tau_2)+J_p\left[\phi,\phi_2\right](\tau_1,\tau_2) \end{aligned}$$

and, for $\delta \le p \le 2-\delta$, using also $\delta_{dec}-2\delta_0>0$, we have

$$\begin{aligned} & J_p\left[\phi, O\left(\frac{\epsilon}{r^2 u_{trap}^{1+\delta_{dec}-2\delta_0}}\right)\mathfrak{d}\phi_1\right](\tau_1,\tau_2)\\ \lesssim\ & \epsilon^2\left(\int_{\tau_1}^{\tau_2}\|\mathfrak{d}\phi_1\|_{L^2({}^{(trap)}\Sigma(\tau))}\frac{d\tau}{\tau^{1+\delta_{dec}-2\delta_0}}\right)^2+\epsilon^2\int_{{}^{(tr\not{a}p)}\mathcal{M}(\tau_1,\tau_2)} r^{\delta-3}|\mathfrak{d}\phi_1|^2\\ & +\epsilon\left|\int_{\mathcal{M}_{\ge 4m_0}(\tau_1,\tau_2)} r^{p-2}\check{e}_4(\psi)\mathfrak{d}\phi_1\right|\\ \lesssim\ & \epsilon^2\sup_{[\tau_1,\tau_2]}\|\mathfrak{d}\phi_1\|^2_{L^2({}^{(trap)}\Sigma(\tau))}+\epsilon^2\int_{{}^{(tr\not{a}p)}\mathcal{M}(\tau_1,\tau_2)} r^{p-3}|\mathfrak{d}^{\le 2}\psi|^2\\ & +\epsilon\left(\int_{{}^{(tr\not{a}p)}\mathcal{M}(\tau_1,\tau_2)} r^{p-3}|\mathfrak{d}\phi|^2\right)^{\frac{1}{2}}\left(\int_{{}^{(tr\not{a}p)}\mathcal{M}(\tau_1,\tau_2)} r^{p-3}|\mathfrak{d}\phi_1|^2\right)^{\frac{1}{2}}\\ \lesssim\ & \epsilon^2\left(\sup_{[\tau_1,\tau_2]}E^1[\phi_1](\tau)+B^1_p[\phi_1](\tau_1,\tau_2)\right)+\epsilon\left(B^1_p[\phi_1](\tau_1,\tau_2)\right)^{\frac{1}{2}}\left(B^1_p[\phi](\tau_1,\tau_2)\right)^{\frac{1}{2}}. \end{aligned}$$

We immediately deduce

$$\begin{aligned} & \sup_{\tau\in[\tau_1,\tau_2]} E_p[\phi](\tau)+B_p[\phi](\tau_1,\tau_2)+F_p[\phi](\tau_1,\tau_2)\\ \lesssim\ & E_p[\phi](\tau_1)+\epsilon^2\left(\sup_{[\tau_1,\tau_2]}E[\phi_1](\tau)+B_p[\phi_1](\tau_1,\tau_2)\right)+J_p[\phi,\phi_2](\tau_1,\tau_2)\\ & +\int_{{}^{(trap)}\mathcal{M}(\tau_1,\tau_2)}\left|1-\frac{3m}{r}\right||\phi|(|\phi|+|R\phi|)+\int_{{}^{(tr\not{a}p)}\mathcal{M}(\tau_1,\tau_2)} r^{p-3}|\phi|(|\phi|+|\mathfrak{d}\phi|). \end{aligned}$$

This concludes the proof of the corollary. □

Finally, we end this section with the following lemma.

Lemma 10.64. *Let ϕ be a reduced k-scalar for $k=1,2$, and let X a vectorfield. We have, for all $\delta\le p\le 2-\delta$,*

$$\begin{aligned} & \int_{{}^{(trap)}\mathcal{M}(\tau_1,\tau_2)}\left|1-\frac{3m}{r}\right||\mathfrak{d}\phi|(|X\phi|+|R(X\phi)|)\\ & +\int_{{}^{(tr\not{a}p)}\mathcal{M}(\tau_1,\tau_2)} r^{p-3}|\mathfrak{d}\phi|(|X\phi|+|\mathfrak{d}(X\phi)|)\\ \lesssim\ & \left(B_p[X\phi](\tau_1,\tau_2)\right)^{\frac{1}{2}}\left(B_p[\phi](\tau_1,\tau_2)\right)^{\frac{1}{2}}. \end{aligned}$$

Proof. The proof follows immediately from the definition of $B_p[\phi](\tau_1,\tau_2)$. □

10.4.5 Proof of Theorem 5.17

We now conclude the proof of Theorem 5.17 for all $0 \leq s \leq k_{small}+30$ by recovering higher derivatives $s \geq 1$ one by one starting from the estimate $s=0$ provided by (10.4.1). As explained in section 10.4.2, it suffices to recover the estimates for $s=1$ from the one for $s=0$ as the procedure to recover the estimate for $s+1$ from the one for s is completely analogous. We now follow the strategy outlined in section 10.4.2.

10.4.5.1 Recovering estimates for $T\psi$

Recall that ψ satisfies

$$\square_2\psi = V\psi + N, \qquad V = -\kappa\underline{\kappa},$$

and recall also from Corollary 10.53 that we have

$$[T, \square_2]\psi \quad = \quad \mathfrak{d}^{\leq 1}(\Gamma_g)\mathfrak{d}^{\leq 2}\psi.$$

We infer

$$\square_2(T\psi) + VT(\psi) \quad = \quad T(N) + \mathfrak{d}^{\leq 1}(\Gamma_g)\mathfrak{d}^{\leq 2}\psi.$$

In view of Corollary 10.63 with $\phi = T(\psi)$, $\phi_1 = \mathfrak{d}^{\leq 1}\psi$ and $\phi_2 = T(N)$, and in view of (10.4.3), we deduce

$$\begin{aligned}
&\sup_{\tau\in[\tau_1,\tau_2]} E_p[T\psi](\tau) + B_p[T\psi](\tau_1,\tau_2) + F_p[T\psi](\tau_1,\tau_2)\\
\lesssim\ & E_p[T\psi](\tau_1) + \epsilon^2\left(\sup_{[\tau_1,\tau_2]} E[\mathfrak{d}^{\leq 1}\psi](\tau) + B_p[\mathfrak{d}^{\leq 1}\psi](\tau_1,\tau_2)\right)\\
&+J_p[T(\psi),T(N)](\tau_1,\tau_2) + \int_{{}^{(trap)}\mathcal{M}(\tau_1,\tau_2)}\left|1-\frac{3m}{r}\right| |T\phi|(|T\phi| + |R(T\phi)|)\\
&+\int_{{}^{(\not{trap})}\mathcal{M}(\tau_1,\tau_2)} r^{p-3}|T\phi|(|T\phi| + |\mathfrak{d}(T\phi)|),
\end{aligned}$$

and hence, using Lemma 10.64 with $X = T$, we infer for any $\delta \leq p \leq 2-\delta$,

$$\begin{aligned}
&\sup_{\tau\in[\tau_1,\tau_2]} E_p[T\psi](\tau) + B_p[T\psi](\tau_1,\tau_2) + F_p[T\psi](\tau_1,\tau_2)\\
\lesssim\ & E_p[T\psi](\tau_1) + J^1_p[\psi,N](\tau_1,\tau_2) + \epsilon^2\left(\sup_{[\tau_1,\tau_2]} E^1[\psi](\tau) + B^1_p[\psi](\tau_1,\tau_2)\right)\\
&+B_p[\psi](\tau_1,\tau_2). \qquad\qquad (10.4.6)
\end{aligned}$$

10.4.5.2 Recovering estimates for $r \,d\!\!/_2\psi$

Recall that ψ satisfies

$$\square_2\psi = V\psi + N, \qquad V = -\kappa\underline{\kappa},$$

and recall also from Corollary 10.55 that we have

$$r\,{\not{d}}_2(\square_2\psi) - (\square_1 - K)(r\,{\not{d}}_2\psi) = -r\eta\square_2\psi + \mathfrak{d}^{\leq 1}(\Gamma_g)\mathfrak{d}^{\leq 2}\psi.$$

We infer

$$\begin{aligned}\square_1(r\,{\not{d}}_2\psi) + (V-K)r\,{\not{d}}_2\psi &= r\eta\square_2\psi + r\,{\not{d}}_2(N) + \mathfrak{d}^{\leq 1}(\Gamma_g)\mathfrak{d}^{\leq 2}\psi \\ &= -r\eta N + r\,{\not{d}}_2(N) + \mathfrak{d}^{\leq 1}(\Gamma_g)\mathfrak{d}^{\leq 2}\psi,\end{aligned}$$

and hence

$$\square_1(r\,{\not{d}}_2\psi) + (V-K)r\,{\not{d}}_2\psi = -r\eta N + r\,{\not{d}}_2(N) + \mathfrak{d}^{\leq 1}(\Gamma_g)\mathfrak{d}^{\leq 2}\psi.$$

In view of Corollary 10.63 with $\phi = r\,{\not{d}}_2\psi$, $\phi_1 = \mathfrak{d}^{\leq 1}\psi$ and $\phi_2 = -r\eta N + r\,{\not{d}}_2(N)$, and in view of (10.4.3), we deduce

$$\begin{aligned}&\sup_{\tau\in[\tau_1,\tau_2]} E_p[r\,{\not{d}}_2\psi](\tau) + B_p[r\,{\not{d}}_2\psi](\tau_1,\tau_2) + F_p[r\,{\not{d}}_2\psi](\tau_1,\tau_2)\\ &\lesssim E_p[r\,{\not{d}}_2\psi](\tau_1) + \epsilon^2\left(\sup_{[\tau_1,\tau_2]} E[\mathfrak{d}^{\leq 1}\psi](\tau) + B_p[\mathfrak{d}^{\leq 1}\psi](\tau_1,\tau_2)\right)\\ &+J_p\Big[r\,{\not{d}}_2\psi, -r\eta N + r\,{\not{d}}_2(N)\Big](\tau_1,\tau_2)\\ &+\int_{{}^{(trap)}\mathcal{M}(\tau_1,\tau_2)}\left|1-\frac{3m}{r}\right| |r\,{\not{d}}_2\phi|(|r\,{\not{d}}_2\phi| + |R(r\,{\not{d}}_2\phi)|)\\ &+\int_{{}^{(tr\not{a}p)}\mathcal{M}(\tau_1,\tau_2)} r^{p-3}|r\,{\not{d}}_2\phi|(|r\,{\not{d}}_2\phi| + |\mathfrak{d}(r\,{\not{d}}_2\phi)|),\end{aligned}$$

and hence, using Lemma 10.64 with $X = r\,{\not{d}}_2$, we infer for any $\delta \leq p \leq 2-\delta$,

$$\begin{aligned}&\sup_{\tau\in[\tau_1,\tau_2]} E_p[r\,{\not{d}}_2\psi](\tau) + B_p[r\,{\not{d}}_2\psi](\tau_1,\tau_2) + F_p[r\,{\not{d}}_2\psi](\tau_1,\tau_2)\\ &\lesssim E_p[r\,{\not{d}}_2\psi](\tau_1) + J^1_p[\psi,N](\tau_1,\tau_2) + \epsilon^2\left(\sup_{[\tau_1,\tau_2]} E^1[\psi](\tau) + B^1_p[\psi](\tau_1,\tau_2)\right)\\ &+B_p[\psi](\tau_1,\tau_2). \qquad (10.4.7)\end{aligned}$$

10.4.5.3 Recovering estimates for $R\psi$ in $r \leq r_0$

We start with the following lemma.

Lemma 10.65. *Let ψ satisfy*

$$\square_2\psi = V\psi + N, \qquad V = -\kappa\underline{\kappa}.$$

Then, $R^2\psi$ satisfies

$$R^2\psi = -\Upsilon N + T^2\psi + O(r^{-2})\,{\not{\mathfrak{d}}}^2\psi + O(r^{-1})\mathfrak{d}\psi + O(r^{-2})\psi.$$

Proof. Recall that we have

$$\square_2\psi = -e_3(e_4(\psi)) + \not{\Delta}_2\psi - \frac{1}{2}\kappa e_3\psi + \left(-\frac{1}{2}\underline{\kappa} + 2\underline{\omega}\right) e_4\psi + 2\eta e_\theta\psi$$

and

$$e_4 = T + R, \quad \Upsilon e_3 = (T - R).$$

We infer

$$\begin{aligned}\Upsilon\square_2\psi &= -(T-R)(T+R)\psi + \Upsilon\not{\Delta}_2\psi - \frac{\Upsilon}{2}\kappa e_3\psi + \Upsilon\left(-\frac{1}{2}\underline{\kappa} + 2\underline{\omega}\right) e_4\psi \\ &\quad +2\Upsilon\eta e_\theta\psi \\ &= -T^2\psi + R^2\psi - [T,R]\psi + O(r^{-2})\not{\mathfrak{d}}^2\psi + O(r^{-1})\mathfrak{d}\psi\end{aligned}$$

and hence

$$\begin{aligned}R^2\psi &= -\Upsilon\square_2\psi + T^2\psi - [T,R]\psi + O(r^{-2})\not{\mathfrak{d}}^2\psi + O(r^{-1})\mathfrak{d}\psi \\ &= -\Upsilon N + T^2\psi - [T,R]\psi + O(r^{-2})\not{\mathfrak{d}}^2\psi + O(r^{-1})\mathfrak{d}\psi + O(r^{-2})\psi\end{aligned}$$

where we used the fact that $\square_2\psi = V\psi + N$ and $V = \kappa\underline{\kappa} = O(r^{-2})$. Also, we have

$$\begin{aligned}[T,R]\psi &= \frac{1}{4}[e_4 + \Upsilon e_3, e_4 - \Upsilon e_3]\psi \\ &= \frac{1}{2}\Big(-e_4(\Upsilon)e_3 + \Upsilon[e_3,e_4]\Big)\psi \\ &= O(r^{-2})\mathfrak{d}\psi\end{aligned}$$

and thus

$$R^2\psi = -\Upsilon N + T^2\psi + O(r^{-2})\not{\mathfrak{d}}^2\psi + O(r^{-1})\mathfrak{d}\psi + O(r^{-2})\psi.$$

This concludes the proof of the lemma. $\square$

We now estimate $R\psi$ in $r \le r_0$ for a fixed $r_0 \ge 4m_0$ that will be chosen large enough. First, in view of the identity of the previous lemma, i.e.,

$$R^2\psi = -\Upsilon N + T^2\psi + O(r^{-2})\not{\mathfrak{d}}^2\psi + O(r^{-1})\mathfrak{d}\psi + O(r^{-2})\psi,$$

we infer

$$\begin{aligned}&\sup_{[\tau_1,\tau_2]}\int_{\Sigma_{r\le r_0}(\tau)}|R^2\psi|^2 + \int_{\mathcal{M}_{r\le r_0}(\tau_1,\tau_2)}\left(1-\frac{3m}{r}\right)^2(R^2\psi)^2 \\ &\lesssim \sup_{[\tau_1,\tau_2]}\left(E[T\psi] + E[r\not{d}_2\psi] + E[\psi] + \int_{\Sigma_{r\le r_0}(\tau)}N^2\right) \\ &\quad + \int_{\mathcal{M}_{r\le r_0}(\tau_1,\tau_2)}N^2 + \mathrm{Morr}[T\psi](\tau_1,\tau_2) \\ &\quad + \mathrm{Morr}[r\not{d}_2\psi](\tau_1,\tau_2) + \mathrm{Morr}[\psi](\tau_1,\tau_2).\end{aligned} \tag{10.4.8}$$

Next, we remove the degeneracy of the above estimate at $r = 3m$. Recall from Corollary 10.57 that we have in the region $r \leq 4m_0$

$$\square_2(R\psi) = \left(1 - \frac{3m}{r}\right)\mathfrak{d}^2\psi + O\left(\frac{\epsilon}{u_{trap}^{1+\delta_{dec}-2\delta_0}}\right)\mathfrak{d}^2\psi + O(1)\mathfrak{d}^{\leq 1}\psi + O(1)\mathfrak{d}^{\leq 1}N.$$

Then,

1. multiplying $R\psi$ with a cut-off function equal to one on $[5/2m_0, 7/2m_0]$ and vanishing on $[9/4m_0, 4m_0]$ and inferring the corresponding wave equation from the above one for $R\psi$,
2. relying on the Morawetz estimate of Proposition 10.12 with the particular choice $f(r) = r - 3m$,
3. adding a large multiple of the energy estimate,
4. and using Proposition 10.32 for the boundary terms,

we easily infer the following estimate:

$$\begin{aligned}&\int_{{}^{(trap)}\mathcal{M}(\tau_1,\tau_2)}(R^2\psi)^2\\ \lesssim\ &\int_{\mathcal{M}_{r\leq 4m_0}(\tau_1,\tau_2)}\left(\left(1-\frac{3m}{r}\right)^2(\mathfrak{d}^2\psi)^2 + (\mathfrak{d}\psi)^2 + (\mathfrak{d}^{\leq 1}N)^2\right)\\ &+E[R\psi](\tau_1) + \epsilon \sup_{[\tau_1,\tau_2]} E^1[\psi](\tau).\end{aligned}$$

Together with (10.4.8), we infer

$$\begin{aligned}&\sup_{[\tau_1,\tau_2]}\int_{\Sigma_{r\leq r_0}(\tau)}|R^2\psi|^2 + \int_{\mathcal{M}_{r\leq r_0}(\tau_1,\tau_2)}|R^2\psi|^2\\ \lesssim\ & E[R\psi](\tau_1) + \sup_{[\tau_1,\tau_2]}\Big(\epsilon E^1[\psi](\tau) + E[T\psi](\tau) + E[r\, \not{d}_2\psi](\tau) + E[\psi](\tau)\Big)\\ &+J_p^1[\psi,N](\tau_1,\tau_2) + \text{Morr}[T\psi](\tau_1,\tau_2) + \text{Morr}[r\,\not{d}_2\psi](\tau_1,\tau_2) + \text{Morr}[\psi](\tau_1,\tau_2).\end{aligned} \tag{10.4.9}$$

10.4.5.4 Recovering estimates for $Y_{\mathcal{H}}\psi$

Recall that ψ satisfies

$$\square_2\psi = V\psi + N, \qquad V = -\kappa\underline{\kappa},$$

and recall also from Lemma 10.59

$$\begin{aligned}[\square_2, Y_{\mathcal{H}}]\psi =\ & d_0 Y_{(0)}(Y_{\mathcal{H}}\psi) + 1_{\Upsilon\leq 2\delta_{\mathcal{H}}}\left(\square_2\psi + \mathfrak{d}T\psi + \mathfrak{d}^{\leq 1}(\Gamma_g)\mathfrak{d}^2\psi + \frac{1}{\delta_{\mathcal{H}}^2}\mathfrak{d}^{\leq 1}\psi\right)\\ &+\frac{1}{\delta_{\mathcal{H}}}1_{\delta_{\mathcal{H}}\leq\Upsilon\leq 2\delta_{\mathcal{H}}}\mathfrak{d}^{\leq 2}\psi\end{aligned}$$

where the scalar function d_0 satisfies the bound

$$d_0 = \frac{1}{2m_0} + O(\delta_{\mathcal{H}}) \text{ on the support of } \kappa_{\mathcal{H}}.$$

We infer

$$\square_2(Y_{\mathcal{H}}\psi) - VY_{\mathcal{H}}(\psi) = d_0 Y_{(0)}(Y_{\mathcal{H}}\psi) + 1_{\Upsilon\leq 2\delta_{\mathcal{H}}}\left(N + \mathfrak{d}T\psi + \epsilon\mathfrak{d}^2\psi + \frac{1}{\delta_{\mathcal{H}}^2}\mathfrak{d}^{\leq 1}\psi\right) + \frac{1}{\delta_{\mathcal{H}}}1_{\delta_{\mathcal{H}}\leq\Upsilon\leq 2\delta_{\mathcal{H}}}\mathfrak{d}^{\leq 2}\psi + Y_{\mathcal{H}}(N).$$

Then,

1. we use the redshift vectorfield $Y_{\mathcal{H}}$ as a multiplier,
2. we rely on Proposition 10.29,
3. we use the fact that $d_0 \geq 0$,
4. we add a large multiple of the energy,
5. and we use Proposition 10.32 for the boundary terms.

We easily infer

$$\begin{aligned}
&\sup_{[\tau_1,\tau_2]} E[Y_{\mathcal{H}}\psi] + \text{Morr}[Y_{\mathcal{H}}\psi](\tau_1,\tau_2) \\
\lesssim\; & E[Y_{\mathcal{H}}\psi](\tau_1) + J_p^1[\psi, N](\tau_1,\tau_2) + \epsilon\text{Morr}[\mathfrak{d}\psi](\tau_1,\tau_2) + \text{Morr}[R\psi](\tau_1,\tau_2) \\
&+\text{Morr}[T\psi](\tau_1,\tau_2) + \text{Morr}[r\,\not{d}_2\psi](\tau_1,\tau_2) + \text{Morr}[\psi](\tau_1,\tau_2). \qquad (10.4.10)
\end{aligned}$$

10.4.5.5 Recovering estimates for $re_4\psi$ in $r \geq r_0$

Recall that ψ satisfies

$$\square_2\psi = V\psi + N, \qquad V = -\kappa\underline{\kappa},$$

and recall also from Lemma 10.60

$$[\square_2, re_4]\psi = \frac{\Upsilon}{r}\left(1 + \frac{2m}{r\Upsilon^2}\right)\check{e}_4(re_4\psi) + \square_2\psi + \Gamma_g\mathfrak{d}^2\psi + \frac{1}{\Upsilon}r^{-2}\mathfrak{d}T\psi + r^{-2}\,\not{\mathfrak{d}}^2\psi + r^{-2}\mathfrak{d}\psi.$$

We infer

$$\square_2(re_4\psi) - Vre_4(\psi) = \frac{\Upsilon}{r}\left(1 + \frac{2m}{r\Upsilon^2}\right)\check{e}_4(re_4\psi) + O\left(\frac{\epsilon}{r^2}\right)\mathfrak{d}^2\psi + \frac{1}{\Upsilon}r^{-2}\mathfrak{d}T\psi + r^{-2}\,\not{\mathfrak{d}}^2\psi + r^{-2}\mathfrak{d}^{\leq 1}\psi + N + re_4(N).$$

Then,

1. as in section 10.2.3, we use the vectorfield $f_p e_4$ as a multiplier, where we have $f_p = \theta_{r_0}(r)r^p e_4$ and the cut-off $\theta_{r_0}(r)$ is equal to one in the region $r \geq r_0$ and vanishes in the region $r \leq r_0/2$,
2. we rely on Proposition 10.43 to control the bulk and the boundary terms,
3. and we use the fact that the prefactor of the term $\check{e}_4(re_4\psi)$ on the right-hand side is positive for $r \geq 4m_0$, i.e.,

$$\frac{\Upsilon}{r}\left(1 + \frac{2m}{r\Upsilon^2}\right) \geq 0 \text{ for } r \geq 4m_0.$$

We easily infer

$$\begin{aligned}
&\sup_{\tau\in[\tau_1,\tau_2]} E_{p,r\geq r_0}[re_4\psi](\tau)+B_{p,r\geq r_0}[re_4\psi](\tau_1,\tau_2)+F_{p,r\geq r_0}[re_4\psi](\tau_1,\tau_2)\\
\lesssim\;& E^1_p[re_4\psi](\tau_1)+J^1_p[\psi,N](\tau_1,\tau_2)+B^1_{p,r_0/2\leq r<r_0}[\psi](\tau_1,\tau_2)+\epsilon B^1_p[\psi](\tau_1,\tau_2)\\
&+B_p[T\psi](\tau_1,\tau_2)+B_p[r\,\not{d}_2\psi](\tau_1,\tau_2)+B_p[\psi](\tau_1,\tau_2). \qquad (10.4.11)
\end{aligned}$$

10.4.5.6 Conclusion of the proof of Theorem 5.17

Gathering the estimates (10.4.6), (10.4.7), (10.4.9), (10.4.10) and (10.4.11), we infer, for any $\delta\leq p\leq 2-\delta$,

$$\begin{aligned}
&\sup_{\tau\in[\tau_1,\tau_2]} E^1_p[\psi](\tau)+B^1_p[\psi](\tau_1,\tau_2)+F^1_p[\psi](\tau_1,\tau_2)\\
\lesssim\;& E^1_p[\psi](\tau_1)+J^1_p[\psi,N](\tau_1,\tau_2)+\epsilon^2\left(\sup_{[\tau_1,\tau_2]} E^1_p[\psi](\tau)+B^1_p[\psi](\tau_1,\tau_2)\right)\\
&+\sup_{[\tau_1,\tau_2]} E_p[\psi](\tau)+B_p[\psi](\tau_1,\tau_2),
\end{aligned}$$

and hence

$$\begin{aligned}
&\sup_{\tau\in[\tau_1,\tau_2]} E^1_p[\psi](\tau)+B^1_p[\psi](\tau_1,\tau_2)+F^1_p[\psi](\tau_1,\tau_2)\\
\lesssim\;& E^1_p[\psi](\tau_1)+J^1_p[\psi,N](\tau_1,\tau_2)+\sup_{[\tau_1,\tau_2]} E_p[\psi](\tau)+B_p[\psi](\tau_1,\tau_2).
\end{aligned}$$

In view of (10.4.1), we deduce

$$\sup_{\tau\in[\tau_1,\tau_2]} E^1_p[\psi](\tau)+B^1_p[\psi](\tau_1,\tau_2)+F^1_p[\psi](\tau_1,\tau_2)\;\lesssim\; E^1_p[\psi](\tau_1)+J^1_p[\psi,N](\tau_1,\tau_2)$$

which is Theorem 5.17 in the case $s=1$. We have thus deduced Theorem 5.17 in the case $s=1$ from the case $s=0$, i.e., (10.4.1). Since going from $s=0$ to $s=1$ is analogous to going from s to $s+1$, higher order derivatives $k\leq k_{small}+30$ are recovered in the same fashion. This concludes the proof of Theorem 5.17.

10.4.6 Proof of Theorem 5.18

We now conclude the proof of Theorem 5.18 for all $0\leq s\leq k_{small}+29$ by recovering higher derivatives $s\geq 1$ one by one starting from the estimate $s=0$ provided by (10.4.2). As explained in section 10.4.2, it suffices to recover the estimate for $s=1$ from the one for $s=0$ as the procedure to recover the estimate for $s+1$ from the one for s is completely analogous. We now follow the strategy outlined in section 10.4.2.

10.4.6.1 Recovering estimates for $T\check{\psi}$

Recall from Proposition 10.47 that $\check{\psi}=f_2\check{e}_4\psi$ satisfies

$$\square\check{\psi}-V\check{\psi} \;=\; \frac{2}{r}\left(1-\frac{3m}{r}\right)e_4\check{\psi}+\check{N}+f_2\left(e_4+\frac{3}{r}\right)N$$

where

$$\check{N} = \begin{cases} O(r^{-2})\mathfrak{d}^{\leq 1}\psi + r\Gamma_b e_4\mathfrak{d}\psi + \mathfrak{d}^{\leq 1}(\Gamma_b)\mathfrak{d}^{\leq 1}\psi \\ +r\mathfrak{d}^{\leq 1}(\Gamma_g)e_3\psi + \mathfrak{d}^{\leq 1}(\Gamma_g)\mathfrak{d}^2\psi, \quad r \geq 6m_0, \\ \\ O(1)\mathfrak{d}^{\leq 2}\psi, \qquad 4m_0 \leq r \leq 6m_0, \end{cases}$$

and recall also from Corollary 10.53 that we have

$$[T, \square_2]\check{\psi} = \mathfrak{d}^{\leq 1}(\Gamma_g)\mathfrak{d}^{\leq 2}\check{\psi}.$$

We infer

$$\square(T\check{\psi}) - VT(\check{\psi}) = \frac{2}{r}\left(1 - \frac{3m}{r}\right)e_4(T\check{\psi}) + N^T + T\left(f_2\left(e_4 + \frac{3}{r}\right)N\right),$$

where we have, in view of the estimates[24] of Lemma 5.1 for $\mathfrak{d}^k\Gamma_g$ and $\mathfrak{d}^k\Gamma_b$ with $k \leq k_{small} + 30$,

$$N^T = \begin{cases} O\left(\frac{1}{\tau^{1+\delta_{dec}-2\delta_0}}\right)\left(e_4\mathfrak{d}^2\psi + r^{-1}\mathfrak{d}^{\leq 2}\psi\right) \\ +O\left(\frac{1}{r\tau^{\frac{1}{2}+\delta_{dec}-2\delta_0}}\right)e_3\mathfrak{d}^{\leq 1}\psi + O\left(\frac{1}{r^2}\right)\left(\mathfrak{d}^{\leq 3}\psi + \epsilon\mathfrak{d}^{\leq 2}\check{\psi}\right), \quad r \geq 6m_0, \\ \\ O(1)\mathfrak{d}^{\leq 3}\psi, \qquad 4m_0 \leq r \leq 6m_0. \end{cases}$$

In view of (10.4.2) with $T\check{\psi}$ instead of $\check{\psi}$ and with

$$N^T + T\left(f_2\left(e_4 + \frac{3}{r}\right)N\right)$$

instead of $\check{N} + f_2\left(e_4 + \frac{3}{r}\right)N$, we deduce

$$\begin{aligned} &\sup_{\tau\in[\tau_1,\tau_2]} E_q[T\check{\psi}](\tau) + B_q[T\check{\psi}](\tau_1,\tau_2) \\ \lesssim\ & E_q[T\check{\psi}](\tau_1) + J_q\left[T\check{\psi}, N^T + T\left(f_2\left(e_4 + \frac{3}{r}\right)N\right)\right](\tau_1,\tau_2) \\ &+E^1_{\max(q,\delta)}[T\psi](\tau_1) + J^1_{\max(q,\delta)}[T\psi, TN] \\ \lesssim\ & E_q[T\check{\psi}](\tau_1) + \check{J}^1_q\left[\check{\psi}, N\right](\tau_1,\tau_2) + J_q\left[T\check{\psi}, N^T\right](\tau_1,\tau_2) \\ &+E^2_{\max(q,\delta)}[\psi](\tau_1) + J^2_{\max(q,\delta)}[\psi, N], \end{aligned}$$

so that it remains to estimate

$$J_q[T\check{\psi}, N^T](\tau_1,\tau_2) = J_{q,4m_0}\left[T\check{\psi}, N^T\right](\tau_1,\tau_2) = \left|\int_{\mathcal{M}_{\geq 4m_0}(\tau_1,\tau_2)} r^q(\check{e}_4(T\check{\psi}))N^T\right|.$$

[24]Here, unlike the proof of Theorem 5.17 above, the non-sharp estimates of section 10.4.1 are not enough, and we need instead to rely on the stronger estimates provided by Lemma 5.1.

We have, in view of the definition of N^T,

$$
\begin{aligned}
& J_q[T\check{\psi}, N^T](\tau_1,\tau_2) \\
\lesssim\; & \left|\int_{\mathcal{M}_{\geq 4m_0}(\tau_1,\tau_2)} r^q(\check{e}_4(T\check{\psi}))\frac{1}{\tau^{1+\delta_{dec}-2\delta_0}}\Big(e_4\mathfrak{d}^2\psi + r^{-1}\mathfrak{d}^{\leq 2}\psi\Big)\right| \\
& + \left|\int_{\mathcal{M}_{\geq 4m_0}(\tau_1,\tau_2)} r^q(\check{e}_4(T\check{\psi}))\frac{1}{r\tau^{\frac{1}{2}+\delta_{dec}-2\delta_0}}e_3\mathfrak{d}^{\leq 1}\psi\right| \\
& + \left|\int_{\mathcal{M}_{\geq 4m_0}(\tau_1,\tau_2)} r^q(\check{e}_4(T\check{\psi}))\frac{1}{r^2}\mathfrak{d}^{\leq 3}\psi\right| + \left|\int_{\mathcal{M}_{\geq 4m_0}(\tau_1,\tau_2)} r^q(\check{e}_4(T\check{\psi}))\frac{\epsilon}{r^2}\mathfrak{d}^{\leq 2}\check{\psi}\right| \\
\lesssim\; & \left(\sup_{[\tau_1,\tau_2]}\int_{\Sigma(\tau)} r^q(\check{e}_4(T\check{\psi}))^2\right)^{\frac{1}{2}}\left\{\left(\sup_{[\tau_1,\tau_2]}\int_{\Sigma(\tau)} r^q\Big((e_4\mathfrak{d}^2\psi)^2 + r^{-2}(\mathfrak{d}^{\leq 2}\psi)^2\Big)\right)^{\frac{1}{2}}\right. \\
& \left. + \left(\int_{\mathcal{M}_{r\geq 4m_0}(\tau_1,\tau_2)} r^{q-2}(e_3\mathfrak{d}^{\leq 1}\psi)^2\right)^{\frac{1}{2}}\right\} + \epsilon\int_{\mathcal{M}(\tau_1,\tau_2)} r^{q-3}(\mathfrak{d}^{\leq 2}\check{\psi})^2 \\
& + \left(\int_{\mathcal{M}(\tau_1,\tau_2)} r^{q-1}(\check{e}_4(T\check{\psi}))^2\right)^{\frac{1}{2}}\left(\int_{\mathcal{M}(\tau_1,\tau_2)} r^{q-3}(\mathfrak{d}^{\leq 3}\psi)^2\right)^{\frac{1}{2}}
\end{aligned}
$$

which yields, using in particular the fact that $q \leq 1-\delta$,

$$
\begin{aligned}
& J_q[T\check{\psi}, N^T](\tau_1,\tau_2) \\
\lesssim\; & \left(\sup_{[\tau_1,\tau_2]} E_q[T\check{\psi}](\tau)\right)^{\frac{1}{2}}\left\{\sup_{[\tau_1,\tau_2]} E^2_{\max(q,\delta)}[\psi](\tau) + B^1_{\max(q,\delta)}[\psi](\tau_1,\tau_2)\right\}^{\frac{1}{2}} \\
& +\epsilon B^1_q[\check{\psi}](\tau_1,\tau_2) + \Big(B_q[T\check{\psi}](\tau_1,\tau_2)\Big)^{\frac{1}{2}}\Big(B^2_{\max(q,\delta)}[\psi](\tau_1,\tau_2)\Big)^{\frac{1}{2}}.
\end{aligned}
$$

We deduce

$$
\begin{aligned}
& \sup_{\tau\in[\tau_1,\tau_2]} E_q[T\check{\psi}](\tau) + B_q[T\check{\psi}](\tau_1,\tau_2) \\
\lesssim\; & E_q[T\check{\psi}](\tau_1) + \check{J}^1_q\left[\check{\psi}, N\right](\tau_1,\tau_2) + \epsilon B^1_q[\check{\psi}](\tau_1,\tau_2) \\
& + \sup_{[\tau_1,\tau_2]} E^2_{\max(q,\delta)}[\psi](\tau) + B^2_{\max(q,\delta)}[\psi](\tau_1,\tau_2) + J^2_{\max(q,\delta)}[\psi, N].
\end{aligned}
$$

Together with Theorem 5.17, this yields

$$
\begin{aligned}
& \sup_{\tau\in[\tau_1,\tau_2]} E_q[T\check{\psi}](\tau) + B_q[T\check{\psi}](\tau_1,\tau_2) \\
\lesssim\; & E_q[T\check{\psi}](\tau_1) + \check{J}^1_q\left[\check{\psi}, N\right](\tau_1,\tau_2) + \epsilon B^1_q[\check{\psi}](\tau_1,\tau_2) + E^{\,2}_{\max(q,\delta)}[\psi](\tau_1) \\
& + J^2_{\max(q,\delta)}[\psi, N]. \qquad (10.4.12)
\end{aligned}
$$

10.4.6.2 Recovering estimates for $r \, d\!\!/_2 \check{\psi}$

Recall from Proposition 10.47 that $\check{\psi} = f_2 \check{e}_4 \psi$ satisfies

$$\square \check{\psi} - V \check{\psi} = \frac{2}{r}\left(1 - \frac{3m}{r}\right) e_4 \check{\psi} + \check{N} + f_2 \left(e_4 + \frac{3}{r}\right) N.$$

Recall also from Corollary 10.55 that we have

$$r \, d\!\!/_2(\square_2 \check{\psi}) - (\square_1 - K)(r \, d\!\!/_2 \check{\psi}) = -r\eta \square_2 \check{\psi} + \mathfrak{d}^{\leq 1}(\Gamma_g)\mathfrak{d}^{\leq 2} \check{\psi}.$$

We infer

$$\begin{aligned} \square_1(r \, d\!\!/_2 \check{\psi}) + (V - K) r \, d\!\!/_2 \check{\psi} &= \frac{2}{r}\left(1 - \frac{3m}{r}\right) e_4(r \, d\!\!/_2 \check{\psi}) + N^{r \, d\!\!/_2} \\ &\quad + (r \, d\!\!/_2 - r\eta)\left(f_2 \left(e_4 + \frac{3}{r}\right) N\right) \end{aligned}$$

where

$$N^{r \, d\!\!/_2} = -r\eta \check{N} + r \, d\!\!/_2(\check{N}) + \mathfrak{d}^{\leq 1}(\Gamma_g)\mathfrak{d}^{\leq 2} \check{\psi}.$$

In view of (10.4.5) with $r \, d\!\!/_2 \check{\psi}$ instead of $\check{\psi}$ and with

$$N^{r \, d\!\!/_2} + (r \, d\!\!/_2 - r\eta)\left(f_2 \left(e_4 + \frac{3}{r}\right) N\right)$$

instead of $\check{N} + f_2 \left(e_4 + \frac{3}{r}\right) N$, we deduce

$$\begin{aligned} &\sup_{\tau \in [\tau_1, \tau_2]} E_q[r \, d\!\!/_2 \check{\psi}](\tau) + B_q[r \, d\!\!/_2 \check{\psi}](\tau_1, \tau_2) \\ &\lesssim E_q[r \, d\!\!/_2 \check{\psi}](\tau_1) + \check{J}_q\left[r \, d\!\!/_2 \check{\psi}, N^{r \, d\!\!/_2} + (r \, d\!\!/_2 - r\eta)\left(f_2 \left(e_4 + \frac{3}{r}\right) N\right)\right](\tau_1, \tau_2) \\ &\quad + E^1_{\max(q,\delta)}[r \, d\!\!/_2 \psi](\tau_1) + J^1_{\max(q,\delta)}[r \, d\!\!/_2 \psi, r \, d\!\!/_2 N] \\ &\quad + \int_{\mathcal{M}(\tau_1, \tau_2)} r^{q-3} |r \, d\!\!/_2 \check{\psi}| (|r \, d\!\!/_2 \check{\psi}| + |\mathfrak{d}(r \, d\!\!/_2 \check{\psi})|) \\ &\lesssim E_q[r \, d\!\!/_2 \check{\psi}](\tau_1) + \check{J}_q\left[r \, d\!\!/_2 \check{\psi}, N^{r \, d\!\!/_2}\right](\tau_1, \tau_2) \\ &\quad + E^2_{\max(q,\delta)}[\psi](\tau_1) + J^2_{\max(q,\delta)}[\psi, N] + \left(B_q[r \, d\!\!/_2 \check{\psi}](\tau_1, \tau_2)\right)^{\frac{1}{2}} \left(B_q[\check{\psi}](\tau_1, \tau_2)\right)^{\frac{1}{2}} \end{aligned}$$

so that it remains to estimate

$$\begin{aligned} \check{J}_q\left[r \, d\!\!/_2 \check{\psi}, N^{r \, d\!\!/_2}\right](\tau_1, \tau_2) &= J_{q, 4m_0}\left[r \, d\!\!/_2 \check{\psi}, N^{r \, d\!\!/_2}\right](\tau_1, \tau_2) \\ &= \left| \int_{\mathcal{M}_{\geq 4m_0}(\tau_1, \tau_2)} r^q (\check{e}_4(r \, d\!\!/_2 \check{\psi})) N^{r \, d\!\!/_2} \right|. \end{aligned}$$

The estimate follows along the same lines as the above one for $J_q[T \check{\psi}, N^T](\tau_1, \tau_2)$ so we leave the details to the reader. In the end, we arrive at the following analog of (10.4.12):

$$\begin{aligned}
&\sup_{\tau\in[\tau_1,\tau_2]} E_q[r\,\not{d}_2\,\check{\psi}](\tau)+B_q[r\,\not{d}_2\,\check{\psi}](\tau_1,\tau_2)\\
&\lesssim\; E_q[r\,\not{d}_2\,\check{\psi}](\tau_1)+B_q[\check{\psi}](\tau_1,\tau_2)+\check{J}^1_q\left[\check{\psi},N\right](\tau_1,\tau_2)+\epsilon B^1_q[\check{\psi}](\tau_1,\tau_2)\\
&\quad+E^{\,2}_{\max(q,\delta)}[\psi](\tau_1)+J^2_{\max(q,\delta)}[\psi,N]. \qquad (10.4.13)
\end{aligned}$$

10.4.6.3 Recovering estimates for $re_4\,\check{\psi}$

Recall from Proposition 10.47 that $\check{\psi}=f_2\check{e}_4\psi$ satisfies

$$\square\check{\psi}-V\check{\psi} \;=\; \frac{2}{r}\left(1-\frac{3m}{r}\right)e_4\check{\psi}+\check{N}+f_2\left(e_4+\frac{3}{r}\right)N.$$

Recall also from Lemma 10.60 that we have

$$\begin{aligned}
[\square_2,re_4]\check{\psi} \;=\;& \frac{\Upsilon}{r}\left(1+\frac{2m}{r\Upsilon^2}\right)\check{e}_4(re_4\check{\psi})+\square_2\check{\psi}+\Gamma_g\mathfrak{d}^2\check{\psi}+\frac{1}{\Upsilon}r^{-2}\mathfrak{d}T\check{\psi}\\
&+r^{-2}\not{\mathfrak{d}}^2\check{\psi}+r^{-2}\mathfrak{d}\check{\psi}.
\end{aligned}$$

We infer[25]

$$\begin{aligned}
\square(re_4\check{\psi})-Vre_4(\check{\psi}) \;=\;& \left(\frac{2}{r}\left(1-\frac{3m}{r}\right)+\frac{\Upsilon}{r}\left(1+\frac{2m}{r\Upsilon^2}\right)\right)e_4(re_4\check{\psi})+N^{re_4}\\
&+re_4\left(f_2\left(e_4+\frac{3}{r}\right)N\right),
\end{aligned}$$

where

$$N^{re_4} \;=\; re_4(\check{N})+\check{N}+\Gamma_g\mathfrak{d}^2\check{\psi}+\frac{1}{\Upsilon}r^{-2}\mathfrak{d}T\check{\psi}+r^{-2}\not{\mathfrak{d}}^2\check{\psi}+r^{-2}\mathfrak{d}\check{\psi}.$$

The rest follows along the same lines as the estimate for $T\check{\psi}$ and we arrive at the following analog of (10.4.12):

$$\begin{aligned}
&\sup_{\tau\in[\tau_1,\tau_2]} E_q[re_4\,\check{\psi}](\tau)+B_q[re_4\,\check{\psi}](\tau_1,\tau_2) \qquad (10.4.14)\\
&\lesssim\; E_q[re_4\,\check{\psi}](\tau_1)+B_q[\check{\psi}](\tau_1,\tau_2)+B_q[T\check{\psi}](\tau_1,\tau_2)+B_q[r\,\not{d}_2\,\check{\psi}](\tau_1,\tau_2)\\
&\quad+\check{J}^1_q\left[\check{\psi},N\right](\tau_1,\tau_2)+\epsilon B^1_q[\check{\psi}](\tau_1,\tau_2)+E^{\,2}_{\max(q,\delta)}[\psi](\tau_1)+J^2_{\max(q,\delta)}[\psi,N].
\end{aligned}$$

[25] Notice that the coefficient in front of the term $e_4(re_4\check{\psi})$ in the RHS of the wave equation for $re_4\check{\psi}$ differs from the one in front of the term $e_4\check{\psi}$ in the RHS of the wave equation for $\check{\psi}$. Nevertheless, we may apply (10.4.2) with $re_4\check{\psi}$ instead of $\check{\psi}$ since the only property of this coefficient which is used is that it is positive on $r\geq 4m_0$, i.e.,

$$\frac{2}{r}\left(1-\frac{3m}{r}\right)+\frac{\Upsilon}{r}\left(1+\frac{2m}{r\Upsilon^2}\right)\geq 0 \text{ on } r\geq 4m_0.$$

10.4.6.4 Conclusion of the proof of Theorem 5.18

Gathering the estimates (10.4.12), (10.4.13) and (10.4.14), we infer, for any q such that $-1+\delta < q \le 1-\delta$,

$$\begin{aligned}&\sup_{\tau\in[\tau_1,\tau_2]} E^1_q[\check{\psi}](\tau) + B^1_q[\check{\psi}](\tau_1,\tau_2)\\ \lesssim\ & E^1_q[\check{\psi}](\tau_1) + \check{J}^1_q[\check{\psi},N](\tau_1,\tau_2) + B_q[\check{\psi}](\tau_1,\tau_2) + \epsilon B^1_q[\check{\psi}](\tau_1,\tau_2)\\ &+E^2_{\max(q,\delta)}[\psi](\tau_1) + J^2_{\max(q,\delta)}[\psi,N]\end{aligned}$$

and hence

$$\begin{aligned}\sup_{\tau\in[\tau_1,\tau_2]} E^1_q[\check{\psi}](\tau) + B^1_q[\check{\psi}](\tau_1,\tau_2) \lesssim{}& E^1_q[\check{\psi}](\tau_1) + \check{J}^1_q[\check{\psi},N](\tau_1,\tau_2) + B_q[\check{\psi}](\tau_1,\tau_2)\\ &+ E^2_{\max(q,\delta)}[\psi](\tau_1) + J^2_{\max(q,\delta)}[\psi,N].\end{aligned}$$

In view of (10.4.2), we deduce

$$\begin{aligned}\sup_{\tau\in[\tau_1,\tau_2]} E^1_q[\check{\psi}](\tau) + B^1_q[\check{\psi}](\tau_1,\tau_2) \ \lesssim\ & E^1_q[\check{\psi}](\tau_1) + \check{J}^1_q[\check{\psi},N](\tau_1,\tau_2)\\ &+E^2_{\max(q,\delta)}[\psi](\tau_1) + J^2_{\max(q,\delta)}[\psi,N]\end{aligned}$$

which is Theorem 5.18 in the case $s=1$. We have thus deduced Theorem 5.18 in the case $s=1$ from the case $s=0$, i.e., (10.4.2). Since going from $s=0$ to $s=1$ is analogous to going from s to $s+1$, higher order derivatives $k \le k_{small}+29$ are recovered in the same fashion. This concludes the proof of Theorem 5.18.

10.5 MORE WEIGHTED ESTIMATES FOR WAVE EQUATIONS

The goal of this section is to derive Theorem 10.67 and Proposition 10.69, see below, which is needed for the proof of Theorem M8 in Chapter 8. Recall that we have used so far in Chapter 10 the global frame of Proposition 3.26. For this last section of Chapter 10, we rely instead on the global frame used in Theorem M8, i.e., the one of Proposition 3.23, as it is more regular and allows us to derive estimates for up to k_{large} derivatives.

Remark 10.66. *Recall that in the frame of Proposition 3.23, we only have*[26] $\eta \in \Gamma_b$. *Note that the assumptions on the frame used in Chapter 10 are all consistent with* $\eta\in\Gamma_b$, *so that all results in this chapter apply for the frame of Proposition 3.23.*

Theorem 10.67. *Let* ψ *a reduced 2-scalar, and* ϕ *a reduced 0-scalar satisfying respectively*

$$(\square_2+V_2)\psi = N_2, \quad (\square_0+V_0)\phi = N_0, \quad V_2 = -\frac{2}{r^2}\left(1+\frac{2m}{r}\right), \quad V_0 = \frac{8m}{r^3}.$$

Also, assume that the Ricci coefficients and curvature components associated to the

[26]Unlike the frame of Proposition 3.26 for which $\eta\in\Gamma_g$.

global null frame we are using satisfy the estimates of section 10.4.1 for $k \leq k_{small}$ derivatives. Then, for any $1 \leq s \leq k_{large} - 1$, we have

$$
\begin{aligned}
&\sup_{\tau\in[\tau_1,\tau_2]} E^s_\delta[\psi](\tau) + B^s_\delta[\psi](\tau_1,\tau_2) + F^s_\delta[\psi](\tau_1,\tau_2) \\
\lesssim\ & E^s_\delta[\psi](\tau_1) + \sup_{\tau\in[\tau_1,\tau_2]} E^{s-1}_\delta[\psi](\tau) + B^{s-1}_\delta[\psi](\tau_1,\tau_2) + F^{s-1}_\delta[\psi](\tau_1,\tau_2) \\
&+D_s[\Gamma]\left(\sup_{\mathcal{M}(\tau_1,\tau_2)} r u_{trap}^{\frac{1}{2}+\delta_{dec}} |\mathfrak{d}^{\leq k_{small}}\psi|\right)^2 \\
&+\int_{\mathcal{M}(\tau_1,\tau_2)} r^{1+\delta}|\mathfrak{d}^{\leq s}N_2|^2 + \left|\int_{{}^{(trap)}\mathcal{M}(\tau_1,\tau_2)} T(\mathfrak{d}^s\phi)\mathfrak{d}^s N_2\right| \qquad (10.5.1)
\end{aligned}
$$

and

$$
\begin{aligned}
&\sup_{\tau\in[\tau_1,\tau_2]} E^s_\delta[\phi](\tau) + B^s_\delta[\phi](\tau_1,\tau_2) + F^s_\delta[\phi](\tau_1,\tau_2) \\
\lesssim\ & E^s_\delta[\phi](\tau_1) + \sup_{\tau\in[\tau_1,\tau_2]} E^{s-1}_\delta[\phi](\tau) + B^{s-1}_\delta[\phi](\tau_1,\tau_2) + F^{s-1}_\delta[\phi](\tau_1,\tau_2) \\
&+D_s[\Gamma]\left(\sup_{\mathcal{M}(\tau_1,\tau_2)} r u_{trap}^{\frac{1}{2}+\delta_{dec}} |\mathfrak{d}^{\leq k_{small}}\phi|\right)^2 + \int_{\Sigma(\tau_2)} \frac{(\mathfrak{d}^{\leq s}\phi)^2}{r^3} \\
&+\int_{\mathcal{M}(\tau_1,\tau_2)} r^{1+\delta}|\mathfrak{d}^{\leq s}N_0|^2 + \left|\int_{{}^{(trap)}\mathcal{M}(\tau_1,\tau_2)} T(\mathfrak{d}^s\phi)\mathfrak{d}^s N_0\right|, \qquad (10.5.2)
\end{aligned}
$$

where $D_s[\Gamma]$ is defined by

$$
\begin{aligned}
D_s[\Gamma] \ :=\ & \int_{{}^{(int)}\mathcal{M}\cup{}^{(ext)}\mathcal{M}(r\leq 4m_0)} (\mathfrak{d}^{\leq s}\check{\Gamma})^2 \\
&+\sup_{r_0\geq 4m_0}\left(r_0\int_{\{r=r_0\}} |\mathfrak{d}^{\leq s}\Gamma_g|^2 + r_0^{-1}\int_{\{r=r_0\}} |\mathfrak{d}^{\leq s}\Gamma_b|^2\right).
\end{aligned}
$$

The proof of Theorem 10.67 relies on the following theorem.

Theorem 10.68. *Let ψ a reduced scalar, and ϕ a reduced 0-scalar satisfying respectively*

$$
(\square_2 + V_2)\psi = N_2, \quad (\square_0 + V_0)\phi = N_0, \quad V_2 = -\frac{2}{r^2}\left(1+\frac{2m}{r}\right), \quad V_0 = \frac{8m}{r^3}.
$$

Then, we have

$$\sup_{\tau\in[\tau_1,\tau_2]} E_\delta[\psi](\tau)+B_\delta[\psi](\tau_1,\tau_2)+F_\delta[\psi](\tau_1,\tau_2)$$
$$\lesssim E_\delta[\psi](\tau_1)+\int_{{}^{(trap)}\mathcal{M}(\tau_1,\tau_2)}\left|1-\frac{3m}{r}\right||\psi|(|\psi|+|R\psi|)$$
$$+\int_{{}^{(trap\!\!\!/)}\mathcal{M}(\tau_1,\tau_2)} r^{\delta-3}|\psi|(|\psi|+|\mathfrak{d}\psi|)$$
$$+\int_{\mathcal{M}(\tau_1,\tau_2)} r^{1+\delta}|N_2|^2+\left|\int_{{}^{(trap)}\mathcal{M}(\tau_1,\tau_2)} T(\psi)N_2\right|, \tag{10.5.3}$$

and

$$\sup_{\tau\in[\tau_1,\tau_2]} E_\delta[\phi](\tau)+B_\delta[\phi](\tau_1,\tau_2)+F_\delta[\phi](\tau_1,\tau_2)$$
$$\lesssim E_\delta[\phi](\tau_1)+\int_{\mathcal{A}(\tau_1,\tau_2)\cup\Sigma(\tau_2)\cup\Sigma_*(\tau_1,\tau_2)}\frac{\phi^2}{r^3} \tag{10.5.4}$$
$$+\int_{{}^{(trap)}\mathcal{M}(\tau_1,\tau_2)}\left|1-\frac{3m}{r}\right||\phi|(|\phi|+|R\phi|)+\int_{{}^{(trap\!\!\!/)}\mathcal{M}(\tau_1,\tau_2)} r^{\delta-3}|\phi|(|\phi|+|\mathfrak{d}\phi|)$$
$$+\int_{\mathcal{M}(\tau_1,\tau_2)} r^{1+\delta}|N_0|^2+\left|\int_{{}^{(trap)}\mathcal{M}(\tau_1,\tau_2)} T(\phi)N_0\right|.$$

Proof. The proof of Theorem 10.68 is analogous to the one of Theorem 10.61. The only differences are

- the treatment of the right-hand sides N_0 and N_2 in the spacetime region ${}^{(trap)}\mathcal{M}$;
- and the boundary term on $\mathcal{A}(\tau_1,\tau_2)\cup\Sigma(\tau_2)\cup\Sigma_*(\tau_1,\tau_2)$ appearing in the right-hand side of[27] (10.5.4).

The treatment of N_0 and N_2 is similar, so we focus on the one of N_2. The only estimate in which N_2 appear in the trapping region is the Morawetz estimate. More precisely, it appears under the form, see (10.1.72),

$$\left|\int_{{}^{(trap)}\mathcal{M}(\tau_1,\tau_2)}\left(f_{\widehat{\delta}}R(\Psi)+\Lambda T(\Psi)+\frac{1}{2}w\Psi\right)N_2\right|,$$

where we recall that Λ is a constant, and $f_{\widehat{\delta}}$, w are functions which are in particular

[27]This boundary term, as discussed below, is due to the fact that V_0 is positive, which explains why no such term is present in (10.5.3) due to the negativity of the potential V_2 for the wave equation satisfied by ψ.

bounded on ${}^{(trap)}\mathcal{M}$. We infer

$$\begin{aligned}
&\left|\int_{{}^{(trap)}\mathcal{M}(\tau_1,\tau_2)}\left(f_{\widehat{\delta}}R(\Psi)+\Lambda T(\Psi)+\frac{1}{2}w\Psi\right)N_2\right|\\
\lesssim\ & \int_{{}^{(trap)}\mathcal{M}(\tau_1,\tau_2)}(|R(\Psi)|+|\Psi|)|N_2|+\left|\int_{{}^{(trap)}\mathcal{M}(\tau_1,\tau_2)}T(\Psi)N_2\right|\\
\lesssim\ & \lambda B_\delta[\psi](\tau_1,\tau_2)+\lambda^{-1}\int_{{}^{(trap)}\mathcal{M}(\tau_1,\tau_2)}|N_2|^2+\left|\int_{{}^{(trap)}\mathcal{M}(\tau_1,\tau_2)}T(\Psi)N_2\right|
\end{aligned}$$

which yields the desired control provided $\lambda > 0$ is chosen small enough so that the term $\lambda B_\delta[\psi](\tau_1,\tau_2)$ can be absorbed by the LHS in (10.5.3).

Concerning the boundary terms on $\mathcal{A}(\tau_1,\tau_2)\cup\Sigma(\tau_2)\cup\Sigma_*(\tau_1,\tau_2)$ appearing in the right-hand side of (10.5.4), the potential V_0 does not appear in the boundary term of the r^p-weighted estimates, but it does appear in the boundary term of the energy estimates.[28] More precisely, it appears in

$$\int_{\mathcal{A}(\tau_1,\tau_2)\cup\Sigma(\tau_2)\cup\Sigma_*(\tau_1,\tau_2)}Q_{34}=\int_{\mathcal{A}(\tau_1,\tau_2)\cup\Sigma(\tau_2)\cup\Sigma_*(\tau_1,\tau_2)}\Big(|\nabla\!\!\!\!/\,\phi|^2+V_0\phi^2\Big).$$

Now, we have in view of the definition of V_0

$$\begin{aligned}
\int_{\mathcal{A}(\tau_1,\tau_2)\cup\Sigma(\tau_2)\cup\Sigma_*(\tau_1,\tau_2)}Q_{34}\ \geq\ & \int_{\mathcal{A}(\tau_1,\tau_2)\cup\Sigma(\tau_2)\cup\Sigma_*(\tau_1,\tau_2)}|\nabla\!\!\!\!/\,\phi|^2\\
&-O(1)\int_{\mathcal{A}(\tau_1,\tau_2)\cup\Sigma(\tau_2)\cup\Sigma_*(\tau_1,\tau_2)}\frac{\phi^2}{r^3}
\end{aligned}$$

and the control of the boundary terms follows. This concludes the proof of 10.68. □

We are now in position to prove Theorem 10.67. Note first that we have

$$\int_{\mathcal{A}(\tau_1,\tau_2)\cup\Sigma_*(\tau_1,\tau_2)}\frac{(\mathfrak{d}^{\leq s}\phi)^2}{r^3}\ \lesssim\ F_\delta^{s-1}[\phi](\tau_1,\tau_2)$$

which explains why the term $\int_{\mathcal{A}(\tau_1,\tau_2)\cup\Sigma_*(\tau_1,\tau_2)}\frac{(\mathfrak{d}^{\leq s}\phi)^2}{r^3}$, that one would a priori expect in view of (10.5.4), is not present on the right-hand side of (10.5.2). Also, the estimates for ψ and ϕ are similar, so we focus on the estimate for ψ.

Proof of Theorem 10.67. The proof of Theorem 10.67 follows along the same lines as the one of Theorem 5.17. More precisely, following the strategy in section 10.4.2, we recover derivatives one by one starting from Theorem 10.68 and use it iteratively in conjonction with the commutator estimates of section 10.4.3. The only difference is the treatment of the derivatives for $s \geq k_{small}+1$ as we assume that the estimates of section 10.4.1 for the Ricci coefficients and curvature components only hold for $k \leq k_{small}$ derivatives. Thus, to conclude, we need to consider the terms for which at least $k_{small}+1$ derivatives fall on the Ricci coefficients and curvature components. Since on the other hand we have $s \leq k_{large}-1$, in view of the definition (3.3.7) of

[28]The boundary term of the r^p-weighted estimates involves only $Q_{44}=(e_4\phi)^2$, while the one of the energy estimate involves also $Q_{34}=|\nabla\!\!\!\!/\,\phi|^2+V_0\phi^2$.

k_{small} in terms of k_{large}, and in view of the commutator estimates of section 10.4.3, one easily checks that these terms are bounded in absolute value from above by

$$\Big(|\mathfrak{d}^{\leq s}(\Gamma_g)| + r^{-1}|\mathfrak{d}^{\leq s}(\Gamma_b)|\Big)|\mathfrak{d}^{\leq k_{small}}\psi|.$$

We thus need, in view of Theorem 10.68, to estimate

$$\begin{aligned}
&\int_{\mathcal{M}(\tau_1,\tau_2)} r^{1+\delta}\Big(|\mathfrak{d}^{\leq s}(\Gamma_g)| + r^{-1}|\mathfrak{d}^{\leq s}(\Gamma_b)|\Big)^2|\mathfrak{d}^{\leq k_{small}}\psi|^2\\
&+\int_{^{(trap)}\mathcal{M}(\tau_1,\tau_2)} |T\mathfrak{d}^s\psi||\mathfrak{d}^{\leq s}(\check{\Gamma})||\mathfrak{d}^{\leq k_{small}}\psi|\\
\lesssim\;& \sup_{\mathcal{M}(\tau_1,\tau_2)}\Big(r^2|\mathfrak{d}^{\leq k_{small}}\psi|^2\Big)\int_{\mathcal{M}(\tau_1,\tau_2)} r^{-1+\delta}\Big(|\mathfrak{d}^{\leq s}(\Gamma_g)| + r^{-1}|\mathfrak{d}^{\leq s}(\Gamma_b)|\Big)^2\\
&+\Bigg(\sup_{^{(trap)}\mathcal{M}(\tau_1,\tau_2)} u^{\frac{1}{2}+\delta_{dec}}|\mathfrak{d}^{\leq k_{small}}\psi|\Bigg)\Bigg(\sup_{\tau\in[\tau_1,\tau_2]} E^s_\delta[\psi](\tau)\Bigg)^{\frac{1}{2}}\\
&\times\Bigg(\int_{^{(trap)}\mathcal{M}(\tau_1,\tau_2)} |\mathfrak{d}^{\leq s}\check{\Gamma}|^2\Bigg)^{\frac{1}{2}}\\
\lesssim\;& \Bigg(\sup_{\mathcal{M}(\tau_1,\tau_2)} ru_{trap}^{\frac{1}{2}+\delta_{dec}}|\mathfrak{d}^{\leq k_{small}}\psi|\Bigg)^2 D_s[\Gamma]\\
&+\Bigg(\sup_{\mathcal{M}(\tau_1,\tau_2)} ru_{trap}^{\frac{1}{2}+\delta_{dec}}|\mathfrak{d}^{\leq k_{small}}\psi|\Bigg)\Bigg(\sup_{\tau\in[\tau_1,\tau_2]} E^s_\delta[\psi](\tau)\Bigg)^{\frac{1}{2}}\sqrt{D_s[\Gamma]}
\end{aligned}$$

where we have used the definition of $D_s[\Gamma]$. We infer

$$\begin{aligned}
&\int_{\mathcal{M}(\tau_1,\tau_2)} r^{1+\delta}\Big(|\mathfrak{d}^{\leq s}(\Gamma_g)| + r^{-1}|\mathfrak{d}^{\leq s}(\Gamma_b)|\Big)^2|\mathfrak{d}^{\leq k_{small}}\psi|^2\\
&+\int_{^{(trap)}\mathcal{M}(\tau_1,\tau_2)} |T\mathfrak{d}^s\psi||\mathfrak{d}^{\leq s}(\check{\Gamma})||\mathfrak{d}^{\leq k_{small}}\psi|\\
\lesssim\;& \lambda^{-1}\Bigg(\sup_{\mathcal{M}(\tau_1,\tau_2)} ru_{trap}^{\frac{1}{2}+\delta_{dec}}|\mathfrak{d}^{\leq k_{small}}\psi|\Bigg)^2 D_s[\Gamma] + \lambda\sup_{\tau\in[\tau_1,\tau_2]} E^s_\delta[\psi](\tau)
\end{aligned}$$

for any $\lambda > 0$ and the last term is then absorbed from the left-hand side of the desired estimate by choosing $\lambda > 0$ small enough which concludes the proof of Theorem 10.67. □

Proposition 10.69. *Let ψ a reduced 2-scalar satisfying*

$$\square_2\psi = f_2(r,m)Y_{(0)}\psi + \widetilde{N}_2,$$

where the function f_2 is smooth and positive, and where the vectorfield $Y_{(0)}$ has been introduced in Proposition 10.29 in connection with the redshift vectorfield and is given by

$$Y_{(0)} := \left(1 + \frac{5}{4m}(r-2m) + \Upsilon\right)e_3 + \left(1 + \frac{5}{4m}(r-2m)\right)e_4.$$

Also, assume that the Ricci coefficients and curvature components associated to the global null frame we are using satisfy the estimates of section 10.4.1 for $k \leq k_{small}$ derivatives. Then, for any $1 \leq s \leq k_{large} - 1$, we have

$$\int_{{}^{(int)}\mathcal{M}(\tau_1,\tau_2)} (\mathfrak{d}^{s+1}\psi)^2 \lesssim E^s_\delta[\psi](\tau_1) + \int_{{}^{(ext)}\mathcal{M}_{r\leq\frac{5}{2}m_0}(\tau_1,\tau_2)} (\mathfrak{d}^{s+1}\psi)^2$$
$$+D_s[\Gamma]\left(\sup_{{}^{(int)}\mathcal{M}(\tau_1,\tau_2)\cup{}^{(ext)}\mathcal{M}_{r\leq\frac{5}{2}m_0}} r|\mathfrak{d}^{\leq k_{small}}\psi|\right)^2$$
$$+\int_{{}^{(int)}\mathcal{M}(\tau_1,\tau_2)\cup{}^{(ext)}\mathcal{M}_{r\leq\frac{5}{2}m_0}} \Big((\mathfrak{d}^{\leq s}\psi)^2 + (\mathfrak{d}^{\leq s+1}\widetilde{N}_2)^2\Big).$$

Proof. Recall from Proposition 10.29 that the redshift vectorfield is given by

$$Y_{\mathcal{H}} := \kappa_{\mathcal{H}} Y_{(0)}, \qquad \kappa_{\mathcal{H}} := \kappa\left(\frac{\Upsilon}{\delta_{\mathcal{H}}^{\frac{1}{10}}}\right),$$

where κ is a positive bump function $\kappa = \kappa(r)$, supported in the region in $[-2, 2]$ and equal to 1 for $[-1, 1]$.

To estimate ψ in ${}^{(int)}\mathcal{M}$, we consider

$$\widetilde{\psi} := \widetilde{\kappa}\left(\frac{r - 2m_0(1+2\delta_{\mathcal{H}})}{2m_0\delta_{\mathcal{H}}}\right)$$

where $\widetilde{\kappa}$ is a positive bump function $\kappa = \kappa(r)$, supported in the region in $(-\infty, 1]$ and equal to 1 for $(-\infty, 0]$. Since ${}^{(int)}\mathcal{M}$ is included in $r \leq 2m_0(1+2\delta_{\mathcal{H}})$, we infer in particular

$$\widetilde{\psi} = \psi \text{ on } {}^{(int)}\mathcal{M}, \qquad \text{supp}(\widetilde{\psi}) \subset {}^{(int)}\mathcal{M}(\tau_1,\tau_2) \cup {}^{(ext)}\mathcal{M}_{r\leq 2m_0(1+3\delta_{\mathcal{H}})}.$$

Also, we have, in view of the wave equation for ψ,

$$\square_2\widetilde{\psi} = f_2(r,m)Y_{(0)}\widetilde{\psi} + \widetilde{N}_2'$$

where $\widetilde{N}_2'$ satisfies

$$\int_{{}^{(int)}\mathcal{M}(\tau_1,\tau_2)\cup{}^{(ext)}\mathcal{M}_{r\leq\frac{5}{2}m_0}} (\mathfrak{d}^{\leq s+1}\widetilde{N}_2')^2 \lesssim \int_{{}^{(ext)}\mathcal{M}_{r\leq\frac{5}{2}m_0}(\tau_1,\tau_2)} (\mathfrak{d}^{\leq s+1}\psi)^2$$
$$+ \int_{{}^{(int)}\mathcal{M}(\tau_1,\tau_2)\cup{}^{(ext)}\mathcal{M}_{r\leq\frac{5}{2}m_0}} (\mathfrak{d}^{\leq s+1}\widetilde{N}_2)^2.$$

Since $\widetilde{\psi} = \psi$ on ${}^{(int)}\mathcal{M}$, it thus suffices to prove for $\widetilde{\psi}$ the following estimate:

$$\begin{aligned}\int_{\mathcal{M}(\tau_1,\tau_2)} (\mathfrak{d}^{s+1}\widetilde{\psi})^2 \quad \lesssim \quad & E^s_\delta[\widetilde{\psi}](\tau_1) \\ & + D_s[\Gamma]\left(\sup_{{}^{(int)}\mathcal{M}(\tau_1,\tau_2)\cup {}^{(ext)}\mathcal{M}_{r\leq\frac{5}{2}m_0}} r|\mathfrak{d}^{\leq k_{small}}\widetilde{\psi}|\right)^2 \\ & + \int_{{}^{(int)}\mathcal{M}(\tau_1,\tau_2)\cup {}^{(ext)}\mathcal{M}_{r\leq\frac{5}{2}m_0}} \Big((\mathfrak{d}^{\leq s}\widetilde{\psi})^2 + (\mathfrak{d}^{\leq s+1}\widetilde{N}'_2)^2\Big).\end{aligned}$$

This estimate follows from first deriving the corresponding estimate for $s = 0$ by using the redshift as a multiplier, and then by recovering derivatives one by one using commutation with T, $\not{\mathfrak{d}}$ and the redshift vectorfield. Note that

- $\widetilde{\psi}$ is supported on $r \leq 2m_0(1 + 2\delta_{\mathcal{H}})$ and hence is estimated on

$$2m_0(1 - 2\delta_{\mathcal{H}}) \leq r \leq 2m_0(1 + 2\delta_{\mathcal{H}})$$

  so that the redshift vectorfield $Y_{\mathcal{H}}$ has good properties, both as a multiplier and as a commutator, on the support of $\widetilde{\psi}$;
- and the term $f_2(r, m)Y_{(0)}$ yields a good sign when using $Y_{\mathcal{H}}$ as a multiplier since the function $f_2(r, m)$ is positive, and since $Y_{\mathcal{H}} = \kappa_{\mathcal{H}} Y_{(0)}$.

This concludes the proof of the proposition. □

Appendix A

Appendix to Chapter 2

A.1 PROOF OF PROPOSITION 2.64

In a neighborhood of a given sphere S, we consider a (u, s, θ, φ) coordinates system, where θ is such that $e_4(\theta) = 0$. Then, in this coordinates system, we have

$$\partial_s = e_4.$$

Since we have

$$\partial_s \left(\int_S f \right) = \int_S \Big(\partial_s f + g(\mathbf{D}_{e_\theta} \partial_s, e_\theta) f + g(\mathbf{D}_{e_\varphi} \partial_s, e_\varphi) f \Big),$$

we infer

$$e_4 \left(\int_S f \right) = \int_S (e_4(f) + \kappa f).$$

In particular, choosing $f = 1$, we deduce

$$\frac{1}{|S|} e_4(|S|) = \overline{\kappa}$$

and since $|S| = 4\pi r^2$,

$$e_4(r) = \frac{r\overline{\kappa}}{2}.$$

Next, let ∂_u the coordinates vectorfield in the (u, s, θ, φ) coordinates system. We have

$$\begin{aligned}
\partial_u \left(\int_S f \right) &= \int_S \Big(\partial_u f + g(\mathbf{D}_{e_\theta} \partial_u, e_\theta) f + g(\mathbf{D}_{e_\varphi} \partial_u, e_\varphi) f \Big) \\
&= \int_S \Big(\partial_u f + g(\mathbf{D}_{e_\theta} \partial_u, e_\theta) f - g(\partial_u, \mathbf{D}_{e_\varphi} e_\varphi) f \Big) \\
&= \int_S \Big(\partial_u f + g(\mathbf{D}_{e_\theta} \partial_u, e_\theta) f + g(\partial_u, D^a(\Phi) e_a) f \Big).
\end{aligned}$$

On the other hand, we have, see (2.2.42),

$$\partial_u = \varsigma \left(\frac{1}{2} e_3 - \frac{1}{2} \underline{\Omega} e_4 - \frac{1}{2} \sqrt{\gamma} \underline{b} e_\theta \right).$$

We infer

$$g(D_{e_\theta} \partial_u, e_\theta) + g(\partial_u, D^a(\Phi) e_a) = \frac{1}{2} \varsigma \underline{\kappa} - \frac{1}{2} \varsigma \underline{\Omega} \kappa - \frac{1}{2} \not{d}_1 (\varsigma \sqrt{\gamma} \underline{b})$$

and thus

$$\varsigma\left(\frac{1}{2}e_3 - \frac{1}{2}\underline{\Omega}e_4\right)\left(\int_S f\right) = \int_S \left(\varsigma\left(\frac{1}{2}e_3 - \frac{1}{2}\underline{\Omega}e_4 - \frac{1}{2}\sqrt{\gamma}\underline{b}e_\theta\right)f + \frac{1}{2}\varsigma\underline{\kappa}f - \frac{1}{2}\varsigma\underline{\Omega}\kappa f - \frac{1}{2}\not{d}_1(\sqrt{\gamma}\underline{b})f\right).$$

We deduce

$$e_3\left(\int_S f\right) = \underline{\Omega}e_4\left(\int_S f\right) + \varsigma^{-1}\int_S\left(\varsigma e_3 f - \varsigma\underline{\Omega}e_4 f + \varsigma\underline{\kappa}f - \varsigma\underline{\Omega}\kappa f - \not{d}_1(\varsigma\sqrt{\gamma}\underline{b}f)\right).$$

Next, we use

$$e_4\left(\int_S f\right) = \int_S (e_4(f) + \kappa f)$$

and

$$\int_S \not{d}_1(\varsigma\sqrt{\gamma}\underline{b}f) = 0.$$

We infer

$$\begin{aligned} e_3\left(\int_S f\right) &= \underline{\Omega}\int_S (e_4(f)+\kappa f) + \varsigma^{-1}\int_S\left(\varsigma e_3 f - \varsigma\underline{\Omega}e_4 f + \varsigma\underline{\kappa}f - \varsigma\underline{\Omega}\kappa f\right)\\ &= \varsigma^{-1}\int_S \varsigma(e_3 f + \underline{\kappa}f) + \left(\check{\underline{\Omega}} + \varsigma^{-1}\overline{\underline{\Omega}}\check{\varsigma}\right)\int_S (e_4 f + \kappa f)\\ &\quad -\varsigma^{-1}\overline{\underline{\Omega}}\int_S \check{\varsigma}(e_4 f + \kappa f) - \varsigma^{-1}\int_S \check{\underline{\Omega}}\varsigma(e_4 f + \kappa f). \end{aligned}$$

We further write

$$\begin{aligned} \varsigma^{-1}\int_S \varsigma(e_3 f + \underline{\kappa}f) &= \varsigma^{-1}\overline{\varsigma}\int_S (e_3 f + \underline{\kappa}f) + \varsigma^{-1}\int_S \check{\varsigma}\,(e_3 f + \underline{\kappa}f)\\ &= \int_S (e_3 f + \underline{\kappa}f) + (\varsigma^{-1}\overline{\varsigma} - 1)\int_S (e_3 f + \underline{\kappa}f)\\ &\quad +\varsigma^{-1}\int_S \check{\varsigma}\,(e_3 f + \underline{\kappa}f)\\ &= \int_S (e_3 f + \underline{\kappa}f) - \varsigma^{-1}\check{\varsigma}\int_S (e_3 f + \underline{\kappa}f) + \varsigma^{-1}\int_S \check{\varsigma}\,(e_3 f + \underline{\kappa}f). \end{aligned}$$

Hence,

$$
\begin{aligned}
e_3\left(\int_S f\right) &= \int_S (e_3 f + \underline{\kappa} f) + \mathrm{Err}\left[e_3\left(\int_S f\right)\right],\\
\mathrm{Err}\left[e_3\left(\int_S f\right)\right] &= -\varsigma^{-1}\check{\varsigma}\int_S (e_3 f + \underline{\kappa} f) + \varsigma^{-1}\int_S \check{\varsigma}\,(e_3 f + \underline{\kappa} f)\\
&+ \left(\check{\underline{\Omega}} + \varsigma^{-1}\overline{\underline{\Omega}}\check{\varsigma}\right)\int_S (e_4 f + \kappa f) - \varsigma^{-1}\overline{\underline{\Omega}}\int_S \check{\varsigma}(e_4 f + \kappa f)\\
&- \varsigma^{-1}\int_S \check{\underline{\Omega}}\varsigma(e_4 f + \kappa f)
\end{aligned}
$$

as desired.

In particular, choosing $f = 1$, we infer

$$
\begin{aligned}
\frac{1}{|S|}e_3(|S|) &= \overline{\underline{\kappa}} - \varsigma^{-1}\check{\varsigma}\,\overline{\underline{\kappa}} + \varsigma^{-1}\,\overline{\check{\varsigma}\underline{\kappa}} + \left(\check{\underline{\Omega}} + \varsigma^{-1}\overline{\underline{\Omega}}\check{\varsigma}\right)\overline{\kappa} - \varsigma^{-1}\overline{\underline{\Omega}}\,\overline{\check{\varsigma}\kappa} - \varsigma^{-1}\overline{\check{\underline{\Omega}}\varsigma\kappa}\\
&= \overline{\underline{\kappa}} - \varsigma^{-1}\check{\varsigma}\,\overline{\underline{\kappa}} + \varsigma^{-1}\,\overline{\check{\varsigma}\check{\underline{\kappa}}} + \left(\check{\underline{\Omega}} + \varsigma^{-1}\overline{\underline{\Omega}}\check{\varsigma}\right)\overline{\kappa} - \varsigma^{-1}\overline{\underline{\Omega}}\,\overline{\check{\varsigma}\check{\kappa}} - \varsigma^{-1}\overline{\check{\underline{\Omega}}\varsigma\kappa}.
\end{aligned}
$$

Hence, since $|S| = 4\pi r^2$, recalling the definition of $\underline{A}$,

$$
\begin{aligned}
\frac{2e_3(r)}{r} &= \overline{\underline{\kappa}} - \varsigma^{-1}\check{\varsigma}\,\overline{\underline{\kappa}} + \varsigma^{-1}\,\overline{\check{\varsigma}\check{\underline{\kappa}}} + \left(\check{\underline{\Omega}} + \varsigma^{-1}\overline{\underline{\Omega}}\check{\varsigma}\right)\overline{\kappa} - \varsigma^{-1}\overline{\underline{\Omega}}\,\overline{\check{\varsigma}\check{\kappa}} - \varsigma^{-1}\overline{\check{\underline{\Omega}}\varsigma\kappa}\\
&= \overline{\underline{\kappa}} + \underline{A}.
\end{aligned}
$$

This concludes the proof of Proposition 2.64.

A.2 PROOF OF PROPOSITION 2.71

We start with the proof for $e_4(m)$. Recall that the Hawking mass m is given by the formula $\frac{2m}{r} = 1 + \frac{1}{16\pi}\int_S \kappa\underline{\kappa}$. Differentiating in the e_4 direction, we deduce

$$
\frac{2e_4(m)}{r} - \frac{2me_4(r)}{r^2} = \frac{1}{16\pi}e_4\left(\int_S \kappa\underline{\kappa}\right) = \frac{1}{16\pi}\int_S\left(e_4(\kappa\underline{\kappa}) + \underline{\kappa}\kappa^2\right).
$$

Now, making use of the e_4 transport equations of Proposition 2.63,

$$
\begin{aligned}
e_4(\kappa\underline{\kappa}) &= \underline{\kappa}\left(-\frac{1}{2}\kappa^2 - \frac{1}{2}\vartheta^2\right) + \kappa\left(-\frac{1}{2}\kappa\underline{\kappa} + 2\rho - 2\,\mathrm{d}\!\!/_1\zeta - \frac{1}{2}\vartheta\underline{\vartheta} + 2\zeta^2\right)\\
&= -\underline{\kappa}\kappa^2 + 2\kappa\rho - 2\kappa\,\mathrm{d}\!\!/_1\zeta - \frac{1}{2}\underline{\kappa}\vartheta^2 - \frac{1}{2}\kappa\vartheta\underline{\vartheta} + 2\kappa\zeta^2.
\end{aligned}
$$

We infer

$$
\begin{aligned}
\frac{2e_4(m)}{r} - \frac{m}{r}\overline{\kappa} &= \frac{1}{16\pi}\int_S\left(2\kappa\rho - 2\kappa\,\mathrm{d}\!\!/_1\zeta - \frac{1}{2}\underline{\kappa}\vartheta^2 - \frac{1}{2}\kappa\vartheta\underline{\vartheta} + 2\kappa\zeta^2\right)\\
&= \frac{1}{8\pi}|S|\overline{\kappa}\,\overline{\rho} + \frac{1}{16\pi}\int_S\left(2\check{\kappa}\check{\rho} + 2e_\theta(\kappa)\zeta - \frac{1}{2}\underline{\kappa}\vartheta^2 - \frac{1}{2}\kappa\vartheta\underline{\vartheta} + 2\kappa\zeta^2\right)\\
&= \frac{r^2}{2}\overline{\kappa}\,\overline{\rho} + \frac{1}{16\pi}\int_S\left(2\check{\kappa}\check{\rho} + 2e_\theta(\kappa)\zeta - \frac{1}{2}\underline{\kappa}\vartheta^2 - \frac{1}{2}\kappa\vartheta\underline{\vartheta} + 2\kappa\zeta^2\right)
\end{aligned}
$$

and hence

$$e_4(m) = \frac{r^3}{4}\overline{\kappa}\left(\overline{\rho} + \frac{2m}{r^3}\right) + \frac{r}{32\pi}\int_S \left(-\frac{1}{2}\underline{\kappa}\vartheta^2 - \frac{1}{2}\kappa\vartheta\underline{\vartheta} + 2\check{\kappa}\check{\rho} + 2e_\theta(\kappa)\zeta + 2\kappa\zeta^2\right).$$

Using the identity $\overline{\rho} = -\frac{2m}{r^3} + \frac{1}{16\pi r^2}\int_S \vartheta\underline{\vartheta}$ (see (2.2.12) of Proposition 2.59), we deduce

$$\begin{aligned} e_4(m) &= \frac{r}{32\pi}\int_S \left(-\frac{1}{2}\underline{\kappa}\vartheta^2 - \frac{1}{2}(\kappa - \overline{\kappa})\vartheta\underline{\vartheta} + 2\check{\kappa}\check{\rho} + 2e_\theta(\kappa)\zeta + 2\kappa\zeta^2\right) \\ &= \frac{r}{32\pi}\int_S \left(-\frac{1}{2}\underline{\kappa}\vartheta^2 - \frac{1}{2}\check{\kappa}\vartheta\underline{\vartheta} + 2\check{\kappa}\check{\rho} + 2e_\theta(\kappa)\zeta + 2\kappa\zeta^2\right) \\ &= \frac{r}{32\pi}\int_S \mathrm{Err}_1 \end{aligned}$$

as desired.

In the same vein,

$$\frac{2e_3(m)}{r} - \frac{2me_3(r)}{r^2} = \frac{1}{16\pi}e_3\left(\int_S \kappa\underline{\kappa}\right) = \frac{1}{16\pi}\int_S \left(e_3(\kappa\underline{\kappa}) + \underline{\kappa}^2\kappa\right) + E_1,$$

with E_1 the error term defined in Proposition 2.64

$$E_1 = \frac{1}{16\pi}\mathrm{Err}\left[e_3\left(\int_{\mathbf{S}} \kappa\underline{\kappa}\right)\right].$$

We make use of the e_3 transport equations of Proposition 2.63,

$$\begin{aligned} e_3(\kappa\underline{\kappa}) &= \underline{\kappa}\left(-\frac{1}{2}\underline{\kappa}\,\kappa + 2\underline{\omega}\kappa + 2\,\not{d}_1\eta + 2\rho - \frac{1}{2}\underline{\vartheta}\,\vartheta + 2\eta^2\right) \\ &+ \kappa\left(-\frac{1}{2}\underline{\kappa}^2 - 2\underline{\omega}\,\underline{\kappa} + 2\,\not{d}_1\underline{\xi} + 2(\eta - 3\zeta)\underline{\xi} - \frac{1}{2}\underline{\vartheta}^2\right) \\ &= -\kappa\underline{\kappa}^2 + 2\underline{\kappa}\,\not{d}_1\eta + 2\kappa\,\not{d}_1\underline{\xi} + 2\rho\underline{\kappa} + \underline{\kappa}\left(2\eta^2 - \frac{1}{2}\vartheta\underline{\vartheta}\right) + 2\kappa(\eta - 3\zeta)\underline{\xi} \\ &\quad -\frac{1}{2}\kappa\underline{\vartheta}^2. \end{aligned}$$

Therefore, setting $E_2 = \frac{1}{16\pi}\int_S \Big(\underline{\kappa}(2\eta^2 - \frac{1}{2}\vartheta\underline{\vartheta}) + 2\kappa(\eta - 3\zeta)\underline{\xi} - \frac{1}{2}\kappa\underline{\vartheta}^2\Big)$,

$$\begin{aligned}
&\frac{2e_3(m)}{r} - \frac{m}{r}(\overline{\underline{\kappa}} + \underline{A})\\
&= \frac{1}{16\pi}\int_S \Big(2\underline{\kappa}\,\not{d}_1\eta + 2\kappa\,\not{d}_1\underline{\xi} + 2\rho\underline{\kappa}\Big) + E_1 + E_2\\
&= \frac{1}{16\pi}\int_S \Big(-2e_\theta(\underline{\kappa})\eta - 2e_\theta(\kappa)\underline{\xi} + 2(\overline{\rho} + \check{\rho})(\overline{\underline{\kappa}} + \check{\underline{\kappa}})\Big) + E_1 + E_2\\
&= \frac{1}{2}r^2\overline{\rho}\,\overline{\underline{\kappa}} + \frac{1}{16\pi}\int_S \Big(-2e_\theta(\underline{\kappa})\eta - 2e_\theta(\kappa)\underline{\xi} + 2\check{\rho}\,\check{\underline{\kappa}}\Big) + E_1 + E_2\\
&= \frac{1}{2}r^2\overline{\underline{\kappa}}\Big(-\frac{2m}{r^3} + \frac{1}{16\pi r^2}\int_S \vartheta\underline{\vartheta}\Big)\\
&+ \frac{1}{16\pi}\int_S \Big(-2e_\theta(\underline{\kappa})\eta - 2e_\theta(\kappa)\underline{\xi} + 2\check{\rho}\,\check{\underline{\kappa}}\Big) + E_1 + E_2.
\end{aligned}$$

We deduce

$$\begin{aligned}
\frac{2e_3(m)}{r} &= \frac{1}{16\pi}\int_S \Big(-2e_\theta(\underline{\kappa})\eta - 2e_\theta(\kappa)\underline{\xi} + 2\check{\rho}\,\check{\underline{\kappa}} + \frac{1}{2}\overline{\underline{\kappa}}\vartheta\underline{\vartheta}\Big) + E_1\\
&+ \frac{1}{16\pi}\int_S \Big(\underline{\kappa}\Big(2\eta^2 - \frac{1}{2}\vartheta\underline{\vartheta}\Big) + 2\kappa(\eta - 3\zeta)\underline{\xi} - \frac{1}{2}\kappa\underline{\vartheta}^2\Big) + \frac{m}{r}\underline{A}\\
&= \frac{1}{16\pi}\int_S \Big(-2e_\theta(\underline{\kappa})\eta + 2\underline{\kappa}\eta^2 - 2e_\theta(\kappa)\underline{\xi} + 2\kappa\eta\underline{\xi} - \frac{1}{2}\kappa\underline{\vartheta}^2\Big)\\
&+ \frac{1}{16\pi}\int_S \Big(2\check{\rho}\,\check{\underline{\kappa}} - 6\kappa\,\zeta\,\underline{\xi} - \frac{1}{2}\check{\underline{\kappa}}\vartheta\,\underline{\vartheta}\Big) + E_1 + \frac{m}{r}\underline{A},
\end{aligned}$$

i.e.,

$$\begin{aligned}
e_3(m) &= \frac{r}{32\pi}\int_S \Big(-2e_\theta(\underline{\kappa})\eta + 2\underline{\kappa}\eta^2 - 2e_\theta(\kappa)\underline{\xi} + 2\kappa\eta\underline{\xi} - \frac{1}{2}\kappa\underline{\vartheta}^2\Big)\\
&+ \frac{r}{32\pi}\int_S \Big(2\check{\rho}\,\check{\underline{\kappa}} - 6\kappa\,\zeta\,\underline{\xi} - \frac{1}{2}\check{\underline{\kappa}}\vartheta\,\underline{\vartheta}\Big) + \frac{r}{2}\Big(E_1 + \frac{m}{r}\underline{A}\Big).
\end{aligned}$$

It remains to calculate $E_1 + \frac{m}{r}\underline{A}$. Using the definitions of E_1 and $\underline{A}$ and grouping similar terms appropriately we find

$$\begin{aligned}
E_1 + \frac{m}{r}\underline{A} &= -\varsigma^{-1}\check{\varsigma}\left[\frac{1}{16\pi}\int_S (e_3(\kappa\underline{\kappa}) + \kappa\underline{\kappa}^2) + \frac{m}{r}\overline{\underline{\kappa}}\right]\\
&+\varsigma^{-1}\left[\frac{1}{16\pi}\int_S \check{\varsigma}\,(e_3(\kappa\underline{\kappa}) + \kappa\underline{\kappa}^2) + \frac{m}{r}\overline{\check{\varsigma}\check{\underline{\kappa}}}\right]\\
&+ (\check{\underline{\Omega}} + \varsigma^{-1}\overline{\underline{\Omega}}\check{\varsigma})\left[\frac{1}{16\pi}\int_S (e_4(\kappa\underline{\kappa}) + \kappa^2\underline{\kappa}) + \frac{m}{r}\overline{\kappa}\right]\\
&-\varsigma^{-1}\overline{\underline{\Omega}}\left[\frac{1}{16\pi}\int_S \check{\varsigma}(e_4(\kappa\underline{\kappa}) + \kappa^2\underline{\kappa}) + \frac{m}{r}\overline{\check{\varsigma}\check{\kappa}}\right]\\
&-\varsigma^{-1}\left[\frac{1}{16\pi}\int_S \check{\underline{\Omega}}\varsigma(e_4(\kappa\underline{\kappa}) + \kappa^2\underline{\kappa}) + \frac{m}{r}\overline{\check{\underline{\Omega}}\varsigma\kappa}\right].
\end{aligned}$$

Now, we have from the above calculations

$$\begin{aligned}
e_4(\kappa\underline{\kappa})+\underline{\kappa}\kappa^2 &= 2\kappa\rho-2\kappa\not{d}_1\zeta+\mathrm{Err}[e_4(\kappa\underline{\kappa})],\\
\mathrm{Err}[e_4(\kappa\underline{\kappa})] &= -\frac{1}{2}\underline{\kappa}\vartheta^2-\frac{1}{2}\kappa\vartheta\underline{\vartheta}+2\kappa\zeta^2,\\
e_3(\kappa\underline{\kappa})+\kappa\underline{\kappa}^2 &= 2\rho\underline{\kappa}+2\underline{\kappa}\not{d}_1\eta+2\kappa\not{d}_1\underline{\xi}+\mathrm{Err}[e_3(\kappa\underline{\kappa})],\\
\mathrm{Err}[e_3(\kappa\underline{\kappa})] &= \underline{\kappa}\left(2\eta^2-\frac{1}{2}\vartheta\underline{\vartheta}\right)+2\kappa(\eta-3\zeta)\underline{\xi}-\frac{1}{2}\kappa\underline{\vartheta}^2.
\end{aligned}$$

We infer

$$\begin{aligned}
E_1+\frac{m}{r}\underline{A} &= -\varsigma^{-1}\check{\varsigma}\left[\frac{1}{16\pi}\int_S(2\rho\underline{\kappa}+2\underline{\kappa}\not{d}_1\eta+2\kappa\not{d}_1\underline{\xi}+\mathrm{Err}[e_3(\kappa\underline{\kappa})])+\frac{m}{r}\overline{\underline{\kappa}}\right]\\
&\quad+\varsigma^{-1}\left[\frac{1}{16\pi}\int_S\check{\varsigma}\,(2\rho\underline{\kappa}+2\underline{\kappa}\not{d}_1\eta+2\kappa\not{d}_1\underline{\xi}+\mathrm{Err}[e_3(\kappa\underline{\kappa})])+\frac{m}{r}\overline{\check{\varsigma}\underline{\check{\kappa}}}\right]\\
&\quad+(\underline{\check{\Omega}}+\varsigma^{-1}\overline{\underline{\Omega}}\check{\varsigma})\left[\frac{1}{16\pi}\int_S(2\kappa\rho-2\kappa\not{d}_1\zeta+\mathrm{Err}[e_4(\kappa\underline{\kappa})])+\frac{m}{r}\overline{\kappa}\right]\\
&\quad-\varsigma^{-1}\overline{\underline{\Omega}}\left[\frac{1}{16\pi}\int_S\check{\varsigma}(2\kappa\rho-2\kappa\not{d}_1\zeta+\mathrm{Err}[e_4(\kappa\underline{\kappa})])+\frac{m}{r}\overline{\check{\varsigma}\check{\kappa}}\right]\\
&\quad-\varsigma^{-1}\left[\frac{1}{16\pi}\int_S\underline{\check{\Omega}}\varsigma(2\kappa\rho-2\kappa\not{d}_1\zeta+\mathrm{Err}[e_4(\kappa\underline{\kappa})])+\frac{m}{r}\overline{\underline{\check{\Omega}}\varsigma\kappa}\right]
\end{aligned}$$

and hence

$$\begin{aligned}
&E_1+\frac{m}{r}\underline{A}\\
=\;& -\varsigma^{-1}\check{\varsigma}\left[\frac{1}{16\pi}\int_S\left(2\check{\rho}\underline{\check{\kappa}}-2e_\theta(\underline{\kappa})\eta-2e_\theta(\kappa)\underline{\xi}+\frac{1}{2}\overline{\underline{\kappa}}\vartheta\underline{\vartheta}+\mathrm{Err}[e_3(\kappa\underline{\kappa})]\right)\right]\\
&+\varsigma^{-1}\left[\frac{1}{16\pi}\int_S\check{\varsigma}\,(2\overline{\rho}\underline{\check{\kappa}}+2\check{\rho}\overline{\underline{\kappa}}+2\check{\rho}\underline{\check{\kappa}}+2\underline{\kappa}\not{d}_1\eta+2\kappa\not{d}_1\underline{\xi}+\mathrm{Err}[e_3(\kappa\underline{\kappa})])+\frac{m}{r}\overline{\check{\varsigma}\underline{\check{\kappa}}}\right]\\
&+(\underline{\check{\Omega}}+\varsigma^{-1}\overline{\underline{\Omega}}\check{\varsigma})\left[\frac{1}{16\pi}\int_S\left(2\check{\kappa}\check{\rho}+2e_\theta(\kappa)\zeta+\frac{1}{2}\overline{\underline{\kappa}}\vartheta\underline{\vartheta}+\mathrm{Err}[e_4(\kappa\underline{\kappa})]\right)\right]\\
&-\varsigma^{-1}\overline{\underline{\Omega}}\left[\frac{1}{16\pi}\int_S\check{\varsigma}(2\overline{\rho}\check{\kappa}+2\check{\rho}\overline{\kappa}+2\check{\rho}\check{\kappa}-2\kappa\not{d}_1\zeta+\mathrm{Err}[e_4(\kappa\underline{\kappa})])+\frac{m}{r}\overline{\check{\varsigma}\check{\kappa}}\right]\\
&-\varsigma^{-1}\left[\frac{1}{16\pi}\int_S\underline{\check{\Omega}}\varsigma(2\overline{\rho}\check{\kappa}+2\check{\rho}\overline{\kappa}+2\check{\rho}\check{\kappa}-2\kappa\not{d}_1\zeta+\mathrm{Err}[e_4(\kappa\underline{\kappa})])+\frac{m}{r}\overline{\underline{\check{\Omega}}\varsigma\kappa}\right].
\end{aligned}$$

We deduce

$$\begin{aligned}
e_3(m) &= (1-\varsigma^{-1}\check{\varsigma})\frac{r}{32\pi}\int_S\underline{\mathrm{Err}}_1+(\underline{\check{\Omega}}+\varsigma^{-1}\overline{\underline{\Omega}}\check{\varsigma})\frac{r}{32\pi}\int_S\mathrm{Err}_1\\
&\quad+\varsigma^{-1}\frac{r}{32\pi}\int_S\check{\varsigma}\,\left(2\overline{\rho}\underline{\check{\kappa}}+2\check{\rho}\overline{\underline{\kappa}}+2\underline{\kappa}\not{d}_1\eta+2\kappa\not{d}_1\underline{\xi}+\underline{\mathrm{Err}}_2\right)\\
&\quad-\varsigma^{-1}\frac{r}{32\pi}\int_S(\overline{\underline{\Omega}}\check{\varsigma}+\underline{\check{\Omega}}\varsigma)\,(2\overline{\rho}\check{\kappa}+2\check{\rho}\overline{\kappa}-2\kappa\not{d}_1\zeta+\mathrm{Err}_2)\\
&\quad-\frac{m}{r}\varsigma^{-1}\left[-\overline{\check{\varsigma}\underline{\check{\kappa}}}+\overline{\underline{\Omega}}\,\overline{\check{\varsigma}\check{\kappa}}+\overline{\underline{\check{\Omega}}\varsigma\kappa}\right],
\end{aligned}$$

where we have introduced

$$\begin{aligned}
\mathrm{Err}_1 &= 2\check{\kappa}\check{\rho} + 2e_\theta(\kappa)\zeta + \frac{1}{2}\overline{\underline{\kappa}}\vartheta\underline{\vartheta} + \mathrm{Err}[e_4(\kappa\underline{\kappa})],\\
\underline{\mathrm{Err}}_1 &= 2\check{\rho}\check{\underline{\kappa}} - 2e_\theta(\underline{\kappa})\eta - 2e_\theta(\kappa)\underline{\xi} + \frac{1}{2}\overline{\underline{\kappa}}\vartheta\underline{\vartheta} + \mathrm{Err}[e_3(\kappa\underline{\kappa})],\\
\mathrm{Err}_2 &= 2\check{\rho}\check{\kappa} + \mathrm{Err}[e_4(\kappa\underline{\kappa})],\\
\underline{\mathrm{Err}}_2 &= 2\check{\rho}\check{\underline{\kappa}} + \mathrm{Err}[e_3(\kappa\underline{\kappa})].
\end{aligned}$$

In view of the definition of $\mathrm{Err}[e_4(\kappa\underline{\kappa})]$ and $\mathrm{Err}[e_3(\kappa\underline{\kappa})]$, this concludes the proof of Proposition 2.71.

A.3 PROOF OF LEMMA 2.72

Recall that we have

$$e_4(\kappa) = -\frac{1}{2}\kappa^2 - \frac{1}{4}\vartheta^2.$$

We infer

$$\begin{aligned}
e_4(\overline{\kappa}) &= \overline{e_4(\kappa)} + \overline{\check{\kappa}^2} = \overline{-\frac{1}{2}\kappa^2 - \frac{1}{4}\vartheta^2} + \overline{\check{\kappa}^2}\\
&= -\frac{1}{2}\overline{\kappa}^2 - \frac{1}{4}\overline{\vartheta^2} + \frac{1}{2}\overline{\check{\kappa}^2}
\end{aligned}$$

and hence

$$\begin{aligned}
e_4\left(\overline{\kappa} - \frac{2}{r}\right) &= -\frac{1}{2}\overline{\kappa}^2 - \frac{1}{4}\overline{\vartheta^2} + \frac{1}{2}\overline{\check{\kappa}^2} + \frac{2}{r}\frac{e_4(r)}{r} = -\frac{1}{2}\overline{\kappa}^2 + \frac{1}{r}\overline{\kappa} - \frac{1}{4}\overline{\vartheta^2} + \frac{1}{2}\overline{\check{\kappa}^2}\\
&= -\frac{1}{2}\overline{\kappa}\left(\overline{\kappa} - \frac{2}{r}\right) - \frac{1}{4}\overline{\vartheta^2} + \frac{1}{2}\overline{\check{\kappa}^2}.
\end{aligned}$$

Next, using

$$e_4(\underline{\omega}) = \rho + \zeta(2\eta + \zeta)$$

we infer that

$$e_4(\overline{\underline{\omega}}) = \overline{e_4(\underline{\omega})} + \overline{\check{\kappa}\check{\underline{\omega}}} = \overline{\rho} + \overline{\zeta(2\eta+\zeta)} + \overline{\check{\kappa}\check{\underline{\omega}}},$$

and hence

$$\begin{aligned}
e_4\left(\overline{\underline{\omega}} - \frac{m}{r^2}\right) &= e_4(\overline{\underline{\omega}}) + \frac{2me_4(r)}{r^3} - \frac{e_4(m)}{r^2}\\
&= \overline{\rho} + \frac{2m}{r^3} + \frac{m}{r^2}\left(\overline{\kappa} - \frac{2}{r}\right) - \frac{e_4(m)}{r^2} + 3\overline{\zeta(2\eta+\zeta)} + \overline{\check{\kappa}\check{\underline{\omega}}}
\end{aligned}$$

as stated.

Next, using

$$e_3(\kappa) + \frac{1}{2}\kappa\underline{\kappa} - 2\underline{\omega}\kappa = 2\,\not{d}_1\eta + 2\rho - \frac{1}{2}\vartheta\underline{\vartheta} + 2\eta^2$$

we deduce

$$
\begin{aligned}
\overline{e_3(\kappa)} &= -\frac{1}{2}\overline{\kappa\underline{\kappa}} + 2\overline{\underline{\omega}\kappa} + 2\overline{\rho} - \frac{1}{2}\overline{\vartheta\underline{\vartheta}} + 2\overline{\eta^2} \\
&= -\frac{1}{2}\overline{\kappa}\,\overline{\underline{\kappa}} + 2\overline{\underline{\omega}}\,\overline{\kappa} + 2\overline{\rho} + 2\overline{\check{\underline{\omega}}\,\check{\kappa}} - \frac{1}{2}\overline{\check{\kappa}\,\check{\underline{\kappa}}} - \frac{1}{2}\overline{\vartheta\underline{\vartheta}} + 2\overline{\eta^2}.
\end{aligned}
$$

Making use of Corollary 2.66

$$
\begin{aligned}
e_3\left(\overline{\kappa}\right) &= \overline{e_3(\kappa)} + \mathrm{Err}[e_3\overline{\kappa}] \\
&= -\frac{1}{2}\overline{\kappa}\,\overline{\underline{\kappa}} + 2\overline{\underline{\omega}}\,\overline{\kappa} + 2\overline{\rho} + 2\overline{\check{\underline{\omega}}\,\check{\kappa}} - \frac{1}{2}\overline{\check{\kappa}\,\check{\underline{\kappa}}} - \frac{1}{2}\overline{\vartheta\underline{\vartheta}} + 2\overline{\eta^2} + \mathrm{Err}[e_3\overline{\kappa}]
\end{aligned}
$$

and

$$
\begin{aligned}
& e_3\left(\overline{\kappa} - \frac{2}{r}\right) \\
= \;& e_3\left(\overline{\kappa}\right) + \frac{2}{r^2}\frac{r}{2}\left(\overline{\underline{\kappa}} + \underline{A}\right) \\
= \;& -\frac{1}{2}\overline{\kappa}\,\overline{\underline{\kappa}} + 2\overline{\underline{\omega}}\,\overline{\kappa} + 2\overline{\rho} + 2\overline{\check{\underline{\omega}}\,\check{\kappa}} - \frac{1}{2}\overline{\check{\kappa}\,\check{\underline{\kappa}}} - \frac{1}{2}\overline{\vartheta\underline{\vartheta}} + 2\overline{\eta^2} + \frac{1}{r}\overline{\underline{\kappa}} + \frac{1}{r}\underline{A} + \mathrm{Err}[e_3\overline{\kappa}] \\
= \;& -\frac{1}{2}\overline{\underline{\kappa}}\left(\overline{\kappa} - \frac{2}{r}\right) + 2\overline{\underline{\omega}}\,\overline{\kappa} + 2\overline{\rho} + 2\overline{\check{\underline{\omega}}\,\check{\kappa}} - \frac{1}{2}\overline{\check{\kappa}\,\check{\underline{\kappa}}} - \frac{1}{2}\overline{\vartheta\underline{\vartheta}} + 2\overline{\eta^2} + \frac{1}{r}\underline{A} + \mathrm{Err}[e_3\overline{\kappa}].
\end{aligned}
$$

Now,

$$
\begin{aligned}
2\overline{\underline{\omega}}\,\overline{\kappa} + 2\overline{\rho} &= 2\overline{\underline{\omega}}\left(\overline{\kappa} - \frac{2}{r}\right) + \frac{4}{r}\overline{\underline{\omega}} + 2\overline{\rho} \\
&= 2\overline{\underline{\omega}}\left(\overline{\kappa} - \frac{2}{r}\right) + \frac{4}{r}\left(\overline{\underline{\omega}} - \frac{m}{r^2}\right) + 2\left(\overline{\rho} + \frac{2m}{r^3}\right).
\end{aligned}
$$

Hence,

$$
\begin{aligned}
e_3\left(\overline{\kappa} - \frac{2}{r}\right) + \frac{1}{2}\overline{\underline{\kappa}}\left(\overline{\kappa} - \frac{2}{r}\right) &= 2\overline{\underline{\omega}}\left(\overline{\kappa} - \frac{2}{r}\right) + \frac{4}{r}\left(\overline{\underline{\omega}} - \frac{m}{r^2}\right) + 2\left(\overline{\rho} + \frac{2m}{r^3}\right) \\
&+ \; 2\overline{\eta^2} + 2\overline{\check{\underline{\omega}}\,\check{\kappa}} - \frac{1}{2}\overline{\check{\kappa}\,\check{\underline{\kappa}}} - \frac{1}{2}\overline{\vartheta\underline{\vartheta}} + \frac{1}{r}\underline{A} + \mathrm{Err}[e_3\overline{\kappa}].
\end{aligned}
$$

In view of Corollary 2.66 the error term $\mathrm{Err}[e_3(\overline{\kappa})]$ is given by

$$
\begin{aligned}
\mathrm{Err}[e_3(\overline{\kappa})] &= -\varsigma^{-1}\check{\varsigma}\left(\overline{e_3\kappa + \underline{\kappa}\kappa} - \overline{\underline{\kappa}\kappa}\right) + \varsigma^{-1}\left(\overline{\check{\varsigma}(e_3\kappa + \underline{\kappa}\kappa)} - \overline{\check{\varsigma}\check{\underline{\kappa}}}\,\overline{\kappa}\right) \\
&+ \left(\check{\underline{\Omega}} + \varsigma^{-1}\overline{\underline{\Omega}}\check{\varsigma}\right)\left(\overline{e_4\kappa + \kappa^2} - \overline{\kappa}^2\right) - \varsigma^{-1}\overline{\underline{\Omega}}\left(\overline{\check{\varsigma}(e_4\kappa + \kappa^2)} - \overline{\check{\varsigma}\check{\kappa}}\,\overline{\kappa}\right) \\
&- \varsigma^{-1}\left(\overline{\check{\underline{\Omega}}\varsigma(e_4\kappa + \kappa^2)} - \overline{\check{\underline{\Omega}}\varsigma\,\kappa}\,\overline{\kappa}\right) + \overline{\check{\underline{\kappa}}\check{\kappa}}.
\end{aligned}
$$

Together with the null structure equations for $e_3(\kappa)$ and $e_4(\kappa)$, we infer

$$
\begin{aligned}
\mathrm{Err}[e_3(\overline{\kappa})] \;=\;& -\varsigma^{-1}\check{\varsigma}\left(\overline{\frac{1}{2}\kappa\underline{\kappa}+2\underline{\omega}\kappa+2\rho+2\,\not{d}_1\eta-\frac{1}{2}\vartheta\underline{\vartheta}+2\eta^2}-\overline{\underline{\kappa}}\,\overline{\kappa}\right)\\
&+\varsigma^{-1}\left(\overline{\check{\varsigma}\left(\frac{1}{2}\kappa\underline{\kappa}+2\underline{\omega}\kappa+2\rho+2\,\not{d}_1\eta-\frac{1}{2}\vartheta\underline{\vartheta}+2\eta^2\right)}-\overline{\check{\varsigma}\check{\underline{\kappa}}}\,\overline{\kappa}\right)\\
&+\left(\check{\underline{\Omega}}+\varsigma^{-1}\overline{\underline{\Omega}}\check{\varsigma}\right)\left(\overline{\frac{1}{2}\kappa^2-\frac{1}{4}\vartheta^2}-\overline{\kappa}^2\right)\\
&-\varsigma^{-1}\overline{\underline{\Omega}}\left(\overline{\check{\varsigma}\left(\frac{1}{2}\kappa^2-\frac{1}{4}\vartheta^2\right)}-\overline{\check{\varsigma}\check{\kappa}}\,\overline{\kappa}\right)\\
&-\varsigma^{-1}\left(\overline{\check{\underline{\Omega}}\varsigma\left(\frac{1}{2}\kappa^2-\frac{1}{4}\vartheta^2\right)}-\overline{\check{\underline{\Omega}}\varsigma\,\kappa}\,\overline{\kappa}\right)+\overline{\check{\underline{\kappa}}\check{\kappa}},
\end{aligned}
$$

and hence

$$
\begin{aligned}
\mathrm{Err}[e_3(\overline{\kappa})] \;=\;& -\varsigma^{-1}\left(-\frac{1}{2}\overline{\underline{\kappa}}\,\overline{\kappa}+2\overline{\underline{\omega}}\,\overline{\kappa}+2\overline{\rho}\right)\check{\varsigma}-\frac{1}{2}\overline{\kappa}^2\left(\check{\underline{\Omega}}+\varsigma^{-1}\overline{\underline{\Omega}}\check{\varsigma}\right)\\
&-\varsigma^{-1}\check{\varsigma}\left(\overline{\frac{1}{2}\check{\kappa}\check{\underline{\kappa}}+2\check{\underline{\omega}}\check{\kappa}-\frac{1}{2}\vartheta\underline{\vartheta}+2\eta^2}\right)\\
&+\varsigma^{-1}\left(\overline{\check{\varsigma}\left(\frac{1}{2}\kappa\underline{\kappa}+2\underline{\omega}\kappa+2\check{\rho}+2\,\not{d}_1\eta-\frac{1}{2}\vartheta\underline{\vartheta}+2\eta^2\right)}-\overline{\check{\varsigma}\check{\underline{\kappa}}}\,\overline{\kappa}\right)\\
&+\left(\check{\underline{\Omega}}+\varsigma^{-1}\overline{\underline{\Omega}}\check{\varsigma}\right)\left(\overline{\frac{1}{2}\check{\kappa}^2-\frac{1}{4}\vartheta^2}\right)-\varsigma^{-1}\overline{\underline{\Omega}}\left(\overline{\check{\varsigma}\left(\frac{1}{2}\kappa^2-\frac{1}{4}\vartheta^2\right)}-\overline{\check{\varsigma}\check{\kappa}}\,\overline{\kappa}\right)\\
&-\varsigma^{-1}\left(\overline{\check{\underline{\Omega}}\varsigma\left(\frac{1}{2}\kappa^2-\frac{1}{4}\vartheta^2\right)}-\overline{\check{\underline{\Omega}}\varsigma\,\kappa}\,\overline{\kappa}\right)+\overline{\check{\underline{\kappa}}\check{\kappa}}
\end{aligned}
\tag{A.3.1}
$$

so that, in view of the definition of $\underline{A}$, we obtain

$$
\begin{aligned}
& e_3\left(\overline{\kappa}-\frac{2}{r}\right)+\frac{1}{2}\overline{\underline{\kappa}}\left(\overline{\kappa}-\frac{2}{r}\right)\\
=\;& 2\overline{\underline{\omega}}\left(\overline{\kappa}-\frac{2}{r}\right)+\frac{4}{r}\left(\overline{\underline{\omega}}-\frac{m}{r^2}\right)+2\left(\overline{\rho}+\frac{2m}{r^3}\right)-\varsigma^{-1}\left(-\frac{1}{2}\overline{\underline{\kappa}}\,\overline{\kappa}+2\overline{\underline{\omega}}\,\overline{\kappa}+2\overline{\rho}\right)\check{\varsigma}\\
&-\frac{1}{2}\overline{\kappa}^2\left(\check{\underline{\Omega}}+\varsigma^{-1}\overline{\underline{\Omega}}\check{\varsigma}\right)-\frac{1}{r}\varsigma^{-1}\overline{\underline{\kappa}}\check{\varsigma}+\frac{1}{r}\overline{\kappa}\left(\check{\underline{\Omega}}+\varsigma^{-1}\overline{\underline{\Omega}}\check{\varsigma}\right)+\mathrm{Err}\left[e_3\left(\overline{\kappa}-\frac{2}{r}\right)\right],
\end{aligned}
$$

with

$$
\begin{aligned}
&\mathrm{Err}\left[e_3\left(\overline{\kappa}-\frac{2}{r}\right)\right]\\
=\ & 2\overline{\eta^2}+2\overline{\underline{\check{\omega}}\,\check{\kappa}}-\frac{1}{2}\overline{\check{\kappa}\,\underline{\check{\kappa}}}-\frac{1}{2}\overline{\vartheta\underline{\vartheta}}+\frac{1}{r}\varsigma^{-1}\overline{\check{\varsigma}\underline{\check{\kappa}}}-\frac{1}{r}\varsigma^{-1}\overline{\underline{\Omega}}\,\overline{\check{\varsigma}\check{\kappa}}-\frac{1}{r}\varsigma^{-1}\overline{\underline{\check{\Omega}}\varsigma\kappa}\\
&-\varsigma^{-1}\check{\varsigma}\,\overline{\left(\frac{1}{2}\check{\kappa}\underline{\check{\kappa}}+2\underline{\check{\omega}}\check{\kappa}-\frac{1}{2}\vartheta\underline{\vartheta}+2\eta^2\right)}\\
&+\varsigma^{-1}\left(\overline{\check{\varsigma}\left(\frac{1}{2}\kappa\underline{\kappa}+2\underline{\omega}\kappa+2\check{\rho}+2\not{d}_1\eta-\frac{1}{2}\vartheta\underline{\vartheta}+2\eta^2\right)}-\overline{\check{\varsigma}\underline{\check{\kappa}}}\,\overline{\kappa}\right)\\
&+\left(\underline{\check{\Omega}}+\varsigma^{-1}\overline{\underline{\Omega}}\check{\varsigma}\right)\overline{\left(\frac{1}{2}\check{\kappa}^2-\frac{1}{4}\vartheta^2\right)}-\varsigma^{-1}\overline{\underline{\Omega}}\left(\overline{\check{\varsigma}\left(\frac{1}{2}\kappa^2-\frac{1}{4}\vartheta^2\right)}-\overline{\check{\varsigma}\check{\kappa}}\,\overline{\kappa}\right)\\
&-\varsigma^{-1}\left(\overline{\underline{\check{\Omega}}\varsigma\left(\frac{1}{2}\kappa^2-\frac{1}{4}\vartheta^2\right)}-\overline{\underline{\check{\Omega}}\varsigma\,\kappa}\,\overline{\kappa}\right)+\overline{\underline{\check{\kappa}}\check{\kappa}}.
\end{aligned}
$$

This concludes the proof of Lemma 2.72.

A.4 PROOF OF PROPOSITION 2.73

In view of Corollary 2.66 applied to

$$
e_4(\kappa)+\frac{1}{2}\kappa^2=-\frac{1}{2}\vartheta^2,
$$

we deduce

$$
e_4\check{\kappa}+\overline{\kappa}\check{\kappa}=-\frac{1}{2}\check{\kappa}^2-\frac{1}{2}\overline{\check{\kappa}^2}-\frac{1}{2}(\vartheta^2-\overline{\vartheta^2}).
$$

In view of Corollary 2.66 applied to

$$
e_4(\underline{\kappa})+\frac{1}{2}\kappa\underline{\kappa}=-2\not{d}_1\zeta+2\rho-\frac{1}{2}\vartheta\underline{\vartheta}+2\zeta^2
$$

we deduce

$$
e_4\underline{\check{\kappa}}+\frac{1}{2}\overline{\kappa}\underline{\check{\kappa}}+\frac{1}{2}\check{\kappa}\overline{\underline{\kappa}}=-\frac{1}{2}\check{\kappa}\underline{\check{\kappa}}-\frac{1}{2}\overline{\check{\kappa}\underline{\check{\kappa}}}+F-\overline{F}
$$

where

$$
\begin{aligned}
F-\overline{F} &= \left(-2\not{d}_1\zeta+2\rho-\frac{1}{2}\vartheta\underline{\vartheta}+2\zeta^2\right)-\overline{\left(-2\not{d}_1\zeta+2\rho-\frac{1}{2}\vartheta\underline{\vartheta}+2\zeta^2\right)}\\
&= -2\not{d}_1\zeta+2\check{\rho}+\left(-\frac{1}{2}\vartheta\underline{\vartheta}+2\zeta^2\right)-\overline{\left(-\frac{1}{2}\vartheta\underline{\vartheta}+2\zeta^2\right)}.
\end{aligned}
$$

Hence,

$$\begin{aligned} e_4\underline{\check{\kappa}} + \frac{1}{2}\overline{\kappa}\underline{\check{\kappa}} + \frac{1}{2}\check{\kappa}\overline{\underline{\kappa}} &= -2\,\not{d}_1\zeta + 2\check{\rho} + \text{Err}[e_4\underline{\check{\kappa}}] \\ \text{Err}[e_4\underline{\check{\kappa}}] : &= -\frac{1}{2}\check{\kappa}\underline{\check{\kappa}} - \frac{1}{2}\overline{\check{\kappa}\underline{\check{\kappa}}} + \left(-\frac{1}{2}\vartheta\underline{\vartheta} + 2\zeta^2\right) - \overline{\left(-\frac{1}{2}\vartheta\underline{\vartheta} + 2\zeta^2\right)}. \end{aligned}$$

In view of Corollary 2.66 applied to $e_4(\underline{\omega}) = \rho + 3\zeta^2$ we deduce

$$e_4\underline{\check{\omega}} = -\overline{\check{\kappa}\underline{\check{\omega}}} + (\rho + 3\zeta^2) - \overline{(\rho + 3\zeta^2)} = \check{\rho} - \overline{\check{\kappa}\underline{\check{\omega}}} + 3(\zeta^2 - \overline{\zeta^2}).$$

In view of Corollary 2.66 applied to

$$e_4(\rho) + \frac{3}{2}\kappa\rho = \not{d}_1\beta - \frac{1}{2}\underline{\vartheta}\alpha - \zeta\beta$$

we deduce

$$e_4\check{\rho} + \frac{3}{2}\overline{\kappa}\check{\rho} + \frac{3}{2}\overline{\rho}\check{\kappa} = -\frac{3}{2}\check{\kappa}\check{\rho} + \frac{1}{2}\overline{\check{\kappa}\check{\rho}} + \not{d}_1\beta - \left(\frac{1}{2}\underline{\vartheta}\alpha + \zeta\beta\right) + \overline{\left(\frac{1}{2}\underline{\vartheta}\alpha + \zeta\beta\right)}.$$

In view of Corollary 2.66 applied to

$$e_4\mu + \frac{3}{2}\kappa\mu = \text{Err}[e_4\mu],$$

we deduce

$$e_4\check{\mu} + \frac{3}{2}\overline{\kappa}\check{\mu} + \frac{3}{2}\overline{\mu}\check{\kappa} = -\frac{3}{2}\check{\kappa}\check{\mu} + \frac{1}{2}\overline{\check{\kappa}\check{\mu}} + \text{Err}[e_4\mu] - \overline{\text{Err}[e_4\mu]}.$$

In view of Corollary 2.66 applied to $e_4(\underline{\Omega}) = -2\underline{\omega}$ we deduce

$$-e_4(\underline{\check{\Omega}}) = 2\underline{\check{\omega}} - \overline{\check{\kappa}\underline{\check{\Omega}}}$$

as stated.

In view of Corollary 2.66 applied to the equation

$$e_3(\kappa) + \frac{1}{2}\underline{\kappa}\kappa = 2\,\not{d}_1\eta + 2\rho + 2\eta^2 + 2\underline{\omega}\kappa - \frac{1}{2}\underline{\vartheta}\vartheta$$

we deduce

$$\begin{aligned} e_3(\check{\kappa}) &= e_3(\kappa) - \overline{e_3(\kappa)} - \text{Err}[e_3(\overline{\kappa})] \\ &= -\frac{1}{2}\underline{\kappa}\kappa + 2\,\not{d}_1\eta + 2\rho + 2\eta^2 + 2\underline{\omega}\kappa - \frac{1}{2}\underline{\vartheta}\vartheta \\ &\quad + \frac{1}{2}\overline{\underline{\kappa}\kappa} - 2\overline{\rho} - 2\overline{\eta^2} - 2\overline{\underline{\omega}\kappa} + \frac{1}{2}\overline{\underline{\vartheta}\vartheta} - \text{Err}[e_3\overline{\kappa}] \\ &= 2\,\not{d}_1\eta + 2\check{\rho} - \frac{1}{2}\left(\underline{\kappa}\check{\kappa} + \kappa\underline{\check{\kappa}}\right) + 2\left(\underline{\omega}\check{\kappa} + \kappa\underline{\check{\omega}}\right) \\ &\quad + 2\left(\eta^2 - \overline{\eta^2}\right) - \frac{1}{2}\overline{\check{\kappa}\underline{\check{\kappa}}} + 2\overline{\underline{\check{\omega}}\check{\kappa}} - \frac{1}{2}\left(\vartheta\underline{\vartheta} - \overline{\vartheta\underline{\vartheta}}\right) - \text{Err}[e_3\overline{\kappa}]. \end{aligned}$$

Now, recall that we have, in view of (A.3.1),

$$
\begin{aligned}
\mathrm{Err}[e_3(\overline{\kappa})] \;=\;& -\varsigma^{-1}\left(-\frac{1}{2}\overline{\underline{\kappa}}\,\overline{\kappa}+2\overline{\underline{\omega}}\,\overline{\kappa}+2\overline{\rho}\right)\check{\varsigma}-\frac{1}{2}\overline{\kappa}^2\left(\underline{\check{\Omega}}+\varsigma^{-1}\underline{\overline{\Omega}}\check{\varsigma}\right)\\
&-\varsigma^{-1}\check{\varsigma}\,\overline{\left(\frac{1}{2}\check{\kappa}\underline{\check{\kappa}}+2\underline{\check{\omega}}\check{\kappa}-\frac{1}{2}\vartheta\underline{\vartheta}+2\eta^2\right)}\\
&+\varsigma^{-1}\left(\overline{\check{\varsigma}\left(\frac{1}{2}\kappa\underline{\kappa}+2\underline{\omega}\kappa+2\check{\rho}+2\,\slashed{d}_1\eta-\frac{1}{2}\vartheta\underline{\vartheta}+2\eta^2\right)}-\overline{\check{\varsigma}\underline{\check{\kappa}}}\,\overline{\kappa}\right)\\
&+\left(\underline{\check{\Omega}}+\varsigma^{-1}\underline{\overline{\Omega}}\check{\varsigma}\right)\overline{\left(\frac{1}{2}\check{\kappa}^2-\frac{1}{4}\vartheta^2\right)}-\varsigma^{-1}\underline{\overline{\Omega}}\left(\overline{\check{\varsigma}\left(\frac{1}{2}\kappa^2-\frac{1}{4}\vartheta^2\right)}-\overline{\check{\varsigma}\check{\kappa}}\,\overline{\kappa}\right)\\
&-\varsigma^{-1}\left(\overline{\underline{\check{\Omega}}\varsigma\left(\frac{1}{2}\kappa^2-\frac{1}{4}\vartheta^2\right)}-\overline{\underline{\check{\Omega}}\varsigma\,\kappa}\,\overline{\kappa}\right)+\overline{\underline{\check{\kappa}}\check{\kappa}}.
\end{aligned}
$$

We deduce

$$
\begin{aligned}
e_3(\check{\kappa}) \;=\;& 2\,\slashed{d}_1\eta+2\check{\rho}-\frac{1}{2}\left(\underline{\kappa}\check{\kappa}+\kappa\underline{\check{\kappa}}\right)+2\left(\underline{\omega}\check{\kappa}+\kappa\underline{\check{\omega}}\right)\\
&+\varsigma^{-1}\left(-\frac{1}{2}\overline{\underline{\kappa}}\,\overline{\kappa}+2\overline{\underline{\omega}}\,\overline{\kappa}+2\overline{\rho}\right)\check{\varsigma}+\frac{1}{2}\overline{\kappa}^2\left(\underline{\check{\Omega}}+\varsigma^{-1}\underline{\overline{\Omega}}\check{\varsigma}\right)\\
&+2\left(\eta^2-\overline{\eta^2}\right)-\frac{1}{2}\overline{\check{\kappa}\underline{\check{\kappa}}}+2\overline{\underline{\check{\omega}}\check{\kappa}}-\frac{1}{2}\left(\vartheta\underline{\vartheta}-\overline{\vartheta\underline{\vartheta}}\right)\\
&+\varsigma^{-1}\check{\varsigma}\,\overline{\left(\frac{1}{2}\check{\kappa}\underline{\check{\kappa}}+2\underline{\check{\omega}}\check{\kappa}-\frac{1}{2}\vartheta\underline{\vartheta}+2\eta^2\right)}\\
&-\varsigma^{-1}\left(\overline{\check{\varsigma}\left(\frac{1}{2}\kappa\underline{\kappa}+2\underline{\omega}\kappa+2\check{\rho}+2\,\slashed{d}_1\eta-\frac{1}{2}\vartheta\underline{\vartheta}+2\eta^2\right)}-\overline{\check{\varsigma}\underline{\check{\kappa}}}\,\overline{\kappa}\right)\\
&-\left(\underline{\check{\Omega}}+\varsigma^{-1}\underline{\overline{\Omega}}\check{\varsigma}\right)\overline{\left(\frac{1}{2}\check{\kappa}^2-\frac{1}{4}\vartheta^2\right)}+\varsigma^{-1}\underline{\overline{\Omega}}\left(\overline{\check{\varsigma}\left(\frac{1}{2}\kappa^2-\frac{1}{4}\vartheta^2\right)}-\overline{\check{\varsigma}\check{\kappa}}\,\overline{\kappa}\right)\\
&+\varsigma^{-1}\left(\overline{\underline{\check{\Omega}}\varsigma\left(\frac{1}{2}\kappa^2-\frac{1}{4}\vartheta^2\right)}-\overline{\underline{\check{\Omega}}\varsigma\,\kappa}\,\overline{\kappa}\right)-\overline{\underline{\check{\kappa}}\check{\kappa}}
\end{aligned}
$$

as desired.

In view of Corollary 2.66 applied to the equation

$$
e_3(\underline{\kappa})+\frac{1}{2}\underline{\kappa}^2=2\,\slashed{d}_1\underline{\xi}-2\underline{\omega}\,\underline{\kappa}+2(\eta-3\zeta)\underline{\xi}-\frac{1}{2}\underline{\vartheta}^2
$$

we deduce

$$
\begin{aligned}
e_3(\underline{\check{\kappa}})+\underline{\kappa}\,\underline{\check{\kappa}} \;=\;& 2\,\slashed{d}_1\underline{\xi}-2\left(\underline{\check{\omega}}\,\underline{\kappa}+\underline{\omega}\,\underline{\check{\kappa}}\right)\\
&-\frac{1}{2}\overline{\underline{\check{\kappa}}^2}-2\overline{\underline{\check{\omega}}\,\underline{\check{\kappa}}}+2(\eta-3\zeta)\underline{\xi}-\overline{2(\eta-3\zeta)\underline{\xi}}-\frac{1}{2}\left(\underline{\vartheta}^2-\overline{\underline{\vartheta}^2}\right)-\mathrm{Err}[e_3\overline{\underline{\kappa}}]
\end{aligned}
$$

where

$$
\begin{aligned}
-\mathrm{Err}[e_3\overline{\underline{\kappa}}] &= \varsigma^{-1}\check{\varsigma}\left(\overline{e_3\underline{\kappa}+\underline{\kappa}^2}-\overline{\underline{\kappa}\underline{\kappa}}\right)-\varsigma^{-1}\left(\overline{\check{\varsigma}(e_3\underline{\kappa}+\underline{\kappa}^2)}-\overline{\check{\varsigma}\check{\underline{\kappa}}}\,\overline{\underline{\kappa}}\right)\\
&\quad-\left(\check{\underline{\Omega}}+\varsigma^{-1}\overline{\underline{\Omega}}\check{\varsigma}\right)\left(\overline{e_4\underline{\kappa}+\kappa\underline{\kappa}}-\overline{\kappa}\,\overline{\underline{\kappa}}\right)+\varsigma^{-1}\overline{\underline{\Omega}}\left(\overline{\check{\varsigma}(e_4\underline{\kappa}+\kappa\underline{\kappa})}-\overline{\check{\varsigma}\check{\kappa}}\,\overline{\underline{\kappa}}\right)\\
&\quad+\varsigma^{-1}\left(\overline{\check{\underline{\Omega}}\varsigma(e_4\underline{\kappa}+\kappa\underline{\kappa})}-\overline{\check{\underline{\Omega}}\varsigma\,\kappa}\,\overline{\underline{\kappa}}\right)-\overline{\check{\underline{\kappa}}^2}.
\end{aligned}
$$

In view of the null structure equations for $e_3(\underline{\kappa})$ and $e_4(\underline{\kappa})$, we infer

$$
\begin{aligned}
-\mathrm{Err}[e_3\overline{\underline{\kappa}}] &= \varsigma^{-1}\check{\varsigma}\left(-\frac{1}{2}\overline{\underline{\kappa}}^2-2\overline{\underline{\omega}}\,\overline{\underline{\kappa}}\right)-\left(\check{\underline{\Omega}}+\varsigma^{-1}\overline{\underline{\Omega}}\check{\varsigma}\right)\left(-\frac{1}{2}\overline{\kappa}\,\overline{\underline{\kappa}}+2\overline{\rho}\right)\\
&\quad+\varsigma^{-1}\check{\varsigma}\left(\overline{\frac{1}{2}\check{\underline{\kappa}}^2-2\check{\underline{\omega}}\,\check{\underline{\kappa}}+2(\eta-3\zeta)\underline{\xi}-\frac{1}{2}\underline{\vartheta}^2}\right)\\
&\quad-\left(\check{\underline{\Omega}}+\varsigma^{-1}\overline{\underline{\Omega}}\check{\varsigma}\right)\left(\overline{\frac{1}{2}\check{\kappa}\,\check{\underline{\kappa}}-\frac{1}{2}\vartheta\,\underline{\vartheta}+2\zeta^2}\right)\\
&\quad-\varsigma^{-1}\left(\overline{\check{\varsigma}\left(\frac{1}{2}\underline{\kappa}^2-2\underline{\omega}\,\underline{\kappa}+2\,\slashed{d}_1\underline{\xi}+2(\eta-3\zeta)\underline{\xi}-\frac{1}{2}\underline{\vartheta}^2\right)}-\overline{\check{\varsigma}\check{\underline{\kappa}}}\,\overline{\underline{\kappa}}\right)\\
&\quad+\varsigma^{-1}\overline{\underline{\Omega}}\left(\overline{\check{\varsigma}\left(\frac{1}{2}\kappa\,\underline{\kappa}-2\,\slashed{d}_1\zeta+2\rho-\frac{1}{2}\vartheta\,\underline{\vartheta}+2\zeta^2\right)}-\overline{\check{\varsigma}\check{\kappa}}\,\overline{\underline{\kappa}}\right)\\
&\quad+\varsigma^{-1}\left(\overline{\check{\underline{\Omega}}\varsigma\left(\frac{1}{2}\kappa\,\underline{\kappa}-2\,\slashed{d}_1\zeta+2\rho-\frac{1}{2}\vartheta\,\underline{\vartheta}+2\zeta^2\right)}-\overline{\check{\underline{\Omega}}\varsigma\,\kappa}\,\overline{\underline{\kappa}}\right)-\overline{\check{\underline{\kappa}}^2}
\end{aligned}
$$

and hence

$$
\begin{aligned}
e_3(\check{\underline{\kappa}})+\underline{\kappa}\,\check{\underline{\kappa}} &= 2\,\slashed{d}_1\underline{\xi}-2\left(\check{\underline{\omega}}\,\underline{\kappa}+\underline{\omega}\,\check{\underline{\kappa}}\right)+\varsigma^{-1}\check{\varsigma}\left(-\frac{1}{2}\overline{\underline{\kappa}}^2-2\overline{\underline{\omega}}\,\overline{\underline{\kappa}}\right)\\
&\quad-\left(\check{\underline{\Omega}}+\varsigma^{-1}\overline{\underline{\Omega}}\check{\varsigma}\right)\left(-\frac{1}{2}\overline{\kappa}\,\overline{\underline{\kappa}}+2\overline{\rho}\right)+\mathrm{Err}[e_3(\check{\underline{\kappa}})],\\
\mathrm{Err}[e_3(\check{\underline{\kappa}})] &= -\frac{1}{2}\overline{\check{\underline{\kappa}}^2}-2\overline{\check{\underline{\omega}}\,\check{\underline{\kappa}}}+2(\eta-3\zeta)\underline{\xi}-\overline{2(\eta-3\zeta)\underline{\xi}}-\frac{1}{2}\left(\underline{\vartheta}^2-\overline{\underline{\vartheta}^2}\right)\\
&\quad-\varsigma^{-1}\left(\overline{\check{\varsigma}\left(\frac{1}{2}\underline{\kappa}^2-2\underline{\omega}\,\underline{\kappa}+2\,\slashed{d}_1\underline{\xi}+2(\eta-3\zeta)\underline{\xi}-\frac{1}{2}\underline{\vartheta}^2\right)}-\overline{\check{\varsigma}\check{\underline{\kappa}}}\,\overline{\underline{\kappa}}\right)\\
&\quad+\varsigma^{-1}\overline{\underline{\Omega}}\left(\overline{\check{\varsigma}\left(\frac{1}{2}\kappa\,\underline{\kappa}-2\,\slashed{d}_1\zeta+2\rho-\frac{1}{2}\vartheta\,\underline{\vartheta}+2\zeta^2\right)}-\overline{\check{\varsigma}\check{\kappa}}\,\overline{\underline{\kappa}}\right)\\
&\quad+\varsigma^{-1}\left(\overline{\check{\underline{\Omega}}\varsigma\left(\frac{1}{2}\kappa\,\underline{\kappa}-2\,\slashed{d}_1\zeta+2\rho-\frac{1}{2}\vartheta\,\underline{\vartheta}+2\zeta^2\right)}-\overline{\check{\underline{\Omega}}\varsigma\,\kappa}\,\overline{\underline{\kappa}}\right)-\overline{\check{\underline{\kappa}}^2}
\end{aligned}
$$

as desired.

In view of Corollary 2.66 applied to the equation

$$
e_3(\rho)+\frac{3}{2}\underline{\kappa}\rho=\slashed{d}_1\underline{\beta}-\frac{1}{2}\vartheta\underline{\alpha}-\zeta\underline{\beta}+2\eta\underline{\beta}+2\underline{\xi}\beta
$$

we deduce

$$
\begin{aligned}
e_3\check{\rho} + \frac{3}{2}\overline{\underline{\kappa}}\check{\rho} + \frac{3}{2}\underline{\check{\kappa}}\overline{\rho} &= \slashed{d}_1\underline{\beta} - \left(\frac{1}{2}\vartheta\underline{\alpha} + \zeta\underline{\beta} - 2\eta\underline{\beta} - 2\underline{\xi}\beta\right) \\
&+ \overline{\left(\frac{1}{2}\vartheta\underline{\alpha} + \zeta\underline{\beta} - 2\eta\underline{\beta} - 2\underline{\xi}\beta\right)} - \frac{3}{2}\underline{\check{\kappa}}\check{\rho} - \mathrm{Err}[e_3\overline{\rho}]
\end{aligned}
$$

where

$$
\begin{aligned}
-\mathrm{Err}[e_3(\overline{\rho})] &= \varsigma^{-1}\check{\varsigma}\,(\overline{e_3\rho + \underline{\kappa}\rho} - \overline{\underline{\kappa}\rho}) - \varsigma^{-1}\left(\overline{\check{\varsigma}(e_3\rho + \underline{\kappa}\rho)} - \overline{\check{\varsigma}}\overline{\underline{\check{\kappa}}}\,\overline{\rho}\right) \\
&- (\underline{\check{\Omega}} + \varsigma^{-1}\overline{\underline{\Omega}}\check{\varsigma})\,(\overline{e_4\rho + \kappa\rho} - \overline{\kappa}\,\overline{\rho}) + \varsigma^{-1}\overline{\underline{\Omega}}\left(\overline{\check{\varsigma}(e_4\rho + \kappa\rho)} - \overline{\check{\varsigma}}\overline{\check{\kappa}}\,\overline{\rho}\right) \\
&+ \varsigma^{-1}\left(\overline{\underline{\check{\Omega}}\varsigma(e_4\rho + \kappa\rho)} - \overline{\underline{\check{\Omega}}\varsigma\,\kappa}\right) - \overline{\underline{\check{\kappa}}\check{\rho}}.
\end{aligned}
$$

In view of the null structure equations for $e_3(\rho)$ and $e_4(\rho)$, we infer

$$
\begin{aligned}
-\mathrm{Err}[e_3(\overline{\rho})] &= -\frac{3}{2}\overline{\underline{\kappa}}\,\overline{\rho}\varsigma^{-1}\check{\varsigma} + \frac{3}{2}\overline{\kappa}\,\overline{\rho}\,(\underline{\check{\Omega}} + \varsigma^{-1}\overline{\underline{\Omega}}\check{\varsigma}) \\
&+ \varsigma^{-1}\check{\varsigma}\,\overline{\left(-\frac{1}{2}\underline{\check{\kappa}}\check{\rho} - \frac{1}{2}\vartheta\,\underline{\alpha} - \zeta\,\underline{\beta} + 2(\eta\,\underline{\beta} + \underline{\xi}\,\beta)\right)} \\
&- \varsigma^{-1}\left(\overline{\check{\varsigma}\left(-\frac{1}{2}\underline{\kappa}\rho + \slashed{d}_1\underline{\beta} - \frac{1}{2}\vartheta\,\underline{\alpha} - \zeta\,\underline{\beta} + 2(\eta\,\underline{\beta} + \underline{\xi}\,\beta)\right)} - \overline{\check{\varsigma}}\overline{\underline{\check{\kappa}}}\,\overline{\rho}\right) \\
&- (\underline{\check{\Omega}} + \varsigma^{-1}\overline{\underline{\Omega}}\check{\varsigma})\,\overline{\left(-\frac{1}{2}\check{\kappa}\check{\rho} - \frac{1}{2}\underline{\vartheta}\,\alpha - \zeta\beta\right)} \\
&+ \varsigma^{-1}\overline{\underline{\Omega}}\left(\overline{\check{\varsigma}\left(-\frac{1}{2}\kappa\rho + \slashed{d}_1\beta - \frac{1}{2}\underline{\vartheta}\,\alpha - \zeta\beta\right)} - \overline{\check{\varsigma}}\overline{\check{\kappa}}\,\overline{\rho}\right) \\
&+ \varsigma^{-1}\left(\overline{\underline{\check{\Omega}}\varsigma\left(-\frac{1}{2}\kappa\rho + \slashed{d}_1\beta - \frac{1}{2}\underline{\vartheta}\,\alpha - \zeta\beta\right)} - \overline{\underline{\check{\Omega}}\varsigma\,\kappa}\right) - \overline{\underline{\check{\kappa}}\check{\rho}}
\end{aligned}
$$

and hence

$$
\begin{aligned}
e_3\check{\rho} + \frac{3}{2}\overline{\underline{\kappa}}\check{\rho} &= -\frac{3}{2}\overline{\rho}\underline{\check{\kappa}} + \not{d}_1\underline{\beta} - \frac{3}{2}\overline{\underline{\kappa}}\,\overline{\rho}\varsigma^{-1}\check{\varsigma} + \frac{3}{2}\overline{\kappa}\,\overline{\rho}\left(\underline{\check{\Omega}} + \varsigma^{-1}\overline{\underline{\Omega}}\check{\varsigma}\right) + \mathrm{Err}[e_3\check{\rho}],\\
\mathrm{Err}[e_3\check{\rho}] &= -\left(\frac{1}{2}\vartheta\underline{\alpha} + \zeta\underline{\beta} - 2\eta\underline{\beta} - 2\underline{\xi}\beta\right) + \overline{\left(\frac{1}{2}\vartheta\underline{\alpha} + \zeta\underline{\beta} - 2\eta\underline{\beta} - 2\underline{\xi}\beta\right)} - \frac{3}{2}\overline{\underline{\check{\kappa}}\check{\rho}}\\
&\quad +\varsigma^{-1}\check{\varsigma}\,\overline{\left(-\frac{1}{2}\underline{\check{\kappa}}\check{\rho} - \frac{1}{2}\vartheta\,\underline{\alpha} - \zeta\,\underline{\beta} + 2(\eta\,\underline{\beta} + \underline{\xi}\,\beta)\right)}\\
&\quad -\varsigma^{-1}\left(\overline{\check{\varsigma}\left(-\frac{1}{2}\underline{\kappa}\rho + \not{d}_1\underline{\beta} - \frac{1}{2}\vartheta\,\underline{\alpha} - \zeta\,\underline{\beta} + 2(\eta\,\underline{\beta} + \underline{\xi}\,\beta)\right)} - \overline{\check{\varsigma}\underline{\check{\kappa}}}\,\overline{\rho}\right)\\
&\quad -\left(\underline{\check{\Omega}} + \varsigma^{-1}\overline{\underline{\Omega}}\check{\varsigma}\right)\overline{\left(-\frac{1}{2}\check{\kappa}\check{\rho} - \frac{1}{2}\underline{\vartheta}\,\alpha - \zeta\beta\right)}\\
&\quad +\varsigma^{-1}\overline{\underline{\Omega}}\left(\overline{\check{\varsigma}\left(-\frac{1}{2}\kappa\rho + \not{d}_1\beta - \frac{1}{2}\underline{\vartheta}\,\alpha - \zeta\beta\right)} - \overline{\check{\varsigma}\check{\kappa}}\,\overline{\rho}\right)\\
&\quad +\varsigma^{-1}\left(\overline{\underline{\check{\Omega}}\varsigma\left(-\frac{1}{2}\kappa\rho + \not{d}_1\beta - \frac{1}{2}\underline{\vartheta}\,\alpha - \zeta\beta\right)} - \overline{\underline{\check{\Omega}}\varsigma\,\kappa}\right) - \overline{\underline{\check{\kappa}}\check{\rho}},
\end{aligned}
$$

which ends the proof of Proposition 2.73.

A.5 PROOF OF PROPOSITION 2.74

In view of the null structure equation for $e_3(\zeta)$, we have

$$
\frac{1}{2}\kappa\underline{\xi} + 2\,\not{d}_1^{\star}\underline{\omega} = e_3(\zeta) + \frac{1}{2}\underline{\kappa}(\zeta + \eta) - 2\underline{\omega}(\zeta - \eta) - \underline{\beta} + \frac{1}{2}\underline{\vartheta}(\zeta + \eta) - \frac{1}{2}\vartheta\underline{\xi}
$$

and hence

$$
\begin{aligned}
\frac{1}{2}\kappa\underline{\xi} + 2\,\not{d}_1^{\star}\underline{\omega} &= \left(\frac{1}{2}\underline{\kappa} + 2\underline{\omega} + \frac{1}{2}\underline{\vartheta}\right)\eta + e_3(\zeta) - \underline{\beta}\\
&\quad + \frac{1}{2}\underline{\kappa}\zeta - 2\underline{\omega}\zeta + \frac{1}{2}\underline{\vartheta}\zeta - \frac{1}{2}\vartheta\underline{\xi}
\end{aligned}
$$

which is the first desired identity.

To prove the second identity we start with

$$
e_3(\kappa) + \frac{1}{2}\underline{\kappa}\,\kappa - 2\underline{\omega}\kappa = 2\,\not{d}_1\eta + 2\rho - \tfrac{1}{2}\underline{\vartheta}\,\vartheta + 2\eta^2.
$$

Applying e_θ,

$$
\begin{aligned}
&e_3(e_\theta(\kappa)) + [e_\theta, e_3]\kappa + \frac{1}{2}\underline{\kappa}e_\theta(\kappa) + \frac{1}{2}\kappa e_\theta(\underline{\kappa}) - 2\underline{\omega}e_\theta(\kappa) - 2\kappa e_\theta(\underline{\omega})\\
= \;& 2e_\theta(\not{d}_1\eta) + 2e_\theta(\rho) - \frac{1}{2}e_\theta(\underline{\vartheta}\,\vartheta) + 2e_\theta(\eta^2).
\end{aligned}
$$

Since $[e_\theta, e_3]\kappa = \frac{1}{2}(\underline{\kappa}+\underline{\vartheta})e_\theta\kappa + (\zeta-\eta)e_3\kappa - \underline{\xi}e_4\kappa$ we deduce

$$\begin{aligned}2e_\theta(\not{d}_1\eta)+\eta e_3(\kappa)+2e_\theta(\eta^2) &= -\underline{\xi}e_4(\kappa)-2\kappa e_\theta(\underline{\omega})+e_3(e_\theta(\kappa))\\ &+ \frac{1}{2}(\underline{\kappa}+\underline{\vartheta})e_\theta(\kappa)+\zeta e_3(\kappa)+\frac{1}{2}\underline{\kappa}e_\theta(\kappa)+\frac{1}{2}\kappa e_\theta(\underline{\kappa})\\ &- 2\underline{\omega}e_\theta(\kappa)-2e_\theta(\rho)+\frac{1}{2}e_\theta(\underline{\vartheta}\,\vartheta),\end{aligned}$$

or, making use of the equations for $e_3\kappa$ and $e_4\kappa$ in Proposition 2.63,

$$\begin{aligned}&2e_\theta(\not{d}_1\eta)+\left(-\frac{1}{2}\kappa\underline{\kappa}+2\underline{\omega}\kappa+2\not{d}_1\eta+2\rho-\frac{1}{2}\vartheta\underline{\vartheta}+2\eta^2\right)\eta+2e_\theta(\eta^2)\\ =\;& -\left(-\frac{1}{2}\kappa^2-\frac{1}{2}\vartheta^2\right)\underline{\xi}-2\kappa e_\theta(\underline{\omega})+e_3(e_\theta(\kappa))\\ &+\frac{1}{2}(\underline{\kappa}+\underline{\vartheta})e_\theta(\kappa)+\left(-\frac{1}{2}\kappa\underline{\kappa}+2\underline{\omega}\kappa+2\not{d}_1\eta+2\rho-\frac{1}{2}\vartheta\underline{\vartheta}+2\eta^2\right)\zeta\\ &+\frac{1}{2}\underline{\kappa}e_\theta(\kappa)+\frac{1}{2}\kappa e_\theta(\underline{\kappa})-2\underline{\omega}e_\theta(\kappa)-2e_\theta(\rho)+\frac{1}{2}e_\theta(\underline{\vartheta}\,\vartheta).\end{aligned}$$

Since $e_\theta = -\not{d}_1^\star$, $\not{d}_1^\star\not{d}_1 = \not{d}_2\not{d}_2^\star + 2K$ and $K = -\rho - \frac{1}{4}\kappa\underline{\kappa} + \frac{1}{4}\vartheta\underline{\vartheta}$, we infer that

$$\begin{aligned}&\left(-2\not{d}_2\not{d}_2^\star+\frac{1}{2}\kappa\underline{\kappa}+2\underline{\omega}\kappa+2\not{d}_1\eta+6\rho-\frac{3}{2}\vartheta\underline{\vartheta}+2\eta^2\right)\eta+2e_\theta(\eta^2)\\ =\;& \kappa\left(\frac{1}{2}\kappa\underline{\xi}+2\not{d}_1^\star\underline{\omega}\right)+e_3(e_\theta(\kappa))\\ &+\frac{1}{2}(\underline{\kappa}+\underline{\vartheta})e_\theta(\kappa)+\left(-\frac{1}{2}\kappa\underline{\kappa}+2\underline{\omega}\kappa+2\not{d}_1\eta+2\rho-\frac{1}{2}\vartheta\underline{\vartheta}+2\eta^2\right)\zeta\\ &+\frac{1}{2}\underline{\kappa}e_\theta(\kappa)+\frac{1}{2}\kappa e_\theta(\underline{\kappa})-2\underline{\omega}e_\theta(\kappa)-2e_\theta(\rho)+\frac{1}{2}e_\theta(\underline{\vartheta}\,\vartheta)+\frac{1}{2}\vartheta^2\underline{\xi}.\end{aligned}$$

Making use of the previously derived identity,

$$\begin{aligned}2\not{d}_1^\star\underline{\omega}+\frac{1}{2}\kappa\underline{\xi} &= \left(\frac{1}{2}\underline{\kappa}+2\underline{\omega}+\frac{1}{2}\underline{\vartheta}\right)\eta+e_3(\zeta)-\underline{\beta}\\ &+\frac{1}{2}\underline{\kappa}\zeta-2\underline{\omega}\zeta+\frac{1}{2}\underline{\vartheta}\zeta-\frac{1}{2}\vartheta\underline{\xi},\end{aligned}$$

we infer that

$$\begin{aligned}&\left(-2\not{d}_2\not{d}_2^\star+\frac{1}{2}\kappa\underline{\kappa}+2\underline{\omega}\kappa+2\not{d}_1\eta+6\rho-\frac{3}{2}\vartheta\underline{\vartheta}+2\eta^2\right)\eta+2e_\theta(\eta^2)\\ =\;& \kappa\left(\left(\frac{1}{2}\underline{\kappa}+2\underline{\omega}+\frac{1}{2}\underline{\vartheta}\right)\eta+e_3(\zeta)-\underline{\beta}\right)\\ &+\kappa\left(\frac{1}{2}\underline{\kappa}\zeta-2\underline{\omega}\zeta+\frac{1}{2}\underline{\vartheta}\zeta-\frac{1}{2}\vartheta\underline{\xi}\right)+e_3(e_\theta(\kappa))\\ &+\frac{1}{2}(\underline{\kappa}+\underline{\vartheta})e_\theta(\kappa)+\left(-\frac{1}{2}\kappa\underline{\kappa}+2\underline{\omega}\kappa+2\not{d}_1\eta+2\rho-\frac{1}{2}\vartheta\underline{\vartheta}+2\eta^2\right)\zeta\\ &+\frac{1}{2}\underline{\kappa}e_\theta(\kappa)+\frac{1}{2}\kappa e_\theta(\underline{\kappa})-2\underline{\omega}e_\theta(\kappa)-2e_\theta(\rho)+\frac{1}{2}e_\theta(\underline{\vartheta}\,\vartheta)+\frac{1}{2}\vartheta^2\underline{\xi},\end{aligned}$$

or

$$
\begin{aligned}
&\left(-2\,\not{d}_2\,\not{d}_2^\star+6\rho+2\,\not{d}_1\eta-\frac{1}{2}\kappa\underline{\vartheta}-\frac{3}{2}\vartheta\underline{\vartheta}+2\eta^2\right)\eta+2e_\theta(\eta^2)\\
=\;&\kappa\left(e_3(\zeta)-\underline{\beta}\right)+e_3(e_\theta(\kappa))\\
&+\kappa\left(\frac{1}{2}\underline{\kappa}\zeta-2\underline{\omega}\zeta+\frac{1}{2}\underline{\vartheta}\zeta-\frac{1}{2}\vartheta\underline{\xi}\right)\\
&+\frac{1}{2}(\underline{\kappa}+\underline{\vartheta})e_\theta(\kappa)+\left(-\frac{1}{2}\kappa\underline{\kappa}+2\underline{\omega}\kappa+2\,\not{d}_1\eta+2\rho-\frac{1}{2}\vartheta\underline{\vartheta}+2\eta^2\right)\zeta\\
&+\frac{1}{2}\underline{\kappa}e_\theta(\kappa)+\frac{1}{2}\kappa e_\theta(\underline{\kappa})-2\underline{\omega}e_\theta(\kappa)-2e_\theta(\rho)+\frac{1}{2}e_\theta(\underline{\vartheta}\,\vartheta)+\frac{1}{2}\vartheta^2\underline{\xi}
\end{aligned}
$$

and hence

$$
\begin{aligned}
&\left(2\,\not{d}_2\,\not{d}_2^\star-2\,\not{d}_1\eta+\frac{1}{2}\kappa\underline{\vartheta}-2\eta^2\right)\eta-2e_\theta(\eta^2)\\
=\;&\kappa\left(-e_3(\zeta)+\underline{\beta}\right)-e_3(e_\theta(\kappa))\\
&-\kappa\left(\frac{1}{2}\underline{\kappa}\zeta-2\underline{\omega}\zeta+\frac{1}{2}\underline{\vartheta}\zeta-\frac{1}{2}\vartheta\underline{\xi}\right)+6\rho\eta-\frac{3}{2}\vartheta\underline{\vartheta}\eta\\
&-\frac{1}{2}(\underline{\kappa}+\underline{\vartheta})e_\theta(\kappa)-\left(-\frac{1}{2}\kappa\underline{\kappa}+2\underline{\omega}\kappa+2\,\not{d}_1\eta+2\rho-\frac{1}{2}\vartheta\underline{\vartheta}+2\eta^2\right)\zeta\\
&-\frac{1}{2}\underline{\kappa}e_\theta(\kappa)-\frac{1}{2}\kappa e_\theta(\underline{\kappa})+2\underline{\omega}e_\theta(\kappa)+2e_\theta(\rho)-\frac{1}{2}e_\theta(\underline{\vartheta}\,\vartheta)-\frac{1}{2}\vartheta^2\underline{\xi}
\end{aligned}
$$

which is the second desired identity.

To prove the third identity we start with

$$
e_3(\underline{\kappa})+\frac{1}{2}\underline{\kappa}^2+2\underline{\omega}\,\underline{\kappa}\;=\;2\,\not{d}_1\underline{\xi}+2(\eta-3\zeta)\underline{\xi}-\frac{1}{2}\underline{\vartheta}^2.
$$

Taking $e_\theta=-\,\not{d}_1^\star$ and using $\not{d}_1^\star\,\not{d}_1=\not{d}_2\,\not{d}_2^\star+2K$ as before,

$$
\begin{aligned}
&e_3(e_\theta(\underline{\kappa}))+[e_\theta,e_3]\underline{\kappa}+\underline{\kappa}e_\theta(\underline{\kappa})+2\underline{\omega}e_\theta(\underline{\kappa})+2\underline{\kappa}e_\theta(\underline{\omega})\\
=\;&-2\,\not{d}_1^\star\,\not{d}_1\underline{\xi}+2e_\theta\Big((\eta-3\zeta)\underline{\xi}\Big)-\frac{1}{2}e_\theta(\underline{\vartheta}^2)\\
=\;&-2(\,\not{d}_2\,\not{d}_2^\star+2K)\underline{\xi}+2e_\theta\Big((\eta-3\zeta)\underline{\xi}\Big)-\frac{1}{2}e_\theta(\underline{\vartheta}^2).
\end{aligned}
$$

Thus, since $[e_\theta,e_3]\underline{\kappa}=\frac{1}{2}(\underline{\kappa}+\underline{\vartheta})e_\theta\underline{\kappa}+(\zeta-\eta)e_3\underline{\kappa}-\underline{\xi}e_4\underline{\kappa}$,

$$
\begin{aligned}
-2(\,\not{d}_2\,\not{d}_2^\star+2K)\underline{\xi}\;=\;&e_3(e_\theta(\underline{\kappa}))+2\underline{\kappa}e_\theta(\underline{\omega})+\frac{1}{2}(\underline{\kappa}+\underline{\vartheta})e_\theta\underline{\kappa}+(\zeta-\eta)e_3\underline{\kappa}-\underline{\xi}e_4\underline{\kappa}\\
+\;&\underline{\kappa}e_\theta(\underline{\kappa})+2\underline{\omega}e_\theta(\underline{\kappa})-2e_\theta\Big((\eta-3\zeta)\underline{\xi}\Big)+\frac{1}{2}e_\theta(\underline{\vartheta}^2).
\end{aligned}
$$

Making use of the equations for $e_3\underline{\kappa}, e_4\kappa$ in Proposition 2.63

$$\begin{aligned}
2(\not{d}_2\not{d}_2^\star+2K)\underline{\xi} &= 2\underline{\kappa}\not{d}_1^\star\underline{\omega}-e_3(e_\theta(\underline{\kappa}))-\frac{1}{2}(\underline{\kappa}+\underline{\vartheta})e_\theta\underline{\kappa}\\
&- (\zeta-\eta)\left(-\frac{1}{2}\underline{\kappa}^2-2\underline{\omega}\,\underline{\kappa}+2\not{d}_1\underline{\xi}+2(\eta-3\zeta)\underline{\xi}-\frac{1}{2}\underline{\vartheta}^2\right)\\
&+ \underline{\xi}\left(-\frac{1}{2}\kappa\,\underline{\kappa}-2\not{d}_1\zeta+2\rho-\frac{1}{2}\vartheta\,\underline{\vartheta}+2\zeta^2\right)\\
&- \underline{\kappa}e_\theta(\underline{\kappa})-2\underline{\omega}e_\theta(\underline{\kappa})+2e_\theta\Big((\eta-3\zeta)\underline{\xi}\Big)-\frac{1}{2}e_\theta(\underline{\vartheta}^2).
\end{aligned}$$

We deduce

$$\begin{aligned}
&2(\not{d}_2\not{d}_2^\star+K)\underline{\xi}\\
=\;& 2\underline{\kappa}\not{d}_1^\star\underline{\omega}+\eta\left(-\frac{1}{2}\underline{\kappa}^2-2\underline{\omega}\,\underline{\kappa}+2\not{d}_1\underline{\xi}+2(\eta-3\zeta)\underline{\xi}-\frac{1}{2}\underline{\vartheta}^2\right)+2e_\theta\Big((\eta-3\zeta)\underline{\xi}\Big)\\
&-e_3(e_\theta(\underline{\kappa}))-\frac{1}{2}e_\theta(\underline{\vartheta}^2)-\frac{1}{2}(\underline{\kappa}+\underline{\vartheta})e_\theta(\underline{\kappa})\\
&-\zeta\left(-\frac{1}{2}\underline{\kappa}^2-2\underline{\omega}\,\underline{\kappa}+2\not{d}_1\underline{\xi}+2(\eta-3\zeta)\underline{\xi}-\frac{1}{2}\underline{\vartheta}^2\right)\\
&+\underline{\xi}\left(-\frac{1}{2}\kappa\,\underline{\kappa}-2K-2\not{d}_1\zeta+2\rho-\frac{1}{2}\vartheta\,\underline{\vartheta}+2\zeta^2\right)-\underline{\kappa}e_\theta(\underline{\kappa})-2\underline{\omega}e_\theta(\underline{\kappa})\\
&-6\eta\zeta\underline{\xi}-6e_\theta(\zeta\underline{\xi}).
\end{aligned}$$

Making use of $K=-\rho-\frac{1}{4}\kappa\underline{\kappa}+\frac{1}{4}\vartheta\underline{\vartheta}$ and reorganizing we deduce

$$\begin{aligned}
&2(\not{d}_2\not{d}_2^\star+K)\underline{\xi}\\
=\;& 2\underline{\kappa}\not{d}_1^\star\underline{\omega}+\eta\left(-\frac{1}{2}\underline{\kappa}^2-2\underline{\omega}\,\underline{\kappa}+2\not{d}_1\underline{\xi}+2\eta\underline{\xi}-\frac{1}{2}\underline{\vartheta}^2\right)+2e_\theta(\eta\underline{\xi})-e_3(e_\theta(\underline{\kappa}))\\
&-\frac{1}{2}e_\theta(\underline{\vartheta}^2)-\frac{1}{2}(\underline{\kappa}+\underline{\vartheta})e_\theta(\underline{\kappa})\\
&-\zeta\left(-\frac{1}{2}\underline{\kappa}^2-2\underline{\omega}\,\underline{\kappa}+2\not{d}_1\underline{\xi}+2(\eta-3\zeta)\underline{\xi}-\frac{1}{2}\underline{\vartheta}^2\right)\\
&+\underline{\xi}\Big(4\rho-\vartheta\underline{\vartheta}-2\not{d}_1\zeta+2\zeta^2\Big)-\underline{\kappa}e_\theta(\underline{\kappa})-2\underline{\omega}e_\theta(\underline{\kappa})-6\eta\zeta\underline{\xi}-6e_\theta(\zeta\underline{\xi}).
\end{aligned}$$

We make use again of the identity

$$2\not{d}_1^\star\underline{\omega}+\frac{1}{2}\kappa\underline{\xi} = \left(\frac{1}{2}\underline{\kappa}+2\underline{\omega}+\frac{1}{2}\underline{\vartheta}\right)\eta+e_3(\zeta)-\underline{\beta}+\frac{1}{2}\underline{\kappa}\zeta-2\underline{\omega}\zeta+\frac{1}{2}\underline{\vartheta}\zeta-\frac{1}{2}\vartheta\underline{\xi},$$

to derive

$$\begin{aligned}
& 2(\not{d}_2 \not{d}_2^\star + K)\underline{\xi} \\
= \; & \underline{\kappa}\left(-\frac{1}{2}\kappa\underline{\xi} + \left(\frac{1}{2}\underline{\kappa} + 2\underline{\omega} + \frac{1}{2}\underline{\vartheta}\right)\eta + e_3(\zeta) - \underline{\beta}\right) \\
& +\underline{\kappa}\left(\frac{1}{2}\underline{\kappa}\zeta - 2\underline{\omega}\zeta + \frac{1}{2}\underline{\vartheta}\zeta - \frac{1}{2}\vartheta\underline{\xi}\right) + \eta\left(-\frac{1}{2}\underline{\kappa}^2 - 2\underline{\omega}\,\underline{\kappa} + 2\not{d}_1\underline{\xi} + 2\eta\underline{\xi} - \frac{1}{2}\underline{\vartheta}^2\right) \\
& +2e_\theta(\eta\underline{\xi}) - e_3(e_\theta(\underline{\kappa})) - \frac{1}{2}e_\theta(\underline{\vartheta}^2) - \frac{1}{2}(\underline{\kappa} + \underline{\vartheta})e_\theta(\underline{\kappa}) \\
& -\zeta\left(-\frac{1}{2}\underline{\kappa}^2 - 2\underline{\omega}\,\underline{\kappa} + 2\not{d}_1\underline{\xi} + 2(\eta - 3\zeta)\underline{\xi} - \frac{1}{2}\underline{\vartheta}^2\right) \\
& +\underline{\xi}\Big(4\rho - \vartheta\underline{\vartheta} - 2\not{d}_1\zeta + 2\zeta^2\Big) - \underline{\kappa}e_\theta(\underline{\kappa}) - 2\underline{\omega}e_\theta(\underline{\kappa}) - 6\eta\zeta\underline{\xi} - 6e_\theta(\zeta\underline{\xi}).
\end{aligned}$$

Grouping terms and using once more the identity $K = -\rho - \frac{1}{4}\kappa\underline{\kappa} + \frac{1}{4}\vartheta\underline{\vartheta}$ we deduce

$$\begin{aligned}
& 2\not{d}_2 \not{d}_2^\star\underline{\xi} \\
= \; & -e_3(e_\theta(\underline{\kappa})) + \underline{\kappa}\left(e_3(\zeta) - \underline{\beta}\right) + \left(2\not{d}_1\underline{\xi} + \frac{1}{2}\underline{\kappa}\,\underline{\vartheta} + 2\eta\underline{\xi} - \frac{1}{2}\underline{\vartheta}^2\right)\eta + 2e_\theta(\eta\underline{\xi}) \\
& -\frac{1}{2}e_\theta(\underline{\vartheta}^2) + \underline{\kappa}\left(\frac{1}{2}\underline{\kappa}\zeta - 2\underline{\omega}\zeta + \frac{1}{2}\underline{\vartheta}\zeta - \frac{1}{2}\vartheta\underline{\xi}\right) - \frac{1}{2}(\underline{\kappa} + \underline{\vartheta})e_\theta(\underline{\kappa}) - \frac{1}{2}\vartheta\underline{\vartheta}\underline{\xi} \\
& -\zeta\left(-\frac{1}{2}\underline{\kappa}^2 - 2\underline{\omega}\,\underline{\kappa} + 2\not{d}_1\underline{\xi} + 2(\eta - 3\zeta)\underline{\xi} - \frac{1}{2}\underline{\vartheta}^2\right) \\
& +\underline{\xi}\Big(6\rho - \vartheta\underline{\vartheta} - 2\not{d}_1\zeta + 2\zeta^2\Big) - \underline{\kappa}e_\theta(\underline{\kappa}) - 2\underline{\omega}e_\theta(\underline{\kappa}) - 6\eta\zeta\underline{\xi} - 6e_\theta(\zeta\underline{\xi})
\end{aligned}$$

which is the third desired identity. This concludes the proof of Proposition 2.74.

A.6 PROOF OF PROPOSITION 2.90

The proof follows by straightforward calculations using the definition of Ricci coefficients and curvature components with respect to the two frames. Recall the transformation (2.3.3)

$$\begin{aligned}
e_4' &= \lambda\left(e_4 + fe_\theta + \frac{1}{4}f^2e_3\right), \\
e_\theta' &= \left(1 + \frac{1}{2}f\underline{f}\right)e_\theta + \frac{1}{2}\underline{f}e_4 + \frac{1}{2}f\left(1 + \frac{1}{4}f\underline{f}\right)e_3, \\
e_3' &= \lambda^{-1}\left(\left(1 + \frac{1}{2}f\underline{f} + \frac{1}{16}f^2\underline{f}^2\right)e_3 + \underline{f}\left(1 + \frac{1}{4}f\underline{f}\right)e_\theta + \frac{1}{4}\underline{f}^2e_4\right).
\end{aligned}$$

We first derive the transformation formulae for κ. We have, under a transformation of type (2.3.3),

$$
\begin{aligned}
\chi' &= g(D_{e'_\theta} e'_4, e_{\theta'}) \\
&= g\left(D_{e'_\theta}\left(\lambda\left(e_4 + fe_\theta + \frac{1}{4}f^2 e_3\right)\right), e'_\theta\right) \\
&= \lambda g\left(D_{e'_\theta}\left(e_4 + fe_\theta + \frac{1}{4}f^2 e_3\right), e'_\theta\right) \\
&= \lambda g\left(D_{e'_\theta} e_4, e'_\theta\right) + \lambda e'_\theta(f)g(e_\theta, e'_\theta) + \frac{\lambda}{4}e'_\theta(f^2)g(e_3, e'_\theta) + \lambda fg\left(D_{e'_\theta} e_\theta, e'_\theta\right) \\
&\quad + \frac{\lambda}{4}f^2 g(D_{e'_\theta} e_3, e'_\theta) \\
&= \lambda g\left(D_{e'_\theta} e_4, e'_\theta\right) + \lambda\left(1 + \frac{1}{2}f\underline{f}\right)e'_\theta(f) - \frac{\lambda}{4}\underline{f}e'_\theta(f^2) + \lambda fg\left(D_{e'_\theta} e_\theta, e'_\theta\right) \\
&\quad + \frac{\lambda}{4}f^2\underline{\chi} + \text{l.o.t.}
\end{aligned}
$$

We recall that the lower order terms we denote by l.o.t., here and throughout the proof, are linear with respect to $\Gamma = \{\xi, \underline{\xi}, \vartheta, \check{\kappa}, \eta, \underline{\eta}, \zeta, \underline{\check{\kappa}}, \underline{\vartheta}\}$ and quadratic or higher order in $f, \underline{f}$, and do not contain derivatives of these latter. We also recall that $\chi = \frac{1}{2}(\kappa + \vartheta)$, $\underline{\chi} = \frac{1}{2}(\underline{\kappa} + \underline{\vartheta})$.

Next, we compute

$$
\begin{aligned}
g\left(D_{e'_\theta} e_4, e'_\theta\right) &= g\left(D_{e'_\theta} e_4, \left(1 + \frac{1}{2}f\underline{f}\right)e_\theta + \frac{1}{2}fe_3\right) + \text{l.o.t.} \\
&= \left(1 + \frac{1}{2}f\underline{f}\right)g\left(D_{\left(1 + \frac{1}{2}f\underline{f}\right)e_\theta + \frac{1}{2}\underline{f}e_4 + \frac{1}{2}fe_3} e_4, e_\theta\right) \\
&\quad + \frac{1}{2}fg\left(D_{e_\theta + \frac{1}{2}\underline{f}e_4 + \frac{1}{2}fe_3} e_4, e_3\right) + \text{l.o.t.} \\
&= \left(1 + f\underline{f}\right)\chi + \underline{f}\xi + f\eta + f\zeta + f\underline{f}\omega - f^2\underline{\omega} + \text{l.o.t.},
\end{aligned}
$$

and

$$
\begin{aligned}
fg\left(D_{e'_\theta} e_\theta, e'_\theta\right) &= \frac{1}{2}f\underline{f}g\left(D_{e'_\theta} e_\theta, e_4\right) + \frac{1}{2}f^2 g\left(D_{e'_\theta} e_\theta, e_3\right) \\
&= -\frac{1}{2}f\underline{f}\chi - \frac{1}{2}f^2\underline{\chi} + \text{l.o.t.}
\end{aligned}
$$

This yields

$$
\begin{aligned}
\chi' &= \lambda g\left(D_{e'_\theta} e_4, e'_\theta\right) + \lambda\left(1 + \frac{1}{2}f\underline{f}\right)e'_\theta(f) - \frac{\lambda}{4}\underline{f}e'_\theta(f^2) + \lambda fg\left(D_{e'_\theta} e_\theta, e'_\theta\right) \\
&\quad + \frac{\lambda}{4}f^2\underline{\chi} + \text{l.o.t.} \\
&= \lambda\left(\chi + \left(1 + \frac{1}{2}f\underline{f}\right)e'_\theta(f) - \frac{1}{4}\underline{f}e'_\theta(f^2) + f(\zeta + \eta) + \underline{f}\xi - \frac{1}{4}f^2\underline{\chi} + \frac{1}{2}f\underline{f}\chi\right. \\
&\quad \left. + f\underline{f}\omega - f^2\underline{\omega} + \text{l.o.t.}\right).
\end{aligned}
$$

Hence,

$$\begin{aligned}\kappa' &= \chi' + e_4'\Phi = \chi' + \lambda\left(e_4 + fe_\theta + \frac{1}{4}f^2 e_3\right)\Phi\\ &= \lambda\Bigg(\kappa + e_\theta'(f) + e_\theta(\Phi)f + \frac{1}{8}(\underline{\kappa} - \underline{\vartheta})f^2 + \frac{1}{2}f\underline{f}e_\theta'(f) - \frac{1}{4}\underline{f}e_\theta'(f^2) + f(\zeta+\eta)\\ &\quad +\underline{f}\xi - \frac{1}{4}f^2\underline{\chi} + \frac{1}{2}f\underline{f}\chi + f\underline{f}\omega - f^2\underline{\omega} + \text{l.o.t.}\Bigg)\\ &= \lambda\Bigg(\kappa + \not{d}_1{}'(f) + \frac{1}{2}f\underline{f}e_\theta'(f) + \frac{1}{4}\underline{f}e_\theta'(f^2) - \frac{1}{4}f^2\underline{\kappa} + f(\zeta+\eta) + \underline{f}\xi + f\underline{f}\omega\\ &\quad -f^2\underline{\omega} + \text{l.o.t.}\Bigg)\end{aligned}$$

and

$$\begin{aligned}\vartheta' &= \chi' - e_4'\Phi = \chi' - \lambda\left(e_4 + fe_\theta + \frac{1}{4}f^2 e_3\right)\Phi\\ &= \lambda\Bigg(\vartheta + e_\theta'(f) - e_\theta(\Phi)f - \frac{1}{8}(\underline{\kappa} - \underline{\vartheta})f^2 + \frac{1}{2}f\underline{f}e_\theta'(f) - \frac{1}{4}\underline{f}e_\theta'(f^2) + f(\zeta+\eta)\\ &\quad +\underline{f}\xi - \frac{1}{4}f^2\underline{\chi} + \frac{1}{2}f\underline{f}\chi + f\underline{f}\omega - f^2\underline{\omega} + \text{l.o.t.}\Bigg)\\ &= \lambda\Bigg(\vartheta - \not{d}_2^\star{}'(f) + \frac{1}{2}f\underline{f}e_\theta'(f) - \frac{1}{4}\underline{f}e_\theta'(f^2) + \frac{1}{4}f\underline{f}\kappa + f(\zeta+\eta) + \underline{f}\xi + f\underline{f}\omega\\ &\quad -f^2\underline{\omega} + \text{l.o.t.}\Bigg).\end{aligned}$$

This yields

$$\begin{aligned}\kappa' &= \lambda\left(\kappa + \not{d}_1{}'(f)\right) + \lambda\mathrm{Err}(\kappa,\kappa'),\\ \mathrm{Err}(\kappa,\kappa') &= \frac{1}{2}f\underline{f}e_\theta'(f) - \frac{1}{4}\underline{f}e_\theta'(f^2) + f(\zeta+\eta) + \underline{f}\xi + \frac{1}{4}f^2\underline{\kappa} + f\underline{f}\omega - f^2\underline{\omega}\\ &\quad +\text{l.o.t.}\\ &= f(\zeta+\eta) + \underline{f}\xi + \frac{1}{4}f^2\underline{\kappa} + f\underline{f}\omega - f^2\underline{\omega} + \text{l.o.t.}\end{aligned}$$

and

$$\begin{aligned}\vartheta' &= \lambda\left(\vartheta - \not{d}_2^\star{}'(f)\right) + \lambda\mathrm{Err}(\vartheta,\vartheta'),\\ \mathrm{Err}(\vartheta,\vartheta') &= \frac{1}{2}f\underline{f}e_\theta'(f) - \frac{1}{4}\underline{f}e_\theta'(f^2) + f(\zeta+\eta) + \underline{f}\xi + \frac{1}{4}f\underline{f}\kappa + f\underline{f}\omega - f^2\underline{\omega}\\ &\quad +\text{l.o.t.}\\ &= f(\zeta+\eta) + \underline{f}\xi + \frac{1}{4}f\underline{f}\kappa + f\underline{f}\omega - f^2\underline{\omega} + \text{l.o.t.}\end{aligned}$$

Next, we derive the transformation formula for $\underline{\kappa}$ and $\underline{\vartheta}$. We have, under a transformation of type (2.3.3),

$$
\begin{aligned}
& \underline{\chi}' \\
= \;& g(D_{e'_\theta} e'_3, e_{\theta'}) \\
= \;& g\left(D_{e'_\theta}\left(\lambda^{-1}\left(\left(1+\frac{1}{2}f\underline{f}+\frac{1}{16}f^2\underline{f}^2\right)e_3+\underline{f}\left(1+\frac{1}{4}f\underline{f}\right)e_\theta+\frac{1}{4}\underline{f}^2e_4\right)\right), e'_\theta\right) \\
= \;& \lambda^{-1}g\left(D_{e'_\theta}\left(\left(1+\frac{1}{2}f\underline{f}+\frac{1}{16}f^2\underline{f}^2\right)e_3+\underline{f}\left(1+\frac{1}{4}f\underline{f}\right)e_\theta+\frac{1}{4}\underline{f}^2e_4\right), e'_\theta\right) \\
= \;& \frac{\lambda^{-1}}{2}e'_\theta\left(f\underline{f}+\frac{1}{8}f^2\underline{f}^2\right)g(e_3,e'_\theta)+\lambda^{-1}e'_\theta\left(\underline{f}\left(1+\frac{1}{4}f\underline{f}\right)\right)g(e_\theta,e'_\theta) \\
& +\frac{\lambda^{-1}}{4}e'_\theta\left(\underline{f}^2\right)g(e_4,e'_\theta)+\lambda^{-1}\left(1+\frac{1}{2}f\underline{f}+\frac{1}{16}f^2\underline{f}^2\right)g\left(D_{e'_\theta}e_3,e'_\theta\right) \\
& +\lambda^{-1}\underline{f}\left(1+\frac{1}{4}f\underline{f}\right)g\left(D_{e'_\theta}e_\theta,e'_\theta\right)+\frac{1}{4}\lambda^{-1}\underline{f}^2g\left(D_{e'_\theta}e_4,e'_\theta\right) \\
= \;& -\frac{\lambda^{-1}}{2}\underline{f}e'_\theta\left(f\underline{f}+\frac{1}{8}f^2\underline{f}^2\right)+\lambda^{-1}\left(1+\frac{1}{2}f\underline{f}\right)e'_\theta\left(\underline{f}\left(1+\frac{1}{4}f\underline{f}\right)\right) \\
& -\frac{\lambda^{-1}}{4}f\left(1+\frac{1}{4}f\underline{f}\right)e'_\theta\left(\underline{f}^2\right)+\lambda^{-1}\left(1+\frac{1}{2}f\underline{f}\right)g\left(D_{e'_\theta}e_3,e'_\theta\right) \\
& +\lambda^{-1}\underline{f}g\left(D_{e'_\theta}e_\theta,e'_\theta\right)+\frac{1}{4}\lambda^{-1}\underline{f}^2\chi+\text{l.o.t.}
\end{aligned}
$$

Then, we easily derive by symmetry from the formula for κ and ϑ

$$
\begin{aligned}
\underline{\kappa}' &= \lambda^{-1}\left(\underline{\kappa}+{\not{d}_1}'(\underline{f})\right)+\lambda^{-1}\text{Err}(\underline{\kappa},\underline{\kappa}'), \\
\text{Err}(\underline{\kappa},\underline{\kappa}') &= -\frac{1}{2}\underline{f}e'_\theta\left(f\underline{f}+\frac{1}{8}f^2\underline{f}^2\right)+\left(\frac{3}{4}f\underline{f}+\frac{1}{8}(f\underline{f})^2\right)e'_\theta(\underline{f}) \\
&\quad +\frac{1}{4}\left(1+\frac{1}{2}f\underline{f}\right)\underline{f}e'_\theta\left(f\underline{f}\right)-\frac{1}{4}f\left(1+\frac{1}{4}f\underline{f}\right)e'_\theta\left(\underline{f}^2\right)+\underline{f}(-\zeta+\underline{\eta}) \\
&\quad +f\underline{\xi}-\frac{1}{4}\underline{f}^2\kappa+f\underline{f}\underline{\omega}-\underline{f}^2\omega+\text{l.o.t.} \\
&= -\frac{1}{4}\underline{f}^2e'_\theta(f)+\underline{f}(-\zeta+\underline{\eta})+f\underline{\xi}-\frac{1}{4}\underline{f}^2\kappa+f\underline{f}\underline{\omega}-\underline{f}^2\omega+\text{l.o.t.}
\end{aligned}
$$

and

$$
\begin{aligned}
\underline{\vartheta}' &= \lambda\left(\underline{\vartheta}-{\not{d}_2^\star}'(\underline{f})\right)+\lambda^{-1}\text{Err}(\underline{\vartheta},\underline{\vartheta}'), \\
\text{Err}(\underline{\vartheta},\underline{\vartheta}') &= -\frac{1}{2}\underline{f}e'_\theta\left(f\underline{f}+\frac{1}{8}f^2\underline{f}^2\right)+\left(\frac{3}{4}f\underline{f}+\frac{1}{8}(f\underline{f})^2\right)e'_\theta(\underline{f}) \\
&\quad +\frac{1}{4}\left(1+\frac{1}{2}f\underline{f}\right)\underline{f}e'_\theta\left(f\underline{f}\right)-\frac{1}{4}f\left(1+\frac{1}{4}f\underline{f}\right)e'_\theta\left(\underline{f}^2\right)+\underline{f}(-\zeta+\underline{\eta}) \\
&\quad +f\underline{\xi}+\frac{1}{4}f\underline{f}\underline{\kappa}+f\underline{f}\underline{\omega}-\underline{f}^2\omega+\text{l.o.t.} \\
&= -\frac{1}{4}\underline{f}^2e'_\theta(f)+\underline{f}(-\zeta+\underline{\eta})+f\underline{\xi}+\frac{1}{4}f\underline{f}\underline{\kappa}+f\underline{f}\underline{\omega}-\underline{f}^2\omega+\text{l.o.t.}
\end{aligned}
$$

Next, we derive the transformation formula for ζ. We have, under a transfor-

mation of type (2.3.3),

$$
\begin{aligned}
2\zeta' &= g(D_{e'_\theta}e'_4, e'_3)\\
&= g\left(D_{e'_\theta}\left(\lambda\left(e_4 + fe_\theta + \frac{1}{4}f^2e_3\right)\right), e'_3\right)\\
&= -2e'_\theta(\log(\lambda)) + \lambda g\left(D_{e'_\theta}\left(e_4 + fe_\theta + \frac{1}{4}f^2e_3\right), e'_3\right)\\
&= -2e'_\theta(\log(\lambda)) + \lambda e'_\theta(f)g\left(e_\theta, e'_3\right) + \frac{1}{4}\lambda e'_\theta(f^2)g\left(e_3, e'_3\right) + \lambda g\left(D_{e'_\theta}e_4, e'_3\right)\\
&\quad +\lambda fg\left(D_{e'_\theta}e_\theta, e'_3\right) + \frac{1}{4}\lambda f^2 g\left(D_{e'_\theta}e_3, e'_3\right)\\
&= -2e'_\theta(\log(\lambda)) + \underline{f}\left(1 + \frac{1}{4}f\underline{f}\right)e'_\theta(f) - \frac{1}{8}\underline{f}^2e'_\theta(f^2) + \lambda g\left(D_{e'_\theta}e_4, e'_3\right)\\
&\quad +\lambda fg\left(D_{e'_\theta}e_\theta, e'_3\right) + \text{l.o.t.}
\end{aligned}
$$

We compute

$$
\begin{aligned}
\lambda g\left(D_{e'_\theta}e_4, e'_3\right) &= g\left(D_{e'_\theta}e_4, e_3 + \underline{f}e_\theta\right) + \text{l.o.t.}\\
&= g\left(D_{e_\theta + \frac{1}{2}\underline{f}e_4 + \frac{1}{2}fe_3}e_4, e_3\right) + \underline{f}g\left(D_{e_\theta + \frac{1}{2}\underline{f}e_4 + \frac{1}{2}fe_3}e_4, e_\theta\right) + \text{l.o.t.}\\
&= 2\zeta + 2\omega\underline{f} - 2\underline{\omega}f + \underline{f}\chi + \text{l.o.t.}
\end{aligned}
$$

and

$$
\begin{aligned}
\lambda fg\left(D_{e'_\theta}e_\theta, e'_3\right) &= fg\left(D_{e'_\theta}e_\theta, e_3\right) + \text{l.o.t.}\\
&= fg\left(D_{e_\theta + \frac{1}{2}\underline{f}e_4 + \frac{1}{2}fe_3}e_\theta, e_3\right) + \text{l.o.t.}\\
&= -f\underline{\chi} + \text{l.o.t.}
\end{aligned}
$$

This yields

$$
\begin{aligned}
2\zeta' &= -2e'_\theta(\log(\lambda)) + \underline{f}\left(1 + \frac{1}{4}f\underline{f}\right)e'_\theta(f) - \frac{1}{8}\underline{f}^2e'_\theta(f^2) + \lambda g\left(D_{e'_\theta}e_4, e'_3\right)\\
&\quad +\lambda fg\left(D_{e'_\theta}e_\theta, e'_3\right) + \text{l.o.t.}\\
&= 2\zeta - 2e'_\theta(\log(\lambda)) + \underline{f}\left(1 + \frac{1}{4}f\underline{f}\right)e'_\theta(f) - \frac{1}{8}\underline{f}^2e'_\theta(f^2) + 2\omega\underline{f} - 2\underline{\omega}f + \underline{f}\chi\\
&\quad -f\underline{\chi} + \text{l.o.t.}
\end{aligned}
$$

and hence

$$
\begin{aligned}
\zeta' &= \zeta - e'_\theta(\log(\lambda)) + \frac{1}{4}(-f\underline{\kappa} + \underline{f}\kappa) + \underline{f}\omega - f\underline{\omega} + \text{Err}(\zeta, \zeta'),\\
\text{Err}(\zeta, \zeta') &= \frac{1}{2}\underline{f}\left(1 + \frac{1}{4}f\underline{f}\right)e'_\theta(f) - \frac{1}{16}\underline{f}^2e'_\theta(f^2) + \frac{1}{4}(-f\underline{\vartheta} + \underline{f}\vartheta) + \text{l.o.t.}\\
&= \frac{1}{2}\underline{f}e'_\theta(f) + \frac{1}{4}(-f\underline{\vartheta} + \underline{f}\vartheta) + \text{l.o.t.}
\end{aligned}
$$

Next, we derive the transformation formulae for η. We have, under a transformation of type (2.3.3),

$$\begin{aligned}
2\eta' &= g\left(D_{e_3'}e_4', e_\theta'\right)\\
&= g\left(D_{e_3'}\left(\lambda\left(e_4+fe_\theta+\frac{1}{4}f^2e_3\right)\right), e_\theta'\right)\\
&= \lambda g\left(D_{e_3'}\left(e_4+fe_\theta+\frac{1}{4}f^2e_3\right), e_\theta'\right)\\
&= \lambda g\left(D_{e_3'}e_4, e_\theta'\right)+\lambda e_3'(f)g\left(e_\theta, e_\theta'\right)+\lambda fg\left(D_{e_3'}e_\theta, e_\theta'\right)+\frac{1}{4}\lambda e_3'(f^2)g\left(e_3, e_\theta'\right)\\
&\quad +\frac{1}{4}\lambda f^2 g(D_{e_3'}e_3, e_\theta')\\
&= \lambda\left(1+\frac{1}{2}f\underline{f}\right)e_3'(f)-\frac{1}{4}\lambda\underline{f}e_3'(f^2)+\lambda g\left(D_{e_3'}e_4, e_\theta'\right)+\lambda fg\left(D_{e_3'}e_\theta, e_\theta'\right)\\
&\quad +\text{l.o.t.}
\end{aligned}$$

We compute

$$\begin{aligned}
\lambda g\left(D_{e_3'}e_4, e_\theta'\right) &= \lambda g\left(D_{e_3'}e_4, e_\theta+\frac{1}{2}\underline{f}e_3\right)+\text{l.o.t.}\\
&= g\left(D_{e_3+\underline{f}e_\theta}e_4, e_\theta\right)+\frac{1}{2}\underline{f}g\left(D_{e_3}e_4, e_3\right)+\text{l.o.t.}\\
&= 2\eta+\underline{f}\chi-2\underline{\omega}f+\text{l.o.t.}
\end{aligned}$$

and

$$\lambda fg\left(D_{e_3'}e_\theta, e_\theta'\right) = \text{l.o.t.}$$

This yields

$$\begin{aligned}
2\eta' &= \lambda\left(1+\frac{1}{2}f\underline{f}\right)e_3'(f)-\frac{1}{4}\lambda\underline{f}e_3'(f^2)+\lambda g\left(D_{e_3'}e_4, e_\theta'\right)+\lambda fg\left(D_{e_3'}e_\theta, e_\theta'\right)\\
&\quad +\text{l.o.t.}\\
&= \lambda\left(1+\frac{1}{2}f\underline{f}\right)e_3'(f)-\frac{1}{4}\lambda\underline{f}e_3'(f^2)+2\eta+\underline{f}\chi-2\underline{\omega}f+\text{l.o.t.}
\end{aligned}$$

and hence

$$\begin{aligned}
\eta' &= \eta+\frac{1}{2}\lambda e_3'(f)+\frac{1}{4}\kappa\underline{f}-f\underline{\omega}+\text{Err}(\eta,\eta'),\\
\text{Err}(\eta,\eta') &= \frac{1}{4}\lambda f\underline{f}e_3'(f)-\frac{1}{8}\lambda\underline{f}e_3'(f^2)+\frac{1}{4}\underline{f}\vartheta+\text{l.o.t.}\\
&= \frac{1}{4}\underline{f}\vartheta+\text{l.o.t.}
\end{aligned}$$

Next, we derive the transformation formulae for $\underline{\eta}$. We have, under a transfor-

mation of type (2.3.3),

$$
\begin{aligned}
& 2\underline{\eta}' \\
= \; & g\left(D_{e_4'} e_3', e_\theta'\right) \\
= \; & g\left(D_{e_4'}\left(\lambda^{-1}\left(\left(1+\frac{1}{2}f\underline{f}+\frac{1}{16}f^2\underline{f}^2\right)e_3+\underline{f}\left(1+\frac{1}{4}f\underline{f}\right)e_\theta+\frac{1}{4}\underline{f}^2 e_4\right)\right), e_\theta'\right) \\
= \; & \lambda^{-1} g\left(D_{e_4'}\left(\left(1+\frac{1}{2}f\underline{f}+\frac{1}{16}f^2\underline{f}^2\right)e_3+\underline{f}\left(1+\frac{1}{4}f\underline{f}\right)e_\theta+\frac{1}{4}\underline{f}^2 e_4\right), e_\theta'\right) \\
= \; & \frac{1}{2}\lambda^{-1} e_4'\left(f\underline{f}+\frac{1}{8}f^2\underline{f}^2\right) g\left(e_3, e_\theta'\right)+\lambda^{-1}\left(1+\frac{1}{2}f\underline{f}+\frac{1}{16}f^2\underline{f}^2\right) g\left(D_{e_4'} e_3, e_\theta'\right) \\
& +\lambda^{-1} e_4'\left(\underline{f}\left(1+\frac{1}{4}f\underline{f}\right)\right) g\left(e_\theta, e_\theta'\right)+\lambda^{-1}\underline{f}\left(1+\frac{1}{4}f\underline{f}\right) g\left(D_{e_4'} e_\theta, e_\theta'\right) \\
& +\frac{1}{4}\lambda^{-1} e_4'(\underline{f}^2) g\left(e_4, e_\theta'\right)+\frac{1}{4}\lambda^{-1}\underline{f}^2 g\left(D_{e_4'} e_4, e_\theta'\right) \\
= \; & -\frac{1}{2}\lambda^{-1}\underline{f} e_4'\left(f\underline{f}+\frac{1}{8}f^2\underline{f}^2\right)+\lambda^{-1}\left(1+\frac{1}{2}f\underline{f}\right) e_4'\left(\underline{f}\left(1+\frac{1}{4}f\underline{f}\right)\right) \\
& -\frac{1}{4}\lambda^{-1} f\left(1+\frac{1}{4}f\underline{f}\right) e_4'(\underline{f}^2)+\lambda^{-1} g\left(D_{e_4'} e_3, e_\theta'\right)+\lambda^{-1}\underline{f} g\left(D_{e_4'} e_\theta, e_\theta'\right)+\text{l.o.t.}
\end{aligned}
$$

We compute

$$
\begin{aligned}
\lambda^{-1} g\left(D_{e_4'} e_3, e_\theta'\right) &= \lambda^{-1} g\left(D_{e_4'} e_3, e_\theta+\frac{1}{2}\underline{f} e_4\right)+\text{l.o.t.} \\
&= 2\underline{\eta}+f\underline{\chi}-2\underline{f}\omega+\text{l.o.t.}
\end{aligned}
$$

and

$$
\lambda^{-1}\underline{f} g\left(D_{e_4'} e_\theta, e_\theta'\right) = \text{l.o.t.}
$$

This yields

$$
\begin{aligned}
2\underline{\eta}' = \; & -\frac{1}{2}\lambda^{-1}\underline{f} e_4'\left(f\underline{f}+\frac{1}{8}f^2\underline{f}^2\right)+\lambda^{-1}\left(1+\frac{1}{2}f\underline{f}\right) e_4'\left(\underline{f}\left(1+\frac{1}{4}f\underline{f}\right)\right) \\
& -\frac{1}{4}\lambda^{-1} f\left(1+\frac{1}{4}f\underline{f}\right) e_4'(\underline{f}^2)+\lambda^{-1} g\left(D_{e_4'} e_3, e_\theta'\right)+\lambda^{-1}\underline{f} g\left(D_{e_4'} e_\theta, e_\theta'\right) \\
& +\text{l.o.t.} \\
= \; & -\frac{1}{2}\lambda^{-1}\underline{f} e_4'\left(f\underline{f}+\frac{1}{8}f^2\underline{f}^2\right)+\lambda^{-1}\left(1+\frac{1}{2}f\underline{f}\right) e_4'\left(\underline{f}\left(1+\frac{1}{4}f\underline{f}\right)\right) \\
& -\frac{1}{4}\lambda^{-1} f\left(1+\frac{1}{4}f\underline{f}\right) e_4'(\underline{f}^2)+2\underline{\eta}+f\underline{\chi}-2\underline{f}\omega+\text{l.o.t.}
\end{aligned}
$$

and hence

$$\begin{aligned}\underline{\eta}' &= \underline{\eta}+\frac{1}{2}\lambda^{-1}e_4'(\underline{f})+\frac{1}{4}\underline{\kappa}f-\underline{f}\omega+\mathrm{Err}(\underline{\eta},\underline{\eta}'),\\ \mathrm{Err}(\underline{\eta},\underline{\eta}') &= \frac{1}{4}\lambda^{-1}f\underline{f}e_4'(\underline{f})-\frac{1}{4}\lambda^{-1}\underline{f}e_4'\left(f\underline{f}+\frac{1}{8}f^2\underline{f}^2\right)\\ &\quad +\lambda^{-1}\frac{1}{8}\left(1+\frac{1}{2}f\underline{f}\right)e_4'\left(f\underline{f}^2\right)-\frac{1}{8}\lambda^{-1}f\left(1+\frac{1}{4}f\underline{f}\right)e_4'(\underline{f}^2)+\frac{1}{4}f\underline{\vartheta}\\ &\quad +\text{l.o.t.}\\ &= -\frac{1}{8}\underline{f}^2\lambda^{-1}e_4'(f)+\frac{1}{4}f\underline{\vartheta}+\text{l.o.t.}\end{aligned}$$

Next, we derive the transformation formulae for ξ. We have, under a transformation of type (2.3.3),

$$\begin{aligned}2\xi' &= g\left(D_{e_4'}e_4',e_\theta'\right)\\ &= g\left(D_{e_4'}\left(\lambda\left(e_4+fe_\theta+\frac{1}{4}f^2e_3\right)\right),e_\theta'\right)\\ &= \lambda g\left(D_{e_4'}\left(e_4+fe_\theta+\frac{1}{4}f^2e_3\right),e_\theta'\right)\\ &= \lambda g\left(D_{e_4'}e_4,e_\theta'\right)+\lambda e_4'(f)g\left(e_\theta,e_\theta'\right)+\lambda fg\left(D_{e_4'}e_\theta,e_\theta'\right)+\frac{1}{4}\lambda e_4'(f^2)g\left(e_3,e_\theta'\right)\\ &\quad +\frac{1}{4}\lambda f^2g\left(D_{e_4'}e_3,e_\theta'\right)\\ &= \lambda\left(1+\frac{1}{2}f\underline{f}\right)e_4'(f)-\frac{1}{4}\lambda\underline{f}e_4'(f^2)+\lambda g\left(D_{e_4'}e_4,e_\theta'\right)+\lambda fg\left(D_{e_4'}e_\theta,e_\theta'\right)\\ &\quad +\text{l.o.t.}\end{aligned}$$

We compute

$$\begin{aligned}\lambda g\left(D_{e_4'}e_4,e_\theta'\right) &= \lambda g\left(D_{e_4'}e_4,e_\theta+\frac{1}{2}fe_3\right)+\text{l.o.t.}\\ &= \lambda^2g\left(D_{e_4+fe_\theta}e_4,e_\theta\right)+\frac{1}{2}\lambda^2fg\left(D_{e_4}e_4,e_3\right)+\text{l.o.t.}\\ &= 2\lambda^2\xi+\lambda^2f\chi+2\lambda^2f\omega+\text{l.o.t.}\end{aligned}$$

and

$$\lambda fg\left(D_{e_4'}e_\theta,e_\theta'\right) = \text{l.o.t.}$$

This yields

$$\begin{aligned}2\xi' &= \lambda\left(1+\frac{1}{2}f\underline{f}\right)e_4'(f)-\frac{1}{4}\lambda\underline{f}e_4'(f^2)+\lambda g\left(D_{e_4'}e_4,e_\theta'\right)+\lambda fg\left(D_{e_4'}e_\theta,e_\theta'\right)\\ &\quad +\text{l.o.t.}\\ &= \lambda\left(1+\frac{1}{2}f\underline{f}\right)e_4'(f)-\frac{1}{4}\lambda\underline{f}e_4'(f^2)+2\lambda^2\xi+\lambda^2f\chi+2\lambda^2f\omega+\text{l.o.t.}\end{aligned}$$

and hence

$$\begin{aligned}\xi' &= \lambda^2\left(\xi+\frac{1}{2}\lambda^{-1}e_4'(f)+\omega f+\frac{1}{4}f\kappa\right)+\lambda^2\mathrm{Err}(\xi,\xi'),\\ \mathrm{Err}(\xi,\xi') &= \frac{1}{4}\lambda^{-1}f\underline{f}e_4'(f)-\frac{1}{8}\lambda^{-1}\underline{f}e_4'(f^2)+\frac{1}{4}f\vartheta+\text{l.o.t.}\\ &= \frac{1}{4}f\vartheta+\text{l.o.t.}\end{aligned}$$

In the particular case when $\lambda=1, \underline{f}=0$, see Remark 2.91, the error term takes the form

$$\mathrm{Err}(\xi,\xi') = \frac{1}{4}f\vartheta+\frac{1}{4}f^2(\eta+2\zeta-\underline{\eta})-\frac{1}{4}f^3\left(\underline{\omega}+\frac{1}{2}\underline{\chi}\right)-\frac{1}{16}f^4\underline{\xi}.$$

Next, we derive the transformation formulae for $\underline{\xi}$. We have, under a transformation of type (2.3.3),

$$\begin{aligned}&2\underline{\xi}'\\ =\;& g\left(D_{e_3'}e_3',e_\theta'\right)\\ =\;& g\left(D_{e_3'}\left(\lambda^{-1}\left(\left(1+\frac{1}{2}f\underline{f}+\frac{1}{16}f^2\underline{f}^2\right)e_3+\underline{f}\left(1+\frac{1}{4}f\underline{f}\right)e_\theta+\frac{1}{4}\underline{f}^2e_4\right)\right),e_\theta'\right)\\ =\;& \lambda^{-1}g\left(D_{e_3'}\left(\left(1+\frac{1}{2}f\underline{f}+\frac{1}{16}f^2\underline{f}^2\right)e_3+\underline{f}\left(1+\frac{1}{4}f\underline{f}\right)e_\theta+\frac{1}{4}\underline{f}^2e_4\right),e_\theta'\right)\\ =\;& \frac{1}{2}\lambda^{-1}e_3'\left(f\underline{f}+\frac{1}{8}f^2\underline{f}^2\right)g\left(e_3,e_\theta'\right)+\lambda^{-1}g\left(D_{e_3'}e_3,e_\theta'\right)\\ &+\lambda^{-1}e_3'\left(\underline{f}\left(1+\frac{1}{4}f\underline{f}\right)\right)g\left(e_\theta,e_\theta'\right)+\lambda^{-1}\underline{f}g\left(D_{e_3'}e_\theta,e_\theta'\right)\\ &+\frac{1}{4}\lambda^{-1}e_3'(\underline{f}^2)g\left(e_4,e_\theta'\right)+\text{l.o.t.}\\ =\;& -\frac{1}{2}\lambda^{-1}\underline{f}e_3'\left(f\underline{f}+\frac{1}{8}f^2\underline{f}^2\right)+\lambda^{-1}\left(1+\frac{1}{2}f\underline{f}\right)e_3'\left(\underline{f}\left(1+\frac{1}{4}f\underline{f}\right)\right)\\ &-\frac{1}{4}\lambda^{-1}f\left(1+\frac{1}{4}f\underline{f}\right)e_3'(\underline{f}^2)+\lambda^{-1}g\left(D_{e_3'}e_3,e_\theta'\right)+\lambda^{-1}\underline{f}g\left(D_{e_3'}e_\theta,e_\theta'\right)+\text{l.o.t.}\end{aligned}$$

We compute

$$\begin{aligned}\lambda^{-1}g\left(D_{e_3'}e_3,e_\theta'\right) &= \lambda^{-1}g\left(D_{e_3'}e_3,e_\theta+\frac{1}{2}\underline{f}e_4\right)+\text{l.o.t.}\\ &= \lambda^{-2}g\left(D_{e_3+\underline{f}e_\theta}e_3,e_\theta\right)+\frac{1}{2}\lambda^{-2}\underline{f}g\left(D_{e_3}e_3,e_4\right)+\text{l.o.t.}\\ &= 2\lambda^{-2}\underline{\xi}+\lambda^{-2}\underline{f}\,\underline{\chi}+2\lambda^{-2}\underline{f}\,\underline{\omega}+\text{l.o.t.}\end{aligned}$$

and

$$\lambda^{-1}\underline{f}g\left(D_{e_3'}e_\theta,e_\theta'\right) = \text{l.o.t.}$$

This yields

$$\begin{aligned}
2\underline{\xi}' &= -\frac{1}{2}\lambda^{-1}\underline{f}e_3'\left(f\underline{f}+\frac{1}{8}f^2\underline{f}^2\right)+\lambda^{-1}\left(1+\frac{1}{2}f\underline{f}\right)e_3'\left(\underline{f}\left(1+\frac{1}{4}f\underline{f}\right)\right)\\
&\quad -\frac{1}{4}\lambda^{-1}f\left(1+\frac{1}{4}f\underline{f}\right)e_3'(\underline{f}^2)+\lambda^{-1}g\left(D_{e_3'}e_3,e_\theta'\right)+\lambda^{-1}\underline{f}g\left(D_{e_3'}e_\theta,e_\theta'\right)\\
&\quad +\text{l.o.t.}\\
&= -\frac{1}{2}\lambda^{-1}\underline{f}e_3'\left(f\underline{f}+\frac{1}{8}f^2\underline{f}^2\right)+\lambda^{-1}\left(1+\frac{1}{2}f\underline{f}\right)e_3'\left(\underline{f}\left(1+\frac{1}{4}f\underline{f}\right)\right)\\
&\quad -\frac{1}{4}\lambda^{-1}f\left(1+\frac{1}{4}f\underline{f}\right)e_3'(\underline{f}^2)+2\lambda^{-2}\underline{\xi}+\lambda^{-2}\underline{f}\,\underline{\chi}+2\lambda^{-2}\underline{f}\,\underline{\omega}+\text{l.o.t.}
\end{aligned}$$

and hence

$$\begin{aligned}
\underline{\xi}' &= \lambda^{-2}\left(\underline{\xi}+\frac{1}{2}\lambda e_3'(\underline{f})+\underline{\omega}\,\underline{f}+\frac{1}{4}\underline{f}\,\underline{\kappa}\right)+\lambda^{-2}\text{Err}(\underline{\xi},\underline{\xi}'),\\
\text{Err}(\underline{\xi},\underline{\xi}') &= -\frac{1}{4}\lambda\underline{f}e_3'\left(f\underline{f}+\frac{1}{8}f^2\underline{f}^2\right)+\frac{1}{4}\lambda f\underline{f}e_3'(\underline{f})+\frac{1}{8}\lambda\left(1+\frac{1}{2}f\underline{f}\right)e_3'\left(f\underline{f}^2\right)\\
&\quad -\frac{1}{8}\lambda f\left(1+\frac{1}{4}f\underline{f}\right)e_3'(\underline{f}^2)+\frac{1}{4}\underline{f}\,\underline{\vartheta}+\text{l.o.t.}\\
&= -\frac{1}{8}\lambda\underline{f}^2e_3'(f)+\frac{1}{4}\underline{f}\,\underline{\vartheta}+\text{l.o.t.}
\end{aligned}$$

Next, we derive the transformation formulae for ω. We have, under a transformation of type (2.3.3),

$$\begin{aligned}
4\omega' &= g\left(D_{e_4'}e_4',e_3'\right)\\
&= g\left(D_{e_4'}\left(\lambda\left(e_4+fe_\theta+\frac{1}{4}f^2e_3\right)\right),e_3'\right)\\
&= -2e_4'(\log\lambda)+\lambda g\left(D_{e_4'}\left(e_4+fe_\theta+\frac{1}{4}f^2e_3\right),e_3'\right)\\
&= -2e_4'(\log\lambda)+\lambda g\left(D_{e_4'}e_4,e_3'\right)+\lambda e_4'(f)g\left(e_\theta,e_3'\right)+\lambda fg\left(D_{e_4'}e_\theta,e_3'\right)\\
&\quad +\frac{1}{4}\lambda e_4'(f^2)g\left(e_3,e_3'\right)+\frac{1}{4}\lambda f^2g\left(D_{e_4'}e_3,e_3'\right)\\
&= -2e_4'(\log\lambda)+\underline{f}\left(1+\frac{1}{4}f\underline{f}\right)e_4'(f)-\frac{1}{8}\underline{f}^2e_4'(f^2)+\lambda g\left(D_{e_4'}e_4,e_3'\right)\\
&\quad +\lambda fg\left(D_{e_4'}e_\theta,e_3'\right)+\text{l.o.t.}
\end{aligned}$$

We compute

$$\begin{aligned}
\lambda g\left(D_{e_4'}e_4,e_3'\right) &= g\left(D_{e_4'}e_4,\left(1+\frac{1}{2}f\underline{f}\right)e_3+\underline{f}e_\theta\right)+\text{l.o.t.}\\
&= \lambda\left(1+\frac{1}{2}f\underline{f}\right)g\left(D_{e_4+fe_\theta+\frac{1}{4}f^2e_3}e_4,e_3\right)\\
&\quad +\lambda\underline{f}g\left(D_{e_4+fe_\theta}e_4,e_\theta\right)+\text{l.o.t.}\\
&= 4\lambda\left(1+\frac{1}{2}f\underline{f}\right)\omega+2\lambda f\zeta-\lambda f^2\underline{\omega}+2\lambda\underline{f}\xi+\lambda f\underline{f}\chi+\text{l.o.t.}
\end{aligned}$$

and

$$
\begin{aligned}
\lambda f g\left(D_{e_4'} e_\theta, e_3'\right) &= f g\left(D_{e_4'} e_\theta, e_3\right) + \text{l.o.t.}\\
&= \lambda f g\left(D_{e_4+f e_\theta} e_\theta, e_3\right) + \text{l.o.t.}\\
&= -2\lambda f \underline{\eta} - \lambda f^2 \underline{\chi} + \text{l.o.t.}
\end{aligned}
$$

This yields

$$
\begin{aligned}
4\omega' &= -2e_4'(\log\lambda) + \underline{f}\left(1+\frac{1}{4}f\underline{f}\right)e_4'(f) - \frac{1}{8}\underline{f}^2 e_4'(f^2) + \lambda g\left(D_{e_4'} e_4, e_3'\right)\\
&\quad +\lambda f g\left(D_{e_4'} e_\theta, e_3'\right) + \text{l.o.t.}\\
&= -2e_4'(\log\lambda) + \underline{f}\left(1+\frac{1}{4}f\underline{f}\right)e_4'(f) - \frac{1}{8}\underline{f}^2 e_4'(f^2) + 4\lambda\left(1+\frac{1}{2}f\underline{f}\right)\omega\\
&\quad +2\lambda f\zeta - \lambda f^2\underline{\omega} + 2\lambda\underline{f}\xi + \lambda f\underline{f}\chi - 2\lambda f\underline{\eta} - \lambda f^2\underline{\chi} + \text{l.o.t.}
\end{aligned}
$$

and hence

$$
\begin{aligned}
\omega' &= \lambda\left(\omega - \frac{1}{2}\lambda^{-1}e_4'(\log(\lambda))\right) + \lambda\text{Err}(\omega,\omega'),\\
\text{Err}(\omega,\omega') &= \frac{1}{4}\underline{f}\left(1+\frac{1}{4}f\underline{f}\right)e_4'(f) - \frac{1}{32}\underline{f}^2 e_4'(f^2) + \frac{1}{2}\omega f\underline{f} - \frac{1}{2}f\underline{\eta} + \frac{1}{2}\underline{f}\xi + \frac{1}{2}f\zeta\\
&\quad -\frac{1}{8}\underline{\kappa}f^2 + \frac{1}{8}f\underline{f}\kappa - \frac{1}{4}\underline{\omega}f^2 + \text{l.o.t.}\\
&= \frac{1}{4}\underline{f}e_4'(f) + \frac{1}{2}\omega f\underline{f} - \frac{1}{2}f\underline{\eta} + \frac{1}{2}\underline{f}\xi + \frac{1}{2}f\zeta - \frac{1}{8}\underline{\kappa}f^2 + \frac{1}{8}f\underline{f}\kappa - \frac{1}{4}\underline{\omega}f^2\\
&\quad +\text{l.o.t.}
\end{aligned}
$$

In the particular case, see Remark 2.91, when $\lambda = 1$, $\underline{f} = 0$ we have the more precise formula

$$
\omega' = \omega + \frac{1}{2}f(\zeta - \underline{\eta}) - \frac{1}{8}f^2\left(2\underline{\omega} + \underline{\kappa} + \underline{\vartheta} + f\underline{\xi}\right).
$$

Next, we derive the transformation formulae for $\underline{\omega}$. We have, under a transfor-

mation of type (2.3.3),

$$
\begin{aligned}
&4\underline{\omega}' \\
= \; & g\left(D_{e_3'}e_3', e_4'\right) \\
= \; & g\left(D_{e_3'}\left(\lambda^{-1}\left(\left(1+\frac{1}{2}f\underline{f}+\frac{1}{16}f^2\underline{f}^2\right)e_3+\underline{f}\left(1+\frac{1}{4}f\underline{f}\right)e_\theta+\frac{1}{4}\underline{f}^2e_4\right)\right), e_4'\right) \\
= \; & 2e_3'(\log(\lambda)) \\
& +\lambda^{-1}g\left(D_{e_3'}\left(\left(1+\frac{1}{2}f\underline{f}+\frac{1}{16}f^2\underline{f}^2\right)e_3+\underline{f}\left(1+\frac{1}{4}f\underline{f}\right)e_\theta+\frac{1}{4}\underline{f}^2e_4\right), e_4'\right) \\
= \; & 2e_3'(\log(\lambda))+\frac{1}{2}\lambda^{-1}e_3'\left(f\underline{f}+\frac{1}{8}f^2\underline{f}^2\right)g\left(e_3, e_4'\right) \\
& +\lambda^{-1}\left(1+\frac{1}{2}f\underline{f}\right)g\left(D_{e_3'}e_3, e_4'\right)+\lambda^{-1}e_3'\left(\underline{f}\left(1+\frac{1}{4}f\underline{f}\right)\right)g\left(e_\theta, e_4'\right) \\
& +\lambda^{-1}\underline{f}g\left(D_{e_3'}e_\theta, e_4'\right)+\frac{1}{4}\lambda^{-1}e_3'(\underline{f}^2)g\left(e_4, e_4'\right)+\text{l.o.t.} \\
= \; & 2e_3'(\log(\lambda))-e_3'\left(f\underline{f}+\frac{1}{8}f^2\underline{f}^2\right)+fe_3'\left(\underline{f}\left(1+\frac{1}{4}f\underline{f}\right)\right)-\frac{1}{8}f^2e_3'(\underline{f}^2) \\
& +\lambda^{-1}\left(1+\frac{1}{2}f\underline{f}\right)g\left(D_{e_3'}e_3, e_4'\right)+\lambda^{-1}\underline{f}g\left(D_{e_3'}e_\theta, e_4'\right)+\text{l.o.t.}
\end{aligned}
$$

We compute

$$
\begin{aligned}
& \lambda^{-1}\left(1+\frac{1}{2}f\underline{f}\right)g\left(D_{e_3'}e_3, e_4'\right) \\
= \; & \left(1+\frac{1}{2}f\underline{f}\right)g\left(D_{e_3'}e_3, e_4+fe_\theta\right) \\
= \; & \lambda^{-1}\left(1+\frac{1}{2}f\underline{f}\right)g\left(D_{(1+\frac{1}{2}f\underline{f})e_3+\underline{f}e_\theta+\frac{1}{4}\underline{f}^2e_4}e_3, e_4+fe_\theta\right)+\text{l.o.t.} \\
= \; & 4\lambda^{-1}\left(1+\frac{1}{2}f\underline{f}\right)\underline{\omega}-2\lambda^{-1}\underline{f}\zeta-\lambda^{-1}\underline{f}^2\omega+2\lambda^{-1}f\underline{f}\underline{\omega} \\
& +2\lambda^{-1}f\underline{\xi}+\lambda^{-1}f\underline{f}\,\underline{\chi}+\text{l.o.t.}
\end{aligned}
$$

and

$$
\begin{aligned}
\lambda^{-1}\underline{f}g\left(D_{e_3'}e_\theta, e_4'\right) &= \underline{f}g\left(D_{e_3'}e_\theta, e_4\right)+\text{l.o.t.} \\
&= \lambda^{-1}\underline{f}g\left(D_{e_3+\underline{f}e_\theta}e_\theta, e_4\right)+\text{l.o.t.} \\
&= -2\lambda^{-1}\underline{f}\eta-\lambda^{-1}\underline{f}^2\chi+\text{l.o.t.}
\end{aligned}
$$

This yields

$$
\begin{aligned}
4\underline{\omega}' = \; & 2e_3'(\log(\lambda))-e_3'\left(f\underline{f}+\frac{1}{8}f^2\underline{f}^2\right)+fe_3'\left(\underline{f}\left(1+\frac{1}{4}f\underline{f}\right)\right)-\frac{1}{8}f^2e_3'(\underline{f}^2) \\
& +4\lambda^{-1}\left(1+\frac{1}{2}f\underline{f}\right)\underline{\omega}-2\lambda^{-1}\underline{f}\zeta-\lambda^{-1}\underline{f}^2\omega+2\lambda^{-1}f\underline{f}\underline{\omega} \\
& +2\lambda^{-1}f\underline{\xi}+\lambda^{-1}f\underline{f}\,\underline{\chi}-2\lambda^{-1}\underline{f}\eta-\lambda^{-1}\underline{f}^2\chi+\text{l.o.t.}
\end{aligned}
$$

and hence

$$\begin{aligned}
\underline{\omega}' &= \lambda^{-1}\left(\underline{\omega}+\frac{1}{2}\lambda e_3'(\log(\lambda))\right)+\lambda^{-1}\mathrm{Err}(\underline{\omega},\underline{\omega}'),\\
\mathrm{Err}(\underline{\omega},\underline{\omega}') &= -\frac{1}{4}e_3'\left(f\underline{f}+\frac{1}{8}f^2\underline{f}^2\right)+\frac{1}{4}fe_3'\left(\underline{f}\left(1+\frac{1}{4}f\underline{f}\right)\right)-\frac{1}{32}f^2e_3'(\underline{f}^2)\\
&\quad +\underline{\omega}f\underline{f}-\frac{1}{2}\underline{f}\eta+\frac{1}{2}f\underline{\xi}-\frac{1}{2}\underline{f}\zeta-\frac{1}{8}\kappa\underline{f}^2+\frac{1}{8}f\underline{f}\underline{\kappa}-\frac{1}{4}\omega\underline{f}^2+\mathrm{l.o.t.}\\
&= -\frac{1}{4}\underline{f}e_3'(f)+\underline{\omega}f\underline{f}-\frac{1}{2}\underline{f}\eta+\frac{1}{2}f\underline{\xi}-\frac{1}{2}\underline{f}\zeta-\frac{1}{8}\kappa\underline{f}^2+\frac{1}{8}f\underline{f}\underline{\kappa}-\frac{1}{4}\omega\underline{f}^2\\
&\quad +\mathrm{l.o.t.}
\end{aligned}$$

Next we derive the formula for α. We have

$$\begin{aligned}
\alpha' &= R(e_4',e_4')=\lambda^2R\left(e_4+fe_\theta+\frac{1}{4}f^2e_3,e_4+fe_\theta+\frac{1}{4}f^2e_3\right)\\
&= \lambda^2\left(R_{44}+2fR_{4\theta}+f^2R_{\theta\theta}+\frac{1}{2}f^2R_{34}\right)+\mathrm{l.o.t.}\\
&= \lambda^2\left(\alpha+2f\beta+\frac{3}{2}f^2\rho\right)+\mathrm{l.o.t.}
\end{aligned}$$

and hence

$$\begin{aligned}
\alpha' &= \lambda^2\alpha+\lambda^2\mathrm{Err}(\alpha,\alpha'),\\
\mathrm{Err}(\alpha,\alpha') &= 2f\beta+\frac{3}{2}f^2\rho+\mathrm{l.o.t.}
\end{aligned}$$

The formula for $\underline{\alpha}$ is easily derived by symmetry from the one on α.

Next we derive the formula for β. We have

$$\begin{aligned}
\beta' &= R(e_4',e_\theta')\\
&= \lambda R\left(e_4+fe_\theta+\frac{1}{4}f^2e_3,\left(1+\frac{1}{2}f\underline{f}\right)e_\theta+\frac{1}{2}(\underline{f}e_4+fe_3)\right)+\mathrm{l.o.t.}\\
&= \lambda\left(R_{4\theta}+fR_{\theta\theta}+\frac{1}{2}\underline{f}R_{44}+\frac{1}{2}fR_{43}\right)+\mathrm{l.o.t.}\\
&= \lambda\left(\beta+\frac{3}{2}f\rho+\frac{1}{2}\underline{f}\alpha\right)+\mathrm{l.o.t.}
\end{aligned}$$

and hence

$$\begin{aligned}
\beta' &= \lambda\left(\beta+\frac{3}{2}f\rho\right)+\lambda\mathrm{Err}(\beta,\beta'),\\
\mathrm{Err}(\beta,\beta') &= \frac{1}{2}\underline{f}\alpha+\mathrm{l.o.t.}
\end{aligned}$$

The formula for $\underline{\beta}$ is easily derived by symmetry from the one on β.

Finally, we derive the formula for ρ. We have

$$\begin{aligned}\rho' &= R(e_4', e_3') = R\left(e_4 + fe_\theta + \frac{1}{4}f^2 e_3, \left(1+\frac{1}{2}f\underline{f}\right)e_3 + \underline{f}e_\theta + \frac{1}{4}\underline{f}^2 e_4\right) + \text{l.o.t.}\\ &= R_{43} + \frac{1}{2}f\underline{f}R_{43} + \underline{f}R_{4\theta} + fR_{\theta 3} + f\underline{f}R_{\theta\theta} + \text{l.o.t.}\\ &= \rho + \frac{3}{2}\rho f\underline{f} + \underline{f}\beta + f\underline{\beta} + \text{l.o.t.}\end{aligned}$$

and hence

$$\begin{aligned}\rho' &= \rho + \mathrm{Err}(\rho, \rho'),\\ \mathrm{Err}(\rho,\rho') &= \frac{3}{2}\rho f\underline{f} + \underline{f}\beta + f\underline{\beta} + \text{l.o.t.}\end{aligned}$$

This concludes the proof of Proposition 2.90.

A.7 PROOF OF LEMMA 2.92

For ξ' and ω', we need a more precise transformation formula than the ones of Proposition 2.90. We have

$$\begin{aligned}2\xi' &= g(D_{e_4'}e_4', e_\theta')\\ &= \lambda^2 g(D_{\lambda^{-1}e_4'}(\lambda^{-1}e_4'), e_\theta')\\ &= \lambda^2 g\left(D_{\lambda^{-1}e_4'}(\lambda^{-1}e_4'), e_\theta + \frac{1}{2}fe_3\right)\\ &\quad +\lambda^2\frac{f}{2}g\left(D_{\lambda^{-1}e_4'}(\lambda^{-1}e_4'), e_4 + fe_\theta + \frac{1}{4}f^2 e_3\right)\\ &= \lambda^2 g\left(D_{\lambda^{-1}e_4'}(\lambda^{-1}e_4'), e_\theta + \frac{1}{2}fe_3\right) + \lambda^2\frac{f}{2}g\left(D_{\lambda^{-1}e_4'}(\lambda^{-1}e_4'), \lambda^{-1}e_4'\right)\\ &= \lambda^2 g\left(D_{\lambda^{-1}e_4'}(\lambda^{-1}e_4'), e_\theta + \frac{1}{2}fe_3\right).\end{aligned}$$

Also, we have

$$\begin{aligned}
&4\omega' \\
=\;& g(D_{e_4'}e_4', e_3') \\
=\;& -2e_4'(\log(\lambda)) + \lambda g(D_{\lambda^{-1}e_4'}(\lambda^{-1}e_4'), \lambda e_3') \\
=\;& -2e_4'(\log(\lambda)) + \lambda g(D_{\lambda^{-1}e_4'}(\lambda^{-1}e_4'), e_3) + \lambda \underline{f} g\left(D_{\lambda^{-1}e_4'}(\lambda^{-1}e_4'), e_\theta + \frac{1}{2}fe_3\right) \\
& +\frac{1}{4}\lambda \underline{f}^2 g\left(D_{\lambda^{-1}e_4'}(\lambda^{-1}e_4'), e_4 + fe_\theta + \frac{1}{4}f^2e_3\right) \\
=\;& -2e_4'(\log(\lambda)) + \lambda g(D_{\lambda^{-1}e_4'}(\lambda^{-1}e_4'), e_3) + \lambda \underline{f} g\left(D_{\lambda^{-1}e_4'}(\lambda^{-1}e_4'), e_\theta + \frac{1}{2}fe_3\right) \\
& +\frac{1}{4}\lambda \underline{f}^2 g\left(D_{\lambda^{-1}e_4'}(\lambda^{-1}e_4'), \lambda^{-1}e_4'\right) \\
=\;& -2e_4'(\log(\lambda)) + \lambda g(D_{\lambda^{-1}e_4'}(\lambda^{-1}e_4'), e_3) + \lambda \underline{f} g\left(D_{\lambda^{-1}e_4'}(\lambda^{-1}e_4'), e_\theta + \frac{1}{2}fe_3\right).
\end{aligned}$$

In view of the change of frame formula for ξ', we infer

$$4\omega' = -2e_4'(\log(\lambda)) + \lambda g(D_{\lambda^{-1}e_4'}(\lambda^{-1}e_4'), e_3) + \lambda^{-1}\underline{f}\xi'.$$

Next, we compute

$$\begin{aligned}
g\left(D_{\lambda^{-1}e_4'}(\lambda^{-1}e_4'), e_\theta\right) &= g\left(D_{\lambda^{-1}e_4'}\left(e_4 + fe_\theta + \frac{1}{4}f^2e_3\right), e_\theta\right) \\
&= g\left(D_{\lambda^{-1}e_4'}e_4, e_\theta\right) + \lambda^{-1}e_4'(f) + \frac{1}{4}f^2 g\left(D_{\lambda^{-1}e_4'}e_3, e_\theta\right) \\
&= g\left(D_{e_4+fe_\theta+\frac{1}{4}f^2e_3}e_4, e_\theta\right) + \left(e_4 + fe_\theta + \frac{1}{4}f^2e_3\right)f \\
&\quad +\frac{1}{4}f^2 g\left(D_{e_4+fe_\theta+\frac{1}{4}f^2e_3}e_3, e_\theta\right) \\
&= 2\xi + f\chi + \frac{1}{2}f^2\eta + \left(e_4 + fe_\theta + \frac{1}{4}f^2e_3\right)f + \frac{1}{2}f^2\underline{\eta} \\
&\quad +\frac{1}{4}f^3\underline{\chi} + \frac{1}{8}f^4\underline{\xi}.
\end{aligned}$$

Also, we have

$$\begin{aligned}
g\left(D_{\lambda^{-1}e_4'}(\lambda^{-1}e_4'), e_3\right) &= g\left(D_{\lambda^{-1}e_4'}\left(e_4 + fe_\theta + \frac{1}{4}f^2e_3\right), e_3\right) \\
&= g\left(D_{\lambda^{-1}e_4'}e_4, e_3\right) + fg\left(D_{\lambda^{-1}e_4'}e_\theta, e_3\right) \\
&= g\left(D_{e_4+fe_\theta+\frac{1}{4}f^2e_3}e_4, e_3\right) + fg\left(D_{e_4+fe_\theta+\frac{1}{4}f^2e_3}e_\theta, e_3\right) \\
&= 4\omega + 2f\zeta - f^2\underline{\omega} - 2\underline{\eta}f - f^2\underline{\chi} - \frac{1}{2}f^3\underline{\xi}.
\end{aligned}$$

We deduce

$$\begin{aligned}
2\xi' &= \lambda^2 g\left(D_{\lambda^{-1}e_4'}(\lambda^{-1}e_4'), e_\theta + \frac{1}{2}fe_3\right)\\
&= \lambda^2\Bigg\{2\xi + f\chi + \frac{1}{2}f^2\eta + \left(e_4 + fe_\theta + \frac{1}{4}f^2e_3\right)f + \frac{1}{2}f^2\underline{\eta} + \frac{1}{4}f^3\underline{\chi}\\
&\quad +\frac{1}{8}f^4\underline{\xi} + \frac{1}{2}f\left(4\omega + 2f\zeta - f^2\underline{\omega} - 2\underline{\eta}f - f^2\underline{\chi} - \frac{1}{2}f^3\underline{\xi}\right)\Bigg\}\\
&= \lambda^2\Bigg\{2\xi + \left(e_4 + fe_\theta + \frac{1}{4}f^2e_3\right)f + f\chi + 2f\omega + \frac{1}{2}f^2\eta - \frac{1}{2}f^2\underline{\eta} + f^2\zeta\\
&\quad -\frac{1}{4}f^3\underline{\chi} - \frac{1}{2}f^3\underline{\omega} - \frac{1}{8}f^4\underline{\xi}\Bigg\}
\end{aligned}$$

and

$$\begin{aligned}
4\omega' &= -2e_4'(\log(\lambda)) + \lambda g(D_{\lambda^{-1}e_4'}(\lambda^{-1}e_4'), e_3) + \lambda^{-1}\underline{f}\xi'\\
&= \lambda\Bigg\{4\omega - 2\left(e_4 + fe_\theta + \frac{1}{4}f^2e_3\right)\log(\lambda) + 2f\zeta - f^2\underline{\omega} - 2\underline{\eta}f - f^2\underline{\chi}\\
&\quad -\frac{1}{2}f^3\underline{\xi}\Bigg\} + \lambda^{-1}\underline{f}\xi'.
\end{aligned}$$

If $\xi' = 0$, we infer

$$\begin{aligned}
2\xi + \left(e_4 + fe_\theta + \frac{1}{4}f^2e_3\right)f + f\chi + 2f\omega + \frac{1}{2}f^2\eta - \frac{1}{2}f^2\underline{\eta} + f^2\zeta - \frac{1}{4}f^3\underline{\chi}&\\
-\frac{1}{2}f^3\underline{\omega} - \frac{1}{8}f^4\underline{\xi} &= 0
\end{aligned}$$

and hence

$$\begin{aligned}
\lambda^{-1}e_4'(f) + \left(\frac{\kappa}{2} + 2\omega\right)f &= -2\xi - \frac{1}{2}\vartheta f - \frac{1}{2}f^2\eta + \frac{1}{2}f^2\underline{\eta} - f^2\zeta + \frac{1}{8}f^3\underline{\kappa} + \frac{1}{2}f^3\underline{\omega}\\
&\quad +\frac{1}{8}f^3\underline{\vartheta} + \frac{1}{8}f^4\underline{\xi}
\end{aligned}$$

which yields the desired transport equation for f

$$\begin{aligned}
\lambda^{-1}e_4'(f) + \left(\frac{\kappa}{2} + 2\omega\right)f &= -2\xi + E_1(f,\Gamma),\\
E_1(f,\Gamma) &= -\frac{1}{2}\vartheta f - \frac{1}{2}f^2\eta + \frac{1}{2}f^2\underline{\eta} - f^2\zeta + \frac{1}{8}f^3\underline{\kappa} + \frac{1}{2}f^3\underline{\omega}\\
&\quad +\frac{1}{8}f^3\underline{\vartheta} + \frac{1}{8}f^4\underline{\xi}.
\end{aligned}$$

Also, if $\xi' = 0$ and $\omega' = 0$, we infer

$$0 = 4\omega - 2\left(e_4 + fe_\theta + \frac{1}{4}f^2e_3\right)\log(\lambda) + 2f\zeta - f^2\underline{\omega} - 2\underline{\eta}f - f^2\underline{\chi} - \frac{1}{2}f^3\underline{\xi}$$

and hence

$$\lambda^{-1}e_4'(\log(\lambda)) = 2\omega + f\zeta - \frac{1}{2}f^2\underline{\omega} - \underline{\eta}f - \frac{1}{4}f^2\underline{\kappa} - \frac{1}{4}f^2\underline{\vartheta} - \frac{1}{4}f^3\underline{\xi}$$

which yields the desired transport equation for $\log(\lambda)$

$$\begin{aligned}\lambda^{-1}e_4'(\log(\lambda)) &= 2\omega + E_2(f,\Gamma),\\ E_2(f,\Gamma) &= f\zeta - \frac{1}{2}f^2\underline{\omega} - \underline{\eta}f - \frac{1}{4}f^2\underline{\kappa} - \frac{1}{4}f^2\underline{\vartheta} - \frac{1}{4}f^3\underline{\xi}.\end{aligned}$$

Finally, we derive the transport equation for $\underline{f}$. In view of the transformation formulas of Proposition 2.90 for ζ' and $\underline{\eta}'$, and the fact that we assume $\zeta' + \underline{\eta}' = 0$, we have

$$\begin{aligned}\frac{1}{2}\lambda^{-1}e_4'(\underline{f}) &= -(\zeta + \underline{\eta}) + e_\theta'(\log(\lambda)) - \frac{1}{4}\underline{f}\kappa + f\underline{\omega} - \frac{1}{2}\underline{f}e_\theta'(f) + \frac{1}{8}\underline{f}^2\lambda^{-1}e_4'(f)\\ &\quad -\frac{1}{4}\underline{f}\vartheta + \text{l.o.t.}\end{aligned}$$

Together with the above identity for $\lambda^{-1}e_4'(f)$, we infer

$$\begin{aligned}\lambda^{-1}e_4'(\underline{f}) + \frac{\kappa}{2}\underline{f} &= -2(\zeta + \underline{\eta}) + 2e_\theta'(\log(\lambda)) + 2f\underline{\omega} + E_3(f, \underline{f}, \Gamma),\\ E_3(f, \underline{f}, \Gamma) &= -\underline{f}e_\theta'(f) - \frac{1}{2}\underline{f}\vartheta + \text{l.o.t.},\end{aligned}$$

which yields the third identity of the statement. This concludes the proof of Lemma 2.92.

A.8 PROOF OF COROLLARY 2.93

In view of Lemma 2.92 and the fact that (e_3, e_4, e_θ) emanates from an outgoing geodesic foliation and hence

$$\xi = 0, \quad \omega = 0, \quad \zeta + \underline{\eta} = 0,$$

we have

$$\begin{aligned}\lambda^{-1}e_4'(f) + \frac{\kappa}{2}f &= E_1(f,\Gamma),\\ \lambda^{-1}e_4'(\log(\lambda)) &= E_2(f,\Gamma),\\ \lambda^{-1}e_4'(\underline{f}) + \frac{\kappa}{2}\underline{f} &= 2e_\theta'(\log(\lambda)) + 2f\underline{\omega} + E_3(f, \underline{f}, \Gamma).\end{aligned}$$

The second equation is the desired identity for $\log(\lambda)$.

We still need to derive the first and the third identities. We start with the first one. We have

$$\begin{aligned}\lambda^{-1}e_4'(rf) &= r\Big(-\frac{\kappa}{2}f + E_1(f,\Gamma)\Big) + \lambda^{-1}e_4'(r)f\\ &= -\frac{r}{2}\left(\kappa - \frac{2\lambda^{-1}e_4'(r)}{r}\right)f + rE_1(f,\Gamma).\end{aligned}$$

Since

$$\lambda^{-1}e_4' = e_4 + fe_\theta + \frac{f^2}{4}e_3,$$

we infer

$$\lambda^{-1}e_4'(r) = \frac{r}{2}\overline{\kappa} + \frac{f^2}{4}e_3(r)$$

and hence

$$\lambda^{-1}e_4'(rf) = -\frac{r}{2}\left(\check{\kappa} - \frac{e_3(r)}{2r}f^2\right)f + rE_1(f,\Gamma)$$

as desired.

Next, we have

$$\begin{aligned}
&\lambda^{-1}e_4'\Big(r\underline{f} - 2r^2e_\theta'(\log(\lambda)) + rf\underline{\Omega}\Big)\\
= \;& r\left(-\frac{\kappa}{2}\underline{f} + 2e_\theta'(\log(\lambda)) + 2f\underline{\omega} + E_3(f,\underline{f},\Gamma)\right) - 2r^2e_\theta'(E_2(f,\Gamma))\\
&+r\underline{\Omega}\left(-\frac{\kappa}{2}f + E_1(f,\Gamma)\right) + \lambda^{-1}e_4'(r)\underline{f} - 2r^2[\lambda^{-1}e_4', e_\theta']\log(\lambda)\\
&-4r\lambda^{-1}e_4'(r)e_\theta'(\log(\lambda)) + r\lambda^{-1}e_4'(\underline{\Omega})f + \lambda^{-1}e_4'(r)f\underline{\Omega}\\
= \;& -\frac{r}{2}\left(\kappa - \frac{2\lambda^{-1}e_4'(r)}{r}\right)\underline{f} + 2r\Big(1 - 2\lambda^{-1}e_4'(r)\Big)e_\theta'(\log(\lambda))\\
&-2r^2\lambda^{-1}[e_4', e_\theta']\log(\lambda) + r\Big(\lambda^{-1}e_4'(\underline{\Omega}) + 2\underline{\omega}\Big)f - \frac{r}{2}\left(\kappa - \frac{2\lambda^{-1}e_4'(r)}{r}\right)\underline{\Omega}f\\
&-2r^2e_\theta'(\log(\lambda))\lambda^{-1}e_4'(\log(\lambda)) + rE_3(f,\underline{f},\Gamma) - 2r^2e_\theta'(E_2(f,\Gamma)) + r\underline{\Omega}E_1(f,\Gamma).
\end{aligned}$$

Since we have

$$\lambda^{-1}e_4' = e_4 + fe_\theta + \frac{f^2}{4}e_3,$$

we infer

$$\begin{aligned}
\lambda^{-1}e_4'(r) &= \frac{r}{2}\overline{\kappa} + \frac{f^2}{4}e_3(r),\\
\lambda^{-1}e_4'(\underline{\Omega}) &= e_4(\underline{\Omega}) + fe_\theta(\underline{\Omega}) + \frac{f^2}{4}e_3(\underline{\Omega})\\
&= -2\underline{\omega} + fe_\theta(\underline{\Omega}) + \frac{f^2}{4}e_3(\underline{\Omega}).
\end{aligned}$$

Together with the transport equation for $\log(\lambda)$ and the commutator identity for

$[e'_4, e'_\theta]$, we infer

$$\begin{aligned}
& \lambda^{-1}e'_4\Big(r\underline{f} - 2r^2 e'_\theta(\log(\lambda)) + rf\underline{\Omega}\Big)\\
= & -\frac{r}{2}\left(\check{\kappa} - \frac{e_3(r)}{2r}f^2\right)\underline{f} + 2r\left(1 - r\overline{\kappa} - \frac{e_3(r)}{2}f^2\right)e'_\theta(\log(\lambda))\\
& + r^2\lambda^{-1}(\kappa' + \vartheta')e'_\theta(\log(\lambda)) + r\left(e_\theta(\underline{\Omega}) + \frac{f}{4}e_3(\underline{\Omega})\right)f^2 - \frac{r}{2}\left(\check{\kappa} - \frac{e_3(r)}{2r}f^2\right)\underline{\Omega}f\\
& -2r^2 e'_\theta(\log(\lambda))E_2(f,\Gamma) + rE_3(f,\underline{f},\Gamma) - 2r^2 e'_\theta(E_2(f,\Gamma)) + r\underline{\Omega}E_1(f,\Gamma).
\end{aligned}$$

Now, recall the following transformation formulas

$$\begin{aligned}
\lambda^{-1}\kappa' &= \kappa + \not{d}'_1(f) + \mathrm{Err}(\kappa,\kappa'),\\
\mathrm{Err}(\kappa,\kappa') &= f(\zeta+\eta) - \frac{1}{4}f^2\underline{\kappa} - f^2\underline{\omega} + \text{l.o.t.}
\end{aligned}$$

We infer

$$\begin{aligned}
& \lambda^{-1}e'_4\Big(r\underline{f} - 2r^2 e'_\theta(\log(\lambda)) + rf\underline{\Omega}\Big)\\
= & -\frac{r}{2}\left(\check{\kappa} - \frac{e_3(r)}{2r}f^2\right)\underline{f} + r^2\left(\check{\kappa} - \left(\overline{\kappa} - \frac{2}{r}\right) - \frac{e_3(r)}{r}f^2\right)e'_\theta(\log(\lambda))\\
& + r^2\Big(\not{d}'_1(f) + \mathrm{Err}(\kappa,\kappa') + \lambda^{-1}\vartheta'\Big)e'_\theta(\log(\lambda))\\
& + r\left(e_\theta(\underline{\Omega}) + \frac{f}{4}e_3(\underline{\Omega})\right)f^2 - \frac{r}{2}\left(\check{\kappa} - \frac{e_3(r)}{2r}f^2\right)\underline{\Omega}f\\
& -2r^2 e'_\theta(\log(\lambda))E_2(f,\Gamma) + rE_3(f,\underline{f},\Gamma) - 2r^2 e'_\theta(E_2(f,\Gamma)) + r\underline{\Omega}E_1(f,\Gamma).
\end{aligned}$$

This concludes the proof of Corollary 2.93.

A.9 PROOF OF LEMMA 2.91

Recall that we have obtained in section A.7

$$\begin{aligned}
2\xi' &= \lambda^2\Bigg\{2\xi + \left(e_4 + fe_\theta + \frac{1}{4}f^2 e_3\right)f + \left(\frac{1}{2}\kappa + 2\omega\right)f + \frac{1}{2}\vartheta f + \frac{1}{2}f^2\eta - \frac{1}{2}f^2\underline{\eta}\\
&\quad + f^2\zeta - \frac{1}{4}f^3\underline{\chi} - \frac{1}{2}f^3\underline{\omega} - \frac{1}{8}f^4\underline{\xi}\Bigg\},\\
4\omega' &= \lambda\Bigg\{4\omega - 2\left(e_4 + fe_\theta + \frac{1}{4}f^2 e_3\right)\log(\lambda) + 2f\zeta - f^2\underline{\omega} - 2\underline{\eta}f - f^2\underline{\chi}\\
&\quad - \frac{1}{2}f^3\underline{\xi}\Bigg\} + \lambda^{-1}\underline{f}\xi'.
\end{aligned}$$

In the case where $\lambda = 1$ and $\underline{f} = 0$, we immediately infer

$$
\begin{aligned}
2\xi' &= 2\xi + \left(e_4 + fe_\theta + \frac{1}{4}f^2 e_3\right) f + \left(\frac{1}{2}\kappa + 2\omega\right) f + \frac{1}{2}\vartheta f + \frac{1}{2}f^2\eta - \frac{1}{2}f^2\underline{\eta} \\
&\quad + f^2\zeta - \frac{1}{4}f^3\underline{\chi} - \frac{1}{2}f^3\underline{\omega} - \frac{1}{8}f^4\underline{\xi}, \\
4\omega' &= 4\omega + 2f\zeta - f^2\underline{\omega} - 2\underline{\eta}f - f^2\underline{\chi} - \frac{1}{2}f^3\underline{\xi},
\end{aligned}
$$

and hence

$$
\begin{aligned}
\xi' &= \xi + \frac{1}{2}e_4'(f) + \left(\frac{1}{4}\kappa + \omega\right) f + \frac{1}{4}\vartheta f + \frac{1}{4}f^2\eta - \frac{1}{4}f^2\underline{\eta} + \frac{1}{2}f^2\zeta - \frac{1}{8}f^3\underline{\chi} \\
&\quad - \frac{1}{4}f^3\underline{\omega} - \frac{1}{16}f^4\underline{\xi}, \\
\omega' &= \omega + \frac{1}{2}f\zeta - \frac{1}{4}f^2\underline{\omega} - \frac{1}{2}\underline{\eta}f - \frac{1}{8}f^2\underline{\kappa} - \frac{1}{8}f^2\underline{\vartheta} - \frac{1}{8}f^3\underline{\xi}.
\end{aligned}
$$

Finally, we compute the change of frame formula for ζ' and η' when $\lambda = 1, \underline{f} = 0$. We have in this case

$$
\begin{aligned}
e_4' &= e_4 + fe_\theta + \frac{1}{4}f^2 e_3, \\
e_\theta' &= e_\theta + \frac{1}{2}fe_3, \\
e_3' &= e_3,
\end{aligned}
$$

and hence

$$
\begin{aligned}
2\zeta' &= g\left(D_{e_\theta'}e_4', e_3'\right) \\
&= g\left(D_{e_\theta'}\left(e_4 + fe_\theta + \frac{1}{4}f^2 e_3\right), e_3\right) \\
&= g\left(D_{e_\theta'}e_4, e_3\right) + fg\left(D_{e_\theta'}e_\theta, e_3\right) \\
&= g\left(D_{e_\theta + \frac{1}{2}fe_3}e_4, e_3\right) + fg\left(D_{e_\theta + \frac{1}{2}fe_3}e_\theta, e_3\right) \\
&= 2\zeta - 2\underline{\omega}f - \underline{\chi}f - \underline{\xi}f^2 \\
&= 2\zeta - \left(\frac{1}{2}\underline{\kappa} + 2\underline{\omega}\right) f - f\left(\frac{1}{2}\underline{\vartheta} + f\underline{\xi}\right)
\end{aligned}
$$

and

$$\begin{aligned}
2\eta' &= g\left(D_{e_3'}e_4', e_\theta'\right)\\
&= g\left(D_{e_3'}\left(e_4 + fe_\theta + \frac{1}{4}f^2 e_3\right), e_\theta'\right)\\
&= g\left(D_{e_3'}e_4, e_\theta'\right) + e_3'(f)g\left(e_\theta, e_\theta'\right) + fg\left(D_{e_3'}e_\theta, e_\theta'\right) + \frac{1}{4}e_3'(f^2)g\left(e_3, e_\theta'\right)\\
&\quad + \frac{1}{4}f^2 g(D_{e_3'}e_3, e_\theta')\\
&= e_3'(f) + g\left(D_{e_3}e_4, e_\theta + \frac{1}{2}fe_3\right) + fg\left(D_{e_3}e_\theta, \frac{1}{2}fe_3\right) + \frac{1}{4}f^2 g(D_{e_3}e_3, e_\theta)\\
&= 2\eta + e_3'(f) - 2f\underline{\omega} - \frac{1}{2}f^2\underline{\xi}
\end{aligned}$$

which yields the desired change of frame formula for ζ' and η'. This concludes the proof of Lemma 2.91.

A.10 PROOF OF PROPOSITION 2.99

Recall that we have

$$\mathfrak{q} = r^4\left(e_3(e_3(\alpha)) + (2\underline{\kappa} - 6\underline{\omega})e_3(\alpha) + \left(-4e_3(\underline{\omega}) + 8\underline{\omega}^2 - 8\underline{\omega}\,\underline{\kappa} + \frac{1}{2}\underline{\kappa}^2\right)\alpha\right),$$

which we write in the form $\mathfrak{q} = r^4 J$ where

$$\begin{aligned}
J &= e_3(e_3(\alpha)) + (2\underline{\kappa} - 6\underline{\omega})e_3(\alpha) + V\alpha,\\
V &= -4e_3(\underline{\omega}) + 8\underline{\omega}^2 - 8\underline{\omega}\,\underline{\kappa} + \frac{1}{2}\underline{\kappa}^2.
\end{aligned}$$

We make use of the general[1] Bianchi equations, see Proposition 2.57,

$$\begin{aligned}
e_3\alpha + \frac{1}{2}\underline{\kappa}\alpha &= -\,{}^{\star}\!\!\not{d}_2\,\beta + 4\underline{\omega}\alpha - \frac{3}{2}\vartheta\rho + \mathrm{Err}[e_3(\alpha)],\\
e_3\beta + \underline{\kappa}\beta &= -\,{}^{\star}\!\!\not{d}_1\rho + 2\underline{\omega}\beta + 3\eta\rho + \mathrm{Err}[e_3(\beta)],\\
e_3\rho + \frac{3}{2}\underline{\kappa}\rho &= \not{d}_1\underline{\beta} + \mathrm{Err}[e_3(\rho)],
\end{aligned}$$

as well as the null structure equations (see Proposition 2.56)

$$\begin{aligned}
e_3\vartheta + \frac{1}{2}\underline{\kappa}\vartheta - 2\underline{\omega}\vartheta &= -2\,{}^{\star}\!\!\not{d}_2\,\eta - \frac{1}{2}\kappa\underline{\vartheta} + \mathrm{Err}[e_3(\vartheta)],\\
e_3(\underline{\kappa}) + \frac{1}{2}\underline{\kappa}^2 + 2\underline{\omega}\,\underline{\kappa} &= 2\not{d}_1\underline{\xi} + \mathrm{Err}[e_3(\underline{\kappa})],
\end{aligned}$$

where $\mathrm{Err}[e_3(\alpha)]$, $\mathrm{Err}[e_3(\beta)]$, $\mathrm{Err}[e_3(\rho)]$, $\mathrm{Err}[e_3(\vartheta)]$, $\mathrm{Err}[e_3(\underline{\kappa})]$ denote the corresponding quadratic terms in each equation. We also make use of the commutation

[1]In an arbitrary **Z**-invariant frame.

formula (see Lemma 2.54)

$$\begin{aligned}
[e_3, \, {\not{d}}_2^\star]\beta &= -\frac{1}{2}\underline{\kappa}\,{\not{d}}_2^\star\beta - \underline{Com}_2^*(\beta),\\
\underline{Com}_2^*(\beta) &= -\frac{1}{2}\underline{\vartheta}\,{\not{d}}_1\beta - (\zeta-\eta)e_3\beta - \eta e_3\Phi\beta + \underline{\xi}(e_4\beta - e_4(\Phi)\beta) - \underline{\beta}\cdot\beta.
\end{aligned}$$

Thus,

$$\begin{aligned}
J &= e_3\left(-\frac{1}{2}\underline{\kappa}\alpha - {\not{d}}_2^\star\beta + 4\underline{\omega}\alpha - \frac{3}{2}\vartheta\rho + \mathrm{Err}[e_3(\alpha)]\right) + (2\underline{\kappa} - 6\underline{\omega})e_3(\alpha) + V\alpha\\
&= \left(\frac{3}{2}\underline{\kappa} - 2\underline{\omega}\right)e_3\alpha + \Big(-\frac{1}{2}e_3\underline{\kappa} + 4e_3(\underline{\omega}) + V\Big)\alpha - {\not{d}}_2^\star e_3\beta - [e_3, \, {\not{d}}_2^\star]\beta - \frac{3}{2}\vartheta e_3\rho\\
&- \frac{3}{2}\rho e_3\vartheta + e_3\mathrm{Err}[e_3(\alpha)]\\
&= \left(\frac{3}{2}\underline{\kappa} - 2\underline{\omega}\right)\left(-\frac{1}{2}\underline{\kappa}\alpha - {\not{d}}_2^\star\beta + 4\underline{\omega}\alpha - \frac{3}{2}\vartheta\rho + \mathrm{Err}[e_3(\alpha)]\right)\\
&+ \left(-\frac{1}{2}e_3\underline{\kappa} + 4e_3(\underline{\omega}) + V\right)\alpha - {\not{d}}_2^\star e_3\beta + \frac{1}{2}\underline{\kappa}\,{\not{d}}_2^\star\beta - \frac{3}{2}\vartheta e_3\rho - \frac{3}{2}\rho e_3\vartheta\\
&+ e_3\mathrm{Err}[e_3(\alpha)] + \underline{Com}_2^*(\beta)\\
&= -{\not{d}}_2^\star e_3\beta + (-\underline{\kappa} + 2\underline{\omega})\,{\not{d}}_2^\star\beta + \Bigg(-\frac{1}{2}e_3\underline{\kappa} + 4e_3(\underline{\omega}) + V\\
&+ \left(\frac{3}{2}\underline{\kappa} - 2\underline{\omega}\right)\left(-\frac{1}{2}\underline{\kappa} + 4\underline{\omega}\right)\Bigg)\alpha - \frac{3}{2}\left(\vartheta e_3\rho + \rho e_3\vartheta + \vartheta\rho\left(\frac{3}{2}\underline{\kappa} - 2\underline{\omega}\right)\right)\\
&+ e_3\mathrm{Err}[e_3(\alpha)] + \left(\frac{3}{2}\underline{\kappa} - 2\underline{\omega}\right)\mathrm{Err}[e_3(\alpha)] + \underline{Com}_2^*(\beta).
\end{aligned}$$

Hence,

$$\begin{aligned}
J &= -{\not{d}}_2^\star e_3\beta + (-\underline{\kappa} + 2\underline{\omega})\,{\not{d}}_2^\star\beta - \frac{3}{2}\left(\vartheta e_3\rho + \rho e_3\vartheta + \vartheta\rho(\frac{3}{2}\underline{\kappa} - 2\underline{\omega})\right)\\
&\quad + W\alpha + E,\\
W &:= -\frac{1}{2}e_3\underline{\kappa} + 4e_3(\underline{\omega}) + V + \left(\frac{3}{2}\underline{\kappa} - 2\underline{\omega}\right)\left(-\frac{1}{2}\underline{\kappa} + 4\underline{\omega}\right),\\
E &:= e_3\mathrm{Err}[e_3(\alpha)] + \left(\frac{3}{2}\underline{\kappa} - 2\underline{\omega}\right)\mathrm{Err}[e_3(\alpha)] + \underline{Com}_2^*(\beta).
\end{aligned} \tag{A.10.1}$$

Now, ignoring cubic and higher order terms,

$$\begin{aligned}
&-{\not{d}}_2^\star e_3\beta + (-\underline{\kappa} + 2\underline{\omega})\,{\not{d}}_2^\star\beta\\
&= -{\not{d}}_2^\star\,(-\underline{\kappa}\beta - {\not{d}}_1^\star\rho + 2\underline{\omega}\beta + 3\eta\rho + \mathrm{Err}[e_3(\beta)]) + (-\underline{\kappa} + 2\underline{\omega})\,{\not{d}}_2^\star\beta\\
&= {\not{d}}_2^\star\,{\not{d}}_1^\star\rho - 3\rho\,{\not{d}}_2^\star\eta + \beta\,{\not{d}}_1^\star(\underline{\kappa} - 2\underline{\omega}) - 3\eta\,{\not{d}}_1^\star\rho - {\not{d}}_2^\star\mathrm{Err}[e_3(\beta)].
\end{aligned}$$

Also,

$$\begin{aligned}
& \vartheta e_3\rho + \rho e_3\vartheta + \vartheta\rho\left(\frac{3}{2}\underline{\kappa} - 2\underline{\omega}\right) \\
= & \ \vartheta\left(-\frac{3}{2}\underline{\kappa}\rho + \not{d}_1\underline{\beta}\right) + \vartheta\rho\left(\frac{3}{2}\underline{\kappa} - 2\underline{\omega}\right) \\
& +\rho\left(-\frac{1}{2}\underline{\kappa}\,\vartheta + 2\underline{\omega}\vartheta - 2\,\not{d}_2^\star\,\eta - \frac{1}{2}\kappa\,\underline{\vartheta} + \mathrm{Err}[e_3(\vartheta)]\right) \\
= & -\frac{1}{2}\underline{\kappa}\rho\vartheta - \frac{1}{2}\kappa\rho\underline{\vartheta} - 2\rho\,\not{d}_2^\star\eta + \vartheta\,\not{d}_1\underline{\beta} + \rho\mathrm{Err}[e_3(\rho)]
\end{aligned}$$

and

$$\begin{aligned}
W = & -\frac{1}{2}e_3\underline{\kappa} + 4e_3(\underline{\omega}) + \left(-4e_3(\underline{\omega}) + 8\underline{\omega}^2 - 8\underline{\omega}\,\underline{\kappa} + \frac{1}{2}\underline{\kappa}^2\right) \\
& + \left(\frac{3}{2}\underline{\kappa} - 2\underline{\omega}\right)\left(-\frac{1}{2}\underline{\kappa} + 4\underline{\omega}\right) \\
= & -\frac{1}{2}e_3\underline{\kappa} + \left(8\underline{\omega}^2 - 8\underline{\omega}\,\underline{\kappa} + \frac{1}{2}\underline{\kappa}^2\right) + \left(-\frac{3}{4}\underline{\kappa}^2 - 8\underline{\omega}^2 + 7\underline{\omega}\underline{\kappa}\right) \\
= & -\frac{1}{2}e_3\underline{\kappa} - \frac{1}{4}\underline{\kappa}^2 - \underline{\omega}\underline{\kappa} \\
= & -\frac{1}{2}\left(-\frac{1}{2}\underline{\kappa}^2 - 2\underline{\omega}\,\underline{\kappa} + 2\,\not{d}_1\underline{\xi} + \mathrm{Err}[e_3(\underline{\kappa})]\right) - \frac{1}{4}\underline{\kappa}^2 - \underline{\omega}\underline{\kappa} \\
= & -\not{d}_1\underline{\xi} - \frac{1}{2}\mathrm{Err}[e_3(\underline{\kappa})].
\end{aligned}$$

Thus, back to (A.10.1),

$$\begin{aligned}
J = & \ \not{d}_2^\star\,\not{d}_1^\star\rho - 3\rho\,\not{d}_2^\star\eta + \beta\,\not{d}_1^\star(\underline{\kappa} - 2\underline{\omega}) - 3\eta\,\not{d}_1^\star\rho - \not{d}_2^\star\mathrm{Err}[e_3(\beta)] \\
& - \frac{3}{2}\left(-\frac{1}{2}\underline{\kappa}\rho\vartheta - \frac{1}{2}\kappa\rho\underline{\vartheta} - 2\rho\,\not{d}_2^\star\eta + \vartheta\,\not{d}_1\underline{\beta} + \rho\mathrm{Err}[e_3(\rho)]\right) - \not{d}_1\underline{\xi}\alpha + E \\
= & \ \not{d}_2^\star\,\not{d}_1^\star\rho + \frac{3}{4}\rho(\underline{\kappa}\vartheta + \kappa\underline{\vartheta}) + \beta\,\not{d}_1^\star(\underline{\kappa} - 2\underline{\omega}) - 3\eta\,\not{d}_1^\star\rho - \frac{3}{2}\vartheta\,\not{d}_1\underline{\beta} - \not{d}_1\underline{\xi}\alpha \\
& -\frac{3}{2}\rho\mathrm{Err}[e_3(\rho)] + E.
\end{aligned}$$

In other words,

$$\begin{aligned}
J = & \ \not{d}_2^\star\,\not{d}_1^\star\rho + \frac{3}{4}\rho(\underline{\kappa}\vartheta + \kappa\underline{\vartheta}) + \mathrm{Err} \\
\mathrm{Err} := & \ \beta\,\not{d}_1^\star(\underline{\kappa} - 2\underline{\omega}) - 3\eta\,\not{d}_1^\star\rho - \frac{3}{2}\vartheta\,\not{d}_1\underline{\beta} - \not{d}_1\underline{\xi}\alpha \\
& - \frac{3}{2}\rho\mathrm{Err}[e_3(\rho)] + e_3\mathrm{Err}[e_3(\alpha)] + \left(\frac{3}{2}\underline{\kappa} - 2\underline{\omega}\right)\mathrm{Err}[e_3(\alpha)] + \underline{Com_2^*}(\beta) + \mathrm{l.o.t.}
\end{aligned}$$

It remains to analyze the lower order terms according to our convention in Definition

2.94. Note that we can write the first line in the expression of Err

$$\begin{aligned}\mathrm{Err}_1 &= r^{-1}\Gamma_b\cdot\beta + r^{-2}\Gamma_g\cdot \not{\mathfrak{d}}\Gamma_g + r^{-2}\Gamma_g\not{\mathfrak{d}}\Gamma_b + r^{-1}\not{\mathfrak{d}}\Gamma_b\cdot\alpha\\ &= r^{-1}\not{\mathfrak{d}}^{\le 1}\Gamma_b\cdot\beta + r^{-2}\Gamma_g\not{\mathfrak{d}}\Gamma_b + \text{l.o.t.}\end{aligned}$$

On the other hand,

$$\begin{aligned}\mathrm{Err}[e_3(\rho)] &= -\frac{1}{2}\vartheta\,\underline{\alpha} - \zeta\,\underline{\beta} + 2(\eta\,\underline{\beta} + \underline{\xi}\,\beta) = \Gamma_g\cdot\Gamma_b + \Gamma_b\cdot\beta,\\ \mathrm{Err}[e_3(\alpha)] &= (\zeta+4\eta)\beta = \Gamma_g\cdot\beta,\\ e_3\mathrm{Err}[e_3(\alpha)] &= e_3(\zeta+4\eta)\beta + (\zeta+4\eta)e_3(\beta)\\ &= e_3(\zeta+4\eta)\cdot\beta + (\zeta+4\eta)(-\underline{\kappa}\beta - \not{d}_1^\star\rho + 2\underline{\omega}\beta + 3\eta\rho)\\ &= e_3(\zeta+4\eta)\cdot\beta + r^{-1}\Gamma_g\beta + r^{-2}\Gamma_g\not{\mathfrak{d}}\Gamma_g + r^{-3}\Gamma_g\cdot\Gamma_g,\end{aligned}$$

$$\begin{aligned}\underline{Com_2^*}(\beta) &= -\frac{1}{2}\underline{\vartheta}\,\not{d}_1\beta - (\zeta-\eta)e_3\beta - \eta e_3\Phi\beta + \underline{\xi}(e_4\beta - e_4(\Phi)\beta) - \underline{\beta}\cdot\beta\\ &= -\frac{1}{2}\underline{\vartheta}\,\not{d}_1\beta - (\zeta-\eta)\left(-\underline{\kappa}\beta - \not{d}_1^\star\rho + 2\underline{\omega}\,\beta + 3\eta\rho\right) - \eta e_3(\Phi)\beta\\ &+ \underline{\xi}\left(-2\kappa\beta + \not{d}_2^\star\alpha - 2\omega\beta\right) - \underline{\xi}e_4\Phi\beta - \beta\cdot\underline{\beta} + \text{l.o.t.}\\ &= r^{-1}\Gamma_b\cdot\not{\mathfrak{d}}^{\le 1}\beta + r^{-2}\Gamma_g\cdot\not{\mathfrak{d}}\Gamma_b + \text{l.o.t.}\end{aligned}$$

Therefore, schematically,

$$\mathrm{Err} = e_3(\zeta+4\eta)\cdot\beta + r^{-1}\Gamma_b\not{\mathfrak{d}}^{\le 1}\beta + r^{-2}\Gamma_g\not{\mathfrak{d}}\Gamma_b + \text{l.o.t.}$$

and therefore,

$$\mathrm{Err}[\mathfrak{q}] = r^4\mathrm{Err} = r^4\left(e_3(\zeta+4\eta)\cdot\beta + r^{-1}\Gamma_b\cdot\not{\mathfrak{d}}^{\le 1}\beta + r^{-2}\Gamma_g\cdot\not{\mathfrak{d}}\Gamma_b\right) + \text{l.o.t.}$$

Since $e_3\zeta \in r^{-1}\mathfrak{d}\Gamma_b$ and $\beta\in r^{-1}\Gamma_g$ we rewrite in the form

$$\mathrm{Err}[\mathfrak{q}] = r^4 e_3\eta\cdot\beta + r^2\mathfrak{d}^{\le 1}(\Gamma_b\cdot\Gamma_g).$$

This concludes the proof of Proposition 2.99.

A.11 PROOF OF PROPOSITION 2.100

We start with the formula (2.3.11)

$$r\mathfrak{q} = r^5\left(\not{d}_2^\star\not{d}_1^\star\rho + \frac{3}{4}\underline{\kappa}\rho\vartheta + \frac{3}{4}\kappa\rho\underline{\vartheta}\right) + r\mathrm{Err}[\mathfrak{q}]$$

with Err[$\mathfrak{q}$] given by (2.3.12). Taking the e_3 derivative we deduce

$$\begin{aligned}e_3(r\mathfrak{q}) &= r^5 L + 5e_3(r)\,\mathfrak{q} + e_3(r\mathrm{Err}[\mathfrak{q}]) - 5e_3(r)\mathrm{Err}[\mathfrak{q}],\\ L &:= e_3\left\{\not{d}_2^\star\not{d}_1^\star\rho + \frac{3}{4}e_3(\underline{\kappa}\rho\vartheta) + \frac{3}{4}e_3(\kappa\rho\underline{\vartheta})\right\}.\end{aligned} \tag{A.11.1}$$

We calculate L as follows:

$$\begin{aligned} L &= e_3 \,\rlap{/}d_2^\star \,\rlap{/}d_1^\star \rho + \frac{3}{4} e_3(\underline{\kappa}\rho\vartheta) + \frac{3}{4} e_3(\kappa\rho\underline{\vartheta}) \\ &= \,\rlap{/}d_2^\star \,\rlap{/}d_1^\star e_3(\rho) + [e_3, \,\rlap{/}d_2^\star \,\rlap{/}d_1^\star]\rho + \frac{3}{4} e_3(\underline{\kappa}\rho\vartheta) + \frac{3}{4} e_3(\kappa\rho\underline{\vartheta}). \end{aligned}$$

Ignoring cubic and higher order terms

$$\begin{aligned} e_3(\underline{\kappa}\rho\vartheta) &= \underline{\kappa}\rho e_3(\vartheta) + e_3(\underline{\kappa})\rho\vartheta + \underline{\kappa} e_3(\rho)\vartheta \\ &= \underline{\kappa}\rho\left(-\frac{1}{2}\underline{\kappa}\,\vartheta + 2\underline{\omega}\vartheta - 2\,\rlap{/}d_2^\star\,\eta - \frac{1}{2}\kappa\,\underline{\vartheta} + \mathrm{Err}[e_3\vartheta]\right) \\ &\quad + \left(-\frac{1}{2}\underline{\kappa}^2 - 2\underline{\omega}\,\underline{\kappa} + 2\,\rlap{/}d_1\underline{\xi} + \mathrm{Err}[e_3\underline{\kappa}]\right)\rho\vartheta \\ &\quad + \underline{\kappa}\left(-\frac{3}{2}\underline{\kappa}\rho + \,\rlap{/}d_1\underline{\beta} - \frac{1}{2}\vartheta\,\underline{\alpha} + \mathrm{Err}[e_3(\rho)]\right)\vartheta \\ &= \underline{\kappa}\rho\left(-\frac{5}{2}\underline{\kappa}\,\vartheta - 2\,\rlap{/}d_2^\star\,\eta - \frac{1}{2}\kappa\,\underline{\vartheta}\right) + \underline{\kappa}\rho\mathrm{Err}[e_3\vartheta] + 2\,\rlap{/}d_1\underline{\xi}\,\rho\vartheta + \underline{\kappa}(\,\rlap{/}d_1\underline{\beta})\vartheta. \end{aligned}$$

We deduce

$$\begin{aligned} e_3(\underline{\kappa}\rho\vartheta) &= \underline{\kappa}\rho\left(-\frac{5}{2}\underline{\kappa}\,\vartheta - 2\,\rlap{/}d_2^\star\,\eta - \frac{1}{2}\kappa\,\underline{\vartheta}\right) + E_1, \\ E_1 &= 2\,\rlap{/}d_1\underline{\xi}\,\rho\vartheta + \underline{\kappa}(\,\rlap{/}d_1\underline{\beta})\vartheta + \underline{\kappa}\rho\mathrm{Err}[e_3(\vartheta)], \end{aligned} \tag{A.11.2}$$

where $\mathrm{Err}[e_3(\vartheta)]$, $\mathrm{Err}[e_3(\underline{\kappa})]$, $\mathrm{Err}[e_3(\rho)]$ denote the quadratic error terms in the corresponding equations. Also,

$$\begin{aligned} e_3(\kappa\rho\underline{\vartheta}) &= \kappa\rho e_3(\underline{\vartheta}) + e_3(\kappa)\rho\underline{\vartheta} + \kappa e_3(\rho)\underline{\vartheta} \\ &= \kappa\rho\Big(-\underline{\kappa}\,\underline{\vartheta} - 2\underline{\omega}\,\underline{\vartheta} - 2\underline{\alpha} - 2\,\rlap{/}d_2^\star\,\underline{\xi} + \mathrm{Err}[e_3(\underline{\vartheta})]\Big) \\ &\quad + \left(-\frac{1}{2}\underline{\kappa}\,\kappa + 2\underline{\omega}\kappa + 2\,\rlap{/}d_1\eta + 2\rho + \mathrm{Err}[e_3(\kappa)]\right)\rho\underline{\vartheta} \\ &\quad + \left(-\frac{3}{2}\underline{\kappa}\rho + \,\rlap{/}d_1\underline{\beta} + \mathrm{Err}[e_3(\rho)]\right)\kappa\underline{\vartheta}. \end{aligned}$$

Hence, ignoring the higher order terms,

$$\begin{aligned} e_3(\kappa\rho\underline{\vartheta}) &= \kappa\rho\Big(-3\underline{\kappa}\,\underline{\vartheta} - 2\underline{\alpha} - 2\,\rlap{/}d_2^\star\,\underline{\xi}\Big) + 2\rho^2\underline{\vartheta} + E_2, \\ E_2 &:= 2\rho\,\rlap{/}d_1\eta\underline{\vartheta} + \kappa\,\rlap{/}d_1\underline{\beta}\underline{\vartheta} + \kappa\rho\,\mathrm{Err}[e_3(\underline{\vartheta})]. \end{aligned} \tag{A.11.3}$$

Also, we have

$$\begin{aligned} \,\rlap{/}d_2^\star \,\rlap{/}d_1^\star e_3(\rho) &= \,\rlap{/}d_2^\star \,\rlap{/}d_1^\star\left(-\frac{3}{2}\underline{\kappa}\rho + \,\rlap{/}d_1\underline{\beta} + \mathrm{Err}[e_3(\rho)]\right) \\ &= \,\rlap{/}d_2^\star \,\rlap{/}d_1^\star \,\rlap{/}d_1\underline{\beta} - \frac{3}{2}\rho\,\rlap{/}d_2^\star \,\rlap{/}d_1^\star\underline{\kappa} - \frac{3}{2}\underline{\kappa}\,\rlap{/}d_2^\star \,\rlap{/}d_1^\star\rho + E_3, \\ E_3 &= \,\rlap{/}d_2^\star \,\rlap{/}d_1^\star\mathrm{Err}[e_3(\rho)] - 3\,\rlap{/}d_1^\star\underline{\kappa}\cdot\,\rlap{/}d_1^\star\rho, \end{aligned}$$

i.e.,

$$
\begin{aligned}
e_3(\underline{\kappa}\rho\vartheta) &= \underline{\kappa}\rho\left(-\frac{5}{2}\underline{\kappa}\,\vartheta - 2\,{\not{d}}_2^\star\,\eta - \frac{1}{2}\kappa\,\underline{\vartheta}\right) + E_1,\\
e_3(\kappa\rho\underline{\vartheta}) &= \kappa\rho\Big(-3\underline{\kappa}\,\underline{\vartheta} - 2\underline{\alpha} - 2\,{\not{d}}_2^\star\,\underline{\xi}\Big) + 2\rho^2\underline{\vartheta} + E_2,\\
{\not{d}}_2^\star\,{\not{d}}_1^\star e_3(\rho) &= {\not{d}}_2^\star\,{\not{d}}_1^\star\,{\not{d}}_1\underline{\beta} - \frac{3}{2}\rho\,{\not{d}}_2^\star\,{\not{d}}_1^\star\underline{\kappa} - \frac{3}{2}\underline{\kappa}\,{\not{d}}_2^\star\,{\not{d}}_1^\star\rho + E_3.
\end{aligned}
\tag{A.11.4}
$$

Now, in view of Lemma 2.54 we have (for $f = {\not{d}}_1^\star\rho \in \mathfrak{s}_{2-1}$),

$$
[e_3, {\not{d}}_2^\star]\,{\not{d}}_1^\star\rho \quad = \quad -\frac{1}{2}\underline{\kappa}\,{\not{d}}_2^\star\,{\not{d}}_1^\star\rho - \underline{Com}_2^*({\not{d}}_1^\star\rho)
$$

and

$$
\begin{aligned}
[e_3, {\not{d}}_1^\star]\rho &= -[e_3, e_\theta]\rho = -\frac{1}{2}\underline{\kappa}\,{\not{d}}_1^\star\rho - \frac{1}{2}\underline{\vartheta}e_\theta\rho + (\zeta-\eta)e_3\rho - \underline{\xi}e_4\rho\\
&= -\frac{1}{2}\underline{\kappa}\,{\not{d}}_1^\star\rho - \frac{1}{2}\underline{\vartheta}e_\theta\rho + (\zeta-\eta)\left(-\frac{3}{2}\underline{\kappa}\rho + {\not{d}}_1\underline{\beta} + \mathrm{Err}[e_3(\rho)]\right)\\
&- \underline{\xi}\left(-\frac{3}{2}\kappa\rho + {\not{d}}_1\beta + \mathrm{Err}[e_4(\rho)]\right)\\
&= -\frac{1}{2}\underline{\kappa}\,{\not{d}}_1^\star\rho - \frac{3}{2}\rho\left[(\zeta-\eta)\underline{\kappa} - \underline{\xi}\kappa\right] + E_{41},\\
E_{41} &= (\zeta-\eta)\,{\not{d}}_1\underline{\beta} - \underline{\xi}\,{\not{d}}_1\beta + (\zeta-\eta)e_3[(\rho)] - \frac{1}{2}\underline{\vartheta}e_\theta\rho - \underline{\xi}\mathrm{Err}[e_4(\rho)].
\end{aligned}
$$

We deduce

$$
\begin{aligned}
{\not{d}}_2^\star[e_3, {\not{d}}_1^\star]\rho &= {\not{d}}_2^\star\Big(-\frac{1}{2}\underline{\kappa}\,{\not{d}}_1^\star\rho - \frac{3}{2}\rho\left[\underline{\kappa}(\zeta-\eta) - \kappa\underline{\xi}\right] + E_{41}\Big)\\
&= -\frac{1}{2}\underline{\kappa}\,{\not{d}}_2^\star\,{\not{d}}_1^\star\rho - \frac{3}{2}\rho\Big(\underline{\kappa}\,{\not{d}}_2^\star(\zeta-\eta) - \kappa\,{\not{d}}_2^\star\underline{\xi}\Big) + E_4,\\
E_4 &= {\not{d}}_2^\star E_{41} - \frac{1}{2}\,{\not{d}}_1^\star\underline{\kappa}\cdot{\not{d}}_1^\star\rho - \frac{3}{2}(\zeta-\eta)\,{\not{d}}_1^\star(\underline{\kappa}\rho) + \frac{3}{2}\underline{\xi}\,{\not{d}}_1^\star(\kappa\rho).
\end{aligned}
$$

Hence, since $[e_3, {\not{d}}_2^\star\,{\not{d}}_1^\star]\rho = [e_3, {\not{d}}_2^\star]\,{\not{d}}_1^\star\rho + {\not{d}}_2^\star[e_3, {\not{d}}_1^\star]\rho$,

$$
[e_3, {\not{d}}_2^\star\,{\not{d}}_1^\star]\rho = -\underline{\kappa}\,{\not{d}}_2^\star\,{\not{d}}_1^\star\rho - \frac{3}{2}\big({\not{d}}_2^\star(\zeta-\eta)\underline{\kappa}\rho - {\not{d}}_2^\star\underline{\xi}\kappa\rho\big) - \underline{Com}_2^*({\not{d}}_1^\star\rho) + E_4. \tag{A.11.5}
$$

We deduce, recalling (A.11.4),

$$
\begin{aligned}
L &= {\not{d}}_2^\star {\not{d}}_1^\star e_3(\rho) + [e_3, {\not{d}}_2^\star {\not{d}}_1^\star]\rho + \frac{3}{4} e_3(\underline{\kappa}\rho\vartheta) + \frac{3}{4} e_3(\kappa\rho\underline{\vartheta})\\
&= {\not{d}}_2^\star {\not{d}}_1^\star {\not{d}}_1 \underline{\beta} - \frac{3}{2}\rho {\not{d}}_2^\star {\not{d}}_1^\star \underline{\kappa} - \frac{3}{2}\underline{\kappa} {\not{d}}_2^\star {\not{d}}_1^\star \rho + E_3\\
&- \underline{\kappa} {\not{d}}_2^\star {\not{d}}_1^\star \rho - \frac{3}{2}\Big({\not{d}}_2^\star(\zeta - \eta)\underline{\kappa}\rho - {\not{d}}_2^\star \underline{\xi}\kappa\rho\Big) - \underline{Com}_2^*({\not{d}}_1^\star \rho) + E_4\\
&+ \frac{3}{4}\left(\underline{\kappa}\rho\left(-\frac{5}{2}\underline{\kappa}\vartheta - 2\,{\not{d}}_2^\star\,\eta - \frac{1}{2}\kappa\underline{\vartheta}\right)\right) + E_1\\
&+ \frac{3}{4}\Big(\kappa\rho\Big(- 3\underline{\kappa}\,\underline{\vartheta} - 2\underline{\alpha} - 2\,{\not{d}}_2^\star\,\underline{\xi}\Big) + 2\rho^2\underline{\vartheta} + \mathrm{Err}_1\Big)\\
&= {\not{d}}_2^\star {\not{d}}_1^\star {\not{d}}_1 \underline{\beta} - \frac{5}{2}\underline{\kappa} {\not{d}}_2^\star {\not{d}}_1^\star \rho - \frac{3}{2}\rho {\not{d}}_2^\star {\not{d}}_1^\star \underline{\kappa} - \frac{3}{2}\ \rho\underline{\kappa}\,{\not{d}}_2^\star \zeta\\
&+ \frac{3}{4}\underline{\kappa}\rho\left(-\frac{5}{2}\underline{\kappa}\vartheta - \frac{1}{2}\kappa\underline{\vartheta}\right) + \frac{3}{4}\kappa\rho\Big(- 3\underline{\kappa}\,\underline{\vartheta} - 2\underline{\alpha}\Big) + \frac{3}{2}\rho^2\underline{\vartheta}\\
&+ E_1 + E_2 + E_3 + E_4 - \underline{Com}_2^*({\not{d}}_1^\star \rho),
\end{aligned}
$$

i.e.,

$$
\begin{aligned}
L &= {\not{d}}_2^\star {\not{d}}_1^\star {\not{d}}_1 \underline{\beta} - \frac{5}{2}\underline{\kappa} {\not{d}}_2^\star {\not{d}}_1^\star \rho - \frac{3}{2}\rho {\not{d}}_2^\star {\not{d}}_1^\star \underline{\kappa} - \frac{3}{2}\ \rho\underline{\kappa}\,{\not{d}}_2^\star \zeta\\
&- \frac{3}{4}\underline{\kappa}\rho\left(\frac{5}{2}\underline{\kappa}\vartheta + \frac{1}{2}\kappa\underline{\vartheta}\right) - \frac{3}{4}\kappa\rho\,(3\underline{\kappa}\,\underline{\vartheta} + 2\underline{\alpha}) + \frac{3}{2}\rho^2\underline{\vartheta} + E,\\
E &= E_1 + E_2 + E_3 + E_4 - \underline{Com}_2^*({\not{d}}_1^\star \rho).
\end{aligned}
\tag{A.11.6}
$$

On the other hand, in view of (2.3.11), writing $e_3 r = \frac{r}{2}\left(\underline{\kappa} + \underline{A}\right)$,

$$
\begin{aligned}
5e_3(r)\mathfrak{q} &= r\frac{5}{2}\underline{\kappa}\mathfrak{q} + 5r\underline{A}\mathfrak{q}\\
&= \frac{5}{2}r^5\underline{\kappa}\left\{ {\not{d}}_2^\star {\not{d}}_1^\star \rho + \frac{3}{4}\underline{\kappa}\rho\vartheta + \frac{3}{4}\kappa\rho\underline{\vartheta} + \mathrm{Err}\right\} + 5r\underline{A}\mathfrak{q}.
\end{aligned}
$$

Hence, in view of (A.11.1) and (A.11.6),

$$
\begin{aligned}
e_3(r\mathfrak{q}) &= r^5 L + 5e_3(r)\mathfrak{q} + e_3(r\mathrm{Err}) - 5e_3(r)\mathrm{Err}\\
&= r^5\left\{\left({\not{d}}_2^\star {\not{d}}_1^\star {\not{d}}_1 \underline{\beta} - \frac{5}{2}\underline{\kappa} {\not{d}}_2^\star {\not{d}}_1^\star \rho - \frac{3}{2}\rho {\not{d}}_2^\star {\not{d}}_1^\star \underline{\kappa} - \frac{3}{2}\ \rho\underline{\kappa}\,{\not{d}}_2^\star \zeta\right)\right.\\
&\left. - \frac{3}{4}\underline{\kappa}\rho\left(\frac{5}{2}\underline{\kappa}\vartheta + \frac{1}{2}\kappa\underline{\vartheta}\right) - \frac{3}{4}\kappa\rho\,(3\underline{\kappa}\,\underline{\vartheta} + 2\underline{\alpha}) + \frac{3}{2}\rho^2\underline{\vartheta}\right\}\\
&+ \frac{5}{2}r^5\underline{\kappa}\left\{ {\not{d}}_2^\star {\not{d}}_1^\star \rho + \frac{3}{4}\underline{\kappa}\rho\vartheta + \frac{3}{4}\kappa\rho\underline{\vartheta}\right\} + r^5 E + e_3(r^5\mathrm{Err}) - 5\frac{e_3 r}{r} r^5\mathrm{Err}\\
&+ 5r\underline{A}\mathfrak{q}\\
&= r^5\left({\not{d}}_2^\star {\not{d}}_1^\star {\not{d}}_1 \underline{\beta} - \frac{3}{2}\rho {\not{d}}_2^\star {\not{d}}_1^\star \underline{\kappa} - \frac{3}{2}\underline{\kappa}\rho\,{\not{d}}_2^\star \zeta - \frac{3}{2}\kappa\rho\underline{\alpha} + \frac{3}{4}(2\rho^2 - \kappa\underline{\kappa}\rho)\underline{\vartheta}\right)\\
&+ \mathrm{Err}[e_3(r\mathfrak{q})]
\end{aligned}
$$

where

$$\begin{aligned}\mathrm{Err}[e_3(r\mathfrak{q})] &= e_3(r\mathrm{Err}[\mathfrak{q}]) - 5e_3 r\mathrm{Err}[\mathfrak{q}] + 5r\underline{A}\mathfrak{q} + r^5 E\\ &= re_3(\mathrm{Err}[\mathfrak{q}]) + \mathrm{Err}[\mathfrak{q}] + 5r\underline{A}\mathfrak{q} + r^5 E \end{aligned} \tag{A.11.7}$$

and

$$E = E_1 + E_2 + E_3 + E_4 - \underline{Com}_2^*(\not{d}_1^\star \rho)$$

with

$$\begin{aligned}
E_1 &= 2\not{d}_1\underline{\xi}\,\rho\vartheta + \underline{\kappa}(\not{d}_1\underline{\beta})\vartheta + \underline{\kappa}\rho\mathrm{Err}[e_3(\underline{\kappa})],\\
E_2 &= 2\rho\not{d}_1\eta\underline{\vartheta} + \kappa\not{d}_1\beta\underline{\vartheta} + \kappa\rho\,\mathrm{Err}[e_3(\underline{\vartheta})],\\
E_3 &= \not{d}_2^\star\not{d}_1^\star\mathrm{Err}[e_3(\rho)] - 3\not{d}_1^\star\underline{\kappa}\cdot\not{d}_1^\star\rho,\\
E_4 &= \not{d}_2^\star E_{41} - \frac{1}{2}\not{d}_1^\star\underline{\kappa}\cdot\not{d}_1^\star\rho - \frac{3}{2}(\zeta-\eta)\not{d}_1^\star(\underline{\kappa}\rho) + \frac{3}{2}\underline{\xi}\not{d}_1^\star(\kappa\rho),\\
E_{41} &= (\zeta-\eta)\not{d}_1\underline{\beta} - \underline{\xi}\not{d}_1\beta - \frac{1}{2}\underline{\vartheta}e_\theta\rho,\\
\underline{Com}_2^*(\not{d}_1^\star\rho) &= -\frac{1}{2}\underline{\vartheta}\not{d}_1\not{d}_1^\star\rho - (\zeta-\eta)e_3\not{d}_1^\star\rho - \eta e_3\Phi\not{d}_1^\star\rho + \underline{\xi}(e_4\not{d}_1^\star\rho - e_4(\Phi)\not{d}_1^\star\rho)\\
&\quad - \underline{\beta}\not{d}_1^\star\rho.
\end{aligned}$$

Note also that

$$\begin{aligned}(\zeta-\eta)e_3\not{d}_1^\star\rho &= (\zeta-\eta)\cdot\not{d}_1^\star e_3\rho\\ &\quad +(\zeta-\eta)\left(-\frac{1}{2}\underline{\kappa}\not{d}_1^\star\rho - \frac{1}{2}\underline{\vartheta}e_\theta\rho + (\zeta-\eta)e_3\rho - \underline{\xi}e_4\check{\rho}\right).\end{aligned}$$

Using our schematic notation

$$\begin{aligned}
\mathrm{Err}[e_3(\underline{\kappa})] &= \Gamma_b\cdot\Gamma_b + \mathrm{l.o.t.},\\
\mathrm{Err}[e_3(\underline{\vartheta})] &= \Gamma_b\cdot\Gamma_b + \mathrm{l.o.t.},\\
\mathrm{Err}[e_3(\rho)] &= \Gamma_g\cdot\underline{\alpha} + \mathrm{l.o.t.} = \Gamma_g\cdot\Gamma_b + \mathrm{l.o.t.},
\end{aligned}$$

and

$$\begin{aligned}
E_1 &= r^{-4}\Gamma_b\cdot\not{\mathfrak{d}}^{\le 1}\Gamma_b + \mathrm{l.o.t.},\\
E_2 &= r^{-4}\Gamma_b\cdot\not{\mathfrak{d}}^{\le 1}\Gamma_b + r^{-2}\Gamma_b\cdot\beta + \mathrm{l.o.t.},\\
E_3 &= r^{-2}\not{\mathfrak{d}}^2(\Gamma_g\cdot\Gamma_b) + r^{-3}(\not{\mathfrak{d}}\Gamma_g)\cdot(\not{\mathfrak{d}}\Gamma_g),\\
E_{41} &= r^{-2}\Gamma_g\cdot(\not{\mathfrak{d}}\Gamma_b) + r^{-1}\Gamma_b\cdot\not{\mathfrak{d}}\beta + r^{-2}\Gamma_b\not{\mathfrak{d}}\cdot\Gamma_g\\
&= r^{-2}\not{\mathfrak{d}}(\Gamma_g\cdot\Gamma_b) + \mathrm{l.o.t.},\\
E_4 &= r^{-3}\not{\mathfrak{d}}^2(\Gamma_g\cdot\Gamma_b) + \mathrm{l.o.t.},\\
\underline{Com}_2^*(\not{d}_1^\star\rho) &= r^{-3}\Gamma_b\cdot\mathfrak{d}^{\le 2}\Gamma_g + r^{-2}\not{\mathfrak{d}}\Gamma_b\cdot\Gamma_g + \mathrm{l.o.t.},
\end{aligned}$$

and, since $r^{-1}\Gamma_b$ can be replaced by Γ_g and $\not{\mathfrak{d}}\beta$ can be replaced by $r^{-1}\Gamma_g$,

$$E = r^{-3}\not{\mathfrak{d}}^2(\Gamma_g\cdot\Gamma_b) + \mathrm{l.o.t.}$$

Taking into account the expression of Err[$\mathfrak{q}$] in Proposition 2.99 we write

$$\begin{aligned} re_3(\text{Err}[\mathfrak{q}]) + \text{Err}[\mathfrak{q}] &= re_3\Big[r^4 e_3\eta\cdot\beta + r^2\mathfrak{d}^{\leq 1}\big(\Gamma_b\cdot\Gamma_g\big)\Big] + r^4 e_3\eta\cdot\beta \\ &\quad + r^2\mathfrak{d}^{\leq 1}\big(\Gamma_b\cdot\Gamma_g\big) \\ &= r^5\mathfrak{d}^{\leq 1}\big(e_3\eta\cdot\beta\big) + r^3\mathfrak{d}^{\leq 2}\big(\Gamma_b\cdot\Gamma_g\big) \end{aligned}$$

and therefore, back to (A.11.7),

$$\begin{aligned} \text{Err}[e_3(r\mathfrak{q})] &= e_3(r\text{Err}[\mathfrak{q}]) + \text{Err}[\mathfrak{q}] + r\underline{A}\mathfrak{q} + r^5 E \\ &= r^5\mathfrak{d}^{\leq 1}\big(e_3\eta\cdot\beta\big) + r^3\mathfrak{d}^{\leq 2}\big(\Gamma_b\cdot\Gamma_g\big) + r\Gamma_b\mathfrak{q} + r^2\,\not{\mathfrak{d}}^2(\Gamma_g\cdot\Gamma_b) + \text{l.o.t.} \\ &= r\Gamma_b\mathfrak{q} + r^5\mathfrak{d}^{\leq 1}\big(e_3\eta\cdot\beta\big) + r^3\mathfrak{d}^{\leq 2}\big(\Gamma_b\cdot\Gamma_g\big). \end{aligned}$$

This concludes the proof of Proposition 2.100.

A.12 PROOF OF THE TEUKOLSKY-STAROBINSKY IDENTITY

According to Proposition 2.100 we have

$$\begin{aligned} e_3(r\mathfrak{q}) &= r^5\left\{\not{d}_2^\star\not{d}_1^\star\not{d}_1\underline{\beta} - \frac{3}{2}\rho\not{d}_2^\star\not{d}_1^\star\underline{\kappa} - \frac{3}{2}\underline{\kappa}\rho\not{d}_2^\star\zeta - \frac{3}{2}\kappa\rho\underline{\alpha} + \frac{3}{4}(2\rho^2 - \kappa\underline{\kappa}\rho)\underline{\vartheta}\right\} \\ &\quad + \text{Err}[e_3\mathfrak{q}]. \end{aligned}$$

We infer that

$$\begin{aligned} e_3(r^2 e_3(r\mathfrak{q})) = r^7 &\left\{ e_3\not{d}_2^\star\not{d}_1^\star\not{d}_1\underline{\beta} - \frac{3}{2}e_3(\rho\not{d}_2^\star\not{d}_1^\star\underline{\kappa}) - \frac{3}{2}e_3(\underline{\kappa}\rho\not{d}_2^\star\zeta) - \frac{3}{2}e_3(\kappa\rho\underline{\alpha})\right. \\ &\left. + \frac{3}{4}e_3\Big((2\rho^2 - \kappa\underline{\kappa}\rho)\underline{\vartheta}\Big)\right\} + 7re_3(r)e_3(r\mathfrak{q}) \\ &+ r^2 e_3\big(\text{Err}[e_3\mathfrak{q}]\big) + r\text{Err}[e_3\mathfrak{q}] + \text{l.o.t.} \end{aligned} \tag{A.12.1}$$

We first compute

$$\begin{aligned} & e_3\not{d}_2^\star\not{d}_1^\star\not{d}_1\underline{\beta} \\ = \;& \not{d}_2^\star\not{d}_1^\star\not{d}_1(e_3\underline{\beta}) + [e_3, \not{d}_2^\star]\not{d}_1^\star\not{d}_1\underline{\beta} + \not{d}_2^\star[e_3, \not{d}_1^\star]\not{d}_1\underline{\beta} + \not{d}_2^\star\not{d}_1^\star[e_3, \not{d}_1]\underline{\beta} \\ = \;& \not{d}_2^\star\not{d}_1^\star\not{d}_1(e_3\underline{\beta}) + \Bigg(-\frac{1}{2}\underline{\kappa}\not{d}_2^\star + \frac{1}{2}\underline{\vartheta}\not{d}_1 + (\zeta-\eta)e_3 + \eta e_3\Phi - \underline{\xi}(e_4 - e_4(\Phi)) \\ & + \underline{\beta}\Bigg)\not{d}_1^\star\not{d}_1\underline{\beta} + \not{d}_2^\star\left(\left(-\frac{1}{2}\underline{\kappa}\not{d}_1^\star + \frac{1}{2}\underline{\vartheta}\not{d}_0 + (\zeta-\eta)e_3 - \underline{\xi}e_4\right)\not{d}_1\underline{\beta}\right) \\ & + \not{d}_2^\star\not{d}_1^\star\left(\left(-\frac{1}{2}\underline{\kappa}\not{d}_1 + \frac{1}{2}\underline{\vartheta}\not{d}_2^\star - (\zeta-\eta)e_3 + \eta e_3\Phi + \underline{\xi}(e_4 + e_4(\Phi)) + \underline{\beta}\right)\underline{\beta}\right). \end{aligned}$$

In view of our general commutation formulas in Lemma 2.54 and our notation convention for error terms,[2] we have[3]

$$
\begin{aligned}
&[e_3, \slashed{d}_2^\star]\, \slashed{d}_1^\star\, \slashed{d}_1 \underline{\beta}\\
&= \left(-\frac{1}{2}\underline{\kappa}\, \slashed{d}_2^\star + \frac{1}{2}\underline{\vartheta}\, \slashed{d}_1 + (\zeta - \eta) e_3 + \eta e_3 \Phi - \underline{\xi}(e_4 - e_4(\Phi)) + \underline{\beta}\right) \slashed{d}_1^\star\, \slashed{d}_1 \underline{\beta}\\
&= -\frac{1}{2}\underline{\kappa}\, \slashed{d}_2^\star\, \slashed{d}_1^\star\, \slashed{d}_1 \underline{\beta} + r^{-4}\Gamma_b \cdot \slashed{\mathfrak{d}}^{\leq 3}\Gamma_b + \text{l.o.t.},
\end{aligned}
$$

$$
\begin{aligned}
\slashed{d}_2^\star [e_3, \slashed{d}_1^\star]\, \slashed{d}_1 \underline{\beta} &= \slashed{d}_2^\star \left(\left(-\frac{1}{2}\underline{\kappa}\, \slashed{d}_1^\star + \frac{1}{2}\underline{\vartheta}\, \slashed{d}_0 + (\zeta - \eta) e_3 - \underline{\xi} e_4\right) \slashed{d}_1 \underline{\beta}\right)\\
&= -\frac{1}{2}\underline{\kappa}\, \slashed{d}_2^\star\, \slashed{d}_1^\star\, \slashed{d}_1 \underline{\beta} + r^{-4}\left(\Gamma_b \cdot \slashed{\mathfrak{d}}^{\leq 3}\Gamma_b + \Gamma_b^{\leq 1} \cdot \slashed{\mathfrak{d}}^{\leq 2}\Gamma_b\right) + \text{l.o.t.},\\
\slashed{d}_2^\star\, \slashed{d}_1^\star [e_3, \slashed{d}_1] \underline{\beta} &= \slashed{d}_2^\star\, \slashed{d}_1^\star \Bigg(\Bigg(-\frac{1}{2}\underline{\kappa}\, \slashed{d}_1 + \frac{1}{2}\underline{\vartheta}\, \slashed{d}_2^\star - (\zeta - \eta) e_3 + \eta e_3 \Phi + \underline{\xi}(e_4 + e_4(\Phi))\\
&\quad + \underline{\beta}\Bigg) \underline{\beta}\Bigg)\\
&= -\frac{1}{2}\underline{\kappa}\, \slashed{d}_2^\star\, \slashed{d}_1^\star\, \slashed{d}_1 \underline{\beta} + r^{-4}\left(\Gamma_b \cdot \slashed{\mathfrak{d}}^{\leq 3}\Gamma_b + \Gamma_b^{\leq 1} \cdot \slashed{\mathfrak{d}}^{\leq 2}\Gamma_b\right) + \text{l.o.t.}
\end{aligned}
$$

Hence, schematically,

$$
e_3\, \slashed{d}_2^\star\, \slashed{d}_1^\star\, \slashed{d}_1 \underline{\beta} = \slashed{d}_2^\star\, \slashed{d}_1^\star\, \slashed{d}_1 (e_3 \underline{\beta}) - \frac{3}{2}\underline{\kappa}\, \slashed{d}_2^\star\, \slashed{d}_1^\star\, \slashed{d}_1 \underline{\beta} + r^{-4}\, \slashed{\mathfrak{d}}^{\leq 3}(\Gamma_b \cdot \Gamma_b).
$$

Using the Bianchi identity $e_3 \underline{\beta} = \slashed{d}_2 \underline{\alpha} - 2(\underline{\kappa} + \underline{\omega})\, \underline{\beta} + (-2\zeta + \eta)\underline{\alpha} + 3\underline{\xi}\rho$, we further deduce

$$
\slashed{d}_2^\star\, \slashed{d}_1^\star\, \slashed{d}_1 (e_3 \underline{\beta}) = \slashed{d}_2^\star\, \slashed{d}_1^\star\, \slashed{d}_1\, \slashed{d}_2 \underline{\alpha} - 2(\underline{\kappa} + \underline{\omega})\, \slashed{d}_2^\star\, \slashed{d}_1^\star\, \slashed{d}_1 \underline{\beta} + 3\rho\, \slashed{d}_2^\star\, \slashed{d}_1^\star\, \slashed{d}_1 \underline{\xi} + r^{-4}\, \slashed{\mathfrak{d}}^{\leq 3}(\Gamma_b \cdot \Gamma_b),
$$

i.e.,

$$
\begin{aligned}
e_3\, \slashed{d}_2^\star\, \slashed{d}_1^\star\, \slashed{d}_1 \underline{\beta} = {}& \slashed{d}_2^\star\, \slashed{d}_1^\star\, \slashed{d}_1\, \slashed{d}_2 \underline{\alpha} - 2(\underline{\kappa} + \underline{\omega})\, \slashed{d}_2^\star\, \slashed{d}_1^\star\, \slashed{d}_1 \underline{\beta} + 3\rho\, \slashed{d}_2^\star\, \slashed{d}_1^\star\, \slashed{d}_1 \underline{\xi}\\
& - \frac{3}{2}\underline{\kappa}\, \slashed{d}_2^\star\, \slashed{d}_1^\star\, \slashed{d}_1 \underline{\beta} + r^{-4}\, \slashed{\mathfrak{d}}^{\leq 3}(\Gamma_b \cdot \Gamma_b).
\end{aligned}
\tag{A.12.2}
$$

We next calculate the second term $e_3(\rho\, \slashed{d}_2^\star\, \slashed{d}_1^\star \underline{\kappa})$ on the right-hand side of (A.12.1)

$$
e_3(\rho\, \slashed{d}_2^\star\, \slashed{d}_1^\star \underline{\kappa}) = \rho\, \slashed{d}_2^\star\, \slashed{d}_1^\star e_3 \underline{\kappa} + \rho [e_3, \slashed{d}_2^\star]\, \slashed{d}_1^\star \underline{\kappa} + \rho\, \slashed{d}_2^\star [e_3, \slashed{d}_1^\star] \underline{\kappa} + e_3(\rho)\, \slashed{d}_2^\star\, \slashed{d}_1^\star \underline{\kappa}.
$$

Using the equation for $e_3 \underline{\kappa}$ in Proposition 2.56 we derive

$$
\begin{aligned}
\rho\, \slashed{d}_2^\star\, \slashed{d}_1^\star e_3 \underline{\kappa} &= \rho\, \slashed{d}_2^\star\, \slashed{d}_1^\star \left(-\frac{1}{2}\underline{\kappa}^2 - 2\underline{\omega}\, \underline{\kappa} + 2\, \slashed{d}_1 \underline{\xi} + \Gamma_b \cdot \Gamma_b\right),\\
&= -\rho\, (\underline{\kappa} + 2\underline{\omega})\, \slashed{d}_2^\star\, \slashed{d}_1^\star \underline{\kappa} - 2\rho \underline{\kappa}\, \slashed{d}_2^\star\, \slashed{d}_1^\star \underline{\omega} + 2\rho\, \slashed{d}_2^\star\, \slashed{d}_1^\star\, \slashed{d}_1 \underline{\xi} + r^{-5}\, \slashed{\mathfrak{d}}^{\leq 2}\Gamma_b \cdot \Gamma_b.
\end{aligned}
$$

Also,

---

[2]In particular we write $\underline{\beta} \in r^{-1}\Gamma_b$.

[3]We also commute once more e_3 and e_4 with $\slashed{d}_1^\star$, $\slashed{d}_1$, $\slashed{d}_2^\star$, $\slashed{d}_2$ and use Bianchi.

$$
\begin{aligned}
[e_3, {\not{d}}_2^\star]\,{\not{d}}_1^\star\underline{\kappa} &= \left(-\frac{1}{2}\underline{\kappa}\,{\not{d}}_2^\star + \frac{1}{2}\underline{\vartheta}\,{\not{d}}_1 + (\zeta-\eta)e_3 + \eta e_3\Phi - \underline{\xi}(e_4 - e_4(\Phi)) + \underline{\beta}\right){\not{d}}_1^\star\underline{\kappa}\\
&= -\frac{1}{2}\underline{\kappa}\,{\not{d}}_2^\star\,{\not{d}}_1^\star\underline{\kappa} + r^{-2}\Gamma_b\cdot{\not{\mathfrak{d}}}^{\leq 2}\Gamma_g,\\
{\not{d}}_2^\star[e_3, {\not{d}}_1^\star]\underline{\kappa} &= {\not{d}}_2^\star\left(\left(-\frac{1}{2}\underline{\kappa}\,{\not{d}}_1^\star - \frac{1}{2}\underline{\vartheta}\,{\not{d}}_1^\star + (\zeta-\eta)e_3 - \underline{\xi}e_4\right)\underline{\kappa}\right)\\
&= -\frac{1}{2}\underline{\kappa}\,{\not{d}}_2^\star\,{\not{d}}_1^\star\underline{\kappa} + {\not{d}}_2^\star(\zeta-\eta)e_3\underline{\kappa} - {\not{d}}_2^\star\underline{\xi}e_4\underline{\kappa} + r^{-2}\,{\not{\mathfrak{d}}}^{\leq 2}(\Gamma_b\cdot\Gamma_g).
\end{aligned}
$$

Using also the Bianchi equation

$$
e_3\rho = -\frac{3}{2}\underline{\kappa}\rho + {\not{d}}_1\underline{\beta} - \frac{1}{2}\vartheta\,\underline{\alpha} - \zeta\,\underline{\beta} + 2(\eta\,\underline{\beta} + \underline{\xi}\,\beta),
$$

we deduce

$$
\begin{aligned}
e_3(\rho\,{\not{d}}_2^\star\,{\not{d}}_1^\star\underline{\kappa}) &= -\rho\,(\underline{\kappa}+2\underline{\omega})\,{\not{d}}_2^\star\,{\not{d}}_1^\star\underline{\kappa} - 2\rho\underline{\kappa}\,{\not{d}}_2^\star\,{\not{d}}_1^\star\underline{\omega} + 2\rho\,{\not{d}}_2^\star\,{\not{d}}_1^\star\,{\not{d}}_1\underline{\xi}\\
&- \rho\underline{\kappa}\,{\not{d}}_2^\star\,{\not{d}}_1^\star\underline{\kappa} + \rho\left({\not{d}}_2^\star(\zeta-\eta)e_3\underline{\kappa} - {\not{d}}_2^\star(\underline{\xi})e_4\underline{\kappa}\right) - \frac{3}{2}\rho\underline{\kappa}\,{\not{d}}_2^\star\,{\not{d}}_1^\star\underline{\kappa}\\
&+ r^{-5}\,{\not{\mathfrak{d}}}^{\leq 2}(\Gamma_b\cdot\Gamma_g),
\end{aligned}
$$

i.e.,

$$
\begin{aligned}
e_3(\rho\,{\not{d}}_2^\star\,{\not{d}}_1^\star\underline{\kappa}) = &-\frac{7}{2}\rho\underline{\kappa}\,{\not{d}}_2^\star\,{\not{d}}_1^\star(\underline{\kappa}) - 2\rho\underline{\omega}\,{\not{d}}_2^\star\,{\not{d}}_1^\star\underline{\kappa} - 2\rho\underline{\kappa}\,{\not{d}}_2^\star\,{\not{d}}_1^\star\underline{\omega} + 2\rho\,{\not{d}}_2^\star\,{\not{d}}_1^\star\,{\not{d}}_1\underline{\xi}\\
&+ \rho\left({\not{d}}_2^\star(\zeta-\eta)e_3\underline{\kappa} - {\not{d}}_2^\star(\underline{\xi})e_4\underline{\kappa}\right) + r^{-5}\,{\not{\mathfrak{d}}}^{\leq 2}(\Gamma_b\cdot\Gamma_g).
\end{aligned}
\tag{A.12.3}
$$

Now,

$$
\begin{aligned}
&{\not{d}}_2^\star(\zeta-\eta)e_3\underline{\kappa} - {\not{d}}_2^\star(\underline{\xi})e_4\underline{\kappa}\\
&= {\not{d}}_2^\star(\zeta-\eta)\left(-\frac{1}{2}\underline{\kappa}^2 - 2\underline{\omega}\,\underline{\kappa} + 2\,{\not{d}}_1\underline{\xi} + \Gamma_b\cdot\Gamma_b\right)\\
&- {\not{d}}_2^\star\underline{\xi}\left(-\frac{1}{2}\kappa\underline{\kappa} + 2\omega\underline{\kappa} + 2\,{\not{d}}_1\underline{\eta} + 2\rho + \Gamma_g\cdot\Gamma_b\right)\\
&= {\not{d}}_2^\star(\zeta-\eta)\left(-\frac{1}{2}\underline{\kappa}^2 - 2\underline{\omega}\,\underline{\kappa}\right) - {\not{d}}_2^\star\underline{\xi}\left(-\frac{1}{2}\kappa\underline{\kappa} + 2\omega\underline{\kappa} + 2\rho\right) + r^{-2}\,{\not{\mathfrak{d}}}^{\leq 1}\Gamma_b\cdot{\not{\mathfrak{d}}}^{\leq 1}\Gamma_b.
\end{aligned}
$$

Therefore, back to (A.12.3),

$$
\begin{aligned}
e_3(\rho\,{\not{d}}_2^\star\,{\not{d}}_1^\star\underline{\kappa}) = &-\frac{7}{2}\rho\underline{\kappa}\,{\not{d}}_2^\star\,{\not{d}}_1^\star(\underline{\kappa}) - 2\rho\underline{\omega}\,{\not{d}}_2^\star\,{\not{d}}_1^\star\underline{\kappa} - 2\rho\underline{\kappa}\,{\not{d}}_2^\star\,{\not{d}}_1^\star\underline{\omega} + 2\rho\,{\not{d}}_2^\star\,{\not{d}}_1^\star\,{\not{d}}_1\underline{\xi}\\
&+ \rho\,{\not{d}}_2^\star(\zeta-\eta)\left(-\frac{1}{2}\underline{\kappa}^2 - 2\underline{\omega}\,\underline{\kappa}\right) - \rho\,{\not{d}}_2^\star\underline{\xi}\left(-\frac{1}{2}\kappa\underline{\kappa} + 2\omega\underline{\kappa} + 2\rho\right)\\
&+ r^{-5}\,{\not{\mathfrak{d}}}^{\leq 2}(\Gamma_b\cdot\Gamma_g).
\end{aligned}
\tag{A.12.4}
$$

We next estimate the third term $e_3(\underline{\kappa}\rho\,{\not{d}}_2^\star\zeta)$ on the right-hand side of (A.12.1),

$$
e_3(\underline{\kappa}\rho\,{\not{d}}_2^\star\zeta) = \underline{\kappa}\rho\,{\not{d}}_2^\star(e_3\zeta) + \underline{\kappa}\rho[e_3, {\not{d}}_2^\star]\zeta + e_3(\underline{\kappa})\rho\,{\not{d}}_2^\star\zeta + \underline{\kappa}e_3(\rho)\,{\not{d}}_2^\star\zeta.
$$

Using again the equations

$$
\begin{aligned}
e_3(\underline{\kappa}) &= -\frac{1}{2}\underline{\kappa}^2 - 2\underline{\omega}\,\underline{\kappa} + 2\not{d}_1\underline{\xi} + \Gamma_b\cdot\Gamma_b,\\
e_3\rho &= -\frac{3}{2}\underline{\kappa}\rho + \not{d}_1\underline{\beta} + \Gamma_g\cdot\Gamma_b,
\end{aligned}
$$

which yields

$$
\begin{aligned}
e_3(\underline{\kappa})\rho\not{d}_2^\star\zeta &= \left(-\frac{1}{2}\underline{\kappa}^2 - 2\underline{\omega}\,\underline{\kappa}\right)\rho\not{d}_2^\star\zeta + r^{-5}\not{\mathfrak{d}}\Gamma_b\cdot\not{\mathfrak{d}}\Gamma_g,\\
\underline{\kappa}e_3\rho\not{d}_2^\star\zeta &= -\frac{3}{2}\underline{\kappa}^2\rho + r^{-4}\not{\mathfrak{d}}\Gamma_b\cdot\not{\mathfrak{d}}\Gamma_g,
\end{aligned}
$$

the equation

$$
e_3\zeta = -\frac{1}{2}\underline{\kappa}(\zeta+\eta) + 2\underline{\omega}(\zeta-\eta) + \underline{\beta} + 2\not{d}_1^\star\underline{\omega} + 2\omega\underline{\xi} + \frac{1}{2}\kappa\underline{\xi} + \Gamma_b\cdot\Gamma_b,
$$

and the commutator formula

$$
\begin{aligned}
[e_3,\not{d}_2^\star]\zeta &= \left(-\frac{1}{2}\underline{\kappa}\not{d}_2^\star + \frac{1}{2}\underline{\vartheta}\not{d}_1 + (\zeta-\eta)e_3 + \eta e_3\Phi - \underline{\xi}(e_4 - e_4(\Phi)) + \underline{\beta}\right)\zeta\\
&= -\frac{1}{2}\underline{\kappa}\not{d}_2^\star\zeta + r^{-1}\Gamma_b\cdot\not{\mathfrak{d}}^{\leq 1}\Gamma_g,
\end{aligned}
$$

we deduce

$$
\begin{aligned}
& e_3(\underline{\kappa}\rho\not{d}_2^\star\zeta)\\
={}& \underline{\kappa}\rho\not{d}_2^\star\left(-\frac{1}{2}\underline{\kappa}(\zeta+\eta) + 2\underline{\omega}(\zeta-\eta) + \underline{\beta} + 2\not{d}_1^\star\underline{\omega} + 2\omega\underline{\xi} + \frac{1}{2}\kappa\underline{\xi} + \Gamma_b\cdot\Gamma_b\right)\\
-{}& \frac{1}{2}\underline{\kappa}^2\rho\not{d}_2^\star\zeta + r^{-5}\Gamma_b\cdot\not{\mathfrak{d}}^{\leq 1}\Gamma_g\\
-{}& \frac{1}{2}\underline{\kappa}^2\rho\not{d}_2^\star\zeta - 2\underline{\omega}\,\underline{\kappa}\rho\not{d}_2^\star\zeta + r^{-5}\not{\mathfrak{d}}^{\leq 1}\Gamma_b\cdot\not{\mathfrak{d}}^{\leq 1}\Gamma_g\\
-{}& \frac{3}{2}\underline{\kappa}^2\rho\not{d}_2^\star\zeta + r^{-4}\not{\mathfrak{d}}\Gamma_b\cdot\not{\mathfrak{d}}\Gamma_g\\
={}& \underline{\kappa}\rho\left(-3\underline{\kappa}\not{d}_2^\star\zeta - \frac{1}{2}\underline{\kappa}\not{d}_2^\star\eta - 2\underline{\omega}\not{d}_2^\star\eta + \not{d}_2^\star\underline{\beta} + 2\not{d}_2^\star\not{d}_1^\star\underline{\omega} + 2\omega\not{d}_2^\star\underline{\xi} + \frac{1}{2}\kappa\not{d}_2^\star\underline{\xi}\right)\\
+{}& r^{-4}\not{\mathfrak{d}}^{\leq 1}\Gamma_b\cdot\not{\mathfrak{d}}^{\leq 1}\Gamma_g + r^{-5}\not{\mathfrak{d}}^{\leq 1}\Gamma_b\cdot\not{\mathfrak{d}}^{\leq 1}\Gamma_b,
\end{aligned}
$$

i.e.,

$$
\begin{aligned}
e_3(\underline{\kappa}\rho\not{d}_2^\star\zeta) ={}& \underline{\kappa}\rho\Bigg(-3\underline{\kappa}\not{d}_2^\star\zeta - \frac{1}{2}\underline{\kappa}\not{d}_2^\star\eta - 2\underline{\omega}\not{d}_2^\star\eta + \not{d}_2^\star\underline{\beta} + 2\not{d}_2^\star\not{d}_1^\star\underline{\omega} + 2\omega\not{d}_2^\star\underline{\xi}\\
&+\frac{1}{2}\kappa\not{d}_2^\star\underline{\xi}\Bigg) + r^{-4}\not{\mathfrak{d}}^{\leq 1}\Gamma_b\cdot\not{\mathfrak{d}}^{\leq 1}\Gamma_g + r^{-5}\not{\mathfrak{d}}^{\leq 1}\Gamma_b\cdot\not{\mathfrak{d}}^{\leq 1}\Gamma_b. \quad \text{(A.12.5)}
\end{aligned}
$$

For the fourth term on the right-hand side of (A.12.1) we have

$$
\begin{aligned}
e_3(\kappa\rho\underline{\alpha}) &= \kappa\rho e_3(\underline{\alpha}) + e_3(\kappa)\rho\underline{\alpha} + \kappa e_3(\rho)\underline{\alpha}\\
&= \kappa\rho e_3(\underline{\alpha}) + \left(-\frac{1}{2}\underline{\kappa}\,\kappa + 2\underline{\omega}\kappa + 2\,\not{d}_1\eta + 2\rho - \frac{1}{2}\underline{\vartheta}\,\vartheta + 2(\underline{\xi}\,\xi + \eta\,\eta)\right)\rho\underline{\alpha}\\
&\quad +\kappa\left(-\frac{3}{2}\underline{\kappa}\rho + \not{d}_1\underline{\beta} - \frac{1}{2}\vartheta\,\underline{\alpha} - \zeta\,\underline{\beta} + 2(\eta\,\underline{\beta} + \underline{\xi}\,\beta)\right)\underline{\alpha}\\
&= \kappa\rho e_3(\underline{\alpha}) + (-2\kappa\underline{\kappa} + 2\underline{\omega}\kappa + 2\rho)\,\rho\underline{\alpha} + \left(r^{-3}\,\mathfrak{d}\Gamma_b + r^{-2}\Gamma_b\cdot\Gamma_b\right)\cdot\underline{\alpha},
\end{aligned}
$$

i.e.,

$$
\begin{aligned}
e_3(\kappa\rho\underline{\alpha}) &= \kappa\rho e_3(\underline{\alpha}) + (-2\kappa\underline{\kappa} + 2\underline{\omega}\kappa + 2\rho)\,\rho\underline{\alpha}\\
&\quad +\left(r^{-3}\,\mathfrak{d}\Gamma_b + r^{-2}\Gamma_b\cdot\Gamma_b\right)\cdot\underline{\alpha}. \qquad \text{(A.12.6)}
\end{aligned}
$$

Finally, for the fifth term on the right-hand side of (A.12.1), using the e_3 equations for $\underline{\vartheta}, \rho, \underline{\kappa}, \kappa$,

$$
\begin{aligned}
e_3\Big((2\rho^2 - \kappa\underline{\kappa}\rho)\underline{\vartheta}\Big) &= (2\rho^2 - \kappa\underline{\kappa}\rho)e_3\underline{\vartheta} + 4\rho e_3(\rho)\underline{\vartheta} - e_3(\kappa)\underline{\kappa}\rho\underline{\vartheta} - \kappa e_3(\underline{\kappa})\rho\underline{\vartheta}\\
&\quad -\kappa\underline{\kappa}e_3(\rho)\underline{\vartheta}\\
&= (2\rho^2 - \kappa\underline{\kappa}\rho)\Big(-\underline{\kappa}\,\underline{\vartheta} - 2\underline{\omega}\,\underline{\vartheta} - 2\underline{\alpha} - 2\,\not{d}_2^\star\,\underline{\xi} + \Gamma_b\cdot\Gamma_b\Big)\\
&\quad +4\rho\left(-\frac{3}{2}\underline{\kappa}\rho + \not{d}_1\underline{\beta} + \Gamma_g\cdot\Gamma_b\right)\underline{\vartheta}\\
&\quad -\left(-\frac{1}{2}\underline{\kappa}\,\kappa + 2\underline{\omega}\kappa + 2\rho + 2\,\not{d}_1\eta + \Gamma_b\cdot\Gamma_b\right)\underline{\kappa}\rho\underline{\vartheta}\\
&\quad -\kappa\left(-\frac{1}{2}\underline{\kappa}^2 - 2\underline{\omega}\,\underline{\kappa} + 2\,\not{d}_1\underline{\xi} + \Gamma_b\cdot\Gamma_b\right)\rho\underline{\vartheta}\\
&\quad -\kappa\underline{\kappa}\left(-\frac{3}{2}\underline{\kappa}\rho + \not{d}_1\underline{\beta} + \Gamma_g\cdot\Gamma_b\right)\underline{\vartheta},
\end{aligned}
$$

i.e.,

$$
\begin{aligned}
e_3\Big((2\rho^2 - \kappa\underline{\kappa}\rho)\underline{\vartheta}\Big) &= (2\rho^2 - \kappa\underline{\kappa}\rho)\Big(-\underline{\kappa}\,\underline{\vartheta} - 2\underline{\omega}\,\underline{\vartheta} - 2\underline{\alpha} - 2\,\not{d}_2^\star\,\underline{\xi}\Big) - 6\underline{\kappa}\rho^2\underline{\vartheta}\\
&\quad - \left(-\frac{1}{2}\underline{\kappa}\,\kappa + 2\underline{\omega}\kappa + 2\rho\right)\underline{\kappa}\rho\underline{\vartheta} + \kappa\left(\frac{1}{2}\underline{\kappa}^2 + 2\underline{\omega}\,\underline{\kappa}\right)\rho\underline{\vartheta} + \frac{3}{2}\underline{\kappa}^2\kappa\rho\\
&\quad + r^{-5}\,\mathfrak{d}^{\leq 1}\Gamma_b\cdot\Gamma_b,
\end{aligned}
$$

from which

$$
\begin{aligned}
& e_3\Big((2\rho^2 - \kappa\underline{\kappa}\rho)\underline{\vartheta}\Big)\\
&= (2\rho^2 - \kappa\underline{\kappa}\rho)\Big(-2\underline{\alpha} - 2\,\not{d}_2^\star\,\underline{\xi}\Big) + \left(\frac{7}{2}\kappa\underline{\kappa}^2\rho - 10\underline{\kappa}\rho^2 + 2\kappa\underline{\kappa}\rho\underline{\omega} - 4\underline{\omega}\rho^2\right)\underline{\vartheta}\\
&\quad +r^{-5}\,\mathfrak{d}^{\leq 1}\Gamma_b\cdot\Gamma_b. \qquad \text{(A.12.7)}
\end{aligned}
$$

Recalling (A.12.1)

$$
\begin{aligned}
r^{-7}e_3(r^2e_3(r\mathfrak{q})) &= e_3\,\not{d}_2^\star\,\not{d}_1^\star\,\not{d}_1\underline{\beta}-\frac{3}{2}e_3(\rho\,\not{d}_2^\star\,\not{d}_1^\star\underline{\kappa})-\frac{3}{2}e_3(\underline{\kappa}\rho\,\not{d}_2^\star\zeta)-\frac{3}{2}e_3(\kappa\rho\underline{\alpha})\\
&\quad+\frac{3}{4}e_3\Big((2\rho^2-\kappa\underline{\kappa}\rho)\underline{\vartheta}\Big)+7r^{-6}e_3(r)e_3(r\mathfrak{q})\\
&\quad+\; r^2e_3(\mathrm{Err}[e_3\mathfrak{q}])+r\mathrm{Err}[e_3\mathfrak{q}]+\mathrm{l.o.t.}
\end{aligned}
$$

and making use of (A.12.4)–(A.12.7) we deduce

$$
\begin{aligned}
&r^{-7}e_3(r^2e_3(r\mathfrak{q}))\\
=&\;\not{d}_2^\star\,\not{d}_1^\star\,\not{d}_1\,\not{d}_2\underline{\alpha}-2(\underline{\kappa}+\underline{\omega})\,\not{d}_2^\star\,\not{d}_1^\star\,\not{d}_1\underline{\beta}+3\rho\,\not{d}_2^\star\,\not{d}_1^\star\,\not{d}_1\underline{\xi}-\frac{3}{2}\underline{\kappa}\,\not{d}_2^\star\,\not{d}_1^\star\,\not{d}_1\underline{\beta}\\
-&\;\frac{3}{2}\Bigg\{-\frac{7}{2}\rho\underline{\kappa}\,\not{d}_2^\star\,\not{d}_1^\star(\underline{\kappa})-2\rho\underline{\omega}\,\not{d}_2^\star\,\not{d}_1^\star\underline{\kappa}-2\rho\underline{\kappa}\,\not{d}_2^\star\,\not{d}_1^\star\underline{\omega}+2\rho\,\not{d}_2^\star\,\not{d}_1^\star\,\not{d}_1\underline{\xi}\\
&\;+\rho\,\not{d}_2^\star(\zeta-\eta)\left(-\frac{1}{2}\underline{\kappa}^2-2\underline{\omega}\,\underline{\kappa}\right)-\rho\,\not{d}_2^\star\underline{\xi}\left(-\frac{1}{2}\kappa\underline{\kappa}+2\omega\underline{\kappa}+2\rho\right)\Bigg\}\\
-&\;\frac{3}{2}\Bigg\{\underline{\kappa}\rho\left(-3\underline{\kappa}\,\not{d}_2^\star\zeta-\frac{1}{2}\underline{\kappa}\,\not{d}_2^\star\eta-2\underline{\omega}\,\not{d}_2^\star\eta+\not{d}_2^\star\underline{\beta}+2\,\not{d}_2^\star\,\not{d}_1^\star\underline{\omega}+2\omega\,\not{d}_2^\star\underline{\xi}+\frac{1}{2}\kappa\,\not{d}_2^\star\underline{\xi}\right)\Bigg\}\\
-&\;\frac{3}{2}\Big\{\kappa\rho e_3(\underline{\alpha})+(-2\kappa\underline{\kappa}+2\underline{\omega}\kappa+2\rho)\,\rho\underline{\alpha}+\big(r^{-3}\,\not{\mathfrak{d}}\Gamma_b+r^{-2}\Gamma_b\cdot\Gamma_b\big)\cdot\underline{\alpha}\Big\}\\
+&\;\frac{3}{4}\Bigg\{(2\rho^2-\kappa\underline{\kappa}\rho)\Big(-2\underline{\alpha}-2\,\not{d}_2^\star\,\underline{\xi}\Big)+\left(\frac{7}{2}\kappa\underline{\kappa}^2\rho-10\underline{\kappa}\rho^2+2\kappa\underline{\kappa}\rho\underline{\omega}-4\underline{\omega}\rho^2\right)\underline{\vartheta}\Bigg\}\\
+&\;7r^{-6}e_3(r)e_3(r\mathfrak{q})+\mathrm{Err}+r^{-7}\Big(r^2e_3(\mathrm{Err}[e_3\mathfrak{q}])+r\mathrm{Err}[e_3\mathfrak{q}]\Big)+\mathrm{l.o.t.},
\end{aligned}
$$

where the error term Err is given by

$$
\begin{aligned}
\mathrm{Err} &= (r^{-3}\,\not{\mathfrak{d}}\Gamma_b+r^{-2}\Gamma_b\cdot\Gamma_b)\cdot\underline{\alpha}+r^{-4}\,\not{\mathfrak{d}}^{\leq 1}\Gamma_b\cdot\not{\mathfrak{d}}^{\leq 1}\Gamma_g\\
&\quad+r^{-5}\,\not{\mathfrak{d}}^{\leq 1}\Gamma_b\cdot\not{\mathfrak{d}}^{\leq 1}\Gamma_b.
\end{aligned}
\tag{A.12.8}
$$

Denoting the expression of the left-hand side of the identity (2.3.15) by I, i.e.,

$$
I := e_3(r^2e_3(r\mathfrak{q}))+2\underline{\omega}r^2e_3(r\mathfrak{q}),
$$

we deduce

$$
\begin{aligned}
r^{-7}I &= \not{d}_2^\star\,\not{d}_1^\star\,\not{d}_1\,\not{d}_2\underline{\alpha}-\left(\frac{7}{2}\underline{\kappa}+2\underline{\omega}\right)\not{d}_2^\star\,\not{d}_1^\star\,\not{d}_1\underline{\beta}\\
&+\;\frac{3}{2}\left(\frac{7}{2}\underline{\kappa}+2\underline{\omega}\right)\rho\,\not{d}_2^\star\,\not{d}_1^\star\underline{\kappa}+\frac{3}{2}\left(\frac{7}{2}\underline{\kappa}+2\underline{\omega}\right)\underline{\kappa}\rho\,\not{d}_2^\star\zeta-\frac{3}{2}\underline{\kappa}\rho\,\not{d}_2^\star\underline{\beta}\\
&-\;\frac{3}{2}\kappa\rho e_3(\underline{\alpha})-\frac{3}{2}\Big(-3\kappa\underline{\kappa}+2\underline{\omega}\kappa+4\rho\Big)\rho\underline{\alpha}\\
&+\;\frac{3}{4}\left(\frac{7}{2}\kappa\underline{\kappa}^2\rho-10\underline{\kappa}\rho^2+2\kappa\underline{\kappa}\rho\underline{\omega}-4\underline{\omega}\rho^2\right)\underline{\vartheta}\\
&+\;r^{-7}\left[7re_3(r)e_3(r\mathfrak{q})+2\underline{\omega}r^2e_3(r\mathfrak{q})\right]+\widetilde{\mathrm{Err}},
\end{aligned}
$$

where the new error term $\widetilde{\mathrm{Err}}$ is given by

$$\widetilde{\mathrm{Err}} = \mathrm{Err} + r^{-7}\Big(r^2 e_3(\mathrm{Err}[e_3\mathfrak{q}]) + r\mathrm{Err}[e_3\mathfrak{q}]\Big) + 2\underline{\omega} r^{-5}\mathrm{Err}[e_3\mathfrak{q}].$$

To calculate the term $J := 7re_3(r)e_3(r\mathfrak{q}) + 2\underline{\omega} r^2 e_3(r\mathfrak{q})$ in the last row we make use once more of the identity of Lemma 2.100 to derive

$$\begin{aligned}
& J \\
= \;& r^2\left(\frac{7}{2}\underline{\kappa} + 2\underline{\omega}\right) e_3(r\mathfrak{q}) + 7r\left(e_3(r) - \frac{r}{2}\underline{\kappa}\right) e_3(r\mathfrak{q}) \\
= \;& r^2\left(\frac{7}{2}\underline{\kappa} + 2\underline{\omega}\right) e_3(r\mathfrak{q}) + r^2\Gamma_b e_3(r\mathfrak{q}) \\
= \;& r^7\left(\frac{7}{2}\underline{\kappa} + 2\underline{\omega}\right)\left\{ \not{d}_2^\star \not{d}_1^\star \not{d}_1 \underline{\beta} - \frac{3}{2}\rho \not{d}_2^\star \not{d}_1^\star \underline{\kappa} - \frac{3}{2}\underline{\kappa}\rho \not{d}_2^\star \zeta - \frac{3}{2}\kappa\rho\underline{\alpha} + \frac{3}{4}(2\rho^2 - \kappa\underline{\kappa}\rho)\underline{\vartheta}\right\} \\
+ \;& r^2\left(\frac{7}{2}\underline{\kappa} + 2\underline{\omega}\right)\mathrm{Err}[e_3(r\mathfrak{q})] + r^2\Gamma_b e_3(r\mathfrak{q}),
\end{aligned}$$

i.e.,

$$\begin{aligned}
& r^{-7}J \\
= \;& \left(\frac{7}{2}\underline{\kappa} + 2\underline{\omega}\right)\left\{ \not{d}_2^\star \not{d}_1^\star \not{d}_1 \underline{\beta} - \frac{3}{2}\rho \not{d}_2^\star \not{d}_1^\star \underline{\kappa} - \frac{3}{2}\underline{\kappa}\rho \not{d}_2^\star \zeta - \frac{3}{2}\kappa\rho\underline{\alpha} + \frac{3}{4}(2\rho^2 - \kappa\underline{\kappa}\rho)\underline{\vartheta}\right\} \\
+ \;& r^{-5}\left(\frac{7}{2}\underline{\kappa} + 2\underline{\omega}\right)\mathrm{Err}[e_3(r\mathfrak{q})] + r^{-5}\Gamma_b e_3(r\mathfrak{q}).
\end{aligned}$$

Combining and simplifying,

$$\begin{aligned}
r^{-7}I &= \not{d}_2^\star \not{d}_1^\star \not{d}_1 \not{d}_2 \underline{\alpha} - \frac{3}{2}\underline{\kappa}\rho \not{d}_2^\star \underline{\beta} - \frac{3}{2}\kappa\rho e_3(\underline{\alpha}) - \frac{3}{2}\left(\frac{1}{2}\kappa\underline{\kappa} + 4\underline{\omega}\kappa + 4\rho\right)\rho\underline{\alpha} \\
&\quad - \frac{9}{4}\underline{\kappa}\rho^2\underline{\vartheta} + \widetilde{\widetilde{\mathrm{Err}}},
\end{aligned}$$

where

$$\widetilde{\widetilde{\mathrm{Err}}} = \widetilde{\mathrm{Err}} + r^{-5}\left(\frac{7}{2}\underline{\kappa} + 2\underline{\omega}\right)\mathrm{Err}[e_3(r\mathfrak{q})] + r^{-5}\Gamma_b e_3(r\mathfrak{q}).$$

Using Bianchi to replace $\not{d}_2^\star\underline{\beta}$, we deduce

$$\begin{aligned}
r^{-7}I &= \not{d}_2^\star \not{d}_1^\star \not{d}_1 \not{d}_2 \underline{\alpha} - \frac{3}{2}\underline{\kappa}\rho\left(-e_4\underline{\alpha} - \frac{1}{2}\kappa\underline{\alpha} + 4\omega\underline{\alpha} - \frac{3}{2}\rho\underline{\vartheta} + r^{-1}\Gamma_g\cdot\Gamma_b\right) \\
&- \frac{3}{2}\kappa\rho e_3(\underline{\alpha}) - \frac{3}{2}\left(\frac{1}{2}\kappa\underline{\kappa} + 4\underline{\omega}\kappa + 4\rho\right)\rho\underline{\alpha} - \frac{9}{4}\underline{\kappa}\rho^2\underline{\vartheta} + r^{-5}\Gamma_g\cdot\Gamma_b + \text{l.o.t.} \\
&= \not{d}_2^\star \not{d}_1^\star \not{d}_1 \not{d}_2 \underline{\alpha} + \frac{3}{2}\underline{\kappa}\rho e_4\underline{\alpha} - \frac{3}{2}\kappa\rho e_3(\underline{\alpha}) - 6\Big(\underline{\kappa}\omega + \underline{\omega}\kappa + \rho\Big)\rho\underline{\alpha} + r^{-5}\Gamma_g\cdot\Gamma_b.
\end{aligned}$$

Hence, in view of the definition of I, we infer

$$\begin{aligned} I &= e_3(r^2e_3(r\mathfrak{q})) + 2\underline{\omega}r^2e_3(r\mathfrak{q}) \\ &= r^7\left\{ \not{d}_2^\star \not{d}_1^\star \not{d}_1 \not{d}_2\underline{\alpha} + \frac{3}{2}\underline{\kappa}\rho e_4\underline{\alpha} - \frac{3}{2}\kappa\rho e_3(\underline{\alpha})\right\} + \mathrm{Err}[ST] \end{aligned}$$

where

$$\begin{aligned} \mathrm{Err}[ST] &= r^7\widetilde{\mathrm{Err}} + r^2\left(\frac{7}{2}\underline{\kappa} + 2\underline{\omega}\right)\mathrm{Err}[e_3(r\mathfrak{q})] + r^2\Gamma_b e_3(r\mathfrak{q}) + r^2\Gamma_g\cdot\Gamma_b \\ &+ r^7\mathrm{Err} + \left(r^2e_3(\mathrm{Err}[e_3\mathfrak{q}]) + r\mathrm{Err}[e_3\mathfrak{q}]\right) + 2\underline{\omega}r^2\mathrm{Err}[e_3\mathfrak{q}] + r^2\Gamma_b e_3(r\mathfrak{q}) \\ &+ r^2\Gamma_g\cdot\Gamma_b. \end{aligned}$$

Recall that, see (A.12.8),

$$\mathrm{Err} = \left(r^{-3}\not{\mathfrak{d}}\Gamma_b + r^{-2}\Gamma_b\cdot\Gamma_b\right)\cdot\underline{\alpha} + r^{-4}\not{\mathfrak{d}}^{\leq 1}\Gamma_b\cdot\not{\mathfrak{d}}^{\leq 1}\Gamma_g + r^{-5}\not{\mathfrak{d}}^{\leq 1}\Gamma_b\cdot\not{\mathfrak{d}}^{\leq 1}\Gamma_b.$$

Hence,

$$\begin{aligned} \mathrm{Err}[ST] &= r^4\left(\not{\mathfrak{d}}\Gamma_b + r\Gamma_b\cdot\Gamma_b\right)\cdot\underline{\alpha} + r^3\not{\mathfrak{d}}^{\leq 1}\Gamma_b\cdot\not{\mathfrak{d}}^{\leq 1}\Gamma_g + r^2\not{\mathfrak{d}}^{\leq 1}\Gamma_b\cdot\not{\mathfrak{d}}^{\leq 1}\Gamma_b \\ &+ r^2\Gamma_b e_3(r\mathfrak{q}) + r^2e_3(\mathrm{Err}[e_3\mathfrak{q}]) + r\mathrm{Err}[e_3\mathfrak{q}] + 2\underline{\omega}r^2\mathrm{Err}[e_3\mathfrak{q}]. \end{aligned}$$

Recall that, see Proposition 2.100,

$$\mathrm{Err}[e_3(r\mathfrak{q})] = r\Gamma_b\mathfrak{q} + r^5\mathfrak{d}^{\leq 1}(e_3\eta\cdot\beta) + r^3\mathfrak{d}^{\leq 2}(\Gamma_b\cdot\Gamma_g).$$

Therefore,

$$\begin{aligned} E &= r^2e_3(\mathrm{Err}[e_3\mathfrak{q}]) + r\mathrm{Err}[e_3\mathfrak{q}] + 2\underline{\omega}r^2\mathrm{Err}[e_3\mathfrak{q}] \\ &= r^2\left(\Gamma_b e_3(r\mathfrak{q}) + e_3(\Gamma_b)r\mathfrak{q}\right) + r^7\mathfrak{d}^{\leq 2}(e_3\eta\cdot\beta) + r^5\mathfrak{d}^{\leq 3}(\Gamma_b\cdot\Gamma_g) \\ &+ r^2\Gamma_b\mathfrak{q} + r^6\mathfrak{d}^{\leq 1}(e_3\eta\cdot\beta) + r^4\mathfrak{d}^{\leq 2}(\Gamma_b\cdot\Gamma_g) \\ &= r^2\left(\Gamma_b e_3(r\mathfrak{q}) + (\mathfrak{d}^{\leq 1}\Gamma_b)r\mathfrak{q}\right) + r^7\mathfrak{d}^{\leq 2}(e_3\eta\cdot\beta) + r^5\mathfrak{d}^{\leq 3}(\Gamma_b\cdot\Gamma_g). \end{aligned}$$

Thus,

$$\begin{aligned} \mathrm{Err}[TS] &= r^4\left(\not{\mathfrak{d}}\Gamma_b + r\Gamma_b\cdot\Gamma_b\right)\cdot\underline{\alpha} + r^2\left(\Gamma_b e_3(r\mathfrak{q}) + (\mathfrak{d}^{\leq 1}\Gamma_b)r\mathfrak{q}\right) \\ &+ r^7\mathfrak{d}^{\leq 2}(e_3\eta\cdot\beta) + r^5\mathfrak{d}^{\leq 3}(\Gamma_b\cdot\Gamma_g) \end{aligned}$$

which ends the proof of Proposition 2.101.

A.13 PROOF OF PROPOSITION 2.107

In this section we give a proof of Proposition 2.107, i.e., we derive the wave equation for the extreme curvature component α,

$$\begin{aligned}
\square_{\mathbf{g}}\alpha &= -4\underline{\omega}e_4(\alpha) + (2\kappa + 4\omega)\, e_3(\alpha) + V\alpha + \mathrm{Err}(\square_{\mathbf{g}}\alpha),\\
V &:= \left(-4e_4(\underline{\omega}) + \frac{1}{2}\kappa\underline{\kappa} - 10\kappa\underline{\omega} + 2\underline{\kappa}\omega - 8\omega\underline{\omega} - 4\rho + 4e_\theta(\Phi)^2\right)\alpha,
\end{aligned} \tag{A.13.1}$$

where

$$\begin{aligned}
\mathrm{Err}(\square_{\mathbf{g}}\alpha) \;=\;& \frac{1}{2}\vartheta e_3(\alpha) + \frac{3}{4}\vartheta^2\rho + e_\theta(\Phi)\vartheta\beta - \frac{1}{2}\kappa(\zeta + 4\eta)\beta - (\zeta + \underline{\eta})e_4(\beta)\\
& -\xi e_3(\beta) + e_\theta(\Phi)(2\zeta + \underline{\eta})\alpha + \beta^2 + e_4(\Phi)\underline{\eta}\beta + e_3(\Phi)\xi\beta\\
& -(\zeta + 4\eta)e_4(\beta) - (e_4(\zeta) + 4e_4(\eta))\beta - 2(\kappa + \omega)(\zeta + 4\eta)\beta\\
& +2e_\theta(\kappa + \omega)\beta - e_\theta((2\zeta + \underline{\eta})\alpha) - 3\xi e_\theta(\rho) + 2\underline{\eta}e_\theta(\alpha)\\
& +\frac{3}{2}\vartheta(e_\theta(\beta) + e_\theta(\Phi)\beta) + 3\rho(\underline{\eta} + \eta + 2\zeta)\xi + (e_\theta(\underline{\eta}) + e_\theta(\Phi)\underline{\eta})\alpha\\
& +\frac{1}{4}\underline{\kappa}\vartheta\alpha - 2\underline{\omega}\vartheta\alpha - \frac{1}{2}\vartheta\underline{\vartheta}\alpha + \xi\underline{\xi}\alpha + \underline{\eta}^2\alpha + \frac{3}{2}\vartheta\zeta\beta + 3\vartheta(\underline{\eta}\beta + \xi\underline{\beta})\\
& -\frac{1}{2}\vartheta(\zeta + 4\eta)\beta.
\end{aligned}$$

The equation for $\underline{\alpha}$ can then be easily inferred by symmetry.

Proof. We make use of the Bianchi identities

$$\begin{aligned}
e_\theta(\beta) - e_\theta(\Phi)\beta &= e_3(\alpha) + \left(\frac{\underline{\kappa}}{2} - 4\underline{\omega}\right)\alpha + \frac{3}{2}\vartheta\,\rho - (\zeta + 4\eta)\beta,\\
e_4(\beta) + 2(\kappa + \omega)\beta &= e_\theta(\alpha) + 2e_\theta(\Phi)\alpha + (2\zeta + \underline{\eta})\alpha + 3\xi\rho,
\end{aligned}$$

to infer that

$$\begin{aligned}
& e_4(e_3(\alpha))\\
=\;& e_4(e_\theta(\beta)) - e_\theta(\Phi)e_4(\beta) - e_4(e_\theta(\Phi))\beta - \left(\frac{\underline{\kappa}}{2} - 4\underline{\omega}\right)e_4(\alpha)\\
& -\left(\frac{e_4(\underline{\kappa})}{2} - 4e_4(\underline{\omega})\right)\alpha - \frac{3}{2}\vartheta e_4(\rho) - \frac{3}{2}e_4(\vartheta)\rho + (\zeta + 4\eta)e_4(\beta)\\
& +(e_4(\zeta) + 4e_4(\eta))\beta\\
=\;& e_4(e_\theta(\beta)) - e_\theta(\Phi)\Big(e_\theta(\alpha) + 2e_\theta(\Phi)\alpha - 2(\kappa + \omega)\beta + (2\zeta + \underline{\eta})\alpha + 3\xi\rho\Big)\\
& -(D_4D_\theta\Phi + D_{D_4e_\theta}\Phi)\beta - \left(\frac{\underline{\kappa}}{2} - 4\underline{\omega}\right)e_4(\alpha) - \left(\frac{e_4(\underline{\kappa})}{2} - 4e_4(\underline{\omega})\right)\alpha\\
& -\frac{3}{2}\vartheta e_4(\rho) - \frac{3}{2}e_4(\vartheta)\rho + (\zeta + 4\eta)e_4(\beta) + (e_4(\zeta) + 4e_4(\eta))\beta.
\end{aligned}$$

Hence,

$$\begin{aligned}
e_4(e_3(\alpha)) &= e_4(e_\theta(\beta)) - e_\theta(\Phi)(e_\theta(\alpha) + 2e_\theta(\Phi)\alpha) + 2e_\theta(\Phi)(\kappa+\omega)\beta - 3e_\theta(\Phi)\xi\rho \\
&\quad + e_4(\Phi)e_\theta(\Phi)\beta - \left(\frac{\underline{\kappa}}{2} - 4\underline{\omega}\right)e_4(\alpha) - \left(\frac{e_4(\underline{\kappa})}{2} - 4e_4(\underline{\omega})\right)\alpha \\
&\quad - \frac{3}{2}\vartheta e_4(\rho) - \frac{3}{2}e_4(\vartheta)\rho - e_\theta(\Phi)(2\zeta+\underline{\eta})\alpha - \beta^2 - e_4(\Phi)\underline{\eta}\beta - e_3(\Phi)\xi\beta \\
&\quad + (\zeta + 4\eta)e_4(\beta) + (e_4(\zeta) + 4e_4(\eta))\beta,
\end{aligned}$$

and

$$\begin{aligned}
& e_\theta(e_\theta(\alpha)) \\
={}& e_\theta(e_4(\beta)) + 2(\kappa+\omega)e_\theta(\beta) + 2e_\theta(\kappa+\omega)\beta - 2e_\theta(\Phi)e_\theta(\alpha) - 2e_\theta(e_\theta(\Phi))\alpha \\
& - e_\theta((2\zeta+\underline{\eta})\alpha) - 3e_\theta(\xi\rho) \\
={}& e_\theta(e_4(\beta)) + 2(\kappa+\omega)\Big(e_\theta(\Phi)\beta + e_3(\alpha) + \left(\frac{\underline{\kappa}}{2} - 4\underline{\omega}\right)\alpha + \frac{3}{2}\vartheta\,\rho - (\zeta+4\eta)\beta\Big) \\
& + 2e_\theta(\kappa+\omega)\beta - 2e_\theta(\Phi)e_\theta(\alpha) - 2(D_\theta D_\theta\Phi + D_{D_\theta e_\theta}\Phi)\alpha - e_\theta((2\zeta+\underline{\eta})\alpha) \\
& - 3e_\theta(\xi\rho) \\
={}& e_\theta(e_4(\beta)) + 2(\kappa+\omega)e_\theta(\Phi)\beta + 2(\kappa+\omega)e_3(\alpha) + 2(\kappa+\omega)\left(\frac{\underline{\kappa}}{2} - 4\underline{\omega}\right)\alpha \\
& + 3(\kappa+\omega)\vartheta\,\rho - 2e_\theta(\Phi)e_\theta(\alpha) - 2\left(\rho - e_\theta(\Phi)^2 + \frac{1}{2}\underline{\chi}e_4(\Phi) + \frac{1}{2}\chi e_3(\Phi)\right)\alpha \\
& - 3e_\theta(\xi)\rho - 2(\kappa+\omega)(\zeta+4\eta)\beta + 2e_\theta(\kappa+\omega)\beta - e_\theta((2\zeta+\underline{\eta})\alpha) - 3\xi e_\theta(\rho).
\end{aligned}$$

In view of Lemma 2.102, we have

$$\begin{aligned}
\square_{\mathbf{g}} f &= -e_4(e_3(f)) + e_\theta(e_\theta(f)) - \frac{1}{2}\underline{\kappa}e_4(f) + \left(-\frac{1}{2}\kappa + 2\omega\right)e_3(f) + e_\theta(\Phi)e_\theta(f) \\
&\quad + 2\underline{\eta}e_\theta(f).
\end{aligned}$$

We infer

$$\begin{aligned}
\square_{\mathbf{g}}\alpha &= -e_4(e_3(\alpha)) + e_\theta(e_\theta(\alpha)) - \frac{1}{2}\underline{\kappa}e_4(\alpha) + \left(-\frac{1}{2}\kappa + 2\omega\right)e_3(\alpha) + e_\theta(\Phi)e_\theta(\alpha) \\
&\quad + 2\underline{\eta}e_\theta(\alpha) \\
&= [e_\theta, e_4](\beta) - e_4(\Phi)e_\theta(\Phi)\beta + \frac{3}{2}\vartheta e_4(\rho) + \frac{3}{2}e_4(\vartheta)\rho + 3(\kappa+\omega)\vartheta\,\rho \\
&\quad - 3(e_\theta(\xi) - e_\theta(\Phi)\xi)\rho - 4\underline{\omega}e_4(\alpha) + \left(\frac{3}{2}\kappa + 4\omega\right)e_3(\alpha) \\
&\quad + \Bigg(\frac{e_4(\underline{\kappa})}{2} - 4e_4(\underline{\omega}) + \kappa\underline{\kappa} - 8\kappa\underline{\omega} + \underline{\kappa}\omega - 8\omega\underline{\omega} - 2\rho + 4e_\theta(\Phi)^2 - \underline{\chi}e_4(\Phi) \\
&\quad - \chi e_3(\Phi)\Bigg)\alpha + e_\theta(\Phi)(2\zeta+\underline{\eta})\alpha + \beta^2 + e_4(\Phi)\underline{\eta}\beta + e_3(\Phi)\xi\beta \\
&\quad - (\zeta+4\eta)e_4(\beta) - (e_4(\zeta) + 4e_4(\eta))\beta - 2(\kappa+\omega)(\zeta+4\eta)\beta \\
&\quad + 2e_\theta(\kappa+\omega)\beta - e_\theta((2\zeta+\underline{\eta})\alpha) - 3\xi e_\theta(\rho) + 2\underline{\eta}e_\theta(\alpha).
\end{aligned}$$

Next, we have

$$\begin{aligned}
[e_\theta, e_4](\beta) &= \chi e_\theta(\beta) - (\zeta + \underline{\eta})e_4(\beta) - \xi e_3(\beta)\\
&= \chi\left(e_\theta(\Phi)\beta + e_3(\alpha) + \left(\frac{\underline{\kappa}}{2} - 4\underline{\omega}\right)\alpha + \frac{3}{2}\vartheta\,\rho - (\zeta + 4\eta)\beta\right)\\
&- (\zeta + \underline{\eta})e_4(\beta) - \xi e_3(\beta)
\end{aligned}$$

and hence

$$\begin{aligned}
\Box_{\mathbf{g}}\alpha &= -4\underline{\omega}e_4(\alpha) + \left(\frac{3}{2}\kappa + \chi + 4\omega\right)e_3(\alpha) + V_1\alpha\\
&+ \frac{3}{2}\vartheta e_4(\rho) + \frac{3}{2}e_4(\vartheta)\rho + 3(\kappa + \omega)\vartheta\,\rho - 3(e_\theta(\xi) - e_\theta(\Phi)\xi)\rho + \frac{3}{2}\chi\vartheta\,\rho\\
&+ \mathrm{Err}_1
\end{aligned}$$

where

$$\begin{aligned}
V_1 &:= \frac{e_4(\underline{\kappa})}{2} - 4e_4(\underline{\omega}) + \kappa\underline{\kappa} - 8\kappa\underline{\omega} + \underline{\kappa}\omega - 8\omega\underline{\omega} - 2\rho + 4e_\theta(\Phi)^2 - \underline{\chi}e_4(\Phi)\\
&- \chi e_3(\Phi) + \chi\frac{\underline{\kappa}}{2} - 4\chi\underline{\omega},
\end{aligned}$$

$$\begin{aligned}
\mathrm{Err}_1 &:= e_\theta(\Phi)\vartheta\beta - \chi(\zeta + 4\eta)\beta - (\zeta + \underline{\eta})e_4(\beta) - \xi e_3(\beta) + e_\theta(\Phi)(2\zeta + \underline{\eta})\alpha\\
&+ \beta^2 + e_4(\Phi)\underline{\eta}\beta + e_3(\Phi)\xi\beta - (\zeta + 4\eta)e_4(\beta) - (e_4(\zeta) + 4e_4(\eta))\beta\\
&- 2(\kappa + \omega)(\zeta + 4\eta)\beta + 2e_\theta(\kappa + \omega)\beta - e_\theta((2\zeta + \underline{\eta})\alpha) - 3\xi e_\theta(\rho) + 2\underline{\eta}e_\theta(\alpha).
\end{aligned}$$

Next, we make use of

$$\begin{aligned}
e_4(\vartheta) + \kappa\vartheta + 2\omega\vartheta &= -2\alpha + 2(e_\theta(\xi) - e_\theta(\Phi)\xi) + 2(\underline{\eta} + \eta + 2\zeta)\xi,\\
e_4(\rho) + \frac{3}{2}\kappa\rho &= e_\theta(\beta) + e_\theta(\Phi)\beta - \frac{1}{2}\underline{\vartheta}\alpha + \zeta\beta + 2(\underline{\eta}\beta + \xi\underline{\beta}),
\end{aligned}$$

to calculate the term

$$\begin{aligned}
I &:= \frac{3}{2}\vartheta e_4(\rho) + \frac{3}{2}e_4(\vartheta)\rho + 3(\kappa + \omega)\vartheta\,\rho - 3(e_\theta(\xi) - e_\theta(\Phi)\xi)\rho + \frac{3}{2}\chi\vartheta\,\rho\\
&= \frac{3}{2}\vartheta\left(-\frac{3}{2}\kappa\rho + \not{d}_1\beta\right) + \frac{3}{2}\rho\left(-\kappa\vartheta - 2\omega\vartheta - 2\alpha + 2(e_\theta(\xi) - e_\theta(\Phi)\xi)\right)\\
&+ 3(\kappa + \omega)\vartheta\,\rho - 3(e_\theta(\xi) - e_\theta(\Phi)\xi)\rho + \frac{3}{2}\frac{\kappa + \vartheta}{2}\vartheta\,\rho + \text{l.o.t.}\\
&= -3\rho\alpha + \frac{3}{2}\vartheta\,\not{d}_1\beta + \frac{3}{4}\vartheta^2\rho.
\end{aligned}$$

Hence,

$$\begin{aligned}
\Box_{\mathbf{g}}\alpha &= -4\underline{\omega}e_4(\alpha) + \left(\frac{3}{2}\kappa + \chi + 4\omega\right)e_3(\alpha) + (V_1 - 3\rho)\,\alpha\\
&+ \mathrm{Err}_1 + \frac{3}{2}\vartheta\,\not{d}_1\beta + \frac{3}{4}\vartheta^2\rho.
\end{aligned}$$

Using, also,

$$e_4(\underline{\kappa}) + \frac{1}{2}\kappa\underline{\kappa} - 2\omega\underline{\kappa} = 2(e_\theta(\underline{\eta}) + e_\theta(\Phi)\underline{\eta}) + 2\rho - \frac{1}{2}\vartheta\underline{\vartheta} + 2(\xi\underline{\xi} + \underline{\eta}^2)$$

and the identities, $2\chi = \kappa + \vartheta$, as well as $2\underline{\chi} = \underline{\kappa} + \underline{\vartheta}$, we finally obtain

$$\square_{\mathbf{g}}\alpha = -4\underline{\omega}e_4(\alpha) + (2\kappa + 4\omega)\, e_3(\alpha) + V\alpha + \mathrm{Err}[\square\alpha]$$

as desired.

We write schematically the error term

$$\begin{aligned}
& \mathrm{Err}[\square\alpha] \\
= & \ \frac{1}{2}\vartheta e_3(\alpha) + \frac{3}{4}\vartheta^2\rho + e_\theta(\Phi)\vartheta\beta - \frac{1}{2}\kappa(\zeta + 4\eta)\beta - (\zeta + \underline{\eta})e_4(\beta) - \xi e_3(\beta) \\
+ & \ e_\theta(\Phi)(2\zeta + \underline{\eta})\alpha + \beta^2 + e_4(\Phi)\underline{\eta}\beta + e_3(\Phi)\xi\beta - (\zeta + 4\eta)e_4(\beta) \\
- & \ (e_4(\zeta) + 4e_4(\eta))\beta - 2(\kappa + \omega)(\zeta + 4\eta)\beta + 2e_\theta(\kappa + \omega)\beta - e_\theta((2\zeta + \underline{\eta})\alpha) \\
- & \ 3\xi e_\theta(\rho) + 2\underline{\eta}e_\theta(\alpha) + \frac{3}{2}\vartheta(e_\theta(\beta) + e_\theta(\Phi)\beta) + 3\rho(\underline{\eta} + \eta + 2\zeta)\xi \\
+ & \ (e_\theta(\underline{\eta}) + e_\theta(\Phi)\underline{\eta})\alpha + \frac{1}{4}\underline{\kappa}\vartheta\alpha - 2\underline{\omega}\vartheta\alpha - \frac{1}{2}\vartheta\underline{\vartheta}\alpha + \xi\underline{\xi}\alpha + \underline{\eta}^2\alpha + \frac{3}{2}\vartheta\zeta\beta \\
+ & \ 3\vartheta(\underline{\eta}\beta + \xi\underline{\beta}) - \frac{1}{2}\vartheta(\zeta + 4\eta)\beta
\end{aligned}$$

as follows:

$$\begin{aligned}
\mathrm{Err}[\square_{\mathbf{g}}\alpha] = & \left(\frac{1}{r}\Gamma_g + \frac{1}{r}\not{\mathfrak{d}}\Gamma_g\right)\alpha + \Gamma_g e_3(\alpha) + \frac{1}{r}\Gamma_g\not{\mathfrak{d}}\alpha \\
+ & \left(\beta + \frac{1}{r}\Gamma_g + \frac{1}{r}\mathfrak{d}\Gamma_g\right)\beta + \frac{1}{r}\Gamma_g\mathfrak{d}(\beta) + \Gamma_g e_3(\beta) \\
+ & (\Gamma_g)^2\rho + \frac{1}{r}\Gamma_g\not{\mathfrak{d}}(\rho) \\
= & \ \Gamma_g e_3(\alpha, \beta) + r^{-1}\Gamma_g^{\leq 1}\cdot\mathfrak{d}^{\leq 1}(\alpha, \beta, \check{\rho}) + \beta^2 + \Gamma_g^2\rho.
\end{aligned}$$

This concludes the proof of Proposition 2.107. □

A.14 PROOF OF THEOREM 2.108

Recall the symbolic notation used in the statement of the theorem.

$$\begin{aligned}
\Gamma_g &= \left\{\vartheta, \eta, \underline{\eta}, \zeta, A\right\}, & \Gamma_b &= \left\{\underline{\vartheta}, \underline{\xi}, \underline{A}\right\}, \\
\mathfrak{d}\Gamma_g &= \left\{\mathfrak{d}\vartheta, re_\theta(\kappa), \mathfrak{d}\eta, \mathfrak{d}\underline{\eta}, \mathfrak{d}\zeta, \mathfrak{d}A\right\}, & \mathfrak{d}\Gamma_b &= \left\{\mathfrak{d}\underline{\vartheta}, e_\theta(\underline{\kappa}), \mathfrak{d}\underline{\xi}, \mathfrak{d}\underline{A}\right\},
\end{aligned}$$

where $A = \frac{2}{r}e_4(r) - \kappa$, $\underline{A} = \frac{2}{r}e_3(r) - \underline{\kappa}$. We also denote, for $s \geq 2$,

$$\mathfrak{d}^s\Gamma_g = \mathfrak{d}^{s-1}\mathfrak{d}\Gamma_g, \qquad \mathfrak{d}^s\Gamma_b = \mathfrak{d}^{s-1}\mathfrak{d}\Gamma_b,$$

for higher derivatives with respect to $\mathfrak{d} = (e_3, re_4, \not{\mathfrak{d}})$ (see definition 2.39 for the notation $\not{\mathfrak{d}}$ and $\not{\mathfrak{d}}^s$).

We also recall Remark 2.95.

Remark A.1. *According to the main bootstrap assumptions* ***BA-E, BA-D*** *(see section 3.4.1.) the terms Γ_b behave worse in powers of r than the terms in Γ_g. Thus, in the symbolic expressions below, we replace the terms of the form $\Gamma_g + \Gamma_b$ by Γ_b. We also replace $r^{-1}\Gamma_b$ by Γ_g. We will denote l.o.t. all cubic and higher error terms in $\check{\Gamma}, \check{R}$. We also include in l.o.t. terms which decay faster in powers of r than those taken into account by the main quadratic terms.*

Recall that

$$\mathfrak{q} = r^4 Q(\alpha), \tag{A.14.1}$$

where Q is the operator

$$Q := e_3 e_3 + (2\underline{\kappa} - 6\underline{\omega})e_3 + W, \qquad W := -4e_3(\underline{\omega}) + 8\underline{\omega}^2 - 8\underline{\omega}\,\underline{\kappa} + \frac{1}{2}\underline{\kappa}^2. \tag{A.14.2}$$

Lemma A.2. *The quantity $\mathfrak{q}$ is fully invariant with respect to the conformal frame transformations*

$$e_3' = \lambda^{-1} e_3, \qquad e_4 = \lambda e_4, \qquad e_\theta' = e_\theta.$$

Proof. The proof is an immediate consequence of Definition A.3 and Lemmas A.5, A.4 below. ☐

We recall that under the above mentioned frame transformation we have

$$\begin{aligned} \alpha' &= \lambda^2 \alpha, \quad \beta' = \lambda\beta, \quad \rho' = \rho, \quad \underline{\kappa}' = \lambda^{-1}\underline{\kappa}, \quad \kappa' = \lambda\kappa, \quad \eta' = \eta, \quad \underline{\eta}' = \underline{\eta}, \\ \underline{\omega}' &= \lambda^{-1}\left(\underline{\omega} + \frac{1}{2}e_3(\log\lambda)\right), \quad \omega' = \lambda\left(\omega - \frac{1}{2}e_4(\log\lambda)\right), \quad \zeta' = \zeta - e_\theta(\log\lambda). \end{aligned}$$

Definition A.3. *We say that a reduced tensor is conformal invariant of type[4] a, i.e., a-conformal invariant, if under the conformal change of frames $e_3' = \lambda^{-1}e_3$, $e_4' = \lambda e_4$, it transforms by*

$$f' = \lambda^a f.$$

Lemma A.4. *Let f be an a-conformal invariant tensor.*

1. *The tensor*

$$\nabla_3 f \quad := \quad e_3 f - 2a\underline{\omega} f \tag{A.14.3}$$

is $a-1$ conformal invariant.

2. *The tensor*

$$\nabla_4 f \quad := \quad e_4 f + 2a\omega f \tag{A.14.4}$$

[4]Note that for a given Ricci or curvature coefficient a coincides with the signature of the component.

is $a+1$ conformal invariant.

3. *The tensor*

$$\nabla\!\!\!/^{(c)}_A f = \nabla\!\!\!/_A f + a\zeta_A f \tag{A.14.5}$$

is a-conformal invariant.

Proof. Immediate verification. □

Lemma A.5. *We have*

$$Q(\alpha) = \nabla_3(\nabla_3\alpha) + 2\underline{\kappa}\nabla_3\alpha + \frac{1}{2}\underline{\kappa}^2\alpha.$$

Proof. We have

$$\begin{aligned}
\nabla_3(\nabla_3\alpha) &= \nabla_3(e_3\alpha - 4\underline{\omega}\alpha) = e_3(e_3\alpha - 4\underline{\omega}\alpha) - 2\underline{\omega}(e_3\alpha - 4\underline{\omega}\alpha)\\
&= e_3e_3\alpha - 4e_3\underline{\omega}\alpha - 4\underline{\omega}e_3\alpha - 2\underline{\omega}e_3\alpha + 8\underline{\omega}^2\alpha.
\end{aligned}$$

Hence,

$$\begin{aligned}
Q(\alpha) &= \nabla_3(\nabla_3\alpha) + 2\underline{\kappa}\nabla_3\alpha + \frac{1}{2}\underline{\kappa}\alpha\\
&= e_3e_3\alpha - 4e_3\underline{\omega}\alpha - 4\underline{\omega}e_3\alpha - 2\underline{\omega}e_3\alpha + 8\underline{\omega}^2\alpha + 2\underline{\kappa}(e_3\alpha - 4\underline{\omega}\alpha) + \frac{1}{2}\underline{\kappa}^2\alpha\\
&= e_3e_3\alpha + (2\underline{\kappa} - 6\underline{\omega})e_3\alpha + \left(-4e_3\underline{\omega} + 8\underline{\omega}^2 - 8\underline{\kappa}\,\underline{\omega} + \frac{1}{2}\underline{\kappa}^2\right)
\end{aligned}$$

as stated. □

Remark A.6. *Using the definitions of ∇_3, ∇_4 the null structure equations for $\kappa, \underline{\kappa}$ take the form*

$$\begin{aligned}
\nabla_3\underline{\kappa} + \frac{1}{2}\underline{\kappa}^2 &= 2\,\not{d}_1\underline{\xi} + \Gamma_b\cdot\Gamma_b = r^{-1}\,\not{\mathfrak{d}}\Gamma_b + l.o.t.,\\
\nabla_4\underline{\kappa} + \frac{1}{2}\kappa\,\underline{\kappa} &= 2\,\not{d}_1\underline{\eta} + 2\rho + \Gamma_g\cdot\Gamma_b = 2\rho + r^{-1}\,\not{\mathfrak{d}}\Gamma_g,\\
\nabla_3\kappa + \frac{1}{2}\underline{\kappa}\,\kappa &= 2\,\not{d}_1\eta + 2\rho + \Gamma_g\cdot\Gamma_b = 2\rho + r^{-1}\,\not{\mathfrak{d}}\Gamma_g,\\
\nabla_4\kappa + \frac{1}{2}\kappa^2 &= 2\,\not{d}_1\xi + \Gamma_g\cdot\Gamma_g = r^{-1}\,\not{\mathfrak{d}}\Gamma_g.
\end{aligned} \tag{A.14.6}$$

Also, since ρ is 0-conformal

$$\nabla_3\rho + \frac{3}{2}\underline{\kappa}\rho = \not{d}_1\underline{\beta} + \Gamma_g\cdot\Gamma_b = r^{-1}\,\not{\mathfrak{d}}\Gamma_g. \tag{A.14.7}$$

Definition A.7. *Given f an a-conformal S-tangent tensor we define its a-conformal Laplacian to be*

$$^{(c)}\Delta\!\!\!/ f = {}^{(c)}\nabla\!\!\!/_A\,{}^{(c)}\nabla\!\!\!/^A f.$$

Lemma A.8. *The following formula holds true for a 2-conformal tensor f*

$$ {}^{(c)}\not\!\Delta f = \not\!\Delta_2 f + 4\zeta\not\nabla f + 2\big(div\,\zeta + 2|\zeta|^2\big)f. $$

In particular, we have

$$ {}^{(c)}\not\!\Delta f = \not\!\Delta_2 f + r^{-1}\not\mathfrak{d}^{\leq 1}(\Gamma_g \cdot f). $$

Proof. Immediate verification. □

The goal of this section is to prove Theorem 2.108 which we recall below for the convenience of the reader.

Theorem A.9. *The invariant scalar quantity $\mathfrak{q}$ defined in* (2.3.10) *verifies the equation*

$$ \square_2\mathfrak{q} + \kappa\underline{\kappa}\,\mathfrak{q} = Err[\square_2\mathfrak{q}] \tag{A.14.8} $$

where, schematically,

$$ Err[\square_2\mathfrak{q}] := r^2\mathfrak{d}^{\leq 2}(\Gamma_g\cdot(\alpha,\beta)) + e_3\Big(r^3\mathfrak{d}^{\leq 2}(\Gamma_g\cdot(\alpha,\beta))\Big) + \mathfrak{d}^{\leq 1}(\Gamma_g\cdot\mathfrak{q}) + l.o.t. $$

Definition A.10. *Given a quadratic or higher order E we say the following:*

1. *$E \in Good$ if r^4E can be expressed in the form* (2.4.8).
2. *$E \in Good_1$ if after applying r^4e_3 or r^3 it can be expressed in the form* (2.4.8).
3. *$E \in Good_2$ if after applying $r^4e_3e_3, r^4e_3$ or r^3 it can be expressed in the form* (2.4.8).

In view of the definition we note that

$$ (e_3 + r^{-1})\text{Good}_1 = \text{Good}, \qquad Q\text{Good}_2 = \text{Good}. $$

To prove the theorem we have to check that $\text{Err}[\square_2\mathfrak{q}] = r^4\text{Good}$.

A.14.1 The Teukolsky equation for α

We recall below Proposition 2.107.

Lemma A.11. *We have*

$$ \begin{aligned} \square_2\alpha &= -4\underline{\omega}e_4(\alpha) + (4\omega + 2\kappa)e_3(\alpha) + V\alpha + Err[\square_{\mathbf{g}}\alpha], \\ V &= -4\rho - 4e_4(\underline{\omega}) - 8\omega\underline{\omega} + 2\omega\,\underline{\kappa} - 10\kappa\,\underline{\omega} + \frac{1}{2}\kappa\,\underline{\kappa}, \end{aligned} $$

where $Err[\square_{\mathbf{g}}\alpha]$ is given schematically by

$$ Err(\square_{\mathbf{g}}\alpha) := \Gamma_g e_3(\alpha) + r^{-1}\mathfrak{d}^{\leq 1}\Big((\eta,\Gamma_g)(\alpha,\beta)\Big) + \xi(e_3(\beta), r^{-1}\mathfrak{d}\check{\rho}). $$

Remark A.12. *Since ξ vanishes for $r \geq 4m_0$, $\eta \in \Gamma_g$ and $e_3\alpha = r^{-1}\mathfrak{d}\alpha$, we deduce*

$$ Err(\square_{\mathbf{g}}\alpha) \in Good_2. $$

Lemma A.13. *The Teukolsky equation for α can be written in the form*

$$\mathcal{L}(\alpha) \quad = \quad Good_2 \tag{A.14.9}$$

where $\mathcal{L}$ is the operator

$$\mathcal{L} \quad := \quad -\nabla_4\nabla_3 + {}^{(c)}\Delta_2 - \frac{5}{2}\kappa\nabla_3 - \frac{1}{2}\underline{\kappa}\nabla_4 - \left(-4\rho + \frac{1}{2}\kappa\underline{\kappa}\right). \tag{A.14.10}$$

We also note that, for a 0-conformal tensor f,

$$\square_2 f \quad = \quad -\nabla_4\nabla_3 f + {}^{(c)}\Delta_2 f - \frac{1}{2}\kappa\nabla_3 f - \frac{1}{2}\underline{\kappa}\nabla_4 f + r^{-1}\Gamma_g \cdot \not{\mathfrak{d}} f. \tag{A.14.11}$$

Proof. Recall that we have (see Definition 2.103)

$$\square_2\alpha \quad = \quad -e_4(e_3(\alpha)) + \Delta_2\alpha - \frac{1}{2}\underline{\kappa}e_4(\alpha) + \left(-\frac{1}{2}\kappa + 2\omega\right) e_3(\alpha) + 2\underline{\eta}e_\theta(\alpha).$$

Therefore,

$$\begin{aligned}
& \mathcal{L}(\alpha) \\
= \quad & -e_4(e_3(\alpha)) + \Delta_2\alpha - \frac{1}{2}\underline{\kappa}e_4(\alpha) + \left(-\frac{1}{2}\kappa + 2\omega\right) e_3(\alpha) + 2\underline{\eta}e_\theta(\alpha) \\
+ \quad & 4\underline{\omega}e_4(\alpha) - (4\omega + 2\kappa)e_3(\alpha) - V\alpha \\
= \quad & -e_4(e_3(\alpha)) + \Delta_2\alpha - \left(\frac{1}{2}\underline{\kappa} - 4\underline{\omega}\right) e_4\alpha - \left(\frac{5}{2}\kappa + 2\omega\right) e_3\alpha + 2\underline{\eta}e_\theta(\alpha) - V\alpha \\
= \quad & -e_4(e_3(\alpha)) + {}^{(c)}\Delta_2\alpha - \left(\frac{1}{2}\underline{\kappa} - 4\underline{\omega}\right) e_4\alpha - \left(\frac{5}{2}\kappa + 2\omega\right) e_3\alpha - V\alpha + \mathrm{Good}_2.
\end{aligned}$$

On the other hand,

$$\begin{aligned}
\nabla_4(\nabla_3(\alpha)) \quad &= \quad \nabla_4(e_3\alpha - 4\underline{\omega}\alpha) = e_4(e_3\alpha - 4\underline{\omega}\alpha) + 2\omega(e_3\alpha - 4\underline{\omega}\alpha) \\
&= \quad e_4e_3\alpha - 4\underline{\omega}e_4\alpha - 4e_4\underline{\omega}\alpha + 2\omega e_3\alpha - 8\omega\underline{\omega}\alpha.
\end{aligned}$$

Hence,

$$\begin{aligned}
-\nabla_4\nabla_3\alpha - \frac{5}{2}\kappa\nabla_3\alpha - \frac{1}{2}\underline{\kappa}\nabla_4\alpha \quad &= \quad -e_4e_3\alpha + 4\underline{\omega}e_4\alpha + 4e_4\underline{\omega}\alpha - 2\omega e_3\alpha + 8\omega\underline{\omega}\alpha \\
&- \quad \frac{5}{2}\kappa(e_3\alpha - 4\underline{\omega}\alpha) - \frac{1}{2}\underline{\kappa}(e_4\alpha + 4\omega\alpha) \\
&= \quad -e_4e_3\alpha - \frac{1}{2}(\underline{\kappa} - 4\underline{\omega})e_4\alpha - \left(\frac{5}{2}\kappa + 2\omega\right) e_3\alpha \\
&+ \quad (4e_4\underline{\omega} + 8\omega\underline{\omega} + 10\kappa\underline{\omega} - 2\omega\underline{\kappa})\alpha.
\end{aligned}$$

We deduce, with $V' = -4\rho + \frac{1}{2}\kappa\underline{\kappa}$,

$$\begin{aligned} & -\nabla_4\nabla_3\alpha - \frac{5}{2}\kappa\nabla_3\alpha - \frac{1}{2}\underline{\kappa}\nabla_4 - V'\alpha \\ = & -e_4e_3\alpha - \frac{1}{2}(\underline{\kappa} - 4\underline{\omega})e_4\alpha - \left(\frac{5}{2}\kappa + 2\omega\right)e_3\alpha \\ + & \left(4e_4\underline{\omega} + 8\omega\underline{\omega} + 10\kappa\underline{\omega} - 2\omega\underline{\kappa} + 4\rho - \frac{1}{2}\kappa\underline{\kappa}\right)\alpha \\ = & -e_4e_3\alpha - \frac{1}{2}(\underline{\kappa} - 4\underline{\omega})e_4\alpha - \left(\frac{5}{2}\kappa + 2\omega\right)e_3\alpha - V\alpha. \end{aligned}$$

Hence,

$$\mathcal{L}\alpha = -\nabla_4\nabla_3\alpha + {}^{(c)}\Delta_2\alpha - \frac{5}{2}\kappa\nabla_3\alpha - \frac{1}{2}\underline{\kappa}\nabla_4\alpha - \left(-4\rho + \frac{1}{2}\kappa\underline{\kappa}\right)\alpha \in \text{Good}_2$$

as desired. The proof of the second part of the lemma follows in the same manner. □

A.14.2 Commutation lemmas

The goal of the following lemmas is to calculate the commutator of Q with $\mathcal{L}$.

Lemma A.14. *Given f an a-conformal tensor we have*

$$[\nabla_3, \nabla_4]f = 2a\rho f + r^{-1}\Gamma_g \not{\mathfrak{d}}^{\leq 1} f. \tag{A.14.12}$$

Proof. We have

$$\begin{aligned} [\nabla_3, \nabla_4]f &= \nabla_3\nabla_4 f - \nabla_4\nabla_3 f \\ &= \big(e_3 - 2(a+1)\underline{\omega}\big)\big(e_4 f + 2a\omega f\big) - \big(e_4 + 2(a-1)\omega\big)\big(e_3 f - 2a\underline{\omega} f\big) \\ &= e_3e_4 f - 2(a+1)\underline{\omega}e_4 f + 2ae_3(\omega f) - 4a(a+1)\underline{\omega}\omega f \\ &- \; e_4e_3 f - 2(a-1)\omega e_3 f + 2ae_4(\underline{\omega} f) + 4a(a-1)\omega\underline{\omega} \\ &= [e_3, e_4]f - 2\underline{\omega}e_4 f + 2\omega e_3(f) + 2a\Big(e_3\omega + e_4\underline{\omega} - 4\omega\underline{\omega}\Big)f. \end{aligned}$$

Recall that

$$\begin{aligned} [e_3, e_4] &= -2\omega e_3 + 2\underline{\omega}e_4 + 2(\underline{\eta} - \eta)e_\theta, \\ e_3\omega + e_4\underline{\omega} - 4\omega\underline{\omega} &= \rho + \Gamma_g \cdot \Gamma_b. \end{aligned}$$

We deduce[5]

$$[\nabla_3, \nabla_4]f = 2a\rho + r^{-1}\Gamma_g \not{\mathfrak{d}}^{\leq 1} f$$

as stated. □

Lemma A.15. *Assume f a-conformal and g is b-conformal. Then fg is $a+b$-*

[5]Recall that $\eta \in \Gamma_g$ in the frame we are using.

conformal and

$$\begin{aligned}\nabla_3(fg) &= f\nabla_3 g + g\nabla_3 f,\\ \nabla_4(fg) &= f\nabla_4 g + g\nabla_4 f.\end{aligned}$$

Proof. Indeed

$$\nabla_3(fg) = e_3(fg) - 2(a+b)\underline{\omega} fg = fe_3 g + ge_3 f - 2(a+b)\underline{\omega} fg = f\nabla_3 g + g\nabla_3 f$$

as stated. □

Lemma A.16. *We have*

$$\begin{aligned}[Q,\nabla_3]\alpha &= \underline{\kappa}^2\nabla_3\alpha + \frac{1}{2}\underline{\kappa}^3\alpha + Good_1,\\ [Q,\nabla_4]\alpha &= \big(2\rho+\kappa\underline{\kappa}\big)\nabla_3\alpha + \frac{1}{2}\kappa\underline{\kappa}^2\alpha + Good_1.\end{aligned} \tag{A.14.13}$$

Also,

$$\begin{aligned}[Q,\nabla_4\nabla_3]\alpha &= \big(-2\rho+\kappa\underline{\kappa}\big)\nabla_3\nabla_3\alpha + \underline{\kappa}^2\nabla_4\nabla_3\alpha + \frac{1}{2}\underline{\kappa}^3\nabla_4\alpha\\ &+ \left(3\rho\underline{\kappa} - \frac{1}{2}\kappa\underline{\kappa}^2\right)\nabla_3\alpha + \frac{3}{2}\underline{\kappa}^2\left(-\frac{1}{2}\kappa\underline{\kappa} + 2\rho\right)\alpha + Good.\end{aligned} \tag{A.14.14}$$

Proof. We have[6]

$$\begin{aligned}[Q,\nabla_3]\alpha &= \left(\nabla_3\nabla_3 + 2\underline{\kappa}\nabla_3 + \frac{1}{2}\underline{\kappa}^2\right)\nabla_3\alpha - \nabla_3\left(\left(\nabla_3\nabla_3 + 2\underline{\kappa}\nabla_3 + \frac{1}{2}\underline{\kappa}^2\right)\alpha\right)\\ &= -2\nabla_3(\underline{\kappa})\nabla_3\alpha - \underline{\kappa}(\nabla_3\underline{\kappa})\alpha\\ &= -2\left(-\frac{1}{2}\underline{\kappa}^2 + r^{-1}\Gamma_b\right)\nabla_3\alpha - \underline{\kappa}\left(-\frac{1}{2}\underline{\kappa}^2 + r^{-1}\Gamma_b\right)\alpha\\ &= \underline{\kappa}^2\nabla_3\alpha + \frac{1}{2}\underline{\kappa}^3\alpha + r^{-1}\Gamma_g\mathfrak{d}^{\le 1}\alpha\end{aligned}$$

and

$$\begin{aligned}&[Q,\nabla_4]\alpha\\ &= \left(\nabla_3\nabla_3 + 2\underline{\kappa}\nabla_3 + \frac{1}{2}\underline{\kappa}^2\right)\nabla_4\alpha - \nabla_4\left(\left(\nabla_3\nabla_3 + 2\underline{\kappa}\nabla_3 + \frac{1}{2}\underline{\kappa}^2\right)\alpha\right)\\ &= (\nabla_3\nabla_3\nabla_4 - \nabla_4\nabla_3\nabla_3)\alpha + 2\underline{\kappa}(\nabla_3\nabla_4 - \nabla_4\nabla_3)\alpha - 2\nabla_4\underline{\kappa}\nabla_3\alpha - \underline{\kappa}(\nabla_4\underline{\kappa})\alpha\\ &= \nabla_3\Big([\nabla_3,\nabla_4]\alpha\Big) + [\nabla_3,\nabla_4]\nabla_3\alpha + 2\underline{\kappa}[\nabla_3,\nabla_4]\alpha - 2\nabla_4\underline{\kappa}\nabla_3\alpha - \underline{\kappa}(\nabla_4\underline{\kappa})\alpha.\end{aligned}$$

In view of Lemma A.14 we have

$$\begin{aligned}[\nabla_3,\nabla_4]\alpha &= 4\rho\alpha + r^{-1}\Gamma_g\cdot \not{\mathfrak{d}}^{\le 1}\alpha,\\ [\nabla_3,\nabla_4]\nabla_3\alpha &= 2\rho\nabla_3\alpha + r^{-1}\Gamma_g\cdot \not{\mathfrak{d}}^{\le 1}\nabla_3\alpha.\end{aligned}$$

[6] Recall that $r^{-1}\Gamma_b = \Gamma_g$.

Hence,

$$\begin{aligned}
[Q, \nabla_4]\alpha &= \nabla_3\Big(4\rho\alpha + r^{-1}\Gamma_g \, \mathfrak{d}^{\leq 1}\alpha\Big)\\
&+ \Big(2\rho + r^{-1}\Gamma_g \, \mathfrak{d}^{\leq 1}\Big)\nabla_3\alpha + 2\underline{\kappa}\Big(4\rho\alpha + r^{-1}\Gamma_g \, \mathfrak{d}^{\leq 1}\alpha\Big) - 2\nabla_4\underline{\kappa}\nabla_3\alpha\\
&- \underline{\kappa}(\nabla_4\underline{\kappa})\alpha\\
&= \big(6\rho - 2\nabla_4\underline{\kappa}\big)\nabla_3\alpha + \big(4\nabla_3\rho + 8\underline{\kappa}\rho - \underline{\kappa}\nabla_4\underline{\kappa}\big)\alpha + \text{Good}_1.
\end{aligned}$$

We now note, using the equations for $\nabla_4\rho$ and $\nabla_4\underline{\kappa}$,

$$\begin{aligned}
4\nabla_3\rho + 8\underline{\kappa}\rho - \underline{\kappa}\nabla_4\underline{\kappa} &= 4\left(-\frac{3}{2}\underline{\kappa}\rho + r^{-1}\,\mathfrak{d}\Gamma_g\right) + 8\underline{\kappa}\rho\\
&\quad -\underline{\kappa}\left(-\frac{1}{2}\kappa\underline{\kappa} + 2\rho + r^{-1}\,\mathfrak{d}\Gamma_g\right)\\
&= \frac{1}{2}\kappa\underline{\kappa}^2 + r^{-1}\,\mathfrak{d}\Gamma_g,\\
6\rho - 2\nabla_4\underline{\kappa} &= 6\rho - 2\left(-\frac{1}{2}\kappa\underline{\kappa} + 2\rho + r^{-1}\,\mathfrak{d}\Gamma_g\right)\\
&= 2\rho + \kappa\underline{\kappa} + r^{-1}\,\mathfrak{d}\Gamma_g.
\end{aligned}$$

Hence,

$$[Q, \nabla_4]\alpha = (2\rho + \kappa\underline{\kappa})\nabla_3\alpha + \frac{1}{2}\kappa\underline{\kappa}^2\alpha + \text{Good}_1$$

as stated.

Also,

$$[Q, \nabla_4\nabla_3]\alpha = [Q, \nabla_4]\nabla_3\alpha + \nabla_4\Big([Q, \nabla_3]\alpha\Big). \tag{A.14.15}$$

We first calculate, as above, for $f = \nabla_3\alpha$

$$\begin{aligned}
&[Q, \nabla_4]f\\
&= \left(\nabla_3\nabla_3 + 2\underline{\kappa}\nabla_3 + \frac{1}{2}\underline{\kappa}^2\right)\nabla_4 f - \nabla_4\left(\left(\nabla_3\nabla_3 + 2\underline{\kappa}\nabla_3 + \frac{1}{2}\underline{\kappa}^2\right)f\right)\\
&= (\nabla_3\nabla_3\nabla_4 - \nabla_4\nabla_3\nabla_3)f + 2\underline{\kappa}(\nabla_3\nabla_4 - \nabla_4\nabla_3)f - 2\nabla_4\underline{\kappa}\nabla_3\alpha - \underline{\kappa}(\nabla_4\underline{\kappa})f\\
&= \nabla_3\Big([\nabla_3, \nabla_4]f\Big) + [\nabla_3, \nabla_4]\nabla_3 f + 2\underline{\kappa}[\nabla_3, \nabla_4]f - 2\nabla_4\underline{\kappa}\nabla_3 f - \underline{\kappa}(\nabla_4\underline{\kappa})f.
\end{aligned}$$

In view of Lemma A.14, since $f = \nabla_3\alpha$ is 1-conformal and $\nabla_3 f$ is 0-conformal, we have

$$\begin{aligned}
[\nabla_3, \nabla_4]f &= 2\rho f + r^{-1}\Gamma_g \, \mathfrak{d}^{\leq 1}f,\\
[\nabla_3, \nabla_4]\nabla_3 f &= r^{-1}\Gamma_g \, \mathfrak{d}^{\leq 1}\nabla_3 f.
\end{aligned}$$

Hence

$$\begin{aligned}
[Q, \nabla_4] f &= \nabla_3\Big(2\rho f + r^{-1}\Gamma_g \not{\mathfrak{d}}^{\leq 1} f\Big) + \Big(r^{-1}\Gamma_g \not{\mathfrak{d}}^{\leq 1}\Big)\nabla_3 f \\
&+ 2\underline{\kappa}\Big(2\rho f + r^{-1}\Gamma_g \not{\mathfrak{d}}^{\leq 1} f\Big) - 2\nabla_4\underline{\kappa}\nabla_3 f - \underline{\kappa}(\nabla_4\underline{\kappa}) f \\
&= \big(2\rho - 2\nabla_4\underline{\kappa}\big)\nabla_3 f + \big(2\nabla_3\rho + 4\underline{\kappa}\rho - \underline{\kappa}\nabla_4\underline{\kappa}\big) f \\
&+ r^{-1}\Gamma_g \not{\mathfrak{d}}^{\leq 1}\nabla_3 f + r^{-2}\Gamma_g \not{\mathfrak{d}}^{\leq 1} f.
\end{aligned}$$

Therefore,

$$[Q, \nabla_4]\nabla_3\alpha = \big(2\rho - 2\nabla_4\underline{\kappa}\big)\nabla_3\nabla_3\alpha + \big(2\nabla_3\rho + 4\underline{\kappa}\rho - \underline{\kappa}\nabla_4\underline{\kappa}\big)\nabla_3\alpha + r^{-2}\Gamma_g \mathfrak{d}^{\leq 2}\alpha.$$

As above,

$$\begin{aligned}
2\rho - 2\nabla_4\underline{\kappa} &= 2\rho - 2\left(-\frac{1}{2}\kappa\underline{\kappa} + 2\rho + r^{-1}\Gamma_g\right) = -2\rho + \kappa\underline{\kappa} + r^{-1}\Gamma_g, \\
2\nabla_3\rho + 4\underline{\kappa}\rho - \underline{\kappa}\nabla_4\underline{\kappa} &= 2\left(-\frac{3}{2}\underline{\kappa}\rho + r^{-1}\Gamma_g\right) + 4\underline{\kappa}\rho - \underline{\kappa}\left(-\frac{1}{2}\kappa\underline{\kappa} + 2\rho + r^{-1}\Gamma_g\right) \\
&= \frac{1}{2}\kappa\underline{\kappa}^2 - \rho\underline{\kappa} + r^{-1}\Gamma_g.
\end{aligned}$$

Hence, since $r^{-1}\Gamma_g(\nabla_3\nabla_3\alpha, \nabla_3\alpha) = r^{-2}\Gamma_g \cdot \mathfrak{d}^{\leq 2}\alpha = \text{Good}$,

$$[Q, \nabla_4]\nabla_3\alpha = \big(-2\rho + \kappa\underline{\kappa}\big)\nabla_3\nabla_3\alpha + \left(\frac{1}{2}\kappa\underline{\kappa}^2 - \rho\underline{\kappa}\right)\nabla_3\alpha + \text{Good}. \quad \text{(A.14.16)}$$

We deduce

$$\begin{aligned}
[Q, \nabla_4\nabla_3]\alpha &= [Q, \nabla_4]\nabla_3\alpha + \nabla_4\Big([Q, \nabla_3]\alpha\Big) \\
&= \big(-2\rho + \kappa\underline{\kappa}\big)\nabla_3\nabla_3\alpha + \left(\frac{1}{2}\kappa\underline{\kappa}^2 - \rho\underline{\kappa}\right)\nabla_3\alpha \\
&+ \nabla_4\left(\underline{\kappa}^2\nabla_3\alpha + \frac{1}{2}\underline{\kappa}^3\alpha + \text{Good}_1\right) + \text{Good} \\
&= \big(-2\rho + \kappa\underline{\kappa}\big)\nabla_3\nabla_3\alpha + \underline{\kappa}^2\nabla_4\nabla_3\alpha + \frac{1}{2}\underline{\kappa}^3\nabla_4\alpha \\
&+ \left(\nabla_4(\underline{\kappa}^2) + \frac{1}{2}\kappa\underline{\kappa}^2 - \rho\underline{\kappa}\right)\nabla_3\alpha + \frac{3}{2}\underline{\kappa}^2\nabla_4\underline{\kappa}\alpha + \text{Good}.
\end{aligned}$$

Note that

$$\begin{aligned}
\nabla_4(\underline{\kappa}^2) + \frac{1}{2}\kappa\underline{\kappa}^2 - \rho\underline{\kappa} &= 2\underline{\kappa}\left(-\frac{1}{2}\kappa\underline{\kappa} + 2\rho + r^{-1}\not{\mathfrak{d}}\Gamma_g\right) + \frac{1}{2}\kappa\underline{\kappa}^2 - \rho\underline{\kappa} \\
&= 3\rho\underline{\kappa} - \frac{1}{2}\kappa\underline{\kappa}^2 + r^{-2}\not{\mathfrak{d}}\Gamma_g, \\
\frac{3}{2}\underline{\kappa}^2\nabla_4\underline{\kappa} &= \frac{3}{2}\underline{\kappa}^2\left(-\frac{1}{2}\kappa\underline{\kappa} + 2\rho + r^{-1}\not{\mathfrak{d}}\Gamma_g\right).
\end{aligned}$$

Hence,

$$
\begin{aligned}
[Q, \nabla_4\nabla_3]\alpha &= \big(-2\rho+\kappa\underline{\kappa}\big)\nabla_3\nabla_3\alpha+\underline{\kappa}^2\nabla_4\nabla_3\alpha+\frac{1}{2}\underline{\kappa}^3\nabla_4\alpha\\
&+ \left(3\rho\underline{\kappa}-\frac{1}{2}\kappa\underline{\kappa}^2\right)\nabla_3\alpha+\frac{3}{2}\underline{\kappa}^2\left(-\frac{1}{2}\kappa\underline{\kappa}+2\rho\right)\alpha+\text{Good}
\end{aligned}
$$

as stated. □

Lemma A.17. *Given f a 2-conformal tensor in $\mathfrak{s}_2$ we have*

$$[\nabla_3, {}^{(c)}\triangle]f = -\underline{\kappa}\,{}^{(c)}\triangle f + r^{-1}\mathfrak{d}^{\leq 2}(\Gamma_b\cdot f).$$

Proof. Recall that for a 2-conformal spacetime tensor f we have

$$ {}^{(c)}\triangle f = \triangle f + r^{-1}\not{\mathfrak{d}}^{\leq 1}(\Gamma_g\cdot f).$$

Hence,

$$[\nabla_3, {}^{(c)}\triangle]f = [\nabla_3, \triangle]f+\nabla_3\big(r^{-1}\not{\mathfrak{d}}^{\leq 1}(\Gamma_g\cdot f)\big)+r^{-1}\not{\mathfrak{d}}^{\leq 1}(\Gamma_g\cdot\nabla_3 f).$$

On the other hand, since $\nabla\!\!\!/\,\underline{\omega}=r^{-1}\not{\mathfrak{d}}\Gamma_b, \nabla\!\!\!/^2\underline{\omega}=r^{-2}\not{\mathfrak{d}}^2\Gamma_b$,

$$[\nabla_3, \triangle]f = [e_3-4\underline{\omega}, \triangle]f = [e_3, \triangle]f + r^{-2}\not{\mathfrak{d}}^{\leq 2}(\Gamma_b\cdot f).$$

We deduce

$$[\nabla_3, {}^{(c)}\triangle]f = [e_3, \triangle]f + r^{-2}\mathfrak{d}^{\leq 2}(\Gamma_b\cdot f)+e_3\big(r^{-1}\not{\mathfrak{d}}^{\leq 1}(\Gamma_g\cdot f)\big).$$

In the reduced form, for an $\mathfrak{s}_2$ tensor f,

$$[\nabla_3, {}^{(c)}\triangle]f = [e_3, \triangle_2]f + r^{-2}\mathfrak{d}^{\leq 2}(\Gamma_b\cdot f)+e_3\big(r^{-1}\not{\mathfrak{d}}^{\leq 1}(\Gamma_g\cdot f)\big).$$

We now recall that $\triangle_2=-\not{d}_2^\star\not{d}_2+2K$. Hence, applying the commutation Lemma[7] 2.54,

$$
\begin{aligned}
[\triangle_2, e_3]f &= [-\not{d}_2^\star\not{d}_2+2K, e_3]f = -\not{d}_2^\star[\not{d}_2, e_3]f-[\not{d}_2^\star, e_3]\not{d}_2 f-2e_3(K)f\\
&= -\not{d}_2^\star\left(\frac{1}{2}\underline{\kappa}\not{d}_2+\underline{Com}_2(f)\right)-\left(\frac{1}{2}\underline{\kappa}\not{d}_2^\star+\underline{Com}_2^*(\not{d}_2 f)\right)-2e_3(K)\\
&= -\underline{\kappa}\not{d}_2^\star\not{d}_2 f-2e_3(K)f+e_\theta(\underline{\kappa})\not{d}_2 f-\not{d}_2^\star\big(\underline{Com}_2(f)\big)-\underline{Com}_2^*(\not{d}_2 f)\\
&= -\underline{\kappa}\not{d}_2^\star\not{d}_2 f-2e_3(K)f+r^{-2}\mathfrak{d}^{\leq 2}(\Gamma_b\cdot f)+r^{-1}\mathfrak{d}^{\leq 1}(\Gamma_g\cdot e_3 f)\\
&= \underline{\kappa}\triangle_2 f-2(e_3K+\underline{\kappa}K)f+r^{-2}\mathfrak{d}^{\leq 2}(\Gamma_b\cdot f)+r^{-1}\mathfrak{d}^{\leq 1}(\Gamma_g\cdot e_3 f).
\end{aligned}
$$

---

[7] Recall that we have

$$
\begin{aligned}
\underline{Com}_2(f) &= -\frac{1}{2}\underline{\vartheta}\not{d}_3^\star f+(\zeta-\eta)e_3 f-2\eta e_3\Phi f-\underline{\xi}(e_4 f+ke_4(\Phi)f)-2\underline{\beta}f,\\
\underline{Com}_2^*(f) &= -\frac{1}{2}\underline{\vartheta}\not{d}_1 f-(\zeta-\eta)e_3 f-\eta e_3\Phi f+\underline{\xi}(e_4 f-e_4(\Phi)f)-\underline{\beta}f.
\end{aligned}
$$

Note that, ignoring the quadratic terms,

$$\begin{aligned}
e_3 K + \underline{\kappa} K &= -e_3\left(\rho + \frac{1}{4}\kappa\underline{\kappa}\right) - \underline{\kappa}\left(\rho + \frac{1}{4}\kappa\underline{\kappa}\right)\\
&= -e_3\rho - \underline{\kappa}\rho - \frac{1}{4}\Big(e_3(\kappa\underline{\kappa}) + \kappa\underline{\kappa}^2\Big)\\
&= \frac{1}{2}\underline{\kappa}\rho - \not{d}_1\underline{\beta}\\
&- \frac{1}{4}\left\{\kappa\left(-\frac{1}{2}\underline{\kappa}^2 - 2\underline{\omega}\,\underline{\kappa}\right) + \underline{\kappa}\left(-\frac{1}{2}\kappa\underline{\kappa} + 2\underline{\omega}\kappa + 2\not{d}_1\eta + 2\rho\right) + \kappa\underline{\kappa}^2\right\}\\
&= -\not{d}_1\underline{\beta} - \frac{1}{2}\underline{\kappa}\not{d}_1\eta.
\end{aligned}$$

We deduce

$$[e_3, \not{\Delta}_2] = -[\not{\Delta}_2, e_3]f = -\underline{\kappa}\not{\Delta}_2 f + r^{-1}\mathfrak{d}^{\leq 2}(\Gamma_b\cdot f).$$

Consequently,

$$[\nabla_3, {}^{(c)}\not{\Delta}]f = -\underline{\kappa}\,{}^{(c)}\not{\Delta}f + r^{-1}\mathfrak{d}^{\leq 2}(\Gamma_b\cdot f)$$

as stated. □

Lemma A.18. *We have*

$$[Q, {}^{(c)}\not{\Delta}]\alpha = -2\underline{\kappa}\nabla_3\,{}^{(c)}\not{\Delta}\alpha - \frac{5}{2}\underline{\kappa}^2\,{}^{(c)}\not{\Delta}\alpha + Good. \tag{A.14.17}$$

Proof. We have

$$\begin{aligned}
[Q, {}^{(c)}\not{\Delta}_2]\alpha &= \left[\nabla_3\nabla_3 + 2\underline{\kappa}\nabla_3 + \frac{1}{2}\underline{\kappa}^2\right]{}^{(c)}\not{\Delta}\alpha - {}^{(c)}\not{\Delta}\left[\nabla_3\nabla_3\alpha + 2\underline{\kappa}\nabla_3\alpha + \frac{1}{2}\underline{\kappa}^2\alpha\right]\\
&= \nabla_3[\nabla_3, {}^{(c)}\not{\Delta}]\alpha + [\nabla_3, {}^{(c)}\not{\Delta}]e_3\alpha + [2\underline{\kappa}\nabla_3, {}^{(c)}\not{\Delta}]\alpha + \left[\frac{1}{2}\underline{\kappa}^2, {}^{(c)}\not{\Delta}_2\right]\alpha.
\end{aligned}$$

Note that

$$\begin{aligned}
[2\underline{\kappa}\nabla_3, {}^{(c)}\not{\Delta}]\alpha &= 2\underline{\kappa}[\nabla_3, {}^{(c)}\not{\Delta}]\alpha + \text{Good},\\
\left[\frac{1}{2}\underline{\kappa}^2, {}^{(c)}\not{\Delta}_2\right]\alpha &= \text{Good}.
\end{aligned}$$

Hence, using the previous commutation lemma,

$$\begin{aligned}
[Q, {}^{(c)}\triangle_2]\alpha &= \nabla_3[\nabla_3, {}^{(c)}\triangle]\alpha + [\nabla_3, {}^{(c)}\triangle]e_3\alpha + 2\underline{\kappa}[\nabla_3, {}^{(c)}\triangle]\alpha + \text{Good}\\
&= \nabla_3\Big(-\underline{\kappa}\,{}^{(c)}\triangle\alpha + r^{-1}\mathfrak{d}^{\leq 2}(\Gamma_b\cdot\alpha)\Big)\\
&+ \Big(-\underline{\kappa}\,{}^{(c)}\triangle\nabla_3\alpha + r^{-1}\mathfrak{d}^{\leq 2}(\Gamma_b\cdot\nabla_3\alpha)\Big)\\
&+ 2\underline{\kappa}\Big(-\underline{\kappa}\,{}^{(c)}\triangle\alpha + r^{-1}\mathfrak{d}^{\leq 2}(\Gamma_b\cdot\alpha)\Big) + \text{Good}\\
&= -\underline{\kappa}(\nabla_3\,{}^{(c)}\triangle\alpha + {}^{(c)}\triangle\nabla_3\alpha) - \big(\nabla_3\underline{\kappa} + 2\underline{\kappa}^2\big)\,{}^{(c)}\triangle\alpha + \text{Good}\\
&= -\underline{\kappa}(2\nabla_3\,{}^{(c)}\triangle\alpha - [\nabla_3, {}^{(c)}\triangle]\alpha) - \big(\nabla_3\underline{\kappa} + 2\underline{\kappa}^2\big)\,{}^{(c)}\triangle\alpha + \text{Good}\\
&= -2\underline{\kappa}\nabla_3\,{}^{(c)}\triangle\alpha - \big(\nabla_3\underline{\kappa} + 3\underline{\kappa}^2\big)\,{}^{(c)}\triangle\alpha + \text{Good}.
\end{aligned}$$

Note that

$$(\nabla_3\underline{\kappa} + 3\underline{\kappa}^2)\,{}^{(c)}\triangle\alpha = \left(\frac{5}{2}\underline{\kappa}^2 + r^{-1}\not{\mathfrak{d}}\Gamma_b\right){}^{(c)}\triangle\alpha = \frac{5}{2}\underline{\kappa}^2 + r^{-2}\not{\mathfrak{d}}\Gamma_g\cdot\not{\mathfrak{d}}^{\leq 2}\alpha.$$

Hence,

$$[Q, {}^{(c)}\triangle_2]\alpha = -2\underline{\kappa}\nabla_3\,{}^{(c)}\triangle\alpha - \frac{5}{2}\underline{\kappa}^2\,{}^{(c)}\triangle\alpha + \text{Good}$$

as stated. □

Lemma A.19. *We have*

$$Q(fg) = Q(f)g + fQ(g) + 2\nabla_3 f\nabla_3 g - \tfrac{1}{2}\underline{\kappa}^2 fg.$$

Also,

$$\begin{aligned}
[Q, fe_4]g &= Q(f)\nabla_4 g + f[Q, e_4]g + 2\nabla_3 f\nabla_3\nabla_4 g - \frac{1}{2}\underline{\kappa}^2 f\nabla_4 g,\\
[Q, f\nabla_3]g &= Q(f)\nabla_3 g + f[Q, \nabla_3]g + 2\nabla_3 f\nabla_3\nabla_3 g - \frac{1}{2}\underline{\kappa}^2 f\nabla_3 g.
\end{aligned}$$

Proof. Recall that

$$Q = \nabla_3\nabla_3 + 2\underline{\kappa}\nabla_3 + \frac{1}{2}\underline{\kappa}^2.$$

Hence,

$$\begin{aligned}
Q(fg) &= \left[\nabla_3\nabla_3 + 2\underline{\kappa}\nabla_3 + \frac{1}{2}\underline{\kappa}^2\right](fg)\\
&= (\nabla_3\nabla_3 f)g + f(\nabla_3\nabla_3 g) + 2\nabla_3 f\nabla_3 g + 2\underline{\kappa}(\nabla_3 f g + f\nabla_3 g) + \frac{1}{2}\underline{\kappa}^2 fg\\
&= \Big(\nabla_3\nabla_3 f + 2\underline{\kappa}\nabla_3 f\Big)g + 2\nabla_3 f\nabla_3 g + fQ(g)\\
&= Q(f)g + fQ(g) + 2\nabla_3 f\nabla_3 g - \frac{1}{2}\underline{\kappa}^2 fg.
\end{aligned}$$

Also,

$$\begin{aligned}[Q, f\nabla_4]g &= Q(f\nabla_4 g) - f\nabla_4 Q(g) = Q(f)\nabla_4 g + fQ\nabla_4(g) + 2\nabla_3 f\nabla_3\nabla_4 g\\ &- \frac{1}{2}\underline{\kappa}^2 f\nabla_4 g - f\nabla_4 Q(g)\\ &= \left(Q(f) - \frac{1}{2}\underline{\kappa} f\right)\nabla_4 g + f[Q, \nabla_4]g + 2\nabla_3 f\nabla_3\nabla_4 g.\end{aligned}$$

Similarly,

$$[Q, f\nabla_3]g = \left(Q(f) - \frac{1}{2}\underline{\kappa}^2 f\right)\nabla_3 g + f[Q, \nabla_3]g + 2\nabla_3 f\nabla_3\nabla_3 g$$

as stated. □

A.14.3 Main commutation

Proposition A.20. *The following identity holds true.*

$$[Q, \mathcal{L}]\alpha = -2\underline{\kappa}\nabla_4 Q(\alpha) + C_Q Q(\alpha) + Good, \tag{A.14.18}$$

where

$$C_Q = -8\rho - \frac{7}{2}\kappa\underline{\kappa}.$$

Proof. In view of Lemma A.13, we have

$$\mathcal{L}\alpha = -\nabla_4\nabla_3\alpha + {}^{(c)}\!\not{\Delta}_2\alpha - \frac{5}{2}\kappa\nabla_3\alpha - \frac{1}{2}\underline{\kappa}\nabla_4\alpha - \left(-4\rho + \frac{1}{2}\kappa\underline{\kappa}\right)\alpha = Good_2.$$

Hence, we infer

$$\begin{aligned}[Q, \mathcal{L}]\alpha &= -[Q, \nabla_4\nabla_3]\alpha + [Q, \not{\Delta}_2]\alpha - \frac{1}{2}[Q, \underline{\kappa}\nabla_4]\alpha - \frac{5}{2}[Q, \kappa\nabla_3]\alpha\\ &+ \left[Q, 4\rho - \frac{1}{2}\kappa\underline{\kappa}\right]\alpha\\ &= I + J + K + L + M\end{aligned} \tag{A.14.19}$$

with I, J, K, L, M, denoting each of the commutators on the left of (A.14.19).

A.14.3.1 Expression for I

In view of Lemma A.16 we have, for $I = -[Q, \nabla_4\nabla_3]\alpha$,

$$\begin{aligned}I &= (2\rho - \kappa\underline{\kappa})\nabla_3\nabla_3\alpha - \underline{\kappa}^2\nabla_4\nabla_3\alpha - \frac{1}{2}\underline{\kappa}^3\nabla_4\alpha - \left(-\frac{1}{2}\kappa\underline{\kappa}^2 + 3\rho\underline{\kappa}\right)\nabla_3\alpha\\ &- \frac{3}{2}\underline{\kappa}^2\left(-\frac{1}{2}\kappa\underline{\kappa} + 2\rho\right)\alpha + \text{Good}.\end{aligned} \tag{A.14.20}$$

A.14.3.2 Expression for J

Using Lemma A.18,

$$J = [Q, {}^{(c)}\Delta\!\!\!\!/]\alpha = -2\underline{\kappa}\nabla_3 {}^{(c)}\Delta\!\!\!\!/\alpha - \frac{5}{2}\underline{\kappa}^2 {}^{(c)}\Delta\!\!\!\!/\alpha.$$

Recalling the definition of $\mathcal{L}$ and the fact that $\mathcal{L}\alpha = \text{Good}_1$ we write

$$\Delta\!\!\!\!/_2\alpha = \nabla_4\nabla_3\alpha + \frac{5}{2}\kappa\nabla_3\alpha + \frac{1}{2}\underline{\kappa}\nabla_4\alpha + \left(-4\rho + \frac{1}{2}\kappa\underline{\kappa}\right)\alpha + \text{Good}_1.$$

Hence,

$$\begin{aligned} J &= -2\underline{\kappa}\nabla_3\left(\nabla_4\nabla_3\alpha + \frac{5}{2}\kappa\nabla_3\alpha + \frac{1}{2}\underline{\kappa}\nabla_4\alpha + \left(-4\rho + \frac{1}{2}\kappa\underline{\kappa}\right)\alpha\right) \\ &- \frac{5}{2}\underline{\kappa}^2\left(\nabla_4\nabla_3\alpha + \frac{5}{2}\kappa\nabla_3\alpha + \frac{1}{2}\underline{\kappa}\nabla_4\alpha + \left(-4\rho + \frac{1}{2}\kappa\underline{\kappa}\right)\alpha\right) \\ &= -2\underline{\kappa}\nabla_3\nabla_4\nabla_3\alpha - 5\kappa\underline{\kappa}\nabla_3\nabla_3\alpha - \underline{\kappa}^2\nabla_3\nabla_4\alpha - 2\underline{\kappa}\left(-4\rho + \frac{1}{2}\kappa\underline{\kappa}\right)\nabla_3\alpha \\ &- 2\underline{\kappa}\left(\frac{5}{2}\nabla_3\kappa\nabla_3\alpha + \frac{1}{2}\nabla_3\underline{\kappa}\nabla_4\alpha + \nabla_3\left(-4\rho + \frac{1}{2}\kappa\underline{\kappa}\right)\alpha\right) \\ &- \frac{5}{2}\underline{\kappa}^2\left(\nabla_4\nabla_3\alpha + \frac{5}{2}\kappa\nabla_3\alpha + \frac{1}{2}\underline{\kappa}\nabla_4\alpha + \left(-4\rho + \frac{1}{2}\kappa\underline{\kappa}\right)\alpha\right). \end{aligned}$$

According to Lemma A.14

$$\begin{aligned} \nabla_3\nabla_4\nabla_3\alpha &= \nabla_4\nabla_3\nabla_3\alpha + [\nabla_3, \nabla_4]\nabla_3\alpha \\ &= \nabla_4\nabla_3\nabla_3\alpha + 2\rho\nabla_3\alpha + r^{-1}\Gamma_g \mathfrak{d}\!\!/^{\le 1}\nabla_3\alpha \\ &= \nabla_4\nabla_3\nabla_3\alpha + 2\rho\nabla_3\alpha + \text{Good}, \\ \nabla_3\nabla_4\alpha &= \nabla_4\nabla_3\alpha + 4\rho\alpha + \text{Good}_1. \end{aligned}$$

We deduce, modulo Good error terms,

$$\begin{aligned} J &= -2\underline{\kappa}(\nabla_4\nabla_3\nabla_3\alpha + 2\rho\nabla_3\alpha) - 5\kappa\underline{\kappa}\nabla_3\nabla_3\alpha - \underline{\kappa}^2(\nabla_4\nabla_3\alpha + 4\rho\alpha) \\ &- 2\underline{\kappa}\left(-4\rho + \frac{1}{2}\kappa\underline{\kappa}\right)\nabla_3\alpha - 5\underline{\kappa}\nabla_3\kappa\nabla_3\alpha - \underline{\kappa}\nabla_3\underline{\kappa}\nabla_4\alpha \\ &- 2\underline{\kappa}\nabla_3\left(-4\rho + \frac{1}{2}\kappa\underline{\kappa}\right)\alpha \\ &- \frac{5}{2}\underline{\kappa}^2\left(\nabla_4\nabla_3\alpha + \frac{5}{2}\kappa\nabla_3\alpha + \frac{1}{2}\underline{\kappa}\nabla_4\alpha + \left(-4\rho + \frac{1}{2}\kappa\underline{\kappa}\right)\alpha\right). \end{aligned}$$

Grouping terms, we rewrite in the form

$$J = -2\underline{\kappa}\nabla_4\nabla_3\nabla_3\alpha - 5\kappa\underline{\kappa}\nabla_3\nabla_3\alpha + J_{43}\nabla_4\nabla_3\alpha + J_4\nabla_4\alpha + J_3\nabla_3\alpha + J_0\alpha.$$

We calculate the coefficients J_{43}, J_4, J_3, J_0 as follows.

$$
\begin{aligned}
J_{43} &= -\underline{\kappa}^2 - \frac{5}{2}\underline{\kappa}^2 = -\frac{7}{2}\underline{\kappa}^2,\\
J_4 &= -\underline{\kappa}\nabla_3\underline{\kappa} - \frac{5}{4}\underline{\kappa}^3 = -\underline{\kappa}\left(-\frac{1}{2}\underline{\kappa}^2 + r^{-1}\not{\mathfrak{d}}\Gamma_b\right) - \frac{5}{4}\underline{\kappa}^3 = -\frac{3}{4}\underline{\kappa}^3 + r^{-2}\not{\mathfrak{d}}\Gamma_b,\\
J_3 &= -4\underline{\kappa}\rho - 2\underline{\kappa}\left(-4\rho + \frac{1}{2}\kappa\underline{\kappa}\right) - 5\underline{\kappa}\nabla_3\kappa - \frac{25}{4}\underline{\kappa}^3\\
&= 4\underline{\kappa}\rho - \frac{29}{4}\underline{\kappa}^3 - 5\underline{\kappa}\left(-\frac{1}{2}\kappa\underline{\kappa} + 2\rho + r^{-1}\not{\mathfrak{d}}\Gamma_g\right)\\
&= -6\underline{\kappa}\rho - \frac{19}{4}\underline{\kappa}^3 + r^{-2}\not{\mathfrak{d}}\Gamma_g,\\
J_0 &= -4\rho\underline{\kappa}^2 - 2\underline{\kappa}\nabla_3\left(-4\rho + \frac{1}{2}\kappa\underline{\kappa}\right) - \frac{5}{2}\underline{\kappa}^2\left(-4\rho + \frac{1}{2}\kappa\underline{\kappa}\right)\\
&= 6\rho\underline{\kappa}^2 - \frac{5}{4}\kappa\underline{\kappa}^3 + 8\underline{\kappa}\nabla_3\rho - \underline{\kappa}\left(\underline{\kappa}\nabla_3\kappa + \kappa\nabla_3\underline{\kappa}\right)\\
&= 6\rho\underline{\kappa}^2 - \frac{5}{4}\kappa\underline{\kappa}^3 + 8\underline{\kappa}\left(-\frac{3}{2}\underline{\kappa}\rho + r^{-1}\not{\mathfrak{d}}\Gamma_g\right)\\
&- \underline{\kappa}\left(\underline{\kappa}\left(-\frac{1}{2}\kappa\underline{\kappa} + 2\rho + r^{-1}\Gamma_g\right) + \kappa\left(-\frac{1}{2}\underline{\kappa}^2 + r^{-1}\not{\mathfrak{d}}\Gamma_b\right)\right)\\
&= -8\underline{\kappa}^2\rho - \frac{5}{4}\kappa\underline{\kappa}^3 + \kappa\underline{\kappa}^3 + r^{-3}\not{\mathfrak{d}}\Gamma_b + r^{-2}\Gamma_g.
\end{aligned}
$$

Hence

$$
\begin{aligned}
J_4\nabla_4\alpha &= -\frac{3}{4}\underline{\kappa}^3 + \text{Good},\\
J_3\nabla_3\alpha &= \left(-6\underline{\kappa}\rho - \frac{19}{4}\underline{\kappa}^3\right)\nabla_3\alpha + \text{Good},\\
J_0\alpha &= -8\underline{\kappa}^2\rho - \frac{1}{4}\kappa\underline{\kappa}^3 + \text{Good}.
\end{aligned}
$$

We finally derive

$$
\begin{aligned}
J = &-2\underline{\kappa}\nabla_4\nabla_3\nabla_3\alpha - 5\kappa\underline{\kappa}\nabla_3\nabla_3\alpha - \frac{7}{2}\underline{\kappa}^2\nabla_4\nabla_3\alpha\\
&- \frac{3}{4}\underline{\kappa}^3\nabla_4\alpha - \left(6\underline{\kappa}\rho + \frac{19}{4}\kappa\underline{\kappa}^2\right)\nabla_3\alpha - \left(8\underline{\kappa}^2\rho + \frac{1}{4}\kappa\underline{\kappa}^3\right)\alpha + \text{Good}.
\end{aligned}
\tag{A.14.21}
$$

A.14.3.3 Expression for K

Also, using Lemma A.19 and Lemma A.16 (according to which we have the identity $[Q, \nabla_4]\alpha = (2\rho + \kappa\underline{\kappa})\nabla_3\alpha + \frac{1}{2}\kappa\underline{\kappa}^2\alpha + \text{Good}_1$)

$$
\begin{aligned}
K &= -\frac{1}{2}\Big[Q, \underline{\kappa}\nabla_4\Big]\alpha = -\frac{1}{2}\left(Q(\underline{\kappa})\nabla_4\alpha + \underline{\kappa}[Q, \nabla_4]\alpha + 2\nabla_3\underline{\kappa}\nabla_3\nabla_4\alpha - \frac{1}{2}\underline{\kappa}^3\nabla_4\alpha\right)\\
&= -\frac{1}{2}\left(Q(\underline{\kappa}) - \frac{1}{2}\underline{\kappa}^3\right)\nabla_4\alpha - \frac{1}{2}\underline{\kappa}\left((2\rho + \kappa\underline{\kappa})\nabla_3\alpha + \frac{1}{2}\kappa\underline{\kappa}^2\alpha\right)\\
&- \nabla_3\underline{\kappa}\nabla_3\nabla_4\alpha + \text{Good}.
\end{aligned}
$$

Hence,

$$K = -\nabla_3\underline{\kappa}\nabla_3\nabla_4\alpha - \frac{1}{2}\left(Q(\underline{\kappa}) - \frac{1}{2}\underline{\kappa}^3\right)\nabla_4\alpha - \frac{1}{2}\underline{\kappa}(2\rho + \kappa\underline{\kappa})\nabla_3\alpha - \frac{1}{4}\kappa\underline{\kappa}^3\alpha + \text{Good}.$$

We calculate the expression,

$$\begin{aligned}
Q(\underline{\kappa}) - \frac{1}{2}\underline{\kappa}^3 &= \nabla_3\nabla_3\underline{\kappa} + 2\underline{\kappa}\nabla_3\underline{\kappa} = \nabla_3\left(-\frac{1}{2}\underline{\kappa}^2 + r^{-1}\not{\mathfrak{d}}\Gamma_b\right)\\
&\quad +2\underline{\kappa}\left(-\frac{1}{2}\underline{\kappa}^2 + r^{-1}\not{\mathfrak{d}}\Gamma_b\right)\\
&= -\underline{\kappa}\left(\nabla_3\underline{\kappa} + \underline{\kappa}^2\right) + \nabla_3(r^{-1}\not{\mathfrak{d}}\Gamma_b) + r^{-2}\not{\mathfrak{d}}\Gamma_b\\
&= -\frac{1}{2}\underline{\kappa}^3 + \nabla_3(r^{-1}\not{\mathfrak{d}}\Gamma_b) + r^{-2}\not{\mathfrak{d}}\Gamma_b.
\end{aligned}$$

Hence,

$$\begin{aligned}
K &= -\nabla_3\underline{\kappa}\nabla_3\nabla_4\alpha + \frac{1}{4}\underline{\kappa}^3\nabla_4\alpha - \frac{1}{2}\underline{\kappa}(2\rho + \kappa\underline{\kappa})\nabla_3\alpha - \frac{1}{4}\kappa\underline{\kappa}^3\alpha\\
&\quad +\nabla_3(r^{-1}\not{\mathfrak{d}}\Gamma_b)\nabla_4\alpha + \text{Good}.
\end{aligned}$$

We note that

$$\begin{aligned}
\nabla_3(r^{-1}\not{\mathfrak{d}}\Gamma_b)\nabla_4\alpha &= \nabla_3(r^{-1}\not{\mathfrak{d}}\Gamma_b\nabla_4\alpha) - r^{-1}\not{\mathfrak{d}}\Gamma_b\nabla_3\nabla_4\alpha\\
&= \nabla_3(r^{-1}\not{\mathfrak{d}}\Gamma_g\mathfrak{d}^{\leq 1}\alpha) - r^{-2}\not{\mathfrak{d}}\Gamma_g\mathfrak{d}^{\leq 2}\alpha = \text{Good}.
\end{aligned}$$

We deduce

$$\begin{aligned}
K &= -\nabla_3\underline{\kappa}\nabla_3\nabla_4\alpha + \frac{1}{4}\underline{\kappa}^3\nabla_4\alpha - \frac{1}{2}\underline{\kappa}(2\rho + \kappa\underline{\kappa})\nabla_3\alpha - \frac{1}{4}\kappa\underline{\kappa}^3\alpha + \text{Good}\\
&= -\left(-\frac{1}{2}\underline{\kappa}^2 + r^{-1}\not{\mathfrak{d}}\Gamma_b\right)\nabla_3\nabla_4\alpha + \frac{1}{4}\underline{\kappa}^3\nabla_4\alpha - \frac{1}{2}\underline{\kappa}(2\rho + \kappa\underline{\kappa})\nabla_3\alpha - \frac{1}{4}\kappa\underline{\kappa}^3\alpha\\
&\quad +\text{Good}\\
&= \frac{1}{2}\underline{\kappa}^2\nabla_3\nabla_4\alpha + \frac{1}{4}\underline{\kappa}^3\nabla_4\alpha - \frac{1}{2}\underline{\kappa}(2\rho + \kappa\underline{\kappa})\nabla_3\alpha - \frac{1}{4}\kappa\underline{\kappa}^3\alpha + \text{Good}.
\end{aligned}$$

In view of Lemma A.14, we have $[\nabla_3, \nabla_4]\alpha = 4\rho\alpha + r^{-1}\Gamma_g\not{\mathfrak{d}}^{\leq 1}\alpha$. Hence

$$\begin{aligned}
K &= \frac{1}{2}\underline{\kappa}^2\left(\nabla_4\nabla_3\alpha + 4\rho\alpha + r^{-1}\Gamma_g\not{\mathfrak{d}}^{\leq 1}\alpha\right) + \frac{1}{4}\underline{\kappa}^3\nabla_4\alpha - \frac{1}{2}\underline{\kappa}(2\rho + \kappa\underline{\kappa})\nabla_3\alpha\\
&\quad -\frac{1}{4}\kappa\underline{\kappa}^3\alpha + \text{Good}\\
&= \frac{1}{2}\underline{\kappa}^2\nabla_4\nabla_3\alpha + \frac{1}{4}\underline{\kappa}^3\nabla_4\alpha - \frac{1}{2}\underline{\kappa}\left(2\rho + \kappa\underline{\kappa}\right)\nabla_3\alpha + \underline{\kappa}^2\left(2\rho - \frac{1}{4}\kappa\underline{\kappa}\right)\alpha + \text{Good}.
\end{aligned}$$

We have thus derived

$$\begin{aligned}
K &= \frac{1}{2}\underline{\kappa}^2\nabla_4\nabla_3\alpha + \frac{1}{4}\underline{\kappa}^3\nabla_4\alpha - \frac{1}{2}\underline{\kappa}(2\rho + \kappa\underline{\kappa})\nabla_3\alpha\\
&\quad +\underline{\kappa}^2\left(2\rho - \frac{1}{4}\kappa\underline{\kappa}\right)\alpha + \text{Good}. \qquad \text{(A.14.22)}
\end{aligned}$$

A.14.3.4 Expression for L

According to Lemma A.19 and Lemma A.16 (according to which we have the identity $[Q, \nabla_3]\alpha = \underline{\kappa}^2\nabla_3\alpha + \frac{1}{2}\underline{\kappa}^3\alpha + \text{Good}_1$)

$$\begin{aligned}
L &= -\frac{5}{2}\Big[Q, \kappa e_3\Big]\alpha = -\frac{5}{2}\left(Q(\kappa)\nabla_3\alpha + \kappa[Q, \nabla_3]\alpha + 2\nabla_3\kappa\nabla_3\nabla_3\alpha - \frac{1}{2}\kappa\underline{\kappa}^2\nabla_3\alpha\right) \\
&= -\frac{5}{2}\left(Q(\kappa)\nabla_3\alpha + \kappa\left(\underline{\kappa}^2\nabla_3\alpha + \frac{1}{2}\underline{\kappa}^3\alpha\right) + 2\nabla_3\kappa\nabla_3\nabla_3\alpha - \frac{1}{2}\kappa\underline{\kappa}^2\nabla_3\alpha\right) \\
&\quad +\text{Good} \\
&= -5\nabla_3\kappa\nabla_3\nabla_3\alpha - \frac{5}{2}\left(Q(\kappa) + \frac{1}{2}\kappa\underline{\kappa}\right)\nabla_3\alpha - \frac{5}{4}\kappa\underline{\kappa}^3\alpha + \text{Good}.
\end{aligned}$$

Note that

$$\begin{aligned}
Q(\kappa) &= \nabla_3\nabla_3\kappa + 2\underline{\kappa}\nabla_3\kappa + \frac{1}{2}\kappa\underline{\kappa}^2 \\
&= \nabla_3\left(-\frac{1}{2}\kappa\underline{\kappa} + 2\rho + r^{-1}\not{\mathfrak{d}}\Gamma_g\right) + 2\underline{\kappa}\left(-\frac{1}{2}\kappa\underline{\kappa} + 2\rho + r^{-1}\not{\mathfrak{d}}\Gamma_g\right) + \frac{1}{2}\kappa\underline{\kappa}^2 \\
&= -\frac{1}{2}(\kappa\nabla_3\underline{\kappa} + \underline{\kappa}\nabla_3\kappa) + 2\left(-\frac{3}{2}\underline{\kappa}\rho + r^{-1}\not{\mathfrak{d}}\Gamma_g\right) - \kappa\underline{\kappa}^2 + 4\rho\underline{\kappa} + \frac{1}{2}\kappa\underline{\kappa}^2 \\
&+ e_3(r^{-1}\not{\mathfrak{d}}\Gamma_g) + r^{-2}\not{\mathfrak{d}}\Gamma_g \\
&= -\frac{1}{2}(\kappa\nabla_3\underline{\kappa} + \underline{\kappa}\nabla_3\kappa) + \rho\underline{\kappa} - \frac{1}{2}\kappa\underline{\kappa}.
\end{aligned}$$

Therefore,

$$\begin{aligned}
Q(\kappa) &= -\frac{1}{2}\kappa\left(-\frac{1}{2}\underline{\kappa}^2 + r^{-1}\not{\mathfrak{d}}\Gamma_b\right) - \frac{1}{2}\underline{\kappa}\left(-\frac{1}{2}\kappa\underline{\kappa} + 2\rho + r^{-1}\not{\mathfrak{d}}\Gamma_g\right) + \underline{\kappa}\rho - \frac{1}{2}\kappa\underline{\kappa}^2 \\
&+ r^{-1}\mathfrak{d}^{\leq 2}\Gamma_g = r^{-1}\mathfrak{d}^{\leq 2}\Gamma_g.
\end{aligned}$$

We deduce

$$\begin{aligned}
L &= -5\nabla_3\kappa\nabla_3\nabla_3\alpha - \frac{5}{4}\kappa\underline{\kappa}^2\nabla_3\alpha + \frac{5}{4}\kappa\underline{\kappa}^3\alpha + \text{Good} \\
&= -5\left(-\frac{1}{2}\kappa\underline{\kappa} + 2\rho\right)\nabla_3\nabla_3\alpha - \frac{5}{4}\kappa\underline{\kappa}^2\nabla_3\alpha + \frac{5}{4}\kappa\underline{\kappa}^3\alpha + \text{Good}.
\end{aligned}$$

Therefore,

$$L = -5\left(-\frac{1}{2}\kappa\underline{\kappa} + 2\rho\right)\nabla_3\nabla_3\alpha - \frac{5}{4}\kappa\underline{\kappa}^2\nabla_3\alpha + \frac{5}{4}\kappa\underline{\kappa}^3\alpha + \text{Good}. \qquad \text{(A.14.23)}$$

A.14.3.5 Expression for M

Similarly, according to Lemma A.19,

$$\begin{aligned} M &= \left[Q, 4\rho - \frac{1}{2}\kappa\underline{\kappa}\right]\alpha \\ &= Q\left(4\rho - \frac{1}{2}\kappa\underline{\kappa}\right)\alpha + 2\nabla_3\left(4\rho - \frac{1}{2}\kappa\underline{\kappa}\right)\nabla_3\alpha - \frac{1}{2}\underline{\kappa}^2\left(4\rho - \frac{1}{2}\kappa\underline{\kappa}\right)\alpha, \end{aligned}$$

i.e.,

$$M = Q\left(4\rho - \frac{1}{2}\kappa\underline{\kappa}\right)\alpha + 2\nabla_3\left(4\rho - \frac{1}{2}\kappa\underline{\kappa}\right)\nabla_3\alpha - \frac{1}{2}\underline{\kappa}^2\left(4\rho - \frac{1}{2}\kappa\underline{\kappa}\right)\alpha.$$

We calculate

$$\begin{aligned} \nabla_3\left(4\rho - \frac{1}{2}\kappa\underline{\kappa}\right) &= 4\nabla_3\rho - \frac{1}{2}\kappa\nabla_3\underline{\kappa} - \frac{1}{2}\underline{\kappa}\nabla_3\kappa \\ &= 4\left(-\frac{3}{2}\underline{\kappa}\rho + r^{-1}\,\mathfrak{d}\!\!/\Gamma_g\right) - \frac{1}{2}\kappa\left(-\frac{1}{2}\underline{\kappa}^2 + r^{-1}\,\mathfrak{d}\!\!/\Gamma_b\right) \\ &- \frac{1}{2}\underline{\kappa}\left(-\frac{1}{2}\kappa\underline{\kappa} + 2\rho + r^{-1}\,\mathfrak{d}\!\!/\Gamma_g\right) \\ &= -7\underline{\kappa}\rho + \frac{1}{2}\kappa\underline{\kappa}^2 + r^{-1}\,\mathfrak{d}\!\!/\Gamma_g. \end{aligned}$$

We deduce

$$M = \left(Q\left(4\rho - \frac{1}{2}\kappa\underline{\kappa}\right) - \frac{1}{2}\underline{\kappa}^2\left(4\rho - \frac{1}{2}\kappa\underline{\kappa}\right)\right)\alpha + \left(-7\underline{\kappa}\rho + \frac{1}{2}\kappa\underline{\kappa}^2\right)\nabla_3\alpha + \text{Good}.$$

It remains to calculate

$$\begin{aligned} M_0 &= Q\left(4\rho - \frac{1}{2}\kappa\underline{\kappa}\right) - \frac{1}{2}\underline{\kappa}^2\left(4\rho - \frac{1}{2}\kappa\underline{\kappa}\right) \\ &= \nabla_3\nabla_3\left(4\rho - \frac{1}{2}\kappa\underline{\kappa}\right) + 2\underline{\kappa}\nabla_3\left(4\rho - \frac{1}{2}\kappa\underline{\kappa}\right) \\ &= \nabla_3\left(-7\underline{\kappa}\rho + \frac{1}{2}\kappa\underline{\kappa}^2 + r^{-1}\,\mathfrak{d}\!\!/\Gamma_g\right) + 2\underline{\kappa}\left(-7\underline{\kappa}\rho + \frac{1}{2}\kappa\underline{\kappa}^2 + r^{-1}\,\mathfrak{d}\!\!/\Gamma_g\right) \\ &= -7\rho\nabla_3\underline{\kappa} - 7\underline{\kappa}\nabla_3\rho + \frac{1}{2}\underline{\kappa}^2\nabla_3\kappa + \kappa\underline{\kappa}\nabla_3\underline{\kappa} + 2\underline{\kappa}\left(-7\underline{\kappa}\rho + \frac{1}{2}\kappa\underline{\kappa}^2\right) \\ &+ \nabla_3(r^{-1}\,\mathfrak{d}\!\!/\Gamma_g) + r^{-2}\,\mathfrak{d}\!\!/\Gamma_g. \end{aligned}$$

Hence,

$$\begin{aligned} M_0 &= -7\rho\left(-\frac{1}{2}\underline{\kappa}^2 + r^{-1}\not{\mathfrak{d}}\Gamma_b\right) - 7\underline{\kappa}\left(-\frac{3}{2}\underline{\kappa}\rho + r^{-1}\not{\mathfrak{d}}\Gamma_g\right) \\ &+ \frac{1}{2}\underline{\kappa}^2\left(-\frac{1}{2}\kappa\underline{\kappa} + 2\rho + r^{-1}\not{\mathfrak{d}}\Gamma_g\right) + \kappa\underline{\kappa}\left(-\frac{1}{2}\underline{\kappa}^2 + r^{-1}\not{\mathfrak{d}}\Gamma_b\right) \\ &+ 2\underline{\kappa}\left(-7\underline{\kappa}\rho + \frac{1}{2}\kappa\underline{\kappa}^2\right) + \nabla_3(r^{-1}\not{\mathfrak{d}}\Gamma_g) + r^{-2}\not{\mathfrak{d}}\Gamma_g \\ &= \underline{\kappa}^2\rho + \frac{1}{4}\kappa\underline{\kappa}^3 + \nabla_3(r^{-1}\not{\mathfrak{d}}\Gamma_g) + r^{-2}\not{\mathfrak{d}}\Gamma_g. \end{aligned}$$

We conclude

$$M = \left(\underline{\kappa}^2\rho + \frac{1}{4}\kappa\underline{\kappa}^3\right)\alpha + 2\left(-7\underline{\kappa}\rho + \frac{1}{2}\kappa\underline{\kappa}^2\right)\nabla_3\alpha + \text{Good}. \tag{A.14.24}$$

Indeed note that

$$\nabla_3(r^{-1}\not{\mathfrak{d}}\Gamma_g)\alpha = \nabla_3(r^{-1}\not{\mathfrak{d}}\Gamma_g\alpha) - r^{-1}\not{\mathfrak{d}}\Gamma_g\nabla_3\alpha = \text{Good}.$$

A.14.3.6 End of the proof of Proposition A.20

Using the equations (A.14.20)–(A.14.24) we deduce, in view of (A.14.19),

$$\begin{aligned} [Q,\mathcal{L}]\alpha &= I + J + K + L + M \\ &= (2\rho - \kappa\underline{\kappa})\nabla_3\nabla_3\alpha - \underline{\kappa}^2\nabla_4\nabla_3\alpha - \frac{1}{2}\underline{\kappa}^3\nabla_4\alpha - \left(-\frac{1}{2}\kappa\underline{\kappa}^2 + 3\rho\underline{\kappa}\right)\nabla_3\alpha \\ &- \frac{3}{2}\underline{\kappa}^2\left(-\frac{1}{2}\kappa\underline{\kappa} + 2\rho\right)\alpha - 2\underline{\kappa}\nabla_4\nabla_3\nabla_3\alpha - 5\kappa\underline{\kappa}\nabla_3\nabla_3\alpha - \frac{7}{2}\underline{\kappa}^2\nabla_4\nabla_3\alpha \\ &- \frac{3}{4}\underline{\kappa}^3\nabla_4\alpha - \left(6\underline{\kappa}\rho + \frac{19}{4}\kappa\underline{\kappa}^2\right)\nabla_3\alpha - \left(8\underline{\kappa}^2\rho + \frac{1}{4}\kappa\underline{\kappa}^3\right)\alpha \\ &+ \frac{1}{2}\underline{\kappa}^2\nabla_4\nabla_3\alpha + \frac{1}{4}\underline{\kappa}^3\nabla_4\alpha - \frac{1}{2}\underline{\kappa}(2\rho + \kappa\underline{\kappa})\nabla_3\alpha + \underline{\kappa}^2\left(2\rho - \frac{1}{4}\kappa\underline{\kappa}\right)\alpha \\ &- 5\left(-\frac{1}{2}\kappa\underline{\kappa} + 2\rho\right)\nabla_3\nabla_3\alpha - \frac{5}{4}\kappa\underline{\kappa}^2\nabla_3\alpha - \frac{5}{4}\kappa\underline{\kappa}^3\alpha \\ &+ \left(\underline{\kappa}^2\rho + \frac{1}{4}\kappa\underline{\kappa}^3\right)\alpha + 2\left(-7\underline{\kappa}\rho + \frac{1}{2}\kappa\underline{\kappa}^2\right)\nabla_3\alpha + \text{Good}. \end{aligned}$$

We deduce

$$[Q,\mathcal{L}]\alpha = -2\underline{\kappa}\nabla_4\nabla_3\nabla_3\alpha + C'_{33}\nabla_3\nabla_3\alpha + C'_{43}\nabla_4\nabla_3\alpha + C'_4\nabla_4\alpha + C'_3\nabla_3\alpha + C'_0\alpha,$$

with

$$
\begin{aligned}
C'_{33} &= (2\rho - \kappa\underline{\kappa}) - 5\kappa\underline{\kappa} - 5\left(-\frac{1}{2}\kappa\underline{\kappa} + 2\rho\right) = -8\rho - \frac{7}{2}\kappa\underline{\kappa},\\
C'_{43} &= -\underline{\kappa}^2 - \frac{7}{2}\underline{\kappa}^2 + \frac{1}{2}\underline{\kappa}^2 = -4\underline{\kappa}^2,\\
C'_4 &= -\frac{1}{2}\underline{\kappa}^3 - \frac{3}{4}\underline{\kappa}^3 + \frac{1}{4}\underline{\kappa}^3 = -\underline{\kappa}^3,\\
C'_3 &= \frac{1}{2}\kappa\underline{\kappa}^2 - 3\rho\underline{\kappa} - \left(6\underline{\kappa}\rho + \frac{19}{4}\kappa\underline{\kappa}^2\right) - \frac{1}{2}\underline{\kappa}(2\rho + \kappa\underline{\kappa}) - \frac{5}{4}\kappa\underline{\kappa}^2\\
&\quad +2\left(-7\underline{\kappa}\rho + \frac{1}{2}\kappa\underline{\kappa}^2\right) = -24\underline{\kappa}\rho - 5\kappa\underline{\kappa}^2,\\
C'_0 &= -\frac{3}{2}\underline{\kappa}^2\left(-\frac{1}{2}\kappa\underline{\kappa} + 2\rho\right) - \left(8\underline{\kappa}^2\rho + \frac{1}{4}\kappa\underline{\kappa}^3\right) + \underline{\kappa}^2\left(2\rho - \frac{1}{4}\kappa\underline{\kappa}\right) - \frac{5}{4}\kappa\underline{\kappa}^3\\
&\quad + \left(\underline{\kappa}^2\rho + \frac{1}{4}\kappa\underline{\kappa}^3\right) = -8\underline{\kappa}^2\rho - \frac{3}{4}\kappa\underline{\kappa}^3.
\end{aligned}
$$

Finally we write, recalling the definition of $Q = \nabla_3\nabla_3 + 2\underline{\kappa}\nabla_3 + \frac{1}{2}\underline{\kappa}^2$,

$$
\nabla_3\nabla_3\alpha = Q(\alpha) - 2\underline{\kappa}\nabla_3\alpha - \frac{1}{2}\underline{\kappa}^2\alpha
$$

and

$$
\nabla_4\nabla_3\nabla_3\alpha = \nabla_4 Q(\alpha) - 2\underline{\kappa}\nabla_4\nabla_3\alpha - \frac{1}{2}\underline{\kappa}^2\nabla_4\alpha - 2\nabla_4\underline{\kappa}\nabla_3\alpha - \underline{\kappa}\nabla_4\underline{\kappa}\alpha.
$$

Hence,

$$
\begin{aligned}
&-2\underline{\kappa}\nabla_4\nabla_3\nabla_3\alpha + C'_{33}\nabla_3\nabla_3\alpha\\
=\;& -2\underline{\kappa}\nabla_4 Q(\alpha) + 4\underline{\kappa}^2\nabla_4\nabla_3\alpha + \underline{\kappa}^3\nabla_4\alpha + 4\underline{\kappa}\nabla_4\underline{\kappa}\nabla_3\alpha\\
+\;& 2\underline{\kappa}^2\nabla_4\underline{\kappa}\alpha + C'_{33}\left(Q(\alpha) - 2\underline{\kappa}\nabla_3\alpha - \frac{1}{2}\underline{\kappa}^2\alpha\right).
\end{aligned}
$$

We deduce

$$
\begin{aligned}
[Q, \mathcal{L}]\alpha &= -2\underline{\kappa}\nabla_4 Q(\alpha) + C'_{33}Q(\alpha) + 4\underline{\kappa}^2\nabla_4\nabla_3\alpha + \underline{\kappa}^3\nabla_4\alpha\\
&+ (4\underline{\kappa}\nabla_4\underline{\kappa} - 2\underline{\kappa}C'_{33})\nabla_3\alpha + \left(2\underline{\kappa}^2\nabla_4\underline{\kappa} - \frac{1}{2}\underline{\kappa}^2 C'_{33}\right)\alpha + C'_{43}\nabla_4\nabla_3\alpha\\
&+ C'_4\nabla_4\alpha + C'_3\nabla_3\alpha + C'_0\alpha.
\end{aligned}
$$

Thus, setting $C_Q = C'_{33}$, we deduce

$$
\begin{aligned}
[Q, \mathcal{L}]\alpha &= -2\underline{\kappa}\nabla_4 Q(\alpha) + C_Q Q(\alpha) + C_{43}\nabla_4\nabla_3\alpha + C_4\nabla_4\alpha + C_3\nabla_3\alpha + C_0\alpha\\
&\quad +\text{Good},
\end{aligned}
$$

where

$$
\begin{aligned}
C_Q &= C'_{33} = -8\rho - \frac{7}{2}\kappa\underline{\kappa},\\
C_{43} &= 4\underline{\kappa}^2 + C'_{43} = 4\underline{\kappa}^2 - 4\underline{\kappa}^2 = 0,\\
C_4 &= \underline{\kappa}^3 + C'_4 = \underline{\kappa}^3 - \underline{\kappa}^3 = 0.
\end{aligned}
$$

Also,

$$
\begin{aligned}
C_3 &= 2\underline{\kappa}(2\nabla_4\underline{\kappa} - C'_{33}) + C'_3\\
&= 2\underline{\kappa}\left(-\kappa\underline{\kappa} + 4\rho + r^{-1}\mathfrak{\not{d}}\Gamma_g + 8\rho + \frac{7}{2}\kappa\underline{\kappa}\right) + C'_3\\
&= 2\underline{\kappa}\left(12\rho + \frac{5}{2}\kappa\underline{\kappa}\right) + \big(-24\underline{\kappa}\rho - 5\kappa\underline{\kappa}^2\big) + r^{-2}\mathfrak{\not{d}}\Gamma_g\\
&= r^{-2}\mathfrak{\not{d}}\Gamma_g,\\
C_0 &= 2\underline{\kappa}^2\nabla_4\underline{\kappa} - \frac{1}{2}\underline{\kappa}^2\, C'_{33} + C'_0\\
&= 2\underline{\kappa}^2\left(-\frac{1}{2}\kappa\underline{\kappa} + 2\rho + r^{-1}\mathfrak{\not{d}}\Gamma_g\right) + \frac{1}{2}\underline{\kappa}^2\left(8\rho + \frac{7}{2}\kappa\underline{\kappa}\right) - 8\underline{\kappa}^2\rho - \frac{3}{4}\kappa\underline{\kappa}^3\\
&= 8\underline{\kappa}^2\rho + \frac{3}{4}\kappa\underline{\kappa}^3 - 8\underline{\kappa}^2\rho - \frac{3}{4}\kappa\underline{\kappa}^3 + r^{-3}\mathfrak{\not{d}}\Gamma_g\\
&= r^{-3}\mathfrak{\not{d}}\Gamma_g.
\end{aligned}
$$

We have therefore checked that

$$
[Q,\mathcal{L}]\alpha = -2\underline{\kappa}\nabla_4 Q(\alpha) + C_Q Q(\alpha) + \text{Good}, \qquad C_Q = -8\rho - \frac{7}{2}\kappa\underline{\kappa},
$$

as stated in Proposition A.20. □

A.14.4 Proof of Theorem 2.108

We start with the following:

Lemma A.21. *We have*

$$
\square_2(fr^4) = r^4\square_2 f - 2r^4\big(\underline{\kappa}e_4 f + \kappa e_3 f\big) + r^4\big(-5\kappa\underline{\kappa} - 4\rho\big)f + O(r^4\mathfrak{d}^{\leq 1}\Gamma_g\cdot f).
$$

We postpone the proof of the lemma to the end of the section and continue below the proof of the theorem. According to Lemma A.13

$$
\mathcal{L}(\alpha) = \text{Good}_2
$$

where $\mathcal{L}$ is the operator

$$
\mathcal{L}\alpha = -\nabla_4\nabla_3\alpha + {}^{(c)}\!\not{\Delta}_2\alpha - \frac{5}{2}\kappa\nabla_3\alpha - \frac{1}{2}\underline{\kappa}\nabla_4\alpha - \left(-4\rho + \frac{1}{2}\kappa\underline{\kappa}\right)\alpha.
$$

Applying Q and recalling the definition of the error terms Good we derive

$$
\mathcal{L}(Q\alpha) = -[Q,\mathcal{L}]\alpha + \text{Good}.
$$

Thus, in view of Proposition A.20,

$$[Q,\mathcal{L}]\alpha = -2\underline{\kappa}\nabla_4 Q(\alpha) + C_Q Q(\alpha), \qquad C_Q = -8\rho - \tfrac{7}{2}\kappa\underline{\kappa}.$$

We deduce

$$\mathcal{L}(Q\alpha) = 2\underline{\kappa}\nabla_4 Q(\alpha) - C_Q Q(\alpha).$$

Therefore, modulo Good terms,

$$\begin{aligned} 2\underline{\kappa}\nabla_4 Q(\alpha) - C_Q Q(\alpha) &= -\nabla_4\nabla_3(Q\alpha) + {}^{(c)}\!\not{\Delta}_2(Q\alpha) - \frac{5}{2}\kappa\nabla_3 Q(\alpha) - \frac{1}{2}\underline{\kappa}\nabla_4 Q(\alpha) \\ &- \left(-4\rho + \frac{1}{2}\kappa\underline{\kappa}\right) Q(\alpha). \end{aligned}$$

We deduce

$$\begin{aligned} -\nabla_4\nabla_3(Q\alpha) + {}^{(c)}\!\not{\Delta}_2(Q\alpha) - \frac{5}{2}\kappa\nabla_3 Q(\alpha) - \frac{5}{2}\underline{\kappa}\nabla_4 Q(\alpha) & \\ + \left(C_Q - \left(-4\rho + \frac{1}{2}\kappa\underline{\kappa}\right)\right) Q(\alpha) &= \text{Good}. \end{aligned}$$

In view of the expression for $\square_2$ in the second part of Lemma A.13, we rewrite in the form

$$\square_2 Q(f) - 2\kappa\nabla_3 Q(\alpha) - 2\underline{\kappa}\nabla_4 Q(\alpha) - \big(4\rho + 4\kappa\underline{\kappa}\big)Q(\alpha) = \text{Good} + r^{-1}\Gamma_g\cdot \not{\mathfrak{d}} Q(\alpha).$$

Finally, making use of Lemma A.21 and recalling that $\mathfrak{q} = r^4 Q(\alpha)$,

$$\begin{aligned} \square_2\mathfrak{q} &= r^4\square_2(Q\alpha) - 2r^4\big(\underline{\kappa}e_4(Q\alpha) + \kappa e_3(Q\alpha)\big) + r^4\big(-5\kappa\underline{\kappa} - 4\rho\big)Qf \\ &\quad + O(r^4\mathfrak{d}^{\le 1}\Gamma_g\cdot Q(\alpha)) \\ &= r^4\Big(2\kappa\nabla_3 Q(\alpha) + 2\underline{\kappa}\nabla_4 Q(\alpha) + \big(4\rho + 4\kappa\underline{\kappa}\big)Q(\alpha) + \text{Good}\Big) \\ &\quad - 2r^4\big(\underline{\kappa}e_4(Q\alpha) + \kappa e_3(Q\alpha)\big) + r^4\big(-5\kappa\underline{\kappa} - 4\rho\big)Qf + O(\mathfrak{d}^{\le 1}\Gamma_g\cdot\mathfrak{q}) \\ &= -\kappa\underline{\kappa}\mathfrak{q} + r^4\text{Good}. \end{aligned}$$

This ends the proof of Theorem 2.108.

A.14.4.1 Proof of Lemma A.21

We have

$$\begin{aligned} \square_2(fr^4) &= \mathbf{D}^\alpha\mathbf{D}_\alpha(fr^4) = \mathbf{D}^\alpha(\mathbf{D}_\alpha f r^4 + f\mathbf{D}_\alpha r^4) \\ &= r^4\square_2 f + 2\mathbf{D}_\alpha(r^4)\mathbf{D}^\alpha f + f\square(r) \\ &= r^4\square_2 f - \big(e_3(r^4)e_4 f + e_4(r^4)e_3 f\big) + f\square(r^4) + r^4\Gamma_g\mathfrak{d}f \\ &= r^4\square_2 f - 4r^3\big(e_3(r)e_4 f + e_4(r)e_3 f\big) + f\square(r^4) + r^4\Gamma_g\mathfrak{d}f \\ &= r^4\square_2 f - 2r^4\Big((\underline{\kappa} + \Gamma_b)e_4 f + (\kappa + \Gamma_g)e_3 f\Big) + f\square(r^4) + r^4\Gamma_g\cdot\mathfrak{d}f \\ &= r^4\square_2 f - 2r^4\big(\underline{\kappa}e_4 f + \kappa e_3 f\big) + f\square(r^4) + r^4\Gamma_g\cdot f. \end{aligned}$$

Also,

$$\begin{aligned}
\square(r^4) &= -e_4(e_3(r^4)) - \frac{1}{2}\underline{\kappa}e_4(r^4) + \left(-\frac{1}{2}\kappa + 2\omega\right)e_3(r^4) + {\not\!\Delta}(r^4) + 2\underline{\eta}e_\theta(r^4)\\
&= -4e_4(r^3e_3(r)) - 2r^3\underline{\kappa}\frac{r}{2}(\kappa + \Gamma_g) + 4r^3\left(-\frac{1}{2}\kappa + 2\omega\right)\frac{r}{2}(\underline{\kappa} + \Gamma_b)\\
&\quad + {\not\!\Delta}(r^4) + 2\underline{\eta}e_\theta(r^4)\\
&= -12r^2(e_4r)(e_3r) - 4r^3e_4e_3r - r^4\underline{\kappa}\kappa + 2r^4\left(-\frac{1}{2}\kappa + 2\omega\right)\underline{\kappa} + O(r^3\Gamma_b)\\
&= -3r^4(\underline{\kappa} + \Gamma_b)(\kappa + \Gamma_g) - 4r^3e_4\left(\frac{r}{2}(\underline{\kappa} + \Gamma_b)\right) - 2r^4\kappa\underline{\kappa} + 4r^4\omega\underline{\kappa}\\
&\quad + O(r^3\Gamma_b).
\end{aligned}$$

Hence,

$$\square(r^4) = -5r^4\kappa\underline{\kappa} + 4r^4\omega\underline{\kappa} - 2r^3e_4(r\underline{\kappa}) + O(r^4\mathfrak{d}^{\leq 1}\Gamma_g).$$

Note that

$$\begin{aligned}
e_4(r\underline{\kappa}) &= re_4(\underline{\kappa}) + \frac{r}{2}\underline{\kappa}(\kappa + \Gamma_g) = r\left(-\frac{1}{2}\kappa\,\underline{\kappa} + 2\omega\underline{\kappa} + 2\,{\not\!d}_1\underline{\eta} + 2\rho\right) + \frac{r}{2}\underline{\kappa}(\kappa + \Gamma_g)\\
&= 2r\rho + 2r\omega\underline{\kappa} + O(\mathfrak{d}^{\leq 1}\Gamma_g).
\end{aligned}$$

Hence,

$$\begin{aligned}
\square(r^4) &= -5r^4\kappa\underline{\kappa} + 4r^4\omega\underline{\kappa} - 2r^3(2r\rho + 2r\omega\underline{\kappa}) + O(r^4\mathfrak{d}^{\leq 1}\Gamma_g)\\
&= r^4\big(-5\kappa\underline{\kappa} - 4\rho\big) + O(r^4\mathfrak{d}^{\leq 1}\Gamma_g).
\end{aligned}$$

We conclude

$$\square_2(fr^4) = r^4\square_2 f - 2r^4\big(\underline{\kappa}e_4 f + \kappa e_3 f\big) + r^4\big(-5\kappa\underline{\kappa} - 4\rho\big)f + O(r^4\mathfrak{d}^{\leq 1}\Gamma_g)$$

as stated.

Appendix B

Appendix to Chapter 8

B.1 PROOF OF PROPOSITION 8.14

Proposition B.1. *The following wave equations hold true.*

1. *The null curvature component* ρ *verifies the identity*

$$\square_{\mathbf{g}}\rho \quad := \quad \underline{\kappa} e_4\rho + \kappa e_3\rho + \frac{3}{2}\Big(\underline{\kappa}\,\kappa + 2\rho\Big)\rho + Err[\square_{\mathbf{g}}\rho],$$

where

$$\begin{aligned} Err[\square_{\mathbf{g}}\rho] \quad = \quad & \frac{3}{2}\rho\left(-\frac{1}{2}\underline{\vartheta}\,\vartheta + 2(\underline{\xi}\,\xi + \eta\,\eta)\right) \\ & + \left(\frac{3}{2}\underline{\kappa} - 2\underline{\omega}\right)\left(\frac{1}{2}\underline{\vartheta}\,\alpha - \zeta\,\beta - 2(\underline{\eta}\,\beta + \xi\,\underline{\beta})\right) \\ & -\frac{1}{2}\underline{\vartheta}\,\not{d}_2^\star\beta + (\zeta - \eta)e_3\beta - \eta e_3(\Phi)\beta - \underline{\xi}(e_4\beta + e_4(\Phi)\beta) - \underline{\beta}\beta \\ & -e_3\left(-\frac{1}{2}\underline{\vartheta}\,\alpha + \zeta\,\beta + 2(\underline{\eta}\,\beta + \xi\,\underline{\beta})\right) \\ & - \not{d}_1^\star(\underline{\kappa})\beta + 2\,\not{d}_1^\star(\underline{\omega})\beta + 3\eta\,\not{d}_1^\star(\rho) - \not{d}_1\Big(-\vartheta\underline{\beta} + \underline{\xi}\alpha\Big) - 2\eta e_\theta\rho. \end{aligned}$$

2. *The small curvature quantity*

$$\tilde{\rho} := r^2\left(\rho + \frac{2m}{r^3}\right)$$

verifies the wave equation

$$\begin{aligned} \square_{\mathbf{g}}(\tilde{\rho}) + \frac{8m}{r^3}\tilde{\rho} \quad = \quad & -6m\frac{\square_{\mathbf{g}}(r) - \left(\frac{2}{r} - \frac{2m}{r^2}\right)}{r^2} - \frac{3m}{r}\left(\kappa\underline{\kappa} + \frac{4\Upsilon}{r^2}\right) \\ & -\frac{3m}{r}\left(A\underline{\kappa} + \underline{A}\kappa\right) + Err[\square_g\tilde{\rho}], \end{aligned}$$

where

$$
\begin{aligned}
Err[\Box_g\tilde{\rho}] \quad := \quad & -\frac{6m}{r}A\underline{A} + \frac{3}{r^2}\tilde{\rho}^2 + \frac{3}{2}\left(\frac{4}{3}A\frac{e_3(r)}{r} + \frac{4}{3}\underline{A}\frac{e_4(r)}{r}\right)\tilde{\rho} \\
& + \left(\frac{3}{2}\left(\kappa\underline{\kappa} - \frac{8m}{r^3} + \frac{2}{3r^2}\Box_{\mathbf{g}}(r^2)\right) + \frac{8m}{r^3}\right)\tilde{\rho} \\
& - Ae_3(\tilde{\rho}) - \underline{A}e_4(\tilde{\rho}) + \frac{2}{r}Ae_3(m) + \frac{2}{r}\underline{A}e_4(m) \\
& + 4D^a(m)D_a\left(\frac{1}{r}\right) + \frac{2}{r}\Box_{\mathbf{g}}(m) + 4r\,{}^{*}\!\!\not{d}_1^{\star}(r)\not{d}_1^{\star}(\rho) + r^2 Err[\Box_{\mathbf{g}}\rho].
\end{aligned}
$$

Proof. We prove the result in the following steps.

Step 1. We start by deriving the wave equation for ρ. From Bianchi, ρ satisfies

$$
e_4\rho + \frac{3}{2}\kappa\rho = \not{d}_1\beta - \frac{1}{2}\underline{\vartheta}\,\alpha + \zeta\,\beta + 2(\underline{\eta}\,\beta + \xi\,\underline{\beta}).
$$

Differentiating with respect to e_3, we obtain

$$
e_3(e_4(\rho)) + \frac{3}{2}\kappa e_3(\rho) + \frac{3}{2}e_3(\kappa)\rho = e_3(\not{d}_1\beta) + e_3\left(-\frac{1}{2}\underline{\vartheta}\,\alpha + \zeta\,\beta + 2(\underline{\eta}\,\beta + \xi\,\underline{\beta})\right).
$$

Also, β satisfies from Bianchi

$$
e_3\beta + \underline{\kappa}\beta \quad = \quad -\not{d}_1^{\star}\rho + 2\underline{\omega}\beta + 3\eta\rho - \vartheta\underline{\beta} + \underline{\xi}\alpha.
$$

Differentiating with respect to $\not{d}_1$, we infer

$$
\begin{aligned}
\not{d}_1(e_3\beta) + \underline{\kappa}\not{d}_1\beta - \not{d}_1^{\star}(\underline{\kappa})\beta \quad = \quad & -\not{d}_1\not{d}_1^{\star}\rho + 2\underline{\omega}\not{d}_1\beta - 2\not{d}_1^{\star}(\underline{\omega})\beta + 3\rho\not{d}_1\eta - 3\eta\not{d}_1^{\star}(\rho) \\
& + \not{d}_1\Big(-\vartheta\underline{\beta} + \underline{\xi}\alpha\Big)
\end{aligned}
$$

and hence

$$
\begin{aligned}
\not{d}_1\not{d}_1^{\star}\rho \quad = \quad & -\not{d}_1(e_3\beta) - \underline{\kappa}\not{d}_1\beta + 2\underline{\omega}\not{d}_1\beta + 3\rho\not{d}_1\eta \\
& + \not{d}_1^{\star}(\underline{\kappa})\beta - 2\not{d}_1^{\star}(\underline{\omega})\beta - 3\eta\not{d}_1^{\star}(\rho) + \not{d}_1\Big(-\vartheta\underline{\beta} + \underline{\xi}\alpha\Big).
\end{aligned}
$$

Next, we add the equation for $\not{d}_1\not{d}_1^{\star}\rho$ from the one for $e_3(e_4(\rho))$. This yields

$$
\begin{aligned}
& e_3(e_4(\rho)) + \not{d}_1\not{d}_1^{\star}\rho + \frac{3}{2}\kappa e_3(\rho) + \frac{3}{2}e_3(\kappa)\rho \\
= \quad & [e_3, \not{d}_1]\beta - \underline{\kappa}\not{d}_1\beta + 2\underline{\omega}\not{d}_1\beta + 3\rho\not{d}_1\eta + e_3\left(-\frac{1}{2}\underline{\vartheta}\,\alpha + \zeta\,\beta + 2(\underline{\eta}\,\beta + \xi\,\underline{\beta})\right) \\
& + \not{d}_1^{\star}(\underline{\kappa})\beta - 2\not{d}_1^{\star}(\underline{\omega})\beta - 3\eta\not{d}_1^{\star}(\rho) + \not{d}_1\Big(-\vartheta\underline{\beta} + \underline{\xi}\alpha\Big).
\end{aligned}
$$

Next, we recall the following commutator identity

$$
[e_3, \not{d}_1]\beta = -\frac{1}{2}\underline{\kappa}\not{d}_1\beta + \frac{1}{2}\underline{\vartheta}\not{d}_2^{\star}\beta - (\zeta - \eta)e_3\beta + \eta e_3(\Phi)\beta + \underline{\xi}(e_4\beta + e_4(\Phi)\beta) + \underline{\beta}\beta.
$$

We infer

$$
\begin{aligned}
& e_3(e_4(\rho)) + \not{d}_1 \not{d}_1^\star \rho + \frac{3}{2}\kappa e_3(\rho) + \frac{3}{2}e_3(\kappa)\rho + \left(\frac{3}{2}\underline{\kappa} - 2\underline{\omega}\right) \not{d}_1\beta - 3\rho \not{d}_1\eta \\
= \ & \frac{1}{2}\underline{\vartheta} \not{d}_2^\star\beta - (\zeta - \eta)e_3\beta + \eta e_3(\Phi)\beta + \underline{\xi}(e_4\beta + e_4(\Phi)\beta) + \underline{\beta}\beta \\
& + e_3\left(-\frac{1}{2}\underline{\vartheta}\,\alpha + \zeta\,\beta + 2(\underline{\eta}\,\beta + \xi\,\underline{\beta})\right) \\
& + \not{d}_1^\star(\underline{\kappa})\beta - 2\not{d}_1^\star(\underline{\omega})\beta - 3\eta \not{d}_1^\star(\rho) + \not{d}_1\Big(- \vartheta\underline{\beta} + \underline{\xi}\alpha\Big).
\end{aligned}
$$

Next, we make use of the Bianchi identities and the null structure equations to compute

$$
\begin{aligned}
& \frac{3}{2}e_3(\kappa)\rho + \left(\frac{3}{2}\underline{\kappa} - 2\underline{\omega}\right) \not{d}_1\beta - 3\rho \not{d}_1\eta \\
= \ & \frac{3}{2}\rho\left(-\frac{1}{2}\underline{\kappa}\,\kappa + 2\underline{\omega}\kappa + 2\not{d}_1\eta + 2\rho - \frac{1}{2}\underline{\vartheta}\,\vartheta + 2(\underline{\xi}\,\xi + \eta\,\eta)\right) \\
& + \left(\frac{3}{2}\underline{\kappa} - 2\underline{\omega}\right)\left(e_4\rho + \frac{3}{2}\kappa\rho + \frac{1}{2}\underline{\vartheta}\,\alpha - \zeta\,\beta - 2(\underline{\eta}\,\beta + \xi\,\underline{\beta})\right) - 3\rho \not{d}_1\eta \\
= \ & \left(\frac{3}{2}\underline{\kappa} - 2\underline{\omega}\right) e_4\rho + \frac{3}{2}\rho\Big(\underline{\kappa}\,\kappa + 2\rho\Big) \\
& + \frac{3}{2}\rho\left(-\frac{1}{2}\underline{\vartheta}\,\vartheta + 2(\underline{\xi}\,\xi + \eta\,\eta)\right) + \left(\frac{3}{2}\underline{\kappa} - 2\underline{\omega}\right)\left(\frac{1}{2}\underline{\vartheta}\,\alpha - \zeta\,\beta - 2(\underline{\eta}\,\beta + \xi\,\underline{\beta})\right).
\end{aligned}
$$

This yields

$$
\begin{aligned}
& e_3(e_4(\rho)) - \not{\Delta}\rho + \frac{3}{2}\kappa e_3(\rho) + \left(\frac{3}{2}\underline{\kappa} - 2\underline{\omega}\right) e_4\rho + \frac{3}{2}\rho\Big(\underline{\kappa}\,\kappa + 2\rho\Big) \\
= \ & -\frac{3}{2}\rho\left(-\frac{1}{2}\underline{\vartheta}\,\vartheta + 2(\underline{\xi}\,\xi + \eta\,\eta)\right) - \left(\frac{3}{2}\underline{\kappa} - 2\underline{\omega}\right)\left(\frac{1}{2}\underline{\vartheta}\,\alpha - \zeta\,\beta - 2(\underline{\eta}\,\beta + \xi\,\underline{\beta})\right) \\
& + \frac{1}{2}\underline{\vartheta} \not{d}_2^\star\beta - (\zeta - \eta)e_3\beta + \eta e_3(\Phi)\beta + \underline{\xi}(e_4\beta + e_4(\Phi)\beta) + \underline{\beta}\beta \\
& + e_3\left(-\frac{1}{2}\underline{\vartheta}\,\alpha + \zeta\,\beta + 2(\underline{\eta}\,\beta + \xi\,\underline{\beta})\right) \\
& + \not{d}_1^\star(\underline{\kappa})\beta - 2\not{d}_1^\star(\underline{\omega})\beta - 3\eta \not{d}_1^\star(\rho) + \not{d}_1\Big(- \vartheta\underline{\beta} + \underline{\xi}\alpha\Big),
\end{aligned}
$$

where we used the fact that $\not{d}_1 \not{d}_1^\star = -\not{\Delta}$.

Next, recall the formula for the wave operator acting on a scalar ψ

$$
\square_{\mathbf{g}}\psi = -e_3e_4\psi + \not{\Delta}\psi + \left(2\underline{\omega} - \frac{1}{2}\underline{\kappa}\right) e_4\psi - \frac{1}{2}\kappa e_3\psi + 2\eta e_\theta\psi.
$$

We infer

$$
\begin{aligned}
& e_3(e_4(\rho)) - \Delta\!\!\!/\,\rho + \frac{3}{2}\kappa e_3(\rho) + \left(\frac{3}{2}\underline{\kappa} - 2\underline{\omega}\right) e_4\rho + \frac{3}{2}\rho\Big(\kappa\underline{\kappa} + 2\rho\Big)\\
= \;& -\square_{\mathbf{g}}\rho + \left(2\underline{\omega} - \frac{1}{2}\underline{\kappa}\right) e_4\rho - \frac{1}{2}\kappa e_3\rho + 2\eta e_\theta\rho\\
& + \frac{3}{2}\kappa e_3(\rho) + \left(\frac{3}{2}\underline{\kappa} - 2\underline{\omega}\right) e_4\rho + \frac{3}{2}\rho\Big(\underline{\kappa}\,\kappa + 2\rho\Big)
\end{aligned}
$$

and hence

$$
\begin{aligned}
\square_{\mathbf{g}}\rho \;=\;& \underline{\kappa} e_4\rho + \kappa e_3\rho + \frac{3}{2}\Big(\underline{\kappa}\,\kappa + 2\rho\Big)\rho\\
& + \frac{3}{2}\rho\left(-\frac{1}{2}\underline{\vartheta}\,\vartheta + 2(\underline{\xi}\,\xi + \eta\,\eta)\right) + \left(\frac{3}{2}\underline{\kappa} - 2\underline{\omega}\right)\left(\frac{1}{2}\underline{\vartheta}\,\alpha - \zeta\,\beta - 2(\underline{\eta}\,\beta + \xi\,\underline{\beta})\right)\\
& - \frac{1}{2}\underline{\vartheta}\, \,{d\!\!\!/}_2^\star\beta + (\zeta - \eta)e_3\beta - \eta e_3(\Phi)\beta - \underline{\xi}(e_4\beta + e_4(\Phi)\beta) - \underline{\beta}\beta\\
& - e_3\left(-\frac{1}{2}\underline{\vartheta}\,\alpha + \zeta\,\beta + 2(\underline{\eta}\,\beta + \xi\,\underline{\beta})\right)\\
& - {d\!\!\!/}_1^\star(\underline{\kappa})\beta + 2\,{d\!\!\!/}_1^\star(\underline{\omega})\beta + 3\eta\,{d\!\!\!/}_1^\star(\rho) - {d\!\!\!/}_1\Big(-\vartheta\underline{\beta} + \underline{\xi}\alpha\Big) - 2\eta e_\theta\rho.
\end{aligned}
$$

Step 2. We derive the following identity

$$
\begin{aligned}
\square_{\mathbf{g}}(r^2\rho) =& -Ae_3(r^2\rho) - \underline{A}e_4(r^2\rho)\\
& + \frac{3}{2}\left(\frac{4}{3}A\frac{e_3(r)}{r} + \frac{4}{3}\underline{A}\frac{e_4(r)}{r} + \kappa\underline{\kappa} + 2\rho + \frac{2}{3r^2}\square_{\mathbf{g}}(r^2)\right)r^2\rho\\
& + 4r\,{d\!\!\!/}_1^\star(r)\,{d\!\!\!/}_1^\star(\rho) + r^2\mathrm{Err}[\square_{\mathbf{g}}\rho].
\end{aligned}
\tag{B.1.1}
$$

Proof. $r^2\rho$ satisfies the following wave equation

$$
\square_{\mathbf{g}}(r^2\rho) \;=\; r^2\square_{\mathbf{g}}\rho + 2D^a(r^2)D_a(\rho) + \rho\square_{\mathbf{g}}(r^2).
$$

On the other hand, recall that we have

$$
\square_{\mathbf{g}}\rho \;=\; \underline{\kappa} e_4\rho + \kappa e_3\rho + \frac{3}{2}\Big(\underline{\kappa}\,\kappa + 2\rho\Big)\rho + \mathrm{Err}[\square_{\mathbf{g}}\rho].
$$

We deduce

$$
\begin{aligned}
\square_{\mathbf{g}}(r^2\rho) \;=\;& \left(r^2\kappa - e_4(r^2)\right) e_3\rho + \left(r^2\underline{\kappa} - e_3(r^2)\right) e_4\rho\\
& + \frac{3}{2}\left(\kappa\underline{\kappa} + 2\rho + \frac{2}{3r^2}\square_{\mathbf{g}}(r^2)\right) r^2\rho + 4r\,{d\!\!\!/}_1^\star(r)\,{d\!\!\!/}_1^\star(\rho) + r^2\mathrm{Err}[\square_{\mathbf{g}}\rho]\\
=\;& -Ae_3(r^2\rho) - \underline{A}e_4(r^2\rho)\\
& + \frac{3}{2}\left(\frac{4}{3}A\frac{e_3(r)}{r} + \frac{4}{3}\underline{A}\frac{e_4(r)}{r} + \kappa\underline{\kappa} + 2\rho + \frac{2}{3r^2}\square_{\mathbf{g}}(r^2)\right)r^2\rho\\
& + 4r\,{d\!\!\!/}_1^\star(r)\,{d\!\!\!/}_1^\star(\rho) + r^2\mathrm{Err}[\square_{\mathbf{g}}\rho]
\end{aligned}
$$

as desired. □

Step 3. We now derive the desired formula for $\square_{\mathbf{g}}\tilde{\rho}$. In view of the definition of $\tilde{\rho}$, we have

$$\begin{aligned}\square_{\mathbf{g}}(\tilde{\rho}) &= \square_{\mathbf{g}}(r^2\rho)+\square_{\mathbf{g}}\left(\frac{2m}{r}\right)\\ &= \square_{\mathbf{g}}(r^2\rho)+2m\square_{\mathbf{g}}\left(\frac{1}{r}\right)+4D^a(m)D_a\left(\frac{1}{r}\right)+\frac{2}{r}\square_{\mathbf{g}}(m).\end{aligned}$$

Together with B.1.1 we deduce

$$\begin{aligned}\square_{\mathbf{g}}(\tilde{\rho}) &= -Ae_3(r^2\rho)-\underline{A}e_4(r^2\rho)\\ &\quad+\frac{3}{2}\left(\frac{4}{3}A\frac{e_3(r)}{r}+\frac{4}{3}\underline{A}\frac{e_4(r)}{r}+\kappa\underline{\kappa}+2\rho+\frac{2}{3r^2}\square_{\mathbf{g}}(r^2)\right)r^2\rho\\ &\quad+4r\,{\not{d}}_1^{\star}(r)\,{\not{d}}_1^{\star}(\rho)+2m\square_{\mathbf{g}}\left(\frac{1}{r}\right)+4D^a(m)D_a\left(\frac{1}{r}\right)+\frac{2}{r}\square_{\mathbf{g}}(m)\\ &\quad+4r\,{\not{d}}_1^{\star}(r)\,{\not{d}}_1^{\star}(\rho)+r^2\mathrm{Err}[\square_{\mathbf{g}}\rho].\end{aligned}$$

Next, we use $r^2\rho=\tilde{\rho}-2mr^{-1}$. This yields

$$\begin{aligned}&\square_{\mathbf{g}}(\tilde{\rho})-\frac{3}{2}\left(\kappa\underline{\kappa}-\frac{8m}{r^3}+\frac{2}{3r^2}\square_{\mathbf{g}}(r^2)\right)\tilde{\rho}\\ =\;& 2m\square_{\mathbf{g}}\left(\frac{1}{r}\right)-\frac{3m}{r}\kappa\underline{\kappa}+\frac{12m^2}{r^4}-\frac{2m}{r^3}\square_{\mathbf{g}}(r^2)-6mA\frac{e_3(r)}{r^2}-6m\underline{A}\frac{e_4(r)}{r^2}\\ &+\frac{3}{r^2}\tilde{\rho}^2+\frac{3}{2}\left(\frac{4}{3}A\frac{e_3(r)}{r}+\frac{4}{3}\underline{A}\frac{e_4(r)}{r}\right)\tilde{\rho}-Ae_3(\tilde{\rho})-\underline{A}e_4(\tilde{\rho})+\frac{2}{r}Ae_3(m)\\ &+\frac{2}{r}\underline{A}e_4(m)+4D^a(m)D_a\left(\frac{1}{r}\right)+\frac{2}{r}\square_{\mathbf{g}}(m)+4r\,{\not{d}}_1^{\star}(r)\,{\not{d}}_1^{\star}(\rho)+r^2\mathrm{Err}[\square_{\mathbf{g}}\rho].\end{aligned}$$

Note that in Schwarzschild

$$\frac{3}{2}\left(\kappa\underline{\kappa}-\frac{8m}{r^3}+\frac{2}{3r^2}\square_{\mathbf{g}}(r^2)\right)=-\frac{8m}{r^3}$$

and hence

$$\begin{aligned}
&\square_{\mathbf{g}}(\tilde{\rho}) + \frac{8m}{r^3}\tilde{\rho}\\
= \;& 2m\square_{\mathbf{g}}\left(\frac{1}{r}\right) - \frac{3m}{r}\kappa\underline{\kappa} + \frac{12m^2}{r^4} - \frac{2m}{r^3}\square_{\mathbf{g}}(r^2) - 6m\frac{e_3(r)}{r^2} - 6m\underline{A}\frac{e_4(r)}{r^2}\\
&+\frac{3}{r^2}\tilde{\rho}^2 + \frac{3}{2}\left(\frac{4}{3}A\frac{e_3(r)}{r} + \frac{4}{3}\underline{A}\frac{e_4(r)}{r}\right)\tilde{\rho}\\
&+\left(\frac{3}{2}\left(\kappa\underline{\kappa} - \frac{8m}{r^3} + \frac{2}{3r^2}\square_{\mathbf{g}}(r^2)\right) + \frac{8m}{r^3}\right)\tilde{\rho}\\
&-Ae_3(\tilde{\rho}) - \underline{A}e_4(\tilde{\rho}) + \frac{2}{r}Ae_3(m) + \frac{2}{r}\underline{A}e_4(m)\\
&+4D^a(m)D_a\left(\frac{1}{r}\right) + \frac{2}{r}\square_{\mathbf{g}}(m) + 4r\,{\not{d}}_1^\star(r)\,{\not{d}}_1^\star(\rho) + r^2\mathrm{Err}[\square_{\mathbf{g}}\rho].
\end{aligned}$$

Also, we have

$$\begin{aligned}
\square_{\mathbf{g}}\left(\frac{1}{r}\right) - \frac{1}{r^3}\square_{\mathbf{g}}(r^2) &= -\frac{\square_{\mathbf{g}}(r)}{r^2} + 2\frac{\mathbf{D}^\alpha(r)\mathbf{D}_\alpha(r)}{r^3} - 2\frac{\square_{\mathbf{g}}(r)}{r^2} - 2\frac{\mathbf{D}^\alpha(r)\mathbf{D}_\alpha(r)}{r^3}\\
&= -3\frac{\square_{\mathbf{g}}(r)}{r^2}
\end{aligned}$$

and hence

$$\begin{aligned}
&\square_{\mathbf{g}}(\tilde{\rho}) + \frac{8m}{r^3}\tilde{\rho} - 6m\frac{\square_{\mathbf{g}}(r)}{r^2} - \frac{3m}{r}\kappa\underline{\kappa} + \frac{12m^2}{r^4}\\
&-6m\frac{e_3(r)}{r^2} - 6m\underline{A}\frac{e_4(r)}{r^2} + \frac{3}{r^2}\tilde{\rho}^2 + \frac{3}{2}\left(\frac{4}{3}A\frac{e_3(r)}{r} + \frac{4}{3}\underline{A}\frac{e_4(r)}{r}\right)\tilde{\rho}\\
&+\left(\frac{3}{2}\left(\kappa\underline{\kappa} - \frac{8m}{r^3} + \frac{2}{3r^2}\square_{\mathbf{g}}(r^2)\right) + \frac{8m}{r^3}\right)\tilde{\rho}\\
&-Ae_3(\tilde{\rho}) - \underline{A}e_4(\tilde{\rho}) + \frac{2}{r}Ae_3(m) + \frac{2}{r}\underline{A}e_4(m)\\
&+4D^a(m)D_a\left(\frac{1}{r}\right) + \frac{2}{r}\square_{\mathbf{g}}(m) + 4r\,{\not{d}}_1^\star(r)\,{\not{d}}_1^\star(\rho) + r^2\mathrm{Err}[\square_{\mathbf{g}}\rho].
\end{aligned}$$

Finally, since

$$-6m\frac{e_3(r)}{r^2} - 6m\underline{A}\frac{e_4(r)}{r^2} = -3mA\frac{\underline{\kappa}}{r} - 3m\underline{A}\frac{\kappa}{r} - \frac{6m}{r}A\underline{A}$$

and

$$-6m\frac{\square_{\mathbf{g}}(r)}{r^2} - \frac{3m}{r}\kappa\underline{\kappa} + \frac{12m^2}{r^4} = -6m\frac{\square_{\mathbf{g}}(r) - \left(\frac{2}{r} - \frac{2m}{r^2}\right)}{r^2} - \frac{3m}{r}\left(\kappa\underline{\kappa} + \frac{4\Upsilon}{r^2}\right),$$

we obtain

$$
\begin{aligned}
\square_{\mathbf{g}}(\tilde{\rho})+\frac{8m}{r^3}\tilde{\rho} \quad = \quad & -6m\frac{\square_{\mathbf{g}}(r)-\left(\frac{2}{r}-\frac{2m}{r^2}\right)}{r^2}-\frac{3m}{r}\left(\kappa\underline{\kappa}+\frac{4\Upsilon}{r^2}\right)\\
& -3mA\frac{\underline{\kappa}}{r}-3m\underline{A}\frac{\kappa}{r}-\frac{6m}{r}A\underline{A}\\
& +\frac{3}{r^2}\tilde{\rho}^2+\frac{3}{2}\left(\frac{4}{3}A\frac{e_3(r)}{r}+\frac{4}{3}\underline{A}\frac{e_4(r)}{r}\right)\tilde{\rho}\\
& +\left(\frac{3}{2}\left(\kappa\underline{\kappa}-\frac{8m}{r^3}+\frac{2}{3r^2}\square_{\mathbf{g}}(r^2)\right)+\frac{8m}{r^3}\right)\tilde{\rho}\\
& -Ae_3(\tilde{\rho})-\underline{A}e_4(\tilde{\rho})+\frac{2}{r}Ae_3(m)+\frac{2}{r}\underline{A}e_4(m)\\
& +4D^a(m)D_a\left(\frac{1}{r}\right)+\frac{2}{r}\square_{\mathbf{g}}(m)+4r\,\not{d}_1^{\star}(r)\,\not{d}_1^{\star}(\rho)+r^2\mathrm{Err}[\square_{\mathbf{g}}\rho].
\end{aligned}
$$

This concludes the proof of the proposition. $\square$

Appendix C

Appendix to Chapter 9

C.1 PROOF OF LEMMA 9.11

We start with the following:

Lemma C.1. *Let $k \geq 0$ an integer and let $f \in \mathfrak{s}_k(\mathbf{S})$. Then, we have*

$$
\begin{aligned}
(\not{d}_k^{\mathbf{S}} f)^\# &= \frac{\sqrt{\gamma}}{\sqrt{\gamma^{\mathbf{S}\,\#}}}\Bigg\{ \mathring{\not{d}}_k(f^\#) + \left(\frac{k}{2}U \int_0^1 \Big(\sqrt{\gamma}(e_\theta(\kappa) - e_\theta(\vartheta))\Big)^{\#_\lambda} d\lambda \right.\\
&+\frac{k}{4}S\int_0^1 \left(\sqrt{\gamma}e_\theta\left(\underline{\kappa} - \underline{\vartheta} - \underline{\Omega}(\kappa - \vartheta) - 2\underline{b}\gamma^{1/2}e_\theta\Phi\right)\right)^{\#_\lambda} d\lambda\\
&\left.+\frac{k}{4}\left(\underline{\kappa} - \underline{\vartheta} - \underline{\Omega}(\kappa - \vartheta) - 2\underline{b}\gamma^{1/2}e_\theta\Phi\right)^\# U' + \frac{k}{2}(\kappa + \vartheta)^\# S'\right) f^\# \Bigg\}
\end{aligned}
$$

where for $0 \leq \lambda \leq 1$, $\#_\lambda$ denotes the pullback by

$$
\psi_\lambda(\mathring{u}, \mathring{s}, \theta) := (\mathring{u} + \lambda U(\theta), \mathring{s} + \lambda S(\theta), \theta).
$$

Proof. For $p \in \mathring{S}$ and f a $\mathbf{Z}$-invariant scalar function on S, we have by definition of the pushforward of a vectorfield

$$
[\Psi_\#(\partial_\theta) f]_{\Psi(p)} = [\partial_\theta(f \circ \Psi)]_p.
$$

We infer

$$
(\not{d}_k^{\mathbf{S}} f)^\# = \frac{1}{\sqrt{\gamma^{\mathbf{S}\,\#}}}\Big(\partial_\theta(f^\#) + k\partial_\theta(\Phi^\#) f^\#\Big)
$$

and hence

$$
\begin{aligned}
(\not{d}_k^{\mathbf{S}} f)^\# &= \frac{\sqrt{\gamma}}{\sqrt{\gamma^{\mathbf{S}\,\#}}}\Big(e_\theta(f^\#) + k e_\theta(\Phi) f^\# + k(e_\theta(\Phi^\#) - e_\theta(\Phi)) f^\#\Big)\\
&= \frac{\sqrt{\gamma}}{\sqrt{\gamma^{\mathbf{S}\,\#}}}\Big(\mathring{\not{d}}_k(f^\#) + k(e_\theta(\Phi^\#) - e_\theta(\Phi)) f^\#\Big).
\end{aligned}
$$

Next, we have

$$
e_\theta(\Phi^\#) - e_\theta(\Phi) = \sqrt{\gamma}^{-1}\Big(\partial_\theta(\Phi^\#) - \partial_\theta\Phi\Big)
$$

and

$$\begin{aligned}\Big(\partial_\theta(\Phi^\#)-\partial_\theta\Phi\Big)(\mathring{u},\mathring{s},\theta) &= \partial_\theta[\Phi(\mathring{u}+U(\theta),\mathring{s}+S(\theta),\theta)]-\partial_\theta\Phi(\mathring{u},\mathring{s},\theta)\\
&= (\partial_\theta\Phi)(\mathring{u}+U(\theta),\mathring{s}+S(\theta),\theta)-\partial_\theta\Phi(\mathring{u},\mathring{s},\theta)\\
&\quad+\Big[(\partial_u\Phi)^\# U'+(\partial_s\Phi)^\# S'\Big](\mathring{u},\mathring{s},\theta)\\
&= \int_0^1\frac{d}{d\lambda}\Big[(\partial_\theta\Phi)(\mathring{u}+\lambda U(\theta),\mathring{s}+\lambda S(\theta),\theta)\Big]d\lambda\\
&\quad+\Big[(\partial_u\Phi)^\# U'+(\partial_s\Phi)^\# S'\Big](\mathring{u},\mathring{s},\theta)\\
&= U(\theta)\int_0^1(\partial_u\partial_\theta\Phi)(\mathring{u}+\lambda U(\theta),\mathring{s}+\lambda S(\theta),\theta)d\lambda\\
&\quad+S(\theta)\int_0^1(\partial_s\partial_\theta\Phi)(\mathring{u}+\lambda U(\theta),\mathring{s}+\lambda S(\theta),\theta)d\lambda\\
&\quad+\Big[(\partial_u\Phi)^\# U'+(\partial_s\Phi)^\# S'\Big](\mathring{u},\mathring{s},\theta)\end{aligned}$$

which we rewrite

$$\partial_\theta(\Phi^\#)-\partial_\theta\Phi = U\int_0^1(\partial_u\partial_\theta\Phi)^{\#_\lambda}d\lambda+S\int_0^1(\partial_s\partial_\theta\Phi)^{\#_\lambda}d\lambda+(\partial_u\Phi)^\# U'+(\partial_s\Phi)^\# S'$$

where $\#_\lambda$ denotes the pullback by the map $\psi_\lambda(\mathring{u},\mathring{s},\theta)=(\mathring{u}+\lambda U(\theta),\mathring{s}+\lambda S(\theta),\theta)$.

Next, recall that

$$\partial_s=e_4,\quad \partial_u=\frac{1}{2}\Big(e_3-\underline{\Omega}e_4-\underline{b}\gamma^{1/2}e_\theta\Big),\quad \partial_\theta=\sqrt{\gamma}e_\theta.$$

Hence,

$$\begin{aligned}\partial_\theta\partial_s\Phi &= \sqrt{\gamma}e_\theta e_4(\Phi),\\
\partial_\theta\partial_u\Phi &= \frac{1}{2}\sqrt{\gamma}e_\theta\Big(e_3\Phi-\underline{\Omega}e_4\Phi-\underline{b}\gamma^{1/2}e_\theta\Phi\Big),\end{aligned}$$

which yields

$$\begin{aligned}\partial_\theta(\Phi^\#)-\partial_\theta\Phi &= U\int_0^1\Big(\sqrt{\gamma}e_\theta e_4(\Phi)\Big)^{\#_\lambda}d\lambda\\
&\quad+S\int_0^1\bigg(\frac{1}{2}\sqrt{\gamma}e_\theta\Big(e_3\Phi-\underline{\Omega}e_4\Phi-\underline{b}\gamma^{1/2}e_\theta\Phi\Big)\bigg)^{\#_\lambda}d\lambda\\
&\quad+\frac{1}{2}\Big(e_3\Phi-\underline{\Omega}e_4\Phi-\underline{b}\gamma^{1/2}e_\theta\Phi\Big)^\# U'+(e_4\Phi)^\# S'\\
&= \frac{1}{2}U\int_0^1\Big(\sqrt{\gamma}(e_\theta(\kappa)-e_\theta(\vartheta))\Big)^{\#_\lambda}d\lambda\\
&\quad+\frac{1}{4}S\int_0^1\Big(\sqrt{\gamma}e_\theta\Big(\underline{\kappa}-\underline{\vartheta}-\underline{\Omega}(\kappa-\vartheta)-2\underline{b}\gamma^{1/2}e_\theta\Phi\Big)\Big)^{\#_\lambda}d\lambda\\
&\quad+\frac{1}{4}\Big(\underline{\kappa}-\underline{\vartheta}-\underline{\Omega}(\kappa-\vartheta)-2\underline{b}\gamma^{1/2}e_\theta\Phi\Big)^\# U'+\frac{1}{2}(\kappa+\vartheta)^\# S'.\end{aligned}$$

We deduce

$$
\begin{aligned}
(\not{d}_k^{\mathbf{S}} f)^{\#} &= \frac{\sqrt{\gamma}}{\sqrt{\gamma^{\mathbf{S}\,\#}}}\Bigg\{ \mathring{\not{d}}_k(f^{\#}) + \Bigg(\frac{k}{2}U\int_0^1 \Big(\sqrt{\gamma}(e_\theta(\kappa)-e_\theta(\vartheta))\Big)^{\#_\lambda} d\lambda \\
&\quad +\frac{k}{4}S\int_0^1 \Big(\sqrt{\gamma}e_\theta\left(\underline{\kappa}-\underline{\vartheta}-\underline{\Omega}(\kappa-\vartheta)-2\underline{b}\gamma^{1/2}e_\theta\Phi\right)\Big)^{\#_\lambda} d\lambda \\
&\quad +\frac{k}{4}\left(\underline{\kappa}-\underline{\vartheta}-\underline{\Omega}(\kappa-\vartheta)-2\underline{b}\gamma^{1/2}e_\theta\Phi\right)^{\#}U' + \frac{k}{2}(\kappa+\vartheta)^{\#}S'\Bigg)f^{\#}\Bigg\}.
\end{aligned}
$$

This concludes the proof of the lemma. □

We are ready to prove the higher derivative comparison Lemma 9.11 which we recall below.

Lemma C.2. *Let $\mathring{S}\subset\mathcal{R}=\mathcal{R}(\mathring{\epsilon},\mathring{\delta})$ as in Definition 9.1 verifying the assumptions* ***A1–A3****. Let $\Psi:\mathring{S}\to\mathbf{S}$ be $\mathbf{Z}$-invariant deformation. Assume the bound*

$$
\|(U',S')\|_{L^\infty_1(\mathring{S})} + \mathring{r}^{-1}\max_{0\le s\le s_{max}-1}\|(U',S')\|_{\mathfrak{h}_s(\mathring{S},\mathring{\not{g}})} \lesssim \mathring{\delta}. \tag{C.1.1}
$$

Then, we have for any reduced scalar h defined on $\mathcal{R}$

$$
\|h\|_{\mathfrak{h}_s(\mathbf{S})}\lesssim \sup_{\mathcal{R}}|\mathfrak{d}^{\le k}h| \quad \text{for } 0\le s\le s_{max}.
$$

Also, if $f\in\mathfrak{h}_s(\mathbf{S})$ and $f^{\#}$ is its pullback by ψ, we have

$$
\|f\|_{\mathfrak{h}_s(\mathbf{S})} = \|f^{\#}\|_{\mathfrak{h}_s(\mathring{S},\not{g}^{\mathbf{S},\#})} = \|f^{\#}\|_{\mathfrak{h}_s(\mathring{S},\mathring{\not{g}})}(1+O(\mathring{\epsilon})) \quad \text{for } 0\le s\le s_{max}-1.
$$

Remark C.3. *Note that the estimates of the lemma are independent of the size $\mathring{r}$ of the sphere $\mathring{S}=S(\mathring{u},\mathring{s})\subset\mathcal{R}$, see Definition 9.1. To simplify the argument below we assume $\mathring{r}\approx 1$. The general case can be easily deduced by a simple scaling argument or making obvious adjustments in the inequalities below.*

Proof. We argue by iteration. We consider the following iteration assumptions:

$$
\text{If (9.2.14) holds, then we have } \|h\|_{\mathfrak{h}_s(\mathbf{S})}\lesssim \sup_{\mathcal{R}}|\mathfrak{d}^{\le s}h|, \tag{C.1.2}
$$

and

$$
\text{if (9.2.14) holds, then we have } \|f^{\#}\|_{\mathfrak{h}_s(\mathring{S},\not{g}^{\mathbf{S},\#})} = \|f^{\#}\|_{\mathfrak{h}_s(\mathring{S},\mathring{\not{g}})}(1+O(\mathring{\delta})). \tag{C.1.3}
$$

First, note that (C.1.2) holds trivially for $s=0$ and (C.1.3) holds for $s=0$ by Lemma 9.8. Thus, from now on, we assume that (C.1.2) and (C.1.3) hold for some s with $0\le s\le s_{max}-2$, and our goal is to prove that it also holds for s replaced by $s+1$.

We start with (C.1.2). We have

$$
\not{d}_k^{\mathbf{S}} h = e_\theta^{\mathbf{S}} h + e_\theta^{\mathbf{S}}(\Phi)h.
$$

Now, recall that we have

$$e^{\mathbf{S}}_\theta = \frac{1}{\sqrt{\gamma^{\mathbf{S}}}}\partial^{\mathbf{S}}_\theta, \quad \partial^{\mathbf{S}}_\theta|_{\Psi(p)} = \left(\left(S' - \frac{1}{2}\underline{\Omega}U'\right)e_4 + \frac{1}{2}U'e_3 + \sqrt{\gamma}\left(1 - \frac{1}{2}\underline{b}U'\right)e_\theta\right)\Big|_{\Psi(p)}.$$

This yields

$$\begin{aligned}(\not{d}^{\mathbf{S}}_k h)_{|\Psi(p)} &= \Bigg\{\frac{1}{\sqrt{\gamma^{\mathbf{S}}}}\Bigg(\left(S' - \frac{1}{2}\underline{\Omega}U'\right)e_4(h) + \left(S' - \frac{1}{2}\underline{\Omega}U'\right)e_4(\Phi)h \\ &\quad + \frac{1}{2}U'e_3(h) + \frac{1}{2}U'e_3(\Phi)h + \sqrt{\gamma}\left(1 - \frac{1}{2}\underline{b}U'\right)\not{d}_k(h)\Bigg)\Bigg\}_{|\Psi(p)}.\end{aligned}$$

Together with the iteration assumption (C.1.3), we infer

$$\begin{aligned}&\|\not{d}^{\mathbf{S}}_k h\|_{\mathfrak{h}_s(\mathbf{S})} \\ &= \|(\not{d}^{\mathbf{S}}_k h)^\#\|_{\mathfrak{h}_s(\overset{\circ}{S}, \overset{\circ}{\not{g}})}(1 + O(\overset{\circ}{\delta})) \\ &\lesssim \Bigg\|\frac{1}{\sqrt{\gamma^{\mathbf{S},\#}}}\Bigg(\left(S' - \frac{1}{2}\underline{\Omega}^\# U'\right)(e_4(h))^\# + \left(S' - \frac{1}{2}\underline{\Omega}^\# U'\right)(e_4(\Phi)h)^\# \\ &\quad + \frac{1}{2}U'(e_3(h))^\# + \frac{1}{2}U'(e_3(\Phi)h)^\# + \sqrt{\gamma^\#}\left(1 - \frac{1}{2}\underline{b}^\# U'\right)(\not{d}_k(h))^\#\Bigg)\Bigg\|_{\mathfrak{h}_s(\overset{\circ}{S}, \overset{\circ}{\not{g}})},\end{aligned}$$

i.e.,

$$\begin{aligned}&\|\not{d}^{\mathbf{S}}_k h\|_{\mathfrak{h}_s(\mathbf{S})} \\ &\lesssim \|(\not{d}_k(h))^\#\|_{\mathfrak{h}_s(\overset{\circ}{S}, \overset{\circ}{\not{g}})} + \left\|\left(\frac{\sqrt{\gamma^\#}}{\sqrt{\gamma^{\mathbf{S},\#}}} - 1\right)(\not{d}_k(h))^\#\right\|_{\mathfrak{h}_s(\overset{\circ}{S}, \overset{\circ}{\not{g}})} \\ &\quad + \Bigg\|\frac{1}{\sqrt{\gamma^{\mathbf{S},\#}}}\Bigg(\left(S' - \frac{1}{2}\underline{\Omega}^\# U'\right)(e_4(h))^\# + \left(S' - \frac{1}{2}\underline{\Omega}^\# U'\right)(e_4(\Phi)h)^\# \\ &\quad + \frac{1}{2}U'(e_3(h))^\# + \frac{1}{2}U'(e_3(\Phi)h)^\# - \frac{1}{2}\sqrt{\gamma^\#}\underline{b}^\# U'(\not{d}_k(h))^\#\Bigg)\Bigg\|_{\mathfrak{h}_s(\overset{\circ}{S}, \overset{\circ}{\not{g}})}.\end{aligned}$$

Together with a non-sharp product rule in $\mathfrak{h}_s(\overset{\circ}{S}, \overset{\circ}{\not{g}})$ and the repeated use of the iteration assumptions (C.1.2), (C.1.3), we can bound the right-hand side of the above inequality by

$$
\begin{aligned}
\lesssim\ & \left(1+\left(\|\sqrt{\gamma}\|_{\mathfrak{h}_s(\mathbf{S})}+\left\|(\sqrt{\gamma})^{\#}\right\|_{\mathfrak{h}_1^\infty(\overset{\circ}{S})}\right)\left(\left\|\frac{1}{\sqrt{\gamma^{\mathbf{S}}}}\right\|_{\mathfrak{h}_s(\mathbf{S})}+\left\|\frac{1}{\sqrt{\gamma^{\mathbf{S},\#}}}\right\|_{\mathfrak{h}_1^\infty(\overset{\circ}{S})}\right)\right)\\
&\times\|\not{d}_k(h)\|_{\mathfrak{h}_s(\mathbf{S})}+\|(U',S')\|_{\mathfrak{h}_s(\overset{\circ}{S},\overset{\circ}{\not{g}})\cap\mathfrak{h}_1^\infty(\overset{\circ}{S})}\left(\left\|\frac{1}{\sqrt{\gamma^{\mathbf{S}}}}\right\|_{\mathfrak{h}_s(\mathbf{S})}+\left\|\frac{1}{\sqrt{\gamma^{\mathbf{S},\#}}}\right\|_{\mathfrak{h}_1^\infty(\overset{\circ}{S})}\right)\\
&\times\left(1+\|(\underline{\Omega},\underline{b}\sqrt{\gamma})\|_{\mathfrak{h}_s(\mathbf{S})}+\|(\underline{\Omega},\underline{b}\sqrt{\gamma})^{\#}\|_{\mathfrak{h}_1^\infty(\overset{\circ}{S})}\right)\\
&\times\left\|\Big((e_3,e_4,\not{d}_k)h,e_3(\Phi)h,e_4(\Phi)h\Big)\right\|_{\mathfrak{h}_s(\mathbf{S})}.
\end{aligned}
$$

Therefore $\|\not{d}_k^{\mathbf{S}}h\|_{\mathfrak{h}_s(\mathbf{S})}$ can be bounded by

$$
\begin{aligned}
\lesssim\ & \left(1+\left(\|\sqrt{\gamma}\|_{\mathfrak{h}_s(\mathbf{S})}+\left\|(\sqrt{\gamma})^{\#}\right\|_{\mathfrak{h}_1^\infty(\overset{\circ}{S})}\right)\left(\left\|\frac{1}{\sqrt{\gamma^{\mathbf{S}}}}\right\|_{\mathfrak{h}_s(\mathbf{S})}+\left\|\frac{1}{\sqrt{\gamma^{\mathbf{S},\#}}}\right\|_{\mathfrak{h}_1^\infty(\overset{\circ}{S})}\right)\right)\\
&\times\sup_{\mathcal{R}}\left|\mathfrak{d}^{\le s}\not{d}_k h\right|+\left(\left\|\frac{1}{\sqrt{\gamma^{\mathbf{S}}}}\right\|_{\mathfrak{h}_s(\mathbf{S})}+\left\|\frac{1}{\sqrt{\gamma^{\mathbf{S},\#}}}\right\|_{\mathfrak{h}_1^\infty(\overset{\circ}{S})}\right)\\
&\times\left(1+\|(\underline{\Omega},\underline{b}\sqrt{\gamma})\|_{\mathfrak{h}_s(\mathbf{S})}+\|(\underline{\Omega},\underline{b}\sqrt{\gamma})^{\#}\|_{\mathfrak{h}_1^\infty(\overset{\circ}{S})}\right)\\
&\times\sup_{\mathcal{R}}\left|\mathfrak{d}^{\le s}(\mathfrak{d}h,e_3(\Phi)h,e_4(\Phi)h)\right|,
\end{aligned}
$$

where we used in the last inequality the assumption (9.2.14) on (U',S'). Together with (9.1.12) and (9.1.15), we infer

$$
\begin{aligned}
&\|\not{d}_k^{\mathbf{S}}h\|_{\mathfrak{h}_s(\mathbf{S})}\\
\lesssim\ & \left\{\left(1+\left(1+\left\|(\sqrt{\gamma})^{\#}\right\|_{\mathfrak{h}_1^\infty(\overset{\circ}{S})}\right)\left(\left\|\frac{1}{\sqrt{\gamma^{\mathbf{S}}}}\right\|_{\mathfrak{h}_s(\mathbf{S})}+\left\|\frac{1}{\sqrt{\gamma^{\mathbf{S},\#}}}\right\|_{\mathfrak{h}_1^\infty(\overset{\circ}{S})}\right)\right)\right.\\
&\left.+\left(1+\left\|\frac{1}{\sqrt{\gamma^{\mathbf{S}}}}\right\|_{\mathfrak{h}_s(\mathbf{S})}+\left\|\frac{1}{\sqrt{\gamma^{\mathbf{S},\#}}}\right\|_{\mathfrak{h}_1^\infty(\overset{\circ}{S})}\right)\left(1+\|(\underline{\Omega},\underline{b}\sqrt{\gamma})^{\#}\|_{\mathfrak{h}_1^\infty(\overset{\circ}{S})}\right)\right\}\\
&\times\sup_{\mathcal{R}}\left|\mathfrak{d}^{\le s+1}h\right|.
\end{aligned}
$$

Also, for a reduced scalar v defined on $\mathcal{R}$, we have in view of the assumption (9.2.14) on (U', S')

$$\begin{aligned}\|v^{\#}\|_{\mathfrak{h}_1^\infty(\overset{\circ}{S})} &= \|v \circ \psi\|_{\mathfrak{h}_1^\infty(\overset{\circ}{S})} \\ &\lesssim \left(1 + \sup_{0\leq\theta\leq\pi} |\psi'(\theta)|\right) \sup_{\mathcal{R}} |\mathfrak{d}^{\leq 1} v| \\ &\lesssim \left(1 + \|(U', S')\|_{\mathfrak{h}_1^\infty(\overset{\circ}{S})}\right) \sup_{\mathcal{R}} |\mathfrak{d}^{\leq 1} v| \\ &\lesssim (1 + \overset{\circ}{\delta}) \sup_{\mathcal{R}} |\mathfrak{d}^{\leq 1} v|. \end{aligned} \tag{C.1.4}$$

Together with (9.1.12) and (9.1.15), we infer

$$\|\not{d}_k^{\mathbf{S}} h\|_{\mathfrak{h}_s(\mathbf{S})} \lesssim \left\{1 + \left\|\frac{1}{\sqrt{\gamma^{\mathbf{S}}}}\right\|_{\mathfrak{h}_s(\mathbf{S})} + \left\|\frac{1}{\sqrt{\gamma^{\mathbf{S},\#}}}\right\|_{\mathfrak{h}_1^\infty(\overset{\circ}{S})}\right\} \sup_{\mathcal{R}} \left|\mathfrak{d}^{\leq s+1} h\right|.$$

Now, recall that

$$\begin{aligned}\gamma^{\mathbf{S}}(\psi(\theta)) &= \gamma(\psi(\theta)) + \left(\underline{\Omega}(\psi(\theta)) + \frac{1}{4}(\underline{b}(\psi(\theta)))^2\gamma(\psi(\theta))\right)(U'(\theta))^2 \\ &- 2U'(\theta)S'(\theta) - \gamma(\psi(\theta))\underline{b}(\psi(\theta))U'(\theta).\end{aligned}$$

Together with a repeated application of the iteration assumptions and a non-sharp product rule in $\mathfrak{h}_s(\overset{\circ}{S}, \overset{\circ}{\not{g}})$ and (C.1.4), this yields

$$\begin{aligned}&\|\gamma^{\mathbf{S}}\|_{\mathfrak{h}_s(\mathbf{S})} + \|\gamma^{\mathbf{S},\#}\|_{\mathfrak{h}_1^\infty(\overset{\circ}{S})} \\ \lesssim\ & \left(1 + \sup_{\mathcal{R}} |\mathfrak{d}^{\leq 1}(\gamma, \underline{\Omega}, \underline{b}^2\gamma, \underline{b}\gamma)| + \sup_{\mathcal{R}} |\mathfrak{d}^{\leq s}(\gamma, \underline{\Omega}, \underline{b}^2\gamma, \underline{b}\gamma)|\right) \\ \times\ & \left(1 + \|(U', S')\|_{\mathfrak{h}_1^\infty(\overset{\circ}{S})} + \|(U', S')\|_{\mathfrak{h}_s(\overset{\circ}{S}, \overset{\circ}{\not{g}})}\right) \\ \lesssim\ & 1\end{aligned}$$

where we used in the last estimate the assumption (9.2.14) on (U', S') and (9.1.15). We infer

$$\left\|\frac{1}{\sqrt{\gamma^{\mathbf{S}}}}\right\|_{\mathfrak{h}_s(\mathbf{S})} + \left\|\frac{1}{\sqrt{\gamma^{\mathbf{S},\#}}}\right\|_{\mathfrak{h}_1^\infty(\overset{\circ}{S})} \lesssim 1$$

and hence

$$\|\not{d}_k^{\mathbf{S}} h\|_{\mathfrak{h}_s(\mathbf{S})} \lesssim \sup_{\mathcal{R}} \left|\mathfrak{d}^{\leq s+1} h\right|$$

which corresponds to the first iteration assumption (C.1.2) with s replaced with $s+1$ for $s \leq s_{max} - 2$.

Next, we focus on recovering the second iteration assumption (C.1.3) with s replaced with $s+1$ for $s \leq s_{max} - 2$. Recall from Lemma C.1 that we have for

$f \in \mathfrak{s}_k(\mathbf{S})$

$$
\begin{aligned}
(\not{d}_k^{\mathbf{S}} f)^\# &= \frac{\sqrt{\gamma}}{\sqrt{\gamma^{\mathbf{S}\,\#}}}\Bigg\{ \mathring{\not{d}}_k(f^\#) + \Bigg(\frac{k}{2}U\int_0^1 \left(\sqrt{\gamma}(e_\theta(\kappa) - e_\theta(\vartheta))\right)^{\#_\lambda} d\lambda \\
&+ \frac{k}{4}S\int_0^1 \left(\sqrt{\gamma}e_\theta\left(\underline{\kappa} - \underline{\vartheta} - \underline{\Omega}(\kappa - \vartheta) - 2\underline{b}\gamma^{1/2}e_\theta\Phi\right)\right)^{\#_\lambda} d\lambda \\
&+ \frac{k}{4}\left(\underline{\kappa} - \underline{\vartheta} - \underline{\Omega}(\kappa - \vartheta) - 2\underline{b}\gamma^{1/2}e_\theta\Phi\right)^\# U' + \frac{k}{2}(\kappa + \vartheta)^\# S'\Bigg) f^\#\Bigg\}
\end{aligned}
$$

where for $0 \le \lambda \le 1$, $\#_\lambda$ denotes the pullback by

$$
\psi_\lambda(\mathring{u}, \mathring{s}, \theta) = (\mathring{u} + \lambda U(\theta), \mathring{s} + \lambda S(\theta), \theta).
$$

For convenience, we rewrite some of the terms as follows:

$$
\begin{aligned}
e_\theta(\kappa) - e_\theta(\vartheta) &= -\not{d}_1^\star(\kappa) - \frac{1}{2}(\not{d}_1\vartheta - \not{d}_2^\star\vartheta), \\
\underline{b}\gamma^{1/2}e_\theta\Phi &= \frac{1}{2}\gamma^{1/2}(\not{d}_1\underline{b} + \not{d}_2^\star\underline{b}),
\end{aligned}
$$

and

$$
\begin{aligned}
& e_\theta\left(\underline{\kappa} - \underline{\vartheta} - \underline{\Omega}(\kappa - \vartheta) - 2\underline{b}\gamma^{1/2}e_\theta\Phi\right) \\
={}& -\not{d}_1^\star(\underline{\kappa}) - \frac{1}{2}(\not{d}_1\underline{\vartheta} - \not{d}_2^\star\underline{\vartheta}) + \not{d}_1^\star(\underline{\Omega}\kappa) - \not{d}_1^\star(\underline{\Omega})\vartheta + \frac{1}{2}\underline{\Omega}(\not{d}_1\underline{\vartheta} - \not{d}_2^\star\underline{\vartheta}) \\
& + \not{d}_1^\star(\gamma^{1/2})(\not{d}_1\vartheta + \not{d}_2^\star\vartheta)\underline{b} - 2\gamma^{1/2}e_\theta(\underline{b}e_\theta\Phi) \\
={}& -\not{d}_1^\star(\underline{\kappa}) - \frac{1}{2}(\not{d}_1\underline{\vartheta} - \not{d}_2^\star\underline{\vartheta}) + \not{d}_1^\star(\underline{\Omega}\kappa) - \not{d}_1^\star(\underline{\Omega})\vartheta + \frac{1}{2}\underline{\Omega}(\not{d}_1\underline{\vartheta} - \not{d}_2^\star\underline{\vartheta}) \\
& + \not{d}_1^\star(\gamma^{1/2})(\not{d}_1\vartheta + \not{d}_2^\star\vartheta)\underline{b} - 2\gamma^{1/2}(-e_\theta(\Phi)\not{d}_2^\star\underline{b} - K\underline{b}) \\
={}& -\not{d}_1^\star(\underline{\kappa}) - \frac{1}{2}(\not{d}_1\underline{\vartheta} - \not{d}_2^\star\underline{\vartheta}) + \not{d}_1^\star(\underline{\Omega}\kappa) - \not{d}_1^\star(\underline{\Omega})\vartheta + \frac{1}{2}\underline{\Omega}(\not{d}_1\underline{\vartheta} - \not{d}_2^\star\underline{\vartheta}) \\
& + \not{d}_1^\star(\gamma^{1/2})(\not{d}_1\vartheta + \not{d}_2^\star\vartheta)\underline{b} + \frac{1}{2}\gamma^{1/2}(\not{d}_2\not{d}_2^\star\underline{b} + \not{d}_3^\star\not{d}_2^\star\underline{b}) + 2\gamma^{1/2}K\underline{b}
\end{aligned}
$$

where we used the identities

$$
\begin{aligned}
e_\theta(e_\theta(\Phi)) &= -(e_\theta(\Phi))^2 - K, \\
2\gamma^{1/2}e_\theta\Phi\,\not{d}_2^\star\underline{b} &= \frac{1}{2}\gamma^{1/2}(\not{d}_2\not{d}_2^\star\underline{b} + \not{d}_3^\star\not{d}_2^\star\underline{b}).
\end{aligned}
$$

This yields

$$
\begin{aligned}
&(\not{d}_k^{\mathbf{S}} f)^\# \\
&= \frac{\sqrt{\gamma}}{\sqrt{\gamma^{\mathbf{S}\,\#}}}\Bigg\{ \mathring{\not{d}}_k(f^\#) + \Bigg(\frac{k}{2}U\int_0^1\Big(\sqrt{\gamma}\Big(-\not{d}_1^\star(\kappa) - \frac{1}{2}(\not{d}_1\vartheta - \not{d}_2^\star\vartheta)\Big)\Big)^{\#_\lambda} d\lambda \\
&+\frac{k}{4}S\int_0^1\Bigg(\sqrt{\gamma}\Bigg(- \not{d}_1^\star(\underline{\kappa}) - \frac{1}{2}(\not{d}_1\underline{\vartheta} - \not{d}_2^\star\underline{\vartheta}) + \not{d}_1^\star(\underline{\Omega}\kappa) - \not{d}_1^\star(\underline{\Omega})\vartheta \\
&+\frac{1}{2}\underline{\Omega}(\not{d}_1\underline{\vartheta} - \not{d}_2^\star\underline{\vartheta}) + \not{d}_1^\star(\gamma^{1/2})(\not{d}_1\vartheta + \not{d}_2^\star\vartheta)\underline{b} \\
&+\frac{1}{2}\gamma^{1/2}(\not{d}_2\not{d}_2^\star\underline{b} + \not{d}_3^\star\not{d}_2^\star\underline{b}) + 2\gamma^{1/2}K\underline{b}\Bigg)\Bigg)^{\#_\lambda} d\lambda \\
&+\frac{k}{4}\Big(\underline{\kappa} - \underline{\vartheta} - \underline{\Omega}(\kappa-\vartheta) - \gamma^{1/2}(\not{d}_1\vartheta + \not{d}_2^\star\vartheta)\underline{b}\Big)^\# U' + \frac{k}{2}(\kappa+\vartheta)^\# S'\Bigg) f^\#\Bigg\}.
\end{aligned}
$$

Next, we take the $\mathfrak{h}_s(\mathring{S}, \mathring{\not{g}})$-norm of this identity, and we use the iteration assumption to replace the norm on the left-hand side with the $\mathfrak{h}_s(\mathring{S}, \not{g}^{\mathbf{S},\#})$-norm. We infer

$$
\begin{aligned}
&\|(\not{d}_k^{\mathbf{S}} f)^\#\|_{\mathfrak{h}_s(\mathring{S},\not{g}^{\mathbf{S},\#})}(1+O(\mathring{\delta})) \\
&= \Bigg\|\frac{\sqrt{\gamma}}{\sqrt{\gamma^{\mathbf{S}\,\#}}}\Bigg\{ \mathring{\not{d}}_k(f^\#) + \Bigg(\frac{k}{2}U\int_0^1\Big(\sqrt{\gamma}\Big(-\not{d}_1^\star(\kappa) - \frac{1}{2}(\not{d}_1\vartheta - \not{d}_2^\star\vartheta)\Big)\Big)^{\#_\lambda} d\lambda \\
&+\frac{k}{4}S\int_0^1\Bigg(\sqrt{\gamma}\Bigg(- \not{d}_1^\star(\underline{\kappa}) - \frac{1}{2}(\not{d}_1\underline{\vartheta} - \not{d}_2^\star\underline{\vartheta}) + \not{d}_1^\star(\underline{\Omega}\kappa) - \not{d}_1^\star(\underline{\Omega})\vartheta \\
&+\frac{1}{2}\underline{\Omega}(\not{d}_1\underline{\vartheta} - \not{d}_2^\star\underline{\vartheta}) + \not{d}_1^\star(\gamma^{1/2})(\not{d}_1\vartheta + \not{d}_2^\star\vartheta)\underline{b} \\
&+\frac{1}{2}\gamma^{1/2}(\not{d}_2\not{d}_2^\star\underline{b} + \not{d}_3^\star\not{d}_2^\star\underline{b}) + 2\gamma^{1/2}K\underline{b}\Bigg)\Bigg)^{\#_\lambda} d\lambda \\
&+\frac{k}{4}\Big(\underline{\kappa} - \underline{\vartheta} - \underline{\Omega}(\kappa-\vartheta) - \gamma^{1/2}(\not{d}_1\vartheta + \not{d}_2^\star\vartheta)\underline{b}\Big)^\# U' \\
&+\frac{k}{2}(\kappa+\vartheta)^\# S'\Bigg) f^\#\Bigg\}\Bigg\|_{\mathfrak{h}_s(\mathring{S},\mathring{\not{g}})}.
\end{aligned}
$$

Next, we use a non-sharp product rule in $\mathfrak{h}_s(\overset{\circ}{S}, \overset{\circ}{\not g})$ to infer

$$
\begin{aligned}
& \|(\not d_k^{\mathbf{S}} f)^\#\|_{\mathfrak{h}_s(\overset{\circ}{S}, \not g^{\mathbf{S},\#})}(1+O(\overset{\circ}{\delta})) \\
= & \left(1+O(1)\left\|\frac{\sqrt{\gamma}}{\sqrt{\gamma^{\mathbf{S}\,\#}}}-1\right\|_{\mathfrak{h}_s(\overset{\circ}{S}, \overset{\circ}{\not g})\cap\mathfrak{h}_1^\infty(\overset{\circ}{S})}\right)\Bigg\{\left\|\overset{\circ}{\not d}_k(f^\#)\right\|_{\mathfrak{h}_s(\overset{\circ}{S}, \overset{\circ}{\not g})} \\
& +O(1)\Bigg(\|U\|_{\mathfrak{h}_{s+1}(\overset{\circ}{S}, \overset{\circ}{\not g})} \\
& \times\int_0^1\left\|\left(\sqrt{\gamma}\left(-\not d_1^\star(\kappa)-\frac{1}{2}(\not d_1\vartheta-\not d_2^\star\vartheta)\right)\right)^{\#_\lambda}\right\|_{\mathfrak{h}_s(\overset{\circ}{S}, \overset{\circ}{\not g})\cap\mathfrak{h}_1^\infty(\overset{\circ}{S})}d\lambda \\
& +\|S\|_{\mathfrak{h}_{s+1}(\overset{\circ}{S}, \overset{\circ}{\not g})}\int_0^1\Bigg\|\Bigg(\sqrt{\gamma}\Bigg(-\not d_1^\star(\underline{\kappa})-\frac{1}{2}(\not d_1\underline{\vartheta}-\not d_2^\star\underline{\vartheta})+\not d_1^\star(\underline{\Omega}\kappa)-\not d_1^\star(\underline{\Omega})\vartheta \\
& +\frac{1}{2}\underline{\Omega}(\not d_1\underline{\vartheta}-\not d_2^\star\underline{\vartheta})+\not d_1^\star(\gamma^{1/2})(\not d_1\vartheta+\not d_2^\star\vartheta)\underline{b} \\
& +\frac{1}{2}\gamma^{1/2}(\not d_2\not d_2^\star\underline{b}+\not d_3^\star\not d_2^\star\underline{b})+2\gamma^{1/2}K\underline{b}\Bigg)\Bigg)^{\#_\lambda}\Bigg\|_{\mathfrak{h}_s(\overset{\circ}{S}, \overset{\circ}{\not g})\cap\mathfrak{h}_1^\infty(\overset{\circ}{S})}d\lambda \\
& +\left\|\left(\underline{\kappa}-\underline{\vartheta}-\underline{\Omega}(\kappa-\vartheta)-\gamma^{1/2}(\not d_1\vartheta+\not d_2^\star\vartheta)\underline{b}\right)^\#\right\|_{\mathfrak{h}_s(\overset{\circ}{S}, \overset{\circ}{\not g})\cap\mathfrak{h}_1^\infty(\overset{\circ}{S})}\|U'\|_{\mathfrak{h}_{s+1}(\overset{\circ}{S}, \overset{\circ}{\not g})} \\
& +\|(\kappa+\vartheta)^\#\|_{\mathfrak{h}_s(\overset{\circ}{S}, \overset{\circ}{\not g})\cap\mathfrak{h}_1^\infty(\overset{\circ}{S})}\|S'\|_{\mathfrak{h}_{s+1}(\overset{\circ}{S}, \overset{\circ}{\not g})}\Bigg)\|f^\#\|_{\mathfrak{h}_{s+1}(\overset{\circ}{S}, \overset{\circ}{\not g})}\Bigg\}.
\end{aligned}
$$

Since $s+1 \le s_{max}-1$, we infer in view of (9.2.14) and the fact that $U(0) = S(0) = 0$,

$$
\begin{aligned}
& \|(\not{d}_k^{\mathbf{S}} f)^{\#}\|_{\mathfrak{h}_s(\overset{\circ}{S}, \not{g}^{\mathbf{S},\#})}(1+O(\overset{\circ}{\delta})) \\
= & \left(1+O(1)\left\|\frac{\sqrt{\gamma}}{\sqrt{\gamma^{\mathbf{S}\,\#}}}-1\right\|_{\mathfrak{h}_s(\overset{\circ}{S}, \overset{\circ}{\not{g}})\cap\mathfrak{h}_1^\infty(\overset{\circ}{S})}\right)\Bigg\{\left\|\overset{\circ}{\not{d}}_k(f^{\#})\right\|_{\mathfrak{h}_s(\overset{\circ}{S}, \overset{\circ}{\not{g}})} \\
& +O(\overset{\circ}{\delta})\left(\int_0^1\left\|\left(\sqrt{\gamma}\left(-\not{d}_1^\star(\kappa)-\frac{1}{2}(\not{d}_1\vartheta-\not{d}_2^\star\vartheta)\right)\right)^{\#_\lambda}\right\|_{\mathfrak{h}_s(\overset{\circ}{S}, \overset{\circ}{\not{g}})\cap\mathfrak{h}_1^\infty(\overset{\circ}{S})} d\lambda \right.\\
& +\int_0^1\left\|\left(\sqrt{\gamma}\left(-\not{d}_1^\star(\underline{\kappa})-\frac{1}{2}(\not{d}_1\underline{\vartheta}-\not{d}_2^\star\underline{\vartheta})+\not{d}_1^\star(\underline{\Omega}\kappa)-\not{d}_1^\star(\underline{\Omega})\vartheta\right.\right.\right. \\
& +\frac{1}{2}\underline{\Omega}(\not{d}_1\underline{\vartheta}-\not{d}_2^\star\underline{\vartheta})+\not{d}_1^\star(\gamma^{1/2})(\not{d}_1\vartheta+\not{d}_2^\star\vartheta)\underline{b} \\
& \left.\left.\left.+\frac{1}{2}\gamma^{1/2}(\not{d}_2\not{d}_2^\star\underline{b}+\not{d}_3^\star\not{d}_2^\star\underline{b})+2\gamma^{1/2}K\underline{b}\right)\right)^{\#_\lambda}\right\|_{\mathfrak{h}_s(\overset{\circ}{S}, \overset{\circ}{\not{g}})\cap\mathfrak{h}_1^\infty(\overset{\circ}{S})} d\lambda \\
& +\left\|\left(\underline{\kappa}-\underline{\vartheta}-\underline{\Omega}(\kappa-\vartheta)-\gamma^{1/2}(\not{d}_1\vartheta+\not{d}_2^\star\vartheta)\underline{b}\right)^{\#}\right\|_{\mathfrak{h}_s(\overset{\circ}{S}, \overset{\circ}{\not{g}})\cap\mathfrak{h}_1^\infty(\overset{\circ}{S})} \\
& \left.+\|(\kappa+\vartheta)^{\#}\|_{\mathfrak{h}_s(\overset{\circ}{S}, \overset{\circ}{\not{g}})\cap\mathfrak{h}_1^\infty(\overset{\circ}{S})}\right)\|f^{\#}\|_{\mathfrak{h}_{s+1}(\overset{\circ}{S}, \overset{\circ}{\not{g}})}\Bigg\}.
\end{aligned}
$$

Next, we have by the iteration assumption (C.1.3)

$$
\begin{aligned}
& \left\|\left(\not{d}^{\le 2}\left(\check{\Gamma}, r^{-2}\gamma-1, \underline{b}, \underline{\Omega}+\Upsilon\right)\right)^{\#}\right\|_{\mathfrak{h}_s(\overset{\circ}{S}, \overset{\circ}{\not{g}})} \\
& +\sup_{0\le\lambda\le 1}\left\|\left(\not{d}^{\le 2}\left(\check{\Gamma}, r^{-2}\gamma-1, \underline{b}, \underline{\Omega}+\Upsilon\right)\right)^{\#_\lambda}\right\|_{\mathfrak{h}_s(\overset{\circ}{S}, \overset{\circ}{\not{g}})} \\
\lesssim & \left\|\left(\not{d}^{\le 2}\left(\check{\Gamma}, r^{-2}\gamma-1, \underline{b}, \underline{\Omega}+\Upsilon\right)\right)^{\#}\right\|_{\mathfrak{h}_s(\overset{\circ}{S}, \not{g}^{\mathbf{S},\#})} \\
& +\sup_{0\le\lambda\le 1}\left\|\left(\not{d}^{\le 2}\left(\check{\Gamma}, r^{-2}\gamma-1, \underline{b}, \underline{\Omega}+\Upsilon\right)\right)^{\#_\lambda}\right\|_{\mathfrak{h}_s(\overset{\circ}{S}, \not{g}^{\mathbf{S},\#_\lambda})} \\
\lesssim & \left\|\not{d}^{\le 2}\left(\check{\Gamma}, r^{-2}\gamma-1, \underline{b}, \underline{\Omega}+\Upsilon\right)\right\|_{\mathfrak{h}_s(S)} \\
& +\sup_{0\le\lambda\le 1}\left\|\not{d}^{\le 2}\left(\check{\Gamma}, r^{-2}\gamma-1, \underline{b}, \underline{\Omega}+\Upsilon\right)\right\|_{\mathfrak{h}_s(\mathbf{S}_\lambda)}
\end{aligned}
$$

where the surface $\mathbf{S}_\lambda$ is the image of $\overset{\circ}{S}$ by ψ_λ. Since $s \le s_{max} - 2$, we infer in view of our iteration assumption (C.1.2) and our assumptions (9.1.12), (9.1.15) on the

(u,s) foliation

$$\begin{aligned}
&\left\|\left(\not{d}^{\leq 2}\left(\check{\Gamma}, r^{-2}\gamma-1, \underline{b}, \underline{\Omega}+\Upsilon\right)\right)^{\#}\right\|_{\mathfrak{h}_s(\overset{\circ}{S}, \overset{\circ}{\not{g}})}\\
&+\sup_{0\leq\lambda\leq 1}\left\|\left(\not{d}^{\leq 2}\left(\check{\Gamma}, r^{-2}\gamma-1, \underline{b}, \underline{\Omega}+\Upsilon\right)\right)^{\#_\lambda}\right\|_{\mathfrak{h}_s(\overset{\circ}{S}, \overset{\circ}{\not{g}})}\\
\lesssim\;& \sup_{\mathcal{R}}\left|\mathfrak{d}^{\leq s}\not{d}^{\leq 2}\left(\check{\Gamma}, r^{-2}\gamma-1, \underline{b}, \underline{\Omega}+\Upsilon\right)\check{\Gamma}\right|\\
\lesssim\;& \sup_{\mathcal{R}}\left|\mathfrak{d}^{\leq s+2}\left(\check{\Gamma}, r^{-2}\gamma-1, \underline{b}, \underline{\Omega}+\Upsilon\right)\check{\Gamma}\right| \lesssim \overset{\circ}{\delta}. \qquad \text{(C.1.5)}
\end{aligned}$$

Also, we have

$$\begin{aligned}
&\left\|\left(\not{d}^{\leq 2}\left(\check{\Gamma}, r^{-2}\gamma-1, \underline{b}, \underline{\Omega}+\Upsilon\right)\right)^{\#}\right\|_{\mathfrak{h}_1^\infty(\overset{\circ}{S})}\\
&+\sup_{0\leq\lambda\leq 1}\left\|\left(\not{d}^{\leq 2}\left(\check{\Gamma}, r^{-2}\gamma-1, \underline{b}, \underline{\Omega}+\Upsilon\right)\right)^{\#_\lambda}\right\|_{\mathfrak{h}_1^\infty(\overset{\circ}{S})}\\
=\;& \left\|\left(\not{d}^{\leq 2}\left(\check{\Gamma}, r^{-2}\gamma-1, \underline{b}, \underline{\Omega}+\Upsilon\right)\right)\circ\psi\right\|_{\mathfrak{h}_1^\infty(\overset{\circ}{S})}\\
&+\sup_{0\leq\lambda\leq 1}\left\|\left(\not{d}^{\leq 2}\left(\check{\Gamma}, r^{-2}\gamma-1, \underline{b}, \underline{\Omega}+\Upsilon\right)\right)\circ\psi_\lambda\right\|_{\mathfrak{h}_1^\infty(\overset{\circ}{S})}\\
\lesssim\;& \left(\sup_{\mathcal{R}}\left|\mathfrak{d}^{\leq 3}\left(\check{\Gamma}, r^{-2}\gamma-1, \underline{b}, \underline{\Omega}+\Upsilon\right)\right|\right)\left(1+\sup_{0\leq\theta\leq\pi}|\psi'(\theta)|\right)\\
\lesssim\;& \left(\sup_{\mathcal{R}}\left|\mathfrak{d}^{\leq 3}\left(\check{\Gamma}, r^{-2}\gamma-1, \underline{b}, \underline{\Omega}+\Upsilon\right)\right|\right)\left(1+\|(U',S')\|_{\mathfrak{h}_1^\infty(\overset{\circ}{S})}\right)\\
\lesssim\;& \overset{\circ}{\epsilon}
\end{aligned}$$

where we used our assumptions (9.1.12), (9.1.15) on the (u,s) foliation and our assumption (9.2.14) on (U', S'). Therefore,

$$\begin{aligned}
\left\|\left(\not{d}^{\leq 2}\left(\check{\Gamma}, r^{-2}\gamma-1, \underline{b}, \underline{\Omega}+\Upsilon\right)\right)^{\#}\right\|_{\mathfrak{h}_1^\infty(\overset{\circ}{S})} &\lesssim \overset{\circ}{\delta},\\
\sup_{0\leq\lambda\leq 1}\left\|\left(\not{d}^{\leq 2}\left(\check{\Gamma}, r^{-2}\gamma-1, \underline{b}, \underline{\Omega}+\Upsilon\right)\right)^{\#_\lambda}\right\|_{\mathfrak{h}_1^\infty(\overset{\circ}{S})} &\lesssim \overset{\circ}{\delta}.
\end{aligned} \qquad \text{(C.1.6)}$$

We deduce

$$\begin{aligned}
&\|(\not{d}_k^{\mathbf{S}} f)^{\#}\|_{\mathfrak{h}_s(\overset{\circ}{S}, \not{g}^{\mathbf{S},\#})}(1+O(\overset{\circ}{\delta}))\\
=\;& \left(1+O(1)\left\|\frac{\sqrt{\gamma}}{\sqrt{\gamma^{\mathbf{S}\,\#}}}-1\right\|_{\mathfrak{h}_s(\overset{\circ}{S}, \overset{\circ}{\not{g}})\cap\mathfrak{h}_1^\infty(\overset{\circ}{S})}\right)\\
&\times\left\{\left\|\overset{\circ}{\not{d}}_k(f^{\#})\right\|_{\mathfrak{h}_s(\overset{\circ}{S}, \overset{\circ}{\not{g}})}+O(\overset{\circ}{\delta})\left\|f^{\#}\right\|_{\mathfrak{h}_{s+1}(\overset{\circ}{S}, \overset{\circ}{\not{g}})}\right\}.
\end{aligned}$$

Next, we estimate the term in the RHS involving γ and $\gamma^{\mathbf{S}\,\#}$. From the proof of

Lemma 9.8, we have

$$\gamma^{\mathbf{S},\#} - \gamma = \frac{1}{2}U\int_0^1 \left(\left(e_3 - \underline{\Omega}e_4 - \underline{b}\gamma^{1/2}e_\theta\right)\gamma\right)^{\#_\lambda} d\lambda + S\int_0^1 (e_4\gamma)^{\#_\lambda}\, d\lambda$$
$$+\left(\underline{\Omega} + \frac{1}{4}\underline{b}^2\gamma\right)^{\#} (U')^2 - 2U'S' - (\gamma\underline{b})^{\#}\, U'.$$

Using a non-sharp product rule, we infer

$$\begin{aligned}
&\left\|\gamma^{\mathbf{S},\#} - \gamma\right\|_{\mathfrak{h}_s(\overset{\circ}{S},\overset{\circ}{\not g})\cap\mathfrak{h}_1^\infty(\overset{\circ}{S})}\\
\lesssim\; & \|U\|_{\mathfrak{h}_s(\overset{\circ}{S},\overset{\circ}{\not g})\cap\mathfrak{h}_1^\infty(\overset{\circ}{S})}\int_0^1 \left\|\left(\left(e_3 - \underline{\Omega}e_4 - \underline{b}\gamma^{1/2}e_\theta\right)\gamma\right)^{\#_\lambda}\right\|_{\mathfrak{h}_s(\overset{\circ}{S},\overset{\circ}{\not g})\cap\mathfrak{h}_1^\infty(\overset{\circ}{S})} d\lambda\\
& \|S\|_{\mathfrak{h}_s(\overset{\circ}{S},\overset{\circ}{\not g})\cap\mathfrak{h}_1^\infty(\overset{\circ}{S})}\int_0^1 \left\|(e_4\gamma)^{\#_\lambda}\right\|_{\mathfrak{h}_s(\overset{\circ}{S},\overset{\circ}{\not g})\cap\mathfrak{h}_1^\infty(\overset{\circ}{S})} d\lambda\\
& +\left\|\left(\underline{\Omega} + \frac{1}{4}\underline{b}^2\gamma\right)^{\#}\right\|_{\mathfrak{h}_s(\overset{\circ}{S},\overset{\circ}{\not g})\cap\mathfrak{h}_1^\infty(\overset{\circ}{S})} \|U'\|_{\mathfrak{h}_s(\overset{\circ}{S},\overset{\circ}{\not g})\cap\mathfrak{h}_1^\infty(\overset{\circ}{S})}\\
& +\|U'\|_{\mathfrak{h}_s(\overset{\circ}{S},\overset{\circ}{\not g})\cap\mathfrak{h}_1^\infty(\overset{\circ}{S})}\|S'\|_{\mathfrak{h}_s(\overset{\circ}{S},\overset{\circ}{\not g})\cap\mathfrak{h}_1^\infty(\overset{\circ}{S})}\\
& +\left\|(\gamma\underline{b})^{\#}\right\|_{\mathfrak{h}_s(\overset{\circ}{S},\overset{\circ}{\not g})\cap\mathfrak{h}_1^\infty(\overset{\circ}{S})}\|U'\|_{\mathfrak{h}_s(\overset{\circ}{S},\overset{\circ}{\not g})\cap\mathfrak{h}_1^\infty(\overset{\circ}{S})}\\
\lesssim\; & \overset{\circ}{\delta}\int_0^1 \left\|\left(\not{d}^{\leq 1}\left(r^{-2}\gamma - 1, \underline{b}, \underline{\Omega} + \Upsilon\right)\right)^{\#_\lambda}\right\|_{\mathfrak{h}_s(\overset{\circ}{S},\overset{\circ}{\not g})\cap\mathfrak{h}_1^\infty(\overset{\circ}{S})} d\lambda\\
& +\overset{\circ}{\delta}\left\|\left(r^{-2}\gamma - 1, \underline{b}, \underline{\Omega} + \Upsilon\right)^{\#}\right\|_{\mathfrak{h}_s(\overset{\circ}{S},\overset{\circ}{\not g})\cap\mathfrak{h}_1^\infty(\overset{\circ}{S})} + \overset{\circ}{\delta}
\end{aligned}$$

where we used our assumption (9.2.14) on (U', S') and $U(0) = S(0) = 0$. Using the estimates (C.1.5), (C.1.6) for $(r^{-2}\gamma - 1, \underline{b}, \underline{\Omega} + \Upsilon)$, we infer

$$\left\|\gamma^{\mathbf{S},\#} - \gamma\right\|_{\mathfrak{h}_s(\overset{\circ}{S},\overset{\circ}{\not g})\cap\mathfrak{h}_1^\infty(\overset{\circ}{S})} \lesssim \overset{\circ}{\delta}.$$

Together with (9.1.15) for γ, we infer

$$\left\|\frac{\sqrt{\gamma}}{\sqrt{\gamma^{\mathbf{S}\,\#}}} - 1\right\|_{\mathfrak{h}_s(\overset{\circ}{S},\overset{\circ}{\not g})\cap\mathfrak{h}_1^\infty(\overset{\circ}{S})} \lesssim \overset{\circ}{\delta}$$

and hence

$$\begin{aligned}
&\|(\not{d}_k^{\mathbf{S}} f)^{\#}\|_{\mathfrak{h}_s(\overset{\circ}{S},\not g^{\mathbf{S},\#})}(1 + O(\overset{\circ}{\delta}))\\
=\; & \left(1 + O(\overset{\circ}{\delta})\right)\left\{\left\|\overset{\circ}{\not{d}}_k(f^{\#})\right\|_{\mathfrak{h}_s(\overset{\circ}{S},\overset{\circ}{\not g})} + O(\overset{\circ}{\delta})\,\|f^{\#}\|_{\mathfrak{h}_{s+1}(\overset{\circ}{S},\overset{\circ}{\not g})}\right\}.
\end{aligned}$$

Now, we have

$$\begin{aligned}\|f^\#\|_{\mathfrak{h}_{s+1}(\overset{\circ}{S}, \not{g}^{\mathbf{S},\#})} &= \|f^\#\|_{L^2(\overset{\circ}{S}, \not{g}^{\mathbf{S},\#})} + \|(\not{d}_k^{\mathbf{S}} f)^\#\|_{\mathfrak{h}_s(\overset{\circ}{S}, \not{g}^{\mathbf{S},\#})},\\ \|f^\#\|_{\mathfrak{h}_{s+1}(\overset{\circ}{S}, \overset{\circ}{\not{g}})} &= \|f^\#\|_{L^2(\overset{\circ}{S}, \overset{\circ}{\not{g}})} + \|\overset{\circ}{\not{d}}_k(f^\#)\|_{\mathfrak{h}_s(\overset{\circ}{S}, \overset{\circ}{\not{g}})}.\end{aligned}$$

Together with Lemma 9.8, this yields

$$\|f^\#\|_{\mathfrak{h}_{s+1}(\overset{\circ}{S}, \not{g}^{\mathbf{S},\#})} = \|f^\#\|_{\mathfrak{h}_{s+1}(\overset{\circ}{S}, \overset{\circ}{\not{g}})}(1 + O(\overset{\circ}{\delta})).$$

This corresponds to our iteration assumption (C.1.3) with s replaced with $s+1$ for $s \leq s_{max} - 2$. Thus, we have finally derived both iteration assumption (C.1.3) and (C.1.2) with s replaced with $s+1$ respectively for $s \leq s_{max} - 2$. Hence, we deduce that they hold for $0 \leq s \leq s_{max} - 1$, i.e.,

$$\|h\|_{\mathfrak{h}_s(\mathbf{S})} \lesssim \sup_{\mathcal{R}} |\mathfrak{d}^{\leq k} h| \text{ for } 0 \leq s \leq s_{max} - 1 \tag{C.1.7}$$

and

$$\|f^\#\|_{\mathfrak{h}_s(\overset{\circ}{S}, \not{g}^{\mathbf{S},\#})} = \|f^\#\|_{\mathfrak{h}_s(\overset{\circ}{S}, \overset{\circ}{\not{g}})}(1 + O(\overset{\circ}{\delta})) \text{ for } 0 \leq s \leq s_{max} - 1.$$

Together with Lemma 9.7, we deduce

$$\|f\|_{\mathfrak{h}_s(\mathbf{S})} = \|f^\#\|_{\mathfrak{h}_s(\overset{\circ}{S}, \not{g}^{\mathbf{S},\#})} = \|f^\#\|_{\mathfrak{h}_s(\overset{\circ}{S}, \overset{\circ}{\not{g}})}(1 + O(\overset{\circ}{\delta})) \text{ for } 0 \leq s \leq s_{max} - 1. \tag{C.1.8}$$

Finally, notice that the restriction $s \leq s_{max} - 2$ for the iteration assumptions (C.1.2), (C.1.3) was only necessary to replace s with $s+1$ in (C.1.3). Indeed, a direct inspection of the proof reveals that to replace s with $s+1$ in (C.1.2), we only need the restriction $s \leq s_{max} - 1$. Thus, running the iteration again, now with $s = s_{max} - 1$, we deduce

$$\|h\|_{\mathfrak{h}_s(\mathbf{S})} \lesssim \sup_{\mathcal{R}} |\mathfrak{d}^{\leq k} h| \text{ for } 0 \leq s \leq s_{max}.$$

This concludes the proof of the lemma. □

Appendix D

Appendix to Chapter 10

D.1 HORIZONTAL S-TENSORS

Consider a null pair e_3, e_4 on $(\mathcal{M}, \mathbf{g})$ and, at every point $p \in \mathcal{M}$, the horizontal space $S = \{e_3, e_4\}^\perp$. Let γ the metric induced on S. By definition, for all $X, Y \in \mathbf{T}_S\mathcal{M}$, i.e., vectors in $\mathcal{M}$ tangent to S,

$$h(X,Y) = \mathbf{g}(X,Y).$$

For any $Y \in T(\mathcal{M})$ we define its horizontal projection,

$$Y^\perp = Y + \frac{1}{2}\mathbf{g}(Y, e_3)e_4 + \frac{1}{2}\mathbf{g}(Y, e_3)e_4. \tag{D.1.1}$$

Definition D.1. *A k-covariant tensor U is said to be S-horizontal, $U \in \mathbf{T}^k_S(\mathcal{M})$, if for any $X_1, \ldots, X_k$, we have*

$$U(Y_1, \ldots, Y_k) = U(Y_1^\perp, \ldots, Y_k^\perp).$$

We define the projection operator,

$$\Pi_\mu^{\ \nu} := \delta_\mu^{\ \nu} - \frac{1}{2}(e_3)_\mu(e_4)^\nu - \frac{1}{2}(e_4)_\mu(e_3)^\nu.$$

Clearly $\Pi_\alpha^{\ \mu}\Pi_\mu^{\ \beta} = \Pi_\alpha^{\ \beta}$. An arbitrary tensor $U_{\alpha_1\ldots\alpha_m}$ is said to be an S-horizontal tensor, or simply S-tensor, if

$$\Pi_{\alpha_1}^{\ \beta_1} \ldots \Pi_{\alpha_m}^{\ \beta_m} U_{\beta_1\ldots\beta_m} = U_{\alpha_1\ldots\alpha_m}.$$

Definition D.2. *Given $X \in \mathbf{T}(\mathcal{M})$ and $Y \in \mathbf{T}_S(\mathcal{M})$ we define*

$$\dot{\mathbf{D}}_X Y \quad := \quad (\mathbf{D}_X Y)^\perp.$$

Remark D.3. *In the particular case when S is integrable and both $X, Y \in \mathbf{T}_S\mathcal{M}$ then $\dot{\mathbf{D}}_X Y$ is the standard induced covariant differentiation on S.*

Definition D.4. *Given a general, covariant, S-horizontal tensorfield U we define its horizontal covariant derivative according to the formula*

$$\dot{\mathbf{D}}_X U(Y_1, \ldots, Y_k) = X(U(Y_1, \ldots, Y_k)) - U(\dot{\mathbf{D}}_X Y_1, \ldots, Y_k) - \ldots - U(Y_1, \ldots, \dot{\mathbf{D}}_X Y_k),$$

where $X \in \mathbf{T}\mathcal{M}$ and $Y_1, \ldots Y_k \in \mathbf{T}_S\mathcal{M}$.

Proposition D.5. *For all $X \in \mathbf{T}\mathcal{M}$ and $Y_1, Y_2 \in \mathbf{T}_S\mathcal{M}$,*

$$Xh(Y_1, Y_2) = h(\dot{\mathbf{D}}_X Y_1, Y_2) + h(Y_1, \dot{\mathbf{D}}_X Y_2).$$

Proof. Indeed, we have

$$\begin{aligned} Xh(Y_1,Y_2) &= X\mathbf{g}(Y_1,Y_2) = \mathbf{g}(\mathbf{D}_X Y_1, Y_2) + \mathbf{g}(Y_1, \mathbf{D}_X Y_2) \\ &= \mathbf{g}(\dot{\mathbf{D}}_X Y_1, Y_2) + \mathbf{g}(Y_1, \dot{\mathbf{D}}_X Y_2) \\ &= h(\dot{\mathbf{D}}_X Y_1, Y_2) + h(Y_1, \dot{\mathbf{D}}_X Y_2) \end{aligned}$$

as desired. □

Given an orthonormal frame e_1, e_2 on S we have

$$\dot{\mathbf{D}}_\mu e_A = \sum_{B=1,2} (\Lambda_\mu)_{AB}\, e_B \qquad A, B = 1, 2,$$

where

$$(\Lambda_\mu)_{\alpha\beta} := \mathbf{g}(\mathbf{D}_\mu e_\beta, e_\alpha).$$

D.1.1 Mixed tensors

We consider tensors $\mathbf{T}^k\mathcal{M} \otimes \mathbf{T}^l_S\mathcal{M}$, i.e., tensors of the form

$$U_{\mu_1\ldots\mu_k, A_1\ldots A_L}$$

for which we define

$$\begin{aligned} \dot{\mathbf{D}}_\mu U_{\nu_1\ldots\nu_k, A_1\ldots A_L} &= e_\mu U_{\nu_1\ldots\nu_k, A_1\ldots A_l} \\ &\quad - U_{\mathbf{D}_\mu \nu_1\ldots\nu_k, A_1\ldots A_l} - \ldots - U_{\nu_1\ldots\mathbf{D}_\mu\nu_k, A_1\ldots A_l} \\ &\quad - U_{\nu_1\ldots\nu_k, \dot{\mathbf{D}}_\mu A_1\ldots A_l} - \ldots - U_{\nu_1\ldots\nu_k, A_1\ldots\dot{\mathbf{D}}_\mu A_l}. \end{aligned}$$

We are now ready to prove the following:

Proposition D.6. *We have the curvature formula*

$$(\dot{\mathbf{D}}_\mu\dot{\mathbf{D}}_\nu - \dot{\mathbf{D}}_\nu\dot{\mathbf{D}}_\mu)\Psi_A = \mathbf{R}_A{}^B{}_{\mu\nu}\Psi_B.$$

More generally,

$$(\dot{\mathbf{D}}_\mu\dot{\mathbf{D}}_\nu - \dot{\mathbf{D}}_\nu\dot{\mathbf{D}}_\mu)\Psi_{\lambda A} = \mathbf{R}_\lambda{}^\sigma{}_{\mu\nu}\Psi_{\sigma A} + \mathbf{R}_A{}^B{}_{\mu\nu}\Psi_{\lambda B}.$$

Proof. Straightforward verification. □

D.1.2 Invariant Lagrangian

We introduce

$$\mathcal{L} = g^{\mu\nu} h_{AB} \dot{\mathbf{D}}_\mu \Psi^A \dot{\mathbf{D}}_\mu \Psi^B + W h_{AB}\Psi^A\Psi^B.$$

Proposition D.7. *The Euler-Lagrange equations are given by*

$$\dot{\square}\Psi^A = W\Psi^A$$

where $\dot{\square}\Psi^A := \mathbf{g}^{\mu\nu}\dot{\mathbf{D}}_\mu\dot{\mathbf{D}}_\nu\Psi^A$.

Proof. The variation of the action is given by

$$\begin{aligned} 0 &= 2\int_{\mathbf{M}} h_{AB}\left(\mathbf{g}^{\mu\nu}\dot{\mathbf{D}}_\mu\Psi^A\dot{\mathbf{D}}_\nu(\delta\Psi)^B + W\Psi^A\delta\Psi^B\right)dv_{\mathbf{g}} \\ &= 2\int_{\mathbf{M}} \mathbf{D}_\nu\left(\mathbf{g}^{\mu\nu}h_{AB}\dot{\mathbf{D}}_\mu\Psi^A(\delta\Psi)^B\right)dv_{\mathbf{g}} \\ &\quad -2\int_{\mathbf{M}} h_{AB}\left(\mathbf{g}^{\mu\nu}\dot{\mathbf{D}}_\nu\dot{\mathbf{D}}_\mu\Psi^A(\delta\Psi)^B - W\Psi^A\delta\Psi^B\right)dv_{\mathbf{g}} \\ &= -2\int_{\mathbf{M}} h_{AB}\left(\mathbf{g}^{\mu\nu}\dot{\mathbf{D}}_\nu\dot{\mathbf{D}}_\mu\Psi^A(\delta\Psi)^B - W\Psi^A\delta\Psi^B\right)dv_{\mathbf{g}} \end{aligned}$$

from which the proposition follows. □

D.1.3 Comparison of the Lagrangians

Let $\Psi \in \mathcal{S}_2(\mathcal{M})$ and $\psi \in \mathfrak{s}_2$ its reduced form. Note that the Lagrangian of the scalar equation

$$\square_{\mathbf{g}}\psi = V\psi + 4(e_\theta\Phi)^2\psi$$

is given by

$$\mathcal{L}(\psi) := \mathbf{g}^{\mu\nu}\partial_\mu\psi\partial_\nu\psi + (V + 4(e_\theta\Phi)^2)\psi^2$$

while the Lagrangian for

$$\dot{\square}_{\mathbf{g}}\Psi = V\Psi$$

is given by

$$\mathcal{L}(\Psi) = \mathbf{g}^{\mu\nu}\dot{\mathbf{D}}_\mu\Psi\cdot\dot{\mathbf{D}}_\nu\Psi + V\Psi\cdot\Psi.$$

Proposition D.8. *We have*

$$\mathcal{L}(\Psi) = 2\mathcal{L}(\psi). \tag{D.1.2}$$

Proof. Observe that

$$\mathbf{g}^{\mu\nu}\dot{\mathbf{D}}_\mu\Psi\dot{\mathbf{D}}_\nu\Psi = -\dot{\mathbf{D}}_3\Psi\cdot\dot{\mathbf{D}}_4\Psi + \dot{\mathbf{D}}_\theta\Psi\cdot\dot{\mathbf{D}}_\theta\Psi + \dot{\mathbf{D}}_\varphi\Psi\cdot\dot{\mathbf{D}}_\varphi\Psi.$$

Now, recalling that

$$\begin{aligned} \nabla\!\!\!/_\varphi e_\varphi &= -e_\theta\Phi e_\theta, & \nabla\!\!\!/_\varphi e_\theta = e_\theta(\Phi)e_\varphi, \\ \nabla\!\!\!/_\theta e_\theta &= 0, & \nabla\!\!\!/_\theta e_\varphi = 0, \end{aligned}$$

we deduce

$$\begin{aligned}
\dot{\mathbf{D}}_3\Psi\cdot\dot{\mathbf{D}}_4\Psi &= e_3\Psi\cdot e_4\Psi = 2e_3\psi e_4\psi\\
\dot{\mathbf{D}}_\theta\Psi\cdot\dot{\mathbf{D}}_\theta\Psi &= \dot{\mathbf{D}}_\theta\Psi_{\theta\theta}\dot{\mathbf{D}}_\theta\Psi_{\theta\theta}+2\dot{\mathbf{D}}_\theta\Psi_{\theta\varphi}\dot{\mathbf{D}}_\theta\Psi_{\theta\varphi}+\dot{\mathbf{D}}_\theta\Psi_{\varphi\varphi}\dot{\mathbf{D}}_\theta\Psi_{\varphi\varphi}\\
&= 2(e_\theta\psi)^2\\
\dot{\mathbf{D}}_\varphi\Psi\cdot\dot{\mathbf{D}}_\varphi\Psi &= \dot{\mathbf{D}}_\varphi\Psi_{\theta\theta}\dot{\mathbf{D}}_\varphi\Psi_{\theta\theta}+2\dot{\mathbf{D}}_\varphi\Psi_{\theta\varphi}\dot{\mathbf{D}}_\varphi\Psi_{\theta\varphi}+\dot{\mathbf{D}}_\varphi\Psi_{\varphi\varphi}\dot{\mathbf{D}}_\varphi\Psi_{\varphi\varphi}\\
&= 2(e_\varphi\psi)^2+2(-\Psi_{\dot{\mathbf{D}}_\varphi\theta\varphi}-\Psi_{\theta\dot{\mathbf{D}}_\varphi\varphi})\cdot(-\Psi_{\dot{\mathbf{D}}_\varphi\theta\varphi}-\Psi_{\theta\dot{\mathbf{D}}_\varphi\varphi})\\
&= 2(e_\varphi\psi)^2+2(-e_\theta(\Phi)\Psi_{\varphi\varphi}+e_\theta(\Phi)\Psi_{\theta\theta})\cdot(-e_\theta(\Phi)\Psi_{\varphi\varphi}+e_\theta(\Phi)\Psi_{\theta\theta})\\
&= 2(e_\varphi\psi)^2+8(e_\theta\Phi)^2\psi^2.
\end{aligned}$$

Hence,

$$\mathbf{g}^{\mu\nu}\dot{\mathbf{D}}_\mu\Psi\dot{\mathbf{D}}_\nu\Psi = -2e_3\psi e_4\psi+2(e_\theta\psi)^2+2(e_\varphi\psi)^2+4(e_\theta\Phi)^2\psi^2$$

and

$$\mathcal{L}(\Psi) = -2e_3\Psi e_4\psi+2(e_\theta\psi)^2+2(e_\varphi\psi)^2+8(e_\theta\Phi)^2\psi^2+2V\psi^2.$$

□

D.1.4 Energy-momentum tensor

Consider the energy-momentum tensor

$$\mathcal{Q}_{\mu\nu} := \dot{\mathbf{D}}_\mu\Psi\cdot\dot{\mathbf{D}}_\nu\Psi-\frac{1}{2}\mathbf{g}_{\mu\nu}\left(\dot{\mathbf{D}}_\lambda\Psi\cdot\dot{\mathbf{D}}^\lambda\Psi+V\Psi\cdot\Psi\right).$$

Lemma D.9. *We have*

$$\mathbf{D}^\nu\mathcal{Q}_{\mu\nu} = \dot{\mathbf{D}}_\mu\Psi\cdot(\dot{\square}\Psi-V\psi)+\dot{\mathbf{D}}^\nu\Psi^A\mathbf{R}_{AB\nu\mu}\Psi^B-\frac{1}{2}\mathbf{D}_\mu V\Psi\cdot\Psi.$$

Proof. We have

$$\begin{aligned}
\mathbf{D}^\nu\mathcal{Q}_{\mu\nu} &= \dot{\mathbf{D}}^\nu\dot{\mathbf{D}}_\nu\Psi\cdot\dot{\mathbf{D}}_\mu\Psi+\dot{\mathbf{D}}^\nu\Psi\cdot\left(\dot{\mathbf{D}}_\nu\dot{\mathbf{D}}_\mu-\dot{\mathbf{D}}_\mu\dot{\mathbf{D}}_\nu\right)\Psi-V\mathbf{D}_\mu\Psi\cdot\Psi\\
&\quad -\frac{1}{2}\mathbf{D}_\mu V\Psi\cdot\Psi\\
&= \dot{\mathbf{D}}_\mu\Psi\cdot\dot{\mathbf{D}}^\nu\dot{\mathbf{D}}_\nu\Psi+\dot{\mathbf{D}}^\nu\Psi^A\mathbf{R}_{AB\nu\mu}\Psi^B-V\mathbf{D}_\mu\Psi\Psi-\frac{1}{2}\mathbf{D}_\mu V\Psi\cdot\Psi\\
&= \dot{\mathbf{D}}_\mu\Psi\left(\dot{\square}\Psi-V\Psi\right)+\dot{\mathbf{D}}^\nu\Psi^A\mathbf{R}_{AB\nu\mu}\Psi^B-\frac{1}{2}\mathbf{D}_\mu V\Psi\cdot\Psi.
\end{aligned}$$

□

Lemma D.10. *Relative to an arbitrary* **Z**-*polarized frame* $e_3, e_4, e_\theta, e_\varphi$ *we have*

$$\begin{aligned}
\mathcal{Q}_{33} &= |e_3\Psi|^2,\\
\mathcal{Q}_{44} &= |e_4\Psi|^2,\\
\mathcal{Q}_{34} &= |\nabla\!\!\!/\,\Psi|^2+V|\Psi|^2.
\end{aligned}$$

If ψ is the reduced form of Ψ,

$$\begin{aligned}\mathcal{Q}_{33} &= 2(e_3\psi)^2,\\ \mathcal{Q}_{44} &= 2(e_4\psi)^2,\\ \mathcal{Q}_{34} &= 2(e_\theta\psi)^2 + 2(e_\varphi\psi)^2 + 2V|\psi|^2 + 8(e_\theta\Phi)^2\psi^2.\end{aligned}$$

Also,

$$\mathbf{g}^{\mu\nu}\mathcal{Q}_{\mu\nu} = -\mathcal{L}(\Psi) - V|\Psi|^2,$$

$$|\mathcal{L}(\Psi)| \lesssim |e_3\Psi|\,|e_4\Psi| + |\nabla\!\!\!/\,\Psi|^2 + V|\Psi|^2,$$

and

$$\begin{aligned}|\mathcal{Q}_{AB}| &\le |e_3\Psi||e_4\Psi| + |\nabla\!\!\!/\,\Psi|^2 + |V||\Psi|^2,\\ |\mathcal{Q}_{A3}| &\le |e_3\Psi||\nabla\!\!\!/\,\Psi|,\\ |\mathcal{Q}_{A4}| &\le |e_4\Psi||\nabla\!\!\!/\,\Psi|.\end{aligned}$$

D.2 STANDARD CALCULATION

Proposition D.11. *Consider an admissible spacetime $\mathcal{M}$ and $\Psi \in \mathcal{S}_2(\mathcal{M})$ and X a vectorfield of the form*

$$X = ae_3 + be_4.$$

1. *The 1-form $\mathcal{P}_\mu = \mathcal{Q}_{\mu\nu}X^\nu$ verifies*

$$\mathbf{D}^\mu\mathcal{P}_\mu = X^\mu\dot{\mathbf{D}}_\mu\Psi\cdot(\dot{\square}\Psi - V\Psi) - X(V)\Psi\cdot\Psi.$$

2. *Let X as above, w a scalar and $\mathbf{M}$ a 1-form. Define*

$$\mathcal{P}_\mu = \mathcal{P}_\mu[X,w,M] = \mathcal{Q}_{\mu\nu}X^\nu + \frac{1}{2}w\Psi\cdot\dot{\mathbf{D}}_\mu\Psi - \frac{1}{4}|\Psi|^2\partial_\mu w + \frac{1}{4}|\Psi|^2M_\mu.$$

Then, with $|\Psi|^2 := \Psi\cdot\Psi$,

$$\begin{aligned}\mathbf{D}^\mu\mathcal{P}_\mu[X,w,M] &= \frac{1}{2}\mathcal{Q}\cdot{}^{(X)}\pi - \frac{1}{2}X(V)\Psi\cdot\Psi + \frac{1}{2}w\mathcal{L}[\Psi] - \frac{1}{4}|\Psi|^2\square_{\mathbf{g}}w\\ &+ \frac{1}{4}|\Psi|^2\mathbf{Div}M + \frac{1}{2}\Psi\cdot\dot{\mathbf{D}}_\mu\Psi\,M^\mu\\ &+ \left(X(\Psi) + \frac{1}{2}w\Psi\right)\cdot(\dot{\square}\Psi - V\Psi).\end{aligned}$$

Proof. Let $\mathcal{P}_\mu[X,0,0] = \mathcal{Q}_{\mu\nu}X^\nu$. Then,

$$\begin{aligned}\mathbf{D}^\mu\mathcal{P}_\mu &= X^\mu\dot{\mathbf{D}}_\mu\Psi\cdot\left(\dot{\mathbf{D}}^\nu\dot{\mathbf{D}}_\nu\Psi - V\Psi\right) + X^\mu\dot{\mathbf{D}}^\nu\Psi^A\mathbf{R}_{AB\nu\mu}\Psi^B - \frac{1}{2}X^\mu\mathbf{D}_\mu V\Psi\cdot\Psi\\ &= X^\mu\dot{\mathbf{D}}_\mu\Psi\cdot(\dot{\square}\Psi - V\Psi) - \frac{1}{2}X(V)|\Psi|^2.\end{aligned}$$

Assume $X = ae_3 + be_4$. Then, since only the middle components of $\mathbf{R}$ are relevant, and recalling that $\mathbf{R}_{AB43} = -\,{}^*\rho \in_{AB} = 0$, we derive

$$X^\mu \dot{\mathbf{D}}^\nu \Psi^A \mathbf{R}_{AB\nu 3}\Psi^B = a\dot{\mathbf{D}}^4\Psi^A \mathbf{R}_{AB43}\Psi^B + b\dot{\mathbf{D}}^3\Psi^A \mathbf{R}_{AB434}\Psi^B = 0.$$

To prove the second part of the proposition we write with $\mathcal{N}[\Psi] := \dot{\square}\Psi - V\Psi$,

$$\begin{aligned}
\mathbf{D}^\mu \mathcal{P}_\mu[X,w,M] &= \frac{1}{2}\mathcal{Q}\cdot {}^{(X)}\pi + X(\Psi)\cdot\mathcal{N}[\Psi] - \frac{1}{2}X(V)\Psi\cdot\Psi + \frac{1}{2}\mathbf{D}^\mu w\,\Psi\cdot\dot{\mathbf{D}}_\mu\Psi \\
&+ \frac{1}{2}w\,\dot{\mathbf{D}}^\mu\Psi\cdot\dot{\mathbf{D}}_\mu\Psi + \frac{1}{2}w\Psi\dot{\square}_{\mathbf{g}}\Psi - \frac{1}{2}\Psi\cdot\dot{\mathbf{D}}^\mu\Psi\partial_\mu w - \frac{1}{4}|\Psi|^2\square_{\mathbf{g}}w \\
&+ \frac{1}{4}|\Psi|^2\mathbf{Div}M + \frac{1}{2}\Psi\cdot\dot{\mathbf{D}}_\mu\Psi\,M^\mu \\
&= \frac{1}{2}\mathcal{Q}\cdot {}^{(X)}\pi - \frac{1}{2}X(V)\Psi\cdot\Psi + \frac{1}{2}w\,\dot{\mathbf{D}}^\mu\Psi\cdot\dot{\mathbf{D}}_\mu\Psi \\
&+ \frac{1}{2}w\Psi\left(V\Psi + \mathcal{N}[\Psi]\right) - \frac{1}{4}|\Psi|^2\square_{\mathbf{g}}w + \frac{1}{4}|\Psi|^2\mathbf{Div}M \\
&+ \frac{1}{2}\Psi\cdot\dot{\mathbf{D}}_\mu\Psi\,M^\mu + X(\Psi)\cdot\Psi\cdot\mathcal{N}[\Psi].
\end{aligned}$$

Hence,

$$\begin{aligned}
\mathbf{D}^\mu \mathcal{P}_\mu[X,w,M] &= \frac{1}{2}\mathcal{Q}\cdot {}^{(X)}\pi - \frac{1}{2}X(V)\Psi\cdot\Psi + \frac{1}{2}w\mathcal{L}[\Psi] - \frac{1}{4}|\Psi|^2\square_{\mathbf{g}}w \\
&+ \frac{1}{4}|\Psi|^2\mathbf{Div}M + \frac{1}{2}\Psi\cdot\dot{\mathbf{D}}_\mu\Psi\,M^\mu + \left(X(\Psi) + \frac{1}{2}w\Psi\right)\cdot\mathcal{N}[\Psi]
\end{aligned}$$

as desired. □

Remark D.12. *As a consequence of the proposition above we deduce that every time we use vectorfields of the form ae_3+be_4 as multipliers, the equation $\square\Psi - V\Psi = \mathcal{N}$ is treated exactly in the same manner as the scalar equation $\square\psi - V\psi = N$.*

Remark D.13. *Note that in Schwarzschild, our potential $V = -\kappa\underline{\kappa} = 4\Upsilon r^{-2}$ verifies*

$$\begin{aligned}
\frac{1}{4}\partial_r V &= \partial_r\left[r^{-2}\left(1-\frac{2m}{r}\right)\right] = -2r^{-3}\left(1-\frac{2m}{r}\right) + \frac{2m}{r^4} \\
&= -2\frac{r-3m}{r^4}.
\end{aligned}$$

D.3 VECTORFIELD X_f

Lemma D.14. *Let*

$$X_f := fe_4, \qquad {}^{(X)}\Lambda := \frac{2f}{r}, \qquad {}^{(X)}\widetilde{\pi} := {}^{(X)}\pi - {}^{(X)}\Lambda\mathbf{g} = {}^{(X)}\pi - \frac{2f}{r}\mathbf{g}.$$

- *We have*

$$\begin{aligned}
{}^{(X)}\widetilde{\pi}_{44} &= 0, \quad {}^{(X)}\pi_{4\varphi} = 0, \quad {}^{(X)}\pi_{3\varphi} = 0,\\
{}^{(X)}\widetilde{\pi}_{43} &= -2e_4 f + 4f\omega + \frac{4f}{r} = -2\left(e_4(f) - \frac{2f}{r}\right) + 4f\omega,\\
{}^{(X)}\widetilde{\pi}_{4\theta} &= 2f\xi,\\
{}^{(X)}\widetilde{\pi}_{AB} &= 2f\,{}^{(1+3)}\chi_{AB} - \frac{2f}{r}\not{g}_{AB} = 2f\left({}^{(1+3)}\chi_{AB} - \frac{1}{r}\delta_{AB}\right),\\
{}^{(X)}\widetilde{\pi}_{3\theta} &= 2f(\eta+\zeta),\\
{}^{(X)}\widetilde{\pi}_{33} &= -8f\underline{\omega} - 4e_3(f).
\end{aligned} \tag{D.3.1}$$

- *In particular, we have*

$$\begin{aligned}
{}^{(X)}\widetilde{\pi}_{43} &= -2f' + \frac{4f}{r} + O(\epsilon)\min\{w_{1,1}, w_{2,1/2}\}\left(|f| + r|f'|\right),\\
{}^{(X)}\widetilde{\pi}_{4A} &= \epsilon\min\{w_{2,1}, w_{3,1/2}\},\\
{}^{(X)}\widetilde{\pi}_{AB} &= O(\epsilon)\min\{w_{1,1}, w_{2,1/2}\}|f|,\\
{}^{(X)}\widetilde{\pi}_{3A} &= O(\epsilon)w_{1,1}|f|,\\
{}^{(X)}\widetilde{\pi}_{33} &= 4f'\Upsilon - 4\Upsilon' + O(\epsilon)w_{1,1}(|f| + r|f'|).
\end{aligned} \tag{D.3.2}$$

- *We have*

$$\square\,{}^{(X)}\Lambda \;=\; \frac{2}{r}f'' + O\left(\frac{m}{r^4} + \epsilon w_{3,1}\right)\left(|f| + r|f'| + r^2|f''|\right). \tag{D.3.3}$$

Proof. We calculate ${}^{(X)}\pi_{\alpha\beta} = \mathbf{g}(D_{e_\alpha}X, e_\beta) + \mathbf{g}(\mathbf{D}_{e_\beta}X, e_\alpha)$,

$$\begin{aligned}
{}^{(X)}\pi_{44} &= 0,\\
{}^{(X)}\pi_{43} &= -2e_4 f + 4f\omega,\\
{}^{(X)}\pi_{4\theta} &= 2f\xi,\\
{}^{(X)}\pi_{AB} &= 2f\,{}^{(1+3)}\chi_{AB},\\
{}^{(X)}\pi_{3\theta} &= 2f(\eta+\zeta),\\
{}^{(X)}\pi_{33} &= -8f\underline{\omega} - 4e_3(f).
\end{aligned}$$

We deduce, for ${}^{(X)}\widetilde{\pi} = {}^{(X)}\pi - {}^{(X)}\Lambda\mathbf{g} = {}^{(X)}\pi - \frac{2f}{r}\mathbf{g}$,

$$\begin{aligned}
{}^{(X)}\widetilde{\pi}_{44} &= 0,\\
{}^{(X)}\widetilde{\pi}_{43} &= -2e_4 f + 4f\omega + \frac{4f}{r} = -2\left(e_4(f) - \frac{2f}{r}\right) + 4f\omega,\\
{}^{(X)}\widetilde{\pi}_{4\theta} &= 2f\xi,\\
{}^{(X)}\widetilde{\pi}_{AB} &= 2f\,{}^{(1+3)}\chi_{AB} - \frac{2f}{r}\not{g}_{AB} = 2f\left({}^{(1+3)}\chi_{AB} - \frac{1}{r}\delta_{AB}\right),\\
{}^{(X)}\widetilde{\pi}_{3\theta} &= 2f(\eta+\zeta),\\
{}^{(X)}\widetilde{\pi}_{33} &= -8f\underline{\omega} - 4e_3(f).
\end{aligned}$$

Under the assumptions (10.2.7)–(10.2.8) on the Ricci coefficients (with respect to the frame (e_3', e_4')), we deduce

$$
\begin{aligned}
{}^{(X)}\widetilde{\pi}_{43} &= -2e_4 f + 4f\omega = -2f' + \frac{4f}{r} - 2f'(e_4(r) - 1) + 4f(\omega - 1) \\
&= -2f' + \frac{4f}{r} + \epsilon \min\{w_{1,1}, w_{2,1/2}\} \left(|f| + r|f'|\right), \\
{}^{(X)}\widetilde{\pi}_{4A} &= \epsilon \min\{w_{2,1}, w_{3,1/2}\}, \\
{}^{(X)}\widetilde{\pi}_{AB} &= \epsilon \min\{w_{1,1}, w_{2,1/2}\}|f|, \\
{}^{(X)}\widetilde{\pi}_{3A} &= \min\{w_{1,1}, w_{2,1/2}\}|f|, \\
{}^{(X)}\widetilde{\pi}_{33} &= -8f\underline{\omega} - 4e_3(f) = -8f\left(\frac{m}{r^2} + \epsilon w_{1,1}\right) - 4f'(-\Upsilon + \epsilon w_{0,1}) \\
&= 4f'\Upsilon - 4\Upsilon' + \epsilon w_{1,1}(|f| + r|f'|).
\end{aligned}
$$

To prove formula (D.3.3) we make use of the following (see also Lemma 10.11).

Lemma D.15. *If $h = h(r)$ then*

$$
\square h = \Upsilon h''(r) + \left(\frac{2}{r} - \frac{2m}{r^2}\right) h' + O(\epsilon) w_{2,1} \left(|h| + r|h'| + r^2|h''|\right).
$$

Proof. For a general scalar h,

$$
\begin{aligned}
\square h &= -\frac{1}{2}(e_3 e_4 + e_4 e_3)h + \triangle\!\!\!\!/ h + \left({}^{(1+3)}\underline{\omega} - \frac{1}{2}\,{}^{(1+3)}\mathrm{tr}\underline{\chi}\right) e_4 h \\
&+ \left({}^{(1+3)}\omega - \frac{1}{2}\,{}^{(1+3)}\mathrm{tr}\chi\right) e_3 h
\end{aligned}
$$

with $\triangle\!\!\!\!/ h = e_\theta e_\theta h + (e_\theta \Phi)^2 e_\theta h = 0$ if h is radial. Thus,

$$
\begin{aligned}
\square h &= -\frac{1}{2}(e_3 e_4 + e_4 e_3)h + \left({}^{(1+3)}\underline{\omega} - \frac{1}{2}\,{}^{(1+3)}\mathrm{tr}\underline{\chi}\right) e_4 h \\
&+ \left({}^{(1+3)}\omega - \frac{1}{2}\,{}^{(1+3)}\mathrm{tr}\chi\right) e_3 h \\
&= -f''(e_3 r)(e_4 r) - \frac{1}{2} h'(e_3 e_4 + e_4 e_3) r \\
&+ h' \left[\left({}^{(1+3)}\underline{\omega} - \frac{1}{2}\,{}^{(1+3)}\mathrm{tr}\underline{\chi}\right) e_4 r + \left({}^{(1+3)}\omega - \frac{1}{2}\,{}^{(1+3)}\mathrm{tr}\chi\right) e_3 r\right] \\
&= -h''(-\Upsilon + O(\epsilon) w_{0,1})(1 + O(\epsilon) w_{1,1}) + \left(\frac{m}{r^2} + O(\epsilon) w_{1,1}\right) h' \\
&+ h' \Bigg[\left(\frac{m}{r^2} + \frac{\Upsilon}{r} + O(\epsilon) w_{1,1}\right)(1 + O(\epsilon) w_{1,1}) \\
&+ \left(-\frac{1}{r} + O(\epsilon) w_{1,1}\right)(-\Upsilon + O(\epsilon) w_{0,1})\Bigg] \\
&= \Upsilon h'' + \left(\frac{2}{r} - \frac{2m}{r^2}\right) h' + O(\epsilon) w_{2,1} \left(|h| + r|h'| + r^2|h''|\right)
\end{aligned}
$$

which concludes the proof of Lemma D.15. □

In view of Lemma D.15,

$$\begin{aligned}\square^{(X)}\Lambda &= \square\left(\frac{2f}{r}\right)\\ &= \Upsilon\left(\frac{2f}{r}\right)'' + \left(\frac{2}{r}-\frac{2m}{r^2}\right)\left(\frac{2f}{r}\right)' + O(\epsilon)w_{3,1}\left(|f|+r|f'|+r^2|f''|\right).\end{aligned}$$

Note that

$$\begin{aligned}&\Upsilon\left(\frac{2f}{r}\right)'' + \left(\frac{2}{r}-\frac{2m}{r^2}\right)\left(\frac{2f}{r}\right)'\\ =\ &\Upsilon\left(\frac{2f''}{r}-\frac{4f'}{r^2}+\frac{4f}{r^3}\right) + \left(\frac{2}{r}-\frac{2m}{r^2}\right)\left(\frac{2f'}{r}-\frac{2f}{r^2}\right)\\ =\ &\frac{2\Upsilon}{r}f'' - (\Upsilon-1)\frac{4f'}{r^2} + (\Upsilon-1)\frac{4f}{r^3} - \frac{2m}{r^2}\left(\frac{2f'}{r}-\frac{2f}{r^2}\right)\\ =\ &\frac{2}{r} + O\left(\frac{m}{r^4}\right)\left(|f|+r|f'|+r^2|f''|\right).\end{aligned}$$

Hence,

$$\square^{(X)}\Lambda = \frac{2}{r}f'' + O\left(\frac{m}{r^4}\epsilon w_{3,1}\right)\left(|f|+r|f'|+r^2|f''|\right)$$

as desired. This concludes the proof of Lemma D.14. □

D.4 PROOF OF PROPOSITION 10.47

In view of the following Leibniz rule which holds for any scalar f,

$$\begin{aligned}-\triangle_2(f\psi) &= \not{d}_2^\star\not{d}_2(f\psi) + 2Kf\psi\\ &= \not{d}_2^\star(f\not{d}_2\psi + e_\theta(f)\psi) + 2Kf\psi\\ &= -f\triangle_2\psi - e_\theta(f)\not{d}_2\psi + e_\theta(f)\not{d}_3^\star\psi - \triangle_0(f)\psi,\end{aligned}$$

we have the following computation

$$\begin{aligned}e_4(\square_2(r\psi)) &= e_4(r\square_2\psi) - e_4(e_3(r)e_4\psi) - e_4(e_4(r)e_3\psi) - 2e_4(e_\theta(r)\not{d}_2\psi)\\ &\quad +2e_4(e_\theta(r)\not{d}_3^\star\psi) + e_4(\square_0(r)\psi)\\ &= e_4(r\square_2\psi) - e_4\left(\frac{r}{2}(\underline{\kappa}+\underline{A})e_4\psi\right) - e_4\left(\frac{r}{2}(\kappa+A)e_3\psi\right)\\ &\quad +e_4(\square_0(r)\psi) + r^{-1}\mathfrak{d}^{\le 1}(\Gamma_g)\mathfrak{d}^{\le 2}\psi\\ &= e_4(r\square_2\psi) - \frac{1}{2}e_4(r\underline{\kappa}e_4\psi) - \frac{1}{2}e_4(r\kappa e_3\psi) + e_4(\square_0(r)\psi) + r^{-1}\mathrm{Err},\end{aligned}$$

where we have introduced the notation, used throughout the proof of Proposition 10.47,

$$\mathrm{Err} := r^2\Gamma_g e_4e_3\psi + r\Gamma_b e_4\mathfrak{d}\psi + \mathfrak{d}^{\le 1}(\Gamma_b)\mathfrak{d}^{\le 1}\psi + r\mathfrak{d}^{\le 1}(\Gamma_g)e_3\psi + \mathfrak{d}^{\le 1}(\Gamma_g)\mathfrak{d}^2\psi.$$

Next, recall that we have

$$\square_2\psi = -e_4e_3\psi + \not{\Delta}_2\psi + \left(2\omega - \frac{1}{2}\kappa\right)e_3\psi - \frac{1}{2}\underline{\kappa}e_4\psi + 2\underline{\eta}e_\theta\psi.$$

We infer

$$\begin{aligned}\square_0(r) &= -e_4(e_3(r)) + \not{\Delta}_2(r) + \left(2\omega - \frac{1}{2}\kappa\right)e_3(r) - \frac{1}{2}\underline{\kappa}e_4(r) + 2\underline{\eta}e_\theta(r)\\ &= -e_4\left(\frac{r}{2}(\underline{\kappa}+\underline{A})\right) + \left(2\omega - \frac{1}{2}\kappa\right)\frac{r}{2}(\underline{\kappa}+\underline{A}) - \frac{1}{2}\underline{\kappa}\frac{r}{2}(\kappa + A) + r^{-1}\mathfrak{d}^{\leq 1}\Gamma_g\\ &= -\frac{1}{2}e_4(r\underline{\kappa}) + \frac{1}{2}\left(2\omega - \frac{1}{2}\kappa\right)r\underline{\kappa} - \frac{1}{4}r\underline{\kappa}\kappa + r\mathfrak{d}^{\leq 1}\Gamma_g\\ &= \frac{2}{r} + O(r^{-2}) + \mathfrak{d}^{\leq 1}\Gamma_b\end{aligned}$$

and hence

$$\begin{aligned}e_4(\square_2(r\psi)) &= e_4(r\square_2\psi) - \frac{1}{2}e_4(r\underline{\kappa}e_4\psi) - \frac{1}{2}e_4\left(r\kappa e_3\psi\right) + e_4(\square_0(r)\psi)\\ &\quad +\mathfrak{d}^{\leq 1}(\Gamma_g)\mathfrak{d}^{\leq 2}\psi\\ &= e_4(r\square_2\psi) - \frac{1}{2}e_4(r\underline{\kappa}e_4\psi) - \frac{1}{2}e_4\left(r\kappa e_3\psi\right) + e_4\left(\frac{2}{r}\psi\right)\\ &\quad +O(r^{-3})\mathfrak{d}^{\leq 1}\psi + r^{-1}\mathrm{Err}\end{aligned}$$

so that

$$\begin{aligned}e_4(r\square_2\psi) &= e_4(\square_2(r\psi)) + \frac{1}{2}e_4(r\underline{\kappa}e_4\psi) + \frac{1}{2}e_4\left(r\kappa e_3\psi\right) - e_4\left(\frac{2}{r}\psi\right)\\ &\quad +O(r^{-3})\mathfrak{d}^{\leq 1}\psi + r^{-1}\mathrm{Err}.\end{aligned}$$

We infer

$$\begin{aligned}\square_2(e_4(r\psi)) - e_4(r\square_2\psi) &= [\square_2, e_4](r\psi) - \frac{1}{2}e_4(r\underline{\kappa}e_4\psi) - \frac{1}{2}e_4\left(r\kappa e_3\psi\right)\\ &\quad +e_4\left(\frac{2}{r}\psi\right) + O(r^{-3})\mathfrak{d}^{\leq 1}\psi + r^{-1}\mathrm{Err}.\end{aligned}$$

Next, using again

$$\square_2\psi = -e_4e_3\psi + \not{\Delta}_2\psi + \left(2\omega - \frac{1}{2}\kappa\right)e_3\psi - \frac{1}{2}\underline{\kappa}e_4\psi + 2\underline{\eta}e_\theta\psi,$$

we infer

$$\begin{aligned}
[\square_2, e_4]\psi &= -e_4[e_3, e_4]\psi + [\not{\Delta}_2, e_4]\psi + \left(2\omega - \frac{1}{2}\kappa\right)[e_3, e_4]\psi \\
&\quad -e_4\left(2\omega - \frac{1}{2}\kappa\right)e_3\psi + \frac{1}{2}e_4(\underline{\kappa})e_4\psi + 2\underline{\eta}[e_\theta, e_4]\psi - 2e_4(\underline{\eta})e_\theta\psi \\
&= -e_4[e_3, e_4]\psi + [\not{\Delta}_2, e_4]\psi + \left(2\omega - \frac{1}{2}\kappa\right)[e_3, e_4]\psi \\
&\quad -\left(2e_4(\omega) - \frac{1}{2}\left(-\frac{1}{2}\kappa^2 - 2\omega\kappa\right)\right)e_3\psi \\
&\quad +\frac{1}{2}\left(-\frac{1}{2}\kappa\underline{\kappa} + 2\omega\underline{\kappa} + 2\rho\right)e_4\psi + 2\underline{\eta}[e_\theta, e_4]\psi + r^{-2}\mathfrak{d}^{\leq 1}(\Gamma_g)\mathfrak{d}\psi.
\end{aligned}$$

Now, recall

$$[e_3, e_4] = 2\underline{\omega}e_4 - 2\omega e_3 + 2(\eta - \underline{\eta})e_\theta,$$

and, in view of Lemma 2.54, the following commutation formulae for reduced scalars:

1. If $f \in \mathfrak{s}_k$,

$$\begin{aligned}
[\not{d}_k, e_4] &= \frac{1}{2}\kappa\,\not{d}_k f + \mathrm{Com}_k(f), \\
\mathrm{Com}_k(f) &= -\frac{1}{2}\vartheta\,\not{d}^\star_{k+1} f - (\zeta + \underline{\eta})e_4 f - k\underline{\eta}e_4\Phi f - \xi(e_3 f + k e_3(\Phi)f) - k\beta f.
\end{aligned}$$

2. If $f \in \mathfrak{s}_{k-1}$,

$$\begin{aligned}
[\not{d}^\star_k, e_4]f &= \frac{1}{2}\kappa\,\not{d}_k f + \mathrm{Com}^*_k(f), \\
\mathrm{Com}^*_k(f) &= -\frac{1}{2}\vartheta\,\not{d}_{k-1} f + (\zeta + \underline{\eta})e_4 f - (k-1)\underline{\eta}e_4\Phi f + \xi(e_3 f - (k-1)e_3(\Phi)f) \\
&\quad - (k-1)\beta f.
\end{aligned}$$

We infer

$$\begin{aligned}
&[\square_2, e_4]\psi \\
&= -e_4\Big((2\underline{\omega}e_4 - 2\omega e_3 + 2\eta e_\theta)\psi\Big) + \kappa\not{\Delta}_2\psi \\
&\quad + \left(2\omega - \frac{1}{2}\kappa\right)\Big(2\underline{\omega}e_4 - 2\omega e_3 + 2\eta e_\theta\Big)\psi - \left(2e_4(\omega) + \frac{1}{4}\kappa^2 + \omega\kappa\right)e_3\psi \\
&\quad +\frac{1}{2}\left(-\frac{1}{2}\kappa\underline{\kappa} + 2\omega\underline{\kappa} + 2\rho\right)e_4\psi + r^{-2}\mathfrak{d}^{\leq 1}(\Gamma_g)\mathfrak{d}^{\leq 2}\psi \\
&= 2e_4(\omega e_3\psi) + \kappa\not{\Delta}_2\psi - \left(2e_4(\omega) + \frac{1}{4}\kappa^2\right)e_3\psi - \frac{1}{4}\kappa\underline{\kappa}e_4\psi + O(r^{-4})\mathfrak{d}^{\leq 1}\psi \\
&\quad + r^{-2}\mathrm{Err} \\
&= 2\omega e_4(e_3\psi) + \kappa\not{\Delta}_2\psi - \frac{1}{4}\kappa^2 e_3\psi - \frac{1}{4}\kappa\underline{\kappa}e_4\psi + O(r^{-4})\mathfrak{d}^{\leq 1}\psi + r^{-2}\mathrm{Err}.
\end{aligned}$$

This implies

$$\begin{aligned}
&[\square_2, e_4](r\psi)\\
= \;& 2\omega e_4(e_3(r\psi)) + \kappa \Delta\!\!\!\!/\,_2(r\psi) - \frac{1}{4}\kappa^2 e_3(r\psi) - \frac{1}{4}\kappa\underline{\kappa} e_4(r\psi) + O(r^{-3})\mathfrak{d}^{\leq 1}\psi\\
& + r^{-1}\mathrm{Err}\\
= \;& 2\omega e_3(e_4(r\psi)) + 2\omega[e_4, e_3]r\psi + \kappa \Delta\!\!\!\!/\,_2(r\psi) - \frac{1}{4}\kappa^2 e_3(r\psi) - \frac{1}{4}\kappa\underline{\kappa} e_4(r\psi)\\
& + O(r^{-3})\mathfrak{d}^{\leq 1}\psi + r^{-1}\mathrm{Err}\\
= \;& 2\omega e_3(e_4(r\psi)) + \kappa \Delta\!\!\!\!/\,_2(r\psi) - \frac{1}{4}\kappa^2 e_3(r\psi) - \frac{1}{4}\kappa\underline{\kappa} e_4(r\psi) + O(r^{-3})\mathfrak{d}^{\leq 1}\psi\\
& + r^{-1}\mathrm{Err}
\end{aligned}$$

and hence

$$\begin{aligned}
&\square_2(e_4(r\psi)) - e_4(r\square_2\psi)\\
= \;& [\square_2, e_4](r\psi) - \frac{1}{2}e_4(r\underline{\kappa} e_4\psi) - \frac{1}{2}e_4\left(r\kappa e_3\psi\right) + e_4\left(\frac{2}{r}\psi\right) + O(r^{-3})\mathfrak{d}^{\leq 1}\psi\\
& + r^{-1}\mathrm{Err}\\
= \;& 2\omega e_3(e_4(r\psi)) - \frac{1}{2}e_4(r\underline{\kappa} e_4\psi) - \frac{1}{2}e_4\left(r\kappa e_3\psi\right) + \kappa \Delta\!\!\!\!/\,_2(r\psi) - \frac{1}{4}\kappa^2 e_3(r\psi)\\
& - \frac{1}{4}\kappa\underline{\kappa} e_4(r\psi) + e_4\left(\frac{2}{r}\psi\right) + O(r^{-3})\mathfrak{d}^{\leq 1}\psi + r^{-1}\mathrm{Err}.
\end{aligned}$$

Next, we compute

$$\begin{aligned}
& -\frac{1}{2}e_4(r\underline{\kappa} e_4\psi) - \frac{1}{2}e_4\left(r\kappa e_3\psi\right) + \kappa \Delta\!\!\!\!/\,_2(r\psi) - \frac{1}{4}\kappa^2 e_3(r\psi) + e_4\left(\frac{2}{r}\psi\right)\\
= \;& -\frac{1}{2}e_4(\underline{\kappa}(e_4(r\psi) - e_4(r)\psi)) - \frac{1}{2}e_3(e_4(r\kappa\psi)) - \frac{1}{2}[e_4, e_3](r\kappa\psi) + \frac{1}{2}e_4(e_3(r\kappa)\psi)\\
& + r\kappa \Delta\!\!\!\!/\,_2\psi - \frac{1}{4}r\kappa^2 e_3\psi - \frac{1}{4}\kappa^2 e_3(r)\psi + \frac{2}{r^2}e_4\left(r\psi\right) + e_4\left(\frac{2}{r^2}\right)r\psi\\
& + r^{-1}\mathfrak{d}^{\leq 1}(\Gamma_g)\mathfrak{d}^{\leq 2}\psi\\
= \;& -\frac{1}{2}e_4(\underline{\kappa}(e_4(r\psi))) + \frac{1}{2}e_4\left(\underline{\kappa}\frac{r}{2}(\kappa + A)\psi\right) - \frac{1}{2}e_3(\kappa e_4(r\psi)) - \frac{1}{2}e_3(e_4(\kappa)r\psi)\\
& - \frac{1}{2}\Big(- 2\underline{\omega} e_4 + 2\omega e_3 - 2(\eta - \underline{\eta})e_\theta\Big)(r\kappa\psi) + \frac{1}{2}e_4(e_3(r\kappa)\psi)\\
& + r\kappa \Delta\!\!\!\!/\,_2\psi - \frac{1}{4}r\kappa^2 e_3\psi - \frac{1}{4}\kappa^2\frac{r}{2}(\underline{\kappa} + \underline{A})\psi + \frac{2}{r^2}e_4\left(r\psi\right) - \frac{4e_4(r)}{r^2}\psi\\
& + r^{-1}\mathfrak{d}^{\leq 1}(\Gamma_g)\mathfrak{d}^{\leq 2}\psi,
\end{aligned}$$

i.e.,

$$\begin{aligned}
& -\frac{1}{2}e_4(r\underline{\kappa}e_4\psi) - \frac{1}{2}e_4\left(r\kappa e_3\psi\right) + \kappa\mathbin{\rlap{/}\Delta}_2(r\psi) - \frac{1}{4}\kappa^2 e_3(r\psi) + e_4\left(\frac{2}{r}\psi\right)\\
= \;& -\frac{1}{2}e_4(\underline{\kappa}(e_4(r\psi))) + \frac{1}{4}e_4\left(r\underline{\kappa}\kappa\psi\right) - \frac{1}{2}e_3(\kappa e_4(r\psi))\\
& -\frac{1}{2}e_3\left(\left(-\frac{1}{2}\kappa^2 - 2\omega\kappa\right)r\psi\right) - \omega e_3(r\kappa\psi) + \frac{1}{2}e_4(e_3(r\kappa)\psi) + r\kappa\mathbin{\rlap{/}\Delta}_2\psi\\
& -\frac{1}{4}r\kappa^2 e_3\psi - \frac{1}{8}r\kappa^2\underline{\kappa}\psi + \frac{2}{r^2}e_4\left(r\psi\right) - \frac{2\kappa}{r}\psi + O(r^{-3})\mathfrak{d}^{\leq 1}\psi + r^{-1}\mathrm{Err}\\
= \;& -\frac{1}{2}e_4(\underline{\kappa}(e_4(r\psi))) + \frac{1}{4}\underline{\kappa}\kappa e_4\left(r\psi\right) + \frac{1}{4}e_4\left(\underline{\kappa}\kappa\right)r\psi - \frac{1}{2}e_3(\kappa e_4(r\psi))\\
& -\frac{1}{2}e_3\left(\left(-\frac{1}{2}\kappa^2 - 2\omega\kappa\right)r\psi\right) - \omega e_3(r\kappa\psi) + \frac{1}{2}r^{-1}e_3(r\kappa)e_4(r\psi)\\
& +\frac{1}{2}e_4(r^{-1}e_3(r\kappa))r\psi + r\kappa\mathbin{\rlap{/}\Delta}_2\psi - \frac{1}{4}r\kappa^2 e_3\psi - \frac{1}{8}r\kappa^2\underline{\kappa}\psi + \frac{2}{r^2}e_4\left(r\psi\right)\\
& -\frac{2\kappa}{r}\psi + O(r^{-3})\mathfrak{d}^{\leq 1}\psi + r^{-1}\mathrm{Err}.
\end{aligned}$$

We infer

$$\begin{aligned}
& \square_2(e_4(r\psi)) - e_4(r\square_2\psi)\\
= \;& 2\omega e_3(e_4(r\psi)) - \frac{1}{2}e_4(\underline{\kappa}(e_4(r\psi))) + \frac{1}{4}e_4\left(\underline{\kappa}\kappa\right)r\psi - \frac{1}{2}e_3(\kappa e_4(r\psi))\\
& -\frac{1}{2}e_3\left(\left(-\frac{1}{2}\kappa^2 - 2\omega\kappa\right)r\psi\right) - \omega e_3(r\kappa\psi) + \frac{1}{2}r^{-1}e_3(r\kappa)e_4(r\psi)\\
& +\frac{1}{2}e_4(r^{-1}e_3(r\kappa))r\psi + r\kappa\mathbin{\rlap{/}\Delta}_2\psi - \frac{1}{4}r\kappa^2 e_3\psi - \frac{1}{8}r\kappa^2\underline{\kappa}\psi + \frac{2}{r^2}e_4\left(r\psi\right)\\
& -\frac{2\kappa}{r}\psi + O(r^{-3})\mathfrak{d}^{\leq 1}\psi + r^{-1}\mathrm{Err}.
\end{aligned}$$

Since $e_4(r\psi) = r\Upsilon\check{e}_4\psi$, this may be rewritten as

$$\begin{aligned}
& \square_2(r\Upsilon\check{e}_4\psi) - e_4(r\square_2\psi)\\
= \;& 2\omega e_3(r\Upsilon\check{e}_4\psi) - \frac{1}{2}e_4(r\Upsilon\underline{\kappa}\check{e}_4\psi) + \frac{1}{4}e_4(\underline{\kappa}\kappa)r\psi - \frac{1}{2}e_3(r\Upsilon\kappa\check{e}_4\psi)\\
& -\frac{1}{2}e_3\left(\left(-\frac{1}{2}\kappa^2 - 2\omega\kappa\right)r\psi\right) - \omega e_3(r\kappa\psi) + \frac{1}{2}e_3(r\kappa)\Upsilon\check{e}_4\psi\\
& +\frac{1}{2}e_4(r^{-1}e_3(r\kappa))r\psi + r\kappa\mathbin{\rlap{/}\Delta}_2\psi - \frac{1}{4}r\kappa^2 e_3\psi - \frac{1}{8}r\kappa^2\underline{\kappa}\psi + \frac{2}{r}\Upsilon\check{e}_4\psi\\
& -\frac{2\kappa}{r}\psi + O(r^{-3})\mathfrak{d}^{\leq 1}\psi + r^{-1}\mathrm{Err}.
\end{aligned}$$

Now, since

$$\square_2\psi = -e_3e_4\psi + \mathbin{\rlap{/}\Delta}_2\psi + \left(2\underline{\omega} - \frac{1}{2}\underline{\kappa}\right)e_4\psi - \frac{1}{2}\kappa e_3\psi + 2\eta e_\theta\psi,$$

we have

$$
\begin{aligned}
r\kappa\not{\Delta}_2\psi &= r\kappa\square_2\psi + r\kappa e_3e_4\psi - r\kappa\left(2\underline{\omega} - \frac{1}{2}\underline{\kappa}\right)e_4\psi + \frac{1}{2}r\kappa^2 e_3\psi \\
&\quad + r^{-1}\mathfrak{d}^{\leq 1}(\Gamma_b)\mathfrak{d}^{\leq 1}\psi \\
&= r\kappa\square_2\psi + r\kappa e_3(r^{-1}e_4(r\psi)) - r\kappa e_3(r^{-1}e_4(r)\psi) + \frac{1}{2}\kappa\underline{\kappa}e_4(r\psi) \\
&\quad - \frac{1}{2}\underline{\kappa}\kappa e_4(r)\psi + \frac{1}{2}r\kappa^2 e_3\psi + r^{-1}\mathfrak{d}^{\leq 1}(\Gamma_b)\mathfrak{d}^{\leq 1}\psi \\
&= r\kappa\square_2\psi + r\kappa e_3(\Upsilon\check{e}_4\psi) - \frac{1}{2}r\kappa e_3(\kappa)\psi + \frac{1}{2}\kappa\underline{\kappa}r\Upsilon\check{e}_4\psi - \frac{r}{4}\kappa^2\underline{\kappa}\psi \\
&\quad + r^{-1}\mathfrak{d}^{\leq 1}(\Gamma_b)\mathfrak{d}^{\leq 1}\psi
\end{aligned}
$$

and hence

$$
\begin{aligned}
&\square_2(r\Upsilon\check{e}_4\psi) - e_4(r\square_2\psi) \\
&= 2\omega e_3(r\Upsilon\check{e}_4\psi) - \frac{1}{2}e_4(r\Upsilon\underline{\kappa}\check{e}_4\psi) + \frac{1}{4}e_4(\underline{\kappa}\kappa)r\psi - \frac{1}{2}e_3(r\Upsilon\kappa\check{e}_4\psi) \\
&\quad - \frac{1}{2}e_3\left(\left(-\frac{1}{2}\kappa^2 - 2\omega\kappa\right)r\psi\right) - \omega e_3(r\kappa\psi) + \frac{1}{2}e_3(r\kappa)\Upsilon\check{e}_4\psi \\
&\quad + \frac{1}{2}e_4(r^{-1}e_3(r\kappa))r\psi + r\kappa\square_2\psi + r\kappa e_3(\Upsilon\check{e}_4\psi) - \frac{1}{2}r\kappa e_3(\kappa)\psi - \frac{r}{4}\kappa^2\underline{\kappa}\psi \\
&\quad - \frac{1}{4}r\kappa^2 e_3\psi - \frac{1}{8}r\kappa^2\underline{\kappa}\psi - \frac{2\kappa}{r}\psi + O(r^{-3})\mathfrak{d}^{\leq 1}\psi + r^{-1}\mathrm{Err}.
\end{aligned}
$$

Next, we compute

$$
\begin{aligned}
&\frac{1}{4}e_4(\underline{\kappa}\kappa)r\psi - \frac{1}{2}e_3\left(\left(-\frac{1}{2}\kappa^2 - 2\omega\kappa\right)r\psi\right) - \omega e_3(r\kappa\psi) + \frac{1}{2}e_4(r^{-1}e_3(r\kappa))r\psi \\
&\quad - \frac{1}{2}r\kappa e_3(\kappa)\psi - \frac{r}{4}\kappa^2\underline{\kappa}\psi - \frac{1}{4}r\kappa^2 e_3\psi - \frac{1}{8}r\kappa^2\underline{\kappa}\psi - \frac{2\kappa}{r}\psi \\
&= \frac{r}{2}\kappa\rho\psi + r^{-1}\mathfrak{d}^{\leq 1}(\Gamma_b)\psi \\
&= O(r^{-3})\psi + r^{-1}\mathfrak{d}^{\leq 1}(\Gamma_b)\psi
\end{aligned}
$$

so that

$$
\begin{aligned}
&\square_2(r\Upsilon\check{e}_4\psi) \\
&= e_4(r\square_2\psi) + r\kappa\square_2\psi + 2\omega e_3(r\Upsilon\check{e}_4\psi) - \frac{1}{2}e_4(r\Upsilon\underline{\kappa}\check{e}_4\psi) - \frac{1}{2}e_3(r\Upsilon\kappa\check{e}_4\psi) \\
&\quad + \frac{1}{2}e_3(r\kappa)\Upsilon\check{e}_4\psi + r\kappa e_3(\Upsilon\check{e}_4\psi) + O(r^{-3})\mathfrak{d}^{\leq 1}\psi + r^{-1}\mathrm{Err}.
\end{aligned}
$$

Since

$$
\begin{aligned}
\square_2(r^2\check{e}_4\psi) &= r\Upsilon^{-1}\square_2(r\Upsilon\check{e}_4\psi) - e_3(r\Upsilon^{-1})e_4(r\Upsilon\check{e}_4\psi) - e_4(r\Upsilon^{-1})e_3(r\Upsilon\check{e}_4\psi) \\
&\quad + \square_0(r\Upsilon^{-1})r\Upsilon\check{e}_4\psi + \mathfrak{d}^{\leq 1}(\Gamma_g)\mathfrak{d}^{\leq 2}\psi,
\end{aligned}
$$

we infer

$$\begin{aligned}\square_2(r^2\check{e}_4\psi) &= r\Upsilon^{-1}e_4(r\square_2\psi)+r^2\Upsilon^{-1}\kappa\square_2\psi+2r\Upsilon^{-1}\omega e_3(r\Upsilon\check{e}_4\psi)\\ &-\frac{1}{2}r\Upsilon^{-1}e_4(r\Upsilon\underline{\kappa}\check{e}_4\psi)-\frac{1}{2}r\Upsilon^{-1}e_3(r\Upsilon\kappa\check{e}_4\psi)+\frac{1}{2}re_3(r\kappa)\check{e}_4\psi\\ &+r^2\Upsilon^{-1}\kappa e_3(\Upsilon\check{e}_4\psi)-e_3(r\Upsilon^{-1})e_4(r\Upsilon\check{e}_4\psi)-e_4(r\Upsilon^{-1})e_3(r\Upsilon\check{e}_4\psi)\\ &+\square_0(r\Upsilon^{-1})r\Upsilon\check{e}_4\psi+O(r^{-2})\mathfrak{d}^{\leq 1}\psi+\mathrm{Err}.\end{aligned}$$

Now, we have

$$\begin{aligned}&2r\Upsilon^{-1}\omega e_3(r\Upsilon\check{e}_4\psi)-\frac{1}{2}r\Upsilon^{-1}e_4(r\Upsilon\underline{\kappa}\check{e}_4\psi)\\ &-\frac{1}{2}r\Upsilon^{-1}e_3(r\Upsilon\kappa\check{e}_4\psi)+\frac{1}{2}re_3(r\kappa)\check{e}_4\psi+r^2\Upsilon^{-1}\kappa e_3(\Upsilon\check{e}_4\psi)\\ &-e_3(r\Upsilon^{-1})e_4(r\Upsilon\check{e}_4\psi)-e_4(r\Upsilon^{-1})e_3(r\Upsilon\check{e}_4\psi)+\square_0(r\Upsilon^{-1})r\Upsilon\check{e}_4\psi\\ =\;& 2r\frac{1-\frac{3m}{r}}{\Upsilon}e_4(\check{e}_4\psi)+\Bigg\{2r\Upsilon^{-1}\omega e_3(r\Upsilon)-\frac{1}{2}r\Upsilon^{-1}e_4(r\Upsilon\underline{\kappa})-\frac{1}{2}r\Upsilon^{-1}e_3(r\Upsilon\kappa)\\ &+\frac{1}{2}re_3(r\kappa)+r^2\Upsilon^{-1}\kappa e_3(\Upsilon)-e_3(r\Upsilon^{-1})e_4(r\Upsilon)-e_4(r\Upsilon^{-1})e_3(r\Upsilon)\\ &+\square_0(r\Upsilon^{-1})r\Upsilon\Bigg\}\check{e}_4\psi+\mathrm{Err}.\end{aligned}$$

Also, we have

$$\begin{aligned}&2r\Upsilon^{-1}\omega e_3(r\Upsilon)-\frac{1}{2}r\Upsilon^{-1}e_4(r\Upsilon\underline{\kappa})-\frac{1}{2}r\Upsilon^{-1}e_3(r\Upsilon\kappa)+\frac{1}{2}re_3(r\kappa)\\ &+r^2\Upsilon^{-1}\kappa e_3(\Upsilon)-e_3(r\Upsilon^{-1})e_4(r\Upsilon)-e_4(r\Upsilon^{-1})e_3(r\Upsilon)+\square_0(r\Upsilon^{-1})r\Upsilon\\ =\;& 4+O(r^{-1})+r\Gamma_b\\ =\;& -r^2\kappa\underline{\kappa}+O(r^{-1})+r\Gamma_b.\end{aligned}$$

We infer

$$\begin{aligned}(\square_2+\kappa\underline{\kappa})(r^2\check{e}_4\psi) &= r\Upsilon^{-1}e_4(r\square_2\psi)+r^2\Upsilon^{-1}\kappa\square_2\psi+2r\frac{1-\frac{3m}{r}}{\Upsilon}e_4(\check{e}_4\psi)\\ &+O(r^{-2})\mathfrak{d}^{\leq 1}\psi+\mathrm{Err}.\end{aligned}$$

In view of the wave equation satisfied by ψ, i.e.,

$$\square_2\psi+\kappa\underline{\kappa}\psi = N,$$

we have

$$\begin{aligned}
& r\Upsilon^{-1}e_4(r\square_2\psi)+r^2\Upsilon^{-1}\kappa\square_2\psi+2r\frac{1-\frac{3m}{r}}{\Upsilon}e_4(\check{e}_4\psi)\\
= \;& r\Upsilon^{-1}e_4(r(N-\kappa\underline{\kappa}\psi))+r^2\Upsilon^{-1}\kappa(N-\kappa\underline{\kappa}\psi)+\frac{2}{r}\frac{1-\frac{3m}{r}}{\Upsilon}e_4(r^2\check{e}_4\psi)\\
& -4\frac{1-\frac{3m}{r}}{\Upsilon}e_4(r)\check{e}_4\psi\\
= \;& r\Upsilon^{-1}e_4(rN)+r^2\Upsilon^{-1}\kappa N+\frac{2}{r}\frac{1-\frac{3m}{r}}{\Upsilon}e_4(r^2\check{e}_4\psi)+\frac{4m}{r}\check{e}_4\psi-2r^2\Upsilon^{-1}\kappa\rho\psi\\
& +\mathfrak{d}^{\leq 1}(\Gamma_b)\mathfrak{d}^{\leq 1}\psi\\
= \;& r^2\left(\Upsilon^{-1}e_4(N)+\frac{3}{r}N\right)+\frac{2}{r}\frac{1-\frac{3m}{r}}{\Upsilon}e_4(r^2\check{e}_4\psi)+O(r^{-2})\mathfrak{d}^{\leq 1}\psi+\mathfrak{d}^{\leq 1}(\Gamma_b)\mathfrak{d}^{\leq 1}\psi,
\end{aligned}$$

from which we deduce

$$\begin{aligned}
(\square_2+\kappa\underline{\kappa})(r^2\check{e}_4\psi) \;=\;& r^2\left(\Upsilon^{-1}e_4(N)+\frac{3}{r}N\right)+\frac{2}{r}\frac{1-\frac{3m}{r}}{\Upsilon}e_4(r^2\check{e}_4\psi)\\
& +O(r^{-2})\mathfrak{d}^{\leq 1}\psi+\mathrm{Err}.
\end{aligned}$$

Since

$$\check{\psi} \;=\; f_2\check{e}_4\psi=\frac{f_2}{r^2}r^2\check{e}_4\psi,$$

we infer

$$\begin{aligned}
(\square_2+\kappa\underline{\kappa})\,\check{\psi} \;=\;& \frac{f_2}{r^2}(\square_2+\kappa\underline{\kappa})(r^2\check{e}_4\psi)-e_3\left(\frac{f_2}{r^2}\right)e_4(r^2\check{e}_4\psi)-e_4\left(\frac{f_2}{r^2}\right)e_3(r^2\check{e}_4\psi)\\
& +e_\theta\left(\frac{f_2}{r^2}\right)\not{d}_2(r^2\check{e}_4\psi)-e_\theta\left(\frac{f_2}{r^2}\right)\not{d}_3^\star(r^2\check{e}_4\psi)+\square_0\left(\frac{f_2}{r^2}\right)r^2\check{e}_4\psi
\end{aligned}$$

and hence

$$\begin{aligned}
& (\square_2+\kappa\underline{\kappa})\,\check{\psi}\\
= \;& f_2\left(\Upsilon^{-1}e_4(N)+\frac{3}{r}N\right)+\frac{f_2}{r^2}\left\{\frac{2}{r}\frac{1-\frac{3m}{r}}{\Upsilon}e_4(r^2\check{e}_4\psi)+O(r^{-2})\mathfrak{d}^{\leq 1}\psi+\mathrm{Err}\right\}\\
& -e_3\left(\frac{f_2}{r^2}\right)e_4(r^2\check{e}_4\psi)-e_4\left(\frac{f_2}{r^2}\right)e_3(r^2\check{e}_4\psi)\\
& +e_\theta\left(\frac{f_2}{r^2}\right)\not{d}_2(r^2\check{e}_4\psi)-e_\theta\left(\frac{f_2}{r^2}\right)\not{d}_3^\star(r^2\check{e}_4\psi)+\square_0\left(\frac{f_2}{r^2}\right)r^2\check{e}_4\psi.
\end{aligned}$$

Now, recall that Err is defined by

$$\mathrm{Err} \;=\; r^2\Gamma_g e_4e_3\psi+r\Gamma_b e_4\mathfrak{d}\psi+\mathfrak{d}^{\leq 1}(\Gamma_b)\mathfrak{d}^{\leq 1}\psi+r\mathfrak{d}^{\leq 1}(\Gamma_g)e_3\psi+\mathfrak{d}^{\leq 1}(\Gamma_g)\mathfrak{d}^2\psi,$$

so that

$$\begin{aligned}
(\square_2 + \kappa\underline{\kappa})\check{\psi} &= f_2\left(\Upsilon^{-1}e_4(N) + \frac{3}{r}N\right) + \frac{f_2}{r^2}\Bigg\{\frac{2}{r}\frac{1-\frac{3m}{r}}{\Upsilon}e_4(r^2\check{e}_4\psi)\\
&+O(r^{-2})\mathfrak{d}^{\leq 1}\psi + r^2\Gamma_g e_4e_3\psi + r\Gamma_b e_4\mathfrak{d}\psi + \mathfrak{d}^{\leq 1}(\Gamma_b)\mathfrak{d}^{\leq 1}\psi\\
&+r\mathfrak{d}^{\leq 1}(\Gamma_g)e_3\psi + \mathfrak{d}^{\leq 1}(\Gamma_g)\mathfrak{d}^2\psi\Bigg\}\\
&-e_3\left(\frac{f_2}{r^2}\right)e_4(r^2\check{e}_4\psi) - e_4\left(\frac{f_2}{r^2}\right)e_3(r^2\check{e}_4\psi)\\
&+e_\theta\left(\frac{f_2}{r^2}\right)\not{d}_2(r^2\check{e}_4\psi) - e_\theta\left(\frac{f_2}{r^2}\right)\not{d}_3^\star(r^2\check{e}_4\psi) + \square_0\left(\frac{f_2}{r^2}\right)r^2\check{e}_4\psi.
\end{aligned}$$

In view of

$$\square_2\psi = -e_4e_3\psi + \not{\Delta}_2\psi + \left(2\omega - \frac{1}{2}\kappa\right)e_3\psi - \frac{1}{2}\underline{\kappa}e_4\psi + 2\underline{\eta}e_\theta\psi,$$

we have

$$\begin{aligned}
r^2\Gamma_g e_4e_3\psi &= r^2\Gamma_g\left(-\square_2\psi + \not{\Delta}_2\psi + \left(2\omega - \frac{1}{2}\kappa\right)e_3\psi - \frac{1}{2}\underline{\kappa}e_4\psi + 2\underline{\eta}e_\theta\psi\right)\\
&= -r^2\Gamma_g N + r\Gamma_g e_3\psi + \Gamma_g\mathfrak{d}^{\leq 2}\psi
\end{aligned}$$

and hence

$$\begin{aligned}
(\square_2 + \kappa\underline{\kappa})\check{\psi} &= f_2\left(\Upsilon^{-1}e_4(N) + \frac{3}{r}N\right) + \frac{f_2}{r^2}\Bigg\{\frac{2}{r}\frac{1-\frac{3m}{r}}{\Upsilon}e_4(r^2\check{e}_4\psi) + O(r^{-2})\mathfrak{d}^{\leq 1}\psi\\
&+r\Gamma_b e_4\mathfrak{d}\psi + \mathfrak{d}^{\leq 1}(\Gamma_b)\mathfrak{d}^{\leq 1}\psi + r\mathfrak{d}^{\leq 1}(\Gamma_g)e_3\psi + \mathfrak{d}^{\leq 1}(\Gamma_g)\mathfrak{d}^2\psi\Bigg\}\\
&-e_3\left(\frac{f_2}{r^2}\right)e_4(r^2\check{e}_4\psi) - e_4\left(\frac{f_2}{r^2}\right)e_3(r^2\check{e}_4\psi)\\
&+e_\theta\left(\frac{f_2}{r^2}\right)\not{d}_2(r^2\check{e}_4\psi) - e_\theta\left(\frac{f_2}{r^2}\right)\not{d}_3^\star(r^2\check{e}_4\psi) + \square_0\left(\frac{f_2}{r^2}\right)r^2\check{e}_4\psi.
\end{aligned}$$

In particular, we have for $r \geq 6m_0$

$$\begin{aligned}
(\square_2 + \kappa\underline{\kappa})\check{\psi} &= r^2\left(\Upsilon^{-1}e_4(N) + \frac{3}{r}N\right) + \frac{2}{r\Upsilon}\left(1 - \frac{3m}{r}\right)e_4\check{\psi} + O(r^{-2})\mathfrak{d}^{\leq 1}\psi\\
&+r\Gamma_b e_4\mathfrak{d}\psi + \mathfrak{d}^{\leq 1}(\Gamma_b)\mathfrak{d}^{\leq 1}\psi + r\mathfrak{d}^{\leq 1}(\Gamma_g)e_3\psi + \mathfrak{d}^{\leq 1}(\Gamma_g)\mathfrak{d}^2\psi
\end{aligned}$$

and for $4m_0 \leq r \leq 6m_0$,

$$(\square_2 + \kappa\underline{\kappa})\check{\psi} = f_2\left(\Upsilon^{-1}e_4(N) + \frac{3}{r}N\right) + O(1)\mathfrak{d}^2\psi.$$

This concludes the proof of Proposition 10.47.

Bibliography

[1] S. Aksteiner, L. Andersson, T. Bäckdahl, A. G. Shah and B. F. Whiting, *Gauge-invariant perturbations of Schwarzschild spacetime*, arXiv.1611.08291.

[2] X. An and J. Luk, *Trapped surfaces in vacuum arising dynamically from mild incoming radiation*, Adv. Theor. Math. Phys. **21** (2017), 1–120.

[3] L. Andersson and P. Blue, *Hidden symmetries and decay for the wave equation on the Kerr spacetime*, Ann. of Math. **182** (2015), 787–853.

[4] L. Andersson, T. Bäckdahl, P. Blue and S. Ma, *Stability for linearized gravity on the Kerr spacetime*, arXiv:1903.03859.

[5] Y. Angelopoulos, S. Aretakis and D. Gajic, *A vectorfield approach to almost-sharp decay for the wave equation on spherically symmetric, stationary spacetimes*, Ann. PDE **4** (2018), Art. 15, 120 pp.

[6] J. M. Bardeen and W. H. Press, *Radiation fields in the Schwarzschild background*, J. Math. Phys. **14** (1973), 719.

[7] L. Bieri, *An extension of the stability theorem of the Minkowski space in general relativity*, Thesis, ETH, 2007.

[8] P. Blue and A. Soffer, *Semilinear wave equations on the Schwarzschild manifold. I. Local decay estimates*, Adv. Differential Equations **8** (2003), 595–614.

[9] P. Blue and A. Soffer, *Errata for "Global existence and scattering for the nonlinear Schrödinger equation on Schwarzschild manifolds," "Semilinear wave equations on the Schwarzschild manifold I: Local Decay Estimates," and "The wave equation on the Schwarzschild metric II: Local Decay for the spin 2 Regge Wheeler equation,"* gr-qc/0608073, 6 pages.

[10] P. Blue and J. Sterbenz, *Uniform decay of local energy and the semi-linear wave equation on Schwarzschild space*, Comm. Math. Phys. **268** (2006), 481–504.

[11] S. Chandrasekhar, *The mathematical theory of black holes*, Oxford Classic Texts in the Physical Sciences, 1983.

[12] S. Chandrasekhar, *On the equations governing the perturbations of the Schwarzschild black hole*, P. Roy. Soc. Lond. A Mat. **343** (1975), 289–298.

[13] Y. Choquet-Bruhat, *Théorème d'existence pour certains systmes d'équations aux dérivèes partielles non linéaires*, Acta Math. **88** (1952), 141–225.

[14] Y. Choquet-Bruhat and R. Geroch, *Global aspects of the Cauchy problem in general relativity*, Commun. Math. Phys. **14** (1969), 329–335.

[15] S. M. Caroll, *Spacetime and geometry. An introduction to general relativity*, Addison-Wesley, 2004.

[16] P.-N. Chen, M.-T. Wang and S.-T. Yau, *Quasilocal angular momentum and center of mass in general relativity*, Adv. Theor. Math. Phys. **20** (2016), 671–682.

[17] D. Christodoulou, *On the global initial value problem and the issue of singularities*, Class. Quant. Gr. (1999), A23–A35.

[18] D. Christodoulou, *Global solutions of nonlinear hyperbolic equations for small initial data*, Comm. Pure Appl. Math. **39** (1986), 267–282.

[19] D. Christodoulou and S. Klainerman, *Asymptotic properties of linear field theories in Minkowski space*, Comm. Pure Appl. Math. **43** (1990), 137–199.

[20] D. Christodoulou and S. Klainerman, *The global nonlinear stability of the Minkowski space*, Princeton University Press, 1993.

[21] D. Christodoulou, *The formation of black holes in General Relativity* , EMS-monographs in mathematics, 2009.

[22] M. Dafermos and I. Rodnianski, *The redshift effect and radiation decay on black hole spacetimes*, Comm. Pure Appl. Math. **62** (2009), 859–919.

[23] M. Dafermos and I. Rodnianski, *A new physical-space approach to decay for the wave equation with applications to black hole spacetimes*, XVIth International Congress on Mathematical Physics, World Sci. Publ., Hackensack, NJ, 2010, 421–432.

[24] M. Dafermos and I. Rodnianski, *A proof of the uniform boundedness of solutions to the wave equation on slowly rotating Kerr backgrounds*, Invent. Math. **185** (2011), 467–559.

[25] M. Dafermos, G. Holzegel and I. Rodnianski, *A scattering theory construction of dynamical black hole spacetimes*, arXiv:1306.5534.

[26] M. Dafermos, G. Holzegel and I. Rodnianski, *Linear stability of the Schwarzschild solution to gravitational perturbations*, Acta Math. **222** (2019), 1–214.

[27] M. Dafermos, G. Holzegel and I. Rodnianski, *Boundedness and decay for the Teukolsky equation on Kerr spacetimes I: The case* $|a| \ll M$, Ann. PDE (2019), 118 pp.

[28] M. Dafermos, I. Rodnianski and Y. Shlapentokh-Rothman, *Decay for solutions of the wave equation on Kerr exterior spacetimes iii: The full subextremal case* $|a| < m$, Ann. of Math. **183** (2016), 787–913.

[29] E. Giorgi, S. Klainerman and J. Szeftel, *A general formalism for the stability of Kerr*, arXiv:2002.02740.

[30] D. Häfner, P. Hintz and A. Vasy, *Linear stability of slowly rotating Kerr black holes*, arXiv:1906.00860.

[31] P. Hintz and A. Vasy, *The global non-linear stability of the Kerr-de Sitter family of black holes*, Acta Math. **220** (2018), 1–206.

[32] P. Hintz, *Non-linear stability of the Kerr-Newman-de Sitter family of charged black holes*, Ann. PDE **4** (2018), Art. 11, 131 pp.

[33] G. Holzegel, *Ultimately Schwarzschildean Spacetimes and the Black Hole Stability Problem*, arXiv:1010.3216.

[34] P. K. Hung, J. Keller and M. T. Wang, *Linear stability of Schwarzschild spacetime: the Cauchy problem of metric coefficients*, arXiv:1702.02843.

[35] A. Ionescu and S. Klainerman, *Rigidity results in general relativity: a review*, Surveys in Differential Geometry **20** (2015), 123–156.

[36] A. D. Ionescu and S. Klainerman, *On the global stability of the wave-map equation in Kerr spaces with small angular momentum*, Ann. PDE **1** (2015), Art. 1, 78 pp.

[37] T. W. Johnson, *The linear stability of the Schwarzschild solution to gravitational perturbations in the generalised wave gauge*, Ann. PDE **5** (2019), no. 2, Art. 13, 92 pp.

[38] S. Klainerman, *Long time behavior of solutions to nonlinear wave equations*, Proceedings of the International Congress of Mathematicians, Vol. 1, 2 (Warsaw, 1983), 1209–1215, PWN, Warsaw, 1984.

[39] S. Klainerman, *Uniform decay estimates and the Lorentz invariance of the classical wave equations*, Comm. Pure Appl. Math. **38** (1985), 321–332.

[40] S. Klainerman, *The Null Condition and global existence to nonlinear wave equations*, Lect. in Appl. Math. **23** (1986), 293–326.

[41] S. Klainerman, *Remarks on the global Sobolev inequalities*, Comm. Pure Appl. Math. **40** (1987), 111–117.

[42] S. Klainerman and F. Nicolo, The evolution problem in general relativity. Progress in Mathematical Physics **25**, Birkhauser, Boston, 2003.

[43] S. Klainerman and F. Nicolo, Peeling properties of asymptotic solutions to the Einstein vacuum equations, Class. Quantum Grav. **20** (2003), 3215–3257.

[44] S. Klainerman, J. Luk and I. Rodnianski, *A fully anisotropic mechanism for formation of trapped surfaces in vacuum*, Inventiones **198** (2014), 1–26.

[45] S. Klainerman and I. Rodnianski, *On the formation of trapped surfaces*, Acta Math. **208** (2012), 211–333.

[46] S. Klainerman, I. Rodnianski and J. Szeftel, *The bounded L^2 curvature conjecture*, Inventiones **202** (2015), 91–216.

[47] S. Klainerman and J. Szeftel, *Construction of GCM spheres in perturbations of Kerr*, arXiv:1911.00697.

[48] S. Klainerman and J. Szeftel, *Effective results on uniformization and intrinsic*

GCM spheres in perturbations of Kerr, arXiv:1912.12195.

[49] H. Lindblad and I. Rodnianski, *The global stability of Minkowski spacetime in harmonic gauge*, Ann. of Math. **171** (2010), 1401–1477.

[50] S. Ma, *Uniform energy bound and Morawetz estimate for extreme component of spin fields in the exterior of a slowly rotating black hole II: Linearized gravity*, arXiv:1708.07385.

[51] M. Mars, A spacetime characterization of the Kerr metric, Class. Quantum Grav. **16** (1999), 2507–2523.

[52] M. Mars and J. Senovilla, *Axial Symmetry and conformal Killing vectorfields*, arXiv:gr-qc/0201045.

[53] Y. Martel and F. Merle, *Asymptotic stability of solitons for subcritical generalized KdV equations*, Arch. Ration. Mech. Anal. **157** (2001), 219–254.

[54] J. Marzuola, J. Metcalfe, D. Tataru and M. Tohaneanu, *Strichartz estimates on Schwarzschild black hole backgrounds*, Comm. Math. Phys. **293** (2010), 37–83.

[55] F. Merle and P. Raphaël, *On universality of blow-up profile for L^2 critical nonlinear Schrödinger equation*, Invent. Math. **156** (2004), 565–672.

[56] V. Moncrief, *Gravitational perturbations of spherically symmetric systems. I. The exterior problem*, Ann. Phys. **88** (1975), 323–342.

[57] G. Moschidis, *The r^p-weighted energy method of Dafermos and Rodnianski in general asymptotically flat spacetimes and applications*, Ann. PDE 2 (2016), no. 1, Art. 6, 194 pp.

[58] T. Regge and J. A. Wheeler, *Stability of a Schwarzschild singularity*, Phys. Rev. (2), 108:1063–1069, 1957.

[59] A. Rizzi, *Angular Momentum in General Relativity: A new Definition*, Phys. Rev. Lett. **81** (1998), 1150–1153.

[60] J. Sbierski, *On the existence of a maximal Cauchy development for the Einstein equations - a dezornification*, Ann. Henri Poincaré **17** (2016), 301–329.

[61] Y. Shlapentokh-Rothman, *Quantitative Mode Stability for the Wave Equation on the Kerr Spacetime*, Ann. Henri Poincaré. **16** (2015), 289–345.

[62] J. Stogin, *Princeton PHD thesis*, 2017.

[63] D. Tataru and M. Tohaneanu, *A local energy estimate on Kerr black hole backgrounds*, Int. Math. Res. Not. (2011), no 2., 248–292.

[64] S. A. Teukolsky, *Perturbations of a rotating black hole. I. Fundamental equations for gravitational, electromagnetic, and neutrino-field perturbations*, Astrophys. J. **185** (1973), 635–648.

[65] C. V. Vishveshwara, *Stability of the Schwarzschild metric*, Phys. Rev. D, 1 (1970), 2870–2879.

[66] R. M. Wald, *General Relativity*, University of Chicago Press, 1984.

[67] R. M. Wald, *Construction of solutions of gravitational, electromagnetic, or other perturbation equations from solutions of decoupled equations*, Phys. Rev. Lett. **41** (1978), 203–206.

[68] B. Whiting, *Mode stability of the Kerr black hole*, J. Math. Phys. **30** (1989), 1301–1305.

[69] F. J. Zerilli, *Effective potential for even-parity Regge-Wheeler gravitational perturbation equations*, Phys. Rev. Lett. **24** (1970), 737–738.